CONCISE ENCYCLOPEDIA OF

BIOMEDICAL POLYMERS AND POLYMERIC BIOMATERIALS

VOLUME I

Adhesives — Medical Devices and Preparative Medicine

CONCISE ENCYCLOPEDIA OF

BIOMEDICAL POLYMERS AND POLYMERIC BIOMATERIALS

VOLUME I

Adhesives — Medical Devices and Preparative Medicine

——————— edited by ———————

Munmaya Mishra

CRC Press
Taylor & Francis Group
Boca Raton London New York

CRC Press is an imprint of the
Taylor & Francis Group, an **informa** business

CRC Press
Taylor & Francis Group
6000 Broken Sound Parkway NW, Suite 300
Boca Raton, FL 33487-2742

© 2018 by Taylor & Francis Group, LLC
CRC Press is an imprint of Taylor & Francis Group, an Informa business

No claim to original U.S. Government works

Printed in Canada on acid-free paper

International Standard Book Number-13: 978-1-4398-9855-0 (Set) 978-1-138-56314-8 (Volume I) 978-1-138-56318-6 (Volume II)

Visit the Taylor & Francis Web site at
http://www.taylorandfrancis.com

and the CRC Press Web site at
http://www.crcpress.com

Volume I

Adhesives: Tissue Repair and Reconstruction 1
Aerogels: Cellulose-Based 19
Anticancer Agents: Polymeric Nanomedicines 58
Anti-Infective Biomaterials 83
Artificial Muscles 91
Bioabsorbable Polymers: Tissue Engineering 101
Bioactive Systems 105
Bioadhesive Drug Delivery Systems 111
Biocompatibility: Testing 129
Biodegradable Polymers: Bioerodible Systems for Controlled Drug Release 139
Biodegradable Polymers: Biomedically Degradable Polymers 153
Biodegradation 162
Biofunctional Polymers 175
BioMEMS 181
Biomimetic Materials 189
Biomimetic Materials: Smart Polymer Surfaces for Tissue Engineering 214
Biorubber: Poly(Glycerol Sebacate) 229
Blood Vessel Substitutes 237
Bone–Implant: Polymer Physicochemical Modification for Osseointegration 248
Cellulose-Based Biopolymers: Formulation and Delivery Applications 270
Colloid Drug Delivery Systems 301
Conducting Polymers: Biomedical Engineering Applications 312
Conjugated Polymers: Nanoparticles and Nanodots of 327
Conjugates: Biosynthetic–Synthetic Polymer Based 340
Contact Lenses: Gas Permeable 362
Corneas: Tissue Engineering 370
Dendritic Architectures: Delivery Vehicles 395
Dendritic Architectures: Theranostic Applications 402
Dental Polymers: Applications 411
Dental Sealants 433
Drug Delivery Systems: Selection Criteria and Use 439
Drugs and Excipients: Polymeric Interactions 451
Electrets 476
Electroactive Polymeric Materials 482
Electrospinning Technology: Polymeric Nanofiber Drug Delivery 491
Electrospinning Technology: Regenerative Medicine 506
Excipients: Pharmaceutical Dosage Forms 547
Fluorescent Nanohybrids: Cancer Diagnosis and Therapy 560
Functionalized Surfaces: Biomolecular Surface Modification with
    Functional Polymers 585
Gels: Fibrillar Fibrin 616
Gene Carriers: Design Elements 623
Gene Delivery 633
Glues 644
Hair and Skin Care Biomaterials 653
Hemocompatible Polymers 663
Hydrogels 674
Hydrogels: Classification, Synthesis, Characterization, and Applications 685
Hydrogels: Multi-Responsive Biomedical Devices 699
In Vitro Vascularization: Tissue Engineering Constructs 723
Latexes: Magnetic 743
Ligament Replacement Polymers: Biocompatability, Technology, and Design 757
Lipoplexes and Polyplexes: Gene Therapy 765
Magnetomicelles: Theranostic Applications 778
Medical Devices and Preparative Medicine: Polymer Drug Application 789

Volume II

Melt Extrusion: Pharmaceutical Applications 827
Melt-Electrospun Fibers 845
Membranes, Polymeric: Biomedical Devices 864
Metal–Polymer Composite Biomaterials 877
Microcomponents: Polymeric 900
Microgels: Smart Polymer and Hybrid 917
Molecular Assemblies 932
Mucoadhesive Polymers: Basics, Strategies, and Trends 941
Mucoadhesive Systems: Drug Delivery 961
Muscles, Artificial: Sensing, Transduction, Feedback Control, and
    Robotic Applications 978
Nanocomposite Polymers: Functional 994
Nanogels: Chemical Approaches to Preparation 1007
Nanomaterials: Conducting Polymers and Sensing 1035
Nanomaterials: Theranostics Applications 1060
Nanomaterials: Therapeutic Applications 1073
Nanomedicine: Review and Perspectives 1088
Nanoparticles: Biological Applications 1109
Nanoparticles: Biomaterials for Drug Delivery 1170
Nanoparticles: Cancer Management Applications 1182
Natural Polymers: Tissue Engineering 1206
Nerve Guides: Multi-Channeled Biodegradable Polymer Composite 1235
Neural Tissue Engineering: Polymers for 1255
Non-Viral Delivery Vehicles 1272
Orthopedic Applications: Bioceramic and Biopolymer Nanocomposite Materials ... 1276
Peptide–Polymer Conjugates: Synthetic Design Strategies 1289
Pharmaceutical Polymers 1304
Polyurethanes: Medical Applications 1318
Rapid Prototyping 1342
Rapid Prototyping: Tissue Engineering 1350
Scaffolds, Polymer: Microfluidic-Based Polymer Design 1362
Scaffolds, Porous Polymer: Tissue Engineering 1374
Semiconducting Polymer Dot Bioconjugates 1382
Sensing and Diagnosis 1393
Shape Memory Polymers: Applications 1400
Skin Tissue Engineering 1408
Smart Polymers: Imprinting 1424
Smart Polymers: Medicine and Biotechnology Applications 1443
Stem Cell: Hematopoietic Stem Cell Culture, Materials for 1453
Stents: Endovascular 1465
Stimuli-Responsive Materials 1480
Stimuli-Responsive Materials: Thermo- and pH-Responsive Polymers for
    Drug Delivery 1493
Sutures 1514
Theranostics: Biodegradable Polymer Particles for 1530
Ultrasound Contrast Agents: Sonochemical Preparation 1545
Vascular Grafts: Biocompatibility Requirements 1560
Vascular Grafts: Polymeric Materials 1575
Vascular Tissue Engineering: Polymeric Biomaterials 1596
Wound Care: Natural Biopolymer Applications 1607
Wound Care: Skin Tissue Regeneration 1620
Wound Healing: Hemoderivatives and Biopolymers 1642
Zwitterionic Polymeric Materials 1661

To my family

Also, to those who made and will make a difference through polymer research for improving the quality of life and for the betterment of human health.

Concise Encyclopedia of Biomedical Polymers and Polymeric Biomaterials

Editor-in-Chief
Munmaya K. Mishra, Ph. D.
c/o Altria Research Center, Richmond, Virginia U.S.A.

Contributors

Carlos A. Aguilar / *Department of Mechanical Engineering, University of Texas at Austin, Austin, Texas, U.S.A.*

M.R. Aguilar / *Spanish National Research Council Institute of Polymer Science and Technology, Madrid, Spain*

Arti Ahluwalia / *E. Piaggio Interdepartmental Research Center, Department of Chemical Engineering, University of Pisa, Pisa, Italy*

Nasir M. Ahmad / *Polymer and Surface Engineering Lab, Department of Materials Engineering, School of Chemical and Materials Engineering (SCME), National University of Sciences and Technology (NUST), Islamabad, Pakistan*

Naveed Ahmed / *Department of Pharmacy, Quaid-i-Azam University, Islamabad, Pakistan*

Tamer A. E. Ahmed / *Medical Biotechnology Department, Genetic Engineering and Biotechnology Research Institute, City of Scientific Research and Technology Applications (SRTA-City), Alexandria, Egypt Department of Cellular and Molecular Medicine, Faculty of Medicine, University of Ottawa, Ottawa, Ontario, Canada*

A.K.M. Moshiul Alam / *Institute of Radiation and Polymer Technology, Bangladesh Atomic Energy Commission, Dhaka, Bangladesh*

A.-C. Albertsson / *Department of Polymer Technology, Royal Institute of Technology (KTH), Stockholm, Sweden*

Carmen Alvarez-Lorenzo / *Department of Pharmacy and Pharmaceutical Technology, University of Santiago de Compostela (USC), Santiago de Compostela, Spain*

Garima Ameta / *Sonochemistry Laboratory, Department of Chemistry, University College of Science, Mohanlal Sukhadia University, Udaipur, India*

Rakshit Ameta / *Department of Chemistry, Pacific College of Basic and Applied Sciences, Pacific Academy of Higher Education and Research Society (PAHER) University, Udaipur, India*

Suresh C. Ameta / *Department of Chemistry, PAHER University, Udaipur, India*

Sambandam Anandan / *Nanomaterials and Solar Energy Conversion Lab, Department of Chemistry, National Institute of Technology, Tiruchirappalli, India*

S. Anandhakumar / *Department of Physics and Nanotechnology, Sri Ramaswamy Memorial (SRM) University, Kattankulathur, Chennai, India*

Fernanda Andrade / *Laboratory of Pharmaceutical Technology (LTFCICF), Faculty of Pharmacy, University of Porto, Porto, Portugal*

Kirk P. Andriano / *APS Research Institute, Redwood City, California, U.S.A.*

Bruce L. Anneaux / *Poly-Med, Incorporated, Anderson, South Carolina, U.S.A.*

Ilaria Armentano / *Materials Science and Technology Center, University of Perugia, Terni, Italy*

Alexandra Arranja / *Research Institute for Medicine and Pharmaceutical Sciences (iMed. UL), School of Pharmacy, University of Lisbon, Lisbon, Portugal*

Hamed Asadi / *Polymer Chemistry Department, School of Science, University of Tehran, Tehran, Iran*

Murat Ates / *Department of Chemistry, Faculty of Arts and Sciences, Namik Kemal University, Degirmenalti Campus, Tekirdag, Turkey*

A.A. Attama / *Department of Pharmaceutics, Faculty of Pharmaceutical Sciences, University of Nigeria, Nsukka, Nigeria*

Vimalkumar Balasubramanian / *Division of Pharmaceutical Technology, University of Basel, Basel, Switzerland*

Florence Bally / *Institute of Materials Science of Mulhouse (IS2M), University of Upper Alsace, Mulhouse, France, and Institute of Functional Interfaces (IFG), Karlsruhe Institute of Technology (KIT), Eggenstein-Leopoldshafen, Germany*

Indranil Banerjee / *Department of Biotechnology and Medical Engineering, National Institute of Technology, Rourkela, India*

Clark E. Barrett / *Department of Materials Science and Metallurgy, University of Cambridge, Cambridge, U. K.*

Beauty Behera / *Department of Biotechnology and Medical Engineering, National Institute of Technology, Rourkela, India*

Andreas Bernkop-Schnürch / *Institute of Pharmacy, University of Innsbruck, Innsbruck, Austria*

Serena M. Best / *Department of Materials Science and Metallurgy, University of Cambridge, Cambridge, U. K.*

Kumudini Bhanat / *Department of Chemistry, University College of Science, Mohanlal Sukhadia University, Udaipur, India*

Maria Cristina Bonferoni / *Department of Drug Sciences, University of Pavia, Pavia, Italy*

Kellye D. Branch / *Biomedical Engineering Program, Agricultural and Biological Engineering Department, Mississippi State University, Mississippi State, Mississippi, U.S.A.*

Chad Brown / *Merck & Co., Inc., West Point, Pennsylvania, U.S.A.*

Toby D. Brown / *Institute for Health and Biomedical Innovation, Queensland University of Technology, Brisbane, Queensland, Australia*

Joel D. Bumgardner / *Biomedical Engineering Program, Agricultural and Biological Engineering Department, Mississippi State University, Mississippi State, Mississippi, U.S.A.*

K. J. L. Burg / *Department of Bioengineering, Clemson University, Clemson, South Carolina, U.S.A.*

Marine Camblin / *Division of Pharmaceutical Technology, University of Basel, Basel, Switzerland*

Ruth E. Cameron / *Department of Materials Science and Metallurgy, University of Cambridge, Cambridge, U. K.*

Carla Caramella / *Department of Drug Sciences, University of Pavia, Pavia, Italy*

Yu Chang / *Department of Obstetrics and Gynecology, Kaohsiung Medical University Hospital, Kaohsiung Medical University, Kaohsiung, Taiwan*

Robert Chapman / *Key Center for Polymers and Colloids, School of Chemistry, University of Sydney, Sydney, New South Wales, Australia*

Jhunu Chatterjee / *Florida A&M University and College of Engineering, Florida State University, Tallahassee, Florida, U.S.A.*

Jyoti Chaudhary / *Department of Polymer Science, University College of Science, Mohanlal Sukhadia University, Udaipur, India*

Narendra Pal Singh Chauhan / *Department of Chemistry, Bhupal Nobles' Post-Graduate (B.N.P.G.) College, Udaipur, India*

Guoping Chen / *Tissue Regeneration Materials Unit, International Center for Materials Nanoarchitectonics, National Institute for Materials Science, Tsukuba, Japan*

Jianjun Chen / *Department of Materials Science and Engineering, University of Illinois at Urbana-Champaign, Urbana, Illinois, U.S.A.*

Shaochen Chen / *Department of Mechanical Engineering, University of Texas at Austin, Austin, Texas, U.S.A.*

Yung Hung Chen / *Department of Obstetrics and Gynecology, Kaohsiung Medical University Hospital, Kaohsiung Medical University, Kaohsiung, Taiwan*

S.A. Chime / *Department of Pharmaceutics, Faculty of Pharmaceutical Sciences, University of Nigeria, Nsukka, Nigeria*

Joon Sig Choi / *Department of Biochemistry, Chungnam National University, Daejeon, South Korea*

Soonmo Choi / *Department of Nano, Medical and Polymer Materials, Yeungnam University, Gyeongsan, South Korea*

Seth I. Christian / *Biomedical Engineering Program, Agricultural and Biological Engineering Department, Mississippi State University, Mississippi State, Mississippi, U.S.A.*

Chih-Chang Chu / *Department of Textiles and Apparel and Biomedical Engineering Program, Cornell University, Ithaca, New York, U.S.A.*

Jeffrey Chuang / *Boston College, Chestnut Hill, Massachusetts, U.S.A.*

Giuseppe Cirillo / *Department of Pharmacy, Health, and Nutritional Sciences, University of Calabria, Rende, Italy*

Angel Concheiro / *Department of Pharmacy and Pharmaceutical Technology, University of Santiago de Compostela (USC), Santiago de Compostela, Spain*

Patrick Crowley / *Callum Consultancy, Devon, Pennsylvania, U.S.A.*

Paul D. Dalton / *Institute for Health and Biomedical Innovation, Queensland University of Technology, Brisbane, Queensland, Australia*

Vinod B. Damodaran / *New Jersey Center for Biomaterials, Rutgers University, Piscataway, New Jersey, U.S.A.*

Martin Dauner / *Denkendorf Forschungsbereich Blomedizintechnik, Institute of Textile and Process Engineering (ITV) Denkendorf*

Sabrina Dehn / *Key Center for Polymers and Colloids, School of Chemistry, University of Sydney, Sydney, New South Wales, Australia*

Claudia Del Fante / *Immunohaematology and Transfusion Service and Cell Therapy Unit, San Matteo University Hospital, Pavia, Italy*

Prashant K. Deshmukh / *Post Graduate Department of Pharmaceutics and Quality Assurance, H. R. Patel Institute of Pharmaceutical Education and Research, Shirpur, India*

James DiNunzio / *Merck & Co., Inc., Summit, New Jersey, U.S.A.*

Ryan F. Donnelly / *School of Pharmacy, Medical Biology Center, Queen's University Belfast, Belfast, U.K.*

Garry P. Duffy / *Tissue Engineering Research Group, Royal College of Surgeons in Ireland, Dublin, Ireland, Trinity Center for Bioengineering, Trinity College, Dublin, Ireland, Advanced Materials and BioEngineering Research Center (AMBER), Dublin, Ireland*

Abdelhamid Elaissari / *University of Lyon, Lyon, France*

Monzer Fanun / *Colloids and Surfaces Research Center, Al-Quds University, East Jerusalem, Palestine*

Miguel Faria / *Pharmacology Laboratory, Multidisciplinary Biomedical Research Unit, Abel Salazar Institute of Biomedical Science, University of Porto (ICBAS-UP), Oporto, Portugal*

Zahoor H. Farooqi / *Institute of Chemistry, University of the Punjab, New Campus, Lahore, Pakistan*

George Fercana / *Biocompatibility and Tissue Regeneration Laboratories, Department of Bioengineering, Clemson University, Clemson, and Laboratory of Regenerative Medicine, Patewood/ CU Bioengineering Translational Research Center, Greenville Hospital System, Greenville, South Carolina, U.S.A.*

M.M. Fernández / *Spanish National Research Council Institute of Polymer Science and Technology, Madrid, Spain*

Franca Ferrari / *Department of Drug Sciences, University of Pavia, Pavia, Italy*

Seth Forster / *Merck & Co., Inc., West Point, Pennsylvania, U.S.A.*

Keertik S. Fulzele / *Biomedical Engineering Program, Agricultural and Biological Engineering Department, Mississippi State University, Mississippi State, Mississippi, U.S.A.*

Bernard R. Gallot / *Laboratoire des Materiaux Organiques, Proprietes Specifi ques, National Center for Scientific Research (CNRS), France*

Miguel Gama / *Center of Biological Engineering (CBE), University of Minho, Braga, Portugal*

L. García-Fernández / *Networking Biomedical Research Center on Bioengineering, Biomaterials, and Nanomedicine (CIBER-BBN), Madrid, Spain*

Surendra G. Gattani / *School of Pharmacy, Swami Ramanand Teerth Marathwada University, Nanded, India*

Trent M. Gause / *Department of Plastic Surgery, University of Pittsburgh, Pittsburgh, Pennsylvania, U.S.A.*

Shyamasree Ghosh / *School of Biological Sciences, National Institute of Science, Education, and Research, Bhubaneswar, India*

Rory L.D. Gibney / *Departments of Materials Engineering, Chemistry, and Mechanical Engineering, KU Leuven, Leuven, Belgium*

Abhijit Gokhale / *Product Development Services, Patheon Pharmaceuticals Inc., Cincinnati, Ohio, U.S.A.*

Aleksander Góra / *Center for Nanofibers and Nanotechnology, Nanoscience and Nanotechnology Initiative, Faculty of Engineering and Department of Mechanical Engineering, Faculty of Engineering, National University of Singapore, Singapore*

Erin Grassl / *University of Minnesota, Minneapolis, Minnesota, U.S.A.*

Alexander Yu. Grosberg / *Department of Physics, University of Minnesota, Minneapolis, Minnesota, U.S.A.*

E. Haimer / *Division of Chemistry of Renewable Resources, University of Natural Resources and Life Sciences Vienna, Tulln, Austria*

Dalia A.M. Hamza / *Department of Biochemistry, Faculty of Science, Ain Shams University, Cairo, Egypt*

Dong Keun Han / *Polymer Chemistry Laboratory, Korea Institute of Science and Technology, Seoul, South Korea*

Sung Soo Han / *Department of Nano, Medical, and Polymer Materials, Polymer Gel Cluster Research Center, Yeungnam University, Daedong, South Korea*

Takao Hanawa / *Institute of Biomaterials and Bioengineering, Tokyo Medical and Dental University, Tokyo, Japan*

Ali Hashemi / *Chemical and Petroleum Engineering Department, Sharif University of Technology, Tehran, Iran*

Jorge Heller / *APS Research Institute, Redwood City, California, U.S.A.*

Akon Higuchi / *Department of Chemical and Materials Engineering, National Central University, Jhongli, Taiwan, and Department of Botany and Microbiology, King Saud University, Riyadh, Saudi Arabia*

Georgios T. Hilas / *Poly-Med, Incorporated, Anderson, South Carolina, U.S.A.*

Maxwell T. Hincke / *Department of Cellular and Molecular Medicine and Department of Innovation in Medical Education, Faculty of Medicine, University of Ottawa, Ottawa, Ontario, Canada*

Allan S. Hoffman / *Center for Bioengineering, University of Washington, Seattle, Washington, U.S.A.*

J.M. Hook / *Mark Wainwright Analytical Centre, University of New South Wales, Sydney, Australia*

Dietmar W. Hutmacher / *Division of Bioengineering, Department of Orthopaedic Surgery, National University of Singapore, Singapore*

Jörg Huwyler / *Division of Pharmaceutical Technology, University of Basel, Basel, Switzerland*

Francesca Iemma / *Department of Pharmacy, Health, and Nutritional Sciences, University of Calabria, Rende, Italy*

Yoshita Ikada / *Department of Clinical Engineering, Faculty of Medical Engineering, Suzuka University of Medical Science, Mie, Japan*

Kazuhiko Ishihara / *Department of Materials Engineering, School of Engineering, University of Tokyo, Tokyo, Japan*

Saeed Jafarirad / *Research Institute for Fundamental Sciences (RIFS), University of Tabriz, Tabriz, Iran*

B. Jansen / *Institute of Medical Microbiology and Hygiene, University of Cologne, Cologne, Germany*

Katrina A. Jolliffe / *School of Chemistry, University of Sydney, Sydney, New South Wales, Australia*

Priya Juneja / *Jubilant Life Sciences, Noida, India*

Rajesh Kalia / *Department of Physics, Maharishi Markandeshwar University, Mullana, India*

Sapna Kalia / *Department of Physics, Maharishi Markandeshwar University, Mullana, India*

Ta-Chun Kao / *Department of Chemical and Materials Engineering, National Central University, Jhongli, Taiwan*

Paridhi Kataria / *Department of Chemistry, University College of Science, Mohanlal Sukhadia University, Udaipur, India*

Harmeet Kaur / *Prabhu Dayal Memorial (PDM) College of Pharmacy, Bahadurgarh, India*

F.C. Kenechukwu / *Department of Pharmaceutics, Faculty of Pharmaceutical Sciences, University of Nigeria, Nsukka, Nigeria*

Josè Maria Kenny / *Materials Science and Technology Center, University of Perugia, Terni, Italy*

Asad Ullah Khan / *Department of Chemical Engineering, Commission on Science and Technology for Sustainable Development in the South (COMSATS) Institute of Information Technology, Lahore, Pakistan*

Sepideh Khoee / *Polymer Chemistry Department, School of Science, University of Tehran, Tehran, Iran*

Arezoo Khosravi / *Institute for Nanoscience and Nanotechnology, Sharif University of Technology, Tehran, Iran*

Kwang J. Kim / *College of Engineering, University of Nevada–Reno, Reno, Nevada, U.S.A.*

Sanghyo Kim / *Department of Bionanotechnology, Gachon University, Gyeonggi-do, South Korea*

Young Ha Kim / *Polymer Chemistry Laboratory, Korea Institute of Science and Technology, Seoul, South Korea*

Yoshihiro Kiritoshi / *Department of Materials Engineering, School of Engineering, University of Tokyo, Tokyo, Japan*

W. Kohnen / *Institute of Medical Microbiology and Hygiene, University of Cologne, Cologne, Germany*

Soheila S. Kordestani / *ChitoTech Inc., Tehran, Iran*

Abhijith Kundadka Kudva / *Departments of Materials Engineering, Chemistry, and Mechanical Engineering, KU Leuven, Leuven, Belgium*

Ashok Kumar / *Department of Biological Sciences and Bioengineering, Indian Institute of Technology, Kanpur, India*

S. Suresh Kumar / *Department of Medical Microbiology and Parasitology, Putra University, Slangor, Malaysia*

Jay F. Künzler / *Department of Chemistry and Polymer Development, Bausch and Lomb Incorporated*

A. Lauto / *Biomedical Engineering and Neuroscience Research Group, The MARCS Institute, Western Sydney University, Penrith, Australia*

L. James Lee / *Department of Chemical Engineering, Ohio State University, Columbus, Ohio, U.S.A.*

Alexandre F. Leitão / *Center of Biological Engineering (CBE), University of Minho, Braga, Portugal*

Kimon Alexandros Leonidakis / *Departments of Materials Engineering, Chemistry, and Mechanical Engineering, KU Leuven, Leuven, Belgium*

Victoria Leszczak / *Department of Mechanical Engineering, Colorado State University, Fort Collins, Colorado, U.S.A.*

Pei-Tsz Li / *Department of Chemical and Materials Engineering, National Central University, Jhongli, Taiwan*

F. Liebner / *Division of Chemistry of Renewable Resources, University of Natural Resources and Life Sciences Vienna, Tulln, Austria*

Cai Lloyd-Griffith / *Tissue Engineering Research Group, Royal College of Surgeons in Ireland, Dublin, Ireland, Trinity Center for Bioengineering, Trinity College, Dublin, Ireland, Advanced Materials and BioEngineering Research Center (AMBER), Dublin, Ireland*

A. Löfgren / *Department of Polymer Technology, Royal Institute of Technology (KTH), Stockholm, Sweden*

M.L. López-Donaire / *Spanish National Research Council Institute of Polymer Science and Technology, Madrid, Spain*

Yi Lu / *Department of Mechanical Engineering, University of Texas at Austin, Austin, Texas, U.S.A.*

Sofia Luís / *Research Institute for Medicine and Pharmaceutical Sciences (iMed.UL), School of Pharmacy, University of Lisbon, Lisbon, Portugal*

Mizuo Maeda / *Graduate School of Engineering, Kyushu University, Fukuoka, Japan*

Bernard Mandrand / *bioMérieux, National Center for Scientific Research (CNRS), Lyon, France*

João F. Mano / *Biomaterials, Biodegradables and Biomimetics (3Bs) Research Group, Department of Polymer Engineering, School of Engineering, European Institute of Excellence on Tissue Engineering and Regenerative Medicine, University of Minho, Braga and Life and Health Sciences Research Institute (ICVS)Portuguese Government Associate Laboratory, Guimaraes, Portugal*

Alexandra A. P. Mansur / *Center of Nanoscience, Nanotechnology, and Innovation (CeNano), Department of Metallurgical and Materials Engineering, Federal University of Minas Gerais, Belo Horizonte, Minas Gerais, Brazil*

Herman S. Mansur / *Center of Nanoscience, Nanotechnology, and Innovation (CeNano), Department of Metallurgical and Materials Engineering, Federal University of Minas Gerais, Belo Horizonte, Minas Gerais, Brazil*

Asghari Maqsood / *Department of Physics, Center for Emerging Sciences, Engineering and Technology (CESET), Islamabad, Pakistan*

Mohana Marimuthu / *Department of Bionanotechnology, Gachon University, Gyeonggi-do, South Korea*

Kacey G. Marra / *Departments of Plastic Surgery and Bioengineering, McGowan Institute for Regenerative Medicine, University of Pittsburgh, Pittsburgh, Pennsylvania, U.S.A.*

Luigi G. Martini / *Institute of Pharmaceutical Sciences, King's College, London, U.K.*

Shojiro Matsuda / *Research and Development Department, Gunze Limited, Ayabe, Japan*

Mohammad Mazinani / *Department of Materials Engineering, Faculty of Engineering, Ferdowsi University of Mashhad, Mashhad, Iran*

Tara M. McFadden / *Tissue Engineering Research Group, Royal College of Surgeons in Ireland, Dublin, Ireland, Trinity Center for Bioengineering, Trinity College, Dublin, Ireland, Advanced Materials and BioEngineering Research Center (AMBER), Dublin, Ireland*

James W. McGinity / *Division of Pharmaceutics, College of Pharmacy, University of Texas at Austin, Austin, Texas, U.S.A.*

Kiran Meghwal / *Department of Chemistry, University College of Science, Mohanlal Sukhadia University, Udaipur, India*

Shahram Mehdipour-Ataei / *Iran Polymer and Petrochemical Institute, Tehran, Iran*

Taíla O. Meiga / *Departments of Materials Engineering, Chemistry, and Mechanical Engineering, KU Leuven, Leuven, Belgium*

Meymanant Sadat Mohsenzadeh / *Department of Materials Engineering, Faculty of Engineering, Ferdowsi University of Mashhad, Mashhad, Iran*

Mehran Mojarrad / *Pharmaceutical Delivery Systems, Eli Lilly and Co., Indianapolis, Indiana, U.S.A.*

Kwang Woo Nam / *Department of Materials Engineering, School of Engineering, University of Tokyo, Tokyo, Japan*

Robert M. Nerem / *Parker H. Petit Institute for Bioengineering and Bioscience, Georgia Institute of Technology, Atlanta, Georgia, U.S.A.*

Fergal J. O'Brien / *Tissue Engineering Research Group, Royal College of Surgeons in Ireland, Dublin, Ireland, Trinity Center for Bioengineering, Trinity College, Dublin, Ireland, Advanced Materials and BioEngineering Research Center (AMBER), Dublin, Ireland*

J.D.N. Ogbonna / *Department of Pharmaceutics, Faculty of Pharmaceutical Sciences, University of Nigeria, Nsukka, Nigeria*

Maryam Oroujzadeh / *Iran Polymer and Petrochemical Institute, Tehran, Iran*

Kunal Pal / *Department of Biotechnology and Medical Engineering, National Institute of Technology, Rourkela, India*

Abhijeet P. Pandey / *Post Graduate Department of Pharmaceutics and Quality Assurance, H. R. Patel Institute of Pharmaceutical Education and Research, Shirpur, India*

Arpita Pandey / *Department of Chemistry, University College of Science, Mohanlal Sukhadia*

Jong-Sang Park / *Department of Chemistry, Seoul National University, Seoul, South Korea*

Ki Dong Park / *Polymer Chemistry Laboratory, Korea Institute of Science and Technology, Seoul, South Korea*

F. Parra / *Spanish National Research Council Institute of Polymer Science and Technology, Madrid, Spain*

Arpit Kumar Pathak / *Department of Chemistry, University College of Science, Mohanlal Sukhadia University, Udaipur, India*

Pravin O. Patil / *Post Graduate Department of Pharmaceutics and Quality Assurance, H. R. Patel Institute of Pharmaceutical Education and Research, Shirpur, India*

Jennifer Patterson / *Departments of Materials Engineering, Chemistry, and Mechanical Engineering, KU Leuven, Leuven, Belgium*

Shawn J. Peniston / *Poly-Med, Incorporated, Anderson, South Carolina, U.S.A.*

Cesare Perotti / *Immunohaematology and Transfusion Service and Cell Therapy Unit, San Matteo University Hospital, Pavia, Italy*

Sébastien Perrier / *Key Center for Polymers and Colloids, School of Chemistry, University of Sydney, Sydney, New South Wales, Australia*

Susanna Piluso / *Departments of Materials Engineering, Chemistry, and Mechanical Engineering, KU Leuven, Leuven, Belgium*

N. Pircher / *Division of Chemistry of Renewable Resources, University of Natural Resources and Life Sciences Vienna, Tulln, Austria*

Stergios Pispas / *Theoretical and Physical Chemistry Institute of the National Hellenic Research Foundation (TPCI-NHRF), Athens, Greece*

Heinrich Planck / *Denkendorf Forschungsbereich Blomedizintechnik, Institute of Textile and Process Engineering (ITV) Denkendorf*

Ketul C. Popat / *Department of Mechanical Engineering and School of Biomedical Engineering, Colorado State University, Fort Collins, Colorado, U.S.A.*

Krishna Pramanik / *Department of Biotechnology and Medical Engineering, National Institute of Technology, Rourkela, India*

Pinki B. Punjabi / *Department of Chemistry, University College of Science, Mohanlal Sukhadia University, Udaipur, India*

Diana Rafael / *Research Institute for Medicine and Pharmaceutical Sciences (iMed.UL), School of Pharmacy, University of Lisbon, Lisbon, Portugal*

Seeram Ramakrishna / *Center for Nanofibers and Nanotechnology, Nanoscience and Nanotechnology Initiative, Faculty of Engineering and Department of Mechanical Engineering, Faculty of Engineering, National University of Singapore, Singapore*

Stanislav Rangelov / *Laboratory of Polymerization Processes, Scientific Council of the Institute of Polymers, Bulgarian Academy of Sciences, Sofia, Bulgaria*

Manish Kumar Rawal / *Department of Chemistry, Vidya Bhawan Rural Institute, Udaipur, India*

Melissa M. Reynolds / *Department of Chemistry and School of Biomedical Engineering, Colorado State University, Fort Collins, Colorado, U.S.A.*

G. Rodríguez / *Spanish National Research Council Institute of Polymer Science and Technology, Madrid, Spain*

L. Rojo / *Department of Materials and Institute of Bioengineering, Imperial College London, London, U.K.*

T. Rosenau / *Division of Chemistry of Renewable Resources, University of Natural Resources and Life Sciences Vienna, Tulln, Austria*

Aftin M. Ross / *Institute of Functional Interfaces (IFG), Karlsruhe Institute of Technology (KIT), Eggenstein-Leopoldshafen, Germany, and Center for Devices and Radiological Health, Food and Drug Administration, Silver Spring, Maryland, U.S.A.*

Silvia Rossi / *Department of Drug Sciences, University of Pavia, Pavia, Italy*

Shaneen L. Rowe / *Department of Biomedical Engineering, Rensselaer Polytechnic Institute, Troy, New York, U.S.A.*

H. Ruprai / *School of Science and Health, Western Sydney University, Penrith, Australia.*

Elizabeth Ryan / *School of Pharmacy, Medical Biology Center, Queen's University of Belfast, Belfast, U.K.*

Sai Sateesh Sagiri / *Department of Biotechnology and Medical Engineering, National Institute of Technology, Rourkela, India*

Soham Saha / *School of Biological Sciences, National Institute of Science, Education, and Research, Bhubaneswar, India*

Rahul Sahay / *Fluid Division, Department of Mechanical Engineering, National University of Singapore, Singapore*

J. San Román / *Institute for Health and Biomedical Innovation, Queensland University of Technology, Kelvin Grove, Queensland, Australia*

Giuseppina Sandri / *Department of Drug Sciences, University of Pavia, Pavia, Italy*

Smritimala Sarmah / *Department of Physics, Girijananda Chowdhury Institute of Management and Technology, Guwahati, India and Raktim Pratim Tamuli, Department of Forensic Medicine, Guwahati Medical College and Hospital, Guwahati, India*

M. Sasidharan / *Department of Physics and Nanotechnology, Sri Ramaswamy Memorial (SRM) University, Kattankulathur, Chennai, India*

C. Schimper / *Division of Chemistry of Renewable Resources, University of Natural Resources and Life Sciences Vienna, Tulln, Austria*

Deepak Kumar Semwal / *Department of Chemistry, Panjab University, Chandigarh, India*

Ravindra Semwal / *Faculty of Pharmacy, Dehradun Institute of Technology, Dehradun, India*

Ruchi Badoni Semwal / *Department of Chemistry, Panjab University, Chandigarh, India*

Mohsen Shahinpoor / *Artificial Muscle Research Institute (AMRI), School of Engineering and School of Medicine, University of New Mexico, Albuquerque, New Mexico, U.S.A.*

Haseeb Shaikh / *Polymer and Surface Engineering Lab, Department of Materials Engineering, School of Chemical and Materials Engineering (SCME), National University of Sciences and Technology (NUST), Islamabad, Pakistan*

Waleed S. W. Shalaby / *Poly-Med Incorporated, Anderson, South Carolina, U.S.A.*

V. Prasad Shastri / *Vanderbilt University, Nashville, Tennessee, U.S.A.*

Mikhail I. Shtilman / *Biomaterials Scientific and Teaching Center, D. I. Mendeleyev University of Chemical Technology, Moscow, Russia*

Quazi T. H. Shubhra / *Doctoral School of Molecular and Nanotechnologies, Faculty of Information Technology, University of Pannonia, Veszprém, Hungary*

Mohammad Siddiq / *Department of Chemistry, Quaid-I-Azam University, Islamabad, Pakistan*

Ivone Silva / *Department of Angiology and Vascular Surgery, Hospital of Porto, Porto, Portugal*

Dan Simionescu / *Department of Bioengineering, Clemson University, Clemson, and Patewood/CU Bioengineering Translational Research Center, Greenville Hospital System, Greenville, South Carolina, U.S.A., and Department of Anatomy, University of Medicine and Pharmacy, Tirgu Mures, Romania*

Deepti Singh / *Department of Nano, Medical, and Polymer Materials, Polymer Gel Cluster Research Center, Yeungnam University, Daedong, South Korea*

Jasbir Singh / *Department of Pharmacy, University of Health Sciences, Rohtak, India*

Thakur Raghu Raj Singh / *School of Pharmacy, Medical Biology Center, Queen's University of Belfast, Belfast, U.K.*

Vinay K. Singh / *Department of Biotechnology and Medical Engineering, National Institute of Technology, Rourkela, India*

Wesley N. Sivak / *Department of Plastic Surgery, University of Pittsburgh, Pittsburgh, Pennsylvania, U.S.A.*

Radhakrishnan Sivakumar / *Nanomaterials and Solar Energy Conversion Lab, Department of Chemistry, National Institute of Technology, Tiruchirappalli, India*

Daniel H. Smith / *Biomedical Engineering Program, Agricultural and Biological Engineering Department, Mississippi State University, Mississippi State, Mississippi, U.S.A.*

Al Halifa Soultan / *Departments of Materials Engineering, Chemistry, and Mechanical Engineering, KU Leuven, Leuven, Belgium*

Umile Gianfranco Spizzirri / *Department of Pharmacy, Health, and Nutritional Sciences, University of Calabria, Rende, Italy*

Jan P. Stegeman / *Department of Biomedical Engineering, Rensselaer Polytechnic Institute, Troy, New York, U.S.A.*

Kaliappa Gounder Subramanian / *Department of Biotechnology, Bannari Amman Institute of Technology, Sathyamangalam, India*

Abhinav Sur / *School of Biological Sciences, National Institute of Science, Education, and Research, Bhubaneswar, India*

Michael Szycher / *Sterling Biomedical, Incorporated, Lynnfield, Massachusetts, U.S.A.*

G. Lawrence Thatcher / *TESco Associates Incorporated, Tyngsborough, Massachusetts, U.S.A.*

Velmurugan Thavasi / *Elam Pte Ltd., Singapore*

Rong Tong / *Department of Materials Science and Engineering, University of Illinois at Urbana-Champaign, Urbana, Illinois, U.S.A.*

Luigi Torre / *Materials Science and Technology Center, University of Perugia, Terni, Italy*

Robert T. Tranquillo / *Department of Chemical Engineering and Materials Science, University of Minnesota, Minneapolis, Minnesota, U.S.A.*

Yasuhiko Tsuchitani / *Faculty of Dentistry, Osaka University, Osaka, Japan Tohru Wada, Kuraray Company Ltd., Tokyo, Japan*

Laurien Van den Broeck / *Departments of Materials Engineering, Chemistry, and Mechanical Engineering, KU Leuven, Leuven, Belgium*

Jan C. M. van Hest / *Organic Chemistry, Institute for Molecules and Materials, Radboud University Nijmegen, Nijmegen, the Netherlands*

Cedryck Vaquette / *Institute for Health and Biomedical Innovation, Queensland University of Technology, Brisbane, Queensland, Australia*

Marcia Vasquez-Lee / *Biomedical Engineering Program, Agricultural and Biological Engineering Department, Mississippi State University, Mississippi State, Mississippi, U.S.A.*

Jason Vaughn / *Product Development Services, Patheon Pharmaceuticals Inc., Cincinnati, Ohio, U.S.A.*

Raphael Veyret / *bioMérieux, National Center for Scientific Research (CNRS), Lyon, France*

Mafalda Videira / *Research Institute for Medicine and Pharmaceutical Sciences (iMed. UL), School of Pharmacy, University of Lisbon, Lisbon, Portugal*

Vediappan Vijayakumar / *Department of Biotechnology, Bannari Amman Institute of Technology, Sathyamangalam, India*

Tanushree Vishnoi / *Department of Biological Sciences and Bioengineering, Indian Institute of Technology, Kanpur, India*

S. Vivekananthan / *Department of Physics and Nanotechnology, Sri Ramaswamy Memorial (SRM) University, Kattankulathur, Chennai, India*

Giovanni Vozzi / *E. Piaggio Interdepartmental Research Center, Department of Chemical Engineering, University of Pisa, Pisa, Italy*

Ernst Wagner / *Department of Pharmacy, Ludwig Maximilians University, Munich, Germany*

Yadong Wang / *Wallace H. Coulter School of Biomedical Engineering and School of Chemistry and Biochemistry, Georgia Institute of Technology, Atlanta, Georgia, U.S.A.*

Junji Watanabe / *Department of Materials Engineering, School of Engineering, University of Tokyo, Tokyo, Japan*

Jennifer L. West / *Department of Bioengineering, Rice University, Houston, Texas, U.S.A.*

David L. Williams / *Biomedical Engineering Program, Agricultural and Biological Engineering Department, Mississippi State University, Mississippi State, Mississippi, U.S.A.*

Thomas Williams / *Product Development Services, Patheon Pharmaceuticals Inc., Cincinnati, Ohio, U.S.A.*

Dominik Witzigmann / *Division of Pharmaceutical Technology, University of Basel, Basel, Switzerland*

A. David Woolfson / *School of Pharmacy, Medical Biology Center, Queen's University of Belfast, Belfast, U.K.*

Tetsuji Yamaoka / *Department of Polymer Science and Engineering, Kyoto Institute of Technology, Kyoto, Japan*

Can Yang / *College of Engineering, Zhejiang Normal University, Zhejiang, China*

Siou-Ting Yang / *Department of Chemical and Materials Engineering, National Central University, Jhongli, Taiwan*

Xiao-Hong Yin / *College of Engineering, Zhejiang Normal University, Zhejiang, China*

Ali Zarrabi / *Department of Biotechnology, Faculty of Advanced Sciences and Technologies, University of Isfahan, Isfahan, Iran*

Seyed Mojtaba Zebarjad / *Department of Materials Engineering, Faculty of Engineering, Ferdowsi University of Mashhad, Mashhad, Iran*

Contents

Contributors xiii
Preface. cv
About the editor cvii

Volume I

Adhesives: Tissue Repair and Reconstruction / A. Lauto, H. Ruprai, and J.M. Hook 1
Aerogels: Cellulose-Based / F. Liebner, N. Pircher, C. Schimper, E. Haimer, and T. Rosenau. 19
Anticancer Agents: Polymeric Nanomedicines / Rong Tong and Jianjun Cheng. 58
Anti-Infective Biomaterials / W. Kohnen and B. Jansen 83
Artificial Muscles / Mohsen Shahinpoor 91
Bioabsorbable Polymers: Tissue Engineering / K. J. L. Burg and Waleed S. W. Shalaby 101
Bioactive Systems / Mikhail I. Shtilman 105
Bioadhesive Drug Delivery Systems / Ryan F. Donnelly and A. David Woolfson 111
Biocompatibility: Testing / Joel D. Bumgardner, Marcia Vasquez-Lee, Keertik S. Fulzele, Daniel H. Smith,
    Kellye D. Branch, Seth I. Christian, and David L. Williams. 129
Biodegradable Polymers: Bioerodible Systems for Controlled Drug Release / Jorge Heller and
    Kirk P. Andriano 139
Biodegradable Polymers: Biomedically Degradable Polymers / A.-C. Albertsson and A. Löfgren 153
Biodegradation / Mikhail I. Shtilman. 162
Biofunctional Polymers / Jennifer. L. West. 175
BioMEMS / L. James Lee 181
Biomimetic Materials / Kimon Alexandros Leonidakis, Al Halifa Soultan, Susanna Piluso,
    Abhijith Kundadka Kudva, Laurien Van den Broeck, Taíla O. Meiga, Rory L.D. Gibney, and
    Jennifer Patterson 189
Biomimetic Materials: Smart Polymer Surfaces for Tissue Engineering / João F. Mano 214
Biorubber: Poly(Glycerol Sebacate) / Yadong Wang 229
Blood Vessel Substitutes / Jan P. Stegeman, Shaneen L. Rowe, and Robert M. Nerem 237
Bone–Implant: Polymer Physicochemical Modification for Osseointegration / Waleed S. W. Shalaby and
    Bruce L. Anneaux 248
Cellulose-Based Biopolymers: Formulation and Delivery Applications / J.D.N. Ogbonna,
    F.C. Kenechukwu, S.A. Chime, and A.A. Attama. 270
Colloid Drug Delivery Systems / Monzer Fanun. 301
Conducting Polymers: Biomedical Engineering Applications / Smritimala Sarmah and
    Raktim Pratim Tamuli 312
Conjugated Polymers: Nanoparticles and Nanodots of / Garima Ameta, Suresh C. Ameta, Rakshit Ameta,
    and Pinki B. Punjabi 327
Conjugates: Biosynthetic–Synthetic Polymer Based / Jan C. M. van Hest 340
Contact Lenses: Gas Permeable / Jay F. Künzler 362
Corneas: Tissue Engineering / Dalia A.M. Hamza, Tamer A. E. Ahmed, and Maxwell T. Hincke. . . . 370
Dendritic Architectures: Delivery Vehicles / Saeed Jafarirad 395
Dendritic Architectures: Theranostic Applications / Saeed Jafarirad 402
Dental Polymers: Applications / Narendra P.S. Chauhan, Kiran Meghwal, Pinki B. Punjabi,
    Jyoti Chaudhary, and Paridhi Kataria. 411
Dental Sealants / Yasuhiko Tsuchitani and Tohru Wada 433

Drug Delivery Systems: Selection Criteria and Use / *Ravindra Semwal, Ruchi Badoni Semwal, and Deepak Kumar Semwal* 439

Drugs and Excipients: Polymeric Interactions / *James C. DiNunzio and James W. McGinity.* 451

Electrets / *Rajesh Kalia and Sapna Kalia.* 476

Electroactive Polymeric Materials / *V. Prasad Shastri* 482

Electrospinning Technology: Polymeric Nanofiber Drug Delivery / *Narendra Pal Singh Chauhan, Kiran Meghwal, Priya Juneja, and Pinki B. Punjabi* 491

Electrospinning Technology: Regenerative Medicine / *Toby D. Brown, Cedryck Vaquette, Dietmar W. Hutmacher, and Paul D. Dalton.* 506

Excipients: Pharmaceutical Dosage Forms / *Luigi G. Martini and Patrick Crowley.* 547

Fluorescent Nanohybrids: Cancer Diagnosis and Therapy / *Herman S. Mansur and Alexandra A. P. Mansur* 560

Functionalized Surfaces: Biomolecular Surface Modification with Functional Polymers / *Florence Bally and Aftin M. Ross* 585

Gels: Fibrillar Fibrin / *Erin Grassl and Robert T. Tranquillo* 616

Gene Carriers: Design Elements / *Jong-Sang Park and Joon Sig Choi.* 623

Gene Delivery / *Tetsuji Yamaoka* 633

Glues / *Shojiro Matsuda and Yoshita Ikada* 644

Hair and Skin Care Biomaterials / *Bernard R. Gallot* 653

Hemocompatible Polymers / *Young Ha Kim, Ki Dong Park, and Dong Keun Han* 663

Hydrogels / *Junji Watanabe, Yoshihiro Kiritoshi, Kwang Woo Nam, and Kazuhiko Ishihara* 674

Hydrogels: Classification, Synthesis, Characterization, and Applications / *Kaliappa Gounder Subramanian and Vediappan Vijayakumar* 685

Hydrogels: Multi-Responsive Biomedical Devices / *Francesca Iemma, Giuseppe Cirillo, and Umile Gianfranco Spizzirri* 699

***In Vitro* Vascularization: Tissue Engineering Constructs** / *Cai Lloyd-Griffith, Tara M. McFadden, Garry P. Duffy, and Fergal J. O'Brien.* 723

Latexes: Magnetic / *Abdel Hamid Elaissari, Raphael Veyret, Bernard Mandrand, and Jhunu Chatterjee* 743

Ligament Replacement Polymers: Biocompatability, Technology, and Design / *Martin Dauner and Heinrich Planck* 757

Lipoplexes and Polyplexes: Gene Therapy / *Diana Rafael, Fernanda Andrade, Alexandra Arranja, Sofia Luís, and Mafalda Videira* 765

Magnetomicelles: Theranostic Applications / *Prashant K. Deshmukh, Abhijeet P. Pandey, Surendra G. Gattani, and Pravin O. Patil* 778

Medical Devices and Preparative Medicine: Polymer Drug Application / *M.R. Aguilar, L. García-Fernández, M.L. López-Donaire, F. Parra, L. Rojo, G. Rodríguez, M.M. Fernández, and J. San Román* 789

Volume II

Melt Extrusion: Pharmaceutical Applications / *James DiNunzio, Seth Forster, and Chad Brown.* 827

Melt-Electrospun Fibers / *Aleksander Góra, Rahul Sahay, Velmurugan Thavasi, and Seeram Ramakrishna* . . 845

Membranes, Polymeric: Biomedical Devices / *Shahram Mehdipour-Ataei and Maryam Oroujzadeh* 864

Metal–Polymer Composite Biomaterials / *Takao Hanawa* 877

Microcomponents: Polymeric / *Can Yang and Xiao-Hong Yin* 900

Microgels: Smart Polymer and Hybrid / *Zahoor H. Farooqi and Mohammad Siddiq* 917

Molecular Assemblies / *Saeed Jafarirad* 932

Mucoadhesive Polymers: Basics, Strategies, and Trends / *Andreas Bernkop-Schnürch* 941

Mucoadhesive Systems: Drug Delivery / *Ryan F. Donnelly, Elizabeth Ryan, Thakur Raghu Raj Singh, and A. David Woolfson* 961

Muscles, Artificial: Sensing, Transduction, Feedback Control, and Robotic Applications /
Mohsen Shahinpoor, Kwang J. Kim, and Mehran Mojarrad 978

Nanocomposite Polymers: Functional / *Radhakrishnan Sivakumar and Sambandam Anandan* 994

Nanogels: Chemical Approaches to Preparation / *Sepideh Khoee and Hamed Asadi* 1007

Nanomaterials: Conducting Polymers and Sensing / *Murat Ates* 1035

Nanomaterials: Theranostics Applications / *M. Sasidharan, S. Anandhakumar, and S. Vivekananthan* 1060

Nanomaterials: Therapeutic Applications / *Dominik Witzigmann, Marine Camblin, Jörg Huwyler, and
Vimalkumar Balasubramanian.* 1073

Nanomedicine: Review and Perspectives / *Ali Zarrabi, Arezoo Khosravi, and Ali Hashemi.* 1088

Nanoparticles: Biological Applications / *Stanislav Rangelov and Stergios Pispas* 1109

Nanoparticles: Biomaterials for Drug Delivery / *Abhijit Gokhale, Thomas Williams, and Jason Vaughn* 1170

Nanoparticles: Cancer Management Applications / *Shyamasree Ghosh, Soham Saha, and
Abhinav Sur* 1182

Natural Polymers: Tissue Engineering / *Kunal Pal, Sai Sateesh Sagiri, Vinay K. Singh, Beauty Behera,
Indranil Banerjee, and Krishna Pramanik.* 1206

Nerve Guides: Multi-Channeled Biodegradable Polymer Composite / *Wesley N. Sivak,
Trent M. Gause, and Kacey G. Marra* 1235

Neural Tissue Engineering: Polymers for / *Ashok Kumar and Tanushree Vishnoi* 1255

Non-Viral Delivery Vehicles / *Ernst Wagner.* 1272

Orthopedic Applications: Bioceramic and Biopolymer Nanocomposite Materials / *Clark E. Barrett,
Ruth E. Cameron, and Serena M. Best.* 1276

Peptide–Polymer Conjugates: Synthetic Design Strategies / *Sabrina Dehn, Robert Chapman,
Katrina A. Jolliffe, and Sébastien Perrier* 1289

Pharmaceutical Polymers / *Narendra Pal Singh Chauhan, Arpit Kumar Pathak, Kumudini Bhanat,
Rakshit Ameta, Manish Kumar Rawal, and Pinki B. Punjabi* 1304

Polyurethanes: Medical Applications / *Michael Szycher* 1318

Rapid Prototyping / *Shaochen Chen, Carlos A. Aguilar, and Yi Lu* 1342

Rapid Prototyping: Tissue Engineering / *Giovanni Vozzi and Arti Ahluwalia* 1350

Scaffolds, Polymer: Microfluidic-Based Polymer Design / *Mohana Marimuthu and Sanghyo Kim.* 1362

Scaffolds, Porous Polymer: Tissue Engineering / *Guoping Chen* 1374

Semiconducting Polymer Dot Bioconjugates / *Garima Ameta* 1382

Sensing and Diagnosis / *Mizuo Maeda.* 1393

Shape Memory Polymers: Applications / *Meymanant Sadat Mohsenzadeh, Mohammad Mazinani, and
Seyed Mojtaba Zebarjad.* 1400

Skin Tissue Engineering / *Ilaria Armentano, Luigi Torre, Josè Maria Kenny.* 1408

Smart Polymers: Imprinting / *Carmen Alvarez-Lorenzo, Angel Concheiro, Jeffrey Chuang, and
Alexander Yu. Grosberg* 1424

Smart Polymers: Medicine and Biotechnology Applications / *Allan S. Hoffman* 1443

Stem Cell: Hematopoietic Stem Cell Culture, Materials for / *Akon Higuchi, Siou-Ting Yang, Pei-Tsz Li,
Ta-Chun Kao, Yu Chang, Yung Hung Chen, and S. Suresh Kumar.* 1453

Stents: Endovascular / *G. Lawrence Thatcher.* 1465

Stimuli-Responsive Materials / *Quazi T. H. Shubhra and A.K.M. Moshiul Alam.* 1480

Stimuli-Responsive Materials: Thermo- and pH-Responsive Polymers for Drug Delivery /
Jasbir Singh and Harmeet Kaur. 1493

Sutures / *Chih-Chang Chu* 1514

Theranostics: Biodegradable Polymer Particles for / *Naveed Ahmed, Nasir M. Ahmad, Asad Ullah Khan,
Haseeb Shaikh, Asghari Maqsood, and Abdul Hamid Elaissari.* 1530

Ultrasound Contrast Agents: Sonochemical Preparation / *Garima Ameta, Kiran Meghwal,
Arpita Pandey, Narendra Pal Singh Chauhan, and Pinki B. Punjabi* 1545

Vascular Grafts: Biocompatibility Requirements / *Shawn J. Peniston and Georgios T. Hilas* 1560

Vascular Grafts: Polymeric Materials / *Alexandre F. Leitão, Ivone Silva, Miguel Faria, and Miguel Gama* . . 1575

Vascular Tissue Engineering: Polymeric Biomaterials / *George Fercana and Dan Simionescu* 1596

Wound Care: Natural Biopolymer Applications / *Soheila S. Kordestani* 1607

Wound Care: Skin Tissue Regeneration / *Soonmo Choi, Deepti Singh, and Sung Soo Han* 1620

Wound Healing: Hemoderivatives and Biopolymers / *Silvia Rossi, Franca Ferrari, Giuseppina Sandri,
Maria Cristina Bonferoni, Claudia Del Fante, Cesare Perotti, and Carla Caramella* 1642

Zwitterionic Polymeric Materials / *Vinod B. Damodaran, Victoria Leszczak, Melissa M. Reynolds, and
Ketul C. Popat* 1661

Topical Table of Contents

Adhesives and Bone Cements

Adhesives: Tissue Repair and Reconstruction / *A. Lauto, H. Ruprai, and J.M. Hook* 1
Bioadhesive Drug Delivery Systems / *Ryan F. Donnelly and A. David Woolfson* 111
Glues / *Shojiro Matsuda and Yoshita Ikada* 644

Bioactive Systems

Anti-Infective Biomaterials / *W. Kohnen and B. Jansen* 83
Bioactive Systems / *Mikhail I. Shtilman* 105

Biocompatibility

Biocompatibility: Testing / *Joel D. Bumgardner, Marcia Vasquez-Lee, Keertik S. Fulzele, Daniel H. Smith,
Kellye D. Branch, Seth I. Christian, and David L. Williams* 129
Hemocompatible Polymers / *Young Ha Kim, Ki Dong Park, and Dong Keun Han* 663
Wound Healing: Hemoderivatives and Biopolymers / *Silvia Rossi, Franca Ferrari, Giuseppina Sandri,
Maria Cristina Bonferoni, Claudia Del Fante, Cesare Perotti, and Carla Caramella* 1642

Biodegradable Polymers, Fibers, and Composites

Biodegradable Polymers: Bioerodible Systems for Controlled Drug Release / *Jorge Heller and
Kirk P. Andriano* 139
Biodegradable Polymers: Biomedically Degradable Polymers / *A.-C. Albertsson and A. Löfgren* 153
Biodegradation / *Mikhail I. Shtilman* 162
Theranostics: Biodegradable Polymer Particles for / *Naveed Ahmed, Nasir M. Ahmad,
Asad Ullah Khan, Haseeb Shaikh, Asghari Maqsood, and Abdul Hamid Elaissari* 1530

Biomimetic Materials

Biomimetic Materials / *Kimon Alexandros Leonidakis, Al Halifa Soultan, Susanna Piluso,
Abhijith Kundadka Kudva, Laurien Van den Broeck, Taíla O. Meiga, Rory L.D. Gibney, and
Jennifer Patterson.* 189
Biomimetic Materials: Smart Polymer Surfaces for Tissue Engineering / *João F. Mano* 214

Biorubber and Elastomers

Biorubber: Poly(Glycerol Sebacate) / *Yadong Wang* 229

Body Systems

Artificial Muscles / *Mohsen Shahinpoor* 91
Blood Vessel Substitutes / *Jan P. Stegeman, Shaneen L. Rowe, and Robert M. Nerem* 237
Bone–Implant: Polymer Physicochemical Modification for Osseointegration /
Waleed S. W. Shalaby and Bruce L. Anneaux. 248
Contact Lenses: Gas Permeable / *Jay F. Künzler* 362

Body Systems (cont'd.)

Corneas: Tissue Engineering / *Dalia A.M. Hamza, Tamer A. E. Ahmed, and Maxwell T. Hincke.* 370

Dental Polymers: Applications / *Narendra P.S. Chauhan, Kiran Meghwal, Pinki B. Punjabi, Jyoti Chaudhary, and Paridhi Kataria* 411

Dental Sealants / *Yasuhiko Tsuchitani and Tohru Wada* 433

Hair and Skin Care Biomaterials / *Bernard R. Gallot* 653

Ligament Replacement Polymers: Biocompatability, Technology, and Design / *Martin Dauner and Heinrich Planck.* 757

Muscles, Artificial: Sensing, Transduction, Feedback Control, and Robotic Applications / *Mohsen Shahinpoor, Kwang J. Kim, and Mehran Mojarrad* 978

Nerve Guides: Multi-Channeled Biodegradable Polymer Composite / *Wesley N. Sivak, Trent M. Gause, and Kacey G. Marra* 1235

Orthopedic Applications: Bioceramic and Biopolymer Nanocomposite Materials / *Clark E. Barrett, Ruth E. Cameron, and Serena M. Best.* 1276

Skin Tissue Engineering / *Ilaria Armentano, Luigi Torre, Josè Maria Kenny.* 1408

Stents: Endovascular / *G. Lawrence Thatcher.* 1465

Vascular Grafts: Polymeric Materials / *Alexandre F. Leitão, Ivone Silva, Miguel Faria, and Miguel Gama.* 1575

Vascular Tissue Engineering: Polymeric Biomaterials / *George Fercana and Dan Simionescu.* 1596

Wound Care: Skin Tissue Regeneration / *Soonmo Choi, Deepti Singh, and Sung Soo Han* 1620

Chemistry, Properties, and Polymer Architectures

Biofunctional Polymers / *Jennifer. L. West* 175

Conjugates: Biosynthetic–Synthetic Polymer Based / *Jan C. M. van Hest.* 340

Dendritic Architectures: Delivery Vehicles / *Saeed Jafarirad.* 395

Dendritic Architectures: Theranostic Applications / *Saeed Jafarirad* 402

Functionalized Surfaces: Biomolecular Surface Modification with Functional Polymers / *Florence Bally and Aftin M. Ross.* 585

Magnetomicelles: Theranostic Applications / *Prashant K. Deshmukh, Abhijeet P. Pandey, Surendra G. Gattani, and Pravin O. Patil* 778

Metal–Polymer Composite Biomaterials / *Takao Hanawa* 877

Molecular Assemblies / *Saeed Jafarirad.* 932

Nanocomposite Polymers: Functional / *Radhakrishnan Sivakumar and Sambandam Anandan* 994

Nanocomposite Polymers: Functional / *Radhakrishnan Sivakumar and Sambandam Anandan* 994

Peptide–Polymer Conjugates: Synthetic Design Strategies / *Sabrina Dehn, Robert Chapman, Katrina A. Jolliffe, and Sébastien Perrier* 1289

Semiconducting Polymer Dot Bioconjugates / *Garima Ameta* 1382

Devices, Implants, Grafts, and Prosthetics

Bone–Implant: Polymer Physicochemical Modification for Osseointegration / *Waleed S. W. Shalaby and Bruce L. Anneaux.* 248

Hydrogels: Multi-Responsive Biomedical Devices / *Francesca Iemma, Giuseppe Cirillo, and Umile Gianfranco Spizzirri.* 699

Medical Devices and Preparative Medicine: Polymer Drug Application / *M.R. Aguilar, L. García-Fernández, M.L. López-Donaire, F. Parra, L. Rojo, G. Rodríguez, M.M. Fernández, and J. San Román* 789

Membranes, Polymeric: Biomedical Devices / *Shahram Mehdipour-Ataei and Maryam Oroujzadeh* . . . 864

Stents: Endovascular / *G. Lawrence Thatcher.* 1465

Vascular Grafts: Biocompatibility Requirements / *Shawn J. Peniston and Georgios T. Hilas.* 1560

Vascular Grafts: Polymeric Materials / *Alexandre F. Leitão, Ivone Silva, Miguel Faria, and Miguel Gama.* 1575

Drugs and Drug Delivery

Anticancer Agents: Polymeric Nanomedicines / *Rong Tong and Jianjun Cheng*. 58
Colloid Drug Delivery Systems / *Monzer Fanun*. 301
Drug Delivery Systems: Selection Criteria and Use / *Ravindra Semwal, Ruchi Badoni Semwal, and
    Deepak Kumar Semwal*. 439
Drugs and Excipients: Polymeric Interactions / *James C. DiNunzio and James W. McGinity*. 451
Electrospinning Technology: Polymeric Nanofiber Drug Delivery / *Narendra Pal Singh Chauhan,
    Kiran Meghwal, Priya Juneja, and Pinki B. Punjabi*. 491
Excipients: Pharmaceutical Dosage Forms / *Luigi G. Martini and Patrick Crowley*. 547
Fluorescent Nanohybrids: Cancer Diagnosis and Therapy / *Herman S. Mansur and
    Alexandra A. P. Mansur*. 560
Mucoadhesive Polymers: Basics, Strategies, and Trends / *Andreas Bernkop-Schnürch*. 941
Mucoadhesive Systems: Drug Delivery / *Ryan F. Donnelly, Elizabeth Ryan,
    Thakur Raghu Raj Singh, and A. David Woolfson*. 961
Nanoparticles: Biomaterials for Drug Delivery / *Abhijit Gokhale, Thomas Williams, and
    Jason Vaughn*. 1170
Nanoparticles: Cancer Management Applications / *Shyamasree Ghosh, Soham Saha,
    Abhinav Sur, and K. V. S. Girish*. 1182
Non-Viral Delivery Vehicles / *Ernst Wagner*. 1272
Pharmaceutical Polymers / *Narendra Pal Singh Chauhan, Arpit Kumar Pathak, Kumudini Bhanat,
    Rakshit Ameta, Manish Kumar Rawal, and Pinki B. Punjabi*. 1304
Stimuli-Responsive Materials: Thermo- and pH-Responsive Polymers for Drug Delivery /
    Jasbir Singh and Harmeet Kaur. 1493

Electroactive and Ionic Materials

Conducting Polymers: Biomedical Engineering Applications / *Smritimala Sarmah and
    Raktim Pratim Tamuli*. 312
Electrets / *Rajesh Kalia and Sapna Kalia*. 476
Electroactive Polymeric Materials / *V. Prasad Shastri*. 482
Nanomaterials: Conducting Polymers and Sensing / *Murat Ates*. 1035
Zwitterionic Polymeric Materials / *Vinod B. Damodaran, Victoria Leszczak, Melissa M. Reynolds,
    and Ketul C. Popat*. 1661

Fabrication Techniques and Processing

BioMEMS / *L. James Lee*. 181
Electrospinning Technology: Regenerative Medicine / *Toby D. Brown, Cedryck Vaquette,
    Dietmar W. Hutmacher, and Paul D. Dalton*. 506
Latexes: Magnetic / *Abdel Hamid Elaissari, Raphael Veyret, Bernard Mandrand, and Jhunu Chatterjee*. 743
Melt Extrusion: Pharmaceutical Applications / *James DiNunzio, Seth Forster, and Chad Brown*. 827
Melt-Electrospun Fibers / *Aleksander Góra, Rahul Sahay, Velmurugan Thavasi, and
    Seeram Ramakrishna*. 845
Rapid Prototyping / *Shaochen Chen, Carlos A. Aguilar, and Yi Lu*. 1342
Ultrasound Contrast Agents: Sonochemical Preparation / *Garima Ameta, Kiran Meghwal,
    Arpita Pandey, Narendra Pal Singh Chauhan, and Pinki B. Punjabi*. 1545

Gels

Aerogels: Cellulose-Based / *F. Liebner, N. Pircher, C. Schimper, E. Haimer, and T. Rosenau*. 19
Gels: Fibrillar Fibrin / *Erin Grassl and Robert T. Tranquillo*. 616
Hydrogels / *Junji Watanabe, Yoshihiro Kiritoshi, Kwang Woo Nam, and Kazuhiko Ishihara*. 674

Gels (cont'd.)

Hydrogels: Classification, Synthesis, Characterization, and Applications /
Kaliappa Gounder Subramanian and Vediappan Vijayakumar 685
Hydrogels: Multi-Responsive Biomedical Devices / *Francesca Iemma, Giuseppe Cirillo, and*
Umile Gianfranco Spizzirri. 699
Microgels: Smart Polymer and Hybrid / *Zahoor H. Farooqi and Mohammad Siddiq* 917
Nanogels: Chemical Approaches to Preparation / *Sepideh Khoee and Hamed Asadi* 1007

Gene Delivery and Gene Therapy

Gene Carriers: Design Elements / *Jong-Sang Park and Joon Sig Choi*. 623
Gene Delivery / *Tetsuji Yamaoka* 633
Lipoplexes and Polyplexes: Gene Therapy / *Diana Rafael, Fernanda Andrade, Alexandra Arranja,*
Sofia Luís, and Mafalda Videira 765

Immobilization and Sensors

Fluorescent Nanohybrids: Cancer Diagnosis and Therapy / *Herman S. Mansur and*
Alexandra A. P. Mansur. 560
Muscles, Artificial: Sensing, Transduction, Feedback Control, and Robotic Applications /
Mohsen Shahinpoor, Kwang J. Kim, and Mehran Mojarrad. 978
Nanomaterials: Conducting Polymers and Sensing / *Murat Ates* 1035
Sensing and Diagnosis / *Mizuo Maeda*. 1393
Theranostics: Biodegradable Polymer Particles for / *Naveed Ahmed, Nasir M. Ahmad,*
Asad Ullah Khan, Haseeb Shaikh, Asghari Maqsood, and Abdul Hamid Elaissari 1530

Micro- and Nanotechnology

Conjugated Polymers: Nanoparticles and Nanodots of / *Garima Ameta, Suresh C. Ameta,*
Rakshit Ameta, and Pinki B. Punjabi. 327
Microcomponents: Polymeric / *Can Yang and Xiao-Hong Yin* 900
Microgels: Smart Polymer and Hybrid / *Zahoor H. Farooqi and Mohammad Siddiq* 917
Nanocomposite Polymers: Functional / *Radhakrishnan Sivakumar and Sambandam Anandan* 994
Nanogels: Chemical Approaches to Preparation / *Sepideh Khoee and Hamed Asadi* 1007
Nanomaterials: Conducting Polymers and Sensing / *Murat Ates* 1035
Nanomaterials: Theranostics Applications / *M. Sasidharan, S. Anandhakumar, and S. Vivekananthan* . . 1060
Nanomaterials: Therapeutic Applications / *Dominik Witzigmann, Marine Camblin, Jörg Huwyler, and*
Vimalkumar Balasubramanian. 1073
Nanomedicine: Review and Perspectives / *Ali Zarrabi, Arezoo Khosravi, and Ali Hashemi* 1088
Nanoparticles: Biomaterials for Drug Delivery / *Abhijit Gokhale, Thomas Williams, and*
Jason Vaughn. 1170
Nanoparticles: Cancer Management Applications / *Shyamasree Ghosh, Soham Saha, and*
Abhinav Sur. 1182

Natural Sources

Aerogels: Cellulose-Based / *F. Liebner, N. Pircher, C. Schimper, E. Haimer, and T. Rosenau*. 19
Cellulose-Based Biopolymers: Formulation and Delivery Applications / *J.D.N. Ogbonna,*
F.C. Kenechukwu, S.A. Chime, and A.A. Attama 270
Natural Polymers: Tissue Engineering / *Kunal Pal, Sai Sateesh Sagiri, Vinay K. Singh, Beauty Behera,*
Indranil Banerjee, and Krishna Pramanik 1206

Natural Sources (cont'd.)

Peptide–Polymer Conjugates: Synthetic Design Strategies / *Sabrina Dehn, Robert Chapman, Katrina A. Jolliffe, and Sébastien Perrier* 1289

Shape Memory Polymers: Applications / *Meymanant Sadat Mohsenzadeh, Mohammad Mazinani, and Seyed Mojtaba Zebarjad* 1400

Smart Polymers: Imprinting / *Carmen Alvarez-Lorenzo, Angel Concheiro, Jeffrey Chuang, and Alexander Yu. Grosberg* 1424

Smart Polymers: Medicine and Biotechnology Applications / *Allan S. Hoffman* 1443

Stimuli-Responsive Materials / *Quazi T. H. Shubhra and A.K.M. Moshiul Alam* 1480

Stimuli-Responsive Materials: Thermo- and pH-Responsive Polymers for Drug Delivery / *Jasbir Singh and Harmeet Kaur* 1493

Synthetic and Custom-Designed Materials

Polyurethanes: Medical Applications / *Michael Szycher* 1318

Tissue Engineering

Adhesives: Tissue Repair and Reconstruction / *A. Lauto, H. Ruprai, and J.M. Hook* 1

Bioabsorbable Polymers: Tissue Engineering / *K. J. L. Burg and Waleed S. W. Shalaby* 101

Corneas: Tissue Engineering / *Dalia A.M. Hamza, Tamer A. E. Ahmed, and Maxwell T. Hincke* 370

***In Vitro* Vascularization: Tissue Engineering Constructs** / *Cai Lloyd-Griffith, Tara M. McFadden, Garry P. Duffy, and Fergal J. O'Brien* 723

Muscles, Artificial: Sensing, Transduction, Feedback Control, and Robotic Applications / *Mohsen Shahinpoor, Kwang J. Kim, and Mehran Mojarrad* 978

Natural Polymers: Tissue Engineering / *Kunal Pal, Sai Sateesh Sagiri, Vinay K. Singh, Beauty Behera, Indranil Banerjee, and Krishna Pramanik* 1206

Neural Tissue Engineering: Polymers for / *Ashok Kumar and Tanushree Vishnoi* 1255

Rapid Prototyping: Tissue Engineering / *Giovanni Vozzi and Arti Ahluwalia* 1350

Scaffolds, Polymer: Microfluidic-Based Polymer Design / *Mohana Marimuthu and Sanghyo Kim* 1362

Scaffolds, Porous Polymer: Tissue Engineering / *Guoping Chen* 1374

Skin Tissue Engineering / *Ilaria Armentano, Luigi Torre, Josè Maria Kenny* 1408

Stem Cell: Hematopoietic Stem Cell Culture, Materials for / *Akon Higuchi, Siou-Ting Yang, Pei-Tsz Li, Ta-Chun Kao, Yu Chang, Yung Hung Chen, and S. Suresh Kumar* 1453

Vascular Tissue Engineering: Polymeric Biomaterials / *George Fercana and Dan Simionescu* 1596

Wound Care: Skin Tissue Regeneration / *Soonmo Choi, Deepti Singh, and Sung Soo Han* 1620

Wound Healing Applications

Sutures / *Chih-Chang Chu* 1514

Wound Care: Natural Biopolymer Applications / *Soheila S. Kordestani* 1607

Wound Care: Skin Tissue Regeneration / *Soonmo Choi, Deepti Singh, and Sung Soo Han* 1620

Wound Healing: Hemoderivatives and Biopolymers / *Silvia Rossi, Franca Ferrari, Giuseppina Sandri, Maria Cristina Bonferoni, Claudia Del Fante, Cesare Perotti, and Carla Caramella* . . 1642

Preface

This compact desk reference includes carefully selected topics from the print and online versions of the multivolume *Encyclopedia of Biomedical Polymers and Polymeric Biomaterials*. This distillation was skillfully assembled in broad subject area of polymer applications in the medical field, which will enable readers to have an enriching experience in general as well as to gain targeted knowledge in this evolving area. This concise encyclopedia is designed to serve as a ready-reference guide for the researchers in the field, and is expected to provide quick access to some recent advances in the area.

This handy reference work is designed for novices to experienced researchers; it caters to engineers and scientists (polymer and materials scientists, biomedical engineers, biochemists, molecular biologists, and macromolecular chemists), pharmacists, doctors, cardiovascular and plastic surgeons, and students, as well as general readers in academia, industry, and research institutions.

I feel honored to undertake the important and challenging endeavor of developing the ***Concise Encyclopedia of Biomedical Polymers and Polymeric Biomaterials*** that will cater to the needs of many who are working in the field. I would like to express my sincere gratitude and appreciation to the authors for their excellent professionalism and dedicated work. I would like to thank the entire management of encyclopedia program of Taylor & Francis Group (T&F - CRC Press) and particularly to Ms. Megan Hilands who made this possible. Her hard work and professionalism made this a wonderful experience for all parties involved.

I take the opportunity to express my appreciation to my wife Bidu Mishra, Ph.D. for her encouragement, sacrifice, and support, during weekends, early mornings, and holidays spent on this project. Without their help and support, this project would have never been started nor completed.

Munmaya K. Mishra
Editor-in-Chief

About the Editor-in-Chief

Munmaya Mishra, Ph.D. is a polymer scientist who has worked in the industry for more than 30 years. He has been engaged in research, management, technology innovations, and product development. He has contributed immensely to multiple aspects of polymer applications, including biomedical polymers, encapsulation and controlled release technologies, etc. He has authored and coauthored hundreds of scientific articles is the author or editor of multiple books, and hold over 50 U.S. patents, over 50 U.S. patent-pending applications, and over 150 world patents. He has received numerous recognitions and awards. He is the editor-in-chief of three renowned polymer journals and the editor-in-chief of the recently published multivolume *Encyclopedia of Biomedical Polymers and Polymeric Biomaterials*. He is also the founder of new scientific organization "International Society of Biomedical Polymers and Polymeric Biomaterials." In 1995 he founded and established a scientific meeting titled "Advanced Polymers via Macromolecular Engineering," which has gained international recognition and is still being held under the sponsorship of the IUPAC organization.

Concise Encyclopedia of Biomedical Polymers and Polymeric Biomaterials

Volume I
Adhesives through Medical
Pages 1–826

Adhesives—Bioabsorbable

Bioactive—BioMEMS

Biomimetic—Colloid

Conducting—Dendritic

Dental—Electrospinning

Excipients—Gels

Gene—Hydrogels

In Vitro—Medical

Adhesives: Tissue Repair and Reconstruction

A. Lauto
Biomedical Engineering and Neuroscience Research Group, The MARCS Institute, Western Sydney University, Penrith, Australia

H. Ruprai
School of Science and Health, Western Sydney University, Penrith, Australia.

J.M. Hook
Mark Wainwright Analytical Centre, University of New South Wales, Sydney, Australia.

Abstract
In recent years, new modalities of tissue repair have emerged replacing traditional wound closure techniques such as suturing. Biocompatible adhesives are at the forefront of this "surgical revolution" as fibrin, polyethylene glycol and cyanoacrylate glues are currently used in clinical procedures. Other experimental adhesives appear to be very promising; in particular, those that mimic natural mechanisms of adhesion and are light-activated or incorporate nanoparticles and nanostructures.

INTRODUCTION

Sutures, staples and clips are the most common clinical techniques for tissue repair as they are generally reliable in preventing tissue dehiscence and stabilizing the wound. However, they have significant drawbacks because they are unable to seal blood vessels or airways effectively[1,2] and are invasive to the host tissue causing scar formation,[3] foreign body reactions [4] and post-operative pain.[5] For these reasons, clinical and experimental studies have recently focused on evaluating whether tissue adhesives, whch are biocompatible, biodegradable and less invasive, can substitute traditional wound closure techniques. Bioadhesives have the potential to repair and seal tissue effectively without sutures and, at the same time, deliver drugs or biological factors to enhance and accelerate wound healing.[6] A variety of bioadhesives have been developed to date, the majority of which are still in an experimental phase. Among these, the catechol derived ones are of particular interest as they can stick strongly to the tissue in wet environments; polyethylene glycol-based glues are also promising because of their degradability and their light curing ability. Polysaccharide adhesives such as dextran or chitosan have been recently explored by researchers because of their excellent biocompatibility, mechanical properties, and bonding strength, especially when light-activated. The most recent technological frontier is the design and fabrication of nanostructured adhesives that incorporate either nanoparticles or surface pillars of nanoscale dimension (as in the Gecko foot) to increase adhesion to tissue. Along with the experimental adhesives mentioned here, a small number of biomaterials are already being used in surgical procedures for wound closure; at the present time the most clinically tested compounds are fibrin, cyanoacrylate, and polyethylene glycol glues. In this report, we review significant experimental and clinical adhesives, highlighting their chemical and photochemical bonding mechanisms. A paragraph is also dedicated to the most recent nanostructured adhesives, including the "Gecko-inspired" bioadhesives. Our concluding remarks outline future directions of research in the field.

CLINICAL STUDIES

Fibrin Glues

Fibrin glue is a plasma-derived adhesive that mimics the properties of physiological blood clots. It is usually made by mixing a solution of thrombin and calcium ions with fibrinogen and factor XIII. In this mixture, thrombin cleaves the fibrinogen chains into fibrin monomers and, in the presence of calcium ions, activates factor XIII that helps polymerize the monomers into a stable clot.[7] The crosslinking of the polymerized fibrin chains in the clot with extracellular proteins, such as collagen, is responsible for tissue bonding. The fibrin polymer is degraded and broken down into smaller polypeptide fragments by plasmin enzymes in ~2 weeks. The accumulation of these degradation products reduces the thrombin action by competitive inhibition and prevents polymerization of fibrin monomers; to slow down this process fibrinolysis inhibitors, such as aprotinin, may also be added.[8] The concentration of components used to make the glue also influence the properties of the clot that is formed. For example, adhesives with a high concentration of fibrinogen tend to produce stronger and slower forming clots than those with less fibrinogen and high concentration of thrombin.[9] Unfortunately, glue components derived from blood can provoke an immune response[10] and carry viruses if originated from human plasma.[11] In the clinical arena fibrin glues are approved colon sealants, adhesives for skin graft and facial flap attachment, and haemostat agents.[12] A meta-analysis of seven randomized controlled clinical trials[13–19] tested the efficacy of fibrin sealants in patients undergoing total knee arthroplasty, a surgery that is prone to large blood loss. This study[20] showed that patients treated with fibrin adhesives had statistically significant reductions in drainage volume (difference = −354.53 mL, n = 131), haemoglobin decline (difference = −0.72 g/dL, n = 169), and risk of wound hematoma (difference = −0.11%, n = 60). Incidences of deep vein thrombosis, pulmonary embolism, and wound infection were similar for both groups.

To further promote the surgical use of fibrin glues, clinical studies have focused on comparing the efficacy of the glue to other surgical techniques. In a prospective randomized clinical study (n = 41), aerosol fibrin glue was applied in one nasal cavity through endoscopic surgery and no morbidity was elicited. Patients also had a higher degree of satisfaction than with polyvinyl acetal sponge packing, which was applied in the opposite cavity.[21] Patients undergoing conjunctival closure with fibrin glues after a trabeculectomy also had less discomfort than with suturing (n = 28), which required more time to complete the procedure.[22] However, dehiscence (n = 2) occurred only in the fibrin group, highlighting the need for a stronger glue.

Comparative studies of fibrin glues for mesh reinforcement in inguinal hernia repairs is a predominant area of clinical research. A meta-analysis of four randomized controlled trials[23–26] and five prospective observational clinical studies[27–31] showed that mesh fixation with fibrin glues in open inguinal hernioplasty resulted in lower incidences of chronic pain and hematoma/seroma when compared to sutures (n = 806).[32] However, the recurrence of urinary problems as a result of the two procedures was comparable. Another meta-analysis that included five studies[33–37] compared mesh reinforcement between tacks/staples and fibrin glue after laparoscopic inguinal hernioplasty. The study concluded that there was no significant difference between the two procedures in relation to surgery time, postoperative complications, risk for hernia recurrence, and hospital stay.[38] Nonetheless, mesh fixation with tacks/staples was associated with a higher risk of developing chronic inguinal pain compared to using fibrin glue (risk ratio = 4.6, n = 306).

Fibrin glues may also enhance early wound healing by reducing inflammation and promoting fibroblast growth:[39] when the glue was applied in the mouth of 10 patients after periodontal flap surgery, it resulted in significantly lower number of inflammatory cells, blood vessel formation, and higher fibroblast density than in suturing.[40] Aprotinin may be responsible for the minimal inflammatory response as it reduces the formation of fibrin degradation products, which in turn activate macrophages.[41]

Cyanoacrylate Glues

Cyanoacrylates are synthetic glues composed of cyanoacrylic esters with the general formula of $CH_2=C(CN)CO_2R$, where R represents an alkyl group. Cyanoacrylates bind to the skin and other organs through exothermic polymerization when it comes into contact with the weak basic surface of tissue. These glues are broken down by hydrolysis; in this respect the short alkyl chained cyanoacrylates (R < 4 carbon chains) are not used clinically because of their quick degradation into cyanacetate and formaldehyde, which leads to cytotoxic or inflammatory effects in the area of application.[42] The long alkyl chained (e.g., butyl, octyl, and hexyl) cyanoacrylates are currently used for tissue repair with applications mostly in dermatology, plastic surgery, and emergency treatments.[43]

Clinical studies are focused on determining the efficacy of long alkyl chained cyanoacrylates; when healing first degree lacerations, these glues can achieve cosmetic and functional results similar to traditional sutures but result in less pain for patients and shorter procedure time (2.3 vs 7.9 min, n = 74).[44] A similar result was observed in patients undergoing open or laparoscopic inguinal hernioplasty, where the glue (n = 50) was quicker than sutures by 12 minutes (n = 52).[45] Cyanoacrylate glues also resulted in less consumption of analgesics and no post-operative morbidities. In spite of these encouraging results, cyanoacrylates are mostly only applied on external wounds at this moment in time. A clinical study (n = 315) showed that infection from skin closure after spinal surgery occurred only in patients (n = 8) with staples rather than in those with cyanoacrylate.[46] The study also noted that it was faster and cheaper to use cyanoacrylate than staples in the closure. However, there have been reports of cases with inflammatory reactions when cyanoacrylate is applied inside the abdominal cavity[47, 48] and reports of dermatologic, neurologic, allergic, and respiratory conditions developed from the use of the glue in the workplace.[49] Long chained cyanoacrylates may also result in non-degradable polymers which can cause infection,[50] raising safety concerns in internal organ applications.

Cyanoacrylate glues have stronger adhesive properties ($\sim$6.8 N/cm^2) than fibrin ($\sim$1.3 N/cm^2) and polyethylene glues (0.4–1.7 N/cm^2).[51] A randomized clinical study found that this glue has a tensile strength comparable to 5-0 sutures applied on facial skin (n = 20).[52] Another study showed the glue resulted in a lower margin separation than absorbable sutures when closing laparoscopic wounds.[53] However, cyanoacrylates are not as effective when wounds are subjected to high tension: a study comparing the wound closure of groin incisions in children with absorbable sutures and cyanoacrylate glue (n = 50) found dehiscence only in the latter group (26%).[54]

Polyethylene Glycol (PEG) Glues

PEG-based glues are highly biocompatible hydrogels that are used for tissue repair; they consist of two main components: a solution with a chemical or photoactive crosslinker that allows tissue bonding and a PEG polymer solution that is responsible for the mechanical properties of the glue. When these solutions are combined and activated, an adherent hydrogel is formed by covalent bonding between the sealant macromers and the crosslinker molecules. These glues degrade by hydrolysis and form products that can be easily cleared by the kidney or locally metabolized. PEG glues have gained clinical interest as they are more biocompatible than cyanoacrylates, avoid any risk of protein immunogenicity associated with fibrin glues, and can be used for drug delivery.[55]

FocalSeal-L is a PEG sealant approved by the Food and Drug Administration (FDA) for air leaks in lung surgery. A multicentre trial found FocalSeal-L to be better at preventing air leaks than standard techniques (control group), which included suturing, stapling, or tissue grafts (n = 117); 92% versus 29% of patients did not develop intraoperative air leaks after lung resections.[56] Also in the sealant group, 39% patients remained free of air leaks between the time of skin closure and hospital discharge, compared with only 11% in the control group. Postoperative morbidities were similar between the two groups. FocalSeal-L is a photopolymerizable PEG-based hydrogel, comprising two components: an eosin-based primer that is brushed into the tissue followed by the second component that consists of PEG molecules, biodegradable polylactic acid, thimethylene carbonate, and a polymerizable acrylic ester.[57] Polymerization of the glue components is initiated by irradiation with blue–green light; the gel usually degrades in the body after ~8 weeks. The main drawbacks of FocalSeal-L are the two-step preparation procedure, which requires a dry environment to prevent washing out the glue, and possible infections as cautioned by the FDA.[58]

Another type of a PEG glue consists of a PEG ester and trilysine amine; when these components are mixed, they crosslink to form a watertight hydrogel seal, which adheres to the tissue mainly by mechanical means. A multicentre clinical study found that low-swell PEG-trilysine hydrogels are safe and effective for watertight dural closure causing no neurological deficits in patients undergoing spinal surgery.[59] The study showed patients (n = 74) treated with this PEG hydrogel had a higher rate of watertight closure (98%) than the controls (n = 24), which comprised sutures and fibrin glue (79%). No statistical differences were seen in postoperative cerebrospinal fluid leakage, infection, and wound healing. The glue was formulated to last in the body 9–12 weeks with 12% of swelling ratio; swelling is a critical factor because it may cause nerve compression in patients.

A third type of PEG-based sealant is the PEG–PEG system that consists of a glutaryl-succinimidyl ester and a thiol-terminated PEG. The thiol groups and the carbonyl groups of the succinimidyl ester form covalent bonds with the PEG molecules after mixing; this compound in turn reacts with amine groups in the tissue matrix and creates tissue adhesion. The PEG glue is usually resorbed in the body within 4 weeks. A study found that PEG–PEG based sealants improved intraoperative and postoperative management of anastomotic bleeding in Bentall procedures.[60] Patients operated with the sealant (n = 48) required fewer transfusions than the control group (n = 54), who did not have this additive treatment to the suture line. The sealant group also had less postoperative drainage loss than the control group. A poly(ethylene glycol) diacrylate hydrogel has been recently developed to support cartilage matrix production, with easy surgical application. In a pilot study, patients with 2–4 cm^2 defects on the medial femoral condyle received the hydrogel implant that was photopolymerized with ultraviolet light for 4 min (λ = 365 nm, I = 5 mW/cm^2). After 6 months, the patients showed integration of repair tissue with host tissue; in particular, 100% integration in 7 of the 14 patients whereas the remaining patients had small gaps (<2 mm). Treated patients also had less pain compared to controls.[61] It should be remarked that this study did not capture possible long-term bone abnormalities. PEG glues are usually stronger than fibrin glues but weaker than cyanoacrylate adhesives; the low mechanical integrity of PEG hydrogels is a further limit that usually confines their application to tissues under low tension (Fig. 1).[62]

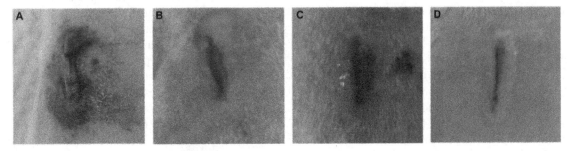

Fig. 1 Skin incisions on back of the rats that were just closed with (**A**) suture, (**B**) fibrin glue, (**C**) Chitosan–PEG–tyramine hydrogel and (**D**) cyanoacrylate glue.
Source: From Lih et al.[94] © 2012, Elsevier. Reprinted with permission

EXPERIMENTAL ADHESIVES

In this part of the review, we focus on building blocks used in the synthesis and chemical activation of materials that have found use as adhesives in biomedical research. Adhesion occurs when two materials are brought into contact, with the aim of keeping them together. Another approach is to design a matrix that allows living cells to inhabit and bind to another bodily tissue, as in the repair of bone, tendon, tooth, or bladder, followed by the degradation of the matrix, after it has delivered its contents, allowing the body to harness regrowth. Many of them remain experimental, with a great deal of research still to be performed before the stage of clinical application. Adhesives have been inspired by the observation of the manner in which particular creatures adhere to surface. At one end of the spectrum is the gecko, capable of sticking to walls and ceilings while moving, whereas at the other end are mussels, barnacles, and oysters, definitely anchored so that they don't move in the tidal zone. So both display adhesion to surfaces but with distinctly different means of attachment, one with nanopatterning pillars and the other with discretely functionalized and carefully orchestrated proteins.

Chemistry and Chemical Activation

One most outstanding observation is of the mussel, which attaches itself to rocks buffeted by waves in a tidal environment. It does so by anchoring itself with strands comprised of a group of proteins rich in 1,2-dihydroxybenzenes derived from the catechol, DOPA. Such a functional group is capable of several distinct chemical actions promoting attachment: acid–base, reacting as a weak acid; oxidation–reduction, as partners with the quinone form; and complexation either to metal ions like aluminium or iron, or to oxidized surfaces, such as silicon dioxide or titanium dioxide. As we shall see, each of these behaviours has been explored in depth so as to develop new materials for applications in biomedicine and tissue engineering. What follows highlights some recent efforts and progress in the use of polymers, copolymers, polymer composites, and nanocomposites bearing the catechol group. Dopamine itself can be polymerized under very mild conditions in aqueous buffer to give polydopamine, which has a tremendous capacity to adhere to a wide variety of materials and surfaces and is the subject of recent reviews,[63,64] including biomaterials applications.[65] For example, polyethylene terephthalate is being investigated as the basis of an artificial graft for ligament reconstruction, but it requires pretreatment with dopamine and calcium phosphate solutions for integration. This forms a polydopamine–hydroxyapatite surface coating suitable for osteogenic compatibility, that gives very promising results in vitro and in vivo.[66] In other research, polydopamine has been used successfully as an anchoring pattern and support for stem and bacterial cells,[67,68] being biostable and biocompatible,

which shows that it has excellent adhesive properties both in relation to the living, implanted cells and the underlying, inert support. Likewise, polydopamine readily adheres to metal used for stents, and this polymer coating can then be further covalently modified with heparin, to reduce thrombosis from the implantation of such a device.[69] Considerations of this kind in the context of tissue repair using living cells supported by an appropriately constructed scaffold, have now appeared[70] and are a definite direction for future research. The incorporation of the catechol group into polymers has been a high priority for many researchers and remains so because of its unique capacity to adhere to a remarkable range of substrates, as already mentioned.[63] Its inclusion in the design of scaffolds for polymeric bioadhesives has been a key feature for promoting attachment, and therefore, adhesives of many varieties. A dental adhesive has been created to bind to the dentin surface of the tooth, utilizing a dopamine-methacrylamide-acrylate copolymer, which offered the desired features of flexibility, immiscibility, stability under wet (saliva) contamination, among other features.[71] Addition of Fe^{3+} to complex the catechol groups in the copolymer appeared to prevent the formation of defects at the tooth–adhesive interface, giving excellent interfacial coverage. Biocompatibility was tested with gingival fibroblast cells, which showed cell adhesion and cell proliferation in the presence of the copolymer, but further research in vivo needs development.[71]

PEG offers water miscibility, but, on its own, has little cohesive structure, or adhesion. Covalent attachment of the catechols group to the PEG, in combination with nanoparticles of Laponite, a synthetic nanosilicate, forms a composite which crosslinks further upon metaperiodate oxidation.[72] The resultant hydrogels have enhanced adhesive and mechanical properties, while remaining sufficiently fluid to be injectable.[72] Although they are yet to be tested in vivo, the hydrogels exhibited biocompatibility, after subcutaneous implantation in rats, and biodegradability. Alternatives to PEG-based catechol adhesives, have been found by using hyperbranched polyglycerol as a carrier for the catechol group.[73] They are readily combined and form crosslinks under mildly oxidizing conditions to produce stable biomaterials, which are of low toxicity that adhere to various metals and oxide surfaces used in biomedical applications, such as implants, that require antifouling surface coatings. Likewise, hyperbranched acrylate[74] and polyoxetane copolymers[75] have also been identified as suitable platforms for pendant catechol groups in bioadhesives. The hyperbranched acrylate features a tertiary amino core and catechol surface, which can function as a tissue adhesive under wet conditions. Activation and curing is required for optimum adhesion, using either fibrinogen or enzymic oxidation.[74] Polyoxetane copolymers can incorporate catechol groups through so-called "click" chemistry for covalently linking the alkyne in the polymer with a catechol azide, producing a stable catechol triazole embedded in the polymer. Activation for adhesion occurs when this polyoxetane copolymer is treated with Fe^{3+}, which may complex with the dihydroxybenzene (catechol) group, or to some extent, in this instance, through the triazole group.[75] The material displays strong adhesion to various substrates including porcine skin, but its compatibility with an aqueous environment needs additional research. Coordination of the catechol groups with metal ions, such as iron, is therefore useful for crosslinking and extending networks for adhesives in biomaterials research. This stems from the observation that the mussel foot, besides having high concentrations of DOPA, is also rich in metal ions, such as iron and copper. Catechol groups can bind to the Fe^{3+} state, which can then undergo reduction to form the ortho-quinone and react further with available amines or thiols or even with itself. It may also remain as the dihydroxybenzene–metal complex, without reduction.[76] The resultant hydrogels exhibit both adhesive and cohesive behaviour, in addition to featuring deformable and self-healing properties, which may be exploited in tissue adhesion and wound sealing.

Natural polymers, such as silk fibroin from the domesticated silk worm, have only recently been modified with groups of two kinds for distinct properties: catechols (as DOPA: 3,4-dihydroxyphenylalanine) to enhance adhesive performance and PEGs, to enhance water miscibility.[77] In this instance, the tendency of the silk fibroin to form ordered β-sheets could also be retained, while periodate oxidation is applied to induce further crosslinking. Lap shear testing clearly demonstrated that the inclusion of DOPA improved adhesion to aluminium, while *in vitro* testing of these materials showed that they could support human mesenchymal stem cells.[77] Deliberate bioengineering of a mussel adhesive protein enzymically enriched in DOPA has been proposed as a flexible and durable sealant for defects in the urinary tract (Fig. 2), when, in combination with hylauronic acid, it forms an example of a

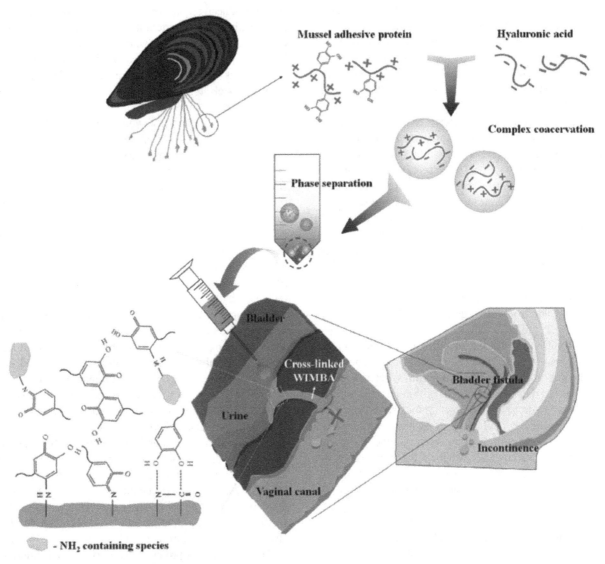

Fig. 2 Schematic representation of the bonding mechanism for the DOPA-based bioadhesive and its application to seal urinary fistula.
Source: From Kim et al.[78] © 2015, Elsevier. Reprinted with permission.

coacervate adhesive.[78] Oxidation with metaperiodate triggered further crosslinking increasing adhesiveness in the material, which demonstrated good underwater adhesion and *ex vivo* sealing of urinary fistulas.[78] Dextran may be oxidised with aqueous periodate to give a di-aldehyde that readily combines with amines, such as ε-poly-lysine (ε-PL) from Streptomyces, to produce hydrogels, dextran-ε-PL adhesives, that exhibits low toxicity—even some antibacterial properties—and binds to collagen more than 10 times stronger than a commercial fibrin glue when tested in a shear bonding strength test.[79] Further investigations have shown that the dextran-ε-PL adhesives undergo self-degradation, a useful property in biomedical applications.[80] While this particular adhesive is yet to be tested *in vivo*, a similar hydrogel-adhesive prepared from oxidised dextran and PEG-amines, was used successfully for sealing and healing corneal incisions in white rabbits.[81] Incorporation of a snake venom, batroxobin, into a dextran-ε-PL adhesive, on the other hand, has allowed the blood clotting effect of the venom to be contained within the discrete area of the wound and to act as a novel hemostatic composite.[82]

Chitin, sourced mainly from crustaceans, and more importantly, its deacetylted version, chitosan, have long been used in biomedical applications and have been the subject of recent reviews,[83] especially as an adhesive.[84] In this regard, the free hydroxyls of chitin have been doubly functionalized with carboxymethyl ether and glyceryl methacrylate groups, and then oxidatively polymerized with ascorbic acid, to give a water-tolerant adhesive, that rivalled cyanoacrylate for performance, after blending with chitin nanofibers.[85] Tensile strength, bursting pressure, and *in vivo* evaluations on a rat model, including lower inflammatory and immune responses, were all reported with positive outcomes, in comparison to cyanoacrylate adhesives. More recently, chitosan has been combined with ε-PL to form an adhesive hydrogel for nerve repair,[86–87] mimicking the polysaccharide protein nature of the tissue for enhanced biocompatibility. The key reaction in hydrogel formation is the Michael addition of a nucleophilic thiol group to a maleimide acceptor. The chitosan portion was functionalized with thiols to serve as nucleophiles, whereas maleimide groups with water miscible PEG spacers were anchored to the ε-PL portion.[86] When the two portions were mixed in aqueous buffer, they reacted rapidly within seconds to form an adhesive hydrogel. Additional adhesive power was derived from the catechol groups that had been incorporated into the ε-PL portion of the system, although it was found that 5% loading of catechol on the ε-PL was optimum, beyond which concentration, the adhesion measured by storage modulus decreased and the cohesion declined. *In vivo* testing of this hydrogel on the sciatic nerve of rats, displayed an improvement over fibrin glue and the suture repair.[86] The reverse approach has also been taken whereby chitosan was set up to carry the catechol groups, whereas the thiol groups were borne by the synthetic copolymer, Pluronic f-147 (polypropylene oxide-polyethylene oxide,

PPO-PEO), abbreviated as Plu-SH.[88] Crosslinking between the oxidsed cathechol/o-quinone groups and the thiols occurred on mixing and setting. The resultant adhesive hydrogel, CHI-C-PluSH, showed thermoresponsive sol-gel properties, and good *in vitro* and in vivo stability and biocompatability. Chitosan decorated with catechol groups, either crosslinked with genepin[89] or formed into nanoparticles[90] has proved effective for adhesion to mucosal linings. In the absence of catechol, there may still be binding to the mucosal lining but, possibly, more through polyionic bonding and coacervation.[91] Although this work was originally conceived as a platform for drug delivery, it demonstrates that adhesion to the type of tissue that is constantly mobile and lubricated can be effectively addressed and holds promise for future research. For example, functionalizing chitosan nanoparticles with rose bengal, produces a material that has great potential in dental health,[92] by reducing biofilm formation and strengthening the underlying dental tissue of dentin and collagen, clearly because of superior adhesive qualities. An inspired approach to cartilage repair has been attempted by employing a catechol-conjugated chitosan adhesive patch to act as a containment barrier for directional release of therapeutic or cellular components towards the repair site.[93] The associated animal model studies show positive results for cartilage regeneration.

Hydrogel adhesives for wound closure and hemostasis are accessible from building blocks based on PEG and the phenol, tyramine. Incorporation of a polyamine, such as chitosan[94] or ε-PL, may then increase its mechanical strength.[95] Cross coupling between the phenolic group of the tyramine with itself, or the amine groups of the chitosan or the ε-PL, can be effected by enzymic oxidative coupling in situ, using horse radish peroxidase and hydrogen peroxide. The hydrogel that forms has excellent adhesive properties with demonstrable applications to hemostasis and wound closure.[94,95] In this particular context, it was found to be superior to fibrin glue, cyanoacrylate, and suturing. Citrate forms the core of a number of polymeric bioadhesives, which have been designed for use in bone regeneration, tissue repair, and wound dressing.[96–100] Combined with the building blocks, PEG for water-miscibility, and DOPA and dopamine for adhesion, a new variety of citrate-based bioadhesives has been created.[96] Inclusion of the bone mineral, hydroxyapatite, in other citrate-based polymers has been shown to enhance the overall performance and integration of the material in animal models.[97,98] Some desirable antifungal/antibacterial/antimicrobial activity can be promoted by inclusion of an undecylenic group, in the citrate–polymer backbone.[96] Oxidation of the catechol groups to promote crosslinking was achieved with meta-periodate or silver nitrate, which also contributed to antimicrobial activity.[96]

The extracellular protein, elastin, rich in the amino-acid, lysine, forms the basis of an injectable hydrogel that has many desirable features as a biomaterial.[101] Careful combination of elastin with a copolymer derived from acryl-amide/hydroxyethyl methacrylate/oligo(ethyleneglycol)/polylactide, afforded a new thermoresponsive hydrogel with tunable properties, such as holding fibroblast cells and supporting their proliferation.[101] A synthetic counterpart to this has been prepared from a combination of two PEG macromonomers, one bearing a vinyl sulphone and the other free thiol groups that react through a Michael addtion to give an amorphous PEG-hydrogel. Fibroblast cells can adhere to this resultant 3-D network with even greater focus, once it has been decorated with suitable peptides or proteins.[102] Extension of this cellular therapy through adherence to suitable 3-D biomaterial networks[70] may form the basis for the delivery of a viable cell group for tissue repair, such as with cardiac tissue implantation,[103] as well as engineering synthetic tissue on a larger scale, such as that mimicking the rudimentary structure of brain tissue.[104]

Light-Activated Adhesives

Chemical activation of glues may be problematic. Cyano-acrylate glues, for example, are particularly awkward when applied to tissue and can result in premature tissue adhe-sion. For this reason, an adhesion trigger that can be adjusted and controlled is of great help to surgeons; light is an optimal trigger being aseptic and allowing surgeons to perform precise tissue bonding. Among light devices, lasers are the most effective as they provide coherent light with spectral purity below 1 nm; laser beams can be deliv-ered to the targets through an optical fiber during open or key-hole surgery. The three main types of laser tissue repair (LTR) are tissue welding, photochemical tissue bonding, and LTR with adhesive biomaterials. Laser tissue welding relies on a photothermal process that heats the tissue above the temperature of collagen denaturation[105] and causes collagen fibres to interlock, forming a bond between the adjoined tissues.[106] The safety and effectiveness of this procedure has been demonstrated in a limited number of clinical studies.[107] However, thermal damage and tissue necrosis,[108] limited bond strength,[109] and excessive scar-ring are possible adverse effects.[110]

Photochemical tissue bonding is usually performed by applying a rose bengal solution between two strips of tissue that are then irradiated with green light. Photochemical reactions are responsible for crosslinking collagen and, thus, bonding tissue at low temperature ($T < 37°C$) [111] and without significant heat production.[112] In particular, the ability of rose bengal to produce singlet oxygen upon green light irradiation facilitates the crosslinking between colla-gen via its amino groups.[113–115] The remarkable advantage of joining tissues photochemically is the elimination of thermal damage that still affects other sutureless tech-niques.[116–118] A variety of tissues have been successfully repaired with this technique including those of the cor-nea,[119] vocal cords,[120] blood vessels,[121] peripheral

nerves,[122] and skin.[123] Repair strength of ~1 N can be achieved when repairing skin incisions in pigs; typical laser irradiance and fluence are 0.5–1.0 W/cm^2 and 25–100 J/cm^2 while the rose bengal concentration ranges between 0.1% and 1% (weight/volume).

The finite depth of light penetration in tissue (1–2 mm) is a serious constraint of photochemical tissue bonding to clinical utility; nonetheless, the use of biocompatible and biodegradable comb-shaped planar waveguides, made of poly(D,L-lactide-co-glycolide), poly(L-lactic acid), and silk, allows the laser beam to penetrate deeper through tis-sue (>10 mm), allowing full thickness wound closure.[124]

Chitosan films have also been combined with rose ben-gal and bonded to tissue (~1.5 N/cm^2) using a green laser ($\lambda = 532$ nm, F ~ 110 J/cm^2); adhesion was completed at ~37°C eliminating detrimental thermal damage.[111,125] Anastomosis of rat median nerves was successfully per-formed *in vivo* and myelinated axons regrew comparable to sutured nerves, whereas the functional recovery of the laser-chitosan technique was superior three months postop-eratively.[126,127] Photocrosslinkable chitosan gels contain-ing azide and lactose moieties were also used in conjunction with UV light ($\lambda = 300$–350 nm, irradiation time ~30 sec-onds) to repair 5 mm incisions in the uterus of pregnant rats. The repair strength of the chitosan gel was signifi-cantly higher than that of fibrin glue (0.090 vs 0.078 N/cm^2) and a histological examination of the wounds showed a mild foreign body reaction 20 days after surgery.[128] Other photochemical techniques for tissue repair have been recently reported; Lang et al. demonstrated LTR with a bio-adhesive patch that was made of poly(glycerol sebacate urethane) and coated with a UV crosslinkable light-activated adhesive.[129] The patches repaired porcine epicar-dial tissue with a bonding strength of ~1.9 N/cm^2 after five seconds of UV light exposure ($\lambda = 320$–390 nm, 0.38 W/cm^2). During the UV irradiation, a preload force of 3 N was applied on the patch to complete tissue bonding. Karp and coworkers developed a nanoparticulate formula-tion of a similar adhesive; after UV irradiation ($\lambda = 365$ nm) at 365 nm and for 10 seconds, the adhesion force of the glue on the epicardium tissue was 1.43 ± 0.30 N/cm^2. Negatively charged alginate was added to the glue to stabi-lize the nanoparticulate surface and significantly reduce its viscosity. Intraocular injection of the glue was then demon-strated *ex vivo* in freshly harvested bovine eye using a 27-gauge syringe.[130]

Photocrosslinkable bioadhesive based on dextran and PEG were tested on glass substrates coated with gelatin. Samples were irradiated with UV light (320–480 nm) for 15 minutes at 40 mW/cm^2 to obtain gelatin–adhesive cross-linking. A maximum adhesion strength of 4.0 ± 0.6 MPa was attained and minimal cytotoxicity was detected on mouse fibroblasts.[131] Other photoactivated tissue adhesive comprising urethane dextran and 2- hydroxyethyl metha-crylate confirmed the bonding capability of dextran hydrogels.[132] PEG-based adhesives, which include hyperbranched polymers, could be crosslinked and bonded to tissue by UV light in a very short time (~15 seconds);[133] nevertheless, UV radiation may cause undesirable side effects such as immunosuppression and carcinogenesis even at longer wavelengths ($\lambda \geq 320$ nm), which may discourage patients and surgeons from adopting these sutureless techniques.[134] The toxicity risks associated with ultraviolet radiation are avoided when photochemical tissue bonding is activated by visible light. Dendritic polymers consisting of glycerol, succinic acid, and poly(ethyleneglycol) were tested to repair porcine corneal wounds. An argon-ion laser ($\lambda = 514$ nm, $P = 200$ mW) photoinitiated the biodendrimers dissolved in a sterile aqueous solution with eosin. After curing for two minutes with one second pulses, a transparent hydrogel adhesive was bonded to the cornea. The adhesive repair withstood a pressure of 3.20 N/cm^2 alone and 1.47 N/cm^2 with 16 sutures. Nonetheless, the application of the liquid solution may be problematic in the wet environment of surgical procedures.[135] A similar hydrogel adhesive was applied to treat osteochondral defects in rabbits and six months postoperatively glycosaminoglycans and collagen II were present in the wounds, and the contralateral unfilled defects healed poorly.[136]

Another study reported a mussel proteins hydrogel that was crosslinked by dityrosine to form tissue bonding. This bioadhesive was photoactivated by blue light ($\lambda = 452$ nm) for ~60 seconds and could close wounds in rat skin effectively with mild and minimal inflammation. The wound breaking strength was almost double (1023 $\pm$ 42 kPa) that of fibrin and cyanoacrylate glues, 14 days after surgery.[137] Hydrogels based on heparin have been recently proposed as bioadhesives; they are cured under visible light using eosin Y as a photoinitiator with triethanolamine (electron donor) to initiate the reaction of thiolated-heparin with acrylate-ended poly(ethylene glycol). Encapsulation of fibroblasts in the hydrogel showed over 96% viability and drug release tests confirmed that the growth factors remain bioactive after irradiation.[138]

Laser tissue repair with nanoparticles

LTR with albumin "solders"[139] was one of the first techniques used to reconnect tissue without sutures; it consisted in applying a solution of concentrated albumin on the wound and then irradiating with a laser to thermally denature the albumin that bonded to tissue. Photosensitive dyes were added to the solder to prevent thermal damage of surrounding tissue and target a specific region. Indocyanine green (ICG) was the chromophore of choice in several procedures, including peripheral nerve repair,[117,118] blood vessel anastomosis,[140] urethral repair,[141] skin repair,[142] and keratoplasty.[143] The photothermal conversion of infrared radiation (upon laser irradiation) by ICG may lead to unnecessary temperature increase (>70°C) that can injure tissue.[144] ICG is highly photosensitive and unstable in

water,[145] and researchers have recently turned to gold nanoshells and nanorods for a more reliable light absorption device.[146] Gold nanoshells are spherical nanoparticles with a dielectric silica core and gold outer coating;[147] they are optically stable and highly tuneable as they absorb radiation strongly in the visible and near infrared regions.[148] A typical silica-gold nanoshell with a diameter of 40 nm has a molar extinction coefficient that is five orders of magnitude larger than indocyanine green (1.08×10^4 M^{-1} cm^{-1} at $\lambda = 778$ nm).[149] Gold nanoshells have therefore the advantage of facilitating laser tissue welding between two thick sections of tissue because near-infrared radiation deeply penetrates tissue and converts effectively into heat to denature collagen and bond tissues together. Gold nanorods are similar to gold nanoshells, but have an ellipsoid shape ranging from 1–100 nm with a typical aspect ratio (length/width) of 3–5.[150] Nanorods are also highly tuneable and their absorption and scattering are controlled by the ratio between the radius of the core and the overall radius;[151] when the gold outer shell varies from 20 to 5 nm, the absorption red-peak shifts ~300 nm.[152] Nanoshells and nanorods convert almost all of the absorbed radiation into heat and for this reason they are used in laser tissue welding and LTR applications.[146, 153] Gold nanoparticles have been used as an exogenous absorber to repair tissue in several studies. Gold nanoshells, for example, were mixed with varying concentrations of bovine serum albumin solder and applied with success to repair full-thickness skin incisions in rats. Wound healing was comparable to that of the sutured group five days postoperatively;[146] the nanoshells were tuned to absorb at 800 nm and were irradiated by a diode laser at 808 nm. Nourbakhsh and Khosroshahi applied a combination of ICG and gold nanoshells (85–115 nm) in conjunction with a 810 nm diode laser to repair sections of sheep skin *in vitro*.[154] The resulting tensile strength was ~1.5 N/cm^2 at irradiance and temperature levels of 24 W/cm^2 and 70°C, respectively.

Recent clinical reports described laser-assisted corneal transplantation in patients using a diode laser and ICG.[155–157] The success of this tissue welding procedure has stimulated the use of gold nanorods instead of ICG with promising results: Rossi et al. transplanted a patch of anterior lens capsule from a donor porcine eye to a recipient animal *ex vivo* using a gold nanorod colloid.[158] The lens capsule was stained with a droplet of the gold colloid before the laser irradiation (100–140 J/cm^2) that caused limited thermal damage within 50–70 μm from the operational site. The reduced thermal damage was also due to the application of 40 ms laser pulses with an energy of 70–100 mJ. Matteini et al. demonstrated in a separate study the sutureless closure of carotid arteries[159] applying gold nanorods, which were embedded in hyaluronan gel after being functionalized with a biopolymeric shell. The gel was applied to a 3 mm incision and irradiated for 50 seconds with a diode laser (810 nm) at an irradiance level of ~30 W/cm^2. No fluid leakage and occlusion occurred immediately after the operation and at 30 days.[160] The authors reported a wavelength shift of the absorption band after the irradiation that worked as a self-terminating procedure.[150]

Nanoparticles can be used for tissue repair and wound closure even without light activation.[161,162] In a recent study, an aqueous solution of SiO_2/Na_2O nanoparticles was applied between two sections of rat liver and held for 30 seconds resulting in a maximum adhesion force of ~300 mN. These silica-based nanoparticles were also put in a 6 mm deep liver incision, holding together the wound edges for approximately one minute; the incisions healed after three days with minimal scarring.[162] When the same technique was applied on the dorsal skin incision, the rat wounds healed and no evidence of inflammation or necrosis appeared seven days postoperatively. The adhesion mechanism involves multiple bonds that are formed between the tissue and nanoparticles because of tissue adsorption. Under a load, these bonds can break and reform dissipating energy and preventing in some degree fractures at the tissue interfaces. This tissue repair modality requires less time than traditional suturing methods; however, it is better suited for soft tissue and may be problematic when applied to hard tissue. These new techniques require further investigations as nanoparticles may bypass or cross the blood–brain barrier and cause permanent damage. Silver nanoparticles, for example, can circulate in the blood and diffuse throughout the brain causing swollen astrocytes and neuronal degeneration in rats.[163] Cell oxidative stress can also result from the nanoparticle production of reactive oxygen that damages lysosomes and mitochondria.[164] In a separate study, it emerged that gold nanoparticles cause abnormally small and underpigmented eyes in the developing zebrafish when functionalized with a cationic ligand.[165]

Researchers have developed matrices to minimize the nanoparticle concentration delivered in situ as they confine and control the nanoparticles' release after tissue repair. Matrices provide an effective barrier against the physiological environment and limit nanoparticle aggregation that can significantly alter the photothermal conversion.[166–168]

Gold nanorods were incorporated in chitosan films[169] that were bonded *in vitro* to porcine carotid arteries and rabbit tendons.[170] The tensile strength of the bond was ~2 N/cm² when a 810 nm laser delivered ~100 ms pulses at a fluence of ~130 J/cm². The pulse regime of the laser helped in circumventing tissue thermal damage. Another porous chitosan film incorporated a mixture of gold nanorods and thermosensitive micelles with a fluorescent dye.[171] The gold nanorods produced heat after being laser-irradiated and caused the release of a fluorescent dye from the micelles in a controlled and localized sequence. Huang et al. developed a nanocomposite solder made of gold nanorods (λ = 780 nm) dispersed in an elastin-like polypeptide that was crosslinked to form a plasmonic compound.[172] This matrix was used to repair intestinal tissue *ex vivo* achieving a repair strength of ~22 N/cm² (T ~ 60°C) when irradiated with a titanium sapphire laser (λ = 800 nm, 20 W/cm², t = 60 seconds). In another study, silica nanoshells with a diameter of 250–270 nm were incorporated inside a porous polycaprolactone scaffold, which was doped with albumin solder.[173] The nanoshells also incorporated ICG, which absorbed a 808 nm laser and consequently bonded the scaffolds to rabbit aortas with a repair strength of ~0.7 N. Thermal damage occurred in the adventitial layer when the external scaffold temperature was 80°C (Irradiance ~14 W/cm²). Redmond and coworkers fabricated a silk nanofiber mat using electrospinning techniques (Fig. 3);[174] this mat was stained with rose bengal solution (0.1% w/v), wrapped around tendons, and photochemically bonded with 532 nm light (Irradiance ~0.3 W/cm², Fluence ~125 J/cm²). No thermal injury affected the tissue; however, adhesion strength was greater in the standard suture-repaired group compared to the laser group.

Gecko-Inspired Adhesives

The discovery of the mechanism that causes the adhesion of Gecko foot to flat surfaces[175] prompted researchers to design and fabricate adhesive films similar to Gecko toes. These comprise high density (5000/mm²) arrays of keratin fibrils (setae) that are ~100 μm long and ~5 μm wide. Setae

Fig. 3 Scanning electron micrographs of electrospun silk mats that were pre-annealed (**A**) and post-annealed (**B**) in 90% methanol for one hour.
Source: From Ni et al.[174] © 2012, John Wiley and Sons. Reprinted with permission

contain numerous nanosized branches named spatulae with a diameter ranging between 200 and 400 nm. Geckos stick to surfaces pushing the setae on a substrate at a very close range (<0.5 nm), thus triggering van der Waals attractive forces.[176] Gecko-inspired adhesives also have arrays of nanostructures, Qu et al. for example, proved that a hierarchical carbon nanotube array can adhere strongly to a dry surface (1000 kPa).[177] Fibrillar adhesives have been integrated with miniature devices for medical diagnostics as they provide an effective and non-invasive system for tissue anchoring, which is otherwise impossible with conventional suturing. A microcapsule robot designed for gastrointestinal monitoring was fixed, for example, to the tissue, using microfibrillar pads that were fabricated on polydimethylsiloxane (PDMS). Tests performed on porcine intestine provided a significant increase in adhesion in the fibrillar pads (1.85 N/cm²) compared to the flat ones (0.86 N/cm²).[178] The design of these pads was modified to create a fibrillar PDMS adhesive with large surface structures (pillar diameter ~140 μm, pillar pitch ~105 μm).[179] Kwak et al.[180] fabricated a similar PDMS adhesive that could stick to dry skin over multiple cycles using mushroom-tipped pillars with a high aspect ratio. The strength of tissue adhesion was ~1.3 N/cm² using 5 μm wide tips with aspect ratio of three and was similar to the strength of patches coated with acrylic adhesive (~3 N/cm²). These fibrillar patches demonstrated repeated adhesion over 30 cycles and incorporated a sensor unit connected to an electrocardiogram device in order to deliver vital signals in real time. The ability of fibrillar adhesives to form strong bonds to a dry surface fails under wet conditions: van der Waals forces are significantly weakened under wet conditions[181] as in the case of wound repair. Indicative in this regard is the study by Vajpayee et al. that showed how a nanostructured film made of PDMS adhered poorly to a hydrophilic surface under water.[182] To overcome this problem, biomedical adhesives have been developed using a multilayered system that is coated with "gluing" polymers to better bond in water. Pereira et al. spin-coated a layer of medical-grade cyanoacrylate adhesive over a micropatterned adhesive made from poly(ε-caprolactone).[183] The bonding of the adhesive that had pillars with diameter and height of 4.9 μm of 19.0 μm, respectively, was significantly higher (~2.5 N/cm²) than the non-patterned adhesive (~1 N/cm²). The application of the adhesive to stomach and colon perforation resulted in minimal inflammatory response in a rat model three weeks postoperatively. Pillars can also be coated with poly(dopamine-co-methoxyethyl acrylate) (poly(DMA-co-MEA)) to produce a reversible wet adhesive.[184] In this instance, the adhesive force per pillar under water was greater than in air (~819 and 120 nN, respectively). A similar increase in adhesion was reported when fibrillar arrays coated with poly(DMA-co-MEA) were used.[185] Mahdavi et al. used oxidized dextran to coat a poly(glycerol sebacate acrylate) (PGSA) fibrillar array enhancing adhesion at the tissue interface.[186] This procedure proved that a strong nonreversible adhesion (~0.7 N/cm²) could be achieved *in vivo* whereas the non-coated array produced weaker bonding (~0.3 N/cm²) two days after implantation (Fig. 4). In a separate study, fibrillar arrays with poly(DMA-co-MEA)[187] had an adhesion strength that was 15 times greater (~33 mN) than the uncoated nanostructures (~2 mN) when fully submerged in water. Adhesives were fabricated using UV photocured

Fig. 4 (**A**) Nanopatterned PGSA polymer after surface spin coating with water as control. (**B** and **C**) Nanopatterned PGSA after surface spin coating with oxidized dextran and aldehyde (DXTA); adhesion of neighbouring pillars occurred. The black arrow indicates how DXTA polymer may cause neighbouring pillar tips to stick together. (**D**) Five percent DXTA were completely obstructed.
Source: From Mahdavi et al.[186] © 2008, PNAS. Reprinted with permission.

PGSA at both the microscale and nanoscale that ranged from 100 nm to 1 μm in width and 0. 8–3.0 μm in height. The micro- and nanopillars were spin-coated with aldehyde functionalized dextran and then crosslinked with tissue to form a nonreversible chemical bond.[186] This adhesive was applied on porcine intestinal tissue and required a force of 4.8 N/cm² to detach it. Furthermore, the adhesive strength was twice that of the samples without the dextran coating two days after implantation.

The extra coating of pillars represents an additional step in the adhesive fabrication that may complicate the manufacturing process; a recent report showed that a simple one-step procedure was devised to fabricate a single-layer chitosan film that had pillars with base diameter in the range of 100 to 600 nm and height of ~70 nm. This chitosan film contained rose bengal and bonded to tissue in a wet environment upon green laser irradiation without pillar coating. In comparison to a "flat" adhesive (without pillars), the nanostructured adhesive bonded significantly stronger to tissue under either stress or pressure (~2.1 N/cm²).[188] Remarkably, the adhesive and tissue temperatures remained below 39°C during photochemical tissue bonding.

The design of fibrillar adhesives is inspired not only by the anatomy of gecko feet but also other micro and nanostructures in nature that exploit ingenious mechanisms for adhesion. Cho et al., for example, developed an adhesive that mechanically interlocked with tissue mimicking the quills of the North American porcupine.[189] Microneedles are present inside the quills with a backward orientation thus providing mechanical lock with tissue upon penetration. Polyurethane quills were fabricated using a replica micromolding process. The barbed quills necessitated 35% less force (30 mN) to penetrate the tissue as compared to the barbless quills. Adhesive patches were then fabricated with and without barbed quills demonstrating that the former patches had a four-fold higher pull-out resistance (0.219 N) than the barbless quill patches (0.063 N). Another fibrillar adhesive that interlocks mechanically with tissue[190] consists of a two-layer system where the backing is made of polystyrene and the functional layer contains polystyrene-*block*-poly(acrylic acid) (PS-*b*-PAA). When the PS-*b*-PAA fibrillar array penetrates the dermis, it takes on water and swells mechanically locking it to the tissue. The adhesive strength was tested on porcine skin and was 7 to 12 times stronger (0.69 ± 0.17 N/cm²) than flat PS-*b*-PAA films. These swellable microneedles could stop fluid leakage and prevent bacteria infiltration across the wound by absorbing water and expanding to fill the entry site. When the adhesive was applied to fix skin grafts on muscle, it was also significantly stronger than stapled skin grafts (0.93 ± 0.23 N/cm² vs 0.28 ± 0.11 N/cm²).[191] Other adhesives consisted of buckypaper with micro- and nanopores made of un-oriented multiwalled carbon nanotubes.[192] The surface topology of the buckypaper is highly randomized with a high surface roughness of ~110 nm, in sharp contrast with the regularly patterned arrays of gecko-based adhesives. The buckypaper was applied to two sections of rabbit muscular fascia and achieved an adhesion strength of ~2 N/cm².

CONCLUSION

Even if a small number of glues are already part of relevant surgical procedures, adhesives have not been embraced by the majority of surgeons. This may be due to several reasons including biocompatibility issues (cyanoacrylates) or lack of adhesion strength that limits the adhesive application to tissue under low tension and in a dry setting (fibrin, PEG). In this respect, sutures are still the best option to surgeons as they guarantee a reliable tool for approximating wound edges that are apart and under tension. For this reason, research needs to focus on increasing the bonding strength of glues in different environmental conditions, under water and tension, for example. When the bonding is strong, less amount of adhesive is required on tissue with the advantage of moderating or avoiding undesirable body reaction and degradability problems. Nanostructured adhesives are relatively new in the field but have already proved that they enhance the strength of tissue bonding; when combined with light-activation they promise to provide a noninvasive technique able to repair tissue in open and laparoscopic surgery.

ACKNOWLEDGMENTS

We would like to thank Dr. Damia Mawad for useful suggestions.

REFERENCES

1. Lumsden, A.B.; Heyman, E.R.; Closure Medical Surgical Sealant Study Group. Prospective randomized study evaluating an absorbable cyanoacrylate for use in vascular reconstructions. J. Vasc. Surg. **2006**, *44* (5), 1002–1009.

2. Downey, D.M.; Harre, J.G.; Pratt, J.W. Functional comparison of staple line reinforcements in lung resection. Ann. Thorac. Surg. **2006**, *82* (5), 1880–1883.

3. Sharma, M.; Wakure, A. Scar revision. Indian. J. Plas. Surg. **2013**, *46* (2), 408–418.

4. Abu Hamdeh, S.; Lytsy, B.; Ronne-Engström, E. Surgical site infections in standard neurosurgery procedures– a study of incidence, impact and potential risk factors. Br. J. Neurosurg. **2014**, *28* (2), 270–275.

5. Carbonell, A.M.; Harold, K.L.; Mahmutovic, A.J.; Hassan, R.; Matthews, B.D.; Kercher, K.W.; Sing, R.F.; Heniford, B.T. Local injection for the treatment of suture site pain after laparoscopic ventral hernia repair. Am. Surg. **2003**, *69* (8), 688–691.

6. Khanlari, S.; Dubé, M.A. Bioadhesives: A review. Macromol. React. Eng. **2013**, *7* (11), 573–587.

7. Currie, L.J.; Sharpe, J.R.; Martin, R. The use of fibrin glue in skin grafts and tissue-engineered skin replacements: A review. Plast. Reconstr. Surg. **2001**, *108* (6), 1713–1726.

8. Busuttil, R.W. A comparison of antifibrinolytic agents used in hemostatic fibrin sealants. J. Am. Coll. Surg. **2003**, *197* (6), 1021–1028.

9. Buchta, C.; Hedrich, H.C.; Macher, M.; Höcker, P.; Redl, H. Biochemical characterization of autologous fibrin sealants produced by CryoSeal and Vivostat in comparison to the homologous fibrin sealant product Tissucol/Tisseel. Biomaterials **2005**, *26* (31), 6233–6241.

10. Kober, B.J.; Scheule, A.M.; Voth, V.; Deschner, N.; Schmid, E.; Ziemer, G. Anaphylactic reaction after systemic application of aprotinin triggered by aprotinin-containing fibrin sealant. Anesth. Analg. **2008**, *107* (2), 406–409.

11. Kawamura, M.; Sawafuji, M.; Watanabe, M.; Horinouchi, H.; Kobayashi, K. Frequency of transmission of human parvovirus B19 infection by fibrin sealant used during thoracic surgery. Ann. Thorac. Surg. **2002**, *73* (4), 1098–1100.

12. Spotnitz, W.D. Fibrin sealant: The only approved hemostat, sealant, and adhesive–a laboratory and clinical perspective. ISRN Surg. **2014**, *2014*, 2039438.

13. Kluba, T.; Fiedler, K.; Kunze, B.; Ipach, I.; Suckel, A. Fibrin sealants in orthopaedic surgery: Practical experiences derived from use of QUIXIL® in total knee arthroplasty. Arch. Orthop. Trauma. Surg. **2012**, *132* (8), 1147–1152.

14. Levy, O.; Martinowitz, U.; Oran, A.; Tauber, C.; Horoszowski, H. The use of fibrin tissue adhesive to reduce blood loss and the need for blood transfusion after total knee arthroplasty. A prospective, randomized, multicenter study. J. Bone. Joint. Surg. Am. **1999**, *81* (11), 1580–1588.

15. McConnell, J.S.; Shewale, S.; Munro, N.; Shah, K.; Deakin, A.H.; Kinninmonth, A.W. Reducing blood loss in primary knee arthroplasty: A prospective randomised controlled trial of tranexamic acid and fibrin spray. Knee **2012**, *19* (4), 295–298.

16. Molloy, D.O.; Archbold, H.A.; McConway, J.; Wilson, R.K.; Beverland, D.E. Comparison of topical fibrin spray and tranexamic acid on blood loss after total knee replacement: A prospective, randomised controlled trial. J. Bone. Joint. Surg. Br. **2007**, *89* (3), 306–309.

17. Notarnicola, A.; Moretti, L.; Martucci, A.; Spinarelli, A.; Tafuri, S.; Pesce, V.; Moretti, B.; Comparative efficacy of different doses of fibrin sealant to reduce bleeding after total knee arthroplasty. Blood. Coagul. Fibrinolysis. **2012**, *23* (4), 278–284.

18. Sabatini, L.; Trecci, A.; Imarisio, D.; Uslenghi, M.D.; Bianco, G.; Scagnelli, R. Fibrin tissue adhesive reduces postoperative blood loss in total knee arthroplasty. J. Orthop. Traumatol. **2012**, *13* (3), 145–151.

19. Wang, G.J.; Hungerford, D.S.; Savory, C.G.; Rosenberg, A.G.; Mont, M.A.; Burks, S.G.; Mayers, S.L.; Spotnitz, W.D. Use of fibrin sealant to reduce bloody drainage and hemoglobin loss after total knee arthroplasty: A brief note on a randomized prospective trial. J. Bone Joint Surg. Am. **2001**, *83* (10), 1503–1505.

20. Li, Z.J.; Fu, X.; Tian, P.; Liu, W.X.; Li, Y.M.; Zheng, Y.F.; Ma, X.L.; Deng, W.M. Fibrin sealant before wound closure in total knee arthroplasty reduced blood loss: A meta-analysis. Knee. Surg. Sports. Traumatol. Arthrosc. **2015**, *23* (7), 2019–2025.

21. Yu, M.S.; Kang, S.H.; Kim, B.H.; Lim, D.J. Effect of aerosolized fibrin sealant on hemostasis and wound healing after endoscopic sinus surgery: A prospective randomized study. Am. J. Rhinol. Allergy. **2014**, *28* (4), 335–340.

22. Martinez-de-la-Casa, J.M.; Rayward, O.; Saenz-Frances, F.; Mendez, C.; Bueso, E.S.; Garcia-Feijoo, J. Use of a fibrin adhesive for conjunctival closure in trabeculectomy. Acta. Ophthalmol. **2013**, *91* (5), 425–428.

23. Testini, M.; Lissidini, G.; Poli, E.; Gurrado, A.; Lardo, D.; Piccinni, G. A single-surgeon randomized trial comparing sutures, N-butyl-2-cyanoacrylate and human fibrin glue for mesh fixation during primary inguinal hernia repair. Can. J. Surg. **2010**, *53* (3), 155–160.

24. Campanelli, G.; Pascual, M.H.; Hoeferlin, A.; Rosenberg, J.; Champault, G.; Kingsnorth, A.; Miserez, M. Randomized, controlled, blinded trial of Tisseel/Tissucol for mesh fixation in patients undergoing lichtenstein technique for primary inguinal hernia repair: Results of the TIMELI trial. Ann. Surg. **2012**, *255* (4), 650–657.

25. Bracale, U.; Rovani, M.; Picardo, A.; Merola, G.; Pignata, G.; Sodo, M.; Di Salvo, E.; Ratto, E.L.; Noceti, A.; Melillo, P.; Pecchia, L. Beneficial effects of fibrin glue (Quixil) versus lichtenstein conventional technique in inguinal hernia repair: A randomized clinical trial. Hernia **2014**, *18* (2), 185–192.

26. Wong, J.U.; Leung, T.H.; Huang, C.C.; Huang, C.S. Comparing chronic pain between fibrin sealant and suture fixation for bilayer polypropylene mesh inguinal hernioplasty: A randomized clinical trial. Am. J. Surg. **2011**, *202* (1), 34–38.

27. Lionetti, R.; Neola, B.; Dilillo, S.; Bruzzese, D.; Ferulano, G.P. Sutureless hernioplasty with light-weight mesh and fibrin glue versus Lichtenstein procedure: A comparison of outcomes focusing on chronic postoperative pain. Hernia **2012**, *16* (2), 127–131.

28. Sözen, S.; Cetinkunar, S.; Emir, S.; Yazar, F.M. Comparing sutures and human fibrin glue for mesh fixation during open inguinal hernioplasty. Ann. Ital. Chir. **2012**, *87* (3), 252–256.29.

29. Hidalgo, M.; Castillo, M.J.; Eymar, J.L.; Hidalgo, A. Lichtenstein inguinal hernioplasty: Sutures versus glue. Hernia **2005**, *9* (3), 242–244.

30. Benizri, E.I.; Rahili, A.; Avallone, S.; Balestro, J.C.; Caï, J.; Benchimol, D. Open inguinal hernia repair by plug and patch: The value of fibrin sealant fixation. Hernia **2006**, *10* (5), 389–394.

31. Negro, P.; Basile, F.; Brescia, A.; Buonanno, G.M.; Campanelli, G.; Canonico, S.; Cavalli, M.; Corrado, G.; Coscarella, G.; Di Lorenzo, N.; Falletto, E.; Fei, L.; Francucci, M.; Fronticelli Baldelli, C.; Gaspari, A.L.; Gianetta, E.; Marvaso, A.; Palumbo, P.; Pellegrino, N.; Piazzai, R; Salvi, P.F.; Stabilini, C.; Zanghì, G.; Open tension-free Lichtenstein repair of inguinal hernia: Use of fibrin glue versus sutures for mesh fixation. Hernia **2011**, *15* (1), 7–14.

32. Liu, H.; Zheng, X.; Gu, Y.; Guo, S. A meta-analysis examining the use of fibrin glue mesh fixation versus suture mesh fixation in open inguinal hernia repair. Dig. Surg. **2014**, *31* (6), 444–451.

33. Boldo, E.; Armelles, A.; Perez de Lucia, G.; Martin, F.; Aracil, J.P.; Miralles, J.M.; Martinez, D.; Escrig, J. Pain

after laparascopic bilateral hernioplasty : Early results of a prospective randomized double-blind study comparing fibrin versus staples. Surg. Endosc. **2008**, *22* (5), 1206–1209.

34. Fortelny, R.H.; Petter-Puchner, A.H.; May, C.; Jaksch, W.; Benesch, T.; Khakpour, Z.; Redl, H.; Glaser, K.S. The impact of atraumatic fibrin sealant vs. staple mesh fixation in TAPP hernia repair on chronic pain and quality of life: Results of a randomized controlled study. Surg. Endos. **2012**, *26* (1), 249–254.

35. Lau, H. Fibrin sealant versus mechanical stapling for mesh fixation during endoscopic extraperitoneal inguinal hernioplasty: A randomized prospective trial. Ann. Surg. **2005**, *242* (5), 670–675.

36. Lovisetto, F.; Zonta, S.; Rota, E.; Mazzilli, M.; Bardone, M.; Bottero, L.; Faillace, G.; Longoni, M. Use of human fibrin glue (Tissucol) versus staples for mesh fixation in laparoscopic transabdominal preperitoneal hernioplasty: A prospective, randomized study. Ann. Surg. **2007**, 245 (2), 222–231.

37. Olmi, S.; Scaini, A.; Erba, L.; Guaglio, M.; Croce, E. Quantification of pain in laparoscopic transabdominal preperitoneal (TAPP) inguinal hernioplasty identifies marked differences between prosthesis fixation systems. Surgery **2007**, *142* (1), 40–46.

38. Sajid, M.S.; Ladwa, N.; Kalra, L.; McFall, M.; Baig, M.K.; Sains, P. A meta-analysis examining the use of tacker mesh fixation versus glue mesh fixation in laparoscopic inguinal hernia repair. Am. J. Surg. **2013**, *206* (1), 103–111.

39. Barbosa, M.D.; Stipp, A.C.; Passanezi, E.; Greghi, S.L. Fibrin adhesive derived from snake venom in periodontal surgery: Histological analysis. J. Appl. Oral. Sci. **2008**, *16* (5), 310–315.

40. Pulikkotil, S.J.; Nath, S. Fibrin sealant as an alternative for sutures in periodontal surgery. J. Coll. Physicians. Surg. Pak. **2013**, *23* (2), 164–165.

41. Levy, L.H. Efficacy and safety of aprotinin in cardiac surgery. Orthopedics **2004**, *27* (6), 659–662.

42. García Cerdá, D.; Ballester, A.M.; Aliena-Valero, A.; Carabén-Redaño, A.; Lloris, J.M. Use of cyanoacrylate adhesives in general surgery. Surg. Today **2014**, *45* (8), 939–956.

43. Martín-Ballester, A.; García-Cerdá, D.; Prieto-Moure, B.; Martín-Martínez, J.M.; Lloris-Carsí, J.M. Use of cyanoacrylate adhesives in dermal lesions: A review. J. Adhes. Sci. and Technol. **2014**, *28* (6), 573–597.

44. Feigenberg, T.; Maor-Sagie, E.; Zivi, E.; Abu-Dia, M.; Ben-Meir, A.; Sela, H.Y.; Ezra, Y. Using adhesive glue to repair first degree perineal tears: A prospective randomized controlled trial. BioMed. Res. Int. **2014**, 2014, 526590.

45. Moreno-Egea, A. Is it possible to eliminate sutures in open (lichtenstein technique) and laparoscopic (totally extraperitoneal endoscopic) inguinal hernia repair? A randomized controlled trial with tissue adhesive (n-hexyl-α-cyanoacrylate). Surg. Innov. **2014**, *21* (6), 590–599.

46. Ando, M.; Tamaki, T.; Yoshida, M.; Sasaki, S.; Toge, Y.; Matsumoto, T.; Maio, K.; Sakata, R.; Fukui, D.; Kanno, S.; Nakagawa, Y.; Yamada, H. Surgical site infection in spinal surgery: A comparative study between 2-octyl-cyanoacrylate and staples for wound closure. Eur. Spine. J. **2014**, *23* (4), 854–862.

47. Chan, R.S.; Vijayananthan, A.; Kumar, G.; Hilmi, I.N. Imaging findings of extensive splenic infarction after cyanoacrylate injection for gastric varices--a case report. Med. J. Malaysia. **2012**, *67* (4), 424–425.

48. Sato, T.; Yamazaki, K.; Toyota, J.; Karino, Y.; Ohmura, T.; Suga, T. Inflammatory tumor in pancreatic tail induced by endoscopic ablation with cyanoacrylate glue for gastric varices. J. Gastroenterol. **2004**, *39* (5), 475–478.

49. Leggat, P.A.; Smith, D.R.; Kedjarune, U. Surgical applications of cyanoacrylate adhesives: A review of toxicity. ANZ J. Surg. **2007**, *77* (4), 209–213.

50. Shalaby, S.W. Cyanoacrylate-Based systems as tissue adhesives. In *Absorbable and Biodegradable Polymers*; Burg, K.J.L., Ed.; CRC Press: Boca Raton, **2004**; 59–76.

51. Lauto, A.; Mawad, D.; Foster, L.J.R. Adhesive biomaterials for tissue reconstruction. J. Chem. Technol. Biot. **2008**, *83* (4), 464–472.

52. Shivamurthy, D.M.; Singh, S.; Reddy, S. Comparison of octyl-2-cyanoacrylate and conventional sutures in facial skin closure. Natl. J. Maxillofac. Surg. **2010**, *1* (1), 15–19.

53. Chen, K.; Klapper, A.S.; Voige, H.; Del Priore, G. A randomized, controlled study comparing two standardized closure methods of laparoscopic port sites. JSLS **2010**, *14* (3), 391–394.

54. van den Ende, E.D.; Vriens, P.W.H.E.; Allema, J.H.; Breslau, P.J. Adhesive bonds or percutaneous absorbable suture for closure of surgical wounds in children. Results of a prospective randomized trial. J. Pediatr. Surg. **2004**, *39* (8), 1249–1251.

55. Banerjee, S.S.; Aher, N.; Patil, R.; Khandare, J. Poly(ethylene glycol)-prodrug conjugates: Concept, design, and applications. J. Drug. Deliv. **2012**, *2012*, 103973.

56. Wain, J.C.; Kaiser, L.R.; Johnstone, D.W.; Yang, S.C.; Wright, C.D.; Friedberg, J.S.; Feins, R.H.; Heitmiller, R.F.; Mathisen, D.J.; Selwyn, M.R. Trial of a novel synthetic sealant in preventing air leaks after lung resection. Ann. Thorac. Surg. **2001**, *71* (5), 1623–1628.

57. Torchiana, D.F. Polyethylene glycol based synthetic sealants. J. Card. Surg. **2003**, *18* (6), 504–506.

58. U.S. Food and Drug Administration, FocalSeal-Synthetic Absorbable Sealant, http://www.fda.gov/MedicalDevices/ProductsandMedicalProcedures/DeviceApprovalsandClearances/Recently-ApprovedDevices/ucm089788.htm (accessed April 2016).

59. Wright, N.M.; Park, J.; Tew, J.M.; Kim, K.D.; Shaffrey, M.E.; Cheng, J.; Choudhri, H.; Krishnaney, A.A.; Graham, R.S.; Mendel, E.; Simmons, N. Spinal sealant system provides better intraoperative watertight closure than standard of care during spinal surgery: A prospective, multicenter, randomized controlled study. Spine (Phila Pa 1976) **2015**, *40* (8), 505–513.

60. Natour, E.; Suedkamp, M.; Dapunt, O.E. Assessment of the effect on blood loss and transfusion requirements when adding a polyethylene glycol sealant to the anastomotic closure of aortic procedures: A case-control analysis of 102 patients undergoing Bentall procedures. J. Cardiothorac. Surg. **2012**, 7, 105.

61. Sharma, B.; Fermanian, S.; Gibson, M.; Unterman, S.; Herzka, D.A.; Cascio, B.; Coburn, J.; Hui, A.Y.; Marcus, N.; Gold, G.E.; Elisseeff, J.H. Human cartilage repair with a photoreactive adhesive-hydrogel composite. Sci. Transl. Med. **2013**, *5* (167), 167ra6.

62. Saunders, M.M.; Baxter, Z.C.; Abou-Elella, A.; Kunselman, A.R.; Trussell, J.C. BioGlue and Dermabond save time, leak less, and are not mechanically inferior to two-layer and modified one-layer vasovasostomy. Fertil. Steril. **2009**, *91* (2), 560–565.

63. Sedó, J.; Saiz-Poseu, J.; Busqué, F.; Ruiz-Molina, D. Catechol-based biomimetic functional materials. Adv. Mater. **2013**, *25* (5), 653–701.

64. Liu, Y.; Ai, K.; Lu, L. Polydopamine and its derivative materials: Synthesis and promising applications in energy, environmental, and biomedical fields. Chem. Rev. **2014**, *114* (9), 5057–5115.

65. Madhurakkat Perikamana, S.K.; Lee, J.; Lee, Y.B.; Shin, Y.M.; Lee, E.J.; Mikos, A.G.; Shin, H. Materials from mussel-inspired chemistry for cell and tissue engineering applications. Biomacromolecules **2015**, *16* (9), 2541–2255.

66. Li, H.; Chen, S.; Chen, J.; Chang, J.; Xu, M.; Sun, Y.; Wu, C. Mussel-inspired artificial grafts for functional ligament reconstruction. ACS. Appl. Mater. Interfaces. **2015**, *7* (27), 14708–14719.

67. Zhou, P.; Wu, F.; Zhou, T.; Cai, X.; Zhang, S.; Zhang, X.; Li, Q.; Li, Y.; Zheng, Y.; Wang, M.; Lan, F.; Pan, G.; Pei, D.; Wei, S. Simple and versatile synthetic polydopamine-based surface supports reprogramming of human somatic cells and long-term self-renewal of human pluripotent stem cells under defined conditions. Biomaterials **2016**, 87, 1–17.

68. Sun, K.; Xie, Y.; Ye, D.; Zhao, Y.; Cui, Y.; Long, F.; Zhang, W.; Jiang, X. Mussel-inspired anchoring for patterning cells using polydopamine. Langmuir **2012**, *28* (4), 2131–2136.

69. Xu, X.; Li, M.; Liu, Q.; Jia, Z.; Shi, Y.; Cheng, Y.; Zheng, Y.; Ruan, L.Q. Facile immobilization of heparin on bioabsorbable iron via mussel adhesive protein (MAPs). Prog. Nat. Sci.: Mater. Int. **2014**, *24* (5), 458–465.

70. Shafiq, M.; Jung, Y.; Kim, S.H. Insight on stem cell preconditioning and instructive biomaterials to enhance cell adhesion, retention, and engraftment for tissue repair. Biomaterials **2016**, *90*, 85–115.

71. Lee, S-B.; González-Cabezas, C.; Kim, K.-M.; Kim, K.-N.; Kuroda, K. Catechol-functionalized synthetic polymer as a dental adhesive to contaminated dentin surface for a composite restoration. Biomacromolecules **2015**, *16* (8), 2265–2275.

72. Liu, Y.; Meng, H.; Konst, S.; Sarmiento, R.; Rajachar, R.; Lee, B.P. Injectable dopamine-modified poly(ethylene glycol) nanocomposite hydrogel with enhanced adhesive property and bioactivity. ACS Appl. Mater. Interfaces **2014**, *6* (19), 16982–16992.

73. Wei, Q.; Becherer, T.; Mutihac, R.-C.; Noeske, P.-L.M.; Paulus, F.; Haag, R.; Grunwald, I. Multivalent anchoring and cross-linking of mussel-inspired antifouling surface coatings. Biomacromolecules **2014**, *15* (8), 3061–3071.

74. Zhang, H.; Bré, L.P.; Zhao, T.; Zheng, Y.; Newland, B.; Wang, W. Mussel-inspired hyperbranched poly(amino ester) polymer as strong wet tissue adhesive. Biomaterials **2014**, *35* (2), 711–719.

75. Jia, M.; Li, A.; Mu, Y.; Jiang, W.; Wan, X. Synthesis and adhesive property study of polyoxetanes grafted with catechols via Cu(I)-catalyzed click chemistry. Polymer **2014**, *55* (5), 1160–1166.

76. Kim, B.J.; Oh, D.X.; Kim, S.; Seo, J.H.; Hwang, D.S.; Masic, A.; Han, D.K.; Cha, H.J. Mussel-mimetic protein-based adhesive hydrogel. Biomacromolecules **2014**, *15* (5), 1579–1585.

77. Burke, K.A.; Roberts, D.C.; Kaplan, D.L. Silk fibroin aqueous-based adhesives inspired by mussel adhesive proteins. Biomacromolecules **2016**, *17* (1), 237–245.

78. Kim, H.J.; Hwang, B.H.; Lim, S.; Choi, B.H.; Kang, S.H.; Cha, H.J. Mussel adhesion-employed water-immiscible fluid bioadhesive for urinary fistula sealing. Biomaterials **2015**, *72*, 104–111.

79. Hyon, S.H.; Nakajima, N.; Sugai, H.; Matsumura, K. Low cytotoxic tissue adhesive based on oxidized dextran and epsilon-poly-L-lysine. J. Biomed. Mater. Res. A **2014**, *102* (8), 2511–2520.

80. Matsumura, K.; Nakajima, N.; Sugai, H.; Hyon, S.H. Self-degradation of tissue adhesive based on oxidized dextran and poly-L-lysine. Carbohy. Polym. **2014**, *113*, 32–38.

81. Chenault, H.K.; Bhatia, S.K.; Dimaio, W.G.; Vincent, G.L.; Camacho, W.; Behrens, A. Sealing and healing of clear corneal incisions with an improved dextran aldehyde-PEG amine tissue adhesive. Curr. Eye. Res. **2011**, *36* (11), 997–1004.

82. You, K.E.; Koo, M.A.; Lee, D.H.; Kwon, B.J.; Lee, M.H.; Hyon, S.H.; Seomun, Y.; Kim, J.T.; Park, J.C. The effective control of a bleeding injury using a medical adhesive containing batroxobin. Biomed. Mat. **2014**, *9* (2), 025002.

83. Dash, M.; Chiellini, F.; Ottenbrite, R.M.; Chiellini, E. Chitosan—A versatile semi-synthetic polymer in biomedical applications. Prog. Polym. Sci. **2011**, *36* (8), 981–1014.

84. Mati-Baouche, N.; Elchinger, P.-H.; de Baynast, H.; Pierre, G.; Delattre, C.; Michaud, P. Chitosan as an adhesive. European Polymer Journal **2014**, *60*, 198–212.

85. Azuma, K.; Nishihara, M.; Shimizu, H.; Itoh, Y.; Takashima, O.; Osaki, T.; Itoh, N.; Imagawa, T.; Murahata, Y.; Tsuka, T.; Izawa, H.; Ifuku, S.; Minami, S.; Saimoto, H.; Okamoto, Y.; Morimoto, M. Biological adhesive based on carboxymethyl chitin derivatives and chitin nanofibers. Biomaterials **2015**, *42*, 20–29.

86. Zhou, Y.; Zhao, J.; Sun, X.; Li, S.; Hou, X.; Yuan, X.; Yuan, X. Rapid gelling chitosan/polylysine hydrogel with enhanced bulk cohesive and interfacial adhesive force: Mimicking features of epineurial matrix for peripheral nerve anastomosis. Biomacromolecules **2016**, *17* (2), 622–630.

87. Nie, W.; Yuan, X.; Zhao, J.; Zhou, Y.; Bao, H. Rapidly in situ forming chitosan/epsilon-polylysine hydrogels for adhesive sealants and hemostatic materials. Carbohydr. Polym. **2013**, *96* (1), 342–348.

88. Ryu, J.H.; Lee, Y.; Kong, W.H.; Kim, T.G.; Park, T.G.; Lee, H. Catechol-functionalized chitosan/pluronic hydrogels for tissue adhesives and hemostatic materials. Biomacromolecules **2011**, *12* (7), 2653–2659.

89. Xu, J.; Strandman, S.; Zhu, J.X.; Barralet, J.; Cerruti, M. Genipin-crosslinked catechol-chitosan mucoadhesive hydrogels for buccal drug delivery. Biomaterials 2015, 37, 395–404.

90. Soliman, G.M.; Zhang, Y.L.; Merle, G.; Cerruti, M.; Barralet, J. Hydrocaffeic acid-chitosan nanoparticles with enhanced stability, mucoadhesion and permeation properties. Eur. J. Pharm. Biopharm. 2014, 88 (3), 1026–1037.

91. Meng-Lund, E.; Muff-Westergaard, C.; Sander, C.; Madelung, P.; Jacobsen, J. A mechanistic based approach for enhancing buccal mucoadhesion of chitosan. Int. J. Pharm. 2014, 461 (1–2), 280–285.

92. Shrestha, A.; Hamblin, M.R.; Kishen, A. Photoactivated rose bengal functionalized chitosan nanoparticles produce antibacterial/biofilm activity and stabilize dentin-collagen. Nanomedicine 2014, 10 (3), 491–501.

93. Lee, J.M.; Ryu, J.H.; Kim, E.A.; Jo, S.; Kim, B.S.; Lee, H.; Im, G.I. Adhesive barrier/directional controlled release for cartilage repair by endogenous progenitor cell recruitment. Biomaterials 2015, 39, 173–181.

94. Lih, E.; Lee, J.S.; Park, K.M.; Park, K.D. Rapidly curable chitosan-PEG hydrogels as tissue adhesives for hemostasis and wound healing. Acta. Biomater. 2012, 8 (9), 3261–3269.

95. Wang, R.; Zhou, B.; Liu, W.; Feng, X.H.; Li, S.; Yu, D.F.; Chang, J.C.; Chi, B.; Xu, H. Fast in situ generated epsilon-polylysine-poly (ethylene glycol) hydrogels as tissue adhesives and hemostatic materials using an enzyme-catalyzed method. J. Biomater. Appl. 2015, 29 (8), 1167–1179.

96. Guo, J.; Wang, W.; Hu, J.; Xie, D.; Gerhard, E.; Nisic, M.; Shan, D.; Qian, G.; Zheng, S.; Yang, J. Synthesis and characterization of anti-bacterial and anti-fungal citrate-based mussel-inspired bioadhesives. Biomaterials 2016, 85, 204–217.

97. Xie, D.; Guo, J.; Mehdizadeh, M.; Tran, R.T.; Chen, R.; Sun, D.; Qian, G.; Jin, D.; Bai, X.; Yang, J. Development of injectable citrate-based bioadhesive bone implants. J. Mater. Chem. B Mater. Biol. Med. 2015, 3, 387–398.

98. Tang, J.; Guo, J.; Li, Z.; Yang, C.; Xie, D.; Chen, J.; Li, S.; Li, S.; Kim, G.B.; Bai, X.; Zhang, Z.; Yang, J. Fast degradable citrate-based bone scaffold promotes spinal fusion. J. Mater. Chem. B Biol. Med. 2015, 3 (27), 5569–5576.

99. Gyawali, D.; Nair, P.; Kim, H.K.; Yang, J. Citrate-based biodegradable injectable hydrogel composites for orthopedic applications. Biomater. Sci. 2013, 1 (1), 52–64.

100. Mehdizadeh, M.; Weng, H.; Gyawali, D.; Tang, L.; Yang, J. Injectable citrate-based mussel-inspired tissue bioadhesives with high wet strength for sutureless wound closure. Biomaterials 2012, 33 (32), 7972–7983.

101. Fathi, A.; Mithieux, S.M.; Wei, H.; Chrzanowski, W.; Valtchev, P.; Weiss, A.S.; Dehghani, F. Elastin based cell-laden injectable hydrogels with tunable gelation, mechanical and biodegradation properties. Biomaterials 2014, 35 (21), 5425–5435.

102. Missirlis, D.; Spatz, J.P. Combined effects of PEG hydrogel elasticity and cell-adhesive coating on fibroblast adhesion and persistent migration. Biomacromolecules 2014, 15 (1), 195–205.

103. Peña, B.; Martinelli, V.; Jeong, M.; Bosi, S.; Lapasin, R.; Taylor, M.R.; Long, C.S.; Shandas, R.; Park, D.; Mestroni, L. Biomimetic polymers for cardiac tissue engineering. Biomacromolecules 2016, 17 (5), 1593–1601.

104. Chwalek, K.; Tang-Schomer, M.D.; Omenetto, F.G.; Kaplan, D.L. In vitro bioengineered model of cortical brain tissue. Nat. Protoc. 2015, 10 (9), 1362–1373.

105. Sun, Y.; Chen, W.-L.; Lin, S.-J.; Jee, S.-H.; Chen, Y.-F.; Lin, L.-C.; So, P.T.C.; Dong, C.-Y. Investigating mechanisms of collagen thermal denaturation by high resolution second-harmonic generation imaging. Biophys. J. 2006, 91 (7), 2620–2625.

106. Schober, R.; Ulrich, F.; Sander, T.; Dürselen, H.; Hessel, S. Laser-induced alteration of collagen substructure allows microsurgical tissue welding. Science 1986, 232 (4756), 1421–1422.

107. Shousha, M.A.; Yoo, S.H.; Kymionis, G.D.; Ide, T.; Feuer, W.; Karp, C.L.; O'Brien, T.P.; Culbertson, W.W.; Alfonso, E. Long-term results of femtosecond laser-assisted sutureless anterior lamellar keratoplasty. Ophthalmology 2011, 118 (2), 315–323.

108. Fung, L.C.; Mingin, G.C.; Massicotte, M.; Felsen, D.; Poppas, D.P. Effects of temperature on tissue thermal injury and wound strength after photothermal wound closure. Lasers. Surg. Med. 1999, 25 (4), 285–290.

109. Huang, T.C.; Blanks, R.H.; Berns, M.W.; Crumley, R.L. Laser vs suture nerve anastomosis. Otolaryngol. Head. Neck. Surg. 1992, 107 (1), 14–20.

110. Fried, N.M.; Walsh, J.T., Jr. Laser skin welding: In vivo tensile strength and wound healing results. Lasers. Surg. Med. 2000, 27 (1), 55–65.

111. Lauto, A.; Mawad, D.; Barton, M.; Gupta, A.; Piller, S.C.; Hook, J. Photochemical tissue bonding with chitosan adhesive films. Biomed. Eng. Online. 2010, 9 (47).

112. Chan, B.P.; Kochevar, I.E.; Redmond, R.W. Enhancement of porcine skin graft adherence using a light-activated process. J. Surg. Res. 2002, 108 (1), 77–84.

113. Verter, E.E.; Gisel, T.E.; Yang, P.; Johnson, A.J.; Redmond, R.W.; Kochevar, I.E. Light-initiated bonding of amniotic membrane to cornea. Invest. Ophthal. Vis. Sci. 2011, 52 (13), 9470–9477.

114. Verweu, H.; Steveninck, J.V. Model studies on photodynamic crosslinking. Photochem. Photobiol. 1982, 35 (2), 265–267.

115. Shen, H.R.; Spikes, J.D.; Kopecková, P.; Kopecek, J. Photodynamic crosslinking of proteins II. Photocrosslinking of a model protein-ribonuclease A. J. Photochem. Photobiol. B. 1996, 35 (3), 213–219.

116. Tal, K.; Strassmann, E.; Loya, N.; Ravid, A.; Kariv, N.; Weinberger, D.; Katzir, A.; Gaton, D.D. Corneal cut closureusing temperature-controlled CO_2 laser soldering system. Lasers. Med. Sci. 2015, 30 (4), 1367–1371.

117. Lauto, A.; Dawes, J.M.; Cushway, T.; Piper, J.A.; Owen, E.R. Laser nerve repair by solid protein band technique I: Identification of optimal laser dose, power, and solder surface area. Microsurgery 1998, 18 (1), 55–59.

118. Lauto, A.; Dawes, J.M.; Piper, J.A.; Owen, E.R. Laser nerve repair by solid protein band technique II: Assessment of long-term nerve regeneration. Microsurgery 1998, 18 (1), 60–64.

119. Wang, Y.; Kochevar, I.E.; Redmond, R.W.; Yao, M. A light-activated method for repair of corneal surface defects. Lasers. Surg. Med. 2011, 43 (6), 481–489.

120. Franco, R.A.; Dowdall, J.R.; Bujold, K.; Amann, C.; Faquin, W.; Redmond, R.W.; Kochevar, I.E. Photochemical repair of vocal fold microflap defects. Laryngoscope 2011, 121 (6), 1244–1251.

121. O'Neil, A.C.; Randolph, M.A.; Bujold, K.E.; Kochevar, I.E.; Redmond, R.W.; Winograd, J.M. Photochemical sealing improves outcome following peripheral neurorrhaphy. J. Surg. Res. 2009, 151 (1), 33–39.

122. Fairbairn, N.G.; Ng-Glazier, J.; Meppelink, A.M.; Randolph, M.A.; Valerio, I.L.; Fleming, M.E.; Winograd, J.M.; Redmond, R.W. Light-activated sealing of nerve graft coaptation sites improves outcome following large gap peripheral nerve injury. Plast. Reconstr. Surg. 2015, 136 (4), 739–750.

123. Xu, N.; Yao, M.; Farinelli, W.; Hajjarian, Z.; Wang, Y.; Redmond, R.W.; Kochevar, I.E. Light-activated sealing of skin wounds. Lasers Surg. Med. 2015, 47 (1), 17–29.

124. Nizamoglu, S.; Gather, M.C.; Humar, M.; Choi, M.; Kim, S.; Kim, K.S.; Hahn, S.K.; Scarcelli, G.; Randolph, M.; Redmond, R.W.; Yun, S.H. Bioabsorbable polymer optical waveguides for deep-tissue photomedicine. Nat. Commun. 2016, 7, 10374.

125. Lauto, A.; Stoodley, M.; Barton, M.; Morley, J.W.; Mahns, D.A.; Longo, L.; Mawad, D. Fabrication and application of rose bengal-chitosan films in laser tissue repair. J. Vis. Exp. 2012, 68 (68), e4158.

126. Barton, M.; Morley, J.W.; Stoodley, M.A.; Ng, K.-S.; Piller, S.C.; Duong, H.; Mawad, D.; Mahns, D.A.; Lauto, A. Laser-activated adhesive films for sutureless median nerve anastomosis. J. Biophotonics. 2013, 6 (11–12), 938–949.

127. Barton, M.J.; Morley, J.W.; Stoodley, M.A.; Shaikh, S.; Mahns, D.A.; Lauto, A. Long term recovery of median nerve repair using laser-activated chitosan adhesive films. J. Biophotonics. 2015, 8 (3), 196–207.

128. Suzuki, K.; Shinya, M.; Kitagawa, M. Basic study of healing of injuries to the myometrium and amniotic membrane using photocrosslinkable chitosan. J. Obstet. Gynaecol. Res. 2006, 32 (2), 140–147.

129. Lang, N.; Pereira, M.J.; Lee, Y.; Friehs, I.; Vasilyev, N.V.; Feins, E.N.; Ablasser, K.; O'Cearbhaill, E.D.; Xu, C.; Fabozzo, A.; Padera, R.; Wasserman, S.; Freudenthal, F.; Ferreira, L.S.; Langer, R.; Karp, J.M.; del Nido, P.J. A blood-resistant surgical glue for minimally invasive repair of vessels and heart defects. Sci. Transl. Med. 2014, 6 (218), 218ra6.

130. Lee, Y.; Xu, C.; Sebastin, M.; Lee, A.; Holwell, N.; Xu, C.; Miranda Nieves, D.; Mu, L.; Langer, R.S.; Lin, C.; Karp, J.M. Bioinspired nanoparticulate medical glues for minimally invasive tissue repair. Adv. Healthc. Mater. 2015, 4 (16), 2587–2596.

131. Li, C.; Wang, T.; Hu, L.; Wei, Y.; Liu, J.; Mu, X.; Nie, J.; Yang, D. Photocrosslinkable bioadhesive based on dextran and PEG derivatives. Mater. Sci. Eng. C Mater. Biol. Appl. 2014, 35, 300–306.

132. Wang, T.; Mu, X.; Li, H.; Wu, W.; Nie, J.; Yang, D. The photocrosslinkable tissue adhesive based on copolymeric dextran/HEMA. Carbohydr. Polym. 2013, 92 (2), 1423–1431.

133. Zhang, H.; Zhao, T.; Duffy, P.; Dong, Y.; Annaidh, A.N.; O'Cearbhaill, E.; Wang, W. Hydrolytically degradable hyperbranched PEG-polyester adhesive with low swelling and robust mechanical properties. Adv. Healthc. Mater. 2015, 4 (15), 2260–2268.

134. Halliday, G.M.; Byrne, S.M.; Damian, D.L. Ultraviolet a radiation: Its role in immunosuppression and carcinogenesis. Semin. Cutan. Med. Surg. 2010, 30 (4), 214–221.

135. Degoricija, L.; Johnson, C.S.; Wathier, M.; Kim, T.; Grinstaff, M.W. Photo cross-linkable Biodendrimers as ophthalmic adhesives for central lacerations and penetrating keratoplasties. Invest. Ophthalmol. Vis. Sci. 2007, 48 (5), 2037–2042.

136. Degoricija, L.; Bansal, P.N.; Söntjens, S.H.; Joshi, N.S.; Takahashi, M.; Snyder, B.; Grinstaff, M.W. Hydrogels for osteochondral repair based on photocrosslinkable carbamate dendrimers. Biomacromolecules 2008, 9 (10), 2863–2872.

137. Jeon, E.Y.; Hwang, B.H.; Yang, Y.J.; Kim, B.J.; Choi, B.H.; Jung, G.Y.; Cha, H.J. Rapidly light-activated surgical protein glue inspired by mussel adhesion and insect structural crosslinking. Biomaterials 2015, 67, 11–19.

138. Fu, A.; Gwon, K.; Kim, M.; Tae, G.; Kornfield, J.A. Visible-light-initiated thiol-acrylate photopolymerization of heparin-based hydrogels. Biomacromolecules 2015, 16 (2), 497–506.

139. Menovsky, T.; Beek, J.F.; van Gemert, M.J. Laser tissue welding of dura mater and peripheral nerves: A scanning electron microscopy study. Lasers. Surg. Med. 1996, 19 (2), 152–158.

140. Xie, H.; Bendre, S.C.; Gregory, K.W.; Furnary, A.P. Laser-assisted end-to-end vascular anastomosis of elastin heterograft to carotid artery with an albumin stent in vivo. Photomed. Laser. Surg. 2004, 22 (4), 298–302.

141. Shumalinsky, D.; Lobik, L.; Cytron, S.; Halpern, M.; Vasilyev, T.; Ravid, A.; Katzir, A. Laparoscopic laser soldering for repair of ureteropelvic junction obstruction in the porcine model. J. Endourol. 2004, 18 (2), 177–181.

142. Simhon, D.; Halpern, M.; Brosh, T.; Vasilyev, T.; Kariv, N.; Argaman, R. In vivo laser soldering of incisions in juvenile pig skins using GaAs or CO_2 lasers and a temperature control system. In Proceedings of SPIE 5312; SPIE: Bellingham, WA, USA, 2004; 162.

143. Menabuoni, L.; Mincione, F.; Mincione, G.; Pini, R. Laser welding to assist penetrating keratoplasty: In vivo studies. In Proceedings of 3195 SPIE; SPIE: Bellingham, WA, USA, 1998; 25.

144. Spector, D.; Rabi, Y.; Vasserman, I.; Hardy, A.; Klausner, J.; Rabau, M.; Katzir, A. In vitro large diameter bowel anastomosis using a temperature controlled laser tissue soldering system and albumin stent. Lasers. Surg. Med. 2009, 41 (7), 504–508.

145. Zhou, J.F.; Chin, M.P.; Schafer, S.A. Aggregation and degradation of indocyanine green. In Proceedings of SPIE 2128; SPIE: Bellingham, WA, USA, 1994; 495.

146. Gobin, A.M.; O'Neal, D.P.; Watkins, D.M.; Halas, N.J.; Drezek, R.A.; West, J.L. Near infrared laser-tissue welding using nanoshells as an exogenous absorber. Lasers. Surg. Med. 2005, 37 (2), 123–129.

147. Erickson, T.A.; Tunnell, J.W. Gold nanoshells in biomedical applications. In Nanotechnologies for the Life Sciences: Mixed Metal Nanomaterials; Challa S.S.R. Kumar., Ed.; Wiley-VCH: Weinheim, Germany, 2009; Vol. 3, 1–44.

148. Duff, D.G.; Baiker, A.; Edwards, P.P. A new hydrosol of gold clusters 1. Formation and particle size variation. Langmuir **1993**, *9* (9), 2301–2309.

149. Jain, P.K.; Lee, K.S.; El-Sayed, I.H.; El-Sayed, M.A. Calculated absorption and scattering properties of gold nanoparticles of different size, shape, and composition: Applications in biological imaging and biomedicine. J. Phys. Chem. B **2006**, *110* (14), 7238–7248.

150. Pérez-Juste, J.; Pastoriza-Santos, I.; Liz-Marzán, L.M.; Mulvaney, P. Gold nanorods: Synthesis, characterization and applications. Coordin. Chem. Rev. **2005**, *249* (17), 1870–1901.

151. Averitt, R.D.; Westcott, S.L.; Halas, N.J. Linear optical properties of gold nanoshells. J. Opt. Soc. Am. B **1999**, *16* (10), 1824–1832.

152. Prodan, E.; Radloff, C.; Halas, N.J.; Nordlander, P. A hybridization model for the plasmon response of complex nanostructures. Science **2003**, *302* (5644), 419–422.

153. Stern, J.M.; Stanfield, J.; Lotan, Y.; Park, S.; Hsieh, J.T.; Cadeddu, J.A. Efficacy of laser-activated gold nanoshells in ablating prostate cancer cells *in vitro*. J. Endourol. **2007**, *21* (8), 939–943.

154. Khosroshahi, M.E.; Nourbakhsh, M.S. Enhanced laser tissue soldering using indocyanine green chromophore and gold nanoshells combination. J. Biomed. Opt. **2011**, *16* (8), 088002.

155. Menabuoni, L.; Canovetti, A.; Rossi, F.; Malandrini, A.; Lenzetti, I.; Pini, R. The 'anvil' profile in femtosecond laser-assisted penetrating keratoplasty. Acta. Ophthalmol. **2013**, *91* (6), e494–e495.

156. Buzzonetti, L.; Capozzi, P.; Petrocelli, G.; Valente, P.; Petroni, S.; Menabuoni, L.; Rossi, F.; Pini, R. Laser welding in penetrating keratoplasty and cataract surgery in pediatric patients: Early results. J. Cataract Refract. Surg. **2013**, *39* (12), 1829–1834.

157. Canovetti, A.; Malandrini, A.; Lenzetti, I.; Rossi, F.; Pini, R.; Menabuoni, L. Laser-assisted penetrating keratoplasty: 1-year results in patients using a laser-welded anvil-profiled graft. Am. J. Ophthalmol. **2014**, *158* (4), 664–670.

158. Ratto, F.; Matteini, P.; Rossi, F.; Menabuoni, L.; Tiwari, N.; Kulkarni, S.K.; Pini, R. Photothermal effects in connective tissues mediated by laser-activated gold nanorods. Nanomedicine **2009**, *5* (2), 143–151.

159. Matteini, P.; Ratto, F.; Rossi, F.; Rossi, G.; Esposito, G.; Puca, A.; Albanese, A.; Maira, G.; Pini, R. *In vivo* carotid artery closure by laser activation of hyaluronan-embedded gold nanorods. J. Biomed. Opt. **2010**, *15* (4), 041508.

160. Esposito, G.; Rossi, F.; Matteini, P.; Ratto, F.; Sabatino, G.; Puca, A.; Albanese, A.; Rossi, G.; Marchese, E.; Maira, G.; Pini, R. Nanotechnology and vascular neurosurgery: An *in vivo* experimental study on microvessels repair using laser photoactivation of a nanostructured hyaluronan solder. J. Biol. Regul. Homeost. Agents **2012**, *26* (3), 447–456.

161. Rose, S.; Prevoteau, A.; Elzière, P.; Hourdet, D.; Marcellan, A.; Leibler, L. Nanoparticle solutions as adhesives for gels and biological tissues. Nature **2014**, *505* (7483), 382–385.

162. Meddahi-Pelle, A.; Legrand, A.; Marcellan, A.; Louedec, L.; Letourneur, D.; Leibler, L. Organ repair, hemostasis, and *in vivo* bonding of medical devices by aqueous solutions of nanoparticles. Angew. Chem. Int. Ed. Engl. **2014**, *53* (25), 6369–6373.

163. Tang, J.; Xiong, L.; Wang, S.; Wang, J.; Liu, L.; Li, J.; Yuan, F.; Xi, T. Distribution, translocation and accumulation of silver nanoparticles in rats. J. Nanosci. Nanotechnol. **2009**, *9* (8), 4924–4932.

164. Yu, K.N.; Yoon, T.J.; Minai-Tehrani, A.; Kim, J.E.; Park, S.J.; Jeong, M.S.; Ha, S.W.; Lee, J.K.; Kim, J.S.; Cho, M.H. Zinc oxide nanoparticle induced autophagic cell death and mitochondrial damage via reactive oxygen species generation. Toxicol. *In Vitro.* **2013**, *27* (4), 1187–1195.

165. Kim, K.T.; Zaikova, T.; Hutchison, J.E.; Tanguay, R.L. Gold nanoparticles disrupt zebrafish eye development and pigmentation. Toxicol. Sci. **2013**, *133* (2), 275–288.

166. Joshi, P.P.; Yoon, S.J.; Hardin, W.G.; Emelianov, S.; Sokolov, K.V. Conjugation of antibodies to gold nanorods through Fc portion: Synthesis and molecular specific imaging. Bioconjug. Chem. **2013**, *24* (6), 878–888.

167. Ungureanu, C.; Kroes, R.; Petersen, W.; Groothuis, T.A.; Ungureanu, F.; Janssen, H.; van Leeuwen, F.W.; Kooyman, R.P.; Manohar, S.; van Leeuwen, T.G. Light interactions with gold nanorods and cells: Implications for photothermal nanotherapeutics. Nano. Lett. **2011**, *11* (5), 1887–1894.

168. Mercatelli, R.; Ratto, F.; Centi, S.; Soria, S.; Romano, G.; Matteini, P.; Quercioli, F.; Pini, R.; Fusi, F. Quantitative readout of optically encoded gold nanorods using an ordinary dark-field microscope. Nanoscale **2013**, *5* (20), 9645–9650.

169. Matteini, P.; Ratto, F.; Rossi, F.; Centi, S.; Dei, L.; Pini, R. Chitosan films doped with gold nanorods as laser-activatable hybrid bioadhesives. Adv. Mater. **2010**, *22* (38), 4313–4316.

170. Matteini, P.; Ratto, F.; Rossi, F.; de Angelis, M.; Cavigli, L.; Pini, R. Hybrid nanocomposite films for laser-activated tissue bonding. J. Biophotonics **2012**, *5* (11–12), 868–877.

171. Matteini, P.; Tatini, F.; Luconi, L.; Ratto, F.; Rossi, F.; Giambastiani, G.; Pini, R. Photothermally activated hybrid films for quantitative confined release of chemical species. Angew. Chem. Int. Ed. Engl. **2013**, *52* (23), 5956–5960.

172. Huang, H.-C.; Walker, C.R.; Nanda, A.; Rege, K. Laser welding of ruptured intestinal tissue using plasmonic polypeptide nanocomposite solders. ACS. Nano. **2013**, *7* (4), 2988–2998.

173. Schöni, D.S.; Bogni, S.; Bregy, A.; Wirth, A.; Raabe, A.; Vajtai, I.; Pieles, U.; Reinert, M.; Frenz, M. Nanoshell assisted laser soldering of vascular tissue. Lasers. Surg. Med. **2011**, *43* (10), 975–983.

174. Ni, T.; Senthil-Kumar, P.; Dubbin, K.; Aznar-Cervantes, S.D.; Datta, N.; Randolph, M.A.; Cenis, J.L.; Rutledge, G.C.; Kochevar, I.E.; Redmond, R.W. A photoactivated nanofiber graft material for augmented Achilles tendon repair. Lasers. Surg. Med. **2012**, *44* (8), 645–652.

175. Autumn, K.; Liang, Y.A.; Hsieh, S.T.; Zesch, W.; Chan, W.P.; Kenny, T.W.; Fearing, R.; Full, R.J. Adhesive force of a single gecko foot-hair. Nature **2000**, *405* (6787), 681–685.

176. London, F. The general theory of molecular forces. Trans. Faraday. Soc. 1937, 33, 8b-26.

177. Qu, L.; Dai, L.; Stone, M.; Xia, Z.; Wang, Z.L. Carbon nanotube arrays with strong shear binding-on and easy normal lifting-off. Science **2008**, *322* (5899), 238–242.

178. Cheung, E.; Karagozler, M.E.; Park, S.; Kim, B.; Sitti, M. In *Proceedings of the 2005 IEEE/ASME*; IEEE: Piscataway, NJ, USA, 2005; 551.

179. del Campo, A.; Arzt, E. Fabrication approaches for generating complex micro and nanopatterns on polymeric surfaces. Chem. Rev. **2008**, *108* (3), 911–945.

180. Kwak, M.K.; Jeong, H.-E.; Suh, K.Y. Rational design and enhanced biocompatibility of a dry adhesive medical skin patch. Adv. Mater. **2011**, *23* (34), 3949–3953.

181. Bergström, L. Hamaker constants of inorganic materials. Adv. Colloid Interface Science **1997**, *70*, 125–169.

182. Vajpayee, S.; Jagota, A.; Hui, C.-Y. Adhesion of a fibrillar interface on wet and rough surfaces. J. Adhesion **2010**, *86* (1), 39–61.

183. Pereira, M.J.; Sundback, C.A.; Lang, N.; Cho, W.K.; Pomerantseva, I.; Ouyang, B.; Tao, S.L.; McHugh, K.; Mwizerwa, O.; Vemula, P.K.; Mochel, M.C.; Carter, D.J.; Borenstein, J.T.; Langer, R.; Ferreira, L.S.; Karp, J.M.; Masiakos, P.T. Combined surface micropatterning and reactive chemistry maximizes tissue adhesion with minimal inflammation. Adv. Healthc. Mater. **2014**, *3* (4), 565–571.

184. Lee, H.; Lee, B.P.; Messersmith, P.B. A reversible wet/dry adhesive inspired by mussels and geckos. Nature **2007**, *448* (7151), 338–341.

185. Glass, P.; Chung, H.; Washburn, N.R.; Sitti. M. Enhanced wet adhesion and shear of elastomeric micro-fiber arrays with mushroom tip geometry and a photopolymerized p(DMA-co-MEA) tip coating. Langmuir **2010**, *26* (22), 17357–17362.

186. Mahdavi, A.; Ferreira, L.; Sundback, C.; Nichol, J.W.; Chan, E.P.; Carter, D.J.; Bettinger, C.J.; Patanavanich, S.; Chignozha, L.; Ben-Joseph, E.; Galakatos, A.; Pryor, H.; Pomerantseva, I.; Masiakos, P.T.; Faquin, W.; Zumbuehl, A.; Hong, S.; Borenstein, J. Vacanti, J.; Karp, J.M. A biodegradable and biocompatible gecko-inspired tissue adhesive. Proc. Natl. Acad. Sci. U. S. A. **2008**, *105* (7), 2307–2312.

187. Glass, P.; Chung, H.; Washburn, N.R.; Sitti, M. Enhanced reversible adhesion of dopamine methacrylamide-coated elastomer microfibrillar structures under wet conditions. Langmuir **2009**, *25* (12), 6607–6612.

188. Frost, S.J.; Mawad, D.; Higgins, M.J.; Ruprai, H.; Kuchel, R.; Tilley, R.; Myers, S.; Hook, J.M.; Lauto, A. Gecko–inspired chitosan adhesive for tissue repair. NPG. Asia. Materials. **2016**, 8, e280.

189. Cho, W.K.; Ankrum, J.A.; Guo, D.; Chester, S.A.; Yang, S.Y.; Kashyap, A.; Campbell, G.A.; Wood, R.J.; Rijal, R.K.; Karnik, R.; Langer, R.; Karp, J.M. Microstructured barbs on the North American porcupine quill enable easy tissue penetration and difficult removal. Proc. Natl. Acad. Sci. USA **2012**, *109* (52), 21289–21294.

190. Yang, S.Y.; O'Cearbhaill, E.D.; Sisk, G.C.; Park, K.M.; Cho, W.K.; Villiger, M.; Bouma, B.E.; Pomahac, B.; Karp, J.M. A bio-inspired swellable microneedle adhesive for mechanical interlocking with tissue. Nat. Comm. **2013**, 4, 1702.

191. Chauvel-Lebret, D.J.; Pellen-Mussi, P.; Auroy, P.; Bonnaure-Mallet, M. Evaluation of the *in vitro* biocompatibility of various elastomers. Biomaterials **1999**, *20* (3), 291–299.

192. Martinelli, A.; Carru, G.A.; D'Ilario, L.; Caprioli, F.; Chiaretti, M.; Crisante, F.; Francolini, I.; Piozzi, A. Wet adhesion of buckypaper produced from oxidized multiwalled carbon nanotubes on soft animal tissue. ACS. Appl. Mater. Interfaces **2013**, *5* (10), 4340–4349.

Aerogels: Cellulose-Based

F. Liebner
N. Pircher
C. Schimper
E. Haimer
T. Rosenau
Division of Chemistry of Renewable Resources, University of Natural Resources and Life Sciences Vienna, Tulln, Austria

Abstract

Aerogels are solids featuring very low density, high specific surface area, and a coherent open-porous network of loosely packed, bonded particles or fibers. Their particular architecture, low weight, and other fascinating properties render aerogels intriguing materials with promising applicability in catalysis, slow-release of bioactive compounds, tissue engineering, high-performance acoustic and thermal insulation, liquid and gas sorption, gas separation, energy storage, kinetic energy absorption, or transportation, among others. Following the development of aerogels from inorganic and petrol-based organic precursors, research on aerogels from biopolymers, in particular from cellulose, has literally been booming since the turn of the millennium. This entry reviews current approaches toward aerogels from naturally occurring cellulose modifications (cellulose Iα and Iβ) and from the regenerated allomorph (cellulose II), also addressing challenges and measures to preserve the hierarchical network structure along the path from hydrogels via solvogels to the aerogels. The impact of various process parameters, such as dissolution, coagulation, and drying conditions on the properties of aerogels, is discussed. This includes also the pros and cons of direct solvents used in the manufacture of cellulose II aerogels, and the opportunities and limitations of reinforcing aerogels by chemical or physical means (cross-linking, interpenetrating networks, *all*-cellulose composites). The potential of cellulosic aerogels for commercial applications will be highlighted in the last part of this entry by some illustrative examples from the fast-developing and fascinating field of cellulosic aerogel research.

INTRODUCTION

The properties of load-bearing natural "construction" materials, such as bone or reed mace, are the result of evolution, adaptation, and optimization processes following the guiding principles of maximizing their stiffness-to-weight ratio and minimizing local stress by efficiently dissipating forces. Most of them are hierarchical, open-porous networks composed of sophisticated natural (composite) materials that are utilized by many living organisms in other life-sustaining functions too, such as accumulation of nutrients from the sea (sponges on reefs), retention and distribution of fluids (wood, bamboo), thermal insulation (cork), or protection of gametes in closed compartments (eggs).

Bacterial cellulose (BC) is another example of a highly porous material that features intriguing properties and is therefore considered to have a broad application potential in particular in biomedicine and cosmetics. It is produced by a variety of bacteria as an integral aspect of their survival strategy as it keeps the aerobic bacteria floating on the interface between aqueous nutrient medium and air. If the interstitial water filling the voids of BC is replaced by air, ultra-lightweight, open-porous aerogels with a large accessible surface are obtained. Inspired by the unique properties of BC, the huge abundance of plant cellulose as nature's most ubiquitous biopolymer, and driven by the everlasting searching for novel, lightweight, functional materials, the development of similar materials based on plant cellulose has been recently moved into the limelight of biomaterial research. The field of potential applications for lightweight, open-porous, biocompatible, and renewable-based materials is very broad and so are the requirements with regard to physical, chemical, or morphological properties, and thus a multitude of approaches to cellulosic aerogels using different source materials and techniques have been developed within the last decade.

This entry reviews the state of research in the relatively young field of cellulosic aerogels, comprising all major types of cellulose source materials. Besides the natural sources rich in cellulose I polymorph (cellulose Iα: BC; cellulose Iβ: plant cellulose), main routes to aerogels from regenerated cellulose (cellulose polymorph II) will be discussed. The respective sections will be complemented with references regarding the tailoring of key properties, such as morphology and mechanical stability, further modification of aerogels, and potential applications.

Concise Encyclopedia of Biomedical Polymers and Polymeric Biomaterials DOI: 10.1081/E-EBPPC-120051062

FROM AQUOGELS TO AEROGELS: THE CHALLENGE OF PRESERVING THE POROUS ARCHITECTURE

Porous natural materials grow and develop in biological environment and are therefore linked to aqueous conditions. If the particular structure of these materials is desired for technical applications in non-aqueous environments (insulation, gas sorption, etc.), water has to be removed beforehand. This has to be accomplished in a way that largely preserves the original network structure.

While the composition of the beams or walls of natural cellular load-bearing construction materials has been adapted to the capillary forces occurring in aqueous media, most of them can be air-dried without significant shrinking, compaction, or loss of porosity. This, however, can be quite different for synthetic inorganic or organic hydrogels in particular when the bulk density of the network forming constituent(s) is very low. Similar to silica aerogels, many biopolymer-based aerogels suffer from extensive shrinking if thermal drying is applied. This is due to capillary forces that occur alongside the capillary walls but mainly adjacent to the solvent menisci. These inward forces at the phase boundaries are most pronounced for thermal drying due to the large differences that exist in the specific energies of the three media, i.e., the void forming walls, solvent, and gas phase. According to the *Young–Laplace* equation, the absolute values of the negative hydrostatic pressure (Ψ_p) are inversely proportional to the capillary radius (r) and increase with the surface tension (σ) of the liquid that fills the pore voids (Eq. 1). Due to the high surface tension of water (72.75 mN m^{-1} at 20°C) in contact with air, hydrogels are particularly sensitive. BC hydrogels, e.g., obtained by static cultivation of *Acetobacter xylinum* AX5 wild type strain, have a density of about 8 mg cm^{-3} only. Assuming an average void radius of about 50 nm and by neglecting the particular impact of the cellulose surface, the hydrostatic pressure (tension) that develops inside the pores of such hydrogels would be in the range of 2.3–2.9 MPa (*cf.* Eq. 1). The occurrence of such strong forces inevitably has the potential to cause pore collapsing

and hence far-reaching destruction of porous materials as demonstrated in Fig. 1.

Freeze-drying of cellulose aquogels based on sublimation of water from solid state is a much better alternative for converting lyogels of low cellulose content into the corresponding aerogels. This is due to the non-existing phase boundary and hence non-existing differences in the specific energies between a liquid and gas phase (γ_{LV}), which turns the numerator of the spherical form of the *Young–Laplace* equation (Eq. 2), and hence the capillary pressure to become zero.

However, as water expands by up to 9 v% during freezing, pore collapsing and crack formation can occur. The reduction in pore volume for BC, e.g., accounts up to 10%.[1] Pore collapsing can be avoided to a certain extent by fast freezing as amorphous, very small ice particles are formed.[2] Replacement of water by solvents of a lower thermal expansion coefficient, such as *tert*-butanol, and/or higher sublimation pressure, such as ethanol, utilization of low-melting liquids, such as butane (−134°C), or the use of cryo-protectants, such as glucose, are further measures to reduce the extent of shrinkage.

Supercritical drying is considered to be the method of choice for drying highly porous, fragile materials, such as low-density cellulose gels. Similarly to freeze-drying, no liquid-gas phase boundaries exist for supercritical fluids. Hence, phenomena such as surface tension or formation of solvent menisci cannot provoke the shrinking of these materials. Furthermore, there is also no liquid-to-solid transition that could alter the open-porous cellular network structure by volume expansion and formation of sharp-edged crystals.

Carbon dioxide is probably the most frequently used supercritical fluid, as it is abundant, cheap, chemically largely inert, incombustible, easily recyclable, environmentally benign, and having a low critical point (30.98°C, 7.375 MPa). While supercritical carbon dioxide (scCO$_2$) has a density and dissolving power similar to that of fluids, it behaves at the same time like a gas, exhibiting a low dynamic viscosity η, and hence a much higher diffusion coefficient. This allows for a rapid mass transport, which is one of the main reasons why scCO$_2$ has found wide use in

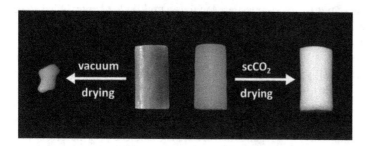

$$\Psi_p = -\frac{2 \cdot \sigma}{r} \qquad (eq.\,1)$$

$$P_{cap} = \frac{2\gamma_{LV}\cos(\theta)}{r} \qquad (eq.\,2)$$

Fig. 1 Impact of solvent and drying method on the preservation of the morphology of hardwood pre-hydrolysis kraft pulp organogels obtained by coagulation of cellulose from 3 w% containing NMMO dopes using DMSO (left set) and 2 w% pulp solutions in Ca(SCN)$_2$ octahydrate using ethanol as an anti-solvent (set on the right). Deformation during drying is due to pore collapse caused by the occurring capillary pressures, depending on the drying method—Eqs. 1 and 2 show the *Young–Laplace* equation in its simple and spherical form, respectively.

extraction protocols.[3] As a non-polar, lipophilic solvent, scCO$_2$ is not miscible with water. Therefore, hydrogels, such as natural BC have to be subject to a solvent exchange step aiming at a quantitative replacement of water by a non-polar or weakly polar organic cellulose anti-solvent that is miscible with scCO$_2$.

Solvent Exchange Prior to scCO$_2$ Drying

Replacement of the interstitial water by a suitable solvent, miscible with scCO$_2$, requires much concern with respect to the experimental protocol as the creation of strong gradients in polarity can lead to a significant reduction in volume and porosity during the solvent exchange. This is due to the different strength of solvent–polymer interactions, which is strongly determined by the surface chemistry of the respective network forming polymer(s), i.e., the abundance of hydroxyl groups in the case of cellulose. According to the *Hansen* model of solvent–polymer interactions, the cohesive energy density (expressed as *Hildebrand* solubility parameter) can be calculated as the sum of a dispersion force component, a polar component, and a hydrogen bonding component. Replacement of water by ethanol, e.g., reduces the cohesive energy density from $\delta_{SI} = 48$ MPa$^{1/2}$ to 26.5 MPa$^{1/2}$, while the total *Hildebrand* parameter of acetone or THF is $\delta_{SI} = 20.0$ MPa$^{1/2}$ and $\delta_{SI} = 19.4$ MPa$^{1/2}$, respectively. The hydrogen bonding component, which, due to the high abundance of OH groups, is supposed to be of particular importance for solvent–polymer interactions, is even more affected and decreases from $\delta_H = 42.3$ MPa$^{1/2}$ (water) to 19.4 MPa$^{1/2}$ (ethanol), 8.0 MPa$^{1/2}$ (THF), and 7.0 MPa$^{1/2}$ (acetone), respectively. The effect of cellulose anti-solvent interactions on the preservation of the fragile cellulose network structure has been demonstrated for the conversion of different BC organogels to the respective aerogels.[4]

Compared to BC aerogels prepared directly from the respective alcogels ($\sigma_{bulk} = 7.8 \pm 0.5$ mg cm^{-3}), replacement of the interstitial ethanol by acetone (9.4 ± 0.6 mg cm^{-3}) or THF (9.6 ± 0.8 mg cm^{-3}) prior to the scCO$_2$ drying step (40°C, 10 MPa) afforded somewhat higher densities reflecting thus the differences in the preceding solvent–polymer interactions. The full miscibility with carbon dioxide at comparatively mild conditions (≥ 8 MPa at 40°C), low viscosity, facileness of recovery, environmental aspects, and the comparatively low price render ethanol one of the most suitable cellulose anti-solvents for the conversion of cellulose aquogels to aerogels.

scCO$_2$ Drying

Besides the type and experimental protocol used to replace the original pore liquid by a scCO$_2$-miscible cellulose anti-solvent, the way of how the scCO$_2$ extraction of the cellulose anti-solvent is subsequently performed can decisively affect the quality of the resulting aerogel in terms of preservation of morphology, shape, porosity, etc. Critical parameters in this respect are the pressure limit beyond which ethanol extraction starts, duration of extraction, rate of depressurization, and residual traces of water.

For scCO$_2$ processes p-x, y diagrams are well suited to describe binary mixtures and their behavior upon pressure variation. Figure 2 shows the phase envelope of the binary mixture CO$_2$/ethanol enclosed by the boiling point and the dew point curves. While the intersect at $c_{CO_2} = 0$ displays the vapor pressure of pure ethanol, that at high CO$_2$ concentrations represents the critical point of the binary mixture for a certain temperature. Beyond the critical temperature of CO$_2$, both the dew point and boiling curve do no longer intersect the y-axis as there is no vapor pressure defined for that state.

Below the critical pressure of the binary system, the solubility of CO$_2$ in ethanol increases with increasing pressure according to the phase envelope (*cf.* Fig. 2, left). During the supercritical drying process, the state of the interstitial liquid filling the voids of the gel must be kept in

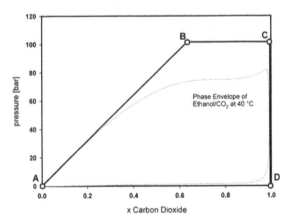

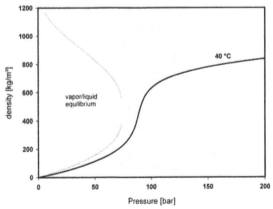

Fig. 2 Phase envelope of the binary system ethanol/CO$_2$ at 40°C (left; capital letters represent the main process steps of scCO$_2$ drying— (**A**) alcogel at ambient conditions, (**B**) critical point of the binary mixture, drying starts; (**C**) extraction of ethanol is concluded; (**D**) return to ambient conditions after depressurization) and specific density of CO$_2$ at 40°C in dependence of pressure (right).

a way that it stays outside this phase envelope at all times. At the beginning of the scCO₂ drying process, all pores of the lyogel are filled with ethanol (Point A). Upon pressurization of the system with carbon dioxide, the amount of dissolved CO₂ within the liquid phase increases forming a CO₂-expanded-liquid phase.[5] By further increasing the concentration of CO₂ in the liquid phase, its surface tension decreases significantly.[6] After reaching pressures above the mixture's critical pressure for the given temperature (Point B), full miscibility of the binary mixture-components is assured. Ethanol can then be removed from the porous matrix by flushing the system with pure CO₂. This flushing step has to be performed until the ethanol content within the pore is low enough to avoid condensation of a liquid phase upon depressurization (Point C). Finally, depressurization to atmospheric pressure gives the final product (Point D).

Below the critical temperature carbon dioxide behaves like a liquid. Starting from low gas-like densities, increasing pressure leads to condensation of CO₂ and separation of two phases, one with high density and one with low density (*cf.* Fig. 2, right). Further compression leads to an increasing volume of the liquid phase until finally all the CO₂ has liquefied. At temperatures above the critical temperature (30.98°C), increasing pressures lead to increased and finally liquid-like densities without crossing the vapor–liquid coexistence region. As no clustering of the molecules and subsequent formulation of droplets is possible at these temperatures, the density of the supercritical liquid can be freely chosen. At the drying temperature of 40°C, the density of CO₂ is gas-like up to pressures of about 7.5 MPa and liquid-like at pressures beyond 12 MPa. In between these pressure limits, the density changes strongly even with small pressure changes. The density of pure CO₂ at drying

conditions (40°C, 10 MPa) is 628 kg m⁻³, which is right between the liquid and the gaseous state. During depressurization of the aerogel, the relatively large density differences occurring in a small pressure range are considered to be one major trigger of gel-shrinking. Thus, depressurization should be performed as slowly as necessary to avoid mechanical over-straining of the material.

Furthermore, the extent of shrinkage is affected by the course of pressurization of the scCO₂ extraction unit, and hence the gels to be dried. The drying process can be divided into three steps, namely: 1) diffusion of CO₂ into the liquid phase within the pores; 2) spillage of the ethanol/CO₂ mixture due to the increased volume of the liquid phase; and 3) convective transport of the spilled mixture within the CO₂-stream out of the matrix (*cf.* Fig. 3).

Starting from a wet gel, CO₂ diffuses into the liquid phase within the pores upon pressurization, causing an increase of the volume of the liquid phase and a slight increase of the liquid phase density. As the volume of the liquid phase increases, diffusion and spillage occur in parallel. Streaming CO₂ outside of the gel and coherent convective transport should be avoided completely until the process of CO₂ diffusion is completed and the CO₂ concentration within the gel-matrix is constant.

As denoted in Fig. 3, surface tension is a function of the density difference of the two involved phases. In order to avoid shrinking of the gel during drying, the interface of the involved phases must always be outside of the gel. Although there are two phases within the porous network upon pressurization (namely, liquid ethanol and the mixture of CO₂ and ethanol), the interfacial tension can be neglected, as the density difference is marginal as long as there is no convective transport at the open end of the pore.

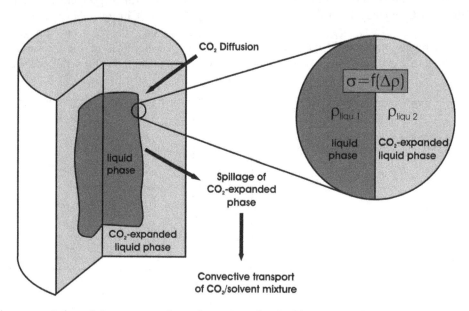

Fig. 3 Schematic representation of the mass transfer pathways associated with pressurization of the alcogel with carbon dioxide. The interfacial tension σ between the liquid ethanol phase and the CO₂/ethanol mixture within the pores of the gel upon pressurization is marginal.

The process of CO_2 diffusion into the liquid phase is rather fast and can be ignored for small gel-dimensions. When drying larger specimen, this effect has imperatively to be taken into account. After reaching uniform CO_2-concentration within the matrix, convective transport can be started.

The presence of water due to incomplete solvent exchange causes high-density differences within the pores, as CO_2 is only poorly soluble in water. In this case, very strong deformation and shrinking of the porous matrix is observed (*cf.* Fig. 4).

The removal of ethanol from the matrix by purging the system with pure carbon dioxide follows pressurization. This step is limited by several mass transfer resistances, which can be summarized to an effective diffusion coefficient (D_{eff}).[7,8] A full model for supercritical drying of aerogels regarding the above-mentioned process steps was proposed by Mukhopadhyay and Rao.[9] Exemplarily, a drying time of 120 min is recommended for cylindrical mesoporous cellulose aerogels with densities of up to about 100 mg cm^{-3}.

Depressurization as the final step in the preparation of aerogels should be started only when the residual content of the $scCO_2$ miscible solvent originally filling the pores of the gel (ethanol in this case) is surely low enough to be outside the phase envelope as shown in Fig. 2 (left). If this is not the case, capillary condensation might cause some pore collapsing, loss of specific surface area, and macroscopic shrinkage of the sample. Furthermore, depressurization should be accomplished at a fairly slow rate, not significantly exceeding 0.2 MPa min^{-1}. This is particularly important at pressures close to the critical pressure of CO_2, as small changes in pressure translate into comparatively significant changes in volume, which can be another reason for pore collapsing especially at very low density of the samples. Cooling below the critical temperature as it can occur by fast depressurization and the provoked *Joule–Thomson* effect can be another reason for volume reduction of the samples during $scCO_2$ extraction.

AEROGELS FROM CELLULOSE Iα-RICH SOURCES

BC is an extracellular natural by-product of the metabolism of various bacteria[10] with *Acetobacter* spp. strains being most frequently used in commercial BC production lines. *Acetobacter xylinum* cultivated in large open tanks using coconut water as nutrient medium affords the well-known *Nata de Coco* (*Spanish: cream of coconut*), which is mainly produced in Southeast Asian countries (Philippines, Indonesia) and commercialized as dietary auxiliary. *Nata de Coco* is a chewy, translucent, colorless, tasteless, fibrous hydrogel consisting of pure BC and water and is used in many Asian desserts and drinks. BC is furthermore produced during the fermentation of sweetened tea (*Kombucha tea*) by the symbiotic activity of *ascomycetes* and *Acetobacter species*. "Mother of vinegar," which is used to convert alcoholic beverages into vinegar and grows on the surface of fermenting wine or cider, is another jelly-like material that is composed of BC and acetic acid.

The formation of BC is considered to be an integral aspect of the survival strategy of many microorganisms (e.g., *Acetobacter xylinum*) as it keeps the aerobic bacteria floating on the surface of the aqueous nutrient medium in direct contact with oxygen. Other bacteria, such as the plant pathogen *Agrobacterium tumefaciens*, use cellulose for better attachment to plants, similar to the symbiotic *Rhizobium* spp. The production of the very lightweight yet comparatively stiff BC networks is accomplished by "biological spinnerets" lined up in BC producing bacteria that release cellulose as elementary fibrils, which subsequently aggregate to ribbons. As the cellulose synthesizing sites are

Cellulose #2 Cellulose #10

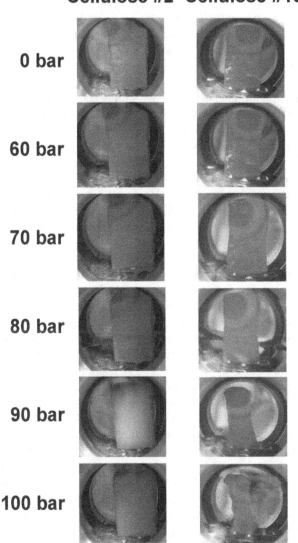

0 bar

60 bar

70 bar

80 bar

90 bar

100 bar

Fig. 4 Effect of water during pressurization of CL alcogels with CO_2: Samples soaked with anhydrous ethanol (left) or ethanol that contained traces of water (right).

duplicated during cell division,[11] mother and daughter cells are each connected to one and the same cellulose ribbon, which forms an interconnected three-dimensional network with a high number of entanglements unsurpassed by aerogels from regenerated cellulose of comparable bulk density.

Batch-wise static cultivation of respective bacteria using large tanks and suitable aqueous growth media is still by far the most frequently applied technology to produce BC. After inoculation of the culture medium with the bacteria strain and a short lag phase, secretion of the exo-polysaccharide on the surface of the growth medium sets in. The thickness of the cellulose layer increases with the cultivation time, which can last up to several weeks. The harvested cellulose is typically purified from the bacteria culture and the remnants of the culture medium by repeated alkaline treatment (e.g., 0.1 M aqueous NaOH at 90°C, 20 min) and thorough washing with (deionized) water.[12] The purified material can be disintegrated to afford nanofibril suspensions, which can be pressed during drying to obtain membranes of uniaxially or uniplanary oriented nanofibrils. They can also easily be cut into sheets or geometric objects of desired size and shape.

Its particular morphology, high purity, molecular weight, crystallinity, fiber strength, water retaining, and moistening capabilities render BC a very interesting biomaterial for an increasing number of applications mainly in biomedicine and cosmetics (cf. section "Applications of Cellulosic Aerogels"). Therefore, current efforts aim to improve existing methods or to develop novel technologies that allow for a more efficient production of BC in terms of production rate, yield, and costs, with the latter being directly linked to energy consumption of respective large-scale production units.[13]

Besides (agitated) static cultivation, airlift reactors, aerosol reactors, or rotary bioreactors are increasingly used. While in aerosol reactors the nutrients are sprayed onto the surface of the growing BC pellicles (thicknesses of up to 7 cm can be reached), rotary bioreactors consist of vertically aligned rotating discs that are alternately exposed segment-wise to the culture medium and air. In membrane bioreactors, the nutrients are supplied through a membrane that separates BC from the circulating culture medium. However, as with shaken or agitated cultures that greatly increase the growth rate of bacteria, the increased risk of mutation can lead to significant reduction in BC yield.[14] In horizontal lift reactors, BC is produced in a continuous way by slowly pulling the BC sheets from respective tanks containing the culture medium.[13]

With respect to future applications, there is also broad consent that successful commercialization of BC will largely depend on its particular properties and hence the targeted special fields of applications where its relatively high price can be justified by the material's performance. Many attempts have been made to tailor the properties of BC (morphology, pore features, strength, surface chemistry, etc.) for particular applications already during its biosynthesis. While the effects of the used bacteria strain, composition of the culture medium, oxygen supply, pH and temperature, and type of bioreactor on the BC yield has been comprehensively reviewed elsewhere,[13] the impact of additives shall be summarized in more detail.

The ultra-structure of cellulose ribbons, their size, intensity of entanglement, and the density of the cellulose pellicles have a large impact on the mechanical properties of BC.[15] Hence, all measures that can alter self-alignment of BC nanofibrils, entanglement of ribbons, and network density during BC production can potentially contribute to control the bulk properties of BC. The addition of xylan to the culture medium, e.g., inhibits ribbon formation by covering the surface of individual microfibrils and decreases both the crystal size and cellulose Iα content of the BC formed.[16] The same effect but at higher intensity is observed when glucomannan is added to Hestrin–Schramm (HS) medium while the addition of pectin has virtually no effect.[17] The crystallinity of cellulose as determined by X-ray diffraction is reduced from 56% to 28% when glucose is replaced by a rice bark extract.[18] The addition of carboxymethyl cellulose (CMC) or hydroxypropyl methyl cellulose (HPMC) to the culture medium has a similar effect, but significantly improves the rehydration ability of BC. Winding of CMC or HPMC around microfibrils creates a larger number of amorphous regions that promote water permeation into the cellulose network and swelling.[19] The addition of microstructuring organic sources (see also the paragraph about porogens in the following text), such as multi-walled carbon nanotubes (MWCNTs) to HS medium inoculated with Acetobacter xylinum, was shown to afford more rigid cellulosic pore walls compared to the reference sample grown on MWCNT-free HS medium.[20] Furthermore, the presence of MWCNT weakened the intermolecular hydrogen bonds of cellulose leading to reduced crystallinity index (CrI), crystal size, and cellulose Iα content. A similar result was observed when wax spheres were added to the growth medium to control the pore size of the formed BC sheets.

The utilization of BC in most applications and in particular in tissue engineering requires effective measures that allow for a far-reaching control of the pore features to provide a cell scaffolding material that supports cell attachment, ingrowth, proliferation, differentiation, diffusion of physiological nutrients and gases to cells, removal of metabolic by-products from cells, cell shaping, reorganization, and gene expression. PEG, e.g., has been demonstrated to be a suitable pore-size modulating culture medium additive, which is, however, not incorporated into the BC network structure-like hemicelluloses or cellulose derivatives (e.g., MC, CMC). Depending on its molecular weight, smaller (PEG, DS 4000) or larger (PEG, DS 400) pores, respectively, compared to the reference material can be obtained. Pore widening has also been observed for β-cyclodextrin.[21] Honeycomb-patterned BC films have been prepared by controlling the movement of cellulose producing bacteria using agarose films equipped with concave honeycomb-patterned grooves.[22] Next to surfactants that can also control the alignment of cellulose fibrils and ribbons,[23] temporary porogens are frequently used too, such as water-soluble salt spheres (e.g., NaCl),[24] carbon nanotubes,[20] paraffin spheres,[25] ice crystals[26] or hydrogel beads from natural polymers such as gelatin,[27] and synthetic polymers such as poly(ethylene glycol).[28] Paraffin spheres that had been suggested as porogens for poly-L-lactic acid (PLLA) and poly(lactic-co-glycolic acid) (PLGA) gels[25] turned out to be also well suited for tuning the porosity of BC toward macroporosity.[29,30]

The conversion of BC hydrogels into the respective aerogels follows the principles as described in the previous section. Water is typically replaced by ethanol prior to the scCO$_2$ extraction step. Depending on the way of BC cultivation, work-up procedure, and scCO$_2$ drying conditions, bulk densities of down to 5.4 mg cm^{-3}[31] and 8.25 ± 0.7 mg cm^{-3} (n = 10), respectively, have been obtained.[12] Drying at temperatures beyond 240°C[32] is not recommended as the chemical integrity of cellulose suffers increasingly from intra- and intermolecular dehydratization reactions.[33–38] Low temperature (40°C) and pressure (10 MPa) have been demonstrated to virtually fully preserve the original cellulose network morphology.[12]

Interestingly, BC resists much better volume reduction during solvent exchange and subsequent scCO$_2$ extraction than it is the case for aquogels of comparable density obtained by coagulation of cellulose from solution state. While aquogels from coagulated commercial pulps three times denser than BC loose at least 25 percent of their original volume,[39] shrinkage is very little for BC (Fig. 5).[12] The full preservation of porosity for BC gels throughout the aerogel preparation is evident from the fact that after soaking the BC aerogels in water, the weight of the original aquogel is virtually fully recovered.[40] BC aerogels were furthermore shown to resist shrinkage during long-term storage even under humid conditions. Whereas cotton linter (CL) aerogels obtained by coagulation (ethanol) of cellulose from respective 3 w% CL containing Lyocell dopes and scCO$_2$ drying suffered from a significant reduction by volume during 14 days of storage (15.5% at 30%, 50.9% at 65%, and 84.2% at 98% relative humidity), BC aerogels of much lower density were confirmed to resist shrinking completely even at 98% relative humidity (r.h.) for at least several days. The almost zero shrinkage during scCO$_2$ drying and open-air storage is supposed to be mainly due to the high portion of crystalline domains, the high number of entanglements, and supposedly higher stiffness of individual BC ribbons.

Its open-porous structure and the full wettability of BC aerogels can be employed for controlled release applications (see following text). The release profiles of D-panthenol and L-ascorbic acid from BC gels of different thickness, e.g., have been shown to be largely independent on the amount of loaded compound due to negligible surface-solute interactions but highly dependent on the thickness of the aerogel layers.[40] It has been furthermore demonstrated

that the mainly diffusion-controlled release of these two compounds can be reliably predicted with the Korsmeyer model that considers both the diffusion of water into an open-porous matrix and simultaneously that of a given organic compound out of it, using an experimentally determined effective diffusion coefficient.[40]

Small-angle X-ray scattering, nitrogen sorption at 77K (calculation of the specific pore surface area by applying the models developed by Brunauer, Emmett, and Teller (BET) and Benjamin, Johnson, and Hui (BJH)), and thermoporosimetry with o-xylene as "confined" solvent confirmed that the dimension of the voids between the nanofibers correspond to interconnected micro-, meso-, and macropores. In particular, smaller macropores of around 100 nm in diameter contribute mostly to the BC aerogel's overall porosity (Fig. 6, left), which is in good agreement with different series of scanning electron microscopy (SEM) and environmental scanning electron microscopy pictures.[4]

Compression tests along the three orthogonal spatial directions revealed that BC aerogels are transversely isotropic (Fig. 6, right). BC obtained by static cultivation features two directions with higher stiffness and strength and a third direction with lower values, the latter being the growth direction of BC, which is perpendicular to the interface between culture medium and air. The latter were obtained by scCO$_2$ drying (40°C, 10 MPa, 2 hr) of respective BC samples after replacing the interstitial water by ethanol. While a Young's modulus of $E = 0.057 ± 0.007$ MPa and yield strength of $R_{P,0.2} = 4.65 ± 0.48$ kPa was observed along the growth direction, the respective values for the other two spatial directions were significantly higher (A: $E = 0.149 ± 0.023$ MPa, $R_{P,0.2} = 7.05 ± 0.55$ kPa; B: $E = 0.140 ± 0.036$ MPa, $R_{P,0.2} = 7.84 ± 1.06$ kPa). The overall smoothness of the curve and the absence of a peak after the linear elastic region indicate that the material deforms in a ductile way on the microscale, in contrast to brittle foams and silica aerogels.

Interestingly, no sample buckling was observed during compression. The Poisson ratio, which describes the change of the cross-section area upon application of mechanical stress, being in the range of 0.1 to 0.3 for silica aerogels, was approximately zero for BC aerogels independent of the loading direction. This is in good agreement with Sescousse et al.[41] who also reported a zero Poisson ratio for cellulosic aerogels from [EMIM][OAc] solution, similar to cork.

Mechanical Reinforcement

Cellulosic aerogels and in particular BC aerogels are comparatively prone to mechanical stress, which is considered a drawback for a couple of applications. Depending on the envisaged type of utilization and the required mechanical stability, different reinforcing approaches are generally applicable, such as the preparation of all-cellulose composite materials, cross-linking, insertion of an interpenetrating

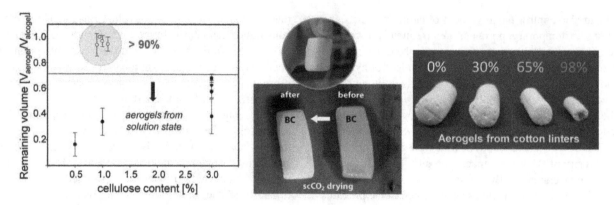

Fig. 5   scCO$_2$ drying at 40°C and 10 MPa virtually fully preserves the cellulosic network structure of BC alcogels and affords ultra-lightweight materials that can be kept in suspense just by surface roughness (left encircled and middle). Shrinkage of aerogels obtained from 3 w% CL containing Lyocell dopes during storage at different levels of relative humidity (right).

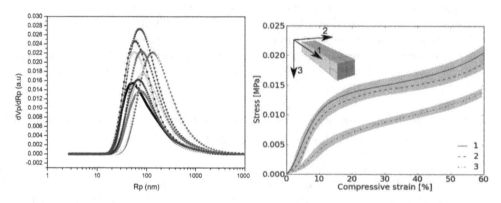

Fig. 6   Pore size distribution (thermoporosimetry) of BC aerogels obtained by scCO$_2$ drying of respective BC pellicles formed by a variety of single or mixed bacteria strains (left). Stress–strain curves of BC aerogels under compression in three orthogonal directions (direction 3: direction of growth; lines are mean values of stress, gray background represents standard deviation, sample size n = 5).
Source: From Liebner et al.[1] © 2012, with permission from American Chemical Society.

network consisting of a second polymer, or the incorporation of network stiffening inorganic or organic particles. However, all of the aforementioned techniques are still in their infancies regarding their adaptation to cellulose aerogels, as controlling the mechanical properties under preservation of the interconnected, high porosity—certainly the most valuable feature of aerogels—is a challenging task.

Interpenetrating networks

Immersion of BC aquogels in silica sols containing different amounts of silica nanoparticles has been shown to be a suitable approach to increase the strength of BC aerogels.[42] However, it turned out to be difficult to load the BC with more than 10 w% of silica in this way. Loadings of up to 50% silica were achieved when the cellulose-producing bacteria strain (*Gluconacetobacter xylinus*) was incubated in a medium that contained the respective silica sol Snowtex 0 (ST 0, pH 2–4) or Snowtex 20 (ST 20, pH 9.5–10.0). Enhanced elastic moduli were observed for silica contents below 4% (ST 20) and 8.7% (ST 0), respectively, whereas higher silica contents led to reduced strength and modulus

of the aerogels. Interpenetrating networks have also been obtained with (derivatized) natural and synthetic polymers. Cationic starch, such as 2-hydroxy-3-trimethyl-ammoniumpropyl starch chloride (TMAP starch) added to the growth medium of *G. xylinus* forms stabilized double-network composites and is incorporated into the wide-mashed BC pre-polymer already during the first 2 days of incubation.[21] Similarly, the addition of 0.5, 1.0, and 2.0% (m/v) CMC or methyl cellulose (MC) to the culture medium have been reported to increase the yield, the fraction of amorphous domains, water retention ability, and ion absorption capacity.[43] Reinforcement with biocompatible polymers, such as polylactic acid (PLA), polycaprolactone (PCL), cellulose acetate (CA), and poly(methyl methacrylate) (PMMA) using scCO$_2$ anti-solvent precipitation as a core technique for forming an interpenetrating secondary polymer network has been recently demonstrated to be a suitable approach to improve the mechanical properties of BC aerogels.[4] BC/CA and BC/PMMA composite aerogels featured the highest gain in specific modulus (density-normalized modulus, E_ρ) compared to pure BC aerogels. For a BC/polymer ratio of 1:8, the respective E_ρ was found to be as high as 50 and 122 MPa cm^{-3} g^{-1}, respectively. The specific modulus of CL aerogels (obtained by coagulation of cellulose from a 1 w% solution in calcium thiocyanate) reinforced with CA at the same BC/polymer ratio exceeded that of the respective BC/CA composite samples by a factor of three.[4] Sleeving of BC fibers with acrylate polymers by in situ atom transfer radical polymerization of methyl methacrylate and *n*-butyl acrylate (BC-*g*-PMMA, BC-*g*-PBA, BC-*g*-PMMA-*co*-PBA)[44] or UV-induced cross-linking radical polymerization of different methacrylate monomer mixtures swollen in BC (glycerol monomethacrylate, 2-hydroxyethyl methacrylate, 2-ethoxyethyl methacrylate) are further examples for the creation of interpenetrating networks in BC-based aerogels.[45]

Cross-linking

Cross-linking of never-dried, microfibrillated, TEMPO-oxidized BC with chitosan (CTS) using 1-ethyl-3-(3-dimethylaminopropyl)carbodiimide (EDC) and *N*-hydroxysuccinimide (NHS) as cross-linking mediator has been recently demonstrated to afford dimensionally stable, highly macroporous cellulosic scaffolds for tissue engineering applications.[46] Following cross-linking at room temperature in slightly acidic aqueous medium (pH 5.5–6),[46,47] subsequent dialysis (removal of excess of EDC and NHS) and concentrating the suspension by dialysis against aqueous polyethylene, the BC/CTS slurry was degassed and cast into molds. The samples were deep frozen (e.g., −30°C) to solidify the solvent and to induce liquid-solid phase separation. BC/CTS aerogels containing 60% of microfibrillated BC, e.g., had an average pore diameter of 284 ± 32 mm, which is more than three orders of magnitude larger than that of native BC and meets the requirements of cell scaffolding materials.[48] Microfibrillation of BC prior to TEMPO-oxidation and subsequent cross-linking with chitosan using the preceding EDC/NHS mediator system can be used to further tune the microstructure (porosity of 120–280 nm) and mechanical properties of respective scaffolds for tissue engineering.[49]

AEROGELS FROM CELLULOSE Iβ-RICH SOURCES

Cellulosic materials can be disintegrated to nanoparticles of different size, shape, morphology, crystallinity, or surface polarity by appropriate mechanical, enzymatic, or chemical treatment or combinations thereof. Depending on the disintegration technique applied, either cellulose microfibrils [microfibrillated cellulose (MFC), also nanofibrillated cellulose (NFC)] or cellulose nanocrystals [(CNC), also cellulose nanowhiskers, nanocrystalline cellulose (NCC)] are obtained. All of these techniques—no matter if MFC or CNC is the desired product—aim at a full preservation of the native cellulose I crystal structure as it is present in the respective parent materials. This is desirable as the mechanical

properties of both cellulose Iα (principal modification in BC) and cellulose Iβ (major type in plant cellulose) are superior to those of cellulose II, which is inevitably obtained if cellulose is recovered from solution-state. The addressed differences are caused by particular molecular features of the cellulose modifications. Based on the molecular mechanics deformation of a number of comprehensively investigated cellulose I and II model structures (Synchrotron X-ray and Neutron Fiber Diffraction),[50–52] stress–strain curves have been derived by Eichhorn and Davies which revealed a chain stiffness of 155 GPa for cellulose Iα, 149 GPa for cellulose Iβ, and 109 GPa for cellulose II, respectively.[53] This value (Ib) is in good agreement with the elastic modulus of single tunicate cellulose microfibrils (151 GPa), as measured by atomic force microscopy.[54]

In homogenous suspension cellulose I microfibrils have a strong tendency to entangle and form stable solvogels. This effect along with the higher stiffness and CrI of cellulose materials that, contrary to cellulose II, have not been subjected to a dissolution/coagulation beforehand is of particular interest for the preparation of aerogels, as the laborious solvent exchange and scCO$_2$ drying steps can be replaced by freeze-drying at a comparably good preservation of the cellulose network structure.[55] Besides drying, the method applied for cellulose disintegration has a strong impact on the properties of MFC or CNC aerogels, as their morphological features, mechanical stability, or void size and shape largely depend on the properties of the respective cellulose I particles (size, shape, morphology, crystallinity, etc.).

While the intriguing mechanical properties of MFC-based materials evoked considerable interest and research activities over the last decade, a major challenge en-route to economically feasible production lines is its energy-intensive production, especially if high-pressure homogenization is applied. However, alternative methods like micro-fluidization or micro-grinding require less energy and enable the reduction or even elimination of refining pretreatment. In any case, due to the ongoing efforts to further develop disintegration techniques, MFC becomes increasingly competitive and available in larger quantities.[56,57]

MFC Aerogels from Mechanically Disintegrated Cellulose

The variety of mechanical treatment techniques and combinations thereof that afford stable aqueous suspensions of cellulose nanofibers is considerably large and includes milling, beating, grinding, high-pressure homogenization, or micro-fluidization. The obtained suspensions can be easily converted into aerogels either by drying in a vacuum oven,[58] by freeze-drying after ultra-sonication,[59] or by freeze-drying after solvent-exchange. Replacement of water by an organic solvent prior to freeze-drying can be a suitable measure to increase the specific surface area of aerogels. This has been demonstrated for aerogels prepared from a MFC sample, which was produced from never dried hardwood pulp by beating the latter in aqueous suspension (2 w%, Valley beater) for 6 hr prior to high-pressure homogenization (500 kg cm^{-2}, 45 min). Compared to freeze-drying from aqueous suspension that afforded aerogels with an internal surface area of 13 m^2 g^{-1}, significantly higher surface areas of up to 100 m^2 g^{-1} were obtained when water was replaced by *tert*-butanol prior to the drying step due to reduced fibril aggregation.[60]

Chen et al.[59] prepared MFC by purification and defibrillation of wood fibers. In the course of the disintegration treatment, cellulose crystallinity increased from 52.0% to 66.3%. Flexible (Fig. 7) and highly porous aerogels (up to >99.9%) were obtained by freeze-drying of aqueous MFC suspensions. Freeze-drying of the supernatant obtained by centrifugation of nanofiber suspensions afforded aerogels of ultra-low densities down to 0.2 mg cm^{-3}.[59]

For an aerogel with a density of 24 mg cm^{-3}, which has been obtained by freeze-drying of homogenized aqueous native birch pulp suspension, a specific compression modulus as high as 21.8 MPa g^{-1} cm^3 was determined,[61] which is in the same range of values obtained for BC aerogels (19–25 MPa g^{-1} cm^3).[4]

MFC Aerogels from Enzymatically/Mechanically Disintegrated Cellulose

Henriksson et al. developed an enzyme-assisted approach to MFC, which is frequently applied to disintegrate cellulosic materials prior to the preparation of aerogels. The sequential process comprised: 1) mechanical pre-treatment of the cellulosic material, e.g., by Papirindustriens Forsknings

Institute (PFI) milling at 1000 revolutions, 2) enzymatic hydrolysis by an endoglucanase preparation (Novozyme 476), 3) mechanical shearing (PFI milling, 4000 revolutions), and 4) high-pressure homogenization of diluted fiber suspensions (e.g., 2 w%, 20 cycles). Commercial bleached Norway Spruce sulfite pulp, e.g., afforded microfibrils with average diameters of 15–30 nm and lengths of several micrometers.[62] Aqueous MFC suspensions have been used to study the impact of different drying techniques on the properties of the obtained aerogels. The drying techniques included: 1) cryogenic freeze-drying (liquid propane, −180°C) followed by vacuum-drying of the sample maintaining its frozen state throughout the entire drying step;[55] 2) conventional freeze-drying;[55] and 3) scCO$_2$-drying of gels that had been subject to a solvent exchange (water was replaced by acetone) beforehand.[63] In a similar way, the effects of different protocols related to a solvent exchange prior to freeze-drying were studied.[64] In summary, it can be concluded that both the general drying technique (freeze-drying vs. scCO$_2$ drying) and side-aspects—such as the implementation of a solvent exchange step, the type of organic solvent used, the increment of increasing its concentration in the gel, the freezing method as well as drying parameters, such as time and temperature—largely affect the properties of aerogels in terms of density, morphology, and porosity. While the homogeneous cellulose network architecture even of fragile, ultra-lightweight aerogels can be largely preserved throughout the scCO$_2$ drying step, freezing prior to vacuum sublimation ("freeze-drying") is accompanied by the formation of ice crystals that can cause significant alterations of the cellulose network morphology depending on the rate of

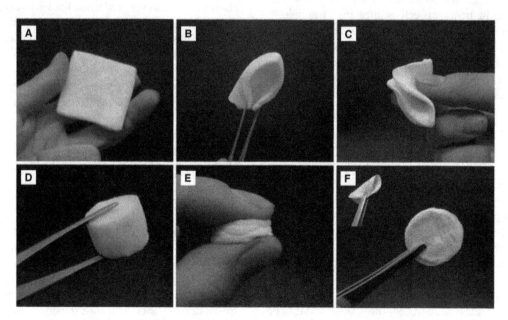

Fig. 7 Deformable and flexible cellulose Iβ aerogels with ultralow densities of 2.6 mg cm^{-3} (**A–C**) and 9.2 mg cm^{-3} (**D**). The sheets can be bent repeatedly (also after compression (**F**)) without destroying the structural integrity.
Source: From Chen et al.,[59] with permission from The Royal Society of Chemistry.

crystal growth and the size of crystallites formed. Deep-freezing of MFC aquogels (never-dried hardwood kraft pulp) using liquid nitrogen ($-196°C$) prior to drying, e.g., affords a rather heterogeneous network morphology that consists of large sheet-like cellulose aggregates interconnected by cellulose fibrils. In contrast, cooling with liquid propane (ca. $-100°C$) affords smaller crystals and prevents fibrils largely from aggregation.[63]

Both partial and full replacement of water by sublimable organic compounds prior to freeze-drying can be beneficial with regard to porosity, specific void surface area, and homogeneity of the formed fibrillary network. This has been demonstrated for an aqueous 2 w% MFC suspension (softwood sulfite pulp) by comparing conventional freeze-drying with approaches where either a major fraction (one-step solvent exchange) or virtually all water (six steps via ethanol) was replaced by t-butanol prior to deep-freezing (liquid nitrogen) and sublimation.[64] While the highest internal surface area of conventionally freeze-dried samples (freezing in liquid propane at $-180°C$)[55] reported to date was 66 m^2 g^{-1} only, replacement of at least a major fraction of water by tert-butanol increased the surface area to 153 m^2 g^{-1} (one-step solvent exchange) and 249 m^2 g^{-1} (six-step solvent exchange via ethanol, freezing in liquid nitrogen prior to vacuum sublimation), respectively.[64]

Besides the drying method (including pretreatment steps, such as solvent exchange or freezing), the solid content of the respective MFC suspensions is another factor that has a significant impact on the morphology and properties of the aerogels. Higher contents result in tighter hydrogen bonding of neighboring microfibrils and the formation of lamellar structures, whereas highly oriented ultrafine fibril networks are obtained from highly diluted suspensions. Other factors promoting self-assembly of cellulose particles are larger particle size, weaker mutual electrostatic repulsion (fewer number of sulfate ester groups), and more surface hydroxyl groups.[65]

A series of MFC aerogels covering a broad range of densities (7–103 mg cm^{-3}) have been prepared from aqueous suspensions of varying MFC content by cryogenic freeze-drying using liquid nitrogen.[66] The cellular materials that suffered only little from shrinkage had porosities of 93.1%–99.5%, compressive moduli (E) between 56 and 5310 kPa, and yield strengths (R_p) ranging from 7.8 to 516 kPa.

The addition of minor amounts of xyloglucan (XG) to the MFC suspension prior to freeze-drying—an attempt to mimic the cellulose–XG interaction in plant cell walls—has been demonstrated to increase the mechanical strength of the obtained aerogels significantly. For example, replacement of 30 w% of MFC by XG in a low-density aerogel ($\rho = 22$ mg cm^{-3}) afforded porous materials of considerably higher mechanical stability as evident from compressive modulus (440 kPa vs. 1470 kPa) and yield strength (31.9 kPa vs. 112.5 kPa) at comparable bulk density ($\rho = 20.2$ mg cm^{-3}).[66]

Aerogels from Chemically/Mechanically Disintegrated Cellulose

Chemical alteration of cellulosic materials prior to mechanical disintegration is considered to be an effective means to 1) reduce the costs of MFC/NFC production and/or 2) afford nanoparticles with interesting properties, such as surface charge, increased/decreased hydrophilicity, increased degree of crystallinity, and lower fibril agglomeration tendency, among others. It can be accomplished by partial hydrolysis, derivatization (e.g., carboxymethylation), or oxidation, such as with 2,2,6,6-tetramethylpiperidine-1-oxyl (TEMPO).

Partially hydrolyzed cellulose

Aerogels from CNC are still rarely found in the literature. However, Heath and Thielemans[67] recently demonstrated that cellulose nanowhiskers—obtained by sulfuric acid hydrolysis of cotton and subsequent surface protonation using ion exchange resins—can afford open-porous materials of extensive nanoscale structure and very high internal surface areas of up to 605 m^2 g^{-1}. This has been accomplished by 1) suspending freeze-dried (liquid nitrogen) nanowhiskers in water; 2) ultrasonication of this suspension until a hydrogel was formed; 3) replacement of water by ethanol; and 4) conversion of alcogels to aerogels by $scCO_2$ drying. The obtained aerogels featured porosities of 91–95% and had a high crystallinity of 88.6%.[67] Particle lengths reported for CNC prepared under similar conditions were in the range of only 100–300 nm length, at an average thickness of 7.3 nm.[68]

Carboxymethyl-MFC aerogels

Chemical introduction of charged moieties onto the surface of cellulose fiber walls prior to disintegration is an approach that can significantly improve the homogenization step as stronger swelling and higher electrostatic repulsion between the liberated particles afford smaller and more uniformly sized fibrils with a typical width of 5–15 nm and length of up to 1 μm.[69,70] It can be accomplished, e.g., by carboxymethylation, and optional conversion of the formed carboxyl groups into the respective sodium carboxylates further promotes delamination of fibers to nanofibrils. High-pressure homogenization of carboxymethylated cellulose has been applied to prepare stable MFC aquogels, which were then converted into aerogels by freeze-drying using liquid nitrogen as a cooling medium.[71] At densities lower than 0.3 mg cm^{-3}, macroporous materials consisting of a largely homogeneous fibril network can be obtained (Fig. 8A). With increasing density (Fig. 8B–F), however, the morphology of the aerogels changes increasingly into sheet-like aggregates of nanofibrils, leading to larger void sizes and significant loss of interconnectivity (Fig. 8C, $\rho_{Bulk} = 5.3$ mg cm^{-3}). At a bulk density of 30 mg cm^{-3}, the pore walls are continuous (Fig. 8F).[71] For low-density carboxymethyl-MFC aerogels (4–14 mg cm⁻³) high porosities of 99.1–99.8% can be obtained at maximum surface area of 42 m² g⁻¹.[70]

TEMPO-oxidized MFC aerogels

2,2,6,6-Tetramethylpiperidine-1-oxyl (TEMPO)-mediated oxidation of cellulose and subsequent mild mechanical treatment to disintegrate the fibrils in water is the most frequently used chemically assisted approach to MFC.[72] The advantage of TEMPO-oxidized MFC is their significantly reduced adhesion tendency in aqueous suspension as the introduction of carboxylate and aldehyde groups largely prevents interfibril hydrogen bonding. Transmission electron microscopic observation of fibrils obtained from TEMPO-mediated oxidation of commercial hardwood bleached kraft pulp (carboxylate content ca. 1.5 mmol g⁻¹) and subsequent stirring of an aqueous 0.1 % (w/v) suspension (1.500 rpm, room temperature, 6–10 days, exclusion of air) revealed an average cellulose fibril width of about 3–5 nm and a length of several microns. The preservation of cellulose crystallinity throughout TEMPO-oxidation is another promising feature of this approach.[73]

Similar as with carboxymethylation, TEMPO-oxidation of cellulose prior to disintegration is a suitable means for tailoring the morphology of aerogels. While freeze-drying (−80°C fridge) of an aqueous 0.8 w% MFC suspension obtained by micro-fluidization of beech sulfite pulp afforded a network of rather inhomogeneous fibril aggregates (Fig. 9A), a significantly more regular macroporous aerogel consisting of pocket-like pores was obtained when the same material had been subject to TEMPO oxidation beforehand (0.1 mmol TEMPO, 10 mml NaClO₂ and 1 mmol NaOCl per 1 g of cellulose, 2 hr of stirring at pH 6.8 and 60°C; Fig. 9B).

The extent of cellulose oxidation seems to play an important role with respect to morphological and mechanical properties of aerogels obtained by freeze-drying of TEMPO-oxidized MFC suspensions. This has been concluded from the observation that doubling the degree of oxidation from 0.1 to 0.2 affords a more homogeneous aerogel morphology and significantly enhanced mechanical strength. Time domain–nuclear magnetic resonance spectroscopy (TD-NMR) indicated that carboxyl groups promote the interaction between cellulose chains and water, thus promoting the formation of larger and more uniform pores.[74]

Sehaqui Zhou, and Berglund reported that aerogels of higher specific surface area can be obtained when the MFC is prepared via TEMPO-mediated oxidation instead of enzymatic/mechanical pretreatment (see earlier). How the interstitial water, filling the pores of the MFC gel, is replaced by t-butanol prior to freeze-drying (liquid nitrogen) is here of minor impact for the specific surface area, as evident from the rather small differences between the variants (one step: 254 m² g⁻¹; six steps via ethanol: 284 m² g⁻¹).[64]

Comparable aerogels but with a higher internal surface area and lower densities have been obtained from mechanically disintegrated, TEMPO-oxidized MFC by drop-wise addition of HCl to aqueous suspensions of two different MFC contents (0.4 and 0.8 w/v%, respectively). After gelation, incremental, consecutive replacement of water by ethanol and t-butanol has been employed prior to the freeze-drying step using liquid nitrogen as well. Protonation of the carboxylate groups was demonstrated to afford aerogels of large specific surface areas (349 and 338 m² g⁻¹). However, the specific compression moduli (12.5 and 13.9 MPa g⁻¹ cm³) of the lightweight materials (r = 5.1 and 9.7 mg cm⁻³) were somewhat lower compared to BC aerogels of similar density (see earlier).[75]

Freeze-drying or solution casting of TEMPO-MFC dispersions that contained small amounts of AgNO₃ (0.2 and 0.5 mmol Ag⁺ · g_{MFC}⁻¹) have been demonstrated to be a suitable approach to MFC-hydrogels, -aerogels, and -films functionalized with silver nanoparticles. The obtained materials are promising wound dressing materials as they show antimicrobial activity against both gram-negative (Escherichia coli) and gram-positive (Staphylococcus aureus) bacteria as exemplarily demonstrated for the respective hydrogels. The addition of silver ions triggered a rapid gelation of the aqueous 1.27% (w/v) TEMPO-MFC suspensions, which is assumed to be due to the strong interaction of MFC-carboxylate groups with Ag⁺ and the associated decrease of negative surface charge. The obtained colorless, strong gels that could not be deformed by shaking (Fig. 10A) turned brownish upon standing for 5 days indicating slow reduction of Ag⁺ by MFC hydroxyl groups. Respective MFC/Ag⁰-aerogel discs (48 mm × 5 mm) were obtained after washing, freeze-drying (ethanol/dry ice bath), and reduction of Ag⁺ to Ag nanoparticles by UV light (λ = 320–395 nm) irradiation from both, top and bottom sides, and had internal surface areas of 8–15 m² g⁻¹. Advantageously, the intensity of the brownish Ag⁰ color pattern (Fig. 10B,C) allows for macroscopic observation of variances in pore size, generated by a topographical variation of the cooling rates that decrease from the gel surface to the core of the samples, where the slow formation of ice crystals affords bigger crystals and hence larger pores.[76]

In a similar approach, Koga et al. replaced sodium on TEMPO carboxylate groups by Cu(I) aiming at the preparation of TEMPO-MFC aerogels that can be used in azide-alkyne [3+2] cycloaddition "click" reactions. Loading of TEMPO-MFC with the copper catalyst was accomplished by the addition of a mixture of aqueous solutions of Cu(NO₃)₂ and sodium L-ascorbate—the latter reduces Cu(II) to Cu(I)—and acetonitrile to an aqueous suspension of the respective cellulose derivative. Following freeze-drying at −30°C, the aerogels were repeatedly rinsed with deionized water and vacuum-dried to remove residual impurities. The porous hybrid materials (11.4 m² g⁻¹) possessed excellent catalytic selectivity and afforded 1,4-disubstituted 1,2,3-triazoles with a turnover frequency of

Adhesives—Bioabsorbable

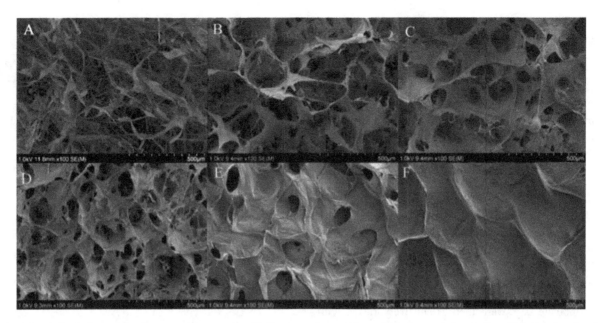

Fig. 8 Morphological transition in terms of progressive fibril aggregation and loss of interconnected porosity with increasing density of carboxymethyl-MFC aerogels from 0.27 (**A**) to 30 mg cm^{-3} (**F**). Applications as wound dressing materials are considered due to their antimicrobial activity.
Source: From Aulin et al.,[71] with permission from The Royal Society of Chemistry.

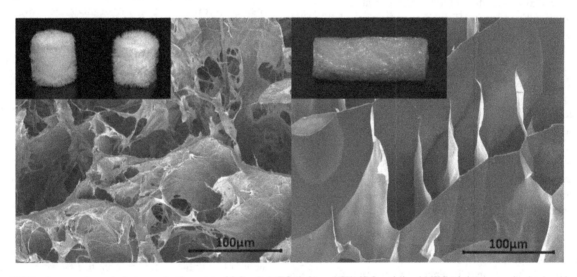

Fig. 9 SEM micrographs and photographs (inserts) of 0.8 w% MFC (left) and TEMPO-oxidized MFC (right) aerogels obtained by freeze drying (−80°C) of aqueous suspensions.

Fig. 10 Images of a TEMPO-MFC aerogel (**A**) and TEMPO-MFC aerogels containing Ag nanoparticles obtained by adding 0.2 (**B**) and 0.5 mM AgNO$_3$ per g MFC (**C**) to the suspension.
Source: From Dong et al.[76] © 2013, with permission from Elsevier.

up to 457 h^{-1}, exceeding that of comparable chitosan-based composite aerogels by about factor 5.[77]

All-Cellulose Composite Aerogels

Partial dissolution of microcrystalline cellulose (MCC, cellulose I) and subsequent coagulation of the dissolved fraction (cellulose II) by a cellulose anti-solvent, such as water affords "*all*-cellulose" composites whose tensile strength can reach that of conventional glass-fiber-reinforced composites.[78–80] This is due to both the particular dissolution behavior of MCC and the strong self-assembly of the two chemically identical constituents, which leads to excellent fiber-matrix adhesion.[81,82] The level of reinforcement achievable for the composites largely depends on the size of the MCC crystals at the stage when coagulation of the dissolved cellulose II fraction sets in. Reinforcement can be hence controlled by dissolution time as large MCC crystals and fiber fragments are progressively split into thinner crystals and cellulose fibrils during dissolution.[78]

Duchemin et al. applied this concept for the preparation of *all*-cellulose composite aerogels.[79] Activated MCC was partially dissolved by stirring a suspension of 5–20 w% MCC for 3 min in *N,N*-dimethylacetamide that contained 8 w% of lithium chloride (DMAc/LiCl). After casting, the mixtures were consecutively exposed to different levels of humidity (33% r.h. for 24 hr, 76% r.h. overnight) to afford slow coagulation of the dissolved cellulose. The *all*-cellulose composite gels were then exhaustively rinsed with water and transformed into aerogels by freeze-drying (−20°C). The densities of the brittle materials ranged from 116 to 350 mg cm^{-3}. Three-point bending tests revealed that maximum flexural strength and stiffness (8.1 and 280 MPa, respectively) is obtained at MCC contents of 10–15 w%. Lowering the cellulose content at otherwise identical conditions resulted in a higher fraction of cellulose II and increased the ductility of the aerogels.[79]

Even though *N,N*-dimethylacetamide/lithium chloride (DMAc/LiCl) is the most frequently used solvent to date for this type of application, other direct solvents, such as ionic liquids[83] or NaOH/urea[84] have been studied too. An example is the room temperature suspension of CNC, obtained by sulfuric acid hydrolysis of high molecular weight cellulose in a solution of low-molecular cellulose in aqueous NaOH/urea obtained by a freezing/thawing approach (*cf.* entry "Aerogels from Cellulose II: Aqueous Salt Solutions"). Gelation caused by thermally induced (40°C) phase separation and subsequent addition of water afforded composite gels of significantly improved dimensional stability and mechanical strength compared to gels that had been prepared without the addition of CNC. Both the gelation behavior and regeneration kinetics furthermore evidenced that cellulose nanowhiskers can act as "bridges" to facilitate crosslinking of cellulose in solution state during gel formation.[84]

Modification of MFC Aerogels

As detailed for BC aerogels, homogeneous and heterogeneous chemical functionalization, cross-linking, surface coating, or formation of interpenetrating networks are typical means to tune the properties of all types of cellulose aerogels including those made from MFC. Aerogels featuring fast shape recovery from compressed state upon dipping in water are just one example for the potential of such facile modification approaches. Respective MFC aerogels

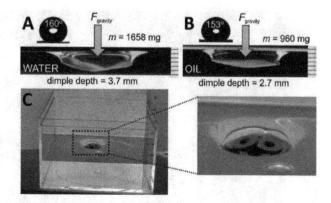

Fig. 11 Cargo carrying fluorinated cellulose aerogels, floating on both water (**A, C**) and oil (**B**) due to surface tension. The membranes can support a weight (m) nearly three orders of magnitude higher than that of the aerogel itself (3.0 mg).
Source: Adapted from Jin et al.[58] © 2011, with permission from American Chemical Society.

can be prepared by cross-linking cellulose with water-borne polyamide-epichlorohydrin resins, such as that of the Kymene™-type, e.g., by adding the latter to an aqueous suspension of MFC, subsequent ultra-sonication and freeze-drying, which affords very lightweight materials with densities as low as 10 mg cm^{-3} at moderately high surface areas (70–110 m^2 g^{-1}).[85] Kymene™-type resins consist of reactive four-membered *N*-heterocyclic groups that can easily undergo crosslinking with carboxyl or hydroxyl groups, largely increasing the wet strength of cellulosic materials.

Hydrophobization of aerogels from mechanically and/or enzymatically disintegrated cellulose suspensions is another modification approach to tune the properties of aerogels for a wide range of applications, and can be accomplished, e.g., by heterogeneous gas-phase esterification using acid halides, such as palmitoyl chloride.[60] It can also be achieved by chemical vapor deposition (CVD) of TiO$_2$, which affords aerogels whose properties can be photo-catalytically switched between water-superabsorbent and water-repellent state[86] or of fluorosilanes that impart cellulose aerogels both hydrophobicity and oleophobicity (Fig. 11).[58] Similar materials of tunable oleophobicity can be obtained by surface silanization, where the variation of their surface texture can be used to control surface wettability.[71] Aerogels from TEMPO-oxidized cellulose that had been surface-modified with octyl-trichlorosilane have been furthermore studied as a separation medium for the selective absorption of oil from water.[70]

A more detailed review of cellulose aerogel modification including MFC aerogels can be found in the entry "Application of Cellulose Aerogels."

AEROGELS FROM ("REGENERATED") CELLULOSE II

The preparation of cellulose II aerogels comprises several processing steps, each one offering a range of parameters for tuning the properties of the final product. Cellulose II aerogels are typically prepared by: 1) molecularly dispersing the dissolution of cellulose in an appropriate direct solvent; 2) shaping; 3) initiation of cellulose coagulation by the addition of a cellulose antisolvent; 4) thorough leaching of any salts including replacement of water by solvents miscible with $scCO_2$; and 5) extraction of the latter with $scCO_2$ aiming at a far-reaching preservation of the highly porous cellulose II network structure, as hierarchical open porosity is a prerequisite to many aerogel applications.

The cellulose dissolution and hence the type of direct cellulose solvent used play a major role with respect to the properties of the final aerogels. Differences in dissolution power, dissolution and gelation mechanisms, solute–solvent interactions, or side reactions between cellulose and solvent are some of the solvent-dependent variables that have the potential to afford considerable variations in the supramolecular cellulose II assembly and can hence be employed to tailor aerogels for specific applications. However, besides their (useful) impact on morphological features, such as surface topography, porosity, void shape and size distribution, or specific surface area, some of the cellulose solvents can negatively affect the chemical integrity of cellulose. Side reactions leading to minor amounts of respective cellulose derivatives can prevent those "impure" aerogels from being used in particular applications, such as in biomedicine.

As the cellulose solvent plays a central role with regard to the properties of cellulose II aerogels, this sub-entry discusses the pros and cons of the most frequently used solvent systems.

Cellulose Dissolution: Pros and Cons of Selected Cellulose Solvents

Virtually all of the comparatively few "direct" cellulose solvents that are able to rearrange the complex network of intramolecular and intermolecular hydrogen bonds in a way that facilitates molecular-dispersing dissolution, but largely maintain the chemical integrity of cellulose, have been studied with regard to their usability for the preparation of cellulose aerogels. However, not all of them turned out to be equally well suited either due to drawbacks with regard to cellulose dissolving performance, handling throughout the processing chain (difficulties with quantitative removal of the solvent from the final aerogel, preparation of shaped monoliths of complex geometry, or laborious and insufficient recovery of expensive solvents), or adverse effects on product properties, such as network homogeneity, surface morphology, porosity, internal surface, mechanical strength, and chemical integrity of cellulose.

The facileness of cellulose dissolution largely depends on molecular and supramolecular features of the respective raw material. In particular, the extensive hydrogen bond network, which is formed between the abundant hydroxyl groups of cellulose, can be seen as a key factor as it largely stabilizes intra- and intermolecular conformations, and promotes cellulose self-assembly. The cellulose dissolving performance of a given solvent is therefore closely related to its capability to weaken, redirect, or cleave these hydrogen bonds in a way that molecular solvation and dispersion of cellulose can occur.

Inorganic Molten Salt Hydrates and Aqueous Salt Solutions

Inorganic molten salt hydrates

Salt hydrates or salt hydrate mixtures, such as zinc chloride ($ZnCl_2 \cdot nH_2O$), calcium thiocyanate ($Ca(SCN)_2 \cdot nH_2O$), lithium thiocyanate ($LiSCN \cdot nH_2O$), or NaSCN/KSCN (eutectic)-$LiSCN_2 \cdot H_2O$, are good cellulose solvents. Besides their ability to swell and dissolve cellulose, they simultaneously support the aggregation of fibrils to colloidal particles and hence gel formation.[87] Depending on the coordination number of the respective metal cation, the crystal lattice of the salt contains constitutional water that essentially contributes to cellulose dissolution. It is assumed that during diffusion of the melt into the cellulose fibers the strong intra- (O3-H...O5'; O6-H...O2') and intermolecular hydrogen bonds (e.g., O6'-H...O3) present in cellulose I are redirected by the salt hydrates in favor of stronger hydrogen bridging between the solvent and cellulose as is the case with most direct solvents. Complexes consisting of single polymer chains and salt melt ions are formed that aggregate to colloidal particles with their cluster size increasing with cooling until a gel is formed.[2]

Calcium thiocyanate tetrahydrate $Ca(SCN)_2 \cdot 4H_2O$ is one of the more frequently used inorganic salt hydrates in cellulose aerogel synthesis.[2,88] It is a solid at room temperature (mp 42.5–46.5°C) and dissolution of cellulose is commonly accomplished at 110–140°C.[89] The addition of water corresponding to a molar $Ca^{2+}:H_2O$ ratio of 1:6 to 1:10 has been found to facilitate cellulose coagulation and gel formation during the cellulose regenerating step without having a negative impact on the cellulose dissolving performance, at least for comparatively low cellulose concentrations. Dissolution of 3 w% of MCC at 110°C, e.g., takes no longer than 30 min even at a $Ca^{2+}:H_2O$ ratio of 1:10.[2] For the preparation of shaped aerogels, the hot cellulose solution is transferred into molds. During cooling cellulose fibrils are assumed to become increasingly aligned to form clusters and fibrillary felts until gel formation sets in at about 80°C.[2] The gels are subsequently immersed first in aqueous ethanol or water to complete cellulose coagulation and to extract the salt hydrate and then in absolute ethanol to prepare them for the final drying step (Fig. 12).

The far-reaching preservation of the cellulose gel morphology throughout consecutive 1) extraction of the inorganic salt hydrate by aqueous ethanol, 2) solvent exchange to absolute ethanol, and 3) $scCO_2$ drying is one of the major advantages of using $Ca(SCN)_2 \cdot nH_2O$. While the volume

Adhesives—Bioabsorbable

reduction was less than 10% when calcium thiocyanate was used for the preparation of CL aerogels, a much stronger shrinkage was observed for aerogels obtained from comparable solutions in TBAF·3H$_2$O/DMSO and NMMO·H$_2$O, respectively.[90] The relatively low cellulose dissolving performance on the other hand is one of the drawbacks of Ca(SCN)$_2$·nH$_2$O, as more than about 4 w% of cellulose cannot be dissolved.[2,91] Jin et al., e.g., prepared aerogels from 0.5–3.0 w% solutions of Whatman CF11 fibrous cellulose powder in a saturated solution of calcium thiocyanate tetrahydrate in water (molar ration of Ca^{2+}:H$_2$O ≈1:6) and subsequent coagulation of cellulose and extraction of Ca(SCN)$_2$ with methanol. Freeze-drying of the obtained alcogels, however, afforded gels that were fissured, partially broken, and had a BET surface area of only 80–160 m^2 g^{-1}. Better aerogels in terms of uniform, open-porous morphology were obtained when water was replaced by *t*-butanol prior to the freeze-drying step. Nitrogen sorption experiments revealed a specific surface area of 160–190 m^2 g^{-1}, slightly increasing with the bulk density of the aerogels (20–100 mg cm^{-3}).[88] The dissolution of MCC in molten Ca(SCN)$_2$·nH$_2$O of varying Ca^{2+}:H$_2$O molar ratio (1:6 to 1:10), cellulose coagulation, as well as extraction of the salt with ethanol and subsequent scCO$_2$ drying afforded aerogels that had specific surface areas of 200–220 m^2 g^{-1} at bulk densities of 10–60 mg cm^{-3}. SEM images confirmed the presence of small macropores (0.1–1 μm) of which the pore size decreased with increasing density.[2]

Aqueous salt solutions

Aqueous NaOH has been intensively studied with regard to cellulose dissolution, coagulation,[92–96] gel formation, and preparation of aerogels.[23,41,97–99] The dissolving ability of aqueous NaOH is supposed to be due to the formation of NaOH "hydrates" such as Na$^+$[OH(H$_2$O)$_n$]$^-$. The latter that are formed by dynamic supramolecular assembly at lower temperature are assumed to be those species capable of re-orienting and cleaving the inter- and intramolecular hydrogen bond network of cellulose.

Cellulose dissolution is typically accomplished by pre-activation at 5°C, vigorous stirring (1000 rpm, 2 hr) in pre-cooled (0–12°C, preferably –6°C),[23] aqueous NaOH (6–12, typically 7.6 w%). Consecutive swelling of cellulose in

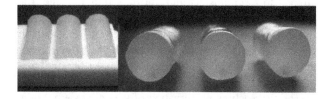

Fig. 12 Cellulose lyogels obtained by dissolution of CL in calcium thiocyanate octahydrate, coagulation with aqueous ethanol (left), washing with water and subsequent solvent exchange to absolute ethanol (right).

8–9 w% aqueous NaOH, freezing the mixture to –20°C, thawing to room temperature, and diluting with water to 5 w% NaOH is another approach that allows for complete dissolution of MCC (Avicel, DPv ≤ 190), CL regenerated from SO$_2$-diethylamine-DMSO (DPv ≤ 850), and regenerated kraft pulps. Native cellulose linters, kraft, or ground wood pulp that consist of cellulose I are only partially soluble.[95,100] Following dissolution, the cellulose solution is kept for a few hours at a temperature sufficiently high (typically 50°C) above the gel point to ensure full gelation. The onset temperature of gel formation depends on the used type of cellulose (e.g., Avicel: 20°C, Solucell: 30°C) and decreases with cellulose concentration. The addition of small amounts of zinc oxide (≤1 w%) can increase the gelation time of cellulose in aqueous NaOH considerably, which can be employed as another means to control the morphology of aerogels but does not improve the thermodynamic quality of the solvent.[101,102] If irreversible gelling (including ripening) has one occurred, the gel is thoroughly rinsed with water or diluted acids until near neutral pH. Direct replacement of aqueous NaOH by ethanol—replacement of water by organic solvents miscible with both water and scCO$_2$ is required prior to scCO$_2$ drying—is not advisable in this case as both increased nucleophilicity and basicity of the hydroxyl ions can promote cellulose degradation.

The cellulose dissolving power of aqueous sodium hydroxide is rather moderate,[23,101] similar to DMAc/LiCl. Complete dissolution is typically only possible for cellulosic materials of comparatively low DS, such as the microcrystalline Avicel, and requires a comparatively high NaOH/anhydrous glucose unit (AGU) molar ratio. For Avicel it has been demonstrated that each AGU has to be closely surrounded by four NaOH molecules at least to accomplish molecular disperse dissolution.[95] Correspondingly, the amount of low DS cellulose that can be dissolved in a 10 w% aqueous sodium hydroxide solution does not significantly exceed 10 w% of cellulose. This, in turn, allows for the preparation of aerogels that have a final apparent density of 100 mg cm^{-3} only, provided that the gels do not shrink during solvent exchange and drying.

However, the cellulose dissolving performance of aqueous solutions of alkali hydroxides (NaOH < LiOH) can be significantly improved by solubilizing additives such as urea and thiourea, as the latter are assumed to form water-soluble inclusion complexes, with the cellulose/[OH(H$_2$O)$_n$]$^-$Na$^+$ system encaged by (thio)urea.[103–105] For example, pre-cooled aqueous solutions of 7 w% NaOH and 12 w% urea can rapidly dissolve cellulose with a molecular mass of up to about 120.000 g mol^{-1}.

Next to films,[106–112] hydrogels,[113,114] microspheres,[115] and spun fibers,[116,117] aqueous alkaline cellulose solutions have also been studied with regard to the manufacture of cellulose II aerogels. Gavillon and Budtova (2008), e.g., prepared aerogels from different types of cellulosic materials (DP 180–950) that had been pre-activated by swelling in water prior to dissolution in 7.6 w% aqueous

NaOH. After stirring at −6°C for 2 hr, the obtained solutions (3–7 w% cellulose) were poured into cylindrical molds and kept for a few hours above the gel point to ensure complete and irreversible gelation. The aquogels were thoroughly rinsed with water and subsequently subjected to a solvent exchange to ethanol or acetone prior to scCO$_2$ drying. The obtained aerogels had heterogeneous pore morphologies, pore diameters of several nanometers to tens of micrometers, bulk densities of 60–300 mg cm^{-3}, and specific surface areas of 200–300 m^2 g^{-1}. The addition of a surfactant (Simulsol SL8, 0.1–1.0 wt % to 5 w% Avicel PH 101 in aq. 7.6 w% NaOH) resulted in an increase of the amount of macropores and, accordingly, a decrease in the sample density with surfactant concentration (Fig. 13).[23] However, strong shrinkage throughout all of the main processing steps has been reported for this approach. Samples prepared from solutions of 5 w% MCC (Avicel PH-101) in 8 w% aqueous NaOH, e.g., largely suffered from strong shrinkage that occurred during cellulose coagulation with water (~40%). Together with the volume reduction that additionally happened throughout replacement of water by acetone (~12%) and scCO$_2$ drying (~20%), the overall shrinkage amounted to about 70 v%.[118]

MWCNT dispersed and immobilized within a cellulose aerogel matrix are promising materials for applications in sensor technology, energy storage, catalysis, and biotechnology as discussed in the following text. Respective aerogels have been obtained from CL that was dissolved within 5 min of vigorous stirring in a pre-cooled (−12°C) mixture that contained varying amounts of MWCNTs (average diameter of 9.5 nm, length of 1.5 μm), 7 w% NaOH, 12 w% urea, and the non-ionic surfactant Brij76 (polyoxyethylene-10-stearyl ether, 1.5-fold excess related to MWCNT). The overall content of MWCNTs in the final aerogels was 1–10 w%. The coagulation of cellulose was accomplished with 5 w% H$_2$SO$_4$. The obtained gels were thoroughly rinsed with water and freeze-dried (liquid nitrogen). As the initial concentration of the constituents (cellulose and MWCNT) was 4 w% for all experiments, both bulk densities (~0.16 g cm^{-3}) and specific surface areas (~150 m^2 g^{-1}) of the aerogels were very similar. Good thermal stability and mechanical properties (up to 90 MPa Young's modulus)

were reported. Cellulose/MWCNT composite aerogels with 3–10 wt% MWCNT have a conductivity of about 2.3 × 10^{-4} to 2.2 × 10^{-2} S cm^{-1}, which is comparable to that of other CNT-based composite aerogels with much higher MWCNT loading. Considering that the porosity of the materials is 90%, the conductivity threshold at MWCNT volume fractions is as low as 3 × 10^{-3}. This is much lower than that of cellulose/MWCNT composites and mostly reported polymer/MWCNT composites and indicates that the used solvent system promotes a very efficient and uniform dispersion of MWCNT in the cellulose matrix.[119]

Aiming at the preparation of aerogels for heat insulation applications, Shi et al. dispersed 2–5 w% CL (M$_n$ = 1.01 × 10^5 mol kg^{-1}) in an aqueous solution of 9.5 w% NaOH and 4.5 w% thiourea. After freezing and subsequent thawing of the suspensions, the received gels were rinsed with water and freeze-dried. To induce hydrophobicity, the aerogels were further treated by cold CCl$_4$ plasma. The aerogels had densities of 0.2–0.4 g cm^{-3} and contained undissolved residues. Their inner morphology was dominated by a heterogeneous macropore structure, while the surface of the aerogel was much more compact.[120] This "skin formation" has previously been reported for aerogels regenerated from aqueous NaOH solution and is supposedly caused by an increase of cellulose concentration on the surface as a result of rapid solvent depletion during regeneration.[23] Shi et al. also prepared cellulose-silica composite aerogels (about 7–16 w% silica) using tetraethoxysilane (TEOS) as a monomeric precursor for the formation of a secondary network penetrating the cellulosic hydrogel. After freeze-drying and hydrophobization by cold plasma treatment (CCl$_4$), comparatively low-porous materials (20–30%) were obtained. For composites of high silica content (e.g., 2 w% cellulose and 7.25 w% SiO$_2$), low thermal conductivity (0.026 W m^{-1} K^{-1}) and hence good insulating properties were reported.[121]

Aiming at applications in tissue engineering, drug delivery, and as high-performance composites, Zhang et al. incorporated small quantities of graphene-oxide-sheets (GOS) into cellulosic aerogels. 0.1 w% of GOS were dispersed in a 10 w% cellulose solution, which had been prepared by dissolving the polysaccharide in a pre-cooled (−10°C) aqueous

Fig. 13 Aerogel obtained from a 5 w% Avicel solution in aqueous NaOH (center). SEM micrographs of aerogel morphologies without Simulsol (left) and with addition of 1 w% Simulsol to the cellulose solution (right), resulting in a higher amount of large macropores.
Source: Adapted from Gavillon & Budtova[23] © 2008, with permission from American Chemical Society.

solution of 9.5 w% NaOH and 4.5 w% thiourea under strong stirring for 5 min. Interestingly, the addition of GOS significantly accelerated the gelation of cellulose, which has been ascribed to their high surface content of oxidized moieties that can undergo close interactions with cellulose via hydrogen bonding or dipole–dipole forces. Crosslinking of cellulose via the incorporated GOS afforded an increase of both compression strength (ca. 30%) and Young's modulus (ca. 90%) for the freeze-dried aerogels compared to respective GO-free reference materials.[122]

The surface polarity of cellulose aerogels prepared from aqueous alkaline solutions largely depends on the anti-solvent used to coagulate cellulose from solution state. This has been demonstrated by Isobe et al. for different aqueous and organic solvents, and a standard cellulose solution prepared from 4.0 w% of filter paper pulp (Advantec MFS, Japan), 4.6 w% LiOH, 15 w% urea and water. Consecutive washing of the coagulated cellulose, replacement of water by ethanol, replacement of ethanol by t-butanol, and freeze-drying afforded gels that were very similar with regard to specific surface area (326–355 $m^2 g^{-1}$) and mesopore size distribution (5–20 nm). However, a significantly higher absorption of Congo red and more pronounced iodine/KI color reaction was observed when organic solvents had been used for the coagulation of cellulose from solution state instead of aqueous media (MeOH > EtOH > acetone > 5 w% aq. H_2SO_4 ≈ 5 w% aq. Na_2SO_4). This is assumed to be due to a higher content of glucopyranoside ring planes exposed on the surface of cellulose fibrils that has been concluded from X-ray diffractometry[123] and supposedly results in enhanced hydrophobicity. Scanning electron microscopy (SEM) revealed on the other hand a membrane-like morphology for aerogels when aqueous cellulose anti-solvents were used. These surfaces are more hydrophilic and seem to have a higher content of hydroxyl groups available for intermolecular cellulose hydrogen bonding.

Low-Melting Organic Salts (Ionic Liquids)

Ionic liquids have been recently moved into the focus of cellulose and lignocelluloses research as they facilitate both direct dissolution[124,125] and homogeneous derivatization.[126–130] They are considered to be potent competitors to NMMO as they are able to dissolve considerable amounts of cellulose, other polysaccharides, and lignin; are not oxidizing, largely stable, and chemically inert; have virtually no vapor pressure; and require lower temperatures for cellulose dissolution. In particular, the 1-butyl-3-methyl-1H-imidazolium (BMIM) and 1-ethyl-3-methyl-1H-imidazolium (EMIM) chlorides and acetates have found wide utilization in cellulose and polysaccharide chemistry.

Next to cellulose fibers,[131,132] films,[133] or beads,[134] ionic liquids have also been used for producing monolithic aerogels[98,135,136] and composite materials.[137] For instance, 15 w% Avicel has been dissolved in 1-ethyl-3-methylimidazolium acetate [EMIM][OAc][138] or 20 w% in 1-butyl-3-methylimidazolium chloride [BMIM][Cl].[139] Applying

the latter under heating in a microwave oven, even a clear 25 w% solution of a dissolving pulp has been obtained.[140] Whether an IL is a good solvent for cellulose depends on several factors. A high hydrogen bond basicity and relatively small size of both cation and anion are usually beneficial. However, ILs containing formate and acetate were also found to facilitate effective cellulose dissolution.[141]

Although ILs are commonly considered to be chemically inert, side reactions may occur with reagents in the course of cellulose derivatization due to the nucleophilic reactivity of the anion. The resulting intermediates could, in turn, react with cellulose. Furthermore, C-2 of 1-alkyl-3-methyl-imidazolium cations reacts with aldehyde/hemi-acetal functions, resulting in minor covalent binding of the solvent to cellulose, which might cause problems, e.g., in biomedical applications. The modification can be suppressed by the absence of catalyzing bases and if contact times are kept below 2 hr, or even completely avoided if an alkyl group is present at the C-2 position of the ILs' imidazolium cation.[142–144]

Aiming at the preparation of aerogels, 15 w% Avicel PH-101 MCC have been dissolved in [EMIM][OAc] (80°C, 48 hr). After coagulation and washing with water, a solvent exchange to acetone was carried out and the obtained gels were dried with scCO2. The resulting aerogels had densities of 60–200 mg cm^{-3} and specific surface areas of 130–230 $m^2 g^{-1}$. The pore size distribution was found to be in the nanometer range, peaking at 10–20 nm, which is in agreement with values obtained for aerogels from Lyocell dopes.[12,41]

Our group achieved complete dissolution of 3 w% CL (DP 880) in a mixture of [EMIM][OAc] and DMSO (30:70, v/v) within 2 hr of stirring at room temperature. The respective aerogels as obtained after casting, cellulose coagulation with water, solvent exchange to ethanol, and scCO2 drying had a density of 57 mg cm^{-3} and a specific surface area of 349 $m^2 g^{-1}$ as calculated from the nitrogen adsorption/desorption isotherms (BET) at 77 K (cf. Fig. 16F).

Dissolution of 1–4 w% MCC in [BMIM][Cl] at 130°C for 3.5 hr, regeneration and washing in water and subsequent freeze-drying using liquid nitrogen, yielded aerogels of high porosity (99%) and specific surface areas of up to 186 $m^2 g^{-1}$.[136] In a similar approach, Aaltonen and Jauhiainen dissolved ~2.9% bleached cellulose pulp in [BMIM][Cl] at 130°C for 4 hr. 90 w% ethanol was used for cellulose regeneration and washing was carried out consecutively by absolute ethanol and liquid carbon dioxide. Finally, drying was achieved by heating above the critical temperature of carbon dioxide. The cellulose aerogels had a density of 48 mg cm^{-3} and a very high surface area of 539 $m^2 g^{-1}$. Besides cellulose, aerogels from soda lignin and composites from mixtures of cellulose, soda lignin, and xylan as well as from spruce wood were prepared by similar procedures. Some composite aerogels had similar or lower densities than their pure cellulose counterpart, even at higher solute concentrations, e.g., 25 mg cm^{-3} for a composite of ~3.8% cellulose and ~1.9% lignin ([BMIM]

[Cl], 130°C, 27 hr), but the specific surface area was always reduced by the presence of other constituents.[135]

If cellulose hydrogels are exposed to certain pressure and temperature conditions, the pore size distribution of the product can be controlled as in classical polymer foaming. Tsioptsias et al. used this effect to obtain microporous cellulose foams from hydrogels by applying several different conditions (80°C, 7.5–17.5 MPa, and 40–80°C, 7.5 MPa respectively) in a high pressure cell for 2 hr, followed by rapid depressurization and freeze-drying. The hydrogels were obtained by dissolving 1.73 and 2.86 w% softwood pulp cellulose (DP 1280) in the IL 1-allyl-3-methylimidazolium chloride [AMIM][Cl] for 5 hr at 80–90°C, allowing the solution to stand for 1–2 hr at <8°C, cellulose regeneration in pre-cooled water and rinsing. A similar gel was prepared from a 1.5 w% solution of the same pulp by regeneration with methanol and subsequent conversion to a nanoporous cellulose aerogel by scCO$_2$ drying (40°C, 20 MPa). The aerogel had a density of 58 mg cm^{-3} and a surface area of 315 m^2 g^{-1}. In contrast to the microporous material, where pores sizes of several hundreds of micrometers were observed by SEM analysis, nitrogen adsorption/desorption isotherms of the aerogel obtained by scCO$_2$ drying revealed that its pore volume was dominated by micropores (<2 nm) and mesopores (2–50 nm).[145]

Aiming at the preparation of hydrophobic aerogels, Granstrom et al. homogeneously esterified MFC in [AMIM][Cl] solution with stearoyl chloride. To preserve unreacted hydroxyl groups, which could take part in the formation of a hydrogen bond network during gelation, they aimed at a very low degree of substitution (DS). After the product was regenerated under stirring, purified and centrifuged, the obtained stearoyl-modified cellulose gels were transferred to ethanol. Thereafter, the ethanol in the 4 w% alcogels was replaced by liquid CO$_2$ in six cycles and finally scCO$_2$ drying was applied. The obtained open porous aerogel structures had an aqueous contact angle of up to 124° at a DS of only 0.07.[146]

An IL was essentially involved in the development of photoluminescent cellulose aerogels that contained covalently immobilized (ZnS)$_x$(CuInS$_2$)$_{1-x}$/ZnS (core/shell) quantum dots (QDs). 1–3 w% eucalyptus pre-hydrolysis kraft pulps were dissolved in 1-hexyl-3-methyl-imidazolium chloride ([HMIM][Cl]) at 100°C for 2 hr. After cooling to 60°C, a suspension of the 3-(trimethoxysilyl)-propyl-functionalized QDs in [HMIM][Cl] was added drop-wise, giving final QD concentrations of 0.01–0.3 w%. After molding, grafting of the QDs onto cellulose via their 3-(trimethoxysilyl)propyl ligands, subsequent cellulose coagulation, and thorough washing with ethanol, the obtained cellulose/QD composite gels were converted to aerogels by scCO$_2$ drying. The resulting hybrid materials had apparent densities of 38–57 mg cm^{-3} and high specific surface areas of 296–686 m^2 g^{-1}. The emission wavelengths of the aerogels were tunable within a wide range of the visible light spectrum mainly by the ratio of the QD core constituents (Fig. 14).[147]

N-Methylmorpholine-N-Oxide (NMMO) Monohydrate

Tertiary amine oxides are good cellulose solvents[148] and in particular N-methylmorpholine-N-oxide (NMMO) monohydrate has found wide use in the large-scale production of Lyocell fibers, which approximates an annual capacity of 200,000 tons. The Lyocell process is considered to be an ecological alternative to the viscose process as the latter requires chemical derivatization (alkalization, xanthation with CS$_2$) and subsequent regeneration (sulfuric acid/sodium sulfate/zinc sulfate) of cellulose prior to spinning with harmful chemicals involved. NMMO as an environmentally friendly, easy-to-recycle solvent has also other

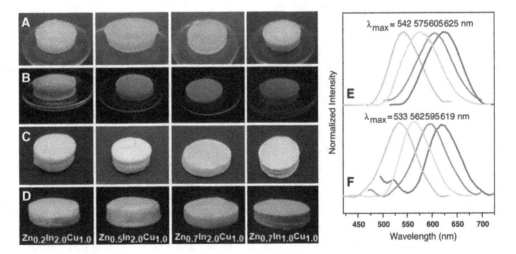

Fig. 14 Pictures of cellulose/QD hybrid alcogels (**A, B**) and aerogels (**C, D**) taken under visible (**A, D**) and UV light (**B, D**). On the right side the fluorescence spectra prior to (**E**) and after (**F**) drying illustrate a considerable influence of the ratio of the QD core constituents on the emission wavelengths within the visible light spectrum.
Source: From Wang et al.[147] © 2013, by the authors.

advantages, such as low toxicity, biodegradability, and high recovery rates that are above 99% even on industrial scale.

The flexibility of the process in terms of cellulosic raw materials is another major advantage. The peculiar solvation power of NMMO (under high shearing, up to 15 w% of cellulose can be dissolved in the NMMO monohydrate at 100–120°C)[149–151] originates in its pronounced ability to disrupt the hydrogen-bond network of cellulose and form solvent complexes by establishing new hydrogen bonds between the macromolecule and the solvent.[152,153] The "chemical" processes during dissolution are thus merely acid–base, i.e., donor–acceptor, interactions finally leading to a far-reaching restructuring of the hydrogen bond network.[154,155]

Cellulose dissolution in the strong oxidant NMMO can be accompanied by a series of negative side effects, such as gradual decomposition of NMMO to products that induce further decomposition of the solvent. In particular, autocatalytic decomposition of NMMO by carbenium-iminium ions can ultimately result in exothermic events, such as deflagrations or blasts. Adverse secondary reactions can furthermore lead to chemical alteration and degradation of cellulose or the formation of chromophores. Fortunately, all of the preceding side reactions can be largely avoided by the addition of appropriate stabilizers. Homolytic (radical) side reactions can be prevented by phenolic antioxidants, such as propyl gallate. Heterolytic side reactions on the other hand can be avoided using N-benzylmorpholine-N-oxide (NBnMO) as a sacrificial stabilizer, which traps both formaldehyde and NMMO-derived carbenium-iminium ions[156,157] and transfers such Mannich intermediates into stable, colorless products.[158]

The use of NMMO·H$_2$O as a cellulose solvent renders some of the process steps to shaped aerogels more simple, which is mainly due to its convenient phase transition: Melting at temperatures above 80°C enables molding of the cellulose solution (Lyocell dope), solidification upon cooling affords shaped specimen that are easy to handle in further processing, and promotes cellulose regeneration. Unfortunately, cooling of Lyocell dopes of low cellulose content (≤4 w%) down to room temperature does not allow for the preparation of homogeneous aerogels, which is due to the particular crystallization behavior of such "diluted" dopes. Solidification here is associated with a strong increase of the dope's density, which happens in the temperature range of 20–30°C and inevitably leads to the formation of cracks.[159,160] Beyond cellulose contents of about 4 w%, this phenomenon diminishes and homogeneous aerogels can be obtained at cellulose contents of ≥8 w%. For dopes of low cellulose content, the coagulation of cellulose at T ≥ 40°C is an appropriate measure to circumvent crack formation, however at the expense of a fibrous morphology (see section in the following text and Gavillon[23]).

The high affinity of NMMO toward cellulose, caused by its strong hydrogen bonding via the exocyclic oxygen atom, is a particular feature of NMMO that can largely impede quantitative removal of the solvent from cellulose

II monoliths. NMMO eluation profiles of Lyocell dopes from different commercial pulps (3 w%) using absolute ethanol as a cellulose anti-solvent revealed that even after repeated and thorough washing residual amounts of NMMO can remain confined in fibrillary interstices of the lyogel.[39] This has to be considered in particular for biomedical applications where high purity of the used materials is demanded. As NMMO is hygroscopic, the remaining traces of the solvent are furthermore supposed to be one of the reasons for the pronounced shrinkage of aerogels by water adsorption and pore collapsing. For aerogels made from respective solutions of 3% CL in NMMO, it is visible already after a few days of storage at comparatively low relative humidity (ca. 40–50%) and has been described in different studies.[39,161,162]

Innerlohinger et al. dissolved a range of pulps at varying concentrations (0.5–13 w%) in NMMO and water, and used propyl gallate as a stabilizer.[162] After molding, cooling, and solidification, the respective Lyocell dopes were immersed in water for simultaneous extraction of NMMO and cellulose coagulation. Prior to scCO$_2$ drying, water was replaced by acetone or ethanol. The obtained materials had densities of 20–200 mg cm^{-3} and specific surface areas of 100–400 m^2 g^{-1}. However, strong shrinkage was observed for both main process steps, i.e., cellulose coagulation and scCO$_2$ drying. In particular, the least dense materials suffered from extensive shrinkage, which was still 60% for aerogels that were obtained from solutions that contained 3 w% of cellulose.[162]

The preparation and detailed characterization of aerogels prepared from Lyocell dopes that contained 3 and 6 w% of different commercial hardwood and softwood pulps, respectively, has been described by Liebner, Haimer et al.[39] Propyl gallate and NBnMO were used as respective stabilizers, simultaneous cellulose coagulation and NMMO extraction was accomplished with ethanol, and scCO$_2$ was employed to convert the respective alcogels to aerogels. Depending on the type of cellulose, the bulk density ranged from 46 to 69 mg cm^{-3} (3 w% cellulose) and from 106 to 137 mg cm^{-3} (6 w% cellulose), respectively. Total shrinkage after coagulation, solvent exchange, and drying was more pronounced for the lighter materials (20–40%), which had average pore sizes of 9–12 nm and specific surface areas of 190–310 m^2 g^{-1} as determined by nitrogen sorption experiments at 77K.[39] The same authors reported furthermore that the extent of shrinkage during the cellulose regenerating step can be significantly reduced from 18 v% to 5.6 v% and 9.2 v%, respectively, when methanol or i-propanol is used for cellulose coagulation/NMMO extraction. Total shrinkage over all the process steps was lowest when i-propanol (21.4 v%) was used instead of ethanol (37.2 v%) while regeneration with water or DMSO resulted in total shrinkage of 36.4 v% and 46.2 v%, respectively.[160] Low depressurization after scCO$_2$ drying at rates ≤ 0.2 MPa min^{-1} is another means to reduce the extent of shrinkage.[1]

Physical, Mechanical, and Morphological Properties of Cellulose II Aerogels: A Comparison

The density of regenerated cellulose aerogels depends on various factors, such as cellulose concentration in the initial solution mixture, constancy of the solvent-to-solute ratio during the dissolution process (as affected by both volatility and stability of the cellulose solvent, and the time required for cellulose dissolution), and the extent of shrinkage of the molded bodies during regeneration, solvent exchange, and drying procedures. All of the common cellulose solvents and anti-solvents have specific limitations and peculiarities with regard to solvent power, cellulose self-assembly and coagulation, and hence, morphology of the formed cellulose II network. The respective solvent/anti-solvent system should therefore be carefully chosen according to the requirements of the respective application. Calcium thiocyanate tetrahydrate, e.g., is an excellent cellulose solvent that can be used to produce aerogels of very low density (≥ 5 mg cm^{-3}) at virtually zero shrinkage. However, if aerogels of high density ($\geq 100...150$ mg cm^{-3}) are required, solvents such as N-methylmorpholine-N-oxide monohydrate or ionic liquids have to be used due to their high cellulose dissolving performance. Similarly, other aspects—such as the impact of the respective cellulose solvent/anti-solvent system or the applied drying technique on morphological features of the aerogels, dimensional stability of the specimen throughout the manufacturing process, homogeneity of the cellulose network (cracks, fissures, etc.), hygroscopicity (shrinkage during storage at elevated humidity), purity, and biocompatibility—have to be considered for a particular application.

Similar to their cellulose I counterparts, cellulose II aerogels are ductile materials. Unlike BC aerogels, they are isotropic in their response toward compression forces. Uniaxial compression stress does not result in a sudden load drop as often observed for brittle foam plastics. On the contrary, the response of cellulose II aerogels toward compression forces is characterized by a pronounced stress plateau (up to about 40% strain) where the compression of the aerogels is accompanied by progressive elastic buckling of the pore walls absorbing a considerable amount of energy, and eventual pore collapsing.[41]

The stiffness of cellulose aerogels as calculated from the elastic region (up to about 5.5% strain) is mainly influenced by their density (Fig. 15A). An increase of the cellulose content of Lyocell dopes from 3 to 6 w% was found to result in an increase of Young's moduli E by about one order of magnitude, due to a higher resistance of the cellulose II structure to cell wall bending and pore collapsing.[39] However, the cellulose solvent also influences the mechanical properties, as can be seen from Fig. 15B. The density-normalized Young's modulus (specific modulus E_ρ) of an aerogel prepared from a solution of Avicel in 8 w% aqueous NaOH is more than twice as high as of a comparable sample derived from an ionic liquid solution ([EMIM][OAC]). Furthermore, the type of cellulose used has an impact. The E_r aerogels from hardwood pulp of different molecular weights and CL, all dissolved in the same solvent (NMMO), show considerable deviations. Interestingly, the specific compression modulus of an MFC aerogel is located only in the middle region and comparable to that of aerogels from Avicel/aqueous NaOH, e.g.

The morphology of cellulose II aerogels can be controlled to some extent by the choice of solvent/anti-solvent used for cellulose coagulation, as the latter can be triggered by different phase separation mechanisms. If cellulose coagulation, e.g., is accomplished from hot, molten Lyocell dopes, phase separation between solvent and solute occurs in one step by spontaneous spinodal decomposition, and globular cellulose aggregate structures are formed (Fig. 16D). A similar morphology most likely caused by the same one-step phase separation mechanism is also obtained when cellulose is coagulated from respective solutions in ionic liquids, such as [BMIM][Cl] or [EMIM][OAc] (Fig. 16E). Networks composed of agglomerated cellulose spheres seem to form only if coagulation is performed from molecular dispersed cellulose solutions.[23,41] If [EMIM][OAc] is diluted with DMSO (30:70, v:v), e.g., the derived aerogels exhibit both globular and fibrillar structural characteristics (Fig. 16F; cf. section "Low-Melting Organic Salts"). This transition suggests the possibility to control the morphology by the ratio of the solvent constituents.

In contrast, solidification of the Lyocell dope leads to a microphase separation of free and bound solvents. Subsequent addition of the antisolvent first dilutes and extracts the free NMMO, before the fraction of cellulose solvent is replaced which is tightly bound to cellulose, leaving behind a network of cellulose fibrils (Fig. 16C).[23] Gelation of cellulose solutions in aqueous NaOH is typically accompanied by partial coagulation of cellulose, whose extent increases with gelation time and temperature. As with NMMO extraction from solidified Lyocell dopes, microphase separation takes place during gelation leading to a fibrillary network structure (Fig. 16A).[41] A similar nanofibrillary morphology is obtained when calcium thiocyanate tetrahydrate is used as cellulose solvent (Fig. 16B), since gelation also occurs prior to regeneration, typically at a temperature of about 80°C.[2]

POTENTIAL APPLICATIONS OF CELLULOSIC AEROGELS: AN OUTLOOK

General Applications of Aerogels

The huge application potential of the various types of cellulosic aerogels discussed in this entry is similar to that of other types of aerogels, such as the "famous" silica aerogels. Silica aerogels obtained by classical sol-gel chemistry from alkoxysilanes are probably the most comprehensively

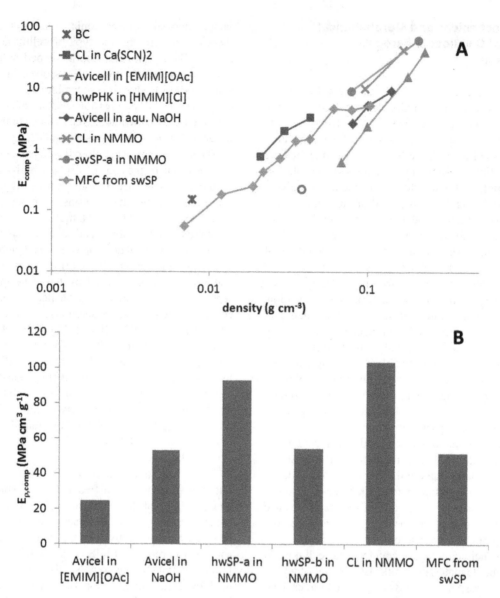

Fig. 15 Young's modulus of a range of cellulosic aerogels under compression stress (**A**). The materials differ in density, cellulosic raw material, crystalline structure (Iα, Iβ, II), and the applied cellulose solvent. Specific modulus (E_ρ) of aerogels of similar densities (94–103 mg cm^{-3}) (**B**). Cellulosic materials: BC in ethanol, CL (CL; DP = 825), Avicel (DP = 185), eucalyptus pre-hydrolysis kraft pulp, totally chlorine-free (TCF) bleached (hwPHK; DP = 495), softwood ammonia sulfite pulp (swSP-a; DP = 915), hardwood calcium sulfite pulp, elemental chlorine-free (ECF) bleached (hwSP-a; DP = 4100); beech magnesium sulfite pulp, TCF bleached (hwSP-b; DP = 1875), MFC from softwood sulfite pulp (MFC from swSP).
Source: Values derived from Sehaqui et al.,[66] Schimper et al.,[90] Wang et al.,[147] Pircher et al.[4] and Wendler et al.[271]

studied subclass of aerogels. Their utilization by the National American Space Agency (NASA) during several space missions and 15 entries in the Guinness Book of Records for their outstanding properties further contributed to their publicity. In particular a high specific surface area (600–1600 m^2 g^{-1}), permittivity (1.007–2), and shock absorption potential on the one hand, and the low bulk density of down to <1 mg cm^{-3} as well as low thermal expansion (α_{lin} = 2·10^{-6} K^{-1}), thermal conductivity (λ = 0.016–0.030 W m^{-1} K^{-1}), sound propagation (v = 100–300 m s^{-1}), and acoustic impedance (Z = 10^3–10^5 kg m^{-2} s^{-1}) on the other hand facilitate a broad range of applications.

Most prominent examples in this respect are the use of silica aerogels for high-performance insulation in cases where the space for placing the insulating material is very limited or the material is exposed to extreme conditions. Silica aerogels have been therefore used by NASA during the Mars Pathfinder expedition in 1996/1997 for protecting electronic devices of the Sojourner Mars rover from the extremely low Mars surface temperature that can be as low as −140°C in polar nights during winter time. The excellent shock absorbing properties have been employed during the NASA stardust project (1999–2006) for trapping high velocity particles without vaporization of the samples for

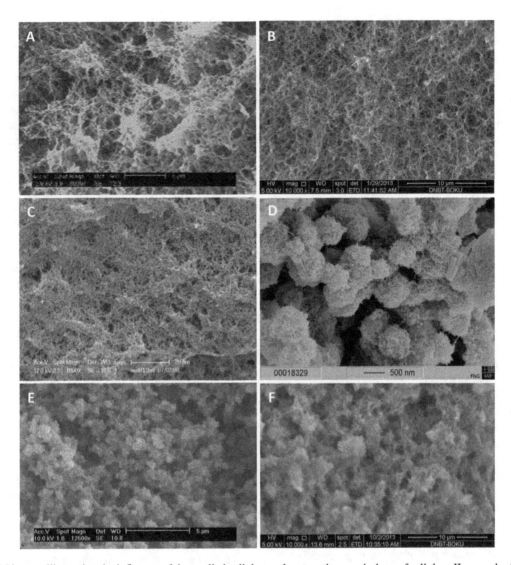

Fig. 16 SEM images illustrating the influence of the applied cellulose solvent on the morphology of cellulose II aerogels. Aerogels from 5 w% Avicel in 8 w% aqueous NaOH (**A**); 1.5 w% CL in calcium thiocyanate hydrate (**B**); 3 w% bleached hardwood sulfite pulp in NMMO monohydrate, solid regeneration; 3 w% cellulose (DP = 950) in NMMO monohydrate, molten regeneration (**D**); 3 w% Avicel in [EMIM][OAc] (**E**); 3 w% CL in [EMIM][OAc]/DMSO (30:70, v/v) (**F**).
Source: (**A**) Adapted from Gavillon & Budtova[23] © 2008, with permission from American Chemical Society, (**C**) From Liebner et al.[39] © 2009, with permission from Walter de Gruyter, (**D, E**) From Sescousse et al.[41] © 2011, with permission from Elsevier.

analytical purposes. Furthermore, the different stopping distance inside the aerogel caused by different speeds of the particles allows for distinguishing copious anthropogenic debris from relatively rare extraterrestrial particles. By advancing this concept, alumina aerogels co-doped with Gd and Tb have been developed, which are able to report the kinetic energy of hypervelocity particles as the latter generate local fluorescence signals due to certain phase transitions, such as the formation of the GdAlO$_3$:Tb perovskite phase.[163]

Aerogels from a wide variety of other materials, such as clay, metals, metal oxides, chalcogens, or silica aerogels, have been developed in the past 20 years. While the open-porous, lightweight structure is a common feature of all aerogels, the properties of the respective source material lead to a broad diversity of properties, such as in terms of surface morphology and chemistry, which are, in turn, attractive for a large variety of applications.[164–168] Aerogels based on silica, carbon, organic polymers, or metals are already commercialized (Aerogel Technologies LLC, Aspen aerogels Inc., Cabot Aerogel Ltd., etc.) for chemisorption, chromatography, low-profile thermal and acoustic insulation, particle detection, optics, chemical spill clean-up, microwave and RF induction heating processes, oil recovery or as catalyst support, diffusion control and battery media, materials with engineered electromagnetic properties, desalination electrodes, and electrodes for battery and supercapacitor applications.[169] Other types of aerogels are supposed to follow soon, such as cobalt-molybdenum-sulfur aerogels that have a high ability for removing mercury from polluted water, for separating hydrogen from other gases, and catalyzing hydro-desulfurization of crude oil to date in an unsurpassed efficiency.[170]

Research on biopolymer-based aerogels has been largely advanced during the past decade. This is mainly due to the generally increasing public and political awareness of the fact that in the future renewable sources will play a much more important role in the production of fine chemicals or functional materials. However, research on biopolymer-based aerogels is still in its infancy even though a series of aerogels with intriguing properties have been obtained from starch, alginates, chitin, chitosan, pectin, agar, carrageenan,[171–176] arabinoxylan,[177] whey protein,[178] or lignin.[179]

APPLICATIONS OF CELLULOSIC AEROGELS

Cellulose and silica aerogels share many similarities with regard to general properties, such as low bulk density (≥ 0.2 mg cm^{-3} vs. 1 mg cm^{-3}),[60,180] low heat transmission (0.029 W m^{-1} K^{-1} vs. ≤ 0.014 W m^{-1} K^{-1}),[120,181] high interconnected porosity (>99.9% vs. $\leq 99.8\%$),[60,182] and void surface area (≤ 605 m^{-2} g^{-1} vs. ≤ 1200 m^{-2} g^{-1}), and target hence a series of similar general applications. However, material-specific differences owing to the different properties of the respective inorganic or organic source material render each of them particularly suited for certain applications.

Technical applications make use of one or several of the following properties of cellulose aerogels: 1) The particular aerogel morphology in terms of accessible pore volume and pore size distribution; 2) stiffness-to-weight ratio; 3) high specific surface area; 4) special surface chemistry resulting from the abundance of hydroxyl groups; 5) biocompatibility; and 6) the characteristics of carbon aerogels obtained under pyrolytic conditions.

Insulation Materials

Their particular pore features in combination with low weight render cellulose aerogels excellent candidates for high-performance thermal insulation, which could greatly contribute to reduce energy consumption for heating, emission of CO_2 and hence global warming, and the overall space consumption of flats or houses at constant effective living space. Just by using super-insulating silica aerogels whose thermal conductivities are lower than that of air ($\lambda = 25$ mW m^{-1} K^{-1}) instead of common glass wool ($\lambda = 40$ mW m^{-1} K^{-1}), the thickness of the insulating layer corresponding to a thermal resistance of R = 4 could be reduced from 16 cm to 6 cm. For a living room of $4 \times 4 \times 2.5$ m with two insulated outer walls, the gain in floor space and volume would be about 0.8 m^2 and 2 m^3, respectively.

However, super-insulation can be only achieved if the respective material virtually has no macroporosity and the diameter of the largest pores is smaller than the mean path length of air, i.e., about 70 nm. While super-insulating silica (Spaceloft®, Pyrogel®XT, Thermablok®) and cross-linked polyimide aerogels[183] are already commercialized, the pore features of polysaccharide-based aerogels do not meet the requirements of super-insulators as reflected by heat conduction values of 29–32 mW K^{-1} m^{-1} as reported by Shi et al.[120] This is due to the comparatively broad pore size distribution of hitherto developed cellulose aerogels, which comprises macropores as well as micro- and mesopores. Current efforts in this particular field include the development of: 1) methods for a more reliable determination of macroporosity; 2) synthetic pathways to suppress the formation of macropores, such as by crosslinking or insertion of interpenetrating networks; and 3) measures to prevent cellulose aerogels from microbial degradation and shrinkage at elevated humidity. Provided these key issues can be solved, cellulose aerogels may become serious competitors of silica-based aerogels as they are superior with regard to economic and environmental issues. Coating of aerogel particles with ethyl cellulose to prevent the adsorption of water[184] or preparation of mesoporous silica/cellulose composite aerogels by sol-gel chemistry using aqueous alkali urea as a cellulose solvent[185] are just two examples of recent progress in this field.

Lightweight Construction Materials

The low bulk density of cellulose aerogels suggests their utilization as lightweight construction materials. However, due to their fragility cellulose aerogels—in particular those of extremely low density—have to be reinforced beforehand to optimize their stiffness-to-weight ratio. Various approaches have been studied in this respect including crosslinking with polyamide-epichlorohydrin resin[85] or of CA with isocyanates,[168,187] the creation of an interpenetrating network composed of methacrylate by loading and subsequent chain-growth polymerization of methacrylic acid,[44] covalent grafting of polymers onto cellulose (e.g., PMMA, PBA, PMMA-co-PBA),[45] dispersion of NFC in a soy protein gel followed by freeze-drying,[188] or anti-solvent precipitation of a secondary polymer (e.g., PLA, PCL, PMMA, CA)[4] on the surface of the fibrous matrix using scCO$_2$, e.g. As with thermal insulation, suitable measures will be required to prevent cellulose-based construction materials from both microbial degradation and shrinkage at higher degrees of humidity. Effective hydrophobization can be achieved, e.g., by heterogeneous (surface) gas-phase esterification with palmitoyl chloride,[60] homogeneous derivatization with stearoyl chloride in ionic liquids,[146] vapor deposition of long-chain perfluoroalkyl or alkyl halogen silanes,[70,71] scCO$_2$ coating with non-polar compounds, such as alkyl keten dimer,[189] CA or PMMA,[4] surface coating with stearoyl esters,[146] or the introduction of an interpenetrating silica network.[42,185,190,191]

Sorption

The high internal surface area of up to 605 m^2 g^{-1}[67] render cellulose aerogels highly suitable for all kinds of applications that rely on surface–substrate interactions. Vapor deposition of hydrophobic octyltrichlorosilane on ultraporous aerogels that were obtained by freeze-drying of a suspension of nanofibrillated, carboxymethylated sulfite softwood-dissolving pulp, e.g., have been demonstrated to be excellent separation media for oil/water mixtures[70] with an oil (hexadecane) absorption capacity of up to 45 g g^{-1}. The materials floating on water are claimed to be recyclable several times without showing significant change in volume upon sorption/desorption. Superoleophobic aerogels can be obtained in a similar way by CVD of 1H,1H,2H,2H-perfluorodecyl-trichlorosilane (PFOTS) e.g.[71] CVD of TiO$_2$ on the surface of MFC has been demonstrated to allow for the preparation of photo-switchable super-adsorbing materials. While the TiO$_2$-coated aerogels have a water-adsorbing capacity of 1.4 g g^{-1} only and are hence essentially water-repellent in native state, UV illumination strongly increases the hydrophilicity of TiO$_2$ as reflected by a strongly increased water adsorbing capacity of 16 g g^{-1}. This has been proposed to be due to defect formation in TiO$_2$ in which the photo-generated oxygen vacancies of TiO$_2$ are presumably favorable for water adsorption. Along with the simultaneously observed photo-catalytic activity being able to decompose organic matter, this type of aerogel opens up new perspectives, such as in air or water purification.[86] Super-adsorbing, shape recovering hydrogels with free swelling capacities of up to 96 g$_{H_2O}$ g$_{gel}$$^{-1}$ have been obtained by cross-linking (120°C, 3 hr, vacuum) of NFC/MFC aerogels with the polyamide-epichlorohydrin resin Kymene™.[85]

Amine-based MFC aerogels have been developed aiming at applications as CO$_2$-adsorbents.[192] The materials exhibited quite low surface areas of 7.1 m^2 g^{-1} and a cyclic CO$_2$ capacity of 0.695 mmol CO$_2$ g^{-1}, which could be maintained over 20 consecutive CO$_2$ adsorption/desorption cycles.

Catalysis

The particular surface chemistry and the high accessible surface area of cellulose aerogels can be also employed for catalysis, slow release applications, or deposition of secondary functional particles (e.g., magnetic particles) or films (e.g., electro-conductive polymers). Topological loading and immobilization of Cu$^+$ ions onto the crystal surfaces of 2,2,6,6-tetramethylpiperidine-1-oxyl (TEMPO)-oxidized cellulose nanofibrils (TOCNs), e.g., were shown to have excellent catalytic efficiency for azide–alkyne [3+2] cyclo-addition reactions.[77] Beyond that it allows for a facile recovery of the catalyst that greatly contributes to economic aspects of the synthesis of 1,4-disubstituted 1,2,3-triazoles under mild conditions.

Magnetic Aerogels

Cellulose microspheres that can be used for the targeted delivery of drugs such as proteins have been prepared by loading CL aerogels with a mixture of FeCl$_3$·6H$_2$O and FeCl$_2$·4H$_2$O at a molar ratio of 2:1 and subsequent situ synthesis of Fe$_3$O$_4$ by adding NH$_4$OH and raising the temperature to 60°C for 30 min.[193] Freeze-dried BC aerogels have been used in a similar way as templates for the preparation of lightweight porous magnetic aerogels, which can be compacted into a stiff magnetic nanopaper (Fig. 17). Different from the earlier approach, a mixture of FeSO$_4$ and CoCl$_2$ (Fe^{2+}/Co^{2+}=2) has been used. The formation of the cobalt ferrite nanoparticles was accomplished by heating on air to 90°C, the addition of a mixture of NaOH/KNO$_3$, and continued heating for another 6 hr. Owing to their flexibility, high porosity, and surface area, these aerogels are expected to be useful in microfluidic devices and as electronic actuators.[194] A comprehensive review on magnetic responsive cellulose nanocomposites and their applications can be found elsewhere.[195]

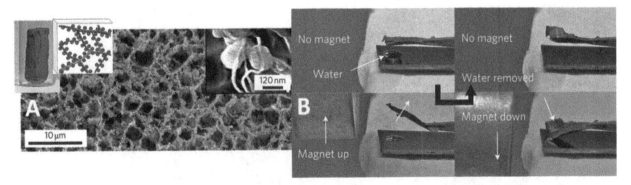

Fig. 17 SEM image of magnetic and highly flexible BC aerogels containing cobalt ferrite nanoparticles (**A**). Image sequence following the black arrow: Aerogel response to the position of the magnet and absorption of a water droplet (**B**).
Source: From Olsson et al.[194] © 2010, with permission from Macmillan Publishers Ltd., Nature Nanotechnology.

Electro-Conductive Aerogels

Conducting polymers, such as poly(*p*-phenylene), poly-thiophene, polyaniline, or polypyrrole, can exist in two potential-dependent redox states, i.e.: 1) an oxidized, conducting state where the polymer is (mostly) positively charged; and 2) a reduced, non-conducting, neutral state. Switching between these two states is coupled to the transport of ions (usually anions) to ensure charge neutrality within the polymer. Due to a reduction in mass transport limitations of the redox reaction, this transport proceeds faster with decreasing thickness of the conducting layer.[196] According to the electrochemically stimulated conformational relaxation model proposed by Otero et al.,[197,198] electro-conductive films are required to have a sufficiently high porosity to ensure fast electrochemical processes as the oxidation rate decreases as more compact the polymeric structure is in its reduced state. Hence, aerogels carrying a thin layer of an electro-conducting polymer are promising materials for all applications that require high specific capacity, specific capacitance, and energy density.

The coating of MFC aerogels with polyaniline (PANI) was one of the first attempts to obtain electro-conducting highly porous cellulose-based materials. Loading of the aerogels was accomplished by dipping them into 6.2 w% polyaniline/dodecyl benzene sulfonic acid doped solution in toluene [PANI(DBSA)$_{1.1}$] and extracting thoroughly the unbound PANI(DBSA). Flexible aerogels with relatively high electrical conductivity of around 1×10^{-2} S cm^{-1} were obtained.[55]

In situ synthesis of the electro-conductive polymer in aqueous suspension with the aerogel forming cellulose source is a different approach that affords functional materials with tunable structural and electrochemical properties. Polymerization of pyrrole in acidic, aqueous suspension (0.25 M HCl) that additionally contained TEMPO-oxidized NFC from softwood sulfite pulp, a small quantity of the surfactant Tween-80 and a slight excess of oxidant (FeCl$_3$·6H$_2$O), afforded aerogels that had specific surface areas of up to 246 m^2 g^{-1}, specific charge capacity of ~220 C g^{-1} at a scan rate of 1 mV s^{-1}. In particular, the high porosity composites that maintained this specific charge capacity also at scan rates of up to 50 mV s^{-1} were considered to be promising candidates for electrode materials in structural batteries.[196]

Layer-by-Layer (LbL) assembling is a third approach for tuning the properties of aerogels. Using this method, carboxymethylated MFC has been layer-wise coated with thin films of either a bioactive compound (hyaluronic acid), a conducting polymer [poly(3,4-ethylenedioxythiophene): poly(styrenesulfonate); PEDOT:PSS], or single-wall carbon nanotubes (SWCNT) using poly(ethyleneimine) (PEI) as an adhesion promoter.[199] The carboxymethylated MFC had been cross-linked by 1,2,3,4-butanetetracarboxylic acid (BTCA) beforehand to further increase the total anionic surface charge (two carboxylate groups per cross-link), which has been claimed to be beneficial for the LbL preparation procedure.[199]

The obtained products were demonstrated to have intriguing features. Thus, a high bulk conductivity of 1.4 10^{-5} S cm^{-1} measured over large electrode distances was obtained for an aerogel consisting of 10 bilayers of PEI/PEDOT:PSS. For the SWCNT-functionalized aerogels, the specific electrode capacitance was calculated to be 419 ± 17 F g^{-1} by accounting only for the measured weight of the active LbL layer. This largely exceeds the best values previously reported for SWCNTs on paper substrates, SWCNT LbL freestanding films, and LbL multiwall-CNT films coated on carbon paper (Fig. 18).[199]

Temporary Templates

Cellulose aerogels can be used as permanent templates to create core-shell structures as shown earlier for the LbL assembly technique. Furthermore, cellulose aerogels can

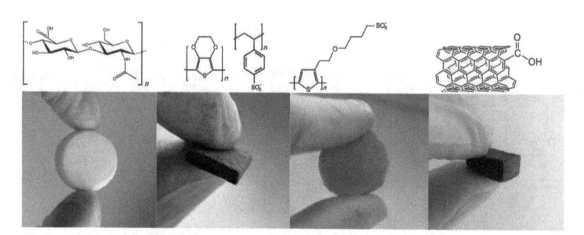

Fig. 18 Layer-by-layer (LbL) functionalized aerogels. LbL coating from left to right: (PAH/HA)$_5$, (PEI/PEDOT:PSS)$_{10}$, (PEI/ADS2000P)$_{10}$, (PEI/SWCNT)$_5$. The functional polyanions are shown above the associated sample images.
Source: From Hamedi et al.[199] © 2013, with permission from Wiley-VCH Verlag GmbH & Co. KGaA, Weinheim.

serve as temporary templates, which allows for a controlled preparation of a secondary inorganic or organic network whose morphology and pore features resemble that of the template that is afterward removed by dissolution or calcination. Following this approach, TiO_2, ZnO, or Al_2O_3 nanotube aerogels have been prepared by atomic layer deposition of the metal oxides onto MFC aerogels and subsequent calcination at 450°C.[63] The feasibility of using cellulosic aerogels as temporary scaffolds for the synthesis of organic aerogels has been demonstrated by Pircher et al. who deposited PMMA on the inner surface of BC cuboids first and removed the temporary template by dissolution in 1-ethyl-3-methyl-imidazolium acetate afterward leaving behind a PMMA aerogel of comparable morphology. The respective composite precursor had been obtained by immersion of the BC template in a solution of PMMA in acetone and subsequent anti-solvent precipitation of PMMA and extraction of the solvent using $scCO_2$.[4]

Biomedical Applications

Bacterial cellulose

BC has become an attractive source material for various applications in biomedicine and cosmetics due to its good biocompatibility and low immunogenic potential,[200,201] the positive effect on skin tissue regeneration,[202] biological degradability,[203,204] high fiber strength, hydrophilicity, moistening capability, and purity. This is reflected by the availability of a series of commercial BC products for topological wound healing (Suprasorb®X, Bioprocess®, XCell® and Biofill®)[205] or skin care (Nanomasque®).[206,207] Furthermore, BC has been studied for hemodialysis,[196,208] vascular grafts[201,209–213] including urinary reconstruction[214] or repair of abdominal wall defects,[215,216] engineering of bone tissue,[217–220] controlled drug release,[40] as artificial blood vessels,[205] semi-permanent artificial skin,[201] cartilage tissue for artificial knee menisci,[221,222] or contact lenses.[223] Figure 19

illustrates selected examples of potential biomedical applications of bacterial cellulose. A more comprehensive review of BC in biomedical applications can be found elsewhere.[224–228]

Plant cellulose

The intriguing properties of BC and its multitude of biomedical applications on the one hand, and the still limited availability for large-scale applications, the time-consuming removal of the bacterial culture and residues of the nutrient medium, the fewer options with regard to shaping and its high price on the other hand have largely stimulated research on comparable materials from native, micro-fibrillated, or regenerated plant cellulose and its derivatives. Even though research on biomedical applications of (non-bacterial) cellulose aerogels is still in its infancy, first results will be summarized in the following text.

Water adsorption

Hydrogels, based on sodium carboxymethyl cellulose (Na-CMC) and hydroxyethyl cellulose (HEC) and cross-linked with divinylsulfone (DVS), were demonstrated to have swelling capabilities comparable to that of polyacrylate-based superadsorbing materials.[229] The high water adsorption capacities of up to 1 L g^{-1} and swelling rates along with the great water retention under centrifugal load or compression have been achieved by phase inversion desiccation in acetone (i.e., a non-solvent for cellulose), which induced the formation of a microporous hydrogel structure and strong capillary effects. Beyond their use in sanitary products, such as diapers or napkins, the novel hydrogels can be used for removal of excess water from the body, in the treatment of some pathological conditions, such as renal failure and diuretic-resistant edemas. In spite of the use of DVS as a crosslinker, the hydrogel formulations tested showed a good biocompatibility for both *in vitro* and *in vivo* experiments.

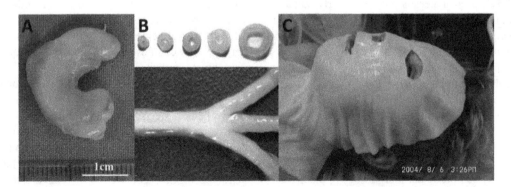

Fig. 19 Potential biomedical applications of BC as meniscus implants (**A**); artificial blood vessels (**B**); and facial masks for wound dressing (**C**).
Source: (**A**) From Bodin et al.[221] © 2007, with permission from John Wiley & Sons, Ltd. (**B**) From Bodin et al.[213] © 2006, with permission from Wiley Periodicals, Inc. (**C**) Adapted from Czaja et al.[227] © 2007, with permission from American Chemical Society.

Controlled drug release

Films consisting of MFC and/or CNC and optionally a hydrophilic polymer, such as HPMC, have been recently proposed as "green" barriers that control the permeability of molecules or active agents, such as drugs, water, pesticides, herbicides, taste-masking agents, flavorings, food additives, nutrients, and vitamins.[230] Commercial films for controlled release applications are frequently composed of one insoluble film-forming polymer and one pore-forming agent. The latter is typically released from the film upon contact with an aqueous solution, leaving pores in the insoluble film that increases the permeability of the film. However, when MFC is used as an insoluble component and HPMC as pore-forming agent, the permeability surprisingly exhibits the opposite effect, i.e., a reduced permeability is observed that opens up new opportunities with regard to applications, such as slowing down the release of active compounds from respective aerogels.

Highly porous cellulose aerogels prepared by freeze-drying of hydrogels from various NFC sources (BC, red pepper, quince seed, TEMPO-oxidized birch pulp) have been recently suggested as nanoparticle reservoirs for oral drug delivery, such as for beclomethasone dipropionate (BDP), which is used in the prophylaxis of asthma, treatment of rhinitis, sinusitis, or severe inflammatory skin disorders unresponsive to less potent steroids. The release profile of the drug from BDP nanoparticles coated with amphiphilic hydrophobin proteins and well dispersed in the NFC aerogels can be largely controlled by the type of cellulosic material used.[231]

Besides (derivatized) bacterial or plant-derived NFC, CMC is another promising candidate in cellulose-based controlled release applications. Radiation-induced copolymerization and crosslinking of CMC and acrylic acid has been shown to afford a material whose swelling behavior is largely dependent on the pH of the aqueous medium. As swelling and hence release of physically bonded drugs confined in the material strongly increases with pH, the preceding CMC-based materials are particularly suitable for orally administered colon-targeted controlled drug delivery that encounter strong pH variations occurring during transitioning from the stomach to the intestine.[232] Thermo- and pH-sensitive films obtained by gamma irradiation of CMC and poly(ethylene oxide) (PEO) blends have been recently proposed for the controlled release of ketoprofen.[233]

Tissue engineering

The high porosity and particular morphology of the hierarchical open-porous biopolymeric network along with the low immunogenic potential[219] and slow degradability of cellulose render cellulosic aerogels potential cell scaffolding matrices for tissue-engineering applications. *In vitro*

generation of bone tissue, e.g., accomplished by combining pluripotent, vital cells with growth factors on a suitable carrier material, is considered to become the method of choice in the near future for repair of larger osseous defects as it has the potential to circumvent the problems related to autogenous or allogeneic bone transplantation and synthetic bone replacement materials. However, even though tissue engineering offers almost unlimited possibilities for providing artificial bone tissue,[219] the required cell scaffolding materials have to meet high demands with regard to sterility, resorbability, biocompatibility, mechanical properties, porosity, and surface topography to provide structural support for cell attachment, spreading, migration, proliferation, and differentiation.[234,235] An interconnected, spread pore network with a highly porous, micro-structured surface is required for the diffusion of physiological nutrients and gases to cells, the removal of metabolic by-products from cells, *in vitro* cell adhesion, ingrowth, and *in vivo* neovascularization.[48] The size, distribution, and shape of pores furthermore control cell survival, signaling, growth, propagation, reorganization, cell shaping, and gene expression.

Resorbable, highly porous cellulose hydrogels containing homogeneously distributed calcium-deficient hydroxyapatite (cd-Hap) nanocrystallites[236] or HAp nanorods[237] have been suggested as promising materials for osseous regeneration, as hydroxyapatite $Ca_{10}(PO_4)_6(OH)_2$ is osteoconductive, biocompatible, and bioactive. Model studies have shown that the presence of hydroxyapatite increases the expression of mRNA encoding the bone matrix proteins osteocalcin, osteopontin, and bone sialoprotein.[238,239] Furthermore, hydroxyapatite increases significantly the mechanical strength of cellulosic scaffolds providing thus improved conditions for cell attachment, spreading, migration, proliferation, and differentiation (Fig. 20).[238,239]

However, despite improved cell attachment, growth, and osteogenic differentiation, the major problem with cellulose aerogels as bone graft material is the insufficient binding and crystallization tendency of hydroxyapatite on the cellulose matrix,[217,218] which impedes a complete biomineralization and osseointegration.[240]

Grafting of negatively charged phosphorous-containing groups onto cellulose is a promising technique for improving the mineralizing properties of cellulose[218] and to support the nucleation of calcium-deficient hydroxyapatite.[49] This has been confirmed by our own work on cellulose phosphate aerogels from different commercial pulps that formed a thin cdHAp surface layer in simulated body fluid similar as reported for poly(ethylene terephthalate)[241] or polysaccharide-based materials, such as chitin fibers,[242] CL,[243] MCC,[240] or BC.[217,218] The creation of surface charges has been known for a while to increase both biodegradation and molecular recognition of cellulose[219,244] and renders it more suitable for bone regeneration.[236] Phosphorylated cellulose has been furthermore confirmed

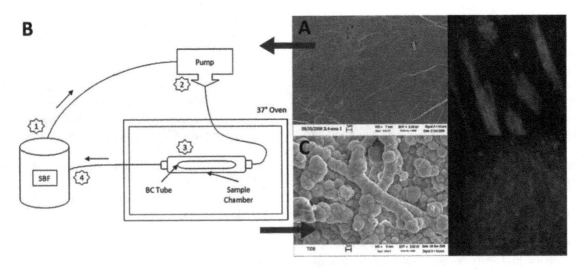

Fig. 20 Scanning electron and fluorescence microscopy images (osteoprogenitor cell morphology) of BC tubes prior to (**A**) and after dynamic biomineralization (**C**). A schematic of the bioreactor system for mineralization of BC tubes is shown (**B**). In contrast to pure BC, were the osteoprogenitor cells cluster together, FM images of mineralized BC show improved adherence to the surface and a confluent layer of cells.
Source: From Zimmermann et al.[49] © 2009, with permission from Elsevier.

to be nontoxic toward human osteoblasts and fibroblasts as demonstrated for MCC.[245]

In a recent study, the authors confirmed that spin-coated layers of phosphorylated CL and hardwood pre-hydrolysis kraft pulp (DS 0.2–0.4) support a robust growth and osteogenic differentiation of human bone-marrow derived mesenchymal stem cells (MSC) similar to clinically used tissue culture polystyrene. Respective cellulose phosphate aerogels (CPA) showed a good hemocompatibility (human whole blood) in terms of hemostasis and inflammatory response. Surprisingly, the low degree of phosphorylation was sufficient to suppress any significant inflammatory response via the alternative pathway for the CP aerogels, which is typically an issue with comparable products of non-derivatized cellulose.[246]

Cellulose-Based Carbon Aerogels

Carbogels of tailored porosity have been confirmed to be promising materials for gas separation and adsorption,[247,248] catalysis,[249–253] hydrogen storage,[254–256] and electrochemical applications where carbogels have been used in proton exchange membrane fuel cells (PEMFCs)[257–259] or electrical double-layer capacitors (EDLCs).[260–263]

PEMFCs utilizing hydrogen as a fuel are being developed to replace batteries in portable electronic devices and internal combustion engines in automobiles on account of their high energy efficiency, low pollutant emission, and low working temperature. PEMFC electrodes that catalyze both of the half-cell reactions, i.e., hydrogen oxidation and oxygen reduction, are typically porous materials covered by a thin film of platinum. For commercial and electrochemical reasons, this platinum film should be as thin as possible.

The latter is due to the fact that the catalyst must have simultaneous access to hydrogen and both of the conducting media (H^+, e^-). If the platinum film would not be thin enough, the rate of proton diffusion within the catalyst layer, the mass transfer rates of the chemical reactants, and products to and from the active sites would result in a loss of energy. This, in turn, can contribute to a significant overpotential or polarization of the electrodes, which can limit the cell performance, particularly at high current densities.[264]

Double-layer capacitors are also referred to as "supercapacitors" that store energy via separation of charges across a polarized electrode/electrolyte interface and bridge the gap between batteries (accumulators) and conventional capacitors.[265] They are able to store more energy than conventional capacitors, release a higher voltage than batteries, store electrical energy almost lossless for a long period of time, and can be (dis)charged very quickly. Potential applications of supercapacitors are uninterruptible power supplies for bridging electrical power outage, short-term supply of high electrical power, such as for starting up industrial machinery and storage of relatively short energy impulses. In addition to voltage, the surface of the interface between electro-conductive solid and surrounding electrolyte is the main criterion determining charge storage.

Even though a multitude of highly porous carbon-based electrode materials, such as activated carbon powders, activated carbon fabrics, or carbon nanotubes, have been developed, none of them are ideal candidates for both PEMFC and EDLC applications.[266] As pore size distribution of most activated carbon materials is not optimum because of poor pore size control in the activation process, the high created surface area cannot be fully exploited to form the double layer. For activated carbon fabrics, such as from rayon or PAN, the production costs are comparatively high, which restricts their use in EDLCs to very specific applications. For purified CNT powders, the specific capacitance values are not impressive and are typically limited to the range of 20–80 F/g.[266] Carbon aerogels that are derived from highly porous cellulosic precursors are therefore of increasing interest and it has been shown that cellulose-based carbon aerogels doped by platinum nanoparticles can compete with state-of-the-art Pt/CB (carbon black) electro-catalysts.[257,258,267] This is mainly due to the better control of the formation of an ordered, regular, and interconnected pore structure that allows for rapid ionic motion[268] and hence high power capabilities of carbon aerogel-based electrodes.[266] High mesoporosity is required for both EDLC and PEMFC applications as a compromise of high surface area and pore wettability, which is of particular importance for polymeric electrolytes or proton conductors, such as the sulfonated tetrafluoroethylene-based fluoropolymer-copolymer Nafion® that is used in PEMFCs. The following examples will demonstrate that mesoporous cellulose-based carbon aerogels for electrochemical applications can be obtained from both cellulose derivatives and MCC.

Cellulose acetate

Pyrolysis of CA (crosslinked beforehand with diphenylmethane diisocyanate using dibutyl tin dilaurate as catalyst) at 800°C in CO_2 atmosphere, subsequent impregnation with a solution of H_2PtCl_6 (0.017 M) in i-propanol/water (1:1, v/v) at room temperature, followed by chemical reduction using an excess of $NaBH_4$ (4×) afforded carbon aerogels that featured a high surface area of up to 400–450 m²/g, which is considered to be appropriate for use in PEMFC active layers.[259] Similarly, a series of mesoporous cellulose-based carbon aerogels with pore surface areas of 160–300 m²/g have been obtained for EDLC applications. Following the classical sol-gel approach, CA organogels were first prepared by crosslinking of CA with polymethylene polyphenylpolyisocyanate (PMDI) in dry acetone using 1.4-diazabicyclo-[2.2.2]octane (DABCO® TMR) as a catalyst. Then, the solvent was extracted using $scCO_2$ (8.5 MPa, 37°C, 4 hr) and the organo-aerogels were subjected to pyrolysis in N_2 atmosphere at constant heating rate of 4°C min⁻¹. The maximum temperature of 1000°C was maintained for 1 hr. The obtained products were further modified by oxidation with H_2O_2 or HNO_3 (48 hr, ambient temperature) and/or ammonization or co-heat-treatment with melamine. Interestingly, both ammonia[269] and carbon dioxide promote the formation of additional micro- and mesopores under pyrolytic conditions and are able to widen existing micropores up to the range of mesoporosity.[269]

Microcrystalline cellulose

Carbogels from commercial MCC (Avicel Ph-101) were obtained by 1) dissolving the cellulose in a pre-cooled aqueous solution of NaOH (–6°C); 2) gel formation at 50°C; 3) coagulation of cellulose using water; 4) replacing water by acetone; 5) scCO₂ drying; and 6) pyrolysis at 830°C (1050°C) in nitrogen atmosphere (Fig. 21).[272]

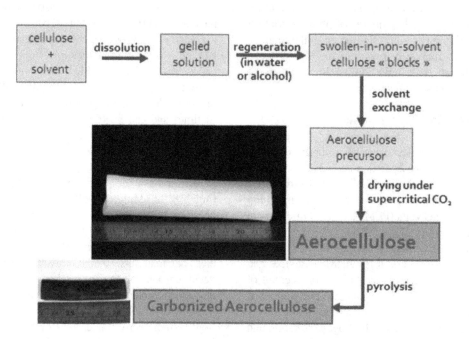

Fig. 21 Scheme of the synthetic route to cellulose-based carbon aerogels.
Source: From Rooke et al.[272] © 2011, with permission from The Electrochemical Society.

Subsequently, the carbogels were doped with platinum particles by 1) thermal activation in CO_2 atmosphere; 2) impregnation with H_2PtCl_6; and 3) platinum salt reduction using either hydrogen (300–400°C)[267] or $NaBH_4$.[257,267] Beyond CA and MCC, other sources of highly crystalline native cellulose, such as bacterial and algal celluloses or ramie fibers, seem to be suitable raw materials for the preparation of carbon aerogels too as the ultrastructure of the parent materials, and in particular their microfibrillar structures are largely retained throughout the carbonization (≥500°C) and subsequent graphitization (≥2000°C) processes.[270]

CONCLUSIONS

Research on cellulose aerogels has experienced a great boom since its early days some 10 years ago. Today, a broad variety of sophisticated approaches is available that allows for the preparation of different types of aerogels from cellulose (polymorph I: BC, microcrystalline cellulose, micro- or NFC; polymorph II: physically or chemically regenerated cellulose; polymorphs I and II: *all*-cellulose composites) and a multitude of cellulose derivatives. Techniques such as layer-by-layer self-assembly, atomic layer deposition, chemical surface modification, crosslinking, templating, creation of interpenetrating networks, etc., are tools that have been further advanced to tailor cellulose aerogels for particular applications by controlling properties such as density, morphology, porosity, pore size distribution and pore surface area, hydrophilicity/hydrophobicity, oleophilicity, biocompatibility, or biodegradability and to provide them particular mechanical, electrical, magnetic, and insulating properties.

In view of the recent efforts and advances in the biorefinery sector, the great availability of cellulose as the most abundant biopolymer on earth and its comparatively low price, it is safe to conclude that the number of commercial applications for cellulose aerogels will further increase. The awakening public interest, activities of respective companies, and the increased funding of research in this field are strongly positive indicators of this development and important signals toward new resource-saving concepts.

ACKNOWLEDGMENT

The financial support by the Austrian Science Fund (FWF: I848-N17), the French *L'Agence Nationale de la Recherché* (ANR-11-IS08-0002*)*, the Austrian Agency for International Cooperation in Education and Research (OeAD: FR10/2010), and the University of Natural Resources and Life Sciences (BOKU-DOC grant to Christian Schimper) is thankfully acknowledged.

REFERENCES

1. Liebner, F.; Aigner, N.; Schimper, C.; Potthast, A.; Rosenau, T. Bacterial cellulose aerogels: From lightweight dietary food to functional materials. In *Functional Materials from Renewable Sources*; Liebner, F., Rosenau, T., Eds.; American Chemical Society: Washington, DC, 2012; 57–74.

2. Hoepfner, S.; Ratke, L.; Milow, B. Synthesis and characterisation of nanofibrillar cellulose aerogels. Cellulose **2008**, *15* (1), 121–129.

3. Mukhopadhyay, M. *Natural Extracts Using Supercritical Carbon Dioxide*; CRC Press: Boca Raton, FL, 2000.

4. Pircher, N.; Veigel, S.; Aigner, N.; Nedelec, J.-M.; Rosenau, T.; Liebner, F. Reinforcement of bacterial cellulose aerogels with biocompatible polymers. Carbohydr. Polym. **2014**, *111*, 505–513.

5. Jessop, P.G.; Subramaniam, B. Gas-Expanded Liquids. Chem. Rev. **2007**, *107* (6), 2666–2694.

6. Dittmar, D.; Bijosono Oei, S.; Eggers, R. Interfacial tension and density of ethanol in contact with carbon dioxide. Chem. Eng. Technol. **2002**, *25* (1), 23–27.

7. Masmoudi, Y.; Rigacci, A.; Ilbizian, P.; Cauneau, F.; Achard, P. Diffusion during the supercritical drying of silica gels. Dry. Technol. **2006**, *24* (9), 1121–1125.

8. Wawrzyniak, P.; Rogacki, G.; Pruba, J.; Bartczak, Z. Effective diffusion coefficient in the low temperature process of silica aerogel production. J. Non-Cryst. Solids **2001**, *285* (1–3), 50–56.

9. Mukhopadhyay, M.; Rao, B.S. Modeling of supercritical drying of ethanol-soaked silica aerogels with carbon dioxide. J. Chem. Technol. Biotechnol. **2008**, *83* (8), 1101–1109.

10. Deinema, M.H.; Zevenhuizen, L.P.T.M. Formation of cellulose fibrils by gram-negative bacteria and their role in bacterial flocculation. Arch. Microbiol. **1971**, *78* (1), 42–57.

11. Brown, R.M., Jr.; Willison, J.H.; Richardson, C.L. Cellulose biosynthesis in *Acetobacter xylinum*: Visualization of the site of synthesis and direct measurement of the *in vivo* process. Proc. Natl Acad. Sci. USA **1976**, *73* (12), 4565–4569.

12. Liebner, F.; Haimer, E.; Wendland, M.; Neouze, M.A.; Schlufter, K.; Miethe, P.; Heinze, T.; Potthast, A.; Rosenau, T. Aerogels from unaltered bacterial cellulose: Application of scCO$_2$ drying for the preparation of shaped, ultralightweight cellulosic aerogels. Macromol. Biosci. **2010**, *10* (4), 349–352.

13. Lee, K.-Y.; Buldum, G.; Mantalaris, A.; Bismarck, A. More than meets the eye in bacterial cellulose: Biosynthesis, bioprocessing, and applications in advanced fiber composites. Macromol. Biosci. **2013**, *14* (1), 10–32.

14. Dudman, W.F. Cellulose production by acetobacter strains in submerged culture. J. General Microbiol. **1960**, *22* (1), 25–39.

15. Yamanaka, S.; Watanabe, K.; Kitamura, N.; Iguchi, M.; Mitsuhashi, S.; Nishi, Y.; Uryu, M. The structure and mechanical properties of sheets prepared from bacterial cellulose. J. Mater. Sci. **1989**, *24* (9), 3141–3145.

16. Iijima, S. Helical microtubules of graphitic carbon. Nature **1991**, *354* (6348), 56–58.

17. Tokoh, C.; Takabe, K.; Sugiyama, J.; Fujita, M. CPMAS 13C NMR and electron diffraction study of bacterial cellulose structure affected by cell wall polysaccharides. Cellulose **2002**, *9* (3), 351–360.

18. Goelzer, F.D.E.; Faria-Tischer, P.C.S.; Vitorino, J.C.; Sierakowski, M.R.; Tischer, C.A. Production and characterization of nanospheres of bacterial cellulose from *Acetobacter xylinum* from processed rice bark. Mater. Sci. Eng. C **2009**, *29* (2), 546–551.

19. Huang, H.-C.; Chen, L.-C.; Lin, S.-B.; Hsu, C.-P.; Chen, H.-H. In situ modification of bacterial cellulose network structure by adding interfering substances during fermentation. Bioresour. Technol. **2010**, *101* (15), 6084–6091.

20. Yan, Z.; Chen, S.; Wang, H.; Wang, B.; Wang, C.; Jiang, J. Cellulose synthesized by *Acetobacter xylinum* in the presence of multi-walled carbon nanotubes. Carbohydr. Res. **2008**, *343* (1), 73–80.

21. Heßler, N.; Klemm, D. Alteration of bacterial nanocellulose structure by in situ modification using polyethylene glycol and carbohydrate additives. Cellulose **2009**, *16* (5), 899–910.

22. Uraki, Y.; Nemoto, J.; Otsuka, H.; Tamai, Y.; Sugiyama, J.; Kishimoto, T.; Ubukata, M.; Yabu, H.; Tanaka, M.; Shimomura, M. Honeycomb-like architecture produced by living bacteria, *Gluconacetobacter xylinus*. Carbohydr. Polym. **2007**, *69* (1), 1–6.

23. Gavillon, R.; Budtova, T. Aerocellulose: New highly porous cellulose prepared from cellulose-NaOH aqueous solutions. Biomacromolecules **2008**, *9* (1), 269–277.

24. Mikos, A.G.; Thorsen, A.J.; Czerwonka, L.A.; Bao, Y.; Langer, R.; Winslow, D.N.; Vacanti, J.P. Preparation and characterization of poly(l-lactic acid) foams. Polymer **1994**, *35* (5), 1068–1077.

25. Ma, P.X.; Choi, J.W. Biodegradable polymer scaffolds with well-defined interconnected spherical pore network. Tissue Eng. **2001**, *7* (1), 23–33.

26. Chen, G.; Ushida, T.; Tateishi, T. Development of biodegradable porous scaffolds for tissue engineering. Mater. Sci. Eng. C **2001**, *17* (1–2), 63–69.

27. Zhou, Q.; Gong, Y.; Gao, C. Microstructure and mechanical properties of poly(L-lactide) scaffolds fabricated by gelatin particle leaching method. J. Appl. Polym. Sci. **2005**, *98* (3), 1373–1379.

28. Kim, J.; Yaszemski, M.J.; Lu, L. Three-dimensional porous biodegradable polymeric scaffolds fabricated with biodegradable hydrogel porogens. Tissue Eng. C **2009**, *15* (4), 583–594.

29. Andersson, J.; Stenhamre, H.; Bäckdahl, H.; Gatenholm, P. Behavior of human chondrocytes in engineered porous bacterial cellulose scaffolds. J. Biomed. Mater. Res. Part A **2010**, *94A* (4), 1124–1132.

30. Bäckdahl, H.; Esguerra, M.; Delbro, D.; Risberg, B.; Gatenholm, P. Engineering microporosity in bacterial cellulose scaffolds. J. Tissue Eng. Regen. Med. **2008**, *2* (6), 320–330.

31. Maeda, H. Preparation and properties of bacterial cellulose aerogel and its application. Cell. Commun. **2006**, *13* (4), 169–172.

32. Maeda, H.; Nakajima, M.; Hagiwara, T.; Sawaguchi, T.; Yano, S. Preparation and properties of bacterial cellulose aerogel. Jap. J. Polym. Sci. Technol. **2006**, *63* (2), 135–137.

33. Tang, M.M.; Bacon, R. Carbonization of cellulose fibers--I. Low temperature pyrolysis. Carbon **1964**, 2 (3), 211–214, IN1, 215–220.

34. Rhee, B.; Yim, H.B. Optimierung des kontinuierlichen thermischen Abbaus zur Graphitierung von Endlos-Zellulose (Transl.). Hwahak Konghak **1975**, *13* (5), 261–268.

35. Cheng, K.-C.; Catchmark, J.M.; Demirci, A. Enhanced production of bacterial cellulose by using a biofilm reactor and its material property analysis. J. Biol. Eng. **2009**, *3*, 12.

36. Yang, C.M.; Chen, C.Y. Synthesis, characterization and properties of polyanilines containing transition metal ions. Synth. Met. **2005**, *153* (1–3), 133–136.

37. Scheirs, J.; Camino, G.; Tumiatti, W. Overview of water evolution during the thermal degradation of cellulose. Eur. Polym. J. **2001**, *37* (5), 933–942.

38. Lampke, T. Beitrag zur Charakterisierung naturfaserverstärkter Verbundwerkstoffe mit hochpolymerer Matrix. Fakultät für Maschinenbau und Verfahrenstechnik. PhD. 2001, Chemnitz, Germany: TU Chemnitz.

39. Liebner, F.; Haimer, E.; Potthast, A.; Loidl, D.; Tschegg, S.; Neouze, M.-A.; Wendland, M.; Rosenau, T. Cellulosic aerogels as ultra-lightweight materials. Part II: Synthesis and properties. Holzforschung **2009**, *63* (1), 3–11.

40. Haimer, E.; Wendland, M.; Schlufter, K.; Frankenfeld, K.; Miethe, P.; Potthast, A.; Rosenau, T.; Liebner, F. Loading of bacterial cellulose aerogels with bioactive compounds by antisolvent precipitation with supercritical carbon dioxide. Macromol. Symp. **2010**, *294* (2), 64–74.

41. Sescousse, R.; Gavillon, R.; Budtova, T. Aerocellulose from cellulose–ionic liquid solutions: Preparation, properties and comparison with cellulose–NaOH and cellulose–NMMO routes. Carbohydr. Polym. **2011**, *83* (4), 1766–1774.

42. Yano, S.; Maeda, H.; Nakajima, M.; Hagiwara, T.; Sawaguchi, T. Preparation and mechanical properties of bacterial cellulose nanocomposites loaded with silica nanoparticles. Cellulose **2008**, *15* (1), 111–120.

43. Seifert, M.; Hesse, S.; Kabrelian, V.; Klemm, D. Controlling the water content of never dried and reswollen bacterial cellulose by the addition of water-soluble polymers to the culture medium. J. Polym. Sci. Part A **2004**, *42* (3), 463–470.

44. Hobzova, R.; Duskova-Smrckova, M.; Michalek, J.; Karpushkin, E.; Gatenholm, P. Methacrylate hydrogels reinforced with bacterial cellulose. Polym. Int. **2012**, *61* (7), 1193–1201.

45. Lacerda, P.S.S.; Barros-Timmons, A.M.M.V.; Freire, C.S.R.; Silvestre, A.J.D.; Neto, C.P. nanostructured composites obtained by ATRP sleeving of bacterial cellulose nanofibers with acrylate polymers. Biomacromolecules **2013**, *14* (6), 2063–2073.

46. Nge, T.T.; Nogi, M.; Yano, H.; Sugiyama, J. Microstructure and mechanical properties of bacterial cellulose/chitosan porous scaffold. Cellulose **2010**, *17* (2), 349–363.

47. Araki, J.; Kuga, S.; Magoshi, J. Influence of reagent addition on carbodiimide-mediated amidation for poly(ethylene glycol) grafting. J. Appl. Polym. Sci. **2002**, *85* (6), 1349–1352.

48. Puppi, D.; Chiellini, F.; Piras, A.M.; Chiellini, E. Polymeric materials for bone and cartilage repair. Progr. Polym. Sci. **2010**, *35* (4), 403–440.

49. Zimmermann, K.A.; LeBlanc, J.M.; Sheets, K.T.; Fox, R.W.; Gatenholm, P. Biomimetic design of a bacterial cellulose/hydroxyapatite nanocomposite for bone healing applications. Mater. Sci. Eng. C **2011**, *31* (1), 43–49.

50. Nishiyama, Y.; Langan, P.; Chanzy, H. Crystal structure and hydrogen-bonding system in cellulose iβ from synchrotron x-ray and neutron fiber diffraction. J. Am. Chem. Soc. **2002**, *124* (31), 9074–9082.

51. Nishiyama, Y.; Sugiyama, J.; Chanzy, H.; Langan, P. Crystal structure and hydrogen bonding system in cellulose iα from synchrotron X-ray and neutron fiber diffraction. J. Am. Chem. Soc. **2003**, *125* (47), 14300–14306.

52. Langan, P.; Nishiyama, Y.; Chanzy, H. A revised structure and hydrogen-bonding system in cellulose II from a neutron fiber diffraction analysis. J. Am. Chem. Soc. **1999**, *121* (43), 9940–9946.

53. Eichhorn, S.J.; Davies, G.R. Modelling the crystalline deformation of native and regenerated cellulose. Cellulose **2006**, *13* (3), 291–307.

54. Iwamoto, S.; Kai, W.; Isogai, A.; Iwata, T. Elastic modulus of single cellulose microfibrils from tunicate measured by atomic force microscopy. Biomacromolecules **2009**, *10* (9), 2571–2576.

55. Pääkko, M.; Vapaavuori, J.; Silvennoinen, R.; Kosonen, H.; Ankerfors, M.; Lindstrom, T.; Berglund, L.A.; Ikkala, O. Long and entangled native cellulose I nanofibers allow flexible aerogels and hierarchically porous templates for functionalities. Soft Matter **2008**, *4* (12), 2492–2499.

56. Trovatti, E.; Carvalho, A.J.F.; Ribeiro, S.J.L.; Gandini, A. Simple green approach to reinforce natural rubber with bacterial cellulose nanofibers. Biomacromolecules **2013**, *14* (8), 2667–2674.

57. Spence, K.; Venditti, R.; Rojas, O.; Habibi, Y.; Pawlak, J. A comparative study of energy consumption and physical properties of microfibrillated cellulose produced by different processing methods. Cellulose **2011**, *18* (4), 1097–1111.

58. Jin, H.; Kettunen, M.; Laiho, A.; Pynnönen, H.; Paltakari, J.; Marmur, A.; Ikkala, O.; Ras, R.H.A. Superhydrophobic and superoleophobic nanocellulose aerogel membranes as bioinspired cargo carriers on water and oil. Langmuir **2011**, *27* (5), 1930–1934.

59. Chen, W.; Yu, H.; Li, Q.; Liu, Y.; Li, J. Ultralight and highly flexible aerogels with long cellulose I nanofibers. Soft Matter **2011**, *7* (21), 10360–10368.

60. Fumagalli, M.; Ouhab, D.; Boisseau, S.M.; Heux, L. Versatile gas-phase reactions for surface to bulk esterification of cellulose microfibrils aerogels. Biomacromolecules **2013**, *14* (9), 3246–3255.

61. Wang, M.; Anoshkin, I.V.; Nasibulin, A.G.; Korhonen, J.T.; Seitsonen, J.; Pere, J.; Kauppinen, E.I.; Ras, R.H.A.; Ikkala, O. Modifying native nanocellulose aerogels with carbon nanotubes for mechanoresponsive conductivity and pressure sensing. Adv. Mater. **2013**, *25* (17), 2428–2432.

62. Henriksson, M.; Henriksson, G.; Berglund, L.A.; Lindström, T. An environmentally friendly method for enzyme-assisted preparation of microfibrillated cellulose (MFC) nanofibers. Eur. Polym. J. **2007**, *43* (8), 3434–3441.

63. Korhonen, J.T.; Hiekkataipale, P.; Malm, J.; Karppinen, M.; Ikkala, O.; Ras, R.H.A. Inorganic hollow nanotube aerogels by atomic layer deposition onto native nanocellulose templates. ACS Nano **2011**, *5* (3), 1967–1974.

64. Sehaqui, H.; Zhou, Q.; Berglund, L.A. High-porosity aerogels of high specific surface area prepared from nanofibrillated cellulose (NFC). Compos. Sci. Technol. **2011**, *71* (13), 1593–1599.

65. Han, J.; Zhou, C.; Wu, Y.; Liu, F.; Wu, Q. Self-assembling behavior of cellulose nanoparticles during freeze-drying: Effect of suspension concentration, particle size, crystal structure, and surface charge. Biomacromolecules **2013**, *14* (5), 1529–1540.

66. Sehaqui, H.; Salajkova, M.; Zhou, Q.; Berglund, L.A. Mechanical performance tailoring of tough ultra-high porosity foams prepared from cellulose I nanofiber suspensions. Soft Matter **2010**, *6* (8), 1824–1832.

67. Heath, L.; Thielemans, W. Cellulose nanowhisker aerogels. Green Chem. **2010**, *12* (8), 1448–1453.

68. Elazzouzi-Hafraoui, S.; Nishiyama, Y.; Putaux, J.-L.; Heux, L.; Dubreuil, F.; Rochas, C. The shape and size distribution of crystalline nanoparticles prepared by acid hydrolysis of native cellulose. Biomacromolecules **2007**, *9* (1), 57–65.

69. Wagberg, L.; Decher, G.; Norgren, M.; Lindstrom, T.; Ankerfors, M.; Axnas, K. The build-up of polyelectrolyte multilayers of microfibrillated cellulose and cationic polyelectrolytes. Langmuir **2008**, *24* (3), 784–795.

70. Cervin, N.; Aulin, C.; Larsson, P.; Wågberg, L. Ultra porous nanocellulose aerogels as separation medium for mixtures of oil/water liquids. Cellulose **2012**, *19* (2), 401–410.

71. Aulin, C.; Netrval, J.; Wågberg, L.; Lindström, T. Aerogels from nanofibrillated cellulose with tunable oleophobicity. Soft Matter **2010**, *6* (14), 3298–3305.

72. Dufresne, A. TEMPO-mediated oxidation. In *Nanocellulose*, Dufresne, A., Ed. Walter de Gruyter: Berlin/Boston, 2012; 162–164.

73. Saito, T.; Kimura, S.; Nishiyama, Y.; Isogai, A. Cellulose nanofibers prepared by TEMPO-mediated oxidation of native cellulose. Biomacromolecules **2007**, *8* (8), 2485–2491.

74. Silva, T.; Habibi, Y.; Colodette, J.; Elder, T.; Lucia, L. A fundamental investigation of the microarchitecture and mechanical properties of tempo-oxidized nanofibrillated cellulose (NFC)-based aerogels. Cellulose **2012**, *19* (6), 1945–1956.

75. Saito, T.; Uematsu, T.; Kimura, S.; Enomae, T.; Isogai, A. Self-aligned integration of native cellulose nanofibrils towards producing diverse bulk materials. Soft Matter **2011**, *7* (19), 8804–8809.

76. Dong, H.; Snyder, J.F.; Tran, D.T.; Leadore, J.L. Hydrogel, aerogel and film of cellulose nanofibrils functionalized with silver nanoparticles. Carbohydr. Polym. **2013**, *95* (2), 760–767.

77. Koga, H.; Azetsu, A.; Tokunaga, E.; Saito, T.; Isogai, A.; Kitaoka, T. Topological loading of Cu(i) catalysts onto crystalline cellulose nanofibrils for the Huisgen click reaction. J. Mater. Chem. **2012**, *22* (12), 5538–5542.

78. Abbott, A.; Bismarck, A. Self-reinforced cellulose nanocomposites. Cellulose **2010**, *17* (4), 779–791.

79. Duchemin, B.J.C.; Staiger, M.P.; Tucker, N.; Newman, R.H. Aerocellulose based on all-cellulose composites. J. Appl. Polym. Sci. **2010**, *115* (1), 216–221.

80. Gindl, W.; Keckes, J. All-cellulose nanocomposite. Polymer **2005**, *46* (23), 10221–10225.

81. Nishino, T.; Matsuda, I.; Hirao, K. All-cellulose composite. Macromolecules **2004**, *37* (20), 7683–7687.

82. Gindl, W.; Schöberl, T.; Keckes, J. Structure and properties of a pulp fibre-reinforced composite with regenerated cellulose matrix. Appl. Phys. A **2006**, *83* (1), 19–22.

83. Yousefi, H.; Nishino, T.; Faezipour, M.; Ebrahimi, G.; Shakeri, A. Direct fabrication of all-cellulose nanocomposite from cellulose microfibers using ionic liquid-based nanowelding. Biomacromolecules **2011**, *12* (11), 4080–4085.

84. Wang, Y.; Chen, L. Impacts of nanowhisker on formation kinetics and properties of all-cellulose composite gels. Carbohydr. Polym. **2011**, *83* (4), 1937–1946.

85. Zhang, W.; Zhang, Y.; Lu, C.; Deng, Y. Aerogels from crosslinked cellulose nano/micro-fibrils and their fast shape recovery property in water. J. Mater. Chem. **2012**, *22* (23), 11642–11650.

86. Kettunen, M.; Silvennoinen, R.J.; Houbenov, N.; Nykänen, A.; Ruokolainen, J.; Sainio, J.; Pore, V.; Kemell, M.; Ankerfors, M.; Lindström, T.; Ritala, M.; Ras, R.H.A.; Ikkala, O. Photoswitchable superabsorbency based on nanocellulose aerogels. Adv. Funct. Mater. **2011**, *21* (3), 510–517.

87. Fischer, S. Anorganische Salzschmelzen—ein unkonventionelles Löse- und Reaktionsmedium für Cellulose. Habilitation thesis. 2003, Freiberg, Germany: TU Bergakademie Freiberg.

88. Jin, H.; Nishiyama, Y.; Wada, M.; Kuga, S. Nanofibrillar cellulose aerogels. Colloids Surf. A **2004**, *240* (1–3), 63–67.

89. Kuga, S. The porous structure of cellulose gel regenerated from calcium thiocyanate solution. J. Colloid Interface Sci. **1980**, *77* (2), 413–417.

90. Schimper, C.; Pircher, N.; Rosenau, T.; Liebner, F. Unpublished results. 2013.

91. Hattori, M.; Shimaya, Y.; Saito, M. Structural changes in wood pulp treated by 55 wt% aqueous calcium thiocyanate solution. Polym. J. **1998**, *30* (1), 37–42.

92. Roy, C.; Budtova, T.; Navard, P.; Bedue, O. Structure of cellulose–soda solutions at low temperatures. Biomacromolecules **2001**, *2* (3), 687–693.

93. Roy, C.; Budtova, T.; Navard, P. Rheological properties and gelation of aqueous cellulose–NaOH solutions. Biomacromolecules **2003**, *4* (2), 259–264.

94. Egal, M. Thèse de doctorat 2006, Sophia-Antipolis: Ecole Nationale Supérieure des Mines de Paris.

95. Egal, M.; Budtova, T.; Navard, P. Structure of aqueous solutions of microcrystalline cellulose/sodium hydroxide below 0°C and the limit of cellulose dissolution. Biomacromolecules **2007**, *8* (7), 2282–2287.

96. Egal, M.; Budtova, T.; Navard, P. The dissolution of microcrystalline cellulose in sodium hydroxide-urea aqueous solutions. Cellulose **2008**, *15* (3), 361–370.

97. Gavillon, R.; Budtova, T. Kinetics of cellulose regeneration from cellulose-NaOH water gels and comparison with cellulose-NMMO-water solutions. Biomacromolecules **2007**, *8* (2), 424–432.

98. Sescousse, R.; Gavillon, R.; Budtova, T. Aerocellulose from cellulose-ionic liquid solutions: Preparation, properties and comparison with cellulose-NaOH and cellulose-NMMO routes. Carbohydr. Polym. **2010**.

99. Sescousse, R.; Gavillon, R.; Budtova, T. Wet and dry highly porous cellulose beads from cellulose–NaOH–water solutions: Influence of the preparation conditions on beads shape and encapsulation of inorganic particles. J. Mater. Sci. **2011**, *46* (3), 759–765.

100. Isogai, A.; Atalla, R.H. Dissolution of cellulose in aqueous NaOH solutions. Cellulose **1998**, *5* (4), 309–319.

101. Liebert, T. Cellulose solvents—Remarkable history, bright future. In *Cellulose Solvents: For Analysis, Shaping and Chemical Modification*; Liebert, T.F., Heinze, T.J., Edgar, K.J., Eds; Oxford University Press: Washington, DC, 2010; 3–54.

102. Liu, W.; Budtova, T.; Navard, P. Influence of ZnO on the properties of dilute and semi-dilute cellulose-NaOH-water solutions. Cellulose **2011**, *18* (4), 911–920.

103. Lue, A.; Zhang, L. Advances in aqueous cellulose solvents. In *Cellulose Solvents: For Analysis, Shaping and Chemical Modification*; Liebert, T.F., Heinze, T.J., Edgar, K.J., Eds.; Oxford University Press: Washington, DC, 2010; 67–89.

104. Cai, J.; Kimura, S.; Wada, M.; Kuga, S.; Zhang, L. Cellulose aerogels from aqueous alkali hydroxide–urea solution. Chem. Sus. Chem. **2008**, *1* (1–2), 149–154.

105. Cai, J.; Zhang, L.; Chang, C.; Cheng, G.; Chen, X.; Chu, B. Hydrogen-bond-induced inclusion complex in aqueous cellulose/LiOH/urea solution at low temperature. Chem. Phys. Chem. **2007**, *8* (10), 1572–1579.

106. Zhou, J.; Zhang, L.; Shu, H.; Chen, F. Regenerated cellulose films from NaOH/urea aqueous solution by coagulating with sulfuric acid. J. Macromol. Sci. Part B **2002**, *41* (1), 1–15.

107. Zhang, L.; Mao, Y.; Zhou, J.; Cai, J. Effects of coagulation conditions on the properties of regenerated cellulose films prepared in NaOH/urea aqueous solution. Ind. Eng. Chem. Res. **2005**, *44* (3), 522–529.

108. Mao, Y.; Zhou, J.; Cai, J.; Zhang, L. Effects of coagulants on porous structure of membranes prepared from cellulose in NaOH/urea aqueous solution. J. Membr. Sci. **2006**, *279* (1–2), 246–255.

109. Zhou, J.; Zhang, L. Structure and properties of blend membranes prepared from cellulose and alginate in NaOH/urea aqueous solution. J. Polym. Sci. Part B **2001**, *39* (4), 451–458.

110. Cai, J.; Wang, L.; Zhang, L. Influence of coagulation temperature on pore size and properties of cellulose membranes prepared from NaOH–urea aqueous solution. Cellulose **2007**, *14* (3), 205–215.

111. Cai, J.; Kimura, S.; Wada, M.; Kuga, S. Nanoporous cellulose as metal nanoparticles support. Biomacromolecules **2008**, *10* (1), 87–94.

112. Qi, H.; Cai, J.; Zhang, L.; Kuga, S. Properties of films composed of cellulose nanowhiskers and a cellulose matrix regenerated from alkali/urea solution. Biomacromolecules **2009**, *10* (6), 1597–1602.

113. Zhou, J.; Chang, C.; Zhang, R.; Zhang, L. Hydrogels prepared from unsubstituted cellulose in NaOH/urea solution. Macromol. Biosci. **2007**, *7* (6), 804–809.

114. Chang, C.; Zhang, L.; Zhou, J.; Zhang, L.; Kennedy, J.F. Structure and properties of hydrogels prepared from cellulose in NaOH/urea aqueous solutions. Carbohydr. Polym. **2010**, *82* (1), 122–127.

115. Luo, X.; Zhang, L. Creation of regenerated cellulose microspheres with diameter ranging from micron to millimeter for chromatography applications. J. Chromatogr. A **2010**, *1217* (38), 5922–5929.

116. Ruan, D.; Zhang, L.; Zhou, J.; Jin, H.; Chen, H. Structure and properties of novel fibers spun from cellulose in NaOH/thiourea aqueous solution. Macromol. Biosci. **2004**, *4* (12), 1105–1112.

117. Cai, J.; Zhang, L.; Zhou, J.; Li, H.; Chen, H.; Jin, H. Novel fibers prepared from cellulose in NaOH/urea aqueous solution. Macromol. Rapid Commun. **2004**, *25* (17), 1558–1562.

118. Sescousse, R.; Budtova, T. Influence of processing parameters on regeneration kinetics and morphology of porous cellulose from cellulose–NaOH–water solutions. Cellulose **2009**, *16* (3), 417–426.

119. Qi, H.; Mader, E.; Liu, J. Electrically conductive aerogels composed of cellulose and carbon nanotubes. J. Mater. Chem. A **2013**, *1* (34), 9714–9720.

120. Shi, J.; Lu, L.; Guo, W.; Sun, Y.; Cao, Y. An environment-friendly thermal insulation material from cellulose and plasma modification. J. Appl. Polym. Sci. **2013**, *130* (5), 3652–3658.

121. Shi, J.; Lu, L.; Guo, W.; Zhang, J.; Cao, Y. Heat insulation performance, mechanics and hydrophobic modification of cellulose–SiO2 composite aerogels. Carbohydr. Polym. **2013**, *98* (1), 282–289.

122. Zhang, J.; Cao, Y.; Feng, J.; Wu, P. Graphene-oxide-sheet-induced gelation of cellulose and promoted mechanical properties of composite aerogels. J. Phys. Chem. C **2012**, *116* (14), 8063–8068.

123. Isobe, N.; Kim, U.-J.; Kimura, S.; Wada, M.; Kuga, S. Internal surface polarity of regenerated cellulose gel depends on the species used as coagulant. J. Colloid Interface Sci. **2011**, *359* (1), 194–201.

124. Swatloski, R.P. Dissolution of cellose with ionic liquids. JACS **2002**, *124* (18), 4974–4975.

125. Zhu, S.; Wu, Y.; Chen, Q.; Yu, Z.; Wang, C.; Jin, S.; Dinga, Y.; Wuc, G. Dissolution of cellulose with ionic liquids and its application: A mini-review. Green Chem. **2006**, *8* (4), 325–327.

126. Barthel, S.; Heinze, T. Acylation and carbanilation of cellulose in ionic liquids. Green Chem. **2006**, *8*, 301.

127. Heinze, T.; Schwikal, K.; Barthel, S. Ionic liquids as reaction medium in cellulose functionalization. Macromol. Biosci. **2005**, *5* (6), 520–525.

128. Heinze, T.; Dorn, S.; Schöbitz, M.; Liebert, T.; Köhler, S.; Meister, F. Interactions of ionic liquids with polysaccharides – 2: Cellulose. Macromol. Symp. **2008**, *262* (1), 8–22.

129. Schöbitz, M.; Meister, F.; Heinze, T. Unconventional reactivity of cellulose dissolved in ionic liquids. Macromol. Symp. **2009**, *280* (1), 102–111.

130. Wu, J.; Zhang, J.; Zhang, H.; He, J.; Ren, Q.; Guo, M. Homogeneous acetylation of cellulose in a new ionic liquid. Biomacromolecules **2004**, *5* (2), 266–268.

131. Quan, S.-L.; Kang, S.-G.; Chin, I.-J. Characterization of cellulose fibers electrospun using ionic liquid. Cellulose **2010**, *17* (2), 223–230.

132. Wendler, F.; Kosan, B.; Krieg, M.; Meister, F. Possibilities for the physical modification of cellulose shapes using ionic liquids. Macromol. Symp. **2009**, *280* (1), 112–122.

133. Turner, M.B.; Spear, S.K.; Holbrey, J.D.; Rogers, R.D. Production of bioactive cellulose films reconstituted from ionic liquids. Biomacromolecules **2004**, *5* (4), 1379–1384.

134. Lin, C.X.; Zhan, H.Y.; Liu, M.H.; Fu, S.Y.; Lucia, L.A. Novel preparation and characterization of cellulose microparticles functionalized in ionic liquids. Langmuir **2009**, *25* (17), 10116–10120.

135. Aaltonen, O.; Jauhiainen, O. The preparation of lignocellulosic aerogels from ionic liquid solutions. Carbohydr. Polym. **2009**, *75* (1), 125–129.

136. Deng, M.; Zhou, Q.; Du, A.; van Kasteren, J.; Wang, Y. Preparation of nanoporous cellulose foams from cellulose-ionic liquid solutions. Mater. Lett. **2009**, *63*, 1851–1854.

137. Zhao, Q.; Yam, R.; Zhang, B.; Yang, Y.; Cheng, X.; Li, R. Novel all-cellulose ecocomposites prepared in ionic liquids. Cellulose **2009**, *16* (2), 217–226.

138. Zhao, H.; Baker, G.A.; Song, Z.; Olubajo, O.; Crittle, T.; Peters, D. Designing enzyme-compatible ionic liquids that can dissolve carbohydrates. Green Chem. **2008**, *10* (6), 696–705.

139. Vitz, J.; Erdmenger, T.; Haensch, C.; Schubert, U.S. Extended dissolution studies of cellulose in imidazolium based ionic liquids. Green Chem. **2009**, *11* (3), 417–424.

140. Swatloski, R.P.; Spear, S.K.; Holbrey, J.D.; Rogers, R.D. Dissolution of cellose with ionic liquids. J. Am. Chem. Soc. **2002**, *124* (18), 4974–4975.

141. Mäki-Arvela, P.; Anugwom, I.; Virtanen, P.; Sjöholm, R.; Mikkola, J.P. Dissolution of lignocellulosic materials and its constituents using ionic liquids—A review. Ind. Crops Prod. **2010**, *32* (3), 175–201.

142. Ebner, G.; Schiehser, S.; Potthast, A.; Rosenau, T. Side reaction of cellulose with common 1-alkyl-3-methyl-imidazolium-based ionic liquids. Tetrahedr. Lett. **2008**, *49* (51), 7322–7324.

143. Liebner, F.; Patel, I.; Ebner, G.; Becker, E.; Horix, M.; Potthast, A.; Rosenau, T. Thermal aging of 1-alkyl-3-methyl-imidazolium ionic liquids and its effect on dissolved cellulose. Holzforschung **2010**, *64* (2), 161–166.

144. Schrems, M.; Ebner, G.; Liebner, F.; Becker, E.; Potthast, A.; Rosenau, T. Side reactions in the system cellulose/1-alkyl-3-methyl-imidazolium ionic liquid. In *Cellulose Solvents: For Analysis, Shaping and Chemical Modification*; Liebert, T.F., Heinze, T.J., Edgar, K.J., Eds.; Oxford University Press: Washington, DC, 2010; 149–164.

145. Tsioptsias, C.; Stefopoulos, A.; Kokkinomalis, I.; Papadopoulou, L.; Panayiotou, C. Development of micro- and nano-porous composite materials by processing cellulose with ionic liquids and supercritical CO2. Green Chem. **2008**, *10* (9), 965–971.

146. Granstrom, M.; nee Paakko, M.K.; Jin, H.; Kolehmainen, E.; Kilpelainen, I.; Ikkala, O. Highly water repellent aerogels based on cellulose stearoyl esters. Polym. Chem. **2011**, *2* (8), 1789–1796.

147. Wang, H.; Shao, Z.; Bacher, M.; Liebner, F.; Rosenau, T. Fluorescent cellulose aerogels containing covalently immobilized (ZnS)x(CuInS2)1–x/ZnS (core/shell) quantum dots. Cellulose **2013**, *20* (6), 3007–3024.

148. Graenacher, C. Cellulose Solutions. US Patent 1943176 A, Sep 16, 1931.

149. Chanzy, H.; Noe, P.; Paillet, M.; Smith, P. Swelling and dissolution of cellulose in amine oxide/water systems, 9. cellulose conference, J.A.P.S.A.P. Symp, Editor 1983: Syracuse, NY, USA. 239–259.

150. Kim, S.O.; Shin, W.J.; Cho, H.; Kim, B.C.; Chung, I.J. Rheological investigation on the anisotropic phase of cellulose–MMNO/H2O solution system. Polymer **1999**, *40* (23), 6443–6450.

151. Liebner, F.; Haimer, E.; Potthast, A.; Rosenau, T. Cellulosic Aerogels. In *Polysaccharide Building Blocks*; John Wiley & Sons, Inc., 2012; 51–103.

152. Ioleva, M.M.; Goikhman, A.S.; Banduryan, S.I.; Papkov, S.P. Characteristics of the interaction of cellulose with N-methylmorpholine-N-oxide. Vysokomol. Soedin. Ser. B **1983**, *25* (11), 803–804.

153. Harmon, K.M.; Akin, A.C.; Keefer, P.K.; Snider, B.L. Hydrogen bonding Part 45. Thermodynamic and IR study of the hydrates of N-methylmorpholine oxide and quinuclidine oxide. Effect of hydrate stoichiometry on strength of H---O---H---O---N hydrogen bonds; Implications for the dissolution of cellulose in amine oxide solvents. J. Mol. Struct. **1992**, *269* (1–2), 109–121.

154. Khanin, V.A.; Bandura, A.V.; Novoselov, N.P. Barriers to rotation of bridging bonds of cellulose molecule in its interaction with N-methylmorpholine N-oxide. Russ. J. Gen. Chem. **1998**, *68* (2), 305–308.

155. Rozhkova, O.V.; Myasoedova, V.V.; Krestov, G.A. Effect of donor-acceptor interactions on solubility of cellulose in methylmorpholine N-oxide-based systems. Khim. Drev. **1985**, *2*, 26–29.

156. Rosenau, T.; Potthast, A.; Adorjan, I.; Hofinger, A.; Sixta, H.; Firgo, H.; Kosma, P. Cellulose solutions in N-methylmorpholine-N-oxide (NMMO)—Degradation processes and stabilizers. Cellulose **2002**, *9* (3–4), 283–291.

157. Rosenau, T.; Potthast, A.; Sixta, H.; Kosma, P. The chemistry of side reactions and by-product formation in the system NMMO/cellulose (Lyocell process). Prog. Polym. Sci. **2001**, *26* (9), 1763–1837.

158. Rosenau, T.; Potthast, A.; Schmid, P.; Kosma, P. On the non-classical course of Polonowski reactions of N-benzyl-morpholine-N-oxide (NBnMO). Tetrahedron **2005**, *61* (14), 3483–3487.

159. Liu, R.-G.; Shen, Y.-Y.; Shao, H.-L.; Wu, C.-X.; Hu, X.-C. An analysis of lyocell fiber formation as a melt–spinning process. Cellulose **2001**, *8* (1), 13–21.

160. Schimper, C.; Haimer, E.; Wendland, M.; Potthast, A.; Rosenau, T.; Liebner, F. The effects of different process parameters on the properties of cellulose aerogels obtained via the Lyocell route. Lenzinger Berichte **2011**, *89*, 109–117.

161. Innerlohinger, J.; Weber, H.K.; Kraft, G. Aerocell: aerogels from cellulosic materials. Lenzinger Ber. **2006**, *86*, 137–143.

162. Innerlohinger, J.; Weber, H.K.; Kraft, G. Aerocellulose: Aerogels and aerogel-like materials made from cellulose. Macromol. Symp. **2006**, *244* (1), 126–135.

163. Domínguez, G.; Westphal, A.J.; Phillips, M.L.F.; Jones, S.M. A fluorescent aerogel for capture and identification of interplanetary and interstellar dust. Astrophys. J. **2003**, *592* (1), 631–635.

164. Hüsing, N.; Schubert, U. *Aerogels, Ullmann's Encyclopedia of Industrial Chemistry*; Wiley VCH: Weinheim, 2006.

165. Akimov, Y.K. Fields of application of aerogels (Review). Instrum. Exp. Techn. **2003**, *46* (3), 287–299.

166. Pierre, A.C.; Pajonk, G.M. Chemistry of aerogels and their applications. Chem. Rev. **2002**, *102* (11), 4243–4266.

167. Fricke, J.; Emmerling, A. Aerogels—Preparation, properties, applications. Struct. Bonding **1992**, *77*, 37–87.

168. Farmer, J.C.; Fix, D.; Mack, G.V.; Pekala, R.W.; Poco, J.F. Capacitive deionization of NaCl and NaNO3 solutions with carbon aerogel electrodes. J. Electrochem. Soc. **1996**, *143* (1), 159–169.

169. Anonymous, http://www.aerogeltechnologies.com/, 2013.

170. Bag, S.; Gaudette, A.F.; Bussell, M.E.; Kanatzidis, M.G. Spongy chalcogels of non-platinum metals act as effective hydrodesulfurization catalysts. Nat. Chem. **2009**, *1* (3), 217–224.

171. Mehling, T.; Smirnova, I.; Guenther, U.; Neubert, R.H.H. Polysaccharide-based aerogels as drug carriers. J. Non-Cryst. Solids **2009**, *355* (50), 2472–2479.

172. García-González, C.A.; Alnaief, M.; Smirnova, I. Polysaccharide-based aerogels—Promising biodegradable carriers for drug delivery systems. Carbohydr. Polym. **2011**, *86* (4), 1425–1438.

173. Kenar, J.A.; Eller, F.J.; Felker, F.C.; Jackson, M.A.; Fanta, G.F. Starch aerogel beads obtained from inclusion complexes prepared from high amylose starch and sodium palmitate. Green Chem. **2013**, *16* (4), 1921–1930.

174. Robitzer, M.; David, L.; Rochas, C.; Di Renzo, F.; Quignard, F. Supercritically-dried alginate aerogels retain the fibrillar structure of the hydrogels. Macromol. Symp. **2008**, *273* (1), 80–84.

175. Tsioptsias, C.; Michailof, C.; Stauropoulos, G.; Panayiotou, C. Chitin and carbon aerogels from chitin alcogels. Carbohydr. Polym. **2009**, *76* (4), 535–540.

176. Chen, H.-B.; Chiou, B.-S.; Wang, Y.-Z.; Schiraldi, D.A. Biodegradable pectin/clay aerogels. ACS Appl. Mater. Interfaces **2013**, *5* (5), 1715–1721.

177. Marquez-Escalante, J.; Carvajal-Millan, E.; Miki-Yoshida, M.; Alvarez-Contreras, L.; Toledo-Guillén, A.R.; Lizardi-Mendoza, J.; Rascón-Chu, A. Water extractable arabinoxylan aerogels prepared by supercritical CO2 drying. Molecules **2013**, *18* (5), 5531–5542.

178. Betz, M.; García-González, C.A.; Subrahmanyam, R.P.; Smirnova, I.; Kulozik, U. Preparation of novel whey protein-based aerogels as drug carriers for life science applications. J. Supercrit. Fluids **2012**, *72* (0), 111–119.

179. Perez-Cantu, L.; Liebner, F.; Smirnova, I. Preparation of aerogels from wheat straw lignin by cross-linking with oligo(alkylene glycol)-α,ω-diglycidyl ethers. J. Micropor. Mesopor. Mater. **2014**, *195* (0), 303–310.

180. Biener, M.M.; Biener, J.; Wang, Y.M.; Shin, S.J.; Tran, I.C.; Willey, T.M.; Pérez, F.N.; Poco, J.F.; Gammon, S.A.; Fournier, K.B.; van Buuren, A.W.; Satcher, J.H.; Hamza, A.V. Atomic layer deposition-derived ultra-low-density composite bulk materials with deterministic density and composition. ACS Appl. Mater. Interfaces **2013**, *5* (24), 13129–13134.

181. Wei, G.; Liu, Y.; Zhang, X.; Yu, F.; Du, X. Thermal conductivities study on silica aerogel and its composite insulation materials. Int. J. Heat Mass Transf. **2011**, *54* (11–12), 2355–2366.

182. Soleimani Dorcheh, A.; Abbasi, M.H. Silica aerogel; Synthesis, properties and characterization. J. Mater. Process. Technol. **2008**, *199* (1–3), 10–26.

183. Anonymous, Mechanically Strong, Flexible Polyimide Aerogels. https://technology.grc.nasa.gov/featured-tech/aerogels.shtm, 2013.

184. Plawsky, J.L.; Littman, H.; Paccione, J.D. Design, simulation, and performance of a draft tube spout fluid bed coating system for aerogel particles. Powd. Technol. **2010**, *199* (2), 131–138.

185. Cai, J.; Liu, S.; Feng, J.; Kimura, S.; Wada, M.; Kuga, S.; Zhang, L. Cellulose-silica nanocomposite aerogels by in situ formation of silica in cellulose gel. Angew. Chem. Int. Ed. **2012**, *51* (9), 2076–2079.

186. Tan, C.; Fung, M.; Newman, J.K.; Vu, C. Organic aerogels with very high impact strength. Adv. Mat. **2001**, *13* (9), 644–646.

187. Fischer, F.; Rigacci, A.; Pirard, R.; Berthon-Fabry, S.; Achard, P. Cellulose-based aerogels. Polymer **2006**, *47* (22), 7636–7645.

188. Arboleda, J.; Hughes, M.; Lucia, L.; Laine, J.; Ekman, K.; Rojas, O. Soy protein–nanocellulose composite aerogels. Cellulose **2013**, *20* (5), 2417–2426.

189. Russler, A.; Wieland, M.; Bacher, M.; Henniges, U.; Miethe, P.; Liebner, F.; Potthast, A.; Rosenau, T. AKD-Modification of bacterial cellulose aerogels in supercritical CO2. Cellulose **2012**, *19* (4), 1337–1349.

190. Litschauer, M.; Neouze, M.A.; Haimer, E.; Henniges, U.; Potthast, A.; Rosenau, T.; Liebner, F. Silica modified cellulosic aerogels. Cellulose **2011**, *18* (1), 143–149.

191. Maeda, H.; Nakajima, M.; Hagiwara, T.; Sawaguchi, T.; Yano, S. Bacterial cellulose/silica hybrid fabricated by mimicking biocomposites. J. Mater. Sci. **2006**, *41* (17), 5646–5656.

192. Gebald, C.; Wurzbacher, J.A.; Tingaut, P.; Zimmermann, T.; Steinfeld, A. Amine-based nanofibrillated cellulose as adsorbent for CO_2 capture from air. Env. Sci. Technol. **2011**, *45* (20), 9101–9108.

193. Luo, X.; Liu, S.; Zhou, J.; Zhang, L. In situ synthesis of Fe_3O_4/cellulose microspheres with magnetic-induced protein delivery. J. Mater. Chem. **2009**, *19* (1), 3538–3545.

194. Olsson, R.T.; Azizi Samir, M.A.S.; Salazar Alvarez, G.; Belova, L.; Strom, V.; Berglund, L.A.; IkkalaO; NoguesJ; Gedde, U.W. Making flexible magnetic aerogels and stiff magnetic nanopaper using cellulose nanofibrils as templates. Nat. Nano. **2010**, *5* (8), 584–588.

195. Liu, S.; Luo, X.; Zhou, J. Magnetic responsive cellulose nanocomposites and their applications. In *Cellulose–Medical, Pharmaceutical and Electronic Applications*; Van de Ven, T., Godbout, L., Eds.; InTech: Rijeka, Croatia, 2013; 105–124.

196. Carlsson, D.O.; Nystrom, G.; Zhou, Q.; Berglund, L.A.; Nyholm, L.; Stromme, M. Electroactive nanofibrillated cellulose aerogel composites with tunable structural and electrochemical properties. J. Mater. Chem. **2012**, *22* (36), 19014–19024.

197. Otero, T.F.; García de Otazo, J.M. Polypyrrole oxidation: Kinetic coefficients, activation energy and conformational energy. Synt. Metals **2009**, *159* (7–8), 681–688.

198. Otero, T.F.; Grande, H.-J.; Rodríguez, J. Reinterpretation of polypyrrole electrochemistry after consideration of conformational relaxation processes. J. Phys. Chem. B **1997**, *101* (19), 3688–3697.

199. Hamedi, M.; Karabulut, E.; Marais, A.; Herland, A.; Nyström, G.; Wågberg, L. Nanocellulose aerogels functionalized by rapid layer-by-layer assembly for high charge storage and beyond. Angew. Chem. Int. Ed. **2013**, *52* (46), 12038–12042.

200. Helenius, G.; Bäckdahl, H.; Bodin, A.; Nannmark, U.; Gatenholm, P.; Risberg, B. *In vivo* biocompatibility of bacterial cellulose. J. Biomed. Mater. Res. Part A **2006**, *76A* (2), 431–438.

201. Klemm, D.; Schumann, D.; Udhardt, U.; Marsch, S. Bacterial synthesized cellulose—Artificial blood vessels for microsurgery. Progr. Polym. Sci. **2001**, *26* (9), 1561–1603.

202. Sutherland, I.W. Novel and established applications of microbial polysaccharides. Trends Biotechnol. **1998**, *16* (1), 41–46.

203. Miyamoto, T.; Takahashi, S.-i.; Ito, H.; Inagaki, H.; Noishiki, Y. Tissue biocompatibility of cellulose and its derivatives. J. Biomed. Mater. Res. **1989**, *23* (1), 125–133.

204. Märtson, M.; Viljanto, J.; Hurme, T.; Laippala, P.; Saukko, P. Is cellulose sponge degradable or stable as implantation material? An *in vivo* subcutaneous study in the rat. Biomaterials **1999**, *20* (21), 1989–1995.

205. Petersen, N.; Gatenholm, P. Bacterial cellulose-based materials and medical devices: Current state and perspectives. Appl. Microbiol. Biotechnol. **2011**, *91* (5), 1277–1286.

206. Amnuaikit, T.; Chusuit, T.; Raknam, P.; Boonme, P. Effects of a cellulose mask synthesized by a bacterium on facial skin characteristics and user satisfaction. Med. Dev. **2011**, *4*, 77–81.

207. Fontana, J.D.; Souza, A.M.; Fontana, C.K.; Torriani, I.L.; Moreschi, J.C.; Gallotti, B.J.; Souza, S.J.; Narcisco, G.P.; Bichara, J.A.; Farah, L.F.X. Acetobacter cellulose pellicle as a temporary skin substitute. Appl. Biochem. Biotechnol. **1990**, *24–25* (1), 253–264.

208. Bowry, S.K.; Rintelen, T. Synthetically modified cellulose (smc) a cellulosic hemodialysis membrane with minimized complement activation. *ASAIO J.* **1998** *44* (5), M579–M583.

209. Schumann, D.; Wippermann, J.; Klemm, D.; Kramer, F.; Koth, D.; Kosmehl, H.; Wahlers, T.; Salehi-Gelani, S. Artificial vascular implants from bacterial cellulose: Preliminary results of small arterial substitutes. Cellulose **2009**, *16* (5), 877–885.

210. Esguerra, M.; Fink, H.; Laschke, M.W.; Jeppsson, A.; Delbro, D.; Gatenholm, P.; Menger, M.D.; Risberg, B. Intravital fluorescent microscopic evaluation of bacterial cellulose as scaffold for vascular grafts. J. Biomed. Mater. Res. Part A **2010**, *93A* (1), 140–149.

211. Fink, H.; Hong, J.; Drotz, K.; Risberg, B.; Sanchez, J.; Sellborn, A. An *in vitro* study of blood compatibility of vascular grafts made of bacterial cellulose in comparison with conventionally-used graft materials. J. Biomed. Mater. Res. Part A **2011**, *97A* (1), 52–58.

212. Bodin, A.; Ahrenstedt, L.; Fink, H.; Brumer, H.; Risberg, B.; Gatenholm, P. Modification of nanocellulose with a xyloglucan–RGD conjugate enhances adhesion and proliferation of endothelial cells: Implications for tissue engineering. Biomacromolecules **2007**, *8* (12), 3697–3704.

213. Bodin, A.; Bäckdahl, H.; Fink, H.; Gustafsson, L.; Risberg, B.; Gatenholm, P. Influence of cultivation conditions on mechanical and morphological properties of bacterial cellulose tubes. Biotechnol. Bioeng. **2007**, *97* (2), 425–434.

Adhesives—Bioabsorbable

Adhesives—Bioabsorbable

214. Bodin, A.; Bharadwaj, S.; Wu, S.; Gatenholm, P.; Atala, A.; Zhang, Y. Tissue-engineered conduit using urine-derived stem cells seeded bacterial cellulose polymer in urinary reconstruction and diversion. Biomaterials 2010, 31 (34), 8889–8901.

215. Suyiene Cordeiro, F.; Antônio Roberto de Barros, C.; Joaquim Evêncio, N. Biomechanical evaluation of microbial cellulose (Zoogloea sp.) and expanded polytetrafluoroethylene membranes as implants in repair of produced abdominal wall defects in rats. Acta Cir. Bras. 2008, 23 (2), 184–191.

216. Suyiene Cordeiro, F.; Joaquim Evêncio, N.; Antônio Roberto de Barros, C. Incorporation by host tissue of two biomaterials used as repair of defects produced in abdominal wall of rats. Acta Cir. Bras./SOBRADPEC 2008, 23 (1), 78–83.

217. Wan, Y.Z.; Hong, L.; Jia, S.R.; Huang, Y.; Zhu, Y.; Wang, Y.L.; Jiang, H.J. Synthesis and characterization of hydroxyapatite-bacterial cellulose nanocomposites. Composites Sci. Technol. 2006, 66 (11–12), 1825–1832.

218. Wan, Y.Z.; Huang, Y.; Yuan, C.D.; Raman, S.; Zhu, Y.; Jiang, H.J.; He, F.; Gao, C. Biomimetic synthesis of hydroxyapatite/bacterial cellulose nanocomposites for biomedical applications. Mater. Sci. Eng. C 2007, 27 (4), 855–864.

219. Zaborowska, M.; Bodin, A.; Bäckdahl, H.; Popp, J.; Goldstein, A.; Gatenholm, P. Microporous bacterial cellulose as a potential scaffold for bone regeneration. Acta Biomater. 2010, 6 (7), 2540–2547.

220. Fang, B.; Wan, Y.; Tang, T.; Gao, C.; Dai, K. Proliferation and osteoblastic differentiation of human bone marrow stromal cells on hydroxyapatite/bacterial cellulose nanocomposite scaffolds. Tissue Eng. Part A 2009, 15 (5), 1091–1098.

221. Bodin, A.; Concaro, S.; Brittberg, M.; Gatenholm, P. Bacterial cellulose as a potential meniscus implant. J. Tissue Eng. Regen. Med. 2007, 1 (5), 406–408.

222. Svensson, A.; Nicklasson, E.; Harrah, T.; Panilaitis, B.; Kaplan, D.L.; Brittberg, M.; Gatenholm, P. Bacterial cellulose as a potential scaffold for tissue engineering of cartilage. Biomaterials 2005, 26 (4), 419–431.

223. Levinson, D.J.; Glonek, T. Microbial cellulose contact lens. US 7832857 B2, August 18, 2008.

224. Hoenich, N. Cellulose for medical applications: Past, present, and future. BioResources 2006, 1 (2), 270–280.

225. Klemm, D.; Heublein, B.; Fink, H.-P.; Bohn, A. Cellulose: Fascinating biopolymer and sustainable raw material. Angew. Chem. Int. Ed. 2005, 44 (22), 3358–3393.

226. Czaja, W.; Krystynowicz, A.; Bielecki, S.; Brown, J.R.M. Microbial cellulose--the natural power to heal wounds. Biomaterials 2006, 27 (2), 145–151.

227. Czaja, W.K.; Young, D.J.; Kawecki, M.; Brown, R.M. The future prospects of microbial cellulose in biomedical applications. Biomacromolecules 2007, 8 (1), 1–12.

228. Klemm, D.; Schumann, D.; Kramer, F.; Heßler, N.; Hornung, M.; Schmauder, H.-P.; Marsch, S. Nanocelluloses as innovative polymers in research and application. In Polysaccharides II, Klemm, D. Ed.; Springer: Heidelberg, 2006; 49–96.

229. Sannino, A.; Mensitieri, G.; Nicolais, L. Water and synthetic urine sorption capacity of cellulose-based hydrogels under a compressive stress field. J. Appl. Polym. Sci. 2004, 91 (6), 3791–3796.

230. Larsson, A.; Larsson, M.; Hjärtstam, J. Microfibrillated cellulose films for controlled release of active agents. WO 2013009253 A1, July 8, 2011.

231. Valo, H.; Arola, S.; Laaksonen, P.; Torkkeli, M.; Peltonen, L.; Linder, M.B.; Serimaa, R.; Kuga, S.; Hirvonen, J.; Laaksonen, T. Drug release from nanoparticles embedded in four different nanofibrillar cellulose aerogels. Eur. J. Pharm. Sci. 2013, 50 (1), 69–77.

232. El-Hag Ali, A.; Abd El-Rehim, H.A.; Kamal, H.; Hegazy, D.E.S.A. Synthesis of carboxymethyl cellulose based drug carrier hydrogel using ionizing radiation for possible use as site specific delivery system. J. Macromol. Sci. Part A 2008, 45 (8), 628–634.

233. El-Din, H.M.N.; El-Naggar, A.W.M.; Abu-El Fadle, F.I. Radiation synthesis of pH-sensitive hydrogels from carboxymethyl cellulose/poly(ethylene oxide) blends as drug delivery systems. Int. J. Polym. Mater. Polym. Biomater. 2013, 62 (13), 711–718.

234. Hollister, S.J. Porous scaffold design for tissue engineering. Nat. Mater. 2005, 4 (7), 518–524.

235. Bonfield, W. Designing porous scaffolds for tissue engineering. Philosophical Trans. R Soc. A 2006, 364 (1838), 227–232.

236. Hutchens, S.A.; Benson, R.S.; Evans, B.R.; Rawn, C.J.; O'Neill, H. A resorbable calcium-deficient hydroxyapatite hydrogel composite for osseous regeneration. Cellulose 2009, 2009 (16), 887–898.

237. Ma, M.-G.; Zhu, J.-F.; Jia, N.; Li, S.-M.; Sun, R.-C.; Cao, S.-W.; Chen, F. Rapid microwave-assisted synthesis and characterization of cellulose-hydroxyapatite nanocomposites in N,N-dimethylacetamide solvent. Carbohydrate Research 2010, 345 (8), 1046–1050.

238. Fang, B.; Wan, Y.-Z.; Tang, T.-T.; Gao, C.; Dai, K.-R. Proliferation and osteoblastic differentiation of human bone marrow stromal cells on hydroxyapatite/bacterial cellulose nanocomposite scaffolds. Tissue Eng. Part A 2009, 15 (5), 1091–1098.

239. Liu, X.; Smith, L.A.; Hu, J.; Ma, P.X. Biomimetic nanofibrous gelatin/apatite composite scaffolds for bone tissue engineering. Biomaterials 2009, 30 (12), 2252–2258.

240. Granja, P.L.; Pouysegu, L.; De Jeso, B.; Rouais, F.; Baquey, C.; Barbosa, M.A. Cellulose phosphates as biomaterials. Mineralisation of chemically modified regenerated cellulose hydrogels. J. Mater. Sci. 2001, 36 (9), 2163–2172.

241. Kato, K.; Eika, Y.; Ikada, Y. Deposition of a hydroxyapatite thin layer onto a polymer surface carrying grafted phosphate polymer chains. J. Biomed. Mater. Res. 1996, 32 (4), 687–691.

242. Yokogawa, Y.; Paz Reyes, J.; Mucalo, M.R.; Toriyama, M.; Kawamoto, Y.; Suzuki, T.; Nishizawa, K.; Nagata, F.; Kamayama, T. Growth of calcium phosphate on phosphorylated chitin fibres. J. Mater. Sci. 1997, 8 (7), 407–412.

243. Mucalo, M.R.; Yokogawa, Y.; Suzuki, T.; Kawamoto, Y.; Nagata, F.; Nishizawa, K. Further studies of calcium phosphate growth on phosphorylated cotton fibres. J. Mater. Sci. 1995, 6 (11), 658–669.

244. Hayashi, T. Biodegradable polymers for biomedical uses. Prog. Polym. Sci. **1994**, *19*, 663–702.

245. Granja, P.L.; Pouységu, L.; Pétraud, M.; De Jéso, B.; Baquey, C.; Barbosa, M.A. Cellulose phosphates as biomaterials. I. Synthesis and characterisation of highly phosphorylated cellulose gels. J. Appl. Polym. Sci. **2001**, *82* (13), 3341–3353.

246. Liebner, F.; Dunareanu, R.; Opietnik, M.; Haimer, E.; Wendland, M.; Werner, C.; Maitz, M.; Seib, P.; Neouze, M.A.; Potthast, A.; Rosenau, T. Shaped hemocompatible aerogels from cellulose phosphates: Preparation and properties. Holzforschung **2011**, *66* (3), 317–321.

247. Carrasco-Marín, F.; Fairén-Jiménez, D.; Moreno-Castilla, C. Carbon aerogels from gallic acid-resorcinol mixtures as adsorbents of benzene, toluene and xylenes from dry and wet air under dynamic conditions. Carbon **2009**, *47* (2), 463–469.

248. Maldonado-Hódar, F.J.; Moreno-Castilla, C.; Carrasco-Marín, F.; Pérez-Cadenas, A.F. Reversible toluene adsorption on monolithic carbon aerogels. J. Hazard. Mater. **2007**, *148* (3), 548–552.

249. Gomes, H.T.; Samant, P.V.; Serp, P.; Kalck, P.; Figueiredo, J.L.; Faria, J.L. Carbon nanotubes and xerogels as supports of well-dispersed Pt catalysts for environmental applications. Appl. Catalysis B **2004**, *54* (3), 175–182.

250. Maillard, F.; Simonov, P.A.; Savinova, E.R. Carbon materials as supports for fuel cell electrocatalysts. In *Carbon Materials for Catalysis*. John Wiley & Sons, Inc., 2008.

251. Maldonado-Hòdar, F.J.; Moreno-Castilla, C.; Rivera-Utrilla, J.; Ferro-Garcla, M.A. Metal-carbon aerogels as catalysts and catalyst supports. In *Studies in Surface Science and Catalysis*; Avelino Corma, F.V.M.S.M., José Luis, G.F., Eds.; Elsevier: 2000; 1007–1012.

252. Moreno-Castilla, C.; Maldonado-Hódar, F.J. Carbon aerogels for catalysis applications: An overview. Carbon **2005**, *43* (3), 455–465.

253. Smirnova, A.; Dong, X.; Hara, H.; Vasiliev, A.; Sammes, N. Novel carbon aerogel-supported catalysts for PEM fuel cell application. Int. J. Hydrogen Energy **2005**, *30* (2), 149–158.

254. Jordá-Beneyto, M.; Suárez-García, F.; Lozano-Castelló, D.; Cazorla-Amorós, D.; Linares-Solano, A. Hydrogen storage on chemically activated carbons and carbon nanomaterials at high pressures. Carbon **2007**, *45* (2), 293–303.

255. Schimmel, H.G.; Nijkamp, G.; Kearley, G.J.; Rivera, A.; de Jong, K.P.; Mulder, F.M. Hydrogen adsorption in carbon nanostructures compared. Mater. Sci. Eng. B **2004**, *108* (1–2), 124–129.

256. Babel, K.; Jurewicz, K. KOH activated lignin based nanostructured carbon exhibiting high hydrogen electrosorption. Carbon **2008**, *46* (14), 1948–1956.

257. Guilminot, E.; Gavillon, R.; Chatenet, M.; Berthon-Fabry, S.; Rigacci, A.; Budtova, T. New nanostructured carbons based on porous cellulose: Elaboration, pyrolysis and use as platinum nanoparticles substrate for oxygen reduction electrocatalysis. J. Power Sources **2008**, *185* (2), 717–726.

258. Guilminot, E.; Fischer, F.; Chatenet, M.; Rigacci, A.; Berthon-Fabry, S.; Achard, P.; Chainet, E. Use of cellulose-based carbon aerogels as catalyst support for PEM fuel cell electrodes: Electrochemical characterization. J. Power Sources **2007**, *166* (1), 104–111.

259. Marie, J.; Chenitz, R.; Chatenet, M.; Berthon-Fabry, S.; Cornet, N.; Achard, P. Highly porous PEM fuel cell cathodes based on low density carbon aerogels as Pt-support: Experimental study of the mass-transport losses. J. Power Sources **2009**, *190* (2), 423–434.

260. Chmiola, J.; Yushin, G.; Dash, R.; Gogotsi, Y. Effect of pore size and surface area of carbide derived carbons on specific capacitance. J. Power Sources **2006**, *158* (1), 765–772.

261. Chmiola, J.; Yushin, G.; Gogotsi, Y.; Portet, C.; Simon, P.; Taberna, P.L. Anomalous Increase in carbon capacitance at pore sizes less than 1 nanometer. Science **2006**, *313* (5794), 1760–1763.

262. Frackowiak, E.; Béguin, F. Electrochemical storage of energy in carbon nanotubes and nanostructured carbons. Carbon **2002**, *40* (10), 1775–1787.

263. Béguin, F.; Frackowiak, E. Nanotextured carbons for electrochemical energy storage. In *Nanomaterials Handbook*; CRC Press: Boca Raton, FL, 2006.

264. Yun, Y.S.; Kim, D.; Tak, Y.; Jin, H.-J. Porous graphene/carbon nanotube composite cathode for proton exchange membrane fuel cell. Synt. Metals **2011**, *161* (21–22), 2460–2465.

265. Zheng, J.P.; Huang, J.; Jow, T.R. The limitation of energy density for electrochemical capacitors. J. Electrochem. Soc. **1997**, *144* (6), 2026–2031.

266. Simon, P.; Burke, A. Nanostructured carbons: Double-layer capacitance and more. Electrochem. Soc. Interface **2008**, *17* (1), 38–43.

267. Rooke, J.; Matos, C.; Chatenet, M.; Sescousse, R.; Budtova, T.; Berthon-Fabry, S.; Mosdale, R.; Maillard, F. Elaboration and characterizations of platinum nanoparticles supported on cellulose-based carbon aerogel. ECS Trans. **2010**, *33* (1), 447–459.

268. Yoon, S.; Lee, J.; Hyeon, T.; Oh, S.M. Electric double-layer capacitor performance of a new mesoporous carbon. J. Electrochem. Soc. **2000**, *147* (7), 2507–2512.

269. Grzyb, B.; Hildenbrand, C.; Berthon-Fabry, S.; Bégin, D.; Job, N.; Rigacci, A.; Achard, P. Functionalisation and chemical characterisation of cellulose-derived carbon aerogels. Carbon **2010**, *48* (8), 2297–2307.

270. Kim, D.-Y.; Nishiyama, Y.; Wada, M.; Kuga, S. Graphitization of highly crystalline cellulose. Carbon **2001**, *39* (7), 1051–1056.

271. Wendler, F.; Schulze, T.; Ciechanska, D.; Wesolowska, E.; Wawro, D.; Meister, F.; Budtova, T.; Liebner, F. Cellulose products from solutions: Film, fibres and aerogels. In *The European Polysaccharide Network of Excellence (EPNOE)*; Navard, P. Ed.; Springer: Vienna, 2013; 153–185.

272. Rooke, J.; de Matos Passos, C.; Chatenet, M.; Sescousse, R.; Budtova, T.; Berthon-Fabry, S.; Mosdale, R.; Maillard, F. Synthesis and properties of platinum nanocatalyst supported on cellulose-based carbon aerogel for applications in PEMFCs. J. Electrochem. Soc. **2011**, *158* (7), B779–B789.

Anticancer Agents: Polymeric Nanomedicines

Rong Tong
Jianjun Cheng
Department of Materials Science and Engineering, University of Illinois at Urbana-Champaign, Urbana, Illinois, U.S.A.

Abstract

Polymers play important roles in the design of delivery nanocarriers for cancer therapies. Polymeric nanocarriers with anticancer drugs conjugated or encapsulated, also known as polymeric nanomedicines, form a variety of different architectures including polymer–drug conjugates, micelles, nanospheres, nanogels, vesicles, and dendrimers. This entry focuses on the current state of the preclinical and clinical investigations of polymer–drug conjugates and polymeric micelles. Recent progress achieved in some promising fields, such as site-specific protein conjugation, pH-sensitive polymer–drug conjugates, polymer nanoparticles for targeted cancer therapy, stimuli-responsive polymeric micelles, polymeric vesicles, and dendrimer-based anticancer nanomedicines, will be highlighted. This entry was originally published as "Anticancer Polymeric Nanomedicines" in the journal *Polymer Reviews*, Vol. 47, No. 3, 345–381.

AN INTRODUCTION TO POLYMERIC NANOMEDICINES IN CANCER DRUG DELIVERY

Nanotechnology is making a significant impact on drug delivery. There is growing interest in integrating nanotechnology with medicine, creating the so-called nanomedicine aiming for disease diagnosis and treatment with unprecedented precision and efficacy.[1] In the past few years, resources allocated to the development of nanomedicine increased dramatically in both the United States and the European Union, highlighting the importance of this evolving field. In drug delivery, nanomedicine is a newly developed term to describe nanometer-sized (1–1000 nm) multicomponent drug or drug delivery systems for disease treatment.[2]

The existing challenge of drug delivery is to design vehicles that can carry sufficient drugs, efficiently cross various physiological barriers to reach disease sites, and cure diseases in a less toxic and sustained manner. As most physiological barriers prohibit the permeation or internalization of particles or drug molecules with large sizes and undesired surface properties, the main input of nanotechnology on nanomedicine is to miniaturize and multifunctionalize drug carriers for improved drug delivery in a time- and disease-specific manner.

Although nanomedicine was conceptualized only recently,[1–5] nanotechnology has been employed in drug delivery for decades.[6,7] For example, nanoparticulate liposomes were first introduced more than 40 years ago.[7] Today, a handful of liposome-based nanoparticulate delivery vehicles have been approved by the FDA for clinical applications.[2,8] Use of colloidal nanoparticles in drug delivery can date back almost 30 years.[2,6] They became clinically promising when long circulating, stealth polymeric nanoparticles were developed.[9] Both micelles and polymer–drug conjugates have been investigated for more than two decades for the treatment of various diseases including cancer.[4,10] The support from both government and industry, the breakthroughs in fundamental nanoscale science and engineering, and the progress of translational science that integrates medicine and nanotechnology have impacted and will continue to impact the development of nanomedicine.

Application of nanotechnology to clinical cancer therapy, also known as cancer nanotechnology, was recently detailed by Ferrari.[3] Cancer is the second leading cause of death in the United States, accounting for 22.7% of total mortality in 2003.[11] Although significant efforts have been devoted to cancer diagnosis and therapy, cancer-induced mortality continues to rise.[11] In cancer drug delivery, delivery strategies can be categorized as either lipid-based or polymer-based. Lipid-based nanomedicines, mainly in the form of liposomes, have been extensively reviewed.[8,12–15] This entry will focus only on various polymer-based nanocarriers that have been developed for cancer therapy. Polymeric-drug nanomedicines to be discussed in detail are polymer–drug conjugates[16–19] and polymeric micelles,[10,20–26] some of which have either been approved for clinic use or currently under clinical investigations.[2,18,27] Other newer delivery systems, such as dendrimers[28–32] and polymeric vesicles[33–39] that have been developed and employed in cancer drug delivery (Fig. 1), will also be discussed.

Concise Encyclopedia of Biomedical Polymers and Polymeric Biomaterials DOI: 10.1081/E-EBPPC-120052058

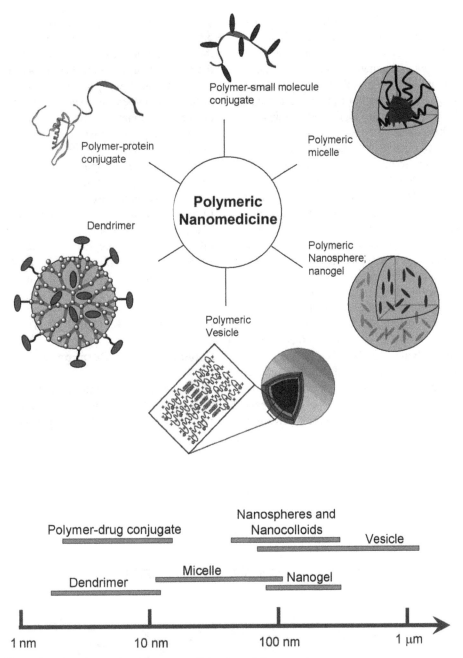

Fig. 1   Schematic illustration of various polymeric nanomedicine drug delivery systems.

DEVELOPMENT OF POLYMER–DRUG NANOMEDICINES: CONJUGATION VERSUS ENCAPSULATION

One of the central themes of drug delivery is to improve the pharmacological and pharmacokinetic profiles of therapeutic molecules. Drug molecules (small molecules or macromolecules) can be either released through the cleavage of a covalent linkage between drug molecules and polymers (conjugation) or through the diffusion from a drug and polymer blended matrix (physical encapsulation).

The covalent conjugation approach was first introduced by Ringsdorf in 1975.[40,41] In his postulated model of polymer–drug conjugates, multiple drug molecules are bound to polymer side chains through covalent, cleavable bonds. The cleavage of the polymer–drug linkage results in the release of the attached drug molecules. This concept received immediate attention since it was introduced. In the late 1970s, Kopecek, Duncan, and others started to develop N-(2-hydroxypropyl)methacrylamide (HPMA) copolymer and designed the first synthetic polymer–drug conjugate.[42] Their efforts led to a handful of HPMA-drug conjugates that later entered several clinical trials.[2,43] Using the same strategy, Maeda and colleagues developed SMANCS conjugate (styrene maleic acid neocarzinostatin) by covalently linking the anticancer drug neocarzinostatin to two styrene maleic anhydride polymer chains.[44] They successfully brought this antitumor protein conjugate to the Japanese market in 1994 as the first polymer–protein conjugate approved for human cancer treatment.[2] Since these early studies, many different polymers have been developed and evaluated as delivery vehicles for both protein and small molecules.[2,17,18,43,45] However, only a limited number of polymeric carriers have reached clinical trials (Fig. 2).[43] Nanosized polymer–drug conjugates based on these polymers as well as the other promising candidates will be discussed in the "Polymer–Drug Conjugates" section.

The physical encapsulation approach controlling drug release from a polymer matrix was originated from the seminal work by Folkman and Long in 1964.[46] They reported that hydrophobic small molecules could diffuse through the wall of silicone tubing at a controlled rate. Later, Langer and Folkman developed the first polymer-based slow-release system.[47] They found that soybean trypsin inhibitor could be encapsulated and released from an ethylene–vinyl acetate copolymer matrix over a 100-day period. This is the first report of sustained release of protein and other macromolecules from the polymer matrix. This concept was extended to the development of Gliadel®, an implantable wafer that can slowly release 1,3-bis (2-chloroethyl)-1-nitrosourea (BCNU) from a degradable poly[bis(p-carboxyphenoxy) propane-sebacic acid] matrix for brain tumor treatment. Through the efforts of Langer, Brem, and others,[48–50] Gliadel was approved by the FDA in 1996 as the first treatment to deliver chemotherapeutics directly to the tumor site using controlled release techniques.

The physical encapsulation approach was also applied to the development of a variety of nanometer-sized delivery vehicles, many of which are based on the aggregation of hydrophobic polymers (polymeric nanoparticles)[51] or the self-assembly of the hydrophobic polymer domain of amphiphilic block copolymers (polymeric micelles and vesicles).[10,33,34,52,53] Compared to polymer–drug conjugates with sizes generally around 10 nm or less, nanoaggregates formed through phase-separation are larger, typically in a range of 20–100 nm for micelles[24] and 100 nm to a few micrometers for polymer vesicles.[35,36,54] Nanocarriers based primarily on physical encapsulation will be covered in the "Polymeric Micelles." section.

POLYMER–DRUG CONJUGATES

Polymer–Protein Conjugates

The application of proteins and peptides as anticancer therapeutics has expanded rapidly in recent years. It is estimated that more than 500 biopharmaceuticals have

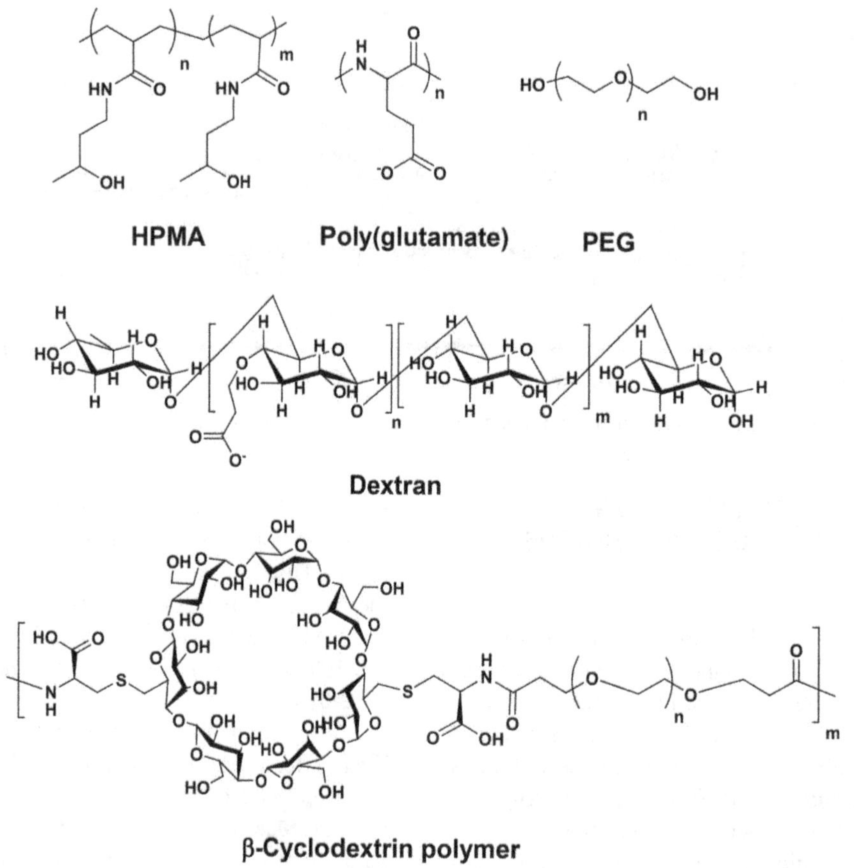

Fig. 2 Polymers in clinical trials as vehicles for conjugated therapeutics.

been developed.[55] Protein and peptide biopharmaceuticals commonly suffer from their pharmacokinetic and pharmacological drawbacks such as short circulating half-lives, immunogenicity, instability against proteolytic degradation, and low solubilities. In addition to the manipulation of amino acid sequence to reduce immunogenicity and improve stability, conjugation of hydrophilic polymers to proteins is frequently employed to overcome these drawbacks. Covalent linking of hydrophilic polymers and protein therapeutics to form polymer–protein conjugates is the most widely adopted strategy. Research on protein modification with polymers started in the late 1960s and early 1970s with dextran as the modifying polymer. However, significant progress in this field was achieved after poly(ethylene glycol) (PEG) was introduced by Frank Davis for protein modification (so-called protein pegylation).[56,57]

PEG is a linear polyether terminated with 1–2 hydroxyl groups (Fig. 2). It is highly flexible, highly water soluble, non-degradable, non-toxic, and non-immunogenic.[58] Conjugation of PEG to a protein or peptide can shield antigenic epitopes of the polypeptide, resulting in significant reduction of recognition by the reticuloendothelial system (RES). Because of the steric effect, pegylation also reduces protein degradation by proteolytic enzymes. In addition, PEG conjugation increases the molecular weight (MW) and hydrodynamic volume of proteins, resulting in decreased blood clearance by renal filtration.

Protein pegylation involves labile biopharmaceutical molecules, therefore coupling reactions are usually carried out under mild conditions. The amino functional groups (or other groups such as thiol and hydroxyl) in proteins are frequently used as the nucleophiles to attack an activated ester of PEG. PEGs are then bound to the ε-amino groups of lysine residues or the N-terminal amino group of the protein. In addition to the amino functional groups on lysine, other conjugation sites include the side chain of cysteine, histidine, tyrosine, and serine.[58] Uncontrollable, multisite pegylation is one of the major drawbacks of pegylation, which leads to pharmaceutical products with heterogeneous structures and reduced activities.[58] For instance, interferon-α2b (IFN-α2b) coupled with an activated 12 kDa mPEG forms as many as 15 different PEG-IFN-α2b products.[58] Less than 10% of bioactivity (relative to the original IFN-α2b) remains after the conjugation of PEG on Lys-83 and Lys-121 of IFN-α2b.[58] Bioactivities of these pegylated IFN-α2b vary dramatically, presumably due to the blocking of certain active sites by PEG. Despite these difficulties, several pegylated systems have received regulatory approvals for clinical applications, such as Oncaspar® (pegylated asparaginase) for the treatment of acute leukemia and Neulasta® (pegfilgrastim) for stimulating neutrophil production that are depleted during chemotherapy.[18] The powerful pegylation techniques have been extended to the delivery of other macromolecules. A branched PEG-anti-VEGF aptamer (pegaptanib sodium injection, Macugen) was approved by the FDA for the treatment of neovascular age-related macular degeneration,[59] which demonstrated the utility of PEG for the systemic delivery of nucleic acids.

The reduction of protein activities of pegylated IFN-α2b is due primarily to uncontrollable PEG conjugation, which suggests the necessity of developing site-specific pegylation. The design of newer generation pegylated proteins has mainly focused on the use of branched or heterodifunctional linear PEG that is capable of controlling site-specific, step-wise conjugation. A unique site-specific pegylation through the formation of a three-carbon bridge was reported by Brocchini, Shaunak, and coworkers.[60] They exploited the chemical reactivity of both thiols in an accessible disulfide bond in a protein molecule for pegylation. An exterior S–S bond in the protein was reduced to a pair of SH groups, both of which subsequently reacted with one PEG monosulfone, a molecule that is specifically designed for interactive bisalkylation with the two SH groups. The "insertion" of PEG into the disulfide bond showed minimum disturbance to the protein structures. This technique can be potentially applied to site-specific pegylation of numerous proteins containing disulfide bonds.

Further development of site-specific conjugation relies on the advancement of new conjugation chemistry. In 2001, click chemistry was introduced by Sharpless and coworkers, which received immediate recognition for its potential in site-specific biological conjugation.[61,62] Click chemistry usually gives very high yields, and proceeds under very mild conditions. Ligand conjugation induced by click chemistry has been successfully carried out both in situ[63] and in vitro.[64,65] One type of click chemistry, the Azide–Alkyne Huisgen cycloaddition, is particularly important for site-specific protein conjugation through the formation of 1,2,3-triazole between an azide and an alkyne.[66] In this reaction, a 1,3-dipolar cycloaddition between an azide and an alkyne gives a 1,2,3-triazole.[62] Conjugation of cellular glycans with fluorescent tags through click chemistry, for example, resulted in rapid, versatile, and site-specific covalent labelings.[66]

Tirrell and coworkers demonstrated that click chemistry can be used for site-specific conjugation of fluorescent tag with genetically engineered proteins containing non-natural homopropargylglycine or ethynylphenylalanine.[67,68] The introduced alkynyl groups on these non-natural amino acids provide sites for the attachment of fluorescent dyes containing azide groups (Fig. 3). Recent advance in protein engineering makes it possible to incorporate many non-natural amino acids at any specific position in a protein. Therefore, this technique may potentially be applied to the site-specific pegylation that gives minimum disturbance to the structure and activity of proteins.

Polymer–Small Molecule Drug Conjugates

Conjugation of hydrophobic small molecule drugs to hydrophilic polymers has been actively pursued for improved pharmacological and pharmacokinetic properties of the

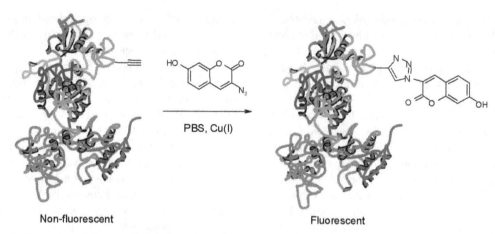

Non-fluorescent   Fluorescent

Fig. 3 Schematic illustration of site-specific labeling of protein through click chemistry.

therapeutic molecules. In general, polymer–drug conjugates have increased aqueous solubility, reduced toxicity, and prolonged plasma circulation half-life compared to free drugs. Polymer–drug conjugation may also change the internalization pathway of small molecules by bypassing P-glycoprotein associated multidrug resistance.[69] Polymers that are particularly important and have track records of preclinical success for small molecule conjugation include HPMA copolymer,[70–73] PEG,[74–78] poly(glutamic acid) (pGlu),[79–82] dextran,[83–86] and cyclodextrin (CD)-based polymer (Fig. 2).[87–90] Conjugates of various anticancer drugs with these polymers are currently in clinical trials (Table 1). Other polymers that have been successfully developed and are currently in clinical trials include polymannopyranose,[91] albumin,[92] and antibody.[93–95]

PEG has been used for the conjugation and delivery of paclitaxel (PTXL),[75] doxorubicin (DOXO),[96] and camptothecin (CPT).[76–78,97,98] Linear PEG only has two terminal hydroxyl groups for conjugation, which limits its drug-carrying capacity. A PEG-CPT conjugate (Prothecan®),[99] for example, only has about 2 wt% CPT linked to PEG.[100] PEG-CPT conjugates showed antitumor efficacy in various preclinical studies,[69,76,98,101] and have also been tested clinically.[9] In a biodistribution study, the plasma half-life of a 20 kDa PEG-DOXO conjugate was found to be less than 10 hours.[96] Protracted antitumor activity was observed with prolonged circulation and improved tumor accumulation due to the enhanced permeability and retention (EPR) effect (Fig. 4).[102] In a phase-I clinical study, PEG-CPT showed a 77-hour plasma clearance half-life,[9] which is much greater than that of a similar system in mice.[96] A study showed that coupling of PEG and CPT through an alanine ester linker can induce apoptosis in tumor and decrease apoptosis in liver and kidney when compared to free CPT.[101] Extended circulation and slow release of CPT may also contribute to the observed neutropenia and thrombocytopenia.[99]

HPMA-drug conjugates are another type of conjugates that have been extensively evaluated in clinic.[42,103,104]

HPMA is very water soluble, biocompatible, and non-degradable, which resembles PEG to some degree. To ensure complete clearance of non-degradable polymers from circulation, polymer MWs have to be maintained at or below 45–50 kDa.[105] Most HPMA copolymers tested in vivo are 30 kDa or shorter.[70,106–108] However, HPMA-drug conjugates with such low MWs showed fast renal clearance, which may adversely affect their antitumor efficacy. Enhanced accumulation through the EPR effect for polymer–drug conjugates with MWs at or around their renal clearance threshold (40–45 kDa) is as effective as their higher MW analogs.[102,109] Compared to PEG, HPMA has a large number of pendent functional groups that allow the conjugation of many hydrophobic small molecules on each HPMA polymer. The drug loading capacity of HPMA is thus significantly larger than that of PEG and is comparable to that of pGlu. HPMA copolymer conjugates with PTXL,[107] CPT,[108,110,111] DOXO,[70,106,112] and platinate[113] have all been evaluated in various clinical trials.

pGlu, a biodegradable polypeptide, has also been used for small molecule drug delivery. pGlu has a large number of pendent carboxyl groups, which makes pGlu extremely water soluble. As much as 30 wt% of PTXL[114,115] or CPT[79] can be conjugated to pGlu, which is much higher than that in PEG conjugates. The resulting pGlu-CPT or pGlu-PTXL still showed sufficiently high water solubility. PTXL molecules linked to pGlu through a degradable ester bond can be released at a controlled hydrolysis rate. The release rate is usually significantly enhanced when the pGlu-PTXL is internalized to cell and exposed to harsh endolysosomal environment. PTXL and CPT conjugated to pGlu showed enhanced preclinical antitumor efficacy in several preclinical tumor models presumably due to the EPR-mediated passive tumor targeting.[81,114,116] Interestingly, pGlu-PTXL also showed positive response in taxane-resistant patients in several Phase I and II studies of various cancers.[117] A completed Phase III trial of pGlu-PTXL in combination with standard chemotherapy against ovarian cancer and non-small-cell lung cancer (NSCLC) suggests that estrogen

Table 1 Polymer–drug conjugates in clinical trials

Name	Polymer	Drug	Linker	Company	Target	Status	References
Prothecan®	PEG (40 kDa)	CPT	Ester	Enzon	SCLC	Phase II	[97,99]
PK1	HPMA (30 kDa)	DOXO	Gly-Phe-Leu-Gly	CRC/Pharmacia	Various cancers	Phase II	[70,106]
PK2	HPMA (30 kDa)	DOXO	Gly-Phe-Leu-Gly	CRC/Pharmacia	Various cancers	Phase I discontinued	[112]
PNU-166945	HPMA (40 kDa)	PTXL	Ester	Pharmacia	Various cancers	Phase I completed	[107]
MAG-CPT	HPMA (30 kDa)	CPT	Gly-6-aminohexanoyl-Gly	Pharmacia	Various cancers	Phase I completed	[108,111]
AP5280 AP5286	HPMA (25 kDa)	Diamine-platinum(II)	Gly-Phe-Leu-Gly	Access Pharmaceuticals	Various cancers	Phase I completed	[217–220]
AP5346	HPMA (25 kDa)	Oxaliplatin	Gly-Gly-Gly	Access Pharmaceuticals	Head and neck cancer	IND approved	[221]
CT-2103 (XYOTAX)	PG (40 kDa)	PTXL	Ester	Cell Therapeutics	Various cancers	Phase III	[114–117,222–224]
CT-2106	PG (50 kDa)	CPT, 5-Fu	Gly-ester	Cell Therapeutics	Various cancers	Phase I	[79]
MTX-HSA	Albumin (67 kDa)	MTX	—	AK St. Georg	Advanced cancers	Phase II	[92,225–228]
DOXO-EMCH	Albumin (67 kDa)	DOXO	Hydrazone	Tumor Biology Center	Various cancers	Phase I	[141]
IT-101	CD polymer	CPT	Gly ester	Insert Therapeutics	Various cancers	Phase I	[87–90]
DAVANAT	Polymannopyranose	5-Fu, AVand LV	—	Propharmaceuticals	Colorectalcancer	Phase II	[91]
AD-70	Dextran (70 kDa)	DOXO	Schiff base	Alpha Therap. GmbH	—	Phase I	[122]
HuC242-DM4	humAb huC242	MTS-DM4	—	ImmunoGen	Various cancers	Phase I	—
BB-10901	humAb N901	MTS-DM1	—	ImmunoGen	SCLC and CD56-SC	Phase II	[93–95,144]

Abbreviations: 5-Fu, 5-fluorouracil; LV, leucovorin; CPT, camptothecin; PTXL, paclitaxel; AV, Avastin; MTS, maytansinoid; SCLC, small-cell lung cancer; DOXO, doxorubicin; MTX, methotrexate; humAb, humanized monoclonal antibody; CD56-SC, CD56-positive SC carcinoma.

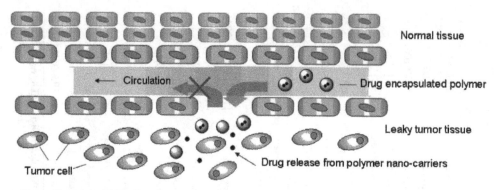

Fig. 4 Schematic illustration of the EPR effect.

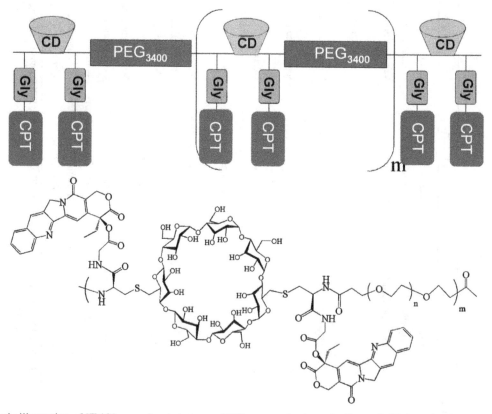

Fig. 5 Schematic illustration of IT-101, a conjugate between 20(S)-camptothecin and a linear, β-CD-based polymer through a glycine ester linker.

may participate in regulating the *in vivo* efficacy of pGlu-PTXL. pGlu-PTXL was found to be efficacious only in certain group of patients, such as premenopausal female NSCLC patients. A pGlu-CPT conjugate (CT-2106) with CPT linked to pGlu through a glycine linker with 33–35 wt% loading is currently in Phase I/II trials.[79]

CD-containing polymer is a new class of hydrophilic biomaterials that has been developed for drug delivery. CDs are cyclic oligomers of glucose that can form water-soluble inclusion complexes with numerous hydrophobic molecules with compatible sizes. CDs are biocompatible, non-immunogenic, and non-toxic, therefore they have

been extensively used in many pharmaceutical applications to improve the bioavailability and solubility of drugs.[118] CD-containing polymers have also been developed and used for decades.[119,120] Because CD has many hydroxyl groups, CD-containing polymers are usually heavily crosslinked with uncontrollable compositions and limited applications. In 1999, Davis and coworkers developed the first linear, β-cyclodextrin polymer (β-CDP)[121] bearing cationic pendant groups for gene delivery.[121–126] CDPs were further modified to introduce pendant carboxyl groups (Fig. 2) for CPT conjugation (IT-101, Fig. 5). CDPs are very water soluble (over 200 mg/mL) and can increase the solubility of CPT by three orders of magnitude after conjugation.[88]

A pharmacokinetic study in rats showed that the half-life of bound CPT in IT-101 is 17–19 hours, which is significantly longer than that of CPT alone.[90] The half-life is also longer than those of PEG-CPT and HPMA-CPT, which may be due in part to the high MW of the β-CDP tested (85 kDa).[90] IT-101 forms large particles (≈50–80 nm) in solution presumably through the interchain interaction between CPT and CD. This unusual nanoaggregation is in sharp contrast to most polymer–drug conjugates reported so far whose sizes typically ranged from 5 to 15 nm. The increase in particle size of IT-101 likely reduces its clearance through glomerular filtration, thus enhances its *in vivo* antitumor efficacy.[89] Protracted antitumor activity was observed in LS174T colon carcinoma tumor-bearing mice[87] as well as in a number of other irinotecan-resistant tumors (MDA-MB-231, Panc-1, and HT29),[89] which is consistent with the hypothesis that polymer–drug conjugates may overcome multidrug resistance. An open-label, dose-escalation Phase I study using IT-101 in patients with inoperable or metastatic solid tumors has been initiated.

Polysaccharides were also developed for the delivery of small-molecule therapeutics. DAVANAT, a [(1–4)-linked-β-D-mannopyranose]-[(1–6)-linked-α-D-galactopyranose] polymer, is currently in the Phase-II trial for colorectal cancer treatment with a combination of 5-fluorouracil (5-FU), avastin, and leucovorin.[91] DAVANAT binds to surface lectins, proteins that are overexpressed in metastatic tumor cells and mediate cell association, apoptosis, and metastasis. The interaction of DAVANAT with lectin may promote transport of 5-FU into the tumor cells. A Phase I open-label study showed that DAVANAT alone or in combination with 5-FU was well tolerated in patients, which facilitated its Phase II clinical trials.[91]

Besides polymannopyranose, other polysaccharides such as dextran and dextran derivatives have also been used for the delivery of small-molecule drugs (Fig. 2). Dextran is biocompatible to some extent and has been approved for certain clinical applications (e.g., as plasma expander). An oxidized form of dextran (70 kDa) was conjugated with DOXO through a Schiff base linker, and the resulting conjugate (AT-70) was subsequently evaluated preclinically and clinically. Severe hepatotoxicity was observed, presumably due to uptake of dextran by the RES.[127] DE-310, another dextran-based conjugate with a 340 kDa carboxymethyldextran polyalcohol conjugated with CPT analog DX-8951 through an Gly-Gly-Phe-Gly linker, was also tested in clinic.[84,128–130] Formation of amide, instead of ester linkages, reduced drug release from DE-310 during systemic circulation. As the peptidyl linker is enzymatically degradable, DX-8951 can presumably only be released after DE-310 is taken up by cells to endolysosomal compartments with active proteinases. Thus drug release can be specifically controlled inside cells.[84] A Phase I study showed dose-limiting toxicities due to thrombocytopenia and neutropenia.[129]

As polymer accumulation in tumor through the EPR effect is usually enhanced with increased polymer MW,[102] it has been actively pursued to develop degradable, high MW polymers using biocompatible building blocks. Duncan and coworkers developed water-soluble and biocompatible polyacetals through the condensation of PEG and tri(ethylene glycol) divinyl ether (Fig. 6).[109,131,132] The acetal moiety was chosen because it can undergo faster hydrolysis under mildly acidic conditions but is stable at

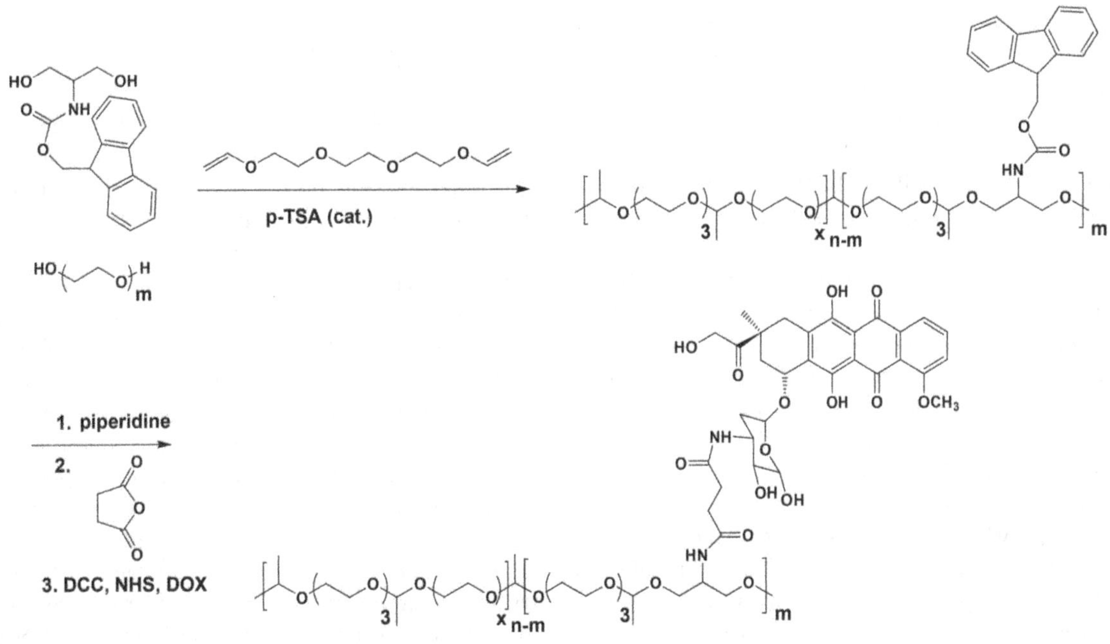

Fig. 6 pH-Sensitive polyacetal for DOXO delivery.

physiological pH. As the main chain of the polyacetals can be hydrolyzed to small, renal-clearable fragments, the polymer can be made significantly larger than 45 kDa for prolonged circulation in blood. One drawback is that the polyacetals were prepared through step-growth polymerization that gave polymers with fairly broad MW distributions (in the range of 1.8–2.6).[109,131,132] The polyacetals displayed remarkable tunability for pH-induced degradation. Enhanced hydrolysis was observed at pH 5.5 (41% M_w loss in 25 hours) when compared with that at pH 7.4 (10% M_w loss in 73 hours). In addition, the polyacetals and their degradation products are non-toxic *in vitro* ($IC_{50} > 5$ mg/mL in B16F10 cells) and *in vivo*. Amine pendant functional groups were incorporated through terpolymerization (Fig. 6), which was used for drug conjugation. A biodistribution study showed no preferential accumulation of polymer in the major organs. In C57 xenograft mice bearing a subcutaneous B16F10 tumor, the pharmacokinetics of intravenously administered polyacetal-DOXO (M_w = 86 kDa, M_w/M_n = 2.6) and HPMA copolymer-GPLG-DOXO (M_w = 30 kDa, M_w/M_n = 1.3–1.5) were compared.[109] Both polyacetal-DOXO and HPMA copolymer-DOXO displayed similar biphasic patterns of plasma clearance with a $t_{1/2}\alpha$ of ~1 hour presumably due to the presence of low MW fragments. But the plasma levels of polyacetal-DOXO were significantly higher than those of HPMA copolymer-DOXO with a $t_{1/2}\beta$ of 19 hours and 3.5 hours for polyacetal-DOXO and HPMA copolymer-DOXO, respectively. The $t_{1/2}\beta$ of polyacetal-DOXO is quite similar to that of the CD polymer with a similar MW (85 kDa, $t_{1/2}\beta$ = 17–19 hours).[90] Because prolonged plasma circulation is the driving force for increased passive tumor targeting,[133,134] polyacetals with higher MWs and lower polydispersities may give improved circulation half-life and tumor accumulation. It is noted that polyacetal-DOXO, although with MW much higher than HPMA copolymer conjugates, showed reduced accumulation in liver and spleen.[131] The high PEG content in polyacetal may contribute to the lower uptake by the RES system.

Polyacetals can also be prepared through selective degradation of polysaccharides. Papisov and coworkers developed acyclic hydrophilic polyals through the lateral cleavage of polyaldoses and polyketoses.[135,136] Polyals obtained through this method consist of acyclic carbohydrate substructures that are potentially biocompatible. The intrachain acetal or ketal groups should enable hydrolytic biodegradation upon cell uptake. In an *in vivo* toxicity study, all mice survived intravenous administration of a 160 kDa polyacetal at a dose as high as 4 g/kg. The polymer gave very low RES response and showed low tissue accumulation even at MW as high as 500 kDa. This class of polymers contains a large number of pendant functional hydroxyl groups, which make it easy for structural modification and drug conjugation. However, it is difficult to control the sites of periodate oxidation, which leads to polymers with poorly controlled compositions.

Albumins have also been evaluated as drug carriers in clinical trials. A methotrexate–human serum albumin conjugate (MTX-HSA) was synthesized by coupling MTX to HSA.[137–139] MTX-HSA showed significant accumulation in rat tumors and displayed high *in vivo* antitumor activity. In a Phase I study, patients with renal cell carcinoma and mesothelioma responded to treatment with MTX-HSA therapy.[139] In a Phase II study of MTX-HSA in combination with cisplatin as first line treatment of advanced bladder cancer,[92] positive response was observed. The combination strategy showed promise for the treatment of urothelial carcinomas with acceptable toxicity. An albumin-DOXO conjugate (DOXO-EMCH) was also developed through an acid-sensitive 6-maleimido-caproyl-hydrazone linker.[140] The covalently linked DOXO prevents its rapid diffusion of DOXO into healthy tissue after intravenous administration and allows passive accumulation of DOXO-EMCH through the EPR effect in solid tumors. DOXO is then released in the acidic environment of tumor tissue through the cleavage of the hydrazone linker. A Phase I study of DOXO-EMCH in 10 patients (6 female, 4 male) showed that DOXO-EMCH could be tolerated up to 40 mg/m².[141]

Antibody has also been extensively used for drug conjugation, creating immunoconjugates as an important group of therapeutics for cancer treatment. For example, BB-10901 (Table 1), a humanized mAb conjugated with cytotoxic maytansinoid DM1 for small-cell lung cancer treatment is currently in Phase I/II clinical trials.[93–95,142] Immunoconjugates for cancer treatment is beyond the coverage of this entry, and has been reviewed elsewhere.[143] It is worth reporting that an alternative strategy of using aptamers for targeted DOXO delivery was developed.[144]

POLYMERIC MICELLES

Amphiphilic block copolymers can self-assemble in aqueous solution to form core–shell micellar nanostructures when the concentrations of the amphiphilic block copolymer are above the critical micellar concentration (Fig. 7). Polymeric micelles have a condensed, compact inner core, which serves as the nanocontainer of hydrophobic compounds. As polymeric micelles are generally more stable than hydrocarbon-based micelles, sustained drug release from polymeric micelles becomes possible.[20,24] Numerous types of amphiphilic copolymers have been employed to form micelles[4,10,20,52,145–147] or other similar architectures such as nanogels[148] and polymer nanoparticles.[51] Details of copolymers structure and drug molecule encapsulated or conjugated are summarized in Table 2.

Polymeric micelles can accumulate in tumors after systemic administration. Their biodistributions are largely determined by their physical and biochemical properties,

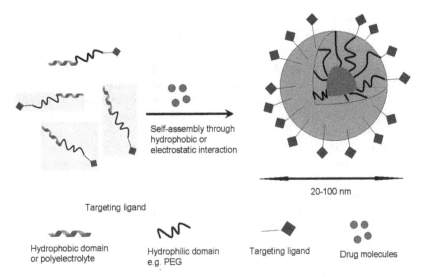

Self-assembly through hydrophobic or electrostatic interaction

20–100 nm

Targeting ligand

Hydrophobic domain or polyelectrolyte

Hydrophilic domain e.g. PEG

Targeting ligand

Drug molecules

Fig. 7 Polymeric micelle core–shell structure and drug encapsulation.

Table 2 Polymeric nanoparticles: Polymer structures and drug incorporated

Block copolymer	Drug (or Dye)[references]
PEG-b-polypeptide	
PEG-b-pAsp	DOXO,[229–233] methotrexate,[234] indomethacin,[235] amphotericin-B,[236–239] KRN 5500,[156] cisplatin,[240,241] and Nile Red[165]
PEG-b-pGlu(Bn)	Clonazepam[242]
PEG-b-pGlu	Cisplatin[158]
PEG-b-pHis/PLA	pH-sensitive micelles; DOXO[243,244]
PEG-b-pLys	Cisplatin[245]
PEG-b-polyester	
PEG-b-PCL	Indomethacin,[246,247] dihydrotestosterone,[248] FK506,[249] L-685,818[249] nimodipine,[250]
PEG-b-PLA	PTXL,[251–253] DOXO[254]
PEG-b-PLGA	DOXO,[255] PTXL,[256] DTXL,[164,152] DOXO/combretastatin[159]
PEG-b-polyether (nanogel)	
Pluronic-P85	Daunorubicin,[257] DOXO,[257,258] vinblastine,[2257] mitomycin,[257] cisplatin,[257] methotrexate,[257] epirubicin,[257] PTXL,[259] etoposide,[259] and digoxin[260]
Pluronic-F127	Nystatin[261]
Pluronic-F68	Nystatin[261]
Other homopolymers and block polymers	
PEG-b-PMA	Pyrene[262] and Nile Red[167]
pLys(EG)-b-pLeu	DiOC$_{18}$ dye[36]
pArg-b-pLeu	Fluorescein[35,54]
pLys-b-pLeu	DiOC$_{18}$ dye[47]
PUA-b-PNIPAAm	N/A[263]
PNIPAAm/PDMAAm-b-PCL/PLA	Pyrene[264]
PLA-PEG-PLA	DOXO[265]
Poly(orthoester)	DNA vaccine[266]
Poly(β-amino ester)	DNA and dye[267,268]
Polyketal	N/A[269]

Abbreviations: PNIPAAm, poly(*N*-isopropylacrylamide); PUA, poly(undecylenic acid); pAsp, poly(aspartate); pGlu(Bn), poly(benzyl-glutamate); pLys, poly(lysine); pHis, poly(histidine); PCL, poly(caprolactone); PLA, poly(D,L-lactide); PLGA, poly(D,L-lactic acid-co-glycolic acid); PMA, polymethacrylate; EG, oligo(ethylene glycol); PDMAAm, poly(*N*,*N*-dimethylacrylamide).

such as particle sizes, hydrophobicity and hydrophilicity of the polymers and drugs, and surface biochemical properties.[149] A major issue that limits the systemic application of micellar nanocarriers is the non-specific uptake by the RES. It is critical to have systems that can circulate for long time without significant accumulation in liver or spleen. The sizes and the surface features of micelles have to be controlled for favored biodistribution and intracellular trafficking.[9] The hydrophilic shells of micelles usually consist of PEGs, which prevent the interaction between the hydrophobic micelle cores and biological membranes, reduce their uptake by the RES, and prevent adsorption of plasma proteins onto nanoparticle surfaces.[22] Micellar nanocontainers are typically in the range of 20–100 nm. The sizes of polymeric micelles resemble that of natural transporting systems (e.g., virus and lipoprotein), which allow efficient cellular uptake via endocytosis.[150] It was also found that the effect of size on polymeric micelle biodistribution is organ specific and non-linear.[151] Therefore, controlling the sizes of micelles in a predefined range can be critical for desired applications. Parameters controlling the size of micelles include relative length of polymer blocks, polymer composition, and the solvent and drug used for encapsulation. A study indicated that the mean volumetric size of PEG-*b*-PLGA micelles correlates linearly with polymer concentration during self-assembly with linear correlation coefficient ≈0.99. Such linear correlation may provide means for preparing polymeric micelles with desirable sizes.[152]

PEG-Polypeptide Micelle

PEG-*b*-poly(aspartic acid) [PEG-*b*-pAsp] micelles and their DOXO conjugates (NK911) were developed by Kataoka and coworkers.[153] This is one of the most intensively investigated micellar drug delivery vehicles. DOXO molecules were conjugated to the copolymers to form micelles with diameters in the range of 15–60 nm. However, DOXO molecules covalently conjugated to the pAsp side chain did not have therapeutic activity. Interestingly, the conjugated DOXO molecules can promote the formation of stable π–π interaction with the encapsulated DOXO molecules.[154] In a Phase I study, the toxicity of NK911 resembled that of free DOXO, and the dose-limiting toxicity was neutropenia.[155] NK911 is currently being evaluated in a Phase II clinical trial.[4]

The compatibility between the core-forming blocks and the drugs to be loaded controls the drug loading capacity and release rate. For example, encapsulation of hydrophobic therapeutic compound KRN5500, a spicamycin derivative with a long-chain fatty acid, requires a hydrophobic core-forming block of pAsp with similar fatty acids side chain.[156] As the micelle core has no interaction with tissue during circulation, drug loading has a minimal effect on the micelle biodistribution.

PEG-*b*-polypeptide micelles have also been used for the delivery of PTXL. For example, PTXL has been incorporated into the 4-phenyl-1-butanolate modified PEG-*b*-pAsp to form polymeric micelles (NK105).[157] An *in vivo* antitumor study revealed that NK105 was more potent than free PTXL, possibly because of the enhanced drug accumulation in tumor tissues through the EPR effect.

Because carboxylate groups can chelate with multivalent metal ions, amphiphilic copolymers containing pAsp and pGlu have been used to complex with anticancer platinum compounds, such as *cis*-dichlorodiammineplatinum(II) (cisplatin).[158] Micelles are formed through the ligand substitution of Cl⁻ on cisplatin with the carboxylate of pAsp or pGlu. *In vivo* studies displayed similar extended plasma half-life and tumor accumulation as reported for other micellar drug delivery vehicles.

PEG-Polyester Micelle

Besides polypeptides, biodegradable polyesters can also be used as a micellar core-forming block. Well-known hydrophobic polyesters include polycaprolactone (PCL), poly(lactic acid) (PLA), poly(glycolic acid), and poly(lactide-co-glycolide) (PLGA), all of which have been approved by the FDA in various clinical applications. These polymers have different degradation profiles, which can be used to tune drug release rates. However, because these polyesters have no pendant functional groups for drug conjugation, drugs are predominantly incorporated into the micellar hydrophobic core through physical encapsulation although conjugation of DOXO through a covalent bond to the terminal hydroxyl group of PLGA has also been tested.[159]

Low MW methoxy-PEG-*b*-PLA was employed to encapsulate PTXL to form copolymer micelles.[160] Evaluation of the *in vivo* antitumor efficacy of this micelle in SKOV-3 human ovarian cancer implanted xenograft mice demonstrated significantly enhanced antitumor activity when compared with free PTXL. At Day 18 after administration, the tumor was undetectable in all mice treated with the micelles at its maximum tolerable dose (60 mg/kg). At the end of the experiment (1 month), all mice remained tumor free. Currently, this PTXL-containing methoxy-PEG-*b*-PLA micellar vehicles are under Phase II clinical evaluation.[161]

The core–shell structures of amphiphilic micelles allow the attachment of targeting ligands to their external surface for active accumulation in tumor tissues. Many small molecules and antibodies have been utilized as such targeting ligands.[162] Recently, aptamers were also developed and used in targeted drug delivery.[163] An A10 2′-fluoropyrimidine RNA aptamer that recognizes the extracellular domain of the prostate-specific membrane antigen was conjugated to a docetaxel (DTXL)-encapsulated COOH-PEG-*b*-PLGA micelle (Fig. 8). The copolymer micelles have terminal

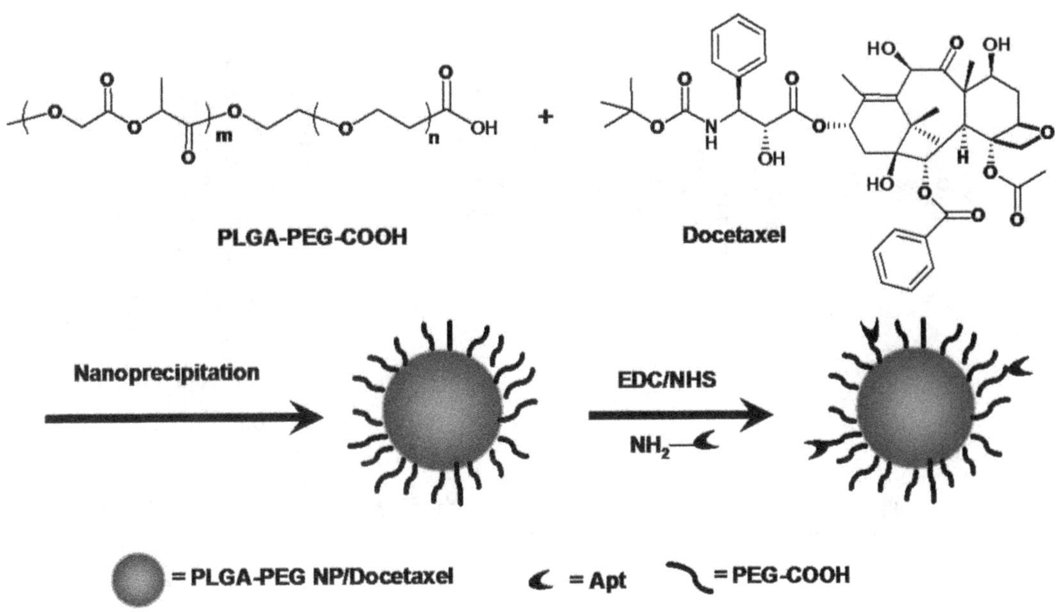

Fig. 8   DTXL-encapsulated, PLGA-b-PEG-COOH micelle and its aptamer conjugate for targeted prostate cancer therapy.

carboxyl groups extruded to the water phase, facilitating the conjugation of aptamer targeting ligands. The aptamer containing micelle displayed enhanced antitumor activity compared to the control group. A single intratumoral injection of DTXL-aptamer nanoparticle resulted in complete tumor remission in five of seven LNCaP xenograft nude mice when compared to tumor remission in two of the seven mice in the control group.[164]

Stimuli-Responsive Polymeric Micelle

Polymeric micelles that are responsive to light, pH, or temperature are potentially exciting nanomedicine modalities for site-specific drug delivery. The mildly acidic pH in tumor and inflammatory tissues (pH ≈ 6.5) as well as in the endosomal intracellular compartments (pH ~4.5–6.5) may trigger drug release from pH-sensitive micelles upon their arrival at the targeted disease sites. Fréchet and coworkers developed a pH-dependent micelle that can release encapsulated cargos significantly faster at pH 5 than at pH 7.4.[165] An amphiphilic copolymer with acid-labile hydrophobic block (Fig. 9) can form micelles at the physiological pH. When exposed to mildly acidic pH, an accelerated hydrolysis of the micelle acetal bonds (Fig. 9) results in the formation of hydroxyl groups in the hydrophobic core, disruption of the micellar assembly, and release of the encapsulated cargos. Another interesting pH-sensitive micellar delivery system was reported by Kataoka and coworkers using an acid-labile hydrazone linker to conjugate DOXO to pAsp.[166] A kinetic study demonstrated pH-dependent release of DOXO, in a manner resembling what was observed in Fréchet's pH-sensitive micelles.

Fréchet and coworkers also reported an alternative release triggering mechanism through the use of infrared light (Fig. 10).[167] The amphiphilic structure has a 2-diazo-1,2-naphthoquinone at the terminal of hydrophobic end and an oligo(ethylene glycol) as the hydrophilic block. When the micelles were exposed to infrared light, 2-diazo-1,2-naphthoquinone undergoes a Wolff rearrangement and forms hydrophilic 3-indenecarboxylate, which destabilizes micelle and causes drug releasing. Because high-wavelength light is safer and has better tissue penetration when compared with low-wavelength light, this design may potentially be used to control drug release in deep tissues harmlessly.

Micelles may not always adopt spherical shapes. Under certain conditions, cylindrical-shaped micelles also called filomicelles can be formed by controlling the fraction of hydrophilic domains.[168] Discher and coworkers studied the biodistribution of a class of filomicelles that are multiple μm long and 22–60 nm in diameters.[169] Surprisingly, these long filomicelles can circulate in rodents for up to one week, which is about 10 times longer than any known synthetic nanoparticles. Various *in vitro* studies suggested that long filomicelles could respond to various biological forces to fragmentize into spheres and short filomicelles that can be taken up by cells more readily than longer filaments.

Other delivery vehicles, such as nanospheres[51,170–172] and nanogels,[173–175] can be prepared using similar methods as micelles by forming nanoaggregates of hydrophobic polymer segments. These systems have been extensively reviewed elsewhere,[176–180] and therefore will not be covered in this entry although some specific systems are highlighted in Table 2.

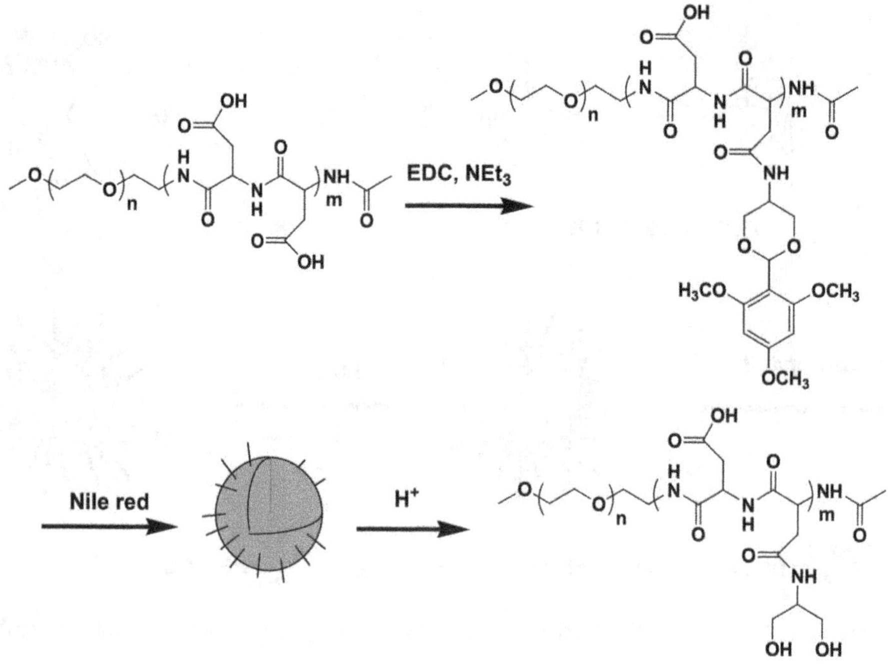

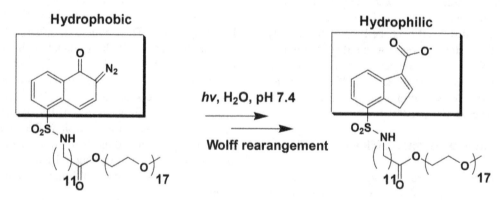

Fig. 9 pH-Sensitive polymeric micelles that can be disrupted at pH 5.

Fig. 10 Formation of IR light-sensitive micelles.

OTHER PROMISING NANOCARRIERS FOR DRUG DELIVERY

Polymeric Vesicles

Besides forming micelles, amphiphilic block copolymers can also form vesicles when the fraction (f) of the hydrophobic domain relative to the hydrophilic domain is controlled within a certain range ($f = 0.2$–0.42).[33,34] Polymeric vesicles form a liposome-like structure with a hydrophobic polymer membrane and hydrophilic inner cavity, therefore they are also called polymersomes.[33,53]

Block copolymers self-assemble into vesicles by forming bilayers through the close packing of lipid-like, amorphous polymer hydrophobic segments in a way similar to phospholipids (Fig. 1). Compared to liposomes, polymeric vesicles are more stable because their membrane-making polymers form much stronger hydrophobic

interactions than the short hydrocarbon segments of liposomes. Polybutadiene (PBD) is a popular bilayer-forming polymer,[33] which can be cross-linked subsequently for enhanced vesicle stability. Other bilayer-forming polymers include biodegradable PLA and PCL for controlled drug release,[168] and polypeptides for conformation-specific vesicle assembly.[35,36] Hydrophilic blocks used in polymeric vesicles include non-ionic PEG or oligo(ethylene-oxide)-modified polypeptide,[36,37,168] and ionic poly(acrylic acid) or polypeptides.[33,54] Triblock[181–184] and tetrablock[185] copolymer vesicles have also been developed and studied.

Polypeptides have more diverse conformations (coils, α-helices, and β-sheets) compared to synthetic polymers, therefore they are very versatile building blocks for polymeric vesicles. Deming and coworkers developed a series of polypeptide-based vesicles.[35,36,54] In addition to the control on the relative length of hydrophilic and hydrophobic segments that are critical to the formation of vesicles, the conformation was found to be another important parameter controlling the formation of peptide vesicles. Conventional uncharged amphiphilic block copolymer vesicles require high hydrophobic contents (approximately 30–60 mol%) to form stable vesicles.[186] However, the block copolypeptides deviate from this trend and can form vesicles with 10–40 mol% hydrophobic domains. This difference is presumably because of the rigid chain conformations of polypeptides and strong intermolecular interactions[187] when compared to PBD-PEG or PLA-PEG vesicles that have more flexible polymer segments. Copolypeptides used in vesicle formation can be designed to adopt rod-like conformations in both hydrophobic and hydrophilic domains due to the strong α-helix-forming tendencies.[188] These rod-like conformations provide a flat interface on hydrophobic association in aqueous solution, thus driving the self-assembly into vesicle structures.

Although polymeric vesicles have only been studied for a few years, they have shown great promise in controlling drug loading, systemic biodistribution, and drug release.[168,169] One of the major challenges in particle-based delivery vehicles is to control drug release kinetics. Polymeric nanoparticles, for example, can release more than 50% of the encapsulated drugs within the first several or tens of hours due to the burst effect.[189] In polymeric vesicles, precise tuning of drug release rates can be achieved through blending vesicle-forming copolymers with a hydrolyzable copolymer (e.g., PLA-PEG). The hydrophilic, hollow interior space of vesicles should also find application in encapsulation and delivery of hydrophilic therapeutics, such as DNA and proteins. A polyarginine–polyleucine copolymer vesicle demonstrated excellent intracellular trafficking properties.[35] The arginine domains not only promote vesicle formation but also mimic the properties of protein transduction domain[190] to enhance cell membrane penetration.

Dendrimer and Dendritic Polymer Nanocarriers

Dendrimers are a class of monodisperse macromolecules with highly branched, symmetric, three-dimensional architectures (Fig. 1). They were first reported in the late 1970s and early 1980s.[191–193] Dendrimers contain layered structures (also known as generations) that extend outwards from a multifunctional core on which dendritic subunits are attached.[194] The sizes of dendrimers are in the range of 1–15 nm.

Syntheses of multigeneration dendrimers involve alternative repetition of a generation-growth and an activation step. Depending on the direction towards which dendrimer grows, the synthetic strategies can be classified as divergent[192,193,195] or convergent.[196,197] Preparation of dendrimers requires alternate and stepwise control on each chain propagation step, which resembles solid-phase peptide synthesis to some extent, therefore synthesis of

dendrimers can be time-consuming and label-intensive, especially for the preparation of monodisperse dendrimers with high generations. The initial efforts in dendrimer research focused primarily on the development of various synthetic methods and the investigation of the physical and chemical properties of dendrimers.[198–204] In the past 10 years, significant efforts have been devoted to explore the potential applications of dendrimers in drug delivery.[28,29,32,45,195,205–214]

Drug molecules can either be conjugated on the surface or encapsulated inside a dendrimer. The periphery of a dendrimer usually contains multiple functional groups for the conjugation of drug molecules or targeting ligands. Surface conjugation is straightforward and easy to control, therefore the majority of dendrimer-based drug delivery is through this covalent conjugation approach. Despite numerous designs of dendrimer-based carriers, only a few of them have been evaluated for their in vivo antitumor activities.[31,102,215]

One early example of dendrimer used as an anticancer carrier in vivo is a sodium carboxyl-terminated G-3.5 polyamidoamine (PAMAM) dendrimer for the conjugation of cisplatin (20–25 wt%).[31] When administered intravenously to treat a subcutaneous B16F10 melanoma, the dendrimer–Pt conjugate displayed significantly enhanced antitumor activity when compared to free cisplatin.[31]

The same type of dendrimer, but with increased size (G-5 PAMAM), was developed and used for the delivery of MTX.[215] The dendrimer surface charge was first reduced by modifying peripheral amines of the G-5 PAMAM dendrimers with acetyl groups. Folate and MTX (≈9 wt%) were subsequently conjugated to PAMAM. Biodistribution study in mice with subcutaneous tumors using radioactively labeled dendrimers displayed internalization and intracellular accumulation in human KB tumors with overexpressed folate receptors.[102] Significant in vivo antitumor activity of the dendrimer-MTX conjugate was also observed.[102]

Szoka and Fréchet developed an asymmetric dendrimer for small molecule delivery.[32] In contrast to the nondegradable PAMAM that forms globular structures, their degradable polyester dendrimers have bow-tie shaped molecular architecture (Fig. 11). The number and size of the PEG chains, and the number of drug conjugation sites can be changed as desired, allowing the formation of a potentially large number of conjugates with variable PEG sizes, branches, and drug loadings. Bow-tie dendrimers with MW over 40 kDa exhibit plasma clearance half-lives greater than 24 hours, which is significantly longer than linear polymer conjugates with similar MW.[102] The branched structure of dendrimers may attribute to the reduced renal clearance and enhanced plasma half-lives as the dendrimers more likely hinder the glomerular filtration in kidney than their linear analogs with similar MWs.[29] Upon intravenous administration to BALB/c mice with subcutaneously implanted C-26 tumors, dendrimer-DOXO

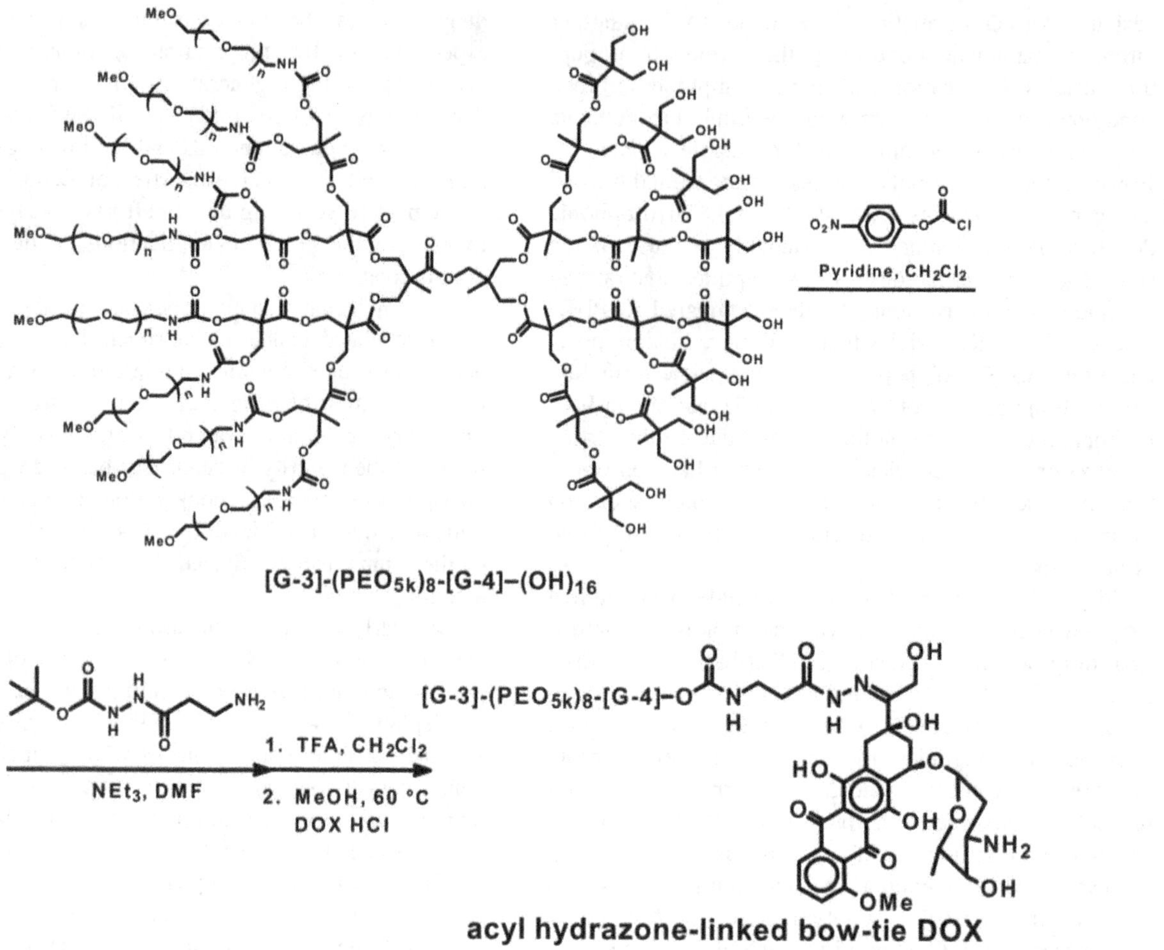

Fig. 11 Functionalization of the [G-3]-(PEO5k)8-[G-4]-(OH)$_{16}$ bow-tie dendrimers for DOXO conjugation through a pH-sensitive acyl hydrazone linker.

was found to be much more efficacious than free DOXO with less toxicity, which was presumably related to enhanced tumor uptake. In fact, dendrimer-DOXO displayed comparable *in vivo* antitumor efficacy as Doxil, an FDA approved, liposome-based DOXO delivery vehicle.

Compared to liposomes and micelles, dendrimer–drug conjugates may be more stable due to their unimolecular structures, and thus are easier to handle (formulation and sterilization). However, in addition to the challenge for the synthesis of monodisperse, high-generation dendrimers, conjugation of a large number of insoluble drugs to the surface of dendrimers may result in significantly increased peripheral hydrophobicity, which may subsequently lead to dendrimer aggregation and increased polydispersity. Although surface hydrophobicity-induced dendrimer aggregation may be reduced by encapsulating drug molecules inside dendrimers and there are some efforts in developing dendritic nanocarriers for encapsulating drugs,[216] this approach is still in the early stage of development with insufficient studies to give a full assessment.

CONCLUSION

Nanotechnology is making a significant impact on cancer drug delivery. In conjunction with the development of lipids-based drug delivery, the advancement of modern polymer chemistry makes it possible for the preparation of a large variety of synthetic polymeric materials with structures tailored to accommodate the specific needs for systemic drug delivery. We reviewed the progress and current state of polymer–drug conjugates and polymeric micelles, the two most extensively investigated polymeric vehicles for drug delivery. We also discussed the exciting progress in some areas that are potentially of importance for controlled drug delivery and cancer therapy. It is anticipated that synergistic integration of the efforts of chemists, materials scientists, chemical and biomedical engineers, and physicians will facilitate the development of polymeric nanomedicine drug delivery at an unprecedented pace, and may eventually allow cancer therapy in a time-, tissue-, or even patient-specific manner.

ACKNOWLEDGMENTS

This work was supported by a start-up grant from University of Illinois, by a seed grant from Siteman Center for Cancer Nanotechnology Excellence (SCCNE, Washington University), Center for Nanoscale Science and Technology (CNST, University of Illinois), and by a Prostate Cancer Foundation Competitive Award. RT acknowledges the support of a student fellowship from SCCNE-CNST.

REFERENCES

1. Farokhzad, O.C.; Langer, R. Nanomedicine: Developing smarter therapeutic and diagnostic modalities. Adv. Drug Deliv. Rev. **2006**, *58* (14), 1456–1459.
2. Duncan, R. Polymer conjugates as anticancer nanomedicines. Nat. Rev. Cancer **2006**, *6* (9), 688–701.
3. Ferrari, M. Cancer nanotechnology: Opportunities and challenges. Nat. Rev. Cancer **2005**, *5* (3), 161–171.
4. Nishiyama, N.; Kataoka, K. Current state, achievements, and future prospects of polymeric micelles as nanocarriers for drug and gene delivery. Pharmacol. Ther. **2006**, *112* (3), 630–648.
5. Moghimi, S.M.; Hunter, A.C.; Murray, J.C. Nanomedicine: Current status and future prospects. FASEB J. **2005**, *19* (3), 311–330.
6. Marty, J.J.; Oppenheim, R.C.; Speiser, P. Nanoparticles—New colloidal drug delivery system. Pharm. Acta Helv. **1978**, *53* (1), 17–23.
7. Bangham, A.D.; Standish, M.M.; Watkins, J.C. Diffusion of univalent ions across the lamellae of swollen phospholipids. J. Mol. Biol. **1965**, *13*, 238–252.
8. Barenholz, Y. Liposome application: Problems and prospects. Curr. Opin. Colloid Interface Sci. **2001**, *6* (1), 66–77.
9. Gref, R.; Minamitake, Y.; Peracchia, M.; Trubetskoy, V.S.; Torchilin, V.P.; Langer, R. Biodegradable long-circulating polymeric nanospheres. Science **1994**, *263* (5153), 1600–1603.
10. Lavasanifar, A.; Samuel, J.; Kwon, G.S. Poly(ethylene oxide)-block-poly(L-amino acid) micelles for drug delivery. Adv. Drug Deliv. Rev. **2002**, *54* (2), 169–190.
11. Jemal, A.; Siegel, R.; Ward, E.; Murray, T.; Xu, J.; Smigal, C.; Thun, M.J. Cancer statistics, 2006. CA. Cancer J. Clin. **2006**, *56* (2), 106–130.
12. Park, J.W.; Benz, C.C.; Martin, F.J. Future directions of liposome- and immunoliposome-based cancer therapeutics. Semin. Oncol. **2004**, *31* (6 Suppl 13), 196–205.
13. Noble, C.O.; Kirpotin, D.B.; Hayes, M.E.; Mamot, C.; Hong, K.; Park, J.W.; Benz, C.C.; Marks, J.D.; Drummond, D.C. Development of ligand-targeted liposomes for cancer therapy. Expert Opin. Ther. Targets **2004**, *8* (4), 335–353.
14. Cattel, L.; Ceruti, M.; Dosio, F. From conventional to stealth liposomes a new frontier in cancer chemotherapy. Tumori **2003**, *89* (3), 237–249.
15. Patel, G.B.; Sprott, G.D. Archaeobacterial ether lipid liposomes (archaeosomes) as novel vaccine and drug delivery systems. Crit. Rev. Biotechnol. **1999**, *19* (4), 317–357.
16. Duncan, R.; Ringsdorf, H.; Satchi-Fainaro, R. Polymer therapeutics—Polymers as drugs, drug and protein conjugates and gene delivery systems: Past, present and future opportunities. J. Drug Target. **2006**, *14* (6), 337–341.
17. Duncan, R.; Ringsdorf, H.; Satchi-Fainaro, R. Polymer therapeutics: Polymers as drugs, drug and protein conjugates and gene delivery systems: Past, present and future opportunities. Adv. Polym. Sci. **2006**, *192*, 1–8.
18. Haag, R.; Kratz, F. Polymer therapeutics: Concepts and applications. Angew. Chem. Int. Ed. **2006**, *45* (8), 1198–1215.
19. Haag, R. Supramolecular drug-delivery systems based on polymeric core-shell architectures. Angew. Chem. Int. Ed. **2004**, *43* (3), 278–282.
20. Torchilin, V.P. Block copolymer micelles as a solution for drug delivery problems. Expert Opin. Ther. Pat. **2005**, *15* (1), 63–75.
21. Torchilin, V.P. Targeted polymeric micelles for delivery of poorly soluble drugs. Cell. Mol. Life Sci. **2004**, *61* (19–20), 2549–2559.
22. Kataoka, K.; Harada, A.; Nagasaki, Y. Block copolymer micelles for drug delivery: Design, characterization and biological significance. Adv. Drug Deliv. Rev. **2001**, *47* (1), 113–131.
23. Jones, M.C.; Leroux, J.C. Polymeric micelles—A new generation of colloidal drug carriers. Eur. J. Pharm. Biopharm. **1999**, *48* (2), 101–111.
24. Kwon, G.S.; Kataoka, K. Block-copolymer micelles as long-circulating drug vehicles. Adv. Drug Deliv. Rev. **1995**, *16* (2), 295–309.
25. Savic, R.; Eisenberg, A.; Maysinger, D. Block copolymer micelles as delivery vehicles of hydrophobic drugs: Micelle-cell interactions. J. Drug Target. **2006**, *14* (6), 343–355.
26. Nishiyama, N.; Kataoka, K. Nanostructured devices based on block copolymer assemblies for drug delivery: Designing structures for enhanced drug function. Adv. Polym. Sci. **2006**, *193*, 67–101.
27. Allen, T.M.; Cullis, P.R. Drug delivery systems: Entering the mainstream. Science **2004**, *303* (5665), 1818–1822.
28. Svenson, S.; Tomalia, D.A. Dendrimers in biomedical applications—Reflections on the field. Adv. Drug Deliv. Rev. **2005**, *57* (15), 2106–2129.
29. Lee, C.C.; MacKay, J.A.; Frechet, J.M.J.; Szoka, F.C. Designing dendrimers for biological applications. Nat. Biotechnol. **2005**, *23* (12), 1517–1526.
30. Quintana, A.; Raczka, E.; Piehler, L.; Lee, I.; Myc, A.; Majoros, I.; Patri, A.K.; Thomas, T.; Mule, J.; Baker, J.R. Design and function of a dendrimer-based therapeutic nanodevice targeted to tumor cells through the folate receptor. Pharm. Res. **2002**, *19* (9), 1310–1316.
31. Malik, N.; Evagorou, E.G.; Duncan, R. Dendrimer-platinate: A novel approach to cancer chemotherapy. Anticancer. Drugs **1999**, *10* (8), 767–776.
32. Lee, C.C.; Gillies, E.R.; Fox, M.E.; Guillaudeu, S.J.; Frechet, J.M.J.; Dy, E.E.; Szoka, F.C. A single dose of doxorubicin-functionalized bow-tie dendrimer cures mice bearing C-26 colon carcinomas. Proc. Natl. Acad. Sci. U.S.A. **2006**, *103* (45), 16649–16654.
33. Discher, D.E.; Ahmed, F. Polymersomes. Annu. Rev. Biomed. Eng. **2006**, *8*, 323–341.

Adhesives—Bioabsorbable

34. Discher, D.E.; Eisenberg, A. Polymer vesicles. Science 2002, 297 (5583), 967–973.

35. Holowka, E.P.; Sun, V.Z.; Kamei, D.T.; Deming, T.J. Polyarginine segments in block copolypeptides drive both vesicular assembly and intracellular delivery. Nat. Mater. 2006, 6 (1), 52–57.

36. Bellomo, E.G.; Wyrsta, M.D.; Pakstis, L.; Pochan, D.J.; Deming, T.J. Stimuli-responsive polypeptide vesicles by conformation-specific assembly. Nat. Mater. 2004, 3 (4), 244–248.

37. Photos, P.J.; Bacakova, L.; Discher, B.; Bates, F.S.; Discher, D.E. Polymer vesicles in vivo: Correlations with PEG molecular weight. J. Control. Release 2003, 90 (3), 323–334.

38. Yu, K.; Zhang, L.F.; Eisenberg, A. Novel morphologies of "crew-cut" aggregates of amphiphilic diblock copolymers in dilute solution. Langmuir 1996, 12 (25), 5980–5984.

39. Vauthey, S.; Santoso, S.; Gong, H.Y.; Watson, N.; Zhang, S.G. Molecular self-assembly of surfactant-like peptides to form nanotubes and nanovesicles. Proc. Natl. Acad. Sci. U.S.A. 2002, 99 (8), 5355–5360.

40. Gros, L.; Ringsdorf, H.; Schupp, H. Polymeric anti-tumor agents on a molecular and on a cellular-level. Angew. Chem. Int. Ed. Engl. 1981, 20 (4), 305–325.

41. Ringsdorf, H. Structure and properties of pharmacologically active polymers. J. Polym. Sci. C Polym. Symp. 1975, 51 (1), 135–153.

42. Duncan, R.; Kopecek, J. Soluble synthetic polymers as potential drug carriers. Adv. Polym. Sci. 1984, 57, 51–101.

43. Duncan, R. The dawning era of polymer therapeutics. Nat. Rev. Drug Discov. 2003, 2 (5), 347–360.

44. Maeda, H. SMANCS and polymer-conjugated macromolecular drugs: Advantages in cancer chemotherapy. Adv. Drug Deliv. Rev. 2001, 46 (1–3), 169–185.

45. Qiu, L.Y.; Bae, Y.H. Polymer architecture and drug delivery. Pharm. Res. 2006, 23 (1), 1–30.

46. Folkman, J.; Long, D.M. The use of silicone rubber as a carrier for prolonged drug therapy. Journal of Surg. Res. 1964, 4 (3), 139–142.

47. Langer, R.; Folkman, J. Polymers for sustained-release of proteins and other macromolecules. Nature 1976, 263 (5580), 797–800.

48. Westphal, M.; Hilt, D.C.; Bortey, E.; Delavault, P.; Olivares, R.; Warnke, P.C.; Whittle, I.R.; Jaaskelainen, J.; Ram, Z. A phase 3 trial of local chemotherapy with biodegradable carmustine (BCNU) wafers (Gliadel wafers) in patients with primary malignant glioma. Neuro Oncol. 2003, 5 (2), 79–88.

49. Brem, H.; Gabikian, P. Biodegradable polymer implants to treat brain tumors. J. Control. Release 2001, 74 (1), 63–67.

50. Brem, H.; Walter, K.A.; Langer, R. Polymers as controlled drug delivery devices for the treatment of malignant brain-tumors. Eur. J. Pharm. Biopharm. 1993, 39 (1), 2–7.

51. Galindo-Rodriguez, S.; Allemann, E.; Fessi, H.; Doelker, E. Physicochemical parameters associated with nanoparticle formation in the salting-out, emulsification-diffusion, and nanoprecipitation methods. Pharm. Res. 2004, 21 (8), 1428–1439.

52. Kakizawa, Y.; Kataoka, K. Block copolymer micelles for delivery of gene and related compounds. Adv. Drug Deliv. Rev. 2002, 54 (2), 203–222.

53. Discher, B.M.; Won, Y.Y.; Ege, D.S.; Lee, J.C.M.; Bates, F.S.; Discher, D.E.; Hammer, D.A. Polymersomes: Tough vesicles made from diblock copolymers. Science 1999, 284 (5417), 1143–1146.

54. Holowka, E.P.; Pochan, D.J.; Deming, T.J. Charged polypeptide vesicles with controllable diameter. J. Am. Chem. Soc. 2005, 127 (35), 12423–12428.

55. Lyczak, J.B.; Morrison, S.L. Biological and pharmacokinetic properties of a novel immunoglobulin-cd4 fusion protein. Arch. Virol. 1994, 139 (1–2), 189–196.

56. Davis, F.; Abuchowski, A.; Van Es, T.; Palczuk, N.; Chen, R.; Savoca, K.; Wieder, K. Enzyme-polyethylene glycol adducts: Modified enzymes with unique properties. In Enzyme Engineering; Springer: US, 1978; Vol. 4, 169.

57. Wieder, K.J.; Palczuk, N.C.; van Es, T.; Davis, F.F. Some properties of polyethylene glycol: Phenylalanine ammonia-lyase adducts. J. Biol. Chem. 1979, 254 (24), 12579–12587.

58. Wang, Y.S.; Youngster, S.; Grace, M.; Bausch, J.; Bordens, R.; Wyss, D.F. Structural and biological characterization of pegylated recombinant interferon alpha-2b and its therapeutic implications. Adv. Drug Deliv. Rev. 2002, 54 (4), 547–570.

59. Wagner, V.; Dullaart, A.; Bock, A.K.; Zweck, A. The emerging nanomedicine landscape. Nat. Biotechnol. 2006, 24 (10), 1211–1217.

60. Shaunak, S.; Godwin, A.; Choi, J.W.; Balan, S.; Pedone, E.; Vijayarangam, D.; Heidelberger, S.; Teo, I.; Zloh, M.; Brocchini, S. Site-specific PEGylation of native disulfide bonds in therapeutic proteins. Nat. Chem. Biol. 2006, 2 (6), 312–313.

61. Kolb, H.C.; Finn, M.G.; Sharpless, K.B. Click chemistry: Diverse chemical function from a few good reactions. Angew. Chem. Int. Ed. 2001, 40 (11), 2004–2021.

62. Demko, Z.P.; Sharpless, K.B. A click chemistry approach to tetrazoles by Huisgen 1,3-dipolar cycloaddition: Synthesis of 5-sulfonyl tetrazoles from azides and sulfonyl cyanides. Angew. Chem. Int. Ed. 2002, 41 (12), 2110–2113.

63. Manetsch, R.; Krasinski, A.; Radic, Z.; Raushel, J.; Taylor, P.; Sharpless, K.B.; Kolb, H.C. In situ click chemistry: Enzyme inhibitors made to their own specifications. J. Am. Chem. Soc. 2004, 126 (40), 12809–12818.

64. Sawa, M.; Hsu, T.L.; Itoh, T.; Sugiyama, M.; Hanson, S.R.; Vogt, P.K.; Wong, C.H. Glycoproteomic probes for fluorescent imaging of fucosylated glycans in vivo. Proc. Natl. Acad. Sci. U.S.A. 2006, 103 (33), 12371–12376.

65. Zhang, K.C.; Diehl, M.R.; Tirrell, D.A. Artificial polypeptide scaffold for protein immobilization. J. Am. Chem. Soc. 2005, 127 (29), 10136–10137.

66. Gopin, A.; Ebner, S.; Attali, B.; Shabat, D. Enzymatic activation of second-generation dendritic prodrugs: Conjugation of self-immolative dendrimers with poly(ethylene glycol) via click chemistry. Bioconjug. Chem. 2006, 17 (6), 1432–1440.

67. Beatty, K.E.; Xie, F.; Wang, Q.; Tirrell, D.A. Selective dye-labeling of newly synthesized proteins in bacterial cells. J. Am. Chem. Soc. 2005, 127 (41), 14150–14151.

68. Kwon, I.; Wang, P.; Tirrell, D.A. Design of a bacterial host for site-specific incorporation of p-bromophenylalanine into recombinant proteins. J. Am. Chem. Soc. 2006, 128 (36), 11778–11783.

69. Minko, T.; Paranjpe, P.V.; Qiu, B.; Lalloo, A.; Won, R.; Stein, S.; Sinko, P.J. Enhancing the anticancer efficacy of camptothecin using biotinylated poly (ethyleneglycol) conjugates in sensitive and multidrug-resistant human ovarian carcinoma cells. Cancer Chemother. Pharmacol. **2002,** *50* (2), 143–150.

70. Thomson, A.H.; Vasey, P.A.; Murray, L.S.; Cassidy, J.; Fraier, D.; Frigerio, E.; Twelves, C. Population pharmacokinetics in phase I drug development: A phase I study of PK1 in patients with solid tumours. Br. J. Cancer **1999,** *81* (1), 99–107.

71. Chandran, S.; Nan, A.; Ghandehari, H.; Denmeade, S.R. An HPMA-prodrug conjugate as a novel strategy for treatment of prostate cancer. Clin. Cancer Res. **2005,** *11* (24), 9004s–9004s.

72. Jelinkova, M.; Rihova, B.; Etrych, T.; Strohalm, J.; Ulbrich, K.; Kubackova, K.; Rozprimova, L. Antitumor activity of antibody-targeted HPMA copolymers of doxorubicin in experiment and clinical practice. Clin. Cancer Res. **2001,** *7* (11), 3677s–3678s.

73. Calolfa, V.R.; Zamai, M.; Ghiglieri, A.; Farao, M.; vande-Ven, M.; Gratton, E.; Castelli, M.G.; Suarato, A.; Geroni, A.C. *In vivo* biodistribution and antitumor activity of novel HPMA-copolymers of camptothecin. Clin. Cancer Res. **2000,** *6,* 4490s–4491s.

74. Greenwald, R.B.; Conover, C.D.; Choe, Y.H. Poly(ethylene glycol) conjugated drugs and prodrugs: A comprehensive review. Crit. Rev. Ther. Drug Carrier Syst. **2000,** *17* (2), 101–161.

75. Pendri, A.; Conover, C.D.; Greenwald, R.B. Antitumor activity of paclitaxel-2'-glycinate conjugated to poly(ethylene glycol): A water-soluble prodrug. Anticancer. Drug Des. **1998,** *13* (5), 387–395.

76. Conover, C.D.; Greenwald, R.B.; Pendri, A.; Gilbert, C.W.; Shum, K.L. Camptothecin delivery systems: Enhanced efficacy and tumor accumulation of camptothecin following its conjugation to polyethylene glycol via a glycine linker. Cancer Chemother. Pharmacol. **1998,** *42* (5), 407–414.

77. Conover, C.D.; Pendri, A.; Lee, C.; Gilbert, C.W.; Shum, K.L.; Greenwald, R.B. Camptothecin delivery systems: The antitumor activity of a camptothecin 20-0-polyethylene glycol ester transport form. Anticancer Res. **1997,** *17* (5A), 3361–3368.

78. Greenwald, R.B.; Pendri, A.; Conover, C.D.; Gilbert, C.W.; Yang, R.; Xia, J. Drug delivery systems. 2. Camptothecin 20-O-poly(ethylene glycol) ester transport forms. J. Med. Chem. **1996,** *39* (10), 1938–1940.

79. Bhatt, R.; Vries, P.D.; Tulinsky, J.; Bellamy, G.; Baker, B.; Singer, J.W.; Klein, P. Synthesis and *in vivo* antitumor activity of poly(L-glutamic acid) conjugates of 20(S)-camptothecin. J. Med. Chem. **2003,** *46* (1), 190–193.

80. Singer, J.W.; Bhatt, R.; Tulinsky, J.; Buhler, K.R.; Heasley, E.; Klein, P.; Vries, P.D. Water-soluble poly-(L-glutamic acid)-gly-camptothecinconjugates enhance camptothecin stability and efficacy *in vivo*. J. Control. Release **2001,** *74* (1–3), 243–247.

81. Li, C. Complete regression of well-established tumors using a novel water-soluble poly (L-glutamic acid)-paclitaxel conjugate. Cancer Res. **1998,** *58* (11), 2404–2409.

82. Zou, Y.; Wu, Q.; Tansey, W.; Chow, D.; Hung, M.; Charnsangavej, C.; Wallace, S.; Li, C. Effectiveness of water soluble poly(L-glutamic acid)-camptothecin conjugate against resistant human lung cancer xenografted in nude mice. Int. J. Oncol. **2001,** *18* (2), 331–336.

83. Mitsui, I.; Kumazawa, E.; Hirota, Y.; Aonuma, M.; Sugimori, M.; Ohsuki, S.; Uoto, K.; Ejima, A.; Terasawa, H.; Sato, K. A new water-soluble camptothecin derivative, Dx-8951f, exhibits potent antitumor-activity against human tumors *in vitro* and *in vivo*. Jpn. J. Cancer Res. **1995,** *86* (8), 776–782.

84. Ochi, Y.; Shiose, Y.; Kuga, H.; Kumazawa, E. A possible mechanism for the long-lasting antitumor effect of the macromolecular conjugate DE-310: Mediation by cellular uptake and drug release of its active camptothecin analog DX-8951. Cancer Chemother. Pharmacol. **2005,** *55* (4), 323–332.

85. Harada, M.; Sakakibara, H.; Yano, T.; Suzuki, T.; Okuno, S. Determinants for the drug release from T-0128, camptothecin analogue-carboxymethyl dextran conjugate. J. Control. Release **2000,** *69* (3), 399–412.

86. Harada, M.; Murata, J.; Sakamura, Y.; Sakakibara, H.; Okuno, S.; Suzuki, T. Carrier and dose effects on the pharmacokinetics of T-0128, a camptothecin analogue-carboxymethyl dextran conjugate, in non-tumor- and tumor-bearing rats. J. Control. Release **2001,** *71* (1), 71–86.

87. Cheng, J.; Khin, K.T.; Davis, M.E. Antitumor activity of beta-cyclodextrin polymer-camptothecin conjugates. Mol. Pharm. **2004,** *1* (3), 183–193.

88. Cheng, J.J.; Khin, K.T.; Jensen, G.S.; Liu, A.J.; Davis, M.E. Synthesis of linear, beta-cyclodextrin-based polymers and their camptothecin conjugates. Bioconjug. Chem. **2003,** *14* (5), 1007–1017.

89. Schluep, T.; Hwang, J.; Cheng, J.J.; Heidel, J.D.; Bartlett, D.W.; Hollister, B.; Davis, M.E. Preclinical efficacy of the camptothecin-polymer conjugate IT-101 in multiple cancer models. Clin. Cancer Res. **2006,** *12* (5), 1606–1614.

90. Schluep, T.; Cheng, J.J.; Khin, K.T.; Davis, M.E. Pharmacokinetics and biodistribution of the camptothecin-polymer conjugate IT-101 in rats and tumor-bearing mice. Cancer Chemother. Pharmacol. **2006,** *57* (5), 654–662.

91. Fuloria, J.; Abubakr, Y.; Perez, R.; Pike, M.; Zalupski, M. A phase I trial of [(1- 4)-linked b-D-mannopyranose]17-[(1-, 6)-linked-a-D-galactopyranose]10 (DAVANAT) co-administered with 5-fluorouracil, in patients with refractory solid tumors. J. Clin. Oncol. **2005,** *23* (16S), 3114.

92. Bolling, C.; Graefe, T.; Lubbing, C.; Jankevicius, F.; Uktveris, S.; Cesas, A.; Meyer-Moldenhauer, W.H.; Starkmann, H.; Weigel, M.; Burk, K.; Hanauske, A.R. Phase II study of MTX-HSA in combination with Cisplatin as first line treatment in patients with advanced or metastatic transitional cell carcinoma. Invest New Drug **2006,** *24* (6), 521–527.

93. Fossella, F.V.; Lippman, S.M.; Tarasoff, P. Phase I/II study of gemcitabine, an active agent for advanced non-small cell lung cancer (NSCLC). Proc. Am. Soc. Clin. Oncol. **1995,** *14,* 371.

94. Tolcher, A.; Forouzesh, B.; McCreery, H. A Phase I and pharmacokinetic study of BB10901, a maytansinoid

immunoconjugate. CD56 expressing tumors. Eur. J. Cancer **2002**, *38* (Suppl. 7), S152–S153.

95. Fossella, F.; McCann, J.; Tolcher, A.; Xie, H.; Hwang, L.L.; Carr, C.; Berg, K.; Fram, R. Phase II trial of BB-10901 (huN901-DM1) given weekly for four consecutive weeks every 6 weeks in patients with relapsed SCLC and CD56-positive small cell carcinoma. J. Clin. Oncol. (2005 ASCO Annual Meeting Proceedings) **2005**, *23* (16S), 7159.

96. Veronese, F.M.; Schiavon, O.; Pasut, G.; Mendichi, R.; Andersson, L.; Tsirk, A.; Ford, J.; Wu, G.F.; Kneller, S.; Davies, J.; Duncan, R. PEG-doxorubicin conjugates: Influence of polymer structure on drug release, *in vitro* cytotoxicity, biodistribution, and antitumor activity. Bioconjug. Chem. **2005**, *16* (4), 775–784.

97. Greenwald, R.B.; Pendri, A.; Conover, C.D.; Lee, C.; Choe, Y.H.; Gilbert, C.; Martinez, A.; Xia, J.; Wu, D.C.; Hsue, M. Camptothecin-20-PEG ester transport forms: The effect of spacer groups on antitumor activity. Bioorg. Med. Chem. **1998**, *6* (5), 551–562.

98. Conover, C.D.; Greenwald, R.B.; Pendri, A.; Shum, K.L. Camptothecin delivery systems: The utility of amino acid spacers for the conjugation of camptothecin with polyethylene glycol to create prodrugs. Anticancer. Drug Des. **1999**, *14* (6), 499–506.

99. Rowinsky, E.K.; Rizzo, J.; Ochoa, L.; Takimoto, C.H.; Forouzesh, B.; Schwartz, G.; Hammond, L.A.; Patnaik, A.; Kwiatek, J.; Goetz, A.; Denis, L.; McGuire, J.; Tolcher, A.W. A phase I and pharmacokinetic study of pegylated camptothecin as a 1-hour infusion every 3 weeks in patients with advanced solid malignancies. J. Clin. Oncol. **2003**, *21* (1), 148–157.

100. Greenwald, R.B.; Choe, Y.H.; McGuire, J.; Conover, C.D. Effective drug delivery by PEGylated drug conjugates. Adv. Drug Deliv. Rev. **2003**, *55* (2), 217–250.

101. Yu, D.S.; Peng, P.; Dharap, S.S.; Wang, Y.; Mehlig, M.; Chandna, P.; Zhao, H.; Filpula, D.; Yang, K.; Borowski, V.; Borchard, G.; Zhang, Z.H.; Minko, T. Antitumor activity of poly(ethylene glycol)-camptothecin conjugate: The inhibition of tumor growth *in vivo*. J. Control. Release **2005**, *110* (1), 90–102.

102. Maeda, H.; Wu, J.; Sawa, T.; Matsumura, Y.; Hori, K. Tumor vascular permeability and the EPR effect in macromolecular therapeutics: A review. J. Control. Release **2000**, *65* (1), 271–284.

103. Seymour, L.; Duncan, R.; Strohalm, J.; Kopecek, J. Effect of molecular weight (mw) of n-(2-hydroxypropyl) methacrylamide copolymers on body distribution and rate of excretion after subcutaneous, intraperitoneal, and intravenous administration to rats. J. Biomed. Mater. Res. **1987**, *21* (11), 1341–1358.

104. Duncan, R. Drug-polymer conjugates: Potential for improved chemotherapy. Anticancer. Drugs **1992**, *3* (3), 175–210.

105. Jorgensen, K.E.; Moller, J.V. Use of flexible polymers as probes of glomerular pore size. Am. J. Physiol. Renal Physiol. **1979**, *236* (2), 103–111.

106. Vasey, P.A.; Kaye, S.B.; Morrison, R.; Twelves, C.; Wilson, P.; Duncan, R.; Thomson, A.H.; Murray, L.S.; Hilditch, T.E.; Murray, T.; Burtles, S.; Fraier, D.; Frigerio, E.; Cassidy, J. Phase I clinical and pharmacokinetic study of PK1 [N-(2-hydroxypropyl)methacrylamide copolymer doxorubicin]: First

member of a new class of chemotherapeutic agents-drug-polymer conjugates. Cancer Research Campaign Phase I/II Committee. Clin. Cancer Res. **1999**, *5* (1), 83–94.

107. Terwogt, J.M.M.; Huinink, W.W.T.; Schellens, J.H.M.; Schot, M.; Mandjes, I.A.M.; Zurlo, M.G.; Rocchetti, M.; Rosing, H.; Koopman, F.J.; Beijnen, J.H. Phase I clinical and pharmacokinetic study of PNU166945, a novel water-soluble polymer-conjugated prodrug of paclitaxel. Anticancer. Drugs **2001**, *12* (4), 315–323.

108. Bissett, D.; Cassidy, J.; de Bono, J.S.; Muirhead, F.; Main, M.; Robson, L.; Fraier, D.; Magne, M.L.; Pellizzoni, C.; Porro, M.G.; Spinelli, R.; Speed, W.; Twelves, C. Phase I and pharmacokinetic (PK) study of MAG-CPT (PNU 166148): A polymeric derivative of camptothecin (CPT). Br. J. Cancer **2004**, *91* (1), 50–55.

109. Tomlinson, R.; Heller, J.; Brocchini, S.; Duncan, R. Poly-acetal-doxorubicin conjugates designed for pH-dependent degradation. Bioconjug. Chem. **2003**, *14* (6), 1096–1106.

110. Wachters, F.M.; Groen, H.J.M.; Maring, J.G.; Gietema, J.A.; Porro, M.; Dumez, H.; de Vries, E.G.E.; van Oosterom, A.T. A phase I study with MAG-camptothecin intravenously administered weekly for 3 weeks in a 4-week cycle in adult patients with solid tumours. Br. J. Cancer **2004**, *90* (12), 2261–2267.

111. Sarapa, N.; Britto, M.R.; Speed, W.; Jannuzzo, M.; Breda, M.; James, C.A.; Porro, M.; Rocchetti, M.; Wanders, A.; Mahteme, H.; Nygren, P. Assessment of normal and tumor tissue uptake of MAG-CPT, a polymer-bound prodrug of camptothecin, in patients undergoing elective surgery for colorectal carcinoma. Cancer Chemother. Pharmacol. **2003**, *52* (5), 424–430.

112. Seymour, L.W.; Ferry, D.R.; Anderson, D.; Hesslewood, S.; Julyan, P.J.; Poyner, R.; Doran, J.; Young, A.M.; Burtles, S.; Kerr, D.J. Hepatic drug targeting: Phase I evaluation of polymer-bound doxorubicin. J. Clin. Oncol. **2002**, *20* (6), 1668–1676.

113. Bouma, M.; Nuijen, B.; Stewart, D.R.; Rice, J.R.; Jansen, B.A.J.; Reedijk, J.; Bult, A.; Beijnen, J.H. Stability and compatibility of the investigational polymer-conjugated platinum anticancer agent AP 5280 in infusion systems and its hemolytic potential. Anticancer Drugs **2002**, *13* (9), 915–924.

114. Singer, J.W.; Shaffer, S.; Baker, B.; Bernareggi, A.; Stromatt, S.; Nienstedt, D.; Besman, M. Paclitaxel poliglumex (XYOTAX; CT-2103): An intracellularly targeted taxane. Anticancer Drugs **2005**, *16* (3), 243–254.

115. Singer, J.W. Paclitaxel poliglumex (XYOTAX, CT-2103): A macromolecular taxane. J. Control. Release **2005**, *109* (1), 120–126.

116. Singer, J.W.; Baker, B.; De Vries, P.; Kumar, A.; Shaffer, S.; Vawter, E.; Bolton, M.; Garzone, P. Poly-(L)-glutamic acid-paclitaxel (CT-2103) [XYOTAX (TM)], a biodegradable polymeric drug conjugate—Characterization, preclinical pharmacology, and preliminary clinical data. Adv. Exp. Med. Biol. **2003**, *519*, 81–99.

117. Sabbatini, P.; Aghajanian, C.; Dizon, D.; Anderson, S.; Dupont, J.; Brown, J.V.; Peters, W.A.; Jacobs, A.; Mehdi, A.; Rivkin, S.; Eisenfeld, A.J.; Spriggs, D. Phase II study of CT-2103 in patients with recurrent epithelial ovarian, fallopian tube, or primary peritoneal carcinoma. J. Clin. Oncol. **2004**, *22* (22), 4523–4531.

118. Davis, M.E.; Brewster, M.E. Cyclodextrin-based pharmaceutics: Past, present and future. Nat. Rev. Drug Discovery **2004**, *3* (12), 1023–1035.

119. Mocanu, G.; Vizitiu, D.; Carpov, A. Cyclodextrin polymers. J. Bioact. Compat. Polym. **2001**, *16* (4), 315–342.

120. Szeman, J.; Fenyvesi, E.; Szejtli, J.; Ueda, H.; Machida, Y.; Nagai, T. Water-soluble cyclodextrin polymers—Their interaction with Drugs. J. Incl. Phenom. **1987**, *5* (4), 427–431.

121. Gonzalez, H.; Hwang, S.J.; Davis, M.E. New class of polymers for the delivery of macromolecular therapeutics. Bioconjug. Chem. **1999**, *10* (6), 1068–1074.

122. Cheng, J.J.; Zeidan, R.; Mishra, S.; Liu, A.; Pun, S.H.; Kulkarni, R.P.; Jensen, G.S.; Bellocq, N.C.; Davis, M.E. Structure—Function correlation of chloroquine and analogues as transgene expression enhancers in nonviral gene delivery. J. Med. Chem. **2006**, *49* (22), 6522–6531.

123. Pun, S.H.; Tack, F.; Bellocq, N.C.; Cheng, J.J.; Grubbs, B.H.; Jensen, G.S.; Davis, M.E.; Brewster, M.; Janicot, M.; Janssens, B.; Floren, W.; Bakker, A. Targeted delivery of RNA-cleaving DNA enzyme (DNAzyme) to tumor tissue by transferrin-modified, cyclodextrin-based particles. Cancer Biol. Ther. **2004**, *3* (7), 641–650.

124. Bellocq, N.C.; Pun, S.H.; Jensen, G.S.; Davis, M.E. Transferrin-containing, cyclodextrin polymer-based particles for tumor-targeted gene delivery. Bioconjug. Chem. **2003**, *14* (6), 1122–1132.

125. Pun, S.H.; Davis, M.E. Development of a nonviral gene delivery vehicle for systemic application. Bioconjug. Chem. **2002**, *13* (3), 630–639.

126. Hwang, S.J.; Bellocq, N.C.; Davis, M.E. Effects of structure of beta-cyclodextrin-containing polymers on gene delivery. Bioconjug. Chem. **2001**, *12* (2), 280–290.

127. Danhauserriedl, S.; Hausmann, E.; Schick, H.D.; Bender, R.; Dietzfelbinger, H.; Rastetter, J.; Hanauske, A.R. Phase-I clinical and pharmacokinetic trial of dextran conjugated doxorubicin (Ad-70, Dox-Oxd). Invest. New Drugs **1993**, *11* (2–3), 187–195.

128. Inoue, K.; Kumazawa, E.; Kuga, H.; Susaki, H.; Masubuchi, N.; Kajimura, T. CM-dextran-polyalcohol-camptothecin conjugate: DE-310 with a novel carrier system and its preclinical data. Adv. Exp. Med. Biol. **2003**, *519*, 145–153.

129. Takimoto, C.H.; Lorusso, P.M.; Forero, L.; Schwartz, G.H.; Tolcher, A.W.; Patnaik, A.; Hammond, L.A.; Syed, S.; Simmons, C.; Ducharme, M.; De Jager, R.; Rowinsky, E.K. A phase I and pharmacokinetic study of DE-310 administered as a 3 hours infusion every 4 weeks (wks) to patients (pts) with advanced solid tumors or lymphomas. Clin. Cancer Res. **2003**, *9* (16), 6105S–6105S.

130. Kumazawa, E.; Ochi, Y. DE-310, a novel macromolecular carrier system for the camptothecin analog DX-8951f: Potent antitumor activities in various murine tumor models. Cancer Sci. **2004**, *95* (2), 168–175.

131. Tomlinson, R.; Klee, M.; Garrett, S.; Heller, J.; Duncan, R.; Brocchini, S. Pendent chain functionalized polyacetals that display pH-dependent degradation: A platform for the development of novel polymer therapeutics. Macromolecules **2002**, *35* (2), 473–480.

132. Vicent, M.J.; Tomlinson, R.; Brocchini, S.; Duncan, R. Polyacetal-diethylstilboestrol: A polymeric drug designed for pH-triggered activation. J. Drug Target. **2004**, *12* (8), 491–501.

133. Seymour, L.W.; Miyamoto, Y.; Maeda, H.; Brereton, M.; Strohalm, J.; Ulbrich, K.; Duncan, R. Influence of molecular-weight on passive tumor accumulation of a soluble macromolecular drug carrier. Eur. J. Cancer **1995**, *31A* (5), 766–770.

134. Noguchi, Y.; Wu, J.; Duncan, R.; Strohalm, J.; Ulbrich, K.; Akaike, T.; Maeda, H. Early phase tumor accumulation of macromolecules: A great difference in clearance rate between tumor and normal tissues. Jpn. J. Cancer Res. **1998**, *89* (3), 307–314.

135. Papisov, M.I.; Hiller, A.; Yurkovetskiy, A.; Yin, M.; Barzana, M.; Hillier, S.; Fischman, A.J. Semisynthetic hydrophilic polyals. Biomacromolecules **2005**, *6* (5), 2659–2670.

136. Yurkovetskiy, A.; Choi, S.W.; Hiller, A.; Yin, M.; McCusker, C.; Syed, S.; Fischman, A.J.; Papisov, M.I. Fully degradable hydrophilic polyals for protein modification. Biomacromolecules **2005**, *6* (5), 2648–2658.

137. Burger, A.M.; Hartung, G.; Stehle, G.; Sinn, H.; Fiebig, H.H. Pre-clinical evaluation of a methotrexate-albumin conjugate (MTX-HSA) in human tumor xenografts *in vivo*. Int. J. Cancer **2001**, *92* (5), 718–724.

138. Stehle, G.; Wunder, A.; Sinn, H.; Schrenk, H.H.; Schutt, S.; Frei, E.; Hartung, G.; Maier-Borst, W.; Heene, D.L. Pharmacokinetics of methotrexate-albumin conjugates in tumor-bearing rats. Anticancer Drugs **1997**, *8* (9), 835–844.

139. Hartung, G.; Stehle, G.; Sinn, H.; Wunder, A.; Schrenk, H.H.; Heeger, S.; Kranzle, M.; Edler, L.; Frei, E.; Fiebig, H.H.; Heene, D.L.; Maier-Borst, W.; Queisser, W. Phase I trial of methotrexate-albumin in a weekly intravenous bolus regimen in cancer patients. Phase I Study Group of the Association for Medical Oncology of the German Cancer Society. Clin. Cancer Res. **1999**, *5* (4), 753–759.

140. Kratz, F.; Warnecke, A.; Scheuermann, K.; Stockmar, C.; Schwab, J.; Lazar, P.; Druckes, P.; Esser, N.; Drevs, J.; Rognan, D.; Bissantz, C.; Hinderling, C.; Folkers, G.; Fichtner, I.; Unger, C. Probing the cysteine-34 position of endogenous serum albumin with thiol-binding doxorubicin derivatives. Improved efficacy of an acid-sensitive doxorubicin derivative with specific albumin-binding properties compared to that of the parent compound. J. Med. Chem. **2002**, *45* (25), 5523–5533.

141. Drevs, J.; Mross, K.; Kratz, F.; Medinger, M.; Unger, C. Phase I dose-escalation and pharmacokinetic (PK) study of a(6-maleimidocaproyl)hydrazone derivative of doxorubicin (DOXO-EMCH) in patients with advanced cancers. J. Clin. Oncol. **2004**, *22* (14), 158S–158S.

142. Fossella, F.V.; Tolcher, A.; Elliott, M.; Lambert, J.M.; Lu, R.; Zinner, R.; Lu, C.; Oh, Y.; Forouzesh, B.; McCreary, H. Phase I trial of the monoclonal antibody conjugate, BB-10901, for relapsed/refractory small cell lung cancer (SCLC) and other neuroendocrine (NE) tumors. Proc. Am. Soc. Clin. Oncol. **2002**, *21*, 309a.

143. Wu, A.M.; Senter, P.D. Arming antibodies: Prospects and challenges for immunoconjugates. Nat. Biotechnol. **2005**, *23* (9), 1137–1146.

144. Bagalkot, V.; Farokhzad, O.C.; Langer, R.; Jon, S. An aptamer-doxorubicin physical conjugate as a novel targeted drug-delivery platform. Angew. Chem. Int. Ed. **2006**, *45* (48), 8149–8152.

Adhesives—Bioabsorbable

145. Huang, H.Y.; Remsen, E.E.; Kowalewski, T.; Wooley, K.L. Nanocages derived from shell cross-linked micelle templates. J. Am. Chem. Soc. **1999**, *121* (15), 3805–3806.

146. Liu, H.B.; Jiang, A.; Guo, J.A.; Uhrich, K.E. Unimolecular micelles: Synthesis and characterization of amphiphilic polymer systems. J. Polym. Sci. A Polym. Chem. **1999**, *37* (6), 703–711.

147. Hagan, S.A.; Coombes, A.G.A.; Garnett, M.C.; Dunn, S.E.; Davis, M.C.; Illum, L.; Davis, S.S.; Harding, S.E.; Purkiss, S.; Gellert, P.R. Polylactide-poly(ethylene glycol) copolymers as drug delivery systems.1. Characterization of water dispersible micelle-forming systems. Langmuir **1996**, *12* (9), 2153–2161.

148. Kabanov, A.V.; Batrakova, E.V.; Alakhov, V.Y. Pluronic (R) block copolymers as novel polymer therapeutics for drug and gene delivery. J. Control. Release **2002**, *82* (2), 189–212.

149. Avgoustakis, K. Pegylated poly(lactide) and poly(lactide-co-glycolide) nanoparticles: Preparation, properties and possible application in drug delivery. Curr. Drug Deliv. **2004**, *1* (4), 321–333.

150. Kabanov, A.V.; Slepnev, V.I.; Kuznetsova, L.E.; Batrakova, E.V.; Alakhov, V.Y.; Melik-Nubarov, N.S.; Sveshnikov, P.G. Pluronic micelles as a tool for low-molecular compound vector delivery into a cell: Effect of Staphylococcus aureus enterotoxin B on cell loading with micelle incorporated fluorescent dye. Biochem. Int. **1992**, *26* (6), 1035–1042.

151. Moghimi, S.M.; Davis, S.S. Innovations in avoiding particle clearance from blood by kupffer cells—Cause for reflection. Crit. Rev. Ther. Drug Carrier Syst. **1994**, *11* (1), 31–59.

152. Cheng, J.; Teply, B.A.; Sherifi, I.; Sung, J.; Luther, G.; Gu, F.X.; Levy-Nissenbaum, E.; Radovic-Moreno, A.F.; Langer, R.; Farokhzad, O.C. Formulation of functionalized PLGA-PEG nanoparticles for *in vivo* targeted drug delivery. Biomaterials **2007**, *28* (5), 869–876.

153. Kataoka, K. Block copolymer micelles as vehicles for drug delivery. J. Control. Release **1993**, *24* (1), 119–132.

154. Yokoyama, M.; Sugiyama, T.; Okano, T.; Sakurai, Y.; Naito, M.; Kataoka, K. Analysis of micelle formation of an adriamycin-conjugated poly(ethylene glycol) poly(aspartic acid) block-copolymer by gel-permeation chromatography. Pharmaceut Res **1993**, *10* (6), 895–899.

155. Matsumura, Y.; Hamaguchi, T.; Ura, T.; Muro, K.; Yamada, Y.; Shimada, Y.; Shirao, K.; Okusaka, T.; Ueno, H.; Ikeda, M.; Watanabe, N. Phase I clinical trial and pharmacokinetic evaluation of NK911, a micelle-encapsulated doxorubicin. Br. J. Cancer **2004**, *91* (10), 1775–1781.

156. Yokoyama, M.; Satoh, A.; Sakurai, Y.; Okano, T.; Matsumura, Y.; Kakizoe, T.; Kataoka, K. Incorporation of water-insoluble anticancer drug into polymeric micelles and control of their particle size. J. Control. Release **1998**, *55* (2), 219–229.

157. Hamaguchi, T.; Matsumura, Y.; Suzuki, M.; Shimizu, K.; Goda, R.; Nakamura, I.; Nakatomi, I.; Yokoyama, M.; Kataoka, K.; Kakizoe, T. NK105, a paclitaxel-incorporating micellar nanoparticle formulation, can extend *in vivo* antitumour activity and reduce the neurotoxicity of paclitaxel. Br. J. Cancer **2005**, *92* (7), 1240–1246.

158. Nishiyama, N.; Okazaki, S.; Cabral, H.; Miyamoto, M.; Kato, Y.; Sugiyama, Y.; Nishio, K.; Matsumura, Y.; Kataoka, K. Novel cisplatin-incorporated polymeric micelles can eradicate solid tumors in mice. Cancer Res. **2003**, *63* (24), 8977–8983.

159. Sengupta, S.; Eavarone, D.; Capila, I.; Zhao, G.L.; Watson, N.; Kiziltepe, T.; Sasisekharan, R. Temporal targeting of tumour cells and neovasculature with a nanoscale delivery system. Nature **2005**, *436* (7050), 568–572.

160. Kim, S.C.; Kim, D.W.; Shim, Y.H.; Bang, J.S.; Oh, H.S.; Kim, S.W.; Seo, M.H. *In vivo* evaluation of polymeric micellar paclitaxel formulation: Toxicity and efficacy. J. Control. Release **2001**, *72* (1), 191–202.

161. Kim, T.Y.; Kim, D.W.; Chung, J.Y.; Shin, S.G.; Kim, S.C.; Heo, D.S.; Kim, N.K.; Bang, Y.J. Phase I and pharmacokinetic study of Genexol-PM, a cremophor-free, polymeric micelle-formulated paclitaxel, in patients with advanced malignancies. Clin. Cancer Res. **2004**, *10* (11), 3708–3716.

162. Brannon-Peppas, L.; Blanchette, J.O. Nanoparticle and targeted systems for cancer therapy. Adv. Drug Deliv. Rev. **2004**, *56* (11), 1649–1659.

163. Farokhzad, O.C.; Jon, S.Y.; Khadelmhosseini, A.; Tran, T.N.T.; LaVan, D.A.; Langer, R. Nanoparticle-aptamer bioconjugates: A new approach for targeting prostate cancer cells. Cancer Res. **2004**, *64* (21), 7668–7672.

164. Farokhzad, O.C.; Cheng, J.J.; Teply, B.A.; Sherifi, I.; Jon, S.; Kantoff, P.W.; Richie, J.P.; Langer, R. Targeted nanoparticle-aptamer bioconjugates for cancer chemotherapy *in vivo*. Proc. Natl. Acad. Sci. U.S.A. **2006**, *103* (16), 6315–6320.

165. Gillies, E.R.; Frechet, J.M.J. A new approach towards acid sensitive copolymer micelles for drug delivery. Chem. Comm. **2003**, (14), 1640–1641.

166. Bae, Y.; Fukushima, S.; Harada, A.; Kataoka, K. Design of environment-sensitive supramolecular assemblies for intracellular drug delivery: Polymeric micelles that are responsive to intracellular pH change. Angew. Chem. Int. Ed. **2003**, *42* (38), 4640–4643.

167. Goodwin, A.P.; Mynar, J.L.; Ma, Y.Z.; Fleming, G.R.; Fréchet, J.M.J. Synthetic micelle sensitive to IR light via a two-photon process. J. Am. Chem. Soc. **2005**, *127* (28), 9952–9953.

168. Ahmed, F.; Discher, D.E. Self-porating polymersomes of PEG-PLA and PEG-PCL: Hydrolysis-triggered controlled release vesicles. J. Control. Release **2004**, *96* (1), 37–53.

169. Geng, Y.; Dalhaimer, P.; Cai, S.; Tsai, R.; Tewari, M.; Minko, T.; Discher, D.E. Shape effects of filaments versus spherical particles in flow and drug delivery. Nat. Nanotechnol. **2007**, *2* (4), 249–255.

170. Bilati, U.; Allemann, E.; Doelker, E. Development of a nanoprecipitation method intended for the entrapment of hydrophilic drugs into nanoparticles. Eur. J. Pharm. Sci. **2005**, *24* (1), 67–75.

171. Peltonen, L.; Koistinen, P.; Hirvonen, J. Preparation of nanoparticles by the nanoprecipitation of low molecular weight poly(l)lactide. S.T.P. Pharm. Sci. **2003**, *13* (5), 299–304.

172. Govender, T.; Stolnik, S.; Garnett, M.C.; Illum, L.; Davis, S.S. PLGA nanoparticles prepared by nanoprecipitation: Drug loading and release studies of a water soluble drug. J. Control. Release **1999**, *57* (2), 171–185.

173. Lemieux, P.; Vinogradov, S.V.; Gebhart, C.L.; Guerin, N.; Paradis, G.; Nguyen, H.K.; Ochietti, B.; Suzdaltseva, Y.G.;

Bartakova, E.V.; Bronich, T.K.; St-Pierre, Y.; Alakhov, V.Y.; Kabanov, A.V. Block and graft copolymers and Nanogel (TM) copolymer networks for DNA delivery into cell. J. Drug Target. **2000**, *8* (2), 91–105.

174. Yu, S.Y.; Hu, J.H.; Pan, X.Y.; Yao, P.; Jiang, M. Stable and pH-sensitive nanogels prepared by self-assembly of chitosan and ovalbumin. Langmuir **2006**, *22* (6), 2754–2759.

175. Vinogradov, S.V.; Batrakova, E.V.; Kabanov, A.V. Nanogels for oligonucleotide delivery to the brain. Bioconjug. Chem. **2004**, *15* (1), 50–60.

176. Vinogradov, S.V.; Bronich, T.K.; Kabanov, A.V. Nanosized cationic hydrogels for drug delivery: Preparation, properties and interactions with cells. Adv. Drug Deliv. Rev. **2002**, *54* (1), 135–147.

177. Kabanov, A.V.; Batrakova, E.V. New technologies for drug delivery across the blood brain barrier. Curr. Pharm. Des. **2004**, *10* (12), 1355–1363.

178. Brigger, I.; Dubernet, C.; Couvreur, P. Nanoparticles in cancer therapy and diagnosis. Adv. Drug Deliv. Rev. **2002**, *54* (5), 631–651.

179. Otsuka, H.; Nagasaki, Y.; Kataoka, K. PEGylated nanoparticles for biological and pharmaceutical applications. Adv. Drug Deliv. Rev. **2003**, *55* (3), 403–419.

180. Panyam, J.; Labhasetwar, V. Biodegradable nanoparticles for drug and gene delivery to cells and tissue. Adv. Drug Deliv. Rev. **2003**, *55* (3), 329–347.

181. Liang, X.M.; Mao, G.Z.; Ng, K.Y.S. Effect of chain lengths of PEO-PPO-PEO on small unilamellar liposome morphology and stability: An AFM investigation. J. Colloid Interface Sci. **2005**, *285* (1), 360–372.

182. Yang, Z.G.; Yuan, J.J.; Cheng, S.Y. Self-assembling of biocompatible BAB amphiphilic triblock copolymers PLL(Z)-PEG-PLL(Z) in aqueous medium. Eur. Polym. J. **2005**, *41* (2), 267–274.

183. Yuan, J.J.; Li, Y.S.; Li, X.Q.; Cheng, S.Y.; Jiang, L.; Feng, L.X.; Fan, Z.Q. The "crew-cut" aggregates of polystyrene-b-poly(ethylene oxide)-b-polystyrene triblock copolymers in aqueous media. Eur. Polym. J. **2003**, *39* (4), 767–776.

184. Liu, F.T.; Eisenberg, A. Preparation and pH triggered inversion of vesicles from poly(acrylic acid)-block-polystyrene-block-poly(4-vinyl pyridine). J. Am. Chem. Soc. **2003**, *125* (49), 15059–15064.

185. Brannan, A.K.; Bates, F.S. ABCA tetrablock copolymer vesicles. Macromolecules **2004**, *37* (24), 8816–8819.

186. Discher, B.M.; Hammer, D.A.; Bates, F.S.; Discher, D.E. Polymer vesicles in various media. Curr. Opin. Colloid Interface Sci. **2000**, *5* (1), 125–131.

187. Fasman, G. *Prediction of Protein Structure and the Principles of Protein Conformation*; Plenum: New York, 1989.

188. Yu, M.; Nowak, A.P.; Deming, T.J.; Pochan, D.J. Methylated mono- and diethyleneglycol functionalized polylysines: Nonionic, alpha-helical, water-soluble polypeptides. J. Am. Chem. Soc. **1999**, *121* (51), 12210–12211.

189. Musumeci, T.; Ventura, C.A.; Giannone, I.; Ruozi, B.; Montenegro, L.; Pignatello, R.; Puglisi, G. PLA/PLGA nanoparticles for sustained release of docetaxel. Int. J. Pharm. **2006**, *325* (1), 172–179.

190. Frankel, A.D.; Pabo, C.O. Cellular uptake of the tat protein from human immunodeficiency virus. Cell **1988**, *55* (6), 1189–1193.

191. Buhleier, E.; Wehner, W.; Vogtle, F. Cascade-chain-like and nonskid-chain-like syntheses of molecular cavity topologies. Synthesis-Stuttgart **1978**, *1978* (2), 155–158.

192. Tomalia, D.A.; Baker, H.; Dewald, J.; Hall, M.; Kallos, G.; Martin, S.; Roeck, J.; Ryder, J.; Smith, P. A New class of polymers—Starburst-dendritic macromolecules. Polym. J. **1985**, *17* (1), 117–132.

193. Newkome, G.R.; Yao, Z.Q.; Baker, G.R.; Gupta, V.K. Micelles.1. Cascade molecules—A New approach to micelles—A [27]-Arborol. J. Org. Chem. **1985**, *50* (11), 2003–2004.

194. Aulenta, F.; Hayes, W.; Rannard, S. Dendrimers: A new class of nanoscopic containers and delivery devices. Eur. Polym. J. **2003**, *39* (9), 1741–1771.

195. Esfand, R.; Tomalia, D.A. Poly(amidoamine) (PAMAM) dendrimers: From biomimicry to drug delivery and biomedical applications. Drug Discov. Today **2001**, *6* (8), 427–436.

196. Hawker, C.J.; Frechet, J.M.J. Preparation of polymers with controlled molecular architecture—A new convergent approach to dendritic macromolecules. J. Am. Chem. Soc. **1990**, *112* (21), 7638–7647.

197. Hawker, C.; Frechet, J.M.J. A new convergent approach to monodisperse dendritic macromolecules. J. Chem. Soc. Chem. Commun. **1990**, *15*, 1010–1013.

198. Zeng, F.W.; Zimmerman, S.C. Dendrimers in supramolecular chemistry: From molecular recognition to self-assembly. Chem. Rev. **1997**, *97* (5), 1681–1712.

199. Zimmerman, S.C. Dendrimers in molecular recognition and self-assembly. Curr. Opin. Colloid Interface Sci. **1997**, *2* (1), 89–99.

200. Matthews, O.A.; Shipway, A.N.; Stoddart, J.F. Dendrimers—Branching out from curiosities into new technologies. Prog. Polym. Sci. **1998**, *23* (1), 1–56.

201. Bosman, A.W.; Janssen, H.M.; Meijer, E.W. About dendrimers: Structure, physical properties, and applications. Chem. Rev. **1999**, *99* (7), 1665–1688.

202. Fischer, M.; Vogtle, F. Dendrimers: From design to application—A progress report. Angew. Chem. Int. Ed. **1999**, *38* (7), 885–905.

203. Liang, C.O.; Frechet, J.M.J. Incorporation of functional guest molecules into an internally functionalizable dendrimer through olefin metathesis. Macromolecules **2005**, *38* (15), 6276–6284.

204. Tomalia, D.A. Birth of a new macromolecular architecture: Dendrimers as quantized building blocks for nanoscale synthetic polymer chemistry. Prog. Polym. Sci. **2005**, *30* (3), 294–324.

205. Yang, H.; Kao, W.Y.J. Dendrimers for pharmaceutical and biomedical applications. J. Biomater. Sci. Polym. Ed. **2006**, *17* (1–2), 3–19.

206. Liu, M.J.; Frechet, J.M.J. Designing dendrimers for drug delivery. Pharm. Sci. Tech. Today **1999**, *2* (10), 393–401.

207. Gillies, E.R.; Frechet, J.M.J. Dendrimers and dendritic polymers in drug delivery. Drug Discov. Today **2005**, *10* (1), 35–43.

208. Florence, A.T.; Hussain, N. Transcytosis of nanoparticle and dendrimer delivery systems: Evolving vistas. Adv. Drug Deliv. Rev. **2001**, *50* (Suppl. 1), S69–S89.

209. Patri, A.K.; Majoros, I.J.; Baker, J.R. Dendritic polymer macromolecular carriers for drug delivery. Curr. Opin. Chem. Biol. **2002**, *6* (4), 466–471.

210. Patri, A.K.; Kukowska-Latallo, J.F.; Baker, J.R. Targeted drug delivery with dendrimers: Comparison of the release kinetics of covalently conjugated drug and non-covalent drug inclusion complex. Adv. Drug Deliv. Rev. **2005**, *57* (15), 2203–2214.

211. Dufes, C.; Uchegbu, I.F.; Schatzlein, A.G. Dendrimers in gene delivery. Adv. Drug Deliv. Rev. **2005**, *57* (15), 2177–2202.

212. D'Emanuele, A.; Attwood, D. Dendrimer-drug interactions. Adv. Drug Deliv. Rev. **2005**, *57* (15), 2147–2162.

213. Kitchens, K.M.; El-Sayed, M.E.H.; Ghandehari, H. Transepithelial and endothelial transport of poly (amido-amine) dendrimers. Adv. Drug Deliv. Rev. **2005**, *57* (15), 2163–2176.

214. Jang, W.D.; Kataoka, K. Bioinspired applications of functional dendrimers. J. Drug Deliv. Sci. Technol. **2005**, *15* (1), 19–30.

215. Kukowska-Latallo, J.F.; Candido, K.A.; Cao, Z.Y.; Nigavekar, S.S.; Majoros, I.J.; Thomas, T.P.; Balogh, L.P.; Khan, M.K.; Baker, J.R. Nanoparticle targeting of anticancer drug improves therapeutic response in animal model of human epithelial cancer. Cancer Res. **2005**, *65* (12), 5317–5324.

216. Kramer, M.; Stumbe, J.F.; Turk, H.; Krause, S.; Komp, A.; Delineau, L.; Prokhorova, S.; Kautz, H.; Haag, R. pH-responsive molecular nanocarriers based on dendritic core-shell architectures. Angew. Chem. Int. Ed. **2002**, *41* (22), 4252–4256.

217. Tibben, M.M.; Rademaker-Lakhai, J.M.; Rice, J.R.; Stewart, D.R.; Schellens, J.H.M.; Beijnen, J.H. Determination of total platinum in plasma and plasma ultrafiltrate, from subjects dosed with the platinum-containing N-(2-hydroxypropyl)methacrylamide copolymer AP5280, by use of graphite-furnace Zeeman atomic-absorption spectrometry. Anal. Bioanal. Chem. **2002**, *373* (4–5), 233–236.

218. Rademaker-Lakhai, J.M.; Terret, C.; Howell, S.B.; Baud, C.M.; De Boer, R.; Beijnen, J.H.; Schellens, J.H. M.; Droz, J.P. A Phase I study of the platinum (Pt) polymer AP5280 as an intravenous infusion once every three weeks in patients with a solid tumor. Br. J. Clin. Pharmacol. **2003**, *56* (4), 469–469.

219. Lin, X.; Zhang, Q.; Rice, J.R.; Stewart, D.R.; Nowotnik, D.P.; Howell, S.B. Improved targeting of platinum chemotherapeutics: The antitumour activity of the HPMA copolymer platinum agent AP5280 in murine tumour models. Eur. J. Cancer **2004**, *40* (2), 291–297.

220. Rademaker-Lakhai, J.M.; Terret, C.; Howell, S.B.; Baud, C.M.; de Boer, R.F.; Pluim, D.; Beijnen, J.H.; Schellens, J.H.M.; Droz, J.P. A phase I and pharmacological study of the platinum polymer AP5280 given as an intravenous infusion once every 3 weeks in patients with solid tumors. Clin. Cancer Res. **2004**, *10* (10), 3386–3395.

221. Rice, J.R.; Gerberich, J.L.; Nowotnik, D.P.; Howell, S.B. Preclinical efficacy and pharmacokinetics of AP5346, a novel diaminocyclohexane-platinum tumor-targeting drug delivery system. Clin. Cancer Res. **2006**, *12* (7), 2248–2254.

222. Nemunaitis, J.; Cunningham, C.; Senzer, N.; Gray, M.; Oldham, F.; Pippen, J.; Mennel, R.; Eisenfeld, A. Phase I study of CT-2103, a polymer-conjugated paclitaxel, and carboplatin in patients with advanced solid tumors. Cancer Invest. **2005**, *23* (8), 671–676.

223. Langer, C.J. CT-2103: A novel macromolecular taxane with potential advantages compared with conventional taxanes. Clin. Lung Cancer **2004**, *6* (Suppl 2), S85–88.

224. Langer, C.J. CT-2103: Emerging utility and therapy for solid tumours. Expert. Opin. Investig. Drugs **2004**, *13* (11), 1501–1508.

225. Graefe, T.; Bolling, C.; Lubbing, C.; Latz, J.; Blatter, J.; Hanauske, A. Pemetrexed in combination with paclitaxel: A phase I clinical and pharmacokinetic trial in patients with solid tumors. J. Clin. Oncol. **2006**, *24* (18), 91S–91S.

226. Graefe, T.; Lubbing, C.; Bolling, C.; Muller-Hagen, S.; Leisner, B.; Fleet, J.; Ludtke, F.E.; Blatter, J.; Suri, A.; Hanauske, A.R. Phase I study of pemetrexed plus paclitaxel in patients with solid tumor. J. Clin. Oncol. **2004**, *22* (14), 152S–152S.

227. Bolling, C.; Lubbing, C.; Graefe, T.; Mack, S.; Von Scheel, J.; Muller-Hagen, S.; Blatter, J.; Depenbrock, H.; Ohnmacht, U.; Hanauske, A.R. Pemetrexed/gemcitabine/cisplatin: Phase I trial in patients with solid tumors. J. Clin. Oncol. **2004**, *22* (14), 218S–218S.

228. Holen, K.D.; Bolling, C.; Saltz, L.B.; Graefe, T.; Ty, V.; Sogo, N.; Starkmann, H.; Burk, K.H.; Hollywood, E.; Hanauske, A.R. Phase I and pharimacokinetic trial of a novel inhibitor of NAD biosynthesis in patients with solid tumors. Clin. Cancer Res. **2003**, *9*, 6158S–6158S.

229. Kwon, G.; Suwa, S.; Yokoyama, M.; Okano, T.; Sakurai, Y.; Kataoka, K. Enhanced tumor accumulation and prolonged circulation times of micelle-forming poly(ethylene oxide-aspartate) block copolymer-adriamycin conjugates. J. Control. Release **1994**, *29* (1), 17–23.

230. Kataoka, K.; Matsumoto, T.; Yokoyama, M.; Okano, T.; Sakurai, Y.; Fukushima, S.; Okamoto, K.; Kwon, G.S. Doxorubicin-loaded poly(ethylene glycol)-poly(beta-benzyl-l-aspartate) copolymer micelles: Their pharmaceutical characteristics and biological significance. J. Control. Release **2000**, *64* (1–3), 143–153.

231. Yokoyama, M.; Kwon, G.S.; Okano, T.; Sakurai, Y.; Naito, M.; Kataoka, K. Influencing factors on *in vitro* micelle stability of adriamycin-block copolymer conjugates. J. Control. Release **1994**, *28* (1), 59–65.

232. Kwon, G.S.; Yokoyama, M.; Okano, T.; Sakurai, Y.; Kataoka, K. Enhanced tumor accumulation and prolonged circulation times of micelle-forming poly(ethylene oxide-aspartate) block copolymer-adriamycin conjugates. J. Control. Release **1994**, *28* (1), 334–335.

233. Yokoyama, M.; Fukushima, S.; Uehara, R.; Okamoto, K.; Kataoka, K.; Sakurai, Y.; Okano, T. Characterization of physical entrapment and chemical conjugation of adriamycin in polymeric micelles and their design for *in vivo* delivery to a solid tumor. J. Control. Release **1998**, *50* (1), 79–92.

234. Li, Y.; Kwon, G.S. Methotrexate esters of poly(ethylene oxide)-block-poly(2-hydroxyethyl-L-aspartamide). Part I: Effects of the level of methotrexate conjugation on the stability of micelles and on drug release. Pharm. Res. **2000**, *17* (5), 607–611.

235. La, S.B.; Okano, T.; Kataoka, K. Preparation and characterization of the micelle-forming polymeric drug indomethacin-incorporated poly(ethylene oxide)-poly(beta-benzyl L-aspartate) block copolymer micelles. J. Pharm. Sci. **1996**, *85* (1), 85–90.

236. Yu, B.G.; Okano, T.; Kataoka, K.; Sardari, S.; Kwon, G.S. *In vitro* dissociation of antifungal efficacy and toxicity for amphotericin B-loaded poly(ethylene oxide)-block-poly(beta-benzyl-L-aspartate) micelles. J. Control. Release **1998**, *56* (1–3), 285–291.

237. Adams, M.L.; Kwon, G.S. Relative aggregation state and hemolytic activity of amphotericin B encapsulated by poly(ethylene oxide)-block-poly(N-hexyl-L-aspartamide)-acyl conjugate micelles: Effects of acyl chain length. J. Control. Release **2003**, *87* (1–3), 23–32.

238. Lavasanifar, A.; Samuel, J.; Kwon, G.S. The effect of fatty acid substitution on the *in vitro* release of amphotericin B from micelles composed of poly(ethylene oxide)-block-poly(N-hexyl stearate-L-aspartamide). J. Control. Release **2002**, *79* (1–3), 165–172.

239. Lavasanifar, A.; Samuel, J.; Kwon, G.S. Micelles self-assembled from poly(ethylene oxide)-block-poly(N-hexyl stearate L-aspartamide) by a solvent evaporation method: Effect on the solubilization and haemolytic activity of amphotericin B. J. Control. Release **2001**, *77* (1–2), 155–160.

240. Nishiyama, N.; Kataoka, K. Preparation and characterization of size-controlled polymeric micelle containing cis-dichlorodiammineplatinum(II) in the core. J. Control. Release **2001**, *74* (1–3), 83–94.

241. Mizumura, Y.; Matsumura, Y.; Hamaguchi, T.; Nishiyama, N.; Kataoka, K.; Kawaguchi, T.; Hrushesky, W.J.M.; Moriyasu, F.; Kakizoe, T. Cisplatin-incorporated polymeric micelles eliminate nephrotoxicity, while maintaining antitumor activity. Jpn. J. Cancer Res. **2001**, *92* (3), 328–336.

242. Jeong, Y.I.; Cheon, J.B.; Kim, S.H.; Nah, J.W.; Lee, Y.M.; Sung, Y.K.; Akaike, T.; Cho, C.S. Clonazepam release from core-shell type nanoparticles *in vitro*. J. Control. Release **1998**, *51* (2–3), 169–178.

243. Lee, E.S.; Na, K.; Bae, Y.H. Doxorubicin loaded pH-sensitive polymeric micelles for reversal of resistant MCF-7 tumor. J. Control. Release **2005**, *103* (2), 405–418.

244. Lee, E.S.; Na, K.; Bae, Y.H. Super pH-sensitive multifunctional polymeric micelle. Nano Lett. **2005**, *5* (2), 325–329.

245. Bogdanov, A.A.; Martin, C.; Bogdanova, A.V.; Brady, T.J.; Weissleder, R. An adduct of cis-diamminedichloroplatinum (II) and poly(ethylene glycol)poly(L-lysine)-succinate: Synthesis and cytotoxic properties. Bioconjug. Chem. **1996**, *7* (1), 144–149.

246. Shin, I.L.G.; Kim, S.Y.; Lee, Y.M.; Cho, C.S.; Sung, Y.K. Methoxy poly(ethylene glycol) epsilon-caprolactone amphiphilic block copolymeric micelle containing indomethacin. I. Preparation and characterization. J. Control. Release **1998**, *51* (1), 1–11.

247. Kim, S.Y.; Shin, I.L.G.; Lee, Y.M.; Cho, C.S.; Sung, Y.K. Methoxy poly(ethylene glycol) and epsilon-caprolactone amphiphilic block copolymeric micelle containing indomethacin. II. Micelle formation and drug release behaviours. J. Control. Release **1998**, *51* (1), 13–22.

248. Allen, C.; Han, J.N.; Yu, Y.S.; Maysinger, D.; Eisenberg, A. Polycaprolactone-b-poly(ethylene oxide) copolymer micelles as a delivery vehicle for dihydrotestosterone. J. Control. Release **2000**, *63* (3), 275–286.

249. Allen, C.; Yu, Y.S.; Maysinger, D.; Eisenberg, A. Polycaprolactone-b-poly(ethylene oxide) block copolymer micelles as a novel drug delivery vehicle for neurotrophic agents FK506 and L-685,818. Bioconjug. Chem. **1998**, *9* (5), 564–572.

250. Ge, H.X.; Hu, Y.; Jiang, X.Q.; Cheng, D.M.; Yuan, Y.Y.; Bi, H.; Yang, C.Z. Preparation, characterization, and drug release behaviors of drug nimodipine-loaded poly(epsilon-caprolactone)-poly(ethylene oxide)-poly(epsilon-caprolactone) amphiphilic triblock copolymer micelles. J. Pharm. Sci. **2002**, *91* (6), 1463–1473.

251. Burt, H.M.; Zhang, X.C.; Toleikis, P.; Embree, L.; Hunter, W.L. Development of copolymers of poly(D,L-lactide) and methoxypolyethylene glycol as micellar carriers of paclitaxel. Colloid Surface B **1999**, *16* (1–4), 161–171.

252. Zhang, X.C.; Jackson, J.K.; Burt, H.M. Development of amphiphilic diblock copolymers as micellar carriers of taxol. Int. J. Pharm. **1996**, *132* (1–2), 195–206.

253. Dong, Y.C.; Feng, S.S. Methoxy poly(ethylene glycol)-poly(lactide) (MPEG-PLA) nanoparticles for controlled delivery of anticancer drugs. Biomaterials **2004**, *25* (14), 2843–2849.

254. Liu, L.; Li, C.X.; Li, X.C.; Yuan, Z.; An, Y.L.; He, B.L. Biodegradable polylactide/poly(ethylene glycol)/polylactide triblock copolymer micelles as anticancer drug carriers. J. Appl. Polym. Sci. **2001**, *80* (11), 1976–1982.

255. Yoo, H.S.; Park, T.G. Biodegradable polymeric micelles composed of doxorubicin conjugated PLGA-PEG block copolymer. J. Control. Release **2001**, *70* (1–2), 63–70.

256. Mu, L.; Feng, S.S. A novel controlled release formulation for the anticancer drug paclitaxel (Taxol(R)): PLGA nanoparticles containing vitamin E TPGS. J. Control. Release **2003**, *86* (1), 33–48.

257. Alakhov, V.Y.; Moskaleva, E.Y.; Batrakova, E.V.; Kabanov, A.V. Hypersensitization of multidrug resistant human ovarian carcinoma cells by pluronic P85 block copolymer. Bioconjug. Chem. **1996**, *7* (2), 209–216.

258. Venne, A.; Li, S.M.; Mandeville, R.; Kabanov, A.; Alakhov, V. Hypersensitizing effect of pluronic L61 on cytotoxic activity, transport, and subcellular distribution of doxorubicin in multiple drug-resistant cells. Cancer Res. **1996**, *56* (16), 3626–3629.

259. Batrakova, E.V.; Li, S.; Miller, D.W.; Kabanov, A.V. Pluronic P85 increases permeability of a broad spectrum of drugs in polarized BBMEC and Caco-2 cell monolayers. Pharm. Res. **1999**, *16* (9), 1366–1372.

260. Batrakova, E.V.; Miller, D.W.; Li, S.; Alakhov, V.Y.; Kabanov, A.V.; Elmquist, W.F. Pluronic P85 enhances the delivery of digoxin to the brain: *In vitro* and *in vivo* studies. J. Pharmacol. Exp. Ther. **2001**, *296* (2), 551–557.

261. Aramwit, P.; Yu, B.G.; Lavasanifar, A.; Samuel, J.; Kwon, G.S. The effect of serum albumin on the aggregation state and toxicity of amphotericin B. J. Pharm. Sci. **2000**, *89* (12), 1589–1593.

262. Jiang, J.Q.; Tong, X.; Zhao, Y. A new design for light-breakable polymer micelles. J. Am. Chem. Soc. **2005**, *127* (23), 8290–8291.

263. Li, Y.Y.; Zhang, X.Z.; Cheng, H.; Kim, G.C.; Cheng, S.X.; Zhuo, R.X. Novel stimuli-responsive micelle self-assembled from Y-shaped P(UA-Y-NIPAAm) copolymer for drug delivery. Biomacromolecules **2006**, *7* (11), 2956–2960.

264. Nakayama, M.; Okano, T.; Miyazaki, T.; Kohori, F.; Sakai, K.; Yokoyama, M. Molecular design of biodegradable polymeric micelles for temperature-responsive drug release. J. Control. Release **2006**, *115* (1), 46–56.

265. Wang, C.H.; Wang, C.H.; Hsiue, G.H. Polymeric micelles with a pH-responsive structure as intracellular drug carriers. J. Control. Release **2005**, *108* (1), 140–149.

266. Wang, C.; Ge, Q.; Ting, D.; Nguyen, D.; Shen, H.R.; Chen, J.Z.; Eisen, H.N.; Heller, J.; Langer, R.; Putnam, D. Molecularly engineered poly(ortho ester) microspheres for enhanced delivery of DNA vaccines. Nat. Mater. **2004**, *3* (3), 190–196.

267. Lynn, D.M.; Amiji, M.M.; Langer, R. pH-responsive polymer microspheres: Rapid release of encapsulated material within the range of intracellular pH. Angew. Chem. Int. Ed. **2001**, *40* (9), 1707–1710.

268. Kim, M.S.; Hwang, S.J.; Han, J.K.; Choi, E.K.; Park, H.J.; Kim, J.S.; Lee, D.S. pH-responsive PEG-poly(beta-amino ester) block copolymer micelles with a sharp transition. Macromol. Rapid. Commun. **2006**, *27* (6), 447–451.

269. Heffernan, M.J.; Murthy, N. Polyketal nanoparticles: A new pH-sensitive biodegradable drug delivery vehicle. Bioconjug. Chem. **2005**, *16* (6), 1340–1342.

Anti-Infective Biomaterials

W. Kohnen
B. Jansen
Institute of Medical Microbiology and Hygiene, University of Cologne, Cologne, Germany

Abstract

Today polymers are essential in many fields of medicine. The benefits of these materials (biomaterials) are beyond doubt. However, aside from complications like mechanical irritation and thrombosis, infection of the materials (foreign-body infection) is the main problem associated with their use.

Today polymers are essential in many fields of medicine. The benefits of these materials (biomaterials) are beyond doubt. However, aside from complications like mechanical irritation and thrombosis, infection of the materials (foreign-body infection) is the main problem associated with their use.[1–3]

The main causative organisms of these so-called foreign-body infections are coagulase-negative *staphylococci*.[4] In most cases, antimicrobial therapy alone cannot cure the infection, and removal of the biomaterial becomes necessary.[5] Along with severe consequences for the individual patient, foreign-body infections lead to prolongation of the hospital stay and thus to an increase in therapy costs. Sugarman and Young[6] has estimated that the cost for therapy of a patient (in the U.S.) with an infected orthopedic implant increases by 400–600%, and more than a billion U.S. dollars are spent each year for the treatment of infected artificial joints. Therefore it is of outstanding importance to prevent this kind of infection.

HOW TO REDUCE FOREIGN-BODY INFECTIONS

Bacterial adhesion is the first step in the pathogenesis of foreign-body infections. After settling to the biomaterial, many bacterial strains (e.g., coagulase-negative staphylococci) are able to produce extracellular substances ("slime") that protect them against host defense mechanisms and antibiotics.[7–10]

Prevention of bacterial adherence or at least significant lowering of the number of viable adhering bacteria seems to be the most effective way to avoid foreign-body infections. There are two principal approaches to preventing foreign-body infections by influencing the interaction between biomaterial and bacterium:

- Development of polymers (polymer surfaces) with anti-adhesive properties

- Development of polymers (polymer surfaces) with anti-microbial properties

For effective development of anti-infective medical devices, detailed knowledge about the interactions between the surface of a biomaterial and bacterial adherence is a prerequisite. Therefore, many scientists have tried to describe bacterial adherence in dependence of surface properties of a biomaterial and to establish physico-chemical models for the adhesion process.

Theoretical Background

From the physico-chemical point of view, bacterial adhesion to polymers can be regarded as the adhesion of a particle to a solid surface in a liquid environment. There are two models describing bacterial adhesion in the case of non-specific interactions:

- A thermodynamic model based on surface tensions
- The DLVO-theory, where additional electrostatic interactions are considered

THERMODYNAMIC MODEL

Provided electrostatic interactions can be neglected, adhesion is controlled by a change of free enthalpy of adhesion (ΔG). Adhesion is thermodynamically favored if the free enthalpy of adhesion becomes negative. With bacteria, ΔG depends on the interfacial tensions between solid and bacteria, γ_{SB}, solid and liquid medium, γ_{SL}, and bacteria and liquid medium, γ_{BL} (Eq. 1).[11]

$$\Delta G = \gamma_{SB} - \gamma_{SL} - \gamma_{SL} \tag{1}$$

There are three approaches to calculating the interfacial tensions: namely, the "equation of state" approach,[12] the

Concise Encyclopedia of Biomedical Polymers and Polymeric Biomaterials DOI: 10.1081/E-EBPPC-120051875

Table 1 Studies showing a correlation between bacterial adhesion and the surface tension or hydrophilicity of the biomaterial

Microorganism	Biomaterial	Reference
Staphylococci	Teflon, PE, PC	[21]
Lactobacillus	FEP, PS, PET, sulfonated PC, silicone	[22]
Pseudomonas	Silicone, glass	[23]
Escherichia	PMMA	[24]

Table 2 Studies showing no correlation between bacterial adhesion and the surface tension or hydrophilicity of the biomaterial

Microorganism	Biomaterial	Reference
Staphylococci	Teflon, PE, PC, PET, silicone, cellulose acetate	[25]
	FEP, PS, PET, sulfonated PC, glass, silicone	[22]

Table 3 Studies showing a correlation between bacterial adhesion and surface charge

Microorganism	Biomaterial	Reference
Escherichia	PMMA with negatice charge	[24]
Staphylococci	PMMA	[31]
Escherichia	Anion exchange resins	[32]

Table 4 Studies showing no correlation between bacterial adhesion and surface charge

Microorganism	Biomaterial	Reference
Mischellaneous bacteria	Negatice charged PS	[33]
Streptococcus	Glass, covered with albumin	[34]
Streptococcus Actinomyces	Apatite minerals	[35]

"geometric mean" approach,[13–16] and the "harmonic mean" approach.[17,18] In the "equation-of-state" approach, interfacial tensions are determined by the surface tensions of the single media.

In the "geometric mean" and the "harmonic mean" approaches, interfacial tensions are calculated from surface tensions and their polar and disperse components.

Above all, the "harmonic mean"-approach is suitable to describe interfaces formed by hydrophilic surfaces.

Absolom et al.[19] and Neumann et al.[20] used the "equation of state" to calculate the adhesion behavior of bacteria. They could distinguish three cases that are dependent on the magnitudes of the individual surface tensions:

- The surface tension of the liquid is greater than the surface tension of the bacterium.⇒bacterial adhesion increases with increasing surface tension of the biomaterial.
- The surface tension of the liquid is lower than the surface tension of the bacterium.⇒bacterial adhesion decreases with increasing surface tension of the biomaterial.
- The surface tension of the liquid is equal to the surface tension of the bacterium.⇒bacterial adhesion is independent of the surface tension of the bacterium.

On the basis of these results, many authors investigated the correlation between adhesion and the surface tension of the biomaterial. In most of the studies, contact angle measurements were taken in order to calculate surface tensions and to describe hydrophilicity. In some cases, a correlation to the adhesion was found (Table 1);[21–24] in others, no correlation could be observed (Table 2).[22,25]

DLVO THEORY

Usually bacteria have a net negative surface charge. Electrostatic interactions must be discussed if the surface of the biomaterial is also charged, e.g., either by formation of an

electric double layer or by a possible surface charge of the polymer itself. In this case, the DLVO theory of colloid stabilization[26,27] can be applied. This theory—originally developed to describe the behavior of colloids in the presence of electrolytes—postulates that the total interaction is the sum of van der Waals interactions and electrostatic interactions. In this approach, the interaction is described as a function of the separation distance. Depending on the electrolyte concentration of the liquid medium, there are two minima in total free energy (i.e., specific distances where the bacterium is in a stable position). The secondary minimum is located at a longer range, typically at a distance of 5–8 µm, while the primary minimum is at a much shorter range. Short-range forces, dominating in the primary minimum, become effective only when the bacterium is able to overcome the energy wall between the secondary and primary minimum.

Bacteria captured in the secondary minimum (dominated by long-range forces) show reversible adhesion.[28] The stability of the minima influences the bacterial adhesion. The DLVO theory can be applied for a qualitative description of bacterial adhesion to biomaterials *in vitro*; quantitative calculations are difficult due to the problems in obtaining the necessary parameters in practice.[3,29,30] Therefore, in most cases only the influence of surface charge on bacterial adhesion was investigated. In some studies a correlation between bacterial adhesion and surface charge could be found (Table 3),[24,31,32] but there are also investigations showing no influence of surface charge (Table 4).[33–35]

POLYMERS WITH ANTIMICROBIAL PROPERTIES

Another strategy to prevent foreign-body infections than the development of "antiadhesive" surfaces by polymer modification is loading polymers with antimicrobial substances. As active agents antibiotics, metal salts, and disinfectants or detergents have been used. The aim is to achieve polymeric drug delivery systems which gradually release the antimicrobial agent and therefore are capable of preventing either primary adhesion of microorganisms or—more likely—colonization of an implant or medical device.

Well known and of clinically documented value is the system polymethylmethacrylate–Gentamicin (Septopal®) for the treatment of chronic osteomyelitis and soft tissue infection.[36,37] Antibiotics have also been incorporated in bone cement,[38,39] vascular prostheses,[40–42] ventricular shunts,[43] as well as in prosthetic heart valves.[44]

For the prevention of vascular graft infection, DACRON® prostheses have been treated with collagen impregnated with amikacin.[40] Rifampin has also been incorporated in vascular grafts exhibiting good performance in animal experiments, e.g., by mixing rifampin with blood used for preclotting of the graft.[41,45] TEFLON® prostheses have been loaded with oxacillin via benzalkoniumchloride.[42,46] Further investigations for the development of infection-resistant vascular grafts include coating polyester with tetracyclines, gentamicin, and rifampin.[47] Tobramycin has been bonded to TEFLON® grafts by simple soaking or via fibrin or cyanoacrylate glues.[48–50] Moore et al.[40] reported on a high infection resistance of vascular prostheses coated by a natural endothelial cell layer.

The number of studies dealing with the incorporation of antimicrobials into i.v. or urinary catheters is large. Diclox-acillin[51] and penicillin have been used, the latter via ionic bonding to tridodecylmethyl ammoniumchloride.[52] Urinary catheters have been loaded with the aminoglyco-sides dibekacin[53] or kanamycin.[54]

Antimicrobials other than "classic" antibiotics have also been investigated as potential candidates to generate anti-microbial activity in medical devices, e.g., chlorhexidine[55] and IRGASAN®.[56] Ruggieri and Abbott[57] have bonded heparin via tridodecylmethyl ammoniumchloride to TEF-LON® or vinyl catheters and found reduced bacterial adherence.

Because of their lack of cross-resistance to "classic" antibiotics and their broad antimicrobial activity against bacteria, fungi, viruses, and parasites, metals or metal salts have been studied. Silver especially has raised the interest of many investigators, perhaps because of its low toxicity and good biocompatibility. Benvenisty et al.[58] and Shah et al.[59] have developed PTFE-prostheses loaded with silver-oxacillin and silver amikacin or silver norfloxacin, respectively. Maki et al.[60] has reported on a central venous "silver" catheter with a better clinical performance than normal control catheters. Today several research groups and companies are working with catheters equipped with silver.

Other investigators have tried to produce polymers with "intrinsic" antimicrobial activity by covalently bonding anti-infectives to polymers.[61–67]

In our group we have also developed polymer systems with antimicrobial activity. In earlier studies, antibiotics like flucloxacillin, clindamycin, and vancomycin have been incorporated into polyurethanes.[68] A commercially available central venous catheter (HYDROCATH®) has been loaded with iodine[69] or teicoplanin.[70] The latter shows good activity to prevent catheter colonization by Gram-positive bacteria in vitro and in vivo. Incorporation of rifampin into silicone catheters used as ventricular shunts leads to catheters with excellent in vitro and in vivo performance.[71,72]

So far, most of the antimicrobial catheter or graft systems mentioned above have been investigated in vitro or in animal experiments. Only a few clinical studies with selected materials have been performed,[73,74] and the routine use of such catheters—if commercially available—is still uncommon. However, it seems that in the near future, more antimicrobial polymer systems will be brought onto the market and into clinical use.

CREATING POLYMERS SUITABLE FOR THE INVESTIGATION OF BACTERIAL ADHESION AND THE DEVELOPMENT OF ANTIMICROBIAL BIOMATERIALS: THE GLOW DISCHARGE TECHNIQUE

In order to investigate correlations between bacterial adhe-sion and biomaterial characteristics and whether the described physico-chemical models are suitable for a description of adhesion, polymers with different surface properties are needed. An elegant and versatile method to achieve polymers with varying surface properties is surface modification by the glow discharge technique, where the surface of a synthetic material is exposed to a glow dis-charge under reduced pressure.[75–78] With this method, the surface of most polymers can be modified in a great vari-ety of ways without changing bulk properties like mechan-ical stability and elasticity. Additionally, any medical device can be equipped with functional groups able to bind antibiotics.

Depending on the nature of the gas that is used in glow discharge treatment, the polymer surface may be modified in different ways. Modifications include etching, cross-linking, functionalization, plasma polymerization, and plasma-induced grafting (Fig. 1).

When non-polymerizable gases (e.g., oxygen, noble gases) are used, etching or crosslinking of the surface will occur. Functionalization with nitrogen or water vapor will cause the formation of new chemical groups in the surface. Treatment with polymerizable gases (e.g., organic com-pounds) results in the formation of a dense coating of a plasma-polymer on the polymer surface, i.e., layers with new properties are generated. Plasma-induced grafting, on the other hand, will lead to the formation of a coating consisting of mobile polymer chains. In this process, glow

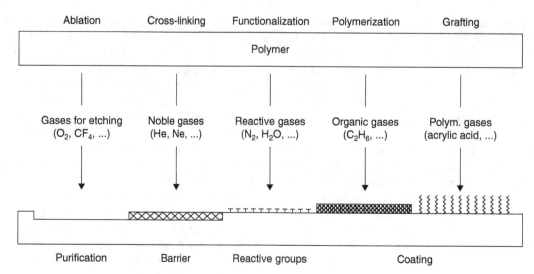

Fig. 1 Possible polymer surface modifications with the glow discharge technique.

discharge is used to generate radicals inside the surface of the synthetic material. Following the glow discharge reaction, a polymerizable monomer is transferred into the system and reacts with the polymer, thereby creating graft chains.

Preparation of New Polymers with the Glow Discharge Technique

As unmodified polymers, polyethylene, polypropylene, polyethyleneterephthalate (both Hoechst AG, Germany), polyetherurethane "Walopur 2201 U" (Wolff Walsrode, Germany), and the poly(tetraflouroethylene-*co*-hexafluoropropylene) "Teflon FEP" (Du Pont) were used.

To obtain modified polymers, the polyurethane was treated with the glow discharge technique. The polyurethane was extracted in ethanol and dried under reduced pressure prior to use. The plasma reactor (Softal, Hamburg, Germany) was designed for the treatment of films with a size of 100 mm × 100 mm. The high frequency generator used for glow discharge operates at a frequency of 20–30 kHz.

Plasma polymerization was done with gas mixtures of CO_2 and C_2H_4 or CO_2 and C_4F_8. In plasma-induced grafting, monomers like acrylic acid, butyl acrylate, and methyl vinyl acetamide were grafted onto polyurethane films that had been treated in a glow discharge with oxygen. The liquid monomers were distilled and stored under reduced pressure prior to use. After modification, the films were cleansed by water and finally dried under reduced pressure.

The surface properties were varied by the change of glow discharge conditions—e.g., gas-flow rates, pressure, or grafting time. For the preparation of silver-loaded antimicrobial polymers, acrylic-acid-grafted polyurethane was incubated for 30 minutes in a saturated aqueous

solution of silver nitrate to obtain a device capable of slowly releasing Ag^+-ions. After incubation, the sample was washed with small portions of water and dried under reduced pressure.

Influence of Surface Properties on Bacterial Adhesion

Contact-angle determination of a water droplet on the surface was used to characterize the hydrophilicity of the dry samples. In doing this, no correlation between bacterial adherence of *Staphylococcus epidermidis* (measured with the bioluminescence assay described by Ludwicka et al.[79]) and the observed water contact angle was found (Fig. 2).

The simple model of a correlation between contact angle (or surface tension) of a biomaterial and bacterial adherence, described previously, could not be proved by these results. Therefore the free enthalpy of adhesion was calculated with the harmonic-mean approach. Thereby a correlation between free enthalpy of adhesion and measured bacterial adherence could be found (Fig. 3). There is high adhesion for negative free enthalpy values decreasing with increasing free enthalpy. In the case of positive free enthalpy—i.e., the bacterium needs energy to adhere—adherence of *S. epidermidis* has a constant value of about 1%*ml/cm². From the physico-chemical standpoint, no adherence should occur if free enthalpy of adhesion is positive. But probably bacteria are able to overcome the energy barrier by specific interactions with the biomaterial. So these results showed that there seems to be a certain *minimum number* of adherent bacteria, independent from the free enthalpy of adhesion and the nature of the polymer surface. Thus, the free enthalpy calculated by the harmonic-mean approach is the parameter to describe bacterial adherence, and not the surface tension of the individual surface.

To investigate the effect of surface charge on bacterial adhesion, the difference between a charged surface and the same surface without charge but the same calculated enthalpy of adhesion has been examined. It was found that a negative surface charge leads to repulsion of bacteria in opposite to uncharged surfaces (Fig. 4). Interestingly, no change of bacterial adhesion in relation to surface charge density, measured by the adsorption of a cationic dye, could be observed (Fig. 5). All charged surfaces had adhesion values of about 1%*ml/cm². This was the same value as the minimum number of adherent bacteria found when studying the influence of free enthalpy of adhesion. Based on the above results, it is concluded that it is obviously impossible to create polymer surfaces by physico-chemical modification to which bacterial adhesion is zero!

To further decrease the number of adherent bacteria, antimicrobial surfaces were developed by binding

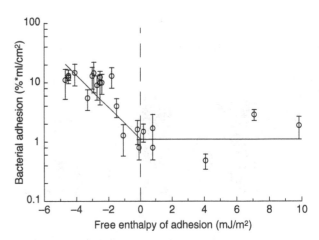

Fig. 3 Correlation between bacterial adhesion and calculated free enthalpy of adhesion.

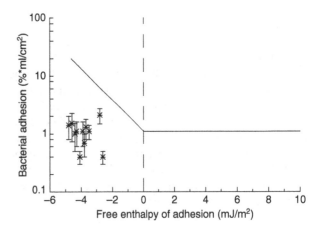

Fig. 4 Influence of surface charge on bacterial adhesion in dependence on calculated free enthalpy of adhesion. The line represents bacterial adhesion of surfaces without charge. The stars show measured bacterial adhesion on surfaces with negative charge.

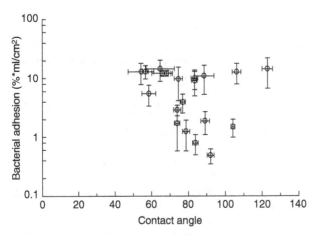

Fig. 2 Influence of the hydrophilicity of dry samples—measured as the contact angle of a water droplet—on bacterial adhesion.

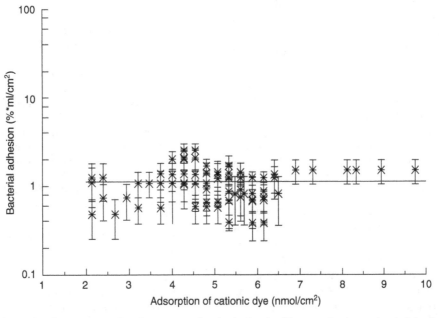

Fig. 5 Influence of charge density—measured as the amount of cationic dye the film can adsorb—on bacterial adhesion.

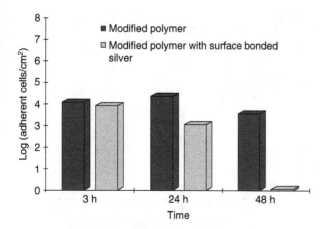

Fig. 6 Adherent viable bacteria in dependence on time of adhesion.

antimicrobial agents, e.g., silver ions, to modified, negatively charged polymers. The maximum silver uptake in the surface was about 23 µg/cm². From this, 3 µg/cm² was absorbed by the polyurethane film, and this part could be washed out with water. In the case of the modified film, the number of adherent viable bacteria did not change with adhesion time (Fig. 6). In contrast, the silver-containing surfaces showed many fewer viable cells after 24 hrs. After 48 hrs, no viable bacterium was detectable. This delay in activity is probably due to the diffusion of the silver ion through the bacterial cell wall and into the cell before it is able to act as an antimicrobial agent. Nevertheless, the silver-containing surfaces reduced adherent viable bacteria from 10^4 to zero.

APPLICATIONS

Modified polymers generated by the glow discharge technique are very helpful to study the interaction between bacteria and biomaterials. By using the glow discharge technique, with the knowledge of the correlation between bacterial adhesion and the surface properties of a biomaterial, it is possible to produce medical devices with reduced adherence. However, in our studies with glow-discharge-modified polyurethanes, we could demonstrate *in vitro* that there exists a certain minimum of adhesion, independent of the free enthalpy of adhesion. Therefore it seems impossible to use only surface modification to obtain polymeric devices to which *no* bacterial adhesion takes place. However, as some of these modified polymers are able to bind antimicrobial agents, it is possible to reduce bacterial adhesion to very low numbers.

Another possibility for creating anti-infective devices by glow discharge treatment results from the different function of serum and tissue proteins in bacterial adherence. After insertion or implantation of a medical device into the body, the material becomes coated by host factors, e.g., serum and tissue proteins.[80] Whereas for some proteins (e.g., fibrinogen, fibronectin, and vitronectin)

adherence-promoting properties are described, it is well known that albumin reduces bacterial adherence.[81–83] Generation of surfaces that selectively bind albumin and thus reduce adherence of bacteria is one of our main interests and is currently under investigation.

REFERENCES

1. Hirshman, H.P.; Schurman, D.J. *Current Clinical Topics in Infectious Diseases*; Remington, J.S., Swartz, M.S., Eds.; McGraw-Hill: New York, 1982; p. 206.
2. Maki, D.G. *Current Clinical Topics in Infectious Diseases*; Remington, J.S., Swartz, M.S., Eds.; McGraw-Hill: New York, 1982; p. 309.
3. Dankert, J.; Hogt, A.H.; Feijen, J. Biomedical polymers: Bacterial adhesion, colonization, and infection. CRC Crit. Rev. Biocompat. **1986**, *2* (1), 219.
4. Jansen, B.; Peters, G.; Pulverer, G. Mechanisms and clinical relevance of bacterial adhesion to polymers. J. Biomater. Appl. **1988**, *2* (4), 520–543.
5. Schoen, F.J. ASAIO Trans. **1987**, *33*, 8.
6. Sugarman, B.; Young, E.J. Infections associated with prosthetic devices: magnitude of the problem. Infect. Dis. Clin. North America **1989**, *3* (2), 187–198.
7. Peters, G.; Locci, R.; Pulverer, G. Adherence and growth of coagulase-negative staphylococci on surfaces of intravenous catheters. J. Infect. Dis. **1982**, *146* (4), 479–482.
8. Christensen, G.D. Simpson, W.A.; Bisno, A.L.; Beachey, E.H. Adherence of slime-producing strains of *Staphylococcus epidermidis* to smooth surfaces. Infect. Immun. **1982**, *37* (1), 318–326.
9. Franson, T.R.; Sheth, N.D.; Rose, H.D.; Sohnle, P.G. Scanning electron microscopy of bacteria adherent to intravascular catheters. J. Clin. Microbiol. **1984**, *20* (3), 500.
10. Marrie, T.J.; Costerton, J.W. Scanning and transmission electron microscopy of in situ bacterial colonization of intravenous and intraarterial catheters. J. Clin. Microbiol. **1984**, *19* (5), 687–693.
11. Gerson, D.F.; Scheer, D. Cell surface energy, contact angles and phase partition III. Adhesion of bacterial cells to hydrophobic surfaces. Biochim. Biophys. Acta **1980**, *602* (3), 505–510.
12. Neumann, A.W.; Good, R.J.; Hope, C.J.; Sejpal, M. An equation-of-state approach to determine surface tensions of low-energy solids from contact angles. J. Colloid Interface Sci. **1974**, *49* (2), 291–304.
13. Fowkes, F. M. Determination of interfacial tensions, contact angles, and dispersion forces in surfaces by assuming additivity of intermolecular interactions in surfaces. J. Phys. Chem. **1962**, *66* (2), 382–382.
14. Owens, D.K.; Wendt, R.C. Estimation of the surface free energy of polymers. J. Appl. Polym. Sci. **1969**, *13* (8), 1741–1747.
15. Kaelble, D.H. Dispersion-polar surface tension properties of organic solids. J. Adhesion **1970**, *2* (2), 66–81.
16. Wu, S. J. Macromol. Sci. Rev. Macromol. Chem. **1974**, *C10*, 1.
17. Wu, S. Calculation of interfacial tension in polymer systems. J. Polym. Sci. **1971**, *34* (1), 19–30.

18. Wu, S. Polar and nonpolar interactions in adhesion. J. Adhesion **1973**, *5* (1), 39–55.

19. Absolom, D.R.; Neumann, A.W.; Zingg, W.; van Oss, C.J. Thermodynamic studies of cellular adhesion. Trans. Amer. Soc. Artif. Int. Org. **1979**, *25* (1), 152–158.

20. Neumann, A.W.; Absolom, D.R.; van Oss, C.J.; Zingg, W. Surface thermodynamics of leukocyte and platelet adhesion to polymer surfaces. Cell Biophys. **1979**, *1* (1), 79–92.

21. Ferreirós, C.M.; Carballo, J.; Criado, M.T.; Sáinz, V.; del Río, M.C. Surface free energy and interaction of *Staphylococcus epidermidis* with biomaterials. FEMS Microbiol. Lett. **1989**, *60* (1), 89–94.

22. Reid, G.; Hawthorn, L.-A.; Eisen, A.; Beg, H.S. Adhesion of *Lactobacillus acidophilus, Escherichia coli* and *Staphylococcus epidermidis* to polymer and urinary catheter surfaces. Coll. Surf. **1989**, *42* (3), 299–311.

23. Mueller, R.F.; Characklis, W.G.; Jones, W.L.; Sears, J.T. Characterization of initial events in bacterial surface colonization by two Pseudomonas species using image analysis. Biotechnol. Bioeng. **1992**, *39* (11), 1161–1170.

24. Harkes, G.; Feijen, J.; Dankert, J. Adhesion of *Escherichia coli* on to a series of poly (methacrylates) differing in charge and hydrophobicity. Biomaterials **1991**, *12* (9), 853–860.

25. Carballo, J.; Fereirós, C.M.; Criado, M.T. Factor analysis in the evaluation of the relationship between bacterial adherence to biomaterials and changes in free energy. J. Biomater. Appl. **1992**, *7* (2), 130–141.

26. Derjaguin, B.V.; Landau, L.T. Acta Phys. Chim. USSR **1941**, *14*, 633.

27. Verwey, E.J.W.; Overbeek, J.T.G. *Thory of Stability of Lyophobic Colloids*; Elsevier: Amsterdam, 1984.

28. van Loosdrecht, M.C.M.; Zehnder, A.J.B. Energetics of bacterial adhesion. *Experientia*, **1990**, *46* (8), 817–822.

29. Berkeley, R.C.W.; Lynch, J.M.; Melling, J.; Rutter, P.R.; Vincent, B.; Eds. *Microbial Adhesion to Surfaces*; Ellis Horwood Lt.: Chichester, 1980; p. 79.

30. Tadros, Th. F. *Microbial Adhesion to Surfaces*; Berkeley, R.C.W., Lynch, J.M., Melling, J., Rutter, P.R., Vincent, B., Eds.; Ellis Horwood Lt.: Chichester, 1980; p. 93.

31. Uyen, H.M.W.; van der Mei, H.C.; Weerkamp, A.H.; Busscher, H.J. Zeta potential and the adhesion of oral streptococci to polymethylmethacrylate. Biomat. Art. Cells Art. Org. **1989**, *17* (4), 385–391.

32. Hattori, R.; Hattori, T. Adsorptive phenomena involving bacterial cells and an anion exchange resin. J. Gen. Appl. Microbiol. **1985**, *31* (2), 147–163.

33. van Loosdrecht, M.C.M.; Lyklema, J.; Norde, W.; Schraa, G.; Zehnder, A.J.B. Electrophoretic mobility and hydrophobicity as a measured to predict the initial steps of bacterial adhesion. Appl. Environ. Microbiol. **1978**, *53* (8), 1898–1901.

34. Abott, A.; Berkeley, R.C.W.; Rutter, P.R. *Microbial Adhesion to Surfaces*; Berkeley, R.C.W., Lynch, J.M., Melling, J., Rutter, P.R., Vincent, B., Eds.; Ellis Horwood Lt.: Chichester, 1980; p. 117.

35. Yelloji Rao, M.K.; Somasundaran, P.; Schilling, K.M.; Carson, B.; Ananthapadmanabhan, K.P. Bacterial adhesion onto apatite minerals—electrokinetic aspects. Colloids Surf. **1993**, *79* (2), 293–300.

36. Marcinko, D.E. J. Foot Surg. **1985**, *24*, 116.

37. Aderhold, L. Dtsch. Z. Mund Kiefer Gesichtschir. **1985**, *9*, 94.

38. Buchholtz, H.W.; Engelbrecht, H. Der Chirurg **1970**, *11*, 511.

39. Welch, A.G. Antibiotics in acrylic bone cement. *In vitro* studies. J. Biomed. Mater Res. **1978**, *12* (5), 679–700.

40. Moore, W.A.; Chrapil, M.; Seiffert, G.; Keown, K. Development of an infection-resistant vascular prosthesis. Arch. Surg. **1981**, *116* (11), 1403–1407.

41. McDougal, E. G.; Burnham, S. J.; Johnson, G. Rifampin protection against experimental graft sepsis. J. Vasc. Surg. **1986**, *4* (1), 5–7.

42. Greco, R.S.; Harvey, R.A. The role of antibiotic bonding in the prevention of vascular prosthetic infections. Ann. Surg. **1982**, *195* (2), 168–171.

43. Bayston, R.; Grove, N.; Siegel, J.; Lawellin, D. Prevention of hydrocephalus shunt catheter colonisation *in vitro* by impregnation with antimicrobials. J. Neurol. Neurosurg. Psych. **1989**, *52* (5), 605–609.

44. Olanoff, L.S.; Anderson, J.M.; Jones, R.D. Sustained release of gentamicin from prosthetic heart valves. ASAIO J. **1979**, *25* (1), 334–338.

45. Powell, T.W.; Burnham, S.J.; Johnson, G. A passive system using rifampin to create an infection-resistant vascular prosthesis. *Surgery*, **1983**, *94* (5), 765.

46. Webb, L.X.; Myers, R.T.; Cordell, R.; Hobgood, C.D.; Costerton, J.W.; Gristina, A.G. Inhibition of bacterial adhesion by antibacterial surface pretreatment of vascular prostheses. J. Vasc. Surg. **1986**, *4* (1), 16–21.

47. Sowinski, A.; Andruszeniec, R.; Lesiakowska, A.; Lipnicki, D.; Okrojek, C.; Noszcyk, W. Polym. Med. **1985**, *15*, 99.

48. Kurata, H. Nippon Ganka Gakkai Zasshi, **1989**, *90*, 1251.

49. Ney, A.L; Kelly, P.H.; Tsukayama, D.T.; Bubrick, M.P. Fibrin glue-antibiotic suspension in the prevention of prosthetic graft infection. J. Trauma **1990**, *30* (8), 1000–1006.

50. Shenk, J.S.; Ney, A.L.; Tsukayama, D.T.; Olson, M.E.; Bubrick, N.P. Tobramycin-adhesive in preventing and treating PTFE vascular graft infections. J. Surg. Res. **1989**, *47* (6), 487–492.

51. Sherertz, R.J.; Forman, D.M.; Soloman, D.D. Efficacy of dicloxacillin-coated polyurethane catheters in preventing subcutaneous Staphylococcus aureus infection in mice. Antimicrob. Agents Chemother. **1989**, *33* (8), 1174–1178.

52. Trooskin, S.Z.; Donetz, A.P.; Baxter, J.; Harvey, R.A.; Greco, R.S. Infection-resistant continuous peritoneal dialysis catheters. Nephron **1987**, *46* (3), 263–267.

53. Sakamoto, I.; Umemura, Y.; Nakano, M.; Nihira, H.; Kitano, T. Efficacy of an antibiotic coated indwelling catheter: a preliminary report. J. Biomed. Mater Res. **1988**, *19* (9), 1031–1041.

54. Cheng, H. Manufacture and clinical employment of an antibiotic silicon-rubber catheter. Eur. Urol. **1988**, *14* (1), 72.

55. Nakano, H.; Seko, S.; Sumii, T.; Nihira, H.; Fujii, M.; Okada, K.; Kitano, T.; Kodama, M. Efficacy of a latex Foley catheter with sustained release of chlorhexidine: 1st report: Clinical trials for prevention of urinary tract infection. Hinyokoika Kiyo **1986**, *32*, 567–574.

56. Kingston, D.; Seal, D.; Hill, I.D. Self-disinfecting plastics for intravenous catheters and prosthetic inserts. J. Hyg. Camb **1986**, *96* (1), 185–198.

57. Ruggieri, M.L.; Abbott, A. J. Gen. Microbiol. **1987**, *138*.

58. Benvenisty, A.I.; Tannenbaum, G.; Ahlborn, T.N.; Fox, C.L. Modak, S.; Sanpath, L.; Reemtsma, K.; Nowygrod, R. Control of prosthetic bacterial infection: evaluation of an easily

incorporated, tightly bound, silver antibiotic PTFE graft. J. Surg. Res. **1988**, *44* (1), 1–7.

59. Shah, P.M.; Modak, S.; Fox, C.L.; Babu, S.C.; Sampath, L. PTFE graft treated with silver norfloxacin (AgNF): drug retention and resistance to bacterial challenge. J. Surg. Res. **1987**, *42* (3), 298–303.

60. Maki, D.G.; Cobb, L.; Garman, J.K.; Shapiro, J.M.; Ringer, M.; Helgerson, R.B. An attachable silver-impregnated cuff for prevention of infection with central venous catheters: a prospective randomized multicenter trial. Am. Med. J. **1988**, *85* (3), 307–314.

61. Solovsky, M.V.; Ulbrich, K.; Kopecek, J. Synthesis of N-(2-hydroxypr opyl) methacrylamide copolymers with antimicrobial activity. Biomaterials **1983**, *4* (1), 44–48.

62. Ghedini, N.; Scapini, G.; Ferruti, P.; Rigidi, P.; Matteuzi, D. Il Farmaco Ed. Sc. **1985**, *40*, 102.

63. Geckler, K.; Bayer, E.; Wolf, H. Polymer-supported organo-metallic compounds as macromolecular antimicrobial agents. Naturwissenschaften **1985**, *72* (2), 88–89.

64. Messinger, P.; Schimpke-Meier, A. Polymere mit antimikrobiell wirksamen Gruppen–Synthese und Wirksamkeit. Arch. Pharm. **1988**, *321* (2), 89–92.

65. Simionescu, C.I.; Dimitriu, S. Antibiotics immobilized on polysaccharides. Makrom. Chem. Suppl. **1985**, *9* (S19851), 179–187.

66. Demitriu, S.; Popa, M.I.; Hăulică, I.; Crîngu, A.; Stratone, A. Bioactive polymers 61. Synthesis and characterization of some retard antibiotics. Colloid Polym. Sci. **1989**, *267* (7), 595–599.

67. Reddy, B.S.R. Int. J. Biochem. Biophys. **1989**, *26*, 80.

68. Jansen, B.; Schareina, S.; Treitz, U.; Peters, G.; Schumacher-Perdreau, F.; Pulverer, G. *Progress in Biomedical Polymers*; Plenum: New York, 1990; p. 347.

69. Jansen, B.; Kristinson, K.; Jansen, S.; Peters, G.; Pulverer, G. *In vitro* efficacy of a central venous catheter complexed with iodine to prevent bacterial colonization. J. Antimicrob. Chemother. **1992**, *30* (2), 135–139.

70. Jansen, B.; Jansen, S.; Peters, G.; Pulverer, G. *In vitro* efficacy of a central venous catheter ('Hydrocath') loaded with teicoplanin to prevent bacterial colonization. J. Hosp. Infect. **1992**, *22* (2), 93–107.

71. Schierholz, J.; Jansen, B.; Jaenicke, W.; Pulverer, G. *In vitro* efficacy of an antibiotic releasing silicone ventricle catheter to prevent shunt infection Biomaterials **1994**, *15* (12), 996–1000.

72. Hampl, J.; Schierholz, J.; Jansen, B.; Aschoff, A. *In vitro* and *in vivo* efficacy of a rifampin-loaded silicone catheter for the prevention of CSF shunt infections. Acta Neuro-chirurgica **1995**, *133* (3–4), 147–152.

73. Kamal, G.D.; Pfaller, M.A.; Remple, L.E.; Jebson, P.J.R. Reduced intravascular catheter infection by antibiotic bonding. JAMA **1991**, *256* (18), 2364.

74. Maki, D.G.; Wheeler, S.J.; Stolz, S.M.; Mermel, L.A. Am. Soc. Microbiol. **1991**, Abstract 461.

75. Gombotz, W.R.; Hoffman, A.S. Crit. Rev. Biocompat. **1986**, *4*, 1, doi:10.3109/07388558609150790.

76. Hoffman, A.S. *Advances in Polymer Science 57*; Dusek, K., Ed.; Springer Verlag: Berlin, 1984; p. 141.

77. Yasuda, H; Gazicki, M. Biomedical applications of plasma polymerization and plasma treatment of polymer surfaces. Biomaterials **1982**, *3* (2), 68–77.

78. Yasuda, H. *Thin Film Processes*; Vossen, J.L., Kern, W., Eds.; Academic: New York, 1978; p. 316.

79. Ludwicka, A.; Switalski, L.M.; Ludin, A.; Pulverer, G.; Wadström, T. Bioluminescent assay for measurement of bacterial attachment to polyethylene. J. Microbiol. Method **1985**, *4* (3), 169–177.

80. Bantjes, A. Clotting phenomena at the blood-polymer interface and development of blood compatible polymer surfaces. Brit. Polym. J. **1978**, *10* (4), 267–274.

81. Vaudaux, P.; Yasuda, H.; Velazco, M.I.; Huggler, E.; Ratti, I.; Waldvogel, F.A.; Lew, D.P. Role of host and bacterial factors in modulating staphylococcal adhesion to implanted polymer surfaces. J. Biomater. Appl. **1990**, *5* (2), 134–153.

82. Vaudaux, P.; Lerch, P.; Velazco, M.; Nydegger, U.E.; Wald-vogel, F.A. Adv. Biomater. **1986**, *6*, 355.

83. Jansen, B.; Ellinghorst, G. Modification of polyetheru-rethane for biomedical application by radiation induced grafting. II. Water sorption, surface properties, and protein adsorption of grafted films. J. Biomed. Mater. Res. **1984**, *18* (6), 655–669.

Artificial Muscles

Mohsen Shahinpoor
Artificial Muscle Research Institute (AMRI), School of Engineering and School of Medicine, University of New Mexico, Albuquerque, New Mexico, U.S.A.

Abstract

This entry presents a brief history of artificial muscles, followed by an overview of the shape memory alloys, polymer-type artificial muscles, and electrical activation of PAN artificial muscles through increasing the conductivity of PAN fibers. The history and properties of ionic polymer-metal composites in artificial muscles are also discussed. It concludes with a brief discussion on liquid crystal elastomer artificial muscles.

INTRODUCTION

The first attempts to create artificial muscles date to the pioneering work of Kuhn and his student Katchalsky and his students in the late 1940s and early 1950s in connection with pH-activated muscles or simply pH muscles.[1-6] Note also the other means of activating artificial muscles such as thermal,[7-9] pneumatic,[10] optical,[11] and magnetic.[12] Chemically stimulated polymers were discovered more than half a century ago when it was shown that collagen filaments could reversibly contract or expand when dipped in acidic or alkaline solutions, respectively.[1] This early work pioneered the development of synthetic polymers that mimic biological muscles.[3,5] However, electrical stimulation has always remained the best means of artificial muscle material actuation and sensing. Shahinpoor[4,5,10,13-19] was one of the pioneers in making electrically active, in sensing and actuation, ionic polymer–conductor composites (IPCCs) and ionic polymer–metal composites (IPMCs). Zhang, Bharti, and Zhao[20] were able to observe a substantial piezoelectric activity in polyvinyledene fluoride trifluoroethylene (PVF2-TrFE) as early as 1998. The greatest progress in artificial muscle materials development has occurred in the last 20 years when effective materials that can induce strains exceeding 100% have emerged.[21]

SPECIFIC ARTIFICIAL MUSCLE MATERIALS

Shape Memory Alloys and Polymers

The history of Shape Memory Alloys (SMA) and Shape Memory Polymers (SMP) artificial muscles is extensive and will not be reported here. However, the pioneering works of Liang and Rogers[22,23] and Liang, Rogers, and Malafeew[24] in developing SMA and SMP actuators are noteworthy. In their work, the load of a spring-biased SMA actuator was modeled as a dead weight. However, many practical applications involve varying loads, as in the cases of SMA rotatory joint actuators.[13,25-27] A general design methodology of various types of bias force SMA actuators has been investigated by Shahinpoor and Wang.[26] Heat-shrink polymers such as polyolefins, poly(vinyl chloride) (PVC), polypropylene, and polyesters are also sometimes considered as heat-activated shape memory polymers and artificial muscles.

Magnetically Activated Artificial Muscles

Magnetically activated gels, so-called ferrogels, are chemically cross-linked polymer networks that are swollen by the presence of a magnetic field.[29] Zrinyi Szabo, and Feher have a pioneers of this technology.[29] Such a gel is a colloidal dispersion of monodomain magnetic nanoparticles. Such magnetic ferrogel materials deform in the presence of a spatially nonuniform magnetic field because its embedded nanoparticles move in reaction to the field. The ferrogel material can be activated to bend, elongate, deform, curve, or contract repeatedly, and it has a response time of less than 100 ms, which is independent of the particle size. Another group of magnetically activated artificial muscles are the magnetostrictive materials such as TERFENOL-D, which are the metallic elements terbium (TER), iron (FE), Naval Ordnance Labs (NOL), and Dysprosium-D[30] and give rise to infinitesimal dimensional changes of the order of 0.1% or so, and magnetic shape memory materials (MSM) such as NiMnGa, which undergo gigantic martensitic-austenitic solid-phase transformation and can give rise to 6–8% dimensional change.[31]

Electronic Electroactive Polymer/Ferroelectric Polymeric Artificial Muscles

The basic phenomenon is called ferroelectricity when a nonconducting crystal or dielectric material exhibits spontaneous electric polarization. These are based on the

Concise Encyclopedia of Biomedical Polymers and Polymeric Biomaterials DOI: 10.1081/E-EBPPC-120014090

phenomenon of piezoelectricity, which is found only in noncentrosymmetric materials such as poly(vinylidene fluoride) (also known as PVDF). These polymers are partly crystalline, with an inactive amorphous phase and a Young's modulus of about 1–10 GPa. This relatively high elastic modulus offers high mechanical energy density. A large applied AC field (–200 MV/m) can induce electrostrictive (nonlinear) strains greater than 1%. Sen, Scheinbeim, and Newman[32] investigated the effect of mixing plasticizers (–65% wt.) with ferroelectric polymers hoping to achieve large strains at reasonable applied fields. However, the plasticizer is also amorphous and inactive, resulting in decreased Young's modulus, permittivity, and electrostrictive strains. As large as 4% electrostrictive strains can be achieved with low-frequency driving fields having amplitudes of about 150 V/μm. As with ceramic ferroelectrics, electrostriction can be considered to be the origin of piezoelectricity in ferroelectric polymers.[33] Unlike electrostriction, piezoelectricity is a linear effect, where not only will the material be strained when voltage is applied, a voltage signal will be induced when a stress is applied. Thus they can be used as sensors, transducers, and actuators. Depoling due to excessive loading, heating, cooling, or electric driving is a problem with these materials.

Dielectric Elastomer Electroactive Polymers

Polymers with low elastic stiffness modulus and high dielectric constant can be packaged with interdigitated electrodes to generate large actuation strain by subjecting them to an electric field. This dielectric elastomer electroactive polymer (EAP) can be represented by a parallel-plate capacitor.[21] The induced strain is proportional to the square of the electric field, multiplied by the dielectric constant and is inversely proportional to the elastic modulus. Dielectric elastomer EAP actuators require large electric fields (100 V/μm) and can induce significant levels of strain (10–200%). Pelrine et al. introduced a new class of polymers that exhibits an extremely high strain response.[34] These acrylic-based elastomers have produced large strains of more than 200% but suffer from the fact that they require gigantic electric fields in the range of hundreds of megavolts per meter.

Liquid Crystal Elastomer Materials

Liquid crystal elastomers (LCE) were pioneered at Albert-Ludwigs Universität (Freiburg, Germany).[35] These materials can be used to form an EAP actuator by inducing isotropic-nematic phase transition due to temperature increase via Joule heating. LCEs are composite materials that consist of monodomain nematic LCE and conductive polymers that are distributed within their network structure.[14,36,37] The actuation mechanism of these materials involves phase transition between nematic and isotropic phases over a period of less

than a second. The reverse process is slower, taking about 10 s, and it requires cooling, causing expansion of the elastomer to its original length. The mechanical properties of LCE materials can be controlled and optimized by effective selection of the liquid crystalline phase, density of cross-linking, flexibility of the polymer backbone, coupling between the backbone and liquid crystal group, and the coupling between the liquid crystal group and the external stimuli.

Ionic EAP/Ionic Polymer Gels

Polymer gels can be synthesized to produce strong actuators having the potential of matching the force and energy density of biological muscles. These materials (e.g., polyacrylonitrile or PAN) are generally activated by a chemical reaction, changing from an acid to an alkaline environment causing the gel to become dense or swollen, respectively. This reaction can be stimulated electrically, as was shown by Shahinpoor et al.[5,10,14–19] Current efforts are directed toward the development of thin layers and more robust techniques of placing electrodes on the surface of the composite. Progress was recently reported by researchers at the University of New Mexico using a mix of conductive and PANS fibers.[38] The mechanism that is responsible for the chemomechanical behavior of ionic gels under electrical excitation is described by Osada and Ross-Murphy,[39] and a model for hydrogel behavior as contractile EAP is described in Gong et al.[40] A significant amount of research and development was conducted at the Hokkaido University, Japan, and applications using ionic gel polymers were explored. These include electrically induced bending of gels[39,41] and electrically induced reversible volume change of gel particles.[42]

Nonionic Polymer Gels/EAPs

Nonionic polymer gels containing a dielectric solvent can be made to swell under a DC electric field with a significant strain. Hirai and his coworkers at Shinshu University in Japan have created bending and crawling nonionic EAPs using a poly(vinyl alcohol) gel with dimethyl sulfoxide.[43,44] A $10 \times 3 \times 2$ mm actuator gel was subjected to an electric field and exhibited bending at angles that were greater than 90° at a speed of 6 cm/s. This phenomenon is attributed to charge injection into the gel and a flow of solvated charges that induce an asymmetric pressure distribution in the gel. Another nonionic gel is PVC, which is generally inactive when subjected to electric fields. However, if PVC is plasticized with dioctyl phthalate (DOP), a typical plasticizer, it can maintain its shape and behave as an elastic nonionic gel.

Ionic Polymer–Metal Composites

Ionic polymer–metal composite is an EAP that bends in response to a small electric field (5–10 V/mm) as a result of mobility of cations in the polymer network. In 1992, IPMC was realized, based on a chemical plating technique developed by Merlet, Pinneri, and coworkers in France and by Kawami and Takanake in Japan in the 1980s. The first working actuators were built by Oguro et al.[45] in Japan and by Shahinpoor[4] in the United States. The first working sensors of this kind were first fabricated by Shahinpoor[4] and Sadeghipour, Salomon, and Neogi[46] in the United States. The operation as actuators is the reverse process of the charge storage mechanism associated with fuel cells[47] and Kim and Shahinpoor.[48,49] A relatively low electric field is required (five orders of magnitude smaller than the fields required for PVDF-TrFE and dielectric elastomers) to stimulate bending in IPMC, where the base polymer provides channels for mobility of positive ions in a fixed network of negative ions on interconnected clusters. In order to chemically deposit electrodes on the polymer films, metal ions (platinum, gold, palladium, or others) are dispersed throughout the hydrophilic regions of the polymer surface and are subsequently reduced to the corresponding zero-valence metal atoms.

Conductive Polymers

Conductive polymers operate under an electric field by the reversible counterion insertion and expulsion that occurs during redox cycling.[6,50] Oxidation and reduction occur at the electrodes, inducing a considerable volume change due mainly to the exchange of ions with an electrolyte. When a voltage is applied between the electrodes, oxidation occurs at the anode and reduction at the cathode. Ions (H^+) migrate between the electrolyte and the electrodes to balance the electric charge. Addition of the ions causes swelling of the polymer and conversely their removal results in shrinkage. As a result, the sandwich assembly bends. Conductive polymer actuators generally require small electric fields in the range of 1–5 V/μm, and the speed increases with the voltage having relatively high mechanical energy densities of over 20 J/cm³, but with low efficiencies at the level of 1%. Several conductive polymers have been reported, including polypyrrole, polyethylenedioxythiophene, poly(p-phenylene vinylene)s, polyanilines, and polythiophenes. Operation of conductive polymers as actuators at the single-molecule level is currently being studied, taking advantage of the intrinsic electroactive property of individual polymer chains.

SMA ARTIFICIAL MUSCLES

Since SMA artificial muscle deformation is temperature-induced and rather small (4–6% linear strain), design means must be incorporated to create larger strains to mimic biological muscles. For example, the human biceps contract more than 30%. Thus, the reader is referred to some design methodologies to create large-strain SMA muscles for biomedical applications.[13,25–27] These papers discuss a new design for a rotatory joint actuator made with shape memory alloy contractile muscle wires. These papers also present a design of SMA rotatory joint actuators using both shape memory effect and pseudoelastic effect. The novelty of such mechanisms is the use of a pseudoelastic SMA wire as a replacement of the bias spring in a bias force type of SMA actuator. Besides the development of a general design methodology of bias force SMA actuators, these efforts by Wang and Shahinpoor intend to investigate a knee and leg muscle exerciser to be used by paraplegics or quadriplegics for exercising their knee and leg muscles. Refer to a dissertation by Guoping Wang[28] for a complete and thorough coverage of SMA artificial muscles.

ELECTROACTIVE POLYACRYLONITRILE FIBERS

Activated PAN fibers, which are suitably annealed, cross-linked, and hydrolyzed, are known to contract and expand when ionically activated with cations and anions, respectively. The key engineering features of PAN fibers are their ability to change their length routinely more than 100% and their comparable strength to biological muscles. Described here are some recent results on electroactive PAN artificial muscles as well as a technique that allows one to electrically control the actuation of active PAN fiber bundles. Increasing the conductivity of PAN fibers by making a composite of them with a conductive medium such as platinum, gold, graphite, carbon nanotubes, and conductive polymers such as polyaniline or polypyrrole has allowed for electric activation of PAN fibers when a conductive polyacrylonitrile (C-PAN) fiber bundle is placed in a chemical electrolysis cell as an electrode. A change in concentration of cations in the vicinity of C-PAN fiber electrode leads to contraction and expansion of C-PAN fibers depending upon the applied electric field polarity. Typically close to 100% change in C-PAN length in a few seconds is observed in a weak electrolyte solution with tens of volts of DC power supply. These results indicate a great potential in developing electrically activated C-PAN muscles and linear actuators, as well as integrated pairs of antagonistic muscles and muscle sarcomere and myosin/actin assembly resembling actual biological muscles.

Activated PAN fibers, which are suitably annealed, cross-linked, and hydrolyzed, are known to contract and expand when ionically activated with cations and anions, respectively. The change in length for these pH-activated fibers is typically greater than 100%, but up to 200% contraction/expansion of PAN fibers has been observed in our laboratories at AMRI. They are comparable in strength to human muscles and have the potential of becoming medically implantable electroactive contractile artificial muscle fibers.[5,19,38,51] Increasing the conductivity of PAN by making a composite with a conductive medium (e.g., a noble metal such as gold, palladium, platinum, or graphite or a conductive polymer such as polypyrrole or polyaniline) has allowed for electrical activation of PAN artificial muscles when it is placed in an electrochemical cell. The electrolysis of water in such a cell produces hydrogen ions at a PAN anode, thus locally decreasing the cationic concentration and causing the PAN muscle to contract. Reversing the electric field allows the PAN muscle to elongate. Typically, close to 100% change in PAN muscle length in a few minutes is observed when it is placed as an electrode in a 10 mM NaCl electrolyte solution and connected to a 20-volt power supply. Recently, the response time has been reduced to a few seconds. These results indicate the potential in developing electrically active PAN artificial muscles and linear actuators.

Activated PAN contracts when exposed to protons (H⁺) in an aqueous medium and elongates when exposed to hydroxyl ion (OH⁻) in a strong alkaline medium. The length of activated PAN fibers can potentially more than double when going from short to long. A possible explanation for the contraction and elongation of activated PAN is the effect carboxylic acid groups have in the molecular geometry. At low cationic concentration, all acid groups are protonated, potentially contracting the polymer. Raw PAN fibers, which are composed of roughly 2000 individual strands of PAN (each about 10 μm in diameter), are first annealed at 220°C for 2 hours. The fibers can be bundled together at this point to form a PAN muscle. The PAN is then placed in a solution of boiling 1 N LiOH for 30 minutes, after which the PAN is elastic like a rubber band. At this point, the PAN can be ionically activated; flooding the activated PAN with a high concentration of cations such as H⁺ induces contraction, while anionic concentration elongates the fibers. An assortment of such PAN muscles is shown in Fig. 1.

PAN has been studied for more than a half century in many institutions and now is popularly used in the textile industry as a fiber form of artificial silk. The useful properties of PAN include the insolubility, thermal stability, and resistance to swelling in most organic solvents. Such properties were thought to be due to the cross-linked nature of the polymer structure. However, a recent finding has shown that a number of strong polar solvents can dissolve PAN and thus has raised a question that its structure could be of the linear zigzag conformations[52,53] with hydrogen bonding between hydrogen and the neighboring nitrogen of the nitrile group. Activated PAN fibers are elastic. The term "activated PAN fibers" refers to the PAN fibers acting as the elastic fibers of which length varies depending upon the ionic concentration of cations in the solution. They contract at low cationic concentration and expand at high cationic concentration. At high cationic concentration they appear to reject water out of the polymer network resulting in shrinkage due to their elastic nature. Conversely, they elongate at low cationic concentration, attracting water from outside into the polymer network. In our laboratory, it has been observed that activated PAN fibers can contract more than 200% relative to that of the expanded PAN. The

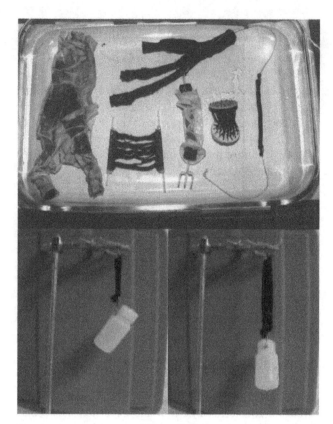

Fig. 1 An assortment of PAN artificial muscles in a variety of configurations.

use of PAN as artificial muscles is promising since they are able to convert chemical energy to mechanical motion, possibly acting as artificial sarcomeres or muscles. Other types of materials that have a capability to be electroactive are polyelectrolyte gels, such as polyacrylamide (PAM), polyvinylalcohol-polyacrylic acid (PAA-PVA), and poly(2-acrylamido-2-methyl propane) sulphonic acid (PAMPS). Under an electric field, these gels are able to swell and de-swell, inducing large changes in the gel volume. Such changes in volume can then be converted to mechanical work. Yet one disadvantage of such polyelectrolyte gels is that they are mechanically weak. Artificial sarcomeres or muscles made from PAN have much greater mechanical strength than polyelectrolyte gels, thus having a greater potential for application as artificial sarcomeres and muscles (or soft actuators).

The strength of activated PAN and its ability to change length up to 100% or more makes it an appealing material for use as linear actuators and artificial muscles. An attractive alternative is electrical activation. Shahinpoor and coworkers[5,19,38,51] have reported considerable efforts on electrical activation of PAN artificial muscles. During the electrolysis of water, hydrogen ions are generated at the anode while hydroxyl ions are formed at the cathode in an electrochemical cell. Electrochemical reactions can then potentially be used to control the length of a PAN artificial muscle. This may be achieved by either locating a PAN muscle near an electrode where the ions are generated, or, if the conductivity of activated PAN can be increased, the PAN muscle can serve as the electrode itself.

The study reported herein takes the second approach, where platinum is deposited on PAN fibers to increase conductivity so the muscle can serve as the electrode directly. This procedure of depositing Pt on the polymer and activating it in an electrochemical cell was initially demonstrated a few years ago with a polyvinyl alcohol–polyacrylic acid copolymer, resulting in about a 5% decrease in length, with both contraction and elongation each taking about 12 minutes. However, our improvements[19,38,51] to electroactive PAN muscles have reduced the activation time to a few seconds with muscle strength approaching human muscles in the range of 2 N per cm^2 of muscle fiber cross section.

In order to manufacture pH-active PAN fibers, the raw fibers are first annealed in air in a computer-controlled oven at an elevated temperature of 220–240°C. It is likely that the annealing process cross-links the polymer and creates a cross-linked structure of pyridine and cyano groups. The preoxidized or annealed PAN fibers show a dark brown or black color depending upon the level of cross-linking. Second, they are subsequently hydrolyzed by boiling in an alkaline solution (2 N NaOH for 20–30 minutes). Following this process, the PAN fibers become extremely pH active (pH muscles). In an approximate pH range of 2–14, the PAN fibers can contract and expand more than 200%.

Based upon the Donnan Theory of ionic equilibrium, the important forces arise from the induced osmotic pressure of free ions between activated PAN fibers and their environment, the ionic interaction of fixed ionic groups, and the charged network itself. Among them, the induced osmotic pressure of free ionic groups is the dominating force. PAN fibers can be activated electrically, by providing a conductive medium or electrode in contact with or within the PAN fibers to act like an anode electrode in a chemical cell with another cathode electrode separated from the PAN fibers. Upon applying an electric field across the electrodes, H$^+$ evolves at the anode electrode via $2H_2O \Rightarrow O_2 + 4H^+ + 4e^+$, and the OH$^-$ evolves at the cathode via $2H_2O + 2e^- \Rightarrow H_2 + 2OH^-$. Upon contact with H^{++} ions in the vicinity of the PAN anode electrode, the decreased pH causes the PAN fibers to contract. If the polarity of the electric field is reversed, then upon contact with OH$^-$ ions in the vicinity of the cathode the increased pH causes the PAN fibers to expand.

IPCCS AND IPMCS AS EAPS

A large number of polyelectrolytes in a composite form with a conductive medium such as a metal (ionic polymer–conductor composites or ionic polymer–metal composites [IPMCs]) can exhibit large dynamic deformation if suitable electrodes are mounted on them and placed in a time-varying

electric field.[4,5,10,13–19] Conversely, dynamic deformation of such polyelectrolytes can produce dynamic electric fields across their electrodes. A recently presented model by de Gennes, Okumura, Shahinpoor, and Kim[54] describes the underlying principle of electrothermodynamics in such polyelectrolytes based upon internal transport phenomena and electrophoresis. It should be pointed out that IPMCs show great potential as soft robotic actuators, artificial muscles, and dynamic sensors in micro to macro size range. An effective way of manufacturing an IPMC starts with making a composite of polyelectrolytes with a metal by means of a chemical reduction process. The current state-of-the-art IPMC manufacturing technique[5,10,14–19] incorporates two distinct preparation processes: initial manufacturing of a polyelectrolyte–metal composite, and the process of placing surface electrodes on the composite. Figure 2 shows illustrative schematics of two different preparation processes (top left and bottom left) and two top-view SEM micrographs for the platinum surface electrode (top right and bottom right).

The initial process of making a polyelectrolyte–metal composite requires an appropriate platinum salt such as Pt(NH$_3$)$_4$HCl or Pd(NH$_3$)$_4$HCl in the context of chemical reduction processes similar to the processes evaluated by a number of investigators including Takenaka et al.[55] and Millet Pinneri, and Durand[56] The principle of the process of making a polyelectrolyte–metal composite is to metalize

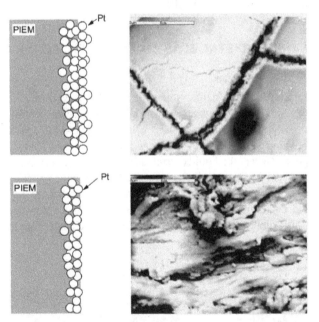

Fig. 2 Schematic diagrams showing two different preparation processes: (top left) a schematic showing initial process of making a polyelectrolyte–metal composite; (top right) its top-view SEM micrograph; (bottom left) a schematic showing the process of placing electrodes on the surface of the composite; and (bottom right) its top-view SEM micrograph where platinum deposited predominately on top of the initial Pt layer. Note that PIEM stands for perfluorinated ion exchange membrane.

the inner surface of the material by a chemical reduction means such as $LiBH_4$ or $NaBH_4$. The polyelectrolyte is soaked in a salt solution to allow platinum-containing cations to diffuse through via the ion-exchange process. Later, a proper reducing agent such as $LiBH_4$ or $NaBH_4$ is introduced to platinize the material.[5,10,14–19] It has been experimentally observed that the platinum particulate layer is buried microns deep (typically 1–20 μm) within the IPMC surface and is highly dispersed. The fabricated IPMCs can be optimized to produce a maximum force density by changing multiple process parameters. These parameters include time-dependent concentrations of the salt and the reducing agents. Applying the Taguchi technique to identify the optimum process parameters for such optimizations has been successful in the past.[19,57] The primary reaction is,

$$LiBH_4 + 4[Pt(NH_3)_4]^{2+} + 8OH^- \Rightarrow 4Pt^\circ \quad (1)$$
$$+ 16NH_3 + LiBO_2 + 6H_2O$$

In the subsequent process of placing electrodes on the surface of the composite, multiple reducing agents are introduced (under optimized concentrations) to carry out the reducing reaction similar to Eq. 1, in addition to the initial platinum layer formed by the initial process of making a polyelectrolyte–metal composite. This is clearly shown in Fig. 2 (bottom right), where the roughened surface disappears because of additional metals deposited on the surface. Figure 3 depicts a typical spectacular steady-state deformation mode of a strip of such ionic polymers under a step voltage.[45,58–73]

LIQUID CRYSTAL ELASTOMER ARTIFICIAL MUSCLES

Liquid single crystal elastomers (LSCEs) have gained a lot of importance in connection with artificial muscles technologies. See Finkelmann et al.,[35,71] Brand and coworkers,[72,73] Shahinpoor,[14] Ratna et al.,[36] and Finkelmann and Shahinpoor[37] where the liquid crystalline phase structure is macroscopically orientated in the network. At the liquid crystalline to isotropic phase transformation temperature, when the liquid crystalline network becomes isotropic like a conventional rubber, the dimensions of the network change. Due to the anisotropy in the phase structure of liquid crystal elastomers, the networks shorten in the direction of the optical axis or director axis. De Gennes and coworkers[74,75] have also alluded to the fact that such nematic networks present a potential as artificial muscles just like the IPMC materials mentioned before. For nematic LSCE synthesized from nematic LC side-chain polymers, the change of this length is normally more than 100% and sometimes up to 400%.[38,51] The speed of this process is determined by the thermal heat conductivity of the network and not by material transport processes. If the sample thickness is limited, calculations indicate that the response time of the LSCE is similar to that of natural muscles and mainly determined by the relaxation behavior of the polymer chains or network strands.

The magnitude of changes in the dimensions of nematic LSCE with temperature is determined by the coupling between the nematic state of order and the conformation of the polymer main chains. In this regard, the conformation of nematic side-chain polymers is only slightly affected. For nematic main-chain polymers, where the mesogenic units are incorporated into the polymer main chain, large conformational changes should occur and be directly expressed in the thermoelastic behavior. The electrical activation of LCEs has also been reported by Shahinpoor[14] and Finkelmann and Shahinpoor.[37]

Making a composite of them with a conductor phase such as a metal or graphite powder does this. Joule heating of some new high-performance nematic LSCE developed by Finkelmann and coworkers[35,71] that consists of a combination of LC side-chain and main-chain polymers has been successful. These materials have been synthesized previously by a hydrosilylation reaction of the monofunctional side-chain mesogen and the bifunctional liquid crystalline main-chain polyether with poly-(methylhydrogensiloxane). Figure 4 displays the main chemical synthesis steps in producing LCEs by hydrosilylation reaction of the monofunctional side-chain mesogen. Note that a weakly cross-linked network is mechanically loaded. Under load, the hydrosilylation reaction is completed. The LSCE has a weight ratio between the mesogenic units in the main chain to those in the side chains of 57/43 and a nematic to isotropic phase transition temperature of $T_{n,i} = 104°C$.

Fig. 3 Typical deformation of strips (10 mm × 80 mm × 0.34 mm) of ionic polymers under a step voltage of 4 volts.

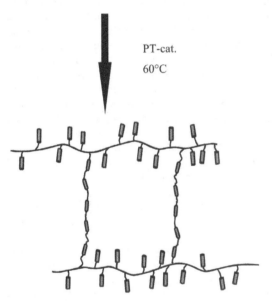

main chain mesogen/crosslinker

PMHS

PT-cat.

60°C

Fig. 4 Synthesis of the nematic coelastomer in a liquid single crystalline coelastomer.

Thus, coelastomers containing network strands comprising LC side- and main-chain polymers exhibit exceptional thermoelastic behavior. These systems might be suitable as model systems for artificial muscles. Thus, liquid crystalline to isotropic phase transition can cause conformational changes of macromolecules, which are sufficient to obtain a large mechanical response of polymer networks.

The samples of monodomain (MD-72/28) nematic liquid side-chain crystal elastomers graphite (LSCE-G) composites under a stress of 100 kPa and contracts up to 400%.[72] The composite of them with a conductor[14] becomes a conductor and has an effective resistivity of about 1.6 ohms/square on its surface and about 0.8 ohm/square across its thickness. Upon applying a DC voltage of 0.5–5 volts to a typical

sample for about 4 seconds, under a stress of about 10 kPa, the sample contracts quickly in about a second to about 18 mm lengthwise with an average linear strain of about 25%. The sample does not contract appreciably in the transverse direction and the thickness direction. The cooling is also quick and it takes the sample about 4.4 seconds to revert back to its initial length under stretching load of 10 kPa.

CONCLUSIONS

We began this entry by discussing a brief history of artificial muscles. We then discussed the shape memory alloys and polymer-type artificial muscles followed by electrical activation of PAN artificial muscles through increasing the conductivity of PAN fibers. The conductivity of PAN was increased by either depositing a coat of metal on the fibers or interweaving it with graphite fibers. Electrochemical reactions were used to generate hydrogen ions or hydroxyl ions for the contraction and elongation, respectively, of these PAN-C muscles. Therefore, by increasing the conductivity of activated PAN, a PAN artificial muscle or linear actuator could be electrically activated in an antagonistic manner (push–pull, biceps, and triceps actions) suitable for industrial and medical applications. Increasing the conductivity of PAN fibers by making a composite of them with a conductive medium such as platinum, gold, graphite, carbon nanotubes, and conductive polymers such as polyaniline or polypyrrole was shown to allow for electric activation of PAN fibers when a C-PAN fiber bundle is placed in a chemical electrolysis cell as an electrode. Typically close to 100% change in C-PAN fiber length in a few seconds is observed in a weak electrolyte solution with tens of volts of DC power supply. These results indicate a great potential in developing electrically activated C-PAN muscles and linear actuators, as well as integrated pairs of antagonistic muscles and muscle sarcomere and myosin/actin assembly. The history of IPMCs in artificial muscles was also discussed. The fundamental properties and characteristics of ionic polymer (polyelectrolyte)–metal composites as biomimetic sensors, actuators, transducers, and artificial muscles were also presented. Finally, a brief discussion was presented on liquid crystal elastomer artificial muscles. It was reported that liquid crystal artificial muscles have the capability of contracting more than 400%. For additional references on artificial muscles, please refer to references 76–91.

REFERENCES

1. Katchalsky, A. Rapid swelling and deswelling of reversible gels of polymeric acids by ionization. Experientia **1949**, 5 (8), 319–320.

2. Kuhn, W.; Hargitay, B.; Katchalsky, A.; Eisenburg, H. Reversible dilatation and contraction by changing the state of ionization of high-polymer acid networks. Nature **1950**, 165 (4196), 514–516.

3. Steinberg, I.Z.; Oplatka, A.; Katchalsky, A. Mechano-chemical engines. Nature **1966**, *210* (5036), 568–571.

4. Shahinpoor, M. Conceptual design, kinematics and dynamics of swimming robotic structures using ionic polymeric gel muscles. Smart Mater. Struc. **1992**, *1* (1), 91–94.

5. Shahinpoor, M.; Bar-Cohen, Y.; Simpon, J.O.; Smith, J. Ionic polymer-metal composites (IPMC) as biomimetic sensors and structures—A review. Smart Mater. Struc. **1998**, *7* (6), 15–30.

6. Otero, T.F.; Grande, H.; Rodriguez, J. A new model for electrochemical oxidation of polypyrrole under conformational relaxation control. J. Electr. Chem. **1995**, *394* (1–2), 211–216.

7. Kishi, R.; Ichijo, H.; Hirasa, O. Thermo-responsive devices using poly(vinyl methyl ether) hydrogels. J. Intell. Mater. Syst. Struct. **1993**, *4* (4), 533–537.

8. Tobushi, H.; Hayashi, S.; Kojima, S. Mechanical properties of shape memory polymer of polyurethane series. JSAE Int. J. Ser. 1 **1992**, *35* (3), 296–302.

9. Li, F.K.; Zhu, W.; Zhang, X.; Zhao, C.T.; Xu, M. Shape memory effect of ethylene-vinyl acetate co-polymers. J. Appl. Polym. Sci. **1999**, *71* (7), 1063–1070.

10. Shahinpoor, M.; Kim, K.J. Ionic polymer-metal composites—I. Fundamentals. Smart Mater. Struc. Int. J. **2001**, *10* (4), 819–833.

11. Van der Veen, G.; Prins, W. Light-sensitive polymers, nature. Phys. Sci. **1971**, *230*, 70–72.

12. Zrinyi, M.; Barsi, L.; Szabo, D.; Kilian, H.G. Direct observation of abrupt shape transition in ferrogels induced by nonuniform magnetic field. J. Chem. Phys. **1997**, *106* (13), 5685–5692.

13. Shahinpoor, M.; Wang, G. In *Design, Modeling and Performance Evaluation of a Novel Large Motion Shape Memory Alloy Actuator*, Proc. SPIE 1995 North American Conference on Smart Structures and Materials, San Diego, CA, February 28–March 2, 1995; Vol. 2447, paper no. 31.

14. Shahinpoor, M. In *Elastically-Activated Artificial Muscles Made with Liquid Crystal Elastomers*, Proceedings of the SPIE's 7th Annual International Symposium on Smart Structures and Materials, EAPAD Conf., Bar-Cohen, Y., Ed.; 2000; Vol. 3987, 187–192.

15. Shahinpoor, M.; Kim, K.J. A solid-state soft actuator exhibiting large electromechanical effect. Appl. Phys. Lett. **2002**, *80* (18), 3445–3447.

16. Shahinpoor, M.; Mojarrad, M. Soft Actuators and Artificial Muscles. U.S. Patent, #6,109,852, April 13, 2000.

17. Shahinpoor, M.; Kim, K.J. The effect of surface-electrode resistance on the performance of ionic polymer metal composites (IPMC) artificial muscles. Smart Mater. Struc. **2000**, *9* (4), 543–551.

18. Shahinpoor, M.; Kim, K.J. Novel ionic polymer-metal composites equipped with physically-loaded particulate electrode as biomimetic sensors, actuators and artificial muscles. Sens. Actuators A Phys. **2002**, *96* (2–3), 125–132.

19. Shahinpoor, M.; Norris, I.D.; Mattes, B.R.; Kim, K.J.; Sillerud, L.O. In *Electroactive Polyacrylonitrile Nanofibers as Artificial Nano-Muscles*, Proceedings of the 2002 SPIE Conference on Smart Mater. Struc., San Diego, CA, March 2002; Vol. 4695 (42), 169–173.

20. Zhang, Q.M.; Bharti, V.; Zhao, X. Giant electrostriction and relaxor ferroelectric behavior in electron-irradiated poly(vinylidene fluoride-trifluorethylene) copolymer. Science **1998**, *280*, 2101–2104.

21. Perline, R.; Kornbluh, R.; Joseph, J.P. Electrostriction of polymer dielectrics with compliant electrodes as a means of actuation. Sens. Actuators A Phys. **1998**, *64* (1), 77–85.

22. Liang, C.; Rogers, C.A. One-dimensional thermometrical constitutive relations of shape memory materials. J. Intell. Mater. Syst. Struct. **1991**, *1* (2), 207–234.

23. Liang, C.; Rogers, C.A. Design of shape memory alloy actuators. ASME J. Mech. Des. **1992**, *114*, 223–230.

24. Liang, C.; Rogers, C.A.; Malafeew, E. Investigation of shape memory polymers and their hybrid composites. J. Intell. Mater. Syst. Struct. **1997**, *8* (4), 380–386.

25. Wang, G.; Shahinpoor, M. Design for shape memory alloy rotary joint actuators using shape memory effect and pseudoelastic effect. Smart Mater. Technol. **1997**, *3040*, 23–30.

26. Wang, G.; Shahinpoor, M. Design, prototyping and computer simulation of a novel large bending actuator made with a shape memory alloy contractile wire. Smart Mater. Struc. Int. J. **1997**, *6* (2), 214–221.

27. Wang, G.; Shahinpoor, M. A new design for a rotatory joint actuator made with shape memory alloy contractile wire. Int. J. Intell. Mater. Syst. Struct. **1997**, *8* (3), 191–279.

28. Wang, G. A general design of Bias Force Shape Memory Alloy (BFSMA) actuators and an electrically-controlled SMA knee and leg muscle exerciser for paraplegics and quadriplegics. In Ph.D.; Department of Mechanical Engineering, University of New Mexico: Albuquerque, NM, 1998.

29. Zrinyi, M.; Szabo, D.; Feher, J. In *Comparative Studies of Electro and Magnetic Field Sensitive Polymer Gels*, Proceedings of the SPIE's 6th Annual International Symposium on Smart Structures and Materials, EAPAD Conf., SPIE Proc., Bar-Cohen, Y., Ed.; 1999; Vol. 3669, 406–413.

30. Dapino, M.; Flatau, A.; Calkins, F. In *Statistical Analysis of Terfenol-D Material Properties*, Proceedings of SPIE 1997 Symposium on Smart Structures and Materials, San Diego, CA, 1997; Vol. 3041, 256–267.

31. O'Handley, R.C. Model for strain and magnetostriction in magnetic shape memory alloys. J. Appl. Phys. **1998**, *83* (6), 3263–3270.

32. Sen, A.; Scheinbeim, J.I.; Newman, B.A. The effect of plasticizer on the polarization of poly(vinylidene fluoride) films. J. Appl. Phys. **1984**, *56* (9), 2433–2439.

33. Furukawa, T.; Seo, N. Electrostriction as the origin of piezoelectricity in ferroetectric polymers. Jpn. J. Appl. Phys. **1990**, *29* (4), 675–680.

34. Pelrine, R.; Kornbluh, R.; Pei, Q.; Joseph, J. High speed electrically actuated elastomers with strain greater than 100%. Science **2000**, *287* (5454), 836–839.

35. Finkelmann, H.; Kock, H.J.; Rehage, G. Investigations on liquid crystalline siloxanes: 3. Liquid crystalline elastomer—A new type of liquid crystalline material. Makromol. Chem. Rapid Commun. **1981**, *2* (4), 317–323.

36. Ratna, B.R.; Selinger, J.V.; Srinivasan, A.; Hong, J.; Naciri, J. In *Namatic Elastomers as Artificial Muscles*, Proceedings of the First World Congress on Biometrics and Artificial Muscles, Albuquerque Conventional Center, Albuquerque, New Mexico, USA, December 9–11, 2002, 2003; 1.

37. Finkelmann, H.; Shahinpoor, M. In *Electrically-Controllable Liquid Crystal Elastomer-Graphite Composites Artificial Muscles*, Proceeding of SPIE 9th Annual International Symposium on Smart Structures and Materials, 2002, San Diego, California, SPIE Publication No. 4695–4653.

38. Schreyer, H.B.; Gebhart, N.; Kim, K.J.; Shahinpoor, M. Electric activation of artificial muscles containing polyacrylonitrite gel fibers. Biomacromolecules **2000**, *1* (4), 642–647.

39. Osada, Y.; Ross-Murphy, S. Intelligent gels. Sci. Am. **1993**, *268*, 82–87.

40. Gong, J.P.; Nitta, T.; Osada, Y. Electrokinetic modeling of the contractile phenomena of polyelectrolyte gels-one dimensional capillary model. J. Phys. Chem. **1994**, *98* (38), 9583–9587.

41. Osada, Y.; Hasebe, M. Electrically activated mechanochemical devices using polyelectrolyte gels. Chem. Lett. **1985**, *12*, 1285–1288.

42. Osada, Y.; Kishi, R. Reversible volume change of microparticles in an electric field. J. Chem. Soc. **1989**, *85* (3), 662–665.

43. Hirai, M.; Hirai, T.; Sukumoda, A.; Nemoto, H.; Amemiya, Y.; Kobayashi, K.; Ueki, T. Electrically induced reversible structural change of a highly swollen polymer gel network. J. Chem. Soc., Faraday Trans. **1995**, *91*, 473–477.

44. Hirai, T.; Zheng, J.; Watanabe, M. In *Solvent-Drag Bending Motion of Polymer Gel Induced by an Electric Field*, Proceedings of the SPIE's 6th Annual International Symposium on Smart Structures and Materials, Bar-Cohen, Y., Ed.; 1999; Vol. 3669, 209–217.

45. Oguro, K.; Kawami, Y.; Takenaka, H. Bending of an ion-conducting polymer film-electrode composite by an electric stimulus at low voltage. Trans. J. Micromach. Soc. **1992**, *5*, 27–30.

46. Sadeghipour, K.; Salomon, R.; Neogi, S. Development of a novel electrochemically active membrane and smart material based vibration sensor/damper. Smart Mater. Struc. **1992**, *1* (2), 172–179.

47. Heitner-Wirguin, C. Recent advances in perfluorinated ionomer membranes: Structure, properties and applications. J. Membr. Sci. **1996**, *120* (1), 1–33.

48. Kim, K.J.; Shahinpoor, M. Application of polyelectrolytes in ionic polymeric sensors, actuators, and artificial muscles. In *Handbook of Polyelectrolytes*; Tripathy, S., Nalwa, H.S., Eds.; Mindspring/Academic Press, 2002.

49. Kim, K.J.; Shahinpoor, M. A novel method of manufacturing three-dimensional ionic polymer-metal composites (IPMC's) biomimetic sensors, actuators and artificial muscle. Polymer **2002**, *43* (3), 797–802.

50. Gandhi, M.R.; Murray, P.; Spinks, G.M.; Wallace, G.G. Mechanisms of electromechanical actuation in polypyrrole. Synth. Met. **1995**, *75* (3), 247–256.

51. Schreyer, H.B.; Shahinpoor, M.; Kim, K.J. In *Electrical Activation of PAN-Pt Artificial Muscles*, Proceedings of SPIE/ Smart Structures and Polymer Actuators and Devices, Newport Beach, CA, March 1999; Vol. 3669, 192–198.

52. Hu, X. Molecular structure of polyacrylonitrile fibers. J. Appl. Polym. Sci. **1996**, *62* (11), 1925–1932.

53. Umemoto, S.; Okui, N.; Sakai, T. *Contraction Behavior of Poly(acrylonitrile) Gel Fibers*, *Polymer Gels*; DeRossi, D., et al., Eds.; Plenum Press: New York, 1991; 257–270.

54. de Gennes, P.G.; Okumura, K.; Shahinpoor, M.; Kim, K.J. Mechanoelectric effects in ionic gels. Lett. **2000**, *50* (4), 513–518.

55. Takenaka, H.; Torikai, E.; Kawami, Y.; Wakabayshi, N. Solid polymer electrolyte water electrolysis. Int. J. Hydrogen Energy **1982**, *7* (5), 397–403.

56. Millet, P.; Pinneri, M.; Durand, R. New solid polymer electrolyte composites for water electrolysis. J. Appl. Electrochem. **1989**, *19*, (2) 162–166.

57. Peace, G.S. *Taguchi Methods: Hands-On Approach*; Addison-Wesley Publishing Company, Inc.: New York, 1993.

58. Adolf, D.; Shahinpoor, M.; Segalman, D.; Witkowski, W. Electrically Controlled Polymeric Gel Actuators. U.S. Patent #5,250,167, October 13, 1993.

59. Brock, D.L. *Review of Artificial Muscle Based on Contractile Polymers*; AI Memo No. 1330, MIT, Artificial Intelligence Lab. 1991.

60. De Rossi, D.E.; Chiarelli, P.; Buzzigoli, G.; Domenici, C.; Lazzeri, L. Contractile behavior of electrically activated mechanochemical polymer actuators. ASAIO Trans. **1986**, *32* (1), 157–163.

61. De Rossi, D.; Suzuki, M.; Osada, Y.; Morasso, P. Pseudo-muscular gel actuators for advanced robotics. J. Intell. Mater. Syst. Struct. **1992**, *3* (1), 75–95.

62. Doi, M.; Matsumoto, M.; Hirose, Y. Deformation of ionic polymer gels by electric fields. Macromolecules **1992**, *25* (20), 5504–5511.

63. Hamlen, R.P.; Kent, C.E.; Shafer, S.N. Electrolytically activated contractile polymer. Nature **1965**, *206* (4989) 1149–1150.

64. Nemat-Nasser, S.; Li, J.Y. Electromechanical response of ionic polymer-metal composites. J. Appl. Phys. **2000**, *87* (7), 3321–3331.

65. Oguro, K.; Kawami, Y.; Takenaka, H. Actuator Element. U.S. Patent #5,268,082, 1993.

66. Oguro, K.; Fujiwara, N.; Asaka, K.; Onishi, K.; Sewa, S. Polymer Electrolyte Actuator with Gold Electrodes, Proceedings of the SPIE's 6th Annual International Symposium on Smart Structures and Materials, SPIE Proc., 1999; Vol. 3669, 64–71.

67. Pei, Q.; Inganas, O.; Lundstrom, I. Bending bilayer strips built from polyaniline for artificial electro-chemical muscles. Smart Mater. Struc. **1993**, *2* (1), 1–5.

68. Segalman, D.; Witkowski, W.; Adolf, D.; Shahinpoor, M. In *Electrically Controlled Polymeric Muscles as Active Materials Used in Adaptive Structures*, Proceedings of ADPA/ AIAA/ASME/SPIE Conference on Active Materials and Adaptive Structures, VA, Nov. 1991.

69. Tanaka, T.; Nishio, I.; Sun, S.T.; Nishio, S.U. Collapse of gels in an electric field. Science **1982**, *218* (4571), 467–469.

70. Woojin, L. Polymer Gel Based Actuator: Dynamic Model of Gel for Real Time Control. In Ph.D. Thesis; Massachusetts Institute of Technology, 1996.

71. Finkelmann, H.; Brand, H.R. Liquid crystalline elastomers— A class of materials with novel properties. Trends Polym. Sci. **1994**, *2*, 222.

72. Brand, H.R.; Finkelmann, H. Physical properties of liquid crystalline elastomers. In *Handbook of Liquid Crystals Vol. 3: High Molecular Weight Liquid Crystals*; Demus, D., Goodby, J., Gray, G.W., Spiess, H.W., Vill, V., Eds.; Wiley-VCH: Weinheim, 1998; 277–289.

73. Brand, H.R.; Pleiner, H. Electrohydrodynamics of nematic liquid crystalline elastomers. Physica **1994**, *208* (3–4), 359.

74. de Gennes, P.G.; Hebert, M.; Kant, R. Artificial muscles based on nematic gels. Macromol. Symp. **1997**, *113* (1), 39.

75. Hebert, M.; Kant, R.; de Gennes, P.G. Dynamics and thermodynamics of artificial muscles based on nematic gels. J. Phys. I. **1997**, *7* (7), 909–918.

76. Shahinpoor, M. Ionic polymer-conductor composites as biomimetic sensors, Robotic actuators and artificial muscle—A review. Electrochimica **2003**, *48* (14–16), 2343–2353.

77. Dusek, K. *Advances in Polymer Science, Vols. 109 and 110, Responsive Gels, Transitions I and II*; Springer-Verlag: Berlin, 1993.

78. Li, J.Y.; Nemat-Nasser, S. Micromechanical analysis of ionic clustering in nafion perfluorinated membrane. Mech. Mater. **2000**, *32* (5), 303–314.

79. Liu, Z.; Calvert, P. Multilayer hydrogens and muscle-like actuators. Adv. Mater. **2000**, *12* (4), 288–291

80. McGehee, M.D.; Miller, E.K.; Moses, D.; Heeger, A.J. Twenty years of conductive polymers: From fundamental science to applications. In *Advances in Synthetic Metal: Twenty Years of Progress in Science and Technology*; Bamier, P., Lefrant, S., Bidan, G., Eds.; Elsevier, 1999; 98–203.

81. Osada, Y.; Matsuda, A. Shape memory in hydrogels. Nature **1995**, *376* (6537), 219.

82. Osada, Y.; Okuzaki, H.; Hori, H. A polymer gel with electrically driven motility. Nature **1992**, *355*, 242–244.

83. Otero, T.F.; Sansifiena, J.M. Soft and wet conducting polymers for artificial muscles. Adv. Mater. **1998**, *10* (6), 491–494.

84. Roentgen, W.C. About the changes in shape and volume of dielectrics caused by electricity. Ann. Phys. Chem. **1880**, *1* (1), 771–786.

85. Sacerdote, M.P. Strains in polymers due to electricity. J. Phys. **1899**, VIII (3 Series, t), 31.

86. Shiga, T. Deformation and viscoelastic behavior of polymer gels in electric fields. Adv. Polym. Sci. **1997**, *134*, 131–136.

87. Shiga, T.; Kurauchi, T. Deformation of polyelectrolyte gels under the influence of electric field. J. Appl. Polym. Sci. **1990**, *39* (11–12), 2305–2320.

88. Shiga, T.; Hirose, Y.; Okada, A.; Kurauchi, T. Bending of ionic polymer gel caused by swelling under sinusoidally varying electric fields. J. Appl. Polym. Sci. **1993**, *47* (1), 113–119.

89. *The Applications of Ferroelectric Polymers*; Wang, T.T., Herbert, J.M., Glass, A.M., Eds.; Chapman and Hall: New York, 1988.

90. Wang, Q.; Du, X.; Xu, B.; Cross, L.E. Electro-mechanical coupling and output efficiency of piezo-electric bending actuators IEEE Trans. Ultrason. Ferroelectr. Freq. Control **1999**, *46* (3), 638–646.

91. Proceedings of the Fall MRS Symposium on Electro-active Polymers (EAP), Warrendale, PA, Zhang, Q.M., Furukawa, T., Bar-Cohen, Y., Scheinbeim, J., Eds.; 1999; 1–336.

Bioabsorbable Polymers: Tissue Engineering

K. J. L. Burg
Department of Bioengineering, Clemson University, Clemson, South Carolina, U.S.A.

Waleed S. W. Shalaby
Poly-Med Incorporated, Anderson, South Carolina, U.S.A.

Abstract

The field of tissue engineering has grown substantially, especially the use of biodegradable polymers as scaffolds or templates. Cells are seeded on an absorbable polymeric or organic matrix, the system is implanted *in vivo*, and the matrix is gradually resorbed as the tissue develops. This entry concentrates on key developments in the field rather than providing a historical overview.

PROPERTIES

The classical field of tissue engineering may be regarded as the development of three-dimensional viable constructs *in vitro*. Constructs can be developed *in vitro* through seeding cells onto an absorbable scaffold or an organic support matrix. This method has an advantage over injecting *cells isolated* directly into the target site in that the cell-seeded constructs have the ability to form a tissue structure—a stable, three-dimensional, site-specific object.[1–5]

Attempts have been made to regenerate such tissues as liver, bone, and cartilage. The advantages of using isolated cells cultured *in vitro* are that cells causing an immune response can be removed, and the surgical expense and complications for a donor are not encountered. Additionally, just those cells supplying a desired function can be isolated. Tissue engineering has these advantages over the more traditional transplantation or prosthetic implantation: the construction of an intricate three-dimensional object is possible, the problem of limited donor tissue *supply* is avoided, and infection and adhesion problems are drastically reduced.

Absorbable polymers that find use in scaffolding include polyglycolide acid,[6–10] 90/10 poly(glycolide-co-L-lactide) or polyglactin-910,[6,11,12] polyanhydrides and polyorthoesters,[11] and poly-L-lactide acid.[10,13–19]

Organic scaffolding materials include collagen gel[20] and hyaluronate gel. Faster-absorbing materials such as polyglycolide acid lead to a relatively unstable tissue because long-term tissue development has not occurred. Various coatings can be applied to the scaffold surface to promote cell adhesion, and cell survival including poly-L-lysine[21] and cell matrix proteins.[22]

Criteria for Success of Hybrid Constructs

The scaffold must maintain structural integrity *in vivo*; the cells must be sufficiently developed and must have adhered to the substrate *in vitro* before their *in vivo* implantation. The construct must be sufficiently malleable for ease of implantation, possibly arthroscopic. The construct's degradation rate must satisfy the cell synthesis rate and not contribute to inflammation, and the scaffold porosity should allow diffusion of nutrients as well as synthesis of extracellular matrix material. The type of polymer used, the relative acidity of its degradation products, and the amount of exposed surface area[11] can all affect inflammatory response.

Bioreactors

The cell culture is also an important factor in a successful construct. A bioreactor is, simply stated, a flask with fluid flow mixing in the form of a fluidized bed reactor or an airlift reactor.[8] There are also rotating microgravity bioreactors in which two concentric cylinders rotate to induce flow. Reactors should allow uniform mixing and control of pH, oxygen, carbon dioxide partial pressures, and nutrient levels, as well as optimal shear stresses. Nutrient, pH, and mass gradients that could be detrimental to cell culture are excluded.

Concise Encyclopedia of Biomedical Polymers and Polymeric Biomaterials DOI: 10.1081/E-EBPPC-120051876

APPLICATIONS

Cartilage Reconstruction

Cartilage degeneration has been a major focus of research because cartilage has minimal ability for self-repair and is therefore a common subject of reconstructive surgery. Attempts to repair cartilage defects have been made using direct chondrocyte transplantation as well as naturally derived or synthetic seeded scaffolds. In direct cartilage transplantation, excised healthy cartilage is used to repair the damaged area. To minimize the need for donor tissue and meet a particular shape requirement, scaffolding or direct chondrocyte transplantation is used. The applications include articular cartilage repair in the knee, reconstruction of nasoseptal cartilage,[5] and cosmetic and reconstructive surgery[3] where a three-dimensional custom-shaped implant is required.

Surface regularity and neocartilage thickness increases when the cell count is high and the chondrocytes are evenly distributed on the scaffold. The cell distribution on the matrix depends on the porosity or fiber spacing, the surface characteristics of the scaffold material and the viscosity of the cell suspension. Varying cell suspension density will change the shape and thickness of the resulting tissue. The density may be optimized in order to use the minimum number of cells necessary to achieve the desired results.[4] Faster-degrading matrices allow a higher cell growth rate because the potential space available for cell growth is increased.[1]

An appropriate cell culture medium is important because seeded systems contain a high concentration of cells over a small area and require frequent media renewal, creating a higher risk of contamination. A continuous perfusion culture system lowers the risk of contamination from handling and better emulates the *in vivo* environment. The cells are distanced from the medium by an agarose gel through which the nutrients must diffuse, thus mimicking the *in vivo* diffusion of nutrients through the collagen and the newly synthesized components into the medium source.

The age of the cartilage donor has an effect on chondrogenesis because younger chondrocytes are more metabolically active. The problem of fixation—the ability to attach the replacement material to the target area[23]—still remains.

Naturally Derived Matrix

The chondrocytes can be embedded in naturally derived matrices such as collagen gel and then cultured *in vitro*.[20] Chondrocytes in this environment tend to retain their ability to synthesize Type II collagen, unlike those in the typical monolayer culture. Although the gel matrix will not maintain a desired shape, it does have the advantage over direct chondrocyte transplantation affording stability, both phenotypically and mechanically, before implantation.

Chondrocytes are a phenotypically unstable cell type *in vitro*. They may dedifferentiate and redifferentiate, potentially resulting in a change in synthesis from Type II to Type I collagen. The retention of a differentiated phenotype is attributed to a malleable cell culture environment, such as a gel, that allows rounding of the cell, while proliferation—the spreading and growth of the cell with loss of specialized function—is attributed to the more rigid substrate of a petri dish.[24] The solid component of articular cartilage contains approximately 50% glycosaminoglycan (GAG) and 50% Type II collagen, so the cell culturing must be carefully planned to induce Type II collagen synthesis.

Liver Reconstruction

Cell transplantation would be ideal for liver replacement because living donors could be used, which would increase the donor pool. Collagenous scaffolds for hepatocytes have been found to lack sufficient mechanical integrity and induce an immune response. Various absorbable foams[18] and wafer discs[11] have also been tested as possible scaffolds. Advanced absorbable foams[25] and gels[26] have been described as suitable matrices for tissue regeneration. In order to seed the cells into porous interconnected structures, the thickness of the implant must be carefully considered so as to ensure the appropriate distribution of cells. The injection of cells into the network may shear or otherwise damage the cells if not performed at the appropriate rate. The rate also affects the distribution of the cells. A surface treatment with a solvent, such as ethanol, can enhance the movement of the cells throughout the foam. Hepatocytes have also been attached to collagen surface treated microcarriers and injected into the peritoneal cavity, thus avoiding a surgical procedure.[27]

Pancreas Reconstruction

Islets are clusters of cells in the pancreas whose damage may lead to diabetes or hypoglycemia. Direct islet allotransplantation has been examined in some detail.[28] Traditionally the islets were transplanted, rather than the complete pancreatic tissue, because it was thought that the exocrine behaved as a contaminant, causing an immunogenic response and implant rejection.[29] To this end, a highly purified islet implant is desired; however, a large amount of pancreatic tissue must be processed to obtain even a small quantity of donor islets. The search for an appropriate immunosuppressive method that would minimize the donor tissue necessary continues. The success of islet viability on absorbable scaffolding *in vitro* has yet to be matched *in vivo*.[11]

Osteoblasts

Limited studies have been directed toward bone repair. Bone regeneration by osteoblast transplantation may be useful in the repair of skeletal defects. Osteoblasts appear to attach and proliferate on a variety of synthetic absorbable polymers.[10]

Skin Reconstruction

Skin replacement research has expanded, largely in response to the needs of burn victims. An effective skin graft is quite difficult to manufacture because of the composite nature of the dermis and epidermis and the difficulty of revascularizing the system quickly enough to support the new tissue growth. Collagen-GAG (col-GAG) grafts seeded with fibroblasts and keratinocytes has been utilized;[6,30] however, they are susceptible to enzymatic degradation, possibly causing the breakdown of newly formed collagen. This problem led to the concept of seeding have fibroblasts onto synthetic, hydrolytically degrading surfaces.[6] Biodegradable synthetic systems are advantageous because they allow precise surgical manipulation and their surface area is large enough to accommodate cell adhesion promoters, angiogenic factors, and cellular attachment.[22]

Vascular Grafts

Ongoing research in this particular area involves implanting a biodegradable scaffold and allowing the tissue to regenerate *in vivo* as the material breaks down.[12–15,31] The typical synthetic nonabsorbable materials are thrombogenic and inelastic and therefore unsuitable for smaller arteries. The degradation time of the graft is important because fast absorbing materials may lead to aneurysm.

CONCLUSIONS

Although progress has been made in the area of tissue engineering, the attainment of major future milestones will depend on the availability of novel forms of absorbable material and rigorous techniques for handling cells and implantable constructs.

REFERENCES

1. Freed, L.E.; Marquis, J.C.; Nohria, A.; Emmanual, J.; Mikos, A.G.; Langer, R. Neocartilage formation *in vitro* and *in vivo* using cells cultured on synthetic biodegradable polymers. J. Biomed. Mater. Res. **1993**, *27* (1), 11–23.
2. Kim, W.S.; Vacanti, J.P.; Cima, L.; Mooney, D.; Upton, J.; Puelacher, W.C.; Vacanti, C.A; Khouri, R.K.; Reddi, A.H. Cartilage engineered in predetermined shapes employing cell transplantation on synthetic biodegradable polymers [Discussion]. Plast. Reconstr. Surg. **1994**, *94* (2), 238–240.
3. Kim, W.S.; Vacanti, J.P.; Cima, L.; Mooney, D.; Upton, J.; Puelacher, W.C.; Vacanti, C.A. Cartilage engineered in predetermined shapes employing cell transplantation on synthetic biodegradable polymers. Plast. Reconstr. Surg. **1994**, *94* (2), 233–237.
4. Puelacher, W.C.; Kim, S.W.; Vacanti, J.P.; Schloo, B.; Mooney, D.; Vacanti, C.A. Tissue-engineered growth of cartilage: The effect of varying the concentration of chondrocytes seeded onto synthetic polymer matrices. Int. J. Oral Maxillofac. Surg. **1994**, *23* (1), 49–53.
5. Puelacher, W.C.; Mooney, D.; Langer, R.; Upton, J.; Vacanti, J.P.; Vacanti, C.A. Design of nasoseptal cartilage replacements synthesized from biodegradable polymers and chondrocytes. Biomaterials **1994**, *15* (10), 774–778.
6. Cooper, M.L.; Hansbrough, J.F.; Spielvogel, R.L.; Cohen, R.; Bartel, R.L.; Naughton, G. *In vivo* optimization of a living dermal substitute employing cultured human fibroblasts on a biodegradable polyglycolic acid or polyglactin mesh. Biomaterials **1991**, *12* (2), 243–248.
7. Atala, A.; Vacanti, J.P.; Peters, C.A.; Mandell, J.; Retik, A.B.; Freeman, M.R. Formation of urothelial structures *in vivo* from dissociated cells attached to biodegradable polymer scaffolds *in vitro*. J. Urol. **1992**, *148* (2 Pt. 2), 658.
8. Freed, L.E.; Vunjak-Novakovic, G.; Langer, R. Cultivation of cell-polymer cartilage implants in bioreactors. J. Cellular Biochem. **1993**, *51* (3), 257–264.
9. Freed, L.E.; Vunjak-Novakovic, G.; Biron, R.J.; Eagles, D.B.; Lesnoy, D.C.; Barlow, S.K.; Langer, R. Biodegradable polymer scaffolds for tissue engineering. Bio/Technology **1994**, *12* (7), 689–693.
10. Ishaug, S.L.; Yaszemski, M.J.; Bizios, R.; Mikos, A.G. Osteoblast function on synthetic biodegradable polymers. J. Biomed. Mater. Res. **1994**, *28* (12), 1445–1453.
11. Vacanti, J.P.; Morse, M.A.; Saltzman, W.M.; Domb, A.J.; Perez-Atayde, A.; Langer, R. Selective cell transplantation using bioabsorbable artificial polymers as matrices. J. Pedia. Surg. **1988**, *23* (1), 3–9.
12. Vacanti, J.P.; Shepard, J.; Retik, A.B. Implantation *in vivo* and retrieval of artificial structures consisting of rabbit and human urothelium and human bladder muscle. J. Urol. **1993**, *150* (2 Pt. 2), 608–612.
13. Lei, B.; Bartels, H.L.; Nieuwenhuis, P.; Wildevuur, C.R.H Microporous, compliant, biodegradable vascular grafts for the regeneration of the arterial wall in rat abdominal aorta. Surgery **1985**, *98* (5), 955–963.
14. Lei, B.; Wildevuur, C.R.H.; Nieuwenhuis, P.; Blaauw, E.H.; Dijk, F.; Hulstaert, C.E.; Molenaar, I. Regeneration of the arterial wall in microporous, compliant, biodegradable vascular grafts after implantation into the rat abdominal aorta. Cell Tissue Res. **1985**, *242* (3), 569.
15. Van der Lei, B.; Wildevuur, C. R.; Dijk, F.; Blaauw, E. H.; Molenaar, I.; Nieuwenhuis, P. Sequential studies of arterial wall regeneration in microporous, compliant, biodegradable small-caliber vascular grafts in rats. J. Thorac. Cardiovasc. Surg. **1987**, *93* (5), 695–707.
16. Beumer, G.J.; van Blitterswijk, C.A.; Bakker, D.; Ponec, M. A new biodegradable matrix as part of a cell seeded skin substitute for the treatment of deep skin defects: A physicochemical characterization. Clin. Mater. **1993**, *14* (1), 21–27.
17. Beumer, G.J.; van Blitterswijk, C.A.; Ponec, M. Degradative behaviour of polymeric matrices in (sub)dermal and muscle tissue of the rat: A quantitative study. Biomaterials **1994**, *15* (7), 551–559.
18. Wald, H.L.; Sarakinos, G.; Lyman, M.D.; Mikos, A.G.; Vacanti, J.P.; Langer, R. Cell seeding in porous transplantation devices. Biomaterials **1993**, *14* (4), 270–278.

19. Wake, M.C.; Patrick, C.W., Jr.; Mikos, A.G. Pore morphology effects on the fibrovascular tissue growth in porous polymer substrates. Cell Transplant. **1994**, *3* (4), 339.

20. Wakitani, S.; Kimura, T.; Hirooka, A.; Ochi, T.; Yoneda, M.; Yasui, N.; Owaki, H.; Ono, K. Repair of rabbit articular surfaces with allograft chondrocytes embedded in collagen gel. J. Bone Joint Surg, **1989**, *71*(1), 74–80.

21. Sittinger, M.; Bujia, J.; Minuth, W.W.; Hammer, C.; Burmester, G.R. Engineering of cartilage tissue using bioresorbable polymer carriers in perfusion culture. Biomaterials **1994**, *15* (6), 451–456.

22. Gilbert, J.C.; Takada, T.; Stein, J.E.; Langer, R.; Vacanti, J.P. Cell transplantation of genetically altered cells on biodegradable polymer scaffolds in syngeneic rats. Transplantation **1993**, *56* (2), 423–426.

23. Freed, L.E.; Grande, D.A.; Lingbin, Z.; Emmanual, J.; Marquis, J.C.; Langer, R. Joint resurfacing using allograft chondrocytes and synthetic biodegradable polymer scaffolds. J. Biomed. Mater. Res. **1994**, *28*, 891–899

24. Ingber, D.E.; Dike, L.; Hansen, L.; Karp, S.; Liley, H.; Maniotis, A.; McNamee, H.; Mooney, D.; Plopper, G.; Sims, J.; Wang, N. Cellular tensegrity: Exploring how mechanical changes in the cytoskeleton regulate cell growth, migration, and tissue pattern during morphogenesis. Int. Rev. Cytol. **1994**, *150* (1), 173–224.

25. Shalaby, S.W.; Roweton, S.L. U.S. Patent 05, 083, 1995.

26. Shalaby, S.W. U.S. Patent application U. S. Patent Application 08/421, 222, 1995.

27. Demetriou, A.A.; Whiting, J.F.; Feldman, D.; Levenson, S.M.; Chowdhury, N.R.; Moscioni, A.D.; Kram, M.; Chowdhury, J.R.; Moscioni, A.D.; Kram, M.; Chowdhury, J.R. Replacement of liver function in rats by transplantation of microcarrier-attached hepatocytes. Science **1986**, *233* (4769), 1190–1192.

28. Kretschmer, G.J.; Sutherland, D.E.R.; Matas, A.J.; Cain, T.L.; Najarian, J.S. The dispersed pancreas: Transplantation without islet purification in totally pancreatectomized dogs. Diabetologia **1977**, *13* (5), 495–502.

29. Gores, P.F.; Sutherland, D.E.R. Pancreatic islet transplantation: Is purification necessary? Am. J. Surg. **1993**, *166* (5), 538–542.

30. Bell, E.; Sher, S.; Hull, B.; Merrill, C.; Rosen, S.; Chamson, A.; Asselineau, D.; Dubertret, L.; Coulomb, B.; Lapiere, C.; Nusgens, B.; Neveux, Y. The reconstitution of living skin. J. Investig. Dermatol. **1983**, *81* (Suppl. 1), 2s–10s.

31. Niu, S.; Kurumatani, H.; Satoh, S.; Kanda, K.; Oka, T.; Watanabe, K. Small diameter vascular prostheses with incorporated bioabsorbable matrices. Am. Soc. Artific. Intern. Organs **1993**, *39* (3), M750–M753.

Bioactive Systems

Mikhail I. Shtilman
Biomaterials Scientific and Teaching Center, D. I. Mendeleyev University of Chemical Technology, Moscow, Russia

Abstract
Polymers are widely used to design various systems that have biological and pharmacological activities. These systems can stimulate or inhibit biological processes; they can also have a biocidal effect. Different aspects of the construction, investigation, and application of these systems are discussed.

HIERARCHY OF POLYMER-CONTAINING BIOLOGICALLY ACTIVE SYSTEMS

There are two main groups of the bioactive systems containing polymer components or macromolecular fragments (Fig. 1).

The first group consists of forms with *biologically active compounds* (BACs) that are not chemically bound to the polymer component (Group 1). The second group comprises polymer compounds that have various biological activities (Group 2).

FORMS CONTAINING POLYMER COMPONENTS (GROUP 1)

Systems of Group 1 are water insoluble, and the active ingredient incorporated in them is delivered to the tissues of the organism by diffusion or as a result of disintegration of the system, which may be caused by different processes, including biodegradation.

Forms of Group 1 can be subdivided into the forms in which the polymer component does not influence the BAC release rate (Group 1.1) and the forms in which the polymer component determines BAC release rate (Group 1.2).

Forms in Which the Polymer Component Does Not Influence the Release Rate of the Active Substance (Group 1.1)

Among polymer components that do not significantly influence BAC release rate are powdered polymer excipients of tablets and components that facilitate their pressing and nonadherence to die molds. Starch, poly(vinyl alcohol), poly(N-vinylpyrrolidone), and other biologically neutral polymers are usually employed as such components.

Forms in Which the Polymer Component Determines the Release Rate of the Active Substance (Group 1.2)

Forms containing non-chemically incorporated BACs in which the polymer component determines the rate of delivery of BACs to living tissues and BACs are released gradually, at a predetermined rate, take on increasing importance. These insoluble systems (as well as polymer carriers that release the BAC due to the gradual disintegration of the chemical bond between the BAC and the carrier, which will be considered in the section "Examples of Forms with Controlled Release of the BAC") are *systems with controlled release of the active ingredient.*

It is well known that the use of these systems can eliminate or mitigate such disadvantages of BACs as the narrow ranges of doses and concentrations that produce beneficial effects. When these doses or concentrations are exceeded, drugs may reach not only the damaged organs but also other parts of the body, causing acute toxic, allergic, and carcinogenic effects. Thus, it is difficult to accurately determine the right dose of the BAC employed.

Moreover, these side effects make it impossible to effectively administer greater amounts of the BAC, which would enable prolonged action of the drug. BACs that are prone to washout, volatilization, biodegradation, and structural changes (e.g., pharmaceutical protein compounds) should be administered at higher, i.e., inhibitory, doses or on a multiple-dosing regimen, and that would significantly increase the cost of the treatment.

Examples of forms with controlled release of the BAC

The controlled release principle is used in many common forms of BACs.

Concise Encyclopedia of Biomedical Polymers and Polymeric Biomaterials DOI: 10.1081/E-EBPPC-120050545

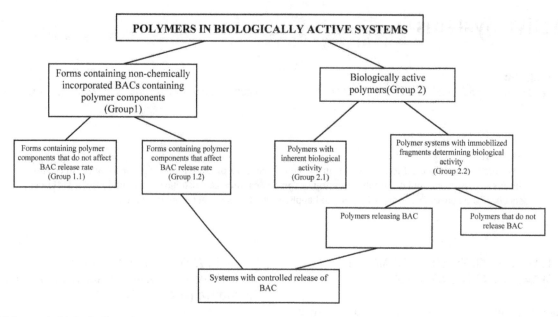

Fig. 1 Polymers in biologically active systems.

Among the most common and well-known forms of the drugs are *tablets*, which release the active ingredient through diffusion or disintegration of the polymer coating. Tablets are usually administered orally, but there are also subcutaneously implanted tablets.

Polymer coatings may be used to deliver the drug to the target region of the gastrointestinal tract. For instance, Eudragit® polymers (Evonik Ind., FRG) are used to produce such coatings containing ionogenic groups that determine polymer solubility at different pH levels. Tablets coated with a polymer containing basic groups, e.g., copolymers of dimethylaminoethyl methacrylate with methyl or butyl methacrylates, degrade in the acidic medium of the stomach. However, tablets coated with polymers containing acid groups, such as copolymers of acrylic acid and ethyl methacrylate or methacrylic acid and methyl methacrylate, are stable in the stomach but degrade in the intestines, in which pH varies from 7.2 to 9.0.[1,2]

Encapsulation of the drug in *micro- and nanocapsules* enables its prolonged action and reduces its side effects such as unpleasant odor and bitter taste.[3,4]

Various macromolecular systems in which BACs are embedded in the polymer and are released through diffusion or gradual dissolution or degradation of the carrier have recently become popular. Pharmaceutical polymer films containing BACs are used in ophthalmology and to prevent ischemic heart disease. Films with embedded BACs can be stored for long periods and readily attach to the mucosal surface of the eye and gum (transmucosal formulations).[5]

The same mechanisms underlie drug delivery by other polymer systems such as coatings of suture materials or sutures incorporating BACs, antiseptic-impregnated catheters or their coatings, stent coatings containing BACs, etc.

Transdermal systems (multi-layer drug-in adhesive patches) are among the most promising polymer-based drug delivery systems as these adhesive patches and special devices are very easy to use.

The main components of a transdermal patch are a backing, drug layer, or reservoir, a polymer membrane that controls drug release by diffusion, an adhesive layer that adheres the patch to the skin, and a liner, which is removed prior to use. The BAC diffuses from the system, penetrates the skin, reaches subcutaneous blood vessels, and spreads through the body.[6]

Nano-sized drug carriers, mainly *liposomes, nanocapsules, nanospheres, and nano-aggregates*, whose sizes vary from several tens to several hundreds of nanometers, have recently begun to attract serious attention.[7]

The optimal routes of administration of nano-sized drug suspensions are injections (e.g., intravenous), inhalation, or oral administration. These formulations are also used as components of eye drops.

Liposomes are effective drug carriers, which usually consist of a lipid bilayer in the form of spherical structures; the BAC may be embedded either inside the liposomes or in the lipid layer. Being composed of natural lipids, liposomes are harmless, and, owing to their small size, compatible with water and biological fluids. A number of liposomal formulations are produced commercially.

Liposomes, however, have some serious drawbacks. They are not stable enough in aqueous systems, and they may aggregate and break up in the presence of different substances such as polycations. Other factors limiting their use are their high cost and storage instability of the lipids used to prepare them.

Various approaches are used to improve the stability of liposomes, but the most promising one is the modification of their membranes by low-molecular-weight amphiphilic water-soluble polymers. The nonpolar moiety of the polymer is introduced into the nonpolar layer of the liposomal membrane, while the polar moiety forms a coating on the surface of the liposome, protecting it from destructive agents. Moreover, this modification prevents the side capture of liposomes by reticuloendothelial cells during injection.

Low-molecular-weight (M = 1500–4000) polymers of ethylene oxide with hydrophobic groups are the amphiphilic polymers that are commonly used to modify liposomes. Polymers of N-vinylpyrrolidone and acrylamide containing long-chain (up to C_{20}) aliphatic groups, which are most often incorporated in the form of end fragments, are also used for the modification of liposomes.[8]

Not only liposomes but also submicron (10–1000 nm) formulations can be used as nanosize drug carriers compatible with aqueous media. These are nanospheres (polymeric spherical solid particles with a drug embedded in them) and nanocapsules (polymeric hollow spheres containing a drug). These polymer systems may be prepared by the polymerization of acrylamide, methyl methacrylate, and esters of cyanoacrylic acid or by using preformed polymers: albumin and polyesters of hydroxycarboxylic acids (lactic and glycolic acids).

Readily synthesized nanoaggregates of amphiphilic polymers, including those capable of enhancing water compatibility of the formulation, are also promising BAC carriers.[9]

Considerable attention has been devoted recently to systems for BAC delivery under preset conditions or in response to environmental conditions. These are so-called smart (intelligent) systems or feedback systems. To some extent, these systems model processes that occur in the body, and they only need certain upgrading to become physiologically optimal therapeutic systems.[10]

Well-known representatives of such systems are hydrogels, whose swelling behavior changes depending on the temperature of the environment, e.g., gels based on N-isopropylacrylamide copolymers. Aqueous solutions of its homopolymers are characterized by the low critical solution temperature (LCST) and form an individual phase at 30–35°C.

Membranes, films, coatings, and fibers based on N-isopropylacrylamide copolymers and other polymers with LCST that behave in the same way have been used to model systems releasing BACs to the environment at a certain temperature.[11]

Similar behavior is characteristic of pH-sensitive hydrogels prepared from ionogenic monomers, whose degree of swelling changes depending on the pH of the medium.[12]

More complex, "smart" polymer systems have been developed for the controlled release of insulin. As an increase in blood glucose level is the main factor stimulating insulin production by the pancreas, glucose-sensitive polymer systems hold promise for the treatment of diabetes. These systems may be implanted to patients and release controlled amounts of insulin in response to an increase in blood glucose level.[13]

Another approach to the delivery of drugs to the target organ includes the incorporation of ferromagnetic substances into the drug followed by the application of a magnetic field on the target organ.[14]

This approach has been developed in a number of studies, but the important issue of elimination of ferromagnetic particles from the organism remains unresolved.

Controlled release polymer formulations of BACs are used not only in medicine. The encapsulation of fertilizers by polymer coatings significantly reduces their amounts applied to the soil as nutrients for plants; fumigant devices and sustained-release pheromone formulations are used in traps for insect pests; polymer antifouling coatings are employed to paint the bottoms of boats and ships.

BIOLOGICALLY ACTIVE POLYMERS

In addition to various forms containing chemically unbound BACs, much attention has been paid to high-molecular-weight compounds that show various types of biological activity. Special consideration should be given to active polymers soluble in water.

Polymers with Inherent Biological Activity

Polymers with inherent biological activity, which is determined by their macromolecular nature, can model naturally occurring polymers and may be used to replace natural blood components.

One of their major uses is to serve as macromolecular components of vital medications—blood substitutes. These polymers perform two important functions of blood proteins: They maintain osmotic properties of blood (components of anti-shock substitutes or hemodynamic blood substitutes) and form complexes with toxic substances in the blood and then remove them from the body (components of detoxifying blood substitutes).

Water-soluble nonionogenic polymers are used as components of blood substitutes.[15]

Biosynthesis-based quantitative recovery of plasma proteins after blood loss takes several days. Administration of hemodynamic blood substitutes expedites this process considerably.

Hemodynamic blood substitutes serve as plasma proteins, primarily serum albumin, which is a major contributor to blood osmotic pressure. In order to perform this function, the polymer must have rather high molecular weight (at least 50–60 kDa), and, thus, naturally occurring (chemically premodified) polymers that can be biodegraded in the body and eliminated from it are used as a basis for this type of medication. These are dextran, partially hydroxyethylated starch, and gelatin—a protein derived by the denaturation of collagen.

Although hemodynamic blood substitutes (such as a dextran formulation with molecular weight about 35,000) are able to form complexes with toxic substances and gradually remove them from the body, this process can be speeded up by using a special group of detoxifying blood substitutes, whose major components are lower molecular weight polymers with molecular weight about 10 kDa. Polymers with this molecular weight are readily excreted from the body with urine, through the kidneys. Therefore, carbochain polymers, such as poly(N-vinylpyrrolidone), poly[N-(2-hydroxypropyl)methacrylamide], and poly(vinyl alcohol), are used in these medications. These polymers have high complexing ability toward various toxicants.

Among pharmaceutical polymers, polymer components of hemodynamic and detoxifying blood substitutes are commercially produced on the largest scale.

Oxygen carriers also belong to blood substitutes. Modified hemoglobin preparations considered in the section "Examples of Forms with Controlled Release of the BAC" are among the polymer forms of oxygen carriers.

Another group of polymers with inherent biological activity comprises water-soluble polyelectrolytes with different types of biological activity. They exert a biocidal effect on microorganisms by mimicking the action of some naturally occurring polymers. Various synthetic polyelectrolytes also show biocidal activity.

Microbicidal activity is exhibited by cationic polyelectrolytes. They are mainly represented by nitrogen-containing polymers, with side- or main-chain primary, secondary, and tertiary amine groups or quaternary ammonia groups, which facilitate the destruction of lipid layers. These polymers interact with cell membranes, causing the breakdown and agglutination of cells or stimulating the absorption of microorganisms by macrophages. Among them are polyethylenimine; homo- and copolymers carrying units of vinylamine and vinylpyridine and their salts; and ionenes, polymers containing main-chain quaternary ammonia groups.

Water-soluble cationites also show other types of biological activity. Polycations are able to form complexes with heparin-natural polysaccharides regulating blood coagulation, which is introduced into the bloodstream to reduce blood clotting before various cardiovascular operations, especially those involving the use of an artificial blood-circulation apparatus. After the surgery, heparin must be removed or neutralized to stop its anticoagulant effect and normalize blood clotting. Excess heparin can be neutralized by polycations.[16]

Biological activity is a characteristic of water-soluble polymer anions (polyanionites). Polyanions containing sulfonate groups may be used as anticoagulant analogs of heparin. Very interesting materials are polycarboxyl polymers, such as hydrolyzed copolymers of divinyl ether and maleic anhydride (the so-called pyran copolymer), which contains substituted pyran and tetrahydrofuran units, and a less toxic copolymer of furan, maleic anhydride, and acrylic acid.

Polyanions may affect the function of absorbing cells of the body, and this may be the reason for their antiviral activity, which makes polyanions promising components of antiviral vaccines. An important aspect of the activity of polyanions is that they activate the formation of interferons—a group of protective proteins generated by vertebrates' cells. Enhanced concentrations of interferons increase the resistance of the body to cancer, and thus polyanions are regarded as substances with anticancer and antiviral activity.[17]

Polymeric N-oxides, e.g., N-oxides of poly (vinylpyridines) exhibiting antisilicosis activity, have inherent biological activity.[18] A polymer containing N-oxide groups, polyoxidonium (the copolymer of N-oxide of 1,4-ethylenepiperazine and (N-carboxyethyl)-1,4-ethylenepiperazinium bromide) has a number of valuable properties. Polyoxidonium exhibits immune modulating activity and enhances the resistance of the body to local and generalized infections; it is used as an adjunct in artificial vaccines.[19]

Polymer Systems with Immobilized Fragments Determining Biological Activity (Group 2.2)

Polymers that do not release BACs

Among biologically active polymers, there are several groups of high-molecular-weight compounds in which a BAC or a group-determining activity is bound to the polymer carrier by the chemical bond that does not break during the functioning of the system.

For instance, *stabilized (or immobilized) enzymes* are used as components of water-soluble drug formulations. The enzyme bound with the polymer carrier or modifier is more resistant to denaturation, which would cause the loss of activity, and insoluble forms acquire the ability to repeatedly take part in enzymatic processes.

The preparation of water-soluble immobilized enzymes is based on the following principle: some functional groups in the protein do not take part in the formation of its active site and can participate in various interactions, including chemical reactions, with functional groups of the polymer modifier (carrier), preserving the water solubility of the system.[20]

An important property of polymer-modified protein in water-soluble systems is that it circulates in the blood stream for longer time periods, which enhances the efficiency of the medication considerably. For instance, dextran-modified streptokinase enzymatic formulations (streptodecase) are effectively used as a fibrinolytic drug.[21]

Insoluble forms of immobilized enzymes are also used in medicine, as components of textile materials for covering wounds and burns.

Formulations of hemoglobin modified by water-soluble polymers and oligomers and used as oxygen-carrying blood substitutes also belong to macromolecular systems containing non-released immobilized fragments. These formulations not only restore oxygen supply to tissues but also facilitate hematopoiesis.[20]

Insoluble forms of immobilized biocatalysts are widely used in biotechnological processes.

Another example of systems with the stably bound (immobilized) ingredient that determines biological activity is a group of immunoactive polymers—conjugates of a polymer carrier and an active, usually low-molecular-weight, group (hapten) causing the stimulation of receptors of immune competent cells.

The function of haptens may be performed by various substances, both similar to the group of antigenic determinants and different from it, such as vitamins, peptides, coenzymes, aromatic nitro compounds, etc. Haptens are used to reveal the structure of the active antigenic determinant and to prepare artificial vaccines in which the polymer modifier enables the interaction of the system with the surface of immune competent cells.

Building artificial vaccines may involve the use of polymer systems containing polymer carriers of antigens (adjuvants), usually polyelectrolyte, which enhance the effect of antigens. Polymer adjuvants stimulate the reproduction and dispersion of precursors of immune competent cells, favor cell differentiation processes, and enhance cell interactions.[22]

Polymers releasing BACs

Such serious disadvantages of low-molecular-weight drug formulations and other biological regulators and biocides as non-optimal doses and concentrations, limited functioning time, rapid unwanted consumption, and insufficient solubility may be eliminated or alleviated if BACs are used as chemical compounds hydrolyzed over time with carriers or modifiers; the most commonly used materials for this are various polymers. These chemical compounds are actually new biologically active polymers whose chemical structure is different from that of the original polymer carriers.

The chemical bond between the BAC and polymer carrier may be disintegrated at a certain rate via hydrolysis, frequently assisted by enzyme systems. The rate of this sustained release may be controlled by the polymer structure or construction of the biologically active system. Thus, these systems, as well as formulations enabling the controlled release of BACs described earlier (in the section "Examples of Forms with Controlled Release of the BAC"), can be classed with controlled release systems.

In contrast to insoluble dosage forms, polymer derivatives of BACs may be prepared as water-soluble formulations with wide applications.

Controlled release of the active ingredient enables extended action of the drug and prevents overdose and, hence, side effects. A very important aspect of developing new systems with the controlled release of the active ingredient is the creation of the polymer for targeted drug delivery.

Numerous studies in this field suggested a schematic structure of the typical medicinal polymer with the controlled release of the active ingredient. This polymer model includes not only the released active ingredient but also a group determining the water solubility of the entire system and a vector group, which determines the affinity of the injected polymer for the tissues of the target organ.[23]

Systems with the gradual or controlled release of the active ingredient form a most important group of biologically active polymer systems. Very many polymer systems with immobilized BACs of different activities have been synthesized and described in a great number of publications such as Plate & Vasilyev,[16] Shtilman,[20] Afinogenov & Panarin,[24] and Duncan.[25] Investigations of immobilization of antitumor antibiotics on poly[N-(2-hydroxypropyl) methacrylamide] derivatives containing side oligopeptide spacers, which provide the enzymatic splitting of the BAC and the monosaccharide vector fragment, and systems using poly(L-glutamic acid) as a carrier are among the most advanced studies in this field.[27–30]

Systems delivering genetic material to the cell also belong to medicinal systems releasing biologically active ingredients. These are complexes of nucleic acids or their fragments and polymer carriers, usually bound together by ionic bonds. These systems fall within the scope of *genetic engineering*.[31]

The most advanced biological objects used in genetic engineering studies are microorganisms and plant cells, but more recent studies report research on genetically engineered animals. It is assumed that genetic engineering may be an effective way to combat human genetic disorders and predisposition to certain diseases.

Polymer carriers are used in genetic engineering to deliver macromolecules carrying genetic information to the cell. On the one hand, they are rapidly destroyed by DNase enzymes in the lymphatic system after introduction into tissue and protect the macromolecules from the immune attack, and, on the other hand, they facilitate the penetration of products of immobilization into the cell via endocytosis and release the immobilized substance at the target site.

In contrast to various carriers used in certain technologies, polymer carriers are much more accessible and, in many cases, more effective. Complexes of polymer carriers and macromolecules carrying genetic information are called biological complexes, bioplexes, or polyplexes.

Polymer carriers intended for the delivery of genetic material to the cell must have reduced toxicity and immunogenicity.

Water-soluble amine-containing polymers—such as polylysine, polyethylenimine, and their derivatives; poly(amino amido) dendrimers; chitosan derivatives, etc.—may be employed as carriers of nucleic acids and their fragments containing the acidic function in the form of phosphoric acid residues.[32]

In addition to polymers with medicinal properties, there are systems showing different bio-regulatory or biocidal activities toward other groups of organisms: regulation of plant growth,[33] fungicidal activity,[34] etc.

CONCLUSION

This brief entry shows that extensive research on employing polymers in therapeutic systems has been carried out and its results have been widely used. The introduction of macromolecular components imparts more useful qualities—prolonged action, predetermined solubility level, targeted delivery of the active ingredient, prevention of its unwanted consumption, and some other properties—to these systems.

Further development of research in this field and practical application of its results should be expected.

REFERENCES

1. Lehmann, K.; Drehe, D. Coating of tablets and small particles with acrylic resins by fluid bed technology. Int. J. Pharm. Technol. Prod. Manuf. **1981**, *2* (4), 31–43.
2. Lehmann, K.M. *Practical Course in Film Coating of Pharmaceutical Dosage Forms with EUDRAGIT*; Rohm Pharma Polymers: Darmstadt, Germany, 2001.
3. Gutcho, M.H. *Capsule Technology and Microencapsulation*; Noyes Data Corporation: Park Ridge, NJ, 1972.
4. Benita, S., Ed. *Microencapsulation: Methods and Industrial Application*, 2nd Ed.; CRC Press: Boca Raton, 2005.
5. Barnhart, S.D.; Sloboda, M.S. The future of dissolvable films. Drug Deliv. Technol. **2007**, *7* (8), 34–37.
6. Langer, R. Transdermal drug delivery: Past progress, current status and future prospects. Adv. Drug Deliv. Rev. **2004**, *56* (5), 557–558.
7. Torchilin, V.P. Recent advances with liposomes as pharmaceutical carriers. Nat. Rev. Drug Discov. **2005**, *4* (2), 145–160.
8. Torchilin, V.P.; Shtilman, M.I.; Trubetskoy, V.S.; Whiteman, K.; Milshtein, A.M. Amphyphylic vinyl polymers effectively prolong liposome circulation time *in vivo*. Biochimica et Biophysica Acta **1994**, *1195* (1), 181–184.
9. Torchilin, V.P.; Levchenko, T.S.; Whiteman, K.R.; Yaroslavov, A.A.; Tsatsakis, A.M.; Rizos, A.K.; Michailova, E.V.; Shtilman, M.I. Amphiphilic poly-N-vinylpyrrolidones: Synthesis, properties and liposome surface modification. Biomaterials **2001**, *22* (22), 3035–3044.
10. Kasagana, V.N.; Karumuri, S.S. Recent advances in smart drug delivery systems. Int. J. Nov. Drug Deliv. Tech. 2011, *1* (3), 201–207.
11. Galaev, I.Y.; Matiason, B. Thermoreactive water-soluble polymers, nonionic surfactants, and hydrogels as reagents in biotechnology. Enzyme Microb. Tech. **1993**, *15* (5), 354–366.
12. Ghandehari, H.; Kopeckova, P.; Yeh, P.-Y.; Kopecek, J. Biodegradable and pH sensitive hydrogels: Synthesis by a polymer-polymer reaction. Macromol. Chem. Phys. **1996**, *197* (3), 965–980.
13. Valuev, L.I.; Valueva, T.A.; Valuev, I.L.; Plate, N.A. Polymeric systems for controlled release of bioactive compounds. Adv. Biol. Chem. **2003**, *43* (2), 307–328.
14. Mosbach, K.; Schroder, U. Preparation and application of magnetic polymers for targeting drugs. FEBS Lett. **1979**, *102* (1), 112–116.
15. Winslow, R., Ed. *Blood Substitutes*; Elsevier: London, 2006.
16. Plate, N.A.; Vasilyev, A.E. *Physiologically Active Polymers*; Chemistry: Moscow, 1986.
17. Seymour, L. Synthetic polymers with intrinsic anti-cancer activity. J. Bioact. Compat. Polym. **1991**, *6* (2), 178–216.
18. Solovsky, M.V.; Panarin, E.F. Polymers in medicinal systems (brief review). In *Synthesis and Transformations of Polymers*, Monakov, Y.B., Ed.; Chemistry: Moscow, 2003; 163–176.
19. Petrov, R.V.; Khaitov, R.M.; Nekrasov, A.V.; Ataullakhanov, R.I.; Pinegin, B.V.; Puchkova, A.S.; Ivanova, A.S. Polyoxidonium – Latest-generation immune modulator: Results of the three-year clinical use. Allergy Asthma Clin. Immunol. **1999**, *3*, 3–6.
20. Shtilman, M.I. *Immobilization on Polymers*; VSP: Utrecht, Tokyo, 1993.
21. Torchilin, V.P., Ed. *Immobilized Enzymes in Medicine*; Springer-Verlag: Berlin, 1991.
22. Petrov, R.V.; Khaitov, R.M. *Artificial Antigens and Vaccines*; Medicina: Moscow, 1988.
23. Ringsdorf, H. Structure and properties of pharmacologically active polymers. J. Polym. Sci. **1975**, *51* (1), 135–153.
24. Afinogenov, G.E.; Panarin, E.F. *Antimicrobial Polymers*; Gippokrat: St. Petersburg, 1993.
25. Duncan, R. The dawning era of polymer therapeutics. Nat. Rev. Drug Discov. **2003**, *2* (5), 347–360.
26. Lu, Z.R.; Shiah, J.C.; Sakuma, S.; Kopechkova, P.; Kopechek, J. Design of novel bioconjugates for targeted drug delivery. J. Control. Release **2002**, *78* (2), 165–173.
27. Kopecek, J. The potential of water soluble polymeric carriers in targeted and site specific drug delivery. J. Control. Release, **1990**, *11* (2), 279–290.
28. Duncan, R.; Vicent, M.J.; Greco, F.; Nicholson, R.I. Polymer-drug conjugates: Towards a novel approach for the treatment of endrocine-related cancer. Endocr. Relat. Cancer **2005**, *12* (Suppl. 1), 189–199.
29. Allen, T.M.; Cullis, P.R. Drug delivery systems: Entering the mainstream. Science **2004**, *303* (5665), 1818–1822.
30. Allen, T.M. Liugand-targeted therapeutics in anticancer therapy. Nat. Rev. Drug Discov. **2002**, *2* (10), 750–763.
31. Anderson, W.F. Human gene therapy. Nature **1998**, *392* (Suppl. 6679), 25–30.
32. Klimenko, O.V.; Shtilman, M.I. Transfection of Kasumi-1 cells with a new type of polymer carriers loaded with miR-155 and antago-miR-155. Cancer Gene Ther. **2013**, *20* (4), 237–241.
33. Tsatsakis, A.M.; Shtilman, M.I. Phytoactive polymers. New synthetic plant growth regulators. In *Plant Morphogenesis*; Angelakis, K.; Tran Thah Van, Eds.; Plenum. Publ. Co.: New York, London, 1993; 255–272.
34. Shtilman, M.I.; Tzatzarakis, M.; Tsatsakis, A.M.; Lotter, M.M. Polymeric fungicides: A review. High Mol. Weight Compd. B **1999**, *41* (8), 1363–1375.

Bioadhesive Drug Delivery Systems

Ryan F. Donnelly
A. David Woolfson
School of Pharmacy, Medical Biology Center, Queen's University Belfast, Belfast, U.K.

Bioactive—BioMEMS

Abstract

Bioadhesive drug delivery systems have been widely studied in the past two decades. Primarily, the application of bioadhesion to drug delivery involves the process of mucoadhesion, a "wet-stick" adhesion requiring, primarily, spreading of the mucoadhesive on a mucin-coated epithelium, followed by interpenetration of polymer chains between the hydrated delivery system and mucin. An alternative mechanism can involve the use of specific bioadhesive molecules, notably lectins, primarily for oral drug delivery applications. The most notable applications of mucoadhesive systems to date remain those involving accessible epithelia, such as in ocular, nasal, buccal, rectal, or intravaginal drug delivery systems. The ancillary effects of certain mucoadhesive polymers, such as polycations, on promoting drug penetration of epithelial tissues, may be increasingly important in the future, particularly for the transmucosal delivery of therapeutic peptides and other biomolecules.

THEORIES OF BIOADHESION

Adhesion can be defined as the bond produced by contact between a pressure-sensitive adhesive (PSA) and a surface.[1] The American Society for Testing and Materials[2] has defined it as the state in which two surfaces are held together by interfacial forces, which may consist of valence forces, interlocking action, or both. Good[3] defined bioadhesion as the state in which two materials, at least one biological in nature, are held together for an extended period of time by interfacial forces. It is also defined as the ability of a material, synthetic or biological, to adhere to a biological tissue for an extended period of time. In biological systems, three main types of bioadhesion can be distinguished:

Type 1: Adhesion between two biological phases, for example, cell fusion, platelet aggregation, wound healing, adhesion between a normal cell and a foreign substance or a pathological cell.

Type 2: Adhesion of a biological phase to an artificial substrate, for example, cell adhesion to culture dishes, platelet adhesion to biomaterials, microbial fouling and barnacle adhesion to ships, biofilm formation on prosthetic devices and inserts.

Type 3: Adhesion of an artificial material to a biological substrate, for example, adhesion of synthetic hydrogels to soft tissues[4] and adhesion of sealants to dental enamel.

Cell-to-cell adhesion can be explained as a balance between nonspecific, repulsive, and attractive physical forces and macromolecular bridges.[5] In contrast to Type 1 adhesion, Types 2 and 3 involve one synthetic phase. When many hard synthetic materials contact a biological fluid, there is formation of an interfacial film of biomolecules. This Type 2 bioadhesion is usually irreversible[6] and has important implications in areas such as artificial kidney hemodialysis where there exists a risk of blood damage. The adhesion of human endothelial cells to polymeric surfaces is dependent on surface wettability.[7] Optimum adhesion occurs when surfaces show moderate wettability, with more hydrophilic and more hydrophobic polymers showing less, or even no, adhesion. Polymers of both natural and synthetic origin comprise the majority of examples that display Type 3 bioadhesion. Polymeric matrices are, of course, the basis of many novel drug delivery systems, including adhesive patches for transdermal and topical indications.[8,9]

For drug delivery purposes, the term bioadhesion implies attachment of a drug carrier system to a specified biological location. The biological surface can be epithelial tissue or the mucus coat on the surface of a tissue. If adhesive attachment is to a mucus coat, the phenomenon is referred to as mucoadhesion. Leung and Robinson[10] described mucoadhesion as the interaction between a mucin surface and a synthetic or natural polymer.

For bioadhesion to occur, a succession of phenomena is required. The first stage involves an intimate contact between a bioadhesive and a biological substrate, either from wetting of the bioadhesive surface or from swelling of the bioadhesive. In the second stage, after contact is established, penetration of the bioadhesive into the crevice of the tissue surface or interpenetration of the chains of the bioadhesive with those of the substrate takes place. Secondary chemical bonds can then establish themselves.[11]

Concise Encyclopedia of Biomedical Polymers and Polymeric Biomaterials DOI: 10.1081/E-EBPPC-120052270

111

One of the most important factors for bioadhesion is tissue surface roughness. Tissue surfaces often display undulations and crevices that provide both attachment points and a key for the deposition and inclusion of polymeric adhesives. These surface irregularities have been analyzed by Merrill,[12] who stated that the surface geography can be expressed in terms of an aspect ratio of maximum depth, d, to maximum width, h, as shown in Fig. 1.

Surfaces that have a ratio of $d/h < 1/20$ are considered too smooth for satisfactory adhesion.[11] For higher ratio values, only adhesives of low viscosity can penetrate into the tissue topography to form mechanical or physical bonds.

Viscosity and wetting are also important factors in bioadhesion.[7] Bioadhesion can be considered to be a wet-stick adhesion process.[9] During wet-stick, polymer chains are released from restraining dry lattice forces by hydration. These chains then move and entangle into the matrix of the substrate. Interaction may then occur on a molecular level between bioadhesive and substrate. The interaction between two molecules is composed of attraction and repulsion. Attractive interactions arise from van der Waal's forces, electrostatic attraction, hydrogen bonding, and hydrophobic interaction. Repulsive interactions occur because of electrostatic and steric repulsion. Attractive and, or, repulsive interactions may occur between the bioadhesive and the lipids, glycolipids, proteins, glycoproteins, and polysaccharides found on epithelial cell membranes. Alternatively the bioadhesive may interact with the glycocalyx, a structure containing polysaccharides on the cell surface. In regions of the body where mucus overlies the epithelial surface, such as the gastrointestinal tract (GIT), the interaction may largely be between bioadhesive and the mucin glycoproteins rather than the epithelial cell itself.[13] For bioadhesion to occur, the attractive interaction should be larger than nonspecific repulsion.[14]

Various theories exist to explain at least some of the experimental observations made during the bioadhesion process. Unfortunately, each theoretical model can only explain a limited number of the diverse range of interactions that constitute the bioadhesive bond.[15] However, four main theories can be distinguished.

Wetting Theory of Bioadhesion

The mechanical or wetting theory is perhaps the oldest established theory of adhesion. It explains adhesion as an embedding process, whereby adhesive molecules penetrate into surface irregularities of the substrate and ultimately harden, producing many adhesive anchors. Free movement of the adhesive on the surface of the substrate means that it must overcome any surface tension effects present at the interface.[16] The wetting theory is best applied to liquid bioadhesives, describing the interactions between the bioadhesive, its angle of contact, and the thermodynamic work of adhesion.

The work done is related to the surface tension of both the adhesive and the substrate, as described by Equation 1, the Dupré equation:[17]

$$\omega_A = \gamma_b + \gamma_t - \gamma_{bt} \tag{1}$$

where

ω_A is the specific thermodynamic work of adhesion

γ_b, γ_t, and γ_{bt} represent the surface tensions of the bioadhesive polymer, the substrate, and the interfacial tension, respectively

The adhesive work done is a sum of the surface tensions of the two adherent phases, less the interfacial tensions apparent between both phases.[18] Figure 2 shows a drop of liquid bioadhesive spreading over a soft tissue surface.

Horizontal resolution of the forces gives the Young equation (Equation 2):

$$\gamma_{ta} = \gamma_{bt} + \gamma_{ba} \cos \theta \tag{2}$$

where

θ is the angle of contact

γ_{bt} is the surface tension between the tissue and polymer

γ_{ba} is the surface tension between polymer and air

γ_{ta} is the surface tension between tissue and air

Equation 3 states that if the angle of contact, θ, is greater than zero, the wetting will be incomplete. If the vector γ_{ta} greatly exceeds $\gamma_{bt} + \gamma_{ba}$, that is

$$\gamma_{ta} \geq \gamma_{bt} + \gamma_{ba} \tag{3}$$

then θ will approach zero and wetting will be complete. If a bioadhesive material is to successfully adhere to a biological surface, it must first dispel barrier substances and then spontaneously spread across the underlying substrate, either tissue or mucus. The spreading coefficient, Sb, can be defined as shown in Equation 4:

$$S_b = \gamma_{ta} - \gamma_{bt} - \gamma_{ba} > 0 \tag{4}$$

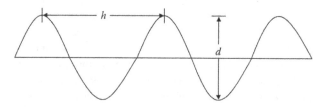

Fig. 1 Schematic representation of the surface roughness of a soft tissue.

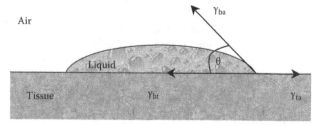

Fig. 2 Liquid bioadhesive spreading over a typical soft tissue surface.

which states that bioadhesion is successful if S_b is positive, thereby setting the criteria for the surface tension vectors, in other words, bioadhesion is favored by large values of γ_{ta} or by small values of γ_{bt} and γ_{ba}.[18]

Electrostatic Theory of Bioadhesion

Separation of two distinct bodies is achieved by overcoming electrostatic forces originating from the establishment of an electrical double layer between an adhesive and a substrate. The electrostatic theory of repulsion proposes the transfer of electrons across the adhesive interface, establishing the electrical double layer and a series of attractive forces responsible for maintaining contact between the two layers.[19] The adhesive interface is considered to be analogous to a parallel-plate condenser, such that work must be done against any electrical charges before separation is achieved. The work of adhesion can be equated to the energy of the condenser, provided that no work is done in overcoming van der Waal's forces.

Diffusion Theory of Bioadhesion

The interpenetration or diffusion theory is currently the most widely accepted physical theory and was first proposed by Voyutskii,[20] who studied the autodiffusion of two contacting, identical polymers. The term autodiffusion was used to describe the diffusion of long-chain molecules across the interface from one polymer into an identical one and is the most likely explanation for the self-tact of rubbers. Many adhesives are applied to the adherents as a solvent or aqueous solution and, on evaporation of the vehicle, the polymer residues are combined. Failure of the joint is not normally expected along the plane of combination because diffusion has made the two layers into one, with a loss of the interface.[18]

While the concept of autodiffusion can be used to explain the autodiffusion of polymers and the heat sealing of thermoplastics, the universality of this theory is not widely accepted. For example, adhesion between different polymers and polymer-to-metal adhesion are not believed to occur as a result of diffusion.[18]

With regard to bioadhesion, polymeric chains from the bioadhesive and the biological substrate, for example the glycoprotein mucin chains found in mucus, intermingle and reach a sufficient depth within the opposite matrix to allow formation of a semipermanent bond.[1] The process can be visualized from the point of initial contact. The existence of concentration gradients will drive the polymer chains of the bioadhesive into the mucus network and the glycoprotein mucin chains into the bioadhesive matrix until an equilibrium penetration depth is achieved as shown in Fig. 3.

The exact depth needed for good bioadhesive bonds is unclear, but is estimated to be in 0.2–0.5 μm range.[21] The

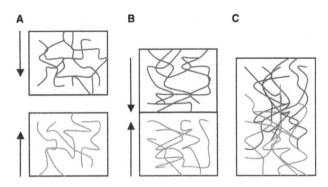

Fig. 3 Schematic representation of the diffusion theory of bioadhesion. Polymer layer and mucus layer before contact (**A**), upon contact (**B**), the interface becomes diffuse after contact for a period of time (**C**).

mean diffusional depth of the bioadhesive polymer segments, s, may be represented by Equation 5:

$$s = \sqrt{2tD} \tag{5}$$

where

    D is the diffusion coefficient

    t is the contact time

Duchêne Touchard, and Peppas[11] adapted Equation 5 to give Equation 6, which can be used to determine the time, t, to bioadhesion of a particular polymer:

$$t = \frac{l^2}{D_b} \tag{6}$$

where

    l represents the interpenetrating depth

    D_b the diffusion coefficient of a bioadhesive through the substrate

Once intimate contact is achieved, the substrate and adhesive chains move along their respective concentration gradients into the opposite phases. Depth of diffusion is obviously dependent on the diffusion characteristics of both phases. Reinhart and Peppas[22] reported that the diffusion coefficient depended on the molecular weight of the polymer strand and that it decreased with increasing cross-linking density. This process of simple physical entanglement has been described as analogous to forcing two pieces of steel wool to intermingle upon contact.[13]

Adsorption Theory of Bioadhesion

According to the adsorption theory, after an initial contact between two surfaces, the materials adhere because of surface forces acting between the chemical structures at the two surfaces.[23] When polar molecules or groups are present, they reorientate at the interface.[18] Chemisorption can occur when adhesion is particularly strong. The theory maintains that adherence to tissue is due to the net result of one or more secondary forces.[13,24,25] The three types of secondary forces are:

1. van der Waal's forces
2. Hydrogen bonding
3. Hydrophobic interaction

The formation of primary chemical bonds, such as those formed between adhesives used in dentistry and surgery and tissue, is undesirable in bioadhesion. This is because the high strength of ionic or covalent bonds may result in permanent adhesion.[23]

BIOADHESIVE MATERIALS

Bioadhesive polymers have numerous hydrophilic groups, such as hydroxyl, carboxyl, amide, and sulfate. These groups attach to mucus or the cell membrane by various interactions such as hydrogen bonding and hydrophobic or electrostatic interactions. These hydrophilic groups also cause polymers to swell in water and, thus, expose the maximum number of adhesive sites.[13]

An ideal polymer for a bioadhesive drug delivery system should have the following characteristics:[1,23]

1. The polymer and its degradation products should be nontoxic and nonabsorbable.
2. It should be nonirritant.
3. It should preferably form a strong non-covalent bond with the mucus or epithelial cell surface.
4. It should adhere quickly to moist tissue and possess some site specificity.
5. It should allow easy incorporation of the drug and offer no hindrance to its release.
6. The polymer must not decompose on storage or during the shelf life of the dosage form.
7. The cost of the polymer should not be high, so that the prepared dosage form remains competitive.

Polymers that adhere to biological surfaces can be divided into three broad categories:[1,23]

1. Polymers that adhere through nonspecific, non-covalent interactions which are primarily electrostatic in nature
2. Polymers possessing hydrophilic functional groups that hydrogen bond with similar groups on biological substrates
3. Polymers that bind to specific receptor sites on the cell or mucus surface

The last category includes lectins, which are generally defined as proteins or glycoprotein complexes of nonimmune origin that are able to bind sugars selectively in a non-covalent manner.[26] Lectins are capable of attaching themselves to carbohydrates on the mucus or epithelial cell surface and have been extensively studied, notably for drug targeting applications.[27,28] These second-generation bioadhesives provide not only for cellular binding but also for subsequent endo- and transcytosis. However, lectins have proved problematic, with recent literature indicating variability in drug absorption associated with their use,[13,29] as well as with their potential toxicity.[30,31] Table 1 shows the chemical structures of several bioadhesive polymers commonly used in modern drug delivery.

FACTORS AFFECTING BIOADHESION

Bioadhesion may be affected by a number of factors, including hydrophilicity, molecular weight, cross-linking, swelling, pH, and the concentration of the active polymer.[1,23,32]

Hydrophilicity

Bioadhesive polymers possess numerous hydrophilic functional groups, such as hydroxyl and carboxyl. These groups allow hydrogen bonding with the substrate, swelling in aqueous media, and allowing maximal exposure of potential anchor sites. Hydrophilic groups also allow polymers to swell in water. In addition, swollen polymers have the maximum distance between their chains leading to increased chain flexibility and efficient penetration of the substrate.

Molecular Weight

There is a certain molecular weight (mw) at which bioadhesion is at a maximum. The interpenetra-tion of polymer molecules is favored by low-molecular-weight polymers, whereas entanglements are favored at higher molecular weights. The optimum molecular weight for the maximum bioadhesion depends on the type of polymer, with bioadhesive forces increasing with the molecular weight of the polymer up to 100,000. Beyond this level, there is no further gain.[33]

Cross-Linking and Swelling

Cross-link density is inversely proportional to the degree of swelling.[34] The lower the cross-link density, the higher the flexibility and hydration rate; the larger the surface area of the polymer, the better the bioadhesion. Interpenetration of chains is easier as polymer chains are disentangled and free of interactions. The force generated by pulling water from the underlying substrate into the swelling polymer also helps polymer strands to penetrate deep into the substrate. To achieve a high degree of swelling, a lightly cross-linked polymer is favored. However, if too much moisture is present and the degree of swelling is too great a slippy mucilage results and this is easily removed from the substrate.[35] The bioadhesion of cross-linked polymers can be

Reason

Table 1 Chemical structures of some bioadhesive polymers used in drug delivery

Chemical name (Abbreviation)	Chemical structure
Poly(ethylene glycol) (PEG)	
Poly(vinyl alcohol) (PVA)	
Poly(vinyl pyrrolidone) (PVP)	
Poly(acrylic acid) (PAA or Carbopol)	
Poly(hydroxyethyl methacrylate) (PHEMA)	
Chitosan	
Hydroxyethylcellulose (HEC): R = H and CH_2CH_2OH Hydroxypropylcellulose (HPC): R = H and $CH_2CH(OH)CH_3$ Hydroxypropylmethylcellulose (HPMC): R = H, CH_3, and $CH_2CH(OH)CH_3$ Methylcellulose: R = H and CH_3 Sodium carboxymethylcellulose (NaCMC): R = H and CH_2COONa	

enhanced by the inclusion in the formulation of adhesion promoters, such as free polymer chains and polymers grafted onto the preformed network.[32]

Spatial Conformation

Besides molecular weight or chain length, spatial conformation of a polymer is also important. Despite a high molecular weight of 19,500,000 for dextrans, they have adhesive strength similar to that of polyethylene glycol (PEG), with a molecular weight of 200,000. The helical conformation of dextran may shield many adhesively active groups, primarily responsible for adhesion, unlike PEG polymers, which have a linear conformation.[1]

pH

The pH at the bioadhesive to substrate interface can influence the adhesion of bioadhesives possessing ionizable groups. Many bioadhesives used in drug delivery are polyanions possessing carboxylic acid functionalities. If the local pH is above the pK_a of the polymer, it will be largely ionized; if the pH is below the pK_a of the polymer, it will be largely unionized. The approximate pK_a for the poly(acrylic acid) (PAA) family of polymers is between 4 and 5. The maximum adhesive strength of these polymers is observed around pH 4–5 and decreases gradually above the pH of 6. A systematic investigation of the mechanisms of bioadhesion clearly showed that the protonated carboxyl groups rather than the ionized carboxyl groups react with mucin molecules, presumably by the simultaneous formation of numerous hydrogen bonds.[36] Fully ionized anionic polymers, despite showing maximal swelling in water, will tend to be repelled by mucus and epithelial surfaces, both of which carry a net negative charge. The pH at the bioadhesive to substrate interface can be influenced by the nature of the tissue or mucus substrate and the composition of the bioadhesive formulation. Polycations, such as poly(lysine), can bind to negatively charged mucus at pH 7.4, presumably by electrostatic interactions. However, cationic polymers tend not to adhere as strongly as anionic ones and the former may cause cell aggregation and other toxic reactions.[1,13]

Concentration of Active Polymer

Ahuja Khar, and Ali[23] stated that there is an optimum concentration of polymer corresponding to the best bioadhesion. In highly concentrated systems, the adhesive strength drops significantly. In concentrated solutions, the coiled molecules become solvent-poor and the chains available for interpenetration are not numerous. This result seems to be of interest only for more or less liquid bioadhesive formulations. It was shown by Duchêne Touchard, and Peppas[11] that, for solid dosage forms, such as tablets, the higher the polymer concentration, the stronger the bioadhesion.

Drug/Excipient Concentration

BlancoFuente et al.[37] showed that the addition of propranolol hydrochloride to Carbopol® (a lightly crosslinked PAA polymer) hydrogels increased adhesion when water was limited in the system, due to an increase in the elasticity caused by complex formation between drug and polymer. However, when large quantities of water were present, the complex precipitated out, leading to a slight decrease in adhesive character. Increasing toluidine blue O concentration in mucoadhesive patches based on Gantrez® (poly(methylvinylether/maleic acid)) significantly increased mucoadhesion to porcine cheek tissue.[38] This was attributed to increased internal cohesion within the patches due to electrostatic interactions between the cationic drug and anionic copolymer. By contrast, miconazole nitrate did not influence bioadhesion up to a concentration of 30% in PAA-based tablets.[39] Voorspoels et al.[40] demonstrated that increasing concentrations of testosterone and its esters produced a decrease in adhesive characteristics of buccal tablets.

For ionic bioadhesive polymers, dissolved salts can have a significant effect on bioadhesion by a variety of mechanisms, including salting out of polymeric components and charge neutralization, preventing expansion of coiled polymer chains and, thus, entanglement with epithelial mucin. Woolfson et al.[41,42] added sodium chloride to polymer blends containing Gantrez and found no effect on bioadhesive character, indicating that the sensitivity of bioadhesive polymers to the presence of dissolved salts varies

considerably, depending on structure. However, sodium chloride can affect bioadhesion through a direct action on the substrate, Thus, sodium chloride, when added to several mucoadhesive polymers, was observed to reduce the stiffness of mucus gel.[43]

Other Factors Affecting Bioadhesion

Bioadhesion may be affected by the initial force of application.[44] Higher forces lead to enhanced interpenetration and high bioadhesive strength.[11] In addition, the greater the initial contact time between bioadhesive and substrate, the greater the swelling and interpenetration of polymer chains.[14] Physiological variables can also affect bioadhesion. The rate of mucus turnover can be affected by disease states and by the presence of a bioadhesive device.[45] In addition, the nature of the surface presented to the bioadhesive can vary significantly depending on the body site and the presence of local or systemic disease.[14]

DETERMINATION OF BIOADHESIVE FORCE OF ATTACHMENT

The evaluation of bioadhesive properties is fundamental to the development of novel bioadhesive delivery systems. Measurement of the mechanical properties of a bioadhesive material after interaction with a substrate is one of the most direct ways to quantify the bioadhesive performance. To measure the force of bioadhesive attachment to tissue involves the application of a stress to the bonding interface. Numerous designs of apparatus have been proposed for this purpose. Since no standard apparatus is available for testing bioadhesive strength, an inevitable lack of uniformity between test methods has arisen. Nevertheless, three main testing modes are recognized: tensile, shear, and peel tests.

When the force of separation is applied perpendicularly to the tissue/adhesive interface, a state of tensile stress is set up. This is, perhaps, the most common configuration used in bioadhesive testing. During shear stress, the direction of the forces is reoriented so that it acts along the joint interface. In both tensile and shear modes, an equal pressure is distributed over the contact area.[46]

The peel test is most applicable to systems involving adhesive tape, where removal of the device is an important parameter. By pulling the two interfaces apart at an acute angle, the force is focused along a single line of contact. This effectively concentrates the applied force at the point of separation and removal of the tape can be achieved easily. The peel test is of limited use in most bioadhesive systems. However, it is of value when the bioadhesive system is formulated as a patch.[47]

Once formed, the interface of the bioadhesive bond is likely to exist in a microenvironment that will inevitably influence its further performance. Factors such as pH, temperature, ionic strength, and water content will all affect bond durability. Experimental rigs have been devised to investigate these important considerations. Although the force used to break the bond can be applied in one of three fashions, as described, the majority of researchers prefer to use the tensile stress method. Irrespective of which configuration is used to apply the force, the measurement of the performance of the bioadhesive bond is generally performed in one of three environments.

Most *in vitro* methods involve measurement of shear or tensile stress. Smart Kellaway, and Worthington[48] described a method to study bioadhesion whereby a polymer-coated glass slide was withdrawn from a mucus solution. The force required to accomplish this was equated to mucoadhesion. A fluorescence probe technique was developed by Park and Robinson[36] to determine bonding between epithelial cells

and a test polymer. Other methods have involved measurement of the force required to detach polymeric materials from excised rabbit corneal endothelium. Mikos and Peppas[49] described a method whereby a polymer particle was blown across a mucus-filled channel. The motion, recorded photographically, gave details regarding the adhesion process.

McCarron et al.[35,47,50] and Donnelly et al.[51] have reported extensively on the use of a commercial apparatus, in the form of a texture profile analyzer (Fig. 4) operating in bioadhesive test mode, to measure the force required to remove bioadhesive films from excised tissue *in vitro*.

The Texture Analyser, operating in tensile test mode and coupled with a sliding lower platform, was also used to determine peel strength of similar formulations (Fig. 5).[47]

The ultimate destination for successful bioadhesive devices is a tissue surface such as the GIT or the buccal cavity. Some investigators have used previous results obtained from their own *in vitro* work to predict and, subsequently test for, *in vivo* performance. Examples of *in vivo* studies in the literature are not as plentiful as *in vitro* work because of the amount and expense of animal trials and the difficulty in maintaining experimental consistency. Commonly measured variables include gastrointestinal transit times of bioadhesive-coated particles and drug release from in situ bioadhesive devices.

Ch'ng et al.[52] studied the *in vivo* transit time for bioadhesive beads in the rat. A ^{51}Cr-radiolabeled bioadhesive was carefully introduced into the stomach. At selected time intervals, the GITs were removed, cut into 20 equal segments, and the radioactivity in each was measured.

Reich et al.[53] described an instrument for measuring the force of adhesion between intraocular lens materials and the endothelium of excised rabbit corneas. It used a metal or

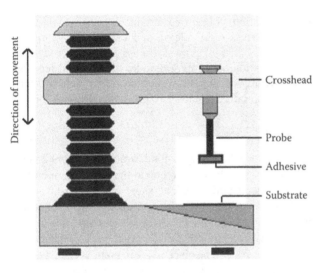

Fig. 4 Texture profile analyzer in bioadhesion test mode.

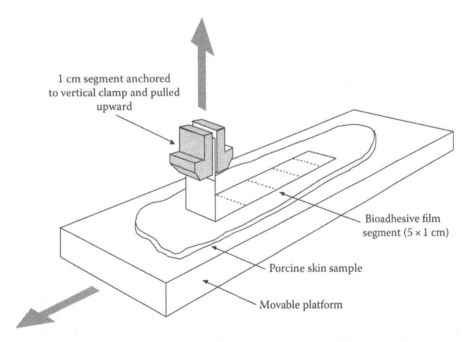

Fig. 5 Simplified representation of a typical test setup used to determine peel strength of bioadhesive films.

glass fiber deflection technique for measuring pressures that were usually lower than 1 g/cm². The contacting surfaces were submerged in saline to eliminate surface tension effects. In contrast to many bioadhesive testing situations, the authors were attempting to find a polymer with a lower adhesion than poly(methylmethacrylate) (PMMA), the material most commonly used to make intraocular lenses. A direct relationship had been shown between the adhesive forces of a polymer to the endothelium and the extent of cell damage. Poly(hydroxyethylmethacrylate) and Duragel®, both hydrophilic materials used to make soft lenses, gave less adhesion and subsequently less cell damage than the hydrophobic PMMA.

Davis[54] described a method to study gastric-controlled release systems. It can be advantageous in certain diseased states for a single dosage unit to be retained in the stomach. This enables released drug to empty from the stomach and have the length of the small intestine available for absorption. Mucoadhesives can help achieve this objective. Therefore, as an alternative to using invasive *in vivo* techniques, a formulation was used containing a gamma-emitting radionuclide. The release characteristics and the position of the device could be monitored using gamma scintigraphy. This technique could measure two different radionuclides simultaneously so that a subject could be given two different formulations together, with their release and transit times then being studied in a single occasion crossover study.

CHARACTERIZATION METHODS FOR BIOADHESIVE DRUG DELIVERY SYSTEMS

A range of techniques are available for the study of bioadhesive delivery systems, with the choice being influenced by site of attachment and substrate tissue, together with the design of the dosage form. Most viable bioadhesive drug delivery applications involve mucoadhesion to accessible (topical) epithelia. Certain tests will be required for regulatory purposes, whilst others, although optional, often constitute an important part of the development of a pharmaceutics package for a novel bioadhesive carrier. Bioadhesive (mucoadhesive) dosage forms vary in bioadhesive performance, with water content being a primary determinant of bioadhesion. Delivery platforms include liquids and gels (low adhesion), viscoelastic semisolids (moderate adhesion), flexible hydrogel films, particulates, and compacts (strong adhesion). Applicable characterization methodologies include:

- Drug release studies
- Drug diffusion (membrane penetration) studies
- Examination of mechanical and textural properties
- Examination of continuous shear properties
- Examination of structural (viscoelastic) properties at defined temperatures
- Evaluation of adhesion to model substrates
- Examination of product/packaging interactions

Texture profile analysis may be used to determine the following characteristics of semisolid bioadhesive delivery systems:[47,55]

- Hardness/compressibility—A measure of the resistance of the formulation to probe depression, which can be used to characterize product spreadability
- Cohesiveness—A measure of the effects of successive deformations on the structural properties of a product
- Adhesiveness—A measure of the work required to remove the probe from the sample, a property related to bioadhesion

Flow rheometry may be used to obtain the following information on bioadhesive semisolids:

- Effects of successive shearing stresses on the rate of deformation of a product
- Information concerning product viscosity which, in turn, affects drug release and ease of application at the site
- Information concerning the rate of structural recovery following deformation (thixotropy) which, in turn, affects product retention at the application site
- Drug release
- Manufacture

Oscillatory rheometry, a nondestructive test (unlike flow rheometry), may be used to obtain the information on bioadhesive semisolids used at sites subjected to variable stresses, for example, in the oral cavity, where chewing, swallowing, and talking all affect the structural rheology of bioadhesive systems.[56,57] Oscillatory rheometry determines strain in the system, resolved into two components: the storage modulus (G′) is the part of the strain which is in phase with the stress and is a measure of the solid character and the loss modulus (G″) is the part of the strain which lags the stress by 90° and is a measure of the liquid character.

Dielectric spectroscopy is an analytical technique which involves the application of an oscillating electric held to a sample and the measurement of the corresponding response over a range of frequencies, from which information on sample structure and behavior may be extrapolated. Craig and Tamburic[58] described studies on sodium alginate gels whereby a model was proposed in order to relate the low frequency response to the gel structure. They also reported on the application of dielectric spectroscopy to the study of cross-linked PAA with respect to the effects of additives such as propylene glycol and chlorhexidine gluconate, the influence of the choice of neutralizing agent, and the effects of ageing on the gel structure.

BIOADHESIVE DRUG DELIVERY SYSTEMS

There has been an increasing interest in the use of bioadhesive polymers in the design of drug delivery systems. One of the advantages of using these materials is that they can maintain contact with mucosal surfaces for much longer periods of time than non-bioadhesive polymers. Since polymers possessing bioadhesive properties can retain drugs in close proximity to membranes rich in underlying vasculature, they may offer a solution to the poor bioavailability of some drugs and a method to avoid enzymatic degradation of others.

Bioadhesive drug delivery systems have in the past been formulated as powders, compacts, sprays, semisolids, or films. For example, compacts have been used for drug delivery to the oral cavity[59] and powders and nanoparticles (NP) have been used to facilitate drug administration to the nasal mucosa.[60,61] Bioadhesive films are generally prepared from solutions of film-forming bioadhesive polymers in appropriate solvents. The final film is traditionally produced by casting of the solution into a mould of defined dimensions. A casting knife may be used to spread the solution and remove excess.[62] Film-forming bioadhesive polymers used in the production of bioadhesive films include the cellulose derivatives,[63] PAA such as Carbopol,[41,42] and Gantrez copolymers such as poly(methylvinylether/maleic anhydride).[9]

Bioadhesive Devices for the Oral Cavity

The oral cavity is a convenient and accessible area, ideally suited to bioadhesive drug delivery. The most commonly used areas are the buccal and sublingual areas. The non-keratinized regions in the oral cavity, such as the soft palate, the mouth floor, the ventral side of the tongue, and the buccal mucosa also offer least resistance to drug absorption.[64] In many instances, these areas are most suitable for locating bioadhesives. Molecular transport across the barrier membrane is achieved by simple diffusion along a concentration gradient from carrier to tissue, possibly aided by some form of penetration enhancement. Larger, hydrophilic molecules encounter greater resistance to diffusion and are believed to cross the oral epithelium by intercellular pathways. The advantages of drug delivery through the oral mucosa have been detailed by Veillard et al.[65] and include bypassing hepatic first-pass metabolism, excellent accessibility, unidirectional drug flux, and improved barrier permeability compared, for example, to intact skin.

Many drugs, such as glyceryl trinitrate, testosterone, and buprenorphine,[66] have been delivered via the buccal route. Absorption is rapid and drains into the reticulated vein. This avoids hepatic first-pass metabolism and intestinal enzymatic attack on the drug species. Because of its accessibility, the buccal area offers excellent patient compliance in comparison to epithelial routes that involve insertion of oleaginous devices, as sometimes found in rectal and vaginal delivery.

The process of drug absorption from conventional oral delivery systems normally occurs from saliva after dissolution from the formulation. Bioadhesive devices differ in that drug diffuses directly through the swollen polymer and then into the membrane. The intimacy of contact concentrates the drug on the epithelial surface. At other mucosal surfaces, the activity of goblet cells builds up a diffusion-limiting mucus layer that can impede drug absorption. This potential problem does not manifest itself at the oral mucosa, which contains no goblet cells, the mouth receiving its mucus primarily from the parotid, submaxillary, and sublingual salivary glands.[65]

Perioli et al.[67] studied the influence of compression force tablet behavior and drug release rate for mucoadhesive buccal tablets. Several tablet batches were produced by varying the compression force and by using hydroxyethyl cellulose (HEC) and Carbopol 940 in a 1:1 ratio as matrix-forming polymers. All the tablets hydrated quickly and their high hydration percentage showed that the compression forces used did not significantly affect water penetration and polymer chain stretching. Mucoadhesion performance and drug release were mainly influenced by compression force; its increase produced higher *ex vivo* and *in vivo* mucoadhesion. *In vitro* and *in vivo* drug release were both observed to decrease with increase of compression force. However, tablets fabricated by using the lowest compression force showed the best *in vivo* mucoadhesive time and hydrated faster when compared to the others. Tablets prepared with the highest forces caused pain during *in vivo* application and gave rise to irritation, needing to be detached by human volunteers. Tablets prepared with the lowest force gave the best results, as they were able to produce the highest drug salivary concentration and no pain. All tablets exhibited an anomalous release mechanism.

The buccal route may be an attractive site for peptide delivery because it is deficient in enzymatic degradation pathways. The cavity is lined with a relatively thick mucous membrane which is highly vascularized and approximately 100 cm^2 in area.[64] Prompted by studies showing rectal absorption of insulin in oleaginous vehicles, various workers tried, largely unsuccessfully, to achieve pharmacologically active blood levels of insulin using bioadhesive buccal patches.[66,68–71] Recently, Cui et al.[72] have described a delivery system based on a mucoadhesive layer (chitosan [CA]-ethylene-diaminetetraacetic acid) containing insulin and an impermeable protective layer made of ethylcellulose. *In vitro* mucoadhesion studies showed that the mucoadhesive force of the hydrogel remained over 17,000 N/m^2 during 4 hr in a simulated oral cavity. The insulin-loaded bilaminate film showed a pronounced hypoglycemic effect following buccal administration to healthy rats, despite only achieving a 17% bioavailability compared with a subcutaneous insulin injection.

Oral mucosal ulceration is a common condition with up to 50% of healthy adults suffering from recurrent minor mouth ulcers (aphthous stomatitis). Shemer et al.[73] evaluated the efficacy and tolerability of a mucoadhesive patch compared with a pain-relieving oral solution for the treatment of aphthous stomatitis. Patients with active aphthous stomatitis were randomly treated either once a day with a mucoadhesive patch containing citrus oil and magnesium salts ($n = 26$) or three times a day with an oral solution containing benzocaine and compound benzoin tincture ($n = 22$). All patients were instructed to apply the medication until pain had resolved, and completed a questionnaire detailing multiple clinical parameters followed by an evaluation of the treatment. The mucoadhesive patch was found to be more effective than the oral solution in terms of healing time and pain intensity after 12 and 24 hr. Local adverse effects 1 hr after treatment were significantly less frequent among the mucoadhesive patch patients compared with the oral solution patients.

Donnelly et al.[38] reported on a mucoadhesive patch containing toluidine blue O (TBO), as a potential delivery system for use in photodynamic antimicrobial chemotherapy of oropharyngeal candidiasis. Patches prepared from aqueous blends of poly(methyl vinyl ether/maleic anhydride) and tripropylene glycol methyl ether possessed suitable properties for use as mucoadhesive drug delivery systems and were capable of resisting dissolution when immersed in artificial saliva. When releasing directly into an aqueous sink, patches containing 50 and 100 mg TBO/cm^2 both generated receiver compartment concentrations exceeding the concentration (2.0–5.0 mg/mL) required to produce high levels of kill (>90%) of both planktonic and biofilm-grown Candida albicans upon illumination. However, the concentrations of TBO in the receiver compartments separated from patches by membranes intended to mimic biofilm structures were an order of magnitude below those inducing high levels of kill, even after 6 hr release. Therefore, the authors concluded that short application times of TBO-containing mucoadhesive patches should allow treatment of recently acquired oropharyngeal candidiasis, caused solely by planktonic cells. Longer patch application times may be required for persistent disease where biofilms are implicated.

Diseases of the oral cavity may be broadly differentiated into two categories, namely, inflammatory and infective conditions. In many cases, the demarcation between these two categories is unclear as some inflammatory diseases may be microbiological in origin.[74] Most frequently, therapeutic agents are delivered into the oral cavity in the form of solutions and gels, as this ensures direct access of the specific agent to the required site in concentrations that vastly exceed those that may be achieved using systemic administration.[74,75] However, the retention of such formulations within the oral cavity is poor, due, primarily, to their inability to interact with the hard and soft tissues and, additionally, to overcome the flushing actions of saliva. Therefore, the use of bioadhesive formulations has been promoted by several authors to overcome these problems and hence improve the clinical resolution of superficial diseases of the oral cavity.[75,76] In particular, within the oral cavity, bioadhesive formulations have been reported for the treatment of periodontal diseases and superficial oral infection, in which specific interactions between the bioadhesive formulations and the oral mucosa may be utilized to "anchor" the formulations to the site of application.

Periodontitis is an inflammatory disease of the oral cavity which results in the destruction of the supporting structures of the teeth.[77] It is characterized by the formation of pockets between the soft tissue of the gingiva and the tooth and which, if untreated, may result in tooth loss.[77,78] Drug delivery problems associated with the periodontal pocket may be overcome by the use of novel, bioadhesive, syringeable semisolid systems.[75,78] Such systems may be formulated to exhibit requisitory flow properties (and hence may be easily administered into the periodontal pocket using a syringe), mucoadhesive properties (ensuring prolonged retention within the pocket), and sustained release of therapeutic agent within this environment. In a series of papers, Andrews et al.,[79] Bruschi et al.,[80] Jones et al.[57,81,82] described the formulation and physicochemical characterization of syringeable semisolid, bioadhesive networks (containing tetracycline, metronidazole, or model protein drugs). Upon contact with mucus, water diffuses into these formulations and controls swelling of the bioadhesive polymers which, in turn, allows interpenetration of the fluidized polymer chains with mucus and, hence, ensures physical and chemical adhesion. The authors concluded that, when used in combination with mechanical treatments, antimicrobial-containing bioadhesive semisolid systems described in these studies would augment periodontal therapy by improving the removal of pathogens and, hence, enhance periodontal health.

Bioadhesive Devices for the GIT

Bioadhesive polymers may provide useful delivery systems for drugs that have limited bioavailability from more conventional dosage forms. To attain a once-daily dosing strategy using peroral bioadhesion, it is desirable for a dosage form to attach itself to the mucosa of the GIT. If attachment is successful in the gastric region, then a steady supply of drug is available to the intestinal tract for absorption. However, gastric motility and muscular contractions will tend to dislodge any such device. Motility in the GIT during the fasted state is known as the interdigestive migrating motor complex (IMMC).[64] During certain phases of the IMMC, a "housekeeper wave" migrates from the foregut to the terminal ileum and is intended to clear all nondigestible items from the gut. The strength of this "housekeeper wave" is such that poorly adhered dosage forms are easily removed and only strong bioadhesives will be of any practical use.

In addition to physical abrasion and erosion exerted on a bioadhesive dosage form, the mucin to which it is attached turns over quickly, especially in the gastric region. Any device that has lodged to surface mucus will be dislodged as newly synthesized mucus displaces the older surface layers. The acid environment will also affect bioadhesion, especially with polyacid polymers, such as PAA, where the mechanism of bond formation is thought to occur chiefly through hydrogen bonding and electrostatic interactions.

Liu et al.[83] prepared amoxicillin mucoadhesive microspheres using ethylcellulose as matrix and Carbopol 934P (C934P) as mucoadhesive polymer for the potential use of treating gastric and duodenal ulcers, which were associated with *Helicobacter pylori*. It was found that amoxicillin stability at low pH was enhanced when entrapped within the microspheres. *In vitro* and *in vivo* mucoadhesive tests showed that the amoxicillin-containing mucoadhesive microspheres adhered more strongly to the gastric mucous layer than nonadhesive amoxicillin microspheres did and could be retained in the GIT for an extended period of time. Amoxicillin-containing mucoadhesive micro-spheres and amoxicillin powder were orally administered to rats. The amoxicillin concentration in gastric tissue was higher in the mucoadhesive microspheres group. *In vivo H. pylori* clearance tests were also carried out by administering, respectively, amoxicillin-containing mucoadhesive microspheres or amoxicillin powder to *H. pylori*–infected BALB/c mice under fed conditions at single or multiple oral dose(s). The results showed that amoxicillin-containing mucoadhesive microspheres had a better clearance effect than amoxicillin powder did. The authors concluded that the prolonged gastrointestinal residence time and enhanced amoxicillin stability observed with the mucoadhesive microspheres of amoxicillin might make a useful contribution to *H. pylori* clearance.

Any swallowed dosage form will be subjected to shear forces as the intestinal contents move. This motility will either prevent the attachment of the bioadhesive or attempt to remove it if bonding to the gut wall has occurred. Previous attempts to reduce the GIT transit time of normal dosage forms have included devices that swelled and floated on the stomach contents, their increasing size preventing them from passing through the pylorus.

Ahmed and Ayres[84] studied gastric retention formulations (GRFs) made of naturally occurring carbohydrate polymers and containing riboflavin *in vitro* for swelling and dissolution characteristics as well as in fasting dogs for gastric retention. The bioavailability of riboflavin, from the GRFs, was studied in fasted healthy humans and compared to an immediate release formulation. It was found that when the GRFs were dried and immersed in gastric juice, they swelled rapidly and released their drug payload in a zero-order fashion for a period of 24 hr. *In vivo* studies in dogs showed that a rectangular shaped GRF stayed in the stomach of fasted dogs for more than 9 hr, then disintegrated and reached the colon in 24 hr. Endoscopic studies in dogs showed that the GRF hydrates and swells back to about 75% of its original size in 30 min. Pharmacokinetic parameters, determined from urinary excretion data from six human subjects under fasting conditions, showed that bioavailability depended on the size of the GRF. Bioavailability of riboflavin from a large size GRF was more than triple than that measured after administration of an immediate release formulation. *In vivo* studies suggested that the large size GRF stayed in the stomach for about 15 hr.

Li et al.[85] investigated distribution, transition, bioadhesion, and release behavior of insulin-loaded pH-sensitive NP in the gut of rats, as well as the effects of a viscosity-enhancing agent. Insulin was labeled with fluorescein isothiocyanate (FITC). The FITC-insulin solution and FITC-insulin nanoparticulate aqueous dispersions, with or without hydro-propylmethylcellulose (HPMC, 0.2%, 0.4%, or 0.8% [w/v]), were orally administered to rats. The amounts of FITC-insulin in both the lumen content and the intestinal mucosa were quantified. The release profiles in the gut were plotted by the percentages of FITC-insulin released versus time. FITC-insulin NP aqueous dispersion showed similar stomach, but lower intestinal, empty rates and enhanced intestinal mucosal adhesion in comparison with FITC-insulin solution. Addition of HPMC reduced the stomach and intestinal empty rates and enhanced adhesion of FITC-insulin to the intestinal mucosa. Release of FITC-insulin from NP in the gut showed an S-shaped profile and addition of HPMC prolonged the release half-life from 0.77 to 1.51 hr. It was concluded that the behaviors of pH-sensitive NP tested in the GIT of rats and the addition of HPMC were favorable to the absorption of the incorporated insulin.

The oral route constitutes the preferred route for drug delivery. However, numerous drugs remain poorly available when administered orally. Drugs associated with bioadhesive polymeric nanoparticulates or small particles in the micrometer size range may be advantageous in this respect due to their ability to interact with the mucosal surface. Targeting applications are possible by this method if there are specific interactions occurring when a ligand attached to the particle is used for the recognition and attachment to a specific site at the mucosal surface.[86]

Salman et al.[87] aimed to develop polymeric nanoparticulate carriers with bioadhesive properties and to evaluate their adjuvant potential for oral vaccination. Thiamine was used as a specific ligand-nanoparticle conjugate (TNP) to target specific sites within the GIT, namely enterocytes and Peyer's patches. The affinity of NP to the gut mucosa was studied in orally inoculated rats. In contrast to conventional noncoated NP, higher levels of TNP were found in the ileum tissue, showing a strong capacity to be captured by Peyer's patches. The adhesion of TNP was found to be three times higher than for control NP. To investigate the adjuvant capacity of TNP, ovalbumin (OVA) was used as a standard antigen. Oral immunization of BALB/c mice with OVA-TNP induced higher serum titers of specific IgG2a and IgG1 and mucosal IgA compared to OVA-NR. This mucosal immune response (IgA) was about four titers higher than that elicited by OVA-NP. The authors concluded that thiamine-coated NP showed promise as particulate vectors for oral vaccination and immunotherapy.

Rastogi et al.[88] formulated spherical microspheres able to prolong the release of isoniazid (INH) by a modified emulsification method, using sodium alginate as the hydrophilic carrier. The release profiles of INH from the microspheres were examined in simulated gastric fluid (SGF, pH 1.2) and simulated intestinal fluid (SIF, pH 7.4). Gamma scintigraphic studies were carried out to determine the location of microspheres on oral administration to rats and the extent of transit through the GIT. The microspheres had smooth surfaces and were found to be discreet and spherical in shape. The particles were heterogeneous with the largest average diameter of 3.7 μm. Results indicated that the mean particle size of the microspheres increased with an increase in the concentration of polymer and cross-linker, as well as cross-linking time. The entrapment efficiency was found to be in the range of 40–91%. Concentrations of the cross-linker up to 7.5% w/w increased entrapment efficiency and the extent of drug release. Optimized INH-alginate microspheres were found to possess good bioadhesion, which resulted in prolonged retention in the small intestine. Microspheres could be observed in the intestinal lumen at 4 hr and were still detectable at lower levels in the intestine 24 hr post-oral administration. Approximately 26% of the INH loading was released in SGIF pH 1.2 in 6 hr and 71.25% in SIF pH 7.4 in 30 hr.

In general, oral applications of bioadhesive technology have been less successful than where the delivery system can be applied directly to an accessible site. Issues such as overhydration, attachment to gastrointestinal mucus, and subsequent shedding and poor resistance to mechanical forces in the GIT have tended to limit the utility of bioadhesive gastric retentive systems to date.

Bioadhesive Devices for Rectal Drug Delivery

The function of the rectum is mostly concerned with removing water. It is only 10 cm in length, with no villi, giving it a relatively small surface area for drug absorption.[64] Drug permeability differs from that found in both the oral cavity and intestinal regions. Most rectal absorption of drugs is achieved by a simple diffusion process through the lipid membrane. However, the rectal route is readily accessible, penetration enhancers can be used, and there is access to the lymphatic system. In contrast, drugs absorbed via the small intestine are transported mostly to the blood with only a small proportion entering the lymphatic system.

Drugs that are liable to extensive first-pass metabolism can benefit greatly if delivered to the rectal area, especially if they are targeted to areas close to the anus. This is because the blood from the lower rectum drains directly into the systemic circulation, whereas blood from the upper regions drain into the portal systems via the superior hemorrhoidal vein and the inferior mesenteric vein. Drug absorbed from this upper site is subjected to liver metabolism. A bioadhesive suppository will attach to the lower rectal area and once inserted will reduce the tendency for migration upward to the upper rectum; this migration can occur with conventional suppositories.

Kim et al.[89] aimed to develop a thermoreversible flurbiprofen liquid suppository base composed of poloxamer and sodium alginate for improvement of rectal

bioavailability of flurbiprofen. Cyclodextrin derivatives, such as α-, β-, γ-cyclodextrin and hydroxypropyl-β-cyclodextrin (HP-β-CD), were used to enhance the aqueous solubility of flurbiprofen. The effects of HP-β-CD and flurbiprofen on the physicochemical properties of liquid suppository were then investigated. Pharmacokinetic studies were performed after rectal administration of flurbiprofen liquid suppositories with and without HP-β-CD or after intravenous administration of a commercially available product (Lipfen®, flurbiprofen axetil-loaded emulsion) to rats, and their pharmacokinetic parameters were compared. HP-β-CD decreased the gelation temperature and reinforced the gel strength and bioadhesive force of liquid suppositories, while the opposite was true for flurbiprofen. Thermoreversible flurbiprofen liquid suppositories demonstrated physicochemical properties suitable for rectal administration. The flurbiprofen liquid suppository with HP-β-CD showed significantly higher plasma levels, AUC, and C_{max} for flurbiprofen than those of the liquid suppository without HP-β-CD, indicating that flurbiprofen could be well-absorbed, due to the enhanced solubility by formation of an inclusion complex. Moreover, the flurbiprofen liquid suppository containing HP-β-CD showed an excellent bioavailability in that the AUC of flurbiprofen after its rectal administration was not significantly different from that after intravenous administration of Lipfen. The authors concluded that HP-β-CD could be a preferable solubility enhancer for the development of liquid suppositories containing poorly water-soluble drugs.

Uchida et al.[90] prepared insulin-loaded acrylic hydrogel formulations containing various absorption enhancers, performed in vitro and in vivo characterization of these formulations, and evaluated the factors affecting insulin availability upon rectal delivery. The acrylic block copolymer of methacrylic acid and methacrylate, Eudispert®, was used to make the hydrogel formulations. As absorption enhancers, 2,6-di-O-methyl-β-cyclodextrin (DM-β-CyD), lauric acid (C_{12}), or the sodium salt of C_{12} (C_{12}Na) was incorporated into the hydrogels. The in vitro release rate of insulin from the hydrogels decreased as polymer concentration was increased. The addition of C_{12}Na further increased release rate, which was greater at higher concentrations of the enhancer. Serum insulin levels were determined at various time points after the administration of insulin solution or insulin-loaded (50 units/kg body weight) Eudispert hydrogels containing 5% w/w of C_{12}, C_{12}Na, or DM-β-CyD to in situ loops in various regions of the rat intestine. The most effective enhancement of insulin release was observed with formulations containing C_{12}Na. The bioavailability of insulin from the hydrogels was lower than that from the insulin solutions, however. Hydrogel formulations containing 7% or 10% w/w Eudispert remained in the rectum for 5 hr after rectal administration. However, the 5% w/w C_{12}Na solution stained with Evan's blue had diffused out and the dye had reached the upper intestinal tract within 2 hr. Finally, the rectal administration of insulin-loaded hydrogels containing 4%, 7%, or 10% w/w Eudispert and 5% w/w enhancer (C_{12}, $C_{12}Na$, or DM-β-CyD) to normal rats was shown to decrease serum glucose concentrations. However, despite such promising results, bioadhesives have not been extensively employed for rectal delivery.

Bioadhesive Devices for Cervical and Vulval Drug Delivery

Nagai[70,71] investigated various bioadhesive forms to treat uterine and cervical cancers. Uterine cancers comprise about 25% of all malignant tumors in Japan, among which carcinoma colli accounts for 95% of this figure.[91] The target cells remain at, or near, the cervical epithelium and can be readily targeted with an appropriate dosage form. Pessaries have been of limited use because drug release is rapid and leakage into the surrounding tissue causes inflammation of vaginal mucosa. In order to meet three important criteria, drug release, swelling of the preparation, and adhesion to the diseased tissue, a bioadhesive disk was prepared. Bleomycin was incorporated into a hydroxypropylcellulose/Carbopol mix and molded into a suitable shape. This could be placed on or into the cervical canal, where it adhered and released the cytotoxic drug. After treatment and following colposcopic examination, areas of necrosis on the lesion were observed, with surrounding normal mucosal cells unaffected. With pessaries, however, this was not the case. In approximately 33% of cases, cancerous foci had completely disappeared.

Woolfson et al.[41,42] described a novel bioadhesive cervical patch containing 5-fluorouracil for the treatment of cervical intraepithelial neoplasia (CIN). The patch was of bilaminar design, with a drug-loaded bioadhesive film cast from a gel containing 2% w/w Carbopol 981 plasticized with 1%w/w glycerin. The casting solvent was ethanol/water 30:70, chosen to give a nonfissuring film with an even particle size distribution. The film, which was mechanically stable on storage under ambient conditions, was bonded directly to a backing layer formed from thermally cured poly(vinyl chloride) emulsion. Bioadhesive strength was independent of drug loading in the bioadhesive matrix over the range investigated but was influenced by both the plasticizer concentration in the casting gel and the thickness of the final film. Release of 5-fluorouracil from the bioadhesive layer into an aqueous sink was rapid but was controlled down to an undetectable level through the backing layer. The latter characteristic was desirable to prevent drug spill from the device onto vaginal epithelium *in vivo*. Despite the relatively hydrophilic nature of 5-fluorouracil, substantial drug release through human cervical tissue samples was observed over approximately 20 hr. Drug release, which was clearly tissue rather than device-dependent, may have been aided by a shunt diffusion route through aqueous pores in the tissue. The bioadhesive and drug release characteristics of the 5-fluorouracil cervical patch indicated that it would be suitable for further clinical investigation as a drug treatment for CIN.[92]

Donnelly et al.[93] described the design, physicochemical characterization, and clinical evaluation of bioadhesive drug delivery systems for photodynamic therapy of difficult-to-manage vulval neoplasias and dysplasias. In photodynamic therapy (PDT), a combination of visible light and a sensitizing drug causes the destruction of selected cells. Aminolevulic acid (ALA) is commonly delivered to the vulva using creams or solutions, which are covered with an occlusive dressing to aid retention and enhance drug absorption. Such dressings are poor at staying in place at the vulva, where shear forces are high in mobile patients. To overcome the problems associated with delivery of ALA to the vulva, the authors produced a bioadhesive patch by a novel laminating procedure. The ALA loading was 38 mg/cm^2. Patches were shown to release more ALA over 6 hr than the proprietary cream (Porphin®, 20% w/w ALA). The ALA concentration in excised tissue at a depth of 2.375 mm following application of the patch was an order of magnitude greater than that found to be cytotoxic to HeLa cells *in vitro*. Clinically, the patch was extensively used in successful PDT of vulval intraepithelial neoplasia, lichen sclerosus, squamous hyperplasia, Paget's disease, and vulvodynia.

Bioadhesive Devices for Vaginal Drug Delivery

Bioadhesives can control the rate of drug release from, and extend the residence time of, vaginal formulations. These formulations may contain drug or, quite simply, act in conjunction with moisturizing agents as a control for vaginal dryness.

Alam et al.[94] developed an acid-buffering bioadhesive vaginal tablet for the treatment of genitourinary tract infections. From bioadhesion experiment and release studies, it was found that polycarbophil and sodium carboxymethylcellulose was a good combination for an acid-buffering bioadhesive vaginal tablet. Sodium monocitrate was used as a buffering agent to provide an acidic pH (4.4), which is an attribute of a healthy vagina. The effervescent mixture (citric acid and sodium bicarbonate) along with a super disintegrant (Ac-Di-sol) was used to enhance the swellability of the bioadhesive tablet. The drugs clotrimazole (antifungal) and metronidazole (antiprotozoal and antibacterial) were used in the formulation along with *Lactobacillus acidophilus* spores to treat mixed vaginal infections. From *ex vivo* retention studies, it was found that the bioadhesive polymers held the tablet for more than 24 hr inside the vagina. The hardness of the acid-buffering bioadhesive vaginal tablet was optimized, at 4–5 kg hardness, the swelling was found to be good and the cumulative release profile of the developed tablet was matched with a marketed conventional tablet (Infa-V®). The *in vitro* spreadability of the swelled tablet was comparable to the marketed gel. In the *in vitro* antimicrobial study, it was found that the acid-buffering bioadhesive tablet produced better antimicrobial action than marketed intravaginal drug delivery systems (Infa-V, Candid-V®, and Canesten®1).

Cevher et al.[95] aimed to prepare clomiphene citrate (CLM) gel formulations possessing appropriate mechanical properties, exhibiting good vaginal retention, and providing sustained drug release for the local treatment of human papilloma virus infections. In this respect, 1% w/w CLM gels including PAA polymers such as C934P, Carbopol 971P (C971P), Carbopol 974P (C974P) in various concentrations, and their conjugates containing thiol groups, were prepared. Based on obtained data, gel formulations containing C934P and its conjugate had appropriate hardness and compressibility to be applied to the vaginal mucosa and showed the highest elasticity and good spreadability. Such gels also exhibited the highest cohesion. The mucoadhesion of the gels changed significantly depending on the polymer type and concentration. Addition of conjugates containing thiol groups caused an increase in mucoadhesion. Gels containing C934P-Cys showed the highest adhesiveness and mucoadhesion. A significant decrease was observed in drug release from gel formulations as the polymer concentration increased.

Bioadhesive Devices for Nasal Drug Delivery

The area of the normal human nasal mucosa is approximately 150 cm². The nasal mucosa and submucosa are liberally populated with goblet cells along with numerous mucous and serous glands. These keep the nasal mucosal surfaces moist. Drug administration to this region is normally reserved for local treatment, such as nasal allergy or inflammation. The nasal mucosa is thin and incorporates a dense vascular network, indicating that drug absorption may be good from this site.[96] Absorbed drugs avoid first-pass metabolism and lumenal degradation associated with the oral route.[64] The nasal mucosa itself is sensitive to drug molecules and to surfactants which are often used to enhance drug absorption. Moreover, the mucociliary escalator travels at 5 mm/min as it drags mucous fluid backward toward the throat. Therefore, if a drug is applied in either a simple powder or liquid formulation it will be quickly cleared from this site of absorption. Thus, any useful dosage form must be nonirritant to the nasal mucosa and be retained for extended periods of time.

Charlton, Davis, and Illum[97] studied the effect of bioadhesive formulations on the direct transport of an angiotensin antagonist drug from the nasal cavity to the central nervous system in a rat model. Three different bioadhesive polymer formulations (3% w/w pectin, 1.0% w/w pectin, and 0.5% w/w CA) containing the drug were administered nasally to rats by inserting a dosing cannula 7 mm into the nasal cavity after which the plasma and brain tissue levels were measured. It was found that the polymer formulations provided significantly higher plasma levels and significantly lower brain tissue levels of drug than a control, in the form of a simple drug solution. Changing the depth of insertion of the cannula from 7 to 15 mm in order to reach the olfactory region in the nasal cavity significantly decreased plasma levels and significantly increased brain tissue levels of drug for the two formulations studied (1.0% w/w pectin and a simple drug solution). There was no significant difference between the drug availability for the bioadhesive formulation and the control in the brain when the longer cannula was used for administration. The authors suggested that the conventional rat model is not suitable for evaluation of the effects of bioadhesive formulations in nose-to-brain delivery.

Nasal delivery of protein and peptide therapeutics can be compromised by the brief residence time at this mucosal surface. Some bioadhesive polymers have been suggested to extend residence time and improve protein uptake across the nasal mucosa. McInnes et al.[98] quantified nasal residence of bioadhesive formulations using gamma scintigraphy and investigated absorption of insulin. A four-way crossover study was conducted in six healthy male volunteers, comparing a conventional nasal spray solution with three lyophilized nasal insert formulations (1–3% w/w HPMC). The conventional nasal spray deposited in the posterior nasal cavity in only one instance, with a rapid clearance half-life of 9.2 min. The nasal insert formulations did not enhance nasal absorption of insulin. However, an extended nasal residence time of 4–5 hr was observed for the 2% w/w HPMC formulation. The 1% w/w HPMC insert initially showed good spreading behavior. However, clearance was faster than for the 2% w/w formulation. The 3% w/w HPMC nasal insert showed no spreading and was usually cleared intact from the nasal cavity within 90 min. The authors concluded that the 2% w/w HPMC lyophilized insert formulation achieved extended nasal residence, demonstrating an optimum combination of rapid adhesion without overhydration.

Coucke et al.[99] studied viscosity-enhancing mucosal delivery systems for the induction of an adaptive immune response against viral antigen. Powder formulations based on spray-dried mixtures of starch (Amioca®) and PAA (C974P) in different ratios were used as carriers of the viral antigen. A comparison of these formulations for intranasal delivery of heat-inactivated influenza virus combined with LTR 192G adjuvant was made in vivo in a rabbit model. Individual rabbit sera were tested for seroconversion against hemagglutinin (HA), the major surface antigen of influenza. The powder vaccine formulations were able to induce systemic anti-HA IgG responses. The presence of C974P improved the kinetics of the immune responses and the level of IgG titers in a dose-dependent way which was correlated with moderately irritating capacities of the formulation. In contrast, mucosal IgA responses were not detected. The authors concluded that the use of bioadhesive carriers based on starch and PAA facilitates the induction of a systemic anti-HA antibody response after intranasal vaccination with a whole virus influenza vaccine.

Overall, the nasal route remains highly promising, particularly for macromolecular absorption, but there are problems regarding effects of bioadhesive delivery systems on nasal cilial beat.

Bioadhesive Devices for Ocular Drug Delivery

Extended drug delivery to the eye is difficult for several reasons. Systemically administered drugs must cross the blood-aqueous humor barrier and to achieve local therapeutic concentrations of drug, high levels of drug in the blood are needed. In addition, lacrimation, blinking, and tear turnover will all reduce the bioavailability of topically administered drug to approximately 1–10%.[100] Conventional delivery methods are not ideal. Solutions and suspensions are readily washed from the cornea and ointments alter the tear refractive index and blur vision.

Sensoy et al.[101] aimed to prepare bioadhesive sulfacetamide sodium microspheres to increase residence time on the ocular surface and to enhance treatment efficacy of ocular keratitis. Microspheres were fabricated by a spray-drying method using a mixture of polymers, such as pectin, polycarbophil, and HPMC at different ratios. A sulfacetamide sodium-loaded polycarbophil microsphere formulation with a polymer/drug ratio of 2:1 was found to be the most suitable for ocular application and used in *in vivo* studies on New Zealand male rabbit eyes with keratitis caused by *Pseudomonas aeruginosa* and *Staphylococcus aureus*. Sterile microsphere suspension in light mineral oil was applied to infected eyes twice a day. Plain sulfacetamide sodium suspension was used as a positive control. On the third and sixth days, the eyes were examined in respect to clinical signs of infection (blepharitis, conjunctivitis, iritis, corneal edema, and corneal infiltrates) and then cornea samples were counted microbiologically. The rabbit eyes treated with microspheres demonstrated significantly lower clinical scores than those treated with sulfacetamide sodium alone. A significant decrease in the number of viable bacteria in eyes treated with microspheres was observed in both infection models when compared to those treated with sulfacetamide sodium alone.

Gene transfer is considered to be a promising alternative for the treatment of several chronic diseases that affect the ocular surface. de la Fuente, Seijo, and Alonso[102] investigated the efficacy and mechanism of action of a bioadhesive DNA nanocarrier made of hyaluronan (HA) and CS, specifically designed for topical ophthalmic gene therapy. The authors first evaluated the transfection efficiency of the plasmid DNA-loaded NP in a human corneal epithelium cell model. Then they investigated the bioadhesion and internalization of the NP in the rabbit ocular epithelia and determined the *in vivo* efficacy of the nanocarriers in terms of their ability to transfect ocular tissues. The results showed that HA-CS NP and, in particular, those made of low molecular weight CS (10–12 kDa) led to high levels of expression of secreted alkaline phosphatase in the human corneal epithelium model. In addition, following topical administration to rabbits, the NP entered the corneal and conjunctival epithelial cells and become assimilated by the cells. More importantly, the NP provided an efficient delivery of the associated plasmid DNA inside the cells, reaching significant transfection levels.

CONCLUSION

Bioadhesive drug delivery systems have been widely studied in the past two decades and a number of interesting, novel drug delivery systems incorporating the use of bioadhesive polymers have been described. Primarily, the application of bioadhesion to drug delivery involves the process of mucoadhesion, a "wet-stick" adhesion requiring, primarily, spreading of the mucoadhesive on a mucin-coated epithelium, followed by interpenetration of polymer chains between the hydrated delivery system and mucin. An alternative mechanism can involve the use of specific bioadhesive molecules, notably lectins, primarily for oral drug delivery applications. Results, however, remain variable by this specific method, as with the use of bioadhesives, generally, for oral systemic drug delivery. The most notable applications of mucoadhesive systems to date remain those involving accessible epithelia, such as in ocular, nasal, buccal, rectal, or intravaginal drug delivery systems. The ancillary effects of certain mucoadhesive polymers, such as polycations, on promoting drug penetration of epithelial tissues, may be increasingly important in the future, particularly for the transmucosal delivery of therapeutic peptides and other biomolecules.

REFERENCES

1. Jiménez-Castellanos, M.R.; Zia, H.; Rhodes, C.T. Mucoadhesive drug delivery systems. Drug Dev. Ind. Pharm. **1993**, *19* (1–2), 143–194.
2. *ASTM Designation D. American Society for Testing and Materials*; West Conshohocken, PA, 1984; 907–977.
3. Good, W.R. Transdermal nitro-controlled delivery of nitroglycerin via the transdermal route. Drug Dev. Ind. Pharm. **1983**, *9*, 647–670.
4. Henriksen, I.; Green, K.L.; Smart, D.; Smistad, G.; Karlsen, G. Bioadhesion of hydrated chitosans: An *in vitro* and *in vivo* study. Int. J. Pharm. **1996**, *145* (1–2), 231–240.
5. Bell, G.I.; Dembo, M.; Bongrand, P. Cell adhesion: Competition between non-specific repulsion and specific bonding. Biophys. J. **1984**, *45* (6), 1051–1064.
6. Larsson, K. Interfacial phenomena: Bioadhesion and biocompatibility. Desalination **1980**, *35*, 105–114.
7. Van Wachem, P.B.; Beugeling, T.; Feijen, J.; Bantjes, A.; Detmers, J.P.; Van Aken, W.G. Interaction of cultured human cells with polymeric surfaces of different wettabilities. Biomaterials **1985**, *6* (6), 403–408.
8. Govil, S.K. Transdermal drug delivery systems. In *Drug Delivery Devices;* Tyle, P., Ed.; Marcel Dekker: New York, 1988.
9. Woolfson, A.D.; McCafferty, D.F.; Moss, G.P. Development and characterisation of a moisture-activated bioadhesive drug delivery system for percutaneous local anaesthesia. Int. J. Pharm. **1998**, *169* (1), 83–94.
10. Leung, S.H.S.; Robinson, J.R. The contribution of anionic polymer structural features related to mucoadhesion. J. Control. Release **1988**, *5* (3), 223–231.
11. Duchêne, D.; Touchard, F.; Peppas, N.A. Pharmaceutical and medical aspects of bioadhesive systems for drug

Bioadhesive–BioMEMS

administration. Drug Dev. Ind. Pharm. **1988**, *14* (2–3), 283–318.

12. Merrill, E.W. Properties of materials affecting the behaviour of blood at their surfaces. Ann. N.Y. Acad. Sci. **1977**, *283*, 6–16.

13. Yang, X.; Robinson, J.R. Bioadhesion in mucosal drug delivery. In *Biorelated Polymers and Gels;* Okano, T., Ed.; Academic Press: London U.K., 1988.

14. Kamath, K.R.; Park, K. Mucosal adhesive preparations. In *Encyclopedia of Pharmaceutical Technology;* Swarbrick, J., Boylan, J.C., Eds.; Marcel Dekker: New York, 1992; 133.

15. Longer, M.A.; Robinson, J.R. Fundamental aspects of bioadhesion. Pharm. Int. **1986**, *7*, 114–117.

16. McBain, J.W.; Hopkins, D.G. On adhesives and adhesive action. J. Phys. Chem. **1925**, *29* (2), 188–204.

17. Pritchard, W.H. The role of hydrogen bonding in adhesion. Aspects Adhes. **1971**, *6*, 11–23.

18. Wake, W.C. *Adhesion and the Formulation of Adhesives;* Applied Science Publishers: London U.K., 1982.

19. Deraguin, B.V.; Smilga, V.P. *Adhesion: Fundamentals and Practice;* McLaren: London U.K., 1969.

20. Voyutskii, S.S. *Autoadhesion and Adhesion of High Polymers;* Wiley: New York, 1963.

21. Peppas, N.A.; Buri, P.A. Surface, interfacial and molecular aspects of polymer bioadhesion on soft tissues. J. Control. Release **1985**, *2*, 257–275.

22. Reinhart, C.P.; Peppas, N.A. Solute diffusion in swollen membranes II. Influence of crosslinking on diffusion properties. J. Membr. Sci. **1984**, *18*, 227–239.

23. Ahuja, A.; Khar, R.K.; Ali, J. Mucoadhesive drug delivery systems. Drug Dev. Ind. Pharm. **1997**, *23*, 489–515.

24. Huntsberger, J.R. Mechanisms of adhesion. J. Paint Technol. **1967**, *39*, 199–211.

25. Kinloch, A.J. The science of adhesion I. Surface and interfacial aspects. J. Mater. Sci. **1980**, *15* (9), 2141–2166.

26. Smart, J.D.; Nicholls, T.J.; Green, K.L.; Rogers, D.J.; Cook, J.D. Lectins in drug delivery: A study of the acute local irritancy of the lectins from *Solanum tuberosum* and *Helixpomatia.* Eur. J. Pharm. Sci. **1999**, *9* (1), 93–98.

27. Naisbett, B.; Woodley, J. The potential use of tomato lectin for oral drug delivery. Int. J. Pharm. **1994**, *107* (3), 223–230.

28. Nicholls, T.J.; Green, K.L.; Rogers, D.J.; Cook, J.D.; Wolowacz, S.; Smart, J.D. Lectins in ocular drug delivery. An investigation of lectin binding sites on the corneal and conjunctival surfaces. Int. J. Pharm. **1996**, *138* (2), 175–183.

29. Lehr, C.M. From sticky stuff to sweet receptors—Achievements, limits and novel approaches to bioad-hesion. Eur. J. Drug Metab. Pharmacokinet. **1996**, *21* (2), 139–148.

30. Banchonglikitkul, C.; Smart, J.D.; Gibbs, R.V.; Donovan, S.J.; Cook, D.J. An *in vitro* evaluation of lectin cytotoxicity using cell lines derived from the ocular surface. J. Drug Target. **2002**, *10* (8), 601–606.

31. Smart, J.D.; Banchonglikitkul, C.; Gibbs, R.V.; Donovan, S.J.; Cook, D.J. Lectins in drug delivery to the oral cavity, *in vitro* toxicity studies. STP Pharm. Sci. **2003**, *13*, 37–40.

32. Peppas, N.A.; Little, M.D.; Huang, Y. Bioadhesive controlled release systems. In *Handbook of Pharmaceutical Controlled Release Technology;* Wise, D.L., Ed.; Marcel Dekker: New York, 2000; 255–269.

33. Gurny, R.; Meyer, J.M.; Peppas, N.A. Bioadhesive intraoral release systems: Design, testing and analysis. Biomaterials **1984**, *5* (6), 336–340.

34. Gudeman, L.; Peppas, N.A. Preparation and characterisation of pH-sensitive, interpenetrating networks of poly(vinyl alcohol) and poly(acrylic acid). J. Appl. Polym. Sci. **1995**, *55* (6), 919–928.

35. McCarron, P.A.; Woolfson, A.D.; Donnelly, R.F.; Andrews, G.P.; Zawislak, A.; Price, J.H. Influence of plasticiser type and storage conditions on the properties of poly(methyl vinyl ether-*co*-maleic anhydride) bioadhesive films. J. Appl. Polym. Sci. **2004**, *91* (3), 1576–1589.

36. Park, H.; Robinson, J.R. Physicochemical properties of water soluble polymers important to mucin/epithelium adhesion. J. Control. Release **1985**, *2*, 47–57.

37. BlancoFuente, H.; AnguianoIgea, S.; OteroEspinar, F.J.; BlancoMendez, J. *In vitro* bioadhesion of Carbopol hydrogels. Int. J. Pharm. **1996**, *142* (2), 169–174.

38. Donnelly, R.F.; McCarron, P.A.; Tunney, M.M.; Woolfson, A.D. Potential of photodynamic therapy in treatment of fungal infections of the mouth. Design and characterisation of a mucoadhesive patch containing toluidine blue O. J. Photochem. Photobiol. B **2007**, *86* (1), 59–69.

39. Bouckaert, S.; Remon, J.P. *In vitro* bioadhesion of a buccal, miconazole slow-release tablet. J. Pharm. Pharmacol. **1993**, *45* (6), 504–507.

40. Voorspoels, J.; Remon, J.P.; Eechaute, W.; DeSy, W. Buccal absorption of testosterone and its esters using a bioadhesive tablet in dogs. Pharm. Res. **1996**, *13* (8), 1228–1332.

41. Woolfson, A.D.; McCafferty, D.F.; McCallion, C.R.; McAdams, E.T.; Anderson, J. Moisture-activated, electrically-conducting bioadhesive hydrogels as interfaces for bio-electrodes: Effect of film hydration on cutaneous adherence in wet environments. J. Appl. Polym. Sci. **1995a**, *58* (8), 1291–1296.

42. Woolfson, A.D.; McCafferty, D.F.; McCarron, P.A.; Price, J.H. A bioadhesive patch cervical drug delivery system for the administration of 5-fluorouracil to cervical tissue. J. Control. Release **1995b**, *35* (1), 49–58.

43. Mortazavi, S.A.; Smart, J.D. Factors influencing gel-strengthening at the mucoadhesive-mucus interface. J. Pharm. Pharmacol. **1994**, *46* (2), 86–90.

44. Smart, J.D. An *in vitro* assessment of some mucoadhesive dosage forms. Int. J. Pharm. **1991**, *73*, 69–74.

45. Lehr, C.M.; Poelma, F.G.J. An estimate of turnover time of intestinal mucus gel layer in the rat in situ loop. Int. J. Pharm. **1991**, *70* (3), 235.

46. Park, K.; Park, H. Test methods of bioadhesion. In *Bioadhesive Drug Delivery Systems;* Lenaerts, V., Gurney, R., Eds.; CRC Press: Boca Raton, FL, 1990.

47. McCarron, P.A.; Donnelly, R.F.; Zawislak, A.; Woolfson, A.D.; Price J.H.; McClelland, R. Evaluation of a water-soluble bioadhesive patch for photodynamic therapy of vulval lesions. Int. J. Pharm. **2005**, *293* (1–2), 11–23.

48. Smart, J.D.; Kellaway, I.W.; Worthington, H.E.C. An *in vitro* investigation of mucosa-adhesive materials for use in controlled drug delivery. J. Pharm. Pharmacol. **1984**, *36* (5), 295–299.

49. Mikos, A.G.; Peppas, N.A. Comparison of experimental techniques for the measurement of the bioadhesive forces of polymeric materials with soft tissues. *Proceedings of the International Symposium on Controlled Release Bioactive Materials* 1986; Vol. 13, 97.

50. McCarron, P.A.; Donnelly, R.F.; Zawislak, A.; Woolfson, A.D. Design and evaluation of a water-soluble bioadhesive patch formulation for cutaneous delivery of 5-aminolevulinic acid to superficial neoplastic lesions. Eur. J. Pharm. Sci. **2006**, *27* (2–3), 268–279.

51. Donnelly, R.F.; McCarron, P.A.; Zawislak, A.A.; Woolfson, A.D. Design and physicochemical characterisation of a bioadhesive patch for dose-controlled topical delivery of imiquimod. Int. J. Pharm. **2006**, *307* (2), 318–325.

52. Ch'ng, H.S., Park, H.; Kelly, P.; Robinson, J.R. Bioadhesive polymers as platforms for oral controlled drug delivery. II Synthesis and evaluation of some swelling, water-insoluble bioadhesive polymers. J. Pharm. Sci. **1985**, *74* (4), 399–405.

53. Reich, S.; Levy, M.; Meshorer, A.; Blumental, M.; Yalon, M.; Sheets, J.W.; Goldberg, E.P. Intraocular-lens endothelial interface-adhesive force measurements. J. Biomed. Mater. Res. **1984**, *18* (7), 737–744.

54. Davis, S.S. The design and evaluation of controlled release systems for the gastro-intestinal tract. J. Control. Release **1985**, *2*, 27–38.

55. Jones, D.S.; Woolfson, A.D. Measuring sensory properties of semi-solid products using texture profile analysis. Pharm. Manuf. Rev. **1997**, *9*, S3–S6.

56. Andrews, G.P.; Laverty, T.P.; Jones, D.S. Mucoadhesive polymeric platforms for controlled drug delivery. Eur. J. Pharm. Biopharm. **2009**, *71* (3), 505–518.

57. Jones, D.S.; Bruschi, M.L.; de Freitas, O.; Gremiao, M.P.D.; Lara, E.H.G.; Andrews, G.P. Rheological, mechanical and mucoadhesive properties of thermoresponsive, bioadhesive binary mixtures composed of Poloxamer 407 and Carbopol 974P designed as platforms for implantable drug delivery systems for use in the oral cavity. Int. J. Pharm. **2009**, *372* (1–2), 49–58.

58. Craig, D.Q.M.; Tamburic, S. Dielectric analysis of bioadhesive gel systems. Eur. J. Pharm. Biopharm. **1997**, *44* (1), 61–70.

59. Ponchel, G.; Touchard, F.; Duchene, D.; Peppas, N.A. Bioadhesive analysis of controlled release systems I. Fracture and interpenetration analysis in poly(acrylic acid)-containing systems. J. Control. Release **1987**, *5* (2), 129–141.

60. Nagai, T.; Konishi, R. Buccal/gingival drug delivery systems. J. Control. Release **1987**, *6* (1), 353–360.

61. Sayin, B.; Somavarapu, S.; Li, X.W.; Thanou, M.; Sesardic, D.; Alpar, H.O.; Senel, S. Mono-*N*-carboxymethyl chitosan (MCC) and *N*-trimethyl chitosan (TMC) nanoparticles for non-invasive vaccine delivery. Int. J. Pharm. **2008**, *363* (1–2), 139–148.

62. Radebaugh, G.W. Film coatings and film-forming materials: Evaluation. In *Encyclopedia of Pharmaceutical Technology;* Swarbrick, J., Boylan, J.C., Eds.; Marcel Dekker: New York, 1992; 1–28.

63. Anders, R.; Merkle, H.P. Evaluation of laminated mucoadhesive patches for buccal drug delivery. Int J. Pharm. **1989**, *49*, 231–240.

64. Leung, S.H.S.; Robinson, J.A. Polyanionic polymers in bioadhesive and mucoadhesive drug delivery. ACS Symp. Series **1992**, *480*, 269–284.

65. Veillard, M.M.; Longer, M.A.; Martens, T.W.; Robinson, J.R. Preliminary studies of oral mucosal delivery of peptide drugs. J. Control. Release **1987**, *6* (1), 123–131.

66. Nagai, T.; Machida, Y. Advances in drug delivery—Mucosal adhesive dosage forms. Pharm. Int. **1985**, *6*, 196–200.

67. Perioli, L.; Ambrogi, V.; Giovagnoli, S.; Blasi, P.; Mancini, A.; Ricci, M.; Rossi, C. Influence of compression force on the behaviour of mucoadhesive buccal tablets. AAPS PharmSciTech **2008**, *9* (1), 274–281.

68. Ishida, M.; Machida, Y.; Nambu, N.; Nagai, T. New mucosal dosage forms of insulin. Chem. Pharm. Bull. (Tokyo) **1981**, *29* (3), 810–816.

69. Nagai, T. Adhesive topical drug delivery systems. J. Control. Release **1985**, *2*, 121–134.

70. Nagai, T. Topical mucosal adhesive dosage forms. Med. Res. Rev. **1986a**, *6* (2), 227–242.

71. Nagai, T. Bioadhesive and mucoadhesive drug delivery systems. *46th International Congress of Pharmaceutical Science of FIP,* Sept 1–5, 1986b, Helsinki.

72. Cui, F.Y.; He, C.B.; He, M.; Tang, C.; Yin, L.C.; Qian, F.; Yin, C.H. Preparation and evaluation of chitosan-ethylene diaminetetraacetic acid hydrogel films for the mucoadhesive transbuccal delivery of insulin. J. Biomed. Mater. Res. A **2009**, *89* (4), 1063–1071.

73. Shemer, A.; Amichai, B.; Trau, H.; Nathansohn, N.; Mizrahi, B.; Domb, A.J. Efficacy of a mucoadhesive patch compared with an oral solution for treatment of aphthous stomatitis. Drugs R D **2008**, *9* (1), 29–35.

74. Addy, M. Local delivery of antimicrobial agents to the oral cavity. Adv. Drug Deliv. Rev. **1994**, *13* (1–2), 123–134.

75. Jones, D.S.; Woolfson, A.D.; Brown, A.F. Textural, viscoelastic and mucoadhesive properties of pharmaceutical gels and polymers. Int. J. Pharm. **1997**, *151* (2), 223–233.

76. Gandhi, R.B.; Robinson, J.R. The oral cavity as a site for bioadhesive drug delivery. Adv. Drug Deliv. Rev. **1994**, *13* (1–2), 43–74.

77. Medlicott, N.J.; Rathbone, M.J.; Tucker, I.J.; Holborow, D.W. Delivery systems for the administration of drugs to the periodontal pocket. Adv. Drug Deliv. Rev. **1994**, *13* (1–2), 181–203.

78. Jones, D.S.; Woolfson, A.D.; Djokic, J.; Coulter, W.A. Development and physical characterization of bioadhesive semi-solid, polymeric systems containing tetracycline for the treatment of periodontal diseases. Pharm. Res. **1996**, *13* (11), 1732–1736.

79. Andrews, G.P.; Jones, D.S.; Redpath, J.M.; Woolfson, A.D. Characterisation of protein-containing binary polymeric gel systems designed for the treatment of periodontal disease. J. Pharm. Pharmacol. **2004**, *56* (Suppl. S), S71–S72.

80. Bruschi, M.L.; de Freitas, O.; Lara, E.H.G.; Panzeri, H.; Gremiao, M.P.D.; Jones, D.S. Precursor system of liquid crystalline phase containing propolis microparticles for the treatment of periodontal disease: Development and characterization. Drug Dev. Ind. Pharm. **2008**, *34* (3), 267–278.

81. Jones, D.S.; Lawlor, M.S.; Woolfson, A.D. Formulation and characterisation of tetracycline-containing bioadhesive polymer networks designed for the treatment of periodontal disease. Curr. Drug Deliv. **2004**, *1* (1), 17–25.

82. Jones, D.S.; Muldoon, B.C.; Woolfson, A.D.; Andrews, G.P.; Sanderson, F.D. Physicochemical characterization of bioactive polyacrylic acid organogels as potential antimicrobial implants for the buccal cavity. Biomacromolecules **2008**, *9* (2), 624–633.

83. Liu, Z.P.; Lu, W.Y.; Qian, L.S.; Zhang, X.H.; Zeng, P.Y.; Pan, J. *In vitro* and *in vivo* studies on mucoadhesive microspheres of amoxicillin. J. Control. Release **2005**, *102* (1), 135–144.

84. Ahmed, I.S.; Ayres, J.W. Bioavailability of riboflavin from a gastric retention formulation. Int. J. Pharm. **2007**, *330* (1–2), 146–154.

85. Li, M.G.; Lu, W.L.; Wang, H.C.; Zhang, X.; Wang, X.Q.; Zheng, A.P.; Zhang, Q. Distribution, transition, adhesion and release of insulin loaded nanoparticles in the gut of rats. Int. J. Pharm. **2007**, *329* (1–2), 182–191.

86. Ponchel, G.; Irache, J.M. Specific and non-specific bioadhesive particulate systems for oral delivery to the gastrointestinal tract. Adv. Drug Deliv. Rev. **1998**, *34* (2–3), 191–219.

87. Salman, H.H.; Gamazo, C.; Agueros, M.; Irache, J.M. Bioadhesive capacity and immunoadjuvant properties of thiamine-coated nanoparticles. Vaccine **2007**, *25* (48), 8123–8132.

88. Rastogi, R.; Sultana, Y.; Aqil, M.; Ali, A.; Kumar, S.; Chuttani, K.; Mishra, A.K. Alginate microspheres of isoniazid for oral sustained drug delivery. Int. J. Pharm. **2007**, *334* (1–2), 71–77.

89. Kim, J.K.; Kim, M.S.; Park, J.S.; Kim, C.K. Thermo-reversible flurbiprofen liquid suppository with HP-beta-CD as a solubility enhancer: Improvement of rectal bioavailability. J. Incl. Phenom. Macro. Chem. **2009**, *64*, 265–272.

90. Uchida, T.; Toida, Y.; Sakakibara, S.; Miyanaga, Y.; Tanaka, H.; Nishikata, M.; Tazuya, K.; Yasuda, N.; Matsuyama, K. Preparation and characterization of insulin-loaded acrylic hydrogels containing absorption enhancers. Chem. Pharm. Bull. (Tokyo) **2001**, *49* (10), 1261–1266.

91. Machida, Y.; Masuda, H.; Fujiyama, N.; Ito, S.; Iwata, M.; Nagai, T. Preparation and phase-II clinical evaluation of topical dosage form for treatment of carcinoma colli containing bleomycin with hydroxypropylcellulose. Chem. Pharm. Bull. (Tokyo) **1979**, *27*, 93–100.

92. Sidhu, H.; Price, J.H.; McCarron, P.A.; McCafferty, D.F.; Woolfson, A.D.; Biggart, D.; Thompson, W. A randomised control trial evaluating a novel cytotoxic drug delivery system for the treatment of cervical intraepithelial neoplasia. Br. J. Obstet. Gynaecol. **1997**, *104* (2), 145–149.

93. Donnelly, R.F.; McCarron, P.A.; Zawislak, A.; Woolfson, A.D. *Photodynamic Therapy of Vulval Neoplasias and Dysplasias: Design and Evaluation of Bioadhesive Photosensitiser Delivery Systems;* VDM Verlag Dr. Mtiller: Saarbrucken, Germany, 2009.

94. Alam, M.A.; Ahmad, F.J.; Khan, Z.I.; Khar, R.K.; Ali, M. Development and evaluation of acid-buffering bioadhesive vaginal tablet for mixed vaginal infections. AAPS PharmSciTech **2007**, *8* (4), 109.

95. Cevher, E.; Taha, M.A.M.; Orlu, M.; Araman, A. Evaluation of mechanical and mucoadhesive properties of clomiphene citrate gel formulations containing carbomers and their thiolated derivatives. Drug Deliv. **2008**, *15* (1), 57–67.

96. Igawa, T.; Maitani, Y.; Machida, Y.; Nagai, T. Intranasal administration of human fibroblast interferon in mice, rats, rabbits and dogs. Chem. Pharm. Bull. (Tokyo) **1990**, *38* (2), 549–551.

97. Charlton, S.T.; Davis, S.S.; Illum, L. Nasal administration of an angiotensin antagonist in the rat model: Effect of bioadhesive formulations on the distribution of drugs to the systemic and central nervous systems. Int. J. Pharm. **2007**, *338* (1–2), 94–103.

98. McInnes, F.J.; O'Mahony, B.; Lindsay, B.; Band, J.; Wilson, C.G.; Hodges, L.A.; Stevens, H.N.E. Nasal residence of insulin containing lyophilised nasal insert formulations, using gamma scintigraphy. Eur. J. Pharm. Sci. **2007**, *31* (1), 25–31.

99. Coucke, D.; Schotsaert, M.; Libert, C.; Pringels, E.; Vervaet, C.; Foreman, P.; Saelens, X.; Remon, J.P. Spray-dried powders of starch and crosslinked poly(acrylic acid) as carriers for nasal delivery of inactivated influenza vaccine. Vaccine **2009**, *27* (8), 1279–1286.

100. Robinson, J.R. Ocular drug delivery mechanism(s) of corneal drug transport and mucoadhesive delivery systems. STP Pharm. Sci. **1989**, *5* (12), 839–846.

101. Sensoy, D.; Cevher, E.; Sarici, A.; Yilmaz, M.; Ozdamar, A.; Bergisadi, N. Bioadhesive sulfacetamide sodium microspheres: Evaluation of their effectiveness in the treatment of bacterial keratitis caused by *Staphylococcus aureus* and *Pseudomonas aeruginosa* in a rabbit model. Eur. J. Pharm. Biopharm. **2009**, *72* (3), 487–495.

102. de la Fuente, M.; Seijo, B.; Alonso, M.J. Bioadhesive hyaluronan-chitosan nanoparticles can transport genes across the ocular mucosa and transfect ocular tissue. Gene Ther. **2009**, *15* (9), 668–676.

103. Nagai, T.; Machida, Y. Bioadhesive dosage forms for nasal administration. In *Bioadhesive Drug Delivery Systems;* Lenaerts, V., Gurney, R., Eds.; CRC Press: Boca Raton, FL, 1990.

Biocompatibility: Testing

Joel D. Bumgardner
Marcia Vasquez-Lee
Keertik S. Fulzele
Daniel H. Smith
Kellye D. Branch
Seth I. Christian
David L. Williams
Biomedical Engineering Program, Agricultural and Biological Engineering Department, Mississippi State University, Mississippi State, Mississippi, U.S.A.

Abstract

This entry provides a general introduction to the types of biocompatibility tests. It is assumed that the physical, mechanical, electrical, and other properties of materials and device designs have been adequately evaluated and determined suitable prior to biocompatibility testing.

INTRODUCTION

A biomaterial may be defined as any material or combination of materials used to replace, augment, or restore function of damaged or diseased tissues.[1] Implicit in this definition is the idea that the materials used in devices must be safe in addition to effective. This characteristic or property of a material is termed biocompatibility. Biocompatibility may generally be regarded as the ability of a material to interact with living cells/tissues or a living system by not being toxic, injurious, or causing immunological reactions while performing or functioning appropriately. The spectrum of biomaterials used in medical applications is quite broad, ranging from cotton pads that stop bleeding of minor cuts within a few minutes; to catheters, dialysis tubes, and contact lenses designed to interact with tissues for a few hours to weeks; to sutures designed to slowly resorb over a few weeks to months; to implant devices like total replacement hips and heart valves intended to last 10–15+ years or the lifetime of the patient. To encompass this diversity of materials and applications, a consensus panel of experts has defined biocompatibility as the ability of a material to perform with an appropriate host response in a specific application.[1] This definition thus accounts for not only the variety of applications of materials in medical devices, but also for the responses of both host tissues to material and material to host tissues. For example, cotton, which may be useful to initiate blood clotting in minor cuts, may not be suitable for use as an artificial blood vessel, because clotting would interfere with or block blood flow through the vessel. The physiological environment causes the release of low levels of metallic ions from

biomedical alloys that may induce host metal hypersensitivity reactions such as Ni-allergy. For heart valves derived from animal tissues, the deposition and buildup of calcium from host fluids eventually increases material stiffness and interferes with valve opening and closing, resulting in failure. It is hence important to realize that: 1) no one material will be appropriate for all medical device applications; 2) the material, its composition, and degradation products may affect host cells and tissues; and 3) the host environment may also affect material properties and device performance.

The determination of the biocompatibility of materials and implant devices involves detailed characterization of the material (e.g., bulk and surface chemical composition; density; porosity; and mechanical, electrical, and degradation properties) and extensive testing, first at the protein/cell/tissue or *in vitro* level, and then in *in vivo* animal models and ultimately in human clinical trials. *In vitro* tests are used to screen materials, their components, and or leachable/soluble/degradation products for cytotoxic, genotoxic, immunological, and hemolytic effects. Animal models are used to evaluate material–host tissue interactions and to predict how the device or prototype may perform in humans. Ultimately, the safety and effectiveness of the device must be evaluated in humans prior to widespread use by physicians and their patients. At each stage, biocompatibility data must be correlated with material properties and with manufacturing, sterilization, packaging, storage, and other handling procedures that also may influence test outcomes.

The design and use of biocompatibility testing protocols are provided by a variety of professional and regulatory

Concise Encyclopedia of Biomedical Polymers and Polymeric Biomaterials DOI: 10.1081/E-EBPPC-120007261

organizations, including the American Society for Testing and Materials International (ASTM), the International Standards Organization (ISO), the American Dental Association (ADA), the National Institutes of Health (NIH), and the Food and Drug Administration (FDA). The use and documentation of biocompatibility tests are required by law in the United States and other countries, and are used to ensure that biomedical devices and their constituent materials are safe and effective under intended use conditions.[2,3] This entry provides a general introduction to the types of biocompatibility tests. It is assumed that the physical, mechanical, electrical, and other properties of materials and device designs have been adequately evaluated and determined suitable prior to biocompatibility testing.

IN VITRO TESTING

As compared with *in vivo* and clinical studies, *in vitro* environments such as cell or bacterial cultures provide a quick and relatively cheap way of estimating a material's biocompatibility. *In vitro* tests allow a great deal of control over the test environment, in contrast to *in vivo* tests, whose animal or human subjects may be influenced by variables such as sex, age, activity, diet, etc. If a material does not perform well *in vitro* it most likely is not a good candidate for implantation. However, it must be noted that a material's *in vitro* characteristics may not necessarily reflect its *in vivo* performance, because dynamic cell/tissue interactions and hormonal and other physiological processes are not reproduced. The case of zinc-oxide eugenol dental cement is a classic example of the contradiction between *in vitro* and *in vivo* performance.[4] In *in vitro* tests, zinc-oxide eugenol cement exhibited severe toxicity, but when used in properly prepared tooth cavities, there is little if any dental pulp tissue reaction. This is because the dentin in the tooth structure acts as a barrier to reduce concentrations of molecules released from the cement to low levels. Therefore, *in vitro* cytotoxicity testing should be used only to supplement *in vivo* experiments.

There are many options for the type of *in vitro* studies that may be performed, and the number is increasing as new technologies (e.g., quantitative polymerase chain reaction, DNA on a chip, cocultures) are developed. Common aspects of host response evaluated *in vitro* are cell or organelle survival, cell proliferation, metabolic and catabolic activity, cell morphology and function (alteration of shape, size, locomotive ability, phagocytic behavior, expression/

response to cytokines/hormones, etc.), and cell damage, including chromosomal damage. Many of these tests are used to investigate and understand basic mechanisms of cell–material interaction that are not possible by other means, as well as to estimate the biocompatibility of materials. This section focuses on several common methods (Table 1) used primarily for estimating biocompatibility *in vitro*.

Cytotoxicity Tests

Cytotoxicity is the ability to cause death or damage at the cellular level by direct cell lysis or by fatally altering cellular metabolism. Inhibition of enzyme activity, changes in cell membrane permeability, and other sublethal effects may also be included in cytotoxic effects. Cytotoxicity assays are used as a screening mechanism to identify candidate materials for further biocompatibility tests. These tests measure the initial reaction of cells to the material and its components, and correlate fairly well with early tissue inflammatory, necrosis, and toxicity reactions at implant sites. Therefore, materials that do not perform well in cytotoxicity studies generally are not considered for human implantation. The three main types of cytotoxicity tests are elution, agar overlay, and direct contact. Specific test protocols may be found in ISO, ASTM, NIH, and related documents.[5–8]

Elution tests

In elution tests, an extract or elute of a test material is taken and added to cells in culture. Serial dilutions of the elute are usually tested to determine toxic doses, and changes in cell growth or proliferation are measured and compared to non-treated cells over 24–78-hr periods. Cell proliferation or growth may be determined through direct cell counting using a hemocytometer or electronic cell counter, incorporation of radiolabeled thymidine (because thymidine is used only in DNA replication), or quantification of DNA material (because the amount of DNA is proportional to cell number). Changes in cell membrane permeability are used to estimate cell damage through the uptake of viability dyes (e.g., trypan blue or neutral red) or the release of intracellular enzymes such as lactate dehydrogenase, because large intracellular enzymes are only released from cells with severely damaged or disrupted cellular membranes. The conversion of tetrazolium salts by intracellular

Table 1 Types of *in vivo* tests for estimating biocompatibility

Cytotoxicity	Hemocompatibility	Mutagenecity	Hypersensitivity
Elution or extract test	Hemolysis assay	Ames test	Lymphocyte transformation test
Agar or agarose overlay test	Clotting and complement activation		Leukocyte migration inhibition test
Direct contact test			

enzymes (primarily mitochondrial enzymes) to colored compounds that are read spectrophotometrically may also be used to estimate viability and proliferation because enzyme activity is retained only in viable cells, and because as cells grow, more dye is converted. Generally, a 50% or greater decrease in the number of cells, in the incorporation of radiolabeled compounds, or in the absorbance values of colored reagents compared to nontreated controls is associated with a severe cytotoxic response. This test is good for evaluating the toxicity of the leachable, soluble components from test materials, although the concentration of test substances must be carefully considered (because almost any material or chemical in high enough concentration will be toxic).

Agar overlay

In agar overlay tests, materials are placed on an agar or agarose layer covering the cells for 24 hours. Components from the test material are allowed to diffuse through the agar or agarose to the cells. Cytotoxicity of the diffusible components is determined by staining the cells with a viability dye and then measuring the zone of dead cells surrounding the test material. Materials that have no or only a limited zone of dead cells (limited to just under the sample) are given low cytotoxic scores, whereas materials with larger zones extending beyond the test sample are given higher cytotoxic scores. This method is good for low-density materials, powders, liquids, and high-density materials that may crush or damage cells if placed in direct contact. Care must be taken in comparing toxicity of different materials because different chemical compounds may have different diffusion rates and solubilities in the agar or agarose layer.

Direct contact tests

In direct contact tests, test materials are placed in direct contact with the cells. This may include simply growing cells on top of the test material or placing the test material on top of the cells in culture. Changes in cell growth or proliferation are measured in similar manner as in the elution tests, and changes in other cellular parameters (including morphology, intracellular ATP, intracellular signaling, chemokine/cytokine release, and total protein production/secretion) may also be measured. Zones of cell death or inhibited cell growth may also be measured in manner similar to the agar overlay tests, although care must be taken to not disturb the sample from its original position. This test is flexible with respect to testing a variety of materials and assessing of toxicity, although care must be taken with bioresorbable/degradable materials because degradation products may build up to nonrelevant toxic levels.

The degree of cytotoxicity of materials and their released components in these tests is determined in comparison to controls. Positive controls such as pure copper or latex rubber are materials that will cause an effect (e.g., cell death). Negative controls such as Teflon® or titanium are materials that cause no or minimal effect.

Other factors affecting *in vitro* tests are the types of cells (e.g., fibroblasts, osteoblasts, macrophage, endothelial, stem cell, etc.), species (e.g., human, rat, pig, etc.), source (normal tissues, tumor tissues, or tissues transformed by viral agents), period of exposure (hours to days), and culture parameters (type of medium, serum additions, number of cells per unit area).[9] These factors must be considered in comparing results and in establishing experimental designs. Generally, cells derived from tumors or transformed cell lines are recommended for cytotoxicity tests because their characteristics are well-defined, they are less variable in genetic makeup and growth, and they are more readily available than normal (noncancerous, nontransformed) cells. Normal cells are more appropriate for investigating questions of specific cell–material interactions (such as cell attachment and spreading, elaboration of extracellular matrix, attachment and growth of osteoblasts, or activation of platelet or macrophage cells). Also, the selection of cell type should be appropriate for the intended application of the implant material.

Hemocompatibility Tests

Hemocompatibility tests are used to evaluate the effect of a material or released compounds on blood coagulation processes, thrombus formation, and hemolysis (the destruction of red blood cells). An upset in blood chemistry or damage to one of its components may be fatal, and therefore most medical devices that come into contact with blood are required to undergo hemocompatibility tests. Materials such as hypodermic needles that make only brief contact with blood need not be tested.

In hemolysis tests, test and control materials or their extracts are incubated with red blood cells, isolated from rabbits, mice, or rats for three hours with intermittent shaking to keep samples mixed and in contact with blood. The amount of hemoglobin released into the supernatant from the cells is determined spectrophotometrically and reported as percent hemolysis with respect to negative controls (materials that do not cause red blood cell lysis). The concentration of extracts and surface-to-volume ratio of solid or particulate materials to volume of red blood cells must be closely controlled to obtain reproducible results.

To evaluate the effects of materials and their surfaces on blood clotting, materials are exposed to whole blood serum. The time for the development of a clot on the material is compared to that of a reference material. Additionally, the serum may be tested for changes in different blood clotting factors after exposure to the test material. However, because changes in blood flow may occur due to device design (e.g., turbulent flow may increase hemolysis and or clotting) the evaluation of test materials under flow conditions should be undertaken. For example, the number of adherent platelets may be determined per unit area after exposure to whole blood under controlled flow conditions.[10] Because of their role in coagulation and thrombus formation, increased adhesion of platelets to the surface may indicate increased clotting rates. *Ex vivo* tests are also popular for evaluating hemocompatibility under more *in vivo*-like blood flow conditions. In *ex vivo* tests, blood is temporarily shunted from an artery and vein of an experimental animal and allowed to flow over or through the test material inserted between inlet and outlet segments.

In addition to the limits of reproducing blood flow dynamics *in vitro*, the variability in blood from donors due to age, diet, health, etc. makes standardization of responses difficult. Hence, repeated tests with the same donor blood and use of appropriate reference materials are important considerations.

Mutagenicity and Genotoxicity

Mutagens are those materials that modify the genome of a host. Materials that have mutagenic effects on their host may therefore be classified as genotoxic. It is widely accepted that carcinogenic behavior proceeds via a mutation in the genome. It is thus expected that all carcinogens will be mutagens; however, the opposite is not necessarily true (i.e., a mutagen is not always a carcinogen).

One common method for studying mutagenicity is the Ames test. Briefly, the Ames test uses a mutant bacterial cell line (*Salmonella typhimurium* or *Escherichia coli*) that must be supplied with histidine to grow. The cells are cultured in a histidine-free environment, and only those materials that mutate the cells back to a state of histidine independence will allow the cells to grow. The Ames test is often used as an early screen for mutagenicity, although it will not detect the full range of mutagens.

Hypersensitivity Tests

Tests such as the leukocyte migration inhibition and lymphocyte transformation tests have been used as *in vitro* models to estimate delayed hypersensitivity reactions to implant materials and their released components.[11–13] The use of these tests in screening the sensitization potential of materials is not well justified because genetic factors control individual sensitivity; however, the tests may be useful in assessing the sensitivity of individuals to particular materials and compounds.

IN VIVO TESTING FOR BIOCOMPATIBILITY

Many systemic physiological processes are complex and cannot be simulated *in vitro*. Therefore *in vivo* animal testing is necessary prior to human clinical testing. *In vivo* tests (Table 2) are used to evaluate the local and systemic interactions of host tissues with implant materials or devices and their leachable/soluble/degradation components. These include physiological effects of the implant and its released components on local and systemic tissues, tissue response to the implant materials and designs, changes in extracellular matrices and regulatory biomolecules, and changes in the material and device due to host physiology.

Animal Welfare Issues

Use of animals for scientific purposes may cause them physical or psychological pain or harm. Thus, it is imperative that scientists use the least painful humane methods accepted in modern veterinary or laboratory practice. *In vivo* animal testing should be done only when absolutely necessary. Whenever possible, a critical survey of the scientific literature and *in vitro* testing should be carried out before *in vivo* tests are undertaken. In the United States, legislative, regulatory, and professional organizations have developed thorough guidelines for using animals in research.[14–18] Similar laws and regulations exist in most other countries that provide protection for lab animals and strongly encourage

- Experimental designs based on the relevance to human or animal health, advancement of knowledge, or the good of society.
- Use of appropriate species, quality, and number of animals.

Table 2 Types of *in vitro* tests for estimating biocompatibility

Short-term implantation tests	Long-term functional tests	Sensitization	Irritation	Other
Subcutaneous, intramuscular, and intraperitoneal implantation tests to evaluate general tissue necrosis, fibrosis, and inflammation	Devices or compositionally identical prototypes are implanted in appropriate animal models to replicate/simulate intended end-use in humans. Functionality of device and histopathological evaluations of tissues/organs are performed.	Guinea pig maximization test	Skin	Genotoxic
		Occluded patch test	Ocular	Carcinogenic
		Open epicutaneous test	Mucosal	Reproductive
				Cerebrospinal
				Hemocompatible

- Avoidance or minimization of discomfort, distress, and pain in concert with sound science.
- Use of appropriate sedation, analgesia, or anesthesia.
- Establishment of experimental end points.
- Conduct of experimentation on living animals only by or under the close supervision of qualified and experienced persons.

Animal and Implant Site Selection

No single species represents an ideal general model for the human species. A variety of sources are devoted to the selection of appropriate animal models.[5–8,18–21] Anatomical, biochemical, physiological, pathological, and/or psychological characteristics must be considered when choosing an animal model.

Animal Tests

Guidelines to carry out animal biocompatibility tests are described by standards organizations and government regulatory agencies such as ASTM, ISO, NIH, FDA, ADA, U.S. and European Pharmacopeia, and others.[5–8] The tests can be divided into functional and nonfunctional tests.

Nonfunctional tests

Nonfunctional tests are generally conducted first to study the direct interactions of the implant with the physiological environment. In these tests, it is assumed that acute toxic or inflammatory reactions are nonspecific and thus may be evaluated in soft tissues (e.g., subcutaneous, intramuscular, intraperitoneal) because implantation in soft tissue requires minor surgery. Specialized sites such as the cornea and cerebral cortex are used for materials intended for those specific applications. After predetermined periods of time, the animals are euthanized and tissue reactions to the test materials are evaluated histologically and compared to controls in the form of blanks or currently accepted implant materials that have a well-characterized known tissue response. Histological analyses may be scored or graded based on degree of tissue necrosis/degeneration, fibrosis, and types and amounts of inflammatory (polymorpho-nuclear leukocytes, macrophages, lymphocytes, etc.) and foreign body giant cells present. Because many factors of the implant device and design (such as functional loading and multicomponent interactions for a total hip) are not taken into account, such tests do not fully measure biocompatibility. Thus, they tend to be of short to intermediate duration, usually a few weeks to 24 months (e.g., ASTM F 763 Standard Practice for Short-Term Screening of Implant Materials or ASTM F 981 Standard Practice for Assessment of Compatibility of Biomaterials for Surgical Implants with Respect to Effect of Materials on Muscle and Bone). Fluid extracts from device materials may be used to determine the acute biological reactivity of possible leachable/soluble/degradation compounds when injected into soft tissues.

Functional tests

After evaluation in soft tissues, the next step is the selection of an animal model and implant site for the device or prototype to simulate the site that ultimately will be utilized in humans (i.e., functional testing). Functional tests require that the implant be placed, at least in some degree, in the functional mode that it would experience in human implant service. Animal model functional tests are obviously of much greater complexity because the design, fabrication, surface treatment, packaging, mechanical testing, and implantation of these devices may be different in the device finally produced for human use. Variation in testing procedures with different animal models and variation in species histology can create difficulties in comparing, assessing, and interpreting results. Changes in animal behavior (for example, in preferentially not loading a limb with an implant) may also be important.

There are also specific *in vivo* physiological assessment tests for genotoxicity, carcinogenicity, reproductive toxicity, delayed-type hypersensitivity, and systemic toxicity. Detailed protocols for these tests may be found in standards documents.[5–8]

Genotoxicity testing

Genotoxic agents or genotoxins cause alterations in DNA or chromosomal structure or other DNA or gene damage that result in permanent inheritable changes in cell function. The major genotoxic effects tested for are gene mutations, chromosomal aberrations, and DNA effects. Gene mutation and chromosomal aberration tests detect actual lesions in the DNA molecule, whereas DNA effects tests detect events that may lead to cell damage. Although *in vitro* tests for genotoxicity/mutagenicity may be conducted using microorganisms or mammalian cells, an *in vivo* animal model is also recommended when the scientific literature or *in vitro* tests indicate potential genotoxicity of the implant or its components.[22]

Gene mutation tests detect base-pair mutations, frameshift mutations, and small deletions. Chromosomal aberration tests detect chromosomal damage induced after one cellular division; structural changes in the chromosomes are evaluated while cells are in the metaphase stage of division. DNA effects tests (generally, mouse bone marrow micronucleus test) detect damage to the chromosomes or the mitotic apparatus of immature red blood cells found in bone marrow.

Carcinogenicity tests

Carcinogenicity testing is done to determine the tumorigenic potential of devices, materials, or extracts (after a single exposure or multiple exposures) over the total life span of the test animal. Specifically, such testing should be considered for a device that will have permanent contact (longer than 30 days) with tissues, either as an implant or as an externally communicating device.

Rodents are invariably chosen as the test species because their relatively short life spans make it practical to carry out lifetime studies. Although most investigators and regulatory agencies agree on undertaking carcinogenicity testing, many available standards lack specific information regarding the number of test animals to be used, the kinds of observations needed, the extent of histopathological evaluations, the number of survivors required at the end of the study, and the type of statistical evaluations to be used. The best scientific practices for carcinogenicity testing have been summarized.[23]

Sensitization tests

Sensitization testing is done to determine whether chemicals or compounds that may be released from specific biomaterials and devices elicit sensitization reactions. Sensitization or hypersensitivity reactions are a result of immunologically medicated reactions resulting in redness (erythema) and swelling (edema). Because most such reactions are of the dermal cell-mediated type rather than the humoral or antigen-antibody type, the skin of laboratory animals (usually guinea pigs because they exhibit dermal sensitivity similar to humans) is used in sensitivity testing.[24] Three common tests are the guinea pig maximization test, the occluded patch test, and the open epicutaneous tests. Typically, the material or material extract is injected subcutaneously. Then one to two weeks after exposure, the test material or solution is reapplied to the skin and the degree of erythema and edema is graded. The use of adjuvants that enhance the immunological response (Freund's adjuvant, sodium lauryl sulfate) may be used during initial exposure to help identify weakly sensitizing materials. These common tests do not detect compounds released from the material that may also act as an adjuvant, or responses to antigens such as the plant proteins found in natural latex. Kimber et al. and Merritt have reviewed sensitization testing.[12,25]

Irritation tests

Irritation is a localized inflammatory response to single, repeated, or continuous exposure to a material without involvement of an immunological mechanism. The intracutaneous, primary skin, and ocular irritation tests are three *in vivo*, nonclinical tests commonly used to evaluate materials for possible contact irritation. The intracutaneous test is similar to the sensitization test. Fresh extract is injected intracutaneously at multiple sites on the shaved backs of albino rabbits and then 24, 48, and 72 hours after injection, the test and control sites are observed and scored for the severity of erythema or edema. Extracts that produce a

significantly greater response than controls are considered irritants. In the primary skin irritation test the portions of the test material itself are simply placed on the shaved backs of rabbits with the aid of an occlusive dressing and the response after 24 hours is evaluated by methods similar to those used in the intracutaneous test.

The ocular irritation test is usually performed for opthalmological materials. A small sample of material (powder or fluid extract) is placed directly into the pocket formed by withdrawing the lower eyelid of a rabbit. The rabbit's other eye is left untreated as a control. After 72 hours, the eye is evaluated for redness and swelling of the conjunctiva, response of the iris to light, corneal opacity, and presence of discharge. Scores are compared with standards to determine the degree of eye irritation. In addition to the intracutaneous, primary skin, and ocular irritation tests, mucosal (oral, rectal, penile, and vaginal) irritation tests may carried out as necessary.[26]

Systemic effects

Systemic toxicity may occur when chemicals and compounds released from implant materials are distributed by the blood and lymphatic system and damage organs and tissues remote from the implant site. The various routes used for sample administration are topical or dermal, inhalation, intravenous, intraperitoneal, and oral. The material may be tested in powder, fluid extract, or solid forms and evaluations may be based on a single or multiple doses. Systemic effects are categorized on the basis of time to initiate adverse effects: acute (within 24 hr), subacute (in 14–28 days), subchronic (10% of an animal's life span), and chronic (longer than 10% of an animal's life span). Test methods used in longer-term systemic toxicity studies are much the same as those used in acute toxicity tests, except that larger groups of animals are used.

Reproduction toxicity tests evaluate the potential effects of devices, materials, or extracts on reproductive function, embryonic development, and prenatal and early postnatal development. Reproductive/development toxicity tests should only be conducted when the device has potential impact on the reproductive system.[26]

The selection of evaluation in *in vivo* protocols depend on the animal used (mice, rabbit, dog, etc.) and the implant material/device. Tests performed during the in-life phase often include measurements of body weight and food consumption, blood and urine analyses, and eye examinations. Recommended hematology test parameters are hematocrit, hemoglobin percentage, erythrocyte counts, and total and differential leukocyte counts. Platelet counts and measurements of prothrombin and thromboplastin concentrations and clotting time are usually performed for blood contacting materials and devices. Clinical pathology protocol should include assays that test proper electrolyte balance, carbohydrate metabolism, and liver and kidney function. Postmortem analyses include gross observations at necropsy, organ weighing, and microscopic examination of selected tissues. Considerable care must be taken in the design of subchronic and chronic studies to make certain they provide assurance of the device's safety.

CLINICAL TRIALS OF BIOMEDICAL IMPLANTS

Clinical trials are designed to test the safety of a new device in humans and are conducted only after they have been extensively evaluated by *in vitro* and *in vivo* animal models. Although clinical trials do provide vital information on the effectiveness of a biomedical implant device, they do not specifically test the biomaterial; rather, they test the device composed of the biomaterial(s) in a specific application. Unlike *in vivo* animal studies, the implants in clinical trials remain exposed to the experimental subject even after the period of observation ends. As time passes and more subjects are exposed to the device without adverse results, confidence in acceptable biological performance increases.

It is standard practice in new drug trials to employ a double-blind study in which a placebo is randomly administered to a portion of the defined group of patients to ensure that neither doctor nor patient know whether the active drug is being administered. In implant testing, however, it is not possible to pair the implanted patient with a placebo-treated patient, because it is not possible to conceal the implantation site from surgeon or patient. For clinical trials of implant materials, comparisons must be made between the condition of the patient before and after implant surgery. An implanted patient must also be compared with a patient with similar implants made of different materials, as well as a nondiseased (control) individual of the same age, sex, and similar home/workplace environment.

Because comparisons must be made between patients, certain standards exist for the selection and treatment of patients. Inclusion/exclusion criteria of the target patient population clarify the intended use of the device. The exclusion criteria serve to eliminate factors that increase operative risk or that may confound the outcome of the study. Medical care administered in a clinical trial must be under the care of a medical professional, and the patients must give informed consent to any experimental procedure.[27–29] Patients must be informed of any possible benefits or risks of the procedure, and the identity of the patients must be protected and confidentiality of their medical records preserved. Also, there must be a reasonable possibility of benefit combined with reasonable assurance of usual risk.

Clinical trials are divided into three phases. Phase I (early trial) involves simply selecting a new treatment from several options for further study. The biomaterial is tested on a small group of people (~60–80). Phase II studies a larger group of subjects (~100–300) to test the effectiveness and further evaluate the safety of the biomaterial. It is divided into phase IIA and phase IIB. If the new treatment

selected in the early trial is not effective, phase IIA (preliminary trial) examines whether further studies should be performed or the treatment abandoned. Otherwise, phase IIB (follow-up trial) estimates the effectiveness of the new treatment if it appears promising. Phase III is a comparison of the effectiveness of the new treatment with a standard of management or some other treatment, and is also divided into phase IIIA and phase IIIB. A large study group of approximately 1000–3000 people is used. If the new implant improves the clinical outcome (phase IIIA), the treatment is refined and examined further in phase IIIB, usually by multiple investigators and institutions.

Clinical trials must be performed under the control of a defined prospective protocol that describes the implant device and outlines the indications of the surgical procedure, the uniform surgical procedure used, and the postoperative treatment and follow-up schedule. The number of patients included and needed to complete the trial to demonstrate benefit or improvement of the device is also critical to the study design, because too few patients will reduce significance, and too many may waste time and resources. Inclusion of and consultation with biostatisticians in developing clinical trial designs will help ensure that the device is adequately evaluated for potential use. Other considerations important in the design of a clinical trial include the study duration, examination schedule, and reporting of adverse events. Adverse events are noted and serious intraoperative and postoperative adverse events reported. A list of possible adverse events should also be included in the protocol of the study. To ensure that a clinical trial is conducted ethically and that the rights of the study participants are protected, an institutional review board or independent committee of physicians, statisticians, and community advocates is selected to help prepare the protocols and review the procedures and safeguards in any experimental program involving human subjects before the clinical trial begins.

The reports of clinical trials should discuss the accuracy and precision of all measurements as well as define a minimum confidence level for all statistical measures of data (usually $p < 0.05$). The reports must also include confidence intervals or other measures of significance associated with all derived parameters, and must indicate the significance of any conclusion arrived at by analysis of the trial.

The complication incidence rate for many biomedical materials/devices is extremely small. A study of the costs and benefits associated with research to improve then current orthopedic prosthetic devices concluded that failure/complication rates, even in the 1970s, were acceptably small and that the investment needed to significantly reduce these rates would be unreasonable with respect to the resulting benefits.[30] In many cases in which the present technology is extremely successful, newer developments cannot demonstrate statistically superior outcomes because of limitations of size and duration of clinical trials. In these cases, the decision to pursue further improvements in the material or device depends on the argument that the human and financial costs to individuals as a result of device malfunction and failure is greater than the financial costs to the manufacturers and research institutions. Because the test of adequate performance described in the legal system is that the device be "safe and effective" and pose no "unreasonable" risk or hazard to the patient, decisions on what may be considered acceptable failure rates remain subjective, and depend on public opinion, expert advice, and administrative action.

Retrieval of implants and associated tissues (due to failure or other complications or at patient death) provides valuable information about the long-term biocompatibility of implants and their materials. Thorough analysis of failed implants provides insights into material design, selection, and manufacture and handling; implant design; surgical protocols; and the interaction between implant and host tissues as they relate to biocompatibility of the device. This information may be used to improve material design, selection, and manufacture, and to revise surgical protocols. Retrieval of implants and tissues is also important because the data provide a true assessment of the safety and efficacy of the device and document long-term host–material interactions. Evaluation of both device failure and success is necessary for improving understanding of the biocompatibility of implant materials and devices.

CONCLUSION

Evaluating the biocompatibility of biomedical materials and devices involves extensive testing at the *in vitro* (protein/cellular/tissue), *in vivo* animal, and human clinical use levels to determine how the materials and devices will perform in their intended applications. The success of a material and device depend on how the materials and device design affect the host and how the host affects device function and material properties. *In vitro* tests have evolved to provide quick and relatively fast biocompatibility screening of candidate biomedical materials, but they are not yet capable of accurately predicting the clinical success of implant devices. However, with societal pressures to develop alternatives to animal testing, and with advances in cell and molecular biology techniques, advances in *in vitro* methods for the prediction of clinical success are continuing. The use of *in vivo* animal models is necessary to evaluate the interaction of an entire physiological system with the device and its components, and to establish safety and performance capabilities prior to use in humans. It is imperative that great care and planning are used in planning *in vivo* animal tests to ensure that relevant and significant biocompatibility data are obtained. Only in clinical studies can the biocompatibility of a material or device and their ability to replace, restore, or augment the function of damaged or diseased human tissues be evaluated. This requires close evaluation of not only patients during the study period, but also of patient health records and of retrieved devices (failed devices and successful devices retrieved at death), even after the planned clinical study time frame to establish epidemiologically successful biomedical materials and devices. Finally, it is noted that methods and evaluation criteria for determining biocompatibility are routinely reviewed and amended as additional information is collected. New methods and standards are being developed as new hybrid devices are developed and refined that involve both biological and synthetic materials (e.g., a patient's stem cells grown on resorbable substrate).

ACKNOWLEDGMENTS

This entry was supported by the Agricultural and Biological Engineering Department and the Mississippi Agriculture and Forestry Experiment Station (MAFES, Ms No. BC 10316) at Mississippi State University.

REFERENCES

1. Williams, D.F. *Definitions in Biomaterials*. Proceedings of a Consensus Conference of the European Society for Biomaterials, Chester, England, Mar 3–5, 1986; Vol. 4, Elsevier: New York.
2. Medical Device Amendment of the U.S. Food and Drug Act of 1976.
3. Safe Medical Devices Act of 1990.
4. Mjör, I.A.; Hensten-Pettersen, A.; Skogedal, O. Biological evaluation of filling materials: A comparison of results using cell culture techniques, implantation tests and pulp studies. Int. Dent. J. **1977**, *27* (2), 124–132.
5. ISO-1099-3: Biological Evaluation of Medical Devices, Part 1—Guidance on Selection of Tests. International Organization for Standardization, Geneva, Switzerland, 1997.
6. ASTM F 748–98 selecting generic biological test methods for materials and devices. In *Annual Book of ASTM Standards Vol. 13.01 Medical Devices*, Emergency Medical Services, ASTM International: West Conshohocken, PA, 2002.
7. U.S. HHS. *Guidelines for Physiochemical Characterization of Bio-materials, Publication 80–2186*; National Institutes of Health, U.S. Department of Health and Human Services: Washington, DC, 1980.
8. AAMI. Standards and Recommended Practices, Vol. 4 Biological Evaluation of Medical Devices; Association for the Advancement of Medical Instrumentation: Arlington, VA, 1994.
9. Bumgardner, J.D.; Gerard, P.D.; Geurtsen, W.; Leyhausen, G. Cytotoxicity of precious and nonprecious alloys—Experimental comparison of *in vitro* data from two laboratories. J. Biomed. Mater. Res. (Appl. Biomater.) **2002**, *63* (2), 214–219.
10. Skarja, G.A.; Kinlough-Rathbone, K.L.; Perry, F.D.; Rubens, F.D.; Brash, J.L. A cone-and-plate device for the investigation of platelet biomaterial interactions. J. Biomed. Mater. Res. **1997**, *34* (4), 427–438.

11. Rae, T. Cell biochemistry in relation to the inflammatory response to foreign materials. In *Fundamental Aspects of Biocompatibility*; Williams, D.F., Ed.; CRC Press, Inc.: Boca Raton, FL, 1981; Vol. 1, 159–181.

12. Merritt, K. Immunological testing of biomaterials. In *Techniques of Biocompatibility Testing*; Williams, D.F., Ed.; CRC Press, Inc.: Boca Raton, FL, 1986; Vol. 2, 123–136.

13. ASTM. F1906–98 Evaluation of Immune Responses in Biocompatibility Testing Using ELISA Tests, Lymphocyte, Proliferation and Cell Migration. In *Annual Book of ASTM Standards Vol. 13.01 Medical Devices*; Emergency Medical Services, ASTM International: West Conshohocken, PA, 2002.

14. Animal Welfare Act of 1996 as amended. United States Code, Title 7, sections 2131 to 2156.

15. NIH. *Guide for the Care and Use of Laboratory Animals, Publication 86–23*; National Institutes of Health, U.S. Department of Health and Human Services: Washington, DC, 1985.

16. Public Health Service. *Policy on Human Care and Use of Laboratory Animals*; Office of Protection from Research Risks, National Institutes of Health: Bethesda, MD, 1986.

17. American Veterinary Medical Associate (AVMA). Report of the AVMA panel on euthanasia. J. Am. Vet. Med. Assoc. **1993**, *202* (2), 229–249.

18. ISO-1099-3: Biological Evaluation of Medical Devices, Part 2—Animal Welfare Requirements. International Organization for Standardization, Geneva, Switzerland, 1992.

19. Mendenhall, H.V. Animal selection. In *Handbook of Biomaterials Evaluation: Scientific, Technical and Clinical Testing of Implant Materials*, 2nd Ed.; von Recum, A.F., Ed.; Macmillan Publishing Company: New York, 1999; 475–480.

20. Schmidt-Nielsen, K. *Scaling: Why Is Animal Size So Important?*; Cambridge University Press: Cambridge, UK, 1985.

21. Vale, B.H.; Willson, J.E.; Niemi, S.M. Animal models. In *Biomaterials Science: An Introduction to Materials in Medicine*; Ratner, B.D., Hoffman, A.S., Schoen, F.J., Lemons, J.E., Eds.; Academic Press: San Diego, CA, 1996; 238–242.

22. Dearfield, K.L.; Cimino, M.C.; McCarroll, N.E.; Mauer, I.; Valcovic, L.R. Genotoxicity risk assessment: A proposed classification strategy. Mutat. Res. **2002**, *521* (1–2), 121–135.

23. Combes, R.; Schechtman, L.; Stokes, W.S.; Blakey, D. The international symposium on regulatory testing and animal welfare: Recommendations on best scientific practives for subchronic/chronic toxicity and carcinogenicity testing. ILAR J. **2002**, *43* (Suppl.), S112–S117.

24. ASTM. F720–81 Testing Guinea Pigs for Contact Allergens: Guinea Pig Maximization Test. In *Annual Book of ASTM Standards vol. 13.01 Medical Devices*; Emergency Medical Services, ASTM International: West Conshohocken, PA, 2002.

25. Kimber, I.; Basketter, D.A.; Berthold, K.; Butler, M.; Garrigue, J.L.; Lea, L.; Newsome, C.; Roggeband, R.; Steiling, W.; Stropp, G.; Waterman, S.; Wieman, C. Skin sensitization testing in potency and risk assessment. Toxicol. Sci. **2001**, *59* (2), 198–208.

26. Thomas, J.A. In-use testing of biomaterials in biomedical devices. In *Handbook of Biomaterials Evaluation: Scientific, Technical and Clinical Testing of Implant Materials*, 2nd Ed.; von Recum, A.F., Ed.; Macmillan Publishing Company: New York, 1999; 313–320.

27. Greenwald, A., Ryan, M., Milvihill, J., Eds. *Human Subject Research: A Handbook for Institutional Review Boards*; Plenum Press: New York, 1982.

28. Federal policy for the protection of human subjects, Department of Health and Human Services regulations, Title 45 Code of Federal Regulations, part 46 (45 CFR46).

29. World Medical Association. Recommendations guiding doctors in clinical research. Br. Med. J. **1964**, *2*, 1119–1129.

30. Piehler, H.R. Risk–benefits of orthopaedic prosthetic devices. Orthop. Rev. **1978**, *7*, 75–88.

BIBLIOGRAPHY

1. American National Standard/American Dental Association (ANSI/ADA) Specification No. 41—Biological Evaluation of Dental Materials; American Dental Association: Chicago, IL, 2001.

2. ASTM. *Annual Book of ASTM Standards Vol. 13.01 Medical Devices*; Emergency Medical Services, ASTM International: West Conshohocken, PA, 2002.

3. Black, J. Biological Performance of Materials—Fundamentals of Biocompatibility; Marcel Dekker, Inc.: New York, 1999.

4. Dee, K.C.; Puleo, D.A.; Bizios, R. *An Introduction to Tissue— Biomaterial Interactions*; John Wiley & Sons, Inc.: Hoboken, NJ, 2002; 173–184.

5. NIH. Guidelines for Blood–Material Interactions, Publication 85–2185; National Institutes of Health: Bethesda, MD, 1985.

6. Hanson, S.; Lalor, P.A.; Niemi, S.M.; Northrup, S.J.; Ratner, B.D.; Spector, M.; Vale, B.H.; Willson, J.E. Testing biomaterials. In *Biomaterials Science—An Introduction to Materials in Medicine*; Ratner, B.D., Hoffman, A.S., Schoen, F.J., Lemons, J.E., Eds.; Academic Press: San Diego, CA, 1996; 215–242.

7. International Standard ISO 7405—Dentistry—Preclinical Evaluation of Biocompatibility of Medical Devices Used in Dentistry—Test Methods for Dental Materials; International Standards Organization: Geneva, Switzerland, 1997.

8. Kirkpatrick, C.J.; Bittinger, F.; Wagner, M.; Kohler, H.; van Kooten, T.G.; Klein, C.L.; Otto, M. Current trends in biocompatibility testing. Proc. Inst. Mech. Eng. H **1998**, *212* (2), 75–84.

9. Mohan, K.; Sargent, H.E. Clinical trials, an introduction. Med. Dev. Diag. Ind. **1996**, *18*, 114–119.

10. Mollnes, T.E. Complement and biocompatibility. Vox Sang. **1998**, *74* (Suppl. 2), 303–307.

11. Technical Committee ISO/TC 94. International Standard ISO 10993—Biological Evaluation of Medical Devices, Part 1–17. International Standards Organization, Geneva, Switzerland, 1997.

12. Thull, R. Physiochemical principles of tissue material interactions. Biomol. Eng. **2002**, *19* (2–6), 43–50.

13. USP XXIII. The Pharmacopeia of the United States of America, 23rd Revision Incorporating The National Formulary, 18th Revision; The United States Pharmacopeial Convention, Inc.: Washington, DC, 1995.

14. von Recum, A.F., Ed. *Handbook of Biomaterials Evaluation—Scienctific, Technical, and Clinical Testing of Implant Materials*, 2nd Ed.; Macmillan Publishing Co.: New York, 1999.

Bioactive—BioMEMS

15.   Walum, E.; Stenberg, K.; Jenssen, D. *Understanding Cell Toxicology—Principles and Practice*; Ellis Horwood Ltd.: New York, 1990.

16.   Wataha, J.C. Biocompatibility of dental materials. In *Restorative Dental Materials*, 11th Ed.; Craig, R.G., Powers, J.M., Eds.; Mosby: St. Louis, 2002; 125–163.

17.   Williams, D.F., Ed. *Fundamental Aspects of Biocompatibiliy*; CRC Press: Boca Raton, FL, 1981; Vols. (1–2).

18.   Williams, D.F., Ed. *Techniques of Biocompatibility Testing*; CRC Press: Boca Raton, FL, 1986; Vols. (1–2).

Biodegradable Polymers: Bioerodible Systems for Controlled Drug Release

Jorge Heller
Kirk P. Andriano
APS Research Institute, Redwood City, California, U.S.A.

Abstract

This entry focuses on delivery systems where a drug has been homogeneously dispersed in a polymer matrix, and divided into hydrophilic systems and hydrophobic systems. The latter class is subdivided into predominantly diffusion-controlled systems and predominantly polymer-hydrolysis-controlled systems.

The development of bioerodible drug-delivery implants is assuming an ever-increasing importance in research on controlled drug-delivery. A major driving force in this development is the need to deliver therapeutic agents directly to the circulatory system, which is important with drugs that undergo significant inactivation by the liver. Another advantage of bioerodible drug-delivery implants is that small, well-tolerated implants can be left in place for very long periods, making delivery regimes lasting one or more years possible.

However, perhaps the major interest in developing bioerodible drug delivery devices arises in connection with protein delivery.[1] Proteins are not active orally and have very short half-lives. In the absence of a suitable controlled systemic delivery device, they must be administered by daily injection—clearly an undesirable therapy.

In discussing such systems, the terms bioerosion and biodegradation are often used as if they were interchangeable. However, this is incorrect because biodegradation signifies changes in polymer structure that occur as a consequence of some chemical reaction, whereas bioerosion signifies solubilization of a solid polymer. Though bioerosion can occur as a consequence of structural changes caused by a chemical reaction, it can also occur without changes in chemical structure; examples are dissolution of a solid water-soluble polymer and dissolution promoted by ionization or protonation of functional groups.[2] Biodegradation, on the other hand, only leads to solubilization if the degradation reaction produces water-soluble structures. Further, although the term "bioerosion" implies that polymer erosion occurs as a consequence of interaction with the biological environment, in actual fact, most polymers erode by a simple hydrolysis or dissolution in the aqueous environment of the tissues.

Although more complex classifications have been proposed,[3,4] in this entry we will only consider delivery systems where a drug has been homogeneously dispersed in a polymer matrix, and we will divide these into hydrophilic systems and hydrophobic systems. The latter class will be further subdivided into predominantly diffusion-controlled systems and predominantly polymer-hydrolysis-controlled systems.

No attempt has been made to provide an exhaustive review. Instead we will emphasize some early developments.

HYDROPHILIC SYSTEMS

When a crosslinked, water-soluble material is placed in water, it will rapidly swell and then retain within its three-dimensional structure a large volume of water. Such materials are known as hydrogels. When they contain water-labile bonds, they can undergo bioerosion, and with time, solubilize to water-soluble fragments. These bioerodible hydrogels can be used as implants to deliver therapeutic agents. However, because these materials are completely permeated by water, they are clearly of little use in delivering small, water-soluble molecules that would rapidly diffuse out. On the other hand, bioerodible hydrogels are of considerable interest in delivering macromolecules that can be physically entangled in the hydrogel structure and, because of their large size, can diffuse out only slowly, if at all. The subject of bioerodible hydrogels has been exhaustively reviewed.[5]

Bioerodible hydrogels can be constructed with water-labile bonds in either the polymer backbone or the crosslink segment.

Water-Labile Bonds in Polymer Backbone

A number of systems involving crosslinked polysaccharides have been described, principally for release of proteins.[6] In these systems, the protein is physically entangled in the hydrogel and is released, principally by diffusion, in proportion to its molecular weight.

A more desirable hydrogel delivery system would be one with a crosslink density high enough so that diffusional

Concise Encyclopedia of Biomedical Polymers and Polymeric Biomaterials DOI: 10.1081/E-EBPPC-120051880

139

Bioactive—BioMEMS

release is prevented and rate of release can be controlled by chemical hydrolysis of the hydrogel. Such a system has been synthesized by crosslinking a water-soluble polyester prepared from fumaric acid and poly(ethylene glycol) with *N*-vinyl pyrrolidone (VP), as shown in Scheme 1.[7]

Because the hydrolysis rate of aliphatic ester linkages at pH 7.4 and 37°C is very slow, release of a protein immobilized in the hydrogel by physical entanglement is also very slow. Hydrolysis rates can be accelerated by placing an electron-withdrawing group adjacent to the ester linkage. Activated polyester hydrogels can be prepared by first preparing a linear, unsaturated, water-soluble polyester based on poly(ethylene glycol), fumaric acid, and a diacid activated by means of an electron-withdrawing group, and then crosslinking by copolymerization with *N*-vinyl pyrrolidone, as before. Examples of activated diacids are diglycolic, ketomalonic, and ketoglutaric, shown in Scheme 2.

Figure 1 shows release of bovine serum albumin (BSA) physically entangled in a hydrogel that was then placed in a pH 7.4 buffer at 37°C. As expected, BSA release from an non-activated ester was very slow, but the use of activated esters enhanced the hydrolytic reactivity in proportion to the reactivity of the activated ester group. Further, there was virtually no early diffusional release of BSA, indicating that the crosslink density of the hydrogel was adequate to completely immobilize the entangled BSA, and that release occurred almost exclusively by hydrogel hydrolysis.

However, these materials hydrolyzed to a nonde-gradable poly (*N*-vinyl pyrrolidone) modified by vicinal hydroxylgroups. For toxicological reasons this may not be desirable, so hydrogels were also prepared by replacing fumaric acid with diacids that contain pendant unsaturation

such as itaconic acid or allylmalonic acids, as shown in Scheme 3.[7]

In this representation, R is poly(ethylene glycol) and the carboxylic acid shown is itaconic acid. Rate of hydrolysis of such hydrogels can be regulated by using mixtures of itaconic or allylmalonic and ketoglutaric acids. These linear, water-soluble polyesters are crosslinked by a free-radical coupling of the pendant unsaturation, and degrade to small molecules.

These materials have been used to release a contraceptive vaccine prepared by conjugating leuteinizing-hormone-releasing hormone (LHRH) to the immunogenic carrier,

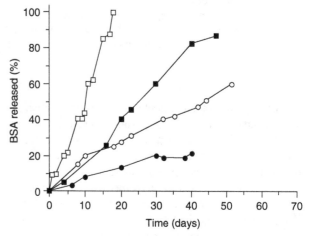

Fig. 1 Release of bovine serum albumin (BSA) at pH 7.4 and 37°C from microparticles prepared from various water-soluble unsaturated polyesters crosslinked with 60 wt% *N*-vinyl pyrrolidone, (□) 4:1 ketomalonic/fumaric, (○) 1:1 diglycolic/fumaric, (■) 1:1 ketoglutaric/fumaric, (●) fumaric.
Source: Data from Heller et al.[7]

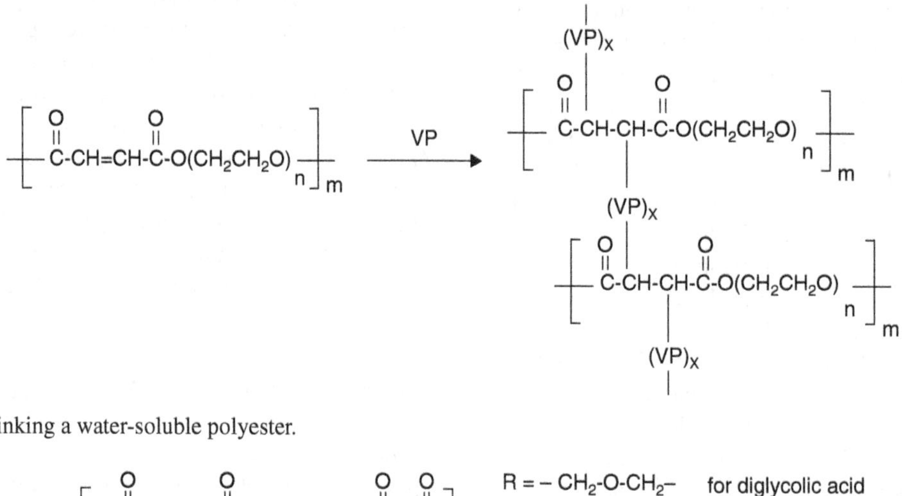

Scheme 1 Crosslinking a water-soluble polyester.

R = − CH₂-O-CH₂−   for diglycolic acid

R = −C− (O)   for ketomalonic acid

R = − CH₂-CH₂-C− (O)   for ketoglutaric acid

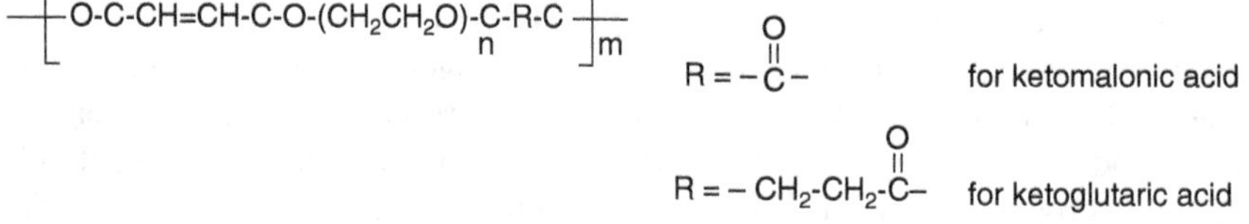

Scheme 2 Activated diacids.

diphtheria toxoid (DT).[8] Immunization with this vaccine generates antibodies to the hapten (LHRH) as well as to the carrier. Release of the vaccine from a hydrogel prepared by crosslinking linear polyesters with pendant unsaturation is shown in Fig. 2. In this particular case, the effect of ketoglutaric acid is relatively minor. However, the effect could be magnified by using higher itaconic to ketoglutaric ratios or by using ketomalonic acid.

To prepare these hydrogels, the water-soluble polyester, the crosslinking monomer (if used), and the protein are dissolved in water, and the crosslinking reaction is carried out at room temperature and pH 7.4, using redox initiation. This method provides a very mild means for incorporating proteins into a matrix, provided that the free radicals are not deleterious to the protein. However, with many proteins, loss of activity can occur because the protein is immobilized in an aqueous environment. Such a loss of activity can often be ascribed to specific chemical interactions, and ways to prevent such reactions can be devised.

Because the use of hydrogels represents one of the mildest means of incorporating proteins into polymers, and because no solvents or elevated temperatures are used, this delivery methodology holds great promise in the continuing search for delivery systems that can be used with sensitive proteins.

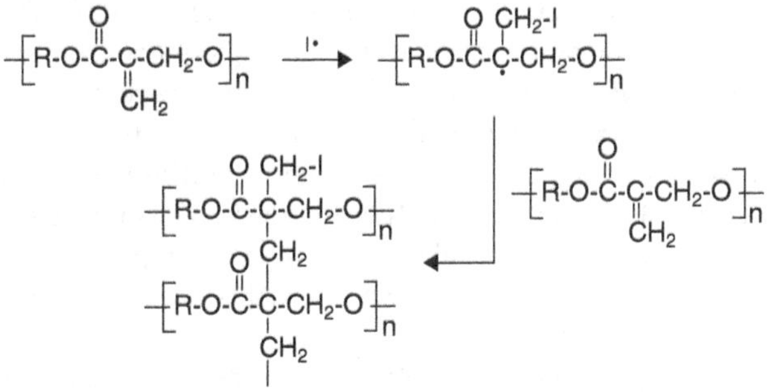

Water-Labile Bonds in Crosslink Segment

One interesting application of such hydrogels is the development of colon-specific delivery systems. There are two major situations where delivery to the colon is important. The first is treatment of inflammatory bowel disease, where in order to reduce systemic toxicity, anti-inflammatory agents need to be delivered directly. The second is enhancement of oral bioavailability of peptides and proteins.

One such hydrogel is shown in Scheme 4.[9,10] It is prepared by reacting the linear precursor with a crosslinker, in this case N,N'-(ω-aminocaproyl)-4-'-diamino azobenzene, which reacts with the activated p-nitrophenyl ester to form crosslinks containing azo linkages. Because the hydrogel contains free carboxylic acid groups, it swells very little in the acidic environment of the stomach, which protects incorporated agents from digestion by enzymes. In the higher pH of the intestines, swelling begins, and when the hydrogel reaches the colon, swelling has increased to a point where azoreductases can cleave the azo bonds, allowing degradation of the hydrogel and release of its contents.

Hydrophobic polymers

The bulk of activity concerning the use of bioerodible polymers for drug delivery centers on hydrophobic polymers.

Drug Release Predominantly Controlled by Diffusion

In this type of system, the drug is homogeneously dispersed in a biodegradable polymer matrix and is mainly released by simple Fickian diffusion. If the rate of polymer hydrolysis is very slow, release can occur entirely by diffusion. On the

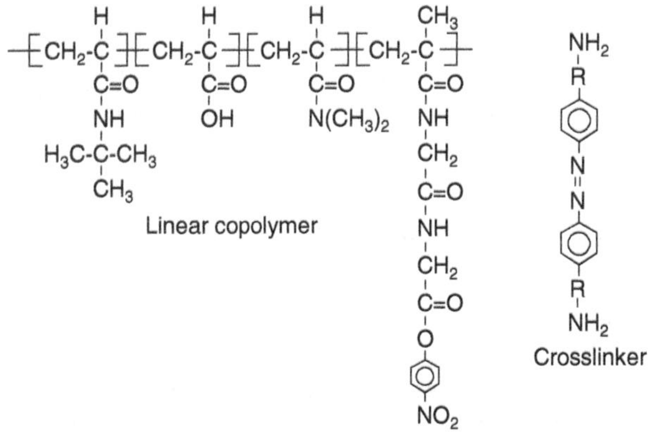

Scheme 3 Replacing fumaric acid with diacids.

Fig. 2 Percent cumulative LHRH-DT release at pH 7.4 and 37°C from microparticles prepared from a crosslinked itaconic (□) and a 2.3:1 itaconic/ketoglutaric (■) copolymer. LHRH-DT loading 0.1 wt%.
Source: Data from Singh et al.[8]

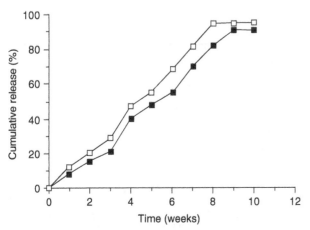

Scheme 4 Reacting the linear precursor with a crosslinker.

other hand, if hydrolysis is relatively rapid, initial release is diffusion-controlled, but as the process continues, polymer hydrolysis becomes an important factor in determining the rate of release.[11] The most extensively investigated bioerodible polymer system in this class consists of poly(lactide-*co*-glycolide) copolymers[12] prepared from the respective glycolides or lactides by heating with an acidic catalyst such as $SnCl_4$. The synthesis is shown in Scheme 5. Copolymers are prepared by polymerizing an appropriate mixture of the lactide and glycolide.[13]

These polymers are extensively used as bioerodible sutures. Poly(glycolic acid) was the first polymer specifically synthesized for such an application.[14] A poly(glycolic acid) suture is available under the trade name Dexon and a 90/10 copolymer of poly(L-lactide-*co*-glycolide) is available under the trade name Vicryl.[15] Because poly(lactic acid) contains a chiral center, it exists as a crystalline L-isomer or D-isomer and as a racemic, noncrystalline DL-isomer. The natural metabolite is L-lactic acid. Another synthetic polymer is poly(*p*-dioxonone), shown in Scheme 6. It is available as a suture material under the trade name PDS.[16]

Initial hydrolysis of these polymers occurs by random chain cleavage without enzymatic intervention, producing low molecular weight fragments that are then attacked by enzymes to produce water-soluble products. Degradation products from lactide and glycolide polymers are eliminated from the body via the Krebs cycle, primarily as carbon dioxide and in the urine. Because of their nontoxic nature and long history of safe use, poly(lactide-*co*-glycolide) copolymers are under intense investigation as bioerodible drug-delivery systems. Their use in drug delivery has been reviewed.[12] The next section will review two applications, the delivery of LHRH analogues and the delivery of antigens in the development of improved vaccination formulations.

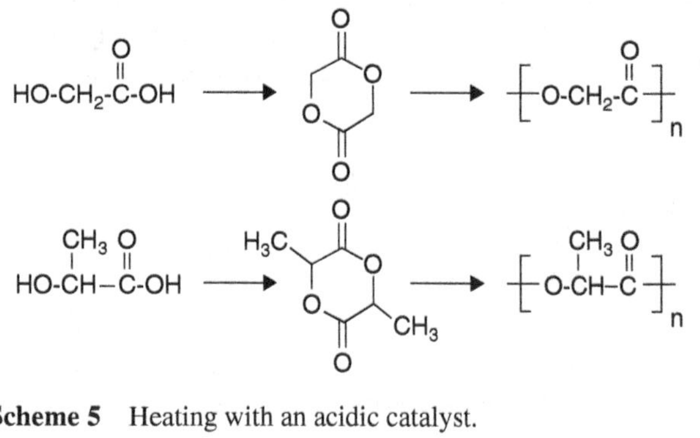

LHRH Analogue Delivery

One of the more interesting current uses of poly(lactide-*co*-glycolide) copolymers is in connection with protein-delivery systems, principally synthetic analogues of the luteinizing-hormone-releasing hormone (LHRH analogues).

LHRH is secreted by the hypothalamus and acts directly on the anterior pituitary to cause secretion of luteinizing hormone (LH) and follicle-stimulating hormone (FSH). These act on the gonads in both females and males, causing the production of estrogen, progesterone, and testosterone, thus initiating the reproductive cycles. When potent LHRH analogues are used, they initially stimulate the production of these hormones. But this is followed by a down-regulation of the receptors and suppression of these events, with a consequent reduction in the levels of estrogen, progesterone, and testosterone. This reduction is clinically important in suppressing prostate cancer in treating endometriosis and precocious puberty, and in male and female contraception.

When the synthetic LHRH analogue Nafarelin was microencapsulated in a 44:56 poly(DL-lactic-*co*-glycolic acid) copolymer followed by its *in vitro* release into an ethanolic phosphate buffer at 37°C, the triphasic release profile shown in Fig. 3 was typically observed.[17] The initial release represents leaching of the hormone from the surface layer. This is followed by a latent period during which the polymer is still intact and the relatively high molecular weight hormone is unable to diffuse from the matrix. The induction period depends on polymer molecular weight and microcapsule size. Proper manipulation of these parameters can achieve a reasonably constant release for about one month.

An important use of LHRH analogues is in the treatment of prostate cancer. Because prostate adenocarcinomas are testosterone-dependent, deprivation of the hormone by

Scheme 5 Heating with an acidic catalyst.

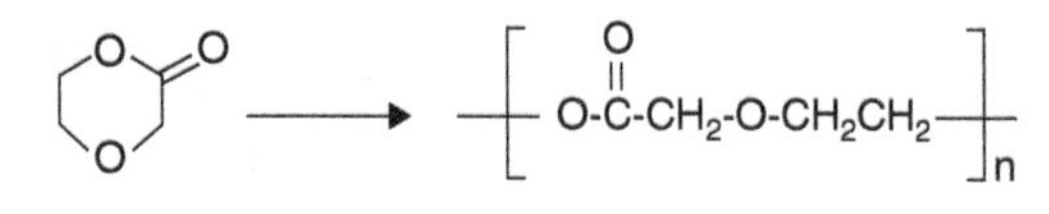

Scheme 6 Poly(*p*-dioxonone).

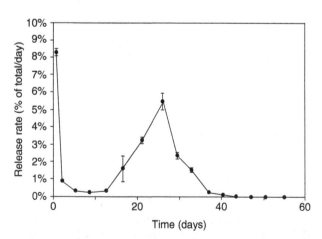

Fig. 3 *In vitro* release of Nafarelin from 44:56 poly(lactide-*co*-glycolide) copolymer microspheres.
Source: Data from Sanders et al.[17]

surgical castration is an effective treatment.[18] The use of LHRH analogues can achieve what has been termed "chemical castration" and is clearly a more appealing therapy than surgical castration. (D-Ser(Bu^i)^6, AzGly^{10}-GnRH) (ICI) and Leuprolide (D-Leu^6, Pro^{9}NEt) LHRH (Takeda-Abbott) have now been approved by the Food and Drug Administration for the treatment of prostate cancer in the United States. Figure 4 compares the effect of surgical castration and the effect of Zoladex administration (chemical castration) on the growth of Dunning R3327H prostate tumor in rats.[19] Clearly, monthly administration of Zoladex from a depot of a 50:50 poly(DL-lactide-*co*-glycolide) copolymer has the same effect as surgical castration.

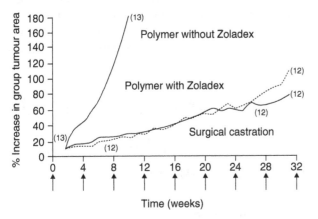

Fig. 4 Growth of androgen-responsive Dunning R 3327H prostate tumors in male rats that were either surgically castrated, or given a single s.c. deposit without Zoladex, or containing 1 mg Zoladex, at 28-day intervals, on 8 occasions, as shown by the arrows. The numbers in parentheses refer to the numbers of animals used.
Source: Data from Furr & Hutchison.[19]

Antigen Delivery

Immunization is the most cost-effective weapon for disease prevention in developing countries, where three-quarters of all people live, and 86% of all births and 96% of all infant deaths occur.[20] However, immunization requires multiple injections, and in developing countries, drop-out rates after the first dose of vaccine can reach levels approximating 70%, depending on the country.[21] The need for patient return visits could be eliminated if the antigen were delivered in a way that would achieve a long-lasting booster effect with a single administration.

The challenge is to develop a pulsed vaccine formulation that mimics conventional therapy by administering the antigen in discrete pulses spaced at desired time intervals. In one study, staphylococcal enterotoxin B (SEB) was encapsulated in a 50:50 poly(dl-lactic acid-*co*-glycolic acid) copolymer having a size distribution of 1–10 μm and 20–125 μm.[22] Mice were then injected with 1–10 μm microspheres, 20–125 μm microspheres, and a mixture of 1–10 μm and 20–125 μm microspheres.

As shown in Fig. 5, mice receiving the 1–10 μm microspheres had a maximum plasma IgG antitoxin titer of 102,400 on days 30 and 40, with a subsequent decrease through day 60 to 25,600. With mice receiving the 20–125 μm microspheres, essentially no antitoxin titer could be detected until day 50, when it reached a value of 51,200 and remained at that value throughout day 60. Co-administration of equal parts of 1–10 μm and 20–125 μm microspheres produced an IgG level identical for the first 30 days to that produced with the 1–10 μm microspheres, but after day 40, the IgG response climbed steadily to 819,200, a level far higher than the additive response to the two microsphere sizes.

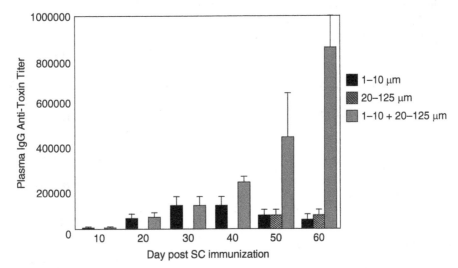

Fig. 5 Controlled vaccine release through alteration in microsphere size. BALB/c mice were immunized with SEB toxoid in 50:50 DL-PG microspheres 1–10 μm in diameter, 20–125 μm in diameter, or in a mixture of 1–10 and 20–125 μm in diameter. Anti-SEB toxin antibodies of the IgG isotype in the plasma were determined by end-point titration in an RIA.
Source: Data from Eldridge et al.[22]

This high IgG response resulted from SEB administration in two distinct pulses. The first begins at day 18 as a result of the 1–10 µm microspheres ability to load a relatively large amount of antigen directly into phagocytic accessory cells, followed by the direct migration of these antigen-containing cells into the draining lymph nodes. Because 20–125 µm spheres are too large to enter phagocytic cells, they release the antigen by the normal hydrolytic degradation of the polymer, so the second pulse begins at day 50. When the two microsphere sizes are combined, two pulses are achieved, with a consequent very large increase in the antitoxin level, similar to that observed when a booster shot is administered.

Drug Release Predominantly Controlled by Polymer Hydrolysis

In contrast to the poly(lactide-*co*-glycolide) copolymers or poly(glycolic acid), which were developed for use as bioerodible sutures and were then later adapted for drug release, two bioerodible polymer systems were specifically designed for drug-delivery applications. These are poly(ortho esters) and polyanhydrides.

Poly(ortho esters)

As of this writing, three major families of poly(ortho esters) have been prepared. These are shown in Scheme 7. Their preparation and applications have been comprehensively reviewed.[23]

Poly(ortho ester) 1, the first such polymer prepared, was developed at the Alza Corporation and described in a series of patents.[24–27] Very little detailed information on the preparation or use of this polymer is available in the scientific literature. It is prepared as shown in Scheme 8.

Because poly(ortho esters) contain acid-sensitive linkages in the polymer backbone, hydrolysis rates can be adjusted between very wide limits by incorporating small amounts of acidic excipients, to accelerate the rate of polymer hydrolysis, or incorporating small amounts of basic excipients, to retard the rate of polymer hydrolysis. However, this is not the case with poly(ortho ester) 1, since hydrolysis of this material, as shown in Scheme 9, produces γ-butyrolactone, which then opens to γ-hydroxybutyric acid. Therefore, to avoid an uncontrolled, autocatalytic hydrolysis reaction, this polymer requires stabilization by a base.

Poly(ortho ester) 2, developed at SRI International, is prepared as shown in Scheme 10.[2,28]

As shown in Scheme 11, this polymer hydrolyzes to initially neutral products. Therefore it is possible to use small amounts of acidic or basic excipients to produce devices having lifetimes of a few weeks to over one year.[29,30] Figure 6 illustrates the use of the acidic excipient suberic acid to control the release of the antimalarial agent pyrimethamine from a polymer prepared from the diketene acetal 3,9-bis(ethylidene) 2,4,8,10-tetraoxaspiro[5,5] undecane and

Scheme 9   Hydrolysis of poly(ortho ester) 1.

Scheme 10   Poly(ortho ester) 2.

Scheme 11   Hydrolysis of poly(ortho ester) 2.

Scheme 7   Three major families of poly(ortho esters).

Scheme 8   Poly(ortho ester) 1.

a mixture of two diols, 1,6-hexanediol and *trans*-cyclohexanedimethanol.[31] Because such polymers are highly hydrophobic, they are quite stable at neutral pH, but relatively small changes in the concentration of the excipient translate to large changes in rates of hydrolysis and release of incorporated agents.

The family of such polymers, poly(ortho ester) 3, also developed at SRI International, is prepared as shown in Scheme 12.[32]

When R is $-(CH_2)_4-$, the polymer is a semisolid at room temperature, even though molecular weights can exceed 35,000 Da. This semisolid consistency provides a number of unique advantages. Dominant among these is the ability to incorporate therapeutic agents into the polymer by simple mixing, without the need to use solvents or elevated temperatures. This is clearly beneficial for the incorporation of heat-sensitive therapeutic agents and proteins.

Because the polymer can be injected using a hypodermic syringe, it is currently under investigation as an injectable 5-flurorouracil delivery system for prevention of failure in glaucoma filtration surgery[33] and in the treatment of periodontal disease.[34] It is also under investigation for the topical treatment of wounds such as deep burns and decubitus ulcers.

This particular polymer hydrolyzes at much faster rates than poly(ortho ester) 2, perhaps because it is not as hydrophobic, although its water uptake has not yet been determined. Drug release rates are controlled using a basic excipient such as $Mg(OH)_2$. The effect of $Mg(OH)_2$ on the rate of release of tetracycline is shown in Fig. 7.[34] Like the suberic acid-controlled devices shown in Fig. 6, $Mg(OH)_2$ is very effective in controlling the rate of hydrolysis and the release of an incorporated drug.

Polyanhydrides

These polymers are prepared as shown in Scheme 13. They were first prepared in 1909,[35] and were subsequently investigated as potential textile fibers, but found unsuitable

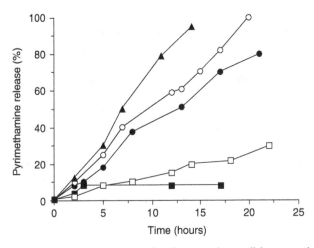

Fig. 6 Release of pyrimethamine from a polymer disk prepared from 3,9-bis(ethylidene) 2,4,8,10-tetraoxaspiro [5,5] undecane and a 60:40 mixture of trans-cyclohexanedimethanol and 1,6-hexanediol, as a function of suberic acid concentration. (■) 0 wt%, (□) 1 wt%, (●) 3 wt%, (○) 5 wt%, (▲) 10 wt%; drug loading 10 wt%; pH 7.4 and 37° C.

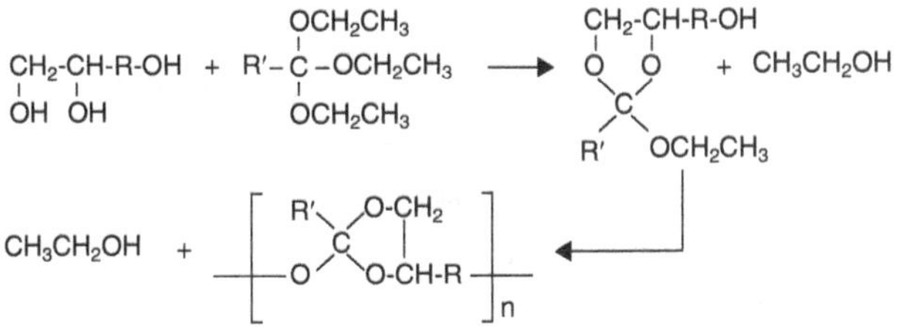

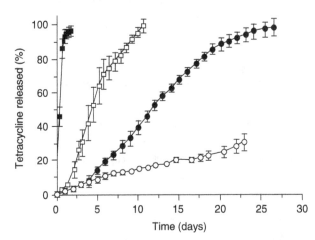

Fig. 7 Cumulative release of tetracycline from a 27 Kdalton propionate polymer at pH 7.4 and 37°C, as a function of $Mg(OH)_2$ content. 0.1 M phosphate buffer, flow rate 1 mL/hr, drug loading 10 wt%. (■) 0%, (□) 0.5%, (●) 1.0%, (○) 2.0%. Error bars are standard deviations, n = 3.

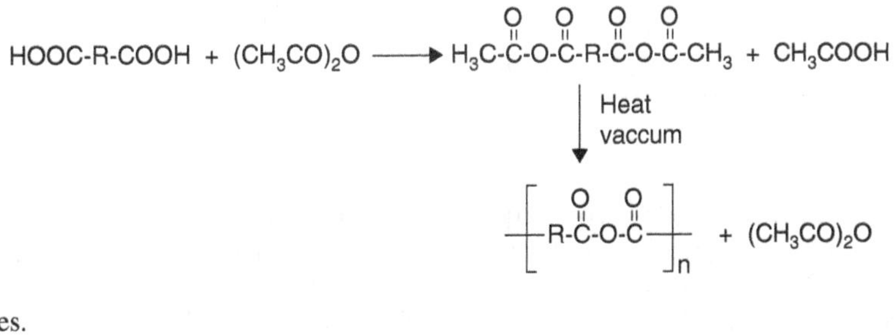

$$CH_2\text{-}CH\text{-}R\text{-}OH + R'\text{-}C\text{-}OCH_2CH_3 \longrightarrow$$

Scheme 12 Poly(ortho ester) 3.

$$HOOC\text{-}R\text{-}COOH + (CH_3CO)_2O \longrightarrow H_3C\text{-}C\text{-}O\text{-}C\text{-}R\text{-}C\text{-}O\text{-}C\text{-}CH_3 + CH_3COOH$$

Scheme 13 Polyanhydrides.

due to their hydrolytic instability.[36] The use of polyanhydrides as bioerodible matrices for the controlled release of therapeutic agents was first reported in 1983.[37] Although anhydride linkages are highly susceptible to hydrolysis, crystalline polymers are very stable because water is unable to penetrate the crystalline regions. However, when crystallinity is disrupted by copolymerization of an aliphatic and aromatic diacid, such as sebacic acid and (p-carboxyphenoxy) propane (shown in Scheme 14), erosion rates that vary from days to projected years can be achieved. Figure 8 shows hydrolysis rates measured as polymer weight loss as a function of copolymer composition.

Polyanhydrides based on aromatic diacids become brittle and eventually fragment after exposure to water, causing water-soluble drugs to be released more rapidly than by polymer erosion. For this reason, a new class of polyanhydrides was prepared from fatty acid dimers derived from naturally occurring oleic and sebacic acids.[39] The structure of these materials is shown in Scheme 15.

When lysozyme, trypsin, ovalbumin, bovine serum albumin, and immunogloulin were incorporated into a 25:75 fatty acid/sebacic acid copolymer at a 2 wt% loading, near constant release of about two weeks was achieved.[40] The fact that these substances are all released at about the same rate suggests the dominance of an erosion-controlled mechanism. Results of the study are shown in Fig. 9.

Use in fracture fixation

Metal implant devices align bone fragments, bring their surfaces into close proximity, and control the relative

motion of the fragments so that union can take place. However, complete healing of the bone depends on its bearing normal loads, which is prevented as long as the device bears part of the load.[41–43] Furthermore, sudden removal of the device can leave the bone temporarily weak and subject to refracture.

The prospect of replacing metal fracture-fixation devices with bioabsorbable polymer or composite devices that have an appropriate combination of initial strength, stiffness, and biocompatibility is of great interest because eventual device absorption has two important advantages. First, as absorption reduces the device's cross-section or the materials elastic modulus, the load is transferred gradually to the healing bone. Second, because the device will be completely absorbed, a second surgical procedure is not necessary.

Only a few materials have been intensively investigated for use in bioabsorbable fracture-fixation devices. These are poly(lactic acid), poly(glycolic acid), polydioxanone, and poly(ortho esters). Comprehensive reviews of the use of biodegradable polymers in surgical applications have been published.[44–47]

Poly(lactic acid)

Because lactic acid is a chiral molecule, poly(L-lactic acid) can exist in four stereoisomeric forms. These are the two

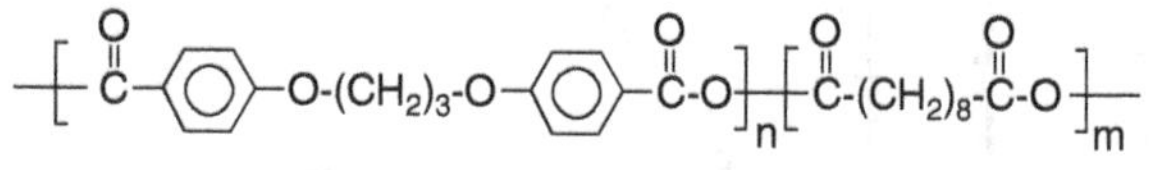

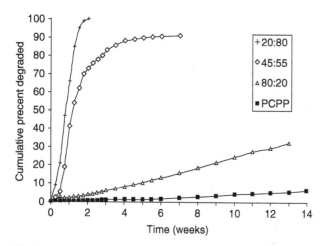

Scheme 14   Crystallinity disrupted by copolymerization of aliphatic and aromatic diacid.

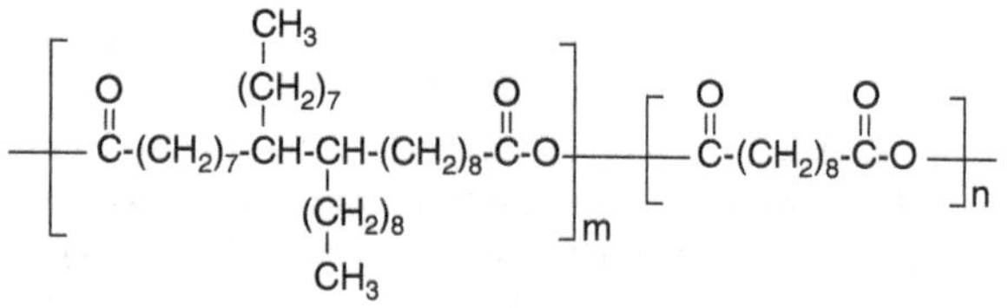

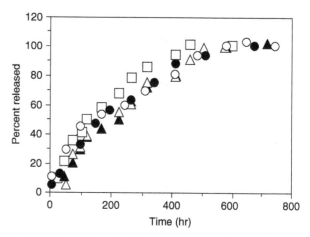

Scheme 15   Polyanhydrides prepared from fatty acid dimers.

Fig. 8   Degradation profiles of compression-molded poly[bis(p-carboxy phenoxy) propane anhydride] (PCPP) and its copolymer with sebacic acid (SA) in 0.1 M pH 7.4 phosphate buffer at 37°C.
Source: Data from Leong, Brott, and Langer.[38]

Fig. 9   Release of proteins from a 42,900 molecular weight 25:75 fatty acid dimer/sebacic polyanhydride in 0.1 M phosphate buffer at 37°C. ($\circ$) lysozyme, ($\bullet$) trypsin, ($\triangle$) ovalbumin, ($\blacktriangle$) BSA, ($\square$) immunoglobulin. Protein loading, 2 wt%.
Source: Data from Tabata & Langer.[39]

stereoregular polymers, poly(D-lactic acid) and poly(L-lactic acid); the racemic mixture of poly(L-lactic acid) and poly(D-lactic acid); and the meso poly(DL-lactic acid). These polymers are obtained by a cationic polymerization of the cyclic lactide, as shown in Scheme 16.[13] Hydrolysis of poly(L-lactic acid) results in l(+) lactic acid, the naturally occurring stereoisomer.

Due to the crystalline nature of poly(L-lactic acid), mechanical and chemical properties of the two enantiomeric forms of the polymer differ significantly. Poly(L-lactic acid) has a T_m = 170–180°C and a T_g = 67°C. Poly(DL-lactic acid) has a T_g = 60°C. Because water penetration into crystalline regions of a polymer is hindered, hydrolysis rates of poly(L-lactic acid) are very slow relative to those of poly(DL-lactic acid).

In comparing various bioabsorbable polymers, it is clear that poly(L-lactic acid) is the stiffest and mechanically strongest available. However, even poly(L-lactic acid) has a tensile modulus of only 3 GPa and a tensile strength of 50 MPa. These values are considerably lower than those of typical metals, which can exhibit tensile moduli in the order of 100–200 GPa.

Mechanical performance of poly(L-lactic acid) can to a certain extent be improved by careful attention to fabrication methods, especially with regard to strength, but major improvement in stiffness can only be achieved by using high-modulus fiber reinforcement. A review covering bioerodible fracture-fixation devices, emphasizing polymer characterization and fabrication procedures, has been published.[48]

Initial work on reinforced materials used nondegradable fibers. Bone plates were fabricated by fusing alternate layers of poly(DL-lactic acid) reinforced with alumina, alumina-boria-silica, and carbon.[49] The alumina composites had a 70 MPa flexural strength and 7.6 GPa flexural modulus, the alumina-boria-silica had a 162 MPa flexural strength and 9.0 GPa flexural modulus, and the carbon composites had a 174 MPa flexural strength and 11.0 GPa flexural modulus. Although these results are encouraging, this is clearly a highly unsatisfactory approach, since polymer bioerosion will leave nondegradable fibers in the tissues. For this reason, only the use of biodegradable reinforcing materials is acceptable.

A number of workers have investigated composites where biodegradable calcium metaphosphate glass fibers were sandwiched between poly(L-lactic acid) and poly(DL-lactic acid) copolymers.[50–53] After maximizing fabrication procedures, a flexural strength and modulus of 193 MPa

and 15.9 GPa, respectively, were obtained. However, as shown in Fig. 10, after the composites were exposed to saline at 37°C, their mechanical properties deteriorated significantly. This serious and general problem was very likely due to water penetration into the polymer, with consequent weakening of the reinforcing interaction between the fiber and the polymer matrix.

Poly(glycolic acid)

Poly(glycolic acid) is prepared by the cationic polymerization of the cyclic glycolide, as shown in Scheme 17.

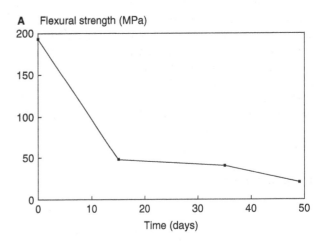

Fig. 10   Effect of exposure to physiological saline at 37°C on the flexural mechanical properties of calcium-metaphosphate-reinforced poly(lactic acid). (**A**) Flexural strength (MPa); (**B**) Flexural modulus (GPa).
Source: Data from Frazza & Schmitt.[14]

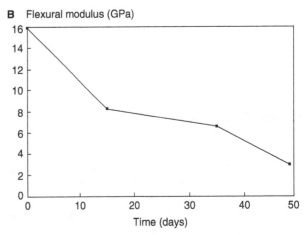

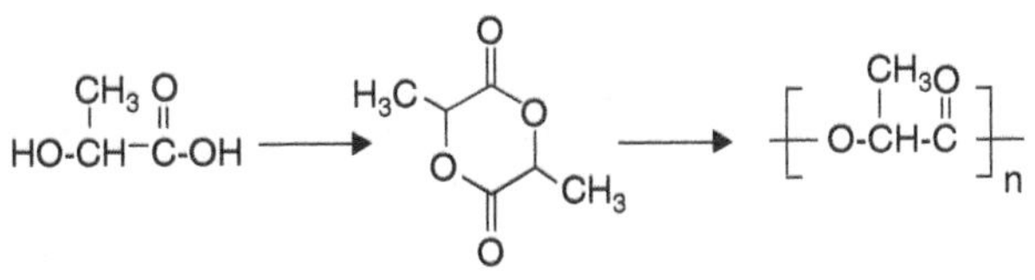

Scheme 16   Cationic polymerization of cyclic lactide.

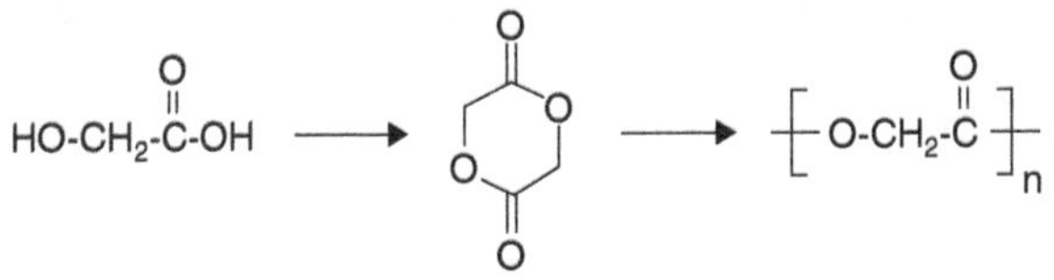

Scheme 17   Poly(glycolic acid).

This polymer is highly crystalline, with a high melting point and very low solubility in common organic solvents. Because it has no chiral carbon, it exists as a single compound. It also lacks the methyl group present in poly(lactic acid), making it significantly more hydrophilic and subject to much faster bioerosion. Typically, the polymer loses most of its strength after being implanted for 2–4 weeks. Poly(glycolic acid) was the first totally synthetic polymer developed by American Cyanamid for absorbable sutures, and was marketed as Dexon.[14] Due to its limited solubility and relatively rapid bioerosion, the only bulk unreinforced property data listed are a tensile strength and modulus of 57 MPa and 6.5 GPa, respectively.[54]

The material is brittle, with only a 0.7% elongation to failure, so it must be reinforced for use in bone fixation. One approach is to sinter commercial Dexon poly(glycolic acid) sutures at high temperature and pressure.[55] In this way, rods containing more than 60% Dexon fibers were prepared. Flexural and shear strength were 370 and 250 MPa respectively, among the highest such strengths achieved with a biodegradable polymer. However, only 5% of the flexural strength was retained after 5 weeks in distilled water at 37°C.

Polydioxanone

Polydioxanone, prepared as shown in Scheme 18, became commercially available in 1981. It is available as a suture material known as PDS,[16] a degradable ligating device,[56] and a degradable bone pin.[57]

Polydioxanone rods made from PDS have been used to repair epiphyseal fractures in growing rabbits.[58,59] The initial strength of the rods was approximately 56 MPa, but dropped to 2.6 MPa after 6 weeks in distilled water at 37°C.

General Toxicological Considerations

Results of numerous clinical studies involving hundreds of patients using self-reinforced poly(glycolic acid) rods, polydioxanone pins, or various other devices prepared from poly(lactic acid) are available.[60] However, only two major products have been approved for routine clinical use in the United States. These are the self-reinforced poly(glycolic acid) rods known as Biofix and a polydioxonone pin known as Orthosorb. A number of other devices based on poly(lactic acid) are in clinical trial and are expected to become commercially available.

However, it has been reported[61,62] that of 516 patients treated with Biofix or a device made from a poly(glycolide-co-lactide) copolymer, 1.2% required reoperation due to

device failure, 1.7% suffered from bacterial infection, and 7–48% developed a late noninfectious inflammatory response that required clinical intervention. The mean interval between device placement and manifestation of an inflammatory response was 12 weeks for poly(glycolic acid), but as long as 3 years for the more slowly degrading poly(L-lactic acid). This is clearly a very serious problem that is not related to device configuration, polymer purity, fabrication method, or surgical technique, but is inherent in the nature of the materials.

One of the hydrolysis products of polyesters is an acid. The poly(lactide) hydrolysis process begins at the outer perimeter of the device with gradual progression into the interior, followed by a catastrophic disintegration.[63–65] There is little doubt that the inflammatory response is due to a massive short-term release of acid that exceeds local tissue-clearance capabilities.[66]

Alternate Polymer Systems

Bringing to market a new biodegradable implant material has been estimated to exceed $100,000,000,[46] yet clinical experience amassed strongly suggests that significant progress can only be achieved by developing new materials. These materials must not only hydrolyze to toxicologically innocuous products., but also, and more importantly, they must have neutral primary degradation products. Ideally they should also be more hydrophobic than are poly(lactide-co-glycolide) and polydioxonone. Two such families of polymers have been under investigation.

Poly(ortho esters)

Poly(ortho esters), which have been under investigation as drug-delivery systems for many years,[23] are highly hydrophobic and completely amorphous. In the absence of incorporated therapeutic agents, hydrolysis occurs predominantly within the surface layers, not the interior, thus avoiding the catastrophic disintegration noticed with poly(lactide-co-glycolide) copolymers and poly(lactic acid) polymers.

These polymers are prepared as shown in Scheme 19.[2,67] They typically have molecular weights on the order of 80–100 kDa, but careful attention to stoichiometry and monomer purity can lead to molecular weights in the 180 kDa range.

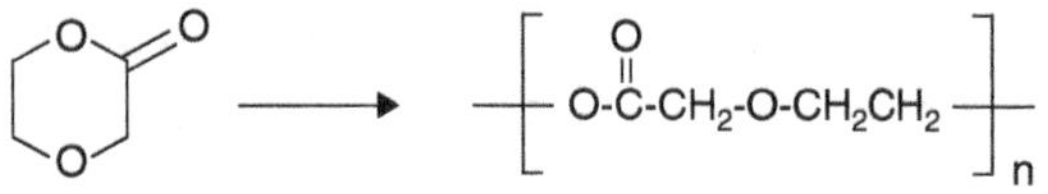

Scheme 18   Polydioxanone.

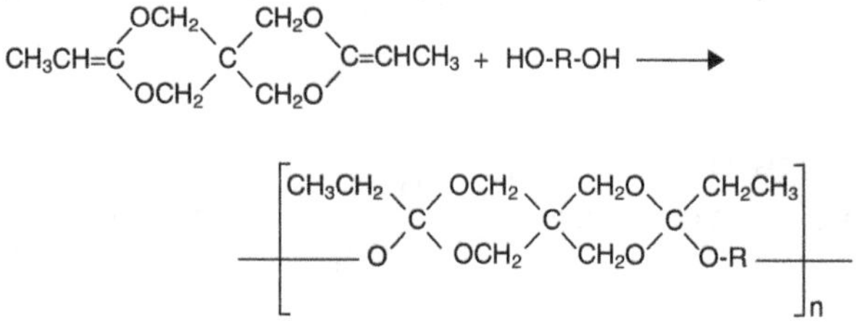

Scheme 19   Poly(ortho esters).

In an aqueous environment, initial hydrolysis occurs at the ortho ester linkages, yielding the original diol or the mixture of diols used in the synthesis, and pentaerythritol dipropionate.[68] The primary degradation products are small neutral molecules that form in the outer layers of the device, so they can diffuse away from the implant site before the second-stage hydrolysis, which does produce an acidic product, takes place (Scheme 20).

To compare the toxicity of accumulated degradation products, six typical biodegradable polymers were investigated using a commercially available acute toxicity assay system based on bioluminescent bacteria.[69] Results of this study[70] showed that poly(glycolic acid), as expected, undergoes rapid hydrolysis and generates toxic products. Of the two poly(lactic acid) samples, the one with a lower molecular weight generated toxic products within about 10 days, while the one with a higher molecular weight did not suppress bioluminescence for the duration of the study. These results reflect the effect of polymer molecular weight on the rate of generation of toxic products. The pH of poly(ortho esters) remained relatively constant throughout the exposure, and no significant toxicity was noted.

Interestingly, weight loss for high molecular weight poly(lactic acid) was less than for poly(ortho ester), but the change in inherent viscosity was greater. This suggests that poly(ortho esters) do indeed undergo surface erosion and that gradual release of degradation products from the surface could minimize complications from the erosion of large fixation devices.

Mechanical properties of poly(ortho esters) indicate that the polymer alone is not suitable for large fracture-fixation devices.[71,72] Successful fracture fixation has been reported for absorbable composites that have initial mechanical properties similar to cortical bone.[44] Totally absorbable composite systems based on poly(ortho ester) have been fabricated using randomly oriented, crystalline microfibers of calcium-sodium-metaphosphate for isotropic reinforcement.[73] Composites prepared with diamine-silane-treated fibers have initial mechanical properties similar to cortical bone—125 MPa flexural strength and 8.3 GPa flexural modulus. Because the polymer matrix is amorphous, poly(ortho ester) composites can be reshaped at temperatures above the polymer's T_g (98°C), but are rigid at body temperature (37°C), a unique and potentially very useful property for certain orthopedic applications.

An outer coating of hot-pressed poly(ortho ester) has been reported to retard influx of fluids into the composite[73] and thus slow degradation of the water-sensitive polymer–fiber interface. The coated composite material retained 70% of its strength and stiffness after 4 weeks of exposure to tris-buffered saline at 37°C (Fig. 11). For comparison, a poly(DL-lactic acid) composite reinforced with continuous calcium-metaphosphate glass fibers and coated with hot films of poly(ε-caprolactone) retained less than 50% of its flexural modulus after 4 weeks of exposure.[74] The superior performance of poly(ortho ester) as a water barrier over poly(ε-caprolactone) is clearly due to its greater hydrophobicity.

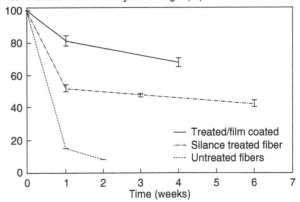

A Retention of flexural yield strength (%)

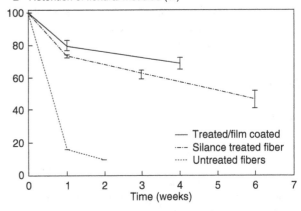

B Retention of flexural modulus (%)

Fig. 11 Percent retention of mechanical properties after exposure to tris-buffered saline, pH 7.4 and 37°C. (**A**) Flexural yield strength (%); (**B**) Flexural modulus (GPa).
Source: Data from Leong Brott, and Langer.[38]

Scheme 20 First-stage and second-stage hydrolysis of poly(ortho esters).

Bioactive—BioMEMS

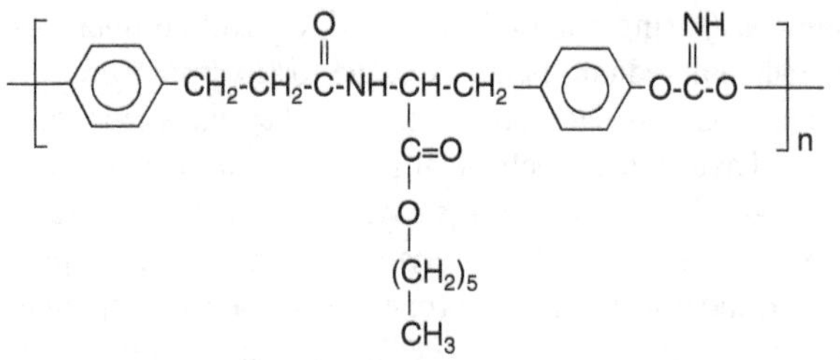

Scheme 21   Pseudo-poly(amino acids).

Tyrosine-Based Polycarbonates

Other hydrophobic degradable polymers include polycarbonates synthesized from derivatives of the natural amino acid l-tyrosine. These polymers are part of a family of materials known as pseudo-poly(amino acids), in which the functional groups on the amino acid side chains are used to link individual amino acids or dipeptides via amide bonds. As illustrated in Scheme 21, this design provides materials in which the amino acid termini assume the position of pendant chains.[46,75]

A series of four polycarbonates derived from the ethyl, butyl, hexyl, and octyl esters of desaminotyrosyl-tyrosine indicated that hydrophobicity, glass-transition temperature, and mechanical strength were related to the length of the alkyl ester pendant chains.[76] The most promising of these polycarbonates for fracture-fixation applications was the one with a hexyl ester side chain. Initial mechanical properties for unoriented films were 62 MPa tensile yield strength, 220 MPa tensile strength at break, and 1.4 GPa tensile modulus. The large difference between tensile yield and strength at break is due to the material's high ductility, 460% elongation to failure. After 40 weeks of exposure to phosphate-buffered saline at 37°C, the material retained 74% of its tensile yield strength, while the high tensile strength at break was maintained throughout the observation period.

The tyrosine-derived polycarbonates, like poly(ortho esters), release pH-neutral degradation products after hydrolytic degradation. The materials were not cytotoxic toward cultured rat lung fibroblasts[76] and showed high tissue compatibility with bone in a rabbit model.[77]

CONCLUSIONS

The development of bioerodible polymers for fracture fixation is a challenging problem receiving increased attention. Ideally, such materials would have the initial biocompatibility, strength, stiffness, and ductility of stainless steel; retain these properties for several weeks or months; and then undergo benign and complete biodegradation and absorption or excretion. The completely bioerodible polymers that have been described do not meet all these requirements, especially for stiffness and ductility. In spite of this, successful clinical fracture fixation has

been reported with poly(lactic acid), poly(glycolic acid), their copolymers, and polydioxanone. The key to this success has been to pick the clinical applications carefully and to design the method of fixation to suit available polymer properties.

The development of tyrosine-based polycarbonates and poly(ortho ester) composite materials were based on rational design processes, rather than using random screening or trial and error to improve biocompatibility performance. These polymers represent new classes of materials that may find many more applications as "second generation" orthopedic fracture-fixation devices.

REFERENCES

1. Heller, J. Polymers for controlled parenteral delivery of peptides and proteins. Adv. Drug Deliv. Rev. **1993**, *10* (2–3), 163–204.

2. Heller, J.; Penhale, D.W.H.; Helwing, R.F. Preparation of poly(ortho esters) by the reaction of diketene acetals and polyols. J. Polym. Sci. Polym. Lett. Ed. **1980**, *18* (9), 619–624.

3. Heller, J. Biodegradable polymers in controlled drug delivery. CRC Crit. Rev. Therap. Drug Carrier Syst. **1984**, *1* (1), 39–90.

4. Heller, J. Controlled drug release from poly(ortho esters)—A surface eroding polymer. J. Control. Release **1985**, *2*, 167–177.

5. Park, K.; Shalaby, W.S.W.; Park, H. *Biodegradable Hydrogels for Drug Delivery*; Technomic: Lancaster, PA, 1993.

6. Edman, P.; Ekman, B.; Sjoholm, I. Immobilization of proteins in microspheres of biodegradable polyacryldextran. J. Pharm. Sci. **1980**, *69* (7), 328–342.

7. Heller, J.; Helwing, R.F.; Baker, R.W.; Tuttle, M.E. Controlled release of water-soluble macromolecules from bioerodible hydrogels. Biomaterials **1983**, *4* (4), 262–266.

8. Singh, M.; Rathi, R.; Singh, A.; Heller, J.; Talwar, G.P.; Kopecek, J. Controlled release of LHRH-DT from bioerodible hydrogel microspheres. Int. J. Pharm. **1991**, *76* (3), R5–R8.

9. Kopeček, J.; Kopečková, P.; Brønsted, H.; Rathi, R.; Ríhová, B.; Yeh, P.-Y; Ikesue, K. Polymers for colon-specific drug delivery. J. Control. Release **1992**, *19* (1–3), 121–130.

10. Kopečková, P.; Rathi, R.; Takada, S.; Ríhová, B.; Berenson, M.M.; Kopeček, J. Bioadhesive N-(2-hydroxypropyl) methacrylamide copolymers for colon-specific drug delivery. J. Control. Release **1994**, *28* (1–3), 211–222.

11. Heller, J.; Baker, R.W. *Controlled Release of Bioactive Materials*; Academic: New York, NY, 1980; Chapter 1.

12. Chasin, M.; Langer, R. *Biodegradable Polymers as Drug Delivery Systems*; Marcel Dekker: New York, NY, 1990; Chapter 1.

13. Kulkarni, R.K.; Moore, E.G.; Hegyelli, A.F.; Leonard, F. Biodegradable poly(lactic acid) polymers. J. Biomed. Mater. Res. **1971**, *5* (3), 169–181.

14. Frazza, E.J.; Schmitt, E.E. A new absorbable suture. J. Biomed. Mater. Res. Symp. **1971**, *5* (2), 43–58.

15. Craig, P.H.; Williams, J.A.; Davis, K.W.; Magoun, A.D.; Levy, A.J.; Bogdansky, S.; Jones, J.P. A biologic comparison

of polyglactin 910 and polyglycolic acid synthetic absorbable sutures. Surg. Gynecol. Obstet. **1975**, *141* (1), 1–10.

16. Ray, J.A.; Doddi, N.; Regula, D.; Williams, J.A.; Melveger, A. Polydioxanone (PDS), a novel monofilament synthetic absorbable suture. Surg. Gynecol. Obstet. **1981**, *153* (4), 497.

17. Sanders, L.M.; McRae, G.I.; Vitale, K.M.; Kell, B.A. Controlled delivery of an LHRH analogue from biodegradable injectable microspheres. J. Control. Release **1985**, *2*, 187–195.

18. Huggins, C.; Hodges, C.V. Studies on prostatic cancer. Cancer Res. **1941**, *1*, 293–297.

19. Furr, B.J.; Hutchison, F.G. A biodegradable delivery system for peptides: Preclinical experience with the gonadotrophin-releasing hormone agonist Zoladex. J. Control. Release **1992**, *21* (1–3), 117–127.

20. Bloom, B.R. Vaccines for the third world. Nature **1989**, *324* (6246), 115–120.

21. Aguado, M.T.; Lambert, P.H. Controlled-release vaccines-biodegradable polylactide/polyglycolide (PL/PG) microspheres as antigen vehicles. Immunobiology **1992**, *184* (2), 113–125.

22. Eldridge, J.H.; Staas, J.K.; Meulbroek, J.A.; McGhee, J.R.; Tice, T.R.; Gilley, R.M. Biodegradable microspheres as a vaccine delivery system. Mol. Immunol. **1991**, *28* (3), 287–294.

23. Heller, J. Poly(ortho esters). Adv. Polym. Sci. **1993**, *107*, 41–92, doi:10.1007/BFb0027551.

24. Choi, N.S.; Heller, J. Poly(carbonates). U.S. Patent 4079038, Mar 5, 1978.

25. Choi, N.S.; Heller, J. Drug Delivery Devices Manufactured from Poly(orthoesters) and Poly(orthocarbonates). U.S. Patent 4093709, Jan 28, 1978.

26. Choi, N.S.; Heller, J. Structured orthoester and orthocarbonate drug delivery devices. U.S. Patent 4131648, Dec 26, 1978.

27. Choi, N.S.; Heller, J. Erodible agent releasing device comprising poly(orthoesters) and poly(orthocarbonates). U.S. Patent 4138344, Feb 6, 1979.

28. Ng, S.Y.; Penhale, D.W.H.; Heller, Poly (ortho esters) by the addition of diols to a diketene acetal. J. Macromol. Synth. **1992**, *11*, 23.

29. Heller, J.; Fritzinger, B.K.; Ng, S.Y.; Penhale, D.W.H. *In vitro* and *in vivo* release of levonorgestrel from poly(ortho esters): I. Linear polymers. J. Control. Release **1985**, *1* (3), 225–232.

30. Heller, J.; Fritzinger, B.K.; Ng, S.Y.; Penhale, D.W.H. *In vitro* and *in vivo* release of levonorgestrel from poly(ortho esters): II. Crosslinked polymers. J. Control. Release **1985**, *1* (3), 233–238.

31. Vandamme, T.F.; Heller, J. Poly(ortho esters) as bioerodible matrices for the controlled delivery of pyrimethamine in chemoprophylaxis of malaria. J. Control. Release **1995**, *36* (3), 209–213.

32. Heller, J.; Ng, S.; Fritzinger, B.K.; Roskos, K.V. Controlled drug release from bioerodible hydrophobic ointments. Biomaterials **1990**, *11* (4), 235–237.

33. Merkli, A.; Heller, J.; Tabatabay, C.; Gurny, R. Semi-solid hydrophobic bioerodible poly (ortho ester) for potential application in glaucoma filtering surgery. J. Control. Release **1994**, *29* (1), 105–112.

34. Roskos, K.V.; Fritzinger, B.K.; Rao, S.S.; Armitage, G.C.; Heller, J. Development of a drug delivery system for the treatment of periodontal disease based on bioerodible poly (ortho esters). Biomaterials **1993**, *16* (4), 313–317.

35. Bucher, J.E.; Slade, W.C. The anhydrides of isophthalic and terephthalic acids. J. Am. Chem. Soc. **1909**, *31* (12), 1319–1321.

36. Conix, A.J. Aromatic polyanhydrides, a new class of high melting fiber-forming polymers. J. Polym. Sci. **1958**, *29* (120), 343–353.

37. Rosen, H.G.; Chang, J.; Wnek, G.E.; Linhardt, G.E.; Langer, R. Biomaterials: New perspectives on their use in the controlled delivery of polypeptides. Biomaterials **1985**, *19* (1), 941–955.

38. Leong, K.W.; Brott, B.C.; Langer, R. Bioerodible polyanhydrides as drug-carrier matrices. I: Characterization, degradation, and release characteristics. J. Biomed. Mater. Res. **1985**, *19* (8), 941–955.

39. Tabata, Y.; Langer, R. Polyanhydride mierospheres that display near-constant release of water-soluble model drug compounds. Pharm. Res. **1993**, *10* (3), 391–399.

40. Tabata, Y.; Gutta, S.; Langer, R. Controlled delivery systems for proteins using polyanhydride microspheres. Pharm. Res. **1993**, *10* (4), 487–496.

41. Bradley, G.W.; McKenna, G.B.; Dunn, H.K.; Daniels, A.U.; Statton, W.O. J. Bone Jt. Surg. **1979**, *61A* (6), 866–872.

42. Terjesen, R.; Apalest, K. The influence of different degrees of stiffness of fixation plates on experimental bone healing. J. Ortho. Res. **1988**, *6* (2), 293–299.

43. Woo, S.L.-Y.; Akeson, W.H.; Coutts, R.D.; Rutherford, L.; Doty, D.; Jemmott, G.F.; Akeson, D.A. A comparison of cortical bone atrophy secondary to fixation with plates with large differences in bending stiffness. J. Bone Joint Surg. **1976**, *58* (2), 190–195.

44. Daniels, A.U.; Chang, M.K.O.; Andriano, K.P.; Heller, J. Mechanical properties of biodegradable polymers and composites proposed for internal fixation of bone. J. Appl. Biomater. **1990**, *1* (1), 57–78.

45. Engelberg, I.; Kohn, J. Physico-mechanical properties of degradable polymers used in medical applications: A comparative study. Biomaterials **1991**, *12* (3), 292–304.

46. Pulapura, S.; Kohn, J. Trends in the development of bioresorbable polymers for medical applications. Biomater. Appl. **1992**, *6* (3), 216–250.

47. Vainionpaa, S.; Rokkanen, P.; Tormala, P. Surgical applications of biodegradable polymers in human tissues. Prog. Polym. Sci. **1989**, *14* (5), 679–716.

48. Hastings, G.W.; Ducheyne, P. *Macromolecular Biomaterials*; CRC: Boca Raton, FL, 1984; 119–142.

49. Lewis, D.; Dunn, R.; Casper, R.; Tipon, A. Trans. Soc. Biomater. **1981**, *4*, 61.

50. Casper, R.A.; Dunn, R.L.; Kelley, B.S. Trans. Soc. Biomater. **1984**, *7*, 278.

51. Casper, R.A.; Kelley, B.S.; Dunn, R.L.; Potter, A.G.; Ellis, D.N. Fiber-reinforced absorbable composite for orthopedic surgery. Polym. Mater. Sci. Eng. **1985**, *53*, 497.

52. Dunn, R.L.; Casper, R.A.; Kelley, B.S. Trans. Soc. Biomater. **1985**, *8*, 213.

53. Gebelein, C.G. *Polymer Science Technology*; Plenum: New York, NY, 1987, Vol. 35, 75–85.

54. Winter, D.G.; Gibbons, D.F.; Plench, J. *Advances in Biomaterials*; John Wiley & Sons: New York, 1982; Vol. 3, 271–280.

55. Vainionpaa, S.; Kilpikari, J.; Laiho, J.; Helevirta, P.; Rokkanen, P.; Tormala, P. Strength and strength retention

vitro, of absorbable, self-reinforced polyglycolide (PGA) rods for fracture fixation. Biomaterials **1987**, *8* (1), 46–48.

56. Hay, D.L.; von Fraunhofer, J.A.; Chegini, N.; Masterson, B.J. A comparative scanning electron microscopic study on degradation of absorbable ligating clips *in vivo* and *in vitro*. J. Biomed. Mater. Res. **1988**, *22* (1), 71–79.

57. Makela, E.A.; Vainionpaa, S.; Vihtonen, K.; Mero, M.; Helevirta, P.; Tormala, P.; Rokkanen, P. The effect of a penetrating biodegradable implant on the growth plate. Clin. Ortho. Rel. Res. **1989**, *241* (1), 300–308.

58. Makela, E.A. Healing of epiphyseal fracture fixed with a biodegradable polydioxanone implant or metallic pins. An experimental study on growing rabbits. Clin. Mater. **1988**, *3* (1), 61–71.

59. Makela, E.A.; Vainionpaa, S.; Vihtonen, K.; Mero, M.; Helevirta, P.; Tormala, P.; Rokkanen, P. The effect of a diagonally placed penetrating biodegradable implant on the epiphyseal plate. An experimental study on growing rabbits with special regard to polydioxanone. Clin. Mater. **1988**, *3* (3), 223–233.

60. Lob, G.; Metzger, T.; Hertlein, H.; Hofmann, G.O. *18th World Congress of the Societe Internationale de Chirurgie Orthopedique et de Traumatologie*; Montreal, 1990, p. 190.

61. Bostman, O.M.; Hirvensalo, E.; Maikinen, J.; Rokkanen, P. Foreign-body reactions to fracture fixation implants of biodegradable synthetic polymers. J. Bone Joint Surg. **1990**, *72* (4), 592–596.

62. Bostman, O.M.J. Bone Joint Surg. **1991**, *73A*, 148.

63. Su Ming, L.; Garreau, H.; Vert, M. Structure-property relationships in the case of the degradation of massive aliphatic poly-(α-hydroxy acids) in aqueous media. J. Mater. Sci. Mater. Med. **1990**, *1* (3), 123–130.

64. Su Ming, L.; Garreau, H.; Vert, M. Structure-property relationships in the case of the degradation of massive poly(α-hydroxy acids) in aqueous media. J. Mater. Sci. Mater. Med. **1990**, *1* (3), 131–139.

65. Su Ming, L.; Garreau, H.; Vert, M. Structure-property relationships in the case of the degradation of massive poly(α-hydroxy acids) in aqueous media. J. Mater. Sci. Mater. Med. **1990**, *1* (4), 198–206.

66. Laurencin, C.; Morris, C.; Pierri-Jacques, H.; Schwartz, E.; Zou, L. Some new optimum Golomb rulers. Trans. Orthop. Res. Soc. **1990**, *36* (1), 183–184.

67. Ng, S.Y.; Penhale, D.W.H.; Heller, J. Macromol. Synth. **1992**, *11*, 23.

68. Heller, J.; Ng, S.Y.; Penhale, D.W.H.; Fritzinger, B.K.; Sanders, L.M.; Burns, R.A.; Gaynon, M.G.; Bhosale, S.S. Use of poly (ortho esters) for the controlled release of 5-fluorouracyl and a LHRH analogue. J. Control. Release **1987**, *6* (1), 217–224.

69. Burton, S.A.; Peterson, R.V.; Dickman, S.N.; Nelson, J.R. Comparison of *in vitro* bacterial bioluminescence and tissue culture bioassays and *in vivo* tests for evaluating acute toxicity of biomaterials. J. Biomed. Mater. Res. **1986**, *20* (6), 827–838.

70. Taylor, M.S.; Daniels, A.U.; Andriano, K.P.; Heller, J. Six bioabsorbable polymers: *in vitro* acute toxicity of accumulated degradation products. J. Appl. Biomater. **1994**, *5* (2), 151–157.

71. Daniels, A.U.; Andriano, K.P.; Smutz, W.P.; Chang, M.K.O.; Heller, J. Evaluation of absorbable poly (ortho esters) for use in surgical implants. J. Appl. Biomater. **1994**, *5* (1), 51–64.

72. Smutz, W.P.; Daniels, A.U.; Andriano, K.P.; France, E.P.; Heller, J. Mechanical test methodology for environmental exposure testing of biodegradable polymers. J. Appl. Biomater. **1991**, *2* (1), 13–22.

73. Andriano, K.P.; Daniels, A.U.; Heller, J. Biocompatibility and mechanical properties of a totally absorbable composite material for orthopaedic fixation devices. J. Appl. Biomater. **1992**, *3* (3), 197–206.

74. Kelley, B.S.; Dunn, R.L.T.; Jackson, E.; Potter, A.S.; Ellis, D.N. *Proc. 3rd World Biomaterials Congress*; Kyoto, Japan, 1988; p471.

75. Kohn, J. Trends Polym. Sci. **1993**, *1* (7), 206–212.

76. Ertel, S.I.; Kohn, J. Evaluation of a series of tyrosine-derived polycarbonates as degradable biomaterials. J. Biomed. Mater. Res. **1994**, *28* (8), 919–930.

77. Ertel, S.I.; Parsons, R.; Kohn, J. Trans. Soc. Biomater. **1993**, *19*, 17.

Biodegradable Polymers: Biomedically Degradable Polymers

A.-C. Albertsson
A. Löfgren
Department of Polymer Technology, Royal Institute of Technology (KTH), Stockholm, Sweden

Abstract

Degradable polymers are well suited for use as temporary aids in wound healing, in tissue replacement, and for use as a drug delivery matrix. Because the material is gradually absorbed by the body, surgical removal of the implant is not needed. Depending on the site and purpose of the implantation, the material must possess different properties, from hard, stiff materials for replacing bone to soft, flexible materials for replacing soft tissues. Other criteria include variations in hydrophilicity to match drugs that are to be incorporated.

Development of degradable polymers is now a well-established field of research, and these materials keep finding increasing numbers of applications. These specialty polymers were almost exclusively intended for use in the biomedical field, mainly because of their relatively high cost. However, with the public's ever-increasing environmental concern, great interest has been shown in the use of the degradable polymers for things like packaging materials. Large-scale production of poly(lactic acid), for example, is a reality in many countries worldwide.

Degradable polymers are well suited for use as temporary aids in wound healing, tissue replacement, and for use as a drug delivery matrix. Because the material is gradually absorbed by the body, surgical removal of the implant is not needed. Depending on the site and purpose of the implantation, the material must possess different properties, from hard, stiff materials for replacing bone to soft, flexible materials for replacing soft tissues. Other criteria include variations in hydrophilicity to match drugs that are to be incorporated.

Biomaterials can be of natural or synthetic origin. To differentiate between natural and synthetic biomaterials, the latter class is often called biomedical materials. Biomedical polymers can be classified as either stable or absorbable. The stable biomedical polymers are made to permanently replace human tissue without losing their properties over time, whereas the absorbable polymers are designed to gradually lose their properties, dissolve or metabolize, and leave the body by natural pathways.

The distinction between these two groups can sometimes be difficult to make. Many so-called stable polymers, when implanted into the body, suffer partial degradation that can cause discomfort or hazards to the host, as in the case of polyurethanes.[1] On the other hand, some supposedly absorbable polymers have in certain cases shown high resistance toward degradation, causing adverse tissue reactions,[2] or have indeed degraded very fast, releasing large amounts of degradation products, with concomitant tissue response.[3]

By far the most common class of synthetic absorbable polymers used today is the poly(α-hydroxy acid)s. These aliphatic polyesters undergo simple hydrolysis *in vivo* and form natural metabolites as degradation products. Vert[4] has proposed that these materials be called "bioresorbable" since they degrade and are further resorbed *in vivo*.

Synthesis of poly(α-hydroxy acid)s involves polycondensation, but generally ring-opening polymerization (ROP) of the cyclic condensation product of two α-hydroxy acid molecules (e.g., lactide or glycolide) is preferred. ROP can be performed with a large number of different initiators to form high molecular weight products.[5–8] Many different applications have been proposed, although only a few have reached commercial practice so far.[4] Examples of commercial products are bone fracture fixation devices, drug-delivery devices, and suture filaments. Some classes of common hydrolyzable polymers are listed in Table 1.

The important property of degradation rate varies a great deal among the different classes of materials listed in Table 1. Aliphatic polyanhydrides are generally hydrolyzed very quickly and are characterized by a surface-erosion type of degradation.[9,10] Aromatic polyanhydrides have also been investigated;[11] they degrade somewhat more slowly due to their more hydrophobic character. At the other end of the spectrum we find the aliphatic polycarbonates, which hydrolyze very slowly; for higher molecular weights, hardly any mass change can be detected even after two years *in vitro*.[12] Poly(ortho ester)s have been proposed as a suitable matrix material in drug-delivery applications. Since the poly(ortho ester) chain contains an acid-sensitive linkage, incorporation of acidic or basic materials in the matrix can accelerate or retard the degradation rate.[13]

Concise Encyclopedia of Biomedical Polymers and Polymeric Biomaterials DOI: 10.1081/E-EBPPC-120051881

153

Table 1 Common class of hydrolyzable aliphatic polymers

Polyesters

Polyanhydrides

Polyorthoesters

Polyesteramides

Polyetheresters

Polyphosphazenes

Polycarbonates

Polyamides

Copolymers including one or several classes make an important contribution to the already existing materials in that they combine the inherent properties of each homopolymer. Examples of commonly used degradable copolymers are PGA/PLA or PGA/PTMC sutures.[14,15]

It was recognized early on that microorganisms had the ability to produce polymers.[16,17] Microorganisms have been used to produce aliphatic polyesters commercially.[18] Microorganisms produce, for example, poly(hydroxy butyrate)/poly(hydroxy valerate) (PHB/PHV) copolymers of perfectly isotactic nature and of high molecular weights.[19]

ROP OF LACTONES

Carothers and colleagues were the first to systematically explore the ROP of various lactones.[20] Due to the increasing interest in the production of degradable materials, many laboratories have since then, been involved in this research area.

The traditional way of synthesizing polyesters has been polycondensation. This method suffers, however, from some major drawbacks, such as long reaction times, high temperatures, the need to remove reaction byproducts, and the need to have a precise stoichiometric balance between reactive acid and hydroxyl groups. Very high conversion is also required to obtain chains with high enough molecular weights to provide useful mechanical properties in the final material. ROP, classified as a polyaddition process, is rid of these shortcomings and allows high molecular weight polyesters to be produced under relatively mild conditions. In addition, ROP can in certain cases be carried out with no or strictly limited side-reactions taking place, making it possible to control properties like molecular weight and molecular weight distribution (MWD).

Investigations reported in the literature have been made using all major polymerization mechanisms to develop the ROP method and make available as many different polymeric structures as possible. The different initiation mechanisms can be divided into anionic, coordinative, cationic, radical, zwitterionic, and active hydrogen processes. The highest yields and molecular weights have, however, been obtained mainly by the anionic and coordinative ROP. Below are some examples.

Coordinative ROP of Lactones

Coordinative initiation differs from its ionic counterpart in that the propagating species consists not of an ionic but of a covalent species. This generally decreases the reactivity, which in some cases leads to fewer side reactions. Living ROP of lactones has indeed been reported in a number of cases. Organometallic derivatives of metals such as aluminum, zinc, and tin have been extensively studied as initiators in ROP of lactones and related compounds.

Cherdron, Ohse, and Korte[21] pioneered the field in the early 1960s when he showed that some Lewis acids, for example, triethyl aluminum added with water or ethanolate of diethyl aluminum, were effective as initiators in lactone polymerization. Aluminum alkoxides have subsequently attracted a lot of interest because of the versatility they offer as initiators.[22–25] Polymerization proceeds according to a "coordination-insertion" mechanism that involves acyl–oxygen bond cleavage of the monomer and insertion into the aluminum–oxygen bond of the initiator (Scheme 1). The coordination of the exocyclic oxygen to the metal results in a polarization of bonds, making the carbonyl carbon of the monomer more susceptible to nucleophilic attack. Transesterification, a side reaction, can also occur in the coordinative ROP of lactones, although this is generally only significant at elevated temperatures (>100°C).[26]

Carboxylates are less nucleophilic than alkoxides and are therefore considered to behave as catalysts rather than as actual initiators. Indeed, metal carboxylates such as stannous-2-ethylhexanoate (stannous octonate) are usually

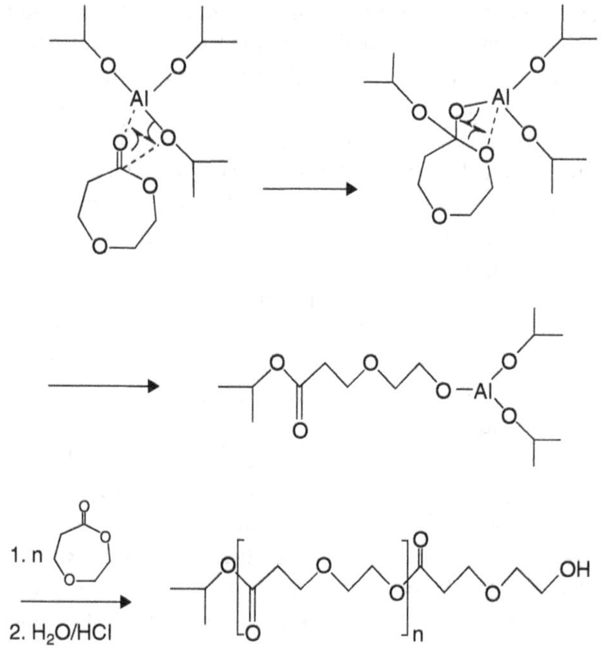

Scheme 1

added with active hydrogen compounds (e.g., alcohols) as coinitiators.[27,28] If no active hydrogen compound is added, many authors have suggested that the actual initiating species consists of hydroxyl-containing impurities.[28–32]

Our research group has synthesized a variety of different degradable polymers including polyesters,[33] polyanhydrides,[10] polycarbonates,[34] poly(ether ester)s,[29] and copolymers thereof.[24] From the very start we realized how important it is that the properties of degradable polymers be modifiable. Early attempts to synthesize degradable elastomeric materials in the form of block copolymers were based on condensation polymerization,[35] but later studies revealed the many advantages of ROP.[29] Poly(β-propiolactone) [PPL (25037-58-5)] was investigated and shown to be formed in high yield and high molecular weights.[36] To make the perfectly alternating copolymer of β-PL and ethylene glycol, a new synthetic route was developed to produce the monomer 1,5-dioxepan-2-one (DXO), a cyclic seven-membered ether-lactone ring.[29] Although the homopolymer [PDXO, (121425-66-9)] of this ether lactone is an amorphous material with a glass-transition at −36 to −39°C, it has proven to yield elastomeric, degradable materials when copolymerized with various lactones.[37,38]

DEGRADABLE COPOLYMERS FROM 1,5-DIOXEPAN-2-ONE

Preparation

Copolymers prepared in bulk were heated at 110–120°C in dry glass bottles, sealed with rubber septums under an inert gas atmosphere. The initiator, stannous-2-ethyl hexanoate, was used as received from the manufacturer. Both monomers (recrystallized twice in dry solvents) and catalyst were

added to the polymerization reactor in a glove-box under strictly anhydrous conditions. The synthesis of 1,5-dioxepan-2-one [35438-57-4] (alternative name: 1,4-dioxepan-5-one) has been described elsewhere.[29,38] Polymerizations were terminated by rapid cooling and subsequent dissolution in CHCl$_3$ and precipitation in cold MeOH.

Solution polymerizations were performed in THF or toluene that had been dried in the presence of sodium/benzophenone complex and distilled immediately prior to use. Polymerization temperatures were kept low, 0 to 25°C, to minimize influences from side reactions like intra- or intermolecular transesterification. Aluminum isopropoxide (distilled and dissolved in dry toluene) was added to the reaction vessel via syringe under anhydrous conditions. Solution polymerizations were terminated by addition of a small amount of dilute HCl (excess to amount of initiator) and subsequent precipitation in hexane. All copolymers were dried in a vacuum oven at ambient temperature until they reached constant weight.

Characterization of the polymers has been described in detail in earlier works.[31,38–40]

Properties

Poly(L- or D,L-lactic acid) (PLLA [26811-96-1] and PDLLA [31587-11-8]) are indeed some of the most common and well-studied degradable polymers on the market today. The high strength and brittleness that characterize these poly(α-hydroxy acid)s can be substantially modified by copolymerization with, for example, 1,5-dioxepan-2-one. The large difference in glass-transition temperatures (T$_g$) of poly(DXO) (−36 to −39°C) and poly(lactide)s (+55 to +58°C) results in copolymers with a wide range of intermediate properties. Table 2 reports the results from the

Table 2 Conditions and results of the copolymerizations of DXO with L- or D,L-lactide in bulk with stannous-2-ethylhexanoate as catalyst

Exp. No.	Mole ratio in feed DXO:D,L-LA	[M]/[C]	M$_w$[a] (g/mol)	Polym. time (h)	Polym. temp. (°C)	M$_w$/M$_n$
1	90:10	600	115,000	16	110	1.7
2	80:20[b]	740	80,000	22	120	1.8
3	60:40	600	90,000	16	110	1.9
4	50:50[b]	860	85,000	22	120	1.8
5	40:60	600	85,000	16	110	2.1
6	23:77[b]	1020	100,000	22	120	1.7
7	10:90 DXO:L-LA	450	35,000	16	110	2.3
8	90:10	800	65,000	9.3	115	1.8
9	80:20[b]	750	95,000	24	120	1.8
10	60:40	900	60,000	9.3	115	2.0
11	50:50[b]	750	110,000	24	120	1.8
12	40:60	450	65,000	9.3	115	1.8
13	20:80[b]	750	140,000	24	120	1.8

[a]Weight-average molecular weights obtained by SEC, polystyrene standard calibration.
[b]Lactide recrystallized twice in dry toluene.

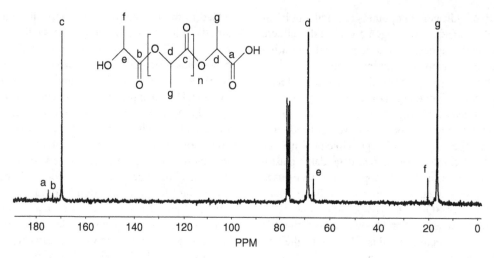

Fig. 1 ^{13}C-NMR spectrum in CDCl$_3$ of a low molecular weight poly(L-lactic acid), showing the presence of carbonyl carbons attached to the carboxyl (C$_a$) and hydroxyl (C$_b$) end-groups (M$_n$ ca. 5000 g/mol).

statistical bulk copolymerization between DXO and L- or D,L-lactide initiated with stannous-2-ethyl hexanoate.

A relatively large difference in reactivity ratios, r$_{dxo}$ = 0.1 and r$_{lactide}$ = 10, results in rather long initial blocks of poly(lactide) being formed, although these are later redistributed by intermolecular transesterification reactions.[38]

The copolymerization of DXO and lactide with Sn(oct)$_2$ is assumed to be initiated by hydroxyl-containing impurities within the monomer and catalyst. Model experiments with DXO and L,L-lactide initiated by Sn(oct)$_2$ have been made using a low monomer-to-initiator molar ratio ([M]/[I] = 5) and a short polymerization time (15–30 min) to obtain polymers with molecular weights of 2000–8000 g/mol. ^{1}H-NMR analyses only show the presence of methyl-(H$_a$) and methine protons (H$_b$) connected to the hydroxyl end-group (-CH(CH$_3$)-OH) formed after hydrolysis of the propagating chain-end, with an intensity corresponding to approximately one hydroxyl end-group per polymer chain. ^{13}C-NMR spectra, on the other hand, show two small peaks (C$_a$ and C$_b$) downfield from the major polymer peak, which corresponds to the carbonyl carbon (C$_c$). The peaks at 175.1 ppm (C$_a$) and 173.6 ppm (C$_b$) have been assigned to the carbonyl carbon attached to the carboxyl and hydroxyl end-groups, respectively (Fig. 1), based on comparison with spectra obtained from lactic acid and oligomers.

In the proposed mechanism, initiation takes place by hydroxyl-containing impurities such as water or lactic acid (Scheme 2). It is clear from this mechanism that the amount of impurities present determines the molecular weight of the resulting polymers. Indeed, many authors have reported that no clear-cut relationship exists between polymer molecular weight and added amount of Sn(oct)$_2$. A good correlation between Sn(oct)$_2$ concentration and polymer molecular weight was, however, obtained by Nijenhuis Grijpma, and Pennings,[30] who assumed that a specific hydroxyl content in monomer and catalyst acted as the true initiator. The effect of additions of hydroxyl or

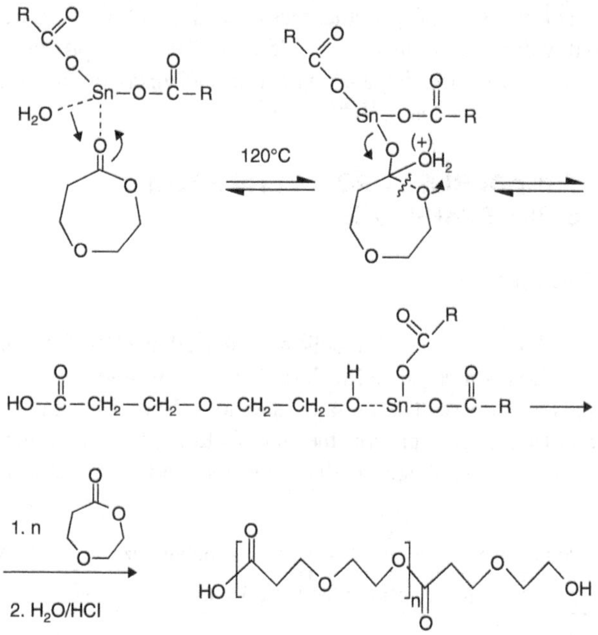

Scheme 2

carboxylic-acid groups in the lactide polymerization with Sn(oct)$_2$ was investigated by Zhang et al.[32]

Conveniently, properties such as crystallinity and glass-transition of the resulting copolymers vary between the respective homopolymer values, depending on the molar composition, as examplified by the poly(DXO-co-D,L-lactide); see Fig. 2. The molar composition also influences the hydrolytic sensitivity of these degradable copolymers, as shown in Fig. 3. The degree of crystallinity observed in the DXO/L,L-lactide copolymers (as opposed to the DXO/D,L-lactide copolymers) greatly influences the hydrolytic stability.

The crystalline material, having a lower water permeability than the amorphous phase, can reside *in vivo* for many years before total absorption.[2,31,41] The *in vivo* study on rats showed differences in the foreign-body response

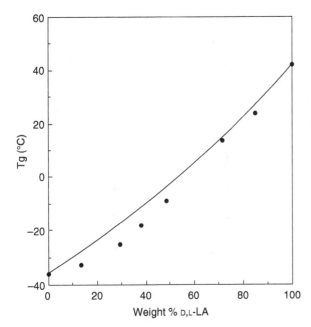

Fig. 2 Glass-transition temperature of poly(DXO-*co*-D,L-LA) copolymers as function of molar content D,L-lactic acid. The continuous line corresponds to expected values calculated from the Fox equation.

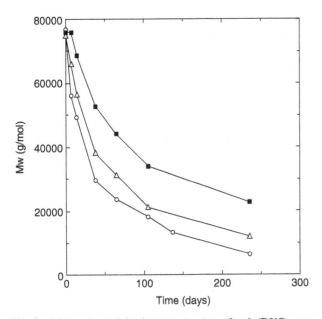

Fig. 3 Molecular weight decrease *in vitro* of poly(DXO-*co*-L-LA) copolymers. Molar ratios DXO/L-LA: 80/20 (■); 50/50 (△); 20/80 (○).

between the amorphous and the crystalline copolymers. Fragmentation at 180 days of the brittle, crystalline DXO/L-lactide copolymer containing about 80 mol% L-lactic acid gave a significantly higher response than did the corresponding DXO/D,L-lactide copolymer.[40] The problem of late tissue-response to highly crystalline debris remaining in the host for many years has been observed in several cases,[2,42,43] and calls for caution in developing new biomedical materials.

If instead of using Sn(oct)₂ we choose to employ aluminum triisopropoxide (Al(O^{i}Pr)₃), the situation becomes quite different. In this case we are dealing with a coordination-insertion mechanism, as proven from the end-group analysis of the formed poly(DXO). This analysis shows the formation of one isopropyl ester and one hydroxyl end-group per polymer chain after termination; see Fig. 4. Some results from the Al(O^{i}Pr)₃-initiated polymerizations are shown in Table 3.

It is obvious from the mechanism that the molecular weight of the polymer is determined by the monomer-to-initiator molar ratio and the monomer conversion, if no side-reactions are present. The narrow MWD indicates that transesterification reactions do not take place and that the polymerization has a living character under the conditions used. This mechanism has earlier been proposed in the case of ε-CL polymerization using the same initiator in toluene[22] which led us to explore the feasibility of block copolymerization of these monomers. Poly(ε-CL) (PCL, [24980-41-4]) is a semicrystalline polymer with a melting point at 60–63°C and a glass transition at −60°C. Block copolymers of DXO and ε-CL would thus be interesting to examine, since combinations of hard and soft blocks in copolymers have been proved to yield valuable materials.[44] The triblock copolymerization is schematized in Scheme 3. In Table 4 we present the results and conditions used.

We now have the ability to control mechanical properties through the respective block lengths in the copolymer, as shown in Table 5. The excellent elastic response displayed by the statistical and particularly the block copolymers is shown by tensile measurements; see Table 6.

Applications

The use of synthetic degradable polymers and copolymers in biomedical devices has been extensively explored. The first commercially successful application was the synthetic absorbable surgical suture.[45] The poly(glycolic acid) used was soon followed by other poly(α-hydroxy acid)s, and copolymers were developed to modify properties like degree of crystallinity and hydrolysis rate.[14,15] Other applications include orthopedic pins, screws, and plates,[46,47] artificial blood vessels,[48] burn dressings,[49,50] nerve guides,[51,52] artificial tendons,[53] and in the regeneration of periodontal tissue.[54] New, more startling ideas also began to emerge, such as "hybrid artificial organs," which may lead to revolutionary advances in transplantation surgery.[55]

Another area of major importance is the controlled release of drugs. The benefits of releasing a predetermined amount of drug at a controllable rate are obvious, as are the needs for various kinds of materials to match different drugs and to provide specific release profiles. The drug can be incorporated into a degradable polymer matrix or be attached via a cleavable bond to a polymer chain.[56] Prospects include "targeted" drug delivery, where the device can be directed by various means to the pathological site of interest.[57]

Bioactive—BioMEMS

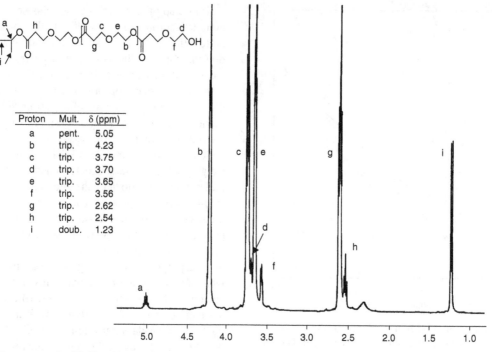

a
h
c
e
d
g
b
f
i
OH
Proton Mult. δ (ppm)
a pent. 5.05
b trip. 4.23
c trip. 3.75
d trip. 3.70
e trip. 3.65
f trip. 3.56
g trip. 2.62
h trip. 2.54
i doub. 1.23
5.0
4.5
4.0
3.5
3.0
2.5
2.0
1.5
1.0

Fig. 4 ^{1}H-NMR spectrum in $CDCl_3$ of low molecular weight poly(DXO), showing an isopropoxide ester end group (H_a and H_i) and methylene protons (H_d and H_f) connected to a hydroxyl chain end (M_n = 1300 g/mol).

Table 3 Polymerization of DXO in THF with $Al(O^iPr)_3$ as initiator

Entry	[M]/[I]	Polym. temp. (°C)	Polym. time (h)	Conv. (%)	$M_{n,th.}$[a] (g/mol)	$M_{n,exp.}$[b] (g/mol)	MWD
1	130	0	2.5	98	15,000	22,700	1.10
2	430	25	16.5	>99	50,000	44,500	1.20
3	517	25	69.5	>99	60,000	60,500	1.30
4	860	25	21.5	97	97,000	126,600	1.15

[a]Theoretical molecular weight – ([M]/[I]) MM x, where [M]/[I] = initial monomer/initiator molar ratio, MM = molecular weight of the monomer, and x = monomer conversion.
[b]Molecular weight measured by SEC (universal calibration with PS standard).

Al(O^iPr)3
Toluene, 0°C
n–1
n
x
y
HCl/H2O

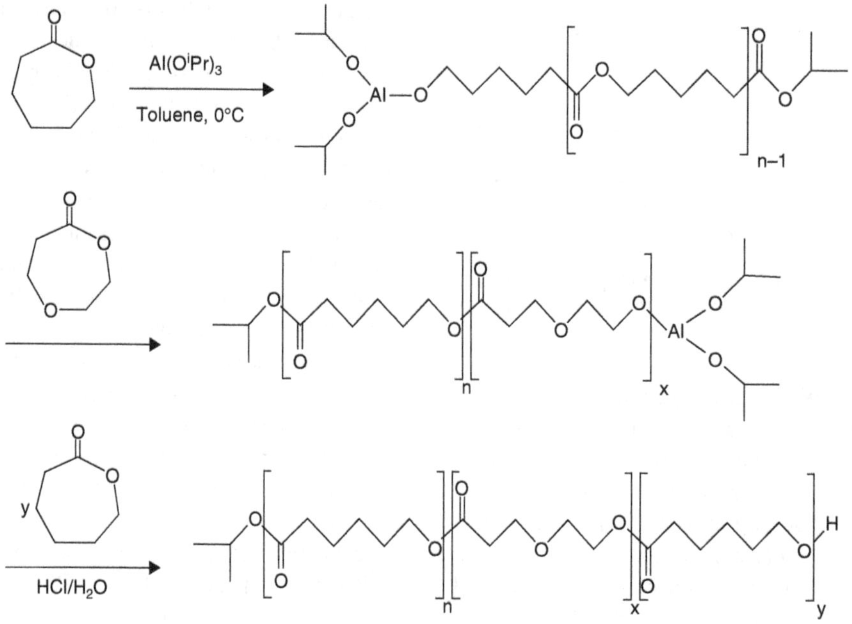

Scheme 3.

Table 4 Triblock copolymerization of ε–CL and DXO initiated by Al(O^{i}Pr)$_3$

Sample	Solvent	Polym. temp. (°C) A/B/A[b]	Conv. (%) A/B/A[b]	Polym. time (h) A/B/A[b]	$M_{n,th}$ (× 10^{-3})[d] (g/mol) mono/di/tri[c]	$M_{n,SEC}$ (× 10^{-3})[d] (g/mol) mono/di/tri[c]	$M_{n,LS}$[e] (g/mol) triblock	MWD mono/di/tri[c]
1	Toluene	0/0/0	>99/>99/37	2.5/2.5/2.0	10/20/24	16/27/32	—	1.18/1.15/1.18
2	THF	0/0/0	>99/>99/15	3.0/5.5/16.5	20/40/43	32/40/44	—	1.13/1.15/1.20
3[a]	THF	0/25/25	>99/93/96	2.0/14/21	10/38/50	17/30/51	40,000	1.13/1.15/1.15
4[a]	THF	0/25/25	>99/95/98	1.3/17/25	10/39/50	22/42/64	55,000	1.10/1.15/1.15
5[a]	THF	0/25/25	>99/96/92	2.0/117/17	15/63/73	28/60/68	—	1.10/1.20/1.20
6[a]	Toluene	0/25/25	>99/97/>99	2.0/36.5/7.0	15/64/80	29/65/90	66,300	1.10/1.20/1.25
7[a]	Toluene	0/25/25	>99/96/>99	1.5/17/16.5	15/82/100	26/73/87	91,300	1/10/1.25/1.30
8	Toluene	0/25/25	>99/96/98	1.5/21/5.0	15/63/80	26/67/77	63,000	1.10/1.25/1.30
9[a]	Toluene	0/25/25	>99/98/>99	1.1/17.5/5.5	15/64/80	26/68/79	79,500	1.10/1.25/1.30

[a]Addition of 1 molar equivalent pyridine compared to amount of initiator.
[b]A/B/A corresponds to data from first, second, and third monomer addition, respectively.
[c]Mono/di/tri corresponds to data from 1st PCL-block, PCL/PDXO-diblock, and PCL/PDXO/PCL-triblock, respectively.
[d]Molecular weights measured by SEC (calibration with PS standards).
[e]Absolute M_n of triblock copolymer obtained by $M_{w,LS}$/MWD, where $M_{w,LS}$ is weight average molecular weight obtained by light—scattering measurements, and MWD = M_w/M_N.

Table 5 Data from DMTA measurements of PCL/PDXO/PCL triblock copolymers

Block lengths A/B/A (g/mol)	10K/30K/10K[a]	15K/50K/15K	15K/70K/15K
Storage modulus at 23°C (MPa)	110	70 [9.1]	23 [12.4]
Loss modulus peak value (MPa)	220	260 [12.4]	285 [11.7]
Loss modulus peak (°C)	−30.8	−31.2 [2.7]	−31.7 [2.9]
Loss tangent peak (°C)	−28.1	−27.9 [2.6]	−27.6 [2.1]
Loss tangent peak value	0.43	0.51 [0.02]	0.84 [0.14]

[a]Single measurement at 1°C/min.
Note: Standard deviation in brackets.

Table 6 Tensile properties of PLLA/PDXO statistical copolymer and PCL/PDXO/PCL triblock polymers

Block lengths A/B/A (g/mol)	15K/50K/15K	15K/70K/15K	PLLA/PDXO 75/25 (mol %)
Tensile modulus (MPa)	31 [8.8]	21 [4.2]	30 [2.7]
Stress at yield (MPa)	3.0 [0.4]	2.1 [0.1]	–
Elongation at yield (%)	15 [4.3]	18 [0.4]	–
Strength at break (MPa)	52 [6.7]	53 [3.0]	21 [1.5]
Elongation at break (%)	1070 [50]	1210 [67]	420 [63]

Note: Standard deviation in brackets.

As mentioned in the introduction, degradable synthetic polymers began to attain a more widespread use because of increasing environmental concern. Great interest has been shown by the agricultural and packaging industries, and although price has been a limiting factor, large-scale production[58] can bring costs down and open the way for new, exciting applications.

REFERENCES

1. Griesser, H.J. Degradation of polyurethanes in biomedical applications—A review. Polym. Degrad. Stab. **1991**, *33* (3), 329–354.
2. Bos, R.R.M.; Rozema, F.R.; Boering, G.; Nijenhuis, A.J.; Pennings, A.J.; Verwey, A.B.; Nieuwenhuis, P.; Jansen, H.W.B. Degradation of and tissue reaction to biodegradable poly(l-lactide) for use as internal fixation of fractures: A study in rats. Biomaterials **1991**, *12* (1), 32–36.
3. Vert, M.; Li, S.M.; Spenlehauer, G.; Guerin, P. Bioresorbability and biocompatibility of aliphatic polyesters. J. Mater. Sci. **1992**, *3* (6), 432–446.
4. Vert, M. Bioresorbability and biocompatibility of aliphatic polyesters. J. Mater. Sci. **1992**, *3* (6), 432–446.
5. Kricheldorf, H.R.; Kreiser-Saunders, I. Polylactones, 19. Anionic polymerization of L-lactide in solution. Makromol. Chem. **1990**, *191* (5), 1057–1066.
6. Kricheldorf, H.R.; Sumbél, M. Polylactones—18. Polymerization of l, l-lactide with Sn (II) and Sn (IV) halogenides. Eur. Polym. J. **1989**, *25* (6), 585–591.
7. Leenslag, J.W.; Pennings, A. Synthesis of high-molecular-weight poly(L-lactide) initiated with tin 2-ethylhexanoate. J. Makromol. Chem. **1987**, *188* (8), 1809–1814.
8. McLain, S.J.; Ford, T.M.; Drysdale, N.E. Polym. Preprint **1992**, *33* (2), 463.
9. Mathiowitz, E.; Staubli, A.; Langer, R. Polym. Preprint **1990**, *32* (1), 431.

10. Albertsson, A.-C.; Lundmark, S. Synthesis, characterization and degradation of aliphatic polyanhydrides. Brit. Polym. J. **1990**, *23* (3), 205–212.

11. Leong, K.W.; Brott, B.C.; Langer, R. Bioerodible polyanhydrides as drug-carrier matrices. I: Characterization, degradation, and release characteristics. J. Biomed. Mater. Res. **1985**, *19* (8), 941–955.

12. Eklund, M.; Albertsson, A.-C. Short methylene segment crosslinks in degradable aliphatic polyanhydride: Network formation, characterization, and degradation. J. Appl. Polym. Sci. **1996**, *34* (8), 1395–1405.

13. Heller, J.; Ng, S.Y.; Fritzinger, B.K. Synthesis and characterization of a new family of poly(ortho esters). Macromolecules **1992**, *25* (13), 3362–3364.

14. Wasserman, D.; Versfelt, C.C. Use of stannous octoate catalyst in the manufacture of L (-) lactide-glycolide copolymer sutures. U.S. Patent 3889297, Oct 1, 1974.

15. Casey, D.J.; Roby, M.S. Eur. Patent 0098394, 1983.

16. Ellar, D.; Lundgren, D.G.; Okamura, K.; Marchessault, R H. Morphology of poly-β-hydroxybutyrate granules. J. Mol. Biol. **1968**, *35* (3), 489–502.

17. Doi, Y.; Tamaki, A.; Kunioka, M.; Soga, K. Biosynthesis of an unusual copolyester (10 mol% 3-hydroxybutyrate and 90 mol% 3-hydroxyvalerate units) in *Alcaligenes eutrophus* from pentanoic acid. J. Chem. Soc. Chem. Commun. **1987**, (21), 1635–1636.

18. Holmes, P.A.; Wright, L.F.; Collins, S.H. (ICI), Eur. Pat. App. 0052459, 1981; Eur. Pat. App. 0069497, 1983.

19. Holmes, P.A. Applications of PHB - A microbially produced biodegradable thermoplastic. Phys. Technol. **1985**, *16* (1), 32–36.

20. Carothers, W.H. Polymers and polyfunctionality. Trans. Faraday Society **1936**, 32, 39–49.

21. Cherdron, H.; Ohse, H.; Korte, F. Die polymerisation von lactonen. Teil 1: Homopolymerisation 4-, 6-und 7-gliedriger lactone mit kationischen initiatoren. Makromol. Chem. **1962**, *56* (1), 179–186.

22. Ouhadi, T.; Stevens, C.; Teyssié, P. Mechanism of ε-caprolactone polymerization by aluminum alkoxides. Makromol. Chem. Suppl. **1975**, *1* (S19751), 191–201.

23. Dubois, P.; Jacobs, C.; Jérôme, R.; Teyssié, P. Macromolecular engineering of polylactones and polylactides. 4. Mechanism and kinetics of lactide homopolymerization by aluminum isopropoxide. Macromolecules **1991**, *24* (9), 2266–2270.

24. Löfgren, A.; Albertsson, A.-C.; Dubois, P.; Jérôme, R.; Teyssié, P. Synthesis and characterization of biodegradable homopolymers and block copolymers based on 1, 5-dioxepan-2-one. Macromolecules **1994a**, *27* (20), 5556–5562.

25. Hofman, A.; Slomkowski, S.; Penczek, S. Polymerization of ε-caprolactone with kinetic suppression of macrocycles. Makromol. Chem. Rapid Commun. **1987**, *8* (8), 387–391.

26. Kricheldorf, H.R.; Berl, M.; Scharnagl, N. Poly(lactones). 9. Polymerization mechanism of metal alkoxide initiated polymerizations of lactide and various lactones. Macromolecules **1988**, *21* (2), 286–293.

27. van der Weij, F.W. The action of tin compounds in condensation-type RTV silicone rubbers. Makromol. Chem. **1980**, *181* (12), 2541–2548.

28. Rafler, G.; Dahlmann, J. Biodegradable polymers. 6th comm. Polymerization of ε-caprolactone. Acta Polym. **1992**, *43* (2), 91–95.

29. Mathisen, T.; Masus, K.; Albertsson, A.-C. Polymerization of 1, 5-dioxepan-2-one. II. Polymerization of 1, 5-dioxepan-2-one and its cyclic dimer, including a new procedure for the synthesis of 1, 5-dioxepan-2-one. Macromolecules **1989**, *22* (10), 3842–3846.

30. Nijenhuis, A.J.; Grijpma, D.W.; Pennings, A.J. Lewis acid catalyzed polymerization of L-lactide. Kinetics and mechanism of the bulk polymerization. Macromolecules **1992**, *25* (24), 6419–6424.

31. Löfgren, A.; Albertsson, A.-C. Copolymers of 1,5-dioxepan-2-one and L- or D,L-dilactide: Hydrolytic degradation behavior. J. Macromol. Sci.-Pure Appl. Chem. **1994**, *52* (9), 1327–1338.

32. Zhang, X.; MacDonald, D.A.; Goosen, M.F.A.; Mcauley, K.B. Mechanism of lactide polymerization in the presence of stannous octoate: The effect of hydroxy and carboxylic acid substances. J. Polym. Sci. Polym. Chem. **1994**, *32* (15), 2965–2970.

33. Albertsson, A.-C.; Ljungquist, O. Degradable Polymers. I. Synthesis, characterization, and long-term *in vitro* degradation of a 14c-labeled aliphatic polyester. J. Macromol. Sci. Chem. **1986**, *A23* (2), 393–409.

34. Albertson, A.-C.; Sjöling, M. Homopolymerization of 1, 8dioxan-2-one to high molecular weight poly(trimethylene carbonate). J. Macromol. Sci. Chem. **1992**, *A29* (1), 43–54.

35. Albertson, A.-C.; Ljungquist, O. Degradable polymers. II. Synthesis, characterization, and degradation of an aliphatic thermoplastic block copolyester. J. Macromol. Sci. Chem. **1986**, *A23* (3), 411–422.

36. Mathisen, T.; Albertsson, A.-C. Hydrolytic degradation of melt-extruded fibers from poly(β-propiolactone). J. Appl. Polym. Sci. **1990**, *39* (3), 591–601.

37. Albertsson, A-C.; Löfgren, A. Copolymers of 1.5-dioxepan-2-one and L-or D, L-dilactide-synthesis and characterization. Makromol. Chem. Macromol. Symp. **1992**, *53* (1), 221–231.

38. Löfgren, A.; Renstad, R.; Albertsson, A.-C. Copolymers of 1,5-dioxepan-2-one and L- or D,L-dilactide: *in vivo* degradation behaviour. J. Appl. Polym. Sci. **1994**, *6* (5), 411–423.

39. Löfgren, A.; Albertsson, A.-C. Copolymers of 1,5-dioxepan-2-one and L- or D,L-dilactide: Hydrolytic degradation behavior. J. Appl. Polym. Sci. **1994**, *52* (9), 1327–1338.

40. Löfgren, A.; Albertsson, A-C. Copolymers of 1,5-dioxepan-2-one and L- or D,L-dilactide: *In vivo* degradation behaviour. J. Biomater. Sci. Polym. Ed. **1995**, *6* (5), 411–423.

41. Pistner, H.; Gutwald, R.; Ordung, R.; Reuther, J.; Mühling, J. Poly(l-lactide): A long-term degradation study *in vivo*. Biomaterials **1993**, *14* (9), 671–677.

42. Lewy, R.B. Responses of laryngeal tissue to granular teflon in situ. Arch. Otolaryng. **1966**, *83* (1), 355–359.

43. Evans, E.J.; Clarke-Smith, E.M.H. Studies on the mechanism of cell damage by finely ground hydroxyapatite particles *in vitro*. Clin. Mater. **1991**, *7* (3), 241–245.

44. Holden, G. *Encyclopedia of Polymer Science and Engineering*, 2nd ed.; Mark, H.F. et al., Eds; Wiley: New York, 1986; Vol. 5, p 416.

45. Schmitt, E.E.; Polistina, R.A. US Patents 32397033, 1967, and 3463158, 1969. In Chem. Abstr **1967**, 66, p. 923826.

46. Vert, M.; Chabot, F.; Leray, J.; Christel, P. Stereoregular bioresorbable polyesters for orthopaedic surgery. Makromol. Chem. Suppl. **1981**, *5* (S19811), 30–41.

47. Vasenius, J.; Vainionpää, S.; Vihtonen, K.; Mero, M.; Mäkelä, P.; Törmälä, P.; Rokkanen, P. A histomorphological study on self-reinforced polyglycolide (SR-PGA) osteosynthesis implants coated with slowly absorbable polymers. J. Biomed. Mater. Res. **1990**, *24* (12), 1615–1635.

48. Hinrichs, W.L.J.; Zweep, H.-P.; Satoh, S.; Feijen, J.; Wildevuur, R.H. Supporting, microporous, elastomeric, degradable prostheses to improve the arterialization of autologous vein grafts. Biomaterials **1994**, *15* (2), 83–91.

49. Beumer, G.J.; Blitterswijk, C.A.; Ponec, M. Biocompatibility of a biodegradable matrix used as a skin substitute: An *in vivo* evaluation. J. Biomed. Mater. Res. **1994**, *28* (5), 545–552.

50. Gatti, A.M.; Pinchiorri, P.; Monari, E. Physical characterization of a new biomaterial for wound management. J. Mater. Sci. Mater. Med. **1994**, *5* (4), 190–193.

51. den Dunnen, W.F.A.; Schakenraad, J.M.; Zondervan, G.J.; Pennings, A.J.; van der Lei, B.; Robinson, P.H. A new PLLA/PCL copolymer for nerve regeneration. J. Mater. Sci. Mater. Med. **1993**, *4* (5), 521–525.

52. Perego, G.; Cella, G.D.; Aldini, N.N.; Fini, M.; Giardino, R. Preparation of a new nerve guide from a poly(l-lactide-co-6-caprolactone). Biomaterials **1994**, *15* (3), 189–193.

53. Davis, P.A.; Huang, S.J.; Ambrosio, L.; Ronca, D.; Nicolais, L. A biodegradable composite artificial tendon. J. Mater. Sci. Mater. Med. **1991**, *3* (5), 359–364.

54. Lundgren, D.; Sennerby, L.; Falk, H.; Friberg, B.; Nyman, S. The use of a new bioresorbable barrier for guided bone regeneration in connection with implant installation. Case reports. Clinical oral implants research **1994**, *5* (3), 177–184.

55. Barrera, D.; Langer, R.S.; Lansbury, P.T.; Vacanti, J.P. WO patent 94/09760, May 1994.

56. Langer, R. Chem. Eng. Commun. **1980**, *6*, 1.

57. Ottenbrite, R.M. *Encyclopedia of Polymer Science and Engineering, Suppl. Vol.*, 2nd Ed.; Mark, H.F., Bikales, N.M., Overberger, C.G., Menges, G., Kroschwitz, J.I., Eds.; Wiley: New York, 1989; p. 164.

58. Gruber, P.R.; Hall, E.S.; Kolstad, J.J.; Iwen, M.L.; Benson, R.D.; Borchardt, R.L. WO patent 93/15127, Aug 1993.

Biodegradation

Mikhail I. Shtilman
Biomaterials Scientific and Teaching Center, D. I. Mendeleyev University of Chemical Technology, Moscow, Russia

Abstract

Biodegradation is the ability of an object to disintegrate in the biological environment accompanied by the loss of its mass and volume. Biodegradation is the result of physical, chemical, and biochemical (i.e., involving enzymes) processes. The ability of an object's sorption and diffusion of water, and hydrolysis of groups of the chain of polymer, from which the object is made, are the important factors affecting biodegradation. Other factors such as degree of crystallinity, the physical form of the object (e.g., porosity, presence of defects), and so on can also affect biodegradation. Simple dissolution of the object is observed if it is made of a polymer soluble in the surrounding biological medium. In this case, biodegradation of the object can occur without any chemical transformations.

DEFINITION

Biodegradation is the process of destroying the physical object under the influence of the biological medium, accompanied by the loss of its mass and volume.

RELATED TERMS

A biodegradable object (e.g., implant)—An object that can undergo biodegradation.

Biotransformation—The process of change of material chemical structure as a result of interaction with biological medium.

Biodestruction—A separate case of biotransformation of material (for macromolecular systems) leading to chemical decay, in particular, of the main chain, that is accompanied by reduction of its molecular weight.

Bioerosion—The process of biodegradation beginning from the surface of an object.

OBSOLETE TERMS

Resorption, bioresorption, absorption, restorability.

INTRODUCTION

Notably, there are groups of materials, which must be resistant to biodegradation in a biological environment, or vice versa undergoing decomposition in biological medium with varying rates:

1. Objects that must be stable under the conditions of functioning in contact with aggressive biological medium over a prolonged period of time (surfaces, which are resistant to biodamage, antifouling coating, implants, which stay for a long time in the body).
2. Objects that are functioning for a long time, but decaying as quickly as possible in the biological environment after use.
3. Objects that are degraded with controlled rate of destruction.

The biodegradation processes are most important for some types of implants, for example, fixing bone systems, bone defect filling compositions, sutures, and adhesives for soft tissue. The biodegradation process is very important for a number of dosage forms (tablets, capsules, microcapsules) and substrates (matrixes) that are used in tissue engineering and packaging products that need to be destroyed after use, which is important from an environmental perspective.

The biological environments in these cases are blood, lymph, interstitial (intercellular) fluid, and other body fluids, for example, gastric and pancreatic fluids, saliva, and others. These aqueous systems contain a variety of inorganic and organic substances, including enzymes, and also different types of cells.

In the case of the biodegradation of objects in the environment, the biological media are soil, composting systems, and aqueous systems, which contain various microorganisms such as soil bacteria, fungi, and microscopic algae, among other substances.

In most cases, the process of biodegradation is observed for polymer objects. Rarely, does it occur in the case of products based on inorganic materials (special types of glass and ceramic, used in medicine), even less in the production of metals making up a medical device. The biodegradation is the result of physical, chemical, and biochemical (i.e., involving enzymes) processes. The biodegradation process has certain peculiarities for different objects.

Concise Encyclopedia of Biomedical Polymers and Polymeric Biomaterials DOI: 10.1081/E-EBPPC-120050403

MECHANISMS OF BIODEGRADATION

In principle, the biodegradation of polymer object may occur.

1. In the surface layer-external kinetic diffusion region, that is, in a layer accessible by diffusion of liquid environment. In this case the properties of the material, including polymer molecular weight outside this range, are not changed before the destruction of the outer layer. This process leads to a *bioerosion* of object surface, that is, the appearance of irregularities, cavities, and other changes on the surface.
2. In the mass of polymeric object, when the penetration rate of a liquid exceeds the rate of polymer degradation, for example, for objects with a high swellability in a biological environment, when the penetration of substances contributing to biodegradation, including enzymes, is facilitated. The first of these mechanisms is the most common.

The biodegradability of an object depends in these cases not only on the chemical structure of the polymer, but also on such factors as product shape and size, the presence of the perforation, the porosity, the presence of filler (for composites), and so on.

The contact of liquid biological medium with the surface of the object and therefore with polymer macromolecules from which it was made, in the case of the appropriate chemical and physical structure of the polymer, causes the start of the process of biodegradation. This stage of biodegradation takes place without the participation of cells (extracellular biodegradation).

Depending on the biological medium condition there are a number of chemical and physicochemical factors which may affect the process of extracellular biodegradation. Depending on the structure and properties of the polymer, all these processes may be either kinetically consistent or proceed simultaneously.

Influence of the Structural Features of the Materials, Undergoing Biodegradation

The role of hydrophilicity of biodegradable object

A very important influence on the biodegradation is the access of the surrounding biological medium to the molecules of the substance of the biodegradable object as a result of sorption and diffusion. These processes begin on the surface after penetration of the object by the surrounding medium into its surface layers. Interactions with the environment is the first step in the complex of chemical, physicochemical, and biological processes, which are involved in the biodegradation of an object. Water sorption and transport of water molecules from surroundings into the interior of an object have a great influence on biodegradability of the object.

Although the penetration of the aqueous solution into the object or at least into its surface layers is a necessary step, high values of sorption capacity of the substances to water and water vapor diffusion themselves do not guarantee a high rate of biodegradation (Table 1).

As is seen in Table 1, the objects made from polysiloxanes (polydimethylsiloxane), which are characterized by high diffusion coefficient of water vapor are not likely to undergo biodegradation. A moderate rate of biodegradation of an object (by hydrolysis of the ester side groups of the polymer) was observed for the products based on poly(2-hydroxyethylmethacrylate) having a high level of water sorption. On other hand, the object made from polyhydroxycarboxylic acids (polylactides and polyglycolides), which are widely used as materials for biodegradable implants, are characterized by relatively moderate values of sorption and diffusion.

Nevertheless, as a rule, the polymers degradable in biological fluids possess a high affinity for water as a result of the presence of polar groups. On the other hand, nonpolar or low polarity polymers (polyhydrocarbons, silicons, and halogen-containing carbon chain polymers) are nonbiodegradable in the body.

Thus, a combination of certain properties of the polymers is required for the realization of the biodegradation process. Primarily, these are the polymer chemical and physical structures of an object allowing penetration of the surrounding liquid into the mass of the polymer object and the presence of functional groups susceptible to hydrolysis in the polymer chain. Other factors such as solubility of the polymer in water, degree of crystallinity, the physical form of the object (e.g., porosity, presence of defects), and so on can also affect the biodegradation of the object.

Table 1 Sorption of water and water vapor diffusion coefficients of some polymers, used for making implants

Polymer	Temperature °C	Sorption H$_2$O/100 g of polymer	Diffusion coefficient cm$^2 \times$ s^{-1}
High rate of biodegradation of object			
Polyglycolide	37	8.0	5.0
Copolymer of glycolic and lactic acids	37	8.0	7.0
Moderate rate of biodegradation of object			
Segmented polyurethanes	37	1.0–1.2	400–700
Low rate of biodegradation of object			
Poly(2-hydroxyethylmethacrylate)	37	40	0.05
High level of bioinertness of object			
Polydimethylsiloxane	35	0.07	70000

Source: Data from Iordanskii, Rudakova, and Zaikov,[1] and Barrie.[2]

Chemical, physicochemical, and biochemical processes accompanying biodegradation

When the object, undergoing the influence of biological environment, enters directly into the biodegradation, the biodegradation is the result of different chemical, physicochemical, and biochemical processes:

1. The dissolution of chemically stable polymers in a biological medium. These processes occur when the biodegraded object is made of a water-soluble polymer.
2. The dissolution by products of chemical and biochemical reactions of the material, into which it enters in a biological environment. These are the products formed as a result of change in the chemical structure of the substance:
   a. Hydrolytic disintegration of the polymer chain:
      i. Hydrolysis of the chemical bonds contained in the main polymer chain.
      ii. Hydrolysis of the groups formed in the polymer chain after previous chemical reactions.
   b. Formation of water-soluble products after reactions of polymer, which has no influence on the length of the polymer chain.
      i. Formation of water-soluble polymer by reactions forming side lyophilic groups.
      ii. Decomposition of polymer–polymer complexes.

DISSOLUTION OF BIODEGRADABLE OBJECT

Simple dissolution of the object is observed if it is made of a polymer soluble in the surrounding biological medium. In this case the biodegradation of the object can be without any chemical transformations. Such polymers should not have a cross-linked structure and must contain polar functional groups in the main chain or side branches, which will help in the solubility of the polymer in to the surrounding aqueous medium.

Dissolution of material proceeds according to the following common polymer dissolution steps: diffusion of medium in the polymer mass, solvation of the polymer macromolecules with solvent molecules (in this case, water), and desorbing of the solvated macromolecules from object into the solution. Dissolution is accompanied by a change in the supramolecular structure of the polymer, the ratio of crystalline and amorphous regions, and other structural changes.

The tendency of the polymer to dissolve can be regulated by introducing various functional groups into the polymer chain. For synthetic polymers it may be hydroxyl (–OH), carboxylate (–COO–cation+), amide (–CO–NH$_2$), and substituted amide groups (–CO–NHR and –CONR$_2$). Such groups are introduced into the polymer to impart water solubility as side groups of the macromolecular chain. In most cases, these are homo- and copolymers of the vinyl

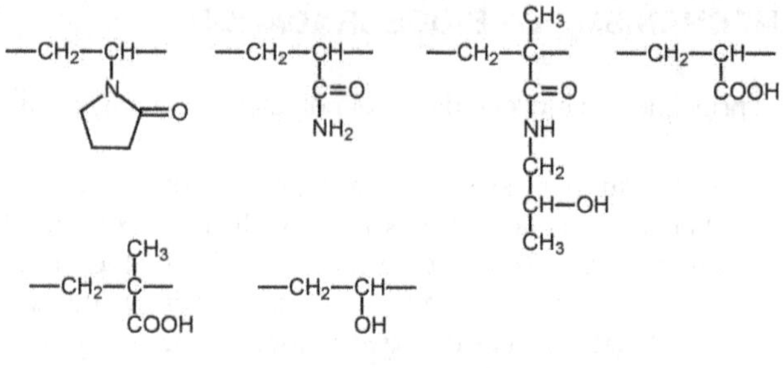

Fig. 1 Homo- and copolymers of vinyl alcohol, acrylamide, N-vinylpyrrolidone, acrylic acids salts, acrylamide, alkyl- and dialkyl-substituted acrylamide, and N-(2-hydroxypropyl)methacrylamide.

alcohol, acrylamide, N-vinylpyrrolidone, acrylic acids salts, acrylamide, alkyl- and dialkyl-substituted acrylamide, and N-(2-hydroxypropyl)methacrylamide (Fig. 1).

Water solubility is also possessed by polymers containing aliphatic ethers groups in the backbone chain, such as polyethyleneoxide.

Soluble materials can be obtained using globular proteins (e.g., albumin) and low-molecular weight fractions of fibrillar proteins (e.g., gelatin), as well as some polysaccharides, such as dextran, starch, mucopolysaccharides, salts of alginic acid, and carboxymethyl cellulose. In most cases, water-soluble polymers are used to make biodegradable drug formulations disintegrating in the organism and for the manufacturing of biodegradable packages.

DISSOLUTION OF THE PRODUCTS OF CHEMICAL TRANSFORMATION OF BIODEGRADABLE OBJECTS

Several distinct types of polymer degradation can be identified for object biotransformation via a biodestruction mechanism (see "Related Terms" section), that is, decomposition of the main chain of linear polymers, reactions in the side branches, and side reactive groups, while the integrity of the main chain is preserved, decomposition of cross-linked polymer maintaining linear chain fragments, and full decomposition of cross-linked polymers up to low–molecular weight products. Of course, in these cases intermediate directions of reactions are possible.

Hydrolytic Break-Up of Polymer Chain

Hydrolysis of groups in the polymer chain

In most of the cases, for hydrolytic degradation to occur, the main chain of polymers must contain hydrolytically active functional groups. The most commonly used hydrolysable groups are ester, carbonate, amide, anhydride, urethane, urea, and semicarbazide groups. Those groups are analogous of groups of many natural low molecular weight substances and polymers such as lipids, peptides, and proteins. Hydrolysis of these groups occurs in accordance with well-known schemes (Fig. 2).

It should be noted that the rate of chemical hydrolysis of these groups depends largely on the structure of the radicals to which they are linked. For example, the rate of hydrolysis of esters is reduced in an order (Fig. 3). Therefore, for example, objects from polyethylene terephthalate in which the ester group is formed by an aliphatic alcohol (ethylene glycol) and an aromatic acid (terephthalic acid) biodegrade slowly in the body, and this polymer is widely used in implants intended for long-term function in the body, such as vascular prostheses and structure components of the artificial heart valves (Fig. 4).

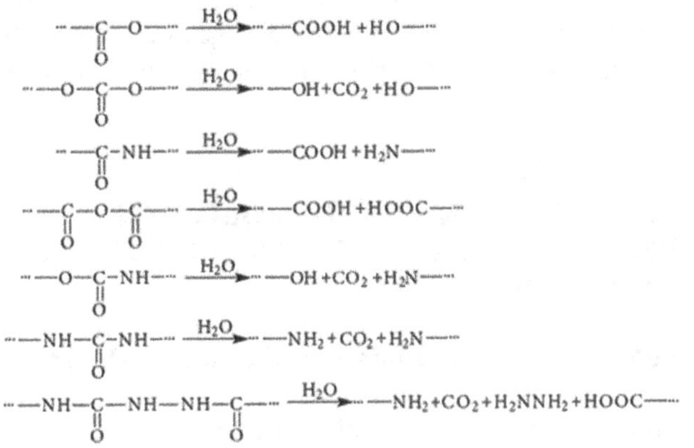

Fig. 2 Schemes of the most commonly used hydrolysable groups.

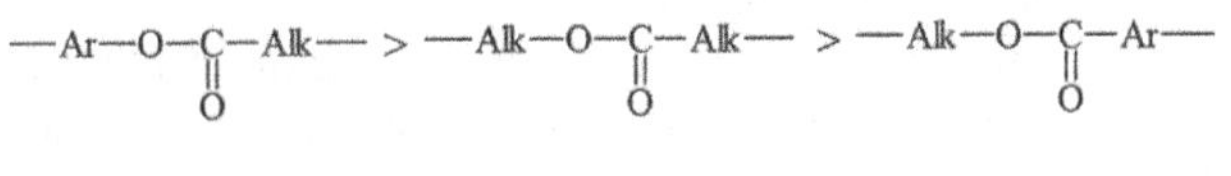

Fig. 3 Order of the rate of ester hydrolysis.

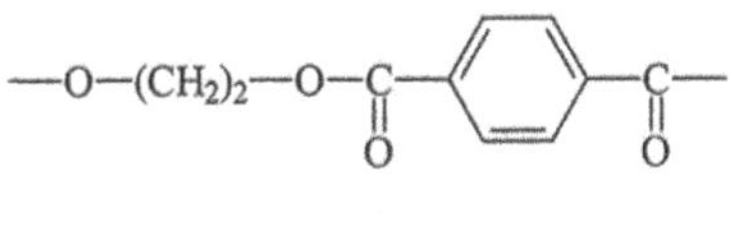

Fig. 4 Structure of a typical polyester.

-O-CH$_2$-CO-   -O-CH(CH$_3$)-CO-   -O-CH(CH$_3$)-CH$_2$-CO-

Fig. 5 Polymer chains of hydroxycarboxylic acid polymers broken down by chemical and enzymatic hydrolysis.

Packaging containers made from the same polymer present serious environmental problems (e.g., widely used bottles for liquids disintegrate in nature very slowly).

On the other hand, the hydroxycarboxylic acid polymers, such as polyglycolides, polylactides, poly-3-hydroxybutirate, and an ester group which is located between aliphatic moieties of the polymer chain, are easy to break down via the chemical and enzymatic hydrolysis (Fig. 5).

Hydrolysis of groups formed in polymer chain after preliminary chemical reactions

In some cases, carbon-chain polymers are biodegradable under the influence of the biological environment. Thus, for certain polymers, which do not contain a hydrolytic bond in the main chain, their biodestruction can occur after the pre-existing transformation of other groups in the polymer, whereby the fragments susceptible to hydrolysis appear in the chain.

For example, objects made from polyvinyl alcohol become degradable in the living tissues after implantation or under the influence of microorganisms.[3] In various cases, it may be polyvinyl alcohol (PVA-oxidase) and β-diketone hydrolase, PVA-dehydrogenase from *Alcaligenes faecalis* KK314, PQQ-dependent PVA-dehydrogenase in symbiotic bacterial culture. A biodestruction of a polyvinyl alcohol chain in symbiotic cultures of *Pseudomonas* sp. VM15C, synthesizing PVA-oxidase and oxidized PVA hydrolase, and *Pseudomonas putida* VM15A, synthesizing methanol dehydrogenase, and in symbiotic cultures of P.SP.M1 and *P. putida* M2, separated from sludge of biological protection stations after the influence of *Phanerochaete chrysosporium*, has been reported. In these cases the decomposition of the polymer chain took place via the formation of diketone structures.

Carbon-chain polymers such as poly-(α-cyanoacrylic acid) are also biodegradable in the body. This process is very important for the application of a variety of drugs, embolizing and bone-filling systems, and adhesives based on polycyanoacrylates.

Susceptibility of polycyanoacrylates to biodegradation is determined by the presence of two strong electron withdrawing groups—nitrile and carbonyl in a cyanoacrylic acid chain, which causes a strong polarization of the electron pair of the C–C bond.[4–6] This process can take place in accordance with the scheme in Fig. 6, where one of the resulting products is formaldehyde, which is not deleterious to the organism.

The process of degradation of a poly-α-cyanoacrylate chain starts at pH 7.0 and is markedly accelerated by increasing the pH to 8.0. A decrease of the length of the alcohol radicals leads to an increase in the rate of hydrolysis. Notably, a polymer of methyl ester undergoes the fastest biodegradation. Polymers which contain long-chain alcohol radicals decompose slowly, accumulating a minimum of formaldehyde in tissues and organs, which allows their use in medical practice.

The objects based on polymers of esters of α-cyanoacrylate acid (e.g., adhesives) degrade in the body at a significant rate of 60%–80% in the first months. The molecular weight of poly-α-cyanoacrylates has a significant effect on their hydrolysis, and the initial rate of hydrolysis is higher for the lower molecular weight polymers. The degradation rate of the object in the body (the observed reduction of its mass) is dependent on the pH and the molecular weight of the polymer length of the alcohol radical, and this reaction can be catalyzed by transition of the metal salts. Furthermore, the biodestruction of poly-α-cyanoacrylates can be accompanied by the hydrolysis of the ester side groups.

Formation of Water-Soluble Products after Reactions of Polymer, Which Has No Influence on the Length of the Polymer Chain

Formation of water-soluble polymer by reactions forming side lyophilic groups

Improved solubility of the polymer and the inclusion of the object, of which it is made in the biodegradation process are observed in a series of transformations of functional groups of the polymer, especially their ionization.

A typical example of such polymers is the programmed dissolving of polymers containing ionic groups and intended to create a coating on dosage forms such as tablets. In particular, such polymers include acrylic copolymers of Eudragit® (Evonik Ind, FRG, Germany) and contain the acid (carboxyl) or base (tertiary amine) groups, which after ionization assist in dissolution of these polymers in the basic environment of the intestine or in an acidic medium such as the stomach, respectively.

Formation of side hydrophilic groups in the polymer can occur as a result of hydrolysis of the functional side groups introduced into the polymer during synthesis, for example, derivatives of esters of vinyl alcohol or acrylic acid, in particular, polymers of vinyl acetate, methyl, ethyl, and butylmethacrylates. In these cases the side hydroxyl or carboxyl groups form in polymers.

Disintegration of polymer–polymer complexes

Polymer–polymer complexes (interpolymer complexes) are systems formed of two or more polymers by a strong cooperative (multipoint) intermolecular (e.g., involving hydrogen binding) or ionic interaction. Such systems are used for the creation of some implants and membrane systems. Their destruction with dissolving formed components may occur as a result of interaction with a competitive complexing agent.

Polymer–polymer complexes can be destroyed by the introduction of low molecular weight agents competitively interacting with the components of the polymer–polymer complex.

FEATURES OF INTERACTION BETWEEN BIODEGRADABLE OBJECT AND BIOLOGICAL ENVIRONMENT

Features of Biodegradation of Different Objects

Implants and the role of inflammatory reactions

The introduction of an implant is a surgical procedure, which injures tissues and leads to the inflammation process. Inflammation begins after surgical trauma to the blood vessels and tissues, and the formation of active medium in the implant zone, comprising a mixture of biological fluids (blood, lymph, interstitial, and intracellular fluid) and fragments of the damaged cells.

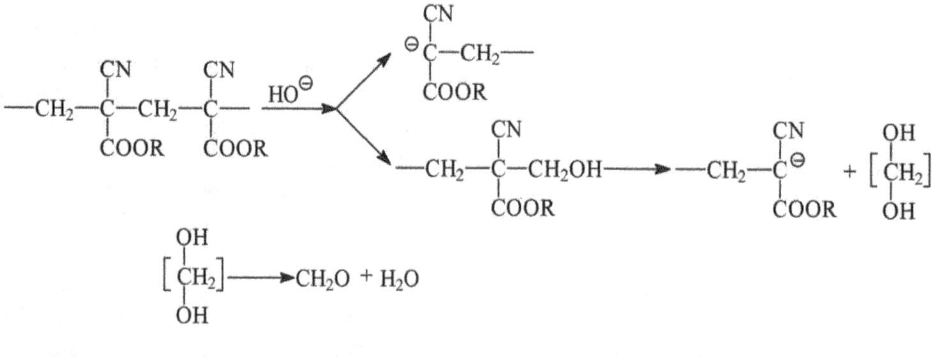

Fig. 6 Degradation of polycyanoacrylates.

All these processes lead to the changes in composition of the implant environment, decreased pH, and increased activity of enzymes, whose activity is different in relation to different polymers. Furthermore, as a result of the protective action of the organism, different cells move into the implant zone.

On the one hand, young cells of new growing tissue appear in the implant zone, which in case of a low ability of the implant to biodegradation form a protective tissue capsule around it. On the other hand, these are protective cells try to metabolize and destroy tissue debris and biopolymers, as well as the implanted foreign object. Among these protective cells monocytes, macrophages, and foreign body giant cells (see section "Intracellular Biodegradation") play an important role. Thus, on the stage of inflammation the organism is protected from the inserted foreign object in two ways—biodegradation and the formation of a tissue capsule.

Prevalence of any of these processes and their specific characteristics are determined by a number of factors, primarily, the chemical and physical structure of the material, the implant form, and its surface characteristics. The place of implantation in the body also plays a significant role because the composition of the active medium, in particular, the composition and amount of enzymes and cells participating in the interaction with the implant, is different for different tissues.

Biodegradation in the gastrointestinal tract

pH, acidic, basic, and enzymatic hydrolysis play an important role in the processes of destruction of object material in the stomach (acidic medium) and in the intestinal system (basic medium).

Dissolution of decomposition products after chemical and biochemical processes flows in the gastrointestinal tract under the influence of acid (stomach) and basic (intestinal system) media and enzymes contained therein. Enzymes, mainly hydrolases involved in digestion— peptidases (primarily pepsin), esterases, lipases, and nucleases—promote disintegration of proteins and other natural polymers belonging to food *in vivo*. Also they can participate in the hydrolysis of synthetic polymers containing functional groups characteristic for natural objects—peptide, ester, and others.

Biodegradation in the environment

The biodegradation of various objects in the environment is of great importance, especially in terms of ecology, since many objects are stable for a long time. For example, paper towels are destroyed in seawater in 2–4 weeks, whereas very popular products such as plastic bags (polyethylene) are destroyed over 10–20 years, the plastic bottles (polyethylenetherephtalate) take over 100 years (data obtained from Mote Marine Laboratory, Sarasota, Florida, USA). This confirms the need to use degradable materials for manufacturing these products (see Section "Practical Approaches to the Design of Materials for Production of Biodegradable Products").

In the case of destruction of the object in the environment, environmental conditions are essential: temperature, pH, and presence of microorganisms and their interaction with the surface of the object, which is to be destroyed. It is possible to distinguish several stages of microbial attack—the adhesive bonding of microorganisms on the surface, their colonization, fouling, surface bioerosion, and destruction (biodegradation) of the object.

In natural conditions in the microbial attack on a biodegradable object, the association of microorganisms is observed—anaerobes and aerobes mutually influence each other in producing the enzymes involved in the destruction of the material. The growth and activity of these microorganisms are influenced by environmental conditions and chemical composition and structure of the material from which the object is made.[7,8]

In the microorganism attack, bacteria, such as *Cytophaga*, *Bacillus*, *Streptomyces*, *Micobacterium*, *Pseudomonas*, and mycelial fungi, and yeasts, such as *Aspergillus*, *Alternaria*, *Penicillium*, *Trichoderma*, *Phanerochaete chrysosporium*, and others, play an important role.

The effect of biodegradability in the environment can exert its artificial acceleration, for example, by acidification of silage and composting waste polymer when placed in the ground.

Extracellular Biodegradation

Low molecular weight substances and enzymes present in the system may have different effects on the course of all the above processes, and the resulting degradation products may enter into further conversion.

It should be noted that a number of chemical and physicochemical factors determined by the biological characteristics of the environment may influence the course of extracellular biodegradation of the object. In particular this relates to biodegradation in a living organism.

In these processes, chemical catalytic hydrolysis occurs with the participation of substances present in the system in accordance with the various mechanisms (basic, acidic, nucleophilic, micellar catalysis, catalysis with the metal ions, and the hydrolysis in the presence of phase transfer catalysts).

For example, the presence of different salts (Table 2) and the pH (Table 3) of biological fluids affect the course of various reactions in the bulk and on the surface of the biodegradable object, and therefore, have a significant effect on the rate of hydrolysis of the polymers.

Auto-oxidation plays a definitive role in the overall process of polymer biodegradation. All organic polymers are prone to the processes of autoxidation–autocatalytic reaction with oxygen. Auto-oxidation is a complex reaction, which is known to proceed via a radical chain mechanism.

Table 2 Average concentrations of various ions in the extracellular fluid

Ion	Concentration, mM	Ion	Concentration, mM
Na^+	140	HCO_3^-	27
K^+	4	PO_3^{2-}	2
Ca^{2+}	2.5	SO_4^{2-}	1.1
Mg^{2+}	1.5	Cl^-	100

Table 3 The pH of some biological fluids

Liquid	pH	Liquid	pH
Blood	7.32–7.43	Saliva	6.8–7.4
Interstitial fluid	7.4	Intracellular fluid:	
		- in muscle	6.1
		- in liver	6.9
Gastric fluid	1.2–3.0	Zone of inflammation	6.8–7.0
Acidity in the intestine	7.3–8.7	Pancreatic fluid	7.8–8.0

In these processes, the products that are formed as a result of the initiating reactions can themselves spontaneously react with the polymer molecules. Metal ions present in the system (catalyst residual, impurities, and others) can play an essential role in the processes of oxidation and auto-oxidation.

Polymers contain a significant number of groups other than those in the backbone, which creates conditions for the occurrence of such reactions. Examples of such groups are hydrocarbon radicals in the branching points of polymer chain.

Finally, the mechanodestruction can play a significant role in the biodegradation of implants that undergo varying dynamic loads (heart valves, endoprostheses of ligaments, tendons, and others). Mechanochemical process can be schematically represented as a combination of the following steps:

- mechanoinitiation—formation of reaction chain;
- growth of the reaction chain—development of chain process in a variety of directions depending on the conditions (temperature, composition of fluids, structure of macroradicals and macrochains);
- termination of the chain reaction, formation of stable endproducts of mechanochemical transformations of the biomaterial.

Overstrain and technological defects (cracks, air bubbles) in the material can occur during the manufacturing process, which, in turn, can initiate and accelerate the hydrolytic and oxidative reactions of biodegradation upon interaction of implant with the living organism.

Initiation of implant biodegradation can occur due to internal and external mechanical stresses. In this case the primary act of destruction is the overstrain of chemical

bonds under the influence of the mechanical forces. In crystalline polymeric materials they occur in amorphous regions and give rise to initial cracks (submicrocracks) with sizes ranging from 10 to 106 Å, followed by the formation of larger microcracks. The permanent static or dynamic load can contribute to the fragmentation of the material with subsequent initiation of intracellular biodegradation. Enzymatic hydrolysis plays an increasingly prominent role in the case of implants made from natural polymers (biopolymers) or synthetic polymers containing biopolymer fragments in the polymer chain. Examples of such polymers are polyurethanes, containing saccharide and oligosaccharide fragments.[9] Enzymes also react with synthetic polymers containing the corresponding groups and molecule fragments surrounding these groups.

Conformational and electrostatic complementarity between the substrate and the enzyme molecules and the unique structure of the active center of the enzyme define a high specificity of enzymes. For example, enzymes with esterase activity are effective for the degradation of polymers containing ester groups. Peptide linkage (–CO–NH–) is the site of action for pepsin, and an ester linkage is the site of action of lipase, which catalyzes the hydrolysis of fats into glycerol and fatty acids. A similar group possesses specificity for trypsin, chymotrypsin, peptidases, and others.

Examples of direction of action of different enzymes (e.g., hydrolases and oxidoreductases, which play an important role in biodegradation) into functional groups of polymers are shown in Table 4. In particular, oxidizing enzymes play an important role, together with the metal ions in the process of autoxidation.

Although primarily a rate of chemical and enzymatic hydrolysis depends on the chemical structure of the hydrolysable group and neighboring groups, other characteristics of the polymer structure also affect the process of implant biodegradation. For example, implants prepared from high

Table 4 Types of reactions, which are catalyzed by hydrolases and oxireductases

Types of enzyme	Types of chemical bonds cleaved by enzyme
Hydrolases	- Ester bond
	- Peptide bond
	- Glycosidic bond
	- C–N-bonds different from peptide bonds
	- Anhydride
Oxidoreductases	–CH–OH
	–C=O
	–CH=CH–
	–CH=NH
	–CH–NH–

molecular weight polymers are more resistant to biodegradation than the objects prepared from the low molecular weight polymers.

The steric hindrances play a very important role in the interaction of the active center of high molecular weight enzymes with the hydrolysable groups of implant material. Therefore, extracellular biodegradation via enzymatic hydrolysis predominantly proceeds in the surface layers of the implant. Thus, the rate of diffusion of the enzyme in the swollen material (hydrogel) is largely determined by steric factors, including cross-linking density of swellable material, hydrophilic macromolecular chain fragments, and the presence of disordered structures.

Thermolability or sensitivity to temperature increase is one of the characteristic properties of enzymes, sharply distinguishing them from inorganic catalysts. For the catalytic action of enzymes, the most optimal temperature is 37°C–40°C for warm-blooded animals. For most enzymes, there is a certain pH value, at which their activity is maximal (Table 5).

Optimal reaction conditions for enzymes are important for *in vitro* experiments mimicking enzymatic hydrolysis of biomaterials. In addition, degradation of implants depends on the implantation site, type of surrounding tissues, and chemical structure of the material.

Enzymes participating in the processes of hydrolytic and oxidative degradation of polymers on the extracellular biodegradation step can be either present in extracellular fluids or excreted from the cells adhered on the surface of the object. In the case of implants, such cells can be, for example, activated macrophages, neutrophils, and others.

Effect of bacterial enzymes on biodegradation. In some cases, degradation of implant is brought about by the action of microorganisms present in the implantation site. Many types of synthetic polymers containing different types of groups—ester, amide, urea, and urethane—undergo degradation under the influence of a large variety of microorganisms. These microorganisms are able to utilize the polymer degradation products, incorporating them in various metabolic cycles.

Because most microorganisms are not capable of endocytosis, microbial biodegradation of implants is associated with extracellular enzymes excreted by microbes into the environment. Typically, these are depolymerases that have low specificity.

In particular, among secreted enzymes are three most common types of proteases—serine protease, metallo, and acid proteases—having optimum pH in an alkaline (pH 8.0–11.0), neutral (pH 6.0–7.8), or acidic (pH 2.5–3.5) medium, respectively. After extracellular degradation of polymeric substrate, the low molecular weight products can be assimilated by microorganisms.

Bacteria infecting the injured surface can significantly alter the degradation rate of the implant. In clinical practice, different rates of degradation were observed for suture materials such as polyglycolic acid and catgut in infected and noninfected wounds.

It was shown that bacterial cells contain enzyme systems that require oxygen, and which are capable of oxidative destruction of the polymer. Thus, the presence of bacteria in the area of the implant can accelerate the biodegradation process via enzymatic and oxidative degradation.

Under appropriate conditions, primarily determined by the structure of the polymer and active environment, all these processes at the stage of noncellular biodegradation lead to surface bioerosion of the object. On the late stages of bioerosion, the porous structures with the dimensions comparable to the size of cells in the surrounding biological environment are formed in the surface layer. This determines the active involvement of the cells in the process of destroying the object and the initial stage of intracellular biodegradation.

Fragmentation of the implant surface is favorable for the transition of formed soluble fragments, colloidal particles, or macromolecules into a surrounding fluid, where they are further subjected to more intensive chemical or enzymatic hydrolysis by dissolved enzymes.[10]

Intracellular Biodegradation

Erosion primarily occurs in the most hydrophilic and disordered areas of the surface. Crystalline defects and supramolecular structures play an important role in biodegradation. Erosion develops over time. Numerous pores on the surface of the implant form a foam layer consisting of a chaotic volume stitches, cavities, and so on.

The beginning of the stage of cell biodegradation is possible when the size of protruding fragments reaches approximately 20–30 microns. As noted above, macrophages and foreign body giant cells that have high metabolizing capacity play a crucial role in the process of cellular degradation.

Electron microscopy observations have shown that macrophages are capable of diffusing and adhering to the surface of the implant. The action of different types of macrophages and foreign body giant cells involves the process of endocytosis. As is known, there are three types of

Table 5 Optimal pH values for certain enzymes

Enzyme	pH	Enzyme	pH
Pepsin	1.5–2.5	Trypsin	7.5–8.5
Catepsin B	4.5–5.0	Arginase	9.5–10.0
Amylase of saliva	6.8–7.0	Papain	3.0–7.5
Catalase	7.3–8.7	Esterase	6.8–7.0
Lipase pancreatic	7.0–8.5	Collagenase	6.8–8.0
Chymotrypsin	7.5–8.5		

endocytosis: phagocytosis, pinocytosis, and specific endocytosis. The first and the second may be termed as nonspecific endocytosis.

In the case of high fragmentation of the implant surface, the cell membrane captures and envelops protruding parts of surface, which thus enters into the cell (endocytosis).[11]

At first, a piece of material is included into a vesicle—a spheroid organelle formed by the plasma membrane fragments—which may be further fused with other intracellular structures.

After the polymer particles enter into the cell, active metabolism of these foreign objects begins with the participation of the enzymes that are inside the special cellular structures—mitochondria and lysosomes. After biodegradation of the absorbed particles with the mechanism of biodestruction with participation of enzymes which are inside the macrophage, decomposition products are derived from the cells (the process of exocytosis) or absorbed by it.

For the different polymers and specific location of the implant defining the composition of the environment, extracellular and intracellular steps may occur for different periods of time.

Summarizing the presently known data, it can be noted that the rate of biodegradation of the objects made from polymer materials depends on many factors, including polymer water solubility and hydrophilicity, chemical and physical structure, molecular weight, biodegradation mechanism, the localization of the implant, and its size, shape, and purpose for introduction as well as the patient's condition.

In a typical case, the sequence of the processes involved is as follows:

- The concentration of macrophages around the implant.
- The appearance of phagocytes and fibroblasts.
- The appearance of giant cells and active macrophages.
- Activation of mitochondria of macrophages and giant cells, contacting with polymer fragments.
- The enveloping of polymer fragments by macrophages and giant cells.
- The beginning of participation in the hydrolysis of polymer functional groups of systems of cells.
- Full biodegradation of implant.
- Active cellular metabolism of the reaction products.

PRODUCTS OF BIODEGRADATION

A set of different products is formed as a result of the extracellular and intracellular processes of biodegradation of polymer implants via the mechanisms of dissolution and biodestruction. Among them are the low and high molecular weight products, differing in chemical structure, molecular weight, and solubility in water, which then diffuse into the blood and lymphatic systems.

Products of biodegradation can be directly cleared from the body through the kidneys, metabolized or accumulated in the body with possible pathological consequences, as well as undergo further chemical transformations. As an example, the breakdown products of some heterochain polymers used in the manufacture of implants are shown in Table 6.

The products of the biodegradation of polymer materials may be combined in several groups:

1. Products that can be directly involved in metabolic processes and are practically nontoxic (such as glycolic acid and lactic acid).
2. Products that can be involved in additional chemical transformations. For example, a number of such transformations are known for ethylene glycol, which is formed during biodegradation of certain types of polyetherurethanes and polyethylene terephthalate. In particular, the toxicity of ethylene glycol is connected to the formation of formic acid and oxalic acid, when the latter leads to deposition of poorly soluble oxalate in the kidney.
3. Insoluble hydrolysis products are cleared from the body unchanged. Such products can be removed as conjugates with blood proteins, primarily with albumin, which are formed via ionic, hydrophobic, or donor–acceptor interactions. 6-Amino-n-dodecanoic acid is an example of such substances.
4. Water-soluble hydrolysis products, which do not enter into the metabolism of the organism, and are excreted unchanged.
5. The products that do not enter into the process of metabolism and are deposited in the tissues.

In the process of gradual biodegradation of polymeric objects, the degradation products must not have a negative effect on the surrounding tissue and the body in general (acute toxic, carcinogenic, mutagenic, allergenic effects,

Table 6 Degradation products of some heterochain polymers, which are used for the manufacture of implants

Polymer	Degradation products	Solubility in water
Polyglycolide	Glycolic acid	Good
$-O-CH_2-CO-$	$HO-CH_2-COOH$	
Copolymer of glycolic and lactic acids	Gycolic and lactic acids	Good
	$HO-CH_2-COOH$	
$-O-CH_2-CO-\ldots-O-$ $CH(CH_3)-CO-$	$HO-CH(CH_3)-COOH$	
Polycaproamide	6-Amino-n-hexanoic acid	Good
$-NH-(CH_2)_5-CO-$	$H_2N-(CH_2)_5-COOH$	
Polydodecaamide	6-Amino-n-dodecanoic acid	Bad
$-NH-(CH_2)_{11}-CO-$	$H2N-(CH_2)_{11}-COOH$	
Polyethylene terephthalate)	Terephthalic acid	Bad
$-O-(CH_2)_2-CO-C_4H_6-$ $CO-$	$HOOC-C_6H_4-COOH$	Good
	Ethylene glycol	
	$HO-(CH_2)_2-OH$	

and so on) or, in any case, this effect should not exceed the permissible limits.

PRACTICAL APPROACHES TO THE DESIGN OF MATERIALS FOR PRODUCTION OF BIODEGRADABLE PRODUCTS

Rational design of biodegradable materials for common applications as biodegradable objects has been of great attention in recent years. Such materials are important not only in medicine, but also for making environmentally degradable packaging products, disposable items, special types of clothing, and so on.

Unlike medical applications where it is possible to use quite low-tonnage polymers (polyanhydrides, polyorthoesters, polydioxanone, and the like), general-purpose materials must be produced in a large-scale technology.

Several groups of such materials can be identified:

1. Natural polymers traditionally continue to be used, primarily cellulose derivatives, including paper products, and cellophane. Based on the data for 2008, this type of packaging has occupied approximately 36% of the market.
2. Composites containing starch as a biodegradable additive are promising as film materials having a high rate of biodegradation. Above all, this applies to compositions based on polyethylene, to which up to 20% starch can be added to maintain a certain level of strength. Films based on these composites can be pretty quickly fragmented in the soil. Also of interest are materials based on starch: Vegemat® (Vegeplast S.A.S., Bazet, France), Solanil® (Rodenburg Biopolymers, Oosterhout, Netherland), EarthShell® (EarthShell Corp., Lutherville Timonium, MD, USA), etc.
3. An important group of biodegradable polymers consists of polymers of hydroxycarboxylic acids (see section "Polyesters Based on Polyhydroxycarboxylic Acids"); large-scale production of these materials is organized in a number of countries.
4. Systems containing additives to promote the interaction of the material with oxygen and UV-light have recently attracted the attention as substances that impart the ability to decay for many known large-tonnage polymers. In particular, the initial decay of macromolecules under the action of atmospheric oxygen and UV-light provokes acceleration of intracellular biodegradation.

These additives (prooxidants) represent derivatives of transition metals, such as cobalt, iron, manganese, and nickel, and may be used to accelerate the biodegradation of carbon- and hetero-chain polymers. In particular, commercially available materials PDQ, PDQ_H, PBA, UV_H (Willow Ridge), and AddiFlex (Vastra Frolunda, Sweden) are known as additives of this type.

EXAMPLES OF POLYMERS USED TO PRODUCE THE BIODEGRADABLE PRODUCTS

Data on the individual types of polymers that can be used to manufacture biodegradable objects are reported in the literature.[12–17]

Polyesters Based on Polyhydroxycarboxylic Acids

Polymers based on polyhydroxycarboxylic acids are used for making biodegradable products, which decompose by the mechanism of biodestruction, such as sutures, fastening devices for connecting bone fragments, drug delivery systems, and, to a lesser degree, for packaging materials. Polymers of hydroxycarboxylic acids may be prepared chemically and microbiologically. Their decomposition occurs due to degradation of the main chain by hydrolysis of the ester group.

Among this group of synthetic polymers to be noted are the polymers and copolymers of glycolic and lactic acids, as well as polymers and copolymers of ε-caprolactone, and dioxanone, and those that are synthesized by ionic polymerization of cyclic diesters—lactide and glycolide, and cyclic esters of low molecular weight ε-caprolactone and dioxanone, respectively (Fig. 7).

Materials based on polylactic acid are produced by a number of companies, for example, Biophan (Trespaphan, England), Ecoloju (Mitsubishi Plastics, Japan), and EcoPla (Cargill Dow Polymers, The Netherlands). Polycaprolactone is produced under the brand name Celgreen (Daicel Chemical, USA). This polymer is a component of suture material (Monocryl, Ethicon Endo-Surgery, Inc., USA), endodontic dental materials (e.g., Resilon), and materials for tissue engineering. One such biodegradable product made from the polylactic acid is NatureWorks™ PLA, developed by Cargill and Dow Chemical, Cargill Dow LLC, and manufactured by BIOTA Brands of America, Inc.

Another group of polyesters of hydroxycarboxylic acids is polyhydroxyalkanoates produced by microbiological synthesis by culturing various microorganisms. Often these are a bacterium of the genus *Ralstonia eutropha*, *Azotobacter vinelandii*, *Pseudomonas oleovorans*, several strains of methylotrophs, or transgenic strains of *Escherichia coli*, *Alcaligenes eutrophus*, and *Klebsiella aerogenes*.

Homopolymers of 3-hydroxybutyric acid (poly-3-hydroxybutyrate), 3-hydroxyvaleric acid, 3-hydroxyhexanoic acid, and so on and their copolymers, the properties of which vary within wide limits, may be prepared by microbiological synthesis (Fig. 8).

Among the polymers of this group, poly(3-hydroxybutyrate) and its copolymers with poly(3-hydroxyvalerate) are frequently used (Fig. 9). Polyhydroxybutyrate is produced, in particular, under the brand names BioMer (Biomer Technology Ltd., Cheshire, England) and Biopol® (Metabolix, Lowell, MA, USA).

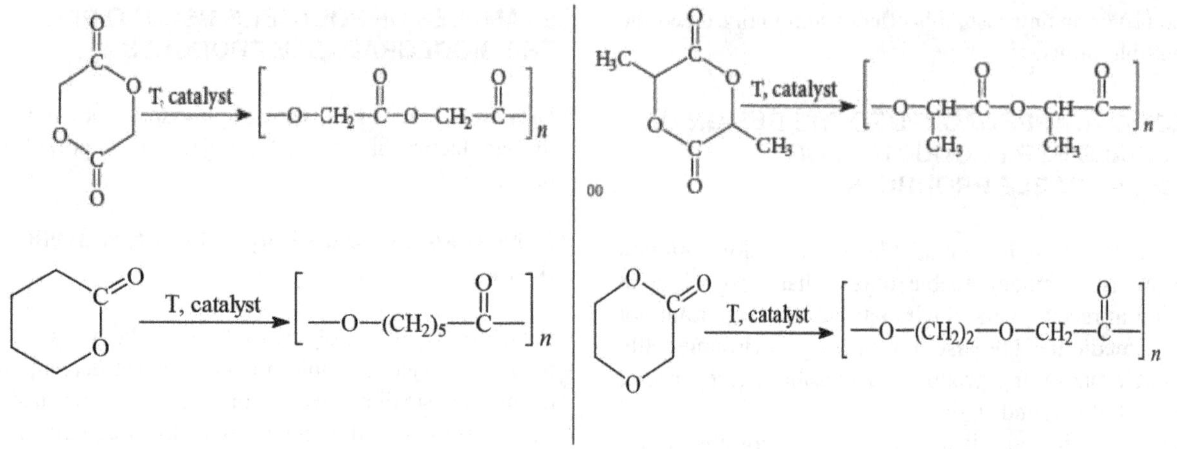

Fig. 7 Synthesis of synthetic polymers by ionic polymerization of cyclic diesters and cyclic esters of low–molecular weight ε-caprolactone and dioxanone.

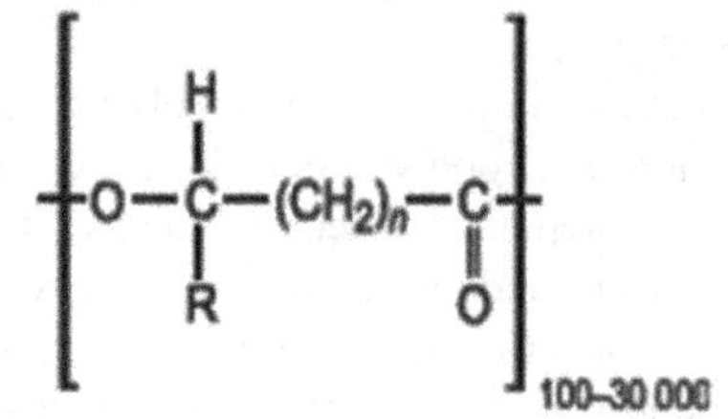

n = 1   R = hydrogen – poly(3-hydroxypropionate),
          R = methyl   – poly(3-hydroxybutyrate),
          R = ethyl   – poly(3-hydroxyvalerate),
          R = propyl   – poly(3-hydroxyhexanoate),
          R = pentyl   – poly(3-hydroxyoctanoate),
          R = nonyl   – poly(3-hydroxydodecanoate),
n = 2   R = hydrogen   – poly(4-hydroxybutyrate),
n = 3   R = hydrogen   – poly(5-hydroxyvalerate).

Fig. 8 Preparation of homopolymers of 3-hydroxybutyric acid (poly-3-hydroxybutyrate), 3-hydroxyvaleric acid, 3-hydroxyhexanoic acid, and so on and their copolymers.

Poly-Orthoesters

Polymeric orthoesters, that is, esters of nonexistent orthocarbonic acid, were synthesized as materials capable of hydrolysis in acidic medium, but they are stable in alkaline medium. Polyorthoesters may be prepared, for example, by transesterification of orthoesters with low molecular weight glycols. In particular, the polymers of the structure given in Fig. 10 were synthesized. The mechanism of hydrolysis of these polymers is associated with the hydrolysis of each ortoester to one carboxyl group and the three alcohol groups.

Polyorthoesters are mainly used in drug delivery systems, but can also be used to create biodegradable implants. In particular, it is possible to use them for the manufacture of fasteners for connecting bone fragments.

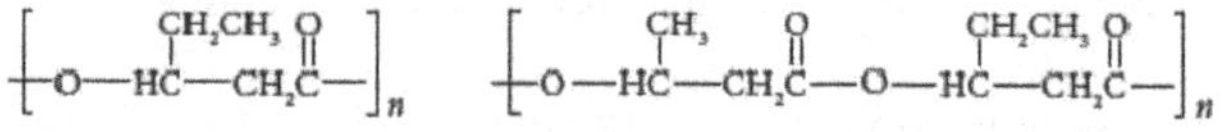

Fig. 9 Poly(3-hydroxybutyrate) and its copolymers with poly(3-hydroxyvalerate).

Polyamides

Polyamides, that is, polymers contained in the backbone chain of amide groups, may degrade by hydrolysis into fragments containing terminal amine and carboxyl groups (Fig. 11). As in the case of polyesters, the tendency of polyamides to hydrolyse depends on the structure of the hydrocarbon radicals surrounding the amide group.

Lower polyamides—polyhexamethylene adipamide (polyamide-66) and poly-ε-caproamide (Polyamide 6)—are unstable in biological fluids and may have applications in implants that gradually decompose in the body, for example, sutures.

On the other hand, polidodecylamide (polyamide-12), containing long hydrocarbon radicals, is more hydrophobic and less prone to hydrolysis, which allows us to consider it as a material for implants of extended use, such as joint prosthesis components.

At the same time, polyamides of α-amino acids having L-conformation are decomposed by peptidases. In particular, they are decomposed in lysosomes.

Segmented Polyurethanes

Segmented polyurethanes (block-*co*-polyurethanes) are widely used to make implants, including those contacting with blood, and have a rather complicated structure.

As a result of detailed study of the effects of chemical and physical structures of segmented polyurethanes, the geometry of the objects on their degradation, a set of

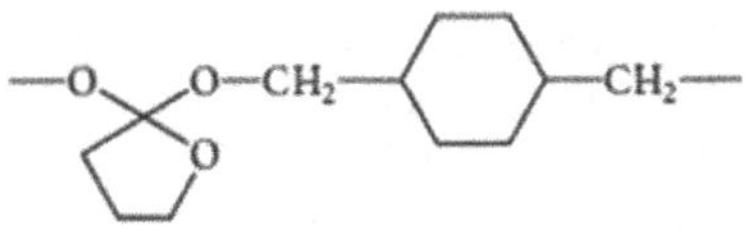

Fig. 10 Structure of synthesized polymeric orthoester.

Fig. 11 Fragments containing terminal amine and carboxyl groups.

individual materials was developed with the decay period in the organism ranging from several weeks to several years. Specifically, the ability to undergo a sufficiently rapid biodegradation is required for the adhesive layers on the basis of polyurethanes. At the same time, the materials used for the replacement devices, for example, a heart prosthesis, must be resistant to degradation and tissue capsule is formed over the device at the early stages after administration into the body.

As is well known, polyurethanes are the polymers containing the urethane (carbamate) group between the hydrocarbon radicals in the main chain, and are usually obtained by reacting monomers with hydroxyl and isocyanate groups.

However, medical materials used under this name have a more complicated structure; their synthesis involves a number of series-parallel reactions with a variety of other reagents. Typically, these polymers comprise two types of fragments—flexible blocks (oligoether, oligoester, and others) and "rigid" fragments containing the aromatic or cycloaliphatic groups and also the groups that can ensure the formation of intermolecular hydrogen bonds—amide and urea groups.

The mechanisms of hydrolysis of polyetherurethanes are similar to those of the corresponding low molecular weight compounds. Thus, polymer degradation can occur in the systems involving both chemical hydrolysis and enzymatic hydrolysis. In the last case, various polymers exhibit differing stability in the presence of different enzymes. Thus, block-*co*-polyurethanes based on polyesters with a molecular weight of 1000 units are degraded by esterase, papain, ficin, chymotrypsin, and trypsin, and at the same time they are relatively stable to collagenase and oxidase.

Collagen

The fibers of native collagen, the main structural proteins of connective tissue, are sufficiently resistant to the action of proteolytic enzymes: pepsin, trypsin, papain, chymotrypsin, and others. However, collagen resistance to enzymatic hydrolysis sharply decreases after its denaturation.

Collagen material implanted into the body is biodegradable, wherein the intensity and rate depends on an implantation site, the shape and structure of the collagen material,

and the degree of intermolecular cross-linking treatment controlled by preservatives and sterilizing. The most rapid biodegradation of collagen samples takes place after intraperitoneal implantation and is slower for subcutaneous implantation. Biodegradation of the implanted collagen material has mainly intracellular character, and macrophages play a major role in this process.

CONCLUSION

The ability for biodegradation is one of the most important properties of the objects used in medicobiological fields. Interest in these products has increased significantly in the last few decades, which is reflected in the significantly enlarged market of materials suitable for the creation of such products. This area is reflected in a number of fundamental marketing research, such as those addressed in.[17–20]

It is believed that in the coming years a key role in the creation of materials suitable for creating biodegradable products will be a study of correlation between the structure and properties of these materials and their ability to decay at a given speed. This will create a biomedical product destination with the specified characteristics, including biodegradability.

REFERENCES

1. Iordanskii, A.L.; Rudakova, T.E.; Zaikov, G.E. *New Concepts in Polymer Science. Interaction of Polymers with Bioactive Aid Corrosive Media*; VSP: Utrecht, 1994.
2. Barrie J.A. Water in polymers. In *Diffusion in Polymers.* Clark, J., Park, G.S., Eds.; Academic Press: London, 1968; 259.
3. Chiellini, E.; Corti, A.; D'Antone, S.; Solaro, R. Biodegradation of poly (vinyl alcohol) based materials. Prog. Polym. Sci. **2003**, *28* (6), 963–1014.
4. Tseng, Y.-C.; Tabata, Y.; Hyon, S.-H.; Ikada. Y. *In vitro* toxicity test of 2-cyanoacrylate polymers by cell culture method. J. Biomed. Mater. Res. **1990**, *24* (10), 1355–1367.
5. Vauther, C.; Couvreur, P.; Dubernet, C. Poly(alkylcyanacrylates). From preparation to real application as drug delivery systems. In *Colloidal Biomolecules, Biomaterials, and Biomedical Application.* Elaissari, A. Ed.; CRC Press: Boca Raton, 2003.
6. Wide C.W.R.; Leonard F. Degradation of poly(methyl-2-cyanoacrylates). J. Biomed. Mater. Res. **1980**, *14* (1), 93–106.
7. Allsopp, D.; Seal, K.J.; Gaylarde, C.C. *Introduction to Biodeterioration*, 2nd Ed.; University Press: Cambridge, 2004.
8. Kuznetsov, A.E., Ed. *Applied Ecobiotechnology*, Vol. 1, 2; BINOM: Moscow, 2010.
9. Phakadze, G.A. *Biodestructable Polymers*; Naukova Dumka: Kiev, 1990.
10. van der Elst, M.; Klein, C.P.A.T.; de Blieck-Hogervorst, J.M.; Patka, P.; Haarman, H.J.T.M. Bone tissue response to biodegradable polymers used for intramedullary fracture

fixation: A long term *in vivo* study in sheep femora. Bioma-terials **1999**, *20* (1), 121–128.

11. Voronov, I.; Santere, J.P.; Hinek, A.; Callham, J.W.; Sandhu, J.; Boynton, E.L. Macrophage phagocytosis of polyethylene particulate *in vitro*. J. Biomed. Mater. Res. **1998**, *39* (1), 40–51.

12. Chiellini, E.; Solaro, R., Eds *Recent Advances in Biodegradable Polymers and Plastics*; Wiley-VCH: Weinheim, 2003.

13. Shalaby, S.W.; Burg, J.L.K., Eds. *Absorbable and Biodegradable Polymers*; CRC Press: Boca Raton, 2004.

14. Bastioli, C., Ed. *Handbook of Biodegradable Polymers*; Rapra Technol. Limited: Novamont SpA, Italy, 2005.

15. Griffin, G.J.L. *Chemistry and Technology of Biodegradable Polymers*; Chapman and Hall: New York, 1994.

16. Halim Hamid, S., Ed. *Handbook of Polymer Degradation*, 2nd Ed.; Dekker: New York, 2000.

17. Platt, D.K. *Biodegradable Polymers (Market Report)*, Smithers Rapra Lim.: Shawbury, UK, 2006.

18. RP_175 (Report) Biodegradable Polymers. Business Communications Co., Inc. www.bccresearch.com.

19. Yezze, I.A. *The Global Market for Bioplastic*; Helmut Kaiser Consultancy: 2008.

20. Bioplastics Market Worldwide 2007–2025, http://www.hkc22.com/bioplastics.html

Biofunctional Polymers

Jennifer L. West
Department of Bioengineering, Rice University, Houston, Texas, U.S.A.

Abstract

The development of bioactive biomaterials is expanding the capabilities and performance of materials used in a variety of biomedical applications. As advances are made in cellular and molecular biology, the functionality that can be built into these hybrid materials should greatly expand, hopefully providing solutions to many currently unmet medical needs, such as synthetic small-diameter vascular grafts.

INTRODUCTION

In many applications utilizing polymeric biomaterials, such as in scaffolds for tissue engineering, investigators can choose from natural materials, such as collagen or fibrin, or synthetic polymers such as poly(lactic-co-glycolic acid) (PLGA). Each choice has its unique advantages and disadvantages. For example, the extracellular matrix (ECM) proteins have evolved over billions of years to have specific cellular interactions and activities, to be remodeled by cellular activity, and even to guide tissue formation processes. However, the isolation, purification, and processing of these materials can be difficult, mechanical properties are often poor, and there are some concerns about the possibility for disease transmission. Synthetic polymers, on the other hand, offer easier handling and processing as well as a greater degree of control of material properties. Unfortunately, when using synthetic materials such as PLGA, cellular interactions with the material are mediated by nonspecifically adsorbed proteins and are generally difficult to control. In order to obtain some of the advantages of natural bioactive compounds, such as ECM proteins, while gaining the ease of processing and safety of synthetic materials, a number of researchers have developed hybrid materials that are primarily composed of standard synthetic polymers but are also grafted to, tethered to, or copolymerized with biologically functional segments such as peptides. Examples discussed in this entry include the development of ECM-mimetic materials for use in tissue engineering applications, as well as the development of pharmacologically active biomaterials.

ECM-MIMETIC MATERIALS

In natural tissue, cells grow within a scaffold consisting of a variety of proteins and polysaccharides—the extracellular matrix (ECM). The ECM does far more than just provide mechanical support for the cells, which is all that many of the tissue engineering scaffolds in current use are capable of doing. Rather, the ECM has direct biological interactions with the entrapped cells, influencing cell growth, migration, morphology, and differentiation. Using natural polymers such as collagen or fibrin gels allows one to reconstitute some of these interactions, but controlling these interactions and other properties of the materials is difficult. Modifying synthetic polymers that have the appropriate mechanical properties and processing conditions with biofunctional moieties that provide some of the ECM interactions may allow one to get the benefits of a synthetic material with many of the advantages normally associated with natural materials.

Cell-Adhesive Materials

In tissues, cell-adhesive interactions occur between cell surface receptors, such as the integrins, and specific ligands on ECM proteins. A number of short peptide sequences derived from adhesive ECM proteins have been identified that are able to bind to cell surface receptors and mediate cell adhesion with affinity and specificity similar to that obtained with intact proteins. Oligopeptides are preferable to intact proteins because they are not subject to denaturation and may be less susceptible to proteolysis. Some of these, along with their biological sources, are shown in Table 1. A number of methods for incorporating peptides into biomaterials have been investigated, including physicochemical adsorption, chemisorption, and covalent attachment. Only covalent attachment will be addressed in this entry. Covalent attachment may be in the form of surface grafting, grafting to polymer chains, or incorporation into polymer backbones. Important considerations in designing these types of bio-hybrid materials include the hydrophobicity of the synthetic substrate, steric hindrance, peptide orientation, and whether interactions need to be two- or three-dimensional. When modifying materials with adhesive peptides, however, one must be aware

Concise Encyclopedia of Biomedical Polymers and Polymeric Biomaterials DOI: 10.1081/E-EBPPC-120013900

Bioactive—BioMEMS

Table 1 Peptide sequences that can mediate cell adhesion.

Peptide sequence	ECM protein
RGD	Fibronectin, vitronectin, laminin, collagen
YIGSR	Laminin
IKVAV	Laminin
REDV	Fibronectin
DGEA	Collagen
VGVAPG	Elastin

that activation of these cell-signaling pathways can alter many other cellular activities, including proliferation, migration, and synthesis of new ECM.[1,2]

Perhaps the most extensively studied cell adhesion peptide is the sequence Arg–Gly–Asp (RGD). It is found in many cell adhesion proteins and binds to integrin receptors on a wide variety of cell types.[3] Modification with RGD peptides may enhance cell adhesion to a biomaterial and activate integrin signaling pathways. A number of studies have been performed with RGD-containing peptides grafted to polymers. In initial studies, RGD peptides were immobilized on a polymer-modified glass substrate, and cell adhesion parameters were correlated with peptide density.[4] A peptide density of 10 fmol/cm^2 was sufficient to support adhesion and spreading of fibroblast cells, clustering of integrin receptors, and organization of actin stress fibers. At 1 fmol/cm^2 the cells were fully spread, but the morphology of the stress fibers was abnormal, and the cells did not form focal contacts.

In addition to grafting RGD peptides to polymeric surfaces, peptides can be incorporated into polymer matrices to allow adhesion throughout the material. For example, a biodegradable copolymer of polylysine and poly(lactic acid) has been used for the attachment of RGD peptides.[5] These types of materials can be processed into highly porous sponges to allow cells to permeate into the matrix and adhere throughout.[6,7] When these copolymers were modified with RGD peptides, significant enhancement of cell adhesion was observed.

When designing peptide-modified materials, it is often desirable to start with intrinsically cell-nonadhesive polymers, effectively providing a blank slate upon which one can build desired biological interactions. Polyethylene glycol (PEG)-based materials, such as photocrosslinked PEG hydrogels, have been particularly interesting for this approach.[2,8,9] Figure 1 shows the interaction of fibroblasts with an unmodified PEG hydrogel and with a RGD-containing PEG hydrogel, demonstrating adhesion and spreading only with inclusion of the adhesion ligand. In addition to providing controlled and specific biological interactions, modification of cell-nonadhesive materials can allow one to generate cell-specific materials, i.e., materials that are adhesive only to certain cell types, if adhesion

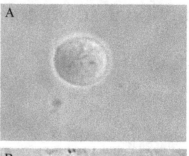
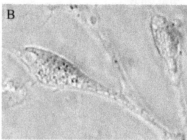

Fig. 1 Fibroblasts were seeded on the surfaces of either PEG hydrogels (**A**) or PEG hydrogels with tethered RGD peptide (**B**). Cells are able to adhere to and spread on these materials only in the presence of an immobilized adhesion ligand. If cell-selective adhesion ligands were used, one could generate a biomaterial that would allow adhesion of only certain cell types.

ligands with appropriate selectivity are utilized. For example, the peptide REDV has been shown to be adhesive to endothelial cells but not to smooth muscle cells, fibroblasts, or platelets,[10] and thus it may be useful in the development of vascular grafts.

Proteolytically Degradable Polymers

The majority of biomaterials currently used in tissue engineering applications are designed to degrade via hydrolysis. With many of these materials, such as poly(lactide-co-glycolide), the degradation rate can be tailored to some degree by altering factors such as the copolymer composition, degree of crystallinity, or initial molecular weight. However, in these types of applications, it can be very difficult to predict a priori what the rate of tissue formation will be and, thus, what the rate of polymer degradation should be. In the ECM, during processes such as wound healing, the rate of scaffold (ECM) degradation is precisely matched to the rate of tissue formation. ECM is degraded by proteolytic enzymes, such as the matrix metalloproteases and plasmin, produced by cells during cell migration and tissue remodeling.[11] Many of the natural polymer scaffolds, such as collagen and fibrin, are degraded by the same cellular mechanisms. Recently, synthetic biomaterials have been designed that are similarly degraded by cellular proteolytic activity.[2,12] These materials are block copolymers of synthetic polymers, such as polyethylene glycol (PEG), and peptides that are substrates for proteolytic enzymes. When exposed to solutions containing the targeted protease, PEG-peptide copolymer hydrogels degraded, whereas they remained stable in the absence of protease or in the presence of a nontargeted protease. Furthermore, cells can migrate through PEG-peptide hydrogels that contain both proteolytically degradable peptide sequences and cell-adhesive sequences.[13] In this case, degradable sequences are required so that the cells can create pathways through the solid structure for migration, and the cell adhesion ligands are required so that cells can develop the traction force required for migration.

PHARMACOLOGICALLY ACTIVE BIOMATERIALS

Pharmacologically active polymers are another class of biofunctional materials, with the goal of being used for prolonged, site-specific pharmacotherapy. Covalent modification of materials by means of pharmacological agents has been investigated for a number of applications. Most commonly, the objective is simply to develop a more biocompatible biomaterial, for instance, to improve blood compatibility. In tissue engineering applications, the goal may be to stimulate or inhibit specific cellular processes, such as proliferation or ECM synthesis. The majority of applications of pharmacologically active biomaterials have been performed by grafting the pharmacological agent to the synthetic polymer substrate, though other material designs are possible. A wide range of pharmacological agents have been employed, including polysaccharides, polypeptide growth factors, enzymes, anti-inflammatory agents, and nitric oxide.

Polysaccharides

Heparin has been the most extensively studied polysaccharide for biomaterials modification. Heparin is a polymer of heterogeneously sulfated L-iduronic acid and D-glucosamine, with molecular weights ranging from 2–50 kDa. Heparin has been commonly used as an anticoagulant and antiplatelet agent due to its binding and potent activation of antithrombin III.[14] Covalent immobilization of heparin onto blood-contacting materials should provide thromboresistance for prolonged periods of time. Numerous synthetic polymers have been covalently heparinized, including polyurethanes,[15,16] polyethylene terephthalate,[17] polyvinyl alcohol,[18,19] poly(dimethylsiloxane),[20] polypyrrole,[21] and polylactide.[22]

Synthesis of covalently heparinized materials often involves the use of a hydrophilic spacer molecule in order to increase the bioavailability of the bound heparin. PEG is often used as a spacer, usually being coupled to the biomaterial and the heparin through di-isocyanate groups. PEG has several advantages for this application. PEG is widely known to be highly resistant to platelet adhesion, so the PEG spacers themselves help to create a more hemocompatible surface. Additionally, when heparin is immobilized

through a PEG spacer, it retains its ability to bind both thrombin and antithrombin III, though to a slightly lesser extent than unmodified heparin.[23] When directly immobilized, however, the heparin is able to bind only to thrombin.

Hyaluronic acid, a linear copolymer of D-glucuronic acid and N-acetylglucosamine, is also generating interest for biomaterials applications. It has been shown to be both antithrombotic and angiogenic.[24–27]

Hirudin

Hirudin is a 65-amino acid polypeptide, originally derived from the salivary glands of the leech, that acts as an anticoagulant through its ability to directly inhibit thrombin. Both recombinant hirudin and synthetic hirudin are now available. Hirudin has been covalently attached to the surfaces of PET and polyurethane vascular grafts, using bovine serum albumin as a spacer, in order to improve their blood compatibility.[28,29] Hirudin immobilized in this manner retains its ability to bind and inhibit thrombin. Hirudin has also been directly coupled to PLGA using glutaraldehyde, demonstrating improvements in several aspects of blood compatibility.[30] Unfortunately, the binding between hirudin and thrombin is irreversible, leading to decreased bioactivity over time.[30]

Plasminogen Activators

During fibrinolysis, circulating plasminogen is converted to the active enzyme plasmin by plasminogen activators, as depicted in Fig. 2. Plasmin is then able to degrade fibrin to lyse the clot structure (Fig. 2). Plasminogen activators include tissue plasminogen activator (tPA), urokinase, streptokinase, and lumbrokinase. Immobilization of plasminogen activators on blood-contacting materials may be a mechanism to impart thromboresistance. Additionally, immobilization of these enzymes may stabilize them, imparting greater resistance to temperature, pH, and proteolytic activity.[31]

Urokinase, a plasminogen activator expressed in many tissue types, has been immobilized to a wide variety of polymeric materials including polyurethanes,[32] poly(tetrafluoroethylene),[33] poly(vinylchloride),[34] and poly(hydroxyethyl methacrylate).[35] Thrombus formation on these materials has been shown to be greatly reduced with

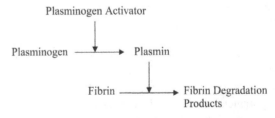

Fig. 2 Plasminogen activators convert circulating plasminogen to the active enzyme, plasmin, which then lyses fibrin clots.

enzyme immobilization. Similarly, lumbrokinase, derived from an earthworm, and streptokinase, derived from streptococcal bacteria, have been immobilized to polymeric materials and shown to improve blood compatibility.[36,37] However, there are significant concerns regarding adverse immunological interactions with materials modified with these foreign proteins.

Growth Factors

Growth factors are polypeptides involved in the regulation of a variety of cellular activities such as growth and differentiation, as well as many metabolic processes. Many different growth factors have been described in the literature, with potential applications ranging from accelerated wound healing to the induction of bone growth into implanted biomaterials. Modification of biomaterials with growth factors may target activity to the desired tissue and may maintain biological function for a prolonged period of time. Growth factors that have been covalently immobilized to polymeric biomaterials include epidermal growth factor (EGF), basic fibroblast growth factor (bFGF), and transforming growth factor-beta (TGF-B).

EGF is a 6 kDa polypeptide that is mitogenic and chemotactic for many of the cell types involved in wound healing. EGF was also the first growth factor that was shown to remain active after covalent immobilization. In that first study, EGF was immobilized to aminated glass via a star PEG spacer molecule and was shown to retain mitogenic activity, whereas physically adsorbed EGF showed no activity.[38] Prior to this work, it was believed that receptor dimerization and internalization were required for growth factor activity, and thus that immobilization would not be effective. However, that does not appear to be true. More recently, EGF has also been immobilized on poly(methyl methacrylate)[39] and polystyrene,[40] with good retention of bioactivity.

bFGF is a 16 kDa polypeptide that is mitogenic for many cell types and that appears to play an important role in angiogenesis. bFGF binds strongly to heparin, so heparinized materials have been widely used for the complexation and presentation of bFGF.[41–43] For example, polyurethane vascular grafts with coimmobilized heparin and bFGF have been shown to support enhanced endothelialization.[42]

Although in many cell types it does not stimulate growth, TGF-B plays a very important role in regulating ECM metabolism. For example, synthesis of collagen and elastin is greatly upregulated in vascular smooth muscle cells upon exposure to TGF-B. In tissue engineered vascular grafts, a potential strategy to improve mechanical properties is to increase ECM deposition. Thus, PEG-based hydrogel scaffolds have been synthesized with TGF-B tethered to the hydrogel network.[44] Tethering TGF-B to the polymeric biomaterial actually enhanced its ability, relative to soluble TGF-B, to stimulate ECM protein production by encapsulated vascular smooth muscle cells.

This effect may be due to the ability of the cells to interact with TGF-B via the appropriate receptors, while not being able to internalize it. In addition, engineered tissues formed with tethered TGF-B in the scaffold formulation were shown to have improved tensile strength.[44]

Anti-Inflammatory Biomaterials

Salicylic acid, the active component of aspirin, is commonly used as an anti-inflammatory agent in a broad spectrum of medical applications. Polymeric materials have been synthesized with salicylic acid incorporated into the backbone of the polymer.[45–47] These materials are poly anhydride esters composed of alkyl chains linked to salicylic acid by ester bonds. Hydrolytic degradation occurs as the ester bonds are cleaved, producing salicylic acid. These materials may be useful in wound closure and wound healing applications, where they could not only release an anti-inflammatory agent but could also serve as a biodegradable scaffold for tissue regeneration. This material could also have use in the treatment of Crohn's disease, a chronic inflammation of the intestine. In this application, the release of salicylic acid following oral ingestion of the polymer would be enhanced in the intestinal tract, due to enhanced polymer degradation at basic pH.

Nitric Oxide–Generating Materials

Nitric oxide (NO) acts on most tissues in the body through its ability to upregulate cGMP.[48] For example, in the vascular system, NO inhibits platelet adhesion and aggregation, increases endothelial cell proliferation, and suppresses smooth muscle cell proliferation. Thus, NO is attractive for use as a therapeutic agent for the prevention of restenosis or for the improvement of vascular graft or stent performance. A number of small-molecule drugs, referred to as NO donors, have been developed that undergo hydrolysis to generate NO in the body.[49] However, systemic NO therapy for applications such as restenosis has generally failed, due to the difficulties in achieving therapeutic dosages at the appropriate site without incurring substantial systemic side effects.[50] Bioactive polymers containing NO donor groups may allow the creation of local delivery vehicles for NO therapy, or even NO-generating medical devices.

Diazeniumdiolates have been attached to pendant groups of a polymer chain to provide NO release for up to five weeks.[51] Diazeniumdiolates attached to crosslinked polyethyleneimine (PEI) inhibited the proliferation of smooth muscle cells *in vitro*. PTFE vascular grafts were also coated with the PEI-NO polymer and implanted in a baboon arteriovenous shunt model. Platelet adhesion to the NO-releasing grafts was reduced almost five-fold relative to control biomaterials. Microspheres can also be formed from the PEI-NO material.[52] Diazeniumdiolate NO donors have also been covalently bound to polyurethane and

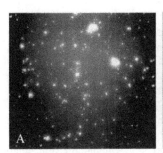

Fig. 3 Pretreatment of mepacrine-labeled whole blood with NO-producing hydrogel materials significantly reduced platelet adhesion to a model thrombogenic surface, collagen-coated glass. (**A**) control hydrogel; (**B**) NO-generating hydrogel.

polyvinyl chloride films.[53] NO release from these materials ranged from 10 to 72 hr, and the NO-modified polymers successfully reduced platelet adhesion *in vitro*.

Hydrogel materials have been formed by photocrosslinking copolymers of PEG with NO donors, such as diazeniumdiolates and S-nitrosothiols, to achieve NO delivery for periods ranging from hours to months.[54] Exposure of blood to NO-generating hydrogels dramatically decreased platelet adhesion to collagen-coated surfaces, as shown in Fig. 3. Furthermore, these materials dramatically reduced smooth muscle cell proliferation. These types of hydrogel materials can be polymerized in situ to form thin coatings on the luminal surface of a blood vessel following angioplasty or stent deployment,[55] which provides highly localized treatment for the prevention of thrombosis and restenosis.

CONCLUSION

The development of bioactive biomaterials is expanding the capabilities and performance of materials used in a variety of biomedical applications. In particular, the development of ECM-mimetic polymers may significantly improve the ability to manipulate cell differentiation and behaviors while growing engineered tissues. As advances are made in cellular and molecular biology, the functionality that can be built into these hybrid materials should greatly expand, hopefully providing solutions to many currently unmet medical needs, such as synthetic small-diameter vascular grafts.

REFERENCES

1. Mann, B.K.; Tsai, A.T.; Scott-Burden, T.; West, J.L. Modification of surfaces with cell adhesion peptides alters extracellular matrix deposition. Biomaterials **1999**, *20* (23–24), 2281–2286.
2. Mann, B.K.; Gobin, A.S.; Tsai, A.T.; Schmedlen, R.H.; West, J.L. Smooth muscle cell growth in photo-polymerized hydrogels with cell adhesive and proteolytically degradable domains: Synthetic ECM analogs for tissue engineering. Biomaterials **2001**, *22* (22), 3045–3051.
3. Humphries, M.J. The molecular basis and specificity of integrin-ligand interactions. J. Cell Sci. **1990**, *97* (Pt. 4), 585–592.
4. Massia, S.P.; Hubell, J.A. An RGD spacing of 440 nm is sufficient for integrin mediated fibroblast spreading and 140 nm for focal contact and stress fiber formation. J. Cell Biol. **1991**, *114* (5), 1089–1100.
5. Barrera, D.A.; Zylestra, E.; Lansbury, P.T.; Langer, R. Synthesis and RGD peptide modification of a new biodegradable polymer—Poly(lactic acid-co-lysine). J. Am. Chem. Soc. **1993**, *115* (23), 11010–11011.
6. Mikos, A.G.; Thorsen, A.J.; Czerwonka, L.A.; Bao, Y.; Langer, R.; Winslow, D.N.; Vacanti, J.P. Preparation and characterization of poly(L-lactide) foams. Polymer **1994**, *35*, 1068–1077.
7. Mooney, D.J.; Park, S.; Kaufmann, P.M.; Sano, K.; McNamara, K.; Vacanti, J.P.; Langer, R. Biodegradable sponges for hepatocyte transplantation. J. Biomed. Mater. Res. **1995**, *29* (8), 959–965.
8. Hern, D.L.; Hubbell, J.A. Incorporation of adhesion peptides into nonadhesive hydrogels useful in tissue resurfacing. J. Biomed. Mater. Res. **1998**, *39* (2), 266–277.
9. Mann, B.K.; West, J.L. Cell adhesion peptides alter smooth muscle cell adhesion, proliferation, and matrix protein synthesis on modified surfaces and in polymer scaffolds. J. Biomed. Mater. Res. **2002**, *60* (1), 86–93.
10. Hubbell, J.A.; Massia, S.P.; Drumheller, P.D. Endothelial cell-selective tissue engineering in the vascular graft via a new receptor. Biotechnology **1991**, *9* (6), 568–572.
11. Rabbani, S.A. Metalloproteases and urokinase in angiogenesis and tumor progression. *In Vivo* **1998**, *12* (1), 135–142.
12. West, J.L.; Hubbell, J.A. Polymeric biomaterials with degradation sites for proteases involved in cell migration. Macromolecules **1999**, *32* (1), 241–244.
13. Gobin, A.S.; West, J.L. Cell migration through defined synthetic ECM analogs. FASEB J. **2002**, *16* (7), 751–753.
14. Barritt, D.W.; Jordan, S.C. Anticoagulant drugs in the treatment of pulmonary embolism: A controlled clinical trial. Lancet **1960**, *1* (7138), 1309–1312.
15. Han, D.K.; Jeong, S.Y.; Kim, Y.H. Evaluation of blood compatibility of PEO grafted and heparin immobilized polyurethanes. J. Biomed. Mater. Res. **1989**, *23* (A2 Suppl.), 211–228.
16. Han, D.K.; Park, K.D.; Ahn, K.D.; Kim, Y.H. Preparation and surface characterization of PEO-grafted and heparin immobilized polyurethanes. J. Biomed. Mater. Res. **1989**, *23* (A1 Suppl.), 87–104.
17. Kim, Y.J.; Kang, I.K.; Muh, M.W.; Yoon, S.C. Surface characterization and *in vitro* blood compatibility of poly(ethylene terephthalate) immobilized with insulin and/or heparin using plasma glow discharge. Biomaterials **2000**, *21* (2), 121–130.
18. Cholakis, C.H.; Sefton, M.V. *In vitro* platelet interactions with a heparin-polyvinyl alcohol hydrogel. J. Biomed. Mater. Res. **1989**, *23* (4), 399–415.
19. Piao, A.Z.; Jacobs, H.A.; Park, K.D.; Kim, S.W. Heparin immobilization by surface amplification. ASAIO J. **1992**, *38* (3), M638–M643.
20. Grainger, D.; Feijen, J.; Kim, S.W. Poly(dimethyl siloxane)–poly(ethylene oxide)–heparin block co-polymers, I: Synthesis and characterization. J. Biomed. Mater. Res. **1988**, *22* (3), 231–242.

21. Garner, B.; Georgevich, B.; Hodgson, A.J.; Liu, L.; Wallace, G.G. Polypyrrole-heparin composites as stimulus-responsive substrates for endothelial cell growth. J. Biomed. Mater. Res. **1999**, *44* (2), 121–129.

22. Seifert, B.; Groth, T.; Hermann, K.; Romaniuk, P. Immobilization of heparin on polylactide for application to degradable biomaterials in contact with blood. J. Biomater. Sci., Polym. Ed. **1995**, *7* (3), 277–287.

23. Byun, Y.; Jacobs, H.A.; Kim, S.W. Heparin surface immobilization through hydrophilic spacers: Thrombin and antithrombin III binding kinetics. J. Biomater. Sci. Polym. Ed. **1994**, *6* (1), 1–13.

24. West, D.C.; Hampson, I.N.; Arnold, F.; Kuman, S. Angiogenesis induced by degradation products of hyaluronic acid. Science **1985**, *228* (4705), 1324–1326.

25. Vercruysse, K.P.; Marecak, D.M.; Marecak, J.F.; Prestwich, G.D. Synthesis and *in vitro* degradation of new polyvalent hydrazide crosslinked hydrogels of hyaluronic acid. Bioconjug. Chem. **1997**, *8* (5), 686–694.

26. Verheye, S.; Markou, C.P.; Salame, M.Y.; Wan, B.; King, S.B.; Robinson, K.A.; Chronos, N.A.; Hanson, S.R. Reduced thrombus formation by hyaluronic acid coating of endovascular devices. Arterioscler. Thromb. Vasc. Biol. **2000**, *20* (4), 1168–1172.

27. Barbucci, R.; Magnani, A.; Rappuoli, R.; Lamponi, S.; Consumi, M. Immobilization of sulfated hyaluronan for improved biocompatibility. J. Inorg. Biochem. **2000**, *79* (1), 119–125.

28. Ito, R.K.; Phaneuf, M.D.; LoGerfo, F.W. Thrombin inhibition by covalently bound hirudin. Blood Coagul. Fibrinolysis **1991**, *2* (1), 77–81.

29. Phaneuf, M.D.; Berceli, S.A.; Bide, M.J.; Quist, W.C.; LoGerfo, F.W. Covalent linkage of recombinant hirudin to poly(ethylene terepthalate) (Dacron): Creation of a novel antithrombin surface. Biomaterials **1997**, *18* (10), 755–765.

30. Seifert, B.; Romaniuk, P.; Groth, T. Covalent immobilization of hirudin improves the haemocompatibility of polylactide-polyglycolide *in vitro*. Biomaterials **1997**, *18* (22), 1495–1502.

31. Torchilin, V.P.; Makismenko, A.V.; Mazaev, A.V. Immobilized enzymes for thrombolytic therapy. Meth. Enzym. **1988**, *137*, 552–566.

32. Kitamoto, Y.; Tomita, M.; Kiyama, S.; Inoue, T.; Yabushita, Y.; Sato, T.; Ryoda, H. Antithrombotic mechanisms of urokinase immobilized polyurethane. Thrombosis **1991**, *65* (1), 73–76.

33. Forster, R.I.; Bernath, F. Analysis of urokinase immobilization on the polytetrafluoroethylene vascular prosthesis. Am. J. Surg. **1988**, *156* (2), 130–132.

34. Sugitachi, A.; Tanaka, M.; Kawahara, T.; Kitamura, N.; Takagi, K. A new type of drain tube. Artif. Organs **1981**, *5*, 69.

35. Liu, L.S.; Ito, Y.; Imanishi, Y. Biological activity of urokinase immobilized to cross linked poly(2-hydroxy-ethyl methacrylate). Biomaterials **1991**, *12* (6), 545–549.

36. Drummond, R.K.; Peppas, N.A. Fibrinolytic behaior of streptokinase-immobilized poly(methacrylic acid-g-ethylene oxide). Biomaterials **1991**, *12* (4), 356–360.

37. Ryu, G.H.; Park, S.; Kim, M.; Han, D.K.; Kim, Y.H.; Min, B. Antithrombogenicity of lumbrokinase-immobilized polyurethane. J. Biomed. Mater. Res. **1994**, *28* (9), 1069–1077.

38. Kuhl, P.R.; Griffith-Cima, L.G. Tethered epidermal growth factor as a paradigm for growth factor-induced stimulation from the solid phase. Nat. Med. **1996**, *2* (9), 1022–1027.

39. Ito, Y.; Li, J.S.; Takahasi, T.; Imanishi, Y.; Okabayashi, Y.; Kido, Y.; Kasuga, M. Enhancement of the mitogenic effect by artificial juxtacrine stimulation using immobilized EGF. J. Biochem. **1997**, *121* (3), 514–520.

40. Ito, Y.; Chen, G.; Imanishi, Y. Micropatterned immobilization of epidermal growth factor to regulate cell function. Bioconjug. Chem. **1998**, *9* (2), 277–282.

41. Bos, G.W.; Scharenborg, N.M.; Poot, A.A.; Engbers, G.H.M.; Beugling, T.; van Aken, W.G.; Feijen, J. Proliferation of endothelial cells on surface-immobilized albumin-heparin conjugate loaded with basic fibroblast growth factor. J. Biomed. Mater. Res. **1999**, *44* (3), 330–340.

42. Doi, K.; Matsuda, T. Enhanced vascularization in a microporous polyurethane graft impregnated with basic fibroblast growth factor and heparin. J. Biomed. Mater. Res. **1997**, *34* (3), 361–370.

43. Wissink, M.J.B.; Beernink, R.; Poot, A.A.; Engbers, G.H.M.; Beugling, T.; van Aken, W.G.; Feijen, J. Improved endothelialization of vascular grafts by local release of growth factor from heparinized collagen matrices. J. Control. Release **2000**, *64* (1–3), 103–114.

44. Mann, B.K.; Schmedlen, R.H.; West, J.L. Tethered TGF-beta increases extracellular matrix production of vascular smooth muscle cells. Biomaterials **2001**, *22* (5), 439–444.

45. Erdmann, L.; Campe, C.; Palms, D.; Uhrich, K. Polymer prodrugs with pharmaceutically active degradation products. Polym. Prepr. **1997**, *41*, 1048–1049.

46. Erdmann, L.; Campe, C.; Bedell, C.; Uhrich, K. Polymeric prodrugs: Novel polymers with bioactive components. ACS Symp. Ser. **1998**, *709*, 83–91.

47. Krogh-Jespersen, E.; Anastasiou, T.; Uhrich, K. Synthesis of a novel aromatic poly anhydride containing amino salicylic acid. Polym. Prepr. **2000**, *41*, 1048–1049.

48. Feldman, P.L.; Griffith, O.W.; Stuehr, D.J. The surprising life of nitric oxide. Chem. Eng. News **1993**, *71* (51), 26–38.

49. Maragos, C.M.; Morley, D.; Wink, D.A.; Dunams, T.M.; Saavedra, J.E.; Hoffman, A.; Bove, A.A.; Isaac, L.; Hrabie, J.A.; Keefer, L.K. Complexes of NO with nucleophiles as agents for the controlled biological re-lease of nitric oxide—Vasorelaxant effects. J. Med. Chem. **1991**, *34* (11), 3242–3247.

50. Loscalzo, J. Nitric oxide and restenosis. Clin. Appl. Thromb. Hemost. **1996**, *2* (1), 7–10.

51. Smith, D.J.; Chakravarthy, D.; Pulfer, S.; Simmons, M.L.; Hrabie, J.A.; Citro, M.E.; Saavedra, J.E.; Davies, K.M.; Hutsell, T.C.; Mooradian, D.L.; Hanson, S.R.; Keefer, L.K. Nitric oxide-releasing polymers containing the [N(O)NO]-group. J. Med. Chem. **1996**, *39* (5), 1148–1156.

52. Pulfer, S.K.; Ott, D.; Smith, D.J. Incorporation of nitric oxide-releasing crosslinked polyethyleneimine microspheres into vascular grafts. J. Biomed. Mater. Res. **1997**, *37* (2), 182–189.

53. Mowery, K.A.; Schoenfisch, M.H.; Saavedra, J.E.; Keefer, L.K.; Meyerhoff, M.E. Preparation and characterization of hydrophobic polymeric films that are thromboresistant via nitric oxide release. Biomaterials **2000**, *21* (1), 9–21.

54. Bohl, K.S.; West, J.L. Nitric oxide-generating polymers reduce platelet adhesion and smooth muscle cell proliferation. Biomaterials **2000**, *21* (22), 2273–2278.

55. Hill-West, J.L.; Chowdhury, S.M.; Slepian, M.J.; Hubbell, J.A. Inhibition of thrombosis and intimal thickening by in situ photopolymerization of thin hydrogel barriers. Proc. Natl. Acad. Sci. U.S.A. **1994**, *91* (13), 5967–5971.

BioMEMS

L. James Lee

Department of Chemical Engineering, Ohio State University, Columbus, Ohio, U.S.A.

Abstract

Miniaturization methods and materials are known as microelectro-mechanical systems (MEMSs). In recent years, MEMS applications have also been extended to the optical communication and biomedical fields. The former are called micro-optic electromechanical systems, while the latter are known as biomicroelectromechanical systems (bioMEMSs). Major potential and existing bioMEMS products are biochips/sensors, drug delivery systems, advanced tissue scaffolds, and miniature bioreactors. In this entry, major bioMEMS applications and microfluidics relevant to bioMEMS applications are briefly introduced.

INTRODUCTION

Miniaturization methods and materials are well developed in the integrated circuit industry. They have been used in other industries to produce microdevices, such as camera and watch components, printer heads, automotive sensors, micro-heat exchangers, micro-pumps, microreactors, etc., in the last 15 years.[1,2] These new processes are known as microelectro-mechanical systems (MEMSs), with a combined international market of over US$ 15 billion in 1998.[3] In recent years, MEMS applications have also been extended to the optical communication and biomedical fields. The former are called micro-optic electromechanical systems (MOEMSs), while the latter are known as biomicroelectromechanical systems (bioMEMSs). Potential MOEMS structures include optical switches, connectors, grids, diffraction gratings, and miniature lenses and mirrors. Major potential and existing bioMEMS products are biochips/sensors, drug delivery systems, advanced tissue scaffolds, and miniature bioreactors.

Future markets for biomedical microdevices for human genome studies, drug discovery and delivery in the pharmaceutical industry, clinical diagnostics, and analytical chemistry are enormous (tens of billions of U.S. dollars).[4] In the following sections, major bioMEMS applications and microfluidics relevant to bioMEMS applications are briefly introduced. Because of the very large volume of publications on this subject, only selected papers or review articles are referenced in this entry.

BIOMEMS APPLICATIONS

Biochips/Biosensors

Chip-based microsystems for genomic and proteomic analysis are the first bioMEMS products to have been commercialized. A large number of articles have been published in this field in recent years. Here, a brief introduction is given based on several recent review articles.[5–9] Biosensors are not necessarily microsystems.[10,11] MEMS techniques, however, may greatly enhance the performance of biosensors and reduce their cost. Microfabricated biosensors can be considered a division of biochips.[10–12]

Most molecular and biological assays and tests are very tedious, as shown in Fig. 1. They include the following steps: 1) obtaining a cellular sample (e.g., blood or tissue); 2) separating the cellular material of interest; 3) lysing the cells to release the crude DNA, RNA, and protein; 4) purifying the crude lysate; 5) performing necessary enzymatic reactions, such as denaturing, cleaving, and amplifying of the lysate by polymerase chain reaction (PCR); 6) sequencing DNA/genes using gel or capillary electrophoresis; and finally, 7) detecting and analyzing data. This process requires skilled technicians working in well-equipped biomedical laboratories, for periods of time ranging from many hours to several days to analyze a single sample. Much of today's diagnostic equipment is costly and bulky. It has limited use in medical diagnostics and is unsuited for emergency response at sites of care. To improve public health services, there is a great need to develop efficient and affordable methods and devices that can simplify the diagnostic process and be used as portable units. In recent years, the concept of integrating many analysis systems into one microdevice has attracted a great deal of interest in industry and academia. Such devices are called "laboratories-on-chips." They combine a number of biological functions (such as enzymatic reactions, antigen-antibody conjugation, and DNA/gene probing) with proper microfluidic techniques (such as sample dilution, pumping, mixing, metering, incubation, separation, and detection in micrometer-sized channels and reservoirs) in a miniaturized device. The integration and automation involved can improve the reproducibility of results and eliminate the

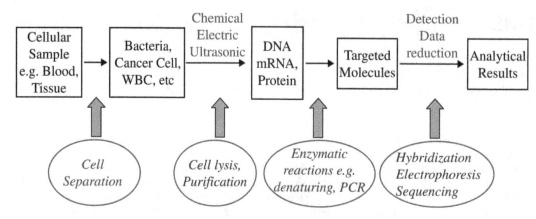

Fig. 1 Schematic of molecular diagnostics.

labor, time, and sample preparation errors that occur in the intermediate stages of an analytical procedure. The miniaturized devices also allow realization of low-energy and "point-of-care," parallel detection from a very small sample size, and easy data storage and transfer through computers and the Internet.

Biochips used for genomic analysis range from those used for separations for DNA sequencing, to those used in microvolume PCR, to complete analysis systems. Sequencing separations of single-stranded DNA fragments on a microchip follow the same principles as in conventional capillary electrophoresis. The process, however, is much faster because a higher electric field can be applied to the micrometer-sized separation channels without Joule heating problems. Automatic injection of a very small sample volume and parallel processing of a large number of separations can also be easily achieved. PCR allows amplification of a specific region of a DNA chain. PCR carried out on microchips is much more efficient than that on commercial PCR thermocyclers. In PCR, a sample solution is mixed, containing DNA, primers (synthesized short oligomers whose sequences flank the region of DNA to be amplified), a thermostable DNA polymerase enzyme, and the individual deoxyribonucleotides. Melting (or denaturization) of the double-stranded DNA molecules to single-stranded ones is done by heating the sample to ~95°C. The system is then quickly cooled to ~60°C for annealing; during this process, the added primers adhere to the single-stranded DNA. Finally, the sample is heated to ~72°C, at which temperature the polymerase is most active. During this extension period, complementary dinucleotide triphosphates are added to the growing strand using the target DNA as template. Each PCR cycle may double the amount of DNA of the required length. In the ideal case, 1 mol of targeted DNA fragments can be produced after 79 cycles. Practically, 20–50 cycles are needed to obtain a measurable quantity. Due to the high surface area/volume ratio associated with microdevices (it is important that PCR-friendly surfaces are produced in these devices), heat transfer is more efficient and temperature control is much easier than

in large systems. PCR time can be easily reduced from hours to minutes, particularly in continuous-flow PCR chips.[6]

Miniaturized proteomic analysis devices include enzyme assays and immunoassays. The enzyme assay chip is mainly a sophisticated incubator and flow-through system. It can perform multiple functions typically required by the biochemist, namely, diluting substrate and buffer, mixing enzyme and substrate, incubating during conversion, and allowing for detection in a flow channel. Immunoassay chips are similar to enzyme assay chips, except that the main focus is antigen-antibody interaction for clinical diagnostics and drug discovery.

There are a small number of commercially available biochips in the market today. Most are microarray-based systems, with biomolecules such as DNA probes, enzymes, and antigens being immobilized on the chip surface (e.g., GenChip® from Affymetrix, NanoChip™ from Nanogen, Inc., and GeneXpert® from Cepheid), or simple microfluidic systems capable of DNA sequencing by either electrophoresis (e.g., LabChip® from Caliper, Inc., and LabCard™ from ACLARA Biosciences, Inc.) or PCR. DNA microarray chips and DNA sequencing microfluidic chips have contributed to the Human Genome Project.

In addition to commercially production, a great deal of research and development work on biochips has been going on both in industry and in academia. Genomic analysis of DNA and RNA continues to be the focus of interest, but more and more effort is being spent on proteomic analysis of proteins and peptides.[5–9] Several enzyme assays and immunoassays designed based on microarray-based systems with simple microfluidic control are close to commercialization. They can be a vital tool in clinic diagnostics, drug discovery, and biomedical research.

Completely integrated micro-total analysis systems (μ-TAS) that can perform all the functions mentioned in Fig. 1 would be very valuable for high-throughput drug screening and personalized healthcare. However, only model systems have been proposed by research groups at present.[9] The mass production of such complicated systems at low cost is a challenging issue. Silicon and glass have been the most popular materials for fabricating microchips, but polymers are increasingly being used because of the availability of flexible, low-cost, high-throughput manufacturing methods for the micro-/nanoscale features needed for these types of applications.

Sensitive detection in microfluidic analytical devices is a challenge because of the extremely small detection volumes available. In conventional capillary electrophoresis, the most commonly employed detection method is ultraviolet absorption. In microscale biochips, laser-induced fluorescence (LIF) in conjunction with optical microscopy is currently the dominant detection technique because of its high sensitivity and noncontact nature. LIF microscopy, however, is costly, and the equipment size is quite large. To ensure wide application of the miniaturized biomedical devices, simple, portable, and low-cost detection methods are essential. Considerable efforts have been made lately to explore electrochemical methods, because the use of electrodes for detection leads to smaller instruments and cost reduction. Amperometry, conductimetry, and electro-chemiluminescence are also likely methods to complement fluorescence detection for on-chip analysis.[13]

Drug Delivery

Self-regulated and controllable drug delivery systems

Most conventional drug delivery systems are based on polymers or lipid vesicles. Drug safety and efficacy can be greatly improved by encapsulating the drug inside or attaching it to a polymer or lipid. The three general mechanisms by which drugs are delivered from polymer or lipid systems are 1) diffusion of the drug species through a polymer membrane, 2) a chemical or enzymatic reaction leading to cleavage of the drug from the system, and 3) solvent activation through swelling or osmosis of the system.[14] A major limitation of currently available delivery devices is that they release drugs at a predetermined rate. Certain disease states, such as diabetes, heart disease, hormonal disorders, and cancer, require drug administration either at a life-threatening moment or repeatedly at a certain critical time of day. Drug delivery technology can be taken to the next level by the fabrication of "smart" polymers or devices that are "responsive" to the individual patient's therapeutic requirements and deliver a certain amount of drug in response to a biological state. Given the miniature size of implantable devices, micromachining techniques will be essential for their manufacture. Currently, there are no commercial products based on the micromachined responsive drug delivery approach, and only some early research activity is seen in this direction.

The controlled release of drugs has been explored by adapting intelligent polymers, such as functional hydrogels, which respond to stimuli such as magnetic fields, ultrasound, electric current, temperature, and pH change.[15–17] These chemically synthesized materials are biocompatible and have good functionality. However, they often lack well-defined properties because of their inherent size and structure distribution resulting from chemical synthesis.[18] On the other hand, microfabrication technology developed for microelectronic applications is capable of mechanically creating devices with more precisely defined features, in a size range similar to that of polymeric and lipid materials.[19] Using hydrogels as switches or gates for controlled drug delivery and microfluidics has been explored recently by several researchers.[20,21] Cao et al.[22] describe the design of a self-regulated drug delivery device based on the integration of both mechanical and chemical methods. A pH-sensitive hydrogel switch is used to regulate the drug release, while a constant release rate is achieved by carefully designing the shape of the drug reservoir.

Biocapsules, membranes, and engineered particles for drug delivery

Immunoisolation is the protection of implanted cells from the host's immune system by the complete prevention of contact of immune molecules with the implanted cells, generally by the use of a semipermeable membrane. To achieve this without preventing nutrients from reaching the cells or waste from being removed, it is necessary to have an absolute pore size just below the minimum size needed to block out the smallest immune molecule, immunoglobulin G (IgG). The polymer and ceramic membranes used currently in bio-medical devices possess nanopores with nonuniform size distributions, which makes it difficult to control the passage of drugs and immunoglobulins (~30–50 nm in size) through these membranes.[23,24] Nonuniform porosity also requires the use of long, torturous flow paths, necessitating the use of thick membranes. Nanoscale resistance to flow in such thick membranes is high, so that high applied pressures (~1–4 MPa) are needed, which further complicates use.[25] These membranes also show incomplete virus retention.

Ferrari and co-workers[19] examined the feasibility of using microfabricated silicon nano-channels for immunoisolation. A suspension of cells was placed between two microfabricated structures with nanoporous membranes to fabricate an immunoisolation biocapsule. Characterization of diffusion through the nanoporous membranes demonstrated that 18 nm channels did not completely block IgG but did provide adequate immunoprotection (immunoprotected cells remained functional *in vitro* in a medium containing immune factors for more than 2 weeks, while unprotected cells ceased to function within 2 days). A major application of biocapsules containing nanochannels is immunoisolation of transplanted cells for the treatment of hormonal and biochemical deficiency diseases, such as diabetes.

Polymer microparticles have attracted much attention for drug delivery applications. Traditional microparticle fabrication protocols, such as phase separation, emulsification, and spray drying, have been successfully used for the production of drug delivery micro-spheres.[26,27] However, due to the surface-driven manufacturing process of these methods, the structural complexity of the resulting particles is limited. These methods are also difficult to apply for producing a monodispersed particle size distribution. Size control of microparticles is an important factor, since there are many routes of drug administration. According to DeLuca et al.[28] very large microparticles (>100 μm) with a broad particle size distribution are acceptable for embolization and drug delivery by implantation. Microparticles in the size range of 10–100 μm can be used for subcutaneous and intramuscular administration. Here, the particle size distribution is not a critical factor. Intravenous administration results in localization in the capillary vasculature and uptake by macro-phages and phagocytes. Microparticles larger than 8 μm lodge predominately in the lung capillaries, whereas those smaller than 8 μm may clear the lung and be localized in the liver and spleen. Therefore, it is most important to control the size of the largest particle.

During inhalation administration to the lung, filtering of particles occurs in the upper airways by inertial impaction, with large particles (aerodynamic diameter d_a > 5 μm) being deposited in the mouth and the first few generations of airways. Very small particles (d_a < 1 μm) are dispersed by diffusion, and a large fraction of these particles remain suspended in the airflow and are exhaled. Microparticles with the optimal size range of 1–5 μm are deposited in the central and peripheral airways and in the alveolar lung region by a combination of inertial impaction and sedimentation.[29] Therefore, the size and distribution of microparticles for inhalation therapies must be closely controlled to achieve high efficiency. Inhalation is a noninvasive drug delivery route and has been used widely for the treatment of diseases such as asthma, cystic fibrosis, and chronic obstructive pulmonary disease. Potential applications of new inhalation products in the near future include the treatment of diseases such as diabetes, pain, and growth deficiency, where proteins and lipids-based drugs will be used. For these biomolecule-based drugs, processing conditions such as high temperature and long solvent contact time may result in drug denaturization, so particle formation methods must avoid such conditions. For certain envisioned functional features of drug delivery vehicles, such as targeted and controlled vector release on cancer tumors, "highly engineered" microparticles (i.e., each particle is essentially a microdevice) may be required. This is another limiting factor for the traditional microparticle fabrication methods.

Compared to conventional polymer microparticle fabrication methods, microfabrication offers greater control of particle features and geometries. The shape and size of the particles can be controlled tightly. Perhaps more importantly, the components and surface properties can be designed to achieve particular functions. Using soft lithography (this fabrication method is explained in a later section), Guan et al.[30] developed a simple method to fabricate nonspherical polymer microparticles of precise shape and size, which can serve as either drug delivery vessels or substrates for further processing to produce functional drug delivery devices. Fig. 2 shows a micrograph of thin, platelike microparticles fabricated using this method. Combining the surface micropatterning and surface-tension self-assembly of autofolding,[31,32] well-defined 3D micropolyhedra, e.g., cubes and pyramids, can be fabricated from metals.[33] In our laboratory, similar micropolyhedra are currently being developed using functional polymers (e.g., biodegradable polymers and functional hydrogels). Large protein and gene molecules may be wrapped in such well-defined microstructures and delivered to targeted sites by either pulmonary delivery or intravenous administration.

Tissue Engineering

Tissue engineering is the regeneration, replacement, or restoration of human tissue function by combining synthetic and living molecules in appropriate configurations and environments.[34] The scaffold, the cells, and the cell-scaffold interactions are the three major components of any tissue-engineered construct. Although many tissue scaffold materials, such as foams and nonwoven fabrics, have been developed and used,[35,36] many challenges must be overcome for the promise of tissue engineering to become a reality. These include 1) low-cost fabrication of well-defined 3D scaffold configurations at both micro- and nanoscale; 2) incorporation of appropriate biocompatibility, bioactivity, and biodegrad-ability in the scaffolding construct to manipulate cellular and subcellular functions; and 3) active control of transport phenomena and cell growth kinetics to mimic microvasculature functions. Micro-/nanofabrication technology of polymers has tremendous potential in this field because it can achieve topographical, spatial, chemical, and immunological control over cells and thus create more functional tissue engineering constructs.[37]

An ideal tissue scaffolding process should be able to produce well-controlled pore sizes and porosity, provide high reproducibility, and use no toxic solvents. This is because these physical factors are associated with nutrient supply and vascularization of the cells in the implant as well as the development of a fibrous tissue layer that may impede nutrient access to the cells. Current processing methods used for polymer scaffolds include solvent casting, plastic foaming, fiber bonding, and membrane lamination.[35] However, precise, reproducible features in the micrometer and nanometer range are difficult to attain using these methods. By combining living cells and microfabricated two dimensions (2D) and 3D scaffolds with carefully controlled surface chemistry, investigators have

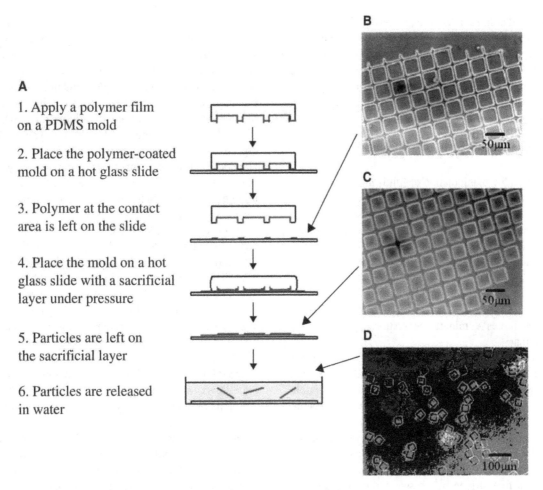

Fig. 2 Thin platelike microparticle fabricated by hot stamping. (**A**) The schematic of the procedure; (**B**) polymer at the contact area left on the glass slide; (**C**) polymeric microparticles left on the sacrificial layer after hot stamping; (**D**) and in water after release.

begun to address fundamental issues such as cell migration, growth, differentiation, apoptosis, orientation, and adhesion, as well as tissue integration and vascularization.

The functioning of tissues such as retinal, cardiac, and vascular tissue is dependent on the controlled orientation of multiple cell types. A key issue in the engineering of these tissues is control of the spatial distribution of cells *in vitro* to recreate a lifelike environment. The current approach to seeding cells is to allow cells to be randomly distributed in the scaffold. Microfabrication techniques, on the other hand, can produce short- and long-range surface patterns to mediate cell distribution and adhesion, biological interaction, and immune responses. Porous scaffolds without integrated blood supply rely solely on diffusion for mass transfer. They are limited to millimeters in size, while normal tissues leverage convection from blood vessels to enable oxygenation of large tissues.[38] Incorporating microfluidic networks in 3D tissue scaffolds for cell culture and implantation can be achieved by microfabrication techniques.[47] This new approach can provide the functional equivalents of microvasculature and enable scale-up of tissue engineering. Many cell-based bioreactors can be designed in a similar manner.[39]

Scientific and commercial work to date in tissue engineering has been largely devoted to clinical needs and focused on physiological aspects. There is a lack of low-cost, solventless, and mass-producible processing methods to fabricate scaffolds with well-defined micro- and nano-structures. In our laboratory, a manufacturing protocol is currently being developed for 3D tissue scaffolds of various shapes. The scaffold can be easily fabricated by combining micropatterned biodegradable polymers and supercritical CO_2 foaming technology.

Depending on the type of bioMEMS application, the polymers used can range from low-cost commodity plastics for disposable biochips to biodegradable and biofunctional polymers for drug delivery and tissue engineering. The feature size in these microsystems can be in either the micrometer or the nanometer scale. For instance, 10–100 μm is the desired microchannel size in microfluidic biochips. Below that, detection is too difficult, and, above that, mixing, heat transfer, and mass transfer are too slow. Particles used in drug delivery and the cell size in tissue scaffolds are also in the micrometer range. On the other hand, nanosized features are essential for immunoisolation in cell-based gene therapy and cell culture in tissue engineering. The enabling processing methods need to cover a broad size range, be mass producible and affordable, and be compatible with the polymers and biomolecules used in the process. Fluid transport in bioMEMS devices is crucial in many applications such as fast DNA sequencing, protein separation, drug delivery, and tissue generation.

Microfluidics

Microfluidics is the manipulation of fluids in channels, with at least 2D at the micrometer or submicrometer scale. This is a core technology in a number of miniaturized systems developed for chemical, biological, and medical applications. Both gases and liquids are used in micro-/nanofluidic applications[40,41] and generally, low-Reynolds-number hydrodynamics is relevant to bioMEMS applications. Typical Reynolds numbers for biofluids flowing in microchannels with linear velocity up to 10 cm/s are less than 30.[42] Therefore, viscous forces dominate the response and the flow remains laminar.

Fluid motion in these small-scale systems can be driven by applied pressure difference, electric fields associated with charged Debye double layers (or electrical double layer—EDL)—common when ionic solutions are present, or capillary driving forces owing to wetting of surfaces by the fluid.[42] Pressure-driven flow is similar to the classic Poiseuille flow. Electrokinetic effects can result in either electro-osmotic flow (EOF) or electrophoretic responses. EOF is a bulk flow driven by stresses induced in the thin EDL (i.e., 1–10 nm) near the channel walls, caused by an electric field imposed across the channel length. The velocity profile in the core of the channel is pluglike, even for a channel height as small as 24 nm.[43] Higher electrical permittivity of the fluid, imposed electric field strength, and zeta potential on the wall surface may all increase the flow rate.

Electrophoretic response, on the other hand, is the motion of charged molecules in a fluid caused by an electric field imposed across the channel length. Positively charged molecules move to the negative electrode, while negatively charged molecules move towards the positive electrode, leading to molecule separation. Electrophoresis is the most widely used separation method in the biotechnology field today. Typically, a buffer solution is chosen such that all biomolecules in the fluids, e.g., DNA/RNA fragments and proteins, are negatively charged. They all migrate from the sampling point to the detection point. Since DNA molecules have the same charge/mass ratio, separation is usually achieved by placing an immobilized gel or a mobile "gel" solution in the separation channel. EOF may cause unwanted washout of the gel solution during electrophoresis, so some sort of channel coating may be necessary for EOF suppression if the channel wall has a high zeta potential (e.g., glass). On the other hand, undesirable electrophoretic separation may occur in EOF if the sample solution contains components with different charges. A high-ionic-strength-plugs method has been developed to facilitate sample transport. The use of solutions at different ionic strengths and therefore different electroosmotic mobility, however, creates a quite complex situation in microfluidics. Electrokinetic flows work very well in microchannels because of the large surface-to-volume ratio, which minimizes the Joule heating problem in this type of flow. Very high electric field strength (i.e., hundreds to thousands of volts per centimeter channel length) can be easily applied in microdevices to speed up the processing time from hours to minutes or even seconds. For nanosized channels, it has been found that very low electric power (e.g., several volts per micrometer channel length) can generate a volume flow rate that is practical for controlled drug delivery.[43]

Capillary separation is also highly favored in microfluidics. This method is simple and low-cost, but a gas–liquid interface must exist. It is mainly used for reagent loading and release in portable biochips and drug delivery systems. The velocity profile is similar to that in pressure-driven flow, but the flow is very sensitive to the surface tension of the fluid, solid surface energy and roughness, and channel shape.[44] Active control of surface tension forces to manipulate flows in microchannels can be achieved by forming gradients in interfacial tension on the channel surface[45] or by electrowetting.[46]

For most cases involving the flow of small-molecule liquids, such as buffer solutions, the standard continuum description of transport processes works very well, except that surface forces (surface tension, electrical effects, van der Waals interactions, and, in some cases, steric effects) play a more important role than usual. Although some discrepancies have been reported between pressure-driven flow measurements made in microchannels and calculations based on the Navier–Stokes equations, most have been found to be experimental errors.[42] This is because the pressure drop, as a function of flow rate, varies as the inverse fourth power of channel radius (or inverse third power of channel height), and a small change in the radius (or channel height) due to manufacturing imperfections or channel-wall contamination produces large changes in the flow. Since the volumetric flow rate varies linearly with channel radius (or height) for electrically driven flow,[43] EOF is a more reliable way than pressure-driven flow to verify microfluidic experiments with calculations. A recent study[43] shows that calculated flow rates from classical EOF analysis agree well with experimental data for channel heights in the range of 10–20 nm.

Retardation of flow of ionic liquids and solutions in microchannels, however, can be significant when the channel walls have either the same static charge[47] (e.g., glass surface is negatively charged) in pressure-driven flow or opposite charges in EOF.[43] In the former case, the flow causes charges inside the EDL to accumulate downstream, while charges on the solid channel wall remain immobile. Such excess charge creates a potential drop in the channel direction, causing a "backflow." For channel height in the range of 100 μm, this electroviscous effect (flow retardation is often counted as an increase in fluid viscosity) is small. But a retardation of 70% is observed when the glass channel diameter is in the range of several micrometers.[47] In the latter case, the backflow can be manipulated by surface micropatterning of opposite charges on the walls of the microchannel to achieve laminar chaotic mixing[48] or controllable membrane permeation.

In many bioMEMS applications, the sample fluid contains molecules and particles of various sizes. Small organic molecules are a few angstroms in size, typical protein molecules are about 2–5 nm, and large DNA molecules and cells are in the range of 1–10 μm. In some cases, the radii of near-spherical fluid droplets or gas bubbles are comparable to that of the channel, but in others, their lengths may be larger.[42] Non-Newtonian fluid and multiphase flow mechanics must be applied. In microchannels, the shear rate can be very high, e.g., 10^7 sec^{-1}, even though the Reynolds number is low. Rheological characterization of polymeric fluids and biofluids in such a flow field has recently been studied in our laboratory.[48] It was found that the standard rheological analysis used at the macroscale also works at the microscale. The high-shear Newtonian plateau can be easily observed. For solutions containing large polymer (or DNA) molecules, polymer degradation and wall slip are substantial when the flow rate is high. Rheology in microchannels needs to be studied further because many biofluids are highly non-Newtonian. One advantage of microfluidics is that a single biomolecule such as DNA can be isolated and analyzed on a biochip containing small channels or wells.[49,50] Since the molecule size is comparable to the channel (well) dimension, understanding and manipulating both the macroscopic and microscopic transport phenomena of the confined molecule undergoing flow is an active area of research.[42]

CONCLUSIONS

The miniaturization of biomedical and biochemical devices for bioMEMSs has gained a great deal of attention in recent years. Products include biochips/biosensors, drug delivery devices, tissue scaffolds, and bioreactors. In the past, MEMS devices have been fabricated almost exclusively in silicon, glass, or quartz because of the comparable technology available in the microelectronics industry. For applications in the biochemistry and biomedical field, polymeric materials are desirable because of their lower cost, good process-ability, and biocompatibility. Polymer microfabrication techniques, however, are still not well developed.

REFERENCES

1. Madou, M.J. *Fundamentals of Microfabrication: The Science of Miniaturization*, 2nd Ed.; CRC Press: Boca Raton, FL, 2002.
2. Jensen, K.F. Microchemical systems: Status, challenges, and opportunities. AIChE J. **1999**, *45* (10), 2051–2054.
3. Freemantle, M. Downsizing chemistry: Chemical analysis and synthesis on microchips promise a variety of potential benefits. Chem. Eng. News **1999**, *77* (7), 27–36.
4. Snyder, M.R. Micromolding technology extends sub-gram part fabrication capability. Mod. Plast. **1999**, *76* (1), 85.
5. Bousse, L.; Cohen, C.; Nikiforov, T.; Chow, A.; Kopf-Sill, A.R.; Dubrow, R.; Parce, J.W. Electrokinetically controlled microfluidic analysis systems. Annu. Rev. Biophys. Biomol. Struct. **2000**, *29* (1), 155–181.
6. Sanders, G.H.W.; Manz, A. Chip-based microsystems for genomic and proteomic analysis. Trends Anal. Chem. **2000**, *19* (6), 364–378.
7. Carrilho, E. DNA sequencing by capillary array electrophoresis and microfabricated array systems. Electrophoresis **2000**, *21* (1), 55–65.
8. Kricka, L.J. Microchips, microarrays, biochips and nanochips: Personal laboratories for the 21st century. Clin. Chim. **2001**, *307* (1–2), 219–223.
9. Krishnan, M.; Namasivayam, V.; Lin, R.; Pal, R.; Burns, M.A. Microfabricated reaction and separation systems. Curr. Opin. Biotechnol. **2001**, *12* (1), 92–98.
10. Vo-Dinh, T.; Cullum, B. Biosensors and biochips: Advances in biological and medical diagnostics. Fresenius J. Anal. Chem. **2000**, *366* (6–7), 540–551.
11. Wang, J. Glucose biosensors: 40 years of advances and challenges. Electroanalysis **2001**, *13* (12), 983–988.
12. Lauks, I.R. Microfabricated biosensors and microanalytical systems for blood analysis. Acc. Chem. Res. **1998**, *31* (5), 317–324.
13. Schwarz, M.A.; Hauser, P.C. Recent developments in detection methods for microfabricated analytical devices. Lab on a Chip Miniaturis. Chem. Biol. **2001**, *1* (1), 1–6.
14. Langer, R. Drug delivery and targeting. Nature **1998**, *392* (6679), 5–10
15. Lowman, A.M.; Peppas, N.A. Analysis of the complexation/decomplexation phenomena in graft copolymer networks. Macromolecules **1997**, *30* (17), 4959–4965.
16. Torres-Lugo, M.; Peppas, N.A. Molecular design and *in vitro* studies of novel pH-sensitive hydrogels for the oral delivery of calcitonin. Macromolecules **1999**, *32* (20), 6646–6651.
17. Traitel, T.; Cohen, Y.; Kost, J. Characterization of glucose-sensitive insulin release systems in simulated *in vivo* conditions. Biomaterials **2000**, *21* (16), 1679–1687.
18. Lanza, R.P.; Chick, W. Encapsulated cell therapy. Sci. Am. Sci. Med. **1995**, *2* (4), 16–25.
19. Desai, T.A.; Hansford, D.; Ferrari, M. Characterization of micromachined silicon membranes for immunoisolation and bioseparation applications. J. Membrane Sci. **1999**, *159* (1), 221–231.
20. Kaetsu, I.; Uchida, K.; Shindo, H.; Gomi, S.; Sutani, K. Intelligent type controlled release systems by radiation techniques. Radiat. Phys. Chem. **1999**, *55* (2), 193–201.
21. Liu, R.H.; Yu, Q.; Bauer, J.M.; Jo, B.-H.; Moore, J.S.; Beebe, D.J. In-channel processing to create autonomous hydrogel microvalues. In *Micro Total Analysis Systems* 2000, Proceedings of the 4th μTAS Symposium, Enschede, Netherlands, May 14–18, 2000; 45–48.
22. Cao, X.; Lai, S.; Lee, L.J. Design of a self-regulated drug delivery device. Biomed. Microdev. **2001**, *3* (2), 109–118.

23. Colton, C.K. Implantable biohybrid artificial organs. Cell Transplant. **1995,** *4* (4), 415–436.

24. Desai, T.A.; Hansford, D.J.; Kulinsky, L.; Nashat, A.H.; Rasi, G.; Tu, J.; Wang, Y.; Zhang, M.; Ferrari, M. Nanopore technology for biomedical applications. Biomed. Microdev. **2000,** *2* (1), 11–40.

25. Kim, K.J.; Stevens, P.V. Hydraulic and surface characteristics of membranes with parallel cylindrical pores. J. Membrane Sci. **1997,** *123* (2), 303–314.

26. Jain, R.A. The manufacturing techniques of various drug loaded biodegradable poly(lactide-co-glycolide) (PLGA) devices. Biomaterials **2000,** *21* (23), 2475–2490.

27. Langer, R. Biomaterials in drug delivery and tissue engineering: One laboratory's experience. Acc. Chem. Res. **2000,** *33* (2), 94–101.

28. DeLuca, P.P.; Mehta, R.C.; Hausberger, A.G.; Thanoo, B.C. Biodegradable polyesters for drug and polypeptide delivery. In *Polymer Delivery Systems, Properties and Applications*; El-Nokaly, M.A., Piatt, D.M., Charpentier, B.A., Eds.; Chapter 4, ACS Symposium Series 520; Amercian Chemical Society: Washington, DC, 1993; 53–79.

29. Edwards, D.A. Delivery of biological agents by aerosols. AlChE J. **2002,** *48* (1), 2–6.

30. Guan, J.; He, H.; Hansford, D. J.; Lee, L. J. Self-folding of three-dimensional hydrogel microstructures. J. Phys. Chem. B **2005,** *109* (49), 23134–23137.

31. Green, P.W.; Syms, R.R.A.; Yeatman, E.M. Demonstration of three-dimensional microstructure self-assembly. J. Microelectromech. Syst. **1995,** *4* (4), 170–176.

32. Harsh, K.F.; Bright, V.M.; Lee, Y.C. Solder self-assembly for three-dimensional microelectrome-chanical systems. Sens. Actuat. **1999,** *77*, 237–244.

33. Gracias, D.H.; Kavthekar, V.; Love, J.C.; Paul, K.E.; Whitesides, G.M. Fabrication of micrometer-scale, patterned polyhedra by self-assembly. Adv. Mater. **2002,** *14* (3), 235–238.

34. Langer, R.; Vacanti, J.P. Tissue engineering: The design and fabrication of living replacement devices for surgical reconstruction and transplantation. Lancet **1999,** *354 (Suppl. 1), 32–34.*

35. Mikos, A.G.; Sarakinos, G.; Leite, S.M.; Vacanti, J.P.; Langer, R. Laminated three-dimensional biodegradable foams for use in tissue engineering. Biomaterials **1993,** *14* (5), 323–330.

36. Li, Y.; Yang, S.-T. Effects of three-dimensional scaffolds on cell organization. Biotechnol. Bioprocess Eng. **2001,** *6* (5), 311–325.

37. Desai, T.A. Micro- and nanoscale structures for tissue engineering constructs. Med. Eng. Phys. **2000,** *22* (9), 595–606.

38. Griffith, L.G.; Noughton, G. Tissue engineering—Current challenges and expanding opportunities. Science **2002,** *295* (5557), 1009–1014.

39. King, K.R.; Terai, H.; Wang, C.C.; Vacanti, J.P.; Borenstein, J.T. Microfluidics for tissue engineering microvasculatuer: Endothelial cell culture. In *Micrototal Analysis Systems*; 2001, Proceedings of the 5th μ_{TAS} 2001 Symposium, Monterey, CA, USA, October 21–25, 2001; 247–249.

40. Gad-el-Hak, M. The fluid mechanics of microdevices. J. Fluids Eng. **1999,** *121* (1), *5–33.*

41. Giordano, N.; Cheng, J.-T. Microfluid mechanics: Progress and opportunities. J. Phys. Condens. Matter **2001,** *13, R271–R295.*

42. Stone, H.A.; Kim, S. Microfluidics: Basic issues, applications, and challenges. AlChE J. **2001,** *47* (6), 1250–1254.

43. Conlisk, A.T.; McFerran, J.; Zheng, Z.; Hansford, D. Mass transfer and flow in electrically charged micro- and nanochannels. Anal. Chem. **2002,** *74* (9), 2139–2150.

44. Kang, K.; Lee, L.J.; Koelling, K.W. High shear microfluidics and its application in rheological measurements. Exp. Fluids **2005,** *38* (2), 222–232.

45. Gallardo, B.; Gupta, V.K.; Eagerton, F.D.; Jong, L.I.; Craig, V.S.; Shah, R.R.; Abbott, N.L. Electrochemical principles for active control of liquids on submillimeter scales. Science **1999,** *283* (5398), 57–60.

46. Pollack, M.G.; Fair, R.B.; Shenderov, A.D. Electrowetting-based actuation of liquid droplets for microfluidic applications. Appl. Phys. Lett. **2000,** *77* (11), 1725–1726.

47. Kulinsky, L.; Wang, Y.; Ferrari, M. Electroviscous effects in microchannels. SPIE Proc. **1999,** *3606*, 158–168.

48. Stroock, A.D.; Weck, M.; Chiu, D.T.; Huck, W.T.S.; Kenis, P.J.A.; Ismagilov, R.F.; Whitesides, G.M. Patterning electro-osmotic flow with patterned surface charge. Phys. Rev. Lett. **2000,** *84* (15), 3314–3317.

49. Smith, D.E.; Babcock, H.P.; Chu, S. Single-polymer dynamics in steady shear flow. Science **1999,** *283* (5408), 1724–1727.

50. Shrewsbury, P.J.; Muller, S.J.; Liepmann, D. Effect of flow on complex biological macromolecules in microfluidic devices. Biomed. Microdev. **2001,** *3* (3), 225–238.

Biomimetic Materials

Kimon Alexandros Leonidakis
Al Halifa Soultan
Susanna Piluso
Abhijith Kundadka Kudva
Laurien Van den Broeck
Taíla O. Meiga
Rory L.D. Gibney
Jennifer Patterson
Departments of Materials Engineering, Chemistry, and Mechanical Engineering, KU Leuven, Leuven, Belgium

Abstract

Biomimetic materials portray biomimesis, which is derived from the ancient Greek bios, meaning "life," and mimesis, meaning "imitation." This field studies the constituents of biological tissues and aims to replicate their form and function. Biomimetic materials are used in regenerative medicine to limit the host response, to promote migration of host cells leading to regeneration, and to support cells as tissue engineering scaffolds. They imitate their biological counterparts in a number of aspects, with structural and mechanical properties being among the most important. They aim to reproduce the extracellular matrix (ECM), which is a complex network of proteins and polysaccharides that gives tissue its structure, supports cell signaling, and influences the presentation of biochemical cues that are vital for tissue morphogenesis. The ECM of each tissue type has a specific chemical composition to support interactions with cells. Materials targeting cell-surface receptors are particularly interesting as receptor–ligand interactions have a great impact on the cellular response to biomimetic materials and their eventual integration. The spatial arrangement and orientation of ligands has a major effect on receptor–ligand interactions and their downstream signaling, and this has led to the development of more complex methods of attaching bioactive molecules to impart specific nanoscale patterns. The development of biomimetic materials is a growing area of research, and understanding of the mechanisms involved in successfully imitating biological entities is evolving. This entry focuses on materials mimicking the structure and mechanical properties, transport, biochemical modifications, and topography of the ECM and their applications.

INTRODUCTION

Biomimetic materials are materials that portray biomimesis or biomimicry. The term biomimesis is derived from the ancient Greek *bios* (βίος), meaning "life," and *mimesis* (μίμησις), meaning "imitation." Millions of years of evolution have allowed nature to generate biological tissues that are optimized for their function and that can be repaired by surrounding cells when damaged, at least to a certain degree. Therefore, it is logical to think that biologically derived materials would be ideal to heal damaged tissues or organs. However, it is often difficult to reproducibly extract and store these components, and there are risks of immunogenicity or disease transfer. The field of biomimetic materials studies these tissues and their biological constituents and aims to replicate their form and function by artificial means. This entry focuses on materials mimicking the extracellular matrix (ECM), which is the native environment of the cell. Important features include its structure and topography, transport and mechanical properties, and biochemical composition. Specific examples of state-of-the-art materials exhibiting these biomimetic properties as well as the chemical and processing methods required for their fabrication are provided. The host response to these materials and their application as tissue engineering scaffolds are also discussed.

OVERVIEW AND PERSPECTIVES

Biomimicry has led to advancements not only for biomedical applications but also for everyday products. Sharkskin-inspired wetsuits supported swimmers in breaking world swimming-records across all strokes at the 2008 Olympics, despite conflicting results in scientific studies looking at the ability of the material to reduce drag.[1] Studies of geckos' feet have led to advancements in adhesive research with an aim to create dry adhesives with superior properties that

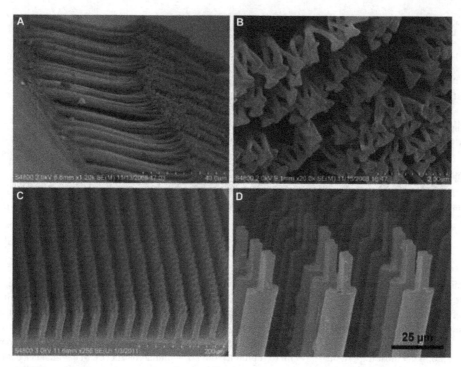

Fig. 1 SEM images of (**A**) setae, the main structures in the adhesive system of a gecko's foot, which split into approximately 1,000 spatulae per seta at their terminus; (**B**) magnified view of an array of spatulae, which generate adhesion through Van der Waals' forces; (**C**) a synthetic dry adhesive surface inspired by the gecko's foot; and (**D**) a magnified view of the adhesive structures which resemble setae.
Source: From Jin et al.[232] © 2012, American Chemical Society. Reprinted with permission.

benefit from Van der Waals' forces (Fig. 1).[2] The term "biomimetic" is used to describe many things from the first flying machines, which imitated bird flight,[3] to skyscraper ventilation that is inspired by termite dens.[4] In the strictest sense, biomimesis is the replication of the form and function of a biological entity, whereas many materials or medical devices portray biomimicry, that is, they superficially imitate a biological system.

Biomimetic materials are of particular importance in regenerative medicine where their ability to imitate natural biological tissues allows for the prevention or reduction of the host response and for the migration of host cells into the implant, leading to improved tissue regeneration. They can further be used as scaffolds to support cells in tissue engineering approaches to create living tissue substitutes or tissue-engineered constructs. In these ways, biomimetic materials could treat a wide range of tissues or organs that are damaged due to trauma, disease, lifestyle, or aging. Most biomimetic materials need to imitate their biological counterparts in a number of aspects to successfully integrate into the host, with structural and mechanical properties being among the most important. For example, bone has a specific architecture to provide strength and structure to the body while allowing for cell activity. Dense cortical bone makes up the exterior of bone giving it strength, and spongy cancellous bone forms the mesoscale porous interior that allows it to withstand more strain than cortical bone[5] and provides a lipid-rich microenvironment for

some of bone's secondary functions, for example, hematopoiesis.[6] Replicating this mesoscale structure is necessary for biomimetic materials that aim to integrate with bone.

More generally, biomimetic materials target to reproduce the local microenvironment of cells—the ECM, which is a complex network of proteins and polysaccharides in the interstitial space that gives tissue its structure. However, the ECM has many functions beyond structure. It supports cell signaling and influences the transport, sequestration, and presentation of biochemical cues that are vital for cell differentiation and tissue morphogenesis. The ECM of each tissue type has a specific chemical composition of proteins and polysaccharides and, therefore, contains different ligands to support interactions with cells. Integrins, transmembrane proteins that have a receptor head in the intercellular space and a short tail in the cytoplasm, are the mediators of most interactions between the ECM and cells.[7] They allow for cell attachment, and different signaling pathways can be initiated by integrin receptors upon binding to an ECM ligand. These signals regulate cytoplasmic kinases, growth factor receptors, and ion channels, and they can also induce changes in the structure of the actin cytoskeleton.[7] As these pathways play a role in the proliferation, migration, and differentiation of cells, materials targeting cell-surface receptors can be used to influence the cellular response to biomimetic materials and their eventual integration.

The spatial arrangement and orientation of ligands has been shown to have a major influence on receptor-ligand interactions and downstream signaling. Integrins form clusters at focal adhesion sites, which are responsible for most ECM-cell interactions,[8] and cell-to-cell interactions also occur on the nanoscale. Growth factors, which include a number of proteins involved in cell signaling and lead to cellular proliferation and differentiation, also interact with cells via cell-surface receptors. Interestingly, integrins are often in close proximity to growth factor receptors, which further enhances ECM-cell signaling as the ECM provides an efficient pathway for growth factor sequestration. In this case, a growth factor binds initially to a matrix protein, which then binds to integrins on the cell surface facilitating activation of growth factor receptors that are in close proximity.[9] By binding to the ECM, many growth factors also significantly increase stability and, hence, the duration of their bioactivity. Nanoscale patterning of materials can imitate these mechanisms of cell signaling and communication. This has inspired the development of more complex methods, such as lithography, for attaching bioactive molecules that impart specific nanoscale patterns of binding motifs and growth factors to enhance cell attachment and growth factor signaling, respectively.[10] In addition to patterned surface modification, the topography of tissue engineering scaffolds and other implants can impact cell interactions. For example, the use of nanoparticles exploits the reaction of a cell to the geometry of these particles or of an array of particles.[10]

The development of biomimetic materials is a relatively new and growing area of research, and understanding of the mechanisms involved in successfully imitating biological entities is continually evolving. Already some excellent reviews, which provide a coherent view of the direction that current research is taking, are available.[11–14] Here, we identify key features of the ECM, namely its structure and mechanical properties, transport, biochemical modifications, and topography, to mimic with these materials. In addition to providing specific examples of such biomimetic materials, the *in vivo* response to these materials and their use as tissue engineering scaffolds are also discussed.

MATERIALS MIMICKING MECHANICAL AND STRUCTURAL PROPERTIES

A high degree of sophistication can be found in natural tissues, which are composed of multiple components organized in elegant and intricate patterns. This complex architecture can improve the mechanical properties of the resulting composite tissues although the performance of the individual constituents is significantly less. For example, bone tissue, a highly mineralized composite consisting of hydroxyapatite and collagen, is simultaneously tough and stiff,[15] whereas individually, hydroxyapatite is brittle and collagen is not as strong. In addition to their compositional diversity, natural tissues also possess anisotropy and porosity. Aligned collagen fibrils in compact bone, for example, guide directional cell growth and

provide the typical anisotropic mechanical characteristics of bone.[15] The porosity of many natural tissues results in reduced weight and allows for mass transport.[16–18] When designing scaffolds, these features of the ECM can be mimicked by engineering the chemical composition and nano- to mesoscale structure of the materials to match the anisotropy and porosity of the target tissue to obtain better mechanical and transport properties.[17] For this purpose, a new set of tools in computational modeling, manufacturing, and nanotechnology has been developed and is also discussed in this section.

A common approach for the design of tissue engineering scaffolds is to approximate the stiffness of the native tissue that is to be regenerated. For example, in a study on the growth of human bladder stromal cells on polycaprolactone (PCL) and poly(lactide-co-glycolide) (PLGA) scaffolds, the cells grown on PCL, which better matches the native tissue in terms of stiffness, showed a greater attachment and rate of growth, whereas the cells on PLGA, which is more rigid, differentiated to a contractile phenotype.[19] A mismatch of mechanical properties and the surrounding tissue can also lead to negative effects during tissue regeneration. For example, intimal hyperplasia occurs at anastomotic sites of small blood vessels with materials of higher stiffness.[20] Matrix stiffness also influences the structure of focal adhesion complexes and the cytoskeleton within cells that are cultured on or in these substrates (Fig. 2).[21–27] This phenomenon is explained by the presence of cellular mechanotransducers that sense the force generated by a cell to deform the matrix, which is then translated into morphological changes and lineage specification.[21,28–30] For instance, matrix stiffness has been shown to influence the phenotype to which naïve mesenchymal stem cells (MSCs) commit. Soft matrices ($E \sim 0.1$–1 kPa) appeared to be neurogenic and induced branched morphologies for cells, whereas matrices that approached a stiffness of $E \sim$ 8–17 kPa were myogenic and promoted a spindle-shaped cell morphology. More rigid matrices ($E \sim 25$–40 kPa) have proven to be osteogenic and promoted a polygonal morphology.[31] Tuning of the matrix stiffness may provide a more suitable environment for differentiated cells including neurons, myoblasts, fibroblasts, and osteoblasts compared to traditional 2D culture where cells adhere to and proliferate on plastic substrates that differ in stiffness by several orders of magnitude compared to their native tissue.[27,32,33]

In addition to bulk mechanical properties, the porosity and pore alignment of biomimetic scaffolds are important to retain flexibility, reduce weight, and/or provide paths for mass transport.[17] Electrospinning has gained attention as a manufacturing technique to create porous structures from long fibers with micro- and nanoscale diameters to mimic the morphology and geometry of native tissue. In this process, first reported in 1934,[34] a high electrical field is applied to a droplet of polymer melt or solution, which leads to the ejection of a charged jet and the formation of continuous fibers.[35] Many biodegradable and biocompatible

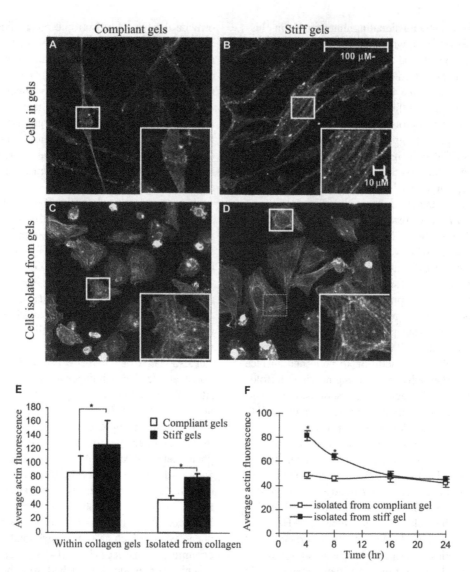

Fig. 2 Qualitative and quantitative assessment of actin, which is an important determinant of the mechanical properties of a cell. (**A**, **B**) Cells encapsulated within gels. (**C**, **D**) Cells after isolation from gels. (**A**, **C**) Compliant gels. (**B**, **D**) Stiff gels. Cells within the compliant gels exhibited less intense actin staining and less prominent stress fibers as compared to cells within stiff gels, which were similar to cells that were isolated from the gels and plated on glass slides. (**E**) Quantitative image analysis of average actin fluorescence. (**F**) Average actin fluorescence for cells isolated from stiff and compliant gels, which decreased as a function of time after plating.
Source: From Byfield et al.[27] © 2009, Elsevier. Reprinted with permission.

polymers, such as collagen, polyethylene oxide, PCL, and fibrinogen, have been electrospun into scaffolds for use in target tissues including cartilage, bone, heart, skin, blood vessels, and nerve.[36–42] For example, aligned biocomposite fibers of electrospun gelatin and PCL for nerve tissue restoration supported neurite outgrowth parallel to the aligned fibers after seeding them with nerve stem cells.[38] Other processing methods for forming porous materials include cryogelation, where scaffold architecture is dictated by the structure of ice crystals that are present during the gelation process. Alginate cryogels showed shape memory properties, large interconnected pores of 150 to 200 µm on average, and enhanced mechanical stability when compared to alginate hydrogels formed at room temperature.[43]

Despite the advantages and biomimetic aspects of porous scaffolds, the addition of pores typically results in a decrease in stiffness. Therefore, addition of a reinforcing phase can be used to stiffen otherwise soft polymeric matrices and create composites. In nature, for example, such reinforcement occurs when cells in bone tissue cause mineral deposition in a process called biomineralization, which has inspired researchers to combine polymeric matrix materials with ceramic particles. For example, bioresorbable foams made of collagen fibers and synthetic apatite nanocrystals were created by freeze drying and crosslinking and have succeeded in mimicking the chemical, structural, and mechanical properties of bone.[44] The collagen/apatite scaffolds showed excellent biocompatibility and osteoconductive properties; they had a compressive stiffness of 37.3 ± 2.2 MPa, similar to the compressive stiffness of cancellous and cortical bone (2–12 MPa and 100–230 MPa, respectively),[16] and successfully healed critical sized defects in pig tibia.[44] Alternatively, directionally freezing bio-inspired ceramic suspensions based on aluminum oxide and polymethyl methacrylate, an approach termed freeze casting, resulted in a yield strength and fracture toughness that were 300 times higher than those of the individual constituents[45] and superior to that of cortical bone.[16,46]

Further, a number of exciting techniques have been developed to refine scaffold architecture at the meso-, micro-, and nanoscales. Casting around a sacrificial material has been used to create patterned networks of channels resembling blood vessels within cellularized tissue constructs to allow for perfusion.[47] For example, an inexpensive, biocompatible, and dissolvable carbohydrate glass was used with 3D printing techniques to create a sacrificial network, which was then embedded in a hydrogel.[48] Cells survived on and near the hydrogel surface and near perfused channels that also showed endothelialization. Using similar approaches, fluidic channels have been created in alginate, agarose, fibrin, poly(ethylene glycol) (PEG), and Matrigel.[47] Another intriguing technique to create scaffolds with unique 3D architectures involves the shaping of carbon nanotubes (CNTs) to form cavities of adjustable size to control the spatial distribution of cells.[49,50] CNTs were grown and vertically aligned by chemical vapor deposition on a silicon substrate. Subsequently, the CNTs were wetted to form regular cavities as a result of capillary and tensile forces (Fig. 3). L929 fibroblasts attached themselves to the bottom of the polygonal cavities and covered the 3D network after 7 days of growth, while the interconnected network retained its shape to support the developing tissue.[49] This kind of structure could be used to mimic tissues such as cancellous bone and, simultaneously, direct cell growth and provide mechanical reinforcement. These are few of the many possibilities to mimic the mechanical and structural properties of target tissue that allow functional support and guide cell growth and differentiation.

DYNAMICALLY RESPONSIVE MATERIALS

In addition to providing appropriate mechanical properties, it can be interesting to develop materials that allow for dynamic processes. Synthetic polymer systems can be used to create responsive interfaces or structures that can adapt to changes in their environment.[51] These materials, typically elastomeric hydrogels, are triggered by external stimuli such as temperature, pH, humidity, or electric fields.[51] Ion and molecule transport can be regulated in this way, and chemical and biochemical signals can be converted into electrical and mechanical responses, which can play an important role in tissue engineering, drug delivery, and biosensing.

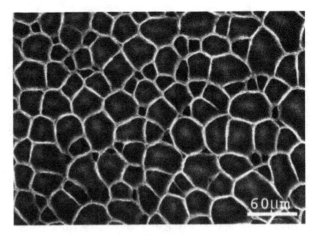

Fig. 3 SEM image of the 3D networks made from multiwall-CNTs of 50 μm length. Scale bar is 60 μm.
Source: From Correa-Duarte et al.[49] © 2004, American Chemical Society. Reprinted with permission.

Conducting and contracting materials are proving useful for applications in nerve, heart, and muscle tissue engineering. For example, melanin, a natural pigment with unique electrical properties, was used to create a conductive collagen film for nerve tissue engineering, which resulted in enhanced Schwann cell growth and neurite extension *in vitro* when compared to reference collagen films.[52,53] CNTs that were organized into bundles also had an advantageous effect on neural signal transmission because of their high electrical conductivity and fractal-like nanostructure.[54–57] An emerging strategy is the combination of conductive and elastic materials to generate electrically responsive scaffolds for heart tissue. Elastic hydrogels with conductive gold nanoparticles led to increased expression of connexin 43 (a gene facilitating electrical signaling) by neonatal rat cardiomyocytes seeded on the hydrogels when compared to those on control hydrogels.[58] Also, alginate cardiac patches incorporating gold nanowires have resulted in synchronous contraction of seeded heart cells after electrical stimulation.[59] The tissue grown on these scaffolds was thicker and better aligned when compared to reference alginate scaffolds, and the cells expressed higher levels of proteins that are involved in electrical coupling and muscle contraction. As a last example, 3D contracting, porous, and anisotropic poly(glycerol sebacate) (PGS) scaffolds with controlled stiffness and an accordion-like honeycomb microstructure were created by excimer laser microablation to resemble the myocardium.[60] The mechanical properties of the scaffolds closely matched native adult rat right ventricular myocardium, electrophysiological assessment demonstrated synchronous contraction, and the scaffolds promoted preferential alignment of neonatal rat cardiomyocytes within the graft. Thus, the use of conductive materials and architecture control can allow conduction and contraction to better mimic the natural function of the target tissue.

Biomimetic—Colloid

EFFECT OF NANOTOPOGRAPHY ON CELL BEHAVIOR

The electrospun materials described earlier form nanofibers, an important step toward mimicking the native environment of the cell. Natural tissues have a very complex architecture, and most of the current biomaterials recapitulate only some of the properties of the natural ECM where cells are surrounded by a substratum of nanometer size,[61] which provides cells with two benefits: a high specific surface area and mechanical support from the reinforcement of the matrix by nanofibers.[62] Moreover, nanostructures play a key role in directing cellular behavior via cell-surface interactions. Although cells have micrometer dimensions, they contain nanoscale features, such as focal adhesion complexes and fine processes (e.g., cilia and filopodia).[63] It is also well-known that cells can sense and respond to a variety of chemical and physical cues, including surface topography and rigidity.[64,65] Topographical sensing is a phenomenon affecting cell morphology, migration, and fate.[66,67] The influence of topographical cues on cell behavior was first observed by Harrison in 1911, when he grew cells on a spider web and the cells followed the fibers of the web in a phenomenon subsequently described as contact guidance.[68] Later, in 1934, Weiss further explored the principle of topographical guidance by culturing neurons on glass covered with oriented fibrin, forming aligned nerve fibers *in vitro*.[69,70]

An emerging approach to mimic this complex nanoscale 3D environment consists of the preparation of nanocomposite hydrogels, which are physically or chemically crosslinked hydrated polymeric networks that incorporate nanoparticles or nanofibers within the matrix.[71] A range of nanostructures such as carbon-based nanomaterials (CNTs, graphene, nanodiamonds), polymeric nanoparticles, inorganic/ceramic nanoparticles (hydroxyapatite, silica, silicates), and metal nanoparticles (gold, silver) have been used for the preparation of nanocomposite hydrogels. Initially, these nanomaterials were mainly incorporated within the hydrogel matrix to reinforce their mechanical properties, as described earlier. However, it is now becoming clear that incorporation of these nanomaterials can result in topographical signals to guide cell behavior. The response of different cell types, including fibroblasts, osteoblasts, endothelial cells, smooth muscle cells, and epithelial cells,[72] to nanostructures has been observed with the effects on cell attachment and spreading being described the most.[62] For instance, fibroblasts and osteoblasts responded differently to engineered topography gradients of three gold nanoparticle sizes (16 nm, 38 nm, and 68 nm) (Fig. 4). For the gradient with the 16 nm nanoparticles, adhesion and spreading of both cell types increased compared to the smooth surface. In the areas with the 38 nm nanoparticles, fibroblasts showed a similar behavior when compared to the smooth surface, whereas the number of adherent osteoblasts and their spreading decreased. Finally, poor cell adhesion was observed in the regions with the 68 nm nanoparticles for both cell types. The study concluded that there was a specific size range of nanotopographical features that promoted cell adhesion and spreading and that it depended on the cell type.[73] In another example, the use of nanopatterned tissue culture substrates significantly increased the proliferation and expression of functional markers (Na^+/K^+-ATPase and tight

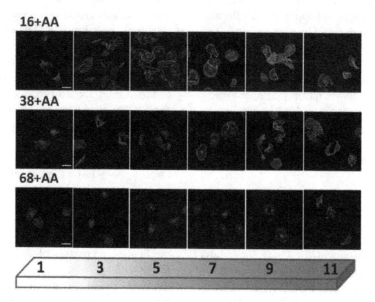

Fig. 4 Confocal microscope images of 3T3 fibroblasts cultured on six positions with increasing density of nanoparticles of three different sizes. The images show a larger number of cells on the gradient of 16 nm nanoparticles. Moreover, cells adhere in large numbers in areas with moderate densities of nanoparticles relative to the smooth end and to the end with the highest density of nanoparticles. Scale bar is 50 μm.

Source: From Goreham et al.[73] © 2013, Royal Society of Chemistry. Reprinted with permission.

junction protein Zona Occludens-1) of human corneal endothelial cells.[74] Nanoscale features have also been reported to enhance the angiogenic capacity of endothelial cells. More specifically, endothelial cells seeded on aligned nanofibrillar scaffolds and implanted in an ischemic limb increased blood perfusion and recovery after 14 days when compared to nonpatterned scaffolds.[75]

The influence of nanotopography on cell function has also been observed when using bioactive nanoceramics, such as hydroxyapatite, bioactive glass (BG), and silicates. Hydroxyapatite nanoparticles (nHA) have been extensively investigated due to their chemical and structural similarity to bone mineral. A recent study showed that nHA could improve proliferation of osteoblasts when compared with microscale hydroxyapatite particles. Specifically, the size effects of 20 nm, 80 nm, and microscale particles were studied using human osteoblast-like cells *in vitro*, and the 20 nm particles had the best effect on cell proliferation.[76] In another study, MSCs seeded on nanofibrous PLGA/nHA scaffolds showed upregulation of alkaline phosphatase (ALP) activity and osteogenic gene expression as well as production of mineralized tissue, when compared with cells seeded on PLGA scaffolds only.[77] This effect was mainly attributed to the osteoconductive property of nHA and increased protein adsorption on nanofibers of PLGA/nHA. PLGA/nHA scaffolds were also investigated for osteochondral regeneration and demonstrated a higher attachment, viability, and proliferation of MSCs than PLGA scaffolds without nHA.[78] Moreover, cartilage defects in rats treated with PLGA/nHA scaffolds with MSCs were completely repaired with regenerated hyaline cartilage after 12 weeks compared to control PLGA scaffolds with MSCs. In combination with naturally derived materials, nHA have also been used as a coating for gelatin scaffolds, resulting in enhanced osteogenic differentiation of marrow-derived MSCs when compared with non-coated gelatin scaffolds.[79]

Nanocomposite hydrogels have also been prepared combining nanoscale bioactive glass (nBG) with synthetic and natural polymers. A bioactive glass, also referred to by its commercial name Bioglass®, is a silicate glass with low silica content, high amount of sodium and calcium, and high calcium/phosphorous ratio. When bioactive glasses are in contact with body fluids, they undergo specific reactions leading to the formation of a hydroxyapatite-like layer, similar to the mineral constituent of bone, on the glass surface.[80,81] In recent years, there has been an increasing interest in the development of nanomaterials comprising bioactive glass due to their good mineralization properties. For instance, nBG and chitosan have been combined to produce nanocomposite membranes, and their interaction with MSCs and periodontal ligament cells (PDLCs) has been investigated.[82,83] The addition of nBG enhanced the proliferation of PDLCs but not that of MSCs. However, both MSCs and PDLCs seeded on chitosan/nBG displayed a higher calcium content than cells seeded on

chitosan membranes, suggesting enhanced mineralized matrix production.[83] Moreover, nanofibrous collagen/nBG scaffolds have been explored for the adhesion, proliferation, migration, and differentiation of MSCs.[84] The addition of nBG to collagen resulted in a significant increase in ALP activity by the cells compared to pure collagen scaffolds. Recently, the addition of silicate nanoparticles to electrospun PCL to induce osteogenic differentiation has been reported.[85] The silicate nanoparticles supported the adhesion and proliferation of human MSCs, and osteogenic differentiation increased as silicate content increased from 0.1% to 10%, as evidenced by enhanced ALP activity and matrix mineralization. Although the mentioned examples suggest the ability of nanostructures to affect the behavior of cells, it is still unclear if the observed effects are due to topography only or to a combined effect of chemical, mechanical, and topographical cues.

Carbon-based nanomaterials, such as CNTs, graphene, and nanodiamonds, have attracted increasing interest due to their unique structural and mechanical properties.[82] The incorporation of CNTs into scaffolds has resulted in enhanced cell attachment, proliferation, and differentiation in several cases. For example, thin films of PEGylated, multiwalled CNTs supported the growth and induced the osteogenic differentiation of MSCs without additional exogenous agents.[82] In another study, poly(acrylic acid)-grafted CNTs led to enhanced viability, adhesion, and neurogenic differentiation of embryonic stem cells when compared to poly-L-ornithine surfaces, which are standard substrates used for neuron culture. This effect was likely because of the topographical cues from CNTs as poly(acrylic acid) alone is known to have negative effects on neuronal cell attachment and differentiation.[82]

As a final example, the effect of nanotopography on cell behavior has also been investigated by developing a high-throughput screening device (TopoChip) that measures cell responses to large libraries of materials with different topographies. Specifically, mathematical algorithms were used to design random surface features, and chips of poly(lactic acid) with 2,176 different topographies were produced. MSCs were seeded on the chips, and by using high content imaging, the surface topographies that were able to enhance proliferation or osteogenic differentiation of the cells were identified.[86,87] In summary, nanoscale topography of synthetic scaffolds has attracted a lot of interest for tissue engineering applications since nanotopography has been shown to affect various cell functions. Nanostructures have been shown to induce a significantly stronger effect than microstructures in terms of morphology, proliferation, and differentiation of cells grown on several materials. Although different hypotheses have been proposed to clarify the mechanism by which nanotopography affects cell behavior, further investigation of these phenomena is needed to support the design of advanced biomaterials.

TRANSPORT AND CONTROLLED RELEASE

Mimicking the transport properties of the native ECM with a natural or synthetic biomaterial that can serve as a carrier for encapsulated cells and/or selectively bind biological molecules of interest is an open challenge in regenerative medicine. The mass transport properties of a carrier biomaterial in a biological context are important in two ways: the transport of soluble molecules through the matrix and the control of binding or release of molecules of interest. The first is relevant for a carrier biomaterial where small molecules including oxygen, glucose, and other nutrients must be able to reach the encapsulated cells via passive diffusion since there are no blood vessels present during *in vitro* culture or immediately after implantation, assuming that fluid flow is absent. The role of oxygen in particular is an important aspect for biomimicry, as different tissues have different needs and measuring oxygen concentration in tissue-engineered constructs is not always straightforward, especially in 3D. However, experiments performed with fluorescent oxygen-sensitive microbeads in cell aggregates[88] and experiments combined with mathematical modeling of hydrogel constructs encapsulating cells[89] have shown that quantitative assessment of oxygen tension was feasible, which could lead to similar approaches for measuring oxygen tension *in vivo*. On the other hand, transport of morphogens is more complex. Several mechanisms have been proposed, from theoretical models which assume that diffusion, which can itself be hindered or facilitated, controls this process to mechanisms based on cellular transport such as transcytosis (transport through the cells) or transport via cytonemes (long cellular extensions).[90]

Quantification of the mass transport of molecules of interest within the native biological environment or in biomimetic materials can be done via a number of techniques. Two particularly interesting methods are fluorescence recovery after photobleaching (FRAP)[91] and raster image correlation spectroscopy (RICS).[92] Both techniques require fluorescently tagged solutes and are performed using a confocal microscope. In FRAP, a region of interest (ROI) is photobleached, and as molecules diffuse into and out of the ROI, the fluorescence intensity is gradually regained. The recovery curve of fluorescence over time yields information about the kinetics of the system from which the diffusion coefficient can be calculated. RICS, on the other hand, is based on a very different principle. In confocal microscopy, there is a known time lag between two consecutive pixels when an image is scanned. RICS takes advantage of this spatial and temporal information contained in the images to calculate the diffusion coefficient using an autocorrelation function. Both FRAP and RICS have advantages and disadvantages, depending on the specific application.[93] FRAP requires a strong fluorescent signal, and thus a high concentration of the fluorescent solute is necessary for a successful measurement. RICS, on the other hand, requires much lower concentrations to quantify molecular transport. Further, RICS can be applied to study the localized properties of a small ROI because of its high spatial resolution, whereas FRAP gives an average view of the system kinetics and, thus, is less influenced by possible outliers. Finally, RICS may yield more information about binding in addition to diffusivity; however, FRAP is more robust when it comes to quantifying transport in unknown systems.

The second important transport property of biomimetic materials is the potential of the substrate to bind molecules of interest, usually growth factors or other proteins, thus preventing them from freely diffusing out of the matrix. This feature is particularly interesting for applications in regenerative medicine, as it allows a tissue-engineered construct to carry not only cells but also growth factors to enhance cell signaling toward a desired differentiation state.[9,94] Growth factors that are not bound to or otherwise retained in the carrier material rapidly diffuse out of the therapeutic region. This requires higher doses, and negative side effects can emerge.[95] Several strategies have been developed for the controlled release and delivery of growth factors from ECM substitutes (Fig. 5).[9,14,96–99] Physical entrapment is the simplest case where a growth factor, or a nano-/microparticle loaded with the growth factor, is kept within the pores of a polymer network (Fig. 5A). Entrapment occurs due to the larger hydrodynamic radius of the molecule or the nano-/microparticle compared to the pore radius of the network. Swelling or degradation of the matrix or the carrier particle dictates the release rate of the molecule of interest. Natural or synthetic hydrogels can also serve as growth factor carriers by inherent or chemically modified affinity binding sites (Fig. 5B). For example, fibrin contains heparin-binding domains where several growth factors can also bind.[100] Alternatively, growth factors can be fused with peptides to bind selectively with the ECM (Fig. 5C) or within biomaterials containing affinity binding domains. For example, a heparin-binding domain was added to IGF-1 that allowed the growth factor to be retained in the cartilage through binding to chondroitin sulfate.[101] Further, heparin can be covalently attached to a polymer matrix so that molecules with an affinity for heparin can bind to the matrix (Fig. 5D).[102] In another study, a heparin-binding peptide was covalently attached to fibrin, and exogenous heparin was used as a bridge to retain nerve growth factor in the modified matrix longer than in fibrin alone.[103] In addition, more sophisticated technologies have been implemented for tethering molecules within hydrogel matrices, for instance by fusion of growth factors with substrates of crosslinking agents like the transglutaminase factor XIIIa (Fig. 5Ei).[104] Alternatively, a growth factor can be tethered directly to a synthetic network (Fig. 5Eii); for example, transforming growth factor-β (TGF-β) was covalently attached to PEG hydrogels.[105] Both strategies result in the covalent attachment of growth factors to the biomaterial network, and they are released according to the

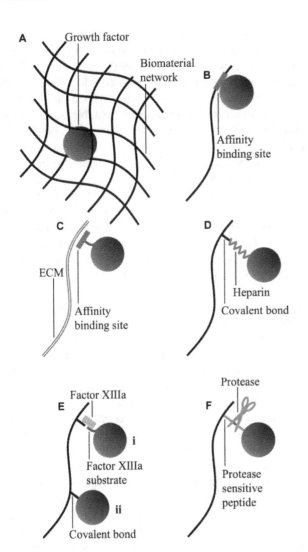

Fig. 5 Binding strategies for the controlled release of growth factors from a biomaterial matrix. (**A**) If the hydrodynamic radius of a growth factor is larger than the pore size of the network, the growth factor can be physically entrapped. (**B**) Growth factors can bind to a biomaterial matrix via affinity binding sites (e.g., BMP-2 to fibrin via the heparin-binding domain). (**C**) Affinity binding domains can be fused with the growth factor to bind to the ECM. (**D**) Heparin can be covalently attached to the biomaterial matrix so that growth factors with an affinity for heparin can bind to the matrix. (**E**) A growth factor can be fused with a substrate for factor XIIIa, so that the latter can mediate the formation of a covalent bond between fibrin and the growth factor (i), or the growth factor can be chemically modified to create a covalent bond directly with the polymer matrix (ii). (**F**) The growth factor can be attached to the matrix via protease sensitive peptides, so that it can be released later in a controlled manner via cell secreted enzymes that show specificity for these peptides.

degradation rate of the matrix. Finally, another approach is the incorporation of protease-sensitive peptides to mediate growth factor release from the polymer network (Fig. 5F). These peptides are selectively cleaved by cell-secreted enzymes, such as collagenase or plasmin,[106] and can be incorporated either in the linker region for chemically

bound molecules or within the hydrogel backbone/cross-links.[107,108]

In addition to the above matrix-mediated approaches, biomimetic controlled release strategies can also be used for intracellular delivery of molecules of interest. Natural transport vesicles are self-assembled structures usually composed of phospholipids, molecules with a hydrophilic head and a hydrophobic tail of fatty acids, and they provide a mechanism for intracellular transport of fatty acids and hydrophobic proteins, protecting them from the aqueous environment of the cytoplasm.[109,110] These vesicles have been replicated by forming micelles or polymersomes from block copolymers with hydrophobic and hydrophilic domains (for example, Pluronic®).[111] Many characteristics, for example, the size of the vesicles or the formation of mono- vs. bilayer membranes or inverted micelles, can be controlled by the molecular weight of and the ratio between blocks of the copolymer. By adding a drug to a solution of these block copolymers, the drug becomes encapsulated in vesicles formed by the block copolymers, just like proteins in transport vesicles.[112] The synthetic vesicles can also be functionalized with ligands to target specific sites/cells, where they release their encapsulated drug either due to degradation or in response to a stimulus such as change in pH.[113] Overall, it is clear that biomimetic materials used as cell carriers and constructs for the controlled release of therapeutic molecules need to be adaptable to their environment. Like their biological counterparts they should not be inert; on the contrary, they need to respond to external stimuli to be successful.

BIOACTIVE MATERIALS AND BIOCHEMICAL MODIFICATIONS

Biomimetic materials that are also bioactive can replace and mimic the native environment of the cell by provoking specific biological responses. These materials can be harvested from natural sources (e.g., ECM),[114] can be created by artificial means (e.g., synthetic self-assembling peptides),[115] or can be semisynthetic or hybrid materials (e.g., collagen/poloxamine matrices).[116] Naturally derived materials retain many of the functions of the native ones and have been used to support tissue reconstruction and/or regeneration, with skeletal muscle as an example.[117,118] One such natural biopolymer is fibrin, which constitutes the main part of the blood clot, the final product of the coagulation cascade. Its primary component is derived from fibrinogen, a 340 kDa protein found in blood plasma. When thrombin, a serine protease, comes into contact with fibrinogen it cleaves off fibrinopeptides A and B to form the fibrin monomer. These monomers spontaneously assemble in a half-staggered fashion to form protofibrils, which then aggregate laterally and form fibers.[119] Ultimately, fibrin consists of a random network of fiber bundles with individual fiber characteristics and the network organization dependent on the polymerization conditions. Several studies have focused on the effect of conditions, such as fibrinogen,[120] thrombin,[121] factor XIII,[122] calcium,[123] and chloride[124] concentrations; ionic strength;[125] and pH,[126] on the structure of fibrin hydrogels. These differences in structure lead to significant changes in other aspects, for example, the mass transport or mechanical properties, of the hydrogels. Since fibrin hydrogel preparation is a process easily recapitulated *in vitro* starting from the isolated natural components, it is a biomaterial that can be tailored to fit the required structural, diffusional, and mechanical properties of a specific application. At the same time, several growth factors have been shown to have an affinity for fibrin, via the heparin-binding domain,[100] which makes fibrin hydrogels a natural biomimetic carrier for these molecules. As such, they have been extensively studied as potential carrier materials in regenerative medicine applications.[127] While the properties of fibrin hydrogels can be easily manipulated, this is not true for all naturally derived materials, and they typically suffer from a potential immunogenic response following implantation, batch-to-batch variability, and difficulty to control their shape, size, and mechanical properties.[128]

On the other hand, synthetic materials, such as polymeric macromolecules, can be designed to be tunable, tailorable, and biodegradable.[129–131] They can also be manipulated to form structures such as fibers in order to mimic the ECM, as described earlier. Nevertheless, they often lack the biological cues required for controlling cell behavior, such as adhesion,[132] proliferation, and differentiation, and cellular processes, such as respiration, glycolysis, and so on. Semisynthetic materials are hybrid materials that contain natural and synthetic components that are covalently or physically linked to each other, and they address many of the limitations given by synthetic and natural materials individually.[133] Hybrid materials can be designed in two different ways. A natural component can be used as the starting material and modified with synthetic components to create 3D scaffolds. Alternatively, the starting material can be synthetic, to which naturally derived biomolecules are grafted. Polymers are often used as the synthetic component because of their ease of modification, whereas peptides and proteins are typically used to provide a specific biological activity to the conjugates.[134–137] Ringsdorf first proposed the concept of polymer biomacromolecule conjugates as a strategy to create bioactive polymeric systems.[138] The methods to synthesize protein polymer conjugates as well as some of their applications are discussed in a thorough review published in 2015.[139]

PEG is one of the most investigated polymers for conjugation to proteins or peptides. On the one hand, cells and proteins do not readily attach to PEG,[140] and on the other hand, the use of PEG reduces immunogenicity and antigenicity compared to other synthetic polymers[141] and decreases protein degradation by mammalian cells and enzymes.[142] Meanwhile, the peptides/proteins will improve the biomimetic aspect of the conjugates due to their biocompatibility and their biological functions. Protein- and peptide-functionalized PEG materials have been widely used to generate hydrogel scaffolds for the growth and differentiation of cells[143] as depicted in Fig. 6A. Because of their biocompatibility and biodegradability, proteins such as albumin, collagen, and elastin-like polypeptides have been conjugated to different synthetic polymers, including PEG.[144–146] Semisynthetic materials have already shown their effectiveness as biocompatible materials with tunable mechanical and physico-chemical properties as well as bioactivity. For example, photocrosslinked thiolated gelatin-PEG diacrylate hydrogels supported spreading of neonatal human dermal fibroblasts and the formation of a cellular network over 28 days, whereas fibroblasts encapsulated in physically crosslinked hydrogels remained spheroidal.[147] In another case, full-length proteins were reversibly patterned within polymeric hydrogels using three orthogonal chemical reactions. The resulting system enabled attachment and subsequent removal of vitronectin (a glycoprotein that binds to integrins and promotes cell adhesion and spreading) with good control in space and time, thus allowing selective differentiation of human MSCs to osteoblasts.[148]

Many materials that are not inherently biomimetic can be made bioactive using synthetic peptides. These peptides can be based on the shortest amino acid sequence from proteins that exhibit the desired biological response or can be identified by screening of peptide libraries for a specific functionality. For example, a peptide mimetic (Ile-Gly-Lys-Tyr-Lys-Leu-Gln-Tyr-Leu-Glu-Gln-Trp-Thr-Leu-Lys or IGKYKLQYLEQWTLK or QK) of vascular endothelial growth factor (VEGF) was attached to decellularized ECM coated on substrate surfaces via a copper(I)-catalyzed alkyne-azide cycloaddition (Fig. 6B). The QK peptide mimics the helical structures at the binding site of VEGF and exhibits a high binding affinity to VEGF receptors, thus promoting angiogenesis.[149] Endothelial cells cultured on the QK-functionalized material showed a significantly enhanced angiogenic response as indicated by the increased formation of branched tubular networks.[150] Further, the integrin-binding motifs from several ECM proteins have been determined and artificially synthesized so that they can be incorporated as short amino acid chains in biomimetic materials.[132] The most widely used of these binding motifs is the Arg-Gly-Asp (RGD) sequence,[151] which is found in both fibronectin and vitronectin. Short peptide sequences containing RGD can be attached to polymeric materials and promote cell attachment and differentiation as full-length proteins do.[152–154] As an example, a peptide containing RGD and Pro-His-Ser-Arg-Asn (PHSRN; a cell adhesive peptide sequence also from fibronectin), linked together via a polyglycine sequence to recapitulate the native spacing within fibronectin, was prepared and attached to PEG. A hydrogel formed from this PEG-peptide conjugate

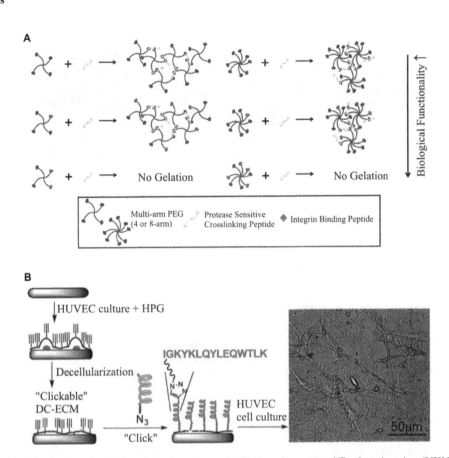

Fig. 6 (**A**) Crosslinking mechanism for RGD-modified multi-arm PEG (4- or 8-arms) and Tyr-Lys-Asn-Arg (YKNR) crosslinking peptides through a Michael-type addition reaction. (**B**) Attachment of the QK peptide onto modified decellularized ECM via a copper-catalyzed click reaction.
Source: (**A**) From Kim et al.[233] © 2016, Royal Society of Chemistry. Reprinted with permission. (**B**) From[150] © 2014, American Chemical Society. Reprinted with permission.

improved the adhesion, spreading, and proliferation of osteoblasts as well as the formation of focal adhesions when compared to controls (PEG with RGD alone and unmodified PEG).[155] The affinity of integrins for ECM proteins has been shown to be influenced by the mechanical properties of the substrate, particularly stiffness, in addition to the biochemical environment.[156] Thus, cell attachment is determined not solely by ligands, but the bulk material properties can also have an influence. Mechanobiology is an emerging area of research[157,158] that should soon lead to a better understanding of the biological response to such biomimetic materials.

In addition to their bioactivity, peptides and proteins have been widely investigated in view of their inherent tendency to self-assemble and form supramolecular architectures as shown in Fig. 7.[159–161] Physical interactions between molecules that self-assemble to form different nanostructures can be used to create scaffolds for tissue engineering.[162,163] These scaffolds have the advantage of forming nanofibrous networks that structurally resemble the collagen fibrils of the native ECM. For example, a peptide containing four repeats of Arg-Ala-Asp-Ala (RADA-16) and its derivatives are some of the

most investigated self-assembling peptides in tissue engineering. RADA-16 self-assembles in aqueous phase into β-sheet structures, leading to the formation of nanofibers (Fig. 7A). Hydrogels based on RADA-16, with additional bioactive peptide functionalization, have been used for several applications such as angiogenesis and bone and cartilage regeneration, and advances have been summarized in recent reviews.[164,165] As an example, Ile-Lys-Val-Ala-Val (IKVAV), a ligand for integrin β1 that is derived from laminin, has been attached to RADA-16 and favored neuronal differentiation of progenitor cells; however, the unaltered RADA-16 allowed less differentiation overall and directed cells nonspecifically toward both neuronal and astroglial lineages.[166,167] Moreover, IKVAV-functionalized RADA-16 also drove pluripotent P19 embryonic carcinoma cells toward neuronal differentiation.[168] Shorter peptides such as Leu-Ile-Val-Ala-Gly-Asp (LIVAGD or LD$_6$) (Fig. 7B) and Gln-Gln-Arg-Phe-Glu-Trp-Glu-Phe-Glu-Gln-Gln (QQR-FEWEFEQQ or P11-4) that also self-assemble into fibers represent alternatives to RADA-16 for creating peptide-based scaffolds because of their ease of synthesis.[161,169–172]

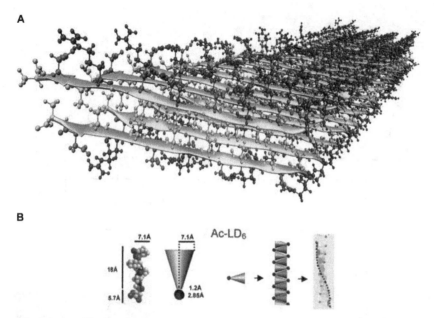

Fig. 7 (**A**) 3D representation of a nanofiber formed from RADA16-I. (**B**) Schematic of the formation of a fiber by stacking of Ac-LIVAGD (Ac-LD$_6$) peptide monomers.
Source: (A) From Cormier et al.[234] © 2013, American Chemical Society. Reprinted with permission. (B) From[235] © 2011, Elsevier. Reprinted with permission.

Designing or engineering new (semi)synthetic biomaterials for biomedical applications represents a promising strategy to overcome the limitations given by the naturally derived materials; however, they could still trigger an immune response. Therefore, the immunogenicity of these materials needs to be thoroughly studied before their clinical use. This aspect of the (semi)synthetic biomaterials has been investigated by several researchers and is discussed in the next section.

THE HOST RESPONSE TO BIOMIMETIC MATERIALS

Every process of tissue repair by a biomaterial implant initiates a local host response. Just after implantation, blood serum proteins are deposited on the implant's surface and activate the coagulation cascade, complement system, platelets, and immune cells, resulting in provisional matrix formation and an inflammatory response (Fig. 8).[173] During the early phases of the host response, there is recruitment of pro-inflammatory neutrophils and/or macrophages, which secrete cytokines and recruit leukocytes and other cell types involved in the inflammatory response and wound healing/remodeling. In some cases, the continued presence of these cells can lead to secondary damage and persistent activation, with macrophage fusion to form foreign body giant cells (FBGCs) that surround the implant in a fibrotic scar tissue. On the other hand, the immune response can contribute strongly to the regeneration of damaged tissues.[174] Biomaterial properties like size, shape, morphology, and physico-chemical and biological surface modifications can

modulate the intensity and duration of the host response to their implantation,[175,176] and these properties are frequently modified when one is trying to make a material more biomimetic.

Biomimetic materials that aim to reproduce features of the ECM may affect the immune response by enhancing or suppressing normal immune cell functions.[175] Natural materials (e.g., collagen, fibrin, and chitosan) are generally considered as biocompatible and can have inherent bioactivity, as described earlier, but the host response to their implantation can vary depending on the species of origin, processing technique, and sterilization method. In most cases, they do not promote chronic inflammation, but they can transmit pathogens and promote an immunogenic response.[177] Synthetic materials like PLGA, PEG, and polyvinyl alcohol (PVA) can be functionalized to promote host tissue interaction, for example, by using surface modification to control cell adhesion[178] or provide resistance to protein adsorption and nonspecific cell attachment,[179] thus affecting the immune response. Semisynthetic materials can also lead to an immune response if the bioactive component acts as an antigen.

The properties that render a material biomimetic, as described throughout this entry, can also influence the host response to these materials. Strategies to limit the intensity of the host response involve modifications of the size of the implant as well as the surface topography. The hydrophilicity/hydrophobicity and charge of the material surface can directly affect biocompatibility, which is defined as the "ability of a material to perform with an appropriate host response in a specific application."[180] For example, a decrease in monocyte and macrophage adhesion and activation has been seen on hydrophilic[181] or anionic[182] surfaces in comparison to hydrophobic or cationic ones, respectively. Aligned nanofibrous scaffolds generated by electrospinning also minimized the host response through a decrease in monocyte adhesion and the formation of a thinner fibrous capsule when compared with random fiber biomaterials.[183] Macrophage fusion to FBGCs can also be affected by material dimensions and topography. For example, spheres (made from hydrogels, ceramic, metal, or plastic) with a diameter bigger than 1.5 mm decreased the degree of macrophage activation while a diameter of 0.5 mm allowed activation and fusion.[184] Mimicking fibrillar orientation in the natural ECM by creating parallel gratings on polymeric surfaces with line widths ranging from 0.5–2 μm affected macrophage morphology and functionality *in vitro*. Cells elongated on the aligned surfaces and presented differing cytokine secretion profiles when cultured on the different topographies. At the same time, macrophage adhesion and fusion on the gratings was reduced *in vivo*.[185] Porosity is another characteristic that can influence macrophage adhesion, activation, and fusion as well as fibrous encapsulation.[186,187] For instance, biomaterials with pore sizes ranging from 30 to 40 μm limited FBGC formation and enhanced tissue repair, when compared with materials with a pore size of 20 μm, and decreased the pro-inflammatory phase, when compared with materials with a pore size of 60 μm or with solid materials.[186]

Biomaterial functionalization by incorporation of bioactive molecules not only promotes specific cell adhesion but also modulates the immune response. For example, materials functionalized with RGD alone resulted in rapid FBGC formation by increasing both the number of adherent macrophages and their fusion rate while materials functionalized with both RGD and PHSRN demonstrated low macrophage adhesion and activity.[188] An interpenetrating polymer network of gelatin and PEGylated RGD containing soluble keratinocyte growth factor-1 (KGF-1), which stimulates endothelial cell growth, improved long term healing of full-thickness skin wounds in a rat model when compared with a control of unmodified gelatin. After 3 weeks of implantation, the engineered material resulted in lower macrophage and fibroblast densities, a reduction in the formation of a foreign body capsule, and a higher extent of ECM organization.[189]

Polymeric biomaterials conjugated with bioactive molecules can also be applied as tools to modulate the immune response as vaccines or in the treatment of cancer by taking advantage of natural transport mechanisms. For example, polymeric micelles modified by a mannose receptor-targeting motif delivered small interfering RNA (siRNA) to primary macrophages at a higher rate than the one obtained for the same carrier without a specific surface target, leading to 87% gene knockdown.[190] The delivery of siRNA-containing micelles via mannose receptors (CD206) modulated the immune response by master gene silencing and is hypothesized to be a potential alternative to cancer treatments as tumor-associated macrophages, which secrete immunosuppressive cytokines [e.g., interleukin 10 (IL-10) and TGF-β] and contribute to tumor progression, also upregulate CD206.[191]

Polymers with controllable mass transport properties have also been studied as a technique for treating type I diabetes by microencapsulation of islets in a semipermeable membrane to avoid the need for immunosuppressive therapy after transplantation. The use of amphiphilic PEG-phosphatidylethanolamine (DPPE) with an immobilized human endoderm cell line on the surface of the engineered islets allowed insulin secretion upon glucose stimulation.[192] A multilayered, ultra-thin membrane of PVA-PEG-DPPE was also developed for long-term encapsulation of islets, because of the stability of PVA, and did not impede cell viability or their ability to release insulin.[193]

Ultimately, the host response is responsible for the acceptance or rejection of a biomaterial and can be tailored by strategies that control different steps in the process, from the initial acute inflammatory phase through the final fibrous encapsulation. Current studies are focused not only on passive or inert biomaterials but also on active biomaterials that combine controllable mechanical properties from synthetic materials with active biological components for an approach where the host response is not avoided but is guided for the desired tissue response and eventual biomaterial acceptance.

TISSUE ENGINEERING AND REGENERATIVE MEDICINE APPLICATIONS

Biomimetic materials play a vital role in the success of tissue engineering and regenerative medicine therapies. Tissue engineering is a field that aims to create a functional construct that combines cells, biological molecules, and a 3D scaffold to heal and restore the function of damaged tissues and organs. At present, organ transplantation and autologous grafts persist as the primary options for several tissue defects, although complications arise. The drive for a better alternative has brought the research of biomimetic materials to the forefront because they can be tailored to accelerate naturally occurring phenomena in healthy subjects or to induce cellular responses that might not be normally present in damaged/diseased tissue.[194] In addition, many biomimetic materials are degraded by the host as a new ECM is generated, which addresses limitations of materials of a more permanent nature, such as metals or ceramics.[195] Although natural materials have been used to create 3D scaffolds as therapies for several tissue types,[196] the desire for greater control over the material properties and the cellular response is leading toward the use of synthetic biomaterials integrated with bioactive components.[11] Biomimetic materials to enhance the regeneration of numerous tissues have been developed, but for the scope of this entry only orthopedic and cardiovascular applications are discussed.

Orthopedic tissue engineering, for tissues such as bone, cartilage, and tendon, is one of the most researched areas in the field. Bone formation during both development and fracture healing follows a similar path where cells undergo sequential events beginning with condensation and proliferation of osteoprogenitor cells, followed by differentiation, matrix formation, and mineralization. While the majority of fractures heal on their own, those that do not are bigger than a critical size, and they often contain a highly compromised environment. Bone tissue engineering scaffolds can support bone tissue regeneration by osteoinduction (the process of initiating bone formation via the recruitment and stimulation of immature, undifferentiated cells) and/or osteoconduction (the growth of bone neo-tissue on the surface)[197] while providing sufficient mechanical and structural stability as well as enabling blood vessel formation and bone marrow production.[198] Naturally derived materials have been used for this application, both clinically and in preclinical research. For example, by designing a porous sponge of collagen type I, which comprises 95% of the collagen content and 80% of all proteins in bone and is responsible for the mechanical properties and toughness of bone,[199] combined with recombinant human bone morphogenetic protein (BMP), a powerful osteoinductive growth factor,[200] researchers were able to successfully regenerate an ulnar defect in rabbits[201] as well as a radius defect in Rhesus monkeys.[202] These positive results led to the commercialization and FDA approval of this combination (Infuse®) for spinal procedures, treatment of tibial shaft fractures, and oral maxillofacial surgeries. As another example, silk-based scaffolds combined with different compositions of calcium phosphate, such as hydroxyapatite (comprising 70% of natural bone mineral),[203] demonstrated increased osteogenic potential *in vitro* as well as *in vivo*, which was hypothesized to be due to their maintenance of mechanical properties, interconnected pore structure, improved particle distribution, and increased nucleation sites for mineral deposition.[204,205] On the other hand, there has been a shift toward using synthetic polymers as the base scaffold material.[195] For instance, following a biomimetic approach (Fig. 9), scaffolds combining a synthetic polymer such as PEG with proteolytically degradable crosslinks in the hydrogel network, growth factors such as VEGF or BMP-2 or TGF-β, and integrin-binding sequences such as RGD or Gly-Phe-Hyp-Gly-Glu-Arg (GFOGER) (Fig. 9A) displayed highly successful bone forming capacities *in vitro* as well as *in vivo* in mouse and rat models[206–208] (Fig. 9B&C).

In contrast to bone, damaged cartilage almost never heals by itself, and biomimetic scaffolds can play an even more crucial role here. Cartilage-mimicking materials need to be able to provide structural and mechanical support in addition to being able to sequester large amounts of water,

which forms part of the native cartilaginous ECM.[209] Early biomimetic materials used in cartilage regeneration were either purely synthetic (e.g., PEG) or purely natural, particularly hyaluronic acid, which is a major component of the cartilage ECM.[210,211] Since then, hybrid scaffolds of synthetic polymers and natural materials have garnered attention as they take advantage of the properties of both components. For example, combining a synthetic polymer with different types of glycosaminoglycans (GAGs) led to varying levels of chondrogenesis, wherein certain combinations led to higher levels of type II collagen production, indicative of superficial cartilage tissue, and other combinations led to higher levels of type X collagen, a sign of hypertrophic cartilage, thus mimicking the different layers of articular cartilage.[212–215] Furthermore, advances in micro- and nanoscale fabrication, via fiber weaving, electrospinning, macromolecular self-assembly, and so on, have enabled researchers to control the scaffold architecture and allow for better spatial distribution of signaling molecules, resulting in scaffolds with varying levels of heterogeneity in composition, again mimicking the native cartilage ECM.[216–218] When it comes to tissue-engineered ligaments, several groups have explored ligament-like scaffolds, but these have demonstrated only a limited success.[219] For example, a biomimetic, 3D, polymer-based, braided scaffold displayed success *in vitro* as well as led to healing of an anterior cruciate ligament (ACL) injury *in vivo* in a rabbit model.[220,221]

Cardiovascular tissue engineering is a highly challenging field of research. This is due to the complexity of blood vessel regeneration, which is initiated by endothelial cell organization into tubes that are later stabilized by smooth muscle cells and pericytes and requires a cascade of growth factors (VEGFs, ephrins, angiopoetins, etc.). For large blood vessels, synthetic grafts made of elastomeric polymers such as polyurethanes have been used as an alternative to autologous vessel grafts. However, these synthetic grafts led to complications such as thrombus formation, mechanical noncompliance with adjacent vessels, and failures in narrow diameter vessels.[222,223] In a biomimetic approach, they were combined with cell adhesion ligands (RGD, Tyr-Ile-Gly-Ser-Arg or YIGSR, and Arg-Glu-Asp-Val or REDV), which allowed for better endothelial cell attachment and spreading, reduced thrombus formation, and enhanced cellular retention on the graft under shear stress.[224,225] To stimulate angiogenesis, PEG scaffolds with covalently bound ephrin-A1 resulted in improved endothelial cell attachment and increased vessel density *in vitro* as well as *in vivo* in a mouse cornea micropocket model.[226] In this model, the biomaterial is placed within a surgically created pocket in the cornea, thus taking advantage of the avascular nature of the cornea to investigate new vessel formation (Fig. 10). In a similar manner, scaffolds with a low sustained release of VEGF led to more stable vessel formation as opposed to scaffolds with a burst release that led to leaky vessels.[227]

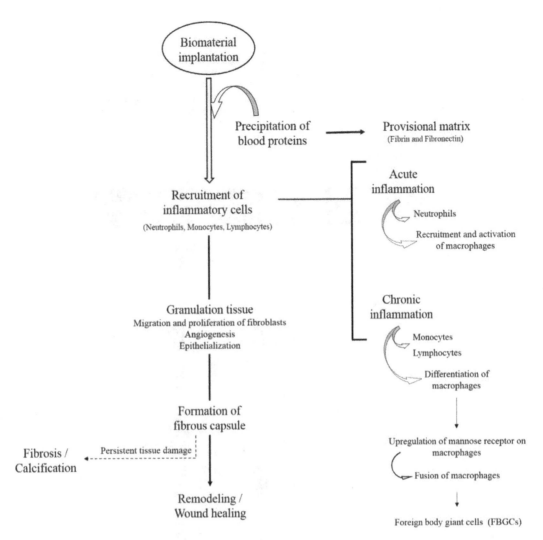

Fig. 8 Scheme of events that happen during the host response to an implanted biomaterial. The process starts with an injury caused during the implantation of the biomaterial that gives rise to a cascade of reactions that recruit and activate related cells at different stages. The host response can lead to both positive (wound healing and tissue remodeling) and negative (fibrosis, calcification, foreign body giant cells) outcomes.

Subsequent release of platelet-derived growth factor BB (PDGF-BB) helped in the recruitment of smooth muscle cells, which favored vessel maturation.[228] Lastly, when it comes to cardiac tissue engineering, a challenge is to mimic the cellular organization of the heart muscle without sacrificing tissue vascularization. The myocardium contains a dense, overlapping array of muscle cells arranged in different circumferential orientations. One of the simplest biomimetic scaffolds was a collagen hydrogel combined with cardiac myocytes that displayed interconnected, beating cells when implanted around rat hearts.[229] To recapitulate the organization of the myocardium, a biodegradable, overlapping nanofibrous PCL mesh allowed for the attached cardiac myocytes to have strong cellular contacts and synchronous beating after two weeks of *in vitro* culture. In addition, the cells displayed an *in vivo*-like cellular morphology during this *in vitro* culture.[230] Further, lanes of cardiomyocytes on degradable polymer films grew up to three cell layers thick and formed an organized, functional tissue.[231] Finally, microstructured honeycomb PGS biodegradable sheets, described in an earlier section here, elicited the directional alignment of neonatal rat heart cells. The cells displayed anisotropic electrical excitation thresholds, which are important for directional contractions and efficient blood transfer.[60]

The potential for biomimetic materials in tissue engineering and regenerative medicine is high. Because of the complexity of each tissue type, it is important to understand the structural, mechanical, and biological properties of the tissue prior to choosing the materials for a given application. Although the aforementioned biomimetic materials have helped to push the boundary of tissue engineering, there still is a lot of research that needs to be done, especially for the purpose of clinical translation.

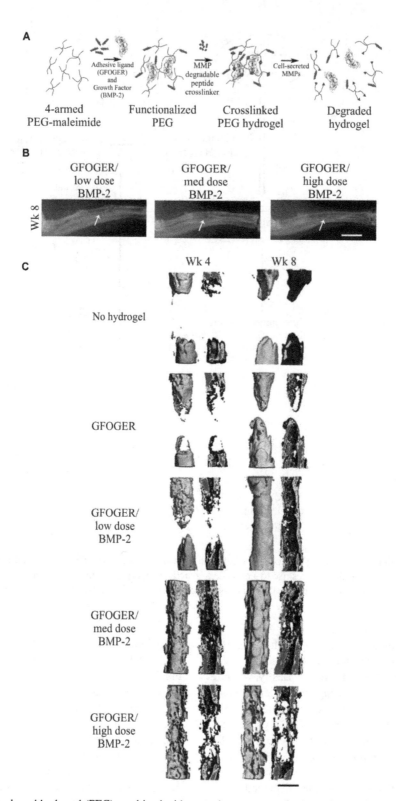

Fig. 9 Synthetic polymer-based hydrogel (PEG) combined with natural components for bone tissue engineering. (**A**) Schematic of the synthesis of a biomimetic, proteolytically degradable, 3D PEG hydrogel that is functionalized with an adhesive ligand (GFOGER) and a growth factor (BMP-2). (**B**) Radiographic images of radial segmental defects in wild type mice that were treated with the biomimetic hydrogels shown in (**A**), leading to normal ulnar structure. Arrows indicate space between ulna and radius, which is not present in the high BMP-2 dose image; scale bar is 2 mm. (**C**) 3D reconstructions of microCT images of radii in sagittal view (left) with mineral density mapping (right) for the various treatment groups; scale bar is 1 mm.
Source: From Shekaran et al.[207] © 2014, Elsevier. Reprinted with permission.

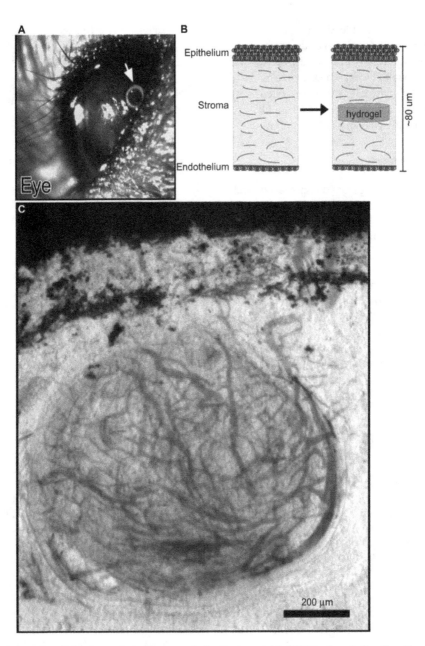

Fig. 10 Newly formed, functional blood vessels in 3D, proteolytically degradable PEG hydrogels functionalized with a pro-angiogenic growth factor (VEGF) and an adhesive ligand (RGD). (**A**) Image showing location of the small incision that was made in the cornea of mice to create a micropocket for hydrogel implantation. (**B**) Schematic of location of hydrogel in relation to host tissue. (**C**) Image showing a dye that was injected intravenously into the mice and perfused through the newly formed blood vessels within the hydrogel, indicating functionality of the newly formed vessels.
Source: From Moon et al.[236] © 2013, Elsevier. Reprinted with permission.

CONCLUSIONS

Biomimetic materials that recapitulate the structure and topography, transport and mechanical properties, and biochemical composition of the native ECM are currently being researched and developed. This entry has reviewed the state of the art for these materials, which are often naturally derived or combine synthetic polymers to allow control of

material properties with bioactive components to influence the cellular response and tissue regeneration. A number of synthesis and fabrication techniques have been developed to allow processing of these materials into complex architectures that more closely mimic the natural tissue structure, to control their mechanical properties and topography, and to allow for controlled presentation and/or release of bioactive molecules. These features have been shown to influence cell behavior *in vitro* and lead to tissue regeneration *in vivo*, which points to the potential of these technologies. The material properties can also affect the host response after implantation. Although few examples are currently in clinical use, biomimetic materials are being extensively researched in the field of tissue engineering and are likely to form the basis of regenerative therapies of the future.

ACKNOWLEDGMENTS

The authors acknowledge funding from the special research fund of the KU Leuven [grant nos. CREA/13/017 (S.P., A.H.S., and J.P.) and IDO/13/016 (S.P., A.H.S., and J.P.)], the Research Foundation Flanders [FWO, grant nos. G.0B39.14 (R.L.D.G. and J.P.), G.0858.12 (K.A.L.), and G.0982.11N (A.K.K.)], the People Programme (Marie Curie Actions) of the European Union's Seventh Framework Programme (FP7/2007–2013) under REA grant agreement no. 608765 (A.H.S.), ERC Grant Agreement nos. 308223 (K.A.L.) and EU-FP7-REJOIND-RTD (A.K.K.), and the Conselho Nacional de Desenvolvimento Científico e Tecnológico of Brazil (T.O.M.).

REFERENCES

1. Oeffner, J.; Lauder, G.V. The hydrodynamic function of shark skin and two biomimetic applications. J. Exp. Biol. **2012**, *215* (5), 785–795.

2. Autumn, K.; Liang, Y.A.; Hsieh, S.T.; Zesch, W.; Chan, W.P.; Kenny, T.W.; Fearing, R.; Full, R.J. Adhesive force of a single gecko foot-hair. Nature **2000**, *405* (6787), 681–685.

3. Mittal, R.; Utturkar, Y.; Udaykumar, H. Computational modeling and analysis of biomimetic flight mechanisms. 40th AIAA Aerospace Sciences Meeting & Exhibit. Aerospace Sciences Meetings: American Institute of Aeronautics and Astronautics; 2002.

4. Zari, M.P. Biomimetic design for climate change adaptation and mitigation. Archit. Sci. Rev. **2010**, *53* (2), 172–183.

5. Hall, S.J. *Basic Biomechanics*. 6th ed: Mc Graw Hill: New York; 2012.

6. Lund, P.K.; Abadi, D.M.; Mathies, J.C. Lipid composition of normal human bone marrow as determined by column chromatography. J. Lipid. Res. **1962**, *3* (1), 95–98.

7. Giancotti, F.G.; Ruoslahti, E. Integrin signaling. Science **1999**, *285* (5430), 1028–1032.

8. Arnold, M.; Cavalcanti-Adam, E.A.; Glass, R.; Blummel, J.; Eck, W.; Kantlehner, M.; Kessler, H.; Spatz, J.P. Activation of integrin function by nanopatterned adhesive interfaces. Chemphyschem **2004**, *5* (3), 383–388.

9. Mitchell, A.C.; Briquez, P.S.; Hubbell, J.A.; Cochran, J.R. Engineering growth factors for regenerative medicine applications. Acta Biomater. **2016**, *30*, 1–12.

10. Kim, D.H.; Lee, H.; Lee, Y.K.; Nam, J.M.; Levchenko, A. Biomimetic nanopatterns as enabling tools for analysis and control of live cells. Adv. Mater. **2010**, *22* (41), 4551–4566.

11. Lutolf, M.P.; Hubbell, J.A. Synthetic biomaterials as instructive extracellular microenvironments for morphogenesis in tissue engineering. Nat. Biotechnol. **2005**, *23* (1), 47–55.

12. Fernandez-Yague, M.A.; Abbah, S.A.; McNamara, L.; Zeugolis, D.I.; Pandit, A.; Biggs, M.J. Biomimetic approaches in bone tissue engineering: Integrating biological and physicomechanical strategies. Adv. Drug. Deliv. Rev. **2015**, *84*, 1–29.

13. Padmanabhan, J.; Kyriakides, T.R. Nanomaterials, inflammation, and tissue engineering. Wiley. Interdiscip. Rev. Nanomed. Nanobiotechnol. **2015**, *7* (3), 355–370.

14. Patterson. J.; Martino. M.M.; Hubbell. J.A. Biomimetic materials in tissue engineering. Materials Today **2010**, *13* (1–2), 14–22.

15. Weiner, S.; Wagner, H.D. The material bone: Structure-mechanical function relations. Ann. Rev. Mat. Sci. **1998**, *28* (1), 271–298.

16. Gibson, L.J.; Ashby, M.F. *Cellular Solids: Structure and Properties*. Cambridge, UK: Cambridge University Press; 1999.

17. Wegst, U.G.; Bai, H.; Saiz, E.; Tomsia, A.P.; Ritchie, R.O. Bioinspired structural materials. Nat. Mater. **2015**, *14* (1), 23–36.

18. Wegst, U.G.K.; Ashby, M.F. The mechanical efficiency of natural materials. Philosophical Magazine **2004**, *84* (21), 2167–2186.

19. Baker, S.C.; Rohman, G.; Southgate, J.; Cameron, N.R. The relationship between the mechanical properties and cell behaviour on PLGA and PCL scaffolds for bladder tissue engineering. Biomaterials **2009**, *30* (7), 1321–1328.

20. Sarkar, S.; Salacinski, H.J.; Hamilton, G.; Seifalian, A.M. The mechanical properties of infrainguinal vascular bypass grafts: Their role in influencing patency. Eur. J. Vasc. Endovasc. **2006**, *31* (6), 627–636.

21. Bershadsky, A.D.; Balaban, N.Q.; Geiger, B. Adhesion-dependent cell mechanosensitivity. Annu. Rev. Cell. Dev. Biol. **2003**, *19*, 677–695.

22. Cukierman, E.; Pankov, R.; Stevens, D.R.; Yamada, K.M. Taking cell-matrix adhesions to the third dimension. Science **2001**, *294* (5547), 1708–1712.

23. Engler, A.J.; Griffin, M.A.; Sen, S.; Bonnemann, C.G.; Sweeney, H.L.; Discher, D.E. Myotubes differentiate optimally on substrates with tissue-like stiffness: Pathological implications for soft or stiff microenvironments. J. Cell. Biol. **2004**, *166* (6), 877–887.

24. Discher, D.E.; Janmey, P.; Wang, Y.L. Tissue cells feel and respond to the stiffness of their substrate. Science **2005**, *310* (5751), 1139–1143.

25. Lo, C.M.; Wang, H.B.; Dembo, M.; Wang, Y.L. Cell movement is guided by the rigidity of the substrate. Biophys. J. **2000**, *79* (1), 144–152.

26. Pelham, R.J.; Wang Y. Cell locomotion and focal adhesions are regulated by substrate flexibility. Proc. Natl. Acad. Sci. U S A **1997**, *94* (25), 13661–13665.

27. Byfield, F.J.; Reen, R.K.; Shentu, T.P.; Levitan, I.; Gooch, K.J. Endothelial actin and cell stiffness is modulated by substrate stiffness in 2D and 3D. J. Biomech. **2009**, *42* (8), 1114–1119.

28. Beningo, K.A.; Dembo, M.; Kaverina, I.; Small, J.V.; Wang, Y.L. Nascent focal adhesions are responsible for

Biomimetic—Colloid

the generation of strong propulsive forces in migrating fibroblasts. J. Cell. Biol. **2001**, *153* (4), 881–887.

29. Tamada, M.; Sheetz, M.P.; Sawada, Y. Activation of a signaling cascade by cytoskeleton stretch. Dev. Cell. **2004**, *7* (5), 709–718.

30. Alenghat. F.J.; Ingber. D.E. Mechanotransduction: All signals point to cytoskeleton, matrix, and integrins. Sci. STKE. **2002**, *2002* (119), pe6.

31. Engler, A.J.; Sen, S.; Sweeney, H.L.; Discher, D.E. Matrix elasticity directs stem cell lineage specification. Cell. **2006**, *126* (4), 677–689.

32. Baker, B.M.; Chen, C.S. Deconstructing the third dimension - how 3D culture microenvironments alter cellular cues. J. Cell. Sci. **2012**, *125* (13), 3015–3024.

33. Zaman, M.H.; Trapani, L.M.; Siemeski, A.; MacKellar, D.; Gong, H.; Kamm, R.D.; Wells, A.; Lauffenburger, D.A.; Matsudaira, P. Migration of tumor cells in 3D matrices is governed by matrix stiffness along with cell-matrix adhesion and proteolysis. Proc. Natl. Acad. Sci. U S A **2006**, *103* (37), 13897–13897.

34. Formhals, A. Process and apparatus for preparing artificial threads. US Patent, 1975504. 1934.

35. Agarwal, S.; Wendorff, J.H.; Greiner, A. Progress in the field of electrospinning for tissue engineering applications. Adv. Mater. **2009**, *21* (32–33), 3343–3351.

36. Buttafoco, L.; Kolkman, N.G.; Engbers-Buijtenhuijs, P.; Poot, A.A.; Dijkstra, P.J.; Vermes, I.; Feijen, J. Electrospinning of collagen and elastin for tissue engineering applications. Biomaterials **2006**, *27* (5), 724–734.

37. Doshi, J.; Reneker, D.H. Electrospinning process and applications of electrospun fibers. J. Electrostat. **1995**, *35* (2–3), 151–160.

38. Ghasemi-Mobarakeh, L.; Prabhakaran, M.P.; Morshed, M.; Nasr-Esfahani, M.-H.; Ramakrishna, S. Electrospun poly(ε-caprolactone)/gelatin nanofibrous scaffolds for nerve tissue engineering. Biomaterials **2008**, *29* (34), 4532–4539.

39. Wnek, G.E.; Carr, M.E.; Simpson, D.G.; Bowlin, G.L. Electrospinning of nanofiber fibrinogen structures. Nano. Lett. **2003**, *3* (2), 213–216.

40. Li, D.; Xia, Y. Direct fabrication of composite and ceramic hollow nanofibers by electrospinning. Nano. Lett. **2004**, *4* (5), 933–938.

41. Yang, L.; Fitié, C.F.C.; van der Werf, K.O.; Bennink, M.L.; Dijkstra, P.J.; Feijen, J. Mechanical properties of single electrospun collagen type I fibers. Biomaterials **2008**, *29* (8), 955–962.

42. Heydarkhan-Hagvall, S.; Schenke-Layland, K.; Dhanasopon, A.P.; Rofail, F.; Smith, H.; Wu, B.M.; Shemin, R.; Beygui, R.E.; MacLellan, W.R. Three-dimensional electrospun ECM-based hybrid scaffolds for cardiovascular tissue engineering. Biomaterials **2008**, *29* (19), 2907–2914.

43. Bencherif, S.A.; Sands, R.W.; Bhatta, D.; Arany, P.; Verbeke, C.S.; Edwards, D.A.; Mooney, D.J. Injectable preformed scaffolds with shape-memory properties. Proc. Natl. Acad. Sci. U S A. **2012**, *109* (48), 19590–19595.

44. Pek, Y.S.; Gao, S.; Arshad, M.S.M.; Leck, K.-J.; Ying, J.Y. Porous collagen-apatite nanocomposite foams as bone regeneration scaffolds. Biomaterials **2008**, *29* (32), 4300–4305.

45. Munch, E.; Launey, M.E.; Alsem, D.H.; Saiz, E.; Tomsia, A.P.; Ritchie, R.O. Tough, bio-inspired hybrid materials. Science **2008**, *322* (5907), 1516–1520.

46. Bayraktar, H.H.; Morgan, E.F.; Niebur, G.L.; Morris, G.E.; Wong, E.K.; Keaveny, T.M. Comparison of the elastic and yield properties of human femoral trabecular and cortical bone tissue. J. Biomech. **2004**, *37* (1), 27–35.

47. Miller, J.S.; Stevens, K.R.; Yang, M.T.; Baker, B.M.; Nguyen, D.-H.T.; Cohen, D.M.; Toro, E.; Chen, A.A.; Galie, P.A.; Yu, X.; Chaturvedi, R.; Bhatia, S.N.; Chen, C.S. Rapid casting of patterned vascular networks for perfusable engineered three-dimensional tissues. Nat. Mater. **2012**, *11* (9), 768–774.

48. Visconti, R.P.; Kasyanov, V.; Gentile, C.; Zhang, J.; Markwald, R.R.; Mironov, V. Towards organ printing: Engineering an intra-organ branched vascular tree. Expert. Opin. Biol. Ther. **2010**, *10* (3), 409–420.

49. Correa-Duarte, M.A.; Wagner, N.; Rojas-Chapana, J.; Morsczeck, C.; Thie, M.; Giersig, M. Fabrication and biocompatibility of carbon nanotube-based 3D networks as scaffolds for cell seeding and growth. Nano. Lett. **2004**, *4* (11), 2233–2236.

50. De Volder M.; Hart A.J. Engineering hierarchical nanostructures by elastocapillary self-assembly. Angew. Chem. Int. Ed. Eng. **2013**, *52* (9), 2412–2425.

51. Stuart, M.A.; Huck, W.T.; Genzer, J.; Muller, M.; Ober, C.; Stamm, M.; Sukhorukov, G.B.; Szleifer, I.; Tsukruk, V.V.; Urban, M.; Winnik, F.; Zauscher, S.; Luzinov, I.; Minko, S. Emerging applications of stimuli-responsive polymer materials. Nat. Mater. **2010**, *9* (2), 101–113.

52. Bettinger, C.J.; Bruggeman, J.P.; Misra, A.; Borenstein, J.T.; Langer, R. Biocompatibility of biodegradable semiconducting melanin films for nerve tissue engineering. Biomaterials **2009**, *30* (17), 3050–3057.

53. Bothma, J.P.; de Boor, J.; Divakar, U.; Schwenn, P.E.; Meredith, P. Device-quality electrically conducting melanin thin films. Adv. Mater. **2008**, *20* (18), 3539–3542.

54. Cellot, G.; Cilia, E.; Cipollone, S.; Rancic, V.; Sucapane, A.; Giordani, S.; Gambazzi, L.; Markram, H.; Grandolfo, M.; Scaini, D.; Gelain, F.; Casalis, L.; Prato, M.; Giugliano, M.; Ballerini, L. Carbon nanotubes might improve neuronal performance by favouring electrical shortcuts. Nat. Nano. **2009**, *4* (2), 126–133.

55. Harrison, B.S.; Atala A. Carbon nanotube applications for tissue engineering. Biomaterials **2007**, *28* (2), 344–353.

56. Mazzatenta, A.; Giugliano, M.; Campidelli, S.; Gambazzi, L.; Businaro, L.; Markram, H.; Prato, M.; Ballerini, L. Interfacing neurons with carbon nanotubes: Electrical signal transfer and synaptic stimulation in cultured brain circuits. J. Neurosci. **2007**, *27* (26), 6931–6936.

57. Giugliano, M.; Prato, M.; Ballerini, L. Nanomaterial/neuronal hybrid system for functional recovery of the CNS. Drug. Disc. Today: Dis. Models **2008**, *5* (1), 38–43.

58. You, J.-O.; Rafat, M.; Ye, G.J.C.; Auguste, D.T. Nanoengineering the heart: Conductive scaffolds enhance connexin 43 expression. Nano. Letters. **2011**, *11* (9), 3643–3648.

59. Dvir. T.; Timko B.P.; Brigham, M.D.; Naik, S.R.; Karajanagi, S.S.; Levy, O.; Jin, H.; Parker, K.K.; Langer, R.; Kohane, D.S. Nanowired three-dimensional cardiac patches. Nat. Nano. 2011, *6* (11), 720–725.

60. Engelmayr, G.C.; Cheng, M.; Bettinger, C.J.; Borenstein, J.T.; Langer, R.; Freed, L.E. Accordion-like honeycombs for tissue engineering of cardiac anisotropy. Nat. Mater. **2008**, *7* (12), 1003–1010.

61. Engel, E.; Michiardi, A.; Navarro, M.; Lacroix, D.; Planell, J.A. Nanotechnology in regenerative medicine: The materials side. Trends Biotechnology **2008**, *26* (1), 39–47.

62. Ng, R.; Zang, R.; Yang, K.K.; Liu, N.; Yang, S.T. Three-dimensional fibrous scaffolds with microstructures and nanotextures for tissue engineering. RSC Adv. **2012**, *2* (27), 10110–10124.

63. Kim, D.H.; Provenzano, P.P.; Smith, C.L.; Levchenko, A. Matrix nanotopography as a regulator of cell function. J. Cell. Biol. **2012**, *197* (3), 351–360.

64. Geiger, B.; Spatz, J.P.; Bershadsky, A.D. Environmental sensing through focal adhesions. Nat. Rev. Mol. Cell. Biol. **2009**, *10* (1), 21–33.

65. Vogel, V.; Sheetz, M. Local force and geometry sensing regulate cell functions. Nat. Rev. Mol. Cell. Biol. **2006**, *7* (4), 265–275.

66. Chen, W.; Villa-Diaz, L.G.; Sun, Y.; Weng, S.; Kim, J.K.; Lam, R.H.W.; Han, L.; Fan, R.; Krebsbach, P.H.; Fu, J. Nanotopography influences adhesion, spreading, and self-renewal of human embryonic stem cells. ACS Nano. **2012**, *6* (5), 4094–4103.

67. Stevens, M.M.; George, J.H. Exploring and engineering the cell surface interface. Science **2005**, *310* (5751), 1135–1138.

68. Harrison R.G. The reaction of embryonic cells to solid structures. J. Exp. Zool. Part A: Ecological Genetics and Physiology **1914**, *17* (4), 521–44.

69. Weiss, P. *In vitro* experiments on the factors determining the course of the outgrowing nerve fiber. J. Exp. Zool. **1934**, *68* (3), 393–448.

70. Kim, H.N.; Jiao, A.; Hwang, N.S.; Kim, M.S.; Kang, D.H.; Kim, D.-H.; Suh, K.-Y. Nanotopography-guided tissue engineering and regenerative medicine. Adv. Drug. Deliv. Rev. **2013**, *65* (4), 536–558.

71. Gaharwar. A.K.; Peppas. N.A.; Khademhosseini. A. Nanocomposite hydrogels for biomedical applications. Biotechnol. Bioeng. **2014**, *111* (3), 441–453.

72. McNamara, L.E.; McMurray, R.J.; Biggs, M.J.P.; Kantawong, F.; Oreffo, R.O.C.; Dalby, M.J. Nanotopographical control of stem cell differentiation. J. Tissue. Eng. **2010**, *2010*, 120623.

73. Goreham, R.V.; Mierczynska, A.; Smith, L.E.; Sedev, R.; Vasilev, K. Small surface nanotopography encourages fibroblast and osteoblast cell adhesion. RSC. Adv. **2013**, *3* (26), 10309–10317.

74. Muhammad, R.; Peh, G.S.L.; Adnan, K.; Law, J.B.K.; Mehta, J.S.; Yim, E.K.F. Micro- and nano-topography to enhance proliferation and sustain functional markers of donor-derived primary human corneal endothelial cells. Acta Biomater. **2015**, *19*, 138–148.

75. Nakayama, K.H.; Hong, G.; Lee, J.C.; Patel, J.; Edwards, B.; Zaitseva, T.S.; Paukshto, M.V.; Dai, H.; Cooke, J.P.; Woo, Y.J.; Huang, N.F. Aligned-braided nanofibrillar scaffold with endothelial cells enhances arteriogenesis. ACS Nano. **2015**, *9* (7), 6900–6908.

76. Shi, Z.; Huang, X.; Cai, Y.; Tang, R.; Yang, D. Size effect of hydroxyapatite nanoparticles on proliferation and apop-

77. tosis of osteoblast-like cells. Acta Biomater. **2009**, *5* (1), 338–345.

77. Lee, J.H.; Rim, N.G.; Jung, H.S.; Shin, H. Control of osteogenic differentiation and mineralization of human mesenchymal stem cells on composite nanofibers containing poly[lactic-co-(glycolic acid)] and hydroxyapatite. Macromol. Biosci. **2010**, *10* (2), 173–182.

78. Xue, D.; Zheng, Q.; Zong, C.; Li, Q.; Li, H.; Qian, S.; Zhang, B.; Yu, L.; Pan, Z. Osteochondral repair using porous poly(lactide-co-glycolide)/nano-hydroxyapatite hybrid scaffolds with undifferentiated mesenchymal stem cells in a rat model. J. Biomed. Mater. Res. **2010**, *94A* (1), 259–270.

79. Zandi, M.; Mirzadeh, H.; Mayer, C.; Urch, H.; Eslaminejad, M.B.; Bagheri, F.; Mivehchi, H. Biocompatibility evaluation of nano-rod hydroxyapatite/gelatin coated with nano-HAp as a novel scaffold using mesenchymal stem cells. J. Biomed. Mater. Res. **2010**, *92A* (4), 1244–1255.

80. Rahaman, M.N.; Day, D.E.; Bal, B.S.; Fu, Q.; Jung, S.B.; Bonewald, L.F.; Tomsia, A.P. Bioactive glass in tissue engineering. Acta Biomater. **2011**, *7* (6), 2355–2373.

81. Fu, Q.; Saiz, E.; Rahaman, M.N.; Tomsia, A.P. Bioactive glass scaffolds for bone tissue engineering: State of the art and future perspectives. Mat. Sci. Eng. C-Mater. **2011**, *31* (7), 1245–1256.

82. Kerativitayanan, P.; Carrow, J.K.; Gaharwar, A.K. Nanomaterials for engineering stem cell responses. Adv. Healthc. Mater. **2015**, *4* (11), 1600–1627.

83. Mota, J.; Yu, N.; Caridade, S.G.; Luz, G.M.; Gomes, M.E.; Reis, R.L.; Jansen, J.A.; Walboomers, X.F.; Mano J.F. Chitosan/bioactive glass nanoparticle composite membranes for periodontal regeneration. Acta Biomater. **2012**, *8* (11), 4173–4180.

84. Hong, S.-J.; Yu, H.-S.; Noh, K.-T.; Oh, S.-A.; Kim, H.-W. Novel scaffolds of collagen with bioactive nanofiller for the osteogenic stimulation of bone marrow stromal cells. J. Biomater. Appl. **2010**, *24* (8), 733–750.

85. Gaharwar, A.K.; Mukundan, S.; Karaca, E.; Dolatshahi-Pirouz, A.; Patel, A.; Rangarajan, K.; Mihaila, S.M.; Iviglia, G.; Zhang, H.; Khademhosseini, A. Nanoclay-enriched poly(ε-caprolactone) electrospun scaffolds for osteogenic differentiation of human mesenchymal stem cells. Tissue Eng. Part A. **2014**, *20* (15–16), 2088–2101.

86. Hulsman, M.; Hulshof, F.; Unadkat, H.; Papenburg, B.J.; Stamatialis, D.F.; Truckenmüller, R.; van Blitterswijk, C.; de Boer, J.; Reinders, M.J.T. Analysis of high-throughput screening reveals the effect of surface topographies on cellular morphology. Acta Biomater. **2015**, *15*, 29–38.

87. Unadkat, H.V.; Hulsman, M.; Cornelissen, K.; Papenburg, B.J.; Truckenmüller, R.K.; Carpenter, A.E.; Wessling, M.; Post, G.F.; Uetz, M.; Reinders, M.J.T.; Stamatialis, D.; Blitterswijk, C.A.V.; Boer J.D. An algorithm-based topographical biomaterials library to instruct cell fate. Proc. Natl. Acad. Sci. U S A **2011**, *108* (40), 16565–16570.

88. Lambrechts, D.; Roeffaers, M.; Kerckhofs, G.; Roberts, S.J.; Hofkens, J.; Van de Putte, T.; Van Oosterwyck, H.; Schrooten, J. Fluorescent oxygen sensitive microbead incorporation for measuring oxygen tension in cell aggregates. Biomaterials **2013**, *34* (4), 922–929.

89. Demol, J.; Lambrechts, D.; Geris, L.; Schrooten, J.; Van Oosterwyck, H. Towards a quantitative understanding of

oxygen tension and cell density evolution in fibrin hydrogels. Biomaterials **2011**, *32* (1), 107–118.

90. Müller, P.; Rogers, K.W.; Yu, S.R.; Brand, M.; Schier, A.F. Morphogen transport. Development **2013**, *140* (8), 1621–1638.

91. Jönsson, P.; Jonsson, M.P.; Tegenfeldt, J.O.; Höök, F. A method improving the accuracy of fluorescence recovery after photobleaching analysis. Biophys. J. **2008**, *95* (11), 5334–5348.

92. Rossow, M.J.; Sasaki, J.M.; Digman, M.A.; Gratton, E. Raster image correlation spectroscopy in live cells. Nat. Prot. **2010**, *5* (11), 1761–1774.

93. Norris, S.C.P.; Humpolíčková, J.; Amler, E.; Huranová, M.; Buzgo, M.; Macháň, R.; Lukáš, D.; Hof, M. Raster image correlation spectroscopy as a novel tool to study interactions of macromolecules with nanofiber scaffolds. Acta Biomater. **2011**, *7* (12), 4195–4203.

94. Malafaya, P.B.; Silva, G.A.; Reis, R.L. Natural-origin polymers as carriers and scaffolds for biomolecules and cell delivery in tissue engineering applications. Adv. Drug. Deliv. Rev. **2007**, *59* (4–5), 207–233.

95. DeVine, J.; Dettori, J.; France, J.; Brodt, E.; McGuire, R. The use of rhBMP in spine surgery: Is there a cancer risk? Evid. Based. Spine. Care. J. **2012**, *3* (2), 35–41.

96. Censi, R.; Di Martino, P.; Vermonden, T.; Hennink, W.E. Hydrogels for protein delivery in tissue engineering. J. Control. Rel. **2012**, *161* (2), 680–692.

97. Briquez, P.S.; Clegg, L.E.; Martino, M.M.; Gabhann, F.M.; Hubbell, J.A. Design principles for therapeutic angiogenic materials. Nat. Rev. Mat. **2016**, *1*, 15006.

98. Martino, M.M.; Briquez, P.S.; Maruyama, K.; Hubbell, J.A. Extracellular matrix-inspired growth factor delivery systems for bone regeneration. Adv. Drug. Deliv. Rev. **2015**, *94*, 41–52.

99. Briquez, P.S.; Hubbell, J.A.; Martino, M.M. Extracellular matrix-inspired growth factor delivery systems for skin wound healing. Adv. Wound. Care. **2015**, *4* (8), 479–489.

100. Martino, M.M.; Briquez, P.S.; Ranga, A.; Lutolf, M.P.; Hubbell, J.A. Heparin-binding domain of fibrin(ogen) binds growth factors and promotes tissue repair when incorporated within a synthetic matrix. Proc. Natl. Acad. Sci. U S A **2013**, *110* (12), 4563–4568.

101. Miller, R.E.; Grodzinsky, A.J.; Cummings, K.; Plaas, A.H.K.; Cole, A.A.; Lee, R.T.; Patwari, P. Intraarticular injection of heparin-binding insulin-like growth factor 1 sustains delivery of insulin-like growth factor 1 to cartilage through binding to chondroitin sulfate. Arthritis. Rheum. **2010**, *62* (12), 3686–3694.

102. Yang, H.S.; La, W.G.; Cho, Y.M.; Shin, W.; Yeo, G.D.; Kim, B.S. Comparison between heparin-conjugated fibrin and collagen sponge as bone morphogenetic protein-2 carriers for bone regeneration. Exp. Mol. Med. **2012**, *44* (5), 350–355.

103. Sakiyama-Elbert, S.E.; Hubbell, J.A. Controlled release of nerve growth factor from a heparin-containing fibrin-based cell ingrowth matrix. J. Control. Rel. **2000**, *69* (1), 149–158.

104. Sacchi, V.; Mittermayr, R.; Hartinger, J.; Martino, M.M.; Lorentz, K.M.; Wolbank, S.; Hofmann, A.; Largo, R.A.; Marschall, J.S.; Groppa, E.; Gianni-Barrera, R.; Ehrbar, M.; Hubbell, J.A.; Redl, H.; Banfi, A. Long-lasting fibrin matri-

ces ensure stable and functional angiogenesis by highly tunable, sustained delivery of recombinant VEGF164. Proc. Natl. Acad. Sci. U S A **2014**, *111* (19), 6952–6957.

105. Mann, B.K.; Schmedlen, R.H.; West, J.L. Tethered-TGF-beta increases extracellular matrix production of vascular smooth muscle cells. Biomaterials **2001**, *22* (5), 439–444.

106. West, J.L.; Hubbell, J.A. Polymeric biomaterials with degradation sites for proteases involved in cell migration. Macromolecules **1999**, *32*, 241–244.

107. Patterson, J.; Hubbell, J.A. Enhanced proteolytic degradation of molecularly engineered PEG hydrogels in response to MMP-1 and MMP-2. Biomaterials **2010**, *31* (30), 7836–7845.

108. Patterson, J.; Hubbell, J.A. SPARC-derived protease substrates to enhance the plasmin sensitivity of molecularly engineered PEG hydrogels. Biomaterials **2011**, *32* (5), 1301–1310.

109. Sprong, H.; van der Sluijs, P.; van Meer, G. How proteins move lipids and lipids move proteins. Nat. Rev. Mol. Cell. Bio. **2001**, *2* (9), 698–698.

110. Thiam, A.R.; Farese, R.V.; Walther, T.C. The biophysics and cell biology of lipid droplets. Nat. Rev. Mol. Cell. Bio. **2013**, *14* (12), 775–786.

111. Batrakova, E.V.; Kabanov, A.V. Pluronic block copolymers: Evolution of drug delivery concept from inert nanocarriers to biological response modifiers. J. Control. Rel. **2008**, *130* (2), 98–106.

112. Rapoport, N. Stabilization and activation of Pluronic micelles for tumor-targeted drug delivery. Colloid. Surface B **1999**, *16* (1–4), 93–111.

113. Wei, H.; Zhuo, R.-X.; Zhang, X.-Z. Design and development of polymeric micelles with cleavable links for intracellular drug delivery. Prog. Poly. Sci. **2013**, *38* (3–4), 503–535.

114. Sreejit, P.; Verma, R.S. Natural ECM as biomaterial for scaffold based cardiac regeneration using adult bone marrow derived stem cells. Stem. Cell. Rev. **2013**, *9* (2), 158–171.

115. Kakiuchi, Y.; Hirohashi, N.; Murakami-Murofushi, K. The macroscopic structure of RADA16 peptide hydrogel stimulates monocyte/macrophage differentiation in HL60 cells via cholesterol synthesis. Biochem. Biophys. Res. Commun. **2013**, *433* (3), 298–304.

116. Sosnik, A.; Sefton, M.V. Semi-synthetic collagen/poloxamine matrices for tissue engineering. Biomaterials **2005**, *26* (35), 7425–7435.

117. Turner, N.J.; Badylak, S.F. Regeneration of skeletal muscle. Cell. Tissue. Res. **2011**, *347* (3), 759–774.

118. Turner, N.J.; Yates, A.J.; Weber, D.J.; Qureshi, I.R.; Stolz, D.B.; Gilbert, T.W.; Badylak, S.F. Xenogeneic extracellular matrix as an inductive scaffold for regeneration of a functioning musculotendinous junction. Tissue Eng. Part A **2010**, *16* (11), 3309–3317.

119. Weisel, J.W.; Litvinov, R.I. Mechanisms of fibrin polymerization and clinical implications. Blood **2013**, *121* (10), 1712–1719.

120. Ryan, E.A.; Mockros, L.F.; Weisel, J.W.; Lorand, L. Structural origins of fibrin clot rheology. Biophys. J. **1999**, *77* (5), 2813–2826.

121. Domingues, M.M.; Macrae, F.L.; Duval, C.; McPherson, H.R.; Bridge, K.I.; Ajjan, R.A.; Ridger, V.C.; Connell, S.D.; Philippou, H.; Ariëns, R.A.S. Thrombin and fibrinogen γ'

Biomimetic—Colloid

impact clot structure by marked effects on intrafibrillar structure and protofibril packing. Blood **2016**, *127* (4), 487–496.

122. Kurniawan, N.A.; Grimbergen, J.; Koopman, J.; Koenderink, G.H. Factor XIII stiffens fibrin clots by causing fiber compaction. J. Thromb. Haemost. **2014**, *12* (10), 1687–1696.

123. Carr, M.E.; Gabriel, D.A.; McDonagh, J. Influence of Ca^{2+} on the structure of reptilase-derived and thrombin-derived fibrin gels. Biochem. J. **1986**, *239* (3), 513–516.

124. Di Stasio, E.; Nagaswami, C.; Weisel, J.W.; Di Cera, E. Cl^- regulates the structure of the fibrin clot. Biophys. J. **1998**, *75* (4), 1973–1979.

125. Yeromonahos, C.; Polack, B.; Caton, F. Nanostructure of the fibrin clot. Biophys. J. **2010**, *99* (7), 2018–2027.

126. Okude, M.; Yamanaka, A.; Akihama, S. The effects of pH on the generation of turbidity and elasticity associated with fibrinogen-fibrin conversion by thrombin are remarkably influenced by sialic acid in fibrinogen. Biol. Pharm. Bull. **1995**, *18* (2), 203–207.

127. Ahmed, T.A.E.; Dare, E.V.; Hincke, M. Fibrin: A versatile scaffold for tissue engineering applications. Tissue Eng. Part B: Rev. **2008**, *14* (2), 199–215.

128. Ige, O.O.; Umoru, L.E.; Aribo S.; Natural products: A minefield of biomaterials. Inter. Sch. Res. Notices, **2012**, *2012*, e983062.

129. Dhandayuthapani, B.; Yoshida, Y.; Maekawa, T.; Kumar, D.S. Polymeric scaffolds in tissue engineering application: A review. Inter. J. Poly. Sci. **2011**, *2011*, e290602.

130. Salgado, A.J.; Coutinho, O.P.; Reis, R.L. Bone tissue engineering: State of the art and future trends. Macromol. Biosci. **2004**, *4* (8), 743–765.

131. Cordonnier, T.; Sohier, J.; Rosset, P.; Layrolle, P. Biomimetic materials for bone tissue engineering – state of the art and future trends. Adv. Eng. Mater. **2011**, *13* (5), B135-B150.

132. Rahmany, M.B.; Van Dyke, M. Biomimetic approaches to modulate cellular adhesion in biomaterials: A review. Acta Biomater. **2013**, *9* (3), 5431–5437.

133. Santambrogio, L. *Biomaterials in Regenerative Medicine and the Immune System*; Springer: US; 2015, 283.

134. Duncan, R. Polymer conjugates as anticancer nanomedicines. Nat. Rev. Cancer. **2006**, *6* (9), 688–701.

135. Jatzkewitz, H. Peptamin (glycyl-L-leucyl-mescaline) bound to blood plasma expander (polyvinylpyrrolidone) as a new depot form of a biologically active primary amine (mescaline). Zeitschrift für Naturforschung B **1955**, *10* (1), 27–31.

136. Shumikhina, K.I.; Panarin, E.F.; Ushakov, S.N. [Experimental study of polymer salts of penicillins]. Antibiotiki **1966**, *11* (9), 767–770.

137. Nicoletti, S.; Seifert, K.; Gilbert, I.H. N-(2-hydroxypropyl)methacrylamide–amphotericin B (HPMA–AmB) copolymer conjugates as antileishmanial agents. Inter. J. Antimicrob. Agents. **2009**, *33* (5), 441–448.

138. Ringsdorf H. Structure and properties of pharmacologically active polymers. J. Polym. Sci. C Polym. Symp. **1975**, *51* (1), 135–153.

139. Cobo, I.; Li, M.; Sumerlin, B.S.; Perrier, S. Smart hybrid materials by conjugation of responsive polymers to biomacromolecules. Nat. Mater. **2015**, *14* (2), 143–159.

140. Gombotz, W.R.; Guanghui, W.; Horbett, T.A.; Hoffman, A.S. Protein adsorption to and elution from polyether surfaces. In *Poly(Ethylene Glycol) Chemistry. Topics in Applied Chemistry*; Harris, J.M., Ed.; Springer: US; 1992. 247–261.

141. Abuchowski, A.; van Es, T.; Palczuk, N.C.; Davis, F.F. Alteration of immunological properties of bovine serum albumin by covalent attachment of polyethylene glycol. J. Biol. Chem. **1977**, *252* (11), 3578–3581.

142. Peter, K.W.; Mary, S.N.; Judy, J.; Joel, B.C. Safety of poly(ethylene glycol) and poly(ethylene glycol) derivatives. Poly(ethylene glycol). In *ACS Symposium Series. 680*: American Chemical Society: Washington, 1997, 45–57.

143. Shu, J.Y.; Panganiban, B.; Xu, T. Peptide-polymer conjugates: From fundamental science to application. Ann. Rev. of Phys. Chem. **2013**, *64* (1), 631–657.

144. Gonen-Wadmany, M.; Oss-Ronen, L.; Seliktar, D. Protein-polymer conjugates for forming photopolymerizable biomimetic hydrogels for tissue engineering. Biomaterials **2007**, *28* (26), 3876–3886.

145. Swierczewska, M.; Hajicharalambous, C.S.; Janorkar, A.V.; Megeed, Z.; Yarmush, M.L.; Rajagopalan, P. Cellular response to nanoscale elastin-like polypeptide polyelectrolyte multilayers. Acta Biomater. **2008**, *4* (4), 827–837.

146. Wu, Y.Z.; Ng, D.Y.W.; Kuan, S.L.; Weil, T. Protein-polymer therapeutics: A macromolecular perspective. Biomater. Sci. **2015**, *3* (2), 214–230.

147. Fu, Y.; Xu, K.; Zheng, X.; Giacomin, A.J.; Mix, A.W.; Kao, W.J. 3D cell entrapment in crosslinked thiolated gelatin-poly(ethylene glycol) diacrylate hydrogels. Biomaterials **2012**, *33* (1), 48–58.

148. DeForest, C.A.; Tirrell, D.A. A photoreversible protein-patterning approach for guiding stem cell fate in three-dimensional gels. Nat. Mater. **2015**, *14* (5), 523–531.

149. D'Andrea, L.D.; Iaccarino, G.; Fattorusso, R.; Sorriento, D.; Carannante, C.; Capasso, D.; Trimarco, B.; Pedone, C. Targeting angiogenesis: Structural characterization and biological properties of a de novo engineered VEGF mimicking peptide. Proc. Natl. Acad. Sci. U S A **2005**, *102* (40), 14215–14220.

150. Wang, L.; Zhao, M.; Li, S.; Erasquin, U.J.; Wang, H.; Ren, L.; Chen, C.; Wang, Y.; Cai, C. "Click" immobilization of a VEGF-mimetic peptide on decellularized endothelial extracellular matrix to enhance angiogenesis. ACS. Appl. Mater. Interfaces. **2014**, *6* (11), 8401–8406.

151. Cook, A.D.; Hrkach, J.S.; Gao, N.N.; Johnson, I.M.; Pajvani, U.B.; Cannizzaro, S.M.; Langer, R. Characterization and development of RGD-peptide-modified poly(lactic acid-co-lysine) as an interactive, resorbable biomaterial. J. Biomed. Mater. Res. **1997**, *35* (4), 513–523.

152. Castelletto, V.; Gouveia, R.M.; Connon, C.J.; Hamley, I.W.; Seitsonen, J.; Nykanen, A.; Ruokolainen, J. Alanine-rich amphiphilic peptide containing the RGD cell adhesion motif: A coating material for human fibroblast attachment and culture. Biomater. Sci. **2014**, *2* (3), 362–369.

153. Grover, G.N.; Lam, J.; Nguyen, T.H.; Segura, T.; Maynard, H.D. Biocompatible hydrogels by oxime click chemistry. Biomacromolecules **2012**, *13* (10), 3013–3017.

154. Harris, B.P.; Kutty, J.K.; Fritz, E.W.; Webb, C.K.; Burg, K.J.L.; Metters, A.T. Photopatterned polymer brushes promoting cell adhesion gradients. Langmuir **2006**, *22* (10), 4467–4471.

155. Benoit, D.S.; Anseth, K.S. The effect on osteoblast function of colocalized RGD and PHSRN epitopes on PEG surfaces. Biomaterials **2005**, *26* (25), 5209–5220.

156. Ingber, D.E. Cellular mechanotransduction: Putting all the pieces together again. Faseb. J. **2006**, *20* (7), 811–827.

157. Mammoto, T.; Mammoto, A.; Ingber, D.E. Mechanobiology and developmental control. Annu. Rev. Cell. Dev. Biol. **2013**, *29*, 27–61.

158. Guilak, F.; Butler, D.L.; Goldstein, S.A.; Baaijens, F.P. Biomechanics and mechanobiology in functional tissue engineering. J. Biomech. **2014**, *47* (9), 1933–1940.

159. Rajagopal, K.; Schneider, J.P. Self-assembling peptides and proteins for nanotechnological applications. Curr. Opin. Struct. Biol. **2004**, *14* (4), 480–486.

160. Nune, M.; Kumaraswamy, P.; Krishnan, U.M.; Sethuraman, S. Self-assembling peptide nanofibrous scaffolds for tissue engineering: Novel approaches and strategies for effective functional regeneration. Curr. Protein. Pept. Sc. **2013**, *14* (1), 70–84.

161. Lakshmanan, A.; Zhang, S.; Hauser, C.A.E. Short self-assembling peptides as building blocks for modern nanodevices. Trends. Biotechnol. **2012**, *30* (3), 155–165.

162. Lu, T.L.; Li, Y.H.; Chen, T. Techniques for fabrication and construction of three-dimensional scaffolds for tissue engineering. Int. J. Nanomed. **2013**, *8*, 337–350.

163. Zhang, S. Building from the bottom up. Materials Today **2003**, *6* (5), 20–27.

164. Arosio, P.; Owczarz, M.; Wu, H.; Butte, A.; Morbidelli, M. End-to-end self-assembly of RADA 16-I nanofibrils in aqueous solutions. Biophys. J. **2012**, *102* (7), 1617–1626.

165. Koutsopoulos, S. Self-assembling peptide nanofiber hydrogels in tissue engineering and regenerative medicine: Progress, design guidelines, and applications. J. Biomed. Mater. Res. **2016**, *104* (4), 1002–1016.

166. Silva, G.A.; Czeisler, C.; Niece, K.L.; Beniash, E.; Harrington, D.A.; Kessler, J.A.; Stupp, S.I. Selective differentiation of neural progenitor cells by high-epitope density nanofibers. Science **2004**, *303* (5662), 1352–1355.

167. Cheng, T.Y.; Chen, M.H.; Chang, W.H.; Huang, M.Y.; Wang, T.W. Neural stem cells encapsulated in a functionalized self-assembling peptide hydrogel for brain tissue engineering. Biomaterials **2013**, *34* (8), 2005–2016.

168. Li, Q.; Cheung, W.H.; Chow, K.L.; Ellis-Behnke, R.G.; Chau, Y. Factorial analysis of adaptable properties of self-assembling peptide matrix on cellular proliferation and neuronal differentiation of pluripotent embryonic carcinoma. Nanomedicine **2012**, *8* (5), 748–756.

169. Hauser, C.A.E.; Deng, R.; Mishra, A.; Loo, Y.; Khoe, U.; Zhuang, F.; Cheong, D.W.; Accardo, A.; Sullivan, M.B.; Riekel, C.; Ying, J.Y.; Hauser, U.A. Natural tri- to hexapeptides self-assemble in water to amyloid beta-type fiber aggregates by unexpected alpha-helical intermediate structures. Proc. Natl. Acad. Sci. U S A **2011**, *108* (4), 1361–1366.

170. Loo, Y.; Wong, Y.-C.; Cai, E.Z.; Ang, C.-H.; Raju, A.; Lakshmanan, A.; Koh, A.G.; Zhou, H.J.; Lim, T.-C.; Moochhala, S.M.; Hauser, C.A.E. Ultrashort peptide nanofibrous hydrogels for the acceleration of healing of burn wounds. Biomaterials **2014**, *35* (17), 4805–4814.

171. Seow, W.Y.; Hauser, C.A.E. Tunable mechanical properties of ultrasmall peptide hydrogels by crosslinking and functionalization to achieve the 3D distribution of cells. Adv. Healthc. Mat. **2013**, *2* (9), 1219–1223.

172. Aggeli, A.; Bell, M.; Carrick, L.M.; Fishwick, C.W.G.; Harding, R.; Mawer, P.J.; Radford, S.E.; Strong, A.E.; Boden, N. pH as a trigger of peptide beta-sheet self-assembly and reversible switching between nematic and isotropic phases. J. Am. Chem. Soc. **2003**, *125* (32), 9619–9628.

173. Wilson, C.J.; Clegg, R.E.; Leavesley, D.I.; Pearcy, M.J. Mediation of biomaterial-cell interactions by adsorbed proteins: A review. Tissue Eng. **2005**, *11* (1–2), 1–18.

174. Boehler, R.M.; Graham, J.G.; Shea, L.D. Tissue engineering tools for modulation of the immune response. Biotechniques **2011**, *51* (4), 239–240.

175. Franz, S.; Rammelt, S.; Scharnweber, D.; Simon, J.C. Immune responses to implants - A review of the implications for the design of immunomodulatory biomaterials. Biomaterials **2011**, *32* (28), 6692–6709.

176. Anderson J.M. Biological responses to materials. Ann Rev Mater Res 2001, 31, 81–110.

177. Valentin, J.E.; Badylak, J.S.; Mccabe, G.P.; Badylak, S.F. Extracellular matrix bioscaffolds for orthopaedic applications - A comparative histologic study. J. Bone Joint Surg. Am. **2006**, *88a* (12), 2673–2686.

178. Shen, M.C.; Pan, Y.V.; Wagner, M.S.; Hauch, K.D.; Castner, D.G.; Ratner, B.D.; Horbett, T.A. Inhibition of monocyte adhesion and fibrinogen adsorption on glow discharge plasma deposited tetraethylene glycol dimethyl ether. J. Biomat. Sci-Polym E. **2001**, *12* (9), 961–978.

179. VandeVondele, S.; Voros, J.; Hubbell, J.A. RGD-Grafted poly-l-lysine-graft-(polyethylene glycol) copolymers block non-specific protein adsorption while promoting cell adhesion. Biotechnol. Bioeng. **2003**, *82* (7), 784–790.

180. Williams, D.F. Williams dictionary of biomaterials: Liverpool University Press: Cambridge. 1999.

181. Hezi-Yamit, A.; Sullivan, C.; Wong, J.; David, L.; Chen, M.F.; Cheng, P.W.; Shumaker, D.; Wilcox, J.N.; Udipi, K. Impact of polymer hydrophilicity on biocompatibility: Implication for DES polymer design. J. Biomed. Mater. Res. **2009**, *90a* (1), 133–141.

182. Brodbeck, W.G.; Patel, J.; Voskerician, G.; Christenson, E.; Shive, M.S.; Nakayama, Y.; Matsuda, T.; Ziats, N.P.; Anderson, J.M. Biomaterial adherent macrophage apoptosis is increased by hydrophilic and anionic substrates *in vivo*. Proc. Natl. Acad. Sci. U S A **2002**, *99* (16), 10287–10292.

183. Cao, H.Q.; Mchugh, K.; Chew, S.Y.; Anderson, J.M. The topographical effect of electrospun nanofibrous scaffolds on the *in vivo* and *in vitro* foreign body reaction. J. Biomed. Mater. Res. **2010**, *93a* (3), 1151–1159.

184. Veiseh, O.; Doloff, J.C.; Ma, M.L.; Vegas, A.J.; Tam, H.H.; Bader, A.R.; Li, J.; Langan, E.; Wyckoff, J.; Loo, W.S.; Jhunjhunwala, S.; Chiu, A.; Siebert, S.; Tang, K.;

Hollister-Lock, J.; Aresta-Dasilva, S.; Bochenek, M.; Mendoza-Elias, J.; Wang, Y.; Qi, M.; Lavin, D.M.; Chen, M.; Dholakia, N.; Thakrar, R.; Lacik, I.; Weir, G.C.; Oberholzer, J.; Greiner, D.L.; Langer, R.; Anderson, D.G. Size- and shape-dependent foreign body immune response to materials implanted in rodents and non-human primates. Nat. Mater. **2015**, *14* (6), 643–U125.

185. Chen, S.L.; Jones, J.A.; Xu, Y.G.; Low, H.Y.; Anderson, J.M.; Leong, K.W. Characterization of topographical effects on macrophage behavior in a foreign body response model. Biomaterials **2010**, *31* (13), 3479–3491.

186. Madden, L.R.; Mortisen, D.J.; Sussman, E.M.; Dupras, S.K.; Fugate, J.A.; Cuy, J.L.; Hauch, K.D.; Laflamme, M.A.; Murry, C.E.; Ratner, B.D. Proangiogenic scaffolds as functional templates for cardiac tissue engineering. Proc. Natl. Acad. Sci. U S A **2010**, *107* (34), 15211–15216.

187. Underwood, R.A.; Usui, M.L.; Zhao, G.; Hauch, K.D.; Takeno, M.M.; Ratner, B.D.; Marshall, A.J.; Shi, X.F.; Olerud, J.E.; Fleckman, P. Quantifying the effect of pore size and surface treatment on epidermal incorporation into percutaneously implanted sphere-templated porous biomaterials in mice. J. Biomed. Mater. Res. **2011**, *98a* (4), 499–508.

188. Kao, W.J.; Liu, Y.P. Utilizing biomimetic oligopeptides to probe fibronectin-integrin binding and signaling in regulating macrophage function *in vitro* and *in vivo*. Front. Biosci. **2001**, *6*, D992-D999.

189. Waldeck, H.; Chung, A.S.; Kao, W.J. Interpenetrating polymer networks containing gelatin modified with PEGylated RGD and soluble KGF: Synthesis, characterization, and application in *in vivo* critical dermal wound. J. Biomed. Mater. Res. **2007**, *82a* (4), 861–871.

190. Yu, S.S.; Lau, C.M.; Barham, W.J.; Onishko, H.M.; Nelson, C.E.; Li, H.M.; Smith, C.A.; Yull, F.E.; Duvall, C.L.; Giorgio, T.D. Macrophage-specific RNA interference targeting via "click", mannosylated polymeric micelles. Mol. Pharm. **2013**, *10* (3), 975–987.

191. Vasievich, E.A.; Huang, L. The suppressive tumor microenvironment: A challenge in cancer immunotherapy. Mol. Pharm. **2011**, *8* (3), 635–641.

192. Teramura, Y.; Iwata, H. Islet encapsulation with living cells for improvement of biocompatibility. Biomaterials **2009**, *30* (12), 2270–2275.

193. Teramura, Y.; Kaneda, Y.; Iwata, H. Islet-encapsulation in ultra-thin layer-by-layer membranes of poly(vinyl alcohol) anchored to poly(ethylene glycol)-lipids in the cell membrane. Biomaterials **2007**, *28* (32), 4818–4825.

194. Hubbell, J.A. Biomaterials in tissue engineering. Biotechnology (N Y) **1995**, *13* (6), 565–576.

195. Liu, X.H.; Ma, P.X. Polymeric scaffolds for bone tissue engineering. Ann. Biomed. Eng. **2004**, *32* (3), 477–486.

196. Mano, J.F.; Silva, G.A.; Azevedo, H.S.; Malafaya, P.B.; Sousa, R.A.; Silva, S.S.; Boesel, L.F.; Oliveira, J.M.; Santos, T.C.; Marques, A.P.; Neves, N.M.; Reis, R.L. Natural origin biodegradable systems in tissue engineering and regenerative medicine: Present status and some moving trends. J. R. Soc. Interface. **2007**, *4* (17), 999–1030.

197. Albrektsson, T.; Johansson, C. Osteoinduction, osteoconduction and osseointegration. Eur. Spine. J. **2001**, *10*, S96-S101.

198. Athanasiou, K.A.; Zhu, C.; Lanctot, D.R.; Agrawal, C.M.; Wang, X. Fundamentals of biomechanics in tissue engineering of bone. Tissue. Eng. **2000**, *6* (4), 361–381.

199. Niyibizi, C.; Eyre, D.R. Structural characteristics of cross-linking sites in type-V collagen of bone - chain specificities and heterotypic links to type-I collagen. Eur. J. Biochem. **1994**, *224* (3), 943–950.

200. Urist, M.R. Bone - formation by autoinduction. Science **1965**, *150* (3698), 893–899.

201. Bouxsein, M.L.; Turek, T.J.; Blake, C.A.; D'Augusta, D.; Li, X.; Stevens, M.; Seeherman, H.J.; Wozney, J.M. Recombinant human bone morphogenetic protein-2 accelerates healing in a rabbit ulnar osteotomy model. J. Bone. Joint. Surg. Am. **2001**, *83-A* (8), 1219–1230.

202. Friess, W.; Uludag, H.; Foskett, S.; Biron, R.; Sargeant, C. Characterization of absorbable collagen sponges as rhBMP-2 carriers. Int. J. Pharm. **1999**, *187* (1), 91–99.

203. Rey, C. Calcium phosphate biomaterials and bone mineral. Differences in composition, structures and properties. Biomaterials **1990**, *11*, 13–15.

204. Zhang, Y.; Wu, C.; Friis, T.; Xiao, Y. The osteogenic properties of CaP/silk composite scaffolds. Biomaterials **2010**, *31* (10), 2848–2856.

205. Bhumiratana, S.; Grayson, W.L.; Castaneda, A.; Rockwood, D.N.; Gil, E.S.; Kaplan, D.L.; Vunjak-Novakovic, G. Nucleation and growth of mineralized bone matrix on silk-hydroxyapatite composite scaffolds. Biomaterials **2011**, *32* (11), 2812–2820.

206. Lutolf, M.R.; Weber, F.E.; Schmoekel, H.G.; Schense, J.C.; Kohler, T.; Muller, R.; Hubbell, J.A. Repair of bone defects using synthetic mimetics of collagenous extracellular matrices. Nat. Biotech. **2003**, *21* (5), 513–518.

207. Shekaran, A.; Garcia, J.R.; Clark, A.Y.; Kavanaugh, T.E.; Lin, A.S.; Guldberg, R.E.; Garcia, A.J. Bone regeneration using an alpha 2 beta 1 integrin-specific hydrogel as a BMP-2 delivery vehicle. Biomaterials **2014**, *35* (21), 5453–5461.

208. Ripamonti, U. Soluble osteogenic molecular signals and the induction of bone formation. Biomaterials **2006**, *27* (6), 807–822.

209. Kisiday, J.; Jin, M.; Kurz, B.; Hung, H.; Semino, C.; Zhang, S.; Grodzinsky, A.J. Self-assembling peptide hydrogel fosters chondrocyte extracellular matrix production and cell division: Implications for cartilage tissue repair. Proc. Natl. Acad. Sci. U S A **2002**, *99* (15), 9996–10001.

210. Solchaga, L.A.; Dennis, J.E.; Goldberg, V.M.; Caplan, A.I. Hyaluronic acid-based polymers as cell carriers for tissue-engineered repair of bone and cartilage. J. Orthop. Res. **1999**, *17* (2), 205–213.

211. Elisseeff, J.; McIntosh, W.; Anseth, K.; Riley, S.; Ragan, P.; Langer, R. Photoencapsulation of chondrocytes in poly(ethylene oxide)-based semi-interpenetrating networks. J. Biomed. Mater. Res. **2000**, *51* (2), 164–171.

212. Park, Y.D.; Tirelli, N.; Hubbell, J.A. Photopolymerized hyaluronic acid-based hydrogels and interpenetrating networks. Biomaterials **2003**, *24* (6), 893–900.

213. Varghese, S.; Hwang, N.S.; Canver, A.C.; Theprungsirikul, P.; Lin, D.W.; Elisseeff, J. Chondroitin sulfate based niches for chondrogenic differentiation of mesenchymal stem cells. Matrix. Biol. **2008**, *27* (1), 12–21.

Biomimetic—Colloid

214. Burdick, J.A.; Chung, C.; Jia, X.; Randolph, M.A.; Langer, R. Controlled degradation and mechanical behavior of photopolymerized hyaluronic acid networks. Biomacromolecules **2005**, *6* (1), 386–391.

215. Nguyen, L.H.; Kudva, A.K.; Guckert, N.L.; Linse, K.D.; Roy, K. Unique biomaterial compositions direct bone marrow stem cells into specific chondrocytic phenotypes corresponding to the various zones of articular cartilage. Biomaterials **2011**, *32* (5), 1327–1338.

216. Moutos, F.T.; Freed, L.E.; Guilak, F. A biomimetic three-dimensional woven composite scaffold for functional tissue engineering of cartilage. Nat. Mater. **2007**, *6* (2), 162–167.

217. Li, W.J.; Danielson, K.G.; Alexander, P.G.; Tuan, R.S. Biological response of chondrocytes cultured in three-dimensional nanofibrous poly(epsilon-caprolactone) scaffolds. J. Biomed. Mater. Res A. **2003**, *67* (4), 1105–1114.

218. Ahmed, M.; Ramos, T.A.D.; Damanik, F.; Le B.Q.; Wieringa, P.; Bennink, M.; van Blitterswijk, C.; de Boer, J.; Moroni, L. A combinatorial approach towards the design of nanofibrous scaffolds for chondrogenesis. Sci. Rep. **2015**, *5*, 14804.

219. Altman, G.H.; Horan, R.L.; Lu, H.H.; Moreau, J.; I M.; Richmond J.C.; Kaplan D.L. Silk matrix for tissue engineered anterior cruciate ligaments. Biomaterials **2002**, *23* (20), 4131–4141.

220. Cooper, J.A.; Lu, H.H.; Ko, F.K.; Freeman, J.W.; Laurencin, C.T. Fiber-based tissue-engineered scaffold for ligament replacement: Design considerations and *in vitro* evaluation. Biomaterials **2005**, *26* (13), 1523–1532.

221. Cooper, J.A.; Sahota, J.S.; Gorum, W.J.; Carter, J.; Doty, S.B.; Laurencin, C.T. Biomimetic tissue-engineered anterior cruciate ligament replacement. Proc. Natl. Acad. Sci. U. S. A. **2007**, *104* (9), 3049–3054.

222. Ratcliffe, A. Tissue engineering of vascular grafts. Matrix. Biol. **2000**, *19* (4), 353–357.

223. Shin, H.; Jo, S.; Mikos, A.G. Biomimetic materials for tissue engineering. Biomaterials **2003**, *24* (24), 4353–4364.

224. Walluscheck, K.P.; Steinhoff, G.; Kelm, S.; Haverich, A. Improved endothelial cell attachment on ePTFE vascular grafts pretreated with synthetic RGD-containing peptides. Eur. J. Vasc. Endovasc. Surg. *12* (3), 321–330.

225. Lin, Y.S.; Wang, S.S.; Chung, T.W.; Wang, Y.H.; Chiou, S.H.; Hsu, J.J.; Chou, N.K.; Hsieh, T.H.; Chu, S.H. Growth of endothelial cells on different concentrations of Gly-Arg-Gly-Asp photochemically grafted in polyethylene glycol modified polyurethane. Artif. Organs **2001**, *25* (8), 617–621.

226. Saik, J.E.; Gould, D.J.; Keswani, A.H.; Dickinson, M.E.; West, J.L. Biomimetic hydrogels with immobilized ephrinA1 for therapeutic angiogenesis. Biomacromolecules **2011**, *12* (7), 2715–2722.

227. von Degenfeld, G.; Banfi, A.; Springer, M.L.; Wagner, R.A.; Jacobi, J.; Ozawa, C.R.; Merchant, M.J.; Cooke, J.P.; Blau, H.M. Microenvironmental VEGF distribution is critical for stable and functional vessel growth in ischemia. Faseb. J. **2006**, *20* (14), 2657–2659.

228. Hao, X.J.; Silva, E.A.; Mansson-Broberg, A.; Grinnemo, K.H.; Siddiqui, A.J.; Dellgren, G.; Wardell, E.; Brodin, L.A.; Mooney, D.J.; Sylven, C. Angiogenic effects of sequential release of VEGF-A(165) and PDGF-BB with alginate hydrogels after myocardial infarction. Cardiovasc. Res. **2007**, *75* (1), 178–185.

229. Zimmermann, W.H.; Melnychenko, I.; Wasmeier, G.; Didie, M.; Naito, H.; Nixdorff, U.; Hess, A.; Budinsky, L.; Brune, K.; Michaelis, B.; Dhein, S.; Schwoerer, A.; Ehmke, H.; Eschenhagen, T. Engineered heart tissue grafts improve systolic and diastolic function in infarcted rat hearts. Nat. Med. **2006**, *12* (4), 452–458.

230. Ishii, O.; Shin, M.; Sueda, T.; Vacanti, J.P. *In vitro* tissue engineering of a cardiac graft using a degradable scaffold with an extracellular matrix-like topography. J. Thorac. Cardiov. Sur. **2005**, *130* (5), 1358–1363.

231. McDevitt, T.C.; Woodhouse, K.A.; Hauschka, S.D.; Murry, C.E.; Stayton, P.S. Spatially organized layers of cardiomyocytes on biodegradable polyurethane films for myocardial repair. J. Biomed. Mater. Res. **2003**, *66a* (3), 586–595.

232. Jin, K.; Tian, Y.; Erickson, J.S.; Puthoff, J.; Autumn, K.; Pesika, N.S. Design and fabrication of gecko-inspired adhesives. Langmuir **2012**, *28* (13), 5737–5742.

233. Kim, J.; Kong, Y.P.; Niedzielski, S.M.; Singh, R.K.; Putnam, A.J.; Shikanov, A. Characterization of the cross-linking kinetics of multi-arm poly(ethylene glycol) hydrogels formed via Michael-type addition. Soft Matter **2016**, *12* (7), 2076–2085.

234. Cormier, A.R.; Pang, X.D.; Zimmerman, M.I.; Zhou, H.X.; Paravastu, A.K. Molecular Structure of RADA16-I designer self-assembling peptide nanofibers. ACS Nano. **2013**, *7* (9), 7562–7572.

235. Mishra, A.; Loo, Y.H.; Deng, R.S.; Chuah, Y.J.; Hee, H.T.; Ying, J.Y.; Hauser, C.A.E. Ultrasmall natural peptides self-assemble to strong temperature-resistant helical fibers in scaffolds suitable for tissue engineering. Nano Today **2011**, *6* (4), 438–438.

236. Moon, J.J.; Saik, J.E.; Poche, R.A.; Leslie-Barbick, J.E.; Lee, S.H.; Smith, A.A.; Dickinson M.E.; West J.L. Biomimetic hydrogels with pro-angiogenic properties. Biomaterials **2010**, *31* (14), 3840–3847.

Biomimetic Materials: Smart Polymer Surfaces for Tissue Engineering

João F. Mano

Biomaterials, Biodegradables and Biomimetics (3Bs) Research Group, Department of Polymer Engineering, School of Engineering, European Institute of Excellence on Tissue Engineering and Regenerative Medicine, University of Minho, Braga and Life and Health Sciences Research Institute (ICVS)Portuguese Government Associate Laboratory, Guimaraes, Portugal

Abstract

Surfaces play a key role in the use of polymeric-based materials to be used in tissue engineering. This contribution reviews the employment of biomimetic methodologies and the use of stimuli-responsive polymers and polymeric surfaces that could have a direct impact on the interaction with cells and on the properties of the substrate.

INTRODUCTION

In the new paradigms of regenerative medicine, the use of materials in contact with biological materials (cells, tissues/organs, physiological fluids, and biomolecules) is a contemporary illustration of the need of interdisciplinary scientific approaches that combine the most recent advances in material science and technology, basic sciences, and life sciences. In tissue engineering (TE), matrices are developed to support cells, promoting their differentiation and proliferation toward the formation of a new tissue. Such strategies allow to produce hybrid structures that can be implanted in patients to induce the regeneration of tissues or replace failing or malfunctioning organs.

Materials applied in medical applications, both *in vivo* and *ex vivo*, have been developed since the early 1950s in different or combined directions: metals, polymers, ceramics, and composites. Increasing attention has been paid to the use of natural-based polymers and biomimetic approaches in a series of biomedical applications; to reasons can explain to explain such interest:

1. The essence of life is essentially "macromolecular" (DNA, proteins, polysaccharides, etc.).
2. Natural materials are intricate structures well adapted to their functions, which have risen from hundreds of millions of years of evolution. During this process, nature was able to conceive structures more complex than manmade materials, forming complex arrays, presenting hierarchical organizations, and exhibiting multifunctional properties.[1,2]

In TE, natural-based polymers can be recognized by the body, which could process them through established metabolic pathways; the metabolic products end up into simple sugars, amino acids, or short macromolecular chains able to be cleared by the kidneys. This reduces the possible cytotoxicity and inflammatory response from the host, a situation occurring, e.g., with wear particles of long-term orthopedic joint prostheses made of metal or synthetic non-degradable polymers.[3] Moreover, natural-based polymers are similar to biological macromolecules that could help in avoiding the stimulation of chronic inflammation or immunological reactions and toxicity, often detected in synthetic polymers. In this sense, nature offers sources of polymers and lessons to synthetize new macromolecules that could have great potential to be used in TE.[4]

Three main families of macromolecules can be considered to be used in TE:

1. From the abovementioned reasons, there are advantages of considering the use of natural-based polymers in the development of devices aimed be used in TE.
2. Synthetic polymers may also present some advantages in some biomedical applications:[6] some of them degrade *in vivo* into safe end products mainly by hydrolysis (e.g., L-lactic acid, a molecule existing in the human body, is able to metabolize into CO_2 and water), in a few weeks to several months, depending on several factors, including molecular structure/morphology, average molecular weight, size, and shape. They may be synthetized with very precise molecular weights and structures and can often be processed by melt-based techniques or using solvents into tailor-made materials for diverse applications.
3. A third class of polymers are artificial protein-based macromolecules prepared using recombinant DNA technology. The biosynthesis of repetitive polypeptides or well-designed sequences of amino acids permits to synthetize complex synthetic amino acid–containing macromolecules with relevant elements that exist in

Concise Encyclopedia of Biomedical Polymers and Polymeric Biomaterials DOI: 10.1081/E-EBPPC-120050615

real proteins, providing special characteristic to the synthetic counterparts, including mechanical strength, cell-binding characteristics, and stimuli-responsiveness or biomineralization promotion; in the primary structure, the macromolecule may combine more than one of these properties, giving rise to multifunctional materials. Examples of such systems are elastin-like recombinamers[7] and chimeric proteins based on silk.[8]

In the field of implantable biomaterials and tissue-engineered constructs, the bulk properties of materials are usually recognized as being important for the overall properties of the device, such as mechanical strength and degradation profile. Surface properties, however, are of utmost importance as they have an influence on subsequent tissue and cellular events, including protein adsorption, cell adhesion, and inflammatory response[9]—all these events should be considered in the process of tissue remodeling. Therefore, in this entry a special focus is given on the design, processing, and properties of surfaces involving the three classes of materials previously mentioned. As many strategies involve the development of products directed to the orthopedic field, we will also consider the inclusion of inorganic elements, especially bioactive inorganic nanoparticles in the systems (see Section "Bioactive materials").

A property that crosses the three classes of macromolecular systems discussed earlier is the ability that some polymers have to respond to external stimuli. Such a phenomenon is inspired by the strong nonlinear behavior observed in some natural macromolecules, especially proteins and polysaccharides: They may be stable long wide ranges of some external variable, e.g., pH or temperature, but undergo drastic conformational or physicochemical changes upon a narrow variation of this variable around a given critical point. Such a phenomenon is based on highly cooperative interactions that allow huge transformations to take place by the cumulative contribution throughout the repeating units of typically small interactions. Such an observation has inspired the development of biomimetic strategies in order to use this concept in a variety of polymers for a series of applications, including biomedicine.[10] Many examples are found in the literature on stimuli-responsive hydrogels or injectable systems (e.g.,[11,12]). However, in this text emphasis is given on the development and use of stimuli-responsive polymeric surfaces that can find applications as substrates to control protein or cell adhesion, in controlled drug delivery, or in biosensors.[13]

The use of biomaterials in TE must take into account how cells react with its environment. Long experience in studying how cells interact with the milieu enables to realize that living cells are complex entities presenting a remarkable, inherent capacity to sense, integrate, and respond to environmental cues at the micro- and nanoscales.[14,15] Their native environment is a three-dimensional (3D) scaffold comprising an insoluble aggregate of several highly organized, multifunctional large proteins

and glycosaminoglycans, collectively known as the extracellular matrix (ECM). By interacting directly with the ECM, cells gather information about the chemical and physical nature of the environment, integrate and interpret it, and then generate an appropriate physiological response.[14,16] Keeping in mind this extreme intelligence and sensitivity of cells, it seems convenient that their behavior can be directed through a precisely designed environment. Biomaterials to be used in TE should take into account the instructive role observed in natural environments to maintain cell viability and control cell behavior. Adult stem cells reside in tissue-specific microenvironments, so-called niches.[17] It is important to mimic such critical biophysical and biochemical native environments in order to understand the relevant factors that instruct stem-cell fate (both *in vitro* and *in vivo*). Such an understanding will ultimately help in creating optimized implantable hybrid constructs that could have real clinical relevance in regenerative medicine strategies. However, this task is extremely complex as one should take into account the multiplicity of factors that influence cell behavior in a natural or artificial milieu (Fig. 1).

In the development of biomaterials and scaffolds for TE, it is then required to test different factors to encounter the best conditions that could allow to direct cells toward the formation of new tissue. New biomaterials designed specifically to be integrated in TE devices should possess adequate characteristics both in terms of physicochemical (e.g., mechanical properties, degradation rate, hydrophilicity, stimuli-responsiveness/injectability, and mineralization inducer) and biochemical (e.g., cell-binding domains,

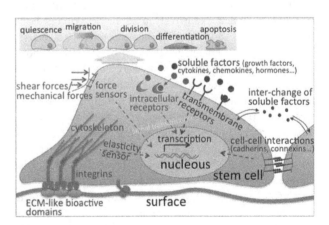

Fig. 1 There are multiple factors influencing stem-cell behavior when embedded in artificial environments: i) signals can come from the liquid medium, such as soluble factors composing the culture medium or released by other cells, and molecules released from micro/nanoparticles; ii) the chemical/biochemical/mechanical/topographic/physical properties of the surface; iii) existence of cell–cell contacts in co-culture conditions; iv) mechanical stimuli (shear stresses, compression forces, etc.); v) architecture of the scaffold (2D vs. 3D, nano/micro features, external, and internal architecture); and vi) the cell phenotype and differentiation state of cells.

osteoinductivity, and differentiation promoters) properties. High-throughput methodologies have started to be employed for first-screening evaluations, rather than for time-consuming and costly traditional one-at-a-time empirical approaches that reveal to be impractical if one tries to address the influence of a great number of combinations of parameters. Much work has been done in settling combinatorial libraries of biomaterials that could quantify cell response to many material properties in a single experiment.[18] Biomimetic platforms were proposed for the development of high-throughput chips in which different combinations of materials/cells/culture media could be screened independently, including a 3D environment (see Section "Superhydrophobic substrates as platforms for high-throughput analysis").

Bioactive Materials

Biomaterials and their surfaces that interact with bone should ideally have specific characteristics that could elicit a positive interfacial response to this tissue. The bone-bonding ability of a biomaterial is a very important property for bone tissue regeneration/replacement applications. Some ceramics and glasses show osteoconductive properties, such as a variety of calcium phosphates (Ca-P)s and some silica glasses. Hench[19] showed for the first time that some glasses, which contain SiO_2, Na_2O, CaO, and P_2O_5 in specific proportions, spontaneously bond to living bone. Since then, many bioactive glasses have been used clinically with the bone-bonding ability. They have been developed in the forms of bulks and particulates with dense and porous structures. For example, the well-known commercially available bioactive glass bioglass in the form of particulates has been extensively used in periodontal bone repair. Such bioactive ceramics and glasses spontaneously form chemical and mechanical strong bonds with the bone through a biologically active hydroxycarbonated apatite layer formed at their surface, which is chemically and structurally similar to the mineral phase of the bone when they are implanted.[20] This calcium phosphate layer is not observed around materials that are not bioactive, like metals and polymers in bone defects, demonstrating that this biologically active bone-like apatite layer is a prerequisite for the bonding between an artificial material and living bone. Ca-P may be integrated into hydrophilic polymeric templates by alternate immersion in solutions containing calcium and phosphate ions. For example, it was possible to produce bioactive poly(L-lactic acid) (PLLA)-chitosan hybrid scaffolds by in situ calcification of the polysaccharide component using such methodology.[21] The analysis of the bioactivity of artificial materials when implanted *in vivo* has been reproduced *in vitro* by immersion experiments using a simulated physiological solution that mimics the typical ion concentrations in body fluids:[22] the simulated body fluid (SBF). Upon immersion of a bioactive substrate in SBF, one observes the formation of an apatite bone-like layer onto the surface of the biomaterial.

Most of the substances used in the biomedical field, including the polymers referred to in Section "Introduction," are not bioactive. Some strategies may be followed to induce bioactivity in biomaterials and have been discussed elsewhere.[23] An obvious possibility is to combine polymers with inorganic bioactive particles: e.g., it was possible to produce a bioactive porous scaffold by combining PLLA and particles of bioglass.[24] However, in such procedures there are advantages of using inorganic particles, such as bioactive glasses, with nanometric sizes. Nanocomposites prepared using such nanoparticles usually exhibit enhanced osteoconductivity, due to the increase the surface area of the inorganic elements, exhibit superior mechanical properties, and may also elicit beneficial biological signals.[25] It should also be noticed that mineralized structures in nature are composed by well-organized combinations of a soft macromolecular fraction and nano-sized (usually calcium-containing) mineral elements—the peculiar structural features of natural nanocomposites have been inspired by novel approaches to develop synthetic materials and coatings, including osteoconductive biomaterials.[26] Synthetic bioactive nanoparticles based on the systems SiO_2-CaO or SiO_2-CaO-P_2O_5 have been prepared by a simple sol-gel method.[27,28] These kinds of nanoparticles could induce the bioactivity of natural-based polymers[29] and also elicit interesting biological effects, such as the stimulation of spreading bone marrow stem cells and the promotion of the formation of endothelial networks that could facilitate the vascularization of biomaterials.[30] Such inorganic nanoparticles may be integrated into biomimetic concepts combining natural-based polymers that could be interesting for TE applications (see, e.g., Section "Nanostructured nacre-like multilayered systems").

Biomimetic Surfaces

The success of polymeric scaffolds and hydrogels in TE is determined by the response they elicit from the surrounding biological environment upon implantation. This response is governed, to a large extent, by the surface properties of the biomaterials. Surfaces of polymeric scaffolds have a significant effect on protein and cell attachment. Multiple approaches have been developed to provide micrometer to nanometer scale alterations in the surface architecture of scaffolds to enable improved protein and cell interactions. The chemical modification of polymeric scaffold surfaces is one of the upcoming approaches that enables improved biocompatibility. Similarly, physical adsorption, radiation-mediated modifications, grafting, and protein immobilization are other methods that have been employed successfully for alterations of surface properties of polymeric scaffolds. Biomimetic approaches may be used in this context in the design and production of special surfaces with interest to be used in TE.

With almost 4 billion years of research and development on its side, nature has already solved problems that human designers and engineers still struggle with. The observation of the materials produced by living organisms allow us to identify some common trends that could help in proposing bio-inspired strategies, which includes: 1) hierarchical organization of the structure—the biologically controlled growth of natural materials leads to high multilevel architectures, exhibiting high control over the orientation of structural elements from the molecular level up to the final macroscopic structure; 2) mild-processing conditions/green chemistry/self-assembly—the production of biomaterials is carried out through the self-assembling of elementary units at mild conditions of ambient temperature, neutral or physiological pH, always in aqueous environment; 3) recurrent use of molecular constituents—widely variable properties are attained from apparently similar elementary units, such as amino acids in the production of proteins or the presence of calcium in many biogenic minerals; 4) energy saving—biology uses information and structure as the main vector for constructing materials, as compared with synthetic technologies based on the larger usage of energy; 5) functionally graded properties—the properties and composition along the material may vary, gradually or abruptly, adapting to the local requirements; 6) self-healing and damage repair—some natural materials have the possibility to self-repair or remodel; 7) biodegradability—natural materials are degraded either by hydrolytic mechanisms or through enzymatic action.

Some emerging surface engineering approaches have focused on creating biomimetic substrates for bio-related applications, which could have at least two mains functions: 1) to provide particular physicochemical features to the surface, such as stimuli-responsive properties, improved mechanical behavior, or special wettability or adhesiveness; 2) to elicit specific biological response to control the interaction of the surface with cells or biomolecules. The last case is especially important in the development of substrates for TE, biological diagnosis, or biosensors. In the particular case of regenerative medicine, biomimetic surfaces may be seen as substrates that elicit specified cellular responses to induce new tissue formation. The biomolecular recognition of materials by cells may be achieved by surface modification via chemical or physical methods with bioactive molecules such as a native long chain of ECM proteins as well as short peptide sequences derived from intact ECM proteins that can interact with cell receptors.[31]

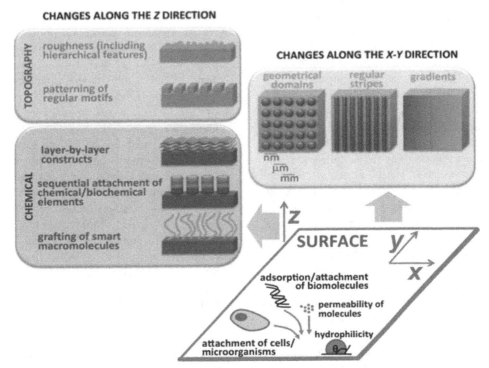

Fig. 2 Schematic description of possible modifications that can be performed onto polymeric substrates. In the X-Y axis confined geometrical domains or more continuous patterns may be produced at different length scales, as well as gradient chemical modifications. Along the Z axis, the deposition of smart macromolecules may be carried out as well as sequential attachment of different chemical/biochemical elements; alternate deposition of macromolecules or other objects may be performed by self-assembling using the layer-by-layer method. Topographic modifications may be performed, either inducing regular patterns or by modifying the roughness properties of the surface. Such modifications envisage the spatial and temporal control of the attachment of cells, microorganisms, or biomolecules, the permeability of molecules through the surface, or the surface energy of the substrate.
Source: Data from Mano[13] and Alves et al.[32]

Many works have been produced proposing a variety of methodologies to modify the surface of materials. Figure 2 shows representative examples on how surfaces may be engineered in both X-Y and Z directions. The spatial control enables to tune the surface properties in certain regions. Microcontact printing or soft-lithography[33,34] are possible techniques that enable to pattern distinct chemical features along the X-Y surface. Those could be used to immobilize proteins, DNA, or antibodies to control specific cell adhesion or to prepare microarrays for throughput analysis. Although many methods have been proposed in this context,[35] improved bio-inspired bottom-up methods based on self-assembly are still to be developed. There are many examples in the nature of functional gradient materials and many methods have been proposed to produce surfaces with gradient properties, especially to be used in the biomedical field;[36] the authors stated that gradient surfaces offer potential not only as models for molecular recognition and interactions in biological systems, and for cell mobility and diagnostic studies, but also for practical applications such as cell separation, drug delivery systems, and sensors in biotechnology.

Surface modifications along the X-Y axis may be combined, or not, with surface adjustment along the vertical direction (Fig. 2). Polymers or other chemical elements can be attached to the surface to offer a reversible stimuli-responsive character (see Section "Smart surfaces"). The most usual procedure in this case is to fabricate polymer brushes using smart macromolecules. The chemical treatment may be also performed in a sequential fashion, in which different molecules or macromolecules are gradually attached along the Z-direction. The use of a self-assembling monolayer is often employed as the first coating onto which the other modifications are executed. Self-assembling is also used to produce multi-layered films of polymers or other objects using the so-called layer-by-layer (LbL) method, which will later be discussed in detail (see Section "Nanostructured surfaces using LbL"), along with some examples; the thickness of such thin coatings can be tuned by adjusting the number of layers and the processing conditions. Besides chemical modifications, topographic features may be created by different techniques. Regular patterns, such as pillars or microgrooves, may be produced using microfabrication and nanofabrication technologies readily available from the semiconductor industry, such as soft-, photo-, and electron-beam lithography. More random textures may be produced by other methods that could also influence cell attachment and morphology, such as the ones produced by the generation of spherulites with different sizes by controlling polymer crystallization.[37] Roughness may also be generated to induce special characteristics to the surface—examples will be given of the fabrication of bio-inspired hierarchical roughness onto polymeric substrates, in order to produce surfaces with extreme wettabilities (see Section "Biomimetic surfaces exhibiting extreme wettabilities").

In vivo, cells inhabit a 3D rather a 2D world. Thus, in regenerative medicine efforts are made to achieve this environment using a variety of supports, including coalescent particles, porous scaffolds, and hydrogels. The structural control at different length scales in 3D systems is much more difficult than in flat surfaces; it is also a challenge to combine different materials, pattern the surface with topographic or chemical/biochemical features, or simply modify in a homogeneous way the inner surface properties in 3D porous scaffolds or hydrogels. Examples will be given on the translation of surface features usually addressed in flat substrates into 3D objects, emphasizing the relevance of being able to design, process, modify, and characterize surfaces in the context of developing 3D structures for TE.

Smart Surfaces

Depending on the application, the surface of a biomaterial should either promote cell adhesion or should avoid the attachment of specific proteins and cells. Much work has been devoted in the immobilization, typically by covalent bonds, of polymers or biopolymers onto the surface of different kinds of substrates.[38,39] Many more possibilities are offered if the coating may respond to external stimuli, where the surface energy or some other kind of property may be switched.[10,40] Usually, smart surfaces are prepared by depositing stimuli-responsive polymers or groups onto the surface of a substrate, either by covalent linkage or by adsorption.

For the case of temperature-responsive polymers, poly(*N*-isopropylacrylamide) (PNIPAAm) has been the most used macromolecule in smart temperature-sensitive systems; the change in properties with temperature in PNIPAAm is based on the existence of a lower-critical solution temperature (LCST), a phenomenon that is thermodynamically similar to that causing temperature-induced protein folding.[41] Above the LCST, a reversible conformational transition occurs from the expanded coil (soluble chains) to the compact globule (insoluble state), at around 32°C in pure water.[42] The solubility is affected because the amphiphilic PNIPAAm chains hide the hydrophilic amide groups and expose the hydrophobic isopropyl groups in the compact globule conformation. During this transition, hydrogen bonds existing between water molecules and the polymeric chain are disrupted, the thermodynamically favorable increase in entropy being the main driving force for the occurrence of the transition. For the case of crosslinking systems, this transition can be seen through an abrupt shrinkage of the gel above the LCST associated with the change in the swelling capability of the material. The grafting of PNIPAAm onto surfaces enables to switch the wettability of the surface around the LCST—this property has been applied to biomedicine due to the fact that cells adhere very differently to hydrophobic and hydrophilic substrates. Such an observation has been the basis of the development of smart temperature-responsive coatings and surfaces designed for cell culture. Okano and co-workers noticed that the ability for the platelet to attach and spread could be switched with temperature in polystyrene tissue culture dishes grafted with PNIPAAm.[43] Such a concept was extended to the production of implantable membranes made of cells: Mammalian cells cultured on such temperature-responsible substrates can be cultured until confluence at 37°C (hydrophobic surface) and can be recovered as confluent cell sheets, while keeping the newly deposited ECM intact, simply by lowering the temperature, where the surface exhibits a hydrophilic character.[44,45] Therefore, the cell-to-cell junctions are maintained, making it possible to produce cell sheets with adequate mechanical integrity. Such technology offers a gentle alternative to harvesting cultured cells as compared to the use of enzymes employed to remove cell monolayers from conventional culture dishes. Cell sheets could be used in a variety of TE applications, some of them already applied clinically. Independent cell sheets have been produced for the regeneration of cornea and epidermis, and showed to adhere easily to host tissues without the need for sutures. However, this technology can also be used to fabricate more complex tissue equivalents, either composed of single or multiple cell phenotypes, by the sequential layering of different cell sheets. 3D cellular organizations have been achieved, enabling the production of layered cardiomyocyte, hepatocytes, and endothelial cell sheets. Therefore, there is the possibility to produce hierarchical biological devices avoiding the use of scaffold materials that could be directly implanted for the regeneration of tissues and organs. Other materials can also be used in this context; it was possible, for example, to graft PNIPAAm onto chitosan membranes and engineer cell sheets onto such flexible substrates.[46] In this case, the surface may be easily processed into different shapes. Moreover chitosan is capable of taking up water and could also permeate small bioactive molecules useful to be delivered to the cells during the culturing period. A different ability for proteins to adsorb hydrophilic and hydrophobic substrates has been also been used to produce smart surfaces for programmed adsorption and the release of proteins in the context of microfluidic devices:[47] PNIPAAm was integrated into a microfluidic hot plate and was shown that protein adsorption and desorption switching could happen in less than 1 sec.

In connection to Section "Bioactive materials," PNIPAAm-based intelligent surfaces have also been used to control/trigger the onset of the biomineralization potential of surfaces when immersed in SBF upon the change of temperature. Such "smart" biomineralization surfaces were produced in PLLA/bioglass composites where PNIPAAm was grafted onto the surface through plasma activation: The precipitation of bone-like apatite occurred at physiological temperature, but was prevented at room temperature.[48] One could also induce a spatial control of the precipitation of apatite by patterning selected regions of the composite with grafted PNIPAAm. Such a concept was extended to pH-responsive systems, where chitosan was grafted onto such composites:[49] In this case calcification occurred just at physiological pH and was prevented in acidic media, where the surface is more hydrophilic.

Other alternatives have been proposed to substitute grafted PNIPAAm in such temperature-responsive surfaces. Our group proposed the use of a recombinant elastin-like polymer containing the cell attachment sequence arginine-glycine-aspartic acid (RGD) to produce smart thin coatings by simple deposition of the polymer dissolved in aqueous-based solutions onto chitosan substrates;[50] contact angle (CA) measurements at room temperature and 50°C showed reversible changes from a moderate hydrophobic behavior to an extremely wettable surface. The existence of RGD sequences in the primary structure of the synthetic protein also seemed to considerably improve cell adhesion—such characteristics suggest that these kinds of biomimetic smart coatings could have the potential to be used in TE and in the controlled delivery of bioactive agents.

Besides temperature, cell adhesion ability onto surfaces may be achieved through other kinds of stimuli that could also control/switch the surface wettability. In this context, a light-responsive copolymer combining PNIPAAm and spiropyran chromophores has been used to tailor cell-adhesion.[51] Enhanced cell adhesion was obtained by irradiation with 365 nm light and "reseted" by visible light irradiation (400–440 nm).[52] Furthermore, living cell patterning could be obtained by the previous projection of ultra violet (UV) light just onto the predefined regions of the surface. In another concept, Nakanishi et al. built self-assembled monolayers (SAMs) having photocleavable 2-nitrobenzyl groups that could be used to template cell adhesion:[53] Bovine serum albumin (BSA), inert to cell adhesion, was adsorbed over this monolayer. Upon exposing defined regions with UV light, BSA was released from such regions, and finally fibronectin, a protein promoting cell adhesion, was then incubated and covered the irradiated regions. This method allowed to produce patterned regions for cell adhesions and by subsequent illumination it could be also possible to induce cell migration and proliferation.

NANOSTRUCTURED SURFACES USING LbL

Fabrication of Polyelectrolyte Multilayered Films

The design of advanced, nanostructured materials at the molecular level is of tremendous interest for the development of materials in the biomedical field. Two techniques related to surface modification (through the Z-direction, according to Fig. 2) dominated research in this area: Langmuir–Blodgett deposition and SAMs. Several intrinsic disadvantages of both methods limit their application in the field of biology. For the Langmuir–Blodgett deposition, the drawbacks are expensive instrumentation and long fabrication periods for the preparation of the biomolecule films, limited types of biomolecules that can be embedded in the film (including the need of the assembly components to be amphiphilic), limitation when complex geometric substrates are intended to be coated, and a considerable instability of biomolecules resulting from weak physical attraction inside the films. For SAMs, disadvantages are a low loading of biological components in the films because of their monolayer nature, a limited number of substrate types, since SAMs are only fabricated by the adsorption of thiols onto noble metal surfaces or by silanes onto silica surfaces, and a limited stability of films under ambient and physiological conditions. Among other available techniques, the LbL assembly method introduced by Decher and co-workers in 1992[54] has attracted extensive attention because it possesses extraordinary advantages for biomedical applications: ease of preparation, versatility, capability of incorporating high loadings of different types of biomolecules in the films, fine control over the material's structure, and robustness of the products under ambient and physiological conditions.

The LbL methodology consists in cyclically depositing polyelectrolytes that assemble and self-organize on the material's surface, leading to the formation of highly tuned, functional polyelectrolyte multilayer films (PEM) with a nanometer-level control of film composition and structure. The procedure is simple and in principle applicable to many different kinds of substrates. In most methodologies, the deposition of the elements onto the surface is performed by dipping the substrate into solutions containing the polyelectrolytes, intercalated with immersions in clear solutions to wash the excess adsorbed material. Other procedures have been proposed, such as the use of spraying to disperse the solutions onto the substrates. It is claimed that the uniformity of the films may be improved and automatization of the process is also possible.[55]

The initial studies involving PEM in the biomedical field were based on more fundamental aspects related to the nanostructure developed during the build-up of the PEM and relationships between the properties of the coating with cell behavior. For example, the system chitosan/alginate was investigated using quartz-crystal microbalance during the construction up of the PEMs.[56] The changes of the viscoelastic properties of the films were monitored upon the crosslinking with glutaraldeyde—it was found that crosslinking mainly takes place in the top chitosan layer. The evolution of the dissipation factor during crosslinking was modelled with first-order kinetics; it was found that more robust films could be produced by crosslinking the intermediate layers of chitosan during their formation. This could have implications in terms of protein adsorption and cell adhesion. Crosslinking of the chitosan/alginate multilayers with glutaraldeyde could enhance the adsorption of albumin at physiological conditions;[57] however, as the chemical modification just took place in the top layer, the change in the mechanical properties of the PEM was not significant and the crosslinking did

not significantly change cell adhesion and proliferation. Other methodologies could be used to crosslink the PEM in a more extensive degree. It was shown earlier that in these cases, crosslinking increases the stiffness of PEMs, which may have a positive response to cells.[58,59]

It would be desirable, in some cases, to process the PEMs at neutral pH and physiological temperature, which is not possible using conventional chitosan. We showed that PEM can be formed in mild conditions using water-soluble chitosan[60]—specifically, using this polymer and alginate, films assembled at a pH of 5.5 exhibited a more rigid behavior and compact state, whereas at a pH of 7.0 the layers displayed more viscoelastic properties and higher thickness, evidencing the key role of the electrostatic interactions during polyelectrolyte adsorption.

Over the past 5 years, possibilities for the spatiotemporal control of cell growth onto PEMs have emerged and the first in vivo studies have been performed. It is only recently that more complex cell processes such as cell differentiation have begun to be explored and controlled by means of PEM. It was shown that poly(sodium-4-styrene-sulfonate)/poly(allylamine hydrochloride) multilayers constitute adequate coatings that permit the differentiation of endothelial progenitor cells into mature endothelial cells and the formation of an endothelium-like confluent cellular monolayer within only 2 weeks (compared with 2 months for classical coatings).[61]

In many TE strategies, it is important to combine therapeutic molecules, including growth factors, with the supporting biomaterials, in order to control cell behavior. PEMs are highly attractive in this context as they can act as ultrathin biologic reservoirs, with the capability of coating difficult geometries, the use of aqueous processing likely to preserve fragile protein function, and the tunability of incorporation and release profiles. The group of Hammond presented LbL films capable of a microgram-scale release of the biologic bone morphogenetic protein 2 (BMP-2), which is capable of directing the host tissue response to create bone from native progenitor cells.[62] The same group proposed the use of PEMs to incorporate more than one molecule that could be released with different time scales.[63] Such a topic is quite relevant and it is often necessary to induce cellular events that take place at different times during the regenerative process. The group of Picart also proposed the use of multilayers of hyaluronic acid (HA) and heparin as polyanions and poly(L-lysine) as a polycation to prepare PEMs as a reservoir to encapsulate and release recombinant human BMP-2.[64]

Overall, the possibilities of using a wide range of polyelectrolytes and nano-objects combined with the advantages offered by PEM coatings, such as spatial confinement and localized delivery, as well as protective effects on exposure to physiological media and external stresses, considerably enrich the biological applications for PEM films. In the following sections, two case studies of flat multilayered films will be discussed, which could have direct interest in the field of biomaterials and TE.

Nanostrucured Nacre-Like Multilayered Systems

The LbL technique is applicable not only to polyelectrolyte/polyelectrolyte systems. Almost any type of charged species, including inorganic molecular clusters, nanoparticles, nanotubes and nanowires, nanoplates, organic dyes, dendrimers, porphyrins, biological macromolecules, and viruses, can be successfully used as components to prepare LbL films.[65] Such a technique can then be used to prepare laminated nanocomposites by intercalated macromolecules (soft component) and inorganic nanoparticles. For example, Tang et al. prepared multilayered films with thicknesses higher than 5 μm by sequential the immersion of a glass slide in solutions of a polycation and anionic montmorillonite clay.[66] The objective of that work was to try to reproduce, suing the LbL technology, coatings and films inspired by the structure of nacre (mother of pearl). Nacre is found in the shells of mollusks and possesses amazing properties that have inspired many researchers to prepare synthetic high-tough materials.[67] The nacreous layer is constituted by aragonite (calcium carbonate) polygonal tablets, with a thickness of about 500 nm, between thin sheets of protein-polysaccharide organic matrix. This organization plays a determinant role in the nacre's structural strength and toughness. Moreover, properties such as interfacial compatibility and adhesion between organic and inorganic layers are also crucial. Between every plate-like aragonite crystal is a tight protein-polysaccharide folding responsible for the inelastic behavior of nacre when loaded in tension and shear, fundamental for notch resistance. In the end, such factors contribute for the remarkable toughness of nacre that has a work of fracture about 3,000 times greater than inorganic aragonite.

Inspired by the extraordinary structure of nacre, biomimetic multilayered nanocomposite films were conceived using LbL, intercalating a polycation (chitosan) and an inorganic element: a dispersion in water of the bioactive nanoparticles that were discussed in Section "Bioactive materials."[68] The concept combined the nacre-like organization achieved by the LbL technique and the osteoconductivity properties of the inorganic component, which could be useful to coat implantable devices with a robust layer and also to provide good bone-bonding ability. Such multilayers deposited onto flat glass surfaces showed to be bioactive *in vitro* as they induced the precipitation of apatite upon immersion in SBF. However, as the deposition of the elements is based on dipping in aqueous-based solutions, it should be possible to coat more complex substrates, including the inner pores of scaffolds for bone TE.

Smart Multilayered Films

As discussed in Section "Smart surfaces," smart surfaces may be useful in a series of biomedical applications including TE and controlled drug delivery. Stimuli responsiveness may also be incorporated into multilayered systems by using, e.g., polymers or other elements that could respond to external parameters, such as temperature, pH, ionic strength, electrical or magnetic fields, or light.[69] As compared to traditional responsive materials such as hydrogels and surfaces grafted with smart polymers, PEMs have the advantages of nanoscale control of film structures and ease in manipulating the chemical composition of the film.

As discussed earlier, temperature-responsive surfaces, using typically polystyrene substrates covalently grafted with PNIPAAm-based macromolecules, have been employed to produce free-standing layers of autologous cells that could be used in a series of regenerative medicinal applications. In a recent work, we showed that small chains of PNIPAAm could be grafted into chitosan and that this copolymer could be used to produce PEMs with the alginate counterpart.[70] This offers the advantage of producing easy-to-build smart coating from multilayers having a versatile structural and chemical control without being covalently linked to the substrate. Moreover, the attachment of the temperature-responsive element to chitosan also permits to have a pH-responsiveness (around the pK_a of chitosan) that could provide to the PEM a dual responsive character. Cells could be cultured over such PEMs until reaching confluence. After reducing the temperature below PNIPAAm LCST, a gradual detachment of cell sheets from these coatings could be observed. The detached cells maintained their viability. Such results open new perspectives of using innovative strategies to prepare novel substrates for cell-sheet technology, in which other features could be envisaged, such as the possibility for releasing active drugs, coating surfaces with complex geometry, or controlling the mechanical properties of the surface.

The use of smart biomimetic polymers obtained by genetic engineering techniques, as refereed in Section "Introduction," may open further possibilities in the context of stimuli-responsive nanostructured multilayers. An elastin-like polymer was biosynthesized, and it was shown that it could be assembled with chitosan into multilayers using LbL.[71] AFM images of the obtained film showed clear differences when taken below or above the inverse temperature transition. Such a concept was extended using more bioactive elastin-like polymers, containing the RGD sequence.[72] The smart properties of the coatings were tested for their wettability by CA measurements as a function of different external stimuli, namely, temperature, pH, and ionic strength. Wettability transitions were observed from a moderate hydrophobic surface (CAs approximately from 62° to 71°) to an extremely wettable one (CA considered as 0°) as the temperature, pH, and ionic strength were raised above 50°C, 11, and 1.25 M, respectively. Enhanced cell adhesion was observed in these coatings, as compared to a coating with a chitosan-ending film and a scrambled arginine–(aspartic acid)–glycine biopolymer. The results from that work suggest that such films could be used in the future as smart biomimetic coatings of biomaterials for different biomedical applications, including those in TE or in controlled delivery systems.

Using LbL for Processing Structures for TE

The passage of the assembly process of multilayered films using LbL from flat surfaces to 3D substrates permits to open new possibilities of preparing or modifying supports to be used in TE. Interconnected 3D porous structures with macroscopic sizes were obtained employing LbL over leachable templates:[73] A perfusion technique allowed to produce the multilayers over a 3D template formed by a free form moldable assembly of paraffin wax spheres held together without any binder—after leaching the spheres, an interconnected porous construct could be formed just composed by the nanostructured assembly of the polyelectrolytes used during the LbL procedure. The final structure exhibited porosities higher than 99% but possessed enough mechanical stability to be handled. The culture of osteoblast-like cells in such scaffolds showed that after 3 days the cells could maintain their viability and could invade the interior of the construct. Similar structures were tested *in vitro* and demonstrated good properties to be used in cartilage TE.[74]

LbL can also be used to link together biocompatible spheres to produce scaffolds prepared by particle aggregation.[75] The multilayers built up over the exposed surface of the particles are used to promote the mechanical fixation of the microspheres assembly and simultaneously control the properties of its surface. Chitosan particles crosslinked with genipin were agglomerated using alginate/chitosan PEMs, deposited using a perfusion procedure. The obtained scaffolds could support the attachment and proliferation of ATDC5 (chondrocyte-like) cells. Oppositely to the previous example, scaffolds prepared by this method present low porosity (higher stiffness), but the interconnectivity of the pores is kept.

Besides the conventional porous scaffolds, other structures could be used to support cell attachment and proliferation. Capsules may be an interesting possibility if one intends to develop injectable systems combining cells and materials in which cells could be protected, e.g., from the immune system. We showed that LbL also could be used to build up the shell of capsules incorporating living cells in their interior.[76] The encapsulation methodology involved three steps: 1) ionotropic gelation to produce alginate beads containing cells; 2) LbL coating of the beads in mild conditions, using water-soluble chitosan and alginate; and 3) core liquefaction using a chelating agent for calcium ions. Cells were found viable up to 3 days. All the capsules exhibited a spherical shape, smooth surface, a liquid-core characteristic, and a mechanical stability that was dependent on the number of layers of the coating. Such a concept was upgraded to liquefied capsules containing solid microparticles dispersed inside the liquid core to provide support to adherent cells.[77]

BIOMIMETIC SURFACES EXHIBITING EXTREME WETTABILITIES

Effect of Wettability on Cell Behavior

It is well known that cell adhesion and protein adsorption onto a substrate are highly affected by the wettability and chemical nature of the surface.[78–80] It was found that cells present a maximum attachment capability onto surfaces with moderate wettability: For the case of surfaces with controlled chemistry produced by SAMs maximum adhesion was reached for water CAs between 40 and 60°. The results suggested that cell attachment is mainly determined by surface wettability, but is also affected by the surface functional groups, their surface density, and of course the nature of cells.

The effect of the wettability of biomaterials has been mainly investigated in flat substrates. However, as discussed earlier, in TE applications cell experience a 3D environment, so it would be more adequate to analyze this effect on 3D samples. It was shown that scaffolds with different wettabilities could be produced by using copolymers containing different compositions of hydrophilic and hydrophobic repeating units, enabling the production of model porous structures with controlled hydrophilicities.[81] In that work, methacrylate-endcapped caprolactone networks with a tailored water sorption ability were processed into 3D porous scaffolds using poly(methyl methacrylate) microspheres as a leachable template. Goat bone marrow stromal cells were seeded and cultured for different days with such structures. Osteoblastic differentiation was followed using typical markers. It was found for this particular case that cell proliferation and the expression of osteogenic markers were higher in the more hydrophobic scaffolds.

Another drawback of conventional wettability studies is that they are limited to surfaces ranging from hydrophilic to hydrophobic, as smooth surfaces have been typically used. There is both fundamental and practical interest in extending such studies toward the superhydrophilic (CAs below 5°) and superhydrophobic (CAs higher than 150°) limits. For example, new insights may be obtained on the influence of such extreme environments on the physiological response of cells, including their contractile characteristics and signaling activity that may influence adhesion, morphology/anisotropy, migration, proliferation, and differentiation. The only possibility to reach such extreme values of CA is to prepare surfaces with special roughness features. By mimicking the effects found in nature such as in the lotus leaf in which the surface exhibits extreme water repellence, PLLA biomimetic surfaces that combined nano- and micro-level roughness were developed, using a phase-separation method.[82,83] Such surfaces were found to be highly hydrophobic, exhibiting water CAs above 150°. The behavior of bone marrow–derived cells in contact with these PLLA surfaces was found to be significantly affected by superhydrophobicity: Such rough surfaces inhibit cell adhesion and proliferation, as compared with a more flat one.[86] Moreover, by using plasma treatments the hydrophilic nature of the polymer could be changed and the wettability of such initial rough substrates could be tailored between the superhydrophobic down to the superhydrophilic regime,[84] opening the possibility of studying cell attachment or protein adsorption within an extremely large range of wettability.

Superhydrophobic polystyrene or chitosan-based surfaces could also be produced using a similar method.[85,86] The *in vitro* performance of three different cell lines (SaOs-2, L929, and ATDC5) was assessed on the superhydrophobic polystyrene surfaces.[85] Moreover, well-defined superhydrophilic spots were patterned on superhydrophobic surfaces by using hollowed masks to localize the UV-ozone irradiation treatment. It was possible to localize the proliferation of SaOs-2 cells on these superhydrophilic spots, demonstrating that random micro/nano roughness and further patterned chemical modification by UV-ozone irradiation may be an elegant and easy method to control spatially the attachment/proliferation of cells in distinct materials with possible uses in the high-throughput analysis or microfluidic systems (see next section).

Superhydrophobic Substrates as Platforms for High-Throughput Analysis

As mentioned in Section "Introduction," high-throughput screening (HTS) approaches permit to correlate the characteristics of materials and surfaces and the corresponding biological response, including cell adhesion, growth and differentiation, or gene expression in a single experiment. Different libraries/methods have been employed to produce such HTS, including surfaces varying in roughness, surface chemistry, energy mechanical properties, and density of biochemical elements. Substrates for this kind of HTS have been fabricated by the robotic DNA spotter; microfabrication masking techniques, such as photolithography, soft-lithography, microfluidics, contact printing, templating, imprint lithography, microelectronics, and magnetic forces; and microfabrication non-contact printing techniques, such as ink-jet printing, electron beam lithography, and dip pen nanolithography. The microarray format enables the rapid synthesis of suitable polymers or the deposition of different materials, thereafter screening a large library of multiple biomaterials and microenvironments. However, in the methodologies all the spots employed in the chip are usually tested in the same biological environment, which means the entire device is immersed in a unique culture medium. Advances in this field should offer the possibility to screen individually but in the same chip different combinations of biomaterials under different conditions, including different cells, culture mediums, or solutions with different proteins or other molecules. We proposed a new method for a rapid, microliter-scale deposition of biomaterials or proteins and

a characterization of their interactions with cells based on the use of patterned superhydrophobic surfaces.[87] The proposed HTS approach uses patterned superhydrophobic substrates with wettable spots to produce flat microarray chips for multiplexing the evaluation of material–cell interactions. The hypothesis is that liquid volumes may be confined in well-defined regions due to strong contrasts in surface tension, enabling to deposit with high control materials, cells, and other substances. To validate the methodology, we investigated the cellular behavior onto patterned polystyrene superhydrophobic substrates having preadsorbed combinations of two proteins, human serum albumin (HSA) and human fibronectin (HFN). After 4 hr of culture, the chip was washed and the cells were fixed, stained, and imaged using confocal microscopy. Image analysis was used to count the number of cells present in each spot allowing to generate an intensity map for the cell number. As a tendency that was useful to proof the concept, more cells were detected in the spots richer in HFN. Such findings were consistent with the fact that HSA is a passivating protein and HFN has cell adhesive properties.

Another similar work used such patterned superhydrophobic chips to test the bioactivity potential of the nanoparticles discussed in Section "Bioactive materials," prepared by a sol-gel procedure.[88] Nanoparticles from different formulations were deposited in the spots of the chip and after immersion in SBF the content and nature of the Ca-P deposited in each spot could be evaluated in a high-throughput way.

The above mentioned works demonstrated that superhydrophobic flat substrates with controlled wettable spots could be used to produce microarray chips as a new low cost platform for high-throughput analysis that permits to screen the biological performance of the combinations of biomaterials, cells, and culture media. Such inexpensive and simple bench-top methods, or simple adaptations from it, could be integrated in tests involving larger libraries of substances that could be tested under distinct biological conditions, constituting a new tool accessible to virtually anyone to be used in the field of TE/regenerative medicine, cellular biology, diagnosis, drug discovery, and drug delivery monitoring.

The idea of the platform used to test cell response to proteins deposited onto patterned superhydrophobic surfaces was transposed to the fabrication of chips for combinatorial cell/3D biomaterial screening assays, which was much more adequate for tests in the context of TE.[89] Again, arrays of hydrophilic regions were patterned in such surfaces using UV/ozone radiation, but in this case the spots were dispensed with liquid precursors also containing cells that gave rise to 3D hydrogels. Three different polymers—chitosan, collagen, and HA—were combined with alginate in different proportions in order to obtain combinatorial binary alginate-based polymeric arrays. The effect of the addition of gelatin to the binary structures was also tested. The gels were chemically analyzed by FTIR microscopic mapping. Cell culture results varied according to the hydrogel composition and encapsulated cell types (L929 fibroblast cells and MC3T3-E1 pre-osteoblast cells). Cell viability and number could be assessed by conventional methods, such as MTS reduction test and dsDNA quantification. Non-destructive image analysis was performed using cytoskeleton and nuclei staining agents and the results were consistent with the ones obtained by conventional sample-destructive techniques. Briefly, L929 cells showed a higher number and viability for higher alginate-content and collagen-containing hydrogels, while MC3T3-E1 showed a higher cell number and viability in lower alginate-content and chitosan-containing hydrogels. In the particular case of the conditions explored in this work, the addition of gelatin did not influence significantly the cell metabolic activity or cell number in any of the encapsulated cell types. The results from this work offer a promising indication that such platforms can have potential to be used to test 3D biomaterial–cell relationships. In this context, similar tests were extended to miniaturized scaffolds obtained by freeze drying after dispensing distinct liquid formulations on the hydrophilic spots of the chips.[90]

Superhydrophobic Substrates as Platforms for Processing Polymeric or Hydrogel Particles

Biomaterials may be processed into different shapes to act as a temporary support for cells in TE strategies. Although porous scaffolds have been the most employed architecture, microparticles have been applied as building blocks and matrices for the delivery of soluble factors or support cell attachment and proliferation, aiming for the construction of TE scaffolds, either by fusion giving rise to porous scaffolds or as injectable systems for in situ scaffold formation, avoiding complicated surgery procedures. These and other less conventional strategies of using microparticles in TE have been reviewed elsewhere;[91] the work also overviewed the different technologies used to process microparticles for TE. In most methodologies employed to fabricate spherical particles, often derived from the pharmaceutical industry, the initial liquid droplets harden into a hydrogel or solid form while immersed in another insoluble liquid substrate, during which a fraction of the molecules that are initially in the liquid phase may be lost. Then, in these "wet" methods, the encapsulation of proteins or other molecules will never be totally efficient and some molecules could not maintain the biological activity. Regarding the internalization of therapeutic substances by diffusion processes, this process could limit the loading capability of the matrix. Another drawback of the most of the presented processes is the use of organic liquids that are required in several steps but are limiting when is required the encapsulation of living cells or microorganisms.

A new processing method, based on the rolling of water drops over superhydrophobic surfaces, proposed the fabrication of hydrogel and polymeric spheres depositing drops

of liquid precursors containing the polymer and other substances onto the surface.[92] This method presents advantages over conventional gelation and emulsion techniques as the contact of the dispensed drops with an outer liquid environment is avoided. Due to the mild conditions associated to this technique, the encapsulation of living cells is also possible strengthening the potential of this technology into TE applications. This processing method also avoids the exposure of the microparticles to stirring forces and allows to obtain precisely shaped and sized structures. In this method, the collection of the particles is also facilitated as the drops can easily be removed as the contact area with the surface is very small.

This methodology permits the encapsulation of therapeutic molecules with high efficiency and in non-aggressive conditions. This hypothesis was proved by the successful encapsulation of proteins (albumin and insulin) in formulations based on photo-crosslinkable methacrylated dextran and PNIPAAm[93]—the work showed that the proteins could be homogeneously distributed in the particle network and their release profile responded to temperature. The method using superhydrophobic platforms can also be used to produce concentric multilayered particles constructed by sequential dispensing of liquid droplets followed by crosslinking and dispensing of another liquid volume over the formed solid spheres.[94]

One concept of using microparticles in TE is in injectable systems in which cells promote the aggregation of the particles. Chitosan particles crosslinked with genipin showed the ability to support the growth of goat bone marrow stromal cells;[95] the cells could also induce the aggregation of the particles in vitro forming a 3D hybrid construct. We also used the superhydrophobic surfaces to prepare particles for these kinds of applications. In this case, biomimetic elastin-like polymers, that have been discussed previously, were processed in the shape of particles, by crosslinking to different extents the droplets of the polymeric solutions deposited in water-repellent polystyrene surfaces using hexamethylene diisocyanate.[96] It was found that the most crosslinked condition was the most favorable for cell proliferation and to form a cell-induced aggregation scaffold.

CONCLUSIONS AND FINAL REMARKS

The case studies presented demonstrated the relevance of surfaces in the field of biomedicine, not only in terms of the interface between biomaterials and cells but also on the use of special substrates exhibiting useful characteristics that permit, e.g., to switch the biomineralization capability or be used to process particles or arrays for high-throughput analysis. For the specific case of regenerative medicine, much work is still needed to provide the adequate cues to control cell behavior, based on the mechanisms by which tissues form and heal. Two difficulties should be surpassed in this context: 1) Much information is still needed on how the natural capacity of the regeneration of tissues takes place or can be stimulated—fundamental information that has been accumulated in the last two decades from developmental biology studies and the combination of molecular and stem cell biology have been most valuable, but more work is still necessary to have a more clear picture on how cells interact with (and be controlled by) their environment. 2) Such basic information to the production of synthetic supports for cells, in which a number of specifications of the interphase should be considered, including topographic aspects of the substrate, the distribution of chemical and biochemical groups throughout the surface, the mechanical properties, and the capability of surface properties to switch over the action of external stimuli, needs to be transposed—in this case, the design of surfaces should bring together the most recent advances in surface chemistry, materials processing and engineering, nanoscience and nanotechnology, and sophisticated characterization tools. Concomitantly, mastering the complexity associated with multiple factors that control cell behavior when in contact with biomaterials will required the development of advanced methodologies to perform combinatory essays in realistic simulated environments (including *in vivo* tests)—it was shown that biomimetic routes may provide cues to develop polymeric surfaces with spatial controlled wettabilities that could be used as platforms for high-throughput analysis. In this context, and in the field of implantable devices, more efforts are necessary to adapt technologies usually used to process or modify flat surfaces into the 3D space. For example, the use of the LbL methodology to produce 3D porous scaffolds using either a leaching technique or by particle agglomeration, or to produce liquefied capsules to immobilize and protect cells, may have potential applicability in regenerative medicine.

Multidisciplinarity has been claimed to be a fundamental tool to develop the field of TE. However, more radical and open-minded non-conventional approaches may also furnish innovative ideas with applicability in this area, and this attitude should ideally be incited to young researchers. Case studies were presented that combined non-evident ideas from different fields toward the development of concepts directly related to TE and regenerative medicine; an example was the use of superhydrophobic biomimetic surfaces as platforms to process in a highly efficient manner polymeric particles or hydrogels that could encapsulate therapeutic molecules or living cells.

The examples shown tended to privilege a more academic interest, but the results revealed potential applicability of some of the concepts. Efforts should be made to translate such fundamental knowledge into the clinical needs, where regulatory issues should be considered, as well as the commercial viability of the products. Those are most likely the biggest obstacles from the application point of view, especially for the case of implantable devices.

REFERENCES

1. Meyers, M.A.; Chen, P.Y.; Lin, A.Y.M.; Seki, Y. Biological materials: Structure and mechanical properties. Prog. Mater. Sci. **2008**, *53*, 1–206.
2. *Biomimetic Approaches for Biomaterials Development*; Mano, J.F., Ed.; Wiley-VCH: Weinheim, 2012.
3. Anderson, J.M. Biological responses to materials. Annu. Rev. Mater. Res. **2001**, *31*, 81–110.
4. Mano, J.F.; Silva, G.A.; Azevedo, H.S.; Malafaya, P.B.; Sousa, R.A.; Silva, S.S.; Boesel, L.F.; Oliveira, J.M.; Santos, T.C.; Marques, A.P.; Neves, N.M.; Reis, R.L. Natural origin biodegradable systems in tissue engineering and regenerative medicine: Present status and some moving trends. J. R. Soc. Interface **2007**, *4* (17), 999–1030.
5. Alves, N.M.; Mano, J.F. Chitosan derivatives obtained by chemical modifications for biomedical and environmental applications. Int. J. Biol. Macromol. **2008**, *43* (5), 401–414.
6. Middleton, J.C.; Tipton, A.J. Synthetic biodegradable polymers as orthopedic devices. Biomaterials **2000**, *21* (23), 2335–2346.
7. Rodríguez-Cabello, J.C.; Martin, L.; Girotti, A.; García-Arévalo, C.; Arias, F.J.; Alonso, M. Emerging applications of multifunctional elastin-like recombinamers. Nanomedicine **2011**, *6* (1), 111–122.
8. Gomes, S.; Leonor, I.B.; Mano, J.F.; Reis, R.L.; Kaplan, D.L. Natural and genetically engineered proteins for tissue engineering. Prog. Polym. Sci. **2012**, *37* (1), 1–17.
9. Thevenot, P.; Hu, W.; Tang, L. Surface chemistry influences implant biocompatibility. Curr. Top. Med. Chem. **2008**, *8* (4), 270–280.
10. Mano, J.F. Stimuli-responsive polymeric systems for biomedical applications. Adv. Eng. Mater. **2008**, *10* (6), 515–527.
11. Couto, D.S.; Hong, Z.; Mano, J.F. Development of bioactive and biodegradable chitosan-based injectable systems containing bioactive glass nanoparticles. Acta Biomater. **2009**, *5* (1), 115–123.
12. Santos, J.R.; Alves, N.M.; Man, J.F. New thermo-responsive hydrogels based on poly(N-isopropylacrylamide)/hyaluronic acid semi-interpenetrated polymer networks: Swelling properties and drug release studies. J. Bioact. Compat. Polym. **2010**, *25* (2), 169–184.
13. Mano, J.F. Biomimetic and smart polymeric surfaces for biomedical and biotechnological applications. Mat. Sci. Forum. **2010**, *3*, 636–637.
14. Parent, C.A.; Devreotes, P.N. A cell's sense of direction. Science **1999**, *284* (5415), 765–770.
15. Girard, P.; Cavalcanti-Adam, E.; Kemkemer, R.; Spatz, J. Cellular chemomechanics at interfaces: Sensing, integration and response. Soft Matter **2007**, *3*(3), 307–326.
16. Spatz, J.P.; Geiger, B. Molecular engineering of cellular environments: Cell adhesion to nano-digital surfaces. Methods Cell Biol. **2007**, *83*, 89–111.
17. Lutolf, M.P.; Gilbert, P.M.; Blau, H.M. Designing materials to direct stem-cell fate. Nature **2009**, *462* (7272), 433–441.
18. Yang, J.; Mei, Y.; Hook, A.L.; Taylor, M.; Urquhart, A.J.; Bogatyrev, S.R.; Langer, R.; Anderson, D.G.; Davies, M.C.; Alexander, M.R. Polymer surface functionalities that control human embryoid body cell adhesion revealed by high throughput surface characterization of combinatorial material microarrays. Biomaterials **2010**, *31* (34), 8827–8838.

Biomimetic—Colloid

19. Hench, L.L. Bioceramics: From concept to clinic. J. Am. Ceram. Soc. **1991**, *74* (7), 1487–1510.

20. Kokubo, T.; Ito, S.; Sakka, S.; Yamamuro, T. Formation of a highstrength bioactive glass-ceramic in the system MgO-CaO-SiO2-P2O5. J. Mater. Sci. **1986**, *21* (2), 536–540.

21. Mano, J.F.; Hungerford, G.; Gómez Ribelles, J.L. Bioactive poly(L-lactic acid)-chitosan hybrid scaffolds. Mater. Sci. Eng. C **2008**, *28* (8), 1356–1365.

22. Kokubo, T.; Takadama, H. How useful is SBF in predicting *in vivo* bone bioactivity? Biomaterials **2006**, *27* (15), 2907–2915.

23. Alves, N.M.; Leonor, I.B.; Azevedo, H.S.; Reis, R.L.; Mano, J.F. Designing biomaterials based on biomineralization of bone. J. Mater. Chem. **2010**, *20* (15), 2911–2921.

24. Ghosh, S.; Viana, J.C.; Reis, R.L.; Mano, J.F. Osteochondral tissue engineering constructs with a cartilage part made of poly(L-lactic acid)/starch blend and a bioactive poly(L-lactic acid) composite layer for subchondral bone. Key Eng. Mat. **2006**, *1109*, 309–311.

25. Boccaccini, A.R.; Erol, M.; Stark, W.J.; Mohn, D.; Hong, Z.; Mano, J.F. Polymer/bioactive glass nanocomposites for biomedical applications: A review. Comp. Sci. Tech. **2010**, *70* (13), 1764–1776.

26. Luz, G.M.; Mano, J.F. Mineralized structures in nature: Examples and inspirations for the design of new composite materials and biomaterials. Comp. Sci. Tech. **2010**, *70* (13), 1777–1788.

27. Hong, Z.; Reis, R.L.; Mano, J.F. Preparation and *in vitro* characterization of scaffolds of poly(L-lactic acid) containing bioactive glass ceramic nanoparticles. Acta Biomater. **2008**, *4* (5), 1297–1306.

28. Luz, G.M.; Mano, J.F. Preparation and characterization of bioactive glass nanoparticles prepared by sol-gel for biomedical applications. Nanotechnology **2011**, *22* (49), 494014.

29. Hong, Z.; Merino, E.G.; Reis, R.L.; Mano, J.F. Novelrice-shaped bioactive glassceramic nanoparticles. Adv. Eng. Mater. **2009**, *11* (5), B25–B29.

30. Hong, Z.; Luz, G.M.; Hampel, P.J.; Jin, M.; Liu, A.; Chen, X.; Mano, J.F. Mono-dispersed bioactive glass nanospheres: Preparation and effects on biomechanics of mammalian cells. J. Biomed. Mater. Res. A **2010**, *95* (3), 747–754.

31. Shin, H.; Jo, S.; Mikos, A.G. Biomimetic materials for tissue engineering. Biomaterials **2003**, *24* (24), 4353–4364.

32. Alves, N.M.; Pashkuleva, I.; Reis, R.L.; Mano, J.F. Controlling cell behavior through the design of polymer surfaces. Small **2010**, *6* (20), 2208–2220.

33. Xia, Y.N.; Whitesides, G.M. Soft lithography. Ann. Rev. Mater. Sci. **1998**, *28*, 153–184.

34. Bernard, A.; Delamarche, E.; Schmid, H.; Michel, B.; Bosshard, H.R.; Biebuyck, H. Printing patterns of proteins. Langmuir **1998**, *14* (9), 2225–2229.

35. Nie, Z.H.; Kumacheva, E. Patterning surfaces with functional polymers. Nat. Mater. **2008**, *7* (4), 277–290.

36. Kim, M.S.; Khang, G.; Bang, H. Gradient polymer surface for biomedical applications. Prog. Polym. Sci. **2008**, *33* (1), 138–164.

37. Costa Marínez, E.; Rodríguez Hernández, J.C.; Machado, M.; Mano, J.F.; Gómez Ribelles, J.L.; Monleón Pradas, M.; Salmerón Sánchez, M. Human chondrocyte morphology, its dedifferentiation, and fibronectin conformation on different

PLLA microtopographies. Tissue Eng. Part A **2008**, *14* (10), 1751–1762.

38. Ikada, Y. Surface modification of polymers for medical applications. Biomaterials **1994**, *15* (10), 725–736.

39. Custódio, C.A.; Alves, C.M.; Reis, R.L.; Mano, J.F. Immobilization of fibronectin in chitosan substrates improves cell adhesion and proliferation. J. Tissue Eng. Regen. Med. **2010**, *4* (4), 316–323.

40. Sun, A.; Lahann, J. Dynamically switchable biointerfaces. Soft Matter **2009**, *5* (8), 1555–1561.

41. Urry, D.W. Molecular machines: How motion and other functions of living organisms can result from reversible chemical changes. Angew. Chem. Int. Ed. Engl. **1993**, *32* (6), 819–841.

42. Schild, H.G. Poly(N-isopropylacrylamide): Experiment, theory and application. Prog. Polym. Sci. **1992**, *17* (2), 163.

43. Uchida, K.; Sakai, K.; Ito, E.; Kwon, O.H.; Kikuchi, A.; Yamato, M.; Okano, T. Temperature-dependent modulation of blood platelet movement and morphology on poly(N-isopropylacrylamide)-grafted surfaces. Biomaterials **2000**, *21* (9), 923–929.

44. Yamato, M.; Akiyama, Y.; Kobayashi, J.; Yang, J.; Kikuchi, A.; Okano, T. Temperature-responsive cell culture surfaces for regenerative medicine with cell sheet engineering. Prog. Polym. Sci. **2007**, *32* (8–9), 1123–1133.

45. da Silva, R.M.P.; Mano, J.F.; Reis, R.L. Smart thermoresponsive coatings and surfaces for tissue engineering: Switching cell-material boundaries. Trends Biotechnol. **2007**, *25* (12), 577–583.

46. da Silva, R.M.P.; Lopez-Peres, P.; Elvira, C.; Mano, J.F.; San Román, J.; Reis, R.L. Poly(N-isopropylacrylamide) surface-grafted chitosan membranes as a new substrate for cell sheet engineering and manipulation. Biotech. Bioeng. **2008**, *101* (6), 1321–1331.

47. Huber, D.L.; Manginell, R.P.; Samara, M.A.; Kim, B.I.; Bunker, B.C. Programmed adsorption and release of proteins in a microfluidic device. Science **2003**, *301* (5631), 352–354.

48. Shi, J.; Alves, N.M.; Mano, J.F. Thermally responsive biomineralization on biodegradable substrates. Adv. Funct. Mater. **2007**, *17* (16), 3312–3318.

49. Dias, C.I.; Mano, J.F.; Alves, N.M. pH-responsive biomineralization onto chitosan grafted biodegradable substrates. J. Mater. Chem. **2008**, *18* (21), 2493–2499.

50. Costa, R.R.; Custódio, C.A.; Testera, A.M.; Arias, F.J.; Rodríguez-Cabello, J.C.; Alves, N.M.; Mano, J.F. Stimuli-responsive thin coatings using elastin-like polymers for biomedical applications. Adv. Funct. Mater. **2009**, *19* (20), 3210–3218.

51. Garcia, A.; Marquez, M.; Cai, T.; Rosário, R.; Hu, Z.; Gust, D.; Hayes, M.; Vail, S.A.; Park, C.D. Photo-, thermally, and pH-responsive microgels. Langmuir **2007**, *23* (1), 224–229.

52. Edahiro, J.I.; Sumaru, K.; Tada, Y.; Ohi, K.; Takagi, T.; Kameda, M.; Shinbo, T.; Kanamori, T.; Yoshimi, Y. In situ control of cell adhesion using photoresponsive culture surface. Biomacromolecules **2005**, *6* (2), 970–974.

53. Nakanishi, J.; Kikuchi, Y.; Takarada, T.; Nakayama, H.; Yamaguchi, K.; Maeda, M. Spatiotemporal control of cell adhesion on a self-assembled monolayer having a photocleavable protecting group. Anal. Chim. Acta **2006**, *578* (1), 100–104.

54. Decher, G.; Hong, J.D.; Schmitt, J. Buildup of ultrathin multilayer films by a self-assembly process: II. Consecutive

adsorption of anionic and cationic bipolar amphiphiles and polyelectrolytes on charged surfaces. Thin Solid Films **1992**, *210* (1–2), 831–835.

55. Krogman, K.C.; Zacharia, N.S.; Schroeder, S.; Hammond, P.T. Automated process for improved uniformity and versatility of layer-by-layer deposition. Langmuir **2007**, *23* (6), 3137–3141.

56. Alves, N.M.; Picart, C.; Mano, J.F. Self assembling and crosslinking of polyelectrolyte multilayer films of chitosan and alginate studied by QCM and IR spectroscopy. Macromol. Biosc. **2009**, *9* (8), 776–785.

57. Martins, G.V.; Merino, E.G.; Mano, J.F.; Laves, N.M. Crosslink effect and albumin adsorption onto chitosan/alginate multilayered systems: An in situ QCM-D study. Macromol. Biosc. **2010**, *10* (12), 1444–1455.

58. Schneider, A.; Francius, G.; Obeid, R.; Schwinté, P.; Frisch, B.; Schaaf, P.; Voegel, J.C.; Senger, B.; Picart, C. Polyelectrolyte multilayers with a tunable Young's modulus: Influence of film stiffness on cell adhesion. Langmuir **2006**, *22* (3), 1193–1200.

59. Ren, K.; Crouzier, T.; Roy, C.; Picart, C. Polyelectrolyte multilayer films of controlled stiffness modulate myoblast cells differentiation. Adv. Funct. Mat. **2008**, *18* (9), 1378–1389.

60. Martins, G.V.; Mano, J.F.; Alves, N.M. Nanostructured self-assembled films containing chitosan fabricated at neutral pH. Carbohydr. Polym. **2010**, 80 (2), 570–573.

61. Berthelemy, N.; Kerdjoudj, H.; Gaucher, C.; Schaaf, P.; Stoltz, J.F.; Lacolley, P.; Voegel, J.C.; Menu, P. Polyelectrolyte films boost progenitor cell differentiation into endothelium-like monolayers. Adv. Mater. **2008**, *20* (14), 2674–2678.

62. Macdonald, M.L.; Samuel, R.E.; Shah, N.J.; Padera, R.F.; Beben, Y.M.; Hammond, P.T. Tissue integration of growth factor-eluting layer-by-layer polyelectrolyte multilayer coated implants. Biomaterials **2011**, *32* (5), 1446–1453.

63. Shah, N.J.; Macdonald, M.L.; Beben, Y.M.; Padera, R.F.; Samuel, R.E.; Hammond, P.T. Tunable dual growth factor delivery from polyelectrolyte multilayer films. Biomaterials **2011**, *32* (26), 6183–6193.

64. Crouzier, T.; Szarpak, A.; Boudou, T.; Auzely-Velty, R.; Picart, C. Polysaccharide-blend multilayers containing hyaluronan and heparin as a delivery system for rhBMP-2. Small **2010**, *6* (5), 651–662.

65. Tang, Z.; Wang, Y.; Podsiadlo, P.; Kotov, N.A. Biomedical applications of layer-by-layer assembly: From biomimetics to tissue engineering. Adv. Mater. **2006**, *18* (24), 3203–3224.

66. Tang, Z.; Kotov, N.A.; Magonov, S.; Ozturk, B. Nanostructured artificial nacre. Nat. Mater. **2003**, *2* (6), 413–418.

67. Luz, G.M.; Mano, J.F. Biomimetic design of materials and biomaterials inspired by the structure of nacre. Philos. Trans. A Math. Phys. Eng. Sci. **2009**, *367* (1893), 1587–1605.

68. Couto, D.S.; Alves, N.M.; Mano, J.F. Nanostructured multilayer coatings combining chitosan with bioactive glass nanoparticles. J. Nanosci. Nanotechnol. **2009**, *9* (3), 1741–1748.

69. Sukhishvili, S.A. Responsive polymer films and capsules via layer-by-layer assembly. Curr. Opin. Colloid Interf. Sci. **2005**, *10* (1–2), 37–44.

70. Martins, G.V.; Mano, J.F.; Alves, N.M. Dual responsive nanostructured surfaces for biomedical applications. Langmuir **2011**, *27* (13), 8415–8423.

71. Barbosa, J.S.; costa, R.R.; Testera, A.M.; Alonso, M.; Rodríguez-Cabello, J.C.; Mano, J.F. Multi-Layered films containing a biomimetic stimuli-responsive recombinant protein. Nanoscale Res. Lett. **2009**, *4* (10), 1247–1253.

72. Costa, R.R.; Custódio, C.A.; Arias, F.J.; Rodríguez-Cabello, J.C.; Mano, J.F. Layer-by-layer assembly of chitosan and recombinant biopolymers into biomimetic coatings with multiple stimuli-responsive properties. Small **2011**, *7* (18), 2640–2649.

73. Sher, P.; Custódio, C.A.; Mano, J.F. Layer-by-layer technique for producing porous nanostructured 3D constructs using moldable freeform assembly of spherical templates. Small **2010**, *6* (23), 2644–2648.

74. Silva, J.M.; Georgi, N.; Costa, R.; Sher, P.; Reis, R.L.; Van Blitterswijk, C.A.; Karperien, M.; Mano, J.F. Nanostructured 3D constructs based on chitosan and chondroitin sulphate multilayers for cartilage tissue engineering. PLoS ONE 2013, *8* (2), e55451.

75. Miranda, E.S.; Silva, T.H.; Reis, R.L.; Mano, J.F. Nanostructured natural-based polyelectrolyte multilayers to agglomerate chitosan particles into scaffolds for tissue engineering. Tissue Eng. Part A **2011**, *17* (21–22), 2663–2674.

76. Costa, N.L.; Sher, P.; Mano, J.F. Liquefied capsules coated with multilayered polyelectrolyte films for cell immobilization. Adv. Eng. Mater. **2011**, *13* (6), B218–B224.

77. Correia, C.R.; Sher, P.; Reis, R.L.; Mano, J.F. Liquified chitosan-alginate multilayer capsules incorporating poly(L-lactic acid) microparticles as cell carriers. Soft Matter 2013, *9* (7), 2125–2130.

78. Arima, Y.; Iwata, H. Effect of wettability and surface functional groups on protein adsorption and cell adhesion using well-defined mixed self-assembled monolayers. Biomaterials **2007**, *28* (20), 3074–3082.

79. Benoit, D.S.W.; Schwartz, M.P.; Durney, A.R.; Anseth, K.S. Small functional groups for controlled differentiation of hydrogel-encapsulated human mesenchymal stem cells. Nat. Mater. **2008**, *7* (10), 816–823.

80. Lim, J.Y.; Shaughnessy, M.C.; Zhou, Z.Y.; Noh, H.; Vogler, E.A.; Donahue, H.J. Surface energy effects on osteoblast spatial growth and mineralization. Biomaterials **2008**, *29* (12), 1776–1784.

81. Ivirico, J.L.E.; Salmeron-Sanchez, M.; Ribelles, J.L.G.; Pradas, M.M.; Soria, J.M.; Gomes, M.E.; Reis, R.L.; Mano, J.F. Proliferation and differentiation of goat bone marrow stromal cells in 3D scaffolds with tunable hydrophilicity. J. Biomed. Mat. Res. B Appl. Biomater. **2009**, *91* (1), 277–286.

82. Shi, J.; Alves, N.M.; Mano, J.F. Towards bioinspired superhydrophobic poly(L-lactic acid) surfaces using phase inversion-based methods. Bioinspir. Biomim. **2008**, *3* (3), 034003.

83. Alves, N.M.; Shi, J.; Oramas, E.; Santos, J.L.; Tomas, H.; Mano, J.F. Bioinspired superhydrophobic poly(L-lactic acid) surfaces control bone marrow derived cells adhesion and proliferation. J. Biomed. Mat. Res. Part A 2009, *91* (2), 480–488.

84. Song, W.L.; Veiga, D.D.; Custodio, C.A.; Mano, J.F. Bioinspired degradable substrates with extreme wettability properties. Adv. Mater. **2009**, *21* (18), 1830–1834.

85. Oliveira, W. Song, N.M.; Alves, Mano, J.F. Chemical modification of bioinspired superhydrophobic polystyrene surfaces to control cell attachment/proliferation. Soft Matter **2011**, *7* (19), 8932–8941.

86. Song, W.; Gaware, V.S.; Rúnarsson, O.V.; Másson, M.; Mano, J.F. Functionalized superhydrophobic biomimetic chitosan-based films. Carbohydr. Polym. **2010**, *81* (1), 140–144.

87. Neto, A.I.; Custódio, C.A.; Song, W.; Mano, J.F. High-throughput evaluation of interactions between biomaterials, proteins and cells using patterned superhydrophobic substrates. Soft Matter **2011**, *7* (9), 4147–4151.

88. Luz, G.M.; Leite, A.J.; Neto, A.I.; Song, W.; Mano, J.F. Wettable arrays onto superhydrophobic surfaces for bioactivity testing of inorganic nanoparticles. Mater. Lett. **2011**, *65* (2), 296–299.

89. Salgado, C.L.; Oliveira, M.B.; Mano, J.F. Combinatorial cell-3D biomaterials cytocompatibility screening for tissue engineering using bioinspired superhydrophobic substrates. Integr. Biol. **2012**, *4* (3), 318–327.

90. Oliveira, M.B.; Salgado, C.L.; Song, W.; Mano, J.F. Combinatorial on-chip study of miniaturized 3D porous scaffolds using a patterned superhydrophobic platform. Small **2013**, *9* (5), 768–778

91. Oliveira, M.B.; Mano, J.F. Polymer-based microparticles in tissue engineering and regenerative medicine. Biotechnol. Prog. **2011**, *27* (4), 897–912.

92. Song, W.; Lima, A.C.; Mano, J.F. Bioinspired methodology to fabricate hydrogel spheres for multi-applications using superhydrophobic substrates. Soft Matter **2010**, *6* (23), 5868–5871.

93. Lima, A.C.; Song, W.; Blanco-Fernandez, B.; Alvarez-Lorenzo, C.; Mano, J.F. Synthesis of temperature-responsive dextran-MA/PNIPAAm particles for controlled drug delivery using superhydrophobic surfaces. Pharm. Res. **2011**, *28* (6), 1294–1305.

94. Lima, A.C.; Custódio, C.A.; Alvarez-Lorenzo, C.; Mano, J.F. Biomimetic methodology to produce polymeric multilayered particles for biotechnological and biomedical applications. Small **2013**, *9* (15), 2487–2492, 2486.

95. Garcia Cruz, D.M.; Ivirico, J.L.E.; Gomes, M.M.; Ribelles, J.L.G.; Gomez Ribelles, J.L.; Salmeron Sanchez, M.S.; Reis, R.L.; Mano, J.F. Chitosan microparticles as injectable scaffolds for tissue engineering. J. Tissue Eng. Regen. Med. **2008**, *2* (6), 378–380.

96. Oliveira, M.B.; Song, W.; Martín, L.; Oliveira, S.M.; Caridade, S.G.; Alonso, M.; Rodríguez-Cabello, J.C.; Mano, J.F. Development of an injectable system based on elastin-like recombinamer particles for tissue engineering applications. Soft Matter **2011**, *7* (14), 6426–6434.

Biorubber: Poly(Glycerol Sebacate)

Yadong Wang
Wallace H. Coulter School of Biomedical Engineering and School of Chemistry and Biochemistry, Georgia Institute of Technology, Atlanta, Georgia, U.S.A.

Abstract
A strong, biodegradable, biocompatible elastomer could be useful in fields such as tissue engineering, drug delivery, and medical devices. Compared with conventional biodegradable polymers, poly-glycerol sebacate (PGS) appears to be tougher, less expensive, and more flexible.

INTRODUCTION

Biodegradable polymers have significant potential in biotechnology and bioengineering. However, for some applications, they are limited by their mechanical properties' mismatch and unsatisfactory compatibility with cells and tissues. A strong, biodegradable, biocompatible elastomer could be useful in fields such as tissue engineering, drug delivery, and medical devices. Poly(glycerol sebacate) (PGS) is a tough biodegradable elastomer designed and synthesized from biocompatible monomers. PGS forms a covalently cross-linked three-dimensional network of random coils with hydroxyl groups attached to its backbone. Both cross-linking and the hydrogen bonding interactions between the hydroxyl groups likely contribute to the unique properties of the elastomer. *In vitro* and *in vivo* studies show the polymer has good biocompatibility. *In vivo* degradation shows a near linear loss of mass with time, and a gradual decrease of mechanical strength with very limited water uptake. Polymer implants under animal skin are absorbed completely within 60 days, followed by restoration of the implantation sites to their normal architectures. PGS supports the growth of many cell types, and may be useful in tissue engineering, medical devices, and other fields in biomedical engineering.

BACKGROUND

Advances in biomaterials have significantly impacted many fields in life sciences and technology, and improved the quality of life of tens of millions of individuals.[1–3] Annual sale of medical systems using biomaterials in the United States alone exceeds $100 billion.[4] An increasingly important category of biomaterials is biodegradable polymers, which have been used widely in fields such as tissue engineering, drug delivery, and medical devices. Many biomedical devices are placed in a mechanically dynamic environment in the body, which requires the implants to sustain and recover from various deformations without mechanical irritations to the surrounding tissues. In many cases, the matrices and scaffolds of these implants would ideally be made of biodegradable polymers whose properties resemble those of extracellular matrices (ECM)—soft, tough, and elastomeric in order to provide mechanical stability and structural integrity to tissues and organs. Hence a soft biodegradable elastomer that readily recovers from relatively large deformations is advantageous for maintaining the implant's proper function without mechanical irritation to the host. One such material is a series of biodegradable elastomers based on glycerol and sebacic acid, hence the name PGS.

These elastomers are biodegradable, biocompatible, tough, and inexpensive. Upon degradation *in vivo*, PGS loses mass almost linearly with time, and its mechanical strength decreases gradually with very limited water uptake. Unmodified PGS supports the growth of many different cell types, including endothelial and smooth muscle cells, Schwann cells, hepatocytes, cardiomyocytes, and chondrocytes. PGS has been fabricated into different physical forms such as flat sheets, foams, and tubes. Microfabricated PGS has also been used as a scaffolding material to make patent endothelialized branching capillary networks that may serve as artificial microvasculature for engineering vital organs. As one of the few biodegradable and biocompatible elastomers, PGS may be useful in soft tissue engineering, medical devices, and other fields in biomedical engineering.

DESIGN OF PGS

PGS is analogous to vulcanized rubber in that it forms a cross-linked three-dimensional network of random coils. Such a strategy to achieve tough and elastomeric materials is also found in nature; collagen and elastin, the major fibrous protein components of ECM, are both cross-linked. In addition to covalent cross-linking, hydrogen bonding interactions through hydroxyproline hydroxyl groups also contribute to the mechanical strength of collagen.

Concise Encyclopedia of Biomedical Polymers and Polymeric Biomaterials DOI: 10.1081/E-EBPPC-120025406

229

Biomimetic—Colloid

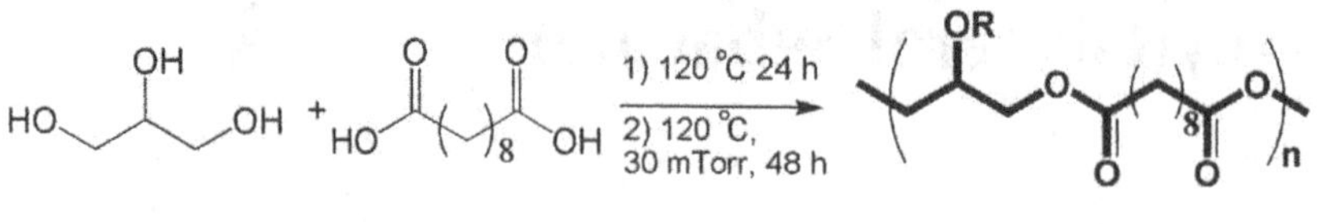

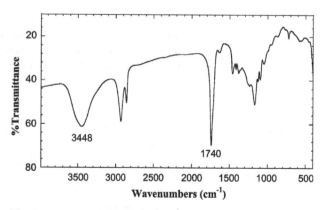

Scheme 1 Polymer synthesized by melt polycondensation of an equimolar amount of glycerol and sebacic acid at 120°C under argon for 24 hours.

The design of our polymer is based on two hypotheses: 1) good mechanical properties can be obtained through covalent cross-linking and hydrogen bonding interactions; and 2) rubberlike elasticity can be obtained by building a three-dimensional network of random coils through copolymerization whereby at least one monomer is trifunctional. To realize this design, the following criteria were considered: 1) degradation mechanism—hydrolysis was preferred to minimize individual differences in degradation characteristics caused by enzymes;[2] 2) hydrolyzable chemical bond—ester was chosen for its established and versatile synthetic methods; 3) cross-link density—low density was preferred, as a high degree of cross-linking usually leads to rigid and brittle polymers; 4) cross-link chemical bonds were chosen to be hydrolyzable and identical to those in the backbone to minimize the possibility of heterogeneous degradation; and 5) specific monomers— they should be nontoxic, at least one should be trifunctional, and at least one should provide hydroxyl groups for hydrogen bonding. Glycerol [$CH_2(OH)CH(OH)CH_2OH$], the basic building block for lipids, satisfies all three requirements and was chosen as the alcohol monomer. From the same toxicological and polymer chemistry standpoints, sebacic acid [$HOOC(CH_2)_8COOH$] was chosen as the acid monomer. Sebacic acid is the natural metabolic intermediate in ω-oxidation of medium- to long-chain fatty acids,[5,6] and has been shown to be safe *in vivo*.[7] Both monomers are inexpensive, an advantage for large-scale applications. This approach was likely to yield biodegradable polymers with improved mechanical properties and biocompatibility.

SYNTHESIS OF PGS

The polymer was synthesized by melt polycondensation of an equimolar amount of glycerol and sebacic acid at 120°C under argon for 24 hr (Scheme 1). The pressure was then reduced from 1 Torr to 40 mTorr over 5 hr. The polymer at this stage is a highly viscous liquid above 80°C, and is soluble in organic solvents such as tetrahydrofuran (THF), *N,N*-dimethylformamide, and ethanol. The polymer was cured at 40 mTorr and 120°C for 48 hr. The resultant polymer is a transparent, almost colorless elastomer. [Rigid, totally crosslinked polymer has been synthesized from glycerol and sebacic acid (glycerol/sebacic acid molar ratio: 2/3) under different conditions.[8]] The resulting polymer features a small amount of cross-links and hydroxyl groups

Fig. 1 FTIR spectrum of PGS. Note the intense C=O and O–H stretches at 1740 cm⁻¹ and 3448 cm⁻¹, respectively.

directly attached to the backbone. PGS is insoluble in water and swells only 2.1 ± 0.33% after soaking in water for 24 hr. As is the case for other hydrolyzable polymers, PGS should be kept in an anhydrous environment for long-term storage.

CHARACTERIZATION OF PGS

The chemical and physical properties of PGS were characterized by various instrumental methods. Fourier transformed infrared spectroscopy (FTIR) on a newly prepared KBr pellet of PGS showed an intense C=O stretch at 1740 cm⁻¹, which confirmed the formation of ester bonds. The absence of C=O stretches corresponding to carboxylic acids suggested a high conversion rate to the polymer. FTIR also showed a broad intense OH stretch at 3448 cm⁻¹, indicating the hydroxyl groups were hydrogen-bonded (Fig. 1).

Differential scanning calorimetery (DSC) measurement showed two crystallization temperatures, −52.14°C and −18.50°C, and two melting temperatures, 5.23°C and 37.62°C, respectively. No glass transition temperature was observed above −80°C, the lower detection limit of the instrument. The DSC results indicate that the polymer was totally amorphous at 37°C.

Elemental analysis on vacuum-dried PGS samples confirmed the composition of the polymer as approximately 1 glycerol:1 sebacic acid. Water-in-air contact angle was measured at room temperature using the sessile drop method and an image analysis of the drop profile on slabs of polymer fixed on glass slides. The polymer surface is very hydrophilic due to the hydroxyl groups attached to its backbone. Its water-in-air contact angle was 32.0°, almost identical to that of a flat 2.7 nm thick type I collagen film (31.9°).[9] The cross-linking density is expressed by n, moles of active network chains per unit volume, which was 38.3 ± 3.40 mol/m³, and M_c, the molecular weight between cross-links, which was $18,300 \pm 1620$, calculated from equation:[10]

$$n = E_0/3R\,T = \rho/M_c$$

where E_0 is Young's modulus, R is the universal gas constant, T is the temperature, and ρ is the density. The low M_c value confirms that the cross-linking density is low in PGS. Depending on the monomer ratio and process parameters, the modulus of PGS can be controlled within at least an order of magnitude. On the low modulus (soft) limit, tensile tests on thin strips of PGS revealed a stress-strain curve characteristic of an elastomeric and tough material (Fig. 2). The nonlinear shape of the tensile stress-strain curve is typical for elastomers, and resembled those of ligament[11–13] and vulcanized rubber.[14] PGS can be elongated repeatedly to at least three times its original length without rupture. The total elongation is unknown, as grip breaks occurred at $267 \pm 59.4\%$ strain. The tensile Young's modulus of the polymer was 0.282 ± 0.0250 MPa with an ultimate tensile strength greater than 0.5 MPa (grip break). The value of the Young's modulus of PGS is between those of ligaments (kPa scale),[11–13] which contain a large amount of elastin in addition to collagen, and tendon (GPa scale),[15–17] which is mainly made of collagen. The strain to failure of PGS is similar to that of arteries and veins (up to 260%),[18] and much larger than that of tendons (up to 18%).[19] On the high modulus limit, PGS can be stretched to about 20%, still much higher than stiff

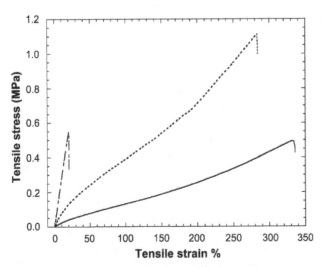

Fig. 2 Stress strain curves of two extremes of PGS, the soft (——) and the stiffer one (..........), and vulcanized rubber (—). Both soft PGS and vulcanized rubber are marked by low modulus and large elongation ratio, indicating elastomeric and tough materials. The modulus of the stiffer PGS is 10 times higher than that of the soft one. Grip break for both PGS.

biodegradable polymers such as poly(lactide-*co*-glycolide) (PLG). Once cross-linked, PGS is insoluble in water or organic solvents. After soaking 24 hr in water, the weight of PGS barely changed and the mechanical properties were virtually the same as for dry polymer.

DEGRADATION OF PGS

The degradation characteristics of PGS were examined both *in vitro* and *in vivo*. Agitation for 60 days in phosphate buffered saline solution (PBS) at 37°C caused the polymer to degrade $17 \pm 6\%$ as measured by change of dry sample weight. In contrast, subcutaneous (SC) implants in rats were totally absorbed in 60 days. For the *in vivo* experiment, enzymes, and perhaps macrophages as well, may have caused differences in degradation rate. *In vivo* degradation thinned the polymer implants, with the explants maintaining their square shape and relatively sharp edges up to at least 35 days. Both the mechanical strength and the mass decreased almost linearly as the polymer degraded. Both results suggest PGS predominantly degrades through surface erosion. The preservation of integrity during the degradation process can be important for certain types of tissue-engineered implants, drug delivery devices, and medical devices.

In Vivo Degradation

Appropriate degradation properties of biodegradable polymers are critical for the success of their clinical applications. The *in vivo* degradation characteristics of PGS were evaluated using a SC model in Sprague-Dawley (SD) rats with PLG as the control material. Carboxyl-terminated PLG (50:50, carboxyl-ended, MW 15,000) powder was pressed into round disks. The surface area/volume ratio was the same for PGS and PLG samples. PGS samples were autoclaved, and PLG samples were sterilized by ethylene oxide (4 hr sterilization, 48 hr ventilation) before implantation. All samples were implanted subcutaneously in the backs of SD rats. The animals were randomly divided into five groups. At days 7, 14, 21, 28, and 35, implants were explanted from one group of animals under general anesthesia. The implants were carefully retrieved and rinsed sequentially with PBS solution and deionized (DI) water. The geometry and surface properties of the explants were examined, and the change of mass, water content, and mechanical strength with time were also monitored. All these parameters are important indications for the mechanism of degradation, and are crucial for the integrity and proper function of an implant. Macroscopically, the PGS explants maintained their geometry throughout the time period tested (Fig. 3). In contrast, the geometry of PLG explants was distorted significantly within 14 days, most likely because of bulk degradation and swelling. They were deformed considerably from transparent disks to white

Biomimetic—Colloid

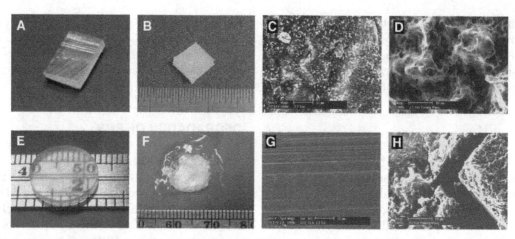

Fig. 3 Photographs and scanning electron micrographs (1000×) of PGS (top) and PLG (bottom) explants at various time points of degradation. PGS implants preserve their geometry and surface integrity throughout the degradation period, whereas PLG implants deform severely. PGS, (**A–D**); PLG, (**E–H**). (**A**) new PGS slab; (**B**) 21 days implantation; (**C**) surface of autoclavated PGS slab—new; (**D**) surface of PGS slab implanted for 21 days. (**E**) new PLG disk implantation; (**F**) PLG after 21 days implantation; (**G**) surface of ethylene oxide sterilized PLG disk—new; (**H**) surface of PLG disk implanted for 21 days.

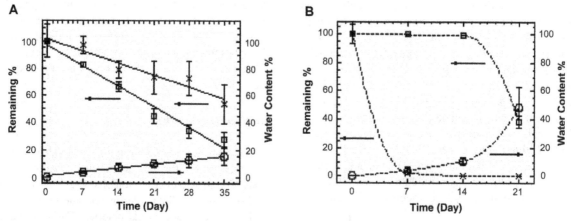

Fig. 4 Comparison of the changes in mass (□), mechanical strength (×), and water content (○) of PGS (solid line) and PLG (dashed line) implants upon degradation. (**A**) Steady, almost linear changes of PGS implants' properties upon degradation. (**B**) Abrupt, nonlinear changes of PLG implants' properties upon degaradation. Data were plotted as mean with standard deviation.
Source: Adapted from Wang et al.[22]

opaque irregular lumps (Fig. 3). No PLG implants were retrieved successfully beyond three weeks due to excessive swelling and the fragile nature of the implants.

The surface integrity of the explants was observed by scanning electron microscopy (SEM). Both pristine sterilized samples and cleaned and dried explants were mounted on aluminum stubs, and their surface morphologies were observed. SEM observation showed that the PGS explant surface maintained its integrity well. Contoured features developed on the surface after autoclaving. Such features remained throughout the course of the experiment; however, no crack formation was observed (Fig. 3). In case of PLG implants (Fig. 3), 20 μm holes developed on the surface 7 days postimplantation; cracks about 20 μm wide formed within 14 days, and by 21 days, networks of both larger cracks wider than 40 μm and smaller cracks could be seen throughout the PLG surface.

The degree of swelling of degradable polymers *in vivo* is a key parameter for proper implant materials. Excessive swelling is usually undesirable for an implant, as it distorts the shape of the implant and softens the polymer.[20,21] The degree of swelling was estimated by measuring the explant weight differences before and after drying. The explants were cleaned and the surface water removed with a Kimwipe before weighing. Each explant was thoroughly dried at 40°C under vacuum (85 mTorr) for 48 hr. Each explant was weighed again before any subsequent testing. The swelling ratio was calculated from explant weight difference before and after drying: $(W_w - W_d)/W_d$, where W_w is the wet sample weight and W_d is the dry sample weight. The water content of PGS implants rose linearly and reached 15% in 35 days, at which time the polymer had degraded >70% (Fig. 4A). In contrast, water uptake of PLG implants showed a time lag followed by a surge of water content, in a pattern similar to its mass loss (Fig. 4B). The water content of PLG implants increased gradually to 11% within 14 days, then increased abruptly, and reached 49% within the next 7 days. One of the key functions of a degradable polymer in an implant is to provide mechanical support. Hence it's critical to know how the mechanical strength changes with degradation. The PGS implants lost mechanical strength gradually and slowly after implantation, about 8% each week. At day 35, when <30% of the PGS implant's mass was left, the modulus was >50% (Fig. 4A). In contrast, PLG implants lost their mechanical strength shortly after implantation (>98% within 7 days). At 14 days, PLG implants' moduli were reduced to 0.25%. At 21 days, with 42% of the mass left, their moduli were reduced to 0.023% (Fig. 4B). This demonstrates that PGS implants maintained their mechanical strength much better than PLG implants. The explants were tested according to ASTM standard D575-91. Briefly, the explants were compressed at a fixed ramp speed of 2 mm/min. PGS explants were compressed to 50% strain, whereas PLG explants were compressed to failure. Pristine samples and PGS explants had regular geometry and were measured with a digital caliper. PLG explants deformed upon degradation, and their dimensions were measured to the best approximation. The rate of mass loss is of fundamental importance in studying the degradation characteristics of a biodegradable polymer. PGS implants lost their weight steadily and linearly over the test period of 35 days, at which time they had lost >70% of their mass (Fig. 4A).[22] The mass loss of PLG was negligible (<1%) within 14 days, then surged abruptly, and reached 61% in the next 7 days (Fig. 4B). The dramatic mass loss after the initial lag for PLG upon degradation was similar to what has been reported in the literature.[23–25] The differences in degradation characteristics between PGS and PLG under identical conditions indicate they probably degrade by different mechanisms. Unlike PLG, which degrades mostly by bulk degradation, *in vivo* degradation of PGS is dominated by surface erosion, as indicated by linear mass loss with time, preservation of implant geometry, better retention of mechanical strength, absence of surface cracks, and minimal water uptake. Upon degradation, PGS implants keep their integrity better than PLG implants, and may prove useful in biomedical applications where other such polymers are unsuccessful.

BIOCOMPATIBILITY OF PGS

PGS appears to be biocompatible both *in vitro* and *in vivo*. Under identical conditions, cultured fibroblasts grew faster on PGS surfaces than on PLG surfaces. SC implantation experiments showed that PGS induced only acute inflammation, whereas PLG implantation sites presented signs of chronic inflammation such as fibrous capsule formation and foreign body giant cells.

In Vitro Biocompatibility

The *in vitro* biocompatibility of PGS was evaluated in comparison with PLG, both in morphology and in cell growth rate. Glass Petri dishes were coated with PGS and PLG, respectively. The coated dishes were sterilized by UV radiation for 15 min. Each dish was soaked in growth medium for 4 hr; the medium was replaced with fresh medium and the dishes soaked for another 4 hr to remove any unreacted monomers or residual solvents before cell seeding. NIH 3T3 fibroblast cells were seeded homogeneously on these coated dishes. The cells in PGS sample wells were viable and showed normal morphology with higher growth rate than the control, as tested by methyl thiazol tetrazolium assay.[26] Cells in PLG wells tended to form clusters, and the number of floating cells was higher; furthermore, most of the attached cells adopted a long thin threadlike morphology. These experiments suggested that PGS is at least as biocompatible as PLG *in vitro*.

In Vivo Biocompatibility

SC implantation in SD rats was used to compare the *in vivo* biocompatibility of PGS and PLG. The surface area/volume ratio was kept the same for both PGS and PLG implants. Autoclaved PGS slabs of approximately 6 × 6 × 3 mm, and ethylene oxide-sterilized PLG (50:50, carboxyl-ended, MW 15,000, Boehringer Ingelheim, Inc., Germany) disks were implanted SC in female SD rats by blunt dissection under general anesthesia. Two implants each of PGS and PLG were implanted symmetrically on the upper and lower back of the same animal. Every implantation site was marked by tattoo marks. The animals were randomly divided into five groups. At each predetermined time point (7, 14, 21, 28, 35 days), one group of rats was sacrificed, and tissue samples surrounding the implants were harvested. The samples were fixed and embedded in paraffin after dehydration. The slides were stained with hematoxylin and eosin (H&E) and Masson's trichrome stain (MTS). The thickness of the inflammatory zone (H&E) and collagen deposition (MTS) for each polymer implant is expressed as the average value of three readings per slide for six slides at each time point.

The inflammatory responses subsided with time for both PGS and PLG implants. In the first three weeks, the inflammatory response of PLG implantation sites was about 16% thinner than that of PGS (Fig. 5). The thickness of the inflammatory zone in both implantation sites was approximately the same at weeks 4 and 5. Fibrous capsules surrounding PLG implants developed within 14 days, and their thickness hovered around 140 μm. Collagen deposition did not appear around PGS implants until 35 days. The collagen layer was highly vascularized and was only about 45 μm thick. Foreign body giant cells were present in the PLG implantation sites 35 days postimplantation. However, no foreign body giant cells were

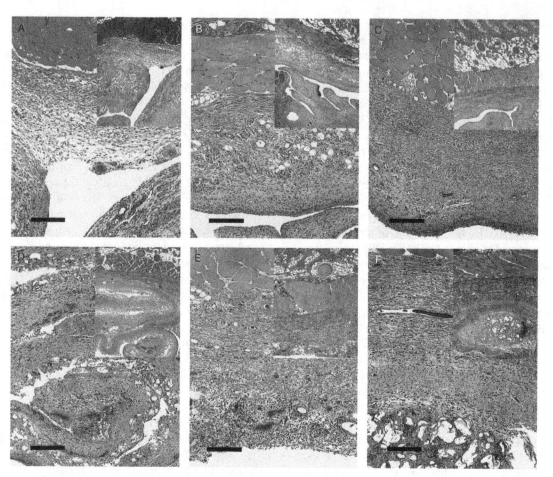

Fig. 5 Photomicrographs of rat skin; comparisons of lumen wall characteristics (H&E, 10×) and fibrous capsule thickness (insets, MTS, 5×) at implantation sites across time. (**A, C, E**) PGS; (**B, D, F**) PLGA. (**A, B**) 7 days postimplantation (pi)—lumenal wall was markedly thickened by a zone of dense vascular proliferation and mild inflammation without detectable collagen deposition. (**C, D**) 21 days pi—lumenal wall was significantly thinner with a modest degenerative inflammatory infiltrate immediately adjacent to the polymer. PLGA implantation site (**D**) was marked by a significant collagen fibrous capsule, which was absent in PGS. (**E, F**) 35 days pi—lumenal wall was reduced to a thin zone of cell debris with no vascular proliferation. Collagen deposition in PGS implantation site was much thinner than that surrounding the fragmented PLGA implant. Top: skin, implantation site: *, scale bar = 200 μ.
Source: Adapted from Wang et al.[26]

observed around PGS implants. The inflammatory response and fibrous capsule formation observed for PLG is similar to those reported in the literature.[27,28] Thick fibrous capsules may block mass transfer between the implants and surrounding tissues, which can impair implant functions. In an *in vivo* study with PGS alone, the SC implantation sites were undetectable despite repeated sectioning of the specimens at multiple levels in 60 days (two implantation sites each in three animals). The implants were completely absorbed without granulation or scar tissues, and the implantation site was restored to its normal histological architecture. Overall, the acute inflammatory response of PGS is similar to that of PLG. However, unlike PLG, PGS barely induces any significant fibrous capsule formation, and the implantation site is free of foreign body giant cells, suggesting the absence of chronic inflammation.

APPLICATIONS OF PGS

Similar to vulcanized rubber, this PGS elastomer is a thermoset polymer. However, the uncross-linked prepolymer can be processed into various shapes, because it can be melted into a liquid and is soluble in common organic solvents. In a typical procedure, a mixture of NaCl particles of appropriate size and an anhydrous THF solution of the prepolymer is poured into a Teflon mold. The polymer is cured in the mold in a vacuum oven at 120°C and 100 mTorr. A porous scaffold that conforms to the shape of the mold is obtained after salt leaching in DI water. So far, the applications of PGS are focused on soft-tissue engineering. The material supports the growth of a wide range of cell types such as endothelial and smooth muscle cells, Schwann cells, cardiomyocytes, hepatocytes, chondrocytes, and fibroblasts.

Artificial Microvasculature

Vital organs maintain dense microvasculatures to sustain the proper function of their cells. For tissue-engineered organs to function properly, artificial capillary networks have to be developed. Due to its elastomeric properties, PGS might be a good candidate for fabricating microvasculature. Standard MicroElectroMechanical Systems techniques were used to etch capillary network patterns onto silicon wafers. The resultant silicon wafers served as micromolds for the devices. The patterned PGS film was bonded with a flat one to create capillary networks. The devices were perfused with a syringe pump at a physiological flow rate, and became endothelialized to near confluence within 7 days. These endothelialized capillary networks could be used in tissue engineering of vital organs.

Cardiac Patch

To evaluate the performance of PGS as a scaffolding material for cardiac tissue engineering, neonatal rat cardiomyocytes were seeded in PGS, collagen, and PLG scaffolds. The PGS constructs beat spontaneously for 3 days after seeding, and they beat synchronously with a DC pulse of 1 Hz until at least 14 days. The collagen constructs behaved similarly up to 6 days. After 6 days, the constructs collapsed, because the collagen matrix was degraded or absorbed so much that it had little mechanical strength to hold the construct in proper shape. PLG scaffolds were also used to build cardiac patches. The construct failed to show any beating at macroscopic level, most likely because of the stiffness of the PLG scaffolds. Compared with collagen foams and PLG, PGS appears to be as good, if not better suited, for tissue engineering of cardiac muscles.

Liver Tissue Engineering

One challenge in liver tissue engineering is to keep the liver-specific functions of the hepatocytes in the constructs. PGS and PLG scaffolds of the same porosity and pore sizes were treated under identical conditions, and seeded with hepatocytes. Hepatocytes on PGS scaffolds showed approximately four times higher ammonia metabolism rate and 40% higher albumin secretion rate than those on PLG.

CONCLUSION

Compared with conventional biodegradable polymers, PGS appears to be tougher, less expensive, and more flexible. In the models tested, the material is biocompatible both *in vitro* and *in vivo*. The degradation of PGS is mostly through surface erosion, which gives a more linear decrease in mass and gradual change in mechanical properties. In addition, the polymer doesn't swell in aqueous media conditions when fresh, and has limited swelling upon degradation. Thus, the geometry and surface integrity of PGS implants are maintained very well. These properties of PGS may be desirable in many applications in biomedical engineering beyond those discussed previously. The polymer's properties—such as hydrophilicity, degradation rate, and pattern—can potentially be tailored by grafting hydrophobic moieties onto the hydroxyl groups. To further control or regulate polymer interaction with cells, biomolecules could be coupled to the hydroxyl groups or integrated into the polymer backbone.

ACKNOWLEDGMENTS

The author would like to thank Drs. Robert Langer, Hiroyuki Ijima, Gordana Vunjak-Novakovic, Milica Radisic, and Hyoungshin Park, and Mrs. Christina Fidkowshki for their contributions to research in PGS.

REFERENCES

1. Peppas, N.A.; Langer, R. New challenges in biomaterials. Science **1994**, *263* (5154), 1715–1720.
2. Langer, R. Biomaterials: Status, challenges, and perspectives. AIChE J. **2000**, *46* (7), 1286–1289.
3. Hench, L.L.; Polak, J.M. Third-generation biomedical materials. Science **2002**, *295* (5557), 1014–1007.
4. Langer, R. Drug delivery and targeting. Nature **1998**, *392* (6679 Suppl.), 5–10.
5. Liu, G.; Hinch, B.; Beavis, A.D. Mechanisms for the transport of α,ω-dicarboxylates through the mitochondrial inner membrane. J. Biol. Chem. **1996**, *271* (41), 25338–25344.
6. Grego, A.V.; Mingrone, G. Dicarboxylic acids, an alternate fuel substrate in parenteral nutrition: An update. Clin. Nutr. **1995**, *14* (3), 143–148.
7. Tamada, J.; Langer, R. The development of polyanhydrides for drug delivery applications. J. Biomater. Sci., Polym. Ed. **1992**, *3* (4), 315–353.
8. Nagata, M.; Machida, T.; Sakai, W.; Tsutsumi, N. Synthesis, characterization, and enzymatic degradation of network aliphatic copolyesters. J. Polym. Sci. A Polym. Chem. **1999**, *37* (13), 2005–2011.
9. Dupont-Gillain, C.C.; Nysten, B.; Rouxhet, P.G. Collagen adsorption on poly(methyl methacrylate): Netlike structure formation upon drying. Polym. Int. **1999**, *48* (4), 271–276.
10. Sperling, L.H. *Introduction to Physical Polymer Science*; John Wiley & Sons: New York, 1992.
11. Yamaguchi, S. Analysis of stress-strain curves at fast and slow velocities of loading *in vitro* in the transverse section of the rat incisor periodontal ligament following the administration of beta-aminopropionitrile. Arch. Oral Biol. **1992**, *37* (6), 439–444.
12. Komatsu, K.; Chiba, M. The effect of velocity of loading on the biomechanical responses of the periodontal ligament in transverse sections of the rat molar *in vitro*. Arch. Oral Biol. **1993**, *38* (5), 369–375.
13. Chiba, M.; Komatsu, K. Mechanical responses of the periodontal ligament in the transverse section of the rat mandibular incisor at various velocities of loading *in vitro*. J. Biomech. **1993**, *26* (4–5), 561–570.

Biomimetic—Colloid

14. Nagdi, K. *Rubber as an Engineering Material: Guideline for Users*; Hanser: Munich, 1993.

15. Fratzl, P.; Misof, K.; Zizak, I.; Rapp, G.; Amenitsch, H.; Bernstorff, S. Fibrillar structure and mechanical properties of collagen. J. Struct. Biol. **1998**, *122* (1–2), 119–122.

16. Misof, K.; Rapp, G.; Fratzl, P. A new molecular model for collagen elasticity based on synchrotron x-ray scattering evidence. Biophys. J. **1997**, *72* (3), 1376–1381.

17. Wang, J.L.; Parnianpour, M.; Shirazi-Adl, A.; Engin, A.E. Failure criterion of collagen fiber: Viscoelastic behavior simulated by using load control data. Theor. Appl. Fract. Mech. **1997**, *27* (1), 1–12.

18. Lee, M.C.; Haut, R.C. Strain rate effects on tensile failure properties of the common carotid artery and jugular veins of ferrets. J. Biomech. **1992**, *25* (8), 925–927.

19. Haut, R.C. The effect of a lathyritic diet on the sensitivity of tendon to strain rate. J. Biomech. Eng. **1985**, *107* (2), 166–174.

20. Yoon, J.J.; Park, T.G. Degradation behaviors of biodegradable macroporous scaffolds prepared by gas foaming of effervescent salts. J. Biomed. Mater. Res. **2001**, *55* (3), 401–408.

21. Kranz, H.; Ubrich, N.; Maincent, P.; Bodmeier, R. Physicomechanical properties of biodegradable poly(D,L-lactide) and poly(D,L-lactide-co-glycolide) films in the dry and wet states. J. Pharm. Sci. **2000**, *89* (12), 1558–1566.

22. Wang, Y.; Kim Yu, M.; Langer, R. *In vivo* degradation characteristics of poly(glycerol sebacate). J. Biomed. Mater. Res. **2003**, *66A* (1), 192–197.

23. Lu, L.; Peter, S.J.; Lyman, M.D.; Lai, H.L.; Leite, S.M.; Tamada, J.A.; Uyama, S.; Vacanti, J.P.; Langer, R.; Mikos, A.G. *In vitro* and *in vivo* degradation of porous poly(DL-lactic-co-glycolic acid) foams. Biomaterials **2000**, *21* (18), 1837–1845.

24. Vert, M.; Li, S.; Garreau, H. More about the degradation of LA/GA-derived matrixes in aqueous media. J. Control. Release **1991**, *16* (1–2), 15–26.

25. Kenley, R.A.; Lee, M.O.; Mahoney, T.R., II; Sanders, L.M. Poly(lactide-co-glycolide) decomposition kinetics *in vivo* and *in vitro*. Macromolecules **1987**, *20* (10), 2398–2403.

26. Wang, Y.; Ameer, G.; Sheppard, B.; Langer, R. A tough biodegradable elastomer. Nat. Biotechnol. **2002**, *20* (6), 602–606.

27. Cadee, J.A.; Brouwer, L.A.; den Otter, W.; Hennink, W.E.; Van Luyn, M.J.A. A comparative biocompati-bility study of microspheres based on crosslinked dextran or poly(lactic-co-glycolic)acid after subcutaneous injection in rats. J. Biomed. Mater. Res. **2001**, *56* (4), 600–609.

28. van der Elst, M.; Klein, C.P.; de BlieckHogervorst, J.M.; Patka, P.; Haarman, H.J. Bone tissue response to biodegradable polymers used for intra medullary fracture fixation: A long-term *in vivo* study in sheep femora. Biomaterials **1999**, *20* (2), 121–128.

Blood Vessel Substitutes

Jan P. Stegeman
Shaneen L. Rowe
Department of Biomedical Engineering, Rensselaer Polytechnic Institute, Troy, New York, U.S.A.

Robert M. Nerem
Parker H. Petit Institute for Bioengineering and Bioscience, Georgia Institute of Technology, Atlanta, Georgia, U.S.A.

Abstract

The need for improved vascular grafts, particularly in small diameter applications, is clear. Preferred materials include biologically appropriate and active conduits, which often incorporate living cells. There has been a great deal of progress in producing such living vascular tissues, as evidenced by the increasing number of implantation studies in both animal and human systems. This entry reviews scaffolding strategies in vascular tissue engineering and describes techniques for producing a blood vessel substitute. The approaches are divided into those using synthetic polymer scaffolds (both permanent and degradable), those using naturally derived polymer scaffolds, and those using cell-secreted scaffolds.

INTRODUCTION

The development of blood vessel substitutes using tissue engineering principles has rapidly progressed in the past decade in response to the clinical need for improved vascular grafts in surgical procedures, especially for small diameter applications. In the United States, around 40% of all deaths are caused by cardiovascular disease, and more than half of these are a result of coronary heart disease. Approximately 500,000 coronary artery bypass grafting procedures are performed annually, each requiring a small diameter vascular conduit to bypass a blocked coronary artery.[1] There has been progress in understanding the biological mechanisms behind vascular disease, and in developing pharmacological and interventional treatments, however, the *in vitro* and preclinical test-beds for evaluating these therapies and for studying vascular biology are still underdeveloped. For these reasons, there has been a strong interest in creating blood vessel analogs for use both *in vivo* as surgical grafts, and *in vitro* as models of the vasculature.

Natural blood vessels are not simply passive conduits for the flow of blood. Their ability to change their dimensions over the short and long term is important in maintaining appropriate hemodynamic conditions, and therefore in preserving patency. In addition, the smaller resistance and capacitance vessels perform an important function in the regulation of blood flow, by constricting and dilating based on neuronal, hormonal and autoregulatory mechanisms.[2] For large diameter (>6 mm inner diameter) vessel replacements, a number of synthetic materials have proven useful for producing vascular conduits. Although these approaches have worked adequately for peripheral grafting and as vascular access shunts, there are currently no acceptable synthetic blood vessel substitutes for small diameter applications. The primary reason is the difficulty in maintaining patency in purely synthetic grafts, due to the lack of an appropriate lumenal surface and the development of intimal hyperplasia at anastomotic sites. For this reason, cell-seeded grafts have been proposed as a more biologically appropriate solution. Initial efforts focused on reconstitution of the intimal layer, through endothelial cell seeding or recruitment, in order to promote hemocompatibility. In addition, more biologically compatible scaffolds were investigated as an alternative to synthetic polymers. These areas of research subsequently widened to include the incorporation of living cells in the vessel wall, in an effort to produce more fully functional grafts that mimic both the passive and active functions of the blood vessel.

The ideal blood vessel substitute would exactly mimic the structure and function of native vessels. Current tissue engineering approaches generally fall short of this ambitious goal, however, important strides have been made. A key requirement of implanted vascular grafts is their ability to remain patent for many years. Hemocompatibility is primarily determined by the lumenal lining of the vessel, and in most tissue engineering approaches the lining is expected to be a monolayer of functional endothelial cells. Another absolute requirement is that the mechanical properties of the engineered tissue must be suitable for long term implantation. This demands not only sufficient mechanical strength (suture retention, burst strength, tensile resistance), but also appropriate compliance and elasticity. In the case of biologically derived scaffolds, a key challenge is achieving appropriate polymer assembly through *in vitro* and *in vivo* processing. Additionally, in

Concise Encyclopedia of Biomedical Polymers and Polymeric Biomaterials DOI: 10.1081/E-EBPPC-120052208

order for an implanted blood vessel substitute to fully replace the function of a native vessel, it must exhibit appropriate short and long term remodeling responses. This is largely a function of the smooth muscle cells that inhabit the vessel wall, and in most tissue engineering approaches these cells are supported in a three-dimensional scaffold material formed in the shape of a tube. Finally, any implanted material or tissue must avoid eliciting a damaging immune response. This is of particular concern in tissue engineering due to the incorporation of living cells in the engineered construct, making cell sourcing a key issue. Determining the right combination of scaffold materials and cell types, and structuring them appropriately, has remained a challenge in vascular tissue engineering.

This review of scaffolding strategies in vascular engineering describes the currently most advanced techniques for producing a blood vessel substitute. The approaches are divided into those using synthetic polymer scaffolds (both permanent and degradable), those using naturally derived polymer scaffolds, and those using cell-secreted scaffolds. In all cases, the discussion is limited to approaches that involve the use of living mammalian cells, whether they are seeded on the construct initially, or recruited into the scaffold after implantation. The focus is on the current state of technology, rather than a historical review of the field, and the goal is to describe how different cell-scaffold combinations seek to address the challenges of creating mechanically sound, thrombosis resistant, immunologically safe blood vessel substitutes.

BLOOD VESSEL STRUCTURE AND FUNCTION

The ultimate goal of vascular tissue engineering is to recreate the function of natural blood vessels, the physical structure of which is shown schematically in Fig. 1. The three main structural layers consist of distinct cell and matrix types, whose composition reflects their function. The intima is composed of a monolayer of endothelial cells with an underlying basement membrane consisting of loosely organized Type IV collagen and laminin. The intima lines the lumen of the blood vessel and provides the hemocompatible blood-contacting surface that is critical for the maintenance of vessel patency. The endothelium is also important in the regulation of vascular tone, by sensing flow conditions and producing signaling molecules in response. The middle layer of the natural blood vessel is the media, and is composed of smooth muscle cells in a matrix of Type I and Type III collagen, elastin and proteoglycans. The collagen fibers of the media are aligned circumferentially or helically along the axis of the vessel, reflecting the need for them to resist tension created by pulsatile blood flow. This layer has been identified as being the most significant mechanically, and is also responsible for the phasic and tonic contraction associated with regulation of blood flow by the vasculature.[3] The interaction between the endothelial cells of the intima and the smooth muscle

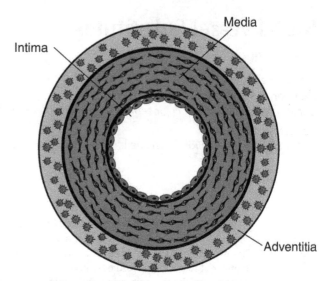

Fig. 1 Schematic diagram of the structure of a normal blood vessel showing the three main tissue layers: the intima, the media, and the adventitia.

cells of the media is a critical component of normal arterial function. The outer layer of the blood vessel, the adventitia, consists of loosely arranged Type I and Type III collagen, along with elastin and fibroblasts. Its structure and thickness varies with vessel size and location, and its function is as an anchoring connective tissue that maintains longitudinal tension on the blood vessel, and in larger vessels provides a bed for the ingrowth of a blood supply and nerves.

The primary function of blood vessels is to act as a conduit for the flow of blood, and they are therefore one of the main transport systems in the body. Compared to complex organs such as the liver and brain, blood vessels have a relatively simple structure and straightforward function. They are therefore good candidates for the application of a tissue engineering approach; using living cells and appropriate matrix materials, combined with defined biochemical and mechanical stimulation, to recreate tissue *in vitro*. It should be noted, however, that even a relatively simple tissue such as a muscular artery has a very specific matrix and cell organization, combined with an important and often complex set of signaling pathways, which lead to its normal biological function. Recapitulating this structure and obtaining the appropriate function is a problem that has been approached in a variety of ways.

The ideal blood vessel substitute will be able to withstand the same mechanical forces that native vessels are constantly exposed to, including cyclic radial stress, longitudinal stress, fluid shear and fluid pressure. However, it is not clear whether engineered blood vessels must have the same properties as the native vessel in order to restore function. It is likely that vessels with properties inferior to native vessels will be able to perform acceptably, at least initially, and that over time they will be further remodeled *in vivo* to better suit the environment. However, inappropriate mechanical matching over the longer term may lead to complications. Therefore, determination of the mechanical properties required for the replacement of vascular function is an important challenge in the field.

Biomechanical analysis of engineered vascular tissues can provide insight into the properties and expected behavior of these materials.[4,5] The burst or rupture strength is a critical property that indicates the pressure required to cause failure of the vessel wall. This can be determined experimentally or can be estimated from the ultimate tensile stress, which is the maximum value of stress on a conventional stress vs. strain diagram. The material modulus is a measure of stiffness that is determined by finding the slope of the linear portion of the stress vs. strain diagram. In native vessels, the major structural protein is Type I collagen, which provides tissue stiffness and high tensile strength.

Vessel compliance and elasticity are also important. Compliance is the ability of a material to dilate under pressure, defined as the change in volume in response to a change in pressure. Elasticity is the capability of a material to regain its original dimensions after deformation. In native vessels, fibrillar collagen and elastin combine to provide a matrix that is both compliant and elastic, allowing dilation without permanent deformation under the pulsatile pressures in the cardiovascular system. Native vessels and blood substitutes generally also exhibit viscoelastic behavior, meaning they exhibit properties of both a viscous fluid and an elastic solid. These materials exhibit a time-dependent response to an applied load, and their mechanical properties change as the strain rate is varied.

The complexity of natural blood vessel mechanics makes it difficult to develop an engineered substitute that fully matches the properties of native tissue. Important strides have been made in this area, mainly in the area of achieving adequate burst strengths, which, at a minimum, allows implantation of the vessel. Achievement of the appropriate elastic and viscoelastic properties has remained a more elusive goal.

ENDOTHELIALIZATION STRATEGIES

Since occlusion due to thrombus formation is one of the main reasons for the failure of synthetic vascular grafts,[6] there has been great emphasis on creating more hemocompatible lumenal linings, designed to shield the blood from the clot-inducing graft material. Early cell-based approaches in this area focused on coating the inner surface of permanent synthetic polymer grafts with a monolayer of living endothelial cells, as shown schematically in Fig. 2. These efforts represented the beginning of vascular tissue engineering, in that they used living cells to provide a more physiological solution to improving graft function.[7,8] It is likely that any eventual engineered vascular tissue will require an intact, functioning endothelium in order to function well over the long term, and therefore endothelialization strategies will remain critical to the success of this field.

Recent evaluations of endothelialized synthetic grafts implanted in the lower extremities of human patients have shown promising results.[9] These studies used autologous endothelial cells harvested from a subcutaneous vein, which were expanded in culture for approximately 21 days before being used to coat the graft lumen. The cells were seeded onto expanded polytetrafluoroethylene (ePTFE) vascular prostheses that had been precoated with fibrin glue to promote cell attachment, and were cultured for a further 9 days to achieve confluence. Using 6 mm inner diameter grafts, the patency rate at 7 years exceeded 60%, and a shorter term evaluation of 7 mm inner diameter grafts showed greater than 80% patency at 4 years. In spite of the relatively recent success of endothelialization in peripheral (large diameter) vascular grafting, this technology has not yet produced similarly promising results in small diameter applications. Encouragingly, recent studies using 4 mm inner diameter ePTFE grafts[10] and decellularized vascular

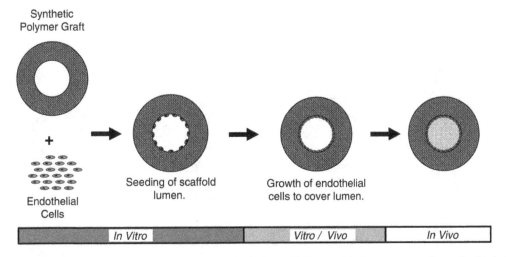

Fig. 2 Schematic diagram of endothelialization of permanent synthetic scaffold materials for use as vascular grafts. Endothelial cells are seeded onto the lumen of the graft and grown to cover the inner surface, in order to provide a hemocompatible lining.
Source: From Herring et al.,[8] Meinhart, Gardner, and Glover[9] and Williams et al.[12]

tissue[11] have suggested that endothelialization can also be used to improve the function of small caliber vessels.

A critical issue in developing practically feasible endothelialization strategies is the sourcing of cells, since endothelial cells are strongly involved in the vascular immune reaction. Initial successes using preseeded autologous endothelial cells have shown that this approach has promise, but in order to provide "off-the-shelf" availability of tissue for implantation it is likely that alternate strategies will be needed. Efforts at decreasing the time required to generate an endothelial cell lining, such that a long *in vitro* culture period is not required, have led to cell "sodding" techniques using microvascular endothelial cells extracted from liposuction fat at the time of graft implant surgery.[12] Recruitment of native endothelial cells to the graft surface after implantation is another possibility, though this is challenging in human patients due to the low proliferation and migration rate of endothelial cells *in vivo*. Immobilization of cell adhesion peptides[13] and release of growth factors[14] have been used to increase the degree graft coverage, as well as the level of endothelial cell function, and the effects of shear forces on endothelial cells seeded onto grafts also are being investigated.[15] In addition, endothelial progenitor cells are known to circulate in the blood, though in very small numbers.[16] This is another potential source of autologous cells for lumenal seeding[17] or recruitment after graft implantation. The ability to use allogeneic endothelial cells to line the lumen of implanted grafts would be a great advantage in terms of cell sourcing. In this case, however, a major challenge would be controlling the immune response in order to provide a functional, hemocompatible graft lining.

SCAFFOLDING STRATEGIES

As the understanding of vascular biology has progressed, it has become increasingly clear that the smooth muscle cells that populate the medial layer of the vessel wall contribute in important ways to both the short and long term function of the blood vessel. This includes generation of contractile forces important in the vasoactive response, longer term remodeling of vessel dimensions in response to changing hemodynamic conditions, as well as pathological responses such as intimal hyperplasia and the generation of atherosclerotic plaques. In addition, the interaction between the endothelial cells of the intima and the smooth muscle cells of the media is receiving more attention as the regulatory effect of each of these cell types on the other is becoming more completely understood.

The generation of appropriate scaffolds for smooth muscle cell growth and function is a challenge that has been approached in a variety of ways. Of key importance is providing an extracellular matrix that is sufficiently strong and compliant to withstand the pulsatile pressures of the vascular system, while at the same time maintaining the cell-specific functions of the vascular smooth muscle cell. The following sections outline three general strategies for achieving these goals.

Degradable Synthetic Scaffolds

The archetypal tissue engineering technology of using a biodegradable synthetic polymer scaffold seeded with isolated cells[18] has also been applied to the vascular system, as shown schematically in Fig. 3. This approach involves seeding cultured cells onto a preformed porous scaffold, which is chemically designed to degrade over time in the physiological environment. In vascular applications, the most commonly used degradable polymer scaffolds are polycaprolactone, polyglycolic acid (PGA), polylactic acid (PLA), as well as derivatives and copolymers of these materials.

The feasibility of this approach was first reported in an ovine model using PGA–polyglactin copolymers seeded in stages with autologous fibroblasts, smooth muscle cells

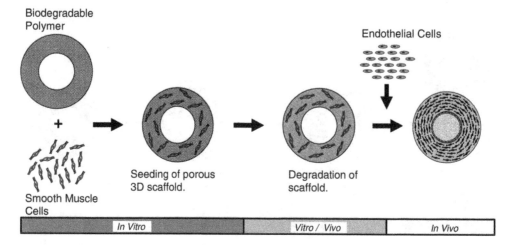

Fig. 3 Schematic representation of the use of biodegradable polymer scaffolds as vascular grafts. Smooth muscle cells are seeded onto a porous degradable scaffold, which is eventually replaced by living tissue and can be lined with endothelial cells.
Source: From Shinoka et al.,[19] Niklason et al.,[21] and Kim et al.[23]

and endothelial cells.[19] The 15 mm diameter constructs were cultured *in vitro* for 7 days, before being used to replace short segments in the low pressure pulmonary circulation. Examination of explanted grafts showed that cell-seeded grafts remained patent for up to 24 weeks, whereas control acellular polymer grafts occluded via thrombus formation. The polymer scaffold had been completely degraded by 11 weeks *in vivo*, but the resulting tissues were not deemed suitable for implantation in the higher pressure systemic circulation. A subsequent study used a modified polymer system (a PGA–polyhydroxyalkanoate composite) in order to augment the longer term mechanical properties of the construct.[20] A mixed culture of autologous ovine endothelial cells, smooth muscle cells and fibroblasts were seeded simultaneously on the lumen of the 7 mm diameter construct, and were allowed to attach and proliferate *in vitro* for 7 days prior to the construct being implanted as an infrarenal aortic graft. These engineered vessels exhibited patency up to 22 weeks, whereas acellular polymer tubes used as controls became occluded at times ranging from 1 day to 14 weeks. Evaluation of the cell seeded constructs showed collagen and elastin fiber formation in the medial layer, as well as staining for endothelial cell markers on the lumenal surface.

A similar approach using bovine and porcine cell-seeded PGA scaffolds implanted into a porcine model[21] also demonstrated very promising results. Notably, this study used a bioreactor system to impart cyclic mechanical strain to the smooth muscle cell-seeded constructs as they developed *in vitro* over a period of 8 weeks. It was found that pulsatile strain promoted homogeneous cell distribution in the construct wall, as well as more complete resorption of the polymer scaffold, leading to enhanced mechanical properties. Endothelial cells were added to the lumen of pulsed constructs, and were cultured for a period of 3 days in the presence of continuous perfusion of the vessel lumen with culture medium. Implantation of one xenogeneic construct, as well as three autologous constructs, showed that these engineered vessels could remain patent for up to 4 weeks, and that application of pulsatile strain *in vitro* improved patency rates as well as graft morphology and function.

Other investigations into the use of degradable synthetic polymer scaffolds have focused on modifying the scaffold to guide cell function and better control the biological response. Modification of PGA–PLA scaffolds has been shown to modulate cell adhesion,[22] as well as cell phenotype,[23] and mechanical stimulation has been shown to enhance tissue properties.[24] Chemical design of other polymers has also been used to produce degradable scaffolds with potential application in the vascular system,[25] including the incorporation of tethered growth factors in order to affect cell function.[26]

Although encouraging results have been obtained using cell-seeded degradable polymer systems, significant challenges remain. Since the scaffold is designed to be replaced by cells and extracellular matrix, a high degree of cell proliferation and matrix synthesis is required to transform these synthetic constructs into fully biological tissue. Complete degradation of the polymer scaffold is difficult to achieve, especially since mass transport limitations can hamper the degradation process as the tissue becomes more densely filled with proliferating cells and new extracellular matrix. In addition, although the degradation products of most of these systems are relatively benign, the hydrolysis reaction involved can change the tissue microenvironment in such a way as to be damaging to nearby cells, for example by decreasing the local pH.[27] The inflammatory and immune responses may also be activated by polymer degradation.[28] These challenges are being approached in a variety of creative ways, by controlling scaffold properties and directing cell function.

Naturally-derived Scaffolds

Another promising strategy to engineering living tissue is to provide cells with a scaffold similar to the one they are exposed to in native tissue, with the expectation that they will recognize and react appropriately to the supplied matrix and produce a competent vascular tissue. The predominant structural components of natural blood vessels are collagen (Type I and Type III) and elastin. These materials have been used to varying degrees in reconstructing vascular tissue *in vitro*. The main approaches using naturally derived matrices can be divided into two subcategories, as shown schematically in Figs. 4 and 5. In the first case, cells are combined with a solubilized form of the matrix proteins and a tissue is formed by subsequent gelation of the scaffold, generally using a molding process that results in the cells being embedded directly in the tissue matrix. In the second case, natural allogeneic or xenogeneic tissue is processed to remove cellular and immunogenic components, leaving an intact decellularized matrix that is subsequently used as a scaffold for repopulation with autologous cells.

Most tissue molding approaches (Fig. 4) have used Type I collagen as the main matrix protein, since it is difficult to obtain sufficient quantities of Type III collagen, and solubilized elastin will generally not reconstitute into a functional fibrillar form. Initial studies using tubular Type I collagen gels[29] established the feasibility of this technique, though the low mechanical strength of the matrix necessitated the use of a reinforcing synthetic polymer sleeve. More recent studies have provided important insight into how collagen gel matrices are formed, how they can be modified and controlled, and how cells interact with and remodel them. Cell-mediated gel compaction has been shown to be modulated by the presence of mechanical constraints during gelation,[30] resulting in increased fiber alignment.[31] The properties of such engineered tissues can also be improved by chemical means, for example, by promoting appropriate collagen cross-linking,[32–34] and through application of cyclic

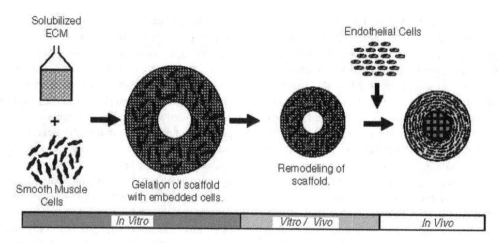

Fig. 4 Schematic diagram of the use of solubilized extracellular matrix proteins, for example, Type I collagen, as scaffolds in vascular tissue engineering. Smooth muscle cells are embedded in a molded hydrogel, which is remodeled by the cellular component to produce a vascular tissue. Endothelial cells can be seeded onto the lumen to provide a hemocompatible surface.
Source: From Weinberg and Bell,[29] Girton, Oegema, and Tranquillo[33] and Seliktar et al.[35]

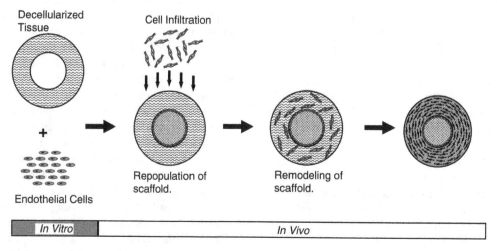

Fig. 5 Schematic diagram of the use of decellularized natural tissues to produce vascular scaffolds. Endothelialization and implantation result in the recruitment of cells from the host, which repopulate and remodel the matrix.
Source: From Hodde et al.,[48] Huynh et al.,[50] and Conklin et al.[51]

mechanical strain during the *in vitro* tissue development process.[35] Remodeling of the collagen matrix by embedded cells through the action of matrix metalloproteinases can affect the mechanical properties of collagen-based engineered tissues, and this process is also affected by the application of mechanical forces.[36,37]

A key challenge in the use of molded Type I collagen matrices as a matrix in vascular tissue engineering is the difficulty in obtaining graft mechanical properties that are appropriate for implantation. One potential solution to this problem is the use of reinforcing sleeves that provide the required mechanical strength while the cells are embedded in a reconstituted collagen matrix. The reinforcing sleeve can be made of a naturally derived polymer that has been strengthened by crosslinking,[38] or it can be a synthetic polymer with superior mechanical properties.[39] In the former

case, the sleeve may be gradually remodeled and resorbed by the cellular component, whereas in the latter case a degradable or permanent polymer can be used. Reinforcement of collagen-based vascular constructs with synthetic polymers has been tested in autologous implantation studies in a canine model, and showed patency up to 27 weeks,[40] depending on the properties of the reinforcing material. Interestingly, this technique has also been used to fabricate branched vessels.[41]

In addition to Type I collagen, other naturally derived polymers that can be reconstituted in fibrillar form have been investigated as potential matrices in tissue engineering. The blood clotting protein fibrin has been suggested as an alternative to collagen matrices,[42,43] and has also been used as an additive to provide enhanced properties to collagen-based engineered tissues.[44] The production of biomimetic peptides using genetic engineering techniques is also being used to produce extracellular matrix proteins with desired mechanical and biochemical properties,[45] and methods for creating scaffolds from these are being developed. For example, electrospinning has been used to produce mesh scaffolds of collagen fibers,[46,47] although the mechanical properties and interaction with cells of these matrices has not yet been extensively characterized.

The use of decellularized natural tissue matrices as scaffolds for vascular tissue engineering also has been investigated. This method is shown schematically in Fig. 5. Isolation and processing of intact extracellular matrix lamina from porcine tissue has been used to produce scaffold materials that are composed entirely of native structural proteins, with the associated attachment proteins also included.[48] The small intestinal submucosa (SIS), in particular, has been used to create vascular conduits, which when tested in xenogeneic implant studies[49] showed patency for up to 9 weeks as well as evidence of remodeling and improvement of graft mechanical properties. SIS has also been used in combination with gelled and crosslinked bovine collagen that has been heparinized to reduce thrombogenicity.[50] This model showed patency after 13 weeks of implantation into a xenogeneic host, as well as cellular infiltration and a recovery of vasomotor activity in response to certain contractile agonists. Decellularized vascular tissue has also been used in an effort to provide new small diameter blood vessels. Porcine carotid arteries have been processed to remove the cellular component, heparinized to reduce thrombosis and implanted in a dog model for up to 9 weeks.[51] Porcine iliac vessels have also been used in decellularized form, and have been lined with autologous endothelial progenitor cells before being implanted in a sheep model.[17] These vessels exhibited patency for up to 29 weeks, as well as a vasoactive response upon removal.

Cell-Secreted Scaffolds

Creation of engineered tissues does not necessarily require that a scaffold be supplied initially, since it can be argued that the most appropriate tissue matrix is that produced by the specialized tissue cells themselves. The resulting construct is fully biological, and the resident cells are well integrated into the tissue, with the ability to produce, degrade, and remodel the matrix in response to environmental cues. Although this approach is simple in concept, directing cell function to recreate appropriate tissues has remained a major challenge in tissue engineering and regenerative medicine.

In the case of vascular tissue engineering, an innovative technology has been developed that uses sequential layering of cultured tissue sheets to produce a tubular blood vessel analog,[52] as shown schematically in Fig. 6. The robust production of extracellular matrix that is required for generation of sheets of tissue is promoted by the addition of ascorbic acid to the culture medium, a method adapted from

the development of engineered dermal tissue. Fibroblasts are cultured for 35 weeks to generate an initial tissue sheet, which is then dehydrated and wrapped around a mandrel to produce an acellular inner sleeve. Smooth muscle cells are cultured for 3 weeks to produce sheets of tissue several cell layers thick. These sheets are then removed from the culture plate and wrapped arounda mandrel to form a tubular medial structure, which is cultured in a flow-through bioreactor for 1 week. A cell-containing fibroblast sheet can then be wrapped around the medial tissue to produce an analog of the adventitia. This construct is cultured for at least an additional 8 weeks before an endothelial cell lining is seeded onto the lumen of the construct. Vascular tissues produced in this manner have exhibited remarkable burst pressures (>2500 mmHg), and have shown some success in xenogeneic implant studies. In addition, these vessels have exhibited evidence of contractility in response to pharmacological stimulation.[53]

The natural foreign body response, and the associated generation of granulation tissue, has also been harnessed to produce fully biological vascular tissue,[54] as shown schematically in Fig. 7. In this approach, silicone tubing is implanted into the peritoneal cavity of the host in order to stimulate fibrous encapsulation. After 2 weeks, the resulting fibrotic response around the tubing produces a vascular structure consisting of dense connective tissue surrounded by cell-rich granulation tissue, with an outer monolayer of mesothelial cells. By removing the silicone and everting the formed tissue, the outer mesothelial cell layer becomes the inner lumenal lining and a structure similar to a blood vessel is formed. The mechanical properties of these engineered vessels have not been extensively characterized, however, autologous implantation studies showed the ability to remain patent for up to 4 months, and explanted vessels exhibited modest contractile responses to vasoactive agents.

Generation of a fully biological blood vessel substitute using only cell-secreted scaffolds has many attractive aspects, although important practical issues still need to be resolved. In the tissue-wrapping approach, the long culture times required for generation of appropriate tissue *in vitro* (at least 13 weeks, not including formation of the initial fibroblast-generated sleeve) and the extensive tissue handling required to wrap successive layers pose a challenge to producing a widely available graft. Use of the fibrotic response to generate new tissue *in vivo* has the potential to produce autologous grafts; however, it is not clear that the human fibrotic response will be appropriate for this technique. In addition, the origin and eventual function of the cells that are recruited in this approach have not been fully characterized.

BIOLOGICAL RESPONSE TO SCAFFOLD

As more fully biological tissue substitutes are developed, an increasing amount of attention is being directed at understanding how living cells interact with their surroundings.

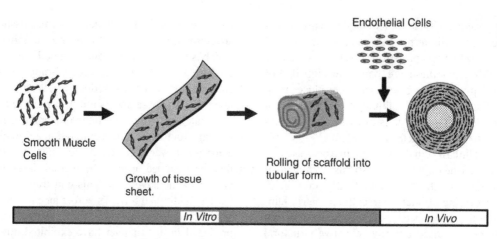

Fig. 6 Schematic representation of using rolled sheets of smooth muscle cell-generated tissue to produce a vascular graft. Rolled tissues are cultured to form a conduit, which can be lined with endothelial cells.
Source: From L'Heureux et al.[52] and Hoerstrup et al.[63]

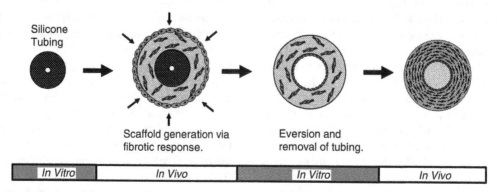

Fig. 7 Schematic diagram of using the production of granulation tissue to generate a tubular conduit. A piece of silicone tubing is implanted into the peritoneal cavity and becomes encapsulated by the fibrotic host response, with an outer layer of mesothelial cells. Upon explantation, the tissue can be everted to yield a vascular graft.
Source: From Campbell, Efendy, and Campbell.[54]

In tissue engineering applications, many of the cell–matrix interactions take place in a complex three-dimensional environment and cell function is further influenced by soluble biochemical as well as mechanical signals. In the case of vascular tissue, the endothelial cells that line the lumen are in the form of a monolayer, which is one of the few physiological situations in which two-dimensional culture systems are representative of the *in vivo* geometry. However, these cells are in contact with the complex biochemical environment of whole blood, and are strongly influenced by fluid shear forces.[55] In addition, it must be recognized that native endothelial cells are found on a three-dimensional substrate populated by smooth muscle cells, and that these cell types can interact directly and through biochemical signaling.[56–58] The ability of the endothelial lining to provide a continuous hemocompatible surface therefore depends on the fluid dynamic environment, as well as the composition and function of the substrate.

The smooth muscle cells of the medial layer are responsible for the agonist-stimulated contraction and dilation of blood vessels, and are also involved in longer

term remodeling of the tissue. These cells are surrounded by a three-dimensional matrix that provides a variety of signals that affect cell function. The matrix itself has biochemical effects on the embedded cells, and is also involved in signaling by growth factors and mechanical forces.[59,60] It is increasingly being recognized that smooth muscle cell function in three-dimensional scaffolds is under complex control, and that studies in two-dimensional culture do not always reflect this complexity.[61] The properties of the scaffold itself can have an effect on cell phenotype,[23] as can the combined effect of the matrix, biochemical factors and mechanical forces.[62] In order to get appropriate tissue structure and remodeling, the effects of these parameters must be understood and ideally harnessed to produce vascular substitutes with the desired properties.

SUMMARY AND FUTURE DIRECTIONS

The need for improved vascular grafts, particularly in small diameter applications, is clear. The trend over the past several decades has been toward more biologically appropriate and active conduits, often incorporating living cells. These engineered tissues are functionally more complex than purely synthetic or acellular materials; however, they offer the potential to more closely replicate the active functions of living native tissues. This includes the anticoagulative properties of an endothelial cell lining, as well as the remodeling and vasoactive functions of a smooth muscle cell-populated vascular wall.

There has been a great deal of progress in producing such living vascular tissues, as evidenced by the increasing number of implantation studies in both animal and human systems. However, a number of key challenges remain to be overcome before a widely usable living vascular substitute is obtained. These include achievement of appropriate mechanical properties, hemocompatibility, tissue remodeling, and vasoactivity. Careful selection and study of scaffolding strategies will certainly play an important role in achieving these goals, as cell–cell and cell–matrix interactions become more completely understood.

ACKNOWLEDGMENTS

The authors gratefully acknowledge the financial support of the National Science Foundation through ERC Award Number EEC-9731643.

REFERENCES

1. American Heart Association (AHA). *Heart Disease and Stroke Statistics—2003 Update*, 2002.

2. Morris, J.L.; Gibbins, I.; Kadowitz, P.J.; Herzog, H.; Kreulen, D.L.; Toda, N.; Claing, A. Roles of peptides and other substances in cotransmission from vascular autonomic and sensory neurons. Can. J. Physiol. Pharmacol. **1995**, *73* (5), 521–532.

3. Griendling, K.K.; Alexander, R.W. Cellular biology of blood vessels. In *Hurst's the Heart*; Schlant, R.C., Alexander, R.W., Eds.; McGraw-Hill Inc.: New York, 1994.

4. Caro, C.G.; Pedley, T.J.; Schroter, R.C.; Seed, W.A. *The Mechanics of the Circulation*; Oxford University Press: Oxford, 1978.

5. Fung, Y.C. *Biomechanics: Mechanical Properties of Living Tissues*; Springer: New York, 1993.

6. Cynamon, J.; Pierpont, C.E. Thrombolysis for the treatment of thrombosed hemodialysis access grafts. Rev. Cardiovasc. Med. **2002**, *3* (Suppl. 2), S84–S91.

7. Graham, L.M.; Vinter, D.W.; Ford, J.W.; Kahn, R.H.; Burkel, W.E.; Stanley, J.C. Cultured autogenous endothelial cell seeding of prosthetic vascular grafts. Surg. Forum **1979**, *30*, 204–206.

8. Herring, M.; Gardner, A.; Glover, J. A single-staged technique for seeding vascular grafts with autogenous endothelium. Surgery **1978**, *84* (4), 498–504.

9. Meinhart, J.G.; Deutsch, M.; Fischlein, T.; Howanietz, N.; Froschl, A.; Zilla, P. Clinical autologous *in vitro* endothelialization of 153 infrainguinal ePTFE grafts. Ann. Thorac. Surg. **2001**, *71* (5 Suppl.), S327–S331.

10. Fields, C.; Cassano, A.; Makhoul, R.G.; Allen, C.; Sims, R.; Bulgrin, J.; Meyer, A.; Bowlin, G.L.; Rittgers, S.E. Evaluation of electrostatically endothelial cell seeded expanded polytetrafluoroethylene grafts in a canine femoral artery model. J. Biomater. Appl. **2002**, *17* (2), 135–152.

11. Carnagey, J.; Hern-Anderson, D.; Ranieri, J.; Schmidt, C.E. Rapid endothelialization of PhotoFix natural biomaterial vascular grafts. J. Biomed. Mater. Res. **2003**, *65B* (1), 171–179.

12. Williams, S.K.; Wang, T.F.; Castrillo, R.; Jarrell, B.E. Liposuction-derived human fat used for vascular graft sodding contains endothelial cells and not mesothelial cells as the major cell type. J. Vasc. Surg. **1994**, *19* (5), 916–923.

13. Krijgsman, B.; Seifalian, A.M.; Salacinski, H.J.; Tai, N.R.; Punshon, G.; Fuller, B.J.; Hamilton, G. An assessment of covalent grafting of RGD peptides to the surface of a compliant poly(carbonate-urea)urethane vascular conduit versus conventional biological coatings: Its role in enhancing cellular retention. Tissue Eng. **2002**, *8* (4), 673–680.

14. Wissink, M.J.B.; Beernink, R.; Poot, A.A.; Engbers, G.H.M.; Beugeling, T.; van Aken, W.G.; Feijen, J. Improved endothelialization of vascular grafts by local release of growth factor from heparinized collagen matrices. J. Control. Release **2000**, *64* (1–3), 103–114.

15. Braddon, L.G.; Karoyli, D.; Harrison, D.G.; Nerem, R.M. Maintenance of a functional endothelial cell monolayer on a fibroblast/polymer substrate under physiologically relevant shear stress conditions. Tissue Eng. **2002**, *8* (4), 695–708.

16. Lin, Y.; Weisdorf, D.J.; Solovey, A.; Hebbel, R.P. Origins of circulating endothelial cells and endothelial outgrowth from blood. J. Clin. Invest. **2000**, *105* (1), 71–77.

17. Kaushal, S.; Amiel, G.E.; Guleserian, K.J.; Shapira, O.M.; Perry, T.; Sutherland, F.W.; Rabkin, E.; Moran, A.M.; Schoen, F.J.; Atala, A.; Soker, S.; Bischoff, J.; Mayer, J.E. Functional small-diameter neovessels created using endothelial progenitor cells expanded *ex vivo*. Nat. Med. **2001**, *7* (9), 1035–1040.

18. Langer, R.; Vacanti, J.P. Tissue engineering. Science **1993**, *260* (5110), 920–926.

19. Shinoka, T.; Shum-Tim, D.; Ma, P.; Tanel, R.; Isogai, N.; Langer, R.; Vacanti, J.; Mayer, J. Creation of viable pulmonary artery autografts through tissue engineering. J. Thorac. Cardiovasc. Surg. **1998**, *115* (3), 536–546.

20. Shum-Tim, D.; Stock, U.; Hrkach, J.; Shinoka, T.; Lien, J.; Moses, M.A.; Stamp, A.; Taylor, G.; Moran, A.M.; Landis, W. Tissue engineering of autologous aorta using a new biodegradable polymer. Ann. Thorac. Surg. **1999**, *68* (6), 2298–2304.

21. Niklason, L.E.; Gao, J.; Abbott, W.M.; Hirschi, K.K.; Houser, S.; Marini, R.; Langer, R. Functional arteries grown *in vitro*. Science **1999**, *284* (5413), 489–493.

22. Nikolovski, J.; Mooney, D.J. Smooth muscle cell adhesion to tissue engineering scaffolds. Biomaterials **2000**, *21* (20), 2025–2032.

23. Kim, B.S.; Nikolovski, J.; Bonadio, J.; Smiley, E.; Mooney, D.J. Engineered smooth muscle tissues: Regulating cell phenotype with the scaffold. Exp. Cell Res. **1999**, *251* (2), 318–328.

24. Kim, B.-S.; Nikolovski, J.; Bonadio, J.; Mooney, D.J. Cyclic mechanical strain regulates the development of engineered smooth muscle tissue. Nat. Biotechnol. **1999**, *17* (10), 979–983.

25. Mann, B.K.; Gobin, A.S.; Tsai, A.T.; Schmedlen, R.H.; West, J.L. Smooth muscle cell growth in photopolymerized hydrogels with cell adhesive and proteolytically degradable domains: Synthetic ECM analogs for tissue engineering. Biomaterials **2001**, *22* (22), 3045–3051.

26. Mann, B.K.; Schmedlen, R.H.; West, J.L. Tethered-TGF-beta increases extracellular matrix production of vascular smooth muscle cells. Biomaterials **2001**, *22* (5), 439–444.

27. Agrawal, C.M.; Athanasiou, K.A. Technique to control pH in vicinity of biodegrading PLA–PGA implants. J. Biomed. Mater. Res. **1997**, *38* (2), 105–114.

28. Athanasiou, K.A.; Niederauer, G.G.; Agrawal, C.M. Sterilization, toxicity, biocompatibility and clinical applications of polylactic acid/polyglycolic acid copolymers. Biomaterials **1996**, *17* (2), 93–102.

29. Weinberg, C.B.; Bell, E. A blood vessel model constructed from collagen and cultured vascular cells. Science **1986**, *231* (4736), 397–400.

30. Barocas, V.H.; Tranquillo, R.T. An anisotropic biphasic theory of tissue-equivalent mechanics: The interplay among cell traction, fibrillar network deformation, fibril alignment, and cell contact guidance. J. Biomech. Eng. **1997**, *119* (2), 137–145.

31. Barocas, V.H.; Girton, T.S.; Tranquillo, R.T. Engineered alignment in media equivalents: Magnetic prealignment and mandrel compaction. J. Biomech. Eng. **1998**, *120* (5), 660–666.

32. Brinkman, W.T.; Nagapudi, K.; Thomas, B.S.; Chaikof, E.L. Photo-cross-linking of type I collagen gels in the presence of smooth muscle cells: Mechanical properties, cell viability, and function. Biomacromolecules **2003**, *4* (4), 890–895.

33. Girton, T.S.; Oegema, T.R.; Tranquillo, R.T. Exploiting glycation to stiffen and strengthen tissue equivalents for tissue engineering. J. Biomed. Mater. Res. **1999**, *46* (1), 87–92.

34. van Wachem, P.B.; Plantinga, J.A.; Wissink, M.J.; Beernink, R.; Poot, A.A.; Engbers, G.H.; Beugeling, T.; van Aken, W.G.; Feijen, J.; van Luyn, M.J. *In vivo* biocompatibility of carbodiimide-crosslinked collagen matrices: Effects of crosslink density, heparin immobilization, and bFGF loading. J. Biomed. Mater. Res. **2001**, *55* (3), 368–378.

35. Seliktar, D.; Black, R.A.; Vito, R.P.; Nerem, R.M. Dynamic mechanical conditioning of collagen-gel blood vessel constructs induces remodeling *in vitro*. Ann. Biomed. Eng. **2000**, *28* (4), 351–362.

36. Asanuma, K.; Magid, R.; Johnson, C.; Nerem, R.M.; Galis, Z.S. Uniaxial strain upregulates matrix-degrading enzymes produced by human vascular smooth muscle cells. Am. J. Physiol. Heart Circ. Physiol. **2003**, *284* (5), H1778–H1784.

37. Seliktar, D.; Nerem, R.M.; Galis, Z.S. The role of matrix metalloproteinase-2 in the remodeling of cell-seeded vascular constructs subjected to cyclic strain. Ann. Biomed. Eng. **2001**, *29* (11), 923–934.

38. Berglund, J.D.; Mohseni, M.M.; Nerem, R.M.; Sambanis, A. A biological hybrid model for collagen-based tissue engineered vascular constructs. Biomaterials **2003**, *24* (7), 1241–1254.

39. Matsuda, T.; He, H. Newly designed compliant hierarchic hybrid vascular grafts wrapped with a microprocessed elastomeric film—I: Fabrication procedure and compliance matching. Cell Transplant. **2002**, *11* (1), 67–74.

40. He, H.; Matsuda, T. Arterial replacement with compliant hierarchic hybrid vascular graft: Biomechanical adaptation and failure. Tissue Eng. **2002**, *8* (2), 213–224.

41. Kobashi, T.; Matsuda, T. Fabrication of branched hybrid vascular prostheses. Tissue Eng. **1999**, *5* (6), 515–524.

42. Grassl, E.D.; Oegema, T.R.; Tranquillo, R.T. Fibrin as an alternative biopolymer to type-I collagen for the fabrication of a media equivalent. J. Biomed. Mater. Res. **2002**, *60* (4), 607–612.

43. Ye, Q.; Zund, G.; Benedikt, P.; Jockenhoevel, S.; Hoerstrup, S.P.; Sakyama, S.; Hubbell, J.A.; Turina, M. Fibrin gel as a three dimensional matrix in cardiovascular tissue engineering. Eur. J. Cardiothorac. Surg. **2000**, *17* (5), 587–591.

44. Cummings, C.L.; Gawlitta, D.; Nerem, R.M.; Stegemann, J.P. Collagen fibrin and collagen–fibrin mixtures as matrix materials for vascular tissue engineering (abstract). In *Proceedings of the 2nd Joint EMBS–BMES Conference*, Houston, Texas, Oct. 23–26 2002; p. 203.

45. Huang, L.; McMillan, R.A.; Apkarian, R.P.; Pourdeyhimi, B.; Conticello, V.P.; Chaikof, E.L. Generation of synthetic elastin-mimetic small diameter fibers and fiber networks. Macromolecules **2000**, *33* (8), 2989–2997.

46. Huang, L.; Apkarian, R.P.; Chaikof, E.L. High-resolution analysis of engineered type I collagen nanofibers by electron microscopy. Scanning **2001**, *23* (6), 372–375.

47. Matthews, J.A.; Wnek, G.E.; Simpson, D.G.; Bowlin, G.L. Electrospinning of collagen nanofibers. Biomacromolecules **2002**, *3* (2), 232–238.

48. Hodde, J.P.; Record, R.D.; Tullius, R.S.; Badylak, S.F. Retention of endothelial cell adherence to porcine-derived extracellular matrix after disinfection and sterilization. Tissue Eng. **2002**, *8* (2), 225–234.

49. Hiles, M.C.; Badylak, S.F.; Lantz, G.C.; Kokini, K.; Geddes, L.A.; Morff, R.J. Mechanical properties of xenogeneic small-intestinal submucosa when used as an aortic graft in the dog. J. Biomed. Mater. Res. **1995**, *29* (7), 883–891.

50. Huynh, T.; Abraham, G.; Murray, J.; Brockbank, K.; Hagen, P.-O.; Sullivan, S. Remodeling of an acellular collagen graft into a physiologically responsive neovessel. Nat. Biotechnol. **1999**, *17* (11), 1083–1086.

51. Conklin, B.S.; Richter, E.R.; Kreutziger, K.L.; Zhong, D.-S.; Chen, C. Development and evaluation of a novel decellularized vascular xenograft. Med. Eng. Phys. **2002**, *24* (3), 173–183.

52. L'Heureux, N.; Paquet, S.; Labbe, R.; Germain, L.; Auger, F.A. A completely biological tissue-engineered human blood vessel. FASEB J. **1998**, *12* (1), 47–56.

53. L'Heureux, N.; Stoclet, J.C.; Auger, F.; Lagaud, G.J.; Germain, L.; Andriantsitohaina, R. A human tissue-engineered vascular media: A new model for pharmacological studies of contractile responses. FASEB J. **2001**, *15* (2), 515–524.

54. Campbell, J.H.; Efendy, J.L.; Campbell, G.R. Novel vascular graft grown within recipient's own peritoneal cavity. Circ. Res. **1999**, *85* (12), 1173–1181.

55. Fisher, A.B.; Chien, S.; Barakat, A.I.; Nerem, R.M. Endothelial cellular response to altered shear stress. Am. J. Physiol. Lung Cell Mol. Physiol. **2001**, *281* (3), L529–L533.

56. Chiu, J.-J.; Chen, L.-J.; Lee, P.-L.; Lee, C.-I.; Lo, L.-W.; Usami, S.; Chien, S. Shear stress inhibits adhesion molecule expression in vascular endothelial cells induced by coculture with smooth muscle cells. Blood **2003**, *101* (7), 2667–2674.

57. Imberti, B.; Seliktar, D.; Nerem, R.M.; Remuzzi, A. The response of endothelial cells to fluid shear stress using a co-culture model of the arterial wall. Endothelium **2002**, *9* (1), 11–23.

58. Powell, R.J.; Carruth, J.A.; Basson, M.D.; Bloodgood, R.; Sumpio, B.E. Matrix-specific effect of endothelial control of smooth muscle cell migration. J. Vasc. Surg. **1996**, *24* (1), 51–57.

59. Boudreau, N.J.; Jones, P.L. Extracellular matrix and integrin signalling: The shape of things to come. Biochem. J. **1999**, *339* (Pt. 3), 481–488.

60. Bottaro, D.P.; Liebmann-Vinson, A.; Heidaran, M.A. Molecular signaling in bioengineered tissue microenvironments. Ann. N.Y. Acad. Sci. **2002**, *961* (1), 143–153.

61. Stegemann, J.P.; Nerem, R.M. Altered response of vascular smooth muscle cells to exogenous biochemical stimulation in two- and three-dimensional culture. Exp. Cell Res. **2003**, *283* (2), 146–155.

62. Stegemann, J.P.; Nerem, R.M. Phenotype modulation in vascular tissue engineering using biochemical and mechanical stimulation. Ann. Biomed. Eng. **2003**, *31* (4), 391–402.

63. Hoerstrup, S.P.; Zund, G.; Cheng, S.; Melnitchouk, S.; Kadner, A.; Sodian, R.; Kolb, S.A.; Turina, M. A new approach to completely autologous cardiovascular tissue in humans. ASAIO J. **2002**, *48* (3), 234–238.

Biomimetic—Colloid

Bone–Implant: Polymer Physicochemical Modification for Osseointegration

Waleed S. W. Shalaby
Bruce L. Anneaux
Poly-Med, Incorporated, Anderson, South Carolina, U.S.A.

Abstract

Biomaterial scientists and engineers are calling for the development of a new generation of implants with improved biocompatibility. This ideology was the driving force for pursuing most of the studies reported in this entry, which deals with the introduction of functional groups capable of positive interaction with osteoblasts to encourage osseointegration. A second aspect of these studies deals specifically with dental and orthopedic implants, which need to meet certain strength and modulus requirements. Accordingly, a segment of the studies reported in this entry describe a new process for enhancing the bulk properties of the pertinent material. A third aspect of the reported studies pertains to the advantage of having microtextured surfaces that support interlocking with bony tissues. A novel process for surface microtexturing is also discussed in the text of the entry.

INTRODUCTION

Traditionally, most biomaterials are developed or selected to meet biological requirements based on their: 1) initial mechanical properties and their retention profiles in the biologic environment; 2) mechanical biocompatibility, such as having practically smooth surface morphology; and 3) chemical biocompatibility, i.e., being free of toxic leachables and inert with no tendency for biological interaction. In fact, inertness was the primary criterion early investigators used for material selection; a biocompatible material was denoted as one that does not elicit toxic, carcinogenic, or significant local inflammatory reactions.[1] An ideal biomaterial for use in long term or permanent (nontransient) implants was typically described by a list of negative adjectives: nontoxic, noncarcinogenic, and nonallergenic.[2] Because our understanding of implantable biomaterials and the biological reactions at the implant–tissue interface have increased significantly over the past three decades, the definition of biocompatible materials has evolved beyond being simply bioinert, and the significance of positive interaction of a biomaterial implant with the surrounding biological environment has been widely acknowledged.[3] Contemporary biomaterial scientists and engineers are calling for the development of a new generation of implants having chemically and/or physically tailored and/or modified surfaces to provide positive interaction with the biological environment and capable of directing pertinent biological events to meet specific functional requirements of such implants.[4]

This ideology was the driving force for pursuing most of the studies reported in this entry, which deal with the introduction of functional groups capable of positive interaction with osteoblasts to encourage osseointegration. A second aspect of these studies deals specifically with dental and orthopedic implants, which need to meet certain strength and modulus requirements. Accordingly, a segment of the studies reported in this entry describe a new process for enhancing the bulk properties of the pertinent material. A third aspect of the reported studies pertains to the advantage of having microtextured surfaces that support interlocking with bony tissues. And a novel process for surface microtexturing is discussed in the text of the entry.

TECHNOLOGY EVOLUTION OF PHYSICOCHEMICAL SURFACE MODIFICATION AND BULK ORIENTATION

Since the beginning of interest in polymers as implantable dental and orthopedic biomaterials, most investigators focused on exploring means to increase the polymer modulus to match or approach those of pertinent bony tissues for optimum biomechanical compatibility. Most of the early studies dealt with solid-state orientation in the tensile mode.[5–7] With the introduction of orthogonal solid-state orientation (OSSO) by Shalaby and coworkers, the application of exceptionally high forces to achieve maximum orientation became possible.[8] However, with the exception of random efforts to modify polymeric surfaces through cold

Concise Encyclopedia of Biomedical Polymers and Polymeric Biomaterials DOI: 10.1081/E-EBPPC-120052219

plasma oxidation radiation grafting of unsaturated monomers and introduction of bioactive molecules of hydrophilic polymers, early efforts on physicochemical surface modification of polymeric implants to optimize their biomechanical and biochemical compatibility were limited.[9–14] More specifically, earlier approaches to the chemical modification of polymeric implant surfaces dealt mostly with blood-contacting surfaces. These and related modifications were addressed by a number of authors and dealt primarily with increasing the surface hydrophilicity of hydrophobic polymeric substrates and entailed 1) surface oxidation using gas plasma; 2) grafting of hydrophilic monomers; and 3) covalently immobilizing hydrophilic or water-soluble polymers on the specific surface.[11–14] In spite of recent emphasis on the implant–tissue interfaces and their key relevance to optimal performance of dental and orthopedic implants, efforts of contemporary investigators in this area were minimal. This led Shalaby and coworkers to pursue part of the studies described in this entry on the physicochemical surface modification of model polymers, such as polyethylene (PE) and polypropylene (PP), as well as high modulus implantable materials, such as polyether ether ketone and its carbon fiber–reinforced composites.[15,16]

Surface Sulfonation of Model Polyolefins

Since its early use in polymer science and technology, PE has been used as a model polymer to investigate new approaches to surface and bulk modification of other polymers because low density polyethylene (LDPE) has the simplest possible chemical structure of all known polymers. Furthermore, the ultrahigh molecular version of PE, that is UHMW-PE, is a key biomaterial used in artificial joints; surface modification is an important aspect as well. On the other hand, PP, which is still a simple polymer, is more reactive than LDPE and can be used as another model polymer for those having methyl side groups. In addition, isotactic polypropylene (i-PP) is currently used in several biomedical implants and most notably in surgical sutures.

LDPE and i-PP as model polymers for studying surface modification; however, these polymers are hydrophobic. Many investigators have attempted to introduce polar moieties, such as carboxylic and sulfonic groups, on the polymer surface to determine their effect on the surface hydrophilicity and the ability to modulate their interaction with the surrounding tissue as part of an implant. For this, introduction of selected types of carboxylic groups is being investigated at Poly-Med, Inc. (Anderson, SC), while efforts at introducing more polar groups, such as sulfonic and phosphonic, are addressed in this entry.

A sulfonation protocol employing fuming sulfuric acid developed at Poly-Med, Inc. (Anderson, South Carolina) was used to surface sulfonate thin films of LDPE and i-PP to render their surfaces more hydrophilic and capable of displaying a negative charge.[17] This study was designed to assess the effect of sulfonation on blood compatibility of

LDPE and i-PP as model surfaces.[17,18] More specifically, the study was designed to determine how changes in surface chemistry of a model implant influence the degree of conformational change of adsorbing proteins and to investigate the correlation between this change and platelet response. Results of the first segment of the study showed that both LDPE and i-PP became more wettable with water in terms of dynamic contact angle measurements—the contact angle of LDPE and i-PP changed from about 88° to 47° and from about 82° to about 52°, respectively.[17] In the study, sulfonated film (LDPE and i-PP) surfaces were treated with poly-D-lysine (PDL) to achieve a positively charged surface resulting from ionically immobilized PDL. To determine the effect of surface modification on the conformation of adsorbed proteins, we used porcine serum albumin and porcine fibrinogen as model proteins. In effect, thin films of LDPE and i-PP were surface modified using sulfonation and immobilized (or preadsorbed) PDL to create a range of surface chemistries. Circular dichroism (CD) studies were then conducted to assess how each surface influenced the secondary structure of adsorbed albumin and fibrinogen as a measure of adsorption-induced conformational changes, and platelet adhesion studies were conducted to investigate how the degree of structural change in the adsorbed proteins influenced the platelet response. From the results of these studies, it was concluded that platelet adhesion to surfaces with a preadsorbed layer of either albumin or fibrinogen is directly related to the degree of adsorption-induced structural change to the protein. It was further concluded that the ability of albumin to serve as a passivation layer to resist platelet adhesion is not a universal property of albumin, but rather is related to albumin's inherent resistance to adsorption-induced structural changes Although controversial, it was suggested that for the same degree of conformational change, nonactivated platelets will adhere to adsorbed albumin as readily as to adsorbed fibrinogen. And collectively, it was concluded that surface treatment of LDPE and i-PP by sulfonation and preadsorption of PDL is an effective means of reducing the degree of surface-induced conformational change of adsorbed albumin and fibrinogen and subsequently to reduce albumin and fibrinogen-mediated platelet adhesion.

The results of this study also suggest that the use of surface chemistry to influence protein adsorption in a manner that minimizes the degree of adsorption-induced structural change may be one of the most important principles for the design of blood-compatible biomaterials to minimize platelet adhesion, subsequent platelet activation, and thrombus formation on biomaterial surfaces.

Surface Phosphonylation of Model Polyolefins and Polyether-Ether Ketone (PEEK)

Treatment of saturated hydrocarbon with phosphorous trichloride in the presence of oxygen was shown to yield alkyl phosphonyl chloride, which can be hydrolyzed to

Biomimetic—Colloid

produce the corresponding alkyl phosphonic acid according to the following reaction scheme:

$$R–H + 2PCl_3 + O_2 \rightarrow R–POCl_2 + POCl_3 + HCl$$

Alkyl phosphonyl chloride

$$R–POCl_2 + 2H_2O \rightarrow R–PO(OH)_2 + 2HCl$$

Alkyl phosphonic acid

The general reaction scheme was later adopted by Shalaby and coworkers to develop new methods for surface activation of preformed polymeric articles, including those made of LDPE, i-PP, and on PEEK.[15,16] The surface treatment scheme was devised as a liquid phase reaction in which a thermoplastic polymer was suspended in a phosphorus trichloride solution in an inert solvent with a continuous flow of oxygen through the system.[16] A gas phase process was also developed in which polymers were suspended directly over a reservoir of phosphorus trichloride in a static oxygen environment.[15] Both schemes led to the introduction of phosphonyl chloride groups, which were subsequently hydrolyzed to phosphonic acid groups at the LDPE, i-PP, and PEEK surfaces, without imparting discernable changes in the bulk properties of these polymers. The phosphonic acid–bearing surfaces were then reacted with calcium ion–containing solutions to produce calcium phosphonate moieties at the polymeric surfaces.[16–20] Such surfaces approximate bioglass and hydroxyapatite, which were recognized earlier for their bone-binding properties.[21,22] Formation of calcium phosphonate as bound moieties PEEK and other polymers of orthopedic significance created the interesting possibility of direct fixation of orthopedic devices to bone without the need for an intermediate grouting material.

Recognizing the simplicity of the gas phase surface phosphonylation of the thermoplastics led to the pursuit of a new study on a modified version.[2,3] In this particular study, surface phosphonylation of LDPE was conducted, and the phosphonylation process was modified using a two-chamber reactor and a dynamic oxygen flow in an effort to secure greater control of the gas phase reaction.[23] Using such a reaction scheme, the study was designed to determine the effect of the physicochemical properties of LDPE surface, as a model for orthopedic materials on bone binding and apposition to phosphonylated surfaces. And, to determine the effect of phosphonylation time and temperature on the surface properties, low density PE films were phosphonylated at both ambient and elevated temperatures for periods ranging from 15 to 60 min. After hydrolyzing the phosphonyl chloride groups, the films were analyzed by scanning electron microscope (SEM), electron dispersive X-ray (EDX) analysis, horizontal ATR-FTIR, surface roughness, and dynamic contact angle measurements. The experimental numerical data are summarized in Table 1.

Horizontal ATR-FTIR (using a Paragon 1000 spectrophotometer) spectra of representative control and films phosphonylated for 15 min were identical and provided no evidence of phosphonylation. For all other groups, the spectra indicated surface modification with the same characteristic group frequencies present in samples treated at 25°C for 30 and 60 min, or 45°C for 60 min. The P–O–H group exhibited four characteristic frequencies at 2525–2725, 2080–2350, 1600–1740, and 917–1040 cm[1]. The first of these frequencies appeared as a slight shoulder on the spectra, while the last three were readily apparent. Coupled with the identification of phosphorus at the surface, these spectra provided conclusive evidence that phosphonic acid moieties exist on the surfaces of films phosphonylated for more than 15 min.

EDX spectra of the control and treated surfaces indicated the absence of phosphorus and chlorine on the control and 15-min treatment groups, while it was present on all other surfaces. More revealing than the EDX spectra is the semiquantitative analysis computed from five such spectra for each treatment group as shown in Table 1. For all cases, the amount of chlorine present is quite low—only a fraction of a percentage—indicating that samples were adequately hydrolyzed following phosphonylation.

After 15 min of phosphonylation time, the surface does not show phosphorus above the amount detected in the control film, again an indication that the LDPE film does not phosphonylate to a detectable degree in this period.

Table 1   Surface analysis data of phosphonylated LDPE

Reaction		Elemental analysis[a]		Roughness[b]		Contact angle (in.)	
Time (min)	Temp (°C)	%P	%Cl	R_q (nm)	R_z (μm)	Water	CH_2I_2
0	–	0.20	0.04	75.59	1.40	96.81	69.32
15	25	0.26	0.04	78.02	1.23	93.49	78.45
30	25	9.59	0.22	127.19	1.90	52.19	67.15
60	25	6.13	0.27	130.43	1.81	56.36	66.47
60	45	13.64	0.35	201.42	2.81	50.70	64.02

[a]Using JEOL JSM-1c 848 electron microscope (JEOL, Peabody, Massachusetts) equipped for electron dispersive X-ray (EDX) analysis.
[b]Using WYKO NT 2000 Profilometer (Veeco Corp., Tuscon, Arizona).

After a 30-min treatment, the film surface is composed of approximately 10% phosphorus. Polyphosphonic acids obtained after hydrolysis are highly hydrophilic, and previous research has shown that these moieties dissolve in water at phosphorus contents in excess of 10%. Therefore, it is no surprise that after a 60-min treatment time the phosphorus content decreases to 6%; isolated surface hydrocarbon chains are excessively phosphonylated and pulled from the bulk during hydrolysis. Solubilization of over-phosphonylated surface molecules upon hydrolysis is consistent with pitting of this film group seen using SEM. Finally, the highest phosphorus incorporation, i.e., 13%, is achieved after 60 min phosphonylation at 45°C, indicating that the material is phosphonylated well below the surface without the concomitant solubilization of overphosphonylated surface molecules.

Surface roughness measurements data are summarized in Table 1. The film treated for 15 min shows essentially the same degree of roughness as the control, reflecting that practically no phosphonylation has taken place. For the 30- and 60-min phosphonylation at room temperature, the surface roughness values were similar to each other and notably rougher than the control. Phosphonylation did appear to cause a physical change in the polymer surface, as also noted from the SEMs in addition to the known chemical modification. Also, while excessively phosphonylated molecules of the 60-min treatment group delaminate from the surface, the overall surface roughness did not change substantially because of this process. Finally, for 60 min treatment at 45°C, both the average roughness and peak-to-valley height were substantially greater than those of other treatment groups. This is consistent with the thesis that the treatment affects the subsurface and is not limited to the surface.

Dynamic contact angle measurements were considered in conjunction with the physicochemical differences of the surfaces described; the contact angles values are summarized in Table 1. As expected, water proved to be a superior probe liquid for discriminating between differences in the treated and untreated surfaces, while methylene iodide (CH_2I_2) was none too revealing. This is because phosphonylation renders the polymer surface hydrophilic, and water, a polar liquid, is quite responsive to this change, while methylene iodide, a nonpolar liquid, is not. Therefore, treatments are discussed in terms of their advancing dynamic contact angle in water.

Results of the Allan et al. study led to the conclusion that using a two-chamber dynamic flow system, phosphonylation of LDPE can be regulated to control the extent and uniformity of surface modification.[23] Under the prevailing reaction conditions, LDPE films do not phosphonylate to a detectable degree at 15 min. Within 30 min, the surface undergoes phosphonylation, and at 60 min, the surface becomes overphosphonylated at localized sites with concomitant surface pitting being observed. Phosphonylation at 45°C results in phosphonylation at the surface as well as

the subsurface. Overall, gas phase phosphonylation of LDPE was noted as being best regulated at 25°C, with reaction times greater than 15 min but less than 60 min providing the most uniform surface treatment.[23]

For preparing surface phosphonylated i-PP films, a protocol similar to the one described for LDPE was used, but the reaction time was limited to 15 and 60 min periods at 45°C. The hydrolysis of the phosphonyl chloride groups of the phosphonylated surface was achieved by sonicating the films in distilled water for 30 min using a Branson Model 3210 Ultrasonic Cleaner (Branson Ultrasonic Corp., Danbury, CT). The films were characterized for surface composition, surface roughness, and contact angle as described previously for LDPE films. Although most of the results were generally similar to those of LDPE, the i-PP films revealed: 1) measurable phosphonylation for the 15-min reaction time; 2) a slightly higher level of surface phosphonylation; 3) a slightly more noticeable decrease in contact angle; and 4) a slightly more pronounced surface roughness or microtexturing.[23]

Attempts to surface phosphonylate PEEK or carbon fiber-reinforced (CFR) PEEK were first made using the protocols applied successfully to LDPE and i-PP. However, it was observed that the aromatic PEEK-based surfaces are far less reactive than their aliphatic counterparts, especially in the case of the gas phase method. Accordingly, a study on the surface phosphonylation of PEEK and CFR-PEEK was conducted on practically unoriented thin films, using the liquid phase method and carbon tetrachloride (CCl_4) as a medium to maximize the effectiveness of the phosphonylation reaction. The phosphonyl chloride groups of the phosphonylated films were hydrolyzed. The films were dried and subjected to the same analytical methods employed for LDPE and i-PP counterparts with the exception of using electron spectroscopy for chemical analysis (ESCA) in addition to EDX for surface elemental analysis. Analytical data obtained using the different methods indicated that phosphonylation occurs mostly at the uppermost layers of the PEEK-based films, while in the polyolefin films, the reaction proceeds well below the surface. This was associated with the finding that the surfaces of PEEK-based films have much lower phosphorus contents than their polyolefin-based counterparts. When EDX was used, it showed comparable and sometimes lower phosphorus concentrations than when the less penetrating ESCA was used. Accordingly, phosphonylation of PEEK-based films appears to be affected by their lower chemical reactivity and high glass transition temperature (T_g) as aromatic substrates.

Surface Microtexturing of Model Polyolefins and PEEK-Based Substrates

The term "surface microtexturing" was initially used relative to improving tissue regeneration and short and long-term mechanical stability of the soft tissue implants. On the other hand, for high modulus dental and orthopedic implants, the term microroughness or microroughening was used as the equivalent of the term surface microtexture, or microtexturing. Contemporary orthopedic and dental investigators, while acknowledging the importance of an implant's chemical inertness, emphasize the importance of bioactivity and surface roughness of such an implant relative to its short and long-term mechanical stability at a bony site.[24,25] Meanwhile, practically all efforts to impart the required surface microroughness or higher level of roughness focused on ceramic and metallic implants to improve their osseointegration with surrounding bony tissue.[25,26] And numerous surface modification schemes have been developed and are currently used to enhance clinical performance of ceramic and metallic implants. Such modifications dealt primarily with blasted, acid-etched, porous sintered, oxidized, plasma-sprayed, and hydroxyapatite-coated surfaces.[26] However, until Shalaby and coworkers developed a process denoted as crystallization-induced microphase separation (CIMS) to produce microporous foam with a continuous cellular structure and an extension thereof, surface microtexturing of crystalline thermoplastic polymers was virtually unknown for dental and orthopedic implants.[27,28] The CIMS process entails

1. Dissolving the crystalline polymer in the melt of a crystalline, low-melting organic compound or diluent to form a one-phase liquid system of the polymer and diluent
2. Quick-quenching the polymer–diluent system to form two bicontinuous crystalline microphases of the polymer and diluent
3. Removing the diluent from the two-phase system by sublimation below the melting temperature of the polymer or extracting the diluent with a solvent that does not dissolve the polymer phase

This process results in a microporous foam with continuous cell structure. The average size of the pores can be modulated primarily by controlling the polymer-diluent ratio, wherein larger pores can be achieved by decreasing this ratio. The CIMS process has been applied successfully for the production of PE and i-PP microporous foams.[29] On the other hand, limiting the CIMS process to the surface of a crystalline polymer results in the formation of a microporous surface that is, in effect, a microtextured surface.[27,28] Application of the CIMS process for surface microtexturing of a preformed implant made of crystalline thermoplastic polymer entails

1. Heating the selected diluent to liquefy and acquire a temperature below the polymer melting temperature (T_m)
2. Dipping the preformed polymeric article or implant in the molten diluent for a specific period to attain a transient, high viscosity, one-phase solution of the polymer with the diluent at the surface

3. Removing the treated implant and quick-quenching it to form a thin, bicontinuous solid coating that is molecularly intermixed with the surface
4. Extracting or subliming the diluent as described for the production of microporous foam

Forming such a coating on the implant surface that yields mostly a microtextured surface that may have a microporous subsurface can be achieved by controlling the: 1) type of diluent in which the polymer exhibits limited solubility; 2) temperature of the molten diluent; and 3) contact time of the polymer with the diluent.

Early attempts in the study of microtexturing dealt with the preparation of microtextured films of i-PP and LDPE using molten naphthalene as a diluent.[27,28] Removal of the naphthalene microphase from the bicontinuous system was achieved by sublimation. Further purification of the microtextured films can be accomplished by extracting residual naphthalene by a solvent such as hexane or methylene chloride. However, surface microtexturing of thin films of PEEK and CFR-PEEK using the CIMS process proved to be more demanding, and the conditions applied successfully for microtexturing polyolefin films were modified substantially in terms of the type of diluent used, reaction temperature, polymer–diluent contact time, and the cooling rate of the treated films. More specifically, the modified CIMS process required the use of: 1) high melting diluents for application at temperatures near the T_g of PEEK; and 2) more polar diluent than naphthalene with higher solubilizing effect relative to PEEK. Specific conditions for microtexturing LDPE, i-PP, PEEK, and CFR-PEEK as implantable devices are outlined in Section "Osseointegration of Phosphonylated Polyolefins and Peek Rods as Tibial Implants" in conjunction with the preparation and evaluation of surface-phosphonylated rods.

OSSO of Crystalline Thermoplastic Polymers

The concept of OSSO of crystalline thermoplastic polymers was first disclosed by Shalaby and coworkers.[30] One aspect of the OSSO process deals with the following steps.[30,31]

1. Placing a sheet or block of practically unoriented crystalline polymer in a U-shaped mold with an adjustable frame to allow precise contact of the polymer in the three sides of the mold
2. Heating the polymer above its T_g but below its T_m to allow for controlled deformation under pressure
3. Applying a compressive force perpendicular to the top surface through a movable plate that fits precisely within the horizontal space of the U-shaped mold
4. Allowing the polymer to deform uniaxially to exit through the open end of the U-shaped mold. The OSSO process was used successfully to produce a number of oriented forms of thermoplastic polymers, such as UHMW-PE and PEEK

Most relevant to the subject of this entry is the study of tailoring the mechanical properties of PEEK implants using the OSSO process.[31,32] In this study, PEEK samples were subjected to the OSSO process at different temperatures. Properties of oriented PEEK were compared with those of unoriented PEEK and carbon fiber–reinforced PEEK (CFR-PEEK). For processing and testing PEEK, dried, molding grade resin (PEEK 450PF, Victrex USA, Inc.) was melt processed into $76 \times 76 \times 32$ mm blocks using a Carver hot press (Model 3895) at 370°C and 9100 kg with a load bearing area of 152×152 mm. The blocks were cut into six bars measuring $76 \times 12.7 \times 32$ mm for orientation. For surgical implants, a solid-state orientation was accomplished at 315°C and 13,600 kg with a final load bearing area of 152×152 mm in a custom designed mold used to achieve uniaxial orientation by applying a compressive force at 90° to the sample surface. The thickness of the oriented material was 10 mm, for a final compression ratio of approximately 3:1. Samples of molded and oriented PEEK, prepared as described above, were machined into rods with a diameter of 5 mm and a length of 10 mm using a JET-1240PD lathe. For comparison purposes, two similar sets of rods were prepared using molded PEEK and CFR-PEEK without implementing the OSSO process. Both sets of rods were used for implantation studies and evaluation of compressive properties. For evaluation of flexural properties, three-point bend samples were prepared as rectangular specimens measuring approximately $28 \times 6 \times 1$ mm. All mechanical testing was accomplished using an MTS 858 MiniBionix universal testing apparatus. Compression tests were conducted at a displacement rate of 1 mm/min. Three-point bend tests were conducted at a rate of 0.54 mm/min with a span of 18 mm. For each test method, force versus displacement curves were captured, and the yield stress and modulus of the materials were calculated.

Results of oriented PEEK, unoriented PEEK, and CFR-PEEK specimens tested for flexible strength and modulus are depicted in Figs. 1 and 2, respectively, and Table 2. PEEK's yield strength and modulus, as determined by compression and three-point bend, were improved by solid state orientation. Table 2 shows compression test results of oriented PEEK and CFR-PEEK rods prepared according to the scheme just described. Values are presented as averages

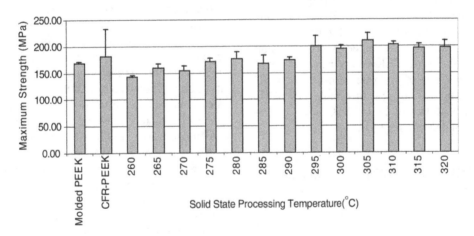

Fig. 1 Maximum strength of oriented PEEK compared to molded and CFR-PEEK using three-point bend.

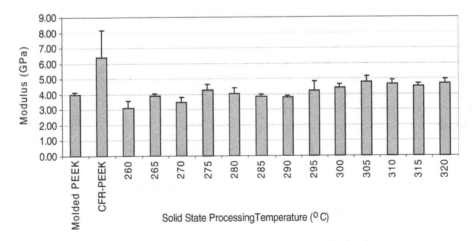

Fig. 2 Modulus of oriented PEEK compared to molded and CFR-PEEK using three-point bend.

Table 2 Compressive properties of typical PEEK-based rods

	Compressive properties	
Sample	Yield strength (MPa)	Modulus (GPa)
Unoriented PEEK	114.2 ± 9.82	2.24 ± 0.16
Oriented PEEK	128.0 ± 4.21	2.77 ± 0.18
Unoriented CFR-PEEK	124.4 ± 21.2	3.20 ± 0.61

with corresponding standard deviations. PEEK orientation at 315°C and a compression ratio of 3 to 1 resulted in an increase in yield strength and modulus of 15 and 24%, respectively, over the molded control. Further, the yield strength of the oriented PEEK was approximately equal to that of CFR-PEEK.

Figures 1 and 2 show results of three-point bend studies conducted on PEEK oriented over a range of temperatures and a compression ratio of 5 to 1. Values on the graphs are shown as averages with corresponding standard deviations. Solid-state orientation of PEEK between 295 and 320°C resulted in increased strength and modulus over the molded PEEK control. Over this range, the strength of the oriented PEEK was equal to or greater than the CFR-PEEK.

The effect of the OSSO process on PEEK T_m as determined by DSC was an increase of 4°C. The results, summarized in Table 2 and Figs. 1 and 2, led to the conclusion that solid state orientation of PEEK results in an increase in the strength and modulus of the material over the unoriented PEEK control. The strength of oriented PEEK is approximately equal to that of CFR-PEEK; however, the modulus is less than that of CFR-PEEK.

INTERACTION OF CELLS WITH CHEMICALLY MODIFIED SURFACES

Interactions of biomaterial surfaces with different types of cells in the biological environment determine, to a great extent, the successful short and long-term use of polymeric implants. The long-term use of polymeric biomaterials in blood is limited by surface-induced thrombosis and biomaterial-associated infections.[14,33–35] Meanwhile, positive interactions and attachment of osteoblasts to biomaterial surfaces are key to the successful application of various dental and orthopedic implants and their osseointegration with surrounding bony tissues.[35–37] In this section, however, discussion of cell interactions with biomaterial surfaces is limited to those related to infection and osseointegration.

Interaction of Bacteria

Biomaterial-associated infection is known to occur as a result of adhesion of bacteria onto the surface.[38] And it is well acknowledged that the biomaterial surface provides a site for bacterial attachment and proliferation.[14] Adherent bacteria can be covered with a biofilm that supports bacterial growth and provides protection against phagocytes and antibiotics.[14,39] Acknowledging the relationship between bacterial cell attachment and device-induced infection, Gerdes and coworkers evoked the strategy of reducing or eliminating the cell attachment as an effective means of reducing the incidence of bacterial infections.[40] To this end, these investigators: 1) adopted the findings of an earlier study where polymers with negatively charged surfaces allowed lower bacterial adherence as compared with chargeless controls; and 2) applied the technology of gas phase phosphonylation discussed in Section "Technology Evolution of Physicochemical Surface Modification and Bulk Orientation" to impart negatively charged functional groups on LDPE as a model implant material.[40,41] It was also suggested that: 1) in addition to the effect of the negative charge, the phosphonylated surface is more hydrophilic than the LDPE control, which rendered the surface unfavorable for bacterial colonization; and 2) phosphonylation-induced microtexturing may increase the charge density and decrease cell adherence further.

In this study, LDPE films were phosphonylated to different degrees in the gas phase by treating the LDPE for 30, 45, and 60 min. Following phosphonylation the samples were sonicated in distilled water for a minimum of 1 hr and dried under reduced pressure. SEM and electron EDX analysis confirmed the presence of phosphorus as well as its relative amounts attached to the surface. Contact angle measurements reconfirmed surface changes with varying intensities of the treatment. Controls included untreated and amine-bearing LDPE. For the latter, the amine-bearing surface was created by phosphonylation of LDPE followed by a 5-hr incubation in a 5% solution of hexane diamine; the final surface exhibited a positive charge. LDPE was treated using the CIMS method at 92°C at intervals from 25 to 300 sec. Following the CIMS treatment, samples were quenched in ice water for a minimum of 2 min. After air drying, the samples were sonicated in toluene for 30 min. Samples were dried under reduced pressure. SEM and noncontacting profilometry was conducted to analyze these surfaces.

Staphylococcus epidermidis, the most common pathogen in device-induced infections, was used as a model bacterium for studying cell attachment and was first characterized for growth. A growth curve was created in tryptic soy broth (TSB) at 37°C. Antibiotic resistance was evaluated to help facilitate the use of the organism in the planned testing.

Results of the study are summarized in Tables 3 and 4. SEM and EDX confirmed the presence of phosphorus on the treated surfaces. The presence of hydrophilic phosphonate groups was verified by contact angle measurements. The gas phase phosphonylation process imparted microtexturing on the surface bearing the functional groups. EDX also showed relative decreases in the carbon

Table 3 Data from EDX and contact angle analysis

Sample treatment	C/P ratio from EDX	Contact angle
LDPE	–	84.46 ± 1.04
LDPE + HD	–	89.88 ± 1.57
LDPE + 30 m P	11.92	49.58 ± 3.48
LDPE + 45 m P	10.21	45.75 ± 3.24
LDPE + 60 m P	4.21	19.83 ± 2.37

Table 4 Data from profilometry analysis

Sample	Roughness average (nm)
LDPE	105.90 ± 7.40
LDPE + 25 s CIMS	161.65 ± 32.71
LDPE + 50 s CIMS	163.30 ± 43.60
LDPE + 160 s CIMS	5688.71 ± 682.32
LDPE + 30 m P	204.49 ± 68.06
LDPE + 45 m P	219.26 ± 9.53
LDPE + 60 m P	297.73 ± 31.27

to phosphorus ratio (C/P) with increases in treatment time. The contact angle data of the phosphonylated surface confirmed increases in hydrophilicity. However, the amine bearing surfaces displayed a slight increase in surface hydrophobicity as compared with the control surface. The examined CIMS samples exhibited microtexture for all periods (25, 50, and 160 sec) used for the treatment. Noncontacting profilometry showed an increase in the roughness average with an increase in CIMS treatment time and phosphonylation reaction time. During the course of this study, the need for reliable methods to assess and quantitate the level of cell attachment was evident. In a subsequent report, the different methods were critically evaluated for their effectiveness in studying the attachment of *Staphylococcus epidermides* on surface activated LDPE, which may be also applicable to similar systems.[42] The methods used and critically assessed were colony count analysis, SEM analysis, dye-elution technique, total DNA isolation, and total protein quantification. Comparative assessment of these methods led to the conclusion that colony counts and SEM are useful methods of monitoring cell attachment under the conditions used in this study, colony counting being the most effective. Colony counts gave repeatable results and enabled the study to be quantitative. The SEM qualitative analysis was helpful and may have been more informative as an extensive study. The dye-elution technique, total DNA, and surface protein methods of quantitating bacterial cells were ineffective in monitoring cell attachment under the conditions used in this study. Because of the active surfaces, both cationogenic and anionogenic, dyeing the cells without binding the dye to the surface was not possible. Total DNA was an ineffective method because of the difficulty in lysing Gram positive species. Although

newer protocols claiming to be capable of penetrating the cell wall were used, no DNA was isolated.[42] The study to quantitate proteins on the cells' surfaces was not sensitive enough to distinguish between ten-fold changes in cell numbers. This was found to be an unacceptable method of quantitating cell numbers.

Interaction of Osteoblasts

It has been shown by Shalaby and coworkers that surface phosphonylated surfaces of LDPE, *i*-PP, and PEEK-based bone implants do encourage bone ingrowth, resulting in osseointegration with surrounding bony tissues, as discussed in more detail in Section "Osseointegration of Phosphonylated Polyolefins and Peek Rods as Tibial Implants."[43–50] This led to the postulate that surfaces capable of immobilizing and chelating calcium ions, as in the case of phosphonylate-bearing surfaces, provide a preferred active substrate for the attachment and proliferation of osteoblasts.[46,47] In an effort to test this hypothesis and demonstrate the relevance of previously noted osseointegration of different implants to osteoblast attachment, a study was conducted on osteoblast attachment to phosphonylated *i*-PP film as a model substrate for typical orthopedic and dental polymeric substrates.[51] In this study, attachment, proliferation, and differentiation of osteoblast to phosphonate-bearing *i*-PP film, pretreated with calcium hydroxide to immobilize Ca^{2+}, was explored. The attachment of osteoblasts and their presence on the activated surface was verified using alkaline phosphatase activity assays and alizarin red staining. Results showed clearly that phosphonylated surfaces do encourage osteoblast attachment, thus providing a preliminary verification of the aforementioned hypothesis. This may also have impact on the growing interest in the in situ tissue engineering, where bioactive scaffolds are placed in defective biological sites to recruit pertinent cells to initiate tissue regeneration within and about the scaffolds. In an extension of this study on phosphonylated *i*-PP, a novel form of surface activated *i*-PP and PEEK-based films were examined for osteoblast attachment.[52] More specifically, the surface activation entailed the carboxylation of these films to produce a special form of carboxylic moieties on the surface, which are capable of binding and/or chelating calcium.[52–54] Osteoblast attachment to these surfaces was measured in terms of cell viability. For conducting the study, human fetal osteoblasts (ATCC, Manassas, VA) were maintained under recommended culture conditions in a 33.5°C, humidified, 5% CO_2–95% air environment in a 1:1 mixture of Dulbecco's modified Eagle's medium and Ham's nutrient mixture F12 supplemented with 10% fetal bovine serum (ATCC), 15 mm HEPES, and 0.3 mg/mL G418 (Invitrogen, Carlsbad, CA). PP and CFR- PEEK films of less than 1 mm thickness were prepared using a heated, automatic hydraulic press (Carver, Wabash, IN) and then surface carboxylated to introduce a special form of carboxylic acid side groups or

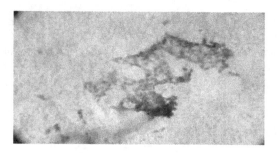

Fig. 3 Control polypropylene.

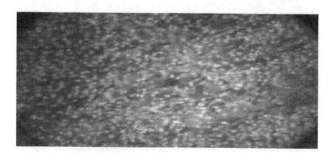

Fig. 5 Control CFR-PEEK.

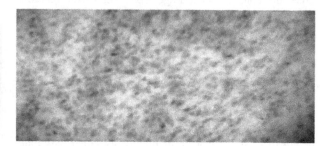

Fig. 4 Carboxylated polypropylene.

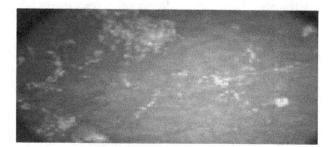

Fig. 6 Carboxylated CFR-PEEK.

left untreated. The individual films were cut into 1 cm² pieces and sterilized by ultraviolet irradiation for 20 min. Prior to cell seeding, films were fixed to the bottom of tissue-culture wells with a small amount of sterile silicone grease and soaked in media for 2 hr. The osteoblasts were seeded onto films at a density of 1.3×10^4 cells/cm². After 7 days of culture, films were transferred to a new tissue-culture plate and visualized with propidium iodide or alizarin red staining. As part of the preliminary outcome of this study, direct microscopic examination of alizarin red–stained specimens are illustrated in Figs. 3 and 4. Compared to untreated PP controls (Fig. 3), there was enhanced osteoblast attachment and proliferation on carboxylated PP films (Fig. 4). Specimens stained with propidium iodide and viewed with fluorescence microscopy revealed that there was osteoblast adhesion on both control and carboxylated CFR-PEEK films (Figs. 5 and 6), but adhesion was hardly enhanced on carboxylated films. Available results of this preliminary study suggest that specially carboxylated surfaces, at least for i-PP films, encourage the attachment and proliferation of osteoblasts.

OSSEOINTEGRATION OF PHOSPHONYLATED POLYOLEFINS AND PEEK RODS AS TIBIAL IMPLANTS

Discussion in the previous sections dealt with demonstrating:

1. The viability of LDPE and i-PP as model substrates for achieving phosphonylation
2. The successful extension of the phosphonylation technology to PEEK and CFR-PEEK

3. The ability to increase the modulus of crystalline polymers of interest in dental and orthopedic applications using the OSSO process so as to approach the moduli of typical bony tissues
4. The ability to achieve microtexturing of key polymers

A logical follow-up, this section deals with the use of simple implants, such as rods, to study their osseointegration in simple animal models, such as the rabbit and goat tibial models.

Osseointegration of Chemically and Physicochemically Modified PE and Propylene Rods as Goat Tibial Implants

In general, the sterilized implants were prepared by: 1) melt extrusion of LDPE and i-PP into rods; 2) surface activation with or without surface microtexturing; and 3) sterilization.[43] More specifically, for the test implants, i-PP and LDPE were extruded using a Randcastle Microtruder laboratory extruder. Extruded polymers were subjected to the types of treatments outlined in Table 5. Implants were phosphonylated in the gas phase and hydrolyzed by sonicating in distilled water for 30 min. Random sets were selected and treated with calcium hydroxide. Rods with microtextured surfaces were produced by immersion into a molten diluent at a temperature between their melting and glass transition temperatures. The i-PP implants were sterilized using a General Purpose Model 2120 autoclave; the LDPE implants were sterilized via ethylene oxide gas at 25°C for 24 hr. Sterilized implants were characterized using SEM, EDX, and surface roughness measurements.

Bone apposition and binding to the implant surface were evaluated using a transcortical plug model in the goat.[43,55] Four rods were implanted through the medial cortex of both tibias in nine goats to give a total of eight implants per surface treatment. The tibial positioning of implants in each group was randomized. *i*-PP and LDPE implants were harvested at 6 and 8 weeks postimplantation, respectively. Six implants per treatment group were used for mechanical tests, and the remaining two were used for histological evaluation. Mechanical push out tests were accomplished using an MTS 858 MiniBionix universal testing apparatus at a displacement rate of 1 mm/min. Samples for histological evaluation were prepared using standard hard tissue techniques and stained with basic fuschin and methylene blue.

In general, results of this study show that *i*-PP and LDPE rods were successfully extruded and surface treated. The EDX spectra of the materials indicated the presence of phosphorus and calcium on all phosphonylated and calcium hydroxide treated implants, respectively. Microtexturing of *i*-PP and LDPE produced the expected irregular surface topography with porosities on the order of 10–100 μm. Microtextured implants exhibited greater surface roughness values than smooth implants. Results of mechanical tests are shown in Figs. 7 and 8. Histological evaluation indicated no significant differences in bone apposition among groups. The biomechanical testing results and histological evaluation data led to the conclusion that formation of a calcium phosphonate surface on thermoplastic polymers through phosphonylation and subsequent treatment with calcium hydroxide resulted in an increased propensity for bone binding in both the *i*-PP and LDPE sets of implants.

Osseointegration of Phosphonylated PEEK Rods as Rabbit Tibial Implants

Successful results of the study on the osseointegration of phosphonylated LDPE and *i*-PP rods using a goat tibial model prompted the pursuit of a limited study on the osseointegration of phosphonylated PEEK rods as rabbit tibial implants.[47] The selection of PEEK as a potential substitute for metallic dental and orthopedic implants was justified following the successful use of the OSSO process for increasing the modulus of the polymer to approach those of typical bony tissues.[31] For preparing and evaluating the test implants, PEEK pellets were compression molded into 32 mm thick blocks using a 30-ton Carver hydraulic press. The blocks were subsequently oriented using the OSSO process.[31] Once oriented, the blocks were machined into rods 3.00 mm in diameter and approximately 5 mm in length.

PEEK rods were divided into two groups, untreated controls and calcium phosphonate bearing surfaces. Control rods were cleaned after machining by sonication, first in acetone and then in water. Rods in the surface treated group were phosphonylated in the liquid phase and then incubated in a saturated solution of calcium hydroxide. The rods were evaluated using EDX analysis to confirm the presence of phosphorus and calcium on the surface of the treated rods and their absence on the control. SEM was used to visualize the material surfaces. Rods were individually packaged in sterilization bags and autoclaved prior to surgery. Sterilized rods were reexamined by EDX and SEM.

Table 5 Types of treatment of LDPE and *i*-PP implants

Material	Group	Treatment
i-PP	I-A	Control, no treatment
i-PP	I-B	Phosphonylated
i-PP	I-C	Microtextured
i-PP	I-D	Microtextured and phosphonylated
i-PP	I-E	Phosphonylated and calcium treated
LDPE	II-A	Control, no treatment
LDPE	II-B	Phosphonylated and calcium treated
LDPE	II-C	Microtextured
LDPE	II-D	Microtextured, phosphonylated, calcium treated

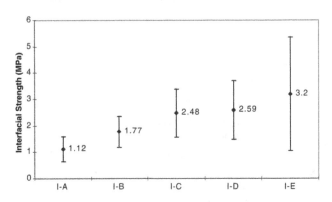

Fig. 7 Strength of the *i*-PP implant–bone interface.

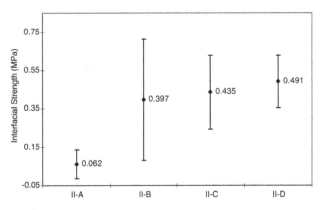

Fig. 8 Strength of the LDPE implant–bone interface.

Bone apposition and binding to the implant surface were evaluated using a transcortical plug model in the rabbit. Rabbits were premedicated and then anesthetized via 2% isofluorane inhalation. The right leg was shaved and scrubbed, and an incision was made along the medial tibia from the tibial crest to the fibula insert. Three holes were drilled into the medial tibia using a 7/64-in. drill bit with an approximate spacing of three diameters between holes. Positioning of the rods in the tibias was randomized. Four rabbits were implanted with three rods each to give a total of six implants per treatment group.

Twelve weeks postimplantation, the rabbits were euthanized, and the implants were harvested with the surrounding bone. The strength of the bone–implant interface was measured via a push out test, which was accomplished using an MTS 858 MiniBionix universal testing apparatus at a displacement rate of 1 mm/min. Explanted rods were stained with a blue marking dye and then examined under a microscope to detect adhering bone fragments.

Results of the push out test are shown in the Fig. 6. The PEEK rods bearing calcium phosphonate surfaces imparted by phosphonylation with calcium hydroxide posttreatment exhibited an interfacial strength with bone that was twice that of the untreated controls. The two groups were compared using a two-tailed, pairwise comparison t-test and were found to be different at a probability level of 99.3%. Observation of explanted rods indicated areas of adhered bone on the calcium phosphonate surfaces. Little to no bone was detected on the control implants. The results of this study led to the conclusion that phosphonylation and calcium hydroxide posttreatment is effective in enhancing the propensity for bone binding and apposition to PEEK tibial implants.

OSSEOINTEGRATION OF CHEMICALLY AND PHYSICOCHEMICALLY MODIFIED PEEK-BASED RODS AS GOAT MANDIBULAR IMPLANTS

Successful results of the study on the osseointegration of phosphonylated PEEK implants as rabbit tibial implants (Section "Osseointegration of Phosphonylated PEEK Rods as Rabbit Tibial Implants") prompted a more detailed study of similar implants in the goat mandible.[32,56]

Preparation of Implantable Sterilized Rods

The PEEK rods were processed by melt extrusion and oriented using the OSSO protocol as described earlier in Section "Osseointegration of Phosphonylated Peek Rods as Rabbit Tibial Implants." On the other hand, CFR-PEEK pellets (RTP 2285LF series, RTP Co.) were melt processed into $76 \times 76 \times 32$ mm blocks using a Carver hot press at 370°C and 9100 kg with a load bearing area of 152×152 mm. The blocks were then cut into 8 mm wide strips for further processing.

For chemical treatment, solid-state oriented PEEK and practically unoriented CFR-PEEK rods were fabricated as described previously with a diameter of 5 mm and a length of 10 mm. Subsequently, the rod surfaces were chemically and morphologically modified as described in Table 6.

To impart the desired surface microtexture and to develop surface microporosity, we treated a fraction of the PEEK and CFR-PEEK cylinders with molten phenylsulfone, rapid cooled them, and then extracted following the general procedure described in Section "Surface Microtexturing of Model Polyolefins and Peek-Based Substrates". Thus, the PEEK and CFR-PEEK rods were immersed in phenyl sulfone at 267°C for 2 min and 305°C for 45 sec, respectively. The rods were then quenched in ice water for 3 min, and extracted (to remove the phenylsulfone phase) by sonicating in toluene for 30 min at room temperature.

To maximize the effectiveness of the phosphonylation process, Allan and coworkers acylated the PEEK and CFR-PEEK with an aliphatic anhydride having a long paraffin chain.[56] This is because the early finding showed that surface phosphonylation of the aliphatic polyolefin was much more effective than in the case of PEEK (Section "Surface Phosphonylation of Model Polyolefins and Polyether-Ether Ketone (PEEK)"). Accordingly, to promote phosphonylation of the PEEK and CFR-PEEK surfaces (both smooth and microtextured), rods marked for phosphonylation were first acylated by immersion in phenyl sulfone in the presence of a catalytic amount of aluminum chloride at 120°C for 2 hours. The samples were then rinsed in a series of steps using distilled water, chloroform, and acetone. Samples were dried under reduced pressure at room temperature for at least 12 hours before further treatment.

Table 6 PEEK and CFR-PEEK implants treatment schemes

Oriented PEEK		CFR-PEEK composite	
Group	Surface treatment	Group	Surface treatment
1	Control, no treatment	5	Control, no treatment
2	Acylated, phosphonylated, calcium post-treatment	6	Acylated, phosphonylated, calcium post-treatment
3	Microtextured using CIMS	7	Microtextured using CIMS
4	Microtextured, acylated, phosphonylated, calcium posttreatment	8	Microtextured, acylated, phosphonylated, calcium posttreatment

Surface phosphonylation was then accomplished as per the liquidphase method (Section "Surface Phosphonylation of Model Polyolefins and Polyether-Ether Ketone (PEEK)"). The rods were rinsed in toluene, and the surface P–Cl groups were hydrolyzed to P–OH groups by sonicating in distilled water. To immobilize calcium ions by surface P–OH groups, the rods were allowed to bind Ca^{2+} through immersion in a saturated solution of calcium oxide in water as described by Campbell.[20] Samples were removed from the solution and dried under reduced pressure at room temperature.

Prior to conducting the *in vivo* evaluation, all implants were sterilized and random samples of were characterized. Accordingly, after all sample preparations and drying were completed, rods were packaged in Chex-all® II Instant Sealing sterilization pouches. Rods were autoclaved prior to surgery. Sterilized implants were characterized using SEM and EDX. Representative sterile rods, which have been surface treated according to the different schemes shown in Table 6, were analyzed by EDX. The EDX data showed the presence of phosphorus and calcium on the surface of these implants.

In Vivo Study of Sterilized Rod Implants Using Goat Mandibles

Animal model and surgical protocol

Bone apposition and binding to the implant surface were evaluated using a recently developed dental implantation model in goats.[45] Sixteen Nubian goats weighing 25–30 kg were procured for use in this study. Prior to surgery, animals were premedicated with glycopyrrolate (0.005–0.01 mg/kg) and buprenex (0.005 mg/kg) via subcutaneous injection. Anesthesia was induced with a mixture of ketamine (3 mg/kg) and xylazine (0.03 mg/kg) intravenously and maintained via inhalation of isofluorane (1.5–3%) in oxygen. A ventral approach was made to the hemimandible, bilaterally just caudal to the last incisor. The subcutaneous tissue and platysma muscle were incised, and the ventral aspect of each hemimandible was isolated. The soft tissues were retracted laterally to expose the mandible caudal to the incisors. On each hemimandible, two holes were drilled in the lateral aspect, avoiding the mental foramen and associated neurovascular bundle. With the gingiva reflected, a 1.9 mm diameter hole was drilled using a 13/64-in. drill bit to a depth of 10 mm. The 5.0 mm diameter cylindrical implants were gently tapped into the prepared holes, and the fascia, subcutaneous, and subcuticular layers were closed using simple interrupted sutures. Eight implants per treatment group were implanted. Buprenex (0.005–0.01 mg/kg) and flunixin meglumine (1.1 mg/kg) were administered via subcutaneous injection every 4–8 hours. and 24–48 hours, respectively, until there were no clinical or behavioral signs of pain.

Harvesting of implants in bone and their biomechanical testing

After 12 weeks healing time, the animals were euthanized via intravenous injection of ketamine (3 mg/kg), xylazine (0.03 mg/kg), and Beuthanasia (1 m/10 lbs. to effect). The implants were harvested along with the surrounding bone. One specimen per goat was immersed in 10% neutral buffered formalin for histological evaluation. The remaining three implants were stored in cups with saline-soaked paper towels for mechanical push out tests. For the biomechanical testing, samples marked for push out tests were prepared by milling the bone on one side to ensure a level support face for testing. Tests were conducted on an MTS 858 MiniBionix universal testing apparatus. Implanted rods were pushed from the bone using the smooth end of a 9/16-in. drill bit at a displacement rate of 0.43 mm/s. Force versus displacement curves were recorded, and the contact area of the implant and bone was measured. The interfacial strength was calculated as the force required to dislodge the rod from the bone divided by the bone–implant contact area.

Histological evaluation

Samples for histological evaluation were fixed in 10% neutral buffered formalin for 1 week and processed using a standard dehydration cycle of alcohols of increasing concentration in a Tissue-Tek VIP. The samples were infiltrated with a proprietary Poly-Med, Inc. methacrylate resin for 1 week under reduced pressure, and then embedded in the resin[56] by curing for 2.0 hours under white light and 6.0 hours under blue light in a Histolux.[56] Embedded samples were sectioned using an Isomet 2000 precision saw and then ground and polished to a thickness of 50 µm using an EXAKT. Finished slides were stained with basic fuschin and methylene blue. Histological slides were evaluated qualitatively and quantitatively to ascertain bone apposition and ingrowth. Images of the histology slides were taken using a Dage-MTI, Inc. 3 CCD camera connected to an Olympus BH-2 microscope at an objective magnification of 2×. The presence of inflammatory cells, fibrous tissue, bone resorption, new bone growth, and other pertinent features of the interface were noted. Image-Pro Plus version 3.0.01.00 for Windows 95/NT was used to measure the length of the implant embedded in bone and the length of bone in direct apposition with the implant. These measurements were expressed as a ratio of implant in contact with bone to the total length of the implant.

Outcome of the *in vivo* study

Implantation surgeries required approximately 30 min for each animal. Within 1 hour of surgery, all goats recovered from anesthesia. All animals healed without significant signs of pain and no incidence of infection. At 12 weeks, the animals were sacrificed and implants were harvested.

Biomimetic—Colloid

The interfacial strength data of the implants are shown in Table 7 and Figs. 9 and 10. Values are presented as averages with corresponding standard deviations. The scatter in the data is attributed to many factors, including straightness of bore upon implantation, variability in healing and bone remodeling rates from animal too animal, and positioning of the specimen in the test grips to ensure central loading of the implant. While care was taken to manage and minimize the variability of each of these and other factors, some deviation occurred. This variability notwithstanding, the trends in mechanical test data are decidedly clear. Groups 2, 6, 7, and 8 showed increases in interfacial strength over their corresponding, untreated controls. Phosphonylation followed by calcium ion immobilization posttreatment resulted in a stronger bond with bone than both PEEK and CFR-PEEK. Further, microtexturing CFR-PEEK also resulted in a stronger interfacial bond. Microtexturing PEEK did not result in an increase in strength, which may be attributed to the difficulty in actually achieving a microtextured surface on the material.

Statistical analysis of the interfacial strength data from the push out test for PEEK and CFR-PEEK dental implants was conducted and results are presented in Table 8. Overall, the phosphonylated PEEK with calcium posttreatment had an interfacial strength greater than that of the untreated control at a probability level of 80% Fig. 9.

All treatment groups of CFR-PEEK showed improvements in interfacial strength over the untreated control at a probability level of 95%.

In the study of bone apposition through histological analysis and interfacial strength measurement, limited correspondence between the two evaluations is common.[57,58] This is due in part to fundamental differences in the methodology of each test. Mechanical evaluation of the bone–implant interface via a push out test assays the entire implant surface as a whole. Conversely, histological analysis provides an image of a singular plane of the interface. In addition, tissue processing and grinding to produce slides for microscopic evaluation can result in implant displacement with respect to the bone, resulting in difficult to interpret or misleading observations regarding tissue interaction with the implanted material. Recognizing this, we selected interfacial strength as determined by mechanical testing as the primary measure of bone ingrowth and apposition in this study. Histology was pursued on a limited basis to explore possible relevance to the mechanical test results. Values obtained from this analysis for fraction of bone in contact with the implant are presented in Table 8 and Fig. 11. Representative histology slides revealed extensive remodeling of the cortex and bone adjacent to the implant. PEEK control slides, group 1, indicated fibrous tissue growth adjacent to the

Table 7   Mechanical and histological evaluation data of rod implants

Implant group	Interfacial strength (MPa)	% Implant in contact with bone
1	0.80 ± 0.70	46.1 ± 14.8
2	2.18 ± 2.07	44.7 ± 16.6
3	0.80 ± 1.24	43.1 ± 17.8
4	0.99 ± 1.23	40.3 ± 4.8
5	0.92 ± 1.23	45.2 ± 8.5
6	4.05 ± 2.71	54.2 ± 18.4
7	7.56 ± 5.59	46.7 ± 21.5
8	7.26 ± 5.67	65.4 ± 24.9

Note: Implant surfaces of groups 2, 4, 6, and 8 contain Ca^{2+} immobilized by the phosphonate functionalities as noted in Table 6.

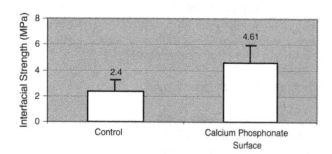

Fig. 9   Interfacial strength comparison of untreated PEEK and calcium phosphonate–bearing PEEK tibial implants.

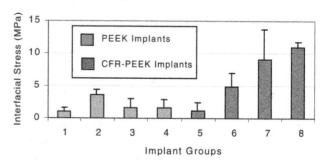

Fig. 10   Implant–bone interfacial strength: the implant surface of groups 2, 4, 6, and 8 contain immobilized Ca^{2+}.

Table 8 Statistical analysis of push-out data

Sample group	Average (MPa)	St. Dev. (MPa)	95% Confidence interval (MPa)	t-Test probability (%)
1	0.80	0.70	0.62	–
2	2.18	2.07	1.82	19.6
3	0.80	1.24	0.99	99.6
4	0.99	1.23	1.12	78.9
5	0.92	1.23	0.98	–
6	4.05	2.71	2.17	2.8
7	7.56	5.59	4.47	1.8
8	7.26	5.67	4.53	2.3

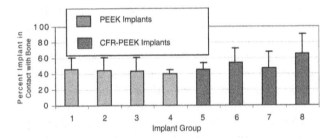

Fig. 11 Histological evaluation of PEEK-based implants.

implant. While group 2, phosphonylated and calcium treated PEEK, also showed extensive remodeling at the bone–implant interface, new bone growth was observed in apposition to isolated portions of the implant surface. CFR-PEEK control implants, group 5, showed a mixed response with both fibrous tissue and new bone growth found at the interface. While groups 6, 7, and 8 of the treated CFR-PEEK materials also exhibited extensive remodeling at the implant–bone interface, new bone was observed in apposition to the implant. In spite of the limited histological findings, there is a correlation between the push out data and bone apposition for groups 6, 7, and 8. Additionally, there is practically no fibrous tissue formation about the surface of the implants corresponding to those groups.

Collectively, the results of the study of PEEK-based implants in the goat mandible led to the conclusion that solid-state orientation of PEEK (using the OSSO process) results in an increase in the strength and modulus of the material over the unoriented PEEK control. The strength of oriented PEEK is approximately equal to that of CFR-PEEK; however, the modulus is less than CFR-PEEK. Phosphonylation and calcium posttreatment of PEEK and CFR-PEEK surfaces led to a stronger interfacial strength over untreated controls as measured by the push out test. Microtexturing CFR-PEEK resulted in an increase of the joint strength at the bone–implant interface. Image analysis of histology slides to assess bone apposition to implants was of limited value in verifying bone ingrowth and providing a strong correlation with the push-out results.

OSSEOINTEGRATION OF PHOSPHONYLATED PEEK AND CFR-PEEK ENDOSTEAL DENTAL IMPLANTS

In a review by Lemon on dental implants, the author noted that modern dental implants began with metallic endosteal systems, such as blades, root forms, endodontic stabilizers, plates, and screws.[59] Most pertinent to the subject of this section are the endosteal screws. An endosteal screw is the major component of a typical prosthesis used to replace a natural tooth. In such a prosthesis, the endosteal screw represents the implant body anchored in the bony tissue of the jaws. To the screw is attached an abutment connector, and the artificial crown is affixed to this connector. Endosteal implants, including those in the form of screws, are usually based on alloys, such as Ti-Al-V and Co-Cr-Mo types. Concerns about the use of metallic implants in bony tissues and associated stress-shielding and subsequent bone resorption directed the attention of contemporary investigators, as discussed earlier in this book, to explore the use of high modulus polymeric alternatives to currently used metallic implants, including those used in dental applications. This provided an incentive to pursue the subject discussed in this section; a second incentive pertains to the limited success associated with the ability of metallic implants to osseointegrate with the surrounding dental bony tissues and attain long-term mechanical stability therein. Among the many early attempts and associated studies to stabilize dental implants through design and surface treatment, those noted below are most pertinent to the subject of this section. Langer and coworkers noted that a 5.0 mm diameter self-tapping metallic dental implant was used successfully for patients who have inadequate bone height, poor bone quality, and who need immediate replacement of nonintegrated or fractured implants.[60] This was followed by a study by Kato and coworkers that showed that the pull-out force for large diameter implants (4.5 mm diameter) was 16% higher than that of the small diameter (3.25 mm diameter) implants.[61] In a study by Mukherjee and coworkers on the fatigue of hydroxyapatitecoated dental metal implants, it was concluded that microcracks widened in some if the samples were fatigued; the surface composition in terms of Ca/P did not change significantly; some P, but not Ca, was released from the samples into the immersion solution during fatigue testing.[62] In a study to determine the effect of surface roughness on TiO$_2$-coated titanium implant screws, Wennerberg and Albrektsson arrived at the following conclusion: Implants with a surface structure that had no dominating pattern, an average surface roughness of 1.4 µm, an average wavelength of 11.1 µm, and a developed area ratio of 1.5 provide the firmest bone fixation.[63]

Contents of this section are designed to take into account the 1) aforementioned discussion pertaining to the growing interest in polymeric alternatives to presently used metallic dental implants while featuring a new design and unique surface physicochemical properties; and 2) discussion in Section "Osseointegration of Chemically and Physico-chemically Modified Peek-Based Rods as Goat Mandibular Implants" outlining the promising properties of phosphonylated materials as dental implants capable of osseointegration. The first segment of this section deals with the development of a new design of a typical endosteal dental implant (EDI) in the form of a screw for evaluation in beagles. This is followed by the development of a suitable animal model and its use in evaluating osseointegration, both histomorphometrically and biomechanically.

Development of a New EDI Design for Evaluation in Beagles

Toward the development of a PEEK-based EDI and its coronal components for evaluation in beagles, a number of requirements were recognized and addressed accordingly in terms of having the following features:[64,65]

1. An EDI with a threaded neck for screwing a metallic abutment connector to which a metallic or ceramic crown is affixed to transfer the load during mastication

2. A screw that can be mechanically placed in the mandibular bone without being distorted
3. A thread design coupled with radially positioned creators and vertical grooves to prevent axial and radial micromotions, respectively, following implantation and during the sought after early EDI bone osseointegration

An illustration of the screw design is depicted in Fig. 12.

In preparation for the animal study using PEEK-based implants and a titanium alloy control for implantation in beagle mandibles, bars of oriented PEEK, unoriented CFR-PEEK, and Ti alloy were formed and micromachined into the design depicted in Fig. 12. More specifically, PEEK and CFR-PEEK pellets were molded into blocks of desired dimensions using a Carver press after which they were cut into bar stock. The PEEK bars were subjected to OSSO using high compressive forces. The oriented PEEK and CFR-PEEK bars were micromachined into the implant forms using a Benchman XT machining center. The machined implants were scoured by sonication in isopropyl alcohol. The implants were surface phosphonylated and processed to immobilize calcium ions on their activated surfaces to allow for osseointegration as discussed earlier. Implants made of Ti alloy (Ti6-6Al-4V) were micromachined and scoured as noted for the PEEK-based implants.[48]

Development of a Beagle Animal Model and Pilot Study of the EDI

During the initial stage of development, a series of EDIs made of PEEK, CFR- PEEK, and Ti alloy having the same preliminary design that was slightly different from the later-optimized design shown in Fig. 12. In the first segment of the pilot study, the three types of EDIs were used. Conventional wisdom was to compare the implants under two modes of placement, immediate and delayed.

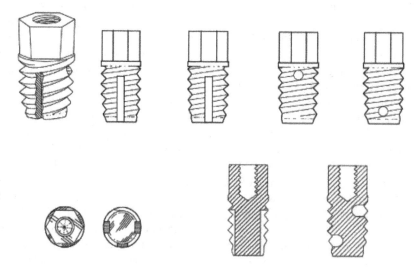

Fig. 12 Optimized endosteal implant design.

In the immediate placement scheme, implants were to be placed in the mandibular bone following extraction of the selected premolars in the canine model. Conversely, the delayed placement scheme was to extract the appropriate premolars, reapproximate the gingiva, and allow a 12-week healing period prior to the placement of the implants in the mandibular bone. However, it was found that the immediate placement presented certain challenges that could not be overcome, at least early on in this study. Namely, the condition of the implant bed following extraction of healthy, multirooted teeth was less than ideal. Given the narrow dimensions of the furcation bone remaining after extraction, immediate placement caused a collapse of the adjacent, empty root cavity. Additionally, success of the immediate placement method is highly dependant on no load being applied to the implant in the early healing period. This led to the focus on delayed placement using an optimized implant design of the EDI in the second segment of the pilot study. More details of the protocol and outcomes of the pilot study are described below.

Beagle mandibular model and placement protocols

In preparation of the pilot study, female beagles of approximately 18 months of age and weighing 15–25 lb were purchased from Harlan (Indianapolis, IN) and held in quarantine for acclimation and observation for not less than 2 weeks. The animals were evaluated for skeletal maturity by radiographic confirmation of closure of the distal femoral and proximal tibial epiphyses. Tooth extraction was performed under anesthesia with the assistance of a certified veterinary surgeon and a registered dental veterinary technician. Complete extraction was verified radiographically. In the immediate placement group, one root cavity was chosen from each of the extracted teeth to receive an implant. This root cavity was prepared to receive an implant by first drilling with a 1/8-in. hole using sterile saline irrigation. The hole was then threaded using a tap custom made to the same thread form as that on the implants. The tapped hole was then irrigated to remove any debris and the implant inserted to a depth that brought the head to the same level as the surrounding gingival tissue followed by gingival closure. In the delayed placement group, the teeth were extracted in the same fashion and the implant bed allowed to heal for a period of 12 weeks. Following the healing period, radiographs were taken to ensure good bone formation and proper healing. The sites for implantation were then selected and the sites drilled and tapped as before. Implants were then placed and the gingiva closed.

Pilot study of the EDIs

To assess the efficacy of the immediate placement approach, we placed EDIs of the original design into two animals on the left side of their mandible in the P2, P3, and P4 (P = premolar) positions. After 2 weeks, several implants were lost and an oral exam revealed that the remaining EDIs were loose. Radiographs were taken and revealed two distinct problems. The first was that the adjacent empty root cavity had collapsed, which eliminated the anchoring of the EDI on one side. This loss of anchoring not only compromised the security of the EDI but also allowed for movement, which prevented any healing at the bone–implant interface. The second source of problems was due to incomplete tooth root extraction. Careful examination of early radiographs showed evidence of root fragments remaining. In subsequent surgeries, radiographic procedures have been modified to improve the quality of the films and aid in the detection of root fragments during the extraction procedure and eliminate this as a source of error. However, the stability issues associated with empty root cavities and thin furcation bone segments have led to suspending the use of the immediate placement approach. After implementing revised extraction techniques and improving the design of the EDIs, a second series of five animals have been implanted with a total of 22 EDIs using the delayed placement method. Of these 22 implants, 21 remain intact with good bone remodeling and implant security as evidenced by radiographic and oral examination. The one implant that was lost was due to incomplete insertion secondary to a sheared head during implantation.

Results of the pilot study led to the conclusion that a refined beagle model, using delayed implant placement, can be used for evaluation of optimally designed surface phosphonylated PEEK-based EDI s to demonstrate their ability to osseointegrate with mandibular bone tissue.

Biomechanical Evaluation of Osseointegration of Metallic and PEEK-Based EDIs

Results of the *in vivo* study discussed in Section "Development of a Beagle Animal Model and Pilot Study of the EDI" on the preliminary evaluation of the EDI design of phosphonylated PEEK-based EDIs reflected the increased propensity of these implants to osseointegrate with mandibular bone following a delayed placement protocol.[48,64] This prompted the pursuit of a more comprehensive study, subject of this section, dealing with the effect of the implantation period on the extent of osseointegration measured in terms of the biomechanical properties of EDI–bone interface.[49] Experimentally, this study was pursued by following the processing protocols described earlier (Section "Development of a Beagle Animal Model and Pilot Study of the EDI") for preparing ready-to-implant EDIs, multiple-member sets were made using solid-state oriented PEEK (by the OSSO process), CFR-PEEK, and titanium alloy.[48,49,64] The number of implants in each set was sufficient to ensure having at least four test specimens per segment of the study and, hence, statistically viable test results. For the animal study, female beagles of approximately 18 months of age and weighing 15–25 lb were used. The study was pursued following the delayed placement protocol and associated preoperative, surgical, and postoperative procedures as discussed earlier.[48] At the conclusion of each study period for different EDIs, animals were euthanized and mandibles removed. Using a band saw, the mandible was cut into proper size blocks (or test specimens) with bone tissue adequately surrounding the test implant to allow conducting reproducible torsional measurements. These were conducted following a protocol similar to one described recently for titanium dental implants.[66] The test specimens were placed immediately in saline and tested within 2 hours of removal. The peak torque (or maximum removal torque, MRT) required to disengage the EDI from surrounding bone tissue was measured using a Multitorque Analyzing System. For this, the individual test specimen was affixed in a metallic holder and rotated at a rate of 16 radians per minute until failure.[49]

Collectively, the study included:

1. Optimization of the direct surface phosphonylation of the PEEK-based EDIs to ensure reproducibility and formation of highly functionalized substrates for immobilizing the calcium ion without compromising the mechanical integrity of the thread.

2. Modification of the EDI head to a hexagonal geometry that is compatible with a specially designed hexagonal applicator to facilitate its insertion without deformation (this was particularly useful for OSSO-PEEK and led to elimination of the notches in the preliminary head design).

3. Use of an acrylic model of the premolar crown to revise the micromachining program to produce the optimum stem design as shown in Fig. 12. The design program modification was implemented directly on the preliminary design program in a stepwise manner following several in vitro attempts to simulate the device assembling steps and to determine the design

requirements for facile insertion of the stem into the dog mandible.

4. Combining the crown-holding post (or abutment connector) and the crown into one component that can be securely attached into the thread receptacle of the EDI head using a commercial dental resin cement (Calibra™). This was pursued to avoid mechanical instability of the post–crown joint and shearing-off of the ceramicor acrylic-based crown shortly after implant assembling at the mandibular site.

5. Use of direct phosphonylation of the EDIs as the method of choice as per the comparative ESCA data of specimens made using the indirect method (i.e., surface acylation following by phosphonylation).

6. Use of heat, radiochemical, or radiation sterilization for preparing sterile EDIs for implantation, depending on the time constraints and coordination with the ESCA verification of surface functionality and testing for implant sterility (see Table 9 for typical ESCA data of randomly selected EDIs from three evaluated sets); standard method for sterility testing employing liquid culture media was used.

7. Use of delayed placement of the EDIs as the only viable protocol for conducting the balance of the studies. This decision was made as a result of many unsuccessful attempts based on immediate placement, which led to loss of time and animals.

8. A brief study on the effect of loading on the biomechanical properties. The results showed that under the prevailing conditions, loading the implant has practically no effect or can lead to minimum improvement in the biomechanical properties in the case of the metallic and OSSO-PEEK or CFR-PEEK EDIs, respectively (Table 10).

Table 9 Typical ESCA[a] data of phosphonylated EDIs, with and without immobilized calcium ion

Treatment	EDI type					
	OSSO-PEEK 1	Specimen 2	No. 3	CFR-PEEK 1	Specimen 2	No. 3
After phosphonylation, atomic phosphorous, %	2.34	2.42	2.46	3.07	3.17	3.61
After phosphonylation and Ca²⁺ immobilization						
Atomic phosphorous, %	1.04	1.40	2.07	1.82	1.86	1.81
Atomic calcium, %	0.50	0.46	0.56	0.41	0.42	0.47

[a] ESCA = electron spectroscopy for chemical analysis.

Table 10 Effect of loading on the biomechanical properties of representative EDI implants

Loading period (weeks)	Maximum removal torque (N-cm)		
	Titanium alloy EDI	Oriented PEEK EDI	CFR-PEEK EDI
0	77.2	96.3	93.2
10	73.6	–	–
20	–	95.4	117.1

Titanium EDIs were implanted for 20 cumulative weeks posthealing following the tooth extraction (loaded + unloaded) while PEEK and CFR-PEEK EDIs were implanted for 30 cumulative weeks posthealing following the tooth extraction.

9. Completion of three sets of studies on the effect of postloading healing period (namely, 10, 15, and 20 weeks following complete assembling of the EDI stem–crown components) on biomechanical stability (Table 11 and Fig. 13).

It is important to note that the data in Table 11 and Fig. 13 dealing with the effect of loading time (namely 10, 15, and 20 weeks following complete assembly of the EDI stem and crown components) on osseointegration–bone apposition, measured in terms of MRT, indicate that

1. At any of the three periods, extent of bone apposition and osseointegration is lowest for titanium alloy and highest for CFR-PEEK EDIs, respectively, with the OSSO-PEEK EDI displaying intermediate values.
2. For both PEEK-based EDIs, the bone apposition–osseointegration values increase progressively with time, thus supporting the postulate that the relatively moderate modulus of the polymeric EDI, as compared with the Ti alloy and surface phosphonylation do support osseointegration and bone formation.
3. For the Ti alloy EDIs, osseointegration and bone apposition or bone formation reached a maximum at 15 weeks, then decreased at 20 weeks, thus supporting the thesis that metallic implants can cause bone resorption upon prolonged presence at implant site.
4. The MRT values for the Ti alloy EDI exceeded those reported by Cho et al. on commercial titanium alloy EDIs with or without surface texturing, thus suggesting that the newly patented proprietary design of the endosteal implant, shown in Fig. 12, contributes positively to the biomechanical properties of the Ti alloy EDI.[65,66]

Results of the study on the biomechanical properties of EDI–bone interface led to the conclusion that surface phosphonylated PEEK-based EDIs and particularly those made of CFR-PEEK, having surface-immobilized calcium ions, should be viewed as clinically preferred alternatives to those made of Ti alloys.

Histomorphometric Evaluation of Osseointegration of Metallic and PEEK-Based EDIs

Extension of the studies described in the preceding sections of the entry on maximizing the effectiveness and clinical relevance of surface phosphonylated implants with hydroxyapatite-like surfaces capable of osseointegration with bone tissue has led to the following:

1. The development of an optimized EDI that minimizes or prevents axial and radial micromotion at the implant site
2. The identification of the delayed placement protocol as the choice approach for introducing mandibular implants after tooth extraction
3. The demonstration that osseointegration, as measured in terms of mechanical properties, of an endosteal implant is time dependent, particularly for PEEK implants
4. The conclusion that both the surface morphology and bulk properties do affect the biomechanical properties, which may be related to the extent of osseointegration[48]

The latter observation provided the incentive to conduct the study, subject of this section, to determine the relevance of the time-dependent mechanical properties of PEEK-based and metallic implants to osseointegration as measured in terms of bone apposition.[48,50,64]

In preparation for conducting the histomorphometric evaluation of osseointegration of the different EDIs,

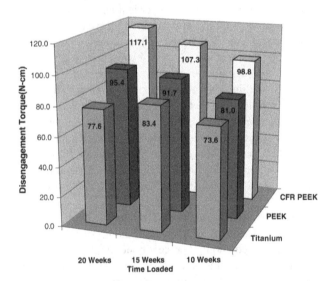

Fig. 13 Effect of postloading period and EDI composition on maximum torque required for disengagement.

Table 11 Effect of post-loading period and composition on biomechanical properties of representative EDI implants

	Maximum removal torque (N-cm)		
Loading period (weeks)	Titanium alloy EDI	Oriented PEEK EDI	CFR-PEEK EDI
10	73.6 ± 4.6	81.0 ± 7.1	98.8 ± 3.7
15	83.4 ± 9.1	91.7 ± 6.5	107.3 ± 8.5
20	77.6 ± 8.9	95.4	117.3 ± 8.6

Measured in terms of maximum torque (N-cm) to disengage the EDI from the mandible.

multiple-member sets of ready-to-implant EDIs were prepared as per the processing protocols described earlier, and test specimens made from OSSO, CFR, and titanium alloy were prepared.[48,50,64] The number of implants in each set was sufficient to provide at least two implants per specimen for individual histological evaluation. For the animal study, female beagles of approximately 18 months of age and weighing 15–25 lb were used. The study was pursued following the delayed placement protocol and associated preoperative, surgical, and postoperative procedures. At the conclusion of each study period, the animals were euthanized and the mandibles were removed and placed in 10% neutral buffered formalin (NBF) for at least 48 hr. Using a band saw, the mandible was cut into properly sized blocks with bone tissue adequately surrounding the test implant. The blocks were then transferred individually to 70% ethanol and stored for histological evaluation. With the exception of a few specimens, histology and histomorphometric evaluations were conducted at SkeleTech (Bothell, WA). The samples were scanned by a Sky-Scan 1076 micro-CT system for a quick turnaround on the three dimensional visualization of bone ingrowth to the implant materials within 2 weeks. The samples were then processed, undecalcified, and embedded in polymethyl methacrylate (PMMA) for sectioning by using the EXAKT system to obtain two vertical and two horizontal sections of the implants and surrounding bone tissue. Sections were stained with toluidine blue. Microscopic evaluation for histopathology and histomorphometry was focused on cell–tissue reaction and percentage bone ingrowth within and surrounding the implant. The bone ingrowth to the grooves of the implanted EDIs was estimated by the groove surface that is in contact with the ingrown bone. Histomorphometric measurements were performed by using the point-counting method for obtaining percent bone ingrowth within and surrounding the implant and by using the OsteoMeasure™ software adjunct to a Nikon Camera Lucida Imaging hardware. Typical histomorphometry data are summarized in Table 12 and illustrate the effect of postloading period and EDI type and composition on percent bone–implant contact.

Histology specimens were subjected to histomorphometric evaluation. No significant differences in the histopathology of the three types of EDIs could be observed.[50,64]

In general the results of the study indicated, specifically, that

1. The osseointegration, or mechanical stability, of the titanium alloy EDI exceeded that reported earlier, which may be attributed to the new design of the EDI depicted in Fig. 12.
2. The OSSO-PEEK functional performance is slightly better in comparison to the Ti-based control in terms of biomechanical stability and osseointegration.
3. CFR-PEEK-based EDI functional performance is superior to both the Ti and OSSO-PEEK-based EDIs.

A representation of a typical histology photomicrograph depicting osseointegration at the EDI–bone interface is shown in Fig. 14. The photomicrograph shows the extent of osseointegration for a CFR-PEEK EDI implanted for a 20-week (10 weeks unloaded and 10 weeks loaded) posthealing period.

Analysis of the histomorphometric data summarized in Table 12 in terms of percentage of bone in contact with different EDIs, which are relevant to the extent of bone apposition and osseointegration at the bone–implant interface, does reflect significant differences among these implants. More specifically, Table 12 outlines the percentage of bone in contact with the surface of different EDIs, which reflects primarily the extent of bone apposition and osseointegration, and can be related to the overall tendency for bone formation at the implant site. Taking this into account, the data in Table 12 and Fig. 14 indicate that

1. At a postloading period of 10 weeks, limited increase in bone apposition and osseointegration takes place for all types of implants, with the EDIs made of CFR-PEEK showing minimum values.
2. At a postloading period of 20 weeks, when sufficient bone formation was allowed to take place, a maximum

Table 12 Effect of postloading period and composition on percentage bone–implant contact for different EDIs: Typical data

Post-loading period	Titanium alloy EDI	OSSO-PEEK EDI	CFR-PEEK EDI
10 weeks	36	35	33
20 weeks	57	36	72

This represents the ratio of implant surface in contact with bone tissue divided by the total surface.

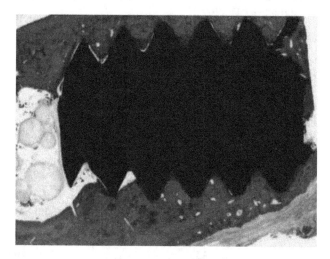

Fig. 14 Histological photomicrograph of a typical longitudinal section of a CFRPEEK EDI implanted for 20 weeks (10 weeks unloaded and 10 weeks loaded) posthealing period.

bone apposition and/or osseointegration was associated with CFR-PEEK EDIs.

3. There is no significant difference between percentage of bone in contact with the titanium and with OSSO-PEEK EDIs.

4. A 30-week posthealing period, including a 20-week loading period, is associated with extensive osseointegration for the CFR-PEEK EDI.

Collective analysis of the results of biomechanical and histomorphometric data of Sections "Biomechanical Evaluation of Osseointegration of Metallic and Peek-Based EDIS" and "Histomorphometric Evaluation of Osseointegration of Metallic and Peek-Based EDIS", respectively, led to the conclusion that

1. The osseointegration and mechanical stability of the Ti alloy EDI exceeded that reported earlier by Cho and Jung,[66] which may be attributed to the newly patented design of the EDI depicted in Fig. 12.

2. The OSSO-PEEK functional performance is slightly better in comparison to the Ti-based control in terms of biomechanical stability and osseointegration.

3. CFR-PEEK-based EDI functional performance is superior to both the Tiand OSSO-PEEK-based EDIs.

CONCLUSION AND PERSPECTIVE ON THE FUTURE

Results of the studies discussed in this entry led to the conclusions that

1. Solid-state orientation of PEEK, among other thermoplastic crystalline polymers, can be used to increase the compression-molded polymer modulus and allow its micromachining to yield useful bone implants, including EDIs with different designs.

2. Direct surface phosphonylation is well-suited as a practical and economical method to create a hydroxyapatite-like surface after the immobilization of calcium ions thereon.

3. Surface microtexturing and/or phosphonylation of high modulus crystalline polymeric bone implants, such as those based on PEEK, can result in their osseointegration and hence, mechanical stability in the bony tissue.

4. CFR-PEEK is the most suitable form of PEEK-based constructs for the production of EDIs. Micromachining is a well-suited method for the production of metallic and PEEK-based EDIs.

5. Using the beagle mandibular model, 10–12 weeks is a sufficient period to allow adequate bone formation following extraction and make available the mandibular site for drilling and preparing the mandibular bone for delayed placement of the EDI.

6. Following the delayed placement of the EDI, most bone formation and osseointegration takes place within a 10-week period followed by variable degrees of bone remodeling and densification after an additional 10, 15, and 20 weeks. The remodeling and densification were minimal in the case of the titanium-based and OSSO-PEEK-based EDIs, but quite discernable in the case of the CFR-PEEK-based EDIs. Under these conditions, the loading of the EDI did not appear to have a significant effect on the amount of bone formation or its quality.

7. Biomechanical properties measured in terms of the torsional force required for disengaging the implant from the surrounding bone is most effective in evaluating the EDI biomechanical stability and the overall extent of osseointegration.

Based on the technical outcome of the studies noted in this entry, the surface activated CFR-PEEK system is recommended for further use in the production of not only screw-type, but also other forms of endosteal implants. Additionally, it is recommended that 1) surface activated OSSO-PEEK be explored for use in maxillofacial applications; 2) surface carboxylation be explored as a facile means of producing substrates capable for osseointegration; and 3) the technological achievement associated with the development of dental implants be explored for suitable applications in the orthopedic area.

REFERENCES

1. von Recum, A.F., Ed. *Handbook of Biomaterials Evaluation*; Macmillan: New York, 1986; p. vii.

2. Williams, D. *Concise Encyclopedia of Medical and Dental Materials*; MIT Press: Cambridge, MA, 1990; p. 52.

3. Williams, D.G. Definitions in Biomaterials: Proceedings of a Consensus Conference of the European Society for Biomaterials, Chester, England, Elsevier, Amsterdam, March 3–5, 1989.

4. Ratner, B.D. New ideas in biomaterials science—A path to engineered biomaterials. J. Biomed. Mat. Res. **1993**, *27* (7), 837–850.

5. Tunc, D. Absorbable Bone Fixation Devices. U.S. Patent 4539981, Sep 10, 1985.

6. Tunc, D. Absorbable Bone Fixation Devices. U.S. Patent 4550449, Nov 5, 1985.

7. Zachariades, A.E. Process for Producing a New Class of Ultra-High Molecular Weight Polyethylene Orthopedic Prostheses with Enhanced Mechanical Properties. U.S. Patent 5030402, Jul 9, 1991.

8. Shalaby, S.W.; Johnson, R.A.; Deng, M. Process of Making a Bone Healing Device, U.S. Patent 5529736, Jun 25, 1996.

9. Kim, S.W.; Feijen, J. Surface modification of polymers for improved blood compatibility. CRC Crit. Revs. Biocomp. **1980**, *1*, 229.

10. Lelah, M.D.; Jordan, C.A.; Pariso, M.E.; Lambarecht, L.K.; Albrecht, R.M.; Cooper, S.L. Blood compatibility of polyethylene and oxidized polyethylene in canine A-V shunt: Relationship to surface properties. In *Polymers as Biomaterials*; Shalaby, S.W., Hoffman, A.S., Ratner, B.D., Horbett, T.A., Eds.; Plenum Press: New York, 1984; p. 257.

11. Hoffman, A.S. Modification of material surfaces to affect how they interact with blood. Ann. N.Y. Acad. Sci. **1987**, *516* (1), 96–101.

12. Ikada, Y. Blood compatible surfaces. Adv. Polym. Sci. **1984**, *57*, 103.

13. Nagaoka, S.; Mori, Y.; Tanzawa, T.; Kikuchi, Y.; Inagaki, F.; Yokota, Y.; Nioshiki, Y. Hydrated dynamic surfaces. Trans. Am. Soc. Artific. Intern. Organs **1987**, *33* (2), 76–77.

14. Amiji, M.; Park, K. Surface modification of polymeric biomaterials with polyethylene oxide. In *Polymers of Biological and Biomedical Significance*, ACS Symposium Series; Shalaby, S.W., Ikada, Y., Langer, R., Williams J., Eds.; American Chemical Society: Washington, D.C., 1992; Chapter 11.

15. Shalaby, S.W.; McCaig, M.S. Process for Phosphonylating the Surface of Polymeric Preforms. U.S. Patent 5491198, 13 Feb, 1996.

16. Shalaby, S.W.; Rogers, K.R. Polymeric Prothesis Having a Phosphonylated Surface. U.S. Patent 5558517, Sep 24, 1996.

17. Hylton, D.M. The Effect of Sulfonation of Polyethylene and Polypropylene on Blood Compatibility, (M.S. thesis, Department of Bioengineering, Clemson University, Clemson, SC, 2002).

18. Hylton, D.M.; Shalaby, S.W.; Latour, R.A., Jr. Direct correlation between adsorption-induced changes in protein structure and platelet adhesion. J. Biomed. Mater. Res. **2005**, *73* (3), 349–358.

19. Clayton, J.O.; Jensen, W.L. Reaction of paraffin hydrocarbons with phosphorus trichloride and oxygen to produce alkanephosphonyl chlorides. J. Am. Chem. Soc. **1948**, *70* (11), 3880–3882.

20. Campbell, C.E. Surface-Phosphonylated Polyethylene and Its Binding Capacity, (Doctoral dissertation, Department of Bioengineering, Clemson University, Clemson, SC, 1996).

21. Hench, L.L.; Paschal, H.W. Histochemical responses at a biomaterial's interface. J. Biomed. Mater. Res. **1974**, *8* (3), 49–64.

22. Holmes, R.E.; Hagler, H.K. Porous hydroxyapatite as a bone graft substitute in mandibular augmentation: A histometric study. J. Oral Maxillofac. Surg. **1987**, *45* (5), 421–429.

23. Allan, J.M.; Dooley, R.L.; Shalaby, S.W. Surface phosphonylation of low-density polyethylene. J. Appl. Polymer Sci. **2000**, *76* (13), 1870–1875.

24. Williams, D.E. The inert-bioactivity conundrum. In *Bio-Implant Interface*; Ellingsen, J.E., Lyngstadaas, S.P., Eds.; CRC Press: Boca Raton, FL, 2003; Chapter 23.

25. Szmukler-Moncler, S.; Zeggel, P.; Perrin, D.; Bernard, J.-P.; Neuman, H.G. From micro-roughness to resorbable bioactive coating. In *Bio-Implant Interface*; Ellingsen, J.E., Lyngstadaas, S.P., Eds.; CRC Press: Boca Raton, FL, 2003; Chapter 5.

26. Esposito, M.; Worthington, H.V.; Coulthard, P.; Wennerberg, A.; Thomsen, P. Role of implant surface properties on the clinical outcome of osseointegrated oral therapy: An evidence-based approach. In *Bio-Implant Interface*; Ellingsen, J.E., Lyngstadaas, S.P., Eds.; CRC Press: Boca Raton, FL, 2003; Chapter 1.

27. Shalaby, S.W.; Roweton, S.L. Continuous Open Cell Polymeric Foam Containing Living Cells, U.S. Patent 5677355, 14 Oct, 1997.

28. Shalaby, S.W.; Roweton, S.L. Microporous Polymeric Foams and Microtextured Surfaces. U.S. Patent 5847012, 8 Dec, 1998.

29. Roweton, S.L.; Shalaby, S.W. Microcellular foams. In *Polymers of Biological and Biomedical Significance*, A.C.S. Symp. Series; Shalaby, S.W., Ikada, Y., Langer, R., Williams, J., Eds.; American Chemical Society: Washington, DC, 1993; Vol. 520.

30. Shalaby, S.W.; Johnson, R.A.; Deng, M. Process of Making a Bone Healing Device. U.S. Patent 5529736, 25 Jun, 1996.

31. Deng, M.; Wrana, J.; Allan, J.M.; Shalaby, S.W. Tailoring mechanical properties of polyether-ether ketone for implants using solid-state orientation. Trans. Cos. Biomater. **1999**, *22*, 477.

32. Shalaby, S.W. Highly Oriented PEEK/PEEK Composites for Dental Implants, Phase I Report, NIH-supported SBIR, Grant No. R43 DE12558-01, March 8, 1999.

33. Andrade, J.D.; Nagaoka, S.; Cooper, S.L.; Okeno, F.; Kim, S.W. Surfaces and blood compatibility, current hypothesis. Trans. Am. Soc. Artif. Intern. Organs **1987**, *33* (2), 75.

34. Hanker, J.S.; Giammara, B.L. Biomaterials and biomedical devices. Science **1988**, *242* (4880), 885–892.

35. Tengvall, P. How surfaces interact with the biological environment. In *Bio-implant Interface*; Ellingsen, J.E., Lyngstadaas, S.P., Eds.; CRC Press: Boca Raton, FL, 2003; Chapter 16.

36. Jansen, J.A.; ter Brugge, P.J.; van der Wal, E.; Vredenberg, A.M.; Wolke, J.G.C. Osteocapacities of calcium phosphate ceramics. In *Bio-implant Interface*; Ellingsen, J.E., Lyngstadaas, S.P., Eds.; CRC Press: Boca Raton, FL, 2003; Chapter 17.

37. Paine, M.L.; Wong, J.C. Osteoblast response to pure titanium and titanium alloy. In *Bio-implant Interface*; Ellingsen, J.E., Lyngstadaas, S.P., Eds.; CRC Press: Boca Raton, FL, 2003; Chapter 7.

38. Gristina, A.G. Biomaterial-centered infection: Microbial adhesion versus tissue integration. Science **1987**, *237* (4822), 1588–1595.

39. Costerton, J.W.; Marrie, T.J.; Change, K.J. Phenomena of bacterial adhesion. In *Bacterial Adhesion Mechanism and Physiological Significance*; Sawage, D.C., Fletcher, M., Eds.; Plenum Press: New York, 1985; p. 3.

40. Gerdes, G.A.; LaBerge, M.L.; Barefoot, S.E.; Dooley, R.L.; Shalaby, S.W. Surface phosphonylated substrate for studying bacterial cell attachment. Trans. Soc. Biomater. **2000**, *23*, 1535.

41. Jansen, B.; Kohnen, W. Prevention of biofilm formation by polymer modification. J. Ind. Microbiol. **1995**, *15* (4), 391–396.

42. Atkins, G.G.; Barefoot, S.G.; LaBerge, M.L.; Dooley, R.E.; Shalaby, S.W. Critical evaluation of methodologies for studying bacterial cell attachment to polymeric material. Trans. Soc. Biomater. **2001**, *24*, 353.

43. Allan, J.M.; Wrana, J.S.; Dooley, R.L.; Budsberg, S.; Shalaby, S.W. Bone in growth into surface phosphonylated polyethylene and polypropylene. Trans Soc. Biomater. **1999**, *22*, 468.

44. Allan, J.M.; Wrana, J.S.; Budsberg, S.C.; Farris, H.M.; Dooley, R.L.; Powers, D.L.; Shalaby, S.W. Surface modified polyolefins for improved adhesion to bone. Proc. 22nd

Annual Meeting, Adhesion Soc., 1999, Virginia Tech: Blacksburg, VA, p. 106.

45. Wrana, J.S.; Allan, J.M.; Budsberg, S.C.; Powers, D.L.; Dooley, R.L.; Shalaby, S.W. Dental implants with modified surfaces for improved bioadhesion. Proc. 22nd Ann. Meeting, Adhesion Soc., 1999, p. 111.

46. Allan, J.M.; Wrana, J.S.; Linden, D.E.; Farris, H.; Budsberg, S.; Dooley, R.L.; Shalaby, S.W. Bone formation into surface phosphonylated polymeric implants. Crit. Rev. Biomed. Eng. **2000**, *28* (3–4), 377.

47. Allan, J.M.; Wrana, J.S.; Kline, J.D.; Gerdes, G.A.; Anneaux, B.L.; Budsberg, S.C.; Farris, H.E.; Shalaby, S.W. Bone ingrowth into phosphonylated PEEK rabbit tibial implants, Sixth World Biomaterials Congress. Trans. Soc. Biomat. **2000**, *23*, 631.

48. Anneaux, B.L.; Hollinger, J.O.; Budsburg, S.C.; Fulton, L.K.; Shalaby, S.W. Surface activated PEEK-based endosteal implants. 7th World Biomaterials Congress, Trans. Soc. Biomater. **2004**, *27*, 967.

49. Anneaux, B.L.; Taylor, M.S.; Johnston, S.A.; Shalaby, S.W. Biomechanical properties of osseointegrated PEEK-based and metallic endosteal implants. Trans. Soc. Biomater. **2005**, *28*, 129.

50. Anneaux, B.L.; Taylor, M.S.; Shih, M.; Fulton, L.K.; Shalaby, S.W. Histomorphometric evaluation of osseointegration of metallic and PEEK-based endosteal dental implants. Trans. Soc. Biomater. **2005**, *28*, 431.

51. Tate, P.L.; Taylor, M.S.; Perry, D.M.; Shalaby, S.W. Attachment of osteoblasts to surface activated biomaterials, Abstract of Second Annual South eastern Tissue Engineering and Biomaterial Conference, Birmingham, Alabama, February 11, 2005.

52. Tate, P.L.; Hucks, M.A.; Nagatomi, S.; Vaughn, M.A.; Shalaby, M.; Shalaby, S.W. Attachment of osteoblasts to surface modified substrates—A preliminary report. Trans. Soc. Biomater. **2006**, *29* (2), 611.

53. Shalaby, S.; Shalaby, S.W. Surface Functionalized Absorbable Medical Devices. U.S. Patent No. 11/378,104, 2005.

54. Shalaby, S.W. Surface Electroconductive Biostable Polymeric Articles. U.S. Patent No. 11/378,178, 2005.

55. Tencer, A.F.; Holmes, R.E.; Johnson, K.D. *Handbook of Biomaterials Evaluation*; von Recum, A.F., Ed.; Macmillan Publishing Co.: New York, 1986; p. 324.

56. Allan, J.M.; Wrana, J.S.; Linden, D.E.; Dooley, R.L.; Farris, H.; Budsberg, S.; Shalaby, S.W. Osseointegration of morphologically and chemically modified polymeric dental implants. Trans. Soc. Biomater. **1999**, *22*, 37.

57. Maxian, S.H.; Zawadsky, J.P.; Dunn, M.G. Effect of Ca/P coating resorption and surgical fit on bone/implant interface. J. Biomed. Mater. Res. **1994**, *28*, 1311.

58. Wei, H.; Hero, H.; Solheim, T.; Kleven, E.; Rorvik, A.M.; Haanaes, H.R. Bonding capacity in bone of HIP-processed HA-coated titanium: Mechanical and histological investigations. J. Biomed. Mater. Res. **1995**, *29* (11), 1443–1449.

59. Lemon, J.E. Dental implants. In *Biomaterials Science*; Ratner, B.D., Hoffman, A.S., Schoen, F.J., Lemons, J.E., Eds.; Academic Press: New York, 1996; Chapter 7.

60. Langer, B.; Langer, L.; Hermann, I.; Jorneus, L. The wide fixture: A solution for special bone situations and a rescue for the compromised implant. Part 1. Int. J. Oral Maxillofac. Impl. **1993**, *8*, 400.

61. Kato, K.; Eika, Y.; Ikada, Y. Depositon of a hydroxyapatite thin layer onto a polymer surface carrying grafted phosphate polymer chains. J. Biomed. Mater. Res. **1996**, *32* (4), 687–691.

62. Mukherjee, D.R.; Wittenberg, J.M.; Rober, S.H.; Kruse, R.N.; Albright, J.A. A fatigue study of hydroxyapatite dental implants. Trans. Soc. Biomater. **1995**, *28*, 283.

63. Wennerberg, A.; Albrektsson, T. The influence of surface roughness on implant take. Trans. Fifth World Biomater. Congr. **1996**, *2*, 459.

64. Shalaby, S.W. Highly Oriented PEEK/PEEK Composites for Dental Implants, Phase II Report, NIH-supported SBIR, Grant No. 2R 44 DE 012558-02A2, August 18, 2004.

65. Shalaby, S.W.; Anneaux, B.L.; Taylor, M.S. Endosteal Dental Implant. U.S. Patent D 503803, Aug 22, 2005.

66. Cho, S.-A.; Jung, S.-K. A removal of torque of the laser-treated titanium implants in rabbit tibia. Biomaterials **2003**, *24* (26), 4859–4863.

Biomimetic—Colloid

Cellulose-Based Biopolymers: Formulation and Delivery Applications

J. D. N. Ogbonna
F. C. Kenechukwu
S. A. Chime
A. A. Attama
Department of Pharmaceutics, Faculty of Pharmaceutical Sciences, University of Nigeria, Nsukka, Nigeria

Abstract

Cellulose is the most abundant naturally occurring polymer of glucose and is usually found as the main constituent of plants and natural fibers. The fundamental unit of cellulose is the microfibril, which are bonded in a special manner, forming a crystalline array. Cellulose derivatives constitute a large class of biopolymers with diverse physicochemical properties and large functional versatility in pharmaceutical application. Cellulose derivatives have been used as drug formulation adjuvants and, recently, microbial cellulose has been shown to be effective in tissue engineering and as a wound-healing device. By modifying the naturally occurring inexpensive renewable resources, both durable and environmentally acceptable materials can be developed. An increasing availability of biodegradable cellulose-based biopolymers will allow many users to choose them on the basis of their user-friendliness. This entry surveys the design and the applications of cellulose-based biopolymers in formulation and delivery of bioactive compounds. These biopolymers have been extensively investigated due to the large availability of cellulose in nature, the intrinsic degradability of cellulose, and the smart behavior possessed by some of the cellulose derivatives in response to physiologically relevant variables such as pH, ionic strength, and temperature.

INTRODUCTION

Biopolymers are polymers produced by living organisms. They contain monomeric units that are covalently bonded to form macromolecules. There are three main classes of biopolymers, classified according to the monomeric units used and the structure of the biopolymer formed:[1] polynucleotides (RNA and DNA), which are long polymers composed of 13 or more nucleotide monomers; polypeptides, which are short polymers of amino acids; and polysaccharides, which are often linear, bonded, polymeric carbohydrate structures. Polysaccharides include cellulose, starch, gums, glycogen, etc.

Cellulose and cellulose derivatives have long been used in the pharmaceutical industry as excipients in many drug device formulations. Devices other than swelling tablets have been developed for controlled drug delivery using these biopolymers. The most recent advances aim not only at the sustained release of a bioactive molecule over a long time period, ranging from hours to weeks, but also at a space-controlled delivery, directly at the site of interest. The need to encapsulate bioactive molecules into a cellulose hydrogel matrix or other delivery devices (e.g., microspheres and nanospheres) is also related to the short half-life displayed by many biomolecules *in vivo*. Smart cellulose-based hydrogels are particularly useful to control the time- and space-release profile of drugs, as swelling–deswelling transitions, which modify the mesh size of the hydrogel network, occur upon changes of physiologically relevant variables, such as pH, temperature, and ionic strength. Controlled release through oral drug delivery is usually based on the strong pH variations encountered when transiting from the stomach to the intestine. Cellulose-based polyelectrolyte hydrogels [e.g., hydrogels containing sodium carboxymethylcellulose (SCMC)] are particularly suitable for this application. The use of cellulose and its derivatives as biomaterials for the design of tissue engineering scaffolds has received increasing attention, due to the excellent biocompatibility of cellulose and its good mechanical properties, and in the form of sponges or fabrics, have been applied for the treatment of severe skin burns, and in studies on the regeneration of cardiac, vascular, neural, cartilage, and bone tissues.[2] This entry focuses on the current design and use of cellulose-based biopolymer hydrogels, which usually couple their biodegradability with a smart stimuli-sensitive behavior. These features, together with the large availability of cellulose in nature and the low cost of cellulose derivatives, make cellulose-based biolpolymers particularly attractive. Both well-established and innovative applications of cellulose-based biopolymers are discussed.

THE ASSEMBLY OF BIOPOLYMERS

Polysaccharides (e.g., cellulose, starch, and glycogen), proteins, and nucleic acids are polymers composed of subunits (monomers) linked to one another in linear sequence by a particular type of bond. The sequence of monomer units in nucleic acids and proteins is always strictly linear; a few polysaccharides (e.g., glycogen) are branched molecules, but such branching is a secondary biosynthetic event, superimposed on a primary linear arrangement of the subunits. Some polysaccharides are homopolymers, consisting of a single, chemically identical repeating subunit. However, many polysaccharides, all proteins and all nucleic acids are heteropolymers, consisting of chemically similar but nonidentical subunits. The polysaccharides that contain more than one kind of subunit show a regular arrangement of the subunits, whereas in proteins and nucleic acids, the sequences of subunits are irregular. The bonds that link together the subunits in some biological polymers such as proteins and polysaccharides are amide and glycosidic bonds, respectively. A characteristic feature of polysaccharide synthesis is the requirement of a primer—a short segment of the polysaccharide in question to act as an acceptor for the monomer units. For instance, in the synthesis of glycogen, it has been found that the primer must contain more than four sugar units to function effectively. Special branching enzymes bring about the branching in these polymers.

Nearly all synthetic polymers and naturally occurring biopolymers possess a range of molecular weights—except proteins and polypeptides. The molecular weight determined is thus an average molecular weight, the value of which depends on the method of measurement: chemical analysis, osmotic pressure, light scattering, viscosimetric method, etc. Small molecules and many biopolymers are monodisperse; i.e., all molecules of a given pure compound have the same molecular weight. In synthetic polymerization reactions, no two chains grow equally fast or for the same length of time. The resultant biopolymers are heterodisperse; i.e., they have different chain lengths and a range of molecular weights, which can be described by an average molecular weight and by a molecular weight distribution. A number of functional properties are attributed to biopolymers for drug delivery.

CELLULOSE

Cellulose is an essential structural component of cell walls in higher plants and remains the most abundant renewable organic polymer on earth. Cellulose is a homo polymer and its monomer is glucose. It consists of long chains of anhydro-D-glucopyranose units (AGUs) with each cellulose molecule having three hydroxyl groups per AGU, with the exception of the terminal ends (Fig. 1). The fundamental unit of cellulose is the microfibril, constituting of a bundle of β-1,4-glucane that are formed by intra- and intermolecular hydrogen bonding of glucan chains, forming a crystalline array. These microfibrils are about 10–38 nm diameter and 30–100 cellulose molecules in extended chains, depending on their origin.[3] The glucose units in cellulose are held together by 1,4-β-glucosidic linkages, which account for the high crystallinity of cellulose and its insolubility in water and other common solvents. However, modification products of cellulose have shown different degrees of solubility in different solvents. Cellulose derivatives constitute a large class of biopolymers with diverse physicochemical properties and large functional versatility in food, cosmetics, and pharmaceutical applications. Cellulose derivatives have been used as diluents, binders in direct compression and wet granulation processes, film coating, viscosity enhancers, and drug-release retardants in controlled-release matrix systems for microparticles and nanoparticles.[4] The excellent biocompatibility of cellulose, cellulosics, and cellulase-mediated degradation products has prompted the large use of cellulose-based devices in biomedical applications.

Both natural and synthetic polymers have been employed for drug delivery purposes. However, cellulose-based biopolymers have continued to gain greater popularity in

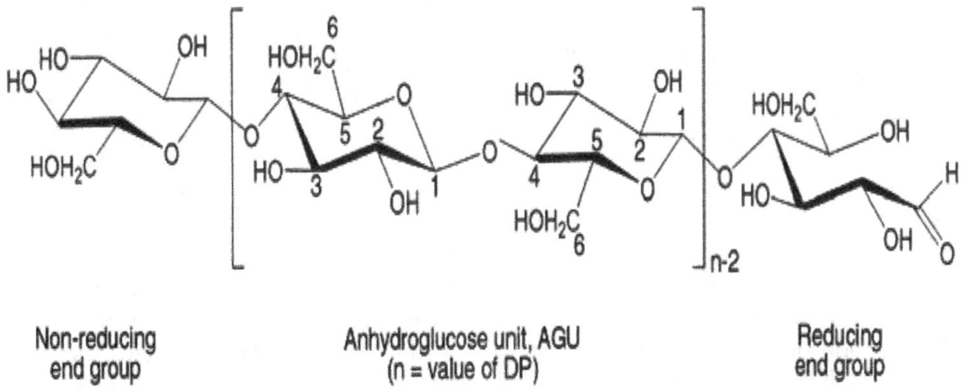

Non-reducing end group   Anhydroglucose unit, AGU (n = value of DP)   Reducing end group

Fig. 1 Molecular structure of cellulose.
Source: From Builders & Attama[5] © 2011, with permission from Nova Science Publishers, Inc.

pharmaceutical manufacture, especially for drug delivery applications because of their diverse and numerous physicochemical and functional properties. In general, the major attraction to the use cellulose-based biopolymers for drug delivery application includes ready availability, low cost, relatively low toxicity, ease of modification, biodegradability, biocompatibility, etc.[5]

Cellulose-Based Biopolymers Used in Formulation and Delivery of Bioactive Compounds

Oxycellulose

Oxidized cellulose (oxycellulose) is cellulose in which some of the terminal primary alcohol groups of the glucose residues have been converted to carboxyl groups.

Microcrystalline cellulose

Since its introduction in the 1960s, microcrystalline cellulose (MCC) has offered great advantages in the formulation of solid dosage forms, but some characteristics have limited its application, such as relatively low bulk density, moderate flowability, loss of compactibility after wet granulation, and sensitivity to lubricants. MCC is purified partially depolymerized cellulose. In many conventional pharmaceutical formulations, it is primarily used as binder and diluents in oral tablets and capsules in both wet granulation and direct compression processes. It is also used as a lubricant and disintegrant in tablets. Because of its desirable physicochemical and functional properties, MCC has been coprocessed with other excipients.[6] It is nonirritant and nontoxic, and is widely used in oral pharmaceutical formulations and food products.[7] Silicification of MCC improves the functionality of MCC with properties such as enhanced density, low moisture content, flowability, lubricity, larger particle size, compactibility, and compressibility. Silicified MCC (SMCC) is manufactured by codrying a suspension of MCC particles and colloidal silicon dioxide such that the dried finished product contains 2% colloidal silicon dioxide.[8] Silicon dioxide simply adheres to the surface of MCC and occurs mainly on the surface of MCC particles; only a small amount was detected in the internal regions of the particles. So, SMCC shows higher bulk density than the common types of MCC.[9]

Cellulose ethers

Cellulose ethers are widely used as important excipients for designing matrix tablets. On contact with water, the cellulose ether swells and forms a hydrogel layer around the dry core of the tablet. The hydrogel presents a diffusional barrier for water molecules penetrating into the polymer matrix and the drug molecules being released.[10–14]

Sodium Carboxymethyl Cellulose. SCMC is a low-cost, soluble, and polyanionic polysaccharide derivative of cellulose that has been employed in medicine as an emulsifying agent in pharmaceuticals and in cosmetics.[15] The many important functions provided by this polymer make it a preferred thickener, suspending aid, stabilizer, binder, and film-former in a wide variety of uses. In biomedicine, it has been employed for preventing postsurgical soft tissue and epidural scar adhesions.

Methylcellulose. Methylcellulose is neutral, odorless, tasteless, and inert. It swells in water to produce a clear to opalescent, viscous, colloidal solution and it is insoluble in most of the common organic solvents. However, aqueous solutions of methylcelluose can be diluted with ethanol. Methylcellulose solutions are stable over a wide range of pH (2–12) with no apparent change in viscosity. They can be used as bulk laxatives, so it can be used to treat constipation, and in nose drops, ophthalmic preparations, burn preparations ointments, etc. Although methylcellulose, when used as a bulk laxative, takes up water quite uniformly, tablets of methylcellulose have caused fecal impaction and intestinal obstruction. Methylcellulose attracts large amounts of water into the colon, producing a softer and bulkier stool so it is used to treat constipation, diverticulosis, hemorrhoids, and irritable bowel syndrome.[16] Methylcellulose dissolves in cold water, but those with higher degree of substitution (DS) result in lower solubility, because the polar hydroxyl groups are masked.

Ethylcellulose. It is the nonionic, pH-insensitive cellulose ether, insoluble in water but soluble in many polar organic solvents. It is used as a nonswellable, insoluble component in matrix or coating systems. When water-soluble binders cannot be used in dosage processing because of water sensitivity of the active ingredient, ethylcellulose is often used. It can also be used to coat one or more active ingredients of a tablet to prevent them from reacting with other materials or with one another. Ethylcellulose can also be used on its own or in combination with water-soluble polymers to prepare sustained-release film coatings that are frequently used for the coating of microparticles, pellets, and tablets.

In addition to ethylcellulose, hydroxyethylcellulose (HEC) is also nonionic, water-soluble, cellulose ether, easily dispersed in cold or hot water to give solutions of varying viscosities and desired properties, yet it is insoluble in organic solvents. It is used as a modified-release tablet matrix, a film former, and a thickener, stabilizer, and suspending agent for oral and topical applications when a nonionic material is desired. Many researchers such as Mura et al.[17] Friedman and Golomb,[18] and Soskolne et al.[19] have demonstrated the ability of ethylcellulose to sustain release of drugs.

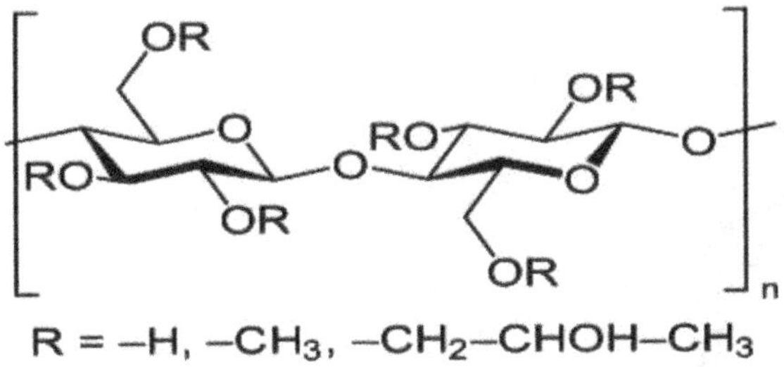

R = −H, −CH₃, −CH₂−CHOH−CH₃

Fig. 2 Structural formula of hydroxypropylmethylcellulose.
Source: From Builders & Attama[5] © 2011, with permission from Nova Science Publishers, Inc.

$C_6H_7O_2(OH)_x(OCH_3)_y[OCH_2CH(CH_3)OH]_z(OCOC_6H_5COOH)_u$

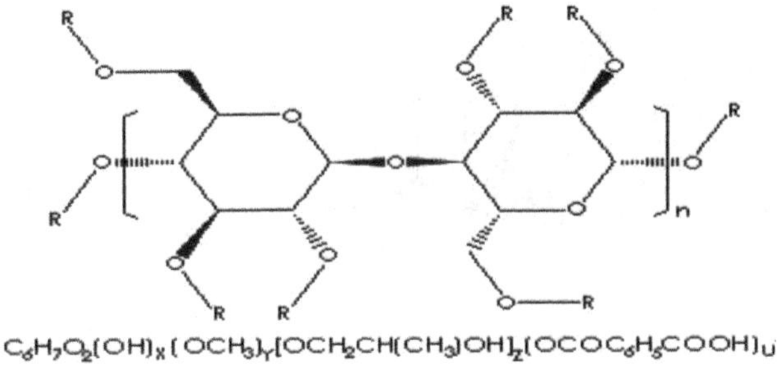

Fig. 3 Structural formula of cellulose acetate phthalate.
Source: From Builders & Attama[5] © 2011, with permission from Nova Science Publishers, Inc.

Hydroxypropyl Cellulose. Hydroxypropyl cellulose (HPC) is nonionic, water-soluble, and pH-insensitive cellulose ether. It can be used as thickening agent, tablet binder, modified-release polymer, and in film coating. By using solid dispersions containing a polymer blend, such as HPC and ethylcellulose, it is possible to precisely control the rate of release of an extremely water-soluble drug, such as oxprenolol hydrochloride.[20–24] In this case, the water-soluble HPC swells in water and is trapped in the water-insoluble EC so that the release of the drug is slowed. These studies have shown that there is a linear relationship between the rate of release of the water-insoluble drug and its interaction with the polymer.[25–27]

Hydroxypropylmethyl Cellulose. Hydroxypropylmethyl cellulose (HPMC) (Fig. 2) is water-soluble cellulose ether and it can be used as hydrophilic polymer for the preparation of controlled-release tablets. Water penetrates the matrix and hydrates the polymer chains, which eventually disentangle from the matrix. Drug release from HPMC matrices follows two mechanisms, drug diffusion through the swelling gel layer and release by matrix erosion of the swollen layer.[28–31]

Cellulose esters

Cellulose esters are part of a large family of cellulose derivatives that are produced by the esterification of the hydroxyl groups of each anhydroglucose monomer. The physical properties of cellulose esters depend on the cellulose chain length and on the type and amount of ester groups attached to the polymer chain. Cellulose acetate phthalate (CAP, Fig. 3) is a partial acetate ester of cellulose that has been reacted with phthalic anhydride. One carboxyl of the phthalic acid is esterified with the cellulose acetate. The finished product contains about 20% acetyl groups and about 35% phthalyl groups. In the acid form, it is soluble in organic solvents and insoluble in water. The nonenteric esters do not show pH-dependent solubility characteristics with the exception of cellulose acetate with low levels of acetyl group. Most nonenteric esters are insoluble in water. The salt form is readily soluble in water. This combination of properties makes it useful in enteric coating of tablets because it is resistant to the acid condition of the stomach but soluble in the more alkaline environment of the intestinal tract.[32]

Hemicellulose

In nature, hemicelluloses are found in the cell walls of woody and annual plants, together with cellulose and lignin.[33] Hemicellulose is made up of a group of complex low-molecular-weight polysaccharides that are bound to the surface of cellulose microfibrils, but their structure prevents them from forming microfibrils by themselves.

The sugar monomers in hemicellulose may include xylose, mannose, galactose, rhamnose, and arabinose. Hemicelluloses contain most of the D-pentose sugars, and occasionally small amounts of l-sugars as well. Their backbones consist mainly of β-1,4-linked D-glycans. Xylose is always the sugar monomer present in the largest amount, but mannuronic acid and galacturonic acid may also be present. Thus, the hemicellulose polysaccharides consist of mannans (Fig. 4), xylan (Fig. 5), and xyloglucan (Fig. 6). These can be extracted from the plant cell wall with a strong alkali. Most of the hemicellulose fraction is soluble in water after alkaline extraction. Xyloglucan has a similar backbone as cellulose, but contains xylose branches on three out of every four glucose monomers and the β→1, 4→linked D-xylan backbone of arabinoxylan contains arabinose branches.[34,35]

Glucomannan is the most commonly used form of the hemicelluloses. The chain structure consists of a linear backbone of (1→4)-linked β-D-mannopyranosyl units to which (1→6)-linked α-D-galactopyranosyl units are substituted. The Man:Gal ratio in the chemical structure of glucomannan shown in Fig. 4 is 2:1. Glucomannan is a water-soluble polysaccharide that is considered a dietary fiber. It is commonly used as a food additive, as an emulsifier, and a thickener. It is also used in drug delivery applications.[36,37] Glucomannans have been specifically derived from softwoods, roots, tubers, and bulbs. Their use in foods and drug delivery applications are primarily based on their viscosity-imparting, gelation, and swelling potentials. Majority have been used as matrices in oral controlled-release formulations. There has been an increase in the

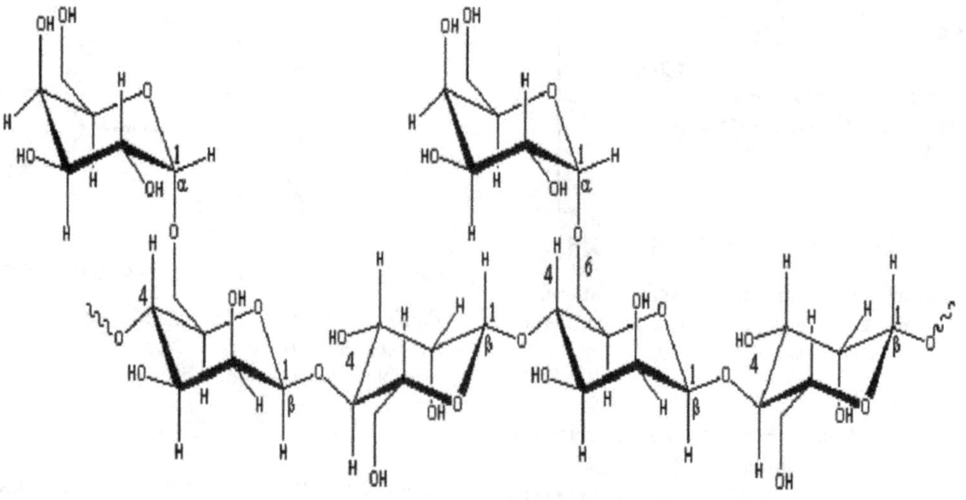

Fig. 4 The chemical structure of a typical galactomannan.
Source: From Builders & Attama[5] © 2011, with permission from Nova Science Publishers, Inc.

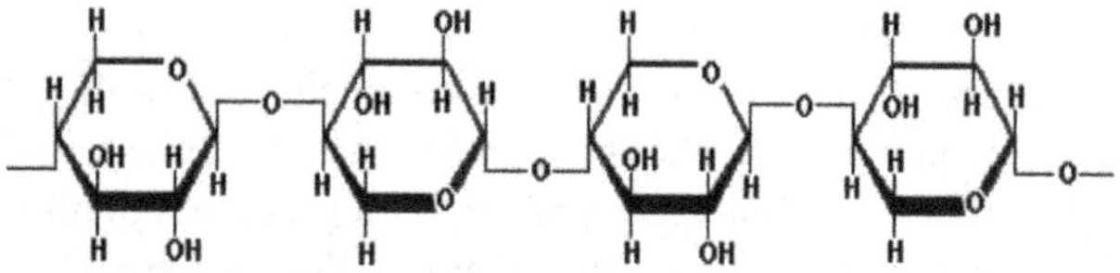

Fig. 5 The chemical structure of a typical xylan.
Source: From Builders & Attama[5] © 2011, with permission from Nova Science Publishers, Inc.

number of commercial glucomannan processed and standardized to meet pharmaceutical grade. The most commonly used types of glucomannan include konjac glucomannan, guar gum, and locust bean gum.[37–39]

FUNCTIONAL PROPERTIES OF CELLULOSE-BASED BIOPOLYMERS IN DRUG FORMULATION AND DELIVERY

A clear understanding of the physical and chemical (physicochemical) properties of biopolymers, especially those to be used for drug delivery application, is very critical to their use for both conventional and modified drug delivery systems. These properties aid in identification and can as well help in modeling the polymer to give a clear understanding how the materials will behave under various conditions. The physical properties of polymers essentially refer to those properties that do not change their chemical nature. The chemical properties refer to those properties that change their chemical nature. To determine these properties, a number of parameters listed here among others are usually evaluated:[5]

Particle Properties

Drug molecules and polymers for formulation consists multiparticulate powders, which are heterogeneous in shape, size, and size distributions. The particle properties of these polymers are critical as many of their physicofunctional properties (both molecular and particulate) as well as the biopharmaceutical properties and the performance of finished products are controlled by their particle characteristics. In the production and processing of cellulose powder into its derivative for direct compression technology, the materials' particle size and size distribution are critical as these affect powder bulkiness, flow, and compaction characteristics. Some particle properties measured during the assessment of solid particulate delivery systems and powdered polymeric excipients include particle size and size distribution, fractal dimension, particle density, and particle strength. These control the bulk powder properties such as flow properties, moisture content, bulk density, water sorption isotherms, compression properties, critical relative humidity, thermal properties, and moisture diffusivity.

Swelling

Swelling is one of the functional properties used to characterize biopolymers required for modified or controlled drug delivery systems. The biopolymers employed for modified drug delivery applications are hydrogels that form three-dimensional polymeric networks when they come into contact with water, they absorb many times their weight of water but they do not dissolve. Swelling of the cellulosics in different biorelevant media must be established.

Bioadhesion

In drug delivery applications, biopolymers with mucoadhesive properties serve to increase the residence time of the bioactive agents at the site of absorption, resulting in a steep concentration gradient to favor drug absorption and

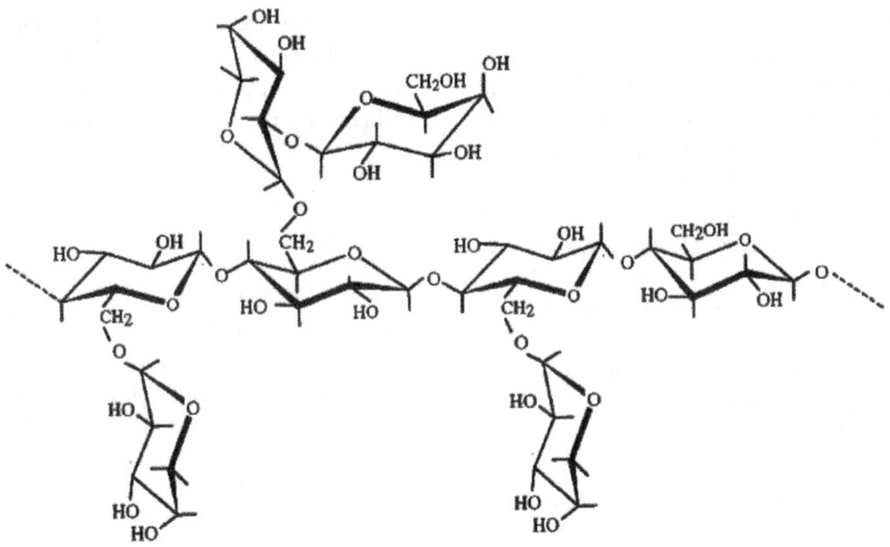

Fig. 6 The chemical structure of a typical xyloglucan.
Source: From Builders & Attama[5] © 2011, with permission from Nova Science Publishers, Inc.

localization in specified regions, thus, improving the bioavailability of the drug. Mucoadhesive polymers are often characterized by certain specific intrinsic properties that have been related to the muco/bioadhesive: These include the presence of strong hydrogen bond–forming group(s) such as carboxylate and hydroxyl groups, presence of a strong anionic charge, high molecular weight, high viscosity, high hydration capacity, sufficient chain flexibility, and high surface energy that favors spreading onto the mucus.

Crystallinity and Solid State Characteristics

Polymer crystallinity refers to the amount of crystalline region in a polymer with respect to its amorphous content. Characterization of polymer crystallinity is based on the molecular chain arrangement. The crystallinity of cellulose partly results from hydrogen bonding between the cellulosic chains, but some hydrogen bonding also occurs in the amorphous phase, although its organization is low. The structural integrity and stability have been monitored using differential scanning calorimetry (DSC). X-ray diffraction has been used to measure the degree of crystallinity of a biopolymer. In the solid state, a polymer may be completely amorphous, perfectly crystalline, or semi-crystalline.

Semi-crystalline polymers are composed of a combination of an amorphous and a crystalline region. Biopolymers are predominantly semi-crystalline as they are composed of varying ratios of crystalline and amorphous regions. The presence of crystalline and amorphous structures forms a material with superior properties in terms of strength and stiffness. The crystallinity of a biopolymer can be used to predict properties such as hardness, modulus, tensile strength, stiffness, moisture sorption, swelling, and melting

characteristics. Highly crystalline biopolymers are rigid, high melting, and less affected by solvent penetration. Amorphous polymers are usually less rigid, weaker, and more easily deformed.

Cellulose is stable to thermal degradation until about 370°C, and then decomposes almost completely over a very short temperature range. At 300°C the cellulose molecule is highly flexible and undergoes depolymerization by transglycosylation to create products such as anhydromonosaccharides, which include levoglucosan (1,6-anhydro-β-D-glyucopyranose) and 1,6-anhydro-β-D-glyucofuranose. Cellulose also produces combustible volatiles such as acetaldehyde, propenal, methanol, butanedione, and acetic acid. These are converted into low-molecular-weight products, randomly linked oligosaccharides and polysaccharides, which lead to carbonized products. Modification of the particle morphology of a biopolymer such as cellulose has been achieved by changing its glass transition temperature.

MODIFICATION OF CELLULOSE-BASED BIOPOLYMERS

The crystalline nature of cellulose originates from intermolecular forces between neighboring cellulose chains over long lengths. All native celluloses show the same crystal lattice structure, called cellulose I. However, various modifications of native cellulose can alter the lattice structure to yield other types of crystals.[40–42] The intermolecular forces in the crystalline domains are mainly hydrogen bonds between adjacent cellulose chains in the same lattice plane, which results in a sheet-like structure of packed cellulose chains. In addition, the sheets are probably connected to one another by hydrogen bonds and/or van der Waal's forces. The organization of cellulose molecules into parallel arrangements is responsible for the formation of crystallites. The length of an elementary crystallite ranges from 12 to 20 nm ($\approx$24–40 glucose units) and the width from 2.5 to 4 nm.[40–42] The crystalline parts of cellulose are rather resistant to degradation by enzymes and a system of several synergistically acting enzymes is necessary to obtain any significant hydrolysis.[43] However, the reactivity of cellulose can be greatly enhanced by various forms of treatment, such as swelling, degradation, or mechanical grinding, which break down the fibrillar aggregations.[40]

To increase their use and to fulfill the various demands for functionality of different cellulose products, they are often modified by physical, chemical, enzymic, or genetic means. Modification leads to changes in the properties and behavior of the polymer and, consequently, improvement of the positive attributes and reduction of the negative characteristics.[42,44,45] The properties of a modified polysaccharide depend on several factors, such as the modification reaction, the nature of the substitution group, the DS, and the distribution of the substitution groups. To direct a modification reaction toward a certain product with the desired properties, it is of importance to have knowledge of the correlations that exist between the modification process, chemical structure, and functional properties of the final product. However, the relationships, if any, between these parameters are still far from fully understood, largely due to difficulties in the elucidation of the modified polymer structure, including the distribution of substituent.[42]

Chemical modification can be used to tune some other cellulose properties, such as hydrophobic or hydrophilic character, elasticity, adsorption, microbial resistance, and heat and mechanical resistance.[46,47] The chemical modification of a surface of an organic polysaccharide follows the same principles as those established for other media, such as for silica gel. However, the hydroxyl groups of cellulose are less reactive and the beginning of the chemical modification takes place in primary hydroxyl found in carbon 6, which may occur for several different routes, highlighting that it is yet to occur also in the secondary hydroxyl groups present on carbons 2 and 3. Cellulose is not modified thermally through melting process,[48] because the decomposition temperature is less than melting temperature.[49] The major modifications of cellulose occur through halogenation, oxidation, etherification, and esterification.[46,50,51] Chemical modification implies the substitution of free hydroxyl groups in the polymer with functional groups, yielding different cellulose derivatives.[42] Chemical modification of cellulose is performed to improve processability and to produce cellulose derivatives (cellulosics) that can be tailored for specific industrial applications.[52,53] Large-scale commercial cellulose ethers include carboxymethyl cellulose (CMC), methyl cellulose, HEC, HPMC, HPC, ethyl hydroxyethyl cellulose (EHEC), and methyl hydroxyethyl cellulose (MHEC).[53]

Pretreatments of the cellulose fiber surface or physical modifications can clean the fiber surface, chemically modify the surface, stop the moisture absorption process, and increase the surface roughness.[54–57]

Physical Modifications

Native cellulose are commonly modified by physical, chemical, enzymic, or genetic means in order to obtain specific functional properties,[42,44,58,59] and to improve some of the inherent properties that limit their utility in certain application. Physical/surface modification of cellulose are performed in order to clean the fiber surface, chemically modify the surface, stop the moisture absorption process, and increase the surface roughness.[54–56] Among the various pretreatment techniques, silylation, mercerization, peroxide, benzoylation, graft copolymerization, and bacterial cellulose treatment are the best methods for surface modification of natural fibers.

Silylation, Mercerization, and Other Surface Chemical Modifications

Silane-coupling agents usually improve the degree of cross-linking in the interface region and offer a perfect bonding. Among the various coupling agents, silane-coupling agents were found to be effective in modifying the natural fiber–matrix interface and may reduce the number of cellulose hydroxyl groups in the fiber–matrix interface. In the presence of moisture, hydrolyzable alkoxy group leads to the formation of silanols. The silanol then reacts with the hydroxyl group of the fiber, forming stable covalent bonds to the cell wall that are chemisorbed onto the fiber surface.[60] Therefore, the hydrocarbon chains provided by the application of silane restrain the swelling of the fiber by creating a cross-linked network due to covalent bonding between the matrix and the fiber.[54] Cellulose fiber treatment with toluene dissocyanate and triethoxyvinyl silane could improve the interfacial properties. Silanes after hydrolysis undergo condensation and bond-formation stage and can form polysiloxane structures by reacting with hydroxyl group of the fibers.[54,56,61]

Mercerization is the common method to produce high-quality fibers.[62] Mercerization leads to fibrillation, which causes the breaking down of the composite fiber bundle into smaller fibers and reduces fiber diameter, thereby increases the aspect ratio, which leads to the development of a rough surface topography that results in better fiber–matrix interface adhesion and an increase in mechanical properties.[56,63] Moreover, mercerization increases the number of possible reactive sites and allows better fiber wetting. Mercerization has an effect on the chemical composition, degree of polymerization, and molecular orientation of the cellulose crystallites due to cementing substances like lignin and hemicellulose, which are removed during the mercerization process.

Peroxide treatment of cellulose fiber has attracted the attention of various researchers due to easy processability and improvement in mechanical properties. Organic peroxides tend to decompose easily to free radicals, which further react with the hydrogen group of the matrix and cellulose fibers.[56]

Polymer Grafting

Desirable and targeted properties can be imparted to the cellulose fibers through graft copolymerization in order to meet the requirement of specialized applications. Graft copolymerization is one of the best methods for modifying the properties of cellulose fibers. Different binary vinyl monomers and their mixtures have been graft-copolymerized onto cellulosic material for modifying the properties of numerous polymer backbones.[54,64] During past decades, several methods have been suggested for the preparation of graft copolymers by conventional chemical techniques. Creation of an active site on the preexisting polymeric backbone is the common feature of most methods for the synthesis of graft copolymers. The active site may be either a free radical or a chemical group that may get involved in an ionic polymerization or in a condensation process. Polymerization of an appropriate monomer onto this activated backbone polymer leads to the formation of a graft copolymer. Ionic polymerization has to be carried out in the presence of anhydrous medium and/or in the presence of considerable quantity of alkali metal hydroxide. Another disadvantage of the ionic grafting is that low-molecular-weight graft copolymers are obtained, while in case of free-radical grafting high-molecular-weight polymers can be prepared. The molecular weight affects the drug delivery applicability of these cellulosics.

Bacterial Modification

The coating of bacterial cellulose onto cellulose fibers provides new means of controlling the interaction between fibers and polymer matrices. Coating of fibers with bacterial cellulose not only facilitates good distribution of bacterial cellulose within the matrix, but also results in an improved interfacial adhesion between the fibers and the matrix. This enhances the interaction between the fibers and the polymer matrix through mechanical interlocking.[55,56,65] Surface modification of cellulose fibers using bacterial cellulose is one of the best methods for greener surface treatment of fibers. Bacterial cellulose has gained attention in the research area for the encouraging properties it possesses; such as its significant mechanical properties in both dry and wet states, porosity, water absorbency, moldability, biodegradability, and excellent biological affinity.[66] Because of these properties, bacteria cellulose has a wide range of

potential applications in drug delivery. *Acetobacter xylinum* (or *Gluconacetobacter xylinus*) is the most efficient producer of bacterial cellulose. Bacteria cellulose is secreted as a ribbon-shaped fibril, less than 100 nm wide, which is composed of much finer 2–4 nm nanofibrils.

Chemical Modifications

Chemical modification is based on reactions of the free hydroxyl groups in the anhydroglucose monomers, resulting in changes in the chemical structure of the glucose units and, ultimately, the production of cellulose derivatives.[42] Chemical modification implies the substitution of free hydroxyl groups in the polymer with functional groups, yielding different cellulose derivatives.[42] Usually, these modifications involve esterification or etherification reactions of the hydroxyl groups. Each AGU is available for up to three sites of substitution; the hydroxyl group on C-2, C-3 or C-6. The DS describes the average number of substituted hydroxyl groups per AGU and ranges from 0 to 3. The term DS applies to derivatives in which the substitution group terminates the reactive hydroxyl sites. Substitution by chemical groups that generate new free hydroxyl groups for further substitution is quantified by the molar substitution (MS). This value is defined as the average number of moles of substituent added per AGU.[67] The MS has no theoretical upper limit. Cellulose derivatives can be characterized by a number of factors, such as type and nature of substitution group, DS, MS, average chain length, and DP. These factors influence the functional properties of the derivative in various ways.[42,67,68]

Chemically modified celluloses were developed primarily in order to overcome the insoluble nature of cellulose, thus extending the range of applications of the polymer. Commercial cellulose derivatives are usually ethers or esters that are soluble in water and/or organic solvents. They are produced by reacting the free hydroxyl groups in the AGUs with various chemical substitution groups. The introduction of substituent disturbs the inter- and intramolecular hydrogen bonds in cellulose, which leads to liberation of the hydrophilic character of the numerous hydroxyl groups and restriction of the chains.[42,68,69] However, substitution with alkyl groups reduces the number of free hydroxyl groups. Cellulose derivatives are used in a wide range of industrial fields and their availability, economic efficiency, easy handling, and low toxicity are reasons for a continuously expanding worldwide market. Common industrial derivatives, properties, and their fields of application are summarized in Table 1. Cellulose ethers are the most widely used derivatives, although there are also some commercial esters, as shown in Table 1.

Preparation of Cellulose Ethers

Etherification of cellulose proceeds under alkaline conditions, generally in aqueous NaOH solutions. Treatment of

Table 1   Some commercially marketed cellulose esters and ethers

Cellulose esters			
Nitrate	1.5–3.0	MeOH, $PhNO_2$, ethanol-ether	Films, fibers, explosives
Acetate	1.0–3.0	Acetone	Films, fibers, coatings, heat- and rot-resistant fabrics
Cellulose ethers			
Methyl	1.5–2.4	Hot H_2O	Food additives, films, cosmetics, greaseproof paper
Carboxymethyl	0.5–1.2	H_2O	Food additives, fibers, coatings, oil-well drilling muds, paper size, paints, detergents
Ethyl	2.3–2.6	Organic solvents	Plastics, lacquers
Hydroxyethyl	Low DS	H_2O	Films
Hydroxypropyl	1.5–2.0	H_2O	Paints
Hydroxypropylmethyl	1.5–2.0	H_2O	Paints
Cyanoethyl	2.0	Organic solvents	Products with high dielectric constants, fabrics with heat and rot resistances

Source: From Varshney & Naithani[71] and Arthur,[72] with permission from Elsevier.

native cellulose with NaOH causes the cellulose to swell, which makes it more readily accessible to the modification reagent.[42,68] Thorough mixing and stirring are of vital importance to ensure uniform swelling and alkali distribution, which are the most important conditions for homogeneous etherification. Uneven distribution of the substituents causes severe loss in solubility due to the unetherified regions in the final product. Two types of reactions dominate cellulose etherification.[42,70]

William etherification

An organic halide is used as the etherification reagent and alkali in amounts that are stoichiometrically equivalent to the reagent are consumed. Unreacted alkali must be washed out of the final product as a salt.[42]

Alkaline-catalyzed oxalkylation

In this reaction, an epoxide is added to the swollen alkali cellulose. Only catalytic amounts of alkali are required; thus, in principle, no alkali is consumed. The reaction may proceed further as new hydroxyl groups are generated during this reaction.

Enzymic Method of Chemical Modification

Enzymic methods for the determination of the substituent distribution in cellulose derivatives are based on the selective degradation of the modified polymer; nonsubstituted regions are easily hydrolyzed compared with low-substituted regions, whereas highly substituted areas are not hydrolysed at all and remain intact. However, one limitation of this enzymic approach is that the hydrolysis of a certain glucosidic linkage does not simply result in cleavage or not, but the hydrolysis rate may differ by some orders of magnitude. The rate of enzymic hydrolysis seems to depend not only on several factors, such as the DS, the nature of the substituent, and position of the substituents in the neighboring glucose unit, but also on the substituent distribution further along the chain, as an oligomeric sequence is usually involved in the formation of the enzyme–substrate complex.[42,43] In addition, the accessibility to enzymic hydrolysis is dependent on the physical features of the polymer, such as crystallinity, degree of swelling, and solubility. In many cases, chemical modification makes the polymer less crystalline and enhances its water solubility, thus increasing the susceptibility of cellulose derivatives to enzymic attack.[42,43]

Cellulose-hydrolyzing enzymes

Cellulose-hydrolyzing enzymes are known as cellulases and they catalyze the hydrolysis of β-D-glucosidic linkages in cellulose and other β-D-glucans. Cellulases are produced by microorganisms, including bacteria and fungi, for example, the fungi *Trichoderma reesei* and *Humicola insolens*.[73]

Endo-1,4-β-glucanase

Endo-1,4-β-glucanase [1,4-(1,3; 1,4)-β-D-glucan 4-glucanohydrolase], often called cellulase, is an endo-enzyme that catalyzes the hydrolysis of (1→4)-β-D-glucosidic linkages in cellulose, lichenin, and cereal β-glucans.[42,73] Endo-1,4-β-glucanase shows low activity on crystalline cellulose, and it is generally believed that the enzyme acts through random hydrolysis of the internal bonds in the amorphous regions of the polymer. Thus, new chain ends that become available for further enzymic hydrolysis are formed.[43] The major hydrolysis products are cellobiose and cellotriose, and also longer cello-oligosaccharides with different chain lengths depending on the source of the enzyme.[42,43]

Biomimetic—Colloid

β-Glucosidase

β-Glucosidase (β-D-glucoside glucohydrolase) is an exo-enzyme that hydrolyzes m-terminal nonreducing β-D-glucose residues in β-D-oligosaccharides with the release of β-glucose. The enzyme is also called cellobiase, as it readily hydrolyzes cellobiose to glucose.[42,73]

Cellobiohydrolase

Cellobiohydrolase (1,4-D-glucan cellobiohydrolase) is an exo-glucanase that catalyzes the hydrolysis of (1→4)-β-D-glucosidic linkages in cellulose with the release of cellobiose. Various forms of this enzyme are thought to hydrolyze cellulose chains either from the reducing or from the nonreducing end.[42,43]

Other Derivatization Processes

Cellulose-based hydrogels and cross-linking strategies

Cellulose-based hydrogels can be obtained via physical or chemical stabilization of aqueous solutions of cellulosics. Additional natural and/or synthetic polymers might be combined with cellulose to obtain composite hydrogels with specific properties.[74–76] Physical, thermoreversible gels are usually prepared from water solutions of methylcellulose and/or hydroxypropyl methylcellulose (in a concentration of 1–10% by weight).[77] The gelation mechanism involves hydrophobic associations among the macromolecules possessing the methoxy group. At low temperatures, polymer chains in solution are hydrated and simply entangled with one another. As temperature increases, macromolecules gradually lose their water of hydration, until polymer–polymer hydrophobic associations take place, thus forming the hydrogel network. The sol–gel transition temperature depends on the DS of the cellulose ethers as well as on the addition of salts. A higher DS of the cellulose derivatives provides them a more hydrophobic character, thus lowering the transition temperature at which hydrophobic associations take place. A similar effect is obtained by adding salts to the polymer solution, since salts reduce the hydration level of macromolecules by recalling the presence of water molecules around themselves. Both the DS and the salt concentration can be properly adjusted to obtain specific formulations gelling at 37°C and thus potentially useful for biomedical applications.[74,78–80] Liquid formulations, either mixed with therapeutic agents or not, are envisaged to be injected in vivo and their cross-linking reaction triggered by the only physiological environment. However, physically cross-linked hydrogels are reversible,[81] thus might flow under given conditions (e.g., mechanical loading) and might degrade in an uncontrollable manner. Owing to such drawbacks, physical hydrogels based on methylcellulose and HPMC are not recommended for use in vivo. In vitro, methylcellulose hydrogels have been recently proposed as novel cell sheet harvest systems.[79] As opposed to physical hydrogels that show flow properties, stable and stiff networks of cellulose can be prepared by inducing the formation of chemical, irreversible cross-links among the cellulose chains. Either chemical agents or physical treatments (i.e., high-energy radiation) can be used to form stable cellulose-based networks. The degree of cross-linking, defined as the number of cross-linking sites per unit volume of the polymer network, affects the diffusive, mechanical, and degradation properties of the hydrogel and can be controlled to a certain extent during the synthesis. Specific chemical modifications of the cellulose backbone might be performed before cross-linking, in order to obtain stable hydrogels with given properties. For instance, silylated HPMC has been developed, which cross-links through condensation reactions upon a decrease of the pH in water solutions. Such hydrogels show potential for the in vivo delivery of chondrocytes in cartilage tissue engineering.[82,83] As a further example, tyramine-modified SCMC has been synthesized to obtain enzymatically gellable formulations for cell delivery.[84] Photocross-linking of water solutions of cellulose derivatives is achievable following proper functionalization of cellulose. Depending on the cellulose derivatives used, a number of cross-linking agents and catalysts can be employed to form hydrogels. Epichlorhydrin, aldehydes and aldehyde-based reagents, urea derivatives, carbodiimides, and multifunctional carboxylic acids are the most widely used cross-linkers for cellulose. However, some reagents, such as aldehydes, are highly toxic in their unreacted state. Although unreacted chemicals are usually eliminated after cross-linking through extensive washing in distilled water, as a rule, toxic cross-linkers should be avoided to preserve the biocompatibility of the final hydrogel, as well as to ensure an environmentally sustainable production process. The cross-linking reactions among the cellulose chains activated by chemical agents might take place in water solution, organic solvents, or even in the dry state (e.g., polycarboxylic acids can cross-link cellulose macromolecules via condensation reactions, which are favored at high temperature and in the absence of water).[74,85–88] Novel superabsorbent cellulose-based hydrogels crosslinked with citric acid have been recently reported, which combine good swelling properties with biodegradability and absolute safety of the production process.[74,88] In light of environmental and health safety concerns, radiation cross-linking of polymers, based on gamma radiation or electron beams, has been receiving increasing attention in the past years as it does not involve additional chemical reagents, is easily controllable, and, in cases of biomedical applications, allows the simultaneous sterilization of the product. High-energy radiation usually leads to chain scission of the polymer and this has been shown also for cellulose.[89]

However, several cellulosics can be cross-linked under relatively mild radiation, both in aqueous solutions and in solid form, because cross-linking prevails over degradation.[90–92] The cross-linking reaction is affected by the irradiation dose as well as by the cellulose concentration in solution.[74]

Conventional Methods for Cellulose Dissolution and Modifications

Cellulose in aqueous phase are modified by introducing functional groups into the cellulose macromolecules to substitute the free hydroxyl groups, either in the heterogeneous phase or in the homogeneous phase.[49,93] There are many solvents used to dissolve and modify cellulose, such as N,N-dimethylacetamide/lithium chloride, dimethylsulfoxide (DMSO)/tetrabutylammonium fluoride, N-methyl-morpholine-N-oxide and other molten salt hydrates, such as $LiClO_4 \cdot 3H_2O$.[94,95] Commercially, two major solvent systems were used for dissolving and regenerating cellulose materials; one is carbon disulfide solvent used in viscose process and the other is N-methylmorpholine-N-oxide (NMMO) solvent used in Lyocell process.[49]

Ionic Liquids for Cellulose Dissolution and Modification

Recently, certain ionic liquids (ILs) have been applied as green solvents that would dissolve cellulose[96–99] and function as inert and homogeneous reaction media. During the past decade, many researches and studies were done on the application of ionic liquid in cellulose dissolution and modification.[97,98,100–105] ILs have been attracting interest because of their wonderful properties such as high thermal stability, lack of inflammability, low volatility, chemical

stability, and excellent solubility with many organic compounds.[106,107] Many kinds of room-temperature ILs,[108] with a variety of structures, have shown a good ability to dissolve cellulose, such as halogen-based ILs, e.g., 1-butyl-3-methyl- and 1-allyl-3methyl-imidazolium, which contain chloride. Also some imidazolium-based ILs containing phosphate, formate, and acetate anion have shown high ability for dissolving cellulose.[106,107] ILs have been regarded as low-melting (<100°C) salts, and hence, it forms liquids that consist of pure cations and anions.[49] In 1934, Graenacher discovered that molten N-ethylpyridinium chloride, in the presence of nitrogen-containing bases, could be used to dissolve cellulose.[99] It was reported that cellulose could be dissolved in ILs without formation of any derivative—a great progress in cellulose dissolution and cellulose modification had been established and interesting results have been reported by various pioneers in the optimization and applications of ILs in cellulose dissolution and functionalization.[49,96,98,108] Table 2 lists some ILs used for cellulose dissolution, while Fig. 7 shows the possible ways of modifying cellulose with other reagents by use of ILs as a medium for cellulose modification.

Cyclic anhydrides like acetic, phthalic, and succinic anhydride have been widely used in cellulose modifications to produce different cellulose derivatives such as cellulose acetate, cellulose butyrate, cellulose benzoate, cellulose phthalate, and cellulose with or without catalyst,[49,97] which have several applications such as water absorbents for soil in agriculture, drug delivery system, and as thermoplastic.

Due to the wide application of cellulose acetate, cellulose acetylation in ILs has taken significant attention. Several investigations of cellulose acetylation with acetic anhydride or acetyl chloride in ILs have been done, and cellulose acetates with a high DS have been obtained.[96,109–111] The acetylation of cellulose in an ionic liquid has been

Table 2 Different types of ILs used for cellulose dissolution

| Cellulose | Ionic liquids | Conditions | | Solubility (%) |
		Temperature (°C)	Time (min)	
MCC pulp cotton (linter)	[Amim]Cl	100–130	40–240	5–14.5
Avicel (MCC)	[Amim]Cl	90	720	5
Pulp (DP = 1000)	[Bmim]Cl	100	—	10
Pulp (DP = 286)	[Bmim]Cl	83	720	18
Avicel (MCC)	[Bmim]Cl	100	120	20
Avicel (MCC)	[Bmim]Cl	100	60	20
Pulp	[Bmim][Cl]	85	—	13.6
Pulp	[Emim]Cl	85	—	15.8
Avicel (MCC)	[Emim]Cl	100	60	10–14
Avicel (MCC)	[Emim]Cl	90	—	5
Pulp	[Emim]OAc	85	—	13.5
Avicel (MCC)	[Emim]OAc	110	—	15

Source: From Magdi et al.,[49] with permission from the International Journal of Engineering Science and Technology.

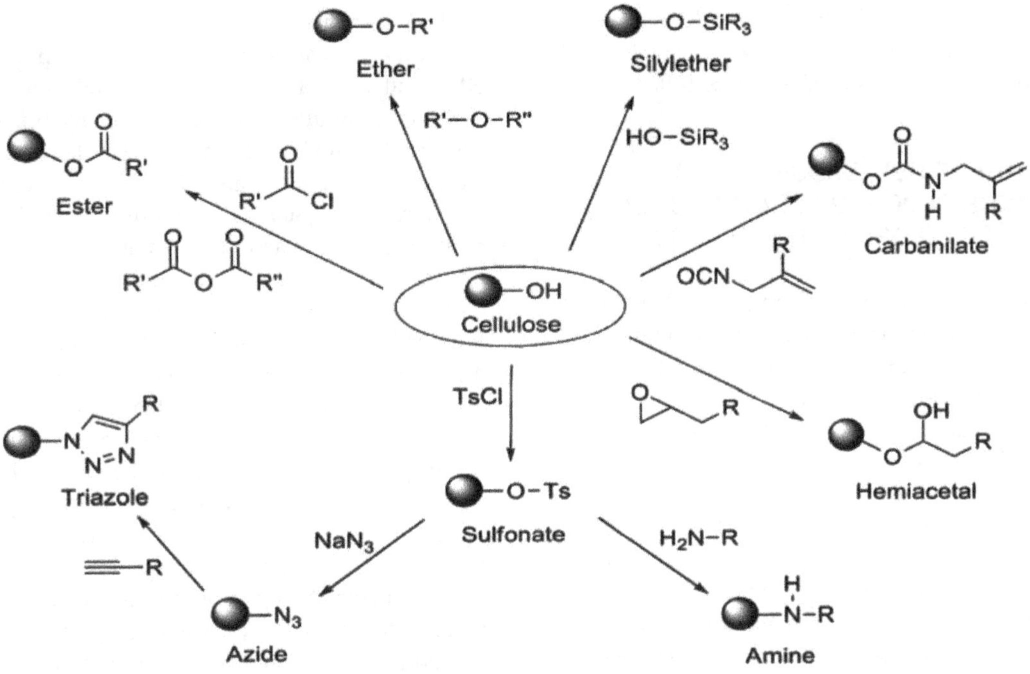

Fig. 7 Functionalization routes for cellulose in ILs medium.
Source: From Magdi et al.,[49] with permission from the International Journal of Engineering Science and Technology.

introduced as a novel process by Zhang and coworkers.[108] They reported that acetylation of cellulose occurs under homogeneous conditions and in the absence of any catalyst by using allylimidazolium-based ionic liquid. Cellulose solution (5%) was prepared by dissolving cellulose with degree of polymerization of 650 into [Amim]Cl ILs at 100°C for 15 min, followed by the addition of acetic anhydrate to acetylate cellulose under a nitrogen atmosphere and temp 80–100°C for 25 min–23 hours. Different degrees of substitution in the range 0.94–2.74 were obtained.[49]

Ionic Liquid as Media for Modification of Cellulose by Grafting Copolymerization and Blends (Composites)

Cellulose has the ability to blend with other biopolymers such as silk, wool, chitin, chitosan, elastin, collagen, keratin, and polyhydroxyalkanoate after dissolution in ILs.[49,99,112] Hence, this part of entry has been directed to overview the cellulose derivatives and their modification by blending graft copolymerization in ILs. Grafting copolymerization is usually achieved by modifying the cellulose molecules through the creation of branches (grafts) of synthetic polymers that impart specific properties onto the cellulose substrate, without destroying its intrinsic properties.[113] Chemical modification of cellulose through graft copolymerization has also been demonstrated to be a promising method for the preparation of new materials, enabling the introduction of special properties into cellulose without destroying their intrinsic characteristics and enlarging their scope for

potential applications.[114] Ring-opening polymerization is a way of grafting copolymers. Generally, it has been accomplished by using cyclic monomer such as oxiranes (epoxides), lactons, amino acid N-carboxy anhydrides (NCAs), and 2-alkyl oxazolines,[115] with cellulose/IL solution. But the formation of homopolymer during the grafting process is unwanted, which should be removed by suitable solvents in extraction process. Zhang et al.[114,116] grafted cellulose in opening ring polymerization. The grafted materials were characterized by FTIR (confirmed the introduction of the side chain into the cellulose backbone via graft copolymerization) and SEM optical microscopy (showed good uniformity). Lin et al.[117] applied microwave heating and obtained similar results.

Cellulose Blends (Composites) in ILs Media

Kadokawa, Masa-Aki, and Akihiko-Kaneko[118] successfully obtained a gel by dissolving a mixture of cellulose–starch in [Bmim]Cl at 100+°C for 24 hours and thereafter, cooling the composites solution to room temperature and kept between two glass plates for five days. ILs were removed by ethanol or acetone. The composite materials were characterized by XRD and SEM, which showed good compatibility with the thermal analysis and TGA exhibiting higher weight loss at higher temperatures.[119] A novel heparin- and cellulose-based biocomposite has been prepared by dissolution of cellulose and heparin in ILs. This study was investigated by Murugesan et al.[120] They obtained a membrane film of cellulose/heparin composite.

Cellulose was dissolved in [Bmim]Cl at 70°C at a concentration of 10 %wt with microwave for 4–5 sec and heparin was dissolved in [Emim]BA at 35°C for 20 min.[49]

APPLICATION OF CELLULOSE-BASED BIOPOLYMERS IN DOSAGE FORM DESIGN AND NOVEL DRUG DELIVERY SYSTEMS

Pharmaceutical formulation development involves various components in addition to the active pharmaceutical ingredients. In recent years, excipient development has become a core area of research in pharmaceutical drug delivery because it influences the formulation development and drug delivery process in various ways. Biopolymers are choice of research as excipient because of their low toxicity, biodegradability, stability, and renewable nature. Cellulose-based biopolymers are naturally obtainable macromolecules and they play an important role in biomedicine with applications in tissue engineering, regenerative medicine, drug-delivery systems, and biosensors. The inherent recyclability, reproducibility, cost-effectiveness, and availability in a wide variety of forms, biocompatibility and biodegradability of these materials make them particularly useful in biomedical applications. Thus, cellulose-based biopolymers are employed pharmaceutically in the formulation and delivery of bioactive compounds as immediate-release and/or sustained/prolonged/extended/modified/controlled-release dosage forms, specialized bioactive carriers as well as nanoparticulate drug delivery systems.

Immediate-Release Dosage Forms

Immediate release refers to the instantaneous availability of drug for absorption or pharmacologic action in which drug products allow drugs to dissolve with no intention of delaying or prolonging dissolution or absorption of the drug. Cellulose-based excipients have been well explored and used in pharmaceutical formulation development of immediate-release dosage forms. Cellulose-based biopolymers are often used as tablet binding, thickening and rheology control agents, for film formation, water retention, improving adhesive strength, and as suspending and emulsifying agents.[8] Cellulose ethers are widely used as important excipients for designing matrix tablets. Large-scale commercial cellulose ethers including CMC, methyl cellulose, HEC, HPMC, HPC, EHEC, and MHEC have been exploited as potential raw materials in the design of immediate-release dosage forms.[48] For instance, methylcellulose has been employed in the formulation of bulk laxatives, as well as solid dispersion granules;[121] MCC has been employed as suspending agent in zinc oxide and sulfadimidine suspensions and is used as diluent and disintegrating agent for immediate-release oral solid dosage forms;[122,123] SCMC is employed as suspending agent

in chalk suspensions and could be used in the formulation of pharmaceutical suspensions for extemporaneous use.[124] HEC and HPC are used in hydrophilic matrix systems, while ethylcellulose can be used in hydrophobic matrix systems.[125] HPC has been employed to formulate buccal tablets containing lidocaine. These tablets were more effective in reducing pain and decreasing the healing time than both experimental and plain tablets.[126] HEC is used as a thickner, stabilizer, and suspending agent for oral and topical applications when a nonionic material is desired. Ethylcellulose can be used to coat one or more active ingredients of a tablet to prevent them from reacting with other materials or with one another, can prevent discoloration of easily oxidizable substances such as ascorbic acid, and allows granulations for easily compressed tablets and other dosage forms.[17] Also, liquid and semi-solid pharmaceutical dosage forms are important physicochemical systems for medical treatment, which require rheological control and stabilizing excipients as essential additives. CMC can be used to adjust the viscosity of syrups.[125] SCMC is a low-cost, soluble, and polyanionic polysaccharide derivative of cellulose that has been employed as an emulsifying agent in pharmaceuticals. The very many important functions provided by this polymer make it a preferred thickening, gelling agent, protective colloid, and film-former in jellies; stabilizer, thickner, and film-former in ointments, creams, lotions, and emulsions; thickner and suspending aid in syrups and suspensions; tablet binder, granulating aid, and tablet-coating film-former.[127] HPMC, a water-soluble cellulose ether, can be used as a hydrophilic polymer for the preparation of immediate-release dosage forms. For instance, immediate-release HPMC matrix tablets of pseudoephedrine, atenolol, and naproxen have been assessed by researchers.[125,128,129]

Immediate-release oral dosage forms (such as tablets and capsules) are the most widely used drug-delivery systems available. These products are designed to disintegrate in the stomach followed by their dissolution in the fluids of the gastrointestinal tract.[130] Dissolution of the drug substance, under physiological conditions, is essential for its systemic absorption. For this reason, dissolution testing is typically performed on solid dosage forms to measure the drug release from the drug product as a test for product quality assurance/product performance and to determine the compliance with the dissolution requirements.[131] Thus, cellulose-based biopolymers, in addition to their usefulness in the design and formulation of immediate-release dosage forms, could also be important excipients for establishing the potential usefulness of the disintegration test as drug product acceptance. A group of researchers has previously assessed verapamil hydrochloride tablets formulated using lactose monohydrate (LMH) as filler, HPMC as binder, SCMC as disintegrating agent, and established that only one formulation (of these tablets) might be suitable for using the disintegration test instead of the dissolution test as the drug product acceptance criteria.[131]

These researchers highlighted the need for systemic studies before using the disintegration test, instead of the dissolution test as the drug acceptance criterion.

Sustained/Modified/Controlled-Release Dosage Forms

Cellulose-based biopolymers are an important class of materials for pharmaceutical and biotechnological applications. There is a great potential in utilizing cellulose-based biopolymers as pharmaceutical adjuvants and in modified/sustained/controlled-release drug delivery systems. Modified-release dosage forms include both delayed and extended-release drug products. Delayed release is defined as the release of a drug at a time other than immediately following administration, while extended-release products are formulated to make the drug available over an extended period after administration. Controlled release includes extended-release and pulsatile-release products. Pulsatile release involves the release of finite amounts (or pulses) of drug at distinct intervals that are programmed into the drug product. Modified-release technologies utilize polymers such as cellulose-based biopolymers to alter the site or time of drug release within the gastrointestinal tract.[125] The need for modified-release technologies arose from an understanding that disintegration of a dosage form in the stomach, resulting in immediate release of drug, is not always desirable. In recent years, there has been an increasing tendency to deliver drug entities as modified-release formulations, and even though it can be appreciated that the unit cost of a modified-release formulation will be greater than the equivalent immediate-release variety, the former version may confer a reduction in overall healthcare costs. This may be in terms of a reduction in the number of doses to be taken to achieve the desired therapeutic effect, therefore reducing overall medication costs, or the subsequent improvement in compliance, negating the implications of ineffective therapy, or the need for further medication required to treat drug-induced side effects of the original treatment.[132]

Different cellulose-based biopolymers can be used either singly or in combination as matrix formers for sustained/modified/controlled drug delivery in different dosage forms ranging from oral to topical drug delivery systems. Hydrophilic cellulose-based biopolymer matrix systems are widely used for designing oral controlled drug delivery dosage forms because of their flexibility to provide a desirable drug-release profile, cost-effectiveness, and broad regulatory acceptance. But owing to rapid diffusion of the dissolved highly water-soluble drugs through the hydrophilic gel network, hydrophobic polymers are usually included in the matrix system to extend the release of highly water-soluble drugs.[133,134] These hydrophobic polymers may or may not be cellulose-based.

Controlled drug delivery remains a research focus for public health to enhance patient compliance and drug efficiency and to reduce the side effects of drugs. Cellulose-based biopolymers are employed in the controlled release of drugs via solid dispersions, a novel drug-delivery system in which compounds are dispersed into water-soluble carriers. This has been generally used to improve the dissolution properties and the bioavailability of drugs that are poorly water soluble.[121,135] Methylcellulose has the hydroxyl group in a structure and is interactive with the carboxylic acid of carboxyvinyl polymer, as well as poly(ethylene oxide) (PEO). Ozeki, Yuasa, and Okada[135] studied the release of phenacetin from the solid dispersion granules containing different ratios of methylcellulose and carboxyvinyl polymer. They found out that it is feasible to control phenacetin release from methylcellulose–carboxyvinyl polymer solid dispersions by controlling the complex formation between methylcellulose and carboxyvinyl polymer, which can be accomplished by varying the methylcellulose–carboxyvinyl polymer ratio and the molecular weight of the methylcellulose. In addition, by using solid dispersions containing a polymer blend, such as HPC and ethylcellulose, it is possible to precisely control the rate of release of an extremely water-soluble drug, such as oxprenolol hydrochloride.[136] In this case, the water-soluble HPC swells in water and is trapped in the water-insoluble ethylcellulose so that the release of the drug is slowed. These studies have shown that there is a linear relationship between the rate of release of the water-insoluble drug and its interaction with the polymer.[137] Furthermore, HPC offers interesting characteristics as controlled-release matrices. Gon et al.[138] observed that graft copolymers could stand alone as an effective matrix for tablets designed for drug-delivery systems. Similarly, buccal delivery formulations containing HPC and polyacrylic acid have been in use for many years,[139,140] with various ratios of the two polymers. More so, mucoadhesive delivery systems based on HPC have been reported for different drugs.[126,141] Furthermore, the ability of ethylcellulose and HEC to sustain the release of drugs has been demonstrated.[127] Additionally, SCMC can be used in the preparation of semi-interpenetrating polymer network microspheres by using glutaraldehyde as a cross-linker. Ketorolac tromethamine, an anti-inflammatory and analgesic agent, was successfully encapsulated into these microspheres and drug encapsulation of up to 67% was achieved. The diffusion coefficients decreased with increasing cross-linking as well as increasing content of SCMC in the matrix, and *in vitro* release studies indicated a dependence of release rate on both the extent of cross-linking and the amount of SCMC used to produce microspheres.[142] Controlled-release preparations of indomethacin could be employed to increase patient compliance and to reduce adverse effects, fluctuation in plasma concentration, and dosing frequency. Waree and Garnpimol[143] prepared a complex of chitosan and CMC and cross-linked by glutaraldehyde to control the release of indomethacin from microcapsules. The membrane of the microcapsules was formed by electrostatic

Biomimetic—Colloid

interaction between positively charged amine on the chitosan chain and the negatively charged hydroxyl group on the CMC chain, and the concentration of CMC affected the formability of chitosan–CMC microcapsules.[144]

Moreover, HPMC, a water-soluble cellulose ether, can be used as hydrophilic polymer for the preparation of controlled-release tablets. Khanvilkar, Ye, and Moore[145] investigated the effects of a mixture of two different grades of HPMC and the apparent viscosity on drug-release profiles of extended-release matrix tablets. The study indicated that lower and higher viscosity grades of HPMC can be mixed uniformly in definite proportions to get the desired apparent viscosity. Also, incorporating a low viscosity grade of HPMC in the formulation would lead to a significantly shorter lag time. However, it imposes minimal impact on the overall dissolution profile. The study showed that the drug release from an HPMC matrix tablet prepared by dry blend and direct compression approach is independent of tablet hardness, is diffusion-controlled, and depends mostly on the viscosity of the gel layer formed.[145] In addition, Ye et al.[146] studied the effect of manufacturing process on the dissolution characteristics of HPMC matrix tablets. These researchers reported that when HPMC matrix tablets were prepared by wet-granulation approach, the tablet hardness, distribution of HPMC within the tablet (intergranular and intragranular), and the amount of water added in the wet granulation step all have a significant impact on dissolution. The results also indicated that incorporating partial amount of HPMC inter-granularly in the dry-blend step, drug-release profiles could be made much less sensitive to the manufacturing process. In a related study, Liu et al.[147] employed alginate as the gelling agent in combination with HPMC (viscosity-enhancing agent) in controlling the release of gatifloxacin. The study showed that the alginate/HPMC solution retained the drug better than the alginate or HPMC solutions alone, indicating that the alginate/HPMC mixture can be used as an in situ gelling vehicle to enhance ocular bioavailability and patient compliance.

Furthermore, owing to the hydration and gel-forming properties of HPMC, it can be used to prolong the release of bioactives. The yahom (a well-known traditional remedy/medicine for treatment of nausea, vomiting, flatulence, and unconsciousness in Thailand) buccal tablet possessed antimicrobial activities that could be able to cure the oral microbial infection and aid in wound healing, but the addition of polyvinyl pyrrolidone (PVP) combined with HPMC could promote the bioadhesivity of yahom tablet.[148] The research by Chantana, Juree, and Thawatchaj[149] indicated that PVP had higher water sorption and erosion, whereas HPMC could prolong the erosion of yahom buccal tablet, and that the tablet containing 50% yahom, which had the polymer mixture of PVP: HPMC 1:2 was suitable for use as buccal tablet. In addition to the cellulose derivatives, cross-linked high amylose starch (CLA) has been successfully used as a controlled-release excipient for the

preparation of solid dosage forms.[150] Rahmouni et al.[151] characterized the gel matrix properties of binary mixtures of CLA/HPMC, and studied the effect of incorporated HPMC on the release kinetics of three model drugs of different solubilities such as pseudoephedrine sulfate (very soluble), diclofenac sodium (sparingly soluble), and prednisone (very slightly soluble). These researchers found out that the swelling characteristics and erosion of granulated cross-linked high amylose starch (CLAgr)/HPMC tablets increased with HPMC concentration and incubation time, and that the presence of HPMC in CLA tablets at concentration 10% protected CLA against α-amylase hydrolysis and reduced the release rate of poorly and moderately water-soluble drugs. However, the release of the highly water-soluble model drug, which occurred predominantly by diffusion, was rapid both in the presence or in the absence of HPMC. In another study, the effect of the concentration of HPMC on naproxen release rate was evaluated. The result showed that an increased amount of HPMC resulted in reduced drug release. The inclusion of buffers to increase the dissolution and to decrease the gastric irritation of weak acid drugs, such as naproxen in the HPMC matrix tablets, enhanced naproxen release. The inclusion of sodium bicarbonate and calcium carbonate in the HPMC matrix improved the naproxen dissolution; however, including sodium citrate did not produce any effect on naproxen dissolution.[152]

Many cellulose-based biopolymer blends have been used in the formulation of bioadhesive/mucoadhesive drug-delivery systems in the form of microparticles, tablets, patches, hydrogels or films. Builders et al.[7] prepared and evaluated mucinated cellulose microparticles for controlled drug-delivery application by mixing of colloidal dispersions of porcine mucin and MCC. The hybrid biopolymer was recovered by precipitating at controlled temperature and pH conditions using acetone. The mucoadhesive property of the new polymer was similar to that of mucin. Scanning electron micrographs (SEMs) showed that the microparticles generated from the hybidization were similar to those of MCC, but with larger and denser particles. The FT-IR spectrum and DSC thermogram of the hybrid polymer were characteristically different from mucin and MCC. The presence of new peaks in the FT-IR spectrum and distinct cold crystallization exotherm, which were absent in both mucin and MCC, confirmed the formation of a new polymer type with synergistic physicochemical and functional properties.[36] The use of admixtures of Carbopols (Carbopols 940 and 941) and SCMC in the formulation of bioadhesive metronidazole tablets has been studied by Ibezim et al.[153] In a related research, ternary cellulose-based biopolymer blends consisting of SCMC, acacia, and Veegum were evaluated for bioadhesive delivery of metronidazole comparing them with the performances of the single biopolymers.[154] Bioadhesive characteristics of tablets prepared with SCMC alone was highest, but blending improved the bioadhesive strengths of acacia and Veegum tablets. Drug release from the single polymers and the ternary biopolymer blends was prolonged, indicating their suitability for the delivery of metronidazole by bioadhesive controlled-release mechanism. Similarly, hydrogel bead-delivery systems of hydrochlorothiazide were studied by Attama and Adikwu.[155] The hydrogel beads consisted of blends of tacca starch and SCMC and Carbopols 940 and 941. The admixtures studied showed improvement on the bioadhesive properties of tacca starch, and confirmed they could be used as bioadhesive motifs for drug delivery into the gastrointestinal tract. More so, the buccoadhesiveness, swelling characteristics, and release profile of hydrochlorothiazide from patches formulated with ethylcellulose and HPMC interpolymer complexes of different ratio were studied to evaluate their applicability in sustained drug delivery.[156] The study indicated that, although the *in vitro* release of hydrochlorothiazide from the patches (prepared by casting) was not appreciably prolonged, the blends with low area swelling ratio are more suitable for the formulation of buccoadhesive drug-delivery systems. In addition, release of diclofenac sodium from bioadhesive hydrophilic matrix tablets composed of polyvinyl pyrrolidone (PVP) and SCMC was also studied.[157] Tablets that satisfied pharmacopoeial standards were obtained, and prolonged release of diclofenac sodium was achieved.

Furthermore, Proddurituri et al.[158] studied a method of improving the physical stability of clotrimazole (CT) and the polymer contained within hot-melt extrusion (HME) films using polymer blends of HPC and poly(ethylene oxide) (PEO). Films containing HPC:PEO:CT in the ratio of 55:35:10 demonstrated optimum physico-mechanical, bioadhesive, and release properties. These researchers concluded that polymer blends of HPC and PEO could be used successfully to tailor the drug release, mechanical and bioadhesive properties, and stability of the HME films, and that the glass transition temperature of the polymers played an important role in determining the physical stability of the solubilized drug. In a study comparing the performance of CMC, HPC, and their admixtures in tableting, tablets containing CMC alone had poor compression characteristics.[159] The hardness values for HPC-containing tablets were the same as or slightly less than those results seen with the HPC/CMC polymer blends. Dissolution test results clearly demonstrate the effects of polymer blending on release rate modification. The tablets that contained 100% CMC reached a T_{80} in 88 min, while 100% HPC tablets had a T_{80} of 224 min. When the tablets that contained the polymer mixture of HPC/CMC (75/25) were tested, the T_{80} value increased to 339 min. Also, combination of HPC/carrageenan (75/25) was tableted and tested. The hardness data showed that the blend is equivalent to or better than the individual polymers demonstrating the advantage of polymer blending in tablets for sustained drug delivery. On the same lines, Kuksal et al.[160] prepared and characterized extended-release matrix tablets of zidovudine (AZT) using hydrophilic

Eudragit® RLPO and Eudragit® PSPO alone or their combination with hydrophobic ethylcellulose. Results of the study demonstrated that a combination of both hydrophilic and hydrophobic polymers could be successfully employed for formulating sustained-release matrix tablets of AZT. The investigated sustained-release matrix tablet was capable of maintaining constant plasma AZT concentration through 12 hours, with good correlation between the dissolution profiles and bioavailability. Mukherjee et al.[161] developed novel transdermal drug delivery system (TDDS) of matrix type containing dexamethasone using blends of two different polymeric combinations, PVP and ethylcellulose, and Eudragit® with PVP. All the formulations were found to be suitable for formulating TDDS in terms of physicochemical characteristics and there was no significant interaction noticed between the drug and polymers used. However, PVP–ethylcellulose polymers performed better than PVP-Eudragit as TDDS for dexamethasone. Yerri-Swamy et al.[162] evaluated interpenetrating polymer network microspheres of HPMC/poly(vinyl alcohol) for controlled release of ciprofloxacin hydrochloride. The HPMC and poly(vinyl alcohol) blend microspheres were prepared by water-in-oil emulsion method and ciprofloxacin hydrochloride was loaded into the interpenetrating polymer network microspheres that were cross-linked with glutaraldehyde. *In vitro* dissolution experiments performed in pH 7.4 buffer medium at 35°C indicates a sustained and controlled release of ciprofloxacin hydrochloride from the interpenetrating polymer network microspheres up to 10 hours.[162]

Iqual et al.[163] studied bacterial cellulose as a promising biopolymer for controlled drug delivery applications. Model tablets were film-coated with bacterial cellulose, using a spray-coating technique, and *in vitro* drug release studies of these tablets were investigated. They concluded that bacterial cellulose could be used as novel aqueous film-coating agent with lower cost and better-film forming properties than existing film-coating agents.[163] Moreover, enteric coating is another important application of cellulose-based biopolymer blends in controlled delivery of bioactives. In this perspective, cellulose acetate phthalate, hydroxypropyl methylcellulose phthalate, and ethylcellulose in blends with other polymers are commonly used for enteric coating.[164–166] Using ethylcellulose as an example, amylose, a plant polysaccharide from starch, can be combined with ethylcellulose to produce a film coating capable of effecting colon-specific drug release from a dosage form through bacterial fermentation of the amylose component. Ethylcellulose is present in the system as a structuring agent in the form of the aqueous dispersion Surelease® grade EA-7100.

Specialized Bioactive Carriers

Cellulose-based biopolymers could be useful in designing specialized drug-delivery devices, such as hydrogels, xerogels/aerogels, osmotic pumps, dual-drug dosage forms

Biomimetic—Colloid

with improved separation of drugs as well as dosage forms combining both immediate-release and prolonged-release modes of drug delivery. Cellulose-based hydrogels are biocompatible and biodegradable materials, which show promise for a number of industrial uses, especially in cases where environmental issues are concerned, as well as biomedical applications. Apart from swelling tablets, more sophisticated hydrogel-based devices have been developed for controlled drug delivery.

The advances in hydrogel-based devices aim not only at the sustained release of a bioactive molecule over a long period of time, ranging from hours to weeks, but also at a space-controlled delivery, directly at the site of interest.[167] Several water-soluble cellulose derivatives can be used, singularly or in combination, to form hydrogel networks possessing specific properties in terms of swelling capability and sensitivity to external stimuli. The trend in the design of cellulose hydrogels is related to the use of nontoxic cross-linking agents or cross-linking treatments to further improve the safety of both the final product and the manufacturing process. Controlled release through oral drug delivery is usually based on the strong pH variations encountered when transiting from the stomach to the intestine. Cellulose-based polyelectrolyte hydrogels (e.g., hydrogels containing SCMC) are particularly suitable for this application. For instance, anionic hydrogels based on CMC have been investigated for colon-targeted drug delivery.[168] The advances in controlled release through a hydrogel matrix deal with the delivery of proteins, growth factors, and genes to specific sites, the need for which has been prompted by tissue engineering strategies. While hydrogel formulations for oral and transdermal delivery can be nondegradable, the direct delivery of drugs or proteins to different body sites requires the hydrogel biodegradation to avoid foreign body reactions and further surgical removal. Injectable hydrogel formulations are particularly appealing and currently under investigation. The cross-linking reaction has to be performed under mild conditions to avoid denaturing the loaded molecule. The microenvironment resulting from degradation of the polymer should be mild as well for the same purpose. With particular regard to cellulose-based hydrogels, injectable formulations, based on HPMC, have been developed to deliver both biomolecules and exogenous cells *in vivo*.[85,86,169] The need to encapsulate bioactive molecules into a hydrogel matrix or other delivery devices (e.g., microspheres) is also related to the short half-life displayed by many biomolecules *in vivo*. When using hydrogels to modulate the drug release, the loading of the drug is performed either after cross-linking or simultaneously during network formation.[167] Moreover, the bioactive molecule can be covalently or physically linked to the polymer network to further tune the release rate. The smart behavior of some cellulose derivatives (e.g., SCMC and HPMC) in response to physiologically relevant variables (i.e., pH, ionic strength, and temperature) makes the resulting hydrogels particularly appealing for *in vivo*

applications. Smart hydrogels are particularly useful to control the time- and space-release profile of the drug as swelling–deswelling transitions, which modify the mesh size of the hydrogel network, occur upon changes of physiologically relevant variables, such as pH, temperature, and ionic strength.[170]

In addition, cellulose-based hydrogel devices could be useful as stomach bulking agents. Novel bulking agents, effective in promoting weight loss, could be developed with cellulose-based superabsorbent hydrogels, since not only can their swelling capacity be properly designed by controlling their chemical composition and physical microstructure, but it can also be modulated by changing the environmental conditions (e.g., pH, ionic strength, and temperature). Here, the notion is that a xerogel-based pill is administered orally before each meal, and that the xerogel powder swells once in the stomach. By so doing, the space available for food intake is reduced, giving a feeling of fullness. Subsequently, the swollen hydrogel is eliminated from the body via the feces. In this direction, the hydrogel is envisaged to pass through the gastrointestinal tract, thus it is supposed to encounter the different pH environments of the stomach and the intestine. Along with superporous acrylate-based hydrogels, which swell very rapidly in aqueous solutions,[171] novel cellulose-based hydrogels, obtained by cross-linking aqueous mixtures of SCMC and HEC, have been shown to be appealing for the production of dietary bulking agents.[172,173] Indeed, such hydrogels possess high biocompatibility, with respect to intestinal tissues, and a high, pH-sensitive water-retention capacity.[173] Although the polyanionic nature of the SCMC network provides higher swelling capabilities at neutral pHs rather than at acid ones, the swelling ratio obtained at acid pHs might still be significant for use of the hydrogel as stomach filler. In particular, cellulose-based hydrogels obtained from nontoxic cross-linking agents are particularly attractive for this kind of application.[91,172,173] Moreover, shaped/silicized cellulosic aerogels have been developed. Reinforced shaped cellulosic aerogels consisting of two interpenetrating networks of cellulose and silica were prepared from shaped cellulose solutions by regenerating (reprecipitating) cellulose with ethanol; subjecting the obtained shaped alcogels to sol–gel condensation with tetraethoxysilane as the principal network-forming compound; and by drying the reinforced cellulose bodies with supercritical carbon dioxide. The influence of different types of cellulose and sol–gel forming parameters on porosity, cellulose integrity, and silica content were studied. The results showed improved functional and physicochemical properties over the normal aerogels in cellulose aerogel applications.[174]

Furthermore, cellulose-based hydrogels hold promise as devices for the removal of excess water from the body (body water retainers) in the treatment of some pathological conditions, such as renal failure and diuretic-resistant edemas. The hydrogel in powder form is envisaged to be administered orally and absorb water in its passage through the intestine, where the pH is about 6–7, without previously swelling in the acid environment of the stomach. The hydrogel is then expelled through the feces, thus performing its function without interfering with body functions. As sensitivity to pH is required, polyelectrolyte cellulose hydrogels based on SCMC and HEC have been investigated for such applications.[175–177] Also, the use of hydrogels in combination with diuretic therapies might be useful in substituting some drugs and in using an intestinal pathway, instead of the systemic one, to remove water from the body.[176]

Additionally, osmotic pump, a specialized drug delivery device developed about 30 years ago, remains an excellent example of the use of polymeric properties to control the delivery of bioactives. The drug-release mechanism in this device is driven by a difference in osmotic pressure between the drug solution and the environment outside the formulation. It represents a family of technologies that have been developed for the extended, optimally zero-order release of pharmaceutical actives.[178] These technologies rely upon the encapsulation of the pharmaceutical active within a membrane (cellulose diacetate is by far the most commonly used membrane polymer), which is highly permeable to water, but is impermeable to salts and to many organics. Upon ingestion of the pill or capsule, water permeates through the membrane and dissolves water-soluble ingredients inside. If the active itself develops sufficient osmotic pressure, it may be used without other osmogents; otherwise inert water-soluble agents (e.g., sodium chloride or other salts) are added to help develop osmotic pressure within the dosage form. That osmotic pressure is very high in comparison to that in the surrounding gut, and powers the ejection of an aqueous drug solution at a constant rate through a small (usually laser-drilled) orifice. Cellulose acetate was one of the first materials used for manufacturing semipermeable membranes in elementary osmotic pumps developed by ALZA Corporation®. These membranes continue to be used in commercial OROS® products, where the semipermeable membrane controls drug-release rate.[178] The release rate of drugs from an OROS is controlled by semipermeable membranes composed typically of cellulose acetate with various flux enhancers. Cellulose acetate butyrate (CAB) was identified as a viable alternative. The CAB membrane matched the cellulose acetate membrane in robustness but had superior drying properties, offering particular advantages for thermolabile formulations.

Many review works has been published on osmotic delivery over the past 40 years, including some recent ones.[179–182] In addition, a research group at the University of Mumbai[183] described a system in which a cellulose diacetate membrane is used for osmotic delivery of pseudoephedrine, with plasticizers used as film dopants in an attempt to create the pores. Good results were obtained

with diethylphthalate (DEP), in some cases containing PEG (both of which have significant miscibility with cellulose acetate). Release rates could be controlled by film thickness and PEG content, and near-zero-order rates were seen in some cases. Reasonable control over release rates was also seen with more hydrophobic (and incompatible with cellulose acetate) plasticizers such as dibutyl sebacate (DBS). It has also been reported that acetaminophen could be microencapsulated mixed with sodium chloride osmogent, using a cellulose acetate coating.[184] The drug, osmogent, and other inert ingredients were extruded, then solution-coated with cellulose acetate. The authors demonstrated zero-order release under certain conditions, and showed that much faster release was obtained in the presence of NaCl than in its absence. There were no holes drilled, nor were pore-forming agents included; the authors speculate that the osmotic pressure tore holes in the cellulose acetate coating of these small particles. One area of more recent concentration has been the development of simpler forms of osmotic delivery systems. In this perspective, Catellani et al.[185] devised a method that avoids laser-drilling entirely, and makes membrane failure unlikely. They created formulations in which a polymer/ drug core tablet is dipped partially in cellulose ester solution, creating a cellulose ester coating on some portion but not the entire pill. For these experiments they used HPMC as the matrix and buflomedil pyridoxal phosphate as the drug. They found out that these systems had an osmotic component to the release rate by virtue of the permeation of water into the system partly via the cellulose ester coating. They could release slowly by using CAP as the coating, or speed it by using increasing amounts of PEG plasticizer in a cellulose acetate coating. The study is interesting but is of unclear practical portent.

In furtherance to that, an interesting application of cellulose-based biopolymers as specialized bioactive carriers has been described.[186] The method involves manufacturing a pharmaceutical tablet for oral administration, the tablet combining both immediate-release and prolonged-release modes of drug delivery and using immediate-release drug that is either insoluble in water or only sparingly soluble and is present in a very small amount compared with the prolonged-release drug. The method involves the use of particles of the immediate-release drug (e.g., glimepiride) that is equal to or less than 10 microns in diameter, applied as a layer or coating over a core of the prolonged-release drug (e.g., metformin hydrochloride), the layer or coating being either the drug particles themselves, applied as an aqueous suspension (e.g., by using HPMC as a suspending agent), or a solid mixture containing the drug, in admixture with a material that disintegrates rapidly in gastric fluid (e.g., lactose, MCC, and combinations of lactose and MCC). The result in both cases is a high degree of uniformity in the proportions of the immediate-release and prolonged-release drugs, uniformity that is otherwise difficult to achieve in view of the insolubility of the immediate-release drug and its relatively small amount compared with the prolonged-released drug.

Another specialized bioactive carrier based on cellulose biopolymers could be seen in dual-drug dosage forms with improved separation of drugs.[187] Drug tablets that include a prolonged-release core and an immediate-release layer or shell are prepared with a thin barrier layer of drug-free polymer between the prolonged-release and immediate-release portions of the tablet. The barrier layer is penetrable by gastrointestinal fluid, thereby providing full access of the gastrointestinal fluid to the prolonged-release core, but remains intact during the application of the immediate-release layer, substantially reducing or eliminating any penetration of the immediate-release drug into the prolonged-release portion. This has been demonstrated using a solid matrix prepared with materials selected from the group consisting of poly(ethylene oxide), HPMC, and combinations of poly(ethylene oxide) and HPMC. It has also been established using a second solid matrix prepared with materials selected from the group consisting of lactose, MCC, and combinations of lactose and MCC.[187]

Cellulose-Based Biopolymer Microparticles for Drug Delivery

By hybridizing mucin and MCC, a novel polymer with a combination of the physicochemical and functional properties characteristic of the two-component polymers was obtained. The new polymer was directly compressible and it possessed mucus membrane protectant. Thus, to produce a novel excipient with membrane-protective, mucoadhesive, and direct compression properties for therapeutics and drug-delivery purposes, a mucin and MCC hybrid was prepared by regenerating colloidal mixtures of mucin and MCC at controlled pH and temperature conditions. Excipients with multiple functional properties confer many advantages, such as reduction of cost of production and the number of steps used during production. MCC–maize starch composite was generated by mixing colloidal dispersions of MCC and chemically gelatinized maize starch at controlled temperature conditions, and this process of polymer composite formation was termed compatibilized reactive polymer blending.[188] This led to direct compression efficiency of the novel polymer formed.

The study of the release kinetic profiles of naproxen from microcapsule compressed as well as matrix tablets using a combination of water-insoluble materials (like beeswax, cetyl alcohol, and stearic acid) with hydrophilic polymers was investigated. The ethylcellulose/HPMC combinations, contributing to an increase in hydrophilic part of blend system, rationally increased the release rate, kinetic constant, and diffusion coefficient thereby, whereas

HPMC/beeswax, HPMC/cetyl alcohol, and HPMC/stearic acid combinations, contributing an increase in hydrophobic part of the blend system, caused a substantial reduction of release.[189]

In several investigations the feasibility of development of a sustained-release form for diclofenac sodium was studied. Matrix-type formulation was designed, which appears to be a very attractive approach from process development and scale up points of view. HPMC is the most important hydrophilic polymer used for the preparation of oral controlled-release drug-delivery systems.[190,191] One of the most important characteristics of HPMC is the high swellability, which has a considerable effect on the release kinetics of the incorporated drug.

Application of Cellulose-Based Biopolymer Nanoparticles in Drug and Bioactive Delivery

Nanotechnology has applications across most economic sectors and allows the development of new enabling science with broad commercial potential. Cellulose-based nanoparticles have received considerable attention in recent years as one of the most promising nanoparticulate drug-delivery systems owing to their unique potentials. Nanoparticle drug-delivery systems are defined as particulate dispersions or solid particles with a size in the range of 10–1000 nm and with various morphologies, including nanospheres, nanocapsules, nanomicelles, nanoliposomes, and nanodrugs. The drug is dissolved, entrapped, encapsulated, or attached to a nanoparticle matrix.[192,193] The nanoparticles take on novel properties and functions such as small size, modified surface, improved solubility, and multifunctionality.[194] Drug-delivery systems of nanoparticles have several advantages, such as high drug-encapsulation efficiency, efficient drug protection against chemical or enzymatic degradation, unique ability to create a controlled release, cell internalization as well as the ability to reverse the multidrug resistance of tumor cells.[195] The use of cellulose-based biopolymer nanoparticles is receiving a significant amount of attention due to the impressive mechanical properties, reinforcing capability, abundance, low weight, low filler load requirements, and biodegradable nature of nanoparticles. These properties make it an ideal candidate for the development of green polymer nanocomposites, especially for drug-delivery applications. Cellulose-based nanoparticles are usually identified with different terminologies such as cellulose nanowhiskers,[196] cellulose nanocrystals,[197] nanocrystalline cellulose (NCC),[198] nanofibrillated cellulose,[199] cellulose nanofibrils,[200] and crystalline nanocellulose.[201] They can be prepared by solvent evaporation method, spontaneous emulsification or solvent diffusion method, self-assembly of hydrophobically modified, and dialysis method.[48,202,203] Application of cellulose nanoparticles in drug delivery is a relatively new research area. Cellulose and lignocellulose have great potential as nanomaterials because they are abundant, renewable, have a nanofibrillar structure, can be made multifunctional, and can self-assemble into well-defined architectures.[204] Thus, the various applications of cellulose-based biopolymer nanoparticles in drug and bioactive delivery cannot be overemphasized.

Cellulose-based biopolymers possess several potential advantages as drug-delivery excipients.[205] They are employed in advanced pelleting systems whereby the rate of tablet disintegration and drug release may be controlled by microparticle inclusion, excipient layering, or tablet coating.[206,207] The very large surface area and negative charge of crystalline nanocellulose suggest that large amounts of drugs might be bound to the surface of this material with the potential for high payloads and optimal control of dosing. The established biocompatibility of cellulose supports the pharmaceutical use of nanocellulose in the controlled delivery of drugs and bioactives. The abundant surface hydroxyl groups on crystalline nanocellulose provide a site for the surface modification of the material with a range of chemical groups by a variety of methods.[208] Surface modification may be used to modulate the loading and release of drugs that would not normally bind to nanocellulose, such as nonionized and hydrophobic drugs. For example, Lönnberg et al.[209] suggested that poly(caprolactone) chains might be conjugated onto NCC for such purpose. Additionally, since crystalline nanocellulose is a low-cost, readily abundant material from a renewable and sustainable resource, its use provides a substantial environmental advantage compared with other nanomaterials.[205]

Recently, the hydrophobically modified cellulose-based biopolymers have received increasing attention because they can form self-assembled nanoparticles for biomedical uses.[210] In the aqueous phase, the hydrophobic cores of polymeric nanoparticles are surrounded by hydrophilic outer shells. Thus, the inner core can serve as a nanocontainer for hydrophobic drugs. For instance, although the successful uses of cellulose-based biopolymers such as CMC for nanoparticle technologies are quite limited due to toxicity issues, Uglea et al.[211] conjugated benzocaine to CMC and oxidized CMC, and tested the effects of these polymers on subcutaneous sarcoma tumors in rat models, and reported some antitumor effect from a single intraperitoneal injection. Also, Sievens-Figueroa et al.[212] prepared and evaluated HPMC films containing stable BCS class II drug nanoparticles for pharmaceutical applications. The ultimate aim of these reserachers was to enhance the dissolution rate of poorly water-soluble drugs. Nanosuspensions produced from wet stirred media milling (WSMM) were transformed into polymer films containing drug nanoparticles by mixing with a low-molecular-weight HPMC (E15V) solution containing glycerin, followed by film casting and drying. Three different BCS class II drugs, naproxen, fenofibrate, and griseofulvin, were studied. The study demonstrated the enhancement in drug dissolu-

tion rate of films due to the large surface area and smaller drug particle size.[212] Similarly, cellulose-based biopolymers have been engineered to deliver a hydrophobic anticancer drug, docetaxel. Here, a compound comprising an acetylated carboxymethylcellulose (CMC-Ac) covalently linked to at least one PEG and at least one hydrophobic drug (docetaxel). The compound was transformed into a functional polymeric self-assembling nanoparticle with the following attributes: it dissolves or transports a hydrophobic drug in an aqueous environment; the drug is protected from metabolism by the particle; the particle is protected from reticulo-endothelial system elimination by PEG or other suitable chemistry; the polymer self-assembles into a suitably scaled nanoparticle due to a balance in hydrophobic and hydrophilic elements; the particle accumulates in the targeted disease compartment through passive accumulation; the link between the drug and particle is reversible, so that the drug can be released; the polymer is biocompatible; and the particle contains an agent to provide imaging contrast or detection in the physiological system.[213]

In addition, a group of researchers has developed methods for the synthesis of cellulose nanocrystals with optimal properties for applications in targeted drug delivery; covalent attachment of fluorescein isothiocyanate molecules to the surface of cellulose nanocrystals for fluorescent labeling; covalent attachment of folic acid molecules to the surface of fluorescently labeled cellulose nanocrystals for cancer targeting; and covalent attachment of doxorubicin molecules to the surface of folic acid-conjugated cellulose nanocrystals for cancer therapy. They assessed the toxicity of cellulose nanocrystals to a variety of human, mouse, and rat cell lines; the cellular uptake of fluorescently labeled cellulose nanocrystals; the cellular uptake of folic acid-conjugated, fluorescently labeled cellulose nanocrystals; and the in vitro efficacy of the targeted drug nanoconjugates on mouth epidermal carcinoma cells (KB cells) as a cancer model. According to the cytotoxicity studies, the cellulose nanocrystals have no cytotoxic effects. The demonstrated lack of cytotoxicity is a necessary prerequisite for the use of cellulose nanocrystals in targeted drug-delivery applications. The cellular uptake studies have demonstrated that targeting of cellulose nanocrystals through folic acid conjugation leads to uptake of cellulose nanocrystals by cancer cells and that nonspecific cellular uptake of cellulose nanocrystals is minimal. These results confirm that cellulose nanocrystals can be targeted for selective uptake by specific cells. The efficacy studies with doxorubicin-conjugated, folate receptor-targeted cellulose nanocrystals have shown that the targeted drug nanoconjugates are more effective in eradicating cancer cells than free doxorubicin. The studies have also shown that doxorubicin-conjugated cellulose nanocrystals that were not targeted to the folate receptor are less toxic to cancer cells than free doxorubicin. The lower toxicity of doxorubicin-conjugated cellulose nanocrystals, compared with that of free doxorubicin, indicates that targeting of the drug nano-conjugates to the folate receptor is crucial for its high efficacy. The findings by these researchers confirmed that cellulose nanocrystals are promising nanoparticles for targeted drug-delivery applications. In addition, a chemical method was developed for covalently attaching molecules of fluorescein isothiocyanate, one of the most widely used fluorescent labels, to the surface of cellulose nanocrystals. The method for fluorescent labeling of cellulose nanocrystals, developed by these researchers, enables the use of fluorescence techniques, such as spectrofluorometry, fluorescence microscopy, and flow cytometry, to study the interaction of cellulose nanocrystals with cells and the biodistribution of cellulose nanocrystals *in vivo*.[214–222]

In a related study, Aswathy et al.[223] developed multifunctional nanoparticles based on CMC. In the study, folate group was attached to nanoparticle for specific recognition of cancerous cells and 5-fluorouracil was encapsulated for delivering cytotoxicity. The whole system was able to be tracked by the semiconductor quantum dots that were attached to the nanoparticle. The multifunctional nanoparticle was characterized by spectroscopic techniques such as ultraviolet-visible spectra and FTIR and microsocopic techniques such as transmission electron microscopy and scanning electron microscopy (SEM) and was targeted to human breast cancer cell, MCF7. The biocompatibility of nanoparticle without drug and cytotoxicity rendered by nanoparticle with drug were studied with MCF7 and L929 cell lines. The epifluorescent images suggest that the folate-conjugated nanoparticles were more internalized by folate receptor positive cell line, MCF7, than the non-cancerous L929 cells.

Another area of application of cellulose-based biopolymer nanoparticles is in the formulation and delivery of bioactive compounds such as curcumin (CUR), a natural diphenol used in the treatment of tumors. Yallapu et al.[224] designed curcumin-loaded cellulose nanoparticles for prostate cancer. They evaluated the comparative cellular uptake and cytotoxicity of β-cyclodextrin, HPMC, poly(lactic-*co*-glycolic acid) (PLGA), magnetic nanoparticles, and dendrimer-based CUR nanoformulations in prostate cancer cells. The study showed that curcumin-loaded cellulose nanoparticles (cellulose-CUR) formulation exhibited the highest cellular uptake and caused maximum ultrastructural changes related to apoptosis (presence of vacuoles) in prostate cancer cells. Secondly, the anticancer potential of the cellulose-CUR formulation was evaluated in cell culture models using cell proliferation, colony formation, and apoptosis (7-AAD staining) assays. In these assays, the cellulose-CUR formulation showed improved anticancer efficacy compared with free curcumin.[224]

Moreover, novel cellulose-based drug-delivery systems were recently developed by Neha,[225] who designed novel cellulosic nanoparticles with potential pharmaceutical and personal care applications. The study involved the synthesis and characterization of polyampholyte nanoparticles composed of chitosan and CMC, a cellulosic ether. 1-Ethyl-3-(3-dimethylaminopropyl)carbodiimide (EDC) chemistry and inverse microemulsion technique was used to produce cross-linked nanoparticles. Chitosan and CMC provided amine and carboxylic acid functionality to the nanoparticles, thereby making them pH responsive. Chitosan and CMC also make the nanoparticles biodegradable and biocompatible, making them suitable candidates for pharmaceutical applications. The synthesis was then extended to chitosan and modified methylcellulose microgel system. The prime reason for using methylcellulose was to introduce thermo-responsive characteristics to the microgel system. Methylcellulose was modified by carboxymethylation to introduce carboxylic acid functionality, and the chitosan-modified methylcellulose microgel system was found to be responsive to pH as well as temperature. Several techniques were used to characterize the two microgel systems. For both systems, polyampholytic behavior was observed in a pH range of 4–9. The microgels showed swelling at low and high pH values and deswelling at isoelectric point (IEP). Zeta potential values confirmed the presence of positive charges on the microgel at low pH, negative charges at high pH, and neutral charge at the IEP. For chitosan-modified methylcellulose microgel system, temperature-dependent behavior was observed with dynamic light scattering. The second study undertaken by the same researcher involved the study of binding interaction between NCC and an oppositely charged surfactant tetradecyl trimethyl ammonium bromide (TTAB). NCC is a crystalline form of cellulose obtained from natural sources like wood, cotton, or animal sources. These rodlike nanocrystals prepared by acid hydrolysis of native cellulose possess negatively charged surface. The interaction between negatively charged NCC and cationic TTAB surfactant was examined, and it was observed that in the presence of TTAB, aqueous suspensions of NCC became unstable and phase separated. A study of this kind is imperative since NCC suspensions are proposed to be used in personal care applications (such as shampoos and conditioners), which also consist of surfactant formulations. Therefore, NCC suspensions would not be useful for applications that employ an oppositely charged surfactant. To prevent destabilization, poly(ethylene glycol) methacrylate (PEGMA) chains were grafted on the NCC surface to prevent the phase separation in presence of a cationic surfactant. Grafting was carried out using the free radical approach. The NCC–TTAB polymer surfactant interactions were studied via isothermal titration calorimetry (ITC), surface tensiometry, conductivity measurements, phase separation, and zeta potential measurements. Grafting of PEGMA on the NCC surface was confirmed using FTIR and ITC experiments. In phase separation experiments, NCC-g-PEGMA samples showed greater stability in the presence of TTAB compared with unmodified NCC. By comparing ITC and phase separation results, an optimum grafting ratio (PEGMA:NCC) for steric stabilization was also proposed.[225]

Applications of cellulose-based biopolymers in formulation and delivery of bioactive compounds discussed here are in no way exhaustive of the works done in the use of cellulose-based biopolymers in the formulation and delivery of bioactive compounds.

Cellulose-Based Biopolymers in Protein and Gene Delivery

Cellulose ethers have long been used in the pharmaceutical industry as excipients in many drug device formulations.[226] Their use in solid tablets allows a swelling-driven release of the drug as physiological fluids come into contact with the tablet itself. The cellulose ether on the tablet surface (e.g., HPMC) starts to swell, forming chain entanglements and physical hydrogel. More sophisticated hydrogel-based devices other than swelling tablets have been developed for controlled drug delivery. Smart hydrogels are particularly useful to control the time- and space-release profile of the drug. The most recent advances aim not only at the sustained release of a bioactive molecule over a long time period, ranging from hours to weeks, but also at a space-controlled delivery, directly at the site of interest. The most recent advances in controlled release through a hydrogel matrix deal with the delivery of proteins, growth factors, and genes to specific sites, the need for which has been prompted by tissue engineering strategies.

Both chitosan and a chitosan oligomer could complex CMC to form stable cationic nanoparticles for subsequent plasmid DNA coating.[227] Chitosan–CMC was subsequently coated with plasmid DNA for genetic immunization.[228] Microcapsules modified with talc and MCC have been shown to exhibit high protein retention in the core in different pH media and could facilitate targeting of protein to the colon.[229,230]

Microparticles containing mixtures of proteins in powder form have been coated with cellulose acetate phthalate using simple preparation techniques based on single emulsion/solvent evaporation. Using aprotinin as a model drug, it was found that these procedures were effective in microencapsulating protein in the solid form without affecting its biological activity. The particles showed adequate *in vitro* release patterns for application to the intestine.[231]

In recent years, a number of polymeric drug/gene-loaded nanoparticles have been developed as drug delivery carriers and their mechanism of circulation in human bodies has been extensively investigated.[232,233] When drug- or gene-loaded nanoparticles are injected into the body, they cross epithelial barriers and circulate in the blood vessels before reaching the target site. Gene therapy has been applied in many different diseases such as cancer, acquired immune deficiency syndrome, and cardiovascular diseases, and is based on the concept that human disease may be treated by the transfer of genetic materials into specific cells of a patient to supply defective genes responsible for disease development.[234] To transfer the genes to the specific site, genes must escape the processes that affect the disposition of macromolecules. Furthermore, the degradation of gene by serum nucleases needs to be avoided. Thus, encapsulation of genes in delivery carrier is necessary to protect the gene until it reaches its target. The delivery carriers must be small enough to internalize into cells and passage to the nucleus. They also need to be capable of escaping endosome–lysosome processing following endocytosis.[234] While both viral and nonviral vectors have been developed for the delivery of genes, nonviral vectors have been studied more actively due to their low immunogenicity and ease of control of their properties.[235,236] Thus, the cationic polymers have a potential for DNA complexation as nonviral vectors for gene therapy applications. To introduce the specificity into the nanoparticle surfaces, the conjugation of cell-specific ligands to the surface of nanoparticles allows for targeted transgene expression. For example, the nanoparticles of genes and the cationic polymers can be modified with proteins (knob, transferrin, or antibodies/antigens) to allow for cell-specific targeting and enhanced gene transfer.[237,238]

Well-defined comb-shaped cationic copolymers composed of long biocompatible HPC backbones and short poly[(2-dimethylamino)ethylmethacrylate][P(DMAEMA)] side chains were prepared as gene vectors via atom transfer radical polymerization from the bromoisobutyryl-terminated HPC biopolymers. The P(DMAEMA) side chains of HPDs was further partially quaternized to produce the quaternary ammonium HPDs (QHPDs). HPDs and QHPDs were assessed *in vitro* for nonviral gene delivery. HPDs exhibit much lower cytotoxicity and better gene transfection yield than high-molecular-weight P(DMAEMA) homopolymers. QHPDs exhibit a stronger ability to complex pDNA, due to increased surface cationic charges. Thus, the approach to well-defined comb-shaped cationic copolymers provided versatile means for tailoring the functional structure of nonviral gene vectors to meet the requirements of strong DNA-condensing ability and high transfection capability.[239]

Cellulose-Based Biopolymers in Wound Healing

Traditional plant-originated cellulose and cellulose-based materials, usually in the form of woven cotton gauze dressings, have been used in medical applications for many years and are mainly utilized to stop bleeding. Even though this conventional dressing is not ideal, its use continues to be widespread. These cotton gauzes consist of an oxidized form of regenerated plant cellulose. In addition, several studies described the implantation of regenerated cellulose hydrogels and revealed their biocompatibility with connective tissue formation and long-term stability. Other *in vitro* studies showed that regenerated cellulose hydrogels promote bone cell attachment and proliferation and are very promising materials for orthopedic applications. Although chemically identical to plant cellulose, the cellulose synthesized by *Acetobacter* is characterized by a unique

Biomimetic—Colloid

fibrillar nanostructure that determines its extraordinary physical and mechanical properties, characteristics that are quite promising for modern medicine and biomedical research. The nonwoven ribbons of microbial cellulose microfibrils closely resemble the structure of native extracellullar matrices, suggesting that it could function as a scaffold for the production of many tissue-engineered constructs. In addition, microbial cellulose membranes, having a unique nanostructure, could have many other uses in wound healing and regenerative medicine, such as guided tissue regeneration, periodontal treatments, or as a replacement for dura mater (a membrane that surrounds brain tissue).[240] Microbial cellulose could function as a scaffold material for the regeneration of a wide variety of tissues, showing that it could eventually become an excellent platform technology for medicine.

CONCLUSIONS

Cellulose derivatives constitute a large class of biopolymers with diverse physicochemical properties and large functional versatility in pharmaceutical application. Cellulose derivatives have been used as diluents, binders in direct compression and wet granulation processes, film coating, viscosity enhancers, drug-release retardants in controlled-release matrix systems for microparticles and nanoparticles. By modifying the naturally occurring inexpensive renewable resources, both durable and environmentally acceptable materials can be developed. The most recent advances in utilization of cellulosics is in the delivery of proteins, growth factors, and genes to specific sites, the need for which has been prompted by tissue engineering strategies. Microbial could be used in many biomedical and biotechnological applications, such as tissue engineering, drug delivery, wound dressings, and medical implants. Different cellulose-based biopolymers can be combined with non-cellulose-based materials to obtain novel polymer composites with unique properties for applications in drug delivery and tissue engineering. Due to the excellent biocompatibility of cellulose and its good mechanical properties, they form good replacement for synthetic materials used as biopolymers. To fully realize the potential of the newly developed celluose-based biopolymers in near future would require coordinated and dedicated multidisciplinary research.

REFERENCES

1. Biopolymer, http://en.wikipedia.org/wiki/Biopolymer (accessed March 2013).
2. Sannino, A.; Demitri, C.; Madaghiele, M. Biodegradable cellulose-based hydrogels: Design and applications. Materials **2009**, *2* (2), 353–373.
3. Baruah, S.D. Biodegradable polymer: The promises and the problems: Science and Culture, Nov, Dec. 2011, 466–470.
4. Attama, A.A.; Builders, P.F. Particulate drug delivery: Recent applications of batural biopolymers. In *Biopolymers in Drug Delivery: Recent Advances and Challenges*; Adikwu, M.U., Esimone, C.O., Eds.; Bentham Science Publishers: UAE, 2009; 63–94.
5. Builders, P.F.; Attama, A.A. Functional properties of biopolymers for drug delivery applications. In *Biodegradable Materials*; Johnson, B.M., Berkel, Z.E., Eds.; Nova Science Pub. Inc.: USA, 2011; ISBN: 978-1-61122-804-5.
6. Rowe, R.C.; Shesky, P.J.; Weoller, P.J. *HandBook of Pharmaceutical Excipients*, 2nd Ed.; Pharm. Press: London, 2003; 120–122, 544–545.
7. Builders, P.F.; Ibekwe, N.; Okpako, L.C.; Attama, A.A.; Kunle, O.O. Preparation and characterization of mucinated cellulose microparticles for therapeutic and drug delivery purposes. Eur. J. Pharm. Biopharm. **2009**, *72* (1), 34–41.
8. Kibbe, A.H. *Handbook of Pharmaceutical Excipients: Cellulose, Silicified Microcrystalline*; American Public Health Association: Washington, 2000.
9. Luukkonen, P.; Schaefer, T.; Hellen, J.; Juppo, A.M.; Yliruusi, J. Rheological characterization of microcrystalline cellulose and silicified microcrystalline cellulose wet masses using a mixer torque rheometer. Int. J. Pharm. **1999**, *188* (2), 181–192.
10. Siepmann, J.; Kranz, H.; Bodmeier, R.; Peppas, N.A. HPMC-matrices for controlled drug delivery: A new model combining diffusion, swelling, and dissolution mechanisms and predicting the release kinetics. Pharm. Res. **1999**, *16* (11), 1748–1756.
11. Colombo, P.; Bettini, R.; Peppas, N.A. Observation of swelling process and diffusion front position during swelling in hydroxypropylmethyl cellulose (HPMC) matrices containing a soluble drug. J. Control. Rel. **1999**, *61* (1–2), 83–91.
12. Lowman, A.M.; Peppas, N.A. Hydrogels. In *Encyclopedia of Controlled Drug Delivery*; Mathiowitz, E., Ed.; Wiley: New York, 2000; 397–417.
13. Le Neel, T.; Morlet-Renaud, C.; Lipart, C.; Gouyette, A.; Truchaud, A.; Merle, C. Image analysis as a new technique for the study of water uptake in tablets. STP Pharma. Sci. **1997**, *7* (2), 117–122.
14. Baumgartner, S.; Šmid-Korbar, J.; Vreèer, F.; Kristl, J. Physical and technological parameters influencing floating properties of matrix tablets based on cellulose ethers. STP Pharma. Sci. **1998**, *8* (5), 182–187.
15. Arion, H. Carboxymethyl cellulose hydrogel-filled breast implants. Our experience in 15 years. Ann. Chir. Plast. Esthet. **2001**, *46* (1), 55–59.
16. Methyl Cellulose, http://en.wikipedia.org/wiki/Methyl_cellulose (accessed March 2013).
17. Mura, P.; Faucci, M.T.; Manderioli, A.; Bramanti, G.; Parrini, P. Thermal behavior and dissolution properties of naproxen from binary and ternary solid dispersion. Drug Dev. Ind. Pharm. **1999**, *25* (3), 257–264.
18. Friedman, M.; Golomb, G. New sustained release dosage form of chlorhexidine for dental use. J. Periodontal Res. **1982**, *17* (3), 323–328.
19. Soskolne, W.A.; Golomb, G.; Friedman, M.; Sela, M.N. New sustained release dosage form of chlorhexidine for dental use. J. Periodontal Res. **1983**, *18* (3), 330–336.

20. Yuasa, H.; Ozeki, T.; Kanaya, Y.; Oishi, K.; Oyake, T. Application of the solid dispersion method to the controlled release of medicine. I. Controlled release of water soluble medicine by using solid dispersion. Chem. Pharm. Bull. **1991**, *39* (2), 465–467.

21. Yuasa, H.; Ozeki, T; Kanaya, Y.; Oishi, K. Application of the solid dispersion method to the controlled release of medicine. II. Sustained release tablet using solid dispersion granule and the medicine release mechanism. Chem. Pharm. Bull. **1992**, *40* (6), 1592–1596.

22. Ozeki, T.; Yuasa, H.; Kanaya, Y.; Oishi, K. Application of the solid dispersion method to the controlled release of medicine. V. Suppression mechanism of the medicine release rate in the three-component solid dispersion system. Chem. Pharm. Bull. **1994**, *42* (2), 337–343.

23. Ozeki, T.; Yuasa, H.; Kanaya, Y.; Oishi, K. Application of the solid dispersion method to the controlled release of medicine. VII. Release mechanism of a highly water-soluble medicine from solid dispersion with different molecular weight of polymer. Chem. Pharm. Bull. **1995**, *43* (4), 660–665.

24. Ozeki, T.; Yuasa, H.; Kanaya, Y.; Oishi, K. Application of the solid dispersion method to the controlled release of medicine. VIII. Medicine release and viscosity of the hydrogel of a water-soluble polymer in a three-component solid dispersion system. Chem. Pharm. Bull. **1995**, *43* (9), 1574–1579.

25. Yuasa, H.; Takahashi, H.; Ozeki, T.; Kanaya, Y.; Ueno, M. Application of the solid dispersion method to the controlled release of medicine. III. Control of the release rate of slightly water soluble medicine from solid dispersion granules. Chem. Pharm. Bull. **1993**, *41* (2), 397–399.

26. Yuasa, H.; Ozeki, T.; Takahashi, H.; Kanaya, Y.; Ueno, M. Application of the solid dispersion method to the controlled release of medicine. VI. Release mechanism of slightly water soluble medicine and interaction between flurbiprofen and hydroxypropyl cellulose in solid dispersion. Chem. Pharm. Bull. **1994**, *42* (2), 354–358.

27. Ozeki, T.; Yuasa, H.; Kanaya, Y. Application of the solid dispersion method to the controlled release of medicine. IX. Difference in the release of flurbiprofen from solid dispersions with poly(ethylene oxide) and hydroxypropyl-cellulose and interaction between medicine and polymers. Int. J. Pharm. **1997**, *115* (2), 209–217.

28. Tahara, K.; Yamamoto, K.; Nishihata, T. Overall mechanism behind matrix sustained release (SR) tablets prepared with hydroxypropylmethyl cellulose. J. Control. Rel. **1995**, *35* (1), 59–66.

29. Skoug, J.W.; Mikelsons, M.V.; Vigneron, C.N.; Stemm, N.L. Qualitative evaluation of the mechanism of release of matrix sustained release dosage forms by measurement of polymer release. J. Control. Rel. **1993**, *27* (3), 227–245.

30. Ford, J.L.; Rubinstein, M.H.; McCaul, F.; Hogan, J.E.; Edgar, P.J. Importance of drug type, tablet shape and added diluents on drug release kinetics from hydroxypropylmethyl cellulose matrix tablets. Int. J. Pharm. **1987**, *40* (3), 223–234.

31. Ranga-Rao, K.V.; Padmalatha, D.; Buri, P. Influence of molecular size and water solubility of the solute on its release from swelling and erosion controlled polymeric matrices. J. Control. Release **1990**, *12* (2), 133–141.

32. Delgado, J.N.; William, A. *Wilson and Gisvold's Textbook of Organic Medicinal and Pharmaceutical Chemistry*; Lippincott-Raven Publishers: Wickford, 1998.

33. Karaaslan, A.M.; Tshabalala, M.A.; Buschle-Diller, G. Wood hemicellulose/chitosan-based semi-interpenetrating network hydrogels: Mechanical, swelling and controlled drug release properties. BioRes **2010**, *5* (2), 1036–1054.

34. Lerouxel, O.; Cavalier, D.M.; Liepman, A.H.; Keegstra, K. Biosynthesis of plant cell wall polysaccharides – a complex process. Curr. Opin. Plant Biol. **2006**, *9* (6), 621–630.

35. Chaa, L.; Joly, N.; Lequart, V.; Faugeron, C.; Mollet, J.; Martin, P.; Morvan, H. Isolation, characterization and valorization of hemicelluloses from *Aristida pungens* leaves as biomaterial. Carbohydr. Polym. **2008**, *74* (3), 597–602

36. Petkowicz, C.L.O.; Reicher, F.; Mazeau, K. Conformational analysis of galactomannans. Carbohydr. Polym. **1998**, *37* (1), 25–39.

37. Chourasia, M.K.; Jain, S.K. Polysaccharides for colon targeted drug delivery. Drug Deliv. **2004**, *11* (2), 129–148.

38. Vendruscolo, C.W.; Andreazza, I.F; Ganter, J.L.M.S.; Ferrero, C.; Bresolin, T.M.B. Xanthan and galactomannan (from *M. scabrella*) matrix tablets for oral controlled delivery of theophylline. Int. J. Pharm. **2005**, *296* (1), 1–11.

39. Beneke, C.E.; Viljoen, A.M.; Hamman, J.H. Polymeric plant-derived excipients in drug delivery. Molecules **2009**, *14* (7), 2602–2620.

40. Krässing, H.; Schurz, J.; Steadman, R.G.; Schliefer, K.; Albrecht, K.X. Cellulose. In *Ullmann's Encyclopedia of Industrial Chemistry*; Campbell, F.T., Pfefferkorn, R., Rounsaville, J.F., Eds.; VCH Verlagsgesellschaft: Weinheim, 1986; 375.

41. Fengel, D.; Wegener, G. *Wood: Chemistry, Ultrastructure, Reactions*; Walter de Gruyter: Berlin, 1989; 26–226.

42. Richardson S, Gorton L. Characterisation of the substituent distribution in starch and cellulose derivatives. Anal. Chim. Acta **2003**, *497* (1), 27–65.

43. Saake, B.; Horner, S.; Puls, J. Progress in the enzymatic hydrolysis of cellulose derivatives. In *Cellulose Derivatives: Modification, Characterization and Nanostructures*; Heinze, T., Glasser G., Eds.; American Chemical Society: Washington, DC, 1998; 201.

44. Guilbot, A.; Mercier, C. Starch. In *The Polysaccharides*; Aspinall, G., Ed.; Academic Press: New York, 1985; 209.

45. BeMiller, J.N. Starch modification: Challenges and prospects. Starch/Stärke **1997**, *49* (7–8), 127.

46. Edson, C.S.F.; Luciano, C.B.L.; Kaline, S.S; Maria, G.F.; Francisco, A.R.P. Calorimetry studies for interaction in solid/liquid interface between the modified cellulose and divalent cation. J. Therm. Anal. Calorim. **2013**, *114* (1), 57–66.

47. McDowall, D.J.; Gupta, B.S.; Stannett, V.T. Grafting of vinyl monomers to cellulose by ceric ion initiation. Prog. Polym. Sci. **1984**, *31* (12), 1–50.

48. Klemm, D.H.; Heublein, B.; Fink, H.P.; Bohn, A. Cellulose: Fascinating biopolymer and sustainable raw material. Angew. Chem. Int. Ed. **2005**, *44* (22), 3358–3393.

49. Magdi, E.G.; Zhang, Y.; Li, X.; Li, H.; Zhong, X.; Li, H.F.; Yu, M. Current status of applications of ionic liquids for cellulose dissolution and modifications: Review. Int. J. Eng. Sci. Tech. **2012**, *4* (7), 3556–3571.

Biomimetic—Colloid

50. Da SilvaFilho, E.C.; De Melo, J.C.P.; Airoldi, C. Preparation of ethylenediamine- anchored cellulose and determination of thermochemical data for the interaction between cations and basic centers at the solid/liquid interface. Carbohyd. Res., **2006**, *341* (17), 2842–2850.

51. De Melo, J.C.P.; Da SilvaFilho, E.C.; Santana, S.A.A.; Airoldi, C. Exploring the favorable ion-exchange ability of phthalylated cellulose biopolymer using thermodynamic data. Carbohydr. Res. **2010**, *345* (13), 1914–1921.

52. Akira, I. Chemical modification of cellulose. In *Wood and Cellulosic Chemistry*; Hon, D.N.S., Shiraishi, N., Ed.; Marcel Dekker: New York, 2001; 599–626.

53. Swati, A.; Achhrish, G.; Sandeep, S. Drug delivery: Special emphasis given on biodegradable polymers. Adv. Polym. Sci. Technol. **2012**, *2* (1), 1–15.

54. Kalia, S.; Kaith, B.S.; Kaur, I. Pretreatments of natural fibers and their application as reinforcing material in polymer composites—A review. Polym. Eng. Sci. **2009**, *49* (7), 1253–1272.

55. Kalia, S.; Kaith, B.S.; Sharma, S.; Bhardwaj, B. Mechanical properties of flax-g-poly(methyl acrylate) reinforced phenolic composites. Fibers Polym. **2008**, *9* (4), 416–422.

56. Kalia, S.; Dufresne, A.; Cherian, B.M.; Kaith, B.S.; Luc, A.; Njuguna, J.; Nassiopoulos, E. Cellulose-based bio- and nanocomposites: A review. Int. J. Poly. Sci. **2011**, Article ID 837875, 35, doi:10.1155/2011/837875.

57. Belgacem, M.N.; Salon-Brochier, M.C.; Krouit, M.; Bras, J. Recent advances in surface chemical modification of cellulose fibres. J. Adh. Sci. Tech. **2011**, *25* (6–7), 661–684.

58. Kennedy, J.F.; Phillips, G.O.; Wedlock, D.J.; Williams, P.A. *Cellulose and its Derivatives: Chemistry, Biochemistry and Applications*; Ellis Horwood: Chichester, 1985; 551.

59. Whistler, R.L.; BeMiller, J.N.; Paschall, E.F. *Starch: Chemistry and Technology*, 2nd Ed.; Academic Press: London, 1984; 26–86.

60. Agrawal, R.; Saxena, N.S.; Sharma, K.B.; Thomas, S.; Sreekala, M.S. Activation energy and crystallization kinetics of untreated and treated oil palm fibre reinforced phenol formaldehyde composites. Mat. Sci. Eng. **2000**, *277* (1–2), 77–82.

61. Sreekala, M.S.; Kumaran, M.G.; Joseph, S.; Jacob, M.; Thomas, S. Oil palm fibre reinforced phenol formaldehyde composites: Influence of fibre surface modifications on the mechanical performance. App. Comp. Mat. **2007**, *7* (5–6), 295–329.

62. Ray, D.; Sarkar, B.K.; Rana, A.K.; Bose, N.R. Effect of alkali treated jute fibres on composite properties. Bull. Mat. Sci. **2001**, *24* (2), 129–135.

63. Joseph, K.; Mattoso, L.H.C.; Toledo, R.D.; Thomas, S.; de Carvalho, L.H. Natural fiber reinforced thermoplastic composites. In *Natural Polymers and Agrofibers Composites*; Frollini, E.; Leão, A.L.; Mattoso, L.H.C.; Eds.; Embrapa, Sãn Carlos: Brazil, 2000; 159–201,

64. Kaith, B.S.; Singha, A.S.; Kumar, S.; Misra, B.N. FASH2O2 initiated graft copolymerization of methylmethacrylate onto flax and evaluation of some physical and chemical properties. J. Polym. Mat. **2005**, 4 (22), 425–432.

65. Pommet, M.; Juntaro, J.; Heng, J.Y.Y.; Athanasios, M.; Adam F.L.; Karen, W.; Gerhard, K.; Milo, S.P.S.; Bismarck, A. Surface modification of natural fibers using

bacteria: Depositing bacterial cellulose onto natural fibers to create hierarchical fiber reinforced nanocomposites. Biomacromolecules **2008**, *9* (6), 1643–1651.

66. Shoda M.; Sugano, Y. Recent advances in bacterial cellulose production. Biotech. Biopr. Eng. **2005**, *10* (1), 1–8.

67. Rutenberg, M.W.; Solarek, D. Starch derivatives: Production and uses. In *Starch Chemistry and Technology*; Whistler, R.L., BeMiller, J.N., Paschall, E.F., Eds.; Academic Press: London, 1984; 311.

68. Brandt, L. Cellulose ethers. In Ullmann's Encyclopedia of Industrial Chemistry; Campbell, F.T., Pfefferkorn, R., Rounsaville, J.F., Ed.; VCH Verlagsgesellschaft: Weinheim, 1986; 461.

69. Marchessault, R.H.; Sundararajan, P.R. Cellulose. In *The Polysaccharides*; Aspinall, G.O., Ed.; Academic Press: New York, 1983; 11.

70. Mondt, J.L. The use of cellulose derivatives in the paint and building industries. In *Cellulose Sources Exploitation*; Kennedy, J.F., Phillips, G.O., Williams, P.A., Ed.; Ellis Horwood: London, 1983; 269.

71. Varshney, V.K.; Naithani, S. Chemical functionalization of cellulose derived from nonconventional sources. In *Cellulose Fibers: Bio- and Nano-Polymer Composites*; Kalia, S., Ed.; Springer-Verlag; Berlin, Heidelberg, 2011; 43–58.

72. Arthur, J.C., Jr. In *Comprehensive Polymer Science*; Allen G., Bevington, J.C., Eds.; Pergamon: Oxford, 1986; 26, 681.

73. Schomburg, D.; Salzmann, M. *Enzyme Handbook*; Springer: Berlin, Heidelberg, 1991; 3.2.1.3.18.

74. Sannino, A.; Demitri, C.; Madaghiele, M. Biodegradable cellulose based hydrogels: Design and applications. Materials **2009**, *2* (2), 353–373.

75. Chen, H.; Fan, M. Novel thermally sensitive pH-dependent chitosan/carboxymethyl cellulose hydrogels. J. Bioact. Compat. Polym. **2008**, *23* (1), 38–48.

76. Chang, C.; Lue, A.; Zhang, L. Effects of crosslinking methods on structure and properties of cellulose/PVA hydrogels. Macromol. Chem. Phys. **2008**, *209* (12), 1266–1273.

77. Sarkar, N. Thermal gelation properties of methyl and hydroxypropyl methylcellulose. J. Appl. Polym. Sci. **1979**, *24* (4), 1073–1087.

78. Tate, M.C.; Shear, D.A.; Hoffman, S.W.; Stein, D.G.; LaPlaca, M.C. Biocompatibility of methylcellulose-based constructs designed for intracerebral gelation following experimental traumatic brain injury. Biomaterials **2001**, *22* (10), 1113–1123.

79. Chen, C.; Tsai, C.; Chen, W.; Mi, F.; Liang, H.; Chen, S.; Sung, H. Novel living cell sheet harvest system composed of thermoreversible methylcellulose hydrogels. Biomacromolecules **2006**, *7* (3), 736–743.

80. Stabenfeldt, S.E.; Garcia, A.J.; LaPlaca, M.C. Thermoreversible laminin-functionalized hydrogel for neural tissue engineering. J. Biomed. Mater. Res. A **2006**, *77* (4), 718–725.

81. Te Nijenhuis, K. On the nature of crosslinks in thermoreversible gels. Polym. Bull. **2007**, *58* (1), 27–42.

82. Vinatier, C.; Magne, D.; Weiss, P.; Trojani, C.; Rochet, N.; Carle, G.F.; Vignes-Colombeix, C.; Chadjichristos, C.; Galera, P.; Daculsi, G.; Guicheux, J. A silanized hydroxypropyl methylcellulose hydrogel for the three-dimensional

culture of chondrocytes. Biomaterials **2005**, *26* (33), 6643–6651.

83. Vinatier, C.; Magne, D.; Moreau, A.; Gauthier, O.; Malard, O.; Vignes-Colombeix, C.; Daculsi, G.; Weiss, P.; Guicheux, J. Engineering cartilage with human nasal chondrocytes and a silanized hydroxypropyl methylcellulose hydrogel. J. Biomed. Mater. Res. A **2007**, *80* (1), 66–74.

84. Ogushi, Y.; Sakai, S.; Kawakami, K. Synthesis of enzimatically-gellable carboxymethylcellulose for biomedical applications. J. Biosci. Bioeng. **2007**, *104* (1), 30–33.

85. Wang, C.; Chen, C. Physical properties of the crosslinked cellulose catalyzed with nanotitanium dioxide under UV irradiation and electronic field. Appl. Catal. A **2005**, *293* (2B), 171–179.

86. Coma, V.; Sebti, I.; Pardon, P.; Pichavant, F.H.; Deschamps, A. Film properties from crosslinking of cellulosic derivatives with a polyfunctional carboxylic acid. Carbohydr. Polym. **2003**, *51* (3), 265–271.

87. Xie, X.; Liu, Q.; Cui, S.W. Studies on the granular structure of resistant starches (type 4) from normal, high amylose and waxy corn starch citrates. Food Res. Int. **2006**, *39* (3), 332–341.

88. Demitri, C.; Del Sole, R.; Scalera, F.; Sannino, A.; Vasapollo, G.; Maffezzoli, A.; Ambrosio, L.; Nicolais, L. Novel superabsorbent cellulose-based hydrogels crosslinked with citric acid. J. Appl.Polym. Sci. **2008**, *110* (4), 2453–2460.

89. Charlesby, A. The degradation of cellulose by ionizing radiation. J. Polym. Sci. **1955**, *15* (79), 263–270.

90. Wach, R.A.; Mitomo, H.; Nagasawa, N.; Yoshii, F. Radiation crosslinking of methylcellulose and hydroxyethylcellulose in concentrated aqueous solutions. Nucl. Instrum. Methods Phys. Res. Sect. B **2003**, *211* (4), 533–544.

91. Pekel, N.; Yoshii, F.; Kume, T.; Guven, O. Radiation crosslinking of biodegradable hydroxypropylmethylcellulose. Carbohydr. Polym. **2004**, *55* (2), 139–147.

92. Liu, P.; Peng, J.; Li, J.; Wu, J. Radiation crosslinking of CMC-Na at low dose and its application as substitute for hydrogels. Rad. Phys. Chem. **2005**, *72* (5), 635–638.

93. Li, W.Y.J.; Liu, A.X.; Sun, C.F.; Zhang, R.C.; Kennedy, J.F. Homogeneous modification of cellulose with succinic anhydride in ionic liquid using 4-dimethylaminopyridine as a catalyst. Carbohydr, Polym, **2009**, *78* (3), 389–395.

94. Gericke, M.L.; TimHeinze, T. Solvent for cellulose chemistry. Nachrichten Aus. Der. Chemie. **2011**, *59* (4), 405–409.

95. Schobitz, M.M.; F.Heinze, T. Unconventional reactivity of cellulose dissolved in ionic liquids. Macromol. Symp. **2009**, *280* (1), 102–111.

96. Heinze, T.D.; Susann, S.; Michael, L.; Tim, K.; Sarah, M.F. Interactions of ionic liquids with polysaccharides - 2: Cellulose. Macromol. Symp. **2008**, *262* (1), 8–22.

97. Pinkert, A.M.; Kenneth, N.P.; Shusheng, S.; Mark, P. Ionic liquids and their interaction with cellulose. Chem. Rev. **2009**, *109* (12), 6712–6728.

98. Swatloski, R.P.S.; Holbrey, J.S.K.; Rogers, R.D. Dissolution of cellulose with ionic liquids. J. Am. Chem. Soc. **2002**, *124* (18), 4974–4975.

99. Zhu, S.; Wu, Y.; Chen, Q.; Yu, Z.; Wang, C.; Jin, S.; Ding, Y.; Wu, G. Dissolution of cellulose with ionic liquids and its application: A mini-review. Green Chem. **2006**, *8* (4), 325–327.

100. Cao, Y.W.; Jin, Z.; Jun, L.; Huiquan, Z.; Yi, H.J. Room temperature ionic liquids (RTILs): A new and versatile platform for cellulose processing and derivatization. Chem. Eng. J. **2009**, *147* (1), 13–21.

101. El Seoud, O.A.K.; A.Fidale, L.; C.Dorn, S.; Heinze, T .Applications of ionic liquids in carbohydrate chemistry: A window of opportunities. Biomacromolecules **2007**, *8* (9), 2629–2647.

102. Zhu, J.; Wang, W.T.; Wang, X.L.; Li, B.; Wang, Y.Z. Green synthesis of a novel biodegradable copolymer base on cellulose and poly(p-dioxanone) in ionic liquid. Carbohydr. Polym. **2009**, *76* (1), 139–144.

103. Przemys, A.; Aw, K. Ionic liquids as solvents for polymerization processes Progress and challenges. Prog. Poly. Sci. **2009**, *34* (12), 1333–1347.

104. Wilpiszewska, K.S.A.T. Ionic liquids: Media for starch dissolution, plasticization and modification. Carbohydr. Polym. **2011**, *86* (2), 424–428.

105. Feng, L.C.; Zhong, I. Research progress on dissolution and functional modification of cellulose in ionic liquids. J. Mol. Liq. **2008**, *142* (1–3), 1–5.

106. Kim, K.W.S.; B.Choi, M.Y.; Kim, M.J. Biocatalysis in ionic liquids: Markedly enhanced enantioselectivity of lipase. Org. Lett. **2001**, *3* (10), 1507–1509.

107. Ren, J.P.; Xinwen, P.; FengSun, R. *Ionic Liquid as Solvent for a Biopolymer: Acetylation of Hemicelluloses*; Sun, R., Fu, S., Eds.; Research Progress in Paper Industry and Biorefinery, South China University of China: Guangzhou, 2010; 57–60.

108. Wu, J.; Zhang, J.; Zhang, H.; He, J.; Ren, Q.; Guo, M. Homogeneous acetylation of cellulose in a new ionic liquid. Biomacromolecules **2004**, *5* (2), 266–268.

109. Koehler, S.L.; Tim, S.; Michael, S.; Jens, M.; Wolfgang, G.F.; Heinze, T. Interactions of ionic liquids with polysaccharides 1. Unexpected acetylation of cellulose with 1-ethyl-3-methylimidazolium acetate. Macromol. Rapid Commun. **2007**, *28* (24), 2311–2317.

110. Cao, Y.L.; Zhang, H.Q.J. Homogeneous synthesis and characterization of cellulose acetate butyrate (CAB) in 1-allyl-3- methylimidazolium chloride (AmimCl) ionic liquid. Ind. Eng. Chem. Res. **2004**, *50* (13), 7808–7814.

111. Abbott, A.P.B.; Handa, S.T.J.; Stoddart, B. O-acetylation of cellulose and monosaccharides using a zinc based ionic liquid. Green Chem. **2005**, *7* (10), 705–707.

112. Turner, M.B.S.; Holbrey, J.D.S.K.; Rogers, R.D. Production of bioactive cellulose films reconstituted from ionic liquids. Biomacromolecules **2004**, *5* (4), 1379–1384.

113. Semsarilar, M.L.; Vincent-Perrier, S. Synthesis of a cellulose supported chain transfer agent and its application to RAFT polymerization. J. Polym. Sci. A **2010**, *48* (19), 4361–4365.

114. Lin, C.X.Z.; Zhang, H.-Y.; Liu, M.-H.; Fu, S.-Y.; Lucia, L.A. Novel preparation and characterization of cellulose microparticles functionalized in ionic liquids. Langmuir **2009**, *25* (17), 10116–10120.

115. Tomasik, P.A. Chemical Modifications of Polysaccharides. In *Chemical and Functional Properties of Food Saccharides*; CRC Press: New York, 2003; 217–229.

116. Yan, C.Z.; Zhang, J.; Lv, Y.; Lu, J.; Wu, J.; Zhang, J.; He, J. Thermoplastic cellulose-graft-poly(L-lactide) copolymers homogeneously synthesized in an ionic liquid with 4 dimethylaminopyridine catalyst. Biomacromolecules **2009**, *10* (8), 2013–2018.

Biomimetic—Colloid

117. Lin, C.X.Z.; Huai-yu, L.; Ming-hua, F.; Shi-yu, H.L. Rapid homogeneous preparation of cellulose graft copolymer in BMIMCL under microwave irradiation. J. Appl. Poly. Sci. **2010**, *118* (1), 399–404.

118. Kadokawa, J.I.M.; Masa-Aki, T.; Akihiko-Kaneko, Y. Preparation of cellulose-starch composite gel and fibrous material from a mixture of the polysaccharides in ionic liquid. Carbohydr. Polym. **2009**, *75* (1), 180–183.

119. Hameed, N.G.; Qipeng, T.; Feng, H.K.; Sergei, G. Blends of cellulose and poly(3-hydroxybutyrate-co-3-hydroxyvalerate) prepared from the ionic liquid 1-butyl-3-methylimidazolium chloride. Carbohydr. Polym. **2011**, *86* (1), 94–104.

120. Murugesan, S.M.; Shaker, V.; Aravind, A.; Pulickel, M.; Linhardt, R.J. Ionic liquid-derived bloodcompatible composite membranes for kidney dialysis. Appl. Biomater. **2006**, *79B* (2), 298–304.

121. Suzuki, H.; Sunada, H. Influence of water-soluble polymers on the dissolution of nifedipine solid dispersions with combined carriers. Chem. Pharm. Bull. **1998**, *46* (3), 482–487.

122. Ofoefule, S.I.; Chukwu, A. Application of blends of microcrystalline cellulose-cissus gum in the formulation of aqueous suspensions. Boll. Chim. Farm. **1999**, *138* (5), 217–222.

123. Hwang, R.-C.; Peck, G.R. A systematic evaluation of the compression and tablets characteristics of various types of microcrystalline cellulose. Pharm. Technol. **2001**, *25* (3), 112–132.

124. Attama, A.A.; Adikwu, M.U.; Esimone, C.O. Sedimentation studies on chalk suspensions containing blends of Veegum and detarium gum as suspending agents. Boll. Chim. Farm. **1999**, *138* (10), 521–525.

125. Kamel, S.; Ali, N.; Jahangir, K.; Shah, S.M.; El-Gendy, A.A. Pharmaceutical significance of cellulose: A review. Expr. Polym. Lett. **2008**, *2* (11), 758–778.

126. Okamoto, H.; Nakamori, T.; Arakawa, Y.; Iida, K.; Danjo, K. Development of polymer film dosage forms of lidocaine for buccal administration. II. Comparison of preparation methods. J. Pharm. Sci. **2002**, *91* (11), 2424–2432.

127. Hercules Incorporated, Aqualon Division, http://www. aqualon.com (accessed July 2008).

128. Katzhendler, I.; Mader, K.; Friedman, M. Structure and hydration properties of hydroxypropyl methylcellulose matrices containing naproxen and naproxen sodium. Int. J. Pharm. **2000**, *200* (2), 161–179.

129. Vazquez, M.-J.; Casalderrey, M.; Duro, R.; Gómez-Amoza, J.-L.; Martinez-Pacheco, R.; Souto, C.; Concheiro, A. Atenolol release from hydrophilic matrix tablets with hydroxypropylmethylcellulose (HPMC) mixtures as gelling agent: Effects of the viscosity of the HPMC mixture. Eur. J. Pharm. Sci. **1996**, *4* (1), 39–48.

130. Gupta, A.; Hunt, R.L.; Shah, R.B.; Sayeed, V.A.; Khah, M.A. Disintegration of highly soluble immediate release tablets: A surrogate for dissolution. AAPS PharmSciTech. **2009**, *10* (2), 495–499.

131. Vogelpoel, H.; Welink, J.; Amidon, G.L.; Junginger, H.E.; Midha, K.K.; Moller, H. Biowaiver monographs for immediate release solid oral dosage forms based on Biopharmaceutics Classification System (BCS) literature data: Verapamil hydrochloride, propranolol hydrochloride, and atenolol. J. Pharm. Sci. **2004**, *93* (8), 1945–1956.

132. Kendall, R.A.; Basit, A.W. The role of polymers in solid oral dosage forms. In *Polymers in Drug Delivery*; Uchegbu, I.F., Ed.; CRC Press: Taylor and Francis, UK, 2006; 280.

133. Alderman, D.A. A review of cellulose ethers in hydrophilic matrices for oral controlled-release dosage forms. Int. J. Pharm. Tech. Prod. Mfr. **1984**, *5* (1), 1–9.

134. Liu, J.; Zhang, F.; McGinity, J.W. Properties of lipophilic matrix tablets containing phenylpropanolamine hydrochloride prepared by hot-melt extrusion. Eur. J. Pharm. Biopharm. **2001**, *52* (2), 181–190.

135. Ozeki, T.; Yuasa, H.; Okada, H. Controlled release of drug via methylcellulose-carboxyvinylpolymer interpolymer complex solid dispersion. AAPS PharmSciTech **2005**, *6* (2), E231–E236.

136. Ozeki, T.; Yuasa, H.; Kanaya, Y.; Oishi, K. Application of the solid dispersion method to the controlled release of medicine. VII. Release mechanism of a highly water-soluble medicine from solid dispersion with different molecular weight of polymer. Chem. Pharm. Bull. **1995**, *43* (4), 660–665.

137. Ozeki, T.; Yuasa, H.; Kanaya, Y. Application of the solid dispersion method to the controlled release of medicine. IX. Difference in the release of flurbiprofen from solid dispersions with poly(ethylene oxide) and hydroxypropylcellulose and interaction between medicine and polymers. Int. J. Pharm. **1997**, *115* (2), 209–217.

138. Gon, M.C.; Ferrero, R.M.; Jimenez, C.; Gurruchaga, M. Synthesis of hydroxypropyl methacrylate/ polysaccharide graft copolymers as matrices for controlled release tablets. Drug Dev. Ind. Pharm. **2002**, *28* (9), 1101–1115.

139. Han, R.-Y.; Fang, J.-Y.; Sung, K.C.; Hu, O.Y.P. Mucoadhesive buccal disks for novel nalbuphine prodrug controlled delivery: Effect of formulation variables on drug release and mucoadhesive performance. Int. J. Pharm. **1999**, *177* (2), 201–209.

140. Park, C.R.; Munday, D.L. Development and evaluation of a biphasic buccal adhesive tablet for nicotine replacement therapy. Int. J. Pharm. **2002**, *237* (1), 215–226.

141. Senel, S.; Hincal, A.A. Drug permeation enhancement via buccal route: Possibilities and limitations. J. Control. Release **2001**, *72* (1), 133–144.

142. Rokhade, A.P.; Agnihotri, S.A.; Patil, S.A.; Mallikarjuna, N.N.; Kulkarni, P.V.; Aminabhavi, T.M. Semi-interpenetrating polymer network microspheres of gelatin and sodium carboxymethyl cellulose for controlled release of ketorolac tromethamine. Carbohyd. Polym. **2006**, *65* (3), 243–252.

143. Waree, T.; Garnpimol, C.R. Development of indomethacin sustained release microcapsules using chitosan-carboxymethyl cellulose complex coacervation. Songklanakarin J. Sci. Technol. **2003**, *25* (2), 245–254.

144. Ritthidej, G.C.; Tiyaboonchai, W. Formulation and drug entrapment of microcapsules prepared from chitosan-carboxymethylcellulose complex coacervation. Thai. J. Pharm. Sci. **1997**, *21* (1), 137–144.

145. Khanvilkar, K.H.; Ye, H.; Moore, A.D. Influence of hydroxypropyl methylcellulose mixture, apparent viscosity, and tablet hardness on drug release using a 23 full factorial design. Drug Dev. Ind. Pharm. **2002**, *28* (5), 601–608.

146. Ye, H.; Khanvilkar, K.H.; Moore, A.D.; Hilliard-Lott, M. Effects of manufacturing process variables on *in vitro* dissolution characteristics of extended-release tablets formulated with hydroxypropylmethyl cellulose. Drug Dev. Ind. Pharm. **2003**, *29* (1), 79–88.

147. Liu, Z.; Li, J.; Nie, S.; Liu, H.; Ding, P.; Pan, W. Study of an alginate/HPMC-based in situ gelling ophthalmic delivery system for gatifloxacin. Int. J. Pharm. **2006**, *315* (1), 12–17.

148. Phaechamu, T.; Vesapun, C.; Kraisit, P. Yahom in dosage form of buccal tablet. In 'Proceedings of the 10th World Congress on Clinical Nutrition. Phuket, Thailand' 193–198, 2004.

149. Chantana, V.; Juree, C.; Thawatchaj, P. Effect of hydroxypropylmethyl cellulose and polyvinyl pyrrolidone on physical properties of Yahom buccal tablets and the antimicrobial activity of Yahom. In *The 4th Thailand Materials Science Technology Conference*; Khong Wan: Thailand, PP01/1–PP01/3, 2006.

150. Lenaerts, V.; Dumoulin, Y.; Mateescu, M.A. Controlled release of theophylline from cross-linked amylose tablets. J. Control. Release **1991**, *15* (1), 39–46.

151. Rahmouni, M.; Lenaerts, V.; Massuelle, D.; Doelker, E.; Johnson, M.; Leroux, J.-C. Characterization of binary mixtures consisting of cross-linked high amylase starch and hydroxypropylmethyl cellulose used in the preparation of controlled release tablets. Pharm. Dev. Technol. **2003**, *8* (4), 335–348.

152. Amaral, M.H.; Sousa-Lobo, J.M.; Ferreira, D.C. Effect of hydroxypropylmethyl cellulose and hydrogenated castor oil on naproxen release from sustained-release tablets. APS PharmSciTech. **2001**, *2* (2), 1–8.

153. Ibezim, E.C.; Attama, A.A.; Dimgba, I.C.; Ofoefule, S.I. Use of Carbopols-sodium carboxymethylcellulose admixtures in the formulation of bioadhesive metronidazole tablets. Acta Pharma. **2000**, *50* (2), 121–130.

154. Ibezim, E.C.; Ofoefule, S.I. *In vitro* evaluation of the bioadhesive properties of sodium carboxymethylcellulose, acacia, Veegum and their admixtures. Afr. J. Pharm. Res. Dev. **2006**, *2* (1), 67–72.

155. Attama, A.A.; Adikwu, M.U. Bioadhesive delivery of hydrochlorothiazide using tacca starch/SCMC and Carbopols 940 and 941 admixtures. Boll. Chim. Farm. **1999**, *138* (7), 329–336.

156. Attama, A.A.; Akpa, P.A.; Onugwu, L.E.; Igwilo, G. Novel buccoadhesive delivery system of hydrochlorothiazide formulated with ethyl cellulose-hydroxy propyl methyl celluolose interpolymer complex. Sci. Res. Essays. **2008**, *3* (6), 343–347.

157. Attama, A.A.; Nnamani, P.O.; Adikwu, M.U. Diclofenac release from bioadhesive hydrophilic matrix tablets formulated with polyvinyl pyrrolidone-sodium carboxymethyl cellulose copolymer. J. Pharm. Appl. Sci. **2003**, *1* (1), 1–7.

158. Proddurituri, S.; Urman, K.L.; Otaigbe, J.U.; Rekpa, M.A. Stabilization of hot-melt extrusion formulations containing solid solutions using polymer blends. AAPS PharmSciTech. **2007**, *8* (2), E1–E10.

159. Hercules Pharmaceutical Technology Report (PTR-016). In Polymer blend matrix for oral sustained drug delivery. 25th International Symposium on Controlled Release of Bioadhesive Materials; Las Vegas, NV, June 21–26, 1998.

160. Kuksal, A.; Tiwary, A.K.; Jain, N.K.; Jain, S. Formulation and *in vitro-in vivo* evaluation of extended-release matrix tablet of zidovudine: Influence of combination of hydrophilic and hydrophobic matrix formers. AAPS PharmSciTech. **2006**, *7* (1), E1–E9.

161. Mukherjee, B.; Mahapatra, S.; Gupta, R.; Patra, B.; Tiwari, A.; Arora, P. A comparison betwen povidone-ethylcellulose and povidone-Eudragit transdermal dexamethasone matrix patches based on *in vitro* skin permeation. Eur. J. Pharm. Biopharm. **2005**, *59* (3), 475–483.

162. Yerri-Swamy, B.; Prasad, C.V.; Reedy, C.L.N.; Mallikarjuma, B.; Rao, K.C.; Subha, M.C.S. Interpenetrating polymer network microspheres of hydroxy propyl methyl cellulose/poly(vinyl alcohol) for control release of ciprofloxacin hydrochloride. Cellulose **2011**, *18* (2), 349–357.

163. Iqual, M.C.; Amin, M.; Abadi, A.G.; Ahmad, N.; Jamia, K.H.; Jamal, A. Bacteria cellulose film caoting as drug delivery system: Thermal and drug release properties. Sains Malays. **2012**, *41* (5), 561–568.

164. Siepmann, F.; Siepmann, J.; Walther, M.; MacRae, R.J.; Bodmeier, R. Blends of aqueous polymer dispersions used for pellet coating: Importance of the particle size. J. Control. Release **2005**, *105* (3), 226–239.

165. Tezuka, Y.; Imai, K.; Oshima, M.; Ito, K. 13C-n.m.r. structural study on an enteric pharmaceutical coating cellulose derivative having ether and ester substituents. Carbohydr. Res. **1991**, *222* (1), 255–259.

166. Wu, S.; Wyatt, D.; Adams, M. *Aqueous Polymeric Coatings for Pharmaceutical Dosage Forms*; McGinity, J., Ed.; Dekker: New York, 1997; 385–418.

167. Drury, J.L.; Mooney, D.J. Hydrogels for tissue engineering: Scaffold design variables and applications. Biomaterials **2003**, *24* (24), 4337–4351.

168. El-Hag Ali, A.; Abd El-Rehim, H.; Kamal, H.; Hegazy, D. Synthesis of carboxymethyl cellulose based drug carrier hydrogel using ionizing radiation for possible use as specific delivery system. J. Macromol. Sci. Pure Appl. Chem. **2008**, *45* (8), 628–634.

169. Trojani, C.; Weiss, P.; Michiels, J.F.; Vinatier, C.; Guicheux, J.; Daculsi, G.; Gaudray, P.; Carle, G.F.; Rochet, N. Three-dimensional culture and differentiation of human osteogenic cells in an injectable hydroxypropylmethylcellulose hydrogel. Biomaterials **2005**, *26* (27), 5509–5517.

170. Peppas, N.A. Hydrogels and drug delivery. Curr. Opin. Colloid Interface Sci. **1997**, *2* (5), 531–537.

171. Chen, J.; Park, H.; Park, K. Synthesis of superporous hydrogels: Hydrogels with fast swelling and superabsorbent properties. J. Biomed. Mater. Res. **1999**, *44* (1), 53–62.

172. Sannino, A.; Pappadà, S.; Madaghiele, M.; Maffezzoli, A.; Ambrosio, L.; Nicolais, L. Crosslinking of cellulose derivatives and hyaluronic acid with water-soluble carbodiimide. Polymer **2005**, *46* (25), 11206–11212.

173. Sannino, A.; Madaghiele, M.; Lionetto, M.G.; Schettino, T.; Maffezzoli, A. A cellulose-based hydrogel as a potential bulking agent for hypocaloric diets: An *in vitro* biocompatibility study on rat intestine. J. Appl. Polym. Sci. **2006**, *102* (2), 1524–1530.

174. Litschauer, M.; Neouze, M.-A.; Haimar, E.; Henniges, U.; Potthast, A.; Rosenau, T.; Liebner, F. Silica modified cellulosic aerogels. Cellulose **2011**, *18* (1), 143–149.

175. Sannino, A.; Esposito, A.; Nicolais, L.; Del Nobile, M.A.; Giovane, A.; Balestrieri, C.; Esposito, R.; Agresti, M. Cellulose-based hydrogels as body water retainers. J. Mater. Sci. **2000**, *11* (4), 247–253.

176. Sannino, A.; Esposito, A.; De Rosa, A.; Cozzolino, A.; Ambrosio, L.; Nicolais, L. Biomedical application of a superabsorbent hydrogel for body water elimination in the treatment of edemas. J. Biomed. Mater. Res. A 2003, *67* (3), 1016–1024.

177. Esposito, A.; Sannino, A.; Cozzolino, A.; Quintiliano, S.N.; Lamberti, M.; Ambrosio, L.; Nicolais, L. Response of intestinal cells and macrophages to an orally administered cellulose-PEG based polymer as a potential treatment for intractable edemas. Biomaterials **2005**, *26* (19), 4101–4110.

178. Shanbhag, A.; Barclay, B.; Koziana, J.; Shivanand, P. Application of cellulose acetate butyrate-based membrane for osmotic drug delivery. Cellulose **2007**, *14* (1), 65–71.

179. Verma, R.K.; Mishra, B.; Garg, S. Osmotically controlled oral drug delivery. Drug Dev. Ind. Pharm. **2000**, *26* (7), 695–708.

180. Theeuwes, F. Oros-osmotic system development. Drug Dev. Ind. Pharm. **1983**, *9* (7), 1331–1357.

181. Santus, G.; Baker, R.W. Osmotic drug delivery: A review of the patent literature. J. Control. Release **1995**, *35* (1), 1–21.

182. Theeuwes, F. Elementary osmotic pump. J. Pharm. Sci. **1975**, *64* (12), 1987–1991.

183. Makhija, S.N.; Vavia, P.R. Controlled porosity osmotic pump-based controlled release systems of pseudoephedrine. I. Cellulose acetate as a semipermeable membrane. J. Control. Release **2003**, *89* (1), 5–18.

184. Schultz, P.; Kleinebudde, P. A new multiparticulate delayed release system. Part I: Dissolution properties and release mechanism. J. Control. Release **1997**, *47* (2), 181–189.

185. Catellani, P.L.; Colombo, P.; Peppas, N.A.; Santi, P.; Bettini, R. Partial permselective coating adds an osmotic contribution to drug release from swellable matrixes. J. Pharm. Sci. **1998**, *87* (6), 726–731.

186. Lim, S. Manufacture of Oral Dosage Forms Delivering Both Immediate-Release and Sustained-Release Drugs. US Patent 0795324;0598309;1330839; WO00/23045. 6682759. Jan 27, 2004.

187. Lee, M. Controlled release of dual drug-loaded hydroxypropylmethylcellulose matrix tablet using drug-containing polymeric coatings. Int. J. Pharm. **1999**, *188* (1), 71–80.

188. Builders, P.F.; Agbo, M.B.; Adelakun, T.; Okpako, L.C.; Attama, A.A. Novel multifunctional pharmaceutical excipients derived from microcrystalline cellulose-starch microparticulate composites prepared by compatibilized reactive polymer blending. Int. J. Pharm. **2010**, *388* (1–2), 159–167.

189. Molla, M.A.K.; Shaheen, S.M.; Rashid, M.; Hossain, A.K.M.M. Rate controlled release of naproxen from HPMC Based sustained release dosage form, I. Microcapsule compressed tablet and matrices. Dhaka Univ. J. Pharm. Sci. **2005**, *4* (1), 588–599.

190. Siepmann, J.; Peppas, N.A. Modeling of drug release from delivery systems based on hydroxypropyl methylcellulose (HPMC). Adv. Drug Deliv. Rev. **2001**, *48* (2–3), 139–157.

191. Siepmann, J.; Streubel, A.; Peppas, N.A. Understanding and predicting drug delivery from hydrophilic matrix tablets using the "sequential layer" model. Pharm. Res. **2002**, *19* (3), 306–314.

192. Kommareddy, S.; Tiwari, S.; Amiji, M. Long-circulating polymeric nanovectors for tumor-selective gene delivery. Technol. Cancer Res. Treat. **2005**, *4* (6), 615–625.

193. Lee, M.; Kim, S. Polyethylene glycol-conjugated copolymers for plasmid DNA delivery. Pharm. Res. **2005**, *22* (1), 1–10.

194. Hamidi, M.; Azadi, A.; Rafiei, P. Hydrogel nanoparticles in drug delivery. Adv. Drug Deliv. Rev. **2008**, *60* (15), 1638–1649.

195. Soma, C.E.; Dubernet, C.; Barratt, G.; Nemati, F.; Appel, M.; Benita, S.; Couvreur, P. Ability of doxorubicin-loaded nanoparticles to overcome multidrug resistance of tumour cells after their capture by macrophages. Pharm. Res. **1999**, *16* (11), 1710–1716.

196. Habibi, Y.; Goffin, A.L.; Schiltz, N.; Duquesne, E.; Dubois, P.; Dufresne, A. Bionanocomposites based on poly(epsilon-caprolactone)-grafted cellulose nanocrystals by ring-opening polymerization. J. Mater. Chem. **2008**, *18* (1), 5002–5010.

197. Paralikar, S.A.; Simonsen, J.; Lombardi, J. Poly(vinyl alcohol)/cellulose nanocrystals barrier membranes. J. Membr. Sci. **2008**, *320* (1), 248–258.

198. Bai, W.; Holbery, J.; Li, K. A techinique for production of nanocrystalline cellulose with a narrow size distribution. Cellulose **2009**, *16* (3), 455–465.

199. Mörseburg, K.; Chinga-Carrasco, G. Assessing the combined benefits of clay and nanofibrillated cellulose in layered TMP-based sheets. Cellulose **2009**, *16* (5), 795–806.

200. Chinga-Carrasco, G.; Syverud, K. Computer-assisted quantification of the multi-scale structure of films made of nanofibrillated cellulose. J. Nanopart. Res. **2010**, *12* (3), 841–851.

201. Moran, J.I.; Alvarez, V.A.; Cyras, V.P.; Vazquez, A. Extraction of cellulose and preparation of nanocellulose from sisal fibers. Cellulose **2008**, *15* (1), 149–159.

202. Aumelas, A.; Serrero, A.; Durand, A.; Dellacherie, E.; Leonard, M. Nanoparticles of hydrophobically modified dextrans as potential drug carrier systems. Colloids Surf. B **2007**, *59* (1), 74–80.

203. Couvreur, P.; Lemarchand, C.; Gref, R. Polysaccharide-decorated nanoparticles. Eur. J. Pharm. Biopharm. **2004**, *58* (2), 327–341.

204. Wegnar, T.H.; Jones, P.E. Advancing cellulose-based nanotechnology. Cellulose **2006**, *13* (2), 115–118.

205. Edgar, K.J. Cellulose esters in drug delivery. Cellulose **2007**, *14* (1), 49–64.

206. Baumann, M.D.; Kang, C.E.; Stanwick, J.C.; Wang, Y.; Kim, H.; Lapitsky, Y.; Shoichet, M.S. An injectable drug delivery platform for sustained combination therapy. J. Control. Release **2009**, *138* (3), 205–213.

207. Watanabe, Y.; Mukai, B.; Kawamura, K.I.; Ishikawa, T.; Namiki, M.; Utoguchi, N.; Fujii, M. Preparation and evaluation of press-coated aminophylline tablet using crystalline cellulose and polyethylene glycol in the outer shell for timed-release dosage forms. Yakugaku Zasshi **2002**, *122* (2), 157–162.

Biomimetic—Colloid

208. Gilberto, S.; Julien, B.; Alain, D. Cellulosic bionanocomposite: A review of preparation, properties and applications. Polymers **2010**, *2* (4), 728–765.

209. Lönnberg, H.; Fogelström, L.; Samir, M.A.S.A.; Berglund, L.; Malmström, E.; Hult, A. Surface grafting of microfibrillated cellulose with poly(ε-caprolactone)—synthesis and characterization. Eur. Polym. J. **2008**, *44* (9), 2991–2997.

210. Hassan, N.; Farzaneh, F.; Abolfazl, H. Nanoparticles based on modified polysaccharides. In *The Delivery of Nanoparticles*; Hashim, A.A., Ed.; InTech: Croatia, 2012; 149–184.

211. Uglea, C.V.; Pary, A.; Corjan, M.; Dumitriu, A.D.; Ottenbrite, R.M. Biodistribution and antitumor activity induced by carboxymethylcellulose conjugates. J. Bioact. Comp. Polym. **2005**, *20* (6), 571–583.

212. Sievens-Figueroa, L.; Bhakay, A.; Jerez-Rozo, J.T.; Pandya, N.; Romanach, R.J.; Michniak-Kohn, B.; Iqbal, Z.; Bilgili, E.; Dave, R.N. Preparation and characterization of hydroxypropylmethyl cellulose films containing stable BCS Class II drug nanoparticles for pharmaceutical applications. Int. J. Pharm. **2012**, *423* (2), 496–508.

213. Ernsting, M.J. Cellulose-Based Nanoparticles for Drug Delivery. US Patent US 20120219508A1, 9 Aug, 2012.

214. Dong, S.; Hirani, A.A.; Lee, Y.W.; Roman, M. *Synthesis of FITC-Labelled, Folate-Targeted Cellulose Nanocrystals*. Technical Programming Archive, 239th ACS National Meeting: San Francisco, CA, Mar 21–25, 2010; CELL-061.

215. Roman, M.; Dong, S.; Hirani, A.; Lee, Y.W. Cellulose nanocrystals for drug delivery. Chapter 12 In *Polysaccharide materials: Performance by design*; Edgar, K.J., Heinze, T., Buchanan, C. Eds.; ACS Symposium Series 1017. American Chemical Society: Washington, DC, 2009.

216. Dong, S.; Roman, M. *Synthesis of Fluorescently-Labelled, Folate-Targeted Cellulose Nanoconjugates*, Proceedings of the 2009 MII Technical Conference and Review, Macromolecules and Interfaces Institute: Blacksburg, VA. Apr, 2009; 73.

217. Dong, S.; Roman, M. Fluorescently labelled cellulose nanocrystals for bioimaging applications. J. Am. Chem. Soc. **2007**, *129* (45), 13810–13811.

218. Roman, M. *Novel Applications of Cellulose Nanocrystals: From Drug Delivery to Micro-Optics*, Proceedings of the MII Technical Conference and Review: Blacksburg, VA, Oct, 2007; 22–24.

219. Dong, S.; Hirani, A.; Lee, Y.W.; Roman, M. *Cellulose Nanocrystals for the Targeted Delivery of Therapeutic Agents*, Proceedings of the MII Technical Conference and Review: Blacksburg, VA, Oct, 2007; 22–24.

220. Dong, S.; Hirani, A.A.; Lee, Y.W.; Roman, M. *Synthesis of FITC-Labelled, Folate-Targeted Cellulose Nanocrystals*. Proceedings of the MII Technical Conference and Review: Blacksburg, VA, Oct, 2010; 11–13.

221. Cho, H.J.; Lee, S.; Dong, S.; Roman, M.; Lee, Y.W. Cellulose nanocrystals as a novel nanocarrier for targeted drug delivery to brain tumor cells. FASEB J. **2011**, *25*, 762.2.

222. Colacino, K.R.; Dong, S.; Roman, M.; Lee, Y.W. Cellulose nanocrystals: A novel biomaterial for targeted drug delivery applications. FASEB J. **2011**, *25*, 762.3.

223. Aswathy, R.G.; Sivakumar, B.; Brahatheeswaran, D.; Raveendran, S.; Ukai, T.; Fukuda, T.; Yoshida, Y.; Maekawa, T.; Sakthikumar, D.N. Multifunctional biocompatible fluorescent carboxymethyl cellulose nanoparticles. J. Biomater. Nanobiotechnol. **2012**, *3* (2), 254–261.

224. Yallapu, M.M.; Dobberpuhl, M.R.; Maher, D.M.; Jaggi, M.; Chauhan, S.C. Design of curcumin loaded cellulose nanoparticles for prostate cancer. Curr. Drug Metab. **2012**, *13* (1), 120–128.

225. Neha, D. Novel cellulose nanoparticles for potential cosmetic and pharmaceutical applications. A Masters' Thesis, Department of Chemical Engineering, University of Waterloo, Ontario, Canada, 2010.

226. Baumgartner, S.; Kristl, J.; Peppas, N.A. Network structure of cellulose ethers used in pharmaceutical applications during swelling and at equilibrium. Pharm. Res. **2002**, *19* (8), 1084–1090.

227. Sachiko, K.N.; Keiji, N. Biopolymer-based nanoparticles for drug/gene delivery and tissue engineering. Int. J. Mol. Sci. **2013**, *14* (1), 1629–1654.

228. Cui, Z.; Mumper, R.J. Chitosan-based nanoparticles for topical genetic immunization. J. Control. Release **2001**, *75* (3), 409–419.

229. Arhewoh, I.M.; Ahonkhai, E.I.; Okhamafe A.O. Optimising oral systems for the delivery of therapeutic proteins and peptides. Afr. J. Biotech. **2005**, *4* (13), 1591–1597.

230. Okhamafe, A.O.; Amsden, B.; Chu, W.; Goosen, M.F.A. Modulation of protein release from chitosan-alginate microcapsules modified with the pH sensitive polymer-hydroxylpropyl methylcellulose acetate succinate (HPMCAS). J. Microencapsul. **1996**, *13* (5), 497–508.

231. Amorim, M.J; Ferreira, J.P. Microparticles for delivering therapeutic peptides and proteins to the lumen of the small intestine. Eur. J. Pharm. Biopharm. **2001**, *52* (1), 39–44.

232. Moghimi, S.M.; Hunter, A.C.; Murray, J.C. Long-circulating and target-specific nanoparticles: Theory to practice. Pharm. Rev. **2001**, *53* (2), 283–318.

233. Adiseshaiah, P.P.; Hall, J.B.; McNeil, S.E. Nanomaterial standards for efficacy and toxicity assessment. Wiley Interdiscip. Rev. **2010**, *2* (1), 99–112.

234. Mansouri, S.; Lavigne, P.; Corsi, K.; Benderdour, M.; Beaumont, E.; Fernandes, J.C. Chitosan-DNA nanoparticles as non-viral vectors in gene therapy: Strategies to improve transfection efficacy. Eur. J. Pharm. Biopharm. **2004**, *57* (1), 1–8.

235. Thomas, M.; Klibanov, A.M. Non-viral gene therapy: Polycation-mediated DNA delivery. Appl. Microbiol. Biotechnol. **2003**, *62* (1), 27–34.

236. Li, S.D.; Huang, L. Gene therapy progress and prospects: Non-viral gene therapy by systemic delivery. Gene Ther. **2006**, *13* (18), 1313–1319.

237. Dang, J.M.; Leong, K.W. Natural polymers for gene delivery and tissue engineering. Adv. Drug Deliv. Rev. **2006**, *58* (4), 487–499.

Biomimetic—Colloid

238. Sachiko, K.N.; Keiji, N. Biopolymer-based nanoparticles for drug/gene delivery and tissue engineering. Int. J. Mol. Sci. **2013**, *14* (1), 1629–1654.

239. Xu, F.J.; Ping, Y.; Ma, J.; Tang, G.P.; Yang, W.T.; Li, J.; Kang, E.T.; Neoh, K.G. Comb-shaped copolymers composed of hydroxypropyl cellulose backbones and cationic poly((2-dimethyl amino) ethyl methacrylate) side chains for gene delivery. Bioconjug. Chem. **2009**, *20* (8), 1449–1458.

240. Czaja, W.K.; Young, D.J.; Kawecki, M.; Malcolm-Brown, R. Jr. The future prospects of microbial cellulose in biomedical applications. Biomacromolecules **2007**, *8* (1), 1–12.

Colloid Drug Delivery Systems

Monzer Fanun
Colloids and Surfaces Research Center, Al-Quds University, East Jerusalem, Palestine

Abstract
Colloids are a dispersion of one phase to another, and their size ranges from 1 nm to 1 μm, and in a system, discontinuities are found at distances of that order. Colloidal properties provide information about absorption, cell membrane interaction, and drug molecule behavior in the surrounding environment. This entry discusses the role of colloids in drug development and delivery.

COLLOIDS

Colloids are a dispersion of one phase to another, and their size ranges from 1 nm to 1 μm, and in a system, discontinuities are found at distances of that order. These nanoscale dispersed systems have considerable application in drug delivery and formulation methods.[1,2] Understanding the properties of colloids can be used as a tool to access rational criteria for formulation development. Colloidal properties provide an idea about absorption, cell membrane interaction, and drug molecule behavior in the surrounding environment. Such properties play a major role in drug-loading capacity and transport mechanisms as charge and colloidal size ranges are set for this capacity. The colloidal particles greatly affect the rate of sedimentation, osmotic pressure as well as the stability and biocompatibility of colloidal drug carrier. The colloidal properties also provide a selection basis for excipients and manufacturing process criteria. These properties include the particle properties studied by the Tyndall effect, turbidity, and dynamic light scattering. Optical properties are widely used to observe the size, shape, and structure of colloidal particles. The kinetic properties, which include Brownian motion, diffusion, osmosis, sedimentation, and viscosity, deal with motion of particles with respect to the dispersion medium. These properties provide a detailed idea on the movement of colloidal carrier within the body as well as the transportation criteria of colloidal carrier across the cell membrane. The physicochemical properties such as physical state, lyophilicity, and lyophobicity are helpful in the selection of colloidal carrier for a particular drug delivery. Electrical properties, which include electrical conductivity and surface charges, are related with migration of particle and give an idea about the selection basis for excipients and the interaction of colloidal carrier with a cell membrane. The magnetic properties of colloids deal with specific delivery of drug inside the body, which helps to increase the efficacy of drug and reduces the toxicity.

The problems experienced by free drugs delivery are instability, poor solubility, toxicity, nonspecificity, and inability to cross blood–brain barrier. The problem experienced by free drug can be overcome by the use of drug delivery system. The development of delivery systems has had a huge impact on our ability to treat numerous diseases. To attain highest pharmacological effects with least side effects of drugs, it should be delivered to target sites without significant distribution to nontarget areas. Colloidal drug delivery carriers are known to improve the solubility of poorly soluble drugs; to provide a microenvironment for the protection of fragile drugs, such as proteins and peptides, that are very often of large size and need to be protected from hydrolysis or enzymatic degradation and absorption through membrane; to increase the drug efficacy and reduce their toxicity; to achieve the targeted delivery of drug by conjugating a specific vector to the carrier; and to have emerged as efficient vehicles for drug delivery, which allow sustained or controlled release for transdermal, topical, oral, nasal, intravenous, ocular, parenteral, and other administration routes of drugs.[1,2] The colloidal delivery systems have been aimed at improving therapeutics by virtue of their submicron size and targetability. The effectiveness of a drug therapy is often governed by the extent to which temporal and distribution control could be achieved. Temporal control is the ability to manipulate the period over which drug release is to take place and/or the possibility of triggering the release at a specified time during the treatment in response to some stimuli such as temperature, pH, and so on. Distribution control is to direct the delivery system to the desired site of action. Colloidal drug delivery systems include micelles, dendrimers, liquid crystals, cubosomes, hexosomes, emulsions, multiple emulsions and self-emulsifying drug delivery systems (SEDDS), nanoemulsions, microemulsions and self-microemulsifying drug delivery systems (SMEDDSs), liposomes, niosomes, polymesomes, nanocapsules, solid–lipid nanoparticles, microspheres, aerosols, and foams. Figure 1 presents a lipid-based drug nanocarrier and its nanoscale dimensional comparison with biological cell, DNA, lipid bilayers, and atoms. In this entry, we have attempted to present a broad

Concise Encyclopedia of Biomedical Polymers and Polymeric Biomaterials DOI: 10.1081/E-EBPPC-120050216

Biomimetic—Colloid

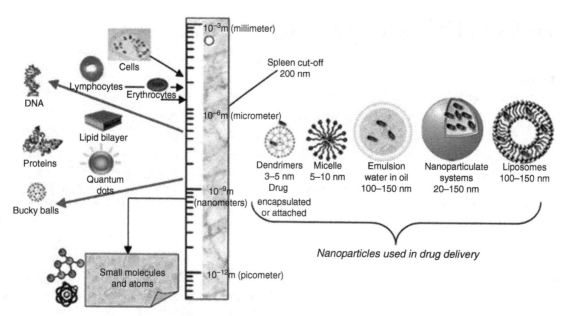

Fig. 1   Lipid-based drug nanocarriers and their nanoscale dimensional comparison with biological cell, DNA, lipid bilayers, and atoms.

view over the past years on these colloidal drug delivery systems as solubilization and dissolution enhancers of poorly soluble drugs, as a medium for generating new drug delivery systems, and as delivery systems themselves.[1,2]

MICELLES

Micelles are the self-assembling nanosized colloidal particles, stealth properties with a hydrophobic core and hydrophilic shell with control size and composition. Micelles are dynamic structures with a liquid core, so they cannot be assumed to be having a definitive rigid shape. For scientific consideration, they are usually regarded as having sphericity. Micelles are formed only when the concentration of the surfactant in the solution increases above critical micellar concentration. Micellar shape may be affected by factors such as concentration, temperature, and presence of added electrolyte. Thus, they may undergo a transition when any of the factors are changed. Micelles also exhibit polydispersity in size. Among the micelle-forming compounds besides surfactants, amphiphilic copolymers, that is, polymers consisting of hydrophobic and hydrophilic blocks, are gaining an increasing attention.[3] The interest has specifically been focused on the potential application of polymeric micelles in the major areas in drug delivery, drug solubilization, controlled drug release, drug targeting, and diagnostic purpose. Increased solubility of a less soluble organic substance in surfactant solution has been applied since long.[1] This increased solubility of a solubilizate in surfactant solution was due to some form of attachment of the solubilizate to the exterior

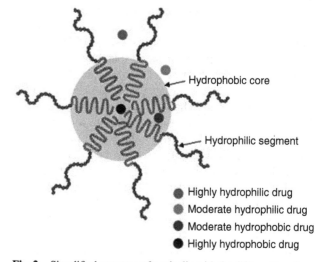

Fig. 2   Simplified structure of a micelle with the different locations of solubilizate.

of the micelle or solution in it. There is a marked difference in the behavior of polar and nonpolar solutes.[1] The solubilizate may be present at different sites in the micelles, dependent on its chemical nature. It is usually observed that the nonpolar solubilizates are accommodated in the hydrocarbon core and the semipolar and polar solubilizates are present in the palisade layer. The location in the palisade layer can be either deep buried or short penetration. Figure 2 depicts the simplified structure of a micelle with the different locations of solubilizate. The application of micelles as drug delivery systems can be employed for enhancing the solubilization of potent but hydrophobic and sparingly soluble drug candidates. A low water concentration inside the micellar core retards the degradation of drugs that are prone to hydrolysis. Micelles facilitate controlled and sustained drug delivery by virtue of partitioning toward it. Apart from these major applications, they may also serve as good excipients, act as adjuncts in vaccines, facilitate taste masking, and so on.[4–6]

DENDRIMERS

Dendrimers have a highly branched, nanoscale architecture with very low polydispersity and high functionality, comprising a central core, internal branches, and a number of reactive surface groups. Because of their unique highly adaptable structures, dendrimers have been extensively investigated for drug delivery and demonstrated great potential for improving therapeutic efficacy.[7–11] Figure 3 presents a schematic of a dendritic structure. To date, dendrimers have been tailored to deliver a variety of drugs for the treatment of various diseases such as cancers. A wide range of drugs can be delivered by dendrimers through either covalent linkages or noncovalent interactions. The presence of a number of end groups on the dendrimer surface enables high drug payload and assembly of multiple functional entities including targeting ligand for targeted drug delivery. Highly adaptable dendritic structures can be engineered to treat various diseases and personalized medicine. In addition to the extensive use of off-the-shelf dendrimers for drug delivery, new dendritic structures are also being developed to obtain additional structural properties to meet specific needs in drug delivery.[7–11]

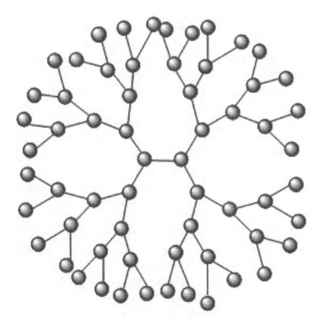

Fig. 3   A schematic of a dendritic structure.

LIQUID CRYSTALS

Liquid crystals are intermediate states of matter or mesophases, halfway between an isotropic liquid and a solid crystal. In nature, some substances, or even mixtures of substances, present these mesomorphic states. These liquid crystalline phases exhibit a local disorder ("liquid-like" behavior) and are dynamic at a molecular level, but a long-range order exists, which endows it with unique rheological, mass transport, and optical properties. Liquid crystals offer a number of useful properties for the drug delivery. Solubilization of drug in the liquid crystalline is similar to the solubilization of drug in micelles. Simultaneously, increase in viscosity of the system helps to provide more localized effects in parenteral (intramuscular), topical, or oral administration. The phase transitions of liquid crystals can be achieved either by temperature or by dilution. The systems can be tailored in such a way the transitions can be achieved at body temperature or in contact with the body fluids. Figure 4 presents a schematic of functional liquid crystal assembly.[12–14]

CUBOSOMES AND HEXOSOMES

Cubosomes, hexosomes, and micellar cubosomes are colloidal aqueous dispersions of surfactant-like lipids with confined inner nanostructures. The formation of these dispersed submicron-sized particles with embedded inverted-type mesophases that display nanostructures closely related to those observed in biological membranes is receiving much attention in pharmaceutical, food, and cosmetical applications. Owing to their unique physicochemical characteristics, they represent an interesting colloidal family that has excellent potential to solubilize bioactive molecules with different physicochemical properties (hydrophilic, amphiphilic, and hydrophobic molecules). In particular, there is an enormous interest in testing the possibility to utilize these dispersions as drug nanocarriers for enhancing the solubilization of poorly water-soluble drugs and for improving their bioavailability. It is important to emphasize that the optimal utilization of these dispersions requires understanding their stability under different conditions (such as presence of salt and pH variation), studying the impact of loading drugs on their internal nanostructures, and fully comprehending the interaction of these drug-loaded dispersed particles with biological interfaces to ensure the efficient transportation of solubilized drugs. Nevertheless, there is a need for further investigations in the future to address the challenges of enhancing the stability of the dispersed particles after administration and of modulating their nanostructure to optimize their interaction with different biological surfaces. It was found that the addition of short charged designer peptide surfactants mimicking biological phospholipids can be used to functionalize the nonlamellar mesophases and to

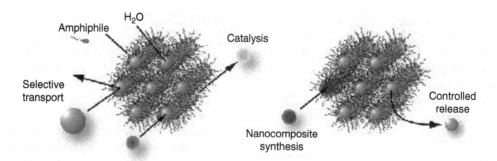

Fig. 4 Functional liquid crystal assembly.

enhance the loading capacity of charged active molecules. The term "functionalization" means to control the solubilization capacity of the liquid crystalline phases by the inclusion of specific anchors such as charged or long-chain amphiphilic molecules.[15–19]

GELS

The use and formulation of three-dimensional gel systems as gelating agents, matrices in patches, and wound dressings for the delivery of dermal and transdermal drugs were investigated.[20–25] Requirements for such a delivery system depend on both the drug characteristics and the conditions in the place of the topical drug application. The properties, composition, duration of drug effect, uses, and side effects of the products that are applicable for the skin are considered. The most applied gels belong to the classes of cellulose derivatives, chitosan, carageenan, polyacrylates, polyvinyl alcohol, polyvinylpyrrolidone, and silicones, which can be further modified to alter the drug delivery or three-dimensional configuration of the gel, including synthesis of stimuli-sensitive polymers.[20–25]

EMULSIONS

Emulsions are thermodynamically unstable system consisting of two immiscible liquids, oil and water, one is dispersed as minute globules in the other continuous medium. The system is stabilized by the presence of surfactants. Emulsions are extensively used as drug delivery systems particularly for topical and oral route of administration. Generally, oil-in-water emulsions are intended for internal use due to obvious reasons. Some of the advantages offered by emulsions in drug delivery are solubilization of hydrophobic drug, high drug release rate, and enhanced chemical stability or protection of drug from hydrolysis in aqueous environment. Pharmaceutical emulsions contain less amount of surfactants (phospholipids, lecithins), which are natural in origin and thus nontoxic in nature. Oil components are always selected from the oils of natural origin such as soya bean oil, cottonseed oil, coconut oil, corn oil, sesame oil, cod liver oil, olive oil, and linseed oil or mixture of different fatty acids in different proportions.[26] SEDDS

are ideally isotropic mixtures of oils and surfactants, and sometimes including cosolvents. They emulsify under mild agitation conditions, just like those encountered in the gastrointestinal tract and produces fine emulsion/lipid droplets, ranging from 100 nm to less than 50 nm. Substantial interfacial disruption and/or ultra-low oil–water interfacial tension are the primary necessities to achieve the self-emulsification. SEDDS have been shown their ability to improve the oral bioavailability of less water-soluble and hydrophobic drugs. Generally, SEDDS are prepared by dissolving drugs in oils containing suitable solubilizing agents. SEDDS are generally formulated by using triglyceride oils and nonionic surfactants at surfactant concentrations more than 25%. As an improvement of conventional liquid-SEDDS, solid-SEDDS (S-SEDDS) may also be developed. S-SEDDS may reduce production cost, having simple manufacturing process, and they are stable as well as patient compliant. S-SEDDS may also be modified to include solid dosage forms for oral as well as for parenteral administration.[26–33]

MULTIPLE EMULSIONS

Multiple emulsions are complex emulsion systems where both oil-in-water (o/w) and water-in-oil (w/o) emulsion types exist simultaneously. The simplest multiple emulsions, popularly known as "double emulsions," are ternary systems consisting of either water-in-oil-in-water (w/o/w) or oil-in-water-in-oil (o/w/o) structures. Multiple emulsions are generally prepared either by a one-step emulsification or by a two-step emulsification process. Multiple emulsions-based formulations have been widely used in the pharmaceutical industry as vaccine adjuvants, sustained release, and parenteral drug delivery systems. However, o/w/o systems have found wide application in the cosmetic and food industries. Many particulate drug delivery systems (nanoparticles, microspheres, liposomes) have been prepared by using multiple emulsions during one of the developmental steps. The poor stability and polydispersibility of multiple emulsions (w/o/w) has been a major challenge for its pharmaceutical applications. Therefore, mechanism of its stability, different formulation methods used to enhance monodispersibility, and evaluation of the finished product ensuring quality were investigated.[34–38] Multiple emulsions are versatile drug carriers, which have a wide range of applications in controlled drug delivery, particulate-based drug delivery systems, targeted drug delivery, taste masking, bioavailability enhancement, enzyme immobilization, overdosage treatment/detoxification, red blood cell substitute, lymphatic delivery, and shear-induced drug release formulations for topical application. Surface-modified fine multiple emulsions containing biopolymers are used as parenteral and transarterial delivery vehicles for various anticancer drugs and antigens employed in cancer chemotherapy and immunotherapy. Some novel applications include encapsulation of a drug by the evaporation of the intermediate phase, leaving behind capsules formed of polymers (polymerosomes), solid particles (colloidosomes), and vesicles.[34–38]

NANOEMULSIONS

Much attention has been paid to potential pharmaceutical uses of nanoemulsions as novel drug delivery systems. Nanoemulsions are transparent or translucent systems that have a dispersed-phase droplet size range of typically 20 to 200 nm; although in earlier cases, these systems have also been called microemulsions. Nanoemulsions are thermodynamically stable, isotropically clear dispersions of oil and water stabilized with the help of surfactant and cosurfactant. These systems are attractive as pharmaceutical formulations, and they are drug carrier systems for oral, topical, and parenteral administration. Especially, these dosage forms have been suggested as carriers for peroral peptide-protein drugs. It was hypothesized that formulating a nanoemulsion of the drug would help to increase the bioavailability of the drug due to the high solubilization capacity as well as the potential for enhanced absorption. On the other hand, the small droplet size, high kinetic stability, and optical transparency of nanoemulsions compared with conventional emulsions give them advantages for their use in many technological applications. The majority of works on nanoemulsion applications deal with the preparation of polymeric nanoparticles using a monomer as the disperse phase (the so-called miniemulsion polymerization method).[39–43]

MICROEMULSIONS

Microemulsions are transparent systems of two immiscible fluids, stabilized by an interfacial film of surfactant or a mixture of surfactants, frequently in combination with a cosurfactant. These systems could be classified as water-in-oil, bicontinuous, or oil-in-water type depending on their microstructure, which is influenced by their physicochemical properties and the extent of their ingredients.[1,2,44–49] SMEDDSs form transparent microemulsions with a droplet size of less than 50 nm. Oil is the most important excipient

in SMEDDSs because it can facilitate self-emulsification and increase the fraction of lipophilic drug transported through the intestinal lymphatic system, thereby increasing absorption from the gastrointestinal tract. Long-chain and medium-chain triglyceride oils, modified or hydrolyzed vegetable oils, have been explored widely. Novel semisynthetic medium-chain triglyceride oils have surfactant properties and are widely replacing the regular medium-chain triglyceride. Microemulsions are characterized by ultra-low interfacial tension between the immiscible phases and offer the advantage of spontaneous formation; thermodynamic stability; simplicity of manufacture; solubilization capacity of lipophilic, hydrophilic, and amphiphilic solutes; improved solubilization and bioavailability of hydrophobic drugs; the large area per volume ratio for mass transfer; and the potential for permeation enhancement. It was demonstrated that microemulsions enhance the solubilization capacity and dissolution efficiency of poorly soluble drugs; the solubilization capacity and dissolution efficiency of drugs are reliant on the microstructure of the microemulsions; the solubilized drugs may influence the boundaries of structural regions and the transition point between different microemulsion's microstructures; drug extent and route of delivery could be influenced by the microstructure of the microemulsions; drug delivery systems generated in microemulsions improved drug release and compatibility; the extent and rate of drug delivery is dependent on the generated system preparation method in microemulsions; and the generated system could influence the selection of a delivery route.[44–49] Figure 5 presents schematically (not to scale) of possible packings of drug along dilution line N60 at the different dilution regions: water-in-oil, bicontinuous, and oil-in-water microemulsions. N60 means that the weight ratio of oil/mixed surfactants equals 4/6.

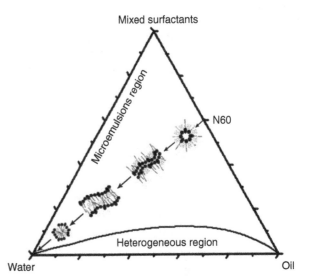

Fig. 5 Schematic presentation (not to scale) of possible packings of drug along dilution line N60 at the different dilution regions: water-in-oil, bicontinuous, and oil-in-water microemulsions. N60 means that the weight ratio of oil/mixed surfactants equals 4/6.

LIPOSOMES

Advances in liposome technologies for conventional and nonconventional drug deliveries have helped pharmacologists with a tool to increase the therapeutic index of several drugs by improving the ratio of the therapeutic effect over the drug's side effect. Thanks to liposome versatility, effective formulations able to deliver hydrophobic drugs, to prevent drug degradation, to alter biodistribution of their associated drugs, to modulate drug release, and, most of all, to selectively target the carrier to specific areas have been obtained. Moreover, by gradually increasing the complexity of the formulation, sophisticated membrane models have been developed, and new landscapes on characterization possibilities of membrane-associated biomacromolecules have been discovered. Basic conventional liposomes are multilamellar or unilamellar vesicles composed of phospholipids, but several different constituents are often included to modulate the fluidity or the permeability of the bilayer. Figure 6 presents a unilamellar liposomal vesicle. With the aim of increasing bioavailability in blood, differently alkylated poly(ethylene glycol)-based polymers are usually added to liposomal formulations. Variations in formulation and in preparation methods may affect the efficacy of the active principle that has to be delivered or the functionality of the biomacromolecule that has to be studied, and this is why a universal liposome preparation protocol cannot be developed.[50–54]

NIOSOMES

Similar to liposomes, nonionic surfactant-based vesicles (niosomes) are vesicularly formed from the self-assembly of nonionic amphiphiles in aqueous media. This self-assembly into closed bilayers is rarely spontaneous and usually requires an energy input in the form of physical agitation or heat. The bilayer assembly ensures minimum contact of hydrophobic parts of the molecule from the aqueous solvent, and the hydrophilic head groups are completely in contact with the same. Like liposomes, niosomes represent a highly flexible platform due to their unique bilayer structure. First, the bilayer structure is composed of two layers of nonionic amphiphiles, with their hydrophobic tails facing one another forming the lipophilic core region while their hydrophilic heads are exposed to the interfacial aqueous environment. Therefore, such a kind of unique structure makes niosomes a promising drug carrier because the interior is able to encapsulate aqueous solutes, and hydrophobic molecules can be incorporated within the lipophilic region of the membrane. Second, unstable drugs (e.g., peptides, proteins, and genetic materials) can be isolated from the adverse environment by being encapsulated into the interior vehicles. Third, the bilayer provides great potentials for surface modification to achieve special functions. The low cost, greater stability, and resultant ease of storage of nonionic surfactants have led to the exploitation of these compounds as alternatives to phospholipids. The niosome formation is affected by surfactant, presence of additives (usually cholesterol), nature of the drug, and physical parameters such as hydration temperature, agitation, size reduction techniques used, and so on. Niosome formation is also based on reduction of interfacial tension as other vesicular assemblies. Typically, vesicles (closed lamellar structures) are formed in the water-rich phase of the binary phase diagram. Almost all the methods used in liposomal surface modification can be easily transferred to the application of niosomal system. For instance, PEGylation has been widely used in liposomal drug delivery systems for the past two decades, which is used to decrease the interaction between liposomes and serum proteins, with a consequent long-circulation effect. Similarly, PEGylated niosomes have also been developed for specific delivery purposes. Niosomal delivery has seen advances, with new applications in biomacromolecular drug delivery. It is believed that surface modification is one of the most important techniques applied in niosomal drug delivery systems, because the niosomal systems can thereby be tailored for some specific purposes—either in physical forms (e.g., stability, drug release) or in biological performance (e.g., biodistribution, half-life, and tissue/cellular penetration). Therefore, multifunctional niosomal delivery systems for biomacromolecules can be developed using some suitable modification techniques.[55–59]

POLYMERSOMES

Polymersomes are polymer-based bilayer vesicles, also termed as nanometer-sized "bags" by scientists. The bilayer structure displayed is similar to liposomes and niosomes as shown in Fig. 7. They can be considered as liposomes but of nonbiological origin. Amphiphilic block copolymers can form various vesicular architectures in solution. They can have different morphologies such as

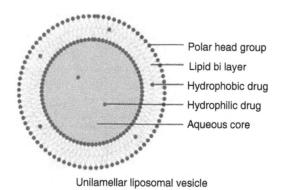

Polar head group
Lipid bi layer
Hydrophobic drug
Hydrophilic drug
Aqueous core

Unilamellar liposomal vesicle

Fig. 6 Unilamellar liposomal vesicle.

uniform common vesicles, large polydisperse vesicles, entrapped vesicles, or hollow concentric vesicles.[60–67]

NANOCAPSULES

Nanocapsules are submicron-sized polymeric colloidal particles with a therapeutic agent of interest encapsulated within their polymeric matrix or adsorbed or conjugated on the surface. They are mainly based on polymeric materials either of natural or of synthetic origin. Natural substances such as proteins, albumin, gelatin, and legumin, and polysaccharides, starch, alginates, and agarose, have been widely studied. Synthetic hydrophobic polymers from the ester class (polylactic acid) and polyglycolic acid copolymers along with ε-caprolactone have been investigated.[68–76] Figure 8 presents a conventional versus surface-functionalized nanocapsules. The theoretical model for a polymeric nanocapsule is a vesicle, where an oily or an aqueous core is surrounded by a thin polymeric wall. Those devices are stabilized by surfactants,

such as phospholipids, polysorbates, poloxamers, and cationic surfactants. Nanocapsules composed of different raw materials have been described, including polyesters and polyacrylates, as polymers, triglycerides, large size alcohols, and mineral oil, as oily cores. Several bioactive molecules have been loaded into these systems, such as antitumorals, antibiotics, antifungals, antiparasitics, anti-inflammatories, hormones, steroids, proteins, and peptides. As a general rule, the type of raw materials used to compound nanocapsules can influence their morphological and functional characteristics, which may influence the *in vitro* release and/or the *in vivo* response.[68–76]

MICROSPHERES

There has been considerable interest of late using protein- or polymer-based microsphere as drug carrier. Several methods of microsphere preparation such as single and double emulsion, phase separation, coacervation, spray drying and congealing, and solvent evaporation techniques are used nowadays. The inclusion of drugs in microparticulate carriers clearly holds significant promise for improvement in the therapy of several disease categories. The microspheres were characterized with respect to physical, chemical, and biological parameters. There are a lot of application of microsphere as drug delivery, which includes dermatology, vaccine adjuvant, ocular delivery, brain targeting, gene therapeutics, cancer targeting, magnetic targeting, and many more. This is confident that microparticulate technology will take its place, along with other drug delivery technologies, in enhancing the effectiveness, convenience, and general utility of new and existing drugs.[77–80]

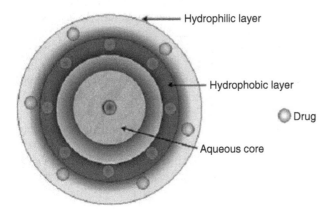

Fig. 7 Drug-loaded bilayer polymerosome.

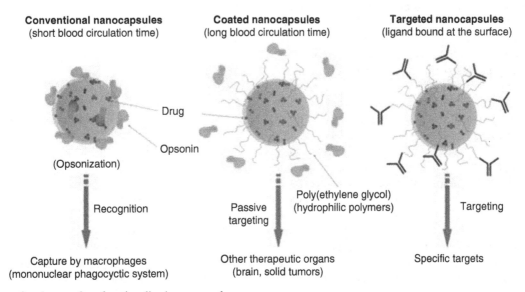

Fig. 8 Conventional vs. surface-functionalized nanocapsules.

AEROSOLS

The term "aerosol" describes systems in which liquids are dispersed in air, similar in principle to emulsions, stabilized by the presence of surfactant. Inhalation drug delivery is gaining increasing popularity in treatment of lung diseases due to its advantages over oral administration such as less side effect and quicker action onset. However, therapeutic results of inhaled medications are dependent upon effective deposition at the target site in the respiratory tract. Determining the regional and local deposition characteristics within the respiratory tract is a critical first step for making accurate predictions of the dose received and the resulting topical and systemic health effects. Furthermore, the diversifying areas of pharmaceutical research and the growing interaction between them make inhalation drug delivery a very much multidisciplinary effort, necessitating inputs from engineering and computer techniques from medicine and physiology. Aerosol drug delivery of pharmaceutical colloidal preparations is a novel mode of drug delivery, which has shown promise in the treatment of various local and systemic disorders. Colloidal preparations delivered through nebulizers, pressurized metered dose inhalers, and dry powder inhalers are the major aerosol delivery systems, and drugs are dispersed directly from the formulations (solutions, suspensions, and dry powders) to the lungs. Noninvasive administration of drugs directly to the lungs by various aerosol delivery techniques results in rapid absorption across bronchopulmonary mucosal membranes. Particle size and size distribution are an important factor in efficient aerosol delivery of medicament into the lungs. A wide variety of medicinal agents has been delivered to the lungs as aerosols for the treatment of diverse diseases; however, currently, most of them are for the management of asthma and chronic obstructive pulmonary diseases. Aerosolized drug delivery into deep lungs is expanding with the increased number of different diseases. Local and systemic delivery of drugs for various diseases is now focused on using aerosol formulations, which have a lot of potential. The future of aerosol delivery of nanoparticles and large molecules for systemic conditions with improved patient compliance is promising. Currently, aerosol therapy is expanding with the advancement of science and technology, especially in the development of nanoparticles, to target the systemic disorders for the delivery of proteins and peptides, gene therapy, pain management, nanotherapeutics, cancer therapy, and vaccines. In addition to current therapeutics for asthma, other drugs such as mucolytics, antituberculosis, antibiotics, drugs for sexual dysfunction, drugs for otitis media, fentanyl for cancer pain, tobramycin, opioids for pain, interferons, alpha-1 antitrypsin, and human growth hormone for lung delivery are in clinical development. For the treatment of specific diseases (lung and systemic) with costly medicines, it is desired to provide medicinal colloidal drugs as aerosol delivery to the targeted area in the human respiratory system. Therefore, pulmonary drug delivery of colloidal drugs would extend the new era of drug delivery research with increased patient compliance, and reduce the total cost of chronic human diseases. With advanced research, it is anticipated that the world will know more about colloidal drug delivery technology as well as the potential applications of deep lung delivery of those drugs. Pulmonary delivery of large molecules for chronic diseases is advancing rapidly and may become successful in the near future. Therefore, deep lung delivery of drugs needs to be focused not only on lung diseases but also for conditions in which rapid onset is desirable such as cancer pains, allergic reactions, brain disorders, and cardiovascular disorders. It is worthwhile to mention that researchers will develop more novel therapeutics, efficient delivery devices, and better formulation to deliver drugs in to the deep lungs for various types of diseases in the near future.[81–85]

FOAMS

A foam is a dispersion of a gas in a liquid or a solid. The formation of foam relies on the surface activity of the surfactants, polymers, proteins, and colloidal particles to stabilize the interface. Thus, the foamability increases with increasing surfactant concentration up to critical micelle concentration because above critical micelle concentration, the unimer concentration in the bulk remains nearly constant. The structure and molecular architecture of the foam is known to influence foamability and its stability. The packing properties at the interface are not excellent for very hydrophilic or very hydrophobic drug. The surfactant promoting a small spontaneous curvature at interface is ideal for foams. Nonionic surfactants are the most commonly used one. The main advantage with foams is its site-specific delivery and multiple dosing of the drug.[86–90]

REFERENCES

1. Fanun, M. Ed. *Colloids in Drug Delivery*; Taylor and Francis: New York, 2010: *Colloids in Biotechnology*; Taylor and Francis: New York, 2010.
2. Burgess, D.J. *Colloids and Colloid Drug Delivery System, Encyclopedia of Pharmaceutical Technology*; 3rd Edn.; Informa: New York, 2007; 636–647.
3. Rapoport, N. Physical stimuli-responsive polymeric micelles for anti-cancer drug delivery. Prog. Polym. Sci. **2007**, *32* (8–9), 962–990.
4. Qiu, L.Y.; Bae, Y.H. Self-assembled polyethylenimine-graft-poly(ε-caprolactone) micelles as potential dual carriers of genes and anticancer drugs. Biomaterials **2007**, *28* (28), 4132–4142.
5. Bae, Y.; Kataoka, K. Intelligent polymeric micelles from functional poly(ethylene glycol)-poly(amino acid) block copolymers. Adv. Drug Deliv. Rev. **2009**, *61* (10), 768–784.
6. Wei, H.; Cheng, S.X.; Zhang, X.Z.; Zhuo, R.X. Thermo-sensitive polymeric micelles based on poly(N-isopropylacrylamide) as drug carriers. Prog. Polym. Sci. **2009**, *34* (9), 893–910.

7. Svenson, S.; Chauhan, A.S. Dendrimers for enhanced drug solubilization. Nanomedicine **2009**, *3* (5), 679–702.

8. Mintzer, M.A.; Grinstaff, M.W. Biomedical applications of dendrimers: a tutorial. Chem. Soc. Rev. **2011**, *40* (1), 173–190.

9. Percec, V.; Wilson, D.A.; Leowanawat, P.; Wilson, C.J.; Hughes, A.D.; Kaucher, M.S.; Hammer, D.A.; Levine, D.H.; Kim, A.J.; Bates, F.S.; Davis, K.P.; Lodge, T.P.; Klein, M.L.; DeVane, R.H.; Aqad, E.; Rosen, B.M.; Argintaru, A.O.; Sienkowska, M.J.; Rissanen, K.; Nummelin, S.; Ropponen, J. Self-assembly of janus dendrimers into uniform dendrimersomes and other complex architectures. Science **2010**, *328* (5981), 1009–1014.

10. Astruc, D.; Boisselier, E.; Ornelas, C. Dendrimers designed for functions: From physical, photophysical, and supramolecular properties to applications in sensing, catalysis, molecular electronics, photonics, and nanomedicine. Chem. Rev. **2010**, *110* (4), 1857–1959.

11. Taratula, O.; Garbuzenko, O.B.; Kirkpatrick, P.; Pandya, I.; Savla, R.; Pozharov, V.P.; He, H.; Minko, T. Surface-engineered targeted PPI dendrimer for efficient intracellular and intratumoral siRNA delivery. J. Control. Release **2009**, *140* (3), 284–293.

12. Fong, W.K.; Hanley, T.; Boyd, B.J. Stimuli responsive liquid crystals provide 'on-demand' drug delivery *in vitro* and *in vivo*. J. Control. Release **2009**, *135* (3), 218–226.

13. Lee, K.W.Y.; Nguyen, T.H.; Hanley, T.; Boyd, B.J. Nanostructure of liquid crystalline matrix determines *in vitro* sustained release and *in vivo* oral absorption kinetics for hydrophilic model drugs. Int. J. Pharm. **2009**, *365* (1–2), 190–199.

14. Boyd, B.J.; Khoo, S.M.; Whittaker, D.V.; Davey, G.; Porter, C.J.H. A lipid-based liquid crystalline matrix that provides sustained release and enhanced oral bioavailability for a model poorly water soluble drug in rats. Int. J. Pharm. **2007**, *340* (1–2), 52–60.

15. Garg, G.; Saraf, S.; Saraf, S. Cubosomes: An overview. Biol. Pharm. Bull. **2008**, *30* (2), 350–353.

16. Yaghmur, A.; Glatter, O. Characterization and potential applications of nanostructured aqueous dispersions. Adv. Colloid Interface Sci. **2009**, *147–148*, 333–342.

17. Rizwan, S.B.; Boyd, B.J.; Rades, T.; Hook, S. Bicontinuous cubic liquid crystals as sustained delivery systems for peptides and proteins. Expert Opin. Drug Deliv. **2010**, *7* (10), 1133–1144.

18. Libster, D.; Aserin, A.; Yariv, D.; Shoham, G.; Garti, N. Soft matter dispersions with ordered inner structures, stabilized by ethoxylated phytosterols. Colloids Surf. B Biointerfaces **2009**, *74* (1), 202–215.

19. Swarnakar, N.K.; Jain, V.; Dubey, V.; Mishra, D.; Jain, N.K. Enhanced oromucosal delivery of progesterone via hexosomes. Pharm. Res. **2007**, *24* (12), 2223–2230.

20. Dash, M.; Chiellini, F.; Ottenbrite, R.M.; Chiellini, E. Chitosan - A versatile semi-synthetic polymer in biomedical applications. Prog. Polym. Sci. **2011**, *36* (8), 981–1014.

21. Ryu, J.H.; Chacko, R.T.; Jiwpanich, S.; Bickerton, S.; Babu, R.P.; Thayumanavan, S. Self-cross-linked polymer nanogels: a versatile nanoscopic drug delivery platform. J. Am. Chem. Soc. **2010**, *132* (48), 17227–17235.

22. Malmsten, M.; Bysell, H.; Hansson, P. Biomacromolecules in microgels – opportunities and challenges for drug

delivery. Curr. Opin. Colloid Interface Sci. **2010**, *15* (6), 435–444.

23. Jayakumar, R.; Menon, D.; Manzoor, K.; Nair, S.V.; Tamura, H. Biomedical applications of chitin and chitosan based nanomaterials—a short review. Carbohydr. Polym. **2010**, *82* (2), 227–232.

24. Saunders, B.R.; Laajam, N.; Daly, E.; Teow, S.; Hu, X.; Stepto, R. Microgels: From responsive polymer colloids to biomaterials. Adv. Colloid Interface Sci. **2009**, *147–148*, 251–262.

25. Hamidi, M.; Azadi, A.; Rafiei, P. Hydrogel nanoparticles in drug delivery. Adv. Drug Deliv. Rev. **2008**, *60* (15), 1638–1649.

26. Malmsten, M. *Surfactants and Polymers in Drug Delivery*; Marcel Dekker, Inc: New York, 2002.

27. Kohli, K.; Chopra, S.; Dhar, D.; Arora, S.; Khar, R.K. Self-emulsifying drug delivery systems: An approach to enhance oral bioavailability. Drug Discov. Today **2010**, *15* (21–22), 958–965.

28. Buyukozturk, F.; Benneyan, J.C.; Carrier, R.L. Impact of emulsion-based drug delivery systems on intestinal permeability and drug release kinetics. J. Control. Release **2010**, *142* (1), 22–30.

29. El Maghraby, G.M. Self-microemulsifying and micro-emulsion systems for transdermal delivery of indomethacin: effect of phase transition. Colloids Surf. B Biointerfaces **2010**, *75* (2), 595–600.

30. Li, Y.; Le Maux, S.; Xiao, H.; McClements, D.J. Emulsion-based delivery systems for tributyrin, a potential colon cancer preventative agent. J. Agric. Food Chem. **2009**, *57* (19), 9243–9249.

31. Balakrishnan, P.; Lee, B.J.; Oh, D.H.; Kim, J.O.; Hong, M.J.; Jee, J.P.; Kim, J.A.; Yoo, B.K.; Woo, J.S.; Yong, C.S.; Choi, H.G. Enhanced oral bioavailability of dexibuprofen by a novel solid Self-emulsifying drug delivery system (SEDDS). Eur. J. Pharm. Biopharm. **2009**, *72* (3), 539–545.

32. Frelichowska, J.; Bolzinger, M.A.; Valour, J.P.; Mouaziz, H.; Pelletier, J.; Chevalier, Y. Pickering w/o emulsions: drug release and topical delivery. Int. J. Pharm. **2009**, *368* (1–2), 7–15.

33. Tang, B.; Cheng, G.; Gu, J.C.; Xu, C.H. Development of solid self-emulsifying drug delivery systems: Preparation techniques and dosage forms. Drug Discov. Today **2008**, *13* (13–14), 606–612.

34. Cohen-Sela, E.; Teitlboim, S.; Chorny, M.; Koroukhov, N.; Danenberg, H.D.; Gao, J.; Golomb, G. Single and double emulsion manufacturing techniques of an amphiphilic drug in PLGA nanoparticles: formulations of mithramycin and bioactivity. J. Pharm. Sci. **2009**, *98* (4), 1452–1462.

35. Cohen-Sela, E.; Chorny, M.; Koroukhov, N.; Danenberg, H.D.; Golomb, G. A new double emulsion solvent diffusion technique for encapsulating hydrophilic molecules in PLGA nanoparticles. J. Control. Release **2009**, *133* (2), 90–95.

36. Hanson, J.A.; Chang, C.B.; Graves, S.M.; Li, Z.; Mason, T.G.; Deming, T.J. Nanoscale double emulsions stabilized by single-component block copolypeptides. Nature **2008**, *455*, 85–88.

37. Ho, M.L.; Fu, Y.C.; Wang, G.J.; Chen, H.T.; Chang, J.K.; Tsai, T.H.; Wang, C.K. Controlled release carrier of BSA made by W/O/W emulsion method containing PLGA and hydroxyapatite. J. Control. Release **2008**, *128* (2), 142–148.

Biomimetic—Colloid

38. Zhang, X.Q.; Intra, J.; Salem, A.K. Comparative study of poly (lactic-co-glycolic acid)-poly ethyleneimine-plasmid DNA microparticles prepared using double emulsion methods. J. Microencapsul. **2008**, *25* (1), 1–12.

39. Kong, M.; Chen, X.G.; Kweon, D.K.; Park, H.J. Investigations on skin permeation of hyaluronic acid based nanoemulsion as transdermal carrier. Carbohydr. Polym. **2011**, *86* (2), 837–843.

40. Shakeel, F.; Ramadan, W. Transdermal delivery of anticancer drug caffeine from water-in-oil nanoemulsions. Colloids Surf. B Biointerfaces **2010**, *75* (1), 356–362.

41. Rapoport, N.Y.; Kennedy, A.M.; Shea, J.E.; Scaife, C.L.; Nam, K.H. Controlled and targeted tumor chemotherapy by ultrasound-activated nanoemulsions/microbubbles. J. Control. Release **2009**, *138* (2), 268–276.

42. Ahmed, M.; Ramadan, W.; Rambhu, D.; Shakeel, F. Potential of nanoemulsions for intravenous delivery of rifampicin. Pharmazie **2008**, *63* (11), 806–811.

43. Kumar, M.; Misra, A.; Babbar, A.K.; Mishra, A.K.; Mishra, P.; Pathak, K. Intranasal nanoemulsion based brain targeting drug delivery system of risperidone. Int. J. Pharm. **2008**, *358* (1–2), 285–291.

44. Fanun, M. Ed. *Microemulsions- Properties and Applications*; Taylor and Francis: New York, 2010.

45. Stubenrauch, C. Ed. *Microemulsions: Background, New Concepts, Applications, Perspectives*; Wiley: New York, 2009.

46. Kumar, P.; Mital, K.L.; Eds. *Handbook of Microemulsion Science and Technology;* Marcel Dekker: New York, 1999.

47. Kunieda, H.; Solans, C. How to prepare microemulsions: temperature-insensitive microemulsions. In *Industrial Applications of Microemulsions*; Solans, C., Kunieda, H., Eds.; Marcel Dekker: New York, 1997; 21–45.

48. Lawrence, M.J.; Rees, G.D. Microemulsion-based media as novel drug delivery systems. Adv. Drug Deliv. Rev. **2000**, *45* (1), 89–121.

49. Kreilgaard, M. Influence of microemulsions on cutaneous drug delivery. Adv. Drug Deliv. Rev. **2002**, *54* (Suppl. 1), S77–S98.

50. Maruyama, K. Intracellular targeting delivery of liposomal drugs to solid tumors based on EPR effects. Adv. Drug Deliv. Rev. **2011**, *63* (3), 161–169.

51. Schäfer, J.; Höbel, S.; Bakowsky, U.; Aigner, A. Liposome-polyethylenimine complexes for enhanced DNA and siRNA delivery. Biomaterials **2010**, *31* (26), 6892–6900.

52. Abu Lila, A.S.; Ishida, T.; Kiwada, H. Recent advances in tumor vasculature targeting using liposomal drug delivery systems. Expert Opin. Drug Deliv. **2009**, *6* (12), 1297–1309.

53. Christensen, D.; Agger, E.M.; Andreasen, L.V. Liposome-based cationic adjuvant formulations (CAF): Past, present, and future. J. Liposome Res. **2009**, *19* (1), 2–11.

54. Schroeder, A.; Kost, J.; Barenholz, Y. Ultrasound, liposomes, and drug delivery: principles for using ultrasound to control the release of drugs from liposomes. Chem. Phys. Lipids **2009**, *162* (1–2), 1–16.

55. Manosroi, J.; Lohcharoenkal, W.; Götz, F.; Werner, R.G.; Manosroi, W.; Manosroi, A. Transdermal absorption enhancement of n-terminal tat-GFP fusion protein (TG) loaded in novel low-toxic elastic anionic niosomes. J. Pharm. Sci. **2011**, *100* (4), 1525–1534.

56. Karim, K.; Mandal, A.; Biswas, N.; Niosome: A future of targeted drug delivery systems. J. Adv. Pharm. Technol. Res. **2010**, *1* (4), 374–380.

57. Azeem, A.; Anwer, M.K.; Talegaonkar, S. Niosomes in sustained and targeted drug delivery: Some recent advances. J. Drug Target **2009**, *17* (9), 671–689.

58. Attia, I.A.; El-Gizawy, S.A.; Fouda, M.A.; Donia, A.M. Influence of a niosomal formulation on the oral bioavailability of acyclovir in rabbits. AAPS PharmSciTech **2007**, *8* (4), 106.

59. Aggarwal, D.; Pal, D.; Mitra, A.K.; Kaur, I.P. Study of the extent of ocular absorption of acetazolamide from a developed niosomal formulation, by microdialysis sampling of aqueous humor. Int. J. Pharm. **2007**, *338* (12), 21–26.

60. Lomas, H.; Johnston, A.P.; Such, G.K.; Zhu, Z.; Liang, K.; van Koeverden, M.P.; Alongkornchotikul, S.; Caruso, F. Polymersome-loaded capsules for controlled release of DNA. Small **2011**, *7* (14), 2109–2019.

61. Brinkhuis, R.P.; Rutjes, F.P.J.T.; Van Hest, J.C.M. Polymeric vesicles in biomedical applications. Polym. Chem. **2011**, *2* (7), 1449–1462.

62. Egli, S.; Nussbaumer, M.G.; Balasubramanian, V.; Chami, M.; Bruns, N.; Palivan, C.; Meier, W. Biocompatible functionalization of polymersome surfaces: A new approach to surface immobilization and cell targeting using polymersomes. J. Am. Chem. Soc. **2011**, *133* (12), 4476–4483.

63. Liu, G.; Ma, S.; Li, S.; Chen, R.; Meng, F.; Jiu, H.; Zhong, Z. The highly efficient delivery of exogenous proteins into cells mediated by biodegradable chimaeric polymersomes. Biomaterials **2010**, *31*, 7575–7585.

64. Upadhyay, K.K.; Bhatt, A.N.; Mishra, A.K.; Dwarakanath, B.S.; Jain, S.; Schatz, C.; Le Meins, J.F.; Farooque, A.; Chandraiah, G.; Jain, A.K.; Misra, A.; Lecommandoux, S. The intracellular drug delivery and anti tumor activity of doxorubicin loaded poly(γ-benzyl l-glutamate)-b-hyaluronan polymersomes. Biomaterials **2010**, *31* (10), 2882–2892.

65. Lo Presti, C.; Lomas, H.; Massignani, M.; Smart, T.; Battaglia, G. Polymersomes: Nature inspired nanometer sized compartments. J. Mater. Chem. **2009**, *19*, 3576–3590.

66. Christian, D.A.; Cai, S.; Bowen, D.M.; Kim, Y.; Pajerowski, J.D.; Discher, D.E. Polymersome carriers: From self-assembly to siRNA and protein therapeutics. Eur. J. Pharm. Biopharm. **2009**, *71* (3), 463–474.

67. Meng, F.; Zhong, Z.; Feijen, J. Stimuli-responsive polymersomes for programmed drug delivery. Biomacromolecules **2009**, *10* (2), 197–209.

68. Yang, X.C.; Samanta, B.; Agasti, S.S.; Jeong, Y.; Zhu, Z.J.; Rana, S.; Miranda, O.R.; Rotella, V.M. Drug delivery using nanoparticle-stabilized nanocapsules. Angew Chem. Int. Ed. Engl. **2011**, *50* (2), 477–481.

69. Mora-Huertas, C.E.; Fessi, H.; Elaissari, A. Polymer-based nanocapsules for drug delivery. Int. J. Pharm. **2010**, *385* (1–2), 113–142.

70. Kamphuis, M.M.J.; Johnston, A.P.R.; Such, G.K.; Dam, H.H.; Evans, R.A.; Scott, A.M.; Nice, E.C.; Heath, J.K.; Caruso, F. Targeting of cancer cells using click-functionalized polymer capsules. J. Am. Chem. Soc. **2010**, *132* (45), 15881–15883.

71. Chen, Y.; Chen, H.; Zeng, D.; Tian, Y.; Chen, F.; Feng, J.; Shi, J. Core/shell structured hollow mesoporous nanocapsules: A potential platform for simultaneous cell imaging and anticancer drug delivery. ACS Nano **2010**, *4* (10), 6001–6013.

72. Delcea, M.; Yashchenok, A.; Videnova, K.; Kreft, O.; Möhwald, H.; Skirtach, A.G. Multicompartmental micro- and nanocapsules: Hierarchy and applications in biosciences. Macromol. Biosci. **2010**, *10* (5), 465–474.

73. Shen, Y.; Jin, E.; Zhang, B.; Murphy, C.J.; Sui, M.; Zhao, J.; Wang, J.; Tang, J.; Fan, M.; Van Kirk, E.; Murdoch, W.J. Prodrugs forming high drug loading multifunctional nanocapsules for intracellular cancer drug delivery. J. Am. Chem. Soc. **2010**, *132* (12), 4259–4265.

74. Matsusaki, M.; Akashi, M. Functional multilayered capsules for targeting and local drug delivery. Expert Opin. Drug Deliv. **2009**, *6* (11), 1207–1217.

75. Huynh, N.T.; Passirani, C.; Saulnier, P.; Benoit, J.P. Lipid nanocapsules: A new platform for nanomedicine. Int. J. Pharm. **2009**, *379* (2), 201–209.

76. Wang, Y.; Bansal, V.; Zelikin, A.N.; Caruso, F. Templated synthesis of single-component polymer capsules and their application in drug delivery. Nano Lett. **2008**, *8* (6), 1741–1745.

77. Mundargi, R.C.; Rangaswamy, V.; Aminabhavi, T.M. pH-Sensitive oral insulin delivery systems using Eudragit microspheres. Drug Dev. Ind. Pharm. **2011**, *37* (8), 977–985.

78. Yang, X.; Chen, L.; Han, B.; Yang, X.; Duan, H. Preparation of magnetite and tumor dual-targeting hollow polymer microspheres with pH-sensitivity for anticancer drug-carriers. Polymer (Guildf) **2010**, *51* (12), 2533–2539.

79. Alagusundaram, M.; Madhu Sudana Chetty, C.; Umashankari, K. Microspheres as a novel drug delivery sysytem - A review. Int. J. Chem. Tech. Res. **2009**, *1*, 526–534.

80. Sun, L.; Zhou, S.; Wang, W.; Li, X.; Wang, J.; Weng, J. Preparation and characterization of porous biodegradable microspheres used for controlled protein delivery. Colloids Surf. A Physicochem. Eng. Asp. **2009**, *345* (1–3), 173–181.

81. Dolovich, M.B.; Dhand, R. Aerosol drug delivery: Developments in device design and clinical use. Lancet **2011**, *377* (9770), 1032–1045.

82. Pilcer, G.; Amighi, K. Formulation strategy and use of excipients in pulmonary drug delivery. Int. J. Pharm. **2010**, *392* (1–2), 1–19.

83. Denyer, J.; Dyche, T. The Adaptive Aerosol Delivery (AAD) technology: past, present, and future. J. Aerosol. Med. Pulm. Drug Deliv. **2010**, 23 (Suppl. 1), S1–S10.

84. Dudley, M.N.; Loutit, J.; Griffith, D.C. Aerosol antibiotics: Considerations in pharmacological and clinical evaluation. Curr. Opin. Biotechnol. **2008**, *19* (6), 637–643.

85. Kleinstreuer, C.; Zhang, Z.; Donohue, J.F. Targeted drug-aerosol delivery in the human respiratory system. Annu. Rev. Biomed. Eng. **2008**, *10*, 195–220.

86. Zhang, Y.; Zhang, J.; Jiang, T.; Wang, S. Inclusion of the poorly water-soluble drug simvastatin in mesocellular foam nanoparticles: drug loading and release properties. Int. J. Pharm. **2011**, *410* (1–2), 118–124.

87. Wu, C.; Wang, Z.; Zhi, Z.; Jiang, T.; Zhang, J.; Wang, S. Development of biodegradable porous starch foam for improving oral delivery of poorly water soluble drugs. Int. J. Pharm. **2011**, *403* (1–2), 162–169.

88. Hegge, A.B.; Andersen, T.; Melvik, J.E.; Bruzell, E.; Kristensen, S.; Tønnesen, H.H. Formulation and bacterial phototoxicity of curcumin loaded alginate foams for wound treatment applications: Studies on curcumin and curcuminoides XLII. J. Pharm. Sci. **2011**, *100* (1), 174–185.

89. Sirsi, S.R.; Borden, M.A. Microbubble compositions, properties and biomedical applications. Bubble Sci. Eng. Technol. **2009**, *1* (1–2), 3–17.

90. Klasse, P.J.; Shattock, R.; Moore, J.P. Antiretroviral drug-based microbicides to prevent HIV-1 sexual transmission. Annu. Rev. Med. **2008**, *59*, 455–71.

Biomimetic—Colloid

Conducting Polymers: Biomedical Engineering Applications

Smritimala Sarmah
Department of Physics, Girijananda Chowdhury Institute of Management and Technology, Guwahati, India

Raktim Pratim Tamuli
Department of Forensic Medicine, Guwahati Medical College and Hospital, Guwahati, India

Abstract

Different procedures of synthesis of conducting polymers are discussed with relation to their applications in biomedical engineering. In addition, applications of conducting polymers in biomolecular sensors, biomolecular actuators, tissue engineering, biomechanical sensing, drug delivery, artificial muscle, neural probe, and optical biosensors are reviewed extensively.

INTRODUCTION

Biomedical engineering is the application of engineering principles and design concepts to medical science and related biology. This ground-breaking field seeks to close the gap between engineering and medicine. It combines the design and problem-solving skills of engineering with medical and biological sciences to improve healthcare diagnosis, monitoring, and therapy. The identification and development of materials that find application in biomedical engineering has a profound outcome on our quality of life. Biomedically engineered materials have seen successful application in the development of highly effective stents,[1,2] bone replacements,[3] pacemakers,[4] bionic ears,[5] and wearable prosthetics.[6] For different applications, the material requirements will vary. However, in all cases they must be compatible with the biological environment in which they are to operate. This compatibility may involve molecular and cellular interactions, be at the skeletal level wherein aspects related to wearability, comfort, lightweight, and esthetics become important. The quest to more effectively monitor and manipulate the biosystems requires the creation of better interfaces between the biological and electronic domains. Materials that bridge this interface are finding utility in the emerging field of bionics. Conducting polymers hold a special position when it comes to biomedical applications, primarily because of the fact that these macromolecules can be chemically or physically engineered to render them with desired properties such as conductivity, reversible oxidation, redox stability, biocompatibility, hydrophobicity, three-dimensional geometry, and required surface topography. Inherently, conducting polymers have been shown to be excellent bionic materials providing utility from the biomolecular to the biomechanical level. The importance and potential impact of this new class of material was recognized by the world scientific community when Shirakawa, Heeger, and MacDiarmid were awarded the Nobel Prize in Chemistry in

2000 for their research in this field.[7–9] Few mostly used conducting polymers are shown in Fig. 1. At the biomolecular level, conducting polymers can be produced in ways that make their integration into implants for tissue engineering (TE) or nerve regeneration possible. At the biomechanical (skeletal) level, fabrication protocols that enable wearable structures containing conducting polymers that function as sensors or mechanical actuators (artificial muscles) have been developed.[10]

Response to stimulus is a basic process of living systems. Based on the lessons from nature, scientists have been designing useful materials that respond to external stimuli such as temperature, pH, light, electric field, chemicals, and ionic strength. These responses are manifested as dramatic changes in one of the following: shape, surface characteristics, solubility, and formation of an intricate molecular self-assembly or a sol-to-gel transition. Applications of stimuli-responsive or "smart" polymers in delivery of therapeutics, TE, bioseparations, sensors, or actuators have been studied extensively and numerous papers and patents are evidence of rapid progress in this area. Understanding the structure–property relationship is essential for the further development and rational design of new functional smart materials. Among various smart materials, conducting polymers are preferred as the best candidates for biomedical applications due to their outstanding electronic and electrochemical switching properties. In this entry, synthesis, processing, fabrication, and applications of conducting polymers in various fields of biomedical engineering have been discussed in detail citing various examples.

SYNTHESIS, PROCESSING, AND FABRICATION OF CONDUCTING POLYMERS

Searching facile and efficient synthetic method is a basic and an important object for the conducting polymer

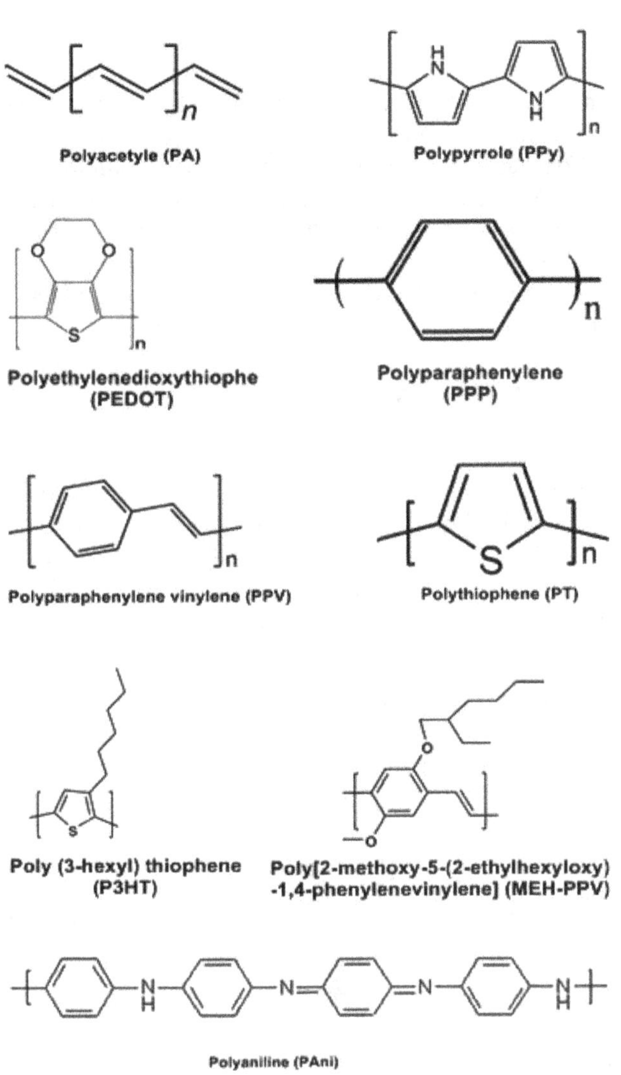

Fig. 1 Structures of few mostly used conducting polymers.

Step 1: Oxidation of monomer

Step 2: Radical coupling

Step 3: Chain propagation

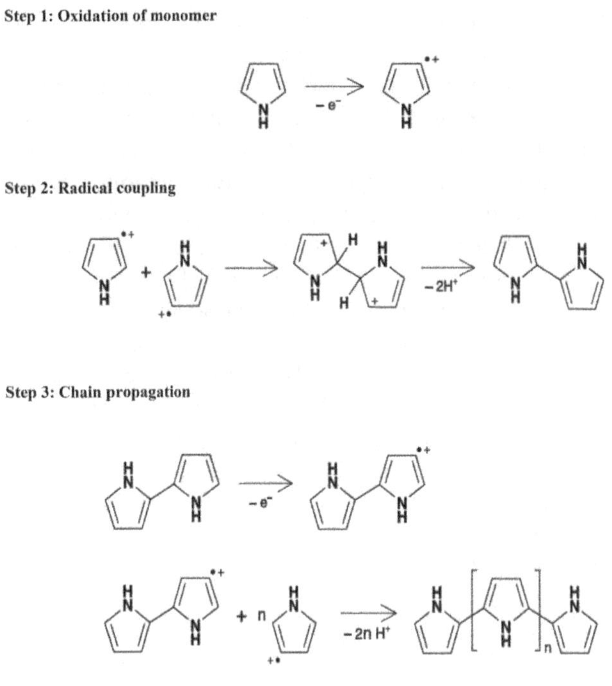

Fig. 2 Chemical polymerization mechanism of PPy.

micro/nanostructures. The preparation of micro-/nanostructured materials with given shapes and sizes requires a controllable and well-confined space for material growth.[11] Different techniques used to control the morphology of application-oriented conducting polymers have been discussed below.

Chemical Synthesis

Chemical oxidation polymerization has been desirable for mass production. Chemical synthesis of conducting polymers is gaining importance not only because of the requirement of bulk quantities but also due to the need for specific morphology of the polymers. The conventional chemical synthesis of conducting polymers in aqueous acids is a typical heterogeneous precipitation polymerization. During chemical synthesis of conducting polymers, a number of factors affect the final morphology such as structure of monomer, dopant, and oxidant as well as the processing parameters. The processing parameters include the concen-

trations of dopant, oxidant, and monomer and the molar ratio of dopant and oxidant to monomer as well as the reaction temperature, time, and stirring. Letheby[12] described chemical oxidative polymerization of aniline into polyaniline in 1862, and Mohilner, Adams, and Argensinger characterized it in 1962,[13] one complete century after its discovery. Polypyrrole was earlier known as pyrrole black and was found on the sides of the container of pyrrole, which was formed by the oxidation of pyrrole in air. Well-defined synthesis of electroactive polyaniline has been reported since 1980.[14] Sun, Park, and Deng[15] studied the morphological evolution of polyaniline (PAni) nanostructures from nanoparticles to nanotubes at different reaction times. Chemical oxidation scheme of polypyrrole (PPy) is shown in Fig. 2.

Electrochemical Synthesis

Electrochemical synthesis, such as chemical polymerization, is a much frequently used technique for obtaining conducting polymers. Electrochemical polymerization is performed using a three-electrode configuration (working, counter, and reference electrodes) in a solution of the monomer, appropriate solvent, and electrolyte (dopant). Current is passed through the solution and electrodeposition occurs at the positively charged working electrode or anode. Monomers at the working electrode surface undergo oxidation to form radical cations that react with other monomers or radical cations, forming insoluble polymer chains on the electrode surface. At a sufficiently high positive (i.e., anodic) electrode potential, some monomers such as aniline or pyrrole undergo electrochemical oxidation yielding cation radicals or other reactive species. Once formed, these species trigger the polymerization process. As a result, oligomers and/or polymers, derived from the corresponding monomers, are formed. Depending on several experimental variables available, such as the monomer concentration, electrolyte used, and electrode potential, different structures of conducting polymers can be obtained, ranging from micrometer or submicrometer (colloid)-sized particles to thick compact deposits on the electrode surface. Yang et al.[16] successfully electropolymerized PPy grains and tubules on stainless steel electrodes via a self-assembly process without and with the addition of methyl orange (MO). At room temperature, the conductivity of the PPy-MO tubule film synthesized by electrochemical polymerization was 76 S/cm. A superhydrophilic PPy nanofiber network was electrochemically synthesized in an aqueous solution using phosphate buffer solution (PBS) in the absence of templates, surfactants, and structure-directing molecules.[17]

Hard Template Synthesis

Hard template method has been used for the one dimensional nanostructures such as nanotubes, nanorods, and nanofibers of conducting polymers and their multifunctional composite nanostructures. In this method, a porous membrane is required as a hard template that guides the growth of the nanostructures within the pore in the membranes. The morphology and diameter of the nanostructures are controlled by the morphology and size of the pores. The length and thickness of the nanostructures are usually adjusted by changing the polymerization time. Martin and his coworkers have synthesized the nanotubes of PPy, PAni, and poly (3-methylthiophene) with hard templates using chemical oxidation and electrochemical polymerization.[18–20] During the polymerization process, the conducting polymer preferentially nucleates and grows on the pore walls of membranes. Resultant polymer tubular structures are tuned by polymerization time. Whereas short polymerization time provides the thin wall of conducting polymer nanotube, long polymerization time produced thick walls. Recently, nanocones of poly (3,4-ethylenedioxythiophene) (PEDOT) on commercially available carbon cloth were prepared by using nanoporous alumina membranes as templates.[21]

Soft Template Synthesis

Soft template method is also called self-assembly technique. The microstructured conducting polymers with desired morphology and sizes are encoded in the original nanostructured host materials formed by nonvalent interactions such as hydrogen bonding, ionic bonding, coordination bonding, or intermolecular forces.

Wan et al.[22–24] developed a micelle template method to synthesize conducting polymer with microstructures for

the first time in 1988. Micro-/nanotubes were chemically synthesized in the aqueous solution of β-naphthalenesulfonic acid (β-NSA), salicylic acid (SA, ortho-hydroxybenzonic acid), camphor sulfonic acid (CSA), p-methylbenzene sulfonic acid (MBSA), or p-dodecylbenzene sulfonic acid (DBSA) as dopant and surfactant, and ammonium persulfate (APS) or ferric chloride ($FeCl_3$) as the oxidant and pyrrole or aniline as the monomers. Transmission electron micrograph of DBSA-doped PPy nanorod and PEDOT nanofibers are shown in Figs. 3 and 4, respectively.

Template-Free Polymerization

Template-free techniques have been extensively studied for the fabrication of conducting polymer nanomaterials. Compared with hard and soft template methods, these methodologies provide a facile and practical route to produce pure, uniform, and high-quality nanofibers. PAni nanofibers were fabricated under ambient environment

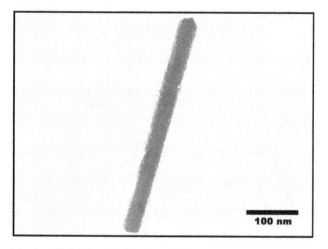

Fig. 3   DBSA-doped PPy nanorod.

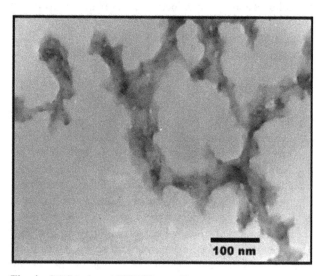

Fig. 4   DBSA-doped PEDOT nanofibers.

using aqueous/organic interfacial polymerization. Despite large-scale production, the diameter of resultant PAni nanofibers ranged from 30 to 50 nm and fiber length varied between 500 nm and several micrometers.[25] Two solutions were prepared before interfacial polymerization of aniline. Aniline monomer was dissolved in an organic solvent (e.g., CCl_4, benzene, toluene, and CS_2). The other solution consisted of oxidant and dopant acid in water. After mixing the two solutions, green PAni formed at the interface between the layers and consecutively diffused into aqueous phase. Dark-green PAni nanofiber was collected in water phase. Polymerization yields ranged from 6 to 10 wt% and approximately 95 vol% of the sample was PAni nanofiber with the diameter of 30–50 nm. Kaner and coworkers have discovered a way to minimize the PAni overgrowth on these nanofibers by undertaking the so-called rapidly mixed reactions. These can be achieved simply by pouring the initiator solution (e.g., ammonium peroxydisulfate) into the aniline solution all at once and rapidly mixing them. As the polymerization begins, the initiator molecules induce the formation of nanofibers by rapidly polymerizing aniline monomers in their vicinity. If the initiator molecules are evenly distributed, then they should be consumed during the formation of nanofibers. Therefore, secondary growth of PAni will be very limited due to the lack of available reactants. The product created in a rapidly mixed reaction is pure nanofibers with a relatively uniform size distribution, comparable to that obtained by interfacial polymerization.[26]

Micro- and Nano-Patterning

The fabrication of organic microelectronics devices and microprocessors usually needs the conducting polymers to be micro-/nano-structured and ordered. And such conducting polymeric materials can be achieved by direct electrodeposition of the polymers on patterned substrates. One important technique is area-selected electropolymerization. The concept behind this technique is to deposit a patterned monolayer, usually an organothiol, onto an electrode, usually gold, so that it selectively blocks electropolymerization of an electroactive monomer. In one of the first examples of its kind, pyrrole was electrochemically polymerized on the exposed regions of a gold electrode patterned with a self-assembled monolayer (SAM). The polypyrrole pattern adheres only weakly to the gold surface and can be transferred to another substrate by contact adhesion. Polymeric patterns having a conductivity of 1 ± 5 S/cm and possessing lateral dimensions of 100 μm for thicknesses up to 1 μm (or 10 μm for thicknesses of 0.25 μm) have been reported.[27] A similar approach has been taken to deposit patterns of PAni.[28]

Electrospinning

Recently, electrospinning has been extensively studied to produce nanofibers of many kinds of soluble or fusible polymers. Electrospinning can consistently produce fibers with the diameter in the sub-micron range.[29] In the electrospinning process, an electric field is applied to a polymer solution held by its surface tension at the end of a capillary tube. The electric field induces charge on the surface of the solution and the mutual charge repulsion causes a force directly opposite to the surface tension. As the electric field is increased, the hemispherical surface of the solution at the tip of the capillary tube elongates to form a conical shape, which is known as the Taylor cone. Above a critical electric field the repulsive electric force overcomes the surface tension and the charged solution jet is ejected from the tip of Taylor cone and split into many jets due to the electric repulsion force. Since the solvent evaporates during travel of the solution jet in air, fibers are formed in the air and randomly deposited on a collecting metal drum electrode, resulting in nonwoven web composed of continuous nanofibers.

MacDiarmid et al.[30] prepared highly conductive sulfuric acid-doped PAni electrospun fibers using a mixture of PAni and different conventional polymers such as PEO, polystyrene, and polyacrylonitrile. Nanobelts and flower-like nanostructures of pure PAni doped with sulfuric acid or hydrochloric acid were prepared via electrospinning by using a coagulation bath as the collector after optimizing the fabrication parameters.[31]

CONDUCTING POLYMERS IN BIOMOLECULAR SENSING

A biomolecular sensor may be considered as a combination of a bioreceptor, the biological component, and a transducer, the detection method. The total effect of a biomolecuar sensor is to transform a biological event into an electrical signal. Biomolecular sensors have found extensive applications in medical diagnostics, environmental pollution control for measuring toxic gases in the atmosphere, and toxic soluble compounds in river water.

Polymers are becoming inseparable from biomolecule immobilization strategies and biosensor platforms. Recent advances in diagnostic chips and microfluidic systems, together with the requirements of mass-production technologies, have raised the need to replace glass by polymeric materials, which are more suitable for production through simple manufacturing processes. Conducting polymers (CPs), in particular, are especially acquiescent for electrochemical biosensor development for providing biomolecule immobilization and for rapid electron transfer. It is expected that not only the combination of known polymer substrates but also new transducing and biocompatible interfaces with nanobiotechnological structures, such as nanoparticles, carbon nanotubes (CNTs), and nanoengineered "smart" polymers, may generate composites with new and interesting properties, providing higher sensitivity and stability of the immobilized molecules, thus constituting the basis for new

Conducting—Dendritic

and improved analytical devices for biomedical and other applications. The use of conducting polymers as biosensors has been extensively reviewed. Gao, Dai, and Wallace highlighted the usefulness of polypyrrole-coated, multiwalled aligned carbon nanotube electrodes as tool for a sensitive and selective glucose sensor.[32] The concentration dependence of glucose on solution-cast poly(aniline) films containing physically absorbed glucose oxidase has been amperometrically measured as a function of temperature (10–90°C). It has been shown that such poly(aniline)-glucose oxidase electrodes are stable up to 77°C and can be efficiently operated up to 25 mM of glucose solution.[33] A biosensor based on a conducting screen-printing ink for the direct amperometric measurement of glucose is described. Carbon electrode prints, containing the work, reference, and auxiliary electrodes, are used as the substrate for the sensor. The active parts of the carbon ink are the redox enzyme glucose oxidase and the conducting polymer poly(pyrrole), which can communicate directly with each other. As a result, the biosensor is largely independent of the oxygen concentration.[34]

Electrochemical glucose biosensor based on conducting microcells was constructed by imprisoning the enzymes in the cell arrays on an electrode without allowing modification of enzyme itself. The microtubular array was fabricated through electrochemical polymerization of EDOT into a template membrane attached on an indium tin oxide (ITO) working electrode. The hollow tubules were loaded with glucose oxidase dispersed in aqueous buffer solution. To imprison the glucose oxidase into the cell, the opening of the tubule was sealed with PEDOT/PSS composite cap. The reinforcing electrochemical polymerization of PEDOT or poly(1,3-phenylenediamine) onto the sealing composite reduce the water solubility of the capping. The steady-state amperometric response to glucose was determined by detecting hydrogen peroxide generated by the action of the glucose oxidase. The sensor could retain good selectivity against interfering substances by capping the opening with electrochemically prepared nonconducting poly(1,3-phenylenediamine) layer. In this work, a glucose sensor with good selectivity and sensitivity was achieved.[35] Similar attempt has been made for fabrication of poly(3-dodecyl thiophene) (P3DT)-based glucose biosensor.[36]

An amperometric enzyme electrode was developed for determination of lactate in serum. To prepare this electrode, commercial lactate oxidase from *Pediococcus* species has been immobilized through glutaraldehyde coupling onto polyaniline-*co*-fluoroaniline film deposited on an ITO-coated glass plate. This plate acted as a working electrode when combined with a Pt electrode as counter electrode to the electrometer for the development of a biosensor. The method is based on generation of electrons from H_2O_2, which is formed from lactic acid by immobilized lactate oxidase. This enzyme electrode was employed for determination of lactate in serum.[37]

A potentiometric biosensor based on bovine serum albumin (BSA)-embedded, surface-modified polypyrrole has been developed for the quantitative estimation of urea in aqueous solution. The enzyme, urease (Urs), was covalently linked to free amino groups present over the BSA embedded modified surface of the conducting polypyrrole film electrochemically deposited onto an ITO-coated glass plate. Potentiometric and spectrophotometric response of the enzyme electrode (Urs/BSA-PPy/ITO) were measured as a function of urea concentration in Tris–HCl buffer (pH 7.0). It has been found that the electrode responds to low urea concentration with wider range of detection. These results indicate an efficient covalent linkage of enzyme to free amino groups of the BSA molecules over the surface of polypyrrole film, which leads to high enzyme loading, an increased lifetime stability of the electrode, and an improved wide range of detection of low urea concentration in aqueous solution.[38]

Cholesterol oxidase, cholesterol esterase, and peroxidase have been co-immobilized onto electrochemically prepared polyaniline films. These polyaniline–enzyme films characterized using spectroscopic techniques have been used to fabricate a cholesterol biosensor.[39]

Self-encapsulation of redox enzyme – glucose oxidase E.C. 1.1.3.4. from Penecillum vitale (GOX) – within conducting polymer polypyrrole (Ppy) has been reported. Polymerization of polypyrrole was initiated by GOX catalytic action product – hydrogen peroxide. The increase in optical absorbance at 460 nm was exploited for the monitoring of polypyrrole polymerization process. The presence of entrapped glucose oxidase within polypyrrole was determined by basic application of polypyrrole-coated glucose oxidase nanoparticles (GOX/Ppy) in electrochemical biosensor design. Further application of nanoparticles based on self-encapsulated glucose oxidase and other oxidases is predicted for various bioanalytical purposes and other biocatalytic applications.[40] Cholesterol oxidase (ChOx) was physically entrapped in poly(3,4-ethylenedioxypyrrole) (PEDOP) to construct an amperometric cholesterol biosensor. The minimum detectable substrate concentration was 0.4 mM, and for a period of 20 days, the biosensor showed the maximum relative activity.[41]

A microbial biosensor based on *Gluconobacter oxydans* cells immobilized on the conducting polymer of 4-(2,5-di(thiophen-2-yl)-1H-pyrrol-1-yl)benzenamine (SNS-NH₂) coated onto the surface of graphite electrode was constructed. The proposed biosensor was characterized using glucose as the substrate. The system was used for ethanol and glucose detection in real samples.[42]

Poly(4,7-di(2,3)-dihydrothienol[3,4-b][1,4]dioxin-5-yl-benzo[1,2,5]thiadiazole) (PBDT) and poly(4,7-di (2,3)-dihydrothienol[3,4-b][1,4]dioxin-5-yl-2,1,3-benzo-selenadiazole) (PESeE) were electrochemically deposited on graphite electrodes and used as immobilization matrices for biosensing studies. After electrochemical deposition of the polymeric matrices, glucose oxidase (GOx) was immobilized on the modified electrodes as the model enzyme. The biosensor was tested on real human blood serum samples.[43]

CONDUCTING POLYMER AS BIOMOLECULAR ACTUATORS

Molecular actuators are materials that can change their physical dimensions when stimulated by an electrical signal. In the case of conducting polymers, the volume change occurs as a result of ion movement in and out of the polymer during redox cycling.[44] The change in volume can be up to 3% (although much higher volume changes have been reported on very small devices). When tested isometrically (at constant length), the stress generated by the volume changes is of the order of 10 MPa. The performance of conducting polymer actuators compares favorably with natural muscle (10% stroke and 0.3 MPa stress) and with piezoelectric polymer (0.1% stroke and 6 MPa). The piezoelectric polymers are driven by high electric fields, usually requiring 100–200 V, whereas the conducting polymers require only 1–5 V to operate. Some of the disadvantages of CEP actuators include a slow response time and limited lifetime, although recent results have shown strain rates of 3%/s (natural muscle can respond at 10%/sec) giving 1% strain for more than 100,000 cycles.

Low et al.[44] have developed the concept of responsive controlled drug release utilizing the actuation capabilities of a conducting polymer, namely polyaniline and a polyaniline composite with poly(2-hydroxyethylmethacrylate) hydrogel in combination with a solid-state silicon-based microvalve structure with metallic electrode contacts. A number of microvalve configurations have been proposed from a sphincter arrangement where the conducting polymer element acts in a constricting fashion, a plunger to a constricting tube arrangement. It was demonstrated that these arrangements acted reversibly, whereas metallic barrier layer valves acted as one-shot devices. Polyaniline–hydrogel blends demonstrated superior actuation properties than the pure polyaniline under the microvalve test conditions. This was primarily due to the hydrogel possessing a higher degree of swelling, which combined with the redox switchability and associated volume changes in the conducting polymer gave clear advantages over the hydrogel and intrinsically conducting polymer (ICP) alone. Potential applications for these device structures for use as a drug-delivery system (reservoir) and biosensor under microprocessor control have been reported.[45]

Massoumi and Entezami[46] reported the controlled release of dexamethasone sodium phosphate (DMP) from a conducting polymer bilayer film consisting of a PPy inner film doped with DMP and poly(N-methylpyrrole)-polystyrene sulfonate (PNMP=PSS) or polyaniline sulfonate (SPANI) outer film. DMP was released from the inner film by an application of less than 0.6 V. It has been demonstrated that it is possible to perform controlled release of biologically active molecules utilizing the actuation of conducting polymer as a flow-gating device.[46]

The electrochemical linear actuation of polyaniline fiber actuators has been studied in a variety of acidic aqueous electrolytes. Experimental results show that the linear strain changes significantly but nonlinearly with the anion volume. For anions smaller than Br^-, a larger strain was obtained for a larger anion, that is, $Br^- > Cl^- > F^-$, while once the anion was larger than Br^-, a larger anion produced a smaller strain, that is, $BF_4^- > ClO_4^- > CF_3SO_3^-$. On the basis of the definition of the ECR (elongation/charge ratio), that is, the contribution of a unit charge to fiber elongation, the maximum linear strain can be estimated by assuming the electrochemical efficiency is 100%. By assuming that ion and solvent insertion contributes mostly to the fiber expansion, a simple mathematical description is developed for the linear strain to show how it is determined by the volume and carried charge of the insert complex and the anisotropicity of the fiber.[47]

The synthesis and characterization of polyaniline integrally skinned asymmetric membranes (PAni-ISAMs) and their use as chemical and electrochemical actuators has been reported. PAni–ISAM cross section showed a thin dense skin and a microporous substructure with a PAni density gradient in scanning electron micrographs. The deformation mechanism of chemical monolithic PAni-ISAM actuation was found to result mainly from asymmetric volume expansion/contraction because of the presence/absence of counterions during PAni doping/dedoping cycles. Actuator performance was affected by acid concentration and film thickness.[48]

Drawn polyaniline films and fibers doped with 2-acrylamido-2-methyl-propane-1-sulfonic acid, PAni-(AMPS), were electrochemically cycled in HCl and their material properties and actuation performance comprehensively characterized. The Young's modulus was obtained as a function of applied voltage. Actuator figures of merit were derived from isotonic and isometric measurements, including strain, stress, work, power, creep, and efficiency. The effects of sample length, solution pH, electrochemical driving method, frequency, and load were studied, as well as the response of current to applied load for sensing applications. This work presents a complete picture of a polyaniline actuator for the first time. The behavior of the actuator is discussed in terms of the changes in the oxidation and protonation states of polyaniline.[49] Actuation of polyaniline (PAni)/Au bending bilayers and stretched polyaniline fibers doped with 2-acrylamido-2-methyl-propane-1-sulfonic acid was studied in aqueous methanesulfonic acid. Electrochemical activity was retained even upon repeated cycling into the pernigraniline state.[50]

Novel thiophene-based conducting polymer molecular actuators, exhibiting electrically triggered molecular conformational transitions, have been reported. In this new class of materials, actuation is the result of conformational rearrangement of the polymer backbone at the molecular level and is not simply due to ion intercalation in the bulk polymer chain upon electrochemical activation. Molecular actuation mechanisms results from pi stacking of thiophene oligomers upon oxidation, producing a reversible molecular displacement that is expected to lead to surprising material properties, such as electrically controllable porosity and large strains.[51] A novel controlled drug-delivery system in which drug release is achieved by electrochemically actuating an array of polymeric valves on a set of drug reservoirs has been developed using gold and PPy bilayers. These valves can be actuated under closed-loop control of sensors responding to a specific biological or environmental stimulus, leading to potential applications in advanced responsive drug-delivery systems.[52]

Intelligent Polymer Research Institute (IPRI) is currently building on a concept introduced by De Rossi to produce a rehabilitation glove in collaboration with North Shore Hospital Service (Sydney, Australia). The actuators will be integrated throughout the wearable glove structure to provide assisted movement during rehabilitation. IPRI is also currently involved in the development of actuators for an electronic Braille screen in collaboration with Quantum Technology (Sydney, Australia). A convenient user interface is the single biggest barrier to blind people's accessing information in the Internet age.[53]

CONDUCTING POLYMERS FOR TE

TE is the use of a combination of cells, engineering and materials methods, and suitable biochemical and physiochemical factors to improve or replace biological functions. According to Langer and Vacanti, TE is "an interdisciplinary field that applies the principles of engineering and life sciences toward the development of biological substitutes that restore, maintain, or improve tissue function or a whole organ."[54] Tailoring specific material properties, bulk as well as surface, could provide novel solutions for tissue-engineered systems, including controlled cell assembly (micro and nanopatterned surfaces), drug release (degradable polymers), tissue release (thermoresponsive polymers), and integrated biosensing [electroactive polymers (EAP)]. In addition, such materials provide a platform for the study of the fundamental underpinning science relating to tissue–material surface interactions.[55] Owing to their capability to electronically control a range of physical and chemical properties, conducting polymers such as polyaniline, polypyrrole, and polythiophene and/or their derivatives and composites provide compatible substrates that promote cell growth, adhesion, and proliferation at the polymer—tissue interface through electrical stimulation. Specific cell responses depend on polymers surface characteristics such as roughness, surface free

energy, topography, chemistry, charge, and other properties as electrical conductivity or mechanical actuation, which depend on the employed synthesis conditions. The biological functions of cells can be dramatically enhanced by biomaterials with controlled organizations at the nanometer scale and in the case of conducting polymers, by the electrical stimulation. The advancement of TE is contingent upon the development and implementation of advanced biomaterials. Conductive polymers have demonstrated potential for use as a medium for electrical stimulation, which has shown to be beneficial in many regenerative medicine strategies, including neural and cardiac TE. Mattioli-Belmonte et al. were the first to demonstrate that conducting polymer is biocompatible *in vitro* and *in vivo*.[56] PAni and its derivatives were found to be able to function as biocompatible substrates, upon which both H9c2 cardiac myoblasts and PC-12 pheochromocytoma cells can adhere, grow, and differentiate.[57,58]

TE of nerve grafts requires synergistic combination of scaffolds and techniques to promote and direct neurite outgrowth across the lesion for effective nerve regeneration. Prabhakaran et al. fabricated a composite polymeric scaffold that is conductive in nature by electrospinning and further performed electrical stimulation of nerve stem cells seeded on the electrospun nanofibers. Poly-L-lactide (PLLA) was blended with polyaniline (PAni) at a ratio of 85:15 and electrospun to obtain PLLA/PAni nanofibers with fiber diameters of 195 ± 30 nm. The electrospun PLLA/PAni fibers showed a conductance of 3×10^{-9} S by two-point probe measurement. *In vitro*, electrical stimulation of the nerve stem cells cultured on PLLA/PAni scaffolds applied with an electric field of 100 mV/mm for a period of 60 min resulted in extended neurite outgrowth compared to the cells grown on nonstimulated scaffolds. Their studies further strengthen the implication of electrical stimulation of nerve stem cells on conducting polymeric scaffolds toward neurite elongation that could be effective for nerve tissue regeneration.[59]

Poly(3-hydroxybutyrate) [P(3HB)] foams exhibiting highly interconnected porosity (85% porosity) were prepared using a unique combination of solvent casting and particulate leaching techniques by employing commercially available sugar cubes as porogen. Bioactive glass (BG) particles of 45S5 Bioglass grade were introduced in the scaffold microstructure, both in micrometer [(m-BG), <5 μm] and in nanometer [(n-BG), 30 nm] sizes. The *in vitro* bioactivity of the P(3HB)/BG foams was confirmed within 10 days of immersion in simulated body fluid and the foams showed high level of protein adsorption. The foams interconnected porous microstructure proved to be suitable for MG-63 osteoblast cell attachment and proliferation. The foams implanted in rats as subcutaneous implants resulted in a nontoxic and foreign body response after 1 week of implantation. In addition to showing bioactivity and biocompatibility, the P(3HB)/BG composite foams also exhibited bactericidal properties, which was tested on the growth of *Staphylococcus aureus*. An attempt was made at developing multifunctional scaffolds by incorporating, in addition to BG, selected concentrations of Vitamin E or/and CNTs. P(3HB) scaffolds with multifunctionalities (viz., bactericidal, bioactive, electrically conductive, and antioxidative behavior) were thus produced, which paves the way for the next generation of advanced scaffolds for bone TE.[60]

Sharma et al. have reported the fabrication and characterization of polyaniline–carbon nanotube/poly(*N*-isopropyl acrylamide-*co*-methacrylic acid) (PAni–CNT/PNIPAm-*co*-MAA) composite nanofibers and PNIPAm-*co*-MAA nanofibers suitable as a three-dimensional (3D) conducting smart tissue scaffold using electrospinning. Cellular response of the nanofibers was studied with mice L929 fibroblasts. Cell viability was checked on 7th day of cell culture by double staining the cells with calcein-AM and PI dye. PAni–CNT/PNIPAm-*co*-MAA composite nanofibers were shown the highest cell growth and cell viability as compared to PNIPAm-*co*-MAA nanofibers. Cell viability in the composite nanofibers was obtained in the order of 98%, which indicates the composite nanofibers provide a better environment as a 3D scaffold for the cell proliferation and attachment suitable for TE applications.[61]

CONDUCTING POLYMER IN BIOMECHANICAL SENSING

Biomechanics is the study of the structure and function of biological systems such as humans, animals, plants, organs, and cells by means of the methods of mechanics.[62] Biomechanical sensors are based on the motion of human body, cellular movement, mechanics of biological system, etc. A review by Engin et al. highlights some of the emergent developments and trends in biomechanical sensors with a specific emphasis on intelligent textiles and wearable electronic devices. Devices such as Georgia Tech's Wearable Motherboard (registered trademark of Georgia Institute of Technology, Atlanta, Georgia) and integrated switches such as the Softswitch (registered trademark of Softswitch Ltd, West Yorkshire, UK) are examples of how function is being embedded into everyday apparel.

Conducting polymers (CPs) being intelligent materials have found applications in biomechanical sensors. Conducting polymer-based electrochemical DNA sensors have shown applicability in a number of areas related to human health, such as diagnosis of infectious diseases, genetic mutations, drug discovery, forensics, and food technology due to their simplicity and high sensitivity.[63] Jensen, Radwin, and Webster have reported the construction and use of a durable and thin force sensor that can be attached to the palmar surface of the fingers and hands for studying the biomechanics of grasp and for use in hand injury rehabilitation. The sensors were used for measuring finger forces during controlled pinching and lifting tasks, and

during ordinary grasping activities, such as picking up a book or a box, where the useful force range and response for these sensors were adequate.[64] A 4 × 4 array of four quad (5 mm × 5 mm) Interlink conductive polymer pressure sensors have been designed and reported by Webster and coworkers. The array was embedded in a shoe insole under the second metatarsal head of a subject. The sensors were calibrated and a microprocessor-based portable data-acquisition system was used to monitor the pressure distribution under the second metatarsal head during normal gait. A center-of-pressure algorithm was used to estimate the maximum metatarsal head movement for one subject.[65]

Conducting polypyrrole-coated textile based on nylon lycra has been prepared using an in situ chemical polymerization process. Factors such as monomer, oxidant, and dopant concentration; fabric sample size; temperature; polymerization time; and convective control have been investigated. The effects of these factors on the conductivity, gauge factor, and stability of the polypyrrole-coated textiles are observed. As the polypyrrole-coated textiles conform to the shape of the human body, they function ideally as wearable biomechanical sensors that can be used in a range of applications to monitor human motion.[66] A molecular template (a sulfonated polyaniline) has been used to facilitate integration of a complementary conductive polymer (polyaniline) into wool-based textiles. The efficiency of the polymerization/coating process is enhanced since the template localizes the reaction within the textile. The presence of the molecular template results in the formation of an adherent, uniform, and stable conducting polymer layer. The coated textiles were found to be suitable for application as wearable strain gauge materials to be used for biomechanical monitoring.[67]

Rajagopalan and coworkers have discovered a new technique to measure urine volume in patients with urinary bladder dysfunction. Polypyrrole was chemically deposited on a highly elastic fabric. This fabric, when placed around a phantom bladder, produced a reproducible change in electrical resistance on stretching. The resistance response to stretching is linear in 20%–40% strain variation. This change in resistance is influenced by chemical fabrication conditions. They also demonstrated the dynamic mechanical testing of the patterned polypyrrole on fabric in order to show the feasibility of passive interrogation of the strain sensor for biomedical sensing applications.[68]

Campbell and his coworkers have studied the breast motion of woman using conducting polymer-coated smart textiles. They used conducting polymer-coated fabric sensors to monitor breast motion, vertical breast motion of two large-breasted women (C+ bra cup) as the subjects walked and ran on a treadmill. They concluded that polymer-coated fabric sensors were able to accurately and reliably represent changes in the amplitude of vertical breast displacement during treadmill gait.[69] A unique textile-based device, the intelligent knee sleeve (IKS), uses conducting polymer technology to provide feedback on knee flexion

Conducting—Dendritic

angle for injury prevention programs. It was concluded that IKS provides valid and reliable feedback on knee flexion angle. Such wearable biofeedback systems have application in a performance enhancement, injury prevention, and rehabilitation of sportspersons.[64]

A study was conducted of socks fitted with thin flexible conductive polymer sensors for the potential use as a smart sock for monitoring foot motion. The thin flexible sensors consisted of a conductive polymer applied on an elastic textile substrate that exhibited a resistance change when strained. Quasi-static response tests of the basic sensor over a static load range of a few newtons were conducted and showed a time-varying response. These sensors were integrated into socks, and preliminary results indicate that distinct responses to different foot motion patterns are detected in sensors placed at different joint locations on the foot. Further processing of strain results from smart socks should provide information about the kinematics and dynamics of the human foot.[70]

An active catheter intended for controllable intravascular maneuvers is presented and initial experimental results have been reported by Shoa et al.[71] A commercial catheter is coated with polypyrrole and laser micromachined into electrodes, which are electrochemically activated, leading to bending of the catheter. The catheters electro-chemomechanical properties are theoretically modeled to design the first prototype device, and used to predict an optimal polypyrrole thickness for the desired degree of bending within 30 sec. They compared the experimental result of catheter bending to the theoretical model with estimated electrochemical strain, showing reasonable agreement. Finally, they used the model to design an encapsulated catheter with polypyrrole actuation for improved intravascular compatibility and performance.[72]

CONDUCTING POLYMERS FOR DRUG DELIVERY

Ion movement to maintain charge neutrality with the mobile species accompanies electrochemical switching of conducting polymers and the direction of ion flux is controlled by polymer–ion interaction.[73] Conducting polymers with immobile high-molecular-weight or multianionic dopant (polyelectrolyte incorporated during electropolymerization) exhibit cation-dominated transport. Recently, there has been a significant effort directed to finding new drug release systems in which bioactive molecules contained in a reservoir can be supplied to a host system while controlling the rate and period of delivery.[74] The optimum mode of administration would be achieved if the drug was delivered to a precise region of the body where it is physiologically required. Polymers have proven especially useful materials for drug carriers since they can be easily processed and their physicochemical properties turned via molecular architecture. Conducting EAPs have been considered for drug delivery due to their unique redox properties, which allow controlled ionic transport through the polymer membrane. Conducting polymer films and coatings are also ideal hosts for the controlled release of chemical substances, including therapeutic drugs and many others. By incorporating the target species as the dopant in the conducting polymer, the redox chemistry of the polymer can be used to release the target species at the desired time.[75]

Valdes-Ramirez et al. reported on the development of a microneedle-based multiplexed drug delivery actuator that enables the controlled delivery of multiple therapeutic agents. Two individually addressable channels on a single microneedle array, each paired with its own reservoir and conducting polymer nanoactuator, are used to deliver various permutations of two unique chemical species. On application of suitable redox potentials to the selected actuator, the conducting polymer is able to undergo reversible volume changes, thereby serving to release a model chemical agent in a controlled fashion through the corresponding microneedle channels. This demonstrates the potential of the drug delivery actuator system to aid in the rapid administration of multiple therapeutic agents and indicates the potential to counteract diverse biomedical conditions.[76]

Svirskis et al. have developed a drug delivery system for implantation featuring electrically controlled release of the antipsychotic drug risperidone from a polypyrrole (PPy) film. PPy-based drug delivery systems have the potential to offer unique benefits to patients where the release rate of drug can be matched to dosing requirements. Changes were observed in risperidone release, the presence of pyrrole, polymer conductivity, surface roughness, and actuation behavior on aging. These changes suggest that the release of risperidone is related to alterations in PPy film morphology during storage.[77] Wadhwa and coworkers proposed a drug delivery system, from conducting polymer (CP) coatings on the electrode sites, to modulate the inflammatory implant–host tissue reaction. In this study, polypyrrole (PPy)-based coatings for electrically controlled and local delivery of the ionic form of an anti-inflammatory drug, dexamethasone (Dex), was investigated.[78] Surface-patterned poly(o-toluidine) (POT) nanofibers and nanotubes with controllable inner diameter has been also reported for controlled drug delivery.[79] Controlled release of an anti-inflammatory drug from PEDOT nanotubes using electrical stimulation is demonstrated by Abidian and his group. The fabrication process includes electrospinning of a biodegradable polymer into which the drug has been incorporated, followed by electrochemical deposition of the conducting polymer around the drug-loaded electrospun nanoscale fibers.[80]

CONDUCTING POLYMERS AS ARTIFICIAL MUSCLE

Natural muscles are one of the most important actuators in biological systems that are larger than a bacterium. EAPs, which emerged in the past fifteen years exhibiting large strain in response to electrical stimulation, are human-made actuators that most closely emulate muscles. For this response, EAPs have earned the moniker "artificial muscles."[81] They are particularly attractive to biomimetic experts since they can be used to mimic the movements of humans, animals, and insects for making biologically inspired mechanisms, devices, and robots.[82]

In 1992, artificial muscles based on conducting polymers (CPs) were constructed with the ability to describe very large and macroscopic angular movements[83,84] when the applied potential is shifted along a potential range lower than 1 V. Baughman et al. and DeRossi et al. suggested that electrochemically induced volume changes in conducting polymers could be applied to the construction of actuator-based artificial muscle. The bending movements obtained by electrochemically induced movements in films of conducting polymer (electro-chemo-mechanical devices) described angles larger than 180 degrees, surpassing all expectations of the stated hypotheses.[85,86] Linear artificial muscles of hydrogel microfibers coated with a conducting polymer can act while working as a sensor of driving current, electrolyte concentration, and temperature. Hybrid conducting polymer/hydrogel microfibers were fabricated from a chitosan solution through wet spinning technique, followed by in situ chemical polymerization of pyrrole in aqueous medium using bis(triflouromethane sulfonyl) imide as dopant. The fiber showed an electrical conductivity of 3.1×10^{-1} S cm^{-1}. The electroactivity was imparted by polypyrrole. An electrochemical linear actuation strain of 0.54% was achieved in aqueous electrolyte. The electrochemical measurements were performed as a function of applied current, temperature, and concentration for a constant charge in 1 M NaCl. A logarithmic dependence of the consumed electrical energy with concentration of the electrolyte during reaction suggested that it can act as a concentration sensor.[87]

Electrically conducting conjugated polymers such as polyaniline (PAni) usually show gel properties with volume changes in the range 1%–100% in response to some external stimuli. The bending beam method provides an effective and sensitive way to detect and make use of these volume changes in bipolymer laminate strips made of a PAni layer and a substrate polyimide layer. These strips bend, corresponding to volume changes in the PAni layer, during electrochemical redox of the PAni in aqueous solutions. The extension and velocity of the movement of the strips can be controlled by applying appropriate potentials and currents.[88] Two types of actuators, viz., "Backbone-type" and "Shell-type" actuators, based on PAni films have been fabricated and reported.[89]

The film made of poly(3,4-ethylenedioxythiophene) doped with poly(4-styrenesulfonate) (PEDOT/PSS) was prepared by casting and electrical conductivity, tensile properties, electromechanical response, and moisture sorption isotherm of the PEDOT/PSS film were investigated. The moisture sorption isotherm of the PEDOT/PSS film indicated that the interaction between water and PEDOT/PSS was superior to that between water molecules, in which isosteric heat of sorption decreased with an increase in the sorption degree and the value reached to the heat of water condensation.[90]

CONDUCTING POLYMER AS NEURAL PROBE

Neural prostheses transduce bioelectric signals to electronic signals at the interface between neural tissue and neural microelectrodes. Conducting polymers may offer the organic, improved bionic interface that is necessary to promote biocompatibility in neural stimulation applications. Factors such as electrode impedance, polymer volume changes under electrical stimulation, charge injection capability, biocompatibility, and long-term stability are of significant importance and may pose as challenges in the future success of conducting polymers in biomedical applications. Microelectrode neural probes facilitate the functional stimulation or recording of neurons in the central nervous system and peripheral nervous system.[10] Controlled neural cell growth can be obtained on poly(3,4-ethylenedioxythiophene) doped with polystyrene sulfonate (PEDOT:PSS) maintaining very low interfacial impedance. Electroadsorbed polylysine enabled long-term neuronal survival and growth on the nanostructured polymer. Neurite extension was strongly inhibited by an additional layer of PSS or heparin, which in turn could be either removed electrically or further coated with spermine to activate cell growth. Binding basic fibroblast growth factor (bFGF) to the heparin layer inhibited neurons but promoted proliferation and migration of precursor cells. This methodology may orchestrate neural cell behavior on EAPs, thus improving cell/electrode communication in prosthetic devices and providing a platform for tissue repair strategie.[91] Abidian and Martin reported electrochemical deposition poly(pyrrole) (PPy) and poly(3,4-ethylenedioxythiophene) (PEDOT) nanotubes on the surface of neural microelectrode sites. An equivalent circuit model comprising a coating capacitance in parallel with a pore resistance and interface impedance in series was developed and fitted to experimental results to characterize the physical and electrical properties of the interface.[92] A synthetically produced, anionically modified, laminin peptides DEDEDYFQRYLI and DCDPGYIGSR were used to dope poly(3,4-ethylenedioxythiophene) (PEDOT) electrodeposited on platinum (Pt) electrodes. Performance of peptide doped films was compared with conventional polymer PEDOT/paratoluene sulfonate (pTS) films using SEM, XPS, cyclic voltammetry, impedance spectroscopy, mechanical hardness, and adherence. Bioactivity of incorporated peptides and their effect on cell growth was assessed using a PC12 neurite outgrowth assay. It was demonstrated that large peptide dopants produced softer PEDOT films with a minimal decrease in electrochemical stability, compared with the conventional dopants.[93]

Conducting—Dendritic

The interface between micromachined neural microelectrodes and neural tissue plays an important role in chronic *in vivo* recording. Electrochemical polymerization was used to optimize the surface of the metal electrode sites. Electrically conductive polymers (polypyrrole) combined with biomolecules having cell adhesion functionality were deposited with great precision onto microelectrode sites of neural probes. The biomolecules used were a silk-like polymer having fibronectin fragments (SLPFs) and nonapeptide CDPGYIGSR.[94] Similar work was reported by Green and his coworkers, which cited the impact of biological inclusions on polymer properties and their ongoing performance in neural prosthetics.[95]

Conductive meshes of PPy and aligned electrospun poly(lactic-*co*-glycolic acid) (PLGA) nanofibers supported the growth and differentiation of rat pheochromocytoma 12 (PC12) cells and hippocampal neurons comparable to noncoated PLGA control meshes, suggesting that PPy–PLGA may be suitable as conductive nanofibers for neuronal tissue scaffold.[96]

Conducting polymers with pendant functionality are advantageous in various bionic and organic bioelectronic applications as they allow facile incorporation of bioregulative cues to provide biomimicry and conductive environments for cell growth, differentiation, and function. Polypyrrole substrates doped with chondroitin sulfate (CS) were electrochemically synthesized and conjugated with type I collagen. Rat pheochromocytoma (nerve) cells showed increased differentiation and neurite outgrowth on the fibrillar collagen, which was further enhanced through electrical stimulation of the underlying conducting polymer substrate.[97]

CONDUCTING POLYMER IN OPTICAL BIOSENSOR

Optical biosensors are based on the measurement of light absorbed or emitted as a consequence of biochemical reaction. Light waves are guided by optical fibers to suitable detectors. Applications of conducting polymers to optical biosensors have aroused much interest. This is because these molecular electronic materials offer control of different parameters such as polymer layer thickness, electrical properties, and bioreagent loading. Conformational changes in conjugated backbone results in changes of color in conducting polymers.

Polythiophenes have been shown to exhibit a variety of optical transitions on external biological stimuli such as proteins opening the way for designing a variety of biochromic sensory devices. As there is a strong correlation between the electronic structure and the backbone conformation in conjugated polymers, any change in the main chain conformation will lead to an alteration of effective conjugation length, coupled with a shift of the absorption in the UV-visible range. The level of urea in blood, as an indicator of several renal diseases, is one of the most important biological parameters, and because of this, many methods have been developed for its determination. An optical biosensor for urea determination has been reported by de Marcos et al.[98] The biosensor is based on the enzymatic reaction with urease, which is first photoimmobilized with polyacrylamide onto a chemically polymerized polypyrrole (PPy) film. The main advantage of this sensor is that no indicator dye or pH indicator is needed, because PPy itself acts as the support and the indicator. These PPy films show an absorbance spectrum in the near IR range, which is pH dependent. The variation of absorbance is thus directly related to the change of pH caused during the enzymatic reaction, which is also dependent on the urea concentration. The linear range of the sensor is from 0.06 to 1 M of urea, which is the common level of urea concentration found in blood and urine samples. Hammarström et al. have developed a novel thiophene-based molecular scaffold, denoted as luminescent conjugated oligothiophenes (LCOs). In this work, use of LCOs as specific ligands for the pathological hallmarks underlying protein misfolding diseases, such as Alzheimer's disease, is described. The use of the conformation-sensitive optical properties of the LCOs for spectral separation of these pathological entities in a diversity of *in vitro*, *ex vivo*, or *in vivo* systems is demonstrated. The protein aggregates are easily identified due to the conformation-dependent emission profile from the LCOs and spectral assignment of protein aggregates can be obtained. Overall, these probes will offer practical research tools for studying protein misfolding diseases and facilitate the study of the molecular mechanism underlying these disorders.[99]

The use of a biochromic conjugated polymer (BCP) sensors for pathogen detection was demonstrated by Song et al.[100] Biologically active cell membrane components were incorporated into conjugated polymers with desirable optical properties. Polydiacetylenic cell membrane-mimicking materials that conveniently report the presence of pathogens with a color change were used for the colorimetric detection of bacterial toxins and influenza virus. It has been suggested that the BCP sensors are convenient for microfabrication and use, since their molecular recognition and signal transduction functionalities are resident in a single functional unit.[101] Su et al.[101] reported the fabrication of mixed phospholipid/polydiacetylene vesicles functionalized with glycolipid for detection of *Escherichia coli*. The paper reported that *E. coli*–glycolipid binding event leads to a visible color change from blue to red, readily seen with the naked eye and quantified by absorption spectroscopy. The biosensor signal amplification has been gained by elevating the pH of the aqueous solutions and increasing the phospholipid content in the mixed lipid vesicles.

Cationic polythiophenes transduce oligonucleotide hybridization into a colorimetric output based on conformational changes of the polymer upon interaction with single-stranded DNAs or double-stranded DNAs. Cationic poly(fluorene-*co*-phenylene) materials serve as donors in fluorescence energy-transfer assays that display signal amplification. Signal transduction in aqueous media is controlled by specific electrostatic interactions.[102]

CONCLUSIONS

Nowadays, from photovoltaic devices to bio-implants, flexible, lightweight, cost-effective conducting polymers (CPs) have found applications. The unique property that ties all these applications together is the conductivity of the CP. CPs are organic in nature, which adds biocompatibility to these materials. Further, the presence of an alternative single and double carbon–carbon conjugated backbone within the polymer endows it with the ability to delocalize π electrons that impart conductivity in the materials. In addition to these highly desirable properties, the ease of preparation and modification of CPs have made them a popular choice for many applications. This is especially true in biomedicine, where many applications benefit from the presence of conductive materials, whether for biosensing or for control over cell proliferation and differentiation. Despite the vast amount of research already conducted on CPs, for biomedical applications, the field is still growing and many questions remain to be answered.

FUTURE PROSPECTS

Although conducting polymers have been found to be biocompatible and have been used for TE, nerve cell regeneration, biosensor, controlled drug release experiments, thus far, only a small number of these (excluding variants of unmodified CPs) have been found to be well suited for biomedical applications. A number of conducting polymers have yet not been explored for biocompatibility and biomedical applications. In near future, many conducting polymers with better functionality have to be fabricated to make biomedical applications cost-effective.

ACKNOWLEDGMENTS

The authors are grateful to the authors of various articles referenced here.

REFERENCES

1. Katchalsky, A.; Zwick. M. Mechanochemistry and ion exchange. J. Polym. Sci. **1955**, *16* (82), 221–234.
2. Katchalsky, A.; Eisenberg, H. Polyvinylphosphate contractile systems. Nature **1950**, *166* (4215), 267.
3. Kuhn, W.; Hargitay, B.; Katchalsky, A.; Eisenberg, H. Reversible dilation and contraction by changing the state of ionization of high-polymer acid networks. Nature **1950**, *165* (4196), 514–516.
4. Katchalsky, A. Rapid swelling and deswelling of reversible gels of polymeric acids by ionization. Experientia **1949**, *5* (8), 319–320.
5. Osada, Y.; Saito, Y. Mechanochemical energy-conversion in a polymer membrane by thermo-reversible polymer–polymer interactions. Makromol. Chem. **1975**, *176* (9), 2761–2764.
6. Tanaka, T.; Fillmore, D.J. Kinetics of swelling of gels. J. Chem. Phys. **1979**, *70* (3), 1214–1218.
7. Shirakawa, H. Nobel lecture: The discovery of polyacetylene film: The dawning of an era of conducting polymers. Rev. Mod. Phys. **2001**, *73* (3), 713–718.
8. Heeger, A.J. Nobel lecture: Semiconducting and metallic polymers: The fourth generation of polymeric materials. Rev. Mod. Phys. **2001**, *73* (3), 681–700.
9. MacDiarmid, A.G. Nobel lecture: "Synthetic metals": A novel role for organic polymers Rev. Mod. Phys. **2001**, *73* (3), 701–712.
10. Innis, P.C.; Moulton, S.E.; Wallace, G.G. Biomedical Applications of Inherently Conducting Polymers (ICPs). In *Handbook of Conducting Polymers*, 3rd Eds.; Skotheim, T.A.; Reynolds, J.R.; CRC press: Boca Raton, 2007; 1–33.
11. Kumar, A.; Sarmah, S.; Nath, C. *Engineering the Properties of Conducting Polymers Through Morphology Control, Polymer Morphology*, 1st Eds.; Yang, H. China, 2012; 79–107.
12. Letheby, H. On the production of a blue substance by the electrolysis of sulphate of aniline. J. Chem. Soc. **1862**, *15* (1), 161–163.
13. Mohilner, D.M.; Adams, R.N.; Argensinger, W.J, Jr. Investigation of the kinetics and mechanism of the anodic oxidation of aniline in aqueous sulfuric acid solution at a platinum electrode. J. Am. Chem. Soc. **1962**, *84* (19), 3618–3622.
14. Diaz, A.F.; Logan, J.A. Electroactive polyaniline films. J. Electroanal. Chem. Interfacial Electrochem. **1980**, *111* (1), 111–114.
15. Sun, Q.; Park, M.C.; Deng, Y. Studies on one-dimensional polyaniline (PANI) nanostructures and the morphological evolution. Mater. Chem. Phys. **2008**, *110* (2–3), 276–279.
16. Yang, X.; Dai, T.; Zhu, Z.; Lu, Y. Electrochemical synthesis of functional polypyrrole nanotubes via a self-assembly process. Polymer **2007**, *48* (14), 4021–4027.
17. Zang, J.; Li, C.M.; Bao, S.J.; Cui, X.; Bao, Q.; Sun, C.Q. Template-free electrochemical synthesis of superhydrophilic polypyrrole nanofiber network, Macromolecules **2008**, *41* (19), 7053–7057.
18. Martin, C.R.; Parthasarathy, R.; Menon, V. Template synthesis of electronically conductive polymers—A new route for achieving higher electronic conductivities. Synth. Met. **1993**, *55* (2–3), 1165–1170.
19. Lei, J.; Cai, Z.; Martin, C.R. Effect of reagent concentrations used to synthesize polypyrrole on the chemical characterristics and optical and electronic properties of the resulting polymer. Synth. Met. **1992**, *46* (1), 53–69.
20. Parthasarathy, R.V.; Martin, C.R. Template-synthesized polyaniline microtubules. Chem. Mater. **1994**, *6* (10), 1627–1632.
21. Rajesh, B.; Thampi, K.R.; Bonard, J.M.; Xanthopoulos, N.; Mathieu, H.J.; Viswanathan, B. Template synthesis of

conducting polymeric nanocones of poly(3-methylthiophene). J. Phys. Chem. B **2004**, *108* (30), 10640–10644.

22. Zhang, L.J.; Wan, M.X. Self-assembly of polyaniline—From nanotubes to hollow microspheres. Adv. Funct. Mater **2003**, *13* (10), 815–820.

23. Yang, Y.S.; Liu, J.; Wan, M.X. Self-assembled conducting polypyrrole micro/nanotubes. Nanotechnology **2002**, *13* (6), 771–774.

24. Zhang, L.J.; Wan, M.X. Synthesis and characterization of self-assembled polyaniline nanotubes doped with D-10-camphorsulfonic acid. Nanotechnology **2002**, *13* (6), 750–755.

25. Huang, J.; Virji, S.; Weiller, B.H.; Kaner, R.B. Polyaniline nanofibers: Facile synthesis and chemical sensors. J. Am. Chem. Soc. **2003**, *125* (2) 314–315.

26. Huang, J.X.; Kaner, R.B. Nanofiber formation in the chemical polymerization of aniline: A mechanistic study. Angew. Chem. Int. Ed. **2004**, *43* (43), 5817–5821.

27. Gorman, C.B.; Biebuyck, H.A.; Whitesides, G.M. Fabrication of patterned, electrically conducting polypyrrole using a self-assembled monolayer: A route to all-organic circuits. Chem. Mater. **1995**, *7* (3) 526–529.

28. Sayre, C.N.; Collard, D.M. Deposition of polyaniline on micro-contact printed self-assembled monolayers of ω-functionalized alkanethiols. J. Mater. Chem. **1997**, *7* (6), 909–912.

29. Doshi, J.; Reneker, D.H. Electrospinning process and applications of electrospun fibers. J. Electrostat. **1995**, *35* (2–3), 151–160.

30. MacDiarmid, A.G.; Jones, W.E., Jr.; Norris, I.D.; Gao, J.; Johnson, A.T., Jr.; Pinto, N.J.; Hone, J.; Han, B.; Ko, F.K.; Okuzaki, H.; Llaguno, M. Electrostatically-generated nanofibers of electronic polymers. Synth. Met. **2001**, *119* (1–3) 27–30.

31. Yu, Q.Z.; Li, Y.; Wang, M.; Chen, H.Z. Polyaniline nanobelts, flower-like and rhizoid-like nanostructures by electro spinning. Chinese Chem. Lett. **2008**, *19* (2), 223–226.

32. Gao, M.; Dai, L.; Wallace, G.G. Biosensors based on aligned carbon nanotubes coated with inherently conducting polymers. Electroanalysis **2003**, *15* (13), 1089–1094.

33. Ramanathan, K.; Annapoorni, S.; Malhotra, B.D. Application of poly(aniline) as a glucose biosensor. Sens. Actuators B **1994**, *21* (3), 165–169.

34. Koopal, C.G.J.; Bos, A.A.C.M.; Nolte, R.J.M. Third-generation glucose biosensor incorporated in a conducting printing ink. Sens. Actuators B **1994**, *18* (1–3), 166–170.

35. Park, J.; Kim, H.K.; Son, Y. Glucose biosensor constructed from capped conducting microtubules of PEDOT. Sens. Actuators B **2008**, *133* (1), 244–250.

36. Singhal, R.; Takashima, W.; Kaneto, K.; Samanta, S.B.; Annapoorni, S.; Malhotra, B.D. Langmuir–Blodgett films of poly(3-dodecyl thiophene) for application to glucose biosensor. Sens. Actuators B **2002**, *86* (1), 42–48.

37. Suman, S.; Singhal, R.; Sharma, A. L.; Malthotra, B.D.; Pundir, C.S. Development of a lactate biosensor based on conducting copolymer bound lactate oxidase. Sens. Actuators B **2005**, *107* (2), 768–772.

38. Ahuja, T.; Mir, I.A.; Kumar, D.; Rajesh. Potentiometric urea biosensor based on BSA embedded surface modified polypyrrole film. Sens. Actuators B **2008**, *134* (1), 140–145.

39. Singh, S.; Solanki, P.R.; Pandey, M.K.; Malhotra, B.D. Cholesterol biosensor based on cholesterol esterase, cholesterol oxidase and peroxidase immobilized onto conducting polyaniline films. Sens. Actuators B **2006**, *115* (1), 534–541.

40. Ramanavicius, A.; Kausaite, A.; Ramanaviciene, A. Polypyrrole-coated glucose oxidase nanoparticles for biosensor design. Sens. Actuators B **2005**, *111–112*, 532–539.

41. Turkarslan, O.; Kayahan, S.K.; Toppare, L. A new amperometric cholesterol biosensor based on poly(3,4-ethylenedioxypyrrole). Sens. Actuators B **2009**, *136* (2), 484–488.

42. Tuncagil, S.; Odaci, D.; Yildiz, E.; Timur, S.; Toppare, L. Design of a microbial sensor using conducting polymer of 4-(2,5-di(thiophen-2-yl)-1H-pyrrole-1-l) benzenamine. Sens. Actuators B **2009**, *137* (1), 42–47.

43. Emre, F.B.; Ekiz, F.; Balan, A.; Emre, S.; Timur, S.; Toppare, L. Conducting polymers with benzothiadiazole and benzoselenadiazole units for biosensor applications. Sens. Actuators B **2011**, *158* (1),117–123.

44. Low, L.M.; Seetharaman, S.; He, K.Q.; Madou. M.J. Microactuators toward microvalves for responsive controlled drug delivery. Sens. Actuators B **2000**, *67* (1–2), 149–160.

45. Jager, E.W.H.; Inganas, O.; Lundstrom, I. Microrobots for micrometer-size objects in aqueous media: Potential tools for single-cell manipulation. Science **2000**, *288* (5475), 2335–2338.

46. Massoumi, B.; Entezami, A. Electrochemically controlled binding and release of dexamethasone from conducting polymer films. J. Bioact. Compat. Polym. **2002**, *17* (1), 51–62.

47. Qi, B.; Lu, W.; Mattes, B. R. Strain and energy efficiency of polyaniline fiber electrochemical actuators in aqueous electrolytes. J. Phys. Chem. B **2004**, *108* (20), 6222–6227.

48. Wang, H.L.; Gao, J.; Sansinena, J.M.; McCarthy, P. Fabrication and characterization of polyaniline monolithic actuators based on a novel configuration: Integrally skinned asymmetric membrane. Chem. Mater. **2002**, *14* (6), 2546–2552.

49. Smela, E.; Lu, W.; Mattes, B.R. Polyaniline actuators: Part 1. PANI (AMPS) in HCl. Synth. Met. **2005**, *151* (1), 25–42.

50. Smela, E.; Mattes, B.R. Polyaniline actuators: Part 2. PANI (AMPS) in methanesulfonic acid. Synth. Met. **2005**, *151* (1), 43–48.

51. Anquetil, P.A.; Yu, H.; Madden, J.D.; Madden, P.G.; Swager, T.M.; Hunter, I.W. Thiophene-based conducting polymer molecular actuators, Smart Structures and Materials 2002: Electroactive Polymer Actuators and Devices (EAPAD), Bar-Cohen, Y., Ed., Proceedings of SPIE, Vol. 4695, 2002.

52. Xu, H.; Wang, C.; Wang, C.; Zoval, J.; Madou, M. Polymer actuator valves toward controlled drug delivery application. Biosens. Bioelectron. **2006**, *21* (11), 2094–2099.

53. Wallace, G.G.; Teasdale, P.R.; Spinks, G.M.; Kane-Maguire, L.A.P. *Conductive Electroactive Polymers: Intelligent Polymer Systems*; 3rd Ed.; CRC Press: Boca Raton: USA, 2008.

54. Langer, R.; Vacanti, J.P. Tissue engineering. Science **1993**, *260* (5110), 920–926.

55. Ateh, D.D.; Navsaria, H.A.; Vadgama, P. Polypyrrole-based conducting polymers and interactions with biological tissues. J. R. Soc. Interface **2006**, *3* (11), 741–752.

56. Mattioli-Belmonte, M.; Giavaresi, G.; Biagini, G.; Virgili, L.; Giacomini, M.; Fini, M.; Giantomassi, F.; Natali, D.; Torricelli, P.; Giardino, R. Tailoring biomaterial compatibility: *In vivo* tissue response versus *in vitro* cell behavior. Int. J. Artif. Organs. **2003**, *26* (12), 1077–1085.

57. Guterman, E.; Cheng Palouian, S.K.; Bide, P.; Lelkes, P.I.; Wei, Y. Peptide-modified electroactive polymers for tissue engineering applications. Polym. Prepr. Am. Chem. Soc. Div. Polym. Chem. **2002**, *43* (1), 766–767.

58. Bidez, P.R.; Li, S.; MacDiarmid, A.G.; Venancio, E.C.; Wei, Y.; Lelkes, P.I. Polyaniline, an electroactive polymer, supports adhesion and proliferation of cardiac myoblasts. J. Biomater. Sci. Polym. Ed. **2006**, *17* (1–2), 199–212.

59. Prabhakaran, M. P.; Ghasemi-Mobarakeh, L.; Jin, G.; Ramakrishna, S. Electrospun conducting polymer nanofibers and electrical stimulation of nerve stem cells. J. Biosci. Bioeng. **2011**, *112* (5), 501–507.

60. Misra, S.K.; Ansari, T.I.; Valappil, S.P.; Mohn, D.; Philip, S.E.; Stark, W.J.; Roy, I.; Knowles, J.C.; Salih, V.; Boccaccini, A.R. Poly(3-hydroxybutyrate) multifunctional composite scaffolds for tissue engineering applications. Biomaterials **2010**, *31* (10), 2806–2815.

61. Sharma, Y.; Tiwari, A.; Hattori, S.; Terada, D.; Sharma, A.K.; Ramalingam, M.; Kobayashi, H. Fabrication of conducting electrospun nanofibers scaffold for three-dimensional cells culture. Int. J. Biol. Macromol. **2012**, *51* (4), 627–631.

62. Alexander, R.M. Mechanics of animal movement. Curr. Biol. **2005**, *15* (16), R616–R619.

63. Peng, H.; Zhang, L.; Soeller, C.; Travas-Sejdic, J. Conducting polymers for electrochemical DNA sensing. Biomaterials **2009**, *30* (11), 2132–2148.

64. Jensen, T.R.; Radwin, R.G.; Webster, J.G.A conductive polymer sensor for measuring external finger forces. J. Biomech. **1991**, *24* (9), 851–858.

65. Webster, J.G.; Tompkins, W.J.; Wretch, J.J. A conductive polymer pressure sensor array, Engineering in Medicine and Biology Society 1989. Images of the Twenty-First Century, Proceedings of the Annual International Conference of the IEEE Engineering, 9–12 Nov, 1989.

66. Wu, J.; Zhou, D.; Too, C.; Wallace, G.G. Conducting polymer coated lycra. Synth. Met. **2005**, *155* (3), 698–701.

67. Wu, J.; Zhou, D.; Looney, M.G.; Waters, P.J.; Wallace, G.G.; Too, C.O. A molecular template approach to integration of polyaniline into textiles. Synth. Met. **2009**, *159* (12), 1135–1140.

68. Rajagopalan, S.; Sawan, M.; Ghafar-Zadeh, E.; Savadogo, O.; Chodavarapu, V.P. A Polypyrrole-based strain sensor dedicated to measure bladder volume in patients with urinary dysfunction. Sensors **2008**, *8* (8), 5081–5095.

69. Campbell, T.E.; Munro, B.J.; Wallace, G.G.; Steele, J.R. Can fabric sensors monitor breast motion? J. Biomech. **2007**, *40* (13), 3056–3059.

70. Munro, B.J.; Campbell, T.E.; Wallace, G.G.; Steele, J.R. The intelligent knee sleeve: A wearable biofeedback device. Sens. Actuators B **2008**, *131* (2), 541–547.

71. Shoa, T.; Madden, J.D.; Niloofar, F.; Munce, N.R.; Victor, Y.X.D. Conducting polymer based active catheter for minimally invasive interventions inside arteries. In Engineering in Medicine and Biology Society, 2008. EMBS 2008. 30th Annual International Conference of the IEEE, pp.2063, 2066, Aug 20–25 2008. doi: 10.1109/IEMBS.2008.4649598

72. Castano, L.M.; Winkelmann, A.E.; Flatau, A.B. A first approach to foot motion monitoring using conductive polymer sensors, Proc. SPIE 7292, Sensors and Smart Structures Technologies for Civil, Mechanical, and Aerospace Systems 2009, 72922O (March 30, 2009); Masayoshi Tomizuka, San Diego, CA, 2009.

73. Ren, X.; Pickup, P.G. Ion transport in polypyrrole and a polypyrrole/polyanion composite. J. Phys. Chem. **1993**, *97* (20), 5356–5362.

74. John, R.; Wallace, G.G. Doping-dedoping of polypyrrole: A study using current-measuring and resistance-measuring techniques. J. Electroanal. Chem. **1993**, *354* (1–2), 145–160.

75. Liang, W.; Martin, C.R. Gas transport in electronically conductive polymers. Chem. Mater. **1991**, *3* (3), 390–391.

76. Valdes-Ramírez, G.; Windmiller, J.R.; Claussen, J.C.; Martinez, A.G.; Kuralay, F.; Zhou, M.; Zhou, N; Polsky, R.; Miller, P.R.; Narayan, R.; Wang, J. Multiplexed and switchable release of distinct fluids from microneedle platforms via conducting polymer nanoactuators for potential drug delivery. Sens. Actuators B **2012**, *161* (1), 1018–1024.

77. Svirskis, D.; Wright, B.E.; Travas-Sejdic, J.; Rodgers, A.; Garg, S. Evaluation of physical properties and performance over time of an actuating polypyrrole based drug delivery system. Sens. Actuators B **2010**, *151* (1), 97–102.

78. Wadhwa, R.; Lagenaur, C.F.; Cui, X.T. Electrochemically controlled release of dexamethasone from conducting polymer polypyrrole coated electrode. J. Control. Release **2006**, *110* (3), 531–541.

79. Han, J.; Wang, L.; Guo, R. Facile synthesis of hierarchical conducting polymer nanotubes derived from nanofibers and their application for controlled drug release. Macromol. Rapid Commun. **2011**, *32* (9–10), 729–735.

80. Abidian, M.R.; Kim, D.H.; Martin, D.C. Conducting-polymer nanotubes for controlled drug release. Adv. Mater. **2006**, *18* (4), 405–409.

81. Bar-Cohen, Y. *Electroactive Polymer (EAP) Actuators as Artificial Muscles - Reality, Potential and Challenges*. 2nd Ed.; SPIE Press: Vol. PM136, 2004; 1–765.

82. Bar-Cohen, Y.; Breazeal, C, Eds. *Biologically-Inspired Intelligent Robots*; SPIE Press: Vol. PM122, 2003; 1–393.

83. Otero, T.F.; Angulo, E.; Rodrıguez, J.; Santamaria, C.J. Electrochemomechanical properties from a bilayer: Polypyrrole/non-conducting and flexible material—Artificial muscle. J. Electroanal. Chem. **1992**, *341* (1), 369–375.

84. Pei, Q.; Inganas, O. Conjugated polymers and the bending cantilever method: Electrical muscles and smart devices. Adv. Mat. **1992**, *4* (4), 277–278.

85. Baughman, R.H.; Shacklette, L.W.; Elsenbaumer, R.L.; Plitchta, E.; Becht, C. *Opportunities in Electronics, Optoelectronics and Molecular Electronics*; Brédas, J.L., Chance, R.R., Eds.; Kuwler Acad. Pub.: Netherlands, 1990; 559–582.

86. DeRossi, D.; Puccio, F.D.; Lorussi, F.; Orsini, P.; Tognetti, A. Feldman's muscle model: Implementation and control

Conducting—Dendritic

of a kinematic chain driven by pseudo-muscular actuators. Acta Bioeng. Biomech. **2004**, *4* (1), 224–225.

87. Ismail, Y.A.; Martínez, J.G.; Al Harrasi, A.S.; Kim, S.J.; Otero, T.F. Sensing characteristics of a conducting polymer/hydrogel hybrid microfiber artificial muscle. Sens. Actuators B **2011**, *160* (1), 1180–1190.

88. Pei, Q.; Inganas, O.; Lundstrom, I. Bending bilayer strips built from polyaniline for artificial electrochemical muscles. Smart Mater. Struct. **1993**, *2* (1), 1–6.

89. Kaneto, K.; Kaneko, M.; Min, Y.; MacDiarmid, A.G. "Artificial muscle" Electromechanical actuators using polyaniline films. Synth. Met. **1995**, *71* (1–3), 2211–2212.

90. Okuzaki, H.; Suzuki, H.; Ito, T. Electrically driven PEDOT/PSS actuators. Synth. Met. **2009**, *159* (21–22), 2233–2236.

91. Collazos-Castro, J.E.; Polo, J.L.; Hernandez-Labrado, G.R.; Padial-Canete, V. Concepción García-Rama, Bioelectrochemical control of neural cell development on conducting polymers. Biomaterials **2010**, *31* (35), 9244–9255.

92. Abidian, M.R.; Martin, D.C. Experimental and theoretical characterization of implantable neural microelectrodes modified with conducting polymer nanotubes. Biomaterials **2008**, *29* (9), 1273–1283.

93. Green, R.A.; Lovell, N.H.; Poole-Warren, L.A. Cell attachment functionality of bioactive conducting polymers for neural interfaces. Biomaterials **2009**, *30* (22), 3637–3644.

94. Cui, X.; Lee, V.A.; Raphael, Y.; Wiler, J.A.; Hetke, J.F.; Anderson, D.J.; Martin, D.C. Surface modification of neural recording electrodes with conducting polymer/biomolecule blends. J. Biomed. Mater. Res. **2001**, *56* (2), 261–272.

95. Green, R.A.; Lovell, N.H.; Wallace, G.G.; Poole-Warren, L.A. Conducting polymers for neural interfaces: Challenges in developing an effective long-term implant. Biomaterials **2008**, *29* (24–25), 3393–3399.

96. Lee, J.Y.; Bashur, C.A.; Goldstein, A.S.; Schmidt C.E. Polypyrrole-coated electrospun PLGA nanofibers for neural tissue applications. Biomaterials **2009**, *30* (26), 4325–4335.

97. Liu, X.; Yue, Z.; Higgins, M.J.; Wallace, G.G. Conducting polymers with immobilised fibrillar collagen for enhanced neural interfacing. Biomaterials **2011**, *32* (30), 7309–7317.

98. de Marcos, S.; Hortigüela, R.; Gatbán, J.A; Castillo, J.R.; Wolfbeis, O.S. Characterization of a urea optical sensor based on polypyrrole. Microchim. Acta **1999**, *130* (4), 267–272.

99. Hammarström, P.; Lindgren, M.; Nilsson, K.P.R. Luminescent conjugated oligothiophenes: optical dyes for revealing pathological hallmarks of protein misfolding diseases. Proc. SPIE 7779, Organic Semiconductors in Sensors and Bioelectronics III, 77790K, August 17, 2010.

100. Song, J.; Cheng, Q.; Zhu, S.; Stevens, R.C. "Smart" materials for biosensing devices: Cell-mimicking supramolecular assemblies and colorimetric detection of pathogenic agents. Biomed. Microdev. **2002**, *4* (3), 213–221.

101. Su, Y.L.; Li, J.R.; Jiang, L.; Cao, J. Biosensor signal amplification of vesicles functionalized with glycolipid for colorimetric detection of Escherichia coli. J. Colloid Interface Sci. **2005**, *284* (1), 114–119.

102. Liu, B.; Bazan, G.C. Homogeneous fluorescence-based DNA detection with water-soluble conjugated polymers. Chem. Mater. **2004**, *16* (23), 4467–4476.

Conjugated Polymers: Nanoparticles and Nanodots of

Garima Ameta
Department of Chemistry, University College of Science, Mohanlal Sukhadia University, Udaipur, India

Suresh C. Ameta
Rakshit Ameta
Department of Chemistry, PAHER University, Udaipur, India

Pinki B. Punjabi
Department of Chemistry, University College of Science, Mohanlal Sukhadia University, Udaipur, India

Abstract

The convergence of nanochemistry and controlled polymerization techniques offer powerful tools for designing hierarchically organized polymer architectures for nanotechnology and nanomedicine. Synthetic polymers are promising building blocks due to their low cost, high processability, and modular functionality. By incorporating small-molecule recognition units into polymer chains, it is possible to target complex and dynamic macromolecule aggregates that may eventually mimic the structure and function of biological entities, from nucleic acids and proteins up to cells, and entire living organisms. Nanoparticles (NPs) and nanodots (d < −30 nm) of conjugated polymers (CPs) have received considerable attention over the last decade (2005–2014) because of their potential to realize brighter luminescence, facile processing, and enhanced device efficiency. In medicine, densely packed CP nanospheres have exhibited higher fluorescence lifetime, emmission rates, and photostability than single molecular dyes. Nanospheres of CPs have enhanced the device performance of organic photovoltaic devices by forming nanoscale domains in thin films. With surface functionalization, bioconjugation, or hybridization, nanospheres have been explored for biological applications, such as labeling, imaging, sensing, and drug delivery. In this entry, we discuss about preparation and applications of various NPs and nanodots of CPs.

INTRODUCTION

Nanoparticles (NPs) based on conjugated polymers (CPs) are emerging as multifunctional nanoscale materials that promise great potential to offer exciting opportunities as imaging agent, biosensor, nanomedicine, and photonic and optoelectronic device materials. These conjugated polymer nanoparticles (CPNs) are desirable for a number of reasons. Their properties can be tuned easily for desired applications through the choice of CPs and surface modification. Additionally, their easy synthesis, tuneable properties, and less toxicity and more biocompatibility compared to the existing inorganic NPs can further make these materials highly attractive in the material choice. These particles consist of a multitude of chromophores confined into small dimensions, often on the same order of magnitude as the size of biological membranes. They are luminescent and their surface can be modified to either tether them to an interface or graft functionalities that allow their insertion into living organisms. Packing a large number of chromophores into a small particle while retaining the ability of the molecule to absorb and emit light results in a unique hybrid that combines the advantages of organic dyes and the high brilliance characteristics of inorganic NPs and quantum dots.

The majority of CPs are relatively rigid and their collapsed state is far from the equilibrium conformation.[1] These macromolecules are often forced into the nanodimension by imposing constraints on the polymer backbone, either cross-linking or physically trapping the chains into a confined space. The most common approach to achieve longlasting nanoconfigurations is to cross-link the polymers, where the resulting dimensions depend on the molecular weight of the polymer, its smallest rigid segment, and the number of cross-links.[2] A new fascinating pathway to form highly luminescent NPs is to confine polymers to a NP without cross-linking.[3] These particles are of particular interest since rearrangements of the chromophores can take place. The conformational freedom of the polymer chains leads to a new class of tunable particles. Their unique photophysics[4] is often determined by the conformation of the polymer backbone.

CPs are macromolecules with π-conjugated backbones, which allow the formation of excitons to facilitate photo- and electroluminescence. Various CPs have been reported to show high extinction coefficient and high

Concise Encyclopedia of Biomedical Polymers and Polymeric Biomaterials DOI: 10.1081/E-EBPPC-120050399

fluorescence.[5] Due to their highly delocalized backbone structures and unique electronic and optical properties, CPs have been widely used for electronics and biosensor applications.[6–9] The application of CPs for cellular imaging has also been reported.[10,11]

Polymer nanocapsules have various useful applications. They can act as drug transporters, constricted reaction vessels, shielding casings for enzymes or cells, gene delivery systems, protective shells for heterogeneous catalysis, dye dispersants, or as mediums for the displacement of contaminated waste. For example, a modified oil-in-water emulsion technique was used to fabricate poly[u-pentadecalactone-co-p-dioxanone; poly(PDL-co-DO)] copolyester NPs that were encapsulated with an anticancer drug, doxorubicin (Dox), or an oligonucleotide, small interfering RNA (siRNA).[12] The poly(PDL-co-DO) copolyesters were being investigated as new materials for biomedical applications. Both the Dox- and the siRNA-encapsulated NPs showed a biphasic release profile over many weeks. It was also found that the physical properties and biodegradation rate could be adjusted over a broad range by varying the copolymer composition. Poly(PDL-co-DO) copolymers that are enzymatically synthesized are promising biomaterials.

Usefulness of CPNs in a clinical setting of drug delivery depends on demonstrating that they are non-toxic and non-immunogenic and possess an appropriate biodistribution profile. Another issue to consider regarding drug delivery using polymer NPs is the amount of time the polymer nanosystem circulates in the blood. Normally, the polymer nanospheres are modified with polyethylene glycol (PEG) to increase their circulation time in the bloodstream.[13] Filamentous polymeric micelles known as filomicelles (which are between 22 and 60 nm in diameter and 2–8 mm in length) stay in the blood longer than PEG vesicles and 10 times longer than spheres of similar chemistry.

CPNs are broadly applied in the form of aqueous dispersions for the preparation of coatings and in paints.[14,15] A key step in these applications is film formation upon evaporation of the dispersing medium, usually water. In comparison to solutions of high-molecular-weight polymers in organic solvents, which possess a very high viscosity even at low polymer concentrations, the particle dispersions retain a low viscosity also at high polymer solids contents. This can be beneficial for handling and processing. Particle dispersions are also useful for the generation of highly disperse heterophase materials. Well-defined and monodispersed colloids of semiconducting polymers may form excellent building blocks for self-assembled materials, but these are not readily available. Highly monodispersed particles of a variety of semiconducting polymers were produced with Suzuki–Miyaura dispersion polymerization.[16] For example, polyacrylate films rendered resistant to soiling by silica NPs, which are prepared from dispersions of organic polymer/inorganic composite NPs.[17] Photoluminescence is considered as a specific property of CPs. This entry focuses on their synthetic

methods and their properties and provides some examples to demonstrate their applications.

SYNTHESIS OF CPNS

Preparation of CPNs can be categorized into two major methods, namely, dispersion of already prepared polymers (postpolymerization) and polymerization from monomers in dispersion media (heterophase polymerizations).

Postpolymerization Dispersion

Postpolymerization generation of polymer particle dispersions is also referred to as secondary dispersion. Dispersion of CPs has employed polymer solutions in an organic solvent as a starting point. Particle formation most commonly occurs either by solvent removal from emulsified solution droplets, which requires a solvent immiscible with the continuous phase of the final particle dispersion, or by precipitation of the polymer upon rapidly adding the polymer solution to an excess of the continuous phase, which requires a solvent miscible with the continuous phase.

Emulsion

This is the most common method used in the synthesis of CPNs. Various CPNs have been prepared by miniemulsion, as shown in Fig. 1.[18–22] To prepare CPNs, the polymer is dissolved in a water immiscible organic solvent and then the resulting solution is injected into an aqueous solution of an appropriate surfactant.[23] The mixture is stirred rapidly by ultrasonicating to form stable miniemulsions containing small droplets of the polymer solution. The organic solvent is evaporated to obtain a stable dispersion of polymer NPs in water. The size of NPs could vary from 30 nm to 500 nm depending on the concentration of the polymer solution. However, the droplets could be destabilized by Ostwald ripening as well as the flocculation caused by the coalescence of droplets. To prevent flocculation, appropriate surfactants are used, while Ostwald ripening can be suppressed by the addition of a hydrophobic agent (a hydrophobe) to the dispersed phase. The hydrophobe promotes the formation of an osmotic pressure inside the droplets that counteracts the Laplace pressure (the pressure difference between the inside and the outside of a droplet), preventing diffusion from one droplet to the surrounding aqueous medium. Many CPNs have been prepared from the preformed polymers using an oil-in-water system. However, synthesis of some polymeric NPs starting from monomers in nonaqueous emulsions (oil-in-oil).[24] Cyclohexane as the continuous phase and acetonitrile as the dispersed phase were used while polyisoprene-block-poly(methyl methacrylate; PI-b-PMMA) was utilized as an emulsifying agent. Poly(3,4-ethylenedioxythiophene; PEDOT), polyacetylene, and

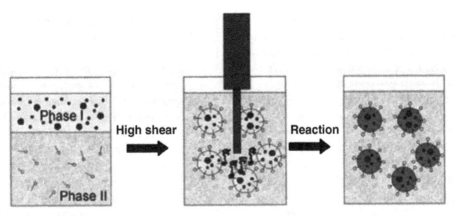

Fig. 1 Miniemulsion method for the preparation of NPs.
Source: From Halls, Walsh, et al.[8] © 2009 Wiley-VCH Verlag GmbH & Co. KGaA.

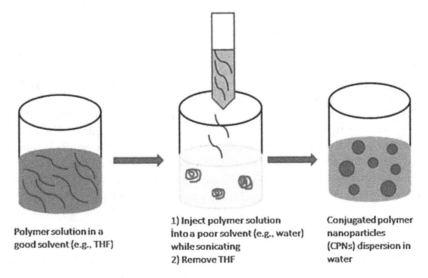

Fig. 2 Nanoprecipitation method for the preparation of polymer NPs.
Source: From Tuncel & Demir[36] © 2010 The Royal Society of Chemistry.

poly(thiphene-3-yl-acetic acid) were prepared as NPs by catalytic and oxidative polymerization using this method. Moreover, after the formation of NPs, the emulsifying agent (PI-b-PMMA) was removed by washing off with tetrahydrofuran (THF). The number-average diameter of the particles is measured as 43 nm (−10 nm). Synthesis of the processable polyacetylene NPs has been carried out by the polymerization of acetylene in an aqueous miniemulsion method.[25] For this purpose, a palladium (Pd) catalyst was dissolved in a minimum amount of a mixture of hexane–ethanol and added into an aqueous solution of surfactant and organic acid (sodium dodecylsulfate and methane sulfonic acid). Subsequently, this mixture was sonicated to form a miniemulsion. An intensely colored, black dispersion of polyacetylene was obtained by stirring the miniemulsion under an acetylene atmosphere. The size of polyacetylene NPs was determined as approximately 20 nm by transmission electron microscopy (TEM).

NPs of poly(arylene diethynylene) (arylene ¼ 2,5-dialkyoxyphenylenes and 9,90-dihexylfluorene) derivatives could also be directly prepared by miniemulsion polymerization of appropriate monomers under Glaser coupling conditions.[26] The molecular weights of these NPs were measured to be in the range of number-average molecular weight 104–105 g mol⁻¹ by gel permeation chromatography and their sizes were determined to be around 30 nm by TEM. They also incorporated covalently 0.1–2 mol% perylene dye and 2–9 mol% fluorenone dye, respectively, to the poly(arylene diethynylene) to obtain conjugated NPs made from these copolymers with the desired emission wavelength through energy transfer.

Nanoprecipitation

In nanoprecipitation method[27–36] presented in Fig. 2, a hydrophobic CP is dissolved in a good solvent (e.g., THF) for the polymer and poured into a poor solvent (e.g., water), which is miscible with the good solvent. The resulting mixture is stirred vigorously usually using a sonicator to assist the formation of NPs. After the NP formation, the organic solvent is removed to leave behind water dispersible NPs. The main driving force for the formation of NPs is the hydrophobic effect. When the solution of polymer in organic solvent is added to water, polymer chains tend to avoid contacting with water and consequently in order to achieve minimum exposure they fold into spherical shapes. The preparation does not involve the use of any additives such as surfactants and hydrophobes and can be applied to a wide variety of CPs that are soluble in organic solvents. Moreover, using this method, it is possible to tune the size of NPs by adjusting the polymer concentration and using polymers with appropriate molecular weights. For example, NPs are produced with 5–10 nm diameter containing single polymer chains.

Synthesis of NPs from poly-(phenylene ethynylene) (PPE) derivatives by phase inversion precipitation[37] involves the dissolution of the polymer in dimethyl sulfoxide and the subsequent addition of this solution into an aqueous SSPE buffer [saline, sodium phosphate, and ethylenediaminetetraacetic acid (EDTA)]. The electron micrographs revealed particles with diameters of 500–800 nm. Dynamic light scattering results show a mean particle size of 400–500 nm. The chemical structure of the polymer, especially the nature and density of the hydrophilic side groups, affected the particles formation. Apparently, the side chain with the protonated amine and the short chain nonionic diethylene oxide moiety stabilize the surface of the forming particle and hinder aggregation and precipitation.

Smaller NPs (as small as 8 nm) from similar hydrophilic CPs can be prepared by tuning the variables in the particle formation process, which include pH, the nature of the acid, salt concentration, and mixing parameters.[38] The resulting NPs were purified by sequential ultrafiltration with acetic acid or tartaric acid, EDTA, and water. Acid-aided removal of metal-ion contamination (Pd and copper) and reduced aromatic backbone aggregation is helpful in the NP formation by generating repulsive forces.

A new drug delivery platform based on a type of NP is self-assembled from a bovine serum albumin (BSA)-PMMA conjugate.[39] BSA, a nutrient to cells, is biodegradable and biocompatible and could be used as a model protein for albumins; PMMA, which is a thermoplastic, is also approved by U.S. Food and Drug Administration for medical applications. Uniform core–shell spherical NPs of BSA-PMMA could be prepared with hydrophobic anticancer drugs being encapsulated in the hydrophobic core (made of PMMA) of the NPs by a simple nanoprecipitation method. The camptothecin-encapsulated BSA–PMMA NPs show enhanced antitumor activity both *in vitro* and in animals. In an animal study, an approximately 79% prohibition of tumor growth was observed when using camptothecin-encapsulated NPs compared with free drug.

Polymerization from Monomers in Dispersion Media

The direct generation in the form of NPs of a polymer during its synthesis from low-molecular-weight monomers, by polymerization in a dispersing medium which is a nonsolvent for the polymer, provides access to a broad scope of NPs in terms of size control and particle structure. Also, polymers entirely insoluble in any solvent are accessible as NPs, which is of particular interest regarding the lack of solubility of many CPs. This low solubility (and infusibility) in general, and the corresponding lack of processability, was indeed a motivation for the first studies of CPNs. During 2000–2010, sterically stabilized NP dispersions of polyacetylene, polypyrrole, and polyaniline were studied intensely, a major aim being the processability to conducting polymeric materials.

The polymerization of EDOT in the presence of polyelectrolyte is employed commercially for the preparation of PEDOT/poly (4-styrenesulfonate) (PSS) dispersions.[40] Some NPs of CPs are of interest for their luminescence, such as poly(phenylenevinylene),[41] poly(phenyleneethynylene),[42] and polyfluorene. Other than polypyrrole, polyaniline, or polythiophene, these polymers cannot be prepared by aqueous oxidative polymerization. Rather, their preparation frequently involves transition metal-catalyzed coupling reactions, which must be compatible with the specific heterophase polymerization for NP synthesis.

In heterophase polymerizations, the miscibility of monomer and polymer decisively impacts the course and outcome of the reaction. For example, in dispersion polymerization, a high solubility of the monomer in the polymer particles, i.e., swelling of the particles by monomer, favors polymerization to occur in the particles once they are formed, rather than in the dispersing medium. In miniemulsion polymerization, a low solubility of the polymer in the monomer will promote phase separation in the droplets. The formation of spherical particles with a thermodynamically favorable minimum surface to volume ratio is favored by miscibility of monomer and polymer, as exemplified by the textbook polystyrene spheres prepared by emulsion polymerization.

A typical feature of the parent unsubstituted representatives of the various classes of CPs is a low miscibility with their monomers, which indeed frequently results in particle shapes other than perfect spheres (note that a nonspherical shape can also result from ordering phenomena such as crystallization). Another feature of many heterophase polymerizations is that the final particles and their number density are determined just as essentially by coagulation of existing particles during polymerization as by nucleation and growth. If the original particles do not completely coalesce in the final particles, this will influence not only their size but also their morphology. This is particularly relevant for polymers not in a rubbery, viscoelastic state during polymerization, which applies to many CPs. Therefore, final particles may consist of more or less strongly bound smaller primary particles, which is not always evident from the analytical data provided.

ENERGY TRANSFER IN CPNS

Energy transfer has been successfully applied in CP blends to enhance the quantum efficiency and to tune the emission color of the light-emitting devices.[43–45] Fluorescence energy transfer [or Förster resonance energy transfer (FRET)] receives increasingly more attention owing to its potential applications in a number of areas such as molecular beacon biosensors and optoelectronic devices. In FRET, the excitation energy of the donor fluorophore is transferred to acceptor fluorophore through a nonradiative pathway. The transfer rate depends on the inverse sixth power of the distance between the donor and acceptor molecules; and for an efficient energy transfer, the emission spectrum of the energy donor should overlap with the absorption spectrum of the energy acceptor.[46] The examples of energy transfer-facilitated sensor applications are DNA and protein sensing.[47–49]

From studies of energy transfer in NPs composed of a blend of different CPs with different band gaps, in addition to FRET, exciton diffusion processes are involved as well.[30] Considering that exciton diffusion lengths of CPs typically range from 5 to 20 nm,[31] which is on the order of the size of the particles studied, such processes can be expected to occur very efficiently. Low amounts (e.g., 1 mol%) of lower energy chromophores, introduced in the form of other CPs or of low-molecular-weight dyes, are sufficient to ensure effective energy transfer from the main component CP, such that emission is only observed from the former.[31] This allows for bathochromic tuning of the emission color of the NPs by blending or covalent incorporation of dye, as shown in Fig. 3.[26,30,31] For a given CP, NPs exhibiting different emission colors can be excited simultaneously with a single laser as a light source, which is of particular interest for simultaneous identification of different analytes and biomedical imaging.

ELECTROCHEMICAL APPLICATIONS

The conductivity of films and other bulk samples generated from NPs of intrinsically conductive polymers is an issue of interest. The ability to detect electrochemical signals from cells, nanopatterned arrays of Cys-(Arg-Gly-Asp)$_4$, RGD-MAP-C (Arg-Gly-Asp-Multi-Armed Peptide-Cys), and Poly-L-Lysine is fabricated on cell chips using nanoporous alumina masks, as shown in Fig. 4.[50] The nanopatterned RGD-MAP-C was the most suitable biomaterial for immobilizing PC12 cells and for enhancing electrochemical signals produced by the cyclic voltammetric and differential pulse voltammetric methods. The redox peak from voltammetric measurement significantly increased. This is very important for sensitive detection of cell viability via electron transfer between cells and the electrode surface. This newly fabricated surface containing RGD-MAP-C nanodots was applied to a cell chip for detecting the effects of polychlorinated biphenyl (PCB), and acute cytotoxicity was sensitively monitored at concentrations of PCB greater than 40 nM. The RGD-MAP-C-patterned surface showed significant superior signal detection compared to that of the other monolayer or patterned peptide surfaces; however, this positive effect needs to be confirmed for a cell line that contains sufficient extracellular matrix proteins and that can attach well on electrode surfaces without additional peptide materials. This peptide nanopatterned electrode on a cell chip can be used for detecting drug effects or for assessing toxicity electrochemically.

For layer-by-layer assemblies[51] of PEDOT NPs stabilized by cationic surfactants with PSS, swelling and shrinking upon oxidation or reduction were studied by electrochemical surface plasmon resonance. In comparison to assemblies of commercial PEDOT/PSS with poly(ethyleneimine), shorter switching times were observed, which was attributed to a better diffusion of charge balancing counterions into the electroactive PEDOT layer.[52]

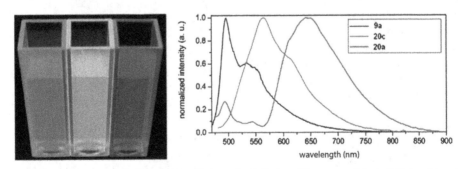

Fig. 3 Dilute aqueous dispersions of NPs: 9a (left) with copolymerized diethynylfluorenone 20c (middle) and copolymerized diethynyl perylene diimide 20a (right) under UV light. Corresponding fluorescence spectra excited at 366 nm.
Source: From Wu, Bull, et al.[29] © 2009 The American Chemical Society.

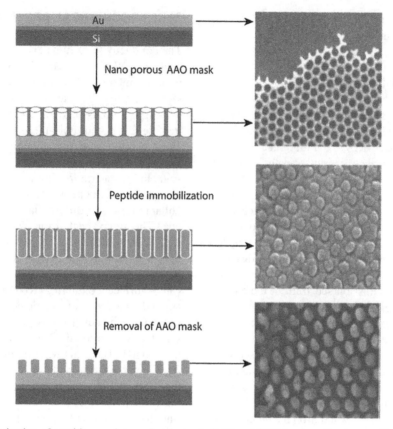

Fig. 4 Mask-assisted fabrication of peptide nanodots on Au electrode (left) and SEM images of corresponding steps (right). **Source:** From Kafi, Kim, et al.[50] © 2010 Elsevier.

OPTOELECTRONIC APPLICATIONS

CPNs made by miniemulsion method have been extensively studied and used in optoelectronic device fabrication. A device that contains a homogeneous single layer of NPs based on a methyl-substituted ladder-type poly(*p*-phenylene) is a light-emitting diode. This NP-based light-emitting diode is reported to exhibit enhanced optoelectronic characteristics (with a lower onset and slightly higher efficiency) compared with organic light emitting diode devices fabricated directly from the CP films.

CP PPE films with well-patterned morphologies can be prepared by tuning the solidification of the crystallizable solvent naphthalene.[53] Controlling the freezing processes play an important role in the preparation of regular morphologies by crystallizable naphthalene. Several strategies have been exploited for the creation of 0-D, 1-D, and 2-D structures, including 0-D nanodots, 1-D-ordered nanodots, 1-D lamellas, and 2-D texture structures through the different freezing direction. The force microscopic (FM) images of the nanodots in Fig. 5 indicated that nanodots are isolated from each other because of the clear fluorescence image of PPE nanodot and black background. The FM images indicate that the size distribution of nanodots prepared at 0°C was more uniform than that prepared at 30°C. CPs film with well-patterned morphologies find applica-

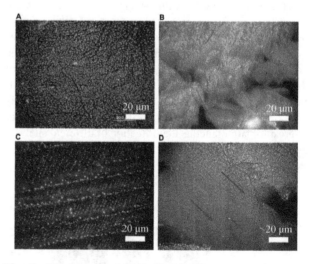

Fig. 5 FM images of PPE films with nanodots structures. (**A**) and (**B**) Film images with disordered nanodots. (**C**) and (**D**) Film images with ordered nanodots. **Source:** From Zuo, Yin, et al.[53] © 2011 Elsevier.

tions in light-energy conversion, catalysis, polymer optoelectronic devices, and so on.

Ferroelectric polymer film with ultrahigh nanodot density has been successfully fabricated through a facile, high-throughput (in half an hour instead of half-day annealing in the case of continuous films), and cost-effective method of imprinting using disposable anodic aluminum oxide mold with orderly arranged nanometer-scale pores, followed by removal of the mold to leave free-standing nanodot arrays.[54]

The nanodots show a large-area smooth surface morphology, and the piezoresponse in each nanodot is strong and uniform. The preferred orientation of the copolymer chains, which are aligned parallel to the substrate, in the nanodot arrays is favorable for polarization switching of single nanodots. This approach allows nanometer electronic feature to be written directly in 2-D by piezoresponse force microscopy (PFM) probe-based technology and can reach a resolution in the order of <10 nm (Fig. 6). Thus, the data storage density of ferroelectric copolymer nanodot arrays can achieve tens or hundreds GB inch^{-2}. The ferroelectric polymer arrays can be operated by a few volts with high writing/erasing speed, which comply with the requirements of integrated circuit.

PHOTOLUMINESCENT TOOL

Brightness and reasonable photostability renders CPNs interesting as probes for cell labeling and bioimaging under both one-photon and especially two-photon excitation. The latter allows for milder deep tissue imaging in the near-infrared spectral range. Obviously, the brightness of a NP when compared to a single low-molecular-weight dye molecule stems in the first place from its larger volume. Utilization of NPs rather than dye molecules does require that the given system under investigation is not disturbed by the probe in an undesirable fashion. As has been demonstrated for NPs of various types of CPs, and different types of cells, the NPs are taken up and accumulated by living cells.[38,55] Toxicity of the NPs was near about zero.[56] This also applies to NPs generated directly by polymerization in an emulsion without intermediate purification of the isolated polymer (Figs. 7 and 8).

Typically, the CPNs accumulate in the cytosol of the cell without penetrating the cell nucleus after being incubated with the living cells for several hours to days (Figs. 7 and 9). Multicolor imaging with different particles varying in emission color allows for ready differentiation of particles located inside and outside the cells.[57] The particles can be imaged by traditional fluorescence microscopy or via multiphoton excitation in the near infrared.[38,57] The photostability of the particles is particularly important considering fluorescence-based imaging techniques, especially for long-term imaging or tracking experiments. The CPNs show no severe photobleaching upon continuous excitation over minutes to hours in a cuvette[55,58] and when imaged in biological systems.[59]

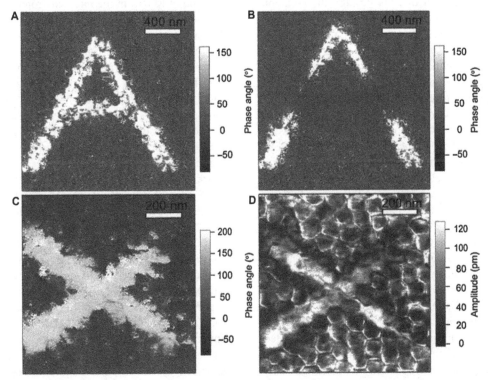

Fig. 6 Demonstration of data memory on the nanoimprinted VDF-TrFE copolymers. (**A**) and (**B**) PFM phase images show that the letter "A" written on the nanodot arrays and then erased by applying a field. (**C**) and (**D**) PFM phase and amplitude images show that the letter "X" written by applying a positive electric field.
Source: From Chen, Li, et al.[54] © 2013 Elsevier.

Conducting—Dendritic

Fig. 7 Confocal fluorescence micrographs of HeLa cells labeled with NPs on the cell surface and inside cells, taken up prior to fixation, excited at 458 nm.
Source: From Chen, Li, et al.[57] © 2010 The American Chemical Society.

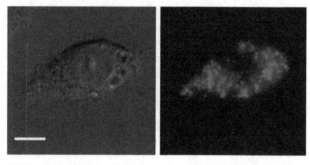

Fig. 8 Differential interference contrast and fluorescence image of a macrophage cell labeled with PPE NPs from reprecipitation.
Source: From Wu, Bull, et al.[58] © 2008 The American Chemical Society.

Longer term tracking of CPN-labeled cells over several days under two-photon excitation has been illustrated in a tissue model. The growth of cells through the collagen gel-based microfluidic device was not stunted by the NPs, proving that they do not affect the cell behavior nor show any toxicity.[38]

Fabrication of dual-modal fluorescent-magnetic NPs with folate receptor-overexpressed have cancer cell targeting ability by co-encapsulation of CP and lipid-coated iron oxides into a mixture of poly(lactic-co-glycolic acid)-poly(ethylene glycol)-folate (PLGA-PEG-FOL) and PLGA. The emission of the CP is designed to fall into far-red/near-infrared (FR/NIR) spectral window. Incorporation of lipid-coated iron oxides into these NPs yielded probes having superparamagnetic properties without sacrificing their fluorescence.[60]

Quantitative analysis using Imaging-Pro Plus software suggests that the average red fluorescence intensity shown in Fig. 9B is approximately 1.4 and 1.3 times brighter as compared to those in Figs. 9A and 10C. 3-D confocal image of cells incubated with Fluorescent-magnetic conjugated polymer nanoparticle (FMCPNPs) shows that the intense fluorescence is mainly from NPs internalized into the cell cytoplasm presented in Fig. 9D. It is noteworthy that no autofluorescence from the cell itself can be detected under the same experimental condition shown in Fig. 9E. The cellular uptake was also quantitatively investigated by flow cytometry. The flow cytometry histograms of MCF-7 cancer cells after incubation with FMCPNPs and MCPNPs (0.25 mg ml^{-1} of NPs) for 4 hrs at 37°C in Fig. 9F. The average fluorescence intensity of each cell incubated with

FMCPNPs is approximately 1.4 times higher as compared to that upon incubation with MCPNPs. It should be noted that more than-99% of the cells are effectively stained by the NPs to show intense fluorescence as compared to the control cells without incubation with NPs.

BIOMEDICAL APPLICATIONS

CP nanosystems will be reviewed in the light of their potential biomedical applications. With high physiological stability, good biodistribution, and long circulation half life, and passive targeting to inflammatory regions, semi-conducting polymer nanoparticle (SPN)-based NIR nano-probe by taking advantage of the reactive oxygen and nitrogen species (RONS) inert property of SPN in conjunction with a RONS-sensitive cyanine derivative allows for detection of RONS in the microenvironment of inflammation using systemic administration. Additionally, its RONS-dependent fluorescence spectral fingerprint in the NIR region permits real-time probe tracking and differentiation of probe activation from accumulation in living mice. This nanoprobe provides more advantages for *in vivo* RONS imaging as compared with "on–off" imaging probes that can only be seen once activated. The current probe does not show specificity over ONOO$^-$, ClO$^-$, and ·OH and is more suitable for detection of the RONS pool at inflammation sites in living animals.[61]

Apply NanoDRONE to detect endogenously generated RONS in cultured cell types relevant to inflammation. RAW264.7, a murine macrophage cell line, showed very weak fluorescence after incubation with NanoDRONE in its resting state in Fig. 10A. To mimic the inflammatory condition that activates resting tissue macrophages, RAW 264.7 cells were successively pretreated with bacterial cell wall lipopolysaccharide (LPS) and phorbol 12-myristate 13-acetate (PMA) to elicit the elevated production of RONS, such as ONOO$^-$ and ClO$^-$.[62,63] With LPS/PMA stimulation, strong fluorescence was observed from the cells as shown in Fig. 10B, indicating the activation of the nanoprobe by RONS under conditions relevant to inflammation. When NAC, a free-radical scavenger with high membrane permeability, was used to treat the cells along with LPS/PMA stimulation, no obvious fluorescence was observed after incubation with NanoDRONE as shown in

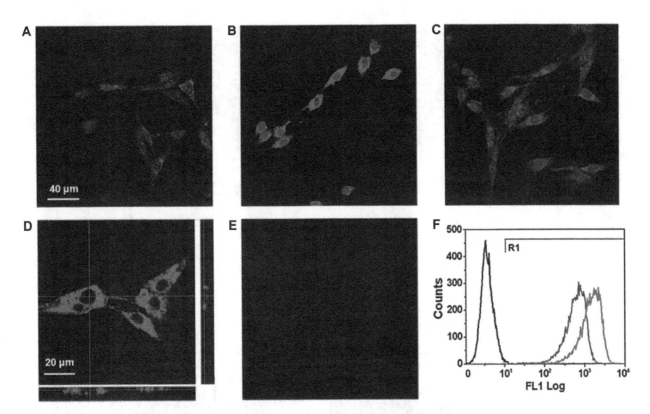

Fig. 9 Confocal images of the MCF-7 breast cancer cells after 4 hour incubation with the MCPNP (**A**) and FMCPNP (**B**) suspension at 0.25 mg ml^{-1} of NPs. (**C**) Confocal image of the free folic acid-pretreated MCF-7 breast cancer cells after 4 hour incubation with 0.25 mg ml^{-1} of FMCPNPs. The nuclei were stained by DAPI. (**D**) 3-D sectional confocal image of the MCF-7 breast cancer cells after 4 hour incubation with 0.25 mg ml^{-1} of FMCPNPs. (**E**) Confocal image of MCF-7 breast cancer cells without incubation with FMCPNPs. (**F**) Flow cytometry line graphs of pure MCF-7 cancer cells (left-most curve) and the cells after 4 hour incubation with the MCPNP (right-most curve) or FMCPNP (middle curve) suspension at 0.25 mg ml^{-1} of NPs. (**A**), (**B**), (**C**), and (**E**) Images have the same scale bar.
Source: From Wu, Bull, et al.[58] © 2008 The American Chemical Society.

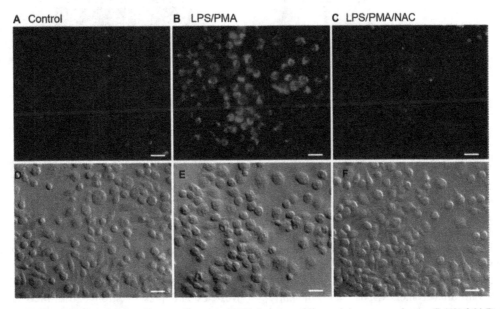

Fig. 10 Fluorescence and differential interference contrast (DIC) images of live murine macrophages (RAW 264.7) incubated with NanoDRONE (1.5 mg ml^{-1}, 3 hours) before imaging: (**A**) nontreated cells, (**B**) cells successively treated with LPS (2 hours) and PMA (0.5 hour), and (**C**) cells pretreated with N-acetylcysteine (NAC) 2 hours before treated with LPS (2 hours) and PMA (0.5 hour), followed with NAC for 1 hour. [LPS] = 1 mg ml^{-1}; [PMA] = 5 mg ml^{-1}; [NAC] = 1 mm. Scale bars: 20 mm.
Source: From Pu, Shuhendler, et al.[61] © 2012 John Wiley and Sons.

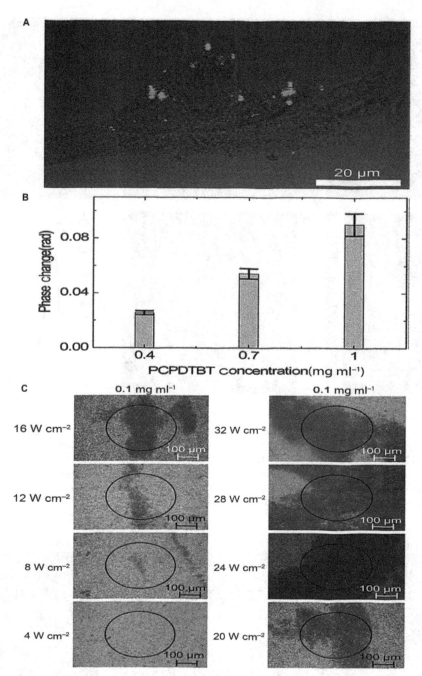

Fig. 11 (**A**) Fluorescence image of HeLa cells after uptaking NPs, (**B**) Optical phase changes with increasing solid contents of PCP-DTBT NPs, and (**C**) photothermal therapeutic effects observed by staining dead HeLa cells with Trypan blue after irradiating them with a laser at 808 nm for 5 min. The circles show the irradiated areas.
Source: From Yoon, Kwag, et al.[64] © 2014 John Wiley and Sons.

Fig. 10C. This indicates that NAC scavenges endogenously generated RONS from macrophage cells and effectively inhibits the activation of NanoDRONE. These *in vitro* data clearly demonstrate that NanoDRONE can efficiently detect RONS produced in stressed cells. Thus, this SPN nanoplatform has the potential to provide real-time, in situ information regarding the RONS status of diseases.

NPs of CPs were prepared from phase-separated nanoassemblies in films of polymers and phospholipids by breaking the films with ultrasonication in water, which is a novel approach for preparing NPs/nanodots of CPs.[64] The resulting lipid-assembled NPs of CPs contain smaller lipid vesicles and present strong absorption at longer wavelengths. The lipid-assembled NPs of CPs can be used as fluorescence imaging probes and photothermal therapeutic agents employing near-infrared light.

The applicability of poly(2,6-(4,4-bis-(2-ethylhexyl)-4-H-cyclopenta[2,1-b;3,A-b']-dithiophene)-alt-4,7-(2,1,3-

benzothiadiazole] (PCPDTBT) NPs was observed by the cellular uptake by confocal microscopy as shown in Fig. 11A, cytotoxicity effect in Fig. 11B, and photothermal therapeutic effects in Fig. 11B and 11C. The cell viability at 0.1 and 0.05 mg ml^{-1} PCPDTBT NP concentrations is comparable to that of the controlled test without NPs. The optical phase differences between the reflections from the top and bottom surfaces varied with increasing NP concentrations as shown in Fig. 11B. These results indicate that heat is generated from the NPs surrounded by water due to radiationless decay and its influence on local densities. *In vitro* experiments using HeLa cell medium and PCPDTBT NPs clearly showed the photothermal therapeutic effect. As shown in Fig. 11C, blue-stained dead cells appeared from a laser power of 8 W cm^{-2} at a NP concentration of 0.1 mg ml^{-1}, proving the photothermal therapeutic effect of PCPDTBT NPs. PCPDTBT NPs show good performance both as a fluorescence imaging agent and as a photothermal therapeutic agent because of their moderate quantum yield.

CPs collapsed into long-lived highly luminescent NPs, or polydots, have opened a new paradigm of tunable organic particles with an immense potential enhancing intracellular imaging and drug delivery.[65]

PLGA NPs have been shown to be an adequate vehicle for the delivery of siRNA.[66] By creating a multifunctional PLGA NP encapsulated with siRNA, target genes could be knocked down and tumor growth could be controlled *in vivo*. Notably, eight separately controlled functions were incorporated in the PLGA NPs. Just as *in vivo* delivery of siRNA is challenging, so is any nucleic acid delivery. Poly(glycidyl methacrylate) is an interesting polymer because its pendant epoxide groups can be opened with different functional groups to fabricate poly(glycerol methacrylate) (PGOHMA) derivatives.[67] Some aminated PGOHMAs readily complexed with an antisense oligonucleotide and show high transfection efficacy. Some water insoluble PGOHMAs could form pH-sensitive nanoassemblies. Due to this property, PGOHMAs represent an alternative to oppositely charged delivery vehicles, such as cationic polymers and lipids traditionally used in nucleic acid delivery.

CONCLUSIONS AND OUTLOOK

NPs based on CPs are highly versatile nanostructured materials. They can find applications in various areas such as optoelectronics, photonics, bioimaging, biosensing, and nanomedicine owing to their straightforward synthesis in desired sizes and properties, biocompatibility, and inherent non-toxicity. A number of different CPs have been used for the formation of CPNs. Although NPs with hydrophilic surface functional groups that can limit the intrinsic hydrophobicity of the CPNs are highly desirable in biological applications, the literature examples are mostly limited to highly hydrophobic CPs carrying no functional groups to

be further modified. However, we believe that this particular problem can be overcome by designing and synthesizing new generation CPNs.

There has also been little effort in developing stable CPNs without using surfactants and hydrophobes and exploring their applications in the area of optoelectronics. Another drawback preventing the exploitation of CPNs is the mechanical instability of these NPs. To this end, the development of CPNs that can be mechanically stable in water as well as in organic solvents and the surface functionalization of these NPs could be highly valuable for many applications.

ACKNOWLEDGMENT

The authors are thankful to Elsevier, John Wiley and Sons, American Chemical Society, and Royal Society of Chemistry for copyright permission.

REFERENCES

1. Maskey, S.; Pierce, F.; Perahia, D.; Grest, G.S. Conformational study of a single molecule of poly para phenylene ethynylenes in dilute solutions. J. Chem. Phys. **2011**, *134* (24), 244906.

2. Beck, J.B.; Killops, K.L.; Kang, T.; Sivanandan, K.; Bayles, A.; Mackay, M.E.; Wooley, K.L.; Hawker, C. Facile preparation of nanoparticles by intramolecular cross-linking of isocyanate functionalized copolymers. J. Macromol. **2009**, *42* (15), 5629–5635.

3. Szymanski, C.; Wu, C.F.; Hooper, J.; Salazar, M.A.; Perdomo, A.; Dukes, A.; McNeill, J. Stabilized core-shell nanoparticles of hydrophobic metal complexes and reprecipation-encapsulation method for preparing same. J. Chem. Phys. B **2005**, *109*, 8543–8546.

4. Wu, C.; Chiu, D.T. Highly fluorescent semiconducting polymer dots for biology and medicine. Angew. Chem. Int. Ed. **2013**, *52* (11), 3086–3109.

5. McQuade, D.T.; Pullen, A.E.; Swager, T.M. Conjugated polymer-based chemical sensors. Chem. Rev. **2000**, *100* (7), 2537–2574.

6. Friend, R.H.; Gymer, R.W.; Holmes, A.B.; Burroughes, J.H.; Marks, R.N.; Taliani, C.; Bradley, D.D.C.; Dos Santos, D.A.; Brédas, J.L.; Logdlund, M.; Salaneck, W.R. Electroluminescence in conjugated polymers. Nature **1999**, *397*, 121–128.

7. Gustafsson, G.; Cao, Y.; Treacy, G.M.; Klavetter, F.; Colaneri, N.; Heeger, A.J. Flexible light-emitting diodes made from soluble conducting polymers. Nature **1992**, *357*, 477–479.

8. Halls, J.J.M.; Walsh, C.A.; Greenham, N.C.; Marseglia, E.A.; Friend, R.H.; Moratti, S.C.; Holmes, A.B. Efficient photodiodes from interpenetrating polymer networks. Nature **1995**, *376* (6540), 498–500.

9. Thomas, S.W.; Joly, G.D.; Swager, T.M. Chemical sensors based on amplifying fluorescent conjugated polymers. Chem. Rev. **2007**, *107*, 1339–1386.

10. Moon, J.H.; McDaniel, W.; MacLean, P.; Hancock, L.F. Live-cell-permeable poly(p-phenylene ethynylene) Angew. Chem. Int. Ed. **2007**, *46* (43), 8223–8225.

11. Kim, I.-B.; Shin, H.; Garcia, A.J.; Bunz, U.H.F. Use of a folate–PPE conjugate to image cancer cells *in vitro*. Bioconjug. Chem. **2007**, *18* (3), 815–820.

12. Liu, J.; Jiang, Z.; Zhang, S.; Liu, C.; Gross, R.A.; Kyriakides, T.R.; Saltzman, W.M. Biodegradation, biocompatibility, and drug delivery in poly(omegapentadecalactone- co-p-dioxanone) copolyesters. Biomaterials **2011**, 32 (27), 6646–6654.

13. Geng, Y.; Dalhaimer, P.; Cai, S.; Tsai, R.; Tewari, M.; Minko, T.; Discher, D.E. Shape effects of filaments versus spherical particles in flow and drug delivery. Nat. Nanotechnol. **2007**, *2* (4), 249–255.

14. Urban, D.; Takamura, K. Polymer Dispersions and Their Industrial Applications; Wiley-VCH: Weinheim, 2002.

15. van Herk, A.M. *Chemistry and Technology of Emulsion Polymerisation*; Blackwell Publishing: Oxford, 2005.

16. Kuehne, A.J.; Gather, M.C.; Sprakel, J. Monodisperse conjugated polymer particles by Suzuki–Miyaura dispersion polymerization. Nature Communications 2012, 3, Article number: 1088. Conjugated Polymers: Nanoparticles and Nanodots 11.

17. Balmer, J.A.; Schmid, A.; Armes, S.P. Colloidal nanocomposite particles: Quo vadis? J. Mater. Chem. **2008**, *18* (47), 5722–5730.

18. Landfester, K. Miniemulsion polymerization and the structure of polymer and hybrid nanoparticles. Angew. Chem. Int. Ed. **2009**, *48* (25), 4488–4507.

19. Landfester, K. Synthesis of colloidal particles in miniemulsions. Annu. Rev. Mater. Res. **2006**, *36*, 231–279.

20. Landfester, K. The generation of nanoparticles in miniemulsions. Adv. Mater. **2001**, *13* (10), 765–768.

21. Landfester, K. Polyreactions in miniemulsions. Macromol. Rapid Commun. **2001**, *22* (12), 896–936.

22. Landfester, K.; Montenegro, R.; Scherf, U.; Guntner, R.; Asawapirom, U.; Patil, S.; Neher, D.; Kietzke, T. Semiconducting polymer nanospheres in aqueous dispersion prepared by a miniemulsion process. Adv. Mater. **2002**, *14* (9), 651–655.

23. Sarrazin, P.; Chaussy, D.; Vurth, L.; Stephan, O.; Beneventi, D. Surfactant (TTAB) role in the preparation of 2,7-poly (9,9-dialkylfluorene-co-fluorenone) nanoparticles by miniemulsion. Langmuir **2009**, *25* (12), 6745–6752.

24. Muller, K.; Klapper, M.; Mullen, K. Synthesis of conjugated polymer nanoparticles in non-aqueous emulsions. Macromol. Rapid Commun. **2006**, *27* (8), 586–593.

25. Berkefeld, A.; Mecking, S. Mechanistic studies of catalytic polyethene chain growth in the presence of water. Angew. Chem. Int. Ed. **2006**, *45* (36), 6044–6046.

26. Baier, M.C.; Huber, J.; Mecking, S. Fluorescent conjugated polymer nanoparticles by polymerization in miniemulsion. J. Am. Chem. Soc. **2009**, *131* (40), 14267–14273.

27. Wu, C.; Szymanski, C.; McNeill, J. Preparation and encapsulation of highly fluorescent conjugated polymer nanoparticles. Langmuir, **2006**, *22* (7), 2956–2960.

28. Wu, C.; Szymanski, C.; Cain, Z.; McNeill, J. Conjugated polymer dots for multiphoton fluorescence imaging. J. Am. Chem. Soc. **2007**, *129* (43), 12904–12905.

29. Wu, C.; Bull, B.; Szymanski, C.; Christensen, K.; McNeill, J. Multicolor conjugated polymer dots for biological fluorescence imaging. ACS Nano **2008**, *2* (11), 2415–2423.

30. Wu, C.; Peng, H.; Jiang, Y.; McNeill, J. Energy transfer mediated fluorescence from blended conjugated polymer nanoparticles. J. Phys. Chem. B **2006**, *110* (29), 14148–14154.

31. Wu, C.; Zheng, Y.; Szymanski, C.; McNeill, J. Energy transfer in a nanoscale multichromophoric system: Fluorescent dye-doped conjugated polymer nanoparticles. J. Phys. Chem. C **2008**, *112* (6), 1772–1781.

32. Wu, C.; McNeill, J. Swelling-controlled polymer phase and fluorescence properties of polyfluorene nanoparticles. Langmuir **2008**, *24* (11), 5855–5861.

33. Peng, H.;Wu, C.; Jiang, Y.; Huang, S.;McNeill, J. Highly luminescent Eu3+ chelate nanoparticles prepared by a reprecipitation–encapsulation method. Langmuir **2007**, *23* (4), 1591–1595.

34. Szymanski, C.;Wu, C.; Hooper, J.; Salazar, M.A.; Perdomo, A.; Dukes, A.; McNeill, J. Single molecule nanoparticles of the conjugated polymer MEH–PPV, preparation and characterization by near-field scanning optical microscopy. J. Phys. Chem. B **2005**, *109* (18), 8543–8546.

35. Kurokawa, N.; Yoshikawa, H.; Hirota, N.; Hyodo, K.; Masuhara, H. Size-dependent spectroscopic properties and thermochromic behavior in poly(substituted thiophene) nanoparticles. Chem. Phys. Chem. **2004**, *5* (10), 1609–1615.

36. Tuncel, D.; Demir, H.V. Conjugated polymer nanoparticles. Nanoscale **2010**, *2* (4), 484–494.

37. Moon, J.H.; Deans, R.; Krueger, E.; Hancock, L.F. Capture and detection of a quencher labeled oligonucleotide by poly(phenylene ethynylene) particles. Chem. Commun. **2003**, (1),104–105.

38. Rahim, N.A.A.; McDaniel, W.; Bardon, K.; Srinivasan, S.; Vickerman, V.; So, P.T.C.; Moon, J.H. Conjugated polymer nanoparticles for two-photon imaging of endothelial cells in a tissue model. Adv. Mater. **2009**, *21* (34), 3492–3496.

39. Ge, J.; Neofytou, E.; Lei, J.; Beygui, R.E.; Zare, R.N. Protein–polymer hybrid nanoparticles for drug delivery. Small **2012**, *8* (23), 3573–3578.

40. Groenendaal, L.B.; Jonas, F.; Freitag, D.; Pielartzik, H.; Reynolds, J.R. Poly(3,4-ethylenedioxythiophene) and its derivatives: Past, present and future. Adv. Mater. **2000**, *12* (7), 481–494.

41. Pecher, J.; Mecking, S. Nanoparticles from step-growth coordination polymerization. Macromolecules **2007**, *40* (22), 7733–7735.

42. Hittinger, E.; Kokil, A.; Weder, C. Synthesis and characterization of cross-linked conjugated polymer milli-, micro-, and nanoparticles. Angew. Chem. Int. Ed. **2004**, 43 (14), 1808–1811.

43. Huebner, C.F.; Roeder, R.D.; Foulger, S.H. Nanoparticle electroluminescence controlling emission color through forster resonance energy transfer in hybrid particles. Adv. Funct. Mater. **2009**, *19* (22), 3604–3609.

44. Luo, J.; Li, X.; Hou, Q.; Peng, J.; Yang, W.; Cao, Y. High-efficiency white-light emission from a single copolymer: Fluorescent blue, green and red chromophores on a conjugated polymer backbone. Adv. Mater. **2007**, *19* (8), 1113–1117.

45. Lee, J.I.; Kang, I.N.; Hwang, D.H.; Shim, H.K.; Jeoung, S.C.; Kim, D. Energy transfer in the blend of electroluminescent conjugated polymers. Chem. Mater. **1996**, *8* (8), 1925.

46. Lakowicz, J.R. Principles of Fluorescence Spectroscopy, 3rd Ed.; Springer: Berlin, 2006.

47. Liu, J.; Xie, Z.Y.; Cheng, Y.X.; Geng, Y.H.; Wang, L.X.; Jing, X.B.; Wang, F.S. Molecular design on highly efficient white electroluminescence from a single-polymer system with simultaneous blue, green, and red emission. Adv. Mater. 2007, 19 (4), 531–535.

48. Chen, L.; McBranch, D.W.; Wang, H.L.; Helgeson, R.; Wudl, F.; Whitten, D.G. Highly sensitive biological and chemical sensors based on reversible fluorescence quenching in a conjugated polymer. Proc. Natl. Acad. Sci. U.S.A. 1999, 96 (22), 12287–12292.

49. Fan, C.H.; Wang, S.; Hong, J.W.; Bazan, G.C.; Plaxco, K.W.; Heeger, A.J. Beyond superquenching: Hyperefficient energy transfer from conjugated polymers to gold nanoparticles. Proc. Natl. Acad. Sci. U.S.A. 2003, 100 (11), 6297–6301.

50. Kafi, M.A.; Kim, T.H.; Yea, C.H.; Kim, H.; Choi, J.W. Effects of nanopatterned RGD peptide layer on electrochemical detection of neural cell chip. Biosens. Bioelectron. 2010, 26 (4), 1359–1365.

51. Decher, G. Fuzzy nanoassemblies: Toward layered polymeric multicomposites. Science 1997, 277 (5330), 1232–1237.

52. Mueller, K.; Park, M.K.; Klapper, M.; Knoll, W.; Muellen, K. Synthesis and layer-by-layer deposition of spherical poly (3,4-ethylenedioxythiophene) nanoparticles—Toward fast switching times between reduced and oxidized states. Macromol. Chem. Phys. 2007, 208 (13), 1394–1401.

53. Zuo, Z.; Yin, X.; Zhou, C.; Chen, N.; Liu, H.; Li, Y.; Li, Y. Organic crystallizable solvent served as template for constructing well-ordered PPE films. J. Colloid Interf. Sci. 2011, 356 (1), 86–91.

54. Chen, X.Z.; Li, Q.; Chen, X.; Guo, X.; Ge, H.X.; Liu, Y.; Shen, Q.D. Nano-imprinted ferroelectric polymer nanodot arrays for high density data storage. Adv. Funct. Mater. 2013, 23 (24), 3124–3129.

55. Howes, P.; Thorogate, R.; Green, M.; Jickells, S.; Daniel, B. Synthesis, characterisation and intracellular imaging of PEG capped BEHP-PPV nanospheres. Chem. Commun. 2009, 18, 2490–2492.

56. Mailander, V.; Landfester, K. Interaction of nanoparticles with cells. Biomacromolecules 2009, 10 (9), 2379–2400.

57. Pecher, J.; Huber, J.; Winterhalder, M.; Zumbusch, A.; Mecking, S. Tailor-made conjugated polymer nanoparticles for multicolor and multiphoton cell imaging. Biomacromolecules 2010, 11 (10), 2776–2780.

58. Wu, C.; Bull, B.; Szymanski, C.; Christensen, K.; McNeill, J. Multicolor conjugated polymer dots for biological fluorescence imaging. ACS Nano 2008, 2 (11), 2415–2423.

59. Moon, J.H.; McDaniel, W.; MacLean, P.; Hancock, L.F. Live-cell-permeable poly(p-phenylene ethynylene). Angew. Chem. Int. Ed. 2007, 46 (43), 8223–8225.

60. Li, K.; Ding, D.; Huo, D.; Pu, K.Y.; Thao, N.N.P.; Yong, H.; Zhi, L.; Liu, B. Conjugated polymer based nanoparticles as dual-modal probes for targeted in vivo fluorescence and magnetic resonance imaging. Adv. Funct. Mater. 2012, 22 (15), 3107–3115.

61. Pu, K.; Shuhendler, A.J.; Rao, J. Semiconducting polymer nanoprobe for in vivo imaging of reactive oxygen and nitrogen species. Angew. Chem. Int. Ed. 2013, 52 (39), 10325–10329.

62. Hashioka, S.; Han, Y.H.; Fujii, S.; Kato, T.; Monji, A.; Utsumi, H.; Sawada, M.; Nakanishi, H.; Kanba, S. Phospholipids modulate superoxide and nitric oxide production by lipopolysaccharide and phorbol 12-myristate-13-acetate-activated microglia. Neurochem. Int. 2007, 50 (3), 499–506.

63. Bergt, C.; Marsche, G.; Panzenboeck, U.; Heinecke, J.W.; Malle, E.; Sattler, W. Human neutrophils employ the myeloperoxidase/hydrogen peroxide/chloride system to oxidatively damage apolipoprotein A-I. Eur. J. Biochem. 2001, 268 (12), 3523–3531.

64. Yoon, J.; Kwag, J.; Shin, T.J.; Park, J.; Lee, Y.M.; Lee, Y.; Park, J.; Heo, J.; Joo, C.; Park, T.J.; Yoo, P.J.; Kim, S.; Park, J. Nanoparticles of conjugated polymers prepared from phase-separated films of phospholipids and polymers for biomedical applications. Adv. Mater. 2014, 26 (26), 4559–4564.

65. Maskey, S.; Osti, N.C.; Perahia, D.; Grest, G.S. Internal correlations and stability of polydots, soft conjugated polymeric nanoparticles. ACS Macro Lett. 2013, 2 (8), 700–704.

66. Zhou, J.; Patel, T.R.; Fu, M.; Bertram, J.P.; Saltzman, W.M. Octa-functional PLGA nanoparticles for targeted and efficient siRNA delivery to tumors. Biomaterials 2012, 33 (2), 583–591.

67. Gao, H.; Elsabahy, M.; Giger, E.V.; Li, D.; Prud'homme, R.E.; Leroux, J.C. Aminated linear and star-shape poly (glycerol methacrylate)s: Synthesis and selfassembling properties. Biomacromolecules 2010, 11 (4), 889–895.

Conducting—Dendritic

Conjugates: Biosynthetic–Synthetic Polymer Based

Jan C. M. van Hest
Organic Chemistry, Institute for Molecules and Materials, Radboud University Nijmegen, Nijmegen, the Netherlands

Abstract

Materials scientists have become increasingly interested in combining natural and synthetic polymers into hybrid polymeric structures. The functionality of (fragments of) biopolymers can be integrated with the synthetic versatility and adaptability of synthetic polymers, thus creating a new class of materials harnessing the best of both worlds. Due to recent developments in synthetic methodologies, the ability to construct well-defined hybrid materials has improved greatly. In this entry, an overview is given of the different approaches used to obtain peptide–polymer hybrids and the applications that are foreseen for this new class of materials. This entry was originally published as "Biosynthetic–Synthetic Polymer Conjugates" in the journal *Polymer Reviews*, Vol. 47, No. 1.

INTRODUCTION

Macromolecules can generally be divided into synthetic and biopolymers, each class with its own specific features. Synthetic polymers are versatile with respect to their composition and architecture. Many different types of monomer building blocks can be used, and with the advance of controlled (radical) polymerization methods a wide variety of topologies, varying from block copolymers to hyperbranched structures, can be prepared in a well-defined manner.[1–4] Synthetic polymers are therefore very adaptable, and can be fine tuned to certain applications. However, although the level of structural control has improved significantly, it is still not possible to have absolute control over molecular weight and monomer sequence.

These specific features are prominently present in biopolymers such as DNA and proteins. The ability of Nature to synthesize perfect monodisperse macromolecules, in which the nucleotide or amino acid sequence is predetermined, makes it possible to store information in these biopolymer chains. In the case of polypeptides this information is translated into a folding process, leading to structural and functional properties attained by the well-defined three-dimensional protein architecture. The subtle relationship between folding and properties makes these biomacromolecules, however, also vulnerable to conformational changes and hence loss of properties during processing of the polymers. A logical development in materials science therefore is to try to design a class of polymers that combines the best of both worlds: the adaptability of synthetic polymers with the structural and functional control of biopolymers.[5–8]

Not only much progress has been made with respect to synthetic polymerization but also in the area of peptide and protein chemistry new techniques have been developed that allow us to build tailor-made (poly)peptides. The use of protein engineering in materials science has become increasingly popular during the last decade.[9] Ligation methods, based on e.g., native chemical ligation or Staudinger ligation, enable us to connect oligopeptide moieties and construct entire proteins.[10–12]

Furthermore, the field of protein or peptide–polymer hybrids has expanded greatly due to the introduction of different bio-orthogonal coupling methods between synthetic and biopolymer moieties. All of these developments have led to a wide variety of peptide–polymer hybrid systems, in which the biological moieties are incorporated in the side chain or main chain of the polymeric species.

This entry will discuss different types of hybrid materials that have been developed. First, polymers with pendant peptide moieties will be discussed, in which a distinction will be made between hybrid polymers with structural and biologically active peptide motifs. The next section will deal with structural and bioactive main chain hybrid systems. The important class of protein–polymer conjugates will be treated separately.

SIDE CHAIN POLYMER HYBRIDS WITH STRUCTURAL MOTIFS

The incorporation of amino acids and peptides in the side chain of synthetic polymers has been a topic of investigation for many years.[13] One of the early examples originates from Morcellet and coworkers, who modified amino acids, such as alanine, glutamic acid, lysine, and asparagine, at the α-amino side with a methacrylamide handle (Fig. 1, **1**).[14] These monomers were subsequently polymerized via a free radical mechanism using AIBN as an initiator. The resulting

Concise Encyclopedia of Biomedical Polymers and Polymeric Biomaterials DOI: 10.1081/E-EBPPC-120052059

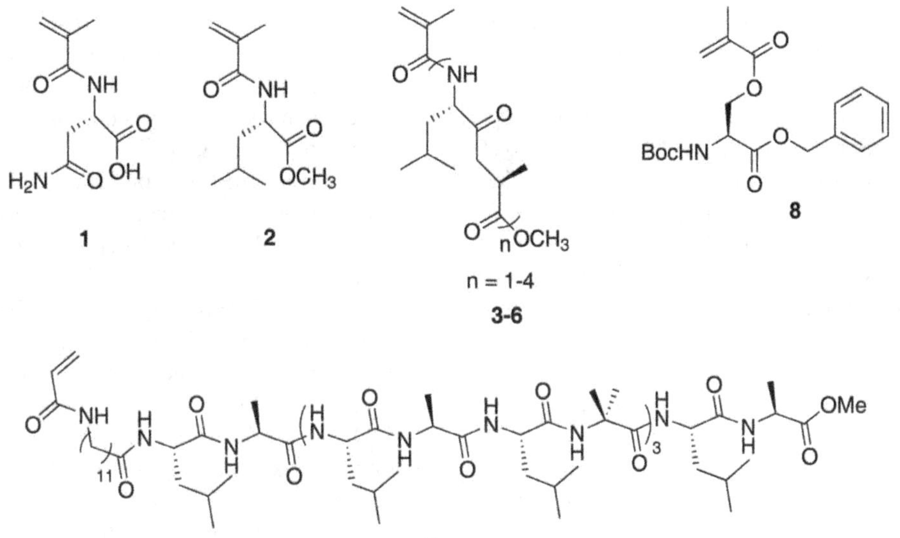

Fig. 1 Amino acid and peptide-based monomers used for free radical polymerization by Morcellet et al.[14] (**1**), Endo et al.[15,16] (**2–7**), and North et al.[17] (**8**).

polymers' ability to complex divalent metal ions, such as Cu(II), was investigated. Motivated by the interesting biological properties of homopolypeptides such as polyleucine, Endo and coworkers used a similar approach to construct polymers containing leucine (**2**), leucyl–alanyl repeats, and leucyl–alanylglycine moieties in the side chain. The effect of peptide length on free radical polymerization was examined by polymerizing one to four repeats of the leucine–alanine sequence (**3–6**).[15] It was observed that all peptides could be conveniently polymerized except for the octapeptide, which only resulted in a degree of polymerization of 3.5. This lower polymerizability was tentatively attributed to the aggregation of the octapeptide monomer, due to hydrogen bonding. A low degree of polymerization (DP) (**3–8**) was also observed for the polymerization of a methacrylamide-functional hexadecapeptide (**7**) based on three repeats of leucine–alanine–leucine–aminoisobutyric acid. This oligopeptide monomer exhibited an α-helical conformation, which was maintained in the oligomer when the polymerization was conducted in chlorobenzene.[16]

North et al. produced amino acid containing polymers by copolymerizing methyl methacrylate with methacrylate-functional serine derivatives.[17] In these experiments, the monomeric handle was connected to the hydroxyl side chain of the amino acids (**8**). They observed a remarkable non-linear dependence of optical rotation of the final polymers, which changed from a positive optical rotation at low incorporation levels of the amino acid monomer to a negative rotation at increased levels.

Introduction of chirality in synthetic polymers was also achieved for polyolefins by Wagener et al. using acyclic diene metathesis polymerization (ADMET) of amino acid and dipeptide containing diene monomers.[18,19] It was observed that the crystallinity and melting temperature of the polymers was increased by increasing the polarity of the amino acid and peptide side chains.

Polyacetylenes and polyisocyanides decorated with amino acids and peptides in the side chain have been prepared with the aim of introducing chirality into the polymer and to affect the conformation of the polymer backbone. Tang and coworkers used leucine-functionalized phenyl acetylene derivatives for the construction of amphiphilic helical polymers (**9, 10**).[20] The polymerization was performed with a rhodium catalyst and resulted in high molecular weight polymers. Only the polymers in which the stereocentre was closely located to the helical backbone (**9**) showed a CD signal and were optically active. The conformations of these polyacetylenes proved to be sensitive to solvent effects. In the case of a polar solvent such as methanol the hydrogen bonding interactions between the leucine moieties were disrupted and the helical structure of the polymer was lost. In the case of a nonpolar solvent intra- and intermolecular interactions remained possible, yielding fiber-like aggregates of helical polymers with an excess of one-handedness. Similar results were found by Masuda and coworkers for polyacetylenes prepared from L-alanine-*N*-propargyl amide (**11**).[21] They also demonstrated that copolymers of D- and L-alanine-*N*-propargyl amide followed the "majority rules" concept, in which the excess of chiral monomers determines the handedness of the helical polymer backbone.[22] Contrary to the previous examples, threonine (**12**) and serine-based polyacetylenes showed an increase in helical content when the MeOH content of the solvent was increased.[23] This unexpected result was explained by the participation of the hydroxyl groups in hydrogen bonding, leading to a helical structure with a shorter pitch. pH-dependent helicity could furthermore be observed in polyacetylenes based on glutamic acid.[24]

Yashima and coworkers prepared poly(phenyl acetylene)s with poly(glutamic acid) (PLGA, **13**) or poly(benzyl glutamate) (PBLG) as side chains.[25] These polypeptides are known to adopt an α-helical or random coil conformation, depending on pH (PLGA) or solvent (PBLG). Switching the pendant polypeptide from one conformation to the other induced a helix-to-helix transition of the polyacetylene backbone (Fig. 2).

Alanyl–alanine dipeptides, modified at the N-terminus with an isocyanide moiety, were polymerized using a Ni catalyst into β-helical polyisocyanopeptides. It was found that these polymers formed rigid rods with extremely long persistence lengths. This rigidity was caused by the formation of β-sheets via hydrogen bonding between the alanines in the side chain.[26,27] The same group also investigated amphiphilic block copolymers containing a poly(styrene)

tail and a charged poly(isocyanide) headgroup, derived from isocyano-L-alanine-L-alanine and isocyano-L-alanine-L-histidine (**14**). This type of rod–coil block copolymers formed micelles, vesicles, and superhelical structures in aqueous solution (Fig. 3).[28,29]

Ayres et al. used the controlled radical polymerization method atom transfer radical polymerization (ATRP) to prepare ABA block copolymers, which contained pendant peptides derived from structural proteins such as silk (**15**) and elastin (**16**) (Fig. 4).[30–32] Regarding the silk-based polymers, the centre B block was prepared using a bifunctional initiator and a methacrylate-functional tetrapeptide (Ala-Gly-Ala-Gly).[30] The poly(methyl methacrylate) (pMMA) A blocks were introduced via an in situ macroinitiation process. The alanylglycine peptide was derived from the antiparallel β-sheet forming domain found in silkworm silk. It was observed that the biomimetic triblock copolymers indeed exhibited similar β-sheet properties to the natural structural protein.

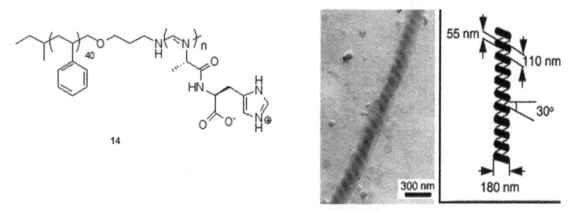

Fig. 2 Poly(phenyl) acetylenes bearing amino acid and oligopeptide side chains.
Source: From Cheuk et al.,[20] Gao, Sanda, and Masuda,[21] Sanda, Araki, and Masuda,[23] and Maeda, Kamiya, and Yashima.[25]

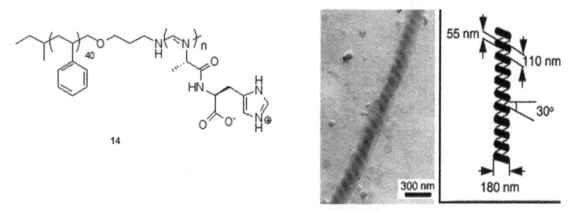

Fig. 3 Helical superstructures formed by amphiphilic block copolymer 14 consisting of polystyrene and polyisocyanide dipeptide.
Source: From Cornelissen et al.[26] © 1998, with permission from AAAS.

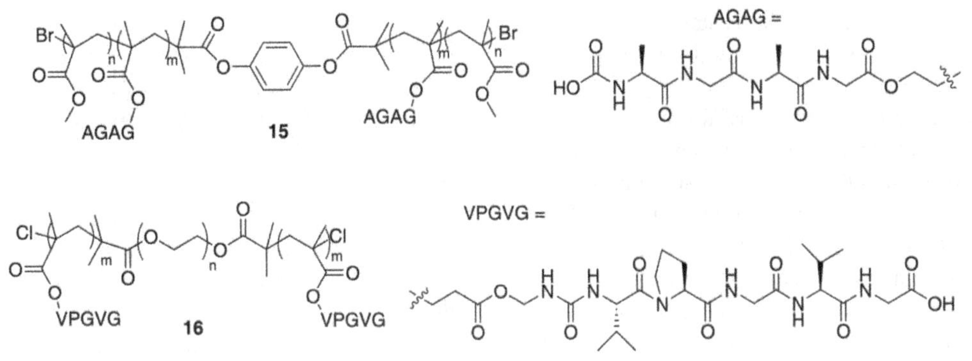

Fig. 4 Side chain peptide block copolymers mimicking structural proteins.
Source: From Ayres et al.[30–32]

Elastin-mimetic triblock copolymers were built up of a central PEG block, flanked by polymethacrylate B blocks, which contained the pentapeptide Val-Pro-Gly-Val-Gly (VPGVG) in the side chains.[31,32] This sequence was derived from the polypeptide poly(VPGVG), which is a mimic of tropoelastin, the precursor protein of elastin. Poly(VPGVG) shows a well-defined reversible lower critical solution temperature (LCST) behavior, which is a result of an entropically driven conformational change of the peptide from random coil to β-spiral upon heating. This transition is an intrinsic property of the pentapeptide sequence. This was demonstrated by the fact that the ABA block copolymer mimic showed similar LCST behavior to poly(VPGVG), with respect to molecular weight and pH dependence. For both silk- and elastin-mimetic polymers, it was therefore observed that the display of multiple peptides along a polymer backbone allowed the introduction of physical properties that are similar to the linear polypeptide analogs.

SIDE CHAIN POLYMER HYBRIDS WITH BIOLOGICALLY ACTIVE MOTIFS

The assembly of multiple peptide moieties along a polymer backbone is interesting from a structural point of view. Also multiple interactions are crucial in many biological processes,[33,34] and the display of bioactive peptides in the polymer side chain therefore can be beneficial for cooperative binding events. Jackson and coworkers have used this concept to create multiple antigenic peptides.[35] A variety of peptide epitopes, derived from species such as the influenza virus and the Malaria parasite, were synthesized in the solid phase and subsequently N-terminally functionalized with an acrylamide monomer unit. The acryloylated peptides were copolymerized with a 50-fold excess of acrylamide in guanidine-HCl using free radical polymerization initiated by ammonium persulfate and tetramethyl ethylene diamine. The concept of these polymeric multiple antigenic peptides was clearly demonstrated. Not only could all polymer-conjugated epitopes still bind their complementary antibody but also the effect of cooperativity resulted in an increase in antigenicity when compared to the free epitopes.

One of the best studied applications of side chain polymer peptide conjugates is polymer drug delivery. This concept was first postulated by Ringsdorf, to overcome problems associated with therapies based on small organic chemotherapeutics, such as lack of selectivity, low solubility, and rapid excretion from the body.[36,37] A polymeric carrier to which the (deactivated) drugs are attached in the side chain can improve solubility and decrease excretion rate, whereas a higher selectivity can be obtained when tumor tissue is targeted by the enhanced permeation and retention (EPR) effect, as discovered by Maeda and coworkers.[38,39] EPR occurs because the leakiness of the tumor cardiovascular system allows large molecules to penetrate tumor tissue more easily than healthy tissue. The ineffective tumor lymphatic drainage furthermore results in subsequent accumulation of these polymers at this specific site.

The release of the drugs from the polymer offers a second option for introduction of selectivity. When, e.g., specific peptide sequences are introduced between the drug and the polymer that can only be cleaved by enzymes present in tumor tissue, release and drug activity are localized at the desired site. Duncan and Kopeček have developed a series of side chain conjugated polymers based on this principle.[40–44] As polymeric carriers, copolymers of N-(2-hydroxypropyl)methacrylamide (HPMA) and methacrylamide-functional peptides were applied, which were prepared by free radical polymerization. Poly(HPMA) is an often employed polymer in biomedical applications, since it is water soluble, non-toxic, and non-immunogenic. The peptide moieties were activated at the C-terminus with a nitrophenyl ester, which made it possible to couple in a post-polymerization reaction the drug doxorubicin to the polymeric carrier. Many peptide sequences were tested and the tetrapeptide glycine–phenylalanine–leucine–glycine (Gly-Phe-Leu-Gly) appeared to be the most suitable degradable spacer, since it was readily cleaved by lysosomal proteases like cathepsin B, which play an important role in

tumor growth and formation of metastasis. These polymeric drug delivery systems are currently in the stage of
clinical trials (Fig. 5A).

Besides passive targeting via the EPR effect, additional
targeting moieties can be attached to enhance the selective
recognition of the conjugate.[45–47] These targeting moieties
range in complexity from carbohydrates and peptides to
hormones and antibodies (Fig. 5B).[48] In targeting cancer
there are two mechanisms, namely the targeting of tumor
cells and the targeting of tumor vasculature. Tumor cell targeting is based on the recognition of receptors or antigens
present in tumor cells. For this purpose, an HPMA-doxorubicin conjugate bearing pendant galactosamine units that
function as targeting ligands for hepatocytes was developed to treat liver cancer. Both drug and saccharide were
connected to the polymer via the abovementioned tetrapeptide spacer. This conjugate, better known as PK2, has also
entered the stage of clinical trials.[49,50]

A second approach is targeting of tumor vasculature.
An important control point in cancer is the formation of
new capillary blood vessels from the pre-existing vasculature, a process referred to as angiogenesis.[51] The process
of angiogenesis involves many growth factors with accompanying receptors, cytokines, proteases, and adhesion
molecules.[52,53] Therefore, many targeting possibilities for
antiangiogenic therapy for cancer exist,[54,55] one of
which is the $\alpha_v\beta_3$ integrin.[56] This marker could be targeted by attachment of a high affinity cyclic peptide ligand
containing the active Arg-Gly-Asp (RGD) sequence to a

copolymer of HPMA. The site of attachment was in this
case a diglycine moiety, which was incorporated into the
polymer via a similar random free radical polymerization
as described for the previous examples. This copolymer
displayed significant tumor localization of the conjugate
in vivo.[57]

The RGD domain is a very attractive peptide, since it is
a generic cell adhesion promoter, which can be found in
extracellular matrix proteins such as fibronectin. It has
therefore been studied extensively for cell culturing and
tissue engineering purposes. Because also in this case
cooperativity plays an important role in binding efficiency,
different strategies have been developed to incorporate
multiple RGD peptides in the side chain of a polymer.
Deng et al., for example, prepared an RGD-modified triblock copolymer of PEG-poly(lactic acid) and PLGA. In
the presence of methoxy PEG, L-lactide was ring opened to
yield the diblock structure.[58] The hydroxy group of the
poly(lactic acid) chain end was converted into a primary
amine by esterification with phenylalanine. This diblock
copolymer was used for the polymerization of the N-carboxy anhydride of γ-benzyl glutamate. After the triblock
polymer was synthesized, the benzyl groups were removed,
and the RGD peptides were connected to the poly(glutamate)
side chains. Coatings were made of blends of this block
copolymer with high molecular weight PLGA. Cell binding studies showed an improved adhesion and proliferation
on RGD-modified surfaces. A similar approach was followed by Hubbell and coworkers, who used a poly(lysine)

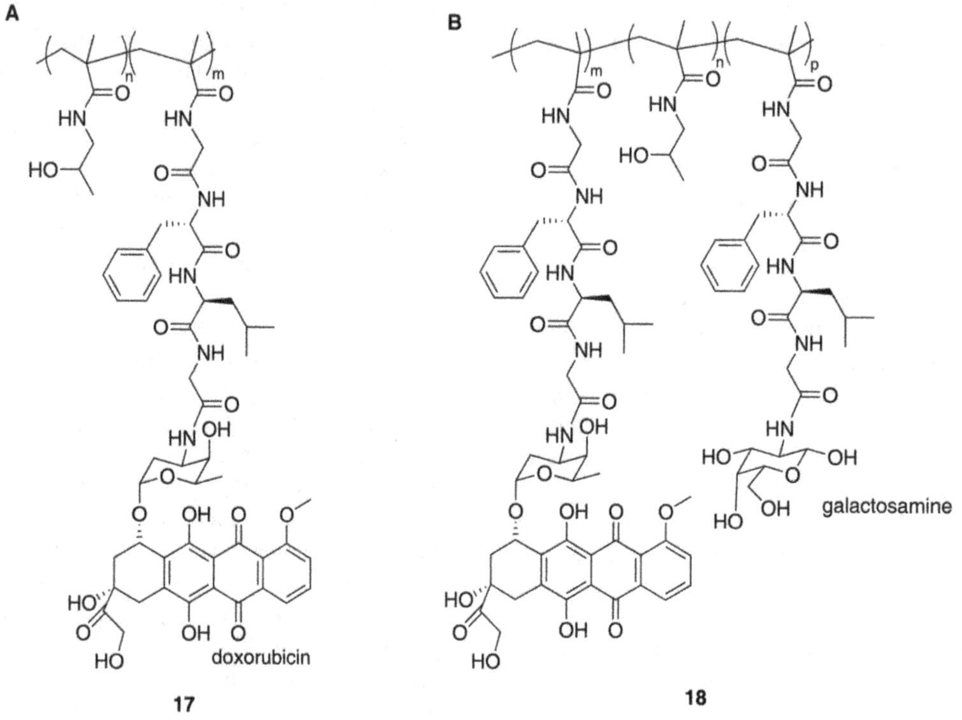

Fig. 5 HPMA-doxorubicin conjugates with cleavable peptide spacers (**A**): 17 (PK1) and (**B**): 18 (PK2), containing the additional liver
targeting moiety galactosamine.

backbone to which short and long PEG chains were connected (Fig. 6, **19**).[59] The longer PEG chains could be further functionalized with RGD domains. The polylysine structure enabled polymer attachment to the surface, the PEG chains prevented non-specific protein adsorption, whereas the RGD domains promoted cell adhesion.

Besides the number of interactions, also the distance and orientation of RGD ligands play an important role in efficient cell binding. Kessler and coworkers investigated these properties by the construction of switchable cyclic RGD domains.[60] The cyclic peptides were coupled to a photoswitchable azobenzene moiety, which was functionalized with an acrylamide monomer handle. The monomers were polymerized photochemically into coatings on pMMA substrates. When the azobenzene moieties had the *E* configuration, enhanced cell binding was observed. When switched to the *Z* isomer, the substrate was more repellent and showed in some cases similar (lack of) adhesion to the untreated pMMA surface.

A third interesting aspect of RGD activity is that cell binding can be enhanced in the presence of a synergistic peptide sequence Pro-His-Ser-Arg-Asn (PHSRN) when these two moieties are in close proximity to each other (approximately 30–40 Å), which could be achieved by incorporating both peptides in the same polymer. The copolymerization of RGD and PHSRN monomers was therefore undertaken by Maynard, Okada, and Grubbs to construct polymers with improved binding efficiency.[61] The polymerization method used was ring-opening metathesis polymerization. For this purpose, both peptides were N-terminally coupled to a norbornene moiety. Studies on inhibition of cell binding to a fibronectin-coated surface were conducted and the synergistic effect of the presence of both multiple RGD copies and both types of peptides along a polymer backbone was demonstrated.

MAIN CHAIN POLYMER HYBRIDS WITH STRUCTURAL MOTIFS

β-Sheet Containing Polymer Hybrids

One of the most common peptide motifs that have been introduced in polymer peptide hybrid materials is the β-sheet. Reasons for this are the synthetic accessibility of β-sheet forming peptides, small sequences already display the desired properties, and their relevance in structural proteins such as silk, in which (anti-parallel) β-sheets are

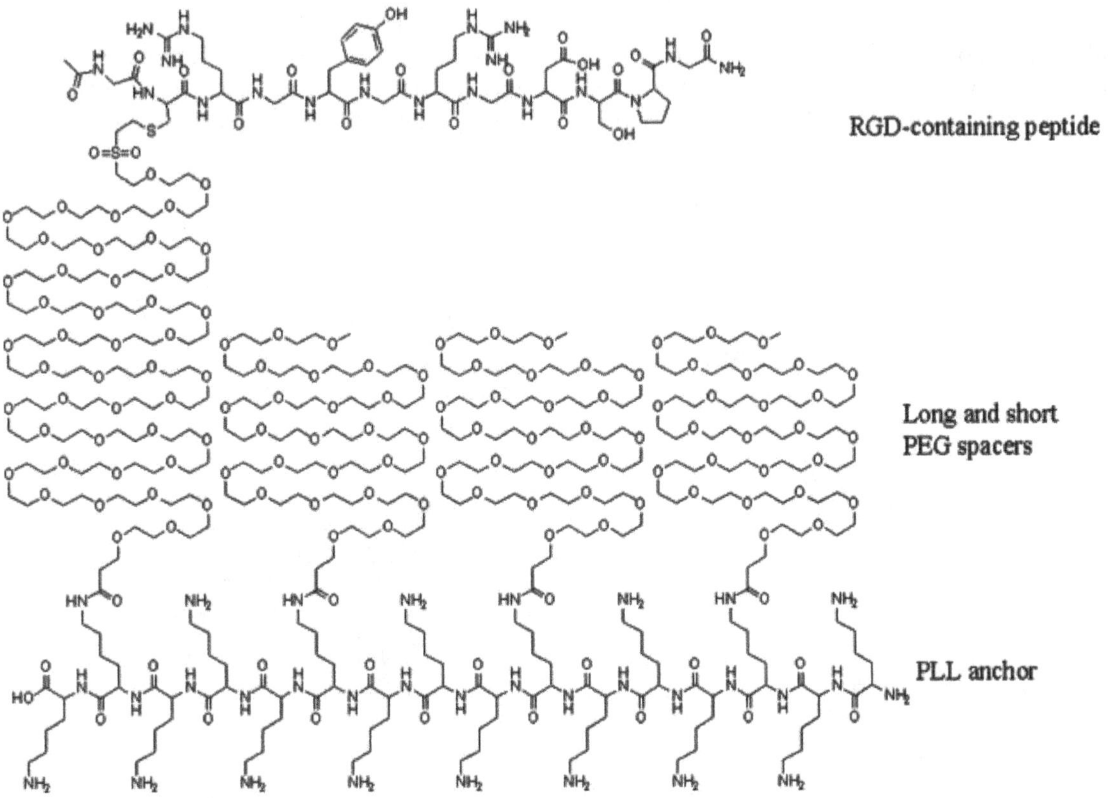

19

Fig. 6 Multifunctional hybrid block copolymer for surface functionalization with RGD domains.
Source: From VandeVondele, Voros, and Hubbell[59] © 2003, with permission from Wiley-VCH.

responsible for crystallinity and hence stiffness of these biopolymers.

An interesting example of a silk-mimetic polymer was developed by Sogah and coworkers.[62,63] They produced hybrid multiblock copolymers, consisting of β-sheet forming peptide sequences, which were connected via PEG chains. Two different types of peptide repeats were used, based on either poly(alanylglycine), derived from silkworm silk, or poly(alanine), commonly found in dragline spider silk. Whereas the peptides mimicked the crystalline part of the silk protein, the PEG chains were introduced to mimic the amorphous, tough protein matrix. Multiblock copolymers were prepared via two approaches. First, two Gly-Ala-Gly-Ala tetrapeptides were coupled via an aromatic hairpin residue to force formation of parallel β-sheets (Fig. 7A). The preformed hairpins were then connected via a condensation reaction with PEG chains. In the second approach, Ala-Gly-Ala-Gly tetrapeptides were coupled to the PEG chains to afford a fully linear system in which the β-sheet segments were free to form intra- and intermolecular parallel or antiparallel β-sheets (Fig. 7B). A linear system was also obtained for oligoalanine-PEG multiblock copolymers. A telechelic hydroxy-functional PEG initiator was used for the anionic ring-opening polymerization of the N-carboxy anhydride of alanine, to a DP of 4–6. The resulting triblock copolymers were connected via telechelic carboxylic acid functional PEGs to afford multiblock polymers.

For all designed polymers, a microphase separated morphology was observed with 20–50 nm β-sheet peptide domains dispersed in a continuous PEG phase. It was also demonstrated that the spider silk-derived oligoalanine containing block copolymers had superior mechanical properties when compared to the silkworm silk-inspired Ala-Gly tetrapeptide containing structures, which reflects the difference in mechanical properties of the natural structural proteins. Similar work was reported by Shao et al.[64]

The β-sheet motif is not only of interest for the introduction of mechanical properties in materials but has also been studied with respect to its potential to direct assembly of polymers. Klok and coworkers applied the Ala-Gly sequence in the construction of a peptide–oligothiophene conjugate.[65] A head-to-tail coupled tetra-3-hexylthiophene was connected to a Gly-Ala-Gly-Ala-Gly pentapeptide via solid-phase peptide chemistry (Fig. 8, 20). This conjugation did not hamper the natural assembly behavior of the peptide, leading to fibrous structures, which could potentially be useful for the construction of organic semiconducting films with a higher level of alignment.

Quite a number of examples have been reported in which PEG is connected to β-sheet forming sequences. An amphiphilic β-sheet peptide, composed of alternating polar and non-polar amino acids, was coupled to PEG on either the C terminus or both C and N termini via a solid-phase peptide chemistry approach (21).[66] In the solid state, superstructures consisting of alternating PEG layers and antiparallel β-sheet ribbons could be detected. Börner and coworkers also used solid-phase peptide chemistry to create a PEG-β-hairpin diblock copolymer (22).[67] The hairpin consisted of two linear peptides composed of a Thr-Val repeat sequence, which were connected to the PEG chain via a hairpin-forming template. It was demonstrated that

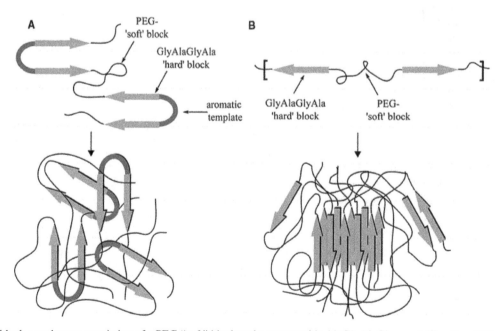

Fig. 7 Multiblock copolymers consisting of a PEG "soft" block and a tetrapeptide AlaGlyAlaGly, crystalline "hard" block in two variants: (**A**) Templated system in which an aromatic hairpin turn is used to force parallel β-sheet formation. (**B**) Non-templated system in which peptide segments are free to form parallel and/or antiparallel β-sheets.

Source: From Rathore & Sogah[63] © 2001, with permission from American Chemical Society.

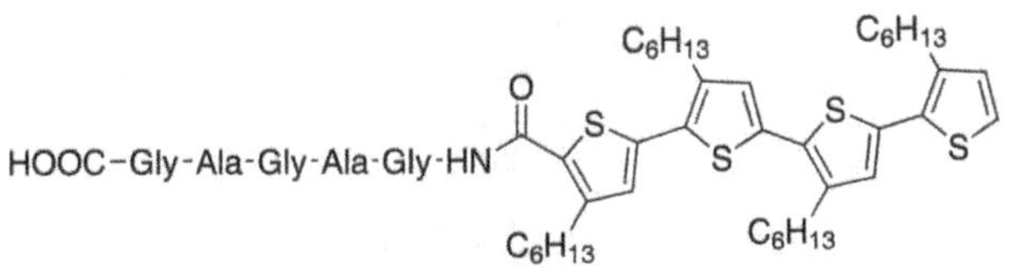

20

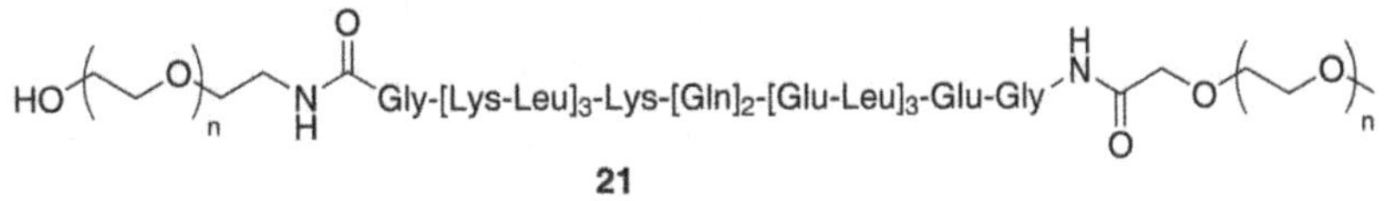

21

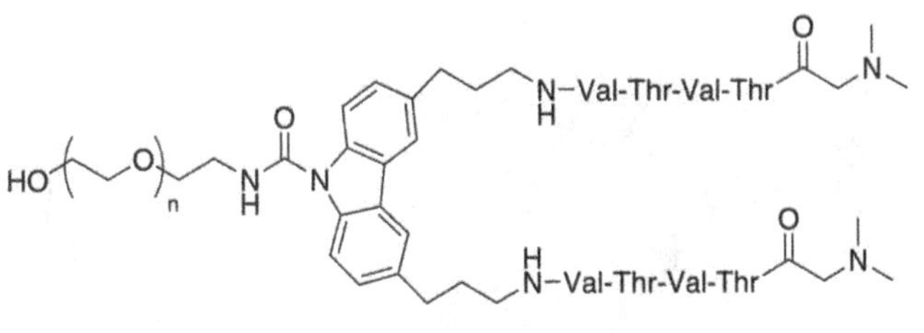

22

Fig. 8 Self-assembling β-sheet hybrid block copolymers.
Source: From Klok et al.,[65] Rösler et al.,[66] and Eckhardt et al.[67]

the preformed hairpin structures had a stronger tendency to aggregate than the linear analogs. In solution micrometer long fibers were formed.

Smeenk et al. used a combination of protein engineering and bioconjugation for the preparation of triblock copolymers consisting of a central β-sheet polypeptide block and PEG end blocks.[68,69] The polypeptide consisted of repeats of the sequence (Ala-Gly)$_3$-Glu-Gly, which was flanked by cysteine residues at both the N- and C-terminal sides. Maleimide-functional PEG was attached to the polypeptide using the cysteine moieties (Fig. 9). The rationale behind attachment of synthetic polymer blocks at the N- and C-termini was to restrict macroscopic crystallization and to preserve translation of the β-sheet design characteristics of width, height, and surface functionality into self-assembled structures. It was shown that the attachment of PEGs of different lengths, varying from 750 to 2000 g/mol, resulted in the formation of well-defined fibrils of multiple micrometers in length, as determined by transmission electron microscopy and atomic force microscopy (Fig. 9B–D). Only when PEGs of 5000 g/mol were connected, the fibers were considerably shortened in length due to steric hindrance effects.

In some cases, aggregation of β-sheet peptides is an undesired feature. This is particularly true for fibril formation of β-amyloid peptide, the primary component of amyloid plaques in Alzheimer's disease. Burkoth et al. investigated the possibility to affect assembly of the C-terminal part of the central domain of β-amyloid peptide (Aβ$_{10-35}$) via modification of this peptide with PEG of 3000 g/mol using standard Fmoc solid-phase peptide synthesis.[70,71] Although aggregation still occurred after attachment of the PEG chains, peptide assembly, unlike in

the case of the native peptide, became completely reversible and single fibers were prevented from forming bundles.

Another type of β-sheet forming peptides is represented by cyclic peptides composed of alternating D and L amino acids (Fig. 10). These peptide rings form tube-like structures through self-assembly via an extended network of hydrogen bonds, preferably via antiparallel β-sheets. In order to affect the surface properties of the tubes, Biesalski and coworkers modified lysine residues in the peptide rings with α-bromo amide moieties, which could be used as an initiator for ATRP of N-isopropyl acrylamide (NIPAAM).[72] Polymerization was performed from the preformed tubes.

The resulting polymer–peptide conjugates yielded distinct rod-shaped objects, which, contrary to the unmodified peptide tubes, did not have the tendency to aggregate into two-dimensional assemblies.

Mussel Adhesive Polymer Hybrids

Another structural protein that has been a source of inspiration for materials scientists is the mussel adhesive protein. This protein enables the blue mussel *Mytilus edulis* to thoroughly adhere itself under aqueous saline conditions to the substrates upon which it resides. Analysis of this protein has shown it to consist of a decapeptide repeat sequence Ala-Lys-Pro-Ser-Tyr-DHP-Hyp-Thr-DOPA-Lys (DHP = dihydroxyproline, Hyp = hydroxyproline, and DOPA = dihydroxyphenylalanine). The presence of DOPA, in particular, seems to be critical for efficient attachment to the surface. Messersmith and coworkers have created polymer–peptide conjugates based on this concept in order to introduce adhesive properties to polymers (Fig. 11, **23**). PEGs of

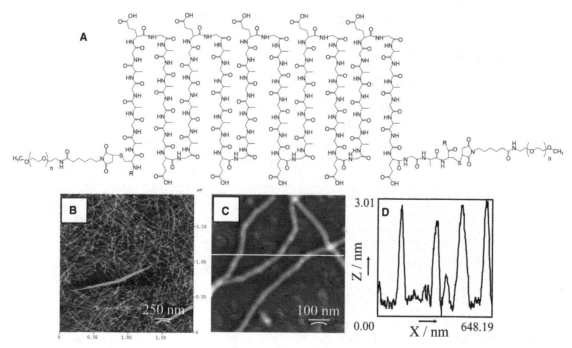

Fig. 9 (**A**) Structure of PEG β-sheet triblock copolymers developed by Smeenk et al.[69] (**B**) and (**C**) AFM images of fibers of a hybrid triblock copolymer with 10 hairpin repeats and PEG 2000 g/mol; (**D**) height image of (**C**).
Source: From Smeenk et al.[69] © 2006, with permission from American Chemical Society.

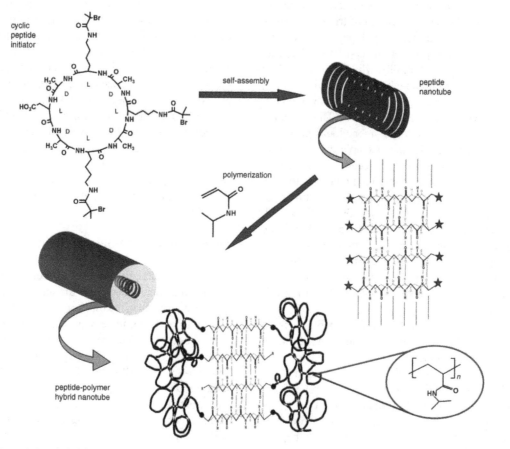

Fig. 10 Peptide–polymer hybrid nanotubes.
Source: From Couet et al.[72] © 2005, with permission from Wiley-VCH.

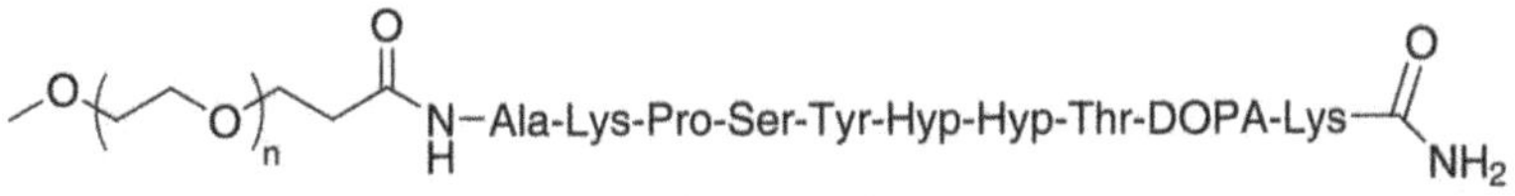

23

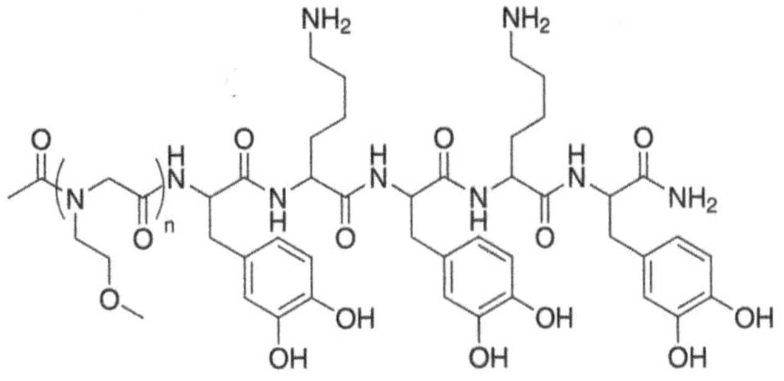

24

Fig. 11 Mussel adhesive protein mimetic polymers used in antifouling applications. Hyp = hydroxyproline, DOPA = dihydroxyphenylalanine.
Source: From Dalsin et al.[73] and Statz et al.[74]

2000 and 5000 g/mol were connected to either the decapeptide (in which DHP was replaced by Hyp) or to a single DOPA residue.[73] Ti and Au surfaces were covered with the modified PEG chains. It was observed that the decapeptide-functional PEGs provided excellent resistance against cellular attachment on both substrates, whereas DOPA-functional PEG only functioned well on Ti. Comparison with non-binding tyrosine-functionalized PEG demonstrated that DOPA was crucial to introduce sufficient adhesion strength in PEG. Similar antifouling properties could be obtained by the peptoid copolymer (**24**) or by first attaching a DOPA-mimetic α-bromo amide ATRP initiator to the substrate, followed by surface-initiated polymerization of oligo ethylene glycol methyl ether methacrylate.[74,75]

Helical Peptide–Polymer Hybrids

The development of controlled polymerization methods for the ring-opening polymerization of N-carboxy anhydrides (NCAs) by Deming, Schlaad, and Hadjichristidis has intensified research in the field of block copolymer hybrids containing homopolypeptide fragments.[4,76,77] Since the α-helical conformation is one of the most frequently observed folding patterns in homopolypeptides such as poly(leucine), (Z-protected) poly(lysine), and poly (γ-benzyl glutamate) (PBLG), the properties of hybrid materials containing this secondary structural element in particular have been extensively studied. Deming and coworkers reported on the synthesis of PBLG-poly(octenamer)-PBLG triblock copolymers, by first polymerizing 1,9-decadiene via ADMET, in the presence of an amine-containing capping agent (Fig. 12). The resulting amine-functional telechelic polymer was modified to the di-Nickel complex (**25**), which was used as a bifunctional initiator for the controlled polymerization of the NCA of γ-benzyl glutamate (**26**). The coordination of the amine

moiety to the Ni metal ion lowered the reactivity of this nucleophile and prevented side reactions from occurring.[78] A similar approach was followed by Steig et al., who developed an initiator capable of performing both Ni-mediated NCA polymerization and ATRP, yielding rod–coil block copolymers,[79] and Kros and Cornelissen, who combined Ni-mediated isocyanide and NCA polymerization to give rod–rod block copolymers.[80]

Schlaad et al. used an alternative approach to obtain control over NCA polymerization.[76,81,82] Well-defined polystyrene (PS) with an amine group was prepared and used as an initiator for NCA polymerization of γ-benzyl glutamate or Z-protected lysine (ε-benzyloxycarbonyl-L-lysine, ZLL). The reactivity of the amine was lowered by the addition of HCl, leading to the formation of the ammonium chloride salt. The properties of these diblock copolymers were examined in the solid phase, both in bulk and in thin films. The effect of DP and polydispersity of the peptide block on bulk assembly was studied with PS-poly(ZLL) block copolymers. In all cases, a lamellar organization was found of alternating polystyrene and polypeptide domains (Fig. 13). The polypeptides showed an interdigitated hexagonal packing and the thickness of the peptide lamellae was linearly proportional to the degree of polymerization. A lower polydispersity index (PDI) led to a more planar interface between the lamellae. A broader size distribution was accommodated in the morphology by the occurrence of kinks in the lamellae. Similar morphologies were detected for PS-PBLG samples, in addition the PBLG segments were folded at least once within the lamellar organization. This folding was not observed in thin films, in which case the PBLG blocks were fully stretched and interdigitated. Lamellar morphologies and hexagonal ordering of the PBLG units were also observed for PBLG-poly(9,9-dihexylfluorene-2,7-diyl)-PBLG triblock copolymers,[83] and polylactide-PBLG

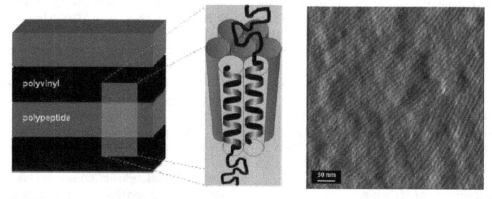

Fig. 12 Hybrid block copolymer synthesis via the Deming NCA polymerization approach. Depe stands for 1,2-bis(diethylphosphino) ethane.
Source: From Brzezinska & Deming.[78]

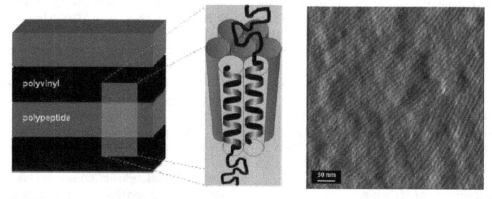

Fig. 13 (**A** and **B**): Schematic representation of hexagonal-in-lamellar solid state morphology of polystyrene–poly(lysine) block copolymers; (**C**) AFM amplitude image of a PS-poly(lysine) film prepared by spin-coating.
Source: From Schlaad, Smarsly, and Losik[81] © 2004, with permission from American Chemical Society.

rod–coil diblock copolymers, of which the PBLG blocks were prepared via classical NCA polymerization.[84]

Oligomeric rod–coil block copolymers composed of oligostyrene (10 units) and either PBLG (10–80 units) or PZLL (20–80 units) have been described by Klok et al.[85,86] The supramolecular organization of these block copolymers was shown to be sensitive to temperature. Depending on the amino acid and the polypeptide length, a partial transition from an α-helical to a β-sheet secondary structure was observed upon increasing the temperature, which prevented the regular organization of the peptide blocks resulting in a change in organization from a hexagonal to a lamellar β-sheet morphology. In solution rod–coil to coil–coil transitions of PS-PBLG block copolymers could be induced by changing the solvent from dioxane to dioxane/trifluoro acetic acid (80/20).[87]

Manners and coworkers produced metallopolymer–polypeptide diblock copolymers poly(ferrocenyldimethylsilane)-PBLG (Fig. 14, **27**) by anionic polymerization of the metallopolymer block, followed by conventional ring-opening NCA polymerization.[88,89] The obtained diblocks showed a lamellar morphology in bulk upon heating to 90°C, due to thermotropic liquid crystalline behavior. When the block copolymer was dissolved in hot toluene, gelation occurred, which was studied in detail and ascribed to the formation of nanoribbons. These ribbons were stabilized by the π–π interactions of the benzyl moieties and solubilized by the metallopolymer units. Similar behavior

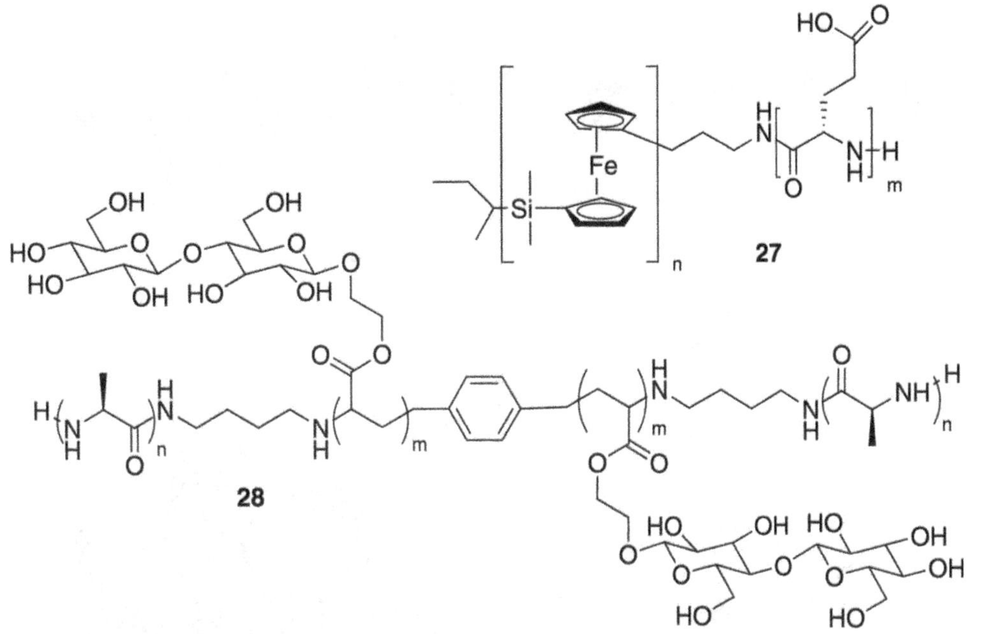

Fig. 14 Amphiphilic hybrid block copolymers, prepared by Manners (**27**) and Chaikof (**28**).
Source: From Kim et al.[88,89] and Dong et al.[90]

was observed for PS-PBLG and PEG-PBLG block copolymers. After the benzyl groups were removed to yield polyglutamate (PLGA), the resulting poly(ferrocenyl-dimethylsilane)-PLGA amphiphilic block copolymer self-assembled into micelles upon dispersion in water.

A glycopolymer–polypeptide triblock copolymer (**28**) was developed by Chaikof et al.[90] Polymerization of a methacrylate-functional, acetyl-protected lactose moiety by ATRP using a bifunctional initiator was followed by polymerization of the NCA of alanine. After removal of the acetyl-protecting groups, the resulting amphiphilic polymer yielded spherical aggregates in water of several hundreds of nanometers in diameter.

Deming and coworkers developed a series of block copolypeptides and investigated their assembly behavior in water.[91,92] Each block copolymer was built up of a polyelectrolyte, either poly(lysine) or poly(glutamate), and a hydrophobic part, poly(valine) or poly(leucine), which adopt a β-sheet and an α-helical conformation, respectively. When parameters such as molecular weight, hydrophobe–hydrophile balance, and most importantly, length of the hydrophobic block were chosen appropriately, stable hydrogels were formed, even to concentrations as low as 0.03 wt% block copolypeptide. The mechanism of hydrogel formation was similar to the abovementioned systems investigated by Manners et al. The helical or β-sheet peptides arrange themselves in ribbon-like structures that are the basis of gel-forming networks. Since poly(leucine) forms more readily secondary (helical) structures than poly(valine), the former polypeptide was more successful in yielding stable gels.

The same block copolymer series was optimized for the preparation of polymeric vesicles, with the block copolypeptide poly(lysine)$_{60}$–poly(leucine)$_{20}$ as the most capable vesicle-forming polymer (Fig. 15).[93] Vesicles prepared from a 1 μM polymer solution could withstand salt concentrations up to 500 mM but were disrupted when placed in serum, due to interaction with anionic proteins. The replacement of the poly(lysine) block with poly(glutamate) of equal length induced stability in the presence of serum.

Vesicles could also be prepared from hybrid polymer–polypeptide systems. Kukula et al. reported the formation of spherical micelles and large vesicles ("peptosomes", diameter ~150 nm) in aqueous solution of the block copolymer poly(butadiene)–poly(glutamate).[94] Chécot et al. showed for vesicular aggregates formed by poly(butadiene)$_{40}$–poly(glutamate)$_{100}$ the occurrence of a transition of the poly(glutamate) block from an α-helical conformation at pH 4.5 to random-coil at elevated pH.[95] This conformational transition was reflected in the hydrodynamic radius of the vesicles, which increased from approximately 100 nm at pH 4.5 to 150 nm at pH 11.5. Furthermore, the addition of salt led to shrinkage of the vesicles as a result of screening of charges.

MAIN CHAIN POLYMER HYBRIDS WITH BIOACTIVE MOTIFS

Compared to polymer hybrids with structural peptide motifs as described in the previous section, only relatively few examples are known in the literature in which bioactive peptides

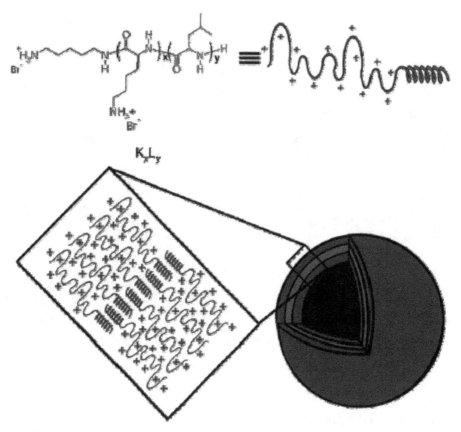

Fig. 15 Proposed model for the self-assembly of poly(leucine)–poly(lysine) block copolymers into vesicular aggregates.
Source: From Holowka, Pochan, and Deming[93] © 2005, with permission from American Chemical Society.

are incorporated in the main chain. The development of controlled radical polymerization methods has led to an increase in activities in this field, since main chain hybrids can be conveniently prepared by polymerizing the synthetic polymer from a peptide-functional initiator. One of the first examples was described by Wooley and coworkers, who synthesized a protein transduction domain and tritrpticin, an antimicrobial peptide, by solid-phase peptide chemistry and introduced at the N-terminal side of the peptides a nitroxide initiator (Fig. 16, **29**).[96,99] Nitroxide-mediated polymerization of, respectively, *tert*-butyl acrylate and methyl acrylate was performed, while the peptide-based initiator was still on the resin. Although polymerization occurred, the final product was difficult to characterize. Tritrpticin was also functionalized with an α-bromo amide initiator for ATRP. Polymerization with this initiator seemed to be better controlled. After hydrolysis of the *tert*-butyl esters, an amphiphilic polymer peptide hybrid was obtained, which formed micelles with the antimicrobial peptides exposed on the surface. The micelles still showed antimicrobial activity. Mei et al. followed a similar approach using ATRP to create poly(hydroxy ethyl methacrylate) functionalized with Gly-Arg-Gly-Asp-Ser (GRGDS), the well-known cell adhesion domain. Moderately low PDIs of 1.3–1.5 were obtained, and the RGD domains still were capable of promoting cell attachment.[100] Much better control over PDI was obtained by Börner and coworkers, who prepared both ATRP and reversible addition fragmentation transfer polymerization (RAFT)-functional peptide initiators (**30, 31**) on the solid support.[97,98] In contrast to the previous examples, the model peptide was cleaved off from the resin prior to polymerization. With both RAFT and ATRP, poly(*n*-butyl acrylate) with low PDI (1.18–1.19) was prepared, demonstrating that the presence of a peptide initiator does not intrinsically hamper controlled radical polymerization.

Besides creating the polymer in situ using a peptide as an initiator, an alternative approach is to couple pre-made peptide and polymer building blocks. This approach was followed by Reynhout et al. for synthesizing via a total solid-phase peptide chemistry approach diblock and triblock copolymers of PS and the peptide Gly-Ala-Asn-Pro-Asn-Ala-Ala-Gly, a peptide sequence often found in the CS protein of the Malaria parasite *Plasmodium falciparum*.[101] Dirks et al. used the efficient [2+3] cycloaddition reaction between azides and alkynes to couple in solution an azide end-functional PS chain, made by ATRP with an acetylene functional tripeptide Gly-Gly-Arg, which is a substrate for the blood clotting enzyme thrombin. The amphiphilic hybrid polymer formed vesicles upon dissolution in water.[102]

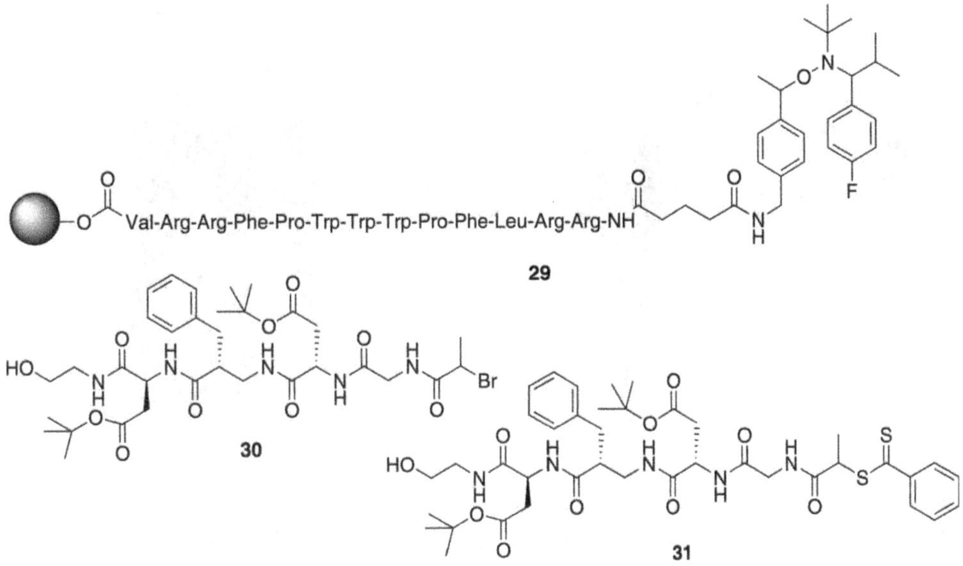

Fig. 16 Peptide initiators for controlled radical polymerization: nitroxide-mediated polymerization (**29**); ATRP (**30**) and RAFT (**31**).
Source: From Becker, Liu, and Wooley,[96] Rettig, Krause, and Börner[97]; and Ten Cate et al.[98]

Klok and coworkers prepared PEG-coiled coil peptide hybrids by solid-phase peptide chemistry.[103] Similar work was published by Kopeček and coworkers.[104] Coiled coil peptides are α-helical peptides that consist of repeats of seven amino acids. Within this heptad repeat sequence apolar amino acids are positioned to create a hydrophobic ribbon along the helical structure. As a result, two to five α-helical peptides intertwine into a coiled coil superhelix. By connecting PEG chains of different lengths to coiled coil peptides it was demonstrated that discrete supramolecular aggregates could be formed and that the number of peptides within the aggregate was dependent on the length of the PEG chain: going from PEG 750 to PEG 2000, the tendency to form unimers and dimers was increased due to steric hindrance (Fig. 17). Another effect of the introduction of PEG chains was an improved thermal stability of the aggregates.

An important biological function of coiled coil proteins is the binding to DNA for regulation of DNA transcription. Klok and Duncan therefore explored the use of coiled coil hybrids for DNA binding and gene delivery[105] and demonstrated the feasibility of this approach.

In traditional non-viral gene delivery, polycationic species are used to condense the (negatively charged) DNA that has to be delivered to the cell. The use of polycationic species, however, has high cytotoxicity as a negative side effect, which severely hampers application of these systems. Furthermore, the cationic charge hinders the escape of DNA from the complex in the endosome. Hybrid polymers have been devised to circumvent these problems, as e.g., shown by Kim and coworkers, who prepared PEG-polypeptide hybrids.[106] Succinic diacid-modified PEG was coupled to poly(lysine), prepared by NCA polymerization. A fraction of the amines in the peptide side chain

was reacted with histidine. The rationale behind this histidine modification was to improve the transfection efficiency of the polymer by facilitating the endosomal escape of the polymer-bound DNA. When 16% of the lysines were functionalized with a histidine moiety an optimal transfection efficiency was observed, combined with a lowered cytotoxicity when compared to poly(lysine).

Another approach to develop efficient and non-toxic gene delivery polymers was undertaken by Guan and coworkers.[107] A hybrid polymer consisting of alternating (oligo)lysine and galactose units was constructed via an interfacial polymerization of the dichloroformate of galactose and two amino groups of a variety of lysine oligomeric structures. Due to the separation of the lysines by the saccharide moiety a strongly reduced toxicity was observed, whereas the ability for transfection was maintained.

POLYMER–PROTEIN CONJUGATES

An important line of research within the area of hybrid biopolymers is the conjugation of synthetic polymers, in particular PEG, to natural proteins, aiming to improve their stability and bioavailability in biomedical applications. The positive effects of attachment of synthetic polymers were already recognized by Abuchowski, Davis, and coworkers in the late 1970s.[108–110] They covalently linked PEG with molecular weights of 1.9 and 5 kDa to the protein bovine liver catalase. The obtained conjugates were injected in mice and exhibited a significantly enhanced circulating half-life, reduced immunogenicity and antigenicity while retaining their bioactivity to a large extent. The rationale behind this stabilization effect is the steric

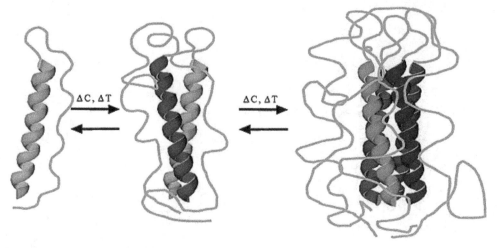

Fig. 17 Assembly behavior of PEG-leucine zipper block copolymers.
Source: From Vandermeulen, Tziatzios, and Klok[103] © 2003, with permission from American Chemical Society.

hindrance of the PEG shell, which prevents reaction of immune cells with the protein and protects it from degrading proteases. The covalent attachment of PEG chains to peptides and proteins has since then been termed *PEGylation* and has been extensively reviewed.[111–115] The PEGylation of peptides and proteins, varying from enzymes to antibodies, nowadays is widely used in the pharmaceutical industry. One of the first PEGylated proteins that appeared in the market is a conjugate of PEG with bovine adenosine deaminase (ADAGEN® by Enzon, Inc.), which received an FDA approval in 1990. The enzyme bovine adenosine deaminase (ADA) is used to treat severe combined immunodeficiency disease,[116] but is cleared rapidly from the plasma. However, the modification of this enzyme with multiple PEG chains with a molecular weight of 5 kDa extends the circulating half-life by a factor of 6.4 in rats.[117]

Although the PEGylation of proteins has proven to be very valuable, many of the first generation PEGylation products suffered from a severe loss in bioactivity, ranging from 20% to 95%. This decrease in activity mainly depends on the chain length of the attached polymers and the site of the protein they are coupled to.[118] Moreover, the attachment of multiple PEG chains leads to mixtures of isomers with different molecular weights, which makes it very difficult to reproduce drug properties from one batch to the next. For these reasons, it is of utmost importance to have control over the conjugation process.

Initial PEGylation processes were based on conjugation to the free amino groups of lysine residues present in proteins, which usually results in statistical multiple site additions. A more specific modification could be achieved by using the difference in pK_a values between the amines of the lysine side chains (10–10.2) and the N-terminal amine of the protein backbone (7.6–8.0). Addition of PEG equipped with an aldehyde moiety to a protein solution at pH 5 in the presence of $NaCNBH_4$ resulted in coupling of

the PEG chain via reductive amination to specifically the protein N terminus.[119] Another approach was to replace lysine residues by other amino acids, thereby limiting the number of conjugation sites.[120]

Site-specific polymer attachment could also be attained via conjugation to thiol groups present in cysteine residues. Free cysteines could be introduced at the surface of proteins either through reduction of disulfide bridges or by introduction of cysteine residues via protein engineering.[121,122] Specific PEGylation using amine-functionalized PEG to the amide group of glutamine could be accomplished under mild conditions using the enzyme transglutaminase.[123]

Recently, methods have been developed to use tyrosine residues as conjugation sites.[124–125] Because only few tyrosines are normally available on the protein surface, this residue is attractive to limit the amount of conjugated polymers. Chymotrypsinogen A, containing three tyrosines on its surface, was subjected to a Mannich-type reaction between the tyrosine residues, formaldehyde, and aniline to yield a modified protein without loss of activity.[124] Another modification involving tyrosine was based on alkylation using π-allylpalladium complexes.[125]

To increase the specificity of the functionalization reaction even further, non-proteinogenic amino acids can be introduced, which carry functionalities in the side chain that can react orthogonally, thus without interference of natural functional groups, with a PEG or polymer chain with complementary functionality. Newly developed protein engineering techniques (see chapter by D.A. Tirrell) enable the site-specific introduction of non-natural amino acids containing, for example, azide moieties.[126,127] This functional group has been successfully employed in the Staudinger ligation with triphenylphosphine derivatives and the Cu-catalyzed click chemistry approach with alkynes for site-specific attachment of PEG (Fig. 18).[128,129]

Site-specific polymer attachment could also be achieved by a complete synthetic build-up of the erythropoiesis

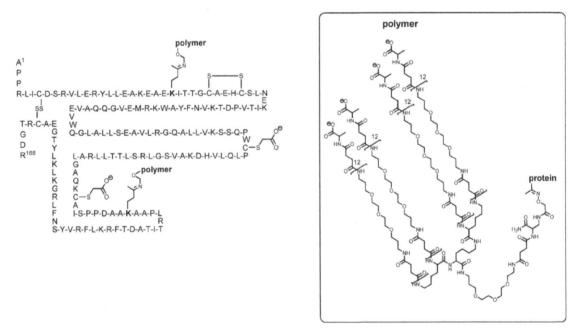

Fig. 18 Site-specific incorporation of azide-functionalized amino acids and the subsequent coupling with PEG exploiting this azido group via (**A**) Staudinger ligation and (**B**) click chemistry.
Source: From Cazalis et al.[128] and Deiters et al.[129]

Fig. 19 Schematic representation of the structure of SEP with two conjugated branched PEG-like polymer moieties.
Source: From Kochendoerfer et al.[130] © 2003, with permission from AAAS.

protein, consisting of a 166 α-amino acid polypeptide chain (Fig. 19).[130] Using peptide chemistry four polypeptide building blocks were created, which were coupled to each other via a native chemical ligation strategy, employing the highly selective and efficient reaction of a C-terminal thioester and an N-terminal cysteine residue. Since the protein was prepared synthetically, PEG-like polymer chains could conveniently be introduced onto two lysine residues (Lys 24 and 126). This synthetic erythropoiesis protein (SEP) hybrid displayed comparable activity *in vitro*

but a prolonged half-life *in vivo* compared to human erythropoietin, which is a glycoprotein hormone controlling the proliferation, differentiation, and maturation of erythroid cells. Kochendoerfer and coworkers extended this methodology to a series of polymers with distinct architectures, showing that the duration of action *in vivo* is strongly dependent on the overall charge and size of the conjugate.[131]

The importance of PEGylation has resulted in the development of a wide variety of different functionalization methods for this specific polymer. Polymer conjugation, however, is not merely limited to PEG. Progress in polymer science has resulted in the controlled synthesis of a myriad of polymers with numerous functional end groups, which can be used for conjugation reactions with proteins. By application of ATRP various polymers with N-succinimidyl ester,[132] aldehyde,[133] N-maleimido,[134] pyridyl disulfide,[135,136] and azide[102] end groups were synthesized and subsequently conjugated to amine, thiol, and acetylene groups, respectively, present in proteins, as depicted in Fig. 20.

In another approach, ATRP initiators were covalently connected to the protein or, via the use of a biotin moiety, non-covalently bound into streptavidin.[137,138] Synthetic polymers were subsequently grown from the protein surface.

The attachment of synthetic polymers other than PEG can be applied for several different purposes. Finn and

coworkers used the [2+3] cycloaddition reaction to couple an azide end-functional neoglycopolymer to a cowpea mosaic virus, surface-residing lysines of which were modified with alkyne-functional moieties.[139] The resulting scaffold with a high saccharide density showed a high binding affinity for lectin.

The conjugation of the thermally responsive synthetic polymer poly(N-isopropyl acrylamide) [poly(NIPAAM)] to proteins has been extensively studied by Hoffman, Stayton, and coworkers.[140] Poly(NIPAAM) is soluble in an aqueous environment below 32°C. When the temperature is increased it precipitates from the solution. This LCST behavior was, for example, used to recover trypsin-poly(NIPAAM) conjugates from a reaction mixture.[141] The enzyme could be recycled efficiently via

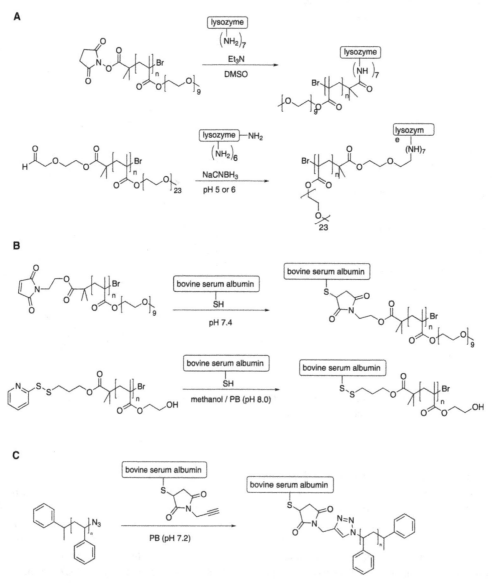

Fig. 20 End-functionalized polymers prepared by ATRP, which are ligated to proteins. (**A**) Coupling to lysines and the N terminus (amino groups); (**B**) ligation to cysteines (thiols); and (**C**) click chemistry onto an acetylene functionalized amino acid.
Source: (**A**) From Lecolley et al.[132] and Tao et al.[133] (**B**) From Mantovani et al.[134] and Bontempo et al.[135] (**C**) From Dirks et al.[102]

this method for at least 14 times, and its activity was increased compared to the native enzyme, probably as a result of a change in the microenvironment of the active site, leading to enhanced partitioning of the substrate in the proximity of the polymer. Poly(NIPAAM) has also been employed as a switch for protein–ligand binding.[142,143] By site-specific conjugation poly(NIPAAM) was positioned close to the protein binding site, which upon heating was closed off due to the collapse of the polymer. This behavior was completely reversible. Similar switching behavior was demonstrated by Schultz et al., who could interfere with protein complex formation, and hence, activity of a protein modified by single-site replacement with an azide moiety, to which a poly(NIPAAM) chain was connected.[144]

The coupling of hydrophobic synthetic polymers to mostly hydrophilic proteins has been investigated in great detail by Nolte and coworkers for the development of giant amphiphiles. Covalent conjugates between *Candida antarctica* lipase B, bovine serum albumin, and PS were constructed via either maleimide–thiol or azide–alkyne cycloaddition chemistry.[145,102] Non-covalent conjugates of horseradish peroxidase and myoglobin by modifying the enzyme cofactors with synthetic polymers could be created.[146] In all cases assemblies in an aqueous environment were observed, varying from cylindrical micelles, spherical micelles to vesicles. The assembled enzymes still retained some of their activity.

CONCLUSIONS

The development of new synthetic tools for the construction of well-defined polymer–peptide hybrid materials has resulted in new opportunities for research in different disciplines. In the area of biochemistry and in particular bioconjugate chemistry, scientists now have the ability to control the position and type of polymer they want to connect to proteins, in order to achieve, for example, improved bioavailability. In the area of materials science and polymer chemistry, researchers are starting to investigate the possibilities to introduce protein-based structure and function into hybrid materials, with potential applications as diverse as nanotechnology and biomedicine. Although the major part of activities has until now been directed to develop the synthetic toolbox, it is to be expected that in the near future materials will be made with a unique set of properties, which cannot be obtained by either natural or purely synthetic polymer materials.

REFERENCES

1. Matyjaszewski, K.; Xia, J.H. Atom transfer radical polymerization. Chem. Rev. **2001**, *101* (9), 2921–2990.
2. Hawker, C.J.; Bosman, A.W.; Harth, E. New polymer synthesis by nitroxide mediated living radical polymerizations. Chem. Rev. **2001**, *101* (12), 3661–3688.
3. Moad, G.; Rizzardo, E.; Thang, S.H. Living radical polymerization by the RAFT process. Aus. J. Chem. **2005**, *58* (6), 379–410.
4. Deming, T.J. Facile synthesis of block copolypeptides of defined architecture. Nature **1997**, *390* (6658), 386–389.
5. Cunliffe, D.; Pennadam, S.; Alexander, C. Synthetic and biological polymers-merging the interface. Eur. Polym. J. **2004**, *40* (1), 5–25.
6. Vandermeulen, G.W.M.; Klok, H.A. Peptide/protein hybrid materials: Enhanced control of structure and improved performance through conjugation of biological and synthetic polymers. Macromol. Biosci. **2004**, *4* (4), 383–398.
7. Alarcon, C.D.H.; Pennadam, S.; Alexander, C. Stimuli responsive polymers for biomedical applications. Chem. Soc. Rev. **2005**, *34* (3), 276–285.
8. Löwik, D.W.P.M.; Ayres, L.; Smeenk, J.M.; Van Hest, J.C.M. Synthesis of bio-inspired hybrid polymers using peptide synthesis and protein engineering. Adv. Polym. Sci. **2006**, *202*, 19–52.
9. van Hest, J.C.M.; Tirrell, D.A. Protein-based materials, toward a new level of structural control. Chem. Comm. **2001**, (19), 1897–1904.
10. Yeo, D.S.Y.; Srinivasan, R.; Chen, G.Y.J.; Yao, S.Q. Expanded utility of the native chemical ligation reaction. Chemistry **2004**, *10* (19), 4664–4672.
11. Muir, T.W. Semisynthesis of proteins by expressed protein ligation. Ann. Rev. Biochem. **2003**, *72* (1), 249–289.
12. Nilsson, B.L.; Soellner, M.B.; Raines, R.T. Chemical synthesis of proteins. Ann. Rev. Biophys. Biomol. Struct. **2005**, *34*, 91–118.
13. Sanda, F.; Endo, T. Syntheses and functions of polymers based on amino acids. Macromol. Chem. Phys. **1999**, *200* (12), 2651–2661.
14. Lekchiri, A.; Morcellet, J.; Morcellet, M. Complex-formation between copper(Ii) and poly(N-methacryloyl-L-asparagine). Macromolecules **1987**, *20* (1), 49–53.
15. Murata, H.; Sanda, F.; Endo, T. Synthesis and radical polymerization behavior of methacrylamides having L-leucyl-L-alanine oligopeptide moieties. Effect of the peptide chain length on the radical polymerizability. Macromol. Chem. Phys. **2001**, *202* (6), 759–764.
16. Murata, H.; Sanda, F.; Endo, T. Synthesis and radical polymerization of a novel acrylamide having an alpha-helical peptide structure in the side chain. J. Polym. Sci. A Polym. Chem. **1998**, *36* (10), 1679–1682.
17. Bush, S.M.; North, M.; Sellarajah, S. Synthesis and chiro-optical properties of copolymers from N-Boc-O-methacry-loyl-(S)-serine benzhydryl ester and methyl methacrylate. Polymer **1998**, *39* (13), 2991–2993.
18. Hopkins, T.E.; Pawlow, J.H.; Koren, D.L.; Deters, K.S.; Solivan, S.M.; Davis, J.A.; Gomez, F.J.; Wagener, K.B. Chiral polyolefins bearing amino acids. Macromolecules **2001**, *34* (23), 7920–7922.
19. Hopkins, T.E.; Wagener, K.B. Amino acid and dipeptide functionalized polyolefins. Macromolecules **2003**, *36* (7), 2206–2214.
20. Cheuk, K.K. L.; Lam, J.W.Y.; Chen, J.; Lai, L.M.; Tang, B.Z. Amino acid-containing polyacetylenes: Synthesis, hydrogen bonding, chirality transcription, and chain helicity of amphiphilic poly(phenylacetylene)s carrying L-leucine pendants. Macromolecules **2003**, *36* (16), 5947–5959.

Conducting—Dendritic

21. Gao, G.Z.; Sanda, F.; Masuda, T. Synthesis and properties of amino acid-based polyacetylenes. Macromolecules **2003**, *36* (11), 3932–3937.

22. Gao, G.Z.; Sanda, F.; Masuda, T. Copolymerization of chiral amino acid-based acetylenes and helical conformation of the copolymers. Macromolecules **2003**, *36* (11), 3938–3943.

23. Sanda, F.; Araki, H.; Masuda, T. Synthesis and properties of serine- and threonine-based helical polyacetylenes. Macromolecules **2004**, *37* (23), 8510–8516.

24. Sanda, F.; Terada, K.; Masuda, T. Synthesis, chiroptical properties, and pH responsibility of aspartic acid- and glutamic acid-based helical polyacetylenes. Macromolecules **2005**, *38* (20), 8149–8154.

25. Maeda, K.; Kamiya, N.; Yashima, E. Poly(phenylacetylene)s bearing a peptide pendant: Helical conformational changes of the polymer backbone stimulated by the pendant conformational change. Chemistry **2004**, *10* (16), 4000–4010.

26. Cornelissen, J.; Fischer, M.; Sommerdijk, N.; Nolte, R.J.M. Helical superstructures from charged poly(styrene)-poly(isocyanodipeptide) block copolymers. Science **1998**, *280* (5368), 1427–1430.

27. Cornelissen, J.J.L.M.; Donners, J.J.J.M.; de Gelder, R.; Graswinckel, W.S.; Metselaar, G.A.; Rowan, A.E.; Sommerdijk, N.A.J.M.; Nolte, R.J.M. Beta-helical polymers from isocyanopeptides. Science **2001**, *293* (5530), 676–680.

28. Vriezema, D.M.; Kros, A.; de Gelder, R.; Cornelissen, J.; Rowan, A.E.; Nolte, R.J.M. Electroformed giant vesicles from thiophene-containing rod-coil diblock copolymers. Macromolecules **2004**, *37* (12), 4736–4739.

29. Vriezema, D.M.; Hoogboom, J.; Velonia, K.; Takazawa, K.; Christianen, P.C.M.; Maan, J.C.; Rowan, A.E.; Nolte, R.J.M. Vesicles and polymerized vesicles from thiophene-containing rod-coil block copolymers. Angew. Chem.-Int. Ed. **2003**, *42* (7), 772–776.

30. Ayres, L.; Adams, H.J.M.P.; Löwik, W.P.M.; van Hest, J.C.M. Beta-sheet side chain polymers synthesized by atom-transfer radical polymerization. Biomacromolecules 2005, *6* (2), 825–831.

31. Ayres, L.; Vos, M.R.J.; Adams, P.J.H.M.; Shklyarevskiy, I.O.; van Hest, J.C.M. Elastin-based side-chain polymers synthesized by ATRP. Macromolecules **2003**, *36* (16), 5967–5973.

32. Ayres, L.; Koch, K.; Adams, P.; van Hest, J.C.M. Stimulus responsive behavior of elastin-based side chain polymers. Macromolecules **2005**, *38* (5), 1699–1704.

33. Kiessling, L.L.; Gestwicki, J.E.; Strong, L.E. Synthetic multivalent ligands as probes of signal transduction. Angew. Chem. Int. Ed. **2006**, *45* (15), 2348–2368.

34. Mammen, M.; Choi, S.K.; Whitesides, G.M. Polyvalent interactions in biological systems: Implications for design and use of multivalent ligands and inhibitors. Angew. Chem. Int. Ed. **1998**, *37* (20), 2755–2794.

35. O'brien Simpson, N.M.; Ede, N.J.; Brown, L.E.; Swan, J.; Jackson, D.C. Polymerization of unprotected synthetic peptides: A view toward synthetic peptide vaccines. J. Am. Chem. Soc. **1997**, *119* (6), 1183–1188.

36. Gros, L.; Ringsdorf, H.; Schupp, H. Polymeric anti-tumor agents on a molecular and on a cellular-level. Angew. Chem. Int. Ed. **1981**, *20* (4), 305–325.

37. Ringsdorf, H. Structure and properties of pharmacologically active polymers. J. Polym. Sci. C Polym. Symp. **1975**, *51* (1), 135–153.

38. Maeda, H. SMANCS and polymer-conjugated macromolecular drugs: Advantages in cancer chemotherapy. Adv. Drug Deliv. Rev. **2001**, *46* (1–3), 169–185.

39. Maeda, H. SMANCS and polymer-conjugated macromolecular drugs—Advantages in cancer-chemotherapy. Adv. Drug Deliv. Rev. **1991**, *6* (2), 181–202.

40. Duncan, R. The dawning era of polymer therapeutics. Nat. Rev. Drug Discov. **2003**, *2* (5), 347–360.

41. Satchi-Fainaro, R.; Duncan, R.; Barnes, C.M. Polymer therapeutics for cancer: Current status and future challenges. Adv. Polym. Sci. **2006**, 193, 1–65.

42. Kopeček, J.; Rejmanova, P.; Duncan, R.; Lloyd, J.B. Controlled release of drug model from N-(2-Hydroxypropyl)-methacrylamide copolymers. Ann. N. Y. Acad. Sci. **1985**, *446* (1), 93–104.

43. Duncan, R.; Cable, H.C.; Lloyd, J.B.; Rejmanova, P.; Kopeček, J. Polymers containing enzymatically degradable bonds.7. Design of oligopeptide side-chains in poly[N-(2-hydroxypropyl)methacrylamide] co-polymers to promote efficient degradation by lysosomal-enzymes. Makromol. Chem. Macromol. Chem. Phys. **1983**, *184* (10), 1997–2008.

44. Kopeček, J.; Kopečkova, P.; Minko, T.; Lu, Z.R. HPMA copolymer-anticancer drug conjugates: Design, activity, and mechanism of action. Eur. J. Pharm. Biopharm. **2000**, *50* (1), 61–81.

45. Kopeček, J.; Kopečkova, P.; Minko, T.; Lu, Z.R.; Peterson, C.M. Water soluble polymers in tumor targeted delivery. J. Control. Release **2001**, *74* (1–3), 147–158.

46. Minko, T.; Dharap, S.S.; Pakunlu, R.I.; Wang, Y. Molecular targeting of drug delivery systems to cancer. Curr. Drug Targets **2004**, *5* (4), 389–406.

47. Allen, T.M. Ligand-targeted therapeutics in anticancer therapy. Nat. Rev. Cancer **2002**, *2* (10), 750–763.

48. Rihova, B. Receptor-mediated targeted drug or toxin delivery. Adv. Drug Deliv. Rev. **1998**, *29* (3), 273–289.

49. Duncan, R.; Kopeček, J.; Rejmanova, P.; Lloyd, J.B. Targeting of N-(2-hydroxypropyl)methacrylamide copolymers to liver by incorporation of galactose residues. Biochim. Biophys. Acta **1983**, *755* (3), 518–521.

50. Rathi, R.C.; Kopečkova, P.; Rihova, B.; Kopeček, J. N-(2-hydroxypropyl) methacrylamide copolymers containing pendant saccharide moieties—Synthesis and bioadhesive properties. J. Polym. Sci. A Polym. Chem. **1991**, *29* (13), 1895–1902.

51. Hanahan, D.; Folkman, J. Patterns and emerging mechanisms of the angiogenic switch during tumorigenesis. Cell **1996**, *86* (3), 353–364.

52. Carmeliet, P. Angiogenesis in life, disease and medicine. Nature **2005**, *438* (7070), 932–936.

53. Carmeliet, P.; Jain, R.K. Angiogenesis in cancer and other diseases. Nature **2000**, *407* (6801), 249–257.

54. Satchi-Fainaro, R.; Barnes, C.M. Drug delivery systems to target the tumour vasculature and the tumour cell. In *The Drug Delivery Companies Report Spring/Summer 2004*; PharmaVentures: Oxford, U.K., 2004; 43–49.

55. Satchi-Fainaro, R.; Puder, M.; Davies, J.W.; Tran, H.T.; Sampson, D.A.; Greene, A.K.; Corfas, G.; Folkman, J.

Targeting angiogenesis with a conjugate of HPMA copolymer and TNP-470. Nat. Med. **2004**, *10* (3), 255–261.

56. Brooks, P.C.; Clark, R.A.F.; Cheresh, D.A. Requirement of vascular integrin alpha(V)beta(3) for Angiogenesis. Science **1994**, *264* (5158), 569–571.

57. Mitra, A.; Mulholland, J.; Nan, A.; McNeill, E.; Ghandehari, H.; Line, B.R. Targeting tumor angiogenic vasculature using polymer-RGD conjugates. J. Control. Release **2005**, *102* (1), 191–201.

58. Deng, C.; Tian, H.Y.; Zhang, P.B.; Sun, J.; Chen, X.S.; Jing, X.B. Synthesis and characterization of RGD peptide grafted poly(ethylene glycol)-b-poly(L-lactide)-b-poly(L-glutamic acid) triblock copolymer. Biomacromolecules **2006**, *7* (2), 590–596.

59. VandeVondele, S.; Voros, J.; Hubbell, J.A. RGD-Grafted poly-l-lysine-graft-(polyethylene glycol) copolymers block non-specific protein adsorption while promoting cell adhesion. Biotech. Bioeng. **2003**, *82* (7), 784–790.

60. Auernheimer, J.; Dahmen, C.; Hersel, U.; Bausch, A.; Kessler, H. Photoswitched cell adhesion on surfaces with RGD peptides. J. Am. Chem. Soc. **2005**, *127* (46), 16107–16110.

61. Maynard, H.D.; Okada, S.Y.; Grubbs, R.H. Inhibition of cell adhesion to fibronectin by oligopeptide-substituted polynorbornenes. J. Am. Chem. Soc. **2001**, *123* (7), 1275–1279.

62. Rathore, O.; Sogah, D.Y. Self-assembly of beta-sheets into nanostructures by poly(alanine) segments incorporated in multiblock copolymers inspired by spider silk. J. Am. Chem. Soc. **2001**, *123* (22), 5231–5239.

63. Rathore, O.; Sogah, D.Y. Nanostructure formation through beta-sheet self-assembly in silk-based materials. Macromolecules **2001**, *34* (5), 1477–1486.

64. Yao, J.M.; Xiao, D.; Chen, X.; Zhou, P.; Yu, T.; Shao, Z. Synthesis and solid-state structure investigation of silk-proteinlike multiblock copolymers. Macromolecules **2003**, *36* (20), 7508–7512.

65. Klok, H.A.; Rösler, A.; Gotz, G.; Mena-Osteritz, E.; Bäuerle, P. Synthesis of a silk-inspired peptide oligothiophene conjugate. Org. Biomol. Chem. **2004**, *2* (24), 3541–3544.

66. Rösler, A.; Klok, H.A.; Hamley, I.W.; Castelletto, V.; Mykhaylyk, O.O. Nanoscale structure of poly(ethylene glycol) hybrid block copolymers containing amphiphilic beta-strand peptide sequences. Biomacromolecules **2003**, *4* (4), 859–863.

67. Eckhardt, D.; Groenewolt, M.; Krause, E.; Börner, H.G. Rational design of oligopeptide organizers for the formation of poly(ethylene oxide) nanofibers. Chem. Comm. **2005**, (22), 2814–2816.

68. Smeenk, J.M.; Otten, M.B.J.; Thies, J.; Tirrell, D.A.; Stunnenberg, H.G.; van Hest, J.C.M. Controlled assembly of macromolecular beta-sheet fibrils. Angew. Chem. Int. Ed. **2005**, *44* (13), 1968–1971.

69. Smeenk, J.M.; Schon, P.; Otten, M.B.J.; Speller, S.; Stunnenberg, H.G.; van Hest, J.C.M. Fibril formation by triblock copolymers of silklike beta-sheet polypeptides and poly(ethylene glycol). Macromolecules **2006**, *39* (8), 2989–2997.

70. Burkoth, T.S.; Benzinger, T.L.S.; Urban, V.; Lynn, D.G.; Meredith, S.C.; Thiyagarajan, P. Self-assembly of A beta(10–35)-PEG block copolymer fibrils. J. Am. Chem. Soc. **1999**, *121* (32), 7429–7430.

71. Burkoth, T.S.; Benzinger, T.L.S.; Jones, D.N.M.; Hallenga, K.; Meredith, S.C.; Lynn, D.G. C-terminal PEG blocks the irreversible step in beta-amyloid(10–35) fibrillogenesis. J. Am. Chem. Soc. **1998**, *120* (30), 7655–7656.

72. Couet, J.; Jeyaprakash, J.D.; Samuel, S.; Kopyshev, A.; Santer, S.; Biesalski, M. Peptide-polymer hybrid nanotubes. Angew. Chem. Int. Ed. **2005**, *44* (21), 3297–3301.

73. Dalsin, J.L.; Hu, B.H.; Lee, B.P.; Messersmith, P.B. Mussel adhesive protein mimetic polymers for the preparation of nonfouling surfaces. J. Am. Chem. Soc. **2003**, *125* (14), 4253–4258.

74. Statz, A.R.; Meagher, R.J.; Barron, A.E.; Messersmith, P.B. New peptidomimetic polymers for antifouling surfaces. J. Am. Chem. Soc. **2005**, *127* (22), 7972–7973.

75. Fan, X.W.; Lin, L.J.; Dalsin, J.L.; Messersmith, P.B. Biomimetic anchor for surface-initiated polymerization from metal substrates. J. Am. Chem. Soc. **2005**, *127* (45), 15843–15847.

76. Dimitrov, I.; Schlaad, H. Synthesis of nearly monodisperse polystyrene-polypeptide block copolymers via polymerisation of N-carboxyanhydrides. Chem. Comm. **2003**, (23), 2944–2945.

77. Aliferis, T.; Iatrou, H.; Hadjichristidis, N. Living polypeptides. Biomacromolecules **2004**, *5* (5), 1653–1656.

78. Brzezinska, K.R.; Deming, T.J. Synthesis of ABA triblock copolymers via acyclic diene metathesis polymerization and living polymerization of alpha-amino acid-N-carboxyanhydrides. Macromolecules **2001**, *34* (13), 4348–4354.

79. Steig, S.; Cornelius, F.; Witte, P.; Staal, B.B.P.; Koning, C.E.; Heise, A.; Menzel, H. Synthesis of polypeptide based rod-coil block copolymers. Chem. Comm. **2005**, (43), 5420–5422.

80. Kros, A.; Jesse, W.; Metselaar, G.A.; Cornelissen, J. Synthesis and self-assembly of rod-rod hybrid poly(gamma-benzyl L-glutamate)-block-polyisocyanide copolymers. Angew. Chem. Int. Ed. **2005**, *44* (28), 4349–4352.

81. Schlaad, H.; Smarsly, B.; Losik, M. The role of chain-length distribution in the formation of solid-state structures of polypeptide-based rod-coil block copolymers. Macromolecules **2004**, *37* (6), 2210–2214.

82. Ludwigs, S.; Krausch, G.; Reiter, G.; Losik, M.; Antonietti, M.; Schlaad, H. Structure formation of a polystyrene-block-poly(gamma-benzyl L-glutamate) in thin films. Macromolecules **2005**, *38* (18), 7532–7535.

83. Kong, X.X.; Jenekhe, S.A. Block copolymers containing conjugated polymer and polypeptide sequences: Synthesis and self-assembly of electroactive and photoactive nanostructures. Macromolecules **2004**, *37* (22), 8180–8183.

84. Caillol, S.; Lecommandoux, S.; Mingotaud, A.F.; Schappacher, M.; Soum, A.; Bryson, N.; Meyrueix, R. Synthesis and self-assembly properties of peptide—Polylactide block copolymers. Macromolecules **2003**, *36* (4), 1118–1124.

85. Lecommandoux, S.; Achard, M.F.; Langenwalter, J.F.; Klok, H.A. Self-assembly of rod-coil diblock oligomers based on alpha-helical peptides. Macromolecules **2001**, *34* (26), 9100–9111.

86. Klok, H.A.; Langenwalter, J.F.; Lecommandoux, S. Self-assembly of peptide-based diblock oligomers. Macromolecules **2000**, *33* (21), 7819–7826.

87. Crespo, J.S.; Lecommandoux, S.; Borsali, R.; Klok, H.A.; Soldi, V. Small-angle neutron scattering from diblock copolymer poly(styrene-d(8))-b-poly(gamma-benzyl L-glutamate) solutions: Rod-coil to coil-coil transition. Macromolecules **2003**, *36* (4), 1253–1256

88. Kim, K.T.; Park, C.; Vandermeulen, G.W.M.; Rider, D.A.; Kim, C.; Winnik, M.A.; Manners, I. Gelation of helical polypeptide-random coil diblock copolymers by a nanoribbon mechanism. Angew. Chem. Int. Ed. **2005**, *44* (48), 7964–7968.

89. Kim, K.T.; Vandermeulen, G.W.M.; Winnik, M.A.; Manners, I. Organometallic-polypeptide block copolymers: Synthesis and properties of poly(ferrocenyl-dimethylsilane)-b-poly-(gamma-benzyl-L-glutamate). Macromolecules **2005**, *38* (12), 4958–4961.

90. Dong, C.M.; Sun, X.L.; Faucher, K.M.; Apkarian, R.P.; Chaikof, E.L. Synthesis and characterization of glyco-polymer-polypeptide triblock copolymers. Biomacromolecules **2004**, *5* (1), 224–231.

91. Nowak, A.P.; Breedveld, V.; Pine, D.J.; Deming, T.J. Unusual salt stability in highly charged diblock co-polypeptide hydrogels. J. Am. Chem. Soc. **2003**, *125* (50), 15666–15670.

92. Nowak, A.P.; Breedveld, V.; Pakstis, L.; Ozbas, B.; Pine, D.J.; Pochan, D.; Deming, T.J. Rapidly recovering hydrogel scaffolds from self-assembling diblock copolypeptide amphiphiles. Nature **2002**, *417* (6887), 424–428.

93. Holowka, E.P.; Pochan, D.J.; Deming, T.J. Charged polypeptide vesicles with controllable diameter. J. Am. Chem. Soc. **2005**, *127* (35), 12423–12428.

94. Kukula, H.; Schlaad, H.; Antonietti, M.; Förster, S. The formation of polymer vesicles or peptosomes by polybuta-diene-block-poly(L-glutamate)s in dilute aqueous solution. J. Am. Chem. Soc. **2002**, *124* (8), 1658–1663.

95. Chécot, F.; Lecommandoux, S.; Gnanou, Y.; Klok, H.A. Water-soluble stimuli-responsive vesicles from peptide-based diblock copolymers. Angew. Chem. Int. Ed. **2002**, *41* (8), 1339–1343.

96. Becker, M.L.; Liu, J.; Wooley, K.L. Functionalized micellar assemblies prepared via block copolymers synthesized by living free radical polymerization upon peptide-loaded resins. Biomacromolecules **2005**, *6* (1), 220–228.

97. Rettig, H.; Krause, E.; Börner, H.G. Atom transfer radical polymerization with polypeptide initiators: A general approach to block copolymers of sequence-defined poly-peptides and synthetic polymers. Macromol. Rapid Comm. **2004**, *25* (13), 1251–1256.

98. Ten Cate, M.G.J.; Rettig, H.; Bernhardt, K.; Börner, H.G. Sequence-defined polypeptide-polymer conjugates utilizing reversible addition fragmentation transfer radical polymerization. Macromolecules **2005**, *38* (26), 10643–10649.

99. Becker, M.L.; Liu, J.Q.; Wooley, K.L. Peptide-polymer bioconjugates: hybrid block copolymers generated via living radical polymerizations from resin-supported peptides. Chem. Comm. **2003**, (2), 180–181.

100. Mei, Y.; Beers, K.L.; Byrd, H.C.M.; Vanderhart, D.L.; Washburn, N.R. Solid-phase ATRP synthesis of peptide-polymer hybrids. J. Am. Chem. Soc. **2004**, *126* (11), 3472–3476.

101. Reynhout, I.C.; Lowik, D.W.P.M.; van Hest, J.C.M.; Cornelissen, J.J.L.M.; Nolte, R.J.M. Solid phase synthesis of bio-hybrid block copolymers. Chem. Comm. **2005**, 602–604.

102. Dirks, A.J.; van Berkel, S.S.; Hatzakis, N.S.; Opsteen, J.A.; van Delft, F.L.; Cornelissen, J.J.L.M.; Rowan, A.E.; Van Hest, J.C.M.; Rutjes, F.p.j.t.; Nolte, R.J.M. Preparation of biohybrid amphiphiles via the copper catalysed Huisgen [3+2] dipolar cylcoaddition reaction. Chem. Comm. **2005**, (33), 4172–4174.

103. Vandermeulen, G.W.M.; Tziatzios, C.; Klok, H.A. Reversible self-organization of poly(ethylene glycol)-based hybrid block copolymers mediated by a De Novo four-stranded alpha-helical coiled coil motif. Macromolecules **2003**, *36* (11), 4107–4114.

104. Pechar, M.; Kopečkova, P.; Joss, L.; Kopeček, J. Associative diblock copolymers of poly(ethylene glycol) and coiled-coil peptides. Macromol. Biosci. **2002**, *2* (5), 199–206.

105. Vandermeulen, G.W.M.; Tziatzios, C.; Duncan, R.; Klok, H.A. PEG-based hybrid block copolymers containing alpha-helical coiled coil peptide sequences: Control of self-assembly and preliminary biological evaluation. Macromolecules **2005**, *38* (3), 761–769.

106. Bikram, M.; Ahn, C.H.; Chae, S.Y.; Lee, M.Y.; Yockman, J.W.; Kim, S.W. Biodegradable poly(ethylene glycol)-co-poly(L-lysine)-g-histidine multiblock copolymers for nonviral gene delivery. Macromolecules **2004**, *37* (5), 1903–1916.

107. Metzke, M.; O'Connor, N.; Maiti, S.; Nelson, E.; Guan, Z.B. Saccharide-peptide hybrid copolymers as biomaterials. Angew. Chem. Int. Ed. **2005**, *44* (40), 6529–6533.

108. Abuchowski, A.; McCoy, J.R.; Palczuk, N.C.; Vanes, T.; Davis, F.F. Effect of covalent attachment of polyethylene-glycol on immunogenicity and circulating life of bovine liver catalase. J. Biol. Chem. **1977**, *252* (11), 3582–3586.

109. Abuchowski, A.; Vanes, T.; Palczuk, N.C.; Davis, F.F. Alteration of immunological properties of bovine serum-albumin by covalent attachment of polyethylene-glycol. J. Biol. Chem. **1977**, *252* (11), 3578–3581.

110. Davis, F.F. Commentary—The origin of pegnology. Adv. Drug Deliv. Rev. **2002**, *54* (4), 457–458.

111. Delgado, C.; Francis, G.E.; Fisher, D. The uses and properties of peg-linked proteins. Crit. Rev. Ther. Drug. Carr. Syst. **1992**, *9* (3–4), 249–304.

112. Zalipsky, S. Chemistry of polyethylene-glycol conjugates with biologically-active molecules. Adv. Drug Deliv. Rev. **1995**, *16* (2–3), 157–182.

113. Kodera, Y.; Matsushima, A.; Hiroto, M.; Nishimura, H.; Ishii, A.; Ueno, T.; Inada, Y. Pegylation of proteins and bioactive substances for medical and technical applications. Prog. Polym. Sci. **1998**, *23* (7), 1233–1271.

114. Veronese, F.M.; Pasut, G. PEGylation, successful approach to drug delivery. Drug Discov. Today **2005**, *10* (21–24), 1451–1458.

115. Caliceti, P.; Veronese, F.M. Pharmacokinetic and biodistribution properties of poly(ethylene glycol)-protein conjugates. Adv. Drug Deliv. Rev. **2003**, *55* (10), 1261–1277.

116. Hershfield, M.S.; Buckley, R.H.; Greenberg, M.L.; Melton, A.L.; Schiff, R.; Hatem, C.; Kurtzberg, J.; Markert, M.L.; Kobayashi, R.H.; Kobayashi, A.L.; Abuchowski, A. Treatment of adenosine-deaminase deficiency with polyethylene-glycol modified adenosine-deaminase. New Engl. J. Med. **1987**, *316* (10), 589–596.

117. Nucci, M.L.; Shorr, R.; Abuchowski, A. The therapeutic value of poly(ethylene glycol)-modified proteins. Adv. Drug Deliv. Rev. **1991**, *6* (2), 133–151.

118. Francis, G.E.; Fisher, D.; Delgado, C.; Malik, F.; Gardiner, A.; Neale, D. PEGylation of cytokines and other therapeutic proteins and peptides: The importance of biological optimisation of coupling techniques. Int. J. Hematol. **1998**, *68* (1), 1–18.

119. Kinstler, O.; Molineux, G.; Treuheit, M.; Ladd, D.; Gegg, C. Mono-N-terminal poly(ethylene glycol)-protein conjugates. Adv. Drug Deliv. Rev. **2002**, *54* (4), 477–485.

120. Yamamoto, Y.; Tsutsumi, Y.; Yoshioka, Y.; Nishibata, T.; Kobayashi, K.; Okamoto, T.; Mukai, Y.; Shimizu, T.; Nakagawa, S.; Nagata, S.; Mayumi, T. Site-specific PEGylation of a lysine-deficient TNF-alpha with full bioactivity. Nat. Biotech. **2003**, *21* (5), 546–552.

121. Wang, J.H.; Tam, S.C.; Huang, H.; Ouyang, D.Y.; Wang, Y.Y.; Zheng, Y.T. Site-directed PEGylation of trichosanthin retained its anti-HIV activity with reduced potency *in vitro*. Biochem. Biophys. Res. Commun. **2004**, *317* (4), 965–971.

122. Doherty, D.H.; Rosendahl, M.S.; Smith, D.J.; Hughes, J.M.; Chlipala, E.A.; Cox, G.N. Site-specific PEGylation of engineered cysteine analogues of recombinant human granulocyte-macrophage colony-stimulating factor. Bioconjug. Chem. **2005**, *16* (5), 1291–1298.

123. Sato, H. Enzymatic procedure for site-specific pegylation of proteins. Adv. Drug Deliv. Rev. **2002**, *54* (4), 487–504.

124. Joshi, N.S.; Whitaker, L.R.; Francis, M.B. A three-component mannich-type reaction for selective tyrosine bioconjugation. J. Am. Chem. Soc. **2004**, *126* (49), 15942–15943.

125. Tilley, S.D.; Francis, M.B. Tyrosine-selective protein alkylation using π-palladium complexes. J. Am. Chem. Soc. **2006**, 128, 1080–1081.

126. Link, J.A.; Tirrell, D.A. Reassignment of sense codons *in vivo*. Methods **2005**, *36*, 292–298.

127. O'Donogue, P.; Ling, J.; Wang, Y.-S.; Söll, D. Upgrading protein synthesis for synthetic biology, Nature Chem. Biol. **2013**, *9*, 594–598.

128. Cazalis, C.S.; Haller, C.A.; Sease-Cargo, L.; Chaikof, E.L. C-terminal site-specific PEGylation of a truncated thrombomodulin mutant with retention of full bioactivity. Bioconj. Chem. **2004**, *15* (5), 1005–1009.

129. Deiters, A.; Cropp, T.A.; Summerer, D.; Mukherji, M.; Schultz, P.G. Site-specific PEGylation of proteins containing unnatural amino acids. Bioorg. Med. Chem. Lett. **2004**, *14* (23), 5743–5745.

130. Kochendoerfer, G.G.; Chen, S.Y.; Mao, F.; Cressman, S.; Traviglia, S.; Shao, H.Y.; Hunter, C.L.; Low, D.W.; Cagle, E.N.; Carnevali, M.; Gueriguian, V.; Keogh, P.J.; Porter, H.; Stratton, S.M.; Wiedeke, M.C.; Wilken, J.; Tang, J.; Levy, J.J.; Miranda, L.P.; Crnogorac, M.M.; Kalbag, S.; Botti, P.; Schindler-Horvat, J.; Savatski, L.; Adamson, J.W.; Kung, A.; Kent, S.B.H.; Bradburne, J.A. Design and chemical synthesis of a homogeneous polymer-modified erythropoiesis protein. Science **2003**, *299* (5608), 884–887.

131. Chen, S.Y.; Cressman, S.; Mao, F.; Shao, H.; Low, D.W.; Beilan, H.S.; Cagle, E.N.; Carnevali, M.; Gueriguian, V.; Keogh, P.J.; Porter, H.; Stratton, S.M.; Wiedeke, M.C.; Savatski, L.; Adamson, J.W.; Bozzini, C.E.; Kung, A.; Kent, S.B. H.; Bradburne, J.A.; Kochendoerfer, G.G. Synthetic erythropoietic proteins: Tuning biological performance by site-specific polymer attachment. Chem. Biol. **2005**, *12* (3), 371–383.

132. Lecolley, F.; Tao, L.; Mantovani, G.; Durkin, I.; Lautru, S.; Haddleton, D.M. A new approach to bioconjugates for proteins and peptides ("pegylation") utilising living radical polymerisation. Chem. Commun. **2004**, (18), 2026–2027.

133. Tao, L.; Mantovani, G.; Lecolley, F.; Haddleton, D.M. alpha-aldehyde terminally functional methacrylic polymers from living radical polymerization: Application in protein conjugation "pegylation". J. Am. Chem. Soc. **2004**, *126* (41), 13220–13221.

134. Mantovani, G.; Lecolley, F.; Tao, L.; Haddleton, D.M.; Clerx, J.; Cornelissen, J.; Velonia, K. Design and synthesis of N-maleimido-functionalized hydrophilic polymers via copper-mediated living radical polymerization: A suitable alternative to PEGylation chemistry. J. Am. Chem. Soc. **2005**, *127* (9), 2966–2973.

135. Bontempo, D.; Heredia, K.L.; Fish, B.A.; Maynard, H.D. Cysteine-reactive polymers synthesized by atom transfer radical polymerization for conjugation to proteins. J. Am. Chem. Soc. **2004**, *126* (47), 15372–15373.

136. Heredia, K.L.; Bontempo, D.; Ly, T.; Byers, J.T.; Halstenberg, S.; Maynard, H.D. In situ preparation of protein—"Smart" polymer conjugates with retention of bioactivity. J. Am. Chem. Soc. **2005**, *127* (48), 16955–16960.

137. Lele, B.S.; Murata, H.; Matyjaszewski, K.; Russell, A.J. Synthesis of uniform protein-polymer conjugates. Biomacromolecules **2005**, *6* (6), 3380–3387.

138. Bontempo, D.; Maynard, H.D. Streptavidin as a macroinitiator for polymerization: In situ protein-polymer conjugate formation. J. Am. Chem. Soc. **2005**, *127* (18), 6508–6509.

139. Sen Gupta, S.; Raja, K.S.; Kaltgrad, E.; Strable, E.; Finn, M.G. Virus-glycopolymer conjugates by copper(I) catalysis of atom transfer radical polymerization and azide-alkyne cycloaddition. Chem. Commun. **2005**, (34), 4315–4317.

140. Hoffman, A.S.; Stayton, P.S. Bioconjugates of smart polymers and proteins: Synthesis and applications. Macromol. Symp. **2004**, *207* (1), 139–151.

141. Ding, Z.L.; Chen, G.H.; Hoffman, A.S. Unusual properties of thermally sensitive oligomer-enzyme conjugates of poly(N-isopropylacrylamide)-trypsin. J. Biomed. Mater. Res. **1998**, *39* (3), 498–505

142. Stayton, P.S.; Shimoboji, T.; Long, C.; Chilkoti, A.; Chen, G.H.; Harris, J.M.; Hoffman, A.S. Control of protein-ligand recognition using a stimuli-responsive polymer. Nature **1995**, *378* (6556), 472–474.

143. Chilkoti, A.; Chen, G.H.; Stayton, P.S.; Hoffman, A.S. Site-specific conjugation of a temperature-sensitive polymer to a genetically-engineered protein. Bioconjug. Chem. **1994**, *5* (6), 504–507.

144. Bose, M.; Groff, D.; Xie, J.M.; Brustad, E.; Schultz, P.G. The incorporation of a photoisomerizable amino acid into proteins in E-coli. J. Am. Chem. Soc. **2006**, *128* (2), 388–389.

145. Velonia, K.; Rowan, A.E.; Nolte, R.J.M. Lipase polystyrene giant amphiphiles. J. Am. Chem. Soc. **2002**, *124* (16), 4224–4225.

146. Boerakker, M.J.; Hannink, J.M.; Bomans, P.H.H.; Frederik, P.M.; Nolte, R.J.M.; Meijer, E.M.; Sommerdijk, N. Giant amphiphiles by cofactor reconstitution. Angew. Chem. Int. Ed. **2002**, *41* (22), 4239–4241.

Conducting—Dendritic

Contact Lenses: Gas Permeable

Jay F. Künzler
Department of Chemistry and Polymer Development, Bausch and Lomb Incorporated

Abstract

There exist three classes of contact lens materials. In each of these material classes there has been an extensive amount of research in the design of new materials with improved oxygen permeability. The main objective of this research is to design contact lens materials for extended wear applications. This entry reviews the currently available high oxygen permeable contact lens materials and their design.

In order to design a successful contact lens material, the candidate polymer must satisfy a number of material requirements.[1–3] The material must be optically transparent, possess chemical and thermal stability, and be biologically compatible with the ocular environment. The material must also possess excellent wettability, durability, and comfort. In addition, it is important that the material can be bulk-polymerized and processed utilizing present contact lens manufacturing techniques.[4] Finally, the material must be permeable to oxygen. Owing to a lack of blood vessels within the corneal framework, the cornea obtains oxygen from the atmosphere. Placing a contact lens over the cornea decreases the exposure to the atmosphere and reduces the supply of oxygen to the cornea. Without an adequate supply of oxygen, corneal edema may occur resulting in a number of adverse physiological responses.[5–8] The two key intrinsic material properties that are a measure of oxygen diffusion are oxygen permeability (DK) and oxygen transmissibility (DK/L) where the actual amount of oxygen reaching the cornea is proportional to the lens thickness L.[9] The units for DK are in barrers where 1 barrer = $[10^{-11}$ cm^3 O$_2$ (STP) cm]/(cm^2 s mmHg). The currently accepted levels of DK/L are 24 barrers for a daily wear lens and 50 barrers for an extended wear lens.[6]

There exist three classes of contact lens materials: the hard, high modulus contact lens materials, the soft, low modulus contact lens materials that possess little or no water. In each of these material classes there has been an extensive amount of research in the design of new materials with improved oxygen permeability. The main objective of this research is to design contact lens materials for extended wear applications. This entry will review the currently available high oxygen permeable contact lens materials and provide a review of advances in the design of contact lens materials possessing high oxygen permeability.

HARD GAS PERMEABLE CONTACT LENSES

Silicone Acrylates

The class of hard contact lenses consists of transparent, crosslinked polymeric materials that possess a high modulus (>100,000 g/mm^2) and contain less than 1% water. The first successful hard contact lens material was based on the rigid polymer poly(methyl methacrylate) (PMMA) [80-62-66] that was developed in the 1950s.[10] PMMA was an ideal candidate for use as a hard contact lens because it possessed a high modulus (200,000 g/mm^2) that resulted in excellent visual acuity, and exhibited a low rate of breakage and excellent dimensional stability. The PMMA lenses were prepared by a bulk, radical polymerization of methyl methacrylate to form rods or buttons from which a lens was obtained by secondary lathing operations.[4] This process has remained essentially unchanged today for all commercially available hard contact lens materials despite research efforts through the years to simplify this process.

Despite its success, the PMMA lens suffered from two major disadvantages: poor initial comfort that required lengthy adaptation wearing times and a low permeability to oxygen. PMMA is essentially a barrier to oxygen with a DK of 0.5 barrers and is not suitable for extended wear. In the early 1970s three new hard contact lens materials with improved levels of oxygen permeability were introduced: lens materials based on cellulose acetate butyrate [9004-36-8], 4-methyl-1-pentene [691-37-2], and t-butylstyrene [26009-55-2].[11–13] All three of these materials posses oxygen permeability values in the 10–20 barrers range, a significant improvement when compared to PMMA; however, these materials suffered from poor wetting characteristics, poor scratch resistance, and poor dimensional stability. A major advance in the design of hard-gas permeable lenses occurred in 1975 with the introduction of the first siloxane based hard contact lens material developed by Gaylord.[14] Gaylord found that by copolymerizing methyl methacrylate with methacrylate functionalized siloxanes, he could blend the excellent stability and processing characteristics of MMA with the high oxygen permeability characteristics of silicone. Silicone based materials, in fact, possess a permeability to oxygen 1000 times higher than that of PMMA. Specifically, Gaylord discovered that by copolymerizing methyl methacrylate with the silicone acrylate, methacryloylpropyl tris(trimethylsiloxy silane) (Scheme 1) (TRIS) [17096-07-0], together with methacrylic acid [79-41-4] as a wetting agent and a crosslinker such as ethylene glycol dimethacrylate [97-90-5], a high modulus lens material possessing excellent scratch resistance, wettability, dimensional stability, and oxygen permeability could be obtained. The bulky tertiary structure of TRIS resulted in high levels of oxygen permeability without a significant drop in modulus, a typical drawback when copolymerizing MMA with long-chain methacrylate functionalized polydimethylsiloxanes (PDMS). A homopolymer film of TRIS is, in fact, a semi-hard material (modulus = 20,000 g/mm^2) possessing an oxygen permeability of >200 barrers. In these TRIS/MMA formulations a ratio of MMA to TRIS exists where optimum levels of oxygen permeability, modulus, dimensional stability, and wettability are achieved. High levels of TRIS result in a reduction in wettability and an increase in lipid deposition owing to the lipophilicity of PDMS-based materials.

With the success of the Gaylord material, an extensive amount of research continued on the design of hard contact lenses possessing improved wetting, deposit resistance, and increased levels of oxygen permeability. The majority of this work focused on the synthesis of new TRIS-like silicone acrylate derivatives. A variety of approaches have been pursued successfully including substitution of the methyl substituent attached to silicone with longer chain silicones, and bulky aliphatic groups and hydrophilic groups.[15–18] In addition, the substitution of linear or cyclic alkyl spacers containing hydrophilic groups between the siloxane and methacrylate and replacement of the methacrylate for other unsaturated polymerizable groups such as acrylamides, itaconates, maleimides, and fumarates has been successfully pursued.[19–26] Several hard lens materials

based on TRIS chemistry are now commercially available (Table 1). The oxygen permeability values for this class of material are in the 10–30 barrers range.

In addition to the TRIS derivatives, an extensive amount of effort into the design of high DK, rigid materials based on methacrylate end capped polydimethylsiloxane telechelics and methacrylate capped siloxane macromers has been reported with no present commercial application.[27–30]

Fluoroacrylates

The design of contact lenses based on fluoro polymers was first reported by Cleaver, Barkdoll, and Girard at Dupont in the late 1960s.[31,32] In this work the high oxygen permeability of fluoromethacrylates and various perfluoro methyl vinyl ether copolymers was described; however, it was not until the mid to late 1980s that contact lenses based on fluoropolymers were actively pursued. The copolymerization of fluoromethacrylates, such as hexafluoroisopropylmethacrylate (HFIM) (Scheme 2), with TRIS and MMA together with suitable wetting agents and crosslinkers, has resulted in a series of new, dimensionally stable, biocompatible, high DK, hard materials possessing DK values in the 30–160 barrers range.[33–36] Several approaches in the design of fluorine-based hard contact lenses have been pursued including the synthesis of polymers based on

Table 1 Representative examples of commercially available hard gas permeable lenses

Manufacturer	Composition	Trade name	DK[a]
Danker Labs	CAB	Meso	12.3
	Silicone acrylate	Dura-sil	18.0
Polymer Technology	Silicone acrylate	Boston IV	28.7
	Fluoro silicone	Equalens[b]	64.0
Cooper Vision	Silicone acrylate	Cooper HGP	14.6
	Fluoro silicone	Fluoroflex	70.0
Menicon	Fluoro silicone	Menicon SF-P	159.0
PBH	Fluoro silicone	Polycon HDK[b]	40.0
Concise Contact Lens	Fluoro silicone	Oxyflow F-92[b]	92.0

[a]DK in units of barrers, 1 barrer = [10^{-11} cm^3 O$_2$ (STP) cm]/(cm^2 s mmHg).
[b]7 day continuous wear approval.

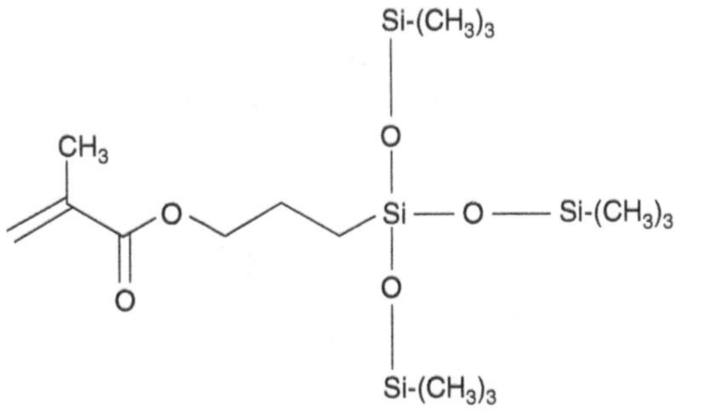

Scheme 1 Methacryloylpropyl tris(trimethylsiloxy silane).

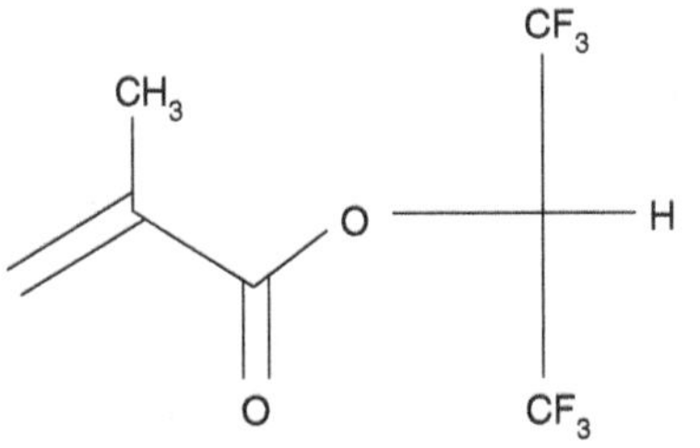

Scheme 2 Hexafluoroisopropylmethacrylate (HFIM).

fluoro-substituted styrenics, fluorosubstituted itaconates, silicone substituted fumarates with fluoromethacrylates, and silicone-fluorine substituted star polymers prepared by group transfer polymerization.[37–40] Several of these fluoro-based materials are now commercially available for seven-day extended wear application (Table 1).

Fluoroethers

A relatively new class of high DK hard materials based on perfluorinated polyethers [25322-68-3] has been developed by Rice and co-workers at 3M.[41] These materials are based on copolymers of methacrylate end-capped, low molecular weight perfluorinated polyethers (Scheme 3). The copolymerization of the perfluorinated polyethers with a variety of wetting agents and additional crosslinkers results in contact lenses possessing an excellent resistance to deposition, and oxygen permeability levels exceeding 100 barrers. These materials are sold through Allergan as the Advent lens.

Polyacetylenes

The preparation of alkyl substituted acetylenes has been extensively studied by the Higashimura and Masuda research group at Kyota University. They have demonstrated that acetylenes containing bulky substituents such as the monomer 1-trimethylsilyl-1-propyne [6224-91-5] (Scheme 4) can be effectively polymerized to high molecular weight polymers using the metal halide-based initiators. Solution casting from these polymers has resulted in stable, clear films possessing oxygen permeability levels 10 times that of polydimethylsiloxane.[42,43] It has been shown that the high levels of oxygen permeability decrease with time, presumably owing to polymer rearrangement; however, the permeability levels remain high (200 barrers).

Initial studies on the use of poly(1-trimethylsilyl-1-propyne) for contact lenses was discouraging. Films of poor wettability could be obtained only by solution-casting

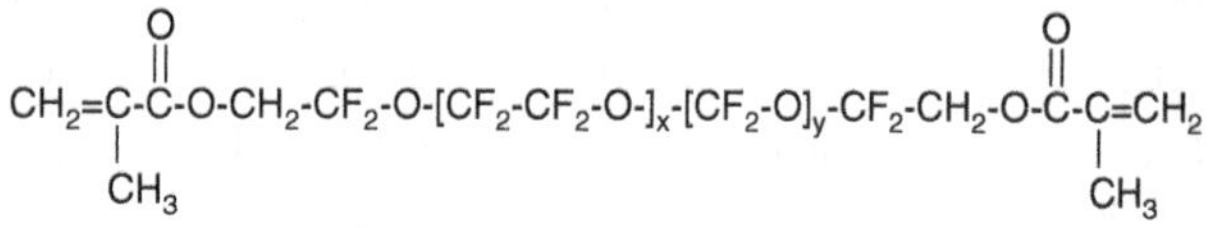

Scheme 3 Methacrylate end-capped, low molecular. weight perfluorinated polyethers.

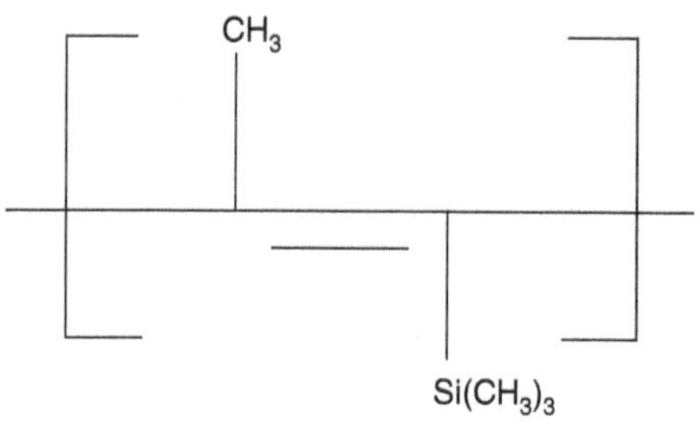

Scheme 4 1-trimethylsilyl-1-propyne [6224-91-5].

techniques. Results by a variety of research groups, however, have shown that the acetylene molecular weight can be controlled and that functionalization with polymerizable groups can be achieved.[44,45] This is encouraging in that lens processing by conventional bulk polymerization techniques may be possible. In addition, it has been reported that wettability can be achieved using surface plasma polymerization techniques.[46] The long-term thermal and oxidative stability of the substituted polyacetylenes remains unknown.

Polyimides

Studies have shown that fluorinated polyimides based on hexafluoroisopropylidene diaryldianhydride (6FDA) and isopropylidenedianiline (Scheme 5) exhibit unexpected high levels of oxygen permeability when compared to other high modulus, transparent polymers. Solution cast films from these polymers show DK levels of 5–16 barrers.[47] In addition, the preparation of aromatic polyimides containing oligodimethydi and tri siloxanes for contact lens use was reported.[48,49] In this work polyimides were synthesized by polycondensation of bis (4-aminophenyl)-functionalized siloxanes with acid dichlorides or dianhydrides. These polymers have reported DK levels of 8–12 barrers.

HYDROGELS

High Water Content

Hydrogels are hydrophilic polymers that absorb water to an equilibrium value and are insoluble in water owing to the presence of a three-dimensional network. The hydrophilicity is due to the presence of hydrophilic groups, such as alcohols, carboxylic acids, amides, sulfonic acids, etc. The swollen equilibrated state results from a balance between the osmotic driving forces that cause the water to enter the hydrophilic polymer and the forces exerted by the polymer chains in resisting expansion.[50] The optimum physical and mechanical properties of a hydrogel for contact lens application include a Youngs modulus between 20 and 200 g/mm^2, a tear strength greater than 2.0 g/mm, water contents in the 35–60% range, and a DK greater than 50 barrers. These physical and mechanical property objectives were chosen based on clinical experience from a variety of commercial and experimental lens materials.

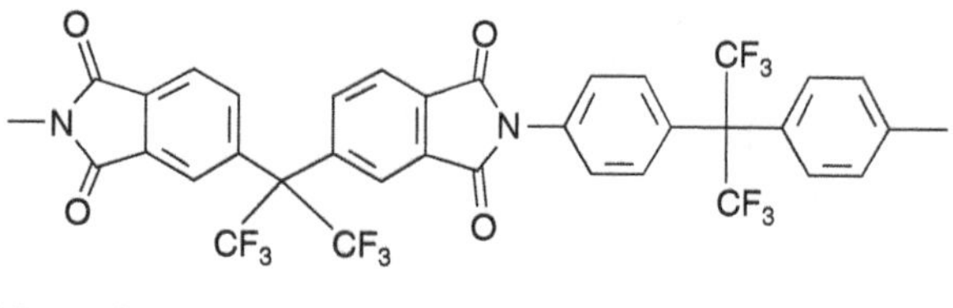

Scheme 5 Isopropylidenedianiline.

The soft hydrogel contact lenses are prepared by a bulk radical polymerization utilizing cast molding and spin-casting techniques.[4]

In the 1960s Wichterle developed the first hydrogel contact lens based on 2-hydroxyethyl methacrylate (HEMA) [868-77-9] which was commercialized in 1971 by Bausch and Lomb.[51] This 38% water-containing hydrogel lens has been extremely successful. Poly(HEMA) possesses excellent wettability, comfort, and deposit resistance. The only significant limitation of poly(HEMA) is that it has a relatively low permeability to oxygen (10 barrers) and is not suited for long-term, extended-wear application. As a result, an extensive amount of research into the design of hydrogels possessing higher levels of oxygen permeability has been undertaken. The first and most obvious route is the design of high water content hydrogels.[9,50] A direct correlation between water content and oxygen permeability exists where the higher the water content, the higher the oxygen permeability.

There have been several approaches to the design of high water-content hydrogels. The first approach has involved the polymerization of highly hydrophilic, non-ionic monomers such as *N*-vinylpyrrolidinone (NVP) [88-12-0], dimethylacrylamide [2680-03-7], polyvinylalcohol [9002-89-5] and glycerol methacrylate [100-92-5].[51-54] The second method involves the co-polymerization of moderately hydrophilic monomers, such as HEMA and hydroxy propyl methacrylate [27813-02-1], with highly hydrophilic, ionic monomers such as methacrylic acid.[55] The ionic functionality in a buffered saline environment dramatically increases the water content of the resultant hydrogel. For example, a formulation consisting of 94% w/w HEMA copolymerized with 6% w/w methacrylic acid results in a hydrogel, following a buffered saline hydration, which contains 70% water (compared to HEMA, which contains 38% water). In addition, a significant amount of work in the design of collagen and other natural polymers has successfully been completed.[56] Several commercially available high water materials are now currently available for 7-day extended wear application (Table 2).

There are several basic limitations of high water-content hydrogels. The first is that high water-content materials typically possess poor tear strength. This issue has been addressed with the development of novel hydrophilic, bulky strengthening agents suitable for high water-content lenses.[57,58] Further, high water-content materials often exhibit a high affinity for protein, particularly for hydrogels possessing an ionic functionality and in dry environments, induce epithelial dehydration, which may cause several adverse physiological responses.[59,60]

Silicone-Based Hydrogels

In an attempt to combine the high oxygen permeability of polydimethylsiloxane materials and the excellent comfort, wetting, and deposit resistance of conventional, non-ionic low water hydrogels, the design of silicone-based hydrogels has been studied for contact lens application. PDMS, owing to its low modulus of elasticity, optical transparency, and high oxygen permeability, is an excellent candidate for use in high DK hydrogels. The design of silicone based hydrogels has primarily involved the copolymerization of methacrylate or vinyl functionalized polydimethylsiloxanes with hydrophilic monomers. The biggest limitation in the design of silicone hydrogels is that silicone-based monomers are hydrophobic and insoluble in hydrophilic monomers. The copolymerization of methacrylate, functionalized silicones with hydrophilic monomers generally results in opaque, phase-separated materials. There have been a variety of approaches to design transparent silicone hydrogels. One approach has been to synthesize siloxanes containing hydrophilic groups incorporated as polymer end caps, blocks, or side chains.[61-63] Keogh prepared methacrylate, end-capped polydimethylsiloxanes containing polyethylene oxide (PEO) side chains. The copolymerization of the PEO side-chain siloxanes with varying concentrations of hydrophilic monomers resulted in transparent hydrogels without the use of a solubilizing agent. These materials exhibited a wide range in water content and high oxygen permeability (DK values in the 50–200 range). In addition, the design of silicone hydrogels based on methacrylate, end-capped polyethylene oxide monomers with methacrylate functionalized silicones has also been pursued.[64,65] Other approaches have

Table 2 Representative examples of commercially available soft contact lens hydrogels

Manufacturer	Composition	Trade name	% Water	DK[a]
Bausch & Lomb	HEMA	Medalist	38	8.4
	MMA/NVP	B&L 70[b]	70	33
Vistacon	HEMA/methacrylic acid	Acuvue	58	28
Wesley-Jessen	HEMA/methacrylic acid	Durasoft[b]	55	17
Ciba Vision	HEMA/methacrylic acid	Softcon[b]	55	16
Cooper Vision	HEMA/NVP	Permaflex[b]	74	38.9
	HEMA/NVP	Permalens[b]	71	34.0

[a]DK in units of barrers, 1 barrer = $[10^{-11} cm^3 O_2 (STP) cm]/(cm^2 s\ mmHg)$.
[b]7 day extended wear approval.

been the design of hydrogels based on silicone-functional-ized urethane prepolymers, methacrylate end capped mac-romers, and vinyl functionalized TRIS derivatives, where compatibility in these systems is achieved either by incor-poration of a hydrophilic group or by the use of a solubiliz-ing agent.[22,30,66–70] One approach has been the design of methacrylate-capped TRIS containing fluorinated side chains (Scheme 6). In this system excellent compatibility with hydrophilic monomers is apparently achieved by the incorporation of the terminal polar -(CF$_2$)-H tail.[71] Finally, the compatibility of hydrophilic monomers with siloxane monomers has been achieved by the trimethylsilyl (TMS) protection of hydrophilic monomers such as HEMA.[72] The TMS protected HEMA is soluble in all proportions with siloxane containing monomers. A contact lens is prepared by curing the monomer mix containing the TMS-protected HEMA and siloxane monomer. The fin-ished lens is then placed in a mild acid solution to remove trimethylsilyl group, organically extracted to remove hexa-methyldisiloxane and then hydrated resulting in transpar-ent hydrogels of high oxygen permeability. At present, silicone hydrogel contact lens materials are not commer-cially available.

Fluorohydrogels

Another approach to prepare oxygen permeable hydrogels has been the design of hydrogels based on fluoropolymers. The preparation of oxygen permeable hydrogels by the copolymerization of fluorinated methacrylates and methac-rylate functionalized fluorinated polyethylene oxides with hydrophilic monomers has been reported.[73–80] The oxygen permeability of these hydrogels is in the 30–50 barrers range, significantly less than the silicone based hydrogels; however, it is claimed that the wetting and biocompatibility characteristics of these silicone-free fluorohydrogels are superior to the silicone-based hydrogels.

ELASTOMERS

Low-Water Soft Silicone

The design of low-water (<1%) silicone-based materials has been extensively studied with limited success. The first

series of silicone-based, low-water elastomers was com-posed of high molecular weight vinyl and hydride func-tionalized polysiloxanes. These systems were polymerized by a platinum-initiated hydrosilation mechanism, and wet-tability was achieved by plasma treatment.[81,82] Initial results from the silicone-based elastomers were mixed. Clinical results showed that there was no change in corneal physiology with long-term wear (owing to the extremely high oxygen permeability of the these materials (500 bar-rers); however, the silicone-based elastomer lenses exhib-ited a high affinity for lipids, poor wetting, and lens adhesion.[83–85]

Some work focused on the design of low-water silicone elastomers using methacrylate functionalized siloxanes. Much of the chemistry described for the design of silicone hydrogels has been applied to the design of low water sil-icone-based lenses.[61–72,86–90] In the work of Deichert low water silicone elastomers were prepared using methacry-late end capped polydimethylsiloxanes (Scheme 7).[87] In these systems the cure was completed either by a thermal or ultraviolet-initiated radical polymerization. Lens wet-tability was achieved by plasma treatment or by the use of an internal polymerizable wetting agent. Another approach was the design of lenses utilizing methacrylate functional-ized fluoro siloxanes and fluoromethacrylates in an attempt to reduce the lipid uptake of the silicone-based lenses.[91,92] One approach consists of designing materials based on a methacrylate end-capped poly (trifluoropro-pylmethylsiloxane) (TFP), octafluoropentylmethacrylate (OFPMA) and a wetting agent, 2-vinyl-4,4-dimethyl-2-oxazolin-5-one (VDMO).[92] The trifluoropropylmethyl siloxane, fluorinated methacrylate methacrylate and VDMO in an optimum concentration provide a lipid resis-tant, oxygen permeable, hydrophilic lens material. The incidence of lens adhesion for these polymers systems was not reported.

Poly(ethylene oxide)-Based Elastomers

The design of low-water content, high-oxygen permeable contact lens materials based on poly(ethylene oxide) poly-mers has been reported by several research groups. These approaches have involved the preparation of methacrylate end-capped polyurethanes containing a poly(ethylene oxide) and propylene oxide soft segments.[93] In this work low modulus, low water, wettable materials possessing DK values approaching 70 barrers have been reported. This high level of oxygen permeability is presumably due to the high chain flexibility of the PEO soft segments. In addition, the preparation of contact lens materials based on sugars

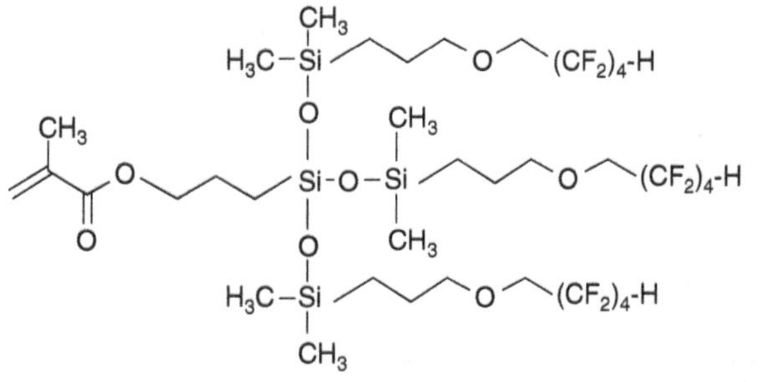

Scheme 6 Methacrylate-capped TRIS.

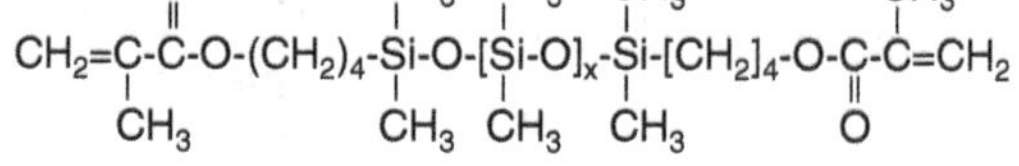

Scheme 7 Low water silicone elastomers.

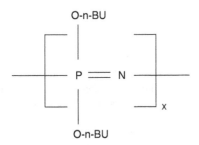

O-n-BU

P ≡ N

O-n-BU

Scheme 8 Poly(n-butyl phosphazene).

containing methacrylate-functionalized polyethylene oxide grafts has also been reported.[94] The long-term hydrolytic stability of the polyethylene oxide-based polymers remains unknown.

Polyphosphazenes

An extensive amount of research on the design of polyorganophosphazenes membranes for use as gas separation membranes and biomedical polymers has been reported by several research groups.[95,96] Although no specific mention in the literature for the use of polyphosphazenes as contact lens materials has been found, the reported oxygen permeability values for several phosphazenes make them excellent candidates for contact lens materials. The reported oxygen permeability of polytrifluoroethoxyphosphazene[28212-50-2] is 40 barrers and poly(n-butyl phosphazene's) (Scheme 8) is 128 barrers. The major limitation of polyphosphazenes for use as contact lens materials is the difficulty in polymer synthesis and polymer processing. Solution casting is presently the only viable method to obtain shaped bodies.

REFERENCES

1. Peppas, N.A.; Yand, W.M. Contact Lens **1981**, *7*, 300.
2. Keogh, P.L. Approaches to the development of improved soft contact lens materials. Artif. Cells Blood Substit. Biotechnol. **1979**, *7* (2), 307–311.
3. Tighe, B.J. The role of permeability and related properties in the design of synthetic hydrogels for biomedical applications. Br. Polym. J. **1986**, *18* (1), 8–13.
4. Ruscio, D.V. Polym. Prep. Am. Chem. Soc. Div. Polym. Mat. Sci. Eng. **1993**, *69*, 221.
5. Holden, B.W.; Mertz, G.W.; McNally, J.J. Corneal swelling response to contact lenses worn under extended wear conditions. Invest. Ophthalmol. Vis. Sci. **1983**, *24* (2), 218–226.
6. Holden, B.A; Mertz, G.W. Critical oxygen levels to avoid corneal edema for daily and extended wear contact lenses. Invest. Ophthalmol. Vis. Sci. **1984**, *25* (10), 1161–1167.
7. Schoessler, J. Corneal endothelial polymegathism associated with extended wear. Int. Contact Lens Clin. **1983**, *10*, 148–155.
8. Sarver, M.; Baggett, D.; Harris, M.; Louie, K. Corneal edema with hydrogel lenses and eye closure, effect of oxygen transmissibility. Am. J. Optom. Physiol. Opt. **1981**, *58* (5), 387–392.
9. Lai, Y.; Wilson, A.; Zantos, S. Contact lenses. In *Kirk-Othmer Encyclopedia of Chemical Technology*, 4th Ed.; 1993; Vol. 7, p. 191.
10. Tuohy, K.M. Contact Lens. U.S. Patent 2510438, Jun 6, 1950.
11. Ivani, E.J. Semi-rigid, Gas Permeable Contact Lenses. U.S. Patent 3900250, Aug 19, 1975.
12. Kamath, P.M. Rigid Gas Permeable Plastic Contact Lens (OCR). U.S. Patent 3551035, Dec 29, 1970.
13. Loschaek, S. Contact Lenses of High Gas Permeability. U.S. Patent 4228269, Oct 14, 1980.
14. Gaylord, N.G. Oxygen-permeable Contact Lens Composition, Methods and Article of Manufacture. U.S. Patent 3808178, Apr 30, 1974.
15. Ellis, E.J. Silicone-containing Contact Lens Material and Contact Lenses Made Thereof. U.S. Patent 4424328, Jan 3, 1984.
16. Novicky, N.N. Oxygen-permeable Contact Lens Compositions, Methods and Articles of Manufacture. U.S. Patent 4216303, Aug 5, 1980.
17. Mueller, K.F.; Harisiades, P. Rigid, Gas-permeable Polysiloxane Contact Lenses. U.S. Patent 4923906, May 8, 1990.
18. Ellis, E.J. Silicone-containing Contact Lens Material and Contact Lenses Made Thereof. U.S. Patent 4463149, Jul 31, 1984.
19. Tanaka, K.; Takahashi, K.; Kanada, M.; Yoshikawa, T. Methyldi(trimethylsiloxy)sylylpropylglycerol Methacrylate. U.S. Patent 4139548, Feb 13, 1979.
20. Cho, E. Siloxane-containing Polymers and Contact Lenses Therefrom. U.S. Patent 4450264, May 22, 1984.
21. Ikari, M. Oxygen Permeable Hard Contact Lens. U.S. Patent 5162391, Nov 10, 1992.
22. Harvey, T.B. Hydrophilic Siloxane Monomers and Dimers for Contact Lens Materials, and Contact Lenses Fabricated Therefrom. U.S. Patent 4711943, Dec 8, 1987.
23. Ellis, E.J.; Salamone, J.C. Silicone-containing Hard Contact Lens Material. U.S. Patent 4152508, May 1, 1979.
24. Kawano, T.; Takehana, J.M.; Yokota, M. Maleimide Polymers and Contact Lenses from Such Polymers. EP0560620, Sep 15, 1993.
25. Amaya, N.; Amami, K.; Egwa, M. Contact Lens Article Made of Silicon- and Fluorine-containing Resin. U.S. Patent 4933406, Jun 12, 1990.
26. Kawaguchi, T. Hard Contact Lens Material Consisting of Alkyl Fumarate and Silicon-alkyl Fumarate Copolymers. U.S. Patent 4868260, Sep 19, 1989.
27. Deichert, W.G.; Su, K.C.; Van Buren, M.F. Polysiloxane Composition and Contact Lens. U.S. Patent 4153641, May 8, 1979.
28. Friends, G.D.; Melpolder, J.B.; Kunzler, J.F.; Park, J.S. Polysiloxane Composition with Improved Surface Wetting Characteristics and Biomedical Devices Made Thereof. U.S. Patent 4495361, Jan 22, 1985.
29. Bany, S.; Koshar, R.; Williams, T. Ophthalmic Devices Fabricated from Urethane Acrylates of Polysiloxane Alcohols. U.S. Patent 4543398, Sep 24, 1985.
30. Mueller, K.F. Reactive Silicone and/or Fluorine Containing Hydrophilic Prepolymers and Polymers Thereof. U.S. Patent 5079319, Jan 7, 1992.
31. Barkdoll, E. Soft, Tough Low Refractive Index Contact Lenses. U.S. Patent 3940207, Feb 24, 1976.

32. Cleaver, C.S. Contact Lens Having an Optimum Combination of Properties. U.S. Patent 3950315, Apr 13, 1976.

33. Kawamura, K.; Yamashita, S.; Yokoyama, Y.; Tsuchiya, M. Oxygen-permeable Hard Contact Lens. U.S. Patent 4540761, Sep 10, 1985.

34. Yamamoto, F.; Suzuki, T.; Ikari, M.; Saito, S.; Ohmori, A.; Yasuhara, T. Contact Lens. U.S. Patent 4684705, Aug 4, 1987.

35. Stoyan, N. Continuous-wear Highly Oxygen Permeable Contact Lenses. U.S. Patent 4829137, May 9, 1989.

36. Kossmehl, G.; Fluthwedel, A.; Schäfer, H. Highly oxygen-permeable copoly(methacrylate/siloxane)s with perfluoroalkyl ester groups. Makromol. Chem. 1992, 193 (1), 157–166.

37. Falcetta, J.J.; Park, J.S. p-(2-Hydroxy hexafluoroisopropyl) styrene [HFIS] Monomer for Ophthalmic Applications. U.S. Patent 4690993, Sep 1, 1987.

38. Ellis, E.J.; Ellis, J.Y. Fluorine Containing Polymeric Compositions Useful in Contact Lenses. U.S. Patent 4686267, Aug 11, 1987.

39. Kawaguchi, T.; Ando, I.; Toyoshima, N.; Yamamoto, Y.; Yoshioka, H.; Yamazaki, T. Ocular Lens Material. U.S. Patent 5250583, Oct 5, 1993.

40. Seidner, L.; Spinelli, H.; Ali, M.; Weintrab, L. Silicone-containing Contact Lens Polymers, Oxygen Permeable Contact Lenses and Methods for Making These Lenses and Treating Patients with Visual Impairment. U.S. Patent 5331067, Jul 19, 1994.

41. Rice, D.; Ihlenfeld, J.V. Contact Lens Containing a Fluorinated Telechelic Polyether. U.S. Patent 4440918, Apr 3, 1984.

42. Masuda, T.; Isobe, E.; Higashimura, T.; Takada, K. Poly[1-(trimethylsilyl)-1-propyne]: A new high polymer synthesized with transition-metal catalysts and characterized by extremely high gas permeability. J. Am. Chem. Soc. 1983, 105 (25), 7473–7474.

43. Masuda, T.; Hamano, T.; Tsuchihara, K.; Higashimura, T. Synthesis and characterization of poly[[o-(trimethylsilyl)phenyl]acetylene]. Macromolecules 1990, 23 (5), 1374–1380.

44. Kunzler, J.F.; Percec, V. The polymerization of alkyl substituted acetylenes using metal halide based initiators: The bulky substituent effect. Polym. Bull. 1992, 29 (3–4), 335–342.

45. Kunzler, J.F.; Percec, V. Oxygen Permeable Polymeric Materials. U.S. Patent 4833262, May 23, 1989.

46. Masuda, T.; Kotoura, M.; Tsuchihara, K.; Higashimura, T. Glow-discharge-induced graft polymerization of acrylic acid onto poly[1-(trimethylsilyl)-1-propyne] film. J. Appl. Polym. Sci. 1991, 43 (3), 423–428.

47. Coleman, M.R.; Koros, W.J. Isomeric polyimides based on fluorinated dianhydrides and diamines for gas separation applications. J. Membr. Sci. 1990, 50 (3), 285–297.

48. Kawakami, Y. Ocular Lens Material. U.S. Patent 5260352, Nov 9, 1993.

49. Kawakami, Y.; Yu, S.; Abe, T. Synthesis and gas permeability of aromatic polyamide and polyimide having oligodimethylsiloxane in main-chain or in side-chain. Polym. J. 1992, 24 (10), 1129–1135.

50. Kudella, V. Hydrogels. In Encyclopedia of Polymer Science and Engineering; Kroschwitz, J.I., Ed.; Wiley-Interscience: New York, 1987; Vol. 7, p. 783.

51. Wichterle, O.; Lim, D. Cross-linked Hydrophilic Polymers and Articles Made Therefrom. U.S. Patent 3220960, Nov 30, 1965.

52. Mancini, W.; Korb, D.; Refojo, M.F. Hydrogels and Articles Made Therefrom. U.S. Patent 3957362, May 18, 1976.

53. Izumitani, T.; Tarumi, N.; Komiya, S.; Sawamoto, T. High-hydration Contact Lens. U.S. Patent 4625009, Nov 25, 1986.

54. Ofstead, R.F. Copolymers Of Poly(vinyl Trifluoroacetate) Or Poly(vinyl Alcohol). U.S. Patent 4618649, Oct 21, 1986.

55. Steckler, R. Anionic Hydrogels Based on Heterocyclic N-vinyl Monomers. U.S. Patent 4036788, Jul 19, 1977.

56. Miyata, T.; Rubin, A.L.; Dunn, M.W.; Stenzel, K.H. Collagen Soft Contact Lens. U.S. Patent 4223984, Sep 23, 1980.

57. Kunzler, J.; Friends, G. Polymer Compositions for Contact Lenses. U.S. Patent 5270418, Dec 14, 1993.

58. Vanderlaan, D.G. Ophthalmic Lens Polymer Incorporating Acyclic Monomer. US5256751, Oct 26, 1993.

59. Minarik, L.; Rapp, J. Contact Lenses (CLOA Journal) 1989, 15, 185.

60. Orsborn, G.; Zantos, S. Contact Lenses (CLOA Journal) 1988, 14, 81.

61. Mueller, K.F.; Kleiner, E.K. Polysiloxane Hydrogels. U.S. Patent 4136250, Jan 23, 1979.

62. Su, K.; Robertson, J.R. Wettable, Flexible, Oxygen Permeable, Substantially Non-swellable Contact Lens Containing Block Copolymer Polysiloxane-Polyoxyalkylene Backbone Units, and Use Thereof. U.S. Patent 4740533, Apr 26, 1988.

63. Keogh, P.L.; Kunzler, J.F.; Niu, G.C. Hydrophilic Contact Lens Made from Polysiloxanes Which are Thermally Bonded to Polymerizable Groups and Which Contain Hydrophilic Sidechains. U.S. Patent 4260725, Apr 7, 1981.

64. Lai, Y.C.; Baccei, L.J. Synthesis and structure–property relationships of UV-curable urethane prepolymers with hard–soft–hard blocks. J. Appl. Polym. Sci. 1991, 42 (7), 2039–2044.

65. Chang, S.H. Hydrophilic, Soft and Oxygen Permeable Copolymer Composition. U.S. Patent 4182822, Jan 8, 1980.

66. Tanaka, K.; Takahashi, K.; Kanada, M.; Kanome, S.; Nakajima, T. Copolymer for Soft Contact Lens, Its Preparation and Soft Contact Lens Made Thereof. U.S. Patent 4139513, Feb 13, 1979.

67. Ratkowski, D.J.; Lue, P.C. Silane Ester Contact Lens Composition, Article and Method of Manufacture. U.S. Patent 4535138, Aug 13, 1985.

68. Bambury, R.E.; Seelye, D.E. Novel Vinyl Carbonate and Vinyl Carbamate Contact Lens Material Monomers. EP0396364, Jul 11, 1990.

69. Mueller, K.F.; Plankl, W.L. Fluorine and/or Silicone Containing Poly(alkylene-oxide)-block Copolymers and Contact Lenses Thereof. U.S. Patent 5115056, May 19, 1992.

70. Braatz, J.; Kehr, C. Contact Lenses Based on Biocompatible Polyurethane and Polyurea-urethane Hydrated Polymers. U.S. Patent 4886866, Dec 12, 1989.

71. Kunzler, J.F.; Ozark, R.E. Fluorosilicone Hydrogels. U.S. Patent 5321108, Jun 14, 1994.

72. Yoshikawa, T.; Shibata, T. Oxygen Permeable Soft Contact Lens Material. U.S. Patent 4649184, Mar 10, 1987.

73. Sawamoto, T.; Nomura, M.; Tarumi, N. Contact Lens. U.S. Patent 5008354, Apr 16, 1991.

74. Mueller, K.F. Dimethylacrylamide-copolymer Hydrogels with High Oxygen Permeability. U.S. Patent 4954587, Sep 4, 1990.

75. Goldenberg, M. Crosslinked Copolymers and Ophthalmic Devices Made from Vinylic Macromers Containing perfluoropolyalkyl Ether and Polyalkyl Ether Segments and Minor Amounts of Vinylic Comonomers. U.S. Patent 4929692, May 29, 1990.

76. Mueller, K.F. Dimethylacrylamide-copolymer Hydrogels with High Oxygen Permeability. U.S. Patent 5011275, Apr 30, 1991.

77. Agou, T.; Sakashita, T.; Shimoda, T.; Sudo, M.; Kuwabara, M.; Tanaka, M. Oxygen-permeable Article of Fluorinated Hydrocarbon Group-grafted (Meth)Acrylate Polymers. U.S. Patent 5057585, Oct 15, 1991.

78. Salamone, J.C. Fluorine Containing Soft Contact Lens Hydrogels. U.S. Patent 4990582, Feb 5, 1991.

79. Kossmehl, J.; Volkheimer, J.; Schaefer, H. Hydrogels based on 2.2.3.3.4.4.4.-heptafluorobutyl methacrylate. Acta Polym. **1992**, *43* (6), 335–342.

80. Futamura, H.; Nomura, M.; Yokoyama, Y. Contact Lens Material and Contact Lens. U.S. Patent 5264465, Nov 23, 1993.

81. Becker, W.E. Corneal Contact Lens Fabricated from Transparent Silicone Rubber. U.S. Patent 3228741, Jan 11, 1966.

82. Burdick, D.F.; Mishler, J.L.; Polmanteer, K.E. Blends of Two Polysiloxane Copolymers with Silica. U.S. Patent 3341490, Sep 12, 1967.

83. Sweeney, D.F.; Holden, B.A. Silicone elastomer lens wear induces less overnight corneal edema than sleep without lens wear. Curr. Eye Res. **1987**, *6* (12), 1391–1394.

84. Josephson, J.; Caffery, B. Int. Contact Lens Clin. **1980**, *7*, 235.

85. Mountford, J. The Wohlk Silflex silicone contact lens: A preliminary clinical evaluation. Aust. J. Optom. **1978**, *61* (6), 197–208.

86. Friends, G.D.; Van Buren, M.F. Contact Lens Made from Polymers of Polysiloxane and Polycyclic Esters of Acrylic Acid or Methacrylic Acid. U.S. Patent 4254248, Mar 3, 1981.

87. Chromecek, R.C.; Deichert, W.G.; Falcetta, J.J.; Van Buren, M.F. Polysiloxane/Acrylic Acid/Polcyclic Esters of Methacrylic Acid Polymer Contact Lens. U.S. Patent 4276402, Jun 30, 1981.

88. Keogh, P.L.; Kunzler, J.F.; Niu, G.C. Hydrophilic Contact Lens Made from Polysiloxanes Containing Hydrophilic Sidechains. U.S. Patent 4259467, Mar 31, 1981.

89. Nakashima, T.; Taniyama, Y.; Sugiyama, A. Contact Lens Material. U.S. Patent 4814402, Mar 21, 1989.

90. Schaefer, H.; Kossmehl, G.; Neumann, W. Modified Silicone Rubber and Its Use As a Material for Optical Lenses and Optical Lenses Made from This Material. U.S. Patent 4853453, Aug 1, 1989.

91. Toyoshima, N.; Shibata, T.; Hirashima, A.; Ando, I.; Iwata, N.; Yoshioka, H.; Itagaki, A.; Yamazaki, T. Soft Ocular Lens Material. U.S. Patent 4954586, Sep 4, 1990.

92. Friends, G.D.; Kunzler, J.F. Polymeric Materials with High Oxygen Permeability and Low Protein Substantivity. U.S. Patent 4810764, Mar 7, 1989.

93. Su, K.; Molock, F.F. Wettable, Flexible, Oxygen Permeable, Substantially Non-swellable Contact Lens Containing Polyoxyalkylene Backbone Units, and Use Thereof. U.S. Patent 4780488, Oct 25, 1988.

94. Nunez, I.M.; Ford, J.D. Soft, High Oxygen Permeability Ophthalmic Lens. U.S. Patent 5196458, Mar 23, 1993.

95. Hirose, T.; Mizogucho, K. Gas transport in poly(alkoxyphosphazenes). J. Appl. Polym. Sci. **1991**, *43* (5), 891–900.

96. Lora, S.; Palma, G.; Bozio, R.; Caliceti, P.; Pezzin, G. Polyphosphazenes as biomaterials: Surface modification of poly(bis(trifluoroethoxy)phosphazene) with polyethylene glycols. Biomaterials **1993**, *14* (6), 430–436.

Conducting–Dendritic

Corneas: Tissue Engineering

Dalia A.M. Hamza
Department of Biochemistry, Faculty of Science, Ain Shams University, Cairo, Egypt

Tamer A.E. Ahmed
Medical Biotechnology Department, Genetic Engineering and Biotechnology Research Institute, City of Scientific Research and Technology Applications (SRTA-City), Alexandria, Egypt
Department of Cellular and Molecular Medicine, Faculty of Medicine, University of Ottawa, Ottawa, Ontario, Canada

Maxwell T. Hincke
Department of Cellular and Molecular Medicine and Department of Innovation in Medical Education, Faculty of Medicine, University of Ottawa, Ottawa, Ontario, Canada

Abstract

Corneal diseases are a major cause of vision loss and blindness worldwide, second only to the impact of cataracts in overall importance. Corneal blindness is a major global health problem and the underlying causes are mainly avoidable either through prevention or early treatment. In this entry, we focus on current treatment strategies to address corneal blindness with a special emphasis on tissue engineering-based approaches. The most widely performed treatment strategy for corneal-associated blindness is corneal transplantation or keratoplasty, including penetrating and lamellar keratoplasty. Because of inadequacies associated with current corneal transplantation strategies, there is an increasing tendency for the ophthalmic community to develop alternative strategies including keratoprosthesis and tissue engineering. Keratoprostheses intended to restore vision of patients with severe corneal disease where corneal transplantation has repeatedly failed or is not an option, however, are associated with several drawbacks including glaucoma and retinal detachment. Tissue-engineered corneal substitutes represent a highly promising treatment option for corneal blindness as they mimic the biomechanical environment of the native cornea and have superior integration capabilities. Currently, a wide range of tissue-engineering-based strategies has been established and is being investigated clinically as an alternative to the routinely used techniques. Tissue engineering- based strategies include reconstruction of a full-thickness corneal equivalent or any of the typical corneal layers (i.e., epithelium, stroma, and endothelium). Transplantation of the tissue engineered full-thickness corneal substitute has shown a satisfactory success rate in human subjects. Tissue-engineered corneal substitutes represent an innovative strategy to replace corneal structure and function with subsequent vision restoration.

INTRODUCTION

Cornea-affecting (damaging) diseases are one of the major causes of vision loss and blindness (14% of global blindness). The gold standard and most widely accepted method for treating typical cases of corneal blindness is the transplantation of human donor corneas. Although corneal transplantation with allogeneic tissue has shown a satisfactory success rate, the major problem in most countries is the shortage of high-quality donor tissue. This shortcoming of corneal transplantation using human donor corneas has prompted intense research to develop alternatives to human donor corneas.

A potential alternative is the replacement of damaged corneas with artificial corneal substitutes. Artificial corneal substitutes are designed to replace partial or full-thickness damaged or diseased corneas. They can include fully synthetic corneal substitutes (keratoprostheses) that solely address replacement of corneal functions, as well as tissue-engineered corneal substitutes that allow regeneration of endogenous cells and nerves to provide fully-integrated biologically functional corneas. Tissue engineering of the functional human corneal implant has recently been presented as a promising solution to overcome the limitations of corneal replacement with allografts and to address the drawbacks of keratoprostheses. Herein, we review the development of the most commonly used strategies to address the replacement of damaged corneas, and discuss the current state of this field, including corneal allotransplantation, corneal xenotransplantation, keratoprothesis, and tissue engineering of corneal substitute.

Concise Encyclopedia of Biomedical Polymers and Polymeric Biomaterials DOI: 10.1081/E-EBPPC-140000003

CORNEAL STRUCTURE AND FUNCTIONS

The human cornea is an optically transparent tissue that constitutes the anteriocentral portion of the eye, with a thickness of approximately 0.5 mm centrally and 0.7 mm peripherally. Its average horizontal diameter is 12 mm.[1] Unlike most tissues in the body, the cornea is avascular and has no nourishing blood vessels; it receives its nourishment (i.e., oxygen and glucose) from tears and aqueous humor.[2] The cornea is densely innervated with sensory nerve fibers and is one of the most sensitive tissues of the body.[3] It provides 75% of the refractive power of the human eye and focuses transmitted light onto the retina. In addition, it has a photo-protective role by absorbing UV radiation. Furthermore, the cornea acts as a thick elastic physical barrier that protects internal ocular structures from external physical, chemical, and microbial insults.[4,5] Moreover, the cornea withstands variations in intraocular pressure (IOP) and curvature changes of the eye.[6] The human cornea consists of five distinct layers, from anterior to posterior as follows: epithelium, Bowman's layer, stroma, Descemet's membrane, and endothelium (Fig. 1).[1,4,7]

Epithelium

It is the outermost layer with a thickness of approximately 50 μm.[8] It is composed of 5–7 stackable cell layers. The main functions of epithelium are: (A) blocking the entry of foreign materials including dust and bacteria into the eye and other corneal layers; and (B) providing a smooth surface that absorbs nutrients from tears for subsequent distribution to the rest of the cornea. The epithelium is a highly innervated structure that makes the cornea extremely sensitive to various physical insults.[4,9,10]

Bowman's Layer

It is a transparent sheet residing directly below the basement membrane of the epithelium having a thickness of approximately 12 μm.[8] It is composed of highly organized collagen type I, III, and V fibers. It maintains corneal structural integrity and facilitates stromal wound healing. Large and central scars of this layer may result in vision loss.[11–13]

Stroma

It resides below Bowman's layer and has a thickness of approximately 500 μm.[8] It comprises about 90% of the entire corneal thickness and does not contain any blood vessels. It consists of aligned arrays of hydrated types I/V heterotypic collagen fibrils (15% wet weight), glycosaminoglycans (GAGs), proteoglycans (PGs), and other proteins. It is interspersed with keratocytes. Due to its unique architecture, the stroma confers upon cornea its resilient shape, transparency, and mechanical properties (i.e., elasticity and strength).[10,14,15]

Descemet's Membrane

It is a strong (tough and highly elastic) sheet residing below the stroma with a thickness of approximately 4μm.[8] It is composed of collagen type VIII fibers, fibronectin, and laminin. Descemet's membrane provides a platform for the adherence of stroma and endothelial cells to provide

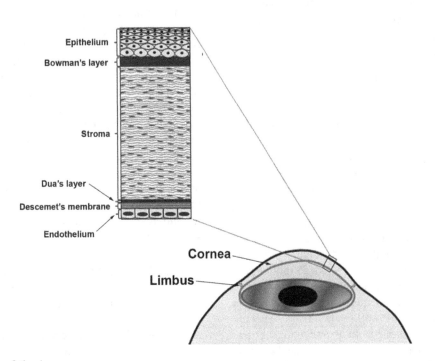

Fig. 1 Structure of the human cornea.

structural integrity to the cornea. In addition, it serves as a protective barrier against infection and injuries. Furthermore, it is regenerated readily after injury and plays a key role in fluid regulation; damage to this membrane can lead to a loss of endothelial cells.[8,10,16] A novel pre-Descemet's layer termed Dua's layer has also been described.[17]

Endothelium

It is the innermost layer with a thickness of approximately 5μm.[8] It is composed of a single layer of mitochondria-rich flattened hexagonal endothelial cells. The corneal endothelium is the most important part of the cornea, since an intact endothelium with sufficient cell density is essential to maintain clarity of the cornea through its dehydrating pump action. Normally, fluid leaks slowly from inside the eye (aqueous humor) to the middle corneal layer (stroma). The primary task of the corneal endothelium is to promote movement of this excess fluid out of the stroma. Without this pumping action, fluid will accumulate inside the stroma leading to stromal swelling, haziness, and finally, opacity.[4,10,18]

Corneal function depends on its transparency which in turn depends on number of anatomical and physiological factors. Anatomical factors affecting corneal transparency include its avascular nature, nonkeratinized epithelium layer, regularity of stromal lamellae, nonmyelinated nerve fibers and precorneal tear film. Physiological factors include corneal dehydration and uniform refractive indices of the corneal tissue.[10,19] Various theories have been postulated by many authors to explain corneal transparency and how corneal function is correlated to its structure.[20] Clarity of cornea is essential for clear vision and therefore irreversible loss of corneal optical clarity results in vision loss and blindness.[21]

PREVALENCE OF CORNEAL BLINDNESS

The World Health Organization (WHO) defines two different types of visual impairments: low vision and blindness.[22] Low vision is defined as visual acuity of less than 6/18 but equal to or better than 3/60 (moderate to severe), or a corresponding visual field loss to less than 20°, in the better eye with the best possible correction. Blindness is defined as visual acuity of less than 3/60 (profound), or a corresponding visual field loss to less than 10°, in the better eye with the best possible correction. According to the WHO report, in 2014, 285 million individuals worldwide are classified as visually impaired, of whom 39 millions are blind.[23] While cataracts are responsible for almost half of the blindness cases, cornea-affecting diseases are the next largest cause of blindness worldwide (14%). Specific causes of corneal blindness include trachoma, corneal opacity, and corneal scarring in children. Corneal scarring is the major cause of childhood blindness in Africa and the

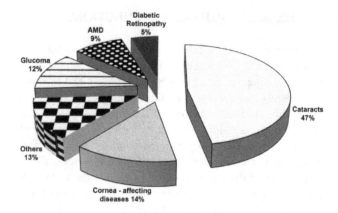

Fig. 2 Worldwide causes of blindness excluding refractive errors.

poorer Asian countries (40%).[24,25] A number of pathologies affect optical clarity of the cornea and may lead to vision loss,[25] including infectious disorders (bacterial and viral keratitis),[26,27] traumatic disorders (chemical, thermal, and mechanical injuries),[28,29] degenerative disorders (keratoconus)[30] and dystrophic disorders (Fuchs' dystrophy, lattice dystrophy, map-dot-fingerprint dystrophy).[31–33] In addition, the cornea can also be damaged secondarily to other eye conditions including tear film abnormalities (dry eye syndrome),[34] eyelid disorders (blepharitis),[35] and glaucoma.[36] Worldwide causes of blindness excluding refractive errors are illustrated in Fig. 2.[22]

TREATMENT STRATEGIES FOR CORNEAL BLINDNESS

Corneal blindness is a major global health problem and 80% of the underlying causes are avoidable either through preventive and basic public health measures or early treatment including available surgical technology and/or drug therapy. Although, preventive measures and early treatment of corneal blindness underlying causes are good strategies to reduce blindness, corneal transplantation is the only valid option for those that are currently blind, and for individuals suffering from diseases that are not currently preventable. Therefore a comprehensive strategy involving both preventive and therapeutic interventions would be the most effective approach to reduce corneal disease-related blindness and would provide both immediate and long-term resolution of the problem.[37,38] The main four treatment strategies for corneal blindness are described in Table 1.

Corneal Transplantation or Keratoplasty

Corneal allotransplantation

Keratoplasty is the medical term that refers to corneal transplantation or corneal grafting, a surgical procedure where a damaged or diseased cornea is replaced by donated

Table 1 Different treatment strategies for corneal blindness

Treatment	Advantages	Disadvantages	Clinical Status	Refs.
Corneal Allotransplantation	1- A fully-integrated biologically functional cornea. 2- One of the most commonly performed and successful type of organ transplant due to the immuno-privileged nature of the cornea.	1- Limited high quality donor corneas 2- Graft rejection and immunosuppression in some recipients 3- Very low but real risk of disease transmission 4- High cost 5- Limited success in some conditions such as corneal nerve damage and in other pathologies or injuries 6- Difficulty in manipulating human donor corneas characteristics to meet specific needs of the recipient	Feasible for clinical practice (the most commonly performed organ transplant)	39–51
Corneal Xenotransplantation	Theoretically, provide unlimited number of corneas, thus, are commercially advantageous	1- Immunogenicity issues 2- Cross-species disease transmission risk 3- Corneal thickness issues 4- Mechanical issues	*In vivo* (Experimental Animals)	53–60
Prosthokeratoplasty	1- Do not opacify and cannot vascularize like human donor corneas. 2- Compensate failure of corneal allotransplantation in some patients 3- Can be produced on large scale and be available on shelf 4- Lower cost than human donor corneas	1- Need specialized centers with skilled surgeons and (complex surgery and lack of trained clinicians) 2- Post-operative complications 3- Improper integration into the host tissue 4- Not applicable to all corneal pathologies 5- Doubtful or unsecured long-term retention or maintenance	The three main keratoprostheses used in clinical practice today are: 1- Osteo-odonto keratoprothesis (OOKP) 2- Boston keratoprosthesis (Boston KPro) 3- AlphaCor Keratoprosthesis	61–83
Tissue Engineering	1- Overcome the limitations of corneal replacement with allografts and drawbacks of keratoprostheses. 2- Allow regeneration of endogenous cells and nerves. 3- Manipulation of tissue engineered corneas characteristics to meet specific needs of the recipient. 4- Addresses or replaces only damaged layer thus it can target a specific disorder so that can be applicable to a wide range of corneal pathologies. 5- *In vitro* preclinical models to reduce the need for animal testing.	1- Difficulty in reproduction of corneal stromal architecture which is considered the backbone of a successful tissue engineered corneal equivalent. 2- Endothelium regeneration issues and the difficulty in populating tissue engineered corneal substitute with viable human corneal endothelial cells 3- Lack of standardized cell sources and optimal culture conditions but research is underway 4- Innervation and integration with host tissue issues 5- Further development in order to produce corneal substitutes with the same tensile strength, curvature, permeability, transparency, and durability as normal cornea	1- No full-thickness tissue engineered corneal replacement at present (cellularized TE corneal equivalent not currently applicable for clinical use). 2- Results of Phase I clinical trials using partial-thickness, acellular corneal matrices, tissue engineered corneal substitute are promising. 3- Tissue engineering methods have already been very successful in the treatment of patients with limbal stem cell deficiency. 4- Favorable results for *in vivo* (Experimental Animals) endothelial cell transplantation.	101–269

Conducting—Dendritic

corneal tissue that has been removed from a recently deceased individual. Careful donor screening is critical to exclude known diseases that might affect the viability of the donated tissue.[39,40] Corneal transplantation is divided into two major types: penetrating keratoplasty (P.K), which is a full thickness corneal grafting (i.e., full-thickness transplantation of the five main layers of the cornea), and lamellar keratoplasty (L.K) which is a partial thickness corneal grafting (partial-thickness transplantation of one, two, or three layers of the cornea).[40] The term keratoplasty was first used by Franz Reisinger when he suggested the use of an animal's eye to provide donor corneal tissue for corneal transplantation in a human;[41] the first successful keratoplasty (allotransplantation) was performed on a patient suffering from bilateral alkali burns in 1906 by Eduard Zirm.[42] Successful corneal transplantation requires all of the following: available high quality corneal tissue, efficient tissue storage and collection systems, skilled surgeons, anti-rejection medications, and appropriate surgical follow-up. In most of the developing world, many or all of these critical components are limited or unavailable. The single most prohibitive factor is limited access to high-quality corneal tissue due to the lack of eye banking infrastructure. From a technical standpoint of view, development of eye banking infrastructure is complex and requires medical educational systems, highly equipped clinical facilities, reliable storage and transportation facilities, and organized distribution networks. In the developed world with access to fresh corneal tissue, corneal transplantation is hindered by prohibitive cost, customs barriers for imported tissue and unreliable compliance by recipients with medications and follow-up instructions. Most importantly, there is an inadequate supply of human donor corneas.[43–46] Major problems encountered during corneal transplantation include: 1) high risk of disease and infection transmission from donor to recipient; 2) difficulty in manipulating the characteristics of human donor corneas to meet the specific needs of the recipient; and 3) graft rejection. However, corneal transplants are the most commonly performed and successful type of organ transplant due to the avascular nature of the cornea that limit corneal antigen exposure to the host immune system.[47–49] In contrast, the success rate of donor corneal transplantation drops dramatically when the host cornea is extensively scarred, deeply vascularized, possessing altered tear film, or in the presence of glaucoma. In addition, poor outcomes are demonstrated when patients are affected by specific conditions such as severe chemical burns, Stevens-Johnson syndrome, trachoma, and severe dry eye syndrome involving limbal stem cell deficiencies.[50,51]

Corneal xenotransplantation

WHO defines xenotransplantation (animal to human) as living cells, tissues or organs of animal origin and human body fluids, cells, tissues or organs that have *ex vivo* with

these living, xenogeneic materials, Xenotransplantation has the potential to constitute an alternative to material of human origin and bridge the shortfall in human material for transplantation.[52] Cross-species transplantation provides new resources of organs and tissues for clinical transplantation with the aim of overcoming the shortage of human material that is continuously increasing the waiting time for patients in need of transplants.[53] The first clinical corneal xenotransplantation (pig-to-blind patient) was carried out by Kissam in 1838, whereas the first corneal allograft (human-to-human) was not carried out until 1906 by Zirm.[54] Attempts to use animal corneas for transplantation in humans have included gibbons, cows, dogs, pigs, and rabbits. Pigs are considered to be the best potential alternative source of organs for humans compared to other species, due to the similarity of anatomical and biomechanical properties of human and porcine corneas; there is currently renewed interest in the possibility of using corneas from pigs.[55,56] Although several xenotransplants are reported to be technically successful, most of the xenografts from widely-disparate species have been rejected due to various immune reactions including activation of the innate and adaptive immune systems, coagulation dysregulation and inflammation.[57] The innate response has largely been overcome by the transplantation of organs from pigs with genetic modifications that shield their tissues from these responses. T cell-mediated rejection can be controlled by immunosuppressive agents that inhibit co-stimulation. Coagulation dysfunction between porcine and primate systems remains problematic but is being overcome by the transplantation of organs from pigs that express human coagulation-regulatory proteins.[57,58] The remaining barriers will be resolved by introducing novel genetically-engineered pigs. Genetic engineering of the pig as an organ source has increased the survival of the transplanted pig corneal graft in non-human primates (NHPs) and may also contribute to resolving any physiological barriers that might be identified, as well as reducing the risk of transfer of potentially infectious micro-organisms. Recent progress in the genetic manipulation of the pig has led to the prospect that the remaining immunological barriers will be overcome and consequently xenotransplantation using pig corneas will become feasible in the near future.[59,60]

Prosthokeratoplasty

Surgical replacement of diseased or scarred corneal tissue with a transparent prosthesis (keratoprosthesis; KPro) is the first engineered artificial corneal substitute designed to replace opacified corneal tissue, thus providing a clear window for the transmission of light.[61,62] The development of keratoprosthesis (KPros) has been ongoing since the 18th century. In 1771, the French ophthalmologist Guillaume Pellier de Quengsy proposed a silver-rimmed glass optic to replace a completely opaque cornea. His suggested porous prosthetic skirt represented a revolutionary concept for current artificial cornea research.[63] In 1856, Nussbaum in Germany manufactured and implanted the first glass KPro into rabbit cornea. Nussbaum discovered that glass would be a suitable biocompatible implant after having implanted several materials including wood, glass, iron, and copper under his own skin to determine the degree of inflammation. The success of his glass KPro in rabbits (maintained for 3 years) led to the first implantation of KPros in humans, which was retained for 7 months.[64] In the1800s, other materials such as quartz, tantalum, vitallium, and celluloid were used unsuccessfully in attempts to create an artificial cornea. With the advent of donor cornea transplantation, the idea of an artificial cornea was abandoned as more attention was given to refining donor keratoplasty techniques.[64–67] It was not until World War II, when polymethylmethacrylate (PMMA) was developed as a potentially implantable material in the eye, that the idea of an artificial cornea was rekindled.[68] The first developed keratoprostheses were one-piece devices consisting of synthetic plastics such as PMMA. This has now been followed by the design of keratoprostheses composed of two pieces, with numerous versions that differ from each other by design and composition.

Types of KPro Designs

There are two basic types of KPro designs. The most popular design (through-and-through or core-and-skirt) consists of a transparent optical core surrounded by a skirt used to anchor the device into the stroma and occasionally covered by transplanted autologous tissue or eyelid skin. The less popular design (nut and bolt or collar button) consists of two plates joined by a central optical stem. It is implanted such that the plates sandwich the cornea between them and are sutured into place like a penetrating keratoplasty graft. [67,69] By 1960s, the first well established PMMA keratoprosthesis was developed (The Cardona KPro) followed by numerous versions. Some KPros are at the R & D stage, several are still under clinical investigation, while others are in clinical use including the following: Osteo-Odonto keratoprothesis (OOKP), Boston KPro (Dohlman-Doane keratoprosthesis), AlphaCor KPro (Chirila keratoprosthesis), Seoul Type keratoprosthesis (S-KPro), Pintucci keratoprosthesis, Legeais BioKPro III, Aachen keratoprosthesis, SupraDescemetic keratoprosthesis, German keratoprosthesis, Stanford Keratoprosthesis and Collagen-based Keratoprosthesis.[70–73]

Main KPros Used in Clinical Practice

1. Osteo-odonto KPros (OOKP): developed by Strampelli in 1964 and consisting of a central PMMA optic surrounded by an osteodental skirt harvested autologously from patient tissues (i.e., tooth root and alveolar bone).[74,75]

2. Dohlman-Doane KPros, also known currently as Boston KPros (Boston KPro), Designed by Dohlman, Harvard Medical school and Massachusetts Eye and Ear Infirmary, Boston. KPros received FDA approval in 1992 and is subdivided into Type I and Type II which are in current clinical use. Type I is more common and consists of a central PMMA optical region, a porous back plate and a titanium locking ring that clamps onto native corneal tissue. Type II is a through-and-through device and has many of the features used by Cardona.[74,76]

3. Chirila Keratoprosthesis, also known commercially as AlphaCor KPros. It consists of a circular optical core of a flexible, transparent poly(2-hydroxyethyl methacrylate) (pHEMA) gel surrounded by a skirt of a porous, opaque pHEMA sponge which allows for cellular invasion and vascularization. AlphaCor KPros received FDA approval in 2003 and two variations of this AlphaCor KPros have been developed for human clinical trials. Type I was a true "artificial corneal button" which was implanted similarly to a penetrating keratoplasty and covered by a conjunctival flap that is later opened to enable a full visual field. The second generation type II device has a smaller diameter and is implanted in a lamellar corneal pocket. The device is available in aphakic and standard powers.[74,77]

Clinical Complications of KPros

Traditional or conventional KPros were developed using synthetic polymers such as PMMA, PHEMA, and PVA. These materials do not promote cell adhesion and are associated with multiple complications. In the last decade, researchers developed KPros with regenerative capacities to improve cell adhesion and migration and to overcome traditional KPro complications.[72,78] Naturally occurring extracellular proteins such as collagen, laminin, and fibronectin, or cell adhesion promoting peptides derived from these proteins (e.g., RGD, IKVAV, YIGSR) have been utilized to develop regenerative KPros.[79–81] It is always preferable to use donor human corneas for purposes of transplantation. However, keratoprostheses are intended to restore vision of patients with severe corneal disease where corneal transplantation has repeatedly failed or is not an treatment option. This includes cases suffering from severe chemical or traumatic injuries and certain immunological conditions including pemphigoid and Stevens-Johnson syndrome. The main complication of prosthokeratoplasty was, and still is, the spontaneous rejection of prosthesis, known as extrusion, which is usually preceded by erosive tissue necrosis (melting) around the prosthetic rim, and dislocation of the prosthesis. Other complications associated with all types of KPros are glaucoma, retinal detachment, and retroprosthetic membrane formation. In order to avoid the aforementioned complications, KPros should meet the following criteria: the device should be transparent and have appropriate curvature, diameter, and refractive power. In addition, it should be flexible, but with sufficient tensile strength to allow further minor surgery. Further, it should be fabricated from a biocompatible material that allows water uptake. Moreover, intraocular pressure (IOP) should be monitored in order to prevent glaucomatous optic nerve damage. Improving the characteristics of current versions of KPros and developing new versions of KPros are the focus of intensive research to fabricate an ideal KPro that is able to replace the natural cornea and help many patients around the world.[74,82,83]

Tissue-Engineered Corneal Substitutes

Tissue engineering is a rapidly growing specialty that emerged in the late 1980s and involves the application of engineering and life sciences principles to develop biological substitutes that restore, maintain or improve tissue functions. It was predicted that tissue engineering - based substitutes would overcome the major limitations of tissue and organ transplantation including donor shortages and graft rejection, along with addressing the limitations of artificial organs and tissues.[84–86] This new and emerging technology relies on three pillars: cellular component (cells), carrier component (supporting/carrier matrix), and bioactive component (stimulating or signaling biomolecules).[87]

The cellular component should consist of healthy viable cells that are accessible, manipulable, and nonimmunogenic. Three sources of cells are available: the patient (autologous cells), another human donor (allogeneic cells) or another species (xenogeneic cells). The selection of appropriate cells, and the conditions for their isolation, purification, and cultivation, is very crucial factor in tissue engineering as it is very important to ensure the most physiologically relevant *in vitro* microenvironments for cells.[88–90] The carrier component has a dual function, since it acts as both a delivery vehicle and as a supporting matrix/scaffold. It is an artificial extracellular matrix in which the cells can attach, proliferate, and differentiate with subsequent new tissue formation. The carrier confers physical, mechanical, and functional properties upon tissues and organs and should possess the following properties: 1) manipulable and possessing sufficient mechanical strength to withstand *in vivo* forces; 2) biodegradable with degradation products that cause minimal immune or inflammatory responses; 3) biocompatible and facilitating oxygen/nutrient transfer; and finally: allow tissue ingrowth. Developing a scaffolding/carrier matrix that possesses all these features is challenging. The composition and structure of the extracellular matrix is specific to each organ or tissue; therefore, a crucial foundation of tissue engineering is the biomaterial from which scaffolds are fabricated. Biomaterials derived from natural, synthetic, and decellularized tissues have been used for various tissue engineering applications.[88,91,92] The bioactive component (stimulating / signaling biomolecules) aims at encouraging functional tissue growth and directing cells to express the desired tissue phenotype. These bioactive components include growth, differentiation, and angiogenic factors. Delivering these factors in the right concentration to the right location is crucial. Strictly regulating the activity of signaling molecules, particularly growth factors, is extremely important to avoid uncontrolled cell growth and proliferation.[88,93,94]

Two main tissue engineering strategies have been adopted by researchers: a) cell-based, where cells are been manipulated to create their own environment or matrix before transplantation to the host. The cells require native-like favorable growth conditions *in vitro* and this achieved by development of defined cell culture conditions, including supplementation of culture medium with specific growth factors; and b) scaffold-based, that are subdivided into either acellular or cell delivery scaffolds. Acellular biointeractive scaffolds are created to mimic the *in vivo* structures and are transplanted into the host to serve as a template for host tissue regeneration, where it induces migration, proliferation, and differentiation of cells from the surrounding tissue. On the other hand, substrates for cell delivery are a three-dimensional supporting biodegradable matrix (scaffold) for transplanting cells to a specific site which becomes capable of replacing the functions of the pathologically altered tissues.[95,96]

Corneal tissue engineering is a promising strategy to restore structure and function for damaged corneas. Successful development of efficient tissue-engineered corneal equivalent requires full understanding of its structure, as well as functional, physical, and biochemical properties. It is crucial to maintain the same biochemical, morphological, physiological, and even genetic components of native corneas. As a multilaminar structure, the cornea affords opportunities to reconstruct each of the layers separately or in combination. Thus, diseases of each layer may be addressed without disruption of the entire structure.[97–99]

Tissue engineering of the full-thickness corneal equivalent

Corneal tissues are mainly populated with three different cell types (epithelium, fibroblasts/keratocytes, and endothelium) that each require optimum culturing conditions if the cornea is to be fabricated by tissue-engineering strategies. Therefore, a highly sophisticated cell transplantation approach must be taken. Several research groups have attempted to fabricate all the cell layers of the cornea using cultured cells. The major limitations of these constructed corneas are stability and proof of function.[100–102] Minami and coworkers reconstructed corneas containing the three main layers using primary bovine corneal cells *in vitro*. Briefly, stromal fibroblasts were resuspended in a type I collagen gel and seeded into a dish containing a nitrocellulose membrane base, and then cultured upside down in a larger culture vessel. Endothelial cells were then seeded onto the bottom of the nitrocellulose base for two days before returning the dish to its up-right position. Epithelial cells were then finally seeded on top of the collagen gel to complete the construction process. When the epithelial layer became confluent, the medium was aspirated to expose the epithelial layer to air to complete epithelial maturation (stratification). The epithelial layer became stratified with a composition ranging from cuboidal to squamous epithelial cells. This is a promising model to study the pathophysiology and diseases of the cornea.[103] In an alternative approach, reconstructed corneas were prepared using primary rabbit epithelial and stromal cells in the presence or absence of an underlying layer of immortalized mouse endothelial cells *in vitro*.[104] The cultures were grown submerged or at a dry or moist interface. Submerged cultures without endothelial cells did not express differentiation markers or basement membrane components, indicating that endothelial cell interaction dramatically enhances the amount and quality of epithelial basement membrane assembly and that epithelial differentiation is influenced by the type of interface between tissue, liquid, and air.[104] A similar approach to construct corneas using cells isolated from fetal pig corneas displayed signs of transition to an organotypic phenotype with the formation of two basement membranes.[105] An equivalent of human cornea was constructed in early 1995 at the Laboratoire d'Organogénèse Expérimentale (LEOX). However, their model was incomplete as it lacked an endothelial layer and no further experiments were done beyond superficial histological analysis.[106] Reconstructed human corneas have been prepared using immortalized human corneal cells.[107] Each cell type was screened for electrophysiological, biochemical, and morphological similarities to the freshly isolated cells (from post-mortem human cornea), before being used in the 3D reconstruction. Collagen–chondroitin sulphate cross-linked with glutaraldehyde was utilized as a tissue matrix. Stromal, epithelial, and endothelial layers were created by mixing cells into, and layering cells above and below, this substrate and then culturing the fabricated constructs. This study represents a promising strategy towards the development of a corneal implant synthesized from human cells, as the resulting construct had some corneal-like properties with respect to morphology, transparency, ion and fluid transport, and gene expression.[107] When different biomatrices were utilized (bovine collagen type I or a mixture of human collagen I and III), Bowman's and Descemet's membrane were clearly visible indicating the development of the three individual corneal layers.[108] Alternatively, a step-by-step procedure showed promise. Corneal endothelial cells were first cultured and then poured over a fibrin–agarose gel embedded with stromal fibroblasts; finally, corneal epithelial cells were cultured on top of the polymerized gel.[109] A complete cornea has been reconstructed on collagen–chondroitin sulfate foam. The scaffold was first seeded with stromal keratocytes and then successively with epithelial and endothelial cells.[110] Tissue-engineered corneal substitutes have been developed using hydrogels based on dendrimer cross-linked collagen or collagen-chitosan.[111,112] A self-assembled transparent corneal equivalent has been developed using the three native corneal cell types to create stroma with fibroblasts, followed by an endothelial cell monolayer which was finally seeded with corneal limbal epithelial cells.[113] Decellularized biological material (combination of amniotic epithelial cells and porcine cornea) in a rabbit lamellar keratoplasty model resulted in degradation of the tissue-engineered cornea due to host rejection.[114] Successful implantation of a full-thickness tissue-engineered corneal replacement into experimental animal models has been always technically challenging. However, a fully engineered corneal replacement does seem to be achievable. In phase I clinical trials a tissue engineered corneal substitute was implanted in 10 patients with keratoconus and corneal scarring using anterior partial keratoplasty surgery.[114–117] Damaged corneal tissue was removed from one eye of each patient and replaced with a tissue engineered corneal substitute composed of recombinant human collagen type III crosslinked with EDC/ NHS. Six-month follow ups showed anchoring of the implants, epithelium regeneration and ingrowth of stromal cells. Two-year follow up showed nerve and tear film regeneration indicating regeneration of corneal epithelial cells over the implant, and stromal cell and nerves growth into the center of the implant. However, the retained surgical sutures of the implants caused a delay in epithelial closure and fibrosis. Four-year postoperative results showed that all implants remained stably integrated without the use of immunosuppression. Nonetheless, these results showed a 60% success rate as vision was improved in six, unchanged in two, and decreased in two patients.[115–117]

Tissue engineering of corneal epithelium

Corneal epithelium is the outermost layer of cornea and the most exposed to external damage. It is constantly renewed through proliferation of different cell types including basal cells, transient amplifying cells, and limbal stem cells (LSCs). Limbus, the circular border of the cornea, was demonstrated to be a reservoir of corneal epithelial stem cells (limbal stem cells). As the LSCs migrate to the central cornea, they differentiate into transient amplifying cells (only dividing a limited number of times) and basal cells (the deepest layer of corneal epithelium).[10,18,118] The reduction or absence of corneal epithelial stem cells due to limbal disorders results in opacification of the cornea and severe visual impairment. In such cases, limbal or corneal allograft transplantation is proposed as a treatment strategy; however, this is limited by the availability of organ donors and requires long-term immunosupression that is usually associated with severe side effects. To overcome these limitations, strategies to develop autologous tissue-engineered corneal epithelium *in vitro* have been developed.[119,120] In patients with unilateral limbal stem-cell deficiency, an autologous limbal transplantation is performed to restore corneal epithelium.

Cultured or Cultivated Limbal Epithelial Transplantation (CLET)

CLET is considered to be a stem cell-based therapy and applied widely in clinical ophthalmology for treatment of patients with unilateral limbal stem-cell deficiency. The first clinical success of CLET was performed on two patients with unilateral limbal stem-cell deficiency.[121] Briefly, after isolation and propagation of cells from a small biopsy of the limbus of the patient healthy eye, a sheet of cells was cultured for 19 days and then transplanted to the prepared patient eye. Long-term followup of both patients revealed the stability of regenerated corneal epithelium and remarkable improvement in visual acuity.[121,122] Although CLET has shown satisfactory outcomes in patients suffering from unilateral limbal-stem cell deficiency, it is technically impossible to apply the same strategy for patients having bilateral limbal stem-cell deficiency as a large number of cells are necessary for transplantation; therefore, alternative cell sources, including autologous oral mucosal epithelial transplantation, are utilized.[123,124]

Cultured or Cultivated Oral Mucosal Epithelial Transplantation (COMET)

COMET is a stem cell-based therapy and widely applied in clinical ophthalmology for treatment of patients with bilateral limbal stem-cell deficiency. Briefly, a biopsy of oral mucosa is harvested from the patient and cultured on different substrates *in vitro*, followed by transplantation of cultured cells to the prepared patient eye.[123,124]

The development of novel and improved strategies for CLET and COMET is the focus of intensive research worldwide. Current research activities in tissue engineering of corneal epithelium focuses on two crucial factors: a) development of optimal cultivation and culturing conditions along with exploration of alternative cell sources for corneal epithelium; and b) development of novel bioactive scaffolds or cell substrates that promote adequate cell proliferation and conditions for long-term maintenance of cells *in vitro* and after transplantation.[118–124]

Cell Source for Tissue-Engineered Corneal Epithelium

Several studies with reasonable success have been reported for culturing of human corneal epithelial cells for corneal tissue engineering, which showed that optimization of culturing conditions is very crucial to reproduce the highly organized and integrated multilaminar architecture of epithelium.[107,108,125–127] Since normal or primary epithelial cells have limited proliferative capacity *in vitro*,

researchers have focused on exploring alternative cell sources.[118] Stem cells are an ideal starting point for tissue engineering of ocular tissues. Corneal epithelial stem cells are an optimal cell source, since physiological renewal occurs via these stem cell populations. Autologous corneal LSCs are now routinely used in the restoration of the injured ocular surface. However, therapeutic use of autologous corneal stem cells is technically challenging. An intact stem cell compartment is not usually available and the number of stem cells in each compartment is very low. Therefore, there is an extended *ex vivo* culture time to generate a sufficient number of cells suitable for transplantation. Furthermore, for patients with only one eye affected, for instance, unilateral limbal stem cell deficiency, the risk of damaging the stem cell compartment of the healthy eye is high. Moreover, in patients with both eyes affected, for instance, bilateral limbal stem deficiency, finding autologous ocular-derived stem cells is technically impossible. Therefore, exploration of non-limbal or extra-ocular cell sources that can be used to construct corneal epithelium is a reasonable strategy. These would include oral mucosal epithelial cells, conjunctival epithelial cells, epidermal stem cells, bone marrow derived mesenchymal stem cells, dental pulp stem cells, hair follicle bulge-derived stem cells and umbilical cord lining stem cells. Among these cell sources, cultured autologous oral mucosal epithelial cells and conjunctival epithelial cells are the only laboratory cultured cell sources that have been explored in humans. Finally, induced pluripotent stem Cells (IPSCs) represent a very promising cell source that might be used in the future for epithelial reconstruction after comprehensive investigation.[128–132]

Cell Substrates or Scaffolds for tissue-engineered Corneal Epithelium

Scaffold utilized for the development of corneal epithelium substitute should be optically transparent, biocompatible, sterilizable, reproducible, suturable and cost-effective. In addition, it should support optimum proliferation and integration of ocular epithelial cells.[107,108,125–127] A wide range of scaffolds have been evaluated for tissue engineering of corneal epithelium:

Thermo-responsive polymers

Thermo-responsive cell culture dishes (carrier-free cell sheets). The majority of tissue engineering strategies to develop corneal epithelium substitutes involve the utilization of scaffold to support cell sheets and facilitate their handling, manipulation, and integration to host corneal stroma after transplantation. However, carrier-free or scaffold-free systems have been also utilized in which intact cell sheets were created using thermo-responsive culture dishes and then transplanted without an underlying supportive scaffold. Thermo-responsive culture dishes are fabricated from synthetic polymers that exhibit reversible hydration and dehydration in response to temperature changes. This type of culture a dish facilitates cell adhesion and growth at 37°C and induces complete detachment of adherent cells in a nondestructive manner upon lowering the temperature to less than 30°C. This culture technique allow detachment of the epithelial sheets from the culture plastic without using proteolytic enzymes that are associated with the destruction of cell-cell junctions and extracellular matrix which are crucial to cell sheet integrity and function. In preclinical settings, carrier-free transplantation of cell sheets ensured optimal preservation of epithelial basement membrane components and cell-matrix adhesion junctions. However, the extended culture time required to generate adequate structures for transplantation, difficult manipulation, and the high batch to batch variability are the major drawbacks of this strategy. This technique promoted the *in vitro* growth of functional stratified epithelium and has been used to prepare cultured autologous limbal and mucosal epithelial cell sheets.[123, 133–136]

Thermoresponsive gel matrix. Thermo-responsive polymers are able to undergo sol-gel transitions at ambient temperatures including Mebiol gel® matrix. This thermo-responsive polymer dissolves in water and forms solutions at temperatures below 20°C which solidifies at the routine incubation temperature of 37°C. This allows either encapsulation of cultured cells by mixing cells with the solution at low temperatures, or cell seeding by placing cells on a solid gel at high temperature.[137] It has been shown that Mebiol gel® was able to support limbal explant proliferation as indicated by the expression of presumed limbal stem cell association markers, transient amplifying cells and cornea phenotype markers.[138]

Amniotic membrane (AM)

AM is the innermost of the three layers forming the fetal membranes and consists of a layer of epithelial cells firmly attached to an underlying basement membrane composed of collagen type IV that is similar to basement membrane of conjunctiva. Autologous limbal epithelial cells (LECs) amniotic membrane has anti-bacterial, anti-inflammatory, and anti-immunogenic properties. It is utilized as a substratum for corneal epithelial cells to grow on. It facilitates epithelialization, reduces inflammation and scarring, and acts as replacement substratum when the underlying stromal tissue is destroyed. Clinically, AM is used as a substrate for LSC transplantation in patients with limbal stem cell deficiency and is considered the most widely used carrier/substrate for epithelium reconstruction.[139–141] In addition, it is used successfully for reconstruction of damaged cornea in several animal models (e.g., rat, rabbit, and goat).[142–146] Furthermore, clinical outcomes of CLET and COMET were promising when AM was utilized as a substrate/carrier matrix. However,

utilization of AM in various clinical settings showed several limitations including the extended time required for preparation along with the absence of a standardized protocol for processing and preservation. In addition, there is a high inter- and intra-donor variability in morphological, chemical and optical properties. Furthermore, an increased risk of bacterial, fungal, and viral transmission is anticipated.[147]

Fibrin-based matrices

Fibrin is a natural biopolymer of the monomer fibrinogen. Fibrin-based matrices have been used as a cell carrier matrix for various tissue engineering applications including corneal tissue due to its satisfactory transparency, adequate mechanical strength, and biodegradability *in vivo*.[95,148] Fibrin gels have been used as a carrier matrix for transplantation of cultured epithelial cells for the treatment of patients suffering from limbal stem cell deficiency, and showed restoration of corneal epithelium as indicated by improvement of transparency in the majority of patient eyes that remained stable for 10 years.[149,150] Encapsulation of corneal epithelial stem cells into a crosslinked autologous fibrin gel enriched with fibronectin is a promising strategy for bioengineering of the ocular surface in terms of flexibility, manipulability, and resistance to the shear forces of the blinking eyelids.[95,151] In patients suffering from limbal stem cell deficiency, fibrin glue has been utilized as a bioadhesive to facilitate the transplantation of reconstructed bioengineered AM-based corneal epithelium and subsequent integration to the underlying stroma in a sutureless procedure.[152] Although fibrin has achieved successful clinical outcomes in tissue engineering of corneal epithelium, the allogeneic rejection and the possibility of disease transmission should be considered.[153]

Collagen-based matrices

Since collagen is the major constituent (predominant extracellular matrix component) of the human cornea, it has been widely evaluated as a reasonable scaffold during tissue engineering of the corneal equivalent[154] Collagen combines some important characteristics making it an ideal natural biomaterial scaffold for various tissue engineering applications including the relatively high abundance of collagen in vertebrate tissues (approximately 30%), biocompatibility, low immunogenicity, and high bioabsorbability. In addition, collagen has been shown to promote cellular proliferation and can be combined with either bioactive components or synthetic polymers to combine the advantages of both. Further, collagen can be easily modified to introduce additional functional groups and can be processed to produce various forms, for instance, films, sheets, fibers, beads, and sponges.[155,156] Dendrimeric crosslinked collagen hydrogel has been used as a scaffold for reconstruction of corneal epithelial cells and has shown superior mechanical and optical properties when compared to thermally cross-linked collagen hydrogels.[111] Collagen hydrogels have been utilized to bioengineer corneal limbal crypts via microcontact printing to generate three-dimensional (3D) human limbal epithelial stem cell niche *in vitro*, which maintained corneal epithelium regeneration *in vivo*[157] Matrigel matrix (laminin, collagen type IV, heparan sulfate, and entactin)[158] has been shown to promote cell proliferation, survival, and expression of stemness markers when used as a substitute for the *in vivo* niche of human LSCs.[159] Similarity, collagen shields (animal scleral collagen, type I) coated with either Matrigl or type IV collagen promoted the attachment of human corneal epithelial cells. However, cells on the shields coated with Matrigel failed to become confluent and subsequently did not form multi-layered stratified tissue as compared to cells grown on collagen shields coated with collagen type IV.[160] Human LSCs can be successfully cultivated *in vitro* on collagen vitrigel membrane (a gel in a stable state produced via a three stage sequence: gelation, vitrification, and rehydration) with subsequent formation of stratified epithelial sheets; these maintained their typical phenotype during the entire culture period.[161] Plastic compressed collagen gels can be used as a potential carrier for limbal epithelial cell (LECs) transplantation. LECs displayed homogenous and smooth morphology after seeding onto compressed collagen gels, in comparison to non-compressed collagen. In addition, cell–cell and cell–matrix attachment, stratification, and cell density were superior in LECs expanded upon compressed collagen gels.[162,163] Plastic compressed collagen produced by a novel tissue engineering approach entitled "Real Architecture for 3D Tissues Production" (RAFT) has allowed the creation of biomimetic epithelial and endothelial tissue equivalents suitable for transplantation and which are ideal for studying cell-cell interactions *in vitro*.[164]

In addition to the aforementioned matrices, a wide range of materials have been evaluated for tissue engineering of corneal epithelium including keratin-based matrices,[165] silk-based matrices,[166] corneal stromas,[167] petrolatum gauze,[121] human keratoplasty lenticules (HKLs),[168] nanofiber scaffolds,[169,170] and electrospun 3D scaffolds.[171,172] The most widely pre-clinically and/or clinically investigated cell carriers/substrata for restoration of corneal epithelium includes AM, collagen, and fibrin along with siloxane—hydrogel contact lenses.[173,174]

Tissue Engineering of Corneal Stroma

Corneal stroma provides the main functions of cornea based on its unique architecture. Stroma comprises 90% of the total corneal thickness and consists of various collagens (mostly type I), glycosaminoglycans (GAGs), proteoglycans (PGs), and other protein constituents including fibronectin and laminin. It is interspersed with keratocytes that regulate the synthesis and degradation of the extracellular

matrix (ECM). The keratocyte population also plays a crucial role in maintaining corneal clarity.[8,10,14,15,175] Stromal collagen fibrils are arranged homogeneously in the form of orthogonal layers (lamellae). This specific spatial arrangement of collagen fibrils provides mechanical strength and transparency to the cornea. The stromal collagen fibrils help transmit the light unscattered due to its unique parallel array alignment of the fibrils which allows light to pass virtually unimpeded. PGs and their associated GAGs contribute to the cornea's mechanical properties including compressive and swelling properties, along with providing uniform spacing to the collagen fibrils. GAGs located in the interfibrillar space have a very strong tendency to attract water. Under normal conditions, the epithelium and the endothelium can maintain stromal hydration at 77% of water per corneal weight.[8,176] Alteration of this corneal stroma function-structure relationship can lead to corneal opacification, visual impairment, and even blindness. Clinically successful tissue engineering - based strategies for reconstruction of corneal substitute should aim at reproducing functions of corneal stroma, which is considered to be a key challenging step.[177]

Cell source for tissue-engineered corneal stroma

Corneal stroma is the secretary product of mesenchymal fibroblastic cells which differentiate into corneal keratocytes. Corneal keratocytes have dendritic morphology (i.e., long cytoplasmic filaments emerging from keratocyte bodies) and form a three-dimensional network of interconnected cells indicating functional interconnection of the entire corneal stroma via keratocyte gap junctions. Following traumatic injury of corneal stroma, keratocytes undergo a remarkable transformation into activated myofibroblasts alongwith disruption of interconnections. After completion of the healing process, myofibroblasts either undergo apoptosis or revert back into quiescent keratocytes.[178,179]

Corneal stroma contains various cell types including mesenchymal fibroblastic cells (which ultimately differentiate into corneal keratocytes), fully-differentiated keratocytes, fibroblasts (usually de-differentiated from corneal keratocytes), myofibroblasts (activated keratocytes and representative of scar tissue during stromal trauma), and other cells observed only under inflammatory conditions including dendritic cells, macrophages, lymphocytes and polymorphonuclear leucocytes.[180] The most abundant stromal cells are keratocytes (also known as dormant fibroblasts). Keratocytes possess the ability to de-differentiate to fibroblasts and to re-differentiate. Understanding the properties and relationships of these cells during development and healing is essential to successfully develop functional cell-derived stromal substitute.[177,180] Under *in vitro* expansion conditions, keratocytes transform from their quiescent to a fibroblastic phenotype. Unlike keratocytes, fibroblasts secrete a disorganized ECM that is typically found in corneal scars. Therefore, maintaining control over the mutable phenotype of the corneal keratocyte when used as a cell source to reconstruct tissue-engineered stroma-like tissue is crucial.[177] It has been shown that injection of the human corneal stromal stem cells into *in vivo* scarred corneal model (mutant mice having opacified cornea due to disruption of stromal organization resembling scarred cornea) results in restoration of corneal thickness and transparency. These promising results suggested that the immune privileged human corneal stromal stem cells is a potential cell source to develop tissue engineering-based strategy for restoration of stromal functions and structures.[181,182]

Scaffolds for tissue-engineered corneal stroma

Self-assembly approach. In this promising approach, the carrier supporting matrix (natural or synthetic) is absent and cell source (for instance fibroblasts) produces its own ECM (mainly collagen) *in vitro* under the effect of bioactive factors, and then organizes this secreted ECM into a structured 3D network. This approach allows the reconstruction of corneal stromas very close physiologically to their *in vivo* counterparts.[106,183] Stimulating fibroblastic cells to produce a functional stroma - like matrix *in vitro* prior to implantation has been reported where corneal fibroblasts were cultured in the presence of ascorbic acid. Under these culture conditions, fibroblasts secrete and organize their own ECM, and after adequate culture time they form sheets that can be assembled into three-dimensional stromal substitutes.[184,185] In one such study, human corneal stromal keratocytes were expanded in the presence of fetal bovine serum and ascorbic acid *in vitro*, and then cells were allowed to build up their ECM for up to 5 weeks.[186] This constructs showed parallel arrays of fibrils alternating in direction in a manner resembling the developing corneal stroma indicating that under the optimum *in vitro* conditions, keratocytes are capable of building a stroma-like array of narrow parallel collagen fibrils with alternating lamellae.[186] However, the major pitfall of this self-assembly approach is the extended time needed to produce enough ECM for transplantation.[106,183]

Stromal cell alignment approach. The effect of matrix surface topography on cellular behavior (development, motility, differentiation, orientation, and alignment) is one of the most investigated cell-matrix interactions. Contact guidance is the ability of the topographical features of the underlying substrate surface to direct or modify the response of cells. The presence of contact guidance platform has been shown to promote specific cell orientation and elongation, and can be utilized to control the development of corneal stromas.[187] The aligned stromal cells secrete aligned collagen fibrils in a pattern mimicking normal human corneal stromal tissue. Therefore, tissue engineering of corneal stroma has focused on the development of functional corneal stroma substrates

through chemical, morphological, and mechanical cues.[188] Stroml cell alignment can be achieved by seeding stromal cells on the following:

Topographically patterned surfaces

A thermoplastic elastomer engraved with a grating period of 4 μm and 1 μm linewidth was used to cultivate corneal fibroblasts and resulted in oriented self-assembled stromal tissue sheets which exhibited a better transparency when compared to non-structured tissue.[189] Similarly, surface-patterned silk films have been shown to support human corneal fibroblast proliferation, and orientation along with ECM deposition and alignment.[190] In addition, RGD surface coupling of groove-patterned silk film, in combination with human corneal stromal stem cells, was an essential factor in enhancing cell attachment, orientation, proliferation, differentiation, and ECM deposition on the silk substratum. This strategy is very promising to develop a highly-ordered collagen fibril-based constructs for corneal regeneration and corneal stromal tissue repair.[191] Further, micropatterned collagen film has been shown to supported proliferation of human corneal keratocytes and oriented secretion of collagen and keratan sulfate.[192] Finally, microgrooved collagen film supported superior stromal fibroblasts alignment when compared to smooth collagen matrices.[193]

Self-assembling peptides (SAPs)

Self-assembling peptides (SAPs) have been widely exploited as potential scaffolds in regenerative medicine and are designed by various strategies.[194,195] Self-assembling peptides include peptide amphiphiles (PAs) that can be functionalized with specific cell bioactive adhesion peptides derived from laminin and fibronectin, such as YIGSR and RGD, which enhance adhesion, proliferation and alignment of human corneal stromal fibroblasts with subsequent corneal stroma regeneration in a rabbit model.[196,197] In addition, the multi-functionalization of PA templates is considered a promising strategy to fabricate native-like tissue sheets *in vitro*. Multi-functionalized PAs have been used not only to direct human corneal fibroblasts to adhere and deposit discreet multiple layers of native ECM, but also to control their own self-directed release.[198]

Aligned Scaffolds

Development of aligned scaffolds/templates/substrates is a very promising tissue engineering strategy to control stromal cell orientation *in vitro*, leading to formation of highly organized typical ECM of corneal stroma. Various approaches have been developed to create aligned scaffolds to promote alignment of stromal cells.[177] Aligned scaffolds can be divided into synthetic such as polyurethane or natural such as collagen.[199,200]

Aligned synthetic substrates

Synthetic polymers have been explored as stromal substrates due their tunable mechanical properties. When seeded onto highly aligned fibrous poly(ester urethane) urea substrate *in vitro*, human corneal stromal stem cells (hCSSCs) were capable of differentiating into keratocyte and deposited multilayered lamellae with orthogonally oriented collagen fibrils in the presence of TGF-β$_3$. The approach of combining a highly aligned fibrous substrate in the presence of appropriate growth factor facilitates the bioengineering of well-organized, collagen-based constructs having suitable nanoscale structure for corneal repair and regeneration.[199] Similarly, 3D constructs fabricated from multiple aligned electrospun poly (L,D lactic acid) nanofiber meshes that were arranged orthogonally (topographical cues) have been shown to influence mechanical, phenotypical and genotypical behaviour of adult human derived corneal stromal (AHDCS) cells. Such highly arranged and aligned nanofiber meshes were capable of permitting cell migration between layers and reverting corneal fibroblasts to a keratocyte phenotype.[188]

Aligned collagen scaffolds

Type I collagen is the most abundant collagen in corneas; in particular, corneal stroma consists mainly of highly ordered collagen produced by keratocytes. Therefore, collagen scaffolds have been the focus of intensive tissue engineering research to develop corneal stromal substitutes. The highly organized collagen ultrastructure is crucial in providing structural support and transparency to the cornea; therefore, collagen-based scaffold fabricated *in vitro* must resemble the natural collagen organization or architecture.[200] Engineering of collagenous matrix to produce a highly organized and functional stroma-like structure is crucial and yet challenging.[177] When combined with human keratocytes, orthogonally aligned collagen promoted superior alignment of cells along the direction of collagen fibers by contact guidance, following the orthogonal design of the collagen template as they penetrate into the bulk of the 3D matrix.[201,202] In addition, the combination of limbal epithelial stem cell and stromal keratocytes with the aligned collagen allowed the reconstruction of human hemicorneas *in vitro* (stroma + epithelium) which is a potential substitute for the anterior region of corneas.[203] In addition to fabrication of corneal stroma using self assembly and stromal cell alignment approaches, various scaffold have been evaluated for development of cornea stromal substitute, including gelain-based matrices,[204] decellularized corneal stromas (matrices),[205,206] synthetic matrices such as polyglycolic acid fibers[207] and poly glycerol sebacate (PGS).[208] However, aligned scaffolds are more promising than other classical scaffolds.[177]

Tissue Engineering of Corneal Endothelium

Corneal endothelium is a single layer of specialized cells that lines the posterior surface of the cornea. In adult human, the average endothelial cell density is ~3000 cells/mm^2 and it decreases steadily with age, from 0.3 to 0.6% per year.

This gradual cell loss cannot be compensated because endothelial cells do not normally divide at a rate sufficient to replace dead or injured cells. Endothelial cells have limited proliferative capacity because they are arrested in Gl phase. Endothelial cell density must remain above a critical threshold, usually between 400 and 500 cells/mm^2, to preserve corneal transparency by regulating water content.[209–211] When the endothelium becomes unable to sustain this important function because of insufficient cell density due to gradual decrease of cell population with age or after endothelial damage, the cornea loses its transparency. Corneal transplantation procedures (penetrating keratoplasty and endothelial keratoplasty) are currently the main treatment options for corneal endothelial damage or dysfunction. However, corneal transplantation is limited by a global donor shortage. Therefore, there is a need to overcome the deficiency of sufficient donor corneal tissue by exploring new approaches to address this problem. Engineering of corneal endothelium by *in vitro* expansion of human corneal endothelial cells followed by transplantation is a promising strategy. Transplantation of corneal endothelial cells offers two important advantages (i) treatment of corneal endothelial damage or dysfunction; (ii) improving the quality of available human donor corneas, which are currently unsuitable for transplantation, by increasing their endothelial cell density. The major challenges of endothelial cell transplantation are optimum culture protocols and delivery methods for bioengineered endothelium to the posterior cornea *in vivo*. Following transplantation, the morphology, cell density and pump function of corneal endothelial cells are critical.[212–215] Therefore, research in tissue engineering of corneal endothelium is focused on developing optimal conditions for the isolation and culturing of corneal endothelial cells and developing optimal scaffolds/ substrates/carriers for transplantation.

Cell source for tissue-engineering corneal endothelium

Endothelial cells retain their proliferative capacity according to telomere length and stop dividing due to inhibitory mechanisms that are activated by stress-induced damage pathways. Therefore culturing of human corneal endothelial cells (HCECs) *in vitro* is technically challenging and is considered a powerful tool for cell or tissue - based regeneration of corneal endothelium. However, like other methods in tissue engineering, transplantation of cultured HCECs may raise ethical concerns that could hinder its use in clinical settings.[216] The proliferative capacity of HCECs is affected by donor age and location from which cells were isolated. HCECs from older donors exhibit an overall reduced proliferative capacity compared to these from young donors. Endothelial cells derived from the peripheral endothelium exhibited greater mitogenic activity (higher replication competence) than cells derived from central endothelium.[211]

Isolation, expansion, and preservation of corneal endothelial cells are essential factors for development of a successful tissue engineering strategy as these cells require optimum conditions to maintain proliferative activity *in vitro*, and this is an active field of research. *In vitro* proliferation of human adult corneal endothelial cells is achievable through development of defined cell culture conditions, including supplementation of culture medium with proper growth factors.[217-220] Besides using central and peripheral corneal endothelial cells,[221] other cell sources have been explored including adult stem cells that reside in the junction between the peripheral corneal endothelium and the anterior part of the trabecular meshwork,[222] as well as corneal stromal derived stem cells,[223] adipose-derived stem cells,[224] umbilical cord blood mesenchymal stem cells[225] and induced pluripotent stem cells (iPS).[226]

Genetic manipulation of human corneal endothelial cells has been reported. However, there is controversy regarding the use of these genetically manipulated endothelial cells in clinical practice due to the unclear fate of the vector and uncertain behavior of the manipulated cells following implantation. New techniques to turn vectors off and the development of vector systems that allow termination of endothelial cell proliferation after the desired cell density are required in order to use these cells effectively in clinical practice.[212]

Scaffolds for tissue-engineered corneal endothelium

A corneal endothelium scaffold provides mechanical support during transplantation and facilitates handling of the fragile cultivated cells. In addition, it provides a favorable microenvironment needed for cellular activity. Ideally, biological, mechanical, chemical, and physiological characteristics of corneal endothelium scaffold should mimic these of Descemet's membrane, the acellular basement membrane which is the natural *in vivo* carrier matrix of corneal endothelial cells. The choice of the scaffold and its suitability for *in vivo* studies are challenges; various criteria including transparency, structural strength, and integration with native tissues must be considered during development of the tissue-engineered endothelial substitute.[219, 220, 227, 228] Different carriers or substrates have been explored for engineering of the corneal endothelial layer including living and devitalized corneal materials, amniotic membrane, chitosan-based membranes,[229] silk fibroin membranes,[230] anterior crystalline lens capsule,[231] collagen based materials (cross-linked collagen)[232] and plastic compressed collagen type I produced by RAFT[164,233]. The most commonly endothelial cell substrates or carriers are:

Thermo-responsive cell culture carrier. Corneal endothelial cells can be transferred to the recipient eye as a cell-sheet membrane using temperature-responsive culture dishes. HCECs are cultivated on culture dishes treated with the temperature-responsive

polymer poly(N-isopropylacrylamide) (PIPAAm). This polymer reversibly alters its hydrophobicity/hydrophilicity dependent on incubation temperature. Under typical culture conditions at 37°C, HCECs adhere and proliferate on hydrophobic PIPAAm-grafted surfaces. After reaching confluence and upon reducing culture temperature below the lower critical solution temperature of 32°C, grafted surfaces become hydrophilic and HCEC sheets detaches. Although HCECs can be detached and harvested as an intact monolayer using this approach, surgical manipulation will be easier if these cell sheets are transferred on top of gelatin hydrogels. Cellular function or bioactivity of cell-sheet membranes obtained from thermoresponsive culture dishes has been comprehensively evaluated in animal models.[234-238]

Living and devitalized, or decellularized, corneal materials. The first carriers used for construction of corneal endothelium were devitalized or decellularized corneal materials. Native or living corneal stroma was used by many investigators as endothelial cell carriers for *in vitro* and *in vivo* studies due to its transparency, physiological curvature, biocompatibility, and mechanical stability. Cultured corneal endothelial cells have been seeded on Descemet's membrane after mechanical removal of its endothelium using a cotton-swab. However, in this approach, cultured corneal endothelial cells are exposed to a high risk of contamination by the proliferating native epithelial and stromal cells within the living carrier, and there is an increased risk of immunological rejection after transplantation. To overcome these limitations, devitalization or decellularization of corneal stroma to remove living cells (epithelial, stromal, and endothelial cells) from the carrier is carried out.[176] Different devitalization or decellularization methods for corneas have been developed including freeze/thaw[239,240] and high-hydrostatic pressurization.[241] The source of these cornea-based carriers are poor quality donated corneas (inferior epithelium or endothelium structure) that were routinely discarded in the past. These devitalized stromal carriers permit good adherence of corneal endothelial cells. However, they display some inherent technical difficulties including possibility of disease transmission as it is a biological tissue, fibroblastic contamination because of resident viable keratocytes, and dependence on donor corneas.[227]

Amniotic membrane (the clinically evaluated carrier). Amniotic membrane has been used as a biological substrate for HECEs in various studies. When rabbit corneas denuded of corneal endothelium and Descemet's membrane were used to host the tissue-engineered endothelial grafts consisting of HCECs sheets on this denuded amniotic membrane, they promoted the formation of endothelial cell density and function comparable to that of normal corneas. The corneas that received transplanted cultivated HCECs sheets had little edema and retained their thinness and transparency.[242] In a related study, cat corneas denuded of Descemet's membrane and endothelium were used to host tissue-engineered endothelial grafts consisting of cat corneal endothelial cells on the basement membrane of amniotic membrane. These displayed typical phenotype of endothelial cells (hexagonal shape) and maintained transparency for six weeks *in vivo*.[243] Similarly, cat corneas denuded of endothelium and part of the Descemet's membrane were used to host tissue-engineered endothelial grafts consisting of HCECs sheet on denuded amniotic membrane in a lamellar keratoplasty model. This reconstructed tissue-engineered human corneal endothelium was able to function as a corneal endothelium equivalent and restore corneal function in cat models.[244]

In addition to the aforementioned reconstruction strategies for corneal endothelium, cultured corneal endothelial cells can be injected directly into the anterior chamber without a supporting carrier. Cultured corneal endothelial cells have been injected into the anterior chamber using the sphere-forming assay; however, directing the cells to attach solely to the posterior stroma and not to the other anterior chamber structures is the main difficulty.[245,246] In order to overcome this limitation, ferromagnetic particles (iron) have been incorporated into cultured corneal endothelial cells in order to control cellular migration and attachment *in vitro* and *in vivo*. This strategy facilitated the guidance of endothelial cells to Descemet's membrane and this can be a method of choice for correcting corneal endothelial decompensation.[247,248] Similarly, superparamagnetic nanoparticles (magnetite oxide) was utilized to facilitate magnetic endothelial cell migration and attachment to Descemet's membrane. Further experiments should be carried out to evaluate biomechanics and the physics required to guide these cells to the desired target. Other points for consideration are the safety aspects of the magnetic nanoparticles-loaded cells within ocular tissues and the possible complications that might arise as a result of injecting cell suspensions into the anterior chamber.[249,250]

Applications and Limitations of Tissue Engineered Corneal Substitutes

The major goal of developing tissue-engineered corneal substitutes for transplantation in humans is to address shortages of high quality donor corneas. However, such materials can also be utilized in pre-clinical and non-clinical settings including (i) drug screening (permeability and toxicity studies), (ii) physiological and pathophysiological studies, and (iii) studying cellular and molecular mechanisms along with basic biological principles. TE corneal substitutes can be used for developing efficient gene therapies through down regulation or overexpression of specific genes in certain cell types *in vitro*.[251–255] A wide range of commercially available corneal preclinical models are

available, including Human Corneal Epithelial (HCE) Model produced by SkinEthic Laboratories (Lyon, France),[256] Clonetics™ Human Corneal Epithelial Culture Model produced by Lonza (Hopkinton, MA),[257] LabCyte Cornea-Model (Japan Tissue Engineering Co., Ltd., Gamagori City, Aichi, Japan),[258] and EpiOcular™ Model (MatTek Corporation, Ashland, MA).[259] These *in vitro* preclinical corneal epithelial models are mainly used in drug permeability and eye irritancy testing to reduce the need for animal testing, as described below:

1. The current drug permeability assessment is performed using bovine or porcine corneal explants. However, they are associated with low reproducibility and species matching problems. Corneal epithelium preclinical models can be utilized as an alternative method to the Bovine Cornea Opacity and Permeability (BCOP) test in which corneas are exposed to test substances and then evaluated for loss of transparency and changes in permeability, to evaluate the potential toxicity of test substances.[260–262]

2. Corneal preclinical models can also used as alternatives to the Draize Test and the Low Volume Eye Test (LVET). These two tests are considered the gold standards for the assessment of ocular irritancy or toxicity in which the cornea of a living rabbit is exposed to test substances followed by monitoring the ocular response of the cornea (e.g., opacification); however, it is difficult to predict human eye response based on responses observed in rabbit eyes, due to species differences.[263–265]

The advantages of tissue-engineered preclinical models over existing models

Commercially available preclinical models overcome scientific and ethical concerns associated with typical forms of testing as they provide more reproducible and efficient testing system, provide complex three-dimensional geometries for cultivation and precise control over cultivation conditions through the use of human cells, and eliminate the need for extensive animal care facilities. However, tissue-engineered preclinical models cannot replace animal models completely as no particular *in vitro* model can currently match the complexity found *in vivo*.[96,266]

Despite recent advancements in corneal tissue engineering, several points should be considered before bringing this technology to clinical settings including: i) Optimum development of corneal substitutes to match tensile strength, curvature, permeability, transparency, and durability of normal corneas, ii) Control of the degradation rate of corneal substitutes and their byproducts, iii) Development of optimal culture conditions for the three different corneal cell types (epithelium, fibroblasts/keratocytes, and endothelium), iv) Control of complex interactions between scaffold, corneal cell types and nerve conduits,

v) The ability of host tissue to support integration of corneal tissue-engineered constructs,[100,117,154] and vi) Innervation in order to create fully functioning corneal tissue, as cornea is one of the most heavily innervated and sensitive tissues in the body with a density of nerve endings greater than skin. The corneal nerves function as mechanical and thermal sensors to maintain the overall cornea health; therefore, that missing innervation results in a reduction of corneal sensitivity and subsequent diffuse corneal ulcers. Having said that, few attempts have been made to induce peripheral nerve proliferation within corneal tissue-engineered constructs.[127, 267–269]

CONCLUSIONS

Vision loss and blindness impose physical, psychological, and financial burdens on the affected individuals; therefore, developing and implementing new strategies to address this problem is the focus of intensive research. Corneal allotransplantation, which is considered the gold standard, requires eye donation with associated eye banking issues and a shortage of high quality donor corneas. Corneal xenotransplantation using genetically modified pigs is still at the investigatory stage and may become a promising solution in the near future. Transplantation using keratoprostheses is associated with many postoperative complications. Various limitations of current strategies have attracted researchers to develop tissue-engineered corneal substitutes. A successful corneal substitute should meet the following criteria. i) It should be transparent and have an appropriate refractive index. ii) It should have sufficient mechanical strength (robust and elastic) to withstand handling and suturing during surgical application as well as daily in vivo mechanical stresses. iii) It should have satisfactory biological functions, including biocompatibility, biodegradability, and the promotion of native tissue growth and repair. iv) It must integrate properly with the host tissue to prevent leakage, infection, and epithelial downgrowth. v) It should be permeable to oxygen and nutrients such as glucose and albumin. vi) Sufficient swelling in aqueous solutions must be exhibited. Corneal tissue engineering includes reconstruction of the three main corneal compartments (epithelium, stroma, and endothelium). The most critical barrier to successfully tissue engineer corneal substitute is the reconstruction of a highly organized, mechanically resilient and transparent stromal matrix. Tissue engineering of corneal epithelium has been most successful and is relatively simple compared to corneal stroma. Tissue engineering strategies have demonstrated satisfactory success in the treatment of patients with limbal stem cell deficiency. Reconstruction of the corneal endothelium is as difficult as that of corneal stroma, and optimizing culture conditions and substrates for corneal endothelial cells will be crucial to develop successful tissue-engineered corneal endothelium substitute.

Development of tissue-engineered corneal substitutes has overcome the limitations of corneal allotransplantations and keratoprostheses and is rapidly gaining importance in the field of preclinical testing. However, further research must be conducted to create tissue engineered corneal substitutes that match the properties of normal corneas and allow the development of new treatment alternatives for corneal diseases in vivo.

ACKNOWLEDGMENTS

We are grateful to Dr. Hincke's lab members and to members of Medical Biotechnology Department, Genetic Engineering and Biotechnology Research Institute, City of Scientific Research and Technology Applications (SRTA-City). The generous assistance and interest of Dr. M. Griffith is acknowledged.

REFERENCES

1. Meek, K.M. The cornea and sclera. In Collagen: Structure and Mechanics; Fratzl, P., Ed.; Springer: New York, 2008; 359–396.
2. Farjo, A.A.; McDermott, M.L.; Soong, H.K. Corneal anatomy, physiology, and wound healing. In Ophthalmology, 3rd Ed.; Yanoff, M., Duker, J.S., Eds.; Mosby Elsevier: Elsevier Inc: Edinburgh, 2009; 203–208.
3. Müller, L.J.; Marfurt, C.F.; Kruse, F.; Tervo, T.M. Corneal nerves: Structure, contents and function. Exp. Eye. Res. 2003, 76 (5), 521–542.
4. Ruberti, J.W.; Roy, A.S.; Roberts, C.J. Corneal structure and function. Annu. Rev. Biomed. Eng. 2011, 13, 269–295.
5. Kolozsvári, L.; Nógrádi, A.; Hopp, B.; Bor, Z. UV absorbance of the human cornea in the 240–400 nm range. Invest. Ophthalmol. Vis. Sci. 2002, 43 (7), 2165–2168.
6. Jue, B.; Maurice, D.M. The mechanical properties of the rabbit and human cornea. J. Biomech. 1986, 19 (10), 847–853.
7. Khaled, M.L.; Helwa, I.; Drewry, M.; Seremwe, M.; Estes, A.; Liu, Y. Molecular and Histopathological Changes Associated with Keratoconus. Biomed. Res. Int. 2017, Epub 2017.
8. Michelacci, Y.M. Collagens and proteoglycans of the corneal extracellular matrix. Braz. J. Med. Biol. Res. 2003, 36 (8), 1037–1046.
9. Pfister R.R. The healing of corneal epithelial abrasions in the rabbit: A scanning electron microscope study. Invest. Ophthalmol. 1975, 14 (9), 648–661.
10. DelMonte, D.W.; Kim, T. Anatomy and physiology of the cornea. J. Cataract. Refract. Surg. 2011, 37 (3), 588–598.
11. Wilson, S.E.; Hong, J.W. Bowman's layer structure and function: critical or dispensable to corneal function? A hypothesis. Cornea 2000, 19 (4), 417–420.
12. Hayashi, S.; Osawa, T.; Tohyama, K. Comparative observations on corneas, with special reference to bowman's layer and descemet's membrane in mammals and amphibians. J. Morphol. 2002, 254 (3), 247–258.

13. Lagali, N.; Germundsson, J.; Fagerholm, P. The role of Bowman's layer in corneal regeneration after phototherapeutic keratectomy: A prospective study using *in vivo* confocal microscopy. Invest. Ophthalmol. Vis. Sci. **2009**, *50* (9), 4192–4198.

14. Meek, K.M.; Leonard, D.W. Ultrastructure of the corneal stroma: A comparative study. Biophys. J. **1993**, *64* (1), 273–280.

15. Müller, L.J.; Pels, L.; Vrensen, G.F. Novel aspects of the ultrastructural organization of human corneal keratocytes. Invest. Ophthalmol. Vis. Sci. **1995**, *36* (13), 2557–2567.

16. Hull, D.S.; Green, K.; Laughter, L. Cornea endothelial rose bengal photosensitization. Effect on permeability, sodium flux, and ultrastructure. Invest. Ophthalmol. Vis. Sci. **1984**, *25* (4), 455–460.

17. Dua, H.S.; Faraj, L.A.; Said, D.G.; Gray, T.; Lowe, J. Human corneal anatomy redefined: A novel pre-Descemet's layer (Dua's layer). Ophthalmology. **2013**, *120* (9), 1778–1785.

18. Beuerman, R.W.; Pedroza, L. Ultrastructure of the human cornea. Microsc. Res. Tech. **1996**, *33* (4), 320–335.

19. Meek, K.M.; Knupp C. Corneal structure and transparency. Prog. Retin. Eye. Res. **2015**, *49*, 1–16.

20. Freegard, T.J. The physical basis of transparency of the normal cornea. Eye (Lond) **1997**, *11* (4), 465–471.

21. Foster, A. Vision 2020--the right to sight. Trop. Doct. **2003**, *33* (4), 193–194.

22. World Health Organization, "Global Initiative for the Elimination of Avoidable Blindness: Action plan 2006–2011." 2007.

23. World Health Organization, Visual impairment and blindness. Fact sheet no. 282, updated August 2014. http://www.who.int/mediacentre/factsheets/fs282/en/ (accessed 15 March 2016).

24. Pascolini, D.; Mariotti, S.P. Global estimates of visual impairment: 2010. Br. J. Ophthalmol. **2012**, *96* (5), 614–618.

25. Whitcher, J.P.; Srinivasan, M.; Upadhyay, M.P. Corneal blindness: A global perspective. Bull. World. Health. Organ. **2001**, *79* (3), 214–221.

26. Derrick, T.; Roberts, Ch.; Last, A.R.; Burr, S.E.; Holland, M.J. Trachoma and ocular chlamydial infection in the era of genomics. Mediators. Inflamm. **2015**, *2015*, 791–847.

27. Shtein, R.M.; Elner, V.M. Herpes simplex virus keratitis: Histopathology and corneal allograft outcomes. Expert. Rev. Ophthalmol. **2010**, *5* (2), 129–134.

28. Scott, R. The injured eye. Philos. Trans R. Soc. Lond B. Biol. Sci. **2011**, *366* (1562), 251–260.

29. Ahmed, F.; House, R.J; Feldman, B.H. Corneal abrasions and corneal foreign bodies. Primary Care. **2015**, *42* (3), 363–375.

30. Ambekar, R.; Toussaint, K.C.; Johnson, A.W. The effect of keratoconus on the structural, mechanical, and optical properties of the cornea. J. Mech. Behav. Biomed. Mater. **2011**, *4* (3), 223–236.

31. Vincent, A.L. Corneal dystrophies and genetics in the International Committee for Classification of Corneal Dystrophies era: A review. Clin. Exp. Ophthalmol. **2014**, *42* (1), 4–12.

32. Wilson, S.E.; Bourne, W.M. Fuchs' dystrophy. Cornea. **1988**, *7* (1), 2–18.

33. Werblin, T.P.; Hirst, L.W.; Stark, W.J.; Maumenee, I.H. Prevalence of map-dot-fingerprint changes in the cornea. Br. J. Ophthalmol. **1981**, *65* (6), 401–409.

34. Yavuz, B.; Bozdağ Pehlivan, S.; Ünlü, N. An overview on dry eye treatment: Approaches for cyclosporin a delivery. Scientific. World. Journal. **2012**, 2012, 194848.

35. Jackson, W.B. Blepharitis: Current strategies for diagnosis and management. Can. J. Ophthalmol. **2008**, *43* (2), 170–179.

36. Clement, C.I.; Goldberg, I. The management of complicated glaucoma. Indian. J. Ophthalmol. **2011**, *59* (Suppl1), S141–S147.

37. WHO, Action plan 2006–2011. Geneva: WHO Press; 2007. Vision 2020 Global initiative for the elimination of avoidable blindness.

38. Goldschmidt, P. Social sciences for the prevention of blindness. Trop. Med. Health. **2015**, *43* (2), 141–148.

39. McColgan, K. Corneal transplant surgery. J. Perioper. Pract. **2009**, *19* (2), 51–54.

40. Tan, D.T., Dart, J.K., Holland, E.J., Kinoshita, S. Corneal transplantation. Lancet **2012**, *379* (9827), 1749–1761.

41. Atallah, M.; Amescua, G . Advancements in anterior lamellar keratoplasty. In *Advances in Medical and Surgical Cornea*; Jeng, B.H., Ed.; Springer Berlin: Heidelberg, 2014, 89–98.

42. Zirm, E.K. Eine erfolgreiche totale Keratoplastik (A successful total keratoplasty). 1906. Refract. Corneal. Surg. **1989**, *5*, 258–261.

43. Garg, P.; Krishna, P.V.; Stratis, A.K.; Gopinathan, U. The value of corneal transplantation in reducing blindness. Eye (Lond). **2005**, *19* (10), 1106–1114.

44. Vieira Silva, J.; Júlio de Faria e Sousa, S.; Mafalda Ferrante, A. Corneal transplantation in a developing country: problems associated with technology transfer from rich to poor societies. Acta Ophthalmol. Scand. **2006**, *84* (3), 396–400.

45. Rao, G.N.; Gopinathan, U. Eye banking: An introduction. Community. Eye. Health. **2009**, *22* (71), 46– 47.

46. Eye Bank Association of America. Eye Banking Statistical Report. Washington DC: EBBA; 2011.

47. Qazi, Y.; Hamrah, P. Corneal allograft rejection: Immunopathogenesis to therapeutics. J. Clin. Cell. Immunol. **2013**, *2013* (Suppl 9).

48. Sellami, D.; Abid, S.; Bouaouaja, G.; Ben Amor, S.; Kammoun, B.; Masmoudi, M.; Dabbeche, K.; Boumoud, H.; Ben Zina, Z.; Feki, J. Epidemiology and risk factors for corneal graft rejection. Transplant. Proc. **2007**, *39* (8), 2609–2611.

49. Fu, H.; Larkin, DF.; George, A.J. Immune modulation in corneal transplantation. Transplant. Rev (Orlando).**2008**, *22* (2),105–115

50. Laibson, P.R. Current concepts and techniques in corneal transplantation. Curr. Opin. Ophth. **2002**, *13* (4), 220–223.

51. Boisjoly, H.M.; Tourigny, R.; Bazin, R.; Laughrea, P.A.; Dubé, I.; Chamberland, G.; Bernier, J.; Roy, R. Risk factors of corneal graft failure. Ophthalmology **1993**, *100* (11), 1728–1735.

52. WHO, Transplantation, Xenotransplantation, http://www.who.int/transplantation/xeno/en/. [accessed 15 March 2015].

53. Cooper, D.K. A brief history of cross-species organ transplantation. Proc (Bayl Univ Med Cent). **2012**, *25* (1), 49–57.

Conducting—Dendritic

54. Ekser, B.; Cooper, D.K.; Tector, A.J. The need for xeno-transplantation as a source of organs and cells for clinical transplantation. Int. J. Surg. **2015**, *23* (Pt B), 199–204.

55. Hara, H.; Cooper, D.K. Xenotransplantation--the future of corneal transplantation? Cornea **2011**, *30* (4),371–378.

56. Kim, M.K.; Wee, W.R.; Park, C.G.; Kim, S.J. Xenocorneal transplantation. Curr. Opin. Organ. Transplant. **2011**, *16* (2), 231–236.

57. Cooper, D.K.; Ezzelarab, M.B.; Hara, H.; Iwase, H.; Lee, W.; Wijkstrom, M.; Bottino, R. The pathobiology of pig-to-primate xenotransplantation: a historical review. Xenotransplantation. **2016**, *23* (2), 83–105.

58. Hara, H.; Cooper, D.K. The immunology of corneal xeno-transplantation: A review of the literature. Xenotransplantation. **2010**, *17* (5), 338–349.

59. Kim, M.K.; Hara, H. Current status of corneal xenotrans-plantation. Int. J. Surg. **2015**, *23* (Pt B), 255–260.

60. Cooper, D.K.; Bottino, R. Recent advances in understand-ing xenotransplantation: Implications for the clinic. Expert. Rev. Clin. Immunol. **2015**, *11* (12), 1379–1390.

61. Castroviejo, R.; Cardona, H.; DeVoe, A.G. The present status of prosthokeratoplasty. Trans. Am. Ophthalmol. Soc. **1969**, *67*, 207–234.

62. Hicks, C.R.; Fitton, J.H.; Chirila, T.V.; Crawford, G.J.; Constable, I.J. Keratoprostheses: Advancing toward a true artificial cornea. Surv. Ophthalmol. **1997**, *42* (2), 175–189.

63. Chirila, T.V.; Hicks, C.R. The origins of the artificial cor-nea: Pellier de Quengsy and his contribution to the modern concept of keratoprosthesis. Gesnerus **1998**, *56* (1–2), 96–106.

64. Chirila, T.V.; Hicks, C.R.; Dalton, P.D.; Vijayasekaran, S.; Lou, X.; Honga, Y.; Claytona, A.B.; Ziegelaara, B.W.; Fittona, J.H.; Plattena, S.; Crawforda, G.J.; Constablea, I.J. Artificial cornea. Prog. Polym. Sci. **1998**, *23* (3), 447–473.

65. Caldwell, D.R. The soft keratoprosthesis. Trans. Am. Ophthalmol. Soc. **1997**, *95*, 751–802.

66. Leibowitz, H.M.; Trinkhaus-Randall, V.; Tsuk, A.G.; Franzbau, C. Progress in the development of a synthetic cornea. Prog. Ret. Eye. Res. **1994**, *13*, 605–621.

67. Khan, B.; Dudenhoefer, E.J.; Dohlman, C.H. Keratopros-thesis: An update. Curr. Opin. Ophthalmol. **2001**, *12* (4), 282–287.

68. Stone, W. Jr.,; Herbert, E. Experimental study of plastic material as replacement for the cornea; A preliminary report. Am. J. Ophthalmol. **1953**, *36* (6), 168–173.

69. Hicks, C.; Crawford, G.; Chirila, T.; Wiffen, S.; Vijayasekaran, S.; Lou, X.; Fitton, J.; Maley, M.; Clayton, A.; Dalton, P.; Platten, S.; Ziegelaar, B.; Hong, Y.; Russo, A.; Constable, I. Development and clinical assessment of an artificial cornea. Prog. Retin. Eye Res. **2000**, *19* (2), 149–170.

70. Avadhanam, V.S.; Smith, H.E.; Liu, C. Keratoprostheses for corneal blindness: A review of contemporary devices. Clin. Ophthalmol. **2015**, *9*, 697–720.

71. Rafat, M.A.; Hackett, J.M.; Fagerholm, P.; Griffith, M. Artificial cornea. In *Ocular Periphery and Disorders*; Dartt, D.A.; Dana, R.; D'Amore, P.; Niederkorn, J., Eds.; Elsiever: Spain, 2011; 311–317

72. Griffith, M.; Fagerholm, P.; Liu, W.; McLaughlin, C.R.; Li, F. Corneal regenerative medicine: corneal substitutes for transplantation. In *Cornea and External Eye Disease*; Reinhard, T.; Larkin, F., Eds.; Springer Berlin: Heidel-berg, 2008; 37–53

73. Princz, M.A.; Sheardown, H.; Griffith, M. Corneal tissue engineering versus synthetic artificial corneas. In *Bioma-terials and Regenerative Medicine in Ophthalmology*; Chirila, T., Ed.; CRC Press/Woodhead Publishing: Cambridge, 2010; 134–149.

74. Gomaa, A.; Comyn, O.; Liu, C. Keratoprostheses in clini-cal practice–a review. Clin. Exp. Ophthalmol. **2010**, *38* (2), 211–224.

75. Liu, C.; Paul, B.; Tandon, R.; Lee, E.; Fong, K.; Mavrikakis, I.; Herold, J.; Thorp, S.; Brittain, P.; Francis, I.; Ferrett, C.; Hull, C.; Lloyd, A.; Green, D.; Franklin, V.; Tighe, B.; Fukuda, M.; Hamada, S. The osteo-odonto-keratopros-thesis (OOKP). Semin. Ophthalmol. **2005**, *20* (2), 113–128.

76. Traish, A.S.; Chodosh, J. Expanding application of the Boston type I keratoprosthesis due to advances in design and improved post-operative therapeutic strategies. Semin. Ophthalmol. **2010**, *25* (5–6), 239–243.

77. Jirásková, N.; Rozsival, P.; Burova, M., Kalfertova, M. AlphaCor artificial cornea: Clinical outcome. Eye. **2011**, *25* (9), 1138–1146.

78. Griffith, M.; Polisetti, N.; Kuffova, L.; Gallar, J.; Forrester, J.; Vemuganti, G.K.; Fuchsluger, T.A. Regenerative approaches as alternatives to donor allografting for restoration of corneal function. Ocul Surf. **2012**, *10* (3), 170–183.

79. Ciolino, J.B.; Dohlman, C.H. Biologic keratoprosthesis materials. Int. Ophthalmol. Clin. **2009**, *49* (1), 1–9.

80. Myung, D.; Duhamel, P.E.; Cochran, J.R.; Noolandi, J.; Ta, C.N.; Frank, C.W. Development of Hydrogel-based keratoprostheses: A materials perspective. Biotechnol. Prog. **2008**, *24* (3), 735–741.

81. Rafat, M.; Xeroudaki, M.; Koulikovska, M.; Sherrell, P.; Groth, F.; Fagerholm, P.; Lagali, N. Composite core-and-skirt collagen hydrogels with differential degradation for corneal therapeutic applications. Biomaterials. **2016**, *83*, 142–155.

82. Ma, J.J.; Graney, J.M.; Dohlman, C.H. Repeat penetrating keratoplasty versus the Boston keratoprosthesis in graft failure. Int. Ophthalmol. Clin. **2005**, *45* (4), 49–59.

83. Ilhan-Sarac, O.; Akpek, E.K. Current concepts and tech-niques in keratoprosthesis. Curr Opin Ophthalmol. **2005**, *16* (4), 246–250.

84. Langer, R.L.; Vacanti, J.P. Tissue Engineering. Science. **1993**, *260* (5110) 920–926.

85. Howard, D.; Buttery, L.D.; Shakesheff, K.M.; Roberts, S.J. Tissue engineering: Strategies, stem cells and scaf-folds. J. Anat. **2008**, *213* (1), 66–72.

86. Jayo, M.J.; Watson, D.D.; Wagner, B.J.; Bertram, T.A. Tissue engineering and regenerative medicine: Role of toxicologic pathologists for an emerging medical technol-ogy. Toxicol. Pathol. **2008**, *36* (1), 92–96.

87. Moreno-Borchart, A. Building organs piece by piece. Accomplishments and future perspectives in tissue engineering. EMBO. Rep. **2004**, *5* (11), 1025–1028.

88. Sipe, J.D. Tissue engineering and reparative medicine. Ann. N Y Acad. Sci. **2002**, *961* (1), 1–9.

89. Tabata, Y. Biomaterial technology for tissue engineering applications. J. R. Soc. Interface.**2009**, *6* (Suppl 3), S311. S324.

90. Parenteau, N.L.; Young, J.H. The use of cells in regenerative medicine. Ann. N Y Acad. Sci. **2002**, *961* (1), 27–39.

91. Chung, H.J.; Park, T.G. Surface engineered and drug releasing pre-fabricated scaffolds for tissue engineering. Adv. Drug. Deliv. Rev. **2007**, *59* (4–5), 249–262.

92. Garg, T.; Singh, O.; Arora, S.; Murthy, R.S. Scaffold: A novel carrier for cell and drug delivery. Crit. Rev. Ther. Drug Carrier Syst. **2012**, *29* (1), 1–63

93. Bottaro, D.P.; Liebmann-Vinson, A.; Heidaran, M.A. Molecular signaling in bioengineered tissue microenvironments. Ann. N Y Acad. Sci. **2002**, *961* (1), 143–153.

94. Rosso, F.; Giordano, A.; Barbarisi, M.; Barbarisi, A. From cell–ECM interactions to tissue engineering. J. Cell. Physiol. **2004**, *199* (2), 174–180.

95. Ahmed, T.A.; Dare, E.V.; Hincke, M. Fibrin: A versatile scaffold for tissue engineering applications. Tissue. Eng. Part B. Rev. **2008**, *14* (2), 199–215.

96. Griffith, L.G.; Naughton, G. Tissue engineering--current challenges and expanding opportunities. Science **2002**, *295* (5557), 1009–1014.

97. Germain, L.; Carrier, P.; Auger, F.A.; Salesse, C.; Guérin, S.L. Can we produce a human corneal equivalent by tissue engineering?. Prog. Retin. Eye Res. **2000**, *19* (5), 497–527.

98. Duan, D.; Klenkler, B.J.; Sheardown, H. Progress in the development of a corneal replacement: keratoprostheses and tissue-engineered corneas. Expert. Rev. Med. Devices. **2006**, *3* (1), 59–72.

99. Karamichos, D. Ocular tissue engineering: Current and future directions. J. Funct. Biomater. **2015**, *6* (1), 77–80.

100. Shah, A.; Brugnano, J.; Sun, S.; Vase, A.; Orwin, E. The development of a tissue-engineered cornea: biomaterials and culture methods. Pediatr. Res. **2008**, *63* (5), 535–544.

101. Griffith, M.; Hakim, M.; Shimmura, S.; Watsky, M.A.; Li, F.; Carlsson, D.; Doillon, C.J.; Nakamura, M.; Suuronen, E.; Nakata, K.; Sheardown, H. Artificial human corneas. Cornea **2002**, *21* (Suppl.2), S54–S61.

102. P De Miguel, M.; L Alio, J.; Arnalich-Montiel, F.; Fuentes-Julian, S.; de Benito-Llopis, L.; Amparo, F.; Bataille, L. Cornea and ocular surface treatment. Curr. Stem. Cell. Res. Ther. **2010**, *5* (2), 195–204.

103. Minami, Y.; Sugihara, H.; Oono, S. Reconstruction of cornea in three-dimensional collagen gel matrix culture. Invest. Ophthalmol. Vis. Sci. **1993**, *34* (7), 2316–24.

104. Zieske, J.D.; Mason, V.S.; Wasson, M.E.; Meunier, S.F.; Nolte, C.J.; Fukai, N.; Olsen, B.R.; Parenteau, N.L. Basement membrane assembly and differentiation of cultured corneal cells: importance of culture environment and endothelial cell interaction. Exp. Cell. Res. **1994**, *214* (2), 621–633.

105. Schneider, A.I.; Maier-Reif, K.; Graeve, T. Constructing an *in vitro* cornea from cultures of the three specific corneal cell types. *In Vitro*. Cell. Dev. Biol. Anim. **1999**, *35* (9), 515–26.

106. Auger, F.A.; Rémy-Zolghadri, M.; Grenier, G.; Germain, L. A truly new approach for tissue engineering: The LOEX self-assembly technique. Ernst. Schering. Res. Found. Workshop. **2002**, *35*, 73–88

107. Griffith, M.; Osborne, R.; Munger, R.; Xiong, X.; Doillon, CJ.; Laycock, N.L.; Hakim, M.; Song, Y.; Watsky, M.A. Functional human corneal equivalents constructed from cell lines. Science **1999**, *286* (5447), 2169–2172

108. Germain, L.; Auger, F.A.; Grandbois, E.; Guignard, R.; Giasson, M.; Boisjoly, H.; GueÂrin, S.L. Reconstructed human cornea produced *in vitro* by tissue engineering. Pathobiol **1999**, *67* (3), 140–147.

109. Alaminos, M.; Sánchez-Quevedo, M.D.; Muñoz -Ávila, J.I.; Serrano, D.; Medialdea, S.; Carreras, I.; Campos, A. Construction of a complete rabbit cornea substitute using a fibrin- agarose scaffold. Invest. Ophthalmol. Vis. Sci. **2006**, *47* (8), 3311–3317.

110. Vrana, N.E.; Builles, N.; Justin, V.; Bednarz, J.; Pellegrini, G.; Ferrari, B.; Damour, O.; Hasirci, V. Development of a reconstructed cornea from collagen–chondroitin sulfate foams and human cell cultures. Invest. Ophthalmol. Vis. Sci. **2008**, *49* (12), 5325–5331.

111. Duan, X.; Sheardown, H. Dendrimer crosslinked collagen as a corneal tissue engineering scaffold: mechanical properties and corneal epithelial cell interactions. Biomaterials. **2006**, *27* (26), 4608–4617.

112. Rafat, M.; Li, F.; Fagerholm, P.; Lagali, N.S.; Watsky, M.A.; Munger, R.; Matsuura, T.; Griffith M. PEG-stabilized carbodiimide crosslinked collagen-chitosan hydrogels for corneal tissue engineering. Biomaterials **2008**, *29* (29), 3960–3972.

113. Uwamaliya, J.; Carrier, P.; Proulx, S.; Deschambeault, A.; Audet, C.; Auger, F.A.; Germain, L. Reconstruction of a human cornea by the self-assembly approach of tissue engineering using the three native cell types. Invest. Ophthalmol. Vis. Sci. **2009**, *50* (13), 1518–1518.

114. Luo, H.; Lu, Y.; Wu, T.; Zhang, M.; Zhang, Y.; Jin, Y. Construction of tissue-engineered cornea composed of amniotic epithelial cells and acellular porcine cornea for treating corneal alkali burn. Biomaterials **2013**, *34* (28), 6748–6759.

115. Fagerholm, P.; Lagali N.S.; Carlsson D.J.; Merrett, K.; Griffith, M. Corneal regeneration following implantation of a biomimetic tissue-engineered substitute. Clin. Transl. Sci. **2009**, *2* (2), 162–164.

116. Fagerholm, P.; Lagali, N.S.; Merrett, K.; Jackson, W.B.; Munger, R.; Liu, Y.; Polarek, J.W.; Söderqvist, M.; Griffith, M. A biosynthetic alternative to human donor tissue for inducing corneal regeneration: 24-month follow-up of a phase 1 clinical study. Sci. Transl. Med. **2010**, *2* (46), 46–61.

117. Fagerholm, P.; Lagali, N.S.; Ong, J.A.; Merrett, K.; Jackson, W.B.; Polarek, J.W.; Suuronen, E.J.; Liu, Y.; Brunette, I.; Griffith, M. Stable corneal regeneration four years after implantation of a cell-free recombinant human collagen scaffold. Biomaterials **2014**, *35* (8), 2420–2427.

118. Takács, L.; Tóth, E.; Berta, A.; Vereb, G. Stem cells of the adult cornea: from cytometric markers to therapeutic applications. Cytometry A. **2009**, *75* (1), 54–66.

119. Ahmad, S. Concise review: Limbal stem cell deficiency, dysfunction, and distress. Stem. Cells. Transl. Med. **2012**, *1* (2), 110–115.

120. Osei-Bempong, C.; Figueiredo, F.C.; Lako, M. The limbal epithelium of the eye–a review of limbal stem cell biology, disease and treatment. Bioessays **2013**, *35* (3), 211–219.

121. 121 Pellegrini, G.; Traverso, C.E.; Franzi, A.T.; Zingirian, M.; Cancedda, R.; De Luca, M. Long-term restoration of damaged corneal surfaces with autologous cultivated corneal epithelium. Lancet **1997**, *349* (9057), 990–993.

122. Haagdorens, M.; Van Acker, S.I.; Van Gerwen, V.; Ní Dhubhghaill, S.; Koppen, C.; Tassignon, M.J.; Zakaria, N. Limbal stem cell deficiency: Current treatment options and emerging therapies. Stem. Cells. Int. **2016**, 2016, 9798374.

123. Nishida, K.; Yamato, M.; Hayashida, Y.; Watanabe, K.; Yamamoto, K.; Adachi, E.; Nagai, S.; Kikuchi, A.; Maeda, N.; Watanabe, H.; Okano, T.; Tano, Y. Corneal reconstruction with tissue-engineered cell sheets composed of autologous oral mucosal epithelium. N. Engl. J. Med. **2004**, *351* (12), 1187–1196.

124. Eslani, M.; Baradaran-Rafii, A.; Ahmad, S. Cultivated limbal and oral mucosal epithelial transplantation. Semin. Ophthalmol. **2012**, *27* (3–4), 80–93.

125. Orwin, E.J.; Hubel, A. *In vitro* culture characteristics of corneal epithelial, endothelial, and keratocyte cells in a native collagen matrix. Tissue. Eng. **2000**, *6* (4), 307–319.

126. Ohji, M.; SundarRaj, N.; Hassell, J.R.; Thoft, R.A. Basement membrane synthesis by human corneal epithelial cells *in vitro*. Invest. Ophthalmol. Vis. Sci. **1994**, *35* (2), 479–485.

127. Li, F.; Carlsson, D.; Lohmann, C.; Suuronen, E.; Vascotto, S.; Kobuch, K.; Sheardown, H.; Munger, R. Nakamura, M.; Griffith, M. Cellular and nerve regeneration within a biosynthetic extracellular matrix for corneal transplantation. Proc.Natl.Acad.Sci.U.S.A. **2003**, *100* (26), 15346–15351.

128. Menzel-Severing, J.; Kruse, F.E.; Schlötzer-Schrehardt, U. Stem cell–based therapy for corneal epithelial reconstruction: Present and future. Can. J. Ophthalmol. **2013**, *48* (1), 13–21.

129. Nakamura, T.; Inatomi, T.; Sotozono, C.; Koizumi, N.; Kinoshita, S. Ocular surface reconstruction using stem cell and tissue engineering. Prog. Retin. Eye. Res. **2016**, *51*, 187–207.

130. Casaroli-Marano, R.P.; Nieto-Nicolau, N.; Martínez-Conesa, E.M.; Edel, M.; B Álvarez-Palomo, A. Potential role of induced pluripotent stem cells (iPSCs) for cell-based therapy of the ocular surface. J. Clin. Med. **2015**, *4* (2), 318–342.

131. Hayashi, R.; Ishikawa, Y.; Ito, M.; Kageyama, T.; Takashiba, K.; Fujioka, T.; Tsujikawa, M.; Miyoshi, H.; Yamato, M.; Nakamura, Y.; Nishida, K. Generation of corneal epithelial cells from induced pluripotent stem cells derived from human dermal fibroblast and corneal limbal epithelium. PLoS One. **2012**, *7* (9), e45435.

132. Sareen, D.; Saghizadeh, M.; Ornelas, L.; Winkler, M.A.; Narwani, K.; Sahabian, A.; Funari, V.A.; Tang, J.; Spurka, L.; Punj, V.; Maguen, E.; Rabinowitz, YS.; Svendsen, C.N.; Ljubimov, A.V. Differentiation of human limbal-derived induced pluripotent stem cells into limbal-like epithelium. Stem. Cells. Transl. Med. **2014**, *3* (9), 1002–1012.

133. Yang, J.; Yamato, M.; Nishida, K.; Ohki, T.; Kanzaki, M.; Sekine, H.; Shimizu, T.; Okano, T. Cell delivery in regenerative medicine: The cell sheet engineering approach. J. Control. Release. **2006**, *116* (2), 193–203.

134. Nishida, K.; Yamato, M.; Hayashida, Y.; Watanabe, K.; Maeda, N.; Watanabe, H.; Yamamoto, K.; Nagai, S.;

Kikuchi, A.; Tano, Y.; Okano, T. Functional bioengineered corneal epithelial sheet grafts from corneal stem cells expanded *ex vivo* on a temperature-responsive cell culture surface. Transplantation. **2004**, *77* (3), 379–385.

135. Kobayashi, T.; Kan, K.; Nishida, K.; Yamato, M.; Okano, T. Corneal regeneration by transplantation of corneal epithelial cell sheets fabricated with automated cell culture system in rabbit model. Biomaterials. **2013**, *34* (36), 9010–9017.

136. Utheim, T.P.; Utheim, Ø.A.; Khan, Q.E.; Sehic, A. Culture of oral mucosal epithelial cells for the purpose of treating limbal stem cell deficiency. J. Funct. Biomater. **2016**, *7* (1), E5.

137. Kataoka, K.; Huh, N. Application of a thermo-reversible gelation polymer, mebiol gel, for stem cell culture and regenerative medicine. J. Stem. Cells. Regen. Med. **2010**, *6* (1), 10–14.

138. Sudha, B.; Madhavan, H.N.; Sitalakshmi, G.; Malathi, J.; Krishnakumar, S.; Mori, Y.; Yoshioka, H.; Abraham, S. Cultivation of human corneal limbal stem cells in mebiol gel—A thermo-reversible gelation polymer. Indian J. Med. Res. **2006**, *124* (6), 655–664.

139. Fukuda, K.; Chikama, T.; Nakamura, M.; Nishida, T. Differential distribution of subchains of the basement membrane components type IV collagen and laminin among the amniotic membrane, cornea, and conjunctiva. Cornea, **1999**, *18* (1), 73–79.

140. Dua, HS.; Gomes, J.A.; King, A.J.; Maharajan, V.S. The amniotic membrane in ophthalmology. Surv. Ophthalmol. **2004**, *49* (1), 51–77.

141. Liu, J.; Sheha, H.; Fu, Y.; Liang, L.; Tseng, S.C. Update on amniotic membrane transplantation. Expert. Rev. Ophthalmol. **2010**, *5* (5), 645–661.

142. Du, Y.; Chen, J.; Funderburgh, J.L.; Zhu, X.; Li, L. Functional reconstruction of rabbit corneal epithelium by human limbal cells cultured on amniotic membrane. Mol. Vis. **2003**, *9*, 635–643

143. Luo, H.; Lu, Y.; Wu, T.; Zhang, M.; Zhang, Y.; Jin, Y. Construction of tissue-engineered cornea composed of amniotic epithelial cells and acellular porcine cornea for treating corneal alkali burn. Biomaterials. **2013**, *34* (28), 6748–6759.

144. Ma, Y.; Xu, Y.; Xiao, Z.; Yang, W.; Zhang, C.; Song, E.; Du, Y.; Li, L. Reconstruction of chemically burned rat corneal surface by bone marrow-derived human mesenchymal stem cells. Stem. Cells. **2006**, *24* (2), 315–321.

145. Mi, S.; Yang, X.; Zhao, Q.; Qu, L.; Chen, S.; Meek, M.K.; Dou, Z. Reconstruction of corneal epithelium with cryopreserved corneal limbal stem cells in a goat model. Mol. Reprod. Dev. **2008**, *75* (11), 1607–1616.

146. Qu, L.; Yang, X.; Wang, X.; Zhao, M.; Mi, S.; Dou, Z.; Wang, H. Reconstruction of corneal epithelium with cryopreserved corneal limbal stem cells in a rabbit model. Vet. J. **2009**, *179* (3), 392–400.

147. Utheim, T.P.; Lyberg, T.; Raeder, S. The culture of limbal epithelial cells. Methods Mol. Biol. **2013**, *1014*, 103–129.

148. Linnes, M.P.; Ratner, B.D.; Giachelli, C.M. A fibrinogen-based precision microporous scaffold for tissue engineering. Biomaterials. **2007**, *28* (35), 5298–5306.

149. Rama, P.; Bonini, S.; Lambiase, A.; Golisano, O.; Paterna, P.; De Luca, M.; Pellegrini, G. Autologous fibrin-cultured

limbal stem cells permanently restore the corneal surface of patients with total limbal stem cell deficiency. Transplantation. **2001**, *72* (9), 1478–1485.

150. Rama, P.; Matuska, S.; Paganoni, G.; Spinelli, A.; De Luca, M.; Pellegrini, G. Limbal stem-cell therapy and long-term corneal regeneration. N. Engl. J. Med. **2010**, *363* (2), 147–155.

151. Han, B.; Schwab, I.R.; Madsen, T.K.; Isseroff, R.R. A fibrin-based bioengineered ocular surface with human corneal epithelial stem cells. Cornea. **2002**, *21* (5), 505–510.

152. Szurman, P.; Warga, M.; Grisanti, S.; Roters, S.; Rohrbach, J.M.; Aisenbrey, S.; Kaczmarek, R.T.; Bartz-Schmidt, K.U. Sutureless amniotic membrane fixation using fibrin glue for ocular surface reconstruction in a rabbit model. Cornea. **2006**, *25* (4), 460–466.

153. Hino, M.; Ishiko, O.; Honda, K.I.; Yamane, T.; Ohta, K.; Takubo, T.; Tatsumi, N. Transmission of symptomatic parvovirus B19 infection by fibrin sealant used during surgery. Br. J. Haematol. **2000**, *108* (1), 194–195.

154. Griffith, M.; Jackson, W.B.; Lagali, N.; Merrett, K.; Li, F.; Fagerholm, P. Artificial corneas: A regenerative medicine approach. Eye (Lond). **2009**, *23* (10), 1985–1989.

155. Bareil, R.P.; Gauvin, R.; Berthod, F. Collagen-based biomaterials for tissue engineering applications. Materials. **2010**, *3* (3), 1863–1887.

156. Lee, H.; Singla, A.; Lee, Y. Biomedical applications of collagen. Int. J. Pharm. **2001**, *221* (1), 1–22.

157. Levis, H.J.; Massie, I.; Dziasko, M.A.; Kaasi, A.; Daniels, J.T. Rapid tissue engineering of biomimetic human corneal limbal crypts with 3D niche architecture. Biomaterials. **2013**, *34* (35), 8860–8868.

158. Benton, G; Kleinman H.K.; George J.; Arnaoutova, I. Multiple uses of basement membrane-like matrix (BME/Matrigel) *in vitro* and *in vivo* with cancer cells. Int. J. Cancer. **2011**, *128* (8), 1751–1757.

159. Ahmadiankia, N.; Ebrahimi, M.; Hosseini, A.; Baharvand, H. Effects of different extracellular matrices and cocultures on human limbal stem cell expansion *in vitro*. Cell. Biol. Int. **2009**, *33* (9), 978–987.

160. He, Y.G.; McCulley, J.P. Growing human corneal epithelium on collagen shield and subsequent transfer to denuded cornea *in vitro*. Curr. Eye. Res. **1991**, *10* (9), 851–863.

161. McIntosh Ambrose, W.; Salahuddin, A.; So, S.; Ng, S.; Ponce Márquez, S.; Takezawa, T.; Schein, O.; Elisseeff, J. Collagen vitrigel membranes for the *in vitro* reconstruction of separate corneal epithelial, stromal, and endothelial cell layers. J. Biomed. Mater. Res B. Appl. Biomater. **2009**, *90* (2), 818–831.

162. 162.Mi, S.; Chen, B.; Wright, B.; Connon, C.J. Plastic compression of a collagen gel forms a much improved scaffold for ocular surface tissue engineering over conventional collagen gels. J. Biomed. Mater. Res A. **2010**, *95* (2), 447–453.

163. Levis, H.J.; Brown, R.A.; Daniels, J.T. Plastic compressed collagen as a biomimetic substrate for human limbal epithelial cell culture. Biomaterials. **2010**, *31* (30), 7726–7737.

164. Levis, H.J.; Kureshi, A.K.; Massie, I.; Morgan, L.;Vernon, A.J.; Daniels, J.T. Tissue engineering the cornea: The evolution of raft. J. Funct. Biomater. **2015**, *6* (1), 50–65.

165. Reichl, S.; Borrelli, M.; Geerling, G. Keratin films for ocular surface reconstruction. Biomaterials. **2011**, *32* (13), 3375–3386.

166. Bray, L.J.; George, K.A.; Suzuki, S.; Chirila, T.V.; Harkin, D.G. Fabrication of a corneal-limbal tissue substitute using silk fibroin. Methods. Mol. Biol. **2013**, *1014*, 165–178.

167. Friend, J.; Kinoshita, S.; Thoft, R.A.; Eliason, J.A. Corneal epithelial cell cultures on stromal carriers. Invest. Ophthalmol. Vis. Sci. **1982**, *23* (1), 41–49.

168. Barbaro, V.; Ferrari, S.; Fasolo, A.; Ponzin, D.; Di Iorio, E. Reconstruction of a human hemicornea through natural scaffolds compatible with the growth of corneal epithelial stem cells and stromal keratocytes. Mol. Vis. **2009**, *15*, 2084–2093.

169. Sharma, S.; Mohanty, S.; Gupta, D.; Jassal, M.; Agrawal, A.K.; Tandon, R. Cellular response of limbal epithelial cells on electrospun poly-epsilon-caprolactone nanofibrous scaffolds for ocular surface bioengineering: A preliminary *in vitro* study. Mol. Vis. **2011**, *17*, 2898–2910.

170. Zajicova, A.; Pokorna, K.; Lencova, A.; Krulova, M.; Svobodova, E.; Kubinova, S.; Sykova, E.; Pradny, M.; Michalek, J.; Svobodova, J.; Munzarova, M.; Holan, V. Treatment of ocular surface injuries by limbal and mesenchymal stem cells growing on nanofiber scaffolds. Cell Transplant. **2010**, *19* (10), 1281–1290.

171. Deshpande, P.; McKean, R.; Blackwood, K.A.; Senior, R.A.; Ogunbanjo, A.; Ryan, A.J.; MacNeil, S. Using poly(lactide-co-glycolide) electrospun scaffolds to deliver cultured epithelial cells to the cornea. Regen. Med. **2010**, *5* (3), 395–401.

172. Sharma, S.; Gupta, D.; Mohanty, S.; Jassal, M.; Agrawal, A.K.; Tandon, R. Surface-modified electrospun poly (ε-Caprolactone) scaffold with improved optical transparency and bioactivity for damaged ocular surface reconstruction. Invest. Ophthalmol. Vis. Sci. **2014**, *55* (2), 899–907.

173. Di Girolamo, N.; Bosch, M.; Zamora, K.; Coroneo, M.T.; Wakefield, D.; Watson, S.L. A contact lens-based technique for expansion and transplantation of autologous epithelial progenitors for ocular surface reconstruction. Transplantation **2009**, *87* (10), 1571–1578.

174. Di Girolamo, N.; Chui, J.; Wakefield, D.; Coroneo, M.T. Cultured human ocular surface epithelium on therapeutic contact lenses. Br. J. Ophthalmol. **2007**, *91* (4), 459–464.

175. Jester, J.V.; Moller-Pedersen, T.; Huang, J.; Sax, C.M.; Kays, W.T.; Cavangh, H D.; Petroll, W.M.; Piatigorsky, J. The cellular basis of corneal transparency: evidence for 'corneal crystallins'. J Cell Sci. **1999**, *112* (5), 613–622.

176. Proulx, S.; Giasson, C.J., Guillemette, M., Gaudreault, M., Carrier, P., Auger, F.A., Guérin, S.L., Germain, L. Tissue engineering of human cornea. In *Biomaterials and Regenerative Medicine in Ophthalmology*; Chirila, T., Ed.; CRC Press/Woodhead Publishing: Cambridge, 2010; 150–192.

177. Ruberti, J.W.; Zieske, J.D. Prelude to corneal tissue engineering–gaining control of collagen organization. Prog. Retin. Eye Res. **2008**, *27* (5), 549–577.

178. Lakshman, N.; Kim, A.; Petroll, W.M. Characterization of corneal keratocyte morphology and mechanical activity

Corneas: Tissue Engineering

391

within 3-D collagen matrices. Exp. Eye Res. **2010**, *90* (2), 350–359.

179. Karamichos, D.; Lakshman, N.; Petroll, W.M. Regulation of corneal fibroblast morphology and collagen reorganization by extracellular matrix mechanical properties. Invest. Ophthalmol. Vis. Sci. **2007**, *48* (11), 5030–5037.

180. Kumar, P.; Pandit, A.; Zeugolis, D.I. Progress in corneal stromal repair: From tissue grafts and biomaterials to modular supramolecular tissue-like assemblies. Adv Mater. **2016**, *28* (27), 5381–5399.

181. Du, Y.; Carlson, E.C.; Funderburgh, M.L.; Birk, D.E.; Pearlman, E.; Guo, N.; Kao, W.W.; Funderburgh, J.L. Stem cell therapy restores transparency to defective murine corneas. Stem. Cells. **2009**, *27* (7), 1635–1642.

182. Pinnamaneni, N.; Funderburgh, J.L. Concise review: Stem cells in the corneal stroma. Stem. Cells. **2012**, *30* (6), 1059–1063.

183. Auger, F.A.; Rémy-Zolghadri, M.; Grenier, G.; Germain, L. Review: The self-assembly approach for organ reconstruction by tissue engineering. e-Biomed: A Journal of Regenerative Medicine. **2004**, *1* (5), 75–86.

184. Ren, R.; Hutcheon, A.E.; Guo, X.Q.; Saeidi, N.; Melotti, S.A.; Ruberti, J.W.; Zieske, J.D.; Trinkaus-Randall, V. Human primary corneal fibroblasts synthesize and deposit proteoglycans in long-term 3-D cultures. Dev. Dyn. **2008**, *237* (10), 2705–2715.

185. Carrier, P.; Deschambeault, A.; Audet, C.; Talbot, M.; Gauvin, R.; Giasson, C.J.; Auger, F.A.; Guérin, S.L.; Germain, L. Impact of cell source on human cornea reconstructed by tissue engineering. Invest. Ophthalmol. Vis. Sci. **2009**, *50* (6), 2645–52.

186. Guo, X.; Hutcheon, A.E.; Melotti, S.A.; Zieske, J.D.; Trinkaus-Randall, V.; Ruberti, J.W. Morphologic characterization of organized extracellular matrix deposition by ascorbic acid-stimulated human corneal fibroblasts. Invest. Ophthalmol. Vis. Sci. **2007**, *48* (9), 4050–4060.

187. Harrison, R.G. The cultivation of tissues in extraneous media as a method of morpho-genetic study. Anat. Rec. **1912**, *6* (4), 181–193.

188. Wilson, S.L.; Wimpenny, I.; Ahearne, M.; Rauz, S.; El Haj, A.J.; Yang, Y. Chemical and topographical effects on cell differentiation and matrix elasticity in a corneal stromal layer model. Adv. Funct. Mater. **2012**, *22* (17), 3641–3649.

189. Guillemette, M.D.; Cui, B.; Roy, E.; Gauvin, R.; Giasson, C.J.; Esch, M.B.; Carrier, P.; Deschambeault, A.; Dumoulin, M.; Toner, M.; Germain, L.; Veres, T.; Auger, F.A. Surface topography induces 3D self-orientation of cells and extracellular matrix resulting in improved tissue function. Integr. Biol (Camb). **2009**, *1* (2), 196–204.

190. Gil, E.S.; Park, S.H.; Marchant, J.; Omenetto, F.; Kaplan, D.L. Response of human corneal fibroblasts on silk film surface patterns. Macromol. Biosci. **2010**, *10* (6), 664–673.

191. Wu, J.; Rnjak-Kovacina, J.; Du, Y.; Funderburgh, M.L.; Kaplan, D.L.; Funderburgh, J.L. Corneal stromal bioequivalents secreted on patterned silk substrates. Biomaterials **2014**, *35* (12), 3744–3755.

192. Vrana, E.; Builles, N.; Hindie, M.; Damour, O.; Aydinli, A.; Hasirci, V. Contact guidance enhances the quality of a tissue engineered corneal stroma. J. Biomed. Mater. Res A. **2008**, *84* (2), 454–463.

193. Crabb, R.A.; Hubel, A. Influence of matrix processing on the optical and biomechanical properties of a corneal stroma equivalent. Tissue. Eng. Part A. **2008**, *14* (1), 173–182.

194. Matson, J.B.; Stupp, S.I. Self-assembling peptide scaffolds for regenerative medicine. Chem Commun (Camb). **2012**, *48* (1), 26–33.

195. Ravichandran, R.; Griffith, M.; Phopase, J. Applications of self-assembling peptide scaffolds in regenerative medicine: The way to the clinic. J. Mater. Chem B. **2014**, *2* (48), 8466–8478.

196. Gouveia, R.M.; Castelletto, V.; Alcock, S.G., Hamley, I.W.; Connon, C.J. Bioactive films produced from self-assembling peptide amphiphiles as versatile substrates for tuning cell adhesion and tissue architecture in serum-free conditions. J. Mater. Chem B. **2013**, *1* (44), 6157–6169.

197. Uzunalli, G.; Soran, Z.; Erkal, T.S.; Dagdas, Y.S.; Dinc, E.; Hondur, A.M.; Bilgihan, K.; Aydin, B.; Guler, M.O.; Tekinay, A.B. Bioactive self-assembled peptide nanofibers for corneal stroma regeneration. Acta. Biomater. **2014**, *10* (3), 1156–1166.

198. Gouveia, R.M.; Castelletto, V.; Hamley, I.W.; Connon, C.J. New self-assembling multifunctional templates for the biofabrication and controlled self-release of cultured tissue. Tissue. Eng. Part A. **2015**, *21* (11–12), 1772–1784.

199. Wu, J.; Du, Y.; Mann, M.M.; Yang, E.; Funderburgh, J.L.; Wagner, W.R. Bioengineering organized, multilamellar human corneal stromal tissue by growth factor supplementation on highly aligned synthetic substrates. Tissue. Eng. Part A. **2013**, *19* (17–18), 2063–2075.

200. Glowacki, J.; Mizuno, S. Collagen scaffolds for tissue engineering. Biopolymers **2008**, *89* (5), 338–344.

201. Torbet, J.; Malbouyres, M.; Builles, N.; Justin, V.; Roulet, M.; Damour, O.; Oldberg, A.; Ruggiero, F.; Hulmes, D.J. Tissue engineering of the cornea: orthogonal scaffold of magnetically aligned collagen lamellae for corneal stroma reconstruction. Con. Proc. IEEE. Eng. Med. Biol. Soc. **2007**, *2007*, 6400.

202. Torbet, J.; Malbouyres, M.; Builles, N.; Justin, V.; Roulet, M.; Damour, O.; Oldberg, A.; Ruggiero, F.; Hulmes, D.J. Orthogonal scaffold of magnetically aligned collagen lamellae for corneal stromal reconstruction. Biomaterials. **2007**, *28* (29), 4268–4276.

203. Builles, N.; Janin-Manificat, H.; Malbouyres, M.; Justin, V.; Rovère M.R.; Pellegrini, G.; Torbet, J.; Hulmes, D.J.; Burillon, C.; Damour, O.; Ruggiero, F. Use of magnetically oriented orthogonal collagen scaffolds for hemi-corneal reconstruction and regeneration. Biomaterials. **2010**, *31* (32), 8313–8322.

204. Mimura, T.; Amano, S.; Yokoo, S.; Uchida, S.; Yamagami, S.; Usui, T.; Kimura, Y.; Tabata, Y. Tissue engineering of corneal stroma with rabbit fibroblast precursors and gelatin hydrogels. Mol. Vis. **2008**, *14*, 1819–1828

205. Diao, J.M.; Pang, X.; Qiu, Y.; Miao, Y.; Yu, M.M.; Fan, T.J. Construction of a human corneal stromal equivalent with non-transfected human corneal stromal cells and acellular porcine corneal stromata. Exp. Eye. Res. **2015**, *132*, 216–224.

206. Fan, T.J.; Hu, X.Z.; Zhao, J.; Niu, Y.; Zhao, W.Z.; Yu, M.M.; Ge, Y. Establishment of an untransfected human corneal stromal cell line and its biocompatibility to

Conducting—Dendritic

acellular porcine corneal stroma. Int. J. Ophthalmol. **2012**, *5* (3), 286–292 .

207. Hu, X.J.; Lui, W.; Cui, L.; Wang, M.; Cao, Y.L. Tissue engineering of nearly transparent corneal stroma. Tissue. Eng. **2005**, *11* (11–12), 1710–1717.

208. Salehi, S.; Fathi, M.; Javanmard, S.H.; Barneh, F.; Moshayedi, M. Fabrication and characterization of biodegradable polymeric films as a corneal stroma substitute. Adv. Biomed. Res. **2015**, *4*, 9.

209. Tuft, S.J.; Coster, D.J. The corneal endothelium. Eye. **1990**, *4* (3), 389–424.

210. Senoo, T.; Joyce, N.C. Cell cycle kinetics in corneal endothelium from old and young donors. Invest. Ophthalmol. Vis. Sci. **2000**, *41* (3), 660–667.

211. Joyce, N.C. Proliferative capacity of corneal endothelial cells. Exp. Eye. Res. **2012**, *95* (1), 16–23.

212. Engelmann, K.; Bednarz, J.; Valtink, M. Prospects for endothelial transplantation. Exp Eye Res. **2004**, *78* (3), 573–578.

213. Peh, G.S.; Beuerman, R.W.; Colman, A.; Tan, D.T.; Mehta, J.S. Human corneal endothelial cell expansion for corneal endothelium transplantation: an overview. Transplantation. **2011**, *91* (8), 811–819.

214. Armitage, W.J. Preservation of human cornea. Transfus. Med. Hemother. **2011**, *38* (2), 143–147.

215. de Araujo, A.L.; Gomes, J.Á. Corneal stem cells and tissue engineering: Current advances and future perspectives. World. J. Stem. Cells. **2015**, *7* (5), 806–814

216. Mimura, T.; Yamagami, S.; Amano, S. Corneal endothelial regeneration and tissue engineering. Prog. Retin. Eye. Res. **2013**, *35*, 1–17.

217. Li, W.; Sabater, A.L.; Chen, Y.T.; Hayashida, Y.; Chen, S.Y.; He, H.; Tseng, S.C. A novel method of isolation, preservation, and expansion of human corneal endothelial cells. Invest. Ophthalmol. Vis. Sci. **2007**, *48* (2), 614–620.

218. Engelmann, K.; Böhnke, M.; Friedl, P. Isolation and long-term cultivation of human corneal endothelial cells. Invest. Ophthalmol. Vis. Sci. **1988**, *29* (11), 1656–1662.

219. Mimura, T.; Yokoo, S.; Yamagami, S. Tissue engineering of corneal endothelium. J. Funct. Biomater. **2012**, *3* (4), 726–744.

220. Zavala, J.; Jaime, G.L.; Barrientos, C.R.; Valdez-Garcia, J. Corneal endothelium: developmental strategies for regeneration. Eye (Lond). **2013**, *27* (5), 579–588.

221. Konomi, K.; Zhu, C.; Harris, D.; Joyce, N.C. Comparison of the proliferative capacity of human corneal endothelial cells from the central and peripheral areas. Invest. Ophthalmol. Vis. Sci. **2005**, *46* (11), 4086–91.

222. Yu, W.Y.; Sheridan, C.; Grierson, I.; Mason, S.; Kearns, V.; Lo, A.C.; Wong, D. Progenitors for the corneal endothelium and trabecular meshwork: A potential source for personalized stem cell therapy in corneal endothelial diseases and glaucoma. J. Biomed. Biotechnol. **2011**, *2011*, 412743.

223. Hatou, S.; Yoshida, S.; Higa, K.; Miyashita, H.; Inagaki, E.; Okano, H.; Tsubota, K.; Shimmura, S. Functional corneal endothelium derived from corneal stroma stem cells of neural crest origin by retinoic acid and wnt/beta-catenin signaling. Stem. Cells. Dev. **2013**, *22* (5), 828–839.

224. Cao, Y.; Sun, Z.; Liao, L.; Meng, Y.; Han, Q.; Zhao, R.C. Human adipose tissue-derived stem cells differentiate into endothelial cells *in vitro* and improve postnatal neovascularization *in vivo*. Biochem. Biophys. Res. Commun. **2005**, *332* (2), 370–379.

225. Joyce, N.C.; Harris, D.L.; Markov, V.; Zhang, Z.; Saitta, B. Potential of human umbilical cord blood mesenchymal stem cells to heal damaged corneal endothelium. Mol. Vis. **2012**, *18*, 547–564.

226. Kim, E.Y.; Song, J.E.; Park, C.H.; Joo, C.K., Khang, G. Recent advances in tissue-engineered corneal regeneration. Inflamm. Regen.. **2014**, *34* (1), 4–14.

227. Navaratnam, J.; Utheim, T.P.; Rajasekhar, V.K.; Shahdadfar, A. Substrates for expansion of corneal endothelial cells towards bioengineering of human corneal endothelium. J. Funct. Biomater. **2015**, *6* (3), 917–945.

228. Proulx, S.; Brunette, I. Methods being developed for preparation, delivery and transplantation of a tissue-engineered corneal endothelium. Exp. Eye Res. **2012**, *95* (1), 68–75.

229. Liang, Y.; Liu, W.; Han, B.; Yang, C.; Ma, Q.; Zhao, W.; Rong, M.; Li, H. Fabrication and characters of a corneal endothelial cells scaffold based on chitosan. J. Mater. Sci. Mater. Med. **2011**, *22* (1), 175–183.

230. Madden, P.W.; Lai, J.N.; George, K.A.; Giovenco, T.; Harkin, D.G.; Chirila, T.V. Human corneal endothelial cell growth on a silk fibroin membrane. Biomaterials **2011**, *32* (17), 4076–4084.

231. Yoeruek, E.; Saygili, O.; Spitzer, M.S.; Tatar, O.; Bartz-Schmidt, K.U.; Szurman, P. Human anterior lens capsule as carrier matrix for cultivated human corneal endothelial cells. Cornea **2009**, *28* (4), 416–420.

232. Mimura, T.; Amano, S.; Usui, T.; Araie, M.; Ono, K.; Akihiro, H.; Yokoo, S.; Yamagami, S. Transplantation of corneas reconstructed with cultured adult human corneal endothelial cells in nude rats. Exp. Eye Res. **2004a**, *79* (2), 231–237.

233. Levis, H.J.; Peh, G.S.; Toh, K.P.; Poh, R.; Shortt, A.J.; Drake, R.A.; Mehta, J.S.; Daniels, J.T. Plastic compressed collagen as a novel carrier for expanded human corneal endothelial cells for transplantation. PLoS One. **2012**, *7* (11), e50993.

234. Teichmann, J.; Valtink, M.; Nitschke, M.; Gramm, S.; Funk, R.H.; Engelmann, K.; Werner, C. Tissue engineering of the corneal endothelium: a review of carrier materials. J. Funct. Biomater. **2013**, *4* (4), 178–208.

235. Lai, J.Y.; Chen, K.H.; Hsiue, G.H. Tissue-engineered human corneal endothelial cell sheet transplantation in a rabbit model using functional biomaterials. Transplantation. **2007**, *84* (10), 1222–1232.

236. Sumide, T.; Nishida, K.; Yamato, M.; Ide, T.; Hayashida, Y.; Watanabe, K.; Yang, J.; Kohno, C.; Kikuchi, A.; Maeda, N.; Watanabe, H.; Okano, T.; Tano, Y. Functional human corneal endothelial cell sheets harvested from temperature-responsive culture surfaces. FASEB J. **2006**, *20* (2), 392–394

237. Hsiue, G.H.; Lai, J.Y.; Chen, K.H.; Hsu, W.M. A novel strategy for corneal endothelial reconstruction with a bioengineered cell sheet. Transplantation. **2006**, *81* (3), 473–476.

238. Watanabe, R.; Hayashi, R.; Kimura, Y.; Tanaka, Y.; Kageyama, T.; Hara, S.; Tabata, Y.; Nishida, K. A novel gelatin hydrogel carrier sheet for corneal endothelial transplantation. Tissue. Eng. Part A. **2011**, *17* (17–18), 2213–2219.

239. Proulx, S.; Bensaoula, T.; Nada, O.; Audet, C.; d'Arc Uwamaliya, J.; Devaux, A.; Allaire, G.; Germain, L.; Brunette, I. Transplantation of a tissue-engineered corneal endothelium reconstructed on a devitalized carrier in the feline model. Investig. Ophthalmol. Vis. Sci. **2009**, *50* (6), 2686–2694.

240. Proulx, S.; Audet, C.; Uwamaliya, J.; Deschambeault, A.; Carrier, P.; Giasson, C.J.; Brunette, I.; Germain, L. Tissue engineering of feline corneal endothelium using a devitalized human cornea as carrier. Tissue, Eng. Part A. **2009**, *15* (7), 1709–1718.

241. Hashimoto, Y.; Funamoto, S.; Sasaki, S.; Honda, T.; Hattori, S.; Nam, K.; Kimura, T.; Mochizuki, M.; Fujisato, T.; Kobayashi, H.; Kishida, A. Preparation and characterization of decellularized cornea using high-hydrostatic pressurization for corneal tissue engineering. Biomaterials. **2010**, *31* (14), 3941–3948.

242. Ishino, Y.; Sano, Y.; Nakamura, T.; Connon, C.J.; Rigby, H.; Fullwood, N.J.; Kinoshita, S. Amniotic membrane as a carrier for cultivated human corneal endothelial cell transplantation. Invest. Ophthalmol. Vis. Sci. **2004**, *45* (3), 800–806.

243. Wencan, W.; Mao, Y.; Wentao, Y.; Fan, L.; Jia, Q.; Qinmei, W.; Xiangtian, Z. Using basement membrane of human amniotic membrane as a cell carrier for cultivated cat corneal endothelial cell transplantation. Curr. Eye Res. **2007**, *32* (3), 199–215.

244. Fan, T.; Ma, X.; Zhao, J.; Wen, Q.; Hu, X.; Yu, H.; Shi, W. Transplantation of tissue-engineered human corneal endothelium in cat models. Mol. Vis. **2013**, *19*, 400–407.

245. Mimura, T.; Yamagami, S.; Yokoo, S.; Yanagi, Y.; Usui, T.; Ono, K.; Araie, M.; Amano, S. Sphere therapy for corneal endothelium deficiency in a rabbit model. Invest. Ophthalmol. Vis. Sci. **2005b**, *46* (9), 3128–3135.

246. Mimura, T.; Yokoo, S.; Araie, M.; Amano, S.; Yamagami, S. Treatment of rabbit bullous keratopathy with precursors derived from cultured human corneal endothelium. Invest. Ophthalmol. Vis. Sci. **2005c**, *46* (10), 3637–3644.

247. Mimura, T.; Shimomura, N.; Usui, T.; Noda, Y.; Kaji, Y.; Yamgami, S.; Amano, S.; Miyata, K.; Araie, M. Magnetic attraction of iron-endocytosed corneal endothelial cells to Descemet's membrane. Exp. Eye Res. **2003**, *76* (6), 745–751.

248. Mimura, T.; Yamagami, S.; Usui, T.; Ishii, Y.; Ono, K.; Yokoo, S.; Funatsu, H.; Araie, M.; Amano, S. Long-term outcome of iron-endocytosing cultured corneal endothelial cell transplantation with magnetic attraction. Exp. Eye Res. **2005**, *80* (2), 149–157.

249. Moysidis, S.N.; Alvarez-Delfin, K.; Peschansky, V.J.; Salero, E.; Weisman, A.D.; Bartakova, A.; Raffa, G.A.; Merkhofer, R.M.; Kador, K.E.; Kunzevitzky, N.J.; Goldberg, J.L. Magnetic field-guided cell delivery with nanoparticle-loaded human corneal endothelial cells. Nanomed.**2015**, *11* (3), 499–509.

250. Patel, S.V.; Bachman, L.A.; Hann, C.R.; Bahler, C.K.; Fautsch, M.P. Human corneal endothelial cell transplantation in a human *ex vivo* model. Invest. Ophthalmol. Vis. Sci. **2009**, *50* (5), 2123–2131.

251. Ghezzi, C.E.; Rnjak-Kovacina, J.; Kaplan, D.L. Corneal tissue engineering: Recent advances and future perspectives. Tissue. Eng. Part B. Rev. **2015**, *21* (3), 278–287.

252. Kitazawa, K.; Kawasaki, S.; Shinomiya, K.; Aoi, K.; Matsuda, A.; Funaki, T.; Yamasaki, K.; Nakatsukasa, M.; Ebihara, N.; Murakami, A.; Hamuro, J.; Kinoshita, S. Establishment of a human corneal epithelial cell line lacking the functional TACSTD2 gene as an *in vitro* model for gelatinous drop-like dystrophy. Invest. Ophthalmol. Vis. Sci. **2013**, *54* (8), 5701–5711.

253. Schneider, A.I.; Maier-Reif, K.; Graeve, T. The use of an *in vitro* cornea for predicting ocular toxicity. *In Vitro.* Toxicol. **1997**, *10*, 309–318..

254. Reichl, S.; Becker, U. Cell culture models of the corneal epithelium and reconstructed cornea equivalents for *in vitro* drug absorption studies. In *Drug Absorption Studies*; Ehrhardt, C.; Kim, K.J., Eds.; Springer US: New York, 2008; 283–306.

255. Curren, R.D.; Harbell, J.W. Ocular safety: A silent (*in vitro*) success story. Altern. Lab. Anim. **2002**, *30* (Suppl 2), 69– 74.

256. Alépée, N.; Adriaens, E.; Grandidier, M.H.; Meloni, M.; Nardelli, L.; Vinall, C J.; Toner, F.; Roper, C.S.; Van Rompay, A.R.; Leblanc, V.; Cotovio, J. Multi-laboratory evaluation of SkinEthic HCE test method for testing serious eye damage/eye irritation using solid chemicals and overall performance of the test method with regard to solid and liquid chemicals testing. Toxicol, *In Vitro.* **2016**, *34*, 55–70.

257. Kulkarni, A.A.; Chang, W.; Shen, J.; Welty, D. Use of Clonetics® human corneal epithelial cell model for evaluating corneal penetration and hydrolysis of ophthalmic drug candidates. Invest. Ophthalmol. Vis. Sci. **2011**, *52* (14), 3259–3259.

258. Katoh, M.; Hamajima, F.; Ogasawara, T.; Hata, K.I. Establishment of a new *in vitro* test method for evaluation of eye irritancy using a reconstructed human corneal epithelial model, LabCyte CORNEA-MODEL. Toxicol. *In Vitro.* **2013**, *27* (8), 2184–2192.

259. Pfannenbecker, U.; Bessou-Touya, S.; Faller, C.; Harbell, J.; Jacob, T.; Raabe, H.; Tailhardat, M.; Alépée, N.; De Smedt, A.; De Wever, B.; Jones, P.; Kaluzhny, Y.; Le Varlet, B.; McNamee, P.; Marrec-Fairley, M.; Van Goethem, F. Cosmetics Europe multi-laboratory pre-validation of the EpiOcular™ reconstituted human tissue test method for the prediction of eye irritation. Toxicol. *In Vitro.* **2013**, *27* (2), 619–626.

260. Gautheron, P.; Dukic, M.; Alix, D.; Sina, J.F. 'Bovine corneal opacity and permeability test: An *in vitro* assay of ocular irritancy'. Fundam. Appl. Toxicol. **1992**, *18* (3), 442–449.

261. Reichl, S.; Bednarz, J.; Müller-Goymann, C.C. Human corneal equivalent as cell culture model for *in vitro* drug permeation studies. Br. J. Ophthalmol. **2004**, *88* (4), 560–565.

262. Reichl, S.; Döhring, S.; Bednarz, J.; Müller-Goymann, C.C. Human cornea construct HCC-an alternative for *in*

vitro permeation studies? A comparison with human donor corneas. Eur. J. Pharm. Biopharm. **2005**, *60* (2), 305–308.

263.   Wilhelmus, K.R. The draize eye test. Surv. Ophthalmol. **2001**, *45* (6), 493–515

264.   Van Goethem, F.; Adriaens, E.; Alepee, N.; Straube, F.; De Wever, B.; Cappadoro, M.; Catoire, S.; Hansen, E.; Wolf, A.; Vanparys, P. Prevalidation of a new *in vitro* reconstituted human cornea model to assess the eye irritating potential of chemicals. Toxicol. *In Vitro*. **2006**, *20*, 1, 1–17.

265.   Yamaguchi, H.; Kojima, H.; Takezawa, T. Vitrigel-eye irritancy test method using HCE-T cells. Toxicol. Sci. **2013**, *135* (2), 347–355.

266.   Gibbons, M.C.; Foley, M.A.; Cardinal, K.O. Thinking inside the box: keeping tissue-engineered constructs *in vitro* for use as preclinical models. Tissue. Eng. Part B. Rev. **2012**, *19* (1), 14–30.

267.   Rozsa, A.J.; Beuerman, R.W. Density and organization of free nerve endings in the corneal epithelium of the rabbit. Pain **1982**, *14* (2), 105–120.

268.   Stern, M.E.; Beuerman, R.W.; Fox, R.I.; Gao, J.; Mircheff, A.K.; Pflugfelder, S.C. A unified theory of the role of the ocular surface in dry eye. Adv. Exp. Med. Biol. **1998**, *438*, 643–651.

269.   Nishida, T. Neurotrophic mediators and corneal wound healing. Ocul. Surf. **2005**, *3* (4), 194–202.

Dendritic Architectures: Delivery Vehicles

Saeed Jafarirad
Research Institute for Fundamental Sciences (RIFS), University of Tabriz, Tabriz, Iran

Abstract

Dendritic polymers including dendrimers, dendrons, hyperbranched polymers, and dendrigrafts are a unique class of synthetic macromolecules. They possess beneficial properties that make them ideal for use in the development of drug delivery systems. This entry briefly discusses the various aspects of dendritic polymers and their use in drug delivery systems. However, the cationic surfaces of their carriers show cytotoxicity that can be improved via different functionalization and nonfunctionalization methods. This entry also describes the potential applications of dendritic delivery vehicles in the field of oral, ocular, transdermal, targeted, pulmonary, and gene delivery systems.

INTRODUCTION

As is well known, the efficiency of a drug carrier with respect to delivery and release depends on some structural parameters such as size, solubility, polydispersity, polyvalency, and shape.[1] In addition, an ideal drug carrier must be biochemically non-toxic and biocompatible and should protect the bioactive agent until it reaches the targeted site of injury. A comparison between the properties of dendritic polymers and their linear counterparts shows that the dendritic architecture supplies some special features in the field of drug delivery such as nanometer size, increased solubility, low polydispersity, controlled polyvalency, and globular shape. Therefore, the application of dendrimers in this area would be more exciting than the linear polymers, in spite of the high effectiveness of linear polymers in the development of drug delivery systems. In this entry, the different aspects of dendritic delivery systems have been highlighted, and their application in a range of delivery systems, including oral, ocular, transdermal, targeted, pulmonary, and gene delivery, has been discussed.

DENDRITIC POLYMERS: DELIVERY ASPECTS

Nanometer Size

Dendritic polymers possess a nanometer-scale range as a result of their step-by-step synthetic methodology. For instance, in the case of polyamido amine (PAMAM) dendrimer with ethylenediamine core, with increasing generation numbers from 1–10, the diameter of the dendrimers grows from 1.1 nm to 12.4 nm.[2] In addition, the generations of G_0–G_3 possess ellipsoidal shapes without suitable dendritic boxes in their interior shells, whereas the generations of G_4–G_{10} possess roughly spherical shapes with well-defined pockets. Accordingly, due to the nanosize and shape of the G_4–G_{10} dendrimers, they can interact efficiently with the majority of cellular nanoelements such as plasma membrane, endosome, mitochondria, enzyme, and nucleus.[3] On the other hand, dendrimer size was a key factor in determining overall uptake in therapeutic use. For instance, polymeric carriers with diameters up to 3 nm can penetrate through the intestinal membranes. Therefore, $G_{2.5}$ and $G_{3.5}$ PAMAM dendrimers could transport across the intestinal tissue. Such dendritic polymers with nanometer size, which were exploited as delivery vehicles, are termed as dendritic nanocarriers (Fig. 1).

Increased Solubility

The solubility of dendritic polymers depends on several factors such as the peripheral groups, generation, polarity of core and polarity of repeated units. The high solubility of dendritic polymers in aqueous or organic media serves as the reason behind their use as solubility enhancers for both hydrophilic and hydrophobic bioactive agents, respectively.[4]

Monodispersity

Monodispersity or very low polydispersity presents a monotonous pharmacokinetic behavior for dendritic polymers, particularly in dendrimers, contrary to that of conventional linear macromolecules with a broad range of molecular weights.

DENDRITIC POLYMERS: CYTOTOXICITY

The solubility of dendritic polymers in aqueous media is as a result of three distinct types of peripheral groups including cationic, anionic, and non-charged hydrophilic moieties.

Concise Encyclopedia of Biomedical Polymers and Polymeric Biomaterials DOI: 10.1081/E-EBPPC-120050064

395

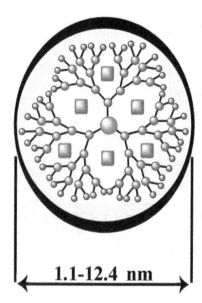

Fig. 1 A schematic representation of PAMAM dendrimer with ethylenediamine core which encapsulates bioactive agents.

Since the cellular surfaces and dendritic polymers possess negative and positive charges, respectively, the cationic surface of the dendritic polymers destabilizes the cell membrane. This unfavored effect which leads to cell lysis has also been reported for the linear type cationic polymers. However, the dendritic polymers with amino-peripheral groups such as PAMAM dendrimers show lower cytotoxicity compared to more linear counterparts containing amine groups. It is owing to the spherical shape of dendrimers, which diminishes adherence to cellular membranes.

The degree of substitution as well as the type of amine groups is significant in which primary amines result in more cytotoxicity than secondary or tertiary amines.[5] However, comparative cytotoxicity data has exhibited that anionic surface groups, like carboxylates, have lower cytotoxicity than cationic surface groups in PAMAM dendrimers. The biocompatibility of dendritic polymers is not merely influenced by the surface groups. Overall, the parameters that influence the hemolysis of the cells may be summarized as:

1. Chemical nature of interior shell and core: It has been shown that the aromatic interior shell of the dendritic polymers containing anionic carboxylate peripheral groups may result in hemolysis via hydrophobic membrane contact.[6] However, the effect of the dendritic core will reduce with increasing number of generations and rigidity of the interior shell. Rigid shells shield the core unit and reduce the interactions between the dendritic core and the cellular surface.
2. Number of generations: Low-generation PAMAM dendrimers possessing carboxylate surface groups show no cytotoxicity up to certain concentrations.[6]

Strategies to Lower the Cytotoxicity

Various strategies have been used to lower the dendritic cytotoxicity and are summarized in this section.

Functionalization strategies

Functionalization strategies can be classified based on the origin of the functionalization reagent based on the chemical conjugation methods such as PEGylation,[7,8] acetylation,[9] and aldehydation[10] and biochemical conjugation methods such as amino acid or peptide conjugating agents,[11] carbohydrate conjugating agents,[12] drug and DNA conjugating agents,[13] antibody conjugating agents,[14] tuftsin conjugating agents,[15] and folate-conjugating agents.[16] Due to the importance of PEGylation of dendritic polymers as a frequently used technique, it is treated as an independent subtitle.

PEGylation: PEGylation is a well-known methodology to decrease the electrostatic interactions between dendritic shell and cellular membranes. Furthermore, a library of PEGylation goals can be listed, such as an increase in solubility of dendritic polymers and circulation half-lives and accordingly, enhancement in the performance of dendritic delivery vehicles, as well as improvement in the other limitations of dendritic nanocarriers such as drug leakage, immunogenicity, biodistribution, bioavailability, pharmacokinetics, and tumor localization.[8,17]

On the other hand, as it is well known that the synthetic preparation of dendritic polymers in high generations is a time- and energy-consuming stage. Therefore, in an alternative approach, the amphiphilic linear-dendritic architectures composed of hydrophobic dendrons and hydrophilic linear blocks have been designed due to their numerous peripheral groups and long circulation half-lives, respectively (Fig. 2). PEG has been selected because of its above-mentioned features and also its low polydispersity index, thus supplying linear-dendritic polymers with low polydispersity.

Nonfunctionalization strategies

Bio-Based Dendrimers: Dendritic bio-nanocarriers meet the necessary criteria of biocompatibility, biodegradability, and non-toxicity in drug delivery systems. Such carriers do not need to be omitted after completion of release because they are hydrolyzed in the body into degradation fragments, which can be metabolized. Therefore, it may be a nice pathway to mitigate the cytotoxicity of dendritic nanocarriers using biodegradable cores and branching units. In recent years, a broad range of bio-based frameworks have been introduced as novel delivery vehicles, including polyether,[18] polyester,[19] polyether imine,[20] polyether-copolyester (PEPE),[21] citric acid,[22] triazine,[23] phosphate,[24] melamine,[25] and peptide dendritic systems.[12]

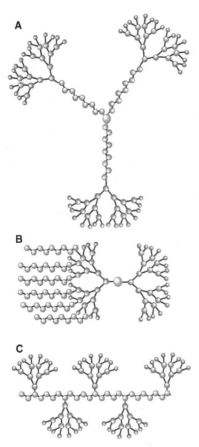

Fig. 2 Different architectures of the amphiphilic linear-dendritic polymers composed of hydrophobic dendrons and hydrophilic linear blocks.

Half-Generation: As was mentioned, the cytotoxicity of dendritic polymers is as a result of their peripheral cationic groups. For instance, PAMAM dendritic polymers have positively charged groups after full generation synthesis. However, their half-generations have anionic groups (carboxylic acid). Thus, half-generation dendritic shells, which possess negatively charged groups, depict less cytotoxicity than their positively charged counterparts.[26]

MECHANISTIC APPROACH TO DELIVERY AND RELEASE VIA DENDRITIC NANOCARRIERS

Delivery Based on Physical Association

Bioactive agents can non-covalently interact with dendritic polymers via either adsorption onto the exterior shell as nanoscaffolding or entrapping in dendritic boxes as nanocontainers, or both.[27] Typically, larger drugs are able to form complexes with dendritic surface groups while small drugs preferably are entrapped in the dendritic boxes. The non-covalent interactions have several limitations with respect to drug incorporation as follows: 1) partial controlled release in spite of their facile establishment and 2) the drug/carrier ratios of incorporation content are low.

Delivery Based on Chemical Conjugation

Alternatively, bioactive agents can covalently be conjugated to dendritic reactive groups (prodrug approach). However, it should be borne in mind that not all the peripheral groups can be conjugated, due to both steric hindrance and backfolding of chains into the dendritic shell. There are two main pathways to create these prodrugs: 1) drug conjugation without spacer (direct route) and 2) drug conjugation to the dendritic shell with a spacer. The spacer or linking agent plays an important role owing to a lack of the desired functional group onto some of the drugs for direct conjugation. Furthermore, it allows a higher degree of conjugation with lowering the steric congestion between drug molecules and dendritic shell. In addition, the spacer may modify solubility profiles and release kinetics.[28]

Triggered Release

The general concept of triggered, and then controlled, release has been established based on both physical association and chemical conjugation approaches (Fig. 3). In the physical association approach, the release can be triggered by structural alterations in the framework of dendritic nanocarriers. These changes may be observed as cleavage or even degradation of dendritic shells as well as protonation of internal reactive groups in the dendritic scaffold. On the other hand, in the chemical conjugation approach, the mechanism of release involves the cleavage of the spacer between the nanocarrier and the bioactive agent.

The external stimuli which cause these changes include, but are not limited to, pH, temperature, enzymes, light, ultrasound, and redox potential. Several of these stimuli, such as pH, occur in different cellular media. However, some other stimuli, like temperature in tumor tissues, are increased by the disease. In the case of the other stimuli, such as light and ultrasound that are not triggered by any biological issue, it is feasible to externally manage the release circumstance. Thus, dendritic nanocarriers are able to participate in the therapeutic and theranostic process, rather than the structurally inert carriers that are currently used in medicine.[29–31]

DENDRITIC ARCHITECTURES AS DELIVERY VEHICLES

Oral Drug Delivery

Some special properties of dendritic polymers make them suitable carriers to transport across the intestine, and hence, are ideal for the development of oral drug delivery systems.

1. They keep the drug concentrations in the favored range of therapeutic dosing schedules.

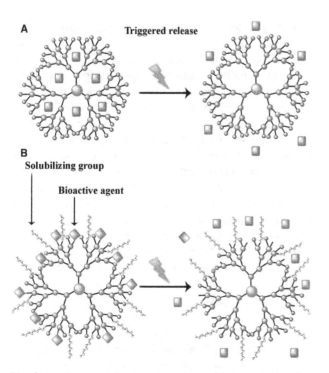

Fig. 3 Different mechanisms for triggered release of bioactive agents from dendritic nanocarriers: (**A**) the release based on physical association and (**B**) the release based on chemical conjugation in which dendritic scaffolds get attached by solubilizing and cleavable spacers for the bioactive groups.

2. They enhance the solubility and, accordingly, the stability of orally administrated drugs in biological media.
3. They possess strong affinity for mucosa due to their bioadhesive properties, so they can prolong the residence time of the drug in contact with the intestinal epithelium.
4. They can increase the oral absorption of low-penetration bioactive agents since they are able to easily penetrate through the tissues in the small intestinal (in accordance to Peyer's patches).[32]
5. They can also penetrate through the intestinal membranes, because of their smaller diameters (less than 3 nm), via either the transcellular or paracellular pathway.[33]

As it was pointed out in the previous section, cationic surface groups of dendritic polymers can interact with negatively charged cell membranes; such strong interactions cause tissue association and consequently a slow transport rate of dendritic nanocarriers. Thus, overall, the three special aspects of dendritic polymers (size, charge, and shape) determine the transport mechanism and rate of dendritic nanocarriers across the intestine.[34]

Ocular Drug Delivery

It is generally agreed that the intraocular bioavailability of topically applied drugs is extremely poor. Therefore,

a large amount of administrated drugs should be used to maintain therapeutic concentrations in the eye, which cause unwanted effects due to accumulation of drugs in all tissues of the body. Ocular drug delivery systems are often applied for systemic drug administration routes in these situations. Generally, ideal ocular drug delivery systems should be nonirritating, biocompatible, biodegradable, sterile, and isotonic. Up to now, some special delivery vehicles of traditional polymers and liposomes have been used in ocular drug delivery systems to overcome these disadvantages. However, most of these carriers lead to unwanted side effects. Dendritic polymers supply better delivery vehicles for ocular drug delivery. For example, PAMAM dendrimers have improved the residence time of pilocarpine in the eye and enhanced pilocarpine bioavailability.[35,36]

On the other hand, dendritic polymers have been used as artificial corneal tissue-engineering scaffolds in corneal diseases, which are the major causes of blindness. These artificial dendritic biomaterials possess much better optical transparency, glucose permeability, mechanical properties, and adhesion ability than their other counterparts without cell cytotoxicity.[37]

Transdermal Drug Delivery

It is well known that nonsteroidal anti-inflammatory drugs (NSAIDs) are very effective in the treatment of chronic rheumatoid arthritis and osteoarthritis. However, clinical practice of NSAIDs is restricted by adverse events such as gastrointestinal and renal side effects when given orally. By using transdermal drug delivery system (TDDS), such adverse effects can be removed. However, because of the slow rate of TDDS, these systems are not successful. It may be assigned to the barrier effect of the skin. The most common method to improve drug penetration through the skin is to use transdermal enhancers. Various transdermal enhancers, like chemical transdermal enhancers, although they increase the permeability, apply an immune response in the skin. Amphiphilic dendritic polymers can form complexes with NSAIDs such as Ketoprofen and Diflunisal and use as penetration enhancers successfully.[38]

Targeted Drug Delivery

Due to the nonselective aspect of conventional chemotherapeutic drugs, they lead to dose-limiting side effects in their treatment of tumors. Therefore, the targeting drug delivery, as an alternative approach, is able to increase the permeability of tumor vasculature to carriers as well as to limit lymphatic drainage. An efficient targeted drug delivery system requires a framework that can couple various components such as targeting, therapeutics, theranostics, as well as combinations of these agents. The polyvalency characters of dendritic polymers facilitate the attachment of various payloads, including targeting molecule, drug, and cancer imaging agent. One of the most efficient cell-specific targeting agents delivered by dendritic nanocarriers is folic acid. The folate-anchored dendritic carriers can enter the cell by receptor-mediated endocytosis, bypassing cellular barriers, and then allow the attached moiety to enter in cancer cells. For example, PAMAM dendrimers have been conjugated with the folic acid and fluorescein isothiocyanate for targeting the tumor cells and imaging, respectively.[39]

Gene Delivery

With regard to the risks of gene delivery by using viruses as the most common vehicles, gene delivery via dendritic nanocarriers, as nonviral vehicles, has been developed as a result of their polyvalency and nanometer size. Dendritic gene delivery systems are based on the electrostatic interaction between multiple positively charged groups, polyca-tions of amine groups, of dendritic polymers with the negative phosphate groups of the genetic material, DNA or RNA strands.[40,41] Therefore, formation of dendriplex (complex of dendritic nanocarrier/genetic material) is the initial step for such nonviral gene transfection. The dendriplexes go through the membrane of the target cells and then are transferred to the endosomal compartment. Finally, the genes escape from the endosomal compartment to the cytosol and internalize to the nucleus. PAMAM and poly(propyleneimine) dendritic polymers are preferably used vehicles due to their commercial availability as well as high affinity to negatively charged backbones of the genes. An outstanding result in gene delivery via PAMAM nano-carriers is that high levels of transfection are obtained by incorporation of defects or partial degradation in the dendritic scaffold. It is because of their increased flexibility, which form more compacted dendriplexes of the gene.[42]

On the other hand, the dendriplexes of plasmid DNA with PEGylated PAMAM nanocarriers possess better flexibility, decreased electrostatic interactions with the DNA helices, and a very low cytotoxicity. In addition, hyperbranched polyether polyols, as the other member of dendritic systems, which partially functionalized with ammonium groups, exhibit a considerable transfection of genetic material.[43]

Pulmonary Drug Delivery

The potential of dendritic polymers in the pulmonary drug administration route needs further research to reach higher biocompatibility. Briefly, it is mentioned that cationic dendritic polymers can be exploited as pulmonary delivery vehicles for large–molecular weight anionic drug. Such nanocarriers interact with anionic drugs electrostatically and hence increase drug absorption via charge neutraliza-tion. In addition, PEGylated dendritic polymers can increase pulmonary absorption and circulation time of the drugs.[17]

Other Drug Delivery Routes

Drug delivery systems based on dendritic polymers are in infancy compared to other delivery vehicles such as lipo-somes and polymeric systems. As mentioned earlier, the penetration of dendritic nanocarriers through the epithelium of intestines is good news for the applications of den-drimers in other delivery routes such as the delivery via rectal, vaginal, and nasal routes. Vivagel™, a dendritic-based microbicide, and other anti-herpes and anti-mycotic drugs for nail fungus are under clinical investigations.[44]

CONCLUSIONS

The focus of this entry is to show the major trends of dendritic nanocarriers in delivery systems. In this direc-tion, diversified pathways of drug administration can exert based on the entrapping/conjugation of into/with dendritc nanocarriers. They are able to offer feasible solutions to drug delivery concerns such as solubility and targeting, due to their nanometer size, globular shape, and polyvalent feature. However, cytotoxicity is a challenging issue for their development. Design of new dendritic polymers with both biocompatible scaffolds and functionalized peripheral groups can present ideal delivery systems with maximum therapeutic efficacy and minimum cytotoxicity. Thus, dendritic delivery systems are able to be commercial vehicles in the field of nanomedicinal applications.

ACKNOWLEDGMENT

The author gratefully acknowledges the University of Tabriz for supporting this project.

REFERENCES

1. Cheng, Y.; Xu, Z.; Ma, M.; Xu T. Dendrimers as drug carri-ers: Applications in different routes of drug administration. J. Pharm. Sci. **2008**, *97* (1), 123–143.
2. Svenson, S.; Tomalia, D.A. Dendrimers in biomedical applications-Reflections on the field. Adv. Drug. Deliv. Rev. **2005**, *57* (15), 2106–2129.
3. Wang, X.L.; Xu, R.; Lu, Z.R. A peptide-targeted delivery system with pHsensitive amphiphilic cell membrane disruption for efficient receptor-mediated siRNA delivery. J. Control. Release **2009**, *134* (3), 207–213.
4. Aulenta, F.; Hayes, W.; Rannard, S. Dendrimers: A new class of nanoscopic containers and delivery devices. Eur. Polym. J. **2003**, *39* (9), 1741–1771.
5. Fischer, D.; Li, Y.; Ahlemeyer, B.; Krieglstein, J.; Kissel, T. *In vitro* cytotoxicity testing of polycations: Influence of polymer structure on cell viability and hemolysis. Biomate-rials **2003**, *24* (7), 1121–1131.

6. Malik, N.; Wiwattanapatapee, R.; Klopsch, R.; Lorenz, K.; Frey, H.; Weener, J.W.; Meijer, E.W.; Paulus, W.; Duncan, R. Dendrimers: Relationship between structure and biocompatibility *in vitro*, and preliminary studies on the biodistribution of I-125-labelled poly(amidoamine) dendrimers *in vivo*. J. Control. Release **2000**, *65* (1–2), 133–148.

7. Namazi, H.; Jafarirad, S. Controlled release of linear-dendritic hybrids of carbosiloxane dendrimer: The effect of hybrid's amphiphilicity on drug-incorporation; hybrid-drug interactions and hydrolytic behavior of nanocarriers. Int. J. Pharm. **2011**, *407* (1–2), 167–173.

8. Stasko, N.A.; Johnson, C.B.; Schoenfisch, M.H.; Johnson, T.A.; Holmuhamedov, E.L. Cytotoxicity of polypropylenimine dendrimer conjugates on cultured endothelial cells. Biomacromolecules **2007**, *8* (12), 3853–3859.

9. Gajbhiye, V.; Kumar, P.V.; Tekade, R.K.; Jain, N.K. PEGylated PPI dendritic architectures for sustained delivery of H2 receptor antagonist. Eur. J. Med. Chem. **2009**, *44* (3), 1155–1166.

10. Hamidi, A.A.; Sharifi, S.; Davaran, S.; Ghasemi S, Omidi, Y.; Rashidi, M.R. Novel aldehyde-terminated dendrimers; synthesis and cytotoxicity assay. BioImpacts **2012**, *2* (2), 97–103.

11. Agashe, H.B.; Dutta, T.D.; Garg, M.; Jain, N.K. Investigations on the toxicological profile of functionalized fifth-generation poly (propylene imine) dendrimer. J. Pharm. Pharmacol. **2006**, *58* (11), 1491–1498.

12. Agrawal, P.; Gupta, U.; Jain, N.K. Glycoconjugated peptide dendrimers-based nanoparticulate system for the delivery of chloroquine phosphate. Biomaterials **2007**, *28* (22), 3349–3359.

13. Lee, C.C.; Gillies, E.R.; Fox, M.E.; Guillaudeu, S.J.; Fréchet, J.M.; Dy, E.E.; Szoka, F.C. A singal dose of doxorubicin-functionalized bow-tie dendrimer cures mice bearing C-26 colon carcinomas. Proc. Natl. Acad. Sci. U.S.A. **2006**, *103* (45), 16649–16654.

14. Wu, G.; Barth, R.F.; Yang, W.; Kawabata, S.; Zhang, L.; Green-Church, K. Targeted delivery of methotrexate to epidermal growth factor receptor-positive brain tumors by means of cetuximab (IMC-C225) dendrimer bioconjugates. Mol. Cancer Ther. **2006**, *5* (1), 52–59.

15. Najjar, V.A.; Nishioka, K. "Tuftsin": A natural phagocytosis stimulating peptide. Nature **1970**, *228* (5272), 672–673.

16. Singh, P.; Gupta, U.; Asthana, A.; Jain, N.K. Folate and folate-PEG-PAMAM dendrimers: Synthesis, characterization, and targeted anticancer drug delivery potential in tumor bearing mice. Bioconjug. Chem. **2008**, *19* (11), 2239–2252.

17. Dufes, C.; Uchegbu, I.F.; Schatzlein, A.G. Dendrimers in gene delivery. Adv. Drug Deliv. Rev. **2005**, *57* (15), 2177–2202.

18. Hawker, C.J.; Frechet, J.M.J. Preparation of polymers with controlled molecular architecture. A new convergent approach to dendritic macromolecules. J. Am. Chem. Soc. **1990**, *112* (21), 7638–7647.

19. Bo, Z.; Zhang, X.; Zhang, C.; Rapid synthesis of polyester dendrimers. J. Chem. Soc. Perkin Trans. **1997**, (19), 2931–2935.

20. Krishna, T.R.; Jain, S.; Tatu, U.S.; Jayaraman, N. Synthesis and biological evaluation of 3-amino-propan-1-ol based poly(ether imine) dendrimers. Tetrahedron **2005**, *61* (17), 4281–4288.

21. Carnahan, M.A.; Grinstaff, M.W. Synthesis and characterization of polyetherester dendrimers from glycerol and lactic acid. J. Am. Chem. Soc. **2001**, *123* (12), 2905–2906.

22. Namazi, H.; Adeli, M. Dendrimers of citric acid and poly (ethylene glycol) as the new drug-delivery agents. Biomaterials **2005**, *26* (10), 1175–1183.

23. Lim, J.; Guo, Y.; Rostollan, C.L.; Stanfield, J.; Hsieh, J.T.; Sun, X.; Simanek, E.E. The role of the size and number of polyethylene glycol chains in the biodistribution and tumor localization of triazine dendrimers. Mol. Pharm. **2008**, *5* (4), 540–547.

24. Domanski, D.M.; Bryszewska, M.; Salamonczyk, G. Preliminary evaluation of the behavior of fifth-generation thiophosphate dendrimer in biological systems. Biomacromolecules **2004**, *5* (5), 2007–2012.

25. Zhang, Z.Y.; Smith, B.D. High-generation polycationic dendrimers are unusually effective at disrupting anionic vesicles: Membrane bending model. Bioconjug. Chem. **2000**, *11* (6), 805–814.

26. Bhadra, D.; Bhadra, S.; Jain, S.; Jain, N.K. PEGylated dendritic nanoparticulate carrier of fluorouracil. Int. J. Pharm. **2003**, *257* (1–2), 111–124.

27. Namazi, H.; Jafarirad, S. Application of hybrid organic/inorganic dendritic ABA type triblock copolymers as new nanocarriers in drug delivery systems. Int. J. Polym. Mater. **2011**, *60* (9), 603–619.

28. Cheng, Y.; Gao, Y.; Rao, T.; Li, Y.; Xu, T. Dendrimer-based prodrugs: Design, synthesis, screening and biological evaluation. Comb. Chem. High Throughput Screen **2007**, *10* (5), 336–349.

29. Ganta, S.; Devalapally, H.; Shahiwala, A.; Amiji, M. A review of stimuli-responsive nanocarriers for drug and gene delivery. J. Control. Release **2008**, *126* (3), 187–204.

30. Calderón, M.; Welker, P.; Licha, K.; Fichtner, I.; Graeser, R.; Haag, R.; Kratz., F. Development of efficient acid cleavable multifunctional prodrugs derived from dendritic polyglycerol with a poly(ethylene glycol) shell. J. Control. Release **2011**, *151* (3), 295–301.

31. Xu, S.; Luo, Y.; Haag, R. Water-soluble pH-responsive dendritic core-shell nanocarriers for polar dyes based on poly(ethylene imine). Macromol. Biosci. **2007**, *7* (8), 968–974.

32. Sakthivel, T.; Toth, I.; Florence, A.T. Distribution of a lipidic 2.5 nm diameter dendrimer carrier after oral administration. Int. J. Pharm. **1999**, *183* (1), 0378–5173.

33. Pantzar, N.; Lundin, S.; Wester, L.; Weström, B.R. Bidirectional small-intestinal permeability in the rat to some common marker molecules *in vitro*. Scand. J. Gastroenterol. **1994**, *29* (8), 703–709.

34. Wiwattanapatapee, R.; Carreno-Gomez, B.; Malik, N.; Duncan, R. Anionic PAMAM dendrimers rapidly cross adult rat intestine *in vitro*: A potential oral delivery system? Pharm. Res. **2000**, *17* (8), 991–998.

35. Vandamme, T.F.; Brobeck, L. Poly(amidoamine) dendrimers as ophthalmic vehicles for ocular delivery of pilocarpine nitrate and tropicamide. J. Control. Release **2005**, *102* (1), 23–38.

36. Yang, H.; Kao, W.J. Dendrimers for pharmaceutical and biomedical applications. J. Biomater. Sci. Polym. Ed. **2006**, *17* (1–2), 3–19.

37. Duan, X.; Sheardown, H. Dendrimer crosslinked collagen as a corneal tissue engineering scaffold: Mechanical properties and corneal epithelial cell interactions. Biomaterials **2006**, *27* (26), 4608–4617.

38. Cheng, Y.; Man, N.; Xu, T.; Fu, R.; Wang, X.; Wang, X.; Wen, L. Transdermal delivery of nonsteroidal anti-inflammatory drugs mediated by polyamidoamine (PAMAM) dendrimers. J. Pharm. Sci. **2007**, *96* (3), 595–602.

39. Choi, Y.; Thomas, T.; Kotlyar, A.; Islam, M.T.; Baker, J.R., Jr. Synthesis and functional evaluation of DNA-assembled polyamidoamine dendrimer clusters for cancer cell-specific targeting. Chem. Biol. **2005**, *12* (1), 35–43.

40. Omidi, Y.; Hollins, A.J.; Drayton, R.M.; Akhtar, S. Polypropylenimine dendrimer-induced gene expression changes: The effect of complexation with DNA, dendrimer generation and cell type. J. Drug Target. **2005**, *13* (7), 431–444.

41. Hollins, A.J.; Benboubetra, M.; Omidi, Y.; Zinselmeyer, B.H.; Schatzlein, A.G.; Uchegbu, I.F.; Akhtar, S. Evaluation of generation 2 and 3 poly (propylenimine) dendrimers for the potential cellular delivery of antisense oligonucleotides targeting the epidermal growth factor receptor. Pharm. Res. **2004**, *21* (3), 458–466.

42. Zinselmeyer, B.H.; Mackay, S.P.; Schatzlein, A.G.; Uchegbu, I.F. The lower-generation polypropylenimine dendrimers are effective gene-transfer agents. Pharmacol. Res. **2002**, *19* (7), 960–967.

43. Tziveleka, L.A.; Psarra, A.M.; Tsiourvas, D.; Paleos, C.M. Synthesis and evaluation of functional hyperbranched polyether polyols as prospected gene carriers. Int. J. Pharm. **2008**, *356* (1–2), 314–324.

44. Gong, E.; Matthews, B.; McCarthy, T.; Chu, J.; Holan, G.; Raff, J.; Sacks, S. Evaluation of dendrimer SPL7013, a lead microbicide candidate against herpes simplex viruses. Int. J. Nanomed. **2005**, *68* (3), 139–146.

Conducting—Dendritic

Dendritic Architectures: Theranostic Applications

Saeed Jafarirad
Research Institute for Fundamental Sciences (RIFS), University of Tabriz, Tabriz, Iran

Abstract

Today's challenge to find safe and innovative nanocarriers for therapeutic as well as diagnostic applications remains as critical as ever. However, over the past decade, various nanomaterial systems have been developed as potent tools to simultaneously monitor and treat disease. In particular, among various kinds of nanomaterial systems such as solid–lipid, magnetic, micellar, and linear macromolecular nanoparticles, dendritic nanostructures have proved to be appropriate nanoparticles in the medical applications. In this entry, we define the term "theranostic" based on dendritic architectures from the early 1990s that afford a promising avenue toward nanomedicine. To define the boundaries of theranostic based on dendritic architectures, this entry first introduces, in brief, the imaging modalities as well as the objective for generating large-molecular-weight contrast agents such as dendritic architectures to modify the pharmacokinetic disadvantages of presently available small-sized agents. Secondly, it highlights the nanopharmaceutical properties of dendritic structures in the growing field of theranostics. And, finally, an overview of select applications that make dendritic architectures optimal candidates for theranostics applications has been mentioned.

INTRODUCTION

Nanomedicine has introduced a number of nanoparticles that have expanded the potential in early detection of the disease. These nanoparticles are very selective to sense the sparse biomarkers of any disease and, accordingly, can be effective in targeted diagnostic imaging. One of the most important nanoparticles that have been used in molecular imaging is dendritic structures.

Conversely, "theranostic" covers two different definitions. We use the term "theranostic" as defined by the combination of therapeutic and diagnostic agents that may simultaneously monitor and treat disease. Typically, the important modalities that are present for monitoring use of dendritic structures include optical imaging, magnetic resonance imaging, X-ray imaging, and computed tomography (CT). In this entry, we first introduce the pharmacological properties of therapeutic techniques. Each imaging modality has relative benefits and drawbacks, which will be discussed. Secondly, we discuss the chemical characteristics and pharmacological properties of polymeric magnetic resonance imaging (MRI) contrast agents (CAs) based on dendritic structures. Lastly, we will discuss their potential clinical applications.

THERANOSTICS APPLICATIONS USING NANOPARTICLES

Recently, opportunities to grow nanomedical systems have developed all effective diagnostic and treatment, such as one-package systems that can target different tissues. A combination of nanotechnology, polymer chemistry, and imaging science approaches has resulted in the development of a broad range of macromolecular bioimaging probes for the diagnosis and therapeutic personalized nanomedicine as well as monitoring theranostics. The concept of theranostics is based on the "find, fight, and follow" tactic. The main parameter in *in vivo* imaging should be mentioned as follows:

1. Generation of nanoparticles that are able to penetrate into cells and crossover biological barriers with special features that prevent them from being cleared before reaching the target tissues.
2. Safety evaluation (*in vitro*/*in vivo* cytotoxicity, haemocompatibility and immunogenicity) as well as biocompatibility with external activation by magnetic field, ultrasound (US), X-ray, or optics that selectively targets diseased cells.

Dendritic nanoparticles, as versatile vehicles, have been designed to develop targeted contrast agents, because:

1. They have a surface that can be functionalized with one or more targeting molecules at a wide range of densities.
2. Their plasma circulation time can be adjusted over several orders of magnitude based on their physicochemical properties.
3. CAs and drugs can be included at predetermined ratios either in the interior or on the surfaces.

Concise Encyclopedia of Biomedical Polymers and Polymeric Biomaterials DOI: 10.1081/E-EBPPC-120050065

The development of effective theranostic nanoparticles will require some knowledge about the modalities available for imaging, including optical imaging, MRI, nuclear imaging, CT, and US.[1]

IMAGING MODALITY

Molecular imaging is a promising technology that allows the characterization of biological processes at the cellular and subcellular levels in organisms. By using specific contrast agents, this powerful technique can detect and characterize early-stage disease. The frequently used molecular imaging modalities include MRI, CT, US, optical imaging (bioluminescence and fluorescence), single photon emission computed tomography, and positron emission tomography. All of these modalities need sufficient amounts of reporter groups to accumulate in a desired situation of tissue.[2]

Optical Imaging

Optical imaging, as a normal technique, exploits photons emitted from bioluminescent or fluorescent probes. It has advantages compared with the other imaging modalities; for instance, in this technique the detection of low-energy photons is inexpensive. Moreover, the spectrum from visible to near-infrared light provides good spatial resolution. However, this modality has some disadvantages such as associated basic background owing to tissue autofluorescence and light absorption by proteins (257–280 nm), heme groups (max absorbance at 560 nm).[3]

Magnetic Resonance Imaging

MRI is one of the well-known, noninvasive detection devices in living systems. This technique is similar to nuclear magnetic resonance (NMR). It is based on the inhomogeneous relaxation time of protons in various tissues. The resonant absorption of energy by the protons will occur when the electromagnetic radiation of the correct radio frequency (RF) is applied to excite the nuclei in the lower-energy state to the higher-energy state. This RF pulse will result in the relaxation of the magnetic spins, which occurs via two different mechanisms referred to as the spin–lattice relaxation time (T_1) and spin–spin relaxation time (T_2). As mentioned, MRI is based on slight differences of environment-sensitive ^1H NMR resonances, mainly of H_2O, in the tissue. Therefore, the signal that is produced contains information on proton density, which will be regenerated by means of a computer from different living systems. However, in many cases, there is an increasing need to produce a much higher differentiation between the MRI images. This differentiation can be developed by addition of CAs to the living system prior to its MRI analysis. These compounds include gadolinium (III) diethyl triamine pentaacetic acid [Gd(III)-DTPA] and gadolinium-tetraazacyclo dodecane tetraacetic acid [Gd(III)-DOTA], which are presently well known as Omniscan™ and Prohance™, respectively. These chelating agents optimize the T_1-relaxation rate of protons in the H_2O molecules in tissue and accordingly the quality of MRI images.[4] Vital features of Gd(III) chelating ligands are high relaxivity, low toxicity and fairly good biocompatibility of the corresponding metal salts.

Computed Tomography

CT technique is a powerful theranostic tool that provides complementary anatomical information without requiring the application of harmful radiation. CT measures the absorption of X-rays as they penetrate through living systems. CT technique is able to distinguish between different tissues depending on distinct degrees of X-ray attenuation. Therefore, differences in absorption between anatomical structures such as fat, bone, and water create high contrast images (Fig. 1).[5]

DENDRITIC CONTRAST AGENTS IN IMAGING

As mentioned, Gd(III) complexes are frequently used as low-molecular-weight CAs in the field of clinical MRI. Each ideal CAs also must possess several vital properties such as good biocompatibility, low toxicity, and high relaxivity. However, the major drawback of currently used low-molecular-weight CAs is their rapid diffusion from blood vessels into the interstitial space and, consequently, relatively high doses and injection rates. In contrast, Gd(III) complexes are highly toxic to serum proteins. Moreover, cardiovascular and oncological MRIs need long blood pool retention. And the long retention in the body can provide a wider time frame for MRI.[6] However, small-molecule CAs are quickly distributed into the extracellular fluid and, accordingly, excreted by glomerular filtration quickly after administration. To improve upon these drawbacks, attachment of low-molecular-weight Gd(III) chelates to linear

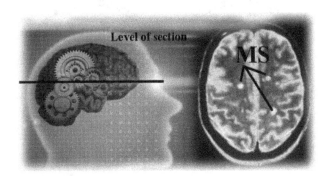

Fig. 1 Schematic diagram showing the places of injured tissues in a multiple sclerosis (MS) patient.

macromolecules such as albumin, dextran, polylysine and poly(ethylene glycol) (PEG), and polysaccharides has been applied. Such polymeric nanoparticles work by reducing the T_1 and T_2 leading to the production of brighter images.[7] They can alter the spin–spin relaxation time of the adjacent molecules of water and, accordingly, they can monitor tissue inflammation and arthritis, and detect tumors.[8]

These CAs based on polymeric nanoparticles can be divided into four classes according to the position of the Gd(III) chelates in the nanocarriers, including (1) block, (2) graft, (3) dendritic, and (4) micellar polymeric nanoparticles, as depicted in Fig. 2. And in order to obtain a wide range of polymeric CAs, critical parameters such as size, shape, surface chemistry, flexibility, and architecture should be designed.

However, the relaxivities of these polymeric nanoparticles were low and independent of temperature at all field strengths. It is due to both flexibility and segmental mobility within the linear polymeric chain or the spacer arms.[9] Alternatively, dendritic CAs are able to improve upon all of the drawbacks described above. Dendritic structures are promising materials to incorporate various components such as therapeutic agents, imaging CAs, and molecular tags. In addition, with respect to the variability of macromolecular architecture, the pharmacokinetics and biodistribution of dendritic CAs can be designed based on the personal nanomedicine. Such a unique aspect arises

from their special characteristics compared to those of equivalent-molecular-weight linear polymers. The defined structure and large number of available functional groups (such as amino groups) of these dendritic structures have resulted in exploiting them as suitable scaffolds. Thus, it is possible to introduce large numbers of chelating agents for generating dendritic MRI contrast agents (DCAs) as well as for adjustment of antibody molecules (Fig. 3).[10]

These dendritic frameworks allow designing several MRI CAs that have similar chemical structures, and meanwhile, coat a wide range of molecular weights and sizes (please see "Dendritic Architectures: Host–Guest Chemistry"). Therefore, by taking advantage of the dendritic structures, many of the difficulties associated with small-molecule CAs can be overcome. Consequently, the possibility of combining diagnosis and therapy in multi-functional dendritic structures can allow the early detection, targeting, and treatment of several diseases (Fig. 4).

In this respect, polyamidoamine (PAMAM), as a water-soluble dendritic nanocarrier, has been one of the most effective vehicles for use as CAs.

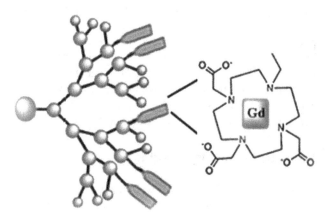

Fig. 3 Schematic diagram showing the preparation of paramagnetic targeted dendritic nanoparticles.

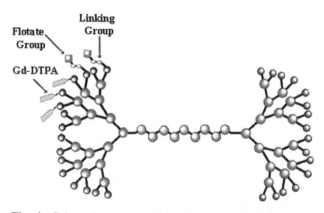

Fig. 4 Schematic showing of the folate-targeting DCAs from PEG-cored structures.

Fig. 2 Diagram showing schematically (**A**) block, (**B**) graft, (**C**) dendritic, and (**D**) micellar polymeric nanoparticles.

Effect of Physicomechanical Factors; Size, Shape, and Flexibility

As is well known the accumulated DCAs may be taken up by cells and metabolized into toxic Gd(III). Therefore, on the one hand, ideal DCAs must be in the body for a sufficient duration to produce the desired effects such as tumor accumulation for oncologic imaging. On the other hand, it must be entirely excreted from the body to decrease unwanted side effects.[11] Target-specific imaging can be realized using DCAs by changing their size, shape, rigidity, and surface modification in order to optimize the pharmacokinetics aspects. For instance, 6-, 8-, and 10-nm nanocarriers of PAMAM, G_4, G_6, and G_{10}-CA are accumulated in the kidneys, blood, and the lymphatic system and, accordingly, are able to act as CAs for renal, blood, and lymphatic imaging, respectively.[12] Moreover, the more hydrophobic polypropylene imine (PPI) G_4-CA, because of longer carbon chains and no amide groups, which has higher accumulation in liver, can be appropriate for liver imaging.[11] Briefly, DCAs larger than 12 nm and smaller than 5.5 nm mainly undergo hepatic and renal clearance, respectively. The effect of shape also significantly affects pharmacokinetics of DCAs. For instance, DCAs modified with short PEG tails have spherical shapes, while those modified with two long PEG chains are nonspherical. Furthermore, the more branched dendritic structures in third generation when conjugated to eight chains of PEG, 5 kDa, depict higher residence time when compared to dendritic structures in second generation conjugated by four chains of PEG, 10 kDa. The DCAs with short PEG tails do not undergo renal clearance, while DCAs with long PEGylated tails show quick renal clearance. In general, dendritic structures with molecular weight (MW) more than 40 kDa normally remain in the blood for a longer period of time than their linear counterparts. Consequently, they show different clearance, suggesting that the shape played a significant role in pharmacokinetics.

Flexibility is also an important factor of pharmacokinetics for DCAs. Normally, increasing flexibility of DCAs can enhance the renal clearance of DCAs.[11]

Effect of Physicochemical Factors; Hydrophilicity and Functionalization

Hydrophilicity also considerably affects the pharmacokinetics of DCAs. For instance, the PPI, G_4-CA has rapid blood clearance and accumulates significantly more in the liver and less in the kidney than does PAMAM, G4-CA.

In addition, functionalization of the peripheral groups of dendritic structures plays a key role in inhibiting or stimulating a patho-physiological response. Thus, conjugating the appropriate ligands to dendritic frameworks will provide increased intracellular trafficking of DCAs in the injured tissues.[13]

THERANOSTIC APPLICATIONS OF DENDRITIC STRUCTURES

As early as 1990, dendritic MRI CAs were exploited by reporting some of the highest known relaxivities for these agents. In 2003, considerable attention was concentrated on the MRI field with the award of the Nobel Prize in Medicine. After the first synthesis of PAMAM dendritic structures, it has been widely used as a nanocarrier for producing polymeric CAs as a result of both the large number of peripheral amino groups and their ordered architectures with minimal amounts of the MRI CAs. Paramagnetic complexes of Gd, including Gd(III)-N, NV, NW, NJ-tetracarboxy methyl-1,4,7,10-tetraazacyclododecane (Gd(III)-DOTA), Gd(III)-diethylenetriamine pentaacetic acid (Gd(III)-DTPA), and their derivatives, enhance the relaxation speed of neighboring H_2O protons and are used as DCAs for MRI (Fig. 5).[14]

As previously mentioned, the clinical uses of polymeric CAs are restricted as a result of their slow excretion rate. Accordingly, they are accumulated within the liver and other organs of the body. The periphery of the dendritic structures can also be functionalized and then complexed by metal ions, which can adjust the relaxivity of the injured tissue. One such commonly employed DCAs for blood pool imaging is marketed, namely, Gadomer-17. *In vivo* experiments with Gadomer-17 exhibit excellent images of long blood circulation times (>100 min) upon intravenous injection. Gadomer-17 has 24 tetraaza cyclododecane-1,4,7,10-triacetic acid functional groups on the surface of a polylysine dendritic structure. However, it exhibits slow renal clearance, probably on account of the globular nature of the dendritic derivative. DCAs also increase the temporal window of dynamic contrast-enhanced MRI, as proved in pharmacokinetic investigations of Gadomer-17.[15]

Dynamic MRI studies suggested that Gadomer-17 can distinguish high-grade infiltrating ductal carcinomas (IDCs). However, it is unable to distinguish low-grade IDCs and fibroadenomas. Therefore, these types of DCAs have depicted great potential for the early diagnosis of several gynecological malignancies.

Blood Pool MRI Using DCAs

DCAs in MRI depict excellent performance for evaluating microvasculature and histological capillary density in tumor tissues.[16] Although postcontrast signal changes in the tumors showed better correlation with histological capillary mass, there are few reports that have studied the lower limits of vessel size that might be visualized by MRI using DCAs. For use as blood pool contrast agents, the G_7 (11 nm) and G_8 (13 nm) dendrimer-based agents were found to be the best candidates for clearly visualizing blood vessels among all the nanosized agents with dendrimer cores described earlier. PEG conjugation appeared to improve the

Conducting—Dendritic

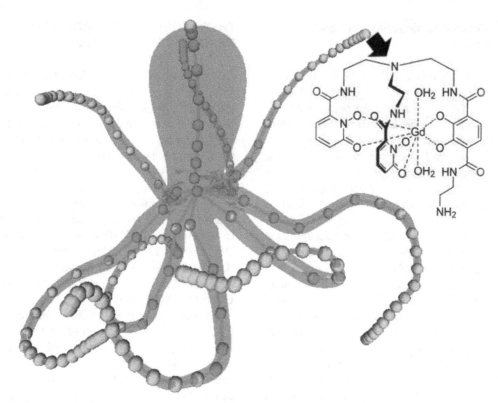

Fig. 5 Schematic line artwork of an octopus-like macromolecular MRI contract agent showing the Gd complexes.

ability of these agents to visualize blood vessels clearly. In addition, dendritic Gd(III) chelates–PAMAM functionalized by 2-(4-isothiocyanatobenzyl)-6-methyl diethylene triamine pentaacetic acid through a thiourea linkage was used to provide flexibility in determining the effect of molecular weight on rotational correlation, relaxivity, and clearance. These DCAs displayed brilliant MRI images of blood vessels as well as longer blood circulation times.[17]

Effect of lysine co-injection and avidin chase

Co-injection of other ligands such as lysine and amino acid mixtures also has been used to moderate renal uptake of radiolabeled peptides. Co-injection of lysine with DCAs of PAMAM, G_4 accelerates urinary excretion of the intact form of the DCAs.[18] Thus, it may be possible to diminish body retention of DCAs potentially lessening any toxicity. It may be as a result of saturating the negative charge on the membrane of the proximal tubules with positively charged lysine molecules or because of blocking the lysine receptor itself. Finally, it permits dendritic sections to pass through without being taken up and retained.

Moreover, the avidin chase system could be exploited to differentiate between extravascular and intravascular fractions of the DCAs. Although, in terms of arriving in clinical applications, a G_4 DCAs functionalized by PEG and lysine co-injection have been confirmed to be a promising option.

Liver and Lymphatic DCAs in MRI

Hydrophobic polymers tend to accumulate rapidly in the liver. For instance, hydrophobic variants of DCAs formed with polypropylenimine diaminobutane (DAB) dendritic cores swiftly accumulate in the liver and have been commercially used as liver CAs. DAB dendritic scaffolds possess a nonpolar polyamine core contrary to the PAMAM frameworks with a polar amide group core. Furthermore, some liver-specific MRI DCAs such as MultiHancek™, $[Gd(BOPTA) (H_2O)]_2$, and Eovistk™, $[Gd(EOB\text{-}DTPA)(H_2O)]_2$ have been applied in clinical uses. Briefly, small changes of chemical architecture can significantly vary the pharmacokinetic features of DCAs.

Assessment of the DACs based on both PAMAM, (G_{2-4}) and DAB demonstrated that all of these nanoparticles are removed from the kidney. These nanoparticles were retained in the blood vessels or urinary tracts with minimal perfusion into the extravascular tissue. Thus, after filtering through the glomerulus, these DACs produced a high-intensity band at the layer of the proximal.[19]

DCAs injected intracutaneously into the lymphatic system have been able to clearly visualize the deep lymphatic system and lymph nodes.[20] The developed resolution of this technique can possess a broad range of application in the field of immunology and cancer in clinical medicine.

Tumor-Specific MRI Using DCAs

With the intention of producing a tumor tissue signal using DCAs, the signal requires to be built by conjugation of Gd(III) atoms on a ligand molecule. In this sense, CAs conjugated to either monoclonal antibodies or avidin is able to play a role of tumor-specific contrasts. Such DCAs have been used by the pharmaceutical industry in commercial developments. For example, the delivery of folate-conjugated DCAs and fluorescent probes to tumor cells that overexpress folate is a potent theranostic method. In this respect, 2-(4-isothiocyanatobenzyl)-6-methyldiethylene triaminepentaacetic acid (TU-DTPA), a chelating ligand of Gd for MRI and fluorescein isothiocyanate for fluorescence is conjugated to PAMAM, G_4-folate systems. Such assemblies reveal drastically amplified signals monitoring rapid cell surface fixation followed by slow internalization.[21]

X-Ray and CT Using DCAs

The X-ray is one of the fundamental diagnostic and noninvasive technologies in medicine. It is used to establish high-resolution images in typical scan times for a few minutes. To obtain a high-resolution X-ray imaging of soft tissues of some diseases require the use of an X-ray CA. Thus, contrast opaque agents with a long-lasting time, providing contrast through tissue vascularization. Similar to MRI contrast agents, low-molecular-weight hydrophilic iodinated X-ray CAs remains limited due to their rapid blood clearance as well as rapidly equilibrate between the intravascular and extracellular fluid compartments of the body. Consequently, high-molecular-weight CAs like DCAs are better for quantitative detection of disease lesions. DCAs of X-ray technology using a variety of organometallic complexes such as bismuth and tin are used to produce a high-resolution X-ray image in several diseases such as arteriosclerotic vasculature, tumors, infarcts, kidneys, or efferent urinary. Alternatively, CT is a medical imaging method used to generate a three-dimensional image of the inside of an object from a large series of two-dimensional X-ray images taken around a single axis of rotation. Usually, CT utilizes iodinated agents and is regarded as a moderate-to-high radiation diagnostic technique. And these iodinated DCAs for CT imaging are synthesized using the "divergent approach." For instance, water-soluble iodinated dendritic nanoparticles, consisting of PAMAM, G4 core, and 3-N-[(N'', N''-dimethylaminoacetyl)amino]-alpha-ethyl-2,4,6-triiodobenzenepropanoic acid (DMAA-IPA) molecules, have been exploited as CT imaging DCA.[22] The high iodine content of these DCA provides a widely available imaging technique with high spatial resolution. In addition, the longer circulatory retention times of such DCAs make it possible to have an extended imaging time scale and potential reduced toxicity that allow for repeated injections of high doses of small iodine molecules.

Dendritic Structures as Molecular Probes and Biosensors

Dendritic structures are attractive scaffolds for use as molecular probes by means of immobilization of sensor units on the surface of dendritic substrate to generate an integrated molecular probe. Such motivating performance returns to the large surface area and the dendritic polyvalency (see the entry "Dendritic Architectures: Host–Guest Chemistry"). To use a dendritic framework as a dendritic sensor in blood media, both a hydrophilic and a chemically inert dendritic structure is vital. A range of DNA-based and nucleobase-based dendritic structures have been produced for signal amplification goals.

Dendritic DNA biosensors

A biosensor is a suitable tool for the recognition of an analyte that merges a biological element with a physicochemical detector element. It consists of four sections: 1) the "sensitive biological component" (for instance, nucleic acids, microorganisms, organelles, cell receptors, tissue, enzymes, and antibodies or biomimic); 2) the "transducer" or the "detector component," which works physicochemically, such as electrochemical, optical, and piezoelectric. (Transducer transforms the acquired signal from the interaction of the analyte with the biological element into another); 3) the amplifier to boost the resulted signals; finally 4) the associated micro- or nanoelectronics or signal processors that are responsible for the representation of the resulted data in a user-friendly manner (Fig. 6).

Nucleic acids have been used as dendritic building blocks, and some of them are commercially available as 3DNA. DNA dendritic systems with up to two million terminal oligonucleotide strands can be produced. As is known, the peripheral polynucleotide strands can be functionalized by various kinds of radioactive or fluorescent labels. Accordingly, these DNA–dendritic structures that commercialized as 3DNA-technology[23] present much potential for single-stranded oligonucleotide detection. For example, as one of the most important applications of DNA dendritic systems, they normally have been utilized as biosensors for DNA hybridization for the rapid diagnosis of genetic and pathogenetic diseases. This biosensor works based on the immobilization of oligonucleotide probes that selectively detect their responsive target sequence through hybridization. Therefore, it can be feasible to possess both increased sensitivity and low detection limits, simultaneously. Quartz–crystal microbalance sensors also make use

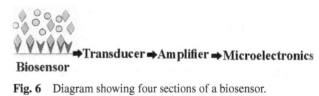

Fig. 6 Diagram showing four sections of a biosensor.

of the amplified hybridization capacity and detection capability of DNA molecules. For instance, DNA dendritic structure in the fourth generation can be immobilized onto a quartz–crystal microbalance. This biosensor is specific to the water-borne pathogen *Cryptosporidium parvum* and can be used as mass-sensitive piezoelectric transducers.[24] In summary, compared to either unimmobilized dendritic structures or the nondendritic systems, immobilized dendritic structures show higher sensitivity in a broad range.

In addition, the oligonucleotide dendritic structures have been applied as a PCR primer in order to design probes with a higher labeling capacity.

Dendritic glucose biosensors

Diabetes is a serious medicinal issue worldwide and enzyme glucose plays a vital role in the development of diabetes biosensors. Therefore, for example, glucose oxidase (GOx) sensors have been successfully designed and used to monitor the glycemia index of diabetic patients.[25] GOx biosensors can also be implanted to detect glucose amounts distantly as well as for applications in insulin pumps. The glucose electrodes utilize oxygen as an electron mediator between GOx and surface of sensor. GOx hydrolyzes glucose and reduces O_2 into hydrogen peroxide (H_2O_2) by the following reaction:

$$H_2O_2 \rightarrow O_2 + 2H^+ + 2e^-$$

The O_2 reduction rate is in direct relation to the glucose levels that can be calculated amperometrically the produced electrical current by increasing and decreasing the amounts of H_2O_2 and O_2, respectively. In this respect, dendritic structures such as ferrocenyl functionalized PPI immobilized onto a platinum biosensor to produce modified electrodes with optimized performance. Moreover, a blend of core-shell SiO_2/Au nanocomposites and PAMAM, G_4 dendric structures has been exploited as flow-through biosensor for assess of immunoglobulin G (IgG) in human serum. The stability, selectivity, and reproducibility of the immunosensor were remarkable.[26] In addition, reduced graphene oxide (RGO)-PAMAM, $G_{3.5}$-Ag nanocomposite was used as a new deposited material for modifying GOx biosensors. Several noticeable properties of this biosensor can be summarized as follows: 1) high sensitivity; 2) low detection limit; 3) satisfactory linear range; 4) negligible interference from ascorbic acid and uric acid that frequently coexisting with glucose in human blood to the glucose biosensor. These characteristics exhibit that the dendritic nanocomposite has outstanding electron transfer properties for GOx and accordingly is a promising matrix for fabricating high-performance glucose biosensors. On the contrary, as mentioned earlier, GOx biosensors also can be implanted to detect glucose amounts distantly. The modified electrode using polyglycerol (PGLD) exhibits a strong and stable amperometric response to glucose. Thus, based on the

electrochemical properties, the bioconjugated PGLD is a good theranostic device in order to generate implantable glucose biosensors.

Dendritic antibody biosensors

A great deal of immunosensor theranostic tools have been established based on the molecular recognition of analytes and immobilized ligands at solid/liquid interfaces. Antibody–theranostic nanomedecene is based on such a solid/liquid interface, which has been concentrated on modifying the sensitivity, selectivity, and sensitivity of immunoassay methods. Immobilization of antibodies onto a substrate at high density with uniform distribution is an important stage in the development of immunosensors. The reversible affinity interactions of immunosensing surfaces are based on biospecific association/displacement reactions between functional antigen ligands and antibody molecules.

Conversely, antibodies can also be used to target a drug into the cell, either by direct conjugation of the drug to the antibody as an immunoconjugate or by conjugation to a carrier molecule. Linear macromolecules, such as N-(2-ydroxypropyl)methacrylamide (HPMA), and poly[N5-(2-hydroxyethyl)-l-glutamine] (PHEG), have been examined as precursors for antibody targeting.[27,28] Dendritic antibody targeting also can overcome several limitations of immunoconjugates based on linear polymers such as reduced drug activity, reduced antibody affinity, and drug resistance. So, theranostic dendritic biosensors play a fundamental role in each of these aspects. As a typical example, a glass surface coated with functionalized PAMAM by carboxylic acid groups (PAMAM COOH) has demonstrated that the modified surface using dendritic precursors has relatively low nonspecific cellular protein adsorption, and can also enable antibody binding. In addition, bifunctional hydroxyl/thiol functionalized PAMAM, G_4 immobilized on the PEG-functionalized assay plate. The dendritic modified plate offers increased sensitivity, decreased nonspecific adsorption, and a better detection limit that were validated in human serum samples from a normal (nonpregnant) woman and pregnant women.[29]

Dendritic central nervous system biosensors

Central nervous system (CNS) disorders affect over 1.5 billion people globally, and theranostics of CNS diseases, disorders, and injuries such as schizophrenia, Parkinson's disease, epilepsy, and stroke is a real challenge. Both glutamate and dopamine are essential neurotransmitters in the mammalian CNS. For instance, amperometric glutamate have been developed by incorporating a nanobiocomposite-dopant based on self-assembling glutamate derivatives and PAMAM-entrapped platinum nanoparticles (PtNPs) on carbon nanotubes. Dopamine concentration, also which is one of the important parameters in determining Parkinson's disease. As a result, the dopamine concentration must be detected. Based on the results of compared analytical performances, the electrochemical quantification with PAMAM-OH-entrapped RhNPs and immobilized on glassy carbon electrodes is feasible.[30]

Dendritic oxygen imaging biosensors

In vivo oxygen imaging is a theranostic method that presents, on the one hand, the possibility of diagnosing complications from disease such as diabetes and peripheral vascular disorders. On the other hand, it can be used as a technique to design an appropriate therapeutic treatment. This method is established by the phosphorescence using oxygen as a quenching agent in the presence of a chromophor. This chromophor, typically Pd chelated by tetrabenzoporphyrins, possesses several features. 1) It is protected from interactions with serum biopolymers like albumin; 2) it is water soluble; and 3) it shows sharp absorption bands in the range of 620–900 nm (near-infrared) to lower the interference of the blood chromophors. In this technique, dendritic carriers with various sizes and structures have an important role. They can physically entrap Pd complexes to regulate and control the oxygen quenching of the phosphorescence. For instance, Pd complexes of tetrabenzoporphyrins having amino acid–based dendritic structures functionalized by PEG segments have been utilized in order to manufacture a bicompatible dendritic biosensor for *in vivo* oxygen imaging.[31]

CONCLUSION

Although dendritic structures used in nanomedicine that can simultaneously detect, image, and treat disease are still the exception rather than the norm, basic research performed from the early 1990s has led to a large number of patents focused in the development of dendritic theranostic systems. Theranostic nanomedicines based on dendritic structures have been a potent tool to confer imaging. Such imaging can present valuable pharmacokinetic and biodistribution data before, during, and after treatment. Therefore, diagnosing techniques such as optical imaging and biosensors have been developed by taking advantages of this revolutionary field. It is as a result of both the polyvalency and the high architectural control of dendritic structures that has increasingly supported the advances in personalized nanomedicine.

ACKNOWLEDGMENT

The author gratefully acknowledges the University of Tabriz for its support of this project.

REFERENCES

1. Debbage, P.; Jaschke, W. Molecular imaging with nanoparticles: Giant roles for dwarf actors. Histochem. Cell Biol. **2008**, *130* (5), 845–875.
2. Massoud, T.F.; Gambhir, S.S. Molecular imaging in living subjects: Seeing fundamental biological processes in a new light. Genes Dev. **2003**, *17* (5), 545–580.
3. Park, K.; Lee, S.; Kang, E.; Kim, K.; Choi, K.; Kwon, I.C. New generation of multifunctional nanoparticles for cancer imaging and therapy. Adv. Funct. Mater. **2009**, *19* (10), 1553–1566.
4. Caravan, P.; Ellison, J.J.; Mc Murry, T.J.; Lauffer, R.B. Gadolinium (III) chelates as MRI contrast agents: Structure, dynamics, and applications. Chem. Rev. **1999**, *99* (9), 2293–2352.
5. Weissleder, R. Scaling down imaging: Molecular mapping of cancer in mice. Nat. Rev. Cancer **2002**, *2* (1), 11–18.
6. Oksendal, A.N.; Hals, P.A. Biodistribution and toxicity of MR imaging contrast media. J. Magn. Reson. Imaging **1993**, *3* (1), 157–165.
7. Yan, G.P.; Robinson, L.; Hogg, P. Magnetic resonance imaging contrast agents: Overview and perspectives. Radiography **2007**, *13* (1), e5–e19.
8. Moghimi, S.M.; Hunter, A.C.; Murray, J.C. Nanomedicine: Current status and future prospects. FASEB J. **2005**, *19* (3), 311–330.
9. Krause, W.; Hackmann-Schlichter, N.; Maier, F.K.; Müllerl, R. Dendrimers in diagnostics. In *Dendrimers II*; Springer: Berlin, Heidelberg, **2000**; 261–308.
10. Kobayashi, H.; Brechbiel, M.W. Nano-sized MRI contrast agents with dendrimer cores. Adv. Drug Deliv. Rev. **2005**, *57* (15), 2271–2286.
11. Longmire, M.R.; Ogawa, M.; Choyke, P.L.; Kobayashi, H. Biologically optimized nanosized molecules and particles: More than just size. Bioconjug. Chem. **2011**, *22* (6), 993–1000.
12. Choyke, P.L.; Kobayashi, H. Functional magnetic resonance imaging of the kidney using macromolecular contrast agents. Abdom. Imaging **2006**, *31* (2), 224–231.
13. Kaminskas, L.M.; Boyd, B.J.; Karellas, P.; Krippner, G.Y.; Lessene, R.; Kelly, B.; Porter, C.J. The impact of molecular weight PEG chain length on the systemic pharmacokinetics of PEGylated poly l-lysine dendrimers. Mol. Pharm. **2008**, *5* (3), 449–463.
14. Sirlin, C.B.; Vera, D.R.; Corbeil, J.A.; Caballero, M.B.; Buxton, R.B.; Mattrey, R.F. Gadolinium-DTPA-dextran: A macromolecular MR blood pool contrast agent. Acad. Radiol. **2004**, *11* (12), 1361–1369.
15. Bumb, A.; Brechbiel, M.W.; Choyke, P. Macromolecular and dendrimer-based magnetic resonance contrast agents. Acta Radiol. **2010**, *51* (7), 751–767.
16. van Dijke, C.F.; Brasch, R.C.; Roberts, T.P.; Weidner, N.; Mathur, A.; Shames, D.M.; Mann, J.S.; Demsar, F.; Lang, P.; Schwickert, H.C. Mammary carcinoma model: Correlation of macromolecular contrast-enhanced MR imaging characterizations of tumor microvasculature and histologic capillary density. Radiology **1996**, *198* (3), 813–818.

17. Brechbiel, M.W.; Xu, H. Method of Preparing Macromolecular Contrast Agents and Uses Thereof. U.S. Patent Application 12/513, 813, Mar 4, 2010.

18. Kobayashi, H.; Sato, N.; Kawamoto, S.; Saga, T.; Hiraga, A.; Ishimori, T.; Konishi, J.; Togashi, K.; Brechbiel, M.W. Novel intravascular macromolecular MRI contrast agent with generation-4 polyamidoamine dendrimer core: Accelerated renal excretion with coinjection of lysine. Magnet. Reson. Med. **2001**, *46* (3), 457–464.

19. Kobayashi, H.; Kawamoto, S.; Jo, S.K.; Sato, N.; Saga, T.; Hiraga, A.; Konishi, J.; Hu, S.; Togashi, K.; Brechbiel, M.W.; Star, R.A. Renal tubular damage detected by dynamic micro-MRI with a dendrimer-based magnetic resonance contrast agent. Kidney Int. **2002**, *61* (6), 1980–1985.

20. Kobayashi, H.; Kawamoto, S.; Sakai, Y.; Choyke, P.L.; Star, R.A.; Brechbiel, M.W.; Sato, N.; Tagaya, Y.; Morris, J.C.; Waldmann, T.A. Lymphatic drainage imaging of breast cancer in mice by micro-magnetic resonance lymphangiography using a nano-size paramagnetic contrast agent. J. Natl. Cancer Inst. **2004**, *96* (9), 703–708.

21. Wiener, E.C.; Konda, S.; Shadron, A.; Brechbiel, M.; Gansow, O. Targeting dendrimer-chelates to tumors and tumor cells expressing the high-affinity folate receptor. Invest. Radiol. **1997**, *32* (12), 748–754.

22. Yordanov, A.T.; Lodder, A.L.; Woller, E.K.; Cloninger, M.J.; Patronas, N.; Milenic, D.; Brechbiel, M.W. Novel iodinated dendritic nanoparticles for computed tomography (CT) imaging. Nano Lett. **2002**, *2* (6), 595–599.

23. Li, Y.; Tseng, Y.D.; Kwon, S.Y.; D'Espaux, L.; Bunchm, J.S.; McEuen, P.L.; Luo, D. Controlled assembly of dendrimer-like DNA. Nat. Mater. **2003**, *3* (1), 38–42.

24. Wang, J.; Jiang, M.; Nilsen, T.W.; Getts, R.C. Dendritic nucleic acid probes for DNA biosensors. J. Am. Chem. Soc. **1998**, *120* (32), 8281–8282.

25. Newman, J.D.; Turner, A.P. Home blood glucose biosensors: A commercial perspective. Biosens. Bioelectron. **2005**, *20* (12), 2435–2453.

26. Tang, D.; Niessner, R.; Knopp, D. Flow-injection electrochemical immunosensor for the detection of human IgG based on glucose oxidase-derivated biomimetic interface. Biosens. Bioelectron. **2009**, *24* (7), 2125–2130.

27. Huang, Y.; Nan, A.; Rosen, G.M. N-(2-Hydroxypropyl) methacrylamide (HPMA) copolymer-linked nitroxides: Potential magnetic resonance contrast agents. Macromol. Biosci. **2003**, *3* (11), 647–652.

28. Thomas, T.P.; Patri, A.K.; Myc, A.; Myaing, M.T.; Ye, J.Y.; Norris, T.B.; Baker, J.R. *In vitro* targeting of synthesized antibody-conjugated dendrimer nanoparticles. Biomacromolecules **2004**, *5* (6), 2269–2274.

29. Han, H.J.; Kannan, R.M.; Wang, S.; Mao, G.; Kusanovic, J.P.; Romero, R. Multifunctional dendrimer-templated antibody presentation on biosensor surfaces for improved biomarker detection. Adv. Funct. Mater. **2010**, *20* (3), 409–421.

30. Bustos, E.B.; Jiménez, M.G.G.; Díaz-Sánchez, B.R.; Juaristi, E.; Chapman, T.W.; Godínez, L.A. Glassy carbon electrodes modified with composites of starburst-PAMAM dendrimers containing metal nanoparticles for amperometric detection of dopamine in urine. Talanta **2007**, *72* (4), 1586–1592.

31. Finikova, O.; Galkin, A.; Rozhkov, V.; Cordero, M.; Hägerhäll, C.; Vinogradov, S. Porphyrin and tetrabenzoporphyrin dendrimers: Tunable membrane-impermeable fluorescent pH nanosensors. J. Am. Chem. Soc. **2003**, *125* (16), 4882–4893.

Conducting—Dendritic

Dental Polymers: Applications

Narendra Pal Singh Chauhan
Department of Polymer Science, University College of Science, Mohanlal Sukhadia University, Udaipur, India

Kiran Meghwal
Pinki B. Punjabi
Department of Chemistry, University College of Science, Mohanlal Sukhadia University, Udaipur, India

Jyoti Chaudhary
Department of Polymer Science, University College of Science, Mohanlal Sukhadia University, Udaipur, India

Paridhi Kataria
Department of Chemistry, University College of Science, Mohanlal Sukhadia University, Udaipur, India

Abstract

Dental composite resins are the common materials used to a esthetically restore the structural integrity of teeth. The most common material used to fabricate the denture base is an acrylic resin made from a mixture of methyl-methacrylate and poly(methyl methacrylate). The matrix of dental composites is mainly based on different methacrylates monomers like 2,2-bis[4-(2-hydroxy-3-methacrylyloxy-propoxy)phenyl] propane, urethanedimethacrylate (UDMA), 2-(HEMA), and triethylene glycol dimethacrylate (TEGDMA). The use of high-pressure/high-temperature (HP/HT) polymerization of urethane dimethacrylate (UDMA) and triethylene glycol dimethacrylate (TEGDMA) to produce resin composite blocks (RCB) is important strategy for dental computer-aided design/manufacture (CAD/CAM) applications. Several techniques such as the micro push-out test for adhesion and failure test by stereomicroscope and scanning electron microscope including nanoindentation have been introduced to evaluate the mechanical performance of resins. The effect of dentin surface moisture and curing mode on microtensile bond strength and nanoindentation characteristics of self-adhesive resin cement will also discussed in this entry. Furthermore, a different chemical surface treatment on the adhesion of resin–core materials to methacrylate resin-based glass fiber posts is explored.

INTRODUCTION

Polymers are a part of everyday life and examples can be found almost anywhere. Many people think of polymers simply as plastics used for packaging, in household objects, and for making fibers, but this is just the tip of the iceberg. Polymers are used in all sorts of applications we might not have thought much about before. Polymers and composites (materials made by combining two or more materials) are vital to modern dentistry. Polymers are used in dentistry to make white fillings for teeth are mainly made from hydroxyapatite (sometimes called hydroxylapatite), which is a type of calcium phosphate with the formula $Ca_{10}(PO_4)_6(OH)_2$. The outside layer of teeth, called enamel, contains over 90% hydroxyapatite. Enamel is the hardest substance in the human body. Bacteria, which live in the mouth, produce acid in the presence of starch and sugar. The acid can cause small holes to form in the tooth's enamel. Eventually these holes become big enough to reach the nerve inside the tooth and it becomes painful. A dentist can fill these holes before they reach the nerve to prevent further decay. More recently, dentists have used a white composite for filling teeth. A composite is a substance made by combining two or more materials. White fillings are a composite of a polymer filled with glass. The filling is put on the tooth in the form of a paste and then a beam of ultraviolet (UV) light is shone on it before they reach the nerve to prevent further decay. The UV light initiates (starts) chemical reactions in the paste, which form a number of cross-links between the polymer chains. The polymer containing the glass filler becomes a solid three-dimensional cross-linked network, which fills the hole in the tooth. Dentistry uses a variety of different polymer materials. Dental polymer materials are based on methacrylate, its polymer, and polyelectrolytes. Methacrylates owe their unique role in dental technology to their amazingly simple processing technology and their excellent properties. Dental polymers are one of the main materials used in modern cosmetic and restorative dentistry.

Features

Dental polymers are dental fillings, tooth coverings, dentures, and any other bisphenol A-glycidyl methacrylate resins. Usually, they are some form of plastic mixed with other materials to create a hard, tooth-like appearance that will last. Most dentists use some form of polymer for at

Concise Encyclopedia of Biomedical Polymers and Polymeric Biomaterials DOI: 10.1081/E-EBPPC-120051131

411

least a few of their cosmetic dental procedures. There are several different types and brands of dental polymers, but the name comes from the fact that several different materials are mixed to create the fillings.

Function

The purpose behind dental polymers is to create replacement teeth, fillings, and dentures with a material that is as hard as a real tooth, is resistant to stains and chipping, is long-lasting, and is a material that has the appearance of a real tooth. In the past, metal fillings or replacements were used which were not cosmetically pleasing and could sometimes cause metal poisoning in the body. Metal fillings also do not attach as well to existing teeth.

Materials

There are several different kinds of dental polymers, and each are created in a slightly different way. Composite polymers are created using a mixture of powdered glass and plastic resin. Resin ionomer cement is a dental polymer that uses a mixture of powdered glass and powdered resin along with organic acid. When exposed to a blue UV light, it hardens. This is used as glue for fillings, crowns, and other cosmetic procedures. A brand that specializes in the creation of dental polymers is Peek.

Benefits

There are many benefits in the use of dental polymers. Since the material is tooth colored, it is easier to make tooth problems appear nonexistent. There is also no danger of metal or mercury poisoning with polymers. Some polymers contain fluoride, which helps in preventing tooth decay. Polymers are extremely durable and resistant to chipping or breaking. Patients are able to resume normal eating and biting habits. Dental polymers usually do not have to be replaced.

METHYL-METHACRYLATE AND POLY(METHYL METHACRYLATE) DENTURE BASE RESIN

The most common material used to fabricate the denture base is an acrylic resin made from a mixture of methylmethacrylate (MMA) and poly(methyl methacrylate) (PMMA). This material possesses favorable working characteristics, polishability, stability in the oral environment, and excellent esthetic appearance.[1] The physical properties exhibited by this material are sensitive to the degree of conversion, processing technique, and conditions presented by the oral environment. High-pressure polymerization is one of the processing technologies that have been developed for the synthesis of high-molecular-weight polymers with well-defined structures. The effect of pressure has been reported for reversible addition fragmentation (chain)

transfer polymerization, resulting in higher polymerization rates and in polymers with higher molecular weights and lower polydispersity.[2] High pressures up to 600 MPa have been reported to enable relatively fast synthesis of well-defined molecular weight polymethacrylates, even at room temperature.[3] It has also been reported that polymerization under pressures up to 500 MPa at 60°C facilitate atom transfer radical polymerization of MMA, resulting in higher polymerization rates and in polymers with higher molecular weights and lower polydispersity indices.[4] A recent report indicated that polymerization under a high pressure of 250 MPa at a high temperature of 180°C increased the flexural strength and hardness of commercially available dental resin composites.[5]

The specimens can be safely polymerized by high pressure up to 980 MPa, and the external appearance of the high-pressure specimens is not distinguishable from that of the control specimens. All the high-pressure specimens can be loaded up to a preset maximum displacement of 8 mm without fracture, except for two of the Acron and one of the PMMA/F(+) samples, all of which fractured into two pieces. The high-pressure group of each material exhibit significantly lower mean values for the 0.2% yield stress ($p < 0.01$), flexural strength ($p < 0.01$), and elastic modulus ($p < 0.01$) in comparison with the control group of the corresponding material (Table 1).

For each material, no statistically significant difference between the two pressure conditions was found in the strain at the maximum stress. No statistically significant difference in the toughness between the different pressures was found in the Acron, where as the high-pressure group revealed significantly higher toughness ($p < 0.01$) than the control in the PMMA.

The scanning electron microscope (SEM) fracture surface microphotographs indicate that the side surface of all the nonfractured high-pressure specimens revealed distributions of crazing on the tensile bottom surface (Fig. 1).

The border between the PMMA beads and the polymer matrix were visible in all the high-pressure specimens of the Acron and the PMMA/F (+) (Fig. 2, upper left). The interpenetrating polymer network (IPN) layer is also present in the high-pressure groups after the tetrahydrofuran (THF) treatment (Fig. 2, lower left). While the border of the PMMA particles is slightly visible in the SEM of the fractured control specimens (Fig. 2, upper right) but not detected for the control groups (Fig. 2, lower right).

The results of the weight measurement of the PMMA/F (−) indicated that there are certain amounts of higher molecular weight polymers in the high-pressure polymerization groups than are present in the control group. In addition, the fraction of the high-molecular-weight polymers increased as the pressure level increased. Thus, the increased toughness of the high-pressure specimens is attributed to the increased average molecular weight of the polymer matrix. In a free radical polymerization, it is explained that high pressure remarkably increases the polymerization rate, with an enhanced propagation rate constant and a reduced termination rate constant.[4] Another potential reason for the increased toughness can be the reduction of the internal voids and defects[7] because the high-pressure environment could reduce the distribution of the voids that are potentially created during the polymer-ization process.[8] It was reported in a study using dental composites that the high pressure/high temperature polym-erization resulted in a reduction in the number and size of defects[9] that led to the improvement of the fracture tough-ness. The high-molecular-weight fraction could increase processing difficulties because of its enormous contribu-tion to the melt viscosity.[10] For these reasons, the low end

Table 1 The mean and the standard deviations of the mechanical properties for the Acron denture base resin and the experimental PMMA with filler in the three-point flexural load test

		Acron		PMMA/F(+)	
		Control	High pressure	Control	High pressure
0.2% yield stress (MPa)	Mean	83.9*	656.3*	85.2*	66.3*
	(S.D.)	−1.8	−1.3	−2.6	−2.3
Flexural strength (MPa)	Mean	124.0*	95.3*	128.0*	97.4*
	(S.D.)	−17.3	−9.9	−4.8	−2.5
Strain at the maximum stress (%)	Mean	6.26	5.81	6.44	6.14
	(S.D.)	−1.78	−1.41	−1.1	−0.29
Elastic modulus (GPa)	Mean	3.36*	2.75*	3.32*	2.79*
	(S.D.)	−0.05	−0.03	−0.10	−0.10
Toughness (N mm)	Mean	0.94	1.56	0.72*	1.76*
	(S.D.)	−0.81	−0.33	−0.18	−0.55

*Asterisk indicates a significant difference between the high pressure and control within the same material groups ($p < 0.01$).
Source: From Natsuko et al.[6] © 2013, with permission from Elsevier.

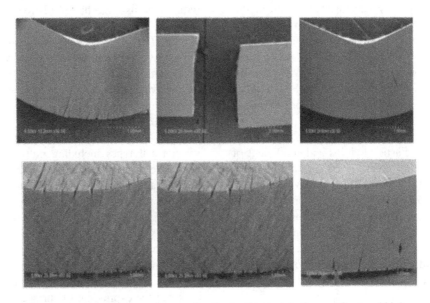

Fig. 1 The SEM images of representative Acron specimens after the loading test. Left: a nonfractured high pressure specimen; Center: a fractured control (ambient pressure) specimen; Right: a nonfractured control specimen. The images on the upper row show the side surface (×30), while the corresponding specimens on the low are viewed from the tensile bottom surface (×50). White arrows indicate the crazing marks on the tensile bottom surface. S: sidesurface, T: tensile bottom surface, C: compressive upper surface. Cleavage-like ratchet marks are evident on the bottom surface, and they created a number of curved crack lines that are originally the straight polishing marks. It is clearly indicated in the Acron and the PMMA/F(+) that the numbers of the cracks are more for the nonfractured high-pressure speci-mens (Left) than those for a few nonfractured control specimens (Right). The cracks generally propagate deeper along the polishing marks at the corner of the bottom plane. The surface cracks and the curved polishing marks are not observed on the side and the bottom surfaces of the fracture specimens (Center).
Source: From Natsuko et al.[6] © 2013, with permission from Elsevier.

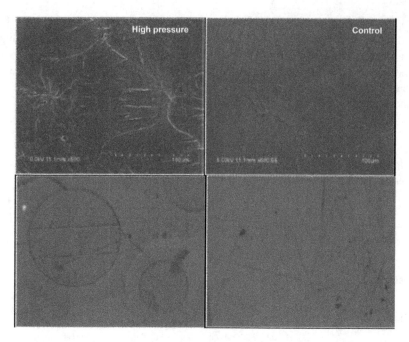

Fig. 2 Typical magnified images of the fractured surfaces of the Acron specimens. The SEM images (upper) and the light microscopic images following the THF treatment (lower) are shown for typical high pressure (left) and control specimens (right). Black arrows indicate the interface between a PMMA particle and the polymer matrix. White triangles indicate the IPN layer (lower left).
Source: From Natsuko et al.[6] © 2013, with permission from Elsevier.

of the distribution might be created and act as a plasticizer, softening the material. The high pressure of 500 MPa during the polymerization process decreased the elastic modulus, the yield, and the maximum flexural strength of the PMMA-based heat curing resin. In the heat-polymerized resin under the ambient pressure, the PMMA particles are likely to swell by absorbing the monomer, resulting in the homogenized appearance in the micrographs (Fig. 2, right). However, in the high-pressure polymerization, the interface became visible in the SEM, presumably because the high pressure increased the interfacial stress between the materials of dissimilar elastic modulus and coefficient of compressibility.[11] During polymerization, the monomers dissolve and diffuse to the surface of the PMMA beads. The high pressure of 500 MPa is employed for the flexural test specimens, although the highest limit of the system used is 980 MPa. Due to reduced mobile capability under an extremely high pressure (above 1 GPa), the monomers are likely to transform into solids and form monomer crystals, thus reducing the polymerization degree.[10]

TRI-ETHYLENE GLYCOL DIMETHACRYLATE OR 2-HYDROXY-ETHYL METHACRYLATE BASED DENTAL RESINS

Despite various modifications in the formulation, the chemical composition of composite resins include inorganic filler particles (quartz, ceramic, or silicia), and additives, which are incorporated into a mixture of an organic resin matrix.[12]

The matrix of dental composites, based on methacrylate chemistry among others, contains strongly viscous major monomers like 2,2-bis [4-(2-hydroxy-3-methacrylyloxy-propoxy) phenyl] propane (Bis-GMA) or urethane dimethacrylate (UDMA), as well as dilutive monomers such as 2-hydroxyethyl methacrylate (HEMA) or the co-monomer triethylene glycol dimethacrylate (TEGDMA) (Fig. 1).[12,13] Physical and chemical properties as well as the clinical performance of composite materials depend on adequate polymerization of resin monomers. Resin monomers like TEGDMA or HEMA-induced cytotoxicity via apoptosis in various cell types including pulp and gingiva cells, and genotoxic or mutagenic effects caused by monomers were reported as well. Most likely as a result of monomer-induced DNA strand breaks, mammalian cells activate functional cell cycle check points through the coordinated activities of regulatory proteins.[14] Monomers also influenced specific cell responses of the innate immune system. TEGDMA and HEMA instantaneously downregulated lipopolysaccharide (LPS)-induced cytokine production in macrophages and inhibited the expression of surface antigens like CD_{14} and other surface markers essential for the controlled interaction of immune cells.[15] Furthermore, low TEGDMA concentrations and chemically related substances like PMMA even inhibited specific odontoblast functions including alkaline phosphatase activity, the matrix mineralizing capability, calcium deposition, and gene expression such as dentin sialoprotein.[16] It has been firmly established that resin monomers deplete the amount of intra-cellular antioxidant glutathione (GSH), the cell's major nonenzymatic antioxidant, while in parallel increasing the formation of reactive oxygen species (ROS).[17]

Monomer-Induced Oxidative Stress and Regulation of Cellular Redox Homeostasis

Nature and sources of reactive oxygen species

Monomer-induced oxidative stress considers the formation of ROS exceeding the capacities of the intracellular nonenzymatic and enzymatic antioxidant system.[17] ROS species are formed physiologically due to an incomplete reduction of molecular oxygen, mostly resulting in superoxide anions (O_2-), hydrogen peroxide (H_2O_2), and highly reactive hydroxyl radicals (HO·).[18] Superoxide anions and H_2O_2, which is mainly a product of the dismutation of O_2-, are known to be incipiently generated ROS in cells. In the course of their decomposition, other radicals are immediately or prevalently formed either enzymatically in reactions that utilize electron transfer during the mitochondrial electron-transport chain, by nicotinamide adenine dinucleotide phosphate-oxidase (NADPH) oxidases, xanthine oxidase, or cytochrome P450s.[19] This biological paradox of ROS function, being toxic but also functioning as signaling molecules, contributes to the integrity of cells and tissues. Specifically, ROS chemically react with atoms of target proteins leading to covalent protein modifications. Thus, ROS are recognized at the atomic and not at the macromolecular level.[20]

Function of the nonenzymatic antioxidant GSH in monomer-exposed cells

Balanced intracellular redox homeostasis, an indication of proper cell function, is tightly controlled by a highly sophisticated antioxidative nonenzymatic and enzymatic system.[21] GSH, a tri-peptide synthesized from glycine, cysteine, and glutamate, is the most important redox regulating nonenzymatic thiol available in cells.[22] The antioxidant function of GSH is due to oxidation of the sulfhydryl group (–SH) in its cysteine residue and has the unique function of sustaining the essential thiol status of proteins by directly scavenging ROS, and by acting as a substrate of glutathione peroxidase (GPx), which catalyzes the reduction of H_2O_2.[22] During normal redox homeostasis, the ratio between GSH and its oxidized form GSSG (glutathione disulfide) in cells is on the order of 100:1, maintained by the reaction of GSH reductase, in which NADPH serves as the substrate.[21] Oxidative stress leads to a quantitative shift in GSH to GSSG ratios and elevates the cellular redox potential, usually within a range of –260 to –200 mV, to a more oxidized level, which may support cell survival and differentiation (mild redox shift), or have general unfavorable consequences on cellular metabolism.[23] Notably, the intracellular amount of GSH was reduced in cells exposed to HEMA or TEGDMA, whereas a contemporaneous increase in GSSG is not detected.[24] A decrease in GSH in monomer-exposed cell cultures is associated with increased formation of ROS and phenomena such as cell death via apoptosis, delayed cell proliferation, or mineralization processes. It is possible that ROS are produced during metabolism of resin monomers or as a result of their effects on ROS-producing enzyme activities. Alternatively, ROS may be produced secondarily due to intracellular GSH depletion through the formation of GSH–monomer adducts.[25]

Functions of antioxidant enzymes in monomer-exposed cells

Besides the nonenzymatic antioxidant GSH, enzymatic antioxidants directly control the steady state of the cellular redox homeostasis by regulating the levels of particular ROS. In the first step, degradation process of ROS, superoxide dismutase eliminates O_2^- by disproportion into H_2O_2 plus O_2.[26] Glutathione peroxidase (GPx1/2), peroxiredoxins, and catalase then decompose the increasing level of H_2O_2 into water, at which point GPx1/2 utilizes GSH as a reducing agent. GSSG, the oxidized form of GSH, is in turn restored to its reduced condition by the expense of NADPH via selective GSH reductase activity.[27] In view of the analysis of dental resin monomer-induced cell responses, presumably based on oxidative stress and redox regulation, the role of GSH content and synthesis, in particular, are considered feasible targets. Moreover, the GSH pathway may be relevant for clinical intervention as well. In this regard, the significance of GSH can be analyzed by lowering its concentration using substances like 1-buthionine sulfoximine (BSO), which selectively inhibits the first step enzyme of GSH synthesis.[28] On the other hand, precursors of the amino acid cysteine like 2-oxothiazolidine-4-carboxylate or N-acetyl cysteine (NAC) should fuel GSH synthesis under stress conditions.[29] Using these tools, it became evident that the intracellular concentration of the nonenzymatic antioxidant GSH directs the expression of enzymatic antioxidants as an adaptive response of cells exposed to monomers like HEMA. HEMA reduced the expression of the H_2O_2 decomposing enzyme GPx1/2, and the presence of BSO, with its associated inhibition of GSH synthesis, decreased GPx1/2 expression even further. This downregulation suggested that expression of GPx1/2 is regulated by GSH levels. It seems that a decreased expression of GPx1/2 in the presence of the resin monomer is the result of GSH depletion, almost certainly caused by adduct formation of GSH with the monomer.[30] HO-1 protein expression can be strongly enhanced by HEMA considering that further antioxidants such as bilirubin are necessary to support enzyme activities that directly balance the cellular redox environment.

The Role of Oxidative Stress and GSH in Monomer-Induced Apoptosis

Dental resin monomers exert cytotoxicity via apoptosis as analyzed in numerous studies.[17] Oxidative stress induced by monomers as a result of elevated ROS generation apparently acts as a signal for the activation of pathways, which control cell survival and death through the redox-sensitive activation of genes and antioxidant proteins. Thus, the monomer-induced disturbance of the intracellular redox homeostasis triggers the onset of apoptosis.[31] Accordingly, substances for scavenging ROS and antioxidants including NAC, ascorbate (vitamin C), or vitamin E are noted to protect cells from monomer-induced cell damage including compensation for monomer-induced apoptosis.[31] The role of NAC in protecting cells from monomer-induced apoptosis appears complex and many faceted. NAC is a known antioxidant for scavenging ROS, but can also act as a surrogate for cysteine after deacetylation to facilitate GSH production.[32] NAC may influence signal transduction pathways leading to the promotion of cell survival and differentiation, and reduction in proliferation and inflammation.[33] Antioxidant pathways as well as differentiation pathways may be intimately related, since protection of NAC may be partly due to its capacity to induce differentiation of cells through the activation of NF-κβ.[34] The nonenzymatic antioxidant GSH is essential to counteract monomer-induced oxidative stress since ROS generation drastically increased when GSH synthesis is inhibited by BSO.[32]

Activation of Mitogen-Activated Protein Kinases and Related Transcription Factors

Principals of mitogen-activated protein kinases signaling and transcription factors downstream

Cellular responses to stimuli of endogenous or exogenous origin are directed by signaling pathways, and the cascade of the family of mitogen-activated protein kinases (MAPK) ranks among those transduction pathways essential for the regulation of cell survival, immune response, or cell differentiation.[35] Initiated by various cellular stresses and often triggered after ligand–receptor binding, signal transduction by MAPKs is consecutively mediated at three levels through phosphorylation-dependent activation of MAPK kinase kinase (MAP3K or MEKK), MAPK kinase (MAP2K or MEK), and finally MAPK. Although the pattern of activation by these three protein kinases appears to function in the same manner, each MAPK, including the extracellular signal-regulated kinases 1 and 2 (ERK1/2), c-Jun N-terminal kinases (JNK), and p38 regulates distinct but convergent processes.[36] The ERK1/2 pathway, which is physiologically, activated via receptor tyrosine kinases after ligand binding, Raf, and MEK1/2, is usually linked to cell proliferation and cell survival.[35] MAPKs JNK1-3, which are also known as stress-activated protein kinases (JNK/SAPK) and isoforms p38α/β/γ/$ kinases, mostly regulate stress-induced apoptosis, but are also associated with differentiation and immune responses. Signal transduction of JNK and p38, initiated by diverse stressors

like UV irradiation, chemotherapeutics, cytokines, or growth factors, and is mediated through multiple MAP3Ks such as the apoptosis signal regulating kinase (ASK), MEKK, or the mixed lineage kinase.[35] The activation of the MAPK cascades finally triggers the expression of immediate early genes to code for transcription factors, which then regulate downstream stress-inducible genes of the cell survival network.[37]

Activation of MAPK signaling and transcription factors downstream by resin monomers

Individual MAPKs are differentially activated by dental resin monomers in various target cells. Long exposure to HEMA and TEGDMA are shown to activate ERK1/2 in salivary gland cells and p38 activation is considered relevant for HEMA-induced apoptosis.[38] It is also described that TEGDMA caused a prolonged activation of ERK1/2 and p38 in a human monocyte cell line possibly linked to a monomer-induced regulation of cell survival, and short-time exposure of primary human pulp cells to HEMA indicated a pro-survival role of ERK1/2.[39]

Likewise, TEGDMA caused a concentration-dependent induction of apoptosis after long exposure periods associated with the simultaneous monomer-induced phosphorylation of MAPK. Both ERK1/2 and p38 can be largely phosphorylated contemporaneously to monomer-induced apoptosis in murine macrophages, and MAPK are also strongly phosphorylated in parallel to the induction of apoptosis in human pulp–derived cells as well. Accordingly, the activation of DUSP5 and DUSP1 gene expression by TEGDMA suggested that the MAPK signal cascade was strongly activated.[31]

DUSP5 and DUSP1 are members of the family of dual specificity protein tyrosine phosphatases (DSPs), which specifically inactivate MAP kinases by dephosphorylation.[40] Enhanced expression of DSPs, which counteract resin phosphorylation of MAPKs, adds to the regulation of resin monomer-induced activation of MAPK cascades. TEGDMA also lead to the upregulation of the protein kinase tribbles homolog 3 (TRiB3), which regulates AKT1, MAPK8, p38, and ERK1/2. Notably, members of the TRiB3 signaling regulator proteins are probably links between "independent" signal processing systems.[41] Furthermore, the monomer TEGDMA inhibited the nuclear expression of transcription factors such as Elk-1, c-Jun, activating transcription factor (ATF-2), and ATF-3 expressed downstream of the MAPK signaling pathways, and this effect is time-related to the induction of apoptosis.[42] The reason for reduced expression of c-Jun and phosphoc-Jun by TEGDMA still remains unknown. It is, however, speculated that a major apoptotic pathway through JNK-c-Jun may be inhibited by degradation of c-Jun, thereby allowing cells to tolerate apoptotic levels of JNK activation.[37] It is also possible that the inhibited expression of c-Jun is linked to a monomer-induced decrease in ATF-2 expression, since c-Jun is a target of ATF-2.[43] Diverse signaling in the activation of ATF-2 is thought to result in various ATF-2 heterodimeric partners activated in a stimulus-specific manner.

The observations on the expression of MAPK and transcription factors indicate that oxidative stress could be dramatically increased beyond the capacities of cellular antioxidants in cells exposed to resin monomers, and the presence of NAC shifted redox homeostasis closer to a physiological balance. Nonetheless, the correlation of the inhibition of MAPK activation by NAC and its influence on the expression of transcription factors downstream with the protection of NAC against apoptosis still demonstrates the relevance of resin monomer-induced oxidative stress at the level of cellular signal transduction pathways. Despite the generation of oxidative stress by resin monomers, MAPK activation is not the primary mechanism of monomer-induced apoptosis but reflects the activation and balance of general redox-sensitive pro-survival and pro-apoptotic pathways.

Signaling Pathways after Monomer-Induced Oxidative DNA Damage

The role of ROS in monomer-induced genotoxicity

ROS are unequally reactive and the extremely reactive HO, catalyzed by transition metal ions in the Haber–Weiss or Fenton reaction, is apparently one of the most hazardous ROS of all.[44] In addition to detrimental effects on lipids and proteins, ROS likely interact with parts of the DNA, such as sugar moieties, chromatin proteins, as well as pyrimidines and purines, which may be rapidly fixed by base or nucleotide excision repair.[45] More severe damage causes when single or double strand breaks and base modifications, and DNA protein cross-links, which consequently impaired the stability of the genome.[46] Various resin monomers including TEGDMA-induced genotoxic and mutagenic effects in vitro.[17] Genotoxicity of monomers could, at least in part, originate from ROS-induced DNA damage since elevated levels of 8-oxoguanine (8-oxoG), a common oxidative DNA lesion, are identified after long exposure periods to TEGDMA in vitro.[39] Correspondingly, it is found that resin monomers induce the expression of 8-hydroxyguanine DNA glycosylase 1, the primary enzyme for the repair of 8-oxoG lesions through base excision.[47] Inhibition of the formation of micronuclei in the presence of NAC indicated that ROS are involved in monomer-induced genotoxicity in vitro.[14] The upregulation of the stress response gene DDIT4 (DNA-damage inducible transcript 4), which enhances oxidative stress-dependent cell death, suggests that ROS may play a role in the formation of genotoxic effects caused by resin monomers.[31] In case the function of the DNA repair machinery is compromised, leaving the cleavage of the DNA backbone unrepaired, oxidative damage may result in DNA strand breaks and deletions of nucleotide sequences as shown with resin monomers.[48]

Signaling pathways originating from monomer-induced DNA damage

The formation of DNA strand breaks is a signal activating sensor, transducer, and effector proteins, which block or delay the cell cycle in order to initiate repair of severe DNA damage or to activate programmed cell death.[49] The current paradigm indicates activation of ataxia-telangiectasia mutated (ATM), a member of the phosphoinositide 3-kinase-like family of serine/threonine protein kinases by autophosphorylation as a first step in response to DNA double-strand breaks (DSB). Activation of ATM is detected in cell cultures exposed to monomers after long exposure periods.[39] ATM, in turn, activates a number of downstream targets at the site of DSB, including histone H2AX, which has been used as a marker of DSBs as well as the cellular checkpoint kinase Chk2 and other downstream targets like p53 that regulate cell cycle checkpoints and apoptosis.[50] The tumor suppressor p53, also known as the "guardian of the genome," is apparently involved in coordination of resin monomer–induced cell responses.[42] Activated upstream through posttranslational modifications like phosphorylation at Ser15 and Ser46 residues via ATM signaling pathway, p53 among others regulates DNA repair, cell cycle progression, and controls apoptosis-related gene expression by interacting with proteins downstream.[51] The resin monomer TEGDMA is reported to slightly enhance the expression of p53 in the nuclear fraction of immune cells and human pulp-derived cells.[42]

USING HIGH-PRESSURE HIGH-TEMPERATURE POLYMERIZATION TO PRODUCE UDMA–TEGDMA BASED RESIN COMPOSITE BLOCKS

Advances in materials science, digital imaging, and computer-assisted design and manufacturing (CAD/CAM) have led to dramatic changes in the production of indirect dental restorations.[52] Developments in manufacturing have been accompanied with corresponding advances in new materials amenable for CAD/CAM. While ceramic blocks, due to their superior esthetics and mechanical properties, remain the material of choice for CAD/CAM applications, resin composite blocks (RCB) seem to gain wider acceptance due to their ease of milling, lower cost, easier through access in case of need (endodontics), and likelihood of easier repair.[52,53] High-pressure high-temperature (HP/HT) polymerization of commercially available resin composites was proposed as a novel methodology to produce RCB suitable for CAD/CAM applications, with superior mechanical properties to those of the commercially available CAD/CAM RCB.[54]

The results of the physical/mechanical characterizations, along with the results of the statistical analysis, are summarized in Table 2.

Table 2 Results (mean ± SD) of mechanical characterization and statistical analysis

Property	Group[b]					
	E55	E60	E65	E60U	Z100	P
σ f (in MPa)	231.9 ± 26.9[ab]	230.5 ± 18.5[ab]	213.7 ± 28.5[b,c]	235.3 ± 17.8[a]	201.1 ± 29[c]	138. ± 24.3[d]
Hardness (in HVN)	80.2 ± 4.2[d]	83.4 ± 4.4[d]	111.2 ± 7.7[b]	92.1 ± 3.7[c]	144 ± 4.8[a]	114.8 ± 4.3[b]
KIC (in MPa m1/2)	1.5 ± 0.33[a,b]	2.06 ± 0.45[a]	1.88 ± 0.41[a]	1.71 ± 0.26[a]	0.99 ± 0.21[b]	0.78 ± 0.21[b]
Weibul modulus	7.8	13.6	8.2	14.5	8.1	5.3
σ 63.21% (in MPa)	246.6	239.0	226.3	243.5	213.4	149.3

[a]Different letters indicate statistically homogeneous subgroups within a tested property.
[b]E55 = experimental 2/3 UDMA 1/3 TEGDMA 55% Vf, HP/HT RCB; E60 = experimental 2/3 UDMA 1/3 TEGDMA 60% Vf, HP/HT RCB; E65 = experimental 2/3 UDMA 1/3 TEGDMA 65% Vf, HP/HT RCB; E60U = experimental 100% UDMA 60% Vf, HP/HT RCB; Z100 = Z100 HP/HT RCB; p = Paradigm MZ100 CAD/CAM RCB.
[c]Weibull characteristic strength.
[d]Weibull modulus (reliability).
Source: From Nguyen et al.[55] © 2013, with permission from Elsevier.

Figure 3 presents Weibull plots of σ_f results. Flexural strength was calculated using the formula:

$$\sigma_f = \frac{3F_1}{2hc^2}$$

where F is the load at fracture, L is the specimen span, h is the specimen width, and c is the specimen height. The results shows that RCB obtained via HP/HT polymerization of UDMA-based experimental resin composites (groups E55, E60, E65, E60U) have superior σ_f, $\sigma_{63.21\%}$ (characteristic Weibull strength), *KIC*, and Weibull modulus (reliability) to those of the commercially available CAD/CAM RCB (group P).[55] The description of the Weibull distribution is given by

$$P_f = 1 - e - \left(\frac{\sigma}{\sigma o}\right) m$$

where P_f is the fracture probability, defined by the relation:

$$P_f = \frac{k}{N+1}$$

where k is the rank in strength from least to greatest, N denotes the total number of specimens in the sample, m is the shape parameter (Weibull modulus), and σ_0 is the scale parameter or characteristic strength, $\sigma_{63.21\%}$.[56] With regard to hardness, only that of E65, the highest filled experimental RCB, came close to the hardness of the commercially available materials (Z100 and P), which contain an equivalent V_f. The properties of the experimental UDMA-based materials are similar or superior to those of HP/HT RCB obtained from the Bis-GMA-based Z100.

Among the UDMA-TEGDMA experimental RCB (E55, E60, and E65), the increase in V_f from 55% to 65% resulted in an increase in hardness but in a decrease in σ_f and $\sigma_{63.21\%}$. With the exception of hardness, the properties of the UDMA-alone experimental RCB (E60U) are similar or superior to those of the UDMA-TEGDMA experimental RCB. The results have also shown that the properties of

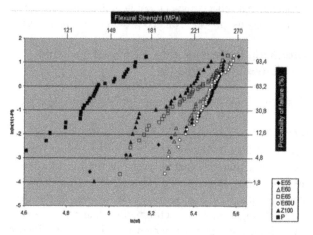

Fig. 3 Weibull plot of the flexural strength (σ_f) results.
Source: From Nguyen et al.[55] © 2013, with permission from Elsevier.

RCB obtained via HP/HT polymerization of Z100 were superior, with exception of *KIC* that was bigger but not statistically different, to those of Paradigm MZ100, a commercial RCB based on Z100.

Figure 4(A–F) shows high magnification (1500×) SEM micrographs of fractured surfaces of representative NTP specimens (with *KIC* close to the average for the group). In the case of the HP/HT experimental RCB (Fig. 4D), crack propagation occurred mainly through the resin, with some failures of larger size filler–resin interfaces present as well, and a higher porosity (not quantified) in the case of E65. The SEM micrographs of Z100 and P groups are very similar in appearance, with a large number of porosities present and with resin–filler failures (Fig. 4E and F).

The first report on the novel method of obtaining dental RCB via HP/HT polymerization of commercially available direct resin composites shows that significant improvements in both mechanical and physical properties could be achieved.[54] The aim of this study is to follow up with the production and characterization of experimental RCB containing only a monomer system and fillers, with no

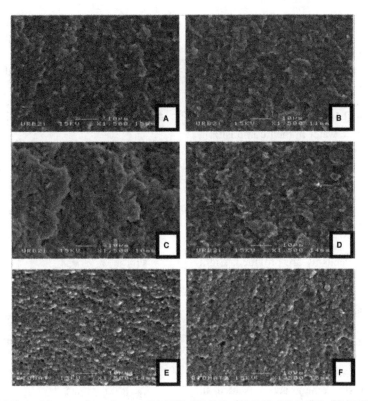

Fig. 4 SEM micrographs (1500×) of fractured specimens: **(A)** E55; **(B)** E60; **(C)** E65; **(D)** E60U; **(E)** Z100; and **(F)** P.
Source: From Nguyen et al.[55] © 2013, with permission from Elsevier.

initiators, accelerators, or any other additives. The filler system used is typical for a hybrid composite. The monomer system selected is based on UDMA, to avoid the use of the controversial Bis-GMA and because it has been shown that the presence of urethanes could improve mechanical/ physical properties.[57] In three of the experimental RCB, UDMA is "diluted" with TEGDMA in a commonly used weight ratio of 2/3 to 1/3, respectively. Three relatively high V_f are incorporated (55%, 60%, and 65%), to investigate if in this range V_f affects or not the physical/mechanical properties of HP/HT RCB. The results obtained are not significantly different between the three groups (E55, E60, and E65), with the exception of hardness (highest for the highest V_f) and Weibull modulus that dropped as V_f increased from 60% to 65%. The latter fact, combined with the relatively higher porosity identified during the SEM characterization of fractured specimens (Fig. 4D), is indicative of practical difficulties in exceeding a certain V_f under the experimental conditions. The formation of polymer networks involving dimethacrylates proceeds from a chemically controlled process to diffusion controlled one[58] to form eventually a heterogeneous network. The cyclization reactions lead to the formation of rather compact, small microgel structures. Further reactions occur during pre- and postgelation periods between the unsaturated bonds present on the surface of these structures. Characterization of laser-ablated surfaces via atomic force microscopy allowed the visualization of network microstructure.[59]

EFFECT OF DENTIN SURFACE MOISTURE AND CURING MODE ON MICROTENSILE BOND STRENGTH AND NANOINDENTATION CHARACTERISTICS OF A SELF-ADHESIVE RESIN CEMENT

The success of an endodontically treated tooth relies not only on the apical sealing but is also highly dependent on coronal sealing. The adhesives used for bonding act as durable barriers hampering coronal micro leakage, thus achieving an effective coronal seal, which is a fundamental step in prevention of bacterial invasion, secondary caries, and de-cementation. It is also suggested that bonding techniques might have significant potential in enhancing the fracture strength of endodontically treated teeth and preventing vertical root fracture.[60] Indirect adhesive procedures represent a substantial portion of esthetic restorative procedures, especially for restorations where a large amount of the natural tooth substance is lacking. Recently, all luting agents required some pretreatment of the dentin, either acid-etching or application of a self-etching primer to prepare the tooth prior to cementation,[61] which resulted in complex and technique-sensitive application procedure. A new type of luting material has been developed that does not require any pretreatment of the tooth surface, the so-called self-adhesive resin cement (SARC). The newly introduced SARCs are expected to offer good esthetics, optimal mechanical properties, dimensional stability, and good adhesion.[62] Dentin is composed of about 50 vol% mineral in the form of a carbonate rich apatite; 30 vol% organic matter, which is largely type I collagen; and about 20 vol% water.[63] Based on the intrinsically moist structure of dentin, bonding has been more complicated compared to enamel, especially with adhesive systems that require moisture control for optimal adhesion.[62] On the other hand, while SARCs contain no water in their composition, water is crucial for their mechanism of action. The mechanism involves ionization of the acidic functional monomer to de-mineralize, penetrate, and establish a chemical bond with calcium ions form dentin apatite, simultaneously allowing for twofold (i.e., micromechanical and chemical) bonding mechanism.[64] The SARCs are categorized as dual-cured resin cements, in which both light-activating and chemical-activating mechanisms are provided. Light activation of the resin cements may increase the degree of conversion when compared to chemical activation alone and enhance physical properties.[65] On the other hand, chemical activation is expected to provide a uniform polymerization at the bottom of deep cavities where access for curing light is limited.

Typical nanoindentation loading cycles with the hold segment for dentin and resin cement are presented in Fig. 5A. Representative curves of C_{IT} during the holding segment are presented in Fig. 5B. Dentin nanoindentation assessment was provided as a reference for a comparison with the resin cement. Nanoindentation creep (C_{IT}) is defined as the relative change of indentation depth while the applied load remained constant during the holding time:

$$\%C_{IT} = \frac{h_2 - h_1}{h_1} \times 100$$

where h_1 and h_2 are penetration depths of indenter at the beginning and the end of hold segment, respectively.

Dentin shows a different creep pattern and is provided for reference. Mean values of nanoindentation hardness and creep are presented in Table 3. Two-way ANOVA indicated that moisture ($p = 0.63$), curing mode ($p = 0.32$), or their interaction ($p = 0.08$) did not significantly affect the hardness of resin cement. However, as for the creep, the data analysis showed that curing mode was a significant factor ($p < 0.01$), while the moisture of dentin surface was not; the chemical curing mode resulted in higher indentation creep ($p = 0.61$). Meanwhile, there was no significant interaction between the two factors ($p = 0.89$).[66]

As shown in Fig. 6, under microscopic examination, a good adaptability is found between the resin cement and the interface before the nanoindentation test. The range of thickness of the resin cement laayer was 30–60 μm.

The evaluation of microtensile bond strength (MTBS) used in the current discussion is an accepted method for laboratory testing of bond strength for dental adhesive materials. It is suggested that the nontrimming microtensile test potentially renders a more realistic expectation of resin–dentin bonding than with conventional bulk

specimens.[67] The variability of MTBS results under the experimental setup is well reflected in the generally low Weibull modulus values ($m < 5.0$), which have been attributed to limitations of bond strength tests prone to high

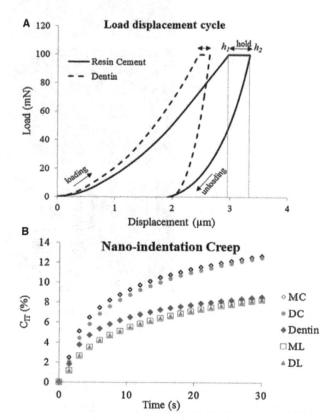

Fig. 5 (A) Nanoindentation loading cycle. Indentation creep data is recorded during the hold-segment (arrows). Two typical cycles are shown for dentin and resin cement. (B) Typical nanoindentation creep plots during the hold segment for each of the substrates tested in this study. The chemical cured cements (MC and DC) apparently show higher creep with a larger curve slope at the end of 30 sec hold segment compared to light-cured (ML and DL).
Source: Reprinted from Moosavi et al.[66] © 2013, with permission from Elsevier.

Table 3 Hardness and creep results

Group	Hardness mean (HM) ± SD (MPa)*	Creep (CIT)mean ± SD (%)**
ML	477 ± 181	9.6 ± 2.9
DL	512 ± 170	9.3 ± 2.9
MC	498 ± 131	10.9 ± 1.9
DC	437 ± 87	10.1 ± 2.7
Dentin***	635 ± 86	8.0 ± 0.8

*Two-way ANOVA showed no significant factors ($p > 0.05$).
**Two-way ANOVA showed that curing mode was a significant factor ($p < 0.05$), but dentin wetness was not ($p > 0.05$).
***Dentin data is provided for reference (not included in the statistical analysis).
Source: From Moosavi et al.[66] © 2013, with permission from Elsevier.

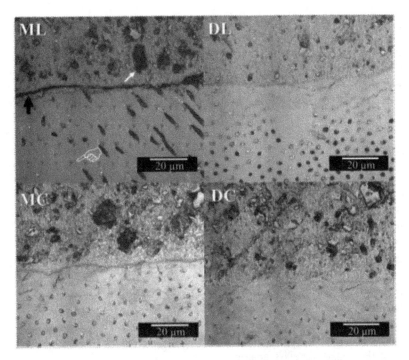

Fig. 6 CLSM images of the interface between resin cement and dentin at magnification of 2500×. ML: Interface of ML after nanoindentation test. Indentation marks are clearly observed on dentin (finger pointer). The interface appears to have developed interfacial gap (black arrow), possibly due to drying and deformation of the specimen during nanoindentation test. Dark areas within resin cement (white arrow) indicate fillers detached during polishing. Indentations located on such defects are excluded from analysis. DL, MC and DC: before nanoindentation, a good adaptation is observed in all groups. No specific difference is observed among groups in terms of interfacial seal.
Source: From Moosavi et al.[66] © 2013, with permission from Elsevier.

scatter from nonuniform stress states and strength controlling flaws present in the specimen.[68] The chemical curing mode of the resin cement appeared to produce bond strength similar to that of light-cured mode, particularly under dry conditions. The only significant difference between dual-cured and chemical-cured groups is in indentation creep. In a polymer matrix, the chemical links established by cross-linking between molecular chains and monomer conversion increase the resistance to plastic flow and creep. In addition, the filler particles may reduce the amount of indentation creep.

RESIN-STRENGTHENING OF A DENTAL PORCELAIN ANALOGUE

Failure of all-ceramic restorations due to fracture, leading to partial or ultimately total restoration loss, continues to compromise clinical treatment outcomes in dentistry.[69] The fracture origin for many classes of all-ceramic dental restorations can be identify using quantitative fractography in failed restorations and finite element approaches, to initiate at the inner surface of the ceramic substrate where the maximum tensile stresses occur in function.[70] Much of the mechanical testing on dental ceramic systems is carried out under standardized conditions, often in a dry environment, and routinely at a fixed crosshead rate, which is unrepresentative of the masticatory cycle.[71] In the context of *in vitro* biaxial flexure strength (BFS) testing and the susceptibility of dental glass–ceramics to slow crack growth, the crosshead rate routinely adopted is 0.75 ± 0.5 mm/min.[72] Water is readily available to the external surfaces of all ceramic restorations from saliva and has been demonstrated as being the "primary rate-dependent mechanism" in the development of cone cracks in monolithic dental crowns.[73] The rate of crack growth in a soda-lime-glass was identified by Charles,[74] using direct observations on loaded four-point-bend specimens, as being reaction rate controlled and dependent on the rate of attack of environmental moisture at the crack tip. Consequently, Charles postulated that water reacted with the molecules at the crack tip, breaking the Si–O–Si bonds to form a hydroxide, which resulted in increased corrosion. The hypothesis tested was that the soda-lime-glass analogue would be susceptible to slow crack growth in air and the magnitude and pattern of resin-strengthening observed would be dependent upon the crosshead test rate and the resin-seating load.

Statistical Analysis

A general linear model univariate analysis demonstrated that BFS can be significantly increased ($p < 0.001$) by resin cementation, by increasing crosshead rate ($p < 0.001$) but not by the seating load used during cementation ($p = 0.095$). However, a significant factorial interactions is observed between cementation seating load crosshead rate ($p = 0.006$). The mean BFS and associated standard deviation of the control soda-lime-glass disc-shaped specimens (alumina particle air abraded and HF acid-etched) tested at crosshead rates of 0.01, 0.1, 1, and 10 mm/min are 54.6 (5.7) MPa (Group A), 64.1 (7.5) MPa (Group B), 80.7 (10.5) MPa (Group C), and 81.4 (8.5) MPa (Group D), respectively (Table 4).

Post hoc analyses demonstrated a significant reduction in mean BFS is identified for Group A specimens (Table 4) compared with Groups B–D (all $p < 0.001$). Further analysis identified a significantly reduced mean BFS for Group B specimens compared with Group C ($p < 0.001$) and Group D ($p < 0.001$) specimens (Table 4), however, no difference in mean BFS values for were identified between Groups C and D ($p = 0.99$). The mean BFS and associated standard deviation data are plotted against the crosshead rate and linear logarithmic regression curve fitted to the raw data for the uncoated control groups (A–D), which demonstrated the static fatigue effects of the soda-lime-glass analogue (Fig. 7).[75]

The mean BFS and associated standard deviation of the Rely-X Veneer resin-coated soda-lime-glass disc-shaped specimens cemented with a 5 N seating load and tested at crosshead rates of 0.01, 0.1, 1, and 10 mm/min are 83.4 (12.8) MPa (Group E), 110.3 (17.7) MPa (Group F), 118.7 (17.8) MPa (Group G) and 123.6 (13.1) MPa (Group H), respectively (Table 4). Increasing the seating load to 30 N prior to testing at crosshead rates of 0.01, 0.1, 1 and 10 mm/min resulted in mean BFS values and associated standard deviations of 74.6 (6.2) MPa (Group I), 96.1 (15.1) MPa (Group J), 121.6 (13.4) MPa (Group K), and 128.7 (22.6) MPa (Group L). Significant decreases in mean BFS are identified for Group I specimens when compared with Group J–L specimens (all $p < 0.001$) and Group J specimens when compared with Group K and L specimens (all $p < 0.001$) but no significant difference in mean BFS is

determined between Group K and L ($p = 0.392$) specimens (Table 4).

Rely-X Veneer resin-coating (Groups E–L) resulted in significant increases (all $p < 0.001$) in the mean BFS data when compared with the uncoated control groups (A–D) at each of the crosshead rates examined (Table 5). Significant increases in mean BFS values are observed between the Rely-X Veneer resin-coated specimens prepared at seating loads of 5 and 30 N and tested at crosshead rates of 0.01 mm/min (Groups E and I; $p = 0.003$) and 0.1 mm/min (Groups F and J; $p = 0.003$). However, no significant differences in mean BFS values are observed at crosshead rates of 1 mm/min (Groups G and K; $p = 0.753$) and 10 mm/min (Groups H and L, $p = 0.51$) for the Rely-X Veneer resin-coated specimens prepared at seating loads of 5 and 30 N.

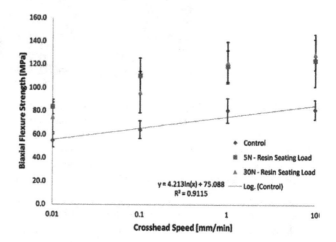

Fig. 7 BFS against crosshead rate for the uncoated disc-shaped controls (Groups A–D) and the resin-coated disc-shaped specimens (Groups E–H with resin-seating load of 5 N, Groups I–L with resin-seating load of 30 N). A linear logarithmic regression curve is fitted to the raw data for the uncoated control groups (A–D) which demonstrated the static fatigue effects of the soda-lime-glass analogue.
Source: From Hooi, Addison, and Fleming[75] © 2013, with permission from Elsevier.

Table 4 Mean BFSs for specimen Groups A–I, where the superscript denominators denote significant differences ($p < 0.05$) between columns for the uncoated control (Group A–D), resin coated with a seating load of 5 N (Groups E–H) and resin coated with a seating load of 30 N (Groups I–L)

Crosshead rate [mm/min]	Uncoated controls [MPa]	Resin-coated at 5 N [MPa]	Resin-coated at 30 N [MPa]
0.01	Group A 54.6 (5.7)	Group E 83.4 (12.8)	Group I 74.6 (6.2)
0.1	Group B 64.1 (7.5)	Group F 110.3 (17.7)	Group J 96.1 (15.1)
1	Group C 80.7 (10.5)	Group G 118.7 (17.8)	Group K 121.6 (13.4)
10	Group D 81.5 (8.5)	Group H 123.6 (13.1)	Group L 128.7 (22.6)

Source: From Hooi, Addison, and Fleming[75] © 2013, with permission from Elsevier.

For the soda-lime-glass dental "porcelain" analogue investigated, when the mean BFS and associated standard deviation data are plotted against the crosshead rate and linear logarithmic regression curve fitted to the raw data for the uncoated controls (Groups A–D) a pattern of strength dependence consistent with the effects of static fatigue is demonstrated (Fig. 7).

Crosshead test rate and resin-cementation load have been shown to significantly modify resin-strengthening mediated by the resin–ceramic hybrid layer in the presence of ambient moisture.

EFFECT OF DIFFERENT CHEMICAL SURFACE TREATMENTS ON THE ADHESION OF RESIN–CORE MATERIALS TO METHACRYLATE RESIN-BASED GLASS FIBER POSTS

Fiber posts are currently widely used in the restoration of endodontically treated teeth.[76] The main advantage of fiber posts is closer elastic modulus of fiber posts ($\approx$20 GPa) to dentin, producing a favorable stress distribution and high success rates without the occurrence of root fractures.[77] It has been shown that the establishments of reliable bonds at the root–post-core interfaces are important for the clinical success of a post-retained restoration.[78] Retention of the composite core to the prefabricated post is affected by various factors, including surface treatment of the post the design of the post head, the post and the composite resin core material.[79] Airborne-particle abrasion with aluminum oxide or silica and hydrofluoric acid etching are techniques used to improve the adhesion between fiber posts and composite resin or resin luting agents.[80] Because these techniques can occasionally damage the glass fibers and affect the integrity of the posts, other chemical treatments have been proposed to improve bonding between fiber posts and composite resin core materials. They included hydrogen peroxide (H_2O_2), potassium permanganate, and sodium ethoxide with varying degree of outcomes. H_2O_2 is commonly used in dental practice, mostly for dental bleaching, and is easy and safe to utilize.[81] Methylene chloride (CH_2Cl_2) has been proposed for use to improve the adhesion between acrylic resin denture base materials and

acrylic resin repair materials by changing the chemical features and surface morphology of denture base resins and increases their repair strength.[82]

Two types of glass fiber posts (Reblida post; VOCO and Rely X post; 3M ESPE) were divided into eight groups according to the surface treatment used; Gr 1 (control; no surface treatment), Gr 2 (silanization for 60 sec), Gr 3 (10% H_2O_2 for 5 min), Gr 4 (10% H_2O_2 for 10 min), Gr 5 (30% H_2O_2 for 5 min), Gr 6 (30% H_2O_2 for 10 min), Gr 7 (CH_2Cl_2 for 5 min), and Gr 8 (CH_2Cl_2 for 10 min).

Three-way ANOVA of the micropush-out bond strength testing data (post type, surface treatment, and core material) revealed that the bond strength is significantly affected by the type of fiber post, by surface treatment, and by core material ($p < 0.001$). There are significant interactions between type of fiber post and core material ($p < 0.001$). However, there were no significant interactions between type of fiber post and surface treatment ($p = 0.917$), the surface treatment and core material ($p = 0.180$), as well as type of fiber post, core material, and surface treatment ($p = 0.957$) as presented in Table 6.

The mean of the micropush-out bond strength values (MPa) and standard deviations are presented in Table 7.

Mean values represented with common or same lowercase letters (row) are not significantly different according to Tukey's test ($p > 0.05$). Mean values represented with common or same uppercase letters (column) are not significantly different according to Tukey's test ($p > 0.05$).

The results of bond strength values achieved with 30% H_2O_2 and CH_2Cl_2 for 5 and 10 min are significantly higher compared with the control and silanization groups for both types of posts with the core materials tested ($p < 0.05$). The type of core material has a significant influence on the bond strength values ($p < 0.05$). The RP/GR (CH_2Cl_2 10 min, CH_2Cl_2 5 min, 30% H_2O_2 10 min and 30% H_2O_2 5 min) groups showed the highest bond strength values (26.4 ± 2.3, 25.9 ± 2.6, 24.4 ± 2.8, and 23.2 ± 2.3 MPa) among the groups. The lowest bond strength values are obtained with F60 composite as a core for both types of posts compared with the other groups.[83]

SEM evaluation of the posts revealed a rather rough surface with some glass fibers exposed for the untreated

Table 5 Significance levels identified when comparing the mean BFS data for the uncoated control (Group A–D) resin coated with a seating load of 5 N (Groups E–H) and resin coated with a seating load of 30 N (Groups I–L) at respective crosshead rates. Analysis of group means was performed utilizing a general linear model univariate analysis where factors where crosshead rates; cementation seating load; and cementation

Crosshead rate [mm/min]	Uncoated vs. resin-coated at 5 N	Uncoated vs. resin-coated at 30 N	Coated at 5 N vs. coated at 30 N
0.01	Groups A and E ($p < 0.001$)	Groups A and I ($p < 0.001$)	Groups E and I ($p = 0.003$)
0.1	Groups B and F ($p < 0.001$)	Groups B and J ($p < 0.001$)	Groups F and J ($p = 0.003$)
1	Groups C and G ($p < 0.001$)	Groups C and K ($p < 0.001$)	Groups G and K ($p = 0.753$)
1	Groups D and H ($p < 0.001$)	Groups D and L ($p < 0.001$)	Groups H and L ($p = 0.510$)

Source: From Hooi, Addison, and Fleming[75] © 2013, with permission from Elsevier.

Table 6 Three-way ANOVA for the post type, surface treatment, core material and the interaction terms according to micropush-out bond strength data ($p < 0.05$)

Source of variation	Sum of squares	D_f	Mean squares	F	p-value*
Post type	262.277	1	262.277	41.471	<0.001
Surface treatment	3,162.585	7	451.798	71.437	<0.001
Core material	1,251.322	1	1,251.322	197.856	<0.001
Post type × surface treatment	16.529	7	2.361	0.373	0.917
Post type × core material	391.196	1	391.196	61.855	<0.001
Surface treatment × core material	65.132	7	9.305	1.471	0.180
Post type × surface treatment × core material	12.825	7	1.832	0.290	0.957
Total	69,223.557	224			

*Statistically significant difference at $p < 0.05$.
Source: From Elsaka[83] © 2013, with permission from Elsevier.

Table 7 Mean (standard deviation) of the micropush-out bond strength values (MPa) of posts to core material combinations with different treatments and Tukey's analysis

Post/core material		Control	Silanization	10% H_2O_2 (5 min)	10% H_2O_2 (10 min)	30% H_2O_2 (5 min)	30% H_2O_2 (10 min)	CH_2Cl_2 (5 min)	CH_2Cl_2 (10 min)
RP	GR	16.2 (2.5)	17.4 (2.7)	18.9 (2.5)	20.8 (2.6)	23.2 (2.3)	24.4 (2.8)	25.9 (2.6)	26.4 (2.3)
	F60	9.4 (2.4)	9.6 (2.7)	10.1 (2.5)	11.1 (2.7)	16.9 (2.5)	17.5 (2.2)	18.3 (2.1)	20.3 (2.8)
RX	GR	12.1 (2.6)	12.2 (2.2)	13.3 (2.8)	16.3 (2.1)A	18.5 (2.8)	20.3 (2.7)	19.8 (2.4)	21.1 (2.3)
	F60	10.0 (2.2)	10.2 (2.4)	11.4 (2.2)	11.9 (2.8)	17.2 (2.2)	18.3 (2.9)	18.5 (2.4)	19.4 (2.4)

Source: From Elsaka[83] © 2013, with permission from Elsevier.

RP, providing potential for micro-mechanical retention compared with the untreated RX posts, which showed a smooth surface (Figs. 8A and 9A). The surface topography of posts can be modified following treatment with H_2O_2 and CH_2Cl_2 (Figs. 8 and 9). The surface treatments dissolved the resin matrix of the posts and exposed the glass fibers of the posts. The dissolution of the resin matrix created retentive areas among the fibers. In addition, the exposed glass fibers are not damaged or fractured by the surface treatments. Nevertheless, treatment with silane showed no changes on the post surface morphology compared with the control group (Figs. 8B and 9B).

MINERAL TRIOXIDE AGGREGATE FILLER

Composite resins have become a mainstay in restorative dentistry. A significant challenge that still remains with dental composite restorations is their inability to prevent formation of secondary caries.[84] One of the modifications to improve the properties of these materials is the use of alternative fillers. One way to incorporate a bioactive calcium-releasing filler in resin would be to use mineral trioxide aggregate (MTA) as a filler. MTA is bioactive; it leaches calcium ions in solution and is capable of forming a bone-like hydroxyapatite layer on its surface when immersed in physiological

solution.[85] This layer provides the benefits of increasing the sealing ability of MTA and promoting remineralization and regeneration of hard tissues. MTA already has numerous applications including pulp-capping, apexification, repair of root perforations, root-end filling, and others.[86]

In this discussion, there is description of characterization and investigation of the chemical properties of composite resins using MTA as filler material.

Characterization of Unhydrated Materials

The X-ray diffractograms of the unhydrated MTA powder and unpolymerized resins are shown in Fig. 10A while the results of FT-IR are shown in Fig. 10B. The XRD plot for the resins showed that there are no crystalline constituents.

The MTA displayed a tri-calcium silicate peak at around 875 cm^{-1} on FT-IR. Two main peaks in the unfilled resin were identified: one at 1730 cm^{-1} and one at 1514 cm^{-1}, attributable to C=O stretching.

Characterization of Hydrated Cements

In the hydrated MTA-W cements, the tri-calcium silicate, di-calcium silicate, and bismuth oxide peak intensities are noticeably reduced for all 28-day samples compared to the

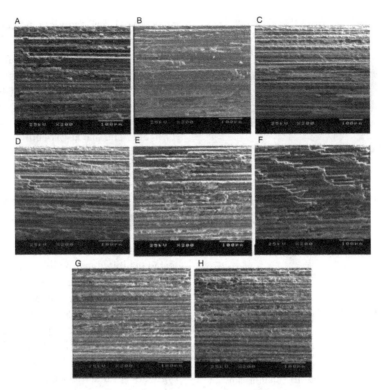

Fig. 8 Representative SEM photomicrographs of Reblida glass fiber post surfaces: **(A)** control, **(B)** silane application for 60 sec, **(C)** etching with 10% H_2O_2 for 5 min, **(D)** etching with 10% H_2O_2 for 10 min, **(E)** etching with 30% H_2O_2 for 5 min, **(F)** etching with 30% H_2O_2 for 10 min, **(G)** etching with CH2Cl2 for 5 min and **(H)** etching with CH2Cl2 for 10 min.
Source: From Elsaka[83] © 2013, with permission from Elsevier.

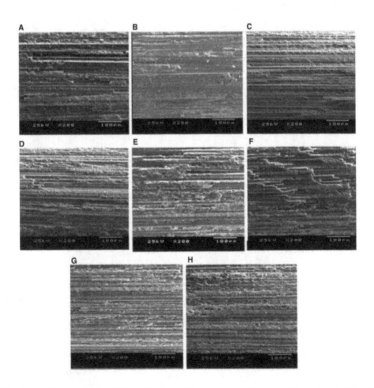

Fig. 9 Representative SEM photomicrographs of RelyX glass fiber post surfaces: **(A)** control, **(B)** silane application for 60 sec, **(C)** etching with 10% H_2O_2 for 5 min, **(D)** etching with 10% H_2O_2 for 10 min, **(E)** etching with 30% H_2O_2 for 5 min, **(F)** etching with 30% H_2O_2 for 10 min, **(G)** etching with CH_2Cl_2 for 5 min and **(H)** etching with CH2Cl2 for 10 min.
Source: From Elsaka[83] © 2013, with permission from Elsevier.

Dental—Electrospinning

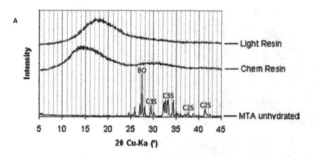

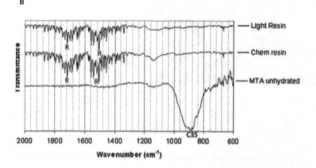

Fig. 10 Characterization cured resins without MTA filler and un-hydrated MTA Plus (**A**) X-ray diffraction and (**B**) Fourier transform infrared spectroscopy showing tri-calcium silicate (C$_3$S), di-calcium silicate (C$_2$S), bismuth oxide (BO) and an absorption peaks due to C = O stretching in the resins (R).

respective 1-day samples, indicating a progression in the hydration reaction (Fig. 11A). The FT-IR plots of the hydrated cement (Fig. 11B) displayed weak peaks of calcium carbonate (at 1400 cm^{-1}) and ettringite (at 900 cm^{-1}) are identified in some of the samples. All materials displayed a tri-calcium silicate peak at around 875 cm^{-1}. Two main peaks in the unfilled resin are identified: one at 1730 cm^{-1} and one at 1514 cm^{-1}, attributable to C=O stretching. These peaks are also detected in the filled resins (except for the 1514 cm^{-1} peak in the chemical curing resin).

pH and Calcium Ion Release in Physiological Solution

The pH results are shown in Fig. 12A. All the materials tested exhibited an alkaline pH, with the pH level generally increasing over time. Statistically significant differences between MTA-Chem. and MTA-Light are only observed at 1 day ($p = 0.025$). Significant differences exist between MTA-W and both composites at all time periods, with the composite materials being slightly less alkaline than MTA-W. The results for calcium ion release are shown in Fig. 12B. All materials leached calcium in solution, with the calcium ion concentration increasing over time. Since the solutions are not changed for the duration of the immersion

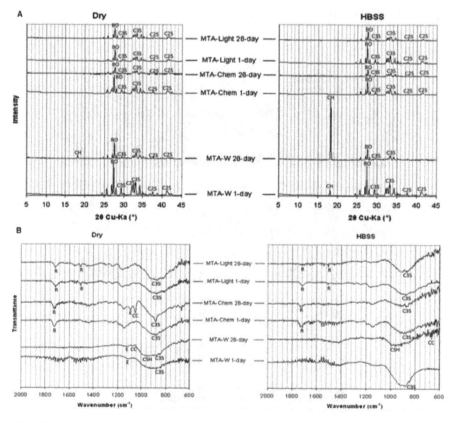

Fig. 11 Characterization of MTA-filled resins, polymerized unfilled resins and MTA mixed with water by (**A**) X-ray diffraction analysis and (**B**) absorption Fourier-transform infrared spectroscopy showing C3S, C2S, **calcium–silicate–hydrate, ettringite (E), calcium carbonate, calcium hydroxide (CH), BO and an absorption peak due to C=O stretching associated with resin (R).
Source: From Formosa, Mallia, and Camilleri[87] © 2009, with permission from Elsevier.

period, the values reported here are cumulative concentrations and thus a decrease over time would be indicative of a reduction in free calcium ions, either due to reaction to form insoluble precipitates, or re-absorption by the test materials. No statistically significant differences are observed between the three materials at 1 day ($p > 0.05$). No significant differences between MTA-W and MTA-Light are observed at 7 days ($p = 0.053$) and 21 days ($p = 0.650$). MTA-Chem. had significantly lower calcium ion release than the other two materials ($p < 0.05$) at all time periods except at 1 day.

MTA was developed at Loma Linda University in the 1990s specifically for use as a root end filling material and as a perforation repair material.[88] It has been patented and has been approved by the FDA in the USA and is commercially available as Pro Root MTA (Tulsa Dental Products, Tulsa, OK, USA), in both gray and white forms. Subsequently, MTA Angelus (Angelus Soluções Odontológicas, Londrina, Brazil) has also become available. In the above discussion, a recently introduced formulation of MTA, "MTA Plus" is used. According to the manufacturer, MTA Plus is similar in composition to Pro Root and MTA Angelus but is ground finer (grain size is approximately 2 μm × 2 μm). The material being proposed would bond to tooth structure, thus providing a hermetic seal, while the MTA would leach calcium hydroxide promoting remineralization and regeneration of hard tissues.[89] In fact, one study

reported that MTA had superior sealing ability to composite resin. It is a well-established fact that MTA releases calcium ions and promotes an alkaline pH in physiological solution.[90] The basis for the biological properties of MTA has been attributed to the production of hydroxyapatite when the calcium ions released by the MTA came into contact with tissue fluid. All materials exhibited an alkaline pH and released calcium ions into solution, indicating all are bioactive. Calcium ion release has been reported for calcium phosphate composite cements and for MTA mixed with light-curing and chemical curing resin.[91]

DUAL CURE DENTAL RESIN

The use of resin-based restorative materials in dentistry has increased exponentially over the last few decades with hardly a single dental procedure today being accomplished without their use. These materials are tooth colored, durable in the oral environment, able to adhere to tooth structure, and have the ability to be placed directly within the prepared tooth. Unfortunately, the issue of initial color, color-matching, and color stability after long-term intraoral exposure still remains.[92] Light-cured resins generally have acceptable color and color stability after polymerization. However, color stability has become more of a problem recently due to the rise in popularity of bleaching and esthetic restorative resins. The inclusion of colorless onium-ion compounds (e.g., *p*-octyloxy phenyl-phenyl iodonium hexafluoroantimonate; OPPI) and other cationic photoinitiators in the photoinitiator system have been suggested as a means to improve degree of cure, lower initial color (to make it less yellow, i.e., whiter), and improve color stability. Such photo-co-initiators would reduce the need for those components that contribute to oxidative color changes such as the ubiquitous blue-light-absorbing photosensitizer camphorquinone (CQ) and an amine initiator. Self-cure and dual-cure (DC) resins have a much darker initial yellow color and a larger color-shift ($^{\wedge}E^*$) to a still darker shade of yellow after polymerization, than do light-cured resins. This causes these resins to be relegated mainly for use as cements and luting agents, where esthetics is not as crucial. Nonetheless, their color-shift is clinically unacceptable especially when used with thin veneers.

DC Resins Using Allylthiourea/Cumyl Hydroperoxide

Figure 13 shows the Rockwell$_{15T}$ (Wilson 3JR Rockwell Hardness Tester) hardness of DC resins formulated using the allylthiourea/cumyl hydroperoxide (T/CH) self-cure initiator system and the PermaFlo light-cure side (PermaFlo LC). PermaFlo DC is significantly harder (80.1 ± 1.7 RHN, $p < 0.05$) than all experimental groups. Permaflo LC is also significantly harder (74.7 ± 5.7 RHN, $p < 0.05$) than all experimental groups except all PermaFlo LC + 1T

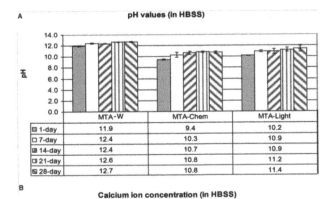

A

pH values (in HBSS)

	MTA-W	MTA-Chem	MTA-Light
1-day	11.9	9.4	10.2
7-day	12.4	10.3	10.9
14-day	12.4	10.7	10.9
21-day	12.6	10.8	11.2
28-day	12.7	10.8	11.4

B

Calcium ion concentration (in HBSS)

	MTA-W	MTA-Chem	MTA-Light
1-day	171.7	29.9	59.8
7-day	390.3	22.6	239.3
14-day	635.0	69.4	471.3
21-day	694.7	107.5	598.0
28-day	841.3	149.7	668.0

Fig. 12 (**A**) pH values over time and (**B**) calcium ion concentration of solution in parts per million of test materials (±SD).
Source: From Formosa, Mallia, and Camilleri[87] © 2009, with permission from Elsevier.

Dental—Electrospinning

groups and PermaFlo LC + 2T:1CH. Very few trends could be identified. At $T = 1$, there are no statistical differences between groups, but as T increases, increase in CH concentration seems to decrease hardness, implying that this DC initiator system is very complex and that the effect of increasing concentrations of initiator components on hardness and possibly degree of cure is not simply additive.[93]

DC Resins Using T/CH and OPPI

Figure 14 shows Rockwell$_{15T}$ hardness of samples 24 hours after light curing with the "Optimal" formulation (77.7 ± 1.4 RHN) is boxed on the graph. Rockwell hardness results show that PermaFlo DC and HSC LC controls had the highest hardness (80.18 ± 1.85 and 79.43 ± 1.23 RHN, respectively). However, the hardness of HSC LC dropped significantly ($p < 0.05$) to 76.05 ± 2.02 RHN when the 1T:5CH self-cure side is added (HSC LC + 1T:5CH) further showing the need to add OPPI. Most experimental groups had hardness values within the range of these controls. The "Optimal" group chosen below for color stability had an above average hardness of 77.7 ± 1.4 RHN and is not significantly lower than the HSC LC control or any other experimental group.

Three-Point Bending Flexural Test

Figure 15 shows the modulus of the control groups and the "Optimal" group. Results show that the use of the low color self-cure initiator system in HSC LC + 1T:5CH decreases the modulus from 10,543 ± 743 to 8737 ± 752 MPa, but this is not a statistically significant difference. However, addition of OPPI increased the modulus of the "Optimal" group (9381 ± 1684 MPa) to levels comparable to that of PermaFlo DC. Thus, OPPI is not only needed to further reduce color shift but also to enhance modulus to a

clinically acceptable level, which is crucial since low modulus can contribute to porcelain veneer fractures. There are no significant differences in ultimate transverse strength among the groups.

Fourier Transform Infrared Spectroscopy Degree of Conversion

Figure 16 shows FTIR DoC of the control and "Optimal" groups following initial cure and after 24-hour cure (DC formulations only). For HSC LC, addition of the self-cure component (HSC LC + 1T:5CH) caused the 60-s DoC to decrease significantly from 46.3 ± 2.7% to 30.1 ± 0.4%, but after 24 hours, it increased significantly to 46.3 ± 0.4%, making it comparable to HSC LC. With the "Optimal" group with OPPI, the initial cure is low (31.1 ± 3.5%), comparable to HSC LC + 1T:5CH, but final cure (49.5 ± 1.5%) after 24 hours is significantly higher than that of

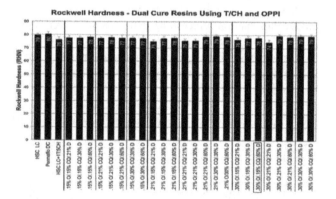

Fig. 14 Rockwell$_{15T}$ hardness of resins made with different concentrations of OPPI, CQ, and DMAEMA on the light-cure side and 5T:1CH on the self-cure side. The "Optimal" formulation is boxed on the graph.
Source: From Oei et al.[93] © 2013, with permission from Elsevier.

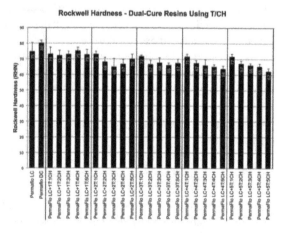

Fig. 13 Rockwell$_{15T}$ hardness of resins made with different concentrations of T and CH.
Source: From Oei et al.[93] © 2013, with permission from Elsevier.

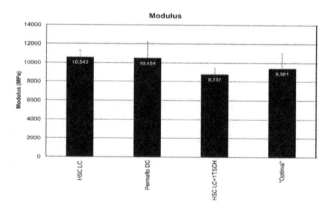

Fig. 15 Modulus of the three control groups and the "Optimal" group.
Source: From Oei et al.[93] © 2013, with permission from Elsevier.

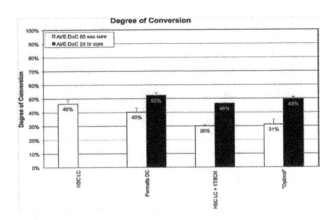

Fig. 16 FTIR degree of conversion of the three controls and the "Optimal" group after initial cure and after 24 hours after cure (DC formulations only).
Source: From Oei et al.[93] © 2013, with permission from Elsevier.

HSC LC + 1T:5CH and comparable to that of PermaFlo DC. Thus, OPPI allowed the use of a different CQ and DMAEMA combination and not only reduced color shift but also restored degree of conversion to levels comparable to that of the commercial resin control.[93]

In short, the feasibility of a novel DC system utilizing an alternative redox system and an additional onium-ion compound to improve color stability is described. The present concept utilizes a resin cement formulation, which could potentially be developed to improve color stability and reduce shrinkage stress in restorative composites, since DC resins are known to have lower shrinkage stress as compared to light-cure resins due to a slower rate of cure.[93]

More recently, we have also prepared some novel resins based on *p*-chloroacetophenone oxime, 8-hydroxyquinoline, vanillinoxime having good thermal stability and antimicrobial activities are found to be an excellent intermediate in dental-based applications.[94–98]

CONCLUSIONS

Dental polymers have found large applications in dental fillings, tooth coverings, dentures, and any other bisphenol A-glycidyl methacrylate resins. The fracture resistance of acrylic denture base, whether for conventional or implant-supported dentures, is one of the important factors in terms of safety, longevity, and patient satisfaction with removable prosthodontics. Dental composite resins are prevalent materials used to esthetically restore the structural integrity of teeth, generally impaired by caries, but also due to attrition, erosion, or fracture. Hence, diverse resinous materials are available, for instance adhesives (primer and bonding agents), flowable and conventional composite resins, fiber-reinforced composites, or resin cements. The compositions of dental formulation are mainly based on different methacrylate and acrlates. Effect of moisture on dentin surface

and curing mode of micro tensile bond strength and nono-indentation of resin cement are very useful characteristics. Recently, the use of HP/HT polymerization to produce UDMA–TEGDMA based RCB are suitable material for dental computer-aided design/manufacture (CAD/CAM) applications. Recent development of DC resins is very important in dental applications of polymers.

ACKNOWLEDGMENT

One of the authors, Dr. Narendra P. S. Chauhan is thankful to Elsevier for copyright permissions. Authors are also thankful to Prof. Suresh C. Ameta, Department of Chemistry, Pacific College of Basic & Applied Sciences, PAHER University, Udaipur (Raj.) for suggestions.

REFERENCES

1. Narva, K.K.; Lassila, L.V.; Vallittu, P.K. The static strength and modulus of fiber reinforced denture base polymer. Dent. Mater. **2005**, *21* (5), 421–428.
2. Rzayev, J.; Penelle, J. HP-RAFT: A free-radical polymerization technique for obtaining living polymers of ultrahigh molecular weights. Angew. Chem. Int. Ed. Engl. **2004**, *43* (13), 1691–1694.
3. Kwiatkowski, P.; Jurczak, J.; Pietrasik, J.; Jakubowski, W.; Mueller, L.; Matyjaszewski, K. High molecular weight polymethacrylates by AGET ATRP under high pressure. Macromolecules **2008**, *41* (4), 1067–1069.
4. Arita, T.; Kayama, Y.; Ohno, K.; Tsujii, Y.; Fukuda, T. High-pressure atom transfer radical polymerization of methyl methacrylate for well-defined ultrahigh molecular-weight polymers. Polym. **2008**, *49* (10), 2426–2429.
5. Rzayev, J.; Penelle, J. Controlled/living free-radical polymerization under very high pressure. Macromolecules **2002**, *35* (5), 1489–1490.
6. Natsuko, M.; Noriyuki, W.; Rie, M.; Akio K.; Yoshimasa, I. Effect of high-pressure polymerization on mechanical properties of PMMA denture base resin. J. Mech. Behav. Biomed. Mater. **2013**, *20*, 98–104.
7. Brosh, T.; Ferstand, N.; Cardash, H.; Baharav, H. Effect of polymerization under pressure on indirect tensile mechanical properties of light-polymerized composites. J. Prosthet. Dent. **2002**, *88* (4), 381–387.
8. Carmai, J.; Dunne, F.P.E. Constitutive equations for densification of matrix-coated fibre composites during hot isostatic pressing. Int. J. Prosthodont. **2003**, *19*, 345–363.
9. Nguyen, J.F.; Migonney, V.; Ruse, N.D.; Sadoun, M. Resin composite blocks via high-pressure high-temperature polymerization. Dent. Mater. **2012**, *28* (5), 529–534.
10. Ahmed, J.; Zhang, J.X.; Song, Z.; Varshney, S.K. Thermal properties of polyactides effect of molecular mass and nature of lactide isomer. J. Therm. Anal. Calorim. **2009**, *95* (3), 957–964.
11. Sasuga, T.; Takehisa, M. Pressure-volume behavior of PMMA-MMA coexistence system as polymerized at high pressure. J. Macromol. Sci. Chem. **1978**, *A12* (9), 1321–1331.

Dental—Electrospinning

12. Cramer, N.B.; Stansbury, J.W.; Bowman, C.N. Recent advances and developments in composite dental restorative materials. J. Dent. Res. **2011**, *90* (4), 402–416.

13. Van Landuyt, K.L.; Snauwaert, J.; De Munck, J.; Peumans, M.; Yoshida, Y.; Poitevin, A. Systematic review of the chemical composition of contemporary dental adhesives. Biomaterials **2007**, *28* (26), 3757–3785.

14. Schweikl, H.; Hartmann, A.; Hiller, K.A.; Spagnuolo, G.; Bolay, C.; Brockhoff, G. Inhibition of TEGDMA and HEMA-induced genotoxicity and cell cycle arrest by N-acetyl cysteine. Dent. Mater. **2007**, *23* (6), 688–695.

15. Bolling, A.K.; Samuelsen, J.T.; Morisbak, E.; Ansteinsson, V.; Becher, R.; Dahl, J.E. Dental monomers inhibit LPS-induced cytokine release from the macrophage cell line RAW264.7. Toxicol. Lett. **2013**, *216* (2–3), 130–138.

16. Galler, K.; Schweikl, H.; Hiller, K.A.; Cavender, A.; Bolay, C.; D'Souza, R. TEGDMA reduces the expression of genes involved in biomineralization. J. Dent. Res. **2011**, *90* (2), 257–262.

17. Schweikl, H.; Spagnuolo, G.; Schmalz, G. Genetic and cellular toxicology of dental resin monomers. J. Dent. Res. **2006**, *85* (10), 870–877.

18. Genestra, M. Oxyl radicals, redox-sensitive signalling cascades and antioxidants. Cell Signal. **2007**, *19* (9), 1807–1819.

19. Pervaiz, S.; Taneja, R.; Ghaffari, S. Oxidative stress regulation of stem and progenitor cells. Antioxid. Redox Signal. **2009**, *11* (11), 2777–2789.

20. D'Autréaux, B.; Toledano, M.B. ROS as signalling molecules: Mechanisms that generate specificity in ROS homeostasis. Nat. Rev. Mol. Cell. Biol. **2007**, *8* (10), 813–824.

21. Circu, M.L.; Aw, T.Y. Reactive oxygen species, cellular redox systems, and apoptosis. Free Radic. Biol. Med. **2010**, *48* (6), 749–762.

22. Lu, S.C. Regulation of glutathione synthesis. Mol. Aspects Med. **2009**, *30* (1), 42–59.

23. Jones, D.P. Redefining oxidative stress. Antioxid. Redox Signal. **2006**, *8* (9–10), 1865–1879.

24. Walther, U.I.; Siagian, I.I.; Walther, S.C.; Reichl, F.X.; Hickel, R. Antioxidative vitamins decrease cytotoxicity of HEMA and TEGDMA in cultured cell lines. Arch. Oral. Biol. **2004**, *49* (2), 125–131.

25. Samuelsen, J.T.; Kopperud, H.M.; Holme, J.A.; Dragland, I.S.; Christensen, T.; Dahl, J.E. Role of thiol-complex formation in 2-hydroxyethyl-methacrylate-induced toxicity *in vitro*. J. Biomed. Mater. Res. A **2011**, *96* (2), 395–401.

26. Miller, A.F. Superoxide dismutases: Ancient enzymes and new insights. Febs. Lett. **2012**, *586* (5), 585–595.

27. Zamocky, M.; Furtmüller, P.G.; Obinger, C. Evolution of catalases from bacteria to humans. Antioxid. Redox Signal. **2008**, *10* (9), 1527–1548.

28. Griffith, O.W.; Meister, A. Potent and specific inhibition of glutathione synthesis by buthionine sulfoximine (S-n-butyl homocysteine sulfoximine). J. Biol. Chem. **1979**, *254* (16), 7558–7560.

29. Williamson, J.M.; Meister, A. Stimulation of hepatic glutathione formation by administration of L-2-oxothiazolidine-4-carboxylate, a 5-oxo-L-prolinase substrate. Proc. Natl. Acad. Sci. USA **1981**, *78* (2), 936–939.

30. Krifka, S.; Hiller, K.A.; Spagnuolo, G.; Jewett, A.; Schmalz, G.; Schweikl, H. The influence of glutathione on redox regulation by antioxidant proteins and apoptosis in macro-phages exposed to 2-hydroxyethyl methacrylate (HEMA). Biomaterials **2012**, *33* (21), 5177–5186.

31. Schweikl, H.; Hiller, K.A.; Eckhardt, A.; Bolay, C.; Spagnuolo, G.; Stempfl, T. Differential gene expression involved in oxidative stress response caused by triethylene glycol dimethacrylate. Biomaterials **2008**, *29* (10), 1377–1387.

32. De Vries, N.; De Flora, S. N-acetyl-l-cysteine. J. Cell. Biochem. Suppl. **1993**, *17F*, 270–277.

33. Zafarullah, M.; Li, W.Q.; Sylvester, J.; Ahmad, M. Molecular mechanisms of N acetyl-cysteine actions. Cell Mol. Life Sci. **2003**, *60* (1), 6–20.

34. Paranjpe, A.; Cacalano, N.A.; Hume, W.R.; Jewett, A. N-acetyl cysteine mediates protection from 2-hydroxyethyl methacrylate induced apoptosis via nuclear factor kappa B-dependent and independent pathways: Potential involvement of JNK. Toxicol. Sci. **2009**, *108* (2), 356–366.

35. Runchel, C.; Matsuzawa, A.; Ichijo, H. Mitogen-activated protein kinases in mammalian oxidative stress responses. Antioxid. Redox Signal. **2011**, *15* (1), 205–218.

36. Keshet, Y.; Seger, R. The MAP kinase signaling cascades: A system of hundreds of components regulates a diverse array of physiological functions. Methods Mol. Biol. **2010**, *661*, 3–38.

37. Whitmarsh, A.J. Regulation of gene transcription by mitogen-activated protein kinase signaling pathways. Biochim. Biophys. Acta **2007**, *1773* (8), 1285–1298.

38. Samuelsen, J.T.; Dahl, J.E.; Karlsson, S.; Morisbak, E.; Becher, R. Apoptosis induced by the monomers HEMA and TEGDMA involves formation of ROS and differential activation of the MAP-kinases p38, JNK and ERK. Dent. Mater. **2007**, *23* (1), 34–39.

39. Eckhardt, A.; Gerstmayr, N.; Hiller, K.A.; Bolay, C.; Waha, C.; Spagnuolo, G. TEGDMA-induced oxidative DNA damage and activation of ATM and MAP Kinases. Biomater. **2009**, *30* (11), 2006–2014.

40. Fox, G.C.; Shafiq, M.; Briggs, D.C.; Knowles, P.P.; Collister, M.; Didmon, M.J. Redox-mediated substrate recognition by Sdp1 defines a new group of tyrosine phosphatases. Nature. **2007**, *447* (7143), 487–492.

41. Hegedus, Z.; Czibula, A.; Kiss-Toth, E. Tribbles: Novel regulators of cell function; evolutionary aspects. Cell Mol. Life Sci. **2006**, *63* (14), 1632–1641.

42. Krifka, S.; Petzel, C.; Bolay, C.; Hiller, K.A.; Spagnuolo, G.; Schmalz, G. Activation of stress-regulated transcription factors by triethylene glycol dimethacrylate monomer. Biomaterials **2011**, *32* (7), 1787–1795.

43. Bhoumik, A.; Lopez-Bergami, P.; Ronai, Z. ATF2 on the double–activating transcription factor and DNA damage response protein. Pigment Cell Res. **2007**, *20* (6), 498–506.

44. Imlay, J.A. Cellular defenses against superoxide and hydrogen peroxide. Annu. Rev. Biochem. **2008**, *77*, 755–776.

45. Wilson, D.M.; Sofinowski, T.M.; McNeill, D.R. Repair mechanisms for oxidative DNA damage. Front Biosci. **2003**, *8*, 963–981.

46. Berquist, B.R.; Wilson, D.M. Pathways for repairing and tolerating the spectrum of oxidative DNA lesions. Cancer Lett. **2012**, *327* (1), 61–72.

47. Blasiak, J.; Synowiec, E.; Tarnawska, J.; Czarny, P.; Poplawski, T.; Reiter, R.J. Dental methacrylates may exert genotoxic effects via the oxidative induction of DNA double strand breaks and the inhibition of their repair. Mol. Biol. Rep. **2012**, *39* (7), 7487–7496.

48. Kleinsasser, N.H.; Schmid, K.; Sassen, A.W.; Harréus, U.A.; Staudenmaier, R.; Folwaczny, M. Cytotoxic and genotoxic effects of resin monomers in human salivary gland tissue and lymphocytes as assessed by the single cell microgel electrophoresis (Comet) assay. Biomaterials 2006, 27 (9), 1762–1770.

49. Jones, R.M.; Petermann, E. Replication fork dynamics and the DNA damage response. Biochem. J. 2012, 443 (1), 13–26.

50. Tanaka, T.; Huang, X.; Halicka, H.D.; Zhao, H.; Traganos, F.; Albino, A.P. Cytometry of ATM activation and histone H2AX phosphorylation to estimate extent of DNA damage induced by exogenous agents. Cytometry A 2007, 71 (9), 648–661.

51. Kruse, J.P.; Gu, W. Modes of p53 regulation. Cell 2009, 137 (4), 609–622.

52. Miyazaki, T.; Hotta, Y.; Kunii, J.; Kuriyama, S.; Tamaki, Y. A review of dental CAD/CAM: Current status and future perspectives from 20 years of experience. Dent. Mater. J. 2009, 28 (1), 44–56.

53. Giordano, R. Materials for chair side CAD/CAM-produced restorations. J. Am. Dent. Assoc. 2006, 137 (Suppl. 1), 14S–21S.

54. Nguyen, J.F.; Migonney, V.; Ruse, N.D.; Sadoun, M. Resin composite blocks via high-pressure high-temperature polymerization. Dent. Mater. 2012, 28 (5), 534–592.

55. Nguyen, J.F.; Migonney, V.; Ruse, N.D.; Sadoun, M. Properties of experimental urethane dimethacrylate-based dental resin composite blocks obtained via thermo-polymerization under high pressure. Dent. Mater. 2013, 29 (5), 535–541.

56. Bona, A.D.; Anusavice, K.J.; DeHoff, P.H. Weibull analysis and flexural strength of hot-pressed core and veneered ceramic structures. Dent. Mater. 2003, 19 (7), 662–669.

57. Barszczewska-Rybarek, I.M. Structure–property relationships in dimethacrylate networks based on Bis-GMA, UDMA and TEGDMA. Dent. Mater. 2009, 25 (9), 1082–1089.

58. Okay, O.; Naghash, H.J.; Capek, I. Free-radical cross-linking copolymerization—Effect of cyclization on diffusion-controlled termination at low conversion. Polymer 1995, 36 (12), 2413–2419.

59. Rey, L.; Duchet, J.; Galy, J.; Sautereau, H.; Vouagner, D.; Carrion, L. Structural heterogeneities and mechanical properties of vinyl/dimethacrylate networks synthesized by thermal free radical polymerisation. Polymer 2002, 43 (16), 4375–4384.

60. Nurrohman, H.; Nikaido, T.; Sadr, A.; Takagaki, T.; Kitayama, S.; Ikeda, M. Long-term regional bond strength of three MMA-based adhesive resins in simulated vertical root fracture. Dent. Mater. J. 2011, 30 (5), 655–663.

61. Hikita, K.; Van Meerbeek, B.; De Munck, J.; Ikeda, T.; Van Landuyt, K.; Maida, T. Bonding effectiveness of adhesive luting agents to enamel and dentin. Dent. Mater. 2007, 23 (1), 71–80.

62. Radovic, I.; Monticelli, F.; Goracci, C.; Vulicevic, Z.R.; Ferrari, M. Self-adhesive resin cements: A literature review. J. Adhes. Dent. 2008, 10 (4), 251–258.

63. Marshall, G.W.; Marshall, S.J.; Kinney, J.H.; Balooch, M. The dentin substrate: Structure and properties related to bonding. J. Dent. 1997, 25 (6), 441–458.

64. Van Landuyt, K.L.; Yoshida, Y.; Hirata, I.; Snauwaert, J.; De Munck, J.; Okazaki, M. Influence of the chemical structure of functional monomers on their adhesive performance. J. Dent. Res. 2008, 87 (8), 757–761.

65. Bolhuis, P.B.; de Gee, A.J.; Kleverlaan, C.J.; El Zohairy, A.A.; Feilzer, A.J. Contraction stress and bond strength to dentin for compatible and incompatible combinations of bonding systems and chemical and light-cured core build-up resin composites. Dent. Mater. 2006, 22 (3), 223–233.

66. Moosavi, H.; Haririb, I.; Sadrb, A.; Thitthaweeratc, S.; Tagami, J. Effects of curing mode and moisture on nanoindentation mechanical properties and bonding of a self-adhesive resin cement to pulp chamber floor. Dent. Mater. 2013, 29 (6), 708–717.

67. Takahashi, R.; Nikaido, T.; Ariyoshi, M.; Foxton, R.M.; Tagami, J. Microtensile bond strengths of a dual-cure resin cement to dentin resin-coated with an all-in-one adhesive system. Dent. Mater. J. 2010, 29 (3), 268–276.

68. Scherrer, S.S.; Cesar, P.F.; Swain, M.V. Direct comparison of the bond strength results of the different test methods: A critical literature review. Dent. Mater. 2010, 26 (2), 78–93.

69. Kelly, J.R. Clinically relevant approach to failure testing of all ceramic restorations. J. Prosthet. Dent. 1999, 81 (6), 652–661.

70. Quinn, J.B.; Quinn, G.D.; Kelly, J.R.; Scherrer, S.S. Fractographic analyses of three ceramic whole crown restoration failures. Dent. Mater. 2005, 21 (10), 920–929.

71. Musanje, L.; Darvell, B.W. Effects of strain rate and temperature on the mechanical properties of resin composites. Dent. Mater. 2004, 20 (8), 750–765.

72. Preis, V.; Behr, M.; Hahnel, S.; Handel, G.; Rosentritt, M. In vitro failure and fracture resistance of veneered and full-contour zirconia restorations. J. Dent. 2012, 40 (11), 921–928.

73. Lee, C.S.; Kim, D.K.; Sanchez, J.; Miranda, P.; Pajares, A.; Lawn, B.R. Rate effects in critical loads for radial cracking in ceramic coatings. J. Am. Ceram. Soc. 2002, 85 (8), 2019–2024.

74. Charles, R.J. Static fatigue of glass II. J. App. Phys. 1958, 29 (11), 1554–1560.

75. Hooi, P.; Addison, O.; Fleming, G.J.P. Testing rate and cementation seating load effects on resin-strengthening of a dental porcelain analogue. J. Dent. 2013, 41 (6), 514–520.

76. Balbosh, A.; Kern, M. Effect of surface treatment on retention of glass-fiber endodontic posts. J. Prosthet. Dent. 2006, 95 (3), 218–223.

77. Vano, M.; Goracci, C.; Monticelli, F.; Tognini, F.; Gabriele, M.; Tay, F.R. The adhesion between fibre posts and composite resin cores: The evaluation of microtensile bond strength following various surface chemical treatments to posts. Int. Endod. J. 2006, 39 (1), 31–39.

78. Monticelli, F.; Grandini, S.; Goracci, C.; Ferrari, M. Clinical behavior of translucent-fiber posts: A 2-year prospective study. Int. J. Prosthodont. 2003, 16 (6), 593–596.

79. Zalkind, M.; Shkury, S.; Stern, N.; Heling, I. Effect of prefabricated metal post-head design on the retention of various core materials. J. Oral Rehabil. 2000, 27 (6), 483–487.

80. Zicari, F.; De Munck, J.; Scotti, R.; Naert, I.; Van Meerbeek, B. Factors affecting the cement-post interface. Dent. Mater. 2012, 28 (3), 287–297.

81. de Sousa Menezes, M.; Queiroz, E.C.; Soares, P.V.; Faria-e-Silva, A.L.; Soares, C.J.; Martins, L.R. Fiber post etching with hydrogen peroxide: Effect of concentration and application time. J. Endod. 2011, 37 (3), 398–402.

82. Minami, H.; Suzuki, S.; Minesaki, Y.; Kurashige, H.; Tanaka, T. *In vitro* evaluation of the influence of repairing condition of denture base resin on the bonding of auto-polymerizing resins. J. Prosthet. Dent. **2004**, *91* (2), 164–170.

83. Elsaka, S.E. Influence of chemical surface treatments on adhesion of fiber posts to composite resin core materials. Dent. Mater. **2013**, *29* (5), 550–558.

84. Sarrett, D.C. Clinical challenges and the relevance of materials testing for posterior composite restorations. Dent. Mater. **2005**, *21* (1), 9–20.

85. Reyes-Carmona, J.F.; Felippe, M.S.; Felippe, W.T. Biomineralization ability and interaction of mineral trioxide aggregate and white Portland cement with dentin in a phosphate-containing fluid. J. Endod. **2009**, *35* (5), 731–736.

86. Parirokh, M.; Torabinejad, M. Mineral trioxide aggregate: A comprehensive literature review. Part I. Chemical, physical, and antibacterial properties. J. Endod. **2010**, *36* (1), 16–27.

87. Formosa, L.M.; Mallia, B.; Camilleri, J. The chemical properties of light- and chemical-curing composites with mineral trioxide aggregate filler. Dent. Mater. **2013**, *29* (2), e11–e19.

88. Lee, S.J.; Monsef, M.; Torabinejad, M. Sealing ability of a mineral trioxide aggregate for repair of lateral root perforations. J. Endod. **1993**, *19* (11), 541–544.

89. Sarkar, N.K.; Caicedo, R.; Ritwik, P.; Moiseyeva, R.; Kawashima, I. Physiochemical basis of the biologic properties of mineral trioxide aggregate. J. Endod. **2005**, *31* (2), 97–100.

90. Duarte, M.A.; Demarchi, A.C.; Yamashita, J.C.; Kuga, M.C.; Fraga Sde, C. pH and calcium ion release of 2 root-end filling materials. Oral Surg. Oral Med. Oral Pathol. Oral Radiol. Endod. **2003**, *95* (3), 345–347.

91. Chung, H.; Kim, M.; Ko, H.; Yang, W. Evaluation of physical and biologic properties of the mixture of mineral trioxide aggregate and 4-META/MMA-TBB resin. Oral Surg. Oral Med. Oral Pathol. Oral Radiol. Endod. **2011**, *112* (5), e6–e11.

92. Sarafianou, A.; Iosifidou, S.; Papadoupoulos, T.; Eliades, G. Color stability and degree of cure of direct composite restoratives after accelerated aging. Oper. Dent. **2007**, *32* (4), 406–411.

93. Oei, J.D.; Mishriky, M.; Barghi, N.; Rawls, H.R.; Cardenas, H.L.; Aguirre, R.; Whang, K. Development of a low-color, color stable, dual cure dental resin. Dent. Mater. **2013**, *29* (4), 405–412.

94. Chauhan, N.P.S. Structural and thermal characterization of macro-branched functional terpolymer containing 8-hydroxyquinoline moieties with enhancing biocidal properties. J. Ind. Eng. Chem. **2013**, *19* (3), 1014–1023.

95. Chauhan, N.P.S.; Ameta, R.; Ameta, S.C. Synthesis, characterization, and thermal degradation of pchloroacetophenone oxime based polymers having biological activities. J Appl. Polym. Sci. **2011**, *122* (1), 573–585.

96. Chauhan, N.P.S.; Ameta, S.C. Preparation and thermal studies of self-crosslinked terpolymer derived from 4-acetyl-pyridine oxime, formaldehyde and acetophenone. Polym. Degrad. Stabil. **2011**, *96* (8), 1420–1429.

97. Chauhan, N.P.S. Terpolymerization of p-acetylpyridine oxime, p-methylacetophenone and formaldehyde, and its thermal studies. J. Therm. Anal. Calorim. **2012**, *110* (3), 1377–1388.

98. Chauhan, N.P.S.; Ameta, R.; Punjabi, P.B.; Ameta, S.C. Synthesis, characterization, and antimicrobial properties of p-chloroacetophenone oxime based furan resins. Int. J. Polym. Mater. **2011**, *61* (14), 57–71.

Dental Sealants

Yasuhiko Tsuchitani
Faculty of Dentistry, Osaka University, Osaka, Japan

Tohru Wada
Kuraray Company Ltd., Tokyo, Japan

Abstract

This entry introduces advancements in the synthetic chemistry and polymer science of dental sealants and discusses those designed for each clinical use, to prolong the lifetime service of natural teeth.

Dental sealants are used to seal high caries–susceptible pits and fissures of the deciduous and permanent molars, and also to seal microspaces between the tooth and restorative materials, enabling those materials to adhere firmly both to prepared cavity walls and to other restorative materials. They provide protection from secondary caries and dental pulp involvements.

Most dental sealants are resinous materials, but glass ionomer dental cements have some limited use as sealing materials.

Prevalence and incidence of dental caries depend on many variables, but incipient caries is frequently detected in three principal locations on teeth: in occlusal pits and fissures (Fig. 1, left), interproximal surfaces under the contact points, and cervical areas (enamel and exposed root dentin). These areas on the teeth are difficult to clean by brushing.

Obturation or sealing of such areas with resinous materials or glass ionomer cements is one modern preventive technology; the sealants used for this purpose are called preventive dental sealants (PDS).

Dental caries that occurs around restoration is clinically called secondary caries. Its prevalence is largely due to the microspaces existing between restorative material and the tooth cavity wall (Fig. 1, right). Sealing the microspaces with adhesive resinous materials is effective in controlling secondary caries; here we call these adhesive materials restorative dental sealants (RDS).

Here we'll introduce advancements in synthetic chemistry and polymer science of dental sealants and discuss those designed for each clinical use, to prolong the lifetime service of natural teeth.

PREVENTIVE DENTAL SEALANTS

Preventive dental sealants used to seal the susceptible areas of teeth are classified into pit and fissure sealants and smooth surface sealants, depending on the part to be sealed. From the viewpoint of material science, pit and fissure sealants can be further classified into resin sealants and glass ionomer cements. Preventive dental sealants are usually placed onto molar teeth of infants and small children who are at higher risk for caries because of inadequate care.

PIT AND FISSURE SEALANTS

Resin Sealants

The chemical nature of resin sealants is a radical-polymerizable monomer mixture that has a viscosity low enough to penetrate into narrow pits and fissures easily and can be cured in them to become a hard and durable sealing material. Bis-phenol-A-diglycidylmethacrylate (bis-GMA), which has bisphenol A structure in its molecule, bis-GMA homologues, and urethane dimethacrylate are very popular as the main monomer for resin sealants, and they are usually used together with other monomers that are less viscous in order to increase their penetration ability (Fig. 2). Pits and fissures are treated with phosphoric acid, citric acid, or other acidic agents prior to application of sealants. This treatment, often called "acid etching," changes the enamel surface to a micro-rugged structure (Fig. 3), which aids penetration of sealants into fissures and results in strong micro-mechanical adhesion between sealants and tooth substances. Acidic monomers have been introduced into the chemical formulation of sealants to further enhance the adhesion and penetration into fissures.

Polymerization of sealants is initiated by redox catalyst or light irradiation. In the former case, benzoylperoxide/ *t*-amine is most common, and sulfinates also are used when acidic monomers are contained.[2] Commercial products are composed of two liquid components: One has

Concise Encyclopedia of Biomedical Polymers and Polymeric Biomaterials DOI: 10.1081/E-EBPPC-120051892

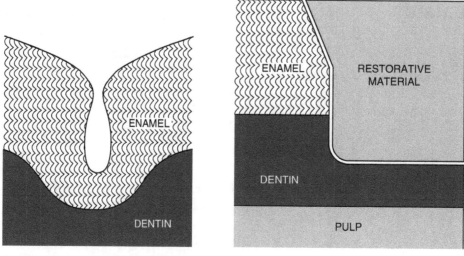

Fig. 1 The typical occlusal fissure of human molar (longitudinal section). Deep invagination is frequently seen (left). Microspaces exist between restored material and the cavity wall (longitudinal section, right).

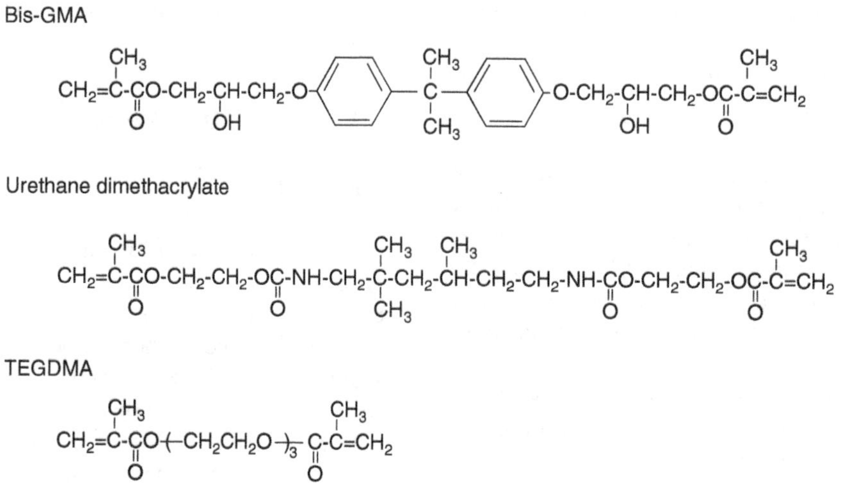

Bis-GMA

Urethane dimethacrylate

TEGDMA

Fig. 2 Monomers used for dental sealants and dental restorative materials. (Bis-GMA is bis-phenol-A-diglycidylmethacrylate; TEGDMA is triethyleneglycol dimethacrylate).

Fig. 3 SEM photographs of unetched (left) and etched (right) enamel surface.

benzoylperoxide and the other has *t*-amine. In the latter case, visible light is irradiated to initiate radical polymerization. Camphorquinone is the most popular photosensitizer and is usually used together with some kinds of amines.[3] Light-cured sealants have one liquid component.

Manipulation of pit and fissure resin sealants is simple. Once one drop of a sealant is placed onto any part of acid-etched pit and fissure, sealant quickly runs along fissure and penetrates into it by capillary action. Photopolymerizable sealants can be cured by 10–30 seconds of visible light irradiation, and self-curing sealants harden within 3–5 minutes after the two liquid components are mixed.

There are some sealants that can release fluoride ion gradually. Fluoride ion converts hydroxyapatite, the major inorganic component of teeth to fluoroapatite, which is more acid resistant than hydroxyapatite, reducing caries susceptability.

Fluoride release technology, which is increasingly used, requires application of microencapsulated sodium fluoride and application of copolymer of methyl methacrylate and methacrylic fluoride.[4] In the latter case, acid fluoride is embedded in the cured sealant; it reacts with water, which slowly permeates into the polymer network of the sealant and releases fluoride ion gradually.

Glass Ionomer Cements

Glass ionomer cements basically consist of the powder of fluoro aluminosilicate glass and the aqueous solution of poly(acrylic acid).[5] Typical glass powder contains SiO_2, Al_2O_3, CaO, Na_2O, and CaF_2. Poly(acrylic acid) sometimes contains maleic acid or itaconic acid as a comonomer.[6] Glass ionomer cements are widely used, and their types vary with the dental treatment. Glass ionomer cements for pit and fissure sealing are characterized by low viscosity of the cement paste and fast setting.

In this process glass powder and polymer solution are mixed by a given proportion and packed into pits and fissures without prior acid etching of tooth surfaces. The cement paste cures within a short while through the ion bridge formation between poly(acrylic acid) and polyvalent cations (Ca^{2+}, Al^{3+}) leaching out from the glass powder. Polyacrylic acid undergoes ionic interaction with calcium ions on the surface of dental enamel and exhibits the bond of 2–4 MPa.

Concentrations of poly(acrylic acid) in solutions are usually about 50% and cured glass ionomers contain a considerable amount of water. Fluoro aluminosilicate glasses slowly release fluoride ion in situ and improve the acid resistance of surrounding dental enamel.

SMOOTH SURFACE SEALANTS

Smooth surface sealants are used to seal caries-susceptible tooth surfaces other than pits and fissures in order to prevent them from bacteria, staining, and physical damage. These surfaces are the cervical area, interproximal area, and exposed root surfaces.

Cured sealants are directly exposed to the oral conditions and therefore strong adhesion to enamel and dentin, good physical properties, good wear resistance, chemical stability, biological stability, and thin film formation are required. Glass ionomer cement cannot fulfill all of these requirements, because of its insufficient physical properties and thick film thickness. Instead, highly crosslinkable resin sealants reinforced with ultrafine silica particles (Aerozil) are used.

RESTORATIVE DENTAL SEALANTS

Amalgam Restoration

Dental amalgam is a restorative material composed of mercury and powder of silver alloy. These two components are mixed and plastic mass is condensed to cavities of molar teeth. The plastic mass hardens through the amalgamation process. There are no chemical interactions between the amalgam and the cavity wall, and irregular microspaces are produced along the interface between them (Fig. 4, top). Amalgam sealant and amalgam bond are used to seal the microspace and control the incidence of secondary caries around the amalgam restoration.

Amalgam Sealant

This sealant was developed by Tsuchitani and co-workers.[8] Chemical composition of this material is as follows:

- monomer: diethyleneglycol dimethacrylate or *o*-mono-methacryloxyethyl phthalate (MEP)
- catalyst: *tert*-butyl hydroperoxide
- accelerator: *o*-sulfobenzimide; and
- inhibitor: hydroquinone.[7]

A primary characteristic of this sealant is that it is composed of only one liquid, which remains uncured before application, but cures quickly once applied to amalgam restoration. *t*-Butyl hydroperoxide and *o*-sulfobenzimide polymerize this sealant anaerobically when it is cut off from air and it contacts with the copper in amalgam. When this sealant is placed between two glass plates (at 25°C), it takes more than 10 minutes to cure. But it cures within 2 minutes between a glass plate and amalgam.

The second characteristic is that methacryloyloxy ethyl acid phthalate (MEP) is combined to enhance penetration of the sealant into microspaces and adhesion both to amalgam and tooth substaces (Fig. 4, bottom). This sealant bonds amalgam and enamel by 1.0 MPa and bonds amalgam and dentin by 2.6 MPa without acid etching.

Dental–Electrospinning

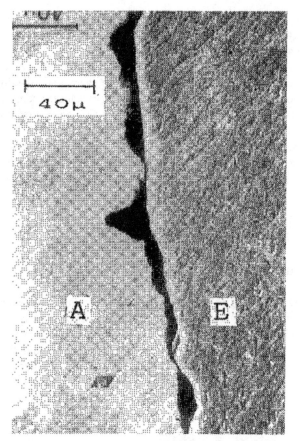

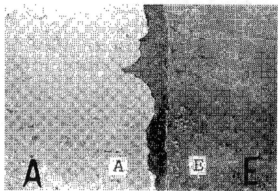

Fig. 4 SEM photographs of the interface between amalgam (A) and enamel (E) (top), and that sealed with amalgam sealants (bottom).

apparatus, in which *Streptococcus mutans*, one caries-inducing bacteria, is cultivated.[8] Development of caries is detected by X-ray photography and histochemical staining. Figure 5 shows results of one study. Decalcification caused by the acids *S. mutans* produces is observed in the surface enamel and around the dento-enamel junction on the tooth without sealant. But such decalcification around the dental enamel junction is absent where the amalgam sealant was applied. The results demonstrate the efficacy of this treatment in caries control.

Amalgam Bonds

Dental adhesives also are used to seal microspaces in amalgam restoration. These adhesives, called amalgam bonds, are of two types: adhesive resin cements, and bonding agents originally developed for composite restorations (which are discussed later).

One commonly used adhesive resin cement is Amalgambond (Sun Medical Company, Japan), which consists of poly(methylmethacrylate) (PMMA) powder, methyl methacrylate (MMA) monomer containing 4-methacryloyloxyethyl trimellitic anhydride[9] (4-META, Fig. 6) and tributylborane oxide (TBB-0) as catalyst. TBB-0 is activated by the water on the surface of adherents and promotes graft polymerization of MMA onto collagen in dentin, resulting in strong adhesion.[10]

Another adhesive cement, Panavia 21 (Kuraray Company, Japan) is a sort of composite cement and is characterized by the adhesive component, methacryloyloxydecyl acid phosphate[11] (MDP, Fig. 6). This adhesive cement consists of a primer and a composite paste. The primer is a monomer solution containing MDP, which develops strong adhesion to enamel and dentin. The composite paste is a mixture of inorganic filler and polymerizable monomers including MDP, which cures through the polymerization caused by the catalyst of BP0/*t*-amine/sulfinate. Panavia 21 exhibits strong adhesion to tooth substances and amalgam, does not deteriorate in wet conditions and provides a reliable seal (Table 1).

The standard procedure for bonded amalgam restoration using Panavia 21 begins with cavity preparation, followed by washing and air drying. Then the entire cavity wall is painted with the primer. Next, the paste is placed onto the primer as a thin and uniform layer, and the cavity is filled tightly with amalgam. Finally, any cement paste that overflowed the cavity is removed.

RDS in Composite Restoration

Dental composite restoration is a method to repair areas deteriorated by caries with powder-reinforced composite materials that consist essentially of dimethacrylate monomers and silanated inorganic fillers. This dental composite is directly placed into cavities, cures there and serves as a

Amalgam sealant applied along the margin of amalgam restoration penetrates into microspaces by capillary action and cures in situ within a few minutes. If the microspace is filled with water, air drying is to be applied. *n*-Butanol is also a good drying agent. The scanning electron micrography (SEM) photographs in Fig. 4 show a section of the interface of amalgam and cavity wall with (bottom) and without (top) sealant. The microspace is well sealed.

The efficacy of this sealant to prevent the secondary caries can be evaluated by an artificial caries production

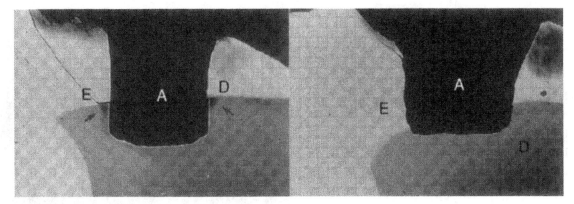

Fig. 5 Microradiographs of amalgam restorations sealed with (right) and without (left) amalgam sealant (A, amalgam; E, enamel; D, dentin). Decalcification in dentin is seen in the unsealed tooth (at arrow). No sign of decalcification is seen in the sealed tooth.

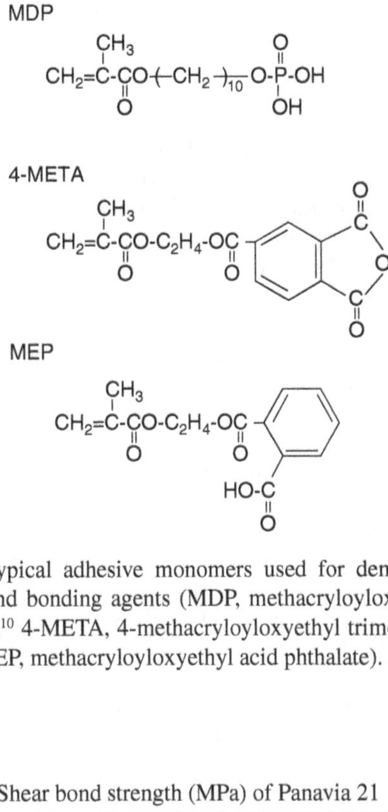

MDP

4-META

MEP

Fig. 6 Typical adhesive monomers used for dental sealants, cements and bonding agents (MDP, methacryloyloxydecyl acid phosphate;[10] 4-META, 4-methacryloyloxyethyl trimetallic anhydride;[11] MEP, methacryloyloxyethyl acid phthalate).

Table 1 Shear bond strength (MPa) of Panavia 21

| | Storage condition | |
Substance	24 h in water (37°C)	3000 TC[a]
Dentin	20.0 (5.8)[b]	20.9 (3.5)
Enamel	37.0 (3.5)	35.0 (3.9)
Amalgam	15.7 (3.3)	12.2 (1.5)
Ni–Cr[c]	47.4 (6.3)	45.1 (5.5)

[a]Thermal cycles between 4°C and 60°C.
[b]Standard deviation.
[c]Sand blasted with 50 μm Al_2O_3.
Source: From Omura et al.[11]

restorative with excellent physical properties. Cured composite is semitranslucent and matches the tooth color well. Wear resistance against tooth brushing and chewing is also very good. Dental composites are the most popular restoratives for anterior and posterior teeth in modern dentistry.

Bis-GMA, Bis-GMA homologues, and urethane dimethacrylate (Fig. 2) are common as monomers, and quartz, glass, and ceramics are used as fillers. The average size of filler particles is 1–10 μm. Redox catalyst (BPO/t-amine) or visible light polymerization catalyst (camphorquinone) is used for curing.

These composites exhibit polymerization shrinkage of 1–5 vol% in the cavity, and microspaces between composite restorations and cavity walls are observed.

The sealants used for restorations (called bonding agents) are applied onto the cavity walls prior to the placement of composites and bond the two substances tightly.

Chemical Nature of Bonding Agent

A bonding agent is a mixture of polymerizable monomers or a solution of monomers in volatile solvents, which cures through radical polymerization on the cavity walls or between cavity wall and composite restorations. Liquid monomers, such as methyl methacrylate, hydroxy ethyl methacrylate, and ethyleneglycol dimethacrylate, exhibit strong adhesion to dry teeth, but cured polymers are easily peeled off of teeth when the specimens are held in water for one day.

Tooth cavity walls are wet, and the application of acidic monomers is common for getting a reliable adhesion.

Monomer composition is selected for a balance of hydrophilicity and hydrophobicity, physical properties of the cured polymers, adhesion to composite restorations, curing time, and so on. Polymerization catalysts are similar to those for pit and fissure sealant.[13]

Table 2 Tensile bond strength of commercial bonding agents to dentin and enamel

Product (Year on market)	Bond strength (MPa)	
	Dentin	Enamel
Clearfil Bond (1978)	7.8 (–)	13.7 (–)
Clearfil New Bond (1984)	12.7 (–)	18.6 (–)
Clearfil Photo-Bond (1985)	11.6 (1.7)	16.0 (5.7)
Clearfil Liner Bond (1991)	17.3 (2.6)	19.0 (4.0)
Clearfil Liner Bond II (1993)	23.3 (2.6)	23.1 (3.8)

Source: From Wada et al.[14]

Clinical Procedures

Pretreatment of cavity walls with bonding agents is used for composite restorations worldwide, and many commercial products are on the market, such as Clearfil Photo Bond (Japan), Scotch Bond (United States), and Prisma Universal Bond (United States).

The typical procedure for composite restoration with a light-cured bonding agent and a light-cured composite is cavity preparation followed by washing and air drying, acid treatment of whole cavity wall, application of bonding agent, visible light irradiation, filling of composite restoration, visible light irradiation, and finishing and polishing.

Researchers have improved this procedure by using acidic monomer in the bonding agent as the acid for etching. Combining the etching and bonding agents reduces the number of steps and gives higher bonding strength to the dentin. Cleafil Liner Bond II, Scotch Bond Multi-Purpose, and Prisma Universal Bond 3 belong to this new category.

Efficacy of Bonding Agent

Bonding agents develop strong bonds between dental composites and tooth substances and some advanced products exhibit long-lasting bond strengths of more than 20 MPa to etched enamel and more than 10 MPa to intact dentin in wet conditions (Table 2).

The sealing effect of the bonding agents can be tested by the percolation test, in which the restored teeth are loaded with a thermal stress of 4°C and 60°C for 1 minute each 100–1000 times in a dye solution. The degree of dye penetration observed on the sectioned specimens determines the strength of the seal. This test was used to evaluate an application of Clearfil Photo Bond (Fig. 7). Composite restoration without bonding agent (on the left in the figure) shows serious dye penetration; while on the right the bonding treatment has maintained a good seal. Sealing microspaces is expected to secure good and long-lasting composite restorations.

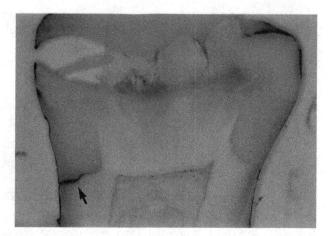

Fig. 7 Cross section of a tooth restored with Photoclearfil A with (right) and without (left) Photobond, prepared after 1000 thermal cycles. Arrow shows dye penetration on the restoration performed without Photobond.

REFERENCES

1. Newbrun, E. *Cariology*; Williams: Baltimore, MD, 1978.
2. Yamauchi, J.; Yamada, K.; Shibatani, K. Adhesive Compositions for the Hard Tissues of the Human Body. U.S. Patent 4182035, Jan 8, 1980.
3. Dart, E.C.; Nemcek, J. Photopolymerisable Composition. Great Britain Patent 1408265, Oct 1, 1975.
4. Kadoma, Y.; Masuhara, E.; Ueda, M.; Imai, Y. Controlled release of fluoride ions from methacryloyl fluoride—Methyl methacrylate copolymers, 1. Synthesis of methacryloyl fluoride—Methacrylate copolymers. Makromol. Chem. **1981**, *182* (1), 273–277.
5. Wilson, A.D.; Kent, B.E. A new translucent cement for dentistry. The glass ionomer cement. Br. Dent. J. **1972**, *132* (4), 133–135.
6. Crisp, S.; Wilson, A.D. Poly(carboxy laic) Cements. Great Britain Patent 1484454, 1977.
7. Fukuda, K. Jpn. J. Cons. Dent. **1978**, *21*, 595.
8. Inoue, K.; Takemura, K.; Tsuchitani, Y. Jpn. J. Cons. Dent. **1969**, *12*, 19.
9. Masuhara, E.; Nakabayashi, N.; Takeyama, J. Curable Composition. U.S. Patent 4148988, Apr 10, 1979.
10. Nakabayashi, N.; Masuhara, E.; Mochida, E.; Ohmori, I. Development of adhesive pit and fissure sealants using a MMA resin initiated by a tri-*n*-butyl borane derivative. J. Biomed. Mater. Res. **1978**, *12* (2), 149–165.
11. Omura, I.; Nagase, Y.; Uemura, F.; Yamauchi, J. U.S. Patent 4539382, Sep 3, 1985.
12. Omura, I.; Kawashima, M. J. Dent. Res. Special Issue. **1994**, *130*.
13. Hino, K.; Nishida, K.; Yamauchi, J. Dental Compositions. U.S. Patent 5321053, Jun 14, 1994.
14. Wada, T. et al. Kuraray Medical Technical Report 940712, 1994.

Drug Delivery Systems: Selection Criteria and Use

Ravindra Semwal
Faculty of Pharmacy, Dehradun Institute of Technology, Dehradun, India

Ruchi Badoni Semwal
Deepak Kumar Semwal
Department of Chemistry, Panjab University, Chandigarh, India

Abstract

Polymers are becoming increasingly important in the field of drug delivery. The advances in polymer science have led to the development of several novel drug-delivery systems (DDSs). The pharmaceutical applications of polymers range from their use as binders in tablets to viscosity and flow controlling agents in liquids, suspensions, and emulsions. Polymers can be used as film coatings to mask the unpleasant taste of a drug, to enhance drug stability, and to modify drug release characteristics. Biodegradable polymers find widespread use in drug delivery as they can be degraded to non-toxic monomers inside the body. Novel supramolecular structures based on polyethylene oxide copolymers and dendrimers are being intensively researched for the delivery of genes and macromolecules. Hydrogels that can respond to a variety of physical, chemical, and biological stimuli hold enormous potential for design of closed-loop DDSs. This entry focuses on the different classes of natural and synthetic pharmaceutical polymers and the significance of these for controlled drug delivery applications. It further focuses on the design and synthesis of novel combinations of polymers that will expand the scope of new DDSs in the future. This entry also covers the selection criteria of a proper polymeric system for different drug delivery system.

INTRODUCTION

The word "polymer" is derived from the Greek roots "poly" and "mer," which mean "many parts." Polymeric substances are composed of many chemical units called monomers, which are joined into large molecular chains consisting of thousands of atoms. The monomers can be connected in linear chains, branched chains, or more complicated structures, each variety yielding interesting and useful properties. Polymers are becoming increasingly important in the field of drug delivery. The pharmaceutical applications of polymers range from their use as binders in tablets to viscosity and flow controlling agents in liquids, suspensions, and emulsions. Polymers can be used as film coatings to disguise the unpleasant taste of a drug, to enhance drug stability, and to modify drug release characteristics. Controlled drug delivery occurs when a polymer, whether natural or synthetic, is judiciously combined with a drug or other active agent in such a way that the active agent is released from the material in a predesigned manner.

Polymers have changed our day-to-day lives over the past several decades. However, the distinction between temporary and permanent biomedical applications of polymers was made only 30 years ago. Subsequently, the amalgamation of polymer science with pharmaceutical sciences led to a quantum leap in terms of "novelty," namely, flexibility in physical state, shape, size, and surface, in design and development of novel drug delivery systems (DDSs). Polymeric delivery systems are mainly intended to achieve either a temporal or spatial control of drug delivery. The introduction of the first synthetic polymer-based (polyglycolic acid) DDS led to an increased interest in the design and synthesis of novel biodegradable polymers that obviated the need to remove the DDS, unlike the nondegradable polymeric systems. Recognizing that intimate contact between a delivery system and an epithelial cell layer will improve the residence time as well as the efficacy of the DDS resulted in the design of bioadhesive polymers. Further advancements in polymer science led to "smart" polymeric hydrogel systems that can self-regulate the delivery of a bioactive agent in response to a specific stimulus. With the availability of a large number of synthetic and natural polymers, this entry discusses the various considerations in the selection and design of polymers for drug-delivery applications.

IDEAL POLYMERS FOR DRUG DELIVERY SYSTEM

The selection and design of a polymer are challenging tasks because of the inherent diversity of structures and require a thorough understanding of the surface and bulk

Concise Encyclopedia of Biomedical Polymers and Polymeric Biomaterials DOI: 10.1081/E-EBPPC-120050409

Table 1 Natural polymers

Polymers	Examples	Advantage	References
Protein based	Collagen	Collagen for its unique structural properties	[7]
	Gelatin		
	Albumin		
Polysaccharides	Chitosan	Good absorption-enhancing and bioadhesive properties	[8]
	Lecithin	Controls the binding uptake and intracellular routing of macromolecules	[9]
	Cyclodextrins	Binds directly to epithelial cells than to the mucus layer	[10]
		For its physiochemical and inclusion capability	
PEO and POP		For its drug targeting and safety	[11]
		Mimics biological transport system like viruses or lipoproteins	
Carbohydrate derivatives Cellulose/PEG blend		Has a controllable phase change property	[12]

properties of the polymer that can give the desired chemical, interfacial, mechanical, and biological functions. The choice of polymer, in addition to its physicochemical properties, is dependent on the need for extensive biochemical characterization and specific preclinical tests to prove its safety. Nowadays, there are different classes of natural and synthetic polymers available for controlled DDSs (Tables 1 and 2). Additionally, there are also many polymers that have been investigated and have applications in different routes of DDSs (Table 3). Surface properties such as hydrophilicity, lubricity, smoothness, and surface energy govern the biocompatibility with tissues and blood, in addition to influencing physical properties such as durability, permeability, and degradability.[1] The surface properties also determine the water sorption capacity of the polymers, i.e., hydrogels, that undergo hydrolytic degradation and swelling. On the other hand, materials for long-term use, e.g., for orthopedic and dental implants, must be water-repellent to avoid degradation or erosion processes that lead to changes in toughness and loss of mechanical strength. Surface properties can be improved by chemical, physical, and biological means to increase their biocompatibility. Grafting of enzymes, drugs, proteins, and antibodies to the polymer surface has led to "polymer therapeutics" for targeting organs and cells.[2] Bulk properties that need to be considered for controlled DDSs include molecular weight, adhesion, solubility based on the release mechanism (diffusion- or dissolution-controlled), and its site of action. Bioadhesiveness needs to be taken into account when DDSs are targeted to mucosal tissues, whereas polymers for ocular devices have to be aqueous or lipid-soluble in addition to having good film-forming ability and mechanical stability for good

retention.[3,4] Structural properties of the matrix, its micromorphology, and pore size are important with respect to mass transfer of water into and out of the polymer-containing drug. For non-biodegradable matrices, drug release in most cases is diffusion-controlled and peptide drugs with low permeability can only be released through the pores and channels created by the dissolved drug phase.[5] With regard to biodegradable polymers, it is essential to recognize that degradation is a chemical process, whereas erosion is a physical phenomenon that depends on the dissolution and diffusion process. Depending on the chemical structure of the polymer backbone, erosion can occur by either surface or bulk erosion. Surface erosion occurs when the rate of erosion exceeds the rate of water permeation into the bulk of the polymer and is desirable because the kinetics (zero order) of erosion and rate of drug release are highly reproducible. Bulk erosion occurs when water molecules permeate into the bulk of the matrix at a faster rate than erosion, thus exhibiting complex degradation/erosion kinetics. Most of the biodegradable polymers used in drug delivery undergo bulk erosion. However, the use of nanoparticle or microparticle formulations possessing massive surface areas results in bulk- and surface-eroding materials that show similar erosion kinetics. Further, the erosion process can be manipulated by modifying the surface area of the DDS or by including hydrophobic monomer units in the polymer. The micro-structural design and chemical composition can be used to adapt the structure–property relationship and tailor improved polymeric matrices. Various polymer architectures, like linear, branched, star-like comb-like polymer, and combinations of polymer species either physically mixed, i.e., polymer blends or interpenetrating networks,

Table 2 Synthetic polymers

Polymers	Examples	Advantage	References
Cellulose derivatives	Methyl cellulose	Releases drugs by attrition since intact hydrated layer is not maintained	[13]
	Anionic polymer	Can interact with cationic drugs and shows increased dissolution in intestinal fluid	
	Carboxymethyl cellulose/carpolene		
	Carboxypolymethylene	Does not adversely hydrate in intestinal fluid	
	HPMC	Best polymer as it does not adversely interact with either acidic or basic drugs and forms gel with water, i.e., more resistant to attrition	
Silicones, colloidal silica	Poly(siloxanes)	For insulating property	[14]
Acrylic polymer, polymethacrylate, polyhydroxy ethylmethacrylate	Polymethylmethacrylate	For physical strength and transparency	
Others	Polyvinyl alcohol	For hydrophilicity and strength	
	Polyethylene	For toughness and lack of swelling	
	Polyvinylpyrrolidone	For suspension capabilities	
Polyester based {Poly(hydroxy butyrate), Poly(beta malic acid), Poly(dioxanone)}	Poly(lactic acid)	Versatility in polymer property for fabrication and bioperformance; represents gold standard	[7]
	Poly(glycolic acid)	Demerit: leads to irritation at site of polymer application	
	Poly(caprolactone)	Monolithic system for drug release	
Polyanhydrides	Poly(sebacic acid)	For their unique property of surface erosion	[15,16]
	Poly(adipic acid)		
	Poly(pterpthalic acid)		
Polyamino acids		Good biocompatibility for the delivery of low molecular weight compounds	[6]
		Limitation: antigenic properties and poor control of release because of dependence of enzymes	
Polyamides	Poly(imino carbonate)	To overcome the limitations of polyaminoacids	
Phosphorus-based polymers	Poly(phosphates)	Uniqueness due to the chemical reactivity of phosphorus	[5,7,17]
	Poly(phosphontes)		
	Poly(phosphazenes)		
Other	Poly(methanes)	For its elasticity	[14] [18,19]
	Poly(orthoesters)	For acid labile linkage, facilitates manipulation rate of hydrolysis	

or chemically bonded, i.e., copolymers, offer tremendous scope in the field of DDS. Although a selection of polymers is a prime concern, especially with regard to compatibility with the drug, the manufacturing process also needs to be considered, because the additives used for polymerization may chemically interact with the drug. Uhrich et al.[6] described a method wherein a protein is first dispersed in a glassy matrix, whose glass-transition temperature is well above the melting point of the polymer, and subsequently is dispersed in the polymer, which then can be fabricated into a DDS of any desired shape, thereby protecting the protein from the harsh manufacturing process. Natural polymers are usually biodegradable and offer excellent biocompatibility, but suffer from batch to batch variation because of difficulties in purification. On the other hand, synthetic polymers are available in a wide variety of compositions with readily adjustable properties.[1] The ideal polymer for drug delivery is that which has

Table 3 Polymers used in different DDSs

Drug delivery system	Polymer used	Attributes	References
Oral transmucosal DDS/ mucoadhesive DDS	Carboxy methyl cellulose	As its viscosity drops during heating, it may be used to improve the volume yield during baking by encouraging gas bubble formation water-holding capacity as this is high even at low viscosity	[20–22]
		The average chain length and degree of substitution are of great importance; the more-hydrophobic lower substituted critical micelle concentrations (CMCs) are thixotropic but more-extended higher substituted CMCs are pseudoplastic	
	Carbopol 934	They swell in water up to 1,000 times their original volume (and 10 times their original diameter) to form a gel when exposed to a pH environment above 4.0 to 6.0. Because the pKa of these polymers is 6.0 to 0.5, the carboxylate groups on the polymer backbone ionize, resulting in repulsion between the negative charges, which adds to the swelling of the polymer. Carbopol polymers have demonstrated zero-order and near zero-order release kinetics. These polymers are effective at low concentrations (less than 10%) and feature extremely rapid and efficient gelation characteristics under both simulated gastric fluid (SGF) and simulated intestinal fluid (SIF) test conditions	[23–26]
	Polycarbophil	Excellent swelling properties without erosion, with chitosan	[25,26]
	Tragacanth	Excellent property to adhere to mucin. Smidsrod parameter B, the empirical stiffness parameter was 0.013, which indicates stiff backbone	[27]
	Polyacrylic acid	pH-sensitiveness of the pluronic-PAA copolymers make them promising excipients for tablets with preferential delivery into a neutral to alkaline pH environment	[28]
	Sodium alginate	Sodium alginate matrices can sustain drug release for at least 8 hours, even for a highly water-soluble drug in the presence of a water-soluble excipient	[26]
Floating DDS	Isapgol	Swelling properties, cheaper, and use in low concentration with respect to HPMC	[23,24]
	Mannitol	Inert polymer does not adversely interact with either acidic or basic drugs	[13]
	HPMC	On contact with water slowly forms a gel, i.e., more resistant to attrition	
Transdermal DDS	Cellulose derivative	Formation of gel in situ. Drug release is controlled by penetration of water through a gel layer produced by hydration of the polymer and diffusion of the drug through the swollen hydrated matrix	
	PVP		
	Sodium CMC		
	Ethyl cellulose		
	Zein	Lengthens dissolution time	
	Shellac	Also lengthens dissolution and disintegration time but on aging because it polymerizes on aging	
	Waxes	Forms controlled release matrix through both pore formation and diffusion.	

(Continued)

Table 3 (*Continued*) Polymers used in different DDSs

Drug delivery system	Polymer used	Attributes	References
	Gums and their derivatives	These polymers are non-reactive, stable with drug, easily fabricated with desired product, and showed inexpensive mucoadhesive properties	[13,14]
	Natural rubber		
	Gelatin		
	Starch	Ensures minimal interaction with polyphenolic activities and sustains drug release	[29]
	Polyvinyl alcohol	They are capable of absorbing large amounts of plasticizers, so that even a 50% plasticizer, 50% PVC resin composition material would result in a non-flowing solid	[30]
	Polyvinyl chloride		
	Polyethylene	The mechanical properties of the drug does not deteriorate when excessive amount of drug is incorporated	
	Polypropylene		
	Polyacrylate		
	Polyamide		
	Polyurea		
	Polyvinylpyrrolidine		
	Polymethyl methacrylate		
	Epoxy		
	Carbopol	Thickening at very low concentrations (less than 1%) to produce a wide range of viscosities and flow properties in topical lotions, creams and gels, oral suspensions, and in transdermal gel reservoirs	[31]
Ocular DDS	Sodium hyaluronate	Viscosity increasing agent, unique physiochemical and polyelectrolyte behavior	[32]
	Chondroitin sulfate		
	Polyvinyl acrylate	Improved ocular contact time, reduces the solution drainage	
	Polyvinyl pyrollidine		
	Methylcellulose		
	Carboxymethyl cellulose hydroxypropyl cellulose		
	Carbopol	Used in the preparation of gel; less blurred vision than ointment	
	Acrylic polymer	Delivers 24-hour pilocarpine dose from single night time replacement in the cul-de-sac	
Osmotic DDS	Sodium carboxymethyl cellulose	Inert polymers do not adversely interact with either acidic or basic drugs	[33]
	Hydroxypropylmethyl cellulose		
	Hydroxy ethyl methyl cellulose, methyl cellulose, polyethylene oxide		
	Polyvinyl pyrollidine		

Dental—Electrospinning

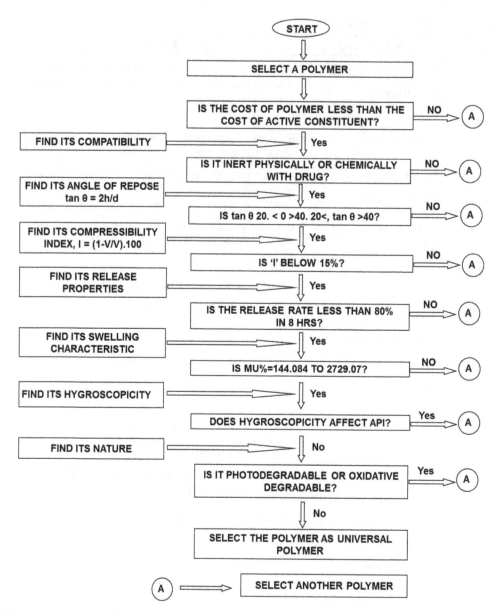

Fig. 1 Typical flow chart for a rationalized polymer selection.

various physicochemical properties such as compatibility with drug and other additives, stability, etc. Before selecting a polymer for drug delivery, we should evaluate various characteristics of the polymer and these are shown by a typical flow chart (Fig. 1).

Natural Polymers

Natural polymers, e.g., gelatin, collagen, and lecithin, have been widely used in pharmaceuticals for many years. These offer various advantages due to their biodegradable nature, excellent biocompatibility, non-toxic, wide applicability as such or after modification, well-known structural and physiological properties, and less immunological properties.[34] The natural polymers are basically of four types, e.g.,

protein based such as collagen and gelatin, polysaccharides based such as lecithin, chitosan polyethylene oxide (PEO) and polyoxypropylene (POP), and some carbohydrate blends, e.g., cellulose/poly(ethylene glycol) (PEG) blend.

Synthetic Polymers

Synthetic polymers have gained a significant attention in recent years due to their wide range of varieties that offer added flexibility in terms of their application in DDSs. These are available in a wide variety of composition with readily adjustable properties. Bulk preparation of synthetic polymers is easy, which also eliminates the additional step of purification in the case of natural polymers.[34] In contrast to natural polymers, these are considered less prone to bacterial contamination. However, synthetic polymers also suffer from some disadvantages,[25] including the following:

(i) Toxicity of the chromium compounds.
(ii) Potential hazards as water pollutants.
(iii) Fairly high temperatures (55–70°C) of operation.
(iv) A system can produce similar etching of plastic articles made from synthetic polymer resins.

The synthetic polymers are basically of two types:

1. Biodegradable such as polyester, poly amide, etc.
2. Non-biodegradable such as cellulose derivatives, acrylic polymer, etc.

BIODEGRADABLE AND NON-DEGRADABLE POLYMERS FOR DRUG DELIVERY

The use of biodegradable and non-biodegradable polymers as drug carriers for various formulations is gaining enormous popularity in recent days. The wide acceptability of these polymers can be appreciated from the fact that the biodegradability can be manipulated by incorporating a variety of labile groups such as ester, orthoester, anhydride, carbonate, amide, urea, and urethane in their basic structure.[5] It is pertinent to mention that Langer[35] has made significant contributions in investigating various polymers for drug delivery and characterizing the release of macromolecules from polymers. Biodegradation can be of enzymatic, chemical, or of microbial origin, and these may operate either separately or simultaneously and are often influenced by many other factors (Table 4). Polyester-based polymers are one of the most widely investigated for the purpose of drug delivery. Poly(lactic acid) (PLA) poly(glycolic acid) (PGA) and their copolymers poly(lactic acid-co-glycolic acid) (PLGA) are some of the well-defined

polymers with regard to design and performance for drug-delivery applications.[36] It is possible to modify the mechanical, thermal, and biological properties of PLA by altering its stereochemistry. Further biodegradability can be tuned by changing the proportion of PLA and PGA in the copolymer. Although PLGA represents the "gold standard" (exemplified by more than 500 patents) of biodegradable polymers, increased local acidity because of degradation can lead to irritation at the site of polymer application.[6] Further, the increased local acidity may also be detrimental to the stability of protein drugs.[37] Polyorthoesters have been under development since the 1970s and most research has focused on the synthesis of polymers by the addition of polyols to diketene acetals. They are unique among all biodegradable polymers, as mechanical properties can be readily varied by choosing appropriate diols or a mixture of diols in their synthesis.[18] A number of applications have been found for polyorthoesters and crosslinked polyorthoesters such as the delivery of 5-flurouracil, periodontal delivery systems of tetracycline, and pH-sensitive polymer systems for insulin delivery. Polyanhydrides are characterized by their fast degradation followed by the rapid erosion of material, but at the same time can be designed to release drugs that last from days to weeks by a suitable choice of monomers.[15,16] By varying the monomer ratio of aliphatic (sebacic acid) and aromatic (carboxyphenoxypropane) polyanhydrides, polymer-carmustine disks were fabricated for the chemotherapy of brain cancer, which was the first US Food and Drug Administration approved polymer-based chemotherapy DDS.[35] Polyamino acids that have good biocompatibility have been investigated for the delivery of low-molecular-weight compounds. However, their widespread use is limited by their antigenic potentials and poor control of release because of the dependence on enzymes for biodegradation. Poly(imino carbonates), which are "pseudo" polyamino acids, have been synthesized from tyrosine dipeptide to overcome the

Table 4 Factors effecting polymeric degradation

S. No.	Factor	S. No.	Factor
1	Chemical structure	11	Processing conditions
2	Chemical composition	12	Annealing
3	Distribution of repeat units	13	Sterilization process
4	Ionic groups	14	Storage history
5	Unexpected units or chain defects	15	Shape
6	Configure structure	16	Site of implantation
7	Molecular weight	17	Adsorbed and absorbed compounds (water, lipids, ions, etc.)
8	Molecular weight distribution	18	Physicochemical factors (ion exchange, ionic strength, pH)
9	Morphology (amorphous/semicrystalline, microstructures, residual stresses)	19	Physical factors (shape and size changes, variations of diffusion coefficients, mechanical stresses, stress- and solvent-induced cracking, etc.
10	Low molecular weight compounds	20	Mechanism of hydrolysis (enzymes vs. water)

Dental—Electrospinning

abovementioned limitations.[6] A relatively new class of biodegradable polymers belonging to polyphosphoesters has a unique backbone consisting of phosphorous atoms attached to either carbon or oxygen. The uniqueness of this class of polymers lies in the chemical reactivity of phosphorous, which enables a wide range of side chains to be attached for manipulating the biodegradation rates and the molecular weight of the polymer.[5] Polyphosphazenes, another new class of polymers, are being investigated for delivery of proteins.[17] Chitosans are promising natural polymers that show good absorption-enhancing, controlled release, as well as bioadhesive properties. The degree of deacetylation and derivatization with various side chains can be a source of manipulation for specific drug-delivery applications.[8] A variety of non-degradable polymers are used in drug delivery of which polysaccharide-based and acrylic-based polymers have found a wide application in the fabrication of peroral dosage forms, transdermal films, and devices.[38] Cyclodextrins are potential high-performance carrier materials that have the ability to alter physical, chemical, and biological properties of guest (drug) molecules through the formation of inclusion complexes both in solution and solid state. The alpha, beta, and gamma cyclodextrins are the most common natural cyclodextrins consisting of six, seven, and eight D-glucopyranose residues, respectively, linked by α-1,4 glycosidic bonds into a macrocycle. There has been an explosion in the number of reports dealing with the practical uses of natural cyclodextrins (more than 16,300 patents and papers). Around 1500 derivatives of cyclodextrins have been discussed in the literature, which include hydrophilic, hydrophobic, and ionic derivatives to expand the physicochemical properties and inclusion capability of natural cyclodextrin as novel drug carriers.[10] Dextrans are being actively investigated for the sustained delivery of therapeutic and imaging agents, particularly for injectables and colon-specific DDSs.[39] PEO and POP copolymers are among the most interesting classes used for nanoparticulate DDSs. They are commercially available as polaxmers in a range of liquids, pastes, and solids. Since their first introduction as non-ionic surfactants in the 1950s, they have found a wide range of applications in the pharmaceutical and biomedical fields.[11] Soluble block copolymers based on PEO–PLA can self-assemble into novel supramolecular structures and are being investigated for the delivery of anti-cancer agents, proteins, and plasmid DNA.[40,41] They are advantageous in terms of drug targeting and safety, and they can mimic biological transport systems, lipoproteins, or viruses. Polysiloxanes are a unique class of non-deformable polymers possessing good low-temperature flexibility, excellent electrical properties, water repellency, and remarkable biocompatibility, features that are not common with hydrocarbon polymers.[38] Because of the ease of fabrication and high permeability, polydimethylsiloxanes are useful for water-soluble drugs and steroids for long-acting DDSs such as subdermal implants and intravagnial systems.[42] Lectins

and lectin-like molecules of plant or microbial origin, as well as biotechnologically generated derivatives of such molecules, have interesting characteristics to control the binding, uptake, and intracellular routing of macromolecules as well as colloidal carrier systems.[9] In contrast with other mucoadhesive polymers, lectin binds directly to epithelial cells rather than to the mucus gel layer. Dendrimers belong to a novel and exciting class of highly branched three-dimensional polymers in which growth emanates from a central core molecule such as ammonia, ethylenediamine, polydiamine, or benzene tricarboxylic acid chloride. Compared with traditional linear polymers, dendrimers have much more accurately controlled structures with a globular shape, a single molecular weight rather than a distribution of molecular weight, and a large number of "controllable" peripheral functionalities.[43] They are truly nanoscale molecules with sizes ranging from 10 to 30Å, and the excitement associated with this class of polymers is mainly because of their applications as synthetic vector systems for gene delivery. Polyaminodiamine is a promising polycationic polymer from this class that can form complexes with the negatively charged nucleic acids; additionally, the surface positive charge can interact and fuse with phospholipids of the cell membrane, thereby facilitating the translocation of DNA into cells.[44] A number of other polymeric systems are also under investigation for DNA delivery.[45] A series of acrylate polymers is marketed under the name of Eudragit (Rohm and Haas Co Inc., Pharma Gmbh, Germany). Eudragit E is a cationic polymer based on dimethyl aminoethyl metha acrylate and other neutral metha acrylic acid ester is freely soluble in gastric fluid. Eudragit L and Eudragit S are resistant to gastric fluid. Eudragit RL and RS (copolymers of acrylic and metha acrylic acid esters) produce films for delayed release and are pH independent.[28] The microspheres were prepared by solvent evaporation method using Eudragit RL or hydroxypropylcellulose as a polymer matrix. In another study, microspheres with a Eudragit RS matrix polymer and different mucoadhesive polymers, i.e., chitosan hydrogen chloride, sodium salt of carboxymethyl cellulose (Na-CMC), and polycarbophil, were prepared and found to be useful as platforms for oral peptide delivery, with a high capacity of binding to bivalent cations, which are essential cofactors for intestinal proteolytic enzymes.[46]

FUTURE GENERATION POLYMERS

The next generation polymers should be the blend of two polymers in order to eliminate the shortcomings of one another. Much of the development of novel polymeric materials in controlled drug delivery is focusing on the preparation and use of these responsive polymers with specifically designed macroscopic and microscopic structure, and chemical features. These systems are described in the following text.

Blend of Natural Polymers and Synthetic Polymers

In order to overcome the biological deficiencies of synthetic polymers and to enhance the mechanical characteristics of natural polymers, the upcoming generation of polymers should be the blend of two. The synthetic polymers, poly(vinyl alcohol) and poly(acrylic acid), were blended in different ratios, with two biological polymers, collagen and hyaluronic acid. These blends were used to prepare films, sponges, and hydrogels that were loaded with growth hormones to investigate their potential use as DDSs. The rate and quantity of growth hormone released were significantly dependent on the collagen or hyaluronic acid content of the polymers.[47]

Chemomechanical Polymers

These new biomaterial-tailor-made copolymers with desirable functional groups are being created by researchers who envision their use not only for innovative DDSs but also as potential linings for artificial organs, as substrates for cell growth or chemical reactors, as agents in drug targeting and immunology testing, as biomedical adhesives and bioseparation membranes, and as substances able to mimic the biological system.[48]

"Cytoadhesion," the next generation mucoadhesive, polymers based on certain materials reversibly bind to cell surfaces in the gastrointestinal tract (GIT).[32] This next generation of mucoadhesives functions with greater specificity because they are based on receptor-ligand-like interactions in which the molecules bind strongly and rapidly directly on the mucosal cell surface rather than the mucus itself. One such class of compounds that has these unique requirements is called lectins. *In vitro* binding and uptake is tomato lectin, which has been shown to bind selectively to the small intestine epithelium.

Complexation Networks Responding Via Hydrogen or Ionic Bonding

A new class of hydrophilic pressure-sensitive adhesives (PSAs) that share the properties of both hydrophobic PSAs and bioadhesives has been developed by Corium Technologies. These Corplex adhesive hydrogels have been prepared by non-covalent (hydrogen bond) cross-linking of a film-forming hydrophilic polymer (e.g., poly-vinylpyrrolidone (PVP)) with a short-chain plasticizer (typically PEG) bearing complementary reactive hydroxyl groups at its chain ends.[49]

Copolymers with Desirable Hydrophilic/Hydrophobic Interactions

Shojaei and Li[50] have designed, synthesized, and characterized a copolymer of poly(amidoamine) (PAA) and PEG monoethylether monomethacrylate (PAA-*co*-PEG) (PEGMM). By adding PEG to these polymers, many of the shortcomings of PAA for mucoadhesion were eliminated. Copolymers of 12 and 16 mole% PEGMM showed higher mucoadhesion than PAA. The effects of hydration on mucoadhesion produce by the copolymers revealed that the film containing lower PEGMM content has higher hydration levels and lower mucoadhesive strengths.[51] The resulting polymer has a lower glass transition temperature than PAA and exists as a rubbery polymer at room temperature.

Block or Graft Copolymers

Block copolymer micelles are among the latest nanoparticles under investigation. Copolymers are polymers composed of several different monomeric units. Block copolymers are defined as the polymers composed of terminally connected structures unlike random polymers. The functions of such polymers can be distinctly designed for each monomeric segment that forms a domain in block copolymer micelle. Block copolymers are divided into three types AB type, ABA type, and $(AB)_n$ multi-segments. In AB type, copolymers are composed of both hydrophilic PEO and hydrophobic blocks such as poly propylene oxide, which allows the polymer to self-assemble as micelles in an aqueous media.[52]

Dendrimers (Star Polymers) as Nanoparticles for Immobilization of Enzymes, Drugs, Peptides, or Other Biological Agents

The term "dendrimer" is derived from "dendron" a Greek word that means trees/branches. In spite of being polymers, they differ from conventional polymers in that they are highly branched and generally have 3D macromolecules with a branch point at each monomer unit. The major focus is on the development of dendrimers based on a new concept of biomimetism/biomorphism with increased plasma stability and compatibility to target cells in various body compartments.[53,54]

Smart Polymers

The concept of "smart" polymers originated from the ability of certain synthetic polymers (hydrogels) to mimic the non-linear response of biopolymers (DNA, proteins, etc.) caused by cooperative interactions between monomers.[55] Because of their excellent water-absorbing capacity, hydrogels resemble natural living tissues more closely than any other class of synthetic polymeric materials.[56] Both the swelling and permeability characteristics of hydrogels and their ability to undergo structural changes in response to a variety of physical, chemical, and biological stimuli (Table 5) have given rise to the concept of "intelligent" or "stimuli"-responsive

Table 5 Factors influencing polymer performance

Stimulus	Hydrogel	Mechanism
pH	Acidic or basic hydrogel	Change in pH—swelling—release of drug
Ionic strength	Ionic hydrogel	Change in ionic strength—change in concentration of ions inside gel—change in swelling—release of drug
Chemical species	Hydrogel containing electron-accepting groups	Electron-donating compounds—formation of charge/transfer complex—change in swelling—release of drug
Enzyme-substrate	Hydrogel containing immobilized enzymes	Substrate present—enzymatic conversion—product changes swelling of gel—release of drug
Magnetic	Magnetic particles dispersed in alginate microspheres	Applied magnetic field—change in pores in gel—change in swelling—release of drug
Thermal	Thermoresponsive hydrogel poly(n-isopropylacrylamide)	Change in temperature—change in polymer–polymer and water–polymer interactions—change in swelling—release of drug
Electrical	Polyelectrolyte hydrogel	Applied electric field—membrane charging—electrophoresis of charged drug—change in swelling—release of drug
Ultrasound irradiation	Ethylene-vinyl alcohol hydrogel	Ultrasound irradiation—temperature increase—release of drug

DDSs.[57,58] Attempts to develop a truly closed-loop-regulated DDS have the eventual goal of delivering insulin in response to blood glucose levels. Typically, these systems have been prepared by incorporating glucose oxidase into hydrogel (cationic, anionic, or neutral polymers) during polymerization, which exhibits a glucose-sensitive swelling behavior. Another approach exploits the competitive binding of glucose and glycosylated insulin to a fixed number of binding sites in concovalin-A immobilized on sepharose beads. The glycosylated insulin, which is biologically active, can be displaced from the concovalin-A in proportion to the amount of glucose that competes for the same binding sites. Polymeric systems that can simultaneously respond to pH and temperature can be achieved by the modification of polyelectrolyte gels with lower critical solution temperature monomers.[59] Recent findings indicate that a hydrogel of polyacrylamide semi-interpenetrating networks can respond to antigen, which could find potential application for the delivery of drugs in response to a specific antigen and has far-reaching implications in the treatment of a variety of immunological-based diseases.[60] Hydrogel systems responsive to microbial infection have been designed based on proteinase specific to a bacteria as a triggering mechanism for the release of antibiotics, and have found application in the localized delivery of antibiotics for wound healing.[61] This can overcome the renal and liver toxicity problems associated with the prolonged use of antibiotics, in addition to reducing the possibility of emergence of drug resistance. The interplay of "innovative" chemistry of coating responsive hydrogel microspheres with a lipid bilayer resulted in emulating the physiological secretory granules that can release the stored drugs on the application of an electroporation pulse, which then allows the fusion of the hydrogel with the lipid bilayer, releasing the drug by an ion-exchange mechanism.[62] At this moment one can say that all these advances in the field of polymer will lead to the golden era in novel drug delivery system.

CONCLUSION

Polymer science has been the backbone for the development of new DDSs for the past few decades. Future advances in polymer science will be based on modifying the chemical and physical properties of the polymer, a novel and "creative" combination of copolymers with targeting and bioresponsive components that can deliver a wide variety of bioactive agents. Further, newer fabrication and manufacturing processes such as molecular imprinting, supercritical fluid technology, and nanoscale engineering are bound to revolutionize the design, development, and performance of polymer-based DDSs. Further stearic modification of certain polymeric structures can pave the way for better incorporation of active pharmaceutical or biological material.

ACKNOWLEDGMENT

Thanks are due to UGC, New Delhi [Grant No. F.4-2/2006(BSR)/13-321/2010(BSR) and F.4-2/2006(BSR)/13-460/2011(BSR)] for financial assistance.

REFERENCES

1. Angelova, N.; Hunkeler, D. Rationalising the design of polymeric biomaterials. Trends Biotechnol. **1999**, *17* (10), 409–421.
2. Brocchini, S.; Duncan, R. Pendant drugs release from polymers. In *Encyclopaedia of Controlled Drug Delivery*; John Wiley and Sons: New York, 1999; Vol. 2, 786–816.

3. Colthrust, M.J.; Williams, R.L.; Hiscott, P.S.; Grierson, I. Biomaterials used in the posterior segment of the eye. Biomaterials **2000**, *21* (7), 649–665.

4. Semwal, R.; Semwal, D.K.; Badoni, R. Chewing gum: A novel approach for drug delivery. J. Appl. Res. **2010**, *10* (3), 115–123.

5. Mao, H.Q.; Kdaiyala, I.; Leong, K.W.; Zhao, Z.; Dhang, W. Biodegradable polymers; poly(phosphoester)s. In *Encyclopedia of Controlled Drug Delivery*; John Wiley and Sons: New York, 1999; Vol. 1, 45–60.

6. Uhrich, K.E.; Cannizzaro, S.M.; Langer, R.S.; Shakesheff, K.M. Polymeric systems for controlled drug release. Chem. Rev. **1999**, *99* (11), 3181–3198.

7. Murthy, R.S.R. Biodegradable polymers. In *Controlled and Novel Drug Delivery*; N.K Jain, Ed.; CBS Publisher: New Delhi, 1997; 27–51.

8. Dodane, V.; Vilivalam, V.D. Pharmaceutical applications of chitosan. Pharm. Sci. Tech. Today **1998**, *1* (16), 246–253.

9. Lehr, C.M. Lectin-mediated drug delivery: The second generation of bioadhesives. J. Control. Release **2000**, *65* (1), 19–29.

10. Szente, L.; Szejtli J. Highly soluble cyclodextrin derivatives, chemistry, properties and trends in development. Adv. Drug Deliv. Rev. **1999**, *36* (1), 17–28.

11. Moghimi, S.M.; Hunter, A.C. Polaxmers and polaxamines in nanoparticles engineering and experimental medicine. Trends Biotechnol. **2000**, *18* (10), 412–420.

12. Qing, S.; Dian, S.L. Cellulose/poly(ethylene glycol) blend and its controllable drug release behaviors *in vitro*. Carbohydr. Polym. **2007**, *69* (2), 293–298.

13. Lachman, L. Polymers used in sustained release matrix tablets. In *The Theory and Practice of Industrial Pharmacy*; Lea & Febiger: Philadelphia, 1991; 453–454.

14. Brannon-Peppas, L. Polymers in controlled drug delivery: Medical Plastics and Biomaterials Magazine. 1997.

15. Gopferich, A. Biodegradable polymers, polyanhydrides. In *Encyclopedia of Controlled Drug Delivery*; John Wiley and Sons: New York, 1999; Vol. 2, 852–874.

16. Gopferich, A. Biodegradable polymers: Polyanhydrides. In *Encyclopaedia of Controlled Drug Delivery*; John Wiley and Sons: New York, 1999; Vol. 1, 60–71.

17. Andarianov, A.K.; Payne, L.G. Protein release from polyphosphazene matrices. Adv. Drug Deliv. Rev. **1998**, *31* (3), 185–196.

18. Heller, J.; Gurny, R. Poly (orthoesters). In *Encyclopaedia of Controlled Drug Delivery*; John Wiley and Sons: New York, 1999; Vol. 2, 852–874.

19. Heller, J. Control of polymer surface erosion by the use of excipients. In *Polymers in Medicine*; Plenum Press: New York II, 1986; pp. 357–358.

20. Padma, V.D.; Manisha, H.A. Oral transmucosal drug delivery. In *Controlled Novel Drug Delivery*; Jain, N.K., Ed.; CBS Publisher: New Delhi, 1997; 68–69.

21. Chaplin, M.F. Structuring and behaviour of water in nanochannels and confined spaces. In *Adsorption and Phase Behaviour in Nanochannels and Nanotubes*; Dunne, L., Manos, G., Eds.; Springer: New York, 2010; 241–255.

22. Sateesh Madhav, N.V.; Semwal, R.; Semwal, D.K.; Semwal, R.B. Recent trends in oral transmucosal drug delivery system: An emphasis on soft palatal route. Expert Opin. Drug Deliv. **2012**, *9* (6), 629–647.

23. Deshpande, A.A.; Rhodes, C.T.; Shah, N.H.; Malick, A.W. Controlled release drug delivery system for prolonged gastric residence an overview. Drug Dev. Ind. Pharm. **1996**, *22* (6), 531–539.

24. Alvisi, V.; Gasparetto, A.; Dentale, A.; Heran, H.; Ambrosi, A. Bioavailability of a controlled release formulation of ursodeoxycholic acid in man. Drugs Exp. Clin. Res. **1996**, *22* (1), 29–33.

25. Chhanda, J.K.; Vidya, B.M. Raft-Forming agents: Antireflux formulations. Drug Dev. Ind. Pharm. **2007**, *33* (12), 1350–1361.

26. Liew, C.V.; Chan, L.W.; Ching, A.L.; Heng, P.W.S. Evaluation of sodium alginate as drug release modifier in matrix tablets. Int. J. Pharm. **2006**, *309* (1), 25–37.

27. Mohammadifar, M.A.; Musavi, S.M.; Kiumarsi, A.; Williams, P.A. Solution properties of targacanthin (water-soluble part of gum tragacanth exudate from *Astragalus gossypinus*). Int. J. Biol. Macromol. **2006**, *38* (1), 31–39.

28. Lachman, L. Acrylate polymers. In *The Theory and Practice of Industrial Pharmacy*; Lea & Febiger: Philadelphia, 1991; 366–368.

29. Pringels, E.; Ameye, D.; Vervaet, C.; Foreman, P.; Remon, J.P. Starch/Carbopol spray-dried mixtures as excipients for oral sustained drug delivery. J. Control. Release **2005**, *103* (3), 635–641.

30. Misra, A.N. Transdermal drug delivery. In *Controlled Novel Drug Delivery*; Jain, N.K., Ed.; CBS Publisher: New Delhi, 1997, 107–109.

31. Guo, J.H. Carbopol polymers for pharmaceutical drug delivery applications. Drug Deliver. Tech. **2003**, *3* (6), 32–36.

32. Lang, J.C. Ocular drug delivery: Conventional ocular formulations. Adv. Drug Deliv. Rev. **1995**, *16* (1), 39–43.

33. Sharma, S. Osmotic controlled drug delivery system. Pharmainfo.net **2008**, *6* (3).

34. Rao, K.P. New concepts in controlled drug delivery. Pure Appl. Chem. **1998**, *70* (6), 1283–1287.

35. Langer, R. Biomaterials in drug delivery and tissue engineering: One laboratory's experience. Acc. Chem. Res. **2000**, *33* (2), 94–101.

36. Li, S.; Vert, M. Biodegradable polymers: Polyesters. In *Encyclopaedia of Controlled Drug Delivery*; Mathowitz, E., John Wiley and Sons: New York, 1999; Vol. 1, 71–93.

37. Fu, K.; Pack, D.W.; Kilbanov, A.M.; Langer, R. Visual evidence of acidic environment within degrading poly(lactic-co-glycolic acid)(PLGA) microspheres. Pharm. Res. **2000**, *17* (1), 100–106.

38. Kumar, M.N.V.; Kumar, N. Polymeric controlled drug delivery systems: Perspective issues and opportunities. Drug Dev. Ind. Pharm. **2001**, *27* (1), 1–30.

39. Mehavar, R. Dextrans for targeted and sustained delivery of therapeutic and imaging agents. J. Control. Release **2000**, *69* (1), 1–25.

40. Kuon, G.S.; Okano, T. Soluble self-assembled block copolymers for drug delivery. Pharm. Res. **1999**, *16* (5), 597–600.

41. Semwal, R.; Semwal, D.K.; Badoni, R.; Gupta, S.; Madan, A.K. Targeted drug nanoparticles: An emphasis on self-assembled polymeric system. J. Med. Sci. **2010**, *10* (5), 130–137.

42. Bodmeir, R.; Siepmann, S. Nondegradable polymers for drug delivery. In *Encyclopaedia of Controlled Drug Delivery*; John Wiley and Sons: New York, 1999; Vol. 1, 664–689.

43. Liu, M.; Frechet, M.J. Designing dendrimers for drug delivery. Pharm. Sci. Tech. Today **1999**, *2* (10), 393–401.

Dental—Electrospinning

44. Eichman, J.D.; Bielinska, A.U.; Latallo, J.F.K.; Baker, J.R.J. The use of PMAM dendrimers in the efficient transfer of genetic material into cells. Pharm. Sci. Tech. Today **2000**, *3* (7), 232–245.

45. Madsen, S.K.; Mooney, D.J. Delivering DNA with polymer matrices: Application in tissue engineering and gene therapy. Pharm. Sci. Tech. Today **2000**, *3* (11), 381–384.

46. Chein, H.; Langer, R. Oral particulate delivery: Status and future trends. Adv. Drug Deliv. Rev. **1998**, *34* (2), 339–350.

47. Maria, G.C.; Bushra, S.; Sandra, D. Blends of synthetic and natural polymers as drug delivery systems for growth hormone. Int. J. Pharm. **1995**, *16* (7), 569–574.

48. Anonymous. Chemomechanical polymers—Smart polymers for drug release, drug screening and cardiovascular applications. Mater. World **2004**, *12* (2), 29–30.

49. Cleary, G.W.; Feldstein, M.M.; Singh, P.; Plate, N.A. A New Polymer Blend Adhesive with Combined Properties to Adhere to Either Skin or Mucosa for Drug Delivery, Podium Abstract, 30th Annual Meeting and Exposition of the Controlled Release Society, Glagsow, Scotland, July 19–23, 2003.

50. Shojaei, A.M.; Li, X. Mechanism of buccal mucoadhesion of novel copolymers of acrylic acid and polyethylene glycol monomethylether monomethacrylate. J. Control. Release **1997**, *47* (2), 151–161.

51. Nagai, T.; Kinishi, R. Buccal/gingival drug delivery systems. J. Control. Release **1987**, *6* (1), 353–360.

52. Yokoyama, M. Block copolymers as drug carriers. Crit. Rev. Ther. Drug Carrier Syst. **1992**, *9* (3–4), 213–248.

53. Cheng, Y.; Xu, T. Dendrimers as potential drug carriers. Part I. Solubilization of non-steroidal anti-inflammatory drugs in the presence of polyamidoamine dendrimers. Eur. J. Med. Chem. **2005**, *40* (11), 1188–1192.

54. Semwal, R.; Semwal, D.K.; Madan, A.K.; Paul, P.; Mujaffer, F.; Badoni, R. Dendrimers: A novel approach for drug targeting. J. Pharm. Res. **2010**, *3* (9), 2238–2247.

55. Galaew, I.Y.; Mathiasson, B. Smart polymer and what they could do in biotechnology and medicine. Trends Biotechnol. **1999**, *17* (8), 335–339.

56. Lowman, A.M.; Peppas, N.A. Hydrogels. In *Encyclopaedia of Controlled Drug Delivery*; John Wiley and Sons: New York, 1999; Vol. 1, 397–418.

57. Kost, J.; Langer, R. Responsive polymeric delivery systems. Adv. Drug. Deliv. Rev. **1991**, *6* (1), 19–50.

58. Dorski, C.M.; Doyle, F.J.; Peppas, N.A. Preparation and characterization of glucose-sensitive P(MAA-g-EG) hydrogels. Polym. Mater. Sci. Eng. Proc. **1997**, *76*, 281–282.

59. Ganorkar, C.R.; Liu, F.; Baudys, M.; Kim, S.W. Modulating insulin release profile from pH/thermosensitive polymeric beads through polymer molecular weight. J. Control. Release **1999**, *59* (3), 287–298.

60. Miyoto, T.; Asami, N.; Uragami, T. A reversibly antigen-responsive hydrogel. Nature **1999**, *399* (6738), 766–769.

61. Tanihara, M.; Suzuki, Y.; Nishimura, Y.; Suzuki, K.; Kakimarau, Y.; Fukunishi, Y. A novel microbial infection responsive drug release systems. J. Pharm. Sci. **1999**, *88* (5), 510–514.

62. Kiser, P.F.; Wilson, G.; Needham, D. A synthetic mimic of the secretory granule for drug delivery. Nature **1998**, *394* (6692), 459–462.

Dental—Electrospinning

Drugs and Excipients: Polymeric Interactions

James C. DiNunzio
Pharmaceutical and Analytical R&D, Hoffmann-La Roche, Incorporated, Basel, Switzerland

James W. McGinity
Division of Pharmaceutics, College of Pharmacy, University of Texas at Austin, Austin, Texas, U.S.A.

Abstract

Polymeric materials play a critical role in the development of solid oral dosage forms and this contribution is likely to grow continuously in the foreseeable future as products continue to become more technologically advanced. Polymers provide a basis for the development of controlled release systems and the stabilization of amorphous forms, presenting two unique delivery platforms which have contributed to the advancement of patient care. As scientists continue to develop a deeper and more complete understanding of the underlying interactions, and the impact of such behaviors on dosage form performance, pharmaceutical systems will continue to become more complex and more efficacious. Polymeric systems are also likely to provide significant contributions in many growing areas of pharmaceutical development, particularly for advanced device manufacturing and bionanotechnology applications. Many of the newest hybrid technologies exploit specific polymeric interactions in order to provide enhanced pharmaceutical efficacy.

INTRODUCTION

Numerous advances in polymer chemistry have brought to market a variety of polymeric materials which have had a substantial impact in numerous industrial fields. Particularly in the pharmaceutical field, these advances have revolutionized not only the type of materials incorporated into drug delivery systems, but also the way in which these systems are produced and the methods for engineering such systems. Examination of the pharmaceutical field shows that prior to the 1970s the vast majority of pharmaceutical systems intended for oral administration were designed to provide a rapid immediate release, with only a few simple matrix systems available at that time for controlled delivery applications. One also notes that the materials utilized for such applications were primarily of natural origin, stemming from cellulose derivatives. In more recent years, synthetic polymers have appeared on the pharmaceutical stage, allowing for significantly greater utility and versatility of polymeric materials for pharmaceutical applications. The emergence of these materials has also been coupled with tremendous advances in processing technology of pharmaceutical systems, particularly the emergence of nanotechnology and the development of amorphous solid dispersions, which have allowed polymeric materials to revolutionize the way in which drug delivery systems are designed, formulated, and manufactured.

As a direct result of the increasing complexity of pharmaceutical systems, polymeric interactions with drugs and other excipients in the formulations play a significant role in determining product attributes. Product properties such as chemical purity, drug release, and physical appearance can all be directly influenced by the type of polymeric material chosen for development of a pharmaceutical system, ultimately leading to success or failure of the product under development. Additionally, because of the regulatory requirements for pharmaceutical systems to provide well-controlled physical and chemical properties throughout the product shelf life, pharmaceutical scientists must consider not only the initial properties of the product but also the long-term properties to ensure accurate and reproducible performance throughout the product life. These properties can be impacted by phenomena occurring over extended timescales. As such, one must think about polymeric interactions with drugs and excipients in terms of the immediate effect and long-term stability of the system in order to develop a suitable pharmaceutical system.

In accordance with the Code of Federal Regulations, the United States Food and Drug Administration (FDA) has been established to determine the safety and efficacy of drug products intended for medicinal use within the country. Similar regulatory bodies exist in other regions of the world, which include the European Medicines Agency and Japanese Ministry of Health, Labour and Welfare. Recently, as the world marketplace has become more global in nature, the major regulatory bodies have worked to harmonize guidelines for regulatory practices under the International Conference on Harmonization. This consortium of regulatory agencies has worked to establish guidelines for the development of pharmaceutical and medical systems, outlining many of the synchronized regulatory policies to which pharmaceutical systems must comply. These policies

Concise Encyclopedia of Biomedical Polymers and Polymeric Biomaterials DOI: 10.1081/E-EBPPC-120052274

cover the properties both for initial product release and for long-term stability. While many of these policies focus primarily on the properties of the active ingredient maintained within the product, growing regulatory discussion is also focusing on the properties of the polymers themselves. Additionally, these regulatory guidelines cover specific drug release properties which are frequently directly related to the chemical and physical stability of the active ingredient and polymers present, such as that found in a controlled release film-coated system. Since these regulations provide a basis for monitoring pharmaceutical product attributes in the context of providing optimized drug delivery systems dependent on the interaction with polymer systems, significant consideration must be given to the testing requirements provided.

Generally speaking, pharmaceutical systems intended for oral drug delivery can be subdivided into three major categories: traditional matrix systems, film-coated systems, and advanced systems. Traditional matrix systems encompass conventional tablet and capsule systems which have been available for over a century. These systems are generally characterized by a rapid drug dissolution profile and are commonly made using well-characterized excipients. A subcategory within this group is the controlled release matrix tablet, where a drug release rate-modifying system regulates the delivery of the active ingredient from the device. In these systems, specific interactions between the rate-controlling polymer and other components in the tablet may impact the observed release behavior from the system. Film-coated systems represent another distinct subclass of oral pharmaceutical systems and consist of dosage forms which have been coated for cosmetic and/or alternative functional reasons, such as immediate release tablets for taste masking, and multiparticulate systems which have been coated to achieve a functional release property, such as an enteric or sustained release. For such systems, particularly for those where the coating has been applied for release-modulating purposes, the interactions of the polymeric material providing the function with other materials present in the dosage form, as well as time, can play a significant role in determining product performance and even patient's safety. Environmental factors also provide significant interactions with the polymers chosen, allowing these interactions to regulate the release behavior from the system. Finally, a third class of oral dosage form is broadly classified as advanced dosage forms, which represent systems prepared using advanced pharmaceutical manufacturing technologies including hot-melt extrusion, spray–drying, and nanoparticle production. Unlike the previous two classifications of pharmaceutical systems, this group typically is employed in pharmaceutical production to enhance the oral bioavailability of the active pharmaceutical ingredient. In recent years, the advances in early-stage drug discovery technologies have led to an increasing number of candidate new chemical entities which exhibit poor aqueous solubility. These advanced technology platforms provide the ability to improve dissolution rate and solubility through increases in specific surface area and modification of thermodynamic properties of the formulation, and while effective at improving oral bioavailability, rely heavily on polymeric interactions to achieve and maintain such properties throughout the product shelf life. Divided into three major subsections based on the type of pharmaceutical production, this entry describes the behavior of pharmaceutical polymers in terms of their interactions with active pharmaceutical ingredients, other commonly used pharmaceutical excipients, and various environmental factors to ultimately yield desired or undesired behavior of the system.

POLYMERIC INTERACTIONS IN TRADITIONAL MATRIX SYSTEMS

Traditional matrix pharmaceutical systems consist of a variety of specific dosage forms however, for oral delivery applications these typically include tablets and capsules. Fundamentally defined as an intimate mixture of drug with additional pharmaceutical adjuvants, these systems can be developed to provide either immediate release, controlled release, or some combination thereof, where drug release is controlled by two unique mechanisms: disintegration and dissolution. For the preparation of such systems, the basic underlying unit operations are similar and have been well-studied over the last century.

For the production of pharmaceutical tablets and capsules, the basic manufacturing scheme is similar regardless of the type of system to be prepared (i.e., immediate release or controlled release) and consists of two major unit operations: blending and compression or capsule filling. During the blending process, the active ingredient is combined with required adjuvants in order to facilitate downstream production of the finished dosage form, with such materials generally consisting of diluents, disintegrants, glidants, and lubricants. In the case of controlled release systems, rate-controlling polymers such as hypromellose hydroxypropyl methylcellulose (HPMC) may be added to the blend. While glidants and lubricants commonly used in pharmaceutical dosage forms are not polymeric in nature, many of the commonly used diluents and disintegrants can be polymeric materials. Following blending, the formulation is then prepared into either a tablet or capsule dosage form via compression or encapsulation, respectively. For compression, rotary compression is the most commonly used method of manufacture and requires a blend capable of meeting the requisite flow and compressibility characteristics. Underlying material properties and component interactions play a significant role in performance, frequently requiring the use of additives or materials exhibiting specific characteristics to enhance performance. Often times, one sees that specific trade-offs must be made to achieve the necessary balance between compressibility and flowability. Additionally, additives to enhance disintegration and dissolution may be incorporated into the formulation, further complicating the delicate balance in designing such systems. Similarly, for encapsulation adequate flow and compressibility properties are necessary to facilitate production. While both flow and compressibility characteristics are important for achieving acceptable production characteristics, the magnitude of each which must be achieved is not the same as for compression. Generally speaking, some degree of flowability and compressibility are necessary for a formulation to function on a dosator or tamping mechanism encapsulator; however, the degree to which these properties must be optimized is significantly less than that required for compression. As such, flow aids such as glidants and lubricants can be minimized in these formulations. Furthermore, because of the loose powder nature of the material superdisintegrant levels may be reduced, although cases have been observed where capsule fill volume and head space have impacted dissolution. Given this, it is clear that the interaction between polymeric components, such as diluents and disintegrants, and other materials can play a significant role on the observed properties of the dosage form.

For optimization of blend properties, the primary product attributes are primarily associated with the physical properties of the materials much more so than the chemical properties. These properties include flowability, compressibility, and uniformity and are substantially influenced by the material morphology and density. Such attributes have been shown to be impacted by material properties such as molecular weight. In a study by Shlieout, Arnold, and Müller[1] the researchers examined the impact of microcrystalline cellulose molecular weight on compressibility, a vital product attribute for the production of pharmaceutical tablets and capsules. Microcrystalline cellulose is a common polymeric pharmaceutical excipient used as both a diluent and disintegrant which is derived from wood pulp. By modifying the degree of polymerization, the researchers showed that flowability and compressibility properties were impacted due to changes induced to the underlying material morphology when celluloses were prepared with different molecular weights, such that increasing molecular weight resulted in decreased flowability with increased compressibility. As such, there is a clear trade-off for production characteristics based on the molecular weight of the material selected to balance compression and flow characteristics. From a formulation perspective, this means that based on the properties of polymeric additives selected, additional components may be required to achieve the requisite compressibility or flowability characteristics. Batch to batch variation of raw materials may also influence performance, depending on supplier quality requirements. The magnitude of these required compensations will be dependent on the inherent properties of other materials in the formulation as well, particularly the active

ingredient and diluent which also generally make up large percentages of the blend formulation.

In addition to the optimization of blend properties to facilitate dosage form production, it is necessary to design a system capable of effective delivery of the active pharmaceutical ingredient. Polymeric interactions which can occur with other materials in the formulation can have a significant impact on oral bioavailability. Such polymeric materials, especially materials such as super-disintegrants and controlled release polymers, which play a critical role in bioavailability and onset of action, can also be subject to specific polymeric interactions with the gastrointestinal (GI) environment encountered upon administration. For example, numerous types of superdisintegrants are commercially available to assist in the disintegration of matrix tablet systems and the vast majority of these systems function by extensive water uptake which leads to a swelling of the polymeric network, thereby exerting a force on the tablet matrix capable of breaking apart interparticulate bonds. The rate at which water is taken up by the material plays a significant role in product performance. This behavior can be impacted by both environmental and compositional considerations. Bussemer, Peppas, and Bodmeier[2] examined the swelling potential of several common disintegrants, evaluating performance using a novel swelling apparatus to measure the force exerted by the materials. Results from the study showed that the swelling energy was proportional to the rate of water uptake, with croscarmellose sodium providing superior performance over a variety of other materials including low substituted hydroxypropyl cellulose (HPC), sodium starch glycolate, crospovidone, and HPMC. Further insight was also provided on the environmental impact to the swelling potential of croscarmellose sodium due to the ionic nature of the polymer. Since croscarmellose sodium is an acidic polymer displaying carboxylic acid functional groups, these groups will remain largely uncharged in acidic conditions, however will become charged in more neutral conditions representative of the later stages of the GI tract. When ionized, the common charge across these groups functions as a repulsive force which complements swelling due to water uptake. Furthermore, the overall rate of water uptake will be increased due to the ionized nature of these materials. Increasing the ionic strength of the bulk fluid, however, showed that there was a decrease in swelling energy due to competition of the ions for free water. Even with this observed behavior however, croscarmellose sodium still provided significantly greater swelling energy than the other materials studied, which indicated that while such polymeric interactions exist they may not have a physiologically relevant impact on the drug product performance. The magnitude of such interactions must therefore be addressed on a case-by-case basis.

Similar polymeric interactions have also been observed in controlled release systems, particularly for controlled release systems based on diffusional release through swollen hydrophilic gels. These systems, which provide one of the main technological platforms for oral controlled release of pharmaceutical compounds, have been extensively characterized and also implemented successfully in numerous commercially available products, due in combination to both the well-understood science and simplicity of manufacture. In many cases, controlled release matrix tablets can be prepared simply by directly blending the active ingredients and other excipients with the controlled release polymer prior to tableting in order to prepare such a system. Other techniques may also be applied to the production of such systems; however, even many of the advanced production techniques such as coprocessing of the drug substance with the controlled release polymer fall well within the realm of traditional pharmaceutical manufacturing.

Controlled release hydrogel matrix tablet technology is based on the incorporation of a swellable polymeric material into the dosage form, which provides a platform for swelling and the development of a gel layer which regulates the release of drug from the device in concert with any competing erosion mechanisms.[3–6] Similar to the behavior of superdisintegrants discussed earlier, controlled release matrix polymers exhibit the ability to absorb water and create a gel layer. Unlike the superdisintegrant class of materials, swelling in these materials is not associated with the same mechanical force that is observed for the later. As noted by Bussemer, Peppas, and Bodmeier[2] HPMC, a commonly used controlled release polymeric material, provided the lowest degree of swelling energy. This behavior allows such materials to form a mechanically intact gel barrier around the dosage form. Within the gel, polymer chains are interconnected through a combination of intermolecular interactions, chemical bonds, and/or physical intermeshing to create a "weblike" structure capable of providing a diffusional barrier. The release of drug substance from such devices is controlled by a series of complex polymer interactions which contribute to the solubilization of the active ingredient, formation of the gel, and transport through the gel layer. Gel formation and resulting layer thickness are governed by three distinct mechanisms: solvent permeation, swelling, and erosion, as shown in Fig. 1. Numerous researchers have developed empirical and first principle models describing release from such systems, with the most complete model to date being the sequential layer method. Simple models have been successfully shown to accurately describe drug release kinetics from such systems as well. Specifically, the Peppas equation illustrates the simple first-order release behavior of such systems which can actually be shown to account for the primary first principle behavior of Fickian diffusion based on drug concentration gradient from a simple geometric shape.[3] In more complex models, such as the sequential layer method,[6] Fickian release of drug from the dosage form is determined in a sequential, stepwise fashion accounting for water permeation,

hydration, solubilization of the active ingredient, and erosion of the gel layer front. Using this model, Siepmann and coworkers were able to use the model to predict dissolution profiles from a variety of HPMC-based controlled release systems having varying polymeric molecular weights and different device geometries. Furthermore, the applicability of this model to a variety of drug substances having a range of physicochemical properties was also demonstrated, providing greater accuracy than could be achieved with simpler earlier generation exponential models.

Given the complex series of competing events which lead to drug release from controlled release matrix systems, it is not surprising that polymeric interactions can also significantly influence release behavior and ultimately pharmaceutical efficacy of the dosage form. Gel layers formed by hydration of the polymeric matrix influence the rate at which drug is released and the environmental factors can significantly influence product performance. It has been well-established that environmental factors such as pH, ionic strength, and temperature can influence the behavior of polymer gels, altering liquid uptake rate, intermolecular chain separation distances, and observed viscosity. Variations in gel behavior due to these interactions can lead to significant changes in release profile, altering the pharmaceutical efficacy of the product produced.

In a study by Johnson and coworkers,[7] they examined the release behavior of phenylpropanolamine from HPC matrix systems having different particle size and molecular weight with respect to ionic strength. Their research demonstrated that at ionic strengths between 0.5 and 1.0 mol/L, disaggregation could occur resulting in dose dumping; however, they also noted that this was not likely to occur under normal physiological conditions where ionic strength is generally not more than 0.2 mol/L. Similar studies have also been undertaken to identify ionic strength–dependent phenomena in HPMC. Xu et al.[8] reported that the ionic strength-dependent behavior of HPMC was also directly related to the ionic competition for free water within the system which could lead to salting out phenomena. Their results mirrored those reported by Johnson and coworkers, showing that at ionic strengths greater than 0.5 mol/L burst release of propanolol from HMPC K15 tablets resulted, while ionic strengths below 0.5 mol/L had only a slight impact on release properties (Fig. 2). In addition to ionic strength, pH conditions experienced along the GI tract can also vary significantly. Although many of the common polymers used for hydrophilic controlled release applications are considered to be pH-independent, it is important to note that many pharmaceutical active ingredients are actually weak acids or weak bases capable of exhibiting pH-dependent solubility and ionization which can impact drug release. Changes in active ingredient solubility, particularly when being present at a high loading within the formulation, can also contribute to significant increases in erosion rates. Furthermore, the viscosity and strength of the gel layer formed may be affected due to the microenvironmental and

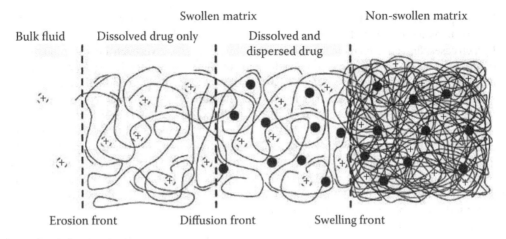

Swollen matrix   Non-swollen matrix

Bulk fluid   Dissolved drug only   Dissolved and dispersed drug

Erosion front   Diffusion front   Swelling front

Fig. 1 Schematic description of gel formulation in hydrophilic controlled release systems.
Source: From Siepmann & Siepmann.[5]

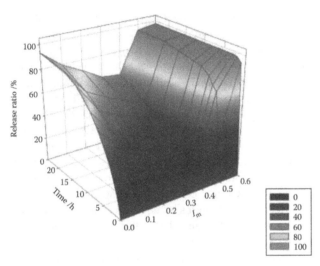

Fig. 2 Dissolution profile of propanolol from HPMC K15M matrix tablets at varying ionic strength.
Source: From Xu et al.[8]

regional pH of the GI tract. While HPMC has been shown to provide relatively independent release in the physiological pH ranges, controlled release tablets composed of carbomer and HPC have been shown to exhibit reduced gel viscosity properties based on pH and can be attributed to the chemical structure of the polymer. HPC, which has nearly 20 times the level of hydroxypropyl substituents than HPMC, providing greater potential for interaction with the environmental media. For carbomer, the carboxylic acids present in the polymer allow for ionization of the material based on environmental conditions, having a reported pK_a of 6.0. This ionization behavior allows for ionic interaction and also can impact intermolecular spacing within the gel based on the state of the material. Changes such as these can significantly impact release by reducing gel layer viscosity in the ionized state, leading to greater release rates in the later stages of the GI tract.

In order to better control the release characteristics from hydrogel-based controlled release matrix systems,

pH-modifying agents can be incorporated into these devices to regulate microenvironmental pH. While some polymeric systems may show susceptibility to environmental changes resulting in variable release profiles, pH-modifying agents are generally incorporated in order to minimize a physical incompatibility of the active moiety or better control its pH-dependent solubility. For example, oral dosage forms of fluvastatin[9] and ifetroban[10] have been successfully stabilized from pH-dependent decomposition by the incorporation of pH-modifying additives into the formulation. With controlled release systems, the design of matrix systems containing active ingredients which exhibit pH-dependent solubility can be better controlled using pH-modifying systems. As with active ingredients however, the pH-modifying agent is also subject to the same mass transport phenomena that occur with the active ingredient therefore requiring some consideration of the material properties such as solubility, as well as potential behavior in the polymer gel layer.[11] Although intuitively, one might think that the ideal modifying agent would have high solubility, this is in fact not the case. High solubility of the pH-modifying agent can lead to increased transport rates through the gel layer as a result of the concentration gradient, which may lead to a depletion of the component over time. Instead, incorporation of lower solubility material such as fumaric acid may provide a more stabilized release profile.[12] Solubility of these additives may also impact water uptake rate within the matrix, yielding a behavior where higher solubility materials facilitate water uptake. Using such an approach, controlled release matrix tablets for ZK 811 752, a novel weakly basic drug indicated in the treatment of autoimmune disease, were prepared by the addition of pH-modifying agents having several different polymeric base materials such as Kollidon® SR, ethyl cellulose, and HPMC.[11] This demonstrated the utility of such an approach, with numerous other examples also available in the literature.

Other interesting approaches to provide constant controlled release of drugs exhibiting a pH-dependent solubility have included the incorporation of pH-sensitive polymers in addition to the use of nonionic polymers to regulate drug release rate.[13] In such cases, the active ingredient exhibits a pH solubility profile opposite of that observed for the pH-sensitive polymer. This behavior creates a system which exhibits greater barrier layer permeability for a reduced concentration-driving force. Vice versa, one also notes a reduced barrier layer permeability corresponding to a higher driving force condition. In a patented composition, sodium alginate and HPMC matrices were prepared to regulate the controlled release of a pH-sensitive active ingredient.[14] At low pH conditions where the active ingredient exhibited greater solubility, the sodium alginate remained undissolved and provided a greater barrier to drug diffusion through the gel layer. Similar approaches have also been successfully implemented using mixtures of ionic and nonionic polymers such as Eudragit® L100-55 and HPMC to provide constant release of pH-dependent active ingredients.

Given the variability of gel properties and the dependence of drug solubility to environmental conditions, coupled with the interaction behavior, it is not surprising that researchers have attempted to better control intermolecular interactions between the polymers and different components in both the environment and dosage form to provide a more controlled drug delivery platform. In recent years, molecular imprinting of polymers has been employed to yield materials having well-controlled physical and chemical properties capable of interacting with the different components of a formulation as well as the environment to regulate drug delivery.[15,16] Generally speaking, molecular imprinting is defined as the process of incorporating specific chemical moieties into a polymer to provide a specific response in the presence of stimuli, such as environmental factors (pH, ionic content, etc.) or specific chemical moieties (drug substance, indicator compounds, etc.). Additionally, the incorporated moieties may also help to refine the polymeric network properties, providing better definition of internal structure, porosity, and tortuosity, thereby regulating the diffusion properties through the materials. Numerous examples of molecularly imprinted systems have been described in the literature covering a range of applications from insulin-responsive systems to improved hydrogel diffusion control. In one recent example where molecular imprinting was used to provide stimuli-responsive control of insulin delivery, a pH-sensitive hydrogel-based polymer was imprinted with glucose oxidase.[17] When exposed to elevated glucose levels, the imprinted glucose oxidase converted glucose to gluconic acid resulting in an elevated pH which triggered swelling of the hydrogel allowing for the release of insulin. Other relevant examples include the development of hydrogels exhibiting a controlled cavity size capable of capturing specific chemical compounds to regulate transport. In a similar fashion, Venkatesh et al.[18] described the use of imprinted hydrogels to control the transport of ketotifen fumarate through molecularly imprinted hydrogels due to both traditional diffusion and

increased drug–polymer interactions resulting the presence of specifically designed chemical moieties intended to elicit specific hydrogen bonding interactions. While still commercially limited in terms of availability, the design of systems capable of providing stimuli-responsive controlled release with a greater degree of reproducibility represents the next generation in drug delivery systems. Control based on both chemical interactions and diffusional limitations also provides significantly greater control over mass transport through nanoscale domains, which represents a substantial advantage for the development of advanced micro- and nano-drug delivery systems.

POLYMERIC INTERACTIONS IN FILM-COATED SYSTEMS

The application of a coated layer to pharmaceutical systems has been extensively utilized over the last century in drug product development. The reasons for such coatings range from cosmetic to functional and can include systems utilizing small molecules, polymers, and different combinations thereof. Beginning with sugar-based coatings to achieve taste masking and provide a cover for cosmetic defects, pharmaceutical coating systems have grown extensively and now rely primarily on polymers applied using both aqueous and organic solvent–based systems to achieve a variety of functional and visual properties for the finished dosage form. Current coating technologies can be divided into two major categories based on the functionality of the system, specifically immediate release coating and modified release coatings. Immediate release coatings are used for both dosage forms and intermediates such as pellets and granules and provide functionality which does not significantly impact oral bioavailability and drug delivery properties. Such systems include cosmetic coatings, moisture protective coatings, and taste masking layers. Conversely, controlled release coatings provide a major function for drug delivery and are typically designed to directly impact oral bioavailability of the dosage form. Examples of controlled release coatings include enteric coatings for delayed release, permeable coating for sustained release, and semipermeable membranes which are designed to facilitate the development of osmotic systems.

In all cases, pharmaceutical coatings must be applied to a substrate which is typically a reservoir of the active ingredient to be delivered. Such coatings, regardless of application, can be applied by several basic unit operations which have been well-described in the literature today. The primary unit operations for such processes include perforated pan coating, fluid bed coating, and rotary coating and the method of which will depend primarily on the substrate size. Conducted in pan coating systems, perforated pan coating is primarily used for the coating of tablets which range in size from approximately 50 mg to upward of 2000 mg and beyond at batch sizes ranging from a few grams to hundreds of kilograms. In such processes, the materials are placed into the pan and rotated under a solution or suspension spray containing the coating system. Dried under heated air to remove the solvent, the coating forms a coherent layer on the substrate which is applicable for immediate and modified release systems, although most commonly used for immediate and delayed release coating due to spray limitations associated with some controlled release systems intended to provide sustained or pulsatile release. Fluid bed coating is commonly used for the coating of multiparticulate dosage forms,[19] although the technology has also been applied to tablets as well.[20] In the case of fluid bed coating, the material is placed inside a product container and large volumes of air are passed through to achieve a fluidized material state. Within the equipment, a Wurster insert separates the upbed and downbed, allowing material to pass upward through the collar and past the coating zone where the coating material is sprayed onto the substrate. Following the pass, material reaches the expansion chamber where the velocity slows before ultimately returning to the product bed via the downbed flow along the outside of the column. For such a system, significantly higher spray rates may be achieved due to the order of magnitude greater drying capacity present in the unit and substantially greater specific surface area of the substrate. Furthermore, multiparticulates present unique advantages over conventional dosage forms, specifically with the large distribution of active ingredient within the product, dose dumping becomes a less substantial concern, and patient-to-patient gastric emptying becomes more uniform.[21] Finally, rotor granulation is the third major technique for applying a coating of a pharmaceutical product. In this unit operation, a modified fluid bed apparatus is used to apply a coating typically onto substrates less than 2 mm in diameter and often times significantly smaller. Furthermore, many rotor granulation processes may be implemented without the need for starting seeds.[22] Unlike a conventional fluid bed however, the rotor granulation unit consists of a traditional tower containing a rotating disk in place of the material screen which imparts an angular velocity to material in the product bed.[23] As the bed of material circulates, it passes a coating zone where a liquid spray is applied to the substrate. Optionally, when configured for dry powder granulation, a second zone will also spray solid micronized material onto the wetted substrate to provide increased coating rates and reduced solvent burden. Due in part to the rotational energy applied to the materials, small powders are frequently coated within these systems for taste masking and controlled release applications.

Since the method of application and conditions during processing for controlled release coatings play a significant role in the performance and behavior of the system, it is important to consider the underlying physics involved in each unit operation. Coating operations, regardless of the specific manufacturing process employed, are thermodynamically similar and based on the drying capacity of the system, where drying capacity is defined as the amount of solvent which can be removed per volume of purge gas.[20] For most applications the purge gas used is air; however, when coatings are applied using potentially explosive organic solvent systems inerting purge gases such as nitrogen are used to prevent potential oxidation of components and also to apply an additional measure of safety. Considering the example of air as a purge gas and the use of an aqueous coating system, one notes that the gas enters the system with a specific temperature, mass flow rate, and quantity of moisture already present within the feed stream. These properties establish the maximum amount of moisture which can be removed from the system per unit time and indirectly determine the coating system spray rate. Additionally, examination of a psychrometric chart allows one to establish the behavior of such an air mass under adiabatic cooling conditions, such as those which can be assumed in most coating systems, in order to model and assist with scale-up of the process. Most importantly, these properties establish the concentration gradient between the bulk air mass traveling through the system and solvent layer formed over the surface of the substrate and ultimately determine the drying rate. Careful control of these conditions is necessary to ensure uniform spreading of materials across the substrate surface and achieve complete drying of the layer to avoid defects such as surface cracking and orange peeling.

Coating systems are not typically single component systems dispersed within a larger solvent phase, but rather they are commonly multicomponent systems, each having different solubilities within the solvent. Coating systems relying on organic solvents are commonly selected to achieve complete solubilization of the majority of materials within the coating system. During processing, the primary polymer for coating is dissolved, as are the majority of additives which may also be present, resulting in a generally simple mechanism for film formation. Upon being sprayed onto the substrate, droplets spread to cover the surface and drying occurs rapidly in many cases because of the commonly volatile nature of many solvents selected. As the solvent is removed, the solution becomes more and more concentrated until ultimately a precipitation point is achieved wherein the components precipitate out. Ensuring similar solubilities within the solvent and a rapid evaporation rate helps to minimize segregation which may occur during the coating process. Generally, because the components are dissolved in solution during the process, one commonly notes a more effective and uniform coating over the surface, with fewer aging-related events during shelf life compared to the aqueous counterpart.

In addition to solvent-based solubilized formulations, many coating systems may actually be colloidal dispersions, such as Aquacoat ECD which is an aqueous dispersion based on the water-insoluble ethyl cellulose. Such systems have gained significant popularity since their introduction in the 1970s.[24] Based on the suspension state of the particles within the system, as well as the number and type of other components, drying rates may have a critical effect on the behavior of the system. Many polymeric coatings, particularly those performed using aqueous systems, are considered to be pseudo-latex systems in which the polymeric particles are actually a true colloidal dispersion, generally prepared by dissolving the polymer within an organic solvent which is then emulsified and the organic phase evaporated to yield a composition containing discrete nanoscale particles.[25] Such systems have tremendous advantages for coating, specifically the ability to achieve a very high polymer content within the solvent without providing a substantial increase of viscosity. This allows for significantly greater solids application rates since there is less solvent per unit solids mass removed during coating. Even still, the drying rate within the system is critical because of the particulate nature of the polymer dispersed within the solvent. As shown in Fig. 3, colloidal polymeric dispersions progress through a series of stages before achieving a final film coating. Following atomization, the spray containing the colloidal dispersion spreads across the surface of the substrate during which time substantial quantities of solvent are removed via evaporation. As the solvent is removed from the system, colloidal polymeric particles experience a gradual coalescence due to decreased void volume. As drying continues the colloidal particles begin to deform leading to the formation of an apparently continuous film which continues to form as water and polymer diffuse within the layer, ultimately yielding a final uniform film. Since the solvent plays a critical role in the coalescence and molecular mobility, the rate

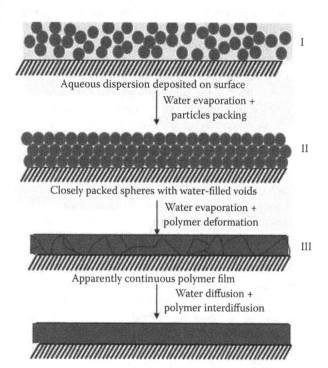

Fig. 3 Stages of film formation from colloidal dispersions.

Aqueous dispersion deposited on surface

Water evaporation + particles packing

Closely packed spheres with water-filled voids

Water evaporation + polymer deformation

Apparently continuous polymer film

Water diffusion + polymer interdiffusion

at which it is removed and extent to which it is removed play a significant role in the formulation of the dosage form. Additionally, incomplete drying may also result in changes in product performance over the storage period, commonly referred to as aging, which is a common problem among aqueous controlled release systems. With manufacturing playing such a critical role in product performance and product shelf life, optimization of the formulation and manufacturing must be investigated and controlled at an early stage to ensure successful development.

Most pharmaceutical systems are complex multicomponent systems containing a combination of polymers and additives all having an intended purpose. Such components include the primary polymer, an opacifying agent, plasticizer, antisticking agent, and stabilizing agent. Each material provides a specific function within the formulation and many of the additives facilitate a desired property of the formulation via a specific interaction with the polymer. For example, many polymeric coatings incorporate plasticizers into the formulation. These materials are added into the system to provide a specific interaction with the primary polymer, reducing the glass transition temperature of the polymer to promote film coalescence and uniformity. Components such as talc may also be included to minimize sticking of substrates during the coating process, with the levels of added material having a strong impact on both processing performance and dosage functionality *in vitro* and *in vivo*. Within this section, focus will be given to explaining polymeric interactions in terms of both controlled release and immediate release coatings, with a strong emphasis placed on controlled release systems.

Film-coated controlled release systems use a combination of erosion and diffusion to govern the release of active ingredient from the dosage form. Divided into two primary categories, delayed release and sustained release, the mechanism for each form of delivery is significantly different. For both forms of controlled release, successful and reliable implementation of these strategies is strongly dependent on polymeric interactions with other components as well as environmental factors.

For delayed release systems, which are designed to protect the contents from the acidic environment of the stomach, the rate of release is generally an erosion-dependent mechanism related to removal of the protective coating. Enteric polymers commonly exhibit acidic functional groups such as carboxylic acid, which remain unionized in the low pH gastric fluid rendering the coating essentially insoluble and impermeable. As the dosage form transits the GI tract, pH increases to above the pK_a of the polymer, resulting in substantial ionization of the polymer and dramatically increasing the solubility of the coating. Due to the increased solubility, rapid erosion of the coating occurs allowing for release of the once protected active ingredient. This interaction between the polymer and environmental pH has been exploited to develop numerous successful products, most notably the blockbuster product omeprazole which is subject to degradation in the presence of acidic gastric fluids.[26]

Uniquely distinct from delayed release systems, sustained delivery systems provide extended durations of release and are commonly used to provide once daily dosing of drug substances which exhibit good absorption properties with a short half life. Additionally, this technique has gained substantial interest in recent years with the development of life cycle management programs for pharmaceutical compounds, which attempt to increase the exclusivity period of a compound by developing more advanced dosage forms which provides improved efficacy and/or reduced dosing. For such systems, release is controlled by a combination of erosion and diffusion, each providing unique release mechanisms and a combined effect which can also contribute to the rate of release. Within the matrix system, drug substance is solubilized at the substrate surface and within the core, establishing a high concentration domain with the film coating functioning as a barrier layer leading to the bulk liquid. This concentration gradient provides a driving force for mass transport through the film layer which can be described by Fick's law.[5,27] Simultaneously, erosion occurring at the surface may reduce the thickness of the film over time, leading to increased mass transfer rates. Unlike matrix systems which are governed by a combination of drug diffusion through the gel and barrier layer erosion, the diffusion properties through the barrier are primarily responsible for governing release of coated systems. From a mechanistic point of view, numerous factors can contribute to the observed diffusion rate through a polymeric film, and include compositional factors as well as environmental factors associated with the conditions during release and storage.

Numerous properties of a polymeric film contribute to the observed diffusion rate of material within the layer, with the type and level of plasticization contributing significantly to this behavior. Plasticizers are generally incorporated into coating systems, particularly aqueous pharmaceutical coatings to provide improved coalescence of colloidal polymeric particles and provide for increased molecular mobility by reducing the glass transition temperature of polymer. This reduction in glass transition temperature occurs due to an interaction between the polymer and plasticizer resulting in an increase in free volume.[28] Selection of the appropriate polymer–plasticizer combination is based on the chemical properties of each material and can generally be predicted based on the solubility parameter of the materials involved.[29] Extensively studied, it has been well established that the selection of a specific plasticizer and given level within a formulation can have a significant impact on the glass transition temperature. Changes in material glass transition temperature can also have a substantial influence on the mechanical and storage properties of controlled release films. In a study conducted by Felton and McGinity,[30] their results highlighted the importance of plasticizer selection for achieving a reduction

in glass transition temperature and also the impact of plasticization on the physicomechanical properties of films, indicating a correlation between adhesional strength of a film and the effectiveness of the plasticizer. Intuitively obvious as well is the impact of changing plasticizer levels within a film. With increasing levels of plasticizer, the polymer will exhibit a lower glass transition temperature and increased molecular mobility. During the coating process, this will result in a film exhibiting greater coalescence, providing a larger diffusional barrier and thereby slower drug release. There is an upper threshold to which plasticizer can be incorporated however, before negative effects are observed, which include the excessive mobility and visual defects of the coating.

Another substantial issue associated with polymeric coatings based on colloidal systems is the apparent instability of the release pattern over time. Due to a combination of molecular mobility and residual colloidal nature, many controlled release films may exhibit aging phenomenon which is a manifestation of continued film coalescence. As highlighted in a study by Amighi and Moes,[31] the rate of theophylline release from Eudragit RS 30D-coated multiparticulates decreased over storage time due to the continued coalescence of the film. Addition of increased levels of plasticizer also showed an increased rate of coalescence on storage due to the increased polymeric mobility resulting from the plasticizer–polymer interaction. Since this behavior can have a significant impact on the therapeutic efficacy of the product, care must be taken to minimize changes in release profile on stability. One technique to avoid such transitions over stability is the intentional curing of material as part of the manufacturing process by maintaining the material under elevated temperature conditions after application of the coating to further improve coalescence. Numerous studies have demonstrated the utility of curing, showing how this process can be utilized to prepare a pharmaceutically stable film.[32]

Excipient interactions between formulations have also been shown to have a destabilizing effect on films during storage. The production of stable sustained release dosage form is predicated on the development of a homogeneous system or the ability to produce in a controlled and reproducible fashion a system with regular heterogeneity. Examples of each type include Eudragit L100-55 films designed to achieve uniform homogeneity and ethyl cellulose films containing low-molecular-weight polyvinylpyrrolidone (PVP) which are designed to have a uniform heterogeneity and drive pore formation to control release. In some cases however, excipients such as plasticizer and surfactants may be incorporated into the formulation which result in dynamic behavior on storage. In many cases, plasticizers may present issues with volatility resulting in a loss of the component from the film during storage. Such volatilization can change the mechanical properties of the formed film, resulting in a more brittle layer which may be subjected to mechanical breakage during the dissolution process,

Dental—Electrospinning

ultimately compromising the release rate.[33] Surfactants, which are also commonly present in aqueous coating dispersions, may be subject to transformations during the storage period. Due to the relatively high level of the materials which are required to support emulsion polymerization, the surfactants may actually exceed the miscibility limit within the polymer. Furthermore, the relatively high melting point of many of these materials may drive recrystallization of the surfactant over the storage period. This behavior creates large crystalline domains within the film that are capable of dissolving more rapidly than the controlled release film, resulting in pore formation that drives a more rapid drug release. An extensively studied example of this behavior is nonoxynol 100 which is used as a surfactant during emulsion polymerization of Eudragit NE30D dispersions. Over time the formation of recrystallized material has been reported, which resulted in a more rapid release from film-coated systems.

Recently, formulation strategies have also been applied to develop more robust systems which can provide sustained release over pharmaceutically relevant timescales through the incorporation of stabilizing materials, ranging from insoluble additives and high glass transition components to immiscible polymers. Under a similar general theory, these materials are incorporated into the films to limit film mobility and prevent further coalescence on storage. By limiting the coalescence which occurs during storage, coated films will be able to maintain a requisite permeability and porosity to provide consistent drug release throughout the pharmaceutical life of the product.

Numerous studies have investigated the use of high glass transition temperature materials in combination with low glass transition temperature-controlled release films in order to stabilize the observed release characteristics. These systems are predicated on the miscibility of the two materials in order to ensure uniform distribution of materials within the film, allowing the higher glass transition material to stabilize the film. One successful example of this was the combination of Eudragit L100-55, a relatively high glass transition temperature polymer, with Eudragit RS 30D, a controlled release polymer having a low glass transition temperature and frequently reported to exhibit physical aging on storage.[34] By incorporating L100-55 into the film, the researchers were able to stabilize the RS 30D and provide a consistent, albeit pH-dependent release of theophylline from the multiparticulate system.

The addition of insoluble additives, particularly at high levels within the formulation, can also be an effective mechanism for stabilizing controlled release films on stability. These materials present a scaffold that mechanically limits the motion of the film to coalesce during storage. Several common pharmaceutical materials such as colloidal silicon dioxide and talc can be used to exert such an effect. By combining extremely high levels of these materials in the formulation it is possible to limit the change in drug release over storage for a pharmaceutical composition.[35]

While drug release is stabilized, it is also important to note that such high percentages can also drive incomplete film formation. Aesthetic properties such as surface roughness may also become exaggerated due to the excess of solid material within the formulation.

Functioning in a similar method to insoluble additives, immiscible materials may also be incorporated into the film formulation to stabilize release properties by providing a physical barrier to film transitions over storage. Numerous materials, including albumin and hydroxyethyl cellulose, have been investigated recently and have shown varying degrees of success in stabilizing films. Researchers have successfully demonstrated the use of albumin in stabilizing Eudragit RS/RL 30D films.[36] Interestingly, initial studies showed that albumin as a 10% additive actually destabilized the dispersion, resulting in a significant change of drug release rates on storage. By acidifying the coating system prior to addition of the albumin, researchers were able to demonstrate that this material could be successfully employed for the stabilization of pharmaceutical coatings. Furthermore, it was demonstrated that this behavior was due to the specific protein-polymer behavior at elevated pH conditions, resulting from interactions caused by the quaternary ammonia group of the polymer and the negatively charged nature of the protein above the isoelectric point. In another study examining the stabilizing properties of hydroxyethyl cellulose as a film stabilizer for Eudragit RS 30D films, researchers showed that incorporation of the material prevented changes in drug release, mechanical properties, and water vapor transmission rates over the storage period.[37] Utilizing atomic force microscopy the researchers were able to characterize the nanostructure of the films produced. As shown in Fig. 4, films formed with hydroxyethyl cellulose showed discrete colloidal particles having a smooth surface while compositions without the stabilizer exhibited no discrete particles. This behavior was ultimately attributed to the coating of the hydrophobic colloidal particles with the hydrophilic stabilizer and that this barrier prevented the coalescence on storage.

Although the design of many multiparticulate systems limits the intermixing of drug substances with the functional coatings, many systems do provide an interfacial region which can result in potential drug–polymer interactions. These interactions can drive the formation of degradation products, thereby lowering the potency of the product and altering the impurity profile of the dosage form. Additionally, interactions may also compromise the drug release characteristics of the dosage form.[38] Studies have shown that active ingredients having alkaline properties can compromise enteric protection, while acidic moieties support the enteric protection by contributing to the microenvironmental pH.[39,40] One also notes that the solubility of the active ingredient can have a significant impact on the performance of the system, not only due to the resulting surface–core concentration gradient, but also due to migration of the active ingredient into the film layer. In a study by Bodmeier and Paeratukul,

Dental—Electrospinning

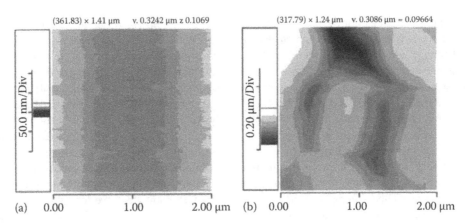

Fig. 4 Atomic force microscopy images showing the difference in film morphology between unstabilized Eudragit RS 30D films and Eudragit RS 30D films containing hydroxyethyl cellulose.
Source: From Zheng et al.[37]

it was shown that this behavior was due to a combination of aqueous solubility and drug–polymer affinity.[41] Such behavior indicated that there is not only an interfacial surface between the layers, but an interfacial volume in which the drug and polymer may be more intimately mixed. This allows for more complex interactions, including decomposition of the drug substance and partial plasticization of the film coating. Similar to the behavior of a traditional plasticizer, many drug substances have been shown to provide plasticizing effects on polymeric materials. In film-coated systems, partial penetration of the active ingredient into the coating can drive such behavior. Similarly, many drug substances will interact with the polymeric coatings to generate complex formation[42] and also drive degradation.[43,44] In general, these interaction behaviors can be limited through the use of subcoatings, which are designed to create a barrier between the layers, minimizing transport of drug into the polymer layer. Such behavior was illustrated by Bruce, Koleng, and McGinity,[38] where the application of a polymeric subcoating improved the enteric protection of Eudragit L100-55 films, as shown in Fig. 5.

Noting the earlier discussion between suspension and solution application of polymeric films, one quickly realizes that there is an inherent difference in solubility for specific polymer–solvent combinations. In many cases, pharmaceutical polymers exhibit significantly higher solubilities in organic solvents. As a result, the presence of such materials in the GI tract can significantly impact drug delivery. While most people do not routinely imbibe many of the pharmaceutically acceptable solvents, alcohol is commonly consumed in many cultures throughout the world. The presence of alcohol can interact with many pharmaceutical compositions, including controlled release coatings, resulting in dose dumping. Dose dumping is defined as the rapid and uncontrolled release of drug from a dosage form. Dose dumping from controlled release systems is a major problem that may raise the likelihood of a fatal exposure due to rapid drug release from the system in

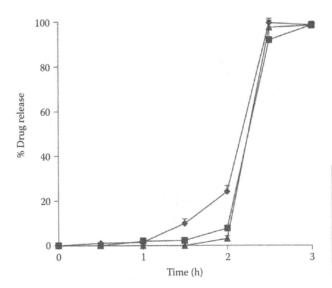

Fig. 5 Drug release profile from Eudragit L100-55 film-coated chlorpheniramine maleate pellets containing 0% (♦); 3% (■); and 5% (▲) Opadry® AMB subcoating. Eudragit L100-55 film coating level is 10%.
Source: From Bruce et al.[38]

combination with the generally higher dose present in the dosage form. In fact, several fatal cases have been reported over the years as a direct result of alcohol-induced dose dumping, with the most dramatic being those which resulted in the FDA-regulated removal of Palldone™ from the market.[45] The behavior of each specific pharmaceutical system varies significantly based on the overall composition and design of the product. Additionally, the concentration of alcohol within the GI tract plays a major role in the observed behavior, further complicating this situation. While many researchers have investigated ways to mitigate the effect of alcohol on controlled release systems, no well-accepted solution has been found to date. As such, products susceptible to dose dumping in the presence of alcohol generally carry a warning label to avoid consuming such products in conjunction with the medication. While an effective temporary solution which does improve safety and compliance, additional research in this field is required to achieve optimum product safety.

Controlled release systems based on modified release coatings are a successful platform for the delivery of pharmaceutical active ingredients, and the successful development and implementation of such products is based on a series of complicated polymeric interactions which help to provide product efficacy. Many of these interactions contribute to the overall success of the product, directly governing the rate of release. Other interactions, however, present a significant barrier to the development of a successful product. In recent years extensive research has been conducted to facilitate the development of successful systems, improving the reliability of release characteristics and minimizing the impact of aging on controlled release systems. Recent advances in immiscible material formulation and insoluble scaffold development present novel opportunities to improve long-term stability, while other areas of research specifically in the area of environmentally induced dose dumping still require significant contributions. Based on this, coated controlled release systems have been shown to be a viable platform for the production of pharmaceutical systems and will continue to play a critical role in the generation of advanced dosage forms designed to improve efficacy and safety.

POLYMERIC INTERACTIONS IN AMORPHOUS SOLID DISPERSIONS AND ADVANCED SYSTEMS

Advances in drug discovery technologies have led to an increasing number of new chemical entities which exhibit low aqueous solubility. This presents a significant limitation to oral absorption and results in a more challenging product development program. Over the last decade, there has been an explosion of new technologies and formulation strategies capable of addressing poor oral bioavailability due to low aqueous solubility, including crystal engineering, co-crystal formation, nanoparticle production, cyclodextrin complexation, and amorphous formation. Each of these processes seeks to maximize the dissolution rate through modification of the intermolecular interactions and/or reduction of particle size which often results in supersaturation, or the ability of a formulation to provide drug concentrations in excess of the equilibrium solubility. Ultimately, however, these formulations eventually return the drug concentration to its equilibrium solubility due to the thermodynamic instability associated with this state which can result in incomplete and variable oral bioavailability. An emerging field of research focuses on the maintenance of supersaturation to provide improved oral bioavailability of pharmaceutical compositions. This technique exploits unique interactions of drug and polymeric

stabilizers to provide longer durations and elevated concentrations of drug in solution to thereby improve oral absorption. Recently, extensive research has been undertaken in this field to establish the utility supersaturating systems for oral bioavailability enhancement and also provide a more detailed scientific understanding of the underlying mechanisms for stabilization. In addition to new formulation technologies pioneered to address aqueous solubility limitations, processing technologies have also been developed to produce stable amorphous compositions providing dissolution and product stability benefits. While many technologies have focused on particle size reduction to provide increased surface area, a select group of processes have been developed to regulate the crystal structure of the active ingredient to alter the thermodynamic properties of the material and enhance dissolution rate. The successful implementation of such technologies for the production of pharmaceutical systems is highly dependent on polymeric interactions with other formulation components, specifically interactions which solubilize drug within the carrier, specific interactions associated with polymer melt behavior, stabilization properties of polymers which lead to amorphous stability, and the behavior of drug–polymer combinations leading to stabilized supersaturation. This section describes the underlying mechanisms for solubility enhancement from supersaturatable systems, as well as current applications of solid dispersion systems for bioavailability enhancement which are dependent on polymeric interactions.

Pharmaceutical materials can be described by a variety of physicochemical properties, including chemical structure, crystal structure, and particle morphology. The chemical structure of a compound refers to the particular atomic composition of the molecule, which can be combined in large clusters of molecules which present a macroscopic three-dimensional structure. Within these clusters, the intermolecular spacing is referred to as crystal structure. In many compounds, this arrangement presents both long and short range order, resulting in specific interactions between the molecules within the ensemble. Such structures are commonly referred to as crystal lattices and are present in seven basic forms, regardless of whether they are organic or ionic crystals. Out of these seven basic structures, organic molecules may also present different packing configurations within the crystal lattice, thereby expanding on the number of possible configurations for arrangement. For a given molecule, multiple configurations may be possible, and such materials are referred to as polymorphic forms. Each polymorphic form also presents a different long and short range order resulting in different intermolecular interactions and variable thermodynamic properties of the system, specifically free energy. As a general rule of thermodynamics, systems always attempt to achieve the lowest free energy state and the variation of free energy within the different crystal forms establishes a most stable polymorphic form and then metastable polymorphic forms having higher free energies.[46] It is important to note that the specific crystal structure which is most stable may only be so under a given set of conditions, and deviation outside of those conditions would establish other forms as the most stable. This behavior is referred to as enantiotropic behavior, while a crystal form which is most stable over all conditions is referred to as monotropic.[47] Free energy of a crystal form is also directly related to the apparent solubility of the compound, true density of the substance, and melting point of the material, such that higher free energy systems exhibit a lower melting point, reduced density, and greater apparent solubility. Application of polymorphic form selection has been well-illustrated throughout the history of the pharmaceutical industry for a variety of applications, including morphological control, stability, and solubility. While a theoretical potential exists for such system, they have generally shown only limited solubility improvements while being hampered by thermodynamic instability.[48] As a result, the pharmaceutical industry has not extensively embraced pharmaceutical polymorph screening for oral bioavailability enhancement.

Material properties of pharmaceutical compounds can be further manipulated to control properties, particularly solubility, by eliminating long and short range order associated with the crystal structure to create an amorphous material. Since amorphous forms lack coherent long range order associated with a crystal form, the material is presented in the highest free energy state and can provide the greatest apparent solubility and dissolution rate, often providing significant increases in such metrics.[46,48] Generation of an amorphous form also results in dramatic changes in the physical properties of such materials. Unlike the crystalline counterparts which are characterized by well-defined melting points, amorphous materials are actually considered supercooled liquids below such transition points and exhibit a glass transition temperature where the equilibrium properties of the material, such as enthalpy and volume, deviate from that of the supercooled glass, as shown in Fig. 6. In such systems, properties such as solubility, molecular mobility, and vapor pressure are enhanced with respect to the crystal state. While improvements in solubility can offer the potential to enhance oral bioavailability, greater molecular mobility and elevated free energy can drive phase transformations and it is this instability that has been viewed by the pharmaceutical industry as one of the greatest drawbacks in the development of such systems.

Resulting from the solubility benefits provided by these systems, extensive research has been conducted investigating the fundamental mechanisms of recrystallization and polymorphic conversion, as well as techniques for improving the overall stability of such compositions. While recrystallization is predicted to occur as a result of the thermodynamic instability of the system, the rates of such transformations are determined by the kinetics associated with such processes, specifically activation energy and molecular mobility which may hinder such phenomena.

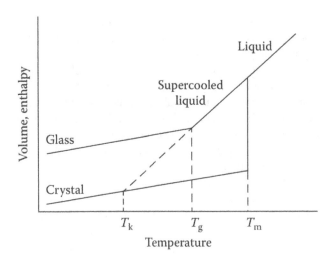

Fig. 6 Schematic diagram illustrating the differences in enthalpy and volume for crystalline and glassy solid as a function of temperature.
Source: From Hancock & Zografi.[118]

Relaxation rates of pharmaceutical systems have been described using the Vogel–Tammann–Fulcher equation which provides an Arrhenius-type relation for molecular relaxation as a function of temperature.[46] Further understanding of molecular motion within such systems is also expressed using the Williams–Landel–Ferry (WLF) equation which describes the temperature dependence of viscosity in such systems.[46] In both relationships, the strong dependence of properties influencing molecular mobility on temperature indicates that controlling the environmental conditions can regulate recrystallization rates. As shown in Fig. 6, as the supercooled liquid line is extrapolated past the glass transition temperature the line will intersect the enthalpy of the crystalline form which denotes the Kauzmann temperature where configurational entropy of the system reaches zero.[46] Utilizing the WLF equation, an approximate temperature of 50 K below the glass transition temperature will provide a situation in which all molecular motion is essentially stopped and recrystallization is no longer possible.[49,50] Using this result, one approach to stabilizing such systems has been to store them under special conditions where this rule can be maintained. It is important to realize, however, that such behavior may be impacted by environmental conditions, such as moisture uptake, which can help to plasticize the material and provide greater molecular motion.

An alternative approach, which is much more common for pharmaceutical products today, is the development of amorphous solid dispersions, which combine the active ingredient and one more carrier materials which can stabilize the amorphous form through specific interactions or a general increase in glass transition temperature.[50] Many of the commonly used pharmaceutical coating polymers and matrix dosage form polymers have also shown utility in such applications. In the most common form, this material creates a solid solution in which the drug substance is homogeneously distributed within the carrier matrix, yielding a single glass transition temperature during solid-state characterization. The behavior of such miscible forms is governed by the Gordon–Taylor equation in a two-component situation which is actually a special case of the multicomponent Fox equation.[51] Interestingly, when the drug and carrier exhibit a synergistic interaction, such as hydrogen bonding, positive deviation in the Gordon–Taylor relationship is observed which results in elevated glass transition temperatures.[52] Several examples have been shown recently, highlighting the ability of pharmaceutical polymers such as PVP and polyvinyl acetate phthalate to interact with drugs such as celecoxib and itraconazole. In addition to the benefits provided by such systems for elevating the glass transition temperature, benefits in dissolution rate can also be achieved through the establishment of polymer dissolution rate-limited release systems, which are described in subsequent sections.

The production of amorphous active ingredients and amorphous solid solutions is generally accomplished by one of four basic techniques, including precipitation from solution, vapor condensation, milling/compaction, and supercooling of a melt.[46] In three of the four schemes, the drug and optional carrier materials are dispersed in a molecular state and then applied through processing to render a solid amorphous form. For milling/compaction processes however, the amorphous form is produced by the introduction of high amounts of energy that drive the formation of molecular disorder within the system to form an amorphous material. While this method has been shown to provide the capability for amorphous material production, this class of unit operation is not commonly used for the production of such systems, but rather serves as a common example for the unintended conversion of crystalline material to amorphous material. For pharmaceutical production methods, amorphous systems are most commonly created using supercooling of a melt and precipitation from solution which are commonly referred to as fusion processing and solvent processing, respectively.[53] For such systems, the interactions between the polymers and other components such as drug substance are critical, helping to regulate uniformity and long-term amorphous stability of the composition.

Solvent processing techniques for the production of amorphous pharmaceutical systems are currently the most popular mode of manufacture for such products and consist of a variety of unit operations including fluid bed coating, spray drying, and particle engineering. While the specifics of each operation vary substantially, the basic principle is the same. Drug and optional carrier materials, most commonly polymers, are dissolved into an appropriate solvent, ranging from a super critical fluid to common organic solvent and mixed to develop a homogeneous solution. Polymers are frequently employed due to their stabilizing properties, high degree of miscibility with many common

active ingredients, and relatively high solubility in volatile organic solvents. That solution is then applied by a process in order to rapidly remove the solvent on a timescale where molecular mobility cannot induce recrystallization. By inhibiting the molecular motion and providing similar solubility characteristics in conjunction with rapid solvent removal, it is possible to prevent recrystallization to develop an amorphous form, capable of providing enhanced oral bioavailability.

Fluid bed coating is the most common commercial method for amorphous form production and several currently marketed products are available based on this technology, including Sporanox®[54] and Prograf®.[55] In both cases, the drug and HPMC carrier are dissolved in an organic solvent system which is layered onto multiparticulate cores. While this technology platform is capable of rendering an amorphous form and also provides benefits in using existing production capabilities, it suffers from requirements for potentially toxic solvents, as well as the need for extended drying periods at elevated temperatures to ensure complete solvent removal. Another technology for solvent-based production of amorphous pharmaceutical systems is spray drying. Similar to fluid bed coating, the drug and appropriate carriers are dissolved in a common solvent and sprayed into a chamber to rapidly remove the liquid phase. Unlike fluid bed coating, the droplets are sprayed through an atomizing air nozzle and directly into a drying chamber, whereas in fluid bed coating, the droplets are sprayed onto the pellet surface forming a film on the substrate. By processing the droplets directly into the air stream, it is possible to control not only the amorphous nature of the particles but the particle morphology. Extensive studies have been conducted to identify the critical variables for controlling such properties and ultimately it has been shown to be related to the relationship of decreasing droplet size and diffusive molecular flow within the system, which can be described by the Peclet number.[56,57] Through careful optimization of the processing conditions it is possible to produce particles having maximized specific surface areas, capable of further improving dissolution, solubility, and oral bioavailability. This technology has also been extensively applied to the development of systems for pulmonary delivery, specifically the production of large porous particles.[56–58] In addition to the obvious drawbacks associated with solvent use during processing, the low bulk density of the resulting material can also have significant issues on downstream processing during compression and encapsulation.[59] When preparing solvent systems for such applications, the loading of material into solution may be a limiting factor due to viscosity or atomization considerations and as a result is generally less than 20% w/w during most conventional processing which results in an inherently low density.[59] While process optimization can facilitate the production of formulations with increased densities and many new spray-drying systems allow for production under pressurized environments, most products are still produced having bulk properties which can negatively impact downstream processing. This presents a need for additional post-processing to increase density using techniques such as roller compaction or slugging. Another critical aspect of spray-drying process control focuses around the optimization of drying rates to prevent phase separation.[60] During the drying process, solvent is removed which results in the formation of concentration gradients for the components dissolved within the particles. Varying drying rates with respect to the differing material solubility can lead to phase separation. Control of such issues during production is another critical area of development, since these issues may result in reduced dissolution rates and poor product stability.

Particle engineering technologies have also been researched extensively, providing another viable platform for the production of pharmaceutical materials having well-controlled physical and morphological properties.[61–63] Three major subclasses for such technologies are supercritical fluid technologies, anti-solvent processes, and cryogenic production, each providing its own unique advantages for the manufacture of such materials.

Supercritical fluid technologies utilize solvents maintained under conditions of temperature and pressure above the critical point, allowing for common materials such as carbon dioxide to be used as a solvent. As a result of the supercritical and often nontoxic nature of the solvent employed, removal is often rapid and complete, only requiring a return of the material to ambient conditions. Unfortunately, supercritical fluids generally exhibit only minimal solubility for pharmaceutical materials, which can limit the applicability of this technology.[64] As a result, many of the technologies using supercritical fluids for pharmaceutical applications rely on it for the extraction of residual solvents, spurring the development of anti-solvent processes. In anti-solvent processes for pharmaceutical applications, the drug and the carrier are dissolved in a solvent that is miscible with a second nontoxic solvent system. The drug–carrier solution is then added into the second solvent, where the miscible toxic solvent diffuses into the solvent, while the drug–carrier mixture which exhibits extremely low solubility precipitates to form a particle. The particles can then be isolated and further dried to remove any residual toxic solvent remaining. This technology has also been effectively used to produce amorphous systems. Kim et al.[65] recently applied supercritical anti-solvent processing techniques to produce amorphous formulations of atorvastatin calcium to provide enhanced oral bioavailability. Their results showed that such technologies could be applied to the production of amorphous systems and provided significant improvements for oral bioavailability in a rat model. In another study by Vaughn et al.[66–68] using evaporative precipitation in aqueous solutions, an anti-solvent technique using an aqueous environment, substantially amorphous danazol particles were produced leading to *in vitro* supersaturation and enhanced oral bioavailability

compared to a physical mixture of drug and carrier in a murine model. One negative attribute of such production techniques is the relatively long timescales for diffusion of the solvent in the bulk anti-solvent and the dynamics of drug and carrier concentration which occur as a result. As highlighted by Vaughn et al., such processes can result in partially crystalline material which can compromise the long-term stability and negatively impact dissolution rates of such systems. During optimization, it is essential to identify processing conditions and formulation variables such as stabilizers which can impact the amorphous nature of the finished product.

Cryogenic processes are a type of solvent process that utilizes significant changes in temperature to produce amorphous systems by exposing the drug–carrier-loaded solvent to significantly reduced temperatures that result in rapid freezing of the material. Rapid freezing is designed to occur on time scales comparable to the precipitation kinetics of the drug and carrier in solution, preventing phase separation. Additionally, drying conducted by lyophilization limits molecular motion as a result of the low temperatures used during processing. Furthermore, the absence of external heat required to drive off the solvent phase presents additional benefits for the production of amorphous particles containing heat-sensitive high-value compounds such as proteins and peptides. Based on this general principle, a variety of specific types of techniques have been utilized to produce amorphous forms, including spray freeze-drying,[69] spray freezing into liquids (SFL),[62] and thin film freezing (TFF).[70] Utilizing such technologies, Vaughn et al.[67,68] produced amorphous itraconazole particles dispersed in a matrix of polyethylene glycol (PEG) 800 and poloxamer 188 by SFL. Similarly, Overhoff et al.[71] demonstrated the applicability of TFF for the production of amorphous danazol–PVP solid dispersion particles which exhibited high specific surface. Application of such technologies to the production of protein formulations was also demonstrated by Engstrom et al.,[69] again using TFF, where engineered particles provided high-specific surface area and also effectively maintain the activity of the model macromolecule.

An alternative to solvent-based processing, fusion processing involves the melting of drug and carrier material, followed by subsequent mixing and cooling to produce an amorphous solid dispersion. Although historically not as accepted as solvent processing, this class of manufacturing processes has become increasingly more common, with the commercially available Kaletra® formulation produced using hot-melt extrusion for oral bioavailability enhancement.[72] Unlike solvent processing technologies, only a few variations of the process are currently used in pharmaceutical research and production, including hot spin mixing, fluidized bed melt granulation, and hot-melt extrusion.[53,73–77]

Fluidized bed melt granulation and hot spin mixing are two forms of fusion processing that have been reported in research literature, however have currently found little application in pharmaceutical production. In a recent study by Walker et al.,[78] fluidized bed melt granulation was utilized to produce amorphous compositions of ibuprofen by granulating for a predetermined time at an elevated temperature of 100°C. While applicable for this particular formulation, compositions containing high melting point drugs or high viscosity polymers may exhibit limitations under the current process design. Modifications to the process for incorporation of melt spraying may also be limited due to the high viscosity of polymer melts or temperatures required for flow and atomization. In another series of studies, hot spin mixing was utilized for the production of amorphous systems. This technology functioned by combining the drug and carrier into a rapidly spinning heated vessel, allowing the material to melt and then ejecting material into a cooling tower. While little information was provided on the exact setup of the process, it was shown to be an effective platform for the production of solid dispersions containing testosterone, dienogest, and progesterone.[79-81]

Hot-melt extrusion is the pharmaceutically preferred method of fusion-based solid dispersion production and is based on polymer melt extrusion which has been in use for over a century. First applied to pharmaceutical compositions in the 1970s, hot-melt extrusion has shown utility in a variety of applications including bioavailability enhancement,[72] device manufacture,[82] and controlled release systems.[83-85]

Based on the equipment utilized for polymer processing, melt extruders typically consist of one or more rotating screws which convey material through a heated barrel, providing intermixing of materials and generation of additional heat due to shear and friction, ultimately forcing material through a shaped die located at the end of the unit to produce a rodlike structure. A schematic diagram of an extruder is presented in Fig. 7. Pharmaceutical extruders are derived from two basic designs based on the number of screws within the unit. Single screw extruders, as the name indicates, contain only one screw and are commonly used for pumping material through the extruder barrel. As a result of the single screw design, minimal flow perturbations are present along the flow pattern and less mixing is achieved than in other configurations. Additionally, these designs also may be subject to flow stagnation along interfacial surfaces, particularly along the melt-screw interface.[86,87] In some cases, specific mixing elements may be included on the screw to increase the frequency of flow discontinuities and improve convective mixing.[86,87] This can also result in prolonged residence times and increased potential for material degradation. Twin screw extruders are the other common form of pharmaceutical extruder and are designed with either corotating or counter rotating screws, each of which provides unique benefits to production. For counter rotating designs, increased convective mixing occurs as a result of the opposite direction of motion which generates intersecting flow patterns inside the barrel. Corotating designs provide less convective mixing than the counter rotating equipment as a result of the similar flow direction at the screw intermeshing; however, this also provides for a material removal action along the screw which is commonly termed "self-wiping."[88] As a result, residence and holdup times inside twin screw extruders can be minimized with respect to the single screw contemporaries. Additional optimization of mixing and material flow can be controlled using different elements for the screw. Screw elements refer to interchangeable segments of the screw which are designed to provide different functions, such as kneading elements for enhanced mixing or conveying elements for greater material throughput. Process optimization of material feed rates, screw speeds, and zone temperatures can all contribute to residence times; however, these values generally range from 30 seconds to 10 minutes based on extruder configuration and scale.

Similar to solvent-based compositions, fusion-based compositions are intended to be produced as a single homogeneous phase to improve dissolution and reduce recrystallization potential due to molecular mobility. Proper preformulation identification of drug–polymer

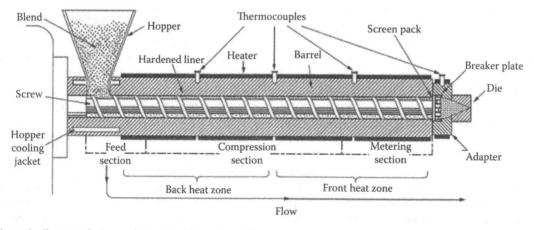

Fig. 7 Schematic diagram of a hot-melt extruder illustrating critical equipment and process aspects.
Source: From Follonier et al.[119]

miscibility is essential for the development of a successful formulation. Solubility parameters provide an indication of interaction energies associated with the mixing of different materials based on the molecular structure of the components. Greenhalgh et al.[29] successfully applied Hildebrand solubility parameters to predict the miscibility of ibuprofen in a variety of hydrophilic carriers and results indicated a strong correlation between observed behavior and that predicted by theory. Similar results have also been obtained for melt-extruded compositions containing indomethacin[89] and lacidipine.[90] Other screening techniques have also proved vital in the development of melt-extruded systems. Applications of differential scanning calorimetry and hot stage microscopy have both been shown to provide valuable information about the behavior of drug–polymer combinations at elevated temperatures, specifically providing information about degradation, miscibility, drug solubility within the carrier, and recrystallization potential upon cooling.[91–93] Extrapolation of results observed in these small-scale experiments can frequently be correlated to fusion processing behavior, although false negatives may occur due to the inability to approximate shear effects present in hot-melt extrusion. Even with careful examination of the formulation requirements, processing of certain types of materials may be difficult by melt extrusion, including the production of solid dispersions using polymers with high melt viscosity and the production of heat-sensitive components.

Unlike the solvent-based systems the use of thermal processes, hot-melt extrusion in particular, is highly dependent on the interaction between the drug and polymer to achieve the requisite processing characteristics. The ability to successfully manufacture compositions by hot-melt extrusion is directly related to the miscibility of the components within the system, melt behavior of the materials, and solubilization affinity of the carrier for the active ingredient. Formulations containing immiscible materials will form inherently unstable solid dispersions and may also yield difficulty during processing due to phase separation. Melt behavior may also drive processing difficulties. Many polymers exhibit a high molecular weight which is directly related to the observed melt viscosity of the material. In such cases, polymeric plasticizers must be included in the formulation or processing modifications must be modified to account for this behavior. Melt behavior also includes the chemical and physical stability of the materials. Many compounds, such as plasticizers, may have a low boiling point which results in venting of the specific material when maintained at temperatures in excess of the vaporization point. Solubilization capacity of the polymer is another critical attribute and is related to the miscibility of the system. It has been well-established in general science that "like dissolves like." In addition to being true for low-molecular-weight systems, it is also true to polymeric systems. Compositions having similar structures with respect to one another generally show some affinity to dissolve the other at elevated temperature. If the polymer shows similar chemical structures with respect to the active ingredient, which is frequently indicated with solubility parameters, solubilization of the drug within the polymer is maintained at temperatures above the glass transition temperature of the polymer. Such specific drug–polymer interactions can be exploited to facilitate lower temperature production and improved overall product quality.

The stability of such formulations is again a function of the glass transition temperature and specific interactions between the drug and carrier. Unlike many solvent processes, melt extrusion frequently requires the addition of a plasticizer to facilitate production by lowering the glass transition temperature to reduce melt viscosity.[74,75] This reduction affects the material under both elevated and room temperature conditions, increasing the molecular mobility of the finished product. In a recent study by Miller et al.,[94,95] a 20% plasticizer loading was required to achieve flow of melt-compounded itraconazole and Eudragit L100-55. Plasticization at this level reduced the glass transition temperature to approximately 50°C, making it unlikely that the formulation would have adequate shelf life upon storage. In another study by Bruce et al.,[96] melt-extruded compositions of guaifenesin and Acryl-Eze, a pre-plasticized form of Eudragit L100-55, exhibited extensive surface recrystallization, due in part to the reduce glass transition temperature and greater molecular mobility. It is important to note that in this case, the polymer was further plasticized by the presence of the drug substance within the solid dispersion. Due to the negative impact of the plasticizer on the solid-state properties of the finished product, studies have also been conducted using temporary plasticizers such as supercritical fluids. Verreck et al. examined the applicability of supercritical carbon dioxide as a temporary plasticizer for a variety of compositions to provide higher finished product glass transition temperatures and also process at reduced temperatures to improve the potency of heat-sensitive compositions.[97–99] Another technique to facilitate processing of temperature-sensitive active ingredients is to control the feedstock characteristics of the active ingredient. Through polymorphic selection it is possible to identify active forms with lower melting points, ultimately achieving the lowest possible processing temperature through use of the amorphous form. Lakshman et al.[100] successfully demonstrated the applicability of such an approach; however, many logistical issues may need to be addressed in order for such an approach to be commercially viable.

Even with the potential drawbacks associated with the addition of processing aids and degradation due to elevated temperatures, hot-melt extrusion has been extensively utilized for the production of amorphous solid dispersions for enhanced oral bioavailability, with some of the earliest uses of this technology for pharmaceutical applications being traced back to the 1970s.[53] To date, a multitude of publications have been presented documenting

Dental–Electrospinning

improvements in dissolution rate as well as bioavailability enhancement in animal models and human subjects. Hulsman et al.[101] utilized hot-melt extrusion for the production of 17β-estradiol in a matrix of PVP and Gelucire® 44/14 which provided a 30-fold increase in dissolution rate. Nimodipine solid dispersion were also produced by melt extrusion and shown to provide improved *in vitro* dissolution rates as well as oral bioavailability in beagle dogs.[102,103] In another example of melt extrusion technology R103757, an experimental compound exhibiting low aqueous solubility, was prepared as an amorphous solid dispersion in HPMC 2910 and provided improved *in vitro* dissolution rates.[104] Oral bioavailability studies in healthy volunteers showed that solid dispersions provided enhanced bioavailability, although the greatest improvement was provided by a cyclodextrin-based complex formulation. Itraconazole, another poorly water-soluble compound, has also been extensively studied using the melt extrusion platform, with results demonstrating that all solid dispersions could provide improved dissolution rates compared to the commercial multiparticulate formulation of Sporanox. Even with this dissolution improvement however, oral bioavailability studies in healthy volunteers showed no statistically significant improvement.[105,106] Subsequent studies by Miller et al.[94,95,107] indicated that the reason for this disparity was the formulation design which targeted supersaturation to the stomach instead of the upper small intestine. Kaletra is another example of a product produced using melt extrusion to provide improved oral bioavailability. This commercially marketed product, indicated in the treatment of human immunodeficiency virus, contains the poorly water-soluble drugs ritonavir and lopinavir[108] and increases oral bioavailability lopinavir through a combination of solubility enhancement and preferential metabolism of ritonavir. Using this platform, a significant reduction in pill burden and removal of strict low temperature storage requirements were obtained when compared to the original soft gelatin capsule formulation. These examples highlight the capabilities of fusion-based solid dispersions to provide improved oral bioavailability.

The current development of formulations capable of achieving supersaturation has been focused primarily on the development of metastable polymorphic forms, nanomaterials, and solid dispersions, each of which is capable of achieving high solubilities due to inherent kinetic and thermodynamic properties of the system. Solid dispersions have recently gained significant popularity for the production of pharmaceuticals and can be defined as an intimate mixture of one or more active ingredients in an inert carrier or matrix at solid state prepared by thermal, solvent, or a combination of processing techniques.[109] By developing these compositions it is possible to create formulations with the smallest possible drug domains, the individual molecules dispersed within a solid carrier, which are termed solid solutions. With solid dispersions, it is possible

to maximize the solubility enhancement while increasing physical stability through proper selection of excipients when compared to polymorphic screening.

One of the most frequently cited properties of these systems is the dissolution rate, which is increased due to the enhancement of surface area. As particles decrease in size, the total surface area required for the same amount of mass increases significantly. Additionally, as described by the Ostwald–Freundlich equation, changes in metastable equilibrium solubility have also been reported due to decreases in particle size. By developing smaller particles, an enhancement in the overall magnitude of kinetic solubility can be achieved in addition to the rate at which that solubility is attained.

Crystal structure of the material is the other major property frequently cited in the literature for solubility enhancement. Pharmaceutical APIs may exist in a variety of crystal structures, commonly referred to as polymorphs, as well as amorphous forms which lack any type of long or short range order associated with a crystalline material. When examining the dissolution process, it can actually be viewed as two separate and discrete steps: dissociation of the solute molecules from the crystal lattice and solvation of solute molecules. Modifications of the crystal structure can be used to reduce the intermolecular interactions and facilitate the dissolution of the drug substance. Ultimately, these forms offer transient solid-state properties and will eventually transition to the thermodynamically stable form of the drug substance. Detailed polymorphic screening and formulation optimization can be used however to provide compositions which are stable for pharmaceutically relevant timescales.

As previously mentioned, an extensive portfolio of technologies has been developed to exploit these mechanisms of solubility enhancement. Particle size reduction processes, including "top-down" and "bottom-up" technologies, maximize surface area to exploit the kinetic and thermodynamic advantages offered by size reduction. Molecular complexation techniques and solid solutions reduce the intermolecular interaction to facilitate dissociation while providing the theoretically smallest possible structure for dissolution, i.e., the individual drug molecule. Additionally, formulation techniques such as the incorporation of hydrophilic polymers and surfactants have been shown to further enhance the dissolution rate through improved wetting of microscopic drug domains within the composition.

Traditional dissolution testing is conducted under sink conditions, meaning that the amount of material added to the vessel is three to five times less than that required to saturate the media within the vessel.[110] By operating under these conditions, it ensures a sufficient driving force for drug release throughout the testing phase in order to mimic conditions found *in vivo* during the dissolution and absorption process. Additionally, operating under sink conditions allows for convenient mathematical assumptions facilitating modeling of the release process. Furthermore, traditionally manufactured crystalline dosage forms lack the requisite thermodynamic and kinetic forces for supersaturation, eliminating the need for examination under these conditions. Solid dispersions are capable of supersaturating their environment, necessitating testing under these conditions. Typically, supersaturated dissolution testing is conducted under similar conditions to sink testing; however, the amount of drug added to the vessel is several fold the amount required for saturation of the media, allowing the formulation, provided it has the underlying properties, to supersaturate the media. Additionally, due to the presence of small particle precipitation which may occur during the testing period filter sizes are frequently smaller than those used under sink conditions. It is generally assumed that the particle size cutoff for cellular uptake is 200 nm, so frequently 0.2 μm polvinylidine fluoride or polytetrafluoroethylene filters are employed to reduce crystallization on the filter membrane while minimizing particle size. Utilizing this testing procedure it becomes possible to ascertain the dissolution rate kinetics associated with a formulation, as well as its ability to provide and maintain supersaturation over prolonged periods of time. In the following sections, examples of *in vitro* supersaturation are presented along with the resulting enhancement in bioavailability to illustrate the utility of solid dispersions and supersaturation for enhanced therapeutic performance.

According to the BCS system, class II compositions exhibit solubility-limited bioavailability making both the compositional equilibrium solubility and the rate at which it is achieved restrictive steps in the oral absorption process. In order to improve the bioavailability of these drugs many formulations, both investigative and commercial, have been developed to provide enhanced dissolution rates. By formulating the solid dispersion with hydrophilic excipients capable of rapid dissolution rates, the drug dissolution rate will become a function of the dissolution rate of the carrier polymer, allowing for enhanced dissolution

rates and the potential for supersaturation. Several commonly used polymeric materials for this application include HPMC, PVP, low-molecular-weight PEG, vinylpyrrolidone vinylacetate (PVPVA), and Eudragit E100. Numerous publications are available in the literature focusing on the dissolution rate enhancement resulting from novel solid dispersion formulations to provide improved bioavailability or faster onset of action; however, only a paucity of these papers examined the ability of these compositions to supersaturate and correlated this behavior to performance in an animal model or human subjects. In recent years, the importance of supersaturation in achieving improved bioavailability has emerged as a critical design factor for formulation development.

Tacrolimus, which is currently marketed under the trade name Prograf, is produced as a solid dispersion using an organic solvent–based coating process.[111] Utilizing a thermal solid dispersion process, it was possible to prepare compositions which exhibited substantial supersaturation and enhanced *in vivo* performance in an animal model that were similar to those produced by compositions using the solvent-based production process, as shown in Fig. 8. The reason for the improved performance of formulations containing HPMC was attributed to the stabilizing interaction between the drug and polymer, which allowed the polymer to function as a recrystallization inhibitor. Similar approaches have also been taken to improve the oral bioavailability of itraconazole[94,95] and nifedipine.[112]

While the enhancement in dissolution rate can provide improved oral bioavailability by achieving the equilibrium solubility faster or providing higher metastable equilibrium solubility values, these formulations may not provide the greatest improvement in bioavailability. Weakly basic drugs, as stated previously, may be ionized at gastric pH and exhibit a higher solubility where hydrophilic polymer-based compositions will dissolve and release the drug. Upon transition to the upper small intestine, the pH rises and the drug may become partially or completely unionized

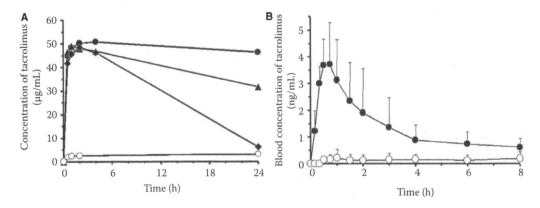

Fig. 8 Supersaturated dissolution profiles of thermally processed tacrolimus solid dispersions (**A**) and *in vivo* plasma profile comparing solvent based and thermally processed solid dispersions (**B**). (•) Tacrolimus: HPMC, (▲) tacrolimus: PVP, (♦) tacrolimus: PEG 6000, (○) crystalline tacrolimus.
Source: From Yamashita et al.[111]

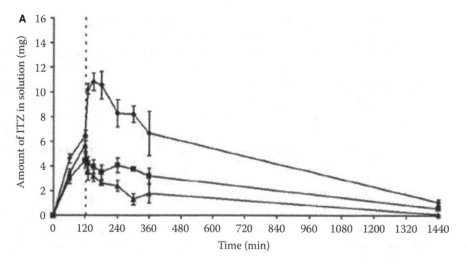

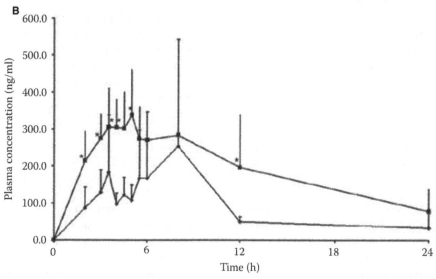

Fig. 9 *In vitro* (**A**) and *in vivo* (**B**) behavior of enteric compositions. (**A**) Supersaturated dissolution profile of ITZ:CAP formulations. Key: 1:2 ITZ:CAP (♦), 1:1 ITZ:CAP (■), 2:1 ITZ:CAP (▲). Each vessel (*n* = 3) contained 37.5 mg ITZ equivalent corresponding to 10 times the equilibrium solubility of ITZ in the acid phase. Testing was conducted for 2 hr in 750 mL of 0.1 N HCl followed by pH adjustment to 6.8 ± 0.5 with 250 mL of 0.2 M tribasic sodium phosphate solution. Dashed vertical line indicates the time of pH change. (**B**) *In vivo* plasma profile. Key: Sporanox pellets (♦), 1:2 ITZ:CAP (■). Formulations were administered by oral gavage at a dose of 15 mg ITZ/kg body weight per rat (*n* = 6). * indicates statistically significant concentration difference between test and reference formulation as determined by one-way ANOVA with Tukey post hoc testing.
Source: From DiNunzio et al.[113]

driving a significant solubility reduction. Furthermore, most drugs are primarily absorbed in the upper small intestine, where the substantial surface area provided by the villi and microvilli facilitate transport across the membrane. Compositions which supersaturate the gastric environment for short durations may also be subject to partial or complete precipitation, achieving only equilibrium solubility prior to entering the upper small intestine and negating the tremendous advantages provided by solid dispersions. In these cases, it would be prudent to target supersaturation to the upper small intestine, which is commonly achieved by using pH-responsive carriers. These carrier materials are insoluble at gastric pH; however upon entering the upper small intestine, the pH change will trigger

ionization of the carboxylic acid functional groups on the polymer chain resulting in dissolution. By providing a range of interactions between the drug and polymer, such formulations are capable of providing extended durations of supersaturation along with site targeting to minimize precipitation of the dissolved drug substance. These techniques have been demonstrated to be highly effective for a range of drugs, including itraconazole,[113] tacrolimus,[114] and HO 221.[115,116] These polymers, which not only maximize the rate of dissolution, but also provide significant stabilization and duration of supersaturation, are termed concentration-enhancing polymers. These materials are generally classified as high-molecular-weight hydrophilic polymers (HPMC E50, PVP K90, etc.) or enteric polymers.

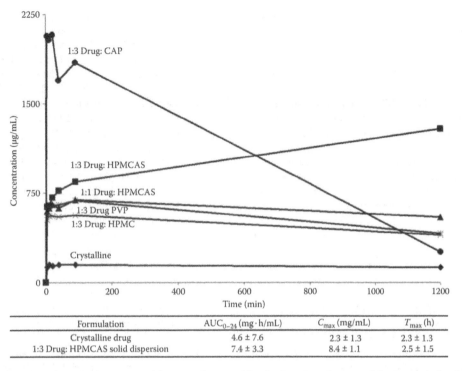

Formulation	AUC$_{0-24}$ (mg·h/mL)	C_{max} (mg/mL)	T_{max} (h)
Crystalline drug	4.6 ± 7.6	2.3 ± 1.3	2.3 ± 1.3
1:3 Drug: HPMCAS solid dispersion	7.4 ± 3.3	8.4 ± 1.1	2.5 ± 1.5

Fig. 10 *In vitro* supersaturation profiles (top) and *in vivo* pharmacokinetic data from human trials (bottom) for developmental solid dispersions.
Source: Adapted from Crew et al.[120]

The behavior for stabilization, although not completely understood, is believed to be the result of steric hinderance associated with molecular weight as well specific drug–polymer interactions such as hydrogen bonding. Further improving stabilization behavior of enteric materials is the charge associated with the polymers when dissolved, which provides a repulsive force limiting colloidal growth.

The application of concentration-enhancing polymers was studied for oral bioavailability enhancement of celecoxib, a poorly water-soluble compound marketed under the trade name Celebrex®.[117] Target concentration-enhancing polymers were identified using a high-throughput screening technique incorporating light scattering and yielded numerous potential candidates. Interestingly, the ability of these materials to inhibit precipitation was related to the CMC of the material. Addition of HPC was shown to enhance stabilization. When lead compositions consisting of TPGS and HPC, as well as formulations with Pluronic F127 and HPC, were dosed in a canine model, rapid and near complete absorption was observed. In a similar study by Overhoff et al.,[114] solid dispersions of tacrolimus were prepared using surfactant-based materials and shown to inhibit precipitation while also providing enhanced oral bioavailability in a rat model. Similar results, highlighted in Fig. 9, were also demonstrated by DiNunzio et al., using engineered amorphous solid dispersions of enteric polymers to provide sustained durations of itraconazole supersaturation to enhance oral bioavailability.[113] Such techniques

have also led to the development of several patented compositions which have been demonstrated to improve oral bioavailability, as shown in Fig. 10.

CONCLUSIONS

Polymeric materials play a critical role in the development of solid oral dosage forms and this contribution is likely to grow continuously in the foreseeable future as products continue to become more technologically advanced. Currently, polymers provide a basis for the development of controlled release systems and the stabilization of amorphous forms, presenting two unique delivery platforms which have contributed to the advancement of patient care. As scientists continue to develop a deeper and more complete understanding of the underlying interactions, and the impact of such behaviors on dosage form performance, pharmaceutical systems will continue to become more complex and more efficacious. Polymeric systems are also likely to provide significant contributions in many growing areas of pharmaceutical development, particularly for advanced device manufacturing and bionanotechnology applications. Many of the newest hybrid technologies exploit specific polymeric interactions in order to provide enhanced pharmaceutical efficacy. Continued learning and a greater understanding of these interactions will provide the basis for developing the next generation of drug delivery systems, providing improved treatment options for a number of disease states.

REFERENCES

1. Shlieout, G.; Arnold, K.; Müller, G. Powder and mechanical properties of microcrystalline cellulose with different degrees of polymerization. AAPS PharmSciTech **2002**, *3* (2), article 11.

2. Bussemer, T.; Peppas, N.A.; Bodmeier, R. Evaluation of the swelling, hydration and rupturing properties of the swelling layer of a rupturable pulsatile drug delivery system. Eur. J. Pharm. Biopharm. **2003**, *56* (2), 261–270.

3. Siepmann, J.; Peppas, N.A. Mathematical modeling of controlled drug delivery. Adv. Drug Deliv. Rev. **2001**, *48* (2–3), 139–157.

4. Siepmann, J.; Podual, K.; Sriwongjanya, M.; Peppas, N.A.; Bodmeier, R. A new model describing the swelling and drug release kinetics from hydroxypropyl methylcellulose tablets. J. Pharm. Sci. **1999**, *88* (1), 65–72.

5. Siepmann, J.; Siepmann, F. Mathematical modeling of drug delivery. Int. J. Pharm. **2008**, *364* (2), 328–343.

6. Siepmann, J.; Streubel, A.; Peppas, N.A. Understanding and predicting drug delivery from hydrophilic matrix tablets using the "sequential layer" model. Pharm. Res. **2002**, *19* (3), 306–314.

7. Johnson, J.L. Influence of ionic strength on matrix integrity and drug release from hydroxypropyl cellulose compacts. Int. J. Pharm. **1993**, *90* (1–2), 151–159.

8. Xu, X.M.; Song, Y.M.; Ping, Q.N.; Wang, Y.; Liu, X.Y. Effect of ionic strength on the temperature-dependent behavior of hydroxy-propyl methylcellulose solution and matrix tablet behavior. J. Appl. Polym. Sci. **2006**, *102* (4), 4066–4074.

9. Kabadi, M.B.; Vivilecchia, R.V. Stabilized pharmaceutical compositions comprising an HMG-CoA reductase inhibitor compound. U.S.P.T. Office, 1994.

10. Nikfar, F.; Serajuddin, A.T.; Jerzewski, R.L.; Jain, N.B. Pharmaceutical compositions having good dissolution properties. U.S. Patent 5506248 A, April 9, 1996.

11. Kranz, H.; Guthmann, C.; Wagner, T.; Lipp, R.; Reinhard, J. Development of a single unit extended release formulation for ZK 811 752, a weakly basic drug. Eur. J. Pharm. Sci. **2005**, *26* (1), 47–53.

12. Siepe, S.; Herrmann, W.; Borchert, H.H.; Lueckel, B.; Kramer, A.; Ries, A., Gurny, R. Microenvironmental pH and microviscosity inside pH-controlled matrix tablets: An EPR imaging study. J. Control. Release **2006**, *112* (1), 72–78.

13. Tatavarti, A.S.; Mehta, K.A.; Augsburger, L.L.; Hoag, S.W. Influence of methacrylic and acrylic acid polymers on the release performance of weakly basic drugs from sustained release hydrophilic matrices. J. Pharm. Sci. **2004**, *93* (9), 2319–2331.

14. Howard, J.R.; Timmins, P. Controlled release formulation. U.S.P.T. Office., 1988.

15. Byrne, M.E.; Park, K.; Peppas, N.A. Molecular imprinting within hydrogels. Adv. Drug Deliv. Rev. **2002**, *54* (1), 149–161.

16. Byrne, M.E.; Salian, V. Molecular imprinting within hydrogels II: Progress and analysis of the field. Int. J. Pharm. **2008**, *364* (2), 188–212.

17. Byrne, M.E.; Hilt, J.Z.; Peppas, N.A. Recognitive biomimetic networks with moiety imprinting for intelligent drug delivery. J. Biomed. Mater. Res. **2008**, *84A* (1), 137–147.

18. Venkatesh, S.; Saha, J.; Pass, S.; Byrne, M.E. Transport and structural analysis of molecular imprinted hydrogels. Eur. J. Pharm. Biopharm. **2008**, *69* (3), 852–860.

19. Jones, D. Air suspension coating for multiparticulates. Drug Dev. Ind. Pharm. **1994**, *20* (20), 3175–3206.

20. Kucharski, J.; Kmiec, A. Hydrodynamics, heat and mass transfer during coating of tablets in a spouted bed. Can. J. Chem. Eng. **1983**, *61* (3), 435–439.

21. Ghebre-Shellasie, I. Pellets: A general overview. In *Pharmaceutical Pelletization Technology;* Ghebre-Shellasie, I., Ed.; Marcel Dekker: New York, 1989; 6–7.

22. Vecchio, C.; Bruni, G.; Gazzaniga, A. Research papers: Preparation of indobufen pellets by using centrifugal rotary fluidized bed equipment without starting seeds. Drug Dev. Ind. Pharm. **1994**, *20* (12), 1943–1956.

23. Vertommen, J.; Kinget, R. The influence of five selected processing and formulation variables on the particle size, particle size distribution, and friability of pellets produced in a rotary processor. Drug Dev. Ind. Pharm. **1997**, *23* (1), 39–46.

24. Banker, G.; Peck, G.E. The new, water based colloidal dispersions. Pharm. Technol. **1981**, *5* (4), 55–61.

25. Carlin, B.; Li, J.-X.; Felton, L.A. Pseudolatex dispersions for controlled delivery applications. In *Aqueous Polymeric Coatings for Pharmaceutical Dosage Forms;* McGinity, J.W., Felton, L.A., Eds.; Informa Healthcare: Hoboken, NJ, 2008; 1–46.

26. Larson, C.; N. Cavuto, J.; Flockhart, D.A.; Weinberg, R.B. Bioavailability and efficacy of omeprazole given orally and by nanogastric tube. Dig. Dis. Sci. **1996**, *41* (3), 475–479.

27. Lecomte, F.; Siepmann, J.; Walther, M.; MacRae, R.J.; Bodmeier, R. pH-Sensitive polymer blends used as coating materials to control drug release from spherical beads: Elucidation of the underlying mass transport mechanism. Pharm. Res. **2005**, *22* (7), 1129–1141.

28. Aharoni, S.M. Increased glass transition temperature in motionally constrained semicrystalline polymers. Polym. Adv. Technol. **1998**, *9* (3), 169–201.

29. Greenhalgh, D.J.; Williams, A.C.; Timmins, P.; York, P. Solubility parameters as predictors of miscibility in solid dispersions. J. Pharm. Sci. **1999**, *88* (11), 1182–1190.

30. Felton, L.A.; McGinity, J.W. Influence of plasticizers on the adhesive properties of an acrylic resin copolymer to hydrophilic and hydrophobic tablet compacts. Int. J. Pharm. **1997**, *154* (2), 167–178.

31. Amighi, K.; Moes, A.J. Influence of plasticizer concentration and storage conditions on the drug release rate from Eudragit® RS 30 D film-coated sustained-release theophylline pellets. Eur. J. Pharm. Biopharm. **1996**, *42* (1), 29–35.

32. Amighi, K.A.; Moes, A.J. Influence of curing conditions in the drug release rate from Eudragit® NE 30 D film coated sustained release theopylline pellets. STP Pharm. Sci. **1997**, *7* (2), 141–147.

33. Anderson, W.; Abdel-Aziz, S.A.M. Ageing effects in cast acrylate-methacrylate film. J. Pharm. Pharmacol. **1976**, (Suppl 22), 28.

34. Wu, C.; McGinity, J.W. Influence of an enteric polymer on drug release rates of theophylline from pellets coated with

Eudragit® RS 30 D. Pharm. Dev. Technol. **2003**, *8* (1), 103–110.

35. Maejima, T.; McGinity, J.W. Influence of film additives on stabilizing drug release rates from pellets coated with acrylic polymers. Pharm. Dev. Technol. **2001**, *6* (2), 211–221.

36. Kucera, S.A.; McGinity, J.W.; Zheng, W.; Shah, N.H.; Malick, A.W.; Infeld, M.H. The use of proteins to minimize the physical aging of Eudragit® sustained release films. Drug Dev. Ind. Pharm. **2007**, *33* (7), 717–726.

37. Zheng, W.; Sauer, D.; McGinity, J.W. Influence of hydroxyethylcellulose on the drug release properties of theophylline pellets coated with Eudragit® RS 30D. Eur. J. Pharm. Biopharm. **2005**, *59* (1), 147–154.

38. Bruce, L.D.; Koleng, J.J.; McGinity, J.W. The influence of polymeric subcoats and pellet formulation on the release of chlorpheniramine maleate from enteric coated pellets. Drug Dev. Ind. Pharm. **2003**, *29* (8), 909–924.

39. Dangel, C.; Kolter, K.; Reich, H.B., et al. Aqueous enteric coatings with methacrylic acid copolymer type C on acidic and basic drugs in tablets and pellets. Part II: Dosage forms containing indomethacin and diclofenac sodium. Pharm. Technol. **2000a**, *24* (4), 36–42.

40. Dangel, C.; Kolter, K.; Reich, H.B., et al. Aqueous enteric coatings with methacrylic acid copolymer type C. On acidic and basic drugs in tablets and pellets. Part I: Acetylsalicylic acid tablets and crystals. Pharm. Technol. **2000b**, *24* (3), 64–70.

41. Bodmeier, R.; Paeratakul, O. The effect of curing on drug release and morphological properties of ethylcellulose pseudolatex-coated beads. Drug Dev. Ind. Pharm. **1994**, *20* (9), 1517–1533.

42. Alvarez-Fuentes, J.; Fernandez-Arevalo, M.; Holgado, M.A.; Caraballo, I.; Llera, J.M.; Rabasco, A.M. Morphine polymeric coprecipitates for controlled release: Elaboration and characterization. Drug Dev. Ind. Pharm. **1994**, *20* (15), 2409–2424.

43. Riedel, A.; Leopold, C.S. Degradation of omeprazole induced by enteric polymer solutions and aqueous dispersions: HPLC investigations. Drug Dev. Ind. Pharm. **2005**, *31* (2), 151–160.

44. Stroyer, A.; McGinity, J.W.; Leopold, C.S. Solid state interactions between the proton pump inhibitor omeprazole and various enteric coating polymers. J. Pharm. Sci. **2006**, *95* (6), 1342–1353.

45. Fadda, H.M.; Mohamed, M.A.M.; Basit, A.W. Impairment of the *in vitro* drug release behaviour of oral modified release preparations in the presence of alcohol. Int. J. Pharm. **2008**, *360* (1–2), 171–176.

46. Hancock, B.C. Amorphous pharmaceutical systems. In *Encyclopedia of Pharmaceutical Technology;* Swarbrick, J., Boylan, J.C., Eds.; Informa Healthcare: Hoboken, NJ, 2007; 83–91.

47. Brittain, H.G. Polymorphism: Pharmaceutical aspects. In *Encyclopedia of Pharmaceutical Technology;* Swarbrick, J., Boylan, J.C., Eds.; Informa Healthcare: Hoboken, NJ, 2007; 2935–2945.

48. Yu, L. Amorphous pharmaceutical solids: Preparation, characterization and stabilization. Adv. Drug Deliv. Rev. **2001**, *48* (1), 27–42.

49. Hancock, B.C.; Christensen, K.; Shamblin, S.L. Estimating the critical molecular mobility temperature (T_K) of amorphous pharmaceuticals. Pharm. Res. **1998**, *15* (11), 1649–1651.

50. Hancock, B.C.; Shamblin, S.L.; Zografi, G. Molecular mobility of amorphous pharmaceutical solids below their glass transition temperatures. Pharm. Res. **1995**, *12* (6), 799–806.

51. Gordon, M.; Taylor, J.S. Ideal copolymers and the second-order transitions of synthetic rubbers. I. Noncrystalline copolymers. J. Appl. Chem. **1952**, *2* (9), 493–500.

52. Gupta, P.; Thilagavathi, R.; Chakraborti, A.K.; Bansal, A.K. Role of molecular interaction in stability of celecoxib-PVP amorphous systems. Mol. Pharm. **2005**, *2* (5), 384–391.

53. Leuner, C.; Dressman, J. Improving drug solubility for oral delivery using solid dispersions. Eur. J. Pharm. Sci. **2000**, *50* (1), 47–60.

54. Gilis, P.M.; De Conde, V.F.V.; Vandecruys, R.P.G. Beads having a core coated with an antifungal and a polymer. U.S.P.T. Office. US Patent 5,633,015, 1997.

55. Letko, E.; Bhol, K.; Pinar, V.; Fosterm C.S.; Ahmed A.R. Tacrolimus (FK 506). Ann. Allergy Asthma Immunol. **1999**, *83* (3), 179–190.

56. Vehring, R. Pharmaceutical particle engineering via spray drying. Pharm. Res. **2008**, *25* (5), 999–1022.

57. Vehring, R.; Foss, W.R.; Lechuga-Ballesteros, D. Particle formation in spray drying. J. Aerosol Sci. **2007**, *38* (7), 728–746.

58. Shoyele, S.A.; Cawthorne, S. Particle engineering techniques for inhaled biopharmaceuticals. Adv. Drug Deliv. Rev. **2006**, *58* (9–10), 1009–1029.

59. Celik, M.; Wendel, S.C. Spray drying and pharmaceutical applications. In *Handbook of Pharmaceutical Granulation Technology;* Parikh, D.M., Ed.; Informa Healthcare: New York, 2005; Vol. 154, 129–158.

60. Patterson, J.E.; James, M.B.; Forster, A.H.; Lancaster, R.W.; Butler, J.M.; Rades, T. Preparation of glass solutions of three poorly water soluble drugs by spray drying, melt extrusion and ball milling. Int. J. Pharm. **2007**, *336* (1), 22–34.

61. Bhardwaj, V.; Hariharan, S.; Bala, I.; Lamprecht, A.; Kumar, N.; Panchagnula, R.; Ravi Kumar, M.N.V. Pharmaceutical aspects of polymeric nanoparticles for oral drug delivery. J. Biomed. Nanotechnol. **2005**, *1* (3), 235–258.

62. Hu, J.; Johnston, K.P.; Williams, R.O., 3rd. Nanoparticle engineering processes for enhancing the dissolution rates of poorly water soluble drugs. Drug Dev. Ind. Pharm. **2004**, *30* (3), 233–245.

63. Jia, L. Nanoparticle formulation increases oral bioavailability of poorly soluble drugs: Approaches, experimental evidences and theory. Curr. Nanosci. **2005**, *1* (3), 237–243.

64. Byrappa, K.; Ohara, S.; Adschiri, T. Nanoparticles synthesis using supercritical fluid technology—Towards biomedical applications. Adv. Drug Deliv. Rev. **2008**, *60* (3), 299–327.

65. Kim, M.-S.; Jin, S.-J.; Kim, J.S.; Park, H.J.; Song, H.S.; Neubert, R.H.; Hwang, S.J. Preparation, characterization and *in vivo* evaluation of atorvastatin calcium nanoparticles using supercritical antisolvent (SAS) process. Eur. J. Pharm. Biopharm. **2008**, *69* (2), 454–465.

66. Vaughn, J.M.; Gao, X.; Yacaman, M.J.; Johnston, K.P.; Williams, R.O., 3rd. Comparison of powder produced by

evaporative precipitation into aqueous solution (EPAS) and spray freezing into liquid (SFL) technologies using novel Z-contrast STEM and complimentary techniques. Eur. J. Pharm. Biopharm. **2005**, *60* (1), 81–89.

67. Vaughn, J.M.; McConville, J.T.; Burgess, D.; Peters, J.I.; Johnston, K.P.; Talbert, R.L.; Williams, R.O., 3rd. Single dose and multiple dose studies of itraconazole nanoparticles. Eur. J. Pharm. Biopharm. **2006a**, *63* (2): 95–102.

68. Vaughn, J.M.; McConville, J.T.; Crisp, M.T.; Johnston, K.P.; Williams, R.O., 3rd. Supersaturation produces high bioavailability of amorphous danazol particles formed by evaporative precipitation into aqueous solution and spray freezing into liquid technologies. Drug Dev. Ind. Pharm. **2006b**, *32* (5), 559–567.

69. Engstrom, J.D.; Simpson, D.T.; Cloonan, C.; Lai, E.S.; Williams, R.O., 3rd; Barrie Kitto, G.; Johnston, K.P. Stable high surface area lactate dehydrogenase particles produced by spray freezing into liquid nitrogen. Eur. J. Pharm. Biopharm. **2007**, *65* (2), 163–174.

70. Overhoff, K.A.; Johnston, K.P.; Tam, J.; Engstrom, J.; Williams, R.O. Use of thin film freezing to enable drug delivery: A review. J. Drug Deliv. Sci. Technol. **2009**, *19* (2), 89–98.

71. Overhoff, K.A.; Engstrom, J.D.; Chen, B.; Scherzer, B.D.; Milner, T.E.; Johnston, K.P.; Williams, R.O., 3rd. Novel ultra-rapid freezing particle engineering process for enhancement of dissolution rates of poorly water-soluble drugs. Eur. J. Pharm. Biopharm. **2007**, *65* (1), 57–67.

72. Breitenbach, J. Melt extrusion can bring new benefits to HIV therapy: The example of Kaletra tablets. Am. J. Drug Deliv. **2006**, *4* (2), 61–64.

73. Andrews, G.P. Advances in solid dosage manufacturing technology. Philos. Trans. R. Soc. Lond. A Math. Phys. Sci. **2007**, *365*, 2935–2949.

74. Crowley, M.M.; Zhang, F.; Repka, M.A.; Thumma, S.; Upadhye, S.B.; Battu, S.K.; McGinity, J.W.; Martin, C. Pharmaceutical applications of hot-melt extrusion: Part I. Drug Dev. Ind. Pharm. **2007**, *33* (9), 909–926.

75. McGinity, J.W.; Repka, M.A.; Koleng, J.J.Jr.; Zhang, F. Hot-melt extrusion technology. In *Encyclopedia of Pharmaceutical Technology;* Swarbrick, J., Boylan, J.C., Eds.; Informa Healthcare: Hoboken, NJ, 2007; 2004–2020.

76. Repka, M.A.; Battu, S.K.; Upadhye, S.B.; Thumma, S.; Crowley, M.M.; Zhang, F.; Martin, C.; McGinity, J.W. Pharmaceutical applications of hot-melt extrusion: Part II. Drug Dev. Ind. Pharm. **2007**, *33* (10), 1043–1057.

77. Repka, M.A.; Majumdar, S.; Kumar Battu, S.; Srirangam, R.; Upadhye, S.B. Applications of hot-melt extrusion for drug delivery. Expert Opin. Drug Deliv. **2008**, *5* (12), 1357–1376.

78. Walker, G.M.; Bell, S.E.J.; Andrews, G.; Jones, D. Co-melt fluidised bed granulation of pharmaceutical powders: Improvements in drug bioavailability. Chem. Eng. Sci. **2007**, *62* (1–2), 451–462.

79. Dittgen, M.; Fricke, S.; Gerecke, H.; Osterwald, H. Hot spin mixing—A new technology to manufacture solid dispersions, part 1: Testosterone. Pharmazie **1995a**, *50* (3), 225–226.

80. Dittgen, M.; Fricke, S.; Gerecke, H.; Osterwald, H. Hot spin mixing—A new technology to manufacture solid dispersions, part 3: Progesterone. Pharmazie **1995b**, *50* (7), 507–508.

81. Dittgen, M.; Graser, T.; Kaufmann, G.; Gerecke, H.; Osterwalf, H.; Oettel, M. Hot spin mixing—A technology to manufacture solid dispersions, part 2: Dienogest. Pharmazie **1995c**, *50* (6), 438–439.

82. Rothen-Weinhold, A.; Oudry, N.; Schwach-Abdellaoui, K.; Frutiger-Hughes, S.; Hughes, G.J.; Jeannerat, D.; Burger, U.; Besseghir, K.; Gurny, R. Formation of peptide impurities in polyester matrices during implant manufacturing. Eur. J. Pharm. Biopharm. **2000**, *49* (3), 253–257.

83. Fukuda, M.; Peppas, N.A.; McGinity, J.W. Properties of sustained release hot-melt extruded tablets containing chitosan and xanthan gum. Int. J. Pharm. **2006**, *310* (1–2), 90–100.

84. Lyons, J.G.; Devine, D.M.; Kennedy, J.E.; Geever, L.M.; O'Sullivan, P.; Higginbotham, C.L. The use of agar as a novel filler for monolithic matrices produced using hot melt extrusion. Eur. J. Pharm. Biopharm. **2006**, *64* (1), 75–81.

85. McGinity, J.W.; Zhang, F. Melt-extruded controlled-release dosage forms. In *Pharmaceutical Extrusion Technology;* Ghebre-Sellassie, I., Martin, C., Eds.; Informa Healthcare: New York, 2003; Vol. 133, 183–208.

86. Kim, S.J.; Kwon, T.H. Enhancement of mixing performance of single-screw extrusion processes via chaotic flows: Part I. Basic concepts and experimental study. Adv. Polym. Technol. **1996a**, *15* (1), 41–54.

87. Kim, S.J.; Kwon, T.H. Enhancement of mixing performance of single-screw extrusion processes via chaotic flows: Part II. Numerical study. Adv. Polym. Technol. **1996b**, *15* (1), 55–69.

88. Thiele, W. Twin-screw extrusion and screw design. In *Pharmaceutical Extrusion Technology;* Ghebre-Sellassie, I., Martin, C., Eds.; Informa Healthcare: New York, 2003; Vol. 133, 69–98.

89. Chokshi, R.J.; Sandhu, H.K.; Iyer, R.M.; Shah, N.H.; Malick, A.W.; Zia, H. Characterization of physico-mechanical properties of indomethacin and polymers to assess their suitability for hot-melt extrusion process as a means to manufacture solid dispersion/solution. J. Pharm. Sci. **2005**, *94* (11), 2463–2474.

90. Forster, A.; Hempenstall, J.; Tucker, I.; Rades, T. Selection of excipients for melt extrusion with two poorly water-soluble drugs by solubility parameter calculation and thermal analysis. Int. J. Pharm. **2001**, *226* (1–2), 147–161.

91. Forster, A.; Hempenstall, J.; Rades, T. Comparison of the Gordon-Taylor and Couchman-Karasz equations for prediction of the glass transition temperature of glass solutions of drug and polyvinylpyrrolidone prepared by melt extrusion. Pharmazie **2003**, *58* (11), 838–839.

92. Van den Brande, J.; Weuts, I.; Verreck, G.; Peeters, J.; Brewster, M.; Van den Mooter, G. DSC analysis of the anti-HIV agent loviride as a preformulation tool in the development of hot-melt extrudates. J. Therm. Anal. Calorim. **2004**, *77* (2), 523–530.

93. Zhou, D.; Zhang, G.G.; Law, D.; Grant, D.J.; Schmitt, E.A. Thermodynamics, molecular mobility and crystallization kinetics of amorphous griseofulvin. Mol. Pharm. **2008**, *5* (6), 927–936.

94. Miller, D.A.; DiNunzio, J.C.; Yang, W.; McGinity, J.W.; Williams, R.O., 3rd. Enhanced *in vivo* absorption of itraconazole via stabilization of supersaturation following

acidic-to-neutral pH transition. Drug Dev. Ind. Pharm. **2008a**, *34* (8), 890–902.

95. Miller, D.A.; DiNunzio, J.C., Yang, W.; McGinity, J.W.; Williams, R.O., 3rd. Targeted intestinal delivery of supersaturated itraconazole for improved oral absorption. Pharm. Res. **2008b**, *25* (6), 1450–1459.

96. Bruce, C.; Fegely, K.A.; Rajabi-Siahboomi, A.R.; McGinity, J.W. Crystal growth formation in melt extrudates. Int. J. Pharm. **2007**, *341* (1–2), 162–172.

97. Verreck, G.; Decorte, A.; Heymans, K.; Adriaensen, J.; Liu, D.; Tomasko, D.; Arien, A.; Peeters, J.; Van den Mooter, G.; Brewster, M.E. Hot stage extrusion of p-amino salicylic acid with EC using CO_2 as a temporary plasticizer. Int. J. Pharm. **2006a**, *327* (1–2): 45–50.

98. Verreck, G.; Decorte, A.; Li, H.; Tomasko, D.; Arien, A.; Peeters, J.; Rombaut, P.; Van den Mooter, G.; Brewster, M.E. The effect of pressurized carbon dioxide as a plasticizer and foaming agent on the hot melt extrusion process and extrudate properties of pharmaceutical polymers. J. Supercrit. Fluids **2006b**, *38* (3), 383–391.

99. Verreck, G.; Decorte, A.; Heymans, K.; Adriaensen, J.; Liu, D.; Tomasko, D.L.; Arien, A.; Peeters, J.; Rombaut, P.; Van den Mooter, G.; Brewster, M.E. The effect of supercritical CO_2 as a reversible plasticizer and foaming agent on the hot stage extrusion of itraconazole with EC 20 cps. J. Supercrit. Fluids **2007**, *40* (1), 153–162.

100. Lakshman, J.P.; Cao, Y.; Kowalski, J.; Serajuddin, A.T. Application of melt extrusion in the development of a physically and chemically stable-energy amorphous solid dispersion of a poorly water-soluble drug. Mol. Pharm. **2008**, *5* (6), 994–1002.

101. Hulsmann, S.; Backensfeld, T.; Keitel, S.; Bodmeier, R. Melt extrusion. An alternative method for enhancing the dissolution rate of 17P-estradiol hemihydrate. Eur. J. Pharm. Biopharm. **2000**, *49* (3), 237–242.

102. Zheng, X.; Yang, R.; Tang, X.; Zheng, L. Part I: Characterization of solid dispersions of nimodipine prepared by hot-melt extrusion. Drug Dev. Ind. Pharm. **2007a**, *33* (7), 791–802.

103. Zheng, X.; Yang, R.; Zhang, Y.; Wang, Z.; Tang, X.; Zheng, L. Part II: Bioavailability in beagle dogs of nimodipine solid dispersions prepared by hot-melt extrusion. Drug Dev. Ind. Pharm. **2007b**, *33* (7), 783–789.

104. Verreck, G.; Vandecruys, R.; De Conde, V.; Baert, L.; Peeters, J.; Brewster, M.E. The use of three different solid dispersion formulations—melt extrusion, film-coated beads, and a glass thermoplastic system—to improve the bioavailability of a novel microsomal triglyceride transfer protein inhibitor. J. Pharm. Sci. **2004**, *93* (5), 1217–1228.

105. Six, K.; Berghmans, H.; Leuner, C.; Dressman, J.; Van Werde, K.; Mullens, J.; Benoist, L.; Thimon, M.; Meublat, L.; Verreck, G.; Peeters, J.; Brewster, M.; Van den Mooter, G. Characterization of solid dispersions of itraconazole and hydroxypropyl-methylcellulose prepared by melt extrusion, part II. Pharm. Res. **2003**, *20* (7), 1047–1054.

106. Six, K.; Daems, T.; de Hoon, J.; Van Hecken, A.; Depre, M.; Bouche, M.P.; Prinsen, P.; Verreck, G.; Peeters, J.; Brewster, M.E.; Van den Mooter, G. Clinical study of solid dispersions of itraconazole prepared by hot-stage extrusion. Eur. J. Pharm. Sci. **2005**, *24* (2–3), 179–186.

107. Miller, D.A.; McConville, J.T.; Yang, W.; Williams, R.O., 3rd; McGinity, J.W. Hot-melt extrusion for enhanced delivery of drug particles. J. Pharm. Sci. **2007**, *96* (2), 361–376.

108. Breitenbach, J.; Lewis, J.. Two concepts, one technology: Controlled-release and solid dispersions with Meltrex. In *Modified-Release Drug Delivery Technology;* Rathbone, M.J., Roberts, J.H.M.S., Eds.; Informa Healthcare: New York, 2002; Vol. 126, 125–134.

109. Chiou, W.L.; Riegelman, S. Pharmaceutical applications of solid dispersions. J. Pharm. Sci. **1971**, *60* (9), 1281–1302.

110. Amidon, G.L.; Lennernas, H.; Shah, V.P.; Crison, J.R. A theoretical basis for a biopharmaceutic drug classification: The correlation of *in vitro* drug product dissolution and *in vivo* bioavailability. Pharm. Res. **1995**, *12* (3), 413–420.

111. Yamashita, K.; Nakate, T.; Okimoto, K.; Ohike, A.; Tokunaga, Y.; Ibuki, R.; Higaki, K.; Kimura, T. Establishment of new preparation method for solid dispersion formulation of tacrolimus. Int. J. Pharm. **2003**, *267* (1–2), 79–91.

112. Ho, H.-O., Su, H.-L., Tsai, T.; Sheu, M.-T. The preparation and characterization of solid dispersions on pellets using a fluidized-bed system. Int. J. Pharm. **1996**, *139* (1–2), 223–229.

113. DiNunzio, J.C.; Miller, D.A.; Yang, W.; McGinity, J.W.; Williams, R.O., 3rd. Amorphous compositions using concentration enhancing polymers for improved bioavailability. Mol. Pharm. **2008**, *5* (6), 968–980.

114. Overhoff, K.A.; McConville, J.T.; Yang, W.; Johnston, K.P.; Peters, J.I.; Williams, R.O., 3rd. Effect of stabilizer on the maximum degree and extent of supersaturation and oral absorption of tacrolimus made by ultra-rapid freezing. Pharm. Res. **2008**, *25* (1), 167–175.

115. Kondo, N.; Iwao, T.; Kikuchi, M.; Shu, H.; Yamanouchi, K.; Yokoyama, K.; Ohyama, K.; Ogyu, S. Pharmacokinetics of a micronized, poorly water-soluble drug, HO-221, in experimental animals. Biol. Pharm. Bull. **1993**, *16* (8), 796–800.

116. Kondo, N.; Iwao, T.; Hirai, K.; Fukuda, M.; Yamanouchi, K.; Yokoyama, K.; Miyaji, M.; Ishihara, Y.; Kon, K.; Ogawa, Y., et al. Improved oral absorption of enteric coprecipitates of a poorly soluble drug. J. Pharm. Sci. **1994**, *83* (4), 566–570.

117. Guzman, H.R.; Tawa, M.; Zhang, Z.; Ratanabanangkoon, P.; Shaw, P.; Gardner, C.R.; Chen, H.; Moreau, J.P.; Almarsson, O.; Remenar, J.F. Combined use of crystalline salt forms and precipitation inhibitors to improve oral absorption of celecoxib from solid oral formulations. J. Pharm. Sci. **2007**, *96* (10), 2686–2702.

118. Hancock, B.C.; Zografi, G. Characteristics and significance of the amorphous state in pharmaceutical systems. J. Pharm. Sci. **1997**, *86* (1), 1.

119. Follonier, N.; Doelker, E.; Cole, E.T. Various way of modulating the release of diltiazem hydrochloride from hot-melt extruded sustained release pellets prepared using polymeric materials. J. Control. Release **1995**, *36* (3), 243.

120. Crew, M.D.; Friesen, D.T.; Hancock, B.C.; Macri, C.; Nightingale, J.A.S.; Shankar, R.M. Pharmaceutical compositions of a sparingly soluble glycogen phosphorylase inhibitor. US Patent 7,235,260 B2, June 26, 2007.

Dental—Electrospinning

Electrets

Rajesh Kalia
Sapna Kalia
Department of Physics, Maharishi Markandeshwar University, Mullana, India

Abstract

This entry aims to reveal the importance of the effects of electrets as biomaterials in the field of biophysical phenomena and hence biomedical applications. The methods of electret formation are reported, which include different methods of charging such as contact electrification; thermal method; corona charging; liquid contact; and irradiation by gamma rays, beta rays, X-rays, and light. The method of investigating charge storage and charge decay by thermally stimulated depolarization current (TSDC) is also discussed in detail. Literature survey of studies by different authors is summarized to examine the choice of materials as bioelectrets and their biomedical applications. The work by different authors on collagen of bones and blood vessels is mentioned, so that a conclusion may be drawn for the choice of compatible biomaterials used in artificial heart vessels and artificial bone devices. This entry also covers the topic of plasma processing on polymers as well as the study on fibrous polystyrene (PS), polyhydroxyethylmethacrylate (PHEMA), polyetheretherketone (PEEK), polyethyleneterephthalate (PET), and polyethylenenaphthalate (PEN). Keratin and bile acid are used for preparation of biomaterials. The dielectric measurements and piezoelectricity of anionic and native collagens are reported and the results are utilized for cardiovascular prostheses, cellular growth, and systems to control drug delivery. Moreover, studies aimed at understanding the microscopic structure of life along with awareness of macroscopic effects through programming are reported. The most recent advancement reported in this entry is the utilization of nanotechnology in the field of bioelectrets. It has been concluded that when carbon nanotubes are introduced into a polymer matrix, there is an increase in electrical conductivity, optical character, and mechanical strength.

INTRODUCTION

On the basis of electrical conductivity, materials are generally divided into three categories: conductors, semiconductors, and insulators. The electrical conductivity of conductors is good, that of insulators is bad, and that of semiconductor lies between conductors and insulators. The insulators when subjected to external electric field develop permanent orientation of molecular dipoles. In fact, the application of external electric field results in trapping of charges inside the insulating material, and these trapped charges remain for a long time within the material; thus, the material termed as an electret.[1]

In ancient times, it was well known that rubbing amber against a fur cloth would attract other things in its vicinity. For more than two millennia, this effect was regarded as a mere scientific curiosity. In the 18th century, Gray[2] investigated the electrostatic attraction of a number of charged materials, such as resins and waxes. A century later, the definition of electrets material was given by Faraday,[3] which is still in use today. Faraday described electrets materials as dielectrics "which retain an electric moment after the externally-applied field has been reduced to zero." The dielectric materials that exhibit a "quasi-permanent" charge storage and/or dipole orientation are called electrets,

which was coined by Heaviside in 1885[4] in analogy to the already established magnet.

The term quasi-permanent indicates that the time constants characteristics for decay of the charge are much longer than the time periods for investigations on the electret.[1] A systematic investigation of electrets was performed by Eguchi in 1919.[5] He introduced thermal charging as an important technique in which the material is exposed to a high electric field while being cooled down from an elevated temperature. It was discovered that the charge adjacent to the electrode may have either the same sign as that on the electrode (named homocharge by Gemant)[6] or the opposite sign (heterocharge). Support for the presence of two different charges was given by Gross.[7,8] The heterocharge was observed due to oriented molecular dipoles, while the homocharge resulted from injected interfacial charging between electrodes and dielectric. Other charging methods, such as high-energy electrons,[9] corona discharge,[10] and liquid contacts,[11,12] were also introduced.

Earlier, research on the electret focused on naturally available materials, such as Carnauba wax. A turning point came with the industrial synthesis of insulating polymers.[13–15]

In polymeric electrets, charge storage and dipole orientation depend on many characteristics, such as chemical

Concise Encyclopedia of Biomedical Polymers and Polymeric Biomaterials DOI: 10.1081/E-EBPPC-120049911

impurities, structure of the polymers, degree of crystallinity, metal–insulator interfaces, interfaces at morphous–crystalline regions in the material, and mechanical stresses.

The properties of polymer electrets were studied extensively by Fukada.[16] Another turning point in the quest for applications of charge-storing materials was the invention of the electret condenser microphone by Sessler and West in 1962.[17] In this, sound causes a charged fluoropolymer membrane to vibrate in front of a static metal-back electrode. Unlike earlier condenser microphones, this design needs no external bias voltage.

The electret formation results in the accumulation of charges in the polymer. The space charge accumulation is accomplished by many methods for example by injecting charge carriers through discharge, by contact electrification, by application of particle beam or by light radiation or by heat. The electret thus formed is known as space charge electret.[1]

The dipolar electret formation is achieved by polarizing the electret material, subjecting it to an electric field. Contact electrification is a method of charging the material. In this method, two polymers are charged by bringing them in contact with each other.[18,19] Here, two phenomena contribute: kinetic effect and equilibrium effect. The kinetic effect originates when one piece of polymer is rubbed against another of the same polymer. Hence, electrification due to the kinetic effect generated as the stationary piece is heated more. The equilibrium effect, i.e., the contact electrification, is due to static contact between different materials.[20–23]

Another important method of charging is the thermal method. In this case, the polymers are subjected to electric field at increased temperature (the temperature should be above glass transition temperature, but it should be below the melting point) and then along with the application of an electric field, allowed to cool. Electrets thus formed are known as thermoelectrets.[24–28]

The charge is transferred to the polymer material because of discharge in the air gap. Generally, corona charging technique is applied for charging of electret by this method. In corona setup, the basic principle of charging is carried out by applying voltage between a pointed upper electrode, which is kept at an appropriate distance from one side of the polymer, and a planar back electrode on the other side. Hence, the ions are deposited on the surface layer and do not penetrate inside the polymer.[29,30]

The polymer can also be charged by liquid contact. In this method, one side of the polymer film is metallized and the other nonmetallized side is put in contact with a wet electrode.[31,32] Hence, a layer of liquid exists between the contact and the polymer. The liquid chosen for this method is water or ethyl alcohol. This method is used to charge a large area of dielectric by just moving the electrode over the surface. When voltage is applied to the electrodes, the liquid evaporates and hence the charge of one polarity is transferred through the liquid to one side of dielectric.

The electrets can be charged by placing them under γ-rays, β-rays, and X-rays. The charging is done by molecular ionization in this case. In fact, irradiation gives rise to an electron–hole pair that is taken with applied field to the electrode and hence results in charge separation.[33–45] A photoelectret is formed by introducing light radiations. Here the polymers that are photoconductors are coated with transparent electrodes and then kept under ultraviolet or visible light.[24,46–49]

TSDC TECHNIQUE

The charge stored inside the electrets and processes of charge decay of electrets in molecular terms are vital for unraveling the important properties of materials. Thermally stimulated depolarization current (TSDC) technique allows fast characterization of the dielectric response of a material under investigation in the temperature domain.[50] The experimental arrangement for TSDC measurement is shown in Fig. 1.

For TSDC measurements, the sample is placed in a furnace with the help of a sample holder. The sample holder consists of two metal electrodes and the sample is kept between these electrodes. The sample is heated from room temperature to the polarizing temperature (T_p) and then dc bias is applied to the sample at this polarizing temperature for a desired time span. Then, the furnace is switched off and the sample is cooled to room temperature under the applied voltage. The sample is cooled after polarization so that the relaxation time increases and polarization can be "frozen in" the material. After that the power supply is switched off. The sample is shorted to remove stray charges for the specified time. The depolarization current is measured at constant heating rate with the help of an electrometer.[51] The various relaxations may be obtained if we plot TSDC as a function of temperature and these relaxations are represented by Fig. 2. The occurrences of these relaxations depend on the material used. The γ-relaxation occurs at a low temperature and is obtained due to impurity centers present in the materials. The β-relaxation occurs mainly due to side-chain disorientation in the polymers.

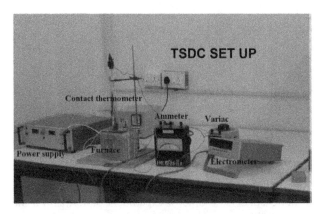

Fig. 1 Experimental arrangement for TSDC measurements.

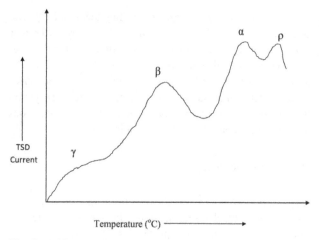

Fig. 2 Different relaxations in polymers.

Disorientation of the main chain segment at glass transition temperature (T_g) is responsible for α-relaxation. There is one more relaxation, i.e., ρ-peak, that arises due to immobilized space charges, which are nonuniformly stored and often reside near the electrodes.

Theoretical Formulations

The rate of decay for the stored polarization (P) and dipolar relaxation time (τ)[52] are related as

$$\frac{\mathrm{d}P}{\mathrm{d}t} = -\frac{P}{\tau} \qquad (1)$$

Polarization (P) is a function of temperature and time. When the dielectric is heated at a uniform rate, Eq. 1 can be expressed as

$$P = P_0 \exp\left(-\int_0^t \frac{\mathrm{d}t}{\tau}\right) \qquad (2)$$

For thermoelectrets, the relaxation time (τ) for decay or growth of polarization (P) is a function of temperature and is given by

$$\tau(T) = \tau_0 \exp\left(\frac{U}{kT}\right) \qquad (3)$$

where τ_0 is, a constant, called the pre-exponential factor, U is activation energy, k is the Boltzmann constant, and T is absolute temperature.

Depending upon activation energy (U), a large variation in $\tau(T)$ can be obtained by varying the temperature of the sample. The electret is defined in terms of its relaxation time as follows: "A material is said to be an electret if the decay times of its stored polarization (P) is very larger as compared to the characteristic time of experiment performed on the material."[53]

REVIEW OF BIOELECTRETS

Murphy and Merchant carried out trials on natural and synthetic polymers to examine the blood compatibility.[54] They concluded that if the surface electronegativity of polymers is increased by electret effect, then it can find a large number of applications in medical devices and also improve blood compatibility. They drew an important conclusion that if artificial heart valves are made of uncharged polymers, then they thrombosed in shorter span of time but negatively charged polymers through electret effect last pretty much longer. Hence they strongly supported the fact that use of electret technique to enhance the surface electronegativity is useful for improvement of blood compatibility and thromboresistant properties. A more revolutionary concept has come with the work of Mascarenhas.[55] The work was carried out on the concept that the bone possesses electrical properties. The investigation was made on bovine, canine, and rodent samples of human; and the results revealed that bone possesses the electret state and is able to store a large amount of polarization. The of electret effect proved to be a great contribution for orthopedics branch of medical science for their utilization in artificial bone devices.

Mascarenhas extended his work on collagen of bones and blood vessels that also contain collagen and elastin. He confirmed that collagen of bone and blood vessel samples exhibit electret polarization curves; furthermore, he calculated the activation energy of these samples that lies between 0.3 and 0.5 eV. This information played an important role for the development of bioelectrical properties of vessels that are thromboresistant in nature.

Fibrous keratin is a class of protein molecules that lies mainly in the structural units of various living tissues. There are major protein components of hair, wool, nail, horn, hoof, and quills of feathers. These molecules possess large electric dipole moment.[56] At a low temperature they are highly ordered but at a higher temperature the helical chain disorient toward the random state. It is similar to electret behavior. So it is confirmed that keratin has total charge production that confirms the uniform dipole orientation within it. This conclusion further opens a new biophysical phenomenon. The studies were carried out for plasma-assisted surface that can be compatible and efficient as artificial polymeric biomaterials. In addition, the molecules that consist of amino acids are employed to form a biomolecule network.

Understanding of resultant complex structure is vital. Studying their structures will enable to control the further development of complex techniques, for example, chemical micropatterning. The techniques were cited first in 1990s and there are ongoing developments on patterning techniques on various biomaterials such as polystyrene (PS), polyetheretherketone (PEEK), and polyethylenenaphthalate (PEN).[57] Schroder et al. studied the effect of plasma processing technique on polymers PS, polyhydroxyethylmethacrylate

(PHEMA), PEEK, polyethyleneterephthalate (PET), and PEN.[58] More developed micropattern was further analyzed using X-ray photoelectron spectroscopy technique and fluorescence staining. An important conclusion of their study reached to the fact that patterning technique is applicable to different hydrocarbon polymers.

Yang, Gu, and Zhu investigated collagen bioelectret, which is studied under various conditions, such as charging voltage, temperature, and time.[59] They concluded that the intracellular calcium level of CHO is increased, when attached to the charged collagen and this attachment also leads to decrease the human cervix uteri tumor HeLa cells.

Wei and Xia worked on the preparation of polymers obtained from bile acids.[60] Bile acids are derivatives cholesterol formed in the liver. They help to bind cholesterol and other fats in the small intestine.

Their work on preparation of biomaterials from bile acids is remarkable as the prepared materials can be utilized for controlled drug release, molecular recognitions, and tissue engineering. In fact, it is convenient to substitute functional groups in bile acids by varying the carboxyl group hydroxyl groups that further leads to polymerization. Moreover, the desired glass transition temperature and the mechanical properties of the materials are obtained by adjusting the ratio of co-monomers. The biomaterials thus obtained have increased compatibility utilized for substitution of soft tissues such as skin and blood vessels.

Blaisten-Barojas and Mascarenhas have found that fibrous keratin constitute the natural bioelectret. Their correlated walk model described the results of thermally stimulated current (TSC) of α-keratin fibers.[61] Their studies, along with the TSC measurements, gave important information regarding hydrogen bond energies and ordering of α-helix molecules that are present in keratin fibers.

Goes et al. had made complex dielectric measurements of anionic and native collagen films.[62] The aim of the studies is to develop a unique biomaterial that can be utilized in coating of cardiovascular prostheses, cellular growth, and systems to control drug delivery. It has been explained by the differential scanning calorimetry curves of collagen films that anionic collagen possess low thermal stability as compared to native collagen. Moreover, anionic collagen has high piezoelectricity than native collagen.

In fact, here alkaline treatment is being done, which results in the hydrolysis of carboxiamide residues of asparagine and glutamine. Hence, there is a presence of net negative change of collagen molecules at neutral pH. Moreover, electrostatic and hydrophobic interactions between amino acid chains of collagen result in the stabilization of the fibers. There is an attraction among opposite-charge amino acids and repulsion among like-charge amino acids. The overall change in charge and ionized groups as pH would result in significant change in fiber physical properties. Finally, they concluded that the variation in the charge and number of ionized groups with the change in pH would change the interaction forces among fiber and hence affect the physical properties of the fiber. They also carried out

the experiment to study the electromechanical resonance that is associated with the piezoelectricity of the sample. In addition, they concluded an important aspect during their calculations that resonance frequency increases with the increase of electrode diameter. Moreover, their results also include the sample film preparation. The collagen film prepared by electrodeposition method shows higher piezoelectric coefficients than does the evaporated films.

Mesquita et al. had emphasized the idea of correlating biology with physics and chemistry.[63] The group believed in the fact that life is an important and natural phenomenon. The phenomena related to life are quite complex to be understood. So the devotion toward physicochemical field needed attention that will facilitate understanding chemical composition of life formation.

Moreover, to understand the microscopic structure of life, one must be aware of macroscopic effects. They mentioned the need of approach of physics that will aid in designing a program that simultaneously deal with the system at microscopic level as well as synergetic aspects. They carried over the same approach to study a large-scale quantum action in brain functioning. According to the group, there is a need for quantum coherence; it means to keep information in an organized form. In studying the quantum effects of a single particle, an atom, or a molecule at a smaller scale, we cannot overlook the quantum system at a larger scale. The group worked on the approach by answering the two questions: one is to investigate the origin of complexity and the second is to use a theoretical approach to deal with the complexity. The work is quite fascinating in the field of biological system undergoing inside the brain. There is ordered water possessing dielectric properties inside the brain. Such ordered state outside of cytoskeletal surfaces and is referred to as an electret state. The group tried to answer theoretically the question of propagation of signals at long distances from brain within the body. Their results include the fact that in a diseased person, there is decrease in the flow of ideas, emotions, and memories. This problem of brain is linked with disruption in the microtubules because of the orientation in their shapes.

Burghate et al. prepared succinic acid-doped glycine (glycine 92% and succinic 8%) from a shell and mortar.[64] Then, the succinic acid-doped glycine pellet was subjected to an electric field and taken in the form of electret. The electric conduction of sample was measured and analyzed using Poole–Frankel, Nordheim, Schottky, $\log(J)$ versus T plots, Richardson and Arrhenius plots. The results reveal that the Schottky–Richardson mechanism of conduction dominates over others in doped samples.

Cao et al. devoted their studies to the utilization of nanotechnology in the field of bioelectrets.[65] The literature survey carried out by them supported the fact that introduction of carbon nanotubes into polymeric matrix results in enhancement of electrical conductivity, optical character, and mechanical strength. The group prepared a nanocomposite through covalent bonding between multiwalled carbon nanotubes (MWCNTs) and grafted collagen matrix. The stress–strain curves of the MWCNTs–grafted collagen nanocomposites show an increase in yield strength of collagen by 92.9% and increase in elastic modulus by 92.3%.

Tanaka et al. prepared ceramic hydroxyapatite (HA) by utilizing proton conductive property.[66] It is concluded in the results that charge storage and dipole relaxation can be controlled by selecting the sintering and poling conditions. Hence, HA electrets can be utilized as tissue engineering material in future.

CONCLUSION

The review described in this entry leads to a conclusion that the study of electrets effect in biomaterials and biophysics is interdisciplinary in nature. Systematic investigations involve the cooperation of the field of biology, physics, and chemistry. The work on bioelectrets is a boon to biomedical applications such as contribution to utilization of electret effect for artificial bone devices and artificial heart valves. Moreover, the applications of electrets in biomedical applications open a new field of systematic investigations in nanotechnology.

Compatible biomaterial for artificial devices can be chosen by investigating through TSDC technique. Moreover, during the formation of a bioelectret, varying polarizing conditions, voltages, times, or heating rates have a significant effect on its properties. Hence, tailor-made changes according to the biomedical applications can be accomplished. Hence, the review of electrets' effects in polymeric biomaterials presents a fascinating field and promises more developments in future.

REFERENCES

1. Turnhout, J.V. Thermally stimulated discharge of electrets. In *Electrets*; Sessler, G.M., Ed.; Springer-Verlag: Berlin, 1980; 33, 81–214.
2. Gray, S.A. Letter from Mr. Stephen to Dr. Mortimer, Secr. R.S. containing a farther account of his experiments concerning electricity. Phil. Trans. R. Soc. London **1732**, Ser. A: 37, 285–291.
3. Faraday, M. *Experimental Researches in Electricity*. Richard and John Edward Taylor: London, 1839; Volume 1, 1234–1250, 1269.
4. Heaviside, O. Electromagnetic induction and its propagation. Electrization and electrification. Natural electrets. Electrician **1885**, 230–231.
5. Eguchi, M. On dielectric polarization. Proc. Phys. Math. Soc. Jpn. **1919**, Series 3, *1*, 326–331.
6. Gemant, A. Electrets. Philos. Mag. **1935**, *20* (136), 929.
7. Gross, B. Experiments on electrets. Phys. Rev. **1944**, *66* (1–2), 26–28.
8. Gross, B. On permanent charges in solid dielectrics. II. Surface charges and transient currents in carnauba wax. J. Chem. Phys. **1949**, *17* (10), 866–872.
9. Gross, B. Irradiation effects in plexiglass. J. Polym. Sci. **1958**, *27* (115), 135–143.

10. Giacometti, J.A.; Fedosov, S.; Costa, M.M. Corona charging of polymers: Recent advances on constant current charging. Braz. J. Phys. **1999**, *29* (2), 269–279.
11. Chudleigh, P.W. Charging of polymer foils using liquid contacts. Appl. Phys. Lett. **1972**, *21* (11), 547–548.
12. Chudleigh, P. W. Mechanism of charge transfer to a polymer surface by a conducting liquid contact. J. Appl. Phys. **1976**, *47* (10), 4475.
13. Sessler, G.M.; Gerhard-Multhaupt, R. *Electrets*, 3rd Ed.; Laplacian Press: Morgan Hill, CA, 1999; Vol. 2, 1.
14. Sperling, L.H. *Introduction to Physical Polymer Science*, 3rd Ed.; Wiley: New York, 2001.
15. Raith, W.; Bergmann, L.; Schäfer, C.; Gobrecht, H. Lehrbuch der Experimentalphysik: Elektromagnetismus, 1999; De Gruyter, Berlin, 2.
16. Fukada, E. Piezoelectricity in polymers and biological materials. Ultrasonics, **1968**, *6* (4), 229.
17. Sessler, G.M.; West, J.E. Self-biased condenser microphone with high capacitance. J. Acoust. Soc. Am. **1962**, *34* (11), 1787–1788.
18. Henry, P.S.H. The role of asymmetric rubbing in the generation of static electricity. Br. J. Appl. Phys. **1953**, *4* (Suppl. 2), S31–S36.
19. Bauser, H. Elektrostatische Aufladung; Verlag Chemie, Weinheim, Germany, 1974; 11–28.
20. Seanor, D. *Electrical Properties of Polymers*; Frisch, K.C., Patsis, A.V., Eds.; Technomic: Westport, 1972; 37–58.
21. Inculet, I.I. Static electrication of dielectrics and at materials' interfaces. In *Electrostatics and Its Applications*, Moore, A.D., Ed.; John Wiley & Sons; New York, 1973; 86–114
22. McCarty, L.S.; Whitesides, G.M. Electrostatic charging due to separation of ions at interfaces: Contact electrification of ionic electrets. Angew. Chem. Int. Ed. Engl. **2008**, *47* (12), 2188–2207.
23. Fuhrmann, H. Contact electrification of dielectric solids. J. Electrostat. **1978**, *4* (2), 109–118.
24. Fridkin, V.M.; Zheludev, I.S. *Photoelectrets and the Electrophotographic Process* Consultants Bureau: New York, 1961; 195.
25. Fridkin, V.M. Springer series in solid state science. In *Photoferroelectrics*; Springer: Berlin, Heidelberg, New York, 1979; Vol. 9, 23–43.
26. Gross, B. *Charge Storage in Solid Dielectrics*; Elsevier: Amsterdam, the Netherlands, 1964, 10–75.
27. Latour, M. Infra-red analysis of poly(vinylidene fluoride) thermoelectrets. Polymer **1977**, *18* (3), 278–280.
28. Pillai, P.K.C.; Jain, K.; Jain, V.K. Thermoelectrets and their applications. Phys. Status Solidi **1972**, *13* (2), 341–357.
29. Tyler, R.W.; Webb, J.H.; York, W.C. Measurements of electrical polarization in thin dielectric materials. J. Appl. Phys. **1955**, *26* (1), 61–68.
30. Carlson, C.F. Electrophotographic Apparatus. US Patent 2588,699, 3rd November 1952.
31. Collins, R.E. Distribution of charge in electrets. Appl. Phys. Lett. **1975**, *26* (12), 675–677.
32. Chudleigh, P.W. Mechanism of charge transfer to a polymer surface by a conducting liquid contact. J. Appl. Phys. **1976**, *47* (10), 4475–4483.
33. Sessler, G.M.; West, J.E. Charging of polymer foils with monoenergetic low-energy electron beams. Appl. Phys. Lett. **1970**, *17* (12), 507–509.

34. Wills, R.F.; Skinner, D.K. Secondary electron emission yield behavior of polymers. Solid State Commun. **1973**, *13* (6), 685–688.

35. Sessler, G.M.; West, J.E. Electret transducers: A review. J. Acoust. Soc. Am. **1973**, *53* (6), 1589–1600.

36. Gross, B.; Sessler, G.M.; West, J.E. Charge dynamics for electron-irradiated polymer-foil electrets. J. Appl. Phys. **1974**, *45* (7), 2841–2851.

37. Perlman, M.M.; Unger, S. Electron bombardment of electret foils. Appl. Phys. Lett. **1974**, *24* (12), 579–580.

38. Gross, B. Compton current and polarization in gamma-irradiated dielectrics. J. Appl. Phys. **1965**, *36* (5), 1635–1641.

39. Sessler, G.M.; West, J.E. Electrets formed by low-energy electron injection. J. Electrostat. **1975**, *1* (2), 111–123.

40. Gross, B. Charge storage effects in dielectrics exposed to penetrating radiation. J. Electrostat. **1975**, *1* (2), 125–140.

41. Murphy, P.V.; Ribeira, S.C.; Milanez, F.; de Moraes, R.J. Effect of penetrating radiation on the production of persistent internal polarization in electret-forming materials. J. Chem. Phys. **1963**, *38* (10), 2400–2404.

42. Fields, D.E.; Moran, P.R. Observation of a radiation-induced thermally activated depolarization in lithium fluoride. Phys. Rev. Lett. **1972**, *29* (11), 721–724.

43. Podgorsak, E.B.; Moran, P.R. Radiation-induced thermally activated polarization transfers in CaF_2. Appl. Phys. Lett. **1974**, *24* (12), 580–583.

44. Gross, B.; Sessler, G.M.; West, J.E. TSC studies of carrier trapping in electron- and γ-irradiated Teflon. J. Appl. Phys. **1976**, *47* (3), 968–975.

45. MacDonald, B.A. Charge transport and storage in the radiation-charged electret ionization chamber. Med. Phys. **1996**, *23* (10), 1819.

46. Freeman, J.R.; Kallmann, H.P.; Silver, M. Persistent internal polarization. Rev. Mod. Phys. **1961**, *33* (4), 553–573.

47. Pillai, P.K.C.; Arya, S.K. The photoelectret state formation and its temperature dependence in evaporated thin-films of CdS. Solid State Electron. **1972**, *15* (11), 1245–1251.

48. Andreichin, R. High-field polarization, photopolarization and photoelectret properties of high-resistance amorphous semiconductors. J. Electrostat. **1975**, *1* (3), 217–230.

49. Kallmann, H.; Rosenberg, B. Persistent internal polarization. Phys. Rev. **1955**, *97* (6), 1596–1610.

50. Kalia, R.; Kalia, S. Investigation of dielectric relaxation parameters of polyetheretherketone (PEEK) films using TSDC technique. J. Polym. Mater. **2012**, *29* (3), 293–300.

51. Kalia, R.; Sharma, V.; Sharma, J.K. Dielectric behavior of polyetheretherketone (PEEK) using TSDC technique. J. Polym. Res. **2012**, *19* (2), doi: 10.1007/s10965-012-9826-4.

52. Vanderschueren, J.; Gasiot, J. Thermally stimulated relaxations in solids. Top. Appl. Phys. **1979**, *37* (4), 135–223, doi: 10.1007/3540095950_10.

53. Mascarenhas, S. Bioelectrets: Electrets in biomaterials and biopolymers. In *Electrets*; Sessler, G.M., Ed.; Springer-Verlag: Berlin, 1987; Vol. 33, 321–346.

54. Murphy, P.V.; Merchant, S. Blood compatibility of polymer electrets. In *Electrets: Charge Storage and Transport Dielectrics*; Perlman, M.M., Ed.; The Electrochemical Society, Inc.: New Jersey, 1973; Vol. 2, 627–649.

55. Mascarenhas, S. Electret behavior of collagen and blood vessel walls. In *Electrets: Charge Storage and Transport Dielectrics*; Perlman, M.M., Ed.; The Electrochemical Society, INC.: New Jersey; 1973; Vol. 2, 657–660.

56. Menefee, E. Thermocurrent from alpha-helix disordering in keratin. In *Electrets: Charge Storage and Transport Dielectrics*; Perlman, M.M., Ed.; The Electrochemical Society, Inc.: New Jersey, 1973; Vol. 2, 661–675.

57. Schroder, K.; Keller, D.; Meyer-Plath, A.; Muller, U.; Ohl, A. Pattern guided cell growth on gas discharge plasma induced chemical microstructured polymer surfaces. In *Materials for Medical Engineering*; Stallforth, H., Revell, P., Eds.; John Wiley-VCH: Weinheim, 2000; Vol. 2, 161.

58. Schroder, K.; Meyer-Plath, A.; Keller, D.; Ohl, A. On the applicability of plasma assisted chemical micropatterning to different polymeric biomaterials. Plasmas Polym. **2002**, *7* (2), 103–125.

59. Yang, X.L.; Gu, J.W.; Zhu, H.S. Preparation of bioelectret collagen and its influence on cell culture *in vitro*. J. Mater. Sci. Mater. Med. **2006**, *17* (8), 767–771.

60. Wei, Z.J.; Xia, Z.X. Biomaterials made of bile acids. Sci. Chin. Ser. B Chem. **2009**, *52* (7), 849–861.

61. Blaisten-Barojas, E.; Mascarenhas, S. A correlated walk model for thermally stimulated depolarization currents in α-keratin. J. Chern. Phys. **1982**, *76* (11), 115643–115645.

62. Goes, J.C.; Figueiro, S.D.; de Paiva, J.A.C.; de Vasconcelos, I.F.; Sombra, A.S.B. On the piezoelectricity of anionic collagen films. J. Phys. Chem. Solids **2002**, *63* (3), 465–470.

63. Mesquita, M.V.; Vasconcellos, A.R.; Luzzi, R.; Mascarenhas, S. Systems biology: An information-theoretic-based thermo-statistical approach. Braz. J. Phys. **2004**, *34* (2A), 459–488.

64. Burghate, D.K.; Deshmukh, S.H.; Joshi, L.; Deogaonkar, V.S.; Deshmukh, P.T. Electrical conduction of succinic acid doped glycine pellets. Ind. J. Pure Appl. Phys. **2004**, *42* (7), 533–538.

65. Cao, Y.; Zhou, Y.M.; Shan, Y.; Ju, H.X.; Xue, X.J. Preparation and characterization of grafted collagen-multiwalled carbon nanotubes composites. J. Nanosci. Nanotechnol. **2007**, *7* (1), 1–5.

66. Tanaka, Y.; Iwasaki, T.; Nakamura, M.; Nagai, A.; Katayama, K.; Yamashita, K. Polarization and microstructural effects of ceramic hydroxyapatite electrets. J. Appl. Phys. **2010**, *107* (1–10), 014107–17.

Electroactive Polymeric Materials

V. Prasad Shastri
Vanderbilt University, Nashville, Tennessee, U.S.A.

Abstract

Electroactive materials may be broadly defined as those materials whose bulk or surface properties may be altered reversibly or irreversibly upon exposure to an electrical stimulus. This entry covers conductive and piezoelectric polymers in detail; overviews of electroresponsive gels and general topics that are material-specific are provided.

INTRODUCTION

Electroactive materials may be broadly defined as those materials whose bulk or surface properties may be altered reversibly or irreversibly upon exposure to an electrical stimulus. In keeping with the scope of this entry, only materials that have potential application in drug delivery, biosensing, tissue engineering, and high throughput screening will be discussed. The entry is divided into four sections. The first two sections cover conductive and piezoelectric polymers in detail. Electroresponsive gels are discussed briefly in the third section and the last section covers more general topics that are material specific. Electrets are not covered in this entry, as examples of these are few and their biomedical applications very limited.

OVERVIEW

Since the early 1980s the application of polymers in biomedical sciences has seen an exponential growth. This growth has been fueled in great part by the demand placed by emerging disciplines, such as drug delivery, tissue engineering, and genetic screening, for novel materials that may be fine-tuned with respect to their bulk and surface characteristics on demand as a function of time. Electroactive materials, by virtue of their ability to predictably respond to electrical stimulus, are therefore ideal candidates.

The role of electrical signals in modulating cellular response is well-recognized.[1] Among various tissues where bioelectricity plays a role, the nervous system and osseous tissue are the most obvious examples. The neural synapse in essence is a biosensor. An action potential (electrical signal) arriving at the synaptic junction is transmitted across the synapse by depolarization of the adjacent synaptic membrane. This depolarization is triggered by acetylcholine, a biomolecule that is secreted in response to the action potential at the synapse.[2] Bone is yet another tissue where electrical activity plays an important role in fulfilling its

function. In the late 1890s, Wolff demonstrated that the bone responds to mechanical stress by growing.[3,4] This relationship between applied mechanical stress and bone remodeling (growth) is known as Wolff's Law. It turns out that bone is piezoelectric in nature. Iwao Yasuda was the first to demonstrate piezoelectricity in bone in the 1950s when he observed that compressive loading along a bone's major axis resulted in the accumulation of electrical charges on the surface along its axis.[5,6] In a seminal study, Bassett and Becker demonstrated that when bone was mechanically loaded, the concave (bone formation) and convex (bone resorption) faces of bone yielded charges of opposing polarity due to compression and tension at the two respective faces.[7] Therefore, cellular architecture in bone is able to sense the mechanical stress due to the localized electrical currents generated upon loading of the inorganic hydroxyapatite phase in bone.

Electroactive materials encompass ceramics and plastics that are capable of electro-mechanical and electrochemical transduction. However, two major considerations for biomedical applications are cellular and tissue compatibility and ease of processing. The library of organic chemical bonds and building blocks available for the synthesis of polymers make it a very versatile system capable of satisfying both of the aforementioned requirements. In emerging applications such as tissue engineering (TE), wherein a scaffold is used to guide the generation of neotissue from dissociated mammalian cells, it is plausible that the use of electroactive scaffolds will offer the possibility of manipulating cellular functions and tissue growth in a localized manner via minimally invasive methodologies.

CONDUCTIVE POLYMERS

Ease of processibility and their insulative properties are two important characteristics that distinguish plastics from metals. In the 1970s, Heeger, MacDiarmid, and Shirakawa, demonstrated that polyacetylene, an organic polymer

Concise Encyclopedia of Biomedical Polymers and Polymeric Biomaterials DOI: 10.1081/E-EBPPC-120013921

characterized by alternating double bonds (Fig. 1A), can exhibit conductivity approaching that of conductors. Conductivities as high as that of copper (10^6 S/cm) have been reported in polyacetylenes doped with a cation or an anion.[8,9] Their seminal work, for which they were awarded the 2000 Nobel Prize in Chemistry, led to the birth of an entirely new class of materials and an exciting field called conductive polymers. The origin of conductivity in this interesting class of polymers lies in the high degree of π-π overlap in a conjugated structure, which reduces the band gap and facilitates carrier mobility.[10] Since then several polymers that exhibit a high degree of extended polyconjugation have been shown to possess electrical conductivity in their doped state. They include poly(aniline), poly(heterocycles), poly(p-phenylene vinylene) and poly(p-phenylene ethynylene).

In 1979, Diaz and coworkers reported the electrochemical synthesis of polypyrrole (Ppy) (Fig. 1B).[11] With the electrochemical approach, freestanding films of Ppy of varying thickness and crystallinity can be obtained by controlling the passage of current during electrochemical deposition and chemistry of the dopant anion. Three years later, Tourillon and Garnier reported the electrochemical polymerization of poly(thiophene) (Pth) from thiophene.[12] Electrochemical polymerization offers several advantages over conventional chemical synthesis: 1) Thin films (membranes) of the polymer can be directly deposited on an electrode of interest or surface of a metal implant; and 2) freestanding thin films can be obtained that can then be fabricated into devices. Both Ppy and Pth exhibit reversible electrochemistry, which enables switching of the polymer from the conductive form (oxidized) to the insulator form (neutral) by the application of a small voltage.[13] Electrical conduction in poly(heterocycles) occur via the interchain hopping and delocalization of radical cations called polarons.[14] The conductivity in Ppy and Pth is significantly influenced by the crystallinity and long-range order in the polymer, which aids in reducing the conduction band gaps, the latter being more important. Among the various conductive polymers, Ppy and Pth (Fig. 1B) are the most promising for biomedical applications due to ease of their

synthesis. However, Ppy is by far the most suited for biosensing and tissue-contacting applications as it can be electrochemically synthesized from aqueous solutions at a relatively low potential of 700 mV versus SEC. Furthermore, it's oxidized form is far more stable in both air and water (physiological conditions of pH 7.2 phosphate buffer, 37°C) in comparison to Pth.

Since the mid-1980s Ppy has been extensively studied as an electrochemical transducer for sensing biomolecules in solution.[15–17] In a Ppy-based biosensor, an oxidative enzyme such as glucose oxidase is entrapped in the film during electrochemical deposition on a platinum or palladium electrode (Fig. 2). When this film is brought into contact with an analyte such as glucose in presence of oxygen, hydrogen peroxide is liberated which is then detected electrochemically at the electrode.[18–22] The advantage of this system is that it allows for the one-step fabrication of a sensor. However, the slow response time, due to the slow diffusion of hydrogen peroxide through the film to the electrode, and low current densities are limitations. These limitations have been overcome by introducing redox couple mediators, such as ferrocene (Fc) (II)↔ferrocinium (Fc⁺) (III), either by coentrapment or by covalent immobilization to the enzyme or polymer, that can act as an electron shunt.[23–26] In this system, the electron transfer occurs directly from enzyme to the oxidized form of the mediator Fc⁺, yielding Fc which is then followed by cyclic voltammetry. In addition to the Fc↔Fc⁺ couple, other charge transfer systems such as tetrathiafulvalene-tetracyanoquinodimethane (TTF-TCNQ)[27] and 4,4'-bipyrridyl mediators have been explored as well.[28] Polymers bearing such electron mediators are said to be wired and they typically yield sensors with faster response times and higher current densities. Besides enzyme-mediated oxidative reactions, changes in local pH have been leveraged in the development of Ppy-based biosensors. Ppy membrane microarrays containing immobilized penicillinase have been used to detect penicillin by monitoring the increase in Ppy membrane conductivity upon the release of acid at the membrane surface due to enzymatic hydrolysis of the penicillin.[29] Ppy membranes/films with immobilized oligonucleotides have been prepared by

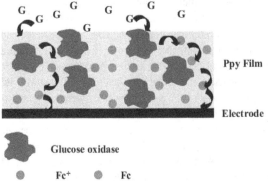

Fig. 2 Enzyme-based conductive polymer biosensor: schematic of a Ppy-based glucose sensor containing glucose oxidase and ferrocene as an electron mediator.

A

B

A⁻

X = NH, Pyrrole

X = S, Thiophene

Fig. 1 (**A**) Chemical structure of oxidized polyacetylene; (**B**) Chemical structure of oxidized poly(heterocycles).

electrochemical polymerization of pyrroles covalently bound to oligonucleotides.[30,31] By individually addressing micro-electrodes in an array, surfaces bearing a spatially well-defined array of oligonucleotides have been fabricated. Such arrays might be useful in the development of high throughput devices for genomics and proteomics that are amperometric rather than colorimetric, thus increasing reliability and response time. In addition to Ppy, Pth and polyanilines have been explored in glucose-sensing applications.[32,33]

In the late 1980s, Miller and coworkers demonstrated that a Ppy film doped with a polyanionic dopant poly(styrenesulfonate) sodium (PSS-Na) can behave as a cation exchanger.[34,35] The immobilization of PSS-Na results in a film that has a net negative charge in its oxidized state. Because these charges are fixed, reduction of the film results in the binding of cations, which can be released upon oxidation of the film. Based on this principle, Miller and co-workers showed that dopamine, a neurotransmitter whose deficiency leads to lesions in the brain such as those found in Parkinson's disease, may be cathodically bond and anodically released from these Ppy films.[34] Prezyna and co-workers later showed that these films can irreversibly bind polycations such as poly(lysine-HCl) and histones, which are responsible for compacting DNA in the nucleus of a cell.[36] However, extensive exploration of Ppy films as drug-delivery vehicles has been limited by unpredictability associated with binding and release kinetics and low drug loading levels. These systems, however, are well suited for highly potent drugs and should not be overlooked especially in the delivery of drugs to the central nervous system. Diseases such as epilepsy and Parkinson's could significantly benefit from self-regulating delivery systems wherein the Ppy electrode serves as a biosensor for disease-predictive markers, thereby reducing the drug delivery response time.

Ppy has been explored in TE applications as well. Wong[37] and Shastri[38,39] have shown that by varying the dopant ion chemistry and oxidation state of Ppy, protein adsorption and conformation can be altered resulting in control over cell cycle progression. This has sparked an interest in using Ppy in cell and tissue contacting applications as interactive interfaces for dictating cell behavior and enhancing cell functions. In a seminal study demonstrating the potential of Ppy as an electroactive interface in manipulating cellular functions, the effect of applying of an electrical stimulus through a Ppy substrate on PC-12 cell differentiation was studied.[40] PC-12 cells respond reversibly to soluble nerve growth factor by differentiation into a neuronal phenotype, a process accompanied by extension of neurites (neuronal processes). In this study, PC-12 cells grown on Ppy-PSS-Na substrates were subjected to an electrical stimulus applied through the polymer film. It was observed that application of a constant potential or current dramatically improved PC-12 differentiation as assessed by an increase in both neurite numbers and lengths. These studies have also shown that tubular scaffolds fabricated from Ppy-PSS-Na can support the guided regeneration of transected sciatic nerve in rats.[41] Furthermore, it has been shown that the

soft-tissue biocompatibility of Ppy film doped with PSS-Na is far superior to that of most commonly used biodegradable polymers such poly(lactic-co-glycolic acid).[40] The response to Ppy films is characterized by the total absence of a fibrous capsule even after 14 weeks (Fig. 3). Schmidt and co-workers have shown that vascularization around Ppy films can be enhanced by doping the film with polysaccharides.[42] Ppy-PSS films have been shown to support the proliferation and osteogenic differentiation of mesenchymal progenitor cells (bone marrow stromal cells).[43] It was observed that the production of alkaline phosphatase, a marker of osteogenic activity in these cells, was enhanced in cells cultured on Ppy-PSS substrates in comparison to those cultured on tissue culture plastic. These observations suggest that Ppy substrates have good cytocompatibility and should be useful as interactive coatings in tissue regeneration paradigms such as TE. One can also envisage the use of Ppy coatings to enhance cellular interactions towards metal implant surfaces.

However, Ppy synthesized from pyrrole or PNpy synthesized from N-methyl pyrrole-yields a highly cross-linked polymer that has low to no solubility in common organic solvents and hence not processible. The cross-linked structure also renders it nondegradable in a biological environment. Imparting degradability in a biological environment should significantly expand the use of Ppy in biomedical uses such as TE and drug delivery. Recently, Shastri and Langer reported the synthesis of conductive bioerodible Ppy from pyrroles substituted in the β-position with ionizable or hydrolyzable side groups.[44] In this system, by controlling the ratio of the ionizable:hydrolyzable group, polymers with erosion times ranging from a few weeks to potentially over a year may be synthesized (Fig. 4). In addition to being bioerosible, Ppy possessing hydrolyzable groups show excellent solubility in common organic solvents such as tetrahydrofuran and methylene

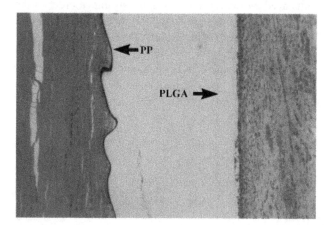

Fig. 3 Hematoxylin-Eosin-stained histological cross-section of Ppy-PSS (PP)-laminated with a biocompatible poly(L-lactic acid-co-glycolic acid) (PLGA) polymer implanted in a intramuscular site in rat for two weeks. Note the absence of fibrous tissue adjacent to the Ppy film. The gap between the Ppy and PLGA phase is an artifact created during processing of the tissue explant.
Source: From Schmidt et al.[40]

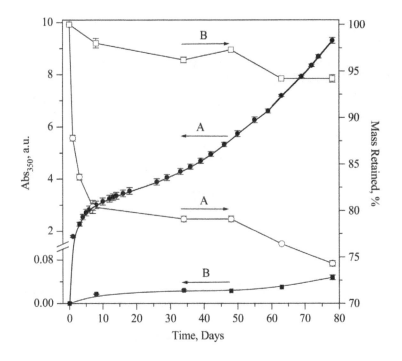

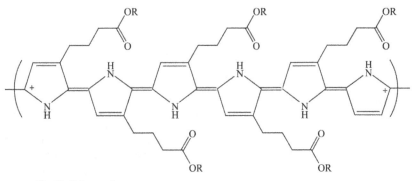

R = H, Polymer A

R = CH₃, Polymer B

Fig. 4 Degradation characteristics of bioerodible Ppy pellets synthesized from pyrroles substituted in the β-position with ionizable or hydrolyzable side groups (polymer) as followed by dissolution (filled symbols) and mass loss (open symbols).
Source: From Zelikin et al.[44]

chloride, thus allowing for solution-based processing of these polymers into coatings and scaffolds for cell and tissue contacting applications. These polymers possess excellent cytocompatibility and have been shown to support the proliferation and differentiation of human marrow-derived mesenchymal progenitor cells into osteoblast lineage.[44] Another approach that has been explored to synthesize degradable Ppy is the coupling of 2,5-bis-(5-(3-hydroxypropoxycarbonyl)-2-pyrrolyl)thiophene with adipoyl chloride to yield a polymer with degradable ester linkages.[45] Although this polymer has good tissue compatibility, it has extremely low conductivity which limits its potential use.

PIEZOELECTRIC POLYMERS

The piezoelectric effect was discovered by Jacques and Pierre Curie in the 1880s, when they observed that certain crystalline materials when subjected to tension or compression became electrically polarized resulting in induced voltages of opposing polarity (compression tension).[46] They also observed that the magnitude of this polarization was directly proportional to the applied stress. Conversely, minute changes in physical dimensions can be introduced in a piezoelectric material by subjecting it to an electric field. This phenomenon is called converse piezoelectricity. Piezoelectricity therefore may be defined as an electrical polarization induced in a solid in response to a mechanical stress (Figs. 5A and B). In order for a material to exhibit piezoelectric behavior, the crystalline domains (or crystal lattice) should be anisotropic (noncentrosymmetric) and polarizable. A piezoelectric material when heated above its Curie temperature loses its piezoelectric properties permanently. Most common examples of piezoelectric materials include naturally occurring substances such as quartz, polycrystalline ceramics such as barium titanate, lead zirconate titanate, and some inorganic salts (e.g., Rochelle's salt). Examples of synthetic and biological polymers that exhibit some form of piezoelectric behavior include

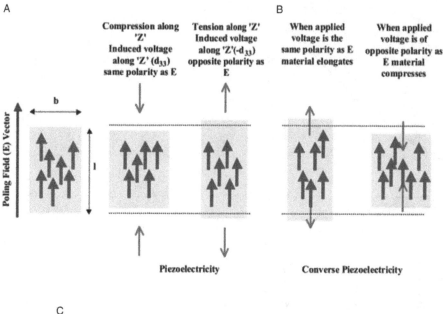

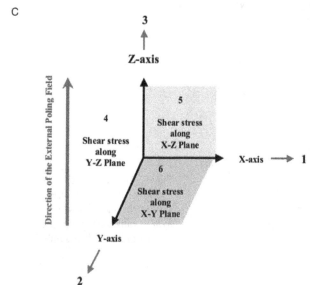

Fig. 5 (**A**) Schematic representation of the piezoelectric phenomenon in a material when subjected to compressive and tensile stress; (**B**) Schematic representation of reverse piezoelectric phenomenon in a material when subjected to an external field of the same and opposing polarity as the poling field; (**C**) Piezoelectric coupling tensors.

poly(vinylidine difluoride) (PVDF), aromatic polyamides, polysulfones, polyparaxylene, poly(lactic acid), and some biopolymers such as collagen and poly(γ-methyl and γ-benzyl-l-glutamate).

In order to induce piezoelectric behavior in a ceramic or polymer, it first has to be poled in an external electric field. During the poling process, the dipoles in the crystalline regions align to form highly ordered domains resulting in a net dipole moment or polarization. The residual polarization after the electric field is removed, termed remnant polarization, is responsible for the piezoelectric behavior in the material. The relationship between the induced polarization and applied stress and between induced strain and applied electrical stimulus is described in terms of coupling constants that relate the two, namely:

- d (C/N) = piezoelectric charge constant, the polarization generated by unit mechanical stress or the induced strain per unit electric field;

- g (Vm/N) = piezoelectric voltage constant, the electric field generated per unit mechanical stress or the induced strain per applied unit displacement of charge;
- C = current in coulombs; N = force in newtons; V = voltage in volts; and m = displacement in meters.

The d and g constants of a piezoelectric material are described using two subscripts, e.g., d_{33} and g_{31}, wherein the first subscript denotes the direction of the induced polarization and the second subscript denotes the direction of the applied stress. The coupling between the induced polarization and the direction of applied stress is described using an X,Y,Z system of notation (Fig. 5C), wherein the Z axis is the direction of the applied external poling field and represents a positive polarization; 1, 2, and 3 represent the stress along X, Y, and Z axes, respectively; and 4, 5, and 6 represent the shear stress about these axes. Therefore, in this system of notation, d_{33} means that the induced polarization is in the direction of the poling (Z axis) (positive polarization) when the stress is applied in that same direction. Similarly, d_{25} means the induced polarization is along the Y axis when shear stress is applied along the X-Z plane.

Piezoelectricity in PVDF was discovered by Kawai in 1969.[47] This discovery was followed by several reports showing that polypeptides and collagen exhibit piezoelectricity when subjected to shear stress.[48–50] PVDF is a semi-crystalline polymer composed of —CH_2-CF_2— repeat units that are linked to each other predominantly in the *head-tail* fashion. The presence of the electronegative fluorine atom makes it responsible for the dipole moment of the CF_2 unit.[51] The crystalline domains, which are responsible for long-range order in the polymer, are structurally anisotropic. The symmetry of the backbone dictates the overall dipole of the unit cell (crystalline domain). The crystalline domains in PVDF exist in four principal forms, called α, β, γ, and δ.[52] Among these phases, only the β phase exhibits an all-*trans* conformation. This imparts a polar characteristic to this phase due to the alignment of the dipoles, something that is absent in the other phases due to a mixture of *trans*, gauche, and anti-gauche conformations. However, in the unpoled state, the dipoles cancel each other out due to random orientation of the crystalline domains resulting in a net dipole moment of zero. It has been observed that head-head —CH_2-CH_2— and tail-tail —CF_2-CF_2— linking of repeat units results in the formation of defects in the crystalline phase of PVDF, a process that favors the formation of the polar β phase.[52] In fact, copolymers of PVDF with tri-fluoroethylene (PVDF-TrFE) and tetrafluoroethylene (PVDF-TeFE) yield polymers with predominantly a β-phase backbone. The formation of β phase can also be induced by orienting the polymer chains through mechanical means such as drawing and stretching of fibers and films during extrusion. To impart persistent polarization in PVDF, the polymer is heated below its Curie temperature which is approximately 110°C and then subjected to an external electric field to align the dipoles in the crystalline domains. The polymer is then cooled in the presence of the field to freeze the orientation of the dipoles. The most likely mechanism for the alignment of dipoles based on the kinetic model proposed by Dvey-Aharon involves a 60° rotation of the polymer chain at the grain boundary resulting in the formation of a kink, which then propagates along the polymer chain and into the phase leading to long-range alignment within the phase.[53] The piezoelectric charge constant, d, for PVDF depends on its configuration (fiber versus film). Films poled across its thickness (Z-axis) show a very strong d_{33} constant of about -30 pC/N and g_{33} constant of around -300 mV-m/N.

Collagen fibrils and other biopolymers exhibit piezoelectric properties when subjected to shear stress. The piezoelectricity in these polymers is thought to originate in the crystalline regions, which are formed when proteins and peptides form ordered structures such as α-helices and β-sheets. Piezoelectricity in these systems is thought to originate in the rotation of the amide bond. The straining of these ordered regions leads to rotation of the polar amide bonds in the backbone, which has a dipole moment of 3.4 debye, resulting in the generation of electrical currents. Biopolymers typically show d_{25} values of a few pC/N. Likewise, in the highly crystalline poly(L-lactic acid) a similar mechanism is presumed to be at play. The d_{25} values in highly oriented films of PLLA can be as large as -10pC/N.

Among polymers, PVDF, poly(lactic acid), and poly(γ-methyl and γ-benzyl L-glutamate) are the most promising with respect to cell and tissue contacting applications. As discussed earlier, it has been shown that mechanical loading of bone leads to stimulation of bone remodeling and growth.[5,6] The interplay between mechanical stress and osteoclastic (bone resorption) and osteoblastic (bone formation) activity occurs via the generation of localized electrical charges that in turn trigger cellular activity.[7] The localized electrical activity is generated by the mechanical deformation of the piezoelectric microcrystalline inorganic hydroxyapatite phase. In addition, collagen fibrils, which are known to exhibit shear piezoelectricity, may play a role as well. The electrical charges generated upon compression and tension-loading in bone are of opposing polarity, which suggests an intricate coupling between the bone resorption and formation processes through the microcrystalline hydroxyapatite-collagen composite matrix. Orthopaedic metal implants coated with various forms of hydroxyapatite have been shown to enhance new bone formation and integration of the implant.[54] Whether the observed effect is due to the piezoelectric nature of the coating or due to the calcium-rich surface is still unclear. Similarly, in wound healing, electrical activity at the injury site due to the depolarization of cell membranes is thought to play a role in the migration of cells to the site. Several studies have shown that electromagnetic stimulation can aid in bone and wound healing.[55–57] Aebischer and coworkers have shown that tubular scaffolds derived from poled PVDF are capable of supporting the regeneration of transected sciatic nerve in rat.[58,59] They observed that the quality of the regenerated sciatic nerve using poled-PVDF guidance channels was superior to that of controls. They have attributed this outcome to localized electrical stimulation of the regenerating neuronal tissue mass by electrical currents generated as a consequence of minute mechanical deformations of the PVDF conduits. So it is entirely conceivable that, in the near future, scaffolds either composed entirely of or coated with piezoelectric polymers will find uses in wound healing and tissue regenerative applications. One can also envision using these polymers as barrier membranes or microvalves in smart drug delivery systems or artificial pores wherein permeability of the drug across the membrane or through a valve is triggered by an external electric field.

ELECTRO-RESPONSIVE GELS

A hydrogel is a cross-linked network of polymer and imbibed water. When the polymer phase is composed of a polyelectrolyte, the gel is called an ionic gel. If the polyelectrolyte is composed of a weakly ionizable group such as a carboxylic acid, the gel will undergo contraction at pHs below the isoelectric point (pIe) of the acid moiety and expansion at pH's above the pIe. This phenomenon is due to the ionization of the acid groups at higher pHs that leads to electrostatic repulsion within the gel resulting in an expansion in gel volume.[60,61]

Pioneering work carried out by Toyoichi Tanaka and later by Osada and coworkers have shown that when an ionic gel, irrespective of the nature of the ionizable group, is placed between two electrodes and subjected to a DC field, it undergoes anisotropic contraction with the concomitant expulsion of water. Upon removal of the electric field, the contracted gel expands by taking up water and returning to its resting volume. This results in the conversion of electrical energy into mechanical strain.[62,63] The basis for this deformation lies in the electrophoretic redistribution of ions within the cross-linked network. For example, in a cross-linked poly(acrylic acid) gel, because the carboxylic acid groups (negative charges) are fixed to the polymer backbone, upon the application of an electric field the positively charged counterions being mobile migrate toward the cathode. This migration of counterions is accompanied by expulsion of water resulting in shrinkage of the gel at the anode.[64,65] The converse is true for a positively charged cross-linked network. Interestingly, when a weak polyelectrolyte gel such as poly(acrylic acid) gel is subjected to a mechanical stress, the pH in the local environment changes in response to this stress.[66] This is thought to occur due to the elongation of polymer chains in response to the stress, resulting in an increase in free energy, which is compensated by ionization of the acid groups to yield protons. It has also been shown that, by a process akin to piezoelectric behavior, mechanical deformation of a weak polyelectrolyte gel can induce an electrical potential. Osada and co-workers have explored this phenomenon to develop tactile sensors that may find uses in robotic devices in remote surgery applications.[66]

When a swollen cross-linked polyelectrolyte gel is exposed to a surfactant bearing opposite charge, the charges in the network are quenched leading to de-swelling of the network. In the presence of an electric field, the de-swelling behavior can yield anisotropic stress within a gel film or fiber, resulting in bending of the gel toward one of the electrodes with concomitant mechanical motion. The gel-bending phenomenon has been leveraged to make a hydrogel-based device called the gel looper that is capable of moving in an electric field.[67] Such smart polymer gels are bound to find applications in drug delivery and tissue engineering as these systems get more integrated with microelectromechanical system (MEMS)-based devices.

OTHER SYSTEMS

Matrices composed of a complex (salt) of polyelectrolytes of opposing charges have been explored as electroerodible systems for the release of proteins.[68] The drug delivery matrix which in essence is a polymer complex can be made to dissociate upon exposure to an electrical potential in aqueous media. The dissociation originates at the face, opposing the counter electrode, and proceeds inward towards the working electrode. Although this system has its limitations with respect to ease of use and thickness of the delivery matrix, truly responsive drug delivery can nevertheless be achieved. Magnetite is an oxide of iron with a chemical composition of Fe_3O_4 wherein iron exists in both $+2$ and $+3$ oxidation states ($Fe_3O_4 = FeO + Fe_2O_3$). Due to the mixed valence state of iron, magnetite has unpaired electrons, hence a net dipole moment, making it ferromagnetic. Ferrofluid is a dispersion of magnetite colloids (nanoparticles) in a hydrophobic medium such as silicone oil. When a ferrofluid is exposed to an external magnetic field, the magnetite nanoparticles become magnetized and align along the magnetic lines of force. This results in a physical transformation of the suspension from a liquid to a solid. Upon removal of the field the solidification is reversed. Ferrofluids can also be made from cobalt, zinc, and manganese oxides. However, magnetite is the most promising with respect to biological applications. One can envisage many applications for ferrofluids in medicine. Some obvious ones are localization media for drug delivery, plugs for coronary embolism, electroresponsive valves in implantable drug delivery systems, and implantable muscle actuators. However, as currently available, ferrofluids are not biocompatible; hence methodologies to improve biocompatibility of magnetite need to be worked out and alternative media for dispersion and delivery of the nanoparticles need to be developed. Iron oxide nanoparticles are already used in cell sorting and are being explored in cell and tissue imaging. Nevertheless, efforts are already underway to explore this system in localization of chemotherapy.[69] Other *ex vivo* applications may include carriers for cell sorting, valves in bio-MEMS devices for high throughput applications, cuffs to prevent vascular embolism, and a battlefield tourniquet. As we gain a better understanding of biomolecular processes, the newly acquired knowledge has enabled us manipulate materials using the very tools of nature.[70] The advent of biomolecular materials engineering as a science and discipline holds immense promise for the future, wherein medical devices are fully integrated with the human body via smart materials and interfaces that can respond to external stimuli such as an electrical signal to deliver drugs on demand[71] or to change a fate of a cell[37,40] or an implant surface[72] to improve clinical outcomes.

REFERENCES

1. Becker, R.O.; Marino, A.A. *Electromagnetism and Life*; State University of New York Press: Albany, New York, 1982.
2. Guyton, E.C. *Text Book of Medical Physiology*, 8th Ed.; W.B. Saunders Company, 1991; Ch. 7.

3. Wolff, J. Das gesetz der transformation der knochen; Hirschwald, A., Ed.; Verlag von August: Berlin, 1891.

4. Forwood, M.R.; Turner, C.H. Skeletal adaptations to mechanical usage. Bone 1995, 17 (4 Suppl.), 197S–205S.

5. Yasuda, I. On the piezoelectric activity of bone. J. Jpn. Orthop. Surg. Soc. 1954, 28, 267–271.

6. Fukada, E.; Yasuda, I. On the piezoelectric effect in bone. J. Physiol. Soc. Jpn. 1957, 12 (10), 1158–1162.

7. Bassett, C.; Becker, R.O. Generation of electric potentials by bone in response to mechanical stress. Science 1962, 137 (3535), 1063–1064.

8. Chiang, C.K.; Druy, M.A.; Gau, S.C.; Heeger, A.J.; Louis, E.J.; MacDiarmid, A.G.; Park, Y.W.; Shirakawa, H. Synthesis of highly conducting films of derivatives of polyacetylene, (CH)x. J. Am. Chem. Soc. 1978, 100 (3), 1013–1015.

9. Kaner, R.B.; MacDiarmid, A.G. Plastics that conduct electricity. Sci. Am. 1988, 258 (2), 106.

10. Skotheim, T.A. Handbook of Conducting Polymers; Marcel Dekker: New York, 1986; Vol. 1.

11. Diaz, A.F.; Kanazawa, K.K.; Garidini, G.P. Electro-chemical polymerization of pyrrole. J. Chem. Soc. Chem. Commun. 1979, 14, 635–636.

12. Tourillon, G.; Garnier, F.J. Electrochemical synthesis of polythiophenes. Electroanal. Chem. 1982, 135, 173.

13. Otero, T.F.; Villanueva, S.; Bengoechea, M.; Brillas, E.; Carrasco, J. Reversible redox switching in polypyrrole and polySNS films. Synth. Met. 1997, 84 (1–3), 183–184.

14. Nalwa, H.R. Handbook of Organic Conductive Molecules and Polymers; John Wiley & Sons, 1997; 2.

15. Cosnier, S.; Innocent, C.; Jouanneau, Y. Amperometric detection of nitrate via nitrate reductase immobilized and electrically wired at the electrode surface. Anal. Chem. 1994, 66 (19), 3198–3201.

16. Trojanowicz, M.; Lewenstam, A.; Krawczynski, T.; Lahdesmaki, I.; Szczepek, W. Flow injection amperometric detection of ammonia using a polypyrrole-modified electrode and its applications in urea and creatinine biosensors. Electroanalysis 1996, 8 (3), 233–243.

17. Garnier, F.; Youssauffi, H.; Srivastava, P.; Yassar, A. Enzyme recognition by polypyrrole functionalized with bioactive peptides. J. Am. Chem. Soc. 1994, 116 (19), 8813–8814.

18. Foulds, N.; Lowe, C. Enzyme entrapment in electrically conducting polymers: Immobilization of glucose oxidase in polypyrrole and its application in amperometric glucose sensor. J. Chem. Soc. 1986, 82, 1259–1264.

19. Bartlett, P.; Whitaker, R. Electrochemical immobilization of enzymes. Part I. Theory. J. Electroanal. Chem. 1987, 224 (1–2), 27–35.

20. Bartlett, P.; Whitaker, R. Electrochemical immobilization of enzymes. Part II. glucose oxidase immobilized in poly(N-methypyrrole). J. Electroanal. Chem. 1987, 224 (1–2), 37–48.

21. Fortier, G.; Brassard, E.; Belanger, D. Optimization of polypyrrole glucose oxidase biosensor. Biosens. Bioelectron. 1990, 5 (6), 473–490.

22. Pandey, P. A new conducting polymer-coated glucose sensor. J. Chem. Soc. Faraday Trans. 1. 1988, 84, 2259–2265.

23. Caglar, P.; Wnek, G.E. Glucose-sensitive polypyrrole/poly(styrenesulfonate) films containing co-immobilized glucose oxidase and (ferrocenylmethyl) trimethyl-ammonium bromide. J. Macromol. Sci. Pure Appl. Chem. 1995, 32 (2), 349–359.

24. Schuhmann, W.; Ohara, T.; Heller, A.; Schmidt, H.L. Electron transfer between glucose oxidase and electrodes via redox mediators bound with flexible chains to the enzyme surface. J. Am. Chem. Soc. 1991, 113 (4), 1394–1397.

25. Foulds, N.; Lowe, C. Immobilization of glucose oxidase in ferrocene modified pyrrole polymers. Anal. Chem. 1988, 60 (22), 2473–2478.

26. Willner, I.; Willner, B. Electrical communication of redox proteins by means of electron relaytethered polymers in photochemical, electrochemical and photoelectrochemical systems. React. Polym. 1994, 22 (3), 267–279.

27. Khan, G.; Ohwa, M.; Wernet, W. Design of stable charge transfer complex electrode for a third generation amperometric glucose sensor. Anal. Chem. 1996, 68 (17), 2939–2945.

28. Schuhmann, W.; Huber, J.; Kranz, C.; Wohlscjalger, H. Conducting polymer based amperometric enzyme electrode: Towards the development of miniaturized reagentless biosensors. Synth. Met. 1994, 61 (1–2), 31–35.

29. Nishizawa, M.; Matsue, T.; Uchida, I. Penicillin sensor based on a microarray electrode coated with pH responsive polypyrrole. Anal. Chem. 1992, 64 (21), 2462–2464.

30. Livache, T.; Roget, A.; Dejean, E.; Barthet, C.; Bidan, G.; Téoule, R. Preparation of a DNA matrix via an electrochemically directed polymerization of pyrrole and oligonucleotides bearing a pyrrole group. Nucleic Acids Res. 1994, 22 (15), 2915–2921.

31. Livache, T.; Roget, A.; Dejean, E.; Barthet, C.; Bidan, G.; Téoule, R. Biosensing effects in functionalized electroconducting conjugated polymer layers: Addressable DNA matrix for the detection of gene mutations. Synth. Met. 1995, 71 (1–3), 2143–2146.

32. Yamato, H.; Ohwa, M.; Wernet, W. Stability of poly-pyrrole and poly(3,4,-ethydioxythiophene) for biosensor application. J. Electroanal. Chem. 1995, 397 (1–2), 163–170.

33. Parente, A.; Marques, E.; Azevedo, W.; Diniz, F.; Melo, E.; Lime Filho, J. Glucose biosensor using glucose oxidase immobilized in polyaniline. Appl. Biochem. Biotechnol. 1992, 37 (3), 267–273.

34. Miller, L.L.; Zhou, Q.X. Poly(N-methypyrrolylium) poly(styrenesulfonate). A conductive, electrically switchable cation exchanger that cathodically binds and anodically releases dopamine. Macromolecules 1987, 20 (7), 1594–1597.

35. Zhou, Q.-X.; Miller, L.L.; Valentine, J.R. Electro-chemically controlled binding and release of protonated dimethyldopamine and other cations from poly (N-methy pyrrole)/polyanion composite redox polymers. J. Electroanal. Chem. 1989, 261 (1), 147–164.

36. Prezyna, L.A.; Qiu, Y.J.; Reynolds, J.R.; Wnek, G.E. Interaction of cationic polypeptides with electroactive polypyrrole/poly(styrenesulfonate) and poly(N-methyl-pyrrole)/poly (styrenesulfonate) films. Macromolecules 1991, 24 (19), 5283–5287.

37. Wong, J.; Ingber, D.E.; Langer, R. Electrically conductive polymers can noninvasively control cell shape. Proc. Natl. Acad. Sci. U. S. A. 1994, 91 (8), 3201–3204.

38. Shastri, V.R.; Wnek, G.E. Effect of dopant ion on cell growth on polypyrrole thin films. ACS Polym. Preprints 1993, 34 (2), 70–71.

39. Shastri, V.R. *Evaluation of Polypyrrole "Thin Films" as Substratum for Mammalian Cell Culture, Doctoral Dissertation*; Rensselaer Polytechnic Institute; Troy, NY, 1995.

40. Schmidt, C.E.; Shastri, V.R.; Vacanti, J.P.; Langer, R. Stimulation of neurite outgrowth using electrically conducting polymer. Proc. Natl. Acad. Sci. U. S. A. **1997**, *94* (17), 8948–8953.

41. Shastri, V.R.; Schmidt, C.E.; Kim, T.H.; Vacanti, J.P.; Langer, R. Polypyrrole-A potential candidate for stimulated nerve regeneration. Mat. Res. Soc. Symp. Proc. **1996**, *414*, 113–118.

42. Collier, J.H.; Camp, J.P.; Hudson, T.W.; Schmidt, C.E. Synthesis and characterization of polypyrrole-hyal-uronic acid composite biomaterials for tissue engineering applications. J. Biomed. Mat. Res. **2000**, *50* (4), 574–584.

43. Shastri, V.P.; Rahman, N.; Martin, I.; Langer, R. Applications of conductive polymers in bone regeneration. Mat. Res. Soc. Symp. Proc. **1999**, *550* (1), 215.

44. Zelikin, A.; Lynn, D.M.; Farhadi, J.; Martin, I.; Shastri, V.P.; Langer, R. Erodible conducting polymers for potential biomedical applications. Angew. Chem. Int. Ed. Engl. **2002**, *41* (1), 141–144.

45. Rivers, T.J.; Hudson, T.W.; Schmidt, C.E. Synthesis of a novel, biodegradable electrically conducting polymer for biomedical applications. Adv. Funct. Mat. **2002**, *12* (1), 33–37.

46. Lines, M.E.; Glass, A.M. *Principles and Applications of Ferroelectric Related Materials*; Clarendon Press: Oxford, 1977.

47. Kawai, H. The piezoelectricity of poly(vinylidine-fluoride). Jpn. J. Appl. Phys. **1969**, *8*, 975–976.

48. Shamos, M.H.; Lavine, L.S. Piezoelectricity as a fundamental property of biological tissue. Nature **1967**, *213* (5073), 267–279.

49. Fakuda, E. Piezoelectricity of biopolymers. Biorheology **1995**, *32* (6), 593–609.

50. Konikoff, J.J. Origins of osseous bioelectric potentials. A review. Ann. Clin. Lab. Sci. **1975**, *5* (5), 330–337.

51. Kepler, R.G.; Anderson, R.A. Ferroelectric polymers. Adv. Phys. **1992**, *41* (1), 1–57.

52. Lovinger, A.J. Poly(Vinylidinefluoride), in developments in crystalline polymers. In *Applied Science*; Bassett, D.C., Ed.; Kluwer Academic Publishers: USA, 1982; 195.

53. Dvey-Aharon, H.; Sluckin, T.J.; Taylor, P.L. Kink propagation as a model for poling in poly(vinylidine-fluoride). Phys. Rev. B. **1980**, *21* (8), 3700–3707.

54. Soballe, K.; Overgaard, S.; Hansen, E.S.; Brokstedt-Rasmussen, H.; Lind, M.; Bunger, C.J. A review of ceramic coatings for implant fixation. J. Long Term Eff. Med. Implants **1999**, *9* (1–2), 131–151.

55. Wahlstrom, O. Stimulation of fracture healing with electromagnetic fields of extremely low frequency. Clin. Orthop. **1984**, *186*, 293–301.

56. Weiss, D.S.; Kirsner, R.; Eaglstein, W.H. Electrical stimulation and wound healing. Arch. Dermatol. **1990**, *126* (2), 222–225.

57. Alvarez, O.M.; Mertz, P.M.; Smerbeck, R.V.; Eaglstein, W.H. The healing of superficial skin wounds is stimulated by external electric fields. J. Invest. Dermatol. **1983**, *81* (2), 144–148.

58. Aebischer, P.; Valentini, R.; Dario, P.; Domencini, C.; Galletti, P. Piezoelectric guidance channels enhance regeneration in the mouse sciatic nerve after axotomy. Brain Res. **1987**, *436* (1), 165–168.

59. Valentini, R.; Vargo, T.; Gardella, J.; Aebischer, P. Electrically charged polymeric substrates enhance nerve fiber outgrowth *in vitro*. Biomaterials **1992**, *13* (3) 183–190.

60. Tanaka, T. Collapse of gels and the critical endpoint. Phys. Rev. Lett. **1978**, *40* (12), 820–823.

61. Tanaka, T.; Fillmore, D.J. Kinetics of swelling of gels. J. Chem. Phys. **1979**, *70* (3), 1214–1218.

62. Tanaka, T.; Nishio, I.; Sun, S.T.; Ueno-Nishio, S. Collapse of gels under an electric field. Science **1982**, *218* (4571), 467–469.

63. Osada, Y.; Ross-Murphy, S. Intelligent gels. Sci. Am. **1993**, *268* (5), 82–87.

64. Osada, Y.; Hasebe, M. Electrically activated mechanochemical devices using polyelectrolyte gels. Chem. Lett. **1985**, *9*, 1285–1288.

65. Gong, J.P.; Nitta, T.; Osada, Y. Electrokinetic modeling of the contractile phenomena of polyelectrolyte gels-one dimensional capillary model. J. Phys. Chem. **1994**, *98* (38), 9583–9587.

66. Osada, Y.; Gong, J.P.; Sawahata, K.J. Synthesis, mechanism, and application of electro-driven chemomechanical system using polymer gels. J. Macromol. Sci. A Chem. **1991**, *28* (11–12), 1189–1205.

67. Osada, Y.; Okuzaki, H.; Hori, H. A polymer gel with electrically driven motility. Nature **1992**, *355*, 242–244.

68. Kwon, I.C.; Bae, Y.H.; Kim, S.W. Electrically erodible polymer gel for controlled release of drugs. Nature **1991**, *354* (6351), 291–293.

69. Alexiou, C.; Arnold, W.; Klein, R.J.; Parak, F.G.; Hulin, P.; Bergemann, C.; Erhardt, W.; Wagenpfeil, S.; Lubbe, A.S. Locoregional cancer treatment with magnetic drug targeting. Cancer Res. **2000**, *60* (23), 6641–6648.

70. Lee, S.W.; Mao, C.; Flynn, C.E.; Belcher, A.M. Ordering of quantum dots using genetically engineered viruses. Science **2002**, *296* (5569), 892–895.

71. Santini, J.T.; Cima, M.J.; Langer, R. A controlled-release microchip. Nature **1999**, *397*, 335–338.

72. Lahann, J.; Mitragotri, S.; Tran, T.H.; Kaido, H.; Sundaram, J.; Choi, I.S.; Hoffer, S.; Somorjai, G.; Langer, R. A reversibly switching surface. Science **2003**, *299* (5605), 371–374.

Electrospinning Technology: Polymeric Nanofiber Drug Delivery

Narendra Pal Singh Chauhan
Department of Polymer Science, University College of Science, Mohanlal Sukhadia University, Udaipur, India

Kiran Meghwal
Department of Chemistry, University College of Science, Mohanlal Sukhadia University, Udaipur, India

Priya Juneja
Jubilant Life Sciences, Noida, India

Pinki B. Punjabi
Department of Chemistry, University College of Science, Mohanlal Sukhadia University, Udaipur, India

Abstract

The employment of electrospun polymeric nanofibers as drug delivery vehicles has been based on their unique functionality and inherent nanoscale morphological characteristics. In addition, due to the flexibility during processing with a variety of structural architectures containing drug molecules, these could be fabricated from monolithic nanofibers to various multiple composition systems. These important benefits allow finely tuned drug-eluting profiles that rely on controlling drug travelling length or modulating the affinity between matrix materials and drugs. In this entry, physical and chemical immobilization methods of bioactive molecules on the surface of various polymeric nanofibers with their applications to drug delivery are described, and various electrospun nanofibers made of poly(ε-caprolactone) (PCL)/hydroxyapatite (HAp), polystyrene (PS), and silk fibroin with improved cell adhesion and proliferation are reviewed. Electrospun nanofibers composed of poly(glycolic acid) (PGA), poly(L-lactic acid) (PLLA), or poly(lactic-co-glycolic acid) (PLGA) were modified with carboxylic acid groups through plasma glow discharge with oxygen and gas-phased acrylic acid. Such hydrophilized nanofibers were shown to enhance fibroblast adhesion and proliferation properties. Drug release mechanism is associated with polymer degradation and complicated diffusion pathway along nano-void spaces within nanofiber mesh. Hollow nanofibrous tubes by coaxial electrospinning also have provided a promising structure for the encapsulation of target drug molecules. This approach succeeded in achieving high drug loading and also facilitation of the solubilization of some insoluble and intractable drugs.

INTRODUCTION

Electrospinning is a technique of producing ultrafine fibers with diameters starting from 10 nm to several microns either in the form of solution or molten liquid by the application of high voltage (30–50 kV). This process is conventional drawing of the fibers where external pressure is used. The features of electrospun nanofibers are high specific surface area, high porosity, and three-dimensional (3D) reticulate structures, which is quite similar to a natural extracellular matrix. Due to these features, electrospun nanofibers have attracted attention for their potential applications in different fields.[1–5] These features allow them to have a wide range of biomedical applications such as tissue engineering,[6] wound dressing,[7] biosensors,[8] and particularly for drug delivery applications.[5,9–12] Electrospinning technique has been used to fabricate nanofibers for drug encapsulation and release. In the late 1500s, Gilbert observed the elctrospraying process. He observed that when a suitably charged piece of amber was brought near to a droplet of water it formed a cone shape and a small droplet ejected from the tip of the cone. Later in the 1950s Cooley and many researchers patented electrospinning process. The various advantages associated with the electrospinning process include that long continuous fibers can be produced and process can be used on laboratory as well as industrial scale. The process also offers cost-effective considerations.

TYPES OF ELECTROSPINNING PROCESS

Chiefly, the electrospinning process may be of two types:

1. Coaxial[12]
2. Emulsion[13,14]

Concise Encyclopedia of Biomedical Polymers and Polymeric Biomaterials DOI: 10.1081/E-EBPPC-120050556

Dental—Electrospinning

Coaxial Electrospinning

A coaxial setup allows for the injection of one solution into another at the tip of the spinneret by using a multiple solution feed system. The sheath fluid acts as a carrier, which draws in the inner fluid at the Taylor Cone of the electrospinning jet.[15] If the solutions are immiscible, then a core shell structure is usually observed. Miscible solutions, however, can result in porosity or a fiber with distinct phases due to phase separation during solidification of the fibers. Taylor cones may be stationary, but they are never static features. Their apices are always the source of emission of charged particles under a rich range of regimens. In this regimen, a steady jet issues continuously from the cone apex, eventually breaking into a spray of charged drops or electrospray (Fig. 1). This steady regimen is not only the best known but also the simplest to analyze, but the limit of high electrical conductivity of the liquid, where the jet radius is typically smaller than 1 μm, enables a division of the problem into two regions. Outer domains (the cone), which are effectively hydrostatic, are slow and have little influence on jet formation, in which the liquid behaves as infinitely conducting, and an inner region, which is dynamic and where a very fine jet carrying a finite current and flow rate forms. Taylor cones of highly conducting liquids offer the only known scheme to produce submicrometer and nanometer jets (down to a diameter of about 10 nm).

Emulsion Electrospinning

Emulsions can be used to create a core shell or composite fibers without modification of the spinneret. However, these fibers are usually more difficult to produce as compared to coaxial spinning due to the greater number of variables, which must be accounted for in creating the emulsion. A water phase and an immiscible solvent phase are mixed in the presence of an emulsifying agent to form the emulsion. Any agent that stabilizes the interface between the immiscible phases can be used for this like surfactants such as sodium dodecyl sulfate, Triton, and nanoparticles.

During the electrospinning process, the emulsion droplets within the fluid are stretched and gradually confined that leads to their coalescence. If the volume fraction of inner fluid is sufficiently high, then a continuous inner core can be formed.[16]

Further, a modified electrospinning method was introduced for the creation of an elastomeric, fibrous sheet [fibrous composite sheet with two distinct submicrometer fiber populations: biodegradable poly(ester urethane) urea (PEUU) and poly(lactide-co-glycolide) (PLGA)], where the PLGA was loaded with the antibiotic tetracycline hydrochloride (PLGA–tet) capable of sustained antibacterial activity *in vitro*. Composite sheets were flexible with breaking strains exceeding 200%, tensile strengths of 5–7 MPa, and high suture-retention capacity. The blending of PEUU fibers markedly reduced the shrinkage ratio observed for PLGA–tet sheets in buffer from 50% to 15%, while imparting elastomeric properties to the composites. In the development of this material, a new approach to two-stream electrospinning (Fig. 2) was used in which one component stream provided for antibiotic release while the other provided mechanical properties deemed essential for the desired application. This material may find applicability in the treatment of temporary abdominal wall closure.

SURFACE MODIFICATION TECHNIQUES OF ELECTROSPUN NANOFIBERS

In order to apply electrospun nanofibers in biomedical uses, their surfaces have been modified chemically and physically with bioactive molecules and cell-recognizable ligands. This subsequently has provided bio-modulating or biomimetic microenvironments for contacting cells and tissues. A variety of functionalization strategies of synthetic electrospun nanofibers with bioactive molecules including proteins, nucleic acids, and carbohydrates have been employed for advanced biological and therapeutic applications.[18]

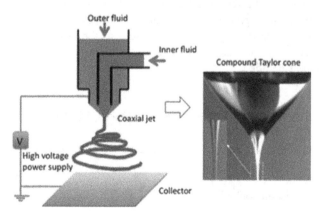

Fig. 1 Diagram showing fiber formation by coaxial electrospinning.

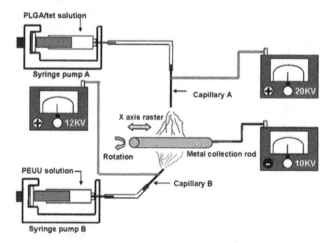

Fig. 2 Two-stream electrospinning setup.
Source: From Agarwal, Wendorff, and Greiner[17] © 2008, with permission from Elsevier.

Synthetic polymers offer easier processability for electrospinning and more controllable nanofibrous morphology than natural polymers. Due to the processing benefits of synthetic polymers, a wide variety of natural polymers having unique biological functions can be immobilized onto the nanofibrous surface of synthetic polymers without compromising bulk properties. Such biologically functionalized synthetic nanofibers can direct enhanced, cell-specific phenotype and organization, because tissue regeneration process is strongly involved with diverse biochemical cues on the cell-contacting surface.[19] For continuous drug delivery applications, the electrospinning process enables a variety of hydrophobic therapeutic agents, to be directly incorporated within the bulk phase of nanoscale fibers for controlled release. For example, a biodegradable polymer solution containing hydrophobic anticancer drugs such as paclitaxel was directly electrospunned to produce drug-releasing nanofibrous mesh.[20] Alternatively, hydrophilic and charged macromolecular drugs such as proteins and nucleic acids were immobilized covalently and physically onto the modified surface of nanofibrous mesh for modulating the cellular functions. The electrospun nanofiber mesh so produced possesses a highly interconnected open nanoporous structure with a high specific surface area that offers an ideal condition for sustained and local drug delivery.[21] Various surface modification techniques for applying synthetic polymer nanofibers to tissue engineering and drug delivery are as follows:

Plasma Treatment

Plasma is a complex energy source for the modification of surface properties of various materials including biomaterials. Plasma is composed of chemically active species that are excited and ionized particles, which may be atomic, and molecular, photons and radicals. These species are highly energetic to induce the chemical reactions.[22] Electrospun nanofibers composed of poly(glycolic acid) (PGA), poly(L-lactic acid) (PLLA), or poly(lactic-co-glycolic acid) (PLGA) have been modified with carboxylic acid groups through plasma glow discharge with oxygen and gas-phased acrylic acid.[23] Such hydrophilized nanofibers were shown to increase fibroblast adhesion and proliferation without compromising with physical and mechanical bulk properties. Air or argon plasma treatment has been widely used as a facile surface modification technique for many biomaterials, because using this technique, the surface hydrophilicity can be easily increased with simultaneous removal of surface impurities. For example, various electrospun nanofibers made of poly(ε-caprolactone) (PCL), PCL/hydroxyapatite (HAp), PS, and silk fibroin were surface modified by air or argon plasma, resulting in an improved cell adhesion and proliferation.[24–27] When PCL nanofibers were modified with argon plasma, rich-quality carboxylic acid groups could be produced on the surface.[28]

When the surface-activated nanofibers were soaked in a simulated body fluid solution, the bone-like calcium phosphate mineralization occurred on the surface of nanofibers. This mineralized nanofibrous scaffold exhibited improved wettability with a bio-mimicking bone structure which indicated potential application for bone grafting (Fig. 3).

Wet Chemical Method

Since the plasma treatment for nanofibrous mesh cannot effectively modify the surface of buried nanofibers deeply located in the mesh due to the limited penetration depth of plasma in the nanopores, wet chemical etching methods can offer the flexibility for surface modification of thick nanofibrous meshes. When biodegradable polymeric nanofibrous meshes are surface modified using the partial hydrolysis method, as shown in Fig. 3, special care must be taken. The duration of the hydrolysis and the concentration of hydrolyzing agents are important to optimally produce surface functional groups only by minimum change in the bulk properties.[29] Since carboxylic acids can chelate calcium ions, surface-induced nucleation and growth of minerals were shown to be enhanced on the surface-modified PLLA electrospun scaffold. PCL electrospun nanofibers were also used to modify the surface of thin PCL membrane for generating nanotopographical surface.[30] When the modified membrane was treated with 5 M NaOH, wettability was dramatically enhanced, showing almost zero water contact angle due to the capillary action on the highly rough surface. When National Institute of Health (NIH) 3T3 cells were cultured on the surface of the modified nanotopographical membrane, favorable cell morphology and adhesion was observed on the modified surface, possibly due to the unique hydrophilic surface topography.[19] NIH 3T3 cells are established from an NIH Swiss mouse embryo. These cells are highly contact inhibited and are sensitive to sarcoma virus focus formation and leukemia virus propagation. The "3T3" designation refers to the abbreviation of 3-day transfer, inoculum 3×10^5 cells.

Surface Graft Polymerization

Almost all types of synthetic biodegradable polymers retain their hydrophobic surface nature, often requiring hydrophilic surface modification for desired cellular responses. Surface graft polymerization has been introduced not only to confer surface hydrophilicity but also to introduce multifunctional groups on the surface for covalent immobilization of bioactive molecules for the purpose of enhanced cell adhesion, proliferation, and differentiation.[31–34] The surface graft polymerization is often initiated with plasma and UV radiation treatment to generate free radicals for the polymerization. Electrospun polyethylene terephthalate nanofibers were modified with poly(methacrylic acid) by graft polymerization in a mild condition without any structural damage in the bulk

Dental—Electrospinning

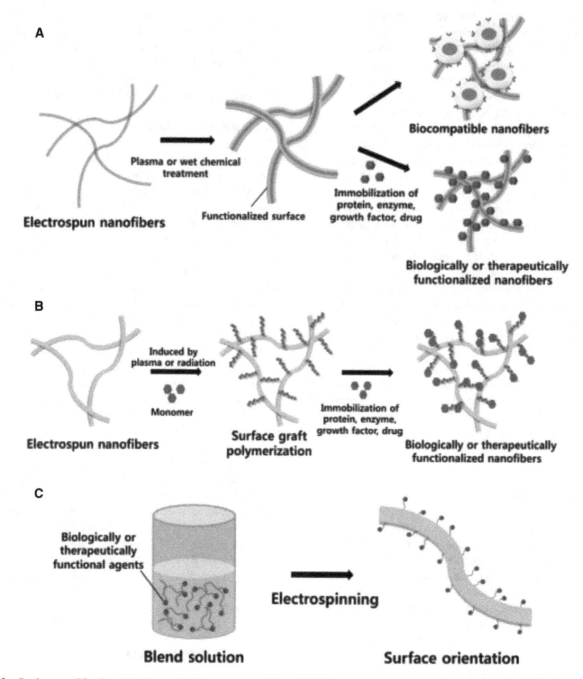

Fig. 3 Surface modification techniques of electrospun nanofibers. (**A**) Plasma treatment or wet chemical method. (**B**) Surface graft polymerization. (**C**) Co-electrospinning.
Source: From Hyuk, Taek, and Tae[19] © 2009, with permission from Elsevier.

phase.[35] For antibacterial applications, electrospun polyurethane (PU) nanofibers were surface modified using poly(4-vinyl-N-hexyl pyridinium bromide).[36] In this study, the PU fibers were first treated with argon plasma, which produced surface oxide and peroxide groups. When the plasma-treated PU fibers were immersed in a 4-vinylpyridine monomer solution with exposure of UV irradiation, poly(4-vinylpyridine)– grafted PU fibers were successfully produced. Through

quaternization of the grafted pyridine groups with hexylbromide, the surface-modified PU fibers were endowed with antibacterial activities. The viability of gram-positive *Staphylococcus aureus* and gram-negative *Escherichia coli* after contact with the PU fibers was measured. The antibacterial efficacy of the modified PU fibers for *S. aureus* and *E. coli* were 99.999% and 99.9%, respectively after 4 hr contact, indicating highly effective antibacterial activities (Fig. 3).

Co-electrospinning of Surface-Active Agents and Polymers

The co-electrospinning process uses two different solutions that are electrospun simultaneously using a spinneret with two coaxial capillaries to produce core/shell nanofibers. The core is then selectively removed and hollow fibers are formed. A well-stabilized co-electrospinning process can be achieved when both the solutions are sufficiently viscous and even spinnable and the solutions are immiscible. When PLLA solution blended with Hap nanocrystals was co-electrospun, HAp was exposed on the surface of the resultant fibers, giving rise to high surface free energy and low water contact angle.[37] These composite fibers exhibited a retarded degradation rate as compared to pure PLLA fibers due to the internal ionic bonding between ester groups in PLLA and calcium ions in HAp. In addition, a novel in situ peptide bio-functionalization method driven by an electric field was developed.[38] Firstly, an antimicrobial peptide, with three repeating units of three anionic amino acids, serine, glutamic acid, and another glutamic acid (SEE)$_3$, was terminally conjugated to polyethylene oxide (PEO). The addition of (SEE)$_3$-PEO conjugate to PEO solution decreased viscosity but increased the solution conductivity. During co-electrospinning of the PEO/(SEE)$_3$-PEO blend solution, electrically polarizable SEE segment had significant influence on fiber morphology. When the collector was connected as an anode, thick and inter-welded fiber morphology could be observed due to the high flow rate of the blend solution under an electric field (Fig. 3).

DRUG LOADING METHOD ON THE NANOFIBER SURFACE FOR BIOMEDICAL APPLICATIONS

By Physical Adsorption

Drug-friendly physical immobilization on the surface can be achieved by using surface-modified prefabricated nanofibrous meshes that have an extremely high surface area to volume ratio which results in higher drug loading amount per unit mass than any other devices. The immediate release of drugs from the nanofiber surface can enable facile dosage control of some therapeutic agents which suits some specific applications such as prevention of bacterial infection occurring within few hours after surgery.[39] In addition to the high drug-loading capacity, hierarchically organized structure such as drug-loaded nanoparticles adsorbed on the nanofibers can allow unique drug-releasing profiles, which nanofiber itself cannot achieve.[40,41] Figure 4 describes three different modes of physical drug loading methods on the surface of electrospun nanofibers for drug delivery application. These are simple physical adsorption, nanoparticles adsorption, and adsorption of multilayer assembly.

Simple physical adsorption

Physical surface adsorption is the simplest approach for loading drug on the nanofibrous mesh. The use of specific interaction between heparin and growth factor is a very typical example. Heparin, a highly sulfated glycosaminoglycan, has strong binding affinity with various growth factors such as fibroblast growth factor, vascular endothelial growth factor, heparin-binding epidermal growth factor, and transforming growth factor-β. This approach offers preservation of biological activity by preventing early degradation of growth factors. Heparin immobilization of biomaterial surface and subsequent attachment of growth factor can be the most efficient way for local delivery of growth factors and consequent mitogenic induction.[42]

The surface-modified nanofibrous meshes have been proven to be an efficient barrier for preventing postsurgical adhesion. As an adhesion barrier, a commercial antibiotic drug, Biteral®, was adsorbed to the electrospun PCL nonwoven sheet.[43] To load the drug, the drug solution was simply dropped on the PCL nonwoven sheet and left to be completely adsorbed. Due to the unique morphological feature of electrospun fibers, such drug loading can be a very straightforward and effective method. A rapid drug release profile is highly desirable for preventing infections at an early stage.

Nanoparticles assembly on the surface

The large interfacial areas of the nanofibers make possible the fabrication of high-performance devices. Any combination of both drug-encapsulated polymeric nanoparticles and nanofibers can be possible to have multifunctional performance. Electrospinning technique is a good method for the formation of such multifunctional devices in which functional nanoparticles were readily embedded within or adsorbed onto nanofibrous mesh.[44,45] An interesting assembly design using nanoparticles and nanofibers was reported for drug delivery applications[46] using electrospinning and

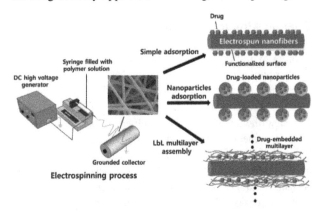

Fig. 4 Three modes of physical drug loading on the surface of electrospun nanofibers.
Source: From Hyuk et al.[19] © 2009, with permission from Elsevier.

electrospraying techniques. The assembly is made up of nanoparticles on nanofibers hierarchical structure having attractive forces between opposite charges. It was reported that poly(methyl methacrylate) (PMMA) solution was electrospun and PS solution was simultaneously electrosprayed from two separate counter-charged nozzles in a side-by-side fashion. Two oppositely charged electrohydrodynamic jets were encountered with neutralization, resulting in a composite structure consisting of electrospun PMMA nano- or microfibers uniformly combined with PS nano- or microparticles. When lipoic acid (antioxidant drug) or gold nanoparticles suspension were electrosprayed, electrospun PMMA fibers were successfully modified with the corresponding composition in an in situ manner.

Layer-by-layer multilayer assembly

The method comprises an alternative layer-by-layer deposition of polyanions and polycations principally driven by an electrostatic force on charged substrates, resulting in self-assembled multilayer coating or free-standing film. This technique has attracted considerable attention due to the ease of its synthesis, universality for any complex structure of substrate, and the possibility of using any composition for the coating layer. Many bioactive agents and chemical drugs have been assembled for topical drug delivery applications mainly on planar substrates such as silicon wafer, quartz slides, and metal oxide due to the ease of their synthesis and analysis. The polyelectrolyte multilayer was deposited on the surface of electrospun PS fibers.[47] PS fibers were initially prepared by electrospinning and then the fiber surface was endowed with negative charges by sulfonation of phenyl groups. Moreover, gold nanoparticles were shown to be homogenously and densely assembled into the PAH/PSS multilayer. It was suggested that this facile surface modification of electrospun fibers with synthetic polymers (PAH and PSS) and biopolymer (DNA) could provide the opportunity for creating a variety of drug-releasing surfaces for biomedical applications.

Chemical Immobilization

In order to immobilize bioactive molecules on the surface of electrospun nanofibers, chemical modification must be done to produce reactive functional groups. Chemical immobilization of bioactive molecules on to the surface of electrospun nanofibers is preferred over physical immobilization in tissue engineering applications. This is so because the immobilized molecules are covalently attached to the nanofibers, and hence, they are not easily leached out from the surface-modified nanofibers when incubated over an extended period. Primary amine and carboxylate groups were most extensively employed to immobilize bioactive molecules onto the surface of nanofibers.[48] Carboxylic groups on the surface of polymeric nanofibers containing different amounts of polyacrylic acid were employed for conjugation with collagen.[23] In other studies, acrylic

acid–immobilized nanofibers were conjugated to amine groups to prepare aminated nanofibers.[49] These nanofibers were further employed for in vitro cultivation of umbilical cord blood cells.[49,50] Hydroxyl groups were also employed for the chemical immobilization of bioactive molecules. Modified PS with a hydroxyl containing initiator was electrospun and alpha-chymotrypsin was covalently attached on the surface.[51] Upon the advancement of electrospinning technology, nanofibrous polymer scaffolds have been diversified in their polymer sources, fiber size, porosity, and texture. Biodegradable synthetic polymers, that is, PLLA, PGA, PLGA, and PCL, as well as natural polymers, such as collagen, silk proteins, and fibrinogen, are fabricated into nanofibers through electrospinning.[52] These nanofiber-based polymer scaffolds have been extensively used for tissue-engineered cartilage or bone, skin, and nerves.[53–56] Transformations of hydrophobic surface either physically or chemically into hydrophilic one have been used to improve poor cell adhesive characteristics. Some bioactive proteins, that is, fibronectin, collagen, and laminin or Arg-Gly-Asp (RGD) peptide can be adsorbed on the polymer surface to increase cellular attachment.[57] Chemicals were also used to change surface characteristics. When PGA mesh scaffolds were treated with sodium hydroxide, the cell seeding density is increased, mainly due to the chemically modified surface with the hydrophilic functional groups.[58] Pre-treatment of electrospun PGA nanofibers in concentrated hydrochloric acid considerably improved cell attachment.[59]

PRINCIPLE OF DRUG DELIVERY

Drug delivery with polymer nanofibers is based on the principle that dissolution rate of a drug particulate increases with increased surface area of both the drug and the corresponding carrier, if necessary. For controlled drug delivery, in addition to their large surface area to volume ratio, polymer nanofibers also have other additional advantages. For example, unlike common encapsulation which involves controlled delivery systems, the polymer nanofibers have been used to improve the therapeutic efficacy and safety of drugs by delivering them to the site of action at a rate dictated by the need of the physiological environment. A wide variety of polymeric materials have been used as delivery matrices and the choice of the delivery vehicle polymer is determined by the requirements of the specific application.[60]

APPLICATION OF DRUG DELIVERY

The applications of some of the polymer nanofibers as drug delivery systems have been described (Table 1).

PLA/Captopril

The nanofiber membranes of PLLA/Captopril were prepared by the electrospinning technique.[61] Captopril is used widely for the treatment of hypertension and congestive

Table 1 Some of the representative electrospun systems used for drug-delivery applications

Electrospun mat	Drug
Poly(caprolactone), PCL	Diclofenac sodium
	Tetracycline hydrochloride
	Resveratrol
	Gentamycin sulfate
	Biteral
Poly(lactic acid), PLA	Tetracycline hydrochloride and mefoxin
Poly(caprolactone-D,L-lactide)	Diclofenac sodium
Poly(vinyl alcohol), PVA	Diclofenac sodium, tetracycline hydrochloride
	Sodium salicylate, naproxen, indomethacin
Poly(maleic anhydride-alt-2-methoxyethylvinylether)	Diclofenac sodium
Poly(lactide–glycolide), PLGA	Paclitaxel (anticancer), tetracycline hydrochloride
Poly(ethylene-*co*-vinylacetate)	Tetracycline hydrochloride
Gelatin	*Centella asiatica*-herbal Extract
Cellulose acetate	Vitamin A and E

Source: From Agarwal, Wendorff, and Greiner[17] © 2008, with permission from Elsevier.

heart failure.[62] Clinically, the general captopril tablets often need to be used three times a day for a long period. The side effects associated are vertigo and headache.[63] To avoid side effects, a novel formulation of captopril using electrospinning technique has been made for promising clinical applications. When the drug is dissolved in the polymer solution and electrospunned into the composite nanofibers, then it can make the drug to be loaded in the carrier of polymer nanofibers. It can not only achieve a relatively high bioavailability of the loaded drug but also minimize their severe side-effects. Further, more than one drug can be encapsulated directly into the electrospun fibers.[61]

Amoxicillin-Loaded Electrospun Nano-HAp/PLA

Drug-loaded halloysite nanotubes (HNTs), a naturally occurring clay material, can be incorporated within PLGA nanofibers by simply electrospinning the mixture solution of PLGA and HNTs drug particles.[64] The hybrid nanofibers formed as a consequence afford the drug with a significantly decreased burst release profile, and simultaneous incorporation of HNTs has greatly improved the mechanical durability of the nanofibers.[64–67] The incorporated HNTs themselves are a kind of drug carrier, which allows drug molecules to be encapsulated within the lumen of the HNTs.

Nano-HAp (n-HA) has been considered as an ideal inorganic drug carrier due to its high surface area to volume ratio, high surface activity, good biocompatibility, porosity, surface hydrophilicity, and strong ability to absorb a variety of chemical species.[68]

A model drug, amoxicillin (AMX) was first loaded onto the n-HA surface via physical adsorption. Then the

AMX-loaded n-HA particles were mixed with PLGA solution for subsequent formation of electrospun AMX/n-HA/PLGA composite nanofibers (Scheme 1A).

The loading of AMX onto n-HA (n-HA/AMX) and the formation of AMX/n-HA/PLGA composite nanofibers were characterized by using different techniques. The antimicrobial activity of the AMX/n-HA/PLGA nanofibers was investigated by using *S. aureus* as a model bacterium both in liquid and in solid medium.

The structural formula of n-HA is $Ca_{10}(PO_4)_6(OH)_2$. It is a principal inorganic ingredient of bone and teeth of the mammal.[70] The loading percentage of AMX onto n-HA was optimized by varying the respective concentration of AMX and n-HA. The apparent density, porosity, and water contact angle data of PLGA, n-HA/PLGA, and AMX/n-HA/PLGA nanofibers (data are representatives of independent experiments and all data are given as mean ± SD, n = 5) are listed in Table 2.

The incorporation of drug-loaded n-HA not only significantly improve the mechanical durability of the nanofibers but also appreciably weaken the initial burst release of the drug. The combination of the two pathways for the AMX dissociation, i.e., first from n-HA surface to PLGA fiber matrix and second from PLGA fiber matrix to the release medium, is proven to be an efficient strategy to slow down the release rate of AMX. This is important for biomedical applications requiring the drug to maintain a long-term antibacterial efficacy.

Wound Dressing

The ideal wound dressing should minimize infection and pain, prevent excessive fluid loss, maintain a moist healing environment, promote epithelial restoration, and be

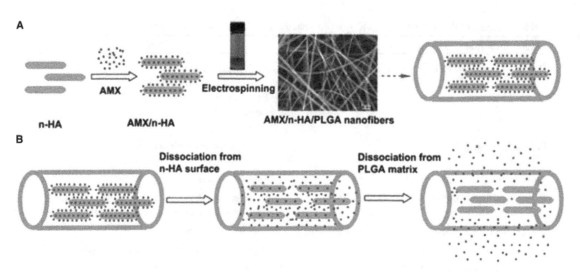

Scheme 1 Schematic illustration of the encapsulation (**A**) and release pathways (**B**) of AMX within n-HA–doped PLGA nanofibers.
Source: From Zheng et al.[69] © 2008, with permission from Elsevier.

Table 2 Apparent density, porosity, and water contact angle of PLGA, n-HA/PLGA, and AMX/n-HA/PLGA nanofibers (data are representative of independent experiments and all data are given as mean ± SD, $n = 5$)

Sample	Apparent density (g/cm³)	Porosity (%)	Water contact angle (°)
PLGA	0.357 ± 0.087	71.5 ± 6.9	139.2 ± 2.1
n-HA/PLGA	0.357 ± 0.067	71.4 ± 5.4	136.3 ± 2.2
AMX/n-HA/PLGA	0.315 ± 0.02	74.8 ± 1.6	137.2 ± 2.9

Source: From Zheng et al.[69] © 2008, with permission from Elsevier.

biocompatible. However, dressings available in the market do not meet all the requirements necessary for an ideal dressing. In addition to the application of dressing, wound treatment includes irrigation of the affected area with an anesthetic solution followed by application of prophylactic antibiotics to prevent wound infection.

Based on these ideas, the concept of a dual drug scaffold wound dressing was put forward. Such a scaffold would offer a unique combination of the inherent properties of electrospun scaffolds such as promoting cell proliferation and simultaneously providing anesthetic and antibiotic activity for pain relief and healing.[71]

In a study, the drugs lidocaine hydrochloride (LH) and mupirocin have been selected to include in drug scaffold.[72] Lidocaine is a routinely used anesthetic in wound-related pain management. It has some antibacterial actions against *E. coli*, *S. aureus*, *Pseudomonas aeruginosa*, and *Candida albicans*. Mupirocin is a commonly used antibiotic in wound care for prophylaxis against cutaneous infection and elimination of carriage of *S. aureus*. It is effective against aerobic gram-positive and some gram-negative flora. The low incidence of adverse effects and rare cases of resistance

add to its attractiveness as an antibiotic of choice. It is formed as a 2% cream or ointment that is applied three times daily from 3 to 10 days as required. The wound is covered with gauze after its application.

To achieve the proposed dual release, a novel electrospinning technique with simultaneous electrospinning from dual spinnerets has been investigated and depicted in Fig. 5.

Characterization of fiber scaffolds

Fiber scaffolds containing fibers of two unique compositions were obtained using the DS electrospinning apparatus. Macroscopically, the scaffold was a conformable and resilient structure having ultrafine cloth appearance. Fluorescence microscopy of the scaffold, which contained one fiber doped with Texas Red and another fiber without Texas Red, showed homogenous distribution of the two fibers (Fig. 6). One can observe two larger diameter fibers: One of which is clearly fluorescent (fiber 1) and another which is not fluorescent (fiber 2).

The DS electrospinning apparatus could be used to electrospin a hybrid mesh of materials of varying degradation rates, mechanical properties, or chemical functionality. Here, the technique was used to create a mesh where one fiber was loaded with an antibiotic and another fiber was loaded with an anesthetic.

Above images are fabricated by DS technique. Fiber 1 is clearly visible by both imaging techniques while Fiber 2 is not visible by fluorescence microscopy.

Cell viability, attachment, and proliferation

Wound-healing scaffolds should be able to support cell proliferation and viability for fast wound healing. Electrospun PLLA has been reported to support the growth of cells such as neural stem cells[73] and cardiac myocytes.[74]

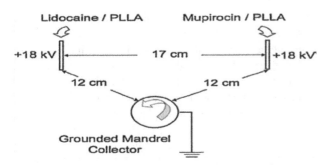

Fig. 5 Schematic of the dual spinneret electrospinning apparatus.
Source: From Thakur et al.[71] © 2008, with permission from Elsevier.

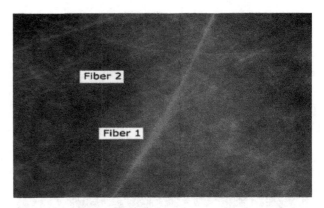

Fig. 6 (**A**) Light microscopy image of electrospun fiber scaffolds. (**B**) Fluorescence image of electrospun fiber scaffolds.
Source: From Thakur et al.[71] © 2008, with permission from Elsevier.

It is possible that inclusion of drugs may alter the cell proliferation *in vivo*.[75] However, it was found that the histopathologic appearance of wounded tissues infiltrated with lidocaine did not vary consistently in relation to collagenization, edema, or acute and chronic inflammatory processes. Lidocaine alone cannot alter wound healing or the breaking strength of the wounds.[76] The dual spinneret electrospinning technique facilitated the fabrication of a polymeric dressing with dual drug release kinetics that could have potential application for wound therapy. An anesthetic and LH were crystallized in the PLLA matrix and eluted through a burst release mechanism. This action would be useful if this mat was used as a wound dressing for immediate relief of pain. Mupirocin, an antibiotic, was released simultaneously with LH through a diffusion-mediated mechanism for extended antibiotic activity. The dual spinneret electrospinning technique achieved the required dual release profiles by allowing LH to crystallize in other PLLA fibers and also by maintaining mupirocin in the non-crystallized form within the PLLA matrix.

Electrospun PU–Dextran Nanofiber

Dextran is a versatile biomacromolecule for preparing electrospun nanofibrous membranes by blending with either water-soluble bioactive agents or hydrophobic biodegradable polymers for biomedical applications. Dextran is a bacterial polysaccharide made up of R-1, 6 linked D-glucopyranose residues with some R-1, 2-, R-1, 3-, or R-1, 4 linked side chains. Due to its biodegradability and biocompatibility, it has been used for various biomedical applications.[77] In comparison to other biodegradable polymers, dextran is inexpensive and readily available. Most importantly, dextran is soluble in both water and some organic solvents. This unique solubility characteristic of dextran put forth the possibility of directly blending it with biodegradable hydrophobic polymers such as PU to prepare composite nanofibrous membranes by using electrospinning method.

PU is a commonly used candidate for wound dressing applications because of its good barrier properties and oxygen permeability.[78] Ciprofloxacin HCl (CipHCl), a fluoroquinolone antibiotic, is one of the most widely used antibiotics in wound healing because of its low minimal inhibitory concentration for both gram-positive and gram-negative bacteria that cause wound infections.[79] The frequency of spontaneous resistance to ciprofloxacin is very low.[80] An electrospun nanofiber membrane containing antibiotic agents has been used as a barrier to prevent the postwound infections. The combination of both these properties can result in a perfect wound dressing material.

Characterization

The morphology of the electrospun PU, PU–dextran, and PU–dextran drug-loaded composite nanofibers was observed by using scanning electron microscopy shown in Fig. 7.

Antimicrobial test of the composite nanofibers

The composite drug-loaded nanofibers have been screened for their antimicrobial activity after an overnight incubation of the agar plate at 37°C. The bactericidal activity was reported by the measurement of zone of inhibition within and around the drug-loaded nanofiber mat for gram-positive and for gram-negative bacteria. As shown in Fig. 8A (1, 2) for gram-positive bacteria *S. aureus* and *Bacillus subtilis* the mean diameter of inhibition ring of drug-loaded PU–dextran composite nanofibers was around 15 and 20 mm, respectively. In the case of gram-negative bacteria, the diameter of the inhibition zone reached 20 mm as shown in Fig. 8B (1, 2, 3, respectively). No bactericidal activity was detected for pristine PU and PU–dextran nanofibrous mats. The drug-loaded PU–dextran composite nanofibers showed excellent bactericidal activity against a wide range of bacteria, thereby avoiding exogenous infections effectively.

It is a known factor that the decontamination of exogenous organisms is a critical factor for a wound-healing material. The antibacterial property plays a crucial role for the electrospun-based wound dressing membranes. As the interconnected nanofibers create perfect blocks and pores

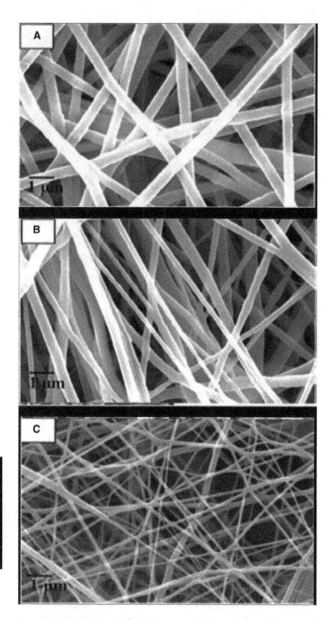

Fig. 7 SEM images of electrospun (**A**) PU, (**B**) PU dextran, and (**C**) PU–dextran–drug nanofibrous mat.
Source: From Unnithan et al.[81] © 2012, with permission from Elsevier.

in a nanofiber membrane, the nanofiber membrane should be able to prevent any bacteria from penetrating and therefore avoiding exogenous infections effectively. The results showed that a composite nanofiber mat is a good antibacterial membrane and it can be applied as a perfect wound dressing material.

Rifampin PLLA Electrospun Fibers

Rifampin, an antituberculosis drug is lipophilic and highly soluble in PLLA/chloroform/acetone solution. Electrospinning of PLLA/chloroform/acetone solutions usually resulted in fibers with a diameter range of 0.3–4.2 μm. The diameter size was greatly decreased by adding rifampin of small

molecular size in the solution. Uniform fibers were obtained as shown in Fig. 9. Red color of the fibers indicated uniform distribution of drug in the fibers. With increasing amount of rifampin, the color of the fibers became deeper. The average diameter was about 700 nm. No rifampin crystals were detected by optical or electronic microscopy, either on the surface of the fibers or outside the fibers, as seen in Fig. 9, indicating that rifampin was perfectly included in the fibers.

When the solution jet was rapidly elongated and the solvent evaporated quickly, the drug remained compatible with PLLA. During the rapid evaporation of the solvent, phase separation took place quickly between the drug and PLLA. Therefore, a reasonable amount of the drug exists on the fiber surface.[82]

Chitosan Nanofiber for Antibacterial Applications

Chitosan (CS) nanofibers with a diameter of 150–200 nm were fabricated from a mixed chitosan/poly (vinyl alcohol) solution by the electrospinning method. Hen egg-white lysozyme was immobilized on electrospun CS nanofibers via cross-linked enzyme aggregates (CLEAs) and used for effective and continuous antibacterial applications. CS has excellent biological properties, such as biodegradability, biocompatibility, antibacterial properties, non-toxicity, hydrophilicity, high mechanical strength, and excellent affinity to proteins. Owing to these characteristics, CS-based materials are used as enzyme immobilization supports. CLEA can be highly efficient biocatalysts with enhanced thermal and environmental stability.

Immobilization of lysozyme-CLEA onto electrospun CS nanofibers

The prepared electrospun CS nanofibers contain a functional group with surface-reactive primary amino groups to allow for the immobilization of proteins. In order to effectively facilitate the covalent coupling and prevent enzyme deactivation, the selection of the optimum coupling time is important. The immobilization efficiency of the lysozyme-CLEA increased with increasing coupling time. The lysozyme-CLEA–immobilized CS nanofibers were confirmed with FE-SEM micrographs. As shown in Fig. 10, after immobilizing the lysozyme, the cross-linked lysozyme aggregates on the CS nanofibers showed a rougher surface morphology (Fig. 10B) than that of the control CS nanofibers (Fig. 10A), confirming formation of strong covalent bonds between the amino groups of the lysozyme molecules and the CS nanofibers. The morphology comparison was useful in confirming the covalent attachments between the CS nanofibers and lysozyme-CLEA molecules.

Antibacterial measurement

The antibacterial activities of the free and immobilized lysozymes were determined by using four pathogenic

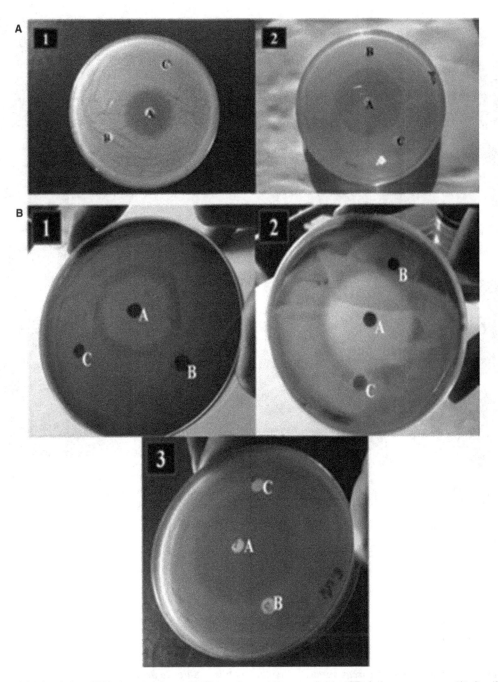

Fig. 8 (**A**) Bactericidal activity of PU–dextran–drug nanofibrous mat with gram-positive *Staphylococcus aureus* (1), *Bacillus subtilis* (2), respectively. PU–dextran–drug, PU–dextran, and pristine PU discs were denoted as A, B, and C, respectively, in the Petri plates. (**B**) Bactericidal activity of PU–dextran–drug nanofibrous mat with gram-negative *Escherichia coli* (1), *Salmonella typhimurium* (2), and *Vibrio vulnificus* (3), respectively. PU–dextran–drug, PU–dextran, and pristine PU discs were denoted as A, B, and C, respectively, in the Petri plates.
Source: From Unnithan et al.[81] © 2012, with permission from Elsevier.

bacteria, viz. *S. aureus*, *B. subtilis*, *Shigella flexneri*, and *P. aeruginosa*. The stability of the antibacterial lysozyme-CLEA–immobilized CS nanofiber was evaluated by multiple reuses. Table 3 shows the effect of reusing the immobilized lysozyme-CLEA on the ratio of bacteriostasis for the four pathogenic bacteria. It can be seen that the antibacterial ratio of the immobilized lysozyme-CLEA decreased with increasing number of reuses, which might be associated with the leakage of the lysozyme from the CS nanofibers upon use. The lysozyme-CLEA–immobilized CS nanofibers retained 82.4, 79.8, 83.4, and 84.1% of their antibacterial ratio after 10 cycles for *S. aureus*, *B. subtilis*, *S. flexneri, and P. aeruginosa*, respectively. These results mean that lysozyme-CLEA–immobilized CS nanofibers could be used as a promising material for enhanced and continuous antibacterial applications.[83]

Fig. 9 SEM photographs of PLLA electrospun fibers containing different amounts of rifampin. (**A**) 30 wt. %; (**B**) 50 wt. %.
Source: From Zeng et al.[82] © 2003, with permission from Elsevier.

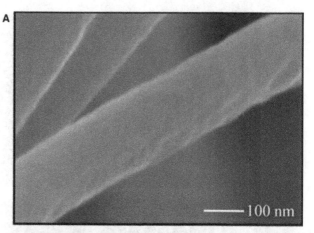

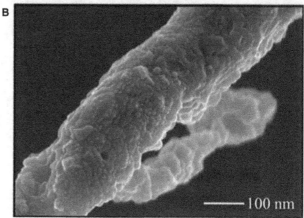

Fig. 10 FE-SEM of the (**A**) control and (**B**) lysozyme-CLEA–immobilized CS nanofibers.
Source: From Park et al.[83] © 2013, with permission from Elsevier.

Table 3 The influence of reusing the lysozyme-CLEA–immobilized CS nanofibers on the bacteriostasis ratio of four different pathogenic bacteria

Number of re-uses	R (%)			
	Staphylococcus aureus	*Bacillus subtilis*	*Shigella flexneri*	*Pseudomonas aeruginosa*
0	100	100	100	100
1	98.3	98.1	97.6	99.2
2	97.6	96.4	98.1	98.1
3	97.5	95.7	96.4	97.8
4	96.2	94.2	94.3	96.1
5	95.5	93.1	92.8	94.9
6	93.2	89.5	91.2	92.1
7	90.5	86.2	89.7	89.6
8	88.4	83.7	88.7	87.5
9	84.7	81.2	86.4	85.2
10	82.4	79.8	83.4	84.1

Source: From Park et al.[83] © 2013, with permission from Elsevier.

CONCLUSION

Electrospun polymeric nanofibers are being extensively used for drug delivery applications because of their ability to produce thin fibers with large surface areas, with high porosity and 3D reticulate structure, simple functionalities, and rich mechanical properties and process ease in applying them to various purposes like tissue engineering, wound dressing, and biosensors. Employing electrospun nanofibers as drug delivery vehicles has found its base in the form of their unique functionality and inherent nanoscale morphological characteristics which makes them compatible to be used as such. Last but not the least, coaxial electrospinning techniques also have succeeded in achieving high drug loading, encapsulation and release and also facilitation of the solubilization of some insoluble and intractable drugs with the aid of hollow nanofibrous tubes.

ACKNOWLEDGMENT

The authors are thankful to Elsevier for copyright permission. Authors are also thankful Prof Suresh C. Ameta, Department of Chemistry, Pacific College of Basic & Applied Sciences, PAHER University, Udaipur (Rajasthan) for suggestions.

REFERENCES

1. Doshi, J.; Reneker, D.H. Electrospinning process and applications of electrospun fibers. J. Electrost. **1995**, *35* (2–3), 151–160.

2. Dzenis, Y. Spinning continuous fibers for nanotechnology. Science. **2004**, *304* (5679), 1917–1919.

3. Huang, Z.M.; Zhang, Y.Z.; Kotaki, M.; Ramakrishna, S.A review on polymer nanofibers by electrospinning and their applications in nanocomposites. Compos. Sci. Technol. **2003**, *63*, 2223–2253.

4. Reneker, D.H.; Chun, I. Nanometre diameter fibres of polymer, produced by electrospinning. Nanotechnology **1996**, *7* (3), 216–223.

5. Qi, R.; Guo, R.; Shen, M.; Cao, X.; Zhang, L.; Xu, J. Electrospun poly (lactic-coglycolic acid)/halloysite nanotube composite nanofibers for drug encapsulation and sustained release. J. Mater. Chem. **2010**, *20*, 10622–10629.

6. Li, W.J.; Laurencin, C.T.; Caterson, E.J.; Tuan, R.S.; Ko, F.K. Electrospunnanofibrous structure : A novel scaffold for tissue engineering. J. Biomed. Mater. Res. **2002**, *60* (4), 613–621.

7. Khil, M.S.; Cha, D.I.; Kim, H.Y.; Kim, I.S.; Bhattarai, N. Electrospunnanofibrous polyurethane membrane as wound dressing. J. Biomed. Mater. Res. **2003**, *67* (3), 675–679.

8. Ding, B.; Wang, M.; Wang, X.; Yu, J.; Sun, G. Electrospunnanomaterials for ultrasensitive sensors. Mater. Today **2010**, *13* (11), 16–27.

9. Zeng, J.; Xu, X.; Chen, X.; Liang, Q.; Bian, X.; Yang, L. Biodegradable electrospun fibers for drug delivery. J. Control. Release. **2003**, *92* (3), 227–231.

10. Luu, Y.; Kim, K.; Hsiao, B.; Chu, B.; Hadjiargyrou, M. Development of a nanostructured DNA delivery scaffold via electrospinning of PLGA and PLA-PEG block copolymers. J. Control. Release. **2003**, *89* (2), 341–353.

11. Kenawy, E.R.; Bowlin, G.L.; Mansfield, K.; Layman, J.; Simpson, D.G.; Sanders, E.H. Release of tetracycline hydrochloride from electrospun poly (ethylene-covinylacetate), poly (lactic acid), and a blend. J. Control. Release **2002**, *81* (1–2), 57–64.

12. Jiang, H.; Hu, Y.; Li, Y.; Zhao, P.; Zhu, K.; Chen, W. A facile technique to prepare biodegradable coaxial electrospunnanofibers for controlled release of bioactive agents. J. Control. Release **2005**, *108* (2–3), 237–243.

13. Xu, X.; Yang, L.; Wang, X.; Chen, X.; Liang, Q.; Zeng, J. Ultrafine medicated fibers electrospun from W/O emulsions. J. Control. Release **2005**, *108* (1), 33–42.

14. Qi, H.; Hu, P.; Xu, J.; Wang, A. Encapsulation of drug reservoirs in fibers by emulsion electrospinning: Morphology characterization and preliminary release assessment. Biomacromolecules **2006**, *7* (8), 2327–2330.

15. Alexander, V.B.; Alexander, L.Y.; Constantine, M.M. Co-electrospinning of core-shell fibers using a single-nozzle technique. Langmuir. **2007**, *23* (5), 2311–2314.

16. Xu, X.; Zhuang, X.; Chen, X.; Wang, X.; Yang, L.; Jing, X. Preparation of core-sheath composite nanofibers by emulsion electrospinning.Macromol.Rapid Commun. **2006**, *27* (19),1637–1642.

17. Agarwal, S.; Wendorff, J.H.; Greiner, A. Use of electrospinning technique for biomedical applications. Polymer **2008**, *49* (26), 5603–5621.

18. Pham, Q.P.; Sharma, U.; Mikos, A.G. Electrospinning of polymeric nanofibers fortissue engineering applications. Tissue Eng. **2006**, *12* (5), 1197–1211.

19. Hyuk, S.Y.; Taek, G.K.; Tae, G.P. Surface-functionalized electrospunnanofibers for tissue engineering and drug delivery. Adv. Drug Deliv. Rev. **2009**, *61* (12), 1033–1042.

20. Xie, J.W.; Wang, C.H. Electrospun micro- and nanofibers for sustained delivery ofpaclitaxel to treat C6 glioma *in vitro*. Pharm. Res. **2006**, *23* (8), 1817–1826.

21. Kim, T.G.; Lee, D.S.; Park, T.G. Controlled protein release from electrospun biodegradable fiber mesh composed of poly(epsilon-caprolactone) and poly (ethylene oxide). Int. J. Pharm. **2007**, *338* (1–2), 276–283.

22. Borcia, C.; Borcia, G.; Dumitrascu, N. Surface treatment of polymers by plasma and uv radiation. Rom. Journ. Phys. **2011**, *56* (1–2), 224–232.

23. Park, K.; Ju, Y.M.; Son, J.S.; Ahn, K.D.; Han, D.K. Surface modification of biodegradable electrospunnanofiber scaffolds and their interaction with fibroblasts. J. Biomater. Sci. Polym. Ed. **2007**, *18* (4), 369–382.

24. Zhu, X.; Chian, K.S.; Park, M.B.E.C.; Lee, S.T. Effect of argon-plasma treatment onproliferation ofhuman-skin-derived fibroblast on chitosan membrane *in vitro*. J. Biomed. Mater. Res. A. **2005**, *73* (3), 264–274.

25. Venugopal, J.; Low, S.; Choon, A.T.; Kumar, A.B.; Ramakrishna, S. Electrospun modified nanofibrous scaffolds for the mineralization of osteoblast cells. J. Biomed. Mater. Res. **2008**, *85* (2), 408–417.

26. Jia, J.; Duan, Y.Y.; Yu, J.; Lu, J.W. Preparation and immobilization of soluble eggshell membrane protein on the

electrospunnanofibers to enhance cell adhesion and growth.J. Biomed. Mater. Res. **2008**, *86* (2), 364–373.

27. Prabhakaran, M.P.; Venugopal, J.; Chan, C.K.; Ramakrishna, S. Surface modified electrospunnanofibrous scaffolds for nerve tissue engineering. Nanotechnology **2008**, *19* (45), 455102.

28. Yang, F.; Wolke, J.G.C.; Jansen, J.A. Biomimetic calcium phosphate coating on electrospun poly (epsilon-caprolactone) scaffolds for bone tissue engineering. Chem. Eng. J. **2008**, *137* (1), 154–161.

29. Croll, T.I.; O'Connor, A.J.; Stevens, G.W.; Cooper-White, J.J. Controllable surface modification of poly(lactic-co-glycolic acid) (PLGA) by hydrolysis or aminolysis I. Physical, chemical, and theoretical aspects. Biomacromolecules **2004**, *5* (2), 463–473.

30. Chen, F.; Lee, C.N.; Teoh, S.H. Nanofibrous modification on ultra-thin poly(epsiloncaprolactone) membrane via electrospinning. Mater. Sci. Eng. **2007**, *27* (2), 325–332.

31. Turmanova, S.; Minchev, M.; Vassilev, K.; Danev, G. Surface grafting polymerization of vinyl monomers on poly(tetrafluoroethylene) films by plasma treatment. J. Polym. Res. **2008**, *15* (4), 309–318.

32. Mori, M.; Uyama, Y.; Ikada, Y. Surface modification of polyethylene fiber by graftpolymerization. J. Polym. Sci. Polym. Chem. **1994**, *32* (9), 1683–1690.

33. Kou, R.Q.; Xu, Z.K.; Deng, H.T.; Liu, Z.M.; Seta, P.; Xu, Y.Y. Surface modification of microporous polypropylene membranes by plasma-induced graft polymerization of alpha-allylglucoside. Langmuir. **2003**, *19*, 6869 – 6875.

34. Liu, Z.M.; Xu, Z.K.; Wang, J.Q.; Wu, J.; Fu, J.J. Surface modification of polypropylene microfiltration membranes by graft polymerization of N-vinyl-2-pyrrolidone. Eur. Polym. J. **2004**, *40* (9), 2077–2087.

35. Ma, Z.W.; Kotaki, M.; Yong, T.; He, W.; Ramakrishna, S. Surface engineering ofelectrospun polyethylene terephthalate (PET) nanofibers towards development of a new material for blood vessel engineering. Biomaterials. **2005**, *26* (15), 2527–2536.

36. Yao, C.; Li, X.S.; Neoh, K.G.; Shi, Z.L.; Kang, E.T. Surface modification and antibacterial activity of electrospun polyurethane fibrous membranes with quaternary ammonium moieties. J. Membr. Sci. **2008**, *320* (1–2), 259–267.

37. Luong, N.D.; Moon, S.; Lee, D.S.; Lee, Y.K.; Nam, J.D. Surface modification of poly (l-lactide) electrospun fibers with nanocrystal hydroxyapatite for engineered scaffold applications. Mater. Sci. Eng. C. **2008**, *28* (8), 1242–1249.

38. Sun, X.Y.; Shankar, R.; Borner, H.G.; Ghosh, T.K.; Spontak, R.J. Field-driven biofunctionalization of polymer fiber surfaces during electrospinning. Adv. Mater. **2007**, *19* (1), 87–91.

39. Chen, C.; Lv, G.; Pan, C.; Song, M.; Wu, C.H.; Guo, D.D.; Wang, X.M.; Chen, B.A.; Gu, Z.Z. Poly(lactic acid) (PLA) based nanocomposites—A novel way of drug-releasing. Biomed. Mater. **2007**, *2*, L1 – L4.

40. Park, C.H.; Kim, K.H.; Lee, J.C.; Lee, J. In-situ nanofabrication via electro hydrodynamic jetting of counter charged nozzles. Polym. Bull. **2008**, *61* (4), 521–528.

41. Ma, Z.W.; He, W.; Yong, T.; Ramakrishna, S. Grafting of gelatin on electrospun poly (caprolactone) nanofibers to improve endothelial cell spreading and proliferation and to control cell orientation. Tissue Eng. **2005**, *11* (7–8), 1149–1158.

42. Joung, Y.K.; Bae, J.W.; Park, K.D. Controlled release of heparin-binding growth factors using heparin-containing particulate systems for tissue regeneration. Expert Opin. Drug Deliv. **2008**, *5* (11), 1173–1184.

43. Bolgen, N.; Vargel, I.; Korkusuz, P.; Menceloglu, Y.Z.; Piskin, E. *In vivo* performance of antibiotic embedded electrospun PCL membranes for prevention of abdominal adhesions. J. Biomed. Mater. Res. B. Appl. Biomater. **2007**, *81* (2), 530–543.

44. Kim, H.W.; Song, J.H.; Kim, H.E. Nanofiber generation of gelatin-hydroxyapatite biomimetics for guided tissue regeneration. Adv. Funct. Mater. **2005**, *15* (12), 1988–1994.

45. Rujitanaroj, P.O.; Pimpha, N.; Supaphol, P. Wound-dressing materials with antibacterial activity from electrospun gelatin fiber mats containing silver nanoparticles. Polymer **2008**, *49* (21), 4723–4732.

46. Park, C.H.; Kim, K.H.; Lee, J.C.; Lee, J. In-situ nanofabrication via electrohydrodynamic jetting of countercharged nozzles. Polym. Bull. **2008**, *61* (4), 521–528.

47. Muller, K.; Quinn, J.F.; Johnston, A.P.R.; Becker, M.; Greiner, A.; Caruso, F. Polyelectrolyte functionalization of electrospun fibers. Chem. Mater. **2006**, *18* (9), 2397–2403.

48. Chua, K.N.; Chai, C.; Lee, P.C.; Tang, Y.N.; Ramakrishna, S.; Leong, K.W.; Mao, H.Q. Surface-aminatedelectrospunnanofibers enhance adhesion and expansion of human umbilical cord blood hematopoietic stem/progenitor cells. Biomaterials **2006**, *27* (36), 6043–6051.

49. Chua, K.N.; Chai, C.; Lee, P.C.; Ramakrishna, S.; Leong, K.W.; Mao, H.Q. Functional nanofiber scaffolds with different spacers modulate adhesion and expansion of cryo preserved umbilical cord blood hematopoietic stem / progenitor cells. Exp. Hematol. **2007**, *35* (5), 771–781.

50. Ye, P.; Xu, Z.K.; Wu, J.; Innocent, C.; Seta, P. Nanofibrous membranes containing reactive groups: Electrospinning from poly(acrylonitrile-co-maleic acid) for lipase immobilization. Macromolecules **2006**, *39* (3), 1041–1045.

51. Ma, Z.; Kotaki, M.; Inai, R.; Ramakrishna, S. Potential of nanofiber matrix as tissue-engineering scaffolds. Tissue Eng. **2005**, *11* (1–2), 101–109.

52. Yoshimoto, H.; Shin, Y.M.; Terai, H.; Vacanti, J.P. A biodegradable nanofiber scaffold by electrospinning and its potential for bone tissue engineering. Biomaterials **2003**, *24* (12), 2077–2082.

53. Li, W.; Tuli, R.; Okafor, C.; Derfoul, A.; Danielson, K.G.; Hall, D.J.; Tuan, R.S. A three-dimensional nanofibrous scaffold for cartilage tissue engineering using human mesenchymal stem cells.Biomaterials **2005**, *26* (6), 599–609.

54. Rho, K.S.; Jeong, L.; Lee, G.; Seo, B.M.; Park, Y.J.; Hong, S.D.; Roh, S.; Cho, J.J.; Park, W.H.; Min, B.M. Electrospinning of collagen nanofibers: Effects on the behavior of normal human keratinocytes and early-stage wound healing. Biomaterials **2006**, *27* (8), 1452–1461.

55. Yang, F.; Murugan, R.; Ramakrishna, S.; Wang, X.; Ma, Y.X.; Wang, S. Fabrication of nano-structured porous PLLA scaffold intended for nerve tissue engineering. Biomaterials **2004**, *25* (10), 1891–1900.

56. Ho, M.H.; Wang, D.M.; Hsieh, H.J.; Liu, H.C.; Hsien, T.Y.; Lai, J.Y.; Hou, L.T. Preparation and characterization of RGD-immobilized chitosan scaffolds. Biomaterials **2005**, *26* (16), 3197–3206.

57. Boland, E.D.; Telemeco, T.A.; Simpson, D.G.; Wnek, G.E.; Bowlin, G.L. Biomimetic nanofibrous scaffolds for

bone tissue engineering. J. Biomed. Mater.Res. **2004**, *71* (1), 144–152.

58. Gao, J.; Niklason, L.; Langer, R. Surface hydrolysis of poly(glycolic acid) meshes increases the seeding density of vascular smooth muscle cells. J. Biomed. Mater. Res. **1998**, *42* (3), 417–424.

59. Mwale, F.; Wang, H.T.; Nelea, V.; Luo, L.; Antoniou, J.; Wertheimer, M.R. The effect of novel nitrogen-rich plasma polymer coatings on the phenotypic profile of notochordal cells.Biomaterials. **2006**, *27*, 2258–2264.

60. Rathinamoorthy, R. Nanofiber for drug delivery system—Principle and application. P. T. J. **2012**, *61*, 45–48.

61. Wei, A.; Wang, J.; Wang, X.; Hou, D.; Wei, Q. Morphology and surface properties of poly(L-lactic acid)/captopril composite nanofiber membranes. J. Eng. Fiber Fabr. **2012**, *7* (1), 129–135.

62. Kalia, K.; Narula, G.D.; Kannan, G.M.; Flora, S.J.S. Effects of combined administration of captopril and DMSA on arsenite induced oxidative stress and blood and tissue arsenic concentration in rats. Comp. Biochem. Physiol. C. Toxicol. Pharmacol. **2007**, *144* (4), 372–379.

63. Abubakr, O.N.; Jun, S.Z. Recent progress in sustained: Controlled oral delivery of Captopril: An overview. Int. J. Pharm. **2000**, *194* (2), 139–146.

64. Qi, R.; Guo, R.; Shen, M.; Cao, X.; Zhang, L.; Xu, J. Electrospun poly (lactic-co-glycolic acid)/halloysite nanotube composite nanofibers for drug encapsulation and sustained release. J. Mater. Chem. **2010**, *20* (47), 10622–10629.

65. Qi, R.; Cao, X.; Shen, M; Guo, R.; Yu, J.; Shi, X. Biocompatibility of electrospunhalloysite nanotube-doped poly (lactic-co-glycolic acid) composite nanofibers. J. Biomater. Sci. Polym. Ed. **2012**, *23* (1–4), 299–313.

66. Qi, R.; Shen, M.; Cao, X.; Guo, R.; Tian, X.; Yu, J. Exploring the dark side of MTT viability assay of cells cultured onto electrospun PLGA-based composite nanofibrous scaffolding materials. Analyst **2011**, *136* (14), 2897–2903.

67. Zhao, Y.; Wang, S.; Guo, Q.; Shen, M.; Shi, X. Hemocompatibility of electrospunhalloysite nanotube and carbon nanotube-doped composite poly (lactic-coglycolic acid) nanofibers. J. Appl. Polym. Sci. **2012**, *127* (1–4), 4825–4832.

68. Zhang, J.; Wang, Q.; Wang, A. In situ generation of sodium alginate/hydroxyapatite nanocomposite beads as drug-controlled release matrices. Acta. Biomater. **2010**, *6* (2), 445–454.

69. Zheng, F.; Wang, S.; Wen, S.; Shen, M.; Zhu, M.; Shi, X. Characterization and antibacterial activity of amoxicillin-loaded electrospunnano-hydroxyapatite/poly(lactic-co-glycolic acid) composite nanofibers. Biomaterials **2013**, *34* (4), 1402–1412.

70. Wang, S.; Wen, S.; Shen, M.; Guo, R.; Cao, X.; Wang, J. Aminopropyltriethoxysilane-mediated surface functionalization of hydroxyapatitenanoparticles: Synthesis, characterization and *in vitro* toxicity assay. Int. J. Nanomed. **2011**, *6*, 3449–3459.

71. Thakur, R.A.; Florek, C.A.; Kohn, J.; Michniak, B.B. Electrospunnanofibrous polymeric scaffold with targeted drug release profiles for potential application as wound dressing. Int. J. Pharm. **2008**, *364* (1), 87–93.

72. Aydin, O.N.; Eyigor, M.; Aydin, N. Antimicrobial activity of ropivacaine and other local anaesthetics. Eur. J. Anaesthesiol. **2001**, *18* (10), 687–694.

73. Yang, F.; Murugan, R.; Wang, S.; Ramakrishna, S. Electrospinning of nano/micro scale poly (l-lactic acid) aligned fibers and their potential in neural tissue engineering. Biomaterials **2005**, *26* (15), 2603–2610.

74. Zong, X.; Bien, H.; Chung, C.Y.; Yin, L.; Fang, D.; Hsiao, B.S.; Chu, B.; Entcheva, E. Electrospun fine-textured scaffolds for heart tissue constructs. Biomaterials **2005**, *26* (26), 5330–5338.

75. Martinsson, T.; Haegerstrand, A.; Dalsgaard, C.J. Ropivacaine and lidocaine inhibit proliferation of non-transformed cultured adult human fibroblasts, endothelial cells and keratinocytes. Agents Actions. **1993**, *40* (1–2), 78–85.

76. Drucker, M.; Cardenas, E.; Arizti, P.; Valenzuela, A.; Gamboa, A. Experimental studies on the effect of lidocaine on wound healing. World J. Surg. **1998**, *22* (4), 394–397.

77. Hennink, W.E.; Van Nostrum, C.F. Novel cross-linking methods to design hydrogels. Adv. Drug Delivery Rev. **2002**, *54* (1), 13–36.

78. Lakshmi, R.L.; Shalumon, K.T.; Sreeja, V.; Jayakumar, R.; Nair, S.V. Preparation of silver nanoparticles incorporated electrospun polyurethane nano-fibrous mat for wound dressing. J. Macromol. Sci. Pure Appl. Chem. **2010**, *47* (10), 1012–1018.

79. Tsou, T.L.; Tang, S.T.; Huang, Y.C.; Wu, J.R.; Young, J.J.; Wang, H.J. Poly(2-hydroxyethyl methacrylate) wound dressing containing ciprofloxacin and its drug release studies. J. Mater. Sci. **2005**, *16* (2), 95–100.

80. Dillen, K.; Vandervoort, J.; Van den Mooter, G.; Verheyden, L.; Ludwig, A. Factorial design, physicochemical characterization and activity of ciprofloxacin-PLGA nanoparticles. Int. J. Pharm. **2004**, *275* (1–2), 171–187.

81. Unnithan, A.R.; Barakat, N.A.M.; Pichiah, P.B.T.; Gnanasekaran, G.; Nirmala, R.; Cha, Y.S.; Jung, C.H.; El-Newehy, M.; Kim, H.Y. Wound-dressing materials with antibacterial activity from electrospun polyurethane–dextran nanofiber mats containing ciprofloxacin HCl. Carbohydr. Polym. **2012**, *90* (4), 1786–1793.

82. Zeng, J.; Xu, X.; Chen, X.; Liang, Q.; Bian, X.; Yang, L.; Jing, X. Biodegradable electrospun fibers for drug delivery. J. Control. Release. **2003**, *92* (3), 227–231.

83. Park, J.M.; Kim, M.; Park, H.S.; Jang, A.; Mind, J.; Kima, Y.H. Immobilization of lysozyme-CLEA onto electrospun chitosan nanofiber for effective antibacterial applications. Int. J.Biol. Macromol. **2013**, *54*, 37–43.

Electrospinning Technology: Regenerative Medicine

Toby D. Brown
Cedryck Vaquette
Institute for Health and Biomedical Innovation, Queensland University of Technology, Brisbane, Queensland, Australia

Dietmar W. Hutmacher
Division of Bioengineering, Department of Orthopaedic Surgery, National University of Singapore, Singapore

Paul D. Dalton
Institute for Health and Biomedical Innovation, Queensland University of Technology, Brisbane, Queensland, Australia

Abstract

Regenerative medicine is an interdisciplinary field combining knowledge of cell biology, materials science, and medicine with the aim to create constructs capable of healing or replacing lost and damaged tissue due to sickness or accidents, eliminating the need for donated tissue. For example, the effective healing of critical size bone defects, as a consequence of injury or disease, is an increasing problem in an aging population. The majority of current treatment strategies involve autografts, where bone is extracted from another part of the patient's body, usually from the iliac crest or the fibula. Although this approach is often successful, there are issues associated with infections and hematomas due to the creation of a secondary injury at the donor site, while the amount of bone available is relatively scarce. Another approach involves using allografts; however, donated bone introduces the risk of transferring diseases between the donor and host, as well as increases the risk of infections and donor site rejection. Tissue engineering offers the possibility to avoid these problems by incorporating three-dimensional constructs, or scaffolds, with osteoinductive and osteoconductive properties to replace the auto/allograft.

INTRODUCTION

Scaffold-based tissue engineering (TE) concepts often involve the combination of viable cells, biomolecules, and a structural scaffold combined into a tissue-engineered construct (TEC) to promote repair or regeneration.[1] The TEC is intended to support cell migration, growth, and differentiation and guide tissue development and organization into a mature and healthy state. Preferably, the scaffold should be absorbed by the surrounding tissue at the same rate as new tissue is formed so that by the time it breaks down the newly formed tissue will take over mechanical loads. Because this field is still in its infancy, the requirements for an ideal scaffold are complex and still not clear and may vary depending on specific tissue types and applications but involve consideration of factors including architecture, structural mechanics, surface properties, degradation properties and by-products, and the changes of these factors with time. However, it is widely accepted that scaffolds in TECs have minimum essential requirements. They should

1. Provide sufficient initial mechanical strength and stiffness to maintain a degree of mechanical function of the diseased or damaged tissue
2. Maintain sufficient structural integrity during the tissue growth and remodeling process in order to achieve stable biomechanical conditions and vascularization
3. Have a scaffold architecture that allows for initial cell attachment and subsequent migration into and through the matrix, mass transfer of nutrients and metabolites, and provide sufficient space for development and later remodeling of organized tissues
4. Have degradation and resorption kinetics based on the relationships between required mechanical properties, mass loss, and tissue development
5. Possess optimized surface properties for the attachment and migration of cell types depending on the desired tissue. Furthermore, the gross size and shape may be customized for a particular application and individual patient[2]

In TE, cellular regeneration requires scaffold support systems to guide the growth of cells and stimulate the development of native extracellular matrix (ECM). Cell proliferation and adhesion are dependent on the specifications of the scaffolds, especially upon the porosity and interconnectivity. Scaffolds for TE are obtained through a variety of fabrication techniques, described elsewhere, each with advantages and disadvantages.[3] These include rapid prototyping, 3D printing; selective laser sintering; extrusion/direct writing; inkjet or organ printing; solid free-form fabrication; and stereo-lithography. However, these techniques have limits to the degree of miniaturization possible. The latest trend in TE has been to use solution

Concise Encyclopedia of Biomedical Polymers and Polymeric Biomaterials DOI: 10.1081/E-EBPPC-120052258

Dental—Electrospinning

electrospun scaffolds, which offer fibers with diameters comparable to those of the fibrils typically found in tissue ECM, rather than other methods for *in vitro* studies. Electrospinning is a technique that fabricates continuous nano- and microscale filaments with potential for regenerative medicine, in particular for growth factor delivery and TE. Even though the production of filamentous material for such biomedical applications has been used for many decades, electrospinning was only recently proposed for TE (in 2002)[4,5] and growth factor delivery (2005).[6,7] Electrospun materials (also described as meshes) have widely been proposed as TE scaffolds, despite the clear evidence that many of these materials do not form a cell-invasive structure, which is a major criterion for a TE scaffold. Exceptions to this include specific strategies that introduce porosity into the electrospun mesh, such as co-electrospraying,[8,9] electrospinning onto ice crystals,[10,11] and the formation of bimodal scaffolds, which deliberately contain differently scaled fibers.[12,13] In comparison, growth factor delivery from these meshes does not necessarily require cell penetration and has been used in surgical paradigms. In this entry, we focus on the approaches to make and modify electrospun scaffolds, rendering the material suitable for applications in regenerative medicine.

FUNDAMENTALS OF ELECTROSPINNING

There are numerous configuration permutations for electrospinning, but a schematic of the most common electrospinning design—using a single collector—is shown in Fig. 1. In electrospinning, an electrical potential difference is generated between a liquid and a collector, which is achieved by applying high voltage to either the liquid,[5,14] collector,[13,15] or both.[16] A spinneret (small orifice or flat-tipped needle) may be used; however, nonspinneret "nozzle-less" systems have also been adopted.[17,18] Typically, as shown in Fig. 1, a (positive or negative) voltage is applied to a polymer solution or melt that is pumped through a spinneret facing a collector at a different electrical potential. As the polymer droplet emerging from the spinneret becomes charged, electrostatic repulsion

counteracts the surface tension and the droplet is stretched. Upon reaching a critical voltage, the surface tension of the polymeric fluid is overcome by the charge buildup, and the droplet elongates into a (Taylor) cone, where a continuous jet is ejected and drawn by the electrostatic force toward the collector. The potential difference between the spinneret and collector typically ranges from 10 to 20 kV, although voltages as low as 0.5 kV[19] and as high as 130 kV[20] have been reported.

The ejected electrospinning polymeric jet, which normally consists of a stable region and an unstable zone (usually called bending or whipping zone), has been subject to significant mathematical analysis. If there are sufficient molecular entanglements in the polymer, the jet does not break up into droplets due to Rayleigh instabilities. Initially, the electrified jet is stable and travels directly toward the collector. The surface charge density of the liquid (which caused the ejection of the jet from the droplet) again increases with decreasing distance to the collector and "twists" the jet, resulting in bending instabilities. In this zone, the instabilities in the jet's path cause it to spiral or "whip," where it rapidly accelerates laterally to the flight path in a conical envelope, leading to further stretching. As the electrified jet passes through the air, the solvent evaporates or the molten polymer cools, depending on whether electrospinning of a polymer solution or melt is occurring, and the jet undergoes magnitudes (up to 5 orders) in diameter reduction by the time it reaches the collector. Electrospinning can be performed with the spinneret(s) and collector(s) configured so that the path of the jet is either vertical or horizontal, since the effects of aerodynamics and gravity are minimal compared to those of the electrostatic forces involved.[21] The path of the polymer solution jet, in particular, is extremely dynamic, and the fiber collection is rapid. Such high speeds, plus long spiraled traveling distances, make accurately controlled deposition of the electrospun fiber technically challenging. Electrospinning commonly results in a nonwoven mat of fibers, although other collection techniques shown here are expanding the morphological nature of the scaffolds.

A variation of electrospinning is melt electrospinning,[22] which avoids issues such as toxicity and solvent removal associated with solution electrospinning. Melt electrospinning behaves differently to solution electrospinning, in that the initial stable region of the jet is relatively much longer and thus there is less lateral movement of the jet[23] resulting in less reduction in the diameter of the jet. However, the larger micron scale fibers that typify melt electrospinning may offer the ability to fabricate thicker structures with greater pore size and porosity, promoting more cell invasiveness. Advances in how melt electrospun fibers can be collected may provide patterned aligned fibrous substrates useful for building complex 3D structures that are of interest in applications such as TE.[22,24]

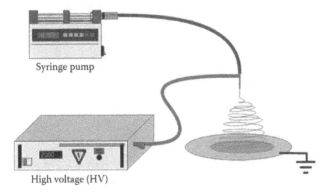

Fig. 1 Schematic of a typical electrospinning setup.

Syringe pump

High voltage (HV)

Dental—Electrospinning

POLYMERS USED FOR ELECTROSPINNING IN REGENERATIVE MEDICINE

There are literally hundreds of different polymers that have been electrospun, and extensive lists of these polymers can be found in other entry.[25–28] Tables 1 through 3 provide a selection of polymers chosen for electrospinning applications in regenerative medicine. Polymers can be categorized into two distinct classes—those derived from biological sources and those that are synthetically produced.

Biologically Derived Polymers

It is understandable that when trying to replace ECM fibrils within the body, the most obvious polymers to use are those that constitute the natural ECM. Such biologically derived polymers (Table 1) provide biological signals to cells and are typically degradable materials. However, the process of dissolving the polymer into a solvent can create problems, particularly denaturation. For example, collagen can be denatured to gelatin during electrospinning, due to both the solvent used (often 1,1,1,3,3,3-hexa-fluoro-2-propanol [HFIP]) and the high voltage applied.[29] Silk fibroin is also used; however, it requires treatment in order to dissolve into a solution capable of being electrospun. A second issue with silk is also a generic issue with the use of aqueous solutions: a degree of post-treatment is needed after generating the electrospun fibers so that they do not redissolve upon placement in aqueous media.[27,30,31] Biologically derived polymers have also been blended with synthetic polymers to improve the processing ability of the fiber.

Numerous studies that blend different biologically derived polymers, such as chitosan and agarose,[46] gelatin and hyaluronic acid, are reported.[47] In addition to the selected entry listed in Table 1, in-depth reviews on the electrospinning of biologically derived polymers areavailable.[27,28]

Synthetic Polymers

The number of different synthetic polymers electrospun (Table 2) is definitely greater than that of those derived from biological sources. Synthesized polymers have widely been electrospun due to the benefits of reproducibility and the ability to tailor important material properties such as molecular weight, which appears to facilitate improved electrospinning than when using biologically derived polymers. There are many specialized reviews on various polymers, with poly(caprolactone) (PCL) and polyethylene glycol (PEG) being two of the most studied polymers for regenerative medicine.

The lack of biological signals provided by synthetic polymers (to trigger and control cellular responses) requires the adsorption of proteins onto the surface, or a surface-engineered strategy: one common approach is to blend biologically derived polymers into a synthetic polymer solution in order to incorporate protein into the fibers. However, denaturation of the protein is still an issue. Furthermore, the two different polymers often phase separate into separate domains because of their intrinsic properties. For example, collagen (hydrophilic) and PCL (hydrophobic) are immiscible and will phase segregate. Table 3 provides a list of some commonly blended "hybrid" polymers.

ELECTROSPINNING PARAMETERS

It is widely appreciated that three broad factors influence the electrospinning process, and thus the morphology of the resulting fibers: the properties of the polymeric fluid used; the instrument conditions (e.g., voltage and flow rate); and the configuration of the instrument. The most widely reported characteristic of electrospun material is the diameter of the fibers. The diameter is known to increase with increasing polymer concentration and also decrease with voltage applied. While the diameter is an important aspect and is discussed later, the morphology and structure

Table 1 Selected biologically derived polymers

Polymer	References
Collagen	[5,29,32]
Gelatin	[33,34]
Silk	[31,35]
Chitosan	[36]
Fibrinogen	[37,38]
Hyaluronic acid	[39,40]
Cellulose	[41]
Laminin	[42,43]
Fibronectin	[44,45]
Chitosan/agarose	[46]

Table 2 Selected synthetic polymers

Polymer	References
PLGA	[13,48,49]
PCL	[50]
PS	[51,52]
PU	[8,53–55]
PNIPAM	[56,57]
PLLA	[58,59]
PLLA-CL	[60,61]
PEG-PCL	[62,63]
PEO	[64,65]
sP(EO-stat-PO)-PLGA	[66]

Table 3 Selected hybrid electrospun systems

Polymer	References
Collagen/PCL	[67]
Gelatin/PLGA	[68]
Chitosan/PVA	[69]
Silk/PEO	[70]
HAp/PLLA	[71]
Hap/Collagen/PVA	[72]

of the electrospun materials significantly have impacts on other important properties such as pore inter-connectivity and cell invasiveness. Due to the large number of parameters that can be altered, and the interdependency of most of these parameters, studying the effects of the processing conditions on scaffold morphology can become excessively difficult. However, through careful experimental design it is possible to deduce definitive relationships.

Polymeric Parameters

When electrospinning polymeric solutions, there are a number of parameters that can be altered to generate different types of structures or different fiber diameters.

Solution viscosity

One of the most important electrospinning parameters is the viscosity of the polymer solution: i.e., the degree of macromolecular entanglements in the polymer that prevents breakup of the liquid jet into droplets. One way by which viscosity can be controlled is by adjusting the concentration of the polymer solution. Chain entanglement can be determined by the molecular weight used and also depends on the polymer type (linear or branched). There is a strong relationship between polymer concentration and fiber diameter, which is well described for a large number of polymers in a large number of studies. Usually highly concentrated solutions produce relatively thick and bead-free fibers. Mckee et al.[73] extensively studied the electrospinnability and resultant fiber morphology of polymer solutions with respect to their rheological properties. The dependence of polymer viscosity on solution concentration can be described by identifying four distinct concentration regimes: the dilute, semidilute unentangled, semidilute entangled, and the concentrated. The boundary between the semidilute un/entangled regimes is named entanglement concentration C_e and corresponds to the point at which there is significant overlap of the polymer chain to constrain the chain motion causing entanglement and couplings.[73] McKee observed that for a concentration below C_e only polymer beads were obtained. Whereas for concentrations between C_e and 2–2.5 times C_e, beaded fibers form, for concentrations above 2.5 times C_e, uniform

bead-free fibers occur. This work showed that there exists a minimum polymer concentration required for solution electrospinning. A fiber diameter prediction model was proposed where the diameter universally scales with the normalized concentration $(C/C_e)^{2.6}$, where C is the concentration of the polymer solution and C_e the entanglement concentration.[73]

Typically, electrospinning of natural polymers in aqueous solutions is challenging due to a number of factors including polymer chain rigidity (which reduces the degree of polymer chain entanglements) and high surface tension in the solution. Increasing the entanglements has been shown to be an efficient way to facilitate the aqueous electrospinning of alginate, a natural polymer:[74] in this strategy better electrospinnability was achieved by using glycerol, a strong polar solvent, which is believed to disrupt the strong intramolecular structure of the polymer. Glycerol also forms new hydrogen bonds with the polymer chains and improves their flexibility. As a result, the viscosity of the solution usually increases due to increased entanglements, whereas surface tension decreases, enhancing the electrospinnability of these polymers. A similar effect was observed when electros-pinning hyaluronic acid in a mixture of water, formic acid, and dimethylformamide.[75] Here again the rigid inner structure of the natural polymer was disrupted by the formic acid, which changed the chain conformation from alpha helix to a coil conformation improving chain flexibility, increasing entanglements and therefore electrospinnability.

Surface tension

During the electrospinning process, a jet forms once the electrostatic forces overcome the surface tension in the droplet emerging from the spinneret. Therefore, the surface tension is a critical parameter and in polymer solutions it can be controlled by blending solvents,[76–78] adding a surfactant or salt,[79] or varying the temperature.[80] It is generally assumed that high surface tension solutions are difficult to electrospin and require higher voltage for the initiation of the jet. Because it tends to minimize the surface area per mass of a fluid, resulting in a spherical shape, surface tension is also thought to participate in the formation of beads on fibers when electrospinning relatively low or moderately viscous solutions. Generally, reduced surface tension produces bead-free smooth fibers.[79]

Solvent conductivity

The forces that tend to elongate the electrospinning jet are created by a buildup of charge within and on the surface of the polymer droplet. For polymer solutions, the net charge carried by the jet is directly bound to the conductivity and the mobility of these charges in the solution. Generally, greater stretching of the jet, and thus thinner fibers, is to be expected when the net charge of the jet is increased, which is facilitated by an increased conductivity. Furthermore, in most cases increasing the conductivity of the solution enhances the bending instability (because it enhances the charge repulsion). Therefore, the jet lengthens and is subject to more whipping, causing more stretching and leading to further reduction in fiber diameter.[81] The significance of this electrospinning parameter is demonstrated by the fact that changes in the conductivity of the same solvent can lead to discrepancies in fiber diameter and scaffold morphology from one experiment to another.[51] For the same conditions and using the same solvent but from different suppliers, Uyar and Besenbacher[51] obtained both beaded and smooth fibers. They correlated these varying fiber morphologies to the differences in the solution conductivities, which varied from 0.7 to 7 µS/cm. The highest conductivity produced the smallest fibers.

The solvent or combination of solvents that are part of electrospun fiber preparation affects the fiber structure, which in turn affects cell behavior.[82] Solvent mixtures have widely been used to alter, generally to increase, the conductivity of the electrospun polymer solution. This approach is valid as long as the viscosity of the solution does not vary too greatly; otherwise, conclusive observations on the effects of conductivity may be difficult. In some cases, adding a highly conductive solvent without a significant variation in the viscosity of the solution is possible.[76] When dimethylformamide (DMF) is added to methylene chloride, the viscosity of a 13 wt% PCL solution does not change, but the conductivity increases up to 10-fold, responsible for a sharp decrease in fiber diameter.[76] The stability of the process may also benefit from the addition of a highly conductive solvent. For example, chloroform or dichloromethane solutions can have an instable Taylor cone, and consequent spinneret clogging can require the operator to clean it frequently. However, upon addition of DMF to the solution the process is stabilized and can be run for several hours without the need to clean the spinneret.

Another way to enhance the conductivity of a polymer solution is to add salt. A substantial amount of literature on this subject has generally shown that thinner fibers are produced when the salt concentration is increased.[81,83,84] It has also been observed that increasing the salt concentration tends to reduce bead formation.[85] Interestingly, while comparing several types of salt, Zong et al.[83] observed that ions with smaller atomic radii produce thinner fibers. This effect was attributed to the higher charge density of smaller ions, providing higher mobility in the solution and therefore producing stronger elongational forces.

Although an increase in polymer solution conductivity generally leads to the production of thinner electrospun fibers, the opposite trend has also been reported.[80] This may be due to additional mass flow created by higher elongational forces.

Instrument Parameters

Flow rate

The effect of mass delivery, or flow rate, on the nature of electrospun scaffolds has been studied extensively.[24,83,86–88] It is widely accepted that increasing the flow rate produces larger fibers: as more material is extruded through the spinneret per unit time, the volume of the Taylor cone increases for the same collection distance. This leads not only to larger diameter fibers but also to the deposition of wet fibers, which may fuse together due to a greater amount of residual solvent. Therefore, more fusion at the points of contact between solution electrospun fibers is generally expected for higher feed rates. This can significantly affect the mechanical properties of the resultant meshes (high fusion density results in tougher scaffolds) and affect the interconnectivity and therefore the cell invasiveness of the construct. For given conditions, increasing the flow rate can induce instability into the Taylor cone. This usually happens when the amount of polymer solution exceeds the amount of polymer that can be drawn by the electrostatic forces. As a result, breaking up of the jet can be observed, and fibers with large beads can also be formed.[89] Therefore, increasing the voltage may be necessary to maintain a stable jet when the flow rate is increased.

Applied voltage

Since an electrical field is inherently required for electrospinning to be possible, the applied voltage plays a major role in the electrospinning process. In most cases, a direct-current (DC) apparatus is utilized to produce the high voltage. However, alternative-current (AC) setups are also possible. The main advantage of using AC voltage resides in the ability to electrospin or electrospray onto non-conductive collectors.[90] Using AC voltage, the alternating polarity induces a neutralizing effect, leading to the reduction of net charge on the jet. Consequently, compared to using DC voltage, the electrostatic forces that produce jet instability and thus the whipping instability are significantly reduced. As a result, fibers obtained using AC voltage are generally larger than those fabricated using DC voltage.[90]

Several phenomenological observations can be made when the DC voltage is increased during electrospinning. Firstly, the shape of the Taylor cone will change from a droplet to a spindle. Several studies agree that there is a reduction in the volume of the Taylor cone with increasing voltage (for a given flow rate).[83,91] As the voltage is increased, the jet acceleration is greater and more volume is drawn from the Taylor cone; therefore, the cone volume and also the jet diameter is reduced.[83] Secondly, increasing the voltage may cause the formation of multiple jets from the Taylor cone:[92] higher voltage favors the formation of thin secondary jets, which may broaden the fiber diameter distribution[81] or produce bimodal fiber distributions.[93] Usually, when secondary jets are not formed the fiber diameter distribution appears to narrow.[94–96] When the voltage is further increased, the Taylor cone can even recede into the spinneret. Thirdly, when the voltage becomes too high corona discharge can be observed, perturbing the deposition.[97]

It is generally accepted that an increase in electrospinning voltage produces a decrease in fiber diameter;[86,94,95] as the jet is increasingly accelerated it is further stretched. Katti et al. observed that after an initial sharp decrease in fiber diameter, a constant diameter is obtained when the voltage is further increased.[95] However, several other studies report no change or increase in the fiber diameter at higher voltages. Therefore, there is no definitive consensus on the effect of the applied voltage; which appears to be dependent on the solution used.

At each voltage extreme (too low or too high voltage), bead formation has frequently been reported. However, once beads appear during fiber deposition, the higher the voltage the higher the bead density.[86,91,98] Lee et al. observed that the morphology of beads from a low viscosity solution could also be controlled by the applied voltage, where they adopted a more pronounced spindle shape at higher voltages.[99]

Spinneret diameter

The diameter of the spinneret through which the polymer is delivered during electrospinning plays a significant role in controlling fiber diameter as it mostly determines the initial size of the Taylor cone. Several studies have shown that smaller spinneret diameters produce smaller fibers.[94,95] A smaller diameter orifice also increases the surface tension of the emerging droplet and thus the critical voltage required to initiate the Taylor cone. It is also believed to reduce bead formation and clogging of the spinneret,[94] while a smaller Taylor cone provides less surface area in contact with the air for solvent evaporation. However, it can prove difficult to extrude highly viscous solutions through small diameter orifices, and the buildup of back pressure in syringe-based systems may affect the feed rate and therefore consistent fiber deposition.

Temperature

The control over the solution/melt temperature can be obtained by developing a jacket-type heat exchanger in which water[80,100,101] or silicone oil[102] is circulated. Other technical solutions can be utilized such as glass fiber heating tape or infrared emitter heating systems.[103] Increasing the temperature generally decreases the fiber diameter since the viscosity of the solution is greatly reduced.[92] Wang et al. systematically studied the effect of elevated temperature over the properties of polyacrylonitrile solution in DMF and on the morphology of the resulting fibers.[102] This study found, as expected, that the viscosity of the solution decreased, but also that the entanglement density was not significantly affected by the elevated temperature. In addition, the surface tension of the solution was found to decrease with temperature, whereas the opposite trend was observed for the conductivity, which increased with temperature. The later phenomenon was attributed to the higher mobility of the molecules caused by the increased temperature. The reduction of the viscosity, surface tension, and the increase in the conductivity lead to smaller fiber diameters.

Besides reducing the fiber diameter, elevated temperatures have other implications for the process of electrospinning. Demir et al. observed improved uniformity in fiber diameter distribution when electrospinning a polyurethane solution at 70°C compared with room temperature.[92] Wang et al. also reported the increase in the Taylor cone volume, which resulted in an increase in the voltage required to electrospin the polymer solution;[102] In Wang's and the work by Givens et al.,[103] a high electrospinning temperature was responsible for a slight decrease in polymer crystallinity. Electrospinning at elevated temperatures can also be used to dissolve polymers that are insoluble at room temperature: linear low density polyethylene (LLDPE) was successfully electrospun at 105–110°C using p-xylene as a solvent.[103] This strategy can also be used to improve the incorporation efficacy of poorly soluble drugs into fibrous polymer matrices, because the solubility of most compounds is improved at high temperatures.[104]

Collection distance

Varying the collection distance, also known as "tip-target distance" or "working distance", strongly impacts the strength of the electrical field and the time of flight of the polymer jet. Although it is generally accepted that a collection distance that is too short will create adverse effects, such as excessive fiber fusion caused by the low solvent evaporation, there is no global consensus on the effects of collection distance on fiber morphology. Several studies have shown very small or insignificant changes in fiber diameter and morphology when increasing the collection distance,[81,105,106] while others have observed significant changes in the fiber diameter.[107] However, the collection distance directly affects the volume of the Taylor cone and plays a significant role in the initial diameter of the jet:[105] small distances increase the field strength and tend to reduce the volume of the Taylor cone. Smaller distances may also lead to greater acceleration of the jet, but not necessarily greater stretching, where induced bead formation has been reported.[86] Longer distances reduce the field strength and have also been associated with the formation of beads or increased bead diameter.[108] Collection distance appears to have less of an effect on fiber diameter and mor-phology than deposition area. As the collector is moved further away from the spinneret, the base of the cone formed by whipping becomes larger, resulting in the depo-sition of the same quantity of material over a larger surface area. Therefore, thinner scaffolds are obtained for the same deposition time. When the working distance is increased, the time of flight is also increased. Consequently, this allows more solvent to evaporate and reduces the fusion density between the fibers. This can have a significant impact upon the mechanical properties of the scaffolds, as more fiber fusion increases the tensile strength of the elec-trospun mat. Therefore, varying the collection distance could also be a means to control the macroscopic mechani-cal properties of the resultant scaffolds. Although the depo-sition of melt electrospun fibers is relatively more focused compared to solution electrospin-ning, a similar effect is experienced where longer collection distances result in less focus of the deposited fibers.[24]

Humidity

Humidity is not always controlled during the electrospin-ning process; however, it seems to play a significant role in the morphology of the fibers obtained. There are several studies that demonstrated the possibility of tailoring the surface morphology of the fiber by controlling the ambient humidity of the electrospinning chamber. It has been shown that increasing the relative humidity leads to the formation of spherical "pores" on the fiber surface, and it is generally accepted that the higher the relative humidity, the larger the pore diameter until the pores fuse together.[109,110] The for-mation of such pores is attributed to the condensation of water molecules onto the fibers during deposition. As a consequence, the specific surface area is greatly increased by the presence of these pores on the fibers.[110] Relative humidity can have significant impact not only on the super-ficial pattern of a single fiber but also on the gross architec-ture of an entire scaffold. When electrospin-ning an aqueous-based solution with various humidity contents, it has been observed that larger diameter fibers are formed at low humidity percentages, due to the higher evaporation rate of the water in the "dry" environment, resulting in less drawing of the jet by the applied voltage. Whereas the fiber diameter was observed to decrease when the relative humidity was increased, up to a point where beads were formed.[111]

Instrument Configuration

This section focuses on the utilization of different configu-rations to generate discrete structures within an electrospun matrix. The variation of instruments described in the litera-ture is far greater than presented in this entry. However, the focus here is on the ability to discretely position and place

fibers in specific locations and patterns. Central to many fibrous systems are woven materials, and therefore creating woven structures or threads using electrospinning is an interesting and relevant component of TE. Initially, static electrospinning instrument configurations are discussed, followed by dynamic collection systems that result in structured electros-pun substrates.

Single collection systems

Using a single, solid collector is the most commonly used electrospinning configuration and results in a nonwoven mesh that is non-cell invasive. It is usually the arrangement that most polymeric materials are initially electrospun onto, since the nonwoven mesh is readily handled and manipulated. In this configuration, the electric field between the spinneret and the collector is uniform or sym-metrical. Although in most cases electrospun fibers are deposited onto a solid collector (Fig. 2A and B), common liquids such as water can be used for fiber collection.[112,113] A straightforward method of collecting a continuous yarn from electrospun fibers by first depositing them onto a liquid medium (as shown in Fig. 2C) was not demonstrated until 2005 by Smit, Buttner, and Sanderson[114] and Khil et al.,[115] even though Formhals had patented various con-tinuous fiber yarn setups in the 1930s.[116,117] This was an important technical breakthrough, because of the potential for continuous yarns of electrospun nanofibers to be woven into textiles, with applications including protective

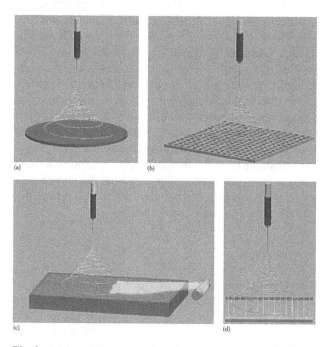

Fig. 2 Various collection systems for electrospinning. (**A**) Single collector; (**B**) Patterned collector; (**C**) Collection onto water with rotating mandrel; and (**D**) Collection across two parallel collectors.

clothing, high performance fabrics, composites, and TE.[118] The method proposed by Smit and colleagues[114] is worthwhile for further discussion. It involved the deposition of randomly oriented electrospun fibers onto the surface of a water bath, which were then drawn to the edge and collected on a rotating mandrel. This caused elongation of the mesh and alignment of the fibers as they were drawn over the water. Surface tension caused the fibrous mesh to collapse into a yarn when it was lifted off the surface of the water onto the rotating collector. As the resultant yarn was rolled onto the mandrel at a rate of 3 m/min, more fibers were deposited on the water surface to feed the drawing process. Electrospun poly(vinyl acetate), poly(vinylidene fluoride), and polyacrylo-nitrile nanofiber yarns were fabricated using this method. Khil et al. employed a similar concept but rotated the mandrel at 30 m/min and used a mixture of water and methanol.[115] However, no comment was made on the difference between yarn collected on water or water/methanol. According to Teo and Ramakrishna,[118] the high surface tension provided by water may be favorable to other liquids with lower surface tension because the electrospun fibers may become submerged, creating a higher drag force as they are drawn to the rotating collector. More studies are required to determine the influence of the liquid bath properties on the yarn.

Dual collector systems

An example of a dual collector system is schematically illustrated in Fig. 2D while a photograph in Fig. 3, shows highly aligned fibers across the separation between two rings placed in parallel, equidistant from the spinneret were deposited.[119] An important advantage of this type of approach, which is also termed the "gap method of alignment," is that the perimeters of the electrode rings where the fibers deposit allow precise and consistent controlled positioning of the fiber bundle (Fig. 4).[120] Such a gap method of alignment has been used to manufacture oriented fibers for *in vitro* use, particularly for applications in the nervous system.[16,67,121–123]

Structured electroconductive collectors

Methods to create patterned 2D and 3D electrospun structures using static electroconductive patterned collectors have also been developed. However, the random formation of honeycombed[65,124] and dimpled structures[125] using this approach has been observed, due to the buildup of electrostatic charges on the deposited fibers and/or the collector, which prevent new deposition directly onto the collector. Furthermore, it is currently reported that the fibers forming 3D structures are very loosely packed and can be easily compressed: rendering them unsuitable for applications where structural integrity is required.[125] Thus, the conditions required to form 3D structures using this method are still unclear.

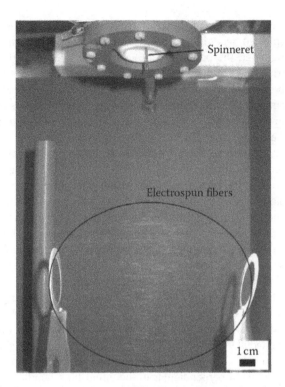

Fig. 3 Oriented electrospun PCL fibers collected using the gap method of alignment. The collectors are spaced 10 cm apart, and the oriented fibers that collect in between are generated in approximately 30 s. (Previously unpublished photograph.)

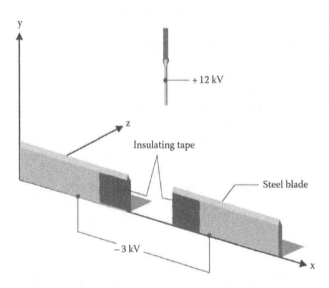

Fig. 4 Experimental setup for precise deposition of electrospun fibers over two known points.
Source: From Teo & Ramakrishna.[120]

Patterned electro-conductive collectors, either in the form of ridges and indentations or grids with pores in between, have been used to form patterned 2D solution electrospun meshes. Ramakrishna et al. reported the use of a collector with conducting ridges and nonconducting indentations where the electrospun fibers were randomly deposited according to time; however, the resultant mesh resembled the pattern of the collector.[125] Aligned fibers formed on the conducting ridges and randomly deposited fibers on the nonconducting regions. Zhang and Chang also demonstrated that protrusions/patterns in the collectors are important parameters that may greatly affect the structures of electrospun meshes (Fig. 5).[126,127] The density of uniaxially aligned fibers deposited between the protrusions on a collector was shown to be dependent on the distance between two protrusions, decreasing with increasing distance between the protrusions. Woven structures were then generated through time-dependent control of the arrangement of a patterned series of electroconductive protrusions: one pair of thin parallel-aligned protrusions, insulated from one another, was mechanically raised higher than the other pairs of protrusions. Fibers were drawn to deposit on the higher pair due to stronger electric attraction forming a parallel-aligned fiber bundle between the protrusions. After a certain time, the first pair was lowered and another pair oriented at 90° was raised and allowed to collect for a similar period. The newly deposited fibers were also parallel but oriented perpendicular to the initial fiber bundle. The process was repeated with additionally located protrusions at 90° to create woven fibrous architectures.[127]

Rotating collectors

Several researchers have shown that it is possible to obtain aligned fibers by using a rotating collector/mandrel.[5,128,129] Matthews et al. demonstrated the effect of the rotating speed of a mandrel on the degree of solution electrospun collagen fiber alignment.[5] At a speed of less than 500 rpm, a random mix of collagen fibers was collected. However, when the rotating speed of the mandrel was increased to 4500 rpm (approximately 1.4 m/s at the surface of the mandrel), the collagen fibers showed significant alignment along the axis of rotation. At such a high rotational speed, the velocity of the electrospinning jet may be slower than the linear velocity of the rotating mandrel. The reported average velocity range of the electrospinning jet is from 2 m/s[130,131] to 186 m/s.[114] A separate study by Zussman,

Theron, and Yarin using a rotating disk collector demonstrated that at high enough rotational speeds, necking of the electrospun fibers occurs.[132] Therefore, the mandrel rotational speed directly influences fiber orientation as well as the material properties of the engineered fibrous structure. In cases where disoriented fibers are collected using a rotating mandrel at sufficient speed, residual charge accumulation on the deposited fibers may interfere with the alignment of incoming fibers. According to Teo and Ramakrishna, reducing this residual charge accumulation on the rotating mandrel as well as reducing the chaotic path of the jet are possible ways to achieve greater fiber alignment, in solution electrospinning applications.[120] Kessick, Fenn, and Tepper used an AC high-voltage supply instead of the typically used DC supply when solution electrospinning polyethylene oxide (PEO)[90] onto a rotating mandrel to improve fiber alignment. Using an AC potential to charge the solution may have a dual function: the electrospinning jet may consist of short segments of alternating polarity, which significantly reduce its chaotic path and allows the fibers to be wound onto the rotating mandrel with greater ease and alignment. Then, given the presence of both positive and negative charges on the surface of the rotating mandrel, there may be a neutralizing effect over the area of the fibers, thus minimizing residual charge accumulation. Another simple approach to reduce charge accumulation is to apply the opposite polarity to the collector than to the spinneret, instead of grounding the collector.[16]

Although the methods incorporating a rotating device described so far have all formed 2D aligned fiber meshes, 3D conduits can be fabricated by depositing fibers onto smaller diameter (<5 mm) collectors. For the construction of vascular grafts, Stitzel et al. electrospun a conduit 120 mm long and 1 mm thick using a polymer solution containing a mixture of collagen type I, elastin, and poly(D,L-lactide-co-glycolide) on a circular mandrel of diameter 4.75 mm.[133] However, in this case the rotational speed was only 500 rpm to facilitate homogeneous deposition of random fibers. Longitudinal and circumferential mechanical testing showed no significant difference between the two directions. Another advantage offered by the electrospinning process when forming small diameter conduits for vascular grafts is the ease of spinning different materials to form layered composites.[134] This way, materials such as collagen, which encourages cell proliferation and attachment but is mechanically weak, can be electrospun as an inner layer, while artificial materials such as polymers with superior mechanical properties can form an outer layer.

Combining electric field manipulation with a rotating collector

The combination of electric field manipulation and a dynamic collector has been used to achieve greatly ordered fiber assemblies. Theron et al. used a knife-edged rotating

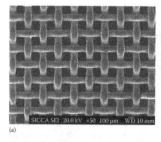

Fig. 5 **(A)** Patterned electro-conductive collector and **(B)** resulting patterned electrospun mesh.
Source: From Zhang & Chang.[127]

disk collector to take advantage of the fiber alignment created by rotary motion as well as the convergence of electric field lines toward the edge of the disk.[135] To collect the fibers, glass coverslips or any nonconducting substrate can be attached around the edge of the discs used in TE studies.[136] Crossed fiber arrays were obtained by Bhattarai et al. using a similar concept to the rotating disk collector, but instead winding copper wire as an electrode on an insulated cylinder and then rotating it at an optimum speed of ~2000 rpm.[137] The size of the fiber bundle was controllable by varying the diameter of the collector. Instead of using static electrodes placed in parallel, Katta et al. used a rotating wire drum,[138] whereas Sundaray et al. used a sharp pin as an electrode on a rotating collector to focus deposition of the electrospinning jet and introduced lateral movement to the collector to obtain cross-bar patterned fibers.[131] Instead of situating an electrode directly below the spinneret to focus electrospun fiber deposition, Teo and Ramakrishna incorporated a negatively charged knife-edged electrode placed at an oblique angle to the spinneret.[120] The electrospinning jet was observed to travel at the same angle as that of the diagonal electric field lines from the tip of the spinneret to the knife edge. Diagonally aligned fibers were then collected on a rotating tube placed in the path of the electrospinning jet. This enabled the fabrication of laminar composites with aligned fibers in different orientations. While the combination of dynamic mechanical devices and electrical field manipulation leads to parallel arrays and grids of electrospun fibers with improved alignment, these methods do not provide sufficient control over the pitch width or the ability to create complex patterns such as circles.[139]

Electrospinning writing

Currently, a limited number of writing approaches using electrospinning are available. Buckling phenomena have been studied where electrically charged solution jets depositing onto collectors moving laterally at a constant velocity produced buckling patterns with limited controllability.[140] While in a different study, the bending and buckling of a molten polymer jet was shown to be close to the collector.[141] There are processes that take advantage of the initially stable region of the solution electrospinning jet, by significantly reducing the spinneret-to-collector distance to below that at which bending instabilities develop along the jet. This results in relatively focused fiber deposition. Then, matching the speed of a translating collector to that of the electrospinning jet enables the ability to write with a continuous filament. Scanning tip electrospinning[128] and near-field electrospinning[19,139,142] processes have demonstrated relatively predictable control over the deposition of submicron fibers using an automated x–y stage as the translating collector. Alternatively, a polymer melt jet will remain

stable over relatively large distances and is well directed toward the collector. Dalton and colleagues utilized this phenomenon in combination with a translating collector, demonstrating the accurate patterning and drawing of a cell adhesive scaffold with a PEG-b-PCL/PCL blended melt across the surface of a microscope slide.[24] Recently Brown, Dalton, and Hutmacher demonstrated a direct writing process using melt electrospun pure PCL, where complex patterns could be drawn with a continuous filament. Layers of melt electrospun fibers were then consistently located on top of each other, to build complex 3D structures up to 1 mm in thickness (Fig. 6). This process also enabled the fabrication of porous TE scaffolds, where control over the location of fiber placement was demonstrated with micron scale resolution.[143]

CELL-INVASIVE ELECTROSPUN MATERIALS

Solution electrospinning offers relatively consistent control over fiber diameters, generally in the sub-micron range. However, due to the nature of the process, it is characterized by disorderly fiber deposition, which has limited its full potential. Despite this, numerous techniques have been developed and are still evolving, which have improved the orderly placement of fibers, including

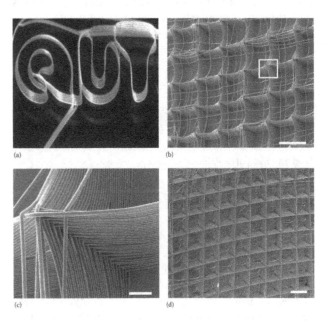

Fig. 6 Electrospinning writing of a polymer melt. Photograph of Queensland University of Technology logo "QUT" written on a microscope slide (**A**), showing stacking of fibers upon each other. SEM image of stacked and interwoven fibers until the stacking becomes affected by either the height of the sample or the uneven z-direction height, the latter caused by differences due to the interweaving of the fibers (**B** and magnified in **C**). The underneath of this box-like structure is shown in (**D**). Scale bars for B, C, and D are 1 mm, 100 μm, and 1 mm, respectively. Unpublished images from the same study as reported in Brown et al.[143]

Dental—Electrospinning

mechanical devices such as rotating collectors, liquid deposition methods, manipulation of the electric field, combining electric field manipulation with a rotating collector, static patterned collectors, and translating collectors. These approaches are important as they progress electrospinning toward true TE scaffold manufacture. Presented here is a selection of techniques developed to produce cell-invasive scaffolds.

Hydrospinninc

In this approach, water is used as a collection medium, and a layer of electrospun fibers forms on the surface. The film of fibers is transferred to a microscope slide and the process repeated, generating multiple layers.[113] The water is then removed with a vacuum and the resulting scaffolds contain considerable volume, are both highly porous, and cell invasive. In this same study, by Levenberg and colleagues, myoblasts were introduced into the process and successfully cultured with an equal density throughout the electrospun scaffold.

Electrospinning onto Ice Crystals

In another study aimed at increasing the porosity of electrospun scaffolds, Simonet et al. electrospun fibers directly onto ice crystals, as they grew off the surface of a rotating drum. The rotating drum was hollow and filled with solid CO_2, causing water vapor in the surrounding air to crystallize on the collector.[10] The ice/fiber mixture was lyophilized, and the remaining scaffold was impressively large, with interconnected pores. Extending on this process, the porous material was mechanically pulled and transformed into a cotton wool-like material, which was then used as a TE scaffold in this form.[11]

Simultaneous Electrospinning and Electrospraying

A similar concept involving building a conductive substrate up from the surface of a collector was devised by researchers trying to manufacture TE constructs. However, instead of using ice crystals, a layer of hydrogel was "grown" from the collector surface: gelatin containing smooth muscle cells was electrosprayed concurrently while electrospinning polyurethane.[8] Similarly, Ekaputra et al. electrosprayed a thiolated hyaluronic acid solution while electrospinning PCL/collagen onto a collector.[9] Human fetal osteoblasts were then observed to migrate up to 200 μm into the scaffold in a mixture of electrospun fibers and gel. In both these instances, a rotating collector was used and the two spinnerets were oriented at 180° to each other.

Bimodal Electrospun Scaffolds

It is recognized that ultrafine electrospun fibers do not readily create a cell-invasive scaffold, because they are too tightly packed and thus do not provide adequate pore size and porosity. However, combining these fibers with a larger structure may allow the benefits of both "nano" and "micro" structures. This approach leads to the generation of scaffolds with high surface area due to the smaller fibers (promoting cell attachment and assisting in higher seeding of the scaffold), as well as suitable volume and pore sizes due to the microfibers. Gentsch et al.[12] showed that solution electrospinning at very low flow rates induced both small and large diameter electrospun materials from a single spinneret. In this instance, PCL was used in conjunction with Chinese hamster ovary (CHO-K1) cells for infiltration experiments with improvements using a bimodal approach. Alternatively, simultaneous melt electrospinning (which produced micron-diameter fibers of 28.0 ± 2.6 μm) and solution electrospinning (which produced submicron diameter fibers of 530 ± 240 nm) of PLGA were performed onto a rotating mandrel, to form a thick cell-invasive bimodal scaffold.

Combining Conventional Technologies with Electrospinning

Electrospun fibers can be mixed with conventionally fabricated fibers to create a scaffold similar in principle to a bimodal electrospun scaffold, where the ultrafine electrospun fibers are mechanically supported by the larger structures. In one case, this was achieved by melt spinning poly(L-lactide-co-epsilon-caprolactone) (PLCL) into a cylindrical structure for vascular TE.[144] In another notable case, Kim and coworkers reported 3D extrusion combined with electrospinning to significantly increase porosity (Fig. 7A). In a different study, an automated scaffold assembly system was used to construct the scaffold: a fused deposition modeling (FDM) process was interrupted between each layer to deposit an electrospun PCL/collagen fiber mesh (Fig. 7B).[145] The resultant constructs contained large pores, while the electrospun fibers provided suitable structures for cell adhesion, effectively increasing the surface area available for penetrating cells to attach. Following this approach, a tubular construct for vascular TE was manufactured, incorporating nanostructured electrospun substrates and FDM fibers for sufficient mechanical strength.[146] Finally, a bimodal scaffold was successfully formed by simultaneously solution electrospinning (low diameter fibers) and melt electrospinning (large fibers) onto a rotating mandrel, using PLGA for both types of electrospinning (Fig. 7C).

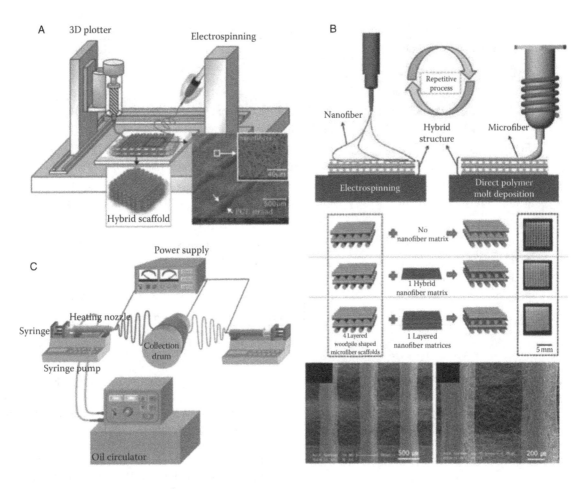

Fig. 7 (**A**) Schematic of a hybrid process combining 3D extrusion and electrospinning. (**B**) Combining Electrospinning with a melt extrusion technique to create an electrospun carrier. (**C**) Combining melt electrospinning and solution electrospinning to create a bimodal scaffold.
Source: (A) From Kim et al.[290] (B) From Park et al.[145] (C) From Kim et al.[13]

Electrospinning Small Diameter Microfibers

There is a reported correlation between fiber diameter and pore size in electrospun meshes: a minimum fiber diameter (2–3 μm) is required for cell penetration (pore size 10–15 μm).[147] Thus, the lack of pore size and interconnectivity resulting from the chaotic deposition of electrospun nanofibers creates a barrier to cell invasion. Interestingly, there are only a handful of studies on cellular interactions on meshes with fiber diameters between 2 and 10 μm, meaning that fibrous scaffolds with both high surface area *and* cell invasiveness are underinvestigated for TE. When such scaffolds are seeded, cells penetrate 1200 μm into the scaffold under flow and 300 μm under static conditions.[147]

GROWTH FACTOR DELIVERY FROM ELECTROSPUN MATERIALS

Electrospun fibers can be used to deliver growth factors and other proteins to their surrounding environment. The growth factor can be included by simple addition to the polymer solution to be electrospun or incorporated using a variety of other means.[148,149] It is important in this case that the bioactivity of the growth factor is retained, since dissolution within a solvent can denature proteins, thus altering their bioactivity. Table 4 provides a summary of different growth factors incorporated within electrospun fibers.

As shown in Table 4, the delivery of growth factors can be achieved through the release of soluble factors or by tethering the molecules to the surface of the electrospun fibers. Drug delivery in the latter approach requires the sustained availability of the growth factor, in a bioactive and natural form. Therefore, the surface conjugation of biomolecules to electrospun fibers is an important area of research and includes single and multistep conjugation approaches for tethering bioactive molecules.

SURFACE MODIFICATION OF ELECTROSPUN FIBERS

Biofunctionalization of the interface between a polymer and a biological entity is an important means to direct cell

Table 4   Selected examples of growth factors delivered from electrospun fibers

Growth factor	Polymer	Comments	References
BMP-2	Silk fibroin		[150]
BMP-2	PLGA/Hap		[151,152]
BMP-2	PLGA/Hap	BMP-2 plasmid delivered	[153]
BDNF	PCL	Surface functionalized	[154]
EGF	PLLA	GF both adsorbed and heparin bound	[155]
FGF-2	PLGA	Two approaches used—normal and coaxial	[156]
FGF-2	Gelatin or collagen	Introduced with perlecan binding domain	[157]
FGF-2	PLLA	Via surface bound heparin	[155,158]
GDNF	PCLEEP		[159]
IGF-1	Polyurethane	Delivered via PLGA microspheres in fibers	[160]
NGF	PLCL	Emulsion electrospinning	[161]
NGF	PCLEEP	NGF co-delivered with BSA	[7]
PDGF	Dextran/PLCL	Coaxial	[162]
VEGF	Gelatin/PCL	Coaxial—VEGF surface bound to heparin	[163]

behavior and/or prevent bacterial infection.[164] The surface of the polymer fiber also influences the properties of the electrospun fiber mesh, such as wettability. Broadly, there are two methods possible to modify the electrospun fiber surface: postprocessing of the fibers, or more recently biofunctionalization was performed in a single electrospinning step.[122,165]

Postprocessinc Approaches

The treatment of electrospun fibers after their manufacture is a common route to biofunctionalization. It is simple, because traditional conjugation chemistry such as 1-ethyl-3-(3-dimethylaminopropyl) carbodiimide hydrochloride (EDC) or N-hydroxysulfosuccinimide (sulfo-NHS) can be adopted. One problem with this approach, however, is that chemical removal and washing in different media is required, and the repetitive transition to different liquids may damage delicate parts of an electrospun scaffold.

Adsorbing proteins onto electrospun fibers

Proteins can also adsorb onto surfaces, particularly hydrophobic surfaces. This is a simple method of introducing biological signals onto electrospun fibers but needs to be compared with including the protein as part of the electrospinning solution. For example, Ramakrishna and colleagues adsorbed laminin onto electrospun fibers. However, when laminin was included within the electrospinning solution, neurite growth was significantly higher compared to electrospun fibers with adsorbed laminin.[40] Some drugs adsorb onto electrospun fibers through simple physical adsorption using electrostatic interactions, hydrogen bonding, and hydrophobic interactions.[166,167]

Chemical binding bioactive functionalities

The direct conjugation of bioactive molecules onto electrospun fibers is an important approach to functionalization. Through various chemistries, brain-derived neurotrophic factor (BDNF),[154] heparin,[158] Cibacron blue F3GA (CB), and bovine serum albumin (BSA)[168] have been covalently attached onto electrospun fibers. By incorporating heparin onto the surface of the fibers, growth factors such as FGF-2 and EGF-1 were bound.[155,158] This set the platform for many different growth factors to be bound using heparin binding sites. Combining adsorption and chemical conjugation, DNA was electrostatically incorporated onto electrospun fibers that have matrix metalloproteinase (MMP) cleavable functionalities.[169]

Protein Resistant Biofunctionalization

There has been some focused work on generating protein resistant electrospun fibers with bioactive surface functionalities. Using amphiphilic block copolymer polymers with the correct hydrophilicity and chemistry, the resulting fibers were hydrophilic.[63] When these amphiphilic polymers were end modified with arginine-glycine-aspartic acid (RGD), biotin, or were thiolated, then they were protein resistant, hydrophilic, and functionalized.[62,63] Interestingly, only when mixed solvents are used (chloroform/methanol instead of chloroform only) there is strong resistance to BSA or streptavidin adsorption. Thiolated electrospun fibers in particular are able to then chemically bind with numerous bioactive molecules.[170] A more simplified approach to that described here has been devised by the same group that uses an amphiphilic and isocyanate star-shaped macromer, a reactive additive to a PCL or

PLGA solution, to create both biofunctionalized and hydrophilic fibers.[122,165] Protein resistant fibers were produced using solutions containing only 7% star-shaped macromer relative to the PLGA or PCL. Functionalization was achieved by adding a bioactive peptide to the macromer/polymer solution prior to electrospinning with RGDS peptide and RDES modified electrospun substrates. The functionalized and hydrophilic fibers exhibited distinct differences in cell adhesion when investigated *in vitro* for neural applications.[122] When such electrospun fibers are placed upon reactive substrates that prevent cell adhesion, cells can only adhere to the fibers (Fig. 8).

IN VITRO STUDIES OF ELECTROSPUN MATERIALS

As previously described, typical electrospinning does not result in cell-invasive scaffolds. Despite this, there are numerous excellent reviews on the interactions between electrospun fibers and cells *in vitro* that address aspects of cell biology[26,171] as well as demonstrate the diversity of polymeric, morphological, and structural properties. In general, electrospun fibers are excellent substrates for cells: except for a few examples, when the fibers are hydrophilic and protein resistant,[62,63,165,170] cells adhere readily onto electrospun material. The cell shape, however, changes with electrospun fiber morphology: random meshes allow cell spreading similar to that seen on flat substrates, whereas cells tend to self-orient onto aligned fibers, particularly if there is no other substrate for them to adhere to Gerardo-Nava et al.[123] In studies where "dilute" quantities of fibers were collected onto reactive substrates, the cells were only permitted to adhere to the fibers, inducing a very oriented structure. Interestingly, phenotypical

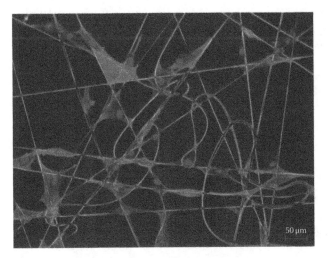

Fig. 8 RGD-modified electrospun fibers deposited onto reactive substrates that prevent adhesion of cell—fibroblasts can only adhere onto the fibers.
Source: From Grafahrend et al.[66]

differences have been observed between cells cultured on oriented fibers rather than random meshes.

Fibroblasts

Fibroblasts are the most abundant cells in connective tissue, and for various reasons, they are by far the most investigated cell type with electrospun material. Fibroblasts excrete ECM, including collagen, and are an important part of the wound-healing process. In addition to collagen, they produce glycosaminoglycans, reticular and elastic fibers, and glycoproteins. Electrospun fibers have often been compared to collagen, which can be fibrillar in structure. Since collagen is widely present in such connective tissue, it is appropriate that electrospun fibers and fibroblasts are so widely studied. For example, there are studies where the integration of fibroblasts into an electrospun scaffold is a crucial component of the final application such as in ligament of skin tissue engineering. Table 5 provides a list of selected references where fibroblasts were cultured with electrospun fibers.

Because wound closure is an important part of fibroblast function, these cells have been used in wound-healing applications. Experiments with electrospun materials have focused on fibroblast adhesion, proliferation, and migration: the majority of electrospinning publications involving fibroblasts are superficial, in that the cells help to determine adhesion and toxicity toward the fiber, rather than as part of a focused wound-healing experiment. However, there are some studies providing in depth analysis of fibroblast function, including upregulation of important genes, such as collagen III production, which is optimal on fibers with diameters between 350 nm and 1.1 μm.[172]

In many instances, fibroblasts are used as a proof of cytotoxicity[173] and biocompatibility.[174] These cells can also be used to ascertain whether a strategy of electrospinning produces feasible fiber morphology,[175] if surface modification of a polymer was successful,[176] or if a drug delivery system is viable.[177] These issues are important, as cells may be the most sensitive measure of changes in bioactivity of a surface, since instrumentation techniques have detection limitations. For example, fibroblasts have successfully been transfected with a nonviral vector incorporated within electrospun fibers so that they express green fluorescent protein.[178] In another study, fibroblasts were used as part of the collection system, to demonstrate that melt electrospun fibers could be deposited directly onto cells without a loss in vitality.[100] Similarly, recent work on making protein resistant and functionalized fibers involved fibroblasts to determine the success of the surface modification approach.[62,63,122,165,170] Although many of the aforementioned electrospinning investigations could be undertaken with almost any other cell type, fibroblasts are inexpensive, readily accessible, and relatively simple to work with. Suspended fibers have also been used with *in*

Table 5 Selected *in vitro* studies with fibroblasts

Polymer	Comments	References
PLGA	Cells adhered and spread according to fiber direction	[4]
PLGA	Bimodal d = 530 nm and 28 μm. Higher cell attachment/spreading in the nano/microfiber than the microfibrous scaffolds without nanofibers	[13]
PCL/PEG-b-PCL blends	Electrospun fibers were drawn onto substrates for discrete cell adhesion	[24]
Fibrinogen	Neonatal rat cardiac fibroblasts readily migrated into and remodel electrospun fibrinogen scaffolds with deposition of native collagen	[37]
PEG-b-PCL	Fibers were both protein resistant and surface functionalized with RGDS	[62,63]
PCL/PEG-b-PCL blends	Direct *in vitro* electrospinning with melt electrospinning	[100]
PCL/sP(EO-stat-PO)-RGDS	Single step functionalization—fibers were both protein resistant and surface functionalized with RGDS	[165]
Thiolated PEG-PDLLA	Adherence and proliferation on RGDC-functionalized fibers with reduced protein adsorption	[170]
Collagen	The aligned scaffold exhibited lower rabbit conjunctiva fibroblast adhesion but higher cell proliferation	[179]
Polyurethane	Rotating cylindrical collector aligned fibers. After mechanical strain, ligament fibroblasts had a spindle shape on aligned fibers and secreted more ECM than random ones	[180]
Polyamide	Cells rearranged their actin cytoskeleton to a more *in vivo*-like morphology	[181]
Collagen and PCL	Supported attachment and proliferation	[182]
Chitosan/silk fibroin	Supported attachment and proliferation	[183]
Ag-containing PVA	The Ag ions and Ag nanoparticles are cytotoxic to cells	[184]
PLLA/silk fibroin/gelatin	After 12 DIV, the scaffold supported attachment, spreading, and growth	[185]
Collagen and hyaluronate	The ratio of foreskin fibroblasts expression of TIMP1 to MMP1 was lower than collagen fibers	[186]
PDLA/PLLA (50:50)	After 5 DIV, the fibroblasts adhered, migrated, and proliferated	[187]
PLGA and PLGA/MWCNTs	MWCNT incorporation promoted attachment, spreading, and proliferation	[188]
PLGA/collagen	Effective as wound-healing accelerators in early stage wound healing	[189]
PEG/PPG/PCL	Supported excellent cell adhesion, comparable to pure PCL	[190]
PCL and PCL/PEO	Proliferated fastest with pores greater than 6 pm. Conformation changed as the pores grew from 12 to 23 pm; cells aligned along single fibers instead of attaching to multiple fibers	[191]
PHB/chitosan	Promoted attachment/proliferation	[192]
pDNA PEI/HA in PCL/ PEG core shell	Complexes of pDNA with PEI-HA released from fiber mesh scaffolds could successfully transfect cells and induce expression of enhanced green fluorescent protein (EGFP)	[178]
PLGA/chitosan and core-shell PLGA/ chitosan	PLGA/chitosan and core-shell PLGA/chitosan showed better adhesion/viability than PLGA	[193]
PCL and ECM proteins	Presence of collagens, tropoelastin, fibronectin, and glycosaminoglycans in scaffolds	[194]
PDLLA/PEG	Dermal fibroblasts interacted and integrated with the fibers containing 20% and 30% PEG	[195]
Plasma silk fibroin	O_2-treated fibers showed higher cellular activities for fibroblasts	[196]
Silk fibroin	L929 fibroblasts showed good cell adhesion after 1, 3, and 7 DIV with confluence at 7 DIV	[197]
Polyurethane/MWNT	MWNT incorporation exhibited highest cell adhesion/proliferation and migration/aggregation	[198]
PVA	Cells grew significantly after 7 DIV but were round-shaped	[199]
Collagen Type I	Qualitative analysis of rabbit corneal fibroblasts morphology and protein expression suggests fibers suitable for engineering a corneal tissue replacement	[200]

(Continued)

Dental—Electrospinning

Table 5 (*Continued*)   Selected *in vitro* studies with fibroblasts

Polymer	Comments	References
Bombyx mori silk fibroin	Attachment/growth with no difference between fibers electrospun with different solvents	[201]
Polyurethane elastomer	Human fibroblasts adhere and migrate, proliferate, and produce components of an ECM	[202]
PLGA	Human skin fibroblasts spread and showed growth. Collagen type III gene expression was upregulated on matrices with fiber diameters in the range of 350–1100 nm	[172]
Hexanoyl chitosan	Fibrous scaffolds supported attachment/proliferation	[174]
Polyethyleneimine	Normal human fibroblasts attached/spread on fibers	[203]
PCL	PCL was electrospun using an auxiliary electrode and chemical blowing agent. Good adhesion with the blown web relative to a normal electrospun mat	[204]

vitro studies, with a minimum separation distance between fibers required for fibroblasts to bridge the gap (Fig. 9).[205]

Epithelial Cells

Simple (one cell thick) or stratified epithelium covers and lines connective tissue (containing fibroblasts) surfaces in the body—from skin, cornea, to the inside of blood vessels, and intestine—thus protecting the underlying tissues. Epithelium also functions to secrete enzymes, hormones, as well as ECM and is separated from connective tissue by a basement membrane. It is substituting the basement membrane that attracts researchers to electrospun fibers. These specialized epithelial cells have been used with electrospun fibers to create layered cellular structures. However, there has been less emphasis on creating cell-invasive substrates, since epithelium should sit on top of electrospun meshes and form a layered construct. Where electrospun fibers fit into epithelium regenerative medicine strategies is not as

well defined as compared to other cell types. Despite this, epithelia are crucial to wound-healing strategies, and they will be increasingly used in electrospun fiber research. Table 6 lists a selection of epithelial studies using electrospun fibers.

Endothelial Cells

Endothelial cells are specialized epithelial cells that line the interior surface of blood vessels to create a barrier between circulating blood and the rest of the vessel wall. Endothelia are of wide interest to electrospinning researchers, which may be due to the structure they comprise: although the most widely used electrospinning collection configuration is a single flat collector, the second most commonly used is a single, rotating collector to generate tubular structures. Since blood vessels are such an important tissue, numerous researchers have looked at incorporating such cells and fibers together to create a functional

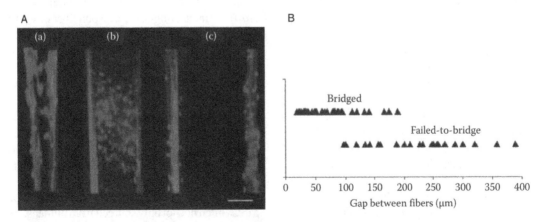

Fig. 9   (**A**) Human dermal fibroblasts cultured on suspended aligned fibers at different distances apart. (**B**) The effect of a gap between suspended fibers and whether cells are able to bridge the gap.
Source: From Sun et al.[205]

Table 6 *In vitro* studies with epithelial cells

Polymer	Comments	References
P(LLA-CL) 75:25	Adhered and proliferated well	[94]
Polyamide	T47D breast epithelial cells underwent morphogenesis to form multicellular spheroids	[181]
Chitosan and alginate	Layer-by-layer scaffolds have biocompatibility with Beas-2B human bronchial epithelial cells	[206]
PDLA	Both protein adsorption and porcine esophageal epithelial cells attachment	[207]
PHBV; PHBV/collagen, and PHBV/gelatin	Co-culture of hair follicular cells with epithelial outer root sheath/ dermal sheath cells 3–5 DIV showed PHBV promoted wound closure and reepithelization more than PHBV/collagen	[208]
Fibronectin-g-PLLCL	Promotes epithelium regeneration, using esophageal epithelial cells	[209]
PCL and PVA co-spinning	Human prostate epithelial cells attached/proliferated more with PVA electrospun PCL mats	[210]

epithelium on electrospun tubes, as shown in Table 7. Desirable mechanical properties of the electrospun tubular mesh are a critical requirement. Electrospun tubes of poly(urethane) and PLLCL in particular have been used, due to their elastic nature and strength.[211–213] In addition, nano and microfibers have been combined to obtain nanoscale structures with sufficient mechanical strength.

Endothelial cells derived from umbilical cords (human umbilical vascular endothelium cells; HUVECs), with the potential to function in regenerative medicine, are an important and increasingly researched cell type, as illustrated in Table 8. HUVECS are popular because they are derived from an unrequired tissue available at birth, and there are few ethical issues with using this form of human tissue.[227]

Smooth Muscle Cells

Blood vessels also consist of nonstriated muscle, and smooth muscle cells (SMCs) are an important cell type in this tissue. SMCs are also found in other tissues, including the aorta, gastrointestinal tract, lymphatic vessels, uterus, urinary bladder, the iris, and the ciliary muscle. Table 9 shows a list of electrospun material cultured with SMCs, where emphasis is placed on seeding cells on tubular constructs mostly for vascular tissue engineering.[234,235] Cell seeding onto 3D constructs creates many challenges and the more anisotropic the scaffold, the more complex and difficult cell seeding becomes. This step is crucial in the *in vitro* culture of cellular implants since homogenous cell distribution appears necessary for successful TE strategies, which explains why there is such an emphasis placed on developing cell seeding techniques in this area. Multilayered vascular constructs have been prepared, demonstrating how complex tissue can be generated using the electrospinning technique.[238] Boland et al.[238] fabricated a three-layered collagen/elastin vascular construct, which

was built with separate media and adventitia seeded with SMCs and fibroblasts for the structural component, while the lumen was seeded with endothelial cells. Stankus et al. also addressed the cell-invasive issue associated with using submicron fibers by simultaneously electrospinning fibers and electrospraying cells in gelatin, to form tubular constructs, which were then included into a bioreactor.[8] The fibers and cells were laid down simultaneously, rather than constructing a scaffold and seeding them in a second step.

Cardiomyocytes, Cardiac Stem Cells, and Cardiac Cell Lines

An excellent article on contractile cells was provided in 2004 by the Vacanti laboratory, where cardiomyocytes were grown on a mesh of PCL, which had been collected and suspended on a metallic ring.[241] This thin sheet began contractile movements, and the metallic wire ring acted as a passive mechanical support for the cells. Important cardiac tissue markers such as myosin, connexin 43, and cardiac troponin I were expressed after 14 DIV (days *in vitro*).[242] Interestingly, electrospun fibers were also embedded into collagen sponges, with an increase in cardiac stem cell attachment on the collagen scaffolds with fibers embedded. Table 10 provides a list of *in vitro* studies with cardiomyocytes, cardiac stem cells, and cardiac cell lines.

Mesenchymal Stem Cells

Mesenchymal stem cells (MSCs) are multipotent stem cells that can differentiate into different cell types, including chondrocytes, osteoblasts, and adipocytes. They are found in bone marrow and can be extracted using a biopsy, with MSCs often but not only sourced from humans. Table 11 provides a list of different publications where MCSs were cultured on electrospun materials and demonstrates that this is one of the most investigated cell types with electrospun

Table 7 *In vitro* studies with endothelial cells

Polymer	Comments	References
PLLCL (70:30)	Human coronary artery endothelial cells were evenly distributed and well spread from 1 to 10 DIV and maintained expression of PECAM-1	[211]
(P(LLA-CL))	Both random and aligned fibers preserved human coronary artery endothelial cell phenotype; expression of PECAM-1, fibronectin, and collagen type IV	[212]
Polyurethane	Small-diameter grafts with microfibers and microgrooves endothelial cells formed confluent monolayers. The cells expressed cadherin	[213]
PCL and CNTs	Good cerebro-microvascular endothelial cell viability	[214]
Collagen/chitosan	Endothelial cells proliferated well	[215]
PCL/collagen	Endothelial cells showed enhanced cellular orientation and focal adhesion	[216]
Coaxial cellulose acetate and chitosan	As mechanical stiffness increased, endothelial cell growth/adhesion and migration promoted. Fibers did not induce platelet aggregation or activation	[217]
Silk fibroin	Iliac endothelial cells attached and proliferated in comparison with cast SF films	[218]
Silk fibroin/PLLCL	Silk fibroin/PLLCL blended fibers increased Iliac endothelial cell growth compared with PLLCL	[219]
Polyurethane/collagen	Iliac endothelial cells migrated inside the scaffold within 24 hr of culture	[220]
Span-80/PLLACL coaxial core/shell	No differences in iliac endothelium cell growth on the fiber types	[221]
HMA:MMA:MAA terpolymer	Cytocompatible allowing human blood outgrowth, endothelial growth, maintained phenotype	[222]
Polyphosphazene	After 16 DIV, neuromicrovascular endothelial cells formed a monolayer on the whole surface	[223]
P(LLA-CL 70:30)	Collagen-coated fibers enhanced spreading, viability, and attachment of human coronary artery endothelial cells and preserved phenotype	[224]
Gelatin-grafted-PET	Gelatin grafting improved the spreading/proliferation of the ECs and preserved phenotype	[225]
PLLA	Endothelial cell function was enhanced on smooth solvent cast rather than on the rough electrospun fibers. Electrospun substrates favored vascular smooth muscle cell behavior	[226]

materials due to their ability to differentiate. With MSCs, the lineage of the cultured cell is of particular interest. Some researchers use differentiation media with electrospun fibers to induce a particular type of cell line with success.[247,248] Therefore, electrospun fibers do not interfere with conditions that are designed to differentiate MSCs into particular lineages. Alternatively, electrospun scaffolds can be used as a drug delivery platform to induce stem cell differentiation into the targeted cell lineage. From a practical viewpoint, it was shown that insulin growth factor (IGF-1) loaded electrospun scaffolds improved MSC survival under hypoxia or nutrient reduced conditions.[160] With such a complex and heterogeneous environment generated during stem cell culture, all conditions and parameters need to be reported in order to allow comparison between studies and to draw definitive conclusions.

Osteoblasts

Ideally, bone TE should provide a substitute for the autologous bone graft, and there is a lot of research to develop a load-bearing material that can integrate with bone defects. Since electrospun fibers do not generate sufficient compressive strength, their use is more to support bone formation in other surgical paradigms, including an artificial periosteum and as a structure to contain hemato-mass between bone fractures.[267] Mineralization is a crucial process in bone formation, and many of the *in vitro* studies with osteoblasts and electrospun fibers measure the alkaline phosphatase activity (ALP), which indicates deposition of mineralized matrix (early bone formation). A distinctive feature of electrospun fibers within bone TE is the inclusion of inorganic substances within the fibers to

Table 8   *In vitro* studies with human umbilical vascular endothelium cells

Polymer	Comments	References
HMA:MMA:MAA terpolymer	Cytocompatible fibers allowed growth, maintained phenotype	[222]
Methacrylic terpolymer	Fiber morphology controlled proliferation, metabolic activity, and morphology of cells	[228]
MWCNT/PU	Fibers activated Rac and Cdc42, while CNT regulated the activation of Rho	[229]
PCL	$D = 0.8$ or 3.6 μm. The larger fibers had higher vitality and improved interaction	[230]
Silk/gelatin	Gelatin improved the mechanical and biological properties	[231]
PLLCL	VE-cadherin expression not detected, but a high degree of tissue integrity was achieved	[232]
PLCL (50:50)	Adhered well and proliferated on small diameter fiber fabrics (0.3 and 1.2 μm; reduced adhesion, cell spreading and no signs of proliferation on the 7.0 μm diameter fibers	[233]

Table 9   *In vitro* studies with smooth muscle cells

Polymer	Comments	References
[P(LLA-CL)] (75:25)	Rotating disk collector aligned fibers. SMCs attached and migrated along aligned fibers; the distribution of cytoskeleton proteins were parallel to the fibers	[136]
Type I and II collagen	Mandrel rotated at approximately 500 rpm. Fibers possessed the typical 67 nm banding pattern observed in native collagen	[5]
PU	Microintegrated cells into a biodegradable elastomer fiber matrix	[8]
[P(LLA-CL)] (75:25)	Adhered and proliferated well	[94]
PCL	Larger fiber diameter improved SMC infiltration into the bilayered vascular scaffolds	[216]
PLGA/COL	Co-culturing of SMCs with ECs, under a pulsatile perfusion system, leads to the enhancement of vascular EC development, as well as the retention of the differentiated cell phenotype	[234]
PU	Cells integrated into a strong, compliant biodegradable tubular matrix	[235]
Collagen-coated PCL	Proliferation on collagen-coated PCL fibers was not different than that on TCP	[236]
PCL	Contains heparin	[237]
Collagen I, II, and elastin	A three-layered vascular construct, integrating fibroblasts and endothelial cells	[238]
PLLA + PLLA/PVP	Morphology/proliferation improved with decreasing of the weight ratio of PLLA/PVP. In platelet adhesion decreased with increment of PVP content in the films	[239]
COL-coated PCL	Same proliferation rate on polycaprolactone (PCL) nanofibrous matrices coated with collagen or tissue culture plates. PCL coated with collagen showed more migration	[240]

improve the biological outcome when using osteoblasts by rendering scaffolds more osteoinductive. Such materials include hydroxyapatite (HAp), silica, titanium dioxide (TiO$_2$), and bioactive glass. The cells used for bone TE can also be MSCs, as they are the stem cell precursor for osteoblasts. Another issue in this type of TE strategy is the degree of cell invasiveness of the electrospun scaffolds, which has been addressed by Ekaputra et al.[9] These authors simultaneously electrospun and electrosprayed different materials, which made the fibrous mesh less dense. For example, when a hyaluronic acid solution is electrosprayed with electrospun PCL onto the same collector, there was increased osteoblast penetration into the scaffold.

A further step in the development of osteoinductive scaffolds is the deposition of a biomimetic calcium

Table 10 *In vitro* studies with cardiomyocytes, cardiac stem cells, and cardiac cell lines

Polymer	Cells	Comments	References
PCL	Cardiomyocytes	A 10 Jim thick mesh is suspended across a wire ring. Myosin, connexin 43, and cardiac troponin I expressed after 14 DIV	[241,242]
PLLA	Cardiomyocytes	Cells with fiber-guided filipodialike protrusions and developed into sarcomeres	[243]
PGA	Cardiac stem cells	Fibers embedded in collagen sponge and more cells attached to the collagen sponge incorporating fibers compared to without. The attachment and proliferation were enhanced by incubation in a bioreactor perfusion system	[244]
Polyurethane	Cardiomyocytes	Cells were elongated, but morphological changes induced by material macrostructure did not directly correlate to functional differences	[245]
PPDL	H9c2	Fibers were not cytotoxic and supported cell proliferation	[246]

phosphate coating on the fibers. This appears to enhance the ALP activity and the osteo-pontin expression of a MC3T3-E1 mouse osteoblastic cell line under osteoblastic induction.[268] In a different study, the same cell line displayed similar behavior onto biomimetically coated electro-spun PCL. The expressions of ALP, RunX2, collagen I, and osteocalcin were upregulated confirming the beneficial effect of a calcium phosphate coating on osteoblast differentiation.[269]

As seen in Table 12, there is significant interest in the interactions between osteoblasts and electrospun fibers. This partial list contains only selected journal articles, and the actual publication list is much greater. The work of Jang, Castano, and Kim is an excellent review on electrospinning for bone regeneration.[285]

Chondrocytes

There is a clear medical need for transplantable cartilage tissue, as cartilage has a limited capacity to repair itself. Cartilage TE research emphasizes regeneration, rather than repair, and an ideal cartilage scaffold should possess similar mechanical properties to the natural tissue. The ECM, which contributes the most to the mechanical properties, is comprised of collagen fibrils (mainly Type II) and proteoglycans and is distributed into zones with distinct mechanical properties.[286,287]

A promising approach to cartilage repair is 3D printing of hydrogels.[288,289] As previously described, this approach can be combined with electrospinning to form a different class of "bimodal scaffold", where the scaffold consists of structures with distinctly different scales. In this instance, the extruded hydrogels form struts approximately 100–200 μm in diameter, while electrospun fibers are incorporated within the larger struts (Fig. 7B).[290] The smaller diameter fibers aid in increased cell seeding efficiency and are included in the printing process by electrospinning separate layers during the printing process. Table 13 summarizes *in vitro* studies with chondrocytes and electrospun fibers.

Neurons and Glia

Electrospun fibers have been investigated for the guidance of neurons and glia, which naturally exist in highly oriented and structured tissues, particularly in the peripheral nerve and the white matter of the central nervous system (CNS). The research with such cells can be divided into two main areas of interest—*in vitro* guidance systems and tubular structures for containing damaged nerves.

Fibers have an inherent guiding property on neuron processes when their diameter is below approximately 250 μm.[298,299] Further reducing the fiber diameter down to submicron levels, as seen in electrospinning, still maintains this guidance effect.[67,121,123] Since nervous tissue is often bundled into organized arrays, the orientation of electrospun fibers is a particularly relevant aspect of the technique. As mentioned previously, electrospun fibers can be aligned by collecting them onto a rapidly rotating mandrel. The oriented electrospun sheet can then be flattened with minimal curvature if the mandrel is of sufficiently large diameter.[300] Alternatively, it is possible to collect electrospun fibers across the gap between two collectors, previously outlined in Section "Instrument Configuration". This has been performed using parallel bars, sharp blades, and metallic rings. When short collection times are used in this approach, the fibers that are suspended across the gap remain "individual", in the sense that they do not adhere to other fibers. However, performing cell culture on such individual fibers can prove difficult, as changes in culture media are difficult to perform.

Recent work on chemically reactive substrates has permitted the study of a single cell on a single oriented fiber, quantifying the effects of changes in surface chemistry by measuring cell elongation and process extension.[67,121,123] Fibers are transferred to a substrate by passing a prepared cover slip through an oriented suspension of fibers. The reactive substrate binds the fibers, but then continues to react with water, so that it is ultimately resistant to both protein and cell adhesion. In this configuration, neurons,

Dental—Electrospinning

Table 11 *In vitro* studies with mesenchymal stem cells

Polymer	Comments	References
PLGA/TCP	Cotton wool-like fibers from collection on ice. Osteocalcin showed osteogenic differentiation	[11]
HAp/TCP/PCL	The solvent or combination for fiber preparation affects properties and cell behavior	[82]
PLLA/PCL	Combined electrospinning with FDM to make a tubular construct	[146]
Silk/PEO/HAp	HAp particles improved bone formation. BMP-2 and HAp resulted in the highest calcium deposition and upregulation of BMP-2 transcript levels	[150]
PLGA/HAp	Encapsulated DNA/chitosan nanoparticles have higher cell attachment, viability, and transfection	[153]
PLGA/HAp	Encapsulation of HAp could enhance cell attachment to scaffolds and lower cytotoxicity	[152]
PLGA	Under hypoxia/nutrient starvation conditions, the IGF-1 loaded scaffolds improved MSC survival	[160]
PLCL and PLCL/coll	Neural differentiation was induced with soluble factors	[247]
PCL	Chondrogenic differentiation was enhanced after incubation in chrondrogenic media	[248]
PCL	Delayed initial attachment and proliferation on meshes, but enhanced mineralization at a later time point	[249]
PLGA and PCL	Supported attachment, proliferation, and differentiation	[250]
PCL	Differentiated to adipogenic, chondrogenic, or osteogenic lineages by specific differentiation media	[251]
PCL	Differentiated to chondrocytic phenotype—a zonal morphology is described	[252]
PCL	Matrix mineralization and collagen type I throughout after 4 weeks	[253]
Collagen	Combined electrospinning with SFF; Enhanced initial cell attachment and cell compactness between pores of the scaffold relative to SFF 3D collagen scaffold only	[254]
HAp/PLGA	Increased ALP activity and expression of osteogenic genes w calcium mineralization of MSCs	[255]
Gelatin/siloxane	Resulted in apatite formation and stimulated proliferation and differentiation	[256]
PCL	Increased osteocalcin and osteopontin 3 weeks of culture	[257]
PCL	Multimodal distribution of fiber size allowed increased cell seeding	[258]
Collagen or PLLA	Early adhesive behavior is affected by texture of surface. Collagen and collagen coated PLLA fiber had similar outcomes	[259]
PCL	Thicker scaffolds provide a better substrate for cell proliferation	[260]
PLCL	MSCs expressed CD44 and CD105 but did not express CD34 or CD14	[261]
P(3HB-co-4HB)	Porcine aortic heart valves were decellularized, coated with electrospun fibers	[262]
Silk fibroin	Fiber alignment affected morphology and orientation	[263]
Chitin/PVA	Fibers supports cell adhesion/attachment and proliferation	[264]
PLA or PLA/DBP	Mineralization with osteogenic supplements was greater PLA after 14 DIV but same level 21 DIV	[265]
PCL or PCL/COL	Multilayer construct, supported cell attachment similar tissue culture polysterene plate	[266]

Dental—Electrospinning

Table 12 *In vitro* studies performed with osteoblasts

Polymer	Cell type	Comments	References
PCL/collagen and PEO; PCL/collagen/ Heprasil	Fetal osteoblasts	All three scaffold types supported attachment/ proliferation; better penetration was with mPCL/ Collagen fibers co-electrosprayed with Heprasil	[9]
PCL/b-TCP	MC3T3-E1 cells	Viscoelastic and biomechanical properties increased with DIV	[270]
Gelatin and HAp	Fetal osteoblasts	Combining of electrospinning of gelatin and electro-spraying of HA increased ALP activity and enhanced mineralization	[271]
PCL-PEG-PCL and HAp	Osteoblasts	Scaffolds could provide a suitable environment for attachment, and MTT revealed the scaffolds had good biocompatibility and nontoxicity	[272]
HA/TiO$_2$	MC3T3-E1 cells	Collagen treated HA/TiO$_2$ fibers possess better cell adhesion and significantly higher proliferation and differentiation than untreated fibrous mats	[273]
PCL-silica	Pre-osteoblast cells	Good proliferation and differentiation of cells	[274]
PLA/bioactive glass	Osteoblasts	Bioactive glass phase enhanced the *in vitro* apatite formation and cells adhered	[275]
PLLA	Osteoblasts	After mechanical expansion of the microfibrous mats, the scaffolds proliferated osteoblasts that actively penetrated the inside of the 3D scaffold	[276]
p(TMC-co-CL)-block-p(p-dioxanone)	MC3T3-E1	Cells proliferated 1.2 times faster at 4 DIV, 1.5 times faster at 7 DIV after seeding; At 28 DIV alkaline phosphatase was four times more than control	[277]
RGD-mod PLCL	Osteoblasts	Adhesion and proliferation was greater on RGD modified fibers meshes up to 7 DIV. ALP activity and calcium content on the RGD-AAc-PLCL meshes were approximately 7.5 and 6.7 times higher than those on the other meshes, respectively. Expression of Cbfa1, ALP, and Osteocalcin (OCN), was upregulated (at least 5–9.7 times greater) on the RGD-AAc-PLCL meshes	[278]
PHBV and cHAp	SaOS-2	Both PHBV and CHA/PHBV supported the prolifera-tion; The alkaline phosphatase was higher than the CHA/PHBV after 14 DIV	[279]
PCL/collagen	Pig bone marrow mesenchymal cells	Osteogenic differentiation markers were more pronounced on PCL/Col fibrous meshes	[280]
PLCL/gelatin/HAp	Fetal osteoblasts	Electrospinning PCL/gelatin and electrospraying of HA nanoparticles made a rough surface with better proliferation, mineralization, and alkaline phospha-tase activity	[281]
Silica	MC3T3-E1 cells	ALP activity and expressions of type I collagen and osteocalcin were higher for cells cultured on nonwoven silica gel fabrics than on TCP	[282]
PLLA and PLLA/HAp	Osteoblasts	n-HA deposition improved seeding efficacy at 20 min time point	[283]
PLA/bioactive glass		As bioglass concentration increased (from 5% to 25%), the bioactivity improved. Cells attached, grew well, and secreted collagen. ALP expression was higher than pure PLA	[284]

Dental—Electrospinning

Table 13 *In vitro* studies with chondrocytes

Polymer	Comments	References
PCL/collagen	Favorable conditions for cell adhesion/proliferation, attributed to nanotopography, chemical composition, and an enlarged surface for cell attachment and growth	[145]
Chitosan	Fibers allowed chondrocytes to proliferate and produce glycosaminoglycan	[291]
PCL and 3D prototyping	This approach is feasible for fabricating 3D scaffolds for chondrocytes	[290]
PCL	Fetal bovine chondrocytes upregulated collagen type IIB expression after 3 weeks	[292]
HA/collagen	Chondrocyte adhesion/proliferation were enhanced maintained chondroblastic morphology	[293]
3D printed PEOT or BPT and electrospun PCL	Combined scaffolds enhanced articular chondrocyte entrapment compared to 3DF scaffolds and higher GAG/DNA ratio after 28 DIV. Spread morphology was observed on 3DF scaffolds, suggesting a direct influence of fiber dimensions on cell differentiation	[294]
PCL on PLA microfibers	Increased cellular infiltration with higher porosity	[295]
PHBV	Rabbit auricular chondrocytes attached more efficiently to electrospun fibers (30% after 2 hr) than cast films (19%)	[296]
Collagen type II	Scaffolds produce a suitable environment for chondrocyte growth	[297]

oligodendrocytes, Schwann cells, astrocytes, and dorsal root ganglia have all been cultured, and changes in the fiber composition (and reactive substrate chemistry) provide distinct differences in migration and adhesion of the cell. Table 14 summarizes these experiments and other investigations where oriented electrospun fibers are cultured with neural derived cells. Co-culture of DRGs on astrocytes/electrospun fibers induced directional growth along the fibers with increased neurites in the presence of astrocytes (Fig. 10).

The other main regenerative medicine approach involving neurons and glia is to manufacture tubular structures to enclose nervous tissue, either the spinal cord or more commonly, the peripheral nerve. Here, tubular electrospun materials have been implanted within animal models, described further in Section "*In Vivo* Results of Electrospun Materials." Such tubular materials must resist compressive forces and be flexible enough to avoid significant mechanical mismatch with the nerve.[311]

Ligament Cells

Ligaments and tendons are soft connective tissues that respectively attach bones to bones or bones to muscles. Their major function is usually to transmit movement and to ensure joint stability. The most predominant proteins in these tissues are collagen type I and III.[312] Ligaments and tendons display a specific structure made of crimped aligned collagen fibers in which cells are embedded, providing high tensile strength and elasticity. Electrospun nano- or microfibers have been mainly used to develop scaffolds for the rotator cuff,[313] periodontal ligament,[314–316] and also for orthopedic applications[317–320] (such as

regeneration of anterior cruciate ligament or tendon). As the structure of the rotator cuff and the anterior cruciate ligament are similar, the scaffolds developed for their regeneration are also alike. In these regenerative strategies, the electrospun fibers were utilized as a cell culture support providing a topographic guide to the cells. There is evidence showing that cell orientation occurs onto aligned electrospun fibers mimicking the morphology of the native tissue. Even though the guidance provided by the aligned fibers appears to be cell specific, it is generally assumed that collagen type I and type III are upregulated as a result of the cell alignment.[180,321] It also seems that aligned fibers enhance the deposition of orientated ECM.[320,322]

As discussed previously, there are a number of means to fabricate scaffolds with aligned fibers. It is also possible to obtain self-crimped aligned electrospun fibers (as long as the polymer used is not too rigid), in order to further mimic the morphology of the native tissue.[323] When aligning fibers onto a rotating mandrel, residual stress is created in the electrospun fibers. Once the fibers are removed from their collection support, spontaneous fiber relaxation occurs, resulting in fiber shrinking producing crimped structures. The crimping process can be controlled by adjusting the temperature at which the fibers are left to relax. This crimped structure has a significant effect on the mechanical properties at low strain (<6% strain) as it results in a J-shape stress-strain curve similar to that observed for native tissue. The particular shape of the tensile curve is explained by the recruitment of the collagen fibers during stretching; as the fibers are being aligned no stress increase is observed (corresponding to the initial toe-region in the J-curve); once the fibers are aligned they take

Table 14 *In vitro* studies with neurons and glia

Polymer	Cell type	Comments	References
Laminin/PLLA	PC-12	Functionalized fibers increased axonal extensions; blended laminin and PLLA is a simple method to modify fibers	[43]
PCL/collagen	Schwann Cells; OECs; DRG explants	Cells aligned on oriented fibers; collagen/PCL blends provided better process growth	[67]
PCL/collagen	Dissociated Schwann Cells; DRGs	Schwann cell processes were longer on collagen containing PCL fibers. When substrate is made adhesive, fibers reduced their guidance	[121]
PCL/sP(EO-stat-PO)-RGDS	DRG	Resistance to protein adsorption with functionalization	[122]
PCL/collagen	U373; hNP-ACs; SH-SY5Y	Glia aligned on the oriented electrospun substrates	[123]
PLGA	C17.2 nerve stem cells	Cells attached and differentiated along the direction of the fibers	[301]
PCL-gelatin	Schwann Cells	Aligned and random PCL/gelatin are better substrates compared with PCL	[302]
Polysialic acid	Schwann Cells	Good viability and directed cell proliferation along the fibers	[303]
Matrigel™	DRG	Supported attachment, elongation of neurites, and migration of Schwann cells similar to electrospun collagen type I fibers	[304]
PLLA	DRG; Schwann Cells	Guided growth along the aligned fibers. Schwann cells had bipolar phenotype, average neurite length was not different from fiber densities	[305]
PPy coated PLLA/PLGA	DRG	Electrical stimulation increased maximum neurite length by 83% and 47%, respectively, for random and aligned samples	[306]
PCL	PC-12	Aligned fibers had similar growth irrespective of diameter. Neurites on aligned fibers were longer than on randomly oriented fibers	[307]
Plasma PCL and PCL/COL	Schwann Cells	17% increased proliferation on plasma modified PCL compared with PCL/collagen. Schwann cells had bipolar elongations	[308]
PCL/chitosan and PCL	Schwann Cells	48% more cell proliferation on PCL/chitosan after 8 DIV	[309]
Polydioxanone	Astrocytes; DRG	Astrocytes had directional growth along the fibers. Increased neurites when co-cultured with DRGs	[310]

over the load applied resulting in a rapid stress increase (known as the linear part of the curve).

Although it is possible to provide topographic cell guidance and to obtain scaffolds with similar stretching mechanisms to ligaments or tendons, one major drawback of the use of aligned electrospun scaffolds comes from the weak mechanical performance of these structures. There is a striking difference between the ultimate tensile force of a human anterior cruciate ligament (around 2500 N[324]) and an electrospun membrane (several tenths of N). Therefore, several groups have developed composite scaffolds in an attempt to overcome this limitation.[317–320] These structures are made of a mechanically resistant knitted construct onto which nano- or microfibers are electrospun. Here again it is also possible to fully mimic the morphology of the ligament and to provide a topographic cell guide by aligning the electrospun fibers onto the knitted structure.[320] These structures displayed a higher ultimate strength of up to 250 N (compared to several tenths of Newtons for electrospun scaffolds only), but this is still far from the ultimate strength of a human ligament. Therefore, research must be pursued in this direction, that is, to increase the mechanical properties of the scaffolds in order to develop functional tissue-engineered constructs.

Dental—Electrospinning

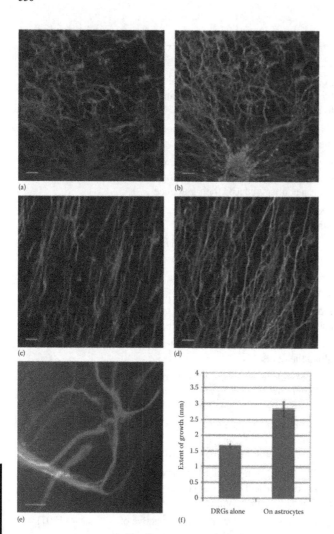

Fig. 10 Images showing DRGs grown on astrocytes and how the directionality of growth is affected by the astrocytes. (**A**) and (**B**) are random electrospun meshes while the cells in (**C**) and (**D**) are grown on oriented fibers. (**E**) shows how neurite processes change direction when encountering an astrocyte. (**F**) is a graph showing that astrocytes increase neurite growth. Scale bars: A–D, 100 μm; E, 30 μm.
Source: From Chow et al.[310]

hydroxyapatite nanoparticles were incorporated in the bulk of the fibers to enhance osteo-integration while maintaining cell occlusiveness.[316] The opposite side of the graded membrane in contact with the root surface can also be loaded with drugs to combat periodontal pathogens and/or favor periodontal fibroblast proliferation. The second strategy using the principles of TE utilizes a cellular scaffold to regenerate the damaged regions.[314] Here, electrospun scaffolds can be used as cell culture supports enabling the fabrication of a multilayered cellular construct upregulating the expression of osteopontin and sialoprotein. Other structures for periodontal tissue regeneration consisted of a complex arrangement of parallel and cross-aligned electrospun fibers, which were drawn onto a rotating mandrel.[326] These structures displayed higher tensile strength than randomly orientated fibers and also enhanced the proliferation and migration of rat periodontal ligament fibroblasts. Cell elongation onto the aligned fibers was also observed to promote a higher level of tissue organization in these scaffolds (Table 15).

Keratinocytes

For effective skin reconstruction, both the dermis and the epidermis need to be reconstructed. This requires the successful culture of keratinocytes. Such culture on a Petri dish presents challenges including the transfer of the cells. Therefore, thin membranes acting as mechanical supports and reinforcements are used during cell culture to aid in handling the cells, thus facilitating their transplantation. Transfer materials such as Laserskin, for example, assist in growing and expanding keratinocytes ready for implantation.[327] In this way, electrospun fibers can be seen as a potential transfer material when fabricated as a sheet and used for *in vitro* culture. Table 16 provides a list of studies where electrospun materials are used as a substrate for keratinocyte culture.

Other Cells

Due to the number of specialized cells, tissue, and organs in the body, it is not possible to list all of the different areas where electrospun fibers are used in regenerative medicine approaches. However, Table 17 provides a selection of these different cell types and electrospun fibers.

IN VIVO RESULTS OF ELECTROSPUN MATERIALS

There are understandably a greater number of *in vitro* experiments using electrospun polymers than there are *in vivo* studies, due to the increased time and effort to perform animal studies. To date, *in vivo* experiments with electrospun

There are two different strategies for periodontal applications. The first strategy concerns the so-called tissue-guided regeneration in which a biodegradable membrane is used to isolate the damaged region from the surrounding connective tissue. In this strategy, the membrane has to be biocompatible but cell-occlusive in order to prevent gingival cell infiltration and to allow the remaining periodontal cell to proliferate onto the root surface. Therefore, electrospun membranes can be considered good candidates due to their intrinsic small pore size that prevents cell infiltration. Gelatin with several levels of crosslinking[315] as well as synthetic biodegradable materials[325] have been used in the past, and they showed promising outcomes in terms of barrier property and stability. Another strategy involved the design of a graded electrospun composite in which

Table 15 *In vitro* studies with ligament cells

Polymer	Cell type	Comments	References
Polyurethane	Human ligament fibroblast	Aligned nanofibers had spindle-shaped, oriented cells; more collagen was synthesized on aligned fibers	[180]
PLGA (85/15)	Human rotator cuff fibroblast	Aligned nanofibers upregulated collagen type I from 3 DIV and type III after 1 DIV, integrin a2 expression was higher for the aligned fibers	[313]
PLGA	Periodontal ligament cells	Osteogenic differentiation and osteopontin, osteocalcin, and bone sialoprotein marker expression. At 3, 6, 9, and 12 DIV cell membrane layers deposited and multilayered structures established	[314]
Gelatin	Periodontal ligament cells	Cells cultured on the membrane *in vitro* exhibited good attachment, growth, and proliferation	[315]
PLGA	Bone marrow stromal cells	Electrospun onto PLGA knitted scaffold. Higher expression of collagen I, decorin, and biglycan genes	[317]
PLGA (65:35)	Porcine bone marrow stromal cells	Nanofiber coated knitted scaffolds could sustain cell growth and proliferation better than woven scaffolds and promoted easy cell seeding	[318]
PLGA(85:15);	Rabbit bone marrow stromal cells	More resistant scaffolds when cell seeded after 21 DIV	[319]
PLCL	Rat bone MSCs	Collagen type III started to form a fibrous network 14 DIV	[320]
Poly(ester urethane)	Bone marrow stromal cells	Spindle-shaped morphology increased with fiber diameter and degree of alignment. Expression of collagen 1 alpha 1, decorin, and tenomodulin was greatest on the smallest fibers	[321]
PLCL (80/20)	Bovine ligament fibroblasts	Immunohistochemistry staining revealed ligament ECM markers collagen type I, collagen type III, and decorin organizing and accumulating along the fibers	[322]
PLCL (80/20)	Bovine ligament fibroblasts	Deposition of ECM onto the fibers	[323]
PLLA/MWNTs/HA	Periodontal ligament cells	PLLA/MWNTs/HA enhanced adhesion/proliferation and inhibited the adhesion and proliferation of gingival epithelial cells	[325]
PLGA (50:50)	Rat periodontal ligament cells	Cells orientation along the fibers, higher level of tissue organization than random scaffold	[326]

Table 16 *In vitro* studies with keratinocytes

Polymer	Comments	References
PLGA	Bimodal d = 530 nm and 28 μm. Higher cell attachment/spreading in the nano/microfiber than the microfibrous scaffolds without nanofibers	[13]
Silk fibroin	Electrospun material was more preferable than SF film and SF microfibers	[30]
Ag-containing PVA	The Ag ions and Ag nanoparticles are cytotoxic to cells	[184]
Silk fibroin	O_2-treated fibers showed higher cellular activities for normal human epidermal keratinocytes	[196]

fibers are all generally positive, with the electrospun material demonstrating potential for the particular application.

Biocompatibility and *In Vivo* Cell Penetration

Electrospun scaffolds do not usually produce any major adverse reaction once implanted. Several studies have demonstrated relatively good biocompatibility with nano- and microfibers.[330–332] Cao et al.[332] compared the foreign body response of PCL randomly orientated nanofibers, aligned nanofibers, and films in a subcutaneous rat model. They found that films and the two types of fibrous scaffold displayed good biocompatibility as shown by a very early resolution of acute and chronic inflammation. Two weeks

Dental—Electrospinning

Table 17 *In vitro* studies with other cell types

Polymer	Cell type	Comments	References
PCL	CHO-K1	Bimodal d = 400 nm and 6 µm; or d = 4 µm. Increased penetration of cells into scaffolds	[12]
Polyamide	Rat kidney cells	Cells rearranged their actin cytoskeleton to a more *in vivo*-like morphology	[181]
PCL	Human amniotic fluid stem cell	Kinetics of osteogenic differentiation were observed between hMSCs and hAFS cells, with the hAFS cells displaying a delayed alkaline phosphatase peak, but elevated mineral deposition, compared with hMSCs	[249]
PCL	CHO	Proliferation was influenced by the thickness of the scaffold (0.1 and 0.6 mm); scaffolds with 0.6 mm thickness were a better substrate for proliferation	[260]
PCL and PCL/ gelatin	Adipose-derived stem cells	Good attachment, although migration was more with PCL/gelatin	[328]
PCL and PCL/HAp	Embryonic stem cells	Cells proliferated at the same rate as cells growing on TCP and maintained pluripotency markers	[329]

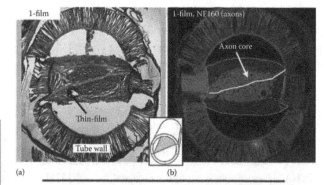

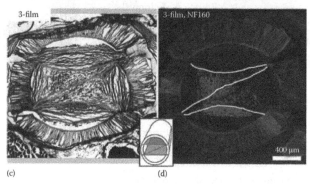

Fig. 11 Nerve guides implanted within a peripheral nerve regeneration model, containing single (**A&B**) or three (**C&D**) sheets of electrospun meshes.
Source: From Clements et al.[335]

post implantation no, or minimal, inflammation was observed. It was also found that aligned nanofibers induced a less severe healing response when compared to random nanofibers. Although no definitive conclusion could be made, it was hypothesized that the aligned nanofibers induced a more moderate inflammation due to better tissue integration as higher cell infiltration was found in this type of scaffold. As opposed to most *in vitro* studies, implanting

electrospun materials generally leads to full[267] or nearly full colonization of the scaffolds.[333] However, increasing the pore size, using for example the sacrificial fiber removal technique, produced better cell infiltration and tissue integration.[333]

Neural Tissue Engineering

In neural TE, most approaches involve attempts at mimicking the natural ECM by fabricating pores, ledges, and fibers so as to present the growing axons with appropriate cues and a permissive environment. Perhaps the most numerous studies involve nerve guides, where cylinders of electrospun material are generated on a rotating mandrel and used to bridge a gap defect in the peripheral nerve. Nerve guidance channels (NGCs) are one of the most frequently implanted electrospun material constructs; however, the *in vivo* results are not particularly compelling so far, when compared with previous nerve guides. For peripheral nerve injuries, a critical gap length for different species dictates what maximum distance regenerates 50% of the nerves.[334] For NGCs from electrospun fibers, regeneration does not appear to extend beyond what would be expected for traditional biomaterials. However, electrospinning is still developing and there are approaches to manufacture electrospun tubes that may provide an advantage over traditional techniques. For example, the Bellamkonda laboratory has been filling NGCs with electrospun meshes, showing encouraging outcomes for repair (Fig. 11).[335]

Bone Tissue Engineering

Electrospun scaffolds were also used in bone TE although for non-load bearing implantation due to the poor intrinsic mechanical properties of the fibrous mats. Generally, electrospun scaffolds do not induce spontaneous and significant bone formation unless surface treated with osteoinductive materials[336,337] or seeded with osteo-induced cells,[338] or loaded with growth factors such as BMP-2.[339] However, when one or several of these conditions are used bone formation can occur *in vivo* even in a nonbony environment such as subcutaneous implantation.[336,337] A recent study on perforated electrospun tubes between bone defects shows some fascinating outcomes, in that the perforated holes were crucial for attaining good bone formation (Fig. 12).

Cartilage Tissue Engineering

TE using electrospun scaffolds offers the potential to develop treatments for defective articular cartilage, where isolated articular chondrocytes or MSCs are seeded onto a scaffold before implantation into a damaged joint. Here, the scaffold architecture is critical, as it should simulate the ECM of cartilage in order to promote cellular adhesion, proliferation, differentiation, and migration, whilst also providing resistance to tensile, compressive, and shear stresses. Such a strategy was used to regenerate a cartilage defect in a mini-pig model. Li et al. observed that MSCs

performed better than chondrocytes seeded onto PCL nanofiber constructs. Fibrous scaffolds seeded with MSCs resulted in the formation of hyaline cartilage-like tissue at 6 months postimplantation, whereas the chondrocytes produced a fibrocartilage tissue.[340] It is also possible to surface treat the nanofibers with cationized gelatin to enhance their tissue integration while maintaining or even promoting the secretion of cartilage ECM such as collagen II (Table 18).[341]

Vascular Tissue Engineering

Alternatively, suitable vascular graft substitutes must be able to withstand pulsation and the high pressure and flow rate of the blood stream. Electrospinning offers the potential to control the composition, mechanical properties, and structure of a graft while making it theoretically possible to match the compliance of the synthetic scaffold to the native artery. Topographically aligned sub-micron fibers have similar circumferential orientations to the cells and fibrils found in the medial layer of the native artery. Because of these similarities between electrospun scaffolds and natural ECM, and because of the large variety of materials that can be used, cell viability is often superior to other scaffold designs. One crucial point in developing vascular grafts is

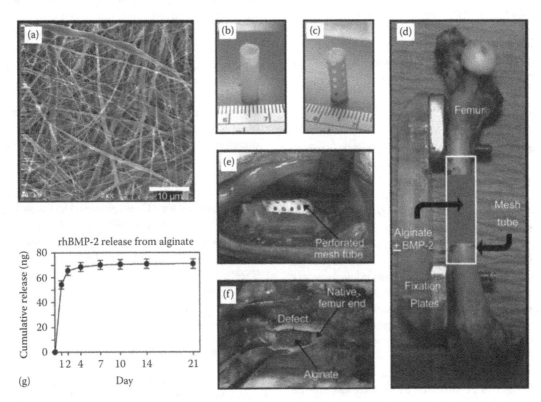

Fig. 12 (**A**) SEM image of electrospun PCL. (**B**) Hollow electrospun tubular implants without perforations and (**C**) with perforations. (**D**) Scheme of implant in segmental bone defect. (**E**) Defect after placement of a perforated mesh tube with alginate seen inside the tube through the perforations. (**F**) One week postsurgery and alginate was present inside the defect, with hematoma present at the bone ends. (**G**) Alginate release kinetics of the rhBMP-2 during the first 4 days.

Table 18 *In vivo* studies with electrospun materials

Polymer	Organ	Comments	References
PCL	Peripheral nerve	Preliminary data on scaffold made using gap method of alignment	[16]
Poly(urethane)	Wound dressing	Rate of epithelialization is increased and dermis is well organized if wounds are covered with electrospun membrane	[53]
P(LLA-CL) (70:30)	Inferior superficial epigastric vein	Tubular scaffolds kept structural integrity for 7 weeks	[211]
PCL/COL	Subcutaneous	Good integration/neovascularization	[266]
PCL-silica	Bone	Biocompatible and bioresorbable	[274]
PLLA/MWNTs/HA	Periodontal ligament cells	Cell/fibers were implanted into the leg muscle pouch	[325]
P(AN-co-MA)	Peripheral nerve	Sheet-like structures in a conventional nerve guide	[335]
PCL	Bone	Woven bone-like appearance, with mineralized matrix	[338]
Chitosan	Peripheral nerve	Similar to autograft	[342]
PLGA/PCL	Peripheral nerve	Modest regeneration over 10 mm gap after 16 weeks	[343]
Bilayered chitosan tube	Peripheral nerve	YIGSR modified tubes regenerated similar to isograft after 5 and 10 weeks	[344]
PCL	Abdominal aortic substitute	Good patency, endothelialization, and cell ingrowth	[345]

the stability of the implants. Any blockage of the inner part of the cylinder due to thrombosis and/or collapse of the electrospun materials should be avoided. PCL-collagen tubular scaffolds have been demonstrated to be very stable after 4 weeks postimplantation under normal hemodynamic conditions in a rabbit aortoiliac by-pass model. This type of scaffold maintained structural integrity and could resist the development of aneurisms.[331]

SUMMARY

With the sudden interest in electrospinning and the desire to control the process to fabricate different assemblies, various setups have been reported. The use of both rotating and translating devices, the manipulation of the electric field profile, and liquid and patterned stationary collectors have proven to be successful in the fabrication of various fibrous assemblies. However, most of the work has focused on solution electrospinning. Here, since increased deposition of fibers will often result in an accumulation of charges, it remains to be seen whether these methods are able to obtain thick, patterned fibrous meshes. The conditions required to form 3D structures are still unclear: currently, the fibers forming reported 3D structures are very loosely packed and easily compressed, rendering them unsuitable for applications where structural integrity is required. Further advances in how scaffolds can be collected using melt electrospinning, coupled with improvements in yield, may provide methods to fabricate aligned fibrous substrates with increased porosity: useful for building complex structures that are of interest in TE applications. By far the most studied of the two types of electrospinning, solution electrospun

substrates are interesting in many applications for regenerative medicine. More importantly, the diversity of materials employed is excellent, and polymers from those both biologically derived and synthetically generated (and blends of both) can be readily used. The ability to generate electrospun scaffolds with a variety of morphologies is also important, with highly oriented, random, and patterned meshes already demonstrated. In this approach, surface segregation of polymeric additives is interesting, since it provides a simple route to both biofunctionalize a fiber and render it protein resistant. Undoubtedly, electrospinning will remain dominated by the use of solutions for the foreseeable future, and advances in controlling the morphology and utility of polymeric scaffold fabrication can advance electrospinning even further. At this point, it is pertinent to consider that electrospinning for TE was proposed less than a decade ago and that the next decade will provide further advances and breakthroughs in the process. We anticipate that electrospinning will remain an important aspect of regenerative medicine research and become clinically important as the field builds on the foundation of the last 10 years.

REFERENCES

1. Hutmacher, D.W.; Woodfield, T.; Dalton, P.D.; Lewis, J. Scaffold design and fabrication. In *Tissue Engineering*; Van Blitterswijk, C., et al., Eds.; Academic Press: New York, 2008; 403–450.
2. Hutmacher, D.W.; Ekaputra, A.K. Design and fabrication principles of electrospinning of scaffolds. In *Biomaterials Fabrication and Processing Handbook*; Chu, P.K., Liu, X., Eds.; Taylor & Francis Group: Boca Raton, FL, 2008; 115–139.

3. Dalton, P.D.; Woodfield, T.; Hutmacher, D.W. Snapshot: Polymer scaffolds for tissue engineering. Biomaterials **2009**, *30* (4), 701–702.

4. Li, W.J.; Laurencin, C.T.; Caterson, E.J.; Tuan, R.S.; Ko, F.K. Electrospun nanofibrous structure: A novel scaffold for tissue engineering. J. Biomed. Mater. Res. **2002**, *60* (4), 613–621.

5. Matthews, J.A.; Wnek, G.E.; Simpson, D.G.; Bowlin, G.L. Electrospinning of collagen nanofibers. Biomacromolecules **2002**, *3* (2), 232–238.

6. Casper, C.L.; Yamaguchi, N.; Kiick, K.L.; Rabolt, J.F. Functionalizing electrospun fibers with biologically relevant macromolecules. Biomacromolecules **2005**, *6* (4), 1998–2007.

7. Chew, S.Y.; Wen, J.; Yim, E.K.F.; Leong, K.W. Sustained release of proteins from electrospun biodegradable fibers. Biomacromolecules **2005**, *6* (4), 2017–2024.

8. Stankus, J.J.; Guan, J.J.; Fujimoto, K.; Wagner, W.R. Microintegrating smooth muscle cells into a biodegradable, elastomeric fiber matrix. Biomaterials **2006**, *27* (5), 735–744.

9. Ekaputra, A.K.; Prestwich, G.D.; Cool, S.M.; Hutmacher, D.W. Combining electrospun scaffolds with electrosprayed hydrogels leads to three-dimensional cellularization of hybrid constructs. Biomacromolecules **2008**, *9* (8), 2097–2103.

10. Simonet, M.; Schneider, O.D.; Neuenschwander, P.; Stark, W.J. Ultraporous 3D polymer meshes by low-temperature electrospinning: Use of ice crystals as a removable void template. Polym. Eng. Sci. **2007**, *47* (12), 2020–2026.

11. Schneider, O.D.; Loher, S.; Brunner, T.J.; Uebersax, L.; Simonet, M.; Grass, R.N.; Merkle, H.P.; Stark, W.J. Cotton wool-like nanocomposite biomaterials prepared by electrospinning: *In vitro* bioactivity and osteogenic differentiation of human mesenchymal stem cells. J. Biomed. Mater. Res. B Appl. Biomater. **2008**, *84B* (2), 350–362.

12. Gentsch, R.; Boysen, B.; Lankenau, A.; Borner, H.G. Single-step electrospinning of bimodal fiber meshes for ease of cellular infiltration. Macromol. Rapid Commun. **2010**, *31* (1), 59–64.

13. Kim, S.J.; Jang, D.H.; Park, W.H.; Min, B.M. Fabrication and characterization of 3-dimensional PLGA nanofiber/microfiber composite scaffolds. Polymer **2010**, *51* (6), 1320–1327.

14. Shin, Y.M.; Hohman, M.M.; Brenner, M.P.; Rutledge, G.C. Electrospinning: A whipping fluid jet generates submicron polymer fibers. Appl. Phys. Lett. **2001**, *78* (8), 1149.

15. Deng, R.J.; Liu, Y.; Ding, Y.; Xie, P.; Luo, L.; Yang, W. Melt electrospinning of low-density polyethylene having a low-melt flow index. J. Appl. Polym. Sci. **2009**, *114* (1), 166–175.

16. Jha, B.S.; Colello, R.J.; Bowman, J.R.; Sell, S.A.; Lee, K.D.; Bigbee, J.W.; Bowlin, G.L.; Chow, W.N.; Mathern, B.E.; Simpson, D.G. Two pole air gap electrospinning: Fabrication of highly aligned, three-dimensional scaffolds for nerve reconstruction. Acta Biomater. **2011**, *7* (1), 203–215.

17. Shimada, N.; Tsutsumi, H.; Nakane, K.; Ogihara, T.; Ogata, N. Poly(ethylene-co-vinyl alcohol) and nylon 6/12 nanofibers produced by melt electrospinning system equipped with a line-like laser beam melting device. J. Appl. Polym. Sci. **2010**, *116* (5), 2998–3004.

18. Zajicova, A.; Pokorna, K.; Lencova, A.; Krulova, M.; Svobodova, E.; Kubinova, S.; Sykova, E.; Pradny, M.; Michalek, J.; Svobodova, J.; Munzarova, M.; Holan, V. Treatment of ocular surface injuries by limbal and mesenchymal stem cells growing on nanofiber scaffolds. Cell Transplant **2011**, *19* (10), 1281–1290.

19. Sun, D.H.; Chang, C.; Li, S.; Lin, L. Near-field electrospinning. Nano Lett. **2006**, *6* (4), 839–842.

20. Malakhov, S.N.; Yu Khomenko, A.; Belousov, S.I.; Prazdnichnyi, A.M.; Chvalun, S.N.; Shepelev, A.D.; Budyka, A.K. Method of manufacturing nonwovens by electrospinning from polymer melts. Fibre Chem. **2009**, *41* (6), 355–359.

21. Reneker, D.H.; Yarin, A.L.; Fong, H.; Koombhongse, S. Bending instability of electrically charged liquid jets of polymer solutions in electrospinning. J. Appl. Phys. **2000**, *87* (9), 4531.

22. Hutmacher, D.W.; Dalton, P.D. Melt electrospinning. Chem. Asian J. **2011**, *6* (1), 44–56.

23. Dalton, P.D.; Grafahrend, D.; Klinkhammer, K.; Klee, D.; Moller, M. Electrospinning of polymer melts: Phenomenological observations. Polymer **2007**, *48* (23), 6823–6833.

24. Dalton, P.D.; Joergensen, N.T.; Groll, J.; Moeller, M. Patterned melt electrospun substrates for tissue engineering. Biomed. Mater. **2008**, *3* (3), 034109.

25. Zahedi, P.; Rezaeian, I.; Ranaei-Siadat, S.O.; Jafari, S.H.; Supaphol, P. A review on wound dressings with an emphasis on electrospun nanofibrous polymeric bandages. Polym. Adv. Technol. **2010**, *21* (2), 77–95.

26. Wang, Y.K.; Yong, T.; Ramakrishna, S. Nanofibres and their influence on cells for tissue regeneration. Aust. J. Chem. **2005**, *58* (10), 704–712.

27. Beglou, M.J.; Haghi, A.K. Electrospun biodegdadable and biocompatible natural nanofibers: A detailed review. Cell. Chem. Technol. **2008**, *42*, 441.

28. Schiffman, J.D.; Schauer, C.L. A review: Electrospinning of biopolymer nanofibers and their applications. Polym. Rev. **2008**, *48* (2), 317–352.v

29. Zeugolis, D.I.; Khew, S.T.; Yew, E.S.; Ekaputra, A.K.; Tong, Y.W.; Yung, L.Y.; Hutmacher, D.W.; Sheppard, C.; Raghunath, M. Electro-spinning of pure collagen nanofibres—Just an expensive way to make gelatin? Biomaterials **2008**, *29*(15), 2293–2305.

30. Min, B.M.; Jeong, L.; Nam, Y.S.; Kim, J.M.; Kim, J.Y.; Park, W.H. Formation of silk fibroin matrices with different texture and its cellular response to normal human keratinocytes. Int. J. Biol. Macromol. **2004**, *34* (5), 281–288.

31. Zhang, X.H.; Reagan, M.R.; Kaplan, D.L. Electrospun silk biomaterial scaffolds for regenerative medicine. Adv. Drug Deliv. Rev. **2009**, *61* (12), 988–1006.

32. Rho, K.S.; Jeong, L.; Lee, G.; Seo, B.M.; Park, Y.J.; Hong, S.D.; Roh, S.; Cho, J.J.; Park, W.H.; Min, B.M. Electrospinning of collagen nanofibers: Effects on the behavior of normal human keratinocytes and early-stage wound healing. Biomaterials **2006**, *27* (8), 1452–1461.

33. Sell, S.A.; McClure, M.J.; Garg, K.; Wolfe, P.S.; Bowlin, G.L. Electrospinning of collagen/biopolymers for

Dental—Electrospinning

regenerative medicine and cardiovascular tissue engineering. Adv. Drug Deliv. Rev. **2009**, *61* (12), 1007–1019.

34. Chen, H.C.; Jao, W.C.; Yang, M.C. Characterization of gelatin nanofibers electrospun using ethanol/ formic acid/ water as a solvent. Polym. Adv. Technol. **2009**, *20* (2), 98–103.

35. Powell, H.M.; Boyce, S.T. Fiber density of electrospun gelatin scaffolds regulates morphogenesis of dermal-epidermal skin substitutes. J. Biomed. Mater. Res. A **2008**, *84A* (4), 1078–1086.

36. Jayakumar, R.; Prabaharan, M.; Nair, S.V.; Tamura, H. Novel chitin and chitosan nanofibers in biomedical applications. Biotechnol. Adv. **2010**, *28* (1), 142–150.

37. McManus, M.C.; Boland, E.D.; Simpson, D.G.; Barnes, C.P.; Bowlin, G.L. Electrospun fibrinogen: Feasibility as a tissue engineering scaffold in a rat cell culture model. J. Biomed.Mater.Res. A **2007**, *81A* (12), 299–309.

38. Wnek, G.E.; Carr, M.E.; Simpson, D.G.; Bowlin, G.L. Electrospinning of nanofiber fibrinogen structures. Nano Lett. **2003**, *3* (2), 213–216.

39. Xu, S.S.; Li, J.; He, A.; Liu, W.; Jiang, X.; Zheng, J.; Han, C.C.; Hsiao, B.S.; Chu, B.; Fang, D. Chemical crosslinking and biophysical properties of electrospun hyaluronic acid based ultrathin fibrous membranes. Polymer **2009**, *50* (15), 3762–3769.

40. Ji, Y.; Ghosh, K.; Shu, X.Z.; Li, B.; Sokolov, J.C.; Prestwich, G.D.; Clark, R.A.; Rafailovich, M.H. Electrospun three-dimensional hyaluronic acid nanofibrous scaffolds. Biomaterials **2006**, *27* (20), 3782–3792.

41. Frey, M.W. Electrospinning cellulose and cellulose derivatives. Polym. Rev. **2008**, *48* (2), 378–391.

42. Neal, R.A.; McClugage, S.G.; Link, M.C.; Sefcik, L.S.; Ogle, R.C.; Botchwey, E.A. Laminin nanofiber meshes that mimic morphological properties and bioactivity of basement membranes. Tissue Eng. Part C Methods **2009**, *15* (1), 11–21.

43. Koh, H.S.; Yong, T.; Chan, C.K.; Ramakrishna, S. Enhancement of neurite outgrowth using nano-structured scaffolds coupled with laminin. Biomaterials **2008**, *29* (26), 3574–3582.

44. Nivison-Smith, L.; Rnjak, J.; Weiss, A.S. Synthetic human elastin microfibers: Stable cross-linked tropoelastin and cell interactive constructs for tissue engineering applications. Acta Biomater. **2010**, *6* (2), 354–359.

45. Ner, Y.; Stuart, J.A.; Whited, G.; Sotzing, G.A. Electrospinning nanoribbons of a bioengineered silk-elastin-like protein (SELP) from water. Polymer **2009**, *50* (24), 5828–5836.

46. Teng, S.H.; Wang, P.; Kim, H.E. Blend fibers of chitosan-agarose by electrospinning. Mater. Lett. **2009**, *63* (28), 2510–2512.

47. Li, J.X.; He, A.; Han, C.C.; Fang, D.; Hsiao, B.S.; Chu, B. Electrospinning of hyaluronic acid (HA) and HA/gelatin blends. Macromol. Rapid Commun. **2006**, *27* (2), 114–120.

48. Puppi, D.; Piras, A.M.; Detta, N.; Dinucci, D.; Chiellini, F. Poly(lactic-co-glycolic acid) electrospun fibrous meshes for the controlled release of retinoic acid. Acta Biomater. **2010**, *6* (4), 1258–1268.

49. Zhao, L.; He, C.; Gao, Y.; Cen, L.; Cui, L.; Cao, Y. Preparation and cytocompatibility of PLGA scaffolds with controllable fiber morphology and diameter using

50. Kim, G.H. Electrospun PCL nanofibers with anisotropic mechanical properties as a biomedical scaffold. Biomed. Mater. **2008**, *3* (2), 025010.

51. Uyar, T.; Besenbacher, F. Electrospinning of uniform polystyrene fibers: The effect of solvent conductivity. Polymer **2008**, *49* (24), 5336–5343.

52. Kim, G.T.; Hwang, Y.J.; Ahn, Y.C.; Shin, H.S.; Lee, J.K.; Sung, C.M. The morphology of electrospun polystyrene fibers.Korean J. Chem. Eng. **2005**, *22* (1), 147–153.

53. Khil, M.S.; Cha, D.I.; Kim, H.Y.; Kim, I.S.; Bhattarai, N. Electrospun nanofibrous polyurethane membrane as wound dressing. J. Biomed. Mater. Res. B Appl. Biomater. **2003**, *67B* (2), 675–679.

54. Pedicini, A.; Farris, R.J. Mechanical behavior of electrospun polyurethane. Polymer **2003**, *44* (22), 6857–6862.

55. Riboldi, S.A.; Sampaolesi, M.; Neuenschwander, P.; Cossu, G.; Mantero, S. Electrospun degradable polyester-urethane membranes: Potential scaffolds for skeletal muscle tissue engineering. Biomaterials **2005**, *26* (22), 4606–4615.

56. Okuzaki, H.; Kobayashi, K.; Yan, H. Non-woven fabric of poly(N-isopropylacrylamide) nanofibers fabricated by electrospinning. Synthetic Metals **2009**, *159* (21–22), 2273–2276.

57. Rockwood, D.N.; Chase, D.B.; Akins, R.E.; Rabolt, J.F. Characterization of electrospun poly(N-isopropyl acrylamide) fibers. Polymer **2008**, *49* (18), 4025–4032.

58. Corey, J.M.; Gertz, C.C.; Wang, B.S.; Birrell, L.K.; Johnson, S.L.; Martin, D.C.; Feldman, E.L. The design of electrospun PLLA nanofiber scaffolds compatible with serum-free growth of primary motor and sensory neurons. Acta Biomater. **2008**, *4* (4), 863–875.

59. Inai, R.; Kotaki, M.; Ramakrishna, S. Structure and properties of electrospun PLLA single nanofibres. Nanotechnology **2005**, *16* (2), 208–213.

60. Liao, G.Y.; Jiang, K.F.; Jiang, S.B.; Xia, H. Synthesis and characterization of biodegradable poly(epsilon-caprolactone)-β-poly(L-lactide) and study on their electrospun scaffolds. J. Macromol.Sci. APure Appl. Chem. **2010**, *47* (11), 1116–1122.

61. Zeng, J.; Chen, X.S.; Liang, Q.Z.; Xu, X.L.; Jing, X.B. Enzymatic degradation of poly(L-lactide) and poly (epsilon-caprolactone) electrospun fibers. Macromol.Biosci. **2004**, *4* (12), 1118–1125.

62. Grafahrend, D.; Calvet, J.L.; Klinkhammer, K.; Salber, J.; Dalton, P.D.; Möller, M.; Klee, D. Control of protein adsorption on functionalized electrospun fibers. Biotechnol.Bioeng. **2008**, *101* (3), 609–621.

63. Grafahrend, D.; Lleixa Calvet, J.; Salber, J.; Dalton, P.D.; Moeller, M.; Klee, D. Biofunctionalized poly(ethylene glycol)-block-poly(epsilon-caprolactone) nanofibers for tissue engineering. J. Mater. Sci. Mater. Med. **2008**, *19* (4), 1479–1484.

64. Arayanarakul, K.; Choktaweesap, N.; Ahtong, D.; Meechaisue, C.; Supaphol, P. Effects of poly(ethylene glycol), inorganic salt, sodium dodecyl sulfate, and solvent system on electrospinning of poly(ethylene oxide). Macromol. Mater. Eng. **2006**, *291* (6), 581–591.

electrospinning method. J. Biomed. Mater. Res. B Appl. Biomater. **2008**, *87B* (1), 26–34.

65. Deitzel, J.M.; Kleinmeyer, J.D.; Hirvonen, J.K.; Tan, N.C.B. Controlled deposition of electrospun poly(ethylene oxide) fibers. Polymer **2001**, *42* (19), 8163–8170.

66. Grafahrend, D.; Heffels, K.H.; Beer, M.V.; Gasteier, P.; Möller, M.; Boehm, G.; Dalton, P.D.; Groll, J. Degradable polyester scaffolds with controlled surface chemistry combining minimal protein adsorption with specific bioactivation. Nat. Mater. **2011**, *10* (1), 67–73.

67. Schnell, E.; Klinkhammer, K.; Balzer, S.; Brook, G.; Klee, D.; Dalton, P.; Mey, J. Guidance of glial cell migration and axonal growth on electrospun nanofibers of poly-epsilon-caprolactone and a collagen/poly-epsilon-caprolactone blend. Biomaterials **2007**, *28* (19), 3012–3025.

68. Meng, Z.X.; Wang, Y.S.; Ma, C.; Zheng, W.; Li, L.; Zheng, Y.F. Electrospinning of PLGA/gelatin randomly-oriented and aligned nanofibers as potential scaffold in tissue engineering. Mater.Sci. Eng. CMater. Biol. Appl. **2010**, *30* (8), 1204–1210.

69. Duan, B.; Yuan, X.; Zhu, Y.; Zhang, Y.; Li, X.; Zhang, Y.; Yao, K. A nanofibrous composite membrane of PLGA-chitosan/PVA prepared by electrospinning. Eur. Polym. J. **2006**, *42* (9), 2013–2022.

70. Jin, H.J.; Fridrikh, S.V.; Rutledge, G.C.; Kaplan, D.L. Electrospinning Bombyx mori silk with poly(ethylene oxide). Biomacromolecules **2002**, *3* (6), 1233–1239.

71. Deng, X.L.; Sui, G.; Zhao, M.L.; Chen, G.Q.; Yang, X.P. Poly(L-lactic acid)/hydroxyapatite hybrid nanofibrous scaffolds prepared by electrospinning. J. Biomater. Sci. Polym. Ed. **2007**, *18* (1), 117–130.

72. Asran, A.S.; Henning, S.; Michler, G.H. Polyvinyl alcohol-collagen-hydroxyapatite biocomposite nanofibrous scaffold: Mimicking the key features of natural bone at the nanoscale level. Polymer **2010**, *51* (4), 868–876.

73. McKee, M.G.; Wilkes, G.L.; Colby, R.H.; Long, T.E. Correlations of solution rheology with electrospun fiber formation of linear and branched polyesters. Macromolecules **2004**, *37* (5), 1760–1767.

74. Nie, H.R.; He, A.; Zheng, J.; Xu, S.; Li, J.; Han, C.C. Effects of chain conformation and entanglement on the electrospinning of pure alginate. Biomacromolecules **2008**, *9* (5), 1362–1365.

75. Liu, Y.; Ma, G.; Fang, D.; Xu, J.; Zhang, H.; Nie, J. Effects of solution properties and electric field on the electrospinning of hyaluronic acid. Carbohydr. Polym. **2011**, *83* (2), 1011–1015.

76. Lee, K.H.; Kim, H.Y.; Khil, M.S.; Ra, Y.M.; Lee, D.R. Characterization of nano-structured poly(epsilon-caprolactone) nonwoven mats via electrospinning. Polymer **2003**, *44* (4), 1287–1294.

77. Lee, K.H.; Kim, H.Y.; Ryu, Y.J.; Kim, K.W.; Choi, S.W. Mechanical behavior of electrospun fiber mats of poly(vinyl chloride)/polyurethane polyblends. J. Polym. Sci. BPolym. Phys. **2003**, *41* (11), 1256–1262.

78. Fong, H.; Liu, W.D.; Wang, C.S.; Vaia, R.A. Generation of electrospun fibers of nylon 6 and nylon 6-montmorillonite nanocomposite. Polymer **2002**, *43* (3), 775–780.

79. Zuo, W.W.; Zhu, M.; Yang, W.; Yu, H.; Chen, Y.; Zhang, Y. Experimental study on relationship between jet instability and formation of beaded fibers during electrospinning. Polym. Eng. Sci. **2005**, *45* (5), 704–709.

80. Mituppatham, C.; Nithitanakul, M.; Supaphol, P. Ultratine electrospun polyamide-6 fibers: Effect of solution conditions on morphology and average fiber diameter. Macromol. Chem. Phys. **2004**, *205* (17), 2327–2338.

81. Zhang, C.X.; Yuan, X.Y.; Wu, L.L.; Han, Y.; Sheng, J. Study on morphology of electrospun poly(vinyl alcohol) mats. Eur. Polym. J. **2005**, *41* (3), 423–432.

82. Patlolla, A.; Collins, G.; Arinzeh, T.L. Solvent-dependent properties of electrospun fibrous composites for bone tissue regeneration. Acta Biomater. **2010**, *6* (1), 90–101.

83. Zong, X.H.; Kim, K.; Fang, D.; Ran, S.; Hsiao, B.S.; Chu, B. Structure and process relationship of electrospun bioabsorbable nanofiber membranes. Polymer **2002**, *43* (16), 4403–4412.

84. You, Y.; Lee, S.J.; Min, B.M.; Park, W.H. Effect of solution properties on nanofibrous structure of electrospun poly(lactic-co-glycolic acid). J. Appl. Polym. Sci. **2006**, *99* (3), 1214–1221.

85. Jun, Z.; Hou, H.Q.; Schaper, A.; Wendorff, J.H.; Greiner, A. Poly-L-lactide nanofibers by electrospinning - Influence of solution viscosity and electrical conductivity on fiber diameter and fiber morphology. E-Polymers Paper 9, 2003.

86. Megelski, S.; Stephens, J.S.; Chase, D.B.; Rabolt, J.F. Micro- and nanostructured surface morphology on electrospun polymer fibers. Macromolecules **2002**, *35* (22), 8456–8466.

87. Zhang, Y.Z.; Wang, X.; Feng, Y.; Li, J.; Lim, C.T.; Ramakrishna S. Coaxial electrospinning of (fluorescein isothiocyanate-conjugated bovine serum albumin)-encapsulated poly(epsilon-caprolactone) nanofibers for sustained release. Biomacromolecules **2006**, *7* (4), 1049–1057.

88. Chen, Z.G.; Wei, B.; Mo, X.M.; Cui, F.Z. Diameter control of electrospun chitosan-collagen fibers. J. Polym. Sci. BPolym. Phys. **2009**, *47* (19), 1949–1955.

89. Zhang, Y.Z.; Huang, Z.M.; Xu, X.J.; Lim, C.T.; Ramakrishna, S. Preparation of core-shell structured PCL-r-gelatin Bi-component nanofibers by coaxial electrospinning. Chem. Mater. **2004**, *16* (18), 3406–3409.

90. Kessick, R.; Fenn, J.; Tepper, G. The use of AC potentials in electrospraying and electrospinning processes. Polymer **2004**, *45* (9), 2981–2984.

91. Deitzel, J.M.; Kleinmeyer, J.; Harris, D.; Tan, N.C.B. The effect of processing variables on the morphology of electrospun nanofibers and textiles. Polymer **2001**, *42* (1), 261–272.

92. Demir, M.M.; Yilgor, I.; Yilgor, E.; Erman, B. Electrospinning of polyurethane fibers. Polymer **2002**, *43* (11), 3303–3309.

93. Ayutsede, J.; Gandhi, M.; Sukigara, S.; Micklus, M.; Chen, H.-E.; Ko, F. Regeneration of Bombyx mori silk by electrospinning. Part 3: Characterization of electrospun nonwoven mat. Polymer **2005**, *46* (5), 1625–1634.

94. Mo, X.M.; Xu, C.Y.; Kotaki, M.; Ramakrishna, S. Electrospun P(LLA-CL) nanofiber: A biomimetic extracellular matrix for smooth muscle cell and endothelial cell proliferation. Biomaterials **2004**, *25* (10), 1883–1890.

95. Katti, D.S.; Robinson, K.W.; Ko, F.K.; Laurencin, C.T. Bioresorbable nanofiber-based systems for wound healing and drug delivery: Optimization of fabrication parameters.

J. Biomed. Mater. Res. B Appl. Biomater. **2004**, *70B* (2), 286–296.

96. Zhou, F.-L.; Gong, R.-H.; Porat, I. Three-jet electrospinning using a flat spinneret. J. Mater. Sci. **2009**, *44* (20), 5501–5508.

97. Zhang, C.; Yuan, X.; Wu, L.; Han, Y.; Sheng, J. Study on morphology of electrospun poly(vinyl alcohol) mats. Eur. Polym. J. **2005**, *41* (3), 423–432.

98. Bölgen, N.; Menceloğlu, Y.Z.; Acatay, K.; Vargel, I.; Pişkin, E. *In vitro* and *in vivo* degradation of non-woven materials made of poly(epsilon-caprolactone) nanofibers prepared by electrospinning under different conditions. J. Biomater. Sci. Polym. Ed. **2005**, *16* (12), 1537–1555.

99. Lee, K.H.; Kim, H.Y.; Bang, H.J.; Jung, Y.H.; Lee, S.G. The change of bead morphology formed on electrospun polystyrene fibers. Polymer **2003**, *44* (14), 4029–4034.

100. Dalton, P.D.; Klinkhammer, K.; Salber, J.; Klee, D.; Moller, M. Direct *in vitro* electrospinning with polymer melts. Biomacromolecules **2006**, *7* (3), 686–690.

101. Detta, N.; Brown, T.D.; Edin, F.K.; Albrecht, K.; Chiellini, F.; Chiellini, E.; Dalton, P.D.; Hutmacher, D.W. Melt electrospinning of polycaprolactone and its blends with poly(ethylene glycol). Polym. Int. **2010**, *59* (11), 1558–1562.

102. Wang, C.; Chien, H.-S.; Hsu, C.-H.; Wang, Y.-C.; Wang, C.-T.; Lu, H.-A. Electrospinning of polyacrylonitrile solutions at elevated temperatures. Macromolecules **2007**, *40* (22), 7973–7983.

103. Givens, S.R.; Gardner, K.H.; Rabolt, J.F.; Chase, D.B. High-temperature electrospinning of polyethylene microfibers from solution. Macromolecules **2007**, *40* (3), 608–610.

104. Yu, D.G.; Gao, L.D.; White, K.; Branford-White, C.; Lu, W.Y.; Zhu, L.M. Multicomponent amorphous nanofibers electrospun from hot aqueous solutions of a poorly soluble drug. Pharm. Res. **2010**, *27* (11), 2466–2477.

105. Wang, C.; Hsu, C.H.; Lin, J.H. Scaling laws in electrospinning of polystyrene solutions. Macromolecules **2006**, *39* (22), 7662–7672.

106. Chen, J.P.; Ho, K.H.; Chiang, Y.P.; Wu, K.W. Fabrication of electrospun poly(methyl methacrylate) nanofibrous membranes by statistical approach for application in enzyme immobilization. J. Membr. Sci. **2009**, *340* (1–2), 9–15.

107. Thompson, C.J.; Chase, G.G.; Yarin, A.L.; Reneker, D.H. Effects of parameters on nanofiber diameter determined from electrospinning model. Polymer **2007**, *48* (23), 6913–6922.

108. Jarusuwannapoom, T.; Hongrojjanawiwat, W.; Jitjaicham, S.; Wannatong, L.; Nithitanakul, M.; Pattamaprom, C.; Koombhongse, P.; Rangkupan, R.; Supaphol, P. Effect of solvents on electro-spinnability of polystyrene solutions and morphological appearance of resulting electrospun polystyrene fibers. Eur. Polym. J. **2005**, *41* (3), 409–421.

109. Casper, C.L.; Stephens, J.S.; Tassi, N.G.; Chase, D.B.; Rabolt, J.F. Controlling surface morphology of electrospun polystyrene fibers: Effect of humidity and molecular weight in the electrospinning process. Macromolecules **2004**, *37* (2), 573–578.

110. Park, J.-Y.; Lee, I.-H. Relative humidity effect on the preparation of porous electrospun polystyrene fibers. J. Nanosci. Nanotechnol. **2010**, *10* (5), 3473–3477.

111. De Vrieze, S.; Van Camp, T.; Nelvig, A.; Hagström, B.; Westbroek, P.; De Clerck, K. The effect of temperature and humidity on electrospinning. J. Mater. Sci. **2009**, *44* (5), 1357–1362.

112. Srinivasan, G.; Reneker, D.H. Structure and morphology of small-diameter electrospun aramid fibers. Polym. Int. **1995**, *36* (2), 195–201.

113. Tzezana, R.; Zussman, E.; Levenberg, S. A layered ultra-porous scaffold for tissue engineering, created via a hydrospinning method. Tissue Eng. PartCMethods **2008**, *14* (4), 281–288.

114. Smit, E.; Buttner, U.; Sanderson, R.D. Continuous yarns from electrospun fibers. Polymer **2005**, *46* (8), 2419–2423.

115. Khil, M.S.; Bhattarai, S.R.; Kim, H.Y.; Kim, S.Z.; Lee, K.H. Novel fabricated matrix via electrospinning for tissue engineering. J. Biomed. Mater. Res. B Appl. Biomater. **2005**, *72B* (1), 117–124.

116. Formhals, A.U. S. P. Office, Ed., 1934.

117. Formhals, A.U. S. P. Office, Ed., 1944.

118. Teo, W.E.; Ramakrishna, S. A review on electrospinning design and nanofibre assemblies. Nanotechnology **2006**, *17* (14), R89.

119. Dalton, P.D.; Klee, D.; Moller, M. Electrospinning with dual collection rings. Polymer **2005**, *46* (3), 611–614.

120. Teo, W.E.; Ramakrishna, S. Electrospun fibre bundle made of aligned nanofibres over two fixed points. Nanotechnology **2005**, *16* (9), 1878.

121. Klinkhammer, K.; Seiler, N.; Grafahrend, D.; Gerardo-Nava, J.; Mey, J.; Brook, G.A.; Möller, M.; Dalton, P.D.; Klee, D. Deposition of electrospun fibers on reactive substrates for *in vitro* investigations. Tissue Eng. Part CMethods **2009**, *15* (1), 77–85.

122. Klinkhammer, K.; Bockelmann, J.; Simitzis, C.; Brook, G.A.; Grafahrend, D.; Groll, J.; Möller, M.; Mey, J.; Klee, D. Functionalization of electrospun fibers of poly(epsilon-caprolactone) with star shaped NCO-poly(ethylene glycol)-stat-poly(propylene glycol) for neuronal cell guidance. J. Mater. Sci. Mater. Med. **2010**, *21* (9), 2637–2651.

123. Gerardo-Nava, J.; Führmann, T.; Klinkhammer, K.; Seiler, N.; Mey, J.; Klee, D.; Möller, M.; Dalton, P.D.; Brook, G.A. Human neural cell interactions with orientated electrospun nanofibers *in vitro*. Nanomedicine (Lond.) **2009**, *4* (1), 11–30.

124. Deitzel, J.M.; Kleinmeyer, J.; Harris, D.; Tan, N.C.B. The effect of processing variables on the morphology of electrospun nanofibers and textiles. Polymer **2001**, *42* (1), 261–272.

125. Ramakrishna, S.; Fujihara, K.; Teo, W.E.; Lim, T.C.; Ma, Z. *An Introduction to Electrospinning and Nanofibers*; World Scientific Publishing: Singapore, 2005.

126. Zhang, D.M.; Chang, J. Electrospinning of three-dimensional nanofibrous tubes with controllable architectures. Nano Lett. **2008**, *8* (10), 3283–3287.

127. Zhang, D.M.; Chang, J. Patterning of electrospun fibers using electroconductive templates. Adv. Mater. **2007**, *19* (21), 3664–3667.

128. Kameoka, J.; Orth, R.; Yang, Y.; Czaplewski, D.; Mathers, R.; Coates, G.W.; Craighead, H.G. A scanning tip electrospinning source for deposition of oriented nanofibres. Nanotechnology 2003, 14 (10), 1124.

129. Subramanian, A.; Vu, D.; Larsen, G.F.; Lin, H.Y. Preparation and evaluation of the electrospun chitosan/PEO fibers for potential applications in cartilage tissue engineering. J. Biomater. Sci. Polym. Ed. 2005, 16 (7), 861–873.

130. Kowalewski, T.A.; Barral, S.; Kowalczyk, T. Modeling electrospinning of nanofibers. Iutam Symp. Model. Nanomater.Nanosyst. 2009, 13, 279.

131. Sundaray, B.; Subramanian, V.; Natarajan, T.S.; Xiang, R.-Z.; Chang, C.-C.; Fann, W.-S. Electrospinning of continuous aligned polymer fibers. Appl. Phys. Lett. 2004, 84 (7), 1222.

132. Zussman, E.; Theron, A.; Yarin, A.L. Formation of nanofiber crossbars in electrospinning. Appl. Phys. Lett. 2003, 82 (6), 973–975.

133. Stitzel, J.; Liu, J.; Lee, S.J.; Komura, M.; Berry, J.; Soker, S.; Lim, G.; Van Dyke, M.; Czerw, R.; Yoo, J.J.; Atala, A. Controlled fabrication of a biological vascular substitute. Biomaterials 2006, 27 (7), 1088–1094.

134. Kidoaki, S.; Kwon, I.K.; Matsuda, T. Mesoscopic spatial designs of nano- and microfiber meshes for tissue-engineering matrix and scaffold based on newly devised multilayering and mixing electrospinning techniques. Biomaterials 2005, 26 (1), 37–46.

135. Theron, A.; Zussman, E.; Yarin, A.L. Electrostatic field-assisted alignment of electrospun nanofibres. Nanotechnology 2001, 12 (3), 384.

136. Xu, C.Y.; Inai, R.; Kotaki, M.; Ramakrishna, S. Aligned biodegradable nanofibrous structure: A potential scaffold for blood vessel engineering. Biomaterials 2004, 25 (5), 877–886.

137. Bhattarai, N.; Edmondson, D.; Veiseh, O.; Matsen, F.A.; Zhang, M.Q. Electrospun chitosan-based nanofibers and their cellular compatibility. Biomaterials 2005, 26 (31), 6176.

138. Katta, P.; Alessandro, M.; Ramsier, R.D.; Chase, G.G. Continuous electrospinning of aligned polymer nanofibers onto a wire drum collector. Nano Lett. 2004, 4 (11), 2215–2218.

139. Chang, C.; Limkrailassiri, K.; Lin, L.W. Continuous near-field electrospinning for large area deposition of orderly nanofiber patterns. Appl. Phys. Lett. 2008, 93 (12), 123111.

140. Han, T.; Reneker, D.H.; Yarin, A.L. Buckling of jets in electrospinning. Polymer 2007, 48 (20), 6064–6076.

141. Zhou, H.J.; Green, T.B.; Joo, Y.L. The thermal effects on electrospinning of polylactic acid melts. Polymer 2006, 47 (21), 7497–7505.

142. Hellmann, C.; Belardi, J.; Dersch, R.; Greiner, A.; Wendorff, J.H.; Bahnmueller, S. High precision deposition electrospinning of nanofibers and nanofiber nonwovens. Polymer 2009, 50 (5), 1197–1205.

143. Brown, T.D.; Dalton, P.D.; Hutmacher, D.W. Direct writing by way of melt electrospinning. Adv. Mater. 2011, 23 (47), 5651–5657.

144. Chung, S.; Moghe, A.K.; Montero, G.A.; Kim, S.H.; King, M.W. Nanofibrous scaffolds electrospun from elastomeric biodegradable poly(L-lactide-co-epsilon-caprolactone) copolymer. Biomedical Materials 2009, 4 (1), 015019.

145. Park, S.H.; Kim, T.G.; Kim, H.C.; Yang, D.Y.; Park, T.G. Development of dual scale scaffolds via direct polymer melt deposition and electrospinning for applications in tissue regeneration. Acta Biomater. 2008, 4 (5), 1198–1207.

146. Centola, M.; Rainer, A.; Spadaccio, C.; De Porcellinis, S.; Genovese, J.A.; Trombetta, M. Combining electrospinning and fused deposition modeling for the fabrication of a hybrid vascular graft. Biofabrication 2010, 2 (1), 014102.

147. Pham, Q.P.; Sharma, U.; Mikos, A.G. Electrospun poly(epsilon-caprolactone) microfiber and multilayer nanofiber/microfiber scaffolds: Characterization of scaffolds and measurement of cellular infiltration. Biomacromolecules 2006, 7 (10), 2796–2805.

148. Ashammakhi, N.; Ndreu, A.; Nikkola, L.; Wimpenny, I.; Yang, Y. Advancing tissue engineering by using electrospun nanofibers. Regen. Med. 2008, 3 (4), 547–574.

149. Goldberg, M.; Langer, R.; Jia, X.Q. Nanostructured materials for applications in drug delivery and tissue engineering. J. Biomater. Sci. Polym. Ed. 2007, 18 (3), 241–268.

150. Li, C.M.; Vepari, C.; Jin, H.J.; Kim, H.J.; Kaplan, D.L. Electrospun silk-BMP-2 scaffolds for bone tissue engineering. Biomaterials 2006, 27 (16), 3115–3124.

151. Fu, Y.C.; Nie, H.; Ho, M.L.; Wang, C.K.; Wang, C.H. Optimized bone regeneration based on sustained release from three-dimensional fibrous PLGA/HAp composite scaffolds loaded with BMP-2. Biotechnol. Bioeng. 2008, 99 (4), 996–1006.

152. Nie, H.; Soh, B.W.; Fu, Y.C.; Wang, C.H. Three-dimensional fibrous PLGA/HAp composite scaffold for BMP-2 delivery. Biotechnol. Bioeng. 2008, 99 (1), 223–234.

153. Nie, H.M.; Wang, C.H. Fabrication and characterization of PLGA/HAp scaffolds for delivery of BMP-2 plasmid composite DNA. J. Control. Release 2007, 120 (1–2), 111–121.

154. Horne, M.K.; Nisbet, D.R.; Forsythe, J.S.; Parish, C.L. Three-dimensional nanofibrous scaffolds incorporating immobilized BDNF promote proliferation and differentiation of cortical neural stem cells. Stem Cells Dev. 2010, 19 (6), 843–852.

155. Lam, H.J.; Patel, S.; Wang, A.J.; Chu, J.; Li, S. In vitro regulation of neural differentiation and axon growth by growth factors and bioactive nanofibers.Tissue Eng. Part A 2010, 16 (8), 2641–2648.

156. Sahoo, S.; Ang, L.T.; Goh, J.C.H.; Toh, S.L. Growth factor delivery through electrospun nanofibers in scaffolds for tissue engineering applications. J. Biomed. Mater. Res. A 2010, 93A (4), 1539–1550.

157. Casper, C.L.; Yang, W.D.; Farach-Carson, M.C.; Rabolt, J.F. Coating electrospun collagen and gelatin fibers with perlecan domain I for increased growth factor binding. Biomacromolecules 2007, 8 (4), 1116–1123.

158. Patel, S.; Kurpinski, K.; Quigley, R.; Gao, H.; Hsiao, B.S.; Poo, M.M.; Li, S. Bioactive nanofibers: Synergistic effects of nanotopography and chemical signaling on cell guidance. Nano Lett. 2007, 7 (7), 2122–2128.

159. Chew, S.Y.; Mi, R.F.; Hoke, A.; Leong, K.W. Aligned protein-polymer composite fibers enhance nerve regeneration: A potential tissue-engineering platform. Adv. Funct. Mater. 2007, 17 (8), 1288–1296.

160. Wang, F.; Li, Z.Q.; Tamama, K.; Sen, C.K.; Guan, J.J. Fabrication and characterization of prosurvival growth

Dental—Electrospinning

factor releasing, anisotropic scaffolds for enhanced mesenchymal stem cell survival/growth and orientation. Biomacromolecules **2009**, *10* (9), 2609–2618.

161. Li, X.Q.; Su, Y.; Liu, S.; Tan, L.; Mo, X.; Ramakrishna, S. Encapsulation of proteins in poly(L-lactide-co-caprolactone) fibers by emulsion electrospinning. Colloids Surf. B Biointerfaces **2010**, *75* (2), 418–424.

162. Li, H.; Zhao, C.; Wang, Z.; Zhang, H.; Yuan, X.; Kong, D. Controlled release of PDGF-bb by coaxial electrospun dextran/poly(L-lactide-co-epsilon-caprolactone) fibers with an ultrafine core/shell structure. J. Biomater. Sci. Polym. Ed. **2010**, *21* (6), 803–819.

163. Lu, Y.; Jiang, H.L.; Tu, K.H.; Wang, L.Q. Mild immobilization of diverse macromolecular bioactive agents onto multifunctional fibrous membranes prepared by coaxial electrospinning. Acta Biomater. **2009**, *5* (5), 1562–1574.

164. Bolgen, N.; Vargel, I.; Korkusuz, P.; Menceloglu, Y.Z.; Piskin, E. *In vivo* performance of antibiotic embedded electrospun PCL membranes for prevention of abdominal adhesions. J. Biomed. Mater. Res. B Appl. Biomater. **2007**, *81B* (2), 530–543.

165. Grafahrend, D.; Heffels, K.-H.; Beer, M.; Gasteier, P.; Moller, M.; Boehm, G.; Dalton, P.D.; Groll, J. Degradable polyester scaffolds with controlled surface chemistry combining minimal protein adsorption with specific bioactivation. Nat. Mater. **2010**, *10* (1), 67–73.

166. Yoo, H.S.; Kim, T.G.; Park, T.G. Surface-functionalized electrospun nanofibers for tissue engineering and drug delivery. Adv. Drug Deliv. Rev. **2009**, *61* (12), 1033–1042.

167. Yoshida, M.; Langer, R.; Lendlein, A.; Lahann, J. From advanced biomedical coatings to multi-functionalized biomaterials. Polym. Rev. **2006**, *46*, 347–375.

168. Ma, Z.W.; Masaya, K.; Ramakrishna, S. Immobilization of Cibacron blue F3GA on electrospun polysulphone ultra-fine fiber surfaces towards developing an affinity membrane for albumin adsorption. J. Membr. Sci. **2006**, *282* (1–2), 237–244.

169. Kim, H.S.; Yoo, H.S. MMPs-responsive release of DNA from electrospun nanofibrous matrix for local gene therapy: *In vitro* and *in vivo* evaluation. J. Control. Release **2010**, *145* (3), 264–271.

170. Losel, R.; Grafahrend, D.; Moller, M.; Klee, D. Bioresorbable electrospun fibers for immobilization of thiol-containing compounds. Macromol. Biosci. **2010**, *10* (10), 1177–1183.

171. Chiu, J.B.; Luu, Y.K.; Fang, D.; Hsiao, B.S.; Chu, B.; Hadjiargyrou, M. Electrospun nanofibrous scaffolds for biomedical applications. J. Biomed. Nanotechnol. **2005**, *1* (2), 115–132.

172. Kumbar, S.G.; Nukavarapu, S.P.; James, R.; Nair, L.S.; Laurencin, C.T. Electrospun poly(lactic acid-co-glycolic acid) scaffolds for skin tissue engineering. Biomaterials **2008**, *29* (30), 4100–4107.

173. Wutticharoenmongkol, P.; Sanchavanakit, N.; Pavasant, P.; Supaphol, P. Preparation and characterization of novel bone scaffolds based on electrospun polycaprolactone fibers filled with nanoparticles. Macromol. Biosci. **2006**, *6* (1), 70–77.

174. Neamnark, A.; Sanchavanakit, N.; Pavasant, P.; Rujiravanit, R.; Supaphol, P. *In vitro* biocompatibility of electrospun hexanoyl chitosan fibrous scaffolds towards human keratinocytes and fibroblasts. Eur. Polym. J. **2008**, *44* (7), 2060–2067.

175. Bashur, C.A.; Dahlgren, L.A.; Goldstein, A.S. Effect of fiber diameter and orientation on fibroblast morphology and proliferation on electrospun poly(D,L-lactic-co-glycolic acid) meshes. Biomaterials **2006**, *27* (33), 5681–5688.

176. Park, K.; Ju, Y.M.; Son, J.S.; Ahn, K.D.; Han, D.K. Surface modification of biodegradable electrospun nanofiber scaffolds and their interaction with fibroblasts. J. Biomater. Sci. Polym. Ed. **2007**, *18* (4), 369–382.

177. Thakur, R.A.; Florek, C.A.; Kohn, J.; Michniak, B.B. Electrospun nanofibrous polymeric scaffold with targeted drug release profiles for potential application as wound dressing. Int. J. Pharm. **2008**, *364* (1), 87–93.

178. Saraf, A.; Baggett, L.S.; Raphael, R.M.; Kasper, F.K.; Mikos, A.G. Regulated non-viral gene delivery from coaxial electrospun fiber mesh scaffolds. J. Control. Release **2010**, *143* (1), 95–103.

179. Zhong, S.P.; Teo, W.E.; Zhu, X.; Beuerman, R.W.; Ramakrishna, S.; Yung, L.Y. An aligned nanofibrous collagen scaffold by electrospinning and its effects on *in vitro* fibroblast culture. J. Biomed. Mater. Res. A **2006**, *79A* (3), 456–463.

180. Lee, C.H.; Shin, H.J.; Cho, I.H.; Kang, Y.M.; Kim, I.A.; Park, K.D.; Shin, J.W. Nanofiber alignment and direction of mechanical strain affect the ECM production of human ACL fibroblast. Biomaterials **2005**, *26* (11), 1261–1270.

181. Schindler, M.; Ahmed, I.; Kamal, J.; Nur-E-Kamal, A.; Grafe, T.H.; Young Chung, H.; Meiners, S. A synthetic nanofibrillar matrix promotes *in vivo*-like organization and morphogenesis for cells in culture. Biomaterials **2005**, *26* (28), 5624–5631.

182. Venugopal, J.; Ramakrishna, S. Biocompatible nanofiber matrices for the engineering of a dermal substitute for skin regeneration. Tissue Eng. **2005**, *11* (5–6), 847–854.

183. Cai, Z.X.; Mo, X.M.; Zhang, K.H.; Fan, L.P.; Yin, A.L.; He, C.L.; Wang, H.S. Fabrication of chitosan/silk fibroin composite nanofibers for wound-dressing applications. Int. J. Mol. Sci. **2010**, *11* (9), 3529–3539.

184. Chun, J.Y.; Kang, H.K.; Jeong, L.; Kang, Y.O.; Oh, J.E.; Yeo, I.S.; Jung, S.Y.; Park, W.H.; Min, B.M. Epidermal cellular response to poly(vinyl alcohol) nanofibers containing silver nanoparticles. Colloids Surf. B Biointerfaces **2010**, *78* (2), 334–342.

185. Gui-Bo, Y.; You-Zhu, Z.; Shu-Dong, W.; De-Bing, S.; Zhi-Hui, D.; Wei-Guo, F. Study of the electrospun PLA/silk fibroin-gelatin composite nanofibrous scaffold for tissue engineering. J. Biomed. Mater. Res. A **2010**, *93A* (1), 158–163.

186. Hsu, F.Y.; Hung, Y.S.; Liou, H.M.; Shen, C.H. Electrospun hyaluronate-collagen nanofibrous matrix and the effects of varying the concentration of hyaluronate on the characteristics of foreskin fibroblast cells. Acta Biomater. **2010**, *6* (6), 2140–2147.

187. Kluger, P.J.; Wyrwa, R.; Weisser, J.; Maierle, J.; Votteler, M.; Rode, C.; Schnabelrauch, M.; Walles, H.; Schenke-Layland, K. Electrospun poly(D/L-lactide-co-L-lactide) hybrid matrix: A novel scaffold material for soft tissue engineering. J. Mater. Sci. Mater. Med. **2010**, *21* (9), 2665–2671.

Dental–Electrospinning

188. Liu, F.J.; Guo, R.; Shen, M.; Cao, X.; Mo, X.; Wang, S.; Shi, X. Effect of the porous microstructures of poly(lactic-co-glycolic acid)/carbon nanotube composites on the growth of fibroblast cells. Soft Mater. **2010**, *8* (3), 239–253.

189. Liu, S.J.; Kau, Y.-C.; Chou, C.-Y.; Chen, J.-K.; Wu, R.-C.; Yeh, W.-L. Electrospun PLGA/collagen nanofibrous membrane as early-stage wound dressing. J. Membr. Sci. **2010**, *355* (1–2), 53–59.

190. Loh, X.J.; Peh, P.; Liao, S.; Sng, C.; Li, J. Controlled drug release from biodegradable thermorespon-sive physical hydrogel nanofibers. J. Control. Release **2010**, *143* (2), 175–182.

191. Lowery, J.L.; Datta, N.; Rutledge, G.C. Effect of fiber diameter, pore size and seeding method on growth of human dermal fibroblasts in electrospun poly(epsilon-caprolactone) fibrous mats. Biomaterials **2010**, *31* (3), 491–504.

192. Ma, G.P.; Yang, D.; Wang, K.; Han, J.; Ding, S.; Song, G.; Nie, J. Organic-soluble chitosan/polyhydroxybutyrate ultrafine fibers as skin regeneration prepared by electrospinning. J. Appl. Polym. Sci. **2010**, *118* (6), 3619–3624.

193. Wu, L.L.; Li, H.; Li, S.; Li, X.; Yuan, X.; Li, X.; Zhang, Y. Composite fibrous membranes of PLGA and chitosan prepared by coelectrospinning and coaxial electrospinning. J. Biomed. Mater. Res. A **2010**, *92A* (2), 563–574.

194. Schenke-Layland,K.; Rofail, F.; Heydarkhan, S.; Gluck, J.M.; Ingle, N.P.; Angelis, E.; Choi, C.H.; MacLellan, W.R.; Beygui, R.E.; Shemin, R.J.; Heydarkhan-Hagvall, S. The use of three-dimensional nanostructures to instruct cells to produce extracellular matrix for regenerative medicine strategies. Biomaterials **2009**, *30* (27), 4665–4675.

195. Cui, W.G.; Zhu, X.L.; Yang, Y.; Li, X.H.; Jin, Y. Evaluation of electrospun fibrous scaffolds of poly(DL-lactide) and poly(ethylene glycol) for skin tissue engineering. Mater. Sci. Eng. CMater. Biol. Appl. **2009**, *29* (6), 1869–1876.

196. Jeong, L.; Yeo, I.S.; Kim, H.N.; Yoon, Y.I.; Jang da, H.; Jung, S.Y.; Min, B.M.; Park, W.H. Plasma-treated silk fibroin nanofibers for skin regeneration. Int. J. Biol. Macromol. **2009**, *44*(3), 222–228.

197. Marelli, B.; Alessandrino, A.; Fare, S.; Tanzi, M.C.; Freddi, G. Electrospun silk fibroin tubular matrixes for small vessel bypass grafting. Mater. Technol. **2009**, *24*, 52.

198. Meng, J.; Kong, H.; Han, Z.; Wang, C.; Zhu, G.; Xie, S.; Xu, H. Enhancement of nanofibrous scaffold of multi-walled carbon nanotubes/polyurethane composite to the fibroblasts growth and biosynthesis. J. Biomed. Mater. Res. A **2009**, *88A* (1), 105–116.

199. Nien, Y.-H.; Chen, Z.-B.; Liang, J.-I.; Yeh, M.-L.; Hsu, H.-C.; Su, F.-C. Fabrication and cell affinity of poly(vinyl alcohol) nanofibers via electrospinning. J. Med. Biol. Eng. **2009**, *29* (2), 98–101.

200. Wray, L.S.; Orwin, E.J. Recreating the microenvironment of the native cornea for tissue engineering applications. Tissue Eng. Part A **2009**, *15* (7), 1463–1472.

201. Zhang, F.; Zuo, B.Q.; Bai, L. Study on the structure of SF fiber mats electrospun with HFIP and FA and cells behavior. J. Mater. Sci. **2009**, *44* (20), 5682–5687.

202. Borg, E.; Frenot, A.; Walkenström, P.; Gisselfält, K.; Gretzer, C.; Gatenholm, P. Electrospinning of degradable elastomeric nanofibers with various morphology and their interaction with human fibroblasts. J. Appl. Polym. Sci. **2008**, *108* (1), 491–497.

203. Khanam, N.; Mikoryak, C.; Draper, R.K.; Balkus, K.J. Electrospun linear polyethyleneimine scaffolds for cell growth. Acta Biomater. **2007**, *3* (6), 1050–1059.

204. Kim, G.; Kim, W. Highly porous 3D nanofiber scaffold using an electrospinning technique. J. Biomed. Mater. Res. B Appl. Biomater. **2007**, *81B* (1), 104–110.

205. Sun, T.; Norton, D.; Mckean, R.J.; Haycock, J.W.; Ryan, A.J.; MacNeil, S.Development of a 3D cell culture system for investigating cell interactions with electrospun fibers. Biotechnol. Bioeng. **2007**, *97* (5), 1318–1328.

206. Deng, H.B.; Zhou, X.; Wang, X.; Zhang, C.; Ding, B.; Zhang, Q.; Du, Y. Layer-by-layer structured polysaccha-rides film-coated cellulose nanofibrous mats for cell culture. Carbohydr. Polym. **2010**, *80* (2), 474–479.

207. Leong, M.F.; Chian, K.S.; Mhaisalkar, P.S.; Ong, W.F.; Ratner, B.D. Effect of electrospun poly(D,L-lactide) fibrous scaffold with nanoporous surface on attachment of porcine esophageal epithelial cells and protein adsorption. J. Biomed. Mater. Res. A **2009**, *89A* (4), 1040–1048.

208. Han, I.; Shim, K.J.; Kim, J.Y.; Im, S.U.; Sung, Y.K.; Kim, M.; Kang, I.K.; Kim, J.C. Effect of poly(3-hydroxybutyrate-co-3-hydroxyvalerate) nanofiber matrices cocultured with hair follicular epithelial and dermal cells for biological wound dressing. Artif. Organs **2007**, *31* (11), 801–808.

209. Zhu, Y.B.; Leong, M.F.; Ong, W.F.; Chan-Park, M.B.; Chian, K.S. Esophageal epithelium regeneration on fibro-nectin grafted poly(L-lactide-co-caprolactone) (PLLC) nanofiber scaffold. Biomaterials **2007**, *28* (5), 861–868.

210. Kim, C.H.; Khil, M.S.; Kim, H.Y.; Lee, H.U.; Jahng, K.Y. An improved hydrophilicity via electrospinning for enhanced cell attachment and proliferation. J. Biomed. Mater. Res. B Appl. Biomater. **2006**, *78B* (2), 283–290.

211. He, W.; Ma, Z.; Teo, W.E.; Dong, Y.X.; Robless, P.A.; Lim, T.C.; Ramakrishna, S. Tubular nanofiber scaffolds for tissue engineered small-diameter vascular grafts. J. Biomed. Mater. Res. A **2009**, *90A* (1), 205–216.

212. He, W.; Yong, T.; Ma, Z.W.; Inai, R.; Teo, W.E.; Ramakrishna, S. Biodegradable polymer nanofiber mesh to maintain functions of endothelial cells. Tissue Eng. **2006**, *12* (9), 2457–2466.

213. Uttayarat, P.; Perets, A.; Li, M.; Pimton, P.; Stachelek, S.J.; Alferiev, I.; Composto, R.J.; Levy, R.J.; Lelkes, P.I. Micropatterning of three-dimensional electrospun poly-urethane vascular grafts. Acta Biomater. **2010**, *6* (11), 4229–4237.

214. Bianco, A.; Del Gaudio, C.; Baiguera, S.; Armentano, I.; Bertarelli, C.; Dottori, M.; Bultrini, G.; Lucotti, A.; Kenny, J.M.; Folin, M. Microstructure and cytocompatibility of electrospun nanocomposites based on poly(epsilon-caprolactone) and carbon nanostructures. Int. J. Artif. Organs **2010**, *33* (5), 271–282.

215. Chen, Z.G.; Wang, P.W.; Wei, B.; Mo, X.M.; Cui, F.Z. Electrospun collagen-chitosan nanofiber: A biomimetic extracellular matrix for endothelial cell and smooth muscle cell. Acta Biomater. **2010**, *6* (2), 372–382.

216. Ju, Y.M.; Choi, J.S.; Atala, A.; Yoo, J.J.; Lee, S.J. Bilayered scaffold for engineering cellularized blood vessels. Biomaterials **2010**, *31* (15), 4313–4321.

Dental—Electrospinning

217. Rubenstein, D.A.; Venkitachalam, S.M.; Zamfir, D.; Wang, F.; Lu, H.; Frame, M.D.; Yin, W. *In vitro* biocompatibility of sheath-core cellulose-acetate-based electrospun scaffolds towards endothelial cells and platelets. J. Biomater. Sci. Polym. Ed. **2010**, *21* (13), 1713–1736.

218. Zhang, K.H.; Mo, X.M.; Huang, C.; He, C.L.; Wang, H.S. Electrospun scaffolds from silk fibroin and their cellular compatibility. J. Biomed. Mater. Res. A **2010**, *93A* (3), 976–983.

219. Zhang, K.H.; Wang, H.; Huang, C.; Su, Y.; Mo, X.; Ikada, Y. Fabrication of silk fibroin blended P(LLA-CL) nanofibrous scaffolds for tissue engineering. J. Biomed. Mater. Res. A **2010**, *93A* (3), 984–993.

220. Chen, R.; Qiu, L.J.; Ke, Q.F.; He, C.L.; Mo, X.M. Electrospinning thermoplastic polyurethane-contained collagen nanofibers for tissue-engineering applications. J. Biomater. Sci. Polym. Ed. **2009**, *20* (11), 1513–1536.

221. Li, X.Q.; Su, Y.; He, C.; Wang, H.; Fong, H.; Mo, X. Sorbitan monooleate and poly(L-lactide-co-epsilon-caprolactone) electrospun nano-fibers for endothelial cell interactions. J. Biomed. Mater. Res. A **2009**, *91A* (3), 878–885.

222. Veleva, A.N.; Heath, D.E.; Johnson, J.K.; Nam, J.; Patterson, C.; Lannutti, J.J.; Cooper, S.L. Interactions between endothelial cells and electrospun methacrylic terpolymer fibers for engineered vascular replacements. J. Biomed. Mater. Res. A **2009**, *91A* (4), 1131–1139.

223. Carampin, P.; Conconi, M.T.; Lora, S.; Menti, A.M.; Baiguera, S.; Bellini, S.; Grandi, C.; Parnigotto, P.P. Electrospun polyphosphazene nanofibers for *in vitro* rat endothelial cells proliferation. J. Biomed. Mater. Res. A **2007**, *80A* (3), 661–668.

224. He, W.; Ma, Z.W.; Yong, T.; Teo, W.E.; Ramakrishna, S. Fabrication of collagen-coated biodegradable polymer nanofiber mesh and its potential for endothelial cells growth. Biomaterials **2005**, *26* (36), 7606–7615.

225. Ma, Z.W.; Kotaki, M.; Yong, T.; He, W. Ramakrishna, S. Surface engineering of electrospun polyethylene terephthalate (PET) nanofibers towards development of a new material for blood vessel engineering. Biomaterials **2005**, *26* (15), 2527–2536.

226. Xu, C.Y.; Yang, F.; Wang, S.; Ramakrishna, S. *In vitro* study of human vascular endothelial cell function on materials with various surface roughness. J. Biomed. Mater. Res. A **2004**, *71A* (1), 154–161.

227. Welin, S. Ethical issues in tissue engineering. In *Tissue Engineering*; Van Blitterswijk, C., et al., Eds.; Academic Press: New York, 2008; 685–703.

228. Heath, D.E.; Lannutti, J.J.; Cooper, S.L. Electrospun scaffold topography affects endothelial cell proliferation, metabolic activity, and morphology. J. Biomed. Mater. Res. A **2010**, *94A* (4), 1195–1204.

229. Meng, J.; Han, Z.; Kong, H.; Qi, X.; Wang, C.; Xie, S.; Xu, H. Electrospun aligned nanofibrous composite of MWCNT/polyurethane to enhance vascular endothelium cells proliferation and function. J. Biomed. Mater. Res. A **2010**, *95A* (1), 312–320.

230. Del Gaudio, C.; Bianco, A.; Folin, M.; Baiguera, S.; Grigioni, M. Structural characterization and cell response evaluation of electrospun PCL membranes: Micrometric versus submicrometric fibers. J. Biomed. Mater. Res. A **2009**, *89A* (4), 1028–1039.

231. Yin, G.; Youzhu, Z.; Weiwei, B.; Jialin, W.; De-bing, S.; Zhu-hui, D.; Wei-guo, F. Study on the properties of the electrospun silk fibroin/gelatin blend nanofibers for scaffolds. J. Appl. Polym. Sci. **2009**, *111* (3), 1471–1477.

232. Inoguchi, H.; Tanaka, T.; Maehara, Y.; Matsuda, T. The effect of gradually graded shear stress on the morphological integrity of a huvec-seeded compliant small-diameter vascular graft. Biomaterials **2007**, *28* (3), 486–495.

233. Kwon, I.K.; Kidoaki, S.; Matsuda, T. Electrospun nano- to microfiber fabrics made of biodegradable copolyesters: Structural characteristics, mechanical properties and cell adhesion potential. Biomaterials **2005**, *26* (18), 3929–3939.

234. Jeong, S.I.; Kim, S.Y.; Cho, S.K.; Chong, M.S.; Kim, K.S.; Kim, H., Lee, S.B.; Lee, Y.M. Tissue-engineered vascular grafts composed of marine collagen and PLGA fibers using pulsatile perfusion bioreactors. Biomaterials **2007**, *28* (6),1115–1122.

235. Stankus, J.J.; Soletti, L.; Fujimoto, K.; Hong, Y.; Vorp, D.A.; Wagner, W.R. Fabrication of cell microintegrated blood vessel constructs through electrohydrodynamic atomization. Biomaterials **2007**, *28* (17), 2738–2746.

236. Venugopal, J.; Ma, L.L.; Yong, T.; Ramakrishna, S. *In vitro* study of smooth muscle cells on polycaprolactone and collagen nanofibrous matrices. Cell Biol. Int. **2005**, *29* (10), 861–867.

237. Luong-Van,E.; Grøndahl, L.; Chua, K.N.; Leong, K.W.; Nurcombe, V.; Cool, S.M. Controlled release of heparin from poly(epsilon-caprolactone) electrospun fibers. Biomaterials **2006**, *27* (9), 2042–2050.

238. Boland, E.D.; Matthews, J.A.; Pawlowski, K.J.; Simpson, D.G.; Wnek, G.E.; Bowlin, G.L. Electrospinning collagen and elastin: Preliminary vascular tissue engineering. Front.Biosci. **2004**, *9*, 1422–1432.

239. Xu, F.; Cui, F.Z.; Jiao, Y.P.; Meng, Q.Y.; Wang, X.P.; Cui, X.Y. Improvement of cytocompatibility of electrospinning PLLA microfibers by blending PVP. J. Mater. Sci. Mater. Med. **2009**, *20* (6), 1331–1338.

240. Venugopal, J.; Ma, L.L.; Yong, T.; Ramakrishna, S. *In vitro* study of smooth muscle cells on polycap-rolactone and collagen nanofibrous matrices. Cell Biol. Int. **2005**, *29* (10), 861–867.

241. Shin, M.; Ishii, O.; Sueda, T.; Vacanti, J.P. Contractile cardiac grafts using a novel nanofibrous mesh. Biomaterials **2004**, *25* (17), 3717–3723.

242. Ishii, O.; Shin, M.; Sueda, T.; Vacanti, J.P. *In vitro* tissue engineering of a cardiac graft using a degrad-able scaffold with an extracellular matrix-like topography. J. Thorac. Cardiovasc. Surg. **2005**, *130* (5), 1358–1363.

243. Zong, X.H.; Bien, H.; Chung, C.Y.; Yin, L.; Fang, D.; Hsiao, B.S.; Chu, B.; Entcheva, E. Electrospun fine-textured scaffolds for heart tissue constructs. Biomaterials **2005**, *26* (26), 5330–5338.

244. Hosseinkhani, H.; Hosseinkhani, M.; Hattori, S.; Matsuoka, R.; Kawaguchi, N. Micro and nano-scale *in vitro* 3D culture system for cardiac stem cells. J. Biomed. Mater. Res. A **2010**, *94A* (1), 1–8.

245. Fromstein, J.D.; Zandstra, P.W.; Alperin, C.; Rockwood, D.; Rabolt, J.F.; Woodhouse, K.A. Seeding bioreactor-produced

embryonic stem cell-derived cardiomyocytes on different porous, degradable, polyurethane scaffolds reveals the effect of scaffold architecture on cell morphology. Tissue Eng. Part A **2008**, *14* (3), 369–378.

246. Focarete, M.L.; Gualandi, C.; Scandola, M.; Govoni, M.; Giordano, E.; Foroni, L.; Valente, S.; Pasquinelli, G.; Gao, W.; Gross, R.A. Electrospun scaffolds of a polyhydroxyalkanoate consisting of omega-hydroxyl-pentadecanoate repeat units: Fabrication and *in vitro* biocompatibility studies. J. Biomater. Sci. Polym. Ed. **2010**, *21* (10), 1283–1296.

247. Prabhakaran, M.P.; Venugopal, J.R.; Ramakrishna, S. Mesenchymal stem cell differentiation to neuronal cells on electrospun nanofibrous substrates for nerve tissue engineering. Biomaterials **2009**, *30* (28), 4996–5003.

248. Wise, J.K.; Yarin, A.L.; Megaridis, C.M.; Cho, M. Chondrogenic differentiation of human mesenchymal stem cells on oriented nanofibrous scaffolds: Engineering the superficial zone of articular cartilage. Tissue Eng. Part A **2009**, *15* (4), 913–921.

249. Kolambkar, Y.M.; Peister, A.; Ekaputra, A.K.; Hutmacher, D.W.; Guldberg, R.E. Colonization and osteogenic differentiation of different stem cell sources on electrospun nanofiber meshes. Tissue Eng. Part A **2010**, *16* (10), 3219–3230.

250. Tuan, R.S.; Boland, G.; Tuli, R. Adult mesenchymal stem cells and cell-based tissue engineering. Arthritis Res. Ther. **2003**, *5*, 32.

251. Li, W.J.; Tuli, R.; Huang, X.X.; Laquerriere, P.; Tuan, R.S. Multilineage differentiation of human mesenchymal stem cells in a three-dimensional nanofibrous scaffold. Biomaterials **2005**, *26* (25), 5158–5166.

252. Li, W.J.; Tuli, R.; Okafor, C.; Derfoul, A.; Danielson, K.G.; Hall, D.J.; Tuan, R.S. A three-dimensional nanofibrous scaffold for cartilage tissue engineering using human mesenchymal stem cells. Biomaterials **2005**, *26* (6), 599–609.

253. Yoshimoto, H.; Shin, Y.M.; Terai, H.; Vacanti, J.P. A biodegradable nanofiber scaffold by electrospinning and its potential for bone tissue engineering. Biomaterials **2003**, *24* (12), 2077–2082.

254. Ahn, S.; Koh, Y.H.; Kim, G. A three-dimensional hierarchical collagen scaffold fabricated by a combined solid freeform fabrication (SFF) and electrospinning process to enhance mesenchymal stem cell (MSC) proliferation. J. Micromech. Microeng. **2010**, *20* (6), 065015.

255. Lee, J.H.; Rim, N.G.; Jung, H.S.; Shin, H. Control of osteogenic differentiation and mineralization of human mesenchymal stem cells on composite nanofibers containing poly lactic-co-(glycolic acid) and hydroxyapatite. Macromol. Biosci. **2010**, *10* (2), 173–182.

256. Ren, L.; Wang, J.; Yang, F.-Y.; Wang, L.; Wang, D.; Wang, T.-X.; Tian, M.-M. Fabrication of gelatin-siloxane fibrous mats via sol-gel and electrospinning procedure and its application for bone tissue engineering. Mater. Sci. Eng. CMater. Biol. Appl. **2010**, *30* (3), 437–444.

257. Ruckh, T.T.; Kumar, K.; Kipper, M.J.; Popat, K.C. Osteogenic differentiation of bone marrow stromal cells on poly(epsilon-caprolactone) nanofiber scaffolds. Acta Biomater. **2010**, *6* (8), 2949–2959.

258. Soliman, S.; Pagliari, S.; Rinaldi, A.; Forte, G.; Fiaccavento, R.; Pagliari, F.; Franzese, O.; Minieri, M.;

Di Nardo, P.; Licoccia, S.; Traversa, E. Multiscale three-dimensional scaffolds for soft tissue engineering via multimodal electrospinning. Acta Biomater. **2010**, *6* (4), 1227–1237.

259. Chan, C.K.; Liao, S.; Li, B.; Lareu, R.R.; Larrick, J.W.; Ramakrishna, S.; Raghunath, M. Early adhesive behavior of bone-marrow-derived mesenchymal stem cells on collagen electrospun fibers. Biomed. Mater. **2009**, *4* (3), 035006.

260. Ghasemi-Mobarakeh,L.; Morshed, M.; Karbalaie, K.; Fesharaki, M.A.; Nematallahi, M.; Nasr-Esfahani, M.H.; Baharvand, H. The thickness of electrospun poly (epsilon-caprolactone) nanofibrous scaffolds influences cell proliferation. Int. J. Artif. Organs **2009**, *32* (3), 150–158.

261. Li, C.M.; Wang, Z.G.; Gu, Y.Q.; Dong, J.D.; Qiu, R.X.; Bian, C.; Liu, X.F.; Feng, Z.G. Preliminary investigation of seeding mesenchymal stem cells on biodegradable scaffolds for vascular tissue engineering *in vitro*. ASAIO J. **2009**, *55* (6), 614–619.

262. Hong, H.; Dong, N.; Shi, J.; Chen, S.; Guo, C.; Hu, P.; Qi, H. Fabrication of a novel hybrid scaffold for tissue engineered heart valve. J. Huazhong Univ. Sci. Technol. Med. Sci. **2009**, *29* (5), 599–603.

263. Meinel, A.J.; Kubow, K.E.; Klotzsch, E.; Garcia-Fuentes, M.; Smith, M.L.; Vogel, V.; Merkle, H.P.; Meinel, L.Optimization strategies for electrospun silk fibroin tissue engineering scaffolds. Biomaterials **2009**, *30* (17), 3058–3067.

264. Shalumon, K.T.; Binulal, N.S.; Selvamurugan, N.; Nair, S.V.; Menon, D.; Furuike, T.; Tamura, H.; Jayakumar, R. Electrospinning of carboxymethyl chitin/poly(vinyl alcohol) nanofibrous scaffolds for tissue engineering applications. Carbohydr. Polym. **2009**, *77* (4), 863–869.

265. Ko, E.K.; Jeong, S.I.; Rim, N.G.; Lee, Y.M.; Shin, H.; Lee, B.K. *In vitro* osteogenic differentiation of human mesenchymal stem cells and *in vivo* bone formation in composite nanofiber meshes. Tissue Eng. Part A **2008**, *14*(12), 2105–2119.

266. Srouji, S.; Kizhner, T.; Suss-Tobi, E.; Livne, E.; Zussman, E. 3-D Nanofibrous electrospun multilayered construct is an alternative ECM mimicking scaffold. J. Mater. Sci. Mater. Med. **2008**, *19* (3), 1249–1255.

267. Kolambkar, Y.M.; Dupont, K.M.; Boerckel, J.D.; Huebsch, N.; Mooney, D.J.; Hutmacher, D.W.; Guldberg, R.E.An alginate-based hybrid system for growth factor delivery in the functional repair of large bone defects. Biomaterials **2011**, *32* (1), 65–74.

268. Mavis, B.; Demirtas, T.T.; Gumusderelioglu, M.; Gunduz, G.; Colak, U. Synthesis, characterization and osteoblastic activity of polycaprolactone nanofibers coated with biomimetic calcium phosphate. Acta Biomater. **2009**, *5* (8), 3098–3111.

269. Yu, H.S.; Jang, J.H.; Kim, T.I.; Lee, H.H.; Kim, H.W. Apatite-mineralized polycaprolactone nanofibrous web as a bone tissue regeneration substrate. J. Biomed. Mater. Res. A **2009**, *88A* (3), 747–754.

270. Erisken, C.; Kalyon, D.M.; Wang, H.J. Viscoelastic and biomechanical properties of osteochondral tissue constructs generated from graded polycaprolactone and beta-tricalcium phosphate composites. J. Biomech. Eng. **2010**, *132* (9), 091013.

271. Francis, L.; Venugopal, J.; Prabhakaran, M.P.; Thavasi, V.; Marsano, E.; Ramakrishna, S. Simultaneous

scaffolds for bone tissue regeneration. Acta Biomater. **2010**, *6* (10), 4100–4109.

272. Fu, S.Z.; Wang, X.H.; Guo, G.; Shi, S.; Liang, H.; Luo, F.; Wei, Y.Q.; Qian, Z.Y. Preparation and characterization of nano-hydroxyapatite/poly(epsilon-caprolactone)-poly(ethylene glycol)-poly(epsilon-caprolactone) composite fibers for tissue engineering. J. Phys. Chem. C **2010**, *114* (43), 18372–18378.

273. Kim, H.M.; Chae, W.P.; Chang, K.W.; Chun, S.; Kim, S.; Jeong, Y.; Kang, I.K. Composite nanofiber mats consisting of hydroxyapatite and titania for biomedical applications. J. Biomed.Mater.Res. BAppl. Biomater. **2010**, *94B* (2), 380–387.

274. Lee, E.J.; Teng, S.H.; Jang, T.S.; Wang, P.; Yook, S.W.; Kim, H.E.; Koh, Y.H. Nanostructured poly(epsilon-caprolactone)-silica xerogel fibrous membrane for guided bone regeneration. Acta Biomater. **2010**, *6* (9), 3557–3565.

275. Noh, K.T.; Lee, H.Y.; Shin, U.S.; Kim, H.W. Composite nanofiber of bioactive glass nanofiller incorporated poly(lactic acid) for bone regeneration. Mater. Lett. **2010**, *64* (7), 802–805.

276. Shim, I.K.; Jung, M.R.; Kim, K.H.; Seol, Y.J.; Park, Y.J.; Park, W.H.; Lee, S.J. Novel three-dimensional scaffolds of poly((L)-lactic acid) microfibers using electros-pinning and mechanical expansion: Fabrication and bone regeneration. J. Biomed. Mater. Res. B Appl. Biomater. **2010**, *95B* (1), 150–160.

277. Shin, T.J.; Park, S.Y.; Kim, H.J.; Lee, H.J.; Youk, J.H.Development of 3-D poly(trimethylenecarbonate-co-epsilon-caprolactone)-block-poly(p-dioxano ne) scaffold for bone regeneration with high porosity using a wet electrospinning method. Biotechnol. Lett. **2010**, *32* (6), 877–882.

278. Shin, Y.M.; Shin, H.; Lim, Y.M. Surface modification of electrospun poly(L-lactide-co-epsilon-caprolactone) fibrous meshes with a RGD peptide for the control of adhesion, proliferation and differentiation of the preosteoblastic cells. Macromol. Res. **2010**, *18*, 472.

279. Tong, H.W.; Wang, M.; Li, Z.Y.; Lu, W.W. Electrospinning, characterization and *in vitro* biological evaluation of nanocomposite fibers containing carbonated hydroxyapatite nanoparticles. Biomed. Mater. **2010**, *5* (5), 054111.

280. Ekaputra, A.K.; Zhou, Y.F.; Cool, S.M.; Hutmacher, D.W. Composite electrospun scaffolds for engineering tubular bone grafts. Tissue Eng. Part A **2009**, *15* (12), 3779–3788.

281. Gupta, D.; Venugopal, J.; Mitra, S.; Dev, V.R.G.; Ramakrishna, S. Nanostructured biocomposite substrates by electrospinning and electrospraying for the mineralization of osteoblasts. Biomaterials **2009**, *30* (11), 2085–2094.

282. Kang, Y.M.; Kim, K.H.; Seol, Y.J.; Rhee, S.H. Evaluations of osteogenic and osteoconductive properties of a nonwoven silica gel fabric made by the electrospinning method. Acta Biomater. **2009**, *5* (1), 462–469.

283. Ngiam, M.; Liao, S.; Patil, A.J.; Cheng, Z.; Yang, F.; Gubler, M.J.; Ramakrishna, S.; Chan, C.K. Fabrication of mineralized polymeric nanofibrous composites for bone graft materials. Tissue Eng. Part A **2009**, *15* (3), 535–546.

284. Kim, H.W.; Lee, H.H.; Chun, G.S. Bioactivity and osteoblast responses of novel biomedical nano-composites of bioactive glass nanofiber filled poly(lactic acid). J. Biomed. Mater. Res. A **2008**, *85A* (3), 651–663.

285. Jang, J.H.; Castano, O.; Kim, H.W. Electrospun materials as potential platforms for bone tissue engineering. Adv. Drug Deliv. Rev. **2009**, *61* (12), 1065–1083.

286. Poole, A.R.; Kojima, T.; Yasuda, T.; Mwale, F.; Kobayashi, M.; Laverty, S. Composition and structure of articular cartilage—A template for tissue repair. Clin. Orthop. Relat.Res. **2001**, *1* (391 Suppl.), S26–S33.

287. Hunziker, E.B. Articular cartilage repair: Basic science and clinical progress. A review of the current status and prospects.Osteoarthr. Cartilage **2002**, *10* (6), 432–463.

288. Klein, T.J.; Malda, J.; Sah, R.L.; Hutmacher, D.W. Tissue engineering of articular cartilage with biomimetic zones. Tissue Eng. Part BRev. **2009**, *15* (2), 143–157.

289. Klein, T.J.; Rizzi, S.C.; Reichert, J.C.; Georgi, N.; Malda, J.; Schuurman, W.; Crawford, R.W.; Hutmacher, D.W. Strategies for zonal cartilage repair using hydrogels. Macromol. Biosci. **2009**, *9* (11), 1049–1058.

290. Kim, G.; Son, J.; Park, S.; Kim, W. Hybrid process for fabricating 3D hierarchical scaffolds combining rapid prototyping and electrospinning. Macromol. Rapid Commun. **2008**, *29* (19), 1577–1581.

291. Shim, I.K.; Suh, W.H.; Lee, S.Y.; Lee, S.H.; Heo, S.J.; Lee, M.C.; Lee, S.J. Chitosan nano-/microfibrous double-layered membrane with rolled-up three-dimensional structures for chondrocyte cultivation. J. Biomed. Mater. Res. A **2009**, *90A* (2), 595–602.

292. Li, W.J.; Danielson, K.G.; Alexander, P.G.; Tuan, R.S. Biological response of chondrocytes cultured in three-dimensional nanofibrous poly(epsilon-caprolactone) scaffolds. J. Biomed. Mater. Res. A **2003**, *67A* (4), 1105–1114.

293. Kim, T.G.; Chung, H.J.; Park, T.G. Macroporous and nanofibrous hyaluronic acid/collagen hybrid scaffold fabricated by concurrent electrospinning and deposition/leaching of salt particles. Acta Biomater. **2008**, *4* (6), 1611–1619.

294. Moroni, L.; Schotel, R.; Hamann, D.; de Wijn, J.R.; van Blitterswijk, C.A. 3D fiber-deposited electrospun integrated scaffolds enhance cartilage tissue formation. Adv. Funct. Mater. **2008**, *18* (1), 53–60.

295. Thorvaldsson, A.; Stenhamre, H.; Gatenholm, P.; Walkenstrom, P. Electrospinning of highly porous scaffolds for cartilage regeneration. Biomacromolecules **2008**, *9*(3), 1044–1049.

296. Lee, I.S.; Kwon, O.H.; Meng, W.; Kang, I.K. Nanofabrication of microbial polyester by electrospinning promotes cell attachment. Macromol. Res. **2004**, *12* (4), 374–378.

297. Shields, K.J.; Beckman, M.J.; Bowlin, G.L.; Wayne, J.S. Mechanical properties and cellular proliferation of electrospun collagen type II. Tissue Eng. **2004**, *10* (9–10), 1510–1517.

298. Smeal, R.M.; Tresco, P.A. The influence of substrate curvature on neurite outgrowth is cell type dependent. Exp. Neurol. **2008**, *213* (2), 281–292.

299. Smeal, R.M.; Rabbitt, R.; Biran, R.; Tresco, P.A. Substrate curvature influences the direction of nerve outgrowth. Ann. Biomed. Eng. **2005**, *33* (3), 376–382.

300. Yang, F.; Murugan, R.; Wang, S.; Ramakrishna, S. Electrospinning of nano/micro scale poly(L-lactic acid) aligned

fibers and their potential in neural tissue engineering. Biomaterials **2005**, *26* (15), 2603–2610.

301. Bini, T.B.; Gao, S.J.; Wang, S.; Ramakrishna, S. Poly(L-lactide-co-glycolide) biodegradable micro-fibers and electrospun nanofibers for nerve tissue engineering: An *in vitro* study. J. Mater. Sci. **2006**, *41* (19), 6453–6459.

302. Gupta, D.; Venugopal, J.; Prabhakaran, M.P.; Dev, V.R.; Low, S.; Choon, A.T.; Ramakrishna, S. Aligned and random nanofibrous substrate for the *in vitro* culture of Schwann cells for neural tissue engineering. Acta Biomater. **2009**, *5* (7), 2560–2569.

303. Assmann, U.; Szentivanyi, A.; Stark, Y.; Scheper, T.; Berski, S.; Dräger, G.; Schuster, R.H. Fiber scaffolds of polysialic acid via electrospinning for peripheral nerve regeneration. J. Mater. Sci. Mater. Med. **2010**, *21* (7), 2115–2124.

304. de Guzman, R.C.; Loeb, J.A.; VandeVord, P.J. Electrospinning of matrigel to deposit a basal lamina-like nanofiber surface. J. Biomater. Sci. Polym. Ed. **2010**, *21* (8–9), 1081–1101.

305. Wang, H.B.; Mullins, M.E.; Cregg, J.M.; Hurtado, A.; Oudega, M.; Trombley, M.T.; Gilbert, R.J. Creation of highly aligned electrospun poly-L-lactic acid fibers for nerve regeneration applications. J. Neural Eng. **2009**, *6* (1), 016001.

306. Xie, J.W.; Macewan, M.R.; Willerth, S.M.; Li, X.; Moran, D.W.; Sakiyama-Elbert, S.E.; Xia, Y. Conductive core-sheath nanofibers and their potential application in neural tissue engineering. Adv. Funct. Mater. **2009**, *19* (14), 2312–2318.

307. Yao, L.; O'Brien, N.; Windebank, A.; Pandit, A. Orienting neurite growth in electrospun fibrous neural conduits. J. Biomed. Mater. Res. B Appl. Biomater. **2009**, *90B* (2), 483–491.

308. Prabhakaran, M.P.; Venugopal, J.; Chan, C.K.; Ramakrishna, S. Surface modified electrospun nanofibrous scaffolds for nerve tissue engineering. Nanotechnology **2008**, *19* (45), 455102.

309. Prabhakaran, M.P.; Venugopal, J.R.; Chyan, T.T.; Hai, L.B.; Chan, C.K.; Lim, A.Y.; Ramakrishna, S. Electrospun biocomposite nanofibrous scaffolds for neural tissue engineering. Tissue Eng. Part A **2008**, *14* (11), 1787–1797.

310. Chow, W.N.; Simpson, D.G.; Bigbee, J.W.; Colello, R.J. Evaluating neuronal and glial growth on electrospun polarized matrices: Bridging the gap in percussive spinal cord injuries. Neuron Glia Biol. **2007**, *3* (2), 119–126.

311. Dalton, P.D.; Flynn, L.; Shoichet, M.S. Manufacture of poly(2-hydroxyethyl methacrylate-co-methyl methacrylate) hydrogel tubes for use as nerve guidance channels. Biomaterials **2002**, *23* (18), 3843–3851.

312. Birk, D.E.; Mayne, R. Localization of collagen types I, III and V during tendon development. Changes in collagen types I and III are correlated with changes in fibril diameter. Eur. J. Cell Biol. **1997**, *72* (4), 352–361.

313. Moffat, K.L.; Kwei, A.S.-P.; Spalazzi, J.P.; Doty, S.B.; Levine, W.N.; Lu, H.H. Novel nanofiber-based scaffold for rotator cuff repair and augmentation. Tissue Eng. Part A **2009**, *15* (1), 115–126.

314. Inanc, B.; Arslan, Y.E.; Seker, S.; Elcin, A.E.; Elcin, Y.M. Periodontal ligament cellular structures engineered with electrospun poly(DL-lactide-co-glycolide) nanofibrous membrane scaffolds. J. Biomed. Mater. Res. A **2009**, *90A* (1), 186–195.

315. Zhang, S.; Huang, Y.; Yang, X.; Mei, F.; Ma, Q.; Chen, G.; Ryu, S.; Deng, X. Gelatin nanofibrous membrane fabricated by electrospinning of aqueous gelatin solution for guided tissue regeneration. J. Biomed. Mater. Res. A **2009**, *90A* (3), 671–679.

316. Bottino, M.C.; Thomas, V.; Janowski, G.M. A novel spatially designed and functionally graded electrospun membrane for periodontal regeneration. Acta Biomater. **2011**, *7* (1), 216–224.

317. Sahoo, S.; Ouyang, H.; Goh, J.C.H.; Tay, T.E.; Toh, S.L. Characterization of a novel polymeric scaffold for potential application in tendon/ligament tissue engineering. Tissue Eng. **2006**, *12* (1), 91–99.

318. Sahoo, S.; Goh, J.C.-H.; Toh, S.L. Development of hybrid polymer scaffolds for potential applications in ligament and tendon tissue engineering. Biomed. Mater. **2007**, *2* (3), 169–173.

319. Sahoo, S.; Toh, S.L.; Goh, J.C.-H. PLGA nanofiber-coated silk microfibrous scaffold for connective tissue engineering. J. Biomed. Mater. Res. B Appl. Biomater. **2010**, *95B* (1), 19–28.

320. Vaquette, C.; Kahn, C.; Frochot, C.; Nouvel, C.; Six, J.L.; De Isla, N.; Luo, L.H.; Cooper-White, J.; Rahouadj, R.; Wang, X. Aligned poly(L-lactic-co-e-caprolactone) electrospun microfibers and knitted structure: A novel composite scaffold for ligament tissue engineering. J. Biomed. Mater. Res. A **2010**, *94A* (4), 1270–1282.

321. Bashur, C.A.; Shaffer, R.D.; Dahlgren, L.A.; Guelcher, S.A.; Goldstein, A.S. Effect of fiber diameter and alignment of electrospun polyurethane meshes on mesenchymal progenitor cells. Tissue Eng. Part A **2009**, *15* (9), 2435–2445.

322. Hayami, J.W.S.; Surrao, D.C.; Waldman, S.D.; Amsden, B.G. Design and characterization of a biodegradable composite scaffold for ligament tissue engineering. J. Biomed. Mater. Res. A **2010**, *92A* (4), 1407–1420.

323. Surrao, D.C.; Hayami, J.W.S.; Waldman, S.D.; Amsden, B.G. Self-crimping, biodegradable, electrospun polymer microfibers. Biomacromolecules **2010**, *11* (12), 3624–3629.

324. Woo, S.L.; Hollis, J.M.; Adams, D.J.; Lyon, R.M.; Takai, S. Tensile properties of the human femur-anterior cruciate ligament complex. The effects of specimen age and orientation. Am. J. Sport Med. **1991**, *19* (3), 217–225.

325. Mei, F.; Zhong, J.; Yang, X.; Ouyang, X.; Zhang, S.; Hu, X.; Ma, Q.; Lu, J.; Ryu, S.; Deng, X. Improved biological characteristics of poly(L-lactic acid) electrospun membrane by incorporation of multiwalled carbon nanotubes/hydroxyapatite nanoparticles. Biomacromolecules **2007**, *8* (12), 3729–3735.

326. Shang, S.H.; Yang, F.; Cheng, X.R.; Walboomers, X.F.; Jansen, J.A. The effect of electrospun fibre alignment on the behaviour of rat periodontal ligament cells. Eur. Cells Mater. **2010**, *19*, 180–192.

327. Lam, P.K.; Chan, E.S.; To, E.W.; Lau, C.H.; Yen, S.C.; King, W.W. Development and evaluation of a new composite Laserskin graft. J. Trauma **1999**, *47* (5), 918–922.

328. Heydarkhan-Hagvall, S.; Schenke-Layland, K.; Dhanasopon, A.P.; Rofail, F.; Smith, H.; Wu, B.M.; Shemin, R.;

Beygui, R.E.; MacLellan, W.R. Three-dimensional electrospun ECM-based hybrid scaffolds for cardiovascular tissue engineering. Biomaterials **2008**, *29* (19), 2907–2914.

329. Bianco, A.; Di Federico, E.; Moscatelli, I.; Camaioni, A.; Armentano, I.; Campagnolo, L.; Dottori, M.; Kenny, J.M.; Siracusa, G.; Gusmano, G. Electrospun poly(epsilon-caprolactone)/Ca-deficient hydroxyapatite nanohybrids: Microstructure, mechanical properties and cell response by murine embryonic stem cells. Mater. Sci. Eng. C Mater. Biol. Appl. **2009**, *29* (6), 2063–2071.

330. Nisbet, D.R.; Rodda, A.E.; Horne, M.K.; Forsythe, J.S.; Finkelstein, D.I. Neurite infiltration and cellular response to electrospun polycaprolactone scaffolds implanted into the brain. Biomaterials **2009**, *30* (27), 4573–4580.

331. Tillman, B.W.; Yazdani, S.K.; Lee, S.J.; Geary, R.L.; Atala, A.; Yoo, J.J. The *in vivo* stability of electrospun polycaprolactone-collagen scaffolds in vascular reconstruction. Biomaterials **2009**, *30* (4), 583–588.

332. Cao, H.Q.; McHugh, K.; Chew, S.Y.; Anderson, J.M. The topographical effect of electrospun nano-fibrous scaffolds on the *in vivo* and *in vitro* foreign body reaction. J. Biomed. Mater. Res. A **2010**, *93A* (3), 1151–1159.

333. Ifkovits, J.L.; Wu, K.; Mauck, R.L.; Burdick, J.A. The influence of fibrous elastomer structure and porosity on matrix organization. PLoS One **2010**, *5* (12), e15717.

334. Dalton, P.D.; Harvey, A.R.; Oudega, M.; Plant, G.W. Tissue engineering of the nervous system. In *Tissue Engineering*; Van Blitterswijk, C., et al., Eds.; Academic Press: New York, 2008; 611–647.

335. Clements, I.P.; Kim, Y.T.; English, A.W.; Lu, X.; Chung, A.; Bellamkonda, R.V. Thin-film enhanced nerve guidance channels for peripheral nerve repair. Biomaterials **2009**, *30* (23–24), 3834–3846.

336. Seyedjafari, E.; Soleimani, M.; Ghaemi, N.; Shabani, I. Nanohydroxyapatite-coated electrospun poly(L-lactide) nanofibers enhance osteogenic differentiation of stem cells and induce ectopic bone formation. Biomacromolecules **2010**, *11* (11), 3118–3125.

337. Nandakumar, A.; Yang, L.; Habibovic, P.; van Blitterswijk, C.Calcium phosphate coated electrospun fiber matrices as scaffolds for bone tissue engineering. Langmuir **2010**, *26* (10), 7380–7387.

338. Shin, M.; Yoshimoto, H.; Vacanti, J.P. *In vivo* bone tissue engineering using mesenchymal stem cells on a novel electrospun nanofibrous scaffold. Tissue Eng. **2004**, *10* (1–2), 33–41.

339. Srouji, S.; Ben-David, D.; Kohler, T.; Müller, R.; Zussman, E.; Livne, E. A model for tissue engineering applications: Femoral critical size defect in immunodeficient mice. Tissue Eng. Part C Methods **2011**, *17* (5), 597–606.

340. Li, W.J.; Chiang, H.; Kuo, T.F.; Lee, H.S.; Jiang, C.C.; Tuan, R.S. Evaluation of articular cartilage repair using biodegradable nanofibrous scaffolds in a swine model: A pilot study. J. Tissue Eng. Regen. Med. **2009**, *3* (1), 1–10.

341. Chen, J.P.; Su, C.H. Surface modification of electrospun PLLA nanofibers by plasma treatment and cationized gelatin immobilization for cartilage tissue engineering. Acta Biomater. **2011**, *7* (1), 234–243.

342. Wang, W.; Itoh, S.; Konno, K.; Kikkawa, T.; Ichinose, S.; Sakai, K.; Ohkuma, T.; Watabe, K. Effects of Schwann cell alignment along the oriented electrospun chitosan nanofibers on nerve regeneration. J. Biomed. Mater. Res. A **2009**, *91A* (4), 994–1005.

343. Panseri, S.; Cunha, C.; Lowery, J.; Del Carro, U.; Taraballi, F.; Amadio, S.; Vescovi, A.; Gelain, F. Electrospun micro- and nanofiber tubes for functional nervous regeneration in sciatic nerve transections. BMC Biotechnol. **2008**, *8*, 39.

344. Wang, W.; Itoh, S.; Matsuda, A.; Aizawa, T.; Demura, M.; Ichinose, S.; Shinomiya, K.; Tanaka, J. Enhanced nerve regeneration through a bilayered chitosan tube: The effect of introduction of glycine spacer into the CYIGSK sequence. J. Biomed. Mater. Res. A **2008**, *85A* (4), 919–928.

345. Nottelet, B.; Pektok, E.; Mandracchia, D.; Tille, J.C.; Walpoth, B.; Gurny, R.; Möller, M. Factorial design optimization and *in vivo* feasibility of poly(epsilon-caprolactone)-micro- and narofiber-based small diameter vascular grafts. J. Biomed. Mater. Res. A **2009**, *89A* (4), 865–875.

Excipients: Pharmaceutical Dosage Forms

Luigi G. Martini
Institute of Pharmaceutical Sciences, King's College, London, U.K.

Patrick Crowley
Callum Consultancy, Devon, Pennsylvania, U.S.A.

Abstract

Virtually all medications contain excipients. They can enhance product performance, patient acceptability, and compliance, providing a more effective or safer medication. This entry is focused on excipients used in parenteral dosage forms and also discusses the current state-of-the-art with respect to commercial products.

INTRODUCTION

Virtually all medications contain excipients. They can enhance product performance, patient acceptability, and compliance, providing a more effective or safer medication. Inclusion in dosage forms may enable or enhance manufacturing operations and product quality. At the same time, they can interact with the active ingredient, with product or packaging components and compromise performance. It is therefore important that during product and process development, manufacture, testing, troubleshooting, or process enhancement account is taken of excipient behaviors. An overview of such possibilities is presented in this entry.

SCOPE OF ENTRY

Oral dosage has traditionally been the most popular mode of administering medication. This is reflected in the number of excipients available for such product forms. Many have been available for some time and have been covered extensively in textbooks, specialized publications like the *Handbook of Pharmaceutical Excipients*, and in the 5th edition of this encyclopedia. In the light of such cover and the relative paucity of any new excipients, this entry does not seek to replicate such information. Rather, there is a greater focus on excipients used in parenteral dosage forms. The content reflects not only the current state-of-the-art with respect to commercial products but also concepts that are at preclinical/proof of concept stage as some of these are likely to progress to commercial products.

Dosage Form Design

Parenteral dosage forms are increasingly being more widely used for delivering medications. This is reflected in the new drug approvals by FDA during 2010. Twelve of the 21 approvals concerned parenteral preparations. Reasons for the increasing prominence and popularity of parenteral dosage include

- The promise and prevalence of biopharmaceuticals, whose peptide nature renders them unsuitable for oral administration.
- There is now ample knowledge that for drugs with a steep dose response, or those that require prolonged plasma presence for optimum efficacy, the gastrointestinal tract can be an inconsistent route of delivery. Parenteral administration can provide more consistent plasma levels and possibly more reliable drug release from modified release systems.
- Development of patient-friendly modes of parenteral administration. These include micro-needles and even needle-free technologies. Consequently, parenteral administration is now less daunting.

Historically, parenteral administration has, with a few exceptions, employed simple formulations, usually aqueous solutions. Excipients were generally restricted to additives such as tonicity modifiers, buffers, and occasionally, stabilizers such as antioxidants and preservatives. Additional excipients are now being used. Drug discovery programs are delivering drug candidates with poor aqueous solubility. This has stimulated interest in agents or vehicles that may provide solution formulations. Targeting specific tissues or organs is also evincing interest. The landscape for parenteral medication is changing accordingly, and excipients are becoming increasingly important as functional aids. Thus, knowledge of excipients, their functions and interactions, is a key requirement for parenteral dosage form design.

Excipient Functionality

A parenteral product needs to possess the following attributes:

- be sterile and endotoxin-free
- be free from animal-derived viral contaminants

Excipients—Gels

- have good appearance, being free from particulate matter if a solution (meets pharmacopeial standards for allowable particulate contamination)
- contain and consistently deliver an accurate dose through an appropriate administration device
- is readily constituted as a liquid for administration if a solid (e.g., a lyophile)
- be well tolerated, not causing pain or irritation on administration
- retain its quality-related attributes through product lifetime and use
- not interact with packaging or administration components
- be prepared by a robust, reliable, and reproducible manufacturing process

In some cases additional requirements may apply, that is,

- be capable of being mixed with appropriate diluents for infusion
- be miscible with some other medications for simultaneous infusion

These requirements can influence and constrain but also be facilitated by excipient selection. Therefore, excipients for parenteral products should ideally meet the following requirements:

- have no biological/pharmacological activity and be safe on parenteral administration
- be nonirritant and decrease any drug-related irritancy at the site of injection
- have high aqueous solubility and dissolve rapidly (if the product is a lyophile and administered as an aqueous preparation)
- be capable of being prepared in sterile form, be free from endotoxins, or be suitable for a product manufacturing process that can deliver such attributes
- be particle-free (if a solution product is being designed)
- have consistent attributes and be free from undesirable residues
- their mode of manufacture (the excipients) is consistent and known to the dosage form design team

Requirements with respect to pharmacological activity and intrinsic safety are self evident and immutable, so are not discussed here. Safety can also be compromised by microbial presence, by endotoxins resulting from contamination, or by viral residues.

Microbial quality

Sterility and feasibility for sterilization are prime requirements for parenteral products. Ideally a product is rendered sterile (and possibly depyrogenated) at the final stage of manufacture, that is, in its final pack. However, lability to heat and to radiation sterilization rules out this

approach for many drugs and excipients. Filtration sterilization, coupled with "clean" processing environments, with associated process checks and precautions is increasingly the norm. Providing finished product with low endotoxin levels means that input ingredients (drug and excipients) must be substantially endotoxin-free as well as of low bioburden.

Most "small molecule" drugs can be isolated (at crystallization) in sterile mode; the mode of manufacture and isolation usually can ensure that they are not pyrogenic. Biopharmaceuticals and some excipients of biological provenance can pose greater challenges. Manufacture may utilize fermentation or other live cell–associated processes or derivatization from animal tissues or natural materials. Downstream processing and purification may reduce or eliminate endotoxins but could also conceivably introduce them. Water of appropriate quality is invariably used in biopharmaceutical downstream processing, usually along with depyrogenations steps. Excipients, however, may be manufactured primarily for use in foodstuffs, cosmetics, or agricultural products where endotoxins may be of less concern. In such cases, sub-batches for use in parenterals may have to be subjected to additional processing for endotoxin reduction. Excipients for parenteral formulation such as sucrose, dextrose, mannitol, cyclodextrins, and trehalose are now available with low endotoxin levels. Apyrogenic buffer salts are also available but are not listed here in the interests of simplicity.

Viral contamination

Viral residues are a concern where materials are of animal or human origin (e.g., gelatin, bovine serum albumin, or human serum albumin). In the case of albumin, virus removal by filtration may be complicated by its propensity to be adsorbed by cellulose-based filters. Heat-sterilization (pasteurization), for example, heating for 10 hours at 60°C, possibly in the presence of low levels of a stabilizer is reportedly effective for viral inactivation of albumin. Use of recombinant albumin eliminates the risk of donor-derived contamination.

Other viral inactivation techniques, mostly applicable to blood-derived products, involve using a detergent such as octylphenyl ethoxylate (Triton × 100) to disrupt the viral lipid coating. Some influenza vaccines contain residues of this material; it is used to disrupt the virus during manufacture.[1]

Solubility and ready dissolution

Excipients in a lyophile or other solid must readily pass into solution and possibly enhance the rate of drug dissolution. Buffering agents may be included for this purpose, providing optimum pH for drug solubility. Other solubilizing techniques are addressed later in this entry.

Rate of dissolution (of drug or excipient) could change during product shelf life, possibly due to solid-state changes (e.g., amorphous to crystalline). It is important therefore to devise a suitable "dissolution rate" test, with a time-defined endpoint for use in stability studies. Verbiage in specifications like "passes readily into solution" or "reconstitutes readily" will not provide evidence that progressive changes to dissolution behavior are occurring over time.

Appearance

Solution products should ideally be "clear and colorless" for aesthetic reasons. Processing and material selection should be designed, where possible to confer clarity. In practice, active ingredients, particularly in concentrated solution, may provide what is euphemistically termed "off-white" (pale yellow) coloration. This may be unavoidable with materials of biological origin, and usually there is little that can be done, additive-wise, to change such appearance. Materials derived from biological sources (drug or excipient) may also have a tendency to foam if shaken vigorously. The silicone, polydimethyl siloxane, is sometimes added in low amounts during downstream processing of biologicals to obviate such effects and is probably carried through to product.

The pH of aqueous solutions, prepared with Water for Injection can increase on prolonged contact with glass. If the formulation contains drug as a salt form, with a steep pH-solubility profile, precipitation of drug may occur. It may be advisable in such circumstances to include a buffer in the formulation, rather than adjusting pH with acid or base. Use of buffers and solubilizers to attain and retain the requisite drug solubility is discussed later.

Dose accuracy may not be an excipient-related issue where product can be formulated or administered as a mobile solution: Accurate dosing in such instances is probably a function of adequacy of vial fill (fill overage) along with accuracy of withdrawal and administration, or of delivery from a more sophisticated dosing device. Suspension products however may pose challenges such as uniformity of content (homogeneity) of the dose for withdrawal and ease of withdrawal and administration ("syringeability"). Vial drainage to facilitate withdrawal is often addressed by coating vials with silicone. An injectable composition of conjugated estrogens (Conjugated Estrogens for Injection, USP) contains 0.2 mg simethicone (dimethyl polysiloxane), but its purpose is not disclosed.[2]

Nanoparticles are evincing much interest as a means of enhancing rate of drug dissolution, or acting as targeting agents, their submicron size allowing intravenous administration.[3] These and other particulate dosage forms need to be homogenous dispersions during administration. However, drugs suspended in liquid vehicles may sediment on standing. This can lead to caking, or can affect withdrawal and delivery accuracy if sedimentation is rapid. Strategies to obviate such issues may be multifaceted. Surfactants may be added at low levels to reduce agglomeration. Minimizing density differences between drug and vehicle, if feasible, may help. A viscosity enhancer or other type of suspending agent could be included. An ideal system should be thixotropic, having high viscosity at low shear to prevent sedimentation on standing or following shaking, but being mobile (low viscosity) at higher shear (shaking) to make homogenous and ensure acceptable syringeability. Agents such as carboxymethyl cellulose sodium, lecithin, and Povidone are listed as additives in parenterals containing insoluble active ingredients.[4,5] Their functions in the referenced products are not disclosed but they are likely to aid dispersibility and enhance viscosity. Gelatin is also used in parenterals and may have a viscosity-enhancing effect. Matching the rheological behaviors of such additives to the characteristics of the drug substance and other formulation components is not a trivial undertaking and has to be done on a case by case basis.

Tolerance at injection site

Irritation, pain, or stinging at the site of administration may be caused by the intrinsic properties of the drug, the tonicity or pH of the formulation, volume or rate of administration, or possibly precipitation of drug. A nonaqueous or partially aqueous vehicle may also cause discomfort. Alleviation using excipients is not common, but benzyl alcohol and lidocaine have been used in high-dose intramuscular preparations, presumably, to provide mild local anesthesia. This practice does not seem to be employed with newer products, possibly because such agents could be categorized as "active ingredients." For spinal injections of therapeutic agents, lidocaine may be administered separately. Lidocaine may also be applied in gel form at the injection site before administration.

If irritation is ascribable to high or low solution pH (necessary for adequate drug solubility) and product is likely to be administered chronically, frequently, or to pediatrics, it may be worth considering an alternative mode of enhancing solubility. Pilocarpine for ophthalmic instillation is unstable at physiological pH. Cyclodextrins (qv) have been used to improve its solubility and stability at neutral pH, providing a better tolerated preparation. Similar strategies could be explored for parenterals.

Salts such as lysine or arginine are reported to reduce pain on intramuscular injection of cephalosporins. Maleic acid reputedly had the same effect with chlordiazepoxide. It is not known whether mechanisms were pH related, for example, avoiding precipitation, or due to some other effect.

Quality retention

Changes that can compromise the quality attributes of a product are manifold. They can concern chemical-, physical-, or microbial-related changes or any combination of these. The breadth of the topic mandates that discussion in this entry be limited to deterioration due to loss of excipient functionality or to drug-excipient interactions. Other entries in this compendium provide a more comprehensive coverage of the topic.

The influence of excipients on quality attributes is the subject of later paragraphs.

Choosing Excipients

Excipients can be used to confer tonicity or act as buffers, preservatives, stabilizers or solubilizers. Agents that modify drug release from the site of administration or target a specific organ or tissue may be included in more sophisticated formulations. Some excipients may serve more than one purpose. In all cases, any additives must be compatible with the active ingredient. Excipient-excipient interactions are also possible and need to be anticipated and avoided.

Tonicity modifiers

Sodium chloride and dextrose are probably the most widely used materials to make solutions for injection isotonic. Sodium chloride can conceivably depress solubility of a sparingly soluble drug, if the drug is a chloride salt, due to a common ion effect or by increasing solution ionic strength. Such possibilities are usually explored and identified at formulation design stage. Dextrose is nonionic, but if the drug product contains electrolyte and is sterilized by autoclaving some isomerization to fructose and 5-hydroxymethyl furfural may occur.[6] These may in turn interact with amino groups in the drug during long-term storage, forming novel molecular constructs. Potential for such occurrence should be explored in dosage form design studies but it may also be advisable to monitor for dextrose isomerization products during longer stability studies, illustrating how it may sometimes be appropriate to monitor "excipient stability" in a program.

Mannitol can also be employed as a tonicity modifier. In lyophilized products, it can also function as a lyophilization aid. Amino acids such as arginine, cysteine, histidine, and glycine have also been used, particularly in biopharmaceutical products where they can also function as aggregation inhibitors.

pH adjustment and retention

Citrates and phosphates are probably the most widely used buffers. Citrate can interact with glass surfaces at neutral pH to form soluble silicon complexes, leading to visible corrosion of the glass.[7] The effect can be accentuated by materials such as gluconate salts or EDTA. Autoclaving is likely to exacerbate the problem, but the issue may sometimes only be noticeable after long-term solution storage.

It is unlikely to be an issue with lyophiles where the product is in solution for a limited time. However, the possibility is worth considering in cases where solutions may be constituted and stored for a period (e.g., over the weekend) as part of a Hospital Pharmacy Services program. Citrate-glass interactions are not likely at solution pH below 5.0.

Ascorbic acid, tartaric acid, and acetate salts have also been used for pH adjustment or as buffers. They do not sequestrate silicon from glass surfaces. Ascorbic acid can also act as a stabilizer (antioxidant) but is susceptible to pH-sensitive degradation on autoclaving and during long-term storage.

Preservation

Preservatives have traditionally been included in multidose containers to guard against contamination on repeated dose withdrawals. Table 1 lists those currently used in parenterals. Most are relatively simple organic compounds and are of long standing. No new preservatives have become available in more than half a century.

Noteworthy properties of some of the materials listed in Table 2 include

- Volatility: Phenols, benzoic acid, benzyl benzoate, parabenzoates, and the listed alkyl/aryl alcohols are volatile or can sublime. Inclusion in a formulation to be lyophilized will result in losses during the secondary drying stage of the cycle. Migration through elastomeric vial seals may also occur.
- Cationic surfactants have a strong positive charge so are likely to interact with negatively charged excipients such as carboxymethyl cellulose sodium.

Preservatives may have pH-solubility profiles that may determine their suitability for the product in question. They can also interact with the drug substance, other excipients, or the container/closure. Benzyl alcohol, 2-phenoxy ethanol, and m-cresol should not be co-formulated by nonionic surfactants. The para-aminobenzoic acids, being esters are susceptible to pH-dependent base catalysis, whereas preservatives that can be volatile (benzoic acid, parabens, phenols) may be depleted due to losses through elastomeric closures or plastic vials. Thus, preservative inclusion levels and product specifications should take account of the minimum effective preservative concentration and losses during product shelf life.

Preservatives have differing modes of action, with various microbial cellular targets such as the microbial cell wall, cytoplasm, or cytoplasmic membrane. The antimicrobial spectrum can depend on factors such as preservative concentration and product pH and presence of other agents. This diversity of targets can also enable the judicious combinations of preservatives to evince synergistic effects and broaden the spectrum of activity.

Table 1 Preservatives used in parenteral products

Class	Preservative	Comment
Alkyl/aryl alcohols	Benzyl alcohol, 2-ethoxy ethanol	Higher concentrations of, benzyl alcohol used as solubilizer
Alkyl/aryl acids	Benzoic acid/Na salt, sorbic Acid	Product pH must be low
Alkyl/aryl ester	Benzyl benzoate	Also used as solubilizer or as non-aqueous solvent
Amino aryl acid esters	Parabens (methyl/ethyl/propyl esters of p-amino benzoic acid)	Combinations usually used. Can sublime/hydrolyze
Biguanide	Chlorhexidine	Can be effective against some viruses
Cationic surfactants	Benzalkonium chloride / Benzethonium chloride	Deactivated by anionic agents; can damage cytoplasmic membranes
Phenols	Phenol, 3-cresol	3-cresol is not suitable for lyophiles
Organic mercurial	Thiome rsal, phenyl-mercuric acid/salts	Usually found in mature products/vaccines

Table 2 Surface active agents/solubilizers

Material	Common name
Phosphatidyl choline	Lecithin
Polyoxy 35 castor oil	Cremophor EL
Polyoxyethylene-650 hydroxystearate	Solutol H15
Polyoxyethylene sorbitan monolaurate	Polysorbate 20/Tween 20
	Polysorbate 40/Tween 40
	Polysorbate 80/Tween 80
Sorbitan monolaurate	Span 20
Sorbitan monopalmitate	Span 40
Sorbitan monooleate	Span 80
Sorbitan trioleate	Span 85
Polyoxyethylene polyoxypropylene block copolymer	Poloxamer 188/Pluronic F68
7-desoxycholic acid sodium	Bile salt

Some preservatives have been linked with adverse reactions and consequently there is a school of thought that parenterals should be single dose and preservative free. It is possible, therefore, that while the materials in Table 2 may all be present in existing products, gaining future approval for use of some of these in new products may be difficult with some regulatory agencies. Such attitudes need to be balanced against other considerations, and countered where appropriate. Single-dose administration may not be practical for economic or logistical reasons in some regions. Additionally, some new drugs only work or work best if administered concurrently with long-established medications. A preservative may be present in the mature product. "Premixing" can involve manipulations akin to multidosing.

Many novel drugs, particularly those "personalized" for patients with a specific genetic makeup/disposition, disease subset, or its stage that can dictate dose level are very expensive. The condition of the patient (who may be seriously ill) can determine the maximum tolerated dose. Dosing may be patient, condition, or disease stage-specific or may reflect patient biomarker measurements. In essence, several considerations and inputs may lead to the decision on dose (volume of product) to be administered. It would be impossible to provide dose units that are "personalized" to cover such possibilities. Multidose presentations may be required in the light of the high cost of (for instance, of anticancer medications) and the complexities of disposing unused material. Furthermore, the prevalence of patient-friendly devices such as auto-injectors, which facilitate parenteral administration to treat chronic conditions, needs to be considered. The potential for undesirable reactions needs to be balanced against the benefits of better modes of dosage, enhanced compliance, and better therapy with such devices, with a consequent need for multidose presentations. In the light of such considerations, common sense, foresight, and risk/benefit considerations may dictate that preservatives continue to be used in parenterals. Interestingly a recently approved product for self-administration by the patient with dose, based on a "biomarker" reading, contains phenol as a preservative.[8]

Lyophilization aids

Some active ingredients, whether of chemical or biological origin may not readily lyophilize. The physical properties of the frozen aqueous plug may cause poor heat transfer, sublimation, or even physical losses due to friability under high vacuum. Inclusion of lyophilization aids such as mannitol, dextrose, or sucrose can remedy such liability. These additives can have dual functions, conferring isotonicity or, in the case of sucrose, acting as an antiaggregant in biopharmaceutical products. Other stabilizers include the amino acids L-arginine and L-cysteine: They may also aid dissolution on constitution.

Solubilizers

Discovery programs regularly select candidate drugs based on potency and specificity. Such materials can be poorly soluble and poorly absorbed on oral dosage. Parenteral administration, if warranted by the severity of the clinical condition, the mechanism of action or (reduced) dose frequency may be a viable option if the drug candidate can be formulated for delivery, preferably as a solution. If the compound is not ionizable and a suitable soluble salt cannot be prepared, nonaqueous solvents (as cosolvents), solubilizing agents, or a combination of these may have to be considered.

Cosolvents. Cosolvents are one of the simpler ways to boost solubility, provided that amounts (inclusion levels) are low, the drug has adequate solubility in the solvent mixture, and drug stability is not affected. Nonaqueous solvents used in parenteral products include ethanol, glycerol, propylene glycol and polyethylene glycol, or mixtures of these. Benzyl alcohol is also used on occasion. They are invariably employed as cosolvents with water or are diluted with an aqueous system before administration. Dose (volume), rate, frequency, and duration of administration as well as severity or intractability of the clinical condition need to be considered when considering the maximum acceptable dose and concentration of cosolvent in an aqueous system.

Many drugs may not have adequate solubility in the limited number of available solvents that can be used, or solubility may decrease markedly on admixture/dilution with aqueous systems. Furthermore, precipitation of drug following injection cannot be discounted, with the potential risk of embolism (IV administration).

Surfactants as Solubilizers. Surfactants may solubilize drugs by forming structures such as micelles with hydrophobic interiors in which hydrophobic drugs dissolve. They are usually derived from vegetable oils or fatty acids/esters that have been rendered more hydrophilic by covalently linking with polyoxethylene or sorbitan moieties to balance hydrophobic and hydrophilic properties. Sterols such as lecithin or bile salt variants have also been used. Table 2 illustrates some examples.

The surfactant inclusion level has to exceed its critical micelle concentration if it is to accommodate drug, and many products are diluted with Water for Injection for administration. A wide range of micelle-forming materials are being evaluated to determine effects of structure, concentration, and polymer combinations to control, target, or otherwise modify drug release and disposition.[9]

Some solubilizing agents have been linked with adverse reactions. Pediatrics and the newborn can be particularly sensitive. Thus, products incorporating them are usually limited to serious clinical conditions, where the therapeutic agent may be of significant benefit, or where dosing is limited or infrequent. The solubilizer Cremophor EL, a polyethoxylated castor oil, illustrates the conundrum that the formulator may face. It is incorporated in parenteral formulations of the anticancer

agents, paclitaxel, docetaxel, and ixapebilone; the immunosuppressant, cyclosporine; and the antifungal miconazole. It is possible that without this material these valuable therapeutic agents would not be available as practical medications. There have been reports of anaphylactoid hypersensitivity, erythrocyte aggregation, and peripheral neuropathy ascribable to its presence. Yet, in the absence of equivalent medication for such severe conditions, there may be little alternative to making decisions on such enabling excipients using risk–benefit considerations.

Hydrotropes. Concentrated aqueous solutions of a wide variety of organic agents are capable of dissolving poorly soluble drugs, possibly due to associations between the drug and hydrotrope in aqueous solution. Solubility can be increased by orders of magnitude. Table 3 lists examples.

Many of the materials in Table 3 are clearly not suitable for parenteral administration (they are included here to exemplify the concept). Furthermore, there are no reports of such solubilization techniques being used to date in commercial parenteral products. A potential complication is that the high concentrations required for solubilization may result in a significant dose of the hydrotrope. Precipitation of drug on dilution is also possible. However, studies are ongoing to link covalently low molecular mass hydrotropes with polyethylene glycol-based hydrophilic

Table 3 Hydrotropic agents and solubility enhancement

Hydrotrope	Drug
Nicotinamide	Allopurinol, diazepam, griseofulvin, indomethacin, moricizine, nifedipine, estradiol, omniquinine, progesterone riboflavin, testosterone
N,N-diethylnicotinamide	Diazepam, griseofulvin, nifedipine
Sodium benzoate	Allopurinol, carbamezapine, etoposide, indomethacin, ketoprofen, nalidixic acid, nifedipine, omniquinine
Sodium p-aminobenzoate	Phenacetin
Sodium salicylate	Indomethacin, etoposide, ketoprofen, nifedipine, piroxicam
Resorcinol	Riboflavin, nalidixic acid
Piperazine	Nimesulide
Sodium butyl monoglycol sulfate	6-amino penicillanic acid
Lysine, urea, gentisic acid, ethanolamide	Acetazolamide

Source: From Kim et al.[10]

Table 4 Characteristics of cyclodextrins

Parent cyclodextrins	α	β	γ
Glucose units	6	7	8
Water solubility (g/100 mL)	14.5	1.85	23.2
Cavity diameter (Å)	4.7–5.3	6.0–6.5	7.5–8.3
Cavity volume (Å)	~174	~262	~472

Table 5 Solubilities of modified cyclodextrins

Form	Degree of substitution	Aqueous solubility (g/100 mL)
Hydroxypropyl	1	0.3
	8.5	50
Sulfobutyl	1	>50
	7	>50

structures to form hydrotropic polymer micelles.[11] The solubility of the anticancer agent Paclitaxel was dramatically improved in such systems.[12] The safety and tolerance of such novel constructs needs to be established but the concept offers possibilities for solubility enhancement along with altered drug disposition and elimination to render them safer and more effective. A comprehensive account of hydrotropes as solubilizers is provided elsewhere in this encyclopedia.

Complexing Agents. If a hydrophobic drug can interact with a complexing agent that is water soluble, such that the complex "disguises" the hydrophobic moiety, while presenting the hydrophilic component of the complexing agent to the aqueous environment, it may be possible to design a soluble drug product. The cholesterol metabolite sodium cholesteryl sulfate has been used successfully as a complexing agent for the antifungal, amphotericin B. Complexation also reputedly alters the distribution and elimination of this drug, reducing systemic toxicity.[13]

Cyclodextrins are cyclic glucose polymers with a hydrophobic inner core and hydrophilic external groups. Table 4 lists the features of the so-called parent (original) cyclodextrins. Cavity diameter and volume is a function of the number of glucose units (although there is no relationship to water solubility).

The dimensions of the annulus, along with the pendant groups within, are such that molecules of molecular mass of about 500 with aromatic structures are readily accommodated and anchored within the cavity. Many small molecule drugs have such structural features. The formed complex essentially acquires the solubility characteristics of the cyclodextrin, being a function of the groups external to the cavity.

The original α, β, and γ forms of cyclodextrins have limited aqueous solubility. This restricts utility to low-dose

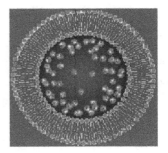

Drug

Phospholipid/cholesterol bilayer

Fig. 1 Schematic of liposome.

drugs. Furthermore, in the case of β cyclodextrin its low solubility was alleged to be associated with renal nephrotoxicity.[14] Modified cyclodextrins in contrast have much better solubilities, being one of the few truly novel excipients with solubilizing potential that have become available in recent decades. Solubility of the hydroxypropyl form is a function of degree of substitution, in contrast to the sulfobutyl form (Table 5). There are no reports of nephrotoxicity on either form and they have been shown to be "clean" in wide ranging animal safety studies. Both are commercially available in low-endotoxin form.

Use of cyclodextrins as excipients is not without complexity. Not only must the guest drug molecule and host cyclodextrin form a stable complex from a solubility and formulation perspective but drug must also dissociate *in vivo* to exert its pharmacologic effect. Such "release" may however occur more readily on parenteral dosage. Endogenous materials like cholesterol and triglycerides are useful displacement agents; release also occurs on dilution. An exception concerns the modified cyclodextrin [(2-Carboxyethyl) thio-γ-cyclodextrin]. This forms a strong complex with the neuromuscular agent, rocuronium, and is used to reverse neuromuscular blockade, postoperatively (this mode of action makes its categorization as an excipient questionable and there are no records of it being used in more conventional medications).

There have been notable successes using modified cyclodextrins as solubilizers in parenteral products (Table 6) and the list of applications is likely to grow.

A wider-ranging account of cyclodextrins, including their utility for other dosage form types, is provided in Entry 54 of this encyclopedia.

Other Solubilizing Agents. If a drug is sufficiently soluble in a lipidic material (or combination of lipids) it may be suited for delivery as an oily solution or as an emulsion. Oil-based solutions are only administered by intramuscular injection and are usually painful, but can provide sustained diffusion (release) to systemic compartments. Solvents may be vegetable oils (e.g.,

Excipients–Gels

Table 6 Parenteral dosage forms utilizing cyclodextrins

Cyclodextrin	Therapeutic compound	Therapeutic area	Route of administration
α Cyclodextrin	Alprostadil PGE$_1$	Erectile dysfunction	Intra-arterial
2-hydroxypropyl-β-cyclodextrin	Itraconazole	Antifungal	IV
	Mitomycin	Oncology	
	Telavancin HCl	Anti-infective	
2-hydroxypropyl-γ-cyclodextrin	TC-99	Oncology	
	Teoboroxime		
	Amiodarone HCl	Antiarrhythmic	
SBE7-β-cyclodextrin	Voriconazole	Antifungal	IM
	Ziprasidone	Antipsychotic	
	Aripiprazole	Mental Disorders	
(2-Carboxyethyl) thio-γ-cyclodextrin	Cyclodextrin only	Drug antidote (neuromuscular blockade)	IV

arachis oil), but benzyl benzoate has also been employed (at lower levels), probably as a cosolvent. Hormones such as progesterone, testosterone, and estradiol have been formulated in this way. Solubility and dose volume constraints may limit utility to such low-dose drugs.

Oil-in water emulsions are better tolerated than oily injections and can be administered intravenously. Oleaginous components can include vegetable oils, fatty acids, their derivatives, or mixtures of these. Esters are also included in this category in the interests of simplicity. Oils derived from marine mammals have been reported as having potential for parenteral administration but they seem to be targeted to parenteral nutrition medications rather than for drug delivery per se.

The fat-soluble Vitamin E derivative (Tocopherol Polyethylene Glycol Succinate) is attracting much interest for oral dosage because of its PgP-inhibiting properties. Its micelle-forming capability may be useful for parenteral dosage forms, and it has been considered as a solubilizer for paclitaxel, potentially replacing polysorbate 80.[15]

Peroxides can be formed in vegetable oils and fatty acids during storage and heating (e.g., sterilization). Antioxidants are often included to retard such formation. These additives can conceivably vary with material source (Vendor). Furthermore, the antioxidant may be depleted over time, as it evinces its stabilizing effect. Thus, there could be batch- or provider-specific effects on drugs susceptible to oxidation. Quality by Design precepts requires that such possibilities are explored using risk-assessment techniques during dosage form design. Suitable vendors, quality standards, storage conditions, and use periods for such oily materials can then be identified and defined. Supply agreements should incorporate change control systems mandating that any changes in additive or manufacturing process is notified to

customers prior to implementation so that potential consequences for product quality are explored and the proposed change is validated.

In summary, there are a number of different strategies and associated materials to enhance solubility. In practice, no single technique may be adequate: it may be necessary to use a combination of approaches[16,17] provide excellent reviews and lists of products that illustrate such multifaceted exercises.

Liposomes as Parenteral Products

Phospholipids and sphingolipids have the unique propensity for self-assembly in aqueous environments. They form uni- or multilamellar structures, with an aqueous interior and lipidic exterior. This property has been exploited to provide "drug carriers" for parenteral administration that can solubilize lipophilic drugs (water-soluble drugs can also be accommodated), target a drug to a specific tissue, or organ or alter its disposition and excretion rate. The drug may be "encapsulated" in the interior of the liposome or be dissolved in the lipid component of the bilayer (or a combination of both).

The membrane formers are usually phospholipids: Lamellar structure and "order" can be enhanced by materials, such as cholesterol or its derivatives, preventing drug leakage (an issue with earlier liposomes). Such components, being endogenous materials and ubiquitous in foodstuffs, can be considered to be safe. Various forms and sources of such primary components afford many opportunities for the design of compositions with different features or performance attributes. The sphingolipid ceramide is reputedly less susceptible to oxidation than other forms. Lecithin, a mixture of phospholipids, triglycerides, glycerol, and fatty acids can be sourced from vegetable oils or egg yolk. Subtle differences in material properties may influence liposome formation, stability, or possibly drug disposition and clearance.[9]

Liposome manufacture is complex. Drug loading, liposome size, and lamellarity (uni or multi), propensity for leakage, and so on can be affected by processing variables, material composition, and quality. Sterility may be a challenge due to the "particulate" nature of the system. Product and process design may be best performed by specialist organizations.

Targeted Drug Delivery

Systems that can localize cytotoxic drugs in organs or tissues have the potential to improve efficacy and reduce general toxicity. The lymphatic system is a good example with respect to delivery of anticancer therapy, being a primary pathway for tumor metastasis. There is evidence that immunomodulators are more active in the lymphatic system rather than in blood, lymphatics being a more integral part of the immuno defense system. It could be an important thoroughfare for delivery of drug treatments. Lymph nodes are essentially "filters" where particulates are trapped. Thus, polymeric microparticles, lipid structures (such as liposomes), or soluble macromolecules associated with a payload of an anticancer drug for delivery to and prolonged residence at lymph nodes should provide better targeted and more effective therapy.[18]

Oral delivery through the lymphatic system requires that the drug be highly lipophilic (Log P > 5) and be reasonably soluble in long-chain triglycerides (>50 mg/mL). Parenteral delivery is less dependent on such requirements. Cancer therapy is used here to exemplify the concept but targeting fungal infections could be another example. Targeting strategies include forming covalent linkages between drug and a lymphotropic agent (effectively a prodrug concept). Other strategies may concern encapsulating or otherwise "anchoring" the drug, for release at the target. There is

Table 7 Lymphatic targeting strategies

Lymphotropic carrier	Model drug
Dextran sulfate	Bleomycin
Dextran	Mitomycin C
Cyclodextrin	1-Hexylcarbamoyl-5-fluorouracil
L-Lactic acid oligomer microsphere	Aclarubicin, cisplatin
Gelatin microsphere	Mitomycin C
Styrene-maleic acid anhydride co-polymer	Neocartinostation
Liposome	Ara-C
Lipid mixed micelle	Interferon, tumor necrosis factor
Chylomicron, LDL	Cyclosporine
Carbon colloid	Mitomycin C

Source: Taken from Muranishi.[18]

evidence that molecules of molecular mass of more than 5–10,000, if injected subcutaneously, are transported mainly through lymph vessels. Appropriately, large particles and/or macromolecules may act as lymphotropic targeting agents. Materials such as oligomers, micellar agents, microspheres, or macromolecules have been considered for such purpose, as listed in Table 7. It is not known at this time whether these concepts are being tested or shown to be effective in clinical studies.

Modified Release Parenterals

Low-dose drugs, for replacement therapy or chronic treatment, are increasingly being developed in sustained release form, for parenteral administration, or as implants. The intrinsic pharmacology of the drug may allow less frequent administration or a prolonged effect may reflect prolonged "residence at site of delivery." This can contrast favorably with oral dosing where GI tract residence time is limited and more variable.

Sustained residence may be achieved by using drug with very low aqueous solubility, possibly linked to controlled crystal size ("macrocrystals" control dissolution of testosterone as the isobutyrate salt). Other modes of prolonging release/dissolution at site of injection include:

- Formation of a drug adsorbate. Aluminum hydroxide is used as an adsorbent/adjuvant in vaccines. There are no reports of its use with other parenterals but its excellent adsorption properties could make it equally useful with selected (low dose) small-molecule drugs or biopharmaceuticals. pH of suspensions are near neutral, particle size can be low, and surface properties mean that it behaves as a colloidal dispersion in aqueous suspension.
- Controlled release formulations akin to those used for oral dosage, that is, release is controlled by rate of permeation across a barrier or diffusion through a matrix. Dextrans, gelatin, polylactic acid, and lactide-glycollide copolymers are being widely evaluated, especially in preclinical models.
- Cross-linked polymeric systems: Biodegradable polymers based on D-L lactide, L-lactide (polylactide), glycolide, and copolymers of these are used as bioresorbable materials in thermosetting scaffolds for implants, stents, surgical plates, and other tissue/joint supports or replacements. They are also being considered as release-modifying agents, particularly where prolonged release (several months) is feasible. Photopolymerized PEG-PLA systems are also reputedly useful for controlling diffusion of hydrophilic molecules such as biopharmaceuticals, providing essentially zero-order release profiles (Fig. 2).[19]

Much of the published work on such modified release systems concern preclinical models but it is very probable

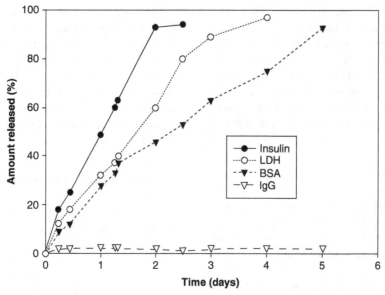

Fig. 2 Protein release from photocrosslinked biodegradabel hydrogel.
Source: From Bari.[19]

Table 8 Residues that may be present in excipients

Excipient	Residue	Comment
Benzyl alcohol	Benzaldehyde	
Lactose	Aldehydes, reducing sugars.	
Mannitol	Sorbitol, maltol, isomaltol	
Povidone, polysorbates	Peroxides, formaldehydes, formic acid	
Polyethylene glycols	Free ethylene oxide, 1-4 dioxane, ethylene glycol, diethylene glycol, peroxides,	Air oxidation can form peroxides, aldehydes and carboxylic acids. Levels can vary with supplier
Polysorbates	Peroxides	Can accumulate over time
Vegetable oils, lipids	Antioxidants, peroxides	Antioxidants added to avoid oxidation

that some of these will progress to clinical evaluation and patient use.

Impurities in Excipients

Many excipients are polymeric in nature or are of natural origin. They are rarely crystalline. Consequently, they are difficult, if not impossible to prepare in highly pure form. They may be complex mixtures or contain low levels of residues, including monomers or low-polymer variants. Levels of such residues may vary with source, between

batches, or may change over time. They may have little impact on functionality, safety, or other attributes but uncertainties as to presence and possible impact make it advisable to acquire knowledge of their presence and consider and explore potential consequences for quality or performance.

Pharmacopeial monographs have historically been deficient in listing low-level impurities or additives in excipients. This may have been understandable when excipients were considered as "inert." Providers of excipients have also, at times been reluctant to divulge such details, stating instead that their materials meet compendial standards, that they possess a Drug Master File (US), or that details of material manufacture and isolation processes are confidential and cannot be divulged. Thankfully, things are changing. Monographs on vegetable oils and lipids now invariably require that the nature and level of any added antioxidant be listed. Residues in polymeric excipients, mostly monomers or structural deviants are also increasingly being named in monographs. This makes it easier to consider the potential for and risks of interaction. In a safety/side effect context groups responsible for dosage form design and quality assessment should acquire or generate knowledge of potential excipient residues. This is particularly important with parenterals, which do not encounter the barriers to systemic entry that are present on oral dosage and which, if liquid may offer greater possibilities for interactions. Residues that may be present in excipients used in parenteral products are listed in Table 8.

Such residues albeit present at low levels could interact with the drug and generate reaction products of unknown biological activity, particularly where the excipient-to-drug ratio is high. Residues may also change during storage or handling. Ethoxylated surfactants such as the Tweens

(Polysorbates) are readily oxidized by atmospheric oxygen to hyperperoxides, peroxides, and carbonyl compounds such as formaldehyde. Peroxides in Polysorbate 20 increased following opening and storage in fluorescent light at ambient conditions: Accumulation is lower in material stored in the dark.[20] Air oxidation of polyethylene glycols (PEGs) can also generate peroxides, aldehydes, and carboxylic acids. Such potential for variable levels could mean that drug-excipient compatibility studies during dosage form design may miss the potential for interactions if fresh or low residue materials are inadvertently used in screening studies.

Biopharmaceutical agents can be particularly susceptible to excipient impurity-related effects. Their structural complexity and multiple modes of degradation offer many possibilities for compositional and conformational changes, and interaction with co-formulated materials. These can compromise potency and also safety if the changes involve the groups on the epitope responsible for antigen/antibody interaction.[21]

The physical state of biopharmaceutical products can also influence interactions. If a freeze-dried solid is amorphous, a potentially destabilizing residue may not be held in a structured form; the amorphous state affords greater molecular mobility and consequent opportunities for reactions. Residual peroxides in polysorbates and PEGs can adversely affect the stability of biopharmaceuticals. Oxidation of Recombinant Human Granulocyte-Colony Stimulating Factor (rhG-CSF) correlated with peroxide content in Polysorbate 80, the effect being more severe than with atmospheric oxygen.[22] Low-aldehyde liquid PEGs and polysorbates are now available commercially. Nevertheless, the effects of storage conditions and usage on their quality and that of product in which they are incorporated should be evaluated as residues can change over time. Appropriate storage, usage periods, and re-test conditions need to be defined.

Reducing sugars in mannitol, an excipient widely used in parenteral lyophiles increased the oxidative degradation of a cyclic heptapeptide.[23] The disaccharides, lactose and sucrose may contain low levels of dextrose, also a reducing sugar.

Drug–Excipient Interactions

Chemical and physical interactions between excipients and drugs that compromise quality can occur with all medicinal products.[24] The risk of such interactions is, arguably greater with parenterals, particularly if formulated as liquids. In the case of biopharmaceuticals their intrinsic fragility during manufacture, transport, storage, and use mandates that excipients be added to maintain and retain their quality. But such agents can have unwanted effects.

- Succinate buffer was shown to crystallize during the freezing stage of a lyophilization cycle, generating lower pH and unfolding of gamma interferon.[25]

- When human growth hormone was lyophilized in the presence of sodium chloride physical changes (aggregation and precipitation) followed, along with chemical change (oxidation and deamidation).[26]
- Phosphate buffer can accelerate the oxidation of methionine structures in proteins. Oxidation can even be accelerated by ascorbic acid, a substance that can be an antioxidant in other scenarios.[27]

The physical state of excipients in lyophilized biopharmaceuticals can also affect stability. Interleukin-6, formulated with sucrose or trehalose, alone or in combination with glycine or mannitol was effectively stabilized during lyophilization, provided that the frozen components were completely amorphous. Long-term stability at ambient and stress conditions was also good, provided that product was stored below the Glass Transition Temperature (Tg) for the lyophile. Trehalose appeared to be the more effective antiaggregant possibly because of its higher Tg. Lyophiles that contained the excipients glycine and mannitol in the crystalline state were less stable.[28]

Excipients as Stabilizers

Environmental stresses such as light and oxygen can degrade drugs, particularly by oxidation mechanisms. It may be possible in many, but not all cases to obviate any effects by environmental controls (during manufacture) or by suitable packaging. Antioxidants can be considered, provided that the mechanism of oxidation and the antioxidant effect of the excipient is elucidated and understood. Photolytic degradation is a complex process, with oxygen often playing a key role. The mechanism of photodegradation of the amino acids, histidine, tryptophan and tyrosine, involves the molecules being raised to an excited state on photoirradiation, excitation energy being transferred to molecular oxygen and converted to highly reactive singlet oxygen atoms.

Use of excipients as photo-stabilizers was pioneered by Thoma, who used materials such as curcumin and riboflavin, with spectra that matched that of the labile drug to provide "spectral overlay" and reduce degradation (Fig. 3).[29]

There are no reports of the approach being utilized with parenteral products (where choice of a suitable stabilizer would be more limited). However, the concept is equally relevant to such presentations, provided that suitable materials for spectral overlay can be used.

Aggregation propensity is a major issue with biopharmaceuticals. It can be induced at any stage, that is, during processing, filling, lyophilization, transport, storage, and use. It can be caused by stirring, agitation, exposure to air, freezing, thawing, and by a host of additives ranging from ionic species to even ostensible stabilizers such as sucrose. Excipients that may help obviate aggregation include low levels (sub-CMC) of nonionic surfactants such as polysorbate (Tween) 80. This is usually added

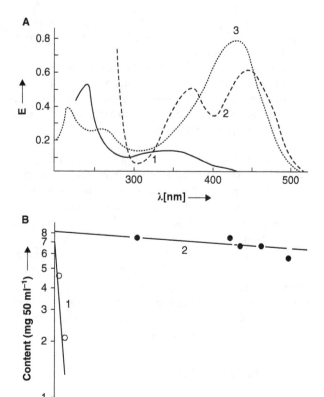

Fig. 3 UV absorption spectra (**A**) and effect on stability of nifedipine (**B**). (**A**) UV spectra of nifedipine 1, riboflavine 2, and curcumin 3. (**B**) Stability in solution of nifedipine with (●) and without (○) curcumin.

during downstream processing of the protein, being carried through to final product. Its widespread presence in commercial biopharmaceutical products suggests that it is an effective antiaggregant. The low inclusion level, coupled with its complex makeup (being essentially a mix of polymeric entities), makes precise and accurate quantitation in product virtually impossible. It is more appropriate therefore to use "Quality by Design" studies, possibly using surrogate measurements, in process validation studies to ensure that it is consistently carried through the process in sufficient quantity to evince its antiaggregant effect.

Mannitol and sucrose are also known aggregation inhibitors. Glucose, formed on hydrolysis of sucrose increased the aggregation of a monoclonal antibody, because of glycosylation. However, the studies were performed at elevated temperature. No effect was seen in material stored under refrigerated conditions. The findings illustrate the hazards of accelerated stability studies to predict performance at less stressful conditions.[30]

Oxidation can be one of the more complex forms of degradation with biopharmaceuticals. It can be induced by light, atmospheric oxygen, oxygen radicals (e.g., peroxides

in solution), or by the presence of metallic ions. Oxidation can alter a protein's folding and subunit association leading in turn to aggregation and fragmentation. Component amino acids in the protein, such as methionine, cysteine, histidine, tryptophan, and tyrosine, are particularly susceptible.

- Oxidation of cysteine can be reduced in biopharmaceuticals by maintaining the redox potential of the formulation, for example, by including the tripeptide glutathione as an excipient. L-methionine and ascorbic acid may also act as antioxidants.[31]
- Cysteine itself may have an antioxidant effect, provided that metal ions are absent.[20]
- Sugars and polyols may prevent metal-catalyzed oxidation because of their capability to complex with metal ions. Glucose, mannitol, and glycerol were shown to protect the reproductive hormone, relaxin against metal-catalyzed oxidation. Chelating agents such as EDTA and citrate can inhibit some metal-catalyzed oxidation.[32]

Future Perspectives

The increased prominence and need for parenteral administration of medicinal agents, especially for controlled delivery has, with a few exceptions not spawned a new generation of excipients to facilitate product development. This contrasts with developments of materials for stents, implants, and replacement joints, which too are "for internal use." Cost of development is clearly a consideration; a novel material would require a level of safety assessment comparable to a new drug, with something akin to a 5-year testing time span. It is encouraging however to note the advances made, using combinations of established excipients and the development of better (purer, more consistent) grades. The synergistic effects (and undesirable interactions) being reported illustrate how excipients now require a level of knowledge of their properties comparable to that of the drugs with which they are formulated.

CONCLUSIONS

This entry was written from an industrial and practical perspective, focusing on the challenges that dosage form design practitioners are likely to encounter in their day-to-day work. The authors did not attempt to provide in-depth information but tried to summarize the salient findings as concisely as possible. There are excellent reviews in other entries in this compendium and in the referenced scientific texts. The reader is encouraged to read these to acquire deeper understanding of the opportunities and challenges presented by excipients in parenteral dosage form design.

REFERENCES

1. Available from: http://www.medicines.org.uk/EMC/medicine/2038/SPC/Fluarix/#COMPOSITION
2. Available from: http://labeling.pfizer.com/showlabeling.aspx?id=467
3. Wong, J.; Brugger, A.; Khare, A.; Chaubal, M.; Papadopoulos, P.; Rabinow, B.; Kipp, J.; Ning, J. Suspensions for intravenous (IV) injection: A review of development, preclinical and clinical aspects. Adv. Drug Deliv. Rev. **2008**, *60* (8), 939–954.
4. Available from: http://www.rxlist.com/bicillin
5. Available from: http://plm.wyeth.com.mx/centroamerica/pdr/html/14580126.htm
6. Buxton, P.C.; Keady, S.; Jahenke, R.W. Degradation of glucose in the presence of electrolytes during heat sterilization. Eur. J. Pharm. Bioopharm. **1994**, *40* (3), 172–175.
7. Bacon, F.R.; Ragoon, F.C. Promotion of attack on glass by citrate and other anions in neutral solution. J. Am. Chem. Ceramic Soc. **1959**, *42* (4), 199–205.
8. Available from: http://www.novo-pi.com/victoza.pdf#guide
9. Giddi, H.S.; Arunagirinathan, M.A.; Bellare, J.R. Self assembled surfactant nano-structures in drug delivery: a review. Indian J. Exp. Biol. **2007**, *45* (2), 133–159.
10. Kim, J.Y.; Kim Lee, S.C.; Ooya, T.; Park K. *Polymeric Delivery Systems for Poorly Soluble Drugs. Encyclopedia of Pharmaceutical Technology*; Informa Healthcare USA, Inc., 2007.
11. Kim, J.Y.; Kim, S.; Pinal, R.; Park K. Hydrotropic polymer micelles as versatile vehicles for delivery of poorly soluble drugs. J. Control. Release **2011**, *152* (1), 13–20.
12. Huh, K.M.; Lee, S.C.; Cho, Y.W.; Lee, J.; Jeong, J.H.; Park, K. Hydrotropic polymer micelle system for delivery of paclitaxel. J. Control. Release **2005**, *101* (1–3), 59–68.
13. Guo, L.S.S.; Fielding, R.M.; Lasic, D.D.; Hamilton, R.L.; Mufson, D. Novel antifungal drug delivery: stable amphotericin B-cholesteryl sulfate discs. Int. J. Pharm. **1991**, *75* (1), 45–54.
14. Brewster, M.E. The potential uses of cyclodextrins in parenteral formulations. J. Parenter. Sci. Technol. **1989**, *43* (5), 231–239.
15. Rubinfeld, R.; Gore, A.Y.; Joshi, R.; Shrotriya, R. In Water Miscible Solubilizer. Oct 24, 2000.
16. Strickley, R.G. Solubilizing excipients in oral and injectable formulations. Pharm. Res. **2004**, *21* (2), 201–230.
17. Vemula, R.V.; Lagishetty, V.; Lingala, S. Solubility enhancement techniques. Int. J. Pharm. Sci. Rev. Res. **2020**, *1*, 41–51.
18. Muranishi, S. Drug targeting towards the lymphatics. Adv. Drug Res. **1991**, *21*, 1–37.
19. Bari, H. A prolonged release parenteral drug delivery system-an overview. Int. J. Pharm. Sci. Rev. Res. **2010**, *3* (1), 1–11.
20. Jaeger, J.; Sorensen, K.; Wolff, S.P. Peroxide accumulation in detergents. J. Biochem. Biophys. Methods **1994**, *29* (1), 77–81.
21. Patel, J.; Kothari, R.; Tunga, R.; Ritter, N.M.; Tunga, B.S. Stability considerations for biopharmaceuticals, part 1. BioProcess Int. **2011**, *9*, 20–31.
22. Herman, A.C.; Boone, T.C.; Lu, H.S. Characterization, formulation and stability of neupogen (Filgrastim), a recombinant human granulocyte-colony stimulating factor. In *Formulation and Stability of Protein Drugs: Case Histories*; Pearlman, R., Wang, Y.J., Eds.; Plenum Press: New York, 1996.
23. Dubost, D.C.; Kaufman, M.J.; Zimmerman, J.A.; Bogusky, M.J.; Coddington, A.B.; Pitzenberger, S.M. Characterization of a solid state reaction product from a lyophilized formulation of a cyclic heptapeptide, a novel example of an excipient-induced oxidation. Pharm. Res. **1996**, *13* (12), 1811–1814.
24. Martini, L.G.; Crowley, P.J. Effects of excipients on the stability of medicinal products. Chimica Oggi **2010**, *5*, vii–xi.
25. Lam, X.M. Replacing succinate with glycollate buffer improves the stability of lyophilized gamma interferon. Int. J. Pharm. **1996**, *142*, 89–95.
26. Pikal, M.K. The effects of formulation variables on the stability of freeze dried human growth hormone. Pharm. Res. **1991**, *8* (4), 427–436.
27. Knepp, V.M.; Whatley, J.L.; Muchnik, A.; Calderwood, T.S. Identification of antioxidants for prevention of peroxide-mediated oxidation of recombinant human ciliary neurotropic factor and recombinant human nerve growth factor. PDA J. Pharm. Sci. Technol. **1996**, *50* (3), 163–171.
28. Lueckel, B.; Helk, B.; Bodmer, D.; Leuenberger, H. Effects of formulation and process variables on the aggregation of freeze dried interleukin-6 (IL-6) after lyophilization and storage. Pharm. Dev. Technol. **1998**, *3* (3), 337–346.
29. Thoma, K.; Klimek, R. Photostabilisation of nifedipine in pharmaceutical formulations. Pharm. Ind. **1981**, *2*, 504–507.
30. Banks, D.D.; Hambly, D.M.; Scavezze, J.L.; Siska, C.C.; Stackhouse, N.L.; Gadgil, H.S. The effect of sucrose hydrolysis on the stability of protein therapeutics during accelerated formulation studies. J. Pharm. Sci. **2009**, *98* (12), 4501–4510.
31. Levine, R.L.; Mosoni, L.; Berlett, B.S.; Stadtman, E.R. Methionine residues as endogenous antioxidants in proteins. Proc. Natl. Acad. Sci. U.S.A. **1996**, *93* (26), 15036–15040.
32. Li, S.; Schoneich, C.; Borchardt, R. Chemical instability of protein pharmaceuticals: mechanisms of oxidation and strategies for stabilization. Biotechnol. Bioeng. **1995**, *48* (5), 490–500.

Excipients—Gels

Fluorescent Nanohybrids: Cancer Diagnosis and Therapy

Herman S. Mansur
Alexandra A. P. Mansur
Center of Nanoscience, Nanotechnology, and Innovation (CeNano), Department of Metallurgical and Materials Engineering, Federal University of Minas Gerais, Belo Horizonte, Minas Gerais, Brazil

Abstract

Material processing technologies combined with fundamental knowledge of the chemistry and physics of colloidal semiconductor nanocrystals or quantum dots (QDs) are advancing at an unprecedented rapid speed and are thus offering new perspectives for the widespread utilization of these nanomaterials in all areas of human society. QDs are a fascinating novel class of fluorescent nanomaterials that promise to revolutionize the emerging applications of nanotechnology in medicine and biology, which is known as nanomedicine. Cancer is one of the main challenges faced by scientists and professionals in the health sciences in the 21st century because it remains one of the world's most devastating diseases with tens of millions of new cases every year. Thus, in this entry, we attempted to survey and present the advances in colloidal semiconductor nanocrystals, focusing on their applications in nanomedicine and, more specifically, in cancer diagnosis and treatment. Nevertheless, because these novel applications are currently drawing the attention of the research communities in the fields of medical and biological sciences, materials and chemical engineering, physics, and chemistry, we endeavored to review these QD-based nanomaterials in order to provide a both comprehensive and accessible report of this information to newcomers to this attractive realm of science.

INTRODUCTION

Since early times, humankind has been fighting against "hostile" animals and all types of diseases in order to survive in the natural environment. From the sixteenth to the nineteenth century, with the birth of some foundations of science, modest advances in longevity of life were gradually achieved by human society. However, in the 20th century, scientists discovered several treatments that have considerably improved the quality of life and increased life expectancy from approximately 35 years in the previous century to more than 80 years in many developed countries. Thus, the major challenge of the current century and millennium relies not only on living longer but also on living healthier. In that sense, new strategies for preventing diseases combined with innovative diagnosis methods for detecting all types of illnesses accurately and rapidly will promote improved quality of life. If not preventable by available means, an earlier detection of a disease will improve the patient's prognosis after treatment and thus increase the possibility of curing the disease. Hence, the emerging areas of nanoscience and nanotechnology will play a decisive role in paving the way to revolutionize all areas of human knowledge, predominantly nanomedicine. Nanomedicine is referred to as the application of nanotechnology and nanoscience in the life sciences for the treatment of diseases, diagnosis, monitoring, and control of biological systems at the dimensional scales of single atoms, molecules, or molecular assemblies.[1] This field integrates a very broad range of theoretical and experimental knowledge, including materials science, chemistry, physics, biochemistry, pharmacy, medicine, nutrition, biology, engineering, and other fields with the aim of offering original solutions to problems faced by health professionals and scientists in medicine and other related biomedical areas. Among the innumerous obstacles to be overcome in medicine, cancer remains one of the world's most devastating diseases with more than 10 million new cases every year. Cancer is the second leading cause of death worldwide, second only to cardiovascular diseases. The statistics show that approximately 33% of women and 50% of men will develop some type of cancer during their lifetimes. Today, millions of patients suffer from cancer due to the late identification and treatment of the disease.[2–4] Consequently, nanomedicine is a promising platform for the development of nanomaterials that may potentially be used for the early diagnosis and treatment of cancer. Quantum dots (QDs) are a new class of nanomaterials that are being intensively studied as novel probes for biomedical imaging, targeting, and drug delivery due to their unique optical, magnetic, and electronic characteristics. QDs possess physicochemical properties, such as size, shape, composition, and surface features, which may be tailored for specific functions and applications. When conjugated with molecular agents of biological interest, such as antibodies, peptides, enzymes, nucleotides, and other molecules,

Excipients—Gels

QD-based probes can be used to target cancer cells with high specificity and sensitivity for molecular imaging and drug delivery to a specific tumor location.[4]

Thus, this entry highlights the promising perspectives for the future use of biologically designed and functionalized QDs in cancer diagnosis and treatment and tumor targeting. We have taken the opportunity to review the recent advancements in a segment of this field, choosing to cover the progress in the chemistry, processing, and some applications of colloidal QDs. Due to the limited restriction of this entry and to a large number of papers and review articles on the subject, we focused our entry on QDs that are directly grown predominately in aqueous media but also considered nanomaterials synthesized through a ligand exchange process to a minor extent. Additionally, the authors would like to emphasize that, in addition to addressing the biomedical applications of QDs in cancer "theranostics" ("fusion of therapeutics and diagnostics"), this entry is written from the materials chemistry perspective and brings together relevant contributions from the broad range of areas involved in this complex and challenging theme.

CANCER

A Brief History of Cancer

Cancer is not a new disease and has afflicted people throughout the world for centuries. The word cancer originated from the Greek words *karkinos*, which was used to describe carcinoma tumors by the physician Hippocrates (460–370 B.C.), who is considered the father of medicine. However, he was not the first to discover this disease because the earliest evidence of human bone cancer was found in mummies in ancient Egypt and in ancient manuscripts that have been dated to approximately 1600 B.C. The Roman physician Celsus (28–50 B.C.) translated the Greek term into *cancer*, which is the Latin word for crab. Later, Galen (130–200 A.D.), another Roman physician, used the word *oncos* (Greek for swelling) to describe tumors (used as a part of the name for cancer specialists—oncologists).[3] Initially, the autopsies performed by Harvey (1628) led to an understanding of the circulation of blood through the heart and body, which had, until then, been a mystery. In 1761, Giovanni Morgagni of Padua was the first to relate a patient's illness to the pathologic findings after death. This finding laid the foundation for scientific oncology, which is the study of cancer. The renowned Scottish surgeon, John Hunter (1728–1793), suggested that some cancers may be cured by surgery. If the tumor had not invaded nearby tissue and was "moveable," he said, "There is no impropriety in removing it." A century later, the development of anesthesia allowed surgery to flourish, and classic cancer operations, such as radical mastectomy, were developed. The 19th century saw the birth of scientific oncology with the use of the modern microscope in the study of diseased tissues. Rudolf Virchow, who is often called the founder of cellular pathology, provided the scientific basis for the modern pathological study of cancer by correlating microscopic pathology to illness. From the 20th century onward, cancer research has experienced an astonishing development, predominantly in the areas of diagnosis, therapeutics, and treatment. This advance was made possible by a combination of new understandings of life sciences, such as cellular and molecular biology, with engineering, physics, and chemistry, which provided novel imaging and diagnostic tools.[5] Figure 1 illustrates the milestones in the evolution of tumor imaging technologies used in oncology.

What Is Cancer?

Cancer is commonly referred to as a diverse group of diseases affecting a variety of tissues but is generally characterized by the uncontrolled proliferation of abnormal cells with the ability to invade surrounding tissues and possibly metastasize. Over the past decades, various studies have attempted to understand the innumerous events and phenomena involved with cancer, such as the regulation of cell adhesion and the dynamics of cancer invasion and progression.[4] However, these studies have largely failed to define the mechanisms that govern cancer, including its initiation, growth, invasion, and progression. There are different types of cancers, and some types of cancer cells often travel to other parts of the body through blood circulation or lymph vessels (metastasis), where they begin to grow. In general, cancer cells develop from normal cells as a result of DNA damage. In most cases of DNA damage, the body is able to repair the damage; however, in cancer cells, the damaged DNA is not repaired. Additionally, individuals can inherit genetic diseases from their parents, which accounts for inherited cancers. Cancer may also be caused or triggered by exposure to some harmful external influences from the environment, such as radiation, pollutants, and social-cultural aspects (nutrition and habits). Cancer generally forms as a solid tumor, but some cancers, such as leukemia (named as "blood cancer"), do not form tumors. Instead, leukemia cells involve the blood and blood-forming organs and circulate through other tissues, where they may invade and grow. Not all tumors are cancerous; in fact, some tumors are benign (noncancerous). Benign tumors do not grow and are not usually life-threatening. If cancer is identified in its early stages, it can be more easily treated, and the patient may have a better chance of the therapy resulting in increased life expectancy.[3] It is believed that the future treatment of cancer relies on the earliest possible detection of cancer lesions and on the efficient and specific delivery of drugs to target the cancer cell site. The detection of cancer at its primary stage (stage 1) is associated with a 5-year survival rate of more than 90%; however, conventional anatomical imaging typically cannot detect

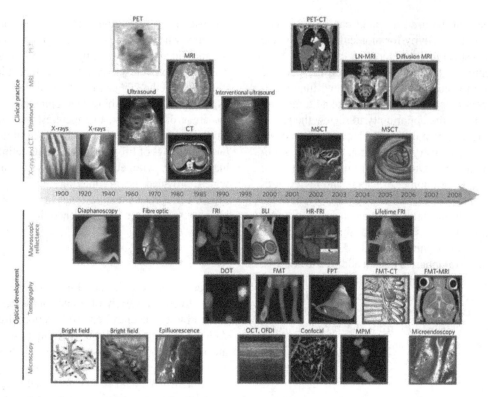

Fig. 1 Illustration of time-frame evolution of tumor imaging technologies used in oncology.
Source: From Weissleder & Pittet[5] © 2008, with permission from Nature Publishing Group.

cancers until they reach a size of more than 1 cm. It is expected that the application of nanotechnology in medicine and biology will offer many exciting possibilities in health care. In fact, nanotechnology has the potential to radically transform the diagnosis and the therapy of several diseases, particularly cancer.[6]

QUANTUM DOTS

What Are QDs? A Short Background on Quantum Physics and Chemistry

For many decades, inorganic solids have been conventionally divided into three distinct classes based on their thermal and electrical conductivities: metals, semiconductors, and insulators, as illustrated in Fig. 2. In fact, this classification may be considered an oversimplified concept of the complexity of the phenomena involved in the mobility of electronic charge carriers in all sorts of materials, particularly at the nanoscale dimension. Nonetheless, this multidisciplinary approach can be useful for addressing the subject. At this point, it is assumed that a short background on the physics and chemistry of semiconductor nanocrystals with some fundamental considerations would be recommended. This background is required to properly understand the number of possibilities associated with exploring the optical and electronic properties of colloidal

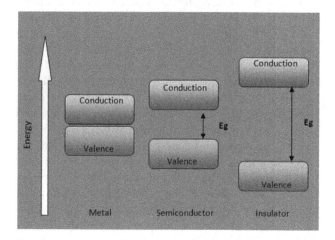

Fig. 2 Schematic representation of the electronic band structure of conductors, insulators, and semiconductors with the energy levels associated with the valence, the conduction band, and the band gap (E_g).

QDs. However, an in-depth theoretical analysis of this topic based on quantum theory is beyond the scope of this entry and has been the subject of various previously published papers and reviews.[7,8] Semiconductors have a filled band and a partially empty band; these are known as the "valence band" and the "conduction band," respectively. From the chemistry perspective, in molecules, these bands are analogous to the transition of excited electrons from the highest occupied molecular orbital to the lowest unoccupied molecular orbital. At nano-order dimensions, the collective electronic properties usually found in solids become extremely distorted, and the electrons at this length scale tend to follow the "particle in-a-box" model to account for the approximated energy band structure.[9] The electronic states are more similar to those found in localized molecular bonds than to those found in the bulk of macroscopic solids, and this finding has significant implications on the total energy of systems and thus on their properties. When the size of a semiconductor crystal becomes sufficiently small that it approaches the size of the material's Bohr radius (a_B), the electron energy levels should be treated as discrete. This phenomenon of discrete energy levels is called "quantum confinement." Consequently, this strong interaction between the pair "hole–electron" (h^+/e^-) is referred to as the "exciton" caused by the dimension quantum confinement.[6,10,11] In other words, after reaching a specific threshold in the nanoparticle size (R = radius), the energy for promoting a band gap transition ($E_{g,QD}$) is larger than that of the original bulk solid material ($E_{g,b}$) due to the bound state of the pair "hole–electron."[12,13] Consider what happens when a semiconductor is irradiated with photons of energy ($h\nu$) higher than the band gap (E_g). An electron (e^-) will be promoted from the valence to the conduction band, leaving a "hole" (h^+) or the "absence of an electron" in the valence band. Thus, this "hole" is assumed to behave as a "particle" with its particular effective mass and positive charge. Brus pioneered the development of studies that relate the particle size to the band gap energy of semiconductor QDs,[12] which is reported by a popular model using Eq. 1 with a reasonable approximation to empirical measurements:

$$E_{g,QD} = E_{g,b} + (h^2/8R^2)(1/m_e + 1/m_h) - 1.8e^2/4\pi\varepsilon_0\,\varepsilon R \qquad (1)$$

where $E_{g,b}$ and $E_{g,QD}$ are the band gap energies of the bulk solid and QD, respectively, R is the QD radius, m_e is the effective mass of the electron in the solid, e is the elementary charge of the electron, h is Planck's constant, m_h is the effective mass of the hole in the semiconductor, and ε is the dielectric constant of the solid. The middle term on the right-hand side of the equation is a "particle-in-a-box-like" term for the exciton, whereas the third term on the right-hand side of the equation represents the electron–hole pair Coulombic attraction mediated by the solid. The quantum confinement of the exciton is the principle that causes the optical and electronic properties of the QDs to be dominated by their dimension. Thus, because the size and optoelectronic properties are correlated, decreasing the QD size results in a higher confinement of the exciton, thereby increasing the band gap energy. The most important consequence of this behavior is that the absorption and emission of the QDs may be adjusted by its size. A smaller nanocrystal will require a higher amount of energy to promote the electron from the valence band to the conduction band, that

is, to generate the excitonic transition. A decrease in the size of the QDs will result in a hypsochromic shift of the absorption and photoluminescence (PL) spectra by increasing the band gap energy. Thus, QDs with the same composition but different sizes can generate the fluorescence of different wavelengths.[14] The representation of this dependence on transition energy (valence to conduction band) and QD size can be observed in Fig. 3A. However, these band structures are simplified approximations of the actual QD systems because these have very high surface-to-volume ratios, which results in imperfections and dangling bonds. This "excess" of atoms at the surface creates electronic trap states, which are usually located within the QD band gap (Fig. 3B).[7,8]

One of the fundamental problems associated with the existence of trap states is related to the optical and electronic properties of QDs that may significantly decrease the fluorescence quantum yield, the broad fluorescence range, and the "blinking" affected by the possible electronic charge recombination pathways. Several studies have investigated how these surface defects can be eliminated or minimized by controlling the nucleation and growth of the QDs using different surfactants and ligands[7,8,15] and surface passivation by the deposition of layers of other semiconductors. Carrillo-Carrión and co-workers published an excellent entry on QD luminescence and the important aspects affecting the photoactivation of QDs.[16] Because defects on the surface of the QDs act as temporary surface traps for charge carriers (i.e., electron/e^- and hole/h^+), these lead to possible nonradiative relaxation and therefore prevent radiative recombination, which greatly reduces the quantum yield. This concept regarding the nature of the trapping state in the colloidal QDs was initially suggested by Efros and Rosen.[17] To minimize or ideally eliminate the photo-optical instability caused by the abundance of "surface states," the combination of two or more semiconductors, which form heterojunctions in which the "core" is surrounded by the "shell" of a wider band-gap semiconductor that passivates the surface states of the inner nanocrystals (these structures are usually called "core–shell" (C-S) nanostructures, has been utilized (Fig. 3C). Moreover, appropriate ligands may noticeably alter the chemical and physical states of the quantum surface.[18–20] In Fig. 4 it is schematically represented the overall optical behavior of QDs as a function of their sizes, affecting the absorption and fluorescent properties.

Synthesis of QDs

The first successful synthesis of QDs was reported approximately three decades ago by Efros and Rosen[17] and Ekimov and Onushchenko,[21] who used glass matrices to host the nucleation and growth of the semiconductor nanocrystals. Since then, a large amount of methods and techniques have been described for the preparation of QDs using different processing routes, including aqueous solution,[10,11,22–34]

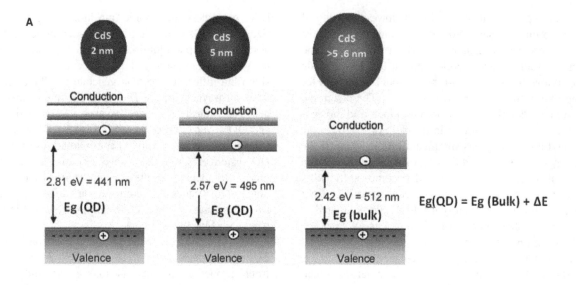

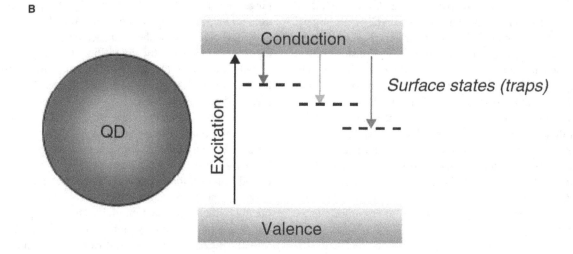

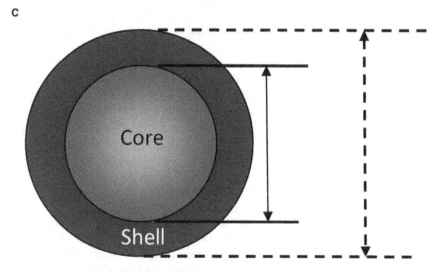

Fig. 3 A representation of the band structure in solids: (**A**) quantum confinement effect on changing QD size; (**B**) surface trap sites with their electronic energy states localized within the QD band gap; (**C**) the electronic band structure of a core–shell QDs made of two semiconductors forming a heterojunction (core–shell).

Source: From Mansur[6] © 2010, with permission from John Wiley & Sons, Inc.

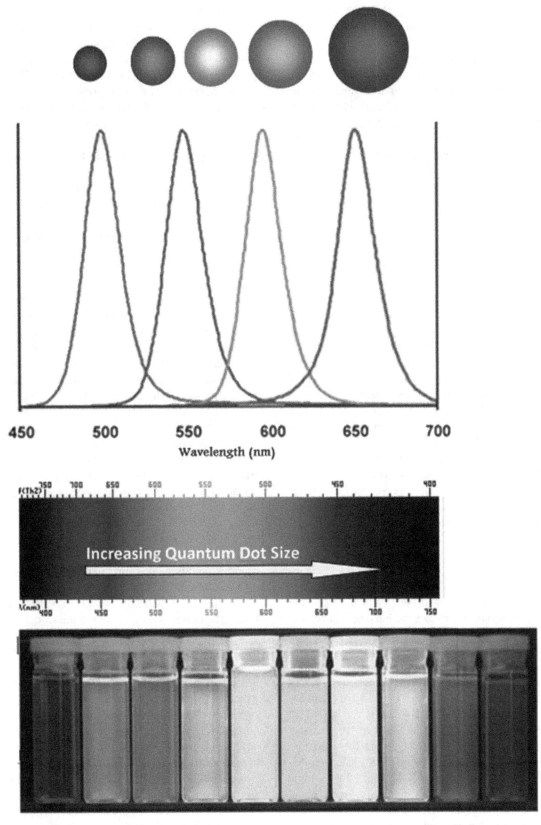

Fig. 4 Schematic drawing representing the changes on optical behavior of nanoparticles associated with their size. Top: Electronic structure of QDs with hypsochromic shift ("blue shift") due to quantum confinement. Medium: A decrease in the size of the QDs results in a hypsochromic shift of the absorption and PL spectra by increasing the band gap. Bottom: Fluorescent emission of QDs in the visible range of electromagnetic spectrum.

Source: From Mansur[6] © 2010, with permission from John Wiley & Sons, Inc.

high-temperature organic solvents,[13,35,36] hydrogels,[37–40] and Langmuir–Blodgett molecular film deposition onto solid substrates.[41–43]

From a broad perspective, semiconductor QDs can be produced by two main processing strategies: one is known as "bottom-up," which is more familiar to chemists, and the other is called "top-down." The first process utilizes molecular or ionic precursors of QDs that are reacted in solution media to produce colloidal nanocrystals. The second approach, which is more familiar to engineers and physics, produces nanostructures typically ranging from 1- to 10-nm scale that are carved out lithographically or electrochemically from a semiconductor substrate. The "top-down" processing methods include molecular beam epitaxy, ion implantation, e-beam lithography, and X-ray lithography.[44,45] In the top-down approaches for synthesizing QDs, the dimension of a bulk semiconductor is significantly reduced. To obtain QDs with a diameter of approximately 30 nm, electron beam lithography, reactive-ion etching, and/or wet chemical etching are commonly used. Alternatively, focused ion or laser beams have also been used to fabricate arrays of zero-dimension dots. The incorporation of impurities into the QDs and structural imperfections by patterning are major disadvantages of these processes.[44] Additionally, a third approach, which is referred to as the hybrid method, has been proposed as an alternative for QD synthesis. This approach combines molecular precursors for the QDs with their chemical reactions in the gas phase and transfers or deposits the QDs as ultrathin films on solid substrates. Thus far, the colloidal route has become the most popular process for QD synthesis and will therefore be the emphasis of this entry. In fact, QDs used in bio-applications are exclusively colloidal nanocrystals. These crystals are commonly synthesized through the introduction of semiconductor precursors under conditions that thermodynamically favor the nucleation and growth of nanocrystals in the presence of stabilizing agents, which function to kinetically limit particle growth and maintain their size within the "quantum-confinement" size regime. The size-dependent optical and electronic properties of QDs can only be engineered if the semiconductor nanocrystals are effectively prepared within very narrow size distributions.

Usually, semiconductor QDs are composed of a combination of elements from groups II–VI, III–V (GaN, GaP, GaAs, InP, InN), or IV–VI (PbSe, PbS) of the periodic table. Group II–VI semiconductor QDs are the most frequently investigated systems based on metal chalcogenides (i.e., MX, M = Cd^{2+}, Zn^{2+}; and X = S^{2-}, Se^{2-}, Te^{2-}) with a band gap, which can be tailored by modifying the nanomaterial composition and size.[46] The most common QDs are synthesized with cadmium, lead, and mercury, which are highly toxic. Thus, recent biological and environmental concerns and regulations restricting the use of toxic metals have increased the interest in the design of QDs containing less toxic or nontoxic metals, such as Zn, Mn, and Cu.[47]

Although it is relatively new, the synthesis of colloidal QD semiconductors is a well-established science, and a large amount of work has been conducted in the chemistry and physics at the nanointerfaces between nanoparticles and ligands in the last two to three decades. For instance, a comprehensive and interesting review on the role of ligands on the synthesis of QDs was published by Mark Green and addresses several relevant aspects associated with the effects of the processing routes and the selection of ligands on the properties of the QDs produced.[48] Nonetheless, the complexity of the dynamic processes occurring at the nano-interfaces is far from been entirely understood. Because semiconductor nanocrystals have a huge surface-to-volume ratio, they are extremely unstable in solution, which causes them to agglomerate and thus form clusters due to the high surface energy. Thus, any route one chooses to synthesize QDs should consider stabilizing the just-formed nanocrystals by minimizing the surface energy through "capping" and avoiding further nanostructure growth. It can be summarized that colloidal QDs are usually produced primarily via two major approaches, i.e., organometallic synthesis and aqueous synthesis.[49,50] It is important to note that in our entry, the "Synthesis of QDs" section surveys methods for both organic and aqueous colloidal chemistries. In fact, in several cases, both approaches can be used, and for some materials, one synthesis route may have an advantage over the other. The reasons for choosing one approach or another may be historical or due to better optical or electronic properties, narrower size distributions, higher quantum efficiency, or some functionality of the ligand that may be advantageous for a specific application. For that reason, because this entry focused on novel nanomaterials for biomedical applications, i.e., cancer diagnosis, targeting, and treatment, this entry predominantly concentrates on relevant contributions in the reported literature that utilize aqueous processing routes.

Therefore, as their major advantage, the organometallic routes for the preparation of QDs have been well established and normally result in the synthesis of monodisperse and chemically stable nanoparticles with excellent optical properties. However, the synthesis is generally performed using high-temperature thermal decomposition of organometallic compounds in high-boiling-point organic solvents, which requires a long reaction time in the presence of surfactants and involves more complex synthesis procedures. Moreover, this type of organic synthesized QDs, which usually use a mixture of trioctylphosphine and trioctylphosphine oxide (TOP/TOPO), are primarily of a hydrophobic nature. This synthesis method produces surfactant-coated nanoparticles, in which the polar surfactant head group is attached to the inorganic surface of the nanocrystals and the hydrophobic chain protrudes into the organic solvent, thereby mediating colloidal stability. At this stage, the particles will be well-dispersed in nonpolar organic solvents (i.e., toluene, hexane, and chloroform), but because of their hydrophobic surface layer, they are not soluble in aqueous media and cannot be directly used in bioapplications.[6,49] A milestone in the synthesis of water-dispersible semiconductor nanocrystals was reached in the published studies conducted by Bruchez et al.[51] and Chan and Nie[52] in 1998. These researchers started a new era in the potential applications of fluorescent water-dispersible QDs in medicine and pharmacy. To alter the originally synthesized hydrophobic QD surface, some additional post-treatment involving an exchange with hydrophilic ligands and/or polymers is needed to render aqueous dispersibility to the QDs, which may cause adverse effects on the optical/physical/chemical properties of the QDs (such as brightness, fluorescent quantum yield, chemical stability, and size). Therefore, a suitable phase transferring agent from the nonaqueous to an aqueous medium would be necessary to retain the optimal properties of the QDs in an aqueous phase.[46,49] Alternatively, the native hydrophobic monolayer of ligands can be retained on the QD surfaces and rendered water-soluble by the adsorption of amphiphilic molecules, such as oligomers and polymers. These molecules contain both a hydrophobic segment (mostly hydrocarbon chains) and a hydrophilic segment (i.e., functional groups, such as alcohol, carboxylic, and amine). In this case, the alkyl chains of the amphiphilic agent interact with the ligands on the QD surfaces (e.g., with the n-octyl groups of TOP or TOPO) through hydrophobic bonds. Simultaneously, the hydrophilic endpoints of the amphiphilic molecules facing outward enable the dispersion of the QDs in an aqueous medium. Thus, the conversion of hydrophobic-capped QDs from the organic phase into an aqueous phase has been intensively studied, and some comprehensive papers and reviews on the subject have been published.[13,35,36] Fundamentally, the conversion of the surface of QDs has been achieved by a large variety of chemical moieties, such as thiol acids, hydrophilic dendrimers, silica-shells, amphiphilic polymers (e.g., PEG), block copolymers, proteins, octylamine-modified polyacrylic acid, amphiphilic polyanhydrides, modified polymaleic acid, and cetyltrimethylammonium bromide.[49,53] This method produces exceptionally stable water-soluble QDs with preserved optical properties because the coating does not directly interact with the nanocrystal surface and does not disturb the surface passivation layer. Nonetheless, a major drawback of these routes using amphiphilic molecules to render water solubility to QDs is associated with the inevitable increase in the "size" of the conjugated systems, which is commonly referred to as the "hydrodynamic diameter" (H_D). Regardless of the processing route one may choose, the final system size must be maintained within a very limited range because the H_D parameter will determine the QD behavior in most biological applications, such as cell targeting, endocytosis, and cytotoxicity.[54] For example, it has been reported that a block copolymer coating increases the diameter of QDs (CdSe/ZnS) from 4 to 8 nm before amphiphilic encapsulation to up to 30 nm (H_D). Such a dramatic increase in the size may be

detrimental for quantitative biomarker detection in some biological environment and may hamper the intracellular penetration of the QD probes. The increased thickness of the polymer coating may also preclude the utilization of the QDs in Forster resonance energy transfer-based applications.[13,55,56] As a consequence, the development of novel methods for producing water-soluble nanoparticles with high-quality optical properties, long-time chemical stability, narrow size distributions, and biocompatibility, which are associated with the lowest possible "hydrodynamic diameter," is still a challenge to be overcome.[57,58]

Water-Soluble QDs

In any case, all approaches and principles used for the synthesis of QDs in nonpolar media that will be potentially used in biological environments must involve the conversion of hydrophobic to hydrophilic surfaces in order to render their compatibility to the physiological environment, in which water is abundant. Henglein and co-workers pioneered the aqueous synthesis of semiconductor QDs in colloidal solutions approximately three decades ago.[59–61] Since then, several developments have demonstrated the great potential of this strategy for the synthesis of semiconductor nanomaterials with specifically designed properties, such as PL, long-term chemical stability, narrow emission range, and biocompatibility, combined with the flexibility of the simultaneously functionalization of the surfaces of the nanoparticles. Because it is relatively straightforward, the aqueous approach provides some advantages over conventional organometallic routes; these advantages include versatility, scalability, environmental friendliness, and cost effectiveness, and lead to very attractive perspectives for a variety of applications. In this sense, innovative facile chemical routes using primarily "one-pot" aqueous media are very promising methods for the synthesis of QDs for biological applications.[62] However, it should be noted that the mild synthetic conditions commonly used in aqueous syntheses often affect the perfect arrangements of the atoms in the crystal lattice of the nanoparticles, which is more easily achieved by syntheses in high-boiling organic solvents. To overcome this problem, the temperature of the media may be increased by the use of microwave irradiation, or the synthesis may be carried out in autoclaves to achieve a higher crystallinity and a narrower size distribution, which results in the synthesis of QDs with improved optical properties.[62–64] Thus, the choice of ligand is crucial to achieving chemical and biological functionalities and to retain the optical properties of the QDs and very often limits the final application range of the synthesized QDs. Several alternatives of ligands may be considered for producing colloidal QDs in aqueous medium, and these include amino acids, organic acids, thiol molecules (i.e., thiol acids), oligomers, polymers, proteins, and nucleotides. In this context, polymers (synthetic or natural) may be used as substitutes for the small organic ligand molecules or as surface

Excipients—Gels

modifiers to avoid the ligand exchange process because these molecules possess some of the most relevant requirements and properties for biomedical applications. Polymers are commercially available, and many are water-soluble, biologically compatible, and have been widely used in the chemistry of colloids as dispersants, chelates, surfactants, food industry, ligands, and so forth. Of the vast number of water-soluble polymers commercially available for preparing colloidal suspensions and films, some have also demonstrated suitable biocompatibility and biodegradability for medicine, biology, and pharmaceutical applications.[65,66] However, surprisingly, despite being a very dynamic and intensive area of research because polymers have been broadly reported in ligand exchange processing routes or as the amphiphilic compatible layer used for preparing semiconductor nanoparticles,[67–77] the use of polymers as direct capping ligands for synthesizing QDs in aqueous colloidal suspensions is relatively unexplored in the literature.[10,78–80] Therefore, polymeric materials offer a broad range of possibilities for directly synthesizing colloidal semiconductor nanocrystals in the aqueous phase that can be potentially engineered with surface functionalities for biomedical applications.[10,11,22–27,81] Most frequently, an aqueous synthesis protocol for preparing QDs consists of the reaction between metal ions and chalcogen-containing precursors in the presence of an appropriate chemical stabilizer and the subsequent nucleation and growth of the nanocrystals, which is typically controlled by processing parameters, such as pH, temperature, reagents concentrations, catalyst, and surfactants. Water-soluble salts, such as nitrates, chlorides, and perchlorates, are commonly used as metal precursors, whereas water-soluble salts or gas sources (H_2S, H_2Se, and H_2Se) are often used as chalcogenide (S, Se and Te) reagents. Sometimes, to control the reaction kinetics, organic compounds that slowly decompose with the controlled release of anions (S^{2-}, Se^{2-}) are utilized for this purpose. For instance, thioacetamide and thiourea are broadly used as sulfur sources for producing QDs, such as CdS, ZnS, and PbS.[62,82]

As mentioned previously, the synthesis route for preparing QDs using aqueous colloidal chemistry with water-soluble ligands is normally performed at low temperatures, which are favorable for biological compatibility and sustainability but usually produce QDs with some disordered crystalline nanostructures and surface defects. Thus, in general, coatings on the QDs with higher-band-gap semiconductor materials is necessary to passivate the surface defects and thus improve the optical properties, such as photostability and luminescence, due to the formation of QDs with core–shell (C-S) nanostructures.[83] For instance, QDs with core–shell structures composed of CdSe/ZnS have been the preferred choice for bioapplications,[13] and their synthesis was first reported by Hines and Guyot-Sionnest.[84] Basically, the ZnS layer grown passivates the surface of the core, protecting it from oxidation and preventing the

leaching of CdSe into the surrounding solution, and also produces a considerable improvement in the quantum yield. Although the production of core–shell structure QDs via the hot-injection organometallic approach has been very successful, these QDs are usually synthesized under high-cost and rigorous experimental conditions, and the QDs are not readily water-dispersible because they are hydrophobic, which seriously limits their use in bioapplications. In fact, the direct synthesis of nanocrystals in aqueous phase is an alternative strategy to obtain water-soluble nanocrystals, and the synthesis is more reproducible, less expensive, and less toxic.[31] Some reviews and reports summarize the major advancements achieved in the last two to three decades with respect to the synthesis of water-dispersible QDs with enhanced optical properties, most of which are based on the building of core–shell nanostructures.[30–33] However, there are only a few successful examples on the improvement of luminescent properties through the growth of an inorganic shell on nanocrystals prepared in aqueous media.[85–87] Depending on the band gaps and the relative position of the electronic energy levels of the involved semiconductors, the shell can have different functions in C–S nanostructures, and these are typically classified into three cases, namely type-I, reverse type-I, and type-II band alignment (schematic drawing presented in Fig. 5). In the first case, the band gap of the shell material is larger than that of the core, and both electrons and holes are confined within the core. In the second, the architecture is the opposite, i.e., the band gap of the semiconductor shell is smaller than that of the core, and depending of the thickness of the shell, the holes and electrons are partially or entirely confined within the shell. In the latter, either the valence-band edge or the conduction band edge of the shell material is located within the band gap range of the core. Thus, upon excitation of the nanostructures, the resulting staggered band alignment leads to a spatial separation of the hole and the electron in different regions of the C–S semiconductor system.[82,88,89] Thus, the combination

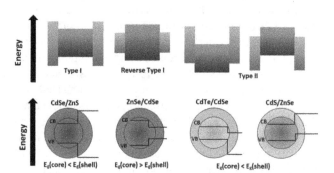

Fig. 5 Schematic representation of core–shell (C–S) nanostructures of QDs. Regarding the band gaps and the relative position of the electronic energy levels of both semiconductors of the C–S structure, they may be classified into three major cases: type-I, reverse type-I, and type-II band alignment.

of different semiconductor heterostructures offer the possibilities for building new systems by engineering the band gap with the selection of the core and shell materials. Additionally, some important requirements must be taken into account for the synthesis of C–S QDs with satisfactory optical and electronic properties, such as the proper matching between the core and shell crystalline lattices and the control of the thickness of the shell, to minimize potential strains that could cause the formation of defect states at the core–shell interface or within the shell. These aspects can act as trap states for photogenerated charge carriers (h^+ or e^-) and significantly deteriorate the photostability, thereby reducing the fluorescence quantum yield.[90]

The number of possibilities for novel capping ligands and routes for synthesizing water-dispersible QDs are continuously increased by new reports. Medintz and co-workers[13] summarized the most-often-used procedures for QD synthesis and divided these into three major classes based on the processing method used. All of these classes share the same goal: the production of QDs with hydrophilic surfaces, narrow size distribution, water-dispersible, and photo and chemical stability in physiological media for biomedical applications. In summary, despite the undeniable progress achieved with respect to improving the optical properties of water-soluble QDs, some refinements with respect to biomedical applications, such as issues of reproducible synthesis, biochemical stability, narrow size distribution, conjugation of affinity ligands, and inorganic passivation, remain to be resolved.[6]

Biofunctionalization of QDs

The major goal of bioconjugation focuses on utilizing the inherent structural, specific recognition, or catalytic properties of biomolecules to assemble (bio)composite and (bio)hybrid nanoscale materials with unique or novel properties. The number of different QD bioconjugates presently being developed is already beyond counting, which amply reflects the breadth of bio-applications and highlights the effervescence of this theme. To avoid ambiguity, the definition of a biomolecule, which was neatly presented by Sapsford et al.[91] is the following: "biological molecules or *biologicals* to include all forms of proteins, peptides, nucleic acids (i.e., DNA/RNA/PNA/LNA as genes, oligomers, aptamers, and ribozymes/DNAzymes), lipids, fatty acids, carbohydrates, etc. whether they are monomeric building blocks (i.e., amino acids, nucleotides, monosaccharides) or fully formed, functional 'polymers,' such as proteins, plasmids, and cellulose. The *biologicals* may be functionally active, offering binding, catalytic, or therapeutic activity (e.g., antibodies or enzymes), or alternatively, may be passively utilized as an inert coating or scaffolding material."

QDs can serve as nanoscale scaffolds with an extremely high surface area and with physicochemical properties and biological activity that can be tailored through interfacial chemistry and functionalization. The conjugation process is usually performed in multiple steps, and the design and execution of each step are critical to the efficacy of the produced QD in its intended application. The functionalization of the surface of QDs is performed to increase its chemical stability and allow the conjugation of biomolecules for specific targeting. Due to their biological applications, QDs must be linked to biomolecules without significantly modifying the biological activity of the conjugated form. Thus, any modification of the outer surface of QDs has to satisfy some requirements, such as the following: 1) render the QDs water-soluble; 2) biocompatibility; 3) surface reactive groups for conjugation; 4) maintain all advantages of QDs with respect to photochemical long-term stability (spectral range, brightness, fluorescence, narrow emission); 5) monodispersed particles (i.e., narrow size distribution); and 6) thin shell and capping layer to avoid increasing the engineered particle size diameter.[6,92,93] In Fig. 6, a schematic drawing of common chemical groups used for the functionalization of QD surfaces for subsequent conjugation with biomolecules is shown. The activities of many types of biomolecules have been found to remain after their conjugation to nanoparticle surfaces, although some decrease in the binding strength and/or affinity may also occur. The optimization of the surface immobilization of biomolecules is currently a very prominent area of research and is beyond the scope of this entry. Moreover, very comprehensive reviews were recently published by Sapsford et al.[91] and Blanco-Canosa et al.,[94] who reported the most relevant aspects involved in the biofunctionalization of nanoparticles, including QDs. Most simply, the conjugation itself may be assumed to be either covalent (i.e., covalent coupling directly to the nanoparticle surface or surface ligand) or noncovalent (i.e., electrostatic attachment, other forms of adsorption, and encapsulation) in nature.[91,94]

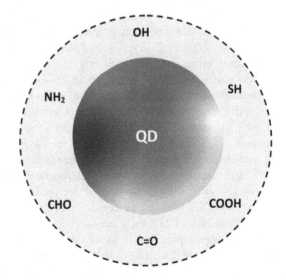

Fig. 6 The most common chemical functional groups on the surface of QDs to be utilized for subsequent bioconjugation.

Noncovalent conjugation

Noncovalent interactions have been exploited by several research groups. Mattousi et al. designed a system for biosensing applications[95] using the protein of interest engineered with a positively charged domain (polyhistidine), which interacts electrostatically with the negatively charged surface of dihydrolipoic acid-capped QDs. The protein–QD conjugates prepared using this approach presented suitable water solubility, and the fluorescence quantum yield was even higher than that of the nonconjugated QDs. Mansur and co-workers have intensively investigated noncovalent conjugation methods using several biomolecules, such as proteins (bovine serum albumin), amino acids (glutamic, aspartic acids), polysaccharides (chitosan and derivatives), for the production of colloidal water-soluble QD bioconjugates.[22,24,26,28,29] Despite the relative success of using biomolecules as direct capping ligands for the synthesis of colloidal QD bioconjugates in aqueous dispersions, there are several important difficulties, including improving the optical properties by reducing surface defects and "blinking," narrowing the nanoparticle size distribution, and achieving synthesis reproducibility, need to be overcome before QDs can be used in the complex biological environment because the electrostatic interactions are generally not sufficiently specific and strong.

Covalent conjugation

Covalent attachment is an effective way of coupling biomolecules to QDs by providing more stable conjugates. In fact, the major goal of covalently capping QDs to biomolecules, including peptides, antibodies, nucleic acids, and small-molecule ligands, is to achieve specific targeting abilities.[51] The availability of such bioconjugation approaches has caused an astonishing growth, mostly in the last decade, in the application of QDs for the imaging of molecular targets in living cells and animal models. The first reports to use QDs for labeling proteins in cells were published approximately 15 years ago and used QDs conjugated to transferring and actin-binding molecules.[51,52] Since then, a myriad of coupling molecules have been utilized for the bio-functionalization of QDs. During synthesis, polymers, organosilane derivatives, or other coatings can include functional groups capable of direct subsequent conjugation. Such linking strategies exploit the functional groups present on both the QD surface and the biomolecule. In Fig. 7 are presented typical chemical functionalities, linkers, and reactions most often utilized for the covalent conjugation of QDs with biomolecules.[96] For example, carbodiimide (1-ethyl-3-(3-dimethylaminopropyl) carbodiimide EDC) compounds are the most frequently used zero-length cross-linkers and are usually applied in conjunction with sulfo-NHS (N-hydroxysulfosuccinimide) or its uncharged analog NHS (N-hydroxysuccinimide) to improve the solubility of the reactive intermediates and increase the

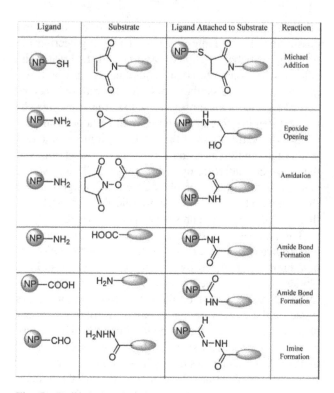

Fig. 7 Typical chemical functionalities, linkers, and reactions most often utilized for the covalent conjugation of QDs with biomolecules.
Source: From Erathodiyil & Ying[96] © 2011, with permission from American Chemical Society.

coupling efficiency of the amino-functionality with carboxyl groups.[94,97] This popularity is essentially because proteins, enzymes, and polypeptides ubiquitously display multiple amines and carboxyls on their amino acid side chains, in addition to their N- and C-termini. Other biomolecules and biopolymers, such as polysaccharides, glycoconjugates, adaptamers, and nucleotides, may also contain amines, carboxyls, thiols, and hydroxyls as chemical functionalities, which make them susceptible for conjugation in order to form amides, acetals, and thioacetal covalent bonds.[11,25] Furthermore, the application of engineered peptides, proteins, and polymers for functional coatings of QDs is a rapidly growing field in nanocrystal modification because these molecules match most of the key requirements of bioconjugates, such as ligand chemical stability, hydrophilic, and tailored bioactive moiety (groups or functionalities), for specific targeting.[11,25] Of the polymer ligands used as capping agents for the preparation of QDs, some have been found to be intrinsically biologically and biochemically active.[98] For instance, some research groups succeeded in functionalizing QDs with amine-containing polymers, such as poly(ethyleneimine),[99] and also being an effective transfection agent. Others used dendrimers based on functionalized poly(amidoamine) (PAMAM) as nanocrystal capping polymers, relying on their important attribute to effectively penetrate cell walls, which makes them

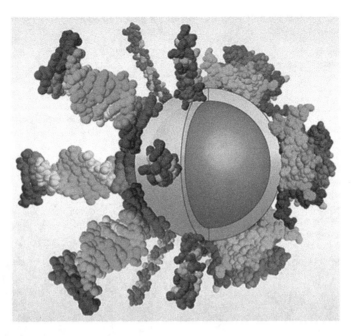

Fig. 8 The general perspective of a typical bioconjugate based on QDs with a core–shell semiconductor inorganic structure and with biomolecules at the surface, resulting in the integrated nanohybrid systems.
Source: From Sapsford et al.[91] © 2013, with permission from American Chemical Society.

potential transfection agents.[100–102] PAMAM polymers possess a large number of primary and tertiary amine groups at the surface and in the interior branches of the molecule, which are known to allow DNA complexation and which can also graft to QD surfaces.[103]

Another common conjugation strategy usually employs the high affinity associated with biotin–streptavidin and biotin–avidin, which requires prior coupling of the QD to streptavidin or avidin, respectively. QD–streptavidin conjugates are useful because a wide range of proteins and other biomolecules can be biotinylated through a relative facile protocol. These conjugates have applications in staining and labeling, live tracking, and drug screening. For a large number of bio-applications, QD conjugates act simply as fluorescent nonfunctional probes and have minimal impact on the experiment with respect to the binding event or the surroundings. Their nonspecific attachment to unintended molecules and the aggregation of QD-conjugates is possible and may negatively impact the results of an experiment. Some strategies have been developed with the aim of minimizing or even eliminating any possible nonspecific protein binding; for instance, the coating of QDs with an inert hydrophilic polymer, such as PEG, which is often referred to as PEGylation, yields uncharged colloidal dispersions. This modification results in a reduced unspecific uptake in cells and also prevents unspecific protein adsorption onto the bioconjugates.[104] As far as designing QD bioconjugates for targeting cancer cells and tissues is concerned, some complications must be taken into account, and these are mainly associated with distinguishing and specifically targeting the malignant tumor and not the normal cells. Moreover, most tumors are highly

heterogeneous and contain a mixture of benign, cancerous, and stroma cells. The heterogeneous nature of tumors makes it very difficult to use some of the aforementioned bioconjugation technologies to accurately reach the disease site.[105] Thus, biomolecules, including antigen-specific antibodies, with high affinity for tumor-associated biomarkers must be used to develop bioconjugates that can effectively detect tumors and tumor-associated tissues both *in vitro* and *in vivo* in animal models and in patients using bio-imaging techniques.[106–109] The innumerous strategies for synthesizing QD nanoconjugates for cancer diagnosis and treatment are presented in more detail in a specific section on this subject. Figure 8 illustrates the overall representation of a typical bioconjugate based on QDs with a core–shell semiconductor inorganic structure and with biomolecules at the surface, resulting in the integrated nanohybrid systems. In Fig. 9 an illustration of some selected surface chemistries and great variety of conjugation strategies that may be applied to make QD bioconjugates designed for biomedical applications is shown.[110]

QDs AND CANCER: A "BRIGHT" FUTURE AHEAD

"The Sooner, The Better": The Importance of Early Diagnosis and Detection of Cancer

Cancer diagnosis and treatment are of great interest due to the widespread occurrence of the disease, which is associated with high death rates and many recurrences after treatment. Despite some major breakthroughs in recent years

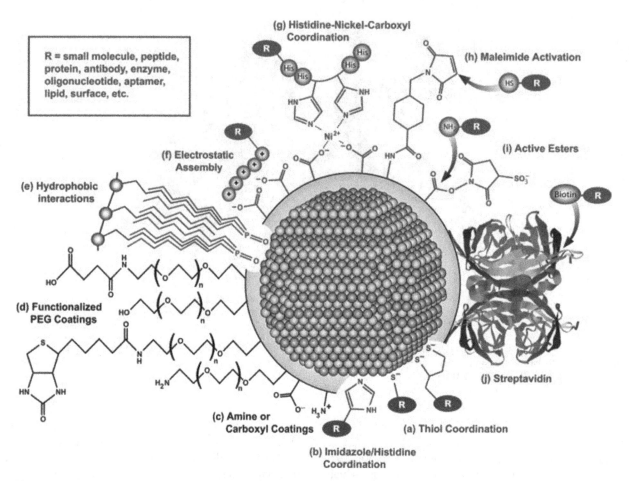

Fig. 9 Illustrative drawing of some selected surface chemistries and the great variety of conjugation strategies that may be applied to make QDs bioconjugates designed for biomedical applications (not to scale).
Source: From Algar et al.[110] © 2010, with permission from Elsevier.

from research on the complexity of the events that cause cancer lesions and potential metastasis, there is no effective treatment for cancers that have spread to other tissues and organs. Thus, cancer is one of the main causes of death in the world, second only to cardiovascular diseases in the current century, and the statistics project that this value will continue to worsen. It has been estimated that cancer will lead to 12 million deaths annually by 2030 compared with the 7.8 million fatal cases observed in 2008, indicating an exponential growth rate.[111] The control of cancer requires substantial public health intervention and, in the long term, must give priority to primary prevention, which would lead to a decrease in cancer incidence. However, in the short term, advances in diagnosis and treatment will be more effective measures. Hence, it is practically a consensus among health professionals and scientists that the prognosis and treatment of cancer greatly relies on the earliest possible detection of cancer lesions and on the efficient and specific delivery of drugs to the cancer site.[49,112] Additionally, the rapid and effective differentiation between normal and cancer cells is an important challenge for the diagnosis and treatment of tumors because the clinical outcome of

cancer diagnosis is strongly related to the stage at which the malignancy is detected. Biomedical imaging is playing an ever-more important role in all phases of cancer management. These include prediction, screening, and biopsy guidance for detection, staging, and metastasis, prognosis, therapy planning, and recurrence. Some of the existing technologies for cancer imaging include noninvasive imaging techniques, such as computed tomography (CT) scans, magnetic resonance, imaging scans, positron emission tomography (PET) scans, single photon emission CT, ultrasound scans, and optical imaging to macroscopically visualize tumors. Unfortunately, most of these imaging techniques are somewhat limited in the detection of abnormalities at the microscopic level. Most solid tumors are currently only detectable once they reach a diameter of approximately ~1 cm, at which point, they comprise millions of cancer cells (threshold of detection ~10^9 cells) that may have already migrated to other sites in metastasis.[49,112] Usually, the growth of human solid tumors may be modeled by "Gompertzian" kinetics, which implies that solid tumors display an initial lag phase starting from the single-cell stage, a log phase heralded by angiogenesis and an escape from diffusion-limited nutrition at approximately the 10^5-cell stage, and a second lag phase culminating in the death of the patient at 10^{12}-cell (~1 kg) stage.[49,112] Thus, the current detection techniques are far behind the goal of cancer imaging (i.e., to detect and image the smallest number of tumor cells) ideally much before the angiogenic switch (<10^5 cells), which is at least fourfold higher than the current resolution. However, it is currently possible to combine some of the existing bioimaging technologies with more sophisticated optical contrast agents for high-resolution *in vivo* cancer imaging to achieve enhancements compared with the conventional methods. In that sense, nanotechnology has emerged as a very promising novel platform in nanomedicine for cancer research and clinical applications by providing groundbreaking new methods and techniques and offering tremendous potential for personalized oncology approaches.[2,113–118] Nanomaterials have the potential to provide solutions to the current obstacles in cancer therapies due to their ultrasmall dimensions (i.e., at least one dimension sized from 1 to 100 nm for medical purposes) and large surface-to-volume ratios. In fact, the size, surface characteristics, and shape of a nanoparticle play a key role in its behavior and performance *in vivo*. Hybrid nanostructures have attracted particular research attention due to their distinctive shape- and composition-dependent properties. Thus, bringing together components with intrinsically different functionalities constitutes a particularly powerful route for the creation of novel multifunctional materials with synergetic properties found in neither of the constituents.[119] These featured properties allow these materials to exhibit potential applications beyond those of the individual components. For instance, the joining of two or more (nano)components to produce a hybrid nanostructure would offer almost countless possibilities in cancer, which may allow the resolution of several challenges, including: 1) improved delivery of poorly water-soluble drugs; 2) site-directed delivery of drugs toward specific biological and molecular targets; 3) development of innovative diagnostic tools; and 4) combination of therapeutic agents with diagnostic probes. Hence, with relevant advances in nanotechnology and a more in-depth understanding of the properties of materials at the nanoscale level, it is possible that several distinct diagnosis and therapeutic systems will be designed, developed, and positively approved (or entered clinical development) for cancer detection and therapy.[120] However, despite some irrefutable progress over the last years, few examples of successful nano-systems for the diagnosis, treatment, and management of cancer have been reported and become effectively clinically available.[121]

Nonetheless, one good example of the improvement introduced by nanotechnology was the development of QDs, which are a heterogeneous class of engineered fluorescent hybrid nanoparticles. QDs are nanometer-sized luminescent semiconductor nanocrystals that possess unique optical properties, such as high brightness, long-term stability, simultaneous detection of multiple signals, and tunable emission spectra, which make them appealing as potential diagnostic and therapeutic systems in oncology (i.e., they can literally "illuminate" cancer tumor lesions). As previously mentioned, one of the most important optical characteristics of QDs is that they are more photostable than conventional organic dyes. For instance, it has been reported that, under the same excitation conditions, 90% of the fluorescence of a normal organic dye fades within 1 min, whereas the fluorescence of QDs remains intact even after 30 min; another feature of these materials is that they may be excited repeatedly.[122] When both QDs and rhodamine green-dextran were exposed to 450-nm light for 80 min, the fluorescence of rhodamine-dextran was lost, whereas the QDs remained fluorescently stable. Moreover, QDs have wide excitation spectra and narrow emission spectra compared with conventional chromophores and contrast agents. These properties of QDs render these materials well suited for multiplex bioimaging.[123–125] Thus, the combination of QD-based nanotechnology with cancer biomarkers may have an encouraging impact in oncology research, such as carcinogenesis and tumor progression, and in clinical oncology, such as cancer diagnosis, treatment, prediction, and monitoring[111,126–130] Obviously, the number of studies on cancer and nanotechnology is enormous and continuously increasing. For that reason, in the following section, the recent progress associated with the promising applications of QDs in cancer diagnosis and therapeutics is summarized from the materials science perspective considering their surface functionalization by biomolecular targets for drug delivery systems and therapy. Whenever possible, the entry focuses on the presentation of hybrid systems involving QDs synthesized with biocompatible polymers and polymer derivatives for the diagnosis and treatment of cancer.

"Seeing Is Believing": QDs as Fluorescent Bioprobes for Cancer Detection and Therapy

To achieve binding specificity or targeting abilities, the polymer-capped or polymer-coated QDs may be linked to affinity ligands, such as antibodies, peptides, polypeptides, oligonucleotides, and synthetic molecules.[49] QD conjugates may be used as cancer bioprobes for labeling and imaging cancer cells *in vitro* or for targeting the diseased organ/tissue *in vivo*. In that sense, the surface-engineered nanohybrids must exhibit specific affinity to the tumor and not to normal cells and must present no cytotoxicity. Thus, biocompatible polymers meet many of the major requirements of ligands for QD bioconjugation because they are water-soluble, chemically stable under physiological conditions, nontoxic, and commercially available at very reasonable prices and can be readily functionalized by chemical and biochemical moieties to assign affinity and specificity properties. For instance, in an elegant study, Gao and co-workers reported a novel class of QD multifunctional bioprobes based on amphiphilic tri-block co-polymers for the simultaneous targeting and imaging of tumors *in vivo* using live animals.[131] Several polymer coatings and ligands, such as polyvinyl alcohol, polymethylmethacrylate (PMMA), and poly-lactide-*co*-glycolide, may be used on the surface of QDs. PEG and derivatives are the most common polymer biocompatible "shell" used with QDs for nanomedicine. These molecules hold great promise for the development of novel advanced multifunctional bioconjugates with QDs for cancer diagnosis and therapeutics. PEG is a coiled polymer of repeating ethylene ether units with dynamic conformations. In both drug delivery and imaging *in vivo* applications, the addition of surface functionalization to nanoparticles by PEG reduces reticuloendothelial system (RES) uptake and increases the circulation time compared with that obtained with the uncoated counterparts.[132] Their aggregation decreases due to passivated surfaces, and their association with nontargeted serum and tissue proteins is diminished, resulting in so-called stealth behavior. The PEG chains reduce the charge-based contact typical of proteins and small-molecule interactions. Their solubility in buffer and serum increases due to the hydrophilic ethylene glycol repeats, and the enhanced permeability and retention (EPR) effect is modulated due to the overall bioconjugated size changes obtained through the addition of a PEG coat. Due to these attributes, PEGylated nanoparticles generally accumulate in the liver at a concentration equal to a half to a third of the amount of accumulated non-PEGylated analogous particles and demonstrate higher tumor accumulation compared

with the background.[133,134] Moreover, PEG is inexpensive, versatile, and FDA-approved for many applications in medicine, nutrition, and biology. Excellent reviews on protein and peptide PEGylation are available in the literature.[135,136] In Fig. 10 is shown a representative drawing of using a PEG coating on nanoparticles for "disguising" the detection by the host immune system and delivering the drug to the cells.[137]

As nanomaterials, two key prominent characteristics distinguish QDs from conventional organic dye compounds to be used in cancer theranostics.[138] First, QDs possess unique optical and physicochemical properties that can be harnessed for diagnostic and therapeutic applications in oncology research. Second, multiple functions can be readily integrated into hybrid nanostructures of QDs and specific biomolecules, which would enable the designed system to perform multimodal image-guided therapy. These novel nanohybrid structures, once introduced into the body, are expected to reach precisely the diseased site and to home in solid tumors either via a passive targeting mechanism (i.e., the EPR effect) or via an active targeting mechanism facilitated by the ligands bound to their surfaces. For that reason, the targeted delivery of QD conjugates to cancer cells is an essential requirement for the selective imaging and effective therapy of cancer. Thus, innovative strategies to incorporate effective targeting ligands in the bioconjugates are of paramount importance to the outcome. The overexpressed receptors in many cancers are ideal targets in cancer cells. The methods for

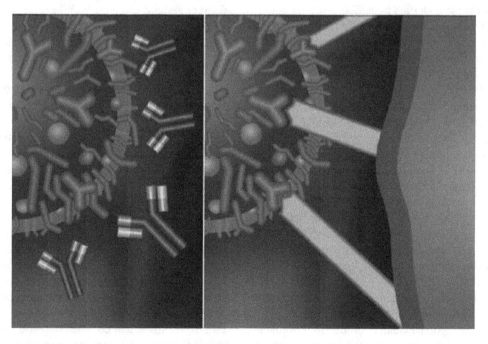

Fig. 10   Representative drawing of the hypothetical mechanism associated with the PEG coating on the surface of nanoparticles for "disguising" the detection by the host immune system and delivering the drug to the cells. The left panel shows that the PEG coating prevents the transient interaction and binding of antibodies with proteins on the nanoparticle surface. The right panel shows that the same PEG coating does not interfere with recognition by the cellular receptors.
Source: From McNeil[137] © 2009, with permission from John Wiley & Sons, Inc.

delivering QDs inside cells include physical and biochemical techniques. Physical techniques, such as electroporation and microinjection, have practical limitations for *in vivo* applications, whereas biochemical techniques, such as peptide-, antibody-, and secondary antibody-based targeting, are promising for the selective labeling of cancer cells with QDs. A comprehensive and excellent review published by Biju and collaborators addresses the most relevant aspects associated with the use of bioconjugated QDs for cancer research.[36] Regarding the targeted imaging of cancer tumors using QDs, most studies in the literature are basically divided into three main groups, namely *in vitro*, *in vivo*, and multifunctional (e.g., imaging and therapy) applications.[14,36,139]

"Shedding Light On" Cancer: QD Bioconjugates for the Imaging of Cancer Cells

In vitro targeted imaging of cancer cells using QDs

One of the most advanced applications of QDs is the *in vitro* imaging of cancer cells. Soon after the introduction of biocompatible QDs for cell imaging by Chan and Nie[52] and Bruchez et al.,[51] several research groups investigated the use of QDs for the imaging of cancer cells.[36] Depending

on the functional group/biomolecule conjugated to the surface of QDs, the methods for targeting and imaging can be classified into antibody-based and ligand-based methods. Usually, antibody-based assays are very accurate, highly specific, and highly sensitive (~pM detection) but are time-consuming and involve complex protocols. In contrast, ligand-based bioconjugates are simpler, faster, and not as precise and reliable as antibodies due to their reduced specificity and affinity toward the receptor to be detected. Thus, cancer biomarker *in vitro* assays are a useful strategy for the screening and diagnosis of cancer. Many proteins are considered useful early diagnostic and prognostic markers of cancer and targets for basic biomedical research. However, these protein biomarkers are only present at very low concentrations; thus, methods with low detection limits are required. Nonetheless, QDs conjugated to cancer-specific ligands/antibodies/peptides are effective for detecting and imaging different human cancer cells, tissues, and organs. QDs have been proven to be applicable for the ultrasensitive detection of protein cancer biomarkers (e.g., ovarian cancer, breast cancer, prostate cancer, pancreatic cancer, gastric adenocarcinoma, metastatic tumor, liver, glioblastoma, and cancers of the bone marrow).[14,49,140–151] An example of targeting and imaging tumor cells using QDs conjugates *in vitro* is illustrated in Fig. 11.[151] It is

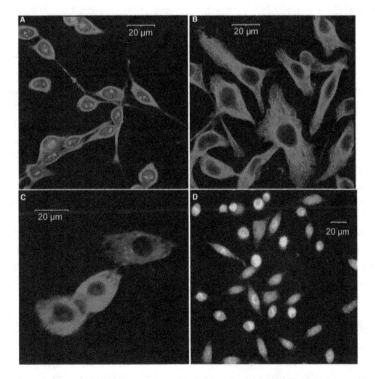

Fig. 11 An example of *in vitro* targeting and imaging tumor cells (cells of a HepG2 cell line, human liver carcinoma) with confocal fluorescence using QDs conjugates staining.[151] The conjugates (glutathione-capped CdTe QDs, GSHCdTe QDs) must present high specificity toward the tumor cells as compared to normal cells ("healthy"). (**A**) Fixed HepG2 cells with nucleoli and cytoplasm stained by GSH-CdTe QDs and GSH-CdTe QDs. (**B**) Fixed NIH 3T3 cells with actin immunostained using biotin-labeled GSH-CdTe618 QDs. (**C**) Live MDA-MB-435 cells incubated with F3-labeled GSH-CdTe618 QDs. (**D**) Live macrophage RAW264.7 cells incubated with GSH-CdTe618 QDs and cell viability calcein dye.
Source: From Zheng et al.[151] © 2007, with permission from Wiley Interscience.

important to highlight that the conjugates must present high specificity toward the tumor cells as compared to normal cells ("healthy").

In vivo targeted imaging of cancer cells using QDs

The long-term chemical and photostability and other outstanding optical properties of QDs make them ideal candidates for *in vivo* targeting and imaging. However, only limited progress has been made in developing QD probes for imaging the intracellular space of living cells. A major problem is the lack of efficient methods for delivering monodispersed (i.e., single) QDs into the cytoplasms of living cells. A common observation is that QDs tend to aggregate inside cells and are often trapped in endocytotic vesicles, such as endosomes and lysosomes. Additionally, the imaging of tumors presents a unique challenge not only because of the urgent need for sensitive and specific imaging agents of cancer but also because of the unique biological attributes inherent to cancerous tissue. In the cancer microenvironment, blood vessels are abnormally formed during tumor-induced angiogenesis, and these have erratic architectures and wide endothelial pores.[139] Thus, the *in vivo* applications of QDs were first tested by Akerman et al.[152] using CdSe/ZnS QDs coated with peptides injected into the tail vein of mice; these researchers found that the injected QDs preferentially distribute in endothelial cells in the lung blood vessels. Thus, in general, the same concepts underlying the *in vitro* targeting of cancer cells may be applied *in vivo*. However, the main challenges for the *in vivo* targeting and imaging of cancers using QDs are the biodistribution of QD bioconjugates, penetration depths of the excitation light and PL, tissue autofluorescence, toxicity, and pharmacokinetics. Bioconjugated QDs were applied *in vivo* either systemically for deep cancers or subcutaneously for peripheral cancers. However, compared with local administration, systemic administration needs more attention due to the possible interactions of QD conjugates with blood components and the stimulation of the immune response. Although it was found that QDs conjugated with various anticancer antibodies were selectively and uniformly distributed in tumor milieu, only a small amount of evidence supports the hypothesis that QDs have the ability to extravasate to reach tumor cells *in vivo*. Indeed, the biodistribution of QDs and their nonspecific uptake by the RES, which includes the liver, spleen, and lymphatic system, is an important issue remaining in the *in vivo* applications of QDs. Nie and coworkers[153] produced QD-based multifunctional nanoparticle probes for cancer targeting and imaging in live animals. These researchers first encapsulated luminescent QDs with an ABC triblock copolymer and then linked this amphiphilic polymer to tumor-targeting ligands. They then applied this probe to achieve both passive tumor imaging and active tumor imaging with high sensitivity and multicolor capabilities. The key features for the success of their study included both the

EPR at tumor sites and the ability of the antibody to recognize cancer-specific cell surface biomarkers. An interesting study was reported using polymer-based QD conjugates for the design and synthesis of folate-poly(ethylene glycol)-polyamidoamine (FPP)-functionalized CdSe/ZnS QDs.[154] The test proved that QDs play a key role in bioimaging, whereas the folate–PEG conjugates of the PAMAM dendrimer serve as a system targeted to folate receptors in tumor cells. Moreover, dendrimer ligands (FPP) were found to encapsulate and solubilize luminescent QDs through direct ligand-exchange reactions. Due to the membrane expression of folic acid receptors in tumor cells, this class of ligand-exchanged QDs is able to target tumor cells. These insights are important for the design and development of polymer-functionalized/QDs bioconjugates for the optical detection of tumor cells and bio-imaging.[119,154]

Multimodal QDs for imaging and therapy

Although relatively few studies on QDs for cancer tumor therapy have been reported, it is conceivable that QDs have the potential for tumor therapy due to their huge surface areas available for modifications with chemical functional groups or therapeutic agents, such as anticancer drugs and photosensitizers.[153,155,156] Thus, QDs may be turned into convenient scaffolds to accommodate multiple imaging (e.g., radionuclide-based or paramagnetic probes) and therapeutic agents (e.g., anticancer drugs), therefore enabling the development of a nearly boundless library of multifunctional nanostructures for multimodality imaging and for the integrated imaging and therapy of cancer. The applications of QDs described in the previous section for *in vivo* imaging are significantly restricted by their tissue penetration depth, quantification problems, and a lack of anatomic resolution and spatial information. Consequently, to address these limitations, some research groups have attempted to couple QD-based optical imaging with other imaging modalities that are not limited by the penetration depth, such as magnetic resonance imaging (MRI), PET, and single photon emission computed tomography.[157,158] For example, Mulder et al.[158] developed a dual modality imaging probe for both optical imaging and MRI by chemically incorporating paramagnetic gadolinium complexes in the lipid-coating layer of QDs.[159] The integration of imaging and therapy functionalities in a single nanohybrid system is a tremendous challenge to be overcome in the cancer research realm. Drug-containing nanoparticles have shown great promise for the treatment of tumors in animal models and even in clinical trials.[160] Both passive and active targeting of nanotherapeutics has been used to increase the local concentration of chemotherapeutics in the tumor. Due to the size and structural similarities between imaging and therapeutic nanoparticles, it is possible that their functions can be integrated to directly monitor their therapeutic biodistribution in order to improve treatment specificity and reduce side effects. This synergy has become the principle foundation for the development of multifunctional hybrid nanocomposites for integrated imaging and cancer treatment. Most studies are still at a very early stage using cultured cancer cells and are not immediately relevant to the *in vivo* imaging and treatment of solid tumors. However, these studies will guide the future design and optimization of multifunctional nanoparticle agents for *in vivo* imaging and therapy.[161–164] As an example, Farokhzad and coworkers reported a ternary system composed of a QD, an aptamer, and the small molecular anticancer drug doxorubicin for *in vitro* targeted imaging, therapy, and sensing of drug release.[162] It is expected that, once the bioconjugates reach the target tissue, their activity can then be turned on using an external stimulus, such as by photothermally converting light energy into heat. As a result, the temperature in the treatment volume would be increased above the thermal damage threshold of the targeted tissue, thereby killing the cells. QDs inside cells are particularly useful for cell tracking to study

anomalous cell division and metastasis. Due to the high stability and multicolor emission of QDs, these particles can act as unique markers for tracking cancer cells *in vivo* during metastasis, which is a critical issue in the development of effective cancer therapies. In particular, the development of near-infrared (NIR) QD conjugates integrated with anticancer drugs may be a powerful strategy that will allow the "whole-body" optical imaging of complex cancer phenomena with minimum invasiveness combined with the therapy of the diseased site. Figure 12 shows a well-designed schematic representation suggested by Sahai[165] of the "whole-body" live imaging process containing several possible events occurring connected to the primary tumor, cell migration, metastases, and the complexity of the microenvironment caused by the cancer disease.[165]

A fine and well-designed study found that QDs in photodynamic therapy (PDT) can act as photo-sensitizers themselves or activate another photo-sensitizer by serving as the energy donor. The energy transfer between QDs and

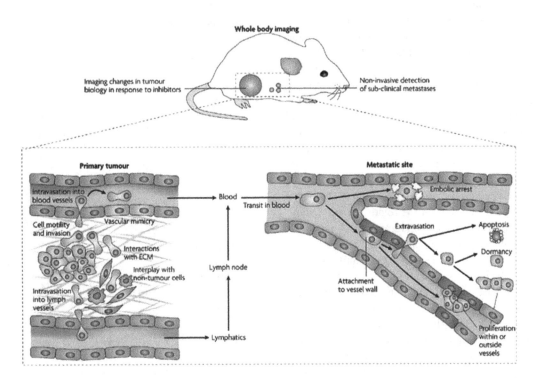

Fig. 12 Whole body imaging can be used to monitor processes in the primary tumor and for the detection of subclinical metastases. Various aspects of tumor cell behavior at the primary site can be imaged including cell motility, cell interactions with extracellular matrix components such as collagen fibres, interplay with nontumor cells including fibroblasts and macrophages, vascular mimicry (cancer cells can sometimes form part of vessel walls), and intravasation into the blood or lymph. Intravital imaging can monitor cells arriving at the metastatic site via transit in the blood and arresting either as emboli in narrow vessels or attaching to vessel walls—darker shading indicates endothelial cells expressing molecules that promote interactions with tumor cells, both processes may be aided through interactions with platelets. Following attachment, cancer cell extravasation, possibly following behind leukocytes, can be observed. Cells will then undergo apoptosis, enter a state of dormancy or begin proliferation within or outside vessels, and these processes can all be imaged (cancer cells, nontumor cells, and endothelial cells).
Source: From Sahai[165] © 2007, with permission from Nature Publishing Group.

cell molecules (such as triple oxygen, reducing equivalents, and pigments) can potentially generate reactive oxygen species to provoke apoptosis in cells. Unlike the visible emission of most conventional photosensitizers, QDs can be tuned to emit in the NIR regions, which can be useful in PDT for deep-seated tumors because NIR is not scattered and absorbed by tissue.[112,166–168]

CONCLUSIONS, CHALLENGES, AND FUTURE OUTLOOK IN CANCER THERANOSTICS

This entry has summarized the essential aspects associated with the synthesis of QD-conjugated nanostructures and their broad emerging applications in cancer diagnosis and therapy. Predominantly, this entry showed the processing routes using colloidal chemistry and focused on the relevant role of polymers as ligands to primarily provide long-term colloidal stability to QDs dispersed in solutions and as surface modifiers with functional groups used in further chemical derivatization for the production of nanohybrid bioconjugates. One major goal of cancer research and treatment is the implementation of multifunctional platforms within a single targeted system that would simultaneously perform diagnosis, targeted delivery, and efficient therapy (i.e., theranostics). Thus, it is hypothesized that the use of polymeric receptor-targeting ligands combined with fluorescent QDs will lead to improved tumor-targeting nanohybrids. Such hybrid nanomaterials would enable the tracking of cells for simultaneous medical therapy and diagnosis. The synergistic combinations of the fluorescent features and bioconjugation of the colloidal hybrid semiconductor nanoparticles have created a series of "tailored" hybrid nanostructures that can potentially behave as bioprobes and nanocarriers for the detection and treatment of tumors. It is anticipated that, in the relatively near future, the research on QDs in cancer imaging will significantly expand their clinical purposes, particularly that of conjugated QDs in targeting metastasis and the quantitative measurement of molecular targets. Currently, conjugated QDs can target essentially solid tumor tissues with mature vasculature; however, the detection of micro-metastasis is challenging. Consequently, the surface of QDs needs to be further engineered to enable efficient extravasation in order to reach micro-metastasis and initiate binding to tumor antigens. One approach to enhance the tumor targeting of QDs is to render them long-circulating in the blood through surface modifications with polymers, such as the introduction of large-molecular-weight PEG molecules to decorate QDs in order to prolong their circulation time and reduce their cellular uptake. However, with all of the advances that have already been achieved in QD nanotechnology, their biocompatibility, distribution, metabolism/excretion, and safety issues, such as cytotoxicity, are major concerns that still need to be addressed. The strategy of adding biocompatible polymer ligands as coating shells have increased the biocompatibility and minimized the toxicity of these ultrasmall particles, resulting in more water-soluble and safer formulations. In that sense, biocompatible and water-soluble polymers and derivatives can be foreseen to be of crucial importance for *in vivo* detection and the subsequent treatment of cancer. In this regard, the combination of QD fluorescence imaging with MRI, PET, and CT will result in the development of powerful techniques for the live diagnosis and therapy of cancer. In Fig. 13 ideally the enormous potential of using QD bioconjugates for cancer detection, targeting, therapy and remission of tumor is displayed, considering an integrated approach for the complete process. Additionally, the development of novel fluorescent NIR and visible QD bioconjugates integrated with photosensitizer molecules may revolutionize oncology because these will be simultaneously utilized for targeting, imaging, and killing cancer cells *in vitro* and *in vivo* through the production of reactive oxygen intermediates specifically at the tumor microenvironment. Based on the aforementioned reasons, a rational scenario in which polymer-based QD nanohybrids have a "bright" future may be envisaged by illuminating the "dark spots" of the existing concerns associated with cancer.

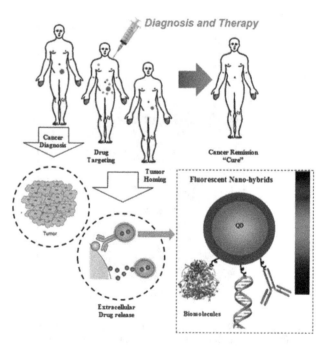

Fig. 13 An idealized strategy for exploiting the enormous potential of using QD bioconjugates for cancer detection, targeting, therapy and remission of tumor, considering it a completely integrated process. The development of fluorescent QD bioconjugates combined with photosensitizer molecules may revolutionize oncology because they may be simultaneously utilized for targeting, imaging, and killing cancer cells *in vitro* and *in vivo* specifically at the tumor microenvironment.

ACKNOWLEDGMENTS

The authors acknowledge the financial support from CAPES, FAPEMIG, and CNPq. This entry was dedicated to all cancer patients, scientists, and health professionals across the world whose daily sacrifices have been the inspiration for combating cancer and the obstinate search for cancer diagnosis, targeting, and treatment.

CONFLICTING INTEREST

The authors have declared that no conflicting interests exist.

REFERENCES

1. Wang, Y.; Chen, L. Quantum dots, lighting up the research and development of nanomedicine. Nanomedicine **2011**, *7* (4), 385–402.

2. Peer, D.; Karp, J.M.; Hong, S.; Farokhzad, O.C.; Margalit, R.; Langer, R. Nanocarriers as an emerging platform for cancer therapy. Nat. Nanotechnol. **2007**, *2* (12), 751–760.

3. Sudhakar, A. History of cancer, ancient and modern treatment methods. J. Cancer Sci. Ther. **2009**, *1* (2), 1–4.

4. Fang, M.; Peng, C.W.; Pang, D.W.; Li, Y. Quantum dots for cancer research: Current status, remaining issues, and future perspectives. Cancer Biol. Med. **2012**, *9* (3), 151–163.

5. Weissleder, R.; Pittet, M.J. Imaging in the era of molecular oncology. Nature **2008**, *452* (7187), 580–589.

6. Mansur, H.S. Quantum dots and nanocomposites. WIREs Nanomed. Nanobiotechnol. **2010**, *2* (2), 113–129.

7. Kumar, S.; Gradzielski, M.; Mehta, S.K. The critical role of surfactants towards CdS nanoparticles: Synthesis, stability, optical and PL emission properties. RSC Adv. **2013**, *3* (8), 2662–2676.

8. Yang, P.; Tretiak, S.; Ivanov, S. Influence of surfactants and charges on cdse quantum dots. J. Clust. Sci. **2011**, *22* (3), 405–431.

9. Knowles, K.E.; Frederick, M.T.; Tice, D.B.; Morris-Cohen, A.J.; Weiss, E.A. Colloidal quantum dots: Think outside the (Particle-in-a-) box. J. Phys. Chem. Lett. **2012**, *3* (1), 18–26.

10. Mansur, H.S.; Mansur, A.A.P. CdSe quantum dots stabilized by carboxylic-functionalized PVA: Synthesis and UV–vis spectroscopy characterization. Mater. Chem. Phys. **2011**, *125* (3), 709–717.

11. Mansur, H.S.; Mansur, A.A.P.; Gonzalez, J.C. Synthesis and characterization of CdS quantum dots with carboxylic-functionalized poly (vinyl alcohol) for bioconjugation. Polymer **2011**, *52* (4), 1045–1054.

12. Brus, L.E. Electron–electron and electron-hole interactions in small semiconductor crystallites: The size dependence of the lowest excited electronic state. J. Chem. Phys. **1984**, *80* (9), 4403–4409.

13. Medintz, I.L.; Uyeda, H.T.; Goldman, E.R.; Mattoussi, H. Quantum dot bioconjugates for imaging, labeling and sensing. Nat. Mater. **2005**, *4* (6), 435–446.

14. Li, J.; Zhu, J.-J. Quantum dots for fluorescent biosensing and bio-imaging applications. Analyst **2013**, *138* (9), 2506–2515.

15. Pilla, V.; Munin, E. Fluorescence quantum efficiency of CdSe/ZnS quantum dots functionalized with amine or carboxyl groups. J. Nanopart. Res. **2012**, *14* (10), 1147.

16. Carrillo-Carrión, C.; Cárdenas, S.; Simonet, B.M.; Valcárcel, M. Quantum dots luminescence enhancement due to illumination with UV/Vis light. Chem. Commun. **2009**, *21* (35), 5214–5226.

17. Efros, A.L.; Rosen, M. Random telegraph signal in the photoluminescence intensity of a single quantum dot. Phys. Rev. Lett. **1997**, *78* (6), 1110–1113.

18. Tit, N.; Obaidat, I.M. Charge confinements in CdSe–ZnSe symmetric double quantum wells. J. Phys. Condens. Matter **2008**, *20* (16), 165205–165214.

19. Coe-Sullivan, S.; Woo, W.; Steckel, J.S.; Bawendi, M.; Bulovi, V. Tuning the performance of hybrid organic/inorganic quantum dot light-emitting devices. Org. Electron. **2003**, *4* (2–3), 123–130.

20. Murphy, C.J.; Coffer, J.L. Quantum dots: A primer. Appl. Spectrosc. **2002**, *56* (1), 16A-27A.

21. Ekimov, A.I.; Onushchenko, A.A. Quantum size effect in the optical-spectra of semiconductor micro-crystals. Sov. Phys. Semicond. **1982**, *16* (7), 775–778.

22. Mansur, H.S.; Mansur, A.A.P.; de Almeida, M.V.; Curti, E. Functionalized-chitosan/quantum dots nano-hybrids for nanomedicine applications: Towards biolabeling and biosorbing phosphate metabolites. J. Mater. Chem. **2013**, *1* (12), 1696–1711.

23. Mansur, A.A.P.; Ramanery, F.P.; Mansur, H.S. Water-soluble quantum dot/carboxylic-poly (vinyl alcohol) conjugates: Insights into the roles of nanointerfaces and defects toward enhancing photoluminescence behavior. Mater. Chem. Phys. **2013**, *141* (1), 223–233.

24. Santos, J.C.C.; Mansur, A.A.P.; Mansur, H.S. One-Step biofunctionalization of quantum dots with chitosan and N-palmitoyl chitosan for potential biomedical applications. Molecules **2013**, *18* (6), 6550–6572.

25. Mansur, H.S.; Mansur, A.A.P. Fluorescent nanohybrids: Quantum dots coupled to polymer recombinant protein conjugates for the recognition of biological hazards. J. Mater. Chem. **2012**, *22* (18), 9006–9918.

26. Mansur, H.S.; Mansur, A.A.P.; Curti, E.; de Almeida, M.V. Bioconjugation of quantum-dots with chitosan and N, N, N-trimethyl chitosan. Carbohydr. Polym. **2012**, *90* (1), 189–196.

27. Mansur, A.; Mansur, H.; González, J. Enzyme-polymers conjugated to quantum-dots for sensing applications. Sensors **2011**, *11* (10), 9951–9972.

28. Mansur, A.A.P.; Saliba, J.B.; Mansur, H.S. Surface modified fluorescent quantum dots with neurotransmitter ligands for potential targeting of cell signaling applications. Colloids Surf. B Biointerfaces **2013**, *111* (1), 60–70.

29. Mansur, H.S.; González, J.C.; Mansur, A.A.P. Biomolecule-quantum dot systems for bioconjugation applications. Colloids Surf. B Biointerfaces **2011**, *84* (2), 360–368.

30. Xu, S.; Wang, C.; Wang, Z.; Zhang, H.; Yang, J.; Xu, Q.; Shao, H.; Li, R.; Lei, W.; Cui, Y. Aqueous synthesis of internally doped Cu: ZnSe/ZnS core–shell nanocrystals with good stability. Nanotechnology **2011**, *22* (27), 275605.

31. Gu, Z.; Zou, L.; Fang, Z.; Zhu, W.; Zhong, X. One-pot synthesis of highly luminescent CdTe/CdS core/shell

Excipients—Gels

nanocrystals in aqueous phase. Nanotechnology **2008**, *19* (13), 135604.

32. Xiao, Q.; Huang, S.; Su, W.; Chan, W.H.; Liu, Y. Facile synthesis and characterization of highly fluorescent and biocompatible N-acetyl-l-cysteine capped CdTe/CdS/ZnS core/shell/shell quantum dots in aqueous phase. Nanotechnology **2012**, *23* (49) 495717.

33. Shu, C.; Huang, B.; Chen, X.; Wang, Y.; Li, X.; Ding, L.; Zhong, W. Facile synthesis and characterization of water soluble ZnSe/ZnS quantum dots for cellar imaging. Spectrochim. Acta A Mol. Biomol. Spectrosc. **2013**, *104* (1), 143–149.

34. da Silva, M.I.N.; Mansur, A.A.P.; Schatkoski, V.; Krambrock, K.W.H.; González, J.; Mansur, H.S. Fluorescent-magnetic nanostructures based on polymer-quantum dots conjugates. Macromol. Symp. **2011**, *319* (1), 114–120.

35. Uyeda, H.T.; Medintz, I.L.; Jaiswal, J.K.; Simon, S.M.; Mattoussi, H. Synthesis of compact multidentate ligands to prepare stable hydrophilic quantum dot fluorophores. J. Am. Chem. Soc. **2005**, *127* (11), 3870–3878.

36. Biju, V.; Mundayoor, S.; Omkumar, R.V.; Anas, A.; Ishikawa, M. Bioconjugated quantum dots for cancer research: Present status, prospects and remaining issues. Biotechnol. Adv. **2010**, *28* (2), 199–213.

37. Jiang, R.; Zhu, H.; Yao, J.; Fu, Y.; Guan, Y. Chitosan hydrogel films as a template for mild biosynthesis of CdS quantum dots with highly efficient photocatalytic activity. Appl. Surf. Sci. **2012**, *258* (8), 3513–3518.

38. Tomczak, N.; Janczewski, D.; Han, M.; Vancso, G.J. Designer polymer–quantum dot architectures. Prog. Polym. Sci. **2009**, *34* (5), 393–430.

39. Liedl, T.; Dietz, H.; Yurke, B.; Simmel, F. Controlled trapping and release of quantum dots in a DNA-switchable hydrogel. Small **2007**, *3* (10), 1688–1693.

40. Gong, Y.; Gao, M.; Wang, D.; Möhwald, H. Incorporating fluorescent CdTe nanocrystals into a hydrogel via hydrogen bonding: Toward fluorescent microspheres with temperature-responsive properties. Chem. Mater. **2005**, *17* (10), 2648–2653.

41. Mansur, H.S.; Grieser, F.; Urquhart, R.S.; Furlong, D.N. Photoelectrochemical behaviour of Q-state CdSxSe1-x particles in arachidic acid LB films. J. Chem. Soc. **1995**, *91* (19), 3399–3404.

42. Mansur, H.S.; Grieser, F.; Marychurch, M.S.; Biggs, S.; Urquhart, R.S.; Furlong, D.N. Photoelectrochemical properties of Q-state CdS particles in arachidic acid LB Films. J. Chem. Soc. **1995**, *91* (4), 665–672.

43. Mansur, H.S.; Vasconcelos, W.L.; Grieser, F.; Caruso, F. Photoelectrochemical behaviour of CdS Q-state semiconductor particles in 10,12 nonacosadiynoic acid polymer LB film. J. Mater. Sci. **1999**, *34* (21), 5285–5291.

44. Valizadeh, A.; Mikaeili, H.; Samiei, M.; Farkhani, S.M.; Zarghami, N.; Kouhi, M.; Akbarzadeh, A.; Davaran, S. Quantum dots: Synthesis, bioapplications, and toxicity. Nanoscale Res. Lett. **2012**, *7* (1), 480.

45. Bera, D.; Qian, L.; Tseng, T.-K.; Holloway, P.H. Quantum dots and their multimodal applications: A review. Materials **2010**, *3* (4), 2260–2345.

46. Gautam, A.; van Veggel, F.C.J.M. Synthesis of nanoparticles, their biocompatibility, and toxicity behavior for bio-

47. Chen, N.; He, Y.; Su, Y.; Li, X.; Huang, Q.; Wang, H.; Zhang, X.; Tai, R.; Fan, C. The cytotoxicity of cadmium-based quantum dots. Biomaterials **2012**, *33* (5), 1238–1244.

48. Green, M. The nature of quantum dot capping ligands. J. Mater. Chem. **2010**, *20* (28), 5797–5809.

49. Smith, A.M.; Dave, S.; Nie, S.; True, L.; Gao, X. Multicolor quantum dots for molecular diagnostics of cancer. Expert Rev. Mol. Diagn. **2006**, *6* (2), 231–244.

50. Kershaw, S.V.; Susha, A.S.; Rogach, A.L. Narrow band-gap colloidal metal chalcogenide quantum dots: Synthetic methods, heterostructures, assemblies, electronic and infrared optical properties. Chem. Soc. Rev. **2013**, *42* (7), 3033–3087.

51. Bruchez, M.; Moronne, M.; Gin, P.; Weiss, S.; Alivisatos, A.P. Semiconductor nanocrystals as fluorescent biological labels. Science **1998**, *281* (5385), 2013–2016.

52. Chan, W.C.W.; Nie, S. Quantum dots bioconjugates for ultrasensitive nonisotopic detection. Science **1998**, *281* (5385), 2016–2018.

53. Geszke-Moritz, M.; Moritz, M. Quantum dots as versatile probes in medical sciences: Synthesis, modification and properties. Mater. Sci. Eng. C **2013**, *33* (3), 1008–1021.

54. Choi, H.S.; Liu, W.; Misra, P.; Tanaka, E.; Zimmer, J.P.; Ipe, Bawendi, M.G.; Frangioni, J.V. Renal clearance of quantum dots. Nat. Biotechnol. **2007**, *25* (10), 1165–1170.

55. Pinaud, F.; Michalet, X.; Bentolila, L.A.; Tsay, J.M.; Doose, S.; Li, J.J.; Iyer, G.; Weiss, S. Advances in fluorescence imaging with quantum dot bio-probes. Biomaterials **2006**, *27* (9), 1679–1687.

56. Zrazhevskiy, P.; Sena, M.; Gao, X. Designing multifunctional quantum dots for bioimaging, detection, and drug delivery. Chem. Soc. Rev. **2010**, *39* (11), 4326–4354.

57. Pons, T.; Uyeda, H.T.; Medintz, I.L.; Mattoussi, H. Hydrodynamic dimensions, electrophoretic mobility, and stability of hydrophilic quantum dots. J. Phys. Chem. B **2006**, *110* (41), 20308–20316.

58. Krueger, K.M.; Al-Somali, A.M.; Mejia, M.; Colvin, V.L. The hydrodynamic size of polymer stabilized nanocrystals. Nanotechnology **2007**, *18* (47), 475709.

59. Henglein, A. Photo-degradation and fluorescence of colloidal-cadmium sulfide in aqueous solution. Ber. Bunsen-Ges. Phys. Chem. **1982**, *86* (4), 301–305.

60. Baral, S.; Fojtik, A.; Weller, H.; Henglein, A. Photochemistry and radiation chemistry of colloidal semiconductors. 12. Intermediates of the oxidation of extremely small particles of CdS, ZnS, and Cd3P2 and size quantization effects. J. Am. Chem. Soc. **1986**, *108* (3), 375–378.

61. Henglein, A. Small-particle research: Physicochemical properties of extremely small colloidal metal and semiconductor particles. Chem. Rev. **1989**, *89* (8), 1861–1873.

62. Lesnyak, V.; Gaponik, N.; Eychmüller, A. Colloidal semiconductor nanocrystals: The aqueous approach. Chem. Soc. Rev. **2013**, *42* (7), 2905—2929.

63. Qian, H.; Qiu, X.; Li, L.; Ren, J. Microwave-assisted aqueous synthesis: A rapid approach to prepare highly luminescent ZnSe(S) alloyed quantum dots. J. Phys. Chem. B **2006**, *110* (18), 9034–9040.

64. He, Y.; Sai, L.M.; Lu, H.T.; Hu, M.; Lai, W.Y.; Fan, Q.L.; Wang, L.H.; Huang, W. Microwave-assisted synthesis of water-dispersed CdTe nanocrystals with high luminescent

Excipients—Gels

efficiency and narrow size distribution. Chem. Mater. **2007**, *19* (3), 359–365.

65. Peppas, N.A.; Simmons, R.E.P. Mechanistic analysis of protein delivery from porous poly(vinyl alcohol) systems. J. Drug Deliv. Sci. Technol. **2004**, *14* (4), 285–289.

66. Costa-Júnior, E.S.; Barbosa-Stancioli, E.F.; Mansur, A. A. P.; Vasconcelos, W.L.; Mansur, H.S. Preparation and characterization of chitosan/poly(vinyl alcohol) chemically crosslinked blends for biomedical applications. Carbohydr. Polym. **2009**, *76* (3), 472–481.

67. Tomczak, N.; Liu, R.; Vancso, J.G. Polymer-coated quantum dots. Nanoscale **2013**, *5* (24), 12018–12032.

68. Liu, L.; Guo, X.H.; Li, Y.; Zhong, X.H. Bifunctional multidentate ligand modified highly stable water-soluble quantum dots. Inorg. Chem. **2010**, *49* (8), 3768–3775.

69. Palui, G.; Na, H.B.; Mattoussi, H. Poly(ethylene glycol)-based multidentate oligomers for biocompatible semiconductor and gold nanocrystals. Langmuir **2012**, *28* (5), 2761–2772.

70. Yildiz, I.; Deniz, E.; McCaughan, B.; Cruickshank, S.F.; Callan, J.F.; Raymo, F.M. Hydrophilic CdSe–ZnS Core–shell quantum dots with reactive functional groups on their surface. Langmuir **2010**, *26* (13), 11503–11511.

71. Wang, X.S.; Dykstra, T.E.; Salvador, M.R.; Manners, I.; Scholes, G.D.; Winnik, M.A. Surface passivation of luminescent colloidal quantum dots with poly-(dimethylaminoethyl methacrylate) through a ligand exchange process. J. Am. Chem. Soc. **2004**, *126* (25), 7784–7785.

72. Potapova, I.; Mruk, R.; Hubner, C.; Zentel, R.; Basche, T.; Mews, A. CdSe/ZnS nanocrystals with dye-functionalized polymer ligands containing many anchor groups. Angew. Chem. Int. Ed. **2005**, *44* (16), 2437–2440.

73. Wang, M.F.; Dykstra, T.E.; Lou, X.D.; Salvador, M.R.; Scholes, G.D.; Winnik, M.A. Colloidal CdSe nanocrystals passivated by a dye-labeled multidentate polymer: Quantitative analysis by size-exclusion chromatography. Angew. Chem. Int. Ed. **2006**, *45* (14), 2221–2224.

74. Wang, M.F.; Oh, J.K.; Dykstra, T.E.; Lou, X.D.; Scholes, G.D.; Winnik, M.A. Surface modification of CdSe and CdSe/ZnS semiconductor nanocrystals with poly(N,N-dimethylaminoethyl methacrylate). Macromolecules **2006**, *39* (10), 3664–3672.

75. Wang, M.F.; Felorzabihi, N., Guerin, G.; Haley, J.C.; Scholes, G.D.; Winnik, M.A. Water-soluble CdSe quantum dots passivated by a multidentate diblock copolymer. Macromolecules **2007**, *40* (17), 6377–6384.

76. Liu, W.H.; Greytak, A.B.; Lee, J.; Wong, C.R.; Park, J.; Marshall, L.F.; Jiang, W.; Curtin, P.N.; Ting, A.Y.; Nocera, D.G.; Fukumura, D.; Jain, R.K.; Bawendi, M.G. Compact biocompatible quantum dots via RAFT-mediated synthesis of imidazole-based random copolymer ligand. J. Am. Chem. Soc. **2010**, *132* (2), 472–483.

77. Zhang, P.; Han, H. Compact PEGylated polymer-caged quantum dots with improved stability. Colloid Surf. A **2012**, *402* (1), 72–79.

78. Ma, X.-D.; Qian, X.-F.; Yin, J.; Xi, H.-A.; Zhu, Z.-K. Preparation and characterization of polyvinyl alcohol-capped CdSe nanoparticles at room temperature. J. Colloid Interface Sci. **2002**, *252* (1), 77–81.

79. Yang, Q.; Tang, K.B.; Wang, C.R.; Zhang, C.J.; Qian, Y.T. Wet synthesis and characterization of MSe (M = Cd, Hg) nanocrystallites at room temperature. J. Mater. Res. **2002**, *17* (5), 1147–1152.

80. Yang, Y.J.; Xiang, B.J. Wet synthesis of nearly monodisperse CdSe nanoparticles at room temperature. J. Cryst. Growth **2005**, *284* (3–4), 453–458.

81. Mansur, H.S.; Pereira, M.M.; Donnici, C.L.; Dias, L.L.S. Synthesis and characterization of chitosan-polyvinyl alcohol-bioactive glass hybrid membranes. Biomatter. **2011**, *1* (1), 114–119.

82. Reiss, P.; Protiere, M.; Li, L. Core/shell semiconductor nanocrystals. Small **2009**, *5* (2), 154–168.

83. Bailey, R.E.; Nie, S. Alloyed semiconductor quantum dots: Tuning the optical properties without changing the particle size. J. Am. Chem. Soc. **2003**, *125* (23), 7100–7106.

84. Hines, M.A.; Guyot-Sionnest, P. Synthesis and characterization of strongly luminescing ZnS-capped CdSe nanocrystals. J. Phys. Chem. **1996**, *100* (2), 468–471.

85. Bao, H.; Gong, Y.; Li, Z.; Gao, M. Enhancement effect of illumination on the photoluminescence of water-soluble CdTe nanocrystals: Toward highly fluorescent CdTe/CdS core-shell structure. Chem. Mater. **2004**, *16* (20), 3853–3859.

86. Wang, Y.; Tang, Z.; Correa-Duarte, M.A.; Pastoriza-Santos, I.; Giersig, M.; Kotov, N.A.; Liz-Marzán, L.M. Mechanism of strong luminescence photoactivation of citrate-stabilized water-soluble nanoparticles with CdSe cores. J. Phys. Chem. B **2004**, *108* (40), 15461–15469;

87. Wang, C.L.; Zhang, H.; Zhang, J.H.; Li, M.J.; Sun, H.Z.; Yang, B. Application of ultrasonic irradiation in aqueous synthesis of highly fluorescent CdTe/CdS core–shell nanocrystals. J. Phys. Chem. C **2007**, *111* (6), 2465–2469.

88. Petryayeva, E.; Algar, W.R.; Medintz, I.L. Quantum dots in bioanalysis: A review of applications across various platforms for fluorescence spectroscopy and imaging. Appl. Spectrosc. **2013**, *67* (3), 215–252.

89. Wei, S.H.; Zunger, A. Calculated natural band offsets of all II–VI and III–V semiconductors: Chemical trends and the role of cation d orbitals. Appl. Phys. Lett. **1998**, *72* (16), 2011–2013.

90. Chen, X.B.; Lou, Y.B.; Samia, A.C.; Burda, C. Coherency strain effects on the optical response of core/shell heteronanostructures. Nano Lett. **2003**, *3* (6), 799–803.

91. Sapsford, K.E.; Algar, W.R; Berti, L.; Gemmill, K.L.; Casey, B.J.; Oh, E.; Stewart, M.H.; Medintz, I.L. Functionalizing nanoparticles with biological molecules: Developing chemistries that facilitate nanotechnology. Chem. Rev. **2013**, *113* (3), 1904–2074.

92. Lee, K.-H. Quantum dots: A Quantum jump for molecular imaging? J. Nucl. Med. **2007**, *48* (9), 1408–1410.

93. Bentolila, L.A.; Michalet, X.; Pinaud, F.F.; Tsay, J.M.; Doose, S.; Li, J.J.; Sundaresan, G.; Wu, A.M.; Gambhir, S.S.; Weiss, S. Quantum dots for molecular imaging and cancer medicine. Discov. Med. **2005**, *5* (26), 213–218;

94. Blanco-Canosa, J.B.; Wu, M.; Susumu, K.; Petryayeva, E.; Jennings, T.L.; Dawson, P.E.; Algar, W.R.; Medintz, I.L. Recent progress in the bioconjugation of quantum dots. Coord. Chem. Rev. **2014**, *263–264*, 101–137.

95. Mattousi, H.; Mauro, J.M.; Goldman, E.R.; Anderson, A.P.; Sunder, V.C.; Mikulec, F.V.; Bawendi, M.G. Self-assembly of CdSe-ZnS quantum dots bioconjugates using an engineered recombinant protein. J. Am. Chem. Soc. **2000**, *122* (49), 12142–12150.

Excipients—Gels

96. Erathodiyil, N.; Ying, J.Y. Functionalization of inorganic nanoparticles for bioimaging applications. Acc. Chem. Res. **2011**, *44* (10), 925–935.

97. Zhang, C.Y.; Yeh, H.C.; Kuroki, M.T.; Wang, T.H. Single quantum-dot-based DNA nanosensor. Nat. Mater. **2005**, *4* (11), 826–831.

98. Hezinger, A.F.; J. Tessmar, J.; Göpferich, A. Polymer coating of quantum dots—A powerful tool toward diagnostics and sensorics. Eur. J. Pharm. Biopharm. **2008**, *68* (1), 138–152.

99. Nann, T. Phase-transfer of CdSe@ZnS quantum dots using amphiphilic hyperbranched polyethyleneimine, Chem. Commun. **2005**, *2005* (13), 1735–1736

100. Pan, B.; Gao, F.; He, R.; Cui, D.; Zhang, Y. Study on interaction between poly(amidoamine) dendrimer and CdSe nanocrystal in chloroform. J. Colloid Interface Sci. **2006**, *297* (1), 151–156.

101. Wisher, A.C.; Bronstein, I.; Chechik, V. Thiolated PAMAM dendrimer-coated CdSe/ZnSe nanoparticles as protein transfection agents. Chem. Commun. **2006**, *2006* (15), 1637–1639.

102. Huang, B.; Tomalia, D.A. Dendronization of gold and CdSe/cdS (core–shell) quantum dots with tomalia type, thiol core, functionalized poly(amidoamine) (PAMAM) dendrons. J. Luminesc. **2005**, *111* (4), 215–223.

103. Talapin, D.V.; Rogach, A.L.; Kornowski, A.; Haase, M.; Weller, H. Highly luminescent monodisperse CdSe and CdSe/ZnS nanocrystals synthesized in a hexadecylamine-trioctylphosphine oxide-trioctylphosphine mixture. Nano Lett. **2001**, *1* (4), 207–211.

104. Bentzen, E.L.; Tomlinson, I.D.; Manson, J.; Gresch, P.; Warnement, M.R.; Wright, D.; Sanders-Bush, E.; Blakely, R.; Rosenthal, S.J. Surface modification to reduce nonspecific binding of quantum dots in live cell assays. Bioconjug. Chem. **2005**, *16* (6), 1488–1494.

105. Xu, H.; Xu, J.; Wang, X.; Wu, D.; Chen, Z.G.; Wang, A.Y. Quantum Dot-based, quantitative, and multiplexed assay for tissue staining. ACS Appl. Mater. Interfaces **2013**, *5* (8), 2901–2907.

106. Sturgis, E.M.; Cinciripini, P.M. Trends in head and neck cancer incidence in relation to smoking prevalence: An emerging epidemic of human papillomavirus-associated cancers? Cancer **2007**, *110* (7), 1429–1435.

107. Cuschieri, K.; Wentzensen, N. Human papillomavirus mRNA and p16 detection as biomarkers for the improved diagnosis of cervical neoplasia cancer epidemiol. Biomarkers Prev. **2008**, *17* (10), 2536–2545.

108. Levenson, R.M.; Mansfield, J.R. Multispectral imaging in biology and medicine: Slices of life. Cytometry A **2006**, 69A (8), 748–758.

109. Mansfield, J.R.; Gossage, K.W.; Hoyt, C.C.; Levenson, R.M. Autofluorescence removal, multiplexing, and automated analysis methods for *in vivo* fluorescence imaging. J. Biomed. Opt. **2005**, *10* (4), 41207.

110. Algar, W.R.; Tavares, A.J.; Krull, U.J. Beyond labels: A review of the application of quantum dots as integrated components of assays, bioprobes, and biosensors utilizing optical transduction. Anal. Chim. Acta **2010**, *673* (1), 1–25.

111. Chen, C.; Peng, J.; Sun, S.-R.; Peng, C.-W.; Li, Y.; Pang, D.-W. Tapping the potential of quantum dots for personalized oncology: Current status and future perspectives. Nanomedicine **2012**, *7* (3), 411–428.

112. Pericleous, P.; Gazouli, M.; Lyberopoulou, A.; Rizos, S.; Nikiteas, N.; Efstathopoulos, E.P. Quantum dots hold promise for early cancer imaging and detection. Int. J. Cancer **2012**, *131* (3), 519–528.

113. Ferrari, M. Cancer nanotechnology: Opportunities and challenges. Nat. Rev. Cancer **2005**, *5* (3), 161–171.

114. Wang, M.D.; Shin, D.M.; Simons, J.W.; Nie, S. Nanotechnology for targeted cancer therapy. Expert Rev. Anticancer Ther. **2007**, *7* (6), 833–837.

115. Davis, M.E.; Chen, Z.G.; Shin, D.M. Nanoparticle therapeutics: An emerging treatment modality for cancer. Nat. Rev. Drug Discov. **2008**, *7* (12), 771–782.

116. Byrne, J.D.; Betancourt, T.; Brannon-Peppas, L. Active targeting schemes for nanoparticle systems in cancer therapeutics. Adv. Drug Deliv. Rev. **2008**, *60* (15), 1615–1526.

117. Farokhzad, O.C.; Langer, R. Impact of nanotechnology on drug delivery. ACS Nano **2009**, *3* (1), 16–20.

118. Petros, R.A.; DeSimone, J.M. Strategies in the design of nanoparticles for therapeutic applications. Nat. Rev. Drug Discov. **2010**, *9* (8), 615–627.

119. Nguyen, T.-D. Portraits of colloidal hybrid nanostructures: Controlled synthesis and potential applications. Colloids Surf. B Biointerfaces **2013**, *113* (1), 326– 344.

120. Bae, K.H; Chung, H.J; Park, T.G. Nanomaterials for cancer therapy and imaging. Mol. Cells **2011**, *31* (4), 295–302.

121. Sanna, V.; Sechi, M. Nanoparticle therapeutics for prostate cancer treatment. Maturitas **2012**, *73* (1), 27–32.

122. Lay, C.L.; Liu, H.Q.; Tan, H.R.; Liu, Y. Delivery of paclitaxel by physically loading onto poly(ethylene glycol) (PEG)-graft-carbon nanotubes for potent cancer therapeutics. Nanotechnology **2010**, *21* (6), 065101.

123. Iga, A.M.; Robertson, J.H.; Winslet, M.C.; Seifalian, A.M. Clinical potential of quantum dots. J. Biomed. Biotechnol. **2007**, *2007* (10), 76087.

124. Dubertret, B.; Skourides, P.; Norris, D.J.; Noireaux, V.; Brivanlou, A.H.; Libchaber, A. *In vivo* imaging of quantum dots encapsulated in phospholipid micelles. Science **2002**, *298* (5599), 1759–1762.

125. Han, M.; Gao, X.; Su, J.Z.; Nie, S. Quantum-dot-tagged microbeads for multiplexed optical coding of biomolecules. Nat. Biotechnol. **2001**, *19* (7), 631–635.

126. Ludwig, J.A.; Weinstein, J.N. Biomarkers in cancer staging, prognosis and treatment selection. Nat. Rev. Cancer **2005**, *5* (11), 845–856.

127. Duffy, M.J.; Crown, J. A personalized approach to cancer treatment: How biomarkers can help. Clin. Chem. **2008**, *54* (11), 1770–1779.

128. Wistuba, I.I.; Gelovani, J.G.; Jacoby, J.J.; Davis, S.E.; Herbst, R.S. Methodological and practical challenges for personalized cancer therapies. Nat. Rev. Clin. Oncol. **2011**, *8* (3), 135–141.

129. Phan, J.H.; Moffitt, R.A.; Stokes, T.H.; Liu, J.; Young, A.N.; Nie, S.; Wang, M.D. Convergence of biomarkers, bioinformatics and nanotechnology for individualized cancer treatment. Trends Biotechnol. **2009**, *27* (6), 350–358.

130. Tan, A.; Yildirimer, L.; Rajadas, J.; de La Peña, H.; Pastorin, G.; Seifalian, A. Quantum dots and carbon nanotubes in oncology: A review on emerging theranostic applications in nanomedicine. Nanomedicine **2011**, *6* (6), 1101–1114.

131. Gao, X.H.; Cui, Y.Y; Levenson, R.M.; Chung, L.W.K.; Nie, S. *In vivo* cancer targeting and imaging with semiconductor quantum dots. Nat. Biotechnol. **2004**, *22* (8), 969–976.

132. Jokerst, J.V.; Lobovkina, T.; Zare, R.N.; Gambhir, S.S. Nanoparticle PEGylation for imaging and therapy. Nanomedicine **2011**, *6* (4), 715–728.

133. Kwon, G.S. Polymeric micelles for delivery of poorly water-soluble compounds. Crit. Rev. Ther. Drug Carrier Syst. **2003**, *20* (5), 357–403.

134. Gref, R.; Minamitake, Y.; Peracchia, M.T.; Trubetskoy, V.; Torchilin, V.; Langer, R. Biodegradable long-circulating polymeric nanospheres. Science **1994**, *263* (5153), 1600–1603.

135. Roberts, M.J.; Bentley, M.D.; Harris, J.M. Chemistry for peptide and protein PEGylation. Adv. Drug Deliv. Rev. **2002**, *54* (4), 459–476.

136. Ballou, B.; Lagerholm, B.C.; Ernst, L.A.; Bruchez, M.P.; Waggoner, A.S. Noninvasive imaging of quantum dots in mice. Bioconjug. Chem. **2004**, *15* (1), 79–86.

137. McNeil, S.E. Nanoparticle therapeutics: A personal perspective. WIREs Nanomed. Nanobiotechnol. **2009**, *1* (3), 264–271.

138. Melancon, M.P.; Zhou, M.; Li C. Cancer theranostics with near-infrared light-activatable multimodal nanoparticles. Acc. Chem. Res. **2011**, *44* (10), 947–956.

139. Smith, A.M.; Duan, H.; Mohs, A.M.; Nie, S. Bioconjugated quantum dots for *in vivo* molecular and cellular imaging. Adv. Drug Deliv. Rev. **2008**, *60* (11), 1226–1240.

140. Srivastava, S.; Srivastava, R.-G. Proteomics in the forefront of cancer biomarker discovery. J. Proteome Res. **2005**, *4* (4), 1098–1103.

141. Cissell, K.A.; Rahimi, Y.; Shrestha, S.; Hunt, E.A.; Deo, S.K. Bioluminescence-based detection of microRNA, miR21 in breast cancer cells. Anal. Chem. **2008**, *80* (7), 2319–2325.

142. Wagner, M.; Li, F.; Li, J.J.; Li, X.-F.; Le, X.C. Quantum dot based assays for cancer biomarkers. Anal. Bioanal. Chem. **2010**, *397* (8), 3213–3224.

143. Wang, H.Z.; Wang, H.Y.; Liang, R.Q.; Ruan, K.C. Detection of tumor marker CA125 in ovarian carcinoma using quantum dots. Acta Biochim. Biophys. Sin. **2004**, *36* (10), 681–686.

144. Chen, C.; Peng, J.; Xia, H.S.; Yang, G.F.; Wu, Q.S.; Chen, L.D.; Zeng, L.B.; Zhang, Z.L.; Pang, D.W.; Li, Y. Quantum dots-based immunofluorescence technology for the quantitative determination of HER2 expression in breast cancer. Biomaterials **2009**, *30* (15), 2912–2918.

145. Barua, S.; Reqe, K. Cancer-cell-phenotype-dependent differential intracellular trafficking of unconjugated quantum dots. Small **2009**, *5* (3), 370–376.

146. Yang, L.; Mao, H.; Wang, Y.A.; Cao, Z.H.; Peng, X.H.; Wang, X.X.; Duan, H.W.; Ni, C.C.; Yuan, Q.G.; Adams, G.; Smith, M.Q.; Wood, W.C.; Gao, X.H.; Nie, S.M. Single chain epidermal growth factor receptor antibody conjugated nanoparticles for *in vivo* tumor targeting and imaging. Small **2009**, *5* (2), 235-243.

147. Voura, E.B.; Jaiswal, J.K.; Mattoussi, H.; Simon, S.M. Tracking metastatic tumor cell extravasation with quantum dot nanocrystals and fluorescence emission-scanning microscopy. Nat. Med. **2004**, *10* (9), 993–998.

148. Stroh, M.; Zimmer, J.P.; Duda, D.G.; Levchenko, T.S.; Cohen, K.S.; Brown, E.B.; Scadden, D.T.; Torchilin, V.P.; Bawendi, M.G.; Fukumura, D.; Jain, R.K. Quantum dots spectrally distinguish multiple species within the tumor milieu *in vivo*. Nat. Med. **2005**, *11* (6), 678–682.

149. Cai, W.B.; Shin, D.W.; Chen, K.; Gheysens, O.; Cao, Q.Z.; Wang, S.X.; Gambhir, S.S.; Chen, X. Peptide-labeled near infrared quantum dots for imaging tumor vasculature in living subjects. Nano Lett. **2006**, *6* (4), 669–676.

150. Zhang, Y.-P.; Sun, P.; Zhang, X.R.; Yang, W.L.; Si, C.S. Synthesis of CdTe quantum dot-conjugated CC49 and their application for *in vitro* imaging of gastric adenocarcinoma cells. Nanoscale Res. Lett. **2013**, *8* (1), 294.

151. Zheng, Y.; Gao, S.; Ying, J.Y. Synthesis and cell-imaging applications of glutathione-capped CdTe quantum dots. Adv. Mater. **2007**, *19* (3), 376–380.

152. Akerman, M.E.; Chan, W.C.W.; Laakkonen, P.; Bhatia, S.N.; Ruoslahti, E. Nanocrystal targeting *in vivo*. Proc. Natl. Acad. Sci. U S A **2002**, *99* (20), 12617–12621.

153. Rhyner, M.N.; Smith, A.M.; Gao, X.H.; Mao, H.; Yang, L.L.; Nie, S.M. Quantum dots and multifunctional nanoparticles: New contrast agents for tumor imaging. Nanomedicine **2006**, *1* (2), 209–217.

154. Zhao, Y.; Liu, S.; Li, Y.; Jiang, W.; Chang, Y.; Pan, S.; Fang, X.; Wang, Y.A.; Wang, J. Synthesis and grafting of folate–PEG–PAMAM conjugates onto quantum dots for selective targeting of folate-receptor-positive tumor cells. J. Colloid Interface Sci. **2010**, *350* (1), 44–50.

155. Liu, L.; Miao, Q.; Liang, G. Quantum dots as multifunctional materials for tumor imaging and therapy. Materials **2013**, *6* (2), 483–499.

156. Juzenas, P.; Chen, W.; Sun, Y.P.; Coelho, M.A.N.; Generalov, R.; Generalova, N.; Christensen, I.L. Quantum dots and nanoparticles for photodynamic and radiation therapies of cancer. Adv. Drug Deliv. Rev. **2008**, *60* (15), 1600–1614.

157. Cai, W.B.; Chen, K.; Li, Z.; Gambhir, S.S.; Chen, X. Dual-functional probe for PET and near-infrared fluorescence imaging of tumor vasculature. J. Nucl. Med. **2007**, *48* (11), 1862–1870.

158. Mulder, W.J.M.; Koole, R.; Brandwijk, R.J.; Storm, G.; Chin, P.T.K.; Strijkers, G.J.; de Mello Donega, C.; Nicolay, K.; Griffioen, A.W. Quantum dots with a paramagnetic coating as a bimodal molecular imaging probe. Nano Lett. **2006**, *6* (1), 1–6.

159. van Tilborg, G.A.F.; Mulder, W.J.M.; Chin, P.T.K.; Storm, G.; Reutelingsperger, C.P.; Nicolay, K.; Strijkers, G.J. Annexin A5-conjugated quantum dots with a paramagnetic lipidic coating for the multimodal detection of apoptotic cells. Bioconjug. Chem. **2006**, *17* (4), 865–868.

160. Cai, W.B.; Chen, X. Nanoplatforms for targeted molecular imaging in living subjects. Small **2007**, *3* (11), 1840–1854.

161. Manabe, N.; Hoshino, A.; Liang, Y.Q.; Goto, T.; Kato, N.; Yamamoto, K. Quantum dot as a drug tracer *in vivo*. IEEE T. Nanobiosci. **2006**, *5* (4), 263–267.

162. Bagalkot, V.; Zhang, L.; Levy-Nissenbaum, E.; Jon, S.; Kantoff, P.W.; Langer, R.; Farokhzad, O.C. Quantum dot–aptamer conjugates for synchronous cancer imaging, therapy, and sensing of drug delivery based on bi-fluorescence resonance energy transfer. Nano Lett. **2007**, *7* (10), 3065–3070.

Excipients—Gels

163. Chen, A.A.; Derfus, A.M.; Khetani, S.R.; Bhatia, S.N. Quantum dots to monitor RNAi delivery and improve gene silencing. Nucleic Acids Res. **2005**, *33* (22), e190.

164. Tan, W.B.; Jiang, S.; Zhang, Y. Quantum-dot based nanoparticles for targeted silencing of HER2/neu gene via RNA interference. Biomaterials **2007**, *28* (8), 1565–1571.

165. Sahai, E. Illuminating the metastatic process. Nat. Rev. Cancer **2007**, *7* (10), 737–749.

166. Hahn, M.A.; Keng, P.C.; Krauss, T.D. Flow cytometric analysis to detect pathogens in bacterial cell mixtures using semiconductor quantum dots. Anal. Chem. **2008**, *80* (3), 864–872.

167. Park, K.; Lee, S.; Kang, E.; Kim, K.; Choi, K.; Kwon, I.C. New generation of multifunctional nanoparticles for cancer imaging and therapy. Adv. Funct. Mater. **2009**, *19* (10), 1553–1566.

168. Nazir, S.; Hussain, T.; Ayub, A.; Rashid, U.; Macrobert, A.J. Nanomaterials in combating cancer: Therapeutic applications and developments. Nanomedicine **2014**, *10* (1), 19–34.

Functionalized Surfaces: Biomolecular Surface Modification with Functional Polymers

Florence Bally
Institute of Materials Science of Mulhouse (IS2M), University of Upper Alsace, Mulhouse, France, and Institute of Functional Interfaces (IFG), Karlsruhe Institute of Technology (KIT), Eggenstein-Leopoldshafen, Germany

Aftin M. Ross
Institute of Functional Interfaces (IFG), Karlsruhe Institute of Technology (KIT), Eggenstein-Leopoldshafen, Germany, and Center for Devices and Radiological Health, Food and Drug Administration, Silver Spring, Maryland, U.S.A.

Abstract

Surface engineering improves biomolecular interactions with biomaterials. Due to the diversity of properties, functional polymer coatings are often deposited on synthetic substrates to tailor abiotic–biological interfaces. Although wet-chemical processes were first developed, increased attention has been given to vapor-based techniques for the fabrication of polymer thin films, mainly due to the ability to easily surface modify 3D geometries. Patterning and gradients provide additional features to mimic the cellular environment. The various surface modification processes described in this entry for the fabrication of (structured) functional polymer coatings are then illustrated via biologically relevant studies aimed at increasing the knowledge of biological mechanisms.

INTRODUCTION

Surface functionalities of materials determine their interaction with their environment. They can be decoupled from the bulk characteristics of materials when surface modification is performed. Mechanical, thermal, and 2D or 3D geometrical properties of the substrate are thus maintained, while the chemical interface between the synthetic coating and surrounding media are tailored. During the last decades, a strong demand for thin polymer films has emerged in the field of biology to improve the interactions of abiotic materials with the biological environment. Cell culture substrates, medical implants, scaffolds for tissue engineering, and biosensors including microfluidic systems for miniaturized bioassays (necessary for the use of small sample volume and for the development of parallel processing) are examples of biomedical devices that may require functional polymer coatings. Tailoring chemical composition of these devices influences biomolecular interactions at their surface (including cell function, morphology, adhesion, migration, proliferation, and differentiation) by mimicking the biological environment. Proteins, polysaccharides, and other recognition ligands are ubiquitous in nature and strongly control numerous biological events. Immobilization of such biomolecules as well as antifouling synthetic polymers, mainly polyethylene glycol (PEG), is a crucial issue for novel biotechnologies. Surface engineering strategies, generic for a multitude of substrates, are therefore required to fabricate functional coatings, enabling the tuning of biomolecular–surface interactions. This entry first describes common surface engineering techniques used for the fabrication of functional polymer coatings. Mimicking the cellular microenvironment requires chemical patterning to address features at the level of cellular actions, i.e., on the micro- and nanoscale. Thus, relevant micro- and nanopatterning processes are reported, including those leading to the creation of compositional gradients. Finally, specific examples of biomolecular interactions with surfaces and their subsequent consequences on cellular or biological behavior illustrate the utility of various techniques used in the design and modification of functional coatings.

FABRICATION OF FUNCTIONAL POLYMER COATINGS

Surface modification by fabrication of functional polymer coatings has been widely used for biomedical applications.[1–3] These polymer thin films can be prepared by physical or chemical coating processes. Physical coating processes are based on adsorption of polymer chains on the surface, whereas chemical coating processes imply covalent bonds between the polymer layer and the substrate. Such classification is illustrated by many examples of surface modification corresponding to wet chemical techniques. In addition, vapor-based techniques have gained increased attention in the last decades due to the possibility to obtain stable, homogeneous functional coatings (even on complex geometries), while limiting the contamination of the substrate with solvent residue. The following section gives an overview of the main techniques reported in the literature for

Concise Encyclopedia of Biomedical Polymers and Polymeric Biomaterials DOI: 10.1081/E-EBPPC-120051690

585

Excipients—Gels

the production of functional polymer coatings with a focus on the advantages and limitations of these engineering techniques for surface modification.

Wet Chemical Methods

Spin coating, dip coating, flow coating, or spray coating are common wet chemical processes used for the fabrication of functional polymer coatings.[4] These physical coating techniques consist of the deposition of a polymer layer on a substrate without generic preliminary treatment. A pre-formed functional polymer or resin is used to prepare the polymer solution or the polymer melt and this liquid solution is deposited on the substrate. Film cohesion is ensured by evaporation of the solvent used to prepare the polymer solution (which usually occurs during the coating process) or by curing of the resin. Film thickness can be tuned by tailoring the polymer concentration in solution and the operating conditions, both of which are influenced by the nature and thus the wettability of the substrate. No particular interaction between the polymer coating and the substrate occurs during such rapid processes, which limits the stability of these systems on the substrate, especially for further use in a biological environment. Electropolymerization is also widely used for the deposition of polymer coatings. Basically, electrochemical polymerization is

performed in the monomer solution by monomer oxidation and the film of the corresponding polymer is progressively deposited on the surface of the electrode, which is often comprised of carbon or metal materials.[5] Deposition occurs due to a decrease in oligomer solubility and coupling of oligomers is still possible after deposition for further chain extension. Conducting and insulating polymers can be fabricated by electropolymerization. An example is the use of the conducting polymer, poly(pyrroles), which is used for subsequent biomolecular immobilization or pyrrole-terminated biomolecules, which may be directly electropolymerized.[6,7] Alternatively, ordered surface coatings may be created by self-assembly of polymers.[8] The Langmuir–Blodgett technique enables the deposition of highly ordered amphiphilic compounds under the form of monolayer on a substrate (Fig. 1).[9,10] A compressed Langmuir monolayer, spread at the air–liquid interface, is transferred to a solid material by immersion of the substrate. For instance, poly[butadiene-*b*-poly(ethylene oxide) (PEO)] monolayer was coated to fabricate a protein-repellant surface.[11] This deposition process can be repeated several times to reach a multilayered surface assembly. Langmuir–Blodgett films have controllable thickness, surface uniformity, and high orientation degree but often lack stability on the substrate since physisorption mechanisms are involved for the immobilization of the first monolayer on the

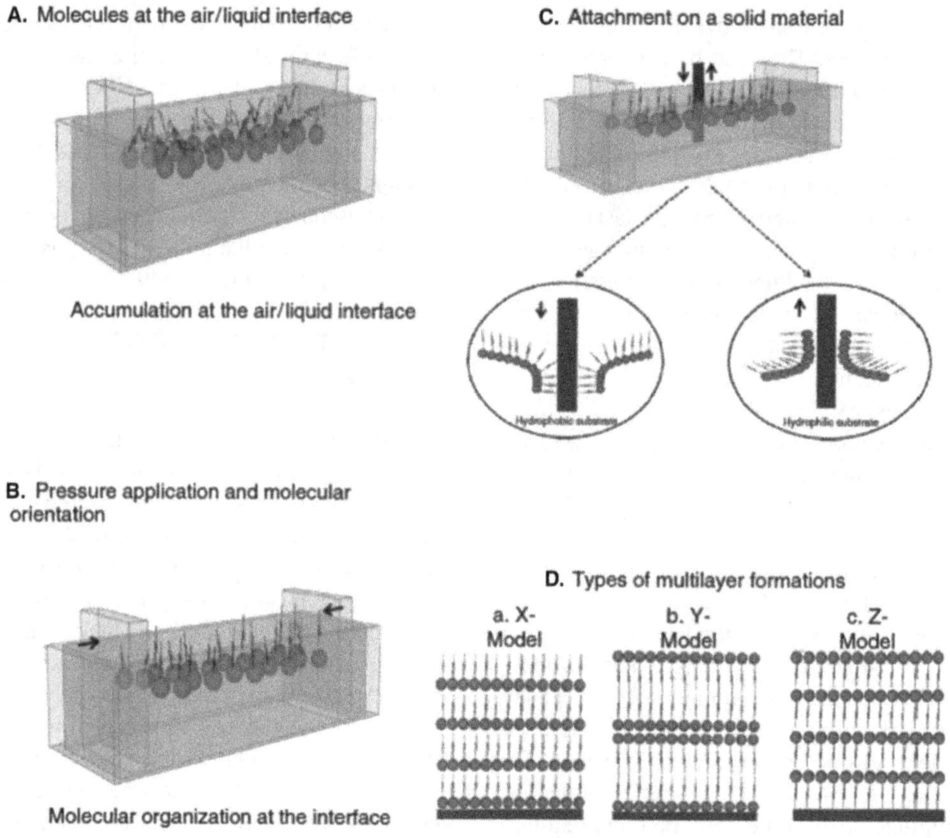

A. Molecules at the air/liquid interface

Accumulation at the air/liquid interface

B. Pressure application and molecular orientation

Molecular organization at the interface

C. Attachment on a solid material

Hydrophobic substrate   Hydrophilic substrate

D. Types of multilayer formations

a. X-Model   b. Y-Model   c. Z-Model

Fig. 1 Schematic representation of Langmuir–Blodgett method for the fabrication of multilayered, thin film.
Source: From Guney et al.[237] © 2013, with permission from Wiley-VCH Verlag GmbH & Co. KGaA, Weinheim.

substrate. Layer-by-layer (LBL) adsorption of polyions has also been the subject of active research for two decades (Fig. 2).[12–14] Multilayered self-assembly of polyelectrolytes are fabricated on charged surfaces, typically a negatively charged surface due to surface oxidation and hydrolysis, by sequential immersion of the surface in a solution of polycations [poly(allylamine hydrochloride), poly(ethyleneimine), and others] and polyanions [poly(styrene sulfonate), poly(vinyl sulfate), or poly(acrylic acid)]. The overcompensation of polyelectrolyte at each step leads to alternative charge inversion within the film. This elegant technique has gained increased interest particularly in the field of biology[15] as it is an easy process that enables control of the assembly at the nanometer scale,

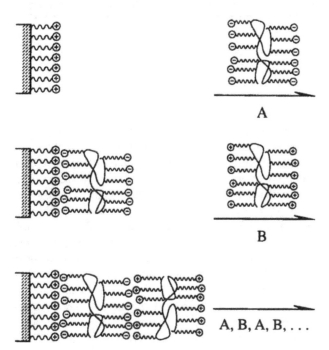

Fig. 2 Simplified schematic for the buildup of multilayer assemblies by consecutive adsorption of anionic and cationic polyelectrolytes.
Source: From Decher[13] © 1992, with permission from Elsevier.

while offering versatility in the materials that may be utilized. Also the stability for LBL films is typically higher than that of Langmuir–Blodgett films.

To improve anchoring of the polymer coating to the substrate, various chemical coating techniques have been developed to covalently graft polymer chains "from" or "to" the surface (Fig. 3).[16–18] As for "grafting to" techniques, pre-formed polymer chains are attached to the surface by direct reaction with the bulk material (which is often a polymeric substrate) by chemical modification or after an activation step by high energy treatment.[19] This method enables precise control of the molecular weight and composition of macromolecules that are grafted on the surface since the polymer is synthesized before its immobilization. For instance, bioactive cellulose derivatives have been grafted onto a poly(vinyl chloride) substrate via a two-step reaction using an isothiocyanate linker.[20] Acidic treatments have been performed to introduce reactive oxygen moieties to poly(ethylene)[21] and poly(propylene),[22] which could subsequently immobilize polyamine chains. Catalyzed hydrolysis or aminolysis of poly(methyl methacrylate) (PMMA) has also been carried out to covalently bond amine-terminated PEG.[23] Similar treatment can also be performed on poly(lactic acid-*co*-glycolic acid) (PLGA)[24] and poly(lactic acid).[25] Many other examples can be found in the literature and describe specific procedures depending on the chemical nature of the substrate.[26–29] Inert surfaces, such as apolar polymer surfaces or polymer surfaces bearing poor reactive functional groups, require a pretreatment to introduce highly reactive groups for subsequent grafting of polymer chains to the substrate in solution. Common surface activation techniques[30] include ionized gas treatments, such as plasma, corona or flame, and light-source irradiation treatments, mainly ultraviolet (UV) irradiation. For example, plasma treatment is commonly used to anchor carboxyl, hydroxyl, amine, and aldehyde groups in the top nanometer of the substrate by ionization of inert or reactive gas such as Ar, N_2, O_2, CO_2, or NH_3.[31] Inert surfaces can be pretreated by such a process[32] to facilitate surface grafting. Materials such as poly(dimethylsiloxane) (PDMS)

Excipients—Gels

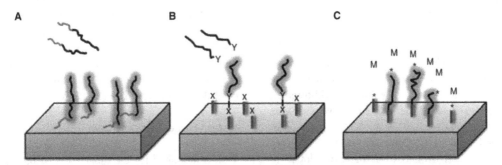

Fig. 3 Schemes of the surface modification methods to obtain tethered polymer layers: (**A**) physical adsorption of polymer chains, (**B**) covalent anchoring via "grafting to," and (**C**) via "grafting from" approaches.
Source: From Zapotoczny[36] © 2013, with permission from Wiley-VCH Verlag GmbH & Co. KGaA, Weinheim.

may also undergo plasma-assisted substitution of chemical functionalities for subsequent derivatization.[33] Alternatively, numerous studies reported UV irradiation–assisted polymer grafting to a substrate as a means of surface modification. For instance, efforts have been made to decrease protein adsorption via covalent grafting (by UV irradiation) of photoreactive-terminated PEG onto polymeric substrates.[34,35] Although, the "grafting to" approach enables good control of the macromolecular characteristics of the immobilized polymer coating, the grafting density is usually lower than the one achieved by the "grafting from" approach. Steric hindrance of the polymer chains covalently bound to the surface progressively hinder the diffusion of additional polymer chains to the surface, limiting the number of polymer chain-ends reacting with the complementary group on the surface. On the contrary, the "grafting from" approach consists in a surface-initiated technique for which a polymerization initiator is covalently bound to the substrate and activates chain extension from the surface. Via this technique, a higher grafting density can be reached and various chain conformations, particularly polymer brushes,[36] can be synthesized on the surface to form the polymer coating. Similar to the "grafting to" approach, direct chemical modification of the bulk material or a preliminary activation step may be necessary to create reactive species at the interface that initiates polymerization.[17] As previously indicated, a redox reaction can be performed on the substrate to generate free-radical sites that initiate polymerization.[16,37] An elegant and well-developed approach consists of functionalization of the substrate with an initiator capable of controlled-radical polymerization initiation.[38–40] For example, surface-initiated atom transfer radical polymerization (ATRP) has been extensively studied in organic as well as in aqueous media[41,42] after previous immobilization of an ATRP starter on silicon wafers. While the grafting of zwitterionic poly(sulfobetaine) brushes has been achieved on cellulose via surface-initiated reversible addition–fragmentation chain-transfer polymerization[43] for improved hemocompatibility and antibiofouling properties of the membrane.[44] Surface-initiated free radical polymerization may also be activated by plasma pretreatment.[31,45,46] In many cases, plasma treatment enables the formation of peroxides that subsequently induce radical polymerization of a monomer in solution. Various other types of activation such as UV irradiation,[47,48] ozone/UV treatment,[49] gamma-rays irradiation,[50] ion beam,[51] and electron beam[52] treatment have also been reported and enable subsequent radical polymerization in solution. Additionally, covalently bonded, self-assembled surfaces have also been widely studied for the fabrication of ordered polymer coatings. Production of self-assembled monolayers (SAMs) of polymers or of polymer initiators has been developed to graft macromolecules onto the surface by simple immersion of the substrate in a surfactant solution. SAMs are molecular assemblies chemisorbed to a surface by the

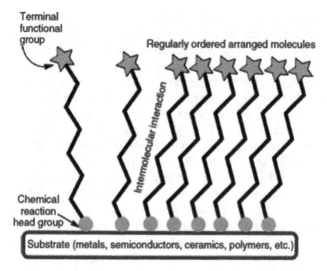

Fig. 4 Schematic representation of self-assembled monolayers. **Source:** From Niepel et al.[238] © 2013, with permission from Wiley-VCH Verlag GmbH & Co. KGaA, Weinheim.

reaction of a head group that provide new physicochemical properties to the substrate by the nature of the chain and end groups (Fig. 4). Commonly, organosilanes have been immobilized to surfaces such as glass, silicon, or other hydroxylated materials via siloxane linkage. Surface modification by silanization is a low-cost technique that may lead to a quasi-crystalline functional monolayer.[53] Thiol- and disulfide-terminated molecules have then preferred to silanes for the fabrication of SAMs[54,55] and have been extensively used for the immobilization of a multitude of (macro)molecules on gold, copper, and silver substrates.[56]

Unfortunately, all these wet chemical methods are not adapted to a wide range of substrates. In particular, techniques involving chemisorptions processes often require a given chemical nature of the bulk material for surface modification. In addition, macromolecules that are immobilized or synthesized on the surface may need to contain specific functional groups to perform their grafting. The density of functional groups at the surface of the coating, and thus available for subsequent reaction, is limited because of steric repulsion. Last but not the least, some of these methods are not adapted to the coating of 3D geometries, and all wet chemical techniques will contain liquid residues in the polymer film (solvent, chemicals, and so on). To overcome these drawbacks, vapor-based processes are elegant alternatives. The primary vapor techniques are described in the following section.

Vapor-Based Deposition

Well established for the fabrication of thin inorganic coatings, an increased interest in chemical vapor deposition (CVD) has emerged during the last decades for the production of polymer thin films.[57–59] The advantages of CVD techniques in comparison with wet chemical techniques are numerous. First of all, vapor-based processes enable the fabrication of conformal and well-adherent coatings, which is a critical issue for their subsequent immersion in biological media. Then, a large variety of reactive, functional polymers can be deposited independently from the solubility of the polymer in a given solvent, which could be limiting for fluoropolymers or crosslinked materials, and also independently from the nature and the geometry of the substrate. Finally, deposition of the polymer film occurs in a solvent-free environment, which not only prevents contamination of the coating but also limits potential swelling or degradation of the substrate. In the following sections, major vapor-based deposition processes are reported with emphasis on chemical system and initiating source of polymerization.

Laser-based process

Laser-based processes developed for the deposition of polymer thin films consist of the ablation of a target material (illuminated by a laser source) and its subsequent deposition on the substrate as a thin film.[60] Pulsed laser deposition (PLD)[61] has been widely used to vaporize polymer fragments that are fragile and an even less harsh desorption process is achieved by matrix-assisted pulsed laser evaporation (MAPLE),[62] for which the target material is previously prepared by freezing a polymer solution (Fig. 5). The latter technique additionally broadens the scope of materials that can be deposited since a multitude of pre-formed polymer targets can be prepared. For instance, biodegradable polymer thin films of PLGA have been fabricated by both the methods.[63] Films produced by PLD were more uniform than the ones made by MAPLE,

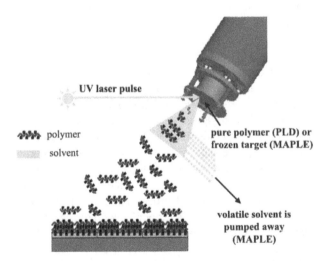

Fig. 5 Experimental setup for the deposition of thin films by conventional pulsed laser deposition (PLD) and matrix-assisted pulsed laser evaporation (MAPLE).
Source: From Mercado et al.,[63] with permission from Springer Science and Business Media.

probably due to a more homogeneous ablation process, thus favoring the transport of individual molecules instead of large ejected target volumes.

Two-compound techniques

Several CVD techniques have been developed for the fabrication of thin, reactive functional coatings and some of them employ the use of a two-compound system: two monomers capable of reacting by step-growth polymerization, called vapor deposition polymerization (VDP), or an initiator and the monomer. In the latter case, techniques are then commonly classified into two categories:[64,65] oxidative and initiated CVD. VDP consists of the evaporation of two complementary monomers that may react by step-growth polymerization and their subsequent deposition on a substrate where a polymer thin film is formed (possibly after film curing or annealing). This technique is operated under high vacuum and film deposition is complex due to the competition between monomer adsorption on the substrate (which is favored by low temperature) and the polymerization reaction, whose kinetics are oppositely influenced by temperature. VDP has been used for the synthesis of aromatic polyamide–polyimide random copolymer films.[66] A variant technique consists of the ring-opening polymerization of amino acid N-carboxy anhydride (NCA), initiated by (γ-aminopropyl) triethoxysilane (APS) molecules present on the substrate, which lead to the formation of a polypeptide film (Fig. 6).[67,68] In this case, ring opening of the newly bonded monomer in the polymer chain creates the amine group required for subsequent monomer addition.

Oxidative CVD (oCVD) consists of step-growth polymerization during which oxidant precursors and monomers, both in the vapor phase, react to grow polymer chains (Fig. 7). Cation radicals are formed by the reaction of an oxidizing agent with monomer followed by a step-growth process, which occurs by oxidative polymerization. Usually performed for the synthesis of conjugated polymers, this technique has also enabled the fabrication of copolymers bearing carboxy and hydroxyl groups suitable for biological applications. For instance, poly(3,4-ethylenedioxythiophene-co-3-thiophene acetic acid) [P(EDOT-co-TAA) has been synthesized by oCVD for subsequent attachment of bovine serum albumin (BSA) via carbodiimide chemistry to fabricate biosensors.[69] When 3-thiopheneethanol (3TE) was copolymerized with 3,4-ethylenedioxythiophene (EDOT), the hydroxyl groups from 3TE were subsequently crosslinked in the presence of avidin proteins (bearing amine groups), by the use of p-mealeimidophenylisocyanate to obtain a chemiresistive biosensor.[70]

On the contrary, initiated CVD (iCVD) consists of a chain-growth polymerization during which polymers are synthesized by chain propagation after free radical initiation.[71] The initiator and the monomer are both in the vapor phase and initiator decomposition may occur via several

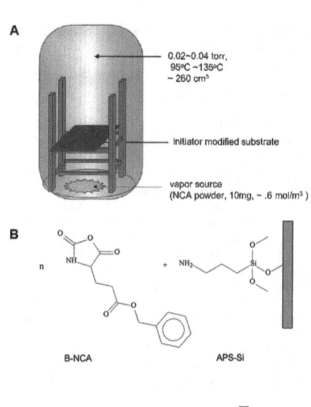

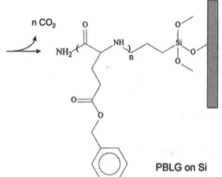

Fig. 6   Vapor deposition polymerization of amino acid *N*-carboxy anhydride (B-NCA) on a (γ-aminopropyl)triethoxysilane-modified silicon substrate (APS-Si): (**A**) experimental setup and (**B**) scheme of the chemical reaction.
Source: From Chang & Frank[67] © 1998, with permission from the American Chemical Society.

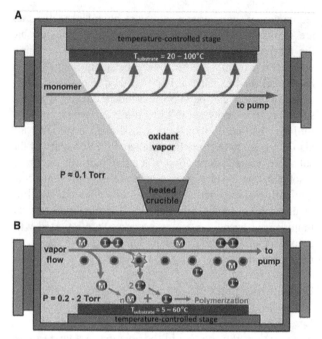

Fig. 7   Reactor configuration for (**A**) oxidative chemical vapor deposition (CVD) and (**B**) initiated CVD processes.
Source: From Coclite et al.[59] © 2013, with permission from Wiley-VCH Verlag GmbH & Co. KGaA, Weinheim.

processes, usually employing rather mild operating conditions. First of all, thermal decomposition of the initiator may be performed. For this purpose, heated filaments are installed in the CVD setup to decompose the initiator. Moderate temperatures such as 180–250°C are enough to break peroxy bonds of *tert*-butyl peroxide,[72] as conducted for the polymerization of poly(glycidyl methacrylate) (PGMA).[73] Filaments heated at 285°C, 310°C, and 360°C were used to initiate polymerization of alkyl acrylates by *tert*-amyl peroxide.[74] For other initiators, higher filament temperatures may be required, as for the polymerization of cyclic vinylmethylsiloxane initiated by perfluorooctane sulfonyl fluoride (350–540°C)[75] or for the polymerization

of methyl methacrylate initiated by triethylamine (550°C).[76] Alternatively, low energy plasma discharge may be used for initiator decomposition and thus polymerization initiation.[77] The initiator source enters the reactor near the plasma source where excitation of the reactant gas occurs and where radicals are formed, while the monomer enters the reaction chamber from another inlet. Radical polymerization takes place on the substrate where initiator radicals and monomers meet each other. This method enables good retention of the functional groups of the monomer in the polymer coating because monomer is not present in the plasma zone. Photoinitiation, particularly by UV light is another technique possibly combined with CVD to initiate polymerization. For instance, PGMA films have been produced by free-radical vapor-phase polymerization using 2,2′-azobis(2-methylpropane), as volatile photoinitiator sensitive to UV irradiation.[78] Such polymerization starts in the vapor phase and leads to the synthesis of linear polymer chains, as demonstrated by the good solubility of the polymer in tetrahydrofuran. This method enables good retention of the functional groups of the monomer since excitation of the initiator is decoupled with the chain propagation. A variant of photoinitiated CVD is grafted CVD for which the initiator is decomposed directly at the surface of the substrate to abstract hydrogen atoms from the substrate material, thus creating radicals that enable chain growth by the supply of monomer vapor in the reactor. The substrate is first pretreated with the photoinitiator by dry or wet processes. For dry processes, vapors of the initiator first enter the polymerization reactor, the initiator adsorbs on the surface and creates radicals during the exposure. For wet processes, a solution of initiator is first cast on the substrate, dried, and the pretreated substrate is introduced in the reactor. The use of benzophenone as a UV-photoinitiator has been reported for polymerization of (dimethylamino)methyl styrene and (diethylamino)ethyl acrylate on PMMA thin films,[79] of acrylamide on PMMA[80] substrate, and of maleic anhydride on poly(ethylene terephtalate) (PTFE).[81] Homogeneous, reactive polymer coatings have also been obtained by various strategies that do not require the use of an initiator. The monomer itself serves as the polymerization precursor and creates reactive species that enable chain growth. These strategies may be classified in two categories: plasma polymerization and thermal-based CVD polymerization, including CVD of [2.2]paracyclophane derivatives.

Plasma polymerization

Plasma polymerization consists of the fabrication of homogeneous, pinhole free polymer films by stimulation through plasma of gaseous monomers, possibly having polymerizable moieties, such as double bonds or cyclic structures. Reactive species (radicals, ions, molecules, and electrons), mainly coming from monomer excitation, fragmentation, or decomposition, react with each other and with the substrate to form a polymer thin film. Several reactor configurations exist and a multitude of operating parameters can be changed to tailor polymer characteristics (functional group retention, architecture, molecular weight, etc.).[82] The final properties of plasma polymers, however, are quite different from conventional polymers, despite numerous efforts to control these characteristics and to understand the formation mechanisms of such coatings.[83–85] Improved control over the architecture of the polymer chains and functional group retention could be achieved by changing the activation mode from continuous-wave plasma to pulsed plasma (Fig. 8).[86] Short plasma pulses serve to activate the monomer to form reactive species ("plasma on" periods) and long "plasma off" periods give time to radicals to initiate polymerization, grow polymer chains, and form the final macromolecular structure. Additionally, plasma polymerization usually leads to well-anchored polymer film because the coating is partially crosslinked with the substrate, generating gradient properties at the interface between the polymer coating and the substrate. Plasma polymerization has been widely used for biomedical applications.[87] For example, in the early 1990s, poly(styrene) and PTFE substrates were modified by radiofrequency plasma polymerization of oxygen-containing molecules (methanol, glutaraldehyde, formic acid, allyl alcohol, and ethylene oxide) to study the influence of the oxygen content of the polymer film on endothelial cell growth on the corresponding surfaces.[88] Fluorocarbon molecules were also polymerized by plasma vapor deposition on silicon rubber vascular grafts to prevent thrombogenesis[89] and nonfouling surfaces were obtained using ethylene oxide derivatives.[90–92] Acrylic acid polymerization was also investigated under various deposition regimes

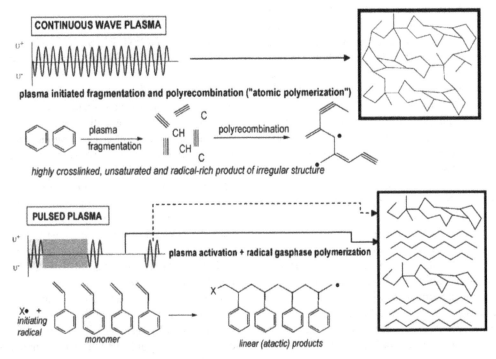

Fig. 8 Scheme of continuous wave plasma polymerization and pulsed plasma polymerization as well as their products.
Source: From Friedrich[85] © 2011, with permission from Wiley-VCH Verlag GmbH & Co. KGaA, Weinheim.

to determine their effect on chemical composition, stability, and morphology of the polymer film for subsequent correlation with cell adhesion properties on the substrate.[93] Nitrogen-rich plasma polymer films were fabricated by atmospheric and low-pressure plasma copolymerization of mixtures of ethylene and ammonia for possible application in orthopedic and cardiovascular research.[94] Additionally, maleic anhydride[95–98] and allylamine[99,100] are commonly polymerized by plasma CVD for subsequent biomedical applications.[31]

Thermal CVD polymerization, including CVD polymerization of [2.2]paracyclophane

Alternatively to plasma excitation, gaseous monomer precursors may also be activated by heat sources in various thermal-based CVD polymerization processes. In this case, a heat source enables radical formation from the monomer and the polymer deposition occurs on a cold substrate. Introduced by Szwarc for the fabrication of parylene coatings by pyrolysis of p-xylene,[101] thermal-based CVD has been widely used and also adapted for the production of other types of polymer thin films. For instance, poly(p-phenylene vinylene)[102–104] or poly(vinylidene fluoride) coatings[105] have been prepared by thermal-based CVD processes. Fabrication of fluorocarbon thin films[106,107] or copolymerization of fluorocarbon–organosilicon monomer, such as hexafluoropropylene oxide and hexamethylcyclotrisiloxane,[108] have also been performed by CVD using a hot filament (550–800°C), located in the reactor chamber to drive pyrolytic decomposition of the gas precursor. This technique has also been named hot-wire CVD,[109] and has been scaled-up for commercialization.[110] Herein, a special focus is made on the fabrication of parylenes, a polymer brand originally developed by Gorham at Union Carbide and commercialized for many applications, including biomedical ones such as the coating of cardiac pacemakers,[111] neural probes,[112] orthodontic devices,[113] or microelectromechanical systems.[114] Numerous examples of the fabrication of (substituted) poly(p-xylylene) coatings can be found in literature, including studies using α,α′-dihyroxy-p-xylylene[115] or α,α′-dibromo-p-xylylene[116] as precursors even as [2.2] paracyclophane precursors gained increased interest.[117] First described in 1966,[118] the Gorham process is based on the use of such dimer precursors to fabricate parylene coatings. It has enabled the fabrication of thin films of poly(p-xylylene) (parylene N), chloro-substituted poly-p-xylylenes (parylene C and parylene D) and has been a source of inspiration for new reactive parylene coatings,[119] particularly substituted [2.2]paracyclophanes. For the fabrication of such coatings, the Lahann group developed a custom-designed CVD polymerization set-up[120] that offers high versatility of operating conditions, which are required for a good retention of substituted functional groups (Fig. 9). In this process, [2.2]paracyclophane derivatives are sublimated (80–120°C) under reduced pressure (<0.1 Torr) before entering a furnace, thermoregulated at 510–800°C where pyrolysis of the precursor occurs. At these temperatures, quinodimethane intermediates are formed by homolytic cleavage of the dimer, whereas no decomposition of substituted functional group is expected. Then reactive species enter the deposition chamber (heated at 80–120°C in order to avoid deposition of the coating on the wall of the chamber), where the substrate is located on a cold sample holder and on which film deposition and polymerization take place. The synthesis of a wide range of functional [2.2]paracyclophanes is now well-established[121] and has led to the preparation of the several substituted poly(p-xylylenes) by CVD polymerization.[122,123] Poly(p-xylylenes) containing amines,[124] esters,[125,126] ketones with fluorinated groups,[127] aldehydes,[128] or alkynes[129] have already been fabricated. In addition, CVD polymerization has also been extended to the production of multifunctional coatings.[130,131] These functional coatings were used for subsequent surface modification, via surface-initiated polymerization[120,132] or via orthogonal click-chemistry reactions.[133,134] This first section reviewed the most commonly used techniques to create functional coatings whose chemical nature may influence biomolecular surface interactions. Patterning or gradients on surfaces provide additional cues to design the cellular environment. Although physical parameters of substrates, such as topography, roughness, and elasticity[135] also have an impact on biomolecular interactions and thus cell behavior, a special focus on spatially controlled chemistry of synthetic coatings are addressed in the following section.

SURFACE ENGINEERING OF FUNCTIONAL COATINGS VIA PATTERNING AND GRADIENTS

The fabrication of homogeneous functional coatings, from a chemical point of view, enables the modification of surface properties of the bulk biomaterial; for instance, to change the hydrophilicity or biomcompatibility of the material and thus the overall response of biological environment to the foreign compound. Additional modulation of surface properties to improve mimicking of environmental cues, to direct biological response, or to fabricate high-throughput screening systems can be provided by spatially controlled surface chemistry.[136,137] Feature sizes as small as tens of nanometers are possible with patterning and/or gradient techniques.[138,139] Currently employed micropatterning methods include soft-lithography,[140,141] photolithography,[142] as well as a range of techniques for the fabrication of nano-scaled patterns such as dip-pen nanolithography (DPN),[143] imprint lithography,[144] or colloidal lithography.[145] These techniques are briefly described in the following section and an emphasis is placed on the combination of these techniques with

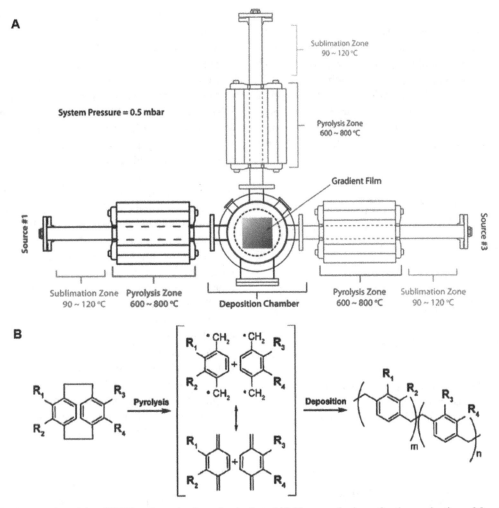

Fig. 9 Chemical vapor deposition (CVD) polymerization of substituted [2.2]paracyclophane for the production of functional parylene coatings: (**A**) custom-designed three-sources CVD setup and (**B**) associated polymerization reaction.
Source: From Chen & Lahann[123] © 2011, with permission from American Chemical Society.

functional coatings obtained via CVD polymerization of [2.2]paracyclophanes[123] because a wide variety of patterning/gradient strategies have been reported in combination with this fabrication technique for generating reactive polymer films.

Chemical Micro- and Nanopatterning

Soft lithography has been used extensively to pattern surfaces. This technique requires the previous fabrication of an elastomeric ("soft") material, subsequently used to create the chemical pattern.[142] As illustrated in Fig. 10, a liquid pre-polymer is cast on a structured master (I). After curing (II), the structured elastomeric stamp (III) is used for surface modification. The popularity of this technique stems from the fact that it is simple to implement and is relatively inexpensive. In addition, micro- to nanopatterns[146] can be generated via this method because the rigidity of the master enables good transfer of the pattern to the stamp as well as a good separation of both materials.

This technology has been derivatized to two main techniques: microcontact printing (μCP) and microfluidic patterning. μCP consists of "inking" the stamp with the molecule of interest and bringing the then inked stamp into physical contact with the substrate to transfer the pattern. For example, μCP has been applied on poly(p-xylylene carboxylic acid pentafluorophenolester-co-p-xylylene) for the selective immobilization of an amino-terminated biotin ligand.[125] This patterning technique has also been used in order to induce selective deposition of parylene coatings on surfaces. Inhibition of parylene-N and parylene-C film growth had been reported on several transition metals such as gold[147] and μCP of alkanethiols on gold has enabled selective deposition of the polymer in the areas pre-treated with this thiol-based SAM.[148] Microfluidic patterning is another widely used soft lithography technique for micropatterning. Microfluidic channels are molded in an elastomeric material by soft lithography and the stencil is put into contact with the substrate. Microfluidic patterning then consists of a transfer of the molecule(s) of interest on the substrate where the soft polymer does not contact the surface.[149] A similar strategy has been combined with CVD polymerization in order to spatially direct the deposition of poly(4-pentafluoropropionyl-*p*-xylylene-*co*-*p*-xylylene) and subsequently the immobilization of the biomolecule (Fig. 11).[150] PDMS stencils, having the desired pattern, were sealed on the substrate and vapor-assisted micropatterning in replica was performed during the CVD polymerization. This technique has also been

used for the fabrication of selective multifunctional polymer films.[134] First, a homogeneous polymer coating of one parylene type is created by CVD polymerization, then a stencil is sealed on the coating and a second CVD polymerization of another parylene type is performed to create a microstructured functional coating during this two-step CVD process. Alternatively, photolithographic techniques are commonly utilized methods for patterning polymeric materials as they are capable of generating large scale patterns and can facilitate patterning in 2D and 3D. Discontinuous patterning is often aided by the use of a mask. Finally, only illuminated areas are allowed to undergo photochemical reaction. For example, PEO has been selectively immobilized to areas of poly(4-benzoyl-*p*-xylylene-*co*-*p*-xylylene) that were exposed to UV irradiation, preventing protein adsorption on these areas (Fig. 12).[151] Using the same functional coating, maskless photolithography that uses programmable digital micromirrors (1024 × 768 pixels) was used to project a pattern on various substrates (Fig. 13).[123,152] Several groups have also begun to use a laser source (or other energy source) to deposit 2D or 3D patterns by direct writing in photoresists.[153] Some examples of this are described in the next section.

The following set of described techniques place an emphasis on patterning on the nanoscale. DPN is a direct writing approach that consists of coating an atomic force microscopy (AFM) probe tip with an ink of interest and

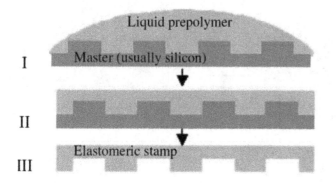

Fig. 10 Schematic representation of the fabrication of elastomeric stamp for soft lithography.
Source: From Falconnet et al.[142] © 2006, with permission from Elsevier.

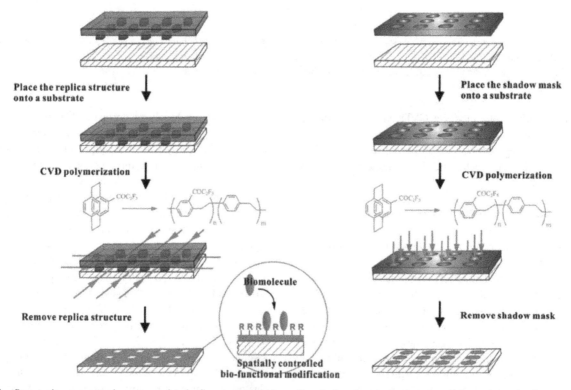

Fig. 11 Structuring process via vapor-assisted microstructuring in replica during chemical vapor deposition polymerization.
Source: From Chen & Lahann[150] © 2007, with permission from Wiley-VCH Verlag GmbH & Co. KGaA, Weinheim.

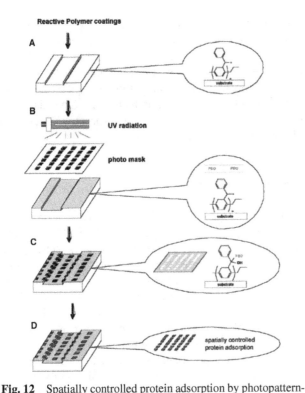

Fig. 12 Spatially controlled protein adsorption by photopatterning of reactive coating deposited in microchannels.
Source: From Chen & Lahann[151] © 2005, with permission from American Chemical Society.

bringing this probe in physical contact with the surface. Printing quality largely depends on the tip geometry, writing speed, contact time between the tip and the substrate, ink properties, and environmental conditions such as humidity.[143] To coat large surfaces with more than one ink type or to increase the production of nanopatterns, an array of DPN probes may be utilized. DPN has been performed on poly(4-ethynyl-*p*-xylylene-*co*-*p*-xylylene) coatings, deposited by CVD polymerization on various substrates via a click chemistry reaction with azide-functionalized inks (Fig. 14).[154] This study also demonstrated that DPN on such functional coatings enables the creation of nanopatterns independently from the substrate. Other techniques have also been applied for the creation of nanopatterns. For instance, nanoimprint lithography (NIL) has been developed to transfer nanopatterned features from a hard mold to a polymeric material by pressing both materials together via physical contact under controlled temperature and pressure (Fig. 15).[144,155,156] Thickness contrast is thus created in the polymeric material and can be subsequently transferred to a substrate. Alternatively, colloidal lithography has emerged for the fabrication of masks that are utilized in sputtering and etching processes to create pattern features (Fig. 16).[145] This inexpensive technique is based on the assembly of nanocolloids at the surface of

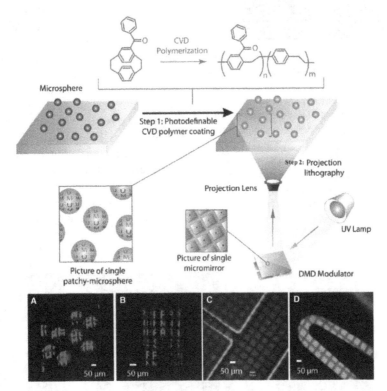

Fig. 13 Schematic description of microstructuring via projection photolithography. Fluorescence micrographs show that this technique can be applied to 3D and complex geometries, including (**A**) microspheres, (**B**) fibers, (**C**) microchannels, and (**D**) stent devices.
Source: From Chen & Lahann[123] © 2011, with permission from American Chemical Society and from Chen et al.[152] © 2007, with permission from National Academy of Sciences, U.S.A.

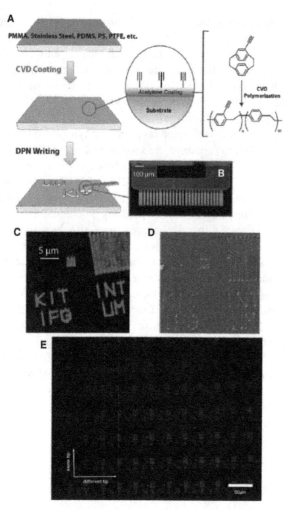

Fig. 14 (**A**) Schematic illustration of the dip-pen nanolithography (DPN) writing process on poly(4-ethynyl-*p*-xylylene-*co*-*p*-xylylene). (**B**) Micrograph of a 1D cantilever array with 26 pens. (**C**) Fluorescence micrograph of nanopatterned coating after DPN writing; features consist of 300-nm lines to form texts and rectangular boxes. (**D**) Atomic force microscopy image (tapping mode topography) showing self-assembled gold-streptavidin particles after DPN writing. (**E**) Fluorescence image of a large-scale DPN pattern, where features in the *x*-direction were written by different tips of the cantilever array and repeated in the *y*-direction.
Source: From Chen et al.[154] © 2010, with permission from American Chemical Society.

a material, thus requiring well-controlled nanoparticle size and appropriate particle physics to form well-defined assemblies. The above-described patterning techniques can be applied to numerous processes reported for the fabrication of functional polymer coatings and additional patterning techniques have also been developed for specific coating processes, including plasma technologies.[157]

Chemical Gradients

Structuring functional coatings by the creation of chemical gradients is an emerging area of interest for biomedical applications.[158,159] Several processes have been reported in the literature to obtain substrates with gradual changes in chemical composition. Many of the patterning methods previously described have also enabled the fabrication of chemical gradients. For example, the exposure time of photochemical techniques can be varied along one direction of the substrate to obtain gradual changes in the photochemical reaction.[160,161] Analogously, the bonding efficiency of the polymer to the substrate follows a similar trend with longer photoexposure times resulting in stronger bonds.[136] Microfluidic pathways have also been used to immobilize species in solution in a gradient fashion via laminar flow inside a network of microchannels (Fig. 17).[162,163] Different gradients have been obtained according to the width of the microchannels, the channel pathway, and operating conditions (flow rate). Experimentally simple and versatile, this low-cost technique only requires an elastomeric relief structure, fabricated by rapid prototyping. Diffusion phenomenon may also direct the formation of compositional gradient via vapor-based processes. This technique has been reported for the fabrication of chemical gradients by plasma polymerization. Briefly, a surface gradient that ranged from hydrophobic plasma polymerized hexane to a more hydrophilic plasma polymerized allylamine were formed with the help of a mask on top of the substrate and subsequent cellular responses to this gradient were investigated (Fig. 18).[164]

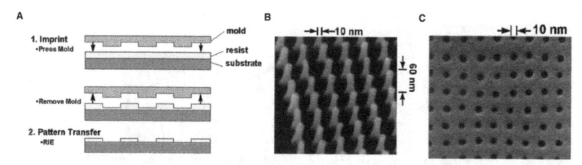

Fig. 15 (**A**) Schematic of the originally proposed nanoimprint lithography process. (**B**) Scanning electron microscopy (SEM) image of a fabricated mold with a 10-nm diameter array. (**C**) SEM image of a hole arrays imprinted in poly(methyl methacrylate) by using such a mold.
Source: From Chou et al.[155] © 1996, permission from American Vacuum Society and from Guo[156] © 2007, permission from Wiley-VCH Verlag GmbH & Co. KGaA, Weinheim.

Compositional gradients may also be created via CVD polymerization of functional parylene. In this approach, a copolymer gradient bearing two reactive chemical groups (trifluoroacetyl and amine moieties, respectively) was created via CVD polymerization in a counterflow

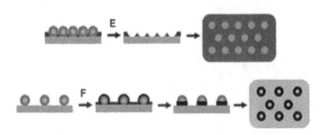

Fig. 16 Examples of colloid assembly used as a template for nanopattern formation: route E shows a templating with a hexagonally packed structure, whereas route F shows the creation of nanorings.
Source: From Yang et al.[145] © 2006, with permission from Wiley-VCH Verlag GmbH & Co. KGaA, Weinheim.

setup (Fig. 19). Specifically two functionalized [2.2] paracyclophanes were fed into a two-source CVD system at a 180° angle and were copolymerized as a gradient. Surface analysis confirmed gradual changes in the copolymer composition along the substrate. The compositional slope of the gradient may be tailored by process parameters such as gas flow rate. The generated gradient was subsequently used for co-immobilization of biomolecules.

BIOMOLECULAR INTERACTIONS— APPLICATIONS

Thus far, the emphasis in this entry has been on the various processes for generating functional polymer coatings. Going forward, the focus is on the use of these coatings for probing and assessing biomolecular interactions in a host of biomedical applications. In particular applications such as the creation of biosensors, cell microenvironments, stem cell culture substrates, and antifouling substrates are highlighted.

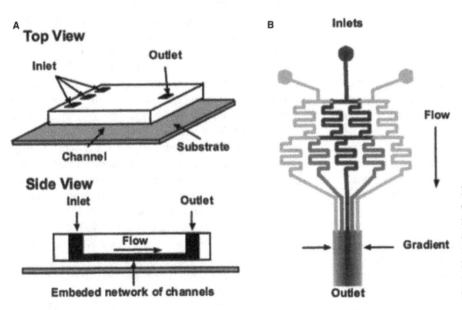

Fig. 17 (**A**) Schematic representations of a poly(dimethylsiloxane) microfluidic gradient generator (**B**) and of the associated gradient-generating microfluidic network.
Source: From Jeon et al.[162] © 2000, with permission from American Chemical Society.

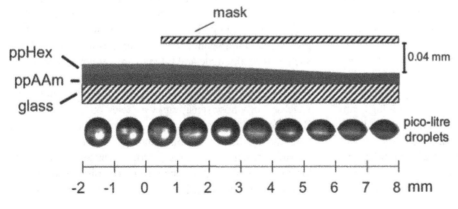

Fig. 18 Schematic representation of the experimental setup used to prepare a shallow diffusion gradient by plasma polymerization.
Source: From Zelzer et al.[164] © 2008, with permission from Elsevier.

Excipients—Gels

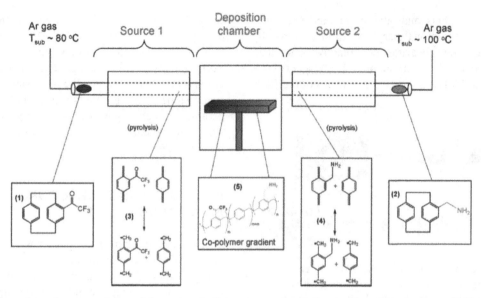

Fig. 19   Side-view schematic representation of the custom-built two-source chemical vapor deposition system used for the fabrication of compositional gradient.

Source: From Elkasabi & Lahann[194] © 2009, with permission from Wiley-VCH Verlag GmbH & Co. KGaA, Weinheim.

Biosensors

Many methods are used to characterize the reactivity and sensitivity of biosensors. However, it is advantageous to use analytical tools and techniques that do not require the labeling of the biomolecule(s) of interest as biomolecular labeling requires additional chemical modification steps and the resulting modification may change the properties of the target molecules.[165] Several label-free methods exist including quartz crystal microbalance (QCM), surface plasmon resonance (SPR), and surface plasmon resonance–enhanced ellipsometry (SPREE) imaging.[166,167] Although the mechanism in which they probe biomolecular interactions is different, electrical means for QCM and optical means for SPR and SPREE, all of the techniques have similar surface requirements for a compatible sensor type. Specifically, the substrates must be stable, reactive, pin-hole free, and sufficiently thin. SAMs of alkanethiols have been utilized for biosensing, but have limited stability and shelf-life.[168] In comparison, vapor-based techniques such as CVD have better stability and shelf-life under ambient conditions. A plethora of chemical reactivities are available for vapor-deposited substrates, and coatings are deposited with good surface coverage such that the resulting coatings are pin-hole free. Sensors must be pin-hole free because the presence of pin-holes results in undefined substrates that can result in inhomgeneities in biomolecular binding. The thickness of CVD coatings can be finely tuned (on the order of nanometers) by controlling the amount of starting material utilized. This is significant as the physical principles on which QCM and SPR are based require thin films as technique sensitivity is hampered by increased coating thickness. A CVD substrate

was used to create a spatially resolved biomolecular sensing array that was compatible with SPREE sensing technologies.[169] In this work, a ketone functionalized CVD coating was spatioselectively patterned with hydrazide-functionalized biotin via μCP. Biotin is a high-affinity reactive partner for the protein streptavidin. Thus the patterned area served as the reactive part of the sensor and the rest of the surface was passivated with hydrazide-terminated PEG as PEG is known to be protein resistant. When placed in a SPREE imaging cell, the CVD sensing array was capable of sensing a cascade of biomolecules consisting of the protein streptavidin with a rhodamine label, a biotinylated antibody for the protein fibrinogen, and the protein fibrinogen with a fluorescein label (Fig. 20). Protein labeling enabled additional confirmation of biomolecular binding via fluorescence microscopy. Not only did this study show the efficacy of CVD coatings as biomolecular sensors for the first time, it also provided insight into the desired coating thickness for said sensors.

The sensing of volatile organic compounds (VOCs) is important in health diagnostics and chemical sensing. For example, the breath is comprised of thousands of VOCs, which may be indicative of various diseases such as breast, prostate, and lung cancers.[170,171] Furthermore, other VOCs such as formaldehyde and toluene are known to be harmful for human health. Microwave plasma-enhanced chemical vapor deposition (MPECVD) has been used to generate a VOC sensor whose sensitivity was assessed using QCM.[172] Because this particular sensor was to be operated under ambient conditions, the influence of moisture on sensor sensitivity had to be considered. Because moisture can limit the sensitivity of a sensor (due to moisture adsorption on the sensor surface), it is imperative that the wettability

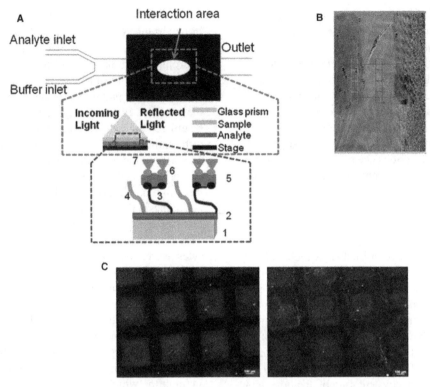

Fig. 20 Surface plasmon resonance–enhanced ellipsometry (SPREE) analysis setup and fluorescence microscopy. (**A**) SPREE imaging was utilized to analyze biomolecular interactions with the surface obtained by chemical vapor deposition (CVD) polymerization. The experimental setup consisted of an inlet for buffer and analyte flow, and the fluids were delivered by syringe pumps to the interaction area. The second inset is a schematic of the sample architecture, with each component represented numerically as follows: (1) Au-coated SPREE slide, (2) CVD coating (PPX-COC$_2$F$_5$), (3) biotin hydrazide long chain, (4) 10K poly(ethylene glycol) hydrazide, (5) TRITC–streptavidin, (6) biotinylated fibrinogen antibody, and (7) FITC–fibrinogen. (**B**) Prior to analysis, portions of the sample patterned with biotin hydrazide and unpatterned areas reacted with poly(ethylene glycol) hydrazide were selected for data collection. (**C**) Secondary confirmation of protein binding via fluorescence microscopy. The left image is the rhodamine channel, which indicates the binding of TRITC–streptavidin, and the right image is the fluorescein channel, which indicates binding of FITC-fibrinogen.
Source: From Ross et al.[169] © 2011, with permission from American Chemical Society.

(hydrophicility and hydrophobicity) of the sensing surface be controlled so as to concentrate the molecules of interest on the sensing surface. In this instance, a QCM sensor was surface modified via the MPECVD of a superhydrophobic coating of *n*-octadecyltrimethoxysilane, which was then used to sense VOCs of formaldehyde and toluene. The surface modification not only increased the hydrophobicity of the sensor (water contact angle of 150°), but also led to the microstructuring of the surface, which increased the effective surface area of the sensor, both of which led to preferential adsorption of toluene and formaldehyde as compared to an uncoated QCM sensor. In addition to VOC sensing, QCM has also been used to evaluate the microbial sensing capabilities of CVD coatings.[173] Here, parylene was used to coat a QCM sensor in order to monitor bacterial cultivation. This parylene-coated QCM sensor (pQCM) was placed into cultivation medium for *Escherichia coli*. First, the impact of medium viscosity, pH, and conductivity on the signaling of the pQCM sensor was assessed to establish a baseline. Then the influence of bacterial cell proliferation on pQCM sensor signaling was evaluated. A satisfactory

correlation was found to exist between the cell number and the response of pQCM sensor. Furthermore, the sensitivity of the sensor to microbial population changes was determined and indicated this method as a promising approach for monitoring microbial cultivation *in vitro* during processes such as fermentation.

Cell-Biomaterial Interactions

In addition to cell–cell interactions, a myriad of environmental cues (including physical and chemical properties of the extracellular matrix (ECM)) signal cellular actions.[174] For example, cellular responses to environmental cues can result in intercellular and intracellular changes in cytoskeletal organization, proliferation, cell differentiation, gene expression, and apoptosis.[175–177] As a result, cell–material interactions are an integral part of *in vitro* cell studies and investigators have used synthetic biomaterials to mimic the cellular microenvironment in terms of its physiochemical properties.[178] Patterning and gradient techniques afford control of physical and chemical surface characteristics on

Excipients—Gels

the level of cellular action, i.e., on the micro- and nanoscale. Generating *in vitro* gradients gives scientists an opportunity to explore distinct surface compositions at the cellular level. Furthermore, the use of surface gradients allows for discrete surface chemistries and morphologies to be assessed on a single sample allowing for more efficient screening of materials. This may prove to be more cost-effective and may enable high-throughput evaluation.[179,180] Here specific consideration is given to the response of numerous cell types to various chemical cues used to recapitulate the cellular microenvironment via the use of chemical patterning and gradients.

Patterning

Patterning provides scientists with a diverse toolbox for the study of cells and biomolecules. Patterning may occur directly, i.e., the attachment of the biomolecule(s) of interest to the substrate via the patterning medium, or indirectly in which a template of the desired pattern is made on the substrate followed by subsequent attachment of the biomolecule(s). The attachment of biomolecular factors (e.g., proteins, growth factors, cytokines) occurs by both physical adsorption and covalent immobilization. However, covalent immobilization affords greater control over the orientation of the biomolecules, and because orientation is known to influence function, covalent immobilization is preferred.

Soft lithography, particularly µCP, has been used extensively to pattern biomolecules to probe cellular interactions. In fact, µCP has served as a platform for patterning neuronal stem cells (NSC) on a PLGA polymer.[181] NSC patterning is potentially beneficial for neural repair and therapy, because it affords control over the spatial distribution and growth of NSCs.[182] However, difficulties in patterning NSCs have limited swift research progress. In this instance, the authors used a PLGA modified with a hydrophobin II protein as the base substrate. Areas of serum were subsequently patterned via µCP and cells adhered to the patterned area, while being repelled from the modified PLGA background. Patterning of NSCs can be utilized in the investigation of cell growth, motility, and differentiation in various environments. µCP was also exploited in the patterning of cardiomyocytes for the study of excitable cell features such as electrical signaling.[183] Here fibronectin (which is known to positively influence cell adhesion) was patterned in 100 µm stripes spaced 100 µm apart onto a multielectrode array (MEA), which can be used to record electrical signals. Rat cardiomyocytes were then cultured on the patterned stripes for a defined time period such that cellular stripes resulted. Aggregation of cardiomyocytes on the stripes leads to the guided excitation and propagation of electrical signals along the cell stripes but not between the cell stripes (Fig. 21). When co-cultured with noncardiomyocytes (NCMs) signaling was attenuated, although the degree of signal attenuation could be controlled via the addition of an inhibitor, which impeded NCM growth. Signal attenuation was also evident in the presence of pattern defects. Results of this work indicate the viability of µCP for investigating factors influencing cardiomyocyte electrical signaling, which has applications in regenerative, developmental, and pharmacological biology.

Transparency-based photolithography was exploited to investigate the impact of exposure time (through a transparent mask) on the concentration of immobilized biomolecules and the influence of patterning on the adhesion behavior of human dermal fibroblasts (HDFs).[184] The

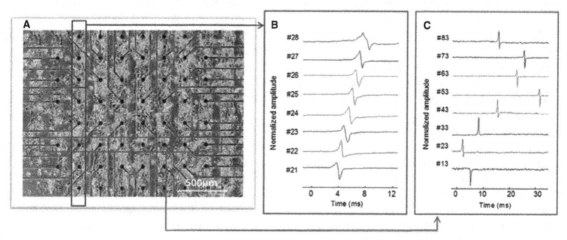

Fig. 21 Time course of field potential of cardiomyocytes after 3 days culture: (**A**) Microphotograph of 100 µm line-and-space cardiomyocytes stripes on the surface of a MEA device; (**B**) data recorded with a linear array of microelectrodes under the same cell stripe, showing a guided propagation; and (**C**) data recorded with another linear array perpendicular to the cell stripes, showing no time–space correlation.
Source: From Wang et al.[183] © 2013, with permission from Elsevier.

biomolecules, a generic adhesion peptide (RGDS) and an endothelial cell–specific adhesion peptide (REDV), were immobilized on the hydrogel by sequential application of the patterning technique. Immobilization occurred by mixing a precursor solution of monoacryloyl–PEG–peptide with a photoinitiator, layering the solution on the hydrogel surface, and then spatioselectively conjugating the acrylated moieties via UV exposure through a photomask. The concentration of biomolecules on the surface varied linearly with exposure time with increased exposure time leading to higher concentrations of immobilized peptides. To ascertain the biological activity of the immobilized peptides, HDFs were seeded onto the patterned surfaces. HDF adhesion was controlled such that cells adhered only to the RGDS patterned regions and not to the REDV areas (Fig. 22). Direct laser writing (DLW) is a variant photolithographic approach that uses two-photon lasers to generate 3D polymer scaffolds and the patterning of these scaffolds by a second DLW step has been used to create composite scaffolds with distinct protein-binding properties.[153] In this work, the two-component scaffold was comprised of poly(ethylene glycol diacrylate) (PEG–DA) and pentaerythritol tetra-acrylate (PETA). Ormocomp square patterns were embedded in a PEG–DA background with increasing concentrations of PETA (during the second DLW step). Patterns and backgrounds with PETA concentrations of 0–100% PETA (w/w) were incubated with the protein fibronectin and then seeded with

chicken fibroblast cells. Below 4.8% PETA (w/w), the scaffold was protein repellant and cells only adhered to the Ormocomp patterns. However, above 4.8% PETA (w/w) cells spread on both the square patterns and the PEG–DA/PETA background. This method provided control over the formation of cell adhesion sites, and consequently, the cell shape in three dimensions. As a result, this work provided evidence that cell growth can be directed in three dimensions.

Electron beam lithography (EBL) is also a direct writing approach, which consists of focusing an electron beam onto a substrate covered with an electron beam photoresist. This patterning technique (which resembles DLW with regard to the manner used to generate the patterns) has been used to create nanoscale patterns of biological ligands in an effort to understand and control cell adhesion on engineered scaffolds. Specifically, EBL is used to fabricate patterns of a cell adhesive ligand (RGD) combined with a growth factor [basic fibroblast growth factor (bFGF)].[185] First, an 8-arm aminooxy-terminated PEG (PEG–AO) was crosslinked with poly(styrene-4-sulfonate-co-poly(ethylene glycol) methacrylate) (pSS-co-PEGMA) to create hydrogels on passivated silicon wafers. Then this hydrogel coating was patterned via EBL such that reactive areas were generated in a nonreactive background. At this juncture, RGD is immobilized by means of oxime bond formation and bFGF by means of electrostatic interactions. Subsequent exposure of the patterned substrate to the appropriate primary antibody followed by a fluorescently labeled secondary antibody, confirmed the colocalization of the RGD and bFGF, respectively, in the patterned areas. Finally, human umbilical vein endothelial cells (HUVECs) are seeded on the patterns to assess their influence on cell adhesion. When both RGD and bFGF were present (Fig. 23), cell areas were significantly larger ($p = 0.009$) as compared to HUVECs grown onto patterns containing only RGD. Thus indicating a synergistic effect of the growth factor on cellular adhesion and suggesting the addition of growth factors to adhesive patterns for cell–material interactions. EBL has also been used to probe the interactions and networks of neural cells.[186] In this instance, a biodegradable hydrogel, poly(amidoamine), was patterned via EBL. After assessing protein adsorption to ensure that proteins adsorbed only on the patterned areas, neuronal cells were seeded onto the patterned substrates. In this instance, the pattern consisted of 10 μm microwells connected by 1 μm microchannels. Cells were provided with neuronal growth factor to induce neurite development. Preferential adhesion of the cells to the patterned microwells and neurite outgrowth into the microchannels resulted in the creation of an intercellular neural network. Moreover, the number of neurites is determined by the number of microchannels and single cell behavior could be observed. Patterning hydrogels with EBL allows for the creation of complex patterns that afford single cell analysis and thus has

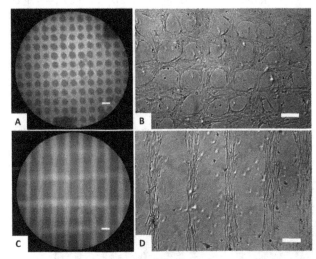

Fig. 22 Fluorescence of PEGDA hydrogels patterned with (**A**) ACRL–PEG–RGDS, a generic adhesion peptide (out of pattern), or with (**C**) ACRL–PEG–REDV, an endothelial cell–specific adhesion peptide (horizontal lines) and ACRL–PEG–RGDS peptide (vertical lines). (**B**, **D**) Phase contrast of human dermal fibroblasts (HDFs) attached to the surface of the ACRL–PEG–RGDS patterned hydrogels (imaged in **A** and **C** respectively). Note that in (**D**), HDFs have bound to RGDS patterned regions but not to REDV-patterned regions, as expected (Scale bar = 200 μm). **Source:** From Hahn et al.[184] © 2006, with permission from Elsevier.

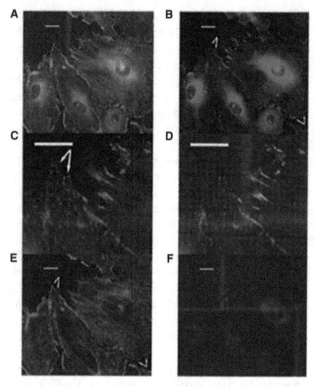

Fig. 23 Fluorescence images of human umbilical vein endothelial cells (HUVECs) adhered on extracellular matrix (ECM)-mimicking substrate containing both a cell adhesive ligand (RGD) and a basic fibroblast growth factor (bFGF). (**A**) Actin filaments terminating at focal adhesions are visible in the composite image. (**B**) Focal adhesions are more clearly visible in the single-channel image corresponding to vinculin staining, and in the zoomed image (**C**), and are highlighted with white arrows. The focal adhesions are colocalized with the ECM features (**D**). A square arrangement of actin filaments termini was also observed as illustrated by the white arrows (**E**), with spacing identical to that of the underlying pattern (**F**). Scale bars = 20 μm.
Source: From Kolodziej et al.[185] © 2011, with permission from American Chemical Society.

applications for multifunctional microdevices as well as the study of single cell behavior in terms of physiological and pharmacological effects.

Other nanoscale patterns have been created via DPN. For example, DPN has been exploited to pattern multiple biomolecules at once by creating a biomimetic lipid membrane pattern for cell culture studies.[187] Specifically, lipids with two different functional groups, biotin and nitrilotriacetic acid, were patterned concurrently using a large array. Then selective binding of proteins via sAV-biotin and histidine tag couplings occurred on the patterned areas. The authors verified the functionality of the proteins under numerous environmental conditions. This multilayer protein/lipid pattern, then served as a substrate for T-cell adhesion and activation. T cells were found to adhere selectively to the curved edges of the patterns and were activated by their interaction with the labeled antibodies in the patterned areas (Fig. 24). Thus far, cells have been patterned indirectly by cell adhesion to pre-patterned ligands. However, cells may also be patterned on substrates directly and researchers have used DPN to directly pattern bacterial cells.[188] Here micron-sized *E. coli* bacteria patterns were generated by DPN onto an amine functionalized substrate using a nanostructured poly(2-methyl-2-oxazoline) (PMeOx) hydrogel-coated tip and carrier agents. To create the PMeOx probe, a silicon oxide tip was chemically modified with 11-iodoundecyltrichlorosilane and then polymerized via ring-opening polymerization of 2-methyl-2 oxazoline monomers. Use of glycerol and tricine as carrier agents facilitated the writing of *E. coli* as they prevented the dehydration of the cells and increased the viscosity and interaction of the cellular ink solution, which enabled the patterning process.[189] Although alive on the tip, the *E. coli* cells died upon patterning likely as a result of membrane disruptions that occurred during the drying process on the substrate.

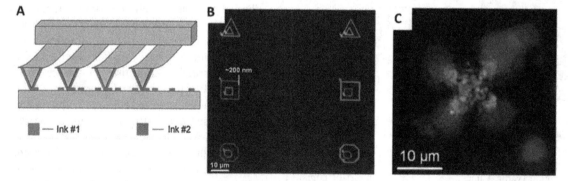

Fig. 24 Lipid writing via dip-pen nanolithography (DPN) and cellular response to said writing. (**A**) Schematic of DPN cantilever array used for writing (**B**) Writing with two fluorophore-labeled lipids (rhodamine and fluorescein). Triangles result from mixing the lipid inks in different concentrations. Fluorescence micrographs of T cells selectively adhered to and activated by functional proteins bound to phospholipid multilayer patterns. (**C**) A three-channel image of T-cells adhering to the corners of lipid protein DPN patterns and activated by functional proteins.
Source: From Sekula et al.[187] © 2008, with permission from Wiley-VCH Verlag GmbH & Co. KGaA, Weinheim.

Gradients

Gradients play a critical role in many biological processes such as embryonic development, immune response, tissue regeneration, and tumor metastasis.[190] Chemical, elasticity, and morphological gradients exist *in vivo* and efforts have been made to mimic these gradients *in vitro*. However, the focus here is on chemical gradients. Biomolecules may be attached to surfaces to create continuous and discrete gradients for investigating cell–material interactions.[191,192] Several methods exist for creating defined chemical gradients. Harris et al. modulated polymer brush density to influence subsequent immobilization of the cell adhesion peptide RGD, which in turn impacted fibroblast adhesion to the polymer brush.[193] In particular, a photoinitiator was deposited onto a silicon wafer and then a methacrylic acid (MAA) monomer was added to the treated substrate. The sample was irradiated with variable exposure times but with constant intensity across the substrate to generate a polymer chain density gradient. Initially, the anionic brush was resistant to cell adhesion and was therefore functionalized to promote fibroblast adhesion. Further functionalization of the PMAA surface involved conjugation of the RGD ligand to the polymer brush via appropriate surface chemistries. Once the gradient was prepared, 3T3 mouse fibroblasts were seeded onto the gradient surfaces and cell adhesion was assessed as a function of ligand density. Results indicate that the RGD surface concentration increased with polymer brush thickness (which increased with increasing irradiation time). Cell adhesion was also found to increase as ligand density increased. Applications of this work include wound healing and nerve regeneration. Compositional gradients may also be created via CVD.[194] CVD copolymer gradients have been used as a platform for modulating cellular transduction.[195] The generated gradient varied compositionally from amine-rich on one end of the substrate to aldehyde-rich on the other end. Then biotinylated ligands reactive for either amine or aldehyde moieties were attached to the substrate followed by immobilization of avidin, which has a strong affinity for biotin. Subsequently, a biotinylated adenovirus was immobilized in a gradient-dependent fashion onto the substrate. Human gingival fibroblasts cultured on the gradients demonstrated asymmetric transduction as a function of the density of the immobilized adenovirus (with decreased transduction in the direction of decreased adenovirus density) as seen in Fig. 25. Although only one immobilization mechanism was shown for both reactive groups, a dual gradient is feasible and can be exploited for the delivery of multiple therapeutic agents.

One of the challenges in utilizing gradients in neuronal cell research is the creation of gradients on the appropriate biological length scale (a few hundred micrometers). Creating substrate-bound chemical gradients would enable the influence of said gradients on neuronal development to be assessed. For example, protein-gradients with complex shapes led to the orientation of neuronal axons.[196] In particular, gradients ranging from pure laminin to pure BSA were generated over a length of 250 µm on a homogenous layer of poly-L-lysine (PLL). Solutions of BSA or laminin were injected into the inlets of the microfluidic network and a gradient was created as the solutions traversed the four or five serpentine streams such that a step-profile was produced in which each stream contained different protein concentrations. Within the serpentine channels, mixing occurs and then all streams exit via a single outlet and were delivered across the PLL-coated microchannel (Fig. 26). Neuronal cells were then cultured on the protein gradient and the direction of their axonal growth assessed. Axonal growth was found to preferentially occur in the direction of increasing laminin surface density.

Few studies have evaluated the cellular response of cells within a 3D gradient hydrogel.[197–199] One such study was used to assess neurite extension for applications in nerve regeneration.[200] In this work, a growth-promoting glycoprotein, laminin-1 (LN-1) was incorporated into a 3D agarose hydrogel. The agarose hydrogel was placed in a chamber where one side was exposed to a concentrated solution of LN-1 (high concentration compartment) and the other side was exposed to phosphate buffered saline (low concentration compartment). Diffusion of the LN-1 across the gel from the high concentration compartment toward the low concentration compartment occurred over the course of 6 hr and then was photocrosslinked via UV light. Because the LN-1 was fluorescently labeled, the persistence of the gradient through the various layers of the 3D network could be assessed. Once a uniform gradient was confirmed via fluorescence microscopy, cellular assessments could begin. The gradient setup for the cell tests was the same as that previously described with the exception that dorsal root ganglia (DRG) from chicken embryos were suspended in the agarose gel prior to LN-1 exposure and that DMEM was used in the low concentration compartment. Neurite outgrowth from the DRGs was assessed over a 4-day period with slopes of the LN-1 concentration gradients ranging from 0.017 to 0.121 µg/mL/mm. Fastest growth rates were seen for the gradient with the mildest slope (0.017 µg/mL/mm). At this point, the fastest neurite outgrowth on gradient gels was compared with that of isotropic gels, and neurite outgrowth was shown to be significantly faster on gradient scaffolds. Results of this work may be used to facilitate peripheral nerve regeneration *in vivo*.

Stem Cell-Culture Substrates and Controlled Stem Cell Differentiation

Although numerous cell types have been highlighted thus far, a discussion of creating stem cell microenvironments was deliberately left until the end of this entry due to the large potential therapeutic impact of this cell type. In

Excipients—Gels

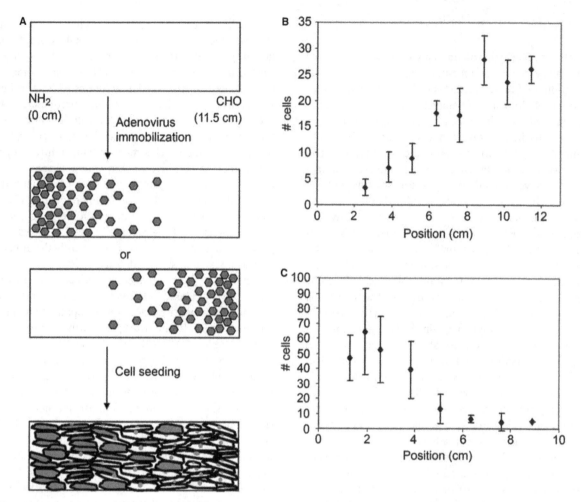

Fig. 25 (A) Schematic of gene delivery and cell transduction gradients. Adenovirus encoding blue fluorescence protein was immobilized onto a cell culture dish treated with the chemical vapor deposition copolymer gradient. Human gingival fibroblasts were then seeded onto the adenovirus gradient. A gradient of transduced cells were generated within the confluent cell population, with the gradient direction determined by (B) biotin hydrazide attachment to aldehyde groups, or (C) sulfo-NHS-LC-biotin attachment to amino groups.
Source: From Elkasabi et al.[195] © 2011, with permission from Elsevier.

addition to proliferation and gene expression, environmental cues are also known to impact the differentiation of stem cells.[201–203] Human pluripotent stem cells (hPSCs), due to their ability to differentiate into multiple cell types, have a myriad of applications in the health care industry.[204,205] Potential uses include cures/treatments for diseases such as heart failure and diabetes as well as drug safety and efficacy testing.[206–208] Feeder layers of mouse embryonic fibroblasts (mEFs) and Matrigel™, an undefined gelatinous protein mixture secreted by mouse carcinoma cells, are currently utilized in hPSC culture.[209] Matrigel is typically placed on the cell culture substrate prior to cell seeding in order to generate a thin monolayer that serves as a complex ECM. Although hPSCs show therapeutic promise, one of the major hurdles inhibiting clinical adoption is the use of this Matrigel culture system as it results in batch-to-batch variation, uses xenogenous factors, which warrant immunogenic concerns, and is not suitable for large-scale hPSC expansion.[210] Synthetic

substrates are a promising alternative for prolonged hPSC culture as they address many of these issues. Because synthetic surfaces are generated from defined materials and processes, there is little variation between batches and thus has greater potential for scaleup.[211] More importantly, eliminating the need for animal byproducts eradicates concerns regarding immunogenicity.[212]

In an attempt to replace poorly defined biological matrices, numerous materials have been studied as potential stem cell substrates, including electroactive polymers, tissue culture plastic, and SAMs of alkanethiols.[213–215] Researchers have focused on the use of hydrogels as a platform for long-term stem cell culture as a large number of stem cells are needed for potential therapeutics. One approach is to use surface-initiated polymerization of chemically defined polymers to generate synthetic cell culture substrates.[216] In particular, various methacrylates were grafted onto tissue culture polystyrene (TCPS) dishes. Synthesis occurs in an oxygen-free glass reaction vessel. Briefly, the reaction vessel

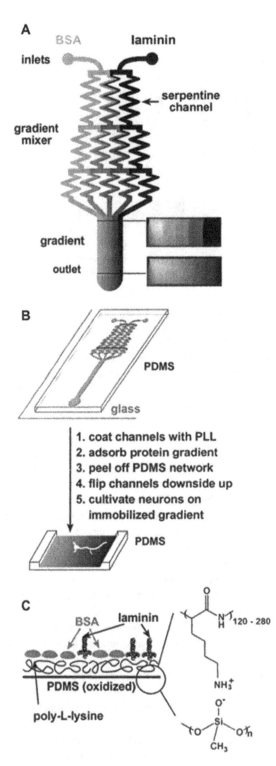

Fig. 26 (**A**) Schematic drawing of the design of a typical microfluidic network in poly(dimethylsiloxane) that we used for the fabrication of immobilized gradients. (**B**) The diagram summarizes the steps from the fabrication of the substrate-bound gradient to the cultivation of neurons on the gradient immobilized on the substrate. (**C**) An idealized schematic diagram of the cross section of the surface composition of a substrate-bound gradient composed of laminin and bovine serum albumin on poly(L-lysine).
Source: From Dertinger et al.[196] © 2002, with permission from National Academy of Sciences, U.S.A.

is degassed via a vacuum–argon purge cycle, which is completed three times. Simultaneously, the solvent, comprised of ethanol and deionized water in a volumetric ratio of 1:4, is degassed via vacuum for 40 min. Then the degassed solvent and the monomer of interest are added to the reaction vessel and heated such that the temperature range is between 76°C and 82°C. Tight temperature control is important for effective polymerization. Prior to monomer synthesis, free radicals must be created on the TCPS dishes by UV ozone plasma treatment. Later, these free radicals will enable the polymerization of the monomers of interest. Once the reaction has reached the desired temperature, TCPS dishes are added to the reaction vessel and polymerization proceeds for 2.5 hr. After the reaction is complete and the reaction vessel has been cooled to at least 60°C, the dishes are removed and rinsed overnight in a 1% saline solution that is maintained at 50°C to remove any excess monomer. Subsequent rinsing steps with 1% saline and deionized water ensure that any unreacted monomer has been eliminated.[217]

Because the influence of hydrogel structure on human embryonic stem cells (hESCs) was unknown, a total of six methacrylate derivatives were generated for cell screening. Specifically, poly(carboxybetaine methacrylate) (PCBMA), poly{[2-(methacryloyloxy)ethyl]trimethylammonium chloride} (PMETAC), poly[poly(ethylene glycol) methyl ether methacrylate] (PPEGMA), poly(2-hydroxyethyl methacrylate) (PHEMA), poly(3-sulfopropyl methacrylate) (PSPMA), and poly[2-(methacryloyloxy)ethyl dimethyl-(3-sulfopropyl)ammonium hydroxide] (PMEDSAH) were characterized in terms of their material properties and by the ability of hESCs to adhere and maintain an undifferentiated state on these surfaces.[216] Of the hydrogels assessed, only the zwitterionic PMEDSAH was able to support undifferentiated hESCs from two cell lines (BG01 and H9) for long-term passage (passage number ≥ 25) as indicated by hESC gene expression, karyotype, and embryoid body formation (Fig. 27). Results from the synthetic substrates were compared with Matrigel, which served as a control and no significant difference was noted. As a proof of concept, a commonly utilized media from animal-derived products, mouse embryonic fibroblast-conditioned media (MEF–CM), was used in this initial screening study. Subsequent studies focused on the behavior of hESCs on PMEDSAH-coated dishes using media which lacked nonhuman animal products or in serum-free, defined media.

Ultimately, hPSCs will be used for applications in regenerative medicine and as such, studies have evaluated the efficacy of the PMEDSAH coating in this context.[218] Here human-induced pluripotent stem cells (hIPSCs) were cultured on PMEDSAH in defined media conditions for 15 passages and then these cells were differentiated into human mesenchymal stem cells (hMSCs), a multipotent stem cell that can differentiate into bone, cartilage, or fat. The resulting hIPSC–MSCs were characterized in terms of their ability to differentiate into the aforementioned lineages and by gene expression. Expression of characteristic MSC

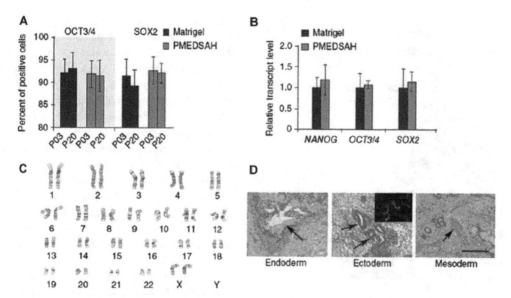

Fig. 27 Cellular characterization of human embryonic stem cells (hESC) cultured on poly(2-(methacryloyloxy)ethyl dimethyl-(3-sulfo-propyl)ammonium hydroxide) (PMEDSAH) substrates in mouse embryonic fibroblast-conditioned media. (**A**) Percentage of hESCs expressing OCT3/4 and SOX2 at passages 3 (P03) and 20 (P20). (**B**) Relative transcript levels of NANOG, OCT3/4, and SOX2 from hESCs cultured on PMEDSAH and Matrigel. (**C, D**) After 25 passages, hESCs cultured on PMEDSAH (**C**) maintained a normal karyotype and (**D**) retained pluripotency as demonstrated by teratoma formation in immunosuppressed mice. Hematoxylin and eosin–stained paraffin sections indicating endoderm (goblet-like cells at arrow), ectoderm (neuroepithelial aggregates at arrow; and cells expressing neuron-restricted protein b-III tubulin in inset) and mesodermal derivatives (cartilage, connective tissue and muscle at arrow). Scale bar, 200 µm.
Source: From Villa-Diaz et al.[216] © 2010, with permission from Macmillan Publishers Ltd.

markers and the facilitation of lineage-specific differentia-tion by these cells indicated successful differentiation to MSCs. The function of hIPSC–MSCs was further assessed via *in vivo* bone regeneration in an animal model. Specifi-cally, these cells were induced into osteoblast differentia-tion over a 4-day period and then transplanted into a craniofacial defect in immunocompromised mice for 8 weeks. Results from MicroCT and histology reveal bone formation in the defects (Fig. 28). These outcomes confirmed that hIPSCs cultured in a defined and xeno-free system have the capability to differentiate into functional MSCs with the ability to form bone *in vivo*.

Anti-Fouling Substrates

Biofouling may be defined as the spontaneous, unwanted buildup of proteins, bacteria, and cells on synthetic sur-faces. Fouling is a significant problem across a variety of industries from health care to defense.[219,220] For example, fouling of biomedical devices impacts device performance via tissue fibrosis, bacterial infection, and so on; often lead-ing to device failure.[221] The range of devices impacted is broad and includes metallic hip replacements, neural pros-thesis, and cardiovascular implants.[222] From a defense per-spective, marine fouling of naval ships and equipment impedes performance by decreasing signal quality and increasing drag and thus fuel inefficiency. Costs associated

with marine fouling are estimated at $1 billion for the US Navy alone.[223] Protein–surface interactions are particu-larly important because biofouling is a multistep process in which protein adsorption is the first step. As a result, pro-tein–surface interactions are a vital part of the fouling pro-cess and inhibiting protein adsorption would prevent fouling altogether. Protein interactions with surfaces are known to be impacted by numerous factors, including wettability and charge. In terms of wettability, nonfouling substrates are typically hydrophilic as surface hydration impedes the hydrophobic interactions of proteins with surfaces that commonly facilitate protein attachment. Therefore, a host of hydrophilic polymers including PEG derivatives such as poly(hydroxypropyl methacrylate) (PHPMA) and PHEMA have been explored as antifouling surface coatings.[224,225] For example, the ability of PHPMA to resist protein adsorp-tion from various media (single protein solutions, 10% human blood plasma and serum, 100% human serum, and 100% human plasma) was evaluated using SPR.[224] The highest resistance to protein adsorption was achieved in the moderate thickness regime of 25–40 nm. Copolymers of PHPMA and PHEMA have also been created to inhibit adsorption of proteins from single protein solutions as well as human blood plasma and serum.[226] Like the homopoly-mers, the degree to which the aforementioned copolymer coatings inhibited protein adsorption was modulated by coating thickness with very short chains and very long

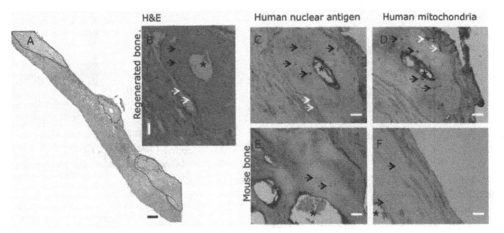

Fig. 28 *In vivo* bone formation by human iPS cell-derived mesenchymal stem cells (hiPSMSCs). (**A**) H and E image of a SCID-mouse skull section where a calvarial defect was created and hiPSMSCs were transplanted. The limits of the mouse skull are delineated by black dashed lines (expanded in **E–F**). The dashed line (expanded in **B–D**) indicates new-formed bone areas with osteocytes (black arrows) surrounded by matrix consistent with woven bone, osteoblasts (white arrows) lining the exterior of the newly formed bone, as well as small bone marrow cavities (indicated by asterisks) within the newly formed bone. Positive immunoreactivity to human nuclear antigen (**C**) and human mitochondria (**D**) antibodies of osteocytes within the newly formed bone, whereas negative in the mouse bone (**E** and **F**) suggest the human origin of the regenerated bone. Scale bar in A = 100 μm, whereas in B–F = 20 μm.
Source: From Villa-Diaz et al.[218] © 2012, with permission from AlphaMed Press.

chains not being as protein resistant (thus revealing a sweet spot for polymer chain length). Poly(*N*-vinylpyrrolidone) (PNVP) is another hydrophilic polymer used in biomedical applications to reduce fouling and membranes comprised of copolymer blends of PNVP have been effective in reducing adsorption of proteins such as BSA, fibrinogen, and human serum albumin.[227,228] However, the long-term use of these blends was prevented by the leaching of PNVP. In order to prevent leaching, PNVP polymer coatings have been generated via covalent grafting, specifically surface-initiated polymerization using ATRP.[229]

Surface charge is also known to play a role in protein adsorption with zwitterionic, or neutrally charged surfaces exhibiting antifouling properties. The antifouling properties of zwitterionic surfaces are commonly attributed to strong electrostatic hydration, which is believed to make protein adsorption unfavorable from an enthalpic viewpoint. Zwitterionic coatings may be generated via ATRP or solution polymerization.[230,231] Of these zwitterionic polymers, poly(carboxy betaine methacrylate) (CBMA) has been widely utilized.[232] In fact, assessment of the fouling characteristics of CBMA via SPR analysis has shown undetectable levels of protein adsorption (0.3 ng/cm^2) when exposed to single protein solutions or complex media.[233–235] Moreover, CBMA has an available functional end group (carboxylateanions), which may be further modified with amine-containing moieties by means of 1-ethyl-3-(3-dimethylaminopropyl)-carbodiimide and *N*-hydroxysuccinimide (EDC/NHS) chemistry thus enabling additional biofunctionalization.[236] This property could be exploited in biomolecular sensing, tissue engineering, or drug delivery.

CONCLUSION

Surface modification of biomaterials via the fabrication of functional polymer coatings influences biomolecular surface interactions. Great technological platforms now exist for the production of functional polymer coatings and for the subsequent modification of the available functional groups for biomedical applications. The development and exploitation of a multitude of these processing techniques have enabled key biological questions to be addressed and aided in our understanding of cellular mechanisms. Moreover, many of these techniques are applicable in the coating of 2D as well as 3D geometries both of which have been successfully combined with micro- and nanopatterning to better mimic features found in biological environments *in vivo*. Fabrication of biomolecular gradients (including those in 3D hydrogels), is an emerging area of study that enhances the potential of chemical modification to mimic biological environments and enables high-throughput screening experiments. In the future, efforts will likely continue in the control and the combination of bioorthogonal chemistries on a single substrate as mimicking the cellular microenvironment is extremely complex and will require precise control in the immobilization of multiple biological compounds. In addition, combinatorial studies including variation in chemical and physical substrate surface properties, such as substrate topography, roughness, elasticity, will be necessary to understand not only the chemical impact on biomolecular surface interactions but environmental effect on biological objects as a whole.

Excipients—Gels

REFERENCES

1. Langer, R.; Tirrell, D.A. Designing materials for biology and medicine. Nature **2004**, *428* (6982), 487–492.

2. Tirrell, M.; Kokkoli, E.; Biesalski, M. The role of surface science in bioengineered materials. Surf. Sci. **2002**, *500* (1–3), 61–83.

3. Lahann, J. Reactive polymer coatings for biomimetic surface engineering. Chem. Eng. Commun. **2006**, *193* (11), 1457–1468.

4. Norrman, K.; Ghanbari-Siahkali, A.; Larsen, N.B. 6 Studies of spin-coated polymer films. Annu. Rep. C Phys. Chem. **2005**, *101* (0), 174–201.

5. Vorotyntsev, M.A.; Zinovyeva, V.A.; Konev, D.V. Mechanisms of electropolymerization and redox activity: Fundamental aspects. In *Electropolymerization: Concepts, Materials and Applications*; Cosnier, S., Karyakin, A., Eds.; Wiley-VCH: Weinheim, 2010; 27–50.

6. Cosnier, S.; Holzinger, M. Biosensors based on electropolymerized films. In *Electropolymerization: Concepts, Materials and Applications*; Cosnier, S., Karyakin, A., Eds.; Wiley-VCH: Weinheim, 2010; 189–213.

7. Sadki, S.; Schottland, P.; Brodie, N.; Sabouraud, G. The mechanisms of pyrrole electropolymerization. Chem. Soc. Rev. **2000**, *29* (5), 283–293.

8. Ariga, K.; Nakanishi, T.; Michinobu, T. Immobilization of biomaterials to nano-assembled films (self-assembled monolayers, Langmuir-Blodgett films, and layer-by-layer assemblies) and their related functions. J. Nanosci. Nanotechnol. **2006**, *6* (8), 2278–2301.

9. Kuhn, H. Present status and future prospects of Langmuir-Blodgett film research. Thin Solid Films **1989**, *178* (1–2), 1–16.

10. Tredgold, R.H. Langmuir-Blodgett films made from preformed polymers. Thin Solid Films **1987**, *152* (1–2), 223–230.

11. Lerum, R.V.; Bermudez, H. Controlled interfacial assembly and transfer of brushlike copolymer films. Chem. Phys. Chem. **2010**, *11* (3), 665–669.

12. Decher, G.; Hong, J.; Schmitt, J. Buildup of ultrathin multilayer films by a self-assembly process. 3. Consecutively alternating adsorption of anionic and cationic polyelectrolytes on charged surfaces. Thin Solid Films **1992**, *210* (1–2), 831–835.

13. Decher, G. Fuzzy nanoassemblies: Toward layered polymeric multicomposites. Science **1997**, *277* (5330), 1232–1237.

14. Caruso, F.; Niikura, K.; Furlong, D.N.; Okahata, Y. Ultrathin multilayer polyelectrolyte films on gold: Construction and thickness determination. Langmuir **1997**, *13* (13), 3422–3426.

15. Tang, Z.; Wang, Y.; Podsiadlo, P.; Kotov, N.A. Biomedical applications of layer-by-layer assembly: From biomimetics to tissue engineering. Adv. Mater. **2006**, *18* (24), 3203–3224.

16. Bhattacharya, A.; Misra, B.N. Grafting: A versatile means to modify polymers: Techniques, factors and applications. Prog. Polym. Sci. **2004**, *29* (8), 767–814.

17. Uyama, Y.; Kato, K.; Ikada, Y. Surface Modification of Polymers by Grafting. In *Grafting/Characterization Techniques/Kinetic Modeling*; Galina, H., Ikada, Y., Kato, K.,

Kitamaru, R., Lechowicz, J., Uyama, Y., Wu, C., Eds.; Advances in Polymer Science; Springer Berlin: Heidelberg, 1998, 1–39.

18. Kato, K.; Uchida, E.; Kang, E.T.; Uyama, Y.; Ikada, Y. Polymer surface with graft chains. Prog. Polym. Sci. **2003**, *28* (2), 209–259.

19. Goddard, J.M.; Hotchkiss, J.H. Polymer surface modification for the attachment of bioactive compounds. Prog. Polym. Sci. **2007**, *32* (7), 698–725.

20. Bigot, S.; Louarn, G.; Kébir, N.; Burel, F. Facile grafting of bioactive cellulose derivatives onto PVC surfaces. Appl. Surf. Sci. **2013**, *283*, 411–416, doi:10.1016/j.apsusc.2013.06.123.

21. Goddard, J.M.; Hotchkiss, J.H. Tailored functionalization of low-density polyethylene surfaces. J. Appl. Polym. Sci. **2008**, *108* (5), 2940–2949.

22. Tao, G.; Gong, A.; Lu, J.; Sue, H.J.; Bergbreiter, D.E. Surface functionalized polypropylene: Synthesis, characterization, and adhesion properties. Macromolecules **2001**, *34* (22), 7672–7679.

23. Patel, S.; Thakar, R.G.; Wong, J.; McLeod, S.D.; Li, S. Control of cell adhesion on poly(methyl methacrylate). Biomaterials **2006**, *27* (14), 2890–2897.

24. Croll, T.I.; O'Connor, A.J.; Stevens, G.W.; Cooper-White, J.J. Controllable surface modification of poly(lactic-*co*-glycolic acid) (PLGA) by hydrolysis or aminolysis I: Physical, chemical, and theoretical aspects. Biomacromolecules **2004**, *5* (2), 463–473.

25. Rasal, R.M.; Janorkar, A.V.; Hirt, D.E. Poly(lactic acid) modifications. Prog. Polym. Sci. **2010**, *35* (3), 338–356.

26. Zhu, Y.B.; Gao, C.Y.; Liu, X.Y.; Shen, J.C. Surface modification of polycaprolactone membrane via aminolysis and biomacromolecule immobilization for promoting cytocompatibility of human endothelial cells. Biomacromolecules **2002**, *3* (6), 1312–1319.

27. Park, K.D.; Kim, Y.S.; Han, D.K.; Kim, Y.H.; Lee, E.H.B.; Suh, H.; Choi, K.S. Bacterial adhesion on PEG modified polyurethane surfaces. Biomaterials **1998**, *19* (7–9), 851–859.

28. Jo, S.; Park, K. Surface modification using silanated poly(ethylene glycol)s. Biomaterials **2000**, *21* (6), 605–616.

29. Kingshott, P.; Thissen, H.; Griesser, H.J. Effects of cloud-point grafting, chain length, and density of PEG layers on competitive adsorption of ocular proteins. Biomaterials **2002**, *23* (9), 2043–2056.

30. Chan, C.M.; Ko, T.M.; Hiraoka, H. Polymer surface modification by plasmas and photons. Surf. Sci. Rep. **1996**, *24* (1–2), 1–54.

31. Siow, K.S.; Britcher, L.; Kumar, S.; Griesser, H.J. Plasma methods for the generation of chemically reactive surfaces for biomolecule immobilization and cell colonization—A review. Plasma Process. Polym. **2006**, *3* (6–7), 392–418.

32. Ren, J.; Hua, X.; Zhang, T.; Zhang, Z.; Ji, Z.; Gu, N. Grafting of telechelic poly(lactic-*co*-glycolic acid) onto O_2 plasma-treated polypropylene flakes. J. Appl. Polym. Sci. **2011**, *121* (1), 210–216.

33. Pinto, S.; Alves, P.; Matos, C.M.; Santos, A.C.; Rodrigues, L.R.; Teixeira, J.A.; Gil, M.H. Poly(dimethyl siloxane) surface modification by low pressure plasma to improve its characteristics towards biomedical applications. Colloids Surf. B Biointerfaces **2010**, *81* (1), 20–26.

34. Thom, V.; Jankova, K.; Ulbricht, M.; Kops, J.; Jonsson, G. Synthesis of photoreactive α-4-azidobenzoyl-ω-methoxy-poly(ethylene glycol)s and their end-on photo-grafting onto polysulfone ultrafiltration membranes. Macromol. Chem. Phys. **1998**, *199* (12), 2723–2729.

35. Lazos, D.; Franzka, S.; Ulbricht, M. Size-selective protein adsorption to polystyrene surfaces by self-assembled grafted poly(ethylene glycols) with varied chain lengths. Langmuir **2005**, *21* (19), 8774–8784.

36. Zapotoczny, S. Surface-grafted polymer brushes. In *Biomaterials Surface Science*; Taubert, A., Mano, J.F., Rodríguez-Cabello, J.C., Eds.; Wiley-VCH: Weinheim, 2013; 27–43.

37. Wang, Y.P.; Yuan, K.; Li, Q.L.; Wang, L.P.; Gu, S.J.; Pei, X.W. Preparation and characterization of poly(*N*-isopropylacrylamide) films on a modified glass surface via surface initiated redox polymerization. Mater. Lett. **2005**, *59* (14–15), 1736–1740.

38. Barbey, R.; Lavanant, L.; Paripovic, D.; Schüwer, N.; Sugnaux, C.; Tugulu, S.; Klok, H.A. Polymer brushes via surface-initiated controlled radical polymerization: Synthesis, characterization, properties, and applications. Chem. Rev. **2009**, *109* (11), 5437–5527.

39. Edmondson, S.; Osborne, V.L.; Huck, W.T.S. Polymer brushes via surface-initiated polymerizations. Chem. Soc. Rev. **2004**, *33* (1), 14–22.

40. Jones, D.M.; Huck, W.T.S. Controlled surface-initiated polymerizations in aqueous media. Adv. Mater. **2001**, *13* (16), 1256–1259.

41. Xu, F.J.; Zhong, S.P.; Yung, L.Y.L.; Kang, E.T.; Neoh, K.G. Surface-active and stimuli-responsive polymer-Si(100) hybrids from surface-initiated atom transfer radical polymerization for control of cell adhesion. Biomacromolecules **2004**, *5* (6), 2392–2403.

42. Ejaz, M.; Ohno, K.; Tsujii, Y.; Fukuda, T. Controlled grafting of a well-defined glycopolymer on a solid surface by surface-initiated atom transfer radical polymerization. Macromolecules **2000**, *33* (8), 2870–2874.

43. Malmstrom, E.; Carlmark, A. Controlled grafting of cellulose fibres—An outlook beyond paper and cardboard. Polym. Chem. **2012**, *3* (7), 1702–1713.

44. Yuan, J.; Huang, X.; Li, P.; Li, L.; Shen, J. Surface-initiated RAFT polymerization of sulfobetaine from cellulose membranes to improve hemocompatibility and antibiofouling property. Polym. Chem. **2013**, *4* (19), 5074–5085.

45. Ulbricht, M.; Belfort, G. Surface modification of ultrafiltration membranes by low temperature plasma. 2. Graft polymerization onto polyacrylonitrile and polysulfone. J. Memb. Sci. **1996**, *111* (2), 193–215.

46. Gupta, B.; Hilborn, J.G.; Bisson, I.; Frey, P. Plasma-induced graft polymerization of acrylic acid onto poly(ethylene terephthalate) films. J. Appl. Polym. Sci. **2001**, *81* (12), 2993–3001.

47. Rohr, T.; Ogletree, D.F.; Svec, F.; Fréchet, J.M.J. Surface functionalization of thermoplastic polymers for the fabrication of microfluidic devices by photoinitiated grafting. Adv. Funct. Mater. **2003**, *13* (4), 264–270.

48. Yang, W.T.; Ranby, B. Bulk surface photografting process and its applications. 2. Principal factors affecting surface photografting. J. Appl. Polym. Sci. **1996**, *62* (3), 545–555.

49. Loh, F.; Tan, K.; Kang, E.; Neoh, K.; Pun, M. Near-UV radiation-induced surface graft-copolymerization of some O₃-pretreated conventional polymer-films. Eur. Polym. J. **1995**, *31* (5), 481–488.

50. Adem, E.; Avalos-Borja, M.; Bucio, E.; Burillo, G.; Castillon, F.F.; Cota, L. Surface characterization of binary grafting of AAc/NIPAAm onto poly(tetrafluoroethylene) (PTFE). Nucl. Instrum. Methods Phys. Res. B **2005**, *234* (4), 471–476.

51. Švorčík, V.; Rybka, V.; Stibor, I.; Hnatowicz, V.; Vacík, J.; Stopka, P. Synthesis of grafted polyethylene by ion beam modification. Polym. Degrad. Stabil. **1997**, *58* (1–2), 143–147.

52. Clochard, M.C.; Bègue, J.; Lafon, A.; Caldemaison, D.; Bittencourt, C.; Pireaux, J.J.; Betz, N. Tailoring bulk and surface grafting of poly(acrylic acid) in electron-irradiated PVDF. Polymer **2004**, *45* (26), 8683–8694.

53. Onclin, S.; Ravoo, B.J.; Reinhoudt, D.N. Engineering silicon oxide surfaces using self-assembled monolayers. Angew. Chem. Int. Ed. **2005**, *44* (39), 6282–6304.

54. Tour, J.M.; Jones, L.; Pearson, D.L.; Lamba, J.J.S.; Burgin, T.P.; Whitesides, G.M.; Allara, D.L.; Parikh, A.N.; Atre, S.V. Self-assembled monolayers and multilayers of conjugated thiols, alpha, omega-dithiols, and thioacetyl-containing adsorbates—Understanding attachments between potential molecular wires and gold surfaces. J. Am. Chem. Soc. **1995**, *117* (37), 9529–9534.

55. Schreiber, F. Structure and growth of self-assembling monolayers. Prog. Surf. Sci. **2000**, *65* (5–8), 151–256.

56. Laibinis, P.; Whitesides, G.M; Omega-terminated alkanethiolate monolayers on surfaces of copper, silver, and gold have similar wettabilities. J. Am. Chem. Soc. **1992**, *114* (6), 1990–1995.

57. Choy, K.L. Chemical vapour deposition of coatings. Prog. Mater. Sci. **2003**, *48* (2), 57–170.

58. Sreenivasan, R.; Gleason, K.K. Overview of strategies for the CVD of organic films and functional polymer layers. Chem. Vap. Deposition **2009**, *15* (4–6), 77–90.

59. Coclite, A.M.; Howden, R.M.; Borrelli, D.C.; Petruczok, C.D.; Yang, R.; Yague, J.L.; Ugur, A.; Chen, N.; Lee, S.; Jo, W.J.; Liu, A.; Wang, X.; Gleason, K.K. 25th anniversary article: CVD polymers: A new paradigm for surface modification and device fabrication. Adv. Mater. **2013**, *25* (38), 5392–5423.

60. Chrisey, D.B.; Pique, A.; McGill, R.A.; Horwitz, J.S.; Ringeisen, B.R.; Bubb, D.M.; Wu, P.K.; Laser deposition of polymer and biomaterial films. Chem. Rev. **2003**, *103* (2), 553–576.

61. Lowndes, D.H.; Geohegan, D.B.; Puretzky, A.A.; Norton, D.P.; Rouleau, C.M. Synthesis of novel thin-film materials by pulsed laser deposition. Science **1996**, *273* (5277), 898–903.

62. Pique, A.; McGill, R.A.; Chrisey, D.B.; Leonhardt, D.; Mslna, T.E.; Spargo, B.J.; Callahan, J.H.; Vachet, R.W.; Chung, R.; Bucaro, M.A. Growth of organic thin films by the matrix assisted pulsed laser evaporation (MAPLE) technique. Thin Solid Films **1999**, *355–356*, 536–541, doi:10.1016/S0257-8972(99)00376-X.

63. Mercado, A.L.; Allmond, C.E.; Hoekstra, J.G.; Fitz-Gerald, J.M. Pulsed laser deposition vs. matrix assisted pulsed laser evaporation for growth of biodegradable polymer thin films. Appl. Phys. A Mater. Sci. Process. **2005**, *81* (3), 591–599.

Excipients—Gels

64. Yang, R.; Asatekin, A.; Gleason, K.K. Design of conformal, substrate-independent surface modification for controlled protein adsorption by chemical vapor deposition (CVD). Soft Matter **2012**, *8* (1), 31–43.

65. Alf, M.E.; Asatekin, A.; Barr, M.C.; Baxamusa, S.H.; Chelawat, H.; Ozaydin-Ince, G.; Petruczok, C.D.; Sreenivasan, R.; Tenhaeff, W.E.; Trujillo, N.J.; Vaddiraju, S.; Xu, J.; Gleason, K.K. Chemical vapor deposition of conformal, functional, and responsive polymer films. Adv. Mater. **2010**, *22* (18), 1993–2027.

66. Miyamae, T.; Tsukagoshi, K.; Matsuoka, O.; Yamamoto, S.; Nozoye, H. Preparation of polyimide-polyamide random copolymer thin film by sequential vapor deposition polymerization. Jpn. J. Appl. Phys. **2002**, 41(Pt. 1, No. 2A), 746–748.

67. Chang, Y.C.; Frank, C.W. Vapor deposition-polymerization of alpha-amino acid *N*-carboxy anhydride on the silicon(100) native oxide surface. Langmuir **1998**, *14* (2), 326–334.

68. Lee, N.H.; Frank, C.W. Surface-initiated vapor polymerization of various alpha-amino acids. Langmuir **2003**, *19* (4), 1295–1303.

69. Bhattacharyya, D.; Gleason, K.K. Single-step oxidative chemical vapor deposition of -COOH functional conducting copolymer and immobilization of biomolecule for sensor application. Chem. Mater. **2011**, *23* (10), 2600–2605.

70. Bhattacharyya, D.; Senecal, K.; Marek, P.; Senecal, A.; Gleason, K.K. High surface area flexible chemiresistive biosensor by oxidative chemical vapor deposition. Adv. Funct. Mater. **2011**, *21* (22), 4328–4337.

71. Yaguee, J.L.; Coclite, A.M.; Petruczok, C.; Gleason, K.K. Chemical vapor deposition for solvent-free polymerization at surfaces. Macromol. Chem. Phys. **2013**, *214* (3), 302–312.

72. Chan, K.; Gleason, K.K. Initiated chemical vapor deposition of linear and cross-linked poly(2-hydroxyethyl methacrylate) for use as thin-film hydrogels. Langmuir **2005**, *21* (19), 8930–8939.

73. Mao, Y.; Gleason, K.K. Hot filament chemical vapor deposition of poly(glycidyl methacrylate) thin films using tert-butyl peroxide as an initiator. Langmuir **2004**, *20* (6), 2484–2488.

74. Lau, K.K.S.; Gleason, K.K. Initiated chemical vapor deposition (iCVD) of poly(alkyl acrylates): An experimental study. Macromolecules **2006**, *39* (10), 3688–3694.

75. Murthy, S.K.; Olsen, B.D.; Gleason, K.K. Initiation of cyclic vinylmethylsiloxane polymerization in a hot-filament chemical vapor deposition process. Langmuir **2002**, *18* (16), 6424–6428.

76. Chan, K.; Gleason, K.K. Initiated CVD of poly(methyl methacrylate) thin films. Chem. Vap. Deposition **2005**, *11* (10), 437–443.

77. Coclite, A.M.; Gleason, K.K. Initiated PECVD of organosilicon coatings: A new strategy to enhance monomer structure retention. Plasma Process. Polym. **2012**, *9* (4), 425–434.

78. Chan, K.; Gleason, K.K. Photoinitiated chemical vapor deposition of polymeric thin films using a volatile photoinitiator. Langmuir **2005**, *21* (25), 11773–11779.

79. Martin, T.P.; Sedransk, K.L.; Chan, K.; Baxamusa, S.H.; Gleason, K.K. Solventless surface photoinitiated polymerization: Grafting chemical vapor deposition (gCVD). Macromolecules **2007**, *40* (13), 4586–4591.

80. Wirsen, A.; Sun, H.; Albertsson, A.C. Solvent free vapour phase photografting of acrylamide onto poly(methyl methacrylate). Polymer **2005**, *46* (13), 4554–4561.

81. Wirsen, A.; Sun, H.; Emilsson, L.; Albertsson, A.C. Solvent free vapor phase photografting of maleic anhydride onto poly(ethylene terephthalate) and surface coupling of fluorinated probes, PEG, and an RGD-peptide. Biomacromolecules **2005**, *6* (4), 2281–2289.

82. Shi, F.F. Recent advances in polymer thin films prepared by plasma polymerization synthesis, structural characterization, properties and applications. Surf. Coat. Technol. **1996**, *82* (1–2), 1–15.

83. Yasuda, H.K. Some important aspects of plasma polymerization. Plasma Process. Polym. **2005**, *2* (4), 293–304.

84. Rau, C.; Kulisch, W. Mechanisms of plasma polymerization of various silico-organic monomers. Thin Solid Films **1994**, *249* (1), 28–37.

85. Friedrich, J. Mechanisms of plasma polymerization—Reviewed from a chemical point of view. Plasma Process. Polym. **2011**, *8* (9), 783–802.

86. Calderon, J.G.; Timmons, R.B. Surface molecular tailoring via pulsed plasma-generated acryloyl chloride polymers: Synthesis and reactivity. Macromolecules **1998**, *31* (10), 3216–3224.

87. Favia, P.; d'Agostino, R. Plasma treatments and plasma deposition of polymers for biomedical applications. Surf. Coat. Technol. **1998**, *98* (1–3), 1102–1106.

88. Ertel, S.I.; Ratner, B.D.; Horbett, T.A. Radiofrequency plasma deposition of oxygen-containing films on polystyrene and poly(ethylene terephthalate) substrates improves endothelial cell growth. J. Biomed. Mater. Res. **1990**, *24* (12), 1637–1659.

89. Yeh, Y.S.; Iriyama, Y.; Matsuzawa, Y.; Hanson, S.R.; Yasuda, H. Blood compatibility of surfaces modified by plasma polymerization. J. Biomed. Mater. Res. **1988**, *22* (9), 795–818.

90. Wu, Y.L.J.; Timmons, R.B.; Jen, J.S.; Molock, F.E. Non-fouling surfaces produced by gas phase pulsed plasma polymerization of an ultra low molecular weight ethylene oxide containing monomer. Colloids Surf. B Biointerfaces **2000**, *18* (3–4), 235–248.

91. Sardella, E.; Gristina, R.; Senesi, G.S.; d'Agostino, R.; Favia, P. Homogeneous and micro-patterned plasma-deposited peo-like coatings for biomedical surfaces. Plasma Process. Polym. **2004**, *1* (1), 63–72.

92. Shen, M.C.; Wagner, M.S.; Castner, D.G.; Ratner, B.D.; Horbett, T.A. Multivariate surface analysis of plasma-deposited tetraglyme for reduction of protein adsorption and monocyte adhesion. Langmuir **2003**, *19* (5), 1692–1699.

93. Detomaso, L.; Gristina, R.; Senesi, G.S.; d'Agostino, R.; Favia, P. Stable plasma-deposited acrylic acid surfaces for cell culture applications. Biomaterials **2005**, *26* (18), 3831–3841.

94. Truica-Marasescu, F.; Girard-Lauriault, P.L.; Lippitz, A.; Unger, W.E.S.; Wertheimer, M.R. Nitrogen-rich plasma polymers: Comparison of films deposited in

atmospheric- and low-pressure plasmas. Thin Solid Films **2008**, *516* (21), 7406–7417.

95. Ryan, M.E.; Hynes, A.M.; Badyal, J.P.S. Pulsed plasma polymerization of maleic anhydride. Chem. Mater. **1996**, *8* (1), 37–42.

96. Jenkins, A.T.A.; Hu, J.; Wang, Y.Z.; Schiller, S.; Foerch, R.; Knoll, W. Pulsed plasma deposited maleic anhydride thin films as supports for lipid bilayers. Langmuir **2000**, *16* (16), 6381–6384.

97. Schiller, S.; Hu, J.; Jenkins, A.T.A.; Timmons, R.B.; Sanchez-Estrada, F.S.; Knoll, W.; Förch, R. Chemical structure and properties of plasma-polymerized maleic anhydride films. Chem. Mater. **2002**, *14* (1), 235–242.

98. Siffer, F.; Ponche, A.; Fioux, P.; Schultz, J.; Roucoules, V.A chemometric investigation of the effect of the process parameters during maleic anhydride pulsed plasma polymerization. Anal. Chim. Acta **2005**, *539* (1–2), 289–299.

99. Fally, F.; Doneux, C.; Riga, J.; Verbist, J. Quantification of the functional-groups present at the surface of plasma polymers deposited from propylamine, allylamine, and propargylamine. J. Appl. Polym. Sci. **1995**, *56* (5), 597–614.

100. Chen, Q.; Förch, R.; Knoll, W. Characterization of pulsed plasma polymerization allylamine as an adhesion layer for DNA adsorption/hybridization. Chem. Mater. **2004**, *16* (4), 614–620.

101. Szwarc, M. Some remarks on the p-xylene molecule. Discuss. Faraday Soc. **1947**, *2* (0), 46–49.

102. Vaeth, K.M.; Jensen, K.F. Chemical vapor deposition of poly(p-phenylene vinylene) based light emitting diodes with low turn-on voltages. Appl. Phys. Lett. **1997**, *71* (15), 2091–2093.

103. Vaeth, K.M.; Jensen, K.F. Chemical vapor deposition of thin polymer films used is polymer-based light emitting diodes. Adv. Mater. **1997**, *9* (6), 490–493.

104. Vaeth, K.M.; Jensen, K.F. Poly(p-phenylene vinylene) prepared by chemical vapor deposition: Influence of monomer selection and reaction conditions on film composition and luminescence properties. Macromolecules **1998**, *31* (20), 6789–6793.

105. Kubono, A.; Kitoh, T.; Kajikawa, K.; Umemoto, S.; Takezoe, H.; Fukuda, A.; Okui, N. 2nd-Harmonic generation in poly(vinylidene fluoride) films prepared by vapor-deposition under an electric-field. Jpn. J. Appl. Phys. Pt 2 Lett. **1992**, *31* (8B), L1195–L1197.

106. Lau, K.K.S.; Caulfield, J.A.; Gleason, K.K. Structure and morphology of fluorocarbon films grown by hot filament chemical vapor deposition. Chem. Mater. **2000**, *12* (10), 3032–3037.

107. Limb, S.J.; Labelle, C.B.; Gleason, K.K.; Edell, D.J.; Gleason, E.F. Growth of fluorocarbon polymer thin films with high CF_2 fractions and low dangling bond concentrations by thermal chemical vapor deposition. Appl. Phys. Lett. **1996**, *68* (20), 2810–2812.

108. Murthy, S.K.; Gleason, K.K. Fluorocarbon-organosilicon copolymer synthesis by hot filament chemical vapor deposition. Macromolecules **2002**, *35* (5), 1967–1972.

109. Lau, K.K.S.; Lewis, H.G.P.; Limb, S.J.; Kwan, M.C.; Gleason, K.K. Hot-wire chemical vapor deposition (HWCVD) of fluorocarbon and organosilicon thin films. Thin Solid Films **2001**, *395* (1–2), 288–291.

110. Lewis, H.G.P.; Bansal, N.P.; White, A.J.; Handy, E.S. HWCVD of polymers: Commercialization and scale-up. Thin Solid Films **2009**, *517* (12), 3551–3554.

111. Iguchi, N.; Kasanuki, H.; Matsuda, N.; Shoda, M.; Ohnishi, S.; Hosoda, S. Contact sensitivity to polychloro-paraxylene-coated cardiac pacemaker. Pacing Clin. Electrophysiol. **1997**, *20* (2), 372–373.

112. Takeuchi, S.; Ziegler, D.; Yoshida, Y.; Mabuchi, K.; Suzuki, T. Parylene flexible neural probes integrated with microfluidic channels. Lab Chip **2005**, *5* (5), 519–523.

113. McKamey, R.P.; Whitley, J.Q.; Kusy, R.P. Physical and mechanical characteristics of a chlorine-substituted poly(para-xylylene) coating on orthodontic chain modules. J. Mater. Sci. Mater. Med. **2000**, *11* (7), 407–419.

114. Weisenberg, B.A.; Mooradian, D.L. Hemocompatibility of materials used in microelectromechanical systems: Platelet adhesion and morphology *in vitro*. J. Biomed. Mater. Res. **2002**, *60* (2), 283–291.

115. Simon, P.; Mang, S.; Hasenhindl, A.; Gronski, W.; Greiner A. Poly(p-xylylene) and its derivatives by chemical vapor deposition: Synthesis, mechanism, and structure. Macromolecules **1998**, *31* (25), 8775–8780.

116. You, L.; Yang, G.; Lang, C.; Moore, J.; Wu, P.; Mcdonald, J.F; Lu, T.M. Vapor-deposition of parylene-F by pyrolysis of dibromotetrafluoro-P-Xylene. J. Vac. Sci. Technol. A **1993**, *11* (6), 3047–3052.

117. Schmidt, C.; Stümpflen, V.; Wendorff, J.H.; Hasenhindl, A.; Gronski, W.; Ishaque, M.; Greiner, A. Structural analysis of PPX prepared by vapor phase pyrolysis of [2.2]para-cyclophane. Acta Polym. **1998**, *49* (5), 232–235.

118. Gorham, W.F. A new, general synthetic method for the preparation of linear poly-p-xylylenes. J. Polym. Sci. A1 Polym. Chem. **1966**, *4* (12), 3027–3039.

119. Lahann, J. Vapor-based polymer coatings for potential biomedical applications. Polym. Int. **2006**, *55* (12), 1361–1370.

120. Lahann, J.; Langer, R. Surface-initiated ring-opening polymerization of epsilon-caprolactone from a patterned poly(hydroxymethyl-p-xylylene). Macromol. Rapid Commun. **2001**, *22* (12), 968–971.

121. Hopf, H. [2.2]Paracyclophanes in polymer chemistry and materials science. Angew. Chem. Int. Ed. **2008**, *47* (51), 9808–9812.

122. Lahann, J.; Klee, D.; Hocker, H. Chemical vapour deposition polymerization of substituted [2.2]paracyclophanes. Macromol. Rapid Commun. **1998**, *19* (9), 441–444.

123. Chen, H.Y.; Lahann, J. Designable biointerfaces using vapor-based reactive polymers. Langmuir **2011**, *27* (1), 34–48.

124. Lahann, J.; Hocker, H.; Langer, R. Synthesis of amino[2.2] paracyclophanes—Beneficial monomers for bioactive coating of medical implant materials. Angew. Chem. Int. Ed. **2001**, *40* (4), 726–728.

125. Lahann, J.; Balcells, M.; Rodon, T.; Lee, J.; Choi, I.S.; Jensen, K.F.; Langer, R. Reactive polymer coatings: A platform for patterning proteins and mammalian cells onto a broad range of materials. Langmuir **2002**, *18* (9), 3632–3638.

126. Lahann, J.; Choi, I.S.; Lee, J.; Jensen, K.F.; Langer, R. A new method toward microengineered surfaces based on

reactive coating. Angew. Chem. Int. Ed. **2001**, *40* (17), 3166–3169.

127. Elkasabi, Y.; Nandivada, H.; Chen, H.Y.; Bhaskar, S.; d'Arcy, J.; Bondarenko, L.; Lahann, J. Partially fluorinated poly-p-xylylenes synthesized by CVD polymerization. Chem. Vap. Deposition **2009**, *15* (4–6), 142–149.

128. Nandivada, H.; Chen, H.Y.; Lahann, J. Vapor-based synthesis of poly [(4-formyl-p-xylylene)-*co*-(p-xylylene)] and its use for biomimetic surface modifications. Macromol. Rapid Commun. **2005**, *26* (22), 1794–1799.

129. Nandivada, H.; Chen, H.Y.; Bondarenko, L.; Lahann, J. Reactive polymer coatings that " click". Angew. Chem. Int. Ed. **2006**, *45* (20), 3360–3363.

130. Elkasabi, Y.; Chen, H.Y.; Lahann, J. Multipotent polymer coatings based on chemical vapor deposition copolymerization. Adv. Mater. **2006**, *18* (12), 1521–1526.

131. Deng, X.; Eyster, T.W.; Elkasabi, Y.; Lahann, J. Bio-orthogonal polymer coatings for co-presentation of biomolecules. Macromol. Rapid Commun. **2012**, *33* (8), 640–645.

132. Jiang, X.; Chen, H.Y.; Galvan, G.; Yoshida, M.; Lahann, J. Vapor-based initiator coatings for atom transfer radical polymerization. Adv. Funct. Mater. **2008**, *18* (1), 27–35.

133. Deng, X.; Friedmann, C.; Lahann, J. Bio-orthogonal "Double-Click" Chemistry based on multifunctional coatings. Angew. Chem. Int. Ed. **2011**, *50* (29), 6522–6526.

134. Bally, F.; Cheng, K.; Nandivada, H.; Deng, X.; Ross, A.M.; Panades, A.; Lahann, J. Co-immobilization of biomolecules on ultrathin reactive chemical vapor deposition coatings using multiple click chemistry strategies. ACS Appl. Mater. Interfaces **2013**, *5* (19), 9262–9268.

135. Ross, A.M.; Jiang, Z.; Bastmeyer, M.; Lahann, J. Physical aspects of cell culture substrates: Topography, roughness, and elasticity. Small **2012**, *8* (3), 336–355.

136. Ross, A.M.; Lahann, J. Surface engineering the cellular microenvironment via patterning and gradients. J. Polym. Sci. B Polym. Phys. **2013**, *51* (10), 775–794.

137. Flemming, R.G.; Murphy, C.J.; Abrams, G.A.; Goodman, S.L.; Nealey, P.F. Effects of synthetic micro-and nano-structured surfaces on cell behavior. Biomaterials **1999**, *20* (6), 573–588.

138. Blattler, T.; Huwiler, C.; Ochsner, M.; Staedler, B.; Solak, H.; Voeroes, J.; Grandin, H.M. Nanopatterns with biological functions. J. Nanosci. Nanotechnol. **2006**, *6* (8), 2237–2264.

139. Arnold, M.; Hirschfeld-Warneken, V.C.; Lohmueller, T.; Heil, P.; Bluemmel, J.; Cavalcanti-Adam, E.A.; López-García, M.; Walther, P.; Kessler, H.; Geiger, B.; Spatz, J.P. Induction of cell polarization and migration by a gradient of nanoscale variations in adhesive ligand spacing. Nano Lett. **2008**, *8* (7), 2063–2069.

140. Kane, R.S.; Takayama, S.; Ostuni, E.; Ingber, D.E.; Whitesides, G.M. Patterning proteins and cells using soft lithography. Biomaterials **1999**, *20* (23–24), 2363–2376.

141. Whitesides, G.M.; Ostuni, E.; Takayama, S.; Jiang, X.Y.; Ingber, D.E. Soft lithography in biology and biochemistry. Annu. Rev. Biomed. Eng. **2001**, *3* (1), 335–373.

142. Falconnet, D.; Csucs, G.; Grandin, H.M.; Textor, M. Surface engineering approaches to micropattern surfaces for cell-based assays. Biomaterials **2006**, *27* (16), 3044–3063.

143. Piner, R.D.; Zhu, J.; Xu, F.; Hong, S.H.; Mirkin, C.A. "Dip-pen" nanolithography. Science **1999**, *283* (5402), 661–663.

144. Chou, S.Y.; Krauss, P.R.; Renstrom, P.J. Imprint lithography with 25-nanometer resolution. Science **1996**, *272* (5258), 85–87.

145. Yang, S.M.; Jang, S.G.; Choi, D.G.; Kim, S.; Yu, H.K. Nanomachining by colloidal lithography. Small **2006**, *2* (4), 458–475.

146. Xia, Y.N.; Whitesides, G.M. Soft lithography. Annu. Rev. Mater. Sci. **1998**, *28* (1), 153–184.

147. Vaeth, K.M.; Jensen, K.F. Transition metals for selective chemical vapor deposition of parylene-based polymers. Chem. Mater. **2000**, *12* (5), 1305–1313.

148. Vaeth, K.M.; Jackman, R.J.; Black, A.J.; Whitesides, G.M.; Jensen, K.F. Use of microcontact printing for generating selectively grown films of poly(p-phenylene vinylene) and parylenes prepared by chemical vapor deposition. Langmuir **2000**, *16* (22), 8495–8500.

149. Delamarche, E.; Bernard, A.; Schmid, H.; Michel, B.; Biebuyck, H. Patterned delivery of immunoglobulins to surfaces using microfluidic networks. Science **1997**, *276* (5313), 779–781.

150. Chen, H.Y.; Lahann, J. Vapor-assisted micropatterning in replica structures: A solventless approach towards topologically and chemically designable surfaces. Adv. Mater. **2007**, *19* (22), 3801–3808.

151. Chen, H.Y.; Lahann, J. Fabrication of discontinuous surface patterns within microfluidic channels using photodefinable vapor-based polymer coatings. Anal. Chem. **2005**, *77* (21), 6909–6914.

152. Chen, H.Y.; Rouillard, J.M.; Gulari, E.; Lahann, J. Colloids with high-definition surface structures. Proc. Natl. Acad. Sci. U.S.A. **2007**, *104* (27), 11173–11178.

153. Klein, F.; Richter, B.; Striebel, T.; Franz, C.M.; von Freymann, G.; Wegener, M.; Bastmeyer, M. Two-component polymer scaffolds for controlled three-dimensional cell culture. Adv. Mater. **2011**, *23* (11), 1341–1345.

154. Chen, H.Y.; Hirtz, M.; Deng, X.; Laue, T.; Fuchs, H.; Lahann, J. Substrate-independent dip-pen nanolithography based on reactive coatings. J. Am. Chem. Soc. **2010**, *132* (51), 18023–18025.

155. Chou, S.Y.; Krauss, P.R.; Renstrom, P.J. Nanoimprint lithography. J. Vac. Sci. Technol. B **1996**, *14* (6), 4129–4133.

156. Guo, L.J. Nanoimprint lithography: Methods and material requirements. Adv. Mater. **2007**, *19* (4), 495–513.

157. Sardella, E.; Favia, P.; Gristina, R.; Nardulli, M.; d'Agostino, R. Plasma-aided micro-and nanopatterning processes for biomedical applications. Plasma Process. Polym. **2006**, *3* (6–7), 456–469.

158. Morgenthaler, S.; Zink, C.; Spencer, N.D. Surface-chemical and -morphological gradients. Soft Matter **2008**, *4* (3), 419–434.

159. Kim, M.S.; Khang, G.; Lee, H.B. Gradient polymer surfaces for biomedical applications. Prog. Polym. Sci. **2008**, *33* (1), 138–164.

160. Hypolite, C.L.; McLernon, T.L.; Adams, D.N.; Chapman, K.E.; Herbert, C.B.; Huang, C.C.; Distefano, M.D.; Hu, W.S. Formation of microscale gradients of protein using heterobifunctional photolinkers. Bioconjug. Chem. **1997**, *8* (5), 658–663.

161. Herbert, C.B.; McLernon, T.L.; Hypolite, C.L.; Adams, D.N.; Pikus, L.; Huang, C.C.; Fields, G.B.; Letourneau, P.C.; Distefano, M.D.; Hu, W.S. Micropatterning gradients and controlling surface densities of photoactivatable biomolecules on self-assembled monolayers of oligo(ethylene glycol) alkanethiolates. Chem. Biol. **1997**, *4* (10), 731–737.

162. Jeon, N.L.; Dertinger, S.K.W.; Chiu, D.T.; Choi, I.S.; Stroock, A.D.; Whitesides, G.M. Generation of solution and surface gradients using microfluidic systems. Langmuir **2000**, *16* (22), 8311–8316.

163. Millet, L.J.; Stewart, M.E.; Nuzzo, R.G.; Gillette, M.U. Guiding neuron development with planar surface gradients of substrate cues deposited using microfluidic devices. Lab Chip **2010**, *10* (12), 1525–1535.

164. Zelzer, M.; Majani, R.; Bradley, J.W.; Rose, F.R.A.J.; Davies, M.C.; Alexander, M.R. Investigation of cell-surface interactions using chemical gradients formed from plasma polymers. Biomaterials **2008**, *29* (2), 172–184.

165. Bally, M.; Halter, M.; Vörös, J.; Grandin, H.M. Optical microarray biosensing techniques. Surf. Interface Anal. **2006**, *38* (11), 1442–1458.

166. Hartwell, S.; Grudpan, K. Flow based immuno/bioassay and trends in micro-immuno/biosensors. Microchim. Acta **2010**, *169* (3–4), 201–220.

167. Fabre, R.M.; Talham, D.R. Stable supported lipid bilayers on zirconium phosphonate surfaces. Langmuir **2009**, *25* (21), 12644–12652.

168. Scarano, S.; Mascini, M.; Turner, A.P.F.; Minunni, M. Surface plasmon resonance imaging for affinity-based biosensors. Biosens. Bioelectron. **2010**, *25* (5), 957–966.

169. Ross, A.M.; Zhang, D.; Deng, X.; Chang, S.L.; Lahann, J. Chemical-vapor-deposition-based polymer substrates for spatially resolved analysis of protein binding by imaging ellipsometry. Anal. Chem. **2011**, *83* (3), 874–880.

170. Peng, G.; Hakim, M.; Broza, Y.Y.; Billan, S.; Abdah-Bortnyak, R.; Kuten, A.; Tisch. U.; Haick, H. Detection of lung, breast, colorectal, and prostate cancers from exhaled breath using a single array of nanosensors. Br. J. Cancer **2010**, *103* (4), 542–551.

171. Phillips, M.; Gleeson, K.; Hughes, J.M.B.; Greenberg, J.; Cataneo, R.N.; Baker, L.; McVay, W.P. Volatile organic compounds in breath as markers of lung cancer: A cross-sectional study. Lancet **1999**, *353* (9168), 1930–1933.

172. Andreeva, N.; Ishizaki, T.; Baroch, P.; Saito, N. High sensitive detection of volatile organic compounds using superhydrophobic quartz crystal microbalance. Sens. Actuators B Chem. **2012**, *164* (1), 15–21.

173. Han, H.C.; Chang, Y.R.; Hsu, W.L.; Chen, C.Y. Application of parylene-coated quartz crystal microbalance for on-line real-time detection of microbial populations. Biosens. Bioelectron. **2009**, *24* (6), 1543–1549.

174. Mark, K.; Park, J.; Bauer, S.; Schmuki, P. Nanoscale engineering of biomimetic surfaces: Cues from the extracellular matrix. Cell Tissue Res. **2010**, *339* (1), 131–153.

175. Geiger, B.; Spatz, J.P.; Bershadsky, A.D. Environmental sensing through focal adhesions. Nat. Rev. Mol. Cell Biol. **2009**, *10* (1), 21–33.

176. Sniadecki, N.J.; Anguelouch, A.; Yang, M.T.; Lamb, C.M.; Liu, Z.; Kirschner, S.B.; Liu, Y.; Reich, D.H.; Chen, C.S. Magnetic microposts as an approach to apply forces to living cells. Proc. Natl. Acad. Sci. U.S.A. **2007**, *104* (37), 14553–14558.

177. Dalby, M.J.; Riehle, M.O.; Sutherland, D.S.; Agheli, H.; Curtis, A.S.G. Use of nanotopography to study mechanotransduction in fibroblasts—Methods and perspectives. Eur. J. Cell Biol. **2004**, *83* (4), 159–169.

178. Lutolf, M.P.; Hubbell, J.A. Synthetic biomaterials as instructive extracellular microenvironments for morphogenesis in tissue engineering. Nat. Biotechnol. **2005**, *23* (1), 47–55.

179. Julthongpiput, D.; Fasolka, M.J.; Zhang, W.; Nguyen, T.; Amis, E.J. Gradient chemical micropatterns: A reference substrate for surface nanometrology. Nano Lett. **2005**, *5* (8), 1535–1540.

180. Potyrailo, R.A.; Hassib, L. Analytical instrumentation infrastructure for combinatorial and high-throughput development of formulated discrete and gradient polymeric sensor materials arrays. Rev. Sci. Instrum. **2005**, *76* (6), 062225.

181. Li, X.; Hou, S.; Feng, X.; Yu, Y.; Ma, J.; Li, L. Patterning of neural stem cells on poly(lactic-*co*-glycolic acid) film modified by hydrophobin. Colloids Surfaces B Biointerfaces **2009**, *74* (1), 370–374.

182. Teng, Y.D.; Lavik, E.B.; Qu, X.; Park, K.I.; Ourednik, J.; Zurakowski, D.; Langer, R.; Snyder, E.Y. Functional recovery following traumatic spinal cord injury mediated by a unique polymer scaffold seeded with neural stem cells. Proc. Natl. Acad. Sci. **2002**, *99* (5), 3024–3029.

183. Wang, L.; Liu, L.; Li, X.; Magome, N.; Agladze, K.; Chen, Y. Multi-electrode monitoring of guided excitation in patterned cardiomyocytes. Microelectron. Eng. **2013**, *111* (0), 267–271.

184. Hahn, M.S.; Taite, L.J.; Moon, J.J.; Rowland, M.C.; Ruffino, K.A.; West, J.L. Photolithographic patterning of polyethylene glycol hydrogels. Biomaterials **2006**, *27* (12), 2519–2524.

185. Kolodziej, C.M.; Kim, S.H.; Broyer, R.M.; Saxer, S.S.; Decker, C.G.; Maynard, H.D. Combination of integrin-binding peptide and growth factor promotes cell adhesion on electron-beam-fabricated patterns. J. Am. Chem. Soc. **2011**, *134* (1), 247–255.

186. Dos Reis, G.; Fenili, F.; Gianfelice, A.; Bongiorno, G.; Marchesi, D.; Scopelliti, P.E.; Borgonovo, A.; Podestà, A.; Indrieri, M.; Ranucci, E.; Ferruti, P.; Lenardi, C.; Milani, P. Direct microfabrication of topographical and chemical cues for the guided growth of neural cell networks on polyamidoamine hydrogels. Macromol. Biosci. **2010**, *10* (8), 842–852.

187. Sekula, S.; Fuchs, J.; Weg-Remers, S.; Nagel, P.; Schuppler, S.; Fragala, J.; Theilacker, N.; Franzreb, M.; Wingren, C.; Ellmark, P.; Borrebaeck, C.A.K.; Mirkin, C.A.; Fuchs, H.; Lenhert, S. Multiplexed lipid dip-pen nanolithography on subcellular scales for the templating of functional proteins and cell culture. Small **2008**, *4* (10), 1785–1793.

188. Kim, J.; Shin, Y.H.; Yun, S.H.; Choi, D.S.; Nam, J.H.; Kim, S.R.; Moon, S.K.; Chung, B.H.; Lee, J.H.; Kim, J.H.; Kim, K.Y.; Kim, K.M.; Lim, J.H. Direct-write patterning of bacterial cells by dip-pen nanolithography. J. Am. Chem. Soc. **2012**, *134* (40), 16500–16503.

Excipients—Gels

189. Senesi, A.J.; Rozkiewicz, D.I.; Reinhoudt, D.N.; Mirkin, C.A. Agarose-assisted dip-pen nanolithography of oligonucleotides and proteins. ACS Nano **2009**, *3* (8), 2394–2402.

190. Keenan, T.M.; Folch, A. Biomolecular gradients in cell culture systems. Lab Chip **2008**, *8* (1), 34–57.

191. Lagunas, A.; Comelles, J.; Martínez, E.; Samitier, J. Universal chemical gradient platforms using poly(methyl methacrylate) based on the biotin–streptavidin interaction for biological applications. Langmuir **2010**, *26* (17), 14154–14161.

192. Genzer, J. Surface-bound gradients for studies of soft materials behavior. Annu. Rev. Mater. Res. **2012**, *42* (1), 435–468.

193. Harris, B.P.; Kutty, J.K.; Fritz, E.W.; Webb, C.K.; Burg, K.J.L.; Metters, A.T. Photopatterned polymer brushes promoting cell adhesion gradients. Langmuir **2006**, *22* (10), 4467–4471.

194. Elkasabi, Y.; Lahann, J. Vapor-based polymer gradients. Macromol. Rapid Commun. **2009**, *30* (1), 57–63.

195. Elkasabi, Y.M.; Lahann, J.; Krebsbach, P.H. Cellular transduction gradients via vapor-deposited polymer coatings. Biomaterials **2011**, *32* (7), 1809–1815.

196. Dertinger, S.K.W.; Jiang, X.; Li, Z.; Murthy, V.N.; Whitesides, G.M. Gradients of substrate-bound laminin orient axonal specification of neurons. Proc. Natil. Acad. Sci. **2002**, *99* (20), 12542–12547.

197. Lühmann, T.; Hänseler, P.; Grant, B.; Hall, H. The induction of cell alignment by covalently immobilized gradients of the 6th Ig-like domain of cell adhesion molecule L1 in 3D-fibrin matrices. Biomaterials **2009**, *30* (27), 4503–4512.

198. Dodla, M.C.; Bellamkonda, R.V. Differences between the effect of anisotropic and isotropic laminin and nerve growth factor presenting scaffolds on nerve regeneration across long peripheral nerve gaps. Biomaterials **2008**, *29* (1), 33–46.

199. Wong, A.P.; Perez-Castillejos, R.; Love, J.C.; Whitesides, G.M. Partitioning microfluidic channels with hydrogel to construct tunable 3-D cellular microenvironments. Biomaterials **2008**, *29* (12), 1853–1861.

200. Dodla, M.C.; Bellamkonda, R.V. Anisotropic scaffolds facilitate enhanced neurite extension *in vitro*. J. Biomed. Mater. Res. A **2006**, *78A* (2), 213–221.

201. Phillips, J.E.; Petrie, T.A.; Creighton, F.P.; García, A.J. Human mesenchymal stem cell differentiation on self-assembled monolayers presenting different surface chemistries. Acta Biomater. **2010**, *6* (1), 12–20.

202. Markiewicz, I.; Sypecka, J.; Domanska-Janik, K.; Wyszomirski, T.; Lukomska, B. Cellular environment directs differentiation of human umbilical cord blood–derived neural stem cells *in vitro*. J. Histochem. Cytochem. **2011**, *59* (3), 289–301.

203. Namgung, S.; Baik, K.Y.; Park, J.; Hong, S. Controlling the growth and differentiation of human mesenchymal stem cells by the arrangement of individual carbon nanotubes. ACS Nano **2011**, *5* (9), 7383–7390.

204. Nelson, T.J.; Martinez-Fernandez, A.; Terzic, A. Induced pluripotent stem cells: Developmental biology to regenerative medicine. Nat. Rev. Cardiol. **2010**, *7* (12), 700–710.

205. Xu, C.; Police, S.; Rao, N.; Carpenter, M.K. Characterization and enrichment of cardiomyocytes derived from human embryonic stem cells. Circ. Res. **2002**, *91* (6), 501–508.

206. Bearzi, C.; Rota, M.; Hosoda, T.; Tillmanns, J.; Nascimbene, A.; De Angelis, A.; Yasuzawa-Amano, S.; Trofimova, I.; Siggins, R.W.; LeCapitaine, N.; Cascapera, S.; Beltrami, A.P.; D'Alessandro, D.A.; Zias, E.; Quaini, F.; Urbanek, C.; Michler, R.E.; Bolli, R.; Kajstura, J.; Leri, A.; Anversa, P. Human cardiac stem cells. Proc. Natl. Acad. Sci. **2007**, *104* (35), 14068–14073.

207. Ramiya, V.K.; Maraist, M.; Arfors, K.E.; Schatz, D.A.; Peck, A.B.; Cornelius, J.G. Reversal of insulin-dependent diabetes using islets generated *in vitro* from pancreatic stem cells. Nat. Med. **2000**, *6* (3), 278–282.

208. Urbán, V.S.; Kiss, J.; Kovács, J.; Gócza, E.; Vas, V.; Monostori, É.; Uher, F. Mesenchymal stem cells cooperate with bone marrow cells in therapy of diabetes. Stem Cells **2008**, *26* (1), 244–253.

209. Nagaoka, M.; Si-Tayeb, K.; Akaike, T.; Duncan, S. Culture of human pluripotent stem cells using completely defined conditions on a recombinant E-cadherin substratum. BMC Develop. Biol. **2010**, *10* (1), 60.

210. Valamehr, B.; Tsutsui, H.; Ho, C.M.; Wu, H. Developing defined culture systems for human pluripotent stem cells. Reg. Med. **2011**, *6* (5), 623–634.

211. Azarin, S.M.; Palecek, S.P. Development of scalable culture systems for human embryonic stem cells. Biochem. Eng. J. **2010**, *48* (3), 378–384.

212. Ilic, D. Culture of human embryonic stem cells and the extracellular matrix microenvironment. Reg. Med. **2006**, *1* (1), 95–101.

213. Wei, Y.; Li, B.; Fu, C.; Qi, H. Electroactive conducting polymers for biomedical applications. Acta Polym. Sin. **2010**, *0* (12), 1399–1405, doi:10.3724/SP.J.1105.2010.10194.

214. Colter, D.C.; Class, R.; DiGirolamo, C.M.; Prockop, D.J. Rapid expansion of recycling stem cells in cultures of plastic-adherent cells from human bone marrow. Proc. Natl. Acad. Sci. U.S.A. **2000**, *97* (7), 3213–3218.

215. Nakaji-Hirabayashi, T.; Kato, K.; Arima, Y.; Iwata, H. Oriented immobilization of epidermal growth factor onto culture substrates for the selective expansion of neural stem cells. Biomaterials **2007**, *28* (24), 3517–3529.

216. Villa-Diaz, L.G.; Nandivada, H.; Ding, J.; Nogueira-de-Souza, N.C.; Krebsbach, P.H.; O'Shea, K.S.; Lahann, J.; Smith, G.D. Synthetic polymer coatings for long-term growth of human embryonic stem cells. Nat. Biotechnol. **2010**, *28* (6), 581–583.

217. Nandivada, H.; Villa-Diaz, L.G.; O'Shea, K.S.; Smith, G.D.; Krebsbach, P.H.; Lahann, J. Fabrication of synthetic polymer coatings and their use in feeder-free culture of human embryonic stem cells. Nat. Protoc. **2011**, *6* (7), 1037–1043.

218. Villa-Diaz, L.G.; Brown, S.E.; Liu, Y.; Ross, A.M.; Lahann, J.; Parent, J.M.; Krebsbach, P.H. Derivation of mesenchymal stem cells from human induced pluripotent stem cells cultured on synthetic substrates. Stem Cells **2012**, *30* (6), 1174–1181.

219. Statz, A.R.; Meagher, R.J.; Barron, A.E.; Messersmith, P.B. New peptidomimetic polymers for antifouling surfaces. J. Am. Chem. Soc. **2005**, *127* (22), 7972–7973.

220. Kirschner, C.M.; Brennan, A.B. Bio-Inspired antifouling strategies. Annu. Rev. Mater. Res. **2012**, *42* (1), 211–229.

221. Statz, A.R.; Barron, A.E.; Messersmith, P.B. Protein, cell and bacterial fouling resistance of polypeptoid-modified surfaces: Effect of side-chain chemistry. Soft Matter **2008**, *4* (1), 131–139.

222. Ratner, B.D. The blood compatibility catastrophe. J. Biomed. Mater. Res. A **1993**, *27* (3), 283–287.

223. Munn, C.B. *Marine Microbiology: Ecology and Applications*; BIOS Scientific Publishers: Abingdon, UK, 2004.

224. Zhao, C.; Li, L.; Zheng, J. Achieving highly effective nonfouling performance for surface-grafted poly(HPMA) via atom-transfer radical polymerization. Langmuir **2010**, *26* (22), 17375–17382.

225. Mrabet, B.; Nguyen, M.N.; Majbri, A.; Mahouche, S.; Turmine, M.; Bakhrouf, A.; Chehimi, M. Anti-fouling poly(2-hydroxyethyl methacrylate) surface coatings with specific bacteria recognition capabilities. Surf. Sci. **2009**, *603* (16), 2422–2429.

226. Zhao, C.; Li, L.; Wang, Q.; Yu, Q.; Zheng, J. Effect of film thickness on the antifouling performance of poly(hydroxy-functional methacrylates) grafted surfaces. Langmuir **2011**, *27* (8), 4906–4913.

227. Ko, M.; Pellegrino, J.; Nassimbene, R.; Marko, P. Characterization of the adsorption-fouling layer using globular-proteins. J. Memb. Sci. **1993**, *76* (2–3), 101–120.

228. Matsuda, M.; Yamamoto, K.; Yakushiji, T.; Fukuda, M.; Miyasaka, T.; Sakai, K. Nanotechnological evaluation of protein adsorption on dialysis membrane surface hydrophilized with polyvinylpyrrolidone. J. Memb. Sci. **2008**, *310* (1–2), 219–228.

229. Wu, Z.; Chen, H.; Liu, X.; Zhang, Y.; Li, D.; Huang, H. Protein adsorption on poly(*N*-vinylpyrrolidone)-modified silicon surfaces prepared by surface-initiated atom transfer radical polymerization. Langmuir **2009**, *25* (5), 2900–2906.

230. Xu, F.J.; Neoh, K.G.; Kang, E.T. Bioactive surfaces and biomaterials via atom transfer radical polymerization. Prog. Polym. Sci. **2009**, *34* (8), 719–761.

231. Futamura, K.; Matsuno, R.; Konno, T.; Takai, M.; Ishihara, K. Rapid development of hydrophilicity and protein adsorption resistance by polymer surfaces bearing phosphorylcholine and naphthalene groups. Langmuir **2008**, *24* (18), 10340–10344.

232. Jiang, S.; Cao, Z. Ultralow-fouling, functionalizable, and hydrolyzable zwitterionic materials and their derivatives for biological applications. Adv. Mater. **2010**, *22* (9), 920–932.

233. Zhang, Z.; Vaisocherová, H.; Cheng, G.; Yang, W.; Xue, H.; Jiang, S. Nonfouling behavior of polycarboxybetaine-grafted surfaces: Structural and environmental effects. Biomacromolecules **2008**, *9* (10), 2686–2692.

234. Ladd, J.; Zhang, Z.; Chen, S.; Hower, J.C.; Jiang, S. Zwitterionic polymers exhibiting high resistance to nonspecific protein adsorption from human serum and plasma. Biomacromolecules **2008**, *9* (5), 1357–1361.

235. Vaisocherová, H.; Yang, W.; Zhang, Z.; Cao, Z.; Cheng, G.; Piliarik, M.; Homola, J.; Jiang, S. Ultralow fouling and functionalizable surface chemistry based on a zwitterionic polymer enabling sensitive and specific protein detection in undiluted blood plasma. Anal. Chem. **2008**, *80* (20), 7894–7901.

236. Zhang, Z.; Chen, S.; Jiang, S. Dual-functional biomimetic materials: Nonfouling poly(carboxybetaine) with active functional groups for protein immobilization. Biomacromolecules **2006**, *7* (12), 3311–3315.

237. Guney, A.; Kara, F.; Ozgen, O.; Aksoy, E.A.; Hasirci, V.; Hasirci, N. Surface modification of polymeric biomaterials. In *Biomaterials Surface Science*; Taubert, A., Mano, J.F., Rodríguez-Cabello, J.C., Eds.; Wiley-VCH: Weinheim, 2013; 89–158.

238. Niepel, M.; Köwitsch, A.; Yang, Y.; Ma, N.; Aggarwal, N.; Guduru, D.; Groth, T. Generic methods of surface modification to control adhesion of cells and beyond. In *Biomaterials Surface Science*; Taubert, A., Mano, J.F., Rodríguez-Cabello J.C., Eds.; Wiley-VCH: Weinheim, 2013; 441–467.

Excipients—Gels

Gels: Fibrillar Fibrin

Erin Grassl
University of Minnesota, Minneapolis, Minnesota, U.S.A.

Robert T. Tranquillo
Department of Chemical Engineering and Materials Science, University of Minnesota, Minneapolis, Minnesota, U.S.A.

Abstract

Fibrin has been, and continues to be widely used in surgical applications as a tissue sealant, although this requires much higher concentrations than that of a clot which cells invade and remodel. More recently, it has been examined as a scaffold for tissue engineering. Fibrin possesses several qualities in addition to those already mentioned that make it ideal for use in tissue engineering. It is biocompatible, biodegradable, and can be produced from human serum, making it possible to use autologous sources. This entry will provide background on the structure and biochemistry of fibrin, as well as an overview of its interactions with cells. We will then finish with a discussion of the tissue engineering applications currently being pursued by researchers.

INTRODUCTION

Collagen, which has seen widespread use in tissue engineering, has several advantages as a scaffold, including the ability to entrap cells directly as it is reconstituted into a gel. However, there are some drawbacks to its use, particularly the suppression of cell proliferation and protein synthesis.[1,2] An alternative biopolymer scaffold that has many of the same features as collagen is fibrin, a protein involved in clotting. It also forms a fibrillar network which can directly entrap the cells and exhibits similar rheology to collagen allowing for cell-mediated compaction and consequent alignment of the fibrils and cells.[3] In addition, it promotes cell proliferation and ECM synthesis and remodeling,[4–6] since its purpose is a temporary scaffold to be remodeled and replaced by new tissue during wound healing.

Fibrin has been, and continues to be widely used in surgical applications as a tissue sealant,[7] although this requires much higher concentrations than the 3.5 mg ml^{-1} of a clot which cells invade and remodel. More recently, it has been examined as a scaffold for tissue engineering. Fibrin possesses several qualities in addition to those already mentioned that make it ideal for use in tissue engineering. It is biocompatible, biodegradable, and can be produced from human serum, making it possible to use autologous sources. In the following sections we will provide background on the structure and biochemistry offibrin, as well as an overview of its interactions with cells. We will then finish with a discussion of the tissue engineering applications currently being pursued by researchers.

STRUCTURE, BIOCHEMISTRY, AND RHEOLOGICAL PROPERTIES OF FIBRIN

Fibrinogen, the monomeric form of fibrin, is a 340 kDa protein made up of three pairs of non-identical polypeptide chains.[8] It consists of three main domains; a central domain containing fibrinopeptide E and two each of fibrinopeptides A and B, and two terminal domains containing fibrinopeptide D and sites that participate in cross-linking, as shown in Fig. 1.

The enzyme thrombin cleaves the fibrinopeptides A and B, allowing the fibrinogen to undergo spontaneous fibrillogenesis, forming a linear fibril. The degree of lateral association of these fibrils depends on the conditions under which the fibrin gels, including pH, ionic strength, and concentration of fibrinogen and thrombin, resulting in fibers with diameters ranging from 10 to 200 nm.[9] At lower pH and ionic strength there is more lateral association, whereas at higher pH and ionic strength there is little lateral association. More lateral association results in a gel comprising thicker fibers which, compared to one with fine fibrils, has a higher modulus, creeps more at short times, and creeps less at long times.[10]

The degradation process of fibrin is termed fibrinolysis, and involves a complex cascade of enzymes. The cascade of enzymes leads to the activation of plasminogen, which becomes the proteolytic enzyme plasmin. Plasmin is a serine protease that attacks not only fibrin, but other plasma proteins as well. *In vivo*, this activity is regulated through the presence of inhibitors of both plasmin and plasminogen activators. The use of these inhibitors *in vitro* will be discussed later.

Concise Encyclopedia of Biomedical Polymers and Polymeric Biomaterials DOI: 10.1081/E-EBPPC-120052191

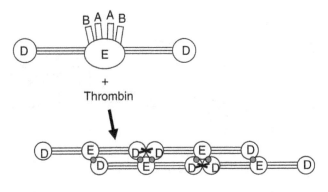

Fig. 1 Simple diagram of fibrinogen molecule and polymerized chain. The gray circles between D and E fibrinopeptides represent the bonds formed once A and B are cleaved. The x between two D peptides represents the covalent amide bonds formed during cross-linking.
Source: Adapted from Grassl.[28]

Greater resistance to fibrinolysis can be achieved through fibrin cross-linking. *In vivo*, enzymes termed transglutaminases participate in this cross-linking through the formation of a bond between primary amines at the γ-carboxamide group of glutamine residues. This forms either ε-(γ-glutamyl) lysine or (γ-glutamyl) polyamine bonds which are covalent, stable, and resistant to proteolysis.[11]

Transglutaminases are found in a variety of tissues. Factor XIII is the enzyme responsible for much of the cross-linking *in vivo*. It is found in both platelets and plasma and participates in cross-linking by mediating the formation of isopeptide bonds between D domains.[12] Factor XIII is composed of two a-chains and two b-chains. It is a proenzyme activated by thrombin to Factor XIIIa. In addition, it facilitates the cross-linking of other proteins, such as collagen and fibronectin, to fibrin.[12] Another transglutaminase that has been suggested to stabilize fibrin to some extent is guinea pig liver transglutaminase.[11,13] Liver transglutaminase has similar binding characteristics, but with a lower affinity for fibrin than factor XIIIa. The binding of liver transglutaminase to fibrin is time- and temperature-dependent and involves binding sites similar to those involved in Factor XIII cross-linking.[13] Liver transglutaminase may be a suitable alternative to Factor XIII for use in the fabrication of cross-linked fibrin tissue equivalents. In addition to these mammalian sources of transglutaminase, several bacterial and fungal transglutaminases have been identified, though only one has been extensively purified and characterized.[14]

For use in tissue engineering applications, fibrin gels are typically prepared by combining fibrinogen and thrombin solutions containing calcium ions. The fibrinogen and thrombin are usually derived from blood plasma including human,[5,15] porcine,[16] or bovine,[17] and several commercial sources exist. Fibrinogen can also be synthesized in a recombinant form by Chinese hamster ovary cells.[18] The resulting gel consists of two component phases: a fibrillar network and an interstitial fluid, often tissue culture medium. The interaction of these two phases determines the response of the gel to an applied force. Fibrin gels respond similarly to collagen gels, which also consist of a fibrillar network and interstitial fluid, exhibiting viscoelastic fluid behavior in shear and compression.[3] The mechanical properties depend on concentration, fibril size, and degree of interaction (cross-linking), among other things. For example, in tensile tests of high and low concentration fibrin sealants, the high concentration sealant exhibited higher values for ultimate tensile strength (UTS) and elastic modulus and less strain hardening than the lower concentration sealant.[19]

INTERACTION OF CELLS WITH FIBRIN

Fibroblasts In Fibrin

The interactions between the cells and matrix, as well as the proliferation and collagen production of fibroblasts, have been explored by several researchers. Compaction of the fibrin matrix by fibroblasts has been shown by the Tranquillo group to generate predictable alignment, as has also been seen in collagen.[20,21] In addition to matrix compaction and alignment, Tuan et al.[6] have demonstrated that fibroblasts proliferate significantly and are very active synthetically, replacing the fibrin matrix with collagen and other unidentified ECM. As early as the second day of culture, collagen was detected, with highly cross-linked collagen detected after 6 days in culture. Coustry et al.[4] also found a significant increase in the production of both collagen and total protein when fibroblasts were cultured in fibrin gels instead of collagen gels. Clark et al.[22] also observed an improvement in collagen production in the presence of the transforming growth factor (TGF)-β. While the collagen-synthetic response of fibroblasts to TGF-β was attenuated when seeded in collagen, the response in fibrin was similar to that seen on tissue culture plastic. The addition of 100 pM TGF-β to the culture medium of fibroblasts in fibrin resulted in a 4.4- and 3.4-fold increase in collagen and non-collagen synthesis, respectively, when compared to fibroblasts in fibrin without the growth factor. More recently, Neidert et al.[23] obtained similar results in a study of the effect of TGF-β and insulin on fibroblasts cultured in either fibrin or collagen. This study, and others looking at the use of fibrin for heart valves, is described in more detail in Section V.E.

Excipients—Gels

Smooth Muscle Cells in Fibrin

Several studies have examined the effect of fibrin and fibrinogen on the adhesion, migration, and proliferation of cultured smooth muscle cells (SMC). Similar to the effect seen with fibroblasts, fibrin promotes the proliferation of SMC in culture. SMC attach to a fibrin clot at 3 hr, begin to proliferate at 6 hr, and show marked proliferation at 24 hr.[5] Naito et al. found that cultured SMC attach to and migrate on fibrinogen- and fibrin-coated dishes, as well as fibrin gels.[5,17,24,25] SMC from explanted tissue also migrate into fibrin gel after an initial lag period.[26]

While there have been many studies examining migration and proliferation of SMC in fibrin, few studies have examined the effect of fibrin on protein synthesis by SMC. Some evidence for the production of ECM can be seen in studies of atherosclerotic fibrous plaques in vessel walls, where SMC appear to proliferate and produce collagen to replace fibrin as they break it down.[27] More recent *in vitro* studies of SMC in fibrin, within the context of a bioartificial artery, have also shown that SMC break down fibrin and replace it with collagen.[28–30] SMC cultured in fibrin gels with complete medium (M199 with 10% fetal bovine serum, 1% penicillin/streptomycin, 1% L-glutamine, and 50 μg ml^{-1} ascorbic acid) produced 3.2–4.9 times the amount of collagen as SMC in collagen gels. They also produced a detectable amount of elastin in fibrin, but not in collagen. This will be discussed in more detail in the section "Tissue Engineering Applications."

Other Cell Types

In addition to fibroblasts and SMC, many other cell types have been grown on and in fibrin. Endo-thelial cells were found to attach and reach confluence. On gels consisting of thicker fibers (and larger pores, presumably) the cells invaded the gel and aligned with the fibrin fibers, whereas on thin fibers they were more randomly distributed on the surface.[31] Other cells that have been cultured in fibrin with some degree of success include chondrocytes, which tend to de-differentiate into a fibroblast-like morphology, periosteal-derived cells, and nucleus pulposus cells.[32] Additional cell types that have been used in fibrin for tissue engineering applications will be discussed in a later section.

CHALLENGES USING FIBRIN

Controlling Fibrinolysis During Formation of New ECM

One of the challenges in using fibrin as a scaffold for engineered tissues is the potentially rapid degradation of the fibrin matrix. The rate and extent of degradation varies with cell type and culture conditions, but in some cases it

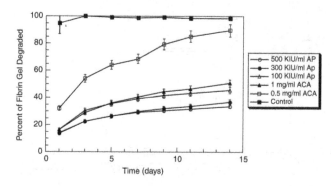

Fig. 2 Degradation of fibrin by neonatal SMC at 4 million cells ml^{-1}, cultured in medium supplemented with various concentrations of aprotinin and ACA.
Source: Taken from Grassl.[28]

is too rapid compared to the rate at which ECM is produced by the cells present to replace it. While fibroblasts slowly degrade the fibrin while replacing it with new ECM, SMCs rapidly degrade the fibrin, resulting in almost complete degradation before significant ECM is produced to replace it.

Rapid degradation of the fibrin matrix can be controlled with several inhibitors. One of these is the serine protease inhibitor aprotinin. Tuan et al.[33] used aprotinin in concentrations of 500 KIU/ml to successfully combat the degradation of fibrin by fibroblasts in the absence of serum. Others have also used aprotinin to prevent degradation during culture of other cell types in fibrin.[7,32,34] Another inhibitor, a lysine analog that competitively inhibits attachment of plasmin and plasminogen to fibrin, is ε-aminocaproic acid (ACA). Herbert et al.[35] used ACA successfully to regulate fibrinolysis in studies of neurite growth in fibrin. Our own studies[29] have shown that ACA can be as effective as aprotinin at appropriate concentrations. Figure 2 compares the inhibition of fibrinolysis by aprotinin and ACA in fibrin gels containing SMC. The fibrin gels contained 5% fluorescently labeled fibrinogen. The fluorescence in the medium above the samples was measured at various time points and used as an indication of fibrin degradation.[29]

Removing (or Masking) Residual Fibrin

It is not clear whether residual fibrin would be prothrombogenic. If so, this may be another consideration in using fibrin as a scaffold, particularly in applications that will have contact with blood. One possibility for removing residual fibrin is to add plasmin or upregulate the activation of plasmin, which should selectively degrade the fibrin without any significant effect on collagen or elastin. However, initial studies with constructs prepared from SMC in fibrin in our lab did not show extensive fibrinolysis. Another possibility is to remove any inhibitor present and allow the cells to degrade it naturally. Our initial studies show that this may be a promising strategy, but needs to be examined further. However, as already mentioned, residual fibrin may not be a problem, particularly in applications where there is little contact with blood, or where the residual fibrin is masked with an endothelium as in the case of blood vessels. In addition, the cells may continue to break down (and remodel) the fibrin once implanted. This is an issue that needs further examination.

Effects of Fibrin Degradation Products

Studies have examined the effect of various fibrin (and fibrinogen) degradation products on several different cell lines under different culture conditions. Most of the work has been done on endothelial cells (EC) and suggests that some Fibrin Degradation Products (FDPs) could be damaging to EC, but results are somewhat conflicting. Dang et al.[36] found that the fibrinogen fragment D, but not E, caused EC to retract from each other and round up. The effect did not appear to be cytotoxic and was reversible since the cells could be replated. However, other studies[16,37] suggest that fragment D does not affect EC, but that certain fractions of low molecular weight FDPs damage EC, causing them to detach from their substrate.

Other cell types have also shown responses to FDPs. Fragment B, which is released during formation of fibrin and may be trapped within the matrix, has been shown to cause direct cell migration of neutrophils and fibroblasts, as well as changes in neutrophil morphology and secretion of enzymes.[38] Fragment D induced proliferation of human hemopoietic cells,[39] while fragment E resulted in IL-6 production by rat peritoneal macrophages[40] and fibrinogen synthesis by cultured rat hepatocytes.[41]

It is not clear whether these results accurately reflect what would be seen in a tissue-engineered construct, since it is a more complex system than the cultured cells in these studies. The potential damage to ECs is of particular concern and warrants a more careful examination using the conditions appropriate for tissue engineering applications.

TISSUE ENGINEERING APPLICATIONS

Fibrin as a Cell Delivery Vehicle

Simple tissue engineering applications of fibrin have stemmed from its use as a sealant in surgery. The use of fibrin glue has been examined for the delivery of cells to a specific site or attachment of cells to another tissue or matrix. An example of this is its use as a delivery vehicle for urothelial cells in urethral reconstruction.[42] Bach et al. suspended cells in fibrin glue and applied the mixture to a connective tissue capsule tube formed *in vivo*. The urothelial cells spread and formed an adherent and confluent cell layer 2 weeks after implantation, at which time the fibrin

clot had already been replaced with other connective tissue. This type of technique has also been used with skin grafts[7] and a composite neotrachea.[43]

Artificial Ocular Surface

Han et al.[44] examined the use of fibrin as a matrix for corneal epithelial stem cells to produce a bioengineered ocular surface. This work was based on the same idea that fibrin provides a favorable matrix environment for epithelial cell growth and differentiation during wound healing. The fibrinogen and thrombin was isolated from blood plasma in the lab and contained a number of other plasma proteins. This demonstrated the possibility of using autologous sources for the fibrin gel. Human corneal stem cells suspended in the fibrin proliferated and exhibited markers of differentiation, expressing keratin 3, as well as keratin 19 in some cases. Keratin 3 is a marker of corneal-type epithelial differentiation, while keratin 19 is proposed as a marker for corneal stem cells. No functional tests were performed, but the tissue was found to be soft, pliable, and elastic.

Skin Grafts

As already mentioned, fibrin has been used in skin graft applications. In addition to its use as a delivery vehicle for cells, it has been used for a more structural approach. In some cases, acellular fibrin was used to induce migration of surrounding cells into the wounded area.[7] More complicated skin grafts have involved fibrin in a composite structure. Meana et al.[34] seeded keratinocytes on a fibrin gel containing fibroblasts. After 15 days of culture a confluent bilayer of keratinocytes had formed, and staining for collagen IV and laminin suggested the formation of a basal membrane between the "dermal" and "epidermal" layers. The fibrin used for these studies was obtained from a cryoprecipitate of human plasma and therefore contained additional components that would not be found in purified commercially available fibrinogen. It was not determined whether the cryoprecipitate contained growth factors necessary for the growth of keratinocytes, and therefore it is not clear whether the commercially available fibrinogen would yield similar results.

Bio-Artificial Arteries

The previous examples have involved fairly simple geometries. However, fibrin gel can be molded into various shapes, and has been examined for the use in more complicated geometries, such as a bio-artificial artery. As in the case of skin grafts, previous work on bioartificial arteries has focused on using collagen as a natural biopolymer.[45–50] However, these constructs typically lack the necessary strength and elasticity (i.e., cell-produced elastic fibers) found in the native artery. Therefore, fibrin was examined as an alternative natural biopolymer because of the tendency of cells to remodel it and replace it with new ECM. In addition, the ability to align the fibrin may be used to provide an aligned template for the newly synthesized matrix, providing the circumferential alignment seen in the native vessel.

Initial studies of 4×10^6 SMC ml^{-1} in 3 mg ml^{-1} fibrin gels prepared from commercially available fibrinogen and thrombin, showed that SMC produced 3.2–4.9 times more collagen when seeded in fibrin than in collagen.[29] The amount of ECM produced was further increased by adding growth factors such as TGF-β and insulin. After a 3 week incubation with 1 mg ml^{-1} ACA (fibrinolysis inhibitor), 5 ng ml^{-1} TGF-β and 2 μg ml^{-1} insulin, the collagen content of the SMC-seeded fibrin gels was 6 times the content of SMC-seeded fibrin gels without TGF-β or insulin. This resulted in a 15- to 20-fold improvement in the mechanical properties (UTS and tangent modulus). After 6 weeks, the collagen content and mechanical properties more than doubled what was seen at 3 weeks.[30] In addition, elastin was found in these samples, though it made up only 2–17% of the protein in the samples.[51] The UTS and modulus of these constructs were on the order of rat aorta, but the burst strength was still quite low, around 120–140 mm Hg, suggesting that further work is needed to produce an adequate artificial artery. More recent work aimed at optimization of the culture conditions has suggested that even greater ECM synthesis can be achieved. When incubated in DMEM supplemented with TGF-β, insulin and 10% FBS, more than 50% of the original fibrin was replaced by elastin and collagen after 4 weeks.[52] With these optimized conditions, burst pressures up to 1100 mm Hg have been achieved (unpublished data). Another exciting result from this work was verification of the hypothesis that an aligned fibrin template would lead to deposition of new ECM with the same alignment. Figure 3 demonstrates the alignment at

1 week when the matrix is mostly fibrin, and alignment in the same direction at 6 weeks when there is substantial collagen produced.

Heart Valves

Another cardiovascular application of fibrin is in the fabrication of a tissue-engineered heart valve or venous valve. Ye et al.[53] and Neidert et al.[23] examined the behavior of fibroblasts in fibrin for use in cardiovascular tissue engineering applications such as the heart valve. Ye et al. combined myofibroblasts with commercially available fibrinogen and thrombin to form a fibrin gel with 750,000 cells ml^{-1} and 3.5 mg ml^{-1} fibrin. They found that the cells would completely degrade the fibrin with 5 μg ml^{-1} or less aprotinin. There was some fibrinolysis with 15 μg ml^{-1}, but no visible fibrinolysis with 20 μg ml^{-1}. They also observed uniform cell distribution and a 20% reduction in thickness of the gel after 1 month. A later study from the same group,[54] examined the effect of different methods of fixing the fibrin to a mold to prevent compaction, and used the results to design a molding technique for forming an aortic valve conduit out of myofibroblasts and fibrin gel. However, the mechanical properties of the myofibroblast-seeded fibrin gel were not examined in either study.

Neidert et al.[23] examined ways to enhance the remodeling and assessed the effect on the mechanical properties. Unlike what was observed with SMC and myofibroblasts, the fibroblasts did not degrade the fibrin rapidly, therefore no inhibitor was necessary. However, the effect of TGF-β and insulin was similar to what was seen with SMC. After 51 days with 5 ng ml^{-1} TGF-β and 2 μg ml^{-1} insulin added to the medium, the fibrin constructs contained nearly 8 times the amount of collagen as those without the additives. In addition, the UTS and modulus were also improved by the addition of these growth factors. Under the optimum conditions, the final constructs consisted of more than 30% collagen and had a modulus and UTS within an order of magnitude of a human heart valve leaflet.

SUMMARY

Fibrin has several characteristics that makes it desirable for tissue engineering. It is biocompatible (able to be made from autologous sources), biodegradable, and capable of entrapping cells directly. In addition, it has adequate mechanical properties and promotes cell growth and remodeling. Studies of cells cultured in fibrin have shown that several different cell types can be successfully grown in fibrin. While there are some challenges to be addressed, fibrin has shown promise as a biopolymer scaffold in several tissue engineering applications including ocular implants, skin, blood vessels, and heart valves. More work is necessary to further develop these applications and examine new ones.

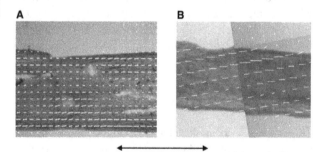

Fig. 3 Alignment maps of a fibrin ME incubated for (**A**) 1 week and (**B**) 6 weeks with 1 mg/ml ACA, 1 ng/ml TGF-β, and 2 mg/ml insulin. The vectors indicate the direction of alignment, with the length indicating the relative strength of alignment. The arrow indicates the circumferential direction.
Source: Adapted from Bromberek et al.[21]

REFERENCES

1. Clark, R.A.F.; Nielsen, L.D.; Welch, M.P.; McPherson, J.M. Collagen matrices attenuate the collagen-synthetic response of cultured fibroblasts to TGF-beta. J. Cell Sci. **1995**, *108* (3), 1251–1261.

2. Thie, M.; Schlumberger, W.; Semich, R.; Rauterberg, J.; Robenek, H. Aortic smooth muscle cells in collagen lattice culture: Effects on ultrastructure, proliferation and collagen synthesis. Eur. J. Cell Biol. **1991**, *55* (2), 295–304.

3. Tranquillo, R.T. Self-organization of tissue-equivalents: The nature and tole of contact guidance. Biochem. Soc. Symp. **1999**, *65*, 27–42.

4. Coustry, F.; Gillery, P.; Maquart, F.X.; Borel, J.P. Effect of transforming growth factor beta on fibroblasts in three-dimensional lattice cultures. FEBS Lett. **1990**, *262* (2), 339–341.

5. Ishida, T.; Tanaka, K. Effects of fibrin and fibrinogen-degradation products on the growth of rabbit aortic smooth muscle cells in culture. Atherosclerosis **1982**, *44* (2), 161–174.

6. Tuan, T.; Song, A.; Chang, S.; Younai, S.; Nimni, M. *In vitro* fibroplasia: Matrix contraction, cell growth, and collagen production of fibroblasts cultured in fibrin gels. Exp. Cell Res. **1996**, *223* (1), 127–134.

7. Currie, L.J.; Sharpe, J.R.; Martin, R. The use of fibrin glue in skin grafts and tissue-engineered skin replacements: A review. Plast. Reconstr. Surg. **2001**, *108* (6), 1713–1726.

8. Doolittle, R.F. Fibrinogen and fibrin. Annu. Rev. Biochem. **1984**, *53* (1), 195–229.

9. Kaibara, M.; Fukada, E.; Sakaoku, K. Rheological study on network structure of fibrin clots under various conditions. Biorheology **1981**, *18* (1), 23–35.

10. Kramer, O., Ed. *Biological and Synthetic Polymer Networks*; Elsevier Applied Science: London, 1988.

11. Greenberg, C.S.; Birckbichler, P.J.; Rice, R.H. Transglutaminases: Multifunctional cross-linking enzymes that stabilize tissues. FASEB J. **1991**, *5* (15), 3071–3077.

12. Ariens, R.A.S.; Lai, T.-S.; Weisel, J.W.; Greenberg, C.S.; Grant, P.J. Role of factor XIII in fibrin clot formation and effects of genetic polymorphisms. Blood **2002**, *100* (3), 743–754.

13. Achyuthan, K E.; Mary, A.; Greenberg, C.S. The binding sites on fibrin(ogen) for guinea pig liver transglutaminase are similar to those of blood coagulation factor XIII. Characterization of the binding of liver transglutaminase to fibrin. J. Biol. Chem. **1988**, *263* (28), 14296–14301.

14. Griffin, M.; Casadio, R.; Bergamini, C.M. Transglutaminases: Nature's biological glues. Biochem. J. **2002**, *368* (Pt. 2), 377–396.

15. Blomback, B.; Carlsson, K.; Hessel, B.; Liljeborg, A.; Procyk, R.; Aslund, N. Native fibrin gel networks observed by 3D microscopy, permeation and turbidity. Biochim. Biophys. Acta **1989**, *997* (1–2), 96–110.

16. Watanabe, K.; Tanaka, K. Influence of fibrin, fibrinogen, and fibrinogen degradation products on cultured endothelial cells. Atherosclerosis **1983**, *48* (1), 57–70.

17. Naito, M.; Nomura, H.; Iguchi, A. Migration of cultured vascular smooth muscle cells into non-crosslinked fibrin gels. Thromb. Res. **1996**, *84* (2), 129–136.

18. Gorkun, O.V.; Veklish, Y.I.; Weisel, J.W.; Lord, S.T. The conversion of fibrinogen to fibrin: Recombinant fibrinogen typifies plasma fibrinogen. Blood **1997**, *89* (12), 4407–4414.

19. Sierra, D.H.; Eberhardt, A.W.; Lemons, J.E. Failure characteristics of multiple-component fibrin-based adhesives. J. Biomed. Mater. Res. **2002**, *59* (1), 1–11.

20. Barocas, V.H.; Girton, T.S.; Tranquillo, R.T. Engineered alignment in media-equivalents: Magnetic prealignment and mandrel compaction. J. Biomech. Eng. **1998**, *120* (5), 660–666.

21. Bromberek, B.A.; Enever, P.A.; Shreiber, D.I.; Caldwell, M.D.; Tranquillo, R.T. Macrophages influence a competition of contact guidance and chemotaxis for fibroblast alignment in a fibrin gel coculture assay. Exp. Cell Res. **2002**, *275* (2), 230–242.

22. Clark, R.A.F.; McCoy, G.A.; Folkvord, J.M.; McPherson, J.M. TGF-beta1 stimulates cultured human fibroblasts to proliferate and produce tissue-like fibroplasia: A fibronectin matrix-dependent event. J. Cell. Physiol. **1997**, *170* (1), 69–80.

23. Neidert, M.R.; Lee, E.S.; Oegema, T.R.; Tranquillo, R.T. Enhanced fibrin remodeling *in vitro* with TGF-beta1, insulin and plasmin for improved tissue-equivalents. Biomaterials **2002**, *23* (17), 3717–3731.

24. Naito, M.; Hayashi, T.; Kuzuya, M.; Funaki, C.; Asai, K.; Kuzuya, F. Effects of fibrinogen and fibrin on the migration of vascular smooth muscle cells *in vitro*. Atherosclerosis **1990**, *83* (1), 9–14.

25. Naito, M.; Funaki, C.; Hayashi, T.; Yamada, K.; Asai, K.; Yoshimine, N.; Kuzuya, F. Substrate-bound fibrinogen, fibrin and other cell attachment-promoting proteins as a scaffold for cultured vascular smooth muscle cells. Atherosclerosis **1992**, *96* (2–3), 227–234.

26. Nomura, H.; Naito, M.; Iguchi, A. Thompson, W.D.; Smith, E.B. Fibrin gel induces the migration of smooth muscle cells from rabbit aortic explants. Thromb. Haemost. **1999**, *82* (4), 1347–1352.

27. Smith, E.B. Fibrinogen, fibrin, and the arterial wall. Eur. Heart J. **1995**, *16* (Suppl. A), 11–15.

28. Grassl, E.D. *Enhancing the Properties of the Medial Layer of a Bioartificial Artery*; Chemical Engineering, University of Minnesota: Minneapolis, 2002; pp. 130.

29. Grassl, E.D.; Oegema, T.R.; Tranquillo, R.T. Fibrin as an alternative biopolymer to type I collagen for fabrication of a media-equivalent. J. Biomed. Mater. Res. **2002**, *60* (4), 607–612.

30. Grassl, E.D.; Oegema, T.R.; Tranquillo, R.T. A fibrin-based arterial media-equivalent. J. Biomed. Mater. Res. **2003**, *66A* (3), 550–561.

31. Shats, E.A.; Nair, C.H.; Dhall, D.P. Interaction of endothelial cells and fibroblasts with modified fibrin networks: Role in atherosclerosis. Atherosclerosis **1997**, *129* (1), 9–15.

32. Perka, C.; Arnold, U.; Spitzer, R.-S.; Lindenhayn, K. The use of fibrin beads for tissue engineering and subsequential transplantation. Tissue Eng. **2001**, *7* (3), 359–361.

33. Tuan, T.L.; Grinnell, F. Fibronectin and fibrinolysis are not required for fibrin gel contraction by human skin fibroblasts. J. Cell. Physiol. **1989**, *140* (3), 577–583.

34. Meana, A.; Iglesias, J.; Del Rio, M.; Larcher, F.; Madrigal, B.; Fresno, M.F.; Martin, C.; San Roman, F.; Tevar, F. Large

surface of cultured human epithelium obtained on a dermal matrix based on live fibroblast-containing fibrin gels. Burns **1998**, *24* (7), 621–630.

35. Herbert, C.B.; Bittner, G.D.; Hubbell, J.A. Effects of fibrinolysis on neurite growth from dorsal root ganglia cultured in two- and three-dimensional fibrin gels. J. Comp. Neurol. **1996**, *365* (3), 380–391.

36. Dang, C.V.; Bell, W.R.; Kaiser, D.; Wong, A. Disorganization of cultured vascular endothelial cell monolayers by fibrinogen fragment D. Science **1984**, *227* (4693), 1487–1490.

37. Lorenzet, R.; Sobel, J.H.; Bini, A.; Witte, L.D. Low molecular weight fibrinogen degradation products stimulate the release of growth factors from endothelial cells. Thromb. Haemost. **1992**, *68* (3), 357–363.

38. Senior, R.M.; Skogen, W.F.; Griffin, G.L.; Wilner, G.D. Effects of fibrinogen derivatives upon the inflammatory response. J. Clin. Invest. **1986**, *77* (3), 1014–1019.

39. Hatzfeld, J.A.; Hatzfeld, A.; Maigne, J. Fibrinogen and its fragment D stimulate proliferation of human hemopoietic cells *in vitro*, Proc. Natl. Acad. Sci. **1982**, *79* (20), 6280–6284.

40. Lee, M.E.; Rhee, K.J.; Nham, S.U. Fragment E derived from both fibrin and fibrinogen stimulates interleukin-6 production in rat peritoneal macrophages. Mol. Cells **1999**, *9* (1), 7–13.

41. Qureshi, G.D.; Guzelian, P.S.; Vennart, R.M.; Evans, H.J. Stimulation of fibrinogen synthesis in cultured rat hepatocytes by fibrinogen fragment E. Biochim. Biophys. Acta **1985**, *844* (3), 288–295.

42. Bach, A.D.; Bannasch, H.; Galla, T.J.; Bittner, K.M.; Stark, G.B. Fibrin glue as a matrix for cultured autologous urothelial cells in urethral reconstruction. Tissue Eng. **2001**, *7* (1), 45–53.

43. Doolin, E.J.; Strande, L.F.; Sheng, X.; Hewitt, C.W. Engineering a composite neotrachea with surgical adhesives. J. Pediatr. Surg. **2002**, *37* (7), 1034–1037.

44. Han, B.; Schwab, I.R.; Madsen, T.K.; Isseroff, R.R. A fibrin-based bioengineered ocular surface with human corneal epithelial stem cells. Cornea **2002**, *21* (5), 505–510.

45. Girton, T.S.; Oegema, T.R.; Grassl, E.D.; Isenberg, B.C.; Tranquillo, R.T. Mechanisms of stiffening and strengthening in media-equivalents fabricated using glycation. J. Biomech. Eng. **2000**, *122* (3), 216–223.

46. Girton, T.S.; Oegema, T.R.; Tranquillo, R.T. Exploiting glycation to stiffen and strengthen tissue-equivalents for tissue engineering. J. Biomed. Mater. Res. **1999**, *46* (1), 87–92.

47. Hirai, J.; Kanda, K.; Oka, T.; Matsuda, T. Highly oriented, tubular hybrid vascular tissue for a low pressure circulatory system. ASAIO J. **1994**, *40* (3), M383–M388.

48. L'Heureux, N.; Germain, L.; Labbe, R.; Auger, F.A. *In vitro* construction of a human blood vessel from cultured vascular cells: A morphological study. J. Vasc. Surg. **1993**, *17* (3), 499–509.

49. Seliktar, D.; Black, R.A.; Vito, R.P.; Nerem, R.M. Dynamic mechanical conditioning of collagen-gel blood vessel constructs induces remodeling *in vitro*. Ann. Biomed. Eng. **2000**, *28* (4), 351–362.

50. Weinberg, C.B.; Bell, E. A blood vessel model constructed from collagen and cultured vascular cells. Science **1986**, *231* (4736), 397–400.

51. Long, J.L.; Tranquillo, R.T. Elastic fiber production in cardiovascular tissue-equivalents. Matrix Biol. **2003**, *22* (4), 339–350.

52. Ross, J.J.; Tranquillo, R.T. ECM gene expression correlates with *in vitro* tissue growth and development in fibrin gel remodeled by neonatal smooth muscle cells. Matrix Biol. **2003**, *22* (6), 477–490.

53. Ye, Q.; Zund, G.; Benedikt, P.; Jockenhoevel, S.; Hoerstrup, S.P.; Sakyama, S.; Hubbell, J.A.; Turina, M. Fibrin gel as a three dimensional matrix in cardiovascular tissue engineering. Eur. J. Cardiothorac. Surg. **2000**, *17* (5), 587–591.

54. Jockenhoevel, S.; Zund, G.; Hoerstrup, S.P.; Chalabi, K.; Sachweh, J.S.; Demircan, L.; Messmer, B.J.; and Turina, M. Fibrin gel—Advantages of a new scaffold in cardiovascular tissue engineering. Eur. J. Cardiothorac. Surg. **2001**, *19* (4), 424–430.

Gene Carriers: Design Elements

Jong-Sang Park
Department of Chemistry, Seoul National University, Seoul, South Korea

Joon Sig Choi
Department of Biochemistry, Chungnam National University, Daejeon, South Korea

Abstract
The focus of this entry is to briefly examine the key elements that are generally required in the field of non-viral gene delivery research. It provides basic criteria that should be addressed before designing novel polymers for powerful, safe, and reliable non-viral vector systems.

INTRODUCTION

Plasmid-based gene therapy is a promising protocol for treating certain human diseases for which other clinical trials are ineffective or unavailable.[1]

During the past decade, intensive research and development in multidisciplinary fields, such as chemistry, molecular biology, pharmaceutics, biochemistry, chemical engineering, and medicine, have been directed towards devising optimized and more effective methods for transferring therapeutic genes into cells and for eventual use in human clinical settings. Several successful clinical trials reported so far involve viral vector systems (retroviruses, adenoviruses) that provide efficient transduction and high levels of gene expression. However, their clinical safety and effectiveness are still hampered by their major drawbacks such as inherent toxicities, short- and long-term risks such as generation of host immune responses, and the possibility of inserted genes combining with activation of oncogenes.[2]

For these reasons, nonviral vector systems that are considered as alternative tools to such risky viral vectors have been introduced and tested for their potential to be safer, and more desirable methods for gene delivery and clinical gene therapy.[3] The nonviral vector systems are generally composed of either naked plasmid DNA or various kinds of DNA-complexing agents such as cationic liposomes and polycationic polymers. However, currently available synthetic nonviral vector systems have been beset with many problems, such as inefficiency, cytotoxicity, and water-solubility problems that limit the many possible applications for their *in vivo* use. Consequently, few nonviral vectors have been so far successful in clinical trials.

The focus of this entry is to examine briefly the key elements that are generally required in the field of nonviral gene delivery research rather than to scrutinize all the polymer-based gene delivery vectors. We wish to provide the basic and newest criteria that should be addressed before designing novel polymers en route to creating powerful, safe, and reliable nonviral vector systems.

TRADITIONAL POLYCATIONIC CHARGED-BASED POLYMERS

Structure of Cationic Polymers

Various types of cationic polymers, i.e., linear, dendritic, cross-linked, branched, and network-type polymers, have been introduced and tested for their potential applicability to the field of gene therapy. The structures of the representative polycationic polymers are shown in Fig. 1. Some linear cationic polymers were found to be promising at the first stage, but unexpected characteristics such as water-solubility of DNA complexes, low level of transfection efficiency, and inherent cytotoxicity, limited their use as *in vivo* gene carriers.[4] However, polycationic dendrimers are still very attractive, because of their well-defined structure and ease of controlling their surface functionality for the design of biomedical applications.[5,6] Already, both polyamidoamine (PAMAM) and polyethylenimine (PEI) dendrimer were tested for their potential utility and these have exhibited relatively high transfection efficiency *in vitro* and *in vivo*.[5,7,8] One of the important features of PEI and PAMAM dendrimers is that they are composed of tertiary amine-containing backbones, which possess pH-sensitive functionality. The so-called "proton sponge effect" or "endosome buffering hypothesis" is the mechanism generally advanced to account for the high transfection efficiency of the polymers.[7] In addition to the

Concise Encyclopedia of Biomedical Polymers and Polymeric Biomaterials DOI: 10.1081/E-EBPPC-120052224

623

endosome buffering functionality, another merit of designing polymers of globular structure rather than linear or branched or flexible structures is that globular polymers show reduced cytotoxicity.[9]

Types of Chargeable Moiety

One of the basic requirements for charge-based complex formation with polyanionic DNA is that the polymer should contain polycationic charge properties. Usually, the types of chargeable moiety are primary, secondary, tertiary, and quaternary amine derivatives. The tertiary amine-containing polymers are less effective at condensation of plasmid DNA than primary or quaternary amine-containing polymers because of the lower degree of protonation at physiological conditions. Interestingly, quaternary amines could bind with DNA effectively, even more strongly than the primary amines–DNA interaction. However, it is noteworthy that a very poor transfection efficiency was observed for the quaternized polycationic polymers.[10]

The condensation of plasmid DNA into nano-sized particles contributes to both physicochemical properties and stability against enzyme action. The formation of polycationic particles with DNA increased transfection efficiency *in vitro* because they could bind to the negatively charged cell membranes and, sometimes, they also could physically come into contact with the cell surface through sedimentation. However, for *in vivo* transfection trials, the systems are inefficient because of the net positive charges of the complexes and formation of large particulates that also reduce the mobility of complexes significantly. The charged particles interact with proteoglycans that are composed of a core protein and sulfated or carboxylic glycosaminoglycans (GAGs) conjugated to the protein. So the transfectivity of the positively charged DNA complexes may be affected by the extracellular polyanionic GAGs that can interact with the complexes, thereby inhibiting their mobility in tissue, and their targeting to some specific cells *in vivo*.

Recently, our group has reported another PAMAM dendrimer-based gene delivery system.[11] PAMAM-OH dendrimer is structurally identical to PAMAM dendrimer except that all terminal functional groups are hydroxyl groups not primary amines. So PAMAM-OH could not form charge-based polyplexes with DNA by itself and shows a deficiency in transfection. The internal quaternary amines were generated by methylation and turned the polymer into a transfection-competent vector preserving the zeta potential of the DNA complexes neutral (Fig. 2).

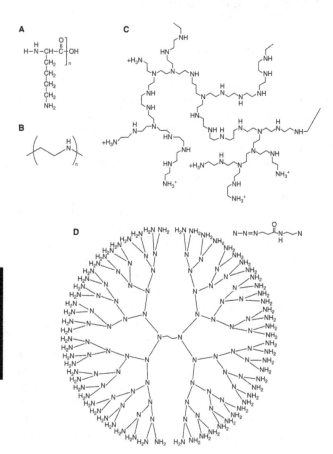

Fig. 1 Structures of the traditional polycationic polymers. (**A**) poly-L-lysine. (**B**) linear polyethylenimine. (**C**) Branched polyethylenimine. (**D**) PAMAM generation 4.

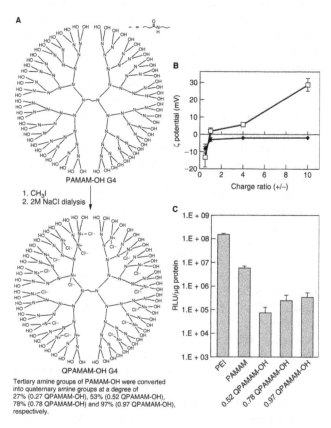

Fig. 2 (**A**) Synthesis of quaternized PAMAM G4 dendrimer. (**B**) Zeta potential values of PAMAM G4/DNA complex (□) and 0.97 QPAMAM-OH/DNA complex (♦). (**C**) Reporter gene expression assay at charge ratio 6 (+/−).

Effect of Charge Density and Molecular Weight

The physicochemical properties of the representative polycationic polymers generally used in gene delivery experiments are presented in Table 1. Poly(L-lysine) (PLL), which was used as a standard basic polymer in the early stages of polycationic polymer-mediated gene delivery experiments, possesses polypeptide backbones and only primary amines at the terminal ends of the side groups of each lysine unit. Due to its deficiency of endosome buffering moiety, the polymer needs an additional chemical, chloroquine, to achieve gene expression. With respect to the charge density (one cationic charge per molecular weight of monomer unit), PLL and PAMAM are smaller in charge density compared to PEI (0.0097 for PLL; 0.0087 for PAMAM, vs. 0.0238 for PEI). As shown in Table 1, PEI has higher transfection efficiency than other transfection reagents due to its endosome buffering and nucleus targeting ability and it shows higher toxicity to cells due to its high charge density and nondegradability. Therefore, cytotoxicity and transfection efficiency are believed to be a function of molecular weight, charge density, degradability, and polymer structure.

Degradability

In view of the cytotoxicity and DNA release in response to specific environmental stimuli, i.e., hydrolysis, enzymatic digestion, pH difference, and reduction potential difference, degradation functionality of the polymer is an essential feature of efficient polymeric gene carriers. The degradable chemical linkages include esters, carbamates, disulfides, ortho esters, acetals, glycosides, and related functional groups. As demonstrated in Fig. 3, the strategy is to construct degradable polyplexes such that the functional particles may attain a lower level of polymer-mediated toxicity and a higher level of gene expression through increased DNA release from the complexes after reaching target sites compared to nondegradable or poorly degradable polymer-based systems.

We and others have designed various degradable polycationic polyester polymers, which can self-assemble with plasmid DNA forming nanoparticles and show gene transfer potency *in vitro*.[12–16] The two major synthesis methods reported are: 1) melting condensation of diols and carboxlic acid derivatives; and 2) polycondensation using Michael addition involving diacrylates and amine groups. The brief polymerization schemes are represented in Fig. 4.

Akinc et al. and Lynn et al. reported the first approach to developing a library of parallel synthesis and screening methods, and suggested effective transfection efficiency by poly(β-amino esters).[17,18] The library was constructed using 140 structurally diverse polymers. Among them, only half of the members, i.e., 70, were soluble in water, which made it possible to characterize them further for DNA condensation and transfection. In addition, 56 polymers of 70 could form complexes with DNA, whereas 14 members could not.

In addition to the relatively low-throughput synthesis and characterization methods, a high-throughput manipulation method was successfully introduced for preparing a library of 2350 structurally diverse, degradable cationic polymers with the aid of liquid handling automation.[19] This provided large amounts of structure-function information.

Another degradable system is composed of disulfide linkages that are sensitive to the reduction potential difference between inside and outside of the cell membranes. The basic idea is that as the concentration of the reduced form of glutathione is 500 times higher than that of the oxidized form in red cells,[20] the polymers composed of disulfide bonds could be degradable due to reduction of the linkages releasing DNA from the complexes. Many groups have studied the redox-triggered DNA releasable polymeric carriers that contain disulfide linkages, and the DNA complexes are susceptible to thiolysis, which influences the multivalent interaction between the cationic polymers and the complexed DNA so that the DNA could migrate into the cytoplasm. Gosselin and colleagues synthesized reversibly cross-linked polyplexes for

Table 1 Physicochemical properties of representative polycationic polymers generally used in gene delivery experiments

Polymer	Molecular weight (kDa)	Order of amines (degree)	Cationic charge/ Monomer (+/Da)[a]	Structure	Degradability	In vitro toxicity[b]	In vitro transfection efficiency[b]
Poly-L-lysine	19.2–36.6	1	0.0097	Linear	Poor, polypeptide	+++	+[c]
PAMAM dendrimer G4	14	1, 3	0.0087	Globular, dendrimer	Poor, polypeptide	+	++
Linear PEI	22	1, 2	0.0238	Linear	Not degradable	++	+++
Branched PEI	25	1, 2, 3	0.0238	Branched, dendrimer	Not degradable	+++	+++

[a]Determined as one cationic charge per molecular weight of monomer unit.
[b]Arbitrary units, compared with branched PEI polymer.
[c]In the presence of chloroquine.

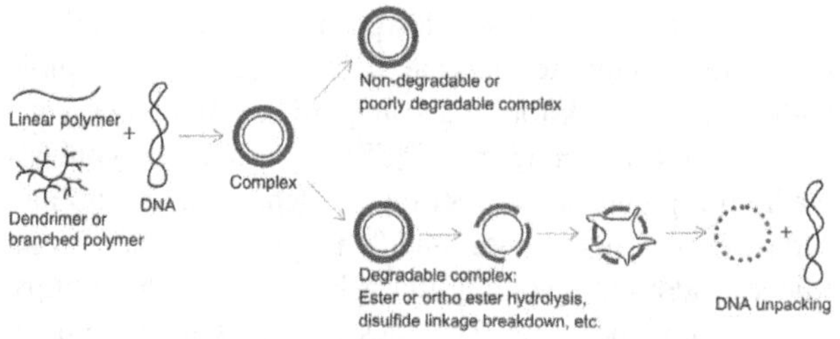

Fig. 3 Scheme of polymer/DNA complex formation and DNA unpacking from the degradable complexes.

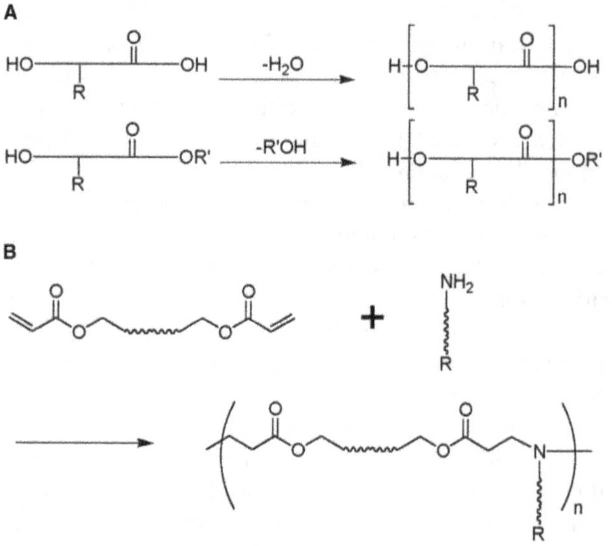

Fig. 4 Synthesis of cationic polyester polymers by (**A**) melting condensation of diols and carboxlic acid derivatives, and by (**B**) polycondensation using Michael addition of diacrylates and amine-containing monomers.

gene delivery.[21,22] They used PEI polymer and employed homobifunctional amine reactive cross-linking reagents to introduce disulfide linkage inside the polyplexes. Some groups have also tried to prepare cysteine-containing polypeptides and evaluated them for polyplex formation and transfection efficiency.[23,24] Despite the disulfide-containing polymers showing enhanced complexation with DNA, as well as increased stability and degradability, the real mechanism is still unclear. One possible explanation could be that the reduced form of glutathione might react with the polyplexes and the DNA released into the cytoplasm is subjected to the intracellular machinery of gene expression. Recently, another hypothesis was also proposed that protein disulfide isomerase, which was reported by Mandel and colleagues in 1993,[25] might play an important role in the thiolysis of biomacromolecules that cannot diffuse through the plasma membranes.[26] Wang and his colleagues have reported novel biodegradable cationic polymers containing a phosphate backbone and positively charged moiety.[27,28]

The polymer backbone was prepared by ring-opening polymerization of 4-methyl-2-oxo-2-hydro-1,3,2-dioxaphospholane using triisobutylaluminum as an initiator. Further, modification of phosphorus atom leads to the possibility of introducing of mono- or multiple-cationic charge functionality for electrostatic interaction with plasmid DNA. Even though the transfection efficiency of the polymers was low and it needs an additional endosome disruptive agent, such as chloroquine, *in vitro* experiments to obtain elevated levels of gene expression, biocompatibility and much lower cytotoxicity are considered to be the key features of the polymers for future *in vivo* application.

HYBRID OR GRAFT POLYMERS FOR MULTIFUNCTIONALITY

Formation of Stealth Complexes: Pegylation

One of the recent strategies attempting to overcome such problems is to link or conjugate polycationic polymers with a hydrophilic polymer, poly(ethylene glycol) (PEG). PEG shows many useful characteristics, such as high solubility in water, non-immunogenicity and improved biocompatibility. PEG has been widely used for delivery of many water-insoluble small molecular weight and proteins drugs. Moreover, PEG is often used as a spacer between targeting ligand and polymeric carriers.[29] For the preparation of synthetic gene delivery carriers, PEG has also been coupled to numerous polycationic polymers, such as PLL, dendrimers, polyspermine, and PEI.[30–38] Therefore, the conjugated PEG helps the reagents to improve their half-life in the bloodstream, to increase solubility, and to reduce the immune reaction of complexes with DNA. In addition, receptor-mediated endocytosis can be realized by introducing specific targeting ligands at the end of PEG. The PEG chain serves as a flexible spacer between ligands and receptors. The formation schemes of PEG-coated polyionic complexes between PEG-conjugated copolymers and DNA are depicted in Fig. 5.

Biocompatibility Issues

Biological evaluation of polymeric gene carriers should be performed to determine the potential risks of toxicity resulting from contact of either the polymer itself or the component materials after degradation with cells and with the body. There are three major considerations proposed by the FDA for medical devices, which are considered to be also applicable to polymeric gene carrier materials (http://www.fda.gov/cdrh/devadvice/pma/special_considerations.html). First, the polymers or the released constituents after degradation of the polymers should not cause any adverse local or systemic effects. Second, they should not be carcinogenic. Finally, they should not produce adverse reproductive and developmental effects. In addition, the biological and chemical characteristics of

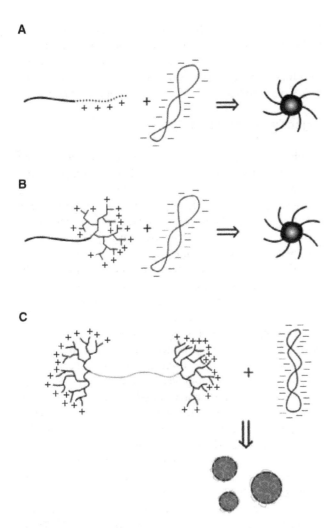

Fig. 5 Polyplex formation of PEG-conjugated hybrid copolymers with DNA. (**A**) AB type copolymer: linear-linear copolymer. (**B**) AB type copolymer: linear-dendrimer copolymer. (**C**) ABA type: dendrimer-linear-dendrimer copolymer.

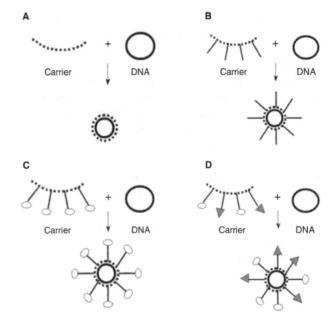

Fig. 6 Types of polymer-mediated complex formation with DNA. (**A**) Homopolymer type. (**B**) Graft copolymer type. (**C**) Functional graft copolymer type. (**D**) Multifunctional copolymer type.

polymers and the nature, degree, frequency, and duration of their exposure to the local cells or the body must be considered. Therefore, careful evaluation of any new polymer intended for clinical use should be based upon sufficient data from systematic testing to ensure that the polymers are safe for *in vivo* use.[39] One of various ways to improve the biocompatibility of polymers is to modify the polymers with hydrophilic polymers, such as PEG, which are reported to reduce the surface adsorption of proteins such as fibrinogen, albumin, or thrombin.[40]

Graft or Surface Functionality

To supply additional functionalities on the polymer, which is monofunctional, difunctional, or multifunctional, is one of the important aspects in the development of smart polymeric gene carriers. As Han and colleagues reported,[41] the construction of polyplexes with various functionalities may be accomplished as presented in Fig. 6.

As an example of a pH-sensitive water-soluble DNA-containing nanoparticle system, Choi et al. reported recently that water-soluble DNA-containing nanoparticles, which are composed of novel pH-sensitive polymeric lipid as well as cationic lipid, showed a much increased level of gene expression compared to that of pH-insensitive nanoparticles.[42] The pH-sensitive polymeric lipid contains an acid-labile linker, ortho ester bond, which is presumed to quickly degrade when the particles are endocytosed by cells and are subjected to low-pH conditions of endosomes. Systematic investigations of the chemical modifications of PEI, the most effective polymeric gene carrier commercially available up until now, and the resulting effects on the characteristics of native PEI have also been reported.[43] As expected, the chemical modification of PEI affected its proton sponge capacity, hydrophobic–hydrophilic balance, lipophilicity, and cytotoxicity and transfection efficiency.

Owing to the polybasic activity of poly-L-histidine, some poly-L-hisitidine polymer-conjugated polymers are also introduced and tested for application to polymeric gene carrier systems.[44,45] Putnam and coworkers have reported that when histidine residue was conjugated to the backbone of the linear poly-L-lysine polymer, the grafted copolymers showed remarkably enhanced gene expression compared to native poly-L-lysine.[46]

As described above, one of the major problems with nonviral gene delivery systems is lower efficiency compared to viral vectors. Many techniques have so far been attempted to overcome such problems by linking or conjugating cell-specific ligands and TAT-derived peptide or oligopeptide, such as oligoarginine derivatives. Recently, some basic peptides known as protein transduction domains or membrane translocation signals were identified, characterized, and used for delivery of drugs, proteins, oligonucleotides, and plasmid DNA.[47,48] These peptide sequences usually contain positively charged amino acid residues, i.e., arginine and lysine. The mechanism of enhanced cellular uptake by these peptide molecules is still unclear and there is debate about whether their entry into cell membranes follows an endocytic pathway or nonendo-cytic pathway or direct penetration into membranes.[49] While most experiments were usually performed by covalently linking these peptides to polymers or lipids for nucleic acid delivery, they have also been simply mixed with nucleic acids for electrostatic interaction.

Arginine-oligopeptides modified with several hydrophobic lipids have been recently reported to be effective gene carriers and, interestingly, those peptides alone did not show a high level of transfection efficacy.[50] In addition, TAT-PEG-PE liposomal systems encapsulating plasmid DNA have been reported to be efficiently incorporated into cells in vitro and in vivo.[51,52] The common characteristic of those systems is thought to be that the arginine residues are rich on the surface of multi-valent liposomal systems. In addtion to these reports, Okuda and colleagues recently reported that the arginine residues located on the surface of poly-L-lysine dendrimer exerted a pronounced influence on the transfection efficiency of the polymer.[53] It was also reported that branched-chain arginine peptides showed a different cellular localization, implying that a liner structure was not necessary, and the formation of a cluster of arginines was suggested to be important for translocation.[49]

Fabrication of Metal Surface with Cationic Moiety for DNA Binding

In line with the development of integrated biological sensors, microelectronic compatible materials such as silicon, glass, and gold can be physically and chemically modified for many applications. Yang and his coworkers reported that DNA could be efficiently and stably conjugated onto the surface of nanocrystalline diamond thin films by a photochemical modification method.[54]

They also demonstrated that DNA-modified ultra-nanocrystalline diamond films showed very high stability and sensitivity compared to other commonly used surfaces, i.e., gold, silicon, glass, and glassy carbon. It is thought that the report provides a basic important principle for the development of injectable medical devices for gene delivery systems as well as drug delivery systems.

Another focus of the design of polymer/metal nanoparticle-mediated DNA delivery is the fabrication of surfaces of gold nanoparticles[55] and bimetallic nanorods (gold-titanium).[56] The use of metal particles for delivery of large sized plasmid DNA through simple chemical modification of the surfaces was successfully implemented.

Such technology involving polymer particles containing metal may open up a new field of polymer-metal hybrid systems for gene delivery. However, the safety problem resulting from the use of metal nanoparticles remains to be thoroughly studied for further clinical application.

Ligands for Targeting

Traditional polycationic charge-based polymers are reported to form complexes with DNA displaying high cationic surface charge. Such surplus surface cationic charge might contribute to increased gene expression for in vitro applications. However, there are many barriers for in vivo applications as reported by many groups, and the positively charged particles could not execute active targeting to some specific target cells or organs. In addition, within minutes after injection into the blood vessel, the serum components adsorb to the particulates regardless of their initial surface charge.[57]

One simple approach to receptor-mediated transfection is to make use of mono- or disaccharides that are recognized and internalized by the asialoglycoprotein receptors of hepatocytes.[58–60] The conjugation of the ligands to PLL in an appropriate configuration caused increased transfection of target cells in vitro and in vivo. Another attempt was to conjugate chemically the synthesized tripeptide Arg-Gly-Asp (RGD), which shows specific interaction with αV- and α5 integrins.[61] It was found that the introduced RGD sequence increased and affected the specific targeting and expression of the DNA complexes.

Hood et al. have developed a targeted gene delivery model that was efficient in in vivo experiments.[62] These authors found the potent αvβ3-targeting ligand candidate from the library of small organic molecules and conjugated the ligand to cationic polymerized lipid nanoparticles for a site-specific gene delivery in vivo. There was pronounced tumor regression in mice after systemic administration of an antiangiogenic gene through specific targeting to the tumor vasculature.

Another approach for targeting cancer cells is to make use of folic acid because the folic acid receptor is generally over-expressed in many cancer cells, especially in ovarian carcinomas.[63] Therefore, many groups have tried to make the ligands tethered onto the surface of polymer/DNA complexes and, more desirably, situated at the end of polymer-conjugated PEG chains for anticancer drug delivery[64–68] or gene delivery.[32,69–74] In general, the polymers used to make folate-decorated polyplexes were pLL, PEI, or poly(dimethylaminomethylmethacrylate (pDMAEMA).

Rebuffat and coworkers have reported an interesting strategy for enhanced gene delivery and expression.[75] They devised a steroid-mediated gene delivery system, which makes use of the interaction between glucocorticoid and glucocorticoid receptors (GRs) that are located in the cytosol. So, once the steroid-decorated DNA is introduced inside cells, it is subsequently transported efficiently to the nucleus by high affinity receptor interaction. They suggested that nuclear receptors in the cytoplasm could be utilized as intracellular gene delivery vehicles. In summary, efficient gene delivery polymers should be equipped with such reagents that facilitate nucleus uptake as well as the specific cell surface targeting ligands.

POLYMERIC NANOPARTICLE- OR INJECTABLE DEPOT-MEDIATED GENE DELIVERY

In addition to the polycationic polymers, physical DNA-trapping methods using degradable polymers have been studied.[76–80] Many kinds of synthetic or natural polymers that are degradable are introduced for controlled release of DNA systems, which usually encompass either micro-, nano-particulate systems or DNA-releasing injectable matrix systems. Such particulate systems are generally produced by a double emulsion-solvent evaporation technique (water-in-oil-in-water system). Due to the degradation of the ester bonds, the DNA trapped inside is released slowly out of the particles. In these systems, sustained and regulated DNA release can be engineered by controlling many factors, and subsequent sustained gene expression may be applicable to the treatment of certain types of localized disease conditions. Compared to traditional charge-based polymers, the particles produced by this system show negative zeta potential values.[81] This may prevent the particles from interacting with the extracellular matrix components, such as proteoglycans, after their *in vivo* administration.

By virtue of the development and accumulated knowledge of sustained release technology in drug delivery systems, it is believed that such technology could be successfully applied to gene therapy, compensating for some of the problems associated with polymeric gene delivery systems. For example, such a DNA-releasing depot system could consist of injectable or implantable matrix types, and nano or micro-particulate formulation types for either localized or targeted gene delivery to the tissue of interest.

Collagen are generally used in clinical settings for tissue reconstruction and have gained the attention of some researchers for their possible utility as an implantable matrix and gene releasing system.[82,83] Further applications have recently been reported as comprising interesting hybrid models for sustained DNA delivery systems composed of collagen loaded with DNA only or DNA with non-viral vectors.[84–90] They are promising attempts and interesting models which combine the merits of different approaches, i.e., nonviral vector systems, biomaterial-based tissue engineering, and controlled drug release systems.

CONCLUDING REMARKS

Virus-mediated gene therapy systems have revealed more unpredictable risks than hitherto known as investigation has intensified, even though their efficiency is still far greater than nonviral vector systems. So, there is no doubt that the demand for safe, efficient and reliable nonviral vectors will continue to increase in contrast to viral vectors in the future. The birth of perfect magic nonviral vectors requires multidisciplinary scientific cooperation and research. Moreover, in line with the development of nonviral vectors that are equipped with multifunctionalities, such as specific targeting, high efficiency, degradability and nontoxicity, research directed toward enhancing knowledge of DNA- or RNA-based therapeutic gene design should also be helpful in setting up essential and promising gene therapy systems for treating certain types of human diseases that would otherwise be fatal.[3,91–93]

Despite recent success in the development of novel nonviral gene delivery systems, there are still many barriers and hurdles to overcome even to the extent that some people still have a pessimistic view of gene therapy and refuse to admit its immense potential and practicality. However, with increasing knowledge of molecular biology, chemistry, biochemistry, pharmaceutics, bioengineering, and medicinal technology, it may no longer be impossible to use genes as therapeutics.

REFERENCES

1. Hunt, K.K.; Vorburger, S.A. Tech.Sight. Gene therapy. Hurdles and hopes for cancer treatment. Science **2002**, *297* (5580), 415–416.
2. Williams, D.A.; Baum, C. Medicine. Gene therapy—New challenges ahead. Science **2003**, *302* (5644), 400–401.
3. Schmidt-Wolf, G.D.; Schmidt-Wolf, I.G. Non-viral and hybrid vectors in human gene therapy: An update. Trends Mol. Med. **2003**, *9* (2), 67–72.
4. Ledley, F.D. Nonviral gene therapy: The promise of genes as pharmaceutical products. Hum. Gene Ther. **1995**, *6* (9), 1129–1144.
5. Stiriba, S.E.; Frey, H.; Haag, R. Dendritic polymers in biomedical applications: From potential to clinical use in diagnostics and therapy. Angew. Chem. Int. Ed. Engl. **2002**, *41* (8), 1329–1334.
6. Eichman, J.D.; Bielinska, A.U.; Kukowska-Latallo, J.F.; Baker, J.R., Jr. The use of PAMAM dendrimers in the efficient transfer of genetic material into cells. Pharm. Sci. Tech. Today **2000**, *3* (7), 232–245.
7. Boussif, O.; Lezoualc'h, F.; Zanta, M.A.; Mergny, M.D.; Scherman, D.; Demeneix, B.; Behr, J.P. A versatile vector for gene and oligonucleotide transfer into cells in culture and *in vivo*: Polyethylenimine. Proc. Natl. Acad. Sci. U.S.A. **1995**, *92* (16), 7297–7301.
8. Kukowska-Latallo, J.F.; Bielinska, A.U.; Johnson, J.; Spindler, R.; Tomalia, D.A.; Baker, J.R. Jr. Efficient transfer of genetic material into mammalian cells using Starburst polyamidoamine dendrimers. Proc. Natl. Acad. Sci. U.S.A. **1996**, *93* (10), 4897–4902.
9. Fischer, D.; Li, Y.; Ahlemeyer, B.; Krieglstein, J.; Kissel, T. *In vitro* cytotoxicity testing of polycations: Influence of

Gene–Hydrogels

polymer structure on cell viability and hemolysis. Biomaterials **2003**, *24* (7), 1121–1131.

10. Wolfert, M.A.; Dash, P.R.; Nazarova, O.; Oupicky, D.; Seymour, L.W.; Smart, S.; Strohalm, J.; Ulbrich, K. Polyelectrolyte vectors for gene delivery: Influence of cationic polymer on biophysical properties of complexes formed with DNA. Bioconjug. Chem. **1999**, *10* (6), 993–1004.

11. Lee, J.H.; Lim, Y.B.; Choi, J.S.; Lee, Y.; Kim, T.I.; Kim, H.J.; Yoon, J.K.; Kim, K.; Park, J.S. Polyplexes assembled with internally quaternized PAMAM-OH dendrimer and plasmid DNA have a neutral surface and gene delivery potency. Bioconjug. Chem. **2003**, *14* (6), 1214–1221.

12. Lim, Y.B.; Han, S.O.; Kong, H.U.; Lee, Y.; Park, J.S.; Jeong, B.; Kim, S.W. Biodegradable polyester, poly[alpha-(4 aminobutyl)-l-glycolic acid], as a non-toxic gene carrier. Pharm. Res. **2000**, *17* (7), 811–816.

13. Lim, Y.B.; Kim, S.M.; Lee, Y.; Lee, W.; Yang, T.; Lee, M.; Suh, H.; Park, J. Cationic hyperbranched poly(amino ester): A novel class of DNA condensing molecule with cationic surface, biodegradable three-dimensional structure, and tertiary amine groups in the interior. J. Am. Chem. Soc. **2001**, *123* (10), 2460–2461.

14. Lim, Y.B.; et al. Cationic hyperbranched polymer as an anionic DNA condensing molecule. Abstracts of Papers of the American Chemical Society, 2001; 221: U417–U417.

15. Lim, Y.B.; Kim, S.M.; Suh, H.; Park, J.S. Biodegradable, endosome disruptive, and cationic network-type polymer as a highly efficient and nontoxic gene delivery carrier. Bioconjug. Chem. **2002**, *13* (5), 952–957.

16. Lynn, D.M.; Langer, R. Degradable poly(beta-amino esters): Synthesis, characterization, and self-assembly with plasmid DNA. J. Am. Chem. Soc. **2000**, *122* (44), 10761–10768.

17. Akinc, A.; Lynn, D.M.; Anderson, D.G.; Langer, R. Parallel synthesis and biophysical characterization of a degradable polymer library for gene delivery. J. Am. Chem. Soc. **2003**, *125* (18), 5316–5323.

18. Lynn, D.M.; Anderson, D.G.; Putnam, D.; Langer, R. Accelerated discovery of synthetic transfection vectors: Parallel synthesis and screening of degradable polymer library. J. Am. Chem. Soc. **2001**, *123* (33), 8155–8156.

19. Anderson, D.G.; Lynn, D.M.; Langer, R. Semi-automated synthesis and screening of a large library of degradable cationic polymers for gene delivery. Angew. Chem. Int. Ed. **2003**, *42* (27), 3153–3158.

20. Stryer, L. *Biochemistry*; W. H. Freeman and Company: New York, 1995; 4th Edn., p. 568.

21. Gosselin, M.A.; Guo, W.; Lee, R.J. Incorporation of reversibly crosslinked polyplexes into LPDII vectors for gene delivery. Bioconjug. Chem. **2002**, *13* (5), 1044–1053.

22. Gosselin, M.A.; Guo, W.; Lee, R.J. Efficient gene transfer using reversibly cross-linked low molecular weight polyethylenimine. Bioconjug. Chem. **2001**, *12* (6), 989–994.

23. McKenzie, D.L.; Smiley, E.; Kwok, K.Y.; Rice, K.G. Low molecular weight disulfide cross-linking peptides as nonviral gene delivery carriers. Bioconjug. Chem. **2000**, *11* (6), 901–909.

24. Oupicky, D.; Parker, A.L.; Seymour, L.W. Laterally stabilized complexes of DNA with linear reducible

polycations: Strategy for triggered intracellular activation of DNA delivery vectors. J. Am. Chem. Soc. **2002**, *124* (1), 8–9.

25. Mandel, R.; Ryser, H.J.; Ghani, F.; Wu, M.; Peak, D. Inhibition of a reductive function of the plasma membrane by bacitracin and antibodies against protein disulfide-isomerase. Proc. Natl. Acad. Sci. U.S.A. **1993**, *90* (9), 4112–4116.

26. Guo, X.; Szoka, F.C. Chemical approaches to triggerable lipid vesicles for drug and gene delivery. Acc. Chem. Res. **2003**, *36* (5), 335–341.

27. Wang, J.; Mao, H.Q.; Leong, K.W. A novel biodegradable gene carrier based on polyphosphoester. J. Am. Chem. Soc. **2001**, *123* (38), 9480–9481.

28. Wang, J.; Zhang, P.C.; Lu, H.F.; Ma, N.; Wang, S.; Mao, H.Q.; Leong, K.W. New polyphosphoramidate with a spermidine side chain as a gene carrier. J. Control. Release **2002**, *83* (1), 157–168.

29. Kataoka, K.; Glenn, K.; Masayuki, Y.; Teruo, O.; Yasuhisa, S. Block copolymer micelles as vehicles for drug delivery. J. Control. Release **1993**, *24* (1–3), 119–132.

30. Kabanov, A.V.; Vinogradov, S.V.; Suzdaltseva, Y.G.; Alakhov, V. Water-soluble block polycations as carriers for oligonucleotide delivery. Bioconjug. Chem. **1995**, *6* (6), 639–643.

31. Kakizawa, Y.; Kataoka, K. Block copolymer micelles for delivery of gene and related compounds. Adv. Drug. Deliv. Rev. **2002**, *54* (2), 203–222.

32. Lee, R.J.; Huang, L. Folate-targeted, anionic liposome-entrapped polylysine-condensed DNA for tumor cell-specific gene transfer. J. Biol. Chem. **1996**, *271* (14), 8481–8487.

33. Nguyen, H.K.; Lemieux, P.; Vinogradov, S.V.; Gebhart, C.L.; Guérin, N.; Paradis, G.; Bronich, T.K.; Alakhov, V.Y.; Kabanov, A.V. Evaluation of polyether-polyethyleneimine graft copolymers as gene transfer agents. Gene Ther. **2000**, *7* (2), 126–138.

34. Wolfert, M.A.; Schacht, E.H.; Toncheva, V.; Ulbrich, K.; Nazarova, O.; Seymour, L.W. Characterization of vectors for gene therapy formed by self-assembly of DNA with synthetic block co-polymers. Hum. Gene Ther. **1996**, *7* (17), 2123–2133.

35. Choi, J.S.; Lee, E.J.; Choi, Y.H.; Jeong, Y.J.; Park, J.S. Poly(ethylene glycol)-block-poly(l-lysine) dendrimer: Novel linear polymer/dendrimer block copolymer forming a spherical water-soluble polyionic complex with DNA. Bioconjug. Chem. **1999**, *10* (1), 62–65.

36. Choi, J.S.; Joo, D.K.; Kim, C.H.; Kim, K.; Park, J.S. Synthesis of a barbell-like triblock copolymer, poly(l-lysine) dendrimer-block-poly(ethylene glycol)-block-poly(l-lysine) dendrimer, and its self-assembly with plasmid DNA. J. Am. Chem. Soc. **2000**, *122* (3), 474–480.

37. Kim, T.I.; Jang, H.-S.; Joo, D.K.; Choi, J.S.; Park, J.-S. Synthesis of diblock copolymer, methoxypoly(ethylene glycol)-block-polyamidoamine dendrimer and its generation-dependent self-assembly with plasmid DNA. Bull. Korean Chem. Soc. **2003**, *24* (1), 123–125.

38. Luo, D.; Haverstick, K.; Belcheva, N.; Han, E.; Sattzman, W.M. Poly(ethylene glycol)-conjugated PAMAM dendrimer for biocompatible, high-efficiency DNA delivery. Macromolecules **2002**, *35* (9), 3456–3462.

39. LaVan, D.A.; McGuire, T.; Langer, R. Small-scale systems for *in vivo* drug delivery. Nat. Biotechnol. **2003**, *21* (10), 1184–1191.

40. Wasiewski, W.; Fasco, M.J.; Martin, B.M.; Detwiler, T.C.; Fenton, J.W., II. Thrombin adsorption to surfaces and prevention with polyethylene glycol 6000. Thromb. Res. **1976**, *8* (6), 881–886.

41. Han, S.; Mahato, R.I.; Sung, Y.K.; Kim, S.W. Development of biomaterials for gene therapy. Mol. Ther. **2000**, *2* (4), 302–317.

42. Choi, J.S.; MacKay, J.A.; Szoka, F.C., Jr. Low-pH-sensitive PEG- stabilized plasmid-lipid nanoparticles: Preparation and characterization. Bioconjug. Chem. **2003**, *14* (2), 420–429.

43. Thomas, M.; Klibanov, A.M. Enhancing polyethylenimine's delivery of plasmid DNA into mammalian cells. Proc. Natl. Acad. Sci. U.S.A. **2002**, *99* (23), 14640–14645.

44. Putnam, D.; Zelikin, A.N.; Izumrudov, V.A.; Langer, R. Polyhistidine- PEG:DNA nanocomposites for gene delivery. Biomaterials **2003**, *24* (24), 4425–4433.

45. Benns, J.M.; Choi, J.S.; Mahato, R.I.; Park, J.S.; Kim, S.W. pH-sensitive cationic polymer gene delivery vehicle: N-Ac-poly(l-histidine)-graft-poly(l-lysine) comb shaped polymer. Bioconjug. Chem. **2000**, *11* (5), 637–645.

46. Putnam, D.; Gentry, C.A.; Pack, D.W.; Langer, R. Polymer-based gene delivery with low cytotoxicity by a unique balance of side-chain termini. Proc. Natl. Acad. Sci. U.S.A. **2001**, *98* (3), 1200–1205.

47. Tung, C.H.; Weissleder, R. Arginine containing peptides as delivery vectors. Adv. Drug. Deliv. Rev. **2003**, *55* (2), 281–294.

48. Henry, C.M. Breaching barriers. Chem. Eng. News **2003**, *81*, 35.

49. Futaki, S. Arginine-rich peptides: Potential for intracellular delivery of macromolecules and the mystery of the translocation mechanisms. Int. J. Pharm. **2002**, *245* (1), 1–7.

50. Futaki, S.; Ohashi, W.; Suzuki, T.; Niwa, M.; Tanaka, S.; Ueda, K.; Harashima, H.; Sugiura, Y. Stearylated arginine-rich peptides: A new class of transfection systems. Bioconjug. Chem. **2001**, *12* (6), 1005–1011.

51. Torchilin, V.P.; Levchenko, T.S.; Rammohan, R.; Volodina, N. Cell transfection *in vitro* and *in vivo* with nontoxic TAT peptide-liposome-DNA complexes. Proc. Natl. Acad. Sci. U.S.A. **2003**, *100* (4), 1972–1977.

52. Torchilin, V.P.; Rammohan, R.; Weissig, V.; Levchenko, T.S. TAT peptide on the surface of liposomes affords their efficient intracellular delivery even at low temperature and in the presence of metabolic inhibitors. Proc. Natl. Acad. Sci. U.S.A. **2001**, *98* (15), 8786–8791.

53. Okuda, T.; Sugiyama, A.; Niidome, T.; Aoyagi, H. Characters of dendritic poly(l-lysine) analogues with the terminal lysines replaced with arginines and histidines as gene carriers *in vitro*. Biomaterials **2004**, *25* (3), 537–544.

54. Yang, W.; Auciello, O.; Butler, J.E.; Cai, W.; Carlisle, J.A.; Gerbi, J.E.; Gruen, D.M.; Hamers, R.J. DNA-modified nanocrystalline diamond thin-films as stable, biologically active substrates. Nat. Mater. **2002**, *1* (4), 253–257.

55. Thomas, M.; Klibanov, A.M. Conjugation to gold nanoparticles enhances polyethylenimine's transfer of plasmid DNA into mammalian cells. Proc. Natl. Acad. Sci. U.S.A. **2003**, *100* (16), 9138–9143.

56. Salem, A.K.; Searson, P.C.; Leong, K.W. Multifunctional nanorods for gene delivery. Nat. Mater. **2003**, *2* (10), 668–671.

57. Juliano, R.L. Factors affecting the clearance kinetics and tissue distribution of liposomes, microspheres and emulsions. Adv. Drug Deliv. Rev. **1988**, *2* (1), 31–54.

58. Perales, J.C.; Gene transfer *in vivo*: Sustained expression and regulation of genes introduced into the liver by receptor-targeted uptake. Proc. Natl. Acad. Sci. U.S.A. **1994**, *91* (9), 4086–4090.

59. Choi, Y.H.; Characterization of a targeted gene carrier, lactose-polyethylene glycol-grafted poly-l-lysine, and its complex with plasmid DNA. Hum. Gene Ther. **1999**, *10* (16), 2657–2665.

60. Choi, Y.H.; Liu, F.; Park, J.S.; Kim, S.W. Lactose-poly(ethylene glycol)- grafted poly-l-lysine as hepatoma cell-tapgeted gene carrier. Bioconjug. Chem. **1998**, *9* (6), 708–718.

61. Hart, S.L. Gene delivery and expression mediated by an integrin-binding peptide. Gene Ther. **1996**, *3* (8), 552–554.

62. Hood, J.D.; Bednarski, M.; Frausto, R.; Guccione, S.; Reisfeld, R.A.; Xiang, R. Tumor regression by targeted gene delivery to the neovasculature. Science **2002**, *296* (5577), 2404–2407.

63. Corona, G.; Giannini, F.; Fabris, M.; Toffoli, G.; Boiocchi, M. Role of folate receptor and reduced folate carrier in the transport of 5-methyltetrahydrofolic acid in human ovarian carcinoma cells. Int. J. Cancer **1998**, *75* (1), 125–133.

64. Sudimack, J.; Lee, R.J. Targeted drug delivery via the folate receptor. Adv. Drug Deliv. Rev. **2000**, *41* (2), 147–162.

65. Stella, B.; Arpicco, S.; Peracchia, M.T.; Desmaële, D.; Hoebeke, J.; Renoir, M.; D'Angelo, J.; Cattel, L.; Couvreur, P. Design of folic acid-conjugated nanoparticles for drug targeting. J. Pharm. Sci. **2000**, *89* (11), 1452–1464.

66. Leamon, C.P.; Low, P.S. Selective targeting of malignant cells with cytotoxin-folate conjugates. J. Drug Target **1994**, *2* (2), 101–112.

67. Leamon, C.P.; DePrince, R.B.; Hendren, R.W. Folate-mediated drug delivery: Effect of alternative conjugation chemistry. J. Drug Target **1999**, *7* (3), 157–169.

68. Lee, R.J.; Low, P.S. Delivery of liposomes into cultured KB cells via folate receptor-mediated endocytosis. J. Biol. Chem. **1994**, *269* (5), 3198–3204.

69. Leamon, C.P.; Weigl, D.; Hendren, R.W. Folate copolymer-mediated transfection of cultured cells. Bioconjug. Chem. **1999**, *10* (6), 947–957.

70. Benns, J.M.; Mahato, R.I.; Kim, S.W. Optimization of factors influencing the transfection efficiency of folate-PEG-folate-graft-polyethylenimine. J. Control. Release **2002**, *79* (1), 255–269.

71. Benns, J.M.; Folate-PEG-folate-graft-polyethylenimine-based gene delivery. J. Drug Target **2001**, *9* (2), 123–139.

72. Mislick, K.A.; Baldeschwieler, J.D.; Kayyem, J.F.; Meade, T.J. Transfection of folate-polylysine DNA complexes: Evidence for lysosomal delivery. Bioconjug. Chem. **1995**, *6* (5), 512–515.

73. Ward, C.M. Folate-targeted non-viral DNA vectors for cancer gene therapy. Curr. Opin. Mol. Ther. **2000**, *2* (2), 182–187.

74. van Steenis, J.H.; Van Maarseveen, E.M.; Verbaan, F.J.; Verrijk, R.; Crommelin, D.J.A.; Storm, G.; Hennink, W.E.

Preparation and characterization of folate targeted pEG-coated pDMAEMA-based polyplexes. J. Control. Release **2003**, *87* (1), 167–176.

75. Rebuffat, A.; Bernasconi, A.; Ceppi, M.; Wehrli, H.; Verca, S.B.; Ibrahim, M.; Frey, B.M.; Frey, F.J.; Rusconi, S. Selective enhancement of gene transfer by steroid-mediated gene delivery. Nat. Biotechnol. **2001**, *19* (12), 1155–1161.

76. Zambaux, M.F.; Bonneaux, F.; Gref, R.; Maincent, P.; Dellacherie, E.; Alonso, M.J.; Labrude, P.; Vigneron, C. Influence of experimental parameters on the characteristics of poly(lactic acid) nanoparticles prepared by a double emulsion method. J. Control. Release **1998**, *50* (1), 31–40.

77. Labhasetwar, V.; Bonadio, J.; Goldstein, S.A.; Levy, J.R. Gene transfection using biodegradable nanospheres: Results in tissue culture and a rat osteotomy model. Colloids Surf. B–Biointerfaces **1999**, *16* (1), 281–290.

78. Wang, D.; Robinson, D.R.; Kwon, G.S.; Samuel, J. Encapsulation of plasmid DNA in biodegradable poly(d,l-lactic-co-glycolic acid) microspheres as a novel approach for immunogene delivery. J. Control Release **1999**, *57* (1), 9–18.

79. Panyam, J.; Labhasetwar, V. Biodegradable nanoparticles for drug and gene delivery to cells and tissue. Adv. Drug Deliv. Rev. **2003**, *55* (3), 329–347.

80. Eliaz, R.E.; Szoka, F.C., Jr. Robust and prolonged gene expression from injectable polymeric implants. Gene Ther. **2002**, *9* (18), 1230–1237.

81. Prabha, S.; Zhou, W.Z.; Panyam, J.; Labhasetwar, V. Size-dependency of nanoparticle-mediated gene transfection: Studies with fractionated nanoparticles. Int. J. Pharm. **2002**, *244* (1), 105–115.

82. Bonadio, J.; Smiley, E.; Patil, P.; Goldstein, S. Localized, direct plasmid gene delivery *in vivo*: Prolonged therapy results in reproducible tissue regeneration. Nat. Med. **1999**, *5* (7), 753–759.

83. Fang, J.; Zhu, Y.Y.; Smiley, E.; Bonadio, J.; Rouleau, J.P.; Goldstein, S.A.; McCauley, L.K.; Davidson, B.L.; Roessler, B.J. Stimulation of new bone formation by direct transfer of osteogenic plasmid genes. Proc. Natl. Acad. Sci. U.S.A. **1996**, *93* (12), 5753–5758.

84. Scherer, F.; Schillinger, U.; Putz, U.; Stemberger, A.; Plank, C. Nonviral vector loaded collagen sponges for sustained gene delivery *in vitro* and *in vivo*. J. Gene Med. **2002**, *4* (6), 634–643.

85. Kyriakides, T.R.; Hartzel, T.; Huynh, G.; Bornstein, P. Regulation of angiogenesis and matrix remodeling by localized, matrix-mediated antisense gene delivery. Mol. Ther. **2001**, *3* (6), 842–849.

86. Tyrone, J.W.; Collagen-embedded platelet-derived growth factor DNA plasmid promotes wound healing in a dermal ulcer model. J. Surg. Res. **2000**, *93* (2), 230–236.

87. Berry, M.; Gonzalez, A.M.; Clarke, W.; Greenlees, L.; Barrett, L.; Tsang, W.; Seymour, L.; Bonadio, J.; Logan, A.; Baird, A. Sustained effects of gene-activated matrices after CNS injury. Mol. Cell. Neurosci. **2001**, *17* (4), 706–716.

88. Pakkanen, T.M.; Laitinen, M.; Hippeläinen, M.; Hiltunen, M.O.; Alhava, E.; Ylä-Herttuala, S. Periadventitial lacZ gene transfer to pig carotid arteries using a biodegradable collagen collar or a wrap of collagen sheet with adenoviruses and plasmid-liposome complexes. J. Gene Med. **2000**, *2* (1), 52–60.

89. Chandler, L.A.; Ma, C.; Gonzalez, A.M.; Doukas, J.; Nguyen, T.; Pierce, G.F.; Phillips, M.L. Matrix-enabled gene transfer for cutaneous wound repair. Wound Repair Regen. **2000**, *8* (6), 473–479.

90. Doukas, J.; Chandler, L.A.; Gonzalez, A.M.; Gu, D.; Hoganson, D.K.; Ma, C.; Nguyen, T.; Pierce, G.F. Matrix immobilization enhances the tissue repair activity of growth factor gene therapy vectors. Hum. Gene Ther. **2001**, *12* (7), 783–798.

91. Patzel, V.; Kaufmann, S.H. Towards simple artificial infectious systems. Trends Mol. Med. **2003**, *9* (11), 479–482.

92. Zuber, G.; Dauty, E.; Nothisen, M.; Belguise, P.; Behr, J.P. Towards synthetic viruses. Adv. Drug Deliv. Rev. **2001**, *52* (3), 245–253.

93. Henry, C.M. Gene delivery—Without viruses. Chem. Eng. News **2001**, *79* (1), 35–41.

Gene Delivery

Tetsuji Yamaoka
Department of Polymer Science and Engineering, Kyoto Institute of Technology, Kyoto, Japan

Abstract

Effective non-viral gene carrier–DNA complexes should achieve well-controlled intracellular trafficking and be recognized by transcription factors after entering the cells. In other words, the complexes should possess supramolecular structures which are biologically active. It has become clear that the physicochemical characteristics of the DNA–carrier complex function through these various steps. The well-controlled and specialized supramolecular structures of the complexes must be the key factors for gene delivery. Detailed information on the structure–function correlation of the supramolecular complexes would permit further molecular design of novel and effective gene carriers. In this entry, various gene carriers are reviewed and the effects of the characteristics of supramolecular complexes on gene delivery efficiency are discussed, with a focus on polymeric gene carriers.

INTRODUCTION

Transfection of foreign genes into mammalian cells *in vitro* and *in vivo* is the most important technique of gene therapy.[1] More than 75% of the clinical protocols involving gene therapy to date employed viral vectors, such as retroviruses (about 50%), adenoviruses (about 20%), adenoassociated viruses, or pox viruses because of their excellent transgene expression. However, despite their high efficiency *in vitro*, their successful use *in vivo* is often limited by toxicity, immunogenicity, inflammatory properties, limited DNA size, and reproduction problems of the viruses.[1]

Various nonviral gene carriers are attracting great attention because they are biologically safe and their chemical structures can be designed.[2–4] These nonviral gene carriers possess positively charged groups, and this permits complex formation with DNA and condensation of the DNA coils into globules.

The most widely studied nonviral gene carrier is the cationic liposome,[3] used in about 20% of clinical trials. The liposomes are accumulated undesirably into the reticuloendothelial systems after systemic injection, resulting in limited *in vivo* applications.[5–7] Polymeric gene carriers also possess cation charges at the side and/or main chains and can condense DNA coils into 10^{-3} to 10^{-4} of their original volume[8] by forming polyion complexes that are conventionally advocated in the field of polymer science.[9–11] In contrast, transgene expression greatly depends on chemical structure and shape, such as the linear or branched structures of the polycations used, suggesting that the formed polyion complex should possess special characteristics for improved gene transfer.

Gene delivery involves control of more complicated factors than the conventional drug delivery system. The introduced gene should pass various barriers before being expressed: passing through the plasma membrane, internalization into cells, inhibited intracellular hydrolysis, controlled intracellular trafficking, and recognition by the transcription factors. Wu and coworkers[9] reported the pioneering work in *in vitro* gene delivery through receptor-mediated endocytosis, using polycations having targeting ligands against hepatocytes, resulting in the improved uptake of the complexes. This approach was based on a similar strategy established in conventional drug delivery systems. However, base carrier materials with poor efficiency cannot lead to sufficient transgene expression even when the DNA molecules are delivered into the cells. This suggests the importance of the nonviral carrier materials. The interaction of DNA and histone, which is based on a similar mechanism, is a biologically active event and many researchers pay much attention to its function. In contrast, the characteristics of gene carrier/DNA complexes affecting the expression of the transgene have not been studied although the characteristic/function correlation in the complexes is apparently important.

Effective nonviral gene carrier/DNA complexes should achieve well controlled intracellular trafficking and be recognized by transcription factors after entering into the cells; in other words, the complexes should possess supramolecular structures which are biologically active. Recently, it has become clear that the physicochemical characteristics of the DNA/carrier complex functions through these various steps. The well controlled and specialized supramolecular structures of the complexes must be the key factors for gene delivery. Detailed information on the structure/function correlation of the supramolecular complexes would permit further molecular design of novel and effective gene carriers. In this entry, various gene

Concise Encyclopedia of Biomedical Polymers and Polymeric Biomaterials DOI: 10.1081/E-EBPPC-120052245

Gene–Hydrogels

carriers are reviewed and the effects of the characteristics of supramolecular complexes on gene delivery efficiency will be discussed, with a focus on polymeric gene carriers.

NONVIRAL GENE CARRIERS

Cationic Liposomes

Cationic liposomes are the most widely used nonviral vectors *in vitro* and *in vivo*.[12] A cationic liposome complex with DNA is known as a lipoplex while the complex of a cationic polymer is known as a polyplex. Among cationic compounds with different chemical structures shown in Fig. 1, 1,2-dioley loxypropyl-3-trimethylammonium bromide (DOTMA) is the most widely used because of its effective gene expression even *in vivo*.[2,3,13] The size[14] and the hydrophilicity[15] of the lipoplex are important factors for effective gene delivery as well as for the linear polycations described later. In spite of their high efficacy, they exhibit poor biocompatibility, unexpected accumulation into the liver or spleen, and rapid degradation of DNA when injected *in vivo*.

Lew et al. reported the pharmacokinetic analysis of plasmid DNA/cationic lipid complexes injected intravenously in mice by Southern blot analyses and showed that the DNA was rapidly degraded in the bloodstream with a half-life of less than 5 min; the intact DNA was no longer detectable at 1 hr post-injection.[16] Serum proteins strongly inhibit the transfer efficiency of the transgene *in vitro* when cationic liposomes are used as carriers, maybe because of the degradation of DNA which is not encapsulated in the liposome membrane by nuclease contained in the serum. In addition, since many lipoplexes are polydisperse in size, charge, and stoichiometry,[17] the effective application *in vivo* is limited to local delivery such as inhalation[5,6] or injection directly into a tumor or tumor-feeding arteries.[7]

Various cationic liposomes with bioderived active ligands on their surfaces were also proposed. For example, fusogenic liposome, which is prepared by fusing unilamellar liposomes with UV-inactivated Sendai virus, produces high gene expression with low cytotoxicity *in vitro* even in the presence of 40% fetal calf serum (FCS) in the medium.[18]

The DNA encapsulated in the membranous structure of the liposome is strongly protected from the digestion by the nuclease, resulting in excellent expression even in *in vivo* gene transfer.[19]

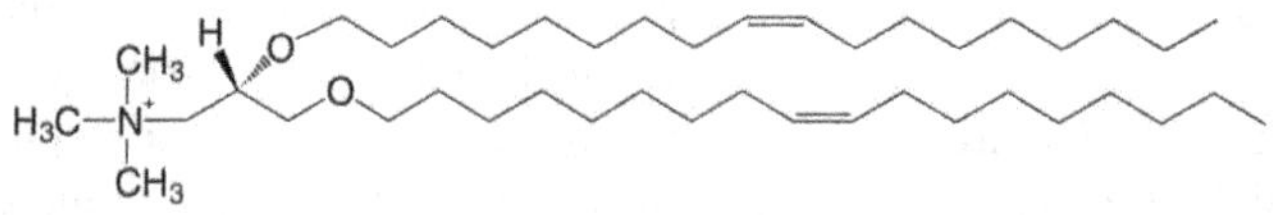

DOTMA [1,2-Dioleyloxypropyl-3-trimethyl ammonium bromide]

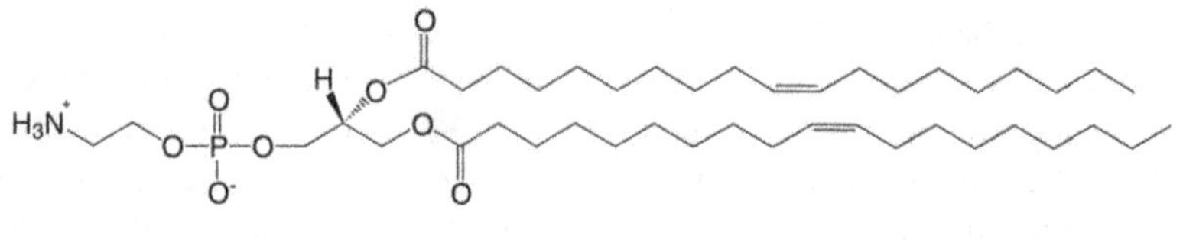

DOPE [Dioleoylphosphatidylethanolamine]

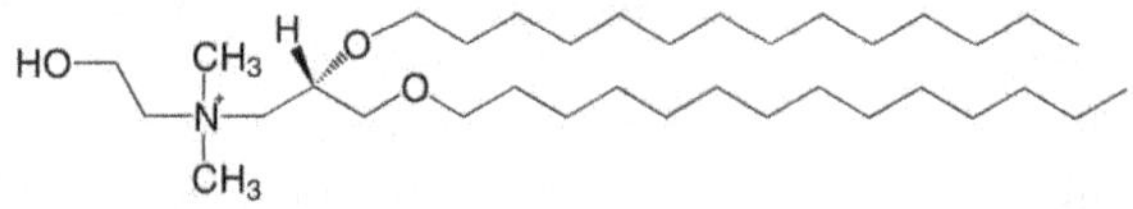

DMRIE [1,2-Dimyristyloxypropyl-3-dimethyl-hydroxyethyl ammonium bromide]

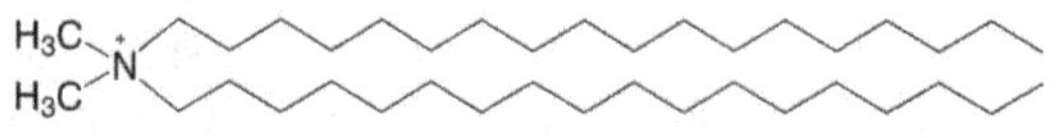

DDAB [Dimethyldioctadecyl ammonium bromide]

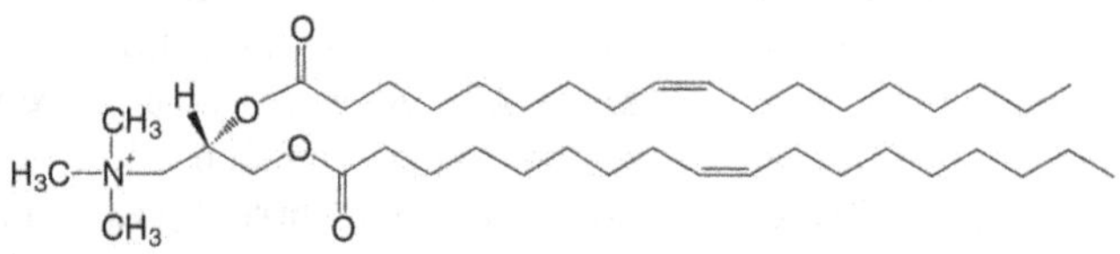

DOTAP [1,2-Dioleoyloxy-3-(trimethylammonio)propane]

Fig. 1 Chemical structures of the compounds used for cationic liposomes.

Linear Polycations

Over the past decade, various cationic polymers have been proposed to act as effective gene carriers *in vitro*. Among them, poly(L-lysine) (PLL) is among the most widely studied polycations because those with various molecular weights are commercially available.[20] Cationic polysaccharides such as diethylaminoethyl–dextran (DEAE–dextran)[21,22] have been utilized in the field of molecular biology. Chitosan[23] was found to be much more effective than DEAE–dextran, and various synthetic polycations such as polyethyleneimine (PEI),[9,11,24–26] polybrene (PB),[27,28] and cationic polymethacrylate derivatives[29,30] were also proposed. Since these polycations have been evaluated separately, little information on the effects of

their chemical structures on gene-introducing efficiency has been offered, resulting in difficulty in designing novel carrier polycations.

In our group, various polycations with different chemical structures as shown in Fig. 2 have been subjected to transfection experiments to determine the required chemical structures of effective polymeric gene carriers. Plasmid DNAs containing functional *lacZ* gene, EGFP gene, or luciferase gene were selected as reporter DNAs and transferred to various types of cells by the osmotic shock procedure reported by Takai et al. (Fig. 3).[31] The osmotic shock procedure enabled us to analyze only the events following the release of the complexes from the endocytotic vesicles into the cytosols (details will be discussed later).

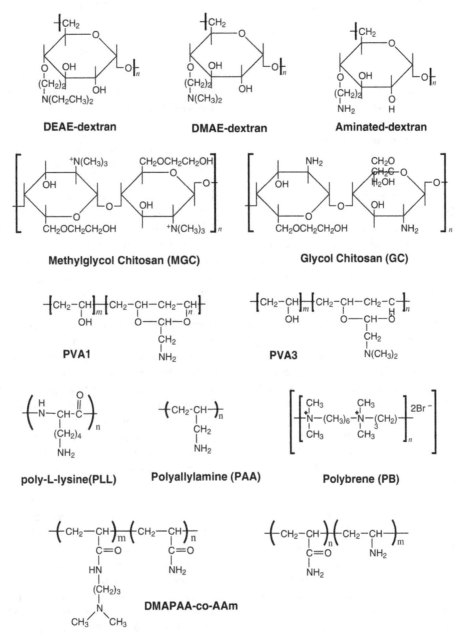

Fig. 2 Chemical structures of polycations used by our group.

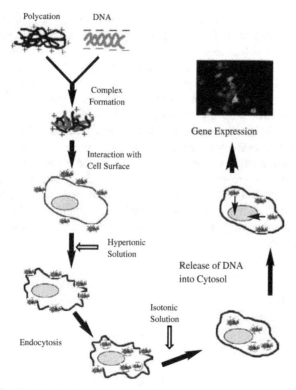

Polycation   DNA

Complex Formation

Interaction with Cell Surface

Hypertonic Solution

Release of DNA into Cytosol

Isotonic Solution

Endocytosis

Fig. 3 Mechanism of the osmotic shock procedure for *in vitro* gene transfer.

Polycation	Main Chain	Hydrophilic Group	Cationic Group	Expression
DEAE-dextran			$-N(C_2H_5)_2$	Excellent
DMAE-dextran			$-N(CH_3)_2$	High
Aminated-dextran	Polysaccharide	-OH	$-NH_2$	Slight
MGC			$-N(CH_3)_2$	High
GC			$-NH_2$	Slight
PLL		—	$-NH_2$	None
DMAPAA-AAm	CH_2-CH_n	$-CONH_2$	$-N(CH_3)_2$	Low
AAm-NH_2	|	$-CONH_2$	$-NH_2$	None
PAA	x or	—	$-NH_2$	None
PVA-1	$(CH_2-CXH)_n$	-OH	$-NH_2$	None
PVA-3		-OH	$-N(CH_3)_2$	Excellent
PB	$R-N(CH_3)_{2n}$	—	$-N(CH_3)_2$	None

Fig. 4 Transient expression of pCH110 plasmid DNA introduced into COS-1 cells using various polycations.

Agarose gel electrophoresis revealed that every polycation form complexes in a similar fashion, depending on the mixing ratio of polycation and DNA. The sizes of the complexes or their zeta potentials evaluated by dynamic light scattering are also similar, irrespective of polycations used. Despite these similarities, the expression of the transgene greatly depended on the chemical structures of the polycations. For all polycations, no gene expression was observed at the C/A ratio (the ratio of cationic groups of polycations to anionic phosphate groups of DNA) less than 1.0, suggesting the importance of the positively charged complexes that would interact effectively with the negatively charged cell surfaces. The qualitative results in gene expression by the polycations and their characteristic structures are summarized in Fig. 4. These results suggest that polycations having tertiary or quaternary ammonium groups led to high gene expression and those with primary ammonium groups were less efficient, indicating that the dissociation behavior of the cationic groups makes an important contribution to the high levels of gene expression. Only polycations with hydroxyl groups led to effective gene expression. According to these results, highly efficient water-soluble gene carriers would seem to have both dissociated cationic groups and sufficient hydroxyl groups. Although the role of the hydroxyl groups is still uncertain, we are focusing on the fact that the hydrophilic groups of polycations strongly maintain the hydrophilic nature of the formed complexes, effectively prevent compaction of the complexes, and allow their disassembling discussed below.

Among the polycations evaluated, PLL and PB, the most widely investigated as base polymers for gene carriers, were found to be the least effective. We have attempted to improve the capacity of PLL. Poly(lysine-*co*-serine), a random copolymer of L-lysine and serine with the unit composition of 3/1 (PLS), was employed as the polycation with hydroxyl groups. The ε-ammonium groups of PLL and PLS were converted into quaternary ammonium groups by their methylation, yielding Poly($N^ε$-trimethyl-L-lysine) (PtmL) and Poly($N^ε$-trimethyl-L-lysine-*co*-serine) (PtmLS), respectively.[32] These polycations have similar molecular weights around 20,000 (Fig. 5). pEGFP-N1 plasmid was transfected into COS-1 cells and the percentage of the cells expressing the EGFP gene was evaluated under a fluorescent microscope. The dependence of the transient expression of the introduced gene using the polypeptides on the C/A ratio is shown in Fig. 5. PL and PtmL showed no transient expression irrespective of the C/A ratio, suggesting that the imparting of high basicity only to PL is not effective in enhancing transfection efficacy. Slightly enhanced expression was observed in the case of PLS with a C/A ratio around 10; PtmLS induced much higher transgene expression. The percent transient expression was rapidly increased at a C/A ratio of 3.0 which is in a good agreement with the fact that the total charge of the PtmLS/DNA complex became positive at C/A > 3.0.

The big difference in the efficacy of PLS and PtmLS indicates an important role of the quaternary ammonium groups in the presence of the hydrophilic residues. The amount of FITC-labeled DNA taken up by cells was in the range of 0.6–1.9 ng/10^4 cells irrespective of the used polypeptides, suggesting that the difference in transient

Gene–Hydrogels

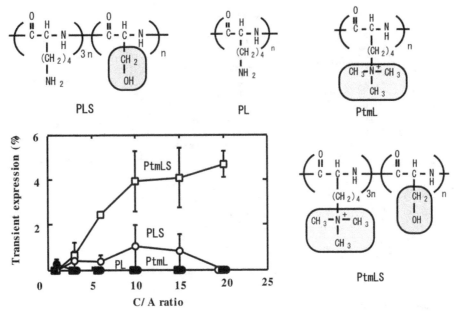

Fig. 5 Transient expression of pEGFP gene introduced into COS-1 cells by the osmotic shock procedure using (●) PL, (○) PLS, (■) PtmL, and (□) PtmLS.

expression was determined at some point following the internalization of the complexes into the cells. When these complexes are subjected to *in vitro* transcription/translation using rabbit reticulocyte lysate, the amount of expressed luciferase for PLS and PtmLS were 3.48 and 2.31 times larger than PL and PtmL, respectively, indicating that the serine residues improve the transcription of the complexes, possibly because of their hydrophilic nature. Erbacher et al. also described the low efficiency of the PLL and the increased efficiency of gluconoylated PLL, and concluded that the decreased strength of the electrostatic interaction between DNA and gluconoylated PLL is a crucial factor in inducing high gene expression.

Branched Polycations

The branched structures of gene carriers are considered to affect gene expression.[9,11,24–26] Plank et al. showed the efficacy of the branched structure,[33] while Ohashi et al. reported more efficient *in vivo* gene transfer using linear PEI than transfer using branched PEI.[34] These results conflict with the viewpoint of the importance of the branched structure, and further research is necessary.

In 1993, Szoka and his coworkers proposed hyperbranched polyamidoamine (PAMAM) cascade polymers, a well-defined class of dendritic polymers from methyl acrylate and ethylenediamine, as novel gene carrier.[10] They successfully achieved very high levels of gene expression using "fractured dendrimers" that were significantly degraded by heat treatment at the amide linkage compared to the expression of intact dendrimers.[35] The great efficiency seems to be based on their high buffering capacity, which results in the rupture of the endocytotic vesicles and accelerates the DNA release from the endosome to the cytosol.[36]

Self-Assembly of Block Copolymer/DNA Complexes

Wolfert et al. reported transgene expression using AB type hydrophilic-cationic block copolymers such as poly(ethylene glycol) (PEG)-PLL and poly-N-(2-hydroxypropyl) methacrylamide-poly(trimethylammonioethyl methacrylate chloride) block copolymers.[37] These copolymers spontaneously form a complex with DNA via the cationic segment resulting in small aggregates with sizes around 100 nm, which are stabilized by the hydrophilic segments PEG or poly-N-(2-hydroxypropyl) methacrylamide, covering their surfaces. The extended structures of the complexes formed revealed high gene expression. Other characteristics of the PEG-PLL/DNA complexes, such as diameter, interexchange reaction with other polyanions, and stability of the DNA structure were also reported.[38,39]

FATE OF INTRODUCED GENES

The mechanism of gene transfer is thought to follow the general endocytotic process shown in Fig. 6. A variety of polycations and other carriers condense DNA through electrostatic interaction. The interactions between the complexes and the cells are normally based on the electrostatic attractive forces between the positive charges of the complexes and the negative charges of the sialic acid at the cell surface. The attached complexes are internalized into the cells and usually localized in the endocytotic vesicles.

The vesicles are then transported in the cells as the internal pH of the vesicles gradually decreases to about 5.5 and fuse with the lysosomes, resulting in secondary lysosomes at which the incorporated DNA is hydrolyzed by the

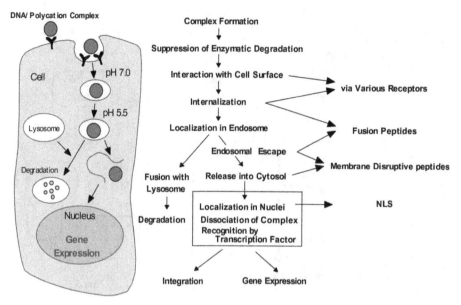

Fig. 6 Proposed mechanisms of gene transfer by nonviral gene carriers.

lysosomal enzymes. Even in the case of receptor-mediated endocytosis, the internalized DNA molecules are encapsulated in the endocytotic vesicles.[40] Therefore, the release of the DNA from the vesicle to the cytoplasm during this process is important for preventing DNA from hydrolysis. Next, the released DNA must be transported to the nucleus by some means and recognized by the transcript factors. This recognition may require the disassembly of the complex to release naked DNAs or somewhat specialized structures of the complexes by which the transcription of the DNA in the complexes is allowed. This transcription is the most important step for gene delivery, which is definitely different from conventional DDS for low molecular weight drugs such as antibiotics or antitumor agents. In the case of conventional DDSs, controlling the drug release rate based on the hydrolysis rate of the covalent bonding is effective and the released drugs can diffuse to the site of action, while the DNA/carrier complexes should be controlled in the adequate structures for effective gene expression. Several research groups are investigating the characteristics of the complexes, the releasing mechanism of the naked DNA, and intracellular trafficking.

Physicochemical Properties of Complexes

Although histones, spermidine, and spermine condense DNA molecules in cells and play an important role in gene transcription,[41,42] they do not necessarily improve the transfection efficiency when used as gene carriers perhaps because of the lower zeta-potentials of the complexes.[36] Polyplexes composed of polycations with sufficient molecular weights exhibit higher zeta potential, which facilitates the internalization of the complexes into the cells.[43] The physicochemical features of the polyplexes such as size,

shape, aggregation properties, thermostability, surface charge, conformational change of DNA, and the ease of interexchange reactions with the other polyanions of transgene expression have attracted great attention.[36,44,45]

Various polycations have been found to form complexes of similar shape and size. Tang and Szoka studied the correlation of these factors and transfection efficiency using linear PLL, intact PAMAM dendrimers, fractured dendrimers, and branched PEI. These polycations formed similar complexes in terms of size and zeta potential, but exhibited different aggregation properties, suggesting the high gene expression induced by fractured dendrimers and branched PEI results from the stable complexes that do not aggregate over time.[36]

Interaction with Cell Surfaces

The complexes, which contain excessive polycation, have positive charges as a whole and then interact with cell surfaces by an electrostatic interaction. This interaction between the complex and the cell surface is not merely the starting stage for the internalization. The negative charges of the cell surface originate from the polysaccharides existing at the cell surface. Recently, Mislick and Baldeschwieler reported an important role of the proteoglycan other than the electrostatic interaction between the complexes and cell surface. DNA molecules in the complexes are replaced by the proteoglycan and released from the complex by interacting with proteoglycan at the surface.[46]

Some other noncovalent bondings are proposed for improving the complex/cell interaction. PEG is one of the candidates for improving cell/complex interaction because it is able to promote cell fusion through the perturbation of the plasma membrane. However, PEG may reduce the interaction because it is also the stabilizing agent for protein or other drugs in preventing endocytosis by the reticuloendothelial system.[47] These two conflicting effects must be carefully considered when designing gene carriers. Zhou et al. reported high gene expression using phosphatidylethanolamine, which has chemical features similar to cell surfaces, as a modification agent for low molecular weight PLL. This carrier also reduces the interaction of the complexes with serum proteins, resulting in effective gene expression *in vitro* even in the presence of 10% FCS.[48]

Some reports deal with the fusion of liposome to the plasma membrane, resulting in the direct gene delivery of DNA into the cytosol. However, DNA that forms complexes with cationic liposomes is not encapsulated in the liposomes but attaches outsides of liposomes. Thus we have no clear proof of the direct delivery of the DNA to the cytoplasm. Even when targeting moieties such as sugar or transferrin that can interact with cell surfaces are used, the complexes follow a similar trafficking pathway after the internalization.

Intracellular Trafficking

Although the efficiency of gene delivery is known to depend greatly on the chemical structures of the gene carriers, the reason for the dependency is unclear. Because the big difference in the gene expression that depends on the carrier molecules used is observed even when the amount of the internalized DNA is taken into consideration, this difference must occur in some steps between the internalization into cells and the recognition by the transcript factors. One possible step is the release of the complexes or free DNA from the endocytotic vesicles to the cytoplasm. Gene delivery using various carriers should be analyzed separately before and after the release. For *in vitro* gene delivery, various introduction methods such as chloroquine treatment,[49] microinjection, and osmotic shock procedure have been employed.

The most widely selected method to enhance transfection efficacy is *in vitro* chloroquine treatment. The detailed mechanism of the chloroquine treatment is still unclear, but chloroquine seems to neutralize the acidic pH of endocytotic vesicles because of their weak acidity, resulting in the inactivation of lysosomal enzymes and inhibition of the fusion of endocytotic vesicles with lysosomes.[50] On the other hand, chloroquine is also known to form a complex with DNA, thus creating some difficulty for analyzing its mechanism.[51] DNA molecules transferred by the osmotic shock procedure were released into the cytosol by osmotic rupture of the endocytotic vesicle. Since the introduced gene should be traced and analyzed in a step-by-step manner in order to evaluate the roles of the carrier polycations, the osmotic shock procedure is a very useful method to deliver DNA into the cytosol more easily than microinjection. The osmotic shock procedure enables us to analyze the events following the release of the complexes from the endocytotic vesicle into the cytosol.

These treatments can be utilized only in *in vitro* experiments and are not applicable for clinical trials. The promising methods for *in vivo* treatment are raising osmotic pressure in the endosome by the action of the carrier polycations and fusion or rupture of the endosomal membrane by the actions of oligopeptides or polypeptides. Godbey et al. showed that PEI internalized by the cells undergoes nuclear localization whether administered with or without DNA.[9] They suggested that the polyplexes come into contact with phospholipids of endosomes and the membranes are permeabilized and burst because of osmotic swelling, resulting in the coating of the polyplexes with the phospholipids. The coated complexes may enter the nucleus via fusion with the nuclear envelope.

The types and arrangements of cationic groups of the carriers are also reported to effect the endosomal escape of the internalized DNA. Cationic species partially protonated at physiological pH may act as proton sponges in the endocytotic vesicles and may result in disruption of the vesicle.[52] For example, dimethylaminoethyl groups improve endosomal escape much more effectively than trimethylaminoethyl groups based on their large buffering capacity.[47] The pK_a of the cationic groups is also influenced by their arrangement, based on the polymer effect of the adjacent charged groups, resulting in large buffering capacity. This seems to be the reason for the high efficacy of dendrimers and PEI.

Another approach is enhanced endosomal release by use of a peptide that mimics the fusogenic activity of the virus[50] and membrane disruptive peptides.[53] This technique may be effective for improving the transgene expression. Midox et al. reported the efficacy of an addition of a 22-residue peptide derived from the influenza viral hemagglutinin HA2N-terminal polypeptide in the culture medium for gene transfection.

DNA molecules released from endocytotic vesicles into the cytosols may be delivered to the nuclei, but the mechanisms are unclear. One possibility is transportation through the nuclear membrane pore but this does not allow the complex, whose size is usually around 100 nm, to pass through. Another possibility is the accumulation during the mitotic event accompanying the nuclear membrane disappearance. To enhance trafficking through the nucleus pore, several nuclear localization signals (NLSs) were utilized.[54,55] NLSs are oligopeptides composed mainly of cationic residues. They are 5–20 amino acids long and have different sequences in many species.

NLS bound to PLL was evaluated by many researchers as a gene carrier and shown to be effective.[56] The importance of the disappearance of the nuclear membrane during the mitotic event is still being debated. The mitotic event was reported to be important for *in vitro* gene transfer in the case of lipoplexes.[57] Pollard et al. studied the nuclear localization of polyplexes and lipoplexes microinjected into the cells and indicated that polycations but not cationic lipids promote gene delivery from the cytoplasm to the nucleus. Moreover, when lipoplexes are injected into the nuclei of oocytes or mammalian cells, gene expression in the nucleus is prevented by the cationic lipids, while lipoplexes injected into the nuclei produce high gene expression.[11] Zauner et al. compared the role of mitosis in the transfection of confluent, contact-inhibited primary human cells using polyplexes and lipoplexes. The lipoplex cannot produce high gene expression at the confluent stage but polyplex can do so.[58]

These results indicate that the intracellular trafficking and gene expression mechanisms for polyplex and lipoplex differ.

Dissociation of Complexes

For transcription of the DNA molecules delivered into the cells, it is considered that the complexes: 1) should be disassembled to release the free DNA; or 2) should have somewhat well controlled structures that allow the transcription of DNA in the complexes without the disassembling.[59]

The ease of the dissociation of DNA/carrier complexes greatly depends on the strength of the electrostatic interaction between the plasmid and the polycations, which is quite important for the subsequent transcription of the transgene. In our experiment using polycations (Fig. 2), only polycations having hydroxyl groups revealed higher gene expression. These hydroxyl groups seemed to impart hydrophilic nature to the complexes, resulting in the easy disassembling of the complexes and in releasing naked DNA in the cells. Indeed, adding other polyanions such as heparan sulfate or heparin, which are reported to play an important role of complex/cell surface interaction, easily disassembles the complexes. We then examined gene expression from the complexes by in vitro transcription/translation experiments. The complexes composed of the polycations having abundant hydroxyl groups showed higher gene expression even in this system. We confirmed that the hydroxyl groups and the amide groups may act similarly as nonionic hydrophilic groups.

Polycations carrying various saccharides were suggested to produce high gene expression due to the enhanced amounts of the internalized complexes and the decreased electrostatic interactions of the complexes based on weakened cationicity.[60,61]

Chloroquine, which is known to form complexes with DNA, was also reported to enhance the dissociation of the complexes in addition to the neutralization and vacuolization of the endosomal and lysosomal vesicles. Erbacher et al. studied the efficacy of chloroquine and some other weak bases which can neutralize the pH inside the endocytotic vesicle and found that only chloroquine enhances transgene expression. They concluded that the strong effect of chloroquine on the gene transfer efficiency is not mainly related to the neutralization of the vesicles, but to the dissociation of the complexes.[62]

Heparan sulfate, which plays the main role in the interaction of the polyplexes with the cell surfaces, has been reported to take part in the disassembly of the polyplexes.[46] When polyanions such as heparan sulfate or heparin are added to the DNA/polycation complex solution, free DNA is released by the interexchange reaction. We recently clarified that this interexchange reaction was strongly improved by the nonionic–hydrophilic groups of carrier polycations such as hydroxyl or amide groups as mentioned above. Toncheva et al. studied gene transfer using PLL with a molecular weight of 1000 or 22,000 and its derivatives grafted with various hydrophilic polymers, including PEG, dextran, and poly [N-(hydroxypropyl)methacrylamide] (pHPMA). By analyzing the size, surface zeta potential, and aggregation properties of the complexes, they confirmed the efficacy of PEG–PLL and pHPMA–PLL.[63] Kabanov reported that polyion interexchange reaction occurred when an adequate amount of polyanion was added to the polycation/DNA complex suspensions. The interexchange reaction of the complexes depended on the kind of polyanion added such as poly(vinyl sulfonate)[39] or poly(aspartic acid).[38] They also reported that the topology of the DNA (linear or super-coiled DNA) affects the stability of the polyplexes based on the phosphate group density of DNA.[45]

The fact that the carrier polycations that improve the interexchange reaction produce effective gene expression does not necessarily suggest the disassembling of the complexes in the cells before the transcription event. Since a weak complex with high hydrophilicity is thought to depress gene expression from the viewpoint of nuclease digestion of DNA, the ease of disassembling is not necessarily an advantage for gene expression. Indeed, the inverse effects of the nuclease resistance and the rate of interexchange reaction of polyplexes were pointed out.[64]

BIODERIVED MOLECULES AS TARGETING MOIETIES

Site-specific gene delivery is attracting great attention, especially for direct in vivo gene transfer using various biologically active moieties such as transferrin[20,65] and sugar.[50,60–62,66–68] These modifications bring DNA/carrier complexes to the target cells or organs and also enhance the internalization of complexes into the cells in a receptor-specific manner, which results in high gene expression.

In 1988, Wu et al.[67,68] developed a system for targeting foreign genes to hepatocytes through receptor-mediated endocytosis for the internalization of DNA/carrier complexes. Hepatocytes possess unique receptors that bind and internalize galactose-terminal asialo glycoproteins. The asialo glycoproteins are internalized by their receptors and delivered into the lysosomes via membrane-limited vesicles.[69]

Asialoorosomucoid-PLL carriers delivered pSV2-CAT plasmid DNA specifically to HepG2 hepatoma cells but not to the other receptor (-) cell lines.[67,68] The advantage of the use of the receptor-mediated endocytosis is not only the cell-type specificity of the gene transfer. Some types of cells, such as nonadherent primary hemopoietic cells, are well known to be difficult or almost impossible to transfect with foreign genes by conventional carriers because their endocytotic activity is quite low. Birnstiel and coworkers developed a system in which transferrin was selected as the ligand and named their system "transferrinfection."[69] They synthesized transferrin–PLL conjugates with various molecular weights of PLL and various transferrin ratios to PLL molecules. They found a strong correlation between DNA condensation evaluated under the electron microscope and cellular DNA uptake. One of the key factors for the high gene expression seems to be the sufficient condensing of DNA into a toroid structure, which may facilitate the endocytotic event.[40]

Other candidates for receptor-mediated gene delivery are the receptors for integlin,[70] insulin,[71] and some growth factors.[72] Interestingly, polycations bound to vascular endothelial growth factor could not deliver DNA into nucleus but basic fibroblast growth factor could. The PEI derivatives conjugated to the integrin-binding peptide CYGGRGDTP via a disulfide bridge produced transgene expression in integrin-expression epithelial (HeLa) cells and fibroblasts (MRC5) at the expression level of 10–100 fold as compared with PEI.

These studies used PLL as a carrier polymer and revealed improved endocytosis and excellent expression. However, the high gene expression did not necessarily result from improved endocytosis because the extent of improved transgene expression is often much larger than that which can be explained only by the enhanced amount of DNA ingested. Even in these cases, the physicochemical changes of the complexes play an important role in enhanced transgene expression.

CONCLUSIONS

Gene delivery is a biological and extremely complicated event. It might be reasonable and clever to use the functional ligands of the living body such as a cell surface receptor, NLSs, fusogenic peptides, and viral proteins. In contrast, the physicochemical character of the well controlled supramolecular complex of polyanionic DNA and polycation has great influence on transgene expression. Interestingly, the factor by which the transcription of DNAs, the final event in gene expression, can be promoted is not a biological one, but the physicochemical behavior of the supramolecular complexes. Development of safer and effective nonviral gene carriers through the cooperative effects of these factors can be expected in the near future.

REFERENCES

1. Smith, A.E. Viral vectors in gene therapy. Annu. Rev. Microbiol. **1995**, *49* (1), 807–837.
2. Ledley, F.D. Nonviral gene therapy: The promise of genes as pharmaceutical products. Hum. Gene Ther. **1995**, *6* (9), 1129–1144.
3. Behr, J.P. Gene transfer with synthetic cationic amphiphiles: Prospects for gene therapy. Bioconjug. Chem. **1994**, *5* (5), 382–389.
4. Luo, D.; Saltzman, W.M. Synthetic DNA delivery systems. Nat. Biotechnol. **2000**, *18* (1), 33–37.
5. Canonico, A.E.; Conary, J.T.; Meyrick, B.O.; Brigham, K.L. Aerosol and intravenous transfection of human alpha 1-antitrypsin gene to lungs of rabbits. Am. J. Respir. Cell Mol. Biol. **1994**, *10* (1), 24–29.
6. Caplen, N.J.; Alton, E.W.; Mddleton, P.G.; Dorin, J.R.; Stevenson, B.J.; Gao, X.; Geddes, D.M. Liposome-mediated CFTR gene transfer to the nasal epithelium of patients with cystic fibrosis. Nat. Med. **1995**, *1* (1), 39–46.
7. Nabel, E.G.; Yang, Z.; Muller, D.; Chang, A.E.; Gao, X.; Huang, L.; Cho, K.J.; Nabel, G.J. Safety and toxicity of catheter gene delivery to the pulmonary vasculature in a patient with metastatic melanoma. Hum. Gene Ther. **1994**, *5* (9), 1089–1094.
8. De Smedt, S.C.; Demeester, J.; Hennink, E. Cationic polymer based gene delivery system. Pharm. Res. **2000**, *17* (2), 113–126.
9. Godbey, W.T.; Wu, K.K.; Mikos, A.G. Tracking the intracellular path of poly(ethylenimine)/DNA complexes for gene delivery. Proc. Natl. Acad. Sci. U.S.A. **1999**, *96* (9), 5177–5181.
10. Haensler, J.; Szoka, F.C. Polyamidoamine cascade polymers mediate efficient transfection of cells in culture. Bioconjug. Chem. **1993**, *4* (5), 372–379.
11. Pollard, H.; Remy, J.S.; Loussouarn, G.; Demolombe, S.; Behr, J.P.; Escande, D. Polyethylenimine but not cationic lipids promotes transgene delivery to the nucleus in mammalian cells. J. Biol. Chem. **1998**, *273* (13), 7507–7511.
12. Fraley, R.; Subramani, S.; Berg, P.; Papahadjopoulos, D. Introduction of liposome-encapsulated SV40 DNA into cells. J. Biol. Chem. **1980**, *255* (21), 10431–10435.
13. Felgner, P.L.; Gadek, T.R.; Holm, M.; Roman, R.; Chan, H.W.; Wenz, M.; Northrop, J.P.; Ringold, G.M.; Danielsen, M. Lipofection: A highly efficient, lipid-mediated DNA-tranfection procedure. Proc. Natl. Acad. Sci. U.S.A. **1987**, *84* (21), 7413–7417.
14. Ross, P.C.; Hui, S.W. Lipoplex size is a major determinant of *in vitro* lipofection efficiency. Gene Ther. **1999**, *6* (4), 651–659.
15. Ross, P.C.; Hui, S.W. Polyethylene glycol enhances lipoplex-cell association and lipofection. Biochim. Biophys. Acta **1999**, *1421* (2), 273–283.
16. Lew, D.; Parker, S.E.; Latimer, T.; Abai, A.M.; Kuwahara-Rundell, A.; Doh, S. G.; Norman, J. Cancer gene therapy using plasmid DNA: Pharmacokinetic study of DNA following injection in mice. Hum. Gene Ther. **1995**, *6* (5), 553–564.
17. Kabanov, A.V. Taking polycation gene delivery systems from *in vitro* to *in vivo*. Pharm. Sci. Technol. Today **1999**, *2* (9), 365–372.
18. Mizuguchi, H.; Nakagawa, T.; Nakanishi, M.; Imazu, S.; Nakagawa, S.; Mayumi, T. Efficient gene transfer into mammalian cells using fusogenic liposome. Biochem. Biophys. Res. Commun. **1996**, *218* (1), 402–407.
19. Kato, K.; Nakanishi, M.; Kaneda, Y.; Uchida, T.; Okada, Y. Expression of hepatitis B virus surface antigen in adult rat

liver. Co-introduction of DNA and nuclear protein by a simplified liposome method. J. Biol. Chem. **1991**, *266* (6), 3361–3364.

20. Wagner, E.; Zenke, M.; Cotten, M.; Beug, H.; Birnstiel, M.L. Transferrin–polycation conjugates as carriers for DNA uptake into cells. Proc. Natl. Acad. Sci. U.S.A. **1990**, *87* (9), 3410–3414.

21. Sheldrick, P.; Laithier, M.; Lando, D.; Ryhiner, M.L. Infectious DNA from herpes simplex virus: Infectivity of doublestrand and single-strand molucules. Proc. Natl. Acad. Sci. U.S.A. **1973**, *70* (12), 3621–3625.

22. Yamaoka, T.; Hamada, N.; Iwata, H.; Murakami, A.; Kimura, Y. Effect of cation content of polycation-tupe gene carriers on *in vitro* gene transfer. Chem. Lett. **1998**, (11), 1171–1172.

23. Lee, K.Y.; Kwon, I.C.; Kim, Y.H.; Jo, W.H.; Jeong, S.Y. Preparation of chitisan self aggregates as a gene delivery system. J. Control. Release **1998**, *51* (2), 213–220.

24. Boussif, O.; Lezoualc'h, F.; Zanta, M.A.; Mergny, M.D.; Scherman, D.; Demeneix, B.; Behr, J.P. A versatile vector for gene and oligonucleotide transfer into cells in culture and *in vivo*: Polyethyleneimine. Proc. Natl. Acad. Sci. U.S.A. **1995**, *92* (16), 7297–7301.

25. Remy, J.S.; Remy, J.S.; Abdallah, B.; Zanta, M.A.; Boussif, O.; Behr, J.P.; Demeneix, B. Gene transfer with lipospermines and polyethylenimines. Adv. Drug Deliv. Rev. **1998**, *30* (1), 85–95.

26. Ferrari, S.; Moro, E.; Pettenazzo, A.; Behr, J.P.; Zacchello, F.; Scarpa, M. ExGen 500 is an effecient vector for gene delivery to lung epithelial cells *in vitro* and *in vivo*. Gene Ther. **1997**, *4* (10), 1100–1006.

27. Aubin, R.A.; Weinfeld, M.; Mirzayans, R.; Paterson, M.C. Polybrene/DMSO-assisted gene transfer. Mol. Biotechnol. **1994**, *1* (1), 29–48.

28. Mita, K.; Zama, M.; Ichmura, S. Effect of charge density of cationic polyelectrolytes on complex formation with DNA. Biopolymers **1977**, *16* (9), 1993–2004.

29. van de Wetering, P.; Cherng, J.Y.; Talsma, H.; Crommelin, D.J.A.; Hennink, W.E. 2-(Dimethylamino.ethyl mathacrylate based copolymers as gene transfer agents. J. Controlled Release **1998**, *53* (1), 145–153.

30. Cherng, J.Y.; van de Wetering, P.; Talsma, H.; Crommelin, D.J.; Hennink, W.E. Effect of size and serum proteins on transfection efficiency of poly ((2-dimethylamino)ethyl methacrylate)-plasmid nanoparticles. Pharm. Res. **1996**, *13* (7), 1038–1042.

31. Takai, T.; Ohmori, H. DNA transfection of mouse lymphoid cells by the combination of DEAE-dextran-mediated DNA uptake and osmotic shock procedure. Biochim. Biophys. Acta **1991**, *1048* (1), 105–109.

32. Yamaoka, T.; Kimura, T.; Iwase, R.; Murakami, A. Enhanced expression of foreign gene transferred to mammalian cells *in vitro* using chemically modified poly(L-lysine) as gene carriers. Chem. Lett. **2000**, *29* (2), 118–119.

33. Plank, C.; Tang, M.X.; Wolfe, A.R.; Szoka, F.C. Branched cationic peptides for gene delivery: Role of type and number of cationic residues in formation and *in vitro* activity of DNA polyplex. Human Gene Ther. **1999**, *10* (2), 319–332.

34. Ohashi, S.; Kubo, T.; Ikeda, T.; Arai, Y.; Takahashi, K.; Hirasawa, Y.; Takigawa, M.; Satoh, E.; Imanishi, J.;

Mazda, O. Cationic polymer-mediated genetic transduction into cultured human chondrosarcoma-derived HCS-2/8 cells. J. Orthop. Sci. **2001**, *6* (1), 75–81.

35. Tang, M.X.; Redemann, C.T.; Szoka, F.C. *In vitro* gene delivery by degraded polyamidoamine dendrimers. Bioconjug. Chem. **1996**, *7* (6), 703–714.

36. Tang, M.X.; Szoka, F.C. The influence of polymer structure on the interactions of cationic polymers with DNA and morphology of the resulting complexes. Gene Ther. **1997**, *4* (8), 823–932.

37. Wolfert, M.A.; Schacht, E.H.; Toncheva, V.; Ulbrich, K.; Nazarova, O.; Seymour, L.W. Characterization of vectors for gene therapy formed by self-assembly of DNA with synthetic block co-polymers. Hum. Gene. Ther. **1996**, *7* (17), 2123–2133.

38. Kabanov, A.V.; Kabanov, V.A. DNA complexes with polycations for the delivery of genetic material into cells. Bioconj. Chem. **1996**, *6*, 7–20.

39. Katayose, S.; Kataoka, K. Remarkable increase in nuclease resistance of plasmid DNA through supramoleclar assembly with poly(ethylene glycol)-poly(L-lysine) block copolymer. J. Pharm. Sci. **1998**, *87* (2), 160–163.

40. Zenke, M.; Steinlein, P.; Wagner, E.; Cotten, M.; Beug, H.; Birnstiel, M.L. Receptor-mediated endocytosis of transferin-polycation conjugates: An efficient way to introduce DNA into hematopoietic cells. Proc. Natl. Acad. Sci. U.S.A. **1991**, *87* (10), 3655–3659.

41. Baeza, I.; Gariglio, P.; Rangel, L.M.; Chavez, P.; Cervantes, L.; Arguello, C.; Wong, C.; Montanez, C. Electron microscopy and biochemical properties of polyamine-compacted DNA. Biochemistry **1987**, *26* (20), 6387–6392.

42. Peng, H.F.; Jackson, V. *In vitro* studies on the maintenance of transcription-induced stress by histones and polyamines. J. Biol. Chem. **2000**, *275* (1), 657–668.

43. Wolfert, M.A.; Dash, P.R.; Nazarova, O.; Oupicky, D.; Seymour, L.W.; Smart, S.; Strohalm, J.; Ulbrich, K. Polyelectrolyte vectors for gene delivery: Influence of cationic polymer on biophysical properties of complexes formed with DNA. Bioconjug. Chem. **1999**, *10* (6), 993–1004.

44. Ward, C.M.; Fisher, K.D.; Seymour, L.W. Turbidometric analysis of polyelectrolyte complexes formed between poly(L-lysine) and DNA. Colloid Surf. B Biointerfaces **1999**, *16* (1), 253–260.

45. Bronich, T.K.; Nguyen, H.K.; Eisenberg, A.; Kabanov, A.V. Recognition of DNA topology in reactions between plasmid DNA and cationic copolymers. J. Am. Chem. Soc. **2000**, *122* (35), 8339–8443.

46. Mislick, K.A.; Baldeschwieler, J.D. Evidence for the role of proteoglycans in cation-mediated gene transfer. Proc. Natl. Acad. Sci. U.S.A. **1996**, *93* (22), 12349–12354.

47. Zuidam, N.J.; Posthumab, G.; Vries, E.D.; Crommelin, D.; Hennink, W.; Storm, G. Effects of physicochemical characteristics of poly(2-(dimethylamino)ethyl methacrylate)-based polyplexes on cellular association and internalization. J. Drug Target. **2000**, *8* (1), 51–66.

48. Zhou, X.; Klibanov, A.L.; Huang L. Lipophilic polylysines mediate efficient DNA transfection in mammalian cells. Biochim. Biophys. Acta **1991**, *1065* (1), 8–14.

49. Yamaoka, T.; et al. Transfection of foreign gene to mammalian cells using low-toxic cationic polymers. Nucleic Acids Symp. Ser. **1994**, *31*, 229–230.

50. Midoux, P.; Mendes, C.; Legrand, A.; Raimond, J.; Mayer, R.; Monsigny, M.; Roche, A.C. Specific gene transfer mediated by lactosylated poly-L-lysine into hepatoma cells. Nucleic Acids Res. **1993**, *21* (4), 871–878.

51. Okada, C.Y.; Rechsteiner, M. Introduction of macromolecule into cultured mammalian cells by osmotic lysis of pinocytic vesicles. Cell **1982**, *29* (1), 33–41.

52. Van de Wetering, P.; Moret, E.E.; Schuurmans-Nieuwenbroek, N.M.E. Structureactivity relationships of water-soluble cationic methacrylate/methacrylamide polymers for nonviral gene delivery. Bioconjug. Chem. **1999**, *10* (4), 589–597.

53. Ohmori, N.; Niidome, T.; Kiyota, T.; Lee, S.; Sugihara, G.; Wada, A.; Hirayama, T.; Aoyagi, H. Importance of hydrophobic region in amphiphilic structures of α-helical peptides for their gene transferability into cells. Biochem. Biophys. Res. Commun. **1998**, *245* (1), 259–265.

54. Garcia-Bustos, J.; Heitman, J.; Hall, M.N. Nuclear protein localization. Biochim. Biophys. Acta **1991**, *1071* (1), 83–101.

55. Yoneda, Y. How proteins are transported from cytoplasm to the nucleus. J. Biochem. **1997**, *121* (5), 811–817.

56. Chan, C.K.; Jans, D.A. Enhancement of polylysine-mediated transferrinfection by nuclear localization sequences: Polylysine does not function as a nuclear localization sequence. Hum. Gene Ther. **1999**, *10* (10), 319–332.

57. Mortimer, I.; Tam, P.; MacLachlan, I.; Graham, R.W.; Saravolac, E.G.; Joshi, P.B. Cationic lipid-mediated transfection of cells in culture requires mitotic activity. Gene Ther. **1999**, *6* (3), 403–411.

58. Zauner, W.; Brunner, S.; Buschle, M.; Ogris, M.; Wagner, E. Differential behaviour of lipid-based and polycation-based gene transfer systems in transfecting primary human fibroblasts: A potential role of polylysine in nuclear transport. Biochim. Biophys. Acta **1999**, *1428* (1), 57–67.

59. Bielinska, A.U.; Kukowska-Latallo, J.F.; Baker, J.R. The interaction of plasmid DNA with polyamidoamine dendrimers: Mechanism of complex formation and analysis of alterations induced in nuclease sensitivity and transcriptional activity of the complexed DNA. Biochim. Biophys. Acta **1997**, *1353* (2), 180–190.

60. Erbacher, P.; Roche, A.C.; Monsigny, M.; Midoux, P. The reduction of the positive charges of polylysine by partial gluconoylation increases the transfection efficiency of polylysine/DNA complexes. Biochim. Biophys. Acta **1997**, *1324* (1), 27–36.

61. Erbacher, P.; Roche, A.C.; Monsigny, M.; Midoux, P. Glycosylated polylysine/DNA complexes: Gene transfer efficiency in relation with the size and the sugar substitution level of glycosylated polylysine and with the plasmid size. Bioconj. Chem. **1995**, *6* (4), 401–410.

62. Erbacher, P.; Roche, A.C.; Monsigny, M.; Midoux, P. Putative role of chloroquine in gene transfer into a human hepatoma cell line by DNA/polylysine complexes. Exp. Cell Res. **1996**, *225* (1), 186–194.

63. Toncheva, V.; Wolfert, M.A.; Dash, P.R.; Oupicky, D.; Ulbrich, K.; Seymour, L.W.; Schacht, E.H. Novel vectors for gene delivery formed by self-assembly of DNA with poly(L-lysine) grafted with hydrophilic polymers. Biochim. Biophys. Acta **1998**, *1380* (3), 354–368.

64. Dash, P.R.; Toncheva, V.; Schacht, E.; Seymour, L.W. Synthetic polymers for vectrial delivery of DNA: Characterisation of polymer-DNA complexes by photon correlation spectroscopy and stability to nuclease degradation and disruption by polyanions *in vitro*. J. Control. Release **1997**, *48* (2–3), 269–276.

65. Wagner, E.; Cotten, M.; Mechtler, K.; Kirlappos, H.; Birnstiel, M.L. DNA-binding transferrin conjugates as functional gene-delivery agents: Synthesis by linkage of polylysine of ethidium homodimer to the transferrin carbohydrate moiety. Bioconjug. Chem. **1991**, *2* (4), 226–231.

66. Wall, D.A.; Wilson, G.; Hubbard A.L. The galactose-specific recognition system of mammalian liver: The route of ligand internalization in rat hepatocytes. Cell **1980**, *21* (1), 79–93.

67. Wu, G.Y.; Wu, C.H. Evidence for targeted gene delivery to Hep G2 hepatoma cells *in vitro*. Biochemistry **1988**, *27* (3), 887–892.

68. Wu, G.Y.; Wu, C.H. Receptor-mediated gene delivery and expression *in vivo*. J. Biol. Chem. **1988**, *263* (29), 4621–14624.

69. Wagner, E.; Cotten, M.; Foisner, R.; Birnstiel, M.L. Transferrin-polycation-DNA complexes: The effect of polycations on the structure of the complex and DNA delivery to cells. Proc. Natl. Acad. Sci. U.S.A. **1991**, *88* (10), 4255–4259.

70. Erbacher, P.; Remy J.S.; Behr, J.P. Gene transfer with synthetic virus-like particles via the integrin-mediated endocytosis pathway. Gene Ther. **1999**, *6* (1), 138–145.

71. Rosenkranz, A.A.; Yachmenev, S.V.; Jans, D.A.; Serebryakova, N.V.; Murav'ev, V.I.; Peters, R.; Sobolev, A.S. Receptor-mediated endocytosis and nuclear transport of a transfecting DNA construct. Exp. Cell Res. **1992**, *199* (2), 323–329.

72. Fisher, K.D.; Ulbrich, K.; Subr, V.; Ward, C.M.; Mautner, V.; Blakey, D.; Seymour, L.W. A versatile system for receptor-mediated gene delivery permits increased entry of DNA into target cells, enhanced delivery to the nucleus and elevated rates of transgene expression. Gene Ther. **2000**, *7* (5), 1337–1343.

Glues

Shojiro Matsuda
Research and Development Department, Gunze Limited, Ayabe, Japan

Yoshita Ikada
Department of Clinical Engineering, Faculty of Medical Engineering, Suzuka University of Medical Science, Mie, Japan

Abstract

Surgical glues are widely used to seal small holes present in diseased soft tissues such as lung (sealants), to stop bleeding from damaged tissues (hemostatic agents), and to bond two separated tissues (surgical adhesives). The focus of this entry is on the different types of surgical glues, their properties, and their applications.

INTRODUCTION

Surgical glues are widely used to seal small holes present in diseased soft tissues such as lung (sealants), to stop bleeding from damaged tissues (hemostatic agents), and to bond two separated tissues (surgical adhesives). The biomaterials currently used for these surgical purposes include fibrin glues, α-cyanoacrylates, and GRF (gelatin-resorcinol-formaldehyde) glues. All of them are applied to tissues in the liquid state (sol), followed by gel formation. This sol–gel transition takes place rapidly on the tissues. Fibrin glue, which comprises fibrinogen and thrombin, is clinically used for the purpose of sealing, hemostasis, and bonding, while cyanoacrylate monomers are utilized mostly as tissue adhesives. GRF with higher gel strength and bonding strength than fibrin glue has a unique application, that is, ceasing of bleeding from acute aortic dissection. Although these currently available glues are greatly contributing to surgical operations, they have disadvantages. For instance, fibrin glues have a potential risk of viral infection in spite of careful screening of plasma, because fibrinogen and thrombin are harvested from human plasma. Low bonding strength of the gelled fibrin to tissues is also a weak point. Cyanoacrylates are instantly allowed to polymerize, forming a gel in the presence of moisture under strong adhesion to tissues, but yield toxic formaldehyde as a result of hydrolytic degradation of the polymer main-chain. As formaldehyde is a major component of the GRF glue, attempts have been made to develop new liquid-type biomaterials that are free of virus, prion, and carcinogen, and can transform into a gel having a high potential of strong tissue adhesion in the presence of water. Glues should be absorbed after having fulfilled their task so as not to induce undesirable reactions as a foreign body.

REQUIREMENTS FOR SURGICAL GLUE

Sutures have been the most widely used method for wound closure, but suturing has shortcomings, such as the requirement of highly skillful procedures and a long time to close wound and the postoperative removal of nonabsorbable suture. It is, therefore, desired to develop surgical glues that can overcome these problems. Various conventional glues are known for home and industrial use, but only three glues, namely, fibrin glues, α-cyanoacrylates, and GRF, are widely available for medical use.

Requirements for these surgical glues are listed in Table 1. As they are applied to the human body, they must be safe, nontoxic, and free from risk of infectious transmission. They should be preferably bioabsorbable and not remain long as a foreign body. Not only must the applying solution and gelled glue be safe, but also the degradation products as well. In addition, they must not hinder the natural healing process.

Most surgical glues are gelled in situ, applicable on rough surfaces, and act as tissue anchors. Appropriate gelling time and mechanical properties of formed gel are important. They should bond rapidly to the surrounding living tissues in the presence of water or aqueous fluids, such as blood, and keep the adhesion strength. Similar to the glue itself, degradation products of the glue must be safe. Besides the chemical nontoxicity, their mechanical properties are also important. If the cured glue is harder than tissue, it may induce undesirable tissue reactions. Good handling (ease of use and gelation speed) is also important.

FIBRIN GLUE

Fibrin glue, which consists of human fibrinogen, thrombin, factor XIII, and bovine aprotinin, has been widely used in

Gene–Hydrogels

Concise Encyclopedia of Biomedical Polymers and Polymeric Biomaterials DOI: 10.1081/E-EBPPC-120007304

Table 1 Requirements for surgical glues

1. In situ curable from liquid-state through polymerization, chemical cross-linking, or solvent evapolation
2. Rapidly curable under wet physiological conditions
3. Strongly bondable to tissues
4. Nontoxic for both agents and their degradation products
5. Tough and pliable as natural tissues
6. Biodegradable

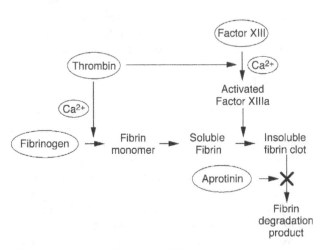

Fig. 1 Mechanism of gelation of fibrin glue.

Europe, Canada, and Japan for about 2 decades and available in the United States since 1998. Figure 1 shows the mechanism of gelation of fibrin glue. When fibrinogen solution and thrombin solution are mixed at the site of injury, fibrin gel formation occurs through the physiological coagulation and the tissue adhesion cascade. Thrombin catalyzes partial hydrolysis of fibrinogen to form fibrin monomers if calcium ion is present. Thrombin also activates factor XIII into factor XIIIa in the presence of calcium ions. Factor XIIIa reacts with fibrin monomers to form insoluble fibrin clots. Aprotinin inhibits fibrinolysis (degradation of fibrin clot). The fibrin clot will be totally resorbed by fibrinolysis and phagocytosis.

Fibrinogen and thrombin are dissolved separately in medium before surgical operation. Each of two solutions is filled in a separate dual syringe, which is set to Y-connector and applying needle or a spray system.

Application of Fibrin Glue

Surgical fields for fibrin glue and its application purposes are shown in Table 2. Mostly they work as sealants, hemostatic agents, and surgical adhesives. Hemostasis is also effectively carried out by suturing, electrocauterization, infrared coagulation, compression, and tamponade. Very small blood vessels, fragile tissues, and small spaces are difficult to apply with these devices, but liquid-type surgical

Table 2 Clinical applications of fibrin glue

Cardiovascular surgery	Hemostasis of bleeding from mediastinum and small blood vessels, reinforcement of suture line, sealing of vascular grafts and anastomoses.
Thoracic surgery	Sealing of air leaks from raw lung surface, and pulmonary and bronchial staple lines.
Gastroenterological surgery	Hemostasis and reinforcement of intestinal anastomosis, prevention of bile leakage, and hemostasis of hepatectomy, pancreatic fistula, and gastroduodenal ulcers.
Neurosurgery	Sealing of cerebrospinal fluid leaks, nonsutured peripheral nerve repair.
Gynecologic surgery	Anastomosis of fallopian tube, adhesion of peritoneum.
Plastic surgery	Closure of skin wound, attachment of skin grafts, fixation of tissue flaps.
Orthopaedic surgery	Fixation of bone fragments, hemostasis during total knee arthroplasty.
Urologic surgery	Anastomosis of urinary tube and adhesion of kidney, pelvis renalis, and ureter.

glues are useful in these cases and also for sealing if they have a proper adhesion strength to tissue, and a proper physical-barrier strength in the presence of water.

The application activity of fibrin glue depends on surgical procedures.[1] Cardiovascular surgery is the most common among them and its major purpose is hemostasis and suture line sealing. Successful local hemostasis of bleeding after open-heart surgery can reduce blood loss, operative time, and the need for resternotomy in these high-risk patients. Fibrin glue is highly effective as a tissue sealant in diffuse, low-pressure bleeding such as coronary vein, right-ventricle/pulmonary-artery conduits, and right-ventricular patches. Fibrin glue is used routinely whenever suturing is impossible, dangerous, or difficult, and for spontaneous hemostasis to save waiting time. Suture holes, staple line, anastomoses, fistulas, and raw surfaces are effectively treated with fibrin glue, similarly to aortic and coronary suture line. It is also used for sealing vascular grafts. Perfect hemostasis is quite important in pediatric cardiac surgery because of the high incidence of coagulopathy after open-heart procedures and increased risks associated with the use of blood and blood products.

Fibrin glue is also very effective for reoperative procedures, because mediastinal bleeding and excessive nonsuturable bleeding from adhesions are highly problematic.

Gene–Hydrogels

The adhesion of tissue induces complications during reoperation, prolonging the operation time and increasing the need for blood transfusion, and hence preventing the formation of adhesion is very important.

In thoracic surgery, fibrin glue is applied to cease hemorrhage and air leaks following pulmonary resection and decortications. It is also used for the treatment of bronchopleural fistulae and pneumothorax. To reduce air leaks after stapled pulmonary resection, several kinds of predget, such as poly(glycolic acid) (PGA) fabric and bovine pericardium, have been clinically applied. Fibrin glue is also used for this purpose with PGA fabric.

In the treatment of bleeding from highly vascularized parenchymal tissues, such as elective hepatic resection and pancreatic fistula, and from anastomotic sites, such as recurrent bleeding from gastroduodenal ulcers, fibrin glue is effectively applied. Many other applications such as dural closure are known to prevent cerebrospinal-fluid (CSF) leaks, burn bleeding after debridement, and bleeding during the total knee arthroplasty (TKA) procedure, and control artery or aneurysm closure in endovascular surgery.

Future and Improvement of Fibrin Glue

Since fibrin glue is made from human and bovine proteins, potential risk of viral transmission cannot be fully denied, but there have been reported no proven cases of viral transmission associated with the use of commercial fibrin sealant. To circumvent this risk, the development of recombinant thrombin and fibrinogen is currently ongoing. Other sources, well-purified materials, and new synthetics are studied as alternatives to fibrin glue.

For the treatment of acute aortic dissection, fibrin glue does not have sufficient gel strength and bonding strength, and GRF and other glues are superior to fibrin glue. It is also claimed that fibrin glue is readily washed away even by weak bleeding, as no compression is possible on the bleeding area. In this case, collagen sheet is effective for the scaffold of fibrin glue.

Other clinical applications under study include the prevention of postoperative adhesion, endoscopic procedure for minimal invasive surgery, drug and cell delivery vehicle, and scaffolding of cell culture.

Fibrin glue sheet

TachoComb (Nycomed Arzneimittel, Germany) is a ready-to-use hemostatic agent.[2] As this spongelike sheet is prepared by coating human fibrinogen, bovine thrombin, and bovine aprotinin on the surface of equine collagen fleece, it does not need to solidify, warm, or set to syringe. This dry sheet can be applied directly to traumatized tissues, in contrast to the liquid application of fibrin glue. TachoComb has mechanical stability, can be effectively used for the treatment of diffuse bleeding of the thoracic wall, and is absorbed within a few weeks. Bleeding site in the thoracic

apex, especially after lysis of adhesion between lung and parietal pleura, is very difficult to treat with fibrin glue and other conventional methods. By using TachoComb, this bleeding is effectively hemostased.[3] Parenchymal leak of the lung after decortication can also be treated with this glue, when the visceral pleura is removed or lacerated. Hemostasis is effective and well tolerated as a secondary procedure in urological, ENT (ear, nose, and throat), hepatic, and vascular surgery when primary hemostatic measures are proved insufficient. TachoComb may be used to control oozing hemorrhage from surgical sites in gynaecological patients effectively and prevent adhesion formation at the application site, thus offering an effective method for preventing adhesion-induced infertility.

Adhesion prevention

Fibrin glue is also expected as an adhesion-prevention material. It is applicable to damaged tissue, fully covering it to avoid contact with another tissue. If the injured tissue comes in contact with another tissue, adhesion formation occurs during wound healing. Since fibrin glue is a kind of hydrogel with a high water content, water-soluble nutrients can pass through this hydrogel without affecting wound healing. The use of fibrin glue results in a significant decrease in postoperative adhesion of a rabbit model compared with the control side in the uterine horns and the ovaries.[4]

Drug release vehicle

Because fibrin glue can form gel under a physiological condition, it can be easily mixed with drugs under a liquid condition and be applied to any tissue. Local delivery of drug has been carried out to avoid complications of systemic administration. The delivery vehicles for local delivery must stay at the applied place without diffusion into any undesired location. Thus, tissue adhesives are effective for sustained release of drugs. Bioabsorbable materials are applicable for sustained release of water-soluble drugs for a relatively short duration because of their high rate of diffusion. Poorly soluble drugs can be released from hydrogels such as fibrin glue as a result of vehicle degradation. Antibiotics for the treatment of large-area burns, antiproliferative drugs for chemotherapeutics, and growth factors such as bone morphogenetic protein (BMP) and basic fibroblast growth factor (bFGF) for tissue regeneration can be incorporated into fibrin glue for this drug delivery system.[5]

Applications in tissue engineering

Fibrin monomer can be polymerized into moldable gels to encapsule cells and to provide three-dimensional scaffold of cell culture. Because fibrin glue is bioabsorbable, it is applicable to tissue engineering. When chondrocytes trapped in fibrin glue gels are transplanted subcutaneously in nude mice, the actively proliferating cells can well form cartilaginous matrix.[6] Urothelial cells, tracheal epithelial cells, and preadipocytes cultured *in vitro* were successfully transplanted onto a prefabricated capsule and tubelike structure using fibrin glue as a delivery vehicle and native cell expansion vehicle.[7–9]

CYANOACRYLATE

When cyanoacrylate monomers come into contact with moisture, they change into a polymer. The detailed mechanism of polymerization of 2-cyanoacrylates is not fully understood, but it seems very probable that they polymerize via an ionic mechanism in which a water molecule acts as co-initiator. The liquid-state monomers polymerize to a solid within a short time period in the presence of water and are stored in one tube as a single component. Very high stiffness of cured glues, low viscosity of monomers, and toxic formaldehyde released during their degradation are drawbacks. However, cyanoacrylates have been used for wound closure, cartilage and bone grafting, coating of corneal ulcers, coating of aphthous ulcers, embolization of gastrointestinal varices, and embolization in neurovascular surgery.

Wound Closure

Surgical glues are preferable to skin sutures owing to good cosmetic outcome and pain reduction, especially in cosmetic surgery and surgery for children. Cyanoacrylate is the only surgical glue that is used for dermal wound closure instead of suturing. Tissue adhesives can save time during wound repair, provide flexible water-resistant protective coating, and eliminate the need for suture removal. Since the 1970s, n-butyl 2-cyanoacrylate (NBCA) with negligible histotoxicity and good bonding strength has been clinically applied and shown to have acceptable cosmetic outcome, but with too high bonding strength and too rapid reaction. Dermabond (Ethicon, Inc., Somerville, New Jersey) is 2-octyl cyanoacrylate that forms a strong bond across wound edges, allowing normal healing. The long-term cosmetic outcome of Dermabond was comparable to that of traditional methods of repair.[10]

Blood Vessel Embolic Material

Endovascular neurosurgery is one of the strategies for treatments of various neurovascular diseases such as cerebral and spinal arteriovenous malformations (AVMs), dural arteriovenous fistulas, and cerebral aneurysms. A microcatheter is advanced into or close to a lesion and then an embolic material is administrated through it to obliterate the lesion. This less invasive method is accepted as an attractive alternative therapy making up a conventional surgical technique. Requirements for blood vessel embolic materials are shown in Table 3.

Table 3 Requirements for blood vessel embolic materials

1. In situ curable from liquid-state (one liquid) through polymerization or solvent evapolation
2. Curable in proper time in the presence of blood
3. Tightly bondable to blood vessel wall tissue and not adhesive to microcatheter
4. Nontoxic
5. Tough and pliable as natural tissues
6. Preferably radiopaque

Various biomaterials have been used for embolization, but each material has some problems. Polymer solutions have ever been used as liquid-type materials for embolization, including poly(ethylene-*co*-vinyl alcohol), cellulose acetate, cationic methacrylate copolymer (Eudragit-E), and poly(vinyl acetate) emulsion. When polymer solution is given into a blood vessel, the organic solvent diffuses into the blood and surrounding tissues, thus forming polymeric precipitates and emboli. However, as most of them contain organic solvents, such as dimethyl sulfoxide and ethanol, they exert toxic effects on the surrounding tissues. Cyanoacrylate monomers that instantly polymerize upon contact with blood are used as liquid embolic material. Among the monomers, NBCA has been preferentially used in Europe and North America. NBCA is injected into the diseased part of the blood vessel through a long thin catheter and then quickly becomes solid, blocking blood flow into the diseased part of the blood vessel.

Although NBCA has been successfully used as embolic material, several troubles have been reported. Its instantaneous polymerization and quite low degradation rate are suitable for embolization of diseased lesions, but its exceptionally high adhesive force sometimes causes adhesion between the tip of the microcatheter and the artery. As a result, it becomes impossible to withdraw the microcatheter from the blood vessel and, in addition, the inner space of the microcatheter is occluded with the rapidly polymerized material when intermittently injected. Therefore, intermittent injection should be avoided and the microcatheter should be withdrawn immediately after completion of embolization. The surgeon is required to accumulate many experiences so that she or he can acquire skillful techniques to carry out embolization safely and effectively when NBCA is used.

To overcome these problems, isostearyl 2-cyanoacrylate (ISCA), which carries a long hydrophobic side isostearyl group with lower reactivity than n-butyl group, was synthesized.[11] Its adhesive force was low enough not to cause these problems. However, as its polymerization rate was too low to obliterate a vascular lesion with a rapid blood flow, ISCA was mixed with NBCA. The adhesive force of this mixture became extremely low, compared with that of NBCA, and adhesion between the tip of microcatheter and artery was not problematic. The viscosity of the mixture was low enough to use as embolic material. Tissue reactions

Gene–Hydrogels

toward the mixture were milder than those of NBCA and radio-angiography became possible by mixing further with Lipiodol. The evaluation of this new embolic material with a rabbit renal artery showed that the obliteration effect of the mixture from ISCA and NBCA was excellent to allow the use of this embolic material for clinical applications.

GRF GLUE

Gelatin is a widely used bioabsorbable material, especially as surgical glue. This natural polymer is extracted from biological tissues, and has many functional groups. Gelatin is soluble in water at temperatures higher than 35°C, and can form gel by lowering temperature below 25°C or by cross-linking. Gelatin gel is as elastic and soft as natural tissue. Braunwald reported that a gelatin-resorcinol mixture cross-linked with a combination of formaldehyde (GRF) and sometimes glutaraldehyde gave satisfactory bonding strength in hepatic and renal tissues of dogs.[12] Formaldehyde undergoes condensation with resorcinol to give a three-dimensionally cross-linked water-insoluble resin. GRF is a two-liquid system glue. One component is aqueous solution of gelatin, resorcinol, and calcium chloride, and the other is an aqueous solution of formaldehyde and glutaraldehyde. After warming the polymer solution to 45°C and mixing 1 ml of polymer solution with two or three drops of aldehyde solution, gel is formed within about one minute.

GRF has largely been used in Europe to treat acute aortic dissections because of its ability to reinforce the delicate structure of the aortic wall.[13] Every acute dissection involving the ascending aorta (Stanford type A) should undergo emergency surgical repair. 75% of Type A AAD patients die within two weeks. The surgical techniques vary depending on the clinical status of patients or the anatomical patterns observed. Furthermore, surgery is generally difficult because of very brittle aortic tissues, especially in advance-aged patients. Bachet et al. applied GRF glue for 212 patients with an emergency operation for type A aortic dissection. Their experience extending over more than 23 years proved that GRF glue is extremely useful, making the procedure much easier and safer. GRF is said to adhere to the false lumen and shows excellent solidity and hemostasis of suture sites.[14]

Since highly concentrated solutions of formaldehyde and glutaraldehyde are directly applied to an injured tissue, its toxicity is of major concern. Formaldehyde will be released from the cured glue when gelatin undergoes enzymatic hydrolysis and, therefore, surgical application of GRF glue is limited to special cases such as where high bonding strength is imperative.

OTHER SURGICAL GLUES

As mentioned earlier, fibrin glue, cyanoacrylates, and GRF glue show risk of viral infection, toxicity, low-adhesion strength, and long curing time. Thus many studies have been carried out to replace these glues. The curing modality of most surgical glues studied is either in situ cross-linking or in situ polymerization of monomer. Less toxic and bioabsorbable polymers such as gelatin, albumin, hyaluronic acid, chitin, poly(ethylene glycol), and poly(L-glutamic acid) have been studied and glutaraldehyde, formaldehyde, water-soluble carbodinimide, N-hydroxysuccinimide, and photoreactive agents have been used as cross-linkers. These agents are mostly multifunctional cross-linkers and are directly used at the application site or immobilized to structural polymers. Several studies will be described below.

Aldehyde System

Aldehyde can react with primary and secondary amines through Schiff's base. Since glutaraldehyde has two aldehyde groups in one molecule, it can react with two amines. If they are from different polymer molecules, glutaraldehyde will produce intermolecular cross-linking. Thus, by mixing the solution of polymer with amines in its side-chain and glutaraldehyde solution, gelation will occur. Natural polypeptides, such as collagen, gelatin, albumin, and enzymes, have amines that are reactive with aldehyde. So, glutaraldehyde is also used as a tissue fixation agent for micrograph observation. As aldehydes with high reactivity show cytotoxicity, too much volume and too high concentration of aldehyde solution may induce untoward tissue reactions.

Aldehyde-polysaccharides and amine-gelatin

Gelatin and polysaccharides were employed to prepare the hemostatic glue, which does not release any low molecular weight substances. To this end, gelatin was modified with ethylenediamine using water-soluble carbodiimide to introduce additional amino groups into the original gelatin, while dextran and hydroxyethyl-starch were oxidized by sodium periodate to convert 1,2-hydroxyl groups into dialdehyde groups.[15] Upon mixing the two polymer components in aqueous solution, Schiff's base was formed between the amino groups in the modified gelatin and the aldehyde groups in the modified polysaccharides, thus resulting in intermolecular cross-linking and gel formation. The fastest gel formation took place within 2 seconds, and its bonding strength to porcine skin was 225 gf/cm^2 when 20 wt% of amino-gelatin (55% amine group) and 10 wt% of aldehyde-HES (>84% dialdehyde) aqueous solutions were mixed. In contrast, the gelation time and bonding strength of fibrin glue was 5 seconds and 120 gf/cm^2, respectively, when the same system was applied for the comparative evaluation.

Albumin glue

Albumin is a natural protein present in blood serum and used as structural polymer for a surgical glue similar to GRF. A glue consisting of 45% bovine serum albumin and

10% glutaraldehyde formed gel within 2 to 3 minutes. Herget et al. reported that this glue was effective as a sealant for bronchial anastomoses and parenchyma lesions in sheep.[16] Bronchial anastomosis and parenchymal tissue repair could be sealed successfully against air leakage with this adhesive. Healing was not complicated much by foreign body reaction and tissue granulation.

Adhesion of aldehyde-gelatin film

It was reported that a glutaraldehyde (GA)-modified dry gelatin film could adhere to tissues.[17] As described above, GA has two reactive aldehyde groups in one molecule and hence can act as a cross-linker of collagen, other proteins, and biological soft tissues. In a similar manner, GA can introduce cross-links into gelatin, making it insoluble even in hot water, where the physical cross-link of gelatin is destroyed. It is highly possible that at the gelation reaction with GA, one aldehyde group of a GA molecule will react with an amino group of gelatin, while the other aldehyde of the same GA molecule will not be able to find an appropriate amino group with which to react, thus remaining free. When a gelatin film was treated with 0.5 M GA solution at 60°C, free aldehyde groups were introduced in the film by up to 150 mmol/g. The dried gelatin film could strongly adhere to the porcine skin as strongly as 250 gf/cm^2, whereas the native gelatin film before GA treatment showed bonding strength of 40 gf/cm^2. The dried film could probably adsorb and remove the water existing between the GA gelatin film and the biological tissue, thus enabling aldehyde groups on the film and the amino groups on the natural tissue to approach each other and result in Schiff's base formation. This reaction scheme was supported by the fact that GA-treated films did not demonstrate high bioadhesion when the remaining aldehydes were quenched with glycine or reduced with NaBH$_4$.

WSC and NHS System

Carbodiimide is a bifunctional cross-linker, similar to GA. 1-Ethyl-3-(3-dimethylaminopropyl) carbodiimide hydrochloride (EDC) and 1-cyclohexyl-3-(2-morpholinoethyl) carbodiimide metho-p-toluenesulfonate (CMC) are available as water-soluble carbodiimide (WSC). WSC is a coupling agent to form an amide bond between amino and carboxyl groups in a short time period in aqueous solution. Therefore, WSC functions as a catalyst to cross-link molecules and no WSC-related fragments are incorporated in the structure of resulting cross-linked molecules. An urea derivative is produced from WSC after the catalytic reaction and is much less toxic than the original WSC.

N-Hydroxysuccinimide (NHS) is a well-known coupling agent that can be immobilized to the carboxyl group of a host polymer through ester bonding. WSC also functions as a catalyst of ester formation. Immobilized NHS-ester can easily react with amines of other molecules through ester-exchange reaction. No NHS-related fragments are incorporated in the reacted molecules as NHS is released after the reaction. Such ester formation is used for the cross-linking of aqueous polymer solution to induce gel formation.

Gelatin, PLGA, and WSC

Gelatin and poly(L-glutamic acid) (PLGA) are natural and synthetic polypeptides, respectively, and are soluble in water and biodegradable. Gelatin is a heat-denatured product of collagen and is obtained commonly by extraction from bovine or porcine bone and skin. This natural polymer contains reactive groups such as $-$COOH and $-$NH$_2$ in the side-chain while PLGA has only $-$COOH groups. An addition of WSC to aqueous solution containing both of these polymers induces gelation, because the amide bond formation between $-$COOH of PLGA and $-$NH$_2$ of gelatin will form intermolecular cross-linking.[18] The gelation time decreased with an increase in the WSC concentration to the comparable level of fibrin glue. When applied to the dog spleen injured by needle pricking, the WSC-modified gelatin-PLGA hydrogel exhibited superior hemostatic capability to fibrin glue. The WSC-modified gelatin-PLGA hydrogel strongly adhered to the surface of the dog spleen, whereas the fibrin hydrogel was easily detached from the spleen surface.

NHS-PLGA and gelatin

WSC is as cytotoxic as glutaraldehyde and the urea, a byproduct of this glue, also has slight toxicity. To circumvent the toxicity of WSC and urea, a new gelatin-PLGA glue that does not use WSC was designed.[19] NHS-ester yields stable products upon reaction with primary or secondary amines and the coupling reaction proceeds efficiently at the physiological pH. An attempt was made to synthesize a new NHS derivative of PLGA, that will induce prompt gelation when mixed with aqueous solution of gelatin. NHS-esters can also react with amines of natural protein, resulting in tissue adhesion. The gelation scheme between gelatin and this PLGA derivative is shown in Fig. 2. Esterification of PLGA with NHS was performed by adding WSC to the reaction mixture. After 20 hr at an ambient temperature, the whole reaction mixture was poured into dehydrated acetone to precipitate the resulting NHS ester of PLGA (NHS-PLGA). The NHS-PLGA could be synthesized at high yields and was found to be stable for an extended time without losing the ability to cross-link with gelatin when stored under a dry-cold condition. This NHS-PLGA could spontaneously form a gel with gelatin in aqueous solution within a short time, comparable to fibrin glue, when gelation was allowed to proceed at pH 8.3. The bonding strength of NHS-PLGA-gelatin glues to natural tissues was higher than that of fibrin glue. With the increase of NHS-ester introduced to PLGA, the gelation time became shorter and the bonding strength increased.

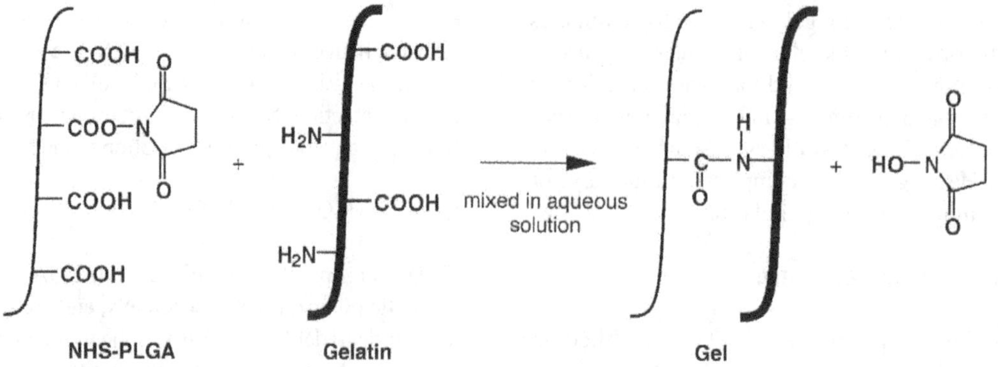

Fig. 2 Gelation scheme between gelatin and NHS-PLGA.

NHS-PEG and SH-PEG

Poly(ethylene glycol) (PEG) is a highly hydrophilic synthetic polymer and has found wide applications in medicine. A new rapidly synthetic tissue sealant can be synthesized from tetra-succinimidyl and tetra-thiol-derivatized PEG. Wallace et al. studied the *in vitro* and *in vivo* efficacy of this glue by dissolving the two reagents in aqueous buffer and subsequently spraying it on a tissue surface.[20] This glue exhibited good adhesion to collagen membrane, PTFE graft, and carotid artery *in vitro*. The gel maintained fluid pressure of 125 mmHg, which is fivefold greater than the capillary blood pressure and one-half that observed in hypertension. Bleeding in rabbit arteries was stopped immediately in five out of six trials. A significant reduction was observed for the time until to hemostasis and blood loss, compared to the control. Carotid artery and subcutaneous implant data in rabbits showed that the formulation was compatible with biological tissues.

NHS-PEG and albumin

Kobayashi et al. reported the ability of an albumin-based hydrogel sealant (ABHS) to prevent air leakage through a suture line after pulmonary surgery.[21] This glue is a hybrid of synthetic and natural products and has two basic components: PEG disuccinimidylsuccinate (cross-linker) and human serum albumin. The terminal reactive ester of the PEG-derived cross-linker and the free amine groups in albumin react to form a gel. When the rat lung was used as an air-leak model, the average burst pressure of the ABHS-treated group was higher than that of fibrin glue throughout the observation period. Histological examination of the incision at day 14 revealed that no sealant was visible at the incision site, without any evidence of adverse tissue reaction.

Photocurable System

With the use of a photosensitizer, UV and visible light can initiate radical polymerization. If dye, azide, and acrylic macromer are used at a suitable combination, polymeriza-

tion is initiated under a mild condition, enabling its application as surgical glue. Water-soluble macromers, such as PEG diacrylate, become insoluble upon light irradiation through polymerization. Optical fiber or other devices are used for light irradiation. UV irradiation should be avoided because of its toxic reaction.

PEG diacrylate

Hubbell and coworkers reported the photoinitiated polymerization of a biodegradable tissue adhesive in 1993.[22] This is made from the macromer that consists of a copolymer of water-soluble PEG with the two terminals capped with biodegradable oligomer such as glycolide and lactide oligomers. The two terminals of this macromer are further capped by acrylates to provide an unsaturated hydrocarbon required for photoinitiation by UV or visible light in the presence of eosin Y, triethanolamine, and 1-vinyl 2-pyrrolidinone. Depending on the type and molecular weight of the capped oligomer and the molecular weight of PEG, the resulting sealant has a wide range of biodegradation kinetics ranging from 1 day to 6 months.

Tanaka et al. evaluated the efficacy of this glue using a canine acute descending aortic dissection model.[23] When the glue was applied to the false cavity for reinforcing and fusing the dissected layers and also to the suture line under a wet condition, hemostasis was easily achieved during surgery. All false cavities were perfectly thrombosed, causing no deleterious effects related to the glue, indicating that this glue can be used easily and effectively for the treatment of acute aortic dissection.

Photoreactive-gelatin and PEG-diacrylate

Nakayama and Matsuda reported on a photocurable glue composed of gelatins partially derivatized with photoreactive xanthene dyes (fluorescein, eosin, and rose bengal) and a hydrophilic difunctional PEG macromer.[24] When a dye-derivatized gelatin (20 wt%), a PEG diacrylate (10 wt%), and ascorbic acid (0.3 wt%) were dissolved in saline solution and irradiated with visible light, a swollen gel was formed within a few tenths of a second due to dye-sensitized photo-cross-linking and photograft polymerization. When a rat liver injured on laparotomy was coated with this hemostatic glue and irradiated with visible light through an optical fiber, the coated viscous solution was immediately converted to a swollen gel and hemostasis was completed. According to a histologic examination at the 7th day after surgery, little gelatin remained in the injured region, scarring with little necrosis occurred, and inflammatory cell infiltration from the surrounding tissue and tissue regeneration proceeded well.

Hyaluronic acid and albumin indocyanine

A photocurable tissue sealant composed of hyaluronic acid, albumin, and indocyanine was studied. For instance, 0.4 ml of hyaluronic acid, 0.2 ml of albumin, and three drops of indocyanine green dye gave a sealant with a peak absorbance of 805 nm, which matches the wavelength (808 nm) of a small, hand-held diode laser. Because tissues do not absorb any light of this wavelength, laser energy is focused in the sealant, minimizing collateral thermal damage. Auteri et al. reported the efficacy of this glue using a canine model of esophageal closure.[25] The tissue sealant was applied to the edges of a hand-sewn closure, and then exposed to diode laser. The end-point was visible shrinking and desiccation of the sealant, which required about 2 minutes. At all time points, closures with this glue had significantly higher bursting pressure than the control closures. Histologic study revealed trace thermal injury with regeneration of intact mucosal lining by 7 days.

Azide chitosan lactose

Ono et al. studied the efficacy of a photo-cross-linkable chitosan glue, to which both azide and lactose moieties were introduced (Az-CH-LA).[26] Introduction of the lactose moieties yielded a much more water-soluble chitosan at the neutral pH. UV irradiation to Az-CH-LA produced an insoluble hydrogel within 60 s. This hydrogel firmly bonded two pieces of sliced ham with each other at appropriate Az-CH-LA concentrations. The bonding strength of the chitosan hydrogel prepared from 30–50 mg/mL of Az-CH-LA was similar to that of fibrin glue. Compared to the fibrin glue, the chitosan hydrogel more effectively sealed air leakage from pinholes on an isolated small intestine and aorta and from incisions on an isolated trachea. Neither Az-CH-LA nor its hydrogel showed any cytotoxicity. Furthermore, all mice studied survived for at least 1 month after implantation of 200 μL of photo-cross-linked chitosan gel and intraperitoneal administration of 1 mL of 30 mg/mL Az-CH-LA solution.

CONCLUSION

Although numerous studies have been carried out to develop surgical glues, as described in the preceding discussion, no perfect glue has been developed yet. Several surgical glues such as fibrin glue, cyanoacrylate, and GRF glue are clinically used, but they have limitations. Requirements for surgical glues such as viscosity, gelation time, adhesion strength, and duration time (biodegradation time), are different depending on the application site and purpose. Handling of glue is also important for the clinical application. Warming and light irradiation may be hazardous and the toxicity of agents should be taken into consideration. Reactive chemicals have some risks of toxicity and natural biomaterials have a risk of virus infection.

Surgical glues were studied first to replace sutures, but they are mostly used also for hemostasis and sealing. In the future, they may be applicable to less invasive surgery and tissue engineering. However, intimate collaboration between material scientists, clinicians, and industries is needed for the development of new good glues.

REFERENCES

1. Sierra, D.H.; Saltz, R. *Surgical Adhesives and Sealants: Current Technology and Applications*; Technomic: Pennsylvania, 1996.
2. Agus, G.B.; Bono, A.V.; Mira, E.; Olivero, S.; Peilowich, A.; Homdrum, E.; Benelli, C. Hemostatic efficacy and safety of TachoComb in surgery. Ready to use and rapid hemostatic agent. Int. Surg. **1996**, *81* (3), 316–319.
3. Hollaus, P.; Pridun, N. The use of Tachocomb in thoracic surgery. J. Cardiovasc. Surg. (Torino). **1994**, *35* (6 Suppl. 1), 169–170.
4. Takeuchi, H.; Toyonari, Y.; Mitsuhashi, N.; Kuwabara, Y. Effects of fibrin glue on postsurgical adhesions after uterine or ovarian surgery in rabbits. J. Obstet. Gynaecol. Res. **1997**, *23* (5), 479–484.
5. Woolverton, C.J.; Fulton, J.A.; Salstrom, S.J.; Hayslip, J.; Haller, N.A.; Wildroudt, M.L.; MacPhee, M. Tetracycline delivery from fibrin controls peritoneal infection without measurable systemic antibiotic. J. Antimicrob. Chemother. **2001**, *48* (6), 861–867.
6. Sims, C.D.; Butler, P.E.; Cao, Y.L.; Casanova, R.; Randolph, M.A.; Black, A.; Vacanti, C.A.; Yaremchuk, M.J. Tissue engineered neocartilage using plasma derived polymer substrates and chondrocytes. Plast. Reconstr. Surg. **1998**, *101* (6), 1580–1585.
7. Wechselberger, G.; Russell, R.C.; Neumeister, M.W.; Schoeller, T.; Piza-Katzer, H.; Rainer, C. Successful transplantation of three tissue-engineered cell types using capsule induction technique and fibrin glue as a delivery vehicle. Plast. Reconstr. Surg. **2002**, *110* (1), 123–129.
8. Bach, A.D.; Bannasch, H.; Galla, T.J.; Bittner, K.M.; Stark, G.B. Fibrin glue as matrix for cultured autologous urothelial cells in urethral reconstruction. Tissue Eng. **2001**, *7* (1), 45–53.
9. Rainer, C.; Wechselberger, G.; Bauer, T.; Neumeister, M.W.; Lille, S.; Mowlavi, A.; Piza, H.; Schoeller, T. Transplantation of tracheal epithelial cells onto a prefabricated capsule pouch with fibrin glue as a delivery vehicle. J. Thorac. Cardiovasc. Surg. **2001**, *121* (6), 1187–1193.

Gene-Hydrogels

10. Bruns, T.B.; Worthington, J.M. Using tissue adhesive for wound repair: A practical guide to dermabond. Am. Fam. Phys. **2000**, *61* (5), 1383–1388.

11. Oowaki, H.; Matsuda, S.; Sakai, N.; Ohta, T.; Iwata, H.; Sadato, A.; Taki, W.; Hashimoto, N.; Ikada, Y. Non-adhesive cyanoacrylate as an embolic material for endovascular neurosurgery. Biomaterials **2000**, *21* (10), 1039–1046.

12. Braunwald, N.S.; Gay, W.J.; Tatooles, C. Evaluation of crosslinked gelatin as a tissue adhesive and hemostatic agent: An experimental study. Surgery **1966**, *59*, 1024–1030.

13. Hata, M.; Shiono, M.; Orime, Y.; Yagi, S.; Yamamoto, T.; Okumura, H.; Kimura, S.; Kashiwazaki, S.; Choh, S.; Negishi, N.; Sezai, Y. The efficacy and mid-term results with use of gelatin resorcin formalin (GRF) glue for aortic surgery. Ann. Thorac. Cardiovasc. Surg. **1999**, *5* (5), 321–325.

14. Bachet, J.; Goudot, B.; Dreyfus, G.; Brodaty, D.; Dubois, C.; Delentdecker, P.; Teimouri, F.; Guilmet, D. Surgery of acute type A dissection: What have we learned during the past 25 years? Z. Kardiol. **2000**, *89* (Suppl. 7), 47–54.

15. Mo, X.; Iwata, H.; Matsuda, S.; Ikada, Y. Soft tissue adhesive composed of modified gelatin and polysaccharides. J. Biomater. Sci. Polym. Ed. **2000**, *11* (4), 341–351.

16. Herget, G.W.; Kassa, M.; Riede, U.N.; Lu, Y.; Brethner, L.; Hasse, J. Experimental use of an albumin-glutaraldehyde tissue adhesive for sealing pulmonary parenchyma and bronchial anastomoses. Eur. J. Cardio-Thorac. Surg. **2001**, *19* (1), 4–9.

17. Matsuda, S.; Iwata, H.; Se, N.; Ikada, Y. Bioadhesion of gelatin films crosslinked with glutaraldehyde. J. Biomed. Mater. Res. **1999**, *45* (1), 20–27.

18. Otani, Y.; Tabata, Y.; Ikada, Y. A new biological glue from gelatin and poly (L-glutamic acid). J. Biomed. Mater. Res. **1996**, *31* (2), 157–166.

19. Iwata, H.; Matsuda, S.; Mitsuhashi, K.; Itoh, E.; Ikada, Y. A novel surgical glue composed of gelatin and N-hydroxysuccinimide activated poly(L-glutamic acid): Part 1. Synthesis of activated poly(L-glutamic acid) and its gelation with gelatin. Biomaterials **1998**, *19* (20), 1869–1876.

20. Wallace, D.G.; Cruise, G.M.; Rhee, W.M.; Schroeder, J.A.; Prior, J.J.; Ju, J.; Maroney, M.; Duronio, J.; Ngo, M.H.; Estridge, T.; Coker, G.C. A tissue sealant based on reactive multifunctional polyethylene glycol. J. Biomed. Mater. Res. **2001**, *58* (5), 545–555.

21. Kobayashi, H.; Sekine, T.; Nakamura, T.; Shimizu, Y. *In vivo* evaluation of a new sealant material on a rat lung air leak model. J. Biomed. Mater. Res. **2001**, *58* (6), 658–665.

22. Sawhney, A.S.; Pathak, C.P.; Hubbell, J.A. Bioerodible hydrogels based on photopolymerized poly (ethylene glycol)-co-Poly (a-hydroxy acid) diacrylate macromers. Macromolecules **1993**, *26* (4), 1–581.

23. Tanaka, K.; Takamoto, S.; Ohtsuka, T.; Kotsuka, Y.; Kawauchi, M. Application of AdvaSeal for acute aortic dissection: Experimental study. Ann. Thorac. Surg. **1999**, *68* (4), 1308–1313.

24. Nakayama, Y.; Matsuda, T. Newly designed hemostatic technology based on photocurable gelatin. ASAIO J. **1995**, *41* (3), M374–M378.

25. Auteri, J.S.; Oz, M.C.; Jeevanandam, V.; Sanchez, J.A.; Treat, M.R.; Smith, C.R. Laser activation of tissue sealant in hand-sewn canine esophageal closure. J. Thorac. Cardiovasc. Surg. **1992**, *103* (4), 781–783.

26. Ono, K.; Saito, Y.; Yura, H.; Ishikawa, K.; Kurita, A.; Akaike, T.; Ishihara, M. Photocrosslinkable chitosan as a biological adhesive. J. Biomed. Mater. Res. **2000**, *49* (2), 289–295.

Hair and Skin Care Biomaterials

Bernard R. Gallot
Laboratoire des Materiaux Organiques, Proprietes Specifiques, National Center for Scientific Research (CNRS), France

Abstract

Cosmetology and dermatology are interrelated and complementary in many ways although they have different purposes. This entry focuses on polymeric biomaterials used in hair and skin applications. Before describing the natural and synthetic polymers used in cosmetology, it briefly defines the products that incorporate them.

Cosmetology and dermatology are interrelated and complementary in many ways although they have different purposes. Cosmetology is concerned with hygiene, beauty enhancement, and the care of hair and skin, whereas dermatology is concerned with preventing and curing diseases of the skin (including the scalp) and its appendages.[1] In this entry we will focus on polymeric biomaterials used in hair and skin applications. The main areas of application of polymeric biomaterials are hair and scalp hygiene (shampoos), hair treatment and care, hair setting, hair coloring and skin care, including makeup.

Before describing the natural and synthetic polymers used in cosmetology, we will briefly define the products that incorporate them.

COSMETIC PRODUCTS INCORPORATING POLYMERS

These can be divided into six categories.

Shampoos

A shampoo is a product available as a clear, opaque or opalescent liquid, gel, cream, foam, aerosol or powder that dissolves in use. A shampoo is formulated using surface-active agents (surfactants and polymers), and exhibits detergent, emulsifying and foaming properties to clean hair and to make hair supple and glossy, and easy to style and disentangle.[1] Therefore, shampoos contain many ingredients, such as detergents, foam stabilizers and softeners, thickeners, opacifiers and agents providing pearlescence, special care products, sequestering agents, preservatives, perfumes and coloring agents.

Hair Care Products

Hair care products restore the natural beauty of hair to give it lightness, volume, spring control, suppleness, softness, and sheen.

Basic compounds

Basic compounds of hair care products are organic acids, fatty compounds and their derivatives, vitamins, protein derivatives, cation–active surfactants, cationic polymers, and hair strengtheners.

Formulation

Different formulations are used to fulfill the various requirements of hair care products: conditioners and deep conditioners, lotions or fluid gels, aerosol foams, hair oils, conditioning shampoos, and hair tonics.

Conditioners are usually fluid creams or gelled emulsions applied after a shampoo and rinsed out after several minutes; whereas deep conditioners are thick creams containing a wide range of conditioning and highly substantive agents.

Lotions or fluid gels are clear or opaque, light-textured and oil-free.

Aerosol foams are easy to handle and contain substantive derivatives that fix selectively on the hair fiber depending on the extent of sheath modification or damage.

Hair oils are used as pre-shampooing or post-shampooing products.

Conditioning shampoos contain conditioning materials to improve combability, manageability and hair properties.

Concise Encyclopedia of Biomedical Polymers and Polymeric Biomaterials DOI: 10.1081/E-EBPPC-120051898

Gene–Hydrogels

Hair tonics are hydroalcoholic lotions containing cationic polymers in combination with nonionic film-forming polymers that make hair easy to style.[1]

Hair–Setting Products

Hair setting products are divided in two families depending on when they are applied: On wet hair after shampooing or on dried, styled hair.

Products for maintaining hair styles

They are applied to wet hair, after shampooing but before combing. They form a flexible film on each hair, provide an overall stiffening of the hair and retard moisture uptake.

The essential components of products for maintaining hairstyle are film-forming polymers; plasticizers for the polymers; disentangling, softening, and glossening agents; solvents; perfumes; and colorants.[1]

Hair sprays

Hair sprays are applied after drying the styled hair. An aerosol formula contains the following components: polymers (described in a following section), plasticizers, softening and glossening agents (often lanolin derivatives), perfume, solvents (ethyl alcohol is the best), and propellants (the nature of which has evolved with environmental legislation and greatly influenced the nature of the polymers used).

Before going further we must briefly discuss the problem of propellants. Propellant gases provide the necessary expelling force but can also act as solvents for the various components of the hair spray formulations. Before environmental regulations were imposed, chlorofluorocarbons (CFCs) were used as propellants as they offer many advantages. They are nonflammable, denser than and immiscible with water, solvents for fatty compounds; compatible with resins, soluble in many organic solvents and generally of low toxicity. Furthermore, they are ideal for obtaining fine, nonwetting sprays.[1] The most widely used CFCs propellants were F.11 (trichlorofluoro methane: CCl_3F) and F12 (dichlorodifluoro methane CCl_2F_2). Unfortunately for CFCs, in 1974 two California researchers, Molina and Rowland, presented the hypothesis that CFCs, under the action of U.V. radiation, would release chlorine atoms that gradually reach the stratosphere and attack the ozone layer. This is not the place to discuss the effect of CFCs on the ozone layer, but the consequence was that United States banned "nonessential use" of CFCs as propellants in consumer products in 1979. Other countries followed suit.

Four alternative propellants have been proposed. Compressible gases (CO_2, N_2O, O_2, N_2) have insufficient solubility in solvents (ethanol, isopropanol) and exhibit a large pressure drop during application so they play a minor role at the present. Difluoro-1,1–ethane ($CH_3–CHF_2$), a non–chlorinated propellant with characteristics similar to those of F12, exhibits low toxicity and a higher flammability than CFCs but much lower than hydrocarbons.

Hydrocarbons (propane, isopropane, butane and isobutane) are non-miscible with water, have low toxicity relatively low cost, but high flammability and low solubility power for resins, requiring the utilization of specially designed polymers.

Dimethyl ether (DME) exhibits vapor pressure and a boiling point similar to those of F12, good dissolving properties for polar and non-polar resins (better than those of CFCs), high miscibility with water, alcohols, and solvents. Unfortunately, it has a greater flammability than hydrocarbons.

Hair Coloring Products

Among the six types of products used in hair coloring, three contain polymers. Coloring conditioners applied after shampooing without subsequent rinsing incorporate cationic polymers with basic dyes in their formulation. Color-setting lotions used for temporary hair coloring are hydroalcoholic solutions of polymers, such as polyvinylpyrrolidone (PVP) or copolymers of vinylacetate and crotonic acid. Coloring hair sprays incorporate dye and a film-forming polymer.

Skin Care Products

Skin care products protect, lubricate and clean the skin. They incorporate a variety of polymers, including polyethylene, polyethylene glycols, ethylene-propylene oxide copolymers, vinyl and acrylic polymer resins, hydrophilic polymers and silicone. Polymers are used as thickeners, film-formers, binders, lubricants, water-resistant agents[2] and in specialized delivery systems such as liposomes, "microsponge," hydrogels.[3]

Eye Makeup Products

The main makeup products are mascaras, shadows, liners brow makeup, creams, false eyelash adhesives and makeup removers.[4] In these products polymers are used as binders, thickeners or film-formers and may provide water resistance, waterproofing and gloss.[5] The main families of polymers used here are cellulose and derivatives, polyethylenes, ethyleneoxide homo and copolymers, vinyl polymers and acrylic polymers such as carbomers[6] and acrylic/acrylates copolymers.

POLYMERS USED IN COSMETOLOGY

The number and characteristics of polymers used in cosmetology and dermatology are rapidly increasing. To put some order in this crowded family of polymers, we have divided them into six classes: natural polymers and derivatives, vinyl and acrylate polymers, cationic polymers, ozone debate polymer bond, water compatible polymers, and siloxanes. We are aware that this classification is arbitrary and that some polymers can easily slide from one class to another; nevertheless, we think our classification will help the reader find his way through the jungle of polymeric biomaterials.

Natural Polymers and Derivatives

Five main types of natural polymers have been used in cosmetics: shellac, cellulose derivatives, xanthan gum, hyaluronic acid and protein derivatives.

Shellac

Shellac is an extract of secretions from the insect *Laccifer lacca* that lives on trees in India and Thailand. It consists primarily of three complex hydroxycarbonic acids: aleuric, shellolic, and jalaric acid.[7] Bleached and dewaxed shellac has good holding ability and excellent weather resistance, but poor solubility in propellants and solvents. It is also very difficult to brush and shampoo away and its films powder easily.

Cellulose Derivatives

Cellulose derivatives sodium carboxymethyl, methyl, hydroxypropylmethyl, hydroxyethyl and hydroxypropyl ether are used in concentrations between 0.1% and 5% in cosmetology, mainly as polymeric thickeners.

Sodium carboxymethyl cellulose is an anionic polymer with extremely high water-retention capacity. It is used in makeup, where it acts as a stabilizer and suspending agent in cream eye shadow, face masks, mascaras, liquid eye liner and liquid rouge.[8]

Hydroxyethyl cellulose is a non-ionic polymer used in emulsion-type mascaras, antiperspirants and deodorants.

Hydroxypropyl cellulose is a non-ionic polymer used in a wide variety of alcohol-based preparations, including shampoos, skin fresheners, and aftershaves.

Methyl cellulose and hydroxypropylmethyl cellulose are multifunctional additives used in makeup products, where they act as binders and film-formers for longer-lasting mascaras, eye-shadow, and eye-liner.[2]

Xanthan Gum

Xanthan gum is a polysaccharide used as emulsion stabilizer and suspending agent in moisturizing lotions and creams.[9]

Hyaluronic Acid

Hyaluronic acid is a mucopolysaccharide that can be used as a transdermal delivery system as it forms a matrix on the skin allowing for increased skin penetration due to skin hydration.[10]

Protein Derivatives

Protein molecules are too large to penetrate hair and fix onto hair keratin, so they are used in the form of partial hydrolysates with molecular weights higher than 1000. Many benefits have been claimed for inclusion of proteins in hair care formulations. When hydrolyzed collagen is used in conditioning rinses and shampoos, it improves body, gloss, manageability, prevents or reduces damage during bleaching, and facilitates a more even penetration of dyes.[11,12] Protein hydrolysates have also been used for substantivity, moisturizing and mildness enhancement in skin-care.[2] The three main sources of protein hydrolysates are collagen, elastin and silk proteins. Collagen films are used for cosmetic masks, where they act as microsponges because, when moistened, they form a gel-like structure that can be molded easily to the contours of skin.[13] Elastin hydrolysates are claimed to improve skin tone and suppleness. Silk protein powder is used to impart a silky feel to the skin.

VINYLIC AND ACRYLIC POLYMERS

Polyvinylpyrrolidone

Among vinylic polymers the best known and most widely used is PVP. It was synthesized first at the I. G. Farben Laboratories and used as a plasma extender during World War II. Since 1950 it has been developed for hair sprays because it exhibits conditioning properties if applied via shampoo. It forms transparent films on hair, providing smoothness and luster. PVP exhibits other useful properties in shampoos: It improves foam stabilization, increases viscosity, reduces eye irritation and is easily removed. Apart from its ability to act as a thickener, dispersing agent, lubricant, or binder, it has a strong affinity for dyes that makes it suitable for use in coloring shampoos and in eye-makeup products. PVP applications in hair and skin formulations result from the following properties: complete solubility in water; high solubility in a wide variety of organic solvents, especially in ethanol propellant mixtures; and absence of odor, taste, and toxicity.

Unfortunately, PVP also has drawbacks, such as hygroscopicity (its films tend to become sticky, dull, and tacky as humidity is absorbed). To overcome such problems formulators introduced small amounts of shellac or silicon oil in their formulations. Typical formulations contain the following constituents in % by weight: PVP, 2; ethoxylated lanolin, 0.6; perfume, 0.2; anhydrous ethanol, 25; propellant F11 and F12, 75; or PVP, 2.6; lanolin, 0.3; shellac, 0.2; perfume, 0.3; alcohol, 16.6; and F12, 80.[14]

Nevertheless, a more efficient way to solve the problem was the copolymerization of the hydrophylic vinylpyrrolidone with a hydrophobic monomer.

Gene–Hydrogels

Copolymers

The first copolymers synthesized were statistic copolymers of vinyl acetate and vinylpyrrolidone, but they were soon followed on the market by many other copolymers. We will briefly describe their main properties.

Vinylpyrrolidone-vinyl acetate copolymers

As vinyl acetate homopolymers present properties complementary to those of PVP; namely, good flexibility in dry film form, good resistance to plastization by humidity, and good hair–spray hold at high humidity, vinyl acetate was copolymerized with vinylpyrrolidone. Since 1956 copolymers PVP/VA with ratios 70/30, 60/40, 50/50, and 30/70 have replaced PVP homopolymers in formulations. An example of formulation is in % by weight: PVP/VA, (70/30), 1.5; laneth, 5, 0.4; perfume, 0.2; alcohol, 27.9; and propellant 11/12 (35/65), 70.[15]

Vinyl acetate-substituted acrylamide copolymers

In 1958 acrylate/acrylamide copolymers appeared that were readily soluble in ethanol and gave clear films with high gloss that was not lost after combing. Variation of the degree of hydrolyzation allowed the modification of the properties.[15]

Vinyl acetate-crotonic acid copolymers

As these polymers contain carboxylic acid groups that can be converted into amine salts by neutralization with bases such as amino alcohols, their solubility in water can be increased and their films softened.[15] The presence of homopoly–vinyl acetate and the variation of the composition of the copolymers in commercial materials gave indesirable properties. To overcome such deficiencies synthesis of homogeneous copolymers was necessary.[15]

Vinyl methy ether-maleic anhydride copolymers

These alternating polymers[16] associate monomers of which homopolymers have complementary properties. Vinyl methyl ether gives rubbery polymers soluble in cold water but insoluble in hot water, whereas maleic anhydride is a hard, polar monomer. These alternating polymers are generally in the form of monoethyl or monobutyl ester.[15] They can be modified by neutralization with alkyl amines or esterification by oleyl alcohol to give more flexibility to their films.[15] PVM/MA copolymers are still popular as hair fixatives owing to their exceptional versatility.

Other copolymers

Hydrophobic monomers were copolymerized at first with vinylpyrrolidone in order to reduce film sensibility to

humidity, and then introduced into a great number of polymers, not only to reduce sensibility problems associated with water but also to ensure good hold by the internal plasticization of the resins and to induce compatibility with propellants. For that purpose two main technics were used: copolymerization of vinylpyrrolidone with different monomers and introduction of different monomers in vinylacetate/crotonic acid copolymers.

PVP was copolymerized with butyl methacrylate and methacrylic acid, alkyl stearate, alkoxyacetic acid, and vinyl stearate, octadecyl vinyl ether, 1-octadecene, alkylacetate, and crotonic acid.[15]

Vinyl acetate/crotonic acid copolymers were modified by including the following monomers: methymethacrylate, hydroxypropyl acrylate, vinyl pivalate, alkyl pivalate, vinyl neodecanoate, vinyl, alkyl and methallyl laurate, stearate, isostearate and behenate, butyraldehyde and B-hydroxyethylcrotonate, bis-2-ethylhexyl maleate, and unsaturated fatty acids.[15]

CATIONIC POLYMERS

Cationic polymers were developed to overcome the disadvantages of cationic surfactants (absence of setting properties, incompatibility with most anionic surfactants used in formulating shampoos, eye irritation, etc.). In cationic polymers, the substantive cationic poles are grafted onto or integrated into a polymeric structure. They are revolutionary in the way that they are substantive for hair and able to cover the hair surface with a continuous film and thereby protect hair against external attack.[1]

The first cationic polymer was a cellulose derivative called Polyquaternium 10. Other polysaccharide derivatives and a number of synthetic polymers were prepared and used in hair and skin care.

Following the CTFA classification cationic polymers are called "Polyquaterniums," as they are polymers that contain a quaternary ammonium radical. Cationic polymers can be divided into four groups: polysaccharide derivatives, acrylamide derivatives, vinylpyrrolidone derivatives, and Ionenes.

Polysaccharide Derivatives

They are derivatives of cellulose, guar gum, or chitosan.

Polyquaternium 10

This polymer is obtained by the reaction of epichlorhydrin on cellulose followed by quaternization by trimethyl amine. It can be described as 2-hydroxypropyl trimethyl ammonium chloride of ethyl cellulose. The liquid-crystalline character of its combinations with anionic surfactants in shampoos imparts good wet combing, curl retention and manageability and mending of split-hair ends.[17] Polyquaternium 10 is also used as additive in traditional or oil-free conditioners and as hair-dye carrier in coloring shampoos.

Polyquaternium 4

This is a copolymer of hydroxyethyl cellulose and diallyldimethylammonium chloride. It is compatible with anionic surfactants such as sodium lauryl sulfate and offers superior curl retention properties, even in high humidity conditions, excellent wet-combing and no build-up.[18]

Guar hydroxypropyltriammonium chloride

This polymer is prepared from guar gum (a galactomannan) by the same procedure as Polyquaternium 10 from cellulose. It combines the thickening properties of guar gum with good sustantivity to hair, and is used in body-building shampoos.

Chitosans

The result from the partial de-acetylation of chitin an insoluble polymer of N-acetylglucosamine extracted from shrimp and crab shells. Chitosans are water soluble cationic polymers used in cosmetic formulations as conditioners to provide set holding, smooth, soft feel, and improved hair luster. In skin care products chitosans are used to form clear protective coatings that retain moisture and are non-allergenic. Chitosans are also useful in after-shave lotions, shaving foams and gels, sunburn lotions, wound healing, and treatment of damaged skin.[19]

Acrylamide Derivatives

Acrylamide-diallylmethylammonium copolymers

These polymers are referred to as Polyquaternium 7. They exhibit excellent substantivity to skin and hair, good film-forming properties and good compatibility with anionic surfactants.

Diallyldimethylammonium homopolymers called Polyquaternium 6 also have excellent substantive properties for hair and skin, but poor compatibility with anionic ingredients.[15]

Acrylamide-quaternized dialkylaminoalcohol acrylate or methacrylate copolymers

These polymers are referred to as polyquaternium 5, and exhibit properties similar to those of polyquaternium 7.

Acrylamide-methacryloyloxyethyltrimethyl ammonium chloride copolymers

These polymers are referred to as polyquaternium 15, and enhance foaming and interact with keratin to give conditioning effects.

Vinylpyrrolidone Derivatives

Dimethylsulfate quaternized copolymers of vinylpyrrolidone and dimethylaminoethyl methacrylate are called Polyquaternium 11. Originally synthesized as a hair fixative polymer[15] they conjugate conditioner and palliative properties for damaged hair with setting and conditioning properties.[15] These properties are related to the uniformity and thickness of the deposit of the polymer on hair.[20–22]

Ionenes

Ionenes are cationic polyelectrolytes formed by the condensation of di(tertiary amines) and dihalides via a quaternization reaction. Ionenes have rather low molecular weight: between 1000 and 20,000[1] and correspond to (Eq. 1):

$$- (- N^+ - A - N^+ - B - N^+ -) - n \qquad (1)$$

with for instance: for Polyquaternium 1: $A = B = CH_2 - CH = CH - CH_2$ with triethanolamine endings. for Polyquaternium 2: $A = (CH_2)_3 - NH - CO - NH - (CH_2)_3$ and $B = CH_2 - CH_2 - O - CH_2 - CH_2$

The advantage of the Ionenes lies in the accuracy with which they can be synthesized to target desired physical properties, taking advantage of the fact that the block lengths, and hence the charge density is accurately predetermined by the choice of the diamine and dihalide reactants.[1]

Ionene polymers are used as conditioners, dyeing supports, wave setting and conditioning lotions, and bleaching agents.[1,15] The setting properties are improved by crosslinking Ionenes[15] and the compatibility with anionic surfactants is improved by including 1,4-piperazine units.[1]

Ozone Controversy Polymer Sons

The U.S. ban of CFCs in aerosols led to the replacement of CFCs by hydrocarbons (propane/butane) and polymers used in hair sprays had to become compatible with those propellants. The increase in the hydrocarbon compatibility of the polymers was achieved by introducing apolar monomers so that the newer resins became terpolymers.

For example, vinyl propionate was introduced in PVP/VA and in VA/Crotonic acid copolymers. In both cases the compatibility with hydrocarbons increased profoundly and formulation of aerosol hair sprays having low cloud points was possible without using a cosolvent. The copolymer PVP/VA/vinyl propionate (30:40:30) forms elastic films that are easily removed by shampoos. The copolymer VA/crotonic acid/vinyl propionate (50:10:40) is used after neutralization until about 75% of the carboxylic functions by aminomethylpropanol.

Excellent propane/butane compatibility (up to 60–70%) is also obtained for PVP/t-butyl acrylate/methacrylic acid copolymers that give films with good curl retention firmness, low water absorption, and are easy to wash off. N-tert-butylacrylamide/ethyl acrylate/acrylic acid copolymers, after neutralization to about 80% by customary aminoalcohols give elastic and permanent hairstyle, even when exposed to an extreme climate.[23]

New problems appeared for hair sprays with the "California Regulation 80% Volatile Organic Compounds (VOCs)" regulation that took effect in January 1993, and that will become VOCs 55% in January 1998. The VOC regulations apply to the presence of water in hair sprays. Nevertheless, two types of terpolymers can already satisfy the 80% VOCs requirement: the terpolymer vinyl caprolactam/vinylpyrrolidone/dimethylaminomethyl methacrylate (VCL/PVP/DMAEMA), and the terpolymer vinylacetate/butylmaleate/isobornyl acrylate.

Terpolymers VCL/PVP/DMAEMA are synthesized by radical polymerization in ethanol.[24] In such polymers, PVP and DMAEMA monomers provide solubility in water or alcohol, while VCL gives excellent compatibility with hydrocarbon propellants. They can be used in aerosol hair sprays, thereby satisfying the 80% VOCs specification and even the zero VOC specification in pump sprays.[25]

Vinyl acetate/N-butylmaleate/Isobornyl acrylate terpolymers combine the properties of vinyl acetate to give relatively flexible segments (homopolymer: T_g = 30°C), isobornylacrylate to produce stiff segments (homopolymer: T_g = 90–100°C), and mono-n-butylmaleate to adjust solubility characteristics by varying the degree and type of neutralization. The terpolymer can be incorporated into aerosol hair sprays with DME or hydrocarbon propellants, propellants satisfying the 80% VOC specification, and pump hair sprays satisfying the 55% VOC specification. The terpolymers give good setting effects without undesirable feel or build-up on repeated applications between shampoos.[26]

WATER COMPATIBLE POLYMERS

An elegant way to fulfill the aerosol regulations consists of replacing aerosol hair sprays by aqueous gels. Aqueous gels are obtained by adding polymers exhibiting one or more of the following properties: thickening, gelling, suspending, emulsifying, and thixotropic.[15]

Classical Polymers

Three types of classical thickening agents are particularly important. They are cellulose derivatives, such as hydroxypropylmethyl cellulose; carbomers obtained by precipitation polymerization of acrylic acid and used in hand and body lotions and creams and hair styling gels;[6,27] and PVM/MA decane crosspolymers used in styling gels, hand and body lotions, and antidandruff shampoos.

Lipopeptide-Based Polymers

A different type of polymer, lipopeptide-based block and comb copolymers, can be used to obtain aqueous gels and stable emulsions.

Block copolymers

Block copolymers are amphiphilic lipopeptides, Cn(AA)p, formed by a hydrophobic lipidic chain, Cn, containing 12–22 carbon atoms linked through an amide bond to a hydrophilic peptidic chain, (AA)p, with a number average degree of polymerization, p, between 1 and 70, where the hydrophilic amino acids are: sacrosine, serine, glutamic acid, lysine, N-hydroxyethylglutamine, or N-hydroxypropylglutamine.[28–30] Lipopeptides with p = 1 are prepared by coupling the amino acid by its α-carboxylic function to the end of a fatty amine. Lipopeptides with p higher than 1 are prepared by polymerization of the N-carboxyanhydride (NCA) of the amino acid, or of the side–chain–protected amino acid at the end of a fatty amine.[28–30]

Lipopeptides exhibit liquid crystalline gels in water solution, and the structure of the gels goes from lamellar to cylindrical hexagonal and to cubic when the degree of polymerization of the amino acid increases.[28,29] Lipopeptides with Cn equal to or greater than C16, and low degrees of polymerization, p, exhibit very interesting emulsifying properties. They give very stable oil in water (O/W) emulsions for very small amounts of lipopeptide (1–2%) with a wide variety of the oils used in the cosmetic industry, such as isopropylmyristate, butylstearate, dodecane, migliol, cosbiol, wheat germ oil, vaseline oil, ricin oil, and silicone oils, as well as stable mini-emulsions of the same oils if 1–3% of cetyl alcohol is added.[31–33] Lipopeptides in C12 with degrees of polymerization between 8 and 20 exhibit good foaming and disentangling properties. Moreover, lipopeptides are highly biocompatible.[33]

Comb-like polymers

Comb-like polymers consist of a polyvinyl main chain of polyacrylamide, polymethacrylamide or polystyrylamide, and lipopeptidic side chains. Comb-like polymers are obtained by radical polymerization of lipopeptides or lipo-amino-acids macromonomers of general formula (Eq. 2):

$$X–CO–NH–\left(CH_2\right)_n–Y–\left(AA\right)_p \qquad (2)$$

with: X=H_2C=CH, H_2C=C(CH$_3$)– or H_2C=CH–C$_6$H$_4$– and Y=NH–CO or CO–NH and AA is one of the following amino acids: Sarcosine, glycine, alanine, serine, aspartic acid, glutamic acid, and tyrosine.[34–38]

Macromonomers are prepared from α,ω-fatty-diamine or α,ω-fatty-amino-acids. In the case of α,ω-diamines one of the amine functions is used to link the polymerizable group and the other amine function to polymerize the NCA of the amino-acid.[34–36] In the case of α,ω-amino-acids the amine function is used to link the polymerizable group and the acid function to link the amino–acid.[37]

Macromonomers are then homopolymerized, or copolymerized with aliphatic acrylamide.[33–38]

In the dry state comb–like polymers exhibit smectic and nematic liquid–crystalline structures, and in the presence of water, lamellar or cylindrical hexagonal liquid–crystalline gels, depending on the nature and the degree of polymerization of the amino acid and the amount of water.[34–38] The hydrophilic–lipophylic balance of comb-like polymers can be regulated by playing on the nature and degree of polymerization of the amino-acid, and the nature and amount of the comonomer.[32]

SILOXANE POLYMERS

The silicone structure is characterized by the repetition of oxygen–silicium bonds in the polymer skeleton: see Eq. 3.

$$CH_3 - \underset{\underset{CH_3}{|}}{\overset{\overset{CH_3}{|}}{Si}} - O - \left(\underset{\underset{CH_3}{|}}{\overset{\overset{CH_3}{|}}{Si}} - O \right)_n - \underset{\underset{CH_3}{|}}{\overset{\overset{CH_3}{|}}{Si}} - CH_3 \tag{3}$$

The silicium–oxygen bonds are responsible for the original properties of the polysiloxanes. The high bond energy (106 Kcal/mole) confers high thermic stability and high oxidation resistance to silicon products. The length of the Si–O bond (1.6 A), and the angle of the Si–O–Si bonds (130°) give very high flexibility and a low glass transition to the siloxane chain. Furthermore, the siloxane blackbone is covered by methyl groups that give weak intermolecular cohesive forces. These structural characteristics give low surface-tension (about 20 dynes/cm), easy spreading on most surfaces, good lubricating properties, and easy formation of uniform, smooth, hydrophobic films to polysiloxanes. As a result of their physical and chemical properties, polysiloxanes are non toxic, non-irritating, and classified as environmentally friendly.[39]

Polydimethylsiloxane

Polydimethylsiloxanes (CTFA name: Dimethicone) are used both for skin and hair care. Skin care products containing Dimethicone are used for protective hand and body lotions and creams, cleansing creams, suntan lotions, sun creams, and aerosol shave lathers. Dimethicones are incorporated in hair care formulations to improve luster and sheen.[2,40]

Cyclic Siloxanes

Cyclic siloxanes (CTFA name: Cyclomethicone) are cyclic low molecular weight siloxanes containing from 3 to 6 Si atoms but generally 4 or 5. They are volatile products compatible with most waxes, fatty esters, and mineral oils. Due to their high volatility they are used in skin lotions, cleansers, sun products, pre- and post-shaving lotions, antiperspirants and deodorants, and in any product requiring a transient effect.[15,18,41,42]

Phenylsiloxanes

Phenylsiloxanes (CTFA name: phenyldimethicone) are polysiloxanes in which phenyl groups replace methyl groups. They exhibit properties similar to Dimethicone but are much more compatible with alcohol and organic solvents.[43]

Organopolysiloxanes

Organopolysiloxanes have been developed to overcome the drawbacks of dimethicone, such as its lack of solubility in water, alcohol, and common solvents. They are prepared by chemically modifying dimethicone.

Organopolysiloxanes are classified in two families depending upon the position of the chemical modification of dimethicones. They are called:

- Terminal or linear organopolysiloxanes when the Dimethicone blackbone is findxb.alpha; ω-modified: see Eq. 4.

$$R - \underset{\underset{CH_3}{|}}{\overset{\overset{CH_3}{|}}{Si}} - O - \left(\underset{\underset{CH_3}{|}}{\overset{\overset{CH_3}{|}}{Si}} - O \right)_n - \underset{\underset{CH_3}{|}}{\overset{\overset{CH_3}{|}}{Si}} - R \tag{4}$$

- Comb organopolysiloxanes when the organic groups are linked as side chains on the skeleton. Comb polymers can contain only one type, R, of organic substituents: Eq. 5, or two types (R1 and R2) of organic substituants: Eq. 6.

$$CH_3 - \underset{\underset{CH_3}{|}}{\overset{\overset{CH_3}{|}}{Si}} - O - \left(\underset{\underset{CH_3}{|}}{\overset{\overset{CH_3}{|}}{Si}} - O \right)_n - \left(\underset{\underset{R}{|}}{\overset{\overset{CH_3}{|}}{Si}} - O \right)_m - \underset{\underset{CH_3}{|}}{\overset{\overset{CH_3}{|}}{Si}} - CH_3 \tag{5}$$

$$CH_3 - \underset{\underset{CH_3}{|}}{\overset{\overset{CH_3}{|}}{Si}} - O - \left(\underset{\underset{CH_3}{|}}{\overset{\overset{CH_3}{|}}{Si}} - O \right)_n - \left(\underset{\underset{R1}{|}}{\overset{\overset{CH_3}{|}}{Si}} - O \right)_{m1} - \left(\underset{\underset{R2}{|}}{\overset{\overset{CH_3}{|}}{Si}} - O \right)_{m2} - \underset{\underset{CH_3}{|}}{\overset{\overset{CH_3}{|}}{Si}} - CH_3 \tag{6}$$

The synthesis of organopolysiloxanes is generally performed in two steps: transformation of Dimethicone into a reactive polysiloxane bearing Si–Cl, Si–OR or Si–H groups followed by esterification of the Si–Cl function or transesterification of the Si–OR function, with both reactions giving Si–O–C linkages, or action of vinyl derivatives in presence of a platinium based catalyst giving more stable Si–C linkages.

The main organopolysiloxanes are: alkyl, alkoxy, ether, amino, betaine, amidoquaternary, and phosphate esters polysiloxanes.

Alkylpolysiloxane copolymers

Their CTFA name is Alkyldimethicone. They are prepared by esterification or transesterification of siloxanes containing hydrolizable Si–Cl or Si–OR groups with fatty alcohols or by hydrosylilation of α-olefins. When the alkyl chains $R=(CH_2)p–CH_3$ in Eq. 5 contain more than 10 carbon atoms they are called silicon waxes and are mostly used in makeup products such as lipsticks and mascaras where they improve the spreadability as well as luster and distribution of pigments.[40] When they contain two types of side chains R1 and R2 in Eq. 6 (alkyl linear chains with two different lengths or a mixture of linear and branched alkyl chains), randomly distributed they enhance the smoothness and softness of the skin and are used in water-in-oil skin creams, water-in-oil protection creams, and oil-in-water moisturizing lotions for sun protection.[44]

The α,ω-dialkoxydimethylpolysiloxanes $R=O–(CH_2)_{17}–CH_3$ in Eq. 4 are used in emulsions for skin care.

Polyetherpolysiloxanes copolymers

Their CTFA name is dimethiconecopolyol. They are generally synthesized by hydrosylilation of allyl polyethers and are non-ionic compounds: $R=(CH_2)_3–O–(C_2H_4O)_x–(C_3H_6O)_y–H$ in Eq. 5.

They are soluble in water when they contain mostly ethylene oxide and soluble in oils when they contain mainly propylene oxide. In these products the polydimethylsiloxane skeleton brings gloss and silky feel, while the ethylene oxide chains increase adherence on polar surfaces. They are used as surface-tension depressants, wetting agents, emulsifiers or foam builders in shampoos, liquid soaps, foaming bath, shower gels, skin care creams, shaving lather, deodorants, and antiperspirants sticks.[40,42,45]

Polyalkylpolyetherpolysiloxane copolymers

When both polyether and α-olefins are grafted on polysiloxanes: $R1=(CH_2)_3–O–(C_2H_4O)_x–(C_3H_6O)_y–H$ and $R2=C_pH_{2\,x+1}$ in Eq. 6.

Very efficient water-in-oil emulsifiers are obtained because polyether chains are dissolved in water and alkyl chains are dissolved in the oil phase while the polysiloxane skeleton fixes the total molecule at the interface.

They also give water in silicon emulsions that are more substantive than the corresponding water-in-oil emulsions.[40,46]

Polyaminopolysiloxane copolymers

They are synthesized by replacing a portion of methyl groups by organic moieties of amine functionalities. They are divided in two families in the CFTA classification: Trimethylsylilamodimethicone (TSA) with $R=(CH_2)_n–NH–(CH_2)_m–NH_2$ in Eq. 5, generally n = 3 and m = 2, and Amodimethicone with the same R side chains but with the α,ω-methylene groups of the main chain replaced by hydroxyl groups.

Silicones with amino functions are strongly substantive to hair, and are ideal conditioners conferring luster and enhancing wet and dry combing.[42,45] Amodimethicones exhibit higher substantivity than TSA because their α,ω–OH groups allow chain extension and crosslinking.[46]

Polyorganobetainepolysiloxane copolymers

They are amphoteric surfactants that can exist with a positive or a negative charge or both in aqueous solution as a function of the pH of the solution. They are of two types: carboxy or phosphobetaines. In carboxybetaines obtained by reaction of epoxysiloxanes with secondary amines at first and then with $ClCH_2–COON_a$ $R=(CH_2)_3–O–CH_2–CH(OH)–N^+(R3,R3)–CH_2–COO^-$ in Eq. 5. In phosphobetaines, in Eq. 5: $R=(CH_2)_3–O–P–O_3–CH_2–CH(OH)–CH_2–N^+(R3,R3)–(CH_2)_3–NR–CO–R4$ with R4=fatty acid derivative, and R3=CH_3 or $CH_2–CH_2–OH$.

Siliconebetaines can be water soluble, insoluble or dispersible. They are non-irritating to skin and eye and are used at concentrations between 1.5% and 5% in conditioning shampoos, bath and shower gels, hair styling gels, facial cleansing, and other kinds of gels.[47]

Aminoquaternarycompoundspolysiloxane copolymers

Aminoquaternary compounds can be of the comb or terminal types. Their CTFA name is Silicon Quaternary. They are prepared by reaction of an alkylamidodimethylamine with dimethiconecopolyol chloroacetic acid ester in aqueous or polar solvent solution. They correspond to Eq. 4 or 5 with: $R=(CH_2)_3–O–(C_2H_4O)_x–(C_3H_6O)_{z\,y}–(C_2H_4O)_z R'$ with: $R'=P–O_3–CH_2–CH(OH)–CH_2–N^+(CH_3,CH_3)–(CH_2)_3–NH–CO–R''$ with R'=fatty acid derivative.

They are called respectively Silicone Quaternium 1, 2 and 3 when their alkylamido group is a cocamidopropyl dimethyl group, a myristamidopropyl dimethyl group and a di-linoleylamidopropyl dimethyl group respectively. Silicone Quaterniums exhibit antistatic properties, improved wet combing and gloss, skin conditioning, and can be formulated into gels, mousses, and virtually any desired product form. They are used at low concentrations (1–2%) in conditioning shampoos, hair conditioners, and bubble baths, where they provide silky feel, conditioning properties, and foam stabilization.[48]

Polyphosphatestersrpolysiloxane copolymers

Polyphosphate esters are anionic surfactants of the terminal or comb type. In Eq. 4 or 5: $R=(CH_2)_3-O-(C_2H_4O)_x$ $(C_3H_6O)_y-(C_2H_4O)_z-P(O)-(OH)_2$.

They are prepared by action of polyphosphoric acid or phosphorous pentoxyde on dimethiconecopolyol. A mixture of mono- and diesters is always obtained but the ratio of the mono and diesters varies with the nature of the phosphating agent used.

The solubility of the polyphosphate esters is easily adjusted by playing on the degree of neutralization of the acid phosphate group and on the nature of the neutralizing agent (tri, di or monoethanol amine, KOH, NaOH).

Silicone-based phosphate esters are emulsifiers producing oil-in-water emulsions. They are used in emollient lotions, light body lotions, sunscreen lotions and creams, at concentrations of about 3%. Incorporated in hair conditioners, they improve curl retention and combability.[48]

Other ionic organomodified silicones

Silicone sulfates and silicone thiosulfates are prepared by the action of $NaHSO_3$ and $Na_2S_2O_3$ on epoxysiloxanes and in Eq. 4: $R=(CH_2)_3-O-CH_2-CH(OH)-SO_3^-$, Na^+ or $S_2O_3^-$, Na^+.

These ionic silicone surfactants are used mainly in hair care products, such as hair rinses and conditioning shampoos where they improve combability, gloss, and handling of hair.[40]

REFERENCES

1. Zviak, C. *The Science of Hair Care*; Marcel Dekker: New York and Basel, 1986.
2. Idson, B. Polymers in skin cosmetics. Cosmet. Toilet. **1988**, *103*, 63.
3. Gans, E.H. Polymer developments of cosmetic interest. Cosmet. Toilet. **1988**, *103*, 94.
4. Wetterhahn, J. *Eye Makeup in Cosmetics Science and Technology*; Bolsam, M.S., Sagarin, E., Eds.; Wiley-Interscience, 1972; Vol. 1.
5. Kapadia, Y.M. Use of polymers in eye makeup. Cosmet. Toilet. **1984**, *99*, 53.
6. Amjad, Z.; Hemker, W.J.; Maiden, C.A.; Rouse, W.H. Sauer, C.E. Carbomer resins: Past, present and future. Cosmet. Toilet. **1992**, *107*, 81.
7. Tannert, U. Shellac: A natural polymer for hair care products. SOFW J. **1992**, *118* (17), 1079.
8. Rutkin, P. Eye makeup. In *The Chemistry and Manufacture of Cosmetics*; de Navarre, M.G., Ed.; Continental, 1981; Vol. 4.
9. Rieger, M. The role of water in performance of hydrophilic gums. Cosmet. Toilet. **1987**, *102*, 101.
10. Balazs, E.A.; Bond, P. Hyaluronic acid: Its structure and use. Cosmet. Toilet. **1984**, *99*, 65.
11. Hunting, A. *Encyclopedia of Shampoo Ingredients*; Micelle: Cranford, NJ, 1983.
12. Hunting, A. *Encyclopedia of Hair Conditioner Ingredients*; Micelle: Cranford, NJ, 1985.
13. Karjala, S.A.; Bouthilet, R.J.; Williamson, J.E. Factors affecting the substantivity of proteins to hair. Proc. Sci. Sec. TGA **1966**, *45*, 6.
14. Kelthler, W.R. *The Formulation of Cosmetics and Cosmetic Specialities*; Drug Cosmet.: New York, 1956.
15. Lochhead, R.Y. The history of polymers in hair care. Cosmet. Toilet. **1988**, *103*, 23.
16. Shildknecht, O.E. *Polymer Processes*; Interscience: New York, 1956.
17. Cannel, D.W. Spilt ends and hair repair. Cosmet. Toilet. **1979**, *94*, 29.
18. Hunting, A.A.L. The functions of polymers in shampoos and conditioners. Cosmet. Toilet. **1984**, *99*, 57.
19. Onsoyen, E.; Dybdahl, M. The application and benefits of chitosan in cosmetics. Cosmet. Toil. **1991**, *106*, 32.
20. Kamath, Y.K.; Dansizer, C.J.; Weigmann, H.D. Surface wettability of human hair. I. Effect of deposition of polymers and surfactants. J. Appl. Polym. Sci. **1984**, *29* (3), 1011–1026.
21. Kamath, Y.K.; Dansizer, C.J.; Weigmann, H.D. Surface wettability of human hair. II. Effect of temperature on the deposition of polymers and surfactants. J. Appl. Polym. Sci. **1985**, *30* (3), 925–936.
22. Kamath, Y.K.; Dansizer, C.J.; Weigmann, H.D. Surface wettability of human hair. III. Role of surfactants in the surface deposition of cationic polymers. J. Appl. Polym. Sci. **1985**, *30* (3), 937–953.
23. Weker, H.U.; Sperling-Vietmeir, K. Modern hair spray formulations: Trends and new polymers. SOFW J. **1990**, *116*, 130.
24. Tazi, M. Progress in water based hair spray. Cosmet. Toilet. **1991**, *106*, 238.
25. Petter, P.J.; Johnson, S.C. Copolymer VC-713: A new resin for hairspray formulation. SOFW J. **1987**, *113*, 427.
26. Patel, D.; Petter, P.J. New polymers for the formulation of hair fixative products for tomorrow's market. SOWF J. **1992**, *118*, 1072.
27. Barzaghi, M.; Tadini, G. Synthalin: The new carbomers of highest purity. Cosmet. Toilet. **1991**, *106*, 100.
28. Douy, A.; Gallot, B. New amphipatic lipopeptides: 1) Synthesis and mesmorphic structures of lipopeptides with polysarcosine peptidic chains. Makromol. Chem. **1986**, *187*, 465.
29. Gallot, B.; Douy, A.; Haj Hassan, H. Synthesis and structural study by X-ray diffraction of lyotropic lipopeptidic block copolymers with polylysine and poly(glutamic acid) peptidic chains. Mol. Cryst. Liq. Cryst. **1987**, *153* (1), 347–356.
30. Gallot, B.; Haj Hassen, H. Lyotropic lipo-amino-acids: Synthesis and structural study. Mol. Cryst. Liq. Cryst. **1989**, *170* (1), 195–214.
31. Gallot, B.; Haj Hassan, H. Liquid crystalline phases and emulsifying properties of block copolymers with hydrophobic aliphatic chains and hydrophilic peptidic chains. In *ACS Symposium Series*; 1989; Vol. 384, 116.
32. Gallot, B. Liposarcosine based polymerizable and polymeric surfactants. In *ACS Symposium Series*; 1991; Vol. 448, 103.
33. Gallot, B. Chapter XVI. In *Polymers in Medicine: Biomedical and Pharmaceutical Applications*; Ottenbrite, A.M., Chiellini, E., Eds.; Technomic: Lancaster, Basel, 1992.

Gene–Hydrogels

34. Gallot, B.; Douy, A. Comb polymers with thermotropic and lyotropic properties: Synthesis and structural study. Mol. Cryst. Liq. Cryst. **1987**, *153* (1), 367–373.

35. Gallot, B.; Douy, A. Synthesis and mesomorphic structures of comb-liquid-crystalline polymers with lipopeptidic side chains. Makromol. Chem. Macromol. Symp. **1989**, *24* (1), 321–329.

36. Gallot, B. Liquid-crystalline comb polymers with polystyrene main chain and liposarcosine side chains. Mol. Cryst. Liq. Cryst. **1991**, *203* (1), 137–148.

37. Gallot, B.; Marchin, B. Liquid-crystalline comb copolymers with lipo-amino-acide side chains: Synthesis and structure study. Liq. Cryst. **1989**, *5* (6), 1719–1727.

38. Gallot, B.; Diao, T. Synthesis and mesomorphic behaviour of comb-like polymers based on lipopeptides with two hydrophobic chains. Liq. Cryst. **1993**, *14* (4), 947–958.

39. Chandra, G.; Disapio, A.; Frye, C.L.; Zellner, D. Silicones for cosmetics and toiletries: An environmental update. Cosmet. Toilet. **1994**, *109*, 63.

40. Schaefer, D. Silicone surfactants: Organo-modified polydimethyl siloxanes as surface active ingredients in cosmetic formulations. Tenside Surf. Det. **1990**, *27*, 154.

41. Roidl, J. Silikone in der Kosmetik als Rezepturbestandteile Aspekte Zur Sicherheit. SOFW J. **1983**, *109*, 91.

42. Roidl, J. Anwendung Der Siliconpolymere und Silicone mit Funktionellen Gruppen in der Kosmetik. Parfumerie und Kosmetik **1986**, *67*, 232.

43. Burczyk, F. Anwendungstechnische Moglichkeiten von Silikonen in der Kosmetik. Parfumerie und Kosmetik **1983**, *109*, 435.

44. Powell, V.; Thimneur, R. New developments in alkyl silicones. Cosmet. Toilet. **1993**, *108*, 87.

45. Roidl, J. Bedeutung der Silicone in der Modernen Kosmetik. Cosmet. Toilet. **1988**, *114*, 51.

46. Disapio, A.; Fridd, P. Silicones: Use of substantive properties on skin and hair. Int. J. Cosmet. Sci. **1988**, *10* (2), 75–89, doi:10.1111/j.1467-2494.1988.tb00004.x.

47. Imperante, J.; O'Lenick, A.; Hannon, J. Silicone phosphobetaines. Cosmet. Toilet. **1994**, *109*, 81.

48. O'Lenick, A.J.; Parkinson, J.K. Silicone quaternary compounds. Cosmet. Toilet. **1994**, *109*, 85.

Hemocompatible Polymers

Young Ha Kim
Ki Dong Park
Dong Keun Han
Polymer Chemistry Laboratory, Korea Institute of Science and Technology, Seoul, South Korea

Abstract
Polymer materials are extending their applications to biomedical uses. In addition to packaging for medical devices and instruments, polymers are widely used as medical devices and instruments, artificial organs and implants, and in drug formulations and delivery systems. Such materials are called biomaterials and are catergorized as metals, ceramics, polymers, or composites.

Polymer materials are extending their applications to biomedical uses. In addition to packaging for medical devices and instruments, polymers are widely used as medical devices and instruments, artificial organs and implants, and in drug formulations and delivery systems. Such materials are called biomaterials and are catergorized as metals, ceramics, polymers, or composites.

In general, biomaterials should fulfill several conditions:[1]

- They should have proper mechanical properties and should not deteriorate in the body. Ceramics are the most stable; metals can be corroded. Many polymers can be degraded by hydrolysis or by oxidation under physiological conditions.
- They should be able to be sterilized. Biomedical products are sterilized by autoclave, ethylene oxide, or γ-ray. Many polymers can be degraded by γ-ray irradiation.
- They should be biocompatible. When materials come into contact with a living organism, there are interactions between the biomaterial and a biologically or chemically active, fluid-based medium. The biomaterials and their degradation products or additives can damage cells, cause cancers, or lead to blood clotting. The biocompatibility of materials can be evaluated in terms of tissue compatibility or blood compatibility. The desired biocompatibility depends on the application site and period. Many researchers have analyzed interfacial phenomena to find a sound basis of biocompatibility between living bodies and artificial materials. However, even now, there are no definite principles of material design. A living organism is made up of numberless constituents, resulting in complicated interactions. Moreover, a standard method of biocompatibility assessment has not been established.

Practical polymer applications include artificial kidneys, membrane oxygenators, and blood vessels. When polymers are used in medical devices contacting blood, coagulation is a serious problem. The blood compatibility of polymers differs depending on their chemical composition and morphology. Despite an enormous amount of research, blood-compatible polymers are still a long way off. That is because the mechanism of thrombus formation is too complicated to find correlations between blood compatibility and material structure.

BLOOD–MATERIAL INTERACTION

The Mechanism of Blood Clotting[2,3]

Blood is composed of liquid medium, plasma, and cellular elements, which are subdivided into red blood cells (erythrocytes), white blood cells (leukocytes), and platelets. The first event that occurs when synthetic materials contact blood, following instantaneous redistribution of interfacially bound water and ions, is the rapid adsorption of plasma protein.

These processes influence subsequent interactions of blood cells, especially platelets and leukocytes, with proteinated surfaces. Adsorption of plasma can lead to the activation of an intrinsic blood coagulation cascade. This activation may be slight, or it may lead to the polymerization and crosslinking of fibrin at the blood interface. Depending on the nature of the surface, adsorbed proteins may initiate different coagulation pathways. The overall scheme of blood coagulation is summarized in Fig. 1.

Protein adsorption

Plasma proteins are immediately adsorbed onto the surface of polymeric materials upon exposure to blood. In the first minutes of contact, there are two modes of plasma protein

Concise Encyclopedia of Biomedical Polymers and Polymeric Biomaterials DOI: 10.1081/E-EBPPC-120051883

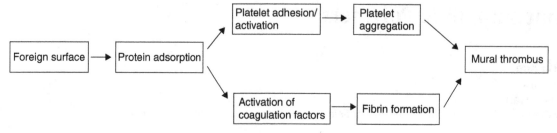

Fig. 1 A schematic diagram for mural thrombus formation on the polymer surfaces.

adsorption at the blood–material interaction. The first occurs as a result of the preferential, competitive adsorption of albumin, fibrinogen, or γ-globulin from the blood plasma; the second mode occurs as platelet and leukocyte adhesion and ultimately results in the polymerization of fibrin onto the surface.

The crucial relationship between adsorption of plasma proteins and the thrombogenicity of biomaterials has been extensively studied. Fibrinogen is thought to enhance platelet adhesion and is also suspected of interacting with leukocytes to contribute for platelet adhesion and aggregation. However, albumin has been repeatedly observed to minimize platelet adhesion and cellular adhesion. If albumin is preferentially adsorbed, there is little or no thrombus formation. γ-Globulins have not been extensively studied, but in some cases are reported to enhance platelet adhesion and activation.

Platelets

Platelets are small disc-type elements containing a variety of granules. Platelets adhere to materials through pseudopodia, which is poorly understood. Following adhesion, platelets undergo a variety of internal changes and release their granules, which are capable of attracting more platelets and initiating thrombus. The important released components are adenonsine diphosphate (ADP) and thromboxane A2, known to be the most potent substance for platelet aggregation. This platelet-release reaction results in platelets sticking to each other, a process known as platelet aggregation. Released platelet granules also stimulate the activation of blood coagulation factors.

Activation of blood coagulation factors

Activation of blood coagulation factors is a complicated sequential reaction, involving several enzymes, lipids, and ions. About 14 coagulation factors have been isolated and their functions characterized. The ultimate procedure is the formation of fibrin. The soluble fibrinogen monomer is polymerized into an insoluble crosslinked fibrin network by the action of thrombin, the ultimate coagulation factor. In reality, fibrin is formed in response to three independent mechanisms, platelet adhesion and activation, extrinsic

pathways, or intrinsic pathways of coagulation factor activation. The extrinsic pathway is activated rapidly by blood exposed to tissue factors (collagen, membrane constituents, lipids, and proteins). In the intrinsic pathway, blood is exposed to artificial surfaces, where various coagulation factors are adsorbed and converted in a chain reaction. The intrinsic and extrinsic pathways differ in initiation mechanism, but stimulate thrombin the same way. The activation of coagulation factors involves platelet phospholipids and Ca^{2+} ions.

Thrombolysis

The body has a defense system against uncontrolled coagulation that is critical for the preservation of the cardiovascular system. The fibrinolytic enzyme plasmin and its proenzyme, plasminogen, help to dissolve fibrin.

Anticoagulants

The principal action of anticoagulants is to prevent the formation of fibrin by inhibiting certain steps. Antithrombin III, a lipoprotein in the body, is the most potent inhibitor of coagulation. It has been suggested for inactivating thrombin and other coagulation factors. Heparin, a carbohydrate containing sulfate and sulfonate groups, functions as a potent anticoagulant by binding to and activating antithrombin to inhibit thrombin. Heparin is now widely used in artificial kidneys and blood oxygenators. Prostaglandins, long-chain hydroxyunsaturated fatty acids, prevent platelets from aggregating and initiating a clot. Several drugs also act as anticoagulants. Aspirin and dipyridamole inhibit platelet aggregation and release reaction. Dicumarol and warfarin interfere with the synthesis of some clotting factors.

Complements and fibronectins

In addition to plasma proteins, platelets, and coagulations factors, two other proteins affect the chronic blood compatibltiy of biomaterials. Complements are the primary humoral mediator of antigen–antibody reactions, and therefore are related to immunity. If complements are activated, leukocytes will adhere to artificial surfaces.

Gene–Hydrogels

These leukocytes and activated complements may also attract platelets and help in the formation of mural thrombus.

Fibronectins are glycoproteins found on the surface of various cells and present in platelet granules. Fibronectins act as adhesive proteins to bind cells to other cells and artificial substrates. They have affinity for fibrinogen, fibrin, and platelets. Fibronectins also appear to be important in mural thrombus.

Rationale: Antithrombogenicity vs. Pseudoneointima

Application of biomaterials for cardiovascular devices varies by site and period. It takes only a few hours for such treatments as artificial kidneys or blood oxygenators. In these cases, anticoagulants such as heparin are used. In devices that contact blood for several days, such as intravenous catheters, the material needs better thromboresistance, as a long administration of heparin can cause an unwanted side effect. When a device is implanted for years (e.g., an artificial heart valve), material design for antithrombogenicity is the most important. The formation of mural thrombus is affected by other factors such as blood flow; constant fast flow decreases the formation of thrombi, whereas irregular or stagnant flow enhances it.

There have been many approaches to preventing or decreasing activation of the thrombogenic pathways by tailoring the physicochemical properties of the polymer either to minimize thrombus formation or to selectively adsorb a passivating albumin layer. These approaches include smooth surfaces, hydrophilic surfaces, microdomained structures, inert surfaces, and incorporation of negative ions. In addition, pharmacologically active agents may be incorporated into the polymer to yield a slow-release system or immobilized onto a surface. From these researches we have made great progress in producing chronically nonthrombogenic cardiovascular devices. However, it appears to be impossible to produce long-lasting, completely nonthrombogenic material. Therefore, the other approaches consist of using rough surfaces, such as velours, flocked, or integrally textured surfaces, to encourage the formation of a living biological lining derived from blood itself. This pseudoneointimal lining, composed of fibrin, fibroblasts, collagen, and pseudoendothelial cells, has been used for years in artificial blood vessels made of polyester or extended poly(tetrafluorethylene). But the lining may be plugged if used in a diameter <4 mm.

Calcification and Biostability[4]

Calcification and biostability are other serious problems when biomaterials are applied in cardiovascular devices over a long period. Many polymers, especially those made by condensation polymerization, may be hydrolyzed under physiological conditions. The hydrolysis of polyesters,

polyamides, polycarbonates, and polyurethanes has been reported to be accelerated by various enzymes and cells (macrophages, leukocytes, etc.) in the body. In many cases even the surface of stable polyolefins is readily oxidized in the body.

Calcification is a deposition of calcium compounds in or onto the implanted material. It is actually a natural process. Extracellular media have 10^{-3} mol concentration of Ca^{2+}, but living cells keep the intracellular calcium concentration of 10^{-7} mol by a metabolic pump that extrudes Ca^{2+}. As artificial materials and dead tissues do not have such a vital function, Ca^{2+} ions and phosphates diffuse into the implanted materials and are precipitated as calcium phosphates. Calcification occurs with a wide spectrum of cardiovascular and noncardiovascular medical devices. It is the leading cause of failure of artificial heart valves. The actual mechanism of calcification is not clearly understood. It is related to host metabolism and to implant chemistry and structure. Adsorbed blood constituents or enzymatic degradation might initiate micro defects on the surface. Complexation of calcium ions to carboxylic groups of tissue proteins has been suggested as a crucial mechanism. Mineralization is enhanced at the sites of the most intense mechanical deformations in a flexing implant. Soft implants made of polymers and modified tissues lose their flexibility and finally fall when minerals are deposited. Although thrombus formation occurs almost instantly on contact with blood, calcification initiates very slowly. The degree of calcification depends on sites and host conditions. The *in vitro* and *in vivo* data do not coincide.

BLOOD COMPATIBILITY OF POLYMERS

Medical Polymers

Biocompatibility in medical polymers appears to be the most crucial factor in long-term safety and performance in the biological environment. Biocompatibility describes the ability of a biomaterial to exist within a living body without adversely and significantly affecting the body, and without the material itself suffering any adverse effects.[5] These two broad aspects of biocompatibility are very much interrelated. It's difficult to define biocompatibility, because it often involves the narrow confines of a specific material prepared for a specific purpose. Hence, a material used as a blood interface is called blood compatible, but the same material, if used as a tissue interface, is called tissue compatible.

Blood compatibility is one of the most important properties of biomedical polymers. In general, blood-compatible polymers should possess two characteristics: they should not induce thrombus formation, immune response, inflammatory reaction, or infection; and they must be nontoxic, noncarcinogenic, and nonmutagenic.[6] Despite the several *in vitro* and *ex vivo/in vivo* methods reported, there is no standard method for evaluating blood compatibility. Furthermore, the differences in the measurement conditions, the forms of the samples, and the individual variance of laboratory animals lead to less reliable or reproducible results.

A National Heart, Lung, and Blood Institute (NHLBI) report recommends techniques for the testing and evaluation of blood compatibility of materials and devices *in vitro* and *in vivo,* considering species effects, fluid dynamic conditions, and so on.[7] Although the report is limited to blood–material interaction, the evaluation standards for actual devices are not discussed. Generally, *in vitro, ex vivo*, and *in vivo* tests are performed successively to evaluate blood compatibility of polymeric materials.[8] *In vitro* experiments are relatively easy to conduct and can be used to screen a number of materials. *In vitro* tests include protein adsorption tests and platelet adhesion and activation test. They also include blood clotting time measurements such as recalcification time, activated partial thromboplastin time, prothrombin time, thrombin time, and whole blood clotting time, including the Lee–White test. The *ex vivo* test method is intermediate between the *in vitro* and *in vivo* tests and includes introducing whole animal blood directly into the test tubes, chambers, or films, together with determination of clot formation and occlusion time. Typical examples of *ex vivo* tests are a canine femoral arteriovenous (A-V) shunt,[9] a canine carotid arteriovenous (A-V) shunt,[10] and a rabbit carotid arterio-arterial (A-A) shunt.[11] *In vivo* tests are useful in that the materials can be tested under conditions similar to clinical use. Widely used

in vivo tests include the vena cava ring test[12] and renal embolus test.[13] Although many approaches have been used to investigate the interactions of blood with biomaterials, because blood coagulation is a broad term that is still incompletely understood, it is likely that biological response tests require proper bulk and surface materials characterization.

Many polymers from natural rubber and cellulose to synthetic elastomers, polyurethanes (PUs), and hydrogel have been used in biomedical applications ranging from disposable syringes to materials for artificial organs.[14] Polymers currently being used are categorized according to their characteristics and end-use applications as follows (Table 1): synthetic nondegradable polymers that were used in most long-term implantable devices and disposables, bioabsorbable or soluble polymers that were used as temporary scaffolding and barrier and drug delivery matrices, and experimental polymers.[15]

Among many polymers, poly(methyl methacrylate) and silicone rubber have been used first as synthetic polymeric biomedical materials. In particular, PUs are widely utilized in biomedical applications because of their excellent physical and mechanical properties and good physiological acceptability.[16] They were found to be relatively blood-compatible materials, which made them suitable for blood filters, pacemaker lead insulation, catheters, heart valves, cardiac-assist devices, and blood oxygenators. Polyester (Dacron) and expanded poly(tetrafluoroethylene) (Gore-Tex), being applied as vascular grafts >6 mm in diameter are relatively thromboresistant as are most hydrogels.

Table 1 Types of polymeric biomaterials

Synthetic nondegradable	Bioabsorbable or soluble	Experimental
Polyformaldehyde	Poly(lactic/glycolic acids)	Poly(alkylene oxalates)
Polyesters	Poly(amino acids)	Poly(alkylene malonates)
Polyamides	Poly(hydroxy butyrate)	Polyetherlactones
Polyolefins	Polycaprolactones	Poly(amino acids)
Poly(tetrafluoroethylene)	Poly(oxyethylene glycolate)	Polyurethanes
Perfluorocarbons	Poly(alkylene oxalates)	Polyphosphazenes
Poly(vinyl chloride)	Poly(ethylene oxide/PET)	Graft copolymers
Polyacrylonitrile	Albumin (crosslinked)	Surface-modified polymers (e.g., heparinized)
Hydrogels	Collagen/gelatin (crosslinked)	Interpenetrating networks
Polyurethanes	Polyanhydrides	Composites
Silicones	Poly(ortho esters)	
Ethylene vinyl-acetate polymers		
Thermoplastic elastomers		
Acrylics		
Polysulfones		
Perfluoroether copolymers		
Poly(2-hydroxy esters)		

Cardiovascular Application of Polymers

Blood compatibilty is of primary importance for those polymers that interact with blood. The polymers used in cardiovascular systems such as blood-contacting devices include a wide variety of commercial materials. There are three kinds of cardiovascular applications:[17] replacements that are permanently implanted in the circulatory system (e.g., artificial hearts, heart valves, and vascular grafts); devices that are inserted into a blood vessel for varying periods of time (e.g., catheters, sensors, fibroscopes, and other imaging agents); and extracorporeal devices that remove and return blood from the body [e.g., blood oxygenators, hemodialysis units (so-called artifical kidneys), cardiopulmonary bypass, and liver perfusion systems].

Table 2 lists biomedical polymers and their main applications and uses in relation to the cardiovascular system or extravascular blood.[18] The applications range from high-volume disposable products such as syringes and blood bags to devices with only blood contact such as intravenous (IV) catheters; high-technology, low-volume implanted organs such as vascular grafts and heart valves; and organs that replace malfunctioned ones such as the artificial heart.[19,20]

Artificial heart devices

Artificial heart devices include percutaneous transluminal coronary angioplasty (PTCA), the intraaortic balloon pump (IABP), ventricular assist devices (VAD), and the total artificial heart (TAH). PTCA, a newer, less invasive method than coronary-artery bypass surgery, is a safe and effective technique of myocardial revascularization. It uses a special balloon dilation catheter made of PVC or PE. IABP is a simpler mechanical circulatory device used to provide temporary cardiac assistance. The balloon material is made of PU such as Cardiothane, Estane, and Biomer. VAD and TAH are used with patients who might die from cardiac failure. Blood pumps of sac or diaphragm type are made from PVC, polyolefin rubber (Hexsyn), silicone rubber (Silastic), and PU such as Avcothane, Biomer, and Pellethane.[19,21]

Heart valves

Commerically available heart valve prostheses include mechanical valves and bioprosthetic heart valves (so-called tissue valves). Polymer valves are currently being studied for temporary VAD. The polymers used in mechanical valves are a Dacron or poly(tetrafluorethylene) (PTFE) sewing cuff and a silicone rubber ball; a poly(ethylene terephtahalate)-covered stent is used for tissue valve. Polymer valves such as seamless trileaflet valves have been made of PUs (e.g., Avcothane and Pellethane),[22] whose blood compatibility and durability are essential for long-term implantation.

Vascular grafts

Vascular grafts made of polymeric materials are among the most successful artificial organs and frequently used in permanent implantations. Only vascular grafts having diameters >6 mm are now clinically used. Their porous structure induces the formation of new pseudoneointima, resulting in the prevention of blood leakage. The polymers most often used are polyester (Dacron) fiber in woven, knitted, and velour types, or porous expanded PTFE (Telfon, Gore-Tex,

Table 2 Cardiovascular polymeric materials and their applications and uses

Material	Application	Medical or surgical use
Knitted dacron or woven dacron	Tubular conduits	Arterial prosthesis
Expanded PTFE	Tubular conduits	Arterial prosthesis (peripheral)
		Arteriorvenous shunt (hemodialysis)
Silicon elastomer (PDMS)	Tubing	Active parts of ECC[a]
		Catheters
Polyurethanes	Tubing	Catheters
	Coatings	Pacemaker leads
	Electrostatically spun tubular conduits	Small-caliber arterial prostheses
	Tubular conduits with a microporous wall	Small-caliber arterial prostheses
	Molding	Total artificial heart or LVAD[b]
Plasticized PVC	Tubing	ECC, catheters
	Molding	Blood bags
Latex	Inflatable balloons fitting intraarterial catheters	Angioplasty
Acrylics	Molding	Housing materials of ECC
Polycarbonates	Molding	Housing materials of ECC

[a]Extracorporeal circulation.
[b]Left ventricular assist devices.
Source: From Baquey.[18]

Impra) fabrics.[23,24] To date, however, vascular grafts <4 mm in diameter are not manufactured because of possible thrombus formation. Nevertheless, significant efforts have been made to develop vascular grafts with various biomaterials. Three promising approaches to vascular substitutes are being studied:[25] application of ultrafine polyester fiber, application of antithrombogenic materials or coatings, and development or hybrid vascular grafts. In particular, this last method, which induces endothelial cell lining, is considered most promising for two reasons: one is the hybridization of polymers with biological components such as peripheral veins or internal elastic lamina.[26,27] The other is the grafting of a cell-adhesive oligopeptide [i.e., Arg-Gly-Asp-Ser(RGDS) or Arg-Glu-Asp-Val (REDV)] on polymer surfaces, in which it is well known to have highly specific adhesivity toward endothelial cells.[28]

Pacemakers

Cardiac pacemakers supply electrical stimulation to the heart in patients whose own myocardial pacemaker activity or electrical conducting pathways are dysfunctional or unreliable. PUs such as Pellethane, taking the place of silicone, which was first employed, are used in the wire lead insulation of pacemakers because of their blood compatibility and durability.[29]

Blood oxygenators

Blood oxygenators, often referred to as extracorporeal heart–lung machines, are used to replace the heart temporarily in open-heart operations. There are two types of blood oxygenators, the bubble type and the membrane type. The former, in which blood is directly contacted with oxygen and carbon dioxide, is composed of soft plasticized PVC film or rigid polycarbonate (PC); in the latter blood is oxygenated via a polymeric membrane such as PP, silicone rubber, PTFE, or polysulfone.[19] Hollowfiber membrane oxygenators are safer and more effective than bubble oxygenators because they reduce damage to blood cells, protein denaturation, and air embolism. In addition, extracorporeal membrane oxygenation (ECMO) is used as an artificial lung to assist patients, especially newborn infants, with respiratory insufficiency.[30]

Catheters

There are many different kinds of intravascular catheters used in intensive care units and diagnostic equipment.

Intravenous (IV)-administration indwelling catheters are widely used and are constructed of PE, PP, silicone rubber, and PU. Arteriographic catheters made of PU and X-ray opaque catheters made of PVC or Teflon-coated Dacron are useful for diagnosing myocardial or cerebral infarction and congenital heart disease, respectively. Balloon-tipped catheters are also used to monitor blood pressure. These devices should possess blood compatibility, but virtually none do.

DESIGN OF BLOOD-COMPATIBLE POLYMERS

As described earlier, polymers have become the material of choice for soft-tissue and blood-contacting implants. However, given the problems plaguing implanted devices, particularly in blood, improving biomedical polymers is extremely important. Although much work has been done on blood-compatible polymers, the results are not conclusive, perhaps because the relationship between surface properties and surface-induced thrombosis has not been completely elucidated.

Every physiochemical factor of polymers influences blood–polymer interactions (hydrophilicity/hydrophobicity balance, surface charge, surface composition, porosity, chemistry/functionalization, surface mobility, etc.) (Table 3).[31] Current approaches to blood-compatible polymers can be divided into three categories: new polymer synthesis, surface modification of existing polymers, and biological approaches.

Tailoring New Polymers

Many surface characteristics that influence biological response can be changed by altering polymer constituent chemistry. In addition, mechanical properties can be tailored for end-use products.

One approach to minimizing thrombus formation is to synthesize nonthrombogenic polymers. These nonthrombogenic polymers prevent activation of the thrombogenic pathway by tailoring the polymer surface to minimize blood interaction.[32]

For many years, microdomain-structured polymers have received much attention for their improved blood compatibility. Several research groups have synthesized and evaluated polymers with microheterogeneous surfaces. Okano et al.[10,33,34] and Shimada et al.[35] reported that block copolymers having hydrophilic–hydrophobic microdomain structures show antithrombogenicity both *in vitro* and *ex vivo/in vivo*. They have showed that hydroxyethylmethacrylate (HEMA)-styrene (ST) block copolymers suppress

Table 3 Physicochemical properties of blood-compatible polymers

Surface composition (polar/apolar, acid/base, H-bonding, ionic charge, and immobilized biomolecules)
Surface hydrophilic/hydrophobic balance (structure, distribution, and sizes)
Surface topography (roughness, porosity, imperfections, etc.)
Surface molecular motions (polymer chain ends, loops, and their flexibility)

Gene–Hydrogels

the platelet adhesion and morphological changes of adhered platelets. The HEMA–ST block copolymer with 0.61 HEMA mole fraction has a microdomain structure in which the morphology is highly ordered, with an alternate lamella having a 20–50 nm width. It has been suggested that an organized protein layer is formed on this block copolymer surface. The hydrophilic domains selectively adsorbed albumin, whereas the hydrophobic domains selectively adsorbed γ-globulin. This protein layer was supposed to regulate the distribution of different binding sites on a molecular level at the platelet and polymer interface, thereby influencing the adhesion and activation of platelets. Similar results were obtained in HEMA-dimethyl siloxane copolymer.

Yui et al.[36,37] investigated the blood compatibility of poly(propylene oxide)-segmented polyamides. They found that the crystallinity of polyamides is closely related to the adhesion behavior of platelets on polymer surfaces, suggesting that microstructure composed of crystalline and amorphous phases suppresses platelet interaction. Platelet adhesion was minimized on copolymer surfaces with a crystalline size of 6–6.5 nm.

Other polymer systems with microdomain structures include segmented polyurethanes (SPUs), which are widely used for medical applications because of their excellent physical and mechanical properties and relatively good blood compatibility. SPUs are segmented heterophase elastomers that exhibit a variety of physical and chemical properties depending on their synthetic conditions. SPUs with various soft and hard segments are summarized in Table 4. Despite extensive research on PUs, their inherent blood compatibility remains a problem, preventing more widespread applications in medicine. Biomedical PUs and their blood compatibility are extensively reviewed by Lelah and Cooper[38] and Ito and Imanishi.[39]

Table 4 Commercial segmented polyurethanes

Trade name	Manufacturer	Structure
Biomer	Ethicon	Linear segmented polyetherurethaneurea
Avocothane (cardiothane)	Avco	Block copolymer of polydimethylsiloxane and segmented polyetherurethaneurea
Pellethane	Dow chemical	Linear segmented polyetherurethane
Texin	Mobay	Linear segmented polyetherurethane
Tecoflex	Thermoelectron	Linear segmented polyetherurethaneurea
TM-3	Toyobo	Linear segmented polyetherurethaneurea
KP-13	Kaneka	Linear segmented polyetherurethaneurea with polydimethylsiloxane segment

Nakabayashi et al. proposed a new blood-compatible polymer-based mimicry of an extracellular surface of the lipid bilayer that forms the matrix of the plasma membrane of blood cells (phosphorylcholine group of phosphatidylcholine and sphingomyelin).[39–42] They designed a methacrylate monomer with a phospholipid polar group, 2-methacryloyloxyethyl phosphorylcholine (MPC). Platelet adhesion and activation were completely suppressed on the surface of MPC copolymers when the MPC composition was above 30 mol%. They suggested that phopholipids adsorbed and organized on the MPC polymer surface make a biomembrane-like surface.

In general, polymer synthesis enables the selection of novel structures designed to improve blood compatibility, but the development of a new polymer may prove too expensive and time consuming to permit clinical use. Therefore, the most common approach improving blood compatibility is polymer modification, mainly surface modification of existing polymers that maintains essential bulk properties of biomaterials. The surface modification of polymeric materials has been reviewed by Ikada.[43]

Surface Modification of Polymers

Another approach of considerable interest is to modify surfaces of existing polymers without changing bulk properties. The modifications are based on factors such as surface free energy, protein adsorption, platelet adhesion, and other blood coagulation factors. Modification techniques have been designed to achieve increased hydrophilicity, chemical modification, and attachment of pharmacologically active agents. Table 5 summarizes surface modification techniques, which may be divided into two categories: physicochemical and biological.

Plasma treatment (plasma gas discharge or rf glow discharge) is an interesting and promising technique.[44–46] Thin polymer films, obtained by plasma polymerization in

Table 5 Surface modification techniques

Physicochemical modification
Plasma treatment
    Plasma gas discharge
    Rf glow discharge
Chemical modification
    Hydrophilic group grafting
    Hydrogel grafting
    Alkyl group grafting (albumin hypothesis)
Physical coating
Biological modification
Preadsorption of protein
Cell seeding
Drug/enzyme immobilization or release

thickness from hundreds of angstroms to several micrometers, change the surface properties of a substrate without changing bulk properties. Plasma treatment techniques have several advantages including ultrathin film deposition, physical durability and chemical stability, and increased adhesion to substrate materials. For instance, polymeric fluorocarbon coatings deposited from a tetrafluoroethylene gas discharge significantly enhance resistance to both thrombotic occlusion and embolization in small-diameter Dacron grafts.[47] In addition, surface modification with γ-rays,[48] electron beams,[49] UV radiation,[50] and ozone treatment are used to improve blood compatibility.[51]

Chemical modification approaches largely consist of hydrophilic group grafting or hydrogel grafting onto the substrate. The potential advantage of hydrogels seems to be the low interfacial free energy exhibited between the hydrogel surface and the blood environment. This might reduce protein adsorption and cell adhesion onto the polymer surfaces.[52,53] Among the hydrophilic polymers are poly(hydroxyethyl methacrylate) (PHEMA), poly(vinyl alcohol) (PVA), poly(ethylene glycol) (PEG), poly(ethylene oxide) (PEO), ethylene oxide-propylene oxide block copolymers, poly(acrylic amide) (PAAm), poly(N-vinyl pyrrolidone) (PVVP), and ionic water-soluble polymers such as poly(acrylic acid) (PAA), as shown in Fig. 2. In general, although protein and cells do not adhere to hydrogel polymers, their mechanical strength is not high enough to allow them to be used as biomaterials. Therefore, these polymers are crosslinked or grafted to other polymeric substrates such as PU, polyethylene, or silicon rubber.

Ratner et al.[46] grafted PHEMA or PNVP hydrogels to silicone rubber, resulting in reduced thrombogenicity as judged by vena cava ring test. They found that hydrogel-grafted surfaces, in general, adsorbed less protein and fewer cells than did ungrafted hydrophobic substrate polymers, which supports the concept of low interfacial free energy. However, later results using a modification of the test with

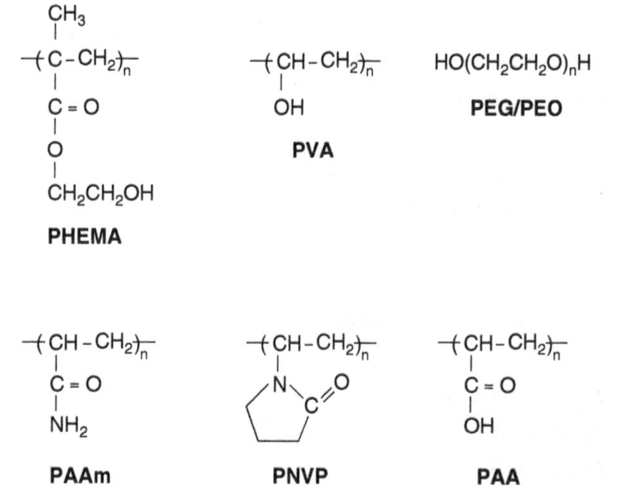

baboons showed thrombus adherence to grafted acrylamides and thromboemboli in the kidney.[54] In addition, polymers containing HEMA were shown to activate complement through hydroxyl groups.[55] The present emphasis is on the use of PEO.

PEO grafting on substrates was shown to improve blood compatibility. Nagaoka et al.[56,57] and many other investigators[58,59] reported that PEO-grafted copolymers were effective for low protein adsorption and low cell adhesion. In addition, PEO can reduce complement activation.[60]

Negatively charged polymer surfaces have proved effective in thrombus formation. This is because cellular materials such as platelets are also negatively charged and therefore electrostatically repelled from the negatively charged surfaces. Cooper et al. synthesized sulfonated polyurethane with propyl sulfonate group grafting on its backbone.[61] Canine ex vivo studies showed that incorporation of sulfonate groups reduced platelet adhesion and activation. Fibrinogen adsorption increased with increasing sulfonate content, despite the low level of platelet activation. Santerre, VanderKamp, and Brash[62] also observed high fibrinogen deposition on a series of PU based on a sulfonate group containing chain extenders. These results suggest that the sulfonate groups interact with fibrinogen, changing its conformation so that functional domains are not recognizable to platelets, resulting in minimized platelet adhesion.

Another property of sulfonated polymers was reported by Jozefowicz and Jozefonvicz[63] They demonstrated that polystyrene sulfonate resins show anticoagulant activity similar to heparin, and bind some amino acids to sulfonated polystyrene resins.

In our laboratory, novel surface modification of PUS has been developed to improve blood compatibility, biostability, and anticalcification. This approach was based on the following concepts:

- The hydrophilic environment of the blood–material interface appears to reduce protein adsorption, platelet adhesion, and activation, and can be achieved by the grafting of hydrophilic polymers, such as PEO. The advantages of PEO include low interfacial free energy,[52] lack of binding sites, highly dynamic motions, and extended chain conformation at the blood–material interface.[56,57,64]
- Sulfonated polymers have shown anticoagulant activity similar to heparin.[63,65]

Hypotheses conceived with this research state that hydrophilic PEO chains are expected to reduce protein adsorption and platelet adhesion because of the unique behaviors of PEO and because the pendent negatively charged sulfonate group provides anticoagulant activity, resulting in prevention of fibrin net formation. This "negative cilia surface" should curtail surface-induced thrombus formation because of the synergistic effect of hydrophilic PEO and the pendent negative charge of sulfonate groups. Sulfonated PEO-grafted PUs (PU-PEO-SO₃) have been prepared by a direct

Fig. 2   Examples of hydrophilic hydrogel polymers.

surface grafting of PU medical devices or by a solution reaction that can be applied as a coating material. The proposed hypotheses have been proven in *in vitro, ex vivo,* and *in vivo* studies.[66–69] In addition, researchers demonstrated reduction in surface crack formation and calcium deposition *in vivo.*[70] The results (improved blood compatibility, biostability, and anticalcification) attest to the usefulness of this negative cilia model for the design of blood-contacting medical devices.

In contrast to the low interfacial free energy concept described above, it has been reported that very hydrophobic surfaces such as PTFE and other fluorinated surfaces are bioinert. As they are water repellent, they should interact minimally with blood. Ito et al.[71] have synthesized L-glutamate copolypeptide containing fluoroalkyl side chains. They have reported that as the content of fluoroalkyl side chains increases, so does blood compatiblility. Han et al.[72] also showed that the blood compatibility of PU was improved by grafting of perfluorodecanoic acid.

Another approach is to design polymers of selectively increased surface affinity toward albumin. Albumin has been shown to adsorb to surfaces and passivate or neutralize surfaces to further protein adsorption (albumin hypothesis). Eberhart et al.[73] coupled alkyl chains of 16 and 18 carbon residues to polyamides, polyester, polyacrylates, and PUs to improve blood compatibility on the basis of the albumin hypothesis.[74,73] They demonstrated that albumin binding is significantly enhanced by ^{16}C and ^{18}C alkylation, because of strong hydrophobic interaction between nonpolar alkyl chains and hydrophobic segments contained within the structure of albumin.

The physical coating techniques use relatively nonthrombogenic polymers over existing blood-contacting devices or substrates without changing bulk properties.[75,76]

Biological Approach

Biological surface modification techniques involve preadsorption or protein,[77,78] cell seeding,[79,80] and release or immobilization of antithrombotic agents to create biologically inert surfaces. Preadsorptions of albumin have been studied because preferential adsorption of albumin was shown to prevent platelet adhesion *in vitro.*[81] Fibronectin coating has been used prior to *in vitro* endothelial cell seeding to improve endothelial cell adhesion to the substrate, which mimics the natural blood vessel.[79,80] Fibrinolytic enzymes [urokinase, streptokinase, and tissue plasminogen activator potent anticoagulants (heparin, hirudin), and their conjugates], various prostaglandins (PGE1-heparin) have been immobilized or released in order to take advantage of their biological activities of fibrin dissolution, antiplatelet aggregation, and antifibrin formation, respectively.[81–88] In addition, albumin–heparin conjugates have been prepared to improve blood compatibility of polymer surfaces either by preadsorption or by covalent coupling onto blood-contacting surfaces.[89]

In summary, a variety of surface modification techniques are being used to improve the blood compatibility of polymeric surfaces. However, many current blood-compatible polymers and devices need improvement, and there are many unfilled needs for new uses of polymeric biomaterials. Understanding the relationship between blood and material would promote better design of new polymers and use of existing polymers for surface modification, and finally the development of improved blood-compatible polymers.

REFERENCES

1. Lyman, D.J.; Rowland, S.M. *Polymers: Biomaterials and Medical Applications*; Kroschwitz, J.I., Ed.; Wiley-Interscience: New York, 1989.
2. Colman, R.W., Eds. *Hemostasis and Thrombosis: Basic Principles and Clinical Practice*; J.B. Lippincott: Philadelphia, 1994.
3. Szycher, M. *Biocompatible Polymers, Metals, and Composites*; Szycher, M., Ed.; Technomic: Lancaster, PA, 1983.
4. Schoen, F.J. et al. J. Biomed. Mater. Res. Appl. Biomater. **1988**, *22* (A-1), 11–36.
5. Williams, D.F., Ed. *Fundamental Aspects of Biocompatibility*; CRC: Boca Raton, FL, 1981.
6. von Recum, A.F. *Handbook of Biomaterials Evaluation: Scientific, Technical, and Clinical Testing of Implant Materials*; Macmillan: New York, 1986.
7. National Institutes of Health Guidelines for Blood–Materials Interactions; Publication No. 85-2185; NIH: Bethesda, MD, 1985.
8. Chandy, T.; Sharma, C.P. *Blood-Compatible Materials and Devices*; Sharma, C.P., Szycher, M., Eds.; Technomic: Lancaster, PA, 1991.
9. Lelah, M.D.; Lambrecht, L.K.; Cooper, S.L. A canine *ex vivo* series shunt for evaluating thrombus deposition on polymer surfaces. J. Biomed. Mater. Res. **1985**, *18* (5), 475–496.
10. Okano, T.; Nishiyama, S.; Shinohara, I.; Akaike, T.; Sakurai, Y.; Kataoka, K.; Tsuruta, T. Effect of hydrophilic and hydrophobic microdomains on mode of interaction between block polymer and blood platelets. J. Biomed. Mater. Res. **1981**, *15* (3), 393–402.
11. Nojiri, C.; Okano, T.; Grainger, D.; Park, K.D.; Nakahama, S.; Suzuki, K.; Kim, S.W. Evaluation of non thrombogenic polymers in a new rabbit A-A shunt model. Trans. Am. Soc. Artif. Intern. Organs **1987**, *33* (3), 596–601.
12. Gott, V.L.; Koepke, D.E.; Daggett, R.L.; Zarnstorff, W.; Young, W.P. The coating of intravascular plastic prostheses with colloidal graphite. Surgery **1961**, *50*, 382–389.
13. Kusserow, B.; Larrow, R.; Nichols, J. Observations concerning prosthesis-induced thromboembolic phenomena made with an *in vivo* embolus test system. Trans. Am. Soc. Artif. Intern. Organs **1970**, *16*, 58–62.
14. Szycher, M. *High-Performance Biomaterials*; Technomic: Lancaster, PA, 1991.
15. Barenberg, S.A. Abridged report of the committee to survey the needs and opportunities for the biomaterials industry. J. Biomed. Mater. Res. **1988**, *22* (12), 1267–1291.
16. Lelah, M.D.; Cooper, S.L. *Polyurethanes in Medicines*; CRC: Boca Raton, FL, 1986.

Gene–Hydrogels

17. Helmus, M.N.; Hubbell, J.A. Chapter 6 Materials selection. Cardiovasc. Pathol. **1993**, *2* (3), 53S–71S.

18. Baquey, C. *Materials Science and Technology, Vol. 14, Medical and Dental Materials*; Williams, D.F. Ed.; VCH: Weinheim, Germany, 1992.

19. Aebischer, P. et al. *Materials Science and Technology, Vol. 14, Medical and Dental Materials*; Williams, D.F., Ed.; VCH: Weinheim, Germany, 1992.

20. Halpren, B.D.; Tong, Y.C. *Polymers: Biomaterials and Medical Applications*; Kroshcwitz, J.I., Ed.; Wiley-Interscience: New York, 1989.

21. Hasircl, N. *High-Performance Biomaterials*; Szycher, M. Ed.; Technomic: Lancaster, PA, 1991; 71–90.

22. Russell, F.B.; Lederman, D.M.; Singh, P.I.; Cumming, R.D.; Morgan, R.A.; Levine, F.H.; Austen, W.G.; Buckley, M.J. Development of seamless tri-leaflet valves. Trans. Am. Soc. Artif. Intern. Organs **1980**, *26*, 66–71.

23. Lee, H.; Neville, K. *Handbook of Biomedical Plastics*; Pasadena Technology: Pasadena, CA, 1971.

24. Lyman, D.J. et al. *Biomedical and Dental Applications, Vol. 14, Polymer Science and Technology*; Plenum: New York, 1981.

25. Okada, H. *Japanese R & D Trend Analysis, Report No. 7, Biomedical Polymers*; KRI International: Tokyo, 1991.

26. Ishii, M. et al. Jpn. J. Artif. Organs **1992**, *21*, 94.

27. Sasajima, T. et al. Jpn. J. Artif. Organs **1991**, *20*, 414.

28. Hubbell, J.A.; Massia, S.P.; Desai, N.P.; Drumheller, P.D. Endothelial cell-selective materials for tissue engineering in the vascular graft via a new receptor. Biotechnology **1991**, *9* (6), 568–572.

29. Devanathan, T.; Sluetz, J.E.; Young, K.A. *In vivo* thrombogenicity of implantable cardiac pacing leads. Biomater. Med. Dev. Artif. Organs **1980**, *8* (4), 369–379.

30. Gille, J.P.; Bagniewski, A.M. Ten years of use of extracorporeal membrane oxygenation (ECMO) in the treatment of acute respiratory insufficiency (ARI). Trans. Am. Soc. Artif. Intern. Organs **1976**, *22*, 102–109.

31. Ishihara, K. *Biomedical Applications of Polymeric Materials*; CRC: Boca Raton, FL, 1993.

32. Cooper, S. L. et al. *The Physics and Chemistry of Protein–Surface Interaction*; Marcel Dekker: New York, 1981.

33. Okano, T. et al. Polymer **1978**, *10*, 233.

34. Okano, T.; Aoyagi, T.; Kataoka, K.; Abe, K.; Sakurai, Y.; Shimada, M.; Shinohara, I. Hydrophilic-hydrophobic microdomain surfaces having an ability to suppress platelet aggregation and their *in vitro* antithrombogenicity. J. Biomed. Mater. Res. **1986**, *20* (7), 919–927.

35. Shimada, M.; Miyahara, M.; Tahara, H.; Shinohara, I.; Okano, T.; Kataoka, K.; Sakurai, Y. Synthesis of 2-hydroxyethyl methacrylate-dimethylsiloxane block copolymers and their ability to suppress blood platelet aggregation. Polym. J. **1983**, *15* (9), 649.

36. Yui, N.; Tanaka, J.; Sanui, K.; Ogata, N.; Kataoka, K.; Okano, T.; Sakurai, Y. Characterization of the microstructure of poly(propylene oxide)–segmented polyamide and its suppression of platelet adhesion. Polym. J. **1984**, *16* (2), 119–128.

37. Yui, N.; Sanui, K.; Ogata, N.; Kataoka, K.; Okano, T.; Sakurai, Y. Effect of microstructure of poly(propylene-oxide)-segmented polyamides on platelet adhesion. J. Biomed. Mater. Res. **1986**, *20* (7), 929–743.

38. Lelah, M.D.; Cooper, S. *Polyurethanes in Medicine*; CRC: Boca Raton, FL, 1986.

39. Ito, Y.; Imanishi, Y. Blood compatibility of polyurethanes. CRC Crit. Rev. Biocompat. **1989**, *5*, 45.

40. Kodama, Y. et al. Jpn. J. Polym. Sci. Technol. **1978**, *35*, 423.

41. Ishihara, K.; Aragaki, R.; Ueda, T.; Watenabe, A.; Nakabayashi, N. Reduced thrombogenicity of polymers having phospholipid polar groups. J. Biomed. Mater. Res. **1990**, *24* (8), 1069–1077.

42. Ishihara, K. et al. Chemtech **1993**, *10*, 19.

43. Ikada, K. Surface modification of polymers for medical applications. Biomaterials **1994**, *15* (10), 725–736.

44. Chinn, J.A.; Horbett, T.A.; Ratner, B.D.; Schway, M.B.; Haque, Y.; Hauschka, S.D. Enhancement of serum fibronectin adsorption and the clonal plating efficiencies of Swiss mouse 3T3 fibroblast and MM14 mouse myoblast cells on polymer substrates modified by radiofrequency plasma deposition. J. Colloid Interface Sci. **1989**, *127* (1), 67–87.

45. Gombotz, W.R.; Hoffman, A.S. Gas discharge technique *for* biomaterial modification. CRC Crit. Rev. Biocompat. **1987**, *4* (1), 1–42.

46. Ratner, B.D.; Chilkoti, A.; Lopez, G.P. *Plasma Deposition and Treatment for Biomaterial Application*; Academic: New York, 1990.

47. Garfinkle, A.M.; Hoffman, A.S.; Ratner, B.D.; Reynolds, L.O.; Hanson, S.R. Effects of a tetrafluoroethylene glow discharge on patency of small diameter dacron vascular grafts. Trans. ASAIO **1984**, *30*, 432–439.

48. Suzuki, M. et al. *Physicochemical Aspects of Polymer Surfaces*; Plenum: New York, 1983.

49. Ellinghorst, G.; Fuehrer, J.; Vierkotten, D. Radiation initiated grafting on fluoro polymers for membrane preparation. Radiat. Phys. Chem. **1981**, *18* (5–6), 889–897.

50. Uyama, Y.; Ikada, Y. Graft polymerization of acrylamide onto UV-irradiated films. J. Appl. Polym. Sci. **1988**, *36* (5), 1087–1096.

51. Fujimoto, K.; Takebayashi, Y.; Inoue, H.; Ikada, Y. Ozone-induced graft polymerization onto polymer surface. J. Polym. Sci. Part A **1993**, *31* (4), 1035–1043.

52. Andrade, J. Med. Instrum. **1976**, *7*, 110.

53. Mori, Y.; Nagaoka, S.; Takiuchi, H.; Kikuchi, T.; Noguchi, N.; Tanzawa, H.; Noishiki, Y. A new antithrombogenic material with long polyethyleneoxide chains. Trans ASAIO **1982**, *28*, 459–463.

54. Hanson, S.R.; Harker, L.A.; Ratner, B.D.; Hoffman, A.S. *In vivo* evaluation of artificial surfaces with a nonhuman primate model of arterial thrombosis. J. Lab. Clin. Med. **1980**, *95* (2), 289–304.

55. Payne, M.S.; Horbett, T.A. Complement activation by hydroxyethylmethacrylate-ethylmethacrylate copolymers. J. Biomed. Mater Res. **1987**, *21* (7), 843–859.

56. Nagaoka, S. et al. *Polymers as Biomaterials*; Plenum: New York, 1984.

57. Nagaoka, S.; Nakao, A. Clinical application of antithrombogenic hydrogel with long poly(ethylene oxide) chains. Biomaterials **1990**, *11* (2), 119–121.

58. Desai, N.P.; Hubbell, J. Biological responses to polyethylene oxide modified polyethylene terephthalate surfaces. J. Biomed. Mater. Res. **1991**, *25* (7), 829–243.

59. Tseng, Y.C.; Park, K. Synthesis of photoreactive poly(ethylene glycol) and its application to the prevention of surface-induced platelet activation. J. Biomed. Mater. Res. **1992**, *26* (3), 373–391.

60. Yu, J.; Sundaram, S.; Weng, D.; Courtney, J.M.; Moran, C.R.; Graham, N.B. Blood interactions with novel polyurethaneurea hydrogels. Biomaterials **1991**, *12* (2), 119–120.

61. Okkema, A.Z.; Yu, X.H.; Cooper, S.L. Physical and blood contacting characteristics of propyl sulphonate grafted biomer. Biomaterials **1991**, *12* (1), 3–12.

62. Santerre, J.P.; VanderKamp, N.H.; Brash, J.L. Effect of solfonation of segmented polyurethanes on the transient adsorption of fibrinogen from plasma: possible correlation with anticoagulant behavior. Trans. Soc. Biomater. **1989**, *12*, 113.

63. Jozefowicz, M.; Jozefonvicz, J. Antithrombogenic polymers. Pure Appl. Chem. **1984**, *56* (10), 1335–1344.

64. Merril, E.W.; Salzman, E.W. ASAIO J. **1983**, *6*, 80.

65. Silver, J.H.; Hart, A.P.; Williams, E.C.; Cooper, S.L.; Charef, S.; Labarre, D.; Jozefowicz, M. Anticoagulant effects of sulphonated polyurethanes. Biomaterials **1993**, *13* (6), 339–344.

66. Han, D.K.; Jeong, S.Y.; Kim, Y.H.; Min, B.G.; Cho, H.I. Negative cilia concept for thromboresistance: Synergistic effect of PEO and sulfonate groups grafted onto polyurethanes. J. Biomed. Mater. Res. **1991**, *25* (5), 561–575.

67. Han, D.K.; Ryu, G.H.; Park, K.D.; Jeong, S.Y.; Kim, Y.H.; Min, B.G. Adsorption behavior of fibrinogen to sulfonated polyethyleneoxide-grafted polyurethane surfaces. J. Biomater. Sci. Polym. Ed. **1993**, *4* (5), 401–413.

68. Han, D.K.; Lee, K.B.; Park, K.D.; Kim, C.S.; Jeong, S.Y.; Kim, Y.H.; Kim, H.M.; Min, B.G. *In vivo* canine studies of a Sinkhole valve and vascular graft coated with biocompatible PU-PEO-SO3. ASAIO J. **1993**, *39* (3), 537–541.

69. Han, D.K.; Lee, N.Y.; Park, K.D. Kim, Y.H.; Cho, H.I.; Min, B.G. Heparin-like anticoagulant activity of sulphonated poly(ethylene oxide) and sulphonated poly(ethylene oxide)-grafted polyurethane. Biomaterials **1995**, *16* (6), 467–471.

70. Han, D.K.; Park, K.D.; Jeong, S.Y.; Kim, Y.H.; Kim, U.Y.; Min, B.G. *In vivo* biostability and calcification-resistance of surface-modified PU-PEO-SO3. J. Biomed. Mater. *Res.* **1993**, *27* (8), 1063–1073.

71. Ito, Y.; Iwata, K.; Kang, I.-K.; Imanishi, Y. Synthesis, blood compatibility and gas permeability of copolypeptides containing fluoroalkyi side groups. Int. J. Biol. Macromol. **1988**, *10* (4), 201–208.

72. Han, D.K.; Jeong, S.Y.; Kim, Y.H.; Min, B.G. Surface characteristics and blood compatibility of polyurethanes grafted by perfluoroalkyl chains. J. Biomater. Sci. Polym. Ed. **1992**, *3* (3), 229–241.

73. Eberhart, R. et al. *Proteins at Interfaces*; ACS Symposium Series 343; American Chemical Society: Washington, DC, 1987.

74. Munro, M.S.; Quattrone, A.J.; Ellsworth, S.R.; Kulkarni, P.; Eberhart, R.C. Alkyl substituted polymers with enhanced albumin affinity. Trans. ASAIO **1981**, *27*, 499–503.

75. Okano, T.; Grainger, D.; Parls, K.D.; Nojiri, C.; Feijen, J.; Kim, S.W. *Artificial Heart II*; Springer Verlag: Tokyo, 1988.

76. Okano, T. et al. Proc. Artif. Org. **1987**, 863.

77. Kim, S.W.; Lee, R.G.; Oster, H.; Coleman, D.; Andrade, J.D.; Lentz, D.J.; Olsen, D. Platelet adhesion to polymer surfaces. Trans. ASAIO **1974**, *20B*, 449–455.

78. Plate, N.A.; Matrosovich, M.M. Akad. Nauk (USSR) **1976**, *220*, 496.

79. Burkel, W.E.; Graham, L.M.; Stanley, J.C. Endothelial linings in prosthetic vascular grafts. Ann. N. Y. Acad. Sci. **1986**, *516* (1), 131–144.

80. Stanley, J.C.; Burkel, W.E.; Graham, L.M.; Lindblad, B. Endothelial cell seeding of synthetic vascular prostheses. Acta. Chir. Scand. Suppl. **1985**, *529*, 17–27.

81. Kim, S.W.; Ebert, C.D.; McRea, J.C.; Briggs, C.; Byun, S.M.; Kim, H.P. The biological activity of antithrombotic agents immobilized on polymer surfaces. Ann. N. Y. Acad. Sci. **1983**, *416* (1), 513–524.

82. Jacoba, H.; Jeong, S.Y.; Kim, S.W; Holmberg, D.L.; McRea, J.C. Self-regulating insulin delivery systems. J. Control. Release **1985**, *2* (1), 143–152.

83. Kim, S.W.; Jacobs, H.; Lin, J.Y.; Nojori, C.; Okano, T. Nonthrombogenic bioactive surfaces. Ann. N. Y. Acad. Sci. **1987**, *516*, 116–130.

84. Park, K.D.; Okano, T.; Nojiri, C.; Kim, S.W. Heparin immobilization onto segmented polyurethane-urea surfaces–effect of hydrophilic spacers. J. Biomed. Mater. Res. **1988**, *22* (11), 977–992.

85. Grainger, D.; Knutson, K.; Kim, S.W.; Feijen, J. Poly(dimethylsiloxane)-poly(ethylene oxide)-heparin block copolymers II: Surface characterization and *in vitro* assessments. J. Biomed. Mater. Res. **1990**, *24* (4), 403–431.

86. Park, K.D.; Kim, S.W. *PEO Chemistry*; Plenum: New York, 1992.

87. Han, D.K.; Park, K.D.; Ahn, K.-D.; Jeong, S.Y.; Kim, Y.H. Preparation and surface characterization of PEO-grafted and heparin-immobilized polyurethanes. J. Biomed. Mater. Res. **1989**, *23* (13), 87–104.

88. Han, D.K.; Jeong, S.Y.; Kim, Y.H. Evaluation of blood compatibility of PEO grafted and heparin immobilized polyurethanes. J. Biomed. Mater. Res. **1989**, *23* (Suppl. A2), 211–228.

89. Hennink, W.E. Ph.D. Dissertation, University of Twente, The Netherlands, 1985.

Hydrogels

Junji Watanabe
Yoshihiro Kiritoshi
Kwang Woo Nam
Kazuhiko Ishihara
Department of Materials Engineering, School of Engineering, University of Tokyo, Tokyo, Japan

Abstract

Hydrogels can be classified as chemically cross-linked hydrogels for use as soft materials, and physically cross-linked hydrogels made of synthetic polymers, natural polymers, and their hybrids. This entry focuses not only on fundamental principles but also on the use of recent anomalous hydrogels as biomaterials.

INTRODUCTION

Hydrogels refer to certain materials that are able to swell under conditions of excess water and hold a large amount of water in the wet state. Hydrogels generally consist of three-dimensional polymer networks that are cross-linked chemically and/or physically. The chemical properties of hydrogels are determined by the polymer backbone, the functional side chain in the monomer unit, and the cross-linking agent. The physical properties—for example, mechanical strength and swelling ratio—are controlled by the cross-link density. In general, hydrogels do not show any biodegradable behavior. Biodegradable hydrogels may be easily designed using natural polymers that are susceptible to enzymatic degradation or synthetic polymers that possess hydrolyzable moieties.

Biomaterials and biomedical engineering practices involve drug delivery carriers, tissue engineering scaffolds, and biomedical devices. Hydrogels provide adequate materials for these purposes. The research area of biomaterials is a multidisciplinary field composed of materials science, chemical engineering, medical engineering, and pharmacology. In particular, a biomaterial can play an important role as a bio-matrix and a bio-interface. Hydrogels provide excellent properties for improving conventional biomaterials, using such elements as copolymerization and polymer blends. As drug carriers, hydrogels can achieve stimuli-responsive drug release (for example, using pH or temperature) synchronized with enzymatic or nonenzymatic degradation. Hydrogels may be loaded with a variety of drugs and solubilizers or excipients. One major advantage of using hydrogels as a three-dimensional tissue reconstruction scaffold is the ability of these highly hydrophilic cross-linked polymer networks to swell and absorb sufficient amounts of cell culture medium. The permeability of the medium is important to the exchange of water-soluble molecules such as nutrients and gases. Widely used hydrogel medical devices include not only disposable materials, such as syringes, sample bags, and sample tubes, but also artificial organs such as blood vessels. The most promising applications may involve lenses, muscle models, and artificial cartilage. Contact lenses are a major application of hydrogel products due to their high transparency.

This entry focuses not only on fundamental principles but also on the use of recent anomalous hydrogels as biomaterials. Hydrogels can be classified as chemically cross-linked hydrogels for use as soft materials, and physically cross-linked hydrogels made of synthetic polymers, natural polymers, and their hybrids. In particular, a series of bioinspired phospholipid polymers show unique properties for application as drug delivery carriers, tissue engineering scaffolds, and biomedical devices.

CHEMICAL GELS

The chemical gels, which are cross-linked by covalent bonds, are synthesized by polymerization of monomers with some difunctional monomers, called cross-linking agents, such as ethylene glycol dimethacrylate (Fig. 1A). The gel is affected by changes in pH, temperature, solvents, and other factors in the surroundings. A chemical gel shrinks or swells, showing a volume phase transition. That is, controlling the polymer network by a change in the surroundings will lead to a change in the gel volume. This phenomenon makes the gel especially useful as a functional material for medical devices.

Soft Contact Lenses

The foreign-body reaction often becomes a problem when artificial materials come in contact with the human body. A significant difference in the surface free energy

Concise Encyclopedia of Biomedical Polymers and Polymeric Biomaterials DOI: 10.1081/E-EBPPC-120007319

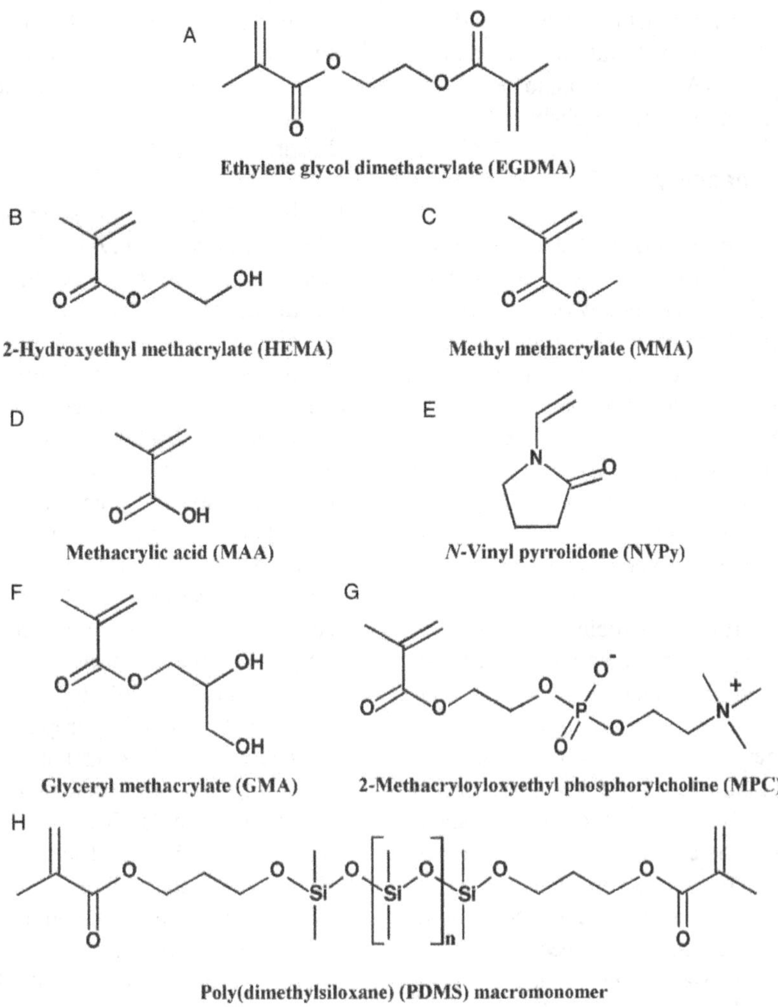

Fig. 1 Chemical structures of various monomers typically used for contact lens material.

between the material and the body is the cause of this undesirable reaction. A hydrogel can avoid this problem because the human body consists of network structures including biomacromolecules and water in a high percentage, which corresponds to the hydrogel structure itself. The surface free energies of the hydrogel and the human body are almost the same, so that the nonspecific interaction between the hydrogel and cell or protein is low, inhibiting protein adsorption following the foreign-body reaction. Therefore, the hydrogel is the most suitable biomaterial.

One of the remarkable properties of a hydrogel is softness, the property of an elastomer such as a rubber. Moreover, hydrogels are characterized by transparency and transmissibility of molecules through water flow. For the hydrogels in practical use, the soft contact lens (SCL) is the ideal manufactured article to take advantage of these properties. The eye naturally needs clarity for vision and the cornea needs oxygen by dermal respiration from the atmosphere. The properties of a hydrogel comply with these requirements.

Background on soft contact lens material

Poly(2-hydroxyethyl methacrylate) (HEMA, Fig. 1B), the first appearance of a soft biomaterial, was a great achievement by Otto Wichterle[1] and cannot be omitted from a history of the SCL. Amazingly, this invention is still the main material used for SCLs some 40 years later. Moreover, poly(HEMA) and its copolymers have widened the range of application to include such products as intraocular lenses. Before poly(HEMA), poly(methyl methacrylate) (MMA, Fig. 1C) was known not to cause critical harm to the human body, a fact discovered when a pilot injured in the eye by a broken aircraft windshield made of poly(MMA) experieced no inflammation. As a consequence of this episode, poly(MMA) was used as the first plastic material for the hard contact lens (HCL). Poly(MMA) lenses not only have good optical properties for clear vision, including greater clarity than glass, but also properties of high strength, light weight, and ease of molding. However, poly(MMA) lenses have limitations of discomfort and prohibited long-term use because of low oxygen permeability.

Wichterle thought that the introduction of hydrophilicity into the polymer could lead to amelioration of the defects of poly(MMA), so poly(HEMA) was the first material with hydrophilicity introduced in the polymer chain.

Increasing oxygen permeability

It is said that in addition to the anterior epithelium of the cornea, the endothelium of the cornea undergoes damage after long-term use of contact lenses having low oxygen permeability. As the physiology of the cornea has become understood, the oxygen permeability of contact lenses has been regarded as the important factor for homeostasis of the cornea. Without an adequate supply of oxygen, adverse events such as corneal edema may occur, followed by a number of possible physiological responses. Therefore, competitive efforts have resulted to develop materials with high performance in terms of regarding oxygen permeability.

The expression DK, where D represents the diffusion coefficient and K the partition coefficient of oxygen—inherent properties for oxygen transport through a material—has become familiar among contact lens manufacturers and users these days. There is a further relationship between oxygen flow and lens thickness; The transmissibility of a particular lens is thickness related and is given by the expression DK/t, where t refers to the thickness of the hydrogel.[2] In the case of SCLs, there are two approaches to increasing oxygen permeability based on DK/t, that is, to reduce the thickness of the SCL or to increase the water content in the hydrogel, because oxygen permeability increases with water content. In the former case, a number of hyperthin poly(HEMA) SCLs (i.e., a center thickness of less than 0.05 mm) have been produced; however, such lenses turned out to be fragile, marked by poor mechanical properties, difficult handling, and difficult manufacture. In the latter case, the hydrophilicity of the polymer was a concern. Actually, the hydrophilicity of poly(HEMA) is too low to provide a water content of about 40%, indicating that not enough water is present to for carry sufficient oxygen to the cornea.

Water content may be adjusted by using various hydrophilic monomers to overcome poor oxygen permeability. For example, methacrylic acid, N-vinyl pyrrolidone, and glyceryl methacrylate are used for this purpose (Fig. 1D–F). However, the DK of pure water is less than necessary to supply the minimum oxygen needed for the cornea. There is naturally a limit to increasing oxygen permeability via increased water content in the hydrogel. The water content in the gel also plays an important role in controlling the mechanical properties of the hydrogel. The higher the water content in the gel, the lower the tensile strength. High water content and high tensile strength are ideal conditions for SCL material; however, a hydrogel cannot have it both ways.

Thus, the same problem that is associated with hyperthin SCLs, fragility, applies. Further, such lenses tend to bind organic materials such as tear proteins, microbes from handling, or inorganic salts, especially those of calcium, all of which can affect oxygen permeability.

Silicone hydrogel

In the HCL field, a silicone-containing lens, poly(dimethylsiloxane) (PDMS, Fig. 1H), is famous for possessing a high DK and many other properties such as optical excellence, softness and flexibility similar to a hydrogel, and high tensile strength. The bulkiness of the siloxane group and the chain mobility are responsible for the high diffusibility of oxygen through siloxane-containing materials.[3] Designing a hydrogel based on PDMS will lead to high performance in terms of oxygen permeability and good elasticity for comfortable wearing; however, the hydrophobicity of the siloxane component makes this difficult to realize. When a PDMS macromonomer is polymerized with some hydrophilic monomers, phase separation occurs and the resultant polymer is opaque.

Several strategies have been considered for improving the hydrophilicity of PDMS. Some macromonomers including PDMS and hydrophilic segments such as polyethylene oxide (PEO) in the molecule were synthesized and polymerized with some hydrophilic monomers such as dimethylacrylamide. The incorporation of a nonpolar fluorosubstituent (CF_2H) had a dramatic effect on the increased solubility with hydrophilic monomers.[4] Those hydrogels based on PDMS macromonomers showed some improvement; however, the siloxane component, which tends to concentrate at the surface of the lens, and the low surface energy of PDMS result in very poor tear wetting, causing binding of tear lipids and lens adhesion to the cornea.

Adverse effects of fouling on the lens surface

With daily wearing of contact lenses, protein or lipid depositions create a biofilm on any kind of lens surface. Proteins have charge distributions, and they attract one another. This could result in the lens losing its ocular properties, and oxygen permeability will be disrupted. Moreover, it is possible for hydrophilic lenses to absorb chemicals into their naturally porous structure. This fouling of the SCL causes adverse events such as adherence of microbes to the lens. Bacteria can colonize in the lens and on its surface, so that the lens itself can become a potential pollutant source, a bacteria bed. This is the first stage in contact lens–associated microbial keratitis, which may ultimately lead to corneal ulceration and to blindness, in the worst case. Thus, lens cleaning and sterilizing are crucial components of daily care, especially for SCLs. However, an inadequately cleaned and sterilized lens could also trigger these adverse events. Also, any sterilizing process, such as heat, peroxide, or chemical disinfection, can harm the lens material, leading to deterioration. Disposable lenses are a way of avoiding these risks. However, there is the problem of the cost and several other risks due to improper usage and mishandling, such as using the lens for more than a certain period.

Requirements for ideal lens material

Surface fouling is inevitably associated with any kind of lens. This fouling has unfavorable effects on both the lens and its user. Unfortunately, there are no SCLs using conventional materials that do not have problems. A matter of primary importance in providing easily handled and problem-free SCLs is biocompatibility, that is, the absence of negative effects due to contact with the constituents of a living organism and the absence of toxic or injurious effects on biological systems.

Poly(2-methacryloyloxyethyl phosphorylcholine) [poly-(MPC), Fig. 1G], a bio-inspired material, was modeled on the surface of a natural biological membrane. The MPC polymers are highly effective in inhibiting the near-instantaneous protein adsorption and subsequent denaturation that can lead to thrombus formation, indicating true biocompatibility.[5,6] Recently, a poly(MPC) hydrogel was made that showed a high swelling ratio, meaning high hydrophilicity.[7] Using this MPC and HEMA, a biomimetic SCL was made.[8] This SCL reduced not only protein adsorption but also the risk of infection by bacteria, exhibiting significantly lower bacterial adhesion *in vitro* compared to the standard poly(HEMA) SCL. Moreover, there has been an attempt to make a hybrid hydrogel having the high oxygen permeability of a silicone hydrogel and the surface of a biomembrane characteristic of the poly(MPC). The potential of a poly(MPC) hydrogel has not been fully realized; however, the poly(MPC) hydrogel assists in widening the range of usage in association with all biomaterials that need true biocompatibility.

Volume Phase Transition

Since Tanaka reported the volume phase transition of a hydrogel in solvent,[9] many research studies on this phenomenon have been initiated, especially for medical devices such as drug delivery systems and tissue engineering. It is desirable to release the drug from the gel by shrinkage or swelling on stimulation. For example, if a hydrogel containing a drug changes volume according to a change in body temperature, patients can automatically receive the drug when they have a fever. The poly(N-isopropyl acrylamide) [poly(NIPAAm)] hydrogel is well qualified for this use (Fig. 2). Poly(NIPAAm) becomes insoluble in water due to dehydration of the polymer chain when the water temperature is over 32°C, which is well known as a lower critical solution temperature (LCST).[10] With this property, the poly(NIPAAm) gel could swell under 32°C but shrink above 32°C, showing a volume phase transition.

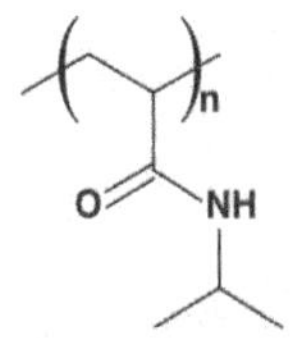

Poly(N-isopropyl acrylamide (NIPAAm))

Fig. 2 Chemical structure of typical polymer used for gel system employing volume phase transition.

Gel with functions

Phenylboronic acid and its derivatives are well known to form covalent complexes with polyol compounds including glucose. Taking advantage of this property, Kataoka et al. have made a hydrogel (NB gel) by polymerization of NIPAAm with 3-acrylamidophenylboronic acid and a small amount of cross-linker, which shows swelling-deswelling behavior in response to a change in glucose concentration (Fig. 3A).[11]

The NB gel showed a volume phase transition at 22°C and pH 9.0; however, the temperature rose with increased glucose concentration from 0 to 5 g/L. In the case of a 1 g/L glucose concentration (the same level as for a healthy human), the NB gel would show a remarkable change in swelling in response to glucose in the range of 25 to 30°C. In fact, the repeated on off regulation of insulin release from the NB gel in synchrony with a change in the concentration of external glucose was confirmed. Although there is still a problem with the difference in pH and temperature in the human body, this type of glucose-responsive gel would be good news for patients under diabetes treatment.

Miyata et al. made a hydrogel that can change its swelling ratio to increase the transmissibility of molecules in response to a specific antigen (Fig. 3B).[12] They first prepared the monomers, the modified antigen and modified antibody, using rabbit immunoglobulin G (rabbit IgG) and goat anti-rabbit IgG (GAR IgG) as the antigen and antibody, respectively. They then synthesized the polymer GAR IgG using vinyl(GAR IgG) and acrylamide (AAm). In the presence of this polymerized GAR IgG, an antigen-antibody semi-interpenetrated polymer network (semi-IPN) hydrogel was continuously prepared by the copolymerization of the vinyl(rabbit IgG), AAm, and some cross-linker. This semi-IPN type of hydrogel showed an abrupt increase in the swelling ratio on the addition of free rabbit IgG; however, it showed no reaction in the case of goat IgG, meaning that the hydrogel recognized a specific antigen. They suggest that the binding of the GAR IgG with rabbit IgG polymerized in the antigen-antibody hydrogel was much weaker than that to native rabbit IgG, due to denaturation of the modified rabbit IgG, causing the intra-chain grafted antigen-antibody binding to be dissociated by exchange of the grafted antigen for free antigen. Although the ratio of the volume change is small, by this

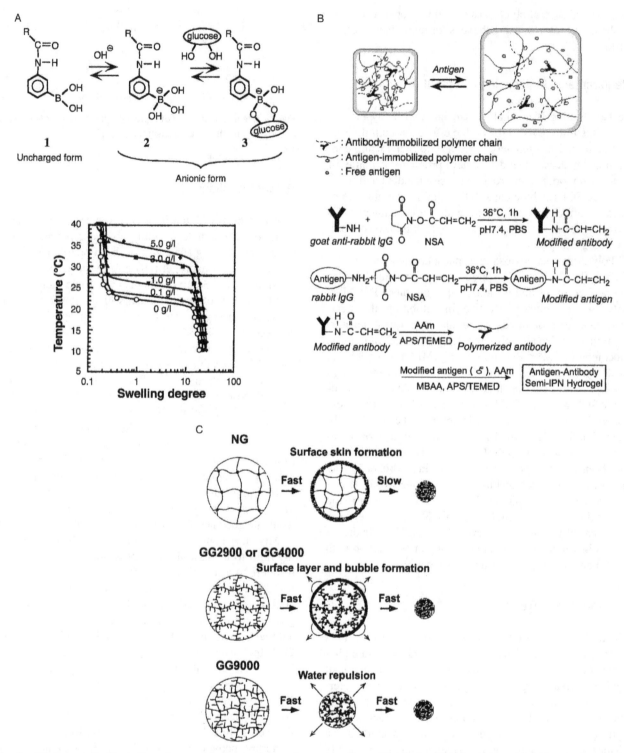

Fig. 3 (**A**) Schematic illustration of reversible change in the glucose bonding with (alkylamido)phenyl boronic acid according to the surroundings, and change in the swelling ratio of NB gel in response to the change in glucose concentration. (**B**) Diagram of a suggested mechanism for the swelling of an antigen–antibody semi-IPN hydrogel in response to a free antigen. Synthesis of the modified antigen and antibody monomers, containing rabbit IgG and GAR IgG as the antigen and antibody, respectively, followed the synthesis of semi-IPN hydrogel using those monomers. (**C**) Schematic illustration of deswelling mechanism of non-grafted gel (NG) and comb-type grafted poly(NIPAAm) gels (GG) having different lengths of grafted chains, with average molecular weights of 2900, 4000, and 9000, above their phase transition temperature.

Source: (A) From Kataoka et al.[11] © 1998, with permission from American Chemical Society. (B) From Miyata et al.[12] © 1999, with permission Macmillan Publishers Ltd. (C) From Kaneko et al.[15] © 1995, with permission from American Chemical Society.

unique antigen-specific type of hydrogel, it is possible to use the hydrogel to release a drug in response to the antigen as a result of the expression of cancer cells.

Quick response hydrogel

For this to be an effectively quick and sharp response to the stimulation, the volume phase transition of the hydrogel is necessary. However, the hydrogel volume change is dominated by the character of hydrogel; that is, the characteristic time of swelling is proportional to the square of the linear dimension of the hydrogel and is also proportional to the diffusion coefficient of the hydrogel network.[13] Several research studies have been undertaken to overcome this limit. Forming pores in the hydrogel is one method to allow water to flow well. There are many ways to form a porous hydrogel—for example, preparing a hydrogel with a water-soluble porosigen, such as sodium chloride, sucrose, PEO, and so on, which can be removed after gelation. However, the pore size of the hydrogel depends on the size of the porosigens.

Chen, Park, and Park successfully created a superporous hydrogel that quickly absorbed water and reached a high swelling equilibrium.[14] They introduced many interconnected pores into the hydrogel that behaved as open channels to allow water flow through a capillary action. They utilized gas bubbles, CO_2, generated by mingling $NaHCO_3$ with acid, to form the pores in the hydrogel. The key point is how to maintain the bubbles during gelation, because the bubbles tend to fade away. They found the answer by mingling the pre-gel solution and a foam stabilizer, [poly(ethylene oxide)-poly(propylene oxide)-poly(ethylene oxide)] tri-block copolymers (PEO–PPO–PEO) together to impart a high viscosity to the pre-gel solution, which helped to stabilize the foam during gelation. They obtained a superporous hydrogel showing up to 200 times faster absorption compared with an ordinary hydrogel.

Okano et al. have created a hydrogel system that quickly shrinks into tiny pieces with a change in temperature (Fig. 3C).[15] They introduced hydrophobic chains into the matrix of poly(NIPAAm) to enhance the aggulutinability, causing a hydrophobic interaction among the polymer chains. First, they made some poly(NIPAAm) macromonomers with different molecular weights by using telomerization of a NIPAAm monomer with 2-aminoethanethiol as a chain transfer agent and a continuous condensation reaction of amino semitelechelic poly(NIPAAm) with N-acryloxysuccinimide. They then synthesized several hydrogels by radical polymerization of the NIPAAm monomer and a NIPAAm macromonomer with a cross-linker, yielding a comb-type grafted polymer hydrogel in which the molecular architecture is different from that of the normal poly(NIPAAm) hydrogel, though the composition is the same. This comb-type grafted polymer gel could shrink much faster than the normal poly(NIPAAm) gel when the surrounding temperature climbs, and the rate of response depended on the length of the grafted polymer chains; that is, a hydrogel having a longer grafted polymer chain shrank faster.

Reentrant volume phase transition

A strange volume phase transition has been reported for the poly(MPC) hydrogel.[6] Generally, the hydrogel swells in a good solvent and shrinks in a poor solvent. Water and ethanol are good solvents for poly(MPC); however, the poly(MPC) hydrogel shrank in certain water/ethanol mixtures. This phenomenon, in which the hydrogel once shrank and then reswelled according to the composition of the solvent, is called reentrant volume phase transition. The water or ethanol condition around polymer chains was altered by the introduction of ethanol or water, which triggered the phase change of the whole system.

The poly(NIPAAm) gel also showed this phenomenon in some water-miscible alcohol-aqueous solutions,[16] but in the case of a poly(MPC) gel, the swelling and shrinking in response to the alcohol composition is dependent on the hydrophobicity of the alcohol. This means that the poly(MPC) gel has a molecular recognition property. By using this unique characteristic of the poly(MPC) gel and in combination with the functions of the hydrogel mentioned above, novel smart biomaterials with high biocompatibility can be realized.

PHYSICAL GELS

Physical gels, which are often called pseudo gels, are continuous, disordered, three-dimensional networks formed by associative forces capable of forming noncovalent cross-links.[17] The noncovalent cross-links are formed by weaker interactions, and the cross-links are reversible. These interactions include hydrogen bonds, ionic interaction, hydrophobic association, stereocomplex formation, coiled-coil interaction, antigen-antibody interaction, crystalline segment cross-linking, and solvent complexation. These interactions would bring the polymer chains together and induce formation of a stable structure by forming a junction zone that would occur as shown in Fig. 4. These junction zones maintain the ordered structure inside the gels. A change in temperature, pH, or the addition of salts, for example, would form or modify the junction zones. The junction zones that had been formed between chain segments tended to stabilize as the segment number increased. Because the cross-linked junctions are not covalently bonded, breakdown of the network may occur due to a change in the environment (reversible phenomena).

Synthetic Polymers

Synthetic polymers that can be utilized for a hydrogel base material have several requirements. They need to be stimuli sensitive, but the most important requirement is biocompatibility, for they are destined to be used for biomaterials.

Gene–Hydrogels

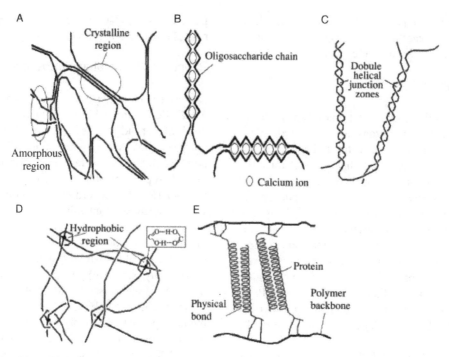

Fig. 4 Physical gels with multiple junctions. (**A**) Aggregation of the chains by crystallization. (**B**) Egg-box model created by specific ionic interaction. (**C**) Helical junction zone. (**D**) Hydrogen bond occurring in hydrophobic domain. (**E**) Hybrid hydrogel formation by coiled-coil proteins.

Poly(vinyl alcohol)

Linear poly(vinyl alcohol) (PVA) is a one of the most actively used synthetic polymers for physical hydrogels. PVA is a polymer that has a thermoreversible property. PVA hydrogel can be prepared by simply lowering the temperature. Due to its thermoreversible property, the hydrogel will form a crystallized zone when the temperature is lowered. Cyclic freeze-thaw or blending a hydrophilic polymer such as poly(sodium L-glutamate) and poly(allylamine) HCl salt together with PVA increases the crystallinity and mechanical properties due to a complex between the PVA and the hydrophilic polymer.[18]

Block copolymers

PEO, PPO, and poly(lactic acid) (PLA) are the materials most actively applied to biomedical and biotechnology because of their excellent physicochemical and biological properties. They are prepared in the form of block or tri-block copolymers. Micelles are formed at low concentrations in aqueous solution, and thermoreversible gels are formed at high concentrations. This block copolymer hydrogel also shows LCST behavior, which allows it to be used for drug delivery. Multiblock copolymers of PEO and poly(butylene terephthalate) (PBT) can be prepared by melt condensation of PEO, butanediol, and dimethyl terephthalate. A phase-separated structure is formed with hard PBT domains, resulting in thermally reversible cross-linking,

dispersed in the soft amorphous phase of hydrophilic PEO and amorphous PBT (Fig. 4A).[19] The PEO–PPO–PEO tri-block, also known as Pluronic, dissolves in cold water, forming a viscous solution. When the temperature is increased, the solution gels due to phase transition as the polymer concentration is raised above a critical value. The increment in temperature leads to an increase in the micelle volume fraction (ϕ_m). When $\phi_m > 0.53$, the system becomes a gel by micelle packing (hard-sphere crystallization). A PLA–PEO–PLA block copolymer hydrogel can be prepared simply by immersing the polymer film or tablet in water. Phase separation of PEO and PLA induces the gelation. It has been reported that when the PEO to PLA ratio is strictly controlled, immersion into an organic solvent followed by the addition of water may produce a softer and more hydrophilic hydrogel.[20]

Acrylamide polymers

One of the most commonly investigated acrylamide polymers for physical gelation is the linear homopolymer poly(NIPAAm). Poly(NIPAAm) is often formed as a gel by chemical cross-linking. However, its precipitation ability above 32°C due to LCST makes subject to the physical gelation process. Generally, an acrylamide polymer including poly(NIPAAm) becomes physically gelled when it is copolymerized with another polymer having a different property. Transparent aqueous polymer solutions of NIPAAm copolymers containing small amounts of acrylic acid are readily polymerized using free radical initiation in benzene and turn opaque as the temperature increases. When the temperature reaches 30–34°C, the solution turns into a gel. As more heat is added to the hydrogel, the gel collapses.[21] Block and star copolymers of PEO and poly(NIPAAm) form aqueous solutions at low temperatures and transform to relatively strong elastic gels upon heating, multiple-arm copolymers appear to form gels via a physical cross-linking mechanism, and diblock copolymers gel due to a micellar aggregation mechanism.

Acrylic polymers

Acrylic polymers possess intra/interchain hydrogen bonds. Poly(acrylic acid) (PAA) and poly(methacrylic acid) (PMAA) form a complex with PEO. Swelling under oscillatory pH conditions reveals the dynamic sensitivity of poly(methacrylic acid-graft-ethylene glycol) gels and shows that network collapse (complexation) occurs more rapidly than network expansion (decomplexation). Hydrophobic interactions are very important for stabilizing the complexes. Poly(organophosphagene) is an organometallic polymer containing double-bonded phosphorus and nitrogen on the main chain with two organic side groups attached to the phosphorus atom. Polyphosphagene hydrogel, like other synthetic hydrogels, can be divided into two major classes: neutral or non-ionic hydrogels and ionic hydrogels. Ionic hydrogels are formed by phosphagene polyelectrolytes, while non-ionic polyphosphagene hydrogels are based on water-soluble polymers ranging from ultimately hydrophilic, such as polyphosphazene containing glycosyl and glyceryl side groups, to less hydrophilic, including poly[bis(methylamino)phosphazene]. The difference in solubility is believed to be partially due to the exposure of skeletal hydrogen bonding to water.[22]

Biomolecules

Oligosaccharides

Oligosaccharides are naturally occurring materials that can be modified for use as suitable biomaterials. Simple modification or a change in the environment may cause gelation phenomena.

Alginate: Alginate is one of most common oligosaccharides used for hydrogel synthesis. The cross-linking is carried out by injecting calcium ions with alignment of each sequence to form an array of cavities called an egg box. In this formation, carboxylate and oxygen atoms are organized into a cavity at room temperature and physiological pH (Fig. 4B). Monovalent cations as well as divalent cations such as magnesium ions do not induce gelation, whereas Ba^{2+} and Sr^{2+} ions induce very strong gelation. High concentrations of α-L-gluconic acid chain sequences lead to brittle gels, while low concentrations lead to more elastic gels.

Dextran: The reaction for the gelation of dextran is due to crystallization caused by the association of chains through hydrogen bonding, but gelation also occurs when potassium ions are added. Although dextran does not have ionic groups on the chain, the cage that is formed by six oxygen atoms from these repeating glucose units yields a cavity that is a perfect fit for potassium ions.[23] However, instability in water makes it unsuitable for biomedical use.

Chitosan: Physical cross-linking has been used to prepare chitosan-based gels by the O- and N-acetylation of chitosan and the attachment of C_{10}-alkyl glycosides to chitosan. Acetylation yields either "rigidly solidified" or "fragile" gels with high- and low-molecular-weight chitosan, respectively, while the attachment of C_{10}-alkyl glycosides forms, gels in acidic media at elevated temperature ($>50°C$).[24]

Gellan and Pectin: Gellan can form either elastic or brittle gels, depending on cation concentration. Gellan gels are soft and easily deformable below the critical calcium, concentration and, brittle above. Using a salt of ions other than calcium, such as Mg^{2+} or K^+, may induce a relatively stronger gellan gel.

Pectins are divided into high-methoxyl (HM) pectins and low-methoxyl (LM) pectins according to the degree of esterification to pectin accomplished in the presence of methanol.[17] In LM pectins, gel formation is influenced by the number and sequence of consecutive carboxyl groups and by the concentration of calcium. LM pectin + Ca^{2+} gels conform to the "egg box" model, while HM pectin gelation is governed by hydrogen bonds and hydrophobic interactions.

Agarose and Carrageenan: Agarose is soluble in water at temperatures above 65°C and, depending on the degree of hydroxyethyl substitution on its side chains, it forms a gel in the range of 17–40°C. Agarose gels are stable and do not swell at constant temperature or reliquefy until heated to 65°C. A commercially available agarose hydrogel is SeaPrep, a polysaccharide derived from red algae that melts at 50°C and gels at 7–18°C.

Carrageenan is gelled by salts such as KCl, but it can form a gel under salt-free conditions also. There are three types of carrageenan: kappa, iota, and lambda. Among these three, only the kappa type shows gel-forming characteristics in water. Potassium and calcium ions induce the gelation, while sodium ions do not. In the iota type, gelation occurs in the presence of calcium ions. At low temperature and high salt concentration, a double helical arrangement of the parallel polysaccharide chains can be formed (Fig. 4C). The temperature for gelation is about 25°C, and remelting occurs at 35°C.

Peptides, proteins, and glycoconjugates

Peptide, protein, and glycoconjugate hydrogels are formed by specific interactions called lock-and-key interactions, but in a different way from polysaccharides. The specific interaction may occur only when the molecules maintain three-dimensional structures such as coiled–coil interaction, or hydrogen-bonded β-sheets or strands.

Fibrin Gels: Fibrin plays an important role in natural wound healing and is used as an adhesive sealant in surgery. It is formed by the enzymatic polymerization of fibrinogen. Cross-linking of the fibrin oligomer can be executed by factor XIIIa. The main application of fibrin is as a scaffold matrix for tissue engineering. The fixation of the gel with poly(L-lysine) during the culturing period was introduced to solve the problems of gel shrinking and low stiffness. Inner tension in the gel as a result of the tendency to shrink and the fixation may have led to the collagen synthesis.

Actin Gels: Globular actin monomer (G-actin) polymerizes to form an actin filament (F-actin), which forms a gel. Cross-linking of the actin network by α-actin can transform isotropic entangled networks into heterogeneous networks of microgels, whose macroscopic viscoelasticity is a complex of contributions from both dense and loose mesh networks.[25] In the case of G-actin *in vitro*, the presence of a salt (K^+, Na^+, Ca^{2+}, Ba^{2+}, Li^+, etc.) may activate the gelation process.

Lectin-Mediated Gels: Lehr elucidated the mechanisms of lectin-cell interactions and explored the pharmaceutical potential of lectin-modified drug carrier systems at epithelial biological barriers.[26] The weak interaction between lectins and glycoconjugates can induce easy dissociation behavior at neutral pH, making lectin suitable for delivery systems.

Biomimetic Polymers

Biomimetic polymers are synthetic polymers that mimic biomolecules. The merit of this approach is that the advantages of synthetic polymers and biomolecules can be achieved at the same time.

Phospholipid polymers

A phospholipid is a substance that comprises the outline of the cell wall and is commonly mimicked by investigators. Its high permeability and biocompatibility have interested investigators for a long time. Ishihara et al. prepared MPC-based polymers, poly(MPC-*co*-methacrylic acid) (PMA) and poly(MPC-*co*-n-butyl methacrylate) (PMB) aqueous solution, to synthesize the hydrogel.[27] The gelation occurs due to hydrogen bonds between the carboxyl groups of the PMA in the hydrophobic domain of the PMB (Fig. 4D). The formation can take place under either high or low aqueous conditions at room temperature. Thermoreversible and pH-reversible phenomena have also been observed, so that the phospholipid hydrogels can be utilized as peptide drug carriers.

Hybrid gels

Hybrid hydrogels refer to hydrogel systems that contain two or more components of distinct classes of molecules. The main purpose of a hybrid polymer is to create stimuli responsiveness to a change in the environment. Numerous research studies are underway on covalently bonding a stimuli-sensitive polymer with double-stranded deoxyribonucleic acid (DNA). DNA forms double or triple strands with its complementary base pair by way of hydrogen bonding. This formation is reversible and is affected by a change in physical or chemical stimuli. Wang, Steward, and Kopecek designed hybrid gels composed of coiled-coil protein domains and a polymer backbone (Fig. 4E). The coiled-coil proteins were engineered by control of amino acid sequences. The coiled-coil formation is regulated by temperature, pH, and ionic strength.[28]

Biomimetic/Synthetic Polymer Hybrid Gels

Biomimetic/synthetic polymer hybrid gels are good candidates for application in biomaterials and biomedical engineering. Most biomimetic polymers are capable of highly sophisticated functions. The hybrid technique is one of the promising approaches for further advancement in the area. In this section, recent biomimetic/synthetic polymer hybrid gels are discussed.

Bio-inspired/biodegradable polymer hybrid

A novel bio-inspired material based on the structure of the cell membrane has been proposed by Ishihara, Ueda, and Nakabayashi They first succeeded with MPC having a phospholipid polar group.[5] To fabricate the cell membrane surface, the MPC was copolymerized with *n*-butyl methacrylate (BMA) as a comonomer, and the resulting copolymer was coated onto a substrate. For a cell culture scaffold, further materials design would be necessary for the MPC polymer. A new concept has been proposed based on a biodegradable polymer hybrid. An enantiomeric poly[L-(D-) lactic acid] [PL(D)LA] macromonomer has been newly synthesized and copolymerized with MPC and BMA.[29] The resulting copolymers would spontaneously form the phospholipid polymer scaffold by stereocomplexation (Fig. 5A). The scaffold is a porous structure formed by a NaCl leaching technique. The enantiomeric PL(D)LA segment was utilized not only as a physical cross-link point but also as a protein adsorption domain in cell culture. One of the advantages was water intrusion into the phospholipid polymer scaffold, which was evaluated by the static contact angle of the water. The contact angle of the phospholipid polymer scaffold quickly decreased. This result suggests that water uptake was enhanced by the MPC unit. This is quite an important property for a porous scaffold to be used in three-dimensional tissue reconstruction.

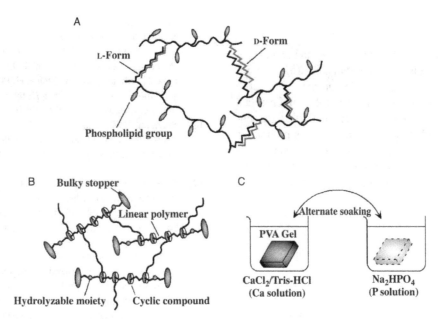

Fig. 5 Schematic illustrations of biomimetic/synthetic polymer hybrid gels. (**A**) Bioinspired/biodegradable polymer hybrid. (**B**) Host/guest supramolecular gels. (**C**) Organic/inorganic composite gels.

Host/guest supramolecular gels

The field of supramolecular chemistry has been explored for constructing new molecular architectures such as molecular association by intermolecular forces and inclusion complexation. Supramolecular science is a new research area, and the study of supramolecular assemblies has progressed due to the study of biomimetic chemistry and molecular recognition in relation to such areas as enzymatic catalysis reactions. One of the major supramolecular structures is a polyrotaxane. Polyrotaxanes, in which many α-cyclodextrins (α-CDs) are threaded onto a PEO chain capped with bulky end groups, have recently been demonstrated to be a molecular necklace. A biodegradable polyrotaxane consisting of α-CDs, PEG, and benzyloxycarbonyl (Z)-L-phenylalanine (L-Phe) introduced via ester linkages has been prepared as an implantable material for tissue engineering.[30] The advantage of the polyrotaxane structure is its rapid and perfect degradation under physiological conditions. The specific structure of the polyrotaxanes may enable us to maintain the shape of the scaffold as a cell support and to achieve complete hydrolysis.

Using these perspectives, Yui et al. have designed PEG hydrogels cross-linked by hydrolyzable polyrotaxanes (Fig. 5B).[31] In the hydrogel design, cyclic molecules in a polyrotaxane can be used as a cross-linking point. The cyclic molecules could be easily dethreaded onto the linear polymeric chain in response to terminal ester hydrolysis; then the cross-link density of the PEG hydrogel would decrease. The expected characteristics of the hydrogels include rapid and perfect disappearance based on supramolecular dissociation via terminal hydrolysis.

PEG hydrogels cross-linked by the polyrotaxane were prepared and characterized to estimate gel erosion and identify the mechanism. It is assumed that terminal hydrolysis of the polyrotaxane cross-links can trigger dethreading of the α-CDs to induce a unique erosion behavior of the hydrogels.

Organic/inorganic composite gels

Mineralized tissues such as bone and dentin are indispensable elements in our body. For example, human bone consists of inorganic salts and organic substances such as collagen. The mineralized tissue has a well-organized structure, with inorganic salts, calcium, and phosphate. For this reason, organic/inorganic composites are good candidates for mineralized tissue reconstruction. In one technique, a hydroxyapatite composite has been produced by conventional flame spraying. The technique is limited to metal implants. In the case of organic substances such as polymeric gels, it is impossible to coat hydroxyapatite because of the heating process. A novel technique for preparation of organic/inorganic composite, especially hydroxyapatite, has been investigated. The hydroxyapatite was prepared by an alternative soaking process between $CaCl_2$/*tris*-HCl (Ca solution) and Na_2HPO_4 (P solution) (Fig. 5C).[32] Polymeric gels, such as PVA and poly(HEMA), were used as a template. The apatite crystal was formed not only on the surface but also in the gel through two subsequent steps.[32] The apatite content and apatite formation rate were correlated with the gel swelling ratio. Three-dimensional gels/hydroxyapatite composites are of great importance in the fabrication of bonelike tissue.

CONCLUSION

This entry summarizes the basic principles and recent topics concerning hydrogels as biomaterials. Hydrogels represent a diverse cross section of biomaterials, e.g., drug delivery carriers, tissue engineering scaffolds, and medical devices. Various chemical and physical properties are controlled by changing chemical reagents such as monomers and polymers and by the preparation conditions of the hydrogels. Hydrogels represent one of the most promising approaches for the fabrication of novel biomaterials.

REFERENCES

1. Wichterle, O.; Lim, D. Hydrophilic gels for biological use. Nature 1960, 185, 117–118.
2. Phillips, A.J.; Speedwell, L. Contact Lenses, 4th Ed.; Butterworth-Heinemann: Oxford, 1997.
3. Nicolson, P.C.; Vogt, J. Soft contact lens polymers: An evolution. Biomaterials 2001, 22 (24), 3273–3283.
4. Künzler, J.; Ozark, R. Methacrylate-capped fluoro side chain siloxanes: Synthesis, characterization, and their use in the design of oxygen-permeable hydrogels. J. Appl. Polym. Sci. 1997, 65 (6), 1081–1089.
5. Ishihara, K.; Ueda, T.; Nakabayashi, N. Preparation of phospholipid polymers and their properties as polymer hydrogel membrane. Polym. J. 1990, 22 (5), 355–360.
6. Ishihara, K.; Nomura, H.; Mihara, T.; Kurita, K.; Iwasaki, Y.; Nakabayashi, N. Why do phospholipid polymers reduce protein adsorption? J. Biomed. Mater. Res. 1998, 39 (3), 323–330.
7. Kiritoshi, Y.; Ishihara, K. Preparation of cross-linked biocompatible poly(2-methacryloyloxyethyl phosphorylcholine) gel and its strange swelling behavior in water/ethanol mixture. J. Biomater. Sci., Polym. Ed. 2002, 13 (2), 213–224.
8. Andrews, C.S.; Denyer, S.P.; Hall, B.; Hanln, G.W.; Lloyd, A.W. A comparison of the use of an ATP-based bioluminescent assay and image analysis for the assessment of bacterial adhesion to standard HEMA and biomimetic soft contact lenses. Biomaterials 2001, 22 (24), 3225–3233.
9. Tanaka, T. Collapse of gels and the critical endpoint. Phys. Rev. Lett. 1978, 40 (12), 820–823.
10. Schild, H.G. Poly(N-Isopropylacrylamide): Experiment, theory and application. Prog. Polym. Sci. 1992, 17 (2), 163–249.
11. Kataoka, K.; Miyazaki, H.; Bunya, M.; Okano, T.; Sakurai, Y. Totally synthetic polymer gels responding to external glucose concentration: Their preparation and application to on–off regulation of insulin release. J. Am. Chem. Soc. 1998, 120 (48), 12694–12695.
12. Miyata, T.; Asami, N.; Uragami, T. A reversibly antigen-responsive hydrogel. Nature 1999, 399 (6738), 766–769.
13. Tanaka, T.; Fillmore, D.J. Kinetics of swelling of gels. J. Chem. Phys. 1979, 70 (3), 1214–1218.
14. Chen, J.; Park, H.; Park, K. Synthesis of superporous hydrogel: Hydrogels with fast swelling and superabsorbent properties. J. Biomed. Mater. Res. 1999, 44 (1), 53–62.
15. Kaneko, Y.; Sakai, K.; Kikuchi, A.; Yoshida, R.; Sakurai, Y.; Okano, T. Influence of freely mobile grafted chain length on dynamic properties of comb-type grafted poly(N-isopropyacrylamide) hydrogel. Macromolecules 1995, 28 (23), 7717–7723.
16. Mukae, K.; Sakurai, M.; Sawamura, S.; Makino, K.; Kim, S.W.; Ueda, I.; Shirahama, K. Swelling of poly(N-isopropylacrylamide) gels in water-alcohol(C1–C4) mixed solvents. J. Phys. Chem. 1993, 97 (3), 737–741.
17. Park, K.; Shalaby, W.S.W.; Park, H. Physical gels. In Biodegradable Hydrogels for Drug Delivery; Technomic Publishing Co.: Lancaster, PA, 1993; 99–140.
18. Shaheen, S.M.; Yamaura, K. Preparation of theophyl-line hydrogels of atatic poly(vinyl alcohol)/NaCl/H2O system for drug delivery system. J. Control. Release 2002, 81 (3), 367–377.
19. Bezemer, J.M.; Radersma, R.; Grijpma, D.W.; Dijkstra, P.J.; Feijen, J.; van Blitterswijk, C.A. Zero-order release of lysozyme from poly(ethylene glycol)/poly (butylene terephthalate) matrices. J. Control. Release 2000, 64 (1–3), 179–192.
20. Molina, I.; Li, S.; Martinez, M.B.; Vert, M. Protein release from physically crosslinked hydrogels of PMA/ PEO/PLA triblock copolymer-type. Biomaterials 2000, 22 (3), 363–369.
21. Bae, Y.H.; Vernon, B.; Han, C.K.; Kim, S.W. Extra-cellular matrix for a rechargeable cell delivery system. J. Control. Release 1998, 53 (1–3), 249–258.
22. Andrianov, A.K.; Payne, L.G. Protein release from polyphosphagene matrices. Adv. Drug Deliv. Rev. 1998, 31 (3), 185–196.
23. Watanabe, T.; Ohtsuka, A.; Murase, N.; Barth, P.; Gersonde, K. NMR studies on water and polymer diffusion in dextran gels. Influence of potassium ions on microstructure formation and gelation mechanism. Magn. Reson. Med. 1996, 35 (5), 697–705.
24. v, K.R.; Hall, L.D. Chitosan derivatives bearing c10-alkyl glycoside branches: A temperature-induced gelling polysaccharide. Macromolecules 1991, 24 (13), 3828–3833.
25. Tempel, M.; Isenberg, G.; Sackmann, E. Temperature-induced sol gel transition and microgel formation in alpha-actinin cross-linked actin networks—A rheological study. Phys. Rev. E. 1996, 54 (2), 1802–1810.
26. Lehr, C.-M. Lectin-mediated drug delivery: The second generation of bioadhesives. J. Control. Release 2000, 65 (1–2), 19–29.
27. Nam, K.; Watanabe, J.; Ishihara, K. Characterization of the spontaneously forming hydrogels composed of water-soluble phospholipid polymers. Biomacromolecules 2002, 3 (1), 100–105.
28. Wang, C.; Steward, R.J.; Kopecek, J. Hybrid hydrogels assembled from synthetic polymers and coiled-coil protein domains. Nature 1999, 397 (6718), 417–420.
29. Watanabe, J.; Eriguchi, T.; Ishihara, K. Stereocomplex formation by enantiomeric poly(lactic acid) graft-type phospholipid polymers for tissue engineering. Biomacromolecules 2002, 3 (5), 1109–1114.
30. Watanabe, J.; Ooya, T.; Yui, N. Effect of acetylation of biodegradable polyrotaxanes on its supramolecular dissociation via terminal ester hydrolysis. J. Biomater. Sci., Polym. Ed. 1999, 10 (12), 1275–1288.
31. Ichi, T.; Watanabe, J.; Ooya, T.; Yui, N. Controllable erosion time and profile in poly(ethylene glycol) hydrogels by supramolecular structure of hydrolyzable polyrotaxane. Biomacromolecules 2001, 2 (1), 204–210.
32. Taguchi, T.; Kishida, A.; Akashi, M. A study on hydroxyapatite formation on/in the hydroxyl groups-bearing nonioinic hydrogels. J. Biomater. Sci. Polym. Ed. 1999, 10 (1), 19–32.

Hydrogels: Classification, Synthesis, Characterization, and Applications

Kaliappa Gounder Subramanian
Vediappan Vijayakumar
Department of Biotechnology, Bannari Amman Institute of Technology, Sathyamangalam, India

Abstract

Polymer hydrogels constitute a group of prospective materials that find extensive use in health care and tissue engineering. They are high-molecular-weight three-dimensional cross-linked hydrophilic polymer networks that can imbibe and hold large quantities of water and biofluids. Their swellability, permeability, viscoelastic features, mechanical properties, etc., are altered by changing their chemical structure, cross-linking, and environmental parameters such as ionic strength, temperature, pH, and presence of electric or magnetic field. This entry is devoted to reviewing the classification, synthesis, structure–property relation, characterization, and medical applications of this potential multifunctional hydrogel in controlled and targeted drug delivery, regenerative medicine, medical device, ophthalmology, wound care, and embolization. The trends in future research of this material are also highlighted.

INTRODUCTION

Hydrogels, characterized by combined properties of both liquids and solids, are cross-linked three-dimensional hydrophilic synthetic or natural polymeric network capable of imbibing water or biofluids many times (>100) its own weight without dissolving and is capable of high flexibility similar to the natural tissue. Cross-linked hydrogel networks show viscoelastic and elastic behaviors unlike aqueous solutions of the linear hydrophilic polymers, which display Newtonian behavior. Their hydrophilic surface has a low interfacial free energy when they come in contact with body fluids, which results in a low tendency for proteins and cells to adhere to these surfaces. Its soft and rubbery nature minimizes irritation of the surrounding tissue. They provide semi-wet three-dimensional environment for cells and tissue interaction and can be linked with biological/therapeutic molecules by covalent or physical binding. Hence, hydrogels have been proposed for a series of biomedical and biological applications, such as tissue engineering, drug release systems, biological sensors, microarrays, imaging and actuators. Hydrogel nanoparticles hold versatility and suitable properties as carriers for efficient delivery of drugs, proteins, peptides, oligosaccharides, vaccines, and nucleic acids. The key to the success of hydrogels in the body is due to their biocompatibility, injectability, relatively low cytotoxicity, biodegradability, mucoadhesiveness, and tunable mechanical, bioadhesive, and other functional properties by chemical modification. The cross-linking density of hydrogels can be modified to control its average pore size, *in vivo*

degradation rate, and thus the rate of drug diffusion from it. The hydrogel network and the thermodynamic features of its components result in their diffusional behavior, variation in molecular mesh sizes, and the molecular stability of the loaded bioactive molecules. The higher the number of the hydrophilic groups, the higher is the water-holding capacity. An increased cross-linking density will decrease the equilibrium swelling due to the decrease in the hydrophilic groups and stretchability of the polymer network. Since polymer hydrogels mimic the behavior of the soft tissues such as cartilage, mucus, vitreous humor, tendons, and blood clots, they can be used for the development of soft-solid biomedical implants for large and small joint applications and they were the first biomaterials to be rationally designed for human use. Gelatin, vitreous humor of the eye, and cartilage are examples of protein hydrogels derived from collagen.

High water content, viscoelasticity, permeability, and perm selectivity of hydrogels make them resemble extracellular matrix and allows them to be populated by cells, and transport nutrients and solutes across spaces in the body. These features are important for 3D cell culture scaffolds and cell-delivery vehicles in cell-based therapy. Hydrogels are used to prevent thrombosis and restenosis in vessels after vascular injury or angioplasty and prevent tissue–tissue adhesion after an operation. They can undergo changes in shape or volume in response to physical or biological conditions such as temperature, pH, and ionic strength. But they have limited resistance to mechanical deformation and poor tear strength due to the plasticizing effect of the water held within the polymer

Concise Encyclopedia of Biomedical Polymers and Polymeric Biomaterials DOI: 10.1081/E-EBPPC-120049894

Gene–Hydrogels

network. Hydrogels can be easily modified chemically to exhibit tunable affinities for target drugs. Injectable thermally responsive hydrogels with a lower critical solution temperature (LCST) below body temperature hold promise as biomaterials for regional tissue mechanical support and cell-delivery applications. The study and understanding of the fundamental aspects and molecular mechanisms associated with the formation of new surfaces and interfaces are necessary to respond to emerging technology problems for polymer hydrogels. The aim of this review is to gather and systemize all information related to classification, design and synthesis, properties, characterization, and applications of hydrogel in biological and biomedical fields. The synthetic hydrogels, which have the advantage of precise control of molar mass, can be tailored to have a wide range of properties and low immunogenicity with minimum risk of biological pathogens or contaminants.

HISTORICAL PROSPECTIVE

Hydrogels are found in nature since life on earth began to evolve. Bacterial biofilm, hydrated living tissue, extracellular matrix components and plant structures are ubiquitous hydrated swollen motifs in nature. Gelatin and agar-agar were also explored in early human history. The base polymer materials used to make the most common hydrogels are known to be inert (nonreactive and nontoxic) and have been around for many years. They were first tested in simple forms as a safe medium for drug and cell delivery and as scaffolds for tissue engineering. But the modern history of hydrogels as a class of biomaterials designed for medical applications can be accurately traced. Hydrogels have been used in numerous biomedical disciplines for more than six decades such as in ophthalmology as contact lenses and in surgery as absorbable sutures, as well as in many other areas of clinical practice to cure illnesses such as diabetes mellitus, osteoporosis, asthma, heart diseases, and neoplasms. In the early 1950s, Wichterle and Lim from Prague in the Czech Republic proposed the design of new biomaterials with features such as shape stability and softness similar to that of the soft tissue, chemical and biochemical stability, absence of extractables, and high permeability for water-soluble nutrients and metabolites for applications in ophthalmology. Based on this rationale, in 1953 they[1] synthesized the first copolymer hydrogels from 2-hydroxyethyl methacrylate (HEMA) and ethylene glycol dimethacrylate (EGDMA) used in contact lens production. Wichterle prepared the first soft hydrogel-based contact lenses by spin casting process and the first investigation on the biocompatibility of hydrogels was published in 1959 and 1960.[2] Design and synthesis of hydrogels for numerous medical applications were also initiated at the same time. In ophthalmology, along with soft contact lenses, glaucoma microcapillary drain fillings for the restoration of detached retina and fillings after enucleation were also studied. Hydrogels have also been designed for augmenting vocal cords.[2]

The emergence of the applications discussed so far had also prompted a detailed study of the relationship between the structure of cross-linked hydrophilic polymers and their biocompatibility.[2,3] These results were translated into successful clinical applications such as the use of HEMA-based hydrogels in rhinoplasty, which produced long-term biocompatibility and excellent cosmetic results.[2] Stimuli-sensitive hydrogels in response to minimal change in environmental conditions had also been made by manipulating their structures.[2,3] Introduction of ionogenic groups into EGDMA cross-linked HEMA hydrogels permitted control of their permeability[2,3] and specific resistance as a function of pH. Hydrogel membranes containing methacrylic acid (MA) units possessed higher permeability of NaCl in alkaline conditions, whereas membranes containing N,N-dimethylaminoethyl methacrylate (DMAEMA) units showed increased permeability under acidic conditions.[2] Permeability of ampholytic membranes, containing both MA and DMAEMA, passed through a minimum at the isoelectric point, and increased as the pH moved from the isoelectric point in either direction. The results for the relationship between the structure of hydrogel membranes and specific resistance revealed similar trends.[4] The main advantage of the biomaterial was its stability under varying pH, temperature, and tonicity conditions. In the 1980s, Lim and Sun obtained calcium alginate microcapsules for cell engineering, and Yannas's group modified synthetic hydrogels with some natural substances, such as collagen and shark cartilage to obtain novel dressings, providing optimal conditions for healing burns.

Temperature-sensitive hydrogels[2,5] usually based on polymers such as poly(N-isopropylacrylamide) (poly(NIPAAm)), exhibit a LCST and collapse as temperature increases. Below the LCST, water molecules form hydrogen bonds with polar groups on the polymer backbone and organize around hydrophobic groups as iceberg water. Above the LCST, bound water molecules are released to the bulk with a large gain in entropy, resulting in collapse of the polymer network. Poly(N,N-diethylacrylamide) and its copolymers with N-$tert$-butylacrylamide demonstrated a similar behavior.[2,5] The sensitivity of NIPAAm-based hydrogels to other stimuli, for example electric current and chemical stimuli, had also been demonstrated and studied in detail. One of the successful strategies to increase the response dynamics of stimuli-sensitive hydrogels was the introduction of short grafts. Comb-type grafted NIPAAm hydrogels demonstrated rapid de-swelling responses to temperature changes. Numerous monomers and cross-linking agents have been used for the synthesis of hydrogels with a wide range of chemical compositions.

CLASSIFICATION OF HYDROGELS: BASED ON SOURCE, STRUCTURE, PROPERTIES, AND APPLICATIONS

Hydrogels are classified as natural (e.g., collagen, agar-agar, alginate, dextran, gelatin, chitosan, etc.) and synthetic (poly(vinyl alcohol) (PVA), poly(ethylene glycol)-poly(lactone)-poly(ethylene glycol) (PEG-PL-PEG)) based on their source. They are categorized as permanent or chemical gels when they feature covalently cross-linked networks[2,6,7] or as reversible or physical gel when the networks are held together by molecular entanglements, and/or secondary forces such as ionic, hydrogen bonding, or hydrophobic interactions. The former attain an equilibrium swelling state, which depends upon the polymer–water interaction parameter and the cross-link density[7] and in the latter gel dissolution is prevented by physical interactions existing between different polymer chains.[2] All these interactions are reversible and can be disrupted by changes in physical conditions or on application of stress.[2,7,8] Some examples of physical hydrogels include poly(vinyl alcohol)–glycine hydrogels, gelatin gels, and agar-agar gels. Hydrogels may also be grouped as homopolymer, copolymer, and multipolymer networks, interpenetrating networks, and double networks based on the nature of the network or as homogeneous (optically transparent) hydrogels, microporous, and macroporous hydrogels based on pores. Based on their fate in an organism they are grouped into degradable and nondegradable hydrogels. Depending upon the properties of synthesized hydrogels, they can be classified into stimuli-sensitive hydrogels (smart, intelligent, or physiologically responsive hydrogels) and non-stimuli responsive hydrogels. Stimuli-sensitive hydrogels are classified further into temperature, pH, glucose, electric signal, light, pressure, specific ion, specific antigen, thrombin-induced infection, protein, solvent, light, pressure, sound, or magnetic field sensitive, hydrogel systems.[9,10] Hydrogels can be classified as neutral or ionic, based on the nature of the side groups. According to their mechanical and structural characteristics, they can be classified as affine or phantom networks. Finally, they can be classified based on the physical structure of the networks as amorphous, semicrystalline, hydrogen-bonded structures, supermolecular structures and hydrocolloidal aggregates.[2,3,6,9,10] They can also be classified as biodegradable and nonbiodegradable hydrogels depending upon their biodegradability. If the hydrogels are absorbing water many hundred times their own weight, they are termed as superabsorbent polymers (SAP).

Superporous hydrogels (SPHs) are a different category of water-absorbing material. Unlike SAPs, SPHs swell fast, within minutes, to the equilibrium swollen state regardless of their size due to absorption through open porous structure by capillary force. The poor mechanical strength of SPHs is overcome by developing the second-generation SPH composites and the third-generation SPH hybrids.[11] If water is removed from these swollen biomaterials they are called xerogels, which are the dried hydrogels. Dehydrated hydrogels are called aerogels as water is removed from them without causing structural deformation. Microgels, however, have small particles with a diameter of 100 nm and also swell in water. This space between molecular chains is often regarded as the molecular mesh or pores. Depending upon the size of these pores, hydrogels can be conveniently classified as (i) macroporous, (ii) microporous, and (iii) nonporous. Macroporous hydrogels have large pores of dimension in the range 0.1–1 μm. These hydrogels release the drug entrapped inside the pores through a mechanism dependent on drug diffusion coefficient. Microporous hydrogels have small pore size of the gel network, usually in the range of 100–1000 Å. In these gels, the loaded drug release occurs by molecular diffusion and convection. However, when drugs and polymers are thermodynamically compatible, partitioning of drug through the hydrogel walls was predominant. Nonporous hydrogels are the mesh-like structures of macromolecular dimension (10–100 Å) and are formed due to cross-linking of monomer chains.

SYNTHESIS

The widely followed approach[2,3,6,12–15] in hydrogel synthesis involves copolymerization/cross-linking of co-monomers using appropriate multifunctional co-monomer as a cross-linking agent. For the formation of hydrogels, the monomers chosen and their corresponding linear polymers should be water-soluble. Polymerization can be achieved in bulk, solution, or suspension and initiated by chemical, photo, thermal, redox modes and gamma ray or microwave irradiation. The second method involves cross-linking of linear preformed polymers by irradiation, or by chemical compounds. For example, the reaction of α,ω-hydroxyl poly(ethylene glycol) with a diisocyanate in the presence of a triol as cross-linker leads to the formation of cross-linked hydrophilic polyurethanes.[6] The monomers used in the preparation of the ionic polymer network contain an ionizable group, a group that can be ionized, or a group that can undergo a substitution reaction after the completion of polymerization. As a result, synthesized hydrogels containing weakly acidic groups like carboxylic acids, or a weakly basic groups like substituted amines, or a strong acidic and basic group like sulfonic acids and quaternary ammonium. Some of the commonly used cross-linking agents include N,N'-methylenebis(acrylamide) (MBAA), divinyl benzene, EGDMA, 1,4-butanediol dimethacrylate, trimethylolpropane triacrylate, poly(ethylene glycol dimetharylate), etc. The widely used monomers in hydrogel synthesis are acrylic acid, methacrylic acid, acrylamide, HEMA, maleic anhydride, glycidyl methacrylate, N,N-dimethyl acrylamide, N-[3-(dimethylamino)propyl]methacrylamide, N-isopropyl acrylamide, N-vinyl pyrrolidone (NVP), itaconic acid (IA), and 2-acrylamido-2-methyl propane sulfonic acid. For synthesis of temperature-sensitive hydrogels, *N*-alkyl and *N*,*N*-dialkyl acrylamides are widely used.[2,5,15,16]

By proper choice of monomers and cross-linking agents, polymer hydrogels with custom-made properties such as biodegradation, mechanical strength, and chemical and biological response to external stimuli can be designed and synthesized for various medical uses. Van den Bulcke et al. have reported on the preparation of gelatin hydrogels cross-linked with partially oxidized dextrans.[17,18] Another approach is the conversion of the hydroxyl end groups of poly(ethylene glycol) into methacrylate, which can then be cross-linked via radical polymerization using co-monomers and a cross-linker. Typical cross-linking between gelatin and chitosan with glutaraldehyde as the cross-linker is shown in Fig. 1. Biodegradable hydrogels are usually synthesized by introducing biodegradable organic moiety in the hydrogel either by modifying an inherently biodegradable preformed polymer or by incorporating biodegradable part either in the backbone or as pendant groups in the hydrogel taking suitable monomers and using appropriate synthetic strategies.[19]

STRATEGIES FOR FABRICATION OF HYDROGELS

The ability to engineer the hydrogels with specific material properties is hampered by lack of good mechanical properties, poor control of molecular weight, chain configuration, and polymerization kinetics. In general, the cross-linked structure of hydrogels is characterized by junctions or tie points to be formed from strong chemical linkages, perma-

nent or temporary physical entanglements, microcrystalline formation, and weak interactions like hydrogen bonds. Covalent cross-links result in stable hydrogels, while other types of cross-links could be used to reverse the gelling properties of the hydrophilic polymers under desired conditions. Novel designs of hydrogels with improved mechanical properties were reviewed by Tanaka, Gong, and Osada.[20] Hydrogels with improved mechanical properties were prepared by introducing sliding cross-linking agents, double network, and nano clay-filling.[20] Hydrogels of well-defined shapes with sufficient mechanical properties for scaffold applications were fabricated by bringing solutions of these polymers into contact with patterned templates of paper wetted with aqueous solutions of multivalent cations.[21,22]

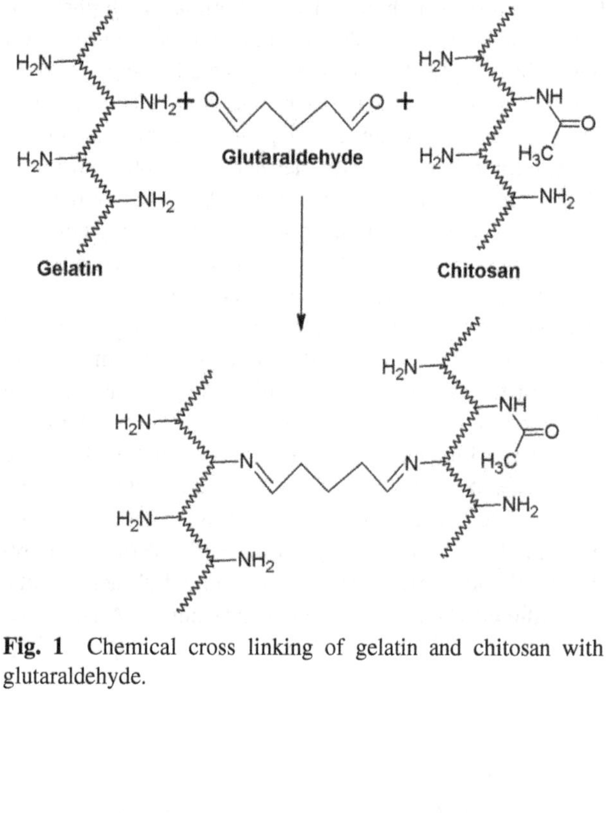

PEG-based hydrogel systems are widely used due to their negligible toxicity and limited immunogenic recognition. Physical (i.e., particle size, shape, surface charge, and hydrophobicity) and chemical properties of hydrogel particles play an important role in cell-particle recognition and response. To fabricate geometrically uniform PEG-based hydrogel structures on cells, electron beam lithography and ultraviolet optical lithography were used. For fabricating hydrogel microspheres, emulsification had historically been used by inserting hydrogel precursors in a hydrophobic medium (such as oil) and breaking up the hydrogel phase into small droplets by agitation. Hydrogels were also fabricated by micro-molding and photolithography to control their size and heterogeneity and directed cell migration into the gel.[23] For tissue engineering micro-fabricated hydrogels were made by bottom-up and top-down approaches.[24,25]

STRUCTURE–PROPERTY CORRELATION

The structure and properties of a specific hydrogel are extremely important in selecting which materials are suitable for specific medical application.[26–34] Knowledge of the structure–property relationship is fundamental to tailor hydrogel properties to the required need. Water inside the hydrogel allows free diffusion of some solute molecules, while the polymer serves as a matrix to hold water together. The semiliquid and semisolid-like properties of hydrogel cause many interesting relaxation behaviors that are not found in either a pure solid or a pure liquid. The hydrogels are characterized by an elastic modulus that exhibits a pronounced plateau extending to times in the order of seconds, and by a viscous modulus that is considerably smaller than the elastic modulus in the plateau region.[26] The equilibrium swelling degree and the elastic modulus of hydrogels depend on the cross-link concentration and charge densities of the polymer network. Increasing the number of ionic groups in hydrogels is known to increase their swelling capacities due to the simultaneous increase in the number of counter-ions inside the gel, which produces an additional osmotic pressure that swells the gel.[27] The most important parameters used to characterize the network

Fig. 1 Chemical cross linking of gelatin and chitosan with glutaraldehyde.

structure of hydrogels are the polymer volume fraction in the swollen state (the amount of polymer in the gel), the average molar mass of the polymer chain between two consecutive cross-linking points ($\overline{M}_c$) and the corresponding mesh size. The mathematical expression for $\overline{M}_c$ between cross-links[49] is given by the Eq. 1

$$\overline{M}_c = -\frac{\left(1 - 2/\phi\right)V_1 v_{2r}^{2/3} v_{2m}^{1/3}}{\overline{v}\left[ln\left(1 - v_{2m}\right) + v_{2m} + \chi v_{2m}^2\right]} \tag{1}$$

where v is the specific volume of the polymer, V_1 is the molar volume of the swelling agent, v_{2m} is the polymer volume fraction in the equilibrium-swollen system, v_{2r} is the polymer volume fraction in the relaxed state (i.e., after cross-linking but before swelling), χ is the Flory's polymer–solvent interaction parameter, and ϕ is the number of branches originating from a cross-linking site. The schematic diagram for network structure in hydrogel is shown in Fig. 2.

The polymer volume fraction in the swollen state is a measure of the amount of fluid imbibed and retained by the hydrogel, $\overline{M}_c$ which can be either chemical or physical in nature, is a measure of the degree of cross-linking of the polymer. The average correlation length (distance between two adjacent cross-links) n, provides a measure of the space available between the macromolecular chains. These parameters are widely determined using equilibrium-swelling theory and rubber-elasticity theory.[29] The structure of hydrogels that do not contain ionic moieties can be analyzed by the Flory–Rehner theory.[29,35] The combination of thermodynamic and elasticity theories states that an equilibrium swelled cross-linked polymer gel in a fluid is subject to two opposing forces, namely, the thermodynamic force of mixing and the retractive force of the polymer chains. At equilibrium these two forces are equal, which can be defined in terms of the Gibbs free energy given by the equation

$$\Delta G_{total} = \Delta G_{elastic} + \Delta G_{mixing} \tag{2}$$

where $\Delta G_{elastic}$ is the contribution from the elastic retractive forces developed inside the gel, and ΔG_{mixing} is the result of the spontaneous mixing of the fluid molecules with the polymer chains and is a measure of the compatibility of the polymer with the molecules of the surrounding fluid and usually expressed by the polymer–solvent interaction parameter. The ability to tailor the molecular structure of hydrogels enables the tailoring of their mechanical, responsive, and diffusive properties.

The nanoscale dispersion of layered silicates or clays in polymer networks is one of the techniques offering significant enhancements in the material properties of hydrogels. Haraguchi et al.[30] prepared such nanocomposite hydrogels starting from acrylamide-based monomers together with Limonite as a physical cross-linker. Limonite, a synthetic clay, when suspended in water, forms disclike particles with a thickness of 1 nm, a diameter of about 25 nm, and a negative surface charge density stabilizing dispersions in water and these nanoparticles act as a multifunctional cross-linker with a large effective functionality.[31–33]

CHARACTERIZATION

Morphology

Morphology of hydrogels are analyzed by particle size, porosity, X-ray diffraction, scanning electron microscopy (SEM), atomic force microscopy (AFM), optical microscopy, and attenuated total reflection Fourier transform infrared spectroscopy.[22,34,36] The volume phase transition of constrained hydrogel layers can be studied by a combination of surface plasma resonance spectroscopy (SPR) and optical waveguide spectroscopy (OWS).[37] SPR and OWS were used to obtain information about the volume phase transition of hydrogel films showing highly anisotropic swelling behavior.[38] A refractive index gradient, perpendicular to the swollen hydrogel film surface, could be analyzed in detail by the application of the reversed Wetzel–Kramer–Brillouin (WKB) approximation to the optical data.[39] This novel approach to analyzing thin-film gradients with the WKB method presents a powerful tool for the characterization of inhomogeneous hydrogels. The SPR/OWS technique has been applied to gain additional insight into the mechanism of the anisotropic swelling of responsive hydrogels. The gel modulus is measured by AFM force–distance curves. Micro-patterned, thermo-responsive hydrogel film were characterized with imaging ellipsometry both on the dry film as well as on a water-swollen sample. Through imaging ellipsometry, it was possible to distinguish the different regions of interest on a micrometer scale and to follow the swelling of the hydrogel part as a function of the temperature.

The swelling transition was probed by a combination of acoustic and optical techniques, namely the quartz crystal microbalance (QCM) and plasmon resonance SPR

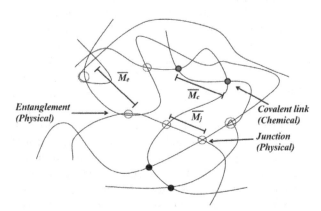

Fig. 2 Schematic network structure of a hydrogel with junctions, entanglements, and covalent linkages ($\overline{M}_c$, $\overline{M}_j$ and $\overline{M}_e$ are the number average molecular weights between crosslinks, junctions and entanglements respectively).

spectroscopy. This combination provides a number of different parameters, which allow for the use of thick layers for sensing. A QCM with dissipation monitoring (QCM-D) was used to assess water and moisture uptake characteristics of plasma-polymerized poly(NIPAAm) hydrogel films[40] as well as the physical properties of interpenetrating polymer networks. The spatial inhomogeneity in hydrogels arising due to uneven distribution of cross-link density is undesirable because it dramatically reduces the optical clarity and strength of hydrogels and this was investigated by light scattering, small angle X-ray scattering, and small angle neutron scattering (SANS). In general, the spatial inhomogeneity increases with the gel cross-link density and it decreases with the ionization degree of gels due to the effects of the mobile counter ions, electrostatic repulsion and the Donnan potential.[41,42] The origin of the toughness and the mechanisms of deformation of tough hydrogels were elucidated by SANS. The density of copolymer hydrogels were found to depend upon the random or blocky sequence of monomers. The transparency of a series of PEG-protein hydrogels synthesized with different proteins was different and this was attributed to the ability of the protein component to self-associate via hydrophobic interactions.[36]

Scanning Electron Microscopy

Scanning electron microscopy (SEM) can be used to provide information about the hydrogel's surface topography, composition, and other properties such as electrical conductivity and to capture the characteristic network structure in hydrogels. Magnification in SEM can be controlled over a range up to six orders of magnitude from about 10 to 500,000 times. The hydrogels were cut with a razor blade and dehydrated by lyophilization overnight to observe their inner morphology. All the samples were kept under vacuum condition and taken out just before the gold sputtering treatment and the gold-coated samples were observed under scanning electron microscopy at 15–20 kV.[43,44]

Porosity

Porosity of swollen hydrogels was determined from the equilibrium water content assuming that the hydrogels prepared without NaCl in polymerization medium will have only a molecular porosity and that prepared with NaCl will have a macroporous structure.[41,42,45] The porosity and mean pore size of dehydarted gels were also determined by mercury porosimetry.[46]

Mechanical Properties

The mechanical property of hydrogel is evaluated by deforming the gel by applying a static compressive loading. The gel resists the continuous load up to the breaking point where it fails. The resistance of the gel to the applied load is measured by the slope of the load deformation curve. The sharper the slope, higher the resistance and hence higher the modulus. The mechanical characteristics of a hydrogels are determined to improve the suitability of the hydrogel for the desirable application. For nonbiodegradable applications, it is essential that the carrier hydrogel matrix maintains physical and mechanical integrity to protect the drugs and other biomolecules from the harmful environments in the body such as extreme pH environment before it is released at the required site. The strength of the material can be increased by incorporating cross-linking agents, co-monomers, and increasing the degree of cross-linking. Cross-linking increases the stability of hydrogels in the wet state. But cross-linking should be at the optimum level because a higher degree of cross-linking leads to brittleness and less elasticity. Elasticity of the gel is important to give flexibility to the cross-linked chains and to facilitate movement of incorporated bioactive agent. Thus a compromise between mechanical strength and flexibility is necessary for appropriate use of these materials. The degree of swelling was also related to mechanical properties. The elastic modulus of hydrogels was determined using a mechanical testing system with load ranging from 0 to 2.5 kg and crosshead speed of 30 mm/min.

Rhelogical Properties

Hydrogels are viscoelastic materials. Their elastic and viscous properties are measured by storage modulus (G') and loss modulus (G''). Viscoelastic properties of swollen gels are measured by a rheometer. The structure of the hydrogels is related to elastic modulus G'[6,47] as per the equation

$$G' = \rho RT / \overline{M}_c \tag{3}$$

where ρ is the density, $\overline{M}_c$ average molecular weight between cross-links. As M_c increases elastic modulus decreases, which can be accounted for by an increase in cross-link density. The rheological properties are very much dependant on the types of structure (i.e., association, entanglement, cross-links) present in the system. Polymer solutions are essentially viscous at low frequencies (ω), tending to fit the scaling laws: $G' \sim \omega^2$ and $G'' \sim \omega$. At high ω, elasticity dominates ($G' > G''$). This corresponds to Maxwell-type behavior with a single relaxation time that may be determined from the crossover point and this relaxation time increases with concentration. For cross-linked microgel dispersions, G' and G'' are almost independent of oscillation frequency.[48] For hydrogels, which are highly cross-linked polymer networks, both G' and G'' are very high and are nearly parallel to each other. The G' and G'' values of a hydrogel is measured in the linear viscosity range. In the case of uncross-linked gel, the point where the G' and G'' intersects each other is known as crossover point and denotes the gel–sol transition temperature.

Swellability

Swellability can be described as a degree of water absorptivity of hydrogel. The water absorptivity of the material is defined as the % absorption expressed by the equation

$$\% \text{ water absorption} = (W_s - W_d/W_d) \times 100 \qquad (4)$$

where W_s and W_d are weights of swollen gel and dried gel, respectively. The swellability of hydrogels depend upon its hydrophilicity, cross-linking ratio (the ratio of moles of cross-linking agent to the moles of polymer repeating units) and chemical structure.[6] The higher the cross-linking ratio, the more is the cross-linking agent incorporated in the hydrogel structure. Highly cross-linked hydrogels have a tighter structure and swell less compared to the same hydrogels with lower cross-linking ratios. Cross-linking hinders the mobility of the polymer chain, hence lowering the swelling ratio. Hydrogels containing hydrophilic groups swell to a higher degree compared to those containing hydrophobic groups. The swelling kinetics of hydrogels can be classified as diffusion-controlled (Fickian) and relaxation-controlled (non-Fickian). When water diffusion into the hydrogel occurs much faster than the relaxation of the polymer chains, the swelling kinetics is diffusion-controlled. The mathematical analysis of the dynamics of swelling of hydrogels was studied by Peppas and Colombo.[47]

Thermal Studies

Differential scanning calorimetry

Differential scanning calorimetry (DSC) is used to determine the influence of cross-linker content and thermal cycling on the volume phase transition temperatures of the hydrogels. The samples were prepared by first allowing the hydrogel discs to swell for 24 hr in distilled water maintained at 15°C. The discs were then removed from water and any excess water on the surfaces was removed by blotting with filter paper. Finally, the discs were coarsely crushed using a mortar and pestle and approximately 5–15 mg samples of hydrogel were weighed into aluminium (Al) pans, which were sealed and then thermal analysis was carried out in a nitrogen gas flow. Empty closed Al pans were used as the reference cell. The temperature was raised from 10°C to 90°C at a rate of 10°C per minute.[49,50] DSC is also used to determine LCST of stimuli-sensitive polymers and to quantify the amount of free and bound water. The endotherm measured when warming the frozen gel represents the melting of the free water, and that value will yield the amount of free water in the hydrogel sample being tested. The bound water is then obtained by the difference of the measured total water content of the hydrogel test specimen, and the calculated free water content. The thermal stability of the hydrogel is evaluated by thermogravimetry.

TOXICOLOGICAL/BIOCOMPATIBILITY STUDIES

Cell culture methods are used to evaluate the toxicity of hydrogels. Three common assays used to assess the toxicity of hydrogels are extract dilution, direct contact, and agar diffusion. Most of the toxic problems associated with hydrogel carriers are residual monomers and initiators and oligomers that may leach out during applications. Hence knowledge on the toxicity of monomers, cross-linker, and initiators used is a must. The relationship between chemical structure and toxicity of monomers, and measures such as high conversion polymerization and thorough washing of the gels have to be followed.[51,52]

Cytotoxicity

Usually biocompatibility is measured in terms of cytotoxicity. It is measured by evaluating the cell viability and cell proliferation in the hydrogel that is in direct contact with the host environmental cells and are evaluated by MTT (tetrazolium salt, 3-[4,5-dimethylthiazol-2-yl]-2,5-diphenyl tetrazolium bromide) assay or visualization through microscopy assay.[52,53] Generally hydrogels are biocompatible and nonirritant in nature. The biocompatibility of the hydrogels is generally associated with the hydrophilic nature of the same, which helps in washing off the toxic and unreacted chemicals during synthesis.[54]

Contact Angle Measurement

There seems to be no relationship between wettability and thrombo resistance of a material. But contact angle is used for the characterization of the nonthrombogenic surfaces. Materials producing small contact angles with blood are considered to be compatible with blood, while materials producing large contact angles have poor blood compatibility.[55] Hydrophilicity, which is inversely related to contact angle, has a proportional relationship with cell attachment.[54–56] Hydrophilicity of the material influences the adsorption of blood proteins, which may promote cellular attachment, onto the material surface. The most common method for measuring the contact angle in a localized region of material surface is the sessile drop method (Fig. 3), which employs an optical arrangement. In this method, the angle between the baseline of the drop and the tangent from the intersection of drop boundary and baseline is measured using a goniometer.

Contact angle is also related with the interfacial free energy and is best expressed by Young's relationship[55,56] given by the Eq. 5

$$\gamma_{SL} = \gamma_{SV} - \gamma_{LV} \cos\theta \qquad (5)$$

where γ_{SL}, γ_{SV}, γ_{LV} and θ denote the solid–liquid interfacial energy, the solid–vapor interfacial energy, the liquid-vapor energy and the contact angle respectively. From

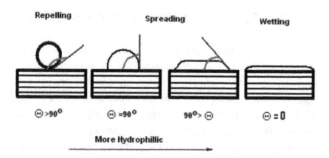

Fig. 3 Schematic representation of hydrophilicity vs. contact angle for blood/hydrogel system.

the Young's relationship, it can be observed that contact angle is directly proportional to the interfacial free energy, i.e., as the contact angle increases, the interfacial free energy also increases. No single *in vitro or in vivo* experiment gives a complete idea of biocompatibility but can reveal useful information. American Standard for Testing of Materials provides an *in vitro* method for the primary evaluation of biomaterials for blood compatibility.[57] The method involves the determination of the percentage of hemolysis of citrated goat blood in the presence of hydrogels. Calves are generally used to evaluate cardiovascular products because of relative ease of handling, convenience, and less complex hematological profile and the evaluation of these products should be done in baboons and/or rhesus monkeys whose hematological profiles closely resemble to the hematological profile of humans. Though *in vitro* tests provide useful information on the biocompatibility of the materials but to get regulatory approval for application on human subjects, rigorous *in vivo* tests in animals are necessary.

Biodegradability

Biodegradability of hydrogels depends upon its chemical structure. Natural polymers such as cellulose, starch, dextran, and gelatin based hydrogels are biodegradable.[58,59] Biodegradability is measured by monitoring the weight loss after incubating the hydrogel in simulated biological fluids (simulated gastric and intestine fluids) or phosphate-buffered saline at 37°C.[60]

APPLICATIONS

Health Care

Pharmaceutical: Drug delivery

Hydrogels have been used for the development of controlled delivery systems.[3,6,61–65] Diffusion attributed to Brownian motion is the main phenomena for diffusion of the drug out of the delivery systems to the surrounding medium. The hydrogel delivery systems for controlled release can be categorized into reservoir and matrix devices. The diffusion of the drug through the hydrogels

may be affected by the property (viz. pH sensitivity, light sensitivity, pressure sensitivity) of the hydrogel, which depends on its chemistry, and has been used successfully to design delivery systems which may release drug at a suitable environment. By controlling the degree of swelling, cross-linking density, pore size and degradation rate, delivery kinetics can be engineered according to the desired drug release schedule. The drug transport mechanisms can be determined by fitting the early time release data[3,6] to the empirical relationship given by the equation.

$$\frac{M_t}{M_\infty} = kt^n \tag{6}$$

where M_t and M_∞ are the amounts of drug released at a given time t and at infinite time respectively, and k and n are the constants (characteristic of drug–polymer system). The diffusional exponent 'n' is dependent on the geometry of the device as well as the physical mechanism of release. The release may be Fickian if $n = 0.5$ and anomalous transport if $n = 0.5–1$. $n = 1$ and >1 indicate case II and super case II transport systems respectively.

Polysaccharide hydrogels such as guar gum are attractive for colon-specific drug delivery because they can have high concentration of polysaccharides enzyme in the colon region of gastrointestinal tract.[66] Photopolymerized hydrogels are attractive for localized drug delivery because they can adhere and conform to targeted tissue when formed in situ. NVP-HEMA-MBAA copolymer, graft-modified chitosan with HEMA-IA, diisocyanate modified gelatin gels, and aloe vera gel were used as drug carriers for controlled release[60,67–69] of cancer and asthma drugs. Stimuli-responsive hydrogels, which respond to environmental stimuli by swelling/deswelling, are extensively used in controlled drug delivery.[70] Bioadhesive hydrogels are used as carriers to localize a delivery device within the body to enhance the drug absorption process in a site-specific manner. Bioadhesion is affected by the synergistic action of the biological environment, the properties of the polymeric-controlled release device, and the presence of the drug itself. The delivery site and the device design are dictated by the drug's molecular structure and its pharmacological behavior.[71,72] They are formed by in situ cross-linking through an enzymatic reaction and possesses excellent biocompatibility, mechanical strength, and excellent tissue adhesiveness.

Wound care

The use of hydrogels in the healing of wounds dates back to late 1970s or early 1980s. Hydrogel is nonadhesive and does not stick to the wound. Hydrogel dressings are seen as an essential component in many different types of wound care as it provides an ideal environment for both cleaning the wound, and allowing the body to rid itself of necrotic tissue. It keeps the wound moist, protects the wound against contamination, and capable of absorbing some exudate and promotes healing.[73,74] Hydrogel treatment promotes the development of new blood vessels and the regeneration of complex layers of skin, including hair follicles and the glands that produce skin oil and this lowers the chance for scarring. It protects the wound from desiccation. The moisture in the wound is also essential in pain management for the patient, and these dressings are very soothing and cooling. With their high moisture content they also help to prevent bacteria and oxygen from reaching the wound, providing a barrier for infections. In addition to this, hydrogels have been found to promote fibroblast proliferation by reducing the fluid loss from the wound surface and protect the wound from external noxae necessary for rapid wound healing. Hydrogels help in maintaining a microclimate for biosynthetic reactions on the wound surface necessary for cellular activities.[75,76] Since hydrogels are transparent it helps to monitor the wound healing without removing the wound dressing. The process of angiogenesis can be initiated by using semiocclusive hydrogel dressings. Angiogenesis of the wound ensures the growth of granulation tissue by maintaining adequate supply of oxygen and nutrients to the wound surface.

Baby diaper and sanitory napkins

Hydrogels (SAP) are finding extensive applications in baby and adult diapers and sanitary napkins to absorb biofluids. For diaper applications the swelling should be high and fast. The commercially available diaper and napkins consist two parts, viz., absorbent core and envelop. The absorbent core contains fluff pulp, SAP, and tissue or nonwoven layer wrap. The mostly used SAP in absorbent core of diaper and napkins were based on cross-linked sodium poly(acrylate). Alla, Sen, and El-Naggar prepared tara gum/acrylic acid (TG/AAc) based SAP by gamma irradiation, in the presence of MBAA as a cross-linking agent[77] for diaper applications.

BIOMEDICAL

Hydrogels have been used as barriers to improve the healing response in tissue injury in order to prevent restenosis or thrombosis due to postoperative adhesion formation.[78–80] Forming a thin hydrogel layer intravascular via interfacial photopolymerization has been reported to prevent restenosis by reducing intima thickening and thrombosis[80,81] as it provides a barrier to prevent platelets, coagulation factors, and plasma proteins from contacting the vascular wall. The contact of these factors to vessel walls stimulates smooth muscle cell proliferation, migration, and matrix synthesis events that lead to restenosis. Hydrogel barriers have additionally been used to prevent postoperative adhesion formation. In one example, poly(ethylene glycol-co-lactic acid) diacrylate hydrogels were formed by bulk photopolymerization on intraperitoneal surfaces. These hydrogel barriers functioned to prevent fibrin deposition and fibroblast attachment at the tissue surface.[78,79]

Tissue Engineering: Scaffold and Body implants

Tissue engineering (TE) is the generation, regeneration, augmentation, or limitation of structure and function of living tissues by the applications of scientific and engineering principles. The principles of TE have been used extensively to restore the function of a traumatized/malfunctioning tissues or organs.[82] Due to the structural similarity of hydrogel to the extracellular matrix of many tissues[83] these are used as scaffolds in regenerative medicine to provide structural integrity and for cellular organization and morphogenic guidance, to serve as tissue barriers and bioadhesives, act as drug depots, deliver bioactive agents that encourage the natural reparative process, and encapsulate and deliver cells.[23,84] Hydrogels induce beneficial tissue responses based on their physical and chemical properties. Due to their high water content they are quite compatible with cell and enter into specific or nonspecific binding with cell receptors. Protein and polysaccharide hydrogels contain ligands to bind with cell and hence form a useful scaffold for cell incorporation. Synthetic hydrogels can be modified to improve cell attachment. Resorbable hydrogels are also being used in TE. Collagen-coated tissue culture inserts are used for growing three-dimensional corneal implant, tracheal gland cells etc.[85]

Porous scaffolding (e.g., filter, swatch of nylon, transwell, biodegradable microcarrier) coated with fibrillar collagen, ideally type III collagen mixed with fibronectin or with matrigel, are used for the culture of the normal mature liver cells (polyploidy liver cells). Various natural biomaterials, including alginate, hyaluronic acid, collagen, fibrin, and agarose, and synthetic polymers, such as PEG and PEG fumigate, have been employed for the preparation of hydrogel scaffolds.[86] Hydrogels that degrade in a manner which matches the rate of new bone formation may be optimal for bone TE. For polymeric degradation products, kidney filtration molecular weight cutoff and possible accumulation of high molecular weight degradation products in the reticule endothelial system have to be considered when applying the polymer scaffolds. Hydrogel-forming polymers can be tailored to exhibit biochemical, cellular, and physical stimuli that guide cellular processes, including cell migration, proliferation and differentiation. Biologically active additives have been added to the hydrogels, including bone morphogenic protein and fibroblast growth factor. Attempts have been made to covalently incorporate cell membrane receptor peptide ligands within the hydrogel matrix to stimulate adhesion, spreading and growth of cells.

Cell-Based TE Approach

Hydrogels can be utilized as cell immobilization matrices to produce various biological products, such as proteins, and to re-create the environments for the damaged or lost tissues. Scaffolds designed to encapsulate cells must be capable of being gelled without damaging the cells, and must be nontoxic to the loaded cells. These hydrogels should allow permeation of biological medium nutrients to promote cell proliferation and/or induce cell differentiation. Hydrogels are applied for bone regeneration and that the modification of hydrogels with bioactive molecules or cell-based approaches resulted in significant increases in new bone formation.[87,88] Cardiomyocytes and stem cells are used to regenerate cardiac tissue and restore heart functions after myocardial infarction. However, the injection of cells directly into the infarcted area involves the problems of low cell retention and engraftment rate. Use of bioactive and biocompatible hydrogel as support matrices can hold cells at the infarcted area initially and further provide support for cell survival and functioning and hence for cardiac tissue regeneration. Both natural and synthetic hydrogels are suitable for cardiac TE because their soft and visco elastic nature mimic the native tissue. Collagen, gelatin, laminin, matrigel, hyaluronic acid (hyaluronan), alginate, and chitosan are typical natural hydrogels having similar or even identical structures to the molecules in biological organisms, thereby reducing the possibility of immune response when implanted *in vivo*. Synthetic hydrogels used for cardiac TE include PEG, poly(lactide-*co*-glycolic acid), and poly(acrylamide). PEG gel has been widely used as a supporting matrix in almost every field of TE (nerve, cartilage, liver, pancreas, bladder, skin) because its low protein adsorption and inert surface reduce the inflammation after implantation. However, low protein affinity is not beneficial for cell adhesion. Methods like conjugating cell adhesive peptides or proteins and incorporating growth factors are used to increase cell adhesion.[89–91] A promising strategy for cardiac tissue engineering lies in hydrogel-based cell therapy. The viscous hydrogel holds the cells in the target place during injection. The gel itself provides mechanical support to the weakened heart wall. Meanwhile, the gel environment may allow the delivered cells to survive and differentiate into cardiomyocytes to regenerate cardiac muscle. To facilitate cell survival, growth, and differentiation, biochemicals can be codelivered with hydrogels.[92,93]

Embolization of Blood Vessels

Therapeutic embolization is the selective transcatheter blockage of blood vessels or diseased vascular structures. Currently the embolization materials under clinical use are permanent. But in some clinical situations temporary embolization is also desirable. Schwarz et al.[94] synthesized degradable hydroxyethyl acrylate (HEA) hydrogel based microspheres and used it for embolization of canine renal arteries and rabbit central auricular arteries, and compared with degradable human serum and albumin (HSA) microspheres. Both HSA and HEA microspheres revealed temporary occlusions. These

micro spheres were recanalized at 1 and 3 weeks, respectively, and led to tissue infarction.

Terumo Medical Corp, USA has launched a hydrogel-coated Pt coil (AZUR TR) as a peripheral hydrocoil embolization system for the occlusion of blood vessels, vascular malformations, and aneurysms. AZUR's hydrogel coating undergoes limited expansion within the first 3 minutes, and fully expands within 20 minutes, resulting in greater filling and mechanical stability with fewer coils compared to platinum coils of the same size. Horák et al.[95,96] had synthesized spherical particles of poly(HEMA) hydrogel and successfully employed it in the embolization of arteriovenous anastomoses, in the suppression of pulmonary haemorrhage and haemoptysis and in the occlusion of some other arteries. Poly(HEMA) embolization particles with enhanced haemostatic properties were prepared by bulk or suspension polymerization of HEMA followed by particle soaking in ethamsylate solution. Alginic acid hydrogel cross-linked with radioactive Ho^{3+} has been used in the antitumor therapy that can be monitored by magnetic resonance imaging (MRI).[81]

Diagnostic Devices

Medical devices

A hydrogel-based microfluidic device capable of generating a steady and long-term linear chemical concentration gradient with no through flow in a microfluidic channel has been developed for successfully monitoring the chemotactic responses of wild-type *Escherichia coli*. This is mainly attributed to the diffusability of proteins and nutrients essential for cell survival into hydrogel. It has the potential of responding to chemical stimuli independently of mechanical stimuli. This device will also be useful in controlling the chemical and mechanical environment during the formation of tissue-engineered constructs.[97–99] Microfluidic mixers are an important component in microfluidic devices. A micromixer that can control mixing with responsive hydrogel actuators to modulate mixing between two adjacent fluids dependant on the chemistries of the fluid has been developed using the pH-responsive hydrogels swelling or contracting under different stimuli, which alters the mixing between the two fluids.

A device that can act as both a microdispensing device and a storable pressure source incorporates a responsive hydrogel valve fluidically isolated from the rest of the device. An array of responsive hydrogel swells to provide the driving pressure to power the dispensing device and power source. Responsive hydrogels have the unique ability to directly transduce a chemical signal into mechanical work and therefore are advantageous for applications where bulky power supplies and control would impede device performance.[100] Interpenetrating network of conducting polymers and hydrogels are used to fabricate bioelectrodes as promising biomaterials for neural interfaces as it minimizes the interfacial impedance between the tissue and the electrode. The premise underlying the choice of these two materials as an electrode coating is that the conducting polymer will act to carry charge whereas the hydrogel will modulate the mechanical properties and enhance the drug-carrying capacity and tissue interfacing of the coatings.

Opthalmology

Hydrogels are widely used to fabricate contact and intraocular lenses (IOLs) because hydrogels render physical, biological, and optical features that make them ideal and suitable for use in foldable IOLs. These characteristics include less dysphotopsia, minimized problems associated with glare, external and internal reflections, and other undesirable visual phenomena. These IOLs have less effect on the blood–aqueous barrier and may be a better option for uveitic and diabetic patients. They possess good optical clarity and resistance to damage during insertion. Hydrophilic acrylic IOLs resist fold marks and forceps damage in contrast to silicone or hydrophobic IOLs. They are less susceptible to biocontamination. Injectable hydrogels are also used as IOLs in ophthalmology.[100–102]

CONCLUSION AND FUTURE DEVELOPMENTS

Hydrogels are hydrophilic biocompatible polymers having intrinsic similarity to soft tissue. They are used extensively in drug delivery, tissue engineering, intraocular lenses, and wound dressing. In this entry, an attempt has been made to give the necessary fundamentals of hydrogel classification, synthesis, fabrication, structure–property relation, biological and medical applications to explain the growing importance of this potential health care material. The use of hydrogels in regenerative medicine has been reviewed briefly covering the most essential concepts. Future advances in TE, and other allied fields, will need considerable integration to realize the true prospective clinical potentials of regenerative medicine. Newer frontiers in tissue engineering may emerge by increased focus in stem cell science, angiogenesis, and molecular biology. Hydrogels synthesized from various newly synthesized or identified nontoxic monomers/cross-linkers, preformed biocompatible polymers, using novel synthetic techniques, may meet the desirable medical requirements. There is ample scope to design synthesis and fabricate new hydrogels fulfilling specific functions for specific needs. Manipulating chemical composition, cross-linking method, or morphological features or fabrication methods may result in new intelligent and smart hydrogels. Only scant reports are available in the literature on hydrogel blends, composites, and interpenetrating networks as smart materials. Investing significant effort in these directions may yield promising biomaterials in the form of laminates or coatings to blends with synthetic hydrophobic polymers with enhanced mechanical properties without sacrificing optical transparency. By proper choice of statistical copolymers with balanced hydrophobicity or hydrophilicity as a matrix, desirable release rates and dissolution profiles can be achieved for controlled oral drug delivery. The significant developments in polymer science and technology have resulted in the synthesis of various stimuli (pH, temperature) sensitive hydrogel carrier for colon-specific delivery of proteins and targeted release of chemotherapeutic agents to tumors. Synthesis of hydrogels that will specifically interact with selective biomolecules may open up newer frontiers in biosensor research. Much research focus on biodegradable hydrogels that are finding increasing use as smart biomaterials may help in the development of nano biotechnology products, which have tremendous applications in controlled and targeted drug delivery. In-depth investigation on the structure–property relationships in hydrogels may help in the synthesis of potential and novel hydrogels hitherto not realized for varied applications in biological and medical fields. The good memory of DNA-based hydrogels[103] may be exploited as carriers for controlled and targeted drug delivery.

REFERENCES

1. Wichterle, O.; Lím, D. Hydrophilic gels for biological use. Nature 1960, 185 (4706), 117–118.
2. Kopeček, J.; Yang, J. Hydrogels as smart biomaterials. Polym. Int. 2007, 56 (9), 1078–1098.
3. Hoffman, A.S. Hydrogels for biomedical applications. Adv. Drug Deliv. Rev. 2012, 64, Suppl. 18–23.
4. Ruel-Gariépy, E.; Leroux, J.C. In situ-forming hydrogels—Review of temperature-sensitive systems. Eur. J. Pharm. Biopharm. 2004, 58 (2), 409–426.
5. You, J.O.; Auguste, D.T. Conductive, physiologically responsive hydrogels. Langmuir 2010, 26 (7), 4607–4612.
6. Peppas, N.A. Hydrogels in Medicine and Pharmacy, Vols. I-III; CRC Press: Boca Raton, 1987.
7. Peppas, N.A.; Bures, P.; Leobandung, W.; Ichikawa, H. Hydrogels in pharmaceutical formulations. Eur. J. Pharm. Biopharm. 2000, 50 (1), 27–46.
8. Ulijn, R.V.; Bibi, N.; Jayawarna, V.; Thornton, P.D.; Todd, S.J.; Mart, R.J.; Smith, A.M.; Gough, J.E. Bioresponsive hydrogels. Mater. Today 2007, 10 (4), 40–48.
9. Gupta, A.K.; Siddiqui, A.W. Environmental responsive hydrogels: A novel approach in drug delivery system. J. Drug Deliv. Ther. 2012, 2 (1), 81–88.
10. Kashyap, N.; Kumar, N.; Ravi Kumar, M.N.V. Hydrogels for pharmaceutical and biomedical applications. Crit. Rev. Ther. Drug Carrier Syst. 2005, 22 (2), 107–150.
11. Omidian, H.; Rocca, J.G.; Park, K. Advances in superporous hydrogels. J. Control. Release 2005, 102 (1), 3–12.
12. Schacht, E.H. Polymer chemistry and hydrogel systems. J. Phys. Conf. Ser. 2004, 3, 22–28.
13. Brannon-Peppas, L.; Harland, R.S.; Eds. Absorbent Polymer Technology; Elsevier: Amsterdam, 1990, pp. 147–169.

14. Bell, C.L.; Peppas, N.A. Biomedical membranes from hydrogels and interpolymer complexes. In *Biopolymers II*; Peppas, N., Langer, R., Eds.; Springer: Berlin, 1995; Vol. 122, 125–175.

15. Peppas, N.A. Physiologically responsive hydrogels. J. Bioact. Compat. Polym. **1991**, *6* (3), 241–246.

16. Hu, Z.; Chen, Y.; Wang, C.; Zheng, Y.; Li, Y. Polymer gels with engineered environmentally responsive surface patterns. Nature **1998**, *393* (6681), 149–152.

17. Schacht, E.; Bogdanov, B.; Bulcke, A.V.D.; De Rooze, N. Hydrogels prepared by crosslinking of gelatin with dextran dialdehyde. React. Funct. Polym. **1997**, *33* (2–3), 109–116.

18. Draye, J.P.; Delaey, B.; Van de Voorde, A.; Van Den Bulcke, A.; Bogdanov, B.; Schacht, E. *In vitro* release characteristics of bioactive molecules from dextran dialdehyde cross-linked gelatin hydrogel films. Biomaterials **1998**, *19* (1–3), 99–107.

19. Chao, G.T.; Qian, Z.Y.; Huang, M.J.; Kan, B.; Gu, Y.C.; Gong, C.Y.; Yang, J.L.; Wang, K.; Dai, M.; Li, X.Y.; Gou, M.L.; Tu, M.J.; Wei, Y.Q. Synthesis, characterization, and hydrolytic degradation behavior of a novel biodegradable pH-sensitive hydrogel based on polycaprolactone, methacrylic acid, and poly(ethylene glycol). J. Biomed. Mater. Res. **2008**, *85A* (1), 36–46.

20. Tanaka, Y.; Gong, J.P.; Osada, Y. Novel hydrogels with excellent mechanical performance. Prog. Polym. Sci. **2005**, *30* (1), 1–9.

21. Panda, P.; Ali, S.; Lo, E.; Chung, B.G.; Hatton, T.A.; Khademhosseini, A.; Doyle, P.S. Stop-flow lithography to generate cell-laden microgel particles. Lab Chip **2008**, *8* (7), 1056–1061.

22. Bracher, P.J.; Gupta, M.; Whitesides, G.M. Shaped films of ionotropic hydrogels fabricated using templates of patterned paper. Adv. Mater. **2009**, *21* (4), 445–450.

23. Slaughter, B.V.; Khurshid, S.S.; Fisher, O.Z.; Khademhosseini, A.; Peppas, N.A. Hydrogels in regenerative medicine. Adv. Mater. **2009**, *21* (32–33), 3307–3329.

24. Ling, Y.; Rubin, J.; Deng, Y.; Huang, C.; Demirci, U.; Karp, J.M.; Khademhosseini, A. A cell-laden microfluidic hydrogel. Lab Chip **2007**, *7* (6), 756–762.

25. Cabodi, M.; Choi, N.W.; Gleghorn, J.P.; Lee, C.S.D.; Bonassar, L.J.; Stroock, A.D. A microfluidic biomaterial. J. Am. Chem. Soc. **2005**, *127* (40), 13788–13789.

26. Almdal, K.; Dyre, J.; Hvidt, S.; Kramer, O. What is a 'gel'? Makromol. Chem. Macromol. Symp. **1993**, *76* (1), 49–51.

27. Flory, P.J. *Polymer Chemistry*; Cornell University Press: New York, 1953; pp. 495–507.

28. Sen, M.; Yakar, A.; Güven, O. Determination of average molecular weight between cross-links (Mc) from swelling behaviours of diprotic acid-containing hydrogels. Polymer **1999**, *40* (11), 2969–2974.

29. Lowman, A.M.; Dziubla, T.D.; Bures, P.; Peppas, N.A. Structural and dynamic response of neutral and intelligent networks in biomedical environments. In *Advances in Chemical Engineering*; Peppas, A., Sefton, M.V., Eds.; Academic Press: New York, 2004; Vol. 29, pp 75–130.

30. Haraguchi, K.; Takehisa, T. Nanocomposite hydrogels: A unique organic–inorganic network structure with extraordinary mechanical, optical, and swelling/de-swelling properties. Adv. Mater. **2002**, *14* (16), 1120–1124.

31. Okay, O.; Oppermann, W. Poly(acrylamide)–clay nanocomposite hydrogels: Rheological and light scattering characterization. Macromolecules **2007**, *40* (9), 3378–3387.

32. Barbucci, R. *Hydrogels: Biological Properties and Applications*; Springer: New York, 2009.

33. Shibayama, M. Structure-mechanical property relationship of tough hydrogels. Soft Matter **2012**, *8* (31), 8030–8038.

34. Lin-Gibson, S.; Jones, R.L.; Washburn, N.R.; Horkay, F. Structure–property relationships of photopolymerizable poly(ethylene glycol dimethacrylate) hydrogels. Macromolecules **2005**, *38* (7), 2897–2902.

35. Peppas, N.A.; Hilt, J.Z.; Khademhosseini, A.; Langer, R. Hydrogels in biology and medicine: From molecular principles to bionanotechnology. Adv. Mater. **2006**, *18* (11), 1345–1360.

36. Kuckling, D.; Harmon, M.E.; Frank, C.W. Photo-cross-linkable PNIPAAm copolymers. 1. Synthesis and characterization of constrained temperature-responsive hydrogel layers. Macromolecules **2002**, *35* (16), 6377–6383.

37. Knoll, W. Interfaces and thin films as seen by bound electromagnetic waves. Annu. Rev. Phys. Chem. **1998**, *49*, 569–638.

38. Kuckling, D. Responsive hydrogel layers—From synthesis to applications. Colloid. Polym. Sci. **2009**, *287* (8), 881–891.

39. Beines, P.W.; Klosterkamp, I.; Menges, B.; Jonas, U.; Knoll, W. Responsive Thin hydrogel layers from photo-cross-linkable poly(N-isopropylacrylamide) terpolymers. Langmuir **2007**, *23* (4), 2231–2238.

40. Tamirisa, P.A.; Hess, D.W. Water and moisture uptake by plasma polymerized thermoresponsive hydrogel films. Macromolecules **2006**, *39* (20), 7092–7097.

41. Kizilay, M.Y.; Okay, O. Effect of hydrolysis on spatial inhomogeneity in poly(acrylamide) gels of various cross-link densities. Polymer **2003**, *44* (18), 5239–5250.

42. Chirila, T.V.; Chen, Y.C.; Griffin, B.J.; Constable, I.J. Hydrophilic sponges based on 2-hydroxyethyl methacrylate. I. effect of monomer mixture composition on the pore size. Polym. Int. **1993**, *32* (3), 221–232.

43. Aouada, F.A.; de Moura, M.R.; Fernandes, P.R.G.; Rubira, A.F.; Muniz, E.C. Optical and morphological characterization of polyacrylamide hydrogel and liquid crystal systems. Eur. Polym. J. **2005**, *41* (9), 2134–2141.

44. El Fray, M.; Pilaszkiewicz, A.; Swieszkowski, W.; Kurzydlowski, K.J. Morphology assessment of chemically modified cryostructured poly(vinyl alcohol) hydrogel. Eur. Polym. J. **2007**, *43*, 2035–2040.

45. Horák, D.; Lednický, F.; Bleha, M. Effect of inert components on the porous structure of 2-hydroxyethyl methacrylate-ethylene dimethacrylate copolymers. Polymer **1996**, *37* (19), 4243–4249.

46. Gemeinhart, R.A.; Park, H.; Park, K. Pore structure of superporous hydrogels. Polym. Adv. Technol. **2000**, *11* (8–12), 617–625.

47. Peppas, N.A.; Colombo, P. Analysis of drug release behavior from swellable polymer carriers using the dimensionality index. J. Control. Release **1997**, *45* (1), 35–40.

48. Omari, A.; Tabary, R.; Rousseau, D.; Calderon, F.L.; Monteil, J.; Chauveteau, G. Soft water-soluble microgel dispersions: Structure and rheology. J. Colloid Interface Sci. **2006**, *302* (2), 537–546.

Gene–Hydrogels

49. Li, S.K.; D'Emanuele, A. Effect of thermal cycling on the properties of thermoresponsive poly(*N*-isopropylacrylamide) hydrogels. Int. J. Pharm. **2003**, *267* (1–2), 27–34.

50. Ma, Z.; Nelson, D.M.; Hong, Y.; Wagner, W.R. Thermally responsive injectable hydrogel incorporating methacrylate-polylactide for hydrolytic lability. Biomacromolecules **2010**, *11* (7), 1873–1881.

51. Yoshii, E. Cytotoxic effects of acrylates and methacrylates: Relationships of monomer structures and cytotoxicity. J. Biomed. Mater. Res. **1997**, *37* (4), 517–524.

52. Schneiderka, P.; Roosova, M.; Jojkova, K.; Prokes, J.; Labsky, J.; Vacik, J. Toxicity of monomers HEMA, DEGMA and AEMA, used in synthesis of hydrogels for medical applications. Sb. Lek. **1996**, *97* (3), 351–367.

53. Yin, H.; Gong, C.; Shi, S.; Liu, X.; Wei, Y.; Qian, Z. Toxicity evaluation of biodegradable and thermosensitive PEG-PCL-PEG hydrogel as a potential in situ sustained ophthalmic drug delivery system. J. Biomed. Mater. Res. B **2010**, *92* (1), 129–137.

54. Horak, D.; Cervinka, M.; Puza, V. Hydrogels in endovascular embolization. VI. Toxicity tests of poly(2-hydroxyethyl methacrylate) particles on cell cultures. Biomaterials **1997**, *18* (20), 1355–1359.

55. De Groot, C.J.; Van Luyn, M.J.A.; Van Dijk-Wolthuis, W.N.E.; Cadée, J.A.; Plantinga, J.A.; Otter, W.D.; Hennink, W.E. *In vitro* biocompatibility of biodegradable dextran-based hydrogels tested with human fibroblasts. Biomaterials **2001**, *22* (11), 1197–1203.

56. van Wachem, P.B.; Hogt, A.H.; Beugeling, T.; Feijen, J.; Bantjes, A.; Detmers, J.P.; van Aken, W.G. Adhesion of cultured human endothelial cells onto methacrylate polymers with varying surface wettability and charge. Biomaterials **1987**, *8* (5), 323–328.

57. Roy Chowdhury, S.K.; Mishra, A.; Pradhan, B.; Saha, D. Wear characteristic and biocompatibility of some polymer composite acetabular cups. Wear **2004**, *256* (11–12), 1026–1036.

58. Sannino, A.; Demitri, C.; Madaghiele, M. Biodegradable cellulose-based hydrogels: Design and applications. Materials **2009**, *2* (2), 353–373.

59. van Dijk-Wolthuis, W.N.E.; Hoogeboom, J.A.M.; van Steenbergen, M.J.; Tsang, S.K.Y.; Hennink, W.E. Degradation and release behavior of dextran-based hydrogels. Macromolecules **1997**, *30* (16), 4639–4645.

60. Subramanian, K.; Vijayakumar, V. Evaluation of isophorone diisocyanate crosslinked gelatin as a carrier for controlled delivery of drugs. Polym. Bull. **2012**, 1–21. doi: 10.1007/s00289-012-0821-z

61. Panyam, J.; Labhasetwar, V. Biodegradable nanoparticles for drug and gene delivery to cells and tissue. Adv. Drug Deliv. Rev. **2003**, *55* (3), 329–347.

62. Seliktar, D. Designing cell-compatible hydrogels for biomedical applications. Science **2012**, *336* (6085), 1124–1128.

63. Li, Y.; Rodrigues, J.; Tomas, H. Injectable and biodegradable hydrogels: Gelation, biodegradation and biomedical applications. Chem. Soc. Rev. **2012**, *41* (6), 2193–2221.

64. Cabral, J.; Moratti, S.C. Hydrogels for biomedical applications. Future Med. Chem. **2011**, *3* (15), 1877–1888.

65. Peppas, N.A. Hydrogels and drug delivery. Curr. Opin. Colloid Interface Sci. **1997**, *2* (5), 531–537.

66. Singh, B.; Sharma, N.; Chauhan, N. Synthesis, characterization and swelling studies of pH responsive psyllium and methacrylamide based hydrogels for the use in colon specific drug delivery. Carbohydr. Polym. **2007**, *69* (4), 631–643.

67. Subramanian, K.; Vijayakumar, V. Synthesis and evaluation of chitosan-graft-poly (2-hydroxyethyl methacrylate-*co*-itaconic acid) as a drug carrier for controlled release of tramadol hydrochloride. Saudi Pharm. J. **2012**, *20* (3), 263–271.

68. Subramanian, K.; Narmadha, S.; Vishnupriya, U.; Vijayakumar, V. Release characteristics of aspirin and paracetamol drugs from tablets with aloe vera gel powder as a drug carrier. Drug Invent. Today **2010**, *2* (9), 424–428.

69. Subramanian, K.; Vijayakumar, V. Characterization of crosslinked poly (2-Hydroxyethyl methacrylate-*co*-N-vinyl-2-pyrrolidone) as a carrier for controlled drug delivery. J. Pharm. Res. **2011**, *4* (3), 743–747.

70. Dumitriu, R.P.; Mitchell, G.R.; Vasile, C. Multi-responsive hydrogels based on N-isopropylacrylamide and sodium alginate. Polym. Int. **2011**, *60* (2), 222–233.

71. Peppas, N.A.; Sahlin, J.J. Hydrogels as mucoadhesive and bioadhesive materials: A review. Biomaterials **1996**, *17* (16), 1553–1561.

72. Collaud, S.; Warloe, T.; Jordan, O.; Gurny, R.; Lange, N. Clinical evaluation of bioadhesive hydrogels for topical delivery of hexylaminolevulinate to Barrett's esophagus. J. Control. Release **2007**, *123* (3), 203–210.

73. Flanagan, M. The efficacy of a hydrogel in the treatment of wounds with non-viable tissue. J. Wound Care **1995**, *4* (6), 264–267.

74. Bale, S.; Banks, V.; Haglestein, S.; Harding, K.G.A comparison of two amorphous hydrogels in the debridement of pressure sores. J. Wound. Care **1998**, *7* (2), 65–68.

75. Witthayaprapakorn, C. Design and preparation of synthetic hydrogels via photopolymerisation for biomedical use as wound dressings. Procedia Eng. **2011**, *8*, 286–291.

76. Jayakumar, R.; Prabaharan, M.; Sudheesh Kumar, P.T.; Nair, S.V.; Tamura, H. Biomaterials based on chitin and chitosan in wound dressing applications. Biotechnol. Adv. **2011**, *29* (3), 322–337.

77. Alla, S.G.A.; Sen, M.; El-Naggar, A.W.M. Swelling and mechanical properties of superabsorbent hydrogels based on Tara gum/acrylic acid synthesized by gamma radiation. Carbohydr. Polym. **2012**, *89* (2), 478–485.

78. Hill-West, J.L.; Dunn, R.C.; Hubbell, J.A. Local release of fibrinolytic agents for adhesion prevention. J. Surg. Res. **1995**, *59* (6), 759–763.

79. Sawhney, A.S.; Pathak, C.P.; van Rensburg, J.J.; Dunn, R.C.; Hubbell, J.A. Optimization of photopolymerized bioerodible hydrogel properties for adhesion prevention. J. Biomed. Mater. Res. **1994**, *28* (7), 831–838.

80. Hill-West, J.L.; Chowdhury, S.M.; Slepian, M.J.; Hubbell, J.A. Inhibition of thrombosis and intimal thickening by in situ photopolymerization of thin hydrogel barriers. Proc. Natl. Acad. Sci. U.S.A. **1994**, *91* (13), 5967–5971.

81. West, J.L.; Hubbell, J.A. Separation of the arterial wall from blood contact using hydrogel barriers reduces intimal thickening after balloon injury in the rat: The roles of medial and luminal factors in arterial healing. Proc. Natl. Acad. Sci. U.S.A. **1996**, *93* (23), 13188–13193.

Gene–Hydrogels

82. Hench, L.L.; Jones, J.R.; Institute of Materials, Minerals and Mining: *Biomaterials, Artificial Organs and Tissue Engineering*; Woodhead Publishing Ltd: Cambridge, England, 2005.

83. Tan, H.; Marra, K.G. Injectable, Biodegradable hydrogels for tissue engineering applications. Materials **2010**, *3* (3), 1746–1767.

84. Lee, K.Y.; Mooney, D.J. Hydrogels for tissue engineering. Chem. Rev. **2001**, *101* (7), 1869–1880.

85. Atala, A.; Lanza, R.P. *Methods of Tissue Engineering*; Academic Press: U.S.A., 2006.

86. Ahn, H.H.; Kim, K.S.; Lee, J.H.; Lee, J.Y.; Kim, B.S.; Lee, I.W.; Chun, H.J.; Kim, J.H.; Lee, H.B.; Kim, M.S. *In vivo* osteogenic differentiation of human adipose-derived stem cells in an injectable in situ-forming gel scaffold. Tissue Eng. A **2009**, *15* (7), 1821–1832.

87. Drury, J.L.; Mooney, D.J. Hydrogels for tissue engineering: Scaffold design variables and applications. Biomaterials **2003**, *24* (24), 4337–4351.

88. Lee, K.Y.; Alsberg, E.; Mooney, D.J. Degradable and injectable poly(aldehyde guluronate) hydrogels for bone tissue engineering. J. Biomed. Mater. Res. **2001**, *56* (2), 228–233.

89. Bhana, B.; Iyer, R.K.; Chen, W.L.K.; Zhao, R.; Sider, K.L.; Likhitpanichkul, M.; Simmons, C.A.; Radisic, M. Influence of substrate stiffness on the phenotype of heart cells. Biotechnol. Bioeng. **2010**, *105* (6), 1148–1160.

90. Motlagh, D.; Senyo, S.E.; Desai, T.A.; Russell, B. Micro-textured substrata alter gene expression, protein localization and the shape of cardiac myocytes. Biomaterials **2003**, *24* (14), 2463–2476.

91. Wang, T.; Jiang, X.J.; Tang, Q.Z.; Li, X.Y.; Lin, T.; Wu, D.Q.; Zhang, X.Z.; Okello, E. Bone marrow stem cells implantation with α-cyclodextrin/MPEG–PCL–MPEG hydrogel improves cardiac function after myocardial infarction. Acta Biomater. **2009**, *5* (8), 2939–2944.

92. Wu, J.; Zeng, F.; Huang, X.P.; Chung, J.C.Y.; Konecny, F.; Weisel, R.D.; Li, R.K. Infarct stabilization and cardiac repair with a VEGF-conjugated, injectable hydrogel. Biomaterials **2011**, *32* (2), 579–586.

93. Singelyn, J.M.; DeQuach, J.A.; Seif-Naraghi, S.B.; Littlefield, R.B.; Schup-Magoffin, P.J.; Christman, K.L. Naturally derived myocardial matrix as an injectable scaffold for cardiac tissue engineering. Biomaterials **2009**, *30* (29), 5409–5416.

94. Schwarz, A.; Zhang, H.; Metcalfe, A.; Salazkin, I.; Raymond, J. Transcatheter embolization using degradable crosslinked hydrogels. Biomaterials **2004**, *25* (21), 5209–5215.

95. Horák, D.; Šeksvec, F.; Kálal, J.; Adamyan, A.A.; Volynskii, Y.D.; Voronkova, O.S.; Kokov, L.S.; Gumargalieva, K.Z. Hydrogels in endovascular embolization. II. Clinical use of spherical particles. Biomaterials **1986**, *7* (6), 467–470.

96. Horák, D.; Galibin, I.; Adamyan, A.; Sitnikov, A.; Dan, V.; Titova, M.; Shafranov, V.; Isakov, Y.; Gumargalieva, K.; Vinokurova, T. Poly(2-hydroxyethyl methacrylate) emboli with increased haemostatic effect for correction of haemorrhage of complex origin in endovascular surgery of children. J. Mater. Sci. Mater. Med. **2008**, *19* (3), 1265–1274.

97. Zielhuis, S.W.; Seppenwoolde, J.H.; Bakker, C.J.G.; Jahnz, U.; Zonnenberg, B.A.; van het Schip, A.D.; Hennink, W.E.; Nijsen, J.F.W. Characterization of holmium loaded alginate microspheres for multimodality imaging and therapeutic applications. J. Biomed. Mater. Res. A **2007**, *82A* (4), 892–898.

98. Cheng, S.Y.; Heilman, S.; Wasserman, M.; Archer, S.; Shuler, M.L.; Wu, M. A hydrogel-based microfluidic device for the studies of directed cell migration. Lab Chip **2007**, *7* (6), 763–769.

99. Deligkaris, K.; Tadele, T.S.; Olthuis, W.; van den Berg, A. Hydrogel-based devices for biomedical applications. Sens. Actuators B **2010**, *147* (2), 765–774.

100. Prettyman, J.B.; Eddington, D.T. Leveraging stimuli responsive hydrogels for on/off control of mixing. Sens. Actuators B **2011**, *157* (2), 722–726.

101. Espandar, L.; Sikder, S.; Moshirfar, M. Softec HD hydrophilic acrylic intraocular lens: Biocompatibility and precision. Clin. Ophthalmol. **2011**, *5*, 65–70.

102. de Groot, J.H.; van Beijma, F.J.; Haitjema, H.J.; Dillingham, K.A.; Hodd, K.A.; Koopmans, S.A.; Norrby, S. Injectable intraocular lens materials based upon hydrogels. Biomacromolecules **2001**, *2* (3), 628–634.

103. Lee, J.B.; Peng, S.; Yang, D.; Roh, Y.H.; Funabashi, H.; Park, N.; Edward J. Rice, E.J.; Chen, L.; Long, R.; Wu, M.; Luo, D. A mechanical metamaterial made from a DNA hydrogel. Nat. Nanotechnol. **2012**, *7* (12), 816–820. doi: 10.1038/nnano.2012.211

Gene–Hydrogels

Hydrogels: Multi-Responsive Biomedical Devices

Francesca Iemma
Giuseppe Cirillo
Umile Gianfranco Spizzirri
Department of Pharmacy, Health, and Nutritional Sciences, University of Calabria, Rende, Italy

Abstract

Pharmaceutical and biomedical application of hydrophilic materials has emerged as one of the most significant trends in the area of nanotechnology. "Intelligent" polymeric devices able to undergo morphological modifications in response to an internal or external stimulus, such as pH, redox balance, temperature, magnetic field, and light have been actively pursued. In an effort to further improve the performances of the biomedical device, novel dual and multistimuli-responsive hydrogels responding to a combination of two or more signals have recently been developed by incorporating different stimulus-responsive elements into a network via polymerization processes. Notably, these combined responses take place either simultaneously at the pathological site, or sequentially from hydrogel preparation, hydrogel transporting pathways, to cellular compartments. These dual and multistimuli-responsive polymeric materials lead to superior *in vitro* and/or *in vivo* therapeutic efficacy, with programmed site-specific feature and remarkable potential for targeted therapy. This entry highlights the recent developments in the synthesis of dual and multistimuli-responsive hydrogels for applications in the biomedical and pharmaceutical fields, with a particular focus on the correlation between the hydrogel physical feature and the precision situ-controlled delivery of bioactive compounds.

INTRODUCTION

Stimuli-responsive polymers mimic biological systems in a crude way where an external stimulus (e.g., changes in pH or temperature) results in a change in properties. This change can be related to conformation, solubility properties, modification of the hydrophilic/hydrophobic balance, or release of a bioactive molecule (e.g,. drug). This often includes a combination of several responses at the same time. In medicine, stimuli-responsive polymers and hydrogels have to show their response properties within the setting of biological conditions. Typical stimuli are temperature, pH, electric and magnetic fields, light, and concentration of specific molecules in solution (e.g., electrolytes or glucose). The potential application of stimulus-responsive hydrogels in drug delivery and tissue engineering has emerged as one of the most significant trends in medicine. This is because stimulus-responsive gels not only possess the required functions of polymers for biomedical application, such as ability to cross biological barriers, protection of the therapeutics from rapid degradation in biological systems, and provision of a large surface area for conjugating targeting ligands, but also show unique characteristics depending on their structural modifications in the different biological environments and conditions.

Multiresponsive hydrogels are hydrophilic materials able to respond to more than one external stimulus.

Multiresponsive hydrogels can be first classified into two main classes: hydrogels responding independently to each of the external stimuli and hydrogels undergoing modification only when all external stimuli exist simultaneously.

Multistimuli-responsive hydrogels are usually prepared by incorporating different stimulus-responsive elements into their network via random copolymerization, graft copolymerization, or core/shell structure methods. Polymerization of monomers and chemical crosslinking of preformed polymers in heterogeneous colloidal environments, particularly in water-in-oil inverse microemulsions, are frequently used techniques for preparing hydrogels. Alternatively, the physical self-assembly of polymers was used to produce various nanogels in mild conditions and in aqueous media, allowing the encapsulation of bioactive macromolecules.

This entry gives a brief overview about some types of multistimuli-responsive hydrogels with the main focus on the strategy involved in the synthesis of materials, their physicochemical characterization and their potential applications in different biological fields.

TEMPERATURE AND pH-RESPONSIVE HYDROGELS

Among the environmentally sensitive polymer systems in drug delivery (Fig. 1),[1–7] synthesized by different

Concise Encyclopedia of Biomedical Polymers and Polymeric Biomaterials DOI: 10.1081/E-EBPPC-120050802

Gene–Hydrogels

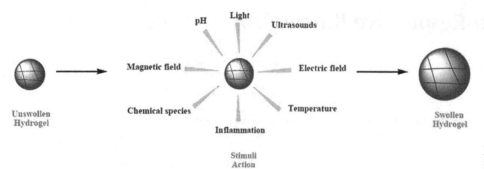

Fig. 1 Stimuli-responsive swelling of hydrogels.

approaches,[8–13] pH- and temperature-responsive hydrogels are the most studied.

The temperature has to be altered externally in most cases, except maybe for hyperthermia therapy, within narrow limits; while the pH changes within the body can therefore be used to direct the response to a certain tissue or cellular compartment. The obvious change in pH along the gastrointestinal tract from acidic in the stomach (pH = 2) to basic in the intestine (pH = 5–8) has to be considered for oral delivery of any kind of drug, but there are also more subtle changes within different tissues.[14]

Most frequently studied negatively temperature-responsive hydrogels are predominantly composed of the polymer p(N-isopropylacrylamide) (pNIPAAM), having a lower critical solution temperature (LCST) in aqueous solution ranging from 30°C to 40°C. Their particle size decreases abruptly when environmental temperature increases above the LCST. The change in the hydration state, which causes the volume phase transition, reflects competing hydrogen-bonding properties, where intra- and intermolecular hydrogen bonding of polymer molecules are favored compared with interaction with water molecules. Thermodynamics can explain this behavior with a balance between entropic effects, due to the dissolution process itself, and to the ordered state of water molecules around the polymer. Enthalpic effects depend on the balance between intra- and intermolecular forces, and on the solvation, e.g., hydrogen bonding and hydrophobic interactions. The transition is then accompanied by coil-to-globule transition. Thermo-responsive behavior of pNIPAAM hydrogel is strongly influenced by polymer–water affinity; at a temperature below the LCST, the hydrophilic groups (amide groups) in the side chains of the PNIPAAM hydrogel interact with water molecules by hydrogen bonds. However, as the external temperature increases, the copolymer–water hydrogen bonds are broken and the water molecules, rigidly structured around the polymer chains, gain more freedom degrees, resulting in their easy and rapid diffusion across the bulk phase. As a result, hydrogen bonds between solvent molecules in the continuous phase are formed, whereas hydrophobic interactions among the isopropyl groups become dominant inside the polymeric network.

An opposite behavior in the volume phase transition is observed for positively temperature-responsive hydrogels, which are based on polymers with an upper critical solution temperature (UCST) in aqueous solution, or interactions between different polymer chains within the hydrogels.[15]

pH-responsive hydrogels are composed of crosslinked polyelectrolytes with weakly acidic (i.e., carboxylic) or basic (i.e., amino) groups, suitable for use as either proton donors or receptors, or through a combination of both. Weak acidic polyelectrolytes, such as p(acrylic acid) (pAA), accept protons in an acidic environment and contribute protons in a basic environment; therefore, a small variation in the environmental pH value can modify the ionization degree of the polyelectrolyte chains.[16]

pH and temperature are the common stimuli in both chemical and biological systems, thus pH and temperature dually responsive hydrogels have been extensively studied (Table 1). A sequential polymerization method is proposed to synthesize dual-responsive nanogels with an interpenetrating polymer network (IPN) structure, in which the stimulus-responsive polymer components are independent of one another. IPN composed of pAA and pNIPAAM have been synthesized by the Jones and Lyon's method, and the IPN particle formation was monitored by measuring the variation of turbidity and the change of the particle hydrodynamic radius as a function of time.[17] IPN aqueous solution with polymer concentration of 2.5% g/mL is found to show an inverse and completely reversible thermoreversible gelation at pH above 5. The IPN gel undergoes the volume phase transition at 34°C, and possesses a pH sensitivity due to the contribution of the pAA. The collected data in the swelling experiments carried out at pH 4.0 and 7.0 indicate that the IPN has a narrower size distribution at acidic pH.

Random radical polymerization of NIPAAM and acrylic acid (AA) was performed in a semibatch surfactant-free emulsion process to obtain transparent p(NIPAAM-co-AA) nanogels with temperature and pH sensitivity, which were addressed to prepare intravenous drug delivery carrier. With respect to the conventional batch method, it has been observed that the absence of surfactant influences both the hydrodynamic diameters, which decrease according to the amount of AA, and the distribution of carboxylic

Table 1 Overview of dual-stimuli–responsive polymer for biomedical applications

Reactive species	Structure	Responsivity	Potential application	References
pAA, pNIPAM	IPN gel	pH/Temperature	Biomedical device	[17]
AA, NIPAAM	Nanogel		Intravenous drug delivery carrier	[18]
NIPAAM, trithiocarbonate-terminated pAA, MEBA	Grafted microgel		Drug delivery	[19]
AA, NIPAAM, modified β-cyclodextrin	Hydrogel		Drug delivery	[20]
Graphene oxide, pNIPAAM-*co*-AA	IPN nanocomposite gel		Drug delivery	[21]
DEAM, MAA	Hydrogel		Drug delivery	[22,23]
Sulfamethazine oligomers, p(ε-caprolactone-*co*-lactide) – p(ethylene glycol) – p(ε-caprolactone-*co*-lactide)	Block copolymer		Drug delivery, cell therapy	[24]
p(vinylcaprolactam-*co*-acetoacetoxy methacrylate), vinylimidazole	Microsphere		Biomedical device	[25]
N,N-(dimethylamino)ethyl methacrylate, p(ethylene glycol) methyl ether methacrylate	Hydrogel		Drug delivery	[26]
NIPAAM, 2-vinylpyridine, MAA, MEBA	Microsphere		Biomedical device	[27]
N-(2-hydroxypropyl) methacrylamide, 2-(dimethylamino) ethyl methacrylate polypropylene glycol dimethacrylate	Hydrogel		Drug delivery	[28]
p(vinyl alcohol)/pNIPAAM, poly(vinyl alcohol), pAA	Covered microcapsule		Drug delivery	[29]
Itaconic acid, oligo(ethylene glycol) acrylates	Hydrogel		Biomedical device	[30]
NIPAAM macromonomer, AA	Grafted microgel		Biomedical device	[31]
NIPAAM, AA, MEBA	Grafted microsphere		Drug delivery	[32]
DEAM, 2-(dimethylamino)ethyl methacrylate	Grafted microgel		Biomedical device	[33]
2-Hydroxyethyl methacrylate, MAA, NIPAAM, EBA	Microsphere		Drug delivery	[34]
Methacrylate polyphosphazene, MAA	Hydrogel		Cell culture media	[35]
Acrylated L-histidine or L-valine, EBA	Hydrogel		Drug delivery	[36]
Acetoacetoxyethyl methacrylate, *N*-vinyl-caprolactam, Fe_3O_4	Composite microgel	Temperature/Magnetic	Biomedical device	[43]
Pluronic F127, Fe_3O_4	Nanosphere		Drug delivery	[44]
Fe_3O_4-undecylenic acid, p (undecylenic acid-*co*-NIPAAM)	Grafted magnetomicelle		Drug delivery	[45]
Fe_3O_4, pNIPAAM	Grafted nanoparticle		Biomedical device	[46]
Pluronic F127, Pluronic F68, magnetite, gelatin, 4-nitrophenyl chloroformate, 1-ethyl-3-(3-dimethylaminopropyl) carbodiimide	Covered spherical nanocapsules		Biomedical device	[47]
pNIPAAM, chitosan, Fe_3O_4	Composite nanohydrogel		Hyperthermia treatment of cancer and targeted drug delivery	[48]
Mesoporous SBA-15, pNIPAAM, Fe_3O_4	Composite nanohydrogel		Drug delivery	[49]
Methoxy p(ethylene glycol)-p(MAA)-p(glycerol monomethacrylate), Fe_3O_4	Composite nanohydrogel	Magnetic/pH	Drug delivery	[50]
p(MEBA-*co*-MAA), Fe_3O_4/SiO_2, 3-(methacryloxy)propyltrimethoxysilane	Composite microsphere		Biomolecule separation	[51]

(*continued*)

Table 1 (*Continued*)   Overview of dual-stimuli–responsive polymer for biomedical applications

Methacryloxypropyltrimethoxysilane-modified Fe$_3$O$_4$, MAA, MEBA	Composite microsphere		Biomolecule separation	[52]
2-(*N,N*-dimethylamino)ethyl methacrylate, *N,N,N′,N′,N″*-pentamethyldiethy lenetriamine, bromide-modified Fe$_3$O$_4$	Composite microsphere		Drug delivery	[53]
Gum arabic-modified Fe$_3$O$_4$	Composite nanospheres		Drug delivery	[54]
p(pentafluorophenylacrylate), NIPAAM, *N*-(2-aminoethyl)-4-(2-phenyldiazenyl) benzamide	Hydrogel	Temperature/ light	Biomedical device	[57]
p(1-Pyrenylmethyl methacrylate), p(dimethylaminoethyl methacrylate)	Nanoparticle	pH/light	Biomedical device	[58]
o-Nitrobenzaldehyde-modified p(NIPAAM-*co*-2-carboxyisopropylacrylamide)	Hydrogel		Drug delivery	[59]
p[(D, L-lactic acid)-*co*-(glycolic acid)]-b-poly(ethylene oxide)-b-p [(D, L-lactic acid)-*co*-(glycolic acid)], polypyrrole	Nanoparticle	Temperature/ electric	Subcutaneous drug delivery	[60]
Multi-walled carbon nanotubes, p(vinyl alcohol), pAA, ethylene glycol dimethacrylate	Composite microcapsules	pH/electric	Drug delivery	[61]
Multi-walled carbon nanotubes, gelatin, MAA, EBA	Composite microspheres		Drug delivery	[62]
Poly(ethyleneimine), poly (styrenesulfonate), pyrolytic graphite	Layer-by-layer film		Bioelectrocatalysis	[63]
Chitin, acrylamide, FeSO$_4$·7H$_2$O	Hydrogel		Drug delivery	[64]
MAA, *N,N*-bis(acryloyl)cystamine	Nanohydrogels	pH/redox	Drug delivery	[65]
Acrylated poly(ferrocenylsilane), MEBA, NIPAAM, silver nanoparticles	Composite hydrogel		Antimicrobial application	[66]
pNIPAAM, *N*-acryloyl-3-aminophenyl boronic acid, 4-(2-acryloyloxyethyl amino)-7-nitro-2,1,3-benzoxadiazole, rhodamine	Microgels	T/glucose	Drug delivery	[67]
Concanavalin A, *N*-(2(dimethylamino) ethyl)-methacrylamide, glucosyl oxyethyl methacrylate	Microgels	pH/glucose	Drug delivery	[68]

Abbreviations: AA, acrylic acid; NIPAAM, *N*-isopropyl acrylamide; MEBA, *N,N″*-methylenebisacrylamide; DEAM, *N,N*-diethylacrylamide; MAA, methacrylic acid; EBA, *N,N″*-ethylenebisacrylamide.

groups located on the exterior of colloidal particles.[18] Polymer networks (pNIPAAM-g-pAA) were prepared via the reversible addition-fragmentation transfer polymerization of NIPAAM with trithiocarbonate-terminated pAA as a macromolecular chain-transfer agent, in the presence of *N,N′*-methylenebisacrylamide (MEBA), as crosslinking agent.[19] Compared with the pNIPAAM, the hydrogels showed a faster response to the external temperature changes due to the presence of water-soluble pAA chains at the surface of hydrogels, which behave as tunnels allowing water molecules to go through, thus accelerating their diffusion.

A series of thermo- and pH-responsive polymeric hydrogels modified with β-cyclodextrin were proposed as controlled delivery devices under external stimuli, using bovine serum albumin as a model drug.[20] The polymers, synthetized by copolymerization of AA and NIPAAM with a modified β-cyclodextrin crosslinker in aqueous solutions, represent an innovative carrier with a particular structure that allows to accommodate a wide variety of molecules into its cavity, from small organic molecules to biomacromolecules, to form a stable inclusion complex (Fig. 2A). A one-step strategy was planned to synthesize graphene oxide interpenetrating pNIPAAM hydrogel networks by covalently bonding graphene oxide sheets and pNIPAAM-*co*-AA microgels directly in water, carrying out to materials that exhibit dual thermal and pH response.[21] The obtained IPN hydrogel combines the mechanical properties of graphene sheets and thermo- and pH-responsivity of pNIPAAM-*co*-AA hydrogel (Fig. 2B).

Crosslinked pH- and temperature-responsive polymers were synthetized by free-radical random copolymerization of *N,N*-diethylacrylamide (DEAM) and methacrylic acid (MAA).[22] The LCST and pH-sensitivity properties are strictly related to the MAA content in the copolymers. An increase in the MAA content, indeed, carries out to increased LCST and the critical phase-transition pH of DEAM–MAA copolymers; the critical phase-transition pH was also found to strictly depend on the aqueous solution temperature. This is due to the intermolecular hydrogen bonding between carboxyl groups and amide groups and to the intramolecular hydrophobic interactions, which allow easier separation of the copolymer from the solution with the gradual increase of the temperature. For the synthesis of p(DEAM-*co*-AA) hydrogel, Hongliang et al. proposed a free radical copolymerization in distilled water, using a "freezing polymerization" technique.[23] The synthesis was conducted by a two-step procedure, an initial polymerization at constant temperature for 15 min, followed by further polymerization at −30°C for 12 hr. This synthetic approach allows a modification of morphology network, leading to the formation of more porous networks compared with those obtained through conventional polymerization methods. Furthermore, faster swelling and deswelling rates were recorded for this kind of hydrogels.

A block copolymer able to form a stable gel under physiological conditions (pH 7.4 and 37°C) has been synthesized by adding pH-sensitive sulfamethazine oligomers to either ends of a thermosensitive p(ε-caprolactone-*co*-lactide)-p(ethylene glycol)-p(ε-caprolactone-*co*-lactide)block copolymer.[24] At pH 8.0, the sulfonamide-modified block copolymer solution maintained its sol phase for about 2 hr at body temperature, but rapidly formed a gel in physiological conditions (pH 7.4 and 37°C) within only 5 min. This sulfonamide-modified block copolymer presented enhanced biocompatibility, both *in vitro* and *in vivo*, making the

sulfonamide-modified block copolymer a good candidate for drug delivery systems and cell therapy.

Microgels based on p(vinylcaprolactam-*co*-acetoacetoxy methacrylate) functionalized with vinylimidazole has been prepared in aqueous medium by a dispersion polymerization procedure.[25] The T- and pH-sensitivity depends on vinyl-imidazole content in the copolymer structure: an increase in the vinylimidazole content within the microgel structure led to an increased swelling in the acidic medium and a strong shift of the volume phase transition temperature to higher temperatures. Considering that the presence of reactive β-diketone groups confers to the microgels the ability to immobilize the proteins, these materials can find applications in the biomedical field.

Reversible addition–fragmentation chain transfer polymerization of *N,N*-(dimethylamino)ethyl methacrylate and p(ethylene glycol) methyl ether methacrylate at 70°C using a radical initiator was employed to synthesize tunable pH- and temperature-sensitive copolymers.[26] The behavior of the dual-sensitive copolymers in aqueous solution linearly depend on the weight percentage of incorporated poly-(ethylene glycol) methyl ether methacrylate in the copolymerization feed.

Radical ter-polymerization of NIPAAM, 2-vinylpyridine and MAA in the presence of MEBA, as a crosslinker agent, carried out to a polyampholyte microgel capable of responding to both temperature and pH changes, and the volume phase transition temperature, ranging from 30°C to 49°C, can be varied by changing the pH value (Fig. 3).[27] This has been explained in terms of the radial gradient distribution of MAA and 2-vinylpyridine components inside the hydrogel structure.

Multifunctional hydrogels used to attain higher site-specific drug delivery and tissue engineering scaffold were synthetized by radical polymerization of *N*-(2-hydroxypropyl) methacrylamide and 2-(dimethylamino) ethyl methacrylate using polypropylene glycol dimethacrylate as the crosslinking agent.[28] The pH responsiveness,

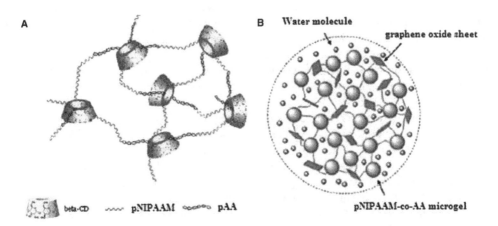

Fig. 2 (A) Thermo- and pH-responsive β-CD–based hydrogel. **(B)** Schematic representation of graphene oxide/pNIPAAM IPN hydrogel. **Source:** **(A)** From Hu et al.[20] © 2012, with permission from Springer Science and Business Media. **(B)** From Sun and Wu.[21] © 2011, with permission from RSC.

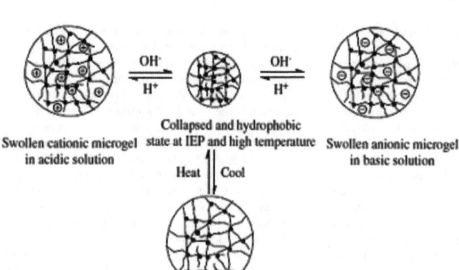

Swollen cationic microgel
in acidic solution

Collapsed and hydrophobic
state at IEP and high temperature

Swollen anionic microgel
in basic solution

Heat ⇅ Cool

Hydrophilic and neutral state
at IEP and low temperature

Fig. 3 Swelling–deswelling behavior for pNIPAAM-*co*-2-vinylpyridine-*co*-MAA (IEP: isoelectric point).
Source: From Li et al.[27] © 2007, with permission from John Wiley and Sons.

coupled to the thermoresponsive behavior, allow the matrix to become active with specific temperature or pH, thus giving a wider window for drug release. The fabricated hydrogel could be used in cancer therapy as the drug release profile (doxorubicin in this case), and the pH-responsive behavior of the gels can contribute to achieve precise targeting of cancer cells and tumor tissues without affecting the healthy surrounding tissues bared on difference in pH system.

Temperature-sensitive p(vinyl alcohol)/pNIPAAM microcapsules were covered by pH-sensitive hydrogel based on poly(vinyl alcohol) and pAA to obtain a biocompatible dual-responsive drug delivery system.[29] The combined effect of pH and temperature was evaluated in vitamin B_{12} release experiments, recording a complete delivery of drug at pH values higher than 7.0 and temperatures below 25°C.

The incorporation of pH-responsive (Itaconic acid) and temperature-responsive (oligo(ethylene glycol) acrylates) monomers allowed to prepare hydrogels with dual responsiveness through gamma radiations.[30] The radiation method has an increased potential for use in biomedical applications because it is an additive-free process, which occurs at room temperature, with an easily controlled crosslinking degree and simultaneous sterilization of the products.

The free-radical copolymerization of NIPAAM macromonomer and AA, utilizing macromonomer technique carried out to graft copolymers, which are both temperature and pH responsive.[31] The LCST of graft copolymer decreases with increasing pNIPAAM content, and the critical phase transition pH value increases with increasing isopropyl units. Similarly, p(NIPAAM-*co*-AA) macromonomer, prepared by radical telomerization of NIPAAM and AA using 2-hydroxyethanethiol, as a chain transfer agent, was employed to prepare graft-type p(NIPAAM-*co*-AA) microgel with linear grafted p(NIPAAM-*co*-AA).[32] The free

radical copolymerization of macromonomer with NIPAAM and AA in the presence of MEBA, as a crosslinker, produces a p(NIPAAM-*co*-AA) backbones with the grafted p(NIPAAM-*co*-AA) freely mobile chains. Due to the mobile nature and the tractive force side chains, the microgels have a synchronously rapid thermo- and pH-response property. The graft-type microgels may thus be very promising for use in many applications in which rapid responses to dual environmental stimuli are required.

Finally, radical telomerization of DEAM and 2-(dimethylamino)ethyl methacrylate with different ratios of the chain transfer agent (2-mercaptoethanol) was performed to prepare macromonomers with three different chain lengths, which have been subsequently employed for dual stimuli–responsive hydrogels preparation.[33] Compared with normal-type hydrogels, the comb-type grafted hydrogels, having the different lengths of the grafted chains, exhibited a macroporous structure and a quite fast reswelling and deswelling behaviors in response to simultaneous dual temperature and pH stimuli. This behavior is due to the side chain free mobility and can be modified by varying the length of the grafted chains.

Dual stimuli–responsive hydrophilic microspheres were prepared by free radical polymerization of MAA, as pH-responsive, hydrophilic monomer, and NIPAAM and *N,N*′-ethylenebisacrylamide (EBA), as thermosensitive monomer and crosslinker, respectively.[34] The LCST values of the microgels, depending on polymerization feed, were in the range 34.6–37.5°C, close to the body temperature and higher than those of NIPAAM homopolymers as a consequence of the increased hydrophilic/hydrophobic balance in the polymeric structure. When in the polymeric chains hydrophilic groups of MAA are randomly inserted, polymer–water interactions significantly increase, and more energy is required to destroy hydrogen bonds and to allow the solvent diffusion and thus the formation of interactions among the isopropyl groups of NIPAAM. The microspheres have been suggested as drug carriers for diclofenac diethylammonium salt and the *in vitro* drug release profiles, recorded at different temperature and pH values, support this theory.

Methacrylate-substituted polyphosphazene and MAA have been used for hydrogel preparation with pH-, thermoresponsive, and ultra-high absorbing behavior (Fig. 4) suggested as cell culture media.[35] The pH response of the hydrogel shows an atypical swelling curve: the network structure collapses not only at lower (<5) but also at high (>8) pH values, whereas the thermoresponsivity was suppressed after the hydrogel annealing and overridden by the pH-response behavior of the pMAA branches.

The free radical crosslinking reaction of acrylated α-aminoacid, such as L-histidine and L-valine, with EBA carried out to hydrogels pH and temperature responsive because of the presence of carboxylic, isopropyl, and amido groups in the monomer structures.[36] Release studies of pilocarpine in physiological conditions showed a burst effect within the first few hours, followed by a sustained release for long time depending on the crosslink density and on the nature of the material. The hydrogels, being nontoxic toward the cell-line 3T3, have been suggested as Ocusert-pilocarpine devices for the treatment of the glaucoma disease.

DUAL-RESPONSIVE HYDROGELS BASED ON MAGNETIC UNITS

Magnetic nanoparticles have been considered as a class of effective supports for guided delivery, which, in combination with an external magnetic field and/or magnetizable implants, allow delivering particles to the desired target area and fixing them at the local site or being taken by the cells while the medication is released and acts locally (magnetic drug targeting).[37,38] For biomedical applications, the biocompatibility of magnetic nanoparticles, especially the shell, is essential. For this purpose, magnetic nanoparticles require to be covered with a shell playing an important role in the interaction with its target.[39] The formation of magnetite/polymer hybrid/composite materials not only stabilized the magnetic nanoparticles, but also endowed the magnetic nanoparticles with functionality. In such a way, magnetic/polymer microspheres have been found wide applications in the fields of biology, medicine, catalysis, and others.[40,41]

Magnetic and Thermoresponsive Hydrogels

Magnetically responsive thermoactive hydrogels offer several potential advantages over other biomaterial systems for the controlled release of drugs and targeted thermal therapy to the cancerous cells. Thermal and magnetic responses of hydrogels can be tuned by varying the size and the concentration of embedded nanoparticles and the magnitude and frequency of the magnetic field. Volumetric changes of the magnetic materials are advisable for controlled drug release and temperature regulation for hyperthermia applications.[42] Gentle magnetic heating causes temperature-responsive polymer to shrink, squeezing drug out from the nanoparticle. Intense magnetic heating additionally ruptures the nanoparticle, triggering a burst-like drug release. Because none of the soft materials suitable for biomedical applications is magnetic, a soft–hard hybrid

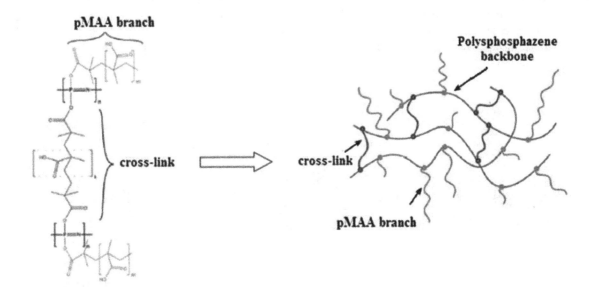

Fig. 4 Thermo- and pH-responsive pMAA/polyphosphazene hydrogel.
Source: From Silva Nykänen et al.[35] © 2011, reprinted with permission from RSC.

construct is required to combine magnetic and thermal sensitivities. The soft temperature-responsive materials of choice form the hydrogel, whereas the hard magnetic material is generally iron oxide, which is relatively safe for biomedical applications and can be readily synthesized in a form of small particles to be embedded into the soft material. Iron oxide can be attracted to a magnet; moreover, using a high-frequency field, remote magnetic heating of iron oxide becomes possible thereby converting a magnetic stimulus to a thermal stimulus.

Hybrid temperature-sensitive microgels including magnetite nanoparticles in their structure have been prepared by surfactant-free emulsion copolymerization of acetoacetoxyethyl methacrylate and N-vinylcaprolactam in water with a water-soluble azo-initiator.[43] The presence of magnetite into microgels lead to the formation of composite particles combining both the temperature-sensitive and magnetic properties. The obtained microgels possess an LCST in water solutions influenced by the amount of loaded magnetite.

Dual-functional nanospheres, composed of magnetic iron oxide nanoparticles embedded in a thermo-sensitive Pluronic F127 matrix were fabricated by an in situ co-precipitation process[44] and studied for the release of anti-cancer drug such as doxorubicin. At a lower temperature (4°C, swelling state), drug diffuses into the nanospheres; after slight heating to 15°C and slight volume shrinkage, the drug gets encapsulated in the nanospheres; finally, under high-frequency magnetic field treatment, a sharp volume shrinkage with accelerative drug release is observed. Nevertheless, the LCST values of these materials, below 37°C, limit applicability as drug carriers (Fig. 5).

Temperature-responsive magnetomicelles were prepared by covering a functionalized Fe_3O_4–undecylenic acid magnetic core with an amphiphilic surface temperature-responsive based on poly(undecylenic acid-co-NIPAAM)[45] and investigated in in vitro controlled delivery of prednisone acetate. The drug release behavior of temperature-responsive magnetomicelles at temperatures around the LCST

shows the potential applicability of this dual stimuli-responsive material as triggered drug release device.

Temperature- and magnetic-responsive hybrid materials were synthesized by coating of ferromagnetic iron oxide nanoparticles with pNIPAAM by a surfactant exchange method.[46] The dynamic light-scattering measurements provide to demonstrate the enhanced agglomeration with increasing temperature in accord to an increasingly hydrophobic interaction. This behavior could be useful for a triggering mechanism for drug release.

Alternatively, magnetic iron oxide nanoparticles can be dispersed in a core of a natural polymer coated with a thermosensitive shell. Self-assembled spherical nanocapsules containing a hydrophilic core and a crosslinked yet thermosensitive shell are prepared by a covalent linkage between p(ethyleneoxide)-p(propylene-oxide)-p(ethylene-oxide) block copolymers and gelatin.[47] The insertion of iron oxide in the core enables the materials to respond to the magnetic field by undergoing irreversible structural changes. Responding to external magnetic induction, magnetic nanocapsules undergo irreversible structural changes, including core/shell disruption, and rapid release of 80% of the core content within 5 min (Fig. 6). Such burst-like response has been proposed for controlled drug release of vitamin B_{12}.

The polymerization of NIPAAM in the presence of chitosan and Fe_3O_4 leads to nanohydrogel formation with optimized LCST of just above 42°C for hyperthermia applications.[48] The amount of Fe_3O_4 magnetic nanoparticles and the ratio of NIPAAM/chitosan into nanohydrogel can modulate the LCST. In particular, the increase in LCST due to the presence of iron oxide in the magnetic nanoparticles may be explained by dipole–dipole interactions, which prevent collapsing of crosslinked polymer segments. The magnetic nanohydrogel shows optimal magnetization, good specific absorption rate (under external magnetic field), and excellent cytocompatibility with L929 cell lines, which suggests a potential application in hyperthermia treatment of cancer and targeted drug delivery.

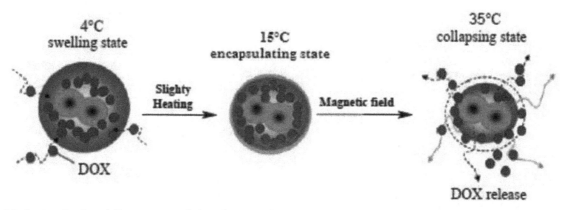

Fig. 5 Mechanism for drug delivery process of pluronic–magnetic nanoparticles (DOX: doxorubicin).
Source: From Liu et al.[44] © 2008, with permission from ACS.

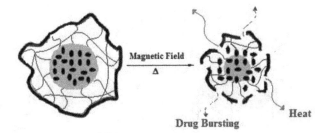

Fig. 6 Event triggered by magnetic heating on the poly(ethyleneoxide)–poly(propylene-oxide)–poly(ethylene-oxide) block copolymers and gelatin: volume shrinkage, core collapse, heat conduction, and drug delivery.
Source: From Liu et al.[47] © 2009, with permission from John Wiley and Sons.

A modified mesoporous SBA-15 combined with maghemite (γ-Fe$_2$O$_3$) nanoparticles and pNIPAAM polymers have been suggested for magnetic and temperature targeting drug release of gentamicin.[49] The phase transition of the temperature-responsive pNIPAAM can control opening and closing of the pores and the drug release can be controlled in response to the environmental temperature. The magnetic γ-Fe$_2$O$_3$ inside the channels of SBA-15 enables the composite to be guided by an external magnetic field. The system exhibits the temperature-responsive controlled release pattern. At lower temperatures, pNIPAAM in the swollen state, blocks the pores preventing gentamicin from release through the channels. On the contrary, at high temperature (>LCST), pNIPAAM is in a collapsed state, resulting in pore enlargement and in an easier diffusion of drug molecules loaded in mesoporous channels.

Magnetic and pH-Responsive Hydrogels

A multifunctional nanocarrier with multilayer core–shell architecture was prepared by alkaline coprecipitation of ferric and ferrous ions in the presence of a triblock copolymer, methoxy p(ethylene glycol)-block-p(MAA)-block-p(glycerol monomethacrylate) in aqueous solution.[50] The core of the nanocarrier is a superparamagnetic Fe$_3$O$_4$ nanoparticle, on which the p(glycerol monomethacrylate) block of the triblock copolymer is attached. The pMAA block forms the inner shell and the p(ethylene glycol) block forms the outermost shell. Adriamycin, as a model of drugs with amine groups and hydrophobic moieties, was loaded into the triblock copolymer-coated Fe$_3$O$_4$ nanoparticles through combined action of ionic bond and hydrophobic interaction by simple mixing with the nanoparticles in aqueous solution at pH 7.4. At endosomal/lysosomal acidic pH (<5.5), the protonation of polycarboxylate anions of pMAA broke the ionic bond between the carrier and Adriamycin, leading to the release of the drug because the hydrophobic interaction alone is very weak due to the relative hydrophilic characteristic of the nanocarrier.

Likewise, Liu et al. describe the preparation of pH-sensitive polymer hollow microspheres with movable magnetic cores via the selective removal of non–cross-linked pMAA midlayer from the magnetite/silica/pMAA/p(MEBA-*co*-MAA) tetra-layer microspheres. The microspheres were synthesized by the distillation precipitation polymerization of monomers in acetonitrile with a radical initiator in the presence of Fe$_3$O$_4$/SiO$_2$/pMAA tri-layer particles as the seeds.[51] The same synthetic approach was employed to fabricate magnetite (Fe$_3$O$_4$)/pMAA composite microspheres through encapsulating γ-methacryloxypropyltrimethoxysilane-modified magnetite colloid nanocrystal clusters with crosslinked pMAA shell.[52] The large amount of carboxyl groups supported by the pMAA shell can be easily functionalized with bioactive molecules, which makes these microspheres to have a great potential in bioseparation and other biomedical applications.

An alternative synthesis of pH- and magnetic-sensitive nanoparticles was performed by coating magnetite beads with p[(2-dimethylamino) ethyl methacrylate], as carriers for targeted drug delivery and controllable release.[53] The magnetite nanoparticles were prepared by alkaline precipitation and modified by α-bromo isobutyric acid to link atom transfer radical polymerization initiator, initiating the polymerization process of p[(2-dimethylamino) ethyl methacrylate] on the surface. The Fe$_3$O$_4$/p[(2-dimethylamino) ethyl methacrylate] hybrid nanoparticles with core–shell structure are able to load drugs into the polymer shell, and the release rate of drug is approximately steady going and can be effectively controlled by altering the pH value.

Coating magnetite nanoparticles with a natural polymer, such as the gum arabic, a multifunctional pH-sensitive magnetic drug nanocarrier for synchronous cancer therapy and sensing was fabricated.[54] The nanocarrier, programmed to display a response to environmental stimuli (pH value), was synthesized by coupling doxorubicin to adipic dihydrazide-grafted gum arabic–modified magnetic nanoparticles via the hydrolytically degradable pH-sensitive hydrazone bond. It exhibited a pH-triggered response characteristic within the endosomal/lysosomal pH range: the drug was released by sensing the change in pH corresponding to the acidic conditions in the intercellular compartment but was relatively stable at physiological pH (7.4). The coupling of doxorubicin to gum arabic–modified magnetic nanoparticles resulted in a reversible self-quenching of fluorescence through the fluorescence resonant energy transfer between the donor gum arabic–modified magnetic nanoparticles and the acceptor Doxorubicin. The release of Doxorubicin from the nanocarrier exposed to acidic media indicated the recovery of fluorescence from both gum arabic–modified magnetic nanoparticles and Doxorubicin. The change in the fluorescence intensity of nanocarrier acts as a potential sensor for the delivery of the drug.

DUAL-RESPONSIVE HYDROGELS CONTAINING THERMO-OR PH-SENSITIVE UNITS

Responsive polymer systems reacting to thermal and light stimuli have been a focus in the biomaterials literature because they have the potential to be less invasive than currently available materials and may perform well in the *in vivo* environment. Photo-responsive isomerization, dimerization, degradation, and triggered reversible and irreversible processes may be used to confer thermal and light responsivity to polymeric systems. Unique wavelengths induce photo-chemical reactions of polymer-bound chromophores to alter the bulk properties of polymer systems. The properties of both thermo- and photo-responsive polymer systems may be taken advantage of to control drug delivery, protein binding, and tissue scaffold architectures.

The polymers responsive to light and temperature contain a light-responsive azobenzene moiety.[55,56] Dual stimuli–responsive copolymers were synthesized by reversible addition–fragmentation chain-transfer copolymerization reaction of the reactive precursor polymer p(pentafluorophenylacrylate) with *N*-(2-aminoethyl)-4-(2-phenyldiazenyl)-benzamide and NIPAAM.[57] The LCST of these materials depends not only on the amount of chromophoric groups, but also on the degree of isomerization trans to cis of the azobenzene induced by ultraviolet (UV) irradiation. The cis-configuration is characterized by an increase of the LCST causing precipitation. Higher LCST values were measured after irradiation and thus, in the temperature region between the LCST of the nonirradiated and the irradiated solution, a light-controlled reversible solubility change was observed.

Feng et al. reported a versatile synthetic method for the preparation of photo-/pH-sensitive nanoparticles using a combined reversible addition–fragmentation chain transfer and atom transfer radical polymerization.[58] The photo-/pH-sensitive nanoparticles composed of p(1-pyrenylmethyl methacrylate) and pH-sensitive p(dimethylaminoethyl methacrylate) exhibited an apparently decreased fluorescence when irradiated by UV light, and their size could be readily tuned through pH variations, which would lead to the swelling or deswelling of the nanoparticles. More recently, Techawanitchai et al. proposed a smart control of an interface movement of proton diffusion in pH-responsive hydrogels using a light-induced spatial pH-jump reaction (Fig. 7).[59] A photoinitiated proton releasing reaction of *o*-nitrobenzaldehyde was integrated into p(NIPAAM-*co*-2-carboxyisopropylacrylamide) hydrogels. The materials demonstrated quick release of proton upon UV irradiation, allowing the pH inside the gel to decrease below the pK_a within a minute. The *o*-nitrobenzaldehyde-integrated gel was shown to shrink rapidly upon UV irradiation without polymer "skin layer" formation due to a uniform decrease of pH inside the gel. The gel was employed for the controlled release of entrapped dextran:

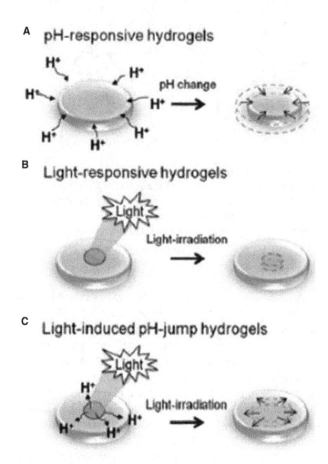

Fig. 7   pH- and light-sensitive p(NIPAAM-*co*-2-carboxyisopropylacrylamide) hydrogels.
Source: From Techawanitchai et al.[59] © 2011, with permission from Elsevier.

dextran was successfully entrapped into the gel and then released into water in a controlled manner under 365 nm UV illumination.

Electric stimuli, easy to generate and control, have been successfully utilized to trigger the release of molecules via conducting polymeric bulk materials or implantable electronic delivery devices. Ge et al. utilize emulsion polymerization techniques to encapsulate drugs in conducting nanoparticles based on polypyrrole dispersed in a block copolymer, p[(D,L-lactic acid)-*co*-(glycolic acid)]-b-poly(ethylene oxide)-b-p[(D,L-lactic acid)-*co*-(glycolic acid)], which is a liquid at low temperature but becomes a gel at body temperature.[60] This mixture can be subcutaneously localized by syringe injection at the place of interest. The subsequent application of a small external electric field releases the drug from the nanoparticles and allows the drug to diffuse through the hydrogel to the surroundings.

Generally, multiwalled carbon nanotubes were applied as active ingredients for enhancing the electroresponsive behavior of hydrogels. Electro- and pH-responsive composite microcapsules were prepared by using multiwalled carbon nanotubes, p(vinyl alcohol), and pAA.[61] Carbon nanotubes were employed to improve the electroresponsive properties of microcapsules and the surface modification of the nanotubes was carried out by oxyfluorination for the improved dispersion in the microcapsule shell. Hydrogels composed of gelatin and multiwalled carbon nanotubes were synthesized by emulsion polymerization in the presence of sodium methacrylate, as pH-sensitive monomer, and EBA, as crosslinker and proposed in the controlled release of diclofenac sodium salt.[62] By a straightforward synthetic strategy based on modified grafting approach, nanosized and uniformly dispersed spherical hybrid hydrogels with enhanced electrical properties, thermal stability and biocompatibility have been prepared (Fig. 8). Drug release experiments demonstrated the ability of the responsive composite to control drug release over time. The electric stimulation resulted in a further increase of the release (+20%) in multiwalled carbon nanotubes containing materials.

Weak polybase branched p(ethyleneimine) and strong polyacid p(styrenesulfonate), assembled into layer-by-layer films on pyrolytic graphite electrodes with spin-coating approach, demonstrated a pH-sensitive "on–off" property toward electroactive probe ferrocenedicarboxylic acid.[63] The mechanism of the pH-dependent permeability of the films toward the probe was attributed to the electrostatic interaction between the two counterparts and provided an example to combine pH-sensitive permeability of layer-by-layer films with bioelectrocatalysis. The pH-sensitive permeability of the films was proposed to control or modulate the electrochemical oxidation of glucose catalyzed by glucose oxidase and mediated by ferrocenedicarboxylic acid.

Chitin, homogeneously functionalized with acrylamide through Michael addition, was employed to synthesize a stimuli-sensitive derivative with water solubility for electrochemically stimulated protein release.[64] Electrical signals to generate the stimuli (pH change or redox state of iron ions) activate the sol–gel transition allowing the release of entrapped drug. Considering the mild conditions for the sol–gel transition, this hydrogel is favorable to release some labile drugs, such as peptides and proteins. The release of protein from hydrogel can be triggered by a cathodic potential–induced pH increase. In the case of cationic ion–crosslinked hydrogel, the conversion of Fe^{2+} to Fe^{3+} was controlled by electrical potentials and the corresponding protein entrapment and release can be achieved based on the fact that Fe^{3+} can crosslink with the hydrogel to form a gel, whereas Fe^{2+} lacks the ability to crosslink.

The introduction of disulfide-functionalized linkages is reversible and responsive to the external redox environment, and usually its redox potential is employed for controlling the biodegradability of nanohydrogels. Pan et al. developed a facile and straightforward way to prepare highly hydrophilic pH/redox dual-responsive biodegradable nanohydrogels with unique properties as drug carriers, such as clean composition, uniform size, pH/redox dual-responses, and quick degradation.[65] The redox/pH dual stimuli–responsive pMAA-based nanohydrogels was prepared from MAA and N,N-bis(acryloyl)cystamine crosslinker via distillation precipitation polymerization. The nanohydrogels could be easily degraded into individual linear short chains in the presence of water-soluble

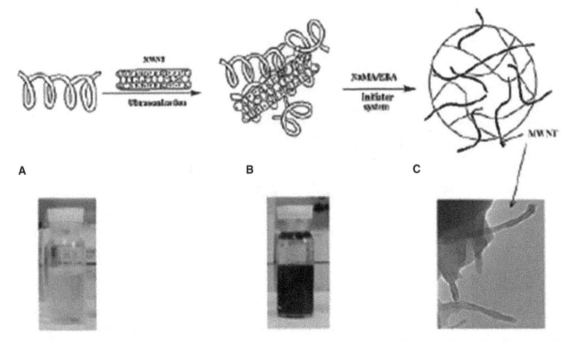

Fig. 8 Schematic representation of the synthesis of hybrid hydrogels. (**A**) Gelatin solution in water; (**B**) multiwalled nanotubes dispersion in gelatin solution; (**C**) hybrid hydrogel SEM.
Source: From Spizzirri et al.[62] © 2013, with permission from Elsevier.

reducing agents such as dithiothreitol or glutathione. Doxorubicin, as a model anticancer drug, was efficiently loaded into the nanohydrogels and the cumulative release profile of the drug-loaded nanohydrogels showed a low level of drug release at pH 7.4, which was significantly accelerated at a lower pH (5.0) and reducing environment, exhibiting an obvious pH/redox dual-responsive controlled drug release capability. In addition, the drug release behavior of the drug-loaded nanohydrogels in the presence of glutathione was very different from the dithiothreitol, as the loaded doxorubicin could be quickly released in the presence of glutathione, but not of dithiothreitol. More recently, Sui et al. proposed a hydrogel prepared by copolymerization of p(ferrocenylsilane) bearing acrylate side groups with NIPAAM and MEBA under UV light-emitting diode irradiation at a wavelength of 365 nm, in the presence of a photoinitiator.[66] Uniform distribution of silver nanoparticles in the hydrogel networks via in situ reduction of silver nitrate with p(ferrocenylsilane) allowed to fabricate a composite showing strong antimicrobial activity while maintaining a high biocompatibility with cells.

Glucose is a particularly interesting target molecule owing to its inherent biological activities and physicochemical properties in living organisms, and thus glucose-responsive materials can play important roles in biomedicines and diabetes therapies. Microspheres containing fluorescence resonance energy transfer pairs, acceptor and donor respectively, and N-acryloyl-3-aminophenylboronic acid have been synthesized via free radical emulsion copolymerization.[67] Because of the thermoresponsive behavior and the presence of ionizable acidic moieties within microgels, they could serve as ratiometric fluorescent pH and temperature probes. Test of cell viability, examined by the cytotoxicity assay, have revealed that this type of thermoresponsive microgels is almost noncytotoxic up to a concentration of 1.6 g·L^{-1}, suggesting their applications for multifunctional purposes such as sensing, imaging, and triggered-release nanocarriers under in vivo conditions. Saccharide-binding affinity of concanavalin A and protonation/deprotonation of tertiary amine groups of N-(2-(dimethylamino) ethyl)-methacrylamide led to a glucose and pH dual-responsive microgels for insulin delivery (Fig. 9).[68] In vitro insulin release in response to different glucose concentrations and small change in pH value were investigated, and the amount of released drug significantly increased with a slight decrease of pH value, which could be favorable for insulin release in the slightly lower physiological pH environment of diabetes mellitus.

DUAL STIMULI–RESPONSIVE HYDROGEL CONTAINING BIOMACROMOLECULES

In the last years, great attention has been paid to the covalent conjugation of a biological macromolecule, like a protein or a polysaccharide, to a stimuli-responsive monomer to produce "smart" hydrogels suitable for pharmaceutical and biomedical applications (Table 2).[69–71]

Microspheres based on albumin are suggested as drug carriers because of the binding ability of albumin to various substances, such as amino acids and drugs. pH and temperature stimuli–responsive microspheres based on functionalized bovine serum albumin have been obtained by radical copolymerization of methacrylate albumin with NIPAAM and MAA sodium salt.[72] The presence of hydrophilic moieties allows to increase the transition

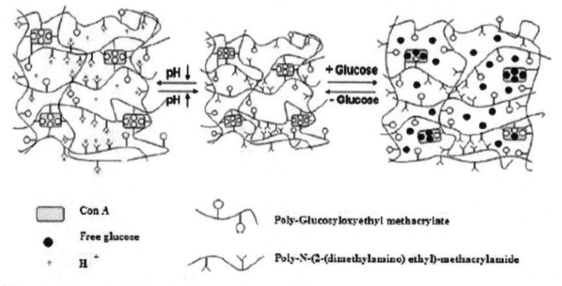

Fig. 9 Schematic of the structural changes in response to glucose and pH.
Source: From Yin et al.[68] © 2011, with permission from Elsevier.

Table 2 Overview of dual-stimuli–responsive polymer based on biomacromolecules for biomedical applications

Reactive species	Structure	Responsivity	Potential application	References
Methacrylate albumin, NIPAAM, MAA	Microspheres	pH/Temperature	Drug delivery	[72]
Trithiocarbonate-modified dextran, NIPAAM	Hydrogel		Drug delivery	[73]
p(NIPAAM-co-acrylamide), pullulan, epichlorohydrin	Microspheres		Drug delivery	[74]
Sodium alginate, NIPAAM, MEBA	Interpenetrating hydrogel		Drug delivery	[75]
Calcium alginate, pNIPAAM, p(sodium acrylate)	Interpenetrating hydrogel		Drug delivery	[76]
Hydroxypropylcellulose, AA, MEBA	Semi-interpenetrating nanoparticles		Drug delivery	[77]
Carboxylated methyl cellulose and p(vinyl alcohol)	Microgels		Drug delivery	[79]
NIPAAM, MEBA, p(ethyleneimine), chitosan	Microgels		Biomedical device	[11]
NIPAAM, chitosan, MAA, methyl methacrylate	Core–shell copolymer		Drug delivery	[80]
Chitosan, maleic anhydride, 2-(dimethylamino)ethyl methacrylate	Graft copolymer		Drug delivery	[81]
Chitosan, pNIPAAM	Graft copolymer		Drug delivery	[82]
Chitosan, AA, pNIPAAM	Core–shell nanoparticles		Biomedical device	[83]
N-phthaloylchitosan, pNIPAAM-b-pAA	Branched copolymer		Biomedical device	[84]
Carboxymethyl chitosan, DEAM	Microgels		Drug delivery	[85]
Carboxymethyl chitosan, carboxy methyl chitosan-g-p(DEAM)	Microgels		Drug delivery	[86]

Abbreviations: AA, acrylic acid; NIPAAM, N-isopropyl acrylamide; MEBA, N,N''-methylenebisacrylamide; DEAM, N,N-diethylacrylamide; MAA, methacrylic acid.

temperature of the microgels close to the body temperature; in addition, the acidic groups confer dual stimuli–responsive characteristic to the microspheres, extending the applicability of these materials. The materials are characterized by high water affinity and a significant volume change in response to both temperature variation and pH changes. This carrier could be useful to reduce the toxicity and increase the pharmacological effect of the nonsteroidal anti-inflammatory drug such as the diclofenac diethyl ammonium salt. The release experiments of a soaked drug on the microspheres varying the temperature and the pH of the releasing media, together with the microgel crosslinking degree showed a dependent stimuli release profiles.

pH- and temperature-sensitive gold nanoparticles have been obtained covalently tethering onto the gold surface by Au–S bonds, a dextran-based dual-sensitive polymers. The functionalized nanoparticles show stability under various conditions, but also a pH-dependent optical response when the temperature is changed from 25°C to 40°C.[73] The dual-sensitive polymer is prepared by radical addition-fragmentation chain transfer polymerization of NIPAAM from trithiocarbonate groups linked to dextran and succinoylation of dextran after polymerization. This smart material presents potential applications in the field of pH and temperature sensing.

Biodegradable pH-/thermo-sensitive microspheres were synthesized by grafting p(NIPAAM-co-acrylamide) onto pullulan microspheres and subsequent insertion of carboxylic groups.[74] The synthetic strategy consists of various steps: firstly the pullulan microspheres were prepared by suspension crosslinking with epichlorohydrin of an aqueous solution of the polymer, then the thermo- and pH-sensitive units were introduced by grafting of thermo-responsive copolymer onto pullulan microspheres, and finally the pH-sensitive units (–COOH) were inserted by reaction between the remaining –OH groups of the pullulan with succinic anhydride. The grafted pullulan microspheres are characterized by an increased hydrophilicity than pullulan microspheres, and by a transition temperature closed to the human body temperature in isotonic phosphate buffer. Loading and thermally controlled release of lysozyme from the pullulan-based microspheres were studied.

Sodium alginate, NIPAAM, and MEBA have been covalently crosslinked to prepare mixed-interpenetrated networks with thermo- and pH-responsive properties. The swelling measurements show a sudden and sharp swelling–deswelling behavior at various temperatures and pH, depending on the composition of the networks. This on–off swelling behavior could make this materials potential candidates as pulsatile drug delivery systems.[75]

Gene–Hydrogels

Interpenetrating hydrogel formed by calcium alginate and pNIPAAM was prepared in the presence of p(sodium acrylate), as a strengthening agent, to improve the mechanical strength of natural hydrogel and to obtain a sustained dual-sensitive drug-delivery device.[76] The mechanical stability, swelling, and drug-release behaviors of hydrogel bead were investigated in phosphate buffered saline at pH 7.4 and HCl solution at pH 1.2 (pH values of intestinal and gastric fluids, respectively) at different temperatures by using indomethacin as a model drug.

The direct polymerization of hydroxypropylcellulose, AA, and MEBA has been exploited to prepare thermo- and pH-responsive semi-interpenetrating nanoparticles.[77] Depending on the chemical composition and the degree of crosslinking, the thermo-responsive property of the gel particles can be shifted from the UCST to the LCST property, and the particle sizes can be changed from 100 to 1 μm in a controllable way. The oxaliplatin-loaded gel particles have high anticancer activity against BCG 823 cell line. Cellulose polymer-based hydrogel with specific dual-responsive absorption properties was prepared from carboxymethyl and hydroxyethyl cellulose in an aqueous solution employing citric acid as a crosslinking agent. The temperature-responsive swelling ability of hydrogel can be formulated based on the hydrogel composition. Application of the selected hydrogels as a thin-film onto cotton knitwear in a durable and stable manner (textile pre-treatment, hydrogels deposition, and grafting), was investigated for the production of surface-functionalized textile materials. It has been confirmed that the stimuli-responsive surface modifying system imparted pH- and temperature-responsiveness to cotton fabric in terms of regulating its water uptake.[78]

Carboxylated methyl cellulose and p(vinyl alcohol) were mixed in aqueous solution to form physical composite hydrogel exhibiting both temperature- and pH-sensitive swelling properties.[79] Rhodamine B is utilized in *in vitro* release profiles, which suggested that hydrogel composite could be a candidate as a carrier for drug controlled release.

Smart polymeric microgels consisting of well-defined temperature-sensitive cores with pH-sensitive shells were obtained directly from aqueous graft copolymerization of NIPAAM and MEBA from water-soluble polymers containing amino groups, such as p(ethyleneimine) and chitosan.[11] The microgels, consisting of well-defined pNIPAAM cores with cationic water-soluble polymer shells, are characterized by a narrow size distribution and a responsiveness to pH and temperature individually tunable.

Thermal-/pH-sensitive core–shell copolymer latex based on crosslinked copolymer of NIPAAM and chitosan was synthesized by soapless dispersion polymerization as the core, whereas a copolymer of MAA and methyl methacrylate was prepared as the shell. To evaluate the protein conjugating ability and the potential of the core–shell copolymer particles being applied on the targeting drug carrier, bovine serum albumin was chosen as model protein

to be conjugated on the surface of copolymer particles by the aid of water-soluble carbodiimide. The amount of albumin conjugated on the surface of particles was mainly influenced by the size of swollen particles or the hydrophobic property of particles.[80]

Chitosan, maleic anhydride, and 2-(dimethylamino) ethyl methacrylate have been combined by grafting and copolymerization to prepare a copolymer with temperature and pH sensitivity.[81] The observed mechanism release of CoA is strictly dependent on pH and temperature of the surrounding medium. By a temperature-dependent self-assembly method, chitosan-g-pNIPAAM nanoparticles were synthesized using cerium ammonium nitrate as the initiator.[82] In the first step, the initiator induces the formation of chitosan-g-pNIPAAM/pNIPAAM blend and then, at an appropriate concentration and temperature, this self-assembled crosslinks to form micelles with thermo- and pH properties. The hydrophobic pNIPAAM was in the interior of chitosan–pNIPAAM micelles and hydrophilic chitosan was on the shell of micelles. Glutaraldehyde was used to crosslink the chitosan chains in the micelles, which could efficiently lock the integrality of micelle nanoparticles. After cooling the reaction system to room temperature, a solution containing chitosan–pNIPAAM porous nanoparticles was prepared. The nanoparticles showed a continuous release of the encapsulated Doxycycline hyclate up to 10 days during an *in vitro* release experiment with environmentally sensitive properties.

Spherical and monodisperse thermo- and pH-sensitive chitosan-based nanoparticles were prepared by "one-pot" polymerization of AA/chitosan in the presence of pNIPAAM macroradical.[83] After the formation of chitosan acrylate salt by complexation of amino groups of chitosan with AA, pNIPAAM is covalently anchored onto chitosan/ pAA nanoparticles as the shell.

Dual stimuli–responsive hydrogel, N-phthaloylchitosan graft copolymer with chitosan backbone and pNIPAAM-b-pAA branch chains was prepared by reversible addition-fragmentation chain transfer polymerization.[84] Copolymer assembles to micelles in aqueous solution in the range of 200–300 nm with narrow size distribution, and the hydrodynamic diameter could be controlled depending on the length of branch chains and temperature. Carboxymethyl chitosan and DEAM carried out to a stable hydrogel beads based on dual physical crosslinking of hydrogen bonds and chelating action.[85] The dual crosslinked method may control the drug release of the vitamin B_2 rate under gastrointestinal tract conditions, which was superior to traditional single crosslinked beads. The same authors prepared microcapsules and microparticles by self-assembly of carboxymethyl chitosan and carboxymethyl chitosan-g-p(DEAM) in aqueous media under mild conditions without the involvement of organic solvent and surfactants.[86] The authors report on a good control over the morphology and the size of the microgels, and also an effectively sustained release of bovine serum albumin encapsulated in the microgels at different of pH and temperature conditions was raised.

TRIPLE-RESPONSIVE HYDROGELS

Although the design of triple stimulus–responsive hydrogels remains a challenge, triple-responsive hydrogels, which could achieve more functionalities and be modulated through more parameters, have emerged as novel drug carriers (Table 3).[87]

A facile way to fabricate novel hybrid particles with multiple sensitivities is to encapsulate magnetic nanoparticles into thermal-/pH-responsive polymers, leading to an enhanced biocompatibility of Fe_3O_4 magnetic nanoparticles. Semi-IPN hydrogel modified magnetic nanoparticles were synthesized by a two-step aqueous polymerization method, where Fe_3O_4 nanoparticles served as the core and a semi-IPN hydrogel consisting of pNIPAAM and pAA was used as the shell.[88] These semi-IPN Fe_3O_4 nanoparticles swelled when raising the temperature or increasing the pH value, with potential applications in the field of intelligent controlled drug delivery. Bilalis et al. proposed the synthesis, by a two-step distillation precipitation polymerization, of core–shell microspheres based on a core of pMAA covered by a shell composed of temperature- and pH-responsive p(NIPAAM-co-MAA) with encapsulated Fe_3O_4 magnetic nanoparticles.[89] Initially, monodispersed pMAA microspheres without crosslinker, used as a template, were synthesized by distillation precipitation polymerization. Successively, in the second stage, the interaction between the acid groups in the core and hydrophilic groups of the monomers leads to the formation of the shell encapsulating the magnetite nanoparticles in the polymerization feed. The drug loading and release behavior of these microcontainers was studied in different conditions, showing a pH and temperature dependence of the daunorubicin hydrochloride release rate. Triple-responsive semi-IPN hydrogels were prepared by radical polymerization, in the presence of the magnetite nanoparticles, of 3-acrylamidephenylboronic acid, N,N-dimethylaminoethylmethacrylate and MEBA, incorporating β-cyclodextrin-epichlorohydrin chains.[90] These semi-IPN hydrogels have a much higher drug-loading ratio for hydrophobic drug and could control the release of therapeutics by adjusting pH value and glucose concentration of release media. Bhattacharya et al. described the synthesis and characterization of multifunctional hybrid microgels by in situ formation of iron oxide nanoparticles in the polymeric structure of microgels exhibiting both temperature and pH sensitivities.[91] The synthesis of the polymeric microgels was performed by batch copolymerization of N-vinylcaprolactam, acetoacetoxyethyl methacrylate, and vinylimidazole, in aqueous medium in the presence of a crosslinking agent (MEBA) and water-soluble azoinitiator. Magnetite can be loaded into the microgels up to 15% (w/w), without any negative impact on the microgel

stability. The authors indicated that the magnetite nanoparticles are presumably located in the thermo-sensitive microgel shell, and lead to the partial shrinkage of the outer microgel layer due to the adsorption of the polymer chains. Despite their high magnetite content, the hybrid microgels possess considerable temperature and pH sensitivity and consequently are able to change their dimensions reversibly depending on the synergic effect of the three effects.

Modified Fe_3O_4 magnetic nanoparticles were proposed in the synthesis of a triple-responsive system consisting of oleic acid–coated magnetite nanoparticles, hydrophilic/thermoresponsive hexa(ethylene glycol) methyl ether methacrylate, hydrophobic/metal binding 2-(acetoacetoxy) ethyl methacrylate, and pH-responsive/thermoresponsive N-diethylaminoethyl methacrylate and N-diethylaminoethyl methacrylate moieties.[92] Single synthetic step free radical copolymerization was proposed for the synthesis of random conetworks, employing EGDMA as crosslinker, and AIBN as radical source. The presence of embedded oleic acid–coated magnetite nanoparticles within the conetworks, leads to nanocomposite materials demonstrating superparamagnetic behavior in the presence of an externally applied magnetic field. The ability of these materials to adsorb and desorb solutes in a controlled manner upon triggering the pH, combined with their tunable superparamagnetic behavior and thermoresponsive properties in aqueous media, allows their employment as drug delivery carriers.

Multiple stimuli–responsive organic/inorganic hybrid hydrogels were prepared via co-precipitation technique by combining dual stimuli–responsive p(2-(2-methoxyethoxy)ethyl methacrylate-co-oligo (ethylene glycol) methacrylate-co-AA) hydrogel with magnetic attapulgite/Fe_3O_4 nanoparticles.[93] Decorating attapulgite, a type of natural fibrillar aluminum silicate with abundant hydroxyl groups on the surface, with Fe_3O_4 magnetic nanoparticles via electrostatic attraction may give birth to novel chemical and physical properties, expecting to be applied in polymeric gels system (Fig. 10). Despite the magnetic functionality, the hydrogels possess considerable temperature/pH sensitivity and excellent mechanical properties, making them interesting candidates for design of artificial muscles, biosensors and actuators, and so on. Chang et al. proposed core–shell composite microspheres prepared by coating of p(NIPAAM-co-MAA) on magnetic mesoporous silica nanoparticles via precipitation polymerization.[94] Doxorubicin hydrochloride was applied as a model drug, and the behaviors of drug storage/release were investigated exhibiting an apparent thermo-/pH-response controlled drug release.

In triple-responsive nanoparticles, thermal and pH units are generally randomly distributed in the polymer chains, usually leading to mutual disturbance. To overcome this limitation, Isojima et al. have prepared two types of thermal/pH/magnetic triple nanoparticles consisting of 5 nm magnetite nanoparticles coated on one side with a

Table 3   Overview of triple-stimuli–responsive polymer for biomedical applications

Reactive species	Structure	Responsivity	Potential application	References
Fe_3O_4 nanoparticles, pNIPAAM, pAA	Semi-IPN hydrogel	Temperature/pH/magnetic	Drug delivery	[88]
Fe_3O_4 nanoparticles, p(NIPAAM-*co*-MAA)	Core–shell microspheres		Drug delivery	[89]
Fe_3O_4 nanoparticles, *N*-vinyl caprolactam, acetoacetoxy ethyl methacrylate, vinylimidazole, MEBA	Composite microgel		Hyperthermia cancer treatment	[91]
Oleic acid-coated Fe_3O_4 nanoparticles, hexa(ethylene glycol) methyl ether methacrylate, 2-(aceto acetoxy)ethyl methacrylate, *N*-diethylaminoethyl methacrylate, 2-(dimethylamino)ethyl methacrylate	Nanocomposite		Biomedical application	[92]
Attapulgite/Fe_3O_4 nanoparticles, p(2-(2-methoxyethoxy)ethyl methacrylate-*co*-oligo(ethylene glycol) methacrylate-*co*-AA)	Organic/inorganic hybrid hydrogels		Artificial muscles, biosensors and actuators	[93]
p(NIPAAM-*co*-MAA), magnetic mesoporous silica nanoparticles	Core–shell composite microsphere		Drug delivery	[94]
Fe_3O_4 nanoparticles, pAA, p-4-styrenesulfonic acid sodium salt hydrate), pNIPAAM	Janus nanoparticles		Biomedical application	[95]
Fe_3O_4 nanoparticles, p(3acrylamide phenylboronic acid-*co*-(2-dimethyl amino) ethyl methacrylate), β-cyclodextrin–epichlorohydrin	Semi-IPN hydrogels	Temperature/pH/magnetic/glucose	Drug delivery	[90]
Fluorescein isothiocyanate-labeled magnetic silica, NIPAAM, MEBA	Core–shell microspheres	Temperature/pH/light	Drug delivery, biomacromolecules separation, cell and protein labeling	[41]
pNIPAAM–allylamine, carboxylate spiropyran	Microgels		Biological applications	[96]
p(NIPAAM-*co*-*N*-hydroxymethyl acrylamide), 2-diazo-1,2-naphthoquinone	Graft copolymer		Drug delivery	[97]
Pluronic F127, 2-hydroxyethyl methacrylate, t-butyl methacrylate, cysteamine	Hydrogel	Temperature/pH/Redox	Biological applications	[98]
Monomethyl oligo(ethylene glycol) acrylate, 2-(5,5-dimethyl-1,3-dioxan-2-yloxy) ethyl acrylate,bis (2-acryloyloxyethyl)disulfide	Nanogels		Hydrophobic antitumor drugs delivery	[100]
pMAA, MAA, p(ethylene glycol) methyl ether methacrylate, MEBA, *N,N*′-bis(acryloyl)-cystamine	Polymer microcontainers	Magnetic/pH/Redox	Biomedical application	[99]
p(Propylene oxide)-containing triamine, 1,3-butadiene diepoxide	Nanogels		Drug delivery	[101]
Concanavalin A, dextran, pDEAM, horseradish peroxidase	Layer-by-layer film	Temperature/pH/salt	Bioelectrocatalysis	[106]
DEAM, 4-vinylpyridine, glucose oxidase	Hydrogel film		Bioelectrocatalysis	[107]
Phenylboronic acid, pAA, pDEAM, horseradish peroxidase	Semi-IPN hydrogel	Temperature/pH/glucose	Electrochemical biosensor, bioelectronic devices	[108]

Abbreviations: AA, acrylic acid; MAA, methacrylic acid; NIPAAM, *N*-isopropyl acrylamide; MEBA, *N,N*′-methylenebisacrylamide; DEAM, *N,N*-diethylacrylamide; IPN, interpenetrating polymer network.

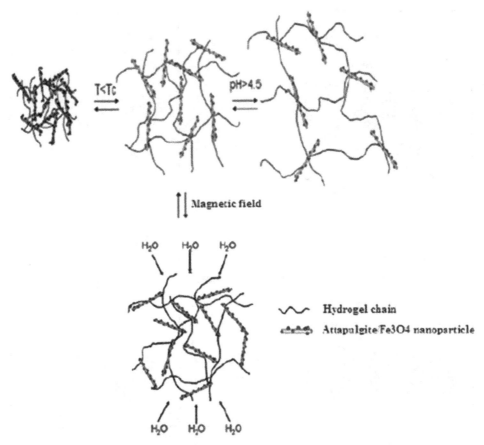

Fig. 10 Schematic of the changes in response to temperature, magnetic field, and pH.
Source: From Wang et al.[93] © 2012, with permission from Elsevier.

pH-dependent but temperature-independent polymer pAA, and functionalized on the other side by a second polymer, which is either pH independent (p-4-styrenesulfonic acid sodium salt hydrate) or temperature-dependent pNIPAAM.[95] Nanoparticles with this special structure, usually denoted as Janus nanoparticles, are dispersed stably as individual particles at high pH values and low temperatures, but can self-assemble at low pH values or at high temperatures (>31°C) to form stable dispersions of clusters of approximately 80–100 nm in hydrodynamic diameter (Fig. 11).

Functional microspheres with multistimuli-responsive properties and a well-defined core–shell structure were prepared employing a core made of magnetic silica labeled with fluorescein isothiocyanate, a popular fluorescent probe, whereas the shell is made of crosslinked pNIPAAM.[41] These systems were proposed as specific drug carriers using doxorubicin as model drug. In addition, by using these microspheres as tracers, the authors studied the magnetic-targeting effect by comparing the *in vivo* distribution of the microspheres in the presence and absence of an external magnetic field. The results suggest that such multistimuli-responsive materials, due to their photophysical and photochemical properties and sensitivity to both external magnetic field and environmental temperature, hold great potential for

the design of controlled drug delivery methods, in which microspheres loaded with a drug could be guided to the tissue of interest by using a magnetic field, followed by the release of the drug as a consequence of changes in the local temperature.

Triple-responsive particles could also be obtained by incorporating photosensitizers into thermal-/pH-responsive polymers. Garcia et al. developed triple-responsive nanoparticles by attaching photo-responsive spiropyran onto the thermal-/pH-responsive microgels based on pNIPAAM–allylamine.[96] Microgels with photo-, thermally, and pH-responsive properties in aqueous suspension have been synthesized first by preparing pNIPAAM–allylamine copolymer and a spiropyran photochrome bearing a carboxylic acid group. Then, the functionalized spiropyran was coupled to the microgel via an amide bond, using 1-ethyl-3-(3-dimethylaminopropyl) carbodiimide), as a condensing agent. After irradiation of visible light, the particle size becomes smaller because spiropyran changes to the relatively nonpolar, closed spiroform. The microgels undergo a volume phase transition in water from a swollen state to a collapsed state with increasing temperature under all light conditions. The microgels in darkness undergo a volume change from a swollen to a collapsed state around to the

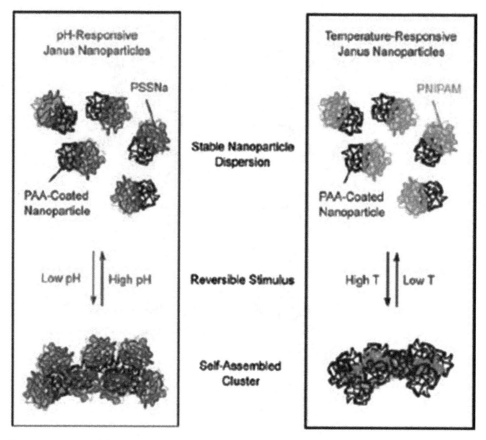

Fig. 11   Schematic representation of the Janus nanoparticles and their self-assembled structures on application of a stimulus.
Source: From Isojima et al.[95] © 2008, with permission from ACS.

transition temperature equal to 34.4°C, whereas under light irradiation, the transition temperature decreases to 32.1°C. This behavior is because the irradiation of the open, hydrophilic form causes photoisomerization to the closed, hydrophobic form, which is similar to the addition of a hydrophobic component to pNIPAAM to reduce its transition temperature. By using a similar method, Yu et al. prepared triple-responsive polymers, combining pH, photo-, and thermoresponsive properties in a single macromolecule, through incorporating photo-responsive 2-diazo-1,2-naphthoquinone into thermal-/pH-responsive polymers.[97] A series of p(NIPAAM-co-N-hydroxy-methylacrylamide) copolymers were firstly synthesized via free radical polymerization. Then, the hydrophobic, photosensitive molecules were partially and randomly grafted onto polymeric backbone through esterification to obtain a triple-stimuli (photo/pH/thermo)-responsive copolymer.

The high redox potential in the cytosol and cell nuclei containing 100–1000 times higher concentration of reducing glutathione tripeptide than body fluids, including blood and extracellular milieu, have recently been exploited for active intracellular release of various drugs. These findings made the insertion in a polymeric structure of redox-sensitive functionalities interesting in the design

of device useful in biomedical and pharmaceutical fields. Nam et al. designed and synthesized a multisensitive injectable hydrogel, which comprises pluronic as thermoresponsive polymer, a benzoic imine bond as acid-sensitive linker, and disulfide bond responsive to redox potential (Fig. 12A).[98] Crosslinked polymers showed the sol–gel phase transition behavior against acidic and redox environment, where a significant change in the behavior of the polymers was found. More recently, the synthesis of novel magnetic-, pH-, and redox-sensitive microcontainers, using a sacrificial template-directed synthesis procedure followed by chemical deposition of magnetic nanocrystals via co-precipitation, was proposed (Fig. 13).[99] The engineering of the nanoscopic device was performed by a two-stage distillation precipitation polymerization procedure, involving the coating on pMAA microspheres, acting as seeds, with MAA, p(ethylene glycol) methyl ether methacrylate, and MEBA and N,N'-bis(acryloyl)-cystamine, as crosslinkers. The fabricated magnetic microcontainers exert operative pH responsiveness, gradual and controlled collapse, once met with highly reducing environment, and efficient magnetic response as drug delivery vectors of daunorubicin hydrochloride. Multiresponsive nanogels were synthesized by miniemulsion copolymerization of monomethyl oligo(ethylene glycol) acrylate and an ortho

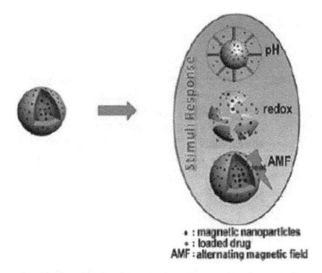

Fig. 12 Schematic representation of the pH-, thermo-, and redox-sensitive hydrogels.

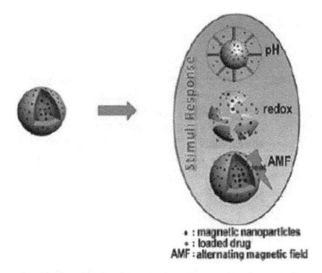

• : magnetic nanoparticles
• : loaded drug
AMF : alternating magnetic field

Fig. 13 Schematic representation of magnetic microcontainer.
Source: From Bilalis et al.[99] © 2012, with permission from RSC.

ester-containing acrylic monomer in the presence of a disulfide-containing crosslinker, bis(2-acryloyloxyethyl) disulfide.[100] These nanogels are thermoresponsive and labile in the weakly acidic or reductive environments. Hydrophobic compounds such as Nile Red, paclitaxel, and doxorubicin were loaded and the analyses of release behaviors of the drug-loaded nanogels showed that the drug release can be greatly accelerated by a cooperative effect of both acid-triggered hydrolysis and dithiothreitol-induced degradation (Fig. 14).

Compared with pH and temperature, oxidation-triggered release is less investigated, yet it is highly relevant in a biomedical field and up-regulation of potent oxidants has been associated with many pathological diseases. Thus, the use of oxidation as a trigger to achieve controlled drug release could be of great interest. In this context, Tang et al. proposed an environmentally friendly process to synthesize a multiple stimuli–responsive (temperature, pH, and

oxidants) polymer consisting of p(propylene oxide) containing triamine and 1,3-butadiene diepoxide (Fig 12B).[101] In principle, the nanogel formation is initiated by the phase separation of the in situ formed thermosensitive intermediate polymer into precursor particles at elevated temperature, followed by aggregation and further crosslinking of precursor particles through amine–epoxide reaction to yield robust nanogels. The applicability of this nanogel system to uptake and release therapeutic chemicals was investigated employing Nile red as a guest molecule, and its release from the carrier was monitored by using a fluorescence spectrophotometer immediately after addition of H_2O_2.

Recently, functional interfaces responsive to environmental stimuli have attracted increasing interest among researchers from various fields, especially bio-related area. These stimuli-sensitive interfaces or surfaces demonstrate great perspective as biosensors, drug delivery, bioseparation, biofuel cells, and so on.[102,103] Bioelectrocatalysis based on enzymatic reactions can provide an important foundation for fabricating electrochemical biosensors and other biodevices.[104] In this regard, the switchable bioelectrocatalysis induced by external stimuli is of interest because the reversible activation/deactivation of bioelectrocatalysis not only finds its application in controllable biosensors, but also presents the basis for information storage, data processing, and signal amplification.[105]

Yao et al. describe a synthetic procedure involving the assembling layer-by-layer of concanavalin A and dextran on pyrolytic graphite electrode surface by lectin–sugar biospecific interactions.[106] The p(N,N-diethylacrylamide) hydrogels containing horseradish peroxidase were then polymerized on their surface, forming the films with the binary architecture. The films demonstrated reversible pH-, thermo-, and salt-responsive on–off behavior toward electroactive probe $Fe(CN)_6^{3-}$ in its cyclic voltammetric responses, and the mechanism of the stimuli–response behavior is different in the inner and

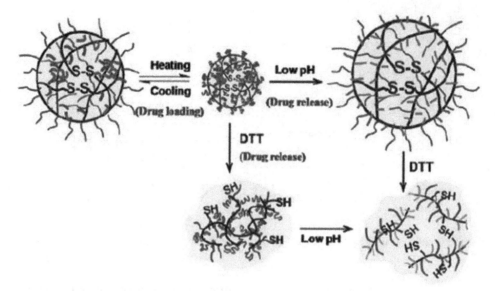

Fig. 14   Schematic representation of the pH-, thermo-, and dithiothreitol-sensitive hydrogels.
Source: From Qiao et al.[100] © 2011, with permission from Elsevier.

outermost layers. For the inner layers, the pH-sensitive property is attributed to the electrostatic interaction between the layers and the probe at different pH; for the p(N,N-diethylacrylamide)–horseradish peroxidase outermost layers, however, the thermo- and salt-sensitive behavior is ascribed to the structure change of the p(N,N-diethylacrylamide) at different temperatures and sulfate concentrations. This multitriggered switch can be used to realize the triply controllable electrochemical reduction of H_2O_2 catalyzed by enzyme immobilized in the films and mediated by $Fe(CN)_6^{3-}$ in solution. Recently, N,N-diethylacrylamide and 4-vinylpyridine were polymerized into copolymer hydrogel films with immobilized glucose oxidase on electrode surface with the simple one-step procedure under mild conditions, to produce a triply switchable bioelectrocatalysis.[107] The pH-sensitive property was attributed to the electrostatic interaction between the 4-vinylpyridine units of the films and the probe at different pH, whereas the thermo- and sulfate-sensitive behavior was ascribed to the structure change of N,N-diethylacrylamide constituent of the films with temperature and Na_2SO_4 concentration, respectively. The same research group proposed a semi-IPN film composed of pAA, phenylboronic acid, and p(N,N-diethylacrylamide) synthesized on electrode surface with entrapped horseradish peroxidase.[108] The films demonstrated reversible pH-, fructose-, and thermo-responsive on–off behavior toward electroactive probe $K_3Fe(CN)_6$ in its cyclic voltammetric response. The pH-responsive property of the system mainly originates from the electrostatic interaction between the pAA moieties of the films and the probe at different pH; the thermo-sensitive behavior is attributed to the structure change of poly(N,N-diethylacrylamide) hydrogel component with temperature whereas the fructose-responsive property is mainly

ascribed to the structure alteration of the films induced by the complexation between the phenylboronic acid constituent and the sugar, where the phenylboronic acid is the recognition unit and the p(N,N-diethylacrylamide) is the switching unit. The films can also be used to realize multiply switchable electrochemical reduction of H_2O_2 catalyzed by horseradish peroxidase entrapped in the films and mediated by $K_3Fe(CN)_6$ in solution.

CONCLUSIONS

The past several years have witnessed a rapid progress in the development of dual and multistimuli-responsive devices for the synthesis of polymeric materials suitable to be used as programmed site-specific drug delivery, scaffold for tissue engineering and in other biological applications. These multifunctional micro- and/or nanoparticles are able to elegantly address the challenging issues of current formulations, including aspects of preparation, *in vivo* stability, specific-targetability. This dual and multistimuli-responsive feature has offered unprecedented control over physicochemical properties of the materials, leading to superior *in vitro* and/or *in vivo* therapeutic effects. It should be noted, however, that research on dual stimuli- and multistimuli-responsive systems is at its infancy, and the devices reported in the literature are mostly proof-of concept studies. To achieve clinical impacts, future efforts shall be directed for the development of dual stimuli- and multistimuli-responsive biodegradable and noncytotoxic hydrogels, which can efficiently load and retain therapeutics in circulation, preferentially accumulate in the specific site, and quickly release drugs at the site of action in response to clinically viable external and/or internal stimuli.

ACKNOWLEDGMENTS

This work was supported by University of Calabria funds. Financial support of Regional Operative Program (ROP) Calabria ESF 2007/2013 – IV Axis Human Capital – Operative Objective M2 - Action D.5 is gratefully acknowledged.

REFERENCES

1. Fleige, E.; Quadir, M.A.; Haag, R. Stimuli-responsive polymeric nanocarriers for the controlled transport of active compounds: Concepts and applications. Adv. Drug Deliv. Rev. **2012**, *64* (9), 866–884.

2. Hoffman, A.S. Hydrogels for biomedical applications. Adv. Drug Deliv. Rev. **2002**, *54* (1), 3–12.

3. Imran, A.B.; Seki, T.; Takeoka, Y. Recent advances in hydrogels in terms of fast stimuli responsiveness and superior mechanical performance. Polym. J. **2010**, *42* (11), 839–851.

4. Deligkaris, K.; Tadele, T.S.; Olthuis, W.; van den Berg, A. Hydrogel-based devices for biomedical applications. Sensor Actuat B Chem. **2010**, *147* (2), 765–774.

5. Bawa, P.; Pillay, V.; Choonara, Y.E.; Du Toit, L.C. Stimuli-responsive polymers and their applications in drug delivery. Biomed. Mater. **2009**, *4* (2), 022001.

6. Tokarev, I.; Minko, S. Stimuli-responsive hydrogel thin films. Soft Matter **2009**, *5* (3), 511–524

7. Klouda, L.; Mikos, A.G. Thermoresponsive hydrogels in biomedical applications. Eur. J. Pharm. Biopharm. **2008**, *68* (1), 34–45.

8. Sanson, N.; Rieger, J. Synthesis of nanogels/microgels by conventional and controlled radical crosslinking copolymerization. Polym. Chem. UK **2010**, *1* (7), 965–977

9. Kuckling, D. Responsive hydrogel layers - From synthesis to applications. Colloid Polym. Sci. **2009**, *287* (8), 881–891.

10. Motornov, M.; Roiter, Y.; Tokarev, I.; Minko, S. Stimuli-responsive nanoparticles, nanogels and capsules for integrated multifunctional intelligent systems. Prog. Polym. Sci. **2010**, *35* (1–2), 174–211.

11. Leung, M.F.; Zhu, J.; Harris, F.W.; Li, P. New route to smart core-shell polymeric microgels: Synthesis and properties. Macromol. Rapid Comm. **2004**, *25* (21), 1819–1823

12. Jiang, J.; Hua, D.; Tang, J. One-pot synthesis of pH- and thermo-sensitive chitosan-based nanoparticles by the polymerization of acrylic acid/chitosan with macro-RAFT agent. Int. J. Biol. Macromol. **2010**, *46* (1), 126–130.

13. Zha, L.; Banik, B.; Alexis, F. Stimulus responsive nanogels for drug delivery. Soft Matter **2011**, *7* (13), 5908–5916.

14. Schmaljohann, D. Thermo- and pH-responsive polymers in drug delivery. Adv. Drug Deliver. Rev. **2006**, *58* (15), 1655–1670.

15. Echeverria, C.; Mijangos, C. Effect of gold nanoparticles on the thermosensitivity, morphology, and optical properties of poly(acrylamide-acrylic acid) microgels. Macromol. Rapid Comm. **2010**, *31* (1), 54–58.

16. Gil, E.S.; Hudson, S.M. Stimuli-reponsive polymers and their bioconjugates. Prog. Polym. Sci. **2004**, *29* (12), 1173–1222.

17. Xia, X.; Hu, Z. Synthesis and light scattering study of microgels with interpenetrating polymer networks. Langmuir **2004**, *20* (6), 2094–2098.

18. Zhang, Q.; Zha, L.; Ma, J.; Liang, B. A novel route to prepare pH- and temperature-sensitive nanogels via a semibatch process. J. Colloid Interface Sci. **2009**, *330* (2), 330–336.

19. Yu, R.; Zheng, S. Poly(acrylic acid)-grafted poly(N-isopropyl acrylamide) networks: Preparation, characterization and hydrogel behavior. J. Biomater. Sci. Polym. Ed. **2011**, *22* (17), 2305–2324.

20. Hu, J.; Zheng, S.; Xu, X. Dual stimuli responsive poly(N-isopropylacrylamide-co-acrylic acid) hydrogels based on a β-cyclodextrin crosslinker: Synthesis, properties, and controlled protein release. J. Polym. Res. **2012**, *19* (11), 9988.

21. Sun, S.; Wu, P. A one-step strategy for thermal- and pH-responsive graphene oxide interpenetrating polymer hydrogel networks. J. Mater. Chem. **2011**, *21* (12), 4095–4097.

22. Liu, S.; Liu, M. Synthesis and characterization of temperature- and pH-sensitive poly(N,N-diethylacrylamide-co-methacrylic acid). J. Appl. Polym. Sci. **2003**, *90* (13), 3563–3568.

23. Liu, H.; Liu, M.; Huang, J.; Ma, L.; Chen, J. A novel method to prepare temperature/pH-sensitive poly (N, N-diethylacrylamide-co-acrylic acid) hydrogels with rapid swelling/deswelling behaviors. Polym. Advan. Technol. **2009**, *20* (12), 1152–1156.

24. Shim, W.S.; Kim, J.-H.; Park, H.; Kim, K.; Chan Kwon, I.; Lee, D.S. Biodegradability and biocompatibility of a pH- and thermo-sensitive hydrogel formed from a sulfonamide-modified poly(ε-caprolactone-co-lactide)-poly(ethylene glycol)-poly(ε-caprolactone-co-lactide) block copolymer. Biomaterials **2006**, *27* (30), 5178–5185.

25. Pich, A.; Tessier, A.; Boyko, V.; Lu, Y.; Adler, H.-J.P. Synthesis and characterization of poly(vinylcaprolactam)-based microgels exhibiting temperature and pH-sensitive properties. Macromolecules **2006**, *39* (22), 7701–7707.

26. Fournier, D.; Hoogenboom, R.; Thijs, H.M.L.; Paulus, R.M.; Schubert, U.S. Tunable pH- and temperature-sensitive copolymer libraries by reversible addition-fragmentation chain transfer copolymerizations of methacrylates. Macromolecules **2007**, *40* (4), 915–920.

27. Li, X.; Zuo, J.; Guo, Y.; Cai, L.; Tang, S.; Yang, W. Volume phase transition temperature tuning and investigation of the swelling-deswelling oscillation of responsive microgels. Polym. Int. **2007**, *56* (8), 968–975.

28. Choi, S.M.; Singh, D.; Cho, Y.W.; Oh, T.H.; Han, S.S. Three-dimensional porous HPMA-co-DMAEM hydrogels for biomedical application. Colloid Polym. Sci. **2013**, *291* (5), 1121–1133.

29. Yun, J.; Kim, H.-I. Dual-responsive release behavior of pH-sensitive PVA/PAAc hydrogels containing temperature-sensitive PVA/PNIPAAm microcapsules. Polym. Bull. **2012**, *68* (4), 1109–1119.

30. Micic, M.; Stamenic, D.; Suljovrujic, E. Radiation-induced synthesis and swelling properties of p(2-hydroxyethyl methacrylate/itaconic acid/oligo (ethylene glycol)

Gene-Hydrogels

acrylate interpolymeric hydrogels. Radiat. Phys. Chem. **2012**, *81* (9), 1451–1455.

31. Liu, S.; Liu, X.; Li, F.; Fang, Y.; Wang, Y.; Yu, J. Phase behavior of temperature- and pH-sensitive poly(acrylic acid-g-N-isopropylacrylamide) in dilute aqueous solution. J. Appl. Polym. Sci. **2008**, *109* (6), 4036–4042.

32. Zhang, J.; Chu, L.-Y.; Cheng, C.-J.; Mi, D.-F.; Zhou, M.-Y.; Ju, X.-J. Graft-type poly(N-isopropylacryl-amide-co-acrylic acid) microgels exhibiting rapid thermo- and pH-responsive properties. Polymer **2008**, *49* (10), 2595–2603.

33. Chen, J.; Dai, P.; Liu, M. Rapid responsive behaviors of the dual stimuli-sensitive poly(DEA-co-DMAEMA) hydrogel via comb-type grafted polymeriziation. Int. J. Polym. Mater. **2012**, *61* (3), 177–198.

34. Spizzirri, U.G.; Iemma, F.; Puoci, F.; Xue, F.; Gao, W.; Cirillo, G.; Curcio, M.; Parisi, O.I.; Picci, N. Synthesis of hydrophilic microspheres with LCST close to body temperature for controlled dual-sensitive drug release. Polym. Adv. Technol. **2011**, *22* (12), 1705–1712.

35. Silva Nykänen, V.P.; Nykänen, A.; Puska, M.A.; Silva, G.G.; Ruokolainen, J. Dual-responsive and super absorbing thermally cross-linked hydrogel based on methacry-late substituted polyphosphazene. Soft Matter **2011**, *7* (9), 4414–4424.

36. Casolaro, M.; Casolaro, I.; Lamponi, S. Stimuli-respon-sive hydrogels for controlled pilocarpine ocular delivery. Eur. J. Pharm. Biopharm. **2012**, *80* (3), 553–561.

37. Jeong, U.; Teng, X.; Wang, Y.; Yang, H.; Xia, Y. Super-paramagnetic colloids: Controlled synthesis and niche applications. Adv. Mater. **2007**, *19* (1), 33–60.

38. Rosengart, A.J.; Kaminski, M.D.; Chen, H.; Caviness, P.L.; Ebner, A.D.; Ritter, J.A. Magnetizable implants and functionalized magnetic carriers: A novel approach for noninvasive yet targeted drug delivery. J. Magn. Magn. Mater. **2005**, *293* (1), 633–638

39. Sun, Q.; Reddy, B.V.; Marquez, M.; Jena, P.; Gonzalez, C.; Wang, Q. Theoretical study on gold-coated iron oxide nanostructure: Magnetism and bioselectivity for amino acids. J. Phys. Chem. C **2007**, *111* (11), 4159–4163.

40. Liu, X.; Guan, Y.; Ma, Z.; Liu, H. Surface modification and characterization of magnetic polymer nanospheres prepared by miniemulsion polymerization. Langmuir **2004**, *20* (23), 10278–10282.

41. Deng, Y.; Wang, C.; Shen, X.; Yang, W.; Jin, L.; Gao, H.; Fu, S. Preparation, characterization, and application of multistimuli-responsive microspheres with fluorescence-labeled magnetic cores and thermoresponsive shells. Chem. Eur. J. **2005**, *11* (20), 6006–6013.

42. Liu, T.-Y.;Hu, S.-H.; Liu, D.-M.; Chen, S.-Y.; Chen, I.-W. Biomedical nanoparticle carriers with combined thermal and magnetic responses. Nano Today **2009**, *4* (1), 52–65.

43. Pich, A.; Bhattacharya, S.; Lu, Y.; Boyko, V.; Adler, H.-J.P. Temperature-sensitive hybrid microgels with magnetic properties. Langmuir **2004**, *20* (24), 10706–10711.

44. Liu, T.-Y.; Hu, S.-H.; Liu, K.-H.; Shaiu, R.-S.; Liu, D.-M.; Chen, S.-Y. Instantaneous drug delivery of magnetic/ther-mally sensitive nanospheres by a high-frequency magnetic field. Langmuir **2008**, *24* (23), 13306–13311.

45. Kim, G.-C.; Li, Y.-Y.; Chu, Y.-F.; Cheng, S.-X.; Zhuo, R.-X.; Zhang, X.-Z. Nanosized temperature-responsive

Fe3O4-UA-g-P(UA-co-NIPAAm) magnetomicelles for controlled drug release. Eur. Polym. J. **2008**, *44* (9), 2761–2767.

46. Kalele, S.; Narain, R.; Krishnan, K.M. Probing temper-ature-sensitive behavior of pNIPAAm-coated iron oxide nanoparticles using frequency-dependent mag-netic measurements. J. Magn. Magn. Mater. **2009**, *321* (10), 1377–1380.

47. Liu, T.-Y.; Liu, K.-H.; Liu, D.-M.; Chen, S.-Y.; Chen, I.-W. Temperature-sensitive nanocapsules for controlled drug release caused by magnetically triggered structural disruption. Adv. Funct. Mater. **2009**, *19* (4), 616–623.

48. Jaiswal, M.K.; Banerjee, R.; Pradhan, P.; Bahadur, D. Thermal behavior of magnetically modalized poly(N-isopropylacrylamide)-chitosan based nanohydrogel. Col-loid Surf. B **2010**, *81* (1), 185–194.

49. Zhu, Y.; Kaskel, S.; Ikoma, T.; Hanagata, N. Magnetic SBA-15/poly(N-isopropylacrylamide) composite: Prepa-ration, characterization and temperature-responsive drug release property. Microporous Mesoporous Mater. **2009**, *123* (1–3), 107–112.

50. Guo, M.; Yan, Y.; Zhang, H.; Yan, H.; Cao, Y.; Liu, K.; Wan, S.; Huang, J.; Yue, W. Magnetic and pH-respon-sive nanocarriers with multilayer core-shell architec-ture for anticancer drug delivery. J. Mater. Chem. **2008**, *18* (42), 5104–5112.

51. Liu, G.; Wang, H.; Yang, X. Synthesis of pH-sensitive hol-low polymer microspheres with movable magnetic core. Polymer **2009**, *50* (12), 2578–2586.

52. Ma, W.; Xu, S.; Li, J.; Guo, J.; Lin, Y.; Wang, C. Hydrophilic dual-responsive magnetite/PMAA core/ shell microspheres with high magnetic susceptibility and pH sensitivity via distillation-precipitation polym-erization. J. Polym. Sci. A Polym. Chem. **2011**, *49* (12), 2725–2733.

53. Zhou, L.; Yuan, J.; Yuan, W.; Sui, X.; Wu, S.; Li, Z.; Shen, D. Synthesis, characterization, and controlla-ble drug release of pH-sensitive hybrid magnetic nanoparticles. J. Magn. Magn. Mater. **2009**, *321* (18), 2799–2804.

54. Banerjee, S.S.; Chen, D.-H. Multifunctional pH-sensitive magnetic nanoparticles for simultaneous imaging, sensing and targeted intracellular anticancer drug delivery. Nano-technology **2008**, *19* (50), 505104.

55. Akiyama, H.; Tamaoki, N. Synthesis and photoinduced phase transitions of poly(N-isopropylacrylamide) deriva-tive functionalized with terminal azobenzene units. Mac-romolecules **2007**, *40* (14), 5129–5132.

56. Luo, C.; Zuo, F.; Ding, X.; Zheng, Z.; Cheng, X.; Peng, Y. Light-triggered reversible solubility of α-cyclodextrin and azobenzene moiety complexes in PDMAA-co-PAPA via molecular recognition. J. Appl. Polym. Sci. **2008**, *107* (4), 2118–2125.

57. Jochum, F.D.; Theato, P. Temperature and light sensitive copolymers containing azobenzene moieties prepared via a polymer analogous reaction. Polymer **2009**. *50* (14), 3079–3085.

58. Feng, H.; Zhao, Y.; Pelletier, M.; Dan, Y.; Zhao, Y. Synthe-sis of photo- and pH-responsive composite nanoparticles using a two-step controlled radical polymerization method. Polymer **2009**, *50* (15), 3470–3477.

Gene–Hydrogels

59. Techawanitchai, P.; Ebara, M.; Idota, N.; Aoyagi, T. Light-induced spatial control of pH-jump reaction at smart gel interface. Colloid Surf. B **2012**, *99*, 53–59.

60. Ge, J.; Neofytou, E.; Cahill, T.J.; Beygui, R.E.; Zare, R.N. Drug release from electric-field-responsive nanoparticles. ACS Nano **2012**, *6* (1), 227–233.

61. Yun, J.; Im, J.S.; Lee, Y.-S.; Bae, T.-S.; Lim, Y.-M.; Kim, H.-I. pH and electro-responsive release behavior of MWCNT/PVA/PAAc composite microcapsules. Colloid Surf. A **2010**, *368* (1–3), 23–30.

62. Spizzirri, U.G.; Hampel, S.; Cirillo, G.; Nicoletta, F.P.; Hassan, A.; Vittorio, O.; Picci, N.; Iemma, F. Spherical gelatin/CNTs hybrid microgels as electro-responsive drug delivery systems. Int. J. Pharm. **2013**, *448* (1), 115–122.

63. Song, S.; Hu, N. pH-controllable bioelectrocatalysis based on "on-off" switching redox property of electroactive probes for spin-assembled layer-by-layer films containing branched poly(ethyleneimine). J. Phys. Chem. B **2010**, *114* (10), 3648–3654.

64. Ding, F.; Shi, X.; Jiang, Z.; Liu, L.; Cai, J.; Li, Z.; Chen, S.; Du, Y. Electrochemically stimulated drug release from dual stimuli responsive chitin hydrogel. J. Mater. Chem. **2013**, *1* (12), 1729–1737.

65. Pan, Y.-J.; Chen, Y.-Y.; Wang, D.-R.; Wei, C.; Guo, J.; Lu, D.-R.; Chu, C.-C.; Wang, C.-C. Redox/pH dual stimuli-responsive biodegradable nanohydrogels with varying responses to dithiothreitol and glutathione for controlled drug release. Biomaterials **2012**, *33* (27), 6570–6579.

66. Sui, X.; Feng, X.; Di Luca, A.; Van Blitterswijk, C.A.; Moroni, L.; Hempenius, M.A.; Vancso, G.J. Poly(N-isopropylacrylamide)-poly(ferrocenylsilane) dual-responsive hydrogels: Synthesis, characterization and antimicrobial applications. Polym. Chem. UK **2013**, *4* (2), 337–342.

67. Wang, D.; Liu, T.; Yin, J.; Liu, S. Stimuli-responsive fluorescent poly(N-isopropylacrylamide) microgels labeled with phenylboronic acid moieties as multifunctional ratiometric probes for glucose and temperatures. Macromolecules **2011**, *44* (7), 2282–2290.

68. Yin, R.; Tong, Z.; Yang, D.; Nie, J. Glucose and pH dual-responsive concanavalin A based microhydrogels for insulin delivery. Int. J. Biol. Macromol. **2011**, *49* (5), 1137–1142.

69. He, C.; Kim, S.W.; Lee, D.S. In situ gelling stimuli-sensitive block copolymer hydrogels for drug delivery. J. Control. Release **2008**, *127* (3), 189–207.

70. Alvarez-Lorenzo, C.; Concheiro, A. Intelligent drug delivery systems: Polymeric micelles and hydrogels. Mini-Rev. Med. Chem. **2008**, *8* (11), 1065–1074.

71. Foss, A.C.; Peppas, N.A. Investigation of the cytotoxicity and insulin transport of acrylic-based copolymer protein delivery systems in contact with caco-2 cultures. Eur. J. Pharm. Biopharm. **2004**, *57* (3), 447–455.

72. Cirillo, G.; Iemma, F.; Spizzirri, U.G.; Puoci, F.; Curcio, M.; Parisi, O.I.; Picci, N. Synthesis of stimuli-responsive microgels for *in vitro* release of diclofenac diethyl ammonium. J. Biomater. Sci. Polym. Ed. **2011**, *22* (4–6), 823–844.

73. Lv, W.; Liu, S.; Fan, X.; Wang, S.; Zhang, G.; Zhang, F. Gold nanoparticles functionalized by a dextran-based pH-And temperature-sensitive polymer. Macromol. Rapid Comm. **2010**, *31* (5), 454–458.

74. Fundueanu, G.; Constantin, M.; Ascenzi, P. Preparation and characterization of pH- and temperature-sensitive pullulan microspheres for controlled release of drugs. Biomaterials **2008**, *29* (18), 2767–2775.

75. Dumitriu, R.P.; Mitchell, G.R.; Vasile, C. Multi-responsive hydrogels based on N-isopropylacrylamide and sodium alginate. Polym. Int. **2011**, *60* (2), 222–233.

76. Sun, X.; Shi, J.; Zhang, Z.; Cao, S. Dual-responsive semi-interpenetrating network beads based on calcium alginate/poly(N-isopropylacrylamide)/poly(sodium acrylate) for sustained drug release. J. Appl. Polym. Sci. **2011**, *122* (2), 729–737.

77. Chen, Y.; Ding, D.; Mao, Z.; He, Y.; Hu, Y.; Wu, W.; Jiang, X. Synthesis of hydroxypropylcellulose-poly(acrylic acid) particles with semi-interpenetrating polymer network structure. Biomacromolecules **2008**, *9* (10), 2609–2614.

78. Gorgieva, S.; Kokol, V. Synthesis and application of new temperature-responsive hydrogels based on carboxymethyl and hydroxyethyl cellulose derivatives for the functional finishing of cotton knitwear. Carbohydr. Polym. **2011**, *85* (3), 664–673.

79. Xiao, C.; Xia, C.; Ma, Y.; He, X. Preparation and characterization of dual sensitive carboxylated methyl cellulose/poly(vinyl alcohol) physical composite hydrogel. J. Appl. Polym. Sci. **2013**, *127* (6), 4750–4755.

80. Lin, C.-L.; Chiu, W.-Y.; Lee, C.-F. Thermal/pH-sensitive core-shell copolymer latex and its potential for targeting drug carrier application. Polymer **2005**, *46* (23), 10092–10101.

81. Guo, B.; Yuan, J.; Gao, Q. Preparation and characterization of temperature and pH-sensitive chitosan material and its controlled release on coenzyme A. Colloid Surf. B **2007**, *58* (2), 151–156.

82. Chuang, C.-Y.; Don, T.-M.; Chiu, W.-Y. Synthesis of chitosan-based thermo- and ph-responsive porous nanoparticles by temperature-dependent self-assembly method and their application in drug release. J. Polym. Sci. A Polym. Chem. **2009**, *47* (19), 5126–5136.

83. Jiang, J.; Hua, D.; Tang, J. One-pot synthesis of pH- and thermo-sensitive chitosan-based nanoparticles by the polymerization of acrylic acid/chitosan with macro-RAFT agent. Int. J. Biol. Macromol. **2010**, *46* (1), 126–130.

84. Zhang, K.; Wang, Z.; Li, Y.; Jiang, Z.; Hu, Q.; Liu, M.; Zhao, Q. Dual stimuli-responsive N-phthaloylchitosan-graft-(poly(N-isopropylacrylamide)-block-poly(acrylic acid)) copolymer prepared via RAFT polymerization. Carbohyd. Polym. **2013**, *92* (1), 662–667.

85. Ma, L.; Liu, M.; Liu, H.; Chen, J.; Gao, C.; Cui, D. Dual crosslinked pH- and temperature-sensitive hydrogel beads for intestine-targeted controlled release. Polym. Advan. Technol. **2010**, *21* (5), 348–355.

86. Ma, L.; Liu, M.; Shi, X. pH- and temperature-sensitive self-assembly microcapsules/microparticles: Synthesis, characterization, *in vitro* cytotoxicity, and drug release properties. J. Biomed. Mater. Res. B **2012**, *100* (2), 305–313.

87. Cheng, R.; Meng, F.; Deng, C.; Klok, H.-A.; Zhong, Z. Dual and multi-stimuli responsive polymeric nanoparticles for programmed site-specific drug delivery. Biomaterials **2013**, *34* (14), 3647–3657.

88. He, F.; Zhang, Y.; Li, J.; Liu, S.; Chi, Z.; Xu J. Preparation and properties of multi-responsive semi-IPN hydrogel modified magnetic nanoparticles as drug carrier. Abstracts/J. Control. Release **2011**, *152* (Suppl. 1), e119–e121.

89. Bilalis, P.; Efthimiadou, E.K.; Chatzipavlidis, A.; Boukos, N.; Kordas, G.C. Multi-responsive polymeric microcontainers for potential biomedical applications: Synthesis and functionality evaluation. Polym. Int. **2012**, *61* (6), 888–894.

90. Huang, Y.; Liu, M.; Chen, J.; Gao, C.; Gong, Q. A novel magnetic triple-responsive composite semi-IPN hydrogels for targeted and controlled drug delivery. Eur. Polym. J. **2012**, *48* (10), 1734–1744.

91. Bhattacharya, S.; Eckert, F.; Boyko, V.; Pich, A. Temperature-, pH-, and magnetic-field-sensitive hybrid microgels. Small **2007**, *3* (4), 650–657.

92. Papaphilippou, P.; Christodoulou, M.; Marinica, O.-M.; Taculescu, A.; Vekas, L.; Chrissafis, K.; Krasia-Christoforou, T. Multiresponsive polymer conetworks capable of responding to changes in pH, temperature, and magnetic field: Synthesis, characterization, and evaluation of their ability for controlled uptake and release of solutes. ACS Appl. Mater. Interfaces **2012**, *4* (4), 2139–2147.

93. Wang, Y.; Dong, A.; Yuan, Z.; Chen, D. Fabrication and characterization of temperature-, pH- and magnetic-field-sensitive organic/inorganic hybrid poly (ethylene glycol)-based hydrogels. Colloid Surf. A **2012**, *415*, 68–76.

94. Chang, B.; Sha, X.; Guo, J.; Jiao, Y.; Wang, C.; Yang, W. Thermo and pH dual responsive, polymer shell coated, magnetic mesoporous silica nanoparticles for controlled drug release. J. Mater. Chem. **2011**, *21* (25), 9239–9247.

95. Isojima, T.; Lattuada, M.; Vander Sande, J.B.; Hatton, T.A. Reversible clustering of pH- and temperature-responsive Janus magnetic nanoparticles. ACS Nano **2008**, *2* (9), 1799–1806.

96. Garcia, A.; Marquez, M.; Cai, T.; Rosario, R.; Hu, Z.; Gust, D.; Hayes, M.; Park, C.-D. Photo-, thermally, and pH-responsive microgels. Langmuir **2007**, *23* (1), 224–229.

97. Yu, Y.Y.; Tian, F.; Wei, C.; Wang, C.C. Facile synthesis of triple-stimuli (photo/pH/thermo) responsive copolymers of 2-diazo-1,2-naphthoquinone-mediated poly(N-isopropylacrylamide-co-n-hydroxymethylacrylamide). J. Polym. Sci. A Polym. Chem. **2009**, *47* (11), 2763–2773.

98. Nam, J.A.; Al-Nahain, A.; Hong, S.; Lee, K.D.; Lee, H.; Park, S.Y. Synthesis and characterization of a multi-sensitive crosslinked injectable hydrogel based on Pluronic. Macromol. Biosci. **2011**, *11* (11), 1594–1602.

99. Bilalis, P.; Chatzipavlidis, A.; Tziveleka, L.-A.; Boukos, N.; Kordas, G. Nanodesigned magnetic polymer containers for dual stimuli actuated drug controlled release and magnetic hyperthermia mediation. J. Mater. Chem. **2012**, *22* (27), 13451–13454.

100. Qiao, Z.-Y.; Zhang, R.; Du, F.-S.; Liang, D.-H.; Li, Z.-C. Multi-responsive nanogels containing motifs of ortho ester, oligo(ethylene glycol) and disulfide linkage as carriers of hydrophobic anti-cancer drugs. J. Control. Release **2011**, *152* (1), 57–66.

101. Tang, S.; Shi, Z.; Cao, Y.; He, W. Facile aqueous-phase synthesis of multi-responsive nanogels based on polyetheramines and bisepoxide. J. Mater. Chem. **2013**, *1* (11), 1628–1634.

102. Cole, M.A.; Voelcker, N.H.; Thissen, H.; Griesser, H.J. Stimuli-responsive interfaces and systems for the control of protein-surface and cell-surface interactions. Biomaterials **2009**, *30* (9), 1827–1850.

103. Nandivada, H.; Ross, A.M.; Lahann, J. Stimuli-responsive monolayers for biotechnology. Prog. Polym. Sci. **2010**, *35* (1–2), 141–154.

104. Murphy, L. Biosensors and bioelectrochemistry. Curr. Opin. Chem. Biol. **2006**, *10* (2), 177–184.

105. Pita, M.; Katz, E. Switchable electrodes: How can the system complexity be scaled up? Electroanalysis **2009**, *21* (3–5), 252–260.

106. Yao, H.; Hu, N. Triply responsive films in bioelectrocatalysis with a binary architecture: Combined layer-by-layer assembly and hydrogel polymerization. J. Phys. Chem. B **2011**, *115* (20), 6691–6699

107. Liang, Y.; Liu, H.; Zhang, K.; Hu, N. Triply switchable bioelectrocatalysis based on poly(N,N-diethylacrylamide-co-4-vinylpyridine) copolymer hydrogel films with immobilized glucose oxidase. Electrochim. Acta **2012**, *60*, 456–463

108. Liu, D.; Liu, H.; Hu, N. pH-, sugar-, and temperature-sensitive electrochemical switch amplified by enzymatic reaction and controlled by logic gates based on semi-interpenetrating polymer networks. J. Phys. Chem. B **2012**, *116* (5), 1700–1708.

In Vitro Vascularization: Tissue Engineering Constructs

Cai Lloyd-Griffith
Tara M. McFadden
Garry P. Duffy
Fergal J. O'Brien
Tissue Engineering Research Group, Royal College of Surgeons in Ireland, Dublin, Ireland
Trinity Center for Bioengineering, Trinity College, Dublin, Ireland
Advanced Materials and BioEngineering Research Center (AMBER), Dublin, Ireland

Abstract

The absence of a sufficient vascular supply within tissue engineering (TE) constructs has been established as one of the major limiting factors in cell survival and implant success in the field of TE to date. Recently, cell-based approaches that involve the engineering of a nascent vasculature within a construct *in vitro*, prior to implantation, has emerged as a potential solution to overcome the issue of implant failure as a result of avascular necrosis. In this entry, we discuss this approach of "*in vitro* vascularization" and detail the numerous methods used to implement it. In addition, we discuss a number of approaches used to promote *in vitro* vascularization of TE constructs, including external stimuli and growth factor release. A number of approaches that involve the promotion of vascularization of scaffolds postimplantation are also discussed.

INTRODUCTION

Tissue engineering (TE) can be defined as a highly interdisciplinary field that aims to restore, maintain, or improve tissue function through the use of a biological substitute.[1] TE constructs show great promise for the regeneration or replacement of both relatively simple organs (e.g., skin)[2,3] and perhaps even more complex structures (e.g., kidney, heart).[4,5] This has been reflected in the work by Raya-Rivera et al., who successfully carried out urethral reconstructions using tissue-engineered urethras[6] and Atala et al., who engineered replacement bladder tissue using collagen–polyglycolic acid scaffolds,[7] both seeded with autologous urothelial and muscle cells. However, the rapid vascularization of implanted tissues is often critical to the success of the implant. In most tissues, cells are capable of surviving only within the range of oxygen and nutrient diffusion (approximately 150–200 μm) of the nearest network of blood vessels.[8] This is especially true in thicker tissues, where proximity to a vascular network is essential for cell survival and homeostasis. This is due to the delayed nature of oxygen diffusion, which occurs more slowly than its consumption. As a result, the absence of a sufficient vascular supply has been established as a major limiting factor in cell survival and implant success in TE.[9,10] The successful engraftment of TE constructs designed to regenerate thick tissues relies on the rapid formation of a stable and functional vasculature postimplantation.[11,12] Therefore, the successful vascularization

of thick constructs for tissue repair remains a major challenge.

A more recent approach of increasing interest, which involves the engineering of a nascent vasculature within a construct *in vitro*, prior to implantation, has emerged as a potential solution to overcome the issue of implant failure as a result of avascular necrosis.[11,13,14] This method of "*in vitro*-vascularization" provides a functional vasculature within the construct, thereby reducing the reliance on vessel invasion from the host, which often occurs too slowly to permit construct viability. Although this system does not provide an instantaneous blood supply, it does allow for quick anastomosis with the host vessels through a process known as "wrapping and tapping," in which the nascent vessels surround nearby host vessels and disrupt underlying host endothelium, leading to the formation of connected, functional vasculature that links the two networks.[15] This connection of vasculature between construct and host significantly enhances the effectiveness of the implanted construct by facilitating the perfusion of oxygen and nutrients throughout.[16] The *in vitro* vascularization approach presents a possible solution to a problem that continues to reduce the therapeutic potential of thick TE constructs.

This entry will focus on some of the advantages and disadvantages of several strategies currently being employed to induce *in vitro* vascularization with the aim of overcoming avascular necrosis in TE constructs. A number of approaches that involve the vascularization of a construct once it is implanted *in vivo* will also be discussed.

In Vitro–Medical

Concise Encyclopedia of Biomedical Polymers and Polymeric Biomaterials DOI: 10.1081/E-EBPPC-120051072

723

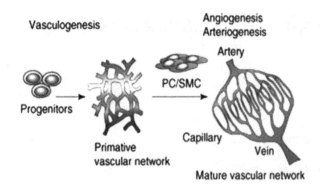

Fig. 1 The process of blood vessel formation. This occurs via several processes: vasculogenesis where the primary networks are formed, followed by angiogenic remodeling that involves the modification of preexisting vasculature. Other stages involved are stabilization and maturation, destabilization, regression, and the sprouting of capillaries from preexisting vessels.
Source: From Rouwkema et al.[245] © 2008, with permission from Elsevier.

VASCULOGENESIS AND ANGIOGENESIS

Blood vessel formation typically occurs via two different processes: 1) angiogenesis, which is the formation of capillaries from the differentiated endothelium of established vasculature; and 2) vasculogenesis, which is the in situ assembly of capillaries from undifferentiated endothelial cells (ECs) (Fig. 1). Until recently, it was thought that postnatal vascularization occurred only by angiogenesis.[17] It is now known, however, that both vasculogenesis and angiogenesis are responsible for the formation of adult blood vessels.[18]

Angiogenesis is a morphogenic process involving the sprouting of capillaries from preexisting vessels and involves six basic steps whereby the new capillaries undergo several remodeling stages to develop into mature vessels: 1) vasodilation of the original vessel; 2) degradation of the basement membrane of the original vessel by proteolytic enzymes; 3) EC migration and proliferation to form the beginnings of a capillary structure; 4) lumen formation and vessel formation; 5) development of the basement membrane; and 6) stabilization of the vasculature through recruitment of perivascular cells.[1,19] Vasculogenesis is the formation of new blood vessels that occurs during embryogenesis or within an avascular region by circulating endothelial precursor cells[20–22] acting in response to particular cues such as growth factors.[23,24] There are five main steps involved in vasculogenesis: 1) ECs are derived from precursor cells; 2) ECs form the vessel primordia and cell–cell arrangements but no lumen is present; 3) an immature tube-like structure is formed containing polarized ECs; 4) a vascular network is established from several EC tubules; and 5) stabilization of the vascular network through recruitment of perivascular cells.

Another process known as arteriogenesis is responsible for the provision of a blood supply to tissues that have suffered from a reduced blood supply. The trigger for arteriogenesis is primarily fluid shear stress and results in the remodeling of arterioles into developed and highly functional arteries.[25–27] However, for the purposes of creating a vascularized construct *in vitro* for TE, it is the process of vasculogenesis that is the primary focus as it does not require the presence of a preexisting vasculature.

BIOMATERIAL-BASED SCAFFOLDS FOR USE IN TISSUE REPAIR

When creating scaffolds for tissue repair, a number of factors must be considered to ensure the manufacture of the necessary structural framework to restore functionality to damaged tissues, as well as the provision of a suitable environment to support and promote cellular attachment, proliferation, and differentiation.[28,29] Implanted constructs must aim to replace the extracellular matrix (ECM), closely mimic the host environment, and function from the time of implantation to the completion of the remodeling process.[29,30] For successful implantation, the biomaterials used to fabricate scaffolds require certain characteristics, which include (but are not limited to): biocompatibility, biodegradability, mechanical properties, and architecture (including porosity, pore size, pore interconnectivity and structure).[29–31]

The most important consideration when choosing a suitable biomaterial for use in the creation of a scaffold is biocompatibility. Biocompatibility is the ability of a scaffold to support cell adherence, growth, and proliferation without eliciting a severe or chronic immune reaction, which could lead to rejection by the body. However, localized inflammation is often required, to some extent, in order to induce healing and tissue repair[32] and to promote the vascularization of an avascular construct. It is essential that this localized inflammation does not become chronic, as this may result in an immune response, which could be detrimental to the healing process.[30,33] Biodegradability is another major consideration in the design of scaffolds. An implanted construct should not be permanent, as it must degrade in order to allow the body to produce its own matrix to replace the damaged tissue, allowing for natural healing to take place.[33] This natural degradation bypasses the need for later removal of the construct via surgery,[34] which could lead to further damage, thus preventing the injured tissue from fully healing. In addition, by-products from the process of degradation must not have a toxic effect on the surrounding cells and should be easily removed via standard metabolic pathways.[31]

The mechanical properties of a scaffold have major effects on its ability to support vessel formation. Substrate stiffness has been shown to affect the type of vasculature formed, for example, Yamamura et al. showed that the stiffness of a collagen gel directly influenced the nature of the three dimensional (3D) capillary-like tubule network formed by bovine pulmonary microvascular ECs. The less-rigid gels facilitated greater cell migration but the vascular branches formed were thin with small lumens, whereas the rigid gels restricted migration but promoted the growth of a thick and continuous network that penetrated deeper into the gel.[35] Despite this thicker network formation, it is commonly believed that stiffer substrates negatively affect the quality of vessels formed.[36] This may be due to the fact that ECs are unable to form stable vascular structures on more rigid substrates as the cells are unable to generate the necessary contractile forces to remodel the substrate and manipulate their environment.[35,37,38]

Architecture is also an important factor in promoting vascularization. Architecture, in particular pore size, has been shown to have a profound effect on a number of important stem cell characteristics. For example, in a study from our own lab, Murphy et al. showed that the mean pore size of collagen–glycosaminoglycan scaffolds can affect cell migration, infiltration, and adhesion of the stem cells seeded within it.[39,40] Along with pore size, interconnectivity between pores is necessary for the undisturbed formation of the developing vascular networks throughout the scaffold.[41–44] There is no consensus on the optimal scaffold architecture to facilitate vascularization but Bai et al., using a β-tricalcium phosphate scaffold with pores that could be accurately adjusted using slip casting, demonstrated that an increase in pore size up to 400 μm led to an increase in both the size and number of blood vessels distributed throughout the scaffold,[45] although this possibly differs between scaffold types. The interconnections between pores can also dictate vessel formation. It was also found that interconnections between pores less than 150 μm produced a "bottleneck" effect, which inhibited vessel growth whereas larger interconnections facilitated larger vessel growth.[42,44,45] These architectural considerations greatly affect the development of scaffolds for vascularization (Fig. 2).

The biomaterials used to fabricate scaffolds for regenerative purposes (the examples used in this case apply mainly to bone) broadly fall into three categories: 1) ceramics; 2) synthetic polymers; and 3) natural polymers, each group with their own benefits and limitations.

Ceramic materials such as calcium phosphate and hydroxyapatite are characterized by their high mechanical stiffness and low elasticity. Despite being composed of artificial material, ceramics have both been shown to be biocompatible and capable of supporting both bone and vascular network formation.[44–46] For example, Choong et al. demonstrated that hydroxyapatite-coated polycaprolactone (PCL) substrates were superior for the attachment and proliferation of a human bone marrow–derived fibroblast/EC co-culture created for the purpose of vascularization, in comparison to untreated PCL substrates.[47] The inherent biocompatibility of ceramics is due to their chemical and structural similarity to the mineral phase of bone.

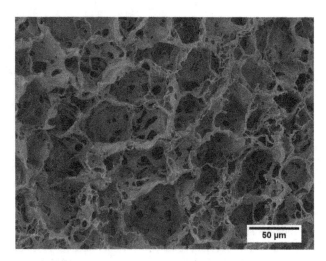

Fig. 2 Scanning electron micrograph (1000×) of a collagen nanohydroxyapatite scaffold depicting the typical porous nature of TE scaffolds.
Source: From Cunniffe et al.[246]

However, the use of ceramic biomaterials has a number of drawbacks, namely that they are brittle, hard to manipulate, and display a slow rate of biodegradation, thus limiting their suitability for vascularization and soft tissue repair.

Synthetic polymers such as poly(lactic-*co*-glycolic acid)[48] polylactide acid, polyglycolic acid, and the previously mentioned PCL[49,50] have all shown promise for use in regenerative medicine. Synthetic polymer scaffold have previously been shown to be capable of supporting vascular network formation.[51–53] For example, Fuchs et al. successfully induced the formation of vessel-like structures on starch PCL fiber meshes using a co-culture of outgrowth endothelial cells (OECs) and human primary osteoblasts.[54] One of the major advantages of synthetic polymer scaffolds is that they can be specifically tailored (via their polymer composition) to meet the requirements for the repair of specific tissues (e.g., degradation rate and physical properties). However, a major disadvantage of synthetic biomaterials is that they have been known to release acidic degradation products that lower the pH at the implantation site, which could affect biocompatibility leading to a foreign body response by the immune system or ultimately lead to tissue necrosis.[55,56]

It is common practice to use natural polymers such as collagen, glycosaminoglycans, elastin, silk, and fibrin for the creation of scaffolds for tissue repair.[14,57–61] These polymers are isolated from the native ECM and as such naturally contain ligands associated with cell adhesion and proliferation. The advantage of using natural biomaterials over other available materials is that they act as a biomimetic support, providing cells with an environment similar to that of native ECM in which they can attach to, proliferate, and differentiate within. These scaffolds are also highly biocompatible, have high porosity, and produce nontoxic degradation products.[62] A large portion of research regarding natural biomaterials has been carried out using

In Vitro–Medical

hydrogel-based materials.[16,20,63–66] Vascularizing a hydrogel is a less complex process than 3D porous scaffolds as the complex architecture can restrict movement of vessel-forming cells that attach to the scaffold struts and are less free to move and form networks than within the fine protein mesh of a hydrogel. This is due to the fact that hydrogels provide an environment that is easily manipulated. Hydrogels can also be fabricated with a tailored architecture and have controllable degradation properties based on polymer composition.[67,68] Much progress has been made vascularizing natural hydrogels such as Matrigel.[65,66,69] Matrigel consists of an assortment of ECM proteins that have been extracted from Englebreth-Holm-Swarm tumors in mice.[70] However, due to its origins, Matrigel could be potentially tumorigenic, rendering it unsuitable for clinical application despite its pro-angiogenic properties,[71,72] although it is typically used only as a research material to better understand vasculogenic/angiogenic processes. A major disadvantage of natural-polymer-based biomaterials in general is that they do not provide high mechanical strength, which may limit their use in load-bearing applications such as bone repair.

In order to overcome such disadvantages, a considerable amount of research has focused on the development of composite biomaterials that consist of a combination of natural, synthetic, and ceramic materials. Composite biomaterials are designed in order to utilize the key properties of each material type while minimizing (or possibly eliminating) the negative attributes of the other. Combinations of all three groups have arisen, with a number of groups utilizing synthetic–natural polymer composites,[73–75] while others have focused on incorporating ceramic material in natural polymer-based scaffolds[76–78] and creating synthetic–ceramic composites[47,56,79] for use in orthopedic repair. The creation of composite biomaterial using components from the same category (e.g., a natural–natural polymer composite) in order to enhance their capacity for tissue repair is also common practice.[58,80,81]

The design of future biomaterials for vascularization will need to consider the balance of the architectural requirements of the biomaterial while still satisfying the biological and mechanical demands of the tissue at the implantation site.

CELL SOURCES TO PROMOTE VASCULARIZATION

Endothelial Cells

To evaluate cell-based model systems for vasculogenesis and angiogenesis, it is crucial to select the appropriate cell type for this purpose. ECs are commonly selected for investigation and have proved highly valuable for cellular therapy, disease models, drug delivery studies, and especially for TE applications such as the generation of prevascularized constructs.[82] ECs can differ in structure and phenotype depending on vessel type[83] and there are a number of EC sources available for TE, for example, microvascular [human dermal microvascular endothelial cells (HDMECs)[11] and macrovascular sources (human coronary artery ECs)].[84] Therefore, selection of an EC source is highly dependent on the specific purpose of investigation and there are associated advantages and disadvantages with every choice. For *in vitro* vascularization of TE constructs, macrovascular human umbilical vein endothelial cells (HUVECs) are a commonly utilized EC type due to the fact that they have been shown to spontaneously self-organize to produce functional vessel-like structures *in vitro* and *in vivo*.[16,85–88] HUVECs are mature, committed ECs, and are often used to study EC function, morphology, and responses to stimuli such as flow and shear stresses for vascular development.[89] Compared to other sources of ECs, HUVECs are easily extracted from an available supply of discarded umbilical cords and can be expanded to large numbers. However, there are associated issues with the use of a macrovascular cell source, as angiogenesis/vasculogenesis typically occurs within the microvasculature. Therefore, the use of a microvascular cell source may be a better representation of the *in vivo* process.[90,91] HDMECs are a commonly used microvascular EC type isolated from juvenile foreskin. These cells are an appealing source of ECs due to their abundance and availability from a tissue, which would otherwise be discarded. HDMECs demonstrate a stable phenotype for a number of passages in culture[92] and have shown to behave similar to HUVECs *in vitro*.[87,89,93] HDMECs have also been shown to be capable of creating vascular networks both *in vitro* and *in vivo*, demonstrating their suitability for vascularization.[11,94,95]

In order to produce larger and more complex TE constructs, a readily available source of ECs are required and isolation of ECs from sources such as the aorta or umbilical vein is not feasible for clinical translation. Furthermore, fully mature ECs have been shown to have a slow expansion rate and limited proliferation capabilities, requiring that a large number of ECs be harvested for therapeutic use.[16] Therefore, there is an increased need to obtain an alternative EC source for vascularization. Recent research has established that multipotent stem cells types may have potential for this purpose.[16]

Endothelial Progenitor Cells

In recent years, endothelial progenitor cells (EPCs) have come to the forefront as an alternative source of vessel forming cells.[96,97] EPCs offer an accessible, autologous source of ECs. They can be easily isolated from peripheral and umbilical cord blood[97] and demonstrate robust proliferation, self-renewal, and superior vessel-forming ability compared with mature ECs.[97] EPCs are also clinically safer to use than, for example, embryonic stem cells, as they differentiate exclusively down an EC lineage. Several studies have demonstrated the vasculogenic ability of EPCs to form functional blood vessels *in vivo*, similar to those formed by HUVECs[98] and have been frequently used in the vascularization of TE constructs.[20,49,51,63,87,99,100] EPCs consist of two different subpopulations, termed early and late EPCs. Although both EPCs are derived from mononuclear cells and express EC markers, they have different morphologies and growth patterns. Early EPCs exhibit a spindle-like morphology and the majority are derived from CD14 positive subpopulations (CD14 positive subpopulations are monocytes capable of differentiating into different cells types). Late EPCs, named because of their late outgrowth potential, exhibit a cobblestone-like morphology and are derived from CD14 negative fractions. In particular, late EPCs are better suited for *in vitro* vascularization as they have been characterized for production of vascular endothelial growth factor (VEGF), the expression of numerous vascular cell surface markers such as CD31, von Willebrand Factor (vWF), and VEGF receptor 2 (VEGFR-2), and they also lack expression of hematopoietic-specific CD45 and CD14 markers.[21,49,98,101,102]

Bone Marrow–Derived Mesenchymal Stem Cells

Bone marrow–derived mesenchymal stem cells (MSCs) are frequently used in TE due to their ability to differentiate along chondrogenic, osteogenic, and adipogenic lineages. As a result of their trilineage potential and high expansion capacity, MSCs have been proposed frequently as an ideal cell source for clinical applications. Of particular interest to this entry, MSCs have been shown to transdifferentiate into vascular ECs both *in vitro* and *in vivo*.[103,104]

Initial studies carried out by Oswald et al. demonstrated that MSCs cultured in VEGF-supplemented media for 7 days acquired major characteristics of mature ECs; expressing vWF, VEGFR-1 and VEGFR-2, VE-cadherin, and VCAM-1. CD31 and CD34 were not shown to be expressed although they are known to occur later in EC differentiation.[103] In some cases, supplementation with VEGF alone has been sufficient to induce expression of CD31 after 10 days in culture whereas other groups have shown that a combination of supplements lead to more advanced differentiation (e.g., VEGF and insulin-like growth factor (IGF-1),[105] or VEGF and basic fibroblast growth factor (bFGF)).[106] MSCs isolated from other tissues such as Wharton's jelly have demonstrated an increased capacity to differentiate down an EC lineage with increased expression of VEGFR-2, vWF, and VE-cadherin compared to bone marrow–derived MSCs.[107] There is still a question over whether ECs derived from MSCs are ideal for *in vitro* and *in vivo* vascularization as there is doubt over their functionality with this acquired phenotype.[108] More recent evidence suggests that MSCs, which are found in perivascular locations, are in fact pericytes[109,110] and have been shown to support *in vitro* vessel formation in co-culture with ECs. Our group has specifi-

cally shown that delayed addition of MSCs to preformed nascent EC-generated vascular networks in a collagen-glycosaminoglycan scaffold results in well developed vasculature, which can then successfully anastomose with host vessels *in vivo*.[111]

Adipose-Derived Stem Cells

Adipose tissue is another readily available source of multipotent adult stem cells for use in TE. Adipose-derived stem cells (hASCs) derived from an aged population revealed that neither advanced age or co-morbidity negatively impacts stem cell harvest or differentiation capacity,[112] which is often an issue with other sources of stem cells.[113,114] There has been some evidence to suggest that hASCs are also capable of differentiating down an EC lineage.[112,115–117] It has been shown that hASCs cultured in EC-supplemented media and exposed to shear stress were capable of forming vessel-like structures, acetylated low-density lipoprotein uptake, and expression of CD31. However, expression of eNOS and vWF was not achieved.[116] More recently, hASCs cultured on an electrospun fiber mesh in endothelial growth media (containing VEGF and bFGF) expressed mature EC markers such as VE-cadherin and vWF after 21 days, thus indicating their potential as an accessible source of ECs for the purposes of *in vitro* vascularization.[115] However, as with bone marrow–derived MSCs, there are certain disadvantages with these stem cells. They require a great amount of manipulation and long culture times to achieve EC differentiation and often do not differentiate fully down this lineage, with the lack of mature marker expression often evident.[108,116]

Amniotic Fluid–Derived Stem Cells

Amniotic fluid–derived stem cells (AFSCs) are a pluripotent stem cell type isolated from backup samples of amniotic fluid retrieved during routine amniocentesis.[118] Unlike a number of multipotent stem cell types, AFSCs are capable of retaining chromosomal stability and stem cell plasticity over long periods of culture.[118,119] AFSCs have been successfully differentiated into functional endothelial-like cells using standard EC growth media supplemented with varying concentrations of VEGF. These differentiated endothelial-like cells display upregulated expression of endothelial markers such as vWF, eNOS, VE-cadherin, and VEGFR-2 and are capable of both tubule formation on Matrigel and uptake of acetylated low-density lipoprotein.[120,121] For these reasons, AFSCs show promise for use as an EC source in the vascularization of a TE construct. However, similar to ASCs and MSCs, AFSCs require a long culture time and a high degree of manipulation to achieve EC differentiation. This cell type has also not been as thoroughly investigated as the others previously mentioned.

In Vitro–Medical

CO-CULTURE SYSTEMS FOR VASCULARIZATION

Even if the ideal EC source for vascularization is determined, it would still be necessary to simulate the relevant cellular and molecular microenvironment in order to replicate the vasculature in a physiologically accurate fashion.[122] This is due to the fact that nascent vessels formed by endothelial progenitors (or by stabilized vessels during angiogenesis) are immature, unstable structures that are susceptible to regression and require stabilization from support cells commonly referred to as pericytes.[123–125] This regression is also seen *in vitro* as vasculature formed using EC monocultures often result in the development of similarly unstable structures.[23,108,65] Larger, more developed blood vessels (veins and arteries) also require support, but from vascular smooth muscle cells (SMCs). A major distinction between SMCs and pericytes is that pericytes cover smaller vessels, like capillaries, one cell at a time, while SMCs cover vessels in either mono or multi-layers.[124] Pericyte-mediated stabilization of the vessel is achieved through two basic mechanisms: the assembly of the vascular basement membrane and the recruitment and integration of perivascular cells into the vessel wall.[123] Stabilization not only relies on physical and mechanical stimulation but on the release of signals that promote EC quiescence and differentiation[126] [angiogenic cytokines secreted by pericytes also play a large role in the support of vessel formation, enhancing the migration, proliferation, and vessel-forming abilities of ECs].[127–130] The mechanism by which pericytes function is not understood in great detail, possibly due their heterogeneous nature and the inherent difficulty in outlining their development. It is suggested that both SMCs and pericytes represent phenotypic variants of a continuous population of mural cells, sharing the same lineage[124] indicating that despite their physical variation, this heterogeneous group of cells can still carry out a similar function.

The creation of a successful co-culture is a complex process. Early attempts at using co-cultures consisting of calf pulmonary artery endothelial veins, retinal fibroblasts, and SMCs in fibrin gel to generate nascent vessels *in vitro* were unsuccessful.[131] Numerous groups have since used various co-culture systems for vascularization.[11,129,132] Similar to the debate over EC sources, there is still no consensus on the ideal cell source for pericytes. Multipotent stem cells are a common choice for use as pericytes due to both their physical and molecular properties. Both ASCs and MSCs have been shown to be capable of enhancing and stabilizing vessel formation by ECs on a number of different substrates both *in vitro* and *in vivo*.[88,63,111,133–136,137] Lin et al. demonstrated that murine MSCs derived from bone marrow, adipose, myocardial and skeletal muscle sources all behaved in a similar perivascular manner when co-implanted with endothelial colony forming cells in Matrigel constructs.[138] AFSCs have also emerged as a

stem cells source suitable for the enhancement of vascularization due to their innate similarity to MSCs in terms of protein and gene expression patterns.[139] However, unlike MSCs, AFSCs retain chromosomal stability and stem cell plasticity over long periods of culture time.[118,119] It has recently been shown that AFSCs significantly improve vessel formation by endothelial colony forming cells in Matrigel plugs implanted in a murine model in comparison to both MSCs and human dermal fibroblasts (HDFs).[140]

Although stem cells clearly have potential for use as a pericyte, non-stem cell types such as human fibroblasts also offer a promising source.[16,47,141] Human fibroblasts have been shown to be as effective as bone marrow–derived MSCs at supporting vessel formation by endothelial colony-forming cells in a collagen scaffold for use as an engineered skin substitute.[142] In our laboratory, rat aortic SMCs have also been proven, both in 2D and in 3D, to be capable of stabilizing an EC-generated capillary network in a biomaterial, in this case a collagen–glycosaminoglycan scaffold.[14] Cell sources for co-culture generated vasculature are also not limited to cells directly associated with the vascular system, for example, a co-culture consisting of human dermal human osteoblasts (HOS) and HDMECs have previously been shown to be capable of vascularizing a number of biomaterials both *in vitro*[95] and *in vivo*.[11] The method of action by which HOS successfully promote vessel formation by their EC counterpart is poorly understood. However, it is theorized that the HOS lay down a bone specific matrix that both supports the nascent vessels formed by the HDMECs and stimulates host inflammatory cells to promote invasion of the construct by host vessels.[143] This invasion of host vessels would be induced partly by the release of pro-angiogenic cytokines such as platelet-derived growth factor (PDGF) and transforming growth factor beta (TGF-β) by platelets from the blood, which come into contact with the implanted biomaterial and later by macrophages recruited to the site of inflammation (Fig. 3).[144]

Recently, a more complex approach to co-culture-induced vascularization has emerged involving a triculture of cell types.[130,145,146] For example, keratinocytes, fibroblasts, and ECs were cultured together to generate a functional vasculature network in a collagen sponge *in vitro*, which was then shown to be capable of anastomosis with murine host vasculature as indicated by the presence of erythrocytes in the nascent vessel network by the fourth day of implantation.[147] Levenberg et al. engineered robustly vascularized muscle tissue constructs using a triculture system composed of myoblasts, embryonic fibroblasts, and ECs. It was found that constructs prevascularized using the triculture system displayed both increased VEGF protein expression and increased vascular density *in vitro* compared to constructs prevascularized using only ECs and myoblasts.[130] These studies indicate a future role for more complex prevascularized constructs in TE.

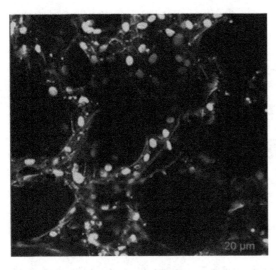

Fig. 3 Photon micrograph demonstrating vessel formation by a delayed addition co-culture model consisting co-culture of human MSCs and HUVECs seeded in a collagen–glycosaminoglycan scaffold *in vitro*. Nuclei are stained using 4′,6-diamidino-2-phenylindole (DAPI).

Generating long-lasting vasculature via co-culture/triculture may be the most viable approach for *in vitro* vascularization, as it is not only more physiologically accurate, but it has already produced vasculature that has demonstrated stability for up to one year post-implantation.[148] However, the complexity of these systems may make them more difficult to implement.

THE USE OF EXTERNAL STIMULI TO ENHANCE *IN VITRO* VASCULARIZATION

One popular approach to *in vitro* vascularization involves mimicry of a more physiologically accurate microenvironment in order to enhance the formation of nascent vasculature in a TE construct. This section covers two of the more common approaches to replicating the natural environments of body, namely the use of fluid shear stress and exposure to low oxygen (hypoxic) conditions in order to stimulate and enhance angiogenesis and by extension, vascularization.

Using Fluid Shear Stress to Enhance *In Vitro* Vascularization

It has previously been demonstrated that fluid shear stress plays a critical role in the maintenance of blood vessels[149,150] and any interruptions to, or the discontinuation of, blood flow results in vessel regression *in vivo*.[151] It is theorized that the regression of vasculature in TE constructs cultured under static conditions could partially be attributed to lack of perfusive flow through the lumen of the scaffolds, resulting in breakdown of the nascent endothelium. It

has also recently become clear that hemodynamic force, that is, fluid shear stress from blood flow, is necessary for proper embryonic blood vessel remodeling.[152,153] This highlights the potential of using shear stress to enhance the formation of *in vitro* vasculature.

Various types of bioreactors have been developed for the application of controllable shear stress to provide a mechanically active microenvironment for cell-seeded scaffolds.[154,155] Spinner flasks are the most basic form of bioreactor, consisting of an anchored cell-seeded scaffold suspended in a culture medium that is continuously agitated by a magnetic stir bar placed at the bottom of the flask. Spinner flasks result in poor mass transfer and non-homogenous cell distribution throughout the construct.[156] Rotating wall bioreactors consist of two concentric cylinders with the cell-seeded scaffold placed in media within the cylinder annulus. The outer cylinder is capable of rotation, thus generating dynamic laminar flow via centrifugal force, reducing the limitation of nutrient and waste diffusion and providing low levels of shear.[157] Hydrostatic pressure bioreactors are used to apply mechanical stimuli whereby constructs are first cultured statically and then transferred to a hydrostatic chamber for loading.[158,159] Flow perfusion bioreactors (Fig. 4) act by forcing media through the interior of a cell seeded scaffold and typically comprise of a perforated scaffold chamber containing the construct, linked with tubing to a media reservoir on one side with a pump to facilitate fluid flow on the other.[160–162]

Shear stress as applied by bioreactor such as those described above has been shown to accelerate differentiation, increase cell proliferation and density, and induce the expression of genes encoding for vasoactivators, adhesion molecules, and growth factors in ECs and their progenitor cells, thus priming them for angiogenesis.[163–166] Shear stress has also been effectively utilized by a number of groups to induce and enhance the endothelial differentiation of a multitude of stem cell types (including embryonic, mesenchymal, and amniotic fluid-derived) for potential use in vascularization.[112,121,167,168] For example, Zhang et al. found that shear stress enhanced the endothelial differentiation of human AFSCs. They saw that shear stress upregulated endothelial markers, increased the ability of the cells to attach to basement membrane components, and induced an antithrombogenic phenotype—all desirable qualities for an endothelial-like cell for use in vascularization.[112]

Shear stress has previously been shown to enhance angiogenesis and consequently vascularization. Lee et al. used a low flow rate (average 4.5 μm/s) to enhance vascular network formation by a co-culture of rat aortic ECs and rat MSCs in a collagen gel. Specifically, shear stress resulted in increased cell proliferation and the formation of denser, wider, and more prominent vascular networks in the gel when compared to a static control. Other groups have also utilized various bioreactor systems to employ shear stress to improve vascular network formation by ECs.[169–171] Shear stress has also been shown to upregulate both

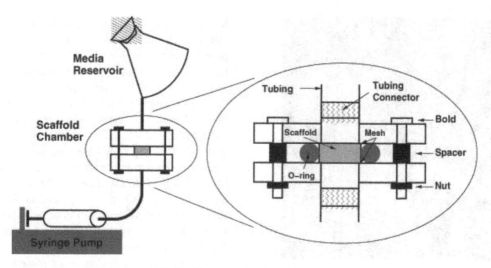

Fig. 4 Diagrammatic representation of a flow perfusion bioreactor developed in our lab. The construct is contained in a perforated scaffold chamber, which is linked with tubing to a media reservoir on one side with a pump to facilitate fluid flow on the other.
Source: From Jungreuthmayer et al.[247] © 2009, with permission from Elsevier.

thrombomodulin and endothelial nitric oxide synthase (eNOS) expression by ECs.[65,172–175] Thrombomodulin is a cell membrane glycoprotein expressed in vascular ECs and is considered the most important endogenous anti-thrombogenic protein.[173,176] eNOS is the enzyme responsible for the production of nitric oxide, an important vasodilator and protective vascular molecule that is stimulated in response cardiovascular risks such as to hypertension, hypercholesterolemia, diabetes mellitus, and chronic smoking.[177] Both thrombomodulin and eNOS are responsible for the maintenance of normal functionality in vasculature, indicating a role for shear stress the homoeostasis of nascent vascular networks by enabling them to present a more quiescent phenotype.

Using Hypoxia to Enhance *In Vitro* Vascularization

Hypoxia is characterized as a state of low oxygen (O_2) that leads to inadequate O_2 supply to the cells and tissues of the body. Hypoxia and the hypoxia-inducible factor (HIF-1) pathway have previously been associated with improved stem cell survival and metabolic enhancement.[178] Hypoxia may also represent another potentially beneficial stimulus for the enhancement of both *in vitro* vascularization and angiogenesis post-implantation. It has been found to be vital in neovascularization of ischemic tissues and the induction of angiogenesis in the healing of hypoxic wounds.[179] The heterodimeric transcription factor known as HIF-1 acts as the central regulator of hypoxia-mediated gene expression (Fig. 5).[180] Activation of the HIF-1 pathway causes the upregulation of a number of angiogenic genes, for example, angiopoietin 2, its receptor Tie2, and most commonly, VEGF.[181–183] These upregulated angiogenic growth factors are known to have an extremely

beneficial effect on vascularization. Hypoxia also represents a more physiologically accurate microenvironment for the stem cells of the body,[184] for example, both hempoetic cells of the bone marrow and fetal membrane cells (both commonly used stem cell sources for vascularization) experience O_2 conditions *in vivo* that are much lower than what is typically encountered in standard cell culture conditions.[185,186]

A number of different methods can be used to activate the HIF-1 pathway in a target cell for the purpose of enhancing angiogenesis and *in vitro* vascularization, each one with its own specific set of advantages and disadvantages. The most commonly used approach is one that most closely mimics physiological hypoxia: the manipulation of environment O_2 levels using a specialized hypoxia chamber. A number of research groups have used varying levels of environmental O_2 levels that can classed as hypoxic, from 5% to as low as 1%, to enhance angiogenesis.[178,187,188] For example, hypoxic environments have been shown to significantly improve the plasticity/potency of human MSCs.[178] Hypoxic exposure has also been associated with an increase in the angiogenic properties of both aged ASCs[189] and murine bone-marrow-derived stromal cells,[188] showing the potential of using hypoxia to prime stem cells for vascularization.

Another method has focused on the exploitation of cell-induced physiological hypoxia to generate angiogenic growth factors within a construct.[190] One such method involves the use of a rolled collagen gel that induces hypoxia via high density cell seeding, which is used to create a "depot" for the release of the angiogenic growth factors that the cells subsequently produce. This hypoxic depot was capable of successfully enhancing the vascularization of 3D spiral acellular collagen scaffold within 7 days of implantation in New Zealand rabbit model.[191]

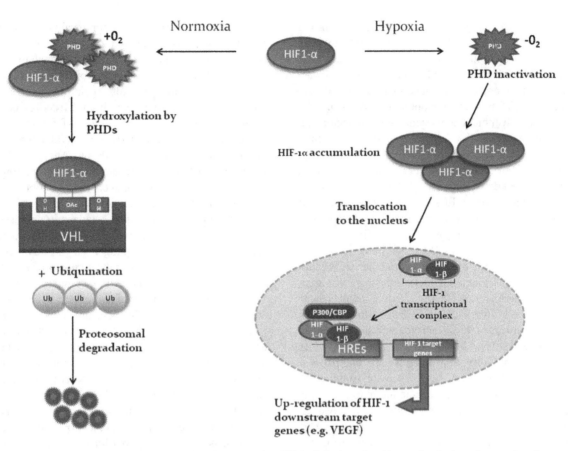

Fig. 5 HIF-1α regulation by proline hydroxylation. In normoxia, HIF-1α is hydroxylated by proline hydroxylases and undergoes degradation. Under hypoxic condition, proline hydroxylation is inhibited. This causes an increase in HIF-1α, which translocates to the nucleus. There, HIF-1α dimerizes with HIF-1β, binds to the HRE causing expression of various metabolic and angiogenic genes.

Pharmacological induction of HIF-1α is also used as a method to induce hypoxic conditions. PHD inhibitors such as desferrioxamine (DFO) and cobalt chloride ($CoCl_2$) have been used to overexpress HIF-1α, leading to HIF-1 pathway activation and angiogenic factor secretion.[192–194] These "hypoxia-mimics" sequester Fe(II) (a necessary cofactor for the HIF-pathway),[195] thus preventing inactivation of HIF-1α. This leads to the formation of the HIF-1 transcription complex. Hypoxia mimics have been shown to have beneficial effects on angiogenesis in a similar manner to hypoxia. DFO treatment has been shown to restore blood flow recovery to ischemic tissues by promoting EC proliferation and reducing oxidative stress[196] and direct injection of DFO at a murine femur fracture site leads to a significant increase in both vessel number and volume of the developing vasculature in the resulting callus.[193] Similarly, $CoCl_2$ pretreated MSCs seeded on a collagen-based scaffold enhanced the vascularization of an engineered periosteal model implanted in both a cranial defect and subcutaneous murine model.[194] However, despite these potential benefits that hypoxia mimics offer, they also present their own unique set of disadvantages when compared to environmental O_2 manipulation, one of which is the requirement for continued application of an effective dosage to maintain HIF-1 activation. This dosage approach could potentially have a detrimental effect on cell viability due to toxicity. It has been found that high levels of $CoCl_2$ and DFO can be toxic to certain cell types and can negatively affect both viability and ability to proliferate.[197–199] Although each method of HIF-1 activation has its own merits and drawbacks, the ideal artificial inducer of HIF-1 has yet to be discovered.

ALTERNATIVE METHODS TO SUPPORT VESSEL FORMATION *IN VITRO* AND *IN VIVO*

The maturation of vessels involves a transition from a series of actively growing vascular branches to a stable, functional network. This process occurs in a spatial and temporal manner and involves pathways that control the proliferation, migration, and morphogenesis of ECs.[200,201] As previously stated, EC–pericyte interactions are of critical importance to newly forming EC sprouts as it has been shown that reduced pericyte recruitment is associated with vessel hypertrophy *in vivo*.[202] There are several bioactive factors implicated in EC–pericyte interactions during vessel formation and these are frequently utilized in order to promote *in vitro* vascularization.

Some of the most important factors associated with vascular development are VEGF, platelet-derived growth factor (PDGF-BB), TGF-β, and angiopoietin-1 (Ang-1).[203,204] VEGF is a highly potent angiogenic growth factor and plays a critical role in the formation of the vasculature. VEGF has been shown to affect EC behavior, morphology, and survival.[65,200,204] Evensen et al. demonstrated that RNA interference knockdown of SMC VEGF expression prevented vessel formation by ECs. Supplemented VEGF did not recover the effects of this knockdown, thus highlighting the importance of pericyte-derived VEGF in driving blood vessel maturation.[204] MSCs have been shown to generate large quantities of VEGF to modulate EC behavior and influence vascular assembly, identifying them as an ideal source of pericytes.[205,206] As a result of its potency as a pro-angiogenic growth factor, VEGF has often been targeted for delivery in order to enhance neovascularization both *in vitro* and *in vivo*. PDGF-BB has also displayed its importance in vessel formation, and this particular isoform is capable of binding to all PDGF receptor isotypes.[203] PDGF-BB is a potent chemoattractant and mitogen that is secreted by ECs to influence the migration of pericytes to the site of vessel assembly.[65] The PDGF-BB/PDGFR-β signaling pathway is the principal pathway responsible for pericyte recruitment, attachment to the vasculature, and the maturation that follows.[127,207] The mitogenic effects of PDGF-BB have previously been demonstrated, as the presence of HUVECs has been shown to increase human MSC migration across a membrane to the EC location. PDGF–BB involvement was confirmed by inhibiting its release, which lead the cessation of MSC migration.[108] The controlled spatiotemporal delivery of VEGF followed by PDGF–BB delivery has also been shown to produce larger, more mature vessels *in vivo*.[208]

Growth factors such as TGF-β1 influence the differentiation of MSCs down a smooth muscle lineage whereby they act as pericytes.[65,200,203,207] TGF-β1 also promotes ECM production, growth arrest, and vessel formation.[23] Ang-1 is a factor predominantly expressed by pericytes and reduces endothelial permeability to enhance stabilization and maturation.[209] Endothelium-specific receptor Tie2 is expressed throughout the developing embryonic endothelium and in the quiescent vasculature of adults. Ang-1 signaling through Tie2 receptor promotes vessel formation and remodeling of blood vessels as well as inducing pericyte recruitment (the alternative Ang-2/Tie2 signaling antagonizes Ang-1 causing detachment and destabilization of EC-pericyte interactions).[203,210] Zacharek et al. demonstrated in a rat model of stroke that delivery of MSCs decreased blood–brain barrier leakage by promoting angiogenesis and vascular stabilization. This was attributed to increased expression of Ang-1, Tie2 and the tight junction protein occludin.[210]

Another target protein that demonstrates great potential in promoting vascularization *in vitro* is the morphogen sonic hedgehog (Shh). Shh is a potent regulator of angiogenesis during early development of the embryo, which is also reactivated during adult repair processes.[211] In cocultures of OECs and primary human osteoblasts, Shh has been shown to increase expression of a number of pro-angiogenic genes, particularly those involved in vessel stabilization (e.g., α SMA, PDGF and Ang-1).[212,213] Dohle et al. demonstrated that in addition to promoting vessel formation *in vitro*, Shh enhanced key factors involved in osteogenesis, for example, increased *alkaline phosphatase,* enhanced mineralization, and osteonectin expression was observed in these co-cultures.[212] These data suggest a synergistic effect of this morphogen on these two processes and thus demonstrates its significant potential for TE applications in particular for bone tissue regeneration.

There are also a number of approaches to generating vasculature that are generally considered in the context of enhancing vascularization postimplantation that may also be useful in promoting *in vitro* vascularization should a cell-based approach be adopted. These approaches generally involve targeting growth factors in order to promote vascularization. The simplest method, which was previously discussed, is the use of co-culture systems. This system exploits what naturally occurs *in vivo* as the cross-talk between ECs and their pericytes is required for functional vessel formation.[86,123,207,209,214,215] Other methods involve a controlled release of bioactive factors to enhance vascularization. In most cases, it has been shown that delivery of single growth factors is not sufficient and that delivery of multiple growth factors produces a faster and longer-lasting repair response.[209,216–220] Such methods include growth factor delivery via nano or microparticles,[221–223] covalent immobilization of bioactive factors onto scaffolds,[49,224] direct incorporation of bioactive factors into scaffolds,[209] targeting other pathways to increase growth factor production (e.g., hypoxia inducible factor pathway),[225,226] or overexpression of key pro-angiogenic growth factors via gene therapy.[227]

The mechanism and timing of biomolecule delivery to cells is also important in enhancing vascularization. Controlled delivery systems aim to sustain the bioactivity of therapeutic molecules in a steady and long-term fashion in order to sufficiently induce and accelerate tissue repair.[16,85,224,228] Hydrogels are often used for co-delivery of cells and bioactive factors in order to promote vascularization. Hydrogels such as alginate,[229,230] hyaluronic acid or chitosan[231,232] have all proven highly successful in the delivery of both genes and growth factors for tissue repair. The use of nanoparticles and microparticles for growth factors delivery is also an area of growing interest due to the fact that it allows for controllable release patterns. The advantages of these materials are that they can easily diffuse into tissues and can be taken up by cells to exert the desired effects.[228] Much research has focused on the delivery of particles to sites of tissue damage (e.g., site of myocardial infarction)[233,234] and in recent years, TE has attempted to incorporate these particles into biomaterials in order to influence cell behavior both *in vitro* and *in vivo*.[235] Direct covalent immobilization of growth factors onto scaffolds has also proved successful both *in vitro* and *in vivo*.[236–238] Chiu and Radisic demonstrated that ECs seeded onto collagen scaffolds with covalently immobilized VEGF and Ang-1 displayed increased proliferation and improved tubule formation *in vitro* compared to scaffolds without these growth factors.[238]

Gene therapy is another promising approach to enhance vascularization. An ideal gene delivery system requires a sustained but transient expression of a transferred piece of genetic material (or transgene) to exert the desired effect and induce rapid tissue repair.[239] To date, the majority of gene therapy has involved the use of viral vectors, but in recent years efforts have been focused on using nonviral vectors, which offer a safer and more clinically relevant alternative.[232,240,241] There are two main approaches that can be adopted for gene therapy: 1) combining gene therapies with cellular therapies; and 2) the use of gene-activated matrices (GAMs) alone, which is a more recent approach. The latter is the preferred approach in terms of clinical translation as it only involves the delivery of a vector on a cell-free scaffold, which can then target the damaged tissue site in order to promote repair. Our group has developed a number of GAMs using nonviral gene vectors such nanohydroxyapatite, polyethyleneimine, and chitosan in combination with collagen-based scaffolds engineered specifically for bone tissue repair, but have also sought to target angiogenesis and vasculogenesis.[78,240–242] Such approaches may also be utilized to enhance vessel formation by cells *in vitro* and as such cell-based gene therapy can be easily applied for *in vitro* vascularization. This involves the transfection of cells to overexpress key angiogenic factors in order to target host cells and enhance neovascularization.[227,243,244] Duffy et al. demonstrated that the overexpression of ephrin B2 in MSCs enhanced their proangiogenic capacity as they were shown to adopt an early EC phenotype under EC culture conditions with increasing expression of vWF and VEGFR-2. These EC-like cells developed an increased ability to form vessel-like structures, produce VEGF, and incorporate into newly formed EC structures.[227] This may have great potential *in vivo* as these cells will target the surrounding host cells to increase angiogenesis in the local environment, thus enhancing tissue repair. Despite the promise presented by gene therapy approaches, issues remain with gene transfer efficacy, safety, and long-term pro-angiogenic expression *in vivo* and thus more research is required in this area.

CONCLUSION

Despite recent advances in the area of vascularization, avascular necrosis in thick TE constructs remains one of the major challenges in regenerative medicine. Generating a nascent vasculature within the construct prior to implantation represents an approach with potential for success, given the small timeframe in which cells can remain viable without the oxygen or nutrients provided by adjacent vasculature. However, to successfully create systems for vascularization using an *in vitro* vascularization technique, one must focus on replicating the inherent cellular complexity of developing vasculature in the body. As a secondary consideration, it is possible that this developing vasculature may be enhanced by mimicking the relevant microenvironment through both physical and biochemical stimuli. The combination of these factors may allow for the creation of more effective vascularized constructs, opening up the way for the development of tissue systems and eventually, whole organs.

ACKNOWLEDGMENTS

The authors would like to thank Rosanne Raftery for providing the image used for Fig. 2. We also acknowledge the following funding bodies: the Health Research Board (PHD/2007/11), the European Research Council (grant agreement number 239685 under EU 7th framework FP7/2007–2013) and the Science Foundation Ireland Short Term Travel Fellowship (11/RFP.1/ENM/3063-STTF 11). Figure 1 is reprinted from Trends in Biotechnology, 26(8): Rouwkema J, Rivron NC and van Blitterswijk CA., Vascularization in TE, 434-41., 2008 with permission from Elsevier. Figure 4 is reprinted from Medical Engineering & Physics, 31(4), Jungreuthmayer C, Jaasma MJ, Al-Munajjed AA, Zanghellini J, Kelly DJ and O'Brien FJ., Deformation simulation of cells seeded on a collagen-GAG scaffold in a flow perfusion bioreactor using a sequential 3D CFD-elastostatics model, 420-7., 2009 with permission from Elsevier.

REFERENCES

1. Ko, H.C.; Milthorpe, B.K.; McFarland, C.D. Engineering thick tissues--the vascularisation problem. Eur. Cell Mater. **2007**, *14*, 1–18.
2. Black, A.F.; Berthod, F.; L'Heureux, N.; Germain, L.; Auger, F.A. *In vitro* reconstruction of a human capillary-like network in a tissue-engineered skin equivalent. FASEB J. **1998**, *12* (13), 1331–1340.
3. Yannas, I.V.; Orgill, D.P.; Burke, J.F. Template for skin regeneration. Plast. Reconstr. Surg. **2011**, 127 (Suppl 1), 60S–70S.
4. Yeh, Y.C.; Lee, W.Y.; Yu, C.L.; Hwang, S.M.; Chung, M.F.; Hsu, L.W.; Chang, Y.; Lin, W.W.; Tsai, M.S.; Wei, H.J.; Sung, H.W. Cardiac repair with injectable cell sheet fragments of human amniotic fluid stem cells in an immune-suppressed rat model. Biomaterials **2010**, *31* (25), 6444–6453.
5. Song, J.J.; Guyette, J.P.; Gilpin, S.E.; Gonzalez, G.; Vacanti, J.P.; Ott, H.C. Regeneration and experimental

orthotopic transplantation of a bioengineered kidney. Nat. Med. **2013**, *19* (5), 646–651.

6. Raya-Rivera, A.; Esquiliano, D.R.; Yoo, J.J.; Lopez-Bayghen, E.; Soker, S.; Atala, A. Tissue-engineered autologous urethras for patients who need reconstruction: An observational study. Lancet **2011**, *377* (9772), 1175–1182.

7. Atala, A.; Bauer, S.B.; Soker, S.; Yoo, J.J.; Retik, A.B. Tissue-engineered autologous bladders for patients needing cystoplasty. Lancet **2006**, *367* (9518), 1241–1246.

8. Folkman, J.; Hochberg, M. Self-regulation of growth in three dimensions. J. Exp. Med. **1973**, *138* (4), 745–753.

9. Cheema, U.; Rong, Z.; Kirresh, O.; Macrobert, A.J.; Vadgama, P.; Brown, R.A. Oxygen diffusion through collagen scaffolds at defined densities: Implications for cell survival in tissue models. J. Tissue. Eng. Regen. Med. **2012**, *6* (1), 77–84.

10. Ishaug-Riley, S.L.; Crane-Kruger, G.M.; Yaszemski, M.J.; Mikos, A.G. Three-dimensional culture of rat calvarial osteoblasts in porous biodegradable polymers. Biomaterials **1998**, *19* (15), 1405–1412.

11. Unger, R.E.; Ghanaati, S.; Orth, C.; Sartoris, A.; Barbeck, M.; Halstenberg, S.; Motta, A.; Migliaresi, C.; Kirkpatrick, C.J. The rapid anastomosis between prevascularized networks on silk fibroin scaffolds generated *in vitro* with cocultures of human microvascular endothelial and osteoblast cells and the host vasculature. Biomaterials **2010**, *31* (27), 6959–6967.

12. Laschke, M.W.; Harder, Y.; Amon, M.; Martin, I.; Farhadi, J.; Ring, A.; Torio-Padron, N.; Schramm, R.; Rücker, M.; Junker, D.; Häufel, J.M.; Carvalho, C.; Heberer, M.; Germann, G.; Vollmar, B.; Menger, M.D. Angiogenesis in tissue engineering: Breathing life into constructed tissue substitutes. Tissue Eng. **2006**, *12* (8), 2093–2104.

13. Montano, I.; Schiestl, C.; Schneider, J.; Pontiggia, L.; Luginbuhl, J.; Biedermann, T.; Böttcher-Haberzeth, S.; Braziulis, E.; Meuli, M.; Reichmann, E. Formation of human capillaries *in vitro*: The engineering of prevascularized matrices. Tissue Eng. Part A **2010**, *16* (1), 269–282.

14. Duffy, G.P.; McFadden, T.M.; Byrne, E.M.; Gill, S.L.; Farrell, E.; O'Brien, F.J. Towards *in vitro* vascularisation of collagen-GAG scaffolds. Eur. Cell. Mater. **2011**, *21* (1), 15–30.

15. Cheng, G.; Liao, S.; Kit Wong, H.; Lacorre, D.A.; di Tomaso, E.; Au, P.; Fukumura, D.; Jain, R.K.; Munn, L.L. Engineered blood vessel networks connect to host vasculature via wrapping-and-tapping anastomosis. Blood **2011**, *118* (17), 4740–4749.

16. Chen, X.; Aledia, A.S.; Ghajar, C.M.; Griffith, C.K.; Putnam, A.J.; Hughes, C.C.; George, S.C. Prevascularization of a fibrin-based tissue construct accelerates the formation of functional anastomosis with host vasculature. Tissue Eng. Part A **2009**, *15* (6), 1363–1371.

17. Dimmeler, S.; Aicher, A.; Vasa, M.; Mildner-Rihm, C.; Adler, K.; Tiemann, M.; Rütten, H.; Fichtlscherer, S.; Martin, H.; Zeiher, A.M. HMG-CoA reductase inhibitors (statins) increase endothelial progenitor cells via the PI 3-kinase/Akt pathway. J. Clin. Invest. **2001**, *108* (3), 391–397.

18. Murasawa, S.; Asahara, T. Endothelial progenitor cells for vasculogenesis. Physiology (Bethesda) **2005**, *20* (1), 36–42.

19. Nomi, M.; Atala, A.; Coppi, P.D.; Soker, S. Principals of neovascularization for tissue engineering. Mol. Aspects Med. **2002**, *23* (6), 463–483.

20. Graupe, D.; Cerrel-Bazo, H.; Kern, H.; Carraro, U. Walking performance, medical outcomes and patient training in FES of innervated muscles for ambulation by thoracic-level complete paraplegics. Neurol. Res. **2008**, *30* (2), 123–130.

21. Hur, J.; Yoon, C.H.; Kim, H.S.; Choi, J.H.; Kang, H.J.; Hwang, K.K.; Oh, B.H.; Lee, M.M.; Park, Y.B. Characterization of two types of endothelial progenitor cells and their different contributions to neovasculogenesis. Arterioscler. Thromb. Vasc. Biol. **2004**, *24* (2), 288–293.

22. Yoder, M.C.; Mead, L.E.; Prater, D.; Krier, T.R.; Mroueh, K.N.; Li, F.; Redefining endothelial progenitor cells via clonal analysis and hematopoietic stem/progenitor cell principals. Blood **2007**, *109* (5), 1801–1809.

23. Kaully, T.; Kaufman-Francis, K.; Lesman, A.; Levenberg, S. Vascularization--the conduit to viable engineered tissues. Tissue Eng. Part B Rev. **2009**, *15* (2), 159–169.

24. Valarmathi, M.T.; Davis, J.M.; Yost, M.J.; Goodwin, R.L.; Potts, J.D. A three-dimensional model of vasculogenesis. Biomaterials **2009**, *30* (6), 1098–1112.

25. Cai, W.; Schaper, W. Mechanisms of arteriogenesis. Acta Biochim Biophys Sin (Shanghai) **2008**, *40* (8), 681–692.

26. Carmeliet, P. Mechanisms of angiogenesis and arteriogenesis. Nat. Med. **2000**, *6* (4), 389–395.

27. Heil, M.; Eitenmuller, I.; Schmitz-Rixen, T.; Schaper, W. rteriogenesis versus angiogenesis: Similarities and differences. J. Cell. Mol. Med. **2006**, *10* (1), 45–55.

28. Lee, E.J.; Kasper, F.K.; Mikos, A.G. Biomaterials for tissue engineering. Ann. Biomed. Eng. **2013**, 13 (6), 565–576.

29. O'Brien, F.J. Biomaterials & scaffolds for tissue engineering. Mater. Today **2011**, *14* (3), 88–95.

30. Hutmacher, D.W. Scaffolds in tissue engineering bone and cartilage. Biomaterials **2000**, *21* (24), 2529–2543.

31. Kim, B.S.; Baez, C.E.; Atala, A. Biomaterials for tissue engineering. World J. Urol. **2000**, *18* (1), 2–9.

32. Arroyo, A.G.; Iruela-Arispe, M.L. Extracellular matrix, inflammation, and the angiogenic response. Cardiovasc. Res. **2010**, *86* (2), 226–235.

33. Mikos, A.G.; McIntire, L.V.; Anderson, J.M.; Babensee, J.E. Host response to tissue engineered devices. Adv. Drug Deliv. Rev. **1998**, *33* (1–2), 111–39.

34. Temenoff, J.S.; Mikos, A.G. Review: Tissue engineering for regeneration of articular cartilage. Biomaterials **2000**, *21* (5), 431–440.

35. Yamamura, N.; Sudo, R.; Ikeda, M.; Tanishita, K. Effects of the mechanical properties of collagen gel on the *in vitro* formation of microvessel networks by endothelial cells. Tissue Eng. **2007**, *13* (7), 1443–1453.

36. Saunders, R.L.; Hammer, D.A. Assembly of human umbilical vein endothelial cells on compliant hydrogels. Cell. Mol. Bioeng. **2010**, *3* (1), 60–67.

37. Califano, J.P.; Reinhart-King, C.A. The effects of substrate elasticity on endothelial cell network formation and traction force generation. Conf. Proc. IEEE Eng. Med. Biol. Soc. **2009**, *2009*, 3343–3345.

38. Kniazeva, E.; Putnam, A.J. Endothelial cell traction and ECM density influence both capillary morphogenesis and maintenance in 3-D. Am. J. Physiol. Cell. Physiol. **2009**, *297* (1), C179–C187.

39. Murphy, C.M.; Haugh, M.G.; O'Brien, F.J. The effect of mean pore size on cell attachment, proliferation and migration in collagen-glycosaminoglycan scaffolds for bone tissue engineering. Biomaterials **2010**, *31* (3), 461–467.

40. Byrne, E.M.; Farrell, E.; McMahon, L.A.; Haugh, M.G.; O'Brien, F.J.; Campbell, V.A.; Prendergast, P.J.; O'Connell, B.C. Gene expression by marrow stromal cells in a porous collagen-glycosaminoglycan scaffold is affected by pore size and mechanical stimulation. J. Mater. Sci. Mater. Med. **2008**, *19* (11), 3455–3463.

41. Kolk, A.; Handschel, J.; Drescher, W.; Rothamel, D.; Kloss, F.; Blessmann, M.; Current trends and future perspectives of bone substitute materials - from space holders to innov-ative biomaterials. J. Craniomaxillofac. Surg. **2012**, *40* (8), 706–718.

42. Lu, J.X.; Flautre, B.; Anselme, K.; Hardouin, P.; Gallur, A.; Descamps, M.; Thierry, B. Role of interconnections in porous bioceramics on bone recolonization *in vitro* and *in vivo*. J. Mater. Sci. Mater. Med. **1999**, *10* (2), 111–120.

43. Kuboki, Y.; Jin, Q.; Kikuchi, M.; Mamood, J.; Takita, H. Geometry of artificial ECM: Sizes of pores controlling phenotype expression in BMP-induced osteogenesis and chondrogenesis. Connect. Tissue Res. **2002**, *43* (2–3), 529–534.

44. Mastrogiacomo, M.; Scaglione, S.; Martinetti, R.; Dolcini, L.; Beltrame, F.; Cancedda, R.; Quarto, R. Role of scaffold internal structure on *in vivo* bone formation in macroporous calcium phosphate bioceramics. Biomaterials **2006**, *27* (17), 3230–3237.

45. Bai, F.; Wang, Z.; Lu, J.; Liu, J.; Chen, G.; Lv, R.; Wang, J.; Lin, K.; Zhang, J.; Huang, X. The correlation between the internal structure and vascularization of controllable porous bioceramic materials *in vivo*: A quantitative study. Tissue Eng. Part A **2010**, *16* (12), 3791–3803.

46. Mirabella, T.; Gentili, C.; Daga, A.; Cancedda, R. Amniotic fluid stem cells in a bone microenvironment: Driving host angiogenic response. Stem Cell. Res. **2013**, *11* (1), 540–551.

47. Choong, C.S.; Hutmacher, D.W.; Triffitt, J.T. Co-culture of bone marrow fibroblasts and endothelial cells on modified polycaprolactone substrates for enhanced potentials in bone tissue engineering. Tissue Eng. **2006**, *12* (9), 2521–2531.

48. Geuze, R.E.; Theyse, L.F.; Kempen, D.H.; Hazewinkel, H.A.; Kraak, H.Y.; Oner, F.C.; Dhert, W.J.; Alblas, J. A differential effect of bone morphogenetic protein-2 and vascular endothelial growth factor release timing on osteogenesis at ectopic and orthotopic sites in a large-animal model. Tissue Eng. Part A **2012**, *18* (19–20), 2052–2062.

49. Singh, S.; Wu, B.M.; Dunn, J.C. Accelerating vascularization in polycaprolactone scaffolds by endothelial progenitor cells. Tissue Eng. Part A **2011**, *17* (13–14), 1819–1830.

50. Navarro, MP.; Planell, J.A. Scaffolds for bone regeneration. Eur. Musculoskel. Rev. **2011**, *6* (4), 1–5.

51. Ghanaati, S.; Fuchs, S.; Webber, M.J.; Orth, C.; Barbeck, M.; Gomes, M.E.; Reis, R.L.; Kirkpatrick, C.J. Rapid vascularization of starch-poly(caprolactone) *in vivo* by outgrowth endothelial cells in co-culture with primary osteoblasts. J. Tissue Eng. Regen. Med. **2011**, *5* (6), e136–e143.

52. Santos, M.I.; Unger, R.E.; Sousa, R.A.; Reis, R.L.; Kirkpatrick, C.J. Crosstalk between osteoblasts and endothelial cells co-cultured on a polycaprolactone-starch scaffold and the *in vitro* development of vascularization. Biomaterials **2009**, *30* (26), 4407–4415.

53. Druecke, D.; Langer, S.; Lamme, E.; Pieper, J.; Ugarkovic, M.; Steinau, H.U.; Homann, H.H. Neovascularization of poly(ether ester) block-copolymer scaffolds *in vivo*: Long-term investigations using intravital fluorescent microscopy. J. Biomed. Mater. Res. A **2004**, *68* (1), 10–18.

54. Fuchs, S.; Ghanaati, S.; Orth, C.; Barbeck, M.; Kolbe, M.; Hofmann, A.; Eblenkamp, M.; Gomes, M.; Reis, R.L.; Kirkpatrick, C.J. Contribution of outgrowth endothelial cells from human peripheral blood on *in vivo* vascularization of bone tissue engineered constructs based on starch polycaprolactone scaffolds. Biomaterials **2009**, *30* (4), 526–534.

55. Yang, S.; Leong, K.F.; Du, Z.; Chua, C.K. The design of scaffolds for use in tissue engineering. Part I. Traditional factors. Tissue Eng. **2001**, *7* (6), 679–689.

56. Liu H.; Slamovich EB.; Webster TJ. Less harmful acidic degradation of poly(lacticco-glycolic acid) bone tissue engineering scaffolds through titania nanoparticle addition. Int J Nanomedicine. **2006**, *1* (4), 541–5.

57. O'Brien FJ.; Harley BA.; Yannas IV.; Gibson L. Influence of freezing rate on pore structure in freeze-dried collagen-GAG scaffolds. Biomaterials. **2004**, *25* (6), 1077–86.

58. Matsiko, A.; Levingstone, T.J.; O'Brien, F.J.; Gleeson, J.P. Addition of hyaluronic acid improves cellular infiltration and promotes early-stage chondrogenesis in a collagen-based scaffold for cartilage tissue engineering. J. Mech. Behav. Biomed. Mater. **2012**, *11*, 41–52.

59. Simionescu, D.T.; Lu, Q.; Song, Y.; Lee, J.S.; Rosenbalm, T.N.; Kelley, C.; Vyavahare, N.R. Biocompatibility and remodeling potential of pure arterial elastin and collagen scaffolds. Biomaterials **2006**, *27* (5), 702–713.

60. Unger, R.E.; Peters, K.; Wolf, M.; Motta, A.; Migliaresi, C.; Kirkpatrick, C.J. Endothelialization of a non-woven silk fibroin net for use in tissue engineering: Growth and gene regulation of human endothelial cells. Biomaterials **2004**, *25* (21), 5137–5146.

61. Unger, R.E.; Wolf, M.; Peters, K.; Motta, A.; Migliaresi, C.; James Kirkpatrick, C. Growth of human cells on a non-woven silk fibroin net: A potential for use in tissue engineering. Biomaterials **2004**, *25* (6), 1069–1075.

62. Lyons, F.; Partap, S.; O'Brien, F.J. Part 1: Scaffolds and surfaces. Technol. Health Care **2008**, *16* (4), 305–317.

63. Allen, P.; Melero-Martin, J.; Bischoff, J. Type I collagen. Fibrin and PuraMatrix matrices provide permissive environments for human endothelial and mesenchymal progenitor cells to form neovascular networks. J. Tissue Eng. Regen. Med. **2010**, *5* (4), e74–e86.

64. Laschke, M.W.; Vollmar, B.; Menger, M.D. Inosculation: Connecting the life-sustaining pipelines. Tissue Eng. Part B Rev. **2009**, *15* (4), 455–465.

65. Lee, E.J.; Niklason, L.E. A Novel flow bioreactor for *in vitro* microvascularization. Tissue Eng. Part C Methods **2010**, *16* (5), 1191–1200.

66. Melero-Martin, J.M.; De Obaldia, M.E.; Allen, P.; Dudley, A.C.; Klagsbrun, M.; Bischoff, J. Host myeloid cells are necessary for creating bioengineered human vascular networks *in vivo*. Tissue Eng. Part A **2010**, *16* (8), 2457–2466.

In Vitro–Medical

67. Lu, L.; Peter, S.J.; Lyman, M.D.; Lai, H.L.; Leite, S.M.; Tamada, J.A.; Uyama, S.;Vacanti, J.P.; Langer, R.; Mikos, A.G. *In vitro* and *in vivo* degradation of porous poly(DL-lactic-co-glycolic acid) foams. Biomaterials. **2000**, *21* (18), 1837–1845.

68. Cipitria, A.; Skelton, A.; Dargaville, T.R.; Dalton, P.D.; Hutmacher, D.W. Design, fabrication and characterization of PCL electrospun scaffolds-a review. J. Mater. Chem. [10.1039/C0JM04502K]. **2011**, *21* (26), 9419–9453.

69. Usami, K.; Mizuno, H.; Okada, K.; Narita, Y.; Aoki, M.; Kondo, T.; Mizuno, D.; Mase, J.; Nishiguchi, H.; Kagami, H.; Ueda, M. Composite implantation of mesenchymal stem cells with endothelial progenitor cells enhances tissue-engineered bone formation. J. Biomed. Mater. Res. A **2009**, *90* (3), 730–741.

70. Hughes, C.S.; Postovit, L.M.; Lajoie, G.A. Matrigel: A complex protein mixture required for optimal growth of cell culture. Proteomics **2010**, *10* (9), 1886–1890.

71. Bonfil, R.D.; Vinyals, A.; Bustuoabad, O.D.; Llorens, A.; Benavides, F.J.; Gonzalez-Garrigues, M.; Fabra, A. Stimulation of angiogenesis as an explanation of Matrigel-enhanced tumorigenicity. Int. J. Cancer **1994**, *58* (2), 233–239.

72. Fridman, R.; Giaccone, G.; Kanemoto, T.; Martin, G.R.; Gazdar, A.F.; Mulshine, J.L. Reconstituted basement membrane (matrigel) and laminin can enhance the tumorigenicity and the drug resistance of small cell lung cancer cell lines. Proc. Natl. Acad. Sci. U S A. **1990**, *87* (17), 6698–6702.

73. Briganti, E.; Spiller, D.; Mirtelli, C.; Kull, S.; Counoupas, C.; Losi, P.; Senesi, S.; Di Stefano, R.; Soldani, G. A composite fibrin-based scaffold for controlled delivery of bioactive pro-angiogenetic growth factors. J. Control. Release **2010**, *142* (1), 14–21.

74. Erggelet, C.; Endres, M.; Neumann, K.; Morawietz, L.; Ringe, J.; Haberstroh, K.; Sittinger, M.; Kaps, C. Formation of cartilage repair tissue in articular cartilage defects pretreated with microfracture and covered with cell-free polymer-based implants. J. Orthop. Res. **2009**, *27* (10), 1353–1360.

75. Vandrovcová, M.; Douglas, T.; Hauk, D.; Grössner-Schreiber, B.; Wiltfang, J.; Bačáková, L.; Warnke, P.H. Influence of collagen and chondroitin sulfate (CS) coatings on poly-(lactide-co-glycolide) (PLGA) on MG 63 osteoblast-like cells. Physiological research/Academia Scientiarum. Bohemoslovaca **2011**, *60* (5), 797–813.

76. Kim, H.W.; Knowles, J.C.; Kim, H.E. Hydroxyapatite and gelatin composite foams processed via novel freeze-drying and crosslinking for use as temporary hard tissue scaffolds. J. Biomed. Mater. Res. A **2005**, *72* (2), 136–145.

77. Yaylaoglu, M.B.; Korkusuz, P.; Ors, U.; Korkusuz, F.; Hasirci, V. Development of a calcium phosphate-gelatin composite as a bone substitute and its use in drug release. Biomaterials **1999**, *20* (8), 711–719.

78. Tierney, E.G.; Duffy, G.P.; Hibbitts, A.J.; Cryan S.A.; O'Brien, F.J. The development of non-viral gene-activated matrices for bone regeneration using polyethyleneimine (PEI) and collagen-based scaffolds. J. Control. Release. **2012**, *158* (2), 304–311.

79. Roether, J.A.; Gough, J.E.; Boccaccini, A.R.; Hench, L.L.; Maquet, V.; Jerome, R. Novel bioresorbable and bioactive composites based on bioactive glass and polylactide foams for bone tissue engineering. J. Mater. Sci. Mater. Med. **2002**, *13* (12), 1207–1214.

80. Tierney, C.M.; Jaasma MJ.; O'Brien FJ. Osteoblast activity on collagen-GAG scaffolds is affected by collagen and GAG concentrations. J. Biomed. Mater. Res. A **2009**, *91* (1), 92–101.

81. Farrell, E.; Byrne, E.M.; Fischer, J.; O'Brien, F.J.; O'Connell, B.C.; Prendergast, P.J.; Campbell, V.A. A comparison of the osteogenic potential of adult rat mesenchymal stem cells cultured in 2-D and on 3-D collagen glycosaminoglycan scaffolds. Technol. Health Care **2007**, *15* (1), 19–31.

82. Kim, S.; von Recum, H. Endothelial stem cells and precursors for tissue engineering: Cell source.; differentiation.; selection.; and application. Tissue Eng. Part B Rev. **2008**, *14* (1), 133–147.

83. Ghitescu, L.; Robert, M. Diversity in unity: The biochemical composition of the endothelial cell surface varies between the vascular beds. Microsc. Res. Tech. **2002**, *57* (5), 381–389.

84. He, W.; Yong, T.; Teo, W.E.; Ma, Z.; Ramakrishna, S. Fabrication and endothelialization of collagen-blended biodegradable polymer nanofibers: Potential vascular graft for blood vessel tissue engineering. Tissue Eng. Part A **2005**, *11* (9–10), 1574–1588.

85. Briganti, E.; Spiller, D.; Mirtelli, C.; Kull, S.; Counoupas, C.; Losi, P.; Senesi, S.; Di Stefano, R.; Soldani, G. A composite fibrin-based scaffold for controlled delivery of bioactive pro-angiogenetic growth factors. J. Control. Release **2009**, *142* (1), 14–21.

86. Correia, C.; Grayson, W.L.; Park, M.; Hutton, D.; Zhou, B.; Guo, XE.; Niklason, L.; Sousa, R.A.; Reis, R.L.; Vunjak-Novakovic, G. *In vitro* model of vascularized bone: Synergizing vascular development and osteogenesis. PLoS One **2011**, *6* (12), e28352.

87. Melero-Martin, J.M.; Khan, Z.A.; Picard, A.; Wu, X.; Paruchuri, S.; Bischoff, J. *In vivo* vasculogenic potential of human blood-derived endothelial progenitor cells. Blood **2007**, *109* (11), 4761–4768.

88. Verseijden, F.; Posthumus-van Sluijs, S.J.; Pavljasevic, P.; Hofer, S.O.; van Osch, G.J.; Farrell, E. Adult human bone marrow- and adipose tissue-derived stromal cells support the formation of prevascular-like structures from endothelial cells *in vitro*. Tissue Eng. Part A **2010**, *16* (1), 101–114.

89. Park, H.J.; Zhang, Y.; Georgescu, S.P.; Johnson, K.L.; Kong, D.; Galper, J.B. Human umbilical vein endothelial cells and human dermal microvascular endothelial cells offer new insights into the relationship between lipid metabolism and angiogenesis. Stem Cell. Rev. **2006**, *2* (2), 93–102.

90. Jackson, C.J.; Nguyen, M. Human microvascular endothelial cells differ from macrovascular endothelial cells in their expression of matrix metalloproteinases. Int. J. Biochem. Cell. Biol. **1997**, *29* (10), 1167–1177.

91. Santos, M.I.; Fuchs, S.; Gomes, M.E.; Unger, R.E.; Reis, R.L.; Kirkpatrick, C.J. Response of micro- and macrovascular endothelial cells to starch-based fiber meshes for bone tissue engineering. Biomaterials **2007**, *28* (2), 240–248.

92. Kraling, B.M.; Bischoff, J. A simplified method for growth of human microvascular endothelial cells results in decreased senescence and continued responsiveness to cytokines and growth factors. *In Vitro* Cell. Dev. Biol. Anim. **1998**, *34* (4), 308–315.

93. Nor, J.E.; Peters, M.C.; Christensen, J.B.; Sutorik, M.M.; Linn, S.; Khan, M.K.; Addison, C.L, Mooney, D.J, Polverini, P.J. Engineering and characterization of functional human microvessels in immunodeficient mice. Lab. Invest. **2001**, *81* (4), 453–463.

94. Supp, D.M.; Wilson-Landy, K.; Boyce, S.T. Human dermal microvascular endothelial cells form vascular analogs in cultured skin substitutes after grafting to athymic mice. FASEB J. **2002**, *16* (8), 797–804.

95. Unger, R.E.; Sartoris, A.; Peters, K.; Motta, A.; Migliaresi, C.; Kunkel, M.; Bulnheim, U.; Rychly, J.; Kirkpatrick, C.J. Tissue-like self-assembly in cocultures of endothelial cells and osteoblasts and the formation of microcapillary-like structures on three-dimensional porous biomaterials. Biomaterials **2007**, *28* (27), 3965–3976.

96. Fuchs, S.; Motta, A.; Migliaresi, C.; Kirkpatrick, C.J. Outgrowth endothelial cells isolated and expanded from human peripheral blood progenitor cells as a potential source of autologous cells for endothelialization of silk fibroin biomaterials. Biomaterials **2006**, *27* (31), 5399–5408.

97. Ingram, D.A.; Mead, L.E.; Tanaka, H.; Meade, V.; Fenoglio, A.; Mortell, K.; Pollok, K.; Ferkowicz, M.J.; Gilley, D.; Yoder, M.C. Identification of a novel hierarchy of endothelial progenitor cells using human peripheral and umbilical cord blood. Blood **2004**, *104* (9), 2752–2760.

98. Mukai, N.; Akahori, T.; Komaki, M.; Li, Q.; Kanayasu-Toyoda, T.; Ishii-Watabe, A.; Kobayashi, A.; Yamaguchi, T.; Abe, M.; Amagasa, T.; Morita, I. A comparison of the tube forming potentials of early and late endothelial progenitor cells. Exp. Cell. Res. **2008**, *314* (3), 430–440.

99. Geuze, R.E.; Wegman, F.; Oner, F.C.; Dhert, W.J.; Alblas, J. Influence of endothelial progenitor cells and platelet gel on tissue-engineered bone ectopically in goats. Tissue Eng Part A. **2009**, *15* (11), 3669–3677.

100. Seebach, C.; Henrich, D.; Kahling, C.; Wilhelm, K.; Tami, A.E.; Alini, M.; Marzi, I. Endothelial progenitor cells and mesenchymal stem cells seeded onto beta-TCP granules enhance early vascularization and bone healing in a critical-sized bone defect in rats. Tissue Eng. Part A **2009**, *16* (6), 1961–1970.

101. Lin, Y.; Weisdorf, D.J.; Solovey, A.; Hebbel, R.P. Origins of circulating endothelial cells and endothelial outgrowth from blood. J. Clin. Invest. **2000**, *105* (1), 71–77.

102. Yoon, C.H.; Hur, J.; Park, K.W.; Kim, J.H.; Lee, C,S.; Oh, I.Y.; Kim, T.Y.; Cho, H.J.; Kang, H.J.; Chae, I.H.; Yang, H.K.; Oh, B.H.; Park, Y.B.; Kim, H.S. Synergistic neovascularization by mixed transplantation of early endothelial progenitor cells and late outgrowth endothelial cells: The role of angiogenic cytokines and matrix metalloproteinases. Circulation **2005**, *112* (11), 1618–1627.

103. Oswald, J.; Boxberger, S.; Jorgensen, B.; Feldmann, S.; Ehninger, G.; Bornhäuser, M.; Werner, C. Mesenchymal stem cells can be differentiated into endothelial cells *in vitro*. Stem Cells. **2004**, *22* (3), 377–384.

104. Silva, G.V.; Litovsky, S.; Assad, J.A.R.; Sousa, A.L.S.; Martin, B.J.; Vela, D.; Coulter, S.C.; Lin, J.; Ober, J.; Vaughn, W.K.; Branco, R.V.; Oliveira, E.M.; He, R.; Geng, Y.J.; Willerson, J.T.; Perin, E.C. Mesenchymal stem cells differentiate into an endothelial phenotypem, enhance vascular density, and improve heart function in a canine chronic ischemia model. Circulation **2005**, *111* (2), 150–156.

105. Jazayeri, M.; Allameh, A.; Soleimani, M.; Jazayeri, S.H.; Piryaei, A.; Kazemnejad, S. Molecular and ultrastructural characterization of endothelial cells differentiated from human bone marrow mesenchymal stem cells. Cell Biol. Int. **2008**, *32* (10), 1183–1192.

106. Liu, D.; Zhang, X.; Li, X.; Zhang, Z.; Guo, G.; Peng, Y. *Differentiation of the Human Marrow Mesenchymal Stem Cells into Vascular Endothelium-like Cells in vitro 7th Asian-Pacific Conference on Medical and Biological Engineering.* Magjarevic, R., Ed.; Springer Berlin Heidelberg; Berlin, 2008; p. 80–83.

107. Chen, M.Y.; Lie, P.C.; Li, Z.L.; Wei, X. Endothelial differentiation of Wharton's jelly-derived mesenchymal stem cells in comparison with bone marrow-derived mesenchymal stem cells. Exp. Hematol. **2009**, *37* (5), 629–640.

108. Au, P.; Tam, J.; Fukumura, D.; Jain, RK. Bone marrow-derived mesenchymal stem cells facilitate engineering of long-lasting functional vasculature. Blood **2008**, *111* (9), 4551–4558.

109. Caplan, A.I. All MSCs are pericytes? Cell Stem Cell. **2008**, *3* (3), 229–230.

110. Crisan, M.; Yap, S.; Casteilla, L.; Chen, C.W.; Corselli, M.; Park, T.S.; Andriolo, G.; Sun, B.; Zheng, B.; Zhang, L.; Norotte, C.; Teng, P.N.; Traas, J.; Schugar, R.; Deasy, B.M.; Badylak, S.; Buhring, H.J.; Giacobino, J.P.; Lazzari, L.; Huard, J.; Péault, B. A perivascular origin for mesenchymal stem cells in multiple human organs. Cell Stem Cell. **2008**, *3* (3), 301–313.

111. McFadden, T.M.; Duffy, G.P.; Allen, A.B.; Stevens, H.Y.; Schwarzmaier, S.M.; Plesnila, N.; Murphy, M.J.; Barry, F.P.; Guldberg, R.E.; O'Brien, F.J. The delayed addition of human MSCs to pre-formed endothelial cell networks results in functional vascularisation of a collagen-GAG scaffold *in vivo*. Acta Biomater. **2013**, *9* (12), 9303–9316.

112. Zhang, P.; Moudgill, N.; Hager, E.; Tarola, N.; Dimatteo, C.; McIlhenny, S.; Tulenko, T.; DiMuzio, P.J. Endothelial differentiation of adipose-derived stem cells from elderly patients with cardiovascular disease. Stem Cell. Dev. **2011**, *20* (6), 977–988.

113. Caplan, A.I. Adult mesenchymal stem cells for tissue engineering versus regenerative medicine. J. Cell. Physiol. **2007**, *213* (2), 341–347.

114. Caplan, A.I. Why are MSCs therapeutic? New data: New insight. J. Pathol. **2009**, *217* (2), 318–324.

115. Zonari, A.; Novikoff, S.; Electo, N.R.P.; Breyner, N.l.M.; Gomes, D.A.; Martins, A.; Neves, N.M.; Reis, R.L.; Goes, A.M. Endothelial differentiation of human stem cells seeded onto electrospun polyhydroxybutyrate/polyhydroxybutyrate-co-hydroxyvalerate fiber mesh. PLoS ONE **2012**, *7* (4), e35422.

116. Fischer, L.J.; McIlhenny, S.; Tulenko, T.; Golesorkhi, N.; Zhang, P.; Larson, R.; Lombardi, J.; Shapiro, I.; DiMuzio,

P.J. Endothelial differentiation of adipose-derived stem cells: Effects of endothelial cell growth supplement and shear force. J. Surg. Res. **2009**, *152* (1), 157–166.

117. Cao, Y.; Sun, Z.; Liao, L.; Meng, Y.; Han, Q.; Zhao, R.C. Human adipose tissue-derived stem cells differentiate into endothelial cells *in vitro* and improve postnatal neovascularization *in vivo*. Biochem. Biophys. Res. Commun. **2005**, *332* (2), 370–379.

118. De Coppi, P.; Bartsch, G.; Siddiqui, M.M.; Xu, T.; Santos, C.C.; Perin, L.; Mostoslavsky, G.; Serre, A.C.; Snyder, E.Y.; Yoo, J.J.; Furth, M.E.; Soker, S.; Atala, A. Isolation of amniotic stem cell lines with potential for therapy. Nat. Biotechnol. **2007**, *25* (1), 100–106.

119. Miranda-Sayago, J.M.; Fernandez-Arcas, N.; Benito, C.; Reyes-Engel, A.; Carrera, J.; Alonso, A. Lifespan of human amniotic fluid-derived multipotent mesenchymal stromal cells. Cytotherapy **2011**, *13* (5), 572–581.

120. Benavides, O.M.; Petsche, J.J.; Moise, K.J.; Johnson, A.; Jacot, JG. Evaluation of endothelial cells differentiated from amniotic fluid-derived stem cells. Tissue Eng. Part A **2012**, *18* (11–12), 1123–1131.

121. Zhang, P.; Baxter, J.; Vinod, K.; Tulenko, T.N.; Di Muzio, P.J. Endothelial differentiation of amniotic fluid-derived stem cells: Synergism of biochemical and shear force stimuli. Stem Cells Dev. **2009**, *18* (9), 1299–1308.

122. James Kirkpatrick, C.; Fuchs, S.; Iris Hermanns, M.; Peters, K.; Unger, RE. Cell culture models of higher complexity in tissue engineering and regenerative medicine. Biomaterials **2007**, *28* (34), 5193–5198.

123. von Tell, D.; Armulik, A.; Betsholtz, C. Pericytes and vascular stability. Exp. Cell. Res. **2006**, *312* (5), 623–629.

124. Gerhardt, H.; Betsholtz, C. Endothelial-pericyte interactions in angiogenesis. Cell Tissue Res. **2003**, *314* (1), 15–23.

125. Darland DC.; D'Amore PA. Blood vessel maturation: Vascular development comes of age. J. Clin. Invest. **1999**, *103* (2), 157–158.

126. Armulik, A.; Abramsson, A.; Betsholtz, C. Endothelial/pericyte interactions. Circ. Res. **2005**, *97* (6), 512–523.

127. Kitahara, T.; Hiromura, K.; Ikeuchi, H.; Yamashita, S.; Kobayashi, S.; Kuroiwa, T.; Kaneko, Y.; Ueki, K.; Nojima, Y. Mesangial cells stimulate differentiation of endothelial cells to form capillary-like networks in a three-dimensional culture system. Nephrol. Dial Transplant. **2005**, *20* (1), 42–49.

128. Roubelakis, M.G.; Tsaknakis, G.; Pappa, K.I.; Anagnou, N.P.; Watt, S.M. Spindle shaped human mesenchymal stem/stromal cells from amniotic fluid promote neovascularization. PLoS One **2013**, *8* (1), e54747.

129. Pedroso, D.C.; Tellechea, A.; Moura, L.; Fidalgo-Carvalho, I.; Duarte, J.; Carvalho, E.; Ferreira, L. Improved survival, vascular differentiation and wound healing potential of stem cells co-cultured with endothelial cells. PLoS One **2011**, *6* (1), e16114.

130. Levenberg, S.; Rouwkema, J.; Macdonald, M.; Garfein, E.S.; Kohane, D.S.; Darland, D.C.; Marini, R.; Van Blitterswijk, C.A.; Mulligan, R.C.; D'amore P.A.; Langer, R. Engineering vascularized skeletal muscle tissue. Nat. Biotechnol. **2005**, *23* (7), 879–884.

131. Nehls, V.; Schuchardt, E.; Drenckhahn, D. The effect of fibroblasts, vascular smooth muscle cells, and pericytes on sprout formation of endothelial cells in a fibrin gel angiogenesis system. Microvasc. Res. **1994**, *48* (3), 349–363.

132. Rouwkema, J.; Westerweel, P.E.; de Boer, J.; Verhaar, M.C.; van Blitterswijk, C.A. The use of endothelial progenitor cells for prevascularized bone tissue engineering. Tissue Eng. Part A **2009**, *15* (8), 2015–2027.

133. Saleh, F.A.; Whyte, M.; Genever, P.G. Effects of endothelial cells on human mesenchymal stem cell activity in a three-dimensional *in vitro* model. Eur. Cell Mater. **2011**, *22*, 242–257.

134. Melero-Martin, J.M.; De Obaldia, M.E.; Kang, S.Y.; Khan, Z.A.; Yuan, L.; Oettgen, P.; Bischoff, J. Engineering robust and functional vascular networks *in vivo* with human adult and cord blood-derived progenitor cells. Circ. Res. **2008**, *103* (2), 194–202.

135. Traktuev, D.O.; Merfeld-Clauss, S.; Li, J.; Kolonin, M.; Arap, W.; Pasqualini, R.; Johnstone, B.H.; March, K.L. A population of multipotent CD34-positive adipose stromal cells share pericyte and mesenchymal surface markers, reside in a periendothelial location, and stabilize endothelial networks. Circ. Res. **2008**, *102* (1), 77–85.

136. Butler, M.J.; Sefton, M,V. Cotransplantation of adipose-derived mesenchymal stromal cells and endothelial cells in a modular construct drives vascularization in SCID/bg mice. Tissue Eng. Part A **2012**, *18* (15–16), 1628–1641.

137. Merfeld-Clauss, S.; Gollahalli, N.; March, K.L.; Traktuev, D.O. Adipose tissue progenitor cells directly interact with endothelial cells to induce vascular network formation. Tissue Eng. Part A **2010**, *16* (9), 2953–2966.

138. Lin, R.Z.; Moreno-Luna, R.; Zhou, B.; Pu, W.T.; Melero-Martin, J.M. Equal modulation of endothelial cell function by four distinct tissue-specific mesenchymal stem cells. Angiogenesis **2012**, *15* (3), 443–455.

139. Roubelakis, M.G.; Pappa, K.I.; Bitsika, V.; Zagoura, D.; Vlahou, A.; Papadaki, HA.; Antsaklis, A.; Anagnou, N.P. Molecular and proteomic characterization of human mesenchymal stem cells derived from amniotic fluid: Comparison to bone marrow mesenchymal stem cells. Stem Cells Dev. **2007**, *16* (6), 931–952.

140. Teodelinda, M.; Michele, C.; Sebastiano, C.; Ranieri, C.; Chiara, G. Amniotic liquid derived stem cells as reservoir of secreted angiogenic factors capable of stimulating neo-arteriogenesis in an ischemic model. Biomaterials **2011**, *32* (15), 3689–3699.

141. Saito, M.; Hamasaki, M.; Shibuya, M. Induction of tube formation by angiopoietin-1 in endothelial cell/fibroblast co-culture is dependent on endogenous VEGF. Cancer Sci. **2003**, *94* (9), 782–790.

142. Athanassopoulos, A.; Tsaknakis, G.; Newey, S.E.; Harris, A.L.; Kean, J.; Tyler, M.P.; Watt, S.M. Microvessel networks [corrected] pre-formed in artificial clinical grade dermal substitutes *in vitro* using cells from haematopoietic tissues. Burns **2012**, *38* (5), 691–701.

143. Ghanaati, S.; Unger, R.E.; Webber, M.J.; Barbeck, M.; Orth, C.; Kirkpatrick, J.A.; Booms, P.; Motta, A.; Migliaresi, C.; Sader, R.A.; Kirkpatrick, C.J. Scaffold vascularization *in vivo* driven by primary human osteoblasts in concert with host inflammatory cells. Biomaterials **2011**, *32* (32), 8150–8160.

144. Diegelmann, R.F.; Evans, M.C. Wound healing: An overview of acute, fibrotic and delayed healing. Front Biosci. **2004**, *9*, 283–289.

145. Lesman, A.; Koffler, J.; Atlas, R.; Blinder, Y.J.; Kam, Z.; Levenberg, S. Engineering vessel-like networks within multicellular fibrin-based constructs. Biomaterials **2011**, *32* (31), 7856–7869.

146. Caspi, O.; Lesman, A.; Basevitch, Y.; Gepstein, A.; Arbel, G.; Habib, I.H.; Gepstein, L.; Levenberg, S. Tissue engineering of vascularized cardiac muscle from human embryonic stem cells. Circ. Res. **2007**, *100* (2), 263–272.

147. Tremblay, P.L.; Hudon, V.; Berthod, F.; Germain, L.; Auger, F.A. Inosculation of tissue-engineered capillaries with the host's vasculature in a reconstructed skin transplanted on mice. Am. J. Transplant. **2005**, *5* (5), 1002–1010.

148. Koike, N.; Fukumura, D.; Gralla, O.; Au, P.; Schechner, J.S.; Jain, R.K. Tissue engineering: Creation of long-lasting blood vessels. Nature **2004**, *428* (6979), 138–139.

149. Meeson, A.; Palmer, M.; Calfon, M.; Lang, R. A relationship between apoptosis and flow during programmed capillary regression is revealed by vital analysis. Development **1996**, *122* (12), 3929–3938.

150. Dimmeler, S.; Haendeler, J.; Rippmann, V.; Nehls, M.; Zeiher, AM. Shear stress inhibits apoptosis of human endothelial cells. FEBS Lett. **1996**, *399* (1–2), 71–74.

151. See, J.R.; Marlon, A.M.; Feikes, H.L.; Cosby RS. Effect of direct revascularization surgery on coronary collateral circulation in man. Am. J. Cardiol. **1975**, *36* (6), 734–738.

152. Lucitti, J.L.; Jones, E.A.; Huang, C.; Chen, J.; Fraser, S.E.; Dickinson ME. Vascular remodeling of the mouse yolk sac requires hemodynamic force. Development **2007**, *134* (18), 3317–3326.

153. Conway, E.M.; Collen, D.; Carmeliet, P. Molecular mechanisms of blood vessel growth. Cardiovasc. Res. **2001**, *49* (3), 507–521.

154. Oragui, E.; Nannaparaju, M.; Khan, W.S. The role of bioreactors in tissue engineering for musculoskeletal applications. Open Orthop. J. **2011**, *5* (Suppl. 2), 267–270.

155. Plunkett, N.; O'Brien, F.J. Bioreactors in tissue engineering. Technol. Health Care. **2011**, *19* (1), 55–69.

156. Goldstein, A.S.; Juarez, T.M.; Helmke, C.D.; Gustin, M.C.; Mikos, A.G. Effect of convection on osteoblastic cell growth and function in biodegradable polymer foam scaffolds. Biomaterials **2001**, *22* (11), 1279–1288.

157. Schwarz, R.P.; Goodwin, T.J.; Wolf, D.A. Cell culture for three-dimensional modeling in rotating-wall vessels: An application of simulated microgravity. J. Tissue Cult. Methods **1992**, *14* (2), 51–57.

158. Darling, E.M.; Athanasiou, K.A. Articular cartilage bioreactors and bioprocesses. Tissue Eng. **2003**, *9* (1), 9–26.

159. Shaikh, F.M.; O'Brien, T.P.; Callanan, A.; Kavanagh, E.G.; Burke, P.E.; Grace, P.A.; McGloughlin, T.M. New pulsatile hydrostatic pressure bioreactor for vascular tissue-engineered constructs. Artif. Organs **2010**, *34* (2), 153–158.

160. Jaasma, M.J.; Plunkett, N.A.; O'Brien, F.J. Design and validation of a dynamic flow perfusion bioreactor for use with compliant tissue engineering scaffolds. J. Biotechnol. **2008**, *133* (4), 490–496.

161. McCoy, R.J.; Widaa, A.; Watters, K.M.; Wuerstle, M.; Stallings, R.L.; Duffy, G.P.; O'Brien, F.J. Orchestrating osteogenic differentiation of mesenchymal stem cells - identification of placental growth factor as a mechanosensitive gene with a pro-osteogenic role. Stem Cells **2013**, *31* (11), 2420–2431.

162. McCoy, R.J.; Jungreuthmayer, C.; O'Brien, F.J. Influence of flow rate and scaffold pore size on cell behavior during mechanical stimulation in a flow perfusion bioreactor. Biotechnol. Bioeng. **2012**, *109* (6), 1583–1594.

163. Chien, S.; Li, S.; Shyy, Y.J. Effects of mechanical forces on signal transduction and gene expression in endothelial cells. Hypertension **1998**, *31* (1 Pt. 2), 162–169.

164. Li, Y.S.; Haga, J.H.; Chien, S. Molecular basis of the effects of shear stress on vascular endothelial cells. J. Biomech. **2005**, *38* (10), 1949–1971.

165. Dewey, C.F.; Bussolari, S.R.; Gimbrone, M.A.; Davies, P.F. The dynamic response of vascular endothelial cells to fluid shear stress. J. Biomech. Eng. **1981**, *103* (3), 177–185.

166. Yamamoto, K.; Takahashi, T.; Asahara, T.; Ohura, N.; Sokabe, T.; Kamiya, A.; Ando, J. Proliferation, differentiation, and tube formation by endothelial progenitor cells in response to shear stress. J. Appl. Physiol. **2003**, *95* (5), 2081–2088.

167. Wang, H.; Riha, G.M.; Yan, S.; Li, M.; Chai, H.; Yang, H.; Shear stress induces endothelial differentiation from a murine embryonic mesenchymal progenitor cell line. Arterioscler. Thromb. Vasc. Biol. **2005**, *25* (9), 1817–1823.

168. Yamamoto K.; Sokabe T.; Watabe T.; Miyazono K.; Yamashita JK.; Obi S.; Ohura, N.; Matsushita, A.; Kamiya, A.; Ando, J. Fluid shear stress induces differentiation of Flk-1-positive embryonic stem cells into vascular endothelial cells *in vitro*. Am. J. Physiol. Heart Circ. Physiol. **2005**, *288* (4), H1915–H1924.

169. Chang, C.C.; Nunes, S.S.; Sibole, S.C.; Krishnan, L.; Williams, S.K.; Weiss, J.A.; Hoying, J.B. Angiogenesis in a microvascular construct for transplantation depends on the method of chamber circulation. Tissue Eng. Part A **2010**, *16* (3), 795–805.

170. Kang, H.; Bayless, K.J.; Kaunas, R. Fluid shear stress modulates endothelial cell invasion into three-dimensional collagen matrices. Am. J. Physiol. Heart Circ. Physiol. **2008**, *295* (5), H2087–H2097.

171. Rotenberg, M.Y.; Ruvinov, E.; Armoza, A.; Cohen, S. A multi-shear perfusion bioreactor for investigating shear stress effects in endothelial cell constructs. Lab. Chip. **2012**, *12* (15), 2696–2703.

172. Takada, Y.; Shinkai, F.; Kondo, S.; Yamamoto, S.; Tsuboi, H.; Korenaga, R.; Ando, J. Fluid shear stress increases the expression of thrombomodulin by cultured human endothelial cells. Biochem. Biophys. Res. Commun. **1994**, *205* (2), 1345–1352.

173. Lee, E.J.; Vunjak-Novakovic, G.; Wang, Y.; Niklason, L.E. A biocompatible endothelial cell delivery system for *in vitro* tissue engineering. Cell Transplant. **2009**, *18* (7), 731–743.

174. Nikmanesh, M.; Shi, Z.D.; Tarbell, JM. Heparan sulfate proteoglycan mediates shear stress-induced endothelial gene expression in mouse embryonic stem cell-derived

endothelial cells. Biotechnol. Bioeng. **2012**, *109* (2), 583–594.

175. Uzarski, J.S.; Scott, E.W.; McFetridge, P.S. Adaptation of endothelial cells to physiologically-modeled, variable shear stress. PLoS One **2013**, *8* (2), e57004.

176. Suzuki, K.; Kusumoto, H.; Deyashiki, Y.; Nishioka, J.; Maruyama, I.; Zushi, M.; Structure and expression of human thrombomodulin.; a thrombin receptor on endothelium acting as a cofactor for protein C activation. EMBO J. **1987**, *6* (7), 1891–1897.

177. Forstermann, U.; Munzel, T. Endothelial nitric oxide synthase in vascular disease: From marvel to menace. Circulation **2006**, *113* (13), 1708–1714.

178. Grayson, W.L.; Zhao, F.; Izadpanah, R.; Bunnell, B.; Ma, T. Effects of hypoxia on human mesenchymal stem cell expansion and plasticity in 3D constructs. J. Cell. Physiol. **2006**, *207* (2), 331–339.

179. Shweiki, D.; Itin, A.; Soffer, D.; Keshet, E. Vascular endothelial growth factor induced by hypoxia may mediate hypoxia-initiated angiogenesis. Nature **1992**, *359* (6398), 843–845.

180. Wang, G.L.; Jiang, B.H.; Rue, E.A.; Semenza, G.L. Hypoxia-inducible factor 1 is a basic-helix-loop-helix-PAS heterodimer regulated by cellular O2 tension. Proc. Natl. Acad. Sci. U S A. **1995**, *92* (12), 5510–5514.

181. Forsythe, J.A.; Jiang, B.H.; Iyer, N.V.; Agani, F.; Leung, S.W.; Koos RD.; Semenza, G.L. Activation of vascular endothelial growth factor gene transcription by hypoxia-inducible factor 1. Mol. Cell. Biol. **1996**, *16* (9), 4604–4613.

182. Willam, C.; Koehne, P.; Jurgensen, JS.; Grafe, M.; Wagner, K.D.; Bachmann, S.; Frei, U.; Eckardt, K.U. Tie2 receptor expression is stimulated by hypoxia and proinflammatory cytokines in human endothelial cells. Circ. Res. **2000**, *87* (5), 370–377.

183. Namiki, A.; Brogi, E.; Kearney, M.; Kim, E.A.; Wu, T.; Couffinhal, T.; Varticovski, L.; Isner, J.M. Hypoxia induces vascular endothelial growth factor in cultured human endothelial cells. J. Biol. Chem. **1995**, *270* (52), 31189–31195.

184. Ivanovic, Z. Hypoxia or in situ normoxia: The stem cell paradigm. J. Cell Physiol. **2009**, *219* (2), 271–275.

185. Chow, D.C.; Wenning, L.A.; Miller, W.M.; Papoutsakis, E.T. Modeling pO(2) distributions in the bone marrow hematopoietic compartment. II. Modified Kroghian models. Biophys. J. **2001**, *81* (2), 685–696.

186. Al-Asmakh, M.; Race, H.; Tan, S.; Sullivan, M.H. The effects of oxygen concentration on *in vitro* output of prostaglandin E2 and interleukin-6 from human fetal membranes. Mol. Hum. Reprod. **2007**, *13* (3), 197–201.

187. Potier, E.; Ferreira, E.; Andriamanalijaona, R.; Pujol, J.P.; Oudina, K.; Logeart-Avramoglou, D.; Petite, H. Hypoxia affects mesenchymal stromal cell osteogenic differentiation and angiogenic factor expression. Bone **2007**, *40* (4), 1078–1087.

188. Annabi, B.; Lee, Y.T.; Turcotte, S.; Naud, E.; Desrosiers, R.R.; Champagne, M.; Eliopoulos, N.; Galipeau, J.; Béliveau, R. Hypoxia promotes murine bone-marrow-derived stromal cell migration and tube formation. Stem Cells. **2003**, *21* (3), 337–347.

189. Efimenko, A.; Starostina, E.; Kalinina, N.; Stolzing, A. Angiogenic properties of aged adipose derived mesenchymal stem cells after hypoxic conditioning. J. Transl. Med. **2011**, *9*, 10.

190. Hadjipanayi, E.; Brown, R.A.; Mudera, V.; Deng, D.; Liu, W.; Cheema, U. Controlling physiological angiogenesis by hypoxia-induced signaling. J. Control. Release **2010**, *146* (3), 309–317.

191. Hadjipanayi, E.; Cheema, U.; Mudera, V.; Deng, D.; Liu, W.; Brown, RA. First implantable device for hypoxia-mediated angiogenic induction. J. Control. Release **2011**, *153* (3), 217–224.

192. Potier, E.; Ferreira, E.; Dennler, S.; Mauviel, A.; Oudina, K.; Logeart-Avramoglou, D.; Desferrioxamine-driven upregulation of angiogenic factor expression by human bone marrow stromal cells. J. Tissue Eng. Regen. Med. **2008**, *2* (5), 272–278.

193. Shen, X.; Wan, C.; Ramaswamy, G.; Mavalli M.; Wang, Y.; Duvall, C.L.; Prolyl hydroxylase inhibitors increase neoangiogenesis and callus formation following femur fracture in mice. J. Orthop. Res. **2009**, *27* (10), 1298–1305.

194. Fan, W.; Crawford, R.; Xiao, Y. Enhancing *in vivo* vascularized bone formation by cobalt chloride-treated bone marrow stromal cells in a tissue engineered periosteum model. Biomaterials **2010**, *31* (13), 3580–3589.

195. Schofield, C.J.; Zhang, Z. Structural and mechanistic studies on 2-oxoglutarate-dependent oxygenases and related enzymes. Curr. Opin. Struct. Biol. **1999**, *9* (6), 722–731.

196. Ikeda, Y.; Tajima, S.; Yoshida, S.; Yamano, N.; Kihira, Y.; Ishizawa, K.; Aihara, K.; Tomita, S.; Tsuchiya, K.; Tamaki, T. Deferoxamine promotes angiogenesis via the activation of vascular endothelial cell function. Atherosclerosis **2011**, *215* (2), 339–347.

197. Milosevic, J.; Adler, I.; Manaenko, A.; Schwarz, S.C.; Walkinshaw, G.; Arend M.; Non-hypoxic stabilization of hypoxia-inducible factor alpha (HIF-alpha): Relevance in neural progenitor/stem cells. Neurotox Res. **2009**, *15* (4), 367–380.

198. Ren, H.; Cao, Y.; Zhao, Q.; Li, J.; Zhou, C.; Liao, L.; Jia, M.; Zhao, Q.; Cai, H.; Han, Z.C.; Yang, R.; Chen, G.; Zhao, R.C. Proliferation and differentiation of bone marrow stromal cells under hypoxic conditions. Biochem. Biophys. Res. Commun. **2006**, *347* (1), 12–21.

199. Karovic, O.; Tonazzini, I.; Rebola, N.; Edstrom, E.; Lovdahl, C.; Fredholm, BB.; Daré, E. Toxic effects of cobalt in primary cultures of mouse astrocytes. Similarities with hypoxia and role of HIF-1alpha. Biochem. Pharmacol. **2007**, *73* (5), 694–708.

200. Holderfield, M.T.; Hughes, C.C. Crosstalk between vascular endothelial growth factor, notch, and transforming growth factor-beta in vascular morphogenesis. Circ. Res. **2008**, *102* (6), 637–652.

201. Jain, R.K. Molecular regulation of vessel maturation. Nat. Med. **2003**, *9* (6), 685–693.

202. Hellstrom, M.; Gerhardt, H.; Kalen, M.; Li, X.; Eriksson, U.; Wolburg, H.; Betsholtz, C. Lack of pericytes leads to endothelial hyperplasia and abnormal vascular morphogenesis. J. Cell. Biol. **2001**, *153* (3), 543–553.

203. Caplan, A.I.; Correa, D. PDGF in bone formation and regeneration: New insights into a novel mechanism involving MSCs. J. Orthop. Res. **2011**, *29* (12), 1795–1803.

204. Evensen, L.; Micklem, D.R.; Blois, A.; Berge, S.V.; Aarsaether, N.; Littlewood-Evans, A.; Wood, J.; Lorens, J.B. Mural cell associated VEGF is required for organotypic vessel formation. PLoS One **2009**, *4* (6), e5798.

205. Kaigler, D.; Krebsbach, P.H.; Polverini, P.J.; Mooney, D.J. Role of vascular endothelial growth factor in bone marrow stromal cell modulation of endothelial cells. Tissue Eng. Part A **2003**, *9* (1), 95–103.

206. Kasper, G.; Dankert, N.; Tuischer, J.; Hoeft, M.; Gaber, T.; Glaeser, J.D.; Zander, D.; Tschirschmann, M.; Thompson, M.; Matziolis, G.; Duda, G.N. Mesenchymal stem cells regulate angiogenesis according to their mechanical environment. Stem Cells **2007**, *25* (4), 903–910.

207. Gaengel, K.; Genove, G.; Armulik, A.; Betsholtz, C. Endothelial-mural cell signaling in vascular development and angiogenesis. Arterioscler. Thromb. Vasc. Biol. **2009**, *29* (5), 630–638.

208. Chen, R.R.; Silva, E.A.; Yuen, W.W.; Mooney, D.J. Spatio-temporal VEGF and PDGF delivery patterns blood vessel formation and maturation. Pharm. Res. **2007**, *24* (2), 258–264.

209. Wakui, S.; Yokoo, K.; Muto, T.; Suzuki, Y.; Takahashi, H.; Furusato, M.; Hano, H.; Endou, H.; Kanai, Y. Localization of Ang-1, -2, Tie-2, and VEGF expression at endothelial-pericyte interdigitation in rat angiogenesis. Lab. Invest. **2006**, *86* (11), 1172–1184.

210. Zacharek, A.; Chen, J.L.; Cui, X.; Li, A.; Li, Y.; Roberts, C.; Feng, Y.; Gao, Q. Chopp M.Angiopoietin1/Tie2 and VEGF/Flk1 induced by MSC treatment amplifies angiogenesis and vascular stabilization after stroke. J. Cereb. Blood Flow Metab. **2007**, *27* (10), 1684–1691.

211. Fuchs, S.; Dohle, E.; Kirkpatrick, C.J. Sonic Hedgehog-mediated synergistic effects guiding angiogenesis and osteogenesis. Vitam. Horm. **2012**, *88*, 491–506.

212. Dohle, E.; Fuchs, S.; Kolbe, M.; Hofmann, A.; Schmidt, H.; Kirkpatrick, C.J. Sonic hedgehog promotes angiogenesis and osteogenesis in a coculture system consisting of primary osteoblasts and outgrowth endothelial cells. Tissue Eng. Part A **2010**, *16* (4), 1235–1237.

213. Dohle, E.; Fuchs, S.; Kolbe, M.; Hofmann, A.; Schmidt, H.; Kirkpatrick, C.J. Comparative study assessing effects of sonic hedgehog and VEGF in a human co-culture model for bone vascularisation strategies. Eur. Cell. Mater. **2011**, *21*, 144–156.

214. Kirkpatrick, C.J.; Fuchs, S.; Unger, R.E. Co-culture systems for vascularization - Learning from nature. Adv. Drug Deliv. Rev. **2011**, *63* (4–5), 291–299.

215. Villars, F.; Bordenave, L.; Bareille, R.; Amedee, J. Effect of human endothelial cells on human bone marrow stromal cell phenotype: Role of VEGF? J. Cell. Biochem. **2000**, *79* (4), 672–685.

216. Saif, J.; Schwarz, T.M.; Chau, D.Y.; Henstock, J.; Sami, P.; Leicht, S.F.; Hermann, P.C.; Alcala, S.; Mulero, F.; Shakesheff, K.M.; Heeschen, C.; Aicher, A. Combination of injectable multiple growth factor-releasing scaffolds and cell therapy as an advanced modality to enhance tissue neovascularization. Arterioscler. Thromb. Vasc. Biol. **2010**, *30* (10), 1897–1904.

217. Geuze, R.; Theyse, L.; Kempen, D.H.; Hazewinkel, H.; Kraak, H.; Oner, F.C.; A differential effect of BMP-2 and VEGF release timing on osteogenesis at ectopic and orthotopic sites in a large animal model. Tissue Eng. Part A **2011**, *18* (19–20), 2052–2062.

218. Kempen, D.H.; Lu, L.; Heijink, A.; Hefferan, T.E.; Creemers, L.B.; Maran, A.; Yaszemski, M.J.; Dhert, W.J. Effect of local sequential VEGF and BMP-2 delivery on ectopic and orthotopic bone regeneration. Biomaterials **2009**, *30* (14), 2816–2825.

219. Xiao, C.; Zhou, H.; Liu, G.; Zhang, P.; Fu, Y.; Gu, P.; Hou, H.; Tang, T.; Fan, X. Bone marrow stromal cells with a combined expression of BMP-2 and VEGF-165 enhanced bone regeneration. Biomed. Mater. **2011**, *6* (1), 015013.

220. Lee, K.; Silva, E.A.; Mooney, D.J. Growth factor delivery-based tissue engineering: General approaches and a review of recent developments. J. R. Soc. Interface **2011**, *8* (55): 153–170.

221. Lim, J.J.; Hammoudi, T.M.; Bratt-Leal, A.M.; Hamilton, S.K.; Kepple, K.L.; Bloodworth, N.C.; McDevitt, T.C.; Temenoff, J.S. Development of nano- and microscale chondroitin sulfate particles for controlled growth factor delivery. Acta Biomater. **2011**, *7* (3), 986–995.

222. Chung, Y.I.; Kim, S.K.; Lee, Y.K.; Park, S.J.; Cho, K.O.; Yuk, S.H.; Tae, G.; Kim, Y.H. Efficient revascularization by VEGF administration via heparin-functionalized nanoparticle-fibrin complex. J. Control. Release **2010**, *143* (3), 282–289.

223. des Rieux, A.; Ucakar, B.; Mupendwa, BP.; Colau, D.; Feron, O.; Carmeliet, P.; Préat, V. 3D systems delivering VEGF to promote angiogenesis for tissue engineering. J. Control. Release **2011**, *150* (3), 272–278.

224. Phelps, E.A.; Landazuri, N.; Thule, P.M.; Taylor, W.R.; Garcia, A.J. Bioartificial matrices for therapeutic vascularization. Proc. Natl. Acad. Sci. U.S.A. **2010**, *107* (8), 3323–3328.

225. Griffith, C.K.; George, S.C. The Effect of hypoxia on *in vitro* prevascularization of a thick soft tissue. Tissue Eng. Part A **2009**, *15* (9), 2423–2434.

226. Yamakawa, M.; Liu, L.X.; Date, T.; Belanger, A.J.; Vincent, K.A.; Akita, G.Y.; Kuriyama, T.; Cheng, S.H.; Gregory, R.J.; Jiang, C. Hypoxia-inducible factor-1 mediates activation of cultured vascular endothelial cells by inducing multiple angiogenic factors. Circ. Res. **2003**, *93* (7), 664–673.

227. Duffy, G.P.; D'Arcy, S.; Ahsan, T.; Nerem, R.M.; O'Brien, T.; Barry, F. Mesenchymal stem cells overexpressing ephrin-b2 rapidly adopt an early endothelial phenotype with simultaneous reduction of osteogenic potential. Tissue Eng. Part A **2010**, *16* (9), 2755–2768.

228. Karal-Yilmaz, O.; Serhatli, M.; Baysal, K.; Baysal, B.M. Preparation and *in vitro* characterization of vascular endothelial growth factor (VEGF)-loaded poly(D,L-lactic-co-glycolic acid) microspheres using a double emulsion/solvent evaporation technique. J. Microencapsul. **2011**, *28* (1), 46–54.

229. Kolambkar, Y.M.; Boerckel, J.D.; Dupont, K.M.; Bajin, M.; Huebsch, N.; Mooney, D.J.; Hutmacher, D.W.; Guldberg, R.E. Spatiotemporal delivery of bone morphogenetic protein enhances functional repair of segmental bone defects. Bone **2011**, *49* (3), 485–492.

230. Kolambkar, Y.M.; Dupont, K.M.; Boerckel, J.D.; Huebsch, N.; Mooney, D.J.; Hutmacher, D.W.; Guldberg, R.E. An alginate-based hybrid system for growth factor delivery in the functional repair of large bone defects. Biomaterials **2010**, *32* (1), 65–74.

231. Hastings, C.L.; Kelly, H.M.; Murphy, M.J.; Barry, F.P.; O'Brien, F.J.; Duffy, G.P. Development of a thermoresponsive chitosan gel combined with human mesenchymal stem cells and desferrioxamine as a multimodal

pro-angiogenic therapeutic for the treatment of critical limb ischaemia. J. Control. Release **2012**, *161* (1), 73–80.

232. Raftery, R.; O'Brien, F.J.; Cryan, S.A. Chitosan for gene delivery and orthopedic tissue engineering applications. Molecules **2013**, *18* (5), 5611–5647.

233. Formiga, F.R.; Pelacho, B.; Garbayo, E.; Abizanda, G.; Gavira, J.J.; Simon-Yarza, T.; Mazo, M.; Tamayo, E.; Jauquicoa, C.; Ortiz-de-Solorzano, C.; Prósper, F.; Blanco-Prieto, M.J. Sustained release of VEGF through PLGA microparticles improves vasculogenesis and tissue remodeling in an acute myocardial ischemia-reperfusion model. J. Control. Release **2010**, *147* (1), 30–37.

234. Simon-Yarza, T.; Formiga, F.R.; Tamayo, E.; Pelacho, B.; Prosper, F.; Blanco-Prieto, M.J. PEGylated-PLGA microparticles containing VEGF for long term drug delivery. Int. J. Pharm. **2013**, *440* (1), 13–18.

235. Jin, Q.; Wei, G.; Lin, Z.; Sugai, J.V.; Lynch, S.E.; Ma, P.X.; Giannobile, W.V. Nanofibrous scaffolds incorporating pdgf-bb microspheres induce chemokine expression and tissue neogenesis *in vivo*. PLoS ONE **2008**, *3* (3), e1729.

236. Shen, YH.; Shoichet, M.S.; Radisic, M. Vascular endothelial growth factor immobilized in collagen scaffold promotes penetration and proliferation of endothelial cells. Acta Biomater. **2008**, *4* (3), 477–489.

237. Steffens, G.C.; Yao, C.; Prevel, P.; Markowicz, M.; Schenck, P.; Noah, EM.; Pallua, N. Modulation of angiogenic potential of collagen matrices by covalent incorporation of heparin and loading with vascular endothelial growth factor. Tissue Eng. **2004**, *10* (9–10), 1502–1509.

238. Chiu, L.L.; Radisic, M. Scaffolds with covalently immobilized VEGF and Angiopoietin-1 for vascularization of engineered tissues. Biomaterials **2010**, *31* (2), 226–241.

239. Melly, L.; Boccardo, S.; Eckstein, F.; Banfi, A.; Marsano, A. Cell and gene therapy approaches for cardiac vascularization. Cells **2012**, *1* (4), 961–975.

240. Curtin, C.M.; Cunniffe, G.M.; Lyons, F.G.; Bessho, K.; Dickson, G.R.; Duffy, G.P.; O'Brien, F.J. Innovative collagen nano-hydroxyapatite scaffolds offer a highly efficient non-viral gene delivery platform for stem cell-mediated bone formation. Adv. Mater. **2012**, *24* (6), 749–754.

241. Tierney, E.G.; Duffy, G.P.; Cryan, S.A.; Curtin, C.M.; O'Brien, F.J. Non-viral gene-activated matrices-next generation constructs for bone repair. Organogenesis **2012**, *9* (1), 22–28.

242. Tierney, E.G.; McSorley, K.; Hastings, C.L.; Cryan, S.A.; O'Brien, T.; Murphy, M.J.; Barry, F.P.; O'Brien, F.J.; Duffy, G.P. High levels of ephrinB2 over-expression increases the osteogenic differentiation of human mesenchymal stem cells and promotes enhanced cell mediated mineralisation in a polyethyleneimine-ephrinB2 gene-activated matrix. J. Control. Release **2013**, *165* (3), 173–182.

243. Jabbarzadeh, E.; Starnes, T.; Khan, Y.M.; Jiang, T.; Wirtel, A.J.; Deng, M.; Laurencin, C.T. Induction of angiogenesis in tissue-engineered scaffolds designed for bone repair: A combined gene therapy-cell transplantation approach. Proc. Natl. Acad. Sci. U.S.A. **2008**, *105* (32), 11099–11104.

244. Peterson, B.; Zhang, J.; Iglesias, R.; Kabo, M.; Hedrick, M.; Benhaim, P.; Lieberman, J.R. Healing of critically sized femoral defects, using genetically modified mesenchymal stem cells from human adipose tissue. Tissue Eng. Part A **2005**, 11 (1–2), 120–129.

245. Rouwkema, J.; Rivron, NC.; van Blitterswijk, CA. Vascularization in tissue engineering. Trends Biotechnol. **2008**, *26* (8), 434–441.

246. Cunniffe, G.M.; Dickson, G.R.; Partap, S.; Stanton, K.T.; O'Brien, F.J. Development and characterisation of a collagen nano-hydroxyapatite composite scaffold for bone tissue engineering. J. Mater. Sci. Mater. Med. **2010**, *21* (8), 2293–2298.

247. Jungreuthmayer, C.; Jaasma, M.J.; Al-Munajjed, A.A.; Zanghellini, J.; Kelly, D.J.; O'Brien, F.J. Deformation simulation of cells seeded on a collagen-GAG scaffold in a flow perfusion bioreactor using a sequential 3D CFD-elastostatics model. Med. Eng. Phys. **2009**, *31* (4), 420–427.

Latexes: Magnetic

Abdul Hamid Elaissari
University of Lyon, Lyon, France

Raphael Veyret
Bernard Mandrand
bioMérieux, National Center for Scientific Research (CNRS), Lyon, France

Jhunu Chatterjee
Florida A&M University and College of Engineering, Florida State University, Tallahassee, Florida, U.S.A.

Abstract

This entry will review the many different applications of magnetic latexes in biomedical diagnostics and therapeutics. The most common utilization is for rapid specific separation of various biological products, often in association with immunological or nucleic acid interactions. Magnetic particles are also utilized in biomolecule concentrations and as a marker in various labeling applications. Most of the criteria required by the magnetic particles are similar in many applications: small particle size to avoid sedimentation, good magnetic response (i.e., high magnetic content in the colloidal particles), high colloidal stability, narrow size distribution, and no residual magnetization when the magnetic field is suppressed (i.e., no remanence).

INTRODUCTION

Magnetic latexes are colloidal composites that combine organic and inorganic materials. The inorganic material may be composed of a metal or iron oxide derivative[1,2] or silica oxide.[3,4] Each of the organic and inorganic components plays a specific role in the properties of the final hybrid material. In inks, cosmetic products, and paints, polymers facilitate compatibility between the pigment and the binder. Thus, the inorganic materials are better dispersed throughout the medium, and there are negligible amounts of aggregated pigments and little aggregation. Consequently, in paints, where aggregate formation is a serious problem, the mechanical properties, durability, and glossiness are improved after the pigment is encapsulated by a polymer layer. The presence of a magnetic material endows the polymer particle with additional properties. For example, iron oxides and silica are used to make conducting polymers,[8] to modify the optical properties of polymer films,[9] and in ink applications, the paper industry,[10,11] high-density recording media,[12] and catalyst carriers.[13]

Many publications in the biomedical field describe the versatility of magnetic polymer particle applications for magnetic separation of biochemical products,[14] cell sorting,[15] magnetic particle guidance for specific drug delivery,[16] and as a contrast agent in magnetic resonance imaging (MRI).[17,18] The polymers used as coating agents generally function to protect inorganic material and induce reactive chemical groups capable of fixing biological molecules (via chemical binding), and magnetic iron oxide ensures the migration under an applied magnetic field. To facilitate adaptation of magnetic colloids, a great variety of magnetic composite particles have been developed commercially and are available; they range from a few nanometers to 10 µm and are available as capsules, microgel, and smooth or porous spheres. In this case, the source of the polymer matrix may be from a natural polymer or biopolymer, such as albumin[19] or starch;[20,21] they are eventually biocompatible and biodegradable particles. These colloidal particles may also contain a synthetic polymer derived from glutaraldehyde and cyanoacrylate.[22] Magnetic microspheres with hydrophilic surfaces are generally synthesized using acrylamides, acrylates, and methacrylate monomer derivatives.[23] Hydrophobic magnetic particles are generally constructed from hydrophobic monomers such as styrene.[24,25]

Given the variety of processes for synthesizing these latexes, they have a wide range of properties (surface charge density, reactive groups, hydrophilic–hydrophobic surface, particle size, size distribution, surface polarity, and magnetic properties) and can be adapted to many biological applications. The aim of this entry is to review the various uses of magnetic particles as solid phase or as particle carrier in the biomedical field.

Magnetic particles are used extensively in pharmacy, biology, and medicine to carry biological compounds. Biomolecules fixed to magnetic particles can be transported and can also be separated quickly from complex mediums; thus they can be used for both therapeutic and diagnostic applications. In the therapeutic domain, magnetic latex can guide a drug to and release it at a specific site,[26] and can extract tumor cells from the organism *in vitro* or *ex vivo*.

In Vitro—Medical

Concise Encyclopedia of Biomedical Polymers and Polymeric Biomaterials DOI: 10.1081/E-EBPPC-120052237

743

It is also used to generate sufficient heat in carcinoma cells to inactivate them, i.e., hyperthermia.[27]

Diagnostic applications involve the use of magnetic latex for biomolecules extraction, separation, and concentration. Both magnetic and nonmagnetic latex particles have long been recognized as good colloid solid supports for antigen detection by immunological reaction or in agglutination diagnostic tests.[28] Characteristics such as a large developed surface area, ease of use in biomolecule adsorption and chemical grafting, fast separation under magnetic fields, and rapid biomolecule kinetics have made magnetic latex an important solid support in immunoassays. Magnetic particles that bear the target molecules (after capturing them) can be easily and rapidly separated by applying a single magnetic field. This section will briefly detail the method of separation and the requirement for magnetic latex particles.

Many magnetic beads with different characteristics have been described in the literature and used in various applications. The first commercialized monodisperse magnetic beads were polystyrene based (e.g., Dynabeads).[29] This type of colloidal particle has been used in numerous processes as a solid support for separating biological compounds. Like other magnetic particles, it is synthesized by the precipitation of magnetic ferric oxide fine particles on porous polymer particles of homogeneous size. After the free iron oxide nanoparticles are removed, the composite beads are encapsulated with a polymer layer that fills the pores and brings to the surface the chemical groups needed to immobilize the biomolecules. Particles may also make up a metal oxide core, usually iron oxide, which is surrounded by a polymer shell bearing reactive groups that can be use to immobilize biomolecules.[30] Other magnetic beads are commercially available: Estapor (Merck Eurolab), Serradyn (Seradyn), Magnisort (Dupont),[30] BioMag M4100 and BioMag M4125 (Advanced Magnetic).[31]

Magnetic beads have been used and evaluated in many different diagnostic processes. Their first area of use is separation process. Magnetic particles have obvious advantages for this, whether or not immunological reactions are used. They are also advantageous as a solid phase for immobilizing various biomolecules and labels. Most of these advantageous characteristics do not vary much from one application to the next, but some of them may be dictated by the intended use.

MAGNETIC SEPARATION

Magnetic particles have been used as support for the separation, selective isolation, and purification of molecules.[32] For example, in biomedical diagnostics, they can replace the cumbersome steps of centrifugation or filtration.[32–34] The major techniques all involve chemical grafting of biomolecules onto magnetic beads to target specific separation of captured biomolecules or of analytes.[29] Since magnetic supports can be separated from solutions containing other species (e.g., suspended solids, cell fragments, and contaminants), magnetic affinity separation is useful for crude samples.[35] Various magnetic particles can be adapted to these kinds of applications, including large particles (above 1 μm),[29] silica magnetic particles,[32] and nanoparticles (below 100 nm).

Therefore, use of a magnetic field to separate composite magnetic particles is, compared with the alternatives, very simple, often cheaper, and above all faster. For example, bacterial control in the food industry requires 10 min of magnetic particle use instead of the 24 hr required with the traditional methods of analysis because the target bacteria concentrations are too weak to be characterized and require culturing. Fast magnetic separation results in direct concentration of the bacteria and therefore eliminates this slow step. The need to apply a powerful magnetic field may, however, be quite expensive, depending on the particle properties. Regardless of the use considered, magnetic particle carriers must have the following properties:

- Colloidal and chemical stability in the separation medium
- Nonmagnetic remanence
- Does not release of iron oxide during biomedical applications
- Low sedimentation velocity compared with magnetic separation
- A surface that is biocompatible with the relevant biomolecules
- Allows complete, rapid, and specific separation

Two specific methods are frequently used in biomedical diagnostics for detecting disease with magnetic particles and specific interaction between biological molecules (e.g., antibody/antigen or nucleic acids):

1. *Direct separation.* The biological sample containing the target molecules is mixed with magnetic particles bearing specific antibodies (generally called sensitive particles). After incubation under given conditions (time, buffer composition, temperature) and the subsequent immunological reaction between the immobilized antibody on the particle and the target antigen,

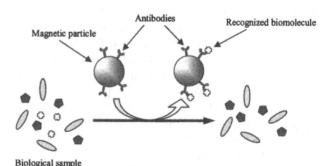

Fig. 1 Schematic illustration of direct specific separation of a target molecule.

the magnetic particles are separated by applying a magnetic field with a single magnet. Such direct capture (Fig. 1) and separation can also be performed with two kinds of antibodies. In this case, the first antibody is fixed on the colloidal particle as a spacer arm for immobilization of the second antibody for specifically capturing the target. The first antibody favors the orientation of the specific antibody immobilized via its Fc part onto the spacer arm–like antibody.

2. *Indirect separation.* The target molecule in a biological sample is first recognized by a specific antibody capable of reacting with the second antibody, which has been chemically grafted onto the surface of magnetic colloidal "carrier" particles. This method requires a thorough knowledge of molecular biology. These indirect methods are often more specific than the direct binding of antibodies (Fig. 2).[36]

On the physical level, the separation of magnetic particles as a function of the magnetic field can be explained and discussed basically on the basis of forces acting (magnetic force F_m and electroviscous force F_v) on the magnetic particles placed under the magnetic field (H), as summarized in the following expression (at equilibrium state):

$$F_v + F_m = 0$$

F_v is the friction force (or electroviscous force), which is basically the resistance to the displacement of the particle in a liquid medium; it is expressed as a function of the medium viscosity by Stokes' equation: $-6\pi\eta RV$, where R is the hydrodynamic radius of the particle, η is the viscosity of the medium, and V is the separation velocity.

F_m is the magnetic force due to the applied magnetic field (magnetic attraction force) and can be expressed as a function of various parameters:

$$F_m = \mu H \nabla H$$

where H is the magnetic field intensity, ∇H is the gradient of the magnetic field H, μ is the magnetic susceptibility ($\mu = \mu_0 \cdot M_s\, 4\pi R^3/3$), M_s is the saturated magnetization of the colloidal magnetic particles (principally related to the nature of iron oxide used), and μ_0 is the permittivity of the vacuum.

The separation speed (V) can therefore be expressed from the previous equations and as a function of saturated magnetization:

$$V = \frac{2\mu_0 M_s \bullet R^2}{9\eta} H \nabla H$$

Consequently, the separation speed increases with both the radius of the particle and the saturation of the magnetization. As expressed by the above equation, the saturated magnetization (M_s) is proportional to the content of magnetic material. In a given medium and in the presence of a fixed magnetic field, separation speed (V) therefore depends on the iron content, the magnetic properties of the particles, and the hydrodynamic size of the final microspheres. The above equation can be used to illustrate the parameters that affect the magnetic velocity of the particles under the magnetic field applied.

IMMUNOMAGNETIC SEPARATION

Immunoseparation involves the use of antibodies with a high affinity with the biomolecules to be separated, such as antigen or cells. Homogeneous magnetic beads are particularly well fitted for these kinds of applications because of their large, standardized surface of immunological reaction, which enables reliable and rapid biomedical diagnostic tests (such as enzyme-linked immunoassay, ELISA),[29] as various papers[33] and patents have reported. The quantification of immunological reactions (i.e., antibody/antigen/antibody, termed sandwich reaction process) is accomplished by the use of labeled antibodies; then the quantification is performed via fluorescence, chemiluminescence, phosphorescence, and radioactivity analysis.[33]

Principle of Immunoassay

The term immunoassay refers to any method for measuring the concentration or amount of analyte in solution based on specific interaction between the antibody and the targeted antigen. This method: 1) requires that the recognized

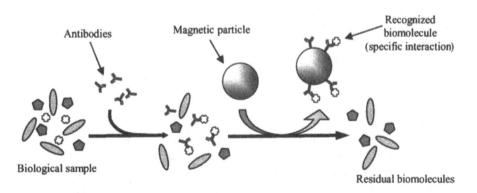

Fig. 2 Schematic illustration of indirect specific separation of a target molecule.

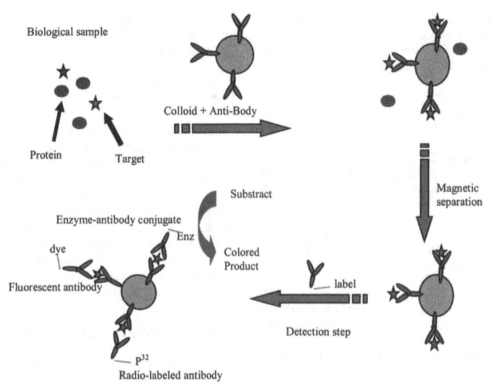

Fig. 3 Schematic illustration of immunoassay using labeled antibody.

analyte be physically separated from the residual biomolecules; and 2) employs a labeled antibody using radiolabel, fluorescent dye, enzyme, chemiluminescent molecule, or other label to measure or estimate the bound targets (Fig. 3). This method is termed competitive when the amount of measurable bound label is inversely proportional to the amount of analyte originally in solution. It is termed noncompetitive when the amount of measurable bound label is directly proportional to the amount of analyte originally present in the biological sample.

Large magnetic particles (10–100 μm) have been used in radioimmunoassays (RIAs) to detect and quantify nortriptyline, methotrexate, digoxin, thyroxine, and human placental lactogen. Such large particles must be stirred carefully to avoid sedimentation phenomena. Smaller particles, such as hollow glass or polypropylene or magnetic core (2–10 μm), have also been used and were evaluated in estradiol RIA, with nondrastic problems related to the particle sedimentation.[32] Nowadays various biomedical kits are available that are based on immunodetection and magnetic beads and either enzyme immunoassays (EIAs) or RIAs.[34]

Immunoaffinity Cell Binding

Magnetic particles are used for negative or positive cell selection. Because hepatocytes require highly specific binding, they are good test cells for developing magnetic particles for cell binding[37] and sorting. Microscopic observation of the interaction between hepatocytes and polystyrene magnetic beads shows that at 20°C the microvillus of the cells enfolds the colloidal particles. The combination of fluorescence microscopy technique analysis and specific immunological interaction between cells and magnetic particles is one of the best methods for visualizing and quantifying cells.

Negative cell selection

Magnetic particles are utilized for negative cell selection, i.e., in lymphocyte depletion, enrichment of monoclonal antibody–producing cells, and selection of nontumorous bone marrow cells.[29] For example, the latter is important in the autologous bone marrow transplantation used because the treatment for malignant disease often kills the bone marrow cells. The bone marrow cells removed and preserved for transplantation may, however, be contaminated with tumor cells. Magnetic beads seem to be one of the most effective methods reported for removing tumor cells from bone marrow.[37]

Allogenic bone marrow transplantation is a treatment for several fatal diseases, including immunodeficiency, leukemia, and aplastic anemia. Graft versus host disease can be avoided by *in vitro* depletion of T lymphocytes from the donor marrow graft.[37] For example, magnetic beads can be used to remove neuroblastoma cells from bone marrow following isolation of the mononuclear cells by gradient centrifugation. Comparison between porous and compact particles indicates that compact particles are interesting in terms of sensitivity. Indirect coupling of antibody to the particles and use of a cocktail of monoclonal antibodies ensures good binding of the tumor cells. Performing this purging operation twice is reported to be more effective.

Magnetic polymer beads of submicrometer size have also been used to remove B-lymphoma cells and T cells from bone marrow. These procedures use direct or indirect coupling of antibodies to the magnetic particles. Purging B cells twice is again more effective, depleting 10^6 tumor cells compared with 10^4 for a one-time purge. Removal of T cells did not result in damage to stem cells.

Positive cell selection

In this case, cells are extracted from a simple biological mixture for specific reuse.[37] Magnetic particles are also utilized to separate eukaryotic cells; to perform HLA typing; to count, culture, and separate rare cells; and to detect or study receptor, genetic, precursor, and progenitor cells for transfixion or autografting (e.g., selection of CD34 precursor cells).[29]

Functionally active cells can be selected with direct or indirect binding on magnetic particles. The ratio of beads to target must be low to maintain cell activity. For example, T8 cells and T and B lymphocytes have been bound to polymer magnetic beads without any interference with the effector functions of the bounded cells. The particles released from the cells by overnight culture can be removed by a suitable magnetic separation with a single magnet.[37]

Polymer magnetic particles are also used to purify CD8 T cells and HLA II cells, which can then be typed for HLA class I and II antigen respectively. The presence of magnetic beads during serological testing has no effect on the results. Compared to the standard method, the isolation is much faster, and a smaller volume of blood is necessary. The cells were isolated directly from the blood with polymer magnetic particles. After isolation, the cells were lysed, stained, and counted with microscopy techniques.[37]

Flux cytometry

This technique allows each kind of cell in a large population to be quantified by processing isolated cells in a detector. In the standard technique, nonmagnetic beads bearing an antibody specific for an antigenic epitope, the antigen, and a labeled antibody (enzyme, fluorescent or radiolabeled compound) are incubated together. Then the excess antibody and beads are washed away. However, the wash step, based on centrifugation, and the redispersion steps are cumbersome. The use of magnetic beads combined with flux cytometry techniques suppresses the drastic problems related to the washing steps. Silica magnetic beads have been explored for this process,[32] as have polymer composite latexes.

Immunomagnetic Separation of Bacteria

Immunomagnetic separation is based on the specific antigen–antibody reaction common to all of the immunoassays presented above. The immobilization of antibodies on a solid phase spurred the development of "sandwich" tests or ELISA.[38] Selective separation is a function of the specific epitope of the biomolecule under consideration: the efficacy (in term of reactivity) of the immobilized antibodies on the surface of the magnetic particles depends primarily on their orientation on the surface of the particles. They are immobilized either via physical adsorption (electrostatic or hydrophobic interactions), or via a chemical grafting process. The latter method results in more stable immobilized antibodies and is thus used more often in various immunological diagnostic tests.[39]

Most bacteria are surrounded by a cellular membrane that protects them from the external environment. They also have a flagella composed of a protein called the fimbria, and the bacteria can be bound to magnetic particles bearing the antifimbria antibody.[40]

The bacteria tested for most often in the food industry are *Salmonella*, *Escherichia coli*, and *Listeria*. Numerous detection processes use magnetic particles to give faster results with a detection limit that is the same as or lower than that of conventional techniques.[41]

Selective magnetic separation is a means of concentrating bacteria populations without having to go through a culture step to enhance the concentration. After magnetic separation, the target bacteria are visualized by physical methods such as scanning electron microscopy, fluorescence, radioactive labeling, or impedance (or conductivity) measurements of the culture medium. The selection of the method depends on the concentration of the bacteria solution.

After the capture step, the bacteria do not need to be released into any medium; they can develop from the surface of the considered colloidal particles. However, in some cases they are released mechanically by incubation or by magnetic agitation for 10–20 hr at 37°C, which takes a long time. Introducing a protein with strong affinity for the magnetic beads or an enzyme capable of destroying the antibody/antigen link permits faster release.[42]

Magnetic particle carriers have proven to be a simple means of detecting viruses difficult to obtain in sufficient quantities by artificial culture. Magnetic beads are then used to separate the bacteria for further applications like ELISA, such as spreading on a solid phase, amplification, and impedance measurement.[29] For example, the use of immunomagnetic beads to capture *Listeria* from environmental samples reduces test time and improves sensitivity, compared with the usual methods. The analysis is performed via commercial tests such as Listertest (Vicam, Watertown, MA, USA). Samples are mixed with immunomagnetic beads coated with anti-*Listeria* antibodies. After binding (i.e., biological recognition), the magnetic beads are isolated by applying a magnetic field; they are then planted on medium and incubated overnight. The next day, a replica is made on a thin plastic, and colorimetric detection is performed with anti-*Listeria* antibodies and anti-antibodies linked to alkaline phosphatase (enzyme). Because the method does not use enrichment, the number of *Listeria* colonies is related to the original level of contamination.[43]

Using magnetic particles makes it possible to avoid some steps in food (disinfection, heating) that modifies the virus envelope and makes polymerase chain reaction (PCR) amplification unsuitable. This system is used, for example, to detect hepatitis A virus in shellfish. In their review, Olsvik et al.[38] describe examples of parasite detection with magnetic microspheres that reduce analysis time from 6 to at less 1 hr and thus permit the diagnosis of a larger number of samples.

Collection of a Substance *In Vivo*

One of the most interesting uses of magnetic particles is for the *in vivo* capture, detection, concentration, or isolation of the target. This has been reported in various papers but only a few patents. One patent[30] describes a technique to isolate and remove an analyte from a body fluid. A molecule ligand, specific to the target, is immobilized on the surface of a magnetic particle. The particle is then introduced *in vivo* (in the body fluid) where it is tolerated long enough to enable the ligand to bind biologically to the target. Once the complex (ligand–target) is formed, it is retrieved from the body fluid by the application of a magnetic field.

The method is particularly fitted for detection of substance in gingival crevices for the diagnosis of periodontal disease. IgG against cachectin or interleukin-1 is chemically grafted to carboxyl groups. The magnetic particles are introduced in the cavity with a microdispenser. A magnetic field applied with a device anatomically compatible with the body cavity and adapted for the magnetic dispersion used then collects the magnetic particles.

Magnetic particles for *in vivo* applications are composed of a magnetic core bearing a hydrophilic-biocompatible biodegradable polymer shell layer or basic silica layer;[32] cells or tissues recognized by the particular bioaffinity adsorbent grafted on the magnetic particle are located, and therapeutic agents immobilized on the particles are delivered by magnetic direction to pathological sites.

SEPARATION AND QUANTIFICATION OF PROTEINS AND ANTIBODIES

The sensitive detection of molecules and biomolecules is one of the most promising research fields. Today various and accessible techniques are used or combined to enhance this sensitivity. To this end, capillary electrophoresis (based on UV or fluorescence analysis), HPLC, fluorescence systems (flux cytometry), PCR, reverse transcriptase (RT) PCR, quantitative PCR, luminescence, phosphorescence, and colorimetric titration have been explored and are widely used. Each technique should be adapted to a given

biomolecules, since the detection of nucleic acids (polyelectrolyte-like substances) and proteins (complex copolymers) is totally different. It is then of paramount interest to develop methods based on both specific and nonspecific isolation and concentration of biomolecules. For example, the detection of a protein in a low concentration of a biological sample has been reported and patented.[44] A small magnetic particle is coated with a given protein specific for the target protein; after a capture step leading to protein–protein interaction, the magnetic particles are concentrated in a small volume. Then, after a protein release process, the magnetic particles are separated from the sample and the protein–protein complexes incubated in the presence of a cleaving agent. The bond between the protein conjugates is then cleaved, and the captured protein is ellipsometrically analyzed. Various kinds of magnetic particles can be used for this detection and specific concentration, with particle sizes ranging from 1 to 10 µm. The magnetic part can be derived from iron oxide (i.e., ferrite). The standard illustration involves the use of particles coated with bovine serum albumin (BSA) to detect anti-BSA antibodies.[44]

Removal of Magnetic Particles

After active cells are selected by immuno affinity magnetic separation, the particles detached from the cells by overnight culture can then be removed by a suitable magnet.[37] This may be due to shedding of the antigen involved in the binding, and this process may not be operative in all cases.

Another method under investigation involves the introduction of a layer of enzymatically degradable polysaccharides between the beads and the cells, or cleavable chemical bonds (S-S), or a Schiff base. Release can also be affected by enzyme digestion with proteinases, such as chymopapain, pronase, or trypsin, or a glycoprotease, to which some antibodies are susceptible, or by the application of ultrasound.[29]

One method for detecting a protein involves coating a small magnetic particle with a protein specific for the target protein; the bond between the two proteins is cleaved after separation from the sample, by introducing the magnetic particle into a cleaving agent solution, composed of weak acid or alkaline solutions.[44]

Anti-Fab antibodies can also be utilized for soft cell release because they inhibit the binding between antibody and antigen. Larger beads may detach more easily, thus leading to the release of captured cells.

It is worth noting that magnetic beads have been tested in many different immunological applications. These beads, which present a large surface area for binding and are easily separated from any medium, seem particularly well fitted to such applications. In addition to the general characteristics of magnetic beads discussed above, the functional groups on the surface of the particles appear very important for binding active antibodies.

NUCLEIC ACID SEPARATIONS

Magnetic separation and isolation of biomolecules always involves interactions between the colloidal particles and the substance to be separated. Interactions in addition to the extremely effective immunological ones have also been explored for separating different biological molecules. Most have been applied to the separation of nucleic acids. Examples of applications include the following:[29]

- Purification and concentration of PCR products (or of single-strand DNA fragments) with biotinylated primers and streptavidin-bearing polymer particles, which are used to detect microorganisms or cellular genes, or to label singlestrand probes
- Specific purification of single-strand DNA or RNA with magnetic particles bearing oligonucleotides (single-strand DNA or RNA fragments of welldefined sequences)
- Purification of proteins associated with nucleic acids, such as transcription factors, regulatory genes, and promoters
- Purification of mRNA with magnetic beads coated with polythymidylic acid for *in vitro* translation or gene expression
- Nonspecific concentration of nucleic acid followed by PCR or RT-PCR amplification

Specific DNA isolation and extraction to obtain purified nucleic acid without the use of any organic solvent as a precipitating agent is required in biomedical diagnostic tests, for which specificity, rapidity, and ease of handling are incontestably necessary. Some applications have been clearly described and discussed more precisely in the literature.

Detection of Mutations

To diagnose leukemia by the detection of mutated mRNA, polymer magnetic beads linked to oligo-dT$_{25}$ are used to purify mRNA after acid guanidium phenol chloroform isolation of crude RNA. Then cDNA extension and PCR are performed in the presence of the magnetic beads.

Mutation detection involves the direct sequencing of the PCR products, obtained with a standard biotinylated primer. They are immobilized by binding onto streptavidin-coated polymer magnetic beads and then by attraction with a permanent magnet during the supernatant removal and washing procedures. Denaturation of the double-stranded PCR product into a single strand is performed in 0.1 M NaOH for 15 min. The single strand, attached to the beads, is then washed and sequenced directly.[45]

Selection of Differentially Expressed mRNA

Biotinylated oligo-dT is used to synthesize a cDNA library, which is mixed and annealed to the mRNA extracted from the cell of interest. The cDNA/mRNA hybrids are removed

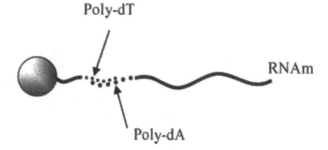

Fig. 4 Selective purification of mRNA with magnetic beads bearing dT oligonucleotide.

with avidin-coated magnetic polymer beads. These two steps are repeated twice, and the remaining mRNA is reverse transcribed and cloned. This process is useful for identifying genes specifically expressed in differentiated cells.[45]

Novagen's mRNA isolation system[46] uses superparamagnetic microspheres covalently coated with oligo-dT25 to selectively extract mRNA from a variety of sources (Fig. 4). After magnetic separation, the purified mRNA is eluted from the magnetic beads for a second round of purification.

Nucleic Acid Extraction and Concentration

The use of cationic colloidal supports (i.e., positively charged particles) in nucleic acid adsorption (i.e., nucleotide multimers) according to size has been explored in numerous studies. The interaction between cationic particles and such negatively charged polyelectrolytes is believed to be based on electrostatic attractive forces. After nucleic acid adsorption and magnetic separation of the colloidal particles, conditions such as pH, salinity, and temperature are adjusted to release the nucleic acids from the support for amplification or direct analysis as schematized in Fig. 5.

For the Human Genome Project the Whitehead Institute at MIT developed a method called solid-phase reversible immobilization (SPRI) wherein DNA is captured onto carboxyl-modified encapsulated superparamagnetic microspheres. After the DNA is bound, the beads are washed with ethanol. The DNA is then eluted from the beads in a low ionic strength solution. This method leads to high-quality DNA template purification and can be used with major templates and enzymes responsible for sequencing.[47] It can also be applied to detecting hybridization between complementary nucleic acid sequences. A long polynucleotide that is partially complementary to a short oligonucleotide is chemically grafted onto the magnetic particles. The short probe is used as a spacer arm and does not bind directly to the particles but rather to the first bonded single-stranded nucleic acid. The specific capture of any nucleic acid sequence can then be performed by a hybridization process on the second, short oligonucleotide, as schematically presented in Fig. 6.

The hybridization can also be performed first in solution. The hybrids are separated from the unbounded probes

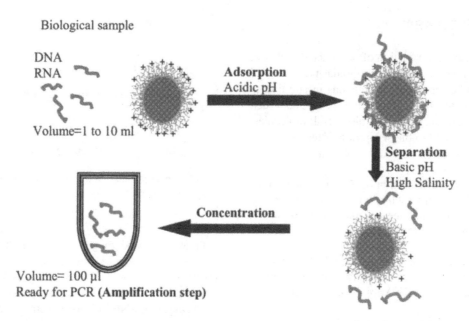

Fig. 5 Illustration of nucleic acids adsorption, extraction, concentration, and amplification.

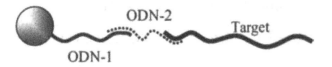

Fig. 6 Schematic illustration of specific capture of nucleic acid fragment by a doublehybridization process onto magnetic polymer particles. ODN-1 oligonucleotide (of a given sequence); ODN-2 oligonucleotides act as a spacer arm and specific to a given part of ODN-1.

by magnetic particles bearing a short oligonucleotide (with a specific sequence complementary to the target) to capture the hybrids. The hybridization yield is 20% in solution and only 5% after binding to the beads.

Another method for capturing free nucleic acids in any biological sample uses positively charged magnetic microspheres for immobilization (via electrostatic interaction) of rRNA as well as DNA molecules. The nucleic acids released are not damaged after the capture of rRNA or DNA and can be amplified via a hybridization process with RT-PCR or PCR. The immobilization of rRNA is favored in the presence of urine combined with acetic acid. Efficient rRNA hybridization is obtained with the proper reagents. RNA/DNA hybrids are recovered from the buffer solution and the unhybridized probes left in solution. Captured polynucleotide probes can then be eluted from the beads with the appropriate conditions. Accordingly, nucleic acids can be purified from such biological samples as cell lysate and sputum. For these applications, various cationic magnetic particles are available; these include quaternary ammonium containing magnetic microsphere and poly-D-lysine-functionalized microspheres. In addition, magnetic particles with amine microspheres have been demonstrated to be compatible with a chemiluminescent nonisotopic

assay[31] or electrochemiluminescence[48] with streptavidin-bearing magnetic particles.

We see from this description of the use of magnetic beads in molecular biology for various applications that various types of magnetic beads appear to be compatible with many enzyme reactions, even PCR or sequencing. Specific biomolecule separation is the application that has been most widely studied, probably because the major advantage of magnetic particles is their easy removability from any liquid simply, just under a magnetic field, even in an unclear and complex solution.

PROTEIN IMMOBILIZATION ONTO REACTIVE MAGNETIC LATEXES

Immobilization of Enzyme

Thermally sensitive magnetic poly(NIPAM/MA/MBA) [N-isopropylacrylamide, NIPAM, methacrylic acid (MA), methylenebisacrylamide (MBA)] latex particles are well adapted to immobilize enzyme or drug carriers. Such composite particles contain small amounts of iron oxide material, which induce a low separation rate under a magnetic field. To enhance the magnetic separation velocity, the particles are thermally aggregated.[49] The thermal sensitive property is attributed to the properties of the poly(NIPAM) polymer. Trypsin has been covalently immobilized (via the well-known carbodiimide method) onto carboxylic thermally sensitive magnetic poly(NIPAM/MA/MBA) latex particles. The content of the carboxyl groups is reported to be an important factor in efficient protein binding. Thus, trypsin activity was higher with a high MA content. However, a high MA content enhances the colloidal stability (electrosteric stabilization) when thermal flocculation of the particles is needed. Because the protein (i.e., enzyme and antibody) activity may be sensitive to temperature changes during thermal flocculation (4°–30°C), the stability of the immobilized proteins has been tested. After repeated thermocycles (10 times), 95% of the initial enzymatic activity was retained.[49]

Immobilization of Antibodies

Different procedures can be used to immobilize a cell-specific antibody to the colloidal particle:[37]

- The monoclonal antibody can be chemically grafted directly onto the particles. This is not very efficient, probably because of the antibody orientation, and the need for a spacer arm between the antibody and the particle surface.
- A polyclonal antibody can be first immobilized on the beads. After washing, the monoclonal antibody is then bound onto it. The polyclonal antibody in this case is used as a spacer arm and corrects the orientation of the monoclonal antibody far from the solid surface.
- In a variant method for cell separation and cell sorting, the monoclonal antibody is first bound to the cells, and then captured onto the particles coated with polyclonal antibodies (indirect method). This technique requires cumbersome washing of the excess antibodies from the cells, leading to cell loss. Washing magnetic particles is much simpler and easy to perform.

The polymer magnetic particles are generally sufficiently hydrophobic to allow a relatively strong physical adsorption of antibodies (or proteinic molecules) via hydrophobic interaction. In addition, the presence of some reactive groups, such as hydroxyl function (–OH), can be used for chemical grafting of antibodies after activation with sulfonyl chlorides. It is also possible to obtain magnetic polymer beads bearing reactive groups such as $-NH_3$, hydrazine, –SH, or –COOH groups on the surface. The covalent coupling of antibody can then be performed via the $-NH_2$, –SH, –COOH, or –CHO (aldehyde) groups of the antibody in question.[37]

Recently, thermally sensitive magnetic latexes have been prepared and evaluated as a support in immunoassay. After covalent coupling of anti-α-fetoprotein onto this stimulus-responsive hydrophilic support (via the carbodiimide activation process), performed in optimal coupling (salinity, temperature, and pH) conditions, the sensitive immunomagnetic colloidal particles were tested with ELISA to detect the α-fetoprotein.[50]

LABELING PROCESS

Accelerated Agglutination

Numerous standard detection methods involve autoagglutination and macroscopic or microscopic visualization.

Agglutination techniques are advantageous because they require only one step, with no washes and no addition of reagents. Use of magnetic particles increases sensitivity and rapidity.[34] In this domain, two interesting patents (JP A 5180842 and WO A 8606493) describe a technique that uses a combination of magnetic and nonmagnetic particles. The non magnetic particles are colored and coated with the same antigen, which are incubated in two samples, one with and one without the antibody specific for the antigen used. Under a magnetic field, decreased color intensity in the nonagglutinated dispersion of the reaction mix indicates the presence of the antibody.[34] Polymers, possibly including magnetic particles, can be used as coloring agents.[51]

Numerous patents describe a method utilizing two kinds of particles—one magnetic and coated with an antigen, the other nonmagnetic and coated with an antiantigen—which are incubated in two samples, one of which does contain and the other of which does not contain the antibody specific for the antigen. After agglutination, the quantity of nonmagnetic and nonagglutinated particles is an indication of the presence of the antibody. This method may not be as sensitive as RIA and EIA methods. The same type of assay has been also patented[34] in a version in which the magnetic and nonmagnetic particles are coated with an antitarget. After incubation of the particles with the biological sample, the magnetic particles are separated under a magnetic field, and the presence of the target is evidenced by observation of the nonagglutinated reaction mix.

For example, the magnetic particle and the nonmagnetic particle can be coated with an antibody against the antigen in question. The evaluation can be done by visual observation, or by measuring the optical density or fluorescence, phosphorescence, luminescence, or chemiluminescence. Nonmagnetic particles can be colored, fluorescent, chemiluminescent, phosphorescent, or luminescent.[34] The technique can identify an antigen in a biological sample (blood, urine, milk, saliva, serum); examples include the degradation product of fibrin, the inhibitor of plasminogen activator, plasminogen activator, C-protein, bacteria, viruses, or antibodies against bacteria, yeast, or viruses. The recommended size of the particles is in between 300 and 800 nm for the magnetic ones, and from 100 to 600 nm for the nonmagnetic.

Accelerated Sedimentation Process

Biomedical diagnostics are now targeting the development of basic and easy techniques for disease and bacteria detection. Various investigators[49,50] have described an immunoquantification method that is one illustration of such an application. This method discriminates reacting and nonreacting colloidal particles by whether they sediment or coat the sensitive walls of a special tube. After the biological sample is introduced into a recipient whose walls are coated with a substance (such as antibody) with a specific immunological affinity for the targeted substance, sensitive magnetic particles bearing captured analyte are then added to the sensitive tube with a specific immunological affinity for the captured substance. If the target is captured by the magnetic particles, they will mainly deposit on the walls of the recipient via specific immunological reaction. If the target is not captured, the sensitive colloidal particles will sediment in the tube. Use of magnetic particles enhances sensitivity via the concentration process, rapid immunological reaction on the walls under magnetic field, and finally high sedimentation of the composite particles. In addition, free magnetic particles can be easily removed magnetically.[51,52]

Contrast Agent in MRI

The resolution and sharpness of the magnetic resonance image in body scanning depend on spin-spin relaxation time, which is reduced by paramagnetic materials.[37,53] A mono-sized polymer particle carrying magnetic iron oxide, given orally, can produce a negative black contrast in MRI, thus eliminating the image of the gastrointestinal track and providing a clearer visualization of the other organs in the abdomen. The particle size varies according to the type of test. Small particle sizes should be used (diameter less than 3 μm) for parenteral use.

CHARACTERISTICS OF THE MAGNETIC MATERIALS FOR DIAGNOSTIC APPLICATIONS

The characteristics of the beads can be modulated according to their intended use. However, for most applications, generally preferred characteristics do not vary.

Particle Size and Size Distribution

Both particle size and size distribution are important in the utilization of magnetic particles in the biomedical field. These parameters may affect the efficiency of their utilization in a given application. For example, light intensity can be explored in flux cytometry, but only if particle size is homogeneous (i.e., narrow size distribution), to make multiple analysis and quantification possible.[29] In any case, monodispersed magnetic particles are particularly appropriate for applications such as immunoseparation because of the large and standardized reaction surface, which makes homogeneous separation reliable and rapid.

Specific Surface Area

The separation rate of magnetic particles under a magnetic field is related to the amount of iron oxide in the composite beads and to the particle size.[33] To favor magnetic separation, the colloidal magnetic particles should contain the minimum needed of the magnetic material (less than 12 wt%) and the size should be sufficient.[35] Therefore very small

magnetizable nanoparticles (such as ferrofluid dispersions with particle size less than 30 nm) are not easy to separate by applying a magnetic field. In fact, the force induced by the thermal agitation (i.e., Brownian diffusion motion) is higher than the magnetic force induced by the magnet.[33] Hence, magnetic separation of small hybrid particles is difficult.

Large magnetic particles (more than 1 μm in diameter) containing less than 12 wt% iron oxide can respond to a weak magnetic field; however, they generally tend to settle rapidly and have a more limited specific surface area (m^2/g) than smaller particles. Accordingly, the quantity of biomolecules immobilized is low and the sensitivity of the biomedical application reduced.[32] The particles containing less than 40 wt% highly magnetic iron oxide must be large enough (0.1–0.3 μm) to obviate high magnetic field intensity for a long period. Thus, the choice of particle size is difficult. The best magnetic separation is obtained with a large particle size, but sedimentation is higher and sensitivity low due to the low available specific surface for biomolecule immobilization. Thus, the choice of the particle size and the amount of iron oxide may be adapted for different applications.

Thermoflocculation of Stimulus-Responsive Magnetic Latexes

Thermally sensitive magnetic latexes have been prepared and evaluated in biomedical domains. Such stimulus-responsive composite particles flocculate as the incubation temperature and salt concentration of the medium increases. The flocculation behavior is related to the reduction of both steric and electrostatic stabilization. Such thermoflocculation is reported to be a reversible process because of the low critical solution temperature of poly[N-alkyl(meth) acrylamide] derivative.[35,50]

The hydrophilic magnetic and thermosensitive poly(NIPAM/MA/MBA) latex particles have a large surface area and are therefore well fitted to immobilize enzymes or antibodies. Moreover, after thermoflocculation, the particles can be quickly separated by a magnetic field. These thermally sensitive particles can satisfy both a large surface area for biomolecule immobilization and rapid magnetic separation.

Thermally sensitive flocculation (Fig. 7) appears to be a very attractive innovation that resolves the antagonism between rapid magnetic separation and surface area. It is worth noting, however, that this property requires a temperature cycle for flocculation and dispersion, which may not be compatible with all biotechnological devices and applications, especially enzymatic activity.

Therapeutic Applications

Magnetic polymer particles must have the appropriate properties for use as a carrier in various in vivo therapeutic applications. The polymer must be biodegradable if the microspheres are to remain inside the organism.[55] The physicochemical and colloidal characteristics of the magnetic carrier and the properties of the continuous phase must be considered. The intensity and distance from the magnetic field applied must be taken into account to check the magnetic particle distribution and displacement before degradation. They must have a biocompatible surface that does not cause an immune system reaction and that generates an efficacious and specific targeting action. Finally, the diameter of the particles is also a crucial parameter. Risks of phagocytosis are limited with particles of diameter greater than 12 μm.[26] On the other hand, small particles (less than 1 μm) can penetrate the capillaries. These act more specifically but do not resist phagocytes.

The total quantity of medicine introduced into the organism is sometimes limited by toxic effects beyond a certain concentration.[26] By injecting the active ingredient in a polymer capsule or matrix, it is possible to control the diffusion and release of the desired product in the organism. This also avoids repetitive administration of the medicine. The targeting is then well controlled by applying a

magnetic field that improves the localization of the chemical product and prevents the particles from being captured by the reticuloendotheliosis system.[58] Because the product is locally concentrated, the total dose administered can be reduced, with a corresponding reduction of toxicity problems. For example, the total dose of doxorubicin, injected freely, is 100 times higher than that introduced with a magnetic carrier for an equivalent concentration at the treatment site.

The active agent is released outside the particles as a function of mechanisms that depend on the properties of the polymer. Adriamycin and doxorubicin in lyophilized albumin magnetic microspheres, conserved at 4°C, diffuse spontaneously at 37°C, in an aqueous solution of 1 g/L NaCl at a rate that depends on the cross-linking of the matrix during synthesis. In the case of certain so-called smart materials, variation of the medium (i.e., pH, temperature, ionic strength) leads to modification of the polymer's properties.[59] For example, certain thermally sensitive polymer-based particles are hydrophilic and expand below their volume phase transition temperature, i.e., poly(NIPAM) particles below 32°C. On the other hand, the particles shrunk above the transition temperature, thus releasing the encapsulated active ingredient, as illustrated in Fig. 8.

Specific Cell Extraction

Generally, the extraction of diseased cells by magnetic particles is based on the chemical difference between healthy and infected cells. Two steps should be considered. The first involves targeting the microspheres to a specific site by applying a magnetic field. Magnetic guidance makes it possible to reach areas that are difficult to access, and its efficacy usually depends on the properties of the carrier, such as particle size and stability in the environment. Secondly, the infected cells are then recognized and immobilized on sensitized particles. Capture yield varies from

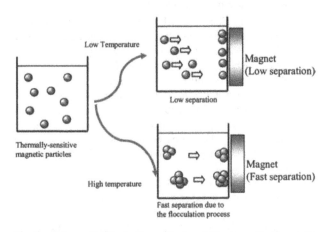

Fig. 7 Schematic illustration of hydrophilic magnetic thermally sensitive latex particles via temperature flocculation process.

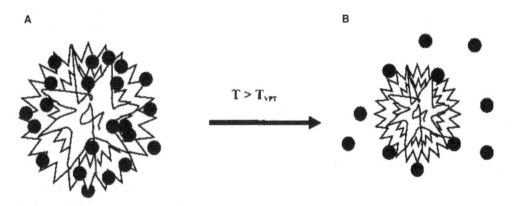

Fig. 8 Release of the active ingredient by controlling incubation temperature. T_{VPT} is the volume phase transition temperature, the T_{VPT} in the case of poly (*N*-isopropylacryl-amide) microgel particles is nearly 35°C. (**A**) Below the T_{VPT} the particles are swollen by the active agent, and (**B**) above the T_{VPT} the active agent is released.

patient to patient since recognition depends on the properties of tumor cells and their affinity vis-à-vis the antibodies immobilized on the colloidal particles. Hancok and Kemshead,[36] Rembaum, Yen, and Molday[59] and Ugelstad et al. [60] initiated the use of magnetic particles to purify marrow. The method was then extended to other tumors such as the treatment of lymphocytes, leukemia and lung cancer. Using chemotherapy and radiation to cure cancerous tumors can affect neighboring healthy cells. To avoid this problem and reduce secondary effects, treatment is carried out outside the organism (*ex vivo*).[61] Specific antigens of tumor cells are fixed on the surface of magnetic particles and make it possible to capture only infected cells.[62] These are then extracted by applying a magnetic field and treated chemically outside the organism (*ex vivo*) before being reinjected into the patient. Another example of this type of promising application is the study of vascular problems by extracting endothelial cells via coating of magnetic particles with lectin.

CONCLUSION

The advantages of colloidal magnetic particles are, on the one hand, the speed of detection and production of the analysis result and, on the other hand, the simplicity of separation, which makes the system easy to automate. Its specificity and reliability are identical to those already proven for nonmagnetic polymer-based particles (reactive latexes).

In view of the applications in the biomedical field, several properties, such as particle size, size distribution, surface polarity, surface charge density, and amount of magnetic material (i.e., iron oxide content), must be considered carefully for magnetic polymer particles.

The real interest of magnetic particles in biomedical field is clearly evidenced by the extensive literature. These particles have been used in many different applications in biomedical diagnostics and therapeutics. The most common utilization is for rapid specific separation of various biological products, often in association with immunological or nucleic acid interactions. Magnetic particles are also utilized in biomolecule concentration and as a marker in various labeling applications. Most of the criteria required by the magnetic particles are similar in many applications: small particle size to avoid sedimentation, good magnetic response (i.e., high magnetic content in the colloidal particles), high colloidal stability, narrow size distribution, and no residual magnetization when the magnetic field is suppressed (i.e., no remanence). For more literature concerning magnetic particles in medicine and biology, the readers can consult Arshady's *Microspheres, Microcapsules and Liposomes*, Volume 3.[63]

REFERENCES

1. Furusawa, K.; Nagashima, K.; Anzai, C. Synthetic process to control the total size and component distribution of multilayer magnetic composite particles. Colloid Polym. Sci. **1994**, *272* (9), 1104–1110.

2. Chen, T.; Somasundaran, P. Preparation of core-shell nanocomposite particles by controlled polymer bridging. J. Am. Ceram. Soc. **1998**, *81* (1), 140–144.

3. Bamnolker, H.; Nitzar, B.; Gura, S.; Margel, S. New solid and hollow, magnetic and non-magnetic, organic-inorganic monodispersed hybrid microspheres: synthesis and characterization. J. Mater. Sci. Lett. **1997**, *16* (16), 1412–1415.

4. Caruso, F.; Mohwald, H. Preparation and characterization of ordered nanoparticle and polymer composite multilayers on colloids. Langmuir **1999**, *15* (23), 8276–8281.

5. Sauzedde, F. Elaboration de latex magnetique hydrophiles fonctionalisés en vue d'applications dans le diagnostic medical. PhD thesis, Claude Bernard University, Lyon, France, 1997.

6. Van Herk, A.M. In *Polymeric Dispersion: Principles and Applications*; Asua, J.M., Ed.; NATO ASI Series E: Applied Sciences Vol. 335, 1996; 435–462.

7. Sugimoto, T. In *Fine Particle: Synthesis, Characterization, and Mechanisms of Growth*; Surfactant Science Series Vol. 92. Marcel Dekker: New York, 2000.

8. Neoh, K.G.; Tan, K.K.; Gob, P.L.; Huang, S.W.; Kang, E.T.; Tan, K.L. Electroactive polymer SiO2 nanocomposites for metal uptake. Polymer **1999**, *40* (4), 887–893.

9. Sohn, B.H.; Cohen, R.E. Processible optically transparent block copolymer films containing superparamagnetic iron oxide nanoclusters. Chem. Mater. **1997**, *9* (1), 264–269.

10. Kommaredi, N.S.; Tata, M.; Vijay, T.J.; McPherson, G.L.; Herman, M.F. Synthesis of superparamagnetic polymer-ferrite composites using surfactant microstructures. Chem. Mater. **1996**, *8* (3), 801–809.

11. Neveu-Prin, S. Synthèse et caractérisation de nanoparticules magnétiques—elaboration de ferrofluides et des capsules magnétiques. PhD thesis, Pierre–Marie Curie University, 1992.

12. Kwon, O.; Soic, J. New interaction effects with a superparamagnetic latex. J. Magnet. Magnet. Mater. **1986**, *54–57* (3), 1699–1700.

13. Tamai, H.; Sakura, H.; Hirota, Y.; Nishiyama, F.; Yasuda, H. Preparation and characteristics of ultrafine meta particles immobilized on fine polymer particles. J. Appl. Polym. Sci. **1995**, *56* (4), 441–449.

14. Patton, W.F.; Kim, J.; Jacobson, B.S. Rapid high-yield purification of cell surface membrane using colloidal magnetite coated with polyvinylamine: sedimentation versus magnetic isolation. Biochim. Biophys. Acta **1985**, *816* (1), 83–92.

15. Kemshead, J.T.; Treleaven, J.G.; Gibson, F.M.; Ugallstad, J.; Rembaum, A.; Philip, T. Removal of malignant cells from marrow using magnetic microsphere and monoclonal antibodies. Prog. Exp. Tumor Res. **1985**, *29*, 249–255.

16. Gupta, P.K.; Hung, C.T. Magnetically controlled targeted micro-carrier system. Life Sci. **1989**, *44* (3), 175–186.

17. Reynold, C.H.; Anan, N.; Beshah, K.; Huber, J.H.; Shaber, S.H.; Lenkinski, R.E.; Wortman, J.A. Gadolinium-loaded nanoparticles: New contrast agents for magnetic resonance imaging. J. Am. Chem. Soc. **2000**, *122* (37), 8940–8945.

18. Ogan, M.; Schmiedl, U.; Moseley, M.; Grodd, W.; Paajenen, H.; Brasch, R.C. Albumin labeled with Gd-DTPA: An intravascular contrast enhancing agent for magnetic resonance blood pool imaging: Preparation and characterization. Invest. Radiol. **1987**, *22* (8), 665.

19. Gupta, P.K.; Hung, C.T. Albumin microspheres 1: physico-chemical characteristics. J. Microencapsul. **1989**, *6* (4), 427–462.

20. Molday, R.S.; Mackenzie, D. Immuno-specific ferromagnetic iron-dextran reagents for labeling and magnetic separation of cells. J. Immunol. Meth. **1982**, *52* (3), 353–367.

21. Kilocoyne, S.H.; Gorisek, A. Magnetic property of iron dextran. J. Magnet. Magnet. Mater. **1998**, *177–181* (2), 1457–1458.

22. Schütt, W.; Grüttner, C.; Häfeli, U.; Zborowski, M.; Teller, J.; Putzar, H.; Schümichen, C. Applications of magnetic targeting in diagnosis and therapy—Possibilities and limitations: A mini-review. Hybridoma **1997**, *16* (1), 109–117.

23. Sauzedde, F.; Elaissari, A.; Pichot, C. Hydrophilic magnetic polymer latexes. 1. Adsorption of magnetic iron oxide nanoparticles onto various cationic latexes. Colloid Polym. Sci. **1999**, *277* (9), 846–855.

24. Charmont, D. Preparation of monodisperse, magnetizable, composite metal/polymer microspheres. Prog. Colloid Polym. Sci. **1989**, *79*, 94–100.

25. Lee, J.; Senna, M. Preparation of monodispersed polystyrene microspheres uniformly coated by magnetite via heterogeneous polymerization. Colloid Polym. Sci. **1995**, *273* (1), 76–82.

26. Langer, R. New methods of drug delivery. Science **1990**, *249* (4976), 1527–1533.

27. Chan, D.C.F.; Kirpotin, D.B.; Bunn, P.A. Synthesis and evaluation of colloidal magnetic iron oxides for the site specific radiofrequency-induced hyperthermia of cancers. J. Magnet. Magnet. Mater. **1993**, *122* (1–3), 374–378.

28. Newman, R.B.; Stevens, R.W.; Gaafar, H.A. Latex agglutination test for the diagnosis of the *Hemophilus influenzae* meningitis. J. Lab. Clin. Med. **1970**, *76*, 107–113.

29. Blondeaux, A.; Caignault, L. La séparation sur bille de taille uniforme. Biofuture **1995**, *147*, 3–8.

30. Rossomando, E.F.; Hadjimichael, J. Method of using magnetic particles for isolating, collecting and assaying diagnostic ligates. US Patent 5158871, University of Connecticut.

31. Lyle, J.A.; Nelson, N.C.; Reynold, M.A.; Waldrop, A.A. Polycationic supports for nucleic acid purification, separation and hybridization. EP 0281390.

32. Chegnon, M.S.; Groman, E.V.; Josephson, L.; Whitehead, R.A. Magnetic particles for use in separations. EP 0125995, Advanced Magnetics Inc.

33. Josephson, L.; Menz, E.; Groman, E. Solvent mediated relaxation assay system. WO 9117428, Advanced Magnetics Inc.

34. Esteve, F.; Amiral, J.; Padula, C.; Solinas, I. Method for assaying an immunological substance using magnetic latex particles and non-magnetic particles. WO 9504279. Societe Diagnostica-Stago.

35. Kondo, A.; Kamura, H.; Higashitani, K. Development and application of thermosensitive magnetic immunomicrospheres for antibody purification. Microbiol. Biotechnol. **1994**, *41* (1), 99–105.

36. Hancok, J.P.; Kemshead, J.T. A rapid and highly selective approach to cell separations using an immunomagnetic colloid. J. Immunol. Meth. **1993**, *164* (1), 51–60.

37. Ugelstad, J.; Berge, A.; Ellingsen, T.; Auno, O.; Kilaas, L.; Nilsen, T.N.; Schmid, R.; Stenstad, P.; Funderud, S.; Kvalheim, G.; Nustad, K.; Lea, T.; Vartdal, F.; Danielsen, H. Monosized magnetic particles and their use in selective cell separation. Macromol. Symp. **1988**, *17* (1), 177–211.

38. Olsvik, O.; Popovic, T.; Skjerve, E.; Cudjoe, K.S.; Hornes, E.; Ugelstad, J.; Uhlen, M. Magnetic separation technique in diagnostic microbiology. Clin. Microbiol. Rev. **1994**, *7* (1), 43–54.

39. Meza, M. In *Scientific and Clinical Applications of Magnetic Carriers*; Häfeli, U., Schütt, W., Teller, J., Zborowski, M., Eds.; Plenum Press: New York, 1997: 303–309.

40. Safarik, I.; Safarikova, M.; Forsythe, S. The application of magnetic separations in applied microbiology. J. Appl. Bacteriol. **1995**, *78* (6), 575–585.

41. Ugelstad, J.; Olsvik, O.; Schmid, R.; Berge, A.; Funderud, S.; Nustad, K. Immunoaffinity separation of cells using monosized magnetic polymer beads. In *Molecular Interactions in Bioseparations*; Springer: US, 1993; Vol. 16, 224–244.

42. Ugelstad, J.; Kilaas, L.; Aune, O. In *Advances in Biomagnetic Separation*; Uhlén, M., Homes, E., Olsvik, O., Eds.; Eaton: Oslo, 1993: 1–19.

43. Mitchell, B.A.; Milbury, J.A.; Brookins, A.M.; Jackson, B.J. Use of immunomagnetic capture on beads to recover *Listeria* from environmental samples. J. Food Prot. **1994**, *57* (8), 743–745.

44. Glaever, I. Diagnostic method and device employing protein-coated magnetic particles. US Patent 4018886. General Electric Company.

45. Mizutani, S.; Asada, M.; Wada, H.; Yamada, A.; Kodama, C. Magnetic separation in molecular studies of human leukemia. In *Advances in Biomagnetic Separation*; Uhlen, M., Homes, E., Olsvik, O., Eds.; Eaton: Oslo, 1994: 127–133.

46. McCormick, M.; Hammer, B. Straight A's mRNA isolation system, rapid, high quality poly(A) + RNA from diverse sources, innovations. 1994, No. 2, Novagen Inc.: Madison, WI.

47. Smith, C.; Ekenberg, S.; McCormick, M. The polyATract magnetic mRNA isolation system: Optimization and performance. Promega Notes 1990, No. 25, Promega: Madison, WI.

48. Kenten, J.H.; Gudibande, S.; Link, J.; Willey, J.; Curfman, B.; Major, E.; Massey, R.J. Improved electrochemiluminescent label for DNA probe assays: Rapid quantitative assays for HIV-1 polymerase chain reaction products. Clin. Chem. **1992**, *38* (6), 873–879.

49. Kondo, A.; Fukuda, H. Preparation of thermo-sensitive magnetic hydrogel microspheres and application to enzyme immobilization. J. Ferment. Bioeng. **1997**, *84* (4), 337–341.

50. Sauzedde, F.; Elaissari, A.; Pichot, C. Thermosensitive magnetic particles as solid phase support in an immunoassay. Macromol. Symp. **2000**, *151* (1), 617–624.

51. Kashlara, A.; Otsuka, C.; Ishikawa, K. Process for preparing particles having monodisperse particle size. EP 0275899.

52. Matte, C.; Muller, A. Procédé et Dispositif Magnétique d'Analyse Immunologique sur Phase Solide. EP 0528708. Pasteur Sanofi Diagnosics.

53. Matte, C.; Muller, A. Procédé et Dispositif Magnétique d'Analyse Immunologique sur Phase Solide. FR 9109242. Diagnostics Pasteurs.

54. Lauterbur, P.C. Magnetic Gels Which Change Volume in Response to Voltage Changes for MRI. US Patent 5532006. The Board of Trustees of the University of Illinois.

55. Ibrahim, A.; Couvreur, P.; Roland, M.; Speiser, P. New magnetic drug carrier. J. Pharma. Phamacol. **1982**, *35* (1), 59–61.

56. Ji, Z.; Pinon, D.I.; Miller, L. Development of magnetic beads for rapid and efficient metal-chelate affinity purifications. J. Anal. Biochem. **1996**, *240* (2), 197–201.

57. Morimoto, Y.; Okurama, M.; Sugibayashi, K.; Kato, Y. Biomedical applications of magnetic fluids. 2. Preparation and magnetic guidance of magnetic albumin microsphere for site specific drug delivery *in vivo*. J. Pharm. Dynam. **1981**, *4* (8), 624–631.

58. Hoffman, A.S. Intelligent polymers in medicine and biotechnology. Macromol. Symp. **1995**, *98* (1), 645–664.

59. Rembaum, A.; Yen, S.P.S.; Molday, R.S. Synthesis and reactions of hydrophilic functional microspheres for immunological studies. J. Macromol. Sci. Phys. **1979**, *A13* (5), 603–632.

60. Treleaven, J.G.; Gibson, F.M.; Ugelstad, J.; Philip, T.; Gibson, F.M.; Rembaum, A.; Caine, G.D.; Kemshead, J.T. Removal of neuroblastoma cells from bonemarrow with monoclonal antibodies conjugated to magnetic microspheres. Lancet **1984**, *323* (8368), 70–73.

61. Lea, T.; Vartdal, F.; Nustad, K.; Funderud, S.; Berge, A.; Ellingsen, T.; Schmid, R.; Stenstad, P.; Ugelstad, J. Monosized, magnetic polymer particles: their use in separation of cells and subcellular components and the study of lymphocyte function *in vitro*. J. Mol. Recog. **1988**, *1* (1), 9–31.

62. Haukanes, B.-I.; Kvam, C. Application of magnetic beads in bioassays. Biotechnology **1993**, *11* (1), 60–63.

63. Arshady, R. Microspheres, Microcapsules and Liposomes. In *Radiolabeled and Magnetic Particles in Medicine & Biology*; Citus Books: London, 2001; Vol. 3.

Ligament Replacement Polymers: Biocompatability, Technology, and Design

Martin Dauner
Heinrich Planck
Denkendorf Forschungsbereich Blomedizintechnik, Institute of Textile and Process Engineering (ITV) Denkendorf

Abstract

All ligament prostheses are produced from fibers that combine the advantage of low diameter, meaning low resulting bending stresses, with the high strength of fibrous materials and uncontested flexibility. The prostheses are designed according to their nature.

One of the first commercial ligament prostheses to undergo the approval tests of the American Food and Drug Administration (FDA) in the late 1970s was the POLYFLEX prosthesis by Richards Manufacturers [U.S. Patent 503990-12331]. It consisted of a 4.76 mm rod in the intra-articular part and 6.35 mm in the osseous bore channels and was made from UHMWPE.

Any quasi-isometric implanted prosthesis is alternatingly bent during knee flexion mainly at the femur at about 60°. Bending results in simultaneous tension/compression stresses, which are proportional to the diameter. As is well known, high alternating stresses lead to fatigue breakage, and respective cases were reported for the POLYFLEX implant, which was finally retracted from the market.[1]

Consequently, today all ligament prostheses are produced from fibers, that combine the advantage of low diameter, meaning low resulting bending stresses, with the high strength of fibrous materials and uncontested flexibility.

With that view, the prostheses are designed according to their nature. Based on the observations of Prockop, Kastelic, and others, who have shown that down to ranges of 10–100 nm the microstructure of the ligaments is formed by fibrils and finally by microfibrils of peptide helices, one can compare that structure with twisted and bundled staple fibers of polymeric origin.[2–4]

CLASSIFICATION

Ligament prostheses can be characterized by their application as a replacement or as a supporting structure, by their proposed way of implantation, and by their intended time of function.

While in the 1970s and early 1980s many prostheses were implanted for total replacement of a destructed ligament, the poor overall performance of most prostheses led in the late 1980s to the use of autologous materials (mostly part of the patellar tendon), which should be supported by an augmentation device temporarily or long-term. Augmentation devices are in use also for primary stabilization of a ligament suture. In the following discussion the ligament prostheses will be described by "ligament replacement," and the supporting structures are named "augmentation devices."

Depending on the application, and also on the cross-section of the implant and the philosophy of the inventors, the devices will be implanted "quasi-isometrical" or "over-the-top." "Quasi-isometrical" means implantation through a tibial and a femoral bore channel that is drilled at the respective insertion sites of the damaged ligament. The "over-the-top" technique, applicable in that way for the anterior cruciate ligament only, also uses a tibial bore channel, while proximally the implant is sited dorsally over the lateral femur condyle.

While a total replacement usually needs a non-degradable, long-term biostable biomaterial, for augmentation a degradable and preferable resorbable material can be used. Some authors consider the self-restoring potency of the organism and propose a long-term, degradable material for "guided tissue regeneration."[5,6]

All these possible classifications cannot be applied strictly, because some implants are used for replacement as well as for augmentation, some are implanted quasi-isometrically and over-the-top, and the biostability of some materials is in question.

For completion, the ligament prostheses of biological origin shall be mentioned mostly as being produced of glutaraldehyde preserved collagen. They don't play an important role today, and therefore a discussion of these implants will not be included in this entry.

In Vitro–Medical

Concise Encyclopedia of Biomedical Polymers and Polymeric Biomaterials DOI: 10.1081/E-EBPPC-120051904

REQUIREMENTS ON PROSTHESES AND AUGMENTATION DEVICES

Different requirements must be considered for ligament replacements and for augmentation devices with respect to their application, for long-term prostheses, and for degradable augmentations.

Biocompatibility

The general requirements for the biocompatibility of any implanted material must be extended for ligaments to their wear particles, which often are produced at the intra-articular entrance of the bore channels. It is reported for Dacron prostheses that wear particles were found to evoke synovitis, while the fiber material itself shows a sufficient biocompatibility.[7–9] Gibbons relates inflammatory responses to relative motion between synthetic fibers and body tissue rather than to the material itself.[10]

Biocompatibility, mild early inflammatory response and no signs of cancerogenicity, and mutagenicity are essentially not only attributed to the implant material and its wear products, but also to the possible degradation products. For totally resorbable materials the observation period can be terminated by the time of total elimination of the material and its degradation products from the body. For other materials the chemical long-term stability must be secured. If chemical (inclusive of enzymatic) degradation occurs, the cancerogenicity and the mutagenicity of the degradation products must be excluded. But their chemical structure is difficult or even impossible to distinguish. Then, relevant concentrations of the degradation products, which may be accumulated in the host body over a couple of years, cannot be generated effectively for *in vitro* cell culture tests. If by theoretical considerations the possible production of cancerogenic or mutagenic substances cannot be excluded for sure, the material has to be considered as potentially not biocompatible. The aggressive and oxidative potency of the intra-articular synovial fluid must be regarded for evaluation of the biostability; subcutaneous implantation tests don't give conclusive evidence.

Fiber materials are generally finished by a preparation to enhance the manufacturing conditions, to reduce electrostatic effects, and to adhere the single fibers together. The biocompatibility of the respective preparation has to be established as well, or the preparation has to be extracted below the measurable quantity. The chemical formulation of the preparation usually is the secret of the fiber manufacturer and will not be revealed. The manufacturer can propose an extraction method for preparation. All relevant mechanical and biological testing must be performed after the washing procedure. For production for clinical use it must be made sure that the fiber manufacturer doesn't change the composition of the finish, at least not without information to the implant producer.

Mechanical Properties

It follows that a high wear resistance of the implant material is advantageous. With the use of high-strength materials, the amount of implanted allogenic material and consequently the size of the prosthesis can be kept small. Unfortunately high modulus materials mostly have a low transversal strength and, therefore, the theoretical longitudinal strength can be used for the prosthesis design in parts only. High creep resistance and low visco-elastic effects are provisions for replacements. For augmentation devices a well determined creep can be part of the functional design to share the load increasingly with the transplant.

To give recommendations for the prosthesis design is rather difficult. It depends mainly on the application and the implantation method. The following two principles may be considered: the implant must stabilize the respective joint enabling the physiological range of motion over the intended time of use, and in any potential case of dysfunction it should not harm the body tissue.

Following these principles and looking at the performance of the natural ligaments in more concrete terms, the mechanical requirements of ligament prostheses are:

- Strength of ligament replacements about 1500 N;
- Initial strength of augmentation devices of 500–1000 N;
- Strength of resorbable augmentations beyond 500 N over the first 2 months;
- Progressive stress–strain curve;[11]
- Elongation according to the implantation and the implant size (see below);
- Stress relaxation asymptotically to 20–25% over 3 hr according to natural ligaments;[12] and
- Augmentation devices can relax to a high degree, but very slowly (in the range of months) in order to share the load increasingly with the restored or replaced ligament.

For prostheses often an "isoelastic behavior" is claimed. This behavior is coupled strongly to the "isometric" implantation site, to its position over the implantation period, and to the integrity of the prosthesis.[4] The structure of the natural ligaments, in particular that of the anterior cruciate ligament, is so complicated that it cannot be reproduced effectively by a synthetic replacement. And the required absolute perfect insertion of such an implant is not realizable. Still, the target is to implant the ligament prostheses in the isometric position, where it will not change its length during unloaded knee flexion, and this can not be fulfilled.

The elongation of the implant must be adjusted to the intended implantation site, based on some experimental data. The osseous integration of the prosthesis is to be considered, depending on the material used and on the design. During the healing period the intra-osseous segments will contribute to the elongation. A complete osseous integration of the prosthesis will restrict the functional length to the intra-articular part, and thereby restrict the effective elongation. Increasing limitation of the knee flexion, damaging of the surrounding bone and cartilage, or even rupture of the prosthesis may result if the prosthesis elongation is inappropriate. Reports that the prosthesis elongation increases owing to ingrowth of collagen or connective tissue refer to specific conditions.[13] They address the possibility that during the increasing osseous integration, stretched intra-osseous parts of the ligament can not relax in the narrowed bore channels because of friction. The elongation of the slackened intra-articular segment is increased due to loss of preload. In that relaxed segment, tissue may possibly ingrow, reducing again the elongation of this part by stretching it to "zero load." Then the ingrown tissue may act as an elastic element, which has a considerable advantage. Yet the total elongation will nevertheless be decreased. The authors don't recommend trusting this appreciable but accidental effect, which depends on many factors such as implant material and design, position of the prosthesis, post-operation procedure, and healing capacity of the host tissue.

One main factor to consider is the preload, which should be low (<20N), to allow the friction-related relaxation effect, which is necessary for an open structure in which tissue can ingrow.

For the prostheses, which should be implanted over-the-top, one has to consider the longer intra-articular segment of the implant, but also the higher elongation during knee flexion. Experiments with a joint model seem expedient.

Post-operative changes of the prosthesis and its position must be considered. The effect of ingrowth of tissue is described above. The properties of the material can change by the overtake of water and of course by any hydrolytical, encymatical, or oxidative degradation. The body will react to overloading, mainly at the entrance to the femur bore channel. There may be loss of bone, affecting the anterior dislocation of the insertion site. More complications result from the growth of osteophytes, which will lead to abrasion and finally rupture of the implant.

Here the second principle for prosthesis design has to be remembered: the implant shall not damage the host tissue, not even in the case of dysfunction. One way to fulfill this is a weak implant, which ruptures at loads below the strength of the surrounding tissue. But experience shows that ruptured prostheses mostly evoke synovitis or even more adverse body reactions.[7–9]

The second way is the use of a prosthesis material that relaxes at too-high stresses. Unfortunately, for most polymeric fiber materials the relaxation coefficient is independent from the applied load in a wide range. That means that the prosthesis will relax at a well-determined percentage from every applied load value, which is specific for the material used. Here the incorporation of collagen or connective tissue into the prosthesis structure can act perfectly as a security element that is pressed out of the structure at too-high loads in case of a traumatic accident.

A third way to avoid damage of the host tissue is the fixation of the implant with strength below the strength of the implant and the damage margin of the host tissue. The loosened implant may be refixed operatively or by tissue integration, which can be an elastic component. This way a limited function of the ligament prosthesis may be maintained. The Marshall and MacIntosh technique addresses this for augmentation devices by suturing the one end to soft (tendon) tissue.[14]

In the case of dysfunction of the ligament prosthesis, rupture, or any adverse body reaction, the removal of the implant is required. An otherwise appreciated osseous integration complicates the pull-out of the ligament and makes the implantation of a new prosthesis difficult.

Design Aspects

From a commercial view the number of sizes and shapes for different application sites of the ligament prostheses should be kept low. The most frequent application of a replacement as well as of an augmentation device is the restorage of the anterior cruciate ligament. But often a combined ACL/MCL-rupture or other combined rupture requires versatility of the implant.

The use of a double-strand ligament for the knee joint respects the complicated morphology of the knee ligaments. The same prosthesis may be used for the shoulder as a single-strand implant. However, the required percentual elongation of the anterior cruciate ligament is at the technological limit, for the lateral ligament the great implant length requires a lower percentual elongation.

With a universally applicable ligament the different sizes can more easily be handled, yet a more sophisticated design regarding fixation and mechanical behavior cannot be considered. A small cross-section of the prosthesis generally facilitates the implantation, enables an isometric position, and enlarges versatility.

TESTING METHODS

Up to now, a standard for testing of ligament prostheses has not been determined. Actually the working group of the CEN TC 285 is involved in the set-up of standard test methods. Degradation tests for Poly-L-Lactide have been discovered in the ISO TC 150 SC 1, which may be used for other resorbable materials as well, changing the procedure where necessary.

For the U.S. FDA, the Division of Surgical and Rehabilitation Devices has given guidance for testing and premarket approval of ligament devices.[15] The document explicitly has no regulatory meaning and should be read as a "scientific positioning paper."

In Vitro-Medical

In 1992, at a symposium in Reisensburg, Germany, and at a subsequent meeting at Ulm, Germany, scientists involved in ligament development and testing set up a proposal for biomechanical testing of ligament prostheses.[16] Both documents shall be reported in brief.

Guidance Document for Testing of Knee Ligament Devices

Nonclinical testing of all implant materials shall be in accordance with the GLP, 21 CFR Part 58. Physical, chemical, biological, and toxicity tests of the implant material are given by the U.S. Pharmacopeia.

All mechanical tests should be performed on treated and sterilized products to include any possible effects in the testing. The prostheses should be tested generally in saline at 37°C if an effect of the elevated temperature and the saline cannot be excluded or calculated by former experience. A soaking of the devices in 37°C saline for 1 month is proposed.

Tensile Testing: at a minimum of three different strain rates:
- Gage length and grips simulating *in vivo* conditions
- 6 devices at each strain rate;
- Data: strain rate (% sec^{-1}), load/stress, and elongation/strain at yield and ultimate load or failure; stiffness/modulus.

Fatigue Testing: load-cycle; elongation-cycle:
- High load-cycles producing failure in less than 10^6 cycles
- Low loads, typical of normal activities;
- Medium-load cycles; maximum 10^7 cycles; then tension test to failure
- Six devices at each load level;
- Data: cycle rate; peak load; number of cycles; total elongation; failure load and elongation; stiffness.

Bending Fatigue Testing: three angles with load of normal activity, or typical bending angle and three-load level:
- Bending angle should be in excess of 90°;
- Bending to failure, but maximum 10^7 cycles; then tension test to failure
- Three devices at each load level or bending angle; total of nine devices;
- Data: cycle rate; peak axial load; bending angle; number of cycles; total elongation; failure load and elongation; stiffness.

Fixation Strength: animal studies are suggested:
- Model implantation on cadaver knees evaluating the pull-out strength of the entire device from its in situ site; the fixation strength of the device from its fixator; and the fixator itself from its bone or soft tissue attachment site
- Data: ultimate pull-out strength; mode of failure.

Abrasion Testing: cadaver bone or "different grades of abrasive materials;" combination with bending.

Bioassay

Long-term animal testing should be performed over at least 1 year; for a new biomaterial it should be performed over at least 2 years. Bioassy on rats is required. Specific high value is set on possible degradation, abrasion, migration of particles to lymph nodes, and examination of synovial fluid. Mechanical tests should include laxity tests, fixation strength and material strength, and stiffness. (A more detailed description of the proposed animal tests and preclinical studies proposed in that guidance exceeds the intention of this entry.)

Comments

Testing conditions

With the use of saline, salt crystallites may act as wear substances. It is advised to investigate the saline stability separately and perform the long-term tests in destilled or deionized water.

Tensile testing

Universal testing machines usually enable a maximum speed of 400 or 500 cm/min or 7–8 mm/sec. That means a strain rate of about 25% sec^{-1} at a gage length of 30 mm (i.e., the typical length of the ACL). For the range from 0.1% sec^{-1} up to 10% sec^{-1} it was found by measurements that commonly used fiber materials don't show a major effect on the strain rate. For essentially higher strain rates, which *in vivo* may occur in accidents, special high-speed or hydraulic machines are needed. At these levels a significant increase of strength and modulus must be expected for synthetic materials as well as for the natural ligaments.[17]

At short-gauge lengths it must be kept in mind that the grips may affect the strain levels. The strong fixation of the testing design cannot be compared with the weak fixation, even with bony ingrowth.

Fatigue testing

No recommendation was given for the cycle rate. Cycling may produce internal heating of the device if it is essentially faster than the physiological one cycle sec^{-1} (1 Hz) and may affect thereby, the lifetime in both directions, depending on the material used.

Bending fatigue test

Fiber materials and devices made of fibers withstand bending without friction usually far beyond the proposed

10^7 cycles. That is diverse from bulk materials. Today all commercial prostheses are made of fibers.

No information was given for the bending radius, which has a great impact on lifetime.

Abrasion test

The abrasion test is the most difficult test to evaluate and compare without standardization. A still imperfect recommendation was set up in Ulm and is described below.

Proposal for the Testing of Ligament Prostheses from the Reisensburg Meeting[16]

Tensile testing

Tying the ligament over a 20 mm cylinder; gauge length 100 mm; cross-head speed 1 mm/sec, i.e., 1% sec^{-1}.

Creep test

Gauge length 100 mm; load 200 N; testing period 48 hr; condition: dry or aqua distillata (Fig. 1).

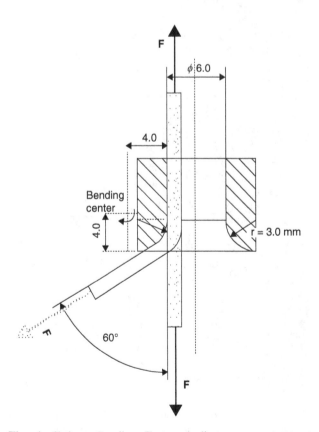

Fig. 1 Fatigue Bending Test according to agreement at Reisensburg.

Fatigue bending test

- Principle: alternating bending with friction around an eccentric radius.
- Load F:40 N and 120 N; cycle rate 2 Hz; bending angle 0° to 60°;
- Condition: aqua distillata, 37°C;
- Wear material: Al_2O_3, diameter 6.0 mm, bending radius 3.0 mm, distance ceramic surface to bending center: 4.0 mm (Fig. 1).

Animal tests should last at least 1 year, at augmentation 2 years, because of the prolonged healing period. A minimum of 7 animals should be used per each test design. The knee stability should be tested at 90° flexion, applying a load of 50 N proximal and distal subsequently, with a speed of 10 mm/min. Rupture test for the anterior cruciate ligament and its replacement is proposed at 90° flexion; the ligament must be in the direction of the load axis. Data reported: maximum load, stiffness, yield load and the respective elongations, and rupture energy.

Comments

Clamps

With special clamps, self-enlarging the holding force in dependence of the applied load, 10% higher tensile strength was measured compared to the proposed cylinder clamps. Yet these clamps are rather complicated.

Gage lengths

The proposed length of 100 mm may not be appropriate for prostheses with change in cross-section in the intra-articular region. The testing speed should be adjusted according to the proportional strain rate of 1% sec^{-1}.

Fatigue bending test

The recommendations do not define the surface roughness of the ceramic. The intention was to use a polished ceramic, which has a highly reproducable surface. In studies it was found that some fiber materials withstand more than 1.5×10^7 cycles without any sign of damage. That is unappropriate for screening tests. Then natural rough ceramics were used, with a roughness of 0.4 μm (arithmetic mean; maximum single value: 3–4 μm). The lifetime of the tested materials was below 1 million cycles. Unfortunately, owing to processing conditions the reproducibility of the natural surface is limited, and the surface profile of each ceramic device must be tested before use. Especially ceramics with high single-surface peaks (>>4 μm) should be excluded from the tests. Yet only a few institutions will be able to measure the surface roughness of these materials.

Manufacturing technologies

The mechanical requirements of the ligament prostheses and the example of the natural ligaments suggest textile processing for the implants. Accordingly, only few devices were produced from bulk material. The above mentioned POLYFLEX-prosthesis [U.S. Patent 503990-12331] is the only one that was considered for clinical application. Experimental designs with a core of PTFE, silicone, and polyurethane, reinforced with fiber materials such as aramide, carbon, and polyester were tested in animal or in preliminary clinical studies.[18–21] Creep of PTFE and abrasion between fibers and the elastomeric materials disqualified the prototypes for further use.

Similar problems must be expected when covering the fibers with elastomeric materials. Wear particles generally provoke synovitis or other adverse tissue reactions. Therefore further discussion will concentrate on textile-processing methods.

Of five principal textile processing methods, the latter three are used for ligament prostheses: nonwovens, weft knits, warp knits, woven fabric, and braids. The interfibrillar strength of a nonwoven is very poor, so that it could be used only as a covering structure. The elastic recovery and the form stability of weft knits are poor.

With woven structures the fibers are regularly entangled. The fiber(s) of a weft knit transfer(s) load from one loop to the next. With the other structures there are considerable numbers of fibers transferring load directly from end to end. The deviation of fibers from the longitudinal axis (parallel to the load direction) implies a structure-specific strain capacity to the implant.

In Fig. 2 the textile structures used for ligament prostheses are shown on the left; on the right their corresponding principle stress–strain curves using two different fiber materials are compared with the curves of the materials itself and a schematic curve of a natural ligament.

Warp Knit (Fig. 2A)

The warp knit is formed by multiple fibers. The fibers are bound together by loops as at the weft knit, yet the main fiber direction is oriented in the longitudinal, or in the load axis. If one fiber breaks the device does not necessarily lose its total strength.

Both of the knitted structures—weft knit and warp knit—have in common that the stretchability in the transversal axis is higher than in the longitudinal axis. Both structures are difficult to process with high modulus fibers because they have very poor transversal strength. Consequently the knitted structures are not used for load-bearing structures, but their high volume offers the application for enhancement of tissue ingrowth.

Woven Fabric (Fig. 2B)

Warp yarns are strictly oriented in the longitudinal axis of the device and are bound by weft yarns in the transversal axis; the cross angle is 90°. The warp yarns are loaded accordingly in their strong longitudinal axis. The elongation of the fabric is determined mainly by the fiber elongation, slightly by the pattern. The transversal elongation is the same when the same fibers, fiber density, and patterns are used for warp and weft.

The low-structure extensibility limits the application to a few prostheses where low elongation is required for application at the shoulder or the pelvis (PDS or Vicryl of Ethicon). The prosthesis acts mainly as a tissue guide (Leeds-Keio), and twisting of the device provides the elongation needed (Trevira-Band of Telos).

Braids (Fig. 2C)

Braids are the best appropriate structures for transfer of high loads providing essential structural elongation.

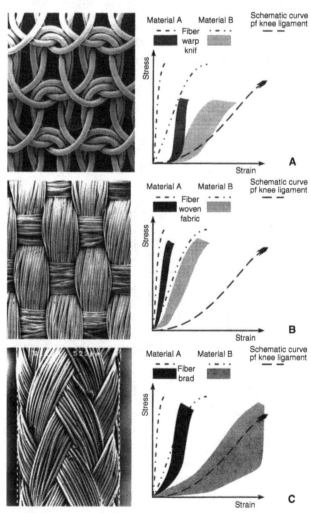

Fig. 2 Textile structures and their principle stress–strain behavior (**A**): warp knit, (**B**): woven fabric, and (**C**): braid.

Two groups of fibers are regularly crossed in a braid angle of 1° to 80° to the longitudinal axis (2 × 1° to 2 × 80° between both groups of fibers; in some literature the angle is defined regarding the transversal axis!). By applying load, the fibers tend to orient in the load axis. The two groups are pressed increasingly against each other. A progressive stress-strain behavior results, more or less comparable to that of the natural ligaments. The transversal pressure of both groups of fibers reduces the overall strength of the device. An increasing braid angle results in increasing extensibility of the braid but reduces the strength.

The braid can be characterized by:

- Tubular braid or flat braid,
- Number of strands (number of carriers),
- Titer of fibers per strand (amount per weight: 1 tex = 1 g/1000 m),
- Pick counts per length (number of fiber crossing, per mm or french, i.e., 27.4 mm),
- Amount of core fibers, (fibers axially oriented inside a tubular braid), and
- Amount of stationary threads (axially oriented fibers bound between the braid fibers).

The authors propose the use of a tubular braid. A middle course should be adopted regarding the titer per carrier and the number of carriers. The pick counts per length results from the other parameters, such as the braid angle, which is adjusted to the required stress–strain curve. The use of core or stationary threads limits the elongation to that of the fiber material. An elastic core in the lumen of a tubular braid enhances the elasticity (reversibility of the extension). Yet friction between the fibers and the core material may lead to abrasion. To support tissue ingrowth a high diameter of single fibers and a high titer per carrier is of advantage.

Other Textile Structures

Plying yarns and orientating them in the loading axis is the way to maintain all the fiber strength in the device. A cover braid can hold together the plied yarns. The elongation of the ply is limited to the fiber's elongation, which is generally low compared with the elongation of the natural ligament. Velour is processed by mixing high and low shrinking yarns in a woven fabric or knitted structure. By thermal treatment of the material above the shrinkage temperature, the high shrinkage yarns contract and the low shrinkage yarns form loops. These loops offer space for tissue ingrowth, used at the STRYKER prosthesis, for example.

More details for design and production of textile structures are reported by the authors elsewhere.[4,12,22–24]

REFERENCES

1. Chen, E.H.; Black, J. Materials design analysis of the prosthetic anterior cruciate ligament. J. Biomed. Mater. Res. **1980**, *14* (5), 567–586.
2. Prockop, D.J.; Guzman, N.A. Collagen diseases and the biosynthesis of collagen. Hosp. Pract. **1977**, *12* (12), 61–68.
3. Kastelic, J.; Galeski, A.; Baer, E. The multicomposite structure of tendon. Connect. Tissue Res. **1978**, *6* (1), 11–23.
4. Dauner, M.; Planck, H.; Syre, I.; Dittel, K.K. In *Medical Textiles for Implantation*; Planck, H., Dauner, M., Renardy, M., Eds.; Springer: Berlin Heidelberg, Germany, 1990; 8, Chapter 2.
5. Alexander, H.; Weiss, A.; Parsons, J.R. Ligament and tendon repair with an absorbable polymer-coated carbon fiber stent. Bull. Hosp. Jt. Dis. Orthop. Inst. **1986**, *46* (2), 155–173.
6. Leandri, J.; Dahhan, P.; Tarragano, O.; Cerol, M.; Rey, P. Geiger, D. In *Implant Materials in Biofunction*; de Putter, C., de Lange, G.L., deGroot, K., Lee, A.J.C., Eds.; Elsevier: Amsterdam, the Netherlands, 1988; p 113.
7. Claes, L.; Dürseleu, L.; Kiefer, H.; Mohr, W. The combined anterior cruciate and medial collateral ligament replacement by various materials: a comparative animal study. J. Biomed. Mater. Res. **1987**, *21* (A3 Suppl.), 319–343.
8. Harth, A.; Meire, D.; Nuyts, R.; Verdonk, R.; Claessens, H. European society for artificial organs. In *XVIth ESAO Congress*; Brussels, Belgium, Sept 13–15, 1989.
9. Barrett, G.R.; Lawrence, L.L.; Shelton, W.R.; Manning, J.O.; Phelps, R. The Dacron ligament prosthesis in anterior cruciate ligament reconstruction. A four-year review. Am. J. Sports Med. **1993**, *21* (3), 367–373.
10. Gibbons, D.F.; Mendenhall, H.V.; Van Kampen, C.L.; Lambrecht, E.G. *Fourth World Biomaterials Congress*; European Society for Biomaterials: Berlin, 1982.
11. Butler, D.L.; Grood, E.S.; Noyes, F.R. Biomechanics of ligaments and tendons. Exerc. Sports Sci. Rev. **1978**, *6* (1), 125–182.
12. Dauner, M.; Planck, H.; Brüning, H.J. In *Die wissenschaftlichen Grundlagen des Bandersatzes*; Claes, L. Ed.; Springer: Berlin Heidelberg, Germany, 1994; p. 25 ff.
13. Claes, L.; Neugebauer, R. *In vivo* and *in vitro* investigation of the long-term behavior and fatigue strength of carbon fiber ligament replacement. Clin. Orthop. Relat. Res. **1985**, *196* (6), 99–111.
14. Marshall, J.L.; Warren, R.F.; Wickiewicz, T.L.; Reider, B. Clin. Orthop. **1979**, *62*, 37.
15. *Guidance Document for the Preparation of Investigational Device Exemptions and Premarket Approval Applications for Intra-Articular Prosthetic Knee Ligament Devices*. U.S. Food and Drug Administration, Rockville, MD, Sept 1, 1987.
16. Claes, L., Ed. *Die wissenschaftlichen Grundlagen des Bandersatzes*; Springer: Berlin Heidelberg, Germany, 1994; p. 205 ff.
17. Noyes, F.R.; Grood, E.S. The strength of the anterior cruciate ligament in humans and Rhesus monkeys. J. Bone Joint Surg. Am. **1976**, *58A* (8), 1074–1082.

In Vitro–Medical

18. James, S.L.; Woods, G.W.; Homsy, C.A.; Prewitt, J.M.; Slocum, D.B. Cruciate ligament stents in reconstruction of the unstable knee. A preliminary report. Clin. Orthop. Relat. Res. **1979**, (143), 90–96.

19. Woods, G.W.; Homsy, C.A.; Prewitt, J.M.; Tullos, H.S. Proplast leader for use in cruciate ligament reconstruction. Am. J. Sports Med **1979**, *7* (6), 314–320.

20. Trembley, G.R.; Laurin, C.A.; Drovin, G. The challenge of prosthetic cruciate ligament replacement. Clin. Orthop. Relat. Res. **1980**, (147), 88–92.

21. Peterson, C.J.; Donachy, J.H.; Kalenak, A. A segmented polyurethane composite prosthetic anterior cruciate ligament *in vivo* study. J. Biomed. Mater. Res. **1985**, *19* (5), 589–594.

22. Planck, H. In *Medical Textiles for Implantation*; Planck, H., Dauner, M., Renardy, M., Eds.; Springer: Berlin Heidelberg, Germany, 1990; 1, Chapter 1.

23. Planck, H. In *Kunststoffe und Elastomere in der Medizin*; Planck, H., Ed.; Kohlhammer: Stuttgart, 1993; 1, Chapter 6.

24. Dauner, M.; Planck, H. In *Kunststoffe und Elastomere in der Medizin*; Planck, H., Ed.; Kohlhammer: Stuttgart, 1993; 1, Chapter 6.3.

In Vitro–Medical

Lipoplexes and Polyplexes: Gene Therapy

Diana Rafael
Research Institute for Medicine and Pharmaceutical Sciences (iMed.UL), School of Pharmacy, University of Lisbon, Lisbon, Portugal

Fernanda Andrade
Laboratory of Pharmaceutical Technology (LTF/CICF), Faculty of Pharmacy, University of Porto, Porto, Portugal

Alexandra Arranja
Sofia Luís
Mafalda Videira
Research Institute for Medicine and Pharmaceutical Sciences (iMed.UL), School of Pharmacy, University of Lisbon, Lisbon, Portugal

Abstract

Gene therapy has gained attention in the last few years because of its efficacy in the treatment of numerous diseases, particularly those caused by oncological and genetic disturbances. The success of gene therapy is completely dependent on the development of nontoxic and highly efficient gene delivery systems. Thus, in order to protect oligonucleotides from damage, facilitate their entry into the target cells, and promote an efficient endosomal release, new lipid- and polymer-based delivery systems have emerged in the past few years. The most promising strategies involving the complexation of nucleic acids with lipids (lipoplexes) and different kinds of polymers (polyplexes) are reviewed in this work. The advantages and disadvantages of each system and the possibility of using polymers and lipids in a same formulation are also explored. Finally, the future perspectives for gene therapy associated with nanotechnology, an approach that seems to be quite promissory, bringing significant new insights into the treatment of numerous diseases are discussed.

INTRODUCTION

Gene therapy is a fascinating and innovative strategy that has caught the attention of several researchers in the last decades as an opportunity to treat both acquired and inherited diseases, since a four-year-old girl with severe combined immunodeficiency caused by adenosine deaminase deficiency was cured in 1990. Delivery and subsequent expression of an exogenous DNA encoding for a missing or defective gene or the silence of a particular gene of interest, such as an oncogene, with the RNA interference technology are different approaches that bring new alternatives and effective solutions for the treatment of several diseases, especially for the dreadful disease cancer. The use of gene therapy in cancer treatment seems to be the best alternative to overcome the well-known adverse effects and drug-resistance problems associated with the conventional pharmacological therapies. However, the difficulty in the delivery of oligonucleotides into the cells has precluded the clinical utilization of gene-based therapies. Therefore, innumerous strategies to deliver the genetic material efficiently have emerged from different research groups all over the world.

In this entry, the most promising strategies regarding the use of different lipids and polymers in the development of efficient gene delivery systems are reviewed. The main differences between them and the advantages and disadvantages of each system are also evaluated. Finally, a brief overview of the innovators for promising that are taking benefit of the advantages of both systems, to create a more complete and effective gene vector is presented.

CHOOSING A VECTOR FOR GENE THERAPY

The vectors for gene delivery can be divided in two main groups, the viral and the nonviral vectors. Despite being very efficient gene transfer vehicles and also the most commonly used in clinical trials nowadays (Fig. 1), the well-established drawbacks related to viral vectors, such as 1) their immunogenicity; 2) mutagenesis; 3) carcinogenesis; 4) limited cargo loading; and 5) time-consuming and high-cost procedures, boosted the development of safer vehicles and explain the enormous outbreak of the nonviral vectors in the last few years.[1–5] Until now, only Glybera® (alipogene tiparvovec using an adeno-associated virus as vector) was granted with marketing authorization by European Commission in late 2012 to treat lipoprotein lipase deficiency.

Among the nonviral vectors, the lipoplexes – complexes formed by nucleic acids and lipids – and the polyplexes – complexes formed by nucleic acids and polymers – (Fig. 2)

Concise Encyclopedia of Biomedical Polymers and Polymeric Biomaterials DOI: 10.1081/E-EBPPC-120050058

765

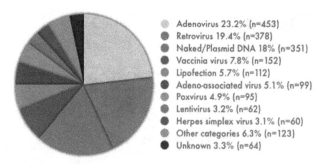

Fig. 1 Type of vectors used in gene-based systems in clinical trials.
Source: From The Journal of Gene Medicine, www.wiley.co.uk/genmed/clinical 2012 with permission from John Wiley and Sons Lda.

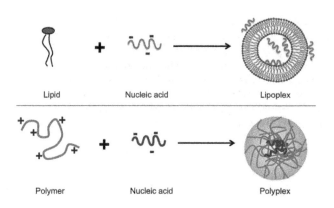

Fig. 2 Schematic representation of lipoplexes and polyplexes.

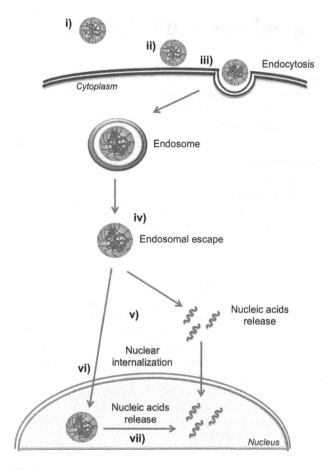

Fig. 3 *In vitro* barriers for nonviral vectors-based gene delivery. (The different barriers here identified as (i) to (vii) are explained and identified with the same nomenclature throughout the Introduction section.)

have attracted the attention of many researchers, especially due to their better safety profile compared with viral vectors. Anionic, neutral, and cationic lipids, as well as neutral, cationic, or amphiphilic polymers have been used to produce gene delivery systems. However, due to an easiest interaction with the negatively charged oligonucleotides and cellular membranes, greater attention is given to the cationic-charged vectors. The use of cationic polymers and lipids in the delivery of genetic material was proposed for the first time in 1987 by Wu and Wu[6] and Felgner et al.,[7] respectively.

Nevertheless, several limitations are also related to the nonviral vectors, namely, their low specificity, cellular toxicity, low biodegradability, rapid blood-circulation clearance, and the limited transfection efficiencies. Moreover, all the steps involved in the gene delivery mediated by these vectors constitute potential barriers to the success of the strategy, namely, i) the nucleic acids complexation, which is dependent on the charge of the lipid or polymer; ii) the binding of lipoplexes and polyplexes to the cellular membrane that, except in the presence of specific targeting ligands, is usually nonspecific; iii) the particles internalization, that usually occurs through one of the five major endocytic pathways

(phagocytosis, macropinocytosis, clathrin-mediated and caveolae-mediated or independent endocytosis) and iv) the endosomal escape, one of the most critical phenomenon since is dependent on the occurrence of the proton sponge effect in the case of polyplexes and the endosomal membrane disruption for the lipoplexes. For the case of DNA there are two addtional barriers, v) the cytoplasmic transportation; vi) the entry to the nucleus; and vii) the decomplexation of DNA–vector within the nucleus (Fig. 3). To overcome the drawbacks usually associated with the use of nonviral vectors for gene delivery, a wide range of different formulations have emerged in the past years aiming for perfect delivery system development.[2,8]

PROMISING LIPOPLEXES DEVELOPED FOR GENE DELIVERY: A STATE OF THE ART

Lipoplexes are usually produced through a simple interaction between nucleic acids and different lipids. Three types of lipids are usually selected to complex the genetic material: the anionic [e.g., cholesteryl hemisuccinate (CHEMS)], neutral [e.g., 1,2-dioleoyl-sn-glycero-3-phosphoethanol-amine (DOPE) and cholesterol (Chol)] and cationic [e.g., *N*-[1-(2,3-dioleyloxy) propyl]-*N*,*N*,*N*-trimethylammonium chloride (DOTMA) or lipofectin, *N*-[1-(2,3-dioleoyloxy)]-*N*,*N*,*N*-trimethylammonium propane methylsulfate (DOTAP), (+)-*N*,*N*-dimethyl-*N*-[2-(sperminecarboxamido)ethyl]-2,3-bis(dioleyloxy)-1-propaniminium pentahydrochloride (DOSPA) or lipofectamine, 1,3-di-oleoyloxy-2-(6-carboxy-spermyl)-propylamid (DOSPER) and oligofectamine] lipids. The final charge of the lipoplex, probably one of the most important factors affecting lipoplexes' stability and transfection efficiency, depends on the ratio between the differently charged lipid vector and the nucleic acid molecules. Besides its charge and composition, the size and morphology of the lipoplexes also tremendously affect their transfection activity. Cationic lipoplexes, for example, have a vast heterogeneity in terms of shape and size and different morphologic forms (such as cubic, beads on a string, spaghettis, meat balls, multilamellar, map-pin, sliding columnar, and inverted hexagonal phase structures) have been reported.[9] Regarding lipoplexes' size, there is still a large controversy about which size range promotes better transfection activities. Some authors, such as Esposito et al.[10] and Massoti et al.[11] advocate that bigger lipoplexes with around 200–400 nm or even 1–2 μm are more efficient, arguing that larger particles have a faster sedimentation process, present a higher contact with cells and form larger intracellular vesicles, which are more easily disrupted. Thus the endosomal escape and the release of genetic material into the cytoplasm are improved.[10,11] However, other authors report that higher transfection efficiencies are obtained using particles with a smaller size of around 100 nm or less.[12,13] Although the optimal size for particles *in vitro* is still unclear, for the *in vivo* applications it is of general consensus that small-diameter nanoparticles are more effective.[9]

Lipoplexes enter the cells through two main mechanisms, in the majority of the cases by endocytosis and in a low percentage of the cases by direct fusion with the cellular membrane. Lipoplexes uptake pathways depend on the type of lipid used and the type of genetic material that is delivered. As example, Rejman, Conese, and Hoekstra[14] demonstrated that DOTAP/DNA lipoplexes uptake occurs through clathrin-mediated endocytosis, but accordingly to Wong, Scales, and Reilly,[15] the uptake of plasmid DNA associated with another cationic lipid, the 1,2-dimyristy-loxypropyl-3-dimethyl- hydroxy ethyl ammonium bromide (DMRIE-C), occurs via the caveolin pathway. For the synthetic siRNAs, another mechanism was reported. Lu, Langer, and Chen[16] showed that besides approximately 95% of siRNA lipoplexes entering the cells by endocytosis, the functional siRNA delivery occurs via minor pathways that involve simply the fusion between siRNA lipoplexes and the plasma membrane. These results are based on the fact that when clathrin-, caveolin- or macropinocytosis-mediated uptake are inhibited, the knockdown of the protein

of interest continues to happen but, on the contrary, when the Chol from the plasma membrane is depleted there is a significant reduction in the knockdown of the protein.[16]

Cationic Lipoplexes

Cationic lipids represent one of the most widely used gene delivery vectors and in the last few years, a number of different lipids have been designed and have become commercially available. Typically, cationic liposomal systems have in their composition a neutral lipid, such as DOPE, Chol, or 1,2-dioleoyl-sn-glycero-3-phosphocho-line (DOPC) that are also called helper lipids, because they play an important role in improving lipoplexes' physiologic stability, genetic material protection from degradation, and transfection efficiency.[2,17,18]

Cationic lipids are amphiphilic molecules formed by a positively charged polar headgroup attached via a linker to a hydrophobic domain, whose hydrocarbon chain lengths vary commonly between C8:0 and C18:1. The characteristics of the headgroup are responsible for the transfection efficiency of the lipoplexes; therefore several synthetic cationic lipids with variations in the headgroup composition have been produced. Headgroups consisting of amine groups with different substitution degrees, guanidinium or pyridinium groups, amino acids, and peptides have been reported.[2–4]

Despite their unquestionable advantages in terms of transfection efficiency, the toxicity of cationic lipids is a well-known problem that needs to be addressed. Therefore, novel cationic lipids-based formulations with minimal toxicity but still with adequate transfection efficiency are continuously being designed and produced. These new cationic lipids are obtained through structural modifications performed in the headgroup, in the linker, and/or in the hydrophobic tail domains that are known to regulate transfection efficiency and cytotoxicity.[19]

The production of such cationic liposomes can be carried out using different techniques on the basis of the type of the entrapped genetic material—it can be used by the hydration of a lipid film, the dehydration–rehydration, the ethanolic injection, the reverse-phase evaporation, or the detergent dialysis technique.[20]

With the final objective to reach an ideal gene delivery system, numerous cationic lipids-based formulations that have been investigated in the last years present promising *in vitro* and *in vivo* results and some of them are already in phase I clinical trials (Table 1).

Anionic and Neutral Lipids

The use of anionic or neutral lipids as gene delivery systems is another strategy that some authors explore with the objective to override the cationic lipids-associated toxicity. However, due to their negative or neutral charge and the consequent impossibility to establish electrostatic interactions

Table 1 Examples of lipoplexes-based system used for *in vitro* and *in vivo* applications as well as in clinical trials

Lipid vector	Genetic material	Utility	State	References
N(4)-linoleoyl-N(9)-oleoyl-1,12-diamino-4,9-diazadodecane/Chol	Enhanced green fluorescent protein (EGFP) siRNA	EGFP delivery and silencing	*In vitro*	[21]
Ethylenediamine/DOPE	pDNA Luciferase (Luc)	Transfection reagent	*In vitro*	[22]
Gd.DOTA.DSA:N1-cholesteryloxycarbonyl-3,7-diazanonane-1,9-diamine:DOPC:DSPE-PEG2000:DOPE-Rhodamine	Surviving siRNA	Tumor reduction	*In vivo*	[12]
Diethanolamine chloride/1-palmitoyl-2-oleoyl-sn-glycero-3-phosphocholine/CHOL/DOPE/ mPEG 2000-DSPE	Argonaute2 siRNA	Antiangiogenesis effect and tumor growth suppression	*In vivo*	[23]
Hexadecyltrimethylammonium bromide	p-ialB and p-omp25 DNA vaccines	Protection against *Brucella melitensis*	*In vivo*	[24]
DOTMA	CpG DNA	Prevent pulmonary metastasis and peritoneal dissemination through NK cell activation	*In vivo*	[25]
PEGylated lipoplex-entrapped alginate scaffolds	Lamin A/C siRNA	Sustained delivery of siRNA to vaginal epithelium	*In vivo*	[26]
AtuFECT	Protein kinase N3 siRNA	Advanced solid tumor	Phase I	[27]
DMRIE/DOPE	pGT-1 gene	Cystic fibrosis	Phase I	[28]
DOTAP/Chol	Fus 1 gene	Non–small-cell lung cancer	Phase I	[29]

with negatively charged molecules, the delivery systems based on anionic or neutral lipids always require a third moiety to promote the interaction with the nucleic acids.[30–32] Therefore, divalent or monovalent cations, such as Ca^{2+}, Na^+, Mg^{2+}, Mn^{2+}, or Ba^{2+} need to be incorporated in the formulation in order to promote the complexation between the nucleic acids and the anionic lipids.[31]

Kapoor and Burgess[32] designed an siRNA delivery system using physiologically occurring anionic lipids [1,2-dioleoyl-sn-glycero-3-[phospho-rac-(1-glycerol)] (DOPG)/DOPE] and calcium ion bridges to promote the oligonucleotide complexation. After formulation optimization, the obtained lipoplexes present high stability in the presence of serum components, efficient cellular uptake and endosomal release, and a significant silencing capacity comparable with the control cationic complexes (Lipofectamine) with a lower toxicity.[32]

Additionally, Mignet et al.[30] designed an innovative system throughout the combination of a cationic lipoplex and pegylated anionic liposomes. The ratio between anionic and cationic lipids was optimized in order to form an anionic pH-sensitive pegylated lipoplex sensitive to pH changes between 5.5 and 6.5. Interestingly, the obtained anionic particles charge was able to reverse to a cationic charge at an adequate pH value. This was observed during endosomal acidification, which leads to the DNA release and the increase of a given gene expression. This result was comparable with the control cationic formulation and more importantly, the ability to

deliver DNA to the tumor was improved relatively to the CHEMS formulations.[30]

Lipoplexes' Surface Decoration

Chemical modification of the cationic lipid structures via attachment of certain functional groups is an approach, which has been widely proposed in order to enhance selectivity.[2] PEGylation is one of the most common modifications to obtain long-circulation lipoplexes. Moreover, decoration with transferrin and different targeting ligands is usually performed to also obtain targeted and multifunctional lipoplexes.

Wojcicki et al. designed hyaluronic acid (HA)-bearing lipoplexes that can target *in vitro* the CD44 cell receptor. Liposomes containing the conjugates between high molecular weight hyaluronic acid and the lipid DOPE (HA-DOPE) were complexed with a reporter plasmid encoding the green fluorescent protein (GFP). The formulation showed a notable transfection efficiency after internalization mediated by the CD44 receptor in CD44-expressing A549 lung cancer cells.[33]

Duarte, Faneca, and de Lima,[34] assessed the effect of noncovalent association of folate to liposomes composed of DOTAP and Chol in lipoplexes stability and transfection efficiency in two different cancer cell lines. Their results show that lipoplexes' functionalization with folate enhance transfection, promote a greater DNA protection by overcoming the serum-related inhibitory effects and significantly increase the lipoplexes' biological activity, which suggests that the folate-associated lipoplexes are promising vectors for *in vivo* gene delivery.[34]

Niu et al.[35] demonstrated *in vitro* and *in vivo* a better efficacy of the lipoplexes formed by complexation of 1,2-distearoyl-sn-glycero-3-phosphoethanolamine-*N*-[PDP(polyethylene glycol)-2000] (PDP-PEG2000-DSPE) and the cytotoxic fusion peptide p14ARF-TAT when the FGF2 targeting peptide was incorporated into the vector. Yonenaga et al.[36] also demonstrated that their DCP-TEPA–based polycation liposomes present higher *in vitro* and *in vivo* gene silencing efficacies when decorated with cyclic arginylglycylaspartic acid (RGD) peptides.

Nakase et al.[37] assessed the efficacy of p53 gene therapy in human osteosarcoma using a cationic liposome decorated with transferrin. They demonstrated *in vitro* a significant growth inhibition and *in vivo* a suppression of tumor growth as well as the reduction of its mean volume.[37]

PROMISING POLYPLEXES DEVELOPED FOR GENE DELIVERY: A STATE OF THE ART

Polyplexes formation usually occurs via a rapid and almost irreversible electrostatic association between the polymer and the nucleic acids.[20,38] The most commonly used polymers for gene delivery are the cationic and the amphiphilic polymers. The ability of amphiphilic polymers as gene delivery systems is explored in other entry of this Encyclopedia entitled "Amphiphilic Polymers in Drug Delivery," thus in this section greater attention is given to the cationic polyplexes. As in the case of lipoplexes, the polyplexes' cell internalization occurs mainly by endocytosis. Several factors such as polymer size, molecular weight, degree of polymer branching, and charge density as well as the composition of the formulation medium and the ratio between the positive and negative charges of polymer and oligonucleotides influence polyplexes' transfection efficiency and biological activity.[39,40]

Cationic Polymers

Cationic polymers include, among others, the oligonucleotide-binding proteins (histones), synthetic polypeptides [e.g., poly(L-lysine) (PLL) and polyarginine (PLA)], polyethylenimine (PEI), cationic dendrimers [e.g., poly(amido amine) (PAMAM)], poly(dimethylaminoethylmethacrylate) and carbohydrate-based polymers (e.g., chitosan).[4,40]

The PLL and PLA are first-generation polymers that are used even now for polyplexes' preparation, but due to its low buffering capacity, PEGylation, functionalization with targeting ligands, or modification by introducing histidine residues in their backbones need to be performed in order to improve their transfection efficiency and circulation times.[4,41,42] Therefore, they have been substituted by other polymers, such as PEI and chitosan, the most widely used nowadays.

In fact, PEI is one of the most commonly used cationic polymers due to its ability to effectively condense nucleic acids into nanoscale particles. These systems proved to induce high transfection efficiencies as well as successfully *in vitro* and *in vivo* gene delivery. PEI-based polyplexes are able to enter the cells via caveolae- or clathrin-dependent routes and easily promote the endosomal release due to its well-known high-buffering properties.[14] PEI is also capable of entering the nucleus and accelerate gene entry into the nucleus from the cytosol.[43,44] However its nonbiodegradability and high amount of positive charges are responsible for the toxicity usually associated with this polymer, namely, the membrane damage, the interaction with blood cells, and activation of the complement system.[45] Therefore, in the last few years, many modifications of the original PEI structure have been done in order to design a polymer with lower toxic effect, maintaining the high transfection efficiencies. PEI can be synthesized in different lengths, branched or linear and can undergo substitution or incorporation of functionalized groups. Generally, branched PEI-based transfection systems are associated with higher ability to nucleic acids complexation and transfection efficiency, but also with higher cytotoxic effects.[45–47] To solve the toxicity problems, researchers have explored the potential of linear PEI-based transfection systems for gene delivery.[48,49] As observed for other cationic polymers, the ability of PEI to condense the DNA and efficiently transfect the cells depends, among the abovementioned factors, particularly on molecular weight, the saline buffer concentration, as well as the N/P ratio.[46,50,51] Generally, higher N/P ratios lead to smaller sizes and increase the zeta-potential of PEI complexes. Higher PEI molecular weights promote an increased buffering capacity and transfection efficiencies due to a higher content of primary, secondary, and tertiary amines but, on the other hand, increase the cytotoxicity due to polymer precipitation and adsorption onto the outer cell membrane.[45,50,52] Concerning the buffer conditions used during preparation, the complexes formed in low ionic strength conditions are usually smaller, physically more stable, and have a lower tendency to aggregate.[39,53]

Regarding PEI modification, one of the most used is the grafting with PEG. This approach, despite leading to polymeric vectors with lower transfection efficiencies is an effective strategy to decrease cytotoxicity, avoid the aggregation problems and nonspecific interactions through the shielding of the particle's surface positive charge.[45,52,54]

To further improve PEI transfection, many other different strategies have been investigated, such as the functionalization with Chol, propionic and succinic acids, alkyl chains, fatty acids, and other lipophilic chains to enhance polyplex interaction with the cellular membrane.[39,55,56]

A well-known alternative to PEI is chitosan, a natural polysaccharide composed of glucosamine and *N*-acetyl glucosamine. This polymer has been extensively proposed as the safer gene delivery vector due to its biodegradability, biocompatibility, low immunogenicity, and strong immune stimulatory ability.[57,58] The ability of chitosan to interact with the negatively charged oligonucleotides is related, among other factors, with its molecular weight and degree of deacetylation. Usually a higher degree of deacetylation enables a better interaction with the genetic material. Chitosan molecular weight should be regulated in order to achieve a balance between protection and release of the oligonucleotides. Higher molecular weight chitosan promotes a better stability of the complexes, whereas lower molecular weight ones lead to a better intracellular oligonucleotides release.[57–59] To improve the transfection efficiency of chitosan-based carriers, different chemical modifications, such as the combination of chitosan with dextran, N-dodecyl, glycyrrhizin, folic acid, or even with PEI have been investigated.[60–66] Additionally, the use of the water-soluble salt forms of chitosan (chloride, acetate, aspartate, glutamate, or lactate salts) is another possible strategy to improve the transfection efficiency.[57,58] Similar to other vectors for gene delivery, the chitosan-based vectors are usually targeted using, for example, monoclonal antibodies, peptides, or sugars with the aim to achieve better particles internalization into specific cells.[58]

Some examples of polymer-based formulations for gene delivery used for *in vitro* and *in vivo* applications, as well as examples of formulations that enrolled clinical trials are given in Table 2.

LIPOPLEXES VERSUS POLYPLEXES: A COMPARATIVE STUDY

Both polyplexes and lipoplexes have the capacity to protect the genetic material from the body environment and undesirable degradation during the administration and transfection process, being a safer alternative to viral vectors. However, the comparative low transfection efficiency is the main drawback of this type of system. Many lipids and polymers with specific properties have been synthesized, modified, and tested in order to enhance their transfection efficiency,[82] including conjugation lipids or polymers with peptides[83] or nucleotides,[84] among others.

Because of their ability to condense genetic material and interact with cellular membrane, cationic lipids and polymers are generally used. The main differences between cationic lipids and cationic polymers rely on the water solubility, the capacity of polymers to more efficiently condense the genetic material, and the mechanism involved in the endosomal escape.[20,82]

Lipoplexes have the advantage of having a platform already developed for industrial production, being of easier scaleup. And because cationic lipoplexes are relatively efficient in gene transfection, they have been largely investigated. Some studies suggested that the presence of fusogenic lipids such as DOPE and derivatives in the lipid formulation enhance their transfection efficiency by promoting the

Table 2 Examples of polyplexes-based system used for *in vitro* and *in vivo* applications as well as in clinical trials

Polymer vector	Genetic material	Utility	State	References
Dendritic PLL-block-poly (L-lactide)-block-dendritic PLL	pEGFP	EGFP expression	*In vitro*	[67]
Chitosan	siRNA luc Gl3	Luciferase (Luc) reporter gene downregulation	*In vitro*	[59]
Derivatives of low molecular weight PEI	pGL3	Luc expression	*In vitro*	[68]
PEI–PEG–TAT peptide	pEGFP-N3	EGFP expression	*In vitro*	[69]
Chitosan oligomers	pCpG-Luc, gWiz-Luc, pCpG-GFP and gWiz-GFP	Luc and GFP expression	*In vivo*	[70]
ARC-520 (Dynamic poly conjugates)	siRNA against apolipoprotein B and peroxisome proliferator-activated receptor-alpha	Hepatitis B	*In vivo*	[71,72]
PEG–PLL	sFlt-1 pDNA	Solid tumors	*In vivo*	[73]
PEG–poly[N′-[N-(2-aminoethyl)-2-aminoethyl]aspartamide]	pGL4.13 and sFlt-1 pDNA	Solid tumors	*In vivo*	[74,75]
Pluronic® P85	pGFP	GFP expression	*In vivo*	[76]
CALAA-01	siRNA against ribonucleotide reductase M2	Melanoma	Phase I	[77,78]
Polyvinylpyrrolidone	pIFN-alpha, pIL-12 and phIGF-1	Cubital tunnel syndrome, oropharyngeal cancer stage unspecified, squamous cell carcinoma and malignant melanoma	Phases I and II	[79–81]

Table 3 Main advantages and disadvantages of lipoplexes and polyplexes

	Advantages	Disadvantages
Lipoplex	Nonimmunogenic	Neutral and anionic – difficult to produce
	Protect genetic material from degradation	Cationic – more toxic
	Cationic – easily form complexes with negatively charged genetic material	Cationic – inactivation in the presence of serum
	Destabilization of the endosomal membrane promoting the endosomal escape	Cationic – cases of low stability upon storage
	Liable to scale-up for industrial production	Lower transfection efficiency compared with viral vectors
	Cationic – interact with the cell membrane facilitating the endocytosis	
	Neutral and anionic – little toxicity	
Polyplex	Nonimmunogenic	Not all polymers used present endosome-lytic characteristics
	Protect genetic material from degradation	Toxicity concerns especially with PEI
	Efficiently condense de genetic material forming complexes – smaller particles	Nonbiodegradability of some polymers (PEI)
	Higher water solubility	Lower transfection efficiency compared with viral vectors
	Tailored in size and composition	
	Easily modified at the surface	

endosomal escape due to its detergent-like destabilization membrane effect.[20,85] Despite the advantages, they interact with opsonins, serum proteins, or enzymes being partially inactivated in the presence of serum.[86] Also, concerns about the cytotoxicity of cationic lipoplexes by interference with cellular ion channels and reports of hepatotoxicity and carcinogenicity via activation of pro-inflammatory cytokines, or generation of reactive oxygen species have been raised.[19] Therefore, lipoplexes with neutral and anionic, more biocompatible and less toxic lipids have been developed, despite being more difficult and time-consuming to produce.[87,88]

Cationic polymers usually complex genetic material more efficiently than lipids, forming complexes with small size that can present higher transfection efficiency, because the size of the complexes is one of the most important factors influencing the cellular uptake. In addition, being mostly synthetic can be tailored with specific size, branching, and composition and easily surface modified.[20]

Conventionally, lipoplexes easily destabilize the endosome by direct interaction with the membrane, whereas in polyplexes, the endosomal release becomes more complicated. Thus, cationic polymers need to have the ability to buffer the acidic interior of endosomes in order to induce their osmotic swelling and disruption, a mechanism that is the basis of the proton sponge effect theory.[51,52,89,90] Although theoretically all cationic polymers present the capacity to complex genetic material, only few such as

PAMAM dendrimers, PEI, chitosan, and some derivatives, can disrupt the endosomal membrane. However, first-generation of cationic polymers are inefficient in terms of endosomal escape and transfection efficiency, requiring the cotransfection of endosome-lytic agents, such as chloroquine, amphipathic peptides, namely, lysine-alanine-leucine-alanine and glutamic acid-alanine-leucine-alanine, or anionic polymers such as poly(ethyl acrylic acid) and poly(propyl acrylic acid).[91]

Nevertheless, the ideal vector that fits all applications has not been designed yet. Each system shows different behaviors depending on the chemical structure of the components, the supramolecular structure of the complexes, the material to deliver, the cell type, and the administration conditions. Therefore, a comparison to choose between lipoplexes and polyplexes is difficult and almost impossible because both present advantages and disadvantages, as described in Table 3.

COMBINATION OF POLYMERS AND LIPIDS: DOES IT MEET THE IDEAL SYSTEM?

To benefit from the advantages of both, some research groups have combined lipids and polymers in the same system. Lipopolyplexes (LPPs) are ternary complexes comprising cationic lipids, a cationic polymer/peptide, and nucleic acids (Fig. 4).[92–95] Although little information is available regarding the structure of these ternary complexes,

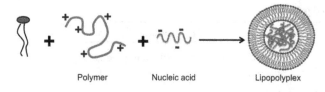

Fig. 4 Schematic representation of lipopolyplexes.

a superior colloidal stability, a reduced cytotoxicity as well as a relatively low immunological response has been reported.[92,96,97] Simultaneously, these novel formulations emerge as the second generation of nonviral gene delivery vectors and have demonstrated an elevated transfection efficiency compared with the first-generation ones, represented by lipoplexes and polyplexes.[92,98–100] As described earlier, these novel ternary complexes combine the delivery capabilities of cationic liposomes with the advantages of cationic polymers/peptides to an efficient complexation and delivery of either DNA,[94,95,98,101–104] siRNA,[105,106] or mRNA.[100,104,107]

Kurosaki et al.[108] has developed an LPP formulation for pulmonary gene delivery composed of PEI, DOTMA, and pDNA. This formulation showed a marked transfection efficiency in HepG2 cells, without any cytotoxicity or aggregation with erythrocytes. *In vivo* studies revealed high levels of transfection efficiency in lung cells after intravenous administration, suggesting that these PEI–DOTMA–pDNA complexes form a safe and effective gene delivery system for lung cancer.[108] Urbiola et al.[92] also evaluated a similar LPP combined with folic acid, which significantly enhanced the transfection efficiency and gene-silencing activity without a significant cytotoxic effect.[92] Pelisek et al.[101] investigated a combination of linear and branched PEI with liposomes (1,3-dioleoyloxy-2-(N(5)-carbamoyl-spermine)-propane, DOSPER, and DOTAP) for gene transfer to slow-proliferating human colon carcinoma cell lines. In all cases, LPP were 5 to 400 times more efficient compared with the corresponding lipoplexes or polyplexes, being therefore promising nanocarriers for *in vitro* and *in vivo* gene transfer in colorectal cancer cells.[101] Garcia et al. developed one of the most efficient LPP composed by PEI/DNA complexes with cationic liposomes formed by DOTAP and Chol.[102,109]

Regarding the LPP complexes compromising liposomes, nucleic acids, and cationic peptides, the most commonly used peptide is protamine,[98,99,103,105,107,110] but lysine,[98,111–113] arginine, and histidine[100,112] have also been reported. Cationic lipid–protamine–DNA complexes have presented an increased DNA protection against enzymatic digestion, a superior physical stability, and higher gene expression in lung cells after intravenous administration.[99,105,110] Li and Huang[105] and Tseng, Mozumdar, and Huang[114] have developed a tumor-targeted LPP for siRNA delivery. The nucleic acid, previously mixed with a carrier DNA, was condensed with protamine and the resulting nanoparticles were coated with DOTAP and Chol.[105,114] Further PEGylation of the LPP complex significantly increased the tumor localization of siRNA by fourfold and increased the silencing effect by two- to threefold in an *in vivo* human lung cancer model.[105,106]

Anionic lipids have also been used for the development of LPP systems, consisting of pH-sensitive liposome and protamine for macrophage gene therapy. *In vitro* tests revealed a high safety profile and increased transfection efficiency, higher than lipofectamine or protamine/DNA complexes.[115]

This strong association between peptides and DNA confers to such LPP particles an enhanced transfection efficiency and gene silencing. Moreover, this complexation promotes the capacity to transfect cells in the presence of serum.[18,116]

Another type of LPP nanocarriers are the pegylated immunolipoplexes (PILPs). In these complexes, the DNA is first compacted by a cationic polymer, and anionic liposomes are added to produce the LPP. Then, the surface is covered with PEG, in order to promote stabilization in the bloodstream, and the PEG strands are functionalized with a targeting monoclonal antibody. The system (mAb-PEG2000/DSPE-PEI/DNA) developed by Hu et al.[117] has been found to be highly effective in protecting DNA from degradation, and to have high efficiency in gene delivery and gene silencing in liver cancer cells without a significant cytotoxic effect. In contrast to the LPP developed by Kurosaki et al.[108] and Li and Huang,[105] which accumulated at the lung, intravenous administration of PILP resulted in liver accumulation with no cytokine production or liver injury.[117–119] PILPs are thus promising gene delivery systems to target liver cancer.

LPP has also been extensively described as an efficient system for mRNA-[93,100,104,107,120] and DNA-based[103,121] cancer vaccines.[104]

Mockey et al.[100] developed an LPP for the delivery of mRNA encoding a tumor antigen (MART-1) against the progression of B16F10 melanoma. MART-1 was condensed in 50 nm particles of PEGylated histidylated PLL and then added to liposomes prepared with L-histidine-(N,N-di-n-hexadecylamine)ethylamide and Chol. Intravenous administration of this formulation induced the production of interferon-γ (IFN-γ) and the activation of cytotoxic T lymphocytes. No effect was observed when the mRNA was complexed with liposomes alone or with the cationic polymer alone.[100] A similar nanosystem for the vaccination of B16F10 melanoma was also developed adding mannosylated and histidylated liposomes to mRNA-PEGylated histidylated PLL polyplexes. Authors were able to enhance the transfection of tumor antigen mRNA in splenic dendritic cells *in vivo* inducing therefore an anticancer immune response.[93,120]

A DNA-based vaccine approach was proposed by Whitmore et al.[122] consisting of DOTAP:Chol cationic liposomes, protamine sulfate, and pDNA with unmethylated CpG motifs. This complex was capable of stimulating a potent Th-1 cytokine response producing tumor necrosis factor-alpha, interleukin (IL)-12, and IFN-γ, and was also able to trigger the antitumor natural killer (NK) activity. The combination of these two innate immune responses was effective in inhibiting the growth of established tumors in mice as they provide an acquired tumor-specific cytotoxic T lymphocytes (CTL) response.[122]

The LPP stability, *in vitro* efficacy, and low toxicity justify its rapid ingoing to clinical trials. In fact, EGEN-001 is an LPP formulation that undergoes phase I and II clinical trials, composed of IL-12 plasmid formulated in PEG–PEI–Chol to treat a variety of cancers such as ovarian, colorectal, and fallopian tube cancer.[123]

CONCLUSION AND FUTURE PERSPECTIVES

Although viral vectors continue to be most widely researched for gene therapy, there is no doubt that the application of nonviral vectors for gene delivery has been an amazing and essential discovery of the last years. Gene therapy mediated by lipoplexes, polyplexes, or lipopolyplexes is a field that is clearly progressing. It is expected that very soon new formulations achieve the clinical trials and enter the market of transfection reagents and especially in clinical applications. Given the chance to promote the silencing of different oncogenes or correct the expression of other absent or mutant genes, among the different health areas to which new nonviral vectors-based delivery systems are being developed, research work is invested and greater applicability is predictable in oncology.

Although lipopolyplexes seem to overcome some of the drawbacks in other systems, the assumption that an ideal nonviral system was not discovered yet, led us to recognize a pressing need for innovative approaches. New discoveries should help to meet quality requirements, which demand a balance between effectiveness, toxicity, and technology, aiming for market approval. To achieve this balance with nonviral vectors is a worthy effort to meet critical patient's needs.

REFERENCES

1. Liu, F.; Huang, L. Development of non-viral vectors for systemic gene delivery. J. Control. Release **2002**, *78* (1–3), 259–266.
2. Wasungu, L.; Hoekstra, D. Cationic lipids, lipoplexes and intracellular delivery of genes. J. Control. Release **2006**, *116* (2), 255–264.
3. Zhang, X.X.; McIntosh, T.J.; Grinstaff, M.W. Functional lipids and lipoplexes for improved gene delivery. Biochimie **2012**, *94* (1), 42–58.
4. Zhang, S.; Xu, Y.; Wang, B.; Qiao, W.; Liu, D.; Li, Z. Cationic compounds used in lipoplexes and polyplexes for gene delivery. J. Control. Release **2004**, *100* (2), 165–180.

5. Scholz, C.; Wagner, E. Therapeutic plasmid DNA versus siRNA delivery: Common and different tasks for synthetic carriers. J. Control. Release **2012**, *161* (2), 554–565.
6. Wu, G.Y.; Wu, C.H. Receptor-mediated *in vitro* gene transformation by a soluble DNA carrier system. J. Biol. Chem. **1987**, *262* (10), 4429–4432.
7. Felgner, P.L.; Gadek, T.R.; Holm, M.; Roman, R.; Chan, H.W.; Wenz, M.; Northrop, J.P.; Ringold, G.M.; Danielsen, M. Lipofection: A highly efficient, lipid-mediated DNA-transfection procedure. Proc. Natl. Acad. Sci. U.S.A. **1987**, *84* (21), 7413–7417.
8. Eliyahu, H.; Joseph, A.; Schillemans, J.P.; Azzam, T.; Domb, A.J.; Barenholz, Y. Characterization and *in vivo* performance of dextran-spermine polyplexes and DOTAP/cholesterol lipoplexes administered locally and systemically. Biomaterials. **2007**, *28* (14), 2339–2349.
9. Ma, B.; Zhang, S.; Jiang, H.; Zhao, B.; Lv, H. Lipoplex morphologies and their influences on transfection efficiency in gene delivery. J. Control. Release **2007**, *123* (3), 184–194.
10. Esposito, C.; Generosi, J.; Mossa, G.; Masotti, A.; Castellano, A.C. The analysis of serum effects on structure, size and toxicity of DDAB-DOPE and DC-Chol-DOPE lipoplexes contributes to explain their different transfection efficiency. Colloids Surf B Biointerfaces **2006**, *53* (2), 187–192.
11. Masotti, A.; Mossa, G.; Cametti, C.; Ortaggi, G.; Bianco, A.; Grosso, N.D.; Malizia, D.; Esposito, C. Comparison of different commercially available cationic liposome-DNA lipoplexes: Parameters influencing toxicity and transfection efficiency. Colloids Surf B Biointerfaces **2009**, *68* (2), 136–144.
12. Kenny, G.D.; Kamaly, N.; Kalber, T.L.; Brody, L.P.; Sahuri, M.; Shamsaei, E.; Miller, A.D.; Bell, J.D. Novel multifunctional nanoparticle mediates siRNA tumour delivery, visualisation and therapeutic tumour reduction *in vivo*. J. Control. Release **2011**, *149* (2), 111–116.
13. Akinc, A.; Goldberg, M.; Qin, J.; Dorkin, J.R.; Gamba-Vitalo, C.; Maier, M.; Jayaprakash, K.N.; Jayaraman, M.; Rajeev, K.G.; Manoharan, M.; Koteliansky, V.; Röhl, I.; Leshchiner, E.S.; Langer, R.; Anderson, D.G. Development of lipidoid-siRNA formulations for systemic delivery to the liver. Mol. Ther. **2009**, *17* (5), 872–879.
14. Rejman, J.; Conese, M.; Hoekstra, D. Gene transfer by means of lipo- and polyplexes: Role of clathrin and caveolae-mediated endocytosis. J. Liposome Res. **2006**, *16* (3), 237–247.
15. Wong, A.W.; Scales, S.J.; Reilly, D.E. DNA internalized via caveolae requires microtubule-dependent, Rab7-independent transport to the late endocytic pathway for delivery to the nucleus. J. Biol. Chem. **2007**, *282* (31), 22953–22963.
16. Lu, J.J.; Langer, R.; Chen, J. A novel mechanism is involved in cationic lipid-mediated functional siRNA delivery. Mol. Pharm. **2009**, *6* (3), 763–771.
17. Ciani, L.; Casini, A.; Gabbiani, C.; Ristori, S.; Messori, L.; Martini, G. DOTAP/DOPE and DC-Chol/DOPE lipoplexes for gene delivery studied by circular dichroism and other biophysical techniques. Biophys. Chem. **2007**, *127* (3), 213–220.

18. Zhang, Y.; Li, H.; Sun, J.; Gao, J.; Liu, W.; Li, B.; Guo, Y.; Chen, J. DC-Chol/DOPE cationic liposomes: A comparative study of the influence factors on plasmid pDNA and siRNA gene delivery. Int. J. Pharm. **2010**, *390* (2), 198–207.

19. Kapoor, M.; Burgess, D.J.; Patil, S.D. Physicochemical characterization techniques for lipid based delivery systems for siRNA. Int. J. Pharm. **2012**, *427* (1), 35–57.

20. Tros de Ilarduya, C.; Sun, Y.; Düzgüneş, N. Gene delivery by lipoplexes and polyplexes. Eur. J. Pharm. Sci. **2010**, *40* (3), 159–170.

21. Metwally, A.A.; Blagbrough, I.S.; Mantell, J.M. Quantitative silencing of EGFP reporter gene by self-assembled siRNA lipoplexes of LinOS and cholesterol. Mol. Pharm. **2012**, *9* (11), 3384–3395.

22. Mochizuki, S.; Kanegae, N.; Nishina, K.; Kamikawa, Y.; Koiwai, K.; Masunaga, H.; Sakurai, K. The role of the helper lipid dioleoylphosphatidylethanolamine (DOPE) for DNA transfection cooperating with a cationic lipid bearing ethylenediamine. Biochim. Biophys. Acta **2012**, *1828* (2), 412–418.

23. Tagami, T.; Suzuki, T.; Matsunaga, M.; Nakamura, K.; Moriyoshi, N.; Ishida, T.; Kiwada, H. Anti-angiogenic therapy via cationic liposome-mediated systemic siRNA delivery. Int. J. Pharm. **2012**, *422* (1–2), 280–289.

24. Commander, N.J.; Brewer, J.M.; Wren, B.W.; Spencer, S.A.; Macmillan, A.P.; Stack, J.A. Liposomal delivery of p-ialB and p-omp25 DNA vaccines improves immunogenicity but fails to provide full protection against B. melitensis challenge. Genet. Vaccines Ther. **2010**, *8* (5), 1–12.

25. Zhou, S.; Kawakami, S.; Higuchi, Y.; Yamashita, F.; Hashida, M. The involvement of NK cell activation following intranasal administration of CpG DNA lipoplex in the prevention of pulmonary metastasis and peritoneal dissemination in mice. Clin. Exp. Metastasis. **2012**, *29* (1), 63–70.

26. Wu, S.Y.; Chang, H.I.; Burgess, M.; McMillan, N.A. Vaginal delivery of siRNA using a novel PEGylated lipoplex-entrapped alginate scaffold system. J. Control. Release **2011**, *155* (3), 418–426.

27. Aleku, M.; Schulz, P.; Keil, O.; Santel, A.; Schaeper, U.; Dieckhoff, B.; Janke, O.; Endruschat, J.; Durieux, B.; Röder, N.; Löffler, K.; Lange, C.; Fechtner, M.; Möpert, K.; Fisch, G.; Dames, S.; Arnold, W.; Jochims, K.; Giese, K.; Wiedenmann, B.; Scholz, A.; Kaufmann, J. Atu027, a liposomal small interfering RNA formulation targeting protein kinase N3, inhibits cancer progression. Cancer Res. **2008**, *68* (23), 9788–9798.

28. Phase I Pilot Study of Gene Therapy for Cystic Fibrosis Using Cationic Liposome Mediated Gene Transfer, http://www.clinicaltrials.gov/ct2/show/NCT00004471 (assessed December 2012).

29. Phase I Study of IV DOTAP: Cholesterol-Fus1 in Non-Small-Cell Lung Cancer, http://clinicaltrials.gov/show/NCT00059605 (assessed December 2012).

30. Mignet, N.; Richard, C.; Seguin, J.; Largeau, C.; Bessodes, M.; Scherman, D. Anionic pH-sensitive pegylated lipoplexes to deliver DNA to tumors. Int. J. Pharm. **2008**, *361* (1–2), 194–201.

31. Srinivasan, C.; Burgess, D.J. Optimization and characterization of anionic lipoplexes for gene delivery. J. Control. Release **2009**, *136* (1), 62–70.

32. Kapoor, M.; Burgess, D.J. Efficient and safe delivery of siRNA using anionic lipids: Formulation optimization studies. Int. J. Pharm. **2012**, *432* (1–2), 80–90.

33. Dufaÿ Wojcicki, A.; Hillaireau, H.; Nascimento, T.L.; Arpicco, S.; Taverna, M.; Ribes, S.; Bourge, M.; Nicolas, V.; Bochot, A.; Vauthier, C.; Tsapis, N.; Fattal, E. Hyaluronic acid-bearing lipoplexes: Physico-chemical characterization and *in vitro* targeting of the CD44 receptor. J. Control. Release **2012**, *162* (3), 545–552.

34. Duarte, S.; Faneca, H.; de Lima, M.C. Non-covalent association of folate to lipoplexes: A promising strategy to improve gene delivery in the presence of serum. J. Control. Release **2011**, *149* (3), 264–272.

35. Niu, G.; Driessen, W.H.; Sullivan, S.M.; Hughes, J.A. *In vivo* anti-tumor effect of expressing p14ARF-TAT using a FGF2-targeted cationic lipid vector. Pharm. Res. **2011**, *28* (4), 720–730.

36. Yonenaga, N.; Kenjo, E.; Asai, T.; Tsuruta, A.; Shimizu, K.; Dewa, T.; Nango, M.; Oku, N. RGD-based active targeting of novel polycation liposomes bearing siRNA for cancer treatment. J. Control. Release. **2012**, *160* (2), 177–181.

37. Nakase, M.; Inui, M.; Okumura, K.; Kamei, T.; Nakamura, S.; Tagawa, T. p53 gene therapy of human osteosarcoma using a transferrin-modified cationic liposome. Mol. Cancer Ther. **2005**, *4* (4), 625–631.

38. Tang, M.X.; Szoka, F.C. The influence of polymer structure on the interactions of cationic polymers with DNA and morphology of the resulting complexes. Gene Ther. **1997**, *4* (8), 823–832.

39. Gunther, M.; Lipka, J.; Malek, A.; Gutsch, D.; Kreyling, W.; Aigner, A. Polyethylenimines for RNAi-mediated gene targeting *in vivo* and siRNA delivery to the lung. Eur. J. Pharm. Biopharm. **2011**, *77* (3), 438–449.

40. Morille, M.; Passirani, C.; Vonarbourg, A.; Clavreul, A.; Benoit, J.P. Progress in developing cationic vectors for non-viral systemic gene therapy against cancer. Biomaterials **2008**, *29* (24–25), 3477–3496.

41. Brown, M.D.; Schätzlein, A.; Brownlie, A.; Jack, V.; Wang, W.; Tetley, L.; Gray, A.I.; Uchegbu, I.F. Preliminary characterization of novel amino acid based polymeric vesicles as gene and drug delivery agents. Bioconjug. Chem. **2000**, *11* (6), 880–891.

42. Pichon, C.; Gonçalves, C.; Midoux, P. Histidine-rich peptides and polymers for nucleic acids delivery. Adv. Drug Deliv. Rev. **2001**, *53* (1), 75–94.

43. Godbey, W.T.; Wu, K.K.; Mikos, A.G. Tracking the intracellular path of poly(ethylenimine)/DNA complexes for gene delivery. Proc. Natl. Acad. Sci. U.S.A. **1999**, *96* (9), 5177–5181.

44. Pollard, H.; Remy, J.S.; Loussouarn, G.; Demolombe, S.; Behr, J.P.; Escande, D. Polyethylenimine but not cationic lipids promotes transgene delivery to the nucleus in mammalian cells. J. Biol. Chem. **1998**, *273* (13), 7507–7511.

45. Roesler, S.; Koch, F.P.; Schmehl, T.; Weissmann, N.; Seeger, W.; Gessler, T.; Kissel, T. Amphiphilic, low molecular weight poly(ethylene imine) derivatives with enhanced stability for efficient pulmonary gene delivery. J. Gene. Med. **2011**, *13* (2), 123–133.

46. Grayson, A.C.; Doody, A.M.; Putnam, D. Biophysical and structural characterization of polyethylenimine-mediated

siRNA delivery *in vitro*. Pharm. Res. **2006**, *23* (8), 1868–1876.

47. Godbey, W.T.; Barry, M.A.; Saggau, P.; Wu, K.K.; Mikos, A.G. Poly(ethylenimine)-mediated transfection: A new paradigm for gene delivery. J. Biomed. Mater. Res. **2000**, *51* (3), 321–328.

48. Breunig, M.; Lungwitz, U.; Liebl, R.; Fontanari, C.; Klar, J.; Kurtz, A.; Blunk, T.; Goepferich, A. Gene delivery with low molecular weight linear polyethylenimines. J Gene Med. **2005**, *7* (10), 1287–1298.

49. Breunig, M.; Lungwitz, U.; Liebl, R.; Klar, J.; Obermayer, B.; Blunk, T.; Goepferich, A. Mechanistic insights into linear polyethylenimine-mediated gene transfer. Biochim. Biophys. Acta **2007**, *1770* (2), 196–205.

50. Godbey, W.T.; Wu, K.K.; Mikos, A.G. Size matters: Molecular weight affects the efficiency of poly(ethylenimine) as a gene delivery vehicle. J. Biomed. Mater. Res. **1999**, *45* (3), 268–275.

51. Akinc, A.; Thomas, M.; Klibanov, A.M.; Langer, R. Exploring polyethylenimine-mediated DNA transfection and the proton sponge hypothesis. J. Gene Med. **2005**, *7* (5), 657–663.

52. Oskuee, R.K.; Philipp, A.; Dehshahri, A.; Wagner, E.; Ramezani, M. T he impact of carboxyalkylation of branched polyethylenimine on effectiveness in small interfering RNA delivery. J. Gene Med. **2010**, *12* (9), 729–738.

53. Goula, D.; Remy, J.S.; Erbacher, P.; Wasowicz, M.; Levi, G.; Abdallah, B.; Demeneix, B.A. Size, diffusibility and transfection performance of linear PEI/DNA complexes in the mouse central nervous system. Gene Ther. **1998**, *5* (5), 712–717.

54. Malek, A.; Czubayko, F.; Aigner, A. PEG grafting of polyethylenimine (PEI) exerts different effects on DNA transfection and siRNA-induced gene targeting efficacy. J. Drug Target. **2008**, *16* (2), 124–139.

55. Masotti, A.; Moretti, F.; Mancini, F.; Russo, G.; Di Lauro, N.; Checchia, P.; Marianecci, C.; Carafa, M.; Santucci, E.; Ortaggi, G. Physicochemical and biological study of selected hydrophobic polyethylenimine-based polycationic liposomes and their complexes with DNA. Bioorg. Med. Chem. **2007**, *15* (3), 1504–1515.

56. Thomas, M.; Klibanov, A.M. Enhancing polyethylenimine's delivery of plasmid DNA into mammalian cells. Proc. Natl. Acad. Sci. U.S.A. **2002**, *99* (23), 14640–14645.

57. Mao, S.; Sun, W.; Kissel, T. Chitosan-based formulations for delivery of DNA and siRNA. Adv. Drug Deliv. Rev. **2010**, *62* (1), 12–27.

58. Rudzinski, W.E.; Aminabhavi, T.M. Chitosan as a carrier for targeted delivery of small interfering RNA. Int. J. Pharm. **2010**, *399* (1–2), 1–11.

59. Holzerny, P.; Ajdini, B.; Heusermann, W.; Bruno, K.; Schuleit, M.; Meinel, L.; Keller, M. Biophysical properties of chitosan/siRNA polyplexes: Profiling the polymer/siRNA interactions and bioactivity. J. Control. Release **2012**, *157* (2), 297–304.

60. Park, I.K.; Park, Y.H.; Shin, B.A.; Choi, E.S.; Kim, Y.R.; Akaike, T.; Cho, C.S.; Park, Y.K.; Park, Y.R. Galactosylated chitosan-graft-dextran as hepatocyte-targeting DNA carrier. J. Control. Release **2000**, *69* (1), 97–108.

61. Kim, T.H.; Ihm, J.E.; Choi, Y.J.; Nah, J.W.; Cho, C.S. Efficient gene delivery by urocanic acid-modified chitosan. J. Control. Release **2003**, *93* (3), 389–402.

62. Kim, T.H.; Kim, S.I.; Akaike, T.; Cho, C.S. Synergistic effect of poly(ethylenimine) on the transfection efficiency of galactosylated chitosan/DNA complexes. J. Control. Release **2005**, *105* (3), 354–366.

63. Wong, K.; Sun, G.; Zhang, X.; Dai, H.; Liu, Y.; He, C.; Leong, K.W. PEI-g-chitosan, a novel gene delivery system with transfection efficiency comparable to polyethylenimine *in vitro* and after liver administration *in vivo*. Bioconjug. Chem. **2006**, *17* (1), 152–158.

64. Lin, A.; Liu, Y.; Huang, Y.; Sun, J.; Wu, Z.; Zhang, X.; Ping, Q. Glycyrrhizin surface-modified chitosan nanoparticles for hepatocyte-targeted delivery. Int. J. Pharm. **2008**, *359* (1–2), 247–253.

65. Gabizon, A.; Shmeeda, H.; Horowitz, A.T.; Zalipsky, S. Tumor cell targeting of liposome-entrapped drugs with phospholipid-anchored folic acid-PEG conjugates. Adv. Drug Deliv. Rev. **2004**, *56* (8), 1177–1192.

66. Moreira, C.; Oliveira, H.; Pires, L.R.; Simões, S.; Barbosa, M.A.; Pêgo, A.P. Improving chitosan-mediated gene transfer by the introduction of intracellular buffering moieties into the chitosan backbone. Acta Biomater. **2009**, *5* (8), 2995–3006.

67. Zhu, Y.; Sheng, R.; Luo, T.; Li, H.; Sun, W.; Li, Y.; Cao, A. Amphiphilic cationic [dendritic poly(L-lysine)]-block-poly(L-lactide)-block-[dendritic poly(L-lysine)]s in aqueous solution: Self-aggregation and interaction with DNA as gene delivery carriers. Macromol. Biosci. **2011**, *11* (2), 174–186.

68. Shum, V.W.; Gabrielson, N.P.; Forrest, M.L.; Pack, D.W. The effects of PVP(Fe(III)) catalyst on polymer molecular weight and gene delivery via biodegradable cross-linked polyethylenimine. Pharm. Res. **2012**, *29* (2), 500–510.

69. Ulasov, A.V.; Khramtsov, Y.V.; Trusov, G.A.; Rosenkranz, A.A.; Sverdlov, E.D.; Sobolev, A.S. Properties of PEI-based polyplex nanoparticles that correlate with their transfection efficacy. Mol. Ther. **2011**, *19* (1), 103–112.

70. Klausner, E.A.; Zhang, Z.; Wong, S.P.; Chapman, R.L.; Volin, M.V.; Harbottle, R.P. Corneal gene delivery: Chitosan oligomer as a carrier of CpG rich, CpG free or S/MAR plasmid DNA. J. Gene Med. **2012**, *14* (2), 100–108.

71. Rozema, D.B.; Lewis, D.L.; Wakefield, D.H.; Wong, S.C.; Klein, J.J.; Roesch, P.L.; Bertin, S.L.; Reppen, T.W.; Chu, Q.; Blokhin, A.V.; Hagstrom, J.E.; Wolff, J.A. Dynamic PolyConjugates for targeted *in vivo* delivery of siRNA to hepatocytes. Proc. Natl. Acad. Sci. U.S.A. **2007**, *104* (32), 12982–12987.

72. Mudd, S.R.; Trubetskoy, V.S.; Blokhin, A.V.; Weichert, J.P.; Wolff, J.A. Hybrid PET/CT for noninvasive pharmacokinetic evaluation of dynamic PolyConjugates, a synthetic siRNA delivery system. Bioconjug. Chem. **2010**, *21* (7), 1183–1189.

73. Itaka, K.; Osada, K.; Morii, K.; Kim, P.; Yun, S.H.; Kataoka, K. Polyplex nanomicelle promotes hydrodynamic gene introduction to skeletal muscle. J. Control. Release **2010**, *143* (1), 112–119.

74. Uchida, S.; Itaka, K.; Chen, Q.; Osada, K.; Miyata, K.; Ishii, T.; Harada-Shiba, M.; Kataoka, K. Combination of chondroitin sulfate and polyplex micelles from

Poly(ethylene glycol)-poly{N'-[N-(2-aminoethyl)-2-aminoethyl]aspartamide} block copolymer for prolonged *in vivo* gene transfection with reduced toxicity. J. Control. Release **2011**, *155* (2), 296–302.

75.   Chen, Q.; Osada, K.; Ishii, T.; Oba, M.; Uchida, S.; Tockary, T.A.; Endo, T.; Ge, Z.; Kinoh, H.; Kano, M.R.; Itaka, K.; Kataoka, K. Homo-catiomer integration into PEGylated polyplex micelle from block-catiomer for systemic anti-angiogenic gene therapy for fibrotic pancreatic tumors. Biomaterials **2012**, *33* (18), 4722–4730.

76.   Chen, Y.C.; Jiang, L.P.; Liu, N.X.; Ding, L.; Liu, X.L.; Wang, Z.H.; Hong, K.; Zhang, Q.P. Enhanced gene transduction into skeletal muscle of mice *in vivo* with pluronic block copolymers and ultrasound exposure. Cell Biochem. Biophys. **2011**, *60* (3), 267–273.

77.   Davis, M.E.; Zuckerman, J.E.; Choi, C.H.; Seligson, D.; Tolcher, A.; Alabi, C.A.; Yen, Y.; Heidel, J.D.; Ribas, A. Evidence of RNAi in humans from systemically administered siRNA via targeted nanoparticles. Nature. **2010**, *464* (7291), 1067–1070.

78.   Zuckerman, J.E.; Hsueh, T.; Koya, R.C.; Davis, M.E.; Ribas, A. siRNA knockdown of ribonucleotide reductase inhibits melanoma cell line proliferation alone or synergistically with temozolomide. J. Invest. Dermatol. **2011**, *131* (2), 453–460.

79.   Phase I Single Dose-Ranging Study Of Formulated hIGF-1 Plasmid In Subjects With Cubital Tunnel Syndrome, http://www.gemcris.od.nih.gov/Contents/GC_CLIN_TRIAL_RPT_VIEW.asp?WIN_TYPE=R&CTID=378 (assessed January 2013).

80.   Phase I/II Multi-Center, Open-Label, Multiple Administration Trial of the Safety, Tolerability, and Efficacy of an IFN-alpha/IL-12 Plasmid-Based Therapeutic, http://www.gemcris.od.nih.gov/Contents/GC_CLIN_TRIAL_RPT_VIEW.asp?WIN_TYPE=R&CTID=116 (assessed January 2013).

81.   A Multicenter, Open-Label, Multiple Administration, Study of the Safety, Tolerability and Efficacy of IFN α/IL-12 Combination Gene Therapy in Patients with Squamous Cell Carcinoma of the Head and Neck (SCCHN). http://www.gemcris.od.nih.gov/Contents/GC_CLIN_TRIAL_RPT_VIEW.asp?WIN_TYPE=R&CTID=184 (assessed January 2013).

82.   Elouahabi, A.; Ruysschaert, J.M. Formation and intracellular trafficking of lipoplexes and polyplexes. Mol. Ther. **2005**, *11* (3), 336–347.

83.   Faham, A.; Herringson, T.; Parish, C.; Suhrbier, A.; Khromykh, A.A.; Altin, J.G. pDNA-lipoplexes engrafted with flagellin-related peptide induce potent immunity and anti-tumour effects. Vaccine **2011**, *29* (40), 6911–6919.

84.   Khiati, S.; Pierre, N.; Andriamanarivo, S.; Grinstaff, M.W.; Arazam, N.; Nallet, F.; Navailles, L.; Barthélémy, P. Anionic nucleotide--lipids for *in vitro* DNA transfection. Bioconjug. Chem. **2009**, *20* (9), 1765–1772.

85.   Fasbender, A.; Marshall, J.; Moninger, T.O.; Grunst, T.; Cheng, S.; Welsh, M.J. Effect of co-lipids in enhancing cationic lipid-mediated gene transfer *in vitro* and *in vivo*. Gene Ther. **1997**, *4* (7), 716–725.

86.   Zhang, Y.; Anchordoquy, T.J. The role of lipid charge density in the serum stability of cationic lipid/DNA complexes. Biochim. Biophys. Acta **2004**, *1663* (1–2), 143–157.

87.   Patil, S.D.; Rhodes, D.G.; Burgess, D.J. Anionic liposomal delivery system for DNA transfection. AAPS J. **2004**, *6* (4), e29.

88.   Mignet, N.; Scherman, D. Anionic pH sensitive lipoplexes. Methods Mol. Biol. **2010**, *605*, 435–444.

89.   Kleemann, E.; Jekel, N.; Dailey, L.A.; Roesler, S.; Fink, L.; Weissmann, N.; Schermuly, R.; Gessler, T.; Schmehl, T.; Roberts, C.J.; Seeger, W.; Kissel, T. Enhanced gene expression and reduced toxicity in mice using polyplexes of low-molecular-weight poly(ethylene imine) for pulmonary gene delivery. J. Drug Target. **2009**, *17* (8), 638–651.

90.   Buyens, K.; Meyer, M.; Wagner, E.; Demeester, J.; De Smedt, S.C.; Sanders, N.N. Monitoring the disassembly of siRNA polyplexes in serum is crucial for predicting their biological efficacy. J. Control. Release **2010**, *141* (1), 38–41.

91.   Cho, Y.W.; Kim, J.D.; Park, K. Polycation gene delivery systems: Escape from endosomes to cytosol. J. Pharm. Pharmacol. **2003**, *55* (6), 721–734.

92.   Urbiola, K.; García, L.; Zalba, S.; Garrido, M.J.; Tros de Ilarduya, C. Efficient serum-resistant lipopolyplexes targeted to the folate receptor. Eur. J. Pharm. Biopharm. **2013**, *83* (3), 358–363.

93.   Perche, F.; Lambert, O.; Berchel, M.; Jaffrès, P.-A.; Pichon, C.; Midoux, P. Gene transfer by histidylated lipopolyplexes: A dehydration method allowing preservation of their physicochemical parameters and transfection efficiency. Int. J. Pharm. **2012**, *423* (1), 144–150.

94.   Whitmore, M.; Li, S.; Huang, L. LPD lipopolyplex initiates a potent cytokine response and inhibits tumor growth. Gene Ther. **1999**, *6*, 1867–1875.

95.   Matsuura, M.; Yamazaki, Y.; Sugiyama, M.; Kondo, M.; Ori, H.; Nango, M.; Oku, N. Polycation liposome-mediated gene transfer *in vivo*. Biochim. Biophys. Acta **2003**, *1612* (2), 136–143.

96.   Rehman, Z.u.; Hoekstra, D.; Zuhorn, I.S. Kinases in cationic lipid/polymer-mediated gene delivery. Drug Discov. Today **2010**, *15* (23–24), 1110.

97.   Ramezani, M.; Khoshhamdam, M.; Dehshahri, A.; Malaekeh-Nikouei, B. The influence of size, lipid composition and bilayer fluidity of cationic liposomes on the transfection efficiency of nanolipoplexes. Colloids Surf B Biointerfaces **2009**, *72* (1), 1–5.

98.   Gao, X.; Huang, L. Potentiation of cationic liposome-mediated gene delivery by polycations. Biochemistry **1996**, *35* (3), 1027–1036.

99.   Li, S.; Huang, L. *In vivo* gene transfer via intravenous administration of cationic lipid-protamine-DNA (LPD) complexes. Gene Ther. **1997**, *4* (9), 891–900.

100.   Mockey, M.; Bourseau, E.; Chandrashekhar, V.; Chaudhuri, A.; Lafosse, S.; Le Cam, E.; Quesniaux, V.F.; Ryffel, B.; Pichon, C.; Midoux, P. mRNA-based cancer vaccine: Prev ention of B16 melanoma progression and metastasis by systemic injection of MART1 mRNA histidylated lipopolyplexes. Cancer Gene Ther. **2007**, *14* (9), 802–814.

101.   Pelisek, J.; Gaedtke, L.; DeRouchey, J.; Walker, G.F.; Nikol, S.; Wagner, E. Optimized lipopolyplex formulations for gene transfer to human colon carcinoma cells under *in vitro* conditions. J. Gene Med. **2006**, *8* (2), 186–197.

102. Garcia, L.; Bunuales, M.; Duzgunes, N.; Tros de Ilarduya, C. Serum-resistant lipopolyplexes for gene delivery to liver tumour cells. Eur. J. Pharm. Biopharm. **2007**, *67*, 58–66.

103. Vangasseri, D.P.; Han, S.J.; Huang, L. Lipid-protamine-DNA-mediated antigen delivery. Curr. Drug Deliv. **2005**, *2* (4), 401–406.

104. Weide, B.; Garbe, C.; Rammensee, H.-G.; Pascolo, S. Plasmid DNA- and messenger RNA-based anti-cancer vaccination. Immunol. Lett. **2008**, *115* (1), 33–42.

105. Li, S.D.; Huang, L. Targeted delivery of antisense oligodeoxynucleotide and small interference RNA into lung cancer cells. Mol. Pharm. **2006**, *3* (5), 579–588.

106. Li, S.D.; Huang, L. Surface-modified LPD nanoparticles for tumor targeting. Ann. N. Y. Acad. Sci. **2006**, *1082* (1), 1–8.

107. Hoerr, I.; Obst, R.; Rammensee, H.G.; Jung, G. *In vivo* application of RNA leads to induction of specific cytotoxic T lymphocytes and antibodies. Eur. J. Immunol. **2000**, *30* (1), 1–7.

108. Kurosaki, T.; Kishikawa, R.; Matsumoto, M.; Kodama, Y.; Hamamoto, T.; To, H.; Niidome, T.; Takayama, K.; Kitahara, T.; Sasaki, H. Pulmonary gene delivery of hybrid vector, lipopolyplex containing N-lauroylsarcosine, via the systemic route. J. Control. Release **2009**, *136* (3), 213–219.

109. Garcia, L.; Urbiola, K.; Duzgunes, N.; Tros de Ilarduya, C. Lipopolyplexes as nanomedicines for therapeutic gene delivery. Methods Enzymol. **2012**, *509*, 327–338.

110. Birchall, J.C.; Kellaway, I.W.; Gumbleton, M. Physical stability and *in vitro* gene expression efficiency of nebulised lipid-peptide-DNA complexes. Int. J. Pharm. **2000**, *197* (1–2), 221–231.

111. Kudsiova, L.; Fridrich, B.; Ho, J.; Mustapa, M.F.; Campbell, F.; Welser, K.; Keppler, M.; Ng, T.; Barlow, D.J.; Tabor, A.B.; Hailes, H.C.; Lawrence, M.J. Lipopolyplex ternary delivery systems incorporating C14 glycerol-based lipids. Mol. Pharm. **2011**, *8* (5), 1831–1847.

112. Welser, K.; Campbell, F.; Kudsiova, L.; Mohammadi, A.; Dawson, N.; Hart, S.L.; Barlow, D.J.; Hailes, H.C.; Lawrence, M.J.; Tabor, A.B. Gene delivery using ternary lipopolyplexes incorporating branched cationic peptides: The role of peptide sequence and branching. Mol. Pharm. **2012**, *10* (1), 127–141.

113. Yan, J.; Berezhnoy, N.V.; Korolev, N.; Su, C.J.; Nordenskiold, L. Structure and internal organization of overcharged cationic-lipid/peptide/DNA self-assembly complexes. Biochim. Biophys. Acta **2012**, *6* (7), 1794–1800.

114. Tseng, Y.C.; Mozumdar, S.; Huang, L. Lipid-based systemic delivery of siRNA. Adv. Drug Deliv. Rev. **2009**, *61* (9), 721–731.

115. Sun, P.; Zhong, M.; Shi, X.; Li, Z. Anionic LPD complexes for gene delivery to macrophage: Preparation, characterization and transfection *in vitro*. J. Drug Target. **2008**, *16* (9), 668–678.

116. Junghans, M.; Loitsch, S.M.; Steiniger, S.C.; Kreuter, J.; Zimmer, A. Cationic lipid-protamine-DNA (LPD) complexes for delivery of antisense c-myc oligonucleotides. Eur. J. Pharm. Biopharm. **2005**, *60* (2), 287–294.

117. Hu, Y.; Li, K.; Wang, L.; Yin, S.; Zhang, Z.; Zhang, Y. Pegylated immuno-lipopolyplexes: A novel non-viral gene delivery system for liver cancer therapy. J. Control. Release **2010**, *144* (1), 75–81.

118. Hu, Y.; Shen, Y.; Ji, B.; Wang, L.; Zhang, Z.; Zhang, Y. Combinational RNAi gene therapy of hepatocellular carcinoma by targeting human EGFR and TERT. Eur. J. Pharm. Sci. **2011**, *42* (4), 387–391.

119. Hu, Y.; Shen, Y.; Ji, B.; Yin, S.; Ren, X.; Chen, T.; Ma, Y.; Zhang, Z.; Zhang, Y. Liver-specific gene therapy of hepatocellular carcinoma by targeting human telomerase reverse transcriptase with pegylated immuno-lipopolyplexes. Eur. J. Pharm. Biopharm. **2011**, *78* (3), 320–325.

120. Perche, F.; Benvegnu, T.; Berchel, M.; Lebegue, L.; Pichon, C.; Jaffrès, P.-A.; Midoux, P. Enhancement of dendritic cells transfection *in vivo* and of vaccination against B16F10 melanoma with mannosylated histidylated lipopolyplexes loaded with tumor antigen messenger RNA. Nanomedicine **2011**, *7* (4), 445–453.

121. Dileo, J.; Banerjee, R.; Whitmore, M.; Nayak, J.V.; Falo, L.D., Jr.; Huang, L. Lipid-protamine-DNA-mediated antigen delivery to antigen-presenting cells results in enhanced anti-tumor immune responses. Mol. Ther. **2003**, *7* (5 Pt. 1), 640–648.

122. Whitmore, M.M.; Li, S.; Falo, L., Jr.; Huang, L. Systemic administration of LPD prepared with CpG oligonucleotides inhibits the growth of established pulmonary metastases by stimulating innate and acquired antitumor immune responses. Cancer Immunol. Immunother. **2001**, *50* (10), 503–514.

123. A Phase I/II Study of Intraperitoneal EGEN-001, http://www.gemcris.od.nih.gov/Contents/GC_CT_RPT.asp?searchfield=polymer&Search.x=0&Search.y=0&Search=Search&browse_=&F_232= (assessed November 2012).

Magnetomicelles: Theranostic Applications

Prashant K. Deshmukh
Abhijeet P. Pandey
Post Graduate Department of Pharmaceutics and Quality Assurance, H. R. Patel Institute of Pharmaceutical Education and Research, Shirpur, India

Surendra G. Gattani
School of Pharmacy, Swami Ramanand Teerth Marathwada University, Nanded, India

Pravin O. Patil
Post Graduate Department of Pharmaceutics and Quality Assurance, H. R. Patel Institute of Pharmaceutical Education and Research, Shirpur, India

Abstract

Magnetic micelles or "magnetomicelles" (MMs) have opened a new avenue for researchers owing to their simultaneous diagnostic and drug-targeting (therapeutic) applications. The present entry highlights the recent diagnostic and therapeutic (theranostic) advances of functional nanostructured micelles with magnetic properties. It explains the potential theranostic use of these materials and the possible future use of this cutting-edge technology in the theranostic field. These MMs possess the capability of diagnosing the delivered NPs, which may be inorganic, organic, or biological in nature. Furthermore, modification of MMs with sensing and therapeutic groups such as ligands, proteins, fluorescent molecules, and monoclonal antibodies into the theranostic nanostructure could prove beneficial.

INTRODUCTION

Today's nanotechnology research is focused on fabrication and development of nanomaterials that are capable to do multiple tasks. Maximum consideration has been given for the development of multifunctional nanoparticles (NPs), which can prove to be useful for diagnostic as well as therapeutic purposes.[1–5] The small size of these NPs make them a suitable candidate for the fabrication of functional nanostructures, which can be used in the medical field such as biomedical imaging and drug delivery.[6–10] Significant attention has been paid for the development of delivery system, which can provide drug localization around the site of action, so as to provide maximum therapeutic effect with minimum toxicity, as in the case of antitumor drugs.[11,12]

Polymeric micelles as drug delivery systems (DDSs) were introduced by researchers in the late 1980s.[13,14] Polymeric micelles are made up of a core and shell. Inner core is the hydrophobic part, which encapsulates the poorly water-soluble drug, whereas the outer shell or corona of the hydrophilic part protects the drug from the aqueous environment and stabilizes the micelles.[15] The main objective of using these micelles for delivering a drug is to modulate or modify the drug disposition in the body, which leads to better therapeutic efficacy of the drug.[16] Polymeric micelles have been conjugated previously with imaging agents such as fluorescent dye[17] and paramagnetic contrast agents like

gadolinium[18] or hydrophobic contrast agents.[19] Polymeric micelles generally consist of two parts, a core and a shell structure. The first part, i.e., the inner core is made up of a hydrophobic part of the block copolymer, in which hydrophobic or water-insoluble drugs are encapsulated. The second part, i.e., the outer shell or corona is made up of the hydrophilic part of block copolymer and it helps in protecting the drug from the aqueous environment and also minimizes the *in vivo* recognition of polymeric micelles by the reticulo-endothelial system (RES). Alternatively, sometimes the core can be made up of a hydrophilic or water-soluble polymer, which is made hydrophobic by the chemical conjugation of water-insoluble drug or complexation of the two oppositely charged polyions, and the formed complex is known as polyions complex (PIC) micelles.[20–22]

Magnetic particle attracted considerable amount of attention in the biomedical field such as bioimaging.[23,24] Even though the first report pertaining to magnetic micelles or "metal-loaded micelles" (MMs) appeared in literature at the beginning of the 21st century,[25] its use in ultrasensitive magnetic resonance imaging (MRI) detection was demonstrated by Ai et al.[19] Since then, a lesser amount of information about MMs was available in the recent few years. From late 2010, research started again in this promising area, and in the year 2011–2012, many investigations were reported, which majorly deal with the potential of these nanostructures in the therapeutic as well as diagnostic fields.[18,26–29]

Concise Encyclopedia of Biomedical Polymers and Polymeric Biomaterials DOI: 10.1081/E-EBPPC-120050754

In Vitro—Medical

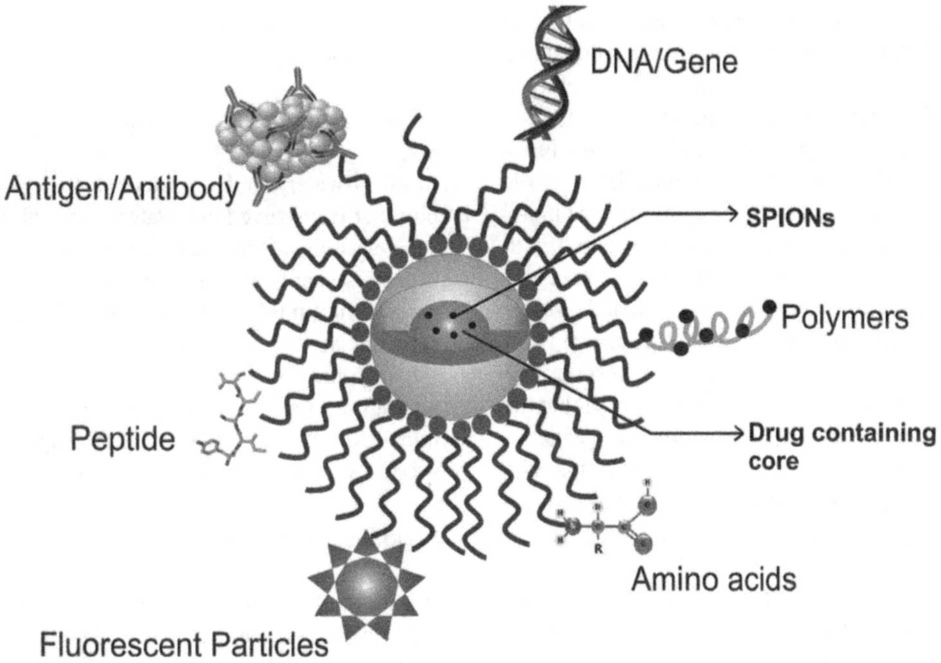

Fig. 1 Schematic representation of some potential theranostic modifications for different applications of magnetic micelles.

Figure 1 summarizes some potential theranostic modifications for the different applications of MMs.

In the present review entry, an attempt has been made to cover various diagnostic, therapeutic, and combined (theranostic) applications of MMs, on which work has either been done or is in progress. The present entry elaborates on the potential applications of these materials and the possible future use of this cutting-edge technology in related fields, with special emphasis on theranostic applications.

BIOLOGICAL SIGNIFICANCE OF POLYMERIC MICELLES

Poor absorption and bioavailability are the most commonly encountered problems with hydrophobic therapeutic agents or poorly water-soluble drugs.[30] Polymeric micelles have been reported to increase the aqueous solubility of the drug by 10–5000 fold, which is sufficient for increasing the bioavailability of hydrophobic agents.[31] When a drug is loaded in polymeric micelles, it gets entrapped into the inner core where it is safe from the aqueous environment as the outer shell of hydrophilic block copolymer reduces the interactions of drugs with the outer aqueous environment. The hydrophilic outer layer of micelle keeps the polymeric micelles stable in blood plasma for a longer duration and prevents their opsonization.[32]

Biodistribution

The objective of using polymeric micelles for drug delivery is to enhance the aqueous solubility and bioavailability, which can further modulate drug disposition in the body and can lead to an increase in the therapeutic efficacy. Avoiding RES recognition is important for increasing the blood circulation of polymeric micelles as RES recognition will lead to elimination of NPs from blood circulation. Surface modification of NPs with hydrophilic and biocompatible polymers, such as PEG, can impart stealth property to NPs which impair or even avoid RES recognition.[33,34] Polymeric micelles may not necessarily dissociate immediately after extreme dilution following administration into the body, and their dissociation is kinetically slow. The kinetically slow dissociation allows the polymeric micelles to circulate in the bloodstream until it reaches the target organ and hence releases the drug at the target site. This type of stable circulation of polymeric micelles in the bloodstream has been confirmed before using gel chromatography assay.[35] The study further revealed that the polymeric micelle avoided RES recognition as well as the entrapment by hepatic sinusoidal capillaries. Furthermore, it was revealed that the constituent block copolymers might be finally excreted into the urine due to their molecular weight being lower than the threshold of glomerular filtration, suggesting the safety of polymeric micelles with a low risk of chronic accumulation in the body.[36,37]

Tumor Accumulation and Tissue Penetration

Long circulating polymeric micelles have been demonstrated to accumulate in tumor which can be explained by the microvascular hyperpermeability to circulating molecules and their impaired lymphatic drainage in tumors which is termed the "Enhanced Permeability and Retention effect".[38,39] It has been demonstrated that polymeric micelle nanocarriers show an enhanced accumulation in solid tumors.[40–43] Much work has been done to examine the relationship between the size of the drug carriers and their circulation time in the blood and tumor accumulation. In a reported study, using polymeric micelles and PEG-coated PIC-based vesicles, it was shown that NPs with the size of 160 nm retained a higher concentration in the blood over 96 hr after systemic administration, compared with NPs having the sizes of 200 and 100 nm. These results give us an idea about the relation between NPs size and circulation time in blood plasma. The result showed that NPs of size approximately 150 nm might possess the best circulation time in the bloodstream. Studies have reported that the micellar formulation achieves significantly stronger pharmacological activity than the other vesicular systems with the aid of TGF-b inhibitors. These results of studies done on tumor penetration and accumulation ability of polymeric micelles demonstrated the feasibility of smaller polymeric micelles for the treatment of intractable hypovascular tumors due to their excellent accumulation and penetration.[44–49]

POLYMERS USED FOR MICELLE PREPARATION

Different types of polymers are used for the preparation of polymeric micelles. Generally, amphiphilic diblock copolymers are used for the preparation of polymeric micelles but triblock copolymers and graft copolymers are also used. Block copolymers like PEO-poly(L-amino acids) are used for the introduction of functional groups that can be derivatized in order to enhance the properties of micelles such as in the case of PEO-poly(L-aspartate); the carboxyl functionality of poly(L-aspartate) block can be used for the conjugation of drugs with the inner core. Other polymers used for the formation of hydrophobic core include poly(L-amino acid), polyesters, and pluronics. In case of triblock copolymers such as pluronics, the chain length of PPO block influences the partitioning of hydrophobic moieties in the micelles. By engineering these block copolymers, temporal and distribution controls can be attained. Temporal control covers the ability to adjust the period of time over which the drug release is supposed to take place or the capability to trigger drug release at a specific time during treatment.[50–54]

Methods of Preparation

Different methods of preparation of NPs have been reported as given below.

Dialysis method

In the dialysis method, small amounts of water are added to the solution of polymer and drug using a water-miscible organic solvent with stirring followed by dialysis against an excess of water using a dialysis bag for the removal of organic solvents.[55,56]

Oil-in-water emulsion solvent evaporation method

In oil-in-water emulsion method, the drug along with the polymer is dissolved in a water-immiscible organic solvent such as chloroform, acetone, or a mixture of different organic solvents like chloroform and ethanol, and the prepared solution is added slowly into the distilled water with vigorous stirring to prepare an emulsion with an internal organic phase and continuous aqueous phase. Under these conditions, the polymer rearranges the polymer to form micelles. This emulsion is then kept open to air with stirring so as to evaporate all the organic solvents.[57–59]

Solid dispersion method

In this method, the drug along with the polymer is dissolved in the organic solvent, and a solid polymer matrix is obtained after the evaporation of solvent under reduced pressure.[60,61]

Microphase separation method

In this method the drug and polymer are dissolved (in an organic solvent) and the solution is added dropwise in water under magnetic stirring. Drug is entrapped in the inner core of the micelles. Organic solvent is removed under reduced pressure.[62] For more insights into the polymeric micelles with respect to drug release, characterization techniques, and drug targeting mechanism for polymeric micelles, authors are suggested to refer the excellent review done by Kedar et al.[15]

SYNTHESIS OF MAGNETIC NPS

Many different methods for the preparation of magnetic NPs (MNPs) have been reported.

In the last decade, increased investigations with several types of iron oxides have been carried out in the field of magnetic NPs including the Fe_3O_4 magnetite, $Fe^{II}Fe^{III}_2O_4$, Fe_2O_3, Fe_2O_3 and FeO, and Fe_2O_3 and Fe_2O_3.[15] The different methods for synthesizing iron oxide NPs include.

Co-precipitation Method

Co-precipitation technique is the most conventional and simple method for preparing iron oxide NPs. In this method, ferric and ferrous ions are taken in a 1:2 molar ratio in a basic solution. The size of the precipitated NPs depends on different factors such as temperature, pH, ratio of ferric and ferrous ions, and the type of salt used, etc.[15] It has been reported previously that for the synthesis of monodispersed and uniformly distributed, small-sized

Fe_3O_4 NPs using co-precipitation method, the reaction should be carried out in an aqueous solution with a molar ratio of $Fe^{II}/Fe^{III} = 0.5$ and a pH between 11 and 12.[63] Apart from this, many publications have described efficient routes to obtain the monodispersed NPs, using surfactants such as dextran or polyvinyl alcohol (PVA).[64,65]

Thermal Decomposition

The second route for synthesis of iron oxide NPs is the thermal decomposition method. In this method, iron complexes such as $Fe(cup)_3$ (cup = N-nitrosophenylhy-droxylamine), $Fe(acac)_3$ (acac = acetylacetonate), or $Fe(CO)_5$ is decomposed at high temperatures followed by oxidation for obtaining high-quality monodispersed iron oxide NPs. A general decomposition approach for the synthesis of small-sized monodispersed magnetic NPs based on high-temperature reactions of $Fe(acac)_3$ in phenyl ether in the presence of alcohol, oleic acid, and oleylamine has also been reported.[66] Direct decomposition of $Fe(Cup)_3$ and $Fe(CO)_5$ single precursors for the preparation of mono-dispersed Fe_2O_3 NPs has also been reported previously.[67,68] The only disadvantage of this method is that, the NPs formed can only be dispersed in nonpolar solvents.

Microemulsion

Microemulsion is a very useful method when it comes to the preparation of NPs of different shapes. In case of binary systems (water/surfactant or oil/surfactant), self-assembling leads to the formation of many different structures ranging, for example, from (inverted) spherical and cylindrical micelles to lamellar phases and bicontinuous microemul-sions.[69] We can prepare small-sized NPs with a narrow size range distribution and high saturation magnetization values.[70]

Hydrothermal Synthesis

The thermal decomposition and microemulsion method leads to a complicated process and require relatively high temperatures for the preparation of iron oxide NPs. Hydrothermal synthesis includes various technologies of crystallizing substances in a sealed container using a high-temperature aqueous solution at high vapour pressure. This technique yields NPs with a better crystallinity than those from other processes, so hydrothermal synthesis is prone to obtain the highly crystalline iron oxide NPs with a narrow size range.

Sonochemical Synthesis

As a competitive alternative, the sonochemical method has been extensively used to generate iron oxide NPs with unusual properties. The chemical effects of ultrasound arise from acoustic cavitation, that is, the formation, growth, and implosive collapse of bubbles in liquid which helps in the formation of small-sized NPs with a narrow size distribution.

THE MMS AS DIAGNOSTIC PLATFORM

Recently, researchers have addressed the issue of toxicity of MNPs and reported that superparamagnetic iron oxide NPs (SPIONs) were biocompatible with enhanced magnetic relaxivity, compared to the gadolinium (Gd) complex.[71] This report demonstrated the superiority of SPIONS as an MRI-contrasting agent. The SPIONs have been used previously as contrast agents for imaging.[71] This justifies the use of these SPIONs in the preparation of MMs, as it is suitable for both the purpose of providing magnetic property to micelles as well as being useful as a contrast agent for imaging purpose.

To date, different methods for loading of hydrophilic SPIONs have been reported in literature. Loading of SPI-ONs into the aqueous hollow interior of cross-linked vesicles fabricated from amphiphilic ABA triblock copolymers have been reported by physical or covalent attachment to the hydrophobic vesicle bilayer.[72–74] Loading of the SPI-ONs using electrostatic interactions between cationic-neutral block polyelectrolytes and negatively charged iron oxide NPs was also documented.[75]

The SPIONs are not very stable at normal physiological conditions because of their high hydrophobic nature and small size, so they need to be stabilized by coating with polymers.[76] Use of polymeric micelles for stabilizing SPIONs has been reported earlier.[17,19]

Pioneering research dealing with polymeric MMs was carried out by Kim et al.,[77] using amphiphilic poly (styrene$_{250}$-block-acrylic acid$_{13}$) components. Fabricated polymeric MMs were further studied for functionalization with organic or biological molecules using NHS (N-hydrox-ylsulfosuccinimide) and EDC (N-ethyl-N'-(3-dimethyl-aminopropyl)carbodiimide methiodide). They lead to the conclusion that MMs could maintain their structural integrity after functionalization and can be applied for biotech-nological protocols.

The SPIONs-conjugated micelles possess better water dispersibility and hence have more colloidal stability. Micelle-encapsulated clustered SPIONs show high T_2 relaxivity and show a better contrast effect for accurate MR imaging, hence acting as a suitable MRI-contrasting probe.[19] Ai and workers were the first to present the proof of using MMs for ultrasensitive MRI detection. They used amphiphilic diblock copolymer of poly(ε-caprolactone)-b-poly(ethylene glycol) (PCL-b-PEG) for the preparation of micelles. Clustering of SPIO particles inside the hydropho-bic core of micelles was responsible for the high SPIO loading, which ultimately increased the MRI sensitivity. They also proposed the use of these MMs for novel diagnostic and therapeutic applications. Lecommandoux and co-workers reported the preparation of MMs by mixing ferrofluid (α-Fe$_2$O$_3$) with PB$_{48}$-b-PGA$_{114}$ (polybutadiene-b-poly(glutamic acid)) or PB$_{48}$-b-PGA$_{145}$ di-block copolymers.[78] When mixed in equal ratio (1:1), a fully dispersed suspension was obtained. AFM images demonstrated high encapsulation efficiency of the MMs formed. In addition to polymeric MMs, the MMs formed from phospholipids have also been investigated. Furthermore, Ai et al. prepared SPIO-loaded DSPE-PEG$_{5000}$ (1, 2-diacyl-*sn*-glycero-3-phosphoethanolamine-*N*-[methoxy (polyethylene glycol)]-5000) lipid micelles. Immobilization of histidine-tagged proteins by MNPs encapsulated with nitrilotriacetic acid (NTA)–phospholipids micelle was reported, where highly monodispersed water-soluble MNPs were encapsulated in phospholipid micelles.[79] It was postulated that such phospholipid MMs can be used for conjugating protein and for bioconjugation of MNPs with various agents for the fabrication of bimodal (magnetic and optical) molecular imaging nanoprobes.

Recently, Lee et al.,[72] fabricated MMs by encapsulating SPION in biodegradable block polymer shells. He also highlighted the effect of molecular weight of block copolymer on the entrapment efficiency of micelles. For the fabrication of SPION-loaded micelle, PLA–PEG of 5–2 kDa MW was selected, as it has more entrapment capacity and forms stable dispersion. The SPIO-loaded micelles showed a higher relaxivity coefficient. The SPIO NPs are known to have a strong effect on T$_2$ relaxation; additionally, SPIO also possesses the advantage of slower kidney clearance and high detection sensitivity as compared to Gd-based contrast agents.

Mi and co-workers reported preparing Gd-DTPA–loaded polymer metal complex micelles using poly(ethyleneglycol)-b-poly{*N*-[*N*0-(2-aminoethyl)-2 aminoethyl] aspartamide} [PEG-b-PAsp(DET)] copolymers for MRI of cancerous cells. The Gd-DTPA–loaded polymeric micelles were having a hydrodynamic diameter of 45 nm. The prepared micelles were stable under physiological conditions, i.e., in 10 mM PBS (pH 7.4) with 150 mM NaCl at 37°C. Nontoxic property of Gd-DTPA/m provides a potential option for *in vivo* imaging. The size, neutral surface charge, stability, and release characteristics of Gd-DTPA/m may contribute to the high penetration and accumulation of Gd-DTPA at the tumor site, which will help in the accumulation of micelles in tumor. The higher tumor accumulation and higher relaxivity of Gd-DTPA/m will enhance the contrast of tumor cells (C-26 tumors) in T$_1$-weighted MR images.

THE MMS AS THERAPEUTIC PLATFORM

Polymeric micelles have been reported previously for their use in drug delivery.[80,81] Preparation of SPION-encapsulated polymeric immunomicelles has also been reported earlier.[76] The immunomicelles were fabricated using polyethylene glycol phosphatidylethanolamine (PEG–PE) and monoclonal antibody 2C$_5$. In this study, T$_2$ relaxation states were higher than that of plain SPION. Epifluorescence microscopy revealed high concentration of SPION in human breast cancer MCF-7 cells, which is important for MRI contrast as well as for drug delivery to tumor cell and hyperthermia therapy.

Boekhorst-te and co-workers worked on similar lines and developed receptor-targeted double-labeled (MR and fluorescent) magnetic immunomicelles.[82] Specific CB$_2$-R (cannabinoid receptor) agonists or antibodies directed to 24p$_3$ were incorporated into di-oleoyl-polyethylene glycol-phosphatidylethanolamine 1000 (DOPE-PEG$_{1000}$) micelles or di-stearoyl-polyethylene glycol-phosphatidylethanolamine 2000 (DSPE-PEG$_{2000}$) micelles to prepare immunomicelles (Fig. 2). That study revealed promising results

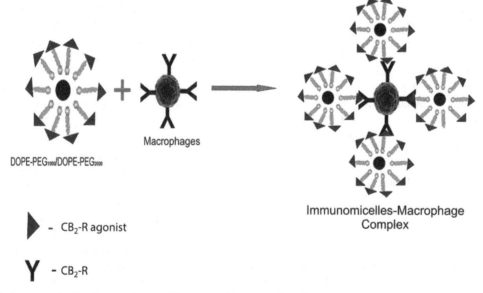

Fig. 2 Magnetic immunomicelles for receptor-specific magnetofluorescent imaging.

for use of these micelles for *in vivo* MRI and florescence imaging.

Bioactivated fluorescent magnetic hybrid micelles were obtained by coencapsulation of hydrophobic CdSe/ZnS quantum dots (QDs) and γ-Fe₂O₃ nanocrystals together in an amphiphilic gallate polymer.[83] These hybrid amphiphilic gallate micelles can be used further for the attachment of proteins, drugs, or any other suitable biomolecules. Biodegradable mPEG*b*-p(HPMAm-Lac2)(poly(ethyleneglycol)-*b*-poly[*N*-(2 hydroxypropyl) methacrylamide dilactate]) was used for the preparation of thermosensitive polymeric MMs by encapsulating hydrophobic oleic acid–coated SPIONs.[84] Sensitivity and stability of the micelles formed were studied in physiological conditions. Loading capacity to about 40% was obtained. The highest concentration of solubilized SPIONs obtained was 0.5 mg/mL, which was further increased by means of magnetic concentration up to 5 mg/mL.

Nikles et al.[85] worked on building micelles for magnetically triggered drug delivery using poly(ethylene glycol)-b-polycaprolactone diblock copolymers (MeO-(EG)₃₆-(CL)₁₆-OH). The drug release capacity of the micelles formed was studied using pyrene (0.010 mM), a fluorescent probe, which was added to an aqueous solution of MeO-(EG)₃₆-(CL)₁₆-OH. They proposed encapsulation of MNPs into the core for determining the interference of NPs with the crystallizing activity of the core.

Park and co-workers reported the synthesis of MMs hybrid NPs.[86] Thermosensitive MMs were also reported by researchers.[87] Poly(*N*-isopropylacrylamide)-block-polylactide (PNIPAAm-b-PLA) copolymers (PNIPAAm-b-PLA) were dialyzed with SPIONs to get composite MMs. They proposed the use of thermosensitive, biodegradable MMs formed in drug delivery.

Yang et al.[88] and Huang et al.[29] fabricated MMs for dual drug-targeted delivery for cancer therapy. Yang et al.

used diblock copolymers of poly(ethylene glycol) (PEG) and poly(3-caprolactone) (PCL) bearing a tumor-targeting ligand, folate (FA) for preparation of micelles. The SPIONs along with doxorubicin (DOX) were encapsulated in these micelles. Attachment of FA group in micelles was done to increase recognition of micelles by tumor cells overexpressing FA receptors. These modifications are important for increasing the efficacy of the delivery system and reducing the drug-related toxicity in case of chemotherapeutic agents by localizing the drug concentration to the site of administration.[11,12,89] Huang also worked on similar platform as that of Yang. Biocompatible pluronic F127 and poly(DL-lactic acid) (F127-PLA) copolymers were conjugated to folic acid for the preparation of micelles. DOX and SPIONs were encapsulated in the micelles for drug targeting. The DOX-loaded MMs showed good *in vitro* and *in vivo* results.

Liao reported the fabrication of DOX-loaded anti-EGFR monoclonal antibody–conjugated magnetic immunomicelles,[90] suggesting that this type of delivery system may be helpful in developing a delivery system for efficient targeting and chemotherapy.

THE MMS AS THERANOSTIC PLATFORM

Park and co-workers provided the first examples of simultaneous targeted drug delivery and dual-mode near-infrared fluorescence imaging and MRI of diseased tissue *in vitro* and *in vivo* using MMs hybrid NPs.[86] They reported the synthesis of micellar hybrid NPs for magnetofluorescent imaging using oleic acid–coated MNPs, synthesized and encapsulated in PEG-modified phospholipids along with trioctylphosphine-coated QDs (Fig. 3). These hybrid micelles were useful for both fluorescence imaging as well

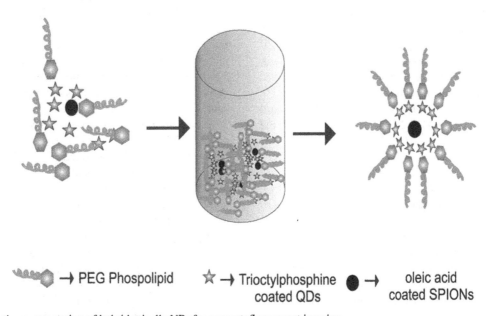

→ PEG Phospolipid   ☆ → Trioctylphosphine coated QDs   ● → oleic acid coated SPIONs

Fig. 3 Schematic representation of hybrid micelle NPs for magnetofluorescent imaging.

as MRI. Use of PEG-modified phospholipids can be replaced with antibody, proteins, or other suitable agents, which can further modify the targeting efficiency of hybrid micelles and would help in developing magnetic immunomicelles.

Hydrotropic MMs were reported recently for their use in MRI along with paclitaxel with higher T_2 relaxivity than HM-SPION alone.[28] It was observed that after systemic administration, HM-SPION-PTX accumulated in tumor cells and allowed the easy detection of tumor cells by MRI. This occurs due to clustering of SPIONs inside micelles. This study also highlights the efficient use of SPION-loaded micelles along with drugs for targeting tumor cells. As mentioned above, the surface modification of MMs, as in the case of immunomicelles can further improve the efficacy of such systems in drug delivery.

The MMs were also explored for their use as gene delivery vehicles in addition to MRI.[27] Gene delivery was achieved by chitosan and polyethyleneimine (PEI)-coated MMs (CP–mag-micelles) by loading monodispersed SPIONs in the core of poly(DL-lactide) (PLA) and monomethoxy polyethylene glycol (mPEG) diblock copolymer. These SPION-loaded mPEG–PLA micelles were coated with chitosan and PEI to form coated MMs. The potential of coated MMs for delivering DNA/gene was tested by performing DNA-binding assays and transfection studies in cell lines such as HEK293, 3T3, and PC3 cells *in vitro*, MR phantom imaging, biodistribution, *in vivo* gene expression, and WST assay. The results revealed that these micelles can be used for gene delivery with simultaneous *in vivo* imaging.

The MMs prepared by self-assembly of fluorine-containing amphiphilic poly(HFMA-g-PEGMA) copolymers was reported recently.[91] Oleic acid–coated SPIONs were encapsulated by poly(HFMA-g-PEGMA) copolymers to form biocompatible MMs with high aqueous stability for MRI of liver (due to high T_2 relaxivity) and spleen (due to good contrast effects). Moreover, the MMs formed were assessed for drug delivery (5-fluorouracil (5-FU)). Hu and co-workers fabricated drug-loaded, SPION–embedded, inorganic/organic, hybrid block copolymer micelles for magnetic imaging and delivery of chemotherapeutic agents.[92] They used PCL-b-PGMA and PCL-b-P(OEGMA-co-FA) diblock copolymers for the fabrication of micelles. The MMs formed showed enhanced T_2 relaxivity due to clustering of SPIO NPs within the micellar coronas.

Oleic acid–coated MNPs were encapsulated in PEG-modified phospholipids along with trioctylphosphine-coated QDs and DOX. The intrinsic fluorescence of DOX enabled the independent imaging of both DOX and QDs encapsulated together in hybrid micelles.

The SPIONs as magnetic contrast agent and DOX as chemotherapeutic agent was used by Yang et al. to fabricate worm-like polymeric vesicles for targeted cancer therapy and MR imaging.[73] Polymer vesicles were formed by

hetero bi-functional amphiphilic triblock copolymers R (R¼ methoxy or FA)-PEG114-PLAx-PEG$_{46}$-acrylate using a double-emulsion method. The SPIONs were attached by crosslinking inner PEG layers of polymeric vesicles. Rapid drug release was observed after the vesicles penetrated into the cells. During drug release study, it was observed that release of DOX from the non-crosslinked polymeric vesicles was faster than that from the crosslinked polymeric vesicles. This might be due to the fact that crosslinking the inner hydrophilic PEG layer helps to maintain the structural integrity of polymer vesicles, hence resulting in a lower drug release rate.

Yang prepared pH-responsive polymer vesicles using hetero functional amphiphilic triblock copolymers, that is, R (FA or methoxy)-poly(ethylene glycol)(MW: 5000)-poly(glutamate hydrozone DOX)-poly(ethylene glycol) acrylate and SPIONs. The SPIONs were encapsulated in aqueous core of vesicles.[74] The high SPIO concentration and the formation of SPIO NP cluster inside the aqueous cores of the polymeric vesicles increased the r_2 relaxivity of the SPIO/DOX-loaded vesicles) and thus the MRI sensitivity. Based on a similar approach, hydrophobically modified ultrasmall γ-Fe$_2$O$_3$ NPs were encapsulated in poly(trimethylene carbonate)-b-poly(L-glutamic acid) (PTMC-b-PGA) block copolymer vesicles by nanoprecipitation method.[93] Loading content of MNPs was observed up to 70% by weight. The release of drug was dependent on the RF oscillating magnetic field. This approach could serve as a platform for imaging and therapy with the help of MMs and RF oscillating magnetic field.

CONCLUSIONS AND FUTURE PROSPECTIVES

In the present entry, we have tried to focus on the recent advancement in MMs and their use in drug delivery and imaging. Polymeric nanomaterials have always gained the attention of researchers, since its discovery. New modification of these nanomaterials has been tried recently, with the intention of increasing the functional capabilities of these NPs. In the recent years, promising investigations are reported about polymeric NPs, proving it as a potential candidate for theranostic purpose.[94] Polymeric micelles have been used in the past for increasing the solubility of poorly soluble drugs[95–97] and in other DDSs for therapeutic application.[80,98,99] Introduction of MMs has opened a new era of research concerned with simultaneous diagnostic and therapeutic targeting. These vesicles are unique in nature due to the properties of SPIONs along with its membrane-bound structure, which can serve as a cargo for multiple applications. Polymeric vesicles have been used before for drug delivery, but the introduction of magnetically active vesicles will make the delivery more efficient. Covalent and non covalent chemistry have been developed to modify the surface of MMs.[77] These MMs possess the capability of diagnosing the delivered NPs, which may be inorganic, organic, or biologic in nature. Further modifications with sensing and introductory groups such as ligands, proteins, fluorescent molecules, monoclonal antibodies, etc. to modify these MMs into theranostic nanostructures could prove beneficial. Conversion of MMs into magnetically active immunomicelles has been reported previously.[76] Slight modification in these nanostructures may be used for the delivery of therapeutics such as siRNA, genes, immunoglobulins, and related materials. This could provide a major breakthrough in the field of targeted theranostic delivery systems. These NPs can be used for sensory purposes in conjugation with other materials such as carbon nanotube and graphene as in the case of chitosan NPs or gold NPs. Immobilization of enzymes or chemical catalysts can be done, thus widening its use in biotechnology and chemical synthesis. Enzyme immobilization can be used for the purpose of modulating a specific reaction at specific sites such as in the case of dopamine deficiency or inhibition of gamma-aminobutyric acid. Chemical or protein immobilization like calcium or thrombin may be useful for blood coagulation in addition to delivering the drug in trauma patient to name a few.

In addition, the stability and multifunctional capability of these micelles can be integrated with microfluidics, which can lead to the development of a miniaturized and high-throughput theranostic platform, which can be used for more target-specific point diagnosis such as specific cells of central nervous system, DNA, and delivery of ssDNA, siRNA, etc. which could prove beneficial for raising the approaches for healthcare. Attachment of antibodies, affibodies, and nanobodies on MMs can serve the purpose of imaging such as MRI. Recently, antibodies and iron oxide NPs have been reported individually to be used in MR imaging of cancer. Conjugation of these antibodies and SPIO NPs can be done on MMs to increase the T_2 relaxivity[19] and enhance the contrasting property in MRI along with increased cell-specific binding of these micelles due to antibodies-mediated targeting of tumor cells. Several studies have been done but still much research is required to establish the potential of these magnetic particles for drug delivery and Imaging. Although not discussed in the entry, the different biomaterials used for fabrication of MMs may also play an important role in its properties and use. It might be possible to extend the use of these MMs for imaging techniques other than MRI and in drug delivery, in addition to cancer therapy. But before that, a lot of research will have to be done in relation to its fabrication and surface modification for truly calling it as an emerging theranostic platform.

ACKNOWLEDGMENTS

The authors are thankful to Dr. Avinash R. Tekade, Professor, Department of Pharmaceutics, Rajarshi Shahu College of Pharmacy and Research, Tathawade, Pune, India for providing valuable suggestions. The authors are also grateful to Mr. Mahendra D. Patil for his meticulous help in typesetting.

REFERENCES

1. Sanvicens, N.; Marco, M.P. Multifunctional nanoparticles—Properties and prospects for their use in human medicine. Trends Biotechnol. **2008**, *26* (8), 425–433.
2. Farokhzad, O.C.; Langer, R. Nanomedicine: Developing smarter therapeutic and diagnostic modalities. Adv. Drug. Deliver. Rev, **2006**, *58* (14), 1456–1459.
3. Torchilin, V.P. Multifunctional nanocarriers. Adv. Drug Deliver. Rev. **2006**, *58* (14), 1532–1555.
4. Torchilin, V.P. Multifunctional and stimuli-sensitive pharmaceutical nanocarriers. Eur. J. Pharm. Biopharm. **2009**, *71* (3), 431–444.
5. Sawant, R.R.; Torchilin, V.P. Multifunctional nanocarriers and intracellular drug delivery. Curr. Opin. Solid. State Mater. **2012**, *16* (6), 269–275.
6. De, M.; Chou, S.S.; Joshi, H.M.; Dravid, V.P. Hybrid magnetic nanostructures (MNS) for magnetic resonance imaging applications. Adv. Drug Deliver. Rev. **2011**, 63 (14–15), 1282–1299.
7. Chacko, A.M.; Hood, E.D.; Zern, B.J.; Muzykantov, V.R. Targeted nanocarriers for imaging and therapy of vascular inflammation. Curr. Opin. Colloid Interface Sci. **2011**, *16* (3), 215–227.
8. Ganta, S.; Devalapally, H.; Shahiwala, A.; Amiji, A. A review of stimuli-responsive nanocarriers for drug and gene delivery. J. Control. Release **2008**, *126* (3), 187-204.
9. Karathanasis, K.; Chan, L.; Balusu, S.R.; D'Orsi, C.J.; Annapragada, A.V.; Sechopoulos, L.; Bellamkonda, R.V. Multifunctional nanocarriers for mammographic quantification of tumor dosing and prognosis of breast cancer therapy. Biomaterials **2008**, *29* (36), 4815–4822
10. Gunaseelan, S.; Gunaseelan, K.; Deshmukh, M.; Zhang, X.; Sinko, P.J. Surface modifications of nanocarriers for effective intracellular delivery of anti-HIV drugs. Adv. Drug. Deliver. Rev **2010**, 62 (4–5), 518–531.
11. Huang, G.; Zhang, N.; Bi, X.; Dou, M. Solid lipid nanoparticles of temozolomide: Potential reduction of cardial and nephric toxicity. Int. J. Pharm. **2008**, *355* (1–2), 314–320.
12. Liu, Q.; Li, R.; Zhu, Z.; Qian, X.; Guan, W.; Yu, L.; Yang, M.; Jiang, X.; Liu, B. Enhanced antitumor efficacy, biodistribution and penetration of docetaxel-loaded biodegradable nanoparticles. Int. J. Pharm. **2012**, *430* (1–2), 350–358.
13. Krovvidi, K.R.; Stroeve, P. A theoretical analysis of micelle-mediated transport through porous membranes. J. Colloid Interface Sci. **1986**, *110* (2), 437–445.
14. Yokoyama, M.; Miyauchi, M.; Yamada, N.; Okano, T.; Sakurai, Y. Polymer micelles as novel drug carrier: Adriamycin-conjugated poly(ethylene glycol)-poly(aspartic acid) block copolymer. J. Control. Release **1990**, *11* (1–3), 269–278.
15. Kedar, U.; Phutane, P.; Shidhaye, S.; Kadam, V. Advances in polymeric micelles for drug delivery and tumor targeting. Nanomed. **2010**, *6* (6), 714–729.
16. Nishiyama, N.; Kataoka, K. Current state, achievements, and future prospects of polymeric micelles as nanocarriers

for drug and gene delivery. Pharmacol. Therapeut. **2006**, *112* (3), 630–648.

17. Nasongkla, N.; Bey, E.; Ren, J.; Ai, H.; Khemtong, C.; Guthi, J.S.; Chin, S.F.; Sherry, A.D.; Boothman, D.A.; Gao, J. Multifunctional polymeric micelles as cancer-targeted, MRI ultrasensitive drug delivery systems. Nano Lett. **2006**, *6* (11), 2427–2430.

18. Mi, P.; Cabral, H.; Kokuryo, D.; Rafi, M.; Terada, Y.; Aoki, I.; Saga, T.; Takehiko, I.; Nishiyama, N.; Kataoka, K. Gd-DTPA-loaded polymer-metal complex micelles with high relaxivity for MR cancer imaging. Biomaterials **2012**, *34* (2), 492–500.

19. Ai, H.; Flask, C.; Weinberg, B.; Shuai, X.; Pagel, M.D.; Farell, D.; Duerk, J.; Gao, J. Magnetite loaded polymeric micelle as ultrasensitive magnetic resonance probe. Adv. Mater. **2005**, *17* (16), 1949–1952.

20. Yokoyama, M.; Okano, T.; Sakurai, Y.; Suwa, S.; Kataoka, K. Introduction of cisplatin into polymeric micelles. J. Control. Release **1996**, *39* (2–3), 351–356.

21. Yokoyama, M.; Miyauchi, M.; Yamada, N.; Okano, T.; Sakurai, Y.; Kataoka, K. Characterization and anticancer activity of the micelle-forming polymeric anticancer drug adriamycin-conjugated poly(ethylene glycol)-poly(aspartic acid) block copolymer. Cancer Res. **1990**, *50* (6), 1693–1700.

22. Kataoka, K.; Togawa, H.; Harada, A.; Yasugi, K.; Matsumoto, T.; Katayose, S. Spontaneous formation of polyion complex micelles with narrow distribution from antisense oligonucleotide and cationic block copolymer in physiological saline. Macromolecules **1996**, *29* (26), 8556–8557.

23. Feldman, A.S.; McDougal, W.S.; Harisinghani, M.G. The potential of NPs-enhanced imaging. Urol. Oncol. **2008**, *26* (1), 65–73.

24. Jain, T.K.; Richey, J.; Strand, M.; Leslie-Pelecky, D.L.; Flask, C.A.; Labhasetwar, A. Magnetic nanoparticles with dual functional properties: Drug delivery and magnetic resonance imaging. Biomaterials **2008**, *29* (29), 4012–4021.

25. Tournier, H.; Hyacinthe, R.; Schneider, M. Gadolinium-containing mixed micelle formulations: A new class of blood pool MRI/MRA contrast agents. Acad. Radiol. **2002**, *9* (Suppl. 1), S20–S28.

26. Chandrasekharan, P.; Maity, D.; Yong, C.X.; Chuang, K.H.; Ding, J.; Feng, S.S. Vitamin E (D-alpha-tocopheryl-co-poly(ethylene glycol) 1000 succinate) micelles-superparamagnetic iron oxide nanoparticles for enhanced thermotherapy and MRI. Biomaterials **2011**, *32* (1), 5663–5672.

27. Wang, C.; Ravi, S.; Martinez, G.V.; Chinnasamy, V.; Raulji, P.; Howell, M.; Davis, Y.; Mallela, J.; Seehra, M.S.; Mohapatra, S. Dual-purpose magnetic micelles for MRI and gene delivery. J. Control. Release **2012**, *163* (1), 82–92.

28. Yoon, Y.; Saravanakumar, G.; Heo, R.; Choi, S.H.; Song, I.C.; Han, M.H.; Kim, K.; Park, J.H.; Choi, K.; Kwon, I.C.; Park, K. Hydrotropic magnetic micelles for combined magnetic resonance imaging and cancer therapy. J. Control. Release **2012**, *160* (3), 692–698.

29. Huang, C.; Tang, Z.; Zhou, Y.; Zhou, X.; Jin, Y.; Li, D.; Yang, Y.; Zhou, S. Magnetic micelles as a potential platform for dual targeted drug delivery in cancer therapy. Int. J. Pharm. **2012**, *429* (1–2), 113–122.

30. Fernandez, A.M.; Van, D.K.; Dasnois, L.; Lebtahi, K.; Dubois, V.; Lobl, T.J. N-succinyl-(β-alanyl-L-leucyl-L-alanyl-L-leucyl) doxorubicin: An extracellularly tumor-activated prodrug devoid of intravenous acute toxicity. J. Med. Chem. **2001**, *44* (22), 3750–3753.

31. Savic, R.; Eisenberg, A.; Maysinger, D. Block copolymer micelles as delivery vehicles of hydrophobic drugs: Micelle-cell interactions. J. Drug Target **2006**, *14* (6), 343–355.

32. Senior, J.H. Fate and behaviour of liposomes *in vivo*: A review of controlling factors. Crit. Rev. Ther. Drug Carrier Syst. **1987**, *3* (2), 123–193.

33. Stolnik, S.; Illum, L.; Davis, S.S. Long circulating microparticulate drug carriers. Adv. Drug Deliv. Rev. **1995**, *16* (2–3), 195–214.

34. Mosqueria, V.C.F.; Legrand, P.; Gulik, A.; Bourdon, O.; Gref, R.; Labarre, D. Relationship between complement activation, cellular uptake and surface physicochemical aspects of novel PEG-modified nanoparticles. Biomaterials **2001**, *22* (22), 2967–2979.

35. Yamamoto, Y.; Nagasaki, Y.; Kato, Y.; Sugiyama, Y.; Kataoka, K. Long-circulating poly(ethylene glycol)-poly(D,L-lactide) block copolymer micelles with modulated surface charge. J. Control. Release **2001**, *77* (1), 27–38.

36. Allen, T.M.; Hansen, C. Pharmacokinetics of stealth versus conventional liposomes: Effect of dose. Biochim. Biophys. Acta. **1991**, *1068* (2), 133–141.

37. Woodle, M.C.; Matthay, K.K.; Newman, M.S.; Hidayat, J.E.; Collins, L.R.; Redemann, C. Versatility in lipid composition showing prolonged circulation with sterically stabilized liposome. Biochim. Biophys. Acta **1992**, *1105* (2), 193–200.

38. Maeda, H. SMANCS and polymer-conjugated macromolecular drugs: Advantages in cancer chemotherapy. Adv. Drug Deliv. Rev. **2001**, *46* (1), 169–185.

39. Matsumura, Y.; Maeda, H. A new concept for macromolecular therapeutics in cancer chemotherapy: Mechanism of tumoritropic accumulation of proteins and the antitumor agent Smancs. Cancer Res. **1986**, *46* (12 Pt. 1), 6387–6392.

40. Kwon, G.S.; Suwa, S.; Yokoyama, M.; Okano, T.; Sakurai, Y.; Kataoka, K. Enhanced tumor accumulation and prolonged circulation times of micelle-forming poly(ethylene oxide-aspartate) block copolymer-Adriamycin conjugate. J. Control. Release **1994**, *29* (1), 17–23.

41. Yokoyama, M.; Okano, T.; Sakurai, Y.; Fukushima, S.; Okamoto, K.; Kataoka, K. Selective delivery of adriamycin to a solid tumor using a polymeric micelle carrier system. J. Drug Target **1999**, *7* (3), 171–186.

42. Nishiyama, N.; Okazaki, S.; Cabral, H.; Miyamoto, M.; Kato, Y.; Sugiyama, Y. Novel cisplatin-incorporated polymeric micelles can eradicate solid tumors in mice. Cancer Res. **2003**, *63* (24), 8977–8983.

43. Hamaguchi, T.; Mastumura, Y.; Suzuki, M.; Shimizu, K.; Goda, R.; Nakamura, I. NK105, a paclitaxel-incorporating micellar NPs formulation, can extend *in vivo* antitumor activity and reduce the neurotoxicity of paclitaxel. Br. J. Cancer **2005**, *92* (7), 1240–1246.

44. Anraku, Y.; Kishimura, A.; Oba, M.; Yamasaki, Y.; Kataoka, K. Spontaneous formation of nanosized unilamellar polyion complex vesicles with tunable size and properties. J. Am. Chem. Soc. **2010**, *132* (5), 1631.

45. Kano, M.R.; Bae, Y.; Iwata, C.; Morishita, Y.; Yashiro, M.; Oka, M.; Fuhii, T.; Komuro, A.; Kiyono, K.; Kaminishi, M.; Hirakawa, K.; Ouchi, Y.; Nishiyama, N.; Kataoka, K.; Miyanozo, K. Improvement of cancer-targeting therapy, using

nanocarriers for intractable solid tumors by inhibition of TGF-β signaling. Natl. Acad. Sci. USA **2007**, *104* (9), 3460.

46. Saito, Y.; Yasunaga, M.; Kuroda, J.; Koga, Y.; Matsumura, Y. Enhanced distribution of NK012, a polymeric micelle-encapsulated SN-38, and sustained release of SN-38 within tumors can beat a hypovascular tumor. Cancer Sci. **2008**, *99*, 1258.

47. Nishiyama, N.; Kataoka, K. Preparation and characterization of size-controlled polymeric micelle containing cis-dichlorodiammineplatinum(II) in the core. J. Control. Release **2001**, *74*, 83–94.

48. Kaida, S.; Cabral, H.; Kumagai, M.; Kishimura, A.; Terada, Y.; Sekino, M.; Aoki, I.; Nishiyama, N., Tani, T.; Kataoka, K. Visible drug delivery by supramolecular nanocarriers directing to single-platformed diagnosis and therapy of pancreatic tumor model. Cancer Res. **2010**, *70* (18), 7031–7041.

49. Mukerjee, A.; Ranjan, A.P.; Vishwanatha, J.K. Combinatorial nanoparticles for cancer diagnosis and therapy. Curr Med Chem. **2012**, *19* (22), 3714–3721.

50. Stepanek, M.; Podhajecka, K.; Tesarova, E.; Prochazka, K. Hybrid polymeric micelles with hydrophobic cores and mixed polyelectrolyte/nonelectrolyte shells in aqueous media. I. Preparation and basic characterization. Langmuir **2001**, *17* (14), 4240–4244.

51. Li, Y.; Kwon, G.S. Methotrexate esters of poly(ethylene oxide)-blockpoly(2-hydroxyethyl-L-aspartamide). I. Effects of the level of methotrexate conjugation on the stability of micelles and on drug release. Pharm. Res. **2000**, *17* (5), 607–611.

52. Torchilin, V.P. Structure and design of polymeric surfactant based drug delivery systems. J. Control. Release **2001**, *73* (2–3), 137–172.

53. Klok, H.A.; Vandermeulen, M.; Guido, W.; Rosler, A. Advanced drug delivery devices via self-assembly of amphiphilic block copolymers. Adv. Drug Del. Rev. **2001**, *53* (1), 95–108.

54. Kozlov, M.Y.; Melik-Nubarov, N.S.; Batrakova, E.V.; Kabanov, A.V. Relationship between Pluronic block copolymer structure, critical micellization concentration and partitioning coefficients of low molecular mass solutes. Macromolecules **2000**, *33* (9), 3305–3313.

55. Kim, J.H.; Emoto, K.; Iijima, M.; Nagasaki, Y.; Aoyagi, T.; Okano, T.Y. Core-stabilized polymeric micelle as potential drug carrier: Increased solubilization of taxol. Polym. Adv. Technol. **1999**, *10* (11), 647–654.

56. Butsele, K.V.; Sibreta, P.; Fustin, C.A.; Gohyb, J.F.; Passirani, C.; Benoitc, J.P. Synthesis and pH-dependent micellization of diblock copolymer mixtures. J Colloid Interface Sci. **2009**, *329* (2), 235–243.

57. Patil, Y.B.; Toti, U.S.; Khdair, A.; Linan, M.; Panyam, J. Single-step surface functionalization of polymeric nanoparticles for targeted drug delivery. Biomaterials **2009**, *30* (5), 859–866.

58. Zhang, J.; Jiang, W.; Zhao, X.; Wang, Y. Preparation and characterization of polymeric micelles from poly(D,L-lactide) and methoxypolyethylene glycol block copolymers as potential drug carriers. Tsinghua Sci. Technol. **2007**, *12* (4), 493–496.

59. Hongyu, C.; Na, L.; Xing, W.; Zhen, J.; Zhiming, C. Morphology and *in vitro* release kinetics of drug-loaded

micelles based on well-defined PMPCb-PBMA copolymer. Int. J. Pharm. **2009**, *371* (1–2), 190–196.

60. Zhang, X.; Jackson, J.K.; Burt, H.M. Development of amphiphilic diblock copolymers as micellar carriers of taxol. Int. J. Pharm. **1996**, *132* (1–2), 195–206.

61. Taillefer, J.; Jones, M.C.; Brasseur, N.; Van-Lier, J.E.; Leroux, J.C. Preparation and characterization of pH-responsive polymeric micelles for the delivery of photosensitizing anticancer drugs. J. Pharm. Sci. **2000**, *89* (1), 52–62.

62. Zhiang, J.; Wu, M.; Yang, J.; Wu, Q.; Jin, Z. Anionic poly(lactic acid)-polyurethane micelles as potential biodegradable drug delivery carriers. Colloids Surfaces A **2009**, *337* (1–3), 200–204.

63. Cornell, R.M.; Schwertmann, U. *The Iron Oxides: Structures, Properties, Reactions, Occurences and Uses*; Wiley-VCH: Weinheim, 2003.

64. Kang, Y.S.; Risbud, S.; Rabolt, J.F.; Stroeve, P. Synthesis and characterization of nanometer-size Fe_3O_4 and γ-Fe_2O_3 particles. Chem. Mater. **1996**, 8 (9), 2209–2211.

65. Novakova, A.A.; Lanchinskaya, V.Y.; Volkov, A.V.; Gendler, T.S.; Kiseleva, T.Y.; Moskvina M.A. Magnetic properties of polymer nanocomposites containing iron oxide nanoparticles. J. Magn. Magn. Mater. **2003**, *258–259*, 354–357.

66. Lee, J.; Isobe, T.; Senna, M. Magnetic properties of ultrafine magnetite particles and their slurries prepared via in-situ precipitation. Colloids Surf. A Physicochem. Eng. Asp. **1996**, *109* (1), 121–127.

67. Sun, S.; Zeng, H.J. Size-controlled synthesis of magnetite nanoparticles. Am. Chem. Soc. **2002**, *124* (28), 8204–8205.

68. Rockenberger, J.; Scher, E.C.; Alivisatos, P.A.J. A new non-hydrolytic single-precursor approach to surfactant-capped nanocrystals of transition metal oxides. Am. Chem. Soc. **1999**, *121* (49), 11595–11596.

69. Woo, K.; Hong, J.; Choi, S.; Lee, H.; Ahn, J.; Kim C.S. Easy synthesis and magnetic properties of iron oxide nanoparticles. Chem. Mater. **2004**, *16* (51), 2814–2818.

70. Hyeon, T.; Lee, S.S.; Park, J.; Chung, Y.; Na, H.B. Synthesis of highly crystalline and monodisperse maghemite nanocrystallites without a size-selection process. J. Am. Chem. Soc. **2001**, *123* (51), 12798–12801.

71. Veiseh, O.; Gunn, J.W.; Zhang, M. Design and fabrication of magnetic nanoparticles for targeted drug delivery and imaging. Adv. Drug. Deliver. Rev. **2010**, *62* (3), 284–304.

72. Lee, E.S.; Lim, C.; Song, H.T.; Yun, J.M.; Lee, K.S.; Lee, B.J.; Youn, Y.S.; Oh, Y.T.; Oh, K.T.A. Nanosized delivery system of superparamagnetic iron oxide for tumor MR imaging. Int. J. Pharm. **2012**, *439* (1–2), 342–348.

73. Yang, X.; Grailer, J.J.; Rowland, I.J.; Javadi, A.; Hurley, S.A.; Steeber, D.A.; Gong, S. Multifunctional SPIO/DOX-loaded wormlike polymer vesicles for cancer therapy and MR imaging. Biomaterials **2010a**, *31* (34), 9065–9073.

74. Yang, X.; Grailer, J.J.; Rowland, I.J.; Javadi, A.; Matson, V.Z.; Steeber, D.A; Gong, S. Multifunctional stable and ph responsive polymer vesicles formed by heterofunctional triblock copolymer for targeted anticancer drug delivery and ultrasensitive MR imaging. ACS Nano. **2010b**, *4* (11), 6805–6817.

75. Berret, J.F.; Schonbeck, H.; Gazeau, F.; Kharrat, D.E.; Sandre, O.; Vacher, A.; Airiau, M. Controlled clustering of superparamagnetic nanoparticles using block copolymers:

In Vitro–Medical

Design of new contrast agents for magnetic resonance imaging. J. Am. Chem. Soc. **2006**, *128* (5), 1755–1761.

76. Sawant, R.M.; Sawant, R.R.; Gultepe, E.; Nagesha, D.; Papahadjopoulos, B.; Sridhar, S.; Torchilin, V. P. Nanosized cancer cell-targeted polymeric immunomicelles loaded with superparamagnetic iron oxide nanoparticles. J. Nanopart. Res. **2009**, *11* (7), 1777–1785.

77. Kim, B.S.; Qiu, J.M.; Wang, J.P.; Taton, T.A. Magnetomicelles: Composite nanostructures from magnetic nanoparticles and cross-linked amphiphilic block copolymers. Nano. Lett. **2005**, *5* (10), 1987–1991.

78. Lecommandoux, S.; Sandre, O.; Checot, F.; Hernandez, J.R.; Perzynski, R. Self-assemblies of magnetic nanoparticles and di-block copolymers: Magnetic micelles and vesicles. J. Magn. Magn. Mater. **2006**, *300* (1), 71–74.

79. Lim, Y.T.; Lee, K.Y.; Lee, K.; Chung, B.H. Immobilization of histidine-tagged proteins by magnetic nanoparticles encapsulated with nitrilotriacetic acid (NTA)-phospholipids micelle. Biochem. Bioph. Res. Commun. **2006**, *344* (3), 926–930.

80. Miyata, K.; Christie, J.R.; Kataoka, K. Polymeric micelles for nano-scale drug delivery. React. Funct. Polym. **2011**, *71* (3), 227–234.

81. Gong, J.; Chen, M.; Zheng, Y.; Wang, S.; Wang, Y. Polymeric micelles drug delivery system in oncology. J. Control. Release **2012**, *159* (3), 312–323.

82. Boekhorst-te, B.C.M.; Bovens, S.M.; Feo, J.R.; Sanders, H.M.H.F.; van-de-Kolk, C.W.A.; De-Kroon, A.I.P.M.; Cramer, M.J.M.; Doevendans, P.A.F.M.; Hove, M.; Pasterkamp.; Echteld, C.J.A.V. Characterization and *in vitro* and *in vivo* testing of cb2-receptor- and ngal-targeted paramagnetic micelles for molecular MRI of vulnerable atherosclerotic plaque. Mol. Imaging Biol. **2010**, *12* (6), 635–651.

83. Roullier, V.; Grasset, F.; Boulmedais, F.; Artzner, F.; Cador, O.; Artzner, V.M. Small bioactivated magnetic quantum dot micelles. Chem. Mater. **2008**, *20* (21), 6657–6665.

84. Talelli, M.; Rijcken, C.J.F.; Lammers, T.; Seevinck, P.R.; Storm, G.; Nostrum, C.F.V.; Hennink, W.E. Superparamagnetic iron oxide nanoparticles encapsulated in biodegradable thermosensitive polymeric micelles: Toward a targeted nanomedicine suitable for image-guided drug delivery. Langmuir **2009**, *25* (4), 2060–2067.

85. Nikles, S.M.; Nikles, J.A.; Hudson, J.S.; Nikles, D.E. Diblock copolymers for magnetically triggered drug delivery systems. Joshua **2010**, *7* (1), 35–39.

86. Park, J.H.; Maltzahn, G.V.; Ruoslahti, E.; Bhatia, S.N.; Sailor, M.J. Micellar hybrid nanoparticles for simultaneous magnetofluorescent imaging and drug delivery. Angew. Chem. Int. Ed. **2008**, *47* (38), 7284–7288.

87. Ren, J.; Jia, M.; Ren, T.; Yuan, W.; Tan, Q. Preparation and characterization of PNIPAAm-b-PLA/Fe$_3$O$_4$ thermo-responsive and magnetic composite micelles. Mater. Lett. **2008**, *62* (29), 4425–4427.

88. Yang, X.; Chen, Y.; Yuan, R.; Chen, G.; Blanco, E.; Gao, J.; Shuai, X. Folate-encoded and Fe$_3$O$_4$-loaded polymeric micelles for dual targeting of cancer cells. Polymer **2008**, *49* (16), 3477–3485.

89. Hong, G.B.; Zhou, J.X.; Yuan, R.X. Folate-targeted polymeric micelles loaded with ultrasmall superparamagnetic iron oxide: Combined small size and high MRI sensitivity. Int. J. Nanomed. **2012**, *7* (1), 2863–2872.

90. Liao, C.; Sun, Q.; Liang, B.; Shen, J.; Shuai, X. Targeting EGFR-overexpressing tumor cells using Cetuximab-immunomicelles loaded with doxorubicin and superparamagnetic iron oxide. Eur. J. Radiol. **2011**, *80* (3), 699–705.

91. Li, X.; Li, H.; Liu, G.; Deng, Z.; Wu, S.; Li, P.; Xu, Z.; Xu, H.; Chu, P.K. Magnetite-loaded fluorine-containing polymeric micelles for magnetic resonance imaging and drug delivery. Biomaterials **2012**, *33* (10), 3013–3024.

92. Hu, J.; Qian, Y.; Wang, X.; Liu, T.; Liu, S. Drug-loaded and superparamagnetic iron oxide NPs surface-embedded amphiphilic block copolymer micelles for integrated chemotherapeutic drug delivery and MR imaging. Langmuir **2012**, *28* (4), 2073–2082.

93. Sanson, C.; Diou, O.; Thevenot, J.; Ibarboure, E.; Soum, A.; Brulet, A.; Miraux, S.; Thiaudiere, E.; Tan, S.; Brisson, A. Doxorubicin loaded magnetic polymersomes: Theranostic nanocarriers for MR imaging and magneto-chemotherapy. ACS Nano. **2011**, *5* (2), 1122–1140.

94. Sahu, S.K.; Maiti, S.; Pramanik, A.; Ghosh, S.K.; Pramanik, P. Controlling the thickness of polymeric shell on magnetic nanoparticles loaded with doxorubicin for targeted delivery and MRI contrast agent. Carbohydr. Polym. **2012**, *87* (4), 2593–2604.

95. Hammad, M.A.; Muller, B.W. Solubility and stability of clonazepam in mixed micelles. Int. J. Pharm. **1998a**, *169* (1), 55–64.

96. Hammad, M.A.; Muller, B.W. Increasing drug solubility by means of bile salt–phosphatidylcholine-based mixed micelles. Eur. J. Pharm. Biopharm. **1998b**, *46* (3), 361–367.

97. Esmaili, M.; Ghaffari, S.M.; Movahedi, Z.M.; Atri, M.S.; Sharifizadeh, A.; Farhadi, M.; Yousefi, R.; Chobert, J.M.; Haertlé, T. Beta casein-micelle as a nano vehicle for solubility enhancement of curcumin; food industry application. Lwt-Food. Sci. Technol. **2011**, *44* (10), 2166–2172.

98. Chung, J.E.; Okoyama, M.; Okano, T. Inner core segment design for drug delivery control of thermo-responsive polymeric micelles. J. Control. Release **2000**, *65* (1–2), 93–103.

99. Liu, D.Z.; Hsieh, J.H.; Fan, X.C.; Yang, J.D.; Chung, T.W. Synthesis, characterization and drug delivery behaviors of new PCP polymeric micelles. Carbohydr. Polym. **2007**, *68* (3), 544–554.

Medical Devices and Preparative Medicine: Polymer Drug Application

M. R. Aguilar
Spanish National Research Council Institute of Polymer Science and Technology, Madrid, Spain

L. García-Fernández
Networking Biomedical Research Center on Bioengineering, Biomaterials, and Nanomedicine (CIBER-BBN), Madrid, Spain

M. L. López-Donaire
F. Parra
Spanish National Research Council Institute of Polymer Science and Technology, Madrid, Spain

L. Rojo
Department of Materials and Institute of Bioengineering, Imperial College London, London, U.K.

G. Rodriguez
M. M. Fernández
Spanish National Research Council Institute of Polymer Science and Technology, Madrid, Spain

Julio San Román
Institute for Health and Biomedical Innovation, Queensland University of Technology, Kelvin Grove, Queensland, Australia

Abstract

Polymer drugs or pharmacologically active macromolecules are one of the most promising choices proposed in the controlled drug delivery field for the development of nanocarriers independently of their origin (natural or synthetic), that can be tailored to give the desired physicochemical properties. These polymeric drugs or polymer–drug conjugates show activity, even though the corresponding monomeric species may or may not be biologically active.

INTRODUCTION

The "perfect" drug presents the precise therapeutic activity and no side effects. Several efforts have been devoted in order to reach these objectives in the last decades. Currently, two different strategies for improving the therapeutic efficacy of drugs are being followed. The first one is based on the development of new agents that modulate the molecular processes and pathways specifically associated with the disease. Genomics and proteomics play a main role in this strategy. The second one consists in the improvement of existing drugs to make them more effective by using nanocarriers that bring more drug molecules to the desired site.

Polymer drugs or pharmacologically active macromolecules are one of the most promising choices proposed in the controlled drug delivery field for the development of nanocarriers independently of their origin (natural or synthetic), that can be tailored to give the desired physicochemical properties. These polymeric drugs or polymer-drug conjugates show activity, even though the corresponding monomeric species may or may not be biologically active. Although there are infinite possibilities for designing polymer-drug conjugates, the most widely accepted model is the one suggested by Ringsdorf in 1975,[1] which considers the covalent binding of low-molecular-weight drugs to a polymeric system must be by means of organic functional groups that can be degraded in the physiological medium.

The desired physicochemical properties of a particular polymer drug can be tailored using the numerous tools offered by the organic chemistry and macromolecular sciences.

Many polymer drugs have been developed for the treatment of different diseases in the last years. In this entry, we report their role in following most important applications:

- Antitumoral polymer drugs
- Antiangiogenic and proangiogenic polymer drugs
- Antibacterial polymer drugs
- Antithrombogenic polymer drugs
- Low friction polymer drugs

APPLICATIONS OF POLYMER DRUGS TO MEDICAL DEVICES AND REPARATIVE MEDICINE

Antitumoral Polymer Drugs

Cancer is considered a leading cause of death worldwide; specifically 7.6 million of deaths were produced in 2008. World Health Organization has estimated to continue

In Vitro–Medical

Concise Encyclopedia of Biomedical Polymers and Polymeric Biomaterials DOI: 10.1081/E-EBPPC-120052267

789

rising to 12 million deaths in 2030 (Http://Www.Who.Int/Cancer/En),[2] The common solid tumors such as breast, prostate, lung, stomach, liver, and colon constitute the most difficult to treat and cause the most cancer deaths.

As mentioned earlier, the efforts devoted to improve anticancer treatment are being concentrated on two different fields: first, by the design and development of new bioactive agents that modulate the molecular processes and pathways associated with the tumor progression and second, by using nanocarriers capable to maintain the anticancer drug in the bloodstream for long periods of time, and to provide a sustained delivery in the tumor tissues, limiting the cytotoxic effect of the drug.

Active Tumor Targeting

This therapy is based on the Ringsdorf's[1] vision about ideal polymer for its use as drug carrier (Fig. 1). The polymer backbone is attached to solubilizing groups that provide the bioavailability of the carrier system, spacers which are bound to macromolecules and whose chemical stability should be unchanged during transport to the target site, and finally, moieties, such as antibodies, saccharides, proteins, and peptides,[3] which are involved in antigens or receptors that are either uniquely expressed or overexpressed on the target cells relative to normal tissues.

One of the first conjugate copolymers bearing a targeting ligand to be tested clinically for cancer treatment was a N-(2-hydroxypropylmethacrylamide) (HPMA) copolymer bearing galactosamine residues, HPMA-Gly-Phe-Leu-Gly-doxorubicin (DOX)-galactosamine (PK2; FCE 28069). In this case, the overhanging galactosamine moieties localize the asialoglycoprotein receptor in hepatocytes.[6,7]

Another ligand which is widely used as target for tumor-specific drug delivery is folate (FOL), an anionic form of folic acid. Its use lies on the fact that one of its receptors, glycosyl phosphati-dylinositol-anchored glypolypeptide, which presents a dissociation constant in the subnanomolar range, is overexpressed in many cancers such as ovary, lung, breast, kidney, brain, endometrium, colon, and hematopoietic cells of myelogenous origin while its presence in healthy cells is limited on the apical (luminal) surface of polarized epithelial cells where it is inaccessible directly through the bloodstream.[8,9] An example of FOL-conjugate systems is based on the amphiphilic block copolymers that self-assemble into spherical micelles FOL-poly (ethylene glycol) (PEG)-poly (aspartate-hydrazone-adriamycin). Adriamycin (ADR) was conjugated to the side chains of the core-forming poly (aspartic acid) block through a hydrazone linkage which can be selectively cleaved under the acidic intracellular environment. FOL-conjugated micelles showed lower in vivo toxicity and higher antitumor activity over a broad range of the dosage from 7.5 to 26.21 mg/kg, which was fivefold broader than free drugs.[10]

Delivery of polymer-conjugated drugs can take place through intra- or extracellular pathways. For intracellular

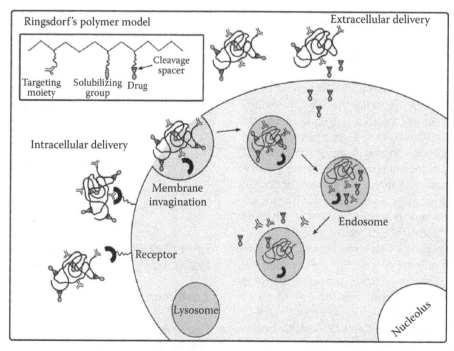

Fig. 1   Endocytic pathway for the cellular uptake of ideal macromolecules carriers defined by Ringsdorf as drug delivery systems.
Source: Adapted from Haag & Kratz[4] and Park et al.[5]

delivery, the lysosometric delivery of the drug defined by De Duve, De Barsy, and Poole[11] could be produced by endocytic pathway in the cytoplasm followed by the trafficking through the endosomal compartments to lysosomes. This point is where the linker plays an important role because it should be selectively cleaved under acidic conditions, based on a significant drop in the pH values existing within the endosomes (pH ~ 5–6.5) and in the primary and secondary lysosomes (pH ~ 4).[12]

Table 1 collects recent references where drugs were coupled to the suitable carriers through acid-sensitive bonds.

Disulfide constitutes other important linkage that is emerging as a fascinating class of reducing sensitive biodegradable polymers. They are exploited for triggered intracellular delivery of a variety of bioactive molecules such as siRNA, DNA, proteins, and low-molecular-weight drugs based on the difference in reducing potential between the intracellular and extracellular environments.[13] In this sense, micelles based on disulfide-linked dextran-fc-poly (ε-caprolactone) diblock copolymer (Dex-SS-PCL) were prepared for an efficient intracellular release of DOX. *In vitro* studies revealed the rapid release of DOX to the cytoplasm as well as to the cell nucleus in comparison with that observed from DOX-loaded reduction insensitive Dex-PCL micelles.[14]

On the other hand, research is also focused on the enzymatic degradation of the linker bond. In this sense, poly (oleyl 2-acetamido-2-deoxy-α-D-glucopyranoside methacrylate-co-N-vinylpyrrolidone) was developed as an active glycopolymer for brain tumor where the drug delivery occurs via ester hydrolysis by the presence of carboxylesterase enzyme.[18]

Passive tumor targeting

Different studies have shown that polymer-drug conjugates without targeting ligands in their structures could interact with different (or a broad number of different types of) antibodies and receptors present on the surface of cancer cells and therefore being entrapped or accumulated in solid tumors and retained for prolonged periods (more than 100 h). This passive tumor targeting was identified by Matsumura and Maeda[19] and is known as enhanced permeability and retention (EPR) effect of macromolecules and lipids in solid tumor (Fig. 2).[20–22]

The EPR effect can be attributed to two factors: the defective vascular architecture based on angiogenesis and hypoxia phenomena that allows macromolecular extravasations not usual in normal tissues, and also, lack of effective tumor lymphatic drainage, preventing clearance of the penetrant macromolecules and promoting their accumulation.[23] The EPR effect is observed for macromolecules with molecular weights greater than 20 kDa, whereas the rate of urinary clearance is inversely related to the tumor uptake. Takakura and Hashida[24] detailed organ distribution of macromolecules and clearance kinetics with regard to targeting liver and kidney tumors.

This phenomenon was observed with HPMA-Gly-Phe-Leu-Gly-DOX copolymer (PK[1]; FCE 28068)[25] which was clinically tested and cumulative doses of HPMA copolymer could be administered without signs of immunogenicity or polymer-related toxicity[26,27] Recently, a new biodegradable graft copolymer-DOX conjugate was designed for passive tumor targeting. HPMA copolymer was grafted to the main chain through degradable,

Table 1 Acid-sensitive bonds copolymers

Benzoic imine		[361]
Hydrazone		[15] [16] [362] [17] [10]
Ketal	or	[363] [364] [365]
Acetal	or	[363] [366] [367]
cis-Aconityl		[368] [369]

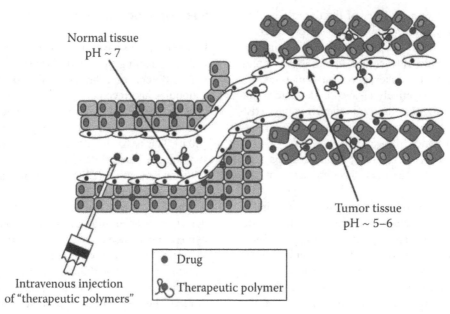

Fig. 2 Passive tumor targeting of polymer therapeutics by the enhanced permeability and retention (EPR) effect.
Source: Adapted from Haag & Kratz[4] and Duncan.[21]

enzymatic, and reductive spacers in order to facilitate the intracellular degradation of the graft polymer carrier after the drug release. The graft polymer-DOX conjugates were tested in mice bearing EL4-T-cells and exhibited a prolongation in the blood circulation and an enhancement in the tumor accumulation compared with its linear analog conjugate.[28]

However, tumor-targeting drug delivery cannot be completely achieved only in base of EPR effect especially for tumors that present hypoxic environment, low vascularization, and permeability.[29] The EPR effect can be facilitated by different factor such as nitric oxide (NO), prostaglandin, peroxynitrite, collagenase, bradykinin, and so on. This artificial augmentation of the EPR effect can be achieved by two different strategies.[22] The first one is based on angiotensin II, which is used to induce hypertension leading to a relative increase in tumor blood flow volume (Fig. 3a). This method was evaluated with the polymeric drug styrene-maleic acid copolymer-conjugated to the protein neocarzinostatin, SMANCS.[19] Clinical results showed an improvement therapy for various advanced solid tumors such as metastatic liver cancer and pancreas.[30]

The second method used to alter the EPR effect is based on the NO generation (a potent mediator of vascular extravasation) from nitroglycerin (NG). NG can be converted into nitrite and then reduced to NO under relatively hypoxic and acidic condition found in cancer tissues (Fig. 3b).[31] The role of NG in the EPR effect was investigated with PEG-conjugated zinc protoporphyrin IX (PZP).[32] The apply of NG exhibited a greater suppression of tumor growth when compared with PZP alone.[33] However, it should be mentioned that these good results are associated with not only the increase of drug delivery but also a

decrease of hypoxic-induced resistance to anticancer drugs in cancer cell lines by means of reduction of hypoxia-inducible factor-1a (HIF-1a).[34]

At present, polymers used in cancer therapy are defined under the umbrella term of "therapeutic polymers" and include different macromolecular architectures such as polymer-drug and polymer-protein conjugates, DNA-polyplexes, polymeric micelles, and dendrimers (Fig. 4). All these systems should be water-soluble, non-immunogenic, nontoxic, and degradable in order to be eliminated from the body after their function.

Polymer-Drugs: Biodistribution, elimination, and metabolism of the polymer-drug conjugate play a crucial role in the design of polymeric systems to be used for drug conjugation, therefore molecular weight and physicochemical properties of the polymeric candidate must be considered.

Despite recent efforts devoted to obtain novel polymer-drug conjugates, at present only three polymers are being tested in clinical phase: PEG, HPMA, and poly(glutamic acid) (PGA).

The most relevant drugs that have been conjugated to HPMA and are under clinical trials are PK1 and PK2.[25–27,35] Both systems have a tetrapeptide spacer, Gly-Phe-Leu-Gly that could be cleaved by lysosomal enzymes present in tumor cells. In the case of PK2, an additional galactosamine group was incorporated onto the main chain as a target ligand in order to promote multivalent targeting of the hepatocyte asialoglycoprotein receptor to treat primary liver cancer.[7] The maximum tolerance dose (MTD) of PK2 in a phase I trial study was 160 mg/m^2 of DOX equivalents, less than the MTD value of PK1 (320 mg/m^2). It is not clear why the PK2 presents

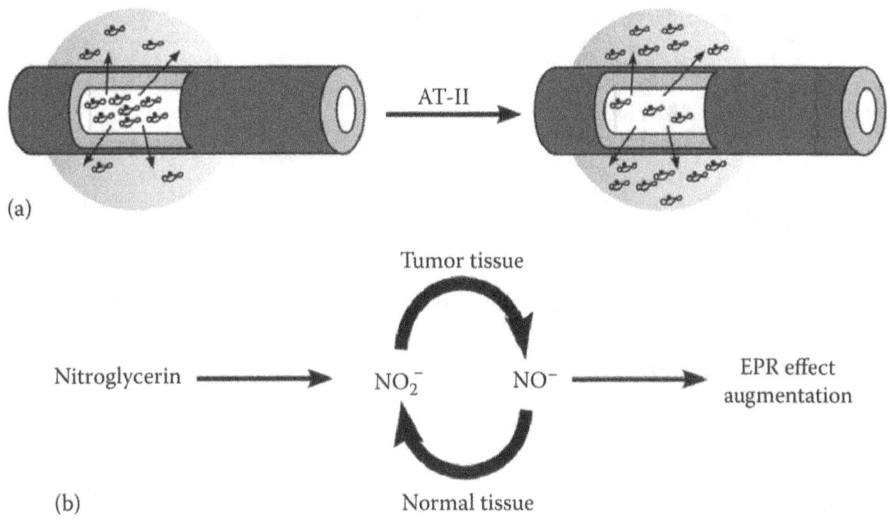

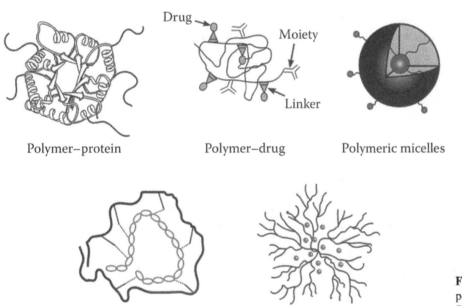

Fig. 3 (a) Blood vasculature in tumor tissue under normotensive (left) and hypertensive (right) conditions. (b) Mechanism of NO generation from nitroglycerin (NG).
Source: (a) Adapted from Nagamitsu et al.[30] and (b) adapted from Seki et al.[33]

Polymer–protein   Polymer–drug   Polymeric micelles

Drug
Moiety
Linker

Polymer–DNA complexes   Dendrimers

Fig. 4 Representative therapeutic polymers used in cancer therapy.
Source: Adapted from Park et al.[5]

higher toxicity than its analog without galactosamine. Table 2 includes other polymer-drug conjugates based on HPMA, where the anticancer drugs are Taxol[36] or camptothecin.[37,38] Based on the good results associated with these systems, current approaches are focused on the HPMA carrier based on pH-sensitive linker,[39,17] enzyme-cleavage linker based on a self-immolative dendritic pro-drug with single or multiple triggering modes of action,[40] and a simultaneous conjugation with multiple chemotherapeutic agents. HPMA was conjugated with DOX through the hydrolytic and degradable hydrazone bond to the drug carrier and additional hydrophobic substituents

(oleic acid, cholesterol, and dodecyl moieties) were introduced into the polymer structure in order to get more flexibility favoring the formation of micelles or nanoparticles (NPs) via self-assembly phenomena.[50] An example of the mentioned system consist of the HPMA conjugation with two therapeutic agents: gemcitabine and DOX.[51]

PGA was the first biodegradable polymer used in polymer–drug conjugates. PGA was linked to paclitaxel via an ester bond to the carboxylic acid of PGA, giving a soluble polymer with a high drug-loading efficiency (37 wt%), *PGA-paclitaxel* (CT-2103; Xyotax). *Paclitaxel* is released

In Vitro–Medical

Table 2 Polymeric drug and polymer–drug conjugates in clinical trials

Name	Drug	Polymer	Indication	References
CT-2103; Xyotax	Paclitaxel	PGA (40 kDa)	Ovarian cancer Non-small cell lung cancer	[41,23]
CT-2106	Camptothecin	PGA (50 kDa)	Various cancer	[42]
PK1: FCE 28068	Doxorubicin	HPMA (30 kDa)	Lung and breast cancer	[25–27]
PK2: FCE 28069	Doxorubicin	HPMA-Galactosamine (30 kDa)	Hepatocellular carcinoma	[7]
PNU 166945	Paclitaxel	HPMA (40 kDa)	Various cancer	[36]
MAG-CPT	Camptothecin	HPMA (40 kDa)	Various cancer	[37,38]
AP5280	Carboplatin platinate	HPMA (28 kDa)	Various cancer	[43]
AP5346 ProLindac	DACH-platinate	HPMA	Various cancer	[23]
AD-70 DOX-OXD	Doxorubicin	Dextran	Various cancer	[44]
DE-310	Camptothecin	Modified dextran	Various cancer	[45]
Prothecan	Camptothecin	PEG	Various cancer	[46]
XMT-1001	Camptothecin	Polyacetal polymer (PHF or Fleximer®)	Advanced cancer	[47]
EZN-2208	SN38	4 arm PEG (40 kDa)	Advanced cancer	[15,48]
NKTR-102	Irinotecan	4 arm PEG	Colorectal breast, ovarian and cervical cancer	
NKTR-105	Docetaxel	4 arm PEG	Solid tumors including hormone refractory	[49]

Abbreviation: DACH, Diaminocyclohexane.

from the polymer backbone by degradation due to the effect of lysosomal cathepsin B after endocytic uptake.[41] In phase I and II clinical trials, *PGA-paclitaxel* was administered every 3 weeks, over 30 min with a MTD of 233 mg/m², higher than that observed for free paclitaxel, and a partial improvement or disease stabilization was observed.[23] In phase III clinical trial, this system was compared with first-line treatments such as *gemcitabine* and *vinorelbine*.[52] The results showed that severe side effects were significantly reduced in comparison with *gemcitabine* and a greater survival was obtained in comparison with that obtained with both components.[23]

Conjugates based on PEG have been obtained with lower drug loads. For example, Prothecan, a *PEG-camptothecin* conjugate, contains only 1.7 wt% of camptothecin because PEG bears only two hydroxyl terminal groups (C-OH) suitable for conjugation. These terminal groups, specifically C20-OH, favor the desired configuration of camptothecin lactone ring. In phase I clinical trial, a prolonged plasma half-life (more than 72 hr) and a raise in the activity were observed.[46]

Despite the desirable properties of PEG, two disadvantages must be overcome such as the low drug payload (two hydroxyl terminal suitable for conjugation) and the nonbiodegradability. In order to overcome these

drawbacks, some researchers have focused on the synthesis of new branched PEG through the end-chain groups or coupling on them small dendron structures with the aim to attain higher drug payloads.[53,54] For the second limitation, researchers developed biodegradable PEGs in response to a specific signal such as changes in pH[55] or reductive conditions.[56]

Recently, polymer-drug conjugates carrying the combination of the anticancer agent epirubicin (EPI) and NO functional groups were reported applying PEG as polymer. The combination of NO and EPI in the same carrier presents two advantages: the improvement of the antitumor activity at the tumor site and the reduction of cardiotoxicity.[57]

Linear polymer containing cyclodextrins (CDs) as part of the backbone (IT-101) were reported for their conjugation with *camptothecin*. IT-101 is linear, highly biocompatible, nonbiodegradable, and has sufficient size to be cleared renally as a single molecule. Their conjugation with *camptothecin* is currently being investigated in human clinical trials.[58]

Several combination therapies have been applied recently with these polymer–drug conjugates. Greco et al.[59] described these types of systems namely: 1) polymer–drug conjugate plus free drug;[60] 2) polymer-drug conjugate plus polymer–drug conjugate;[61] 3) single polymeric carrier carrying a combination of drug;[62] and 4) polymer-directed enzyme prodrug therapy (PDEPT)[63] and polymer–enzyme liposome therapy (PELT).[64]

PDEPT approach requires initial administration of the polymeric drug in order to allow its arrival at the tumor site before it could be activated for the polymer-enzyme conjugate. One example is the coadministration of both HPMA-DOX (PK$_1$) (copolymer–drug conjugate) and HPMA-cathepsin B (enzyme conjugate).[65] When the PDEPT conjugates were applied in the treatment of B16F10 melanoma tumors, the antimitotic activity was 168%, noticeably higher than that obtained when PK$_1$ was administered alone (152%).

Polymer–Protein: Polymer–protein conjugates technique is used to reduce protein immunogenicity and increase protein solubility and stability. When the polymer used is PEG, the technique is known as "PEGylation technology."[66]

Linear or branched PEG derivatives are used by coupling to the surface of the proteins. PEGylations produce a drop in the biological activity of proteins. However, they prolong plasma half-life and avoid receptor-mediated protein uptake by cells of the reticuloendothelial system.[67] Furthermore, it is worth noting that the design of the parameters for PEGylation should be based on the type of anticancer proteins that want to be released. Thus, heterologous proteins, whose administration is limited by their immunogenicity, require low-molecular-weight PEGs and random amine coupling while endogenous proteins, which need a prolonged half-life, are conjugated with high-molecular-weight PEGs.[49]

PEG-L-asparaginase (ONCASPAR) was the first polymer-protein conjugate to be tested clinically in oncology. The clinical status of other PEGylated conjugates is recorded in recent reviews.[49,68] Recently the PEGylation of recombinant human arginase (rhArg-PEG 5000 MW), based on the enzyme rhArg that can produce the depletion of arginase and consequently the inhibition of the tumor growth, angiogenesis, and NO synthesis have showed similar half maximal inhibitory concentration (IC50) with respect to the native enzyme in several human hepatocellular carcinoma cells line.[69]

Despite the great variety of polymer-protein conjugates that are in clinical trials, recent studies have shown the immunogenic response to PEG. It should be suggested that this behavior could be associated with the excessive use of this polymer in the therapeutic applications.[70] The immunogenic response of PEG imposes to develop novel polymers with similar advantages of it/PEG. In this sense, poly (2-ethyl-2-oxazoline) has been suggested as a possible alternative. In fact, this polymer presents similar features than PEG such as high water solubility, amphiphile, flexibility, and nontoxicity being also produced at low polydispersity. Properties of poly (2-ethyl-2-oxazoline) have been determined by its conjugation with a model protein

(tryp-sin) showing a preservation of the enzyme activity and protein rejection properties similar with those of the analogous PEGylated conjugate.[71]

Additionally, novel dextrin-protein conjugates[72,73] have been reported in a new concept named "polymer masked-unmasked protein therapy" which is based on the conjugation of a biodegradable polymer in order to mask a protein activity, and subsequently triggered degradation of the polymer is used to regenerate the protein bioactivity in a controlled manner. Dextrin-phospholipase A2 crotoxin (PLA2) conjugate is an example of this type of systems. Dextrin-PLA2 showed a reduction of its enzymatic activity in comparison with free PLA2. This activity was restored mostly in the target site by the action of a-amylase, an enzyme present in the extracellular fluid but overexpressed in the tumor site. In this sense, the conjugate displayed enhanced *in vitro* cytotoxicity in the tumor site.[73]

Polyplexes: Human Genome Project has received widespread attention in Cancer Gene Therapies during the last decade. This advantageous therapy offers the possibility to kill cancer cells selectively due to the direct action on a deficient gene causing its block or replacement.[74,5] Genes have been applied by direct injection into tumor tissues showing good results.[75] However, the efficiency is improved when genes are administered by a carrier delivery system avoiding their degradation before reaching the target site.[76]

Until now, viral and nonviral vectors are the two delivery systems used in gene therapy. In spite of the excellent transfection capacity, viral vector (retrovirus, adenovirus, and adeno-associated viruses) therapies have been abandoned due to some disadvantages such as their low ability carrying high-molecular-weight DNA molecules and their associated immunogenicity and oncogenic potential effect in the body. In comparison, nonviral polymeric vector-based therapies show lower side effect risks and high affinity for large DNA molecules although they present low transfection efficiency.[77,78]

Many nonviral vectors are based on polymer-DNA polyplexes. Optimal transfection efficiencies require positive-charged polymer species in order to promote the electrostatic interactions needed with the negative charges of DNA. Figure 5 shows the most common cationic polymers used as nonviral vectors for DNA delivery.

Linear and branched polymers based on polyethylenimine (PEI) constitute the most important nonviral vectors used due to their high affinity for high-molecular-weight DNA molecules.[79–81] The affinity and transfection efficiency of DNA-PEI polyplexes increase with the PEI molecular weight due to the proton sponge effect associated to the high density of primary, secondary, and tertiary amino groups.[77] However, the high number of positively charged amino groups leads to an increased toxicity constituting this fact a real bottleneck of these PEI-based polyplexes.

In Vitro–Medical

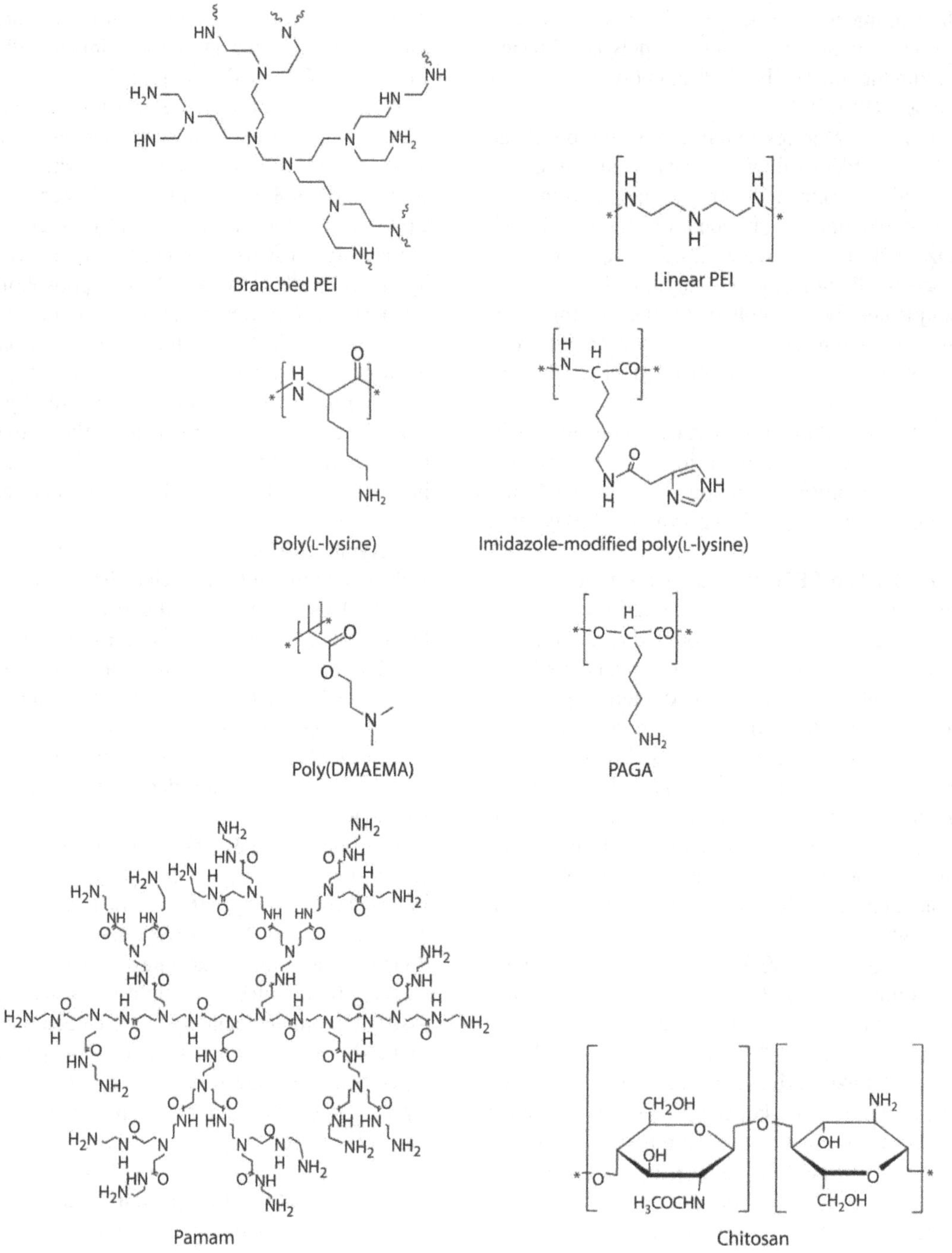

Fig. 5 Nonviral vectors based on cationic polymers for DNA delivery.

To overcome the high toxicity of PEI and of PEGylated PEI (PEG-PEI),[82] FOL was linked on PEG and then grafted the FOL-PEG onto the linear[83] and branched PEI 25 kDa.[84]

Poly (L-lysine) (PLL), a cationic lineal polypeptide, differently from PEI presents an additional desirable feature, its biodegradation. This polymer presents lower toxicity than PEI but also, lower transfection efficiency,[85] which is due to the lack of amino groups present in the polymer structure. However, recent reports using PEGylated PLL[86]

and biocompatible dendrimers such as poly (L-lysine octa (3-aminopropyl)silsesquioxane) dendrimers[87] have shown an increase complex stability and transfection efficiency. These recent studies may lead to a renaissance of PLL as gene delivery agents.

PLL modified with imidazole groups provides systems with enhanced transfection efficiencies, without increasing toxicity, when compared to unmodified PLL[88,89] as a consequence of the imidazole heterocycles buffering capacity in the range of endoly-sosomal pH.

Poly (2-dimethylaminoethyl methacrylate) (PDMAEMA) has also been studied for gene delivery therapies showing good transfection activities. However, these systems also show high cytotoxicity levels and lack of biodegradability limiting their potential in gene delivery.[90] Lately, novel reducible PDMAEMA-based systems combined with biodegradable polymers such as phosphazene[91] were reported showing minimal cytotoxicity levels and enhanced transfection activity when compared with PDMAEMA.

Different chitosans with molecular weight between 30 and 170 kDa are used in this field, showing transfection efficiency values of the same magnitude as PEI[92] and in combination with other polymers.[93]

In the case of the dendrimers, the most used one is the polyamidoamine (PAMAM)[94–97] due to their molecular architecture which shows some unique physical and chemical properties and makes them particularly interesting for gene delivery applications.[98] Recent studies have shown the use of other type of dendrimers (triazine derivatives) with easier synthesis and higher transfection efficiencies in comparison with PAMAM.[99]

Other polymers used are poly (α-(4-aminobutyl)-L-glycolic acid),[100,101] poly (carbonic acid 2-dimethylaminoethyl ester 1-methyl-2-(2-methacryloylamino) ethyl ester) (pHPMA-DMAE),[102,103] and CDs.[104–107]

The high toxicity of these cationic polymers is associated with their interaction with different plasma proteins such as albumin, fibronectin, immunoglobulin, complement factors, or fibrinogen, through their surface charge[108,109] so that they can activate the complement system[110] leading to removal by the reticuloendothelial system. In order to avoid this phenomenon, the surface of these polymers is modified with hydrophilic polymers like HPMA[111] or PEG[112] before or after poly-plex formation. In the case of PEG, a half-life rise in the plasma was observed but it was still far from the desired values.[113,82] These polymer surfaces were also modified with moieties such as transferring for targeting of cancer cells that express transferrin receptor, leading in an increase of the expression levels within the tumor tissues.[114–116]

Studies based on nonviral DNA gene therapy have been ongoing for years and will continue toward improving systemic delivery and transfection efficiencies to the levels required of *in vivo* clinical trials. In the meantime, scientists have focused on a novel therapeutic pathway which uses RNA interference (RNAi) by which harmful genes can be "silenced" by delivering complementary short interfering RNA (siRNA) to target cells.[117]

For this purpose, the same nonviral DNA vectors have been used to deliver siRNA into cancer cells. As examples, linear and branched PEI,[118] poly (D,L-lactide-α-glycolide) (PLGA),[119] PAMAM,[120] poly (isobutyl cyanoacrylate),[121] polycations consisting of histidine, and polylysine residues,[122] chitosan,[123,124] PEI-PEG[125] were used as polycation-based siRNA delivery systems. However, this novel approach shows the same limitations as DNA-based therapies in terms of toxicity and delivery efficiency.

Particulate Systems: Particulate system is a term that comprises liposomes, surfactants, polymeric micelles, polymer-somes, and NPs based on synthetic polymers. The present section is an introduction of the different advances of the last three systems.

Micellar systems are very interesting in the encapsulation of amphiphilic and low solubility drugs.[126,5] The administration of different drugs in form of micelles offers different advantages such as improvement of drug solubility and it avoids its environmental degradation. Also, due to their low size (10–100 nm) a targeting effect at the tumor site and a high half-life in blood are provided.[127,22,128] These kinds of systems are based in liposomes, self-aggregates, microparticles and NPs, and polymer micelles.[5]

Polymer micelles are caused from a spontaneous self-assembling of amphiphilic block copolymers when the critical micellar concentration (CMC) is achieved.[129] The size of the micelle depends on the geometry of the constituent monomers, intermolecular interactions, and conditions of the bulk solution (i.e., concentration, ionic strength, pH, and temperature).[130] The surface properties of polymeric micelles are important factors since they determine their biological fate.[131]

More common polymeric micellar systems are copolymers based on PEG and polyesters such as poly (D,L-lactic acid) (PDLLA), PLGA, and PCL. These systems were widely studied because they do not require removal after their administration due to their biodegradation in the body.[132–137] Most of these systems present a physical encapsulation of the drug. PDMAEMA-*b*-PCL-*b*-PDMAEMA is an example of micellar system loaded with paclitaxel. These micelles are based on the self-assembling of these triblock copolymers which present positive charge on their surface (+29.3 to +35.5 mV) allowing the complexation with siRNA. The combinatorial delivery of siRNA and paclitaxel showed a reduction on the vascular endothelial growth factor (VEGF) expression.[138]

Polymeric micelles where the drug is conjugated were also reported. For example, a block copolymer PEG-*block*-PCL bearing DOX was synthesized and its efficient control over the rate of DOX release in physiological medium was compared with micelles PEG-*block*-poly(α-benzyl carboxylate-epsilon-caprolactone) (PEG-*b*-PBCL) where DOX was encapsulated. Both systems could maintain the cytotoxicity effect over cancer cells.[139]

Despite the success of polymeric micelles as drug carriers *in vitro* and in animal studies, clinical trials revealed a number of problems associated with premature drug release from micelles in the circulation or an absence of adequate drug release upon micelle accumulation in the tumor interstitium. The premature drug release can be prevented by appropriate micelle stabilization[140] or enhancing drug interaction with the hydrophobic blocks.[132,133] On the other hand, the excessive drug retention in micelle cores can be solved by the application of external stimuli[141] that cause micelle destabilization in a specially controlled manner thus increasing the selectivity and efficiency of drug delivery to target cells. The intracellular signals include mainly pH[142–146] and glutathione.[147]

pH-responsive biodegradable micelles based on a block copolymer consisting of a novel acid-labile hydrophobic polycarbonate and PEG were applied for the encapsulation of paclitaxel and DOX at the same time (13% and 11.7%, respectively). The release was clearly pH-dependent being higher at acid pH. This phenomenon is associated with the hydrolysis of the acetal groups of the polycarbonate.[148]

Two different strategies are proposed for the biodegradation of micelles under reductive environment. One is based on the reduction-sensitive cross-linking of micelles[149] and the other is based on diblock copolymer containing a single disulfide linkage such as PEG-SS-PCL.[150]

External factors including heat, ultrasound, and light are expected to trigger micellar degradation. In particular, thermoresponsive polymeric micelles that would show a remarkable change of properties responding to minimum changes in temperature are receiving attention, in particular for cancer therapies as the slightly higher temperature shown in tumor tissue (2–5°C) compared to healthy tissues.[151] In this sense, polymers showing lower and upper critical solution temperatures (LCST and UCST, respectively) such as poly (A-isopropylacrylamide) (PNIPA)[152–154] were reported for the delivery of different anticancer drugs.

Polymersomes or polymeric vesicles constitute another type of self-assembling nanostructures that, unlike micelles, can encapsulate both hydrophilic and hydrophobic molecules in the hydrophilic fluid-filled core or in the hydrophobic bilayer, respectively.[151] Furthermore, unlike liposomes and phospholipids, the features of the polymersome membrane such as thickness, chemical functionality, and stability can be tailored in function of the molecular weight and type of copolymer.[155] However, the main methods to produce vesicles cannot be efficiently controlled in terms of size and morphology and also the final membrane will always contain an amount of organic solvents. For these reasons, recent studies are based on the design of new methods of preparation of new vesicle systems in order to overcome these drawbacks.[156] Polymersomes are prepared mainly from amphiphilic diblock, triblock, graft and dendritic polymers, preferentially biodegradable.[157] Like micelles, recent efforts are focused on the design of stimuli-responsive polymersomes such as redox, pH,[148,158] and temperature.[159]

NPs based on synthetic polymers constitute another type of particulate system which can enhance the intracellular concentration of drugs in cancer cells while avoiding toxicity in normal cells. As examples of these systems, the preparation of Dex-DOX encapsulated in chitosan NPs[160] and the encapsulation of curcumin in biodegradable

nanoparticulated formulations based on PLGA were reported in order to improve the bioavailability and the cellular uptake.[161]

Additionally, new reversible cross-linked dextrans were reported for an efficient intracellular drug delivery within reductive environments that mimic those of the intracellular compartments, showing high drug-loading efficiency and reduction-triggered release of DOX *in vitro* as well as inside tumor cells, in particular within cell nucleus.[162]

Recently, the oral bioavailability of paclitaxel was increased by its encapsulation through complex formation with CDs in poly (anhydride)-based NPs, as a consequence this system produced a synergistic effect associated to the combination of bioadhesive properties of poly (anhydride) and the inhibitory effect of CD.[163] Self-assembled NPs based on hydrophobically modified glycol chitosan have also been reported as a carrier for paclitaxel and camptothecin to display fast cellular and tissue internalization into tumors.[164,165]

Like micelles and polymersomes, temperature can also be exploited as an external factor to tailoring the drug release profiles from NPs. In this sense, a novel drug platform comprising a magnetic core and biodegradable thermoresponsive shell of triblock copolymer was synthesized. Oleic acid-coated Fe_3O_4 NPs and *DOX* were encapsulated with PEG-*b*-PLGA-*b*-PEG triblock copolymer.[166]

Furthermore, NPs can be covalently linked to different moieties that are specifically recognized by cancer cell membrane receptors. Farokhzad et al.[167,168] developed a bioconjugate system based on PLGA-*b*-PEG or PLA-*b*-PEG-COOH and a specific aptamer for targeted delivery to prostate cancer cells.[167,168] Galactosyl was also linked to low-molecular-weight chitosan (Gal-LMWC) NPs for hepatocyte targeting and specifically delivering DOX.[169] Paclitaxel and DOX-loaded PEGylated PLGA-based NPs were also grafted with RGD or RGD-peptidomimetic moieties in order to target tumor vessels and enhance the antitumor efficacy and tumor growth retardation.[170]

Dendrimers: Dendrimers constitute a family of nanostructured macromolecules with a highly regular branching, globular architecture, multivalency, and well-defined molecular weight charge and surface area. Dendrimeric drug delivery carriers present several advantages compared with linear polymers due to their capacity to cross cell barriers through paracellular and transcellular pathways.[171] On the other hand, they have an exceptionally high drug-loading capacity by chemical bonding or physical encapsulation (Fig. 6a and b).

On the other hand, dendrimers, unlike polymer micelles based on amphiphilic block copolymers, can maintain their globular structure regardless of their concentration. Their stability is based on the presence of covalent bounds and does not depend on the CMC.[130] Different types of dendrimers such as PAMAM, polypropylene imine (PPI), and poly (ether hydroxylamine) are being commercialized for biomedical applications.

A particular example of dendrimer drug carrier with antitumoral activity consist of the biocom-patible polyester composed by glycerol and succinic acid dendrimers where the anticancer drug camptothecin was physically encapsulated within intrinsic empty cavities of the dendrimer. The cytotoxicity of this complex was evaluated for several human cancer cell lines showing an increase in cellular uptake and drug retention in human breast adenocarcinoma (MCF-7).[172] A recent publication describes the synthesis of thermoresponsive highly branched PAMAM-PEG-PDLLA dendritic NPs as nanocontainers for camptothecin. These constructed NPs show a significantly improved cytocompatibility and encapsulation efficiency, which was significantly improved in comparison with that observed for PAMAM-based dendrimers.[173]

One of the first dendrimer conjugates in which the drug was covalently bonded to the periphery was PAMAM-cisplatin. This system offered excellent water solubility, selective accumulation in tumor tissues, and reduced toxicity in comparison with the native cisplatin.[174]

Different strategies are developed to reduce the cytotoxicity associated to cationic dendrimers based on partial surface derivatization using chemically inert entities. Acetylated PAMAM dendrimer was conjugated to methotrexate or tritium and analyzed in vivo in animal models of human epithelial cancer and the acetylation improved the pharmacokinetics.[175] Nevertheless, such modifications may limit the number of targeting molecules and/or drugs. In this sense, PAMAM dendrimers were modified by glutamylation in order to increase the number of available functional groups (-COOH and -NH2) for further biologically active molecules immobilization.[176] Moreover, the conversion of primary amine groups to amphoteric ones may prevent toxicity.

Another strategy widely used in order to decrease cytotoxicity and increase half-life of the systems consists of the increase of the molecular weight of the dendrimer by PEGylation (Fig. 6c).[177–179]

Zhu et al. reported recently the conjugation of DOX to different PEGylated PAMAM dendrimers by acid-sensitive linkage (cis-aconityl) or -insensitive linkage (succinic) to produce PPCD and PPSD, respectively. The effect of PEGylation degree and the short of linkage have been investigated against ovarian cancer cell (SKOV-3). Highest PEGylation degree of PPCD conjugate showed greater tumor accumulation in mice inoculated with SKOV-3 cells.[180] Another PEGylate PAMAM dendrimer conjugated with the anticancer drug adriamycin by amide or hydra-zone bond was also reported.[181]

As in the linear polymer, dendrimers can also be conjugated with specific ligands for tumor targeting. Based on this concept, PAMAMG5 dendrimer was conjugated with biotin-targeting moieties. The good biocompatibility and biodistribution made them interesting for further applications as drug delivery systems.[182]

FOL and FOL-PEG-PAMAM dendrimers based on the ligand-mediated targeted FOL loading with the anticancer drug 5-fluorouracil were analyzed and the result showed a reduction in hemolytic toxicity, sustained drug release, as well as a higher accumulation in the tumor area than without PEG-FOL.[183]

Drug-dendrimer conjugation is based usually on covalent bonds but could also be possible by ionic interaction as is the case of the complexes formed between the anticancer drug 7-ethyl-10-hydroxy-camptothecin (SN-38) and PAMAM. The complexes were stable at pH 7.4 and drug was released at pH 5.[184]

It is worth mentioning the carbohydrate-installed polymers (Fig. 6d) as targeted anticancer drug carriers, such as glycodendrimers, linear glycopolymers, and spherical glycopolymers. These systems present a specific sugar-protein biomolecular recognition in living systems and the EPR effect. Methotrexate-loaded polyether-copolyester dendrimers (PEPE) conjugated with d-glucosamine as ligand is an example of glycodendrimers. Results showed that glucosamine can be used as an effective ligand not only for targeting glial tumors but also for a permeability enhancement across blood-brain barrier. Thus, glucosylate PEPE dendrimers can serve as potential delivery system for the treatment of gliomas.[185]

Most of dendrimer vehicles used for anticancer drug delivery are focused on PAMAM but new promising systems are based on triazine dendrimers.[179]

Antiangiogenic and Proangiogenic Polymer Drugs

Angiogenesis consist of a natural process that occurs in the human body, especially during normal growth of organs and during wound healing and is characterized by the sprouting of new blood vessels from existing ones.

Angiogenesis is regulated by the balance between proangiogenic and antiangiogenic signals. This balance is influenced by the biological function of the heparan sulfate proteoglycans (HSPGs) and their capacity to modulate the biological activity of several growth factors involved in the angio-genesis process, mainly fibroblast growth factors (FGF) and VEGF (Fig. 7).[186,187] Therefore, the interaction of VEGFs or FGFs with these cell surface HSPGs seems to be an attractive target to modulate angiogenesis.

Angiogenesis Inhibitors

Up-regulated angiogenic processes play an important role in the development of various pathological processes such as psoriasis,[188] rheumatoid arthritis,[189] tumor growth,[190] diabetic retinopathy,[191] etc. The uncontrolled neofor-mation of blood vessels contributes to the progression of these diseases, especially in the development of solid malignant tumors. The search of synthetic or

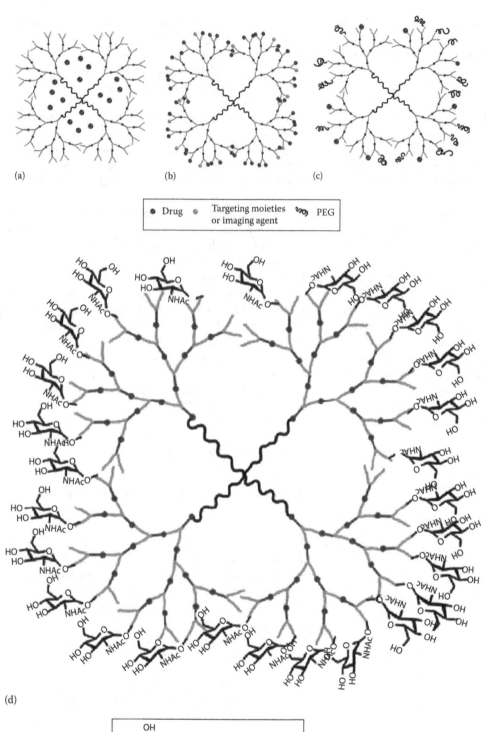

(a)   (b)   (c)

| ● Drug | ● Targeting moieties or imaging agent | ∿ PEG |

(d)

N-acetylglucosamine

Fig. 6 (a) Encapsulation of the drugs in the dendritic interiors. (b) Dendrimer-drug conjugates, den-drimers linked to targeting moieties and imaging agents. (c) PEGylated dendrimer and (d) glycodendrimer.

natural polymers that interfere on the angiogenic process could be a good approach as angiogenic inhibitors. Figure 8 shows different pathways to inhibit the angiogenic process:

Polymer drugs are being used as angiogenic factor (AF) inhibitors, blocking the interaction between the AF and the HSPGs. There are two main strategies in order to avoid the binding between growth factors and HSPG, one consisting on the use of heparin-binding polycations and the other on the use of heparin-like polycations.

Some polycationic compounds, such as protamine, and PLL-based dendrimers are able to compete with growth

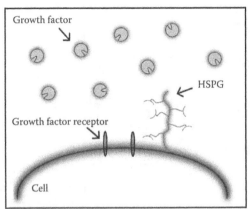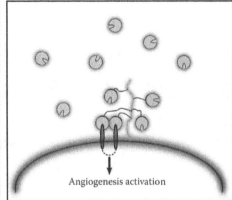

Fig. 7   Interaction between growth factor and its receptors by HSPGs.

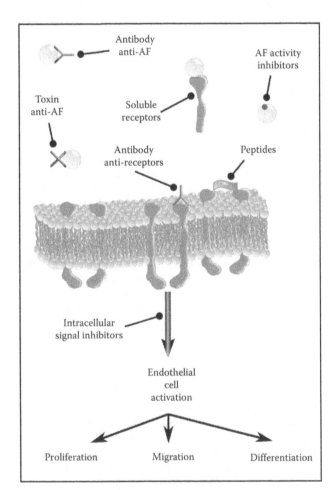

Fig. 8   Mechanisms of inhibition of the HSPG-dependent interaction between growth factors and their receptors.

factors for HSPG interaction. Protamine is a natural polycationic protein that binds to HSPGs and competes with heparin-binding growth factors for its interaction[192] (Fig. 9). The antiangiogenic activity of protamine has been demonstrated in several *in vitro* tests.[193,194] Nowadays, different types of dendrimers have been used as heparin-binding inhibitors such as those synthesized from PLL or

arginine showing good inhibition activity of the angiogenic process without toxic effect.[195,196]

Polyanionic compounds similar to HSPG compete with HSPGs for the interaction with growth factor (Fig. 9). In this way, polysulfonate polymers structurally similar to heparin such as sur-amins block the activity of growth factors by inhibiting their binding to HSPGs.[197] The limitation of suramin is due to the toxic side effects at high doses. Similar compounds have been also developed. 5-Amino-2-naphthalenesulfonic acid (ANSA) showed good antian-giogenic activity and lower toxicity.[198,199] Recent studies on ANSA-based polymer systems showed good results *in vitro* and *in vivo*.[200,201] Other polycationic polymers such as poly (2-acrylamido-2-methyl-1-propanesulfonic acid), poly (anetholesulfonic acid), and poly (4-styrenesulfonic acid) have demonstrated a potent inhibition of neovascularization.[202,203]

In both cases, the binding of AF to endothelial cell surface would be hampered with a consequent inhibition of their angiogenic capacity.

Non-polysulfonate-based polymer compounds with different mechanism of actions have also been investigated. Polyacetylene derivatives exhibit significant and potent antiangiogenic activity. This ability is possible through induction of regulators and cell cycle mediator production. Polyacetylenes induced the formation of cyclin-dependent kinase inhibitors that inhibit the formation of tubelike structures.[204] Other polymers are being used as growth factor receptors inhibitors drugs. CEP-7055 (*N*, *N*-dimethyl glycine ester) or CEP-5210 (a C3-(isopropylmethoxy) fused pyrrolocarbazole) has shown a clear inhibition of cancer angiogenesis.

Angiogenesis Inducers

Therapeutic induction of angiogenesis is a potential treatment to induce neovascularization after vessel wall injury (ischemia, angioplasty). As mentioned earlier, HSPGs play an important role by their interactions with proangiogenic

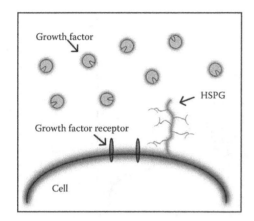

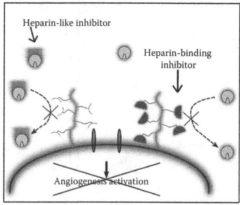

Fig. 9 Interaction between growth factor and growth factor receptors by fucoidan.

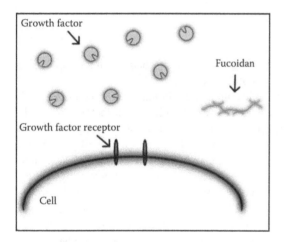

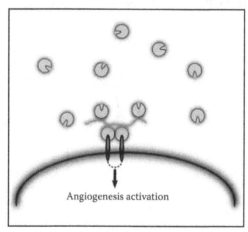

Fig. 10 Possible strategies to obtain antibacterial polymers.

growth factors such as VEGF or FGF. Therefore, those molecules that mimic some biological activities of HSPGs can be used to induce angiogenesis (Fig. 10). In this way, natural sulfated polysaccharides such as fucoidan have been studied as angiogenic modulators.[205] Fucoidans are high-molecular-weight sulfated poly (L-fucopyranose) of marine plants origin. Reduced into low-molecular-weight fraction, the action of fucoidans is similar to HSPGs, binding some proangiogenic growth factors and protecting them from proteolysis and enhancing their angiogenic activity.[105,205,206]

The problem of HSPGs-like polymers is that the same polymer can show different behaviors depending on its molecular weight. Numerous studies support the hypothesis that low-molecular-weight species of fucoidan have a generally positive effect on angiogenesis, whereas high-molecular-weight fucoidans present inhibitory effects.[105,205-207]

A similar effect occurs with heparins. Recent studies demonstrated that the angiogenesis capacity also depends on the heparin molecular weight showing inhibition effects for low-molecular-weight fractions and stimulating angiogenesis systemically for the high-molecular-weight fractions.[208-210]

Another way to induce angiogenesis is based on the preparation of scaffolds with bioactive polymers that stimulate the angiogenic process. Generally, the use of tissue scaffold has been limited by the low level of revascularization.

The use of polymers that promote angiogenesis is a useful method to obtain scaffolds for the treatment of complicated tissue defects. For example, the preparation of scaffolds based on poly-N-acetyl glucosamine nanofibers showed good results for the treatment of diabetic wound healing.[211] This scaffold simulates the extracellular matrix (ECM) and provides to the cell a growth media that enhance the cell metabolism, migration, and therefore the angiogenesis. The problem is the limitation of their use due to the inability to provide sufficient supply of oxygen and nutrients to the inner part of the cell-matrix constructs.[212] The superficial modification with ECM proteins such as FN, collagen, or sequences corresponding to cell adhesion domains contributes to the angiogenic process.[213-215] An example is the preparation of chito-san scaffolds with different degrees of acetylation (DA) coated with FN.[216] The DA affects cell adhesion and growth on FN-coated chitosan porous scaffolds due to the influence on the adsorbed FN. The selection of suitable DA will therefore be highly important for further vascularization strategies.

Antibacterial Polymer Drugs

Contamination and infections caused by pathogen microorganisms constitute a serious problem and a great concern in hospitals, healthcare products, medical devices, and many other biomedi-cal applications. In fact, infections are the most common cause of biomaterial implant failure and represent a significant challenge to the more widespread application of biomedical implants.[217–219] Generally, an infected implant needs to be removed as the formation of a bacterial biofilm hampers the treatment with antibiotics. Therefore, emphasis has been placed on different ways to prevent the infections caused by bacteria that account for 1 million implant-associated infections and 3 billion dollars of healthcare costs annually.[220] Therefore, there is a definitive need of new materials resistant to microbial colonization and polymeric drugs are good candidates for this purpose.

There are many factors than can affect the biological activity of antibacterial polymers such as hydrophilic-hydrophobic balance, spacer length between active site and polymer, molecular weight or interaction between polymer and pathogen. Consequently, an appropriate balance between all these features is necessary to reach an optimum result.

The basic requirements for antibacterial polymers in biomedical applications are the following:

- Does not decompose to and/or emit toxic products
- Stable in long-term usage and storage at the temperature of its intended application
- Biocidal to a broad spectrum of pathogenic microorganisms in brief times of contact

From the technological point of view, and without considering the encapsulation of low-molecular-weight active agents to give rise to leachable polymers, the antimicrobial polymers can be obtained following two different and complementary strategies (Fig. 11):

1. Surface functionalization of natural or synthetic polymers by grafting of antimicrobial agents
2. Polymerization of pharmaceutically active compounds conjugated to polymerizable reactive groups

From the chemical point of view, there are two great groups of bactericidal (kill bacteria) or bacteriostatic (halt bacterial growth) agents that can be included into polymeric systems: cationic biocides and bioactive low-molecular-weight compounds. Cationic biocides are represented to a large extent by guanidines,[221] biguanidines,[222] and

salts of sulfonium,[223] phosphonium,[224] pyridinium,[225] and ammonium,[226] being this last type of salt widely used since the 1930s in domestic and public hygiene for surface disinfection, topical antisepsis, and to control biofouling and microbial contamination in industry. Their mode of action normally involves interaction with the cell envelope, displacing divalent cations. Subsequent interactions with membrane proteins and lipid bilayer depend upon the specific nature of the biocide but generally cationic biocide exposure results in membrane disruption and lethal leakage of cytoplasmic materials.[227]

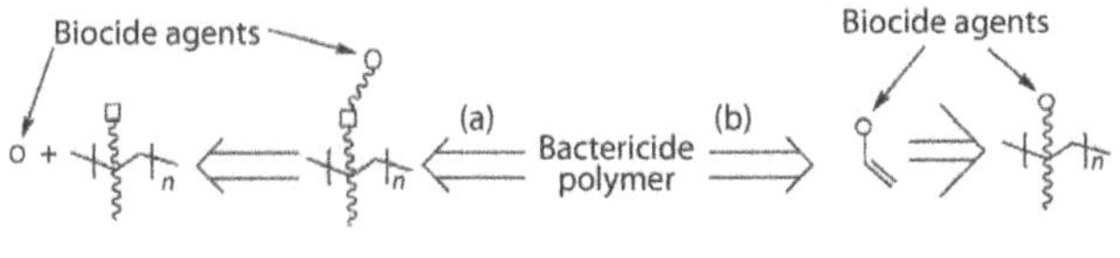

The other great group of substances that can be included into polymeric systems are the low-molecular-weight antibiotic agents widely established in the pharmaceutical industry and the sanitary system from World War II, such as aminoglucosides, ansamicines, carbacephens, carbap-enems, cefalosporins, glycopeptides, macrolides, monobactams, penicillins, polypeptides, quinolones, sulfonamides, tetracyclins, and others. Low-molecular-weight antimicrobial agents present some disadvantages, such as short-term antimicrobial activity, toxicity to the environment, and bioresistance. To overcome all these drawbacks, antimicrobial functional groups can be incorporated into polymer molecules if the antimicrobial agent contains reactive functional groups such as amino, carbonyl, or hydroxyl groups that can be covalently linked to a wide variety of polymerizable derivatives.[228–230]

Considering the effectiveness of bactericides, a distinction needs to be made between two types of bacteria: Gram-positive and Gram-negative. Their essential difference lies in the absence of the cell wall in the Gram-positive bacteria while that the Gram-negative bacteria have an outer membrane structure in the cell wall forming an additional barrier for foreign molecules which determines a lesser sensitivity to chemicals. For the determination of bactericide efficiency, it is important to carry out tests with bacteria representative of both types, being *Staphylococcus aureus* and *Bacillus subtilis* typical representatives for Gram-positive bacteria and *Escherichia coli* and *Pseudomonas aeruginosa* for Gram-negative bacteria.

The membrane of Gram-negative bacteria is composed of 70–80% of phosphatidylethanolamine (PEA) and 20–25% of negatively charged lipids, such as phosphatidylglycerol (PG) or cardiolipin (CL),[231] whereas the membrane of the Gram-positive bacteria are formed mainly by anionic lipids, such as 70% of PG, 12% of PEA, and 4% of CL.[232]

However, the membrane of eukaryotic cells like human erythrocytes is composed by a lipidic bilayer which has in its internal layer 10% of the negatively charged lipid phosphatidylserine (PS), 25% of PEA, 10% of phosphatidylcholine (PC), and 5% of sphingomyelin (SM), whereas its external part is composed of 25% of cholesterol, 33% of PC, 18% SM, and 9% of PEA.[233] Therefore, there is clear difference between bacterial and eukaryotic cells, being the

Fig. 11 Antimicrobial polymers obtained by surface functionalization of natural or synthetic polymers (a) or polymerization of pharmaceutical active compounds conjugated to polymerizable groups (b).

outer wall of bacteria more negatively charged than eukaryotic membrane.

Biocidal cationic polymers present a clear tendency to interact with the bacterial membrane (negatively charged) instead of the eukaryotic membrane due to ionic affinity.[234,235] This along with the fact that cholesterol is a membrane-stabilizing agent in mammalian cells but absent in bacterial cell membranes protects the cells from antimicrobial pep-tides attack, and along with the well-known transmembrane potential that affect peptide-lipid interactions makes bacterial membrane highly vulnerable against positively charged cationic polymers.

In contrast to many conventional antibiotics, the exact mechanism of killing bacteria by cationic polymers is not well-established. These polymers appear to be bactericidal (bacteria killer) instead of bacteriostatic (bacteria growth inhibitor). However, it is known that biocidal cationic polymers interact with the anionic bacteria membrane disrupting the membrane integrity and inducing the cytoplasmic constituents leakage and death of the bacteria.[236,237]

Surface Functionalization of Natural or Synthetic Polymers by Grafting of Antimicrobial Agents

Surface modification by grafting polymerization can be achieved by different coupling mechanisms such as coordination, ionic, or free-radical mechanism.[238] However, most recent techniques for surface modification are based on radical living polymerization, reversible addition-fragmentation chain transfer polymerization (RAFT),[239–241] and atom transfer radical polymerization reactions (ATRP).[242,95,243]

In this context, Matyjaszewski et al.[95] prepared polypropylene (PP)-based surfaces grafted with non-leachable biocide by chemically binding poly (quaternary ammonium) (PQA). The well-defined PDMAEMA, a precursor

of PQA, was grown from the surface of PP via ATRP and the tertiary amine groups in PDMAEMA were consequently converted to the quaternary ammonium in the presence of ethyl bromide (Scheme 1).

The biocidal activity test against *E. coli* of the resultant surfaces depended on the number of available quaternary ammonium units (QA density). With the same grafting density, the surface grafted with relatively high-molecular-weight polymers (Mn > 10,000 g/mol) showed almost 100% killing efficiency, whereas a low biocidal activity (85%) was observed for the surface grafted with shorter PQA chains (Mn = 1500 g/mol).

Cheng et al.[244] prepared a switchable polymer surface coating which combined the advantages of both nonfouling and cationic antimicrobial materials overcoming their individual disadvantages. In this system, poly (*N*,*N*-dimethyl-*N*-(ethoxycarbonylmethyl)-*N*-[2'-(methacryloyloxy)ethyl]-ammonium bromide) was grafted by surface-initiated ATRP onto a gold surface with initiators, giving rise to a surface that killed more than 99.9% of *E. coli* K12 in 1 h and released more than 98% of the dead bacterial cells when the cationic derivative was hydrolyzed to nonfouling zwitterionic polymers.

Natural polymers, such as chitosan and cellulose, offer another source of macromolecules to obtain of antibacterial materials. Nowadays, many attempts have been carried out in order to improve the antibacterial properties of chitosan, a semisynthetic polymer obtained from partial deacetylation of chitin,[245] and that has many interesting properties, such as antimicrobial activity and nontoxicity[246] (Scheme 2).

N-Alkyl chitosan derivatives were prepared by introducing alkyl groups into the amine residue of chitosan, reduction of the intermediate Schiff's base with NaBH4, and subsequent quaternization of the obtained *N*-akyl chitosan derivative with methyl iodide[247] (Scheme 3).

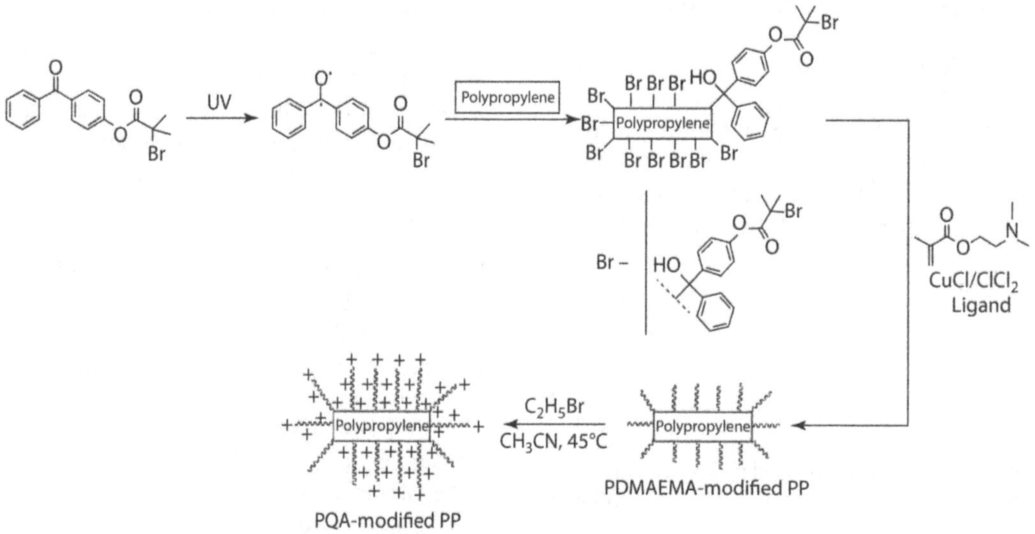

Scheme 1   PP surface modification by PDMAEMA grafting via ATRP.

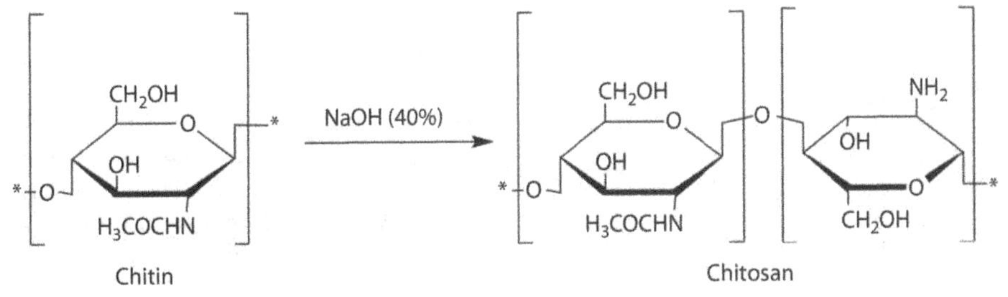

Scheme 2 Transformation of chitin into chitosan.

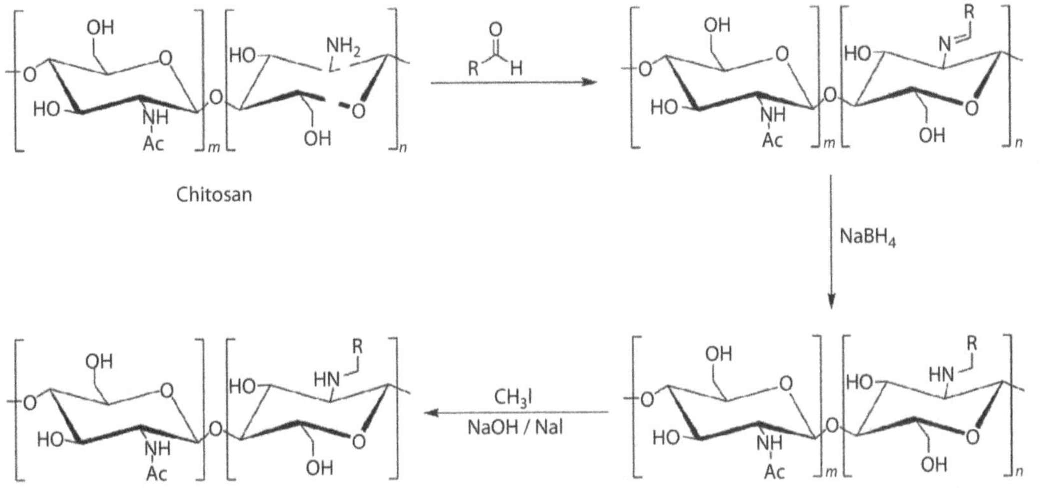

Scheme 3 Synthetic route to N-alkyl chitosan derivatives.

The chitosan quaternary ammonium salts obtained turned out to be better antibacterial than the original chitosan, and it was found that the antimicrobial activities of the chitosan quaternary ammonium salts increased with the increase in the chain length of the alkyl substituent, which was ascribed to the contribution of the increased lipophilic character of the derivatives.

Another way to obtain chitosan quaternary ammonium salts consist of the reaction of chitosan with glycidyltrimethylammonium chloride leading to modified chitosan that showed excellent antimicrobial activity[248] (Scheme 4).

Phenolic derivatives have antiseptic properties damaging cell membranes and causing release of intracellular constituents, although they also cause intracellular coagulation of cytoplasmic constituents leading to cell death or inhibition of cell growth. Therefore, phenolic derivatives have been conjugated to chitosan through its amino group at C2 leading to chitosan derivatives that were highly active against fungi such as *Aspergillus flavus*, *Candida albicans*, and *Fusarium oxysporium*, as well as against bacteria such as *E. coli*, *S. aureus*, and *B. subtilis*[249] (Scheme 5).

Cellulose constitutes another natural polymer commonly used for the preparation of antibacterial materials. Perrier and coworkers grafted PDMAEMA into cellulosic fibers via RAFT polymerization (Scheme 6).[250]

Subsequent quaternization of the tertiary amino groups of the grafted PDMAEMA chains with different chain length alkyl bromides (C8–C16) leads to a large concentration of quaternary ammonium groups on the cellulose surface. The antibacterial activity depends on the alkyl chain length and on the degree of quaternization. PDMAEMA-grafted cellulose with the highest degree of quaternization and quaternized with the shortest alkyl chains exhibited the highest activity against *E. coli*.

Gademann and coworkers created antibacterial titanium oxide surfaces allowing the generation of stable, protein-resistant, nonfouling surfaces thanks to a novel biomimetic strategy for surface modification that exploits the evolutionary optimized strong binding affinities of iron chelators such as anachelin, chromophore that contains a catechol moiety as the anchoring group, which is structurally similar to key elements of mussel-adhesive protein sequences that are thought to be responsible for the very strong wet adhesion of mussels to surfaces.[230,251–253]

In his approximation, Gademann employed an anachelin chromophore-PEG conjugate linked to vancomycin, where the anachelin chromophore enables the immobilization of the hybrid on the surface; the vancomycin interferes with cell-wall biosynthesis and inhibits the growth of bacteria; and the long PEG-3000 linker makes the modified surfaces resistant to proteins and cells, and ensures the optimal positioning of the antibiotic on the surface.

Antimicrobial Polymers by Polymerization of Pharmaceutically Active Compounds Conjugated with Polymerizable Groups

The second strategy to overcome the disadvantages of the low-molecular-weight antimicrobial agents consist of their covalently linking onto polymerizable molecules.

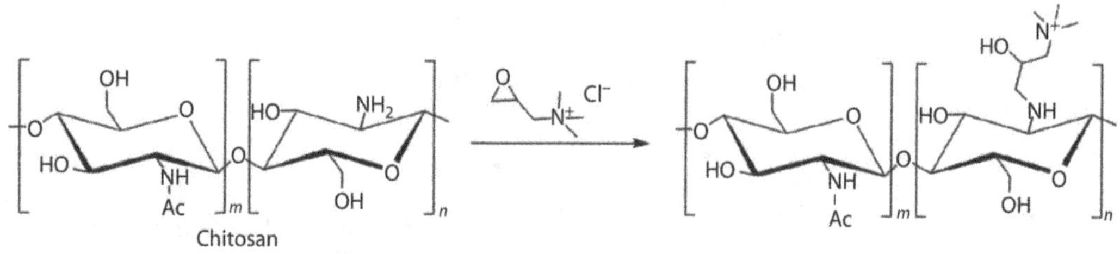

Scheme 4 Reaction of chitosan with glycidyl-trimethylammonium chloride.

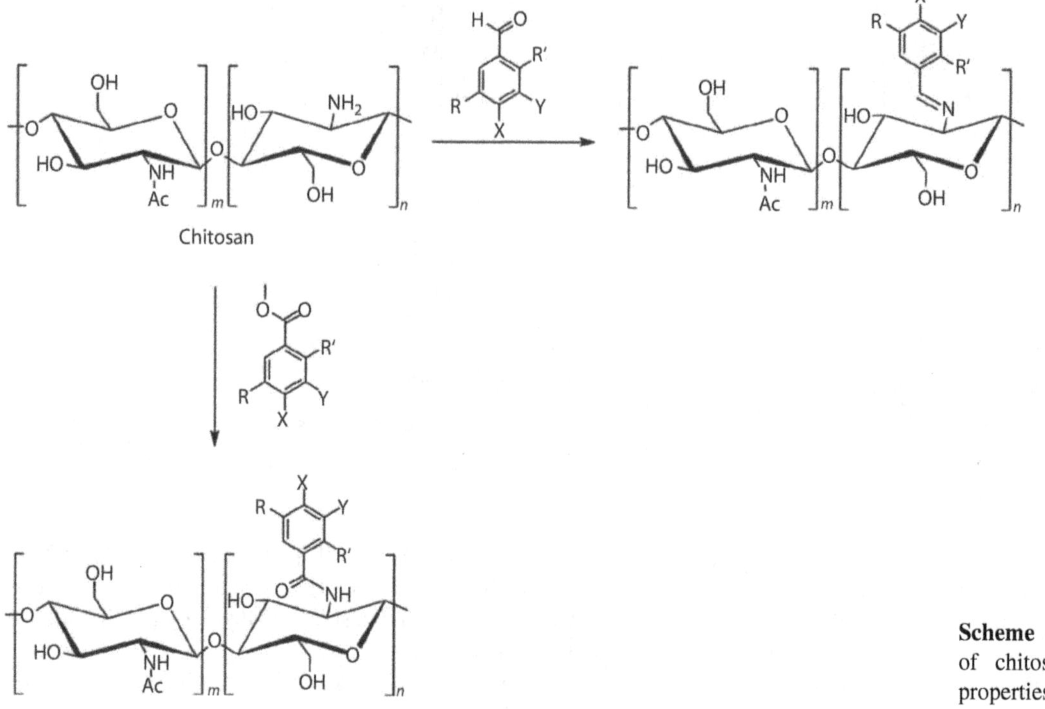

Scheme 5 Phenolic derivatives of chitosan with antimicrobial properties.

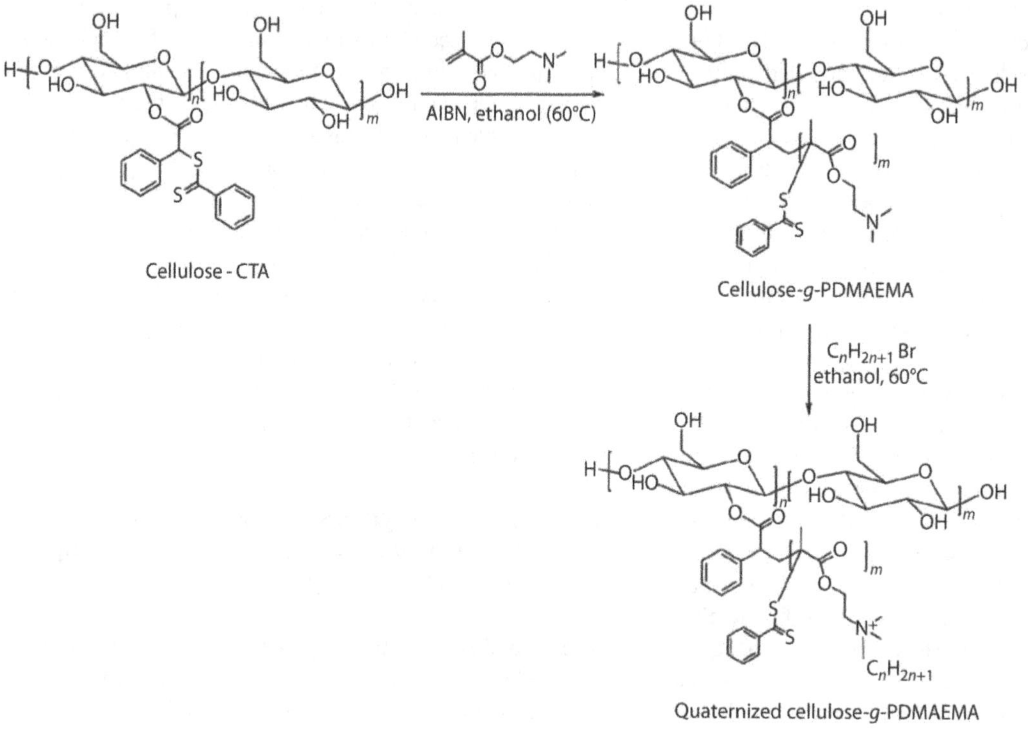

Scheme 6 PDMAEMA grafting into cellulose via RAFT polymerization.

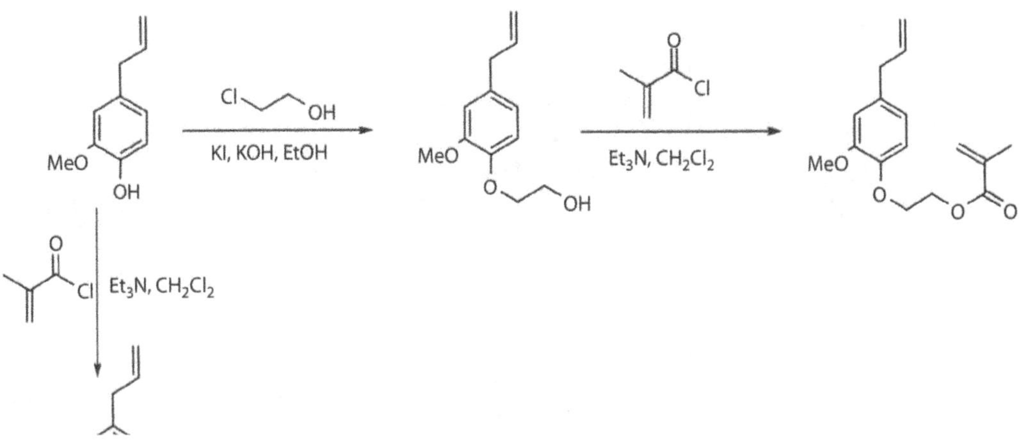

Scheme 7 Synthetic route to polymerizable eugenol derivatives.

Acrylic derivatives bearing pharmaceutically active compounds are one of the most important groups of polymeric drugs. These acrylic drug-conjugated monomers have the advantage that they can be copolymerized with a wide type of compounds and tune the drug concentration or the hydrophilic/hydrophobic functionalities present in the copolymer.

In this context,[254] described the preparation of antimicrobial copolymers by radical copolymerization of 2-hydroxyethyl methacrylate (HEMA) with methacryloyl derivatives of eugenol, an essential oil used in medicine as local antiseptic and anesthetic, or mixed with zinc oxide as temporary pulp capping agent and as filling system in root canals. Eugenol derivatives were obtained by reacting methacryloyl chloride and eugenol or 2-eugenyl ethanol in the presence of triethylamine (Scheme 7).

The copolymers obtained were active against Gram-negative *E. coli* strains and also against *Streptococcus* mutants, Gram-positive bacteria found in the oral cavity and significant contributor to the tooth decay, and were biocompatible with human fibroblast which makes these kinds of compounds suitable for biomedical systems in the field of dentistry and orthopedic applications.[255]

Parra et al.[256] reported the synthesis of polymerizable quaternary ammonium salts with high refractive index and their copolymerization with different methacrylic monomers for the preparation of bactericide copolymers for ophthalmic applications (Scheme 8).

The bactericide quaternary ammonium salts monomers showed inhibition halos of 23–25 mm in antibiogram tests against *Staphylococcus epidermidis* and *P. aeruginosa*, strains found in the ocular cavity and responsible for most postsurgical endolphthalmitis. Biocompatibility of the systems was evaluated in cell cultures using human fibroblasts and in all cases the cellular viability was higher than 90%, and close to 100% in many cases, for the extracts of selected formulations collected at different periods of time. The materials obtained by copolymerization with HEMA and 2-(benzothiazolylthio)ethyl methacrylate presented high refractive index values ($nD0 \approx 1.51$) and good wettability (15–37%), which make these systems good

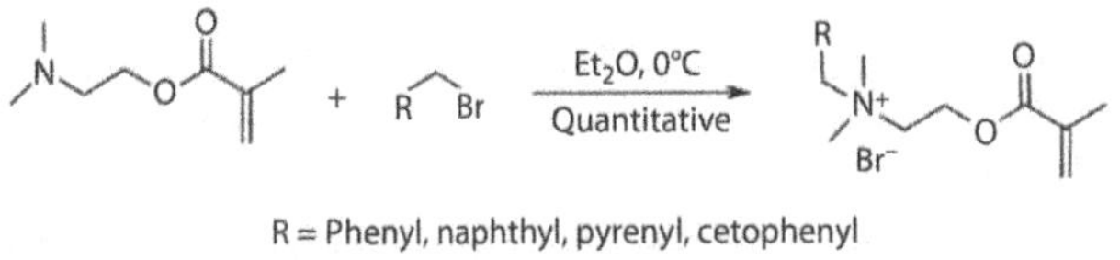

R = Phenyl, naphthyl, pyrenyl, cetophenyl

Scheme 8 Polymerizable quaternary ammonium salts.

candidates for the fabrication of bac-teriostatic and foldable intraocular lenses (IOLs), and can be considered as a good alternative to the currently used foldable hydrogel IOLs.

It is known that the biocidal cationic polymers can mimic the biological activity of the natural host defense peptides (HDPs) that are an evolutionarily conserved component of the innate immune response and are found among all classes of life. HDPs have demonstrated their broad spectrum antibiotic activity, killing Gram-negative and Gram-positive bacteria (including strains that are resistant to conventional antibiotics), mycobacteria (including Mycobacterium tuberculosis), enveloped viruses, fungi, and even transformed or cancerous cells.[257] HDPs are generally between 12 and 50 amino acids and have a large portion of hydrophobic residues, generally >50%, and two or more positively charged residues provided by histidine in acidic environments, or by arginine and lysine. With the aim of mimicking HDPs and taking advantage of the guanidines properties, which are positively charged at physiological pH and have been used widely as antibacterial agents,[258] Gabriel et al. synthesized poly (guanidinium oxanorbornene) (PGON) from norbornene monomers via ring-opening metathesis polymerization[259] (Scheme 9).

The synthesized polymer was not membrane-disruptive indicating that it also had properties similar to polyarginine and other cell-penetrating peptides (CPPs) like HIV-TAT,[260–264] and represents an exciting and fundamentally novel entry in the expanding field of synthetic mimics of antimicrobial peptides like HDPs. Moreover, PGON possess a remarkable combination of antimicrobial activity against both Gram-negative and Gram-positive bacteria as well as low hemolytic activity against human red blood

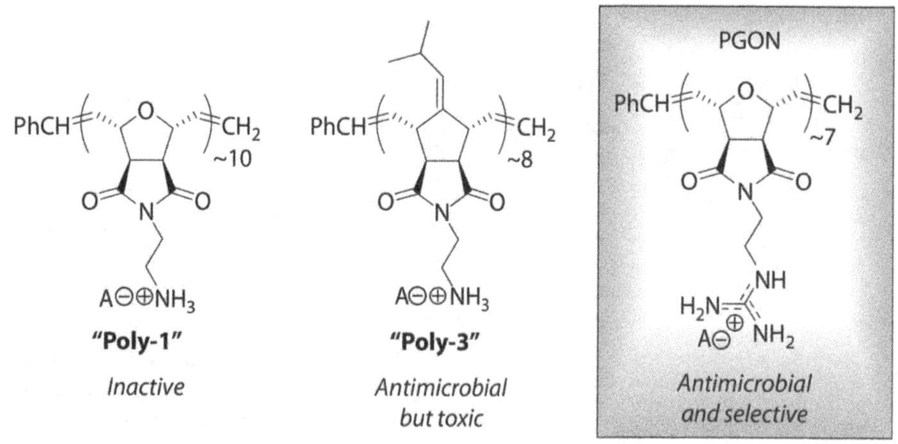

"Poly-1"

Inactive

"Poly-3"

*Antimicrobial
but toxic*

PGON

*Antimicrobial
and selective*

Scheme 9　Synthesis of polyguanidinium oxanorbornene (PGON).

cells, encouraging the development of powerful, nontoxic, antibacterial materials designed to prevent biofilm formation.

Vancomycin is the prototypical glycopeptide antibiotic, and it remains potent against Gram-positive organisms (e.g., *Staphylococcus* spp.) commonly encountered in association with indwelling medical devices such as orthopedic hardware, although resistance can occur.[265] Its activity derives from binding D-Ala-D-Ala sequences found at the terminal end of peptido-glycan precursors, that blocks the action of both transglycosylases and transpeptidases by complex-ing with their substrates and preventing proper cross-linking of peptidoglycan structures, reason why the internalization is not required for its activity.[266] Following this rationale, Lawson and coworkers[267] synthesized polymerizable vancomycin derivatives bearing either acrylamide or PEG-acrylate and were tethered from a surface pendant to a polyacrylate backbone through a living radical polymerization, demonstrating that the vancomycin-PEG-acrylate derivatives showed a significant reduction in bacterial colony-forming units (CFU) with respect to nonfunctionalized control surfaces.

Antithrombogenic Polymer Drugs

The use of a cardiovascular device represents the introduction of a foreign surface into blood circulation. Under normal conditions (intact blood vessels), blood contacts an endothelium with anticoagulant and antithrombogenic properties. However, when artificial surfaces are placed in contact with blood, the hemostatic mechanism is activated; this is a physiological process that involves a complex set of interdependent reactions between the surface, platelets, and coagulation proteins, resulting in the formation of a clot or thrombus, which can be removed by fibrinolysis.[268] The first event that occurs when a biomaterial comes in contact with blood is the adsorption of proteins and other molecules; this protein layer has an impact on further biological processes such as cell adhesion or activation of enzyme cascades of coagulation or inflammation.[269]

In this sense, thrombogenicity is defined as the ability of a biomaterial to induce or promote the formation of thromboemboli.[270] Thrombogenicity is one aspect of hemocompatibility, which can be defined, according to ISO standards,[271] as the conjunction of thrombosis, coagulation, blood platelets, hematology, and immunology. The first three categories may be summarized as hemostasis.

Hemostasis can be divided into plasmatic (soluble) and cellular aspects. The plasmatic part of coagulation consists of a cascade-like activation of inactive proteases, finally leading to the activation of thrombin, which is the key enzyme for the formation of fibrin. Fibrin spontaneously assembles to fibrils, forming a fibrin clot.

Blood platelets (thrombocytes) present the cellular component of the coagulation/thrombotic system. They are 3 μm sized spherical to disk-shaped anuclear cells, which can adhere to surfaces, spread by forming pseudopodia, adhere to each other, release growth factors and cytokines, and enhance the humoral coagulation system.

Immunoreactions to foreign materials are mainly nonspecific. The leading effector of the soluble nonspecific immune system is the complement system. Similar to the coagulation cascade, it consists of a number of proteases, which activate each other in a cascade-like process. Consequences are the opsonization of the target for facilitated phagocytosis, activation of nonspecific inflammatory cells (granulocytes and monocytes), and attraction of these cells (chemotaxis).

Polymers, from natural or synthetic origin, are the biomaterials most widely used for manufacturing medical devices and disposable clinical apparatus which can be used in contact with blood, such as vascular prostheses, artificial kidney, blood pumps, artificial heart, dialyzers, and plasma separators.[272] The polymeric materials used for these applications are conventional materials like poly (tetrafluoroethylene) (PTFE), poly(vinyl chloride) (PVC), segmented poly-etherurethane (SPU), polyethylene (PE), or silicone rubber. These polymers are usually selected because of their technological properties like mechanical and chemical stability, processing and ease sterilization, combined with nontoxicity, and a reasonable hemocompatibility. For this reason, in most of the cases, when the devices are placed in contact with blood they initiate the formation of clots by the activation of platelets and other components of blood coagulation system. These phenomena are harmful for maintaining well-balanced function or even life in patients, being necessary the administration of lifelong antithrombogenic therapy. For example, synthetic vascular grafts are successfully used in the treatment of the pathology of large arteries but when they are used in the replacement of small diameter blood vessels (internal diameter <6 mm) are known to be highly thrombogenic and need special treatments to improve their patency after implantation.[273]

Therefore, the improvement of hemocompatibility of artificial polymer surfaces constitutes one of the main issues in biomaterials science. The main strategies to reduce thrombogenicity in polymeric materials can be resumed in three different approaches:[274]

1. Design of bioactive polymeric materials that prevent the reactions involved in blood-material interaction: activation and aggregation of platelets and/or activation of blood coagulation cascade
2. Preparation of inert polymers that do not trigger blood reactions
3. Promotion of endothelial cell growth (re-endothelization)

Surface modification by bioactive polymers

The inner wall of natural blood vessel is mainly composed of a monolayer of endothelial cells. These cells posses several active anticoagulant mechanisms, for this reason, blood does not coagulate in normal blood vessels.[275] As an attempt to reproduce the behavior of this layer, the immobilization of active molecules onto polymeric surfaces represents one of the most popular approaches to improve the hemocompatibility of cardiovascular devices.

Heparin, as the analog to the cell surface anticoagulant heparan sulfate, has been the first anticoagulant successfully immobilized onto materials in order to reduce its thrombogenicity.[276]

Heparins constitute a family of glycosaminoglycans of various molecular weights with anticoagulant activity due to a high-affinity binding to antithrombin III (AT III) and therefore able to act as indirect thrombin inhibitors. AT III is the molecular target of heparin and, when it is activated, it binds to either thrombin and/or factor X from the coagulation cascade. Many researchers have been able to immobilize heparin by covalent or ionic bonding.[277–281]

The main concern of using heparin is based on the fact that, once immobilized, heparin should be able to adopt its native conformation for its interaction with AT III. It has been demonstrated that binding of AT III is more efficient when heparin is coupled by end-point attachment to a polymeric chain. This strategy was followed by Larm

et al.[282] for the development of Carmeda® Bioactive Surface, a heparin coating technique which has been licensed for its use on vascular grafts, coronary stents, oxygenation systems, and extracorporeal devices. Heparin is first partially depolymerized by deaminate cleavage to produce heparin fragments terminating in an aldehyde group. These heparin fragments are then covalently linked to the primary amino groups of PEI.

Heparin has also been covalently attached onto poly(carbonate-urea)urethane graft (Myolink®) using spacer arm technology. This is a grafting technique consisting on the use of spacer arms to reduce steric hindrance by the proximity of the ligand to the rigid surface of the polymeric backbone.[283] Following this rationale, the combination of heparin with an RGD peptide moieties confers antithrombogenic properties to the surface as well as improves endothelial cell adhesion in comparison with native Myolink.[284]

Another approach for surface heparinization consist of the immobilization of heparin macro-molecules by ionic bonding onto polymeric surfaces. The strong anionic character of this molecule facilitates the anchorage on surfaces previously treated with a cationic substance. In this sense,[285] described a heparin bioconjugate based on a branched thermoresponsive star-shaped cationic polymer. The system combines a PNIPA-based surface adsorption domain with a cationic polymer as a heparin-binding domain. Upon mixing the polymer with heparin, NPs are formed resulting from the formation of polyionic complexes. These particles can be adsorbed onto hydrophobic polymers commonly used as materials for medical devices, such as silicone, PE, polystyrene, or poly(ethylene terephthalate) (PET).

Another ionic approach has been developed by Hydromer[286] and consist of a polysaccharide-based polymer (F202TM). This polymer has been evaluated as antithrombogenic coating on polyurethanes as well as on medical grade electropolished stainless steel, showing in both cases a reduction in platelet adhesion and thrombus formation.

A commercial procedure for ionically bounding of heparin (Duraflo II, Baxter Bentley Healthcare Systems, Irvine, CA) has been used to coat cardiopulmonary bypass circuits and other medical devices. This coating has proved to reduce surface thrombus formation as well as a significant reduction of C3 and C4 complement activation.[287–289]

Technologies based on the preparation of albumin–heparin coatings have also been described.[290,291] These systems are prepared by the layer-by-layer assembly technique consisting on the sequential adsorption of albumin and heparin onto a surface under certain conditions where the constituents bear opposite charges.[292] Albumin–heparin multilayers have been demonstrated to improve hemocompatibility of cardiovascular devices; moreover, the increasing number of layers enhanced the coating bioactivity, suggesting that heparin maintains its activity inside the assemblies.[293]

In Vitro–Medical

As mentioned earlier, platelet response to a foreign material is one of the most important parameters to take into consideration for the evaluation of blood compatibility of a biomaterial, since coagulation on the surface starts with platelet aggregation and the formation of a fibrin network, giving raise to thrombus formation, from the combination of mutually fused platelets plus the insoluble fibrin and the cells that are trapped from blood.[272]

There are several drugs that are used to inhibit platelet aggregation; however, only aspirin, dipyridamole, and thienopyridines are currently approved by the Food and Drug Administration (FDA) for use in patients. Attending to the mechanisms of action, these drugs can be classified into the following four groups:[294]

1. Cyclooxygenase inhibitors: Aspirin and aspirin-like drugs
2. ADP receptor blockers: Thienopyridine derivatives (ticlopidine and clopidogrel)
3. Adenosine uptake inhibitor: Dipyridamole
4. Inhibitors of platelet glycoprotein IIb/IIIa (GP IIb/IIIa): Abciximab, eptifibatide, tirofiban

Several authors have described the direct immobilization of antiplatelet agents onto polymeric surfaces. For example,[274][295] described dipyridamole immobilization onto polyurethane surfaces, in which the drug is linked to a photoreactive moiety via an spacer group. These systems showed reduced platelet adhesion *in vitro* and improved patency in a goat model.

Aspirin and aspirin derivatives have also been extensively incorporated into blood-contacting polymers. In this sense, Paul et al.[296] developed poly (vinyl alcohol) (PVA) hemodialysis membranes loaded with acetylsalicylic acid.[297] described the application of random copolymers of HEMA and a methacrylic derivative of salicylic acid as coating of Dacron vascular prostheses. The presence of salicylic acid in the coating produced a noticeable decrease in the average number of platelets adhered to the graft.

Polyacrylic derivatives of (2-acetyloxy-4-trifluoromethyl)benzoic acid (Triflusal), an antiplatelet drug with a chemical structure closely related to aspirin, have also been described. These systems are based on a Triflusal derivative synthesized by an esterification reaction with HEMA (THEMA). When the homopolymer bearing Triflusal was applied as coating for commercial vascular pros-theses of PTFE, an improvement of the hemocompatibility of these devices was shown, since there was a clear reduction in platelet adhesion when the prostheses were coated with the polymer.[298]

Copolymerization of THEMA with other hydrophilic biocompatible monomers gave rise to materials that not only present activity in its macromolecular form, but also can be applied as drug delivery systems.[299,300] In this sense, copolymers of THEMA and *N,N'*-dimethylacrylamide (DMAA) were prepared. These systems showed good cell

biocompatibility and *in vitro* antiaggregant behavior. Furthermore, copolymer released Triflusal in a sustained way during several months.

A different family of hydrophilic copolymers was prepared by copolymerization of THEMA with AMPS. These systems combined the presence of the antiaggregant drug with a heparin-like behavior due to the presence of sulfonic group from AMPS monomer. THEMA-AMPS copolymers prevented platelet adhesion and aggregation and also showed a zero-order release for several months.[301,302]

Recently, new polymeric systems prepared from THEMA and butyl acrylate (BA) have been described.[303] These systems have been adequate for the development of stent coatings, since they exhibit good adhesion and crack-bridging properties, requirements that are necessary to withstand the mechanical deformation produced during the application of vascular stents. Furthermore, these polymers present an antithrombogenic behavior as well as good biocompatibility. Poly(THEMA-*co*-BA) copolymers have been loaded with antiproliferative drugs such as simv-astatin and paclitaxel in order to develop drug-eluting stents; these coatings simultaneously bear an antiaggregant drug for the prevention of thrombosis and antiproliferative drug to inhibit restenosis postimplantation. The coated stents have been licensed for use in human and commercialized by Iberhospitex (Spain) as IRIST® and ACTIVE®.

Another molecule that prevents blood coagulation and thrombus formation is NO, a signal molecule produced in endothelial cells by the endothelial isoform of nitric oxide synthase (eNOS), a calcium-cadmodulin-sensitive enzyme. In the cardiovascular system, NO triggers a cascade of events leading to smooth muscle relaxation and a subsequent decrease in blood pressure. Furthermore, it prevents platelet adhesion and activation. NO is also involved in the immune response and serves as a potent neurotransmitter at the neuron synapses. In normal physiological conditions, NO has an anti-inflammatory effect, but it could also be a proinflammatory mediator in abnormal situations. On the other hand, NO serves as a central regulator of oxidant reactions and diverse free radical-related disease processes.[304]

Researches in tissue engineering have worked to develop new materials which could release and generate NO.[305] described a NO-releasing polyurethane-PEG copolymer that incorporates the cell adhesive peptide sequence YIGSR. This system showed a decreasing in platelet adhesion as well as an increasing in endothelial cell proliferation when compared to control polyurethane.

Wu et al.[306] have described the preparation of multifunctional bilayer polymeric coatings prepared from commercial silicon rubber-polyurethane copolymers in which thrombomodulin and NO are incorporated in separate layers. By this approach, blood-compatible biomedical surfaces are designed by mimicking the highly thromboresistant endothelium layer.

Surface passivation by inert polymers

Biomaterials researchers have made a great effort in the last decades for the preparation and development of polymeric systems that are inert or passive with respect to blood reactions, and display thromboresistant behavior by minimizing the interaction with proteins and cells. This strategy is known as surface passivation.

Both synthetic and natural polymers have been investigated in the development of strategies to passivate a biomaterial surface, by reducing or eliminating enthalpic or entropic effects that drive protein and cell absorption on a molecular level.[307] In this sense, brushes of long-chained hydrophilic molecules like poly(ethylene oxide) (PEO) or the related molecules PEG or tetraethylene glycol dimethyl ether (tetraglyme) and others are known to be biologically inert and display excellent biocompatibility, and therefore have been suggested for hemocompatible surface modification.[308–312] Several methods have been established with this purpose, including bulk modification, covalent grafting, and physical adsorption. These systems has found application as coatings for dialysis membranes[313,314] as well as for vascular stents,[315,316] microencapsulation of cells,[317,318] or drug delivery systems.[319,320]

Another strategy for the development of inert thromboresistant surfaces consist of the preparation of albumin-coated polymers. Albumin is the most abundant protein in blood plasma. Several studies have demonstrated that this protein induces less platelet adhesion and activation than other plasma protein like fibrinogen or γ-globulin. Surface coating with this molecule is regarded to mask the device from the immune system and coagulation processes; however, this method has limitations due to conformational changes of the adsorbed molecules, exchange processes, physiological degradation of the protein, difficulties during the sterilization process, and risk of transmission of infection diseases due to the human origin of this molecule. Some investigations are directed to the covalent grafting of albumin onto surfaces, while others are based on surface modification with long aliphatic chains in order to enhance the affinity of albumin.

In a similar approach to surface masking with albumin, phosphorylcholine surfaces have been used in order to mimic the cell membrane. Several investigators have hypothesized that a surface similar to the external phospholipid membrane of cells should be nonthrombogenic. Since phosphorylcholine, the main lipid head group present on the external surface of blood cells, is inert in coagulation assays, it has been proposed for incorporation into polymeric surfaces in order to improve its hemocompatibility.

Planar-supported lipid bilayers composed of PC have shown to inhibit protein and cell adhesion *in vivo*. It has been proposed that this phenomenon is due to the zwitterionic nature of the phosphorylcholine head group that while carrying both positive and negative charges is electrically neutral at physiological pH. Application of supported lipid films as coatings for implantable devices has been limited by the inherent instability of a coating that is formed by individual molecules that self-assemble as a monolayer or bilayer film through relatively weak hydrophobic van der Waals interactions. Therefore, several approaches have been focused on the development of stable membrane-mimetic films through protein anchors, heat stabilization, and in situ polymerization of synthetically modified polymerizable phospholipids.[321–324] The protein and cell-resistant properties of the exposed PC layer were retained in all the studies. Jordan et al.[325] described a polymerized membrane-mimetic film prepared by in situ photopolymerization of an acrylate-functionalized phospholipid assembly at a solid-liquid interface. These systems were applied as coating for small diameter vascular grafts and demonstrated a reduction in platelet adhesion using a baboon femoral arteriovenous shunt model.

A different approach is based on the direct chemical grafting of PC head group to metal or polymer surfaces.[326–330]

Promotion of endothelial cells growth

An intact endothelial cell layer is the most hemocompatible blood-contacting surface; therefore, stimulation of growth of this cell type onto a biomaterial surface allows the preparation of a permanently active hemocompatible surface. The underlying assumption of this strategy is that when cultured on artificial biomaterials, a confluent layer of endothelial cells maintain their nonthrombogenic phenotype. Endothelial cells inhibit thrombosis through three interconnected regulatory systems: at the coagulation cascade level, the cellular components of blood such as leukocytes and platelets, at the complement cascade level, and also through effects on fibrinolysis and vascular tone.[331]

Combination of endothelial cells with biomaterials have been carried out in a great number of applications in order to prevent thrombotic or inflammatory reactions, or in other words, to improve the integration of the artificial device. For example, the lumen of vascular grafts has been seeded with endothelial cells as a strategy to create an interface that "hides" the biomaterial, enabling blood contact without significant inflammation and thrombosis, and therefore, maintaining graft patency.[332,333] A different approach consist of the fabrication of tissue-engineered constructs, in which blood must be supplied to cells within the construct without contacting the biomaterial scaffold. In this sense, different strategies have been developed in order to encourage blood vessel formation within the biomaterial scaffolds, involving the seeding of the construct with endothelial cells prior to implantation[334–336] or the immobilization of biomolecules onto the biomaterial, such as specific antibodies[337] or growth factors[338] in order to encourage endothelial cell attachment.

In Vitro–Medical

Low Friction Polymer Drugs

Polymeric materials with slippery or low friction surfaces are valuable in biomedical technologies.[339] Lubricity is required for medical devices involving moving parts such as artificial joints as well as tubular devices, e.g., catheters or endoscopes that are inserted into blood vessels, urethra, or other parts of the body which have mucous membranes. Therefore, low frictional surface properties of these medical devices are desirable in order to reduce the pain accompanied by their introduction into the body and also to limit the risk of damage to the mucous membranes or the intima of blood vessels, which may lead to infectious diseases or mural thrombus formation.[340] The use of lubricants to reduce friction and wear between rubbing surfaces has been documented since antiquity.[341,342] In the last decades, even though many surface modification techniques have been developed in attempt to reduce the friction coefficients (μ) of characteristic medical device surfaces, only a few of them are applicable in the body environment (Fig. 12) (Table 3).

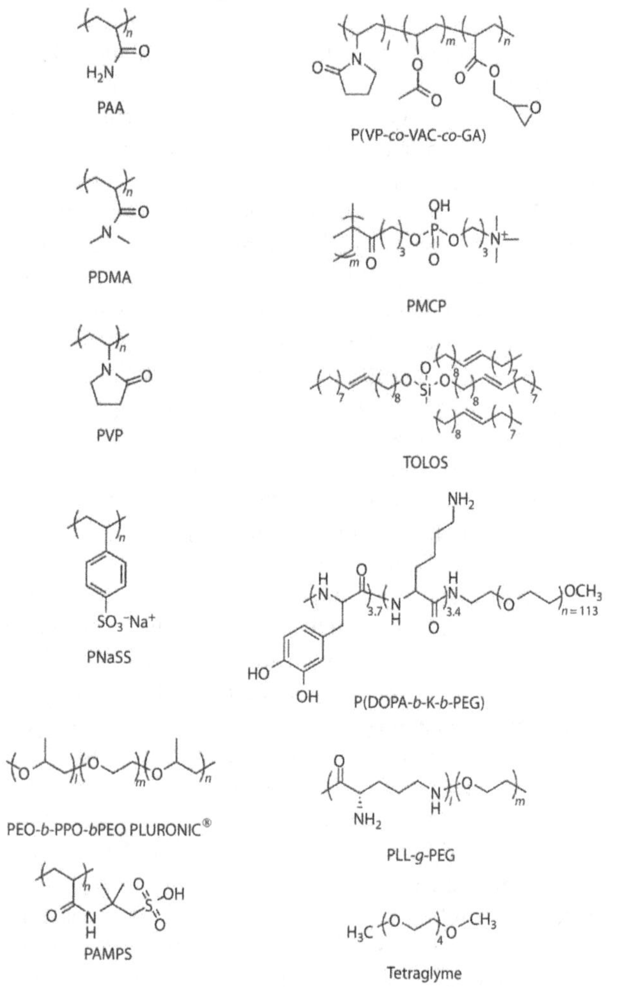

Several methods of decreasing surface friction are known. End-grafting of hydrophilic polymer chains through surface-initiated polymerization of monomers has been extensively investigated as an effective means to impart hydrophilicity and lubricity to various polymer-based medical devices. Nagaoka and Akashi[340] described a methodology of binding N-vinyl pyrrolidone (VP) copolymers with epoxy groups, which were covalently coated on polyurethane catheter substrates. Uyama et al. also reported a methodology to prepare catheter devices possessing low μ values.[343,344] Surface modification of ethylene-vinyl acetate (EVA) copolymers and plasticized PVC, both used as catheter manufacturing polymers, through graft polymerization with nonionic water-soluble monomers such as acrylamide (AAm) and DMAA, was demonstrated useful to reduce their μ values in hydrate states to such an extent as becoming slippery.

2-Methacryloyloxyethyl phosphorylcholine (MPC)-based polymers constitute other example of a common biocompatible and hydrophilic polymers studied so far with high lubricity and low friction and antiprotein adsorption. Kyomoto and coworkers[345] developed an artificial hip joint by using MCP grafted onto the surface of cross-linked PE (PMPC-g-cLPE). This device was designed to reduce wear between the ultrahigh-molecular-weight PE component of the prosthesis and the metallic surface that articulates against, suppressing therefore the progressively bone resorption by osteolysis, leading to aseptic loosing of the artificial joint after a number of years, which is recognized as a serious problem.[346]

Ratner and Hoffman[347] described a process to graft VP-based hydrogels onto organic polymeric substrates using UV radiation for forming biocompatible coatings. Fan and Lawrance[348] developed a polymeric complex for forming biocompatible coatings, some of which could render the surface lubricious when exposed to aqueous or body fluids. Hu et al. achieved the same goal by plasma treatment only.[349,350] The lubricous coating is composed

Fig. 12 Representative low-friction polymers used in medical devices.

Table 3 Friction coefficients (μ) determined for different modified polymeric surfaces commonly used in medical devices

Polymeric Coating/ Polymeric Substrate	Coated Surface Friction Coefficient	Substrate Friction Coefficient
PVP/polyurethane	0.035	0.32
PDMA/poly (vinyl acetate)	<0.1	0.7
PMPC/polyethylene	0.026[a]	0.075[a]
TOLOS/tetrapropoxysilane	0.029	0.531
P(DOPA-co-K)PEG/PDMS	0.03	0.98
PEO-b-PPO-b-PEO/PDMS	0.05	0.90
PLL-g-PEG/PDMS	0.028	0.90
PNaSS/PDMS	10^{-4}	

[a]Dynamic friction coefficient.

of a polyelectrolyte molecular film, along with the hydrophilic, lubricant molecules on a plasma-treated plastic surface. Then the molecular film was further cross-linked with aldehyde-functionalized molecules such as Healon® to form an interpenetrating network, which directly adheres to polymeric matrices fully employed in the development of IOL cartridges.[350] To minimize friction within the cartridge tip and ease IOL deployment, a lubricious coating is necessary.[351]

The use of self-lubricant elastomers has gained widespread medical acceptance as an alternative approach to the hydrogel-coated biomaterials since they present a low risk of unfavorable biological reactions and provide improved patient comfort compared to other biomaterials used for drug delivery, urinary catheters, and other implantable devices. However, silicone lacks inherent lubricity and has a relatively high μ, Woolfson et al. reported on the development of novel self-lubricating silicone elastomers produced by condensation of cure systems employing higher-molecular-weight tetra (alkyloxysilane) cross-linking agents.[352] These silane cross-linkers are simple to synthesize in a one-step process from their low-molecular-weight propoxy analog. The tetraoleyloxysilane (TOLOS) derivative represents a successful example of this low toxicity lubricant agent with a μ near zero to be used in indwelling devices such as prostheses, contact lenses, intravaginal rings, and so on.

Previous studies involving grafting of PEG chains onto elastomeric biomedical devices indicated a dramatic reduction of friction forces under aqueous conditions.[353,16] Chawla and coworkers achieved the modification of poly (dimethylsiloxane) (PDMS) elasto-meric substrates with enhanced tribological properties.[354] Synthetic PEG-based polymers containing 1-3,4-dihydroxyphenylalanine (DOPA) and lysine (K) (DOPA-b-K-b-PEG) were grafted through noncovalent interactions between hydrophobic anchoring groups of PEG-based copolymers and the PDMS surfaces in aqueous media.

An alternative to enhance the tribological properties of medical devices consists in tethering polymer chains to the surfaces by one end which act as molecular "brushes" and facilitate sliding when they are swollen by aqueous solvents. This biomimetic approach is based on the architecture of biological lubricant additives, usually glycoproteins, in which large number of sugar chains are bound along a protein backbone (Fig. 13). For example, mucins are found in most parts of the human body that need lubricating, such as knees and eyes.[355] The characteristic bottlebrush structure of these supramolecular complexes plays a crucial role in the mechanism of aid lubrication. The hydrophilic sugars immobilize large amounts of water within the contact region, while the backbone interconnects to other bottlebrushes or to a surface.[356] Other important characteristic of natural tribological systems is that they usually involve soft surfaces and deform elastically in response to

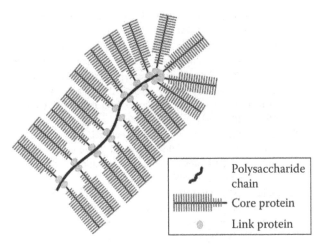

Fig. 13 Hierarchical bottlebrush like architecture of biolubricating systems.

external loads, increasing the contact area resulting in a relatively low contact pressure. The synergic combination of hairy polymers and soft surfaces has been extensively studied and had led to applications in biomedical implants. Lee and Spencer recently reported enhanced water lubrication of brush-like grafted PEO-*block*-PEO-*block*-PEO (PEO-*b*-PPO-*b*-PEO, Pluronic®) and PLL-*graft*-PEG (PLL-*g*-PEG) onto silicone-rubber surfaces,[357] while[358] have gone a step further in biomimicry by using polyelectrolyte brushes hydrogels based on poly (sodium 4-styrene sulfonate) (PNaSS) for aqueous lubrication. These hieratical systems lead to superior lubrication properties acting in the same way than biolubricating systems,[359] where the charged hydrophilic surfaces provide an electrostatic double-layer repulsion, in addition to the steric repulsion of any protruding polyelectrolytes and the hydration layer of tightly bound water molecules.[360]

REFERENCES

1. Ringsdorf, H. Structure and properties of pharmacologically active polymers. J. Polym. Sci. Polym. Symp. **1975**, *51* (1), 135–153.
2. WHOC. TC 194 Biological Evaluation of Medical Devices, ISO 10993–3:2002, Biological Evaluation of Medical Devices—Part 4: Selection of Tests for Interactions with Blood, Int. Organization for Standardisation. http://www.who.int/cancer/en 2009.
3. Allen, T.M. Ligand-targeted therapeutics in anticancer therapy. Nat. Rev. Cancer **2002**, *2* (10), 750–763.
4. Haag, R.; Kratz, F. Polymer therapeutics: Concepts and applications. Angew. Chem. Int. Ed. **2006**, *45*, 1198–1215.
5. Park, J.H.; Lee, S.; Kim, J.H.; Park, K.; Kim, K.; Kwon, I. CPolymeric nanomedicine for cancer therapy. Progr. Polym. Sci. **2008**, *33* (1), 113–137.
6. Duncan, R.; Seymour, L.C.W.; Scarlett, L. Fate of N-(2-hydroxypropyl)methacrylamide copolymers with pendent galactosamine residues after intravenous

administration to rats. Biochim. Biophys. Acta Gen. Subj. **1986**, *880* (1), 62–71.

7. Seymour, L.W.; Ferry, D.R.; Anderson, D.; Hesslewood, S.; Julyan, P.J.; Poyner, R.; Doran, J.; Young, A.M.; Burtles, S.; Kerr, D.J.; Cancer Research Campaign Phase I/II Clinical Trials committee. Hepatic drug targeting: Phase I evaluation of polymer-bound doxorubicin. J. Clinical Oncol. **2002**, *20* (6), 1668–1676.

8. Low, P.S.; Henne, W.A.; Doorneweerd, D.D. Discovery and development of folic-acid-based receptor targeting for imaging and therapy of cancer and inflammatory diseases. Acc. Chem. Res. **2008**, *41* (1), 120–129.

9. Salazar, M.D.; Ratnam, M. The folate receptor: What does it promise in tissue-targeted therapeutics? Cancer and Metastasis Reviews, **2007**, *26*, 141–152.

10. Bae, Y.; Nishiyama, N.; Kataoka, K. *In vivo* antitumor activity of the folate-conjugated pH-sensitive polymeric micelle selectively releasing adriamycin in the intracellular acidic compartments. Bioconj. Chem. **2007**, *18* (4), 1131–1139.

11. De Duve, C.; De Barsy, T.; Poole, B. Lysosomotropic agents. Biochem. Pharmacol. **1974**, *23* (18), 2495–2531.

12. Bareford, L.M.; Swaan, P.W. Endocytic mechanisms for targeted drug delivery. Adv. Drug Deliv. Rev. **2007**, *59* (8), 748–758.

13. Meng, F.; Hennink, W.E.; Zhong, Z. Reduction-sensitive polymers and bioconjugates for biomedical applications. Biomaterials **2009**, *30*, 2180–2198.

14. Sun, H.; Guo, B.; Li, X.; Cheng, R.; Meng, F.; Liu, H.; Zhong, Z. Shell-sheddable micelles based on dextran-SS-poly (e-caprolactone) diblock copolymer for efficient intracellular release of doxorubicin. Biomacromolecules **2010**, *11* (4), 848–854.

15. Guo, K.; Li, J.; Tang, J.P.; Tan, C.P.; Wang, H.; Zeng, Q. Monoclonal antibodies target intracellular PRL phosphatases to inhibit cancer metastases in mice. Cancer Biol. Ther. **2008**, *7*, 752–759.

16. Lee, S.; Spencer, N.D. Poly (L-lysine)-graft-poly (ethylene glycol): A versatile aqueous lubricant additive for tribosystems involving thermoplastics. Lub. Sci. **2008**, 20, 21–34.

17. Sirova, M.; Mrkvan, T.; Etrych, T.; Chytil, P.; Rossmann, P.; Ibrahimova, M.; Kovar, L.; Ulbrich, K.; Rihova, B. Preclinical evaluation of linear HPMA-doxorubicin conjugates with pH-sensitive drug release: Efficacy, safety, and immunomodulating activity in murine model. Pharm. Res. **2010**, *27* (1), 200–208.

18. Lopez Donaire, M.L.; Parra-Caceres, J.; Vazquez-Lasa, B.; García-Alvarez, I.; Fernández-Mayoralas, A.; López-Bravo, A.; San Román, J. Polymeric drugs based on bioactive glycosides for the treatment of brain tumours. Biomaterials **2009**, *30* (8), 1613–1626.

19. Matsumura, Y.; Maeda, H. A new concept for macromolecular therapeutics in cancer chemotherapy: Mechanism of tumoritropic accumulation of proteins and the antitumor agent SMANCS. Cancer Res. **1986**, *46* (12 Pt 1), 6387–6392.

20. Duncan, R. Polymer conjugates for tumour targeting and intracytoplasmic delivery. The EPR effect as a common gateway? Pharm. Sci. Technol. Today **1999**, *2* (11), 441–449.

21. Duncan, R. The dawning era of polymer therapeutics. Nat. Rev. Drug Dis. **2003**, *2* (5), 347–360.

22. Maeda, H.; Wu, J.; Sawa, T.; Matsumura, Y.; Hori, K. Tumor vascular permeability and the EPR effect in macromolecular therapeutics: A review. J. Control. Release, **2000**, *65* (1-2), 271–284.

23. Duncan, R. Polymer conjugates as anticancer nanomedicines. Nat. Rev. Cancer, **2006**, *6* (9), 688–701.

24. Takakura, Y.; Hashida, M. Macromolecular carrier systems for targeted drug delivery: Pharmacokinetic considerations on biodistribution. Pharm. Res. **1996**, *13* (6), 820–831.

25. Duncan, R.; Seymour, L.W.; O'Hare, K.B.; Flanagan, P.A.; Wedge, S.; Hume, I.C.; Suarato, A. Preclinical evaluation of polymer-bound doxorubicin. J. Control. Release **1992**, *19* (1-3), 331–346.

26. Duncan, R. Development of HPMA copolymer-anticancer conjugates: Clinical experience and lessons learnt. Adv. Drug Del. Rev. **2009**, *61* (13), 1131–1148.

27. Vasey, P.A.; Kaye, S.B.; Morrison, R.; Twelves, C.; Wilson, P.; Duncan, R.; Thomson, A.H.; Murray, L.S.; Hilditch, T.E.; Murray, T.; Burtles, S.; Fraier, D.; Frigerio, E.; Cassidy, J. Phase I clinical and pharmacokinetic study of PK1 [N-(2-hydroxypropyl)methacrylamide copolymer doxorubicin]: First member of a new class of chemotherapeutic agents—Drug-polymer conjugates. Clin. Cancer Res. **1999**, *5* (1), 83–94.

28. Etrych, T.; Chytil, P.; Mrkvan, T.; Sírová, M.; Ríhová, B.; Ulbrich, K. Conjugates of doxorubicin with graft HPMA copolymers for passive tumor targeting. J. Control Release **2008**, *132* (3), 184–192.

29. Minchinton, A.I.; Tannock, I.F. Drug penetration in solid tumours. Nat. Rev. Cancer **2006**, *6* (8), 583–592.

30. Nagamitsu, A.; Greish, K.; Maeda, H. Elevating blood pressure as a strategy to increase tumor-targeted delivery of macromolecular drug SMANCS: Cases of advanced solid tumors. Jpn. J. Clin. Oncol. **2009**, *39* (11), 756–766.

31. Maeda, H.; Noguchi, Y.; Sato, K.; Akaike, T. Enhanced vascular permeability in solid tumor is mediated by nitric oxide and inhibited by both new nitric oxide scavenger and nitric oxide synthase inhibitor. Jpn J. Cancer Res. **1994**, *85* (4), 331–334.

32. Fang, J.; Sawa, T.; Akaike, T.; Greish, K.; Maeda, H. Enhancement of chemotherapeutic response of tumor cells by a heme oxygenase inhibitor, pegylated zinc protoporphyrin. Int. J. Cancer **2004**, *109* (1), 1–8.

33. Seki, T.; Fang, J.; Maeda, H. Enhanced delivery of macromolecular antitumor drugs to tumors by nitroglycerin application. Cancer Sci., **2009**, *100* (12), 2426–2430.

34. Yasuda, H.; Nakayama, K.; Watanabe, M. Nitroglycerin treatment may enhance chemosensitivity to docetaxel and carboplatin in patients with lung adenocarcinoma. Clin. Cancer Res. **2006**, *12* (22), 6748–6757.

35. Seymour, L.W.; Ferry, D.R.; Kerr, D.J.; Rea, D.; Whitlock, M.; Poyner, R.; Boivin, C.; Hesslewood, S.; Twelves, C.; Blackie, R.; Schatzlein, A.; Jodrell, D.; Bissett, D.; Calvert, H.; Lind, M.; Robbins, A.; Burtles, S.; Duncan, R.; Cassidy, J. Phase II studies of polymer-doxorubicin (PK1, FCE28068) in the treatment of breast, lung and colorectal cancer. Int. J. Oncol. **2009**, *34* (6), 1629–1636.

36. Atkins, J.H.; Gershell, L.J. Selective anticancer drugs. Market indicators. Nat. Rev. Drug Discov. **2002**, *1* (7), 491–492.

37. Bissett, D.; Cassidy, J.; De Bono, J.S.; Muirhead, F.; Main, M.; Robson, L.; Fraier, D.; Magnè, M.L.; Pellizzoni, C.; Porro, M.G.; Spinelli, R.; Speed, W.; Twelves, C. Phase I and pharmacokinetic (PK) study of MAG-CPT (PNU 166148): A polymeric derivative of camptothecin (CPT). Br. J. Cancer **2004**, *91* (1), 50–55.

38. Sarapa, N.; Britto, M.R.; Speed, W.; Jannuzzo, M.; Breda, M.; James, C.A.; Porro, M.; Rocchetti, M.; Wanders, A.; Mahteme, H.; Nygren, P. Assessment of normal and tumor tissue uptake of MAG-CPT, a polymer-bound prodrug of camptothecin, in patients undergoing elective surgery for colorectal carcinoma. Cancer Chemother. Pharmacol. **2003**, *52* (5), 424–430.

39. Mrkvan, T.; Sirova, M.; Etrych, T.; Chytil, P.; Strohalm, J.; Plocova, D.; Ulbrich, K.; Rihova, B. Chemotherapy based on HPMA copolymer conjugates with pH-controlled release of doxorubicin triggers anti-tumor immunity. J. Control Release, **2005**, *110* (1), 119–129.

40. Erez, R.; Segal, E.; Miller, K.; Satchi-Fainaro, R.; Shabat, D. Enhanced cytotoxicity of a polymer-drug conjugate with triple payload of paclitaxel. Bioorg. Med. Chem. **2009**, *17* (13), 4327–4335.

41. Auzenne, E.; Donato, N.J.; Li, C.; Leroux, E.; Price, R.E.; Farquhar, D.; Klostergaard, J. Superior therapeutic profile of poly-L-glutamic acid-paclitaxel copolymer compared with Taxol in xenogeneic compartmental models of human ovarian carcinoma. Clin. Cancer Res. **2002**, *8*, 573–581.

42. Bhatt, R.; de Vries, P.; Tulinsky, J.; Bellamy, G.; Baker, B.; Singer, J.W.; Klein, P. Synthesis and *in vivo* antitumor activity of poly (l-glutamic acid) conjugates of 20S-camptothecin. J. Med. Chem. **2003**, *46* (1), 190–193.

43. Rademaker-Lakhai, J.M.; Terret, C.; Howell, S.B.; Baud, C.M.; de Boer, R.F.; Pluim, D.; Beijnen, J.H.; Schellens, J.H.; Drozs J.-P. A phase I and pharmacological study of the platinum polymer AP5280 given as an intravenous infusion once every 3 weeks in patients with solid tumors. Clin. Cancer Res. **2004**, *10* (10), 3386–3395.

44. Danhauser-Riedl, S.; Hausmann, E.; Schick, H.; Bender, R.; Dietzfelbinger, H.; Rastetter, J.; Hanauske, A.R. Phase I clinical and pharmacokinetic trial of dextran conjugated doxorubicin (AD-70, doxorubicin-OXD). Invest New Drugs **1993**, *11* (2-3), 187–195.

45. Kumazawa, E.; Ochi, Y.DE-310, a novel macromolecular carrier system for the camptothecin analog DX-8951f: Potent antitumor activities in various murine tumor models. Cancer Sci. **2004**, *95* (2), 168–175.

46. Rowinsky, E.K.; Rizzo, J.; Ochoa, L.; Takimoto, C.H.; Forouzesh, B.; Schwartz, G.; Hammond, L.A.; Patnaik, A.; Kwiatek, J.; Goetz, A.; Denis, L.; McGuire, J.; Tolcher, A.W. A phase I and pharmacokinetic study of pegylated camptoth-ecin as a 1-hour infusion every 3 weeks in patients with advanced solid malignancies. J. Clin. Oncol. **2003**, *21* (1), 148–157.

47. Yurkovetskiy, A.V.; Fram, R.J.XMT-1001, a novel polymeric camptothecin pro-drug in clinical development for patients with advanced cancer. Adv. Drug Deliv. Rev., **2009**, *61* (13), 1193–1202.

48. Sapra P.; Zhao H.; Mehlig, M.; Malaby, J.; Kraft, P.; Longley, C.; Greenberger, L.M.; Horak, I.D. Novel delivery of SN38 markedly inhibits tumor growth in xeno-grafts, including a camptothecin-11- refractory model. Clin. Cancer Res. **2008**, *14* (6), 1888–1896.

49. Pasut, G.; Veronese, F.M. PEG conjugates in clinical development or use as anticancer agents: An overview. Adv. Drug Deliv. Rev. **2009**, *61* (13), 1177–1188.

50. Chytil, P.; Etrych, T.; Konak, C.; Sírová, M.; Mrkvan, T.; Boucek, J.; Ríhová, B.; Ulbrich, K. New HPMA copolymer-based drug carriers with covalently bound hydrophobic substituents for solid tumour targeting. J Control Release **2008**, *127* (2), 121–130.

51. Lammers, T.; Subr, V.; Ulbrich, K. et al. Simultaneous delivery of doxorubicin and gemcitabine to tumors *in vivo* using prototypic polymeric drug carriers. Biomaterials **2009**, *30*, 3466–3475.

52. Langer, C.J. CT- 2103, A novel macromolecular taxane with potential advantages compared with conventional taxanes. Clin. Lung Cancer **2004**, *6*, S85–S88.

53. Choe, Y.H.; Conover, C.D.; Wu, D.; Royzen, M.; Gervacio, Y.; Borowski, V.; Mehlig, M.; Greenwald, R.B. Anticancer drug delivery systems: Multi-loaded N4-acylpoly (ethylene glycol) prodrugs of ara-C. II. Efficacy in ascites and solid tumors. J. Control. Release **2002**, *79*, 55–70.

54. Pasut, G.; Canal, F.; Dalla Via, L.; Arpicco, S.; Veronese, F.M.; Schiavon, O. Antitumoral activity of PEG-gemcitabine prodrugs targeted by folic acid. J. Control Release **2008**, *127* (3), 239–248.

55. Pechar, M.; Braunova, A.; Ulbrich, K.; Jelinkova, M.; Rihova, B. Poly (ethylene glycol)-doxorubicin conjugates with pH-controlled activation. J. Bioactive and Compatible Polymers, **2005**, *20* (4), 319–341.

56. Braunova, A.; Pechar, M.; Laga, R.; Ulbrich, K. Hydrolytically and reductively degradable high-molecular-weight poly (ethylene glycol)s. Macromol. Chem. Phys. **2007**, *208*, 2642–2653.

57. Pasut, G.; Greco, F.; Mero, A.; Mendichi, R.; Fante, C.; Green, R.J.; Veronese, F.M. Polymer-drug conjugates for combination anticancer therapy: Investigating the mechanism of action. J. Medi. Chem. **2009**, *52* (20), 6499–6502.

58. Davis, M.E. Design and development of IT-101, a cyclodextrin-containing polymer conjugate of camptothecin. Adv. Drug Deliv. Rev. **2009**, *61* (13), 1189–1192.

59. Greco, F.; Vicent, M.J. Combination therapy: Opportunities and challenges for polymer-drug conjugates as anticancer nanomedicines. Adv. Drug Deliv. Rev. **2009**, *61* (13), 1203–1213.

60. Verschraegen, C.F.; Skubitz, K.; Daud, A.; Kudelka, A.P.; Rabinowitz, I.; Allievi, C.; Eisenfeld, A.; Singer, J.W.; Oldham, F.B. A phase I and pharmacokinetic study of paclitaxel poliglumex and cisplatin in patients with advanced solid tumors. Cancer Chemother. Pharmacol. **2009**, *63* (5), 903–910.

61. Hongrapipat, J.; Kopeckova, P.; Liu, J.; Prakongpan, S.; Kopecek, J. Combination chemotherapy and photodynamic therapy with Fab' fragment targeted HPMA copolymer conjugates in human ovarian carcinoma cells. Mol. Pharmacol. **2008**, *5* (5), 696–709.

62. Chadna, P.; Saad, M.; Wang, Y. A novel targeted proapoptotic drug delivery system for efficient anticancer therapy.

Proceedings 35th Annual Meeting of the Controlled Release Society, New York, 2008.

63. Satchi-Fainaro, R.; Hailu, H.; Davies, J.W.; Summerford, C.; Duncan, R.PDEPT: Polymer-directed enzyme prodrug therapy. 2. HPMA copolymer-b-lactamase and HPMA copolymer-C-Dox as a model combination. Bioconjug. Chem. **2003**, *14*, 797–804.

64. Duncan, R.; Gac-Breton, S.; Keane, R.; Musila, R.; Sat, Y.N.; Satchi, R.; Searle, F. Polymer-drug conjugates, PDEPT and PELT: Basic principles for design and transfer from the laboratory to clinic. J. Control Release **2001**, *74* (1-3), 135–146.

65. Satchi, R.; Connors, T.A.; Duncan, R.PDEPT: Polymer-directed enzyme prodrug therapy: I. HPMA copolymer-cathepsin B and PK1 as a model combination. Br. J. Cancer **2001**, *85* (7), 1070–1076.

66. Davis, F.F. The origin of pegnology. Adv. Drug Deliv. Rev. **2002**, *54* (4), 457–458.

67. Caliceti, P.; Veronese, F.M. Pharmacokinetic and biodistribution properties of poly (ethylene glycol)-protein conjugates. Adv. Drug Deliv. Rev. **2003**, *55* (10), 1261–1277.

68. Vicent, M.J.; Duncan, R. Polymer conjugates: Nanosized medicines for treating cancer. Trends Biotechnol. **2006**, *24* (1), 39–47.

69. Cheng, P.N.M.; Lam, T.L.; Lam, W.M.; Tsui, S.M.; Cheng, A.W.; Lo, W.H.; Leung, Y.C. Pegylated recombinant human arginase (rhArg-peg5,000mw) inhibits the *in vitro* and *in vivo* proliferation of human hepatocellular carcinoma through arginine depletion. Cancer Res. **2007**, *67* (1), 309–317.

70. Armstrong, J.K.; Hempel, G.; Koling, S.; Chan, L.S.; Fisher, T.; Meiselman, H.J.; Garratty, G. Antibody against poly (ethylene glycol) adversely affects PEG-asparaginase therapy in acute lymphoblastic leukemia patients. Cancer **2007**, *110* (1), 103–111.

71. Mero, A.; Pasut, G.; Via, L.D.; Fijten, M.W.; Schubert, U.S.; Hoogenboom, R.; Veronese, F.M. Synthesis and characterization of poly (2-ethyl 2-oxazoline)-conjugates with proteins and drugs: Suitable alternatives to PEG-conjugates? J. Control. Release, **2008**, *125*, 87–95.

72. Duncan, R.; Gilbert, H.R.P.; Carbajo, R.J.; Vicent, M.J. Polymer masked-unmasked protein therapy. 1. Bioresponsive dextrin-trypsin and -melanocyte stimulating hormone conjugates designed for a-amylase activation. Biomacromolecules **2008**, *9*, 1146–1154.

73. Ferguson, E.L.; Duncan, R. Dextrin-phospholipase A Synthesis and evaluation as a bioresponsive anticancer conjugate. Biomacromolecules **2009**, *10* (6), 1358–1364.

74. Merdan, T.; Kopecek, J.; Kissel, T. Prospects for cationic polymers in gene and oligonucleotide therapy against cancer. Adv. Drug Deliv. Rev. **2002**, *54* (5), 715–758.

75. Walther, W.; Stein, U.; Fichtner, I. Intratumoral low-volume jet-injection for efficient nonviral gene transfer. Applied Biochem. Biotechnol. Part B Mol. Biotechnol. **2002**, *21*, 105–115.

76. Kawabata, K.; Takakura, Y.; Hashida, M. The fate of plasmid DNA after intravenous injection in mice: Involvement of scavenger receptors in its hepatic uptake. Pharm. Res. **1995**, *12* (6), 825–830.

77. Boussif, O.; Lezoualc'h, F.; Zanta, M.A.; Mergny, M.D.; Scherman, D.; Demeneix, B.; Behr, J. P. A versatile vector for gene and oligonucleotide transfer into cells in culture and *in vivo*: Polyethylenimine. Proc. Nat. Acad. Sci. USA **1995**, *92*, 7297–7301.

78. Collins, L.; Fabre, J.W. A synthetic peptide vector system for optimal gene delivery to corneal endothelium. J. Gene Med. **2004**, *6* (2), 185–194.

79. Campeau, P.; Chapdelaine, P.; Seigneurin-Venin, S.; Massie, B.; Tremblay, J.P. Transfection of large plasmids in primary human myoblasts. Gene Ther. **2001**, *8* (18), 1387–1394.

80. Fischer, D.; Bieber, T.; Li, Y.; Elsasser, H.P.; Kissel, T. A novel non-viral vector for DNA delivery based on low molecular weight, branched polyethylenimine: Effect of molecular weight on transfection efficiency and cytotoxicity. Pharm. Res. **1999**, *16* (8), 1273–1279.

81. Marschall, P.; Malik, N.; Larin, Z. Transfer of YACs up to 2.3 Mb intact into human cells with poly-ethylenimine. Gene Therapy, **1999**, *6* (9), 1634–1637.

82. Nguyen, H.K.; Lemieux, P.; Vinogradov, S.V.; Gebhart, C.L.; Guérin, N.; Paradis, G.; Bronich, T.K.; Alakhov, V.Y.; Kabanov, A.V. Evaluation of polyether-polyethyleneimine graft copolymers as gene transfer agents. Gene Ther. **2000**, *7* (2), 126–138.

83. Benns, J.M.; Mahato, R.I.; Kim, S.W. Optimization of factors influencing the transfection efficiency of folate-PEG-folate-graft-polyethylenimine. J. Control Release **2002**, *79* (1-3), 255–269.

84. Liang, B.; He, M.L.; Xiao, Z.P.; Li, Y.; Chan, C.Y.; Kung, H.F.; Shuai, X.T.; Peng, Y. Synthesis and characterization of folate-PEG-grafted-hyperbranched-PEI for tumor-targeted gene delivery. Biochem. Biophys. Res. Commun. **2008**, *367* (4), 874–880.

85. Wolfert, M.A.; Dash, P.R.; Nazarova, O. Polyelectrolyte vectors for gene delivery: Influence of cationic polymer on biophysical properties of complexes formed with DNA. Bioconjug. Chem. **1999**, *10* (6), 993–1004.

86. Mannisto, M.; Vanderkerken, S.; Toncheva, V.; Elomaa, M.; Ruponen, M.; Schacht, E.; Urtti, A. Structure-activity relationships of poly (L-lysines): Effects of pegylation and molecular shape on physicochemical and biological properties in gene delivery. J. Control. Release **2002**, *83* (1), 169–182.

87. Kaneshiro, T.L.; Wang, X.; Lu, Z.R.Synthesis, characterization, and gene delivery of poly-L-lysine octa (3-aminopropyl)silsesquioxane dendrimers: Nanoglobular drug carriers with precisely defined molecular architectures. Mol. Pharm. **2007**, *4* (5), 759–768.

88. Benns, J.M.; Choi, J.S.; Mahato, R.I.; Park, J.S.; Sung Wan, K. pH-Sensitive cationic polymer gene delivery vehicle: N-Ac-poly (L-histidine)-graft-poly (L-lysine) comb shaped polymer. Bioconj. Chem. **2000**, *11*, 637–645.

89. Yang, Y.; Xu, Z.; Jiang, J. Poly (imidazole/DMAEA)phosphazene/DNA self-assembled nanoparticles for gene delivery: Synthesis and *in vitro* transfection. J. Control. Release **2008**, *127* (3), 273–279.

90. Verbaan, F.J.; Klouwenberg, P.K.; Van Steenis, J.H.; Snel, C.J.; Boerman, O.; Hennink, W.E.; Storm, G. Application of poly (2-(dimethylamino) ethyl methacrylate)-based polyplexes for gene transfer into human ovarian carcinoma cells. Int. J. Pharm. **2005**, *304* (1-2), 185–192.

In Vitro–Medical

91. De Wolf, H.K.; De Raad, M.; Snel, C.; van Steenbergen, M.J.; Fens, M.H.; Storm, G.; Hennink, W.E. Biodegradable poly (2-dimethylamino ethylamino)phosp-hazene for *in vivo* gene delivery to tumor cells. Effect of polymer molecular weight. Pharm. Res. **2007**, *24* (8), 1572–1580.

92. Koping-Hoggard, M.; Tubulekas, I.; Guan, H.; Edwards, K.; Nilsson, M.; Vårum, K.M.; Artursson, P. Chitosan as a nonviral gene delivery system. Structure-property relationships and characteristics compared with polyethylenimine *in vitro* and after lung administration *in vivo*. Gene Ther. **2001**, *8* (14), 1108–1121.

93. Zhao, J.; Duan, J.; Zhang, Y. Cationic polybutyl cyanoacrylate nanoparticles for DNA delivery. J. Biomedicine and Biotechnology, Hindawi Publishing Corporation, Volume 2009, Article ID 149254, 9 pp. doi: 10.1155/**2009**/*14924*.

94. Choi, J.S.; Nam, K.; Park, J.Y.; Kim, J.B.; Lee, J.K.; Park, J.S. Enhanced transfection efficiency of PAMAM dendrimer by surface modification with L-arginine. J. Control Release **2004**, *99* (3), 445–456.

95. Huang, J.; Murata, H.; Koepsel, R.R.; Russell, A.J.; Matyjaszewski, K. Antibacterial polypropylene via surface-initiated atom transfer radical polymerization. Biomacromolecules **2007**, *8* (5): 1396–1399.

96. Lin, C.; Blaauboer, C.J.; Timoneda, M.M.; Lok, M.C.; van Steenbergen, M.; Hennink, W.E.; Zhong, Z.; Feijen, J.; Engbersen, J.F. Bioreducible poly (amido amine)s with oligoamine side chains: Synthesis, characterization, and structural effects on gene delivery. J. Control Release, **2008**, *126* (2), 166–174.

97. Patil, M.L.; Zhang, M.; Betigeri, S.; Taratula, O.; He, H.; Minko, T. Surface-modified and internally cationic polyamidoamine dendrimers for efficient siRNA delivery. Bioconjug. Chem. **2008**, *19* (7), 1396–1403.

98. Navarro, G.; Tros De Ilarduya, C. Activated and non-activated PAMAM dendrimers for gene delivery *in vitro* and *in vivo*. Nanomed. Nanotechnol. Biol. Med. **2009**, *5* (3), 287–297.

99. Merkel, O.M.; Mintzer, M.A.; Sitterberg, J.; Bakowsky, U.; Simanek, E.E.; Kissel, T. Triazine dendrimers as nonviral gene delivery systems: Effects of molecular structure on biological activity. Bioconjug. Chem. **2009**, *20* (9), 1799–1806.

100. Lim, Y.B.; Han, S.O.; Kong, H.U.; Lee, Y.; Park, J.S.; Jeong, B.; Kim, S.W. Biodegradable polyester, poly[a-(4-aminobutyl)-l-glycolic acid], as a non-toxic gene carrier. Pharm. Res. **2000**, *17* (7), 811–816.

101. Maheshwari, A.; Mahato, R.I.; Mcgregor, J.; Han, S.; Samlowski, W.E.; Park, J.S.; Kim, S.W. Soluble biodegradable polymer-based cytokine gene delivery for cancer treatment. Mol. Ther. **2000**, *2* (2), 121–130.

102. De Wolf, H.K.; Luten, J.; Snel, C.J.; Storm, G.; Hennink, W.E. Biodegradable, cationic methacrylamide-based polymers for gene delivery to ovarian cancer cells in mice. Mol. Pharm. **2008**, *5* (2), 349–357.

103. Luten, J.; Akeroyd, N.; Funhoff, A.; Lok, M.C.; Talsma, H.; Hennink, W.E. Methacrylamide polymers with hydrolysis-sensitive cationic side groups as degradable gene carriers. Bioconjug. Chem. **2006**, *17* (4), 1077–1084.

104. Bellocq, N.C.; Kang, D.W.; Wang, X.; Jensen, G.S.; Pun, S.H.; Schluep, T.; Zepeda, M.L.; Davis, M.E. Synthetic biocompatible cyclodextrin-based constructs for local gene delivery to improve cutaneous wound healing. Bioconj. Chem. **2004**, *15*, 1201–1211.

105. Lake, A.C.; Vassy, R.; Di Benedetto, M. et al. Low molecular weight fucoidan increases VEGF165- induced endothelial cell migration by enhancing VEGF165 binding to VEGFR-2 and NRP1. J. Biolog. Chem. **2006**, *281*, 37844–37852.

106. Li, J.; Loh, X.J. Cyclodextrin-based supramolecular architectures: Syntheses, structures, and applications for drug and gene delivery. Adv. Drug Deliv. Rev. **2008**, *60* (9), 1000–1017.

107. Pun, S.H.; Tack, F.; Bellocq, N.C.; Cheng, J.; Grubbs, B.H.; Jensen, G.S.; Davis, M.E.; Brewster, M.; Janicot, M.; Janssens, B.; Floren, W.; Bakker, A. Targeted delivery of RNA-cleaving DNA enzyme (DNAzyme) to tumor tissue by transferrin-modified, cyclodextrin-based particles. Cancer Biol. Ther. **2004**, *3* (7), 641–650.

108. Ogris, M.; Brunner, S.; Schuller, S.; Kircheis, R.; Wagner, E. PEGylated DNA/transferrin-PEI complexes: Reduced interaction with blood components, extended circulation in blood and potential for systemic gene delivery. Gene Therapy, **1999**, *6* (4), 595–605.

109. Oupicky, D.; Konak, C.; Dash, P.R.; Seymour, L.W.; Ulbrich, K. Effect of albumin and polyanion on the structure of DNA complexes with polycation containing hydrophilic nonionic block. Bioconjug. Chem. **1999**, *10* (5), 764–772.

110. Plank, C.; Mechtler, K.; Szoka, F.C.Jr.; Wagner, E. Activation of the complement system by synthetic DNA complexes: A potential barrier for intravenous gene delivery. Hum. Gene Ther. **1996**, *7* (12), 1437–1446.

111. Kopecek, J.; Kopeckova, P.; Minko, T.; Lu, Z.R. HPMA copolymer-anticancer drug conjugates: Design, activity, and mechanism of action. Eur. J. Pharm. Biopharm. **2000**, *50* (1), 61–81.

112. Kainthan, R.K.; Gnanamani, M.; Ganguli, M.; Ghosh, T.; Brooks, D.E.; Maiti, S.; Kizhakkedathu, J.N. Blood compatibility of novel water soluble hyper-branched polyglycerol-based multivalent cationic polymers and their interaction with DNA. Biomaterials **2006**, *27* (31), 5377–5390.

113. Kwoh, D.Y.; Coffin, C.C.; Lollo, C.P.; Jovenal, J.; Banaszczyk, M.G.; Mullen, P.; Phillips, A.; Amini, A.; Fabrycki, J.; Bartholomew, R.M.; Brostoff, S.W.; Carlo, D.J. Stabilization of poly-L-lysine/DNA polyplexes for *in vivo* gene delivery to the liver. Biochim. Biophysic. Gene Struct. Expres. **1999**, *1444* (2), 171–190.

114. Bellocq, N.C.; Pun, S.H.; Jensen, G.S.; Davis, M.E.Transferrin-containing, cyclodextrin polymer-based particles for tumor-targeted gene delivery. Bioconj. Chem. **2003**, *14*, 1122–1132.

115. Kichler, A. Gene transfer with modified polyethylenimines. J. Gene. Med. **2004**, *6* (Suppl 1), S3–S10.

116. Kircheis, R.; Wightman, L.; Schreiber, A.; Robitza, B.; Rössler, V.; Kursa, M.; Wagner, E. Polyethylenimine/DNA complexes shielded by transferrin target gene expression to tumors after systemic application. Gene Ther. **2001**, *8* (1), 28–40.

117. Gary, D.J.; Puri, N.; Won, Y.Y. Polymer-based siRNA delivery: Perspectives on the fundamental and phenomenological distinctions from polymer-based DNA delivery. J. Control Release **2007**, *121* (1–2), 64–73.

118. Richards Grayson, A.C.; Doody, A.M.; Putnam, D. Biophysical and structural characterization of polyethylenimine-mediated siRNA delivery *in vitro*. Pharm. Res. **2006**, *23* (8), 1868–1876.

119. Khan, A.; Benboubetra, M.; Sayyed, P.Z.; Ng, K.W.; Fox, S.; Beck, G.; Benter, I.F.; Akhtar, S. Sustained polymeric delivery of gene silencing antisense ODNs, siRNA, DNA-zymes and ribozymes: *In vitro* and *in vivo* studies. J. Drug Target. **2004**, *12* (6), 393–404.

120. Waite, C.L.; Roth, C.M. PAMAM-RGD conjugates enhance siRNA delivery through a multicellular spheroid model of malignant glioma. Bioconjug. Chem. **2009**, *20* (10), 1908–1916.

121. Toub, N.; Bertrand, J.R.; Tamaddon, A.; Elhamess, H.; Hillaireau, H.; Maksimenko, A.; Maccario, J.; Malvy, C.; Fattal, E.; Couvreur, P. Efficacy of siRNA nanocapsules targeted against the EWS-Fli1 oncogene in Ewing sarcoma. Pharm. Res. **2006**, *23* (5), 892–900.

122. Read, M. L.; Singh, S.; Ahmed, Z.; Stevenson, M.; Briggs, S. S.; Oupicky, D.; Barrett, L. B.; Spice, R.; Kendall, M.; Berry, M.; Preece, J. A.; Logan, A.; Seymour, L. W. A versatile reducible polycation-based system for efficient delivery of a broad range of nucleic acids. Nucleic Acids Res. **2005**, *33* (1), 1–16.

123. Howard, K.A.; Rahbek, U.L.; Liu, X.; Damgaard, C.K.; Glud, S.Z.; Andersen, M.Ø.; Hovgaard, M.B.; Schmitz, A.; Nyengaard, J.R.; Besenbacher,. F; Kjems, J. RNA interference *in vitro* and *in vivo* using a novel chitosan/sirna nanoparticle system. Mol. Ther. **2006**, *14* (4), 476–484.

124. Pille, J.Y.; Li, H.; Blot, E.; Bertrand, J.R.; Pritchard, L.L.; Opolon, P.; Maksimenko, A.; Lu, H.; Vannier, J.P.; Soria, J.; Malvy, C.; Soria, C. Intravenous delivery of anti-RhoA small interfering RNA loaded in nanoparticles of chitosan in mice: Safety and efficacy in xenografted aggressive breast cancer.Hum. Gene Ther., **2006**, *17* (10), 1019–1026.

125. Mao, S.; Neu, M.; Germershaus, O.; Merkel, O.; Sitterberg, J.; Bakowsky, U.; Kissel, T. Influence of polyethylene glycol chain length on the physico-chemical and biological properties of poly (ethylene imine)-graft-poly (ethylene glycol) block copolymer/SiRNA polyplexes. Bioconjug. Chem. **2006**, *17* (5), 1209–1218.

126. Lukyanov, A.N.; Torchilin, V.P. Micelles from lipid derivatives of water-soluble polymers as delivery systems for poorly soluble drugs. Adv. Drug Deliv. Rev. **2004**, *56* (9), 1273–1289.

127. Kataoka, K.; Harada, A.; Nagasaki, Y. Block copolymer micelles for drug delivery: Design, characterization and biological significance. Adv. Drug Deliv. Rev. **2001**, *47* (1), 113–131.

128. Torchilin, V.P. Structure and design of polymeric surfactant-based drug delivery systems. J. Control. Release **2001**, *73* (2–3), 137–172.

129. Kwon, G.S.; Naito, M.; Kataoka, K.; Yokoyama, M.; Sakurai, Y.; Okano, T. Block copolymer micelles as vehicles for hydrophobic drugs. Colloids Surf. B **1994**, *2* (4), 429–434.

130. Svenson, S.; Tomalia, D.A. Dendrimers in biomedical applications—Reflections on the field. Adv. Drug Deliv. Rev. **2005**, *57* (15), 2106–2129.

131. Rijcken, C.J.F.; Soga, O.; Hennink, W.E.; Nostrum, C.F.V. Triggered destabilisation of polymeric micelles and vesicles by changing polymers polarity: An attractive

132. Forrest, M.L.; Won, C.Y.; Malick, A.W.; Kwon, G.S. *In vitro* release of the mTOR inhibitor rapamycin from poly (ethylene glycol)-b-poly (ε-caprolactone) micelles. J. Control. Release **2006**, *110* (2), 370–377.

133. Forrest, M.L.; Zhao, A.; Won, C.Y.; Malick, A.W.; Kwon, G.S. Lipophilic prodrugs of Hsp90 inhibitor geldanamycin for nanoencapsulation in poly (ethylene glycol)-b-poly (ε-caprolactone) micelles. J. Control. Release **2006**, *116* (2), 139–149.

134. Lin, W.J.; Chen, Y.C.; Lin, C.C.; Chen, C.F.; Chen, J.W. Characterization of pegylated copolymeric micelles and *in vivo* pharmacokinetics and biodistribution studies. J. Biomed. Mater. Res. Part B Appl. Biomater. **2006**, *77* (1), 188–194.

135. Wang, Y.C.; Tang, L.Y.; Sun, T.M. Self-assembled micelles of biodegradable triblock copolymers based on poly (ethyl ethylene phosphate) and poly (ε-caprolactone) as drug carriers. Biomacromolecules **2008**, *9* (1), 388–395.

136. Xiong, X.B.; Mahmud, A.; Uludag, H.; Lavasanifar, A. Multifunctional polymeric micelles for enhanced intracellular delivery of doxorubicin to metastatic cancer cells. Pharm. Res. **2008**, *25* (11), 2555–2566.

137. Yang, X.; Zhu, B.; Dong, T. Interactions between an anticancer drug and polymeric micelles based on biodegradable polyesters. Macromol. Biosci. **2008**, *8* (12), 1116–1125.

138. Zhu, C.; Jung, S.; Luo, S. Co-delivery of siRNA and paclitaxel into cancer cells by biodegradable cationic micelles based on PDMAEMA-PCL-PDMAEMA triblock copolymers. Biomaterials **2010**, *31* (8), 2408–2416.

139. Mahmud, A.; Xiong, X.B.; Lavasanifar, A. Development of novel polymeric micellar drug conjugates and nano-containers with hydrolyzable core structure for doxorubicin delivery. European J. Pharmaceutics and Biopharmaceutics, **2008**, *69* (3), 923–934.

140. Bae, K.H.; Choi, S.H.; Park, S.Y.; Lee, Y.; Park, T.G. Thermosensitive Pluronic micelles stabilized by shell cross-linking with gold nanoparticles. Langmuir **2006**, *22* (14), 6380–6384.

141. Rapoport, N. Physical stimuli-responsive polymeric micelles for anti-cancer drug delivery. Progress in Polymer Science (Oxford), **2007**, *32*, 962–990.

142. Bae, Y.; Jang, W.D.; Nishiyama, N.; Fukushima, S.; Kataoka, K. Multifunctional polymeric micelles with folate-mediated cancer cell targeting and pH-triggered drug releasing properties for active intracellular drug delivery. Mol. Biosyst. **2005**, *1* (3), 242–250.

143. Bae, Y.; Nishiyama, N.; Fukushima, S.; Koyama, H.; Yasuhiro, M.; Kataoka, K. Preparation and biological characterization of polymeric micelle drug carriers with intracellular pH-triggered drug release property: Tumor permeability, controlled subcellular drug distribution, and enhanced *in vivo* antitumor efficacy. Bioconj. Chem. **2005**, *16* (1), 122–130.

144. Min, K.H.; Kim, J.H.; Bae, S.M.; Fijten, M.W.; Schubert, U.S.; Hoogenboom, R.; Veronese, F.M. Tumoral acidic pH-responsive MPEG-poly (b-amino ester) polymeric micelles for cancer targeting therapy. J. Control Release **2010**, *144* (2), 259–266.

tool for drug delivery. J. Control Release, **2007**, *120* (3), 131–148.

145. Nasongkla, N.; Bey, E.; Ren, J.; Ai, H.; Khemtong, C.; Guthi, J.S.; Chin, S.F.; Sherry, A.D.; Boothman, D.A.; Gao, J. Multifunctional polymeric micelles as cancer-targeted, MRI-ultrasensitive drug delivery systems. Nano Lett. **2006**, *6* (11), 2427–2430.

146. Wu, X.L.; Kim, J.H.; Koo, H. Tumor-targeting peptide conjugated pH-responsive micelles as a potential drug carrier for cancer therapy. Bioconjug. Chem. **2010**, *21* (2), 208–213.

147. Oishi, M.; Nagasaki, Y.; Itaka, K.; Nishiyama, N.; Kataoka, K. Lactosylated poly (ethylene glycol)-siRNA conjugate through acid-labile P-thiopropionate linkage to construct pH-sensitive polyion complex micelles achieving enhanced gene silencing in hepatoma cells. J. Am. Chem. Soc. **2005**, *127* (6), 1624–1625.

148. Chen, W.; Meng, F.; Li, F.; Ji, S.J.; Zhong, Z. pH-responsive biodegradable micelles based on acid-labile polycarbonate hydrophobe: Synthesis and triggered drug release. Biomacromolecules **2009**, *10* (7), 1727–1735.

149. Xu, Y.; Meng, F.; Cheng, R.; Zhong, Z. Reduction-sensitive reversibly crosslinked biodegradable micelles for triggered release of doxorubicin. Macromol. Biosci. **2009**, *9* (12), 1254–1261.

150. Sun, H.; Guo, B.; Cheng, R.; Meng, F.; Liu, H.; Zhong, Z. Biodegradable micelles with sheddable poly(ethylene glycol) shells for triggered intracellular release of doxorubicin. Biomaterials **2009**, *30* (31), 6358–6366.

151. Meng, F.; Zhong, Z.; Feijen, J. Stimuli-responsive polymersomes for programed drug delivery. Biomacromolecules **2009**, *10*, 197–209.

152. Li, G.; Song, S.; Guo, L.; Ma, S. Self-assembly of thermo- and pH-responsive poly (acrylic acid)-b-poly (A-isopropylacrylamide) micelles for drug delivery. J. Polym. Sci. Part A Polym. Chem. **2008**, *46* (15), 5028–5035.

153. Liu, B.; Yang, M.; Li, R.; Ding, Y.; Qian, X.; Yu, L.; Jiang, X. The antitumor effect of novel docetaxel-loaded thermosensitive micelles. Eur. J. Pharm. Biopharm. **2008**, *69* (2), 527–534.

154. Wei, H.; Cheng, S.X.; Zhang, X.Z.; Zhuo, R.X. Thermosensitive polymeric micelles based on poly (N-isopropylacrylamide) as drug carriers. Progr. Polym. Sci. **2009**, *34* (9), 893–910.

155. Smart, T.P.; Mykhaylyk, O.O.; Ryan, A.J.; Battaglia, G. Polymersomes hydrophilic brush scaling relations. Soft Matter **2009**, *5* (19), 3607–3610.

156. Howse, J.R.; Jones, R.A.L.; Battaglia, G.; Ducker, R.E.; Leggett, G.J.; Ryan, A.J. Templated formation of giant polymer vesicles with controlled size distributions. Nat Mater. **2009**, *8* (6), 507–511.

157. Wang, F.; Wang, Y.C.; Yan, L.F.; Wang, J. Biodegradable vesicular nanocarriers based on poly (ecaprolactone)-bloc£-poly (ethyl ethylene phosphate) for drug delivery. Polymer **2009**, *50* (21), 5048–5054.

158. Chen, W.; Meng, F.; Cheng, R.; Zhong, Z. pH-Sensitive degradable polymersomes for triggered release of anticancer drugs: A comparative study with micelles. J. Control Release **2010**, *142* (1), 40–46.

159. Xu, H.; Meng, F.; Zhong, Z. Reversibly crosslinked temperature-responsive nano-sized polymersomes: Synthesis and triggered drug release. J. Mater. Chem. **2009**, *19* (24), 4183–H90.

160. Mitra, S.; Gaur, U.; Ghosh, P.C.; Maitra, A.N. Tumour targeted delivery of encapsulated dextran-doxorubicin conjugate using chitosan nanoparticles as carrier. J. Control. Release, **2001**, *74* (1-3), 317–323.

161. Anand, P.; Nair, H.B.; Sung, B.; Kunnumakkara, A.B.; Yadav, V.R.; Tekmal, R.R.; Aggarwal, B.B. Design of curcumin-loaded PLGA nanoparticles formulation with enhanced cellular uptake, and increased bioactivity *in vitro* and superior bioavailability *in vivo*. Biochem. Pharmacol. **2010**, *79* (3), 330–338.

162. Li, Y.L.; Zhu, L.; Liu, Z.; Cheng, R.; Meng, F.; Cui, J.H.; Ji, S.J.; Zhong, Z. Reversibly stabilized multifunctional dextran nanoparticles efficiently deliver doxorubicin into the nuclei of cancer cells. Angew. Chem. Int. Ed. **2009**, *48*, 9914–9918.

163. Agueros, M.; Zabaleta, V.; Espuelas, S.; Campanero, M.A.; Irache, J.M. Increased oral bioavailability of paclitaxel by its encapsulation through complex formation with cyclodextrins in poly (anhydride) nanoparticles. J. Control Release **2010**, *145* (1), 2–8.

164. Kim, J.H.; Kim, Y.S.; Kim, S.; Park, J.H.; Kim, K.; Choi, K.; Chung, H.; Jeong, S.Y.; Park, R.W.; Kim, I.S.; Kwon, I.C. Hydrophobically modified glycol chitosan nanoparticles as carriers for paclitaxel. J. Control Release **2006**, *111* (1-2), 228–234.

165. Min, K.H.; Park, K.; Kim, Y.S.; Bae, S.M.; Lee, S.; Jo, H.G.; Park, R.W.; Kim, I.S.; Jeong, S.Y.; Kim, K.; Kwon, I.C. Hydrophobically modified glycol chitosan nanoparticles-encapsulated camptothecin enhance the drug stability and tumor targeting in cancer therapy. J. Control. Release **2008**, *127* (3), 208–218.

166. Andhariya, N.; Chudasama, B.; Mehta, R.V.; Upadhyay, R.V. Biodegradable thermoresponsive polymeric magnetic nanoparticles: A new drug delivery platform for doxorubicin. J. Nanopart. Res. **2010**, *12* (1), 1–12.

167. Farokhzad, O.C.; Jon, S.; Khademhosseini, A.; Tran, T.N.T.; LaVan, D.A.; Langer, R. Nanoparticle-aptamer bioconjugates: A new approach for targeting prostate cancer cells. Cancer Res. **2004**, *64* (21), 7668–7672.

168. Farokhzad, O.C.; Cheng, J.; Teply, B.A.; Sherifi, I.; Jon, S.; Kantoff, P.W.; Richie, J.P.; Langer, R. Targeted nanoparticle-aptamer bioconjugates for cancer chemotherapy *in vivo*. Proc. Nat. Acad. Sci. USA **2006**, *103* (16), 6315–6320.

169. Jain, N.K.; Jain, S.K. Development and *in vitro* characterization of galactosylated low molecular weight chitosan nanoparticles bearing doxorubicin. AAPS Pharm. Sci. Tech. **2010**, *11* (2), 686–697.

170. Danhier, F.; Vroman, B.; Lecouturier, N.; Crokart, N.; Pourcelle, V.; Freichels, H.; Jérôme, C.; Marchand-Brynaert, J.; Feron, O.; Préat, V. Targeting of tumor endothelium by RGD-grafted PLGA-nanoparticles loaded with paclitaxel. J. Control. Release **2009**, *140* (2), 166–173.

171. Menjoge, A.R.; Kannan, R.M.; Tomalia, D.A. Dendrimer-based drug and imaging conjugates: Design considerations for nanomedical applications. Drug Discov. Today **2010**, *15* (5-6), 171–185.

172. Morgan, M.T.; Nakanishi, Y.; Kroll, D.J.; Griset, A.P.; Carnahan, M.A.; Wathier, M.; Oberlies, N.H.; Manikumar, G.; Wani, M.C.; Grinstaff, M.W. Dendrimer-encapsulated camptothecins: Increased solubility, cellular

uptake, and cellular retention affords enhanced anticancer activity *in vitro*. Cancer Res. **2006**, *66* (24), 11913–11921.

173. Kailasan, A.; Yuan, Q.; Yang, H. Synthesis and characterization of thermoresponsive polyamidoamine-polyethylene glycol-poly (D, L-lactide) core-shell nanoparticles. Acta Biomater. **2010**, *6* (3), 1131–1139.

174. Malik, N.; Evagorou, E.G.; Duncan, R.Dendrimer-platinate: A novel approach to cancer chemotherapy. Anti-Cancer Drugs, **1999**, *10* (8), 767–776.

175. Kukowska-Latallo, J.F.; Candido, K.A.; Cao, Z.; Nigavekar, S.S.; Majoros, I.J.; Thomas, T.P.; Balogh, L.P.; Khan, M.K.; Baker, J.R.Jr. Nanoparticle targeting of anticancer drug improves therapeutic response in animal model of human epithelial cancer. Cancer Res. **2005**, *65* (12), 5317–5324.

176. Uehara, T.; Ishii, D.; Uemura, T.; Suzuki, H.; Kanei, T.; Takagi, K.; Takama,. M.; Murakami, M.; Akizawa, H.; Arano, Y. Gamma-glutamyl PAMAM dendrimer as versatile precursor for dendrimer-based targeting devices. Bioconjug. Chem. **2010**, *21* (1), 175–181.

177. Ihre, H.R.; Padilla De Jesus, O.L.; Szoka Jr, F.C.; Frechet. Polyester dendritic systems for drug delivery applications: Design, synthesis, and characterization. Bioconjug. Chem. **2002**, *13*, 443–452.

178. Kaminskas, L.M.; Boyd, B.J.; Karellas, P.; Krippner, G.Y.; Lessene, R.; Kelly, B.; Porter, C.J. The impact of molecular weight and PEG chain length on the systemic pharmacokinetics of pegylated poly L-lysine dendrimers. Mol. Pharm. **2008**, *5* (3), 449–463.

179. Lim, J.; Simanek, E.E. Synthesis of water-soluble dendrimers based on melamine bearing 16 pacli-taxel groups. Org. Lett. **2008**, *10* (2), 201–204.

180. Zhu, S.; Hong, M.; Zhang, L. PEGylated PAMAM dendrimer-doxorubicin conjugates: *In vitro* evaluation and *in vivo* tumor accumulation. Pharm. Res. **2010**, *27* (1), 161–174.

181. Kono, K.; Kojima, C.; Hayashi, N.; Nishisaka, E.; Kiura, K.; Watarai, S.; Harada, A. Preparation and cytotoxic activity of poly (ethylene glycol)-modified poly (amidoamine) dendrimers bearing adriamycin. Biomaterials **2008**, *29* (11), 1664–1675.

182. Yang, W.; Cheng, Y.; Xu, T.; Wang, X.; Wen, L.P. Targeting cancer cells with biotin-dendrimer conjugates. Eur J Med Chem. **2009**, *44* (2), 862–868.

183. Singh, P.; Gupta, U.; Asthana, A.; Jain, N.K. Folate and folate-PEG-PAMAM dendrimers: Synthesis, characterization, and targeted anticancer drug delivery potential in tumor bearing mice. Bioconjug. Chem. **2008**, *19* (11), 2239–2252.

184. Kolhatkar, R.B.; Swaan, P.; Ghandehari, H. Potential oral delivery of 7-ethyl-10-hydroxy-camptothecin (SN-38) using poly (amidoamine) dendrimers. Pharm. Res. **2008**, *25* (7), 1723–1729.

185. Dhanikula, R.S.; Argaw, A.; Bouchard, J.F.; Hildgen, P. Methotrexate loaded polyether-copolyester dendrimers for the treatment of gliomas: Enhanced efficacy and intratumoral transport capability. Mol. Pharm. **2008**, *5* (1), 105–116.

186. Iozzo, R.V.; Zoeller, J.J.; Nystrom, A. Basement membrane proteoglycans: Modulators Par Excellence of cancer growth and angiogenesis. Mol. Cells **2009**, *27* (5), 503–513.

187. Johnson, D.E.; Williams, L.T. Structural and functional diversity in the FGF receptor multigene family. Adv. Cancer Res. **1993**, *60*, 1–41.

188. Arbiser, J.L. Angiogenesis and the skin: A primer. J. Am. Acad. Dermatol. **1996**, *34* (3), 486–497.

189. Paleolog, E.M.; Miotla, J.M. Angiogenesis in arthritis: Role in disease pathogenesis and as a potential therapeutic target. Angiogenesis **1998**, *2* (4), 295–307.

190. Folkman, J. What is the evidence that tumors are angiogenesis dependent? J. Nat. Cancer Instit. **1990**, *82* (1), 4–6.

191. Xu, J.H.; Li, R.X.; Zhang, W.; Qi, F.Res. progress of the mechanism on intraocular neovascularization. Int. J. Ophthalmol. **2008**, *8*, 2496–2498.

192. Neufeld, G.; Gospodarowicz, D. Protamine sulfate inhibits mitogenic activities of the extracellular matrix and fibroblast growth factor, but potentiates that of epidermal growth factor. J. Cell. Physiol. **1987**, *132* (2), 287–294.

193. Jackson, C.J.; Giles, I.; Knop, A.; Nethery, A.; Schrieber, L. Sulfated polysaccharides are required for collagen-induced vascular tube formation. Exp. Cell Res. **1994**, *215* (2), 294–302.

194. Kersten, J.R.; Pagel, P.S.; Warltier, D.C. Protamine inhibits coronary collateral development in a canine model of repetitive coronary occlusion. Am. J. Physiol. Heart Circ. Physiol. **1995**, *268*, H720-H728.

195. Kasai, S.; Nagasawa, H.; Shimamura, M.; Uto, Y.; Hori, H. Design and synthesis of antiangiogenic/heparin-binding arginine dendrimer mimicking the surface of endostatin. Bioorg. Med. Chem. Lett. **2002**, *12* (6), 951–954.

196. Al-Jamal, K.T.; Al-Jamal, W.T.; Akerman, S.; Podesta, J.E.; Yilmazer, A.; Turton, J.A.; Bianco, A.; Vargesson, N.; Kanthou, C.; Florence, A.T.; Tozer, G.M.; Kostarelos, K. Systemic antiangiogenic activity of cationic poly-L-lysine dendrimer delays tumor growth. Proc. Natl. Acad. Sci. USA **2010**, *107* (9), 3966–3971.

197. Rusnati, M.; Dell'era, P.; Urbinati, C.; Tanghetti, E.; Massardi, M.L.; Nagamine, Y.; Monti, E.; Presta, M. A distinct basic fibroblast growth factor (FGF-2)/FGF receptor interaction distinguishes urokinase-type plasminogen activator induction from mitogenicity in endothelial cells. Mol. Biol. Cell **1996**, *7* (3), 369–381.

198. Lange, C.; Ehlken, C.; Martin, G.; Konzok, K.; Moscoso Del Prado, J.; Hansen, L.L.; Agostini, HH. T. Intravitreal injection of the heparin analog 5-amino-2-naphthalenesulfonate reduces retinal neovascularization in mice. Exp. Eye Res. **2007**, *85*, 323–327.

199. Lozano, R.M.; Jimenez, M.A.; Santoro, J.; Rico, M.; Gimenez-Gallego, G. Solution structure of acidic fibroblast growth factor bound to 1,3,6-naphthalenetrisulfonate: A minimal model for the anti-tumoral action of suramins and suradistas. J. Mol. Biol. **1998**, *281* (5), 899–915.

200. Garcia-Fernandez, L.; Aguilar, M.R.; Fernandez, M.M.; Lozano, R.M.; Giménez, G.; Valverde, S.; San Román, J. Structure, morphology, and bioactivity of biocompatible systems derived from functionalized acrylic polymers based on 5-amino-2-naphthalene sulfonic acid. Biomacromolecules, **2010**, *11* (7), 1763–1772.

201. Garcia-Fernandez, L.; Halstenberg, S.; Unger, R.E.; Aguilar, M.R.; Kirkpatrick, C.J.; San Román, J. Anti-angiogenic activity of heparin-like poly-sulfonated polymeric

drugs in 3D human cell culture. Biomaterials **2010**, *31* (31), 7863–7872.

202. Liekens, S.; Neyts, J.; Degreve, B.; De Clercq, E. The sulfonic acid polymers PAMPS [poly (2-acryl-amido-2-methyl-1-propanesulfonic acid)] and related analogues are highly potent inhibitors of angiogenesis. Oncol. Res. **1997**, *9* (4), 173–181.

203. Liekens, S.; Leali, D.; Neyts, J.; Esnouf, R.; Rusnati, M.; Dell'Era, P.; Maudgal, P.C.; De Clercq, E.; Presta, M. Modulation of fibroblast growth factor-2 receptor binding, signaling, and mitogenic activity by heparin-mimicking polysulfonated compounds. Mol. Pharmacol. **1999**, *56* (1), 204–213.

204. Wu, L.W.; Chiang, Y.M.; Chuang, H.C.; wang, S.Y.; Yang, G.W.; Chen, Y.H.; Lai, L.Y.; Shyur, L.F. Polyacetylenes function as anti-angiogenic agents. Pharm. Res. **2004**, *21* (11), 2112–2119.

205. Chabut, D.; Fischer, A.-M.; Colliec-Jouault, S.; Laurendeau, I.; Matou, S.; Le Bonniec, B.; Helley, D. Low molecular weight fucoidan and heparin enhance the basic fibroblast growth factor-induced tube formation of endothelial cells through heparan sulfate-dependent alpha6 overexpression. Mol. Pharmacol. **2003**, *64* (3), 696–702.

206. Matou, S.; Helley, D.; Chabut, D.; Bros, A.; Fischer, A.-M. Effect of fucoidan on fibroblast growth factor-2-induced angiogenesis *in vitro*. Thrombosis Res., **2002**, *106* (4-5), 213–221.

207. Soeda, S.; Kozako, T.; Iwata, K.; Shimeno, H. Oversulfated fucoidan inhibits the basic fibroblast growth factor-induced tube formation by human umbilical vein endothelial cells: Its possible mechanism of action. Biochim. Biophys. Acta Mol. Cell Res. **2000**, *1497* (1), 127–134.

208. Marchetti, M.; Vignoli, A.; Russo, L.; Balducci, D.; Pagnoncelli, M.; Barbui, T.; Falanga, A. Endothelial capillary tube formation and cell proliferation induced by tumor cells are affected by low molecular weight heparins and unfractionated heparin. Thrombosis Res. **2008**, *121* (5), 637–645.

209. Norrby, K. Heparin and angiogenesis: A low-molecular-weight fraction inhibits and a high-molecular-weight fraction stimulates angiogenesis systemically. Haemostasis, **1993**, *23*, 141–149.

210. Powers, C.J. Fibroblast growth factors, their receptors and signaling. Endocr. Relat. Cancer, **2000**, *7* (3), 165–197.

211. Scherer, S.S.; Pietramaggiori, G.; Matthews, J.; Perry, S.; Assmann, A.; Carothers, A.; Demcheva, M.; Muise-Helmericks, R.C.; Seth, A.; Vournakis, J.N.; Valeri, R.C.; Fischer, T.H.; Hechtman, H.B.; Orgill, D.P. Poly-A-acetyl glucosamine nanofibers: A new bio-active material to enhance diabetic wound healing by cell migration and angiogenesis. Ann. Surg. **2009**, *250* (2), 322–330.

212. Moon, J.J.; West, J.L. Vascularization of engineered tissues: Approaches to promote angiogenesis in biomaterials. Curr. Top. Med. Chem. **2008**, *8* (4), 300–310.

213. Calonder, C.; Matthew, H.W.T.; Van Tassel, P.R. Adsorbed layers of oriented fibronectin: A strategy to control cell-surface interactions. J. Biomed. Mater. Res. Part A **2005**, *75* (2), 316–323.

214. Kouvroukoglou, S.; Dee, K.C.; Bizios, R.; Mcintire, L.V.; Zygourakis, K. Endothelial cell migration on surfaces modified with immobilized adhesive peptides. Biomaterials **2000**, *21* (17), 1725–1733.

215. Unger, R.E.; Peters, K.; Wolf, M.; Motta, A.; Migliaresi, C.; Kirkpatrick, C.J. Endothelialization of a non-woven silk fibroin net for use in tissue engineering: Growth and gene regulation of human endothelial cells. Biomaterials **2004**, *25* (21), 5137–5146.

216. Amaral, I.F.; Unger, R.E.; Fuchs, S.; Mendonça, A.M.; Sousa, S.R.; Barbosa, M.A.; Pêgo, A.P.; Kirkpatrick, C.J. Fibronectin-mediated endothelialisation of chitosan porous matrices. Biomaterials **2009**, *30* (29), 5465–5475.

217. Kurt, P.; Wood, L.; Ohman, D.E.; Wynne, K.J. Highly effective contact antimicrobial surfaces via polymer surface modifiers. Langmuir, **2007**, *23* (9), 4719–4723.

218. Nishi, K.K.; Antony, M.; Mohanan, P.V.; Anilkumar, T.V.; Loiseau, P.M.; Jayakrishnan, A. Amphotericin B-gum Arabic conjugates: Synthesis, toxicity, bioavailability, and activities against Leishmania and fungi. Pharm. Res. **2007**, *24* (5), 971–980.

219. Waschinski, C.J.; Zimmermann, J.; Salz, U. Design of contact-active antimicrobial acrylate-based materials using biocidal macromers. Adv. Mater. **2008**, *20* (1), 104–108.

220. Hetrick, E.M.; Schoenfisch, M.H. Reducing implant-related infections: Active release strategies. Chem. Soc. Rev. **2006**, *35* (9), 780–789.

221. Wei, D.; Ma, Q.; Guan, Y. Structural characterization and antibacterial activity of oligoguanidine (polyhexamethylene guanidine hydrochloride). Mater. Sci. Eng. C **2009**, *29* (6), 1776–1780.

222. Allen, M.J.; Morby, A.P.; White, G.F. Cooperativity in the binding of the cationic biocide poly hexamethylene biguanide to nucleic acids. Biochem. Biophys. Res. Commun. **2004**, *318* (2), 397–404.

223. Kanazawa, A.; Ikeda, T.; Endo, T. Antibacterial activity of polymeric sulfonium salts. J. Polym. Sci. Part A **1993**, *31* (11), 2873–2876.

224. Nonaka, T.; Hua, L.; Ogata, T.; Kurihara, S. Synthesis of water-soluble thermosensitive polymers having phosphonium groups from methacryloyloxyethyl trialkyl phosphonium chlorides-N-isopropylacrylamide copolymers and their functions. J. Appl. Polym. Sci. **2002**, *87* (3), 386–393.

225. Imazato, S.; Ebi, N.; Tarumi, H.; Russell, R.R.; Kaneko, T.; Ebisu, S. Bactericidal activity and cytotoxicity of antibacterial monomer MDPB. Biomaterials **1999**, *20* (9), 899–903.

226. Kebir, N.; Campistron, I.; Laguerre, A.; Pilard, J.F.; Bunel, C.; Jouenne, T. Use of telechelic cis-1,4-polyisoprene cationomers in the synthesis of antibacterial ionic polyurethanes and copolyurethanes bearing ammonium groups. Biomaterials **2007**, *28* (29), 4200–4208.

227. Gilbert, P.; Moore, L.E. Cationic antiseptics: Diversity of action under a common epithet. J. Appl. Microbiol. **2005**, *99* (4), 703–715.

228. Dizman, B.; Elasri, M.O.; Mathias, L.J. Synthesis, characterization, and antibacterial activities of novel methacrylate polymers containing norfloxacin. Biomacromolecules **2005**, *6* (1), 514–520.

229. Kenawy, E.R.; Worley, S.D.; Broughton, R. The chemistry and applications of antimicrobial polymers: A state-of-the-art review. Biomacromolecules **2007**, *8* (5), 1359–1384.

In Vitro–Medical

230. Wach, J.Y.; Malisova, B.; Bonazzi, S. Protein-resistant surfaces through mild dopamine surface functionalization. Chem. A Eur. J. **2008**, *14* (34), 10579–10584.

231. Glukhov, E.; Stark, M.; Burrows, L.L.; Deber, C.M. Basis for selectivity of cationic antimicrobial peptides for bacterial versus mammalian membranes. J. Biolog. Chem. **2005**, *280* (40), 33960–33967.

232. Epand, R. F.; Schmitt, M. A.; Gellman, S. H.; Epand, R. M. Role of membrane lipids in the mechanism of bacterial species selective toxicity by two a/P-antimicrobial peptides. Biochim. Biophys. Acta Biomembr. **2006**, *1758*, 1343–1350.

233. Verkleij, A.J.; Zwaal, R.F.A.; Roelofsen, B. The asymmetric distribution of phospholipids in the human red cell membrane. A combined study using phospholipases and freeze etch electron microscopy. Biochim. Biophys. Acta **1973**, *323* (2), 178–193.

234. Hancock, R.E.W.; Rozek, A. Role of membranes in the activities of antimicrobial cationic peptides. FEMS Microbiol. Lett. **2002**, *206* (2), 143–149.

235. Zasloff, M. Antimicrobial peptides of multicellular organisms.Nature **2002**, *415* (6870), 389–395.

236. Bagheri, M.; Beyermann, M.; Dathe, M. Immobilization reduces the activity of surface-bound cat-ionic antimicrobial peptides with no influence upon the activity spectrum. Antimicrob. Agents Chemother. **2009**, *53* (3), 1132–1141.

237. Brogden, K.A. Antimicrobial peptides: Pore formers or metabolic inhibitors in bacteria? Nat. Rev. Microbiol. **2005**, *3*, 238–250.

238. Desmet, T.; Morent, R.; De Geyter, N.; Leys, C.; Schacht, E.; Dubruel, P. Nonthermal plasma technology as a versatile strategy for polymeric biomaterials surface modification: A review. Biomacromolecules **2009**, *10* (9), 2351–2378.

239. Iwasaki, Y.; Takamiya, M.; Iwata, R.; Yusa, S.I.; Akiyoshi, K. Surface modification with well-defined biocompatible triblock copolymers. Improvement of biointerfacial phenomena on a poly (dimethylsiloxane) surface. Colloids Surf B Biointerfaces **2007**, *57* (2), 226–236.

240. Li, D.; Luo, Y.; Zhang, B.; Li, B.; Zhu, S. Raft grafting polymerization of MMA/St from surface of silicon wafer. Acta Polymerica Sinica **2007**, 699–704.

241. Ranjan, R.; Brittain, W.J. Tandem RAFT polymerization and click chemistry: An efficient approach to surface modification. Macromolecular Rapid Communications, **2007**, *28*, 2084–2089.

242. Edmondson, S.; Osborne, V.L.; Huck, W.T.S. Polymer brushes via surface-initiated polymerizations. Chem. Soc. Rev. **2004**, *33* (1), 14–22.

243. Lee, B.S.; Lee, J.K.; Kim, W.J.; Jung, Y.H.; Sim, S.J.; Lee, J.; Choi, I.S. Surface-initiated, atom transfer radical polymerization of oligo (ethylene glycol) methyl ether methacrylate and subsequent click chemistry for bioconjugation. Biomacromolecules **2007**, *8* (2), 744–749.

244. Cheng, G.; Xue, H.; Zhang, Z.; Chen, S.; Jiang, S. A switchable biocompatible polymer surface with self-sterilizing and nonfouling capabilities. Angew. Chem. Int. Ed. **2008**, *47*, 8831–8834.

245. Cho, Y.W.; Cho, Y.N.; Chung, S.H.; Yoo, G.; Ko, S.W. Water-soluble chitin as a wound healing accelerator. Biomaterials **1999**, *20* (22), 2139–2145.

246. Rabea, E.I.; Badawy, M.E.T.; Stevens, C.V.; Smagghe, G.; Steurbaut, W. Chitosan as antimicrobial agent: Applications and mode of action. Biomacromolecules, **2003**, *4* (6), 1457–1465.

247. Kim, C.H.; Choi, J.W.; Chun, H.J.; Choi, K.S. Synthesis of chitosan derivatives with quaternary ammonium salt and their antibacterial activity. Polym. Bull. **1997**, *38* (4), 387–393.

248. Nam, C.W.; Kim, Y.H.; Ko, S.W. Modification of polyacrylonitrile (PAN) fiber by blending with N-(2-hydroxy) propyl-3-trimethyl-ammonium chitosan chloride. J. Appl. Polym. Sci. **1999**, *74*, 2258–2265.

249. Kenawy, E.R.; Abdel-Hay, F.I.; El-Magd, A.A.; Mahmoud, Y. Biologically active polymers: Modification and anti-microbial activity of chitosan derivatives. J. Bioact. Comp. Polym. **2005**, *20* (1), 95–111.

250. Roy, D.; Knapp, J.S.; Guthrie, J.T.; Perrier, S. Antibacterial cellulose fiber via RAFT surface graft polymerization. Biomacromolecules, **2008**, *9* (1), 91–99.

251. Deming, T.J. Mussel byssus and biomolecular materials. Curr Opin Chem Biol. **1999**, *3* (1), 100–105.

252. Gademann, K.; Kobylinska, J.; Wach, J.Y.; Woods, T.M. Surface modifications based on the cyanobacterial siderophore anachelin: From structure to functional biomaterials design. Biometals **2009**, *22* (4), 595–604.

253. Zurcher, S.; Wackerlin, D.; Bethuel, Y. Biomimetic surface modifications based on the cyanobacterial iron chelator anachelin. J. Am. Chem. Soc. **2006**, *128* (4), 1064–1065.

254. Rojo, L.; Barcenilla, J.M.; Vazquez, B.; Gonzalez, R.; San Roman, J. Intrinsically antibacterial materials based on polymeric derivatives of eugenol for biomedical applications. Biomacromolecules **2008**, *9* (9), 2530–2535.

255. Rojo, L.; Vazquez, B.; Parra, J.; López Bravo, A.; Deb, S.; San Roman, J. From natural products to polymeric derivatives of "eugenol": A new approach for preparation of dental composites and orthopedic bone cements. Biomacromolecules **2006**, *7* (10), 2751–2761.

256. Parra, F.; Vazquez, B.; Benito, L.; Barcenilla, J.; San Roman, J. Foldable antibacterial acrylic intraocular lenses of high refractive index. Biomacromolecules **2009**, *10* (11), 3055–3061.

257. Gabriel, G.J.; Som, A.; Madkour, A.E.; Eren, T.; Tew, G.N. Infectious disease: Connecting innate immunity to biocidal polymers. Mater. Sci. Eng. R Rep. **2007**, *57* (1–6), 28–64.

258. Broxton, P.; Woodcock, P.M.; Gilbert, P. A study of the antibacterial activity of some polyhexamethylene biguanides towards *Escherichia coli* ATCC 8739. J. Appl. Bacteriol. **1983**, *54* (3), 345–353.

259. Gabriel, G.J.; Pool, J.G.; Som, A.; Dabkowski, JM.; Coughlin, E.B.; Muthukumar, M.; Tew, G.N. Interactions between antimicrobial polynorbornenes and phospholipid vesicles monitored by light scattering and microcalorimetry. Langmuir **2008**, *24* (21), 12489–12495.

260. Futaki, S.; Suzuki, T.; Ohashi, W.; Yagami, T.; Tanaka, S.; Ueda, K.; Sugiura, Y. Arginine-rich peptides. An abundant source of membrane-permeable peptides having potential as carriers for intracellular protein delivery. J. Biol. Chem. **2001**, *276* (8), 5836–5840.

261. Henriques, S.T.; Melo, M.N.; Castanho, M.A.R.B. Cell-penetrating peptides and antimicrobial peptides: How different are they? Biochem. J. **2006**, *399*, 1–7.

262. Mitchell, D.J.; Steinman, L.; Kim, D.T.; Fathman, C.G.; Rothbard, J.B. Polyarginine enters cells more efficiently than other polycationic homopolymers. J. Peptide Res. **2000**, *56* (5), 318–325.

263. Miyatake, T.; Nishihara, M.; Matile, S. A cost-effective method for the optical transduction of chemical reactions. Application to hyaluronidase inhibitor screening with polyarginine-counteranion complexes in lipid bilayers. J. Am. Chem. Soc. **2006**, *128* (38), 12420–12421.

264. Rothbard, J.B.; Kreider, E.; Vandeusen, C.L.; Wright, L.; Wylie, B.L.; Wender, P.A. Arginine-rich molecular transporters for drug delivery: Role of backbone spacing in cellular uptake. J. Med. Chem. **2002**, *45* (17), 3612–3618.

265. Boneca, I.G.; Chiosis, G. Vancomycin resistance: Occurrence, mechanisms and strategies to combat it. Expert Opin. Ther. Targets **2003**, *7* (3), 311–328.

266. Loll, P.J.; Axelsen, P.H. The structural biology of molecular recognition by vancomycin. Ann. Rev. Biophys. Biomol. Struct. **2000**, *29* (1), 265–289.

267. Lawson, M.C.; Shoemaker, R.; Hoth, K.B.; Bowman, C.N.; Anseth, K.S. Polymerizable vancomycin derivatives for bactericidal biomaterial surface modification: Structure-function evaluation. Biomacromolecules **2009**, *10* (8), 2221–2234.

268. Xue, L.; Greisler, H.P. Biomaterials in the development and future of vascular grafts. J. Vascular Surgery, **2003**, *37* (2), 472–480.

269. Gorbet, M.B.; Sefton, M.V. Biomaterial-associated thrombosis: Roles of coagulation factors, complement, platelets and leukocytes. Biomaterials **2004**, *25* (26), 5681–5703.

270. Sefton, M.V.; Gemmell, C.H.; Gorbet, M.B. What really is blood compatibility? J. Biomaterials Science, Polymer Edition, **2000**, *11* (11), 1165–1182.

271. Seyfert, U.T.; Biehl, V.; Schenk, J. *In vitro* hemocompatibility testing of biomaterials according to the ISO 10993–4. Biomol. Eng. **2002**, *19* (2-6), 91–96.

272. Mao, C.; Qiu, Y.; Sang, H.; Mei, H.; Zhu, A.; Shen, J.; Lin, S. Various approaches to modify biomaterial surfaces for improving hemo compatibility. Adv. Colloid Interface Sci. **2004**, *110* (1-2), 5–17.

273. Boura, C.; Menu, P.; Payan, E.; Picart, C.; Voegel, J.C.; Muller, S.; Stoltz, J.F. Endothelial cells grown on thin polyelectrolyte multilayered films: An evaluation of a new versatile surface modification. Biomaterials **2003**, *24* (20), 3521–3530.

274. Aldenhoff, Y.B.J.; Blezer, R.; Lindhout, T.; Koole, L.H. Photo-immobilization of dipyridamole (Persantin®) at the surface of polyurethane biomaterials: Reduction of *in vitro* thrombogenicity. Biomaterials **1997**, *18* (2), 167–172.

275. Werner, C.; Maitz, M.F.; Sperling, C. Current strategies towards hemocompatible coatings. J. Mater. Chem. **2007**, *17* (32), 3376–3384.

276. Gott, V.L.; Whiffen, J.D.; Dutton, R.C. Heparin bonding on colloidal graphite surfaces. Science **1963**, *142* (3597), 1299–1300.

277. Alferiev, I.S.; Connolly, J.M.; Stachelek, S.J.; Ottey, A.; Rauova, L.; Levy, R.J. Surface heparinization of polyurethane via bromoalkylation of hard segment nitrogens. Biomacromolecules **2006**, *7* (1), 317–322.

278. Chen, H.; Chen, Y.; Sheardown, H.; Brook, M.A. Immobilization of heparin on a silicone surface through a heterobifunctional PEG spacer. Biomaterials **2005**, *26* (35), 7418–7424.

279. Chen, M.C.; Wong, H.S.; Lin, K.J.; Chen, H.L.; Wey, S.P.; Sonaje, K.; Lin, Y.H.; Chu, C.Y.; Sung, H.W. The characteristics, biodistribution and bioavailability of a chitosan-based nanoparticulate system for the oral delivery of heparin. Biomaterials **2009**, *30* (34), 6629–6637.

280. Huang Ly, Y.M. Hemocompatibility of layer-by-layer hyaluronic acid/heparin nanostructure coating on stainless steel for cardiovascular stents and its use for drug delivery. J. Nanosci. Nanotechnol. **2006**, *6* (9), 3163–3170.

281. Luong-Van, E.; Gr0ndahl, L.; Chua, K. N.; Leong, K. W.; Nurcombe, V.; Cool, S. M. Controlled release of heparin from poly (epsilon-caprolactone) electrospun fibers. Biomaterials **2006**, *27*, 2042–2050.

282. Larm, O.; Larsson, R.; Olsson, P. A new non-thrombogenic surface prepared by selective covalent binding of heparin via a modified reducing terminal residue. Biomater. Med. Dev. Artif. Organs **1983**, *11*, 161–173.

283. Krijgsman, B.; Seifalian, A.M.; Salacinski, H.J.; Tai, N.R.; Punshon, G.; Fuller, B.J.; Hamilton, G. An assessment of covalent grafting of RGD peptides to the surface of a compliant poly (carbonate-urea)urethane vascular conduit versus conventional biological coatings: Its role in enhancing cellular retention. Tissue Eng. **2002**, *8* (4), 673–680.

284. Tiwari, A.; Salacinski, H.J.; Punshon, G.; Hamilton, G.; Seifalian, A.M. Development of a hybrid cardiovascular graft using a tissue engineering approach. FASEB J. **2002**, *16* (8), 791–796.

285. Nakayama, Y.; Okahashi, R.; Iwai, R.; Uchida, K. Heparin bioconjugate with a thermoresponsive cationic branched polymer: A novel aqueous antithrombogenic coating material. Langmuir, **2007**, *23* (15), 8206–8211.

286. Vicario, P.P.; Lu, Z.; Wang, Z.; Merritt, K.; Buongiovanni, D.; Chen, P. Antithrombogenicity of hydromer's polymeric formula F202™ immobilized on polyurethane and electropolished stainless steel. J. Biomed. Mater. Res. Part B Appl. Biomater. **2008**, *86* (1), 136–144.

287. Fosse, E.; Thelin, S.; Svennevig, J.L.; Jansen, P.; Mollnes, T.E.; Hack, E.; Venge, P.; Moen, O.; Brockmeier, V.; Dregelid, E.; Haldén, E.; Hagman, L.; Videm, V.; Pedersen, T.; Mohr, B. Duraflo II coating of cardiopulmonary bypass circuits reduces complement activation, but does not affect the release of granulocyte enzymes in fully heparinized patients: A European multicentre study. Eur. J. Cardiothor. Surg. **1997**, *11* (2), 320–327.

288. Mangoush, O.; Purkayastha, S.; Haj-Yahia, S.; Kinross, J.; Hayward, M.; Bartolozzi, F.; Darzi, A.; Athanasiou, T. Heparin-bonded circuits versus nonheparin-bonded circuits: An evaluation of their effect on clinical outcomes. Eur. J. Cardio-Thor. Surg. **2007**, *31* (6), 1058–1069.

289. Wildevuur, C.R.H.; Jansen, P.G.M.; Bezemer, P.D. Clinical evaluation of Duraflo II heparin treated extracorporeal circulation circuits (2nd version). The European working group on heparin coated extracorporeal circulation circuits. Eur. J. Cardio-Thor. Surg. **1997**, *11* (4), 616–623.

290. Brynda, E.; Houska, M.; Jirouskova, M.; Dyr, J.E. Albumin and heparin multilayer coatings for blood-contacting

medical devices. J. Biomed. Mater. Res. **2000**, *51* (2), 249–257.

291. Sperling, C.; Houska, M.; Brynda, E.; Streller, U.; Werner, C. *In vitro* hemocompatibility of albu-min-heparin multi-layer coatings on polyethersulfone prepared by the layer-by-layer technique. J. Biomed. Mater. Res. Part A **2006**, *76* (4), 681–689.

292. Brynda, E.; Houska, M. Preparation of organized protein multilayers. Macromol. Rapid Commun. **1998**, *19*, 173–176.

293. Houska, M.; Brynda, E.; Solovyev, A.; Brouckova, A.; Krízová, P.; Vanícková, M.; Dyr, J.E. Hemocompatible albumin-heparin coatings prepared by the layer-by-layer technique. The effect of layer ordering on thrombin inhibition and platelet adhesion. J. Biomed. Mater. Res. Part A **2008**, *86* (3), 769–778.

294. Kidane, A.G.; Salacinski, H.; Tiwari, A.; Bruckdorfer, K.R.; Seifalian, A.M. Anticoagulant and antiplatelet agents: Their clinical and device application (s) together with usages to engineer surfaces. Biomacromolecules **2004**, *5* (3), 798–813.

295. Aldenhoff, Y.B.J.; Koole, L.H.; Curtis, A.; Descouts, P. Platelet adhesion studies on dipyridamole coated polyurethane surfaces. Eur. Cells Mater. **2003**, *5*, 61–67.

296. Paul, W.; Sharma, C.P. Acetylsalicylic acid loaded poly (vinyl alcohol) hemodialysis membranes: Effect of drug release on blood compatibility and permeability. J. Biomaterials Science, Polymer Edition, **1997**, *8* (10), 755–764.

297. San Roman, J.; Bujan, J.; Bellon, J.M.; Gallardo, A.; Escudero, M.C.; Jorge, E.; de Haro, J.; Alvarez, L.; Castillo-Olivares, J.L. Experimental study of the antithrombogenic behavior of Dacron vascular grafts coated with hydrophilic acrylic copolymers bearing salicylic acid residues. J. Biomed. Mater. Res. **1996**, *32* (1), 19–27.

298. Rodriguez, G.; Gallardo, A.; San Roman, J.; Rebuelta, M.; Bermejo, P.; Buján, J.; Bellón, J.M.; Honduvilla, N.G.; Escudero, C. New resorbable polymeric systems with antithrombogenic activity. J. Mater. Sci. Mater Med. **1999**, *10* (12), 873–878.

299. Gallardo, A.; Rodriguez, G.; Fernandez, M.; Aguilar, M.R.; San Roman, J. Polymeric drugs with prolonged sustained delivery of specific anti-aggregant agents for platelets: Kinetic analysis of the release mechanism. J. Biomater. Sci. Polym. Ed. **2004**, *15*, 917–928.

300. Rodriguez, G.; Gallardo, A.; Fernandez, M.; Rebuelta, M.; Buján, J.; Bellón, J.M.; Honduvilla, N.G.; Escudero, C.; San Román, J. Hydrophilic polymer drug from a derivative of salicylic acid: Synthesis, controlled release studies and biological behavior. Macromol Biosci. **2004**, *4* (6), 579–586.

301. Aguilar, M.R.; Rodriguez, G.; Fernandez, M.; Gallardo, A.; San Roman, J. Polymeric active coatings with functionality in vascular applications. J. Mater. Sci. **2002**, *13*, 1099–1104.

302. Gallardo, A.; Rodriguez, G.; Aguilar, M.R.; Fernandez, M.; San Roman, J. A kinetic model to explain the zero-order release of drugs from ionic polymer drug conjugates: Application to AMPS-triflusal-derived polymeric drugs. Macromolecules **2003**, *36* (23), 8876–8880.

303. San Roman, J.; Rodriguez, G.; Fernandez, M. Triflusal-Containing Polymers for Stent Coating. WIPO Patent Application WO/2007/014787; Application Number: EP2006/009156; February 08, 2007.

304. Moncada, S.; Radomski, M.W.; Palmer, R.M.J. Endothelium-derived relaxing factor. Identification as nitric oxide and role in the control of vascular tone and platelet function. Biochem. Pharmacol., **1988**, *37* (13), 2495–2501.

305. Taite, L.J.; Yang, P.; Jun, H.W.; West, J.L. 2008. Nitric oxide-releasing polyurethane-PEG copolymer containing the YIGSR peptide promotes endothelialization with decreased platelet adhesion. J. Biomedical Materials Res.—Part B Applied Biomaterials, 1973, 84, 108–116.

306. Wu, Y.; Zhou, Z.; Meyerhoff, M. E. *in vitro* platelet adhesion on polymeric surfaces with varying fluxes of continuous nitric oxide release. J. Biomed. Mater. Res. Part A, **2007**, *81* (4), 956–963.

307. Jordan, S.W.; Chaikof, E.L. Novel thromboresistant materials. J. Vasc. Surg. **2007**, *45* (6), 104–115.

308. Cao, L.; Chang, M.; Lee, C.Y.; Castner, D.G.; Sukavaneshvar, S.; Ratner, B.D.; Horbett, TA. Plasma-deposited tetraglyme surfaces greatly reduce total blood protein adsorption, contact activation, platelet adhesion, platelet procoagulant activity, and *in vitro* thrombus deposition. J. Biomed. Mater. Res. Part A **2007**, *81* (4), 827–837.

309. Chen, H.; Zhang, Z.; Chen, Y.; Brook, M.A.; Sheardown, H. Protein repellant silicone surfaces by covalent immobilization of poly (ethylene oxide). Biomaterials **2005**, *26* (15), 2391–2399.

310. Gorbet, M.B.; Sefton, M.V. Leukocyte activation and leukocyte procoagulant activities after blood contact with polystyrene and polyethylene glycol-immobilized polystyrene beads. J. Lab. Clin. Med. **2001**, *137* (5), 345–355.

311. Hansson, K.M.; Tosatti, S.; Isaksson, J.; Wetterö, J.; Textor, M.; Lindahl, T.L.; Tengvall, P. Whole blood coagulation on protein adsorption-resistant PEG and peptide functionalised PEG-coated titanium surfaces. Biomaterials **2005**, *26* (8), 861–872.

312. Shen, M.; Martinson, L.; Wagner, M.S.; Castner, D.G.; Ratner, B.D.; Horbett, T.A. PEO-like plasma polymerized tetraglyme surface interactions with leukocytes and proteins: *In vitro* and *in vivo* studies. J. Biomater. Sci. Polym. Ed. **2002**, *13* (4), 367–390.

313. Fushimi, F.; Nakayama, M.; Nishimura, K.; Hiyoshi, T. Platelet adhesion, contact phase coagulation activation, and C5a generation of polyethylene glycol acid-grafted high flux cellulosic membrane with varieties of grafting amounts. Artif. Org. **1998**, *22* (10), 821–826.

314. Sirolli, V.; Di Stante, S.; Stuard, S.; Di Liberato, L.; Amoroso, L.; Cappelli, P.; Bonomini, M. Biocompatibility and functional performance of a polyethylene glycol acid-grafted cellulosic membrane for hemodialysis. Int. J. Artif. Organs **2000**, *23* (6), 356–364.

315. Okner, R.; Domb, A.J.; Mandler, D. Electrochemically deposited poly (ethylene glycol)-based sol-gel thin films on stainless steel stents. New J. Chemistry, **2009**, *33* (7), 1596–1604.

316. Thierry, B.; Merhi, Y.; Silver, J.; Tabrizian, M. Biodegradable membrane-covered stent from chitosan-based polymers. J. Biomed. Mater. Res. A, **2005**, *75* (3), 556–566.

317. Arifin, D.R.; Palmer, A.F. Polymersome encapsulated hemoglobin: A novel type of oxygen carrier. Biomacromolecules **2005**, *6* (4), 2172–2181.

318. Haque, T.; Chen, H.; Ouyang, W.; Martoni, C.; Lawuyi, B.; Urbanska, A.; Prakash, S. Investigation of a new microcapsule membrane combining alginate, chitosan, polyethylene glycol and poly-L-lysine for cell transplantation applications. Int. J. Artif. Organs **2005**, *28* (6), 631–637.

319. Arica, M.Y.; Bayramoglu, G.; Arica, B.; Yalçin, E.; Ito, K.; Yagci, Y. Novel hydrogel membrane based on copoly (hydroxyethyl methacrylate/p-vinylbenzyl-poly (ethylene oxide)) for biomedical applications: Properties and drug release characteristics. Macromol. Biosci. **2005**, *5*, 983–992.

320. Arica, M.Y.; Tuglu, D.; Basar, M.M.; Kiliç, D.; Bayramoğlu, G.; Batislam, E. Preparation and characterization of infection-resistant antibiotics-releasing hydrogels rods of poly[hydroxyethyl methacrylate-co-poly (ethylene glycol)-methacrylate]: Biomedical application in a novel rabbit penile prosthesis model. J. Biomed. Mater. Res. B Appl. Biomater. **2008**, *86* (1), 18–28.

321. Kazuhiko, I.; Hidenori, F.; Toshikazu, Y.; Yasuhiko, I. Antithrombogenic polymer alloy composed of 2-methacryloyloxyethyl phosphorylcholine polymer and segmented polyurethane. J. Biomater. Sci. Polym. Ed. **2000**, *11*, 1183–1195.

322. Kobayashi, K.; Ohuchi, K.; Hoshi, H.; Morimoto, N.; Iwasaki, Y.; Takatani, S. Segmented polyurethane modified by photopolymerization and cross-linking with 2-methacryloyloxyethyl phosphorylcholine polymer for blood-contacting surfaces of ventricular assist devices. J. Artif. Organs **2005**, *8* (4), 237–244.

323. Xu, J.; Yuan, Y.; Shan, B.; Shen, J.; Lin, S. Ozone-induced grafting phosphorylcholine polymer onto silicone film grafting 2-methacryloyloxyethyl phosphorylcholine onto silicone film to improve hemo-compatibility. Colloids Surf. B **2003**, *30* (3), 215–223.

324. Yang, Z.M.; Wang, L.; Yuan, J.; Shen, J.; Lin, S.C. Synthetic studies on nonthrombogenic biomaterials, Synthesis and characterization of poly (ether-urethane) bearing a zwitterionic structure of phosphorylcholine on the surface. J. Biomater. Sci. Polym. Ed. **2003**, *14* (7), 707–718.

325. Jordan, S.W.; Faucher, K.M.; Caves, J.M.; Apkarian, R.P.; Rele, S.S.; Sun, X.L.; Hanson, S.R.; Chaikof, E.L. Fabrication of a phospholipid membrane-mimetic film on the luminal surface of an ePTFE vascular graft. Biomaterials **2006**, *27* (18), 3473–3481.

326. Chandy, T.; Das, G.S.; Wilson, R.F.; Rao, G.H.R. Use of plasma glow for surface-engineering biomolecules to enhance blood compatibility of Dacron and PTFE vascular prosthesis. Biomaterials **2000**, *21* (7), 699–712.

327. Chang Chung, Y.; Hong Chiu, Y.; Wei Wu, Y.; Tai Tao, Y. Self-assembled biomimetic monolayers using phospholipid-containing disulfides. Biomaterials **2005**, *26* (15), 2313–2324.

328. Feng, W.; Zhu, S.; Ishihara, K.; Brash, J.L. Adsorption of fibrinogen and lysozyme on silicon grafted with poly (2-methacryloyloxyethyl phosphorylcholine) via surface-initiated atom transfer radical polymerization. Langmuir **2005**, *21* (13), 5980–5987.

329. Huang, X.J.; Xu, Z.K.; Wan, L.S.; Wang, Z.G.; Wang, J.L. Surface modification of polyacrylonitrile-based membranes by chemical reactions to generate phospholipid moieties. Langmuir **2005**, *21* (7), 2941–2947.

330. Nam, K.; Kimura, T.; Kishida, A. Physical and biological properties of collagen-phospholipid polymer hybrid gels. Biomaterials **2007**, *28* (20), 3153–3162.

331. Mcguigan, A.P.; Sefton, M.V. The influence of biomaterials on endothelial cell thrombogenicity. Biomaterials **2007**, *28* (16), 2547–2571.

332. Heyligers, J.M.M.; Arts, C.H.P.; Verhagen, H.J.M.; De Groot, P.G.; Moll, F.L. Improving small-diameter vascular grafts: From the application of an endothelial cell lining to the construction of a tissue-engineered blood vessel. Ann Vasc Surg. **2005**, *19* (3), 448–456.

333. Kidd, K.R.; Patula, V.B.; Williams, S.K. Accelerated endothelialization of interpositional 1-mm vascular grafts. J. Surg. Res. 2003, 113, 234–242.

334. Feinberg, A.W.; Schumacher, J.F.; Brennan, A.B. Engineering high-density endothelial cell mono-layers on soft substrates. Acta Biomater. **2009**, *5* (6), 2013–2024.

335. Lu, A.; Sipehia, R. Antithrombotic and fibrinolytic system of human endothelial cells seeded on PTFE: The effects of surface modification of PTFE by ammonia plasma treatment and ECM protein coatings. Biomaterials **2001**, *22* (11), 1439–1446.

336. Pawlowski, K.J.; Rittgers, S.E.; Schmidt, S.P.; Bowlin, G.L. Endothelial cell seeding of polymeric vascular grafts. Front. Biosci. **2004**, *9* (1-3), 1412–1421.

337. Aoki, J.; Serruys, P.W.; Van Beusekom, H.; Ong, A.T.; McFadden, E.P.; Sianos, G.; van der Giessen, W.J.; Regar, E.; de Feyter, P.J.; Davis, H.R.; Rowland, S.; Kutryk, M.J. Endothelial progenitor cell capture by stents coated with antibody against CD20, 0534, The HEALING-FIM (Healthy Endothelial Accelerated Lining Inhibits Neointimal Growth-First in Man) registry. J. Am. Coll. Cardiol. **2005**, *45* (10), 1574–1579.

338. Richardson, T.P.; Peters, M.C.; Ennett, A.B.; Mooney, D.J. Polymeric system for dual growth factor delivery.Nat. Biotechnol., **2001**, *19* (11), 1029–1034.

339. Singer, I.L.; Ollock, H.M.; Eds. *Fundamentals of Friction: Microscopic and Macroscopic Processes*; Kluwer: Dordrecht, the Netherlands, 1992.

340. Nagaoka, S.; Akashi, R. Low-friction hydrophilic surface for medical devices. Biomaterials **1990**, *11*, 419-K4.

341. Dowson, D. *History of Tribology*; Longmans: London U. K., 1979.

342. Tabor, D. (Ed.). *Friction*. Doubleday: New York, 1973.

343. Uyama, Y.; Tadokoro, H.; Ikada, Y. Surface lubrication of polymer-films by photoinduced graft polymerization. J. Appl. Polym. Sci. **1990**, *39* (3), 489–498.

344. Uyama, Y.; Tadokoro, H.; Ikada, Y. Low-frictional catheter materials by photoinduced graft-polymerization. Biomaterials **1991**, *12* (1), 71–75.

345. Kyomoto, M.; Moro, T.; Miyaji, F.; Hashimoto, M.; Kawaguchi, H.; Takatori, Y.; Nakamura, K.; Ishihara, K. Effects of mobility/immobility of surface modification by 2-methacryloyloxyethyl phosphorylcholine polymer on the durability of polyethylene for artificial joints. J. Biomed. Mater. Res. A **2009**, *90A*, 362–371.

346. Sochart, D.H. Relationship of acetabular wear to osteolysis and loosening in total hip arthroplasty. Clin. Orthopaed. Relat. Res. 1999, 135–150.

347. Ratner, B.D.; Hoffman, A.S. Process for Radiation Grafting Hydrogels onto Organic Polymeric Substrates. The United States of America as represented by the United States Energy, Washington, DC, 1976.

348. Fan, Y.-L.; Marlin, L. Biocompatible Hydrophilic Complexes and Process for Preparation and Use; Union Carbide Chemicals & Plastics Technology Corporation: Danbury, CT, 1994.

349. Hu, C.B.; Solomon, D.D.; Williamitis, V.A. *Method for Preparing Lubricated Surfaces*; Becton, Dickinson and Company: Franklin Lakes, NJ, 1989.

350. Hu, C.B.; Gwon, A.; Lowery, M.; Makker, H.; Gruber, L. Preparation and evaluation of a lubricious treated cartridge used for implantation of intraocular lenses. J. Biomater. Sci. Polym. Ed. 2007, 18, 179–191.

351. Kohnen, T.; Kasper, T. Incision sizes before and after implantation of foldable intraocular lenses with 6 mm optic using Monarch and Unfolder injector systems. Ophthalmology, 2005, 112 (1), 58–66.

352. Woolfson, A.D.; Malcolm, R.K.; Gorman, S.P. Self-lubricating silicone elastomer biomaterials. J. Mater. Chem. 2003, 13 (10), 2465–2470.

353. Lee, S.; Iten, R.; Muller, M.; Spencer, N.D. Influence of molecular architecture on the adsorption of poly (ethylene oxide)-poly (propylene oxide)-poly (ethylene oxide) on PDMS surfaces and implications for aqueous lubrication. Macromolecules 2004, 37 (22), 8349–8356.

354. Chawla, K.; Lee, S.; Lee, B.P.; Dalsin, J.L.; Messersmith, P.B.; Spencer, N.D. A novel low-friction surface for biomedical applications: Modification of poly (dimethylsiloxane) (PDMS) with polyethylene glycol (PEG)-DOPA-lysine. J. Biomed. Mater. Res. A 2009, 90A, 742–749.

355. Bansil, R.; Stanley, E.; Lamont, J.T. Mucin biophysics. Ann. Rev. Physiol. 1995, 57, 635–657.

356. Lee, S.; Spencer, N. D. Materials science—Sweet, hairy, soft, and slippery. Science 2008, 319, 575–576.

357. Lee, S.; Spencer, N.D. Aqueous lubrication of polymers: Influence of surface modification. Tribol. Int. 2005, 38 (11–12), 922–930.

358. Gong, J.P. Friction and lubrication of hydrogels—Its richness and complexity. Soft Matter 2006, 2 (7), 544–552.

359. Raviv, U.; Giasson, S.; Kampf, N.; Gohy, J.F.; Jérôme, R.; Klein, J. Lubrication by charged polymers. Nature 2003, 425 (6954), 163–165.

360. Urbakh, M.; Klafter, J.; Gourdon, D.; Israelachvili, J. The nonlinear nature of friction. Nature 2004, 430 (6999), 525–528.

361. Ding, C.; Gu, J.; Qu, X.; Yang, Z. Preparation of multifunctional drug carrier for tumor-specific uptake and enhanced intracellular delivery through the conjugation of weak acid labile linker. Bioconjug. Chem. 2009 20(6), 1163–1170.

362. Etrych, T.; Mrkvan, T.; Říhová, B.; Ulbrich, K. Star-shaped immunoglobulin-containing HPMA-based conjugates with doxorubicin for cancer therapy. J. Control. Release 2007, 122(1), 31–38.

363. Jain, R.; Standley, S.M.; Frechet, J.M. Synthesis and degradation of pH-sensitive linear poly (amidoamine) s. Macromolecules 2007, 40(3), 452–457.

364. Heffernan, M.J.; Murthy, N. Polyketal nanoparticles: a new pH-sensitive biodegradable drug delivery vehicle. Bioconjug. Chem. 2005, 16(6), 1340–1342.

365. Shim, M.S.; Kwon, Y.J. Controlled delivery of plasmid DNA and siRNA to intracellular targets using ketalized polyethylenimine. Biomacromolecules 2008, 9(2), 444–455.

366. Chan, Y.; Bulmus, V.; Zareie, M.H.; Byrne, F.L.; Barner, L.; Kavallaris, M. Acid-cleavable polymeric core–shell particles for delivery of hydrophobic drugs. J. Control. Release 2006, 115(2), 197–207.

367. Knorr, V.; Allmendinger, L.; Walker, G.F.; Paintner, F.F.; Wagner, E. An acetal-based PEGylation reagent for pH-sensitive shielding of DNA polyplexes. Bioconjug. Chem. 2007, 18(4), 1218–1225.

368. Yoo, H.S.; Lee, E.A.; Park, T.G. Doxorubicin-conjugated biodegradable polymeric micelles having acid-cleavable linkages. J. Control. Release 2002, 82(1), 17–27.

369. Lavignac, N.; Nicholls, J.L.; Ferruti, P.; Duncan, R. Poly (amidoamine) Conjugates Containing Doxorubicin Bound via an Acid-Sensitive Linker. Macromol. Biosci. 2009, 9(5), 480–487.

Index

ABHS, *see* Albumin-based hydrogel sealant (ABHS)
Abrasion testing, 760, 761
Absorbable suture
  characteristics, 1527
  commercial structure, 1526–1527
  commercial suture materials, 1515–1518
  comparison, 1520, 1523
  degradation products, 1528
  mechanical properties, 1520–1521
  scanning electron images, 1515, 1519, 1527–1528
  tensile breaking force, 1527
Acid etching, 433
Actin gels, 682
Active pharmaceutical ingredients (APIs), 1076, 1465
Adenosine diphosphate (ADP), 1560–1561
Adhesives
  catechol derived, 1, 4–7
  cyanoacrylate glues, 2–3
  fibrin glue, 1–2
  Gecko-inspired, 9–11
  light-activated, 7–9
  nanostructured adhesives, 1
  polyethylene glycol-based glue, 1, 3
  polysaccharide adhesives, 1
Adipose-derived stem cells (ADSCs), 727, 1410
Adriamycin (ADR), 790
Adsorption theory
  bioadhesion, 113–114
  mucoadhesion, 963–964
Aerogels
  alumina, 41
  biopolymer-based, 42
  cellulosic (*see* Cellulosic aerogel)
  cobalt–molybdenum–sulfur, 42
  silica, 39–41
Aerosols, 308
Agar overlay tests, 131
Agarose, 681, 1218–1219, 1259
Agglutination techniques, 751
AGUs, *see* Anhydro-D-glucopyranose units (AGUs)
Albumin
  glue, 648–649
  small-molecule therapeutics delivery, 66
Albumin-based hydrogel sealant (ABHS), 650
Alcohol-resistant polyurethane, 1326
Aldehyde system
  albumin glue, 648–649
  aldehyde–gelatin film, 649
  gelatin and polysaccharides, 648
  glutaraldehyde, 649
Alginate, 681, 1259
Alginic acid
  applications, 1610
  manufacture, 1610

  structure, 1610
  wound care, 1609–1610
Alkaline phosphatase activity (ALP), 523, 525
Alkaline-catalyzed oxalkylation, 278
AM, *see* Amniotic membrane (AM)
Amalgam
  bonds, 436, 437
  restoration, 435, 436
  sealant, 435–437
Ambiphilic mucoadhesive polymers, 950
Amine-based MFC aerogels, 43
Amino acid *N*-carboxy anhydride (B-NCA), 589–590
(γ-Aminopropyl) triethoxysilane (APS) molecules, 589–590
Amniotic fluid-derived stem cells (AFSCs), 727
Amniotic membrane (AM), 1635
  corneal endothelium, 383–384
  corneal epithelium, 379
Amorphous solid dispersion system
  melt extrusion, 828
    bioavailability enhancement applications, 829
    controlled release products, 835, 836
    dissolution-enhanced products, 833–835
    miscibility, 831
    plasticizer, 829
    spring and parachute performance, 832
    surfactants, 831
  polymeric interactions
    anti-solvent process, 465
    basic techniques, 464
    cryogenic production, 465
    crystalline and glassy solid enthalpy and volume, 463
    dissolution rate, 468
    enteric compositions, 470, 471
    fluid bed coating, 464
    hot-melt extrusion, 466–468
    physicochemical properties, 462–463
    supercritical fluid technologies, 465
    tacrolimus solid dispersions, 469
    in vitro supersaturation profiles, 471
Amoxicillin-loaded electrospun nano-HAp/PLA (AMX), 497, 498
Amoxicillin mucoadhesive microspheres, 121
Amphiphilic diblock copolymer self-assembly, PNSs
  elongated micelles, 934
  micellar unimolecules and aggregates, 933
  micellization and gelation, 933
  spherical micelles, 933–934
  vesicles, 934
Amphiphilic hybrid block copolymers, 350–351
AMX, *see* Amoxicillin-loaded electrospun nano-HAp/PLA (AMX)
Amyl acetate (AA), 1551

AN69®, 868
Angiogenic process, 724, 1215
  angiogenic factor, 799–800
  ANSA, 801
  antibacterial polymers, 802
  fucoidan, 801–802
  HSPGs, 800–801
  interaction, 801
  mechanisms, 801
  pathological processes, 799
  therapeutic induction, 801–802
Angiography catheters, 1324–1325
Anhydro-D-glucopyranose units (AGUs), 271, 277
Anionic mucoadhesive polymers, 947, 948
Antibacterial polymers
  angiogenic process, 802
  biomedical applications, 802–803
  cationic biocides, 803
  cationic polymers, 804
  eukaryotic cells, 803–804
  phosphatidylethanolamine, 803
  requirements, 803
  strategies, 803
  surface functionalization, 803
Antibody
  biosensors, dentritic, 408
  small-molecule therapeutics delivery, 66
Antibody-targeted micelles, 1555–1556
Anticoagulants, 664
Anti-fouling substrates, 606–607
Antigen delivery, 143–144
Anti-infective biomaterials
  antimicrobial incorporation, 85
  bacterial adhesion (*see* Bacterial adhesion)
  DACRON® prostheses, 85
  Septopal®, 85
  silicone catheters, 85
  silver catheter, 85
  TEFLON® prostheses, 85
Anti-inflammatory biomaterials, 178
Antimicrobial polyurethanes
  co-extruded antimicrobial catheter, 1333
  fluorinated, 1335
  nosocomial infections, 1329
  polycarbonate-based, 1333
  polyethylene vinyl acetate, 1331
  silver and, 1330–1331
  thermoplastic medical-grade, 1334–1335
  wound dressings, 1333–1334
  zone of inhibition, 1331–1332
Antithrombogenicity *vs.* pseudoneointima, 665
Aprotinin, 2
Aptamers
  drug-encapsulated controlled-release, 1117–1118
  targeted drug delivery, 68–69
Arg-Gly-Asp (RGD) sequence, 344–345

Artificial muscle, 320–321, 1489
  conductive polymers, 93
  electroactive polymer (*see* Electroactive
     polymer (EAP))
  electronic electroactive polymer/ferroelectric
     polymeric, 91–92
  ionic polymer-metal nanocomposites (*see*
     Ionic polymer–metal nanocomposites
     (IPMNCs))
  liquid crystal elastomers, 92, 96–97
  magnetically activated, 91
  polyacrylonitrile, 93–95
  shape memory alloys, 91
  shape memory polymers, 91
  SMA, 93
Atom transfer radical polymerization (ATRP),
     342–343, 588, 922, 1019–1020, 1290,
     1291
Atomic force microscopy (AFM), 689
ATRP, *see* Atom transfer radical
     polymerization (ATRP)
Augmentation devices and prostheses
  biocompatibility, 758
  design aspects, 759
  mechanical properties, 758–759
  principles, 758
Avanace, 1266
Axo guard nerve
  connector, 1266
  protector, 1266
Axogen, 1265–1266
Azide chitosan lactose, 651
AZT, *see* Zidovudine (AZT)

BACs, *see* Biologically active compounds
     (BACs)
Bacterial adhesion, 83
  DLVO theory, 84
  polymer surface modification, 85–86
  polymer surface properties, 86–88
  preventive approaches, 83
  thermodynamic model, 83–84
Bacterial cellulose (BC), 1583, 1219–1220
Bacterial cellulose (BC) aerogels
  Acetobacter spp. strains, 23
  aerosol reactors, 24
  airlift reactors, 24
  batch-wise static cultivation, 24
  biological spinnerets, 23
  biomedical applications, 45
  compression test, 25
  mechanical reinforcement, 25–27
  membrane bioreactors, 24
  open-porous structure and wettability, 25
  paraffin spheres, 25
  pore size distribution, 25, 26
  pore widening, 24–25
  porosity preservation, 25
  rotary bioreactors, 24
  tissue engineering, 24
  ultra-structure, 24
Balloon-delivered stent
  bioabsorbable therapeutics, 1474–1475
  BMS and metal DES platforms, 1473–1474
  finite element analysis, 476
  NSAID polymer, 1475
  OrbusNeich R stent, 1475

  OrbusNeich stent, 1476–1477
  PCL-PGA copolymer, 1474
  pharmacokinetics, 1476
  PLGA copolymer, 1474
  sheet, 1473
  tubes, 1473–1476
  ultraviolet, 1474
Bare metal stents (BMSs), 1466
Basic fibroblast growth factor (bFGF), 178
BCOP, *see* Bovine cornea opacity and
     permeability (BCOP) test
BCP, *see* Biochromic conjugated polymer
     (BCP)
Beam modeling process, 989
Bending fatigue testing, 760–761
β-glucosidase, 279
β-sheet containing polymer hybrids, 345–348
BFS, *see* Biaxial flexure strength (BFS)
BG, *see* Bioactive glass (BG)
Biaxial flexure strength (BFS), 421, 422
Bimodal electrospun scaffolds, 516
Bioabsorbable polymers
  cartilage degeneration, 102
  liver reconstruction, 102
  naturally derived matrix, 102
  osteoblast transplantation, 102–103
  pancreas reconstruction, 102
  skin replacement, 103
  vascular grafts, 103
Bioactive glass (BG), 318–319
Bioadhesion
  adsorption theory, 113–114
  bioadhesive polymers, 114, 115
  concentration of active polymer, 116
  cross-linking and swelling, 114–115
  diffusion theory, 113
  drug/excipient concentration, 116
  electrostatic theory, 112–113
  higher forces, 116
  hydrophilicity, 114
  molecular weight, 114
  pH, 115–116
  physiological variables, 116
  tissue surface roughness, 112
  types of, 111
  viscosity and wetting, 112
  wetting theory, 112–113
Bioadhesive drug delivery systems
  bioadhesive force of attachment, 116–118
  cervical and vulval drug delivery, 123
  characterization methodologies, 118
  compacts, 119
  delivery platforms, 118
  dielectric spectroscopy, 118
  film-forming bioadhesive polymers, 119
  flow rheometry, 118
  GIT, 120–122
  nasal drug delivery, 124
  ocular drug delivery, 125
  oral cavity, 119–120
  oscillatory rheometry, 118
  rectal drug delivery, 122–123
  vaginal drug delivery, 123–124
Bioadhesive force of attachment
  bioadhesive bond, 116–117
  gastric-controlled release systems, 118
  gastrointestinal transit times, 117

  mucoadhesives, 118
  peel test, 116
  shear stress, 116
  tensile stress, 116
  texture profile analyzer, 117, 118
Bio-based dendrimers, 396
Biochips/biosensors
  DNA sequencing separations, 182
  laboratories-on-chip, 181
  laser-induced fluorescence, 183
  microfluidic analytical devices, 183
  micro-total analysis systems, 182–183
  miniaturized device, 181–182
  molecular and biological assays, 181
  tissue engineering, 185
Biochromic conjugated polymer (BCP), 322
Biocompatibility, polymeric membranes
  biological reactions, 864
  polymeric surface interactions, 864
  surface charge adjustment, 865
  surface free energy, 865
  surface hydrophobicity and hydrophilicity,
     865
Biocompatibility tests, PGS
  clinical trials, 135–136
  in vitro tests, 130–132
  in vivo animal testing, 132–135
Bioconjugates
  confocal fluorescence, 575
  in vitro image, 575–576
  in vivo image, 576
  monitor processes, 577
  multimodal QDs, 576–578
  near-infrared, 1389–1390
  semiconducting polymer dot (*see*
     Semiconducting polymer dot
     bioconjugates)
Biodegradability, 692
Biodegradable polymers, 445–446; *see also*
     Bioerodible drug-delivery implants;
     Biomedically degradable polymers
  natural, 871
  synthetic, 871–873, 1378–1379
  theranostic nanoparticles, 1534–1536
  tissue engineering, 1374–1375
Biodegradation
  biochemical process, 164
  biological environments, 162
  chemical process, 164
  collagen, 173
  composites, 171
  decomposition, 162
  definition, 162
  dissolution of object, 164
  dosage forms, 162
  environmental conditions, 167
  ethylene glycol, 170
  extracellular, 167–169
  in gastrointestinal tract, 167
  heterochain polymers products, 170
  hydrolytic degradation, 164–166
  hydrophilicity, 163
  implants, 162
  inflammatory reactions, 166–167
  insoluble hydrolysis products, 170
  intracellular, 169–170
  mechanism, 163

natural polymers, 171
object biotransformation, 164
physicochemical process, 164
polyamides, 172, 173
polyhydroxycarboxylic acids, 171–172
poly-orthoesters, 172, 173
practically nontoxic products, 170
prooxidants, 171
segmented polyurethanes, 172–173
water-soluble hydrolysis products, 170
Biodestruction, 162
Bioelectrets
bile acids, 479
blood compatibility, 478
brain functioning, 479
chemical micropatterning, 478–479
collagen, 478
collagen films dielectric measurements, 479
fibrous keratin, 478
hydroxyapatite, 480
nanotechnology, 479–480
succinic acid-doped glycine, 479
Bioengineered skin equivalents (BSEs), 1408
Bioerodible drug-delivery implants
advantage, 139
hydrogels, 139–141
hydrophobic polymers (see Hydrophobic polymers)
structural changes, 139
Bioerosion, 162
Biofunctional polymers
anti-inflammatory biomaterials, 178
ECM-mimetic materials, 175–177
growth factors, 178
hirudin, 177
nitric oxide–generating materials, 178–179
plasminogen activators, 177–178
polysaccharides, 177
Bio-inspired/biodegradable polymer hybrid, 682–683
Biological macromolecule
carboxymethyl chitosan and DEAM, 712–713
chitosan-pNIPAAM, 712
dual-stimuli-responsive polymer, 710–711
NIPAAM and MAA sodium salt, 710
p(NIPAAM-co-acrylamide), 711
polymerization, 712
pulsatile drug delivery systems, 711
smart polymeric microgels, 712
Biologically active compounds (BACs)
encapsulation, 106
ferromagnetic substances, 107
macromolecular systems, 106
nano-sized drug carriers, 106–107
polyanionites, 108
polymer coatings, 106
tablets, 106
transdermal systems, 106
water-soluble cationites, 108
water-soluble polyelectrolytes, 108
Biologically active polymers
biologically active compounds, 105–107
hemodynamic blood substitutes, 107–108
immobilized fragments, 108–110
microbicidal activity, 108
polycations and polyanions, 108
polymer carriers, 109

water-soluble amine-containing polymers, 109–110
water-soluble cationites, 108
Biologically derived polymers, 508
Biomaterial-based scaffolds
architecture, 725
collagen-glycosaminoglycan scaffolds, 725
extracellular matrix, 724–726
fabricate scaffolds, 724
hydrogel-based materials, 726
mechanical properties, 724–725
PCL, 725
synthetic polymers, 725
tissue repair, 724–726
vasculogenic/angiogenic process, 725–726
Biomechanical sensing, 319–320
Biomedically degradable polymers
aliphatic polyanhydrides, 153
aromatic polyanhydrides, 153
copolymers, 154
degradation rate, 153
1,5-dioxepan-2-one, 156–159
microorganisms, 154
poly(ortho ester), 153
ring-opening polymerization of lactones, 154–155
stable, 153
supposedly absorbable polymers, 153
synthetic absorbable polymers, 153
Biomicroelectromechanical systems (bioMEMSs)
biochips/biosensors, 181–183
controllable drug delivery systems, 183
immunoisolation, 183
microfluidics, 186–187
polymer microparticles, 184
self-regulated drug delivery systems, 183
soft lithography, 184
tissue engineering, 184–186
Biomimetic materials
bioactive materials and biochemical modifications, 197–200
bio-inspired strategies, 217
biomimicry, 189–190
cardiovascular tissue engineering, 202–203, 205
cartilage-mimicking materials, 202
chemical modification, 216
extracellular matrix, 190–201
functions, 217
high-throughput screening approaches, 223–224
nanostrucured nacre-like multilayered systems, 221
orthopedic tissue engineering, 202, 204
polyelectrolyte multilayered films, 219–220
polymeric/hydrogel particles, 224
regenerative medicine, 190
smart multilayered films, 221–222
smart surfaces, 218–219
superhydrophobic substrates, 223–224
surface modifications, 218
wettability, 222–223
Biomimetic polymers, 682
Biomimetic/synthetic polymer hybrid gels, 682–683
Biomolecular actuators, 317–318

Biomolecular sensing, 315–317
Biomolecular surface modification
biomolecular interactions, 585
functional polymer coatings, 585–592
polyethylene glycol, 585
Biomolecule immobilization, metal–polymer composite
hydrogel and gelatin, 893
organic coatings, 889–890
osseointegration, 891
peptides, 891–892
protein and collagen, 892–893
self-assembled monolayers, 891
Biopolymers/hemoderivatives association
AgSD, 1655–1656
CHOS and PL components, 1654
CSG formulation, 1654
HA-based dressing, 1655
normal human dermal fibroblasts, 1656
PDGF AB content, 1654
photograph, 1654
SLNs, 1656
therapeutic platforms, 1653–1654
Bioreactors, 24, 101, 729, 730
Biorubber, see Poly(Glycerol Sebacate) elastomer
Biosensors, 408–409, 598–599; see also Sensors and biosensors
Biosynthetic-synthetic polymer based conjugates, 340
future prospects, 357
main chain polymer hybrids with bioactive motifs, 351–354
main chain polymer hybrids with structural motifs
β-sheet containing polymer hybrids, 345–348
helical peptide–polymer hybrids, 349–352
mussel adhesive polymer hybrids, 347, 349
polymer–protein conjugates
end-functionalized polymers synthesis, 356
poly(N-isopropyl acrylamide) [poly(NIPAAM)], 356–357
positive effects, 353–354
site-specific incorporation, 354–355
synthetic erythropoiesis protein, 355
side chain polymer hybrids with biologically active motifs, 343–345
side chain polymer hybrids with structural motifs
amphiphilic helical polymer construction, 341–342
atom transfer radical polymerization, 342–343
chirality introduction, 341
elastin-mimetic triblock copolymers, 343
free radical polymerization, 340–341
poly(phenyl) acetylenes, 342
Biotransformation, 162
Bis-glycidyldimethacrylate (Bis-GMA), 884
Bismuth oxychloride, 1337–1339
Block copolymer/DNA complexes, 637
Blood clotting
anticoagulants, 664
blood coagulation factors, 664
complements, 664–665
fibronectins, 665

mural thrombus formation, 663–664
plasma proteins, 663–664
platelets, 664
thrombolysis, 664
Blood oxygenator, 668
bubble oxygenator, 868
disk oxygenator, 868
hollow fiber oxygenator, 869
membrane oxygenator, 868
required oxygen transfer, 869–870
sheet membrane, 869
Blood vessel substitutes
biological response, 243–244
cell-secreted scaffolds, 243
degradable synthetic scaffolds, 240–241
endothelialization strategies, 239–240
naturally-derived scaffolds, 241–243
structure and function, 238–239
BODI® inserts, 970
Bone marrow (BM), 1453
Bone marrow mesenchymal stem cells
(BMSCs), 1410
Bone morphogenetic protein-2 (BMP-2), 892
Bone tissue engineering, 532–533
Bone–implant
carbon fiber-reinforced PEEK (see Carbon
fiber-reinforced PEEK)
low density polyethylene (see Low density
polyethylene)
physicochemical surface modification,
248–249
polyether-ether ketone (see Polyether-ether
ketone)
Bovine cornea opacity and permeability
(BCOP) test, 384
Bovine serum albumin (BSA), 316, 589, 748
Bowman's layer, 371
Braids, 762–763
Branched polycations, 637
BSA, see Bovine serum albumin (BSA)
Bulk degradation, 870–871

Caco 2 cell-actin, 1310, 1311
CAD/CAM, see Computer-assisted design and
manufacturing (CAD/CAM)
Calcium phosphate, 1278
Cancer diagnosis and therapy
atoms, molecules/molecular assemblies, 560
bioconjugates (see Bioconjugates)
cancer tumor lesions, 573
conventional methods, 573
diagnosis and detection, 571–573
DNA damage, 561
fluorescent bioprobes (see Fluorescent
bioprobes)
history, 561
invasion and progression, 561
leukemia cells (blood cancer), 561
micro-metastasis, 578
nanomedicine (see Nanomedicine)
nanoparticles (see Nanoparticles (NPs))
QD bioconjugates, 578
QDs (see Quantum dots (QDs))
theranostics, 578
time-frame evolution, 561–562
CAP, see Cellulose acetate phthalate (CAP)
Captopril, 496–497

Carbogels
cellulose acetate, 48
electrical double-layer capacitors, 47–48
microcrystalline cellulose, 48–49
proton exchange membrane fuel cells, 47–48
Carbon fiber (CF) microelectrodes, 1036–1037
Carbon fiber-reinforced (CFR) PEEK
endosteal dental implants, 261–267
goat mandibular implants, 258–261
orthogonal solid-state orientation process,
252–254
osteoblast interaction, 255–256
surface microtexturing, 251–252
surface phosphonylation, 251
Carbon nanostructures and composite
structures, 1035
Carbon nanotubes (CNTs), 315–316,
1037–1038
Carbopol®, 116
Carbothane hemodialysis catheters, 1323
Carboxy betaine methacrylate (CBMA), 607
Carboxybetaines (CB)
anti-ALCAM, 1662–1663
NMFRP, 1662
non-fouling characteristics, 1662–1663
poly-CBMA, 1662–1663
SI-PIMP technique, 1663
zwitterionic polymers, 1661–1662
Carboxymethyl cellulose (CMC), 24, 46, 1313
Carboxymethyl-MFC aerogels, 29–30
Carcinogenicity testing, 133–134
Cardiomyocytes, 522, 525
Cardiomyocyte-targeting peptide (PCM), 1135
Cardiovascular application of polymers, 667
artificial heart devices, 667
blood oxygenators, 668
catheters, 668
heart valves, 667
pacemakers, 668
vascular grafts, 667–668
Cardiovascular catheters
alcohol vapor effects, 1328
alcohol-resistant polyurethane, 1326
angiography, 1324–1325
elastic modulus, 1329
molecular weight effect on kink resistance,
1326–1327
thrombolysis, 1325
transluminal angioplasty, 1325–1326
Cardiovascular disease (CVD)
activation and amplification pathways, 1579
analysis and testing, 1578
arteriovenous grafts, 1578
atherosclerosis, 1575–1576
coagulation, 1578–1579
complement activation, 1578
dacron graft, 1576–1577
expanded polytetrafluoroethylene, 1575–1577
hematology, 1578
long saphenous veins, 1576
mechanical properties, 1579
nondegradable polymers, 1580–1583
nonprotein polymer monomers, 1580–1581
peripheral arterial disease, 1576
platelet activation, 1579
polymeric grafts, 1587
polytetrafluoroethylene, 1577

requirements, 1578
smooth muscle cells, 1576
thrombogenicities, 1579
tissue-engineering approaches, 1583–1587
vein graft, 1576–1577
in vivo tests, 1580
Cardiovascular therapeutics
advantages, 1550
amyl acetate, 1551
capsules and chitosan, 1550
coadministration, 1549–1550
drug therapy, 1549–1550
ethylcellulose, 1550
gene therapy, 1549
sonochemical encapsulation process,
1550–1551
stem cell delivery, 1550
Carrageenan, 681
Cartilage degeneration, 102
Cartilage tissue engineering, 533, 534
Catechol derived bioadhesives, 1
bonding mechanism, 5
chitin, 6
dextran-ε-PL adhesives, 6
dopamine-methacrylamide-acrylate
copolymer, 4
elastin, 6–7
Fe³⁺ addition, 4
hydrogel, 6
PEG-based catechol adhesives, 4
polydopamine, 4
polyoxetane copolymers, 4–5
silk fibroin, 5
Cationic lipoplexes, 767
Cationic liposomes, 634
Cationic mucoadhesive polymers, 947–949
Cationic polymers, 656–658
Cell interactions
cells and matrix, 617
fibroblasts, 617
proliferation and collagen production, 617
smooth muscle cells, 618
TGF-β, 617
types, 618
Cell therapy, 1263
Cell-adhesive materials, 175–176
Cell-based TE approach, 693–694
Cell-biomaterial interactions
gradients, 603
patterning, 600–602
physical and chemical surface, 599–600
Cell-invasive electrospun materials, 515–517
Cellobiohydrolase, 279
Cell-penetrating peptides (CPPs), 807
Cellulose
applications, 1607–1608
bacterial cellulose, 1607
derivatives, 655
ethers, 273
ethylcellulose, 272
hemicellulose, 273–275
hydroxypropyl cellulose, 273
hydroxypropylmethyl cellulose, 273
manufacture, 1608
methyl cellulose, 272
microcrystalline cellulose, 272
molecular structure, 271

oxycellulose, 272
sodium carboxymethyl cellulose, 272
structure of, 1607–1608
wound care application, 1607–1608
Cellulose acetate, 48
Cellulose acetate phthalate (CAP), 273
Cellulose II aerogels
aqueous salt solutions, 34–36
cellulose dissolution, 33
inorganic molten salt hydrates, 33–34
ionic liquids, 36–37
morphological properties, 39
NMMO monohydrate, 37–38
physical and mechanical properties, 39
preparation, 32–33
Cellulose-based biopolymers, 1312–1313
assembly, 271
bacterial modification, 277
bioadhesion, 274–275
cellulose, 273–275
cellulose ethers preparation, 277–278
cellulose-based hydrogels and cross-linking
strategies, 279–280
chemical modification, 277, 278
classes and applications, 270
crystallinity and solid state characteristics,
275
enzymic methods, 278–279
future prospects, 292
ILs (see ionic liquids (ILs))
immediate-release dosage forms, 282–283
microparticles, 288
nanoparticles, 288–291
particle, 274
physical modification, 276
polymer grafting, 277
protein and gene delivery, 291
silylation, mercerization and other surface
chemical, 276–277
specialized bioactive carriers (see Specialized
bioactive carriers)
sustained/modified/controlled-release dosage
forms, 283–285
swelling, 274
wound healing, 291–292
Cellulosic aerogel
carbogels, 47–49
catalysis, 43
from cellulose Iα-rich sources (see Bacterial
cellulose (BC) aerogels)
from cellulose Iβ-rich sources (see
Microfibrillated cellulose (MFC)
aerogels)
cellulose II aerogels, 32–39
electro-conductive aerogels, 44
freeze-drying, 20
insulation materials, 42
low bulk density, 42
magnetic aerogels, 43
pore collapsing, 20
solvent exchange, 21
sorption, 43
supercritical carbon dioxide drying, 21–23
temporary templates, 44–45
Cellulosic membranes, 866–867
Ceramics, 1277–1279
Cerebrospinal-fluid (CSF), 646

Cervical drug delivery, 123
Cervical mucoadhesive delivery systems, 971
Chemical gradients
cellular responses, 596–597
laminar flow, 596
side-view schematic representation, 597–598
Chemical immobilization, 496
Chemical micro- and nanopatterning, 478–479
azide-functionalized inks, 595–596
inexpensive technique, 595–596
liquid pre-polymer, 593
μCP and microfluidic patterning, 593
microstructuring, projection
photolithography, 594–595
nanoimprint lithography process, 595–596
nanopattern formation, 595, 597
preventing protein adsorption, 594
vapor-assisted microstructuring, 594
Chemical vapor deposition (CVD)
advantages, 588–589
laser-based process, 589
plasma polymerization, 591–592
thermal-based CVD polymerization, 592
two-compound techniques, 589–591
vapor-based processes, 589
Chemical-responsive polymer microgels, 921
Chemomechanical polymers, 447
Chirila keratoprosthesis, 375
Chitin, 6, 1645–1646
antibacterial action, 1609
applications, 1609
manufacture, 1609
N-acetyl-glucosamine, 1608
N-glucosamine units, 1608
structure of, 1609
wound care, 1608–1609
Chitosan (CS), 657, 681, 947–949, 1259,
1645–1646
nanofiber, 500–502
scaffolds, 1217–1218
wound care, 1608–1609
Chitosan-gelatin scaffolds, 1215
Chitosan-gelatin sponge wound dressing
(CGSWD), 1489
Chlorofluorocarbons (CFCs), 654
Chondrocytes, 525, 528
Chondroitin sulfates (CHOS), 1649–1650
CLA, see Cross-linked high amylose starch
(CLA)
Classical polymers, 658
CLEAs, see Cross-linked enzyme aggregates
(CLEAs)
CLET, see Cultivated limbal epithelial
transplantation (CLET)
Cloud point (CP), 1443, 1445–1446
CMC, see Carboxymethyl cellulose (CMC)
CNTs, see Carbon nanotubes (CNTs)
Coaxial electrospinning, 492
Cobalt chloride (CoCl$_2$), 731
Cobalt-molybdenum-sulfur aerogel, 42
Co-culture systems (vascularization), 728–729
Co-extruded antimicrobial catheter, 1333
Cold plasma oxidation radiation grafting, 249
Collagen
applications, 1612
biodegradation, 173
bioelectrets, 478, 479

biomolecule immobilization, 892–893
manufacture, 1612–1613
matrices, 379–380
scaffolds, 1213–1214
structure of, 1612
tissue engineering, 1584–1586
triple helix, 1612
wound care, 1611–1612
Collagen-glycosaminoglycan scaffolds, 725
Colloid drug delivery systems
aerosols, 308
colloids, 301–302
cubosomes and hexosomes, 303–304
dendrimers, 303
emulsions, 304
foams, 308
gels, 304
liposomes, 306
liquid crystals, 303, 304
micelles, 302–303
microemulsions, 305
microspheres, 307
multiple emulsions, 304–305
nanocapsules, 307
nanoemulsions, 305
nibosomes, 306
polymersomes, 306–307
Colloids, 301–302
Colony-forming unit-granulocyte monocytes
(CFU-GM), 1456
COMET, see Cultivated oral mucosal epithelial
transplantation (COMET)
Computed tomography (CT), 403, 407, 572
Computer-assisted design and manufacturing
(CAD/CAM), 417
Conducting polymers (CPs), 1262–1263
artificial muscle, 320–321
bioerodible Ppy degradation characteristics,
484–485
biomechanical sensing, 319–320
biomolecular actuators, 317–318
biomolecular sensing, 315–317
chemical synthesis, 313
drug delivery, 320
electrochemical synthesis, 313–314
electrospinning, 315
enzyme-based biosensor, 483–484
future prospects, 323
hard template synthesis, 314
micro- and nano-patterning, 315
neural probe, 321–322
optical biosensor, 322–323
polyacetylene, 482–483
Ppy electrochemical synthesis, 483
soft template synthesis, 314
structures, 312, 313
TE applications, 484
template-free polymerization, 314–315
tissue engineering, 318–319
Conductive polyacrylonitrile (C-PAN) fiber, 93
Conjugated polymer nanoparticles (CPNs)
advantages and properties, 327
biomedical applications, 334–337
coatings and paints preparation, 328
drug delivery and, 328
electrochemical applications, 331–332
energy transfer, 331

future prospects, 337
optoelectronic applications, 332–333
photoluminescent tool, 333–335
polymerization from monomers, dispersion media, 330–331
postpolymerization dispersion, 328–330
Conjugates, *see* Biosynthetic-synthetic polymer based conjugates
Contact angle measurement, 691–692
Contact electrification, 477
Contact lenses, *see* Gas permeable contact lenses
Contrast-enhanced ultrasound (CEUS)
bovine serum albumin (BSA), 1547
gas core, 1547
magnetic resonance imaging (MRI), 1546
microbubbles, 1545
multilayered microbubble, 1547
perfluorocarbon, 1547
shell materials, 1547
systemic circulation, 1545–1546
targeting ligands, 1546–1547
untargeted microbubbles, 1546
Controlled release products, melt extrusion
abuse-deterrent formulations, 835
abuse-deterrent products, 837
amorphous solid dispersion, 836
crystalline solid suspension, 836
delayed release, 835
ethylcellulose, 836
extended release, 835
hot-melt extrusion, 835–836
hypromellose, 836–837
poly(methylmethacrylate), 837
polyethylene oxide, 836
surfactant and plasticizers, 836
Conventional injection molding (CIM), 908–909
Convergent synthesis, peptide-polymer conjugates, 1292
Copolymers
acrylate/acrylamide, 656
carboxylic acid, 656
hydrophobic monomers, 656
methy ether-maleic anhydride, 656
vinyl acetate homopolymers, 656
Co-precipitation method, 780–781
Core-shell nanoparticles, 1535–1536
Core-shell (C-S) nanostructures, 563–564, 568
Core–shell type hybrid microgels, 923, 924
Cornea, 370
Bowman's layer, 371
corneal blindness
corneal allotransplantation, 372, 374
corneal xenotransplantation, 374
prevalence, 372
prosthokeratoplasty (*see* Prosthokeratoplasty)
tissue-engineered corneal substitutes (*see* Tissue engineering (TE))
treatment strategies, 372, 373
Descemet's membrane, 371–372
endothelium, 372
epithelium, 371
stroma, 371
Corneal transplantation
Co-rotating extruder, 828

Cosmetology, 653–655
Counter-rotating extruders, 827
Covalent bioconjugation, PNPs, 59–60, 570–571
bacterial targeting, 1385
biotin-streptavidin labeling system, 1384
bovine serum albumin, 1383
conjugation approaches, 1383
EpCAM receptors, 1384
green fluoresent protein, 1385
Pdot-IgG probes, 1384
PEI and PLA, 1386–1387
polyethylene glycol, 1383
porphyrin-polymer conjugates, 1386
POSS-PFV-loaded NPs, 1386
transmission electron microscopy images, 1385–1386
CPNs, *see* Conjugated polymer nanoparticles (CPNs)
CPs, *see* Conducting polymers (CPs)
Critical micelle concentration (CMC), 797, 1074, 1555–1556
Critical solution temperature (CST), 1493
NIPAAm, 1498
pH-responsive polymers, 1498–1501
poly(vinylcaprolactum) (PVCL), 1498
thermo-responsive polymers, 1496–1497
Cross-linked enzyme aggregates (CLEAs), 500–502
Cross-linked high amylose starch (CLA), 284
Crystalline solid dispersions, melt extrusion, 828
characterization, 832
controlled release products, 835, 837
crystallinity assessment, 832
dissolution-enhanced products, 834
formulation and process, 831
Crystallization-induced microphase separation (CIMS), 252
CS, *see* Chitosan (CS)
CT, *see* Computed tomography (CT)
Cubosomes, 303–304
Cultivated limbal epithelial transplantation (CLET), 378
Cultivated oral mucosal epithelial transplantation (COMET), 378
CUR, *see* Curcumin (CUR)
Curcumin (CUR), 290
Cyanoacrylate glues, 2–3
Cyclic siloxanes, 659
Cyclodextrins, 446, 553
Cyclodextrins (CDs), 794
Cytotoxic T lymphocytes (CTL), 773
Cytotoxicity, 691

DACRON® prostheses, 85, 1576–1577
DC, *see* Dual cure (DC) dental resin
DCAs, *see* Dendritic contrast agents (DCAs)
DDSs, *see* Drug delivery systems (DDSs)
Decellularization
adipose, 1603
application of, 1602–1603
components, 1602
dermis, 1603
penta-galloyl-glucose, 1603–1604
pericardium, 1603
protocols, 1602–1603

scaffold fabrication, 1602
tannic acid, 1603–1604
top-down approach, 1602
trachea, 1603
Degradable synthetic polymer scaffolds, 240–241
Dendrimer-based drug delivery systems, 303, 446, 1099
colloid drug delivery, 303
formulations, 1178
manufacturing, 1177–1178
nanocarriers, 71–72
Dendritic architectures
as delivery vehicles
chemical conjugation, 397
future prospects, 399
gene delivery, 399
increased solubility, 395
monodispersity, 395
nanometer size, 395, 396
ocular drug delivery, 398
oral drug delivery, 397–398
physical association, 397
polymer cytotoxicity, 395–397
pulmonary drug delivery, 399
targeted drug delivery, 398–399
transdermal drug delivery, 398
triggered release, 397, 398
theranostic applications, 402
biosensors and molecuar probes, 407–409
computed tomography, 403
dendritic contrast agents, 403–407
magnetic resonance imaging, 403
nanoparticles, 402–403
optical imaging, 403
Dendritic contrast agents (DCAs), 403–405
blood pool MRI, 405–406
liver and lymphatic, 406
octopus-like macromolecular MRI contract agent, 405, 406
tumor-specific MRI, 407
X-ray and CT, 407
Dental polymers
benefits, 412
DC (*see* Dual cure (DC) dental resin)
dental porcelain analogue resin-strengthening, 421–423
features, 411–412
function, 412
hydroxyapatite, 411
materials, 412
methacrylate resinbased glass fiber posts, 423–425
MMA and PMMA denture base resin, 412–414
MTA (*see* Mineral trioxide aggregate (MTA) filler)
self-adhesive resin cement (SARC), 419–421
TEGDMA (*see* Tri-ethylene glycol dimethacrylate (TEGDMA))
urethane dimethacrylate-TEGDMA based resin composite blocks production, 417–419
Dental sealants
glass ionomer cements, 435
preventive dental sealants (PDS), 433
RDS (*see* Restorative dental sealants (RDS))

resin, 433–435
  smooth surface, 435
  typical occlusal fissure, 433, 434
Dentistry, metal–polymer composite
  adhesive reagents, 884
  bonding to alloys, 885–886
  bonding to base metal alloys, 884–885
  durability in water, 885
Deoxyribonucleic acid (DNA)
  carrying nanoparticles, 1394–1395
  colloidal dispersion, 1396
  colloidal properties, 1396–1398
  cross-linking DNA, 1395
  dependent assembly, 1395–1396
  dispersion, 1396–1397
  free duplex DNA, 1397
  graft copolymer, 1396
  NaCl concentration, 1397–1398
  sequence and chain length-selective
      aggregations, 1398
  temperature dependence, 1396–1397
Descemet's membrane, 371–372
Desferrioxamine (DFO), 731
Dexamethasone sodium phosphate (DMP), 317
Dextran, 6, 681
Dextran hydrogel scaffolds, 1220–1221
Diacetylenic phosphatidylcholine (DAPC),
      1665
Dialkylaminoalcohol/methacrylate copolymers,
      657
Diallylmethylammonium copolymers, 657
Dialysis membranes, 868
Diazeniumdiolates, 178
Dielectric elastomer electroactive polymers, 92
Dielectric spectroscopy, 118
Differential scanning calorimetry (DSC), 691
Diffusion theory
  of bioadhesion, 113
  of mucoadhesion, 963
Diffusion-controlled drug release, 870
  antigen delivery, 143–144
  Fickian diffusion, 141
  LHRH analogue delivery, 142–143
  poly(glycolic acid), 142
  poly(lactic acid), 142
  poly(p-dioxonone), 142
Dimethylformamide (DMF), 510
1,2-Dioley loxypropyl-3-trimethylammonium
      bromide (DOTMA), 634
1,5-Dioxepan-2-one
  applications, 157, 159
  copolymerizations, 155–156
  crystallinity and glass transition, 156, 157
  hydroxyl-containing impurities, 156
  molar composition, 156, 157
  preparation, 155
  triblock copolymerization, 157–159
Dip-pen nanolithography (DPN), 592, 602
Direct laser writing (DLW), 601
Distribution control, 301
Divergent synthesis, peptide-polymer
      conjugates
  grafting through technique, 1291
  limitations, 1291–1292
  peptide synthesis from polymers, 1291
  polymer synthesis from peptides, 1291, 1292
  schematic representation, 1290

DK, see Oxygen permeability (DK)
DK/L, see Oxygen transmissibility (DK/L)
DLVO theory, 84
DMF, see Dimethylformamide (DMF)
DMP, see Dexamethasone sodium phosphate
      (DMP)
DNA biosensors, dentritic, 407–408
Dodecylbenzene sulfonic acid (DBSA), 314
Dohlman-Doane KPros, 375
Donnan theory, 95
Double emulsions, 304
Drug delivery systems (DDSs), 439
  bioadhesive (see Bioadhesive drug delivery
      systems)
  biodegradable and non-degradable polymers,
      445–446
  biodegradable and pH-sensitive hydrogels,
      1488
  blend of natural and synthetic polymers, 447
  chemomechanical polymers, 447
  complexation networks, ionic bonding, 447
  copolymers with hydrophilic/hydrophobic
      interactions, 447
  delivery and release mechanisms, 1487
  dendrimers, 447
  doxorubicin, 1667
  DSPE-PCB, 1667
  future prospects, 448
  goal of, 1487
  graft copolymers, 447
  HBPO-PCB, 1667
  magnetic micelles/magnetomicelles, 778
  microgels, 925
  mucoadhesive (see Mucoadhesive drug
      delivery systems)
  nanomedicine (see Nanomedicine)
  nanoparticles (see Nanoparticles (NPs))
  natural polymers, 440, 444
  P(MAA-g-EG), 1487–1488
  PAA/PMAA, 1488
  PAMAM dendrimers, 1668
  PCB-PLGA, 1667
  PHPMA, 1667–1668
  polyethylene glycol (PEG), 1487
  polymeric membrane
    bulk degradation, 870–871
    diffusion-controlled systems, 870
    natural biodegradable polymers, 871
    osmotic pumps, 871
    surface erosion, 870
    synthetic biodegradable polymers, 871–873
    transdermal drug delivery, 872–874
  polymers used, 440, 442–443
  polyMPC-CPT, 1668
  rationalized polymer selection, 444
  rhodamine B, 1667
  smart polymers, 447–448
  synthetic polymers, 440, 441, 444–445
  theranostic nanoparticles, 1540–1541
  thermo- and pH-responsive (see Thermo- and
      pH-responsive polymers)
Drug eluting stents (DESs), 886–887, 1466
Drug packaging, polymers, 1306–1307
Drug-excipient interactions, 557
Dual collector systems, 513
Dual cure (DC) dental resin, 427
  allylthiourea/cumyl hydroperoxide, 427–428

Fourier transform infrared spectroscopy
      (FTIR) degree of conversion,
      428–429
  T/CH and OPPI, 428
  three-point bending flexural test, 428
Dual spinneret electrospinning apparatus,
      498, 499

Embrionic stem cells (ESCs), 1409–1410
EC, see Ethylcellulose (EC)
ECM, see Extracellular matrix (ECM)
Elastic modulus, 1329
Elastomers
  low-water soft silicone, 366
  poly(glycerol sebacate) (see Poly(glycerol
      sebacate) (PGS) elastomer)
  poly(ethylene oxide)-based, 366–367
  polyphosphazenes, 367
Electrets
  bioelectrets (see Bioelectrets)
  charging methods, 477
  definition and discovery, 476–477
  TSDC technique, 477–478
Electrical double-layer capacitors (EDLCs),
      47–48
Electroactive polymer (EAP)
  conducting (see Conducting polymers (CPs))
  dielectric elastomer, 92
  electro-responsive gels, 487–488
  ferrofluid, 488
  future prospects, 488
  ionic, 92
  ionic polymer–conductor composites, 95
  ionic polymer–metal composites, 92–93,
      95–96
  magnetite, 488
  nonionic polymer, 92
  piezoelectric polymers, 485–487
Electrochemical polymerization, 313–314
Electro-conductive aerogels, 44
Electrodeposition, 888–891
Electron beam lithography (EBL), 601
Electronic electroactive polymer/ferroelectric
      polymeric artificial muscles, 91–92
Electro-osmotic flow (EOF), 186
Electropolymerization, 1042
Electro-responsive gels, 487–488
Electrospinning technology, 315, 1418–1419,
      1632
  polymeric nanofiber drug delivery
    amoxicillin-loaded electrospun nano-HAp/
        PLA, 497, 498
    chemical immobilization, 496
    chitosan nanofiber, antibacterial
        applications, 500–502
    coaxial electrospinning, 492
    definition and discovery, 491
    electrospun PU-dextran nanofiber,
        499–501
    electrospun systems used, 496–497
    emulsion electrospinning, 492
    layer-by-layer multilayer assembly, 496
    nanoparticles assembly on the surface,
        495–496
    PLA/captopril, 496–497
    plasma treatment, 493, 494
    principle, 496

rifampin PLLA electrospun fibers, 500, 502
simple physical adsorption, 495
surface graft polymerization, 493–494
surface-active agents and polymers co-electrospinning, 495
wet chemical method, 493
wound dressing, 497–499
regenerative medicine
applied voltage, 510–511
biocompatibility and *in vivo* cell penetration, 531–532
biologically derived polymers, 508
bone tissue engineering, 532–533
cartilage tissue engineering, 533, 534
cell-invasive electrospun materials, 515–517
collection distance, 511–512
dual collector systems, 513
electric field manipulation, rotating collector, 514–515
electrospinning writing, 515
electrospun fibers surface modification, 517–519
flow rate, 510
fundamentals, 507
future prospects, 534
growth factor delivery, 517, 518
humidity, 512
hybrid electrospun systems, 508, 509
neural tissue engineering, 532
rotating collectors, 514
scaffold support systems, 506–507
single collection systems, 512–513
solution viscosity, 509
solvent conductivity, 509–510
spinneret diameter, 511
structured electroconductive collectors, 513–514
surface tension, 509
synthetic polymers, 508
temperature, 511
tissue-engineered construct (TEC), 506
vascular tissue engineering, 533–534
in vitro studies (*see* Electrospun materials, *in vitro* studies)
Electrospun fibrinogen scaffolds, 1216
Electrospun materials, *in vitro* studies
cardiomyocytes, cardiac stem cells and cardiac cell lines, 522, 525
cell types and electrospun fibers, 530, 532
chondrocytes, 525, 528
endothelial cells, 521–524
epithelial cells, 521, 522
fibroblasts, 519–521
keratinocytes, 530, 531
ligament cells, 528–531
mesenchymal stem cells, 522–523, 526
neurons and glia, 525, 528–530
osteoblasts, 523–525, 527
smooth muscle cells, 522, 524
Electrospun nanofibers, 1261
Electrospun PU-dextran nanofiber, 499–501
Electrostatic theory
of bioadhesion, 113
of mucoadhesion, 963
Elution tests, 130–131

Emulsion electrospinning, 492
Emulsions, 304
Endo-1,4-β-glucanase, 278
Endocytosis, 169–170, 637–638
Endosteal dental implant (EDI), PEEK and CFR-PEEK
beagle mandibular model, 263
biomechanical evaluation, 263–264
delayed implant placement, 263
design, 262
endosteal screw, 261
histomorphometric evaluation, 265–267
hydroxyapatitecoated dental metal implants, 261
immediate implant placement, 263
pilot study, 263
self-tapping metallic dental implant, 261
surface structure, 262
Endothelial cells (ECs), 521–524, 619, 726, 1560–1561
Endothelial progenitor cells (EPCs), 726–727
Endovascular stents
AAA repair, 1471
active pharmaceutical ingredients, 1465
balloon-delivered stent, 1473
bare metal stents, 1466
clinical prelude, 1465–1467
design envelope, 1469–1470
DESs, 1476–1477
drug-eluting stents, 1466
ideal stent design, 1467
optical coherence tomography, 1465–1466
OrbusNeich stent, 1465–1466
poly-L-lactide, 1467–1469
requirement groups, 1469–1470
revascularization, 1471
self-expanding stent, 1472
stent grafts, 1465
vascular restoration therapy, 1467, 1471
End-stage kidney disease (ESKD), 1318–1319
Enhanced permeability and retention (EPR), 64, 343, 574, 791, 1097–1098
Enteral feeding, 1319
Environment-sensitive hydrogels (ESH), 1208
Epidermal growth factor (EGF), 178
Epidermis
basale/germinativum, 1622
basement membrane, 1622
corneum/horned layer, 1621
granulosum/granular layer, 1621
spinosum, 1621
Epineural repair, 1265
Epithelial cells, 521, 522
EPR, *see* Enhanced permeability and retention (EPR)
Equation-of-state approach, 84
ESKD, *see* End-stage kidney disease (ESKD)
Ethylcellulose (EC), 272, 282, 1312
as an insoluble matrix polymer, 836
cardiovascular therapeutics, 1550
Ethylene glycol dimethacrylate (EGDMA), 686
Ethylene oxide (EtO), 1472
3,4-Ethylenedioxythiophene (EDOT), 589
Ethylene-vinyl alcohol (EVOH), 1308
Eudragit® RL PO patches, 968
EVOH, *see* Ethylene-vinyl alcohol (EVOH)
Ex vivo expansion

CFU-GM expansion, 1456–1457
conventional materials, 1456–1457
culture materials, 1458–1459
ECM molecules, 1458
fibronectin, 1460
HSPCs/stem cell, 1455–1456
materials evaluation, 1457
nanotechnology, 1457–1458
natural polymeric materials, 1456–1457
PES nanofibers, 1458
polymeric materials, 1458–1461
protein and oligopeptide, 1458–1459
stem cells, 1457
surface modification, 1460–1461
Exocytosis, 170
Expanded polytetrafluoroethylene (ePTFE), 886, 1575–1577
Extracellular biodegradation
auto-oxidation, 167–168
chemical catalytic hydrolysis, 167
conformational and electrostatic complementarity, 168
enzymatic hydrolysis, 168, 169
ions concentrations, 167, 168
mechanodestruction, 168
pH of biological fluids, 167, 168
steric hindrances, 169
thermolability, 169
Extracellular matrix (ECM), 380, 381, 599–600, 724–726, 1561
ε-aminocaproic acid, 618
epidermis, 1621–1622
fibrillar fibrin, 618
function, 1620–1621
HSPCs, 1454
innermost layer, 1621
skin tissue engineering, 1409
skin tissue regeneration, 1620
sructure of, 1621
tissue engineering, 1568–1569
Extracellular matrix (ECM)-mimetic materials
cell signaling, 191
cell-adhesive interactions, 175–176
dynamically responsive materials, 193, 194–195
functions, 190
host response, 200–201
hydroxyapatite and collagen, 191
interactions, 190–191
matrix stiffness, 191, 192
nanoscale patterning, 191
porosity and pore alignment, 191–192
proteolytically degradable polymers, 177
transport and controlled release, 196–197

Fabrication
biomolecular interations-application, 597–607
chemical coating processes, 585
electrospinning technique, 1632
freeze-drying, 1632
functional polymer coatings, 585–586
functional via patterning and gradients, 592–597
gas foaming method, 1631
particle leaching, 1630–1631
physical coating processes, 585
salt-leaching methods, 1631

solvent casting, 1630
thermally induced phase separation,
1631–1632
vapor-based deposition (*see* Chemical vapor
deposition (CVD))
wet chemical methods, 586–588
Fascicle, 1256
Fatigue testing, 760
FDM, *see* Fused deposition modeling (FDM)
Ferrofluid, 488
Fetal calf serum (FCS), 634
Fibrillar adhesives, 9–11
Fibrillar fibrin
cells (*see* Cell interactions)
degradation process, 616
extracellular matrix, 618
FDP effects, 619
fibrinogen molecule, 616–617
residual fibrin, 618–619
rheological properties, 616–617
structure and biochemistry offibrin, 616–617
tissue engineering applications, 619–620
tissue sealant, 616
transglutaminases, 617
Fibrin degradation products (FDPs), 619
Fibrin gels, 682
Fibrin glue
adhesion prevention, 646
application of, 645–646
clinical applications, 645
colon sealants, 2
concentration, 1–2
drug release vehicle, 646
fibrinogen, 645
gelation, 644–645
spongelike sheet, 646
surgical use, 2
vs. sutures, 2
vs. tacks/staples, 2
tissue engineering, 646–647
wound healing, 2
Fibrin-based matrices, 379
Fibroblasts, 519–521
Fibronectin, 665, 892, 1460
Fibrous keratin, 478
Film-coated systems, polymeric interactions
categories, 456
complex multicomponent systems, 458
delayed release systems, 458–459
dose dumping, 461–462
drug release profile, Eudragit L100-55 films,
461
film formation stages, colloidal dispersions,
457–458
fluid bed coating, 457
hydroxyethyl cellulose, 460–461
perforated pan coating, 456–457
plasticizer, 459
release pattern, 459–460
rotary coating, 457
solvent-based solubilized formulations, 457
sustained release systems, 459
Filtration sterilization, 548
Finite element analysis (FEA), 476
Flow rheometry, 118
Fluid bed coating, 457
Fluid shear stress, 729–730

Fluid-based microfabrication
FDM, 1354, 1355
organ printing, 1354–1355
pressure-assisted microsyringe system,
1353–1354
Fluorescent bioprobes
cancer theranostics, 574
physical and biochemical techniques, 575
poly-lactide-*co*-glycolide, 574
polymer-capped/polymer-coated QDs, 573
polymethylmethacrylate, 574
Fluorescent labeling, 289–290
Fluoride release technology, 435
Fluorinated polyurethane, 1335
Fluoroacrylates, 363–364
Fluoroethers, 364
Fluorohydrogels, 366
Flux cytometry, 747
Foam extrusion
calcium phosphate dihydrate, 839
chemical agents, 838
dosage, 838
gastric retentive dosage forms, 839
high pressure pumps, 838
nucleating agents, 839
physical blowing agents, 838
sodium bicarbonate, 838
Foams, 308
Folate-poly(ethylene glycol)-polyamidoamine
(FPP), 576
Food and Drug Administration (FDA),
451, 1277
Foreign-body infections
anti-infective biomaterials (*see* Anti-infective
biomaterials)
bacterial adhesion, 83–84
biomaterial and bacterium interaction, 83
polymer surface modifications, 85–86
Förster resonance energy transfer (FRET), 331
Fourier transform infrared spectroscopy (FTIR),
428–429
Fracture theory of mucoadhesion, 964
Free radical polymerization, 922
Freeze-drying technique, 1632
FRET, *see* Förster resonance energy transfer
(FRET)
FTIR, *see* Fourier transform infrared
spectroscopy (FTIR)
Fucoidan, 801–802
Fused deposition modeling (FDM), 516
Fused deposition molding (FDM), 1347–1348
Future generation polymers, 446
blend of natural polymers and synthetic, 447
block or graft copolymers, 447
chemomechanical, 447
complexation networks responding via
hydrogen or ionic bonding, 447
copolymers with desirable hydrophilic/
hydrophobic interactions, 447
dendrimers, 447
smart polymers, 447–448

Gantrez®, 116
Gas foaming method, 1631
Gas permeable contact lenses, 362
elastomers, 366–367
hard gas permeable, 362–364

hydrogels, 364–366
Gas sensors, 1043–1044
Gastric retention formulations (GRFs), 121
Gastrointestinal mucoadhesive delivery
systems, 971–972
Gastrointestinal tract (GIT) drug delivery
amoxicillin mucoadhesive microspheres, 121
bond formation, 120
gastric retention formulations, 121
insulin-loaded pH-sensitive NP, 121
interdigestive migrating motor complex, 120
spherical microspheres, 121–122
swallowed dosage form, 121
thiamine and ovalbumin, 121
GDNF, *see* Glial cell-derived nerve growth
factor (GDNF)
Gecko-inspired adhesives, 9–11
Gelatin, 648, 649, 893, 1259
Gelatin-based scaffolds, 1214–1215
Gelatinresorcinol-formaldehyde (GRF) glues,
644
Gellan, 681
Gelling, 1308
Gene delivery, 399
cellular entry, 1121–1123
DNA packaging, 1120–1121
fundamental criteria, 1120
genome insertion, 1123
microgels, 925–926
polymer-based nonviral gene carriers (*see*
Polymer-based nonviral gene carriers)
Gene therapy
bioderived molecules, 640–641
biological evaluation, 626–627
cationic polymers, 623–624
charge-based complex formation, 624
degradable system, 625–626
dissociation, 640
DNA binding, 628
DNA-carrier complex function, 633–634
endocytotic process, 637–638
graft/surface function, 627–628
interaction, 638–639
intracellular trafficking, 639–640
ligand targets, 628–629
lipoplexes (*see* Lipoplexes/lipids)
nonviral vectors, 634–637, 766
nucleic acids and polymers, 765–766
pegylation, 626
physicochemical properties, 625, 638
polymer/DNA complex formation, 626
polymeric nanoparticle, 629
polymer-mediated complex formation, 627
polyplexes (*see* Polyplexes)
protein transduction domains, 628
vectors, 765–766
Gene-activated matrices (GAMs), 733
Genotoxicity, 132, 133
Geometric mean approach, 84
Glass ionomer cements, 435
Gliadel®, 60
Glial cell-derived nerve growth factor (GDNF),
1264
Glial cells, 1263
Glow discharge technique, 85, 86
Glucocorticoid receptors (GRs), 628–629
Glucomannan, 273–274

Glucose biosensors, dentritic, 408
Glucose oxidase (GOX) sensors, 316, 408
Glucose-responsive microgels, 921
Glutaraldehyde (GA), 649
Glutathione (GSH), 414–416
Glutathione peroxidase (GPx), 415
Glycosaminoglycan (GAG), 624, 1374, 1632
Goat mandibular implants, PEEK and
    CFR-PEEK rods
  aliphatic anhydride, 258
  animal model and surgical protocol, 259
  biomechanical testing, 259
  chemical treatment, 258
  implant harvesting, 259
  implant–bone interfacial strength, 260
  interfacial strength data, 260
  mechanical and histological evaluation data,
    259–261
  melt extrusion, 258
  OSSO protocol, 258
  statistical analysis, 260, 261
  sterilization, 259
  surface microtexturing, 258
  surface phosphonylation, 259
Gold nanoparticles, 1038–1039
Gompertzian kinetics, 572
GOX, see Glucose oxidase (GOX)
GPx, see Glutathione peroxidase (GPx)
Graphene, 1036
Green fluorescent protein (GFP), 768
Growth factors, 178
GSH, see Glutathione (GSH)
Guar hydroxypropyltriammonium chloride, 657

HA, see Hydroxyapatite (HA)
Hair and skin care biomaterials
  cationic polymers, 656–658
  coloring products, 654
  compounds, 653
  cosmetology, 653–655
  eye makeup products, 654
  formulations, 653–654
  hair sprays, 654
  hair-care products, 653
  ionenes, 657
  ozone controversy polymer sons, 657–658
  products, 654
  shampoos, 653
  silicone structure, 659–661
  skin-care products, 654
  vinylic and acrylic polymers, 655–656
  vinylpyrrolidone, 657
  water compatible polymers, 658–659
Halloysite nanotubes (HNTs), 497
Hard gas permeable contact lenses
  fluoroacrylates, 363–364
  fluoroethers, 364
  polyacetylenes, 364
  polyimides, 364
  silicone acrylates, 362–363
Hard template synthesis, 314
Harmonic mean approach, 84
HCECs, see Human corneal endothelial cells
    (HCECs)
Health care
  baby diaper and sanitory napkins, 693
  pharmaceutical drug delivery, 692

  wound care, 692–693
Heart valves, 667
HEC, see Hydroxyethylcellulose (HEC)
Helical peptide-polymer hybrids, 349–352
HEMA, see 2-Hydroxyethyl methacrylate
    (HEMA)
Hematopoietic stem and progenitor cells
    (HSPCs)
  bone marrow, 1453–1454
  co-culture systems, 1454–1455
  culture materials (see Ex vivo expansion)
  3D culture, 1461
  ECM molecules, 1454
  ex vivo expansion, 1454
  mesenchymal stem cell, 1454
  umbilical cord blood, 1453
Hemicellulose, 273–275
Hemocompatible polymers
  antithrombogenicity vs. pseudoneointima,
    665
  biocompatibility, 665–668
  biomaterials, 663
  blood clotting, 663–665
  blood-compatible polymers, 668–671
  calcification and biostability, 665
  polymeric biomaterials, 666
Hemoderivatives
  platelet gel, 1651–1652
  platelet lysate, 1652–1653
  platelet-rich plasma, 1650–1651
  tear break up time, 1653
Hemodialysis
  cellulosic membranes, 866–867
  dialyzers, 866
  synthetic membranes, 867–868
Heparin, 177, 495
Hexafluoroisopropylmethacrylate (HFIM), 363
Hexosomes, 303–304
High water content hydrogels, 364–365
High-performance polyurethanes, 1318
Hildebrand solubility parameter, 21
Hirudin, 177
HNTs, see Halloysite nanotubes (HNTs)
HOB, see Human osteoblast-like (HOB) cells
Hollow fiber oxygenator, 869
Honeycomb-patterned bacterial cellulose (BC)
    films, 24
Host defense peptides (HDPs), 807
Host/guest supramolecular gels, 683
Hot embossing, 908
Hot-melt extrusion (HME), 466–468, 835–836
HPC, see Hydroxypropyl cellulose (HPC)
HPMA, see N-(2-hydroxypropyl)
    methacrylamide (HPMA)
HPMC, see Hydroxypropylmethyl cellulose
    (HPMC)
Human corneal endothelial cells (HCECs), 382
Human dermal fibroblasts (HDFs), 600–601
Human dermal human osteoblasts (HOS), 728
Human embryonic stem cells (hESC), 605–606
Human mucins, 942, 943
Human osteoblast-like (HOB) cells, 1283
Human serum albumin (HSA), 694, 1547
Human umbilical vein endothelial cells
    (HUVECs), 601
Humidity, 512
Hyaluronan (HA)

  application, 1614–1615
  manufacture, 1615
  structure of, 1613–1614
  wound care application, 1613
Hyaluronic acid (HA), 177, 651, 655, 1258
  alginate, 1415
  carboxyl groups, 1647
  cellulose, 1416
  chemical modifications, 1649
  chemical structure, 1647
  corneal cell migration, 1648
  fibrin, 1415
  hyaluronan synthases, 1648
  phacoemulsification, 1649
  proliferative phase, 1648
  streptococcus bacterium, 1415
  trabeculectomy, 1648–1649
Hybrid constructs, 101
Hybrid electrospun systems, 508, 509
Hybrid hydrogels, 682
Hybrid microgels
  surface plasmon resonance, 924
  synthesis, 923
  types, 923–924
HYDROCATH®, 85
Hydrochlorothiazide, 285
Hydrocolloids, 1310–1311
Hydrodynamic diameter (H_D), 567
Hydrogel, 685–686, 1259–1260
  adhesives, 6–7
  agarose, 1218–1219
  applications, 692–693
  biomedical, 693–695
  biomolecule immobilization, 893
  characterization, 689–691
  chemical gels, 674–679
  classification, 687
  in crosslink segment, 141
  cross-linked hydrogels, 674
  Flory-Rehner theory, 689
  fluorohydrogels, 366
  high water content, 364–365
  history, 686
  LCST, 686
  multi-responsive biomedical devices,
    699–718
  pH and temperature-responsive (see pH and
    temperature-responsive hydrogels)
  physical gels, 679–683
  in polymer backbone, 139–141
  silicone-based, 365–366
  strategies, 687
  structure and properties, 688–689
  superhydrophobic substrates, 224
  synthesis, 687–688
  temperature-sensitive (see Temperature-
    sensitive imprinted hydrogels)
  toxicological/biocompatibility studies,
    691–692
  triple-stimuli-responsive (see Triple-stimuli-
    responsive hydrogels)
Hydrolysis-controlled drug release
  poly(glycolic acid), 147–148
  poly(lactic acid), 146–147
  poly(ortho esters), 144–145, 148–150
  polyanhydrides, 145–146
  polydioxanone, 148

tyrosine-based polycarbonates, 150
Hydrolytic degradation
  carbon-chain polymers, 165–166
  hydrolysable groups, 164, 165
  hydroxycarboxylic acid polymers, 165
  rate of hydrolysis of esters, 165
  water-soluble products, 166
Hydrophilicity
  bioadhesion, 114
  biodegradation, 163
  vs. contact angle, 691–692
  mucoadhesion, 964
Hydrophilicity to lipophilicity balance (HLB),
    934
Hydrophobic polymers
  diffusion-controlled drug release, 141–144
  hydrolysis-controlled drug release, 144–150
Hydrospinning, 516
Hydrothermal synthesis, 781
Hydroxyapatite (HA), 411, 480, 1277
Hydroxyethyl acrylate (HEA), 694
2-Hydroxyethyl methacrylate (HEMA),
    365, 686
Hydroxyethylcellulose (HEC), 272, 282
Hydroxyethylmethacrylate (HEMA)-styrene
    (ST), 668–669
Hydroxypropyl cellulose (HPC), 273,
    282, 1313
Hydroxypropylmethyl cellulose (HPMC), 273,
    282, 284, 1308, 1649
Hypersensitivity tests, 132
Hypoxia-inducible factor (HIF-1), 730–731
Hypromellose (HPMC), 836

Ice particulates
  aqueous collagen solution, 1376
  collagen concentration, 1376–1377
  compression, 1376
  funnel-like scaffolds, 1378
  photomicrograph, 1377–1378
  PLGA and PLLA sponges, 1375
  porogen material, 1375
  preparation procedure, 1375
  SEM photomicrograph, 1377
  synthetic polymers, 1376
Igaki-Tamai stent, 1472
IKS, *see* Intelligent knee sleeve (IKS)
ILs, *see* Ionic liquids (ILs)
Imaging applications
  computed tomography, 1541–1542
  MRI, 1541–1542
  OI, 1541–1542
  positron emission tomography (PET),
    1541–1542
  SPECT, 1541–1542
  ultrasmall superparamagnetic iron oxide,
    1541–1543
  US, 1541–1542
Immunomagnetic separation
  antibodies, 751
  bacteria, 747–748
  bound targets, 746
  cell binding, 746–747
  enzyme, 750–751
  enzyme immunoassays (EIAs), 746
  principle of, 745–746
  in vivo, 748

In vitro biocompatibility tests
  agar overlay tests, 131
  direct contact tests, 131
  elution tests, 130–131
  genotoxicity, 132
  hemocompatibility tests, 131–132
  hypersensitivity tests, 132
  mutagenicity, 132
  zinc-oxide eugenol cement, 130
In vitro vascularization
  fluid shear stress, 729–730
  hypoxia, 730–731
In vivo animal testing
  animal and implant site selection, 133
  animal welfare issues, 132–133
  carcinogenicity testing, 133–134
  functional tests, 133
  genotoxicity testing, 133
  irritation tests, 134
  nonfunctional tests, 133
  sensitization tests, 134
  systemic toxicity, 134–135
In vivo tissue regeneration
  acellular human dermis substitute, 1635
  amniotic membranes, 1635
  bilayer system, 1635
  bioengineered skin, 1635
  epithelial cell seeding techniques, 1634
  multilayer system, 1635
  prospects, 1635–1636
  restricted success, 1635
  rhEPO, 1636
  skin progress, 1634
  tissue-engineered substitute, 1635
Indium tin oxide (ITO), 316
Inducible Pluripotent stem cells (iPSCs), 1410
Infa-V®, 970
Infusion pumps, 1322–1324, 1340
Inorganic nanoparticles-filled hybrid microgels,
    923, 924
In-situ gelation, 1260–1261
Insulin-dependent diabetes, 1323
Insulin-loaded pH-sensitive NP, 121
Intelligent knee sleeve (IKS), 319–320
Intelligent Polymer Research Institute (IPRI),
    318
Intelligent polymer system
  biomolecules, 1446–1447
  copolymers, 1443, 1445–1446
  CP/LCST, 1443, 1445
  environmental stimuli, 1443–1444
  kinetics, 1450
  molecular mechanisms, 1443–1444
  natural/synthetic biomolecules, 1446–1447
  NIPAAm and AAc, 1445
  polymers and surfactants, 1444
  random *vs.* graft copolymers, 1445
  responses, 1443–1444
  soluble polymers/hydrogels, 1443–1444
  stimuli-responsive polymers, 1446–1450
  systems, 1443–1444
  temperature-sensitive, 1443, 1445–1446
Intercellular adhesion molecule-1 (ICAM-1)-
    targeted nanocarriers, 1144, 1146
Interpenetrating polymer network (IPN), 701
Interpolymer networks (IPNs), 1500–1501
Intracellular biodegradation, 169–170

Ionenes, 657
Ionic liquids (ILs)
  cellulose blends, 281–282
  cellulose dissolution and modification,
    280–281
  grafting copolymerization and blends, 281
Ionic polymer-metal nanocomposites
    (IPMNCs), 92–93, 95–96
  AC impedance characteristics, 979
  bending response, 991–992
  chronoamperometry responses, 980, 981
  dynamic deformation, 978, 979
  dynamic sensing response, 978
  equivalent circuit model, 980, 981
  intrinsic electric field, 979
  ionoelastic beam dynamic deflection model,
    986–991
  low-and high-frequency responses, 978
  mechanoelectric behaviors, 978
  mechanoelectric effect, 979
  near-DC mechanical sensors (*see* Near-DC
    mechanical sensors)
  quasistatic DC sensing data, 981, 982
  SEM micrograph, 981, 982
  sensing response, 978, 979
  surface resistance, 980
Ionoelastic beam dynamic deflection model
  dynamic case, 988–989
  extension, 990
  flapping beam, 989, 990
  moment modification, 990
  polyelectrolyte membrane, 986
  Simulink simulation, 986
  static deflection, 986–988
  validation, 990–991
Iontophoresis, 874
IPRI, *see* Intelligent Polymer Research Institute
    (IPRI)
IR light-sensitive micelles, 69, 70
Irritation tests, 134
Isopropylidenedianiline, 364
ITO, *see* Indium tin oxide (ITO)

Keratin
  applications, 1614
  α-Helical structure, 1614–1615
  intermediate filaments (IFs), 1614
  manufacture, 1614
  wound care, 1614–1615
Keratinocytes, 530, 531
Keratoplasty
  corneal allotransplantation, 372, 374
  corneal xenotransplantation, 374
Keratoprosthesis (KPros), 375–376
Kinetics, 1450
Kink resistance, 1326–1327
Knee ligament devices, 760
KPros, *see* Keratoprosthesis (KPros)

Lactate oxidase, 316
Langmuir-Blodgett method, 586
Laser melting procedure, 848
Laser tissue repair (LTR)
  laser tissue welding, 7
  photochemical tissue bonding, 7–8
  silica-based nanoparticles, 9
  silver nanoparticles, 9

Laser-based process, 589
Laser-induced fluorescence (LIF), 183
Layer-by-layer capsules
    core–shell particle, 1081
    formulation characteristics, 1081–1082
    therapeutic applications, 1082–1083
Layer-by-layer (LbL) functionalized aerogels, 44
Layer-by-layer multilayer assembly, 496
LCOs, see Luminescent conjugated
    oligothiophenes (LCOs)
LCST, see Lower critical solution temperature
    (LCST)
LECs, see Limbal epithelial cell (LECs)
Lectin-mediated gels, 682
Lectins, 446
Leukemia cells (blood cancer), 561
Lidocaine hydrochloride (LH), 498, 499
Ligament cells, 528–531
Ligament prostheses
    augmentation devices, 757
    clamps, 761
    classification, 757
    creep test, 761
    fatigue bending test, 761
    gage lengths, 761
    over-the-top technique, 757
    manufacturing technologies, 762
    POLYFLEX, 757
    quasi-isometrical technique, 757
    requirements (see Augmentation devices and
        prostheses)
    tensile testing, 761
    textile structures, 762, 763
Ligament prosthesis testing
    textile structures, 762
Ligand-targeted micelles, 1555–1556
Light-activated adhesives, 7–9
Light-responsive polymers, 1485–1486
Limbal epithelial cell (LECs), 380
Limbal stem cells (LSCs), 377
Linear peptide conjugates, 1293
Linear polycations
    chemical structures, 635
    COS-1 cells, 636–637
    osmotic shock procedure, 635–636
    poly(L-lysine) (PLL), 635
    polycations, 636
Lipid-based nanomedicines, 58
Lipopeptide-based polymers
    block copolymers, 658
    comb-like polymers, 658–659
Lipoplexes/lipids
    advantages and disadvantages, 771
    anionic and neutral lipids, 767–768
    cationic lipids, 767
    combination, 771–773
    DOTAP/DOTMA, 767
    genetic materials, 766–767
    lipopolyplexes, 771–772
    vs. polyplexes, 770–771
    surface decoration, 768–769
    in vitro and in vivo applications, 767–768
Lipopolyplexes (LPPs), 771–772
Liposome-based nanoparticulate delivery
    vehicles, 58
Liposomes, 306, 1100
    formulations, 1174–1175

manufacturing, 1174
    ultrasound contrast agents, 1554–1555
Liquid crystals, 303, 304
Liquid single crystal elastomers (LSCEs),
    96–97
Liver tissue engineering
    bioabsorbable polymer, 102
    poly(glycerol sebacate), 235
Low density polyethylene (LDPE)
    bacterial interaction, 254–255
    circular dichroism studies, 249
    dynamic contact angle measurements,
        251, 255
    EDX spectra, 250, 255
    horizontal ATR-FTIR, 250
    osseointegration, 256–257
    osteoblast interaction, 255–256
    phosphonylation, 250–251
    physicochemical properties, 250
    profilometry analysis, 255
    sulfonation, 249
    surface roughness measurements, 250
    two-chamber dynamic flow system, 251
Low volume eye test (LVET), 384
Lower critical solution temperature (LCST),
    343, 686, 700, 1443, 1445–1446,
    1495–1496
Low-molecular-weight drug delivery, block
    copolymer micelles
    acetal linkers, 1114
    acetal-terminated-PEO-c-PCL block
        copolymers, 1116
    amino acid-based core cross-linked star-block
        copolymers, 1115
    amphiphilic 6-arm star-block copolymers,
        1118
    camptothecin incorporation, 1113
    carbohydrates, 1115
    complex block copolymer, 1118
    dendron-like poly (ε-benzyloxycarbonyl-l-
        lysine)/linear PEO block copolymer,
        1118
    dibucaine incorporation, 1113
    diminazene diaceturate release, 1113
    drug loading, 1111, 1114
    drug solubilization, 1109, 1110
    drug-conjugated triblock, 1110
    encapsulation, 1113, 1114
    folate-PEO-b-PCL micelles, 1115
    hollow core spherical micelles, 1120
    hydrophilic hyperbranched PEO-hb-PG
        copolymer, 1110
    hydrophobic blocks, 1111–1112
    interaction parameter, 1111
    linear triblock copolymer, 1120
    micellar aggregates, 1119
    micelle-like nanoaggregates, 1113
    miktoarm-based carrier, 1118–1119
    monoclonal antibodies, 1117
    nucleic acid ligands, 1117–1118
    onion-type micelles, 1120
    PEG-b-P(HPMAm-Lac$_n$), 1117
    PEO-b-PAsp (ADR), 1109–1110
    PEO-b-PCL block copolymer, 1114
    PHEMA-b-PHis micelles, 1114–1115
    pH-responsive drug conjugate micelles, 1114
    physical entrapment, 1111

physico-chemical affinity, 1111
poly (β-lactam-isoprene-b-ethylene oxide)
    copolymers, 1112
poly (ethylene glycol)-&-poly[N-(2-hydroxy-
    propyl) methacrylamide-lactate],
    1110
polycaprolactone-g-dextran polymers, 1116
polymeric micelle-like nanoparticles,
    1116–1117
shell cross-linked, knedel-like polymer
    nanoparticles, 1119
small peptide sequences and proteins, 1116
spatial and temporal control of drug delivery,
    1109, 1111
transferrin, 1117
Low-water soft silicone elastomers, 366
LSCs, see Limbal stem cells (LSCs)
Luminescent conjugated oligothiophenes
    (LCOs), 322
LVET, see Low volume eye test (LVET)
Lyocell process, 37–38

Macrogels, 917
Macromolecular drugs delivery
    gene delivery
        cellular entry, 1121–1123
        engineered viruses, 1120
        packaging of DNA, 1120–1121
        polymer-based nonviral gene carriers,
            1123–1140
        synthetic gene delivery system, 1120
    protein therapy, 1140–1146
Macrophages, 1263
Magnetic aerogels, 43
Magnetic latexes
    cell extraction, 753–754
    characteristics of, 752
    diagnostic applications, 744
    direct separation, 744–745
    hydrophilic-hydrophobic processes, 743
    immunoseparation, 745–748
    indirect separation, 745
    labeling process, 751–752
    nucleic acid separations, 749–750
    organic and inorganic materials, 743
    particle size and size distribution, 752
    protein immobilization, 750–751
    proteins and antibodies, 748
    removal of, 748
    sensitive latex particles via temperature
        flocculation process, 753
    separation, 744–745
    surface area, 752
    therapeutic applications, 752–753
    thermoflocculation, 752
Magnetic micelles/magnetomicelles (MMs)
    diagnostic platform, 781–782
    dialysis method, 780
    drug delivery systems, 778
    magnetofl uorescent imaging, 782
    micelle preparation, 780
    oil-in-water emulsion method, 780
    polymeric micelles, 779–780
    potential theranostic modifications, 779
    synthesis, 780–781
    theranostic platform, 783–784
    therapeutic platform, 782–783

Magnetic resonance imaging (MRI), 403, 576, 743, 752, 778
Magnetic targeting gene delivery systems, 1133–1134
Magnetic units, 705–707
Magnetically activated artificial muscles, 91
MAPK, *see* Mitogen-activated protein kinases (MAPK)
Matrix-assisted pulsed laser evaporation (MAPLE), 589
Maximum tolerance dose (MTD), 792–793
MC, *see* Methyl cellulose (MC)
MCC, *see* Microcrystalline cellulose (MCC)
MDP, *see* Methacryloyloxydecyl acid phosphate (MDP)
Mebiol gel®, 379
Medical polymers, 665–666
Melt electrospinning, 507
  biomedical applications, 853–855
  core–shell setup, 847
  electric field effect, 849–850
  electric field simulations, 849
  energy applications, 855–856
  external gaseous fluid, 848
  fiber diameters, 846, 849
  filtration and separation process, 856–857
  heating elements, 847
  high-voltage direct current power supply, 847
  industrial perspective, 857–861
  laser melting procedure, 848
  melt electrospun jet, 846–847
  microfibers, 846
  "needle-less" electrospinning method, 849
  parametric analysis, 845
  polymer concentration, 852
  polypropylene melt, 845–846
  rotating disk, 849
  vs. solution electrospinning, 845, 846
  stainless steel tweezers, 849
  temperature profile, 847
  viscosity and conductivity, 850–852
Melt extrusion technology, solid dispersion
  amorphous solid dispersions (*see* Amorphous solid dispersion system)
  amorphous solid solution, 828
  controlled release products, 835–837
  directly shaped products, 839–840
  dissolution-enhanced products, 833–835
  extruders, 827
  foam extrusion, 838–839
  formulation design, 831
  functional excipient roles, 828, 829
  marketed pharmaceutical products, 828, 829
  miscibility, 831
  plasticizer, 829
  povidone and methacrylic acid copolymer, 829
  solubilization regime, 831–832
  spring and parachute performance, 832, 833
  surfactants, 831
Melt flow index (MFI), 850
Membrane-bound mucins, 942, 962
Membranes, polymeric
  biocompatibility, 864–865
  blood oxygenators, 868–870
  drug delivery, 870–874
  hemodialysis, 865–868
  modification, 865

MEP, *see* Methacryloyloxy ethyl acid phthalate (MEP)
Mercerization, 276–277
Mesenchymal stem cells (MSCs), 522–523, 526, 727, 1454
Metallic biomaterial surface
  atoms, 880
  composition, 880–881
  gas molecules, 880
  immediate reaction, 880
  molecule immobilization, 880
  surface active hydroxyl group, 883–884
  surface oxide film, 881–883
Metal–polymer composite biomaterials
  advantages, 877
  biomolecule immobilization, 889–893
  dentistry, 884–886
  design, 877
  metallic biomaterial surface, 880–884
  metallization of polymers, 894–895
  metals *vs.* polymer, 877–880
  PEG immobilization, 888–889
  porous titanium, 894, 895
  SCA layer, 893–894
  stents and stent grafts, 886–888
Metals *vs.* polymers
  advantages, 877
  cohesive and noncohesive bodies, 878, 880
  dentistry, 878
  disadvantages, 878
  implant, 879
  mechanical and chemical properties, 878, 880
  mechanical reliabilit, 878
  medical devices, 878–880
  orthopedic implant, 878
  stents and stent grafts, 878
  stress–strain curves, 878
  surface modification, 878
Methacrylate resinbased glass fiber posts, 423–425
Methacrylic acid (MAA), 603
4-Methacryloxyethyl-trimellitic anhydride (4-META), 884
  bonding to alloys, 885–886
  bonding to base metal alloys, 884–885
  durability in water, 885
Methacryloyloxy ethyl acid phthalate (MEP), 435
Methacryloyloxyalkyl phosphorylcholines (MAPC), 1665–1666
Methacryloyloxydecyl acid phosphate (MDP), 436, 437
Methacryloyloxyethyltrimethyl ammonium chloride, 657
Methacryloylpropyl tris(trimethylsiloxy silane), 363
Methotrexate–human serum albumin conjugate (MTX-HSA), 66
Methyl cellulose (MC), 272, 283, 1312
Methyl-methacrylate (MMA) and poly(methyl methacrylate) (PMMA) denture base resin, 412–414
Micelles, 302–303, 1099–1100
  nanoparticles, 1535
  polymeric (*see* Polymeric micelles)
  ultrasound contrast agents, 1555–1556
Micro- and nano-patterning, 315

Microbubbles, 1536
Microcomponents, polymeric
  aluminum mold insert, 902, 903
  biodegradable microstent, 901, 902
  biomaterials, 906–907
  biomedical industry, 902–903
  case study, 911–914
  dimensional effects, 901, 903
  microangioplasty balloon, 901, 902
  microfeatured parts, 901
  micromolding, 903–906
  microneedles, 901, 902
  microparts, 901
  microprecision parts, 901
  mold defining approach, 908–911
  photo defining method, 907–908
  physical parameters, 901, 903
  surface roughness, 901–902
  surface-to-volume ratio, 901
Microcontact printing (µCP), 593
Microcrystalline cellulose (MCC), 31–32, 48–49, 272, 282, 284, 453
Microemulsions, 305, 781
Microextrusion, 909–911
Microfibrillated cellulose (MFC) aerogels
  carboxymethyl-MFC aerogels, 29–30
  controlled drug release, 46
  enzymatically/mechanically disintegrated cellulose, 28–29
  high-pressure homogenization, 27
  homogenous suspension, 27
  mechanically disintegrated cellulose, 27–28
  micro-fluidization, 27
  modification, 32
  partially hydrolyzed cellulose, 29
  TEMPO-oxidized, 30–31
Microfluidics
  advantage, 187
  capillary separation, 186
  fluid motion, 186
  non-Newtonian fluid, 187
  retardation of flow, 186–187
  scaffolding design
    advantages, disadvantages and applications, 1362–1363
    bioreactors, 1367–1369
    cell culture systems, 1367, 1368
    fabrication, 1366–1367
    limitations and future prospects, 1369
    polymeric scaffolds used, 1363–1365
Microgels
  atom transfer radical polymerization, 922
  biotechnology, 924–925
  cell culturing, 925
  characteristics, 917
  colloidal dimensions, 917
  cross-linking, 917–918
  drug delivery, 925–927
  free radical polymerization, 922
  gene delivery, 925–926
  hybrid, 923–924
  from macrogels, 922
  from monomers, 921–922
  from polymers, 922
  reversible addition-fragmentation chain transfer polymerization, 922–923
  smart polymer microgels, 918–921

thermo-responsive, 918–919
in vivo diagnosis and therapy, 926
Micromolding, 903–906
Micro-optic electromechanical systems
        (MOEMSs), 181
Microphase separation method, 780
Microreaction injection molding (μRIM), 909
Microspheres, 307
    co-precipitation method, 1533–1534
    emulsification process, 1533–1534
    linear polymers, 1533
    molecule/hydrophilic drug nanoparticles,
        1533, 1535
    polymerization, 1532–1533
    solvent evaporation method, 1533
Microtensile bond strength (MTBS), 419–421
Microwave plasma-enhanced chemical vapor
        deposition (MPECVD), 598–599
Mineral trioxide aggregate (MTA) filler
    hydrated cements characterization, 424, 426
    pH and calcium ion release, 426–427
    unhydrated materials characterization,
        424, 426
Mitogen-activated protein kinases (MAPK),
        416–417
MMA, *see* Methyl-methacrylate (MMA) and
        poly(methyl methacrylate) (PMMA)
        denture base resin
Molar substitution (MS), 277
Mold defining approach
    hot embossing, 908, 909
    injection-based process, 908–910
    microextrusion, 909–912
    precision/ultra-precision machining
        technology, 911, 912
Molecular imprinting technology
    approaches, 1425–1426
    covalent approach, 1425–1426
    gels, 1426–1427
    intuitive view, 1425
    memorization, 1424
    noncovalent/self-assembly approach, 1425
    pH-sensitive gels, 1438–1440
    Tanaka (*see* Tanaka equation)
    temperature-sensitive hydrogels, 1432–1438
Molecular probes, dendritic structures, 408–409
Molecular self-assembly of copolymers
    drug delivery, nanocarriers, 936–937
    Langmuir-Blodgett technique, 932
    polymeric nanoparticles, 936, 938
    polymeric nanospheres, 933–938
    polymeric nanotubes, 936, 938
Molecularly imprinted polymers (MIPs),
        1104, 1425
Monodispersity, 395
Morphology, 689–690
Mouse embryonic fibroblasts (mEFs), 604
MRI, *see* Magnetic resonance imaging (MRI)
mRNA encoding a tumor antigen (MART-1),
        772
MS, *see* Molar substitution (MS)
MSCs, *see* Mesenchymal stem cells (MSCs)
MTA, *see* Mineral trioxide aggregate (MTA)
        filler
MTBS, *see* Microtensile bond strength (MTBS)
Mucoadhesion
    adsorption theory, 963–964

buffer system, 955–956
consolidation stage, 942
contact stage, 942
controlled drug release, 956
covalent bond formation, 942
cross-linking and swelling, 964
diffusion theory, 963
drug/excipient concentration, 965
efflux pump inhibition, 955
electrostatic theory, 963
enzyme inhibition, 951, 953–955
factors influencing, 942, 943
hydrophilicity, 964
initial contact time, 965
interpenetration, 943–944
mucus dehydration, 944
non-covalent bond formation, 942
peel strength, 966
permeation enhancement, 953–955
pH, 964
physiological variables, 965
polymer chain entanglement, 944
polymer concentration, 964–965
polymer gel systems, 965–966
polymer molecular weight, 964
spatial conformation of polymers, 964
tensile and shear tests, 966
wetting theory, 962–963
Mucoadhesive drug delivery systems
    cervical and vulval, 971
    gastrointestinal, 971–972
    nasal mucosa, 969
    ocular route, 969–970
    oral mucosa, 968–969
    rectal, 971
    vagina, 970–971
Mucoadhesive polymers
    advantageous, 941
    ambiphilic polymers, 950
    anionic polymers, 947, 948
    cationic polymers, 947–949
    chemical structures, 972
    covalent binding polymers, 950–952
    hydrogel, 962
    lectins, 972
    mucoadhesion (*see* Mucoadhesion)
    mucus gel composition, 941–942
    non-ionic polymers, 948–949
    PMVE/MA, 973–974
    PMVE/MAH, 972–973
    rheological techniques, 946
    spectroscopic techniques, 946
    tensile tests, 945–946
    thiolated polymers, 972
    visual tests, 944–945
    in vivo methods, 946–947
Mucus, 962
Multichanneled nerve conduits
    axon infiltration, 1245
    bilayer multichannel conduit, 1240–1241
    biocompatibility and biodegradation, 1244
    biomaterial-based nerve conduits, 1244
    biomechanics, 1243–1244
    chitosan and CAD, 1241–1242
    cross-linked collagen based conduit, 1242
    electroplating, 1240
    electrospinning, 1240

fabrication and characterization, 1238
hydrogel-based conduit, 1242
injection molding, 1237, 1239
lyophilizing and wire heating process,
        1238–1239
mandrel coating, 1237–1239
microchannels, 1241
permeability, 1243
stereolithography, 1240
unidirectional freezing technique, 1239–1240
in vitro analysis, 1245–1247
in vivo analysis, 1247–1250
Multielectrode array (MEA), 600
Multiple emulsions, 304–305
Multi-responsive biomedical devices
    biomacromolecules, 710–713
    magnetic units, 705–707
    temperature and pH-responsive hydrogels,
        699–705
    thermo-or pH-sensitive units, 708–710
    triple-responsive hydrogels, 713–718
Multi-screw extruders, 827
Multislice spiral computed tomography
        (MSCT), 1465
Multiwalled carbon nanotubes (MWCNTs),
        479–480
Mupirocin, 498
Muscles, artificial
    ionic polymer-metal nanocomposites (*see*
        Ionic polymer-metal nanocomposites
        (IPMNCs))
Mussel adhesive polymer hybrids, 347, 349
Mutagenicity, 132
MWCNTs, *see* Multiwalled carbon nanotubes
        (MWCNTs)
Myelomeningocele
    gelatin-based injectable scaffolds, 1215
Myosin light chain (MLC), 1419

NAC, *see* N-acetyl cysteine (NAC)
N-acetyl cysteine (NAC), 415, 416
NAFTA, *see* North American Free Trade
        Agreement (NAFTA)
Nanocapsules, 307, 1076
Nanocarriers; *see also* Polymer-based
        nanocarriers
    core length and crystalizability, 936
    dendrimer and dendritic polymer, 71–72
    drug/carrier concentration ratio, 936–937
    drug–core compatibility, 936
    polymeric vesicles, 70–71
    solvent effect, 937
Nanocomposite polymers, *see* Silver/
        polymer-based nanocomposites
Nanocrystalline cellulose (NCC), 288, 289
NanoDRONE, 334–336
Nanoemulsions, 305
Nanofiber drug delivery, *see* Electrospinning
        technology, polymeric nanofiber drug
        delivery
Nanogels
    amine-based cross-linking, 1009–1011
    applications, 1023–1025
    atom-transfer radical polymerization,
        1019–1020
    click chemistry–based cross-linking,
        1011–1015

disulfide-based cross-linking, 1008–1009
heterogeneous controlled/living radical polymerization, 1019
heterogeneous free radical polymerization, 1017
imine bonds–induced cross-linking, 1013, 1015
inverse microemulsion polymerization, 1018–1019
inverse miniemulsion polymerization, 1018, 1019
photo-induced cross-linking, 1014–1017
physical cross-linking, 1016–1017
precipitation polymerization, 1017–1018
properties, 1007
release mechanisms, 1025
reversible addition–fragmentation chain transfer, 1020–1023
Nanoimprint lithography (NIL), 595
Nanoindentation loading cycle, 420
Nanomaterials
carbon fiber microelectrodes, 1036–1037
carbon nanotubes, 1037–1038
carbon surfaces, 1035
conducting polymers, 1042
gold nanoparticles, 1038–1039
graphene, 1036
layer-by-layer capsules, 1081–1083
metal oxides and nanoparticles, 1040–1041
polymeric micelles, 1073–1076
polymeric nanoparticles, 1076–1078
polymersomes, 1078–1081
sensors and biosensors, 1042–1045
silver nanoparticules, 1039–1040
theranostics (see Theranostics)
Nanomedicine; see also Polymeric nanomedicine
advantages, 1104
computational approaches, 1103–1104
drug delivery applications
bulk erosion, 1091
cancer drug delivery, 1100–1101
combination therapy, 1104
molecularly imprinted polymers, 1104
nanocarriers, 1099–1100
phosphorus-containing polymers, 1093–1094
poly(amino acids), 1093
poly(anhydrides), 1092–1093
poly(esters), 1091–1092
poly(ortho esters), 1092
polymer erosion, 1091
polymer–drug nanomedicine, 1101–1102
smart polymers (see Smart polymers)
subcellular drug delivery, 1104
evolution, 1089
polymer therapeutic products, 1089–1090
polymeric
colloidal nanoparticles, 58
covalent conjugation approach, 59–60
dendrimer and dendritic polymer nanocarriers, 71–72
liposome-based nanoparticulate delivery vehicles, 58
micelles, 66–69
physical encapsulation approach, 60
polymeric vesicles, 70–71

polymer–protein conjugates, 60–61
polymer–small molecule drug conjugates, 61–66, 62–66
polymeric carriers, 1090
polymeric sequestrants, 1102–1103
Nanoparticles (NPs), 327
Nanostructured lipid carriers (NLCs), 1173–1174
Nasal drug delivery, 124
Nasal mucoadhesive delivery systems, 969
National Institute of Health (NIH), 493
Natural polymers, 440, 444, 655, 1258
N-carboxy anhydride (NCA), 589–590
NCC, see Nanocrystalline cellulose (NCC)
Near-DC mechanical sensors
accelerometer implementations, 982–983
complex impedance, 983
dry IPMNC impedance, 983–984
frequency response magnitude, 984, 985
IPMNC mounting apparatus, 983
ling 5-lb shaker, 983
long response time, 983
machined aluminum block, 983
material testing system, 986
piezoelectric elements, 983
voltage magnitude, 985
voltage/current output and power output, 985, 986
wet IPMNC impedance, 984
Near-infrared (NIR), 577, 1389–1390
Needle-less electrospinning method, 849
Negative cell selection, 746–747
Nerve guidance channel (NGC), 532, 1265–1266
Nerve guides
autografting, 1236
biodegradable nerve guide implants, 1236
multichanneled conduits, 1236–1237
multichanneled nerve conduits (see Multichanneled nerve conduits)
normal nerve architecture, 1236
trichanneled, 1236
Neulasta®, 61
Neuragen, 1265
Neural probe, 321–322
Neural regeneration
affecting factors, 1266
biomaterial ideal properties, 1257–1258
electrospun nanofibers, 1261
hydrogels, 1259–1260
in-situ gelation, 1260–1261
strategies, 1263, 1264
Neural stem cell (NSC), 1263, 1264
Neural tissue engineering, 532, 1255
agarose, 1259
alginate, 1259
bioactive molecules delivery, 1263–1265
chitosan, 1259
conducting polymers, 1262–1263
future prospects, 1267
gelatin, 1259
hyaluronic acid, 1258
natural polymers, 1258
nerve guidance channel (NGC), 1265–1266
nerve regeneration (see Neural regeneration)
nervous system, 1256
neurodegenerative diseases, 1256–1257

PEG and PLGA, 1259
peripheral nerve injury, 1256, 1257
poly (2-hydroxyethyl methacrylate), 1259
synthetic polymers, 1259
traumatic brain injury and spinal injury, 1256, 1257
Neurodegenerative diseases, 1256–1257
Neuronal stem cells (NSC), 600
Neurons and glia, 525, 528–530
NGC, see Nerve guidance channel (NGC)
N-(2-hydroxypropylmethacrylamide) (HPMA), 343–344, 790
N-Hydroxysuccinimide (NHS), 649–650
Nibosomes, 306
NIH, see National Institute of Health (NIH)
Niosome-based drug delivery, 1176
N-isopropyl acrylamide (NIPAAm), 1664
Nitric oxide–generating materials, 178–179
Nitroxide-mediated free radical polymerization (NMFRP), 1662
N-Methylmorpholine-N-oxide (NMMO) monohydrate, 37–38
NMR, see Nuclear magnetic resonance (NMR)
N,N-dimethylaminoethyl methacrylate (DMAEMA), 686
Nonabsorbable suture materials
biodegradation and absorption properties, 1526–1528
commercial suture materials, 1515
comparison, 1520, 1524
mechanical properties, 1520, 1522
Noncardiomyocytes (NCMs), 600
Non-covalent approaches, 1383
Non-ionic mucoadhesive polymers, 948–949
Nonsteroidal anti-inflammatory drugs (NSAIDs), 398
Non-viral delivery vehicles
future prospects, 1274
polycations for DNA complex formation, 1272
polyplexes
in vivo applications, 1272–1274
therapeutic approaches, 1274
Non-viral vectors, 634–637, 1088
North American Free Trade Agreement (NAFTA), 1329
NPs, see Nanoparticles (NPs)
NSAIDs, see Nonsteroidal anti-inflammatory drugs (NSAIDs)
NSC, see Neural stem cell (NSC)
Nuclear magnetic resonance (NMR), 403
Nucleic acid separations
amplification/direct analysis, 749–750
applications, 749
doublehybridization process, 749–750
extraction and concentration, 749–750
mRNA hybrids, 749
mutation detection, 749

Ocular drug delivery, 125, 398
Ocular irritation test, 134
OECs, see Olfactory ensheathing cells (OECs)
Oil-in-water emulsion method, 780
Olfactory ensheathing cells (OECs), 1263
Oligosaccharides, 681
Oncaspar®, 61
OOKP, see Osteo-odonto KPros (OOKP)

Opthalmology, 695
Optical biosensor, 322–323
Optical coherence tomography (OCT),
    1465–1466
Optical imaging, 403
Oral drug delivery, 397–398
    buccal administration, 119
    compression force tablet behavior, 119
    drug absorption, 119
    drug release rate, 119
    inflammatory and infective conditions, 120
    mucoadhesive, 968–969
    mucoadhesive patch, 120
    oral mucosa, 119–120
    periodontitis, 120
Organic/inorganic composite gels, 683
Organopolysiloxanes, 659–660
    alkylpolysiloxane, 660
    aminoquaternarycompoundspolysiloxane, 660
    ionic organomodified silicones, 661
    polyalkylpolyetherpolysiloxane, 660
    polyaminopolysiloxane, 660
    polyetherpolysiloxanes, 660
    polyorganobetainepolysiloxane, 660
    polyphosphatesterspolysiloxane, 661
OROS®, 287
Orthodontics, 1404–1405
Orthopedics
    PMNCs (see Polymer-ceramic nanocomposite
        (PMNC))
    shape memory polymers, 1404, 1405
    stress shielding, 894
Oscillatory rheometry, 118
Osmotic pumps, 287, 871
Osseointegration
    goat mandibular implants, 258–261
    LDPE and i-PP implants, goat model,
        256–257
    metallic and PEEK-based EDIs, 262–263
    phosphonylated PEEK rods, rabbit model,
        257–258
Osteoblasts, 523–525, 527
    interaction, 255–256
    transplantation, 102–103
Osteo-odonto KPros (OOKP), 375
Oxycellulose, 272
Oxygen imaging biosensors, dentritic, 409
Oxygen permeability (DK), 362
Oxygen transmissibility (DK/L), 362
Ozurdex delivery system, 839–840

[P(3HB)], see Poly(3-hydroxybutyrate)
    [P(3HB)]
p(N-isopropylacrylamide) (pNIPAAM), 700
PAA, see Poly(amidoamine) (PAA)
Pacemaker, 668, 1319–1320
PAMAM, see Polyamidoamine (PAMAM)
Panavia 21, 436, 437
Pancreas reconstruction, 102
PAni-CNT/PNIPAm-co- MAA, see Polyaniline-
    carbon nanotube/poly(N-isopropyl
    acrylamide-co-methacrylic acid)
    (PAni-CNT/PNIPAm-co- MAA)
PAni-ISAMs, see Polyaniline integrally skinned
    asymmetric membranes (PAni-
    ISAMs)
Paraffin spheres, 25

Parenteral dosage forms
    active ingredients, 549
    aggregation, 557–558
    appearance, 549
    attributes, 547–548
    biopharmaceutical agents, 557
    commercial products, 547
    cysteine, 558
    dissolution, 549
    dose accuracy, 549
    drug approvals, 547
    drug-excipient interactions, 557
    environmental stresses, 557
    histroy, 547
    impurities, 556–557
    injection, 555
    liposomes, 554–555
    low-dose drugs, 555
    lymphatic targeting strategies, 555
    lyophilization aids, 551
    microbial quality, 548
    oxidation, 558
    parenteral products, 551
    pH adjustment and retention, 550
    pharmacopeial monographs, 556
    photocrosslinked biodegradabel hydrogel,
        555–556
    photo-stabilizers, 557
    preservation, 550–551
    quality retention, 549–550
    requirements, 548
    residues, 556
    small molecule, 548
    sodium chloride, 550
    solubilizers, 551–554
    solution, 548
    sugars and polyols, 558
    surface agents/solubilizers, 551
    tolerance (injection site), 549
    UV absorption spectra, 557–558
    viral contamination, 548
Parenteral drug delivery (pH-responsive
        polymers)
    acid sulfonamide, 1504
    anionic polymers, 1505
    butylmethacrylate, 1505
    cancerous tissue, 1504
    DMAEMA/HEMA, 1504–1505
    ethyl acrylic acid, 1505
    ODNs model, 1505–1506
    PDEAEMA, 1505
    PEG layer, 1506
    poloxamer copolymer micelles, 1507
    poly(β-amino ester), 1506
    polycationic/polybases, 1504
    PSD-b-PEG, 1504
    pyridyldisulfide acrylate, 1505
    siRNA, 1505
Parkinson's disease, 1320
Parylene-coated QCM sensor (pQCM), 599
Passive tumor targeting
    chemical bonding/physical encapsulation,
        798, 800
    critical micellar concentration, 797
    cyclodextrins, 794
    encapsulation, 798, 800
    EPR effect, 791–792

FOL-PEG-PAMAM, 799
    glycodendrimer, 799–800
    HPMA and PGA, 793
    HPMA-Gly-Phe-Leu-Gly-DOX copolymer,
        791
    maximum tolerance dose, 792–793
    nanoparticles, 793–794
    nitroglycerin, 792
    nonviral vectors, 795–796
    PDEPT approach, 795
    PDMAEMA, 797
    PEGylated PEI, 796
    pH and glutathione, 798
    poly(glutamic acid), 792
    poly(L-lysine), 796–797
    polyamidoamine, 797
    polyether-copolyester dendrimers, 799
    polyethylenimine, 795
    polymer–drug conjugates, 792–794
    polymersomes/polymeric vesicles, 798
    polyplexes, 795
    protein–polymer conjugates, 795
    RGD, 798
    siRNA, 797
    systems, 794–795
    VEGF expression, 797
Patterning and gradients, functional coatings
    cardiomyocytes, 600
    chemical gradients, 596–597
    chemical micro- and nanopatterning, 593–596
    CVD polymerization, 592–593
    dip-pen nanolithography, 592, 602
    direct laser writing, 601
    electron beam lithography, 601
    fabrication, 592
    fluorescence images, 601–602
    human dermal fibroblasts, 600–601
    HUVECs, 601–602
    influence function, 600
    NSC and NCM, 600
    PEGDA hydrogels, 601
    PLGA polymer, 600
    PMeOx, 602
    RGD and bFGF, 601
    RGDS and REDV, 601
PBLG, see Poly(γ-benzyl glutamate) (PBLG)
PCL, see Poly(ε-caprolactone) (PCL)
PCPDTBT, see Poly(2,6-(4,4-bis-(2-
        ethylhexyl)-4-H-cyclopenta[2,1-
        b;3,A-b ′]-dithiophene)-alt-4,7-
        (2,1,3-benzothiadiazole] (PCPDTBT)
PDG, see Polydimethylaminoethylmethacrylate
        (PDG)
PDMS, see Polydimethylsiloxane (PDMS)
PDS, see Preventive dental sealants (PDS)
Pectin, 681
PEDOT, see Poly (3,4-ethylenedioxythiophene)
        (PEDOT)
PEDOT/PSS, see Poly(3,4-
        ethylenedioxythiophene) doped with
        poly(4-styrenesulfonate) (PEDOT/
        PSS)
Peel test, 116
PEG, see Polyethylene glycol (PEG)
PEGMM, see Polyethylene glycol
        monoethylether monomethacrylate
        (PEGMM)

Pegylated immunolipoplexes (PILPs), 772
PEGylated nanoparticles, 574
PEGylation, 396, 1309–1311
PEI, *see* Polyethyleneimine (PEI)
Pendant peptide conjugates, 1292–1293
Penetration enhancers, 873–874
PEO, *see* Polyethylene oxide (PEO)
Peptide–polymer conjugates, 1289–1290
  alkyne functionality introduction, 1296–1298
  azide functionality introduction, 1295–1296
  convergent synthesis, 1290, 1292
  coupling reaction, 1297
  Cu(I)-catalyzed ligation catalytic cycle, 1294–1295
  divergent synthesis, 1290–1292
  efficient coupling requirements, 1293
  linear peptide conjugates, 1293
  pendant peptide conjugates, 1292–1293
  purification and analysis, 1297–1299
  synthesis schematic representation, 1293–1294
Peptide–polymer hybrid nanotubes, 347, 348
Peptides
  biomolecule immobilization, 891–892
  hydrogels, 681–682
Perineural repair, 1265
Periodontitis, 120
Peripheral arterial disease (PAD), 1576
Peripheral nerve injury, 1256, 1257
Peripheral nervous system (PNS), 1255, 1256
Peripherally inserted central catheters (PICC), 1326, 1327
Peritoneal dialysis, 1318–1319
PEVA, *see* Polyethylene vinyl acetate (PEVA)
PFM, *see* Piezoresponse force microscopy (PFM)
PGA, *see* Poly(glycolic acid) (PGA)
PGLD, *see* Polyglycerol (PGLD)
PGOHMA, *see* Poly(glycerol methacrylate) (PGOHMA)
pH and temperature-responsive hydrogels
  β-cyclodextrin, 702–703
  dual-stimuli-responsive polymer, 700–702
  IPN structure, 701
  LCST, 700, 703
  MAA, 704–705
  magnetic units, 707
  multifunctional hydrogels, 703–704
  p(NIPAAM-*co*-AA), 704
  pNIPAAM-*co*-AA hydrogel, 703
  radiation method, 704
  stimuli-responsive swelling, 699–700
  swelling-deswelling model, 703–704
pH-active PAN fibers, 95
Pharmaceutical polymers
  applications, 1304–1306
  barrier properties, 1308
  cellulose-based polymer, 1312–1313
  controlled release dosage forms, 1306
  conventional dosage forms, 1306
  drug packaging, 1306–1307
  examples, 1304
  film coating, 1313, 1314
  future prospects, 1315
  gelling, 1308
  hydrocolloids, 1310–1311
  plastics and rubbers, 1311–1312

polydimethylaminoethylmethacrylate, 1314–1315
  polyethylene glycol, 1309–1311
  polyvinyl alcohol and polyvinyl pyrrolidone, 1310
  reverse micelles, 1313
  rheology modifiers, 1307
  taste masking, 1307
Pharmacologically active polymers
  anti-inflammatory biomaterials, 178
  growth factors, 178
  nitric oxide–generating materials, 178–179
  plasminogen activators, 177–178
  polysaccharides, 177
PHEA, *see* Poly(hydroxyethyl acrylate) (PHEA)
pHEMA, *see* Poly (2-hydroxyethyl methacrylate) (pHEMA)
Phenylsiloxanes, 659
Phosphobetaines (PB)
  bio-membrane mimicry, 1665
  Co-Cr-Mo surface, 1666–1667
  cross-linked polyethylene, 1667
  diacetylenic phosphatidylcholine, 1665
  fluorescent micrograph images, 1666
  macroscopics, 1666
  methacryloyloxyalkyl phosphorylcholines, 1665–1666
  MPC, MHPC and MDPC, 1665–1666
  poly(PB-*co*-HEMA), 1666
  poly-DMS, 1666
  SCL-biosensor, 1666
Phospholipid polymers, 682
Photo defining method, 907–908
Photochemical tissue bonding
  bioadhesive patch, 7
  chitosan films, 7
  depth of light penetration, 7
  photocrosslinkable bioadhesive, 7–8
  rose bengal solution, 7
Photocurable system, 650–651
Photodynamic therapy (PDT), 577, 1067–1068, 1151–1153
Photoluminescence (PL), 563
Photoresponsive microgels, 920–921
pH-responsive microgels, 919–920
pH-responsive polymers; *see also* pH and temperature-responsive hydrogels; Thermo- and pH-responsive polymers
  categories, 1498
  cellulose acetate phthalate, 1498
  EHHKCO, 1500
  interpolymer networks, 1500–1501
  parenteral drug delivery, 1504–1507
  poly(ethyleneimine), 1499
  poly(L-histidine), 1500
  poly(L-lysine), 1500
  poly(vinyl acetate phthalate), 1498
  polyelectrolytes, 1484, 1498
  SRPs, 1484
  structures of, 1499
pH-sensitive imprinted gels
  amylose-based polymer, 1438–1439
  bovine serum albumin, 1439–1440
  network formation, 1438–1439
  polyethylene glycol, 1438
  types, 1438
pH-sensitive polymeric micelles, 69, 70

pH-sensitive 2,4,6-trimethoxybenzylidene-tris (hydroxymethyl)ethane (TMB-THME) hydrophobe, 1126, 1128
PHSRN, *see* Pro-His-Ser-Arg-Asn (PHSRN)
Physical encapsulation approach, 60
Physical gels, 679–683
Physicochemical surface modification
  microtexturing, 251–252
  OSSO process, 252–254
  phosphonylation, 249–251
  sulfonation, 249
PICC, *see* Peripherally inserted central catheters (PICC)
Piezoelectric polymers, 485–487
Piezoresponse force microscopy (PFM), 333
Pit and fissure sealants
  glass ionomer, 435
  resin, 433–435
PLA, *see* Poly(lactic acid) (PLA)
Plasma polymerization, 591–592
Plasminogen activators, 177–178
Plastics and rubbers, 1311–1312
Platelet gel (PG), 1651–1652
Platelet lysate (PL), 1652–1653
Platelet-rich plasma (PRP), 1650–1651
Platelets, 664
Plexiglas®, 837
PLGA, *see* Poly(lactic acid-*co*-glycolic acid) (PLGA)
PLLA, *see* Poly-L-lactide (PLLA)
Pluronic 123, 1273
PMMA, *see* Poly(methyl methacrylate) (PMMA)
PMNC, *see* Polymer-ceramic nanocomposite (PMNC)
PNIPAM, *see* poly(*N*-isopropyl acrylamide) (PNIPAM)
PNS, *see* Peripheral nervous system (PNS)
Poke technique, 1336
Poly(2,6-(4,4-bis-(2-ethylhexyl)-4-H-cyclopenta[2,1-b;3,A-b ′]-dithiophene)-alt-4,7-(2,1,3-benzothiadiazole] (PCPDTBT), 337
Poly(2-hydroxyethyl methacrylate) (HEMA), 675
Poly(2-methyl-2-oxazoline) (PMeOx), 602
Poly(3-hexylthiophenes) (P3HT), 1042
Poly(acrylic acid) (PAA), 435, 681
Poly(acryloyl-*N*-propylpiperazine) (PAcrNPP), 1482
Poly(amidoamine) (PAA), 447
Poly(carboxybetaine methacrylate) (poly-CBMA), 1662
Poly(D,L lactide-*co*-glycolide) (PDLGA), 1567–1568
Poly(D,L -lactide) (PLA), 784
Poly(ε-caprolactone) (PCL), 493, 1261
Poly(ethylene terephthalate) (PET)
  cardiovascular disease, 1580–1581
  thermoplastic polymer, 1580
Poly(γ-benzyl glutamate) (PBLG), 349, 350
Poly(glycerol methacrylate) (PGOHMA), 337
Poly(glycidyl methacrylate) (PGMA), 590
Poly(glycolic acid) (PGA), 445, 1416
  diffusion-controlled drug release, 142
  hydrolysis-controlled drug release, 147–148
  porous scaffolds, 1374

Poly(HFMA-γ-PEGMA), 784
Poly(hydroxyethyl acrylate) (PHEA),
    1297, 1298
Poly(hydroxyethyl methacrylate) (pHEMA),
    893
Poly(hydroxypropyl methacrylate) (PHPMA),
    606
Poly(L-glutamic acid) (PLGA), 649
Poly(lactic acid) (PLA), 445
    diffusion-controlled drug release, 142
    hydrolysis-controlled drug release, 146–147
    polymeric materials, 1416
    porous scaffolds, 1374
Poly(lactic acid-co-glycolic acid) (PLGA), 322,
    445, 1259, 1277
    polymeric materials, 1416
    porous scaffolds, 1374
    wet chemical methods, 587
Poly(lactide-co-caprolactone) (PLCL), 1567
Poly(methacrylic acid) (PMAA), 681
Poly(methyl methacrylate) (PMMA), 362, 574
    controlled drug release, 837
    wet chemical methods, 587
Poly(N-isopropyl acrylamide) (PNIPAM),
    218, 1297, 1298
Poly(N-vinylpyrrolidone) (PNVP), 607
poly(NIPAAM), see Poly(N-isopropyl
    acrylamide) [poly(NIPAAM)]
Poly(ortho esters) (POEs), 445
    biodegradation, 172, 173
    drug delivery, 871–872
    hydrolysis-controlled drug release, 144–145,
        148–150
Poly(p-dioxonone)
    diffusion-controlled drug release, 142
    hydrolysis-controlled drug release, 148
Poly(phosphazenes), 1093–1094
Poly(phosphoesters), 1094
Poly(sulfobetaine methacrylate) (polySBMA),
    1663
Poly(vinyl alcohol) (PVA), 680
Poly(vinylidine difluoride) (PVDF), 486,
    487, 1664
Poly (2-aminoethyl methacrylate) (PAEM),
    1126
Poly (2-dimethylaminoethyl methacrylate)
    (PDMAEMA), 797
Poly (2-hydroxyethyl methacrylate) (PHEMA),
    1114–1115, 1259
Poly (3,4-ethylenedioxythiophene) (PEDOT),
    314
Poly (ethylene oxide-b-aspartate) block
    copolymer-adriamycin [PEO-b-PAsp
    (ADR)] conjugates, 1109–1110
Poly (ethylene oxide)-b-poly (N-hexyl
    stearate-l-aspartamide) (PEO-b-
    PHSA) block copolymer micelles,
    1114
Poly (ethylene oxide)-hyperbranched-
    polyglycerol (PEO-hb-PG), 1110
Poly [N-(hydroxypropyl)methacrylamide]
    (pHPMA), 640
Poly (methyl caprolactone)-O-nitrobenzyl-poly
    (acrylic acid) diblock copolymer,
    1144, 1145
Polyacetylene, 364, 482–483
Poly(phenyl) acetylenes, 342

Polyacrylonitrile (PAN) artificial muscles
    anode electrode, 95
    cationic concentration, 94
    conductivity, 93
    contraction and elongation, 94
    electrical activation, 94
    engineering features, 93
    pH-active PAN fibers, 95
    platinum deposition, 95
    strength and length, 94–95
Polyamidoamine (PAMAM), 71, 395, 396, 404,
    623–624
Polyaminodiamine, 446
Polyanhydrides, 445
    drug delivery, 872, 1092–1093
    hydrolysis-controlled drug release,
        145–146
Polyaniline (PANI), 313, 315, 1042
Polyaniline integrally skinned asymmetric
    membranes (PAni-ISAMs), 317
Polyaniline-carbon nanotube/poly(N-isopropyl
    acrylamide-co-methacrylic acid)
    (PAni-CNT/PNIPAm-co- MAA), 319
Poly(ethylene oxide)-based elastomers,
    366–367
Poly(ethylene glycol) block copolymers, 1092
Polybutylcyanoacrylate nanoparticles
    (PBCA-NP) loaded with mitomycin C
    (MMC-PBCA–NP), 1196
Polybutylcyanoacrylate NPs of diallyl trisulfide
    (DATS–PBCA–NP), 1195
Polycaprolactone (PCL)
    polymeric materials, 1416–1417
    substrates, 725
Polycarbonate-based polyurethane, 1333
Polycyanoacrylate, 165–166
Poly(methyl methacrylate) (PMMA) denture
    base resin, 412–414
Polydimethylaminoethylmethacrylate (PDG),
    1314–1315
Polydimethylsiloxane (PDMS), 10, 363,
    587–588, 659, 813, 1563–1564
Polydimethylsiloxane–chitosan–clay (PDMS/
    Cs/clay) composites, 998
Polydioxanone, 148
Poly-D-lactide (PDLA), 1476
Poly(3,4-ethylenedioxythiophene) doped with
    poly(4-styrenesulfonate) (PEDOT/
    PSS), 321
Poly(glycerol sebacate) (PGS) elastomer
    artificial microvasculature, 235
    cardiac patch, 235
    characterization, 230–231
    cross-linking, 229
    design, 229–230
    hydrogen bonding interactions, 229
    in vitro biocompatibility, 233
    in vivo biocompatibility, 233–234
    in vivo degradation, 231–233
    liver tissue engineering, 235
    matrices and scaffolds, 229
    microfabricated, 229
    synthesis, 230
Polyether-copolyester dendrimers (PEPE), 799
Polyether-ether ketone (PEEK)
    endosteal dental implants, 261–267
    goat mandibular implants, 258–261

orthogonal solid-state orientation process,
    252–254
    osseointegration, 257–258
    osteoblast interaction, 255–256
    surface microtexturing, 251–252
    surface phosphonylation, 251
Poly-ethersulfone (PES) nanofibers, 1458
Polyethylene glycol (PEG), 557, 626, 783,
    1259, 1309–1311
    biomolecular surface modification, 585
    chemical immobilization, 888
    chemical structure, 888
    diacrylate, 650
    electrodeposition, 888–891
    glues, 1, 3
    NHS-PEG and albumin, 650
    NHS-PEG and SH-PEG, 650
    PEG-CPT conjugate, 62
    PEG-DOXO conjugate, 62
    PEG-polyester micelle, 68–69
    PEG-polypeptide micelle, 68
    photoreactive-gelatin, 650–651
    protein conjugation, 61
    spacer, 177
Polyethylene glycol conjugated zinc
    protoporphyrin IX (PZP), 792
Polyethylene glycol monoethylether
    monomethacrylate (PEGMM), 447
Poly-ethylene glycol phosphatidylethanolamine
    (PEG-PE), 782
Polyethylene oxide (PEO), 365, 836
Polyethylene terephthalate (PET), 1467,
    1561–1562
Polyethylene vinyl acetate (PEVA), 1331
Polyethyleneimine (PEI), 623–624, 769, 795,
    1272, 1273, 1386
Polyflux S®, 868
Polyglycerol (PGLD), 408
Polyguanidinium oxanorbornene (PGON),
    807–808
Polyhydroxyalkanoates, 1417
Polyhydroxybutyrate valerate (PHBV), 1476
Polyimides, 364
Polyions complex (PIC), 778
Poly-L-lactide (PLLA), 318, 1467–1469
    biodegradable test polymers, 1467
    Cordis-TESco hybrid, 1468
    PLA/TMC intravascular prototype stent,
        1468–1469
    polyethylene terephthalate, 1467
    polymeric stent development, 1467
    residual monomers/solvents, 1468
Poly-L-lysine (PLL), 603
Polylysine polyplexes, 1272–1273
Polymer drug application
    active tumor targeting, 790–791
    angiogenic processes, 799–801
    antiangiogenic and proangiogenic polymer
        drugs, 799
    antibacterial polymer drugs, 802–804
    antimicrobial properties, 805–806
    antithrombogenic polymer drugs, 808–809
    antitumoral polymer drugs, 789–790
    applications, 789
    bioactive polymers, 809–811
    biolubricating systems, 813
    cellulose, RAFT polymerization, 805–806

colony-forming units, 808
endocytic pathway, 790
endothelial cells growth, 811
glycidyl-trimethylammonium chloride, 805–806
low friction polymer drugs, 812–813
N-akyl chitosan, 804–805
ophthalmic applications, 807
passive tumor targeting, 791–799
PDMAEMA grafting via ATRP, 804
polymerizable molecules, 805, 807–808
representative low-friction polymers, 812
surface modification, 804–805
transformation, 804–805
triethylamine, 807
Polymer gel
hydrogel, 917 (*see also* Hydrogel)
macrogels, 917
microgels (*see* Microgels)
Polymer grafting, 277
Polymer physicochemical modification
cold plasma oxidation radiation grafting, 249
orthogonal solid-state orientation, 248
phosphonylation, 249–251
sulfonation, 249
Polymer surface modifications
antibiotics, 85
antimicrobial incorporation, 85
glow discharge technique, 85, 86
polymethylmethacrylate–gentamicin system, 85
Polymerase chain reaction (PCR), 748–749
Polymer-based nanocarriers
dendrimers, 1099
liposomes, 1100
micelles, 1099–1100
polymersomes, 1100
polysaccharide-based nanoparticles, 1100
vesicles, 1100
Polymer-based nonviral gene carriers
arginine-based polycations, 1135, 1136
chitosan-graft-(PEI-β-cyclodextrin) cationic copolymers, 1137–1139
(Coixan polysaccharide)-graft-PEI-folate, 1139
cyclodextrin-modified PEIs, 1137
histone H3/PEI hybrid polyplexes, 1136
HPMA-oligolysine copolymers, 1136
hyperbranched cationic glycopolymers, 1136, 1137
linear (co)polymers
linear PEG-PEI diblock copolymers, 1123–1124
linear poly (2-aminoethyl methacrylate), 1126
micellar delivery system, 1124
N-ethyl pyrrolidine methacrylamide polymers, 1124
oligomers, 1124
plasmid DNA packaging, 1126
polyphosphonium polymers, 1124–1125
PSOAT system, 1126, 1127
QNPHOS, 1124
nonlinear (co)polymers
boronic acid moieties in polyamidoamines, 1131

cationic polymeric amphiphiles, 1130, 1131
graft densities, 1130–1131
polyamidoamine-PEG-poly(L-lysine) copolymer, 1129, 1130
poly(cyclooctene-graft-oligolysine)s, 1129
star copolymers, 1128–1129
stimuli-sensitive systems, targeting, 1131–1135
water-soluble PEI derivatives, 1126, 1128
nonviral transfection systems, 1123
polypropyleneimine dendrimer polyplexes, 1135
RAFT polymerization, 1136
ternary complexes
aminoterminated polyamidoamine dendrimers, 1140, 1141
DOTAP/pDNA complexes, 1139–1140
homo-catiomer integration, 1139
hyaluronic acid, 1140
magnetoplexes, 1140
trehalose-pentaethylenehexamine glycopolymer, 1134–1135
Polymer-ceramic nanocomposite (PMNC), 1276
bioactivity, 1283–1284
ceramics, 1277–1279
future prospects, 1286
glass transition behavior, 1281–1283
mechanical behavior, 1280–1281
nanocomposite degradation behavior and acidity regulation, 1279–1280
polymers, 1276–1277
silanes, 1279
theoretical approach, 1284–1285
Polymer-directed enzyme prodrug therapy (PDEPT), 795
Polymer-enzyme liposome therapy (PELT), 795
Polymeric interactions
amorphous solid dispersions and advanced systems, 462–471
film-coated systems, 456–462
future prospects, 471
oral drug delivery categories, 452
regulatory guidelines, 451–452
traditional matrix systems, 452–456
Polymeric materials
biodegradable polymer properties, 1411–1413
chitosan, 1414
clinical and preclinical models, 1411
collagen, 1413
gelatin, 1413–1414
hyaluronic acid, 1415
natural and synthetic polymers, 1411
natural polymers, 1411–1416
synthetic materials, 1416–1417
Polymeric micelles
amphiphilic block copolymers, 1073–1074
biodistribution, 779
biodistributions, 66, 68
core–shell structure, 66, 67
crew-cut micelles, 1074
drug encapsulation, 67
formulation characteristics, 1074–1075
hydrophobic agents, 779
morphologies, 1074
nanocontainers, 68
PEG-polyester micelle, 68–69

PEG-polypeptide micelle, 68
preparation, 1074
size effect, 68
sizes and surface features, 68
stimuli-responsive, 69–70
stimuli-responsive polymeric micelle, 69–70
therapeutic applications, 1075–1076
tumor accumulation and tissue penetration, 779–780
Polymeric nanofiber drug delivery, *see* Electrospinning technology, polymeric nanofiber drug delivery
Polymeric nanomedicine
colloidal nanoparticles, 58
covalent conjugation approach, 59–60
dendrimer and dendritic polymer nanocarriers, 71–72
liposome-based nanoparticulate delivery vehicles, 58
micelles, 66–69
physical encapsulation approach, 60
polymeric vesicles, 70–71
polymer–protein conjugates, 60–61
polymer–small molecule drug conjugates, 61–66, 62–66
Polymeric nanoparticles (PNPs)
bioconjugation
covalent approaches, 1383–1386
non-covalent approaches, 1383
cancer research and management
CD40 and toll-like receptor stimulation, 1196–1197
chemopreventive combination polymers, 1196
cisplatin NPs, 1196
clinical trials, 1197–1198
DATS–PBCA–NP, 1195
echogenic PNPs, 1196
immunoliposomes, 1196
MMC-PBCA–NP, 1196
N-acetyl penicillamine-chitosan, 1196
noscapine, 1195
PLGA–DDAB/DNA, 1195
poly(L-γ-glutamylglutamine)–docetaxel conjugate, 1196
superparamagnetic iron oxide NPs, 1196
thiolated chitosan, 1196
zoledronate-conjugated PLGA NPs, 1195
as delivery vehicles, 938
formulation characteristics, 1076–1077
multifunctional, 1153–1159
PDT, 1151–1153
preparation, 936, 1076
as theranostic agents, 938
therapeutic applications, 1077–1078
Polymeric nanospheres (PNSs)
advantages, 1076
amphiphilic diblock copolymer self-assembly, 933–934
as diagnostics agents, 938
disadvantages, 1076
drug incorporation, 937
hyperbranched polymers self-assembly, 935–936
intermolecular interactions, 935
nucleic acid incorporation, 937–938

polypeptide-based copolymers self-assembly, 935
rod–coil block copolymer self-assembly, 934–935
self-assembly components, 933
stimuli-responsive copolymer self-assembly, 934
Polymeric nanotubes (PNTs)
as delivery vehicles, 938
preparation, 936
Polymeric sequestrants
bile acid sequestrants, 1102–1103
iron sequestrants, 1102
phosphate sequestrants, 1102
toxin sequestrants, 1103
Polymer/metal nanocomposites, see Silver/polymer-based nanocomposites
Polymer–protein conjugates, 60–61
end-functionalized polymers synthesis, 356
poly(N-isopropyl acrylamide), 356–357
positive effects, 353–354
site-specific incorporation, 354–355
synthetic erythropoiesis protein, 355
Polymer–small molecule drug conjugates
cyclodextrin (CD)-based polymer, 64–65
HPMA-drug conjugates, 62
PEG conjugates, 62
poly(glutamic acid) (pGlu), 62–64
polysaccharides, 65–66
Polymersomes, 306–307, 1100
bilayer membrane, 1078
drug delivery, 1075
formulation characteristics, 1079
preparation, 1078–1079
therapeutic applications, 1079–1081
Poly[2-(methacryloyloxy)ethyl dimethyl-(3-sulfopropyl)ammonium hydroxide] (PMEDSAH), 605–606
Poly(ethylene glycol) methyl ether methacrylate (PEGMA), 1125–1126
Poly(3-hydroxybutyrate) [P(3HB)], 318–319
Polyphosphazenes, 367, 872
Polyplexes
advantages and disadvantages, 771
cationic polymers, 769–770
combination, 771–773
in vitro applications, 770
in vivo applications, 770, 1272–1274
vs. lipoplexes, 770–771
PEI, 769
PLL and PLA, 769
therapeutic approaches, 1274
Poly(N-isopropyl acrylamide) [poly(NIPAAM)], 356–357
Polypropylene
osseointegration, 256–257
osteoblast interaction, 255–256
Polypropylene melt, 845–846
Polypyrrole (PPy), 313, 319, 320, 483–485, 1042
Polyquaternium 4, 657
Polyquaternium 10, 656
Polysaccharides, 648
adhesives, 1
chemical structure, 1646
chitin and chitosan, 1645–1646
chondroitin sulfates, 1649–1650

coagulation process and angiogenesis, 1646–1647
heparin, 177
human dermal fibroblasts, 1646
hyaluronic acid, 177
hyaluronic acid, 1647–1649
matrix MMPs, 1647
microorganisms, 1647
nitric oxide, 1646
peritoneal macrophages, 1646
polymorphonuclear neutrophils, 1646
poly-N-acetylglucosamine, 1647
Polysiloxanes, 446
Polysorbitol-based osmotically active transporter (PSOAT) system, 1126
Polytetrafluoroethylene (PTFE), 1562–1563, 1577, 1581–1582
Polyurethanes (PUs)
antimicrobial (see Antimicrobial polyurethanes)
bismuth oxychloride as radiopacifier, 1337–1339
cardiovascular catheters (see Cardiovascular catheters)
cardiovascular disease, 1583
enteral feeding, 1319
feeding tube, 1335–1337
high-performance, 1318
infusion pumps, 1322–1324, 1340
neurological leads, 1320–1322
pacemaker leads, 1319–1320
peritoneal dialysis applications, 1318–1319
polymeric materials, 1417
purpose and advantages, 1339–1341
Polyvinyl alcohol (PVA), 1310, 1583
Polyvinylpyrrolidone (PVP), 654, 655, 1310
Porosity, 690
Porous scaffolds
ice particulates, 1375–1377
porogen leaching, 1375
surface pore structures, 1377–1378
synthetic biodegradable polymer, 1378–1379
Porphyrin-polymer conjugates (PPCs), 1386
Positive cell selection, 747
Positron emission tomography (PET) scans, 572
Postpolymerization dispersion
emulsion, 328–329
nanoprecipitation, 329–330
Ppy, see Polypyrrole (Ppy)
Precision extruding deposition (PED), 911
Precision/ultra-precision machining technology, 911–913
Pressure-sensitive adhesives (PSAs), 447
Preventive dental sealants (PDS), 433
Printing head and powder-based microfabrication
3-DP, 1355–1356
laser sintering, 1356–1357
membrane lamination, 1355
photopolymerization, 1357
Pro-His-Ser-Arg-Asn (PHSRN), 345
Prooxidants, 171
Propanolol, 454, 455
Prosthokeratoplasty
development, 374–375
keratoprosthesis (KPro)
clinical complications, 375–376

clinical practice use in, 375
types, 375
Prostin E2® suppository, 971
Protein delivery systems
albumin and tetanus toxoid, 1143–1144
amphiphilic photocleavable block copolymer, 1144, 1145
antibody delivery, 1142
blend formulations, 1143
bovine serum albumin, 1146
chitosan-carrageenan nanoparticles, 1143
cholesterol-bearing pullulan, 1142
core/corona nanoparticles, 1143
enzymatic conversion, 1140
ICAM-1-targeted nanocarriers, 1144, 1146
insulin-loaded nanoparticles, 1143
micrometer/nanometer-size polymer particles, 1142
PEGDMA and MAA-based nanoparticles, 1142
PEO-PPO-PEO block copolymer, 1146
pH-responsive polymeric nanocarriers, 1142
PIC micelles, 1144
polyelectrolyte complexe, 1146
polymer-protein complexes, 1142
polymersomes, 1143, 1144
weak non-covalent interactions, 1142
Protein resistant biofunctionalization, 518–519
Proteolytically degradable polymers, 176–177, 177
Prothecan®, 62
Proton exchange membrane fuel cells (PEMFCs), 47–48
PSAs, see Pressure-sensitive adhesives (PSAs)
Pseudo gels, see Physical gels
Pull technique, 1336
Pulmonary drug delivery, 399
Pulsed laser deposition (PLD), 589
Push technique, 1336
PVA, see Polyvinyl alcohol (PVA)
PVDF, see Poly(vinylidine difluoride) (PVDF)

Quantitative coronary angiography (QCA), 1465
Quantum dots (QDs)
band gap, 563–564
biofunctionalization, 569
biotin-streptavidin, 571
bottom-up and top-down, 566
cancer, 560–561
conduction band, 563–564
conduction band and valence band, 562
core-shell (C-S) nanostructures, 563–564
covalent conjugation, 570–571
electronic band structure, 562
empirical measurements, 563
hole-electron, 563
hydrodynamic diameter, 567
noncovalent interactions, 570
PAMAM polymers possess, 571
particle in-a-box model, 563
quantum confinement, 563
synthesis, 563, 566–567
thermal and electrical conductivities, 562
water-soluble, 567–569
Quartz crystal microbalance (QCM), 598–599, 689–690

Quaternized poly[3,5-bis
(dimethylaminomethylene)-p-
hydroxyl styrene] homopolymer
(QNPHOS), 1124

Rabbit corneal epithelial cells (RCE),
1649–1650
Radiation sterilization, 548
Radio frequency (RF), 403
Radiopacifier, 1337–1339
RAFT, see Reversible addition fragmentation
transfer (RAFT)
Rapid prototyping (RP)
biodegradable microfluidic devices, 1348
case studies, 1342–1343
classification, 1342, 1343
fused deposition molding (FDM), 1347–1348
hydrogels photopattern, 1348
phases, 1342
selective laser sintering, 1344–1345
stereolithography, 1343–1344
three-dimensional printing, 1345–1347
tissue engineering, 1350
commercial systems, 1359
electrospinning, 1358
fluid-based microfabrication (see
Fluid-based microfabrication)
future prospects, 1360
integration, 1358–1359
limitations and critiques, 1359–1360
materials used, 1351–1352
printing head and powder-based
microfabrication (see Printing head
and powder-based microfabrication)
RTM ratio and geometry, 1352–1353
sacrificial molds, 1357–1358
three-dimensional structures
microfabrication, 1350–1351
RDRP, see Reversible-deactivation radical
polymerization (RDRP)
RDS, see Restorative dental sealants (RDS)
Reactive oxygen and nitrogen species (RONS),
334
Reactive oxygen species (ROS), 415
Recombinant Human Granulocyte-Colony
Stimulating Factor (rhG-CSF), 557
Rectal drug delivery, 122–123
Rectal mucoadhesive delivery systems, 971
Regenerative medicine, see Electrospinning
technology, regenerative medicine
Resin
cements, 884
sealants, 433–435
Restorative dental sealants (RDS), 433
amalgam bonds, 436, 437
amalgam restoration, 435, 436
amalgam sealant, 435–437
bonding agent chemical nature and efficacy,
437–438
clinical procedures, 438
composite restoration, 436–437
Reticuloendothelial system (RES), 574, 778
Reverse micelles (RMs), 1313
Reversible addition fragmentation transfer
(RAFT), 1290–1292
nanogels, 1020–1023
polymerization, 352, 922–923

Reversible-deactivation radical polymerization
(RDRP), 1290, 1291
RF, see Radio frequency (RF)
RGD, see Arg-Gly-Asp (RGD) sequence
Rheology modifiers, 1307
Rifampin PLLA electrospun fibers, 500, 502
Ringsdorf model of synthetic polymer drugs,
1171
Rinsed channel, 945
RMs, see Reverse micelles (RMs)
RONS, see Reactive oxygen and nitrogen
species (RONS)
ROS, see Reactive oxygen species (ROS)
Rotating cylinder, 944–945

SAM, see Self-assembled monolayer (SAM)
SAPs, see Self-assembling peptides (SAPs)
SARC, see Self-adhesive resin cement (SARC)
Scaffolds, 1567, 1630
agarose, 1218–1219
alginate, 1216–1217
bacterial cellulose, 1219–1220
biomaterial research, 1410
blend, 1417–1418
cellulosic scaffolds, 1219
chitosan, 1217–1218
collagen-based scaffolds, 1213–1214
composites/nanocomposite, 1418
dextran, 1220–1221
electrospinning, 1212–1213, 1418–1419
fibrin and fibrinogen, 1216
freeze-drying method, 1211, 1212
freeze-extraction method, 1212
gas foaming, 1210–1211
gelatin-based scaffolds, 1214–1215
gellan gum, 1220
hyaluronic acid, 1218
mechanical property, 1208
microfluidic-based polymer design
advantages, disadvantages and
applications, 1362–1363
bioreactors, 1367–1369
cell culture systems for tissue regeneration,
1367, 1368
limitations and future prospects, 1369
polymeric scaffold fabrications, 1366–1367
polymeric scaffolds used, 1363–1365
physicochemical characterization, 1213
polymeric materials, 1411–1417
pore diameter, 1208
porous architecture, 1410
preparation, 1209–1210
processing conditions, 1418
silk fibroin scaffolds, 1215–1216
starch-based scaffolds, 1220
TE, 1410–1411
thermally induced phase separation
technique, 1210
Scanning electron microscopy (SEM), 412, 413,
595–596, 690
Schrom MG, 1321–1322
SCMC, see Sodium carboxymethyl cellulose
(SCMC)
Sealants, see Dental sealants
Secretory mucins, 941–942
SEDDS, see Self-emulsifying drug delivery
systems (SEDDS)

Sedimentation process, 751–752
Self-adhesive resin cement (SARC),
419–421
Self-assembled monolayer (SAM), 315, 588
Self-assembling peptides (SAPs), 381
Self-assembly approach, 381
Self-emulsifying drug delivery systems
(SEDDS), 301, 304
Self-expanding stent, 1472
Self-microemulsifying drug delivery systems
(SMEDDSs), 301, 305
SEM, see Scanning electron microscopy (SEM)
Semiconducting polymer dot bioconjugates
biomedical research, 1386–1387
cellular imaging, 1388
dynamic light scattering measurements,
1387–1388
lyophilization, 1386–1387
MCF-7 cells, 1388–1389
near-infrared, 1389–1390
3-NIR Pdot-SA probes, 1389–1390
PEG lipid PFBT nanoparticles targeting,
1389
polymeric nanoparticles, 1382–1386
PPVseg-COOH, 1388–1389
Qdots, 1382
Sensing and diagnosis
colloidal properties, 1396–1398
diagnostic methods, 1393
DNA-carrying nanoparticles, 1394–1395
DNA-dependent assembly, 1395–1396
dynamic light scattering, 1394
LCST, 1393
molecular parameters, 1395
static light scattering, 1394
types of, 1393–1394
Sensitization tests, 134
Sensors and biosensors
alkanethiol-modified gold electrodes, 1042
ascorbic acid and UA, 1044–1045
dopamine oxidation, 1043
electrochemical DNA sensor, 1042
gas sensors, 1043–1044
glucose biosensors, 1043
SEP, see Synthetic erythropoiesis protein (SEP)
Septopal®, 85
Shape memory alloy (SMA) artificial muscles,
91
Shape memory polymers (SMPs)
artificial muscles, 91
biomedical applications, 1401–1405
brain application, 1405
controlled drug release system, 1405
properties, 1400–1401
Shear stress, 116
Shell cross-linked, knedel-like (SCK) polymer
nanoparticles, 1119
Shellac, 655
Silica aerogel, 39–41
Silicified MCC (SMCC), 272
Silicone acrylates, 362–363
Silicone polymers, 659–661
Silicone-based hydrogels, 365–366
Silk fibroin (SF), 4, 1215–1216
Silkworm silk, 1613
Silver, 1330–1331
Silver nanoparticules, 1039–1040

Silver/polymer-based nanocomposites
  antibacterial activities, 994
  antibacterial efficacy, 998–1000
  antifungicidal activity, 1000–1001, 1003
  cellulose–silver nanocomposites, 1001–1003
  gentamicin-loaded cement, 997
  hydrogel matrix, 1000
  minimum inhibitory concentration, 998
  PA6/Ag nanocomposites, 995–996
  PDMS/Cs/clay composites, 998
  PMMA nanofibers, 997–998
  poly(methyl methacrylate) (PMMA) bone
    cement, 997
  polyether-type waterborne polyurethane,
    996–997
  polyethylene glycol–PU–titania
    nanocomposite polymer films,
    1003–1005
  poly(lactic acid) nanocomposite, 1004
  polyrhodanine nanofibers, 1002–1004
  polystyrene (PS) matrix, 998
  polyvinylpyrrolidone, 1000–1002
  ZOI analysis, 999–1000
Silylation, 276–277
Simple physical adsorption, 495
Single collection systems, 512–513
Single screw extruder, 827
Skin replacement, 103
Skin tissue engineering
  bioengineered skin equivalents, 1408
  biomedical applications, 1419–1420
  cells, 1409–1410
  cell-scaffold interaction, 1408
  fibroblasts and keratinocytes, 1409
  multidisciplinary approach, 1409
  myosin light chain, 1419
  paradigms, 1408–1409
  PLLA-collagen/PLLA-gelatin scaffolds,
    1419–1420
  scaffold, 1410–1419
Skin tissue regeneration
  adipose stem cells, 1629–1630
  allogeneic grafts, 1627
  autologous skin grafts, 1626–1627
  biomechanics, 1622
  conventional strategies, 1626
  deoxyribonucleic acid technology,
    1627–1628
  dermis, 1622
  ECM component, 1620
  epidermis, 1621–1622
  epithelialization and remodeling phase, 1627
  induced pluripotent stem cells, 1628–1629
  inflammatory phase, 1627
  proliferative phase, 1627
  regenerative medicine, 1628
  split-thickness skin graft, 1626
  stem cell role, 1628–1630
  subcutaneous tissue/hypodermis, 1621
  TE approaches, 1630–1634
  treatment strategies, 1626–1630
  in vivo tissue regeneration, 1634–1636
  wounded skin, 1622–1626
  Xenografts, 1627
Small angle neutron scattering (SANS), 690
SMART®, 886
Smart polymers, 447–448

active targeting strategy, 1099
  imprinting (see Molecular imprinting
    technology)
  medicine and biotechnology (see Intelligent
    polymer system)
  microgel
    glucose-responsive microgels, 921
    monomers, 918
    photoresponsive microgels, 920–921
    pH-responsive microgels, 919–920
    swelling ability, 918
    thermo-responsive microgels, 918–919
  passive targeting, 1097–1098
  stimuli-responsive polymers
    dual stimuli, 1097
    electro-responsive polymers, 1095
    enzyme-responsive polymers, 1096–1097
    glucose-responsive polymers, 1096
    inflammation-responsive polymers, 1097
    ion-responsive polymers, 1096
    magneto-responsive polymers, 1095, 1096
    photo-responsive polymers, 1095
    pH-responsive polymers, 1095–1096
    redox-responsive polymers, 1096
    temperature-responsive polymers,
      1094–1095
SMCC, see Silicified MCC (SMCC)
SMCs, see Smooth muscle cells (SMCs)
SMEDDSs, see Self-microemulsifying drug
    delivery systems (SMEDDSs)
Smooth muscle cells (SMCs), 522, 524, 618,
    728, 1576
Smooth surface sealants, 435
Sodium carboxymethyl cellulose (SCMC), 270,
    272, 282, 1313
Soft contact lenses, 674–675
  adverse effects, 676–677
  ideal lens material, 677
  monomers, 675–676
  oxygen permeability, 676
  polyethylene oxide (PEO), 676
  silicone hydrogel, 676
Soft template synthesis, 314
Soft-tissue engineering, 234–235
Solid dispersion, melt extrusion
  amorphous solid dispersions (see Amorphous
    solid dispersion system)
  amorphous solid solution, 828
  controlled release products, 835–837
  crystalline solid dispersions (see Crystalline
    solid dispersions, melt extrusion)
  directly shaped products, 839–840
  dissolution-enhanced products, 833–835
  extruders, 827
  foam extrusion, 838–839
  formulation design, 831
  functional excipient roles, 828, 829
  marketed pharmaceutical products, 828, 829
  miscibility, 831
  plasticizer, 829
  povidone and methacrylic acid copolymer,
    829
  solubilization regime, 831–832
  spring and parachute performance, 832, 833
  surfactants, 831
Solid dispersion method, 780
Solid lipid nanoparticles (SLNs)

excipients, 1172
  high pressure homogenization technique,
    1172
  lipids, 1172–1173
  microemulsion method, 1172, 1173
  proteins and peptides delivery, 1173
  targeted brain drug delivery, 1173
Solid-phase reversible immobilization (SPRI),
    749–750
Solubilizers, 551–552
  cosolvents, 552
  cyclodextrins, 553
  hydrotropes, 552–553
  liposome, 553
  oil-based solutions, 553
  peroxides, 554
  solvents, 553–554
  surfactants, 552
Solution viscosity, 509
Solvent conductivity, 509–510
Sonochemical synthesis, 781
Space charge electret, 477
Specialized bioactive carriers, 285
  dual-drug dosage forms, 288
  hydrogels, 286
  osmotic pump, 287
  xerogels, 286
Split-thickness skin graft (STSG), 1626
Star copolymers, 1128–1129
Starch, 1310
  amylose and amylopectin, 1611
  applications, 1611
  manufacture, 1611
  scaffolds, 1220
  wound care, 1611
Stem cell-culture substrates and controlled
    differentiation
  hIPSC-MSCs, 606
  hMSCs, 605
  hPSCs, 604
  human embryonic stem cells, 605–606
  mouse embryonic fibroblasts, 604
  PMEDSAH, 605–606
  TCPS, 604–605
Stem cells (SCs), 1263
  HSPCs (see Hematopoietic stem and
    progenitor cells (HSPCs))
  iPSCs, 1410
  skin tissue engineering, 1409–1410
Stents and stent grafts
  balloon angioplasty, 886
  drug eluting stents, 886–887
  elasticity/plasticity, 886
  pitting and corrosion, 888
  self-expandable stent, 886
  vascular grafts, 887–888
Stimuli-responsive polymeric materials
    (SRPMs)
  health care applications, 1487–1489
  hydrogels, 1486–1487
  radiation sources, 1481–1482
  traditional method, 1481
  wound dressing materials, 1489–1490
Stimuli-sensitive systems, targeting
  anti-DF3/Mucin1 (MUC1) nanobody, 1134
  cardiomyocyte-targeting peptide, 1135
  folate-conjugated tercopolymers, 1134

hydroxyethyl starch, 1134
light-regulated host-guest interaction, 1133
magnetic targeting gene delivery systems, 1133–1134
poly[oligo (ethylene glycol) methacrylate], 1132
reduction-sensitive reversibly shielded DNA polyplexes, 1134
ternary polyplexes, 1133–1134
thermoresponsive nonviral gene carrier, 1131–1132
Stimulus-responsive magnetic latexes, 752
Stimulus/stimuli-responsive polymers (SRPs)
aqueous solutions, 1446–1447
biomolecule conjugates, 1446, 1448
chemical stimuli, 1480
classification, 1480–1481
electrosensitive, 1485
hydrogels, 1449–1450
hydrophobic surface compositions, 1448
leuco derivative molecule, 1486
light-responsive polymers, 1485–1486
medicine and biotechnology, 1449–1450
micelle, 69–70
novel membrane, 1448–1449
pH-sensitive polymers, 1449–1450, 1485
physical stimuli, 1480
PNIPAAm, 1482
poly(acryloyl-N-propylpiperazine), 1482
polymer adsorption-desorption, 1448–1449
radiation-induced SRPs, 1480
SRPMs (see Stimuli-responsive polymeric materials (SRPMs))
surfaces, 1447–1448
temperature-sensitive polymers, 1482–1484
Stroma, 371
Stromal cell alignment approach, 381–382
Stromal vascular fraction (SVF), 1410
Structured electroconductive collectors, 513–514
Sulfobetaine (SB)
ATRP technique, 1663–1664
human serum albumin (HSA), 1663
LCST and UCST, 1664
PECH-DMAPS, 1664–1665
poly(CBAA), 1663–1664
polySBMA, 1663–1664
PVDF surfaces, 1664
zwitterionic materials, 1663
Super paramagnetic iron oxide NPs (SPIONs), 1183
Superabsorbent polymers (SAP), 687
Supercritical carbon dioxide (scCO$_2$) drying
binary system, 21–22
depressurization, 23
mass transfer pathways, 22
pressurization, 22–23
supercritical liquid, 22
Superoleophobic aerogels, 43
Superparamagnetic iron oxide–based theranostics systems, 1061–1062
Superporous hydrogels (SPHs), 687
Surface erosion, 870
Surface graft polymerization, 493–494
Surface modification techniques
albumin, 671
biological techniques, 671

chemical modification approaches, 670
concepts, 670
hydrophilic hydrogel polymers, 670
hydrophilic PEO chains, 670–671
modification techniques, 669–670
PEO, 670
PHEMA/PNVP hydrogels, 670
plasma treatment techniques, 670
Surface oxide layer
cobalt–chromium alloy, 883
dental precious alloys, 883
passive film, 881
stainless steels, 883
structure model, 881
titanium, 881–882
titanium alloy, 882–883
Surface plasmon resonance (SPR), 598–599, 924
Surface plasmon resonance-enhanced ellipsometry (SPREE), 598–599
Surface tension, 509
Surface-initiated photoinitiator-mediated polymerization (SI-PIMP), 1663
Surgical glues
Aldehyde system, 648–649
cyanoacrylate, 647–648
fibrin glue, 644–647
GRF glue, 648
GRF glues, 644
photocurable system, 650–651
requirements, 644–645
seal small holes, 644
WSC and NHS system, 649–650
Sutures
absorbable and nonabsorbable, 1515
biodegradation and absorption properties, 1526–1528
biological properties, 1525–1526
characteristics, 1520–1521
commercial sutures, 1516–1518
comparison, 1520, 1523
essential properties, 1520–1528
handling properties, 1525
healing process, 1514
history, 1514
knot tie-down and security, 1520
light histologic photomicrographs, 1526
mechanical properties, 1520–1522
needle selection, 1514–1515
nonabsorbable sutures, 1520, 1522, 1524
numerous materials, 1514
physical and mechanical properties, 1520–1525
physical configurations, 1515–1520
scanning electron images, 1515, 1519
steel wire and synthetic nonabsorbable fibers, 1514
tissue reactivities, 1520, 1522
trade names and manufacturers, 1516–1518
USP suture size classification, 1515, 1520
Swellability, 691
Synthetic biodegradable polymer
biodegradable synthetic polymers, 1379
hybrid sponges and meshes, 1378
PGA, PLA and PLGA, 1378
PLGA-collagen hybrid sponge, 1378–1379
Synthetic erythropoiesis protein (SEP), 355

Synthetic polymers, 440, 441, 444–445, 508, 679, 1259
acrylamide polymers, 680–681
acrylic polymers, 680–681
block copolymers, 680
crystallization, 680
egg-box model, 680
helical junction zone, 680
hybrid hydrogel formation, 680
poly(vinyl alcohol) (PVA), 680

Tacrolimus, 469
Tailoring polymers, 668–669
Tanaka equation
binding site possesses, 1430–1431
cross-linker concentration, 1432–1433
experimental assessment, 1431–1432
fixed-point model, 1428–1429
MAPTAC, 1431–1432
N-isopropylacrylamide (NIPA), 1427
power-law dependence, 1429
Py-3 (or Py-4) molecule, 1431
replacement ion concentration, 1432
theoretical considerations, 1427–1431
volume phase transition, 1427
Taste masking, 1307
Taylor cone, 315
TBI, see Traumatic brain injury (TBI)
TCP, see Tricalcium phosphate (TCP)
TDDS, see Transdermal drug delivery system (TDDS)
TE, see Tissue engineering (TE)
TEC, see Tissue-engineered construct (TEC)
TEFLON® prostheses, 85
TEGDMA, see Tri-ethylene glycol dimethacrylate (TEGDMA)
TEM, see Transmission electron microscopy (TEM)
Temperature-sensitive imprinted hydrogels
conformational memory, 1434
dibenzothiophenes (DBT), 1437
equilibrium swelling ratios, 1436
frustrations, 1434
functional monomer Vb-EDA, 1436
Imprinter-Q monomer, 1437
interpenetrated system, 1436
methacrylic acid (MAA), 1434–1435
nonimprinted and imprinted gels, 1435
PNIPA hydrogels, 1438
polymer chains, 1433
post-cross-linking approach, 1434
Py-4 function, 1434
random gels, 1435
thermodynamic phases, 1432
Temperature-sensitive polymers; see also Thermo- and pH-responsive polymers
cloud point, 1483
LCST polymers, 1483
PIPAAm, 1483–1484
structure, 1482–1483
thermosensitive polymers, 1483
Template-free polymerization, 314–315
Temporal control, 301
Tensile stress, 116
Tensile tests, 760
dry polymer compacts, 945
hydrated polymers, 945–946

microspheres, 946
Tetradecyl trimethyl ammonium bromide
    (TTAB), 290
2,2,6,6-Tetramethylpiperidine-1-oxyl
    (TEMPO)-oxidized microfibrillated
    cellulose aerogels, 30–31
Tetrathiafulvalene-tetracyanoquinodimethane
    (TTF-TCNQ), 483
Theranostic nanoparticles (TNPs), 1530–1531
    aliphatic polyesters, 1537
    applications, 1539–1543
    biocompatibilities, 1538–1539
    biodegradable polymers, 1534–1536
    biomedical applications, 1532
    biomedical-biodegradation factors,
        1536–1538
    biopolymers possess, 1533–1535
    citric acid cycle (CTA), 1538
    core-shell nanoparticles, 1535–1536
    diagnostic and therapeutic actions, 1530
    drug delivery applications, 1540–1541
    encapsulation, 1539
    functionalized nanoparticle, 1534, 1536
    hydrolytic biodegradation, 1537–1538
    imaging applications, 1541–1543
    magnetic resonance imaging (MRI), 1532
    materials and methods, 1532
    micellar nanoparticles, 1535
    microbubbles, 1536
    microspheres, 1532–1533
    photo-induced reversible transitions, 1533,
        1535
    PLGA nanoparticles, 1538
    vesicles, 1534
Theranostics
    gadolinium and manganese based contrast
        agents, 1066–1067
    imaging and gene therapy, 1069–1071
    imaging and photodynamic therapy,
        1067–1068
    imaging and photothermal therapy,
        1068–1069
    multimodal imaging probes, 1062–1064
    organic dyes for optical imaging, 1064–1065
    polymeric nanoparticles, 1061
    superparamagnetic iron oxide–based,
        1061–1062
Therapeutic embolization, 694
Thermal decomposition, 781
Thermal-based CVD polymerization, 592
Thermally induced phase separation (TIPS),
        1631–1632
Thermally stimulated current (TSC), 479
Thermally stimulated depolarization current
        (TSDC), 477–478
Thermo- and pH-responsive polymers
    AMPEG, 1496
    critical solution temperature, 1493
    CSTs, 1496–1497
    gastrointestinal tract (GIT), 1493–1494
    LCST and UCST, 1495–1496
    NIPAAm gels, 1495
    oral protein/peptide and drug delivery,
        1503–1504
    parenteral drug delivery, 1504–1507
    PEG-PLGA-PEG aqueous solution, 1493
    phase transition, 1494–1495

pH-responsive polymers, 1498–1501
    PNIPAAm, 1501
    polymeric micelles, 1501–1503
    smart polymers, 1493
    stimulus-sensitive polymers, 1493
    temperature-responsive polymers, 1496–1498
    thermoresponsive hydrogels, 1501
    transitions/transformations, 1493–1494
Thermoelectrets, 477
Thermo-or pH-sensitive units
    dithiothreitol/glutathione, 709–710
    glucose and pH, 710
    hybrid hydrogels, 709
    layer-by-layer films, 709
    LCST, 708
    microcapsules, 708–709
    responsive polymer systems, 708
    sol-gel transition, 709
    UV irradiation, 708
    versatile synthetic method, 708
Thermoplastic medical-grade polyurethane,
        1334–1335
Thermoplastic polymers, 1206
Thermoresponsive and magnetic hydrogels
    block copolymers and gelatin, 706–707
    drug delivery process, 706
    dual-functional nanospheres, 706
    hyperthermia applications, 705–706
    iron oxide nanoparticles, 706
    mesoporous SBA-15, 707
    pluronic-magnetic nanoparticles, 706
    poly(undecylenic acid-co-NIPAAM), 706
Thermo-responsive cell culture
    carrier, 383
    dishes, 378–379
Thermoresponsive gel matrix, 379
Thermo-responsive microgels, 918–919
Thermoresponsive polymeric micelles
    amphiphilic polymers, 1501
    critical micelle temperature, 1502
    micellar structures, 1502–1503
    micelle formation via radical polymerization,
        1503
    NIPAAm, 1502
    PEO-PPO-PEO, 1501–1502
    PNIPAAm, LCST and PEG, 1502
    shell cross-linked (SCL), 1503
Thermosetting polymers, 1206
Thiolated mucoadhesive polymers, 951, 952
3-thiopheneethanol (3TE), 589
Three-point bending flexural test, 428
Thrombolysis, 664
Tissue engineering (TE), 318–319, 693
    adhesion, 1632–1633
    advantages, 101, 723–724
    angiogenesis and vasculogenesis, 724
    approaches, 1630
    artificial ocular surface, 619
    bioabsorbable polymers, 101–103
    bioactive materials, 216, 1207
    bio-artificial arteries, 619–620
    biochips, 185
    biodegradable polymers, 1374–1375
    bioinert polymers, 1206–1207
    biomaterial-based scaffolds, 724–726
    biomimetic materials (see Biomimetic
        materials)

biomimetic surfaces, 216–224
bioresorbable materials, 1207
blood vessel formation, 724
BMMCs, 1568
cell delivery vehicle, 619
cells and scaffolds interaction, 1632–1634
challenges, 184
collagen, 1584–1586
cornea
    advantages, 384–385
    applications and limitations, 384
    carrier properties, 376
    corneal endothelium, 382–384
    corneal epithelium, 377–380
    corneal stroma, 380–382
    criteria, 385
    full-thickness corneal equivalent, 376–377
    future prospects, 385
    strategies, 376
cytokine immobilization, 1633–1634
definition, 723
elastin, 1586
fabrication techniques, 1630–1632
fibrin, 1586–1587
goal of, 1488
heart valves, 619–620
hybrid constructs, 101
hydrogel films, 1208
hydrogels
    mechanical properties, 1208
    physical hydrogel, 1207–1208
    preparation, 1207
    swelling kinetics, 1208
macromolecules, 214–215
macroscale characterization, 1584–1585
matrices, 214
microfabrication techniques, 185
micro-/nanofabrication technology, 184
natural origin, 1568–1569
natural-based polymers, 214
PEG-based hydrogels, 1489
PLCL, 1567
PLGA, 1567
polycaprolactone (PCL), 1584–1585
porous scaffolds (see Porous scaffolds)
scaffolding materials, 1488–1489
scaffolding process, 184
scaffolds, 1567, 1630
    agarose, 1218–1219
    alginate, 1216–1217
    bacterial cellulose, 1219–1220
    cellulosic scaffolds, 1219
    chitosan, 1217–1218
    collagen-based scaffolds, 1213–1214
    dextran, 1220–1221
    electrospinning method, 1212–1213
    fibrin and fibrinogen, 1216
    freeze-drying method, 1211, 1212
    freeze-extraction method, 1212
    gas foaming, 1210–1211
    gelatin-based scaffolds, 1214–1215
    gellan gum, 1220
    hyaluronic acid, 1218
    mechanical property, 1208
    physicochemical characterization, 1213
    pore diameter, 1208
    preparation, 1209–1210

silk fibroin scaffolds, 1215–1216
starch-based scaffolds, 1220
thermally induced phase separation technique, 1210
silk, 1587
skin (*see* Skin tissue engineering; Skin tissue regeneration)
skin grafts, 619
stem-cell behavior, 215
synthetic polymers, 1567–1568
vascular (*see* Vascular tissue engineering)
vascular grafts, 1567–1569
vascularization, 724 (*see also* Vascularization)
wrapping and tapping, 723
Tissue engineering, neural, *see* Neural tissue engineering
Tissue-engineered construct (TEC), 506
TNF alpha, *see* Tumor necrosis factor alpha (TNF alpha)
Traditional matrix systems, polymeric interactions
blending and compression, 452–453
controlled release systems, 453–454
gel formulation, 454, 455
microcrystalline cellulose, 453
molecular imprinting, 456
pH-modifying agents, 455–456
propanolol dissolution profile, 454, 455
swelling energy, 453
Transdermal drug delivery system (TDDS), 285, 398
chemical/physical approach, 872
iontophoresis, 874
penetration enhancers, 873–874
stratum corneum layer, 872
Transforming growth factor-beta (TGF-β), 178, 617
Transluminal angioplasty catheters, 1325–1326
Transmission electron microscopy (TEM), 329
Traumatic brain injury (TBI), 1256, 1257
Traumatic spinal injury (TSI), 1256, 1257
Tricalcium phosphate (TCP)
ceramics, 1277–1279
theoretical approach, 1284–1285
Tri-ethylene glycol dimethacrylate (TEGDMA), 414–415
antioxidant enzymes function, 415
mitogen-activated protein kinases and transcription factors, 416–417
nonenzymatic antioxidant GSH function, 415
oxidative stress and GSH, 415–416
reactive oxygen species, 415
signaling pathways, monomer-induced oxidative DNA damage, 417
Trioctylphosphine oxide (TOP/TOPO), 566
Tri(n-butyl) phosphate (TBP), 1568–1569
Triple-stimuli-responsive hydrogels
biomedical applications, 713–714
1,3-butadiene diepoxide, 717
dithiothreitolinduced degradation, 717–718
hydrodynamic diameter, 715
Janus nanoparticles, 715–716
magnetic microcontainer, 716–717
N,N-diethylacrylamide, 718
p(NIPAAM-*co*-MAA), 713

pH-, thermo-, and redox-sensitive hydrogels, 716–717
pNIPAAM-allylamine, 715
temperature, magnetic field and pH, 713, 715
TSC, *see* Thermally stimulated current (TSC)
TSDC, *see* Thermally stimulated depolarization current (TSDC)
TSI, *see* Traumatic spinal injury (TSI)
TTAB, *see* Tetradecyl trimethyl ammonium bromide (TTAB)
TTF-TCNQ, *see* Tetrathiafulvalene-tetracyanoquinodimethane (TTF-TCNQ)
Tumor necrosis factor alpha (TNF alpha), 1274
Tumor-homing chitosan-based NPs (CNPs), 1195
Twin screw extruders, 827
Two-compound techniques (CVD)
B-NCA and APS-Si, 589–590
oxidative and initiated CVD, 589
PMMA, 590–591
PTFE, 591
reactor configuration, 589–590
tert-butyl peroxide, 590
UV irradiation, 590
VDP, 589
Tyrosine-based polycarbonates, 150

UDMA, *see* Urethane dimethacrylate (UDMA)- TEGDMA based resin composite blocks
Ultimate tensile strength (UTS), 617
Ultrasmall superparamagnetic iron oxide (USPIO), 1541–1543
Ultrasound contrast agents (UCAs)
advantages of, 1556
BSA-AA containers, 1551, 1556
bubble-cell collisions, 1548–1549
cancer treatment, 1553
cardiovascular therapeutics, 1549–1551
CEUS (*see* Contrast-enhanced ultrasound (CEUS))
diagnostic imaging tool, 1548
drug/gene delivery, 1556
drug-loaded microbubbles, 1554
encapsulation materials, 1552
fight cancer, 1554
fluorescence microscopic image, 1553
liposomes, 1554–1555
liquid-filled lysozyme microspheres, 1552–1553
lysozyme microbubbles, 1551–1553
micelles, 1555–1556
microbubbles, 1554
morphology, 1552–1553
nanoemulsions, 1554
Nile red molecules, 1553
PFC-containing nanoemulsions, 1554–1555
protein containers, 1551
protein microspheres, 1545
release profile measurements, 1551
scanning electron micrograph, 1548
sonochemical synthesis, 1547–1548
targeted microbubbles, 1554
temperature-based actuation mechanisms, 1554–1555
therepeutics, 1548

ultrasound, 1548–1549
ultrasound-targeted microbubble destruction, 1549
X-ray and MRI, 1545
Ultrasound-targeted microbubble destruction (UTMD), 1549
Umbilical cord blood (UCB), 1453
Unilamellar liposomal vesicle, 306
Upper critical solution temperature (UCST), 700, 1495–1496
Urethane dimethacrylate (UDMA)- TEGDMA based resin composite blocks, 417–419
Urogenital applications, 1403–1404
Urokinase, 177
US Food and Drug Administration, 445

Vaginal drug delivery, 123–124
Vaginal mucoadhesive delivery systems, 970–971
Val-Pro-Gly-Val-Gly (VPGVG), 343
Vapor deposition polymerization (VDP), 589
Vascular applications
PVA, PLA and PEG, 1401–1402
shape recovery, 1401
SMP thrombectomy device, 1402
treating aneurysm, 1402
Vascular endothelial growth factor (VEGF), 727, 732
Vascular grafts, 103, 667–668; *see also* Blood vessel substitutes
adenosine diphosphate, 1560–1561
artery, 1560–1561
biocompatibility requirements, 1561–1562
blood compatibility, 1565–1566
clinical relevance, 1560
CVD (*see* Cardiovascular disease (CVD))
endothelial cells, 1560–1561
ePTFE and PET, 1564
functional requirement, 1561
inflammatory and immune systems, 1566–1567
materials and natural tissue, 1563
mechanical compatibilities, 1563–1564
modulation, 1564–1565
[N(O)NO]_, 1566
poly(dimethyl siloxane), 1563–1564
polyethylene terephthalate, 1561–1562
polytetrafl uoroethylene, 1562–1563
polyurethanes, 1563
porosity, 1564–1565
surface characteristics, 1564–1565
tissue engineering, 1567–1569
wall shear stress, 1563–1564
Vascular prostheses, 85
Vascular tissue engineering, 533–534
allure, 1599–1600
biological and synthetic scaffolds, 1599
blood vessels, 1597, 1599
characteristics and history of, 1598
collagen, 1600–1601
current biopolymer options, 1600–1602
decellularization, 1602–1603
elastin, 1601–1602
extracellular matrix (ECM), 1600
measurement, 1597, 1599
overview of, 1596–1597

properties, 1596–1597
replacements, 1596
scanning electron micrographs, 1601
small-caliber grafts, 1596
small-diameter prosthetic grafts, 1597, 1599
translational research, 1603–1604
Vascularization; *see also In vitro
vascularization*
adipose tissue, 727
amniotic fluid-derived stem cells, 727
angiogenesis, 724
bone marrow-derived mesenchymal stem
cells, 727
co-culture systems, 728–729
endothelial cells, 726
endothelial progenitor cells, 726–727
vessel formation *in vitro* and *in vivo*,
731–733
Vinylic and acrylic polymers, 655–656
Vinylpyrrolidone, 657
Viscosity-enhancing mucosal delivery systems,
969
Visual analogical scale (VAS), 1652
Vivagel™, 399
Volatile organic compounds (VOCs), 598–599
Volume phase transition
antigen-antibody semi-IPN hydrogel,
677–678
functions, 677
gel system, 677
glucose concentration, 677–678
non-grafted gel (NG), 677–679
quick response hydrogel, 679
reentrant volume phase transition, 679
VPGVG, *see* Val-Pro-Gly-Val-Gly (VPGVG)
Vulval drug delivery, 123
Vulval mucoadhesive delivery systems, 971

Wall shear stress (WSS), 1563–1564
Wallstent®, 886
Warp knit, 762
Water compatible polymers, 658–659
Water-soluble carbodiimide (WSC), 649–650
Water-soluble QDs
aqueous synthesis protocol, 568
biological applications, 567
CdSe/ZnS layer, 568
core-shell (C-S) nanostructures, 568

C-S semiconductor system, 568–569
fluorescence quantum yield, 569
hydrophilic surfaces, 567
polymers (synthetic or natural), 567–568
salts/gas sources, 568
semiconductor nanomaterials, 567
Water-vapor transmission rate (WVTR), 1308
WelChol™, 1103
Wet chemical methods, 493
carbon/metal materials, 586
consecutive adsorption, 587
electropolymerization, 586
grafting from approach, 588
Langmuir-Blodgett method, 586
Layer-by-layer (LBL), 587
polydimethylsiloxane, 587–588
polymer solution, 586
self-assembled monolayers, 588
surface modification methods, 587
ultraviolet (UV) irradiation, 587–588
Wetting theory
bioadhesion, 112–113
mucoadhesion, 962–963
Wetzel–Kramer–Brillouin (WKB) method, 689
WHO, *see* World Health Organization (WHO)
William etherification, 278
Williams–Landel–Ferry (WLF) equation, 463
World Health Organization (WHO), 372
Wound care application
alginate, 1609–1610
cellulose, 1607–1608
chitin and chitosan, 1608–1609
collagen, 1611–1613
hyaluronic acid, 1613–1614
keratin, 1614–1615
oxidized regenerated cellulose (ORC), 1607
raw silk fiber, 1613
silkworm silk, 1613
skin tissue regeneration (*see* Skin tissue
regeneration)
starch, 1611
Wound dressing, 497–499, 1333–1334
Wound healing
acute wounds, 1642–1643
biopolymers/hemoderivatives association,
1653–1656
cellulose-based biopolymers, 291–292
chronic wounds, 1643

corneal epithelial defects, 1643
ECM and signaling compounds, 1643
exogenous factors, 1645
factors, 1645
growth factors, 1642, 1644
hemoderivatives, 1650–1653
inert dressings, 1642
inflammatory phase, 1643
keratocyte apoptosis, 1645
matrix metalloproteinase, 1644
maturation phase, 1643–1644
platelet-rich preparations, 1642
polysaccharides, 1645–1650
reepithelialization, 1644–1645
transforming growth factor beta, 1645
tumor necrosis factor alpha, 1643
Wounded skin tissue
contraction, 1622
healing process (*see* Wound healing)
hemostasis, 1623–1624
hemostatic stage, 1624
impact of, 1622–1623
inflammation, 1624
phases of, 1625
platelet-derived growth factor (PDGF),
1624
proliferative phase, 1624–1626
remodeling phase, 1626
type of, 1622
Woven fabric, 762
WVTR, *see* Water-vapor transmission rate
(WVTR)

Xanthan gum, 655
Xenogenic skin grafts, 1627
Xerogels, 287
Xylan, 273–274
Xyloglucan, 273, 275

Zidovudine (AZT), 285
Zone of inhibition (ZOI), 1331–1332
Zwitterionic polymeric materials
drug delivery, 1667–1668
fouling properties, 1661–1667
miscellaneous applications, 1668
nitric oxide, 1661
poly(ethylene glycol), 1661